ASM Handbook®

Formerly Ninth Edition, Metals Handbook

Volume 8
Mechanical Testing

Prepared under the direction of
the ASM Handbook Committee

John R. Newby, Coordinator

Joseph R. Davis, Senior Technical Editor
Sunniva K. Refsnes, Technical Editor

Deborah A. Dieterich, Production Editor
Heather J. Frissell, Editorial Assistant
Diane M. Jenkins, Word Processing Specialist

William H. Cubberly, Director of Publications
Robert L. Stedfeld, Assistant Director of Publications
Kathleen Mills, Manager of Editorial Operations

Editorial Assistance
Bonnie R. Sanders
Terri L. Weintraub

ASM INTERNATIONAL®

**The Materials
Information Society**

First printing, June 1985
Second printing, June 1987
Third printing, April 1989
Fourth printing, February 1992

ASM Handbook is a collective effort involving thousands of technical specialists. It brings together in one book a wealth of information from world-wide sources to help scientists, engineers, and technicians solve current and long-range problems.

Great care is taken in the compilation and production of this volume, but it should be made clear that no warranties, express or implied, are given in connection with the accuracy or completeness of this publication, and no responsibility can be taken for any claims that may arise.

Nothing contained in the ASM Handbook shall be construed as a grant of any right of manufacture, sale, use, or reproduction, in connection with any method, process, apparatus, product, composition, or system, whether or not covered by letters patent, copyright, or trademark, and nothing contained in the ASM Handbook shall be construed as a defense against any alleged infringement of letters patent, copyright, or trademark, or as a defense against any liability for such infringement.

Comments, criticisms, and suggestions are invited, and should be forwarded to ASM International.

Library of Congress Cataloging in Publication Data

American Society for Metals

Metals handbook.

Includes bibliographical references and indexes.
Contents: v. 1. Properties and selection—v. 2.
Properties and selection—nonferrous alloys and pure
metals—[etc.]—v. 8. Mechanical testing.

1. Metals—Handbooks, manuals, etc. 2. American
Society for Metals. Handbook Committee.
TA459.M43   1978   669   78-14934
ISBN 0-87170-007-7 (v. 1)

SAN 204-7586

Printed in the United States of America

Foreword to the Fourth Printing

With the fourth printing of this Volume, it takes its place in the new *ASM Handbook* series. The *ASM Handbook* was established to build upon the proud tradition of the *Metals Handbook* and to position the series to meet the needs of future engineers, researchers, technicians, and students. The *ASM Handbook* series will encompass volumes from both the 9th and 10th Editions of *Metals Handbook*—as well as new and revised Volumes as they are released—in order to establish one comprehensive set of reference materials. This will allow a much more flexible approach to updating information: technological advances will be the impetus behind Volume revisions or the addition of new Volumes to the series. The title of the new *ASM Handbook* series reflects the increasingly interrelated nature of materials technology and emphasizes the position of ASM International as the premier source of authoritative materials information.

The Editors

Foreword to the First through Third Printings

Advances in the field of materials science have resulted in the improvement of existing materials and the development of new materials—along with a variety of new applications. Most major branches of engineering, particularly those involved in the design and construction of new mechanical or structural elements, depend on the results of mechanical tests for measurements of material properties.

For accurate analysis and use of test data, it is important that engineers, even those not engaged in actual testing work, possess a general knowledge of common test methods and an understanding of the applicability, advantages, and limitations of a given test procedure. This latest addition to the distinguished *Metals Handbook* series is intended to provide such an understanding; in it is detailed the maturing technical development of mechanical testing.

The production of this Handbook is a major accomplishment for all those involved. Although the 1948 and 1985 single-volume editions of *Metals Handbook* each include a section on mechanical testing, and Volumes 10 and 11 of the 8th edition series touch on various aspects of the subject, this marks the first time an entire Handbook volume has been devoted exclusively to this important technology.

The Society owes a debt of gratitude to the more than 200 authors and reviewers who volunteered their time and expertise to this undertaking. Under the guidance of John R. Newby of Armco Inc. and the ASM Handbook Committee, and in collaboration with the Handbook's editorial staff, they have produced for the technical community an authoritative reference work.

M. Brian Ives
President

Edward L. Langer
Managing Director

The Ninth Edition of Metals Handbook
is dedicated to the memory of
TAYLOR LYMAN, A.B. (Eng.), S.M., Ph.D.
(1917—1973)
Editor, Metals Handbook, 1945—1973

Preface

Mechanical testing of materials is a relatively new area of metallurgy. Machines for pulling specimens to determine a material's breaking load were developed in the mid-18th century. One such machine designed by Benjamin Franklin in 1765 is on exhibit at the Franklin Institute in Philadelphia, adjacent to the headquarters of the American Society for Testing and Materials. Franklin, as every schoolchild knows, was also interested in electricity—another important aspect of metallurgy.

Both "test" and "electricity" have ancient origins. Test was the name for the vessels in which 17th-century alchemists assayed metals. The word electricity is derived from the Greek term for amber *(electros),* which was used by Pliny the Younger around 100 A.D. to produce static charges.

However, it was not until the 20th century that the use of either mechanical testing or electricity was fully developed. Tests of mechanical properties were not performed until structures began failing under conditions not previously encountered—such as the famous galloping cast iron bridge in Scotland that collapsed as a train was crossing a firth. It was discovered that holes in the castings had been filled with a black material to make them appear sound. Modern testing machines could not operate without powerful electric motors that can be precisely controlled. As will be seen from the procedures described in this Handbook, electrical resistance strain gages, computers, and other devices are the basis for the most recent advances in mechanical testing.

Mechanical testing is a game of numbers. Numbers are assigned to everything from grains to ultimate tensile strength. Users want to know the various strengths of the materials they are considering and how those strengths relate to the practical problems of forming parts. Designers want a reliable number so that structures will be neither underdesigned nor overdesigned. A number by itself, however, is of no value unless it is correctly related to some test parameter.

Visual observation is the primary mechanical test. A bent beam is easily detected, but questions arise. Is it supposed to be bent? Is it static or moving? What is its function? The next step is to do something to the object, such as strike it, and observe whether it moves, dents, or breaks. This usually requires a tool; the two most common mechanical testing tools are the universal testing machine and the hardness testing machine. A hardness test is much less expensive to perform. Universal testing machines, with their associated equipment and their need for skilled operators, are so costly that most fabricating shops cannot afford them. The numbers that can be generated from such machines, however, are extensively used in selecting materials. For this reason, tension testing is covered in the first major section of this Handbook, as it is the most frequently employed material evaluation test method.

Some of the test methods described later in the Handbook are so difficult to carry out that only a limited number of laboratories are equipped to perform them, and fabricators must rely on their results. In certain structures the properties determined by these methods are critical, but one cannot use a numerical value unless the basis for it is understood. We have asked experts in these fields to describe these tests with the user of the resulting values in mind.

The reader will note that the emphasis is on metals. This does not preclude the use of other materials such as organic plastics or elastomers as specimens. The principles and requirements in determining a mechanical property remain unchanged.

Standards and test methods established by ASTM and other more specialized societies for welding, nondestructive testing, and specific materials are referenced. This Handbook does not attempt to preempt these methods or to rewrite the procedures. The authors have described the tests, why they are made, and how the results are used to evaluate materials. Variations from standard procedures and special techniques to solve problems associated with obtaining valid results are included. Special conditions to simulate in-service environments or external forces that may influence a test are considered, and suggestions are made as to how critical these may be for design and function in service.

In summary, this is a book designed to assist a trained investigator in performing mechanical tests and in selecting the most cost-effective equipment to obtain valid results. For those not equipped to perform mechanical tests of a specific nature critical to their needs, a description of the procedures and interpretation of results is given with adequate reference to in-depth literature on these subjects.

John R. Newby
Coordinator

Policy on Units of Measure

By a resolution of its Board of Trustees, the American Society for Metals has adopted the practice of publishing data in both metric and customary U.S. units of measure. In preparing this Handbook, the editors have attempted to present data primarily in metric units based on Système International d'Unités (SI), with secondary mention of the corresponding values in customary U.S. units. The decision to use SI as the primary system of units was based on the aforementioned resolution of the Board of Trustees, the widespread use of metric units throughout the world, and the expectation that the use of metric units in the United States will increase substantially during the anticipated lifetime of this Handbook.

For the most part, numerical engineering data in the text and in tables are presented in SI-based units with the customary U.S. equivalents in parentheses (text) or adjoining columns (tables). For example, pressure, stress, and strength are shown both in SI units, which are pascals (Pa) with a suitable prefix, and in customary U.S. units, which are pounds per square inch (psi). To save space, large values of psi have been converted to kips per square inch (ksi), where one kip equals 1000 pounds. Some strictly scientific data are presented in SI units only.

On graphs and charts, grids correspond to SI-based units, which appear along the left and bottom edges; where appropriate, corresponding customary U.S. units appear along the top and right edges.

Data pertaining to a specification published by a specification-writing group may be given in only the units used in that specification or in dual units, depending on the nature of the data. For example, the typical yield strength of aluminum sheet made to a specification written in customary U.S. units would be presented in dual units, but the thickness specified in that specification might be presented only in inches.

Dual units of measure also are not provided in some articles dealing with statistical analyses of material properties. For example, stress levels utilized in fatigue or creep numerical data analysis are given in either SI or customary U.S. units, depending on the measurement system used in the original test program.

Data obtained according to standardized test methods for which the standard recommends a particular system of units are presented in the units of that system. For example, the diameters of steel ball indenters used in Rockwell hardness testing are given in inches, followed by the equivalent diameters in millimeters. Conversion factors for these and other units of measure can be found in the Appendix "Metric and Conversion Data for Mechanical Testing" in this Volume.

Conversions and rounding have been done in accordance with ASTM Standard E 380, with careful attention to the number of significant digits in the original data. For example, a test temperature of 2050 °F contains three significant digits. In this instance, the equivalent temperature would be given as 1120 °C; the exact conversion to 1121.11 °C would not be appropriate. For an invariant physical phenomenon that occurs at a precise temperature (such as the melting of pure silver), it would be appropriate to report the temperature as 961.93 °C or 1763.5 °F. In many instances (especially in tables and data compilations), temperature values in °C and °F are alternatives rather than conversions.

The policy on units of measure in this Handbook contains several exceptions to strict conformance to ASTM E 380; in each instance, the exception has been made to improve the clarity of the Handbook. The most notable exception is the use of $MPa\sqrt{m}$ rather than $MPa \cdot m^{-3/2}$ as the SI unit of measure for fracture toughness. Other examples of such exceptions are the use of ''L'' rather than ''l'' as the abbreviation for liter, and the use of g/cm^3 rather than kg/m^3 as the unit of measure for density (mass per unit volume).

SI practice requires that only one virgule (diagonal) appear in units formed by combination of several basic units. Therefore, all of the units preceding the virgule are in the numerator and all units following the virgule are in the denominator of the expression; no parentheses are required to prevent ambiguity.

Authors and Reviewers

Donald F. Adams
University of Wyoming
Albert Anctil
Army Materials & Mechanics Research Center
Peter L. Andresen
General Electric Company
Philip E. Armstrong
Los Alamos National Laboratory
John A. Aurentz
Briles Rivet Corporation
John A. Bailey
North Carolina State University
Gordon Baker
NewAge Industries Inc.
Robert G. Baur
General Electric Company
Raymond G. Bayer
IBM Corporation
William T. Becker
University of Tennessee
Alan P. Berens
University of Dayton Research Institute
S.J. Bless
University of Dayton Research Institute
Bruce Boardman
Deere & Company
M.K. Booker
Oak Ridge National Laboratory
James P. Bosscher
Calvin College
William G. Brazier
Ford Motor Company
Kevin R. Brown
Kaiser Aluminum & Chemical Corporation .
John T. Cammett, III
Metcut Research Associates, Inc.
James E. Campbell
Technical Consultant
Norman L. Carroll
Applied Test Systems, Inc.
Thomas E. Caton
David Taylor Naval Ship Research & Development Center
J.A. Chivinsky
Allegheny Ludlum Steel Corporation
Bruce W. Christ
National Bureau of Standards
Rodney J. Clifton
Brown University

Louis Coffin
General Electric Company
Joseph B. Conway
Mar-Test, Inc.
Carl Cross
Colorado School of Mines
William H. Cullen
Materials Engineering Associates
Ernest J. Czyryca
David Taylor Naval Ship Research & Development Center
P. Dadras
Wright State University
Craig B. Dallam
Colorado School of Mines
Isaac M. Daniels
Illinois Institute of Technology
Robert E. Davies
Alcoa Laboratories
R.F. Davis
North Carolina State University
Anthony DeBellis
United Calibration Corporation
William Degman
Sikorsky Aircraft Division
United Technologies Corporation
Robert J. Dexter
Southwest Research Institute
George E. Dieter
University of Maryland
Gordon B. Dudder
Battelle Pacific Northwest Laboratories
J. Duffy
Brown University
John L. Duncan
McMaster University
D.J. Duquette
Rensselaer Polytechnic Institute
T. Elverso
Satec Systems, Inc.
Richard C. Erickson
Battelle Columbus Laboratories
David C. Erlich
SRI International
Andrew R. Fee
Page-Wilson Corporation
W.N. Findley
Brown University
Stephan A. Finnegan
Naval Weapons Center

Iain Finnie
University of California at Berkeley
Paul S. Follansbee
Los Alamos National Laboratory
F. Peter Ford
General Electric Company
William L. Fourney
University of Maryland
G.R. Fowles
Washington State University
Alfred Fox
AT&T Bell Laboratories
Charles Fry
Satec Systems, Inc.
William J. Galbraith
Shore Instrument & Manufacturing Company, Inc.
Richard P. Gangloff
Exxon Research and Engineering Company
Harold L. Gegel
AFWAL Materials Laboratory
William W. Gerberich
University of Minnesota
Peter P. Gillis
University of Kentucky
Jacques Henri Joseph Giovanola
SRI International
R.P. Goel
AMP, Inc.
R.J. Goode
Naval Research Laboratories
Todd S. Gross
University of Kentucky
A. Guha
Brush Wellman Inc.
George T. Hahn
Vanderbilt University
Duncan L. Hammon
Los Alamos National Laboratory
John Harding
Oxford University
Clayton L. Harmsworth
AFWAL Materials Laboratory
K.A. Hartley
Brown University
R.H. Hawley
Brown University
Thomas G. Herberling
Armco, Inc.

John L. Herron
Herron Testing Laboratories, Inc.
B.M. Hillberry
Purdue University
Richard G. Hoagland
Ohio State University
John E. Hockett
John E. Hockett, Inc.
Dennis Hogan
Satec Systems, Inc.
William F. Hosford
University of Michigan
J. Hoyt
AMP, Inc.
Steve J. Hudak, Jr.
Southwest Research Institute
Charles G. Interrante
National Bureau of Standards
George R. Irwin
University of Maryland
David Jablonski
Instron Corporation
Lee A. James
Westinghouse Hanford Company
Ken Jerina
Washington University in St. Louis
John J. Jonas
McGill University
J. Wayne Jones
University of Michigan
J.F. Kalthoff
Fraunhofer-Institut für
Werkstoffmechanik
S. Keeler
National Steel Corporation
Ran Y. Kim
University of Dayton Research Institute
Richard W. Klopp
Brown University
Rodney H. Klundt
Latrobe Steel Company
Karen L. Knight
IBM Corporation
Lawrence J. Korb
Rockwell International
E. Krempl
Rensselaer Polytechnic Institute
Howard A. Kuhn
University of Pittsburgh
G.D. Lahoti
The Timken Company
Carl W. Lawton
Combustion Engineering, Inc.
Gail E. Leese
NASA Lewis Research Center
William F. Leyda
Technical Consultant
Peter K. Liaw
Westinghouse Research and Development
Center
Ulric S. Lindholm
Southwest Research Institute
M.R. Louthan, Jr.
Virginia Polytechnic Institute

Kenneth C. Ludema
University of Michigan
Douglas P. Lutz
Krautkramer Branson
Lorenzo C. Majno
Instron Corporation
Frank N. Mandigo
Olin Corporation
Douglas Mann
MTS Systems Corporation
Charles W. Marschall
Battelle Columbus Laboratories
Francis J. Marsh
Bethlehem Steel Corporation
James J. Martin
Instron Corporation
I.A. Martorell
Wright Field Operations
Westinghouse Electric Corporation
David K. Matlock
Colorado School of Mines
Frank A. McClintock
Massachusetts Institute of Technology
Robert McDemus
Carpenter Technology Corporation
Paul A. Meyer
Krautkramer Branson
Richard A. Meyer
MTS Systems Corporation
Terry Morton
David Taylor Naval Ship Research &
Development Center
Lawrence E. Murr
Oregon Graduate Center
Michael S. Nagorka
Colorado School of Mines
Michael V. Nathal
NASA Lewis Research Center
Mary G. Natrella
National Bureau of Standards
John R. Newby
Armco, Inc.
Theodore Nicholas
AFWAL Materials Laboratory
Harmon D. Nine
General Motors Research Laboratories
James L. Oess
Dana Corporation
Neal R. Ontko
AFWAL Materials Laboratory
Peter S. Pao
McDonnell Douglas Research
Laboratories
Ralph Papirno
Army Materials & Mechanics Research
Center
Won J. Park
Wright State University
Cecil E. Parsons
Boeing Commercial Airplane Company
Jerry Petrak
AFWAL Materials Laboratory
Alan Porter
Rocketdyne Company

L. Raymond
METTEK Laboratories
Richard C. Rice
Battelle Columbus Laboratories
Randy Riddle
Satec Systems, Inc.
Robert O. Ritchie
University of California at Berkeley
Blaine W. Roberts
Combustion Engineering
Alan R. Rosenfield
Battelle Columbus Laboratories
Lewis D. Roth
Westinghouse Electric Research &
Development Center
Paul E. Ruff
Battelle Columbus Laboratories
Steven W. Rust
Battelle Columbus Laboratories
John A. Schey
University of Waterloo
Thomas V. Schill
Huntington Alloys, Inc.
James E. Scott
Bethlehem Steel Corporation
Peter M. Scott
U.K. Atomic Energy Authority
Robert Segabache
Page-Wilson Corporation
M. Semchyshen
Climax Molybdenum Company
S.L. Semiatin
Battelle Columbus Laboratories
William L. Server
EG&G Idaho, Inc.
Eugene Shapiro
Olin Corporation
Michael M. Shea
General Motors Research Laboratories
Donald H. Sherman
Caterpillar Tractor Company
C.F. Shih
Brown University
Donald A. Shockey
SRI International
G.W. Simmons
Lehigh University
Lars H. Sjodahl
Mar-Test, Inc.
Darryl F. Socie
University of Illinois at Urbana
Carlo Sonnino
Emerson Electric Company
T. Spalvins
NASA Lewis Research Center
Donald O. Sprowls
Alcoa Laboratories (retired)
M.R. Staker
Army Materials & Mechanics Research
Center
R.H. Stentz
Mar-Test, Inc.
Robert S. Strimel
Tinius Olsen Testing Machine Company,
Inc.

Jeffrey W. Swegle
Sandia Laboratories

Robert W. Swindeman
Union Carbide Corporation
Oak Ridge National Laboratory

Brian Taylor
General Motors Corporation

John K. Tien
Columbia University

David G. Tipton
Laque Center for Corrosion Technology,
Inc.

Edward L. Tobolski
Page-Wilson Corporation

R.C. Tucker
Union Carbide Corporation

A. Turnbull
National Physical Laboratory (U.K.)

David A. Utah
General Electric Company

W.A. Van der Sluys
Babcock & Wilcox Research Center

Howard R. Voorhees
Materials Technology Corporation

R.P. Wei
Lehigh University

Edward T. Wessel
Technical Consultant

J. Daniel Whittenberger
NASA Lewis Research Center

Roy Williams
General Electric Company

William Wood
Oregon Graduate Center

Clyde E. Work
Western New England College

Peter J. Wray
Inland Steel Research Laboratory

George R. Yoder
Naval Research Laboratories

Robert E. Zinkham
Reynolds Metals Company

Contents

Terms and Definitions

A

abrasion. Roughening or scratching of a surface due to abrasive wear.

abrasive wear. The removal or displacement of material from a surface when hard particles slide or roll across the surface under pressure. The particles may be loose or may be part of another surface in contact with the surface being worn. Contrast with *adhesive wear*.

accepted reference value. A value that serves as an agreed-on reference for comparison and which is derived as: (1) a theoretical or established value, based on scientific principles, (2) an assigned value, based on experimental work of some national or international standards organization, or (3) a consensus value, based on collaborative experimental work under the auspices of a scientific or engineering group. When the accepted reference value is the theoretical value, it is sometimes referred to as the "true" value.

accuracy. (1) The agreement or correspondence between an experimentally determined value and an accepted reference value for the material undergoing testing. The reference value may be established by an accepted standard (such as those established by ASTM), or in some cases the average value obtained by applying the test method to all the sampling units in a lot or batch of the material may be used. (2) The extent to which the result of a calculation or the reading of an instrument approaches the true value of the calculated or measured quantity.

adhesive wear. The removal or displacement of material from a surface by the welding together and subsequent shearing of minute areas of two surfaces that slide across each other under pressure. In advanced stages, may lead to galling. Contrast with *abrasive wear*.

aging. A change in a material's properties that occurs at ambient or elevated temperatures after hot working, cold working, or a heat treatment. See also *strain aging*.

alligator skin. See preferred term *orange peel*.

alligatoring. The longitudinal splitting of flat slabs in a plane parallel to the rolled surface. Also known as fishmouthing.

all-weld-metal test specimen. A test specimen wherein the portion being tested is composed wholly of weld metal.

angle of bend. The angle between the two legs of the specimen after bending is completed. It is measured before release of the bending force, unless otherwise specified.

anisotropy. A variation in the mechanical properties of a material relative to direction. See also *planar anisotropy*.

apparent area of contact. In *tribology*, the area of contact between two solid surfaces defined by the boundaries of their macroscopic interface. Contrast with *real area of contact*.

arbitration bar. A test bar, cast with a heat of material, used to determine chemical composition, hardness, tensile strength, and deflection and strength under transverse loading in order to establish the state of acceptability of the casting.

asperity. In *tribology*, a protuberance in the small-scale topographical irregularities of a solid surface.

average linear strain. See *engineering strain*.

axial strain. Increase (or decrease) in length resulting from a stress acting parallel to the longitudinal axis of the specimen.

B

batch. A definite quantity of some product or material produced under conditions that are considered uniform. A batch is usually smaller than a *lot*.

Bauschinger effect. The phenomenon by which plastic deformation increases yield strength in the direction of plastic flow and decreases it in other directions.

beach marks. Progression marks on a fracture surface that indicate successive position of the advancing crack front. The classic appearance is of irregular elliptical or semielliptical rings, radiating outward from one or more origins. Beach marks (also known as clamshell marks or tide marks) are typically found on service fractures where the part is loaded randomly, intermittently, or with periodic variations in mean stress or alternating stress.

bearing area. The product of the pin diameter and specimen thickness. See also *bearing stress*.

bearing strain. The ratio of the bearing deformation of the bearing hole, in the direction of the applied force, to the pin diameter.

bearing strength. The maximum bearing stress that a material is capable of sustaining.

bearing stress. The force per unit of *bearing area*.

bearing test. A method of determining the response to stress (load) of sheet products that are subjected to riveting, bolting, or a similar fastening procedure. The purpose of the test is to determine the *bearing strength* of the material and to measure the *bearing stress* versus the deformation of the hole created by a pin or rod of circular cross section that pierces the sheet perpendicular to the surface.

bearing yield strength. The *bearing stress* at which a material exhibits a specified limiting deviation from the proportionality of *bearing stress* to *bearing strain*.

Beilby layer. A layer of metal disturbed by mechanical working presumed to be without regular crystalline structure (amorphous); originally applied to grain boundaries.

bend radius. (1) The inside radius of a bent section. (2) The radius of a tool around which metal is bent during fabrication.

bend test. A test for determining the relative ductility of metal that is to be formed (usually sheet, strip, plate, or wire) or for determining soundness and toughness of metal (after welding, for example). The specimen is usually bent over a specified diameter through a specified angle for a specified number of cycles. There are four general types of bend tests, named according to the manner in which the forces are applied to the specimen to make the bend: *free bend*, *guided bend*, *semiguided bend*, and *wrap-around bend*.

bias. A systematic error that contributes to the difference between a population mean of the measurements or test results and an accepted reference or true value.

brale. An indenter of specified spheroconical shape used with a Rockwell hardness tester. This indenter is used for the A, C, D, and N scales for testing hard metals.

breaking load. The maximum load (or force) applied to a test specimen or structural member loaded to rupture.

breaking stress. See *rupture stress*.

Brinell hardness number, HB. A number related to the applied load and to the surface area of the permanent impression made by a ball indenter computed from the equation:

$$HB = \frac{2P}{\pi D(D - \sqrt{D^2 - d^2})}$$

where P is applied load, kgf; D is diameter of ball, mm; and d is mean diameter of the impression, mm.

Brinell hardness test. A test for determining the hardness of a material by forcing a hard steel or carbide ball of specified diameter into it under a specified load. The result is expressed as the *Brinell hardness number*.

Brinelling. Damage to a solid bearing surface characterized by one or more plastically formed indentations brought about by overload. This term is often applied in the case of rolling element bearings.

brittle crack propagation. A very sudden propagation of a crack with the absorption of no energy except that stored elastically in the body. Microscopic examination may reveal some deformation not noticeable to the unaided eye.

brittle erosion behavior. Erosion behavior having characteristic properties (e.g., little or no plastic flow, the formation of cracks) that can be associated with brittle fracture of the exposed surface. The maximum volume removal occurs at an angle near 90°, in contrast to approximately 25° for *ductile erosion behavior*.

brittle fracture. Separation of a solid accompanied by little or no macroscopic plastic deformation. Typically, brittle fracture occurs by rapid crack propagation with less expenditure of energy than for ductile fracture.

buckling. A compression phenomenon that occurs when, after some critical level of load, a bulge, bend, bow, kink, or other wavy condition is produced in a beam, column, plate, bar, or sheet product form.

bulk modulus. See *bulk modulus of elasticity*.

bulk modulus of elasticity, K. The measure of resistance to change in volume; the ratio of hydrostatic stress to the corresponding unit change in volume. This elastic constant can be expressed by the equation:

$$K = \frac{\sigma_m}{\Delta} = -\frac{p}{\Delta} = \frac{1}{\beta}$$

where K is bulk modulus of elasticity, σ_m is hydrostatic or mean stress tensor, p is hydrostatic pressure, and β is compressibility. Also known as bulk modulus, compression modulus, hydrostatic modulus, and volumetric modulus of elasticity.

button. That part of a weld, including all or part of the weld nugget, that tears out in the destructive testing of a spot, seam, or projection welded specimen.

c

calibrate. To determine, by measurement or comparison with a standard, the correct value of each scale reading on a measuring (test) instrument.

calibration. Determination of the values of the significant parameters by comparison with values indicated by a reference instrument or by a set of reference standards.

capacity. In tensile testing machines, the maximum load and/or displacement for which a machine is designed. Some testing machines have more than one load capacity. These are equipped with accessories that allow the capacity to be modified as desired.

catastrophic wear. Rapidly occurring or accelerating surface damage, deterioration, or change of shape caused by wear to such a degree that the service life of a part is appreciably shortened or its function is destroyed.

caustic cracking. A form of stress corrosion cracking most frequently encountered in carbon steels or iron-chromium-nickel alloys that are exposed to concentrated hydroxide solutions at temperatures of 200 to 250 °C (400 to 480 °F). Also known as caustic embrittlement.

caustic embrittlement. See preferred term *caustic cracking*.

cavitation damage. Erosion of a solid surface through the formation and collapse of cavities in an adjacent liquid.

cavitation erosion. See preferred term *cavitation damage*.

chafing fatigue. Fatigue initiated in a surface damaged by rubbing against another body. See also *fretting*.

characteristic. A property of items in a sample or population that when measured, counted, or otherwise observed helps to distinguish between the items.

Charpy test. An impact test in which a V-notched, keyhole-notched, or U-notched specimen, supported at both ends, is struck behind the notch by a striker mounted at the lower end of a bar that can swing as a pendulum. The energy that is absorbed in fracture is calculated from the height to which the striker would have risen had there been no specimen and the height to which it actually rises after fracture of the specimen. Contrast with *Izod test*.

chevron pattern. A fractographic pattern of radial marks (shear ledges) that looks like nested letters "V"; sometimes called a herringbone pattern. Chevron patterns are typically found on brittle fracture surface in parts whose widths are considerably greater than their thicknesses. The points of the chevrons can be traced back to the fracture origin.

chord modulus. The slope of the chord drawn between any two specific points on a stress-strain curve. See also *modulus of elasticity*.

clamshell marks. See *beach marks*.

cleavage. The tendency to cleave or split along definite crystallographic planes.

cleavage fracture. A fracture, usually of a polycrystalline metal, in which most of the grains have failed by cleavage, resulting in bright reflecting facets. It is one type of crystalline fracture and is associated with low-energy brittle fracture. Contrast with *shear fracture*.

coefficient of compressibility. See *bulk modulus of elasticity*.

coefficient of friction, μ. The ratio of the force resisting tangential motion between two bodies to the normal force pressing these bodies together.

component of variance. A part of a total variance identified with a specified source of variability.

compression modulus. See *bulk modulus of elasticity*.

compression test. A method for assessing the ability of a material to withstand compressive loads.

compressive strength. Maximum compressive stress a material is capable of developing. With a brittle material that fails in compression by fracturing, the compressive strength has a definite value. In the case of ductile, malleable, or semiviscous materials (which do not fail in compression by a shattering fracture), the value obtained for compressive strength is an arbitrary value dependent on the degree of distortion that is regarded as effective failure of the material.

compressive stress. A stress that causes an elastic body to deform (shorten) in the direction of the applied load. Contrast with *tensile stress*.

compressometer. Instrument for measuring change in length over a given gage length caused by application or removal of a force. Commonly used in compression testing of metal specimens.

constant life fatigue diagram. A plot (usually on rectangular coordinates) of a family of curves, each of which is for a single fatigue life (number of cycles), relating alternating stress, maximum stress, minimum stress, and mean stress. The constant life fatigue diagram is generally derived from a family of *S-N* curves, each of which represents a different stress ratio for a 50% probability of survival. See also *nominal stress, maximum stress, minimum stress, S-N curve, fatigue life,* and *stress ratio.*

constraint. Any restriction to the deformation of a body.

conventional strain. See *engineering strain.*

conventional stress. See *engineering stress.*

correction. In the case of a testing machine, the difference obtained by subtracting the indicated load from the correct value of the applied load.

corrosion fatigue. Cracking produced by the combined action of repeated or fluctuating stress and a corrosive environment at lower stress levels or fewer cycles than would be required in the absence of a corrosive environment.

corrosive wear. Wear in which chemical or electrochemical reaction with the environment is significant.

coupon. A piece of metal from which a test specimen is to be prepared—often an extra piece (as on a casting or forging) or a separate piece made for test purposes (such as a test weldment).

crack extension, Δa. An increase in *crack size.* See also *physical crack size, effective crack size, original crack size,* and *crack length.*

crack-extension force, G. The elastic energy per unit of new separation area that would be made available at the front of an ideal crack in an elastic solid during a virtual increment of forward crack extension. This definition is useful for either static cracks or running cracks. From past usage, crack extension force is commonly associated with linear-elastic methods of analysis. See also *J-integral.*

crack-extension resistance, K_R. A measure of the resistance of a material to *crack extension* expressed in terms of the *stress-intensity factor,* the *crack-extension force,* or values of *J* derived using the *J-integral* concept.

crack length (depth), a. In fatigue and stress corrosion cracking, the *physical crack size* used to determine the crack growth rate and the *stress-intensity factor.*

For the compact-type specimen, crack length is measured from the line connecting the bearing points of load application. For the center-cracked-tension specimen, crack length is measured from the perpendicular bisector of the central crack.

crack mouth opening displacement (CMOD). See *crack opening displacement.*

crack opening displacement (COD). On a K_{Ic} specimen, the opening displacement of the notch surfaces at the notch and in the direction perpendicular to the plane of the notch and the crack. The displacement at the tip is called the crack tip opening displacement (CTOD); at the mouth, it is called the crack mouth opening displacement (CMOD).

crack plane orientation. An identification of the plane and direction of a fracture in relation to product geometry. This identification is designated by a hyphenated code, the first letter(s) representing the direction normal to the crack plane and the second letter(s) designating the expected direction of crack propagation.

crack size, a. A lineal measure of a principal planar dimension of a crack. This measure is commonly used in the calculation of quantities descriptive of the stress and displacement fields. In practice, the value of crack size is obtained from procedures for measurement of *physical crack size, original crack size,* or *effective crack size,* as appropriate to the situation under consideration. See also *crack length.*

crack tip opening displacement (CTOD). See *crack opening displacement.*

crack-tip plane strain. A stress-strain field, near a crack tip, that approaches *plane strain* to the degree required by an empirical criterion.

creep. Time-dependent strain occurring under stress. The *creep strain* occurring at a diminishing rate is called primary or transient creep; that occurring at a minimum and almost constant rate, secondary or steady-rate creep; that occurring at an accelerating rate, tertiary creep.

creep rate. The slope of the creep-time curve at a given time determined from a Cartesian plot.

creep recovery. The time-dependent decrease in strain in a solid, following the removal of force. Recovery is usually determined at constant temperature.

creep-rupture strength. The stress that will cause fracture in a creep test at a given time in a specified constant environment. Also known as stress-rupture strength.

creep-rupture test. A test in which progressive specimen deformation and the time for rupture are both measured. In general, deformation is much greater than that de-

veloped during a creep test. Also known as stress-rupture test.

creep strain. The time-dependent total strain (extension plus initial gage length) produced by applied stress during a creep test.

creep strength. The stress that will cause a given creep strain in a creep test at a given time in a specified constant environment.

creep stress. The constant load divided by the original cross-sectional area of the specimen.

creep test. A method of determining the extension of metals under a given load at a given temperature. The determination usually involves the plotting of time-elongation curves under constant load; a single test may extend over many months. The results are often expressed as the elongation (in millimeters or inches) per hour on a given gage length (e.g., 25 mm or 1 in.).

crushing test. (1) A radial compressive test applied to tubing, sintered-metal bearings, or other similar products for determining radial crushing strength (maximum load in compression). (2) An axial compressive test for determining quality of tubing, such as soundness of weld in welded tubing.

crystalline fracture. A pattern of brightly reflecting crystal facets on the fracture surface of a polycrystalline metal, resulting from cleavage fracture of many individual crystals. Contrast with *fibrous fracture* and *silky fracture.* See also *granular fracture.*

crystallographic cleavage. The separation of a crystal along a plane of fixed orientation relative to the three-dimensional crystal structure within which the separation process occurs, with the separation process causing the newly formed surfaces to move away from one another in directions containing major components of motion perpendicular to the fixed plane.

cup fracture (cup-and-cone fracture). A mixed-mode fracture, often seen in tensile test specimens of a ductile material, where the central portion undergoes plane-strain fracture and the surrounding region undergoes plane-stress fracture. One of the mating fracture surfaces looks like a miniature cup; it has a central depressed flat-face region surrounded by a shear lip. The other fracture surface looks like a miniature truncated cone.

cupping. The fracture of severely cold worked rods or wire where one end has the appearance of a cup and the other that of a cone.

cupping test. A mechanical test used to determine the ductility and drawing properties of sheet metal. It consists of measuring the maximum depth of bulge that can be formed before fracture. The test is commonly carried out by drawing the test piece into a circular die by means of a punch with a hemispherical end. See also *cup fracture*, *Erichsen test*, and *Olsen test*.

cycle. In fatigue, one complete sequence of values of applied load that is repeated periodically. The symbol N represents the number of cycles.

cyclic loads. Loads that change value by following a regular repeating sequence of change.

D

decarburization. Loss of carbon from the surface layer of a carbon-containing alloy due to reaction with one or more chemical substances in a medium that contacts the surface.

deformation. A change in the form of a body due to stress, thermal change, change in moisture, or other causes. Measured in units of length.

deformation bands. Parts of a crystal that have rotated differently during deformation to produce bands of varied orientation within individual grains.

dendrite. A crystal that has a treelike branching pattern, being most evident in cast metals slowly cooled through the solidification range.

density, absolute. The mass per unit volume of a solid material, expressed in g/cm^3 or kg/m^3.

dezincification. Corrosion in which zinc is selectively leached from zinc-containing alloys. Most commonly found in copper-zinc alloys containing less than 85% copper after extended service in water containing dissolved oxygen.

diamond pyramid hardness test. See *Vickers hardness test*.

dilatometer. An instrument for measuring length or volume changes in a solid metal during heating and subsequent cooling or isothermal holding.

dimple rupture. A fractographic term describing ductile fracture that occurs through the formation and coalescence of microvoids along the fracture path. The fracture surface of such a ductile fracture appears dimpled when observed at high magnification and usually is most clearly resolved when viewed in a scanning electron microscope. See also *ductile fracture*.

discontinuous yielding. The nonuniform plastic flow of a metal exhibiting a yield point in which plastic deformation is inhomogeneously distributed along the gage length. Under some circumstances, it may occur in metals not exhibiting a distinct

yield point, either at the onset of or during plastic flow.

displacement. The distance that a chosen measurement point on the specimen displaces normal to the crack plane. See also *crack opening displacement* and *crack plane orientation*.

distortion. Any deviation from an original size, shape, or contour that occurs because of the application of stress or the release of residual stress.

ductile crack propagation. Slow crack propagation that is accompanied by noticeable plastic deformation and requires energy to be supplied from outside the body.

ductile erosion behavior. Erosion behavior having characteristic properties (i.e., considerable plastic deformation) that can be associated with ductile fracture of the exposed solid surface. A characteristic ripple pattern forms on the exposed surface at low values of angle of attack. Contrast with *brittle erosion behavior*.

ductile fracture. Fracture characterized by tearing of metal accompanied by appreciable gross plastic deformation and expenditure of considerable energy.

ductility. The ability of a material to deform plastically before fracturing. Measured by elongation or reduction in area in a tensile test, by height of cupping in a cupping test, or by the radius or angle of bend in a bend test. See also *cupping*.

E

edge distance. The distance from the edge of a bearing specimen to the center of the hole in the direction of applied force. See also *bearing test*.

edge distance ratio. The ratio of the edge distance to the pin diameter in a bearing test. See also *bearing test*.

edge strain. Transverse strain lines or Lüders lines ranging from 25 to 300 mm (1 to 12 in.) in from the edges of cold rolled steel sheet or strip. See also *Lüders lines*.

effective crack size, a_e. The *physical crack size* augmented for the effects of cracking plastic deformation. Sometimes the effective crack size is calculated from a measured value of a *physical crack size* plus a calculated value of a *plastic-zone adjustment*. A preferred method for calculation of effective crack size compares compliance from the secant of a load-deflection trace with the elastic compliance from a calibration for the type of specimen.

effective yield strength. An assumed value of uniaxial yield strength that represents the influence of plastic yielding on fracture test parameters.

elastic calibration device. A device for use in verifying the load readings of a testing machine consisting of an elastic member(s) to which loads may be applied, combined with a mechanism or device for indicating the magnitude (or a quantity proportional to the magnitude) of deformation under load.

elastic constants. The factors of proportionality that relate elastic displacement of a material to applied forces. See also *modulus of elasticity*, *bulk modulus of elasticity*, *shear modulus*, and *Poisson's ratio*.

elastic deformation. A change in dimensions directly proportional to and in phase with an increase or decrease in applied force.

elastic energy. The amount of energy required to deform a material within the elastic range of behavior, neglecting small heat losses due to internal friction. The energy absorbed by a specimen per unit volume of material contained within the gage length being tested. It is determined by measuring the area under the stress-strain curve up to a specified elastic strain. See also *modulus of resilience* and *strain energy*.

elastic hysteresis. See preferred term *mechanical hysteresis*.

elastic limit. The maximum stress which a material is capable of sustaining without any permanent strain (deformation) remaining upon complete release of the stress. See also *proportional limit*.

elastic recovery. In hardness testing, the shortening of the original dimensions of the indentation upon release of the applied load.

elastic resilience. The amount of energy absorbed in stressing a material up to the elastic limit; or, the amount of energy that can be recovered when stress is released from the elastic limit.

elastic strain. See *elastic deformation*.

elastic strain energy. The energy expended by the action of external forces in deforming a body elastically. Essentially all the work performed during elastic deformation is stored as elastic energy, and this energy is recovered upon release of the applied force.

elasticity. The property of a material by virtue of which deformation caused by stress disappears upon removal of the stress. A perfectly elastic body completely recovers its original shape and dimensions after release of stress.

elongation. A term used in mechanical testing to describe the amount of extension of a test piece when stressed. See also *elongation, percent*.

elongation, percent. The extension of a uniform section of a specimen expressed as percentage of the original gage length:

$$\text{elongation, } \% = \frac{(L_x - L_o)}{L_o} \times 100$$

where L_o is original gage length, and L_x is final gage length.

end-quench hardenability test. A laboratory procedure for determining the hardenability of a steel or other ferrous alloy; widely referred to as the Jominy test. Hardenability is determined by heating a standard specimen above the upper critical temperature, placing the hot specimen in a fixture so that a stream of cold water impinges on one end, and, after cooling to room temperature is completed, measuring the hardness near the surface of the specimen at regularly spaced intervals along its length. The data are normally plotted as hardness versus distance from the quenched end.

endurance. The capacity of a material to withstand repeated application of stress.

endurance limit. The maximum stress below which a material can presumably endure an infinite number of stress cycles. The value of the *maximum stress*, and the *stress ratio* also should be stated. Compare with *fatigue limit*.

engineering strain, e. A term sometimes used for average linear strain or conventional strain in order to differentiate it from *true strain*. In tension testing it is calculated by dividing the change in the gage length by the original gage length.

engineering stress, s. A term sometimes used for conventional stress in order to differentiate it from *true stress*. In tension testing, it is calculated by dividing the breaking load applied to the specimen by the original cross-sectional area of the specimen.

equiaxed grain structure. A structure in which the grains have approximately the same dimensions in all directions.

Erichsen test. A *cupping test* used for assessing the ductility of sheet metal. The method consists of forcing a conical or hemispherical-ended plunger into the specimen and measuring the depth of the impression at fracture.

erosion. Progressive loss of original material from a solid surface due to mechanical interaction between that surface and a fluid, a multicomponent fluid, or impinging liquid or solid particles.

error. Deviation from the correct value. In the case of a testing machine, the difference obtained by subtracting the load indicated by the calibration device from the load indicated by the testing machine.

estimate. The particular value, or values, of a parameter computed by an estimation procedure for a given sample.

estimation. A procedure for making a statistical inference about the numerical values of one or more unknown population parameters from the observed values in a sample.

extensometer. An instrument for measuring changes in length over a given gage length caused by application or removal of a force. Commonly used in tension testing of metal specimens.

F

false Brinelling. Damage to a solid bearing surface characterized by indentations not caused by plastic deformation resulting from overload but thought to be due to other causes such as *fretting corrosion*. See also *Brinelling*.

fatigue. The phenomenon leading to fracture under repeated or fluctuating stresses having a maximum value less than the ultimate tensile strength of the material. *Fatigue failure* generally occurs at loads which applied statically would produce little perceptible effect. Fatigue fractures are progressive, beginning as minute cracks that grow under the action of the fluctuating stress.

fatigue crack growth rate, da/dN. The rate of crack extension caused by constant-amplitude fatigue loading, expressed in terms of crack extension per cycle of load application.

fatigue failure. Failure that occurs when a specimen undergoing fatigue completely fractures into two parts or has softened or been otherwise significantly reduced in stiffness by thermal heating or cracking.

fatigue life, N. The number of cycles of stress or strain of a specified character that a given specimen sustains before failure of a specified nature occurs.

fatigue life for p% survival. An estimate of the fatigue life that $p\%$ of the population would attain or exceed at a given stress level. The observed value of the median fatigue life estimates the fatigue life for 50% survival. Fatigue life for $p\%$ survival values, where p is any number, such as 95, 90, etc., may also be estimated from the individual fatigue life values.

fatigue limit. The maximum stress that presumably leads to fatigue fracture in a specified number of stress cycles. The value of the *maximum stress*, and the *stress ratio* also should be stated. See also *endurance limit*.

fatigue limit for p% survival. The limiting value of fatigue strength for $p\%$ survival as N becomes very large; p may be any number, such as 95, 90, etc.

fatigue notch factor, K_f (actual). The ratio

of the fatigue strength of an unnotched specimen to the fatigue strength of a notched specimen of the same material and condition; both strengths are determined at the same number of stress cycles.

fatigue notch sensitivity, q. An estimate of the effect of a notch or hole of a given size and shape on the fatigue properties of a material; measured by $q = (K_f - 1)/(K_t - 1)$, where K_f is the fatigue notch factor and K_t is the stress concentration factor. A material is said to be fully notch sensitive if q approaches a value of 1.0; it is not notch sensitive if the ratio approaches 0. See also *fatigue notch factor* and *stress concentration factor*.

fatigue strength. The maximum stress that can be sustained for a specified number of cycles without failure, the stress being completely reversed within each cycle unless otherwise stated.

fatigue strength at N cycles, S_N. A hypothetical value of stress for failure at exactly N cycles as determined from an S-N diagram. The value of S_N thus determined is subject to the same conditions as those which apply to the S-N diagram. The value of S_N that is commonly found in the literature is the hypothetical value of maximum stress, S_{max}, minimum stress, S_{min}, or stress amplitude, S_a, at which 50% of the specimens of a given sample could survive N stress cycles in which the mean stress, S_m, $= 0$. This is also known as the *median fatigue strength at N cycles*.

fatigue strength for p% survival at N cycles. An estimate of the stress level at which $p\%$ of the population would survive N cycles; p may be any number, such as 95, 90, etc. The estimates of the fatigue strengths for $p\%$ survival values are derived from particular points of the fatigue life distribution since there is no test procedure by which a frequency distribution of fatigue strengths at N cycles can be directly observed.

fatigue striations. Parallel lines frequently observed in electron microscope fractographs or fatigue fracture surfaces. The lines are transverse to the direction of local crack propagation; the distance between successive lines represents the advance of the crack front during the one cycle of stress variation.

fatigue test. A method for determining the range of alternating (fluctuating) stresses a material can withstand without failing.

fatigue wear. Wear of a solid surface caused by fracture arising from material fatigue.

ferrite banding. Parallel bands of free ferrite aligned in the direction of working. Sometimes referred to as ferrite streaks.

ferrograph. An instrument used to determine the size distribution of wear particles in lubricating oils of mechanical systems.

fibrous fracture. A gray and amorphous fracture that results when a metal is sufficiently ductile for the crystals to elongate before fracture occurs. When a fibrous fracture is obtained in an impact test, it may be regarded as definite evidence of toughness of the metal. Contrast with *crystalline fracture* and *silky fracture*.

fibrous structure. (1) In forgings, a structure revealed as laminations, not necessarily detrimental, on an etched section or as a ropy appearance on a fracture. It is not to be confused with the *ductile fracture* of a clean metal. (2) In wrought iron, a structure consisting of slag fibers embedded in ferrite. (3) In rolled steel plate stock, a uniform, fine-grained structure on a fractured surface, free of laminations or shale-type discontinuities. As contrasted with (1) above, it is virtually synonymous with *ductile fracture*.

file hardness. Hardness as determined by the use of a file of standardized hardness on the assumption that a material that cannot be cut with the file is as hard as, or harder than, the file. Files covering a range of hardnesses may be employed.

fish eyes. A discontinuity found on the fracture surface of a weld in steel that consists of small pore or inclusion surrounded by an approximately round, bright area.

fishmouthing. See *alligatoring*.

fixed-load or fixed-displacement crack extension force curves. Curves obtained from a fracture mechanics analysis for the test configuration, assuming a fixed applied load or displacement and generating a curve of *crack extension force* versus the *effective crack size* as the independent variable.

flare test. A test applied to tubing, involving a tapered expansion over a cone. Similar to *pin expansion test*.

flattening test. A quality test for tubing in which a specimen is flattened to a specified height between parallel plates.

flexibility. The quality or state of a material that allows it to be flexed or bent repeatedly without undergoing rupture. See also *flexure*.

flexure. A term used in the study of strength of materials to indicate the property of a body, usually a rod or beam, to bend without fracture. See also *flexibility*.

flow. When essentially parallel planes within an element of a material move (slip or slide) in parallel directions; occurs under the action of shearing stress. Continuous action in this manner, at constant volume and without disintegration of the material, is termed *yield*, *creep*, or *plastic deformation*.

flow lines. Texture showing the direction of metal flow during hot or cold working. Flow lines often can be revealed by etching the surface or a section of a metal part.

flow stress. The stress required to produce plastic deformation in a solid metal.

forged roll Scleroscope hardness number (HFRSc or HFRSd). A number related to the height of rebound of a diamond-tipped hammer dropped on a forged steel roll. It is measured on a scale determined by dividing into 100 units the average rebound of a hammer from a forged steel roll of accepted maximum hardness. See also *Scleroscope hardness number* and *Scleroscope hardness test*.

formability. The ease with which a metal can be shaped through plastic deformation. The evaluation of the formability of a metal involves measurement of strength and ductility, as well as the amount of deformation required to cause fracture. *Workability* is used interchangeably with formability; however, formability refers to the shaping of sheet metal, while workability refers to shaping materials by bulk deformation (i.e., forging or rolling).

forming limit diagram. A diagram on which the major strains at the onset of necking in sheet metal are plotted vertically and the corresponding minor strains are plotted horizontally. The onset-of-failure line divides all possible strain combinations into two zones: the "safe" zone, in which failure during forming is not expected, and the "failure" zone, in which failure during forming is expected.

fractography. Descriptive explanation of a fracture process, especially in metals, with specific reference to photographs of the fracture surface. Macrofractography involves photographs at low magnification; microfractography, at high magnification.

fracture. The irregular surface produced when a piece of metal is broken. See also *fracture test*, *granular fracture*, *fibrous fracture*, *silky fracture*, and *crystalline fracture*.

fracture mechanics. See *linear elastic fracture mechanics*.

fracture stress. The true normal stress on the minimum cross-sectional area at the beginning of fracture. This term usually applies to tension tests of unnotched specimens.

fracture test. Breaking a specimen and examining the fractured surface with the unaided eye or with a low-power microscope to determine such things as composition, grain size, case depth, or internal defects.

fracture toughness. A generic term for measures of resistance to extension of a crack. The term is sometimes restricted to results of fracture mechanics tests, which are directly applicable in fracture control. However, the term commonly includes results from simple tests of notched or precracked specimens not based on fracture mechanics analysis. Results from tests of the latter type are often useful for fracture control, based on either service experience or empirical correlations with fracture mechanics tests. See also *stress intensity factor*.

fragmentation. The subdivision of a grain into small, discrete crystallites outlined by a heavily deformed network of intersecting slips as a result of cold working. These small crystals or fragments differ from one another in orientation and tend to rotate to a stable orientation determined by the slip systems.

frame. A list, compiled for sampling purposes, that designates the items (units) of a population or universe to be considered in a study.

free bend. The bend obtained by applying forces to the ends of a specimen without the application of force at the point of maximum bending. In making a free bend, lateral forces first are applied to produce a small amount of bending at two points. The two bends, each a suitable distance from the center, are both in the same direction.

fretting. A type of wear that occurs between tight-fitting surfaces subjected to oscillation at very small amplitude. This type of wear can be a combination of *oxidative wear* and *abrasive wear*. See also *fretting corrosion*.

fretting corrosion. The deterioration at the interface between contacting surfaces as the result of corrosion and slight oscillatory slip between the two surfaces.

fretting fatigue. Fatigue fracture that initiates at a surface area where *fretting* has occurred.

frequency distribution. The way in which the frequencies of occurrence of members of a population, or a sample, are distributed according to the values of the variable under consideration.

G

gage length. The original length of that portion of the specimen over which strain or change of length is determined.

galling. A condition whereby excessive friction between high spots results in localized welding with subsequent spalling and a further roughening of the rubbing surfaces of one or both of two mating parts. See also *spalling*.

grain growth. An increase in the average

size of the grains in polycrystalline metal, usually as a result of heating at elevated temperature. See also *grain size*.

grain size. A measure of the areas or volumes of grains in a polycrystalline material, usually expressed as an average when the individual sizes are fairly uniform. In metals containing two or more phases, the grain size refers to that of the matrix unless otherwise specified. Grain size is reported in terms of number of grains per unit area or volume, average diameter, or as a grain-size number derived from area measurements.

granular fracture. A type of irregular surface produced when metal is broken that is characterized by a rough, grainlike appearance, rather than a smooth or fibrous one. It can be subclassified as transgranular or intergranular. This type of fracture is frequently called *crystalline fracture*; however, the inference that the metal broke because it "crystallized" is not justified, because all metals are crystalline when in the solid state. Contrast with *fibrous fracture* and *silky fracture*.

group. The specimens tested at one time, or consecutively, at one stress level. A group may comprise one or more specimens.

guided bend. The bend obtained by use of a plunger to force the specimen into a die in order to produce the desired contour of the outside and inside surfaces of the specimen.

guided bend test. A test in which the specimen is bent to a definite shape by means of a jig.

H

hardness. A measure of the resistance of a material to surface indentation or abrasion; may be thought of as a function of the stress required to produce some specified type of surface deformation. There is no absolute scale for hardness; therefore, to express hardness quantitatively, each type of test has its own scale of arbitrarily defined hardness. Indentation hardness can be measured by *Brinell, Rockwell, Vickers, Knoop,* and *Scleroscope hardness tests*.

Hartmann lines. See *Lüders lines*.

herringbone pattern. See *chevron pattern*.

hole expansion test. A simulative test in which a flat sheet specimen with a circular hole in its center is clamped between annular die plates and deformed by a punch, which expands and ultimately cracks the edge of the hole.

Hooke's Law. A material in which the stress is linearly proportional to strain is said to obey Hooke's law. See also *modulus of elasticity*.

hydrogen damage. A general term for the embrittlement, cracking, blistering, and hydride formation that can occur when hydrogen is present in some metals.

hydrogen embrittlement. A condition of low ductility or hydrogen-induced cracking in metals resulting from the presence of hydrogen.

hydrogen-induced delayed cracking. A term sometimes used to identify a form of hydrogen embrittlement in which a metal appears to fracture spontaneously under a steady stress less than the yield stress. There is usually a delay between the application of stress (or exposure of the stressed metal to hydrogen) and the onset of cracking. Also referred to as static fatigue.

hydrostatic modulus. See *bulk modulus of elasticity*.

hysteresis. The phenomenon of permanently absorbed or lost energy that occurs during any cycle of loading or unloading when a material is subjected to repeated loading.

I

ideal crack. A simplified model of a crack used in elastic stress analysis. In a stress-free body, the crack has two smooth surfaces that are coincident and join within the body along a smooth curve called the crack front; in two-dimensional representations, the crack front is called the crack tip.

ideal-crack-tip stress field. The singular stress field, infinitesimally close to the crack front, that results from the dominant influence of an ideal crack in an elastic body that is deformed. In a linear-elastic homogeneous body, the significant stress components vary inversely as the square root of the distance from the crack tip. In a linear-elastic body, the crack-tip stress field can be regarded as the superposition of three component stress fields called *modes*.

impact energy. The amount of energy required to fracture a material, usually measured by means of an *Izod test* or *Charpy test*. The type of specimen and test conditions affect the values and therefore should be specified.

impact loads. Especially severe shock loads such as those caused by instantaneous arrest of a falling mass, shock meeting of two parts (in a mechanical hammer, for example), or by explosive impact, in which there can be an exceptionally rapid build-up of stress.

impact strength. See *impact energy*.

impact test. A test for determining the energy absorbed in fracturing a test piece at high velocity, as distinct from static test. The test may be carried out in tension, bending, or torsion, and the test bar may be notched or unnotched. See also *impact energy, Charpy test,* and *Izod test*.

indentation hardness. The resistance of a material to indentation as determined by hardness testing. The indenter, which may be spherical or diamond shaped, is pressed into the surface of a metal under specified load for a given time.

initial recovery. The decrease in strain in a solid during the removal of force before any creep recovery takes place, usually determined at constant temperature. Sometimes referred to as instantaneous recovery.

initial strain. The strain in a solid immediately upon achieving the given loading conditions in a creep test. Sometimes referred to as instantaneous strain.

initial stress. The stress in a specimen immediately upon achieving the given constraint conditions in a stress-relaxation test. Sometimes referred to as instantaneous stress.

initial tangent modulus. The slope of the stress-strain curve at the beginning of loading. See also *modulus of elasticity*.

initiation of stable crack growth. The initiation of slow stable crack advance from the blunted crack tip.

instrumented impact test. An impact test in which the load on the specimen is continually recorded as a function of time and/or specimen deflection prior to fracture.

interval estimate. The estimate of a parameter given by two statistics, defining the end points of an interval.

isotropy. A term indicating equal physical or mechanical properties in all directions within a material.

item. (1) An object or quantity of material on which a set of observations can be made. (2) An observed value or test result obtained from an object or quantity of material.

Izod test. A type of impact test in which a V-notched specimen, mounted vertically, is subjected to a sudden blow delivered by the weight at the end of a pendulum arm. The energy required to break off the free end is a measure of the impact strength or toughness of the material. Contrast with *Charpy test*.

J

J-integral. A mathematical expression, a line or surface integral that encloses the crack front from one crack surface to the other, used to characterize the fracture toughness of a material having appreciable plasticity before fracture. The *J*-integral eliminates the need to describe the behavior of the material near the crack tip by considering the local stress-strain field around the crack front. J_{Ic} is the critical value of the J-integral required to initiate crack extension from a preexisting crack.

Jominy test. See *end-quench hardenability test*.

K

keel block. A standard test casting, for steel and other high-shrinkage alloys, consisting of a rectangular bar that resembles the keel of a boat, attached to the bottom of a large riser, or shrinkhead. Keel blocks that have only one bar are often called Y-blocks; keel blocks having two bars, double keel blocks. Test specimens are machined from the rectangular bar, and the shrinkhead is discarded.

keyhole specimen. A type of specimen containing a hole-and-slot notch, shaped like a keyhole, usually used in impact and bend tests. See *Charpy test*.

Knoop hardness number, HK. A number related to the applied load and to the projected area of the permanent impression made by a rhombic-based pyramidal diamond indenter having included edge angles of 172° 30′ and 130° 0′ computed from the equation:

$$HK = \frac{P}{0.07028d^2}$$

where P is applied load, kgf; and d is long diagonal of the impression, mm. In reporting Knoop hardness numbers, the test load is stated.

Knoop hardness test. An indentation hardness test using calibrated machines to force a rhombic-based pyramidal diamond indenter having specified edge angles, under specified conditions, into the surface of the material under test and to measure the long diagonal after removal of the load.

L

least count. The smallest value that can be read from an instrument having a graduated scale. Except on instruments provided with a *vernier*, the least count is that fraction of the smallest division which can be conveniently and reliably estimated; this fraction is ordinarily one fifth or one tenth, except where the graduations are very closely spaced. Also known as least reading.

least reading. See *least count*.

limiting dome height (LDH) test. A mechanical test, usually performed unlubricated on sheet metal, that simulates the fracture conditions in a practical press-forming operation. The results are dependent on the sheet thickness.

linear-elastic fracture mechanics. A method of fracture analysis that can determine the stress (or load) required to induce fracture instability in a structure containing a crack-like flaw of known size and shape. See also *stress-intensity factor*.

linear (tensile or compressive) strain. The change per unit length due to force in an original linear dimension. An increase in length is considered positive.

load. In the case of testing machines, a force applied to a test piece that is measured in units such as pound-force, newton, or kilogram-force.

load range, P. In fatigue, the algebraic difference between the maximum and minimum loads in a fatigue cycle.

load ratio, R. In fatigue, the algebraic ratio of the minimum to maximum load in a fatigue cycle, that is, $R = P_{min}/P_{max}$. Also known as *stress ratio*.

longitudinal direction. The principal direction of flow in a worked metal.

long transverse. See *transverse*.

lot. A definite quantity of a product or material accumulated under conditions that are considered uniform for sampling purposes. Compare with *batch*.

lubricant. Any substance interposed between two surfaces for the purpose of reducing the friction or wear between them.

Lüders lines. Elongated surface markings or depressions, often visible with the unaided eye, that form along the length of a tension specimen at an angle of approximately 45° to the loading axis. Caused by localized plastic deformation, they result from discontinuous (inhomogeneous) yielding. Also known as Lüders bands, Hartmann lines, Piobert lines, or stretcher strains.

M

macrostrain. The mean strain over any finite gage length of measurement large in comparison with interatomic distances. Macrostrain can be measured by several methods, including electrical-resistance strain gages and mechanical or optical extensometers. Elastic macrostrain can be measured by x-ray diffraction.

malleability. The characteristic of metals that permits plastic deformation in compression without fracture. See also *ductility*.

maximum load, P_{max}. (1) The load having the highest algebraic value in the load cycle. Tensile loads are considered positive and compressive loads negative. (2) Used to determine the strength of a structural member; the load that can be borne before failure is apparent.

maximum strength. See *ultimate strength*.

maximum stress, S_{max}. The stress having the highest algebraic value in the stress cycle, tensile stress being considered positive and compressive stress negative. The *nominal stress* is used most commonly.

maximum stress-intensity factor, K_{max}. The maximum value of the *stress-intensity factor* in a fatigue cycle.

mean stress (or steady component of stress), S_m. The algebraic average of the maximum and minimum stresses in one cycle, that is, $S_m = (S_{max} + S_{min})/2$.

mechanical hysteresis. Energy absorbed in a complete cycle of loading and unloading within the elastic limit and represented by the closed loop of the stress-strain curves for loading and unloading.

mechanical metallurgy. The science and technology dealing with the behavior of metals when subjected to applied forces.

mechanical properties. The properties of a material that reveal its elastic and inelastic behavior when force is applied or that involve the relationship between the intensity of the applied stress and the strain produced. The properties included under this heading are those that can be recorded by *mechanical testing*—for example, *modulus of elasticity*, *tensile strength*, *elongation*, *hardness*, and *fatigue limit*. Compare with *physical properties*.

mechanical testing. The methods by which the *mechanical properties* of a metal are determined.

median fatigue life. The middle value when all of the observed fatigue life values of the individual specimens in a group tested under identical conditions are arranged in order of magnitude. When an even number of specimens are tested, the average of the two middlemost values is used. Use of the sample median rather than the arithmetic mean (that is, the average) is usually preferred.

median fatigue strength at N cycles. An estimate of the stress level at which 50% of the population would survive N cycles. The estimate is derived from a particular point of the fatigue life distribution, since there is no test procedure by which a frequency distribution of fatigue strengths at N cycles can be directly observed. Also known as *fatigue strength at N cycles*.

microhardness. The hardness of a material as determined by forcing an indenter into the surface of a material under very light load; usually the indentations are so small that they must be measured with a microscope. Capable of determining hardnesses of different microconstituents within a structure, or of measuring steep hardness gradients such as those encountered in case hardening. See also *microhardness test*.

microhardness test. A microindentation hardness test using a calibrated machine to force a diamond indenter of specific geometry, under a test load of 1 to 1000 gram-force, into the surface of the test material and to measure the diagonal or diagonals optically.

microstrain. The strain over a gage length comparable to interatomic distances. These are the strains being averaged by the *macrostrain* measurement. Microstrain is not measurable by existing techniques. Variance of the microstrain distribution can, however, be measured by x-ray diffraction.

minimum load, P_{min}. In fatigue, the least algebraic value of applied load in a cycle.

minimum stress, S_{min}. In fatigue, the stress having the lowest algebraic value in the cycle, tensile stress being considered positive and compressive stress negative.

minimum stress-intensity factor, K_{min}. In fatigue, the minimum value of the *stress-intensity factor* in a cycle. This value corresponds to the *minimum load* when the *load ratio* > 0 and is taken to be zero when the *load ratio* is ≤ 0.

mode. One of the three classes of crack (surface) displacements adjacent to the crack tip. These displacement modes are associated with stress-strain fields around the crack tip and are designated one, two, and three. See also *crack-tip plane strain* and *crack opening displacement*.

modulus of elasticity, E. The measure of rigidity or stiffness of a metal; the ratio of stress, below the proportional limit, to the corresponding strain. In terms of the *stress-strain diagram*, the modulus of elasticity is the slope of the stress-strain curve in the range of linear proportionality of stress to strain. Also known as *Young's modulus*. For materials that do not conform to *Hooke's law* throughout the elastic range, the slope of either the tangent to the stress-strain curve at the origin or at low stress, the secant drawn from the origin to any specified point on the stress-strain curve, or the chord connecting any two specific points on the stress-strain curve is usually taken to be the modulus of elasticity. In these cases, the modulus is referred to as the *tangent modulus*, *secant modulus*, or *chord modulus*, respectively.

modulus of resilience. The amount of energy stored in a material when loaded to its elastic limit. It is determined by measuring the area under the stress-strain curve up to the elastic limit. See also *elastic energy*, *resilience*, and *strain energy*.

modulus of rigidity. See *shear modulus*.

modulus of rupture. Nominal stress at fracture in a bend test or torsion test. In bending, modulus of rupture is the bending moment at fracture divided by the section modulus. In torsion, modulus of rupture is the torque at fracture divided by the polar section modulus. See also *modulus of rupture in bending* and *modulus of rupture in torsion*.

modulus of rupture in bending, S_b. The value of maximum tensile or compressive stress (whichever causes failure) in the extreme fiber of a beam loaded to failure in bending computed from the flexure equation:

$$S_b = \frac{Mc}{I}$$

where M is maximum bending moment, computed from the maximum load and the original moment arm; c is initial distance from the neutral axis to the extreme fiber where failure occurs; and I is initial moment of inertia of the cross section about the neutral axis. See also *modulus of rupture*.

modulus of rupture in torsion, S_s. The value of maximum shear stress in the extreme fiber of a member of circular cross section loaded to failure in torsion computed from the equation:

$$S_s = \frac{Tr}{J}$$

where T is maximum twisting moment, r is original outer radius, and J is polar moment of inertia of the original cross section. See also *modulus of rupture*.

modulus of toughness. The amount of work per unit volume of a material required to carry that material to failure under static loading. See also *toughness*.

Mohs scale. A scratch hardness test for determining comparative hardness using 10 standard minerals—from talc (the softest) to diamond (the hardest).

monotron hardness test. A method of determining *indentation hardness* by measuring the load required to force a spherical penetrator into a metal to a specified depth. Now obsolete.

m-value. See *strain-rate sensitivity*.

N

natural strain. See *true strain*.

necking. (1) Reducing the cross-sectional area of metal in a localized area by stretching. (2) Reducing the diameter of a portion of the length of a cylindrical shell or tube.

nominal strain. See *strain*.

nominal strength. See *ultimate strength*.

nominal stress. The stress at a point calculated on the net cross section by simple elasticity theory without taking into account the effect on the stress produced by stress raisers such as holes, grooves, fillets, etc.

normal stress. The stress component perpendicular to a plane on which forces act. Normal stress may be either *tensile* or *compressive*.

notch brittleness. Susceptibility of a material to *brittle fracture* at points of stress concentration. For example, in a notch tensile test, the material is said to be notch brittle if the notch strength is less than the tensile strength of an unnotched specimen. Otherwise, it is said to be notch ductile.

notch depth. The distance from the surface of a test specimen to the bottom of the notch. In a cylindrical test specimen, the percentage of the original cross-sectional area removed by machining an annular groove.

notch ductility. The percentage reduction in area after complete separation of the metal in a tensile test of a notched specimen.

notch strength. The maximum load on a notched tensile-test specimen divided by the minimum cross-sectional area (the area at the root of the notch). Also known as notch tensile strength.

n-value. See *strain hardening exponent*.

O

observed value. The particular value of a characteristic determined as a result of a test or measurement.

offset. The distance along the strain coordinate between the initial portion of a stress-strain curve and a parallel line that intersects the stress-strain curve at a value of stress (commonly 0.2%) that is used as a measure of the *yield strength*. Used for materials that have no obvious *yield point*.

offset yield strength. The stress at which the strain exceeds by a specified amount (the *offset*) an extension of the initial proportional portion of the stress-strain curve. Expressed in force per unit area.

Olsen ductility test. A *cupping test* in which a piece of sheet metal, restrained except at the center, is deformed by a standard steel ball until fracture occurs. The height of the cup at time of fracture is a measure of the ductility.

orange peel. A surface roughening in the form of a pebble-grained pattern where a metal of unusually coarse grain is stressed beyond its elastic limit. Also known as pebbles and alligator skin.

original crack size, a_o. The *physical crack size* at the start of testing.

oxidative wear. A type of wear resulting from the sliding action between two metallic components that generates oxide films on the metal surfaces. These oxide films prevent the formation of a metallic bond between the sliding surfaces, resulting in fine wear debris and low wear rates.

P

parameter. In statistics, a constant (usually unknown) defining some property of the frequency distribution of a population, such as a population median or a population standard deviation.

pebbles. See preferred term *orange peel*.

percent error. In the case of a testing machine, the ratio, expressed as a percentage, of the *error* to the correct value of the applied load.

permanent set. The deformation or strain remaining in a previously stressed body after release of load.

permissible variation. In the case of testing machines, the maximum allowable error in the value of the quantity indicated. It is convenient to express permissible variation in terms of the *percent error*. See also *tolerance*.

physical crack size, a_p. The distance from a reference plane to the observed crack front. This distance may represent an average of several measurements along the crack front. The reference plane depends on the specimen form, and it is normally taken to be either the boundary or a plane containing either the load line or the centerline of a specimen or plate.

physical properties. Properties of a metal or alloy the determination of which does not involve the deformation or destruction of the specimen—for example, density, electrical conductivity, coefficient of thermal expansion, magnetic permeability, and lattice parameter. Does not include chemical reactivity or properties more appropriately regarded as *mechanical properties*.

physical testing. Methods used to determine the entire range of a material's *physical properties*. In addition to density and thermal, electrical, and magnetic properties, physical testing methods may be used to assess simple fundamental physical properties such as color, crystalline form, and melting point.

pin expansion test. A test for determining the ability of a tube to be expanded or for revealing the presence of cracks or other longitudinal weaknesses, made by forcing a tapered pin into the open end of the tube.

pin or mandrel. The plunger or tool used in making semiguided, guided, or wraparound bend tests to apply the bending force to the inside surface of the bend. In free bends or semiguided bends to an angle of 180°, a shim or block of the proper thickness may be placed between the legs of the specimen as bending is completed. This shim or block is also referred to as a pin or mandrel.

Piobert lines. See *Lüders lines*.

pitting. In *tribology*, a type of wear characterized by the presence of surface cavities formed by processes such as fatigue, local adhesion, or cavitation.

planar anisotropy. A variation in physical and/or mechanical properties with respect to direction within the plane of material in sheet form. See also *plastic strain ratio*.

plane strain. The stress condition in linear elastic fracture mechanics in which there is zero strain in a direction normal to both the axis of applied tensile stress and the direction of crack growth (i.e., parallel to the crack front); most nearly achieved in loading thick plates along a direction parallel to the plate surface. Under plane-strain conditions, the plane of fracture instability is normal to the axis of the principal tensile stress.

plane-strain fracture toughness, K_{Ic}. The crack extension resistance under conditions of crack-tip plane strain. See also *stress-intensity factor*.

plane stress. The stress condition in linear elastic fracture mechanics in which the stress in the thickness direction is zero; most nearly achieved in loading very thin sheet along a direction parallel to the surface of the sheet. Under plane-stress conditions, the plane of fracture instability is inclined 45° to the axis of the principal tensile stress.

plane-stress fracture toughness, K_c. The value of the *crack-extension resistance* at the instability condition determined from the tangency between the *R*-curve and the critical *crack-extension force* curve of the specimen. See also *stress intensity factor* and *R-curve*.

plastic deformation. The permanent (inelastic) distortion of metals under applied stresses that strain the material beyond its *elastic limit*.

plastic flow. See *plastic deformation*.

plastic instability. The stage of deformation in a tensile test where the plastic flow becomes nonuniform and *necking* begins.

plasticity. The property that enables a material to undergo permanent deformation without rupture.

plastic strain. Dimensional change that does not disappear when the initiating stress is removed. Usually accompanied by some *elastic deformation*.

plastic strain ratio (*r*-value). The ratio of the true width strain to the true thickness strain in a tensile test, $r = \epsilon_w/\epsilon_t$. Because of the difficulty in making precise measurement of thickness strain in sheet material, it is more convenient to express r in terms of initial and final length and width dimensions. It can be shown that $r = (\ln w_o w_f) - (\ln l_f w_f/l_o w_o)$, where l_o and w_o are initial length and width of gage section, respectively; and l_f and w_f are final length and width, respectively.

plastic-zone adjustment, r_Y. An addition to the *physical crack size* to account for plastic crack-tip deformation enclosed by a linear-elastic stress field.

plowing. In *tribology*, the formation of grooves by plastic deformation of the softer of two surfaces in relative motion.

point estimate. The estimate of a *parameter* given by a single statistic.

Poisson's ratio, ν. The absolute value of the ratio of transverse (lateral) strain to the corresponding axial strain resulting from uniformly distributed axial stress below the *proportional limit* of the material.

population. The hypothetical collection of all possible test specimens that could be prepared in the specified way from the material under consideration. Also known as universe.

precision. The closeness of agreement between randomly selected individual measurements or test results. The standard deviation of the error of measurement may be used as a measure of "imprecision."

principal stress (normal). The maximum or minimum value of the normal stress at a point in a plane considered with respect to all possible orientations of the considered plane. On such principal planes the shear stress is zero. There are three principal stresses on three mutually perpendicular planes. The state of stress at a point may be: (1) uniaxial, a state of stress in which two of the three principal stresses are zero; (2) biaxial, a state of stress in which only one of the three principal stresses is zero; or (3) triaxial, a state of stress in which none of the principal stresses is zero. Multiaxial stress refers to either biaxial or triaxial stress.

proof stress. (1) The stress that will cause a specified small *permanent set* in a material. (2) A specified stress to be applied to a member or structure to indicate its ability to withstand service loads.

proportional limit. The greatest stress a material is capable of developing without a deviation from straight-line proportionality between stress and strain. See also *elastic limit* and *Hooke's law*.

R

radial marks. Lines on a fracture surface

that radiate from the fracture origin and are visible to the unaided eye or at low magnification. Radial lines result from the intersection and connection of brittle fractures propagating at different levels. Also known as shear ledges. See also *chevron pattern*.

radius of bend. The radius of the cylindrical surface of the pin or mandrel that comes in contact with the inside surface of the bend during bending. In the case of free or semiguided bends to 180° in which a shim or block is used, the radius of bend is one half the thickness of the shim or block.

range of stress, S_r. The algebraic difference between the maximum and minimum stress in one cycle—that is,

$$S_r = S_{max} - S_{min}$$

ratchet marks. Lines on a fatigue fracture surface that result from the intersection and connection of fatigue fractures propagating from multiple origins. Ratchet marks are parallel to the overall direction of crack propagation and are visible to the unaided eye or at low magnification.

rate of creep. See *creep rate*.

R-curve. A plot of crack-extension resistance as a function of stable crack extension, which is either the difference between the *physical crack size*, a_p, or the *effective crack size*, a_e, and the *original crack size*, a_o. R-curves normally depend upon specimen thickness and, for some materials, upon temperature and strain rate.

real area of contact. In *tribology*, the sum of the local areas of contact between two solid surfaces, formed by contacting asperities, that transmit the interfacial force between the two surfaces. Contrast with *apparent area of contact*.

reduction in area. The difference between the original cross-sectional area of a tensile specimen and the smallest area at or after fracture as specified for the material undergoing testing. Also known as reduction of area.

relaxation curve. A plot of either the remaining or relaxed stress as a function of time. See also *relaxation rate*.

relaxation rate. The absolute value of the slope of a relaxation curve at a given time.

relaxed stress. The initial stress minus the remaining stress at a given time during a stress-relaxation test. See also *stress relaxation*.

remaining stress. The stress remaining at a given time during a stress-relaxation test. See also *stress relaxation*.

repeatability. A term used to refer to the test result variability associated with a limited set of specifically defined sources of variability within a single laboratory.

reproducibility. A term used to describe test result variability associated with specifically defined components of variance obtained both from within a single laboratory and between laboratories.

residual stress. Stresses that remain within a body as the result of plastic deformation.

resilience. The ability of a material to absorb energy when deformed elastically and return to its original shape upon release of load. See also *modulus of resilience*.

response curve for N cycles. A curve fitted to observed values of percentage survival at N cycles for several stress levels, where N is a preassigned number such as 10^6, 10^7, etc. It is an estimate of the relationship between applied stress and the percentage of the population that would survive N cycles.

Rockwell hardness number, HR. A number derived from the net increase in the depth of impression as the load on an indenter is increased from a fixed minor load to a major load and then returned to the minor load. Rockwell hardness numbers are always quoted with a scale symbol representing the penetrator, load, and dial used.

Rockwell hardness test. An indentation hardness test using a calibrated machine that utilizes the depth of indentation, under constant load, as a measure of hardness. Either a 120° diamond cone with a slightly rounded point, or a ¹⁄₁₆- or ⅛-in.-diam steel ball is used as the indenter.

Rockwell superficial hardness number. See *Rockwell hardness number*.

Rockwell superficial hardness test. Same as *Rockwell hardness test*, except that smaller minor and major loads are used.

rosette. Strain gages arranged to indicate, at a single position, strain in three different directions.

rupture stress. The stress at failure. Also known as *breaking stress* or *fracture stress*.

r-value. See *plastic strain ratio*.

s

sample. One or more units of product (or a relatively small quantity of a bulk material) that are withdrawn from a lot or process stream, and that are tested or inspected to provide information about the properties, dimensions, or other quality characteristics of the lot or process stream. Not be confused with *specimen*.

sample average. The sum of all the observed values in a sample divided by the sample size. It is a point estimate of the population mean. Also known as arithmetic mean.

sample median. The middle value when all observed values in a sample are arranged in order of magnitude. If an even number of samples are tested, the average of the two middlemost values is used. It is a point estimate of the population median, or 50% point.

sample percentage. The percentage of observed values between two stated values of the variable under consideration. It is a point estimate of the percentage of the population between the same two stated values. (One stated value may be $-\infty$ or $+\infty$.)

sample standard deviation, s. The square root of the sample variance. It is a point estimate of the population standard deviation, a measure of the "spread" of the frequency distribution of a population. This value of s provides a statistic that is used in computing interval estimates and several test statistics. For small sample sizes, s underestimates the population standard deviation.

sample variance, s^2. The sum of the squares of the differences between each observed value and the sample average divided by the sample size minus one. It is a point estimate of the population variance.

Scleroscope hardness number (HSc or HSd). A number related to the height of rebound of a diamond-tipped hammer dropped on the material being tested. It is measured on a scale determined by dividing into 100 units the average rebound of the hammer from a quenched (to maximum hardness) and untempered high-carbon water-hardening tool steel test block of AISI W-5.

Scleroscope hardness test. A dynamic indentation hardness test using a calibrated instrument that drops a diamond-tipped hammer from a fixed height onto the surface of the material being tested. The height of rebound of the hammer is a measure of the hardness of the material.

scoring. In *tribology*, a severe form of wear characterized by the formation of extensive grooves and scratches in the direction of sliding.

scratch hardness. The hardness of a metal determined by the width of a scratch made by a cutting point drawn across the surface under a given pressure.

scratching. In *tribology*, the mechanical removal or displacement, or both, of material from a surface by the action of abrasive particles or protuberances sliding across the surfaces. See also *plowing*.

scuffing. A form of adhesive wear that produces superficial scratches or a high polish on the rubbing surfaces. It is observed most often on inadequately lubricated parts.

secant modulus. The slope of the secant drawn from the origin to any specified point on the stress-strain curve. See also *modulus of elasticity*.

semiguided bend. The bend obtained by applying a force directly to the specimen in the portion that is to be bent. The specimen is either held at one end and forced around a pin or rounded edge, or is supported near the ends and bent by a force applied on the side of the specimen opposite the supports and midway between them. In some instances, the bend is started in this manner and finished in the manner of a *free bend*.

set. See *permanent set*.

sharp-notch strength, σ_s. The notch tensile strength measured using specimens with very small notch root radii (approaching the limit for machining capability); values of sharp-notch strength usually depend on notch root radius.

shear fracture. A ductile fracture in which a crystal (or a polycrystalline mass) has separated by sliding or tearing under the action of shear stresses.

shear lip. A narrow, slanting ridge along the edge of a fracture surface. The term sometimes also denotes a narrow, often crescent-shaped, fibrous region at the edge of a fracture that is otherwise of the cleavage type, even though this fibrous region is in the same plane as the rest of the fracture surface.

shear modulus, G. The ratio of shear stress to the corresponding shear strain for shear stresses below the proportional limit of the material. Values of shear modulus are usually determined by torsion testing. Also known as modulus of rigidity.

shear strain. The tangent of the angular change, due to force, between two lines originally perpendicular to each other through a point in a body.

shear strength. The maximum shear stress that a material is capable of sustaining. Shear strength is calculated from the maximum load during a shear or torsion test and is based on the original dimensions of the cross section of the specimen.

shear stress. (1) A stress that exists when parallel planes in metal crystals slide across each other. (2) The stress component tangential to the plane on which the forces act. Also known as tangential stress.

shelf roughness. Roughness on upward-facing surfaces where undissolved solids have settled on parts during a plating operation.

shock loads. The sudden application of an external force that results in a very rapid build-up of stress—for example, piston loading in internal combustion engines.

Shore hardness test. See *Scleroscope hardness test*.

short transverse. See *transverse*.

significance level. The stated probability (risk) that a given test of significance will reject the hypothesis that a specified effect is absent when the hypothesis is true.

significant. Statistically significant. An effect of difference between populations is said to be present if the value of a test statistic is significant, that is, lies outside predetermined limits.

silky fracture. A metal fracture in which the broken metal surface has a fine texture, usually dull in appearance. Characteristic of tough and strong metals. Contrast with *crystalline fracture* and *granular fracture*.

size effect. Effect of the dimensions of a piece of metal on its mechanical and other properties and on manufacturing variables such as forging reduction and heat treatment. In general, the mechanical properties are lower for a larger size.

slant fracture. A type of fracture appearance, typical of plane-stress fractures, in which the plane of metal separation is inclined at an angle (usually about 45°) to the axis of the applied stress.

slenderness ratio. The effective unsupported length of a uniform column divided by the least radius of gyration of the cross-sectional area.

slip. Plastic deformation by the irreversible shear displacement (translation) of one part of a crystal relative to another in a definite crystallographic direction and usually on a specific crystallographic plane. Sometimes called glide. See also *flow*.

slip band. A group of parallel slip lines so closely spaced as to appear as a single line when observed under an optical microscope. See also *slip line*.

slip line. The trace of the slip plane on the viewing surface; the trace is usually observable only if the surface has been polished before deformation. The usual observation on metal crystals (under the light microscope) is of a cluster of slip lines known as a slip band.

S-N curve. A plot of stress, S, against the number of cycles to failure, N. The stress can be the maximum stress, S_{max}, or the alternating stress amplitude S_a. The stress values are usually nominal stresses; i.e., there is no adjustment for stress concentration. The diagram indicates the S-N relationship for a specified value of the mean stress, S_m, or the stress ratio, A, or R and a specified probability of survival. For N a log scale is almost always used. For S a linear scale is used most often, but a log scale is sometimes used. Also known as S-N diagram.

S-N curve for 50% survival. A curve fitted to the median value of fatigue life at each of several stress levels. It is an estimate of the relationship between applied stress and the number of cycles-to-failure that 50% of the population would survive.

S-N curve for p% survival. A curve fitted to the fatigue life for p% survival values at each of several stress levels. It is an estimate of the relationship between applied stress and the number of cycles-to-failure that p% of the population would survive. p may be any number, such as 95, 90, etc.

spalling. The cracking and flaking of particles out of a surface.

specimen. A test object, often of standard dimensions or configuration, that is used for destructive or nondestructive testing. One or more specimens may be cut from each unit of a sample.

springback. The degree to which metal tends to return to its original shape or contour after undergoing a forming operation.

standard deviation. The most usual measure of the dispersion of observed values or results expressed as the positive square root of the variance.

statistic. A summary value calculated from the observed values in a sample.

steady loads. Loads that do not change in intensity, or change so slowly that they may be regarded as steady.

stiffness. (1) The ability of a metal or shape to resist elastic deflection. (2) The rate of stress with respect to strain; the greater the stress required to produce a given strain, the stiffer the material is said to be.

strain. The unit of change in the size or shape of a body due to force. Also known as nominal strain. See also *engineering strain*, *linear strain*, and *true strain*.

strain-age embrittlement. A loss in ductility accompanied by an increase in hardness and strength that occurs when low-carbon steel (especially rimmed or capped steel) is aged following plastic deformation. The degree of embrittlement is a function of aging time and temperature, occurring in a matter of minutes at about 200 °C (400 °F), but requiring a few hours to a year at room temperature.

strain aging. The changes in ductility, hardness, yield point, and tensile strength that occur when a metal or alloy that has been cold worked is stored for some time. In steel, strain aging is characterized by a loss of ductility and a corresponding increase in hardness, yield point, and tensile strength.

strain energy. A measure of the energy absorption characteristics of a material determined by measuring the area under the stress-strain diagram. See also *elastic energy*, *resilience*, and *toughness*.

strain gage. A device for measuring small amounts of strain produced during tensile and similar tests on metal. A coil of fine wire is mounted on a piece of paper, plastic, or similar carrier matrix (backing material), which is rectangular in shape and usually about 25 mm (1.0 in.) long. This is glued to a portion of metal under test. As the coil extends with the specimen, its electrical resistance increases in direct proportion. This is known as bonded resistance-strain gage. Other types of gages measure the actual deformation. Mechanical, optical, or electronic devices are sometimes used to magnify the strain for easier reading. See also *rosette*.

strain hardening. An increase in hardness and strength caused by plastic deformation at temperatures below the recrystallization range. Also known as work hardening.

strain-hardening coefficient. See *strain-hardening exponent*.

strain-hardening exponent. The value n in the relationship $\sigma = K\epsilon^n$, where σ is the *true stress*, ϵ is the *true strain*, and K, which is called the "strength coefficient," is equal to the true stress at a true strain of 1.0. The strain hardening exponent, also called "n-value," is equal to the slope of the true stress/true strain curve up to maximum load, when plotted on log-log coordinates. The n-value relates to the ability of a material to be stretched in metalworking operations. The higher the n-value, the better the formability (stretchability).

strain rate. The time rate of straining for the usual tensile test. Strain as measured directly on the specimen gage length is used for determining strain rate. Because strain is dimensionless, the units of strain rate are reciprocal time.

strain-rate sensitivity (*m* value). The increase in stress (σ) needed to cause a certain increase in plastic strain rate ($\dot{\epsilon}$) at a given level of plastic strain (ϵ) and a given temperature (T).

$$\text{Strain-rate sensitivity} = m = \left(\frac{\Delta \log \sigma}{\Delta \log \dot{\epsilon}} \right)_{\epsilon T}$$

strength. The maximum nominal stress a material can sustain. Always qualified by the type of stress (tensile, compressive, or shear).

strength coefficient. See *strain-hardening exponent*.

stress. The intensity of the internally distributed forces or components of forces that resist a change in the volume or shape of a material that is or has been subjected to external forces. Stress is expressed in force per unit area and is calculated on the basis of the original dimensions of the cross section of the specimen. Stress can be either direct (tension or compression) or shear. See also *engineering stress*, *mean stress*, *nominal stress*, *normal stress*, *residual stress*, and *true stress*.

stress amplitude. One half the algebraic difference between the maximum and minimum stress in one cycle of a repetitively varying stress.

stress-concentration factor (K_t). A multiplying factor for applied stress that allows for the presence of a structural discontinuity such as a notch or hole; K_t equals the ratio of the greatest stress in the region of the discontinuity to the nominal stress for the entire section. Also known as theoretical stress concentration factor.

stress-corrosion cracking (SCC). A time-dependent process in which a metallurgically susceptible material fractures prematurely under conditions of simultaneous corrosion and sustained loading at lower stress levels than would be required in the absence of a corrosive environment. Tensile stress is required at the metal surface and may be a residual stress resulting from heat treatment or fabrication of the metal or the result of external loading. Cracking may be intergranular or transgranular, depending on the combination of alloy and environment.

stress cycle. The smallest segment of the stress-time function that is repeated periodically.

stress cycles endured, *n*. The number of cycles of a specified character (that produce fluctuating stress and strain) that a specimen has endured at any time in its stress history.

stress-intensity calibration. A mathematical expression, based on empirical or analytical results, that relates the stress-intensity factor to load and crack length for a specific specimen planar geometry. Also known as K calibration.

stress-intensity factor. A scaling factor, usually denoted by the symbol K, used in linear-elastic fracture mechanics to describe the intensification of applied stress at the tip of a crack of known size and shape. At the onset of rapid crack propagation in any structure containing a crack, the factor is called the critical stress-intensity factor, or the fracture toughness. Various subscripts are used to denote different loading conditions or fracture toughnesses:

K_c. Plane-stress fracture toughness. The value of stress intensity at which crack propagation becomes rapid in sections thinner than those in which plane-strain conditions prevail.

K_I. Stress-intensity factor for a loading condition that displaces the crack faces in a direction normal to the crack plane (also known as the opening mode of deformation).

K_{Ic}. Plane-strain fracture toughness. The minimum value of K_c for any given material and condition, which is attained when rapid crack propagation in the opening mode is governed by plane-strain conditions.

K_{Id}. Dynamic fracture toughness. The fracture toughness determined under dynamic loading conditions; it is used as an approximation of K_{Ic} for very tough materials.

K_{Iscc}. Threshold stress intensity for stress corrosion cracking when loading conditions meet plane-strain requirements.

K_Q. Provisional value for plane-strain fracture toughness.

K_{th}. Threshold stress intensity for stress corrosion cracking. A value of stress intensity characteristic of a specific combination of material, material condition, and corrosive environment above which stress corrosion crack propagation occurs and below which the material is immune from stress corrosion cracking.

ΔK. The range of the stress-intensity factor during a fatigue cycle.

stress-intensity factor range, *K*. In fatigue, the variation in the stress-intensity factor in a cycle, that is, $K_{\max} - K_{\min}$.

stress raisers. Changes in contour or discontinuities in structure that cause local increases in stress.

stress ratio, *A* or *R*. The algebraic ratio of two specified stress values in a stress cycle. Two commonly used stress ratios are: the ratio of the alternating stress amplitude to the mean stress, $A = S_a/S_m$; and the ratio of the minimum stress to the maximum stress, $R = S_{\min}/S_{\max}$.

stress relaxation. The time-dependent decrease in stress in a solid under constant constraint at constant temperature. The stress-relaxation behavior of a metal is usually shown in a *stress-relaxation curve*.

stress-relaxation curve. A plot of the remaining or relaxed stress as a function of time. The relaxed stress equals the initial stress minus the remaining stress. Also known as stress-time curve.

stress-rupture strength. See *creep-rupture strength*.

stress-rupture test. See *creep-rupture test*.

stress-strain curve. See *stress-strain diagram*.

stress-strain diagram. A graph in which corresponding values of stress and strain are plotted against each other. Values of stress are usually plotted vertically (ordinates or *y* axis) and values of strain horizontally (abscissas or *x* axis). Also known as deformation curve and stress-strain curve.

stretch-bending test. A simulative test for sheet metal formability in which a strip of sheet metal is clamped at its ends in lock beads and deformed in the center by a punch. Test conditions are chosen so that fracture occurs in the region of punch contact.

stretcher strains. See *Lüders lines*.

striation. A fatigue fracture feature, often observed in electron micrographs, that indicates the position of the crack front after each succeeding cycle of stress. The distance between striations indicates the advance of the crack front across that crystal during one stress cycle, and a line normal to the striation indicates the direction of local crack propagation.

Swift cup test. A simulative test in which circular blanks of various diameters are clamped in a die ring and deep drawn into cups by a flat-bottomed cylindrical punch.

T

tangent modulus. The slope of the stress-strain curve at any specified stress or strain. See also *modulus of elasticity*.

temper brittleness. Brittleness that results when certain steels are held within, or are cooled slowly through, a certain range of temperature below the transformation range. The brittleness is manifested as an upward shift in ductile-to-brittle transition temperature, but only rarely produces a low value of reduction in area in a smooth-bar tension test of the embrittled material.

tensile strength. In tensile testing, the ratio of maximum load to original cross-sectional area. Also known as ultimate strength. Compare with *yield strength*.

tensile stress. A stress that causes two parts of an elastic body, on either side of a typical stress plane, to pull apart. Contrast with *compressive stress*.

tensile testing. See *tension testing*.

tension. The force or load that produces elongation.

tension testing. A method of determining the behavior of materials subjected to uniaxial loading, which tends to stretch the metal. A longitudinal specimen of known length and diameter is gripped at both ends and stretched at a slow, controlled rate until rupture occurs. Also known as tensile testing.

testing machine (load-measuring type). A mechanical device for applying a load (force) to a specimen.

theoretical stress concentration factor. See *stress concentration factor*.

thermal fatigue. Fracture resulting from the presence of temperature gradients that vary with time in such a manner as to produce cyclic stresses in a structure.

thermocouple. A device for measuring temperature, consisting of lengths of two dissimilar metals or alloys that are electrically joined at one end and connected to a voltage-measuring instrument at the other end. When one junction is hotter than the other, a thermal electromotive force is produced that is roughly proportional to the difference in temperature between the hot and cold junctions.

threshold stress for stress-corrosion cracking, σ_{th}. An experimentally determined gross section stress below which stress-corrosion cracking will not occur under specified test conditions.

tolerance. The amount by which a quantity, such as a dimension, property, or composition, is allowed to vary; the tolerance is the difference between the maximum allowable limit or limits.

tolerance limits. The extreme values (upper and lower) that define the range of permissible variation in size or other quality characteristic of a part.

torsion. A twisting deformation of a solid body about an axis in which lines that were initially parallel to the axis become helices.

torsion test. A test designed to provide data for the calculation of the *shear modulus*, *modulus of rupture in torsion*, and *yield strength* in shear.

torsional stress. The shear stress on a transverse cross section resulting from a twisting action.

total elongation. A total amount of permanent extension of a test piece broken in a tensile test. See also *elongation, percent*.

total-extension-under-load yield strength. See *yield strength*.

toughness. The ability of a metal to absorb energy and deform plastically before fracturing.

transition temperature. (1) An arbitrarily defined temperature that lies within the temperature range in which metal fracture characteristics (as usually determined by tests of notched specimens) change rapidly, such as from primarily fibrous (shear) to primarily crystalline (cleavage) fracture. (2) Sometimes used to denote an arbitrarily defined temperature within a range in which the ductility changes rapidly with temperature.

transverse. Literally, "across," usually signifying a direction or plane perpendicular to the direction of working. In rolled plate or sheet, the direction across the width is often called long transverse, and the direction through the thickness, short transverse.

transverse strain. Linear strain in a plane perpendicular to the axis of the specimen.

tribology. The science and technology concerned with interacting surfaces in relative motion.

true strain. (1) The ratio of the change in dimension, resulting from a given load increment, to the magnitude of the dimension immediately prior to applying the load increment. (2) In a body subjected to axial force, the natural logarithm of the ratio of the gage length at the moment of observation to the original gage length. Also known as natural strain.

true stress. The value obtained by dividing the load applied to a member at a given instant by the cross-sectional area over which it acts.

U

ultimate strength. The maximum stress (tensile, compressive, or shear) a material can sustain without fracture, determined by dividing maximum load by the original cross-sectional area of the specimen. Also known as nominal strength or maximum strength.

uncertainty. An indication of the variability associated with a measured value that takes into account two major components of error. (1) bias, and (2) the random error attributed to the imprecision of the measurement process.

uniaxial strain. See *axial strain*.

uniform elongation. The elongation at maximum load and immediately preceding the onset of necking in a tensile test.

uniform strain. The strain occurring prior to the beginning of localization of strain (necking); the strain to maximum load in the tension test.

V

variance. A measure of the squared dispersion of observed values or measurements expressed as a function of the sum of the squared deviations from the population mean or sample average.

verification. Checking or testing the instrument to ensure conformance with the specification.

verified loading range. In the case of testing machines, the range of indicated loads for which the testing machine gives results within the permissible variation specified.

vernier. A short auxiliary scale that slides along the main instrument scale to permit more accurate fractional reading of the least main division of the main scale. See also *least count*.

Vickers hardness number, HV. A number related to the applied load and the surface area of the permanent impression made by a square-based pyramidal diamond indenter having included face angles of 136°, computed from the equation:

$$HV = 2P \sin \frac{\dfrac{\alpha}{2}}{d^2} = \frac{1.8544P}{d^2}$$

where P is applied load, kgf; d is mean diagonal of the impression, mm; and α is face angle of diamond, 136°.

Vickers hardness test. An indentation hardness test employing a 136° diamond pyramid indenter (Vickers) and variable loads, enabling the use of one hardness scale for all ranges of hardness—from very soft lead to tungsten carbide. Also know as diamond pyramid hardness test.

volumetric modulus of elasticity. See *bulk modulus of elasticity*.

W

Wallner lines. A distinct pattern of intersecting sets of parallel lines, usually producing a set of V-shaped lines, sometimes observed when viewing brittle fracture surfaces at high magnification in an electron microscope. Wallner lines are attributed to interaction between a shock wave and a brittle crack front propagating at high velocity. Sometimes Wallner lines are misinterpreted as fatigue striations.

wear. Damage to a solid surface, generally involving progressive loss of material, due to relative motion between that surface and a contacting surface or substance.

wear rate. The rate of material removal or dimensional change due to wear per unit of exposure parameter—for example, quantity of material removed (mass, volume, thickness) in unit distance of sliding or unit time.

welding. In *tribology*, the bonding between metallic surfaces in direct contact, at any temperature.

Widmanstätten structure. A structure characterized by a geometrical pattern resulting from the formation of a new phase along certain crystallographic planes of the parent solid solution. The orientation of the lattice in the new phase is related crystallographically to the orientation of the lattice in the parent phase. The structure is readily produced in many alloys by appropriate heat treatment.

workability. See *formability*.

wrap-around bend. The bend obtained when a specimen is wrapped in a closed helix around a cylindrical mandrel. This term is sometimes applied to a semiguided bend of 180° or less.

wrinkling. A wavy condition obtained in drawing in the area of the metal that passes over the draw radius. Wrinkling may also occur in other forming operations when unbalanced compressive forces are set up.

Y

yield. Evidence of plastic deformation in structural materials. Also known as plastic flow or creep. See also *flow*.

yield point. The first stress in a material, usually less than the maximum attainable stress, at which an increase in strain occurs without an increase in stress. Only certain metals—those which exhibit a localized, heterogeneous type of transition from elastic to plastic deformation—produce a yield point. If there is a decrease in stress after yielding, a distinction may be made between upper and lower yield points. The load at which a sudden drop in the flow curve occurs is called the upper yield point. The constant load shown on the flow curve is the lower yield point.

yield point elongation. The amount of strain that is required to complete the yielding process. It is measured from the onset of

yielding to the beginning of strain hardening.

yield strength. The stress at which a material exhibits a specified deviation from proportionality of stress and strain. An offset of 0.2% is used for many metals. Compare with *tensile strength*.

yield stress. The stress level of highly ductile materials, such as structural steels, at which large strains take place without further increase in stress.

Young's modulus. A term used synonymously with modulus of elasticity. The ratio of tensile or compressive stresses to the resulting strain. See also *modulus of elasticity*.

Z

zero time. The time when the given loading or constraint conditions are initially obtained in creep or stress-relaxation tests, respectively.

REFERENCES

- *Annual Book of ASTM Standards, Section 3, Metal Test Methods and Analytical Procedures*, Vol 03.01 and 03.02, ASTM, Philadelphia, 1984
- *Compilation of ASTM Standard Definitions*, 5th ed., ASTM, Philadelphia, 1979
- H.E. Davis, G.E. Troxell, and G.F.W. Hauck, *The Testing of Engineering Materials*, 4th ed., McGraw-Hill, New York, 1982
- G.E. Dieter, *Mechanical Metallurgy*, 2nd ed., McGraw-Hill, New York, 1976
- *Glossary of Metallurgical Terms and Engineering Tables*, American Society for Metals, 1979
- D.N. Lapedes, Ed., *Dictionary of Scientific and Technical Terms*, 2nd ed., McGraw-Hill, New York, 1974
- A.D. Merriman, *A Dictionary of Metallurgy*, Pitman Publishing, London, 1958
- J.G. Tweeddale, *Mechanical Properties of Metals*, American Elsevier, New York, 1964

Tension Testing

Introduction

TENSION TESTING consists of subjecting a prepared specimen of specified size and shape, or a full-size specimen, to a gradually increasing (static) uniaxial load (stress) until failure occurs. The operation is accomplished by gripping opposite ends of the workpiece and pulling it, which results in elongation of the test specimen in a direction parallel to the applied load.

Much attention has been given to tension testing, and great confidence has been placed in the value and significance of the test procedure. The chief advantages of the test are that the stress state is well established, the procedure has been carefully standardized, and it is relatively easy and inexpensive to perform.

One of the primary aims in conducting tension tests, or alternate mechanical property tests, is to determine conformance or nonconformance to specifications. The data may thus serve as an index of the quality of a product in comparison with data obtained previously or from other sources. Such data may also be used to compare a given material with other materials and, in conjunction with other factors, may assist in the replacement or improvement of a material. Manufacturers of metal products frequently use tension test-

ing data for control of manufacturing methods, as well as for guidance in the development of new materials.

The tension test is also used to establish a basis for the selection of values for engineering design. When properly conducted on suitable test specimens, the tension test can be used to evaluate fundamental mechanical properties for use in design, although it should be noted that tensile properties alone are not sufficient to predict the performance of materials under all loading conditions. When standard test methods are used, the results provide assurance that the material or part in question will exhibit satisfactory behavior in service.

Tension testing is not restricted to prepared specimens; full-size fabricated parts and structural members are also tested. Properly conducted tests on representative parts are extremely valuable in determining the performance of structural components under in-service loads and in observing the development of localized weaknesses as well as critical loads. As with any test procedure, however, caution should be exercised when using the tension test as a quality-level indicator, because the significance of the test is limited by its correlation with performance.

This section does not describe step-by-step procedures for conducting tension tests; this information can be obtained by referring to the appropriate ASTM standard or by consulting with test equipment manufacturers. Rather, the interpretation and limitations of test results will be discussed. Articles that describe the mechanical behavior of materials under tension and the engineering stress-strain curve are followed by an article on the factors that influence the reliability and utility of tensile data. These factors, which include specimen preparation, test setup, and test procedures, are generally taken for granted. If not properly accounted for, however, these variables can significantly influence test results.

Articles on the effect of temperature (both high and low) and strain rate on the strength and ductility of engineering materials are also presented. The Section concludes with a review of the current technology of tension testing equipment and extensometry. Additional information on tension testing can be found in the articles "High Strain Rate Tension Testing," "Sheet Formability Testing," and "Bulk Workability Testing" in this Volume.

Mechanical Behavior of Materials Under Tension*

By George E. Dieter
Dean of Engineering
University of Maryland

THE ENGINEERING TENSION TEST is widely used to provide basic design information on the strength of materials and as an acceptance test for the specification of materials. In this test procedure, a specimen is subjected to a continually increasing uniaxial load (force), while simultaneous observations are made of the elongation of the specimen. In this article, emphasis is placed on the interpretation of these observations and on the effect of metallurgical variables on mechanical behavior rather than on the procedures for conducting the tests. Subsequent articles in this Section discuss the influence of test procedure variables, temperature, and strain rate on test results and material flow properties.

Engineering Stress-Strain Curve

In the conventional engineering tension test, an engineering stress-strain curve is constructed from the load-elongation measurements made on the test specimen (Fig. 1). The engineering stress, s, used in this stress-strain curve is the average longitudinal stress in the tensile specimen. It is obtained by dividing the load, P, by the original area of the cross section of the specimen, A_0:

$$s = \frac{P}{A_0} \qquad \text{(Eq 1)}$$

The strain, e, used for the engineering stress-strain curve is the average linear strain, which is obtained by dividing the elongation of the gage length of the specimen, δ, by its original length, L_0:

$$e = \frac{\delta}{L_0} = \frac{\Delta L}{L_0} = \frac{L - L_0}{L_0} \qquad \text{(Eq 2)}$$

*Reprinted from *Mechanical Metallurgy*, 2nd ed., McGraw-Hill, New York, 1976, p 329-348. With permission.

Because both the stress and the strain are obtained by dividing the load and elongation by constant factors, the load-elongation curve has the same shape as the engineering stress-strain curve. The two curves frequently are used interchangeably.

The shape and magnitude of the stress-strain curve of a metal depend on its composition, heat treatment, prior history of plastic deformation, and the strain rate, temperature, and state of stress imposed during the testing. The parameters that are used to describe the stress-strain curve of a metal are the tensile strength, yield strength or yield point, percent elongation, and reduction in area. The first two are strength parameters; the last two indicate ductility.

The general shape of the engineering stress-strain curve (Fig. 1) requires further explanation. In the elastic region, stress is linearly proportional to strain. When the stress exceeds a value corresponding to the yield strength, the specimen undergoes gross plastic deformation. If the load is subsequently reduced to zero, the specimen will remain permanently deformed. The stress required to produce continued plastic deformation increases with increasing plastic strain; i.e., the metal strain hardens. The volume of the specimen (area × length) remains constant during plastic deformation, $AL = A_0L_0$, and as the specimen elongates, its cross-sectional area decreases uniformly along the gage length.

Initially, the strain hardening more than compensates for this decrease in area, and the engineering stress (proportional to load P) continues to rise with increasing strain. Eventually, a point is reached where the decrease in specimen cross-sectional area is greater than the increase in deformation load

arising from strain hardening. This condition will be reached first at some point in the specimen that is slightly weaker than the rest. All further plastic deformation is concentrated in this region, and the specimen begins to neck or thin down locally. Because the cross-sectional area now is decreasing far more rapidly than the deformation load is increased by strain hardening, the actual load required to deform the specimen falls off and the engineering stress defined in Eq 1 continues to decrease until fracture occurs.

The tensile strength, or ultimate tensile strength, s_u, is the maximum load divided by the original cross-sectional area of the specimen:

$$s_u = \frac{P_{max}}{A_0} \qquad \text{(Eq 3)}$$

The tensile strength is the value most frequently quoted from the results of a tension test. Actually, however, it is a value of little fundamental significance with regard to the strength of a metal. For ductile metals, the tensile strength should be regarded as a measure of the maximum load that a metal can withstand under the very restrictive conditions of uniaxial loading. This value bears little relation to the useful strength of the metal under the more complex conditions of stress that usually are encountered.

For many years, it was customary to base the strength of members on the tensile strength, suitably reduced by a factor of safety. The current trend is to the more rational approach of basing the static design of ductile metals on the yield strength. However, because of the long practice of using the tensile strength to describe the strength of materials, it has become a familiar property, and as such, it is a useful identification of a material in the same sense that the chemical composition serves to identify a metal or

Fig. 1 Engineering stress-strain curve

Intersection of the dashed line with the curve determines the offset yield strength. See also Fig. 2 and corresponding text.

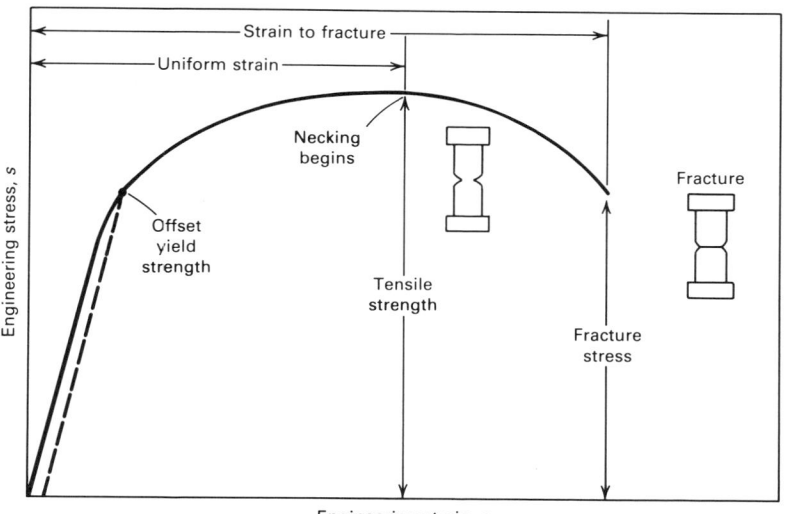

Fig. 2 Typical tension stress-strain curve for ductile metal indicating yielding criteria

Point A, elastic limit; point A', proportional limit; point B, yield strength; line C-B, offset yield strength; 0, intersection of the stress-strain curve with the strain axis

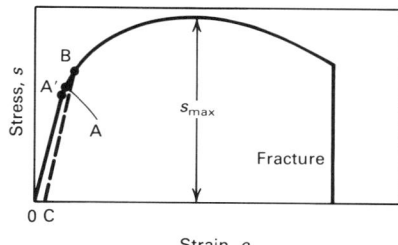

alloy. Furthermore, because the tensile strength is easy to determine and is a reproducible property, it is useful for the purposes of specification and for quality control of a product. Extensive empirical correlations between tensile strength and properties such as hardness and fatigue strength are often useful. For brittle materials, the tensile strength is a valid design criterion.

Measures of Yielding. The stress at which plastic deformation or yielding is observed to begin depends on the sensitivity of the strain measurements. With most materials, there is a gradual transition from elastic to plastic behavior, and the point at which plastic deformation begins is difficult to define with precision. In tests of materials under uniaxial loading, three criteria for the initiation of yielding have been used: the elastic limit, the proportional limit, and the yield strength.

Elastic limit, shown at point A in Fig. 2, is the greatest stress the material can withstand without any measurable permanent strain remaining after the complete release of load. With increasing sensitivity of strain measurement, the value of the elastic limit is decreased until it equals the true elastic limit determined from microstrain measurements. With the sensitivity of strain typically used in engineering studies (10^{-4} in./in.), the elastic limit is greater than the proportional limit. Determination of the elastic limit requires a tedious incremental loading-unloading test procedure. For this reason, it is often replaced by the proportional limit.

Proportional limit, shown at point A' in Fig. 2, is the highest stress at which stress is directly proportional to strain. It is obtained by observing the deviation from the straight-line portion of the stress-strain curve.

The yield strength, shown at point B in Fig. 2, is the stress required to produce a small specified amount of plastic deformation. The usual definition of this property is the offset yield strength determined by the stress corresponding to the intersection of the stress-strain curve and a line parallel to the elastic part of the curve offset by a specified strain (see Fig. 1 and 2). In the United States, the offset is usually specified as a strain of 0.2% or 0.1% ($e = 0.002$ or 0.001):

$$s_0 = \frac{P_{(\text{strain offset} = 0.002)}}{A_0} \qquad \text{(Eq 4)}$$

Offset yield strength determination requires a specimen that has been loaded to its 0.2% offset yield strength and unloaded so that it is 0.2% longer than before the test. The offset yield strength is often referred to in Great Britain as the proof stress, where offset values are either 0.1% or 0.5%. The yield strength obtained by an offset method is commonly used for design and specification purposes, because it avoids the practical difficulties of measuring the elastic limit or proportional limit.

Some materials have essentially no linear portion to their stress-strain curve, for example, soft copper or gray cast iron. For these materials, the offset method cannot be used, and the usual practice is to define the yield strength as the stress to produce some total strain, for example, $e = 0.005$.

Many metals, particularly annealed low-carbon steel, show a localized, heterogeneous type of transition from elastic to plastic deformation that produces a yield point in the stress-strain curve. Rather than having a flow curve with a gradual transition from elastic to plastic behavior, such as in Fig. 1 and 2, metals with a yield point produce a flow curve or a load-elongation diagram similar to Fig. 3. The load increases steadily with elastic strain, drops suddenly, fluctuates about some approximately constant value of load, and then rises with further strain.

The load at which the sudden drop occurs is called the upper yield point. The constant load is called the lower yield point, and the elongation that occurs at constant load is called the yield-point elongation. The deformation occurring throughout the yield-point elongation is heterogeneous. At the upper yield point, a discrete band of deformed metal, often readily visible, appears at a stress concentration such as a fillet. Coincident with the formation of the band, the load drops to the lower yield point. The band then propagates along the length of the specimen, causing the yield-point elongation.

In typical cases, several bands form at several points of stress concentration. These bands are generally at approximately 45° to the tensile axis. They are usually called Lüders bands, Hartmann lines, or stretcher strains, and this type of deformation is sometimes referred to as the Piobert effect. When several Lüders bands are formed, the flow curve during the yield-point elongation is irregular, each jog corresponding to the formation of a new Lüders band. After the Lüders bands have propagated to cover the entire length of the specimen test section, the flow will increase with strain in the typical manner. This marks the end of the yield-point elongation. Lüders bands formed on a rimmed 1008 steel are shown in Fig. 4.

Measures of Ductility. Currently, ductility is considered a qualitative, subjective property of a material. In general, measurements of ductility are of interest in three respects (Ref 1):

Fig. 3 Typical yield point behavior of low-carbon steel

The slope of the initial linear portion of the stress-strain curve, designated by E, is the modulus of elasticity.

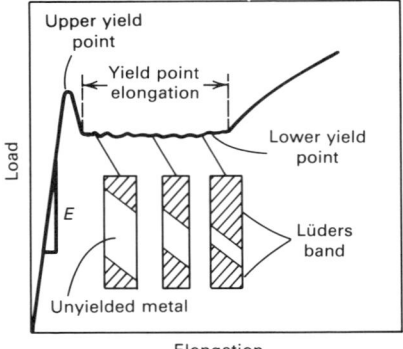

- To indicate the extent to which a metal can be deformed without fracture in metal-working operations such as rolling and extrusion.
- To indicate to the designer the ability of the metal to flow plastically before fracture. A high ductility indicates that the material is "forgiving" and likely to deform locally without fracture should the designer err in the stress calculation or the prediction of severe loads.
- To serve as an indicator of changes in impurity level or processing conditions. Ductility measurements may be specified to assess material quality, even though no direct relationship exists between the ductility measurement and performance in service.

The conventional measures of ductility that are obtained from the tension test are the engineering strain at fracture e_f (usually called the elongation) and the reduction in area at fracture q. Elongation and reduction in area usually are expressed as a percentage. Both of these properties are obtained after fracture by putting the specimen back together and taking measurements of the final length, L_f, and final specimen cross section, A_f:

$$e_f = \frac{L_f - L_0}{L_0} \qquad \text{(Eq 5)}$$

$$q = \frac{A_0 - A_f}{A_0} \qquad \text{(Eq 6)}$$

Because an appreciable fraction of the plastic deformation will be concentrated in the necked region of the tension specimen, the value of e_f will depend on the gage length L_0 over which the measurement was taken (see the section of this article on ductility measurement in tension testing). The smaller

Fig. 4 Rimmed 1008 steel with Lüders bands on the surface as a result of stretching the sheet just beyond the yield point during forming

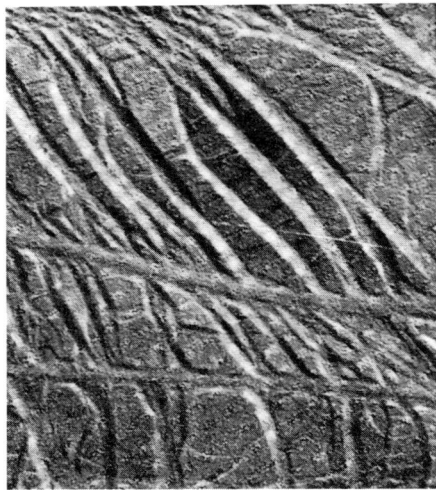

the gage length, the greater the contribution to the overall elongation from the necked region and the higher the value of e_f. Therefore, when reporting values of percentage elongation, the gage length, L_0, should always be given.

Reduction in area does not suffer from this difficulty. These values can be converted into an equivalent zero-gage-length elongation, e_0. From the constancy of volume relationship for plastic deformation, $AL = A_0L_0$:

$$\frac{L}{L_0} = \frac{A_0}{A} = \frac{1}{1 - q}$$

$$e_0 = \frac{L - L_0}{L_0} = \frac{A_0}{A} - 1 = \frac{1}{1 - q} - 1$$

$$= \frac{q}{1 - q} \qquad \text{(Eq 7)}$$

This represents the elongation based on a very short gage length near the fracture.

Another way to avoid the complications resulting from necking is to base the percentage elongation on the uniform strain out to the point at which necking begins. The uniform elongation, e_u, correlates well with stretch-forming operations. Because the engineering stress-strain curve often is quite flat in the vicinity of necking, it may be difficult to establish the strain at maximum load without ambiguity. In this case, the method suggested in Ref 2 is useful.

Modulus of Elasticity. The slope of the initial linear portion of the stress-strain curve is the modulus of elasticity, or Young's modulus, as shown in Fig. 3. The modulus of

elasticity, E, is a measure of the stiffness of the material. The greater the modulus, the smaller the elastic strain resulting from the application of a given stress. Because the modulus of elasticity is needed for computing deflections of beams and other members, it is an important design value.

The modulus of elasticity is determined by the binding forces between atoms. Because these forces cannot be changed without changing the basic nature of the material, the modulus of elasticity is one of the most structure-insensitive of the mechanical properties. Generally, it is only slightly affected by alloying additions, heat treatment, or cold work (Ref 3). However, increasing the temperature decreases the modulus of elasticity. At elevated temperatures, the modulus is often measured by a dynamic method (Ref 4). Typical values of modulus of elasticity for common engineering metals at different temperatures are given in Table 1.

Resilience. The ability of a material to absorb energy when deformed elastically and to return it when unloaded is called resilience. This property usually is measured by the modulus of resilience, which is the strain energy per unit volume, U_0, required to stress the material from zero stress to the yield stress, σ_0. The strain energy per unit volume for uniaxial tension is:

$$U_0 = \frac{1}{2} \sigma_x e_x \qquad \text{(Eq 8)}$$

From the above definition, the modulus of resilience, U_R, is:

$$U_R = \frac{1}{2} s_0 e_0 = \frac{1}{2} s_0 \frac{s_0}{E} = \frac{s_0^2}{2E} \qquad \text{(Eq 9)}$$

This equation indicates that the ideal material for resisting energy loads in applications where the material must not undergo permanent distortion, such as mechanical springs, is one having a high yield stress and a low modulus of elasticity.

For various grades of steel, the modulus of resilience ranges from 100 to 4500 kJ/m³ (14.5 to 650 lbf-in./in.³), with the higher values representing steels with higher carbon or alloy contents (Ref 5). The cross-hatched regions in Fig. 5 indicate the modulus of resilience for two steels. Because of its higher yield strength, the high-carbon spring steel has the greater resilience.

The toughness of a material is its ability to absorb energy in the plastic range. The ability to withstand occasional stresses above the yield stress without fracturing is particularly desirable in parts such as freight-car couplings, gears, chains, and crane hooks. Toughness is a commonly used concept that is difficult to precisely define. Toughness may be considered to be the total area under the stress-strain curve. This area, which is

Table 1 Typical values of modulus of elasticity at different temperatures

Material	Room temperature	Modulus of elasticity, GPa (10^6 psi), at:			
		250 °C (400 °F)	425 °C (800 °F)	540 °C (1000 °F)	650 °C (1200 °F)
Carbon steel	207 (30.0)	186 (27.0)	155 (22.5)	134 (19.5)	124 (18.0)
Austenitic					
stainless steel	193 (28.0)	176 (25.5)	159 (23.0)	155 (22.5)	145 (21.0)
Titanium alloys	114 (16.5)	96.5 (14.0)	74 (10.7)	70 (10.0)	. . .
Aluminum alloys	72 (10.5)	65.5 (9.5)	54 (7.8)	. . .	. . .

Fig. 5 Comparison of stress-strain curves for high- and low-toughness steels

Cross-hatched regions in this curve represent the modulus of resilience, U_R, of the two materials. The U_R is determined by measuring the area under the stress-strain curve up to the elastic limit of the material. Point A represents the elastic limit of the spring steel; point B that of the structural steel.

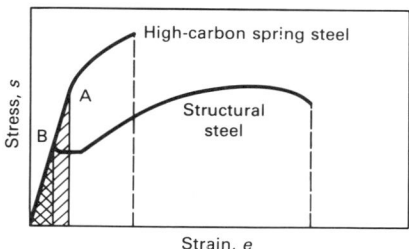

referred to as the modulus of toughness, U_T, is an indication of the amount of work per unit volume that can be done on the material without causing it to rupture.

Figure 5 shows the stress-strain curves for high- and low-toughness materials. The high-carbon spring steel has a higher yield strength and tensile strength than the medium-carbon structural steel. However, the structural steel is more ductile and has a greater total elongation. The total area under the stress-strain curve is greater for the structural steel; therefore, it is a tougher material. This illustrates that toughness is a parameter that comprises both strength and ductility.

Several mathematical approximations for the area under the stress-strain curve have been suggested. For ductile metals that have a stress-strain curve like that of the structural steel, the area under the curve can be approximated by:

$$U_T \approx s_u e_f \qquad \text{(Eq 10)}$$

or

$$U_T \approx \frac{s_0 + s_u}{2} e_f \qquad \text{(Eq 11)}$$

For brittle materials, the stress-strain curve is sometimes assumed to be a parabola, and the area under the curve is given by:

$$U_T \approx \frac{2}{3} s_u e_f \qquad \text{(Eq 12)}$$

True Stress/True Strain Curve

The engineering stress-strain curve does not give a true indication of the deformation characteristics of a metal, because it is based entirely on the original dimensions of the specimen, and these dimensions change continuously during the test. Also, ductile metal that is pulled in tension becomes unstable and necks down during the course of the test. Because the cross-sectional area of the specimen is decreasing rapidly at this stage in the test, the load required to continue deformation falls off.

The average stress based on the original area likewise decreases, and this produces the fall-off in the engineering stress-strain curve beyond the point of maximum load. Actually, the metal continues to strain harden to fracture, so that the stress required to produce further deformation should also increase. If the true stress, based on the actual cross-sectional area of the specimen, is used, the stress-strain curve increases continuously to fracture. If the strain measurement is also based on instantaneous measurement, the curve that is obtained is known as true stress/true strain curve. This is also known as a flow curve, because it represents the basic plastic-flow characteristics of the material.

Any point on the flow curve can be considered the yield stress for a metal strained in tension by the amount shown on the curve. Thus, if the load is removed at this point and then reapplied, the material will behave elastically throughout the entire range of reloading. The true stress, σ, is expressed in terms of engineering stress, s, by:

$$\sigma = \frac{P}{A_0}(e + 1) = s(e + 1) \qquad \text{(Eq 13)}$$

The derivation of Eq 13 assumes both constancy of volume and a homogeneous distribution of strain along the gage length of the tension specimen. Thus, Eq 13 should be used only until the onset of necking. Beyond the maximum load, the true stress should be determined from actual measurements of load and cross-sectional area.

$$\sigma = \frac{P}{A} \qquad \text{(Eq 14)}$$

The true strain, ϵ, may be determined from the engineering or conventional strain, e, by:

$$\epsilon = \ln(e + 1) = \ln \frac{L}{L_0} \qquad \text{(Eq 15)}$$

This equation is applicable only to the onset of necking for the reasons discussed above. Beyond maximum load, the true strain should be based on actual area or diameter, D, measurements:

$$\epsilon = \ln \frac{A_0}{A}$$

$$= \ln \frac{\left(\frac{\pi}{4}\right) D_0^2}{\left(\frac{\pi}{4}\right) D^2} = 2 \ln \frac{D_0}{D} \qquad \text{(Eq 16)}$$

Figure 6 compares the true stress/true strain curve with its corresponding engineering stress-strain curve. Note that, because of the relatively large plastic strains, the elastic region has been compressed into the y axis. In agreement with Eq 13 and 15, the true stress/true strain curve is always to the left of the engineering curve until the maximum load is reached.

However, beyond maximum load, the high, localized strains in the necked region that are used in Eq 16 far exceed the engineering strain calculated from Eq 2. Frequently, the flow curve is linear from maximum load to fracture, while in other cases its slope continuously decreases to fracture. The formation of a necked region or mild notch introduces triaxial stresses that make it difficult to determine accurately the longitudinal tensile stress from the onset of necking until fracture occurs. This concept is discussed in greater detail in the section of this article on instability in tension. The following parameters usually are determined from the true stress/true strain curve.

Fig. 6 Comparison of engineering and true stress/true strain curves

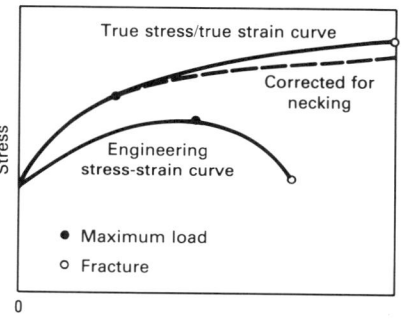

The true stress at maximum load corresponds to the true tensile strength. For most materials, necking begins at maximum load at a value of strain where the true stress equals the slope of the flow curve. Let σ_u and ϵ_u denote the true stress and true strain at maximum load when the cross-sectional area of the specimen is A_u. The ultimate tensile strength can be defined as:

$$s_u = \frac{P_{max}}{A_0} \qquad \text{(Eq 17)}$$

and

$$\sigma_u = \frac{P_{max}}{A_u} \qquad \text{(Eq 18)}$$

Eliminating P_{max} yields:

$$\sigma_u = s_u \frac{A_0}{A_u} \qquad \text{(Eq 19)}$$

and

$$\sigma_u = s_u e^{\epsilon_u} \qquad \text{(Eq 20)}$$

The true fracture stress is the load at fracture divided by the cross-sectional area at fracture. This stress should be corrected for the triaxial state of stress existing in the tensile specimen at fracture. Because the data required for this correction frequently are not available, true fracture stress values are frequently in error.

The true fracture strain, ϵ_f, is the true strain based on the original area, A_0, and the area after fracture, A_f:

$$\epsilon_f = \ln \frac{A_0}{A_f} \qquad \text{(Eq 21)}$$

This parameter represents the maximum true strain that the material can withstand before fracture and is analogous to the total strain to fracture of the engineering stress-strain curve. Because Eq 15 is not valid beyond the onset of necking, it is not possible to calculate ϵ_f from measured values of e_f. However, for cylindrical tensile specimens, the reduction in area, q, is related to the true fracture strain by:

$$\epsilon_f = \ln \frac{1}{1 - q} \qquad \text{(Eq 22)}$$

The true uniform strain, ϵ_u, is the true strain based only on the strain up to maximum load. It may be calculated from either the specimen cross-sectional area, A_u, or the gage length, L_u, at maximum load. Equation 15 may be used to convert conventional uniform strain to true uniform strain. The uniform strain frequently is useful in estimating the formability of metals from the results of a tension test:

$$\epsilon_u = \ln \frac{A_0}{A_u} \qquad \text{(Eq 23)}$$

More detailed information on the use of the uniaxial tension test for estimating formability can be found in the article "Sheet Formability Testing" in this Volume.

The true local necking strain, ϵ_n, is the strain required to deform the specimen from maximum load to fracture:

$$\epsilon_n = \ln \frac{A_u}{A_f} \qquad \text{(Eq 24)}$$

The flow curve of many metals in the region of uniform plastic deformation can be expressed by the simple power curve relation:

$$\sigma = K\epsilon^n \qquad \text{(Eq 25)}$$

where n is the strain-hardening exponent, and K is the strength coefficient. A log-log plot of true stress and true strain up to maximum load will result in a straight line if Eq 25 is satisfied by the data (Fig 7). The linear slope of this line is n, and K is the true stress at $\epsilon = 1.0$ (corresponds to $q = 0.63$). As shown in Fig. 8, the strain-hardening exponent may have values from $n = 0$ (perfectly plastic solid) to $n = 1$ (elastic solid). For most metals, n has values between 0.10 and 0.50 (see Table 2).

The rate of strain hardening $d\sigma/d\epsilon$ is not identical to the strain-hardening exponent. From the definition of n:

$$n = \frac{d(\log \sigma)}{d(\log \epsilon)} = \frac{d(\ln \sigma)}{d(\ln \epsilon)} = \frac{\epsilon}{\sigma}\frac{d\sigma}{d\epsilon}$$

or

$$\frac{d\sigma}{d\epsilon} = n\frac{\sigma}{\epsilon} \qquad \text{(Eq 26)}$$

Deviations from Eq 25 frequently are observed, often at low strains (10^{-3}) or high strains ($\epsilon \approx 1.0$). One common type of deviation is for a log-log plot of Eq 25 to result in two straight lines with different slopes. Sometimes data that do not plot according to Eq 25 will yield a straight line according to the relationship:

$$\sigma = K(\epsilon_0 + \epsilon)^n \qquad \text{(Eq 27)}$$

ϵ_0 can be considered to be the amount of strain hardening that the material received prior to the tension test (Ref 8). Another

Fig. 7 Log-log plot of true stress/true strain curve

n is the strain-hardening exponent; K is the strength coefficient.

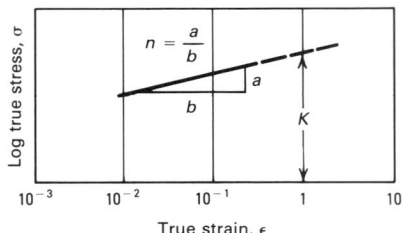

Fig. 8 Various forms of power curve $\sigma = K\epsilon^n$

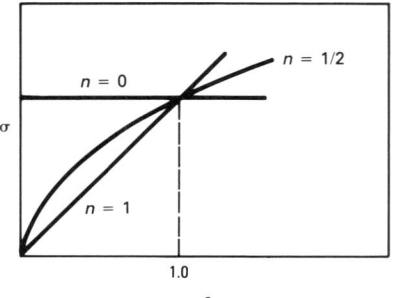

common variation on Eq 25 is the Ludwik equation:

$$\sigma = \sigma_0 + K\epsilon^n \qquad \text{(Eq 28)}$$

where σ_0 is the yield stress, and K and n are the same constants as in Eq 25. This equation may be more satisfying than Eq 25, because the latter implies that at zero true strain the stress is zero. It has been shown that σ_0 can be obtained from the intercept of the strain-hardening portion of the stress-strain curve and the elastic modulus line by (Ref 9):

$$\sigma_0 = \left(\frac{K}{E^n}\right)^{1/1-n} \qquad \text{(Eq 29)}$$

The true stress/true strain curve of metals such as austenitic stainless steel, which deviate markedly from Eq 25 at low strains (Ref 10), can be expressed by:

Table 2 Values for n and K for metals at room temperature

Metal	Condition	n	K MPa	K ksi	Ref
0.05% carbon steel	Annealed	0.26	530	77	6
SAE 4340 steel	Annealed	0.15	641	93	6
0.6% carbon steel	Quenched and tempered at 540 °C (1000 °F)	0.10	1572	228	7
0.6% carbon steel	Quenched and tempered at 705 °C (1300 °F)	0.19	1227	178	7
Copper	Annealed	0.54	320	46.4	6
70/30 brass	Annealed	0.49	896	130	7

$$\sigma = K\epsilon^n + e^{K_1} e^{n_1\epsilon} \qquad \text{(Eq 30)}$$

where e^{K_1} is approximately equal to the proportional limit, and n_1 is the slope of the deviation of stress from Eq 25 plotted against ϵ. Other expressions for the flow curve are available (Ref 11, 12). The true strain term in Eq 25 to 28 properly should be the plastic strain, $\epsilon_p = \epsilon_{\text{total}} - \epsilon_E = \epsilon_{\text{total}} - \sigma/E$, where ϵ_E represents elastic strain.

Instability in Tension

Necking generally begins at maximum load during the tensile deformation of a ductile metal. An exception to this is the behavior of cold rolled zirconium tested at 200 to 370 °C (390 to 700 °F), where necking occurs at a strain of twice the strain at maximum load (Ref 13). An ideal plastic material in which no strain hardening occurs would become unstable in tension and begin to neck as soon as yielding occurred. However, an actual metal undergoes strain hardening, which tends to increase the load-carrying capacity of the specimen as deformation increases.

This effect is opposed by the gradual decrease in the cross-sectional area of the specimen as it elongates. Necking or localized deformation begins at maximum load, where the increase in stress due to decrease in the cross-sectional area of the specimen becomes greater than the increase in the load-carrying ability of the metal due to strain hardening. This condition of instability leading to localized deformation is defined by the condition $dP = 0$:

$$P = \sigma A \qquad \text{(Eq 31)}$$

$$dP = \sigma\, dA + A\, d\sigma = 0 \qquad \text{(Eq 32)}$$

From the constancy-of-volume relationship:

$$\frac{dL}{L} = -\frac{dA}{A} = d\epsilon \qquad \text{(Eq 33)}$$

and from the instability condition:

$$-\frac{dA}{A} = \frac{d\sigma}{\sigma} \qquad \text{(Eq 34)}$$

so that at a point of tensile instability:

$$\frac{d\sigma}{d\epsilon} = \sigma \qquad \text{(Eq 35)}$$

Therefore, the point of necking at maximum load can be obtained from the true stress/true strain curve by finding the point on the curve having a subtangent of unity (Fig. 9a), or the point where the rate of strain hardening equals the stress (Fig. 9b). The necking criterion can be expressed more explicitly if engineering strain is used. Starting with Eq 35:

Fig. 9 Graphical interpretation of necking criterion

The point of necking at maximum load can be obtained from the true stress/true strain curve by finding (a) the point on the curve having a subtangent of unity or (b) the point where $d\sigma/d\epsilon = \sigma$.

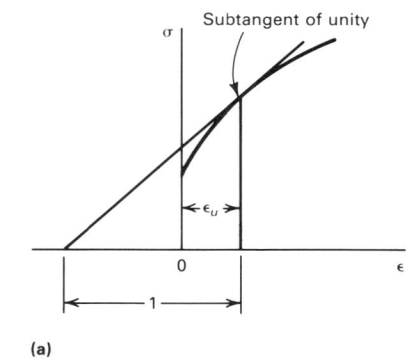

(a)

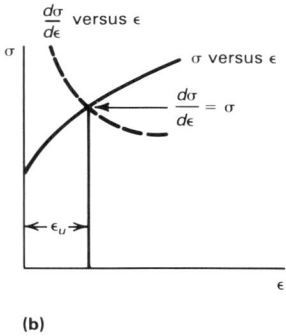

(b)

$$\frac{d\sigma}{d\epsilon} = \frac{d\sigma}{de}\frac{de}{d\epsilon} = \frac{d\sigma}{de}\frac{\dfrac{dL}{L_0}}{\dfrac{dL}{L}} = \frac{d\sigma}{de}\frac{L}{L_0}$$

$$= \frac{d\sigma}{de}(1 + e) = \sigma$$

$$\frac{d\sigma}{de} = \frac{\sigma}{1 + e} \qquad \text{(Eq 36)}$$

Equation 36 permits an interesting geometrical construction for the determination of the point of maximum load (Ref 14). In Fig. 10, the stress-strain curve is plotted in terms of true stress against conventional linear strain. Let point A represent a negative strain of 1.0. A line drawn from point A, which is tangent to the stress-strain curve, will establish the point of maximum load, because according to Eq 36, the slope at this point is $\sigma/(1 + e)$.

By substituting the necking criterion given in Eq 35 into Eq 26, a simple relationship for the strain at which necking occurs is obtained. This strain is the true uniform strain, e_u:

$$\epsilon_u = n \qquad \text{(Eq 37)}$$

Although Eq 26 is based on the assumption that the flow curve is given by Eq 25, it has been shown that $\epsilon_u = n$ does not depend on this power law behavior (Ref 15).

Stress Distribution at the Neck

The formation of a neck in the tensile specimen introduces a complex triaxial state of stress in that region. The necked region is in effect a mild notch. A notch under tension produces radial stress, σ_r, and transverse stress, σ_t, which raise the value of longitudinal stress required to cause the plastic flow. Therefore, the average true stress at the neck, which is determined by dividing the axial tensile load by the minimum cross-sectional area of the specimen at the neck, is higher than the stress that would be required to cause flow if simple tension prevailed.

Figure 11 illustrates the geometry at the necked region and the stresses developed by this localized deformation. R is the radius of curvature of the neck, which can be measured either by projecting the contour of the necked region on a screen or by using a tapered, conical radius gage.

Bridgman made a mathematical analysis that provides a correction to the average axial stress to compensate for the introduction of transverse stresses (Ref 16). This analysis was based on the following assumptions:

- The contour of the neck is approximated by the arc of a circle.
- The cross section of the necked region remains circular throughout the test.
- The von Mises criterion for yielding applies.

Fig. 10 Considére's construction for the determination of the point of maximum load

Source: Ref 14

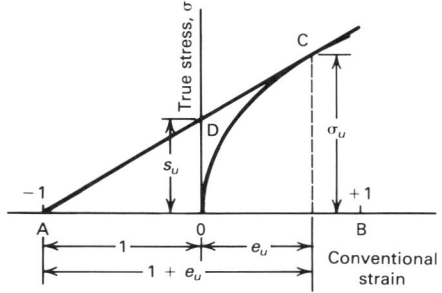

Fig. 11 Stress distribution at the neck of a tensile specimen

(a) Geometry of necked region. R is the radius of curvature of the neck; a is the minimum radius at the neck. (b) Stresses acting on element at point O. σ_x is the stress in the axial direction; σ_r is the radial stress; σ_t is the transverse stress.

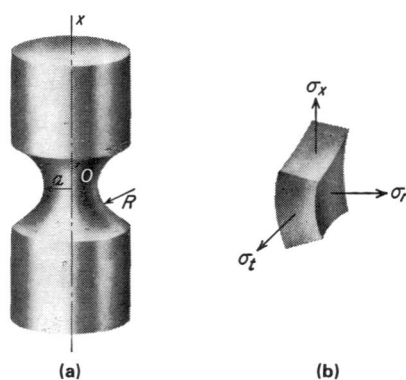

(a)   (b)

- The strains are constant over the cross section of the neck.

According to this analysis, the uniaxial flow stress corresponding to that which would exist in the tension test if necking had not introduced triaxial stresses is:

$$\sigma = \frac{(\sigma_x)_{avg}}{\left(\dfrac{1 + 2R}{a}\right)\left[\ln\left(1 + \dfrac{a}{2R}\right)\right]} \quad \text{(Eq 38)}$$

where $(\sigma_x)_{avg}$ is the measured stress in the axial direction (load divided by minimum cross section). Figure 6 shows how the application of the Bridgman correction changes the true stress/true strain curve. A correction for the triaxial stresses in the neck of a flat tensile specimen has been considered (Ref 17). The values of a/R needed for the analysis can be obtained either by straining a specimen a given amount beyond necking and unloading to measure a and R directly, or by measuring these parameters continuously past necking using photography or a tapered ring gage (Ref 18).

To avoid these measurements, Bridgman presented an empirical relation between a/R and the true strain in the neck. Figure 12 shows that this gives close agreement for steel specimens, but not for other metals with widely different necking strains. A much better correlation is obtained between the Bridgman correction and the true strain in the neck minus the true strain at necking, ϵ_u (Ref 20).

Ductility Measurement in Tension Testing

The measured elongation from a tension specimen depends on the gage length of the specimen, or the dimensions of its cross section. This is because the total extension consists of two components: the uniform extension up to necking and the localized extension once necking begins. The extent of uniform extension depends on the metallurgical condition of the material (through n) and the effect of specimen size and shape on the development of the neck.

Figure 13 illustrates the variation of the local elongation, as defined in Eq 7, along the gage length of a prominently necked tensile specimen. The shorter the gage length, the greater the influence of localized deformation at the neck on the total elongation of the gage length. The extension of a specimen at fracture can be expressed by:

$$L_f - L_0 = \alpha + e_u L_0 \quad \text{(Eq 39)}$$

where α is the local necking extension, and $e_u L_0$ is the uniform extension. The tensile elongation is then:

$$e_f = \frac{L_f - L_0}{L_0} = \frac{\alpha}{L_0} + e_u \quad \text{(Eq 40)}$$

This clearly indicates that the total elongation is a function of the specimen gage length. The shorter the gage length, the greater the percent elongation.

Numerous attempts have been made to rationalize the strain distribution in the tension test. Perhaps the most general conclusion that can be drawn is that geometrically similar specimens develop geometrically similar necked regions. According to Barba's law (Ref 21), $\alpha = \beta\sqrt{A_0}$, and the elongation equation becomes:

$$e_f = \beta \frac{\sqrt{A_0}}{L_0} + e_u \quad \text{(Eq 41)}$$

where β is a coefficient of proportionality.

To compare elongation measurements of different sized specimens, the specimens must be geometrically similar. Equation 41 shows that the critical geometrical factor for which similitude must be maintained is $L_0/\sqrt{A_0}$ for sheet specimens, or L_0/D_0 for round bars. In the United States, the standard round tensile specimen has a 12.8-mm (0.505-in.) diam and a 50-mm (2-in.) gage length. Sub-size specimens have the following respective diameter and gage length: 9.06 mm and 35.6 mm (0.357 in. and 1.4 in.), 6.4 mm and 25 mm (0.252 in. and 1.0 in.), and 4.06 mm and 16.1 mm (0.160 in. and 0.634 in.). Different values of $L_0/\sqrt{A_0}$ are specified for sheet specimens by the standardizing agencies in different countries. In the United States, ASTM recommends a $L_0/\sqrt{A_0}$ value of 4.5 for sheet specimens and a L_0/D_0 value of 4.0 for round specimens.

Generally, a given elongation will be produced in a material if $\sqrt{A_0}/L_0$ is maintained

Fig. 12 Relationship between Bridgman correction factor $\sigma/(\sigma_x)_{avg}$ and true tensile strain

Source: Ref 19

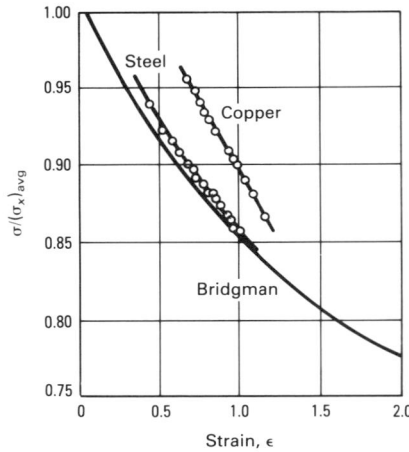

Fig. 13 Variation of local elongation with position along gage length of tensile specimen

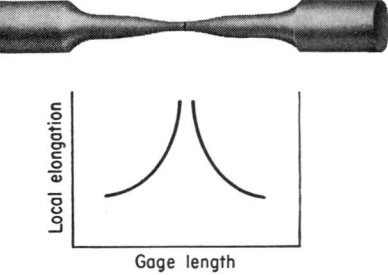

constant as predicted by Eq 41. Thus, at a constant value of elongation $\sqrt{A_1}/L_1 = \sqrt{A_2}/L_2$, where A and L are the areas and gage lengths of two different specimens, 1 and 2, of the same metal. To predict elongation using gage length L_2 on a specimen with area A_2 by means of measurements on a specimen with area A_1, it only is necessary to adjust the gage length of specimen 1 to conform with $L_1 = L_2\sqrt{A_1/A_2}$. For example, suppose that a 3.2-mm (0.125-in.) thick sheet is available, and one wishes to predict the elongation with a 50-mm (2-in.) gage length for the identical material but in 2.0-mm (0.080-in.) thickness. Using 12.7-mm (0.5-in.) wide sheet specimens, a test specimen with a gage length $L = 50$ mm$(3.2$ mm/2.0 mm)$^{1/2}$ = 63 mm, or 2 in.$(0.125$ in./0.080 in.)$^{1/2}$ = 2.5 in., made from the 3.2-mm (0.125-in.) sheet would be predicted to give the same elongation as a 50-mm (2-in.) gage length in 2.0-mm (0.080-in.) thick sheet. Experimental verification for this procedure has been shown in Ref 22.

The occurrence of necking in the tension test, however, makes any quantitative conversion between elongation and reduction in area impossible. Although elongation and reduction in area usually vary in the same way—for example, as a function of test temperature, tempering temperature, alloy content, etc.—this is not always the case. Generally, elongation and reduction in area measure different types of material behavior. Provided the gage length is not too short, percent elongation is primarily influenced by uniform elongation, and thus it is dependent on the strain-hardening capacity of the material.

Reduction in area is more a measure of the deformation required to produce fracture, and its chief contribution results from the necking process. Because of the complicated stress state in the neck, values of reduction in area are dependent on specimen geometry and deformation behavior, and they should not be taken as true material properties. However, reduction in area is the most structure-sensitive ductility parameter, and as such, it is useful in detecting quality changes in the material.

Notch Tensile Test

Ductility measurements on standard smooth tensile specimens do not always reveal metallurgical or environmental changes that lead to reduced local ductility. The tendency for reduced ductility in the presence of a triaxial stress field and steep stress gradients (such as a rise at a notch) is called notch sensitivity. A common way of evaluating notch sensitivity is a tension test using a notched specimen.

The notch tensile test has been used extensively for investigating the properties of high-strength steels, for studying hydrogen embrittlement in steels and titanium, and for investigating the notch sensitivity of high-temperature alloys. More recently, notched tension specimens have been used for fracture mechanics measurements (see the Section on "Fracture Mechanics" in this Volume). Notch sensitivity can also be investigated with the notched-impact test.

The most common notch tensile specimen uses a 60° notch with a root radius 0.025 mm (0.001 in.) or less introduced into a round (circumferential notch) or flat (double-edge notch) tensile specimen. Usually, the depth of the notch is such that the cross-sectional area at the root of the notch is one half of the area in the unnotched section. The specimen is aligned carefully and loaded in tension until fracture occurs. The notch strength is defined as the maximum load divided by the original cross-sectional area at the notch. Because of the plastic constraint at the notch,

this value will be higher than the tensile strength of an unnotched specimen if the material possesses some ductility. Therefore, the common way of detecting notch brittleness (or high notch sensitivity) is by determining the notch-strength ratio, NSR:

$$NSR = \frac{s_{net} \text{ (for notched specimen at maximum load)}}{s_u \text{ (tensile strength for unnotched specimen)}}$$

(Eq 42)

If the NSR is less than unity, the material is notch brittle. The other property that is measured in the notch tensile test is the reduction in area at the notch.

As strength, hardness, or some metallurgical variable restricting plastic flow increases, the metal at the root of the notch is less able to flow, and fracture becomes more likely. Notch brittleness may be considered to begin at the strength level where the notch strength begins to fall or, more conventionally, at the strength level where the NSR becomes less than unity.

The sensitivity of notch strength for detecting metallurgical embrittlement is illustrated in Fig. 14. Note that the conventional elongation measured on a smooth specimen was unable to detect the fall in notch strength produced by tempering in the 330 to 480 °C (600 to 900 °F) range. For a more detailed review of notch tensile testing, see Ref 24.

REFERENCES

1. G.E. Dieter, Introduction to Ductility, in *Ductility*, American Society for Metals, 1968
2. P.G. Nelson and J. Winlock, *ASTM Bull.*, Vol 156, Jan 1949, p 53
3. D.J. Mack, *Trans. AIME*, Vol 166, 1946, p 68-85
4. P.E. Armstrong, Measurement of Elastic Constants, in *Techniques of Metals Research*, Vol V, R.F. Brunshaw, Ed., Interscience, New York, 1971
5. H.E. Davis, G.E. Troxell, and G.F.W. Hauck, *The Testing of Engineering Materials*, McGraw-Hill, New York, 1964, p 33
6. J.R. Low and F. Garofalo, *Proc. Soc. Exp. Stress Anal.*, Vol 4 (No. 2), 1947, p 16-25
7. J.R. Low, *Properties of Metals in Materials Engineering*, American Society for Metals, 1949
8. J. Datsko, *Material Properties and Manufacturing Processes*, John Wiley & Sons, New York, 1966, p 18-20
9. W.B. Morrison, *Trans. Am. Soc. Met.*, Vol 59, 1966, p 824

10. D.C. Ludwigson, *Met. Trans.*, Vol 2, 1971, p 2825-2828
11. H.J. Kleemola and M.A. Nieminen, *Met. Trans.*, Vol 5, 1974, p 1863-1866
12. C. Adams and J.G. Beese, *Trans. ASME*, Series H, Vol 96, 1974, p 123-126
13. J.H. Keeler, *Trans. Am. Soc. Met.*, Vol 47, 1955, p 157-192, and discussion by A.J. Opinsky, p 189-190
14. A. Considere, *Ann. Ponts Chaussées*, Vol 9, 1885, p 574-775
15. G.W. Geil and N.L. Carwile, *J. Res. Nat. Bur. Stand.*, Vol 45, 1950, p 129
16. P.W. Bridgman, *Trans. Am. Soc. Met.*, Vol 32, 1944, p 553
17. J. Aronofsky, *J. Appl. Mech.*, Vol 18, 1951, p 75-84
18. T.A. Trozera, *Trans. Am. Soc. Met.*, Vol 56, 1963, p 280-282
19. E.R. Marshall and M.C. Shaw, *Trans. Am. Soc. Met.*, Vol 44, 1952, p 716
20. W.J.McG. Tegart, *Elements of Mechanical Metallurgy*, Macmillan, New York, 1966, p 22
21. M.J. Barba, *Mem. Soc. Ing. Civils*, Part 1, 1880, p 682
22. E.G. Kula and N.N. Fahey, *Mat. Res. Stand.*, Vol 1, 1961, p 631
23. G.B. Espey, M.H. Jones, and W.F. Brown, Jr., *Am. Soc. Test. Mat. Proc.*, Vol 59, 1959, p 837
24. J.D. Lubahn, *Trans. ASME*, Vol 79, 1957, p 111-115

Fig. 14 Notched and unnotched tensile properties of an alloy steel as a function of tempering temperature
Source: Ref 23

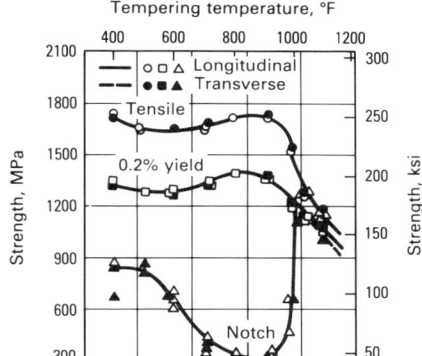

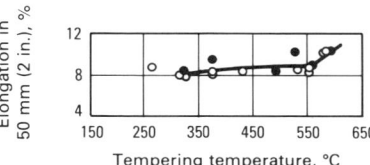

Effect of Specimen Preparation, Setup, and Test Procedures on Test Results

By Bruce W. Christ
Fracture and Deformation Division
Center for Materials Science
National Bureau of Standards

THE UNIAXIAL TENSION TEST is one of the most widely performed mechanical property tests. Analysis of test results provides data about elastic and plastic deformation: modulus of elasticity or Young's modulus, Poisson's ratio, yield and tensile strengths, elongation and reduction in area, and strain-hardening behavior. These data are used for quality control in production, for ranking performance of structural materials, for evaluation of newly developed alloys, and for dealing with the static strength requirements of design.

Specifications or reports often give tensile properties of structural materials without accompanying background information. These tensile data usually are accepted as accurate and reliable. However, experienced technologists know that a detailed description of the test material, test specimen preparation, test setup, and test procedures is needed to evaluate the reliability and utility of such data. This article discusses how the results of a tension test can be affected by each of these important factors.

The results of any mechanical properties test are influenced by the thermal/mechanical history of a test specimen and the details of specimen shape and applied load. For example, a heat treatment applied to a low-alloy steel changes its microstructure from a pearlite-ferrite mixture to tempered bainite, with an accompanying increase in yield and tensile strengths. Introduction of a notch into the gage length of a steel tensile test specimen increases the tensile strength above that measured on an identical steel test specimen without a notch. The increased tensile strength of the notched specimen is caused by a subtle distortion of the applied uniaxial stress, resulting in a localized triaxial stress state at the root of the notch.

If the type of applied load is changed, test results can vary greatly. For example, a low-carbon steel test specimen tested in tension at liquid nitrogen temperature fails in a brittle manner. However, the same type of test specimen tested at liquid nitrogen temperature and under pure shear loading fails in a ductile manner (Ref 1). Furthermore, a low-alloy steel tested under uniaxial tension will exhibit ductility of about 15 to 30% as measured by linear elongation, whereas the same material tested under shear loading will exhibit ductility of about 150 to 250% as measured by rotational elongation (Ref 2). Stress state clearly influences test results for a given material.

Production tension tests of commercial materials are conducted most often at room temperature on smooth (i.e., unnotched) test specimens. The basic information obtained is load-extension data. Average strain rates for most tension tests range between 10^{-2} and 10^{-5} s^{-1}. Some tension tests evaluate performance of notched or fatigue-cracked test specimens because such conditions often occur in actual engineering structures, and other tension tests are carried out on wires and metallic foils.

Standardized tension testing methods have been developed under the auspices of ASTM and are published in the *Annual Book of ASTM Standards* (Ref 3-10). ASTM standard methods specify important details about test specimen selection, size, and shape, test setup and procedures, and data analysis. However, only a few provide guidance in material description and reporting of test results.

Test Specimen Preparation

Test specimens used to evaluate commercial structural materials are often of a size and shape defined by a standard test method. Evaluation of research materials may be performed using standard or nonstandard test specimens, depending on the amount, size, and shape of material available. Typical manufactured product forms from which test specimens are made include sheet, plate, pipe and tubular products, beams, rails, foils, and wires. Fabricated structures from which test specimens might be taken range from bicycles and surgical instruments to cylindrical or spherical pressure vessels and bridges. Sometimes test specimens are taken from welded structures, with the intention of evaluating only weld metal or the metal in the heat-affected zone.

Material Description. A description of the material being tested is useful for understanding the measured tensile properties and comparing them with the properties and microstructures of other materials. It also helps determine whether the material will maintain an acceptably stable microstructure under the conditions of intended applications. Ideally, a description should include chemical composition, thermal and mechanical history, and information on microstructure. Chemical composition can be estimated from product specifications and can be determined more specifically by chemical analysis. Thermal and mechanical history usually can be determined from manufacturing specifications. Details about microstructure (e.g., grain size in an alloy or decarburization of a steel during heat treatment) can be determined by examining suitably prepared metallographic cross sections. This information is useful in all further aspects of test specimen preparation.

The orientation and location of the test specimen in a product can influence measured tensile properties. Many ASTM standards, such as A 370 (Ref 3), E 8 (Ref 5), and B 557 (Ref 6), provide guidance in the selection of test specimen orientation relative to the rolling direction of plate or the major forming axes of other types of products and in the selection of test specimen location relative to the surface of the product.

Orientation is important to standardize test results relative to the directionality of properties that often develops in the microstructure of materials during processing. Some causes of directionality include the fibering of inclusions in steels, the formation of crystallographic textures in most metals and alloys, and the alignment of molecular chains in polymers.

The location from which a test specimen is taken from the initial product form is important because the manner in which a material is processed influences the uniformity of microstructure along the length of the product as well as through its thickness. For example, the properties of metal cut from castings are influenced by the rate of cooling and by shrinkage stresses at changes in section. Generally, specimens taken from near the surface of iron castings are stronger. To standardize test results relative to location, ASTM A 370 recommends that tension specimens be taken from midway between the surface and the center of round, square, hexagon, or octagonal bars (Ref 3). ASTM E 8 recommends that test specimens be taken from the thickest part of a forging from which a test coupon can be obtained, from a prolongation of the forging, or, in some cases, from separately forged coupons representative of the forging (Ref 5).

Size and Shape. Nomenclature for a typical tension specimen is indicated in Fig. 1. The cross section of the specimen may be round (Fig. 2a), rectangular (Fig. 2b and c), or curved. For example, specimens taken from pipe or compressed gas cylinders have a curved cross section in the gage length.

The geometry of a test specimen is often influenced by the product form from which it is taken. For example, only flat specimens can be obtained from sheet products. Test specimens taken from thick plate may be either flat or round; according to ASTM E 8 (Ref 5), the standard diameter of a round tensile specimen must be 12.5 mm (0.5 in.).

Fig. 1 Nomenclature for a typical tension test specimen

The ends of round specimens (see Fig. 2a) may be plain, shouldered, or threaded. Plain ends should be long enough to accommodate some type of wedge grip. Rectangular specimens (see Fig. 2b and c) are generally made with plain ends, but may be shouldered to contain a hole for a pin bearing.

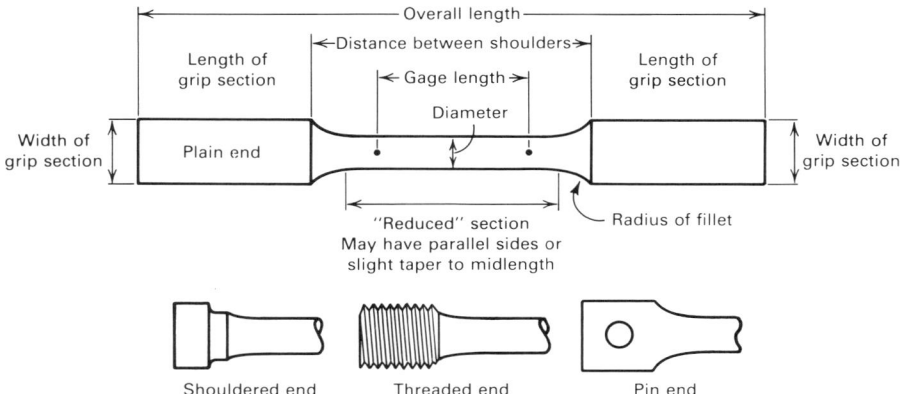

Hence, standard-size round specimens cannot be taken from plate less than 12.5 mm (0.5 in.) thick.

Gage lengths of 50 and 200 mm (2 and 8 in.) are common for standard tension test specimens (see Fig. 2). Most ASTM tension test methods also specify small or subsize specimens that are proportional to the standard specimens. An acceptable small-size specimen can have a gage length of only 11 mm (0.45 in.).

Test specimens received from a machine shop are expected to meet size specifications provided to the shop. To ensure dimensional accuracy, however, each test specimen should be measured prior to testing. Gage length, fillet radius, and cross-sectional dimensions are measured easily. Cylindrical test specimens should be measured for concentricity. Maintaining acceptable concentricity is extremely important in minimizing unintended bending stresses on materials in a brittle state. For example, in a testing program in which beryllium oxide, alumina, and graphite are routinely tested in uniaxial tension, the deviation from true concentricity at the center of the gage length is held to 0.13 mm (0.005 in.) or less, as measured by rotating the gripped tensile specimen 360° about the tensile axis (Ref 11).

Test specimen size can affect the ductility measured in a tension test (Ref 12). For example, in a test on a cylindrical test specimen, elongation to fracture normally increases with decreasing gage length (Fig. 3a). This effect is due to the localization of strain during necking, which occurs after maximum load (Fig. 3b).

Size is also a factor with compact tension test specimens (Fig. 4) used for plane-strain fracture toughness evaluation (Ref 13). In tests to determine the plane-strain stress-

intensity factor, K_{Ic}, for a valid measurement, both the specimen thickness, B, and the crack length, a, must exceed 2.5 $(K_{Ic}/\sigma_{ys})^2$, where σ_{ys} is the 0.2% offset yield strength of the material in tension as determined according to ASTM E 8 (Ref 5) and as discussed in the article "Mechanical Behavior of Materials Under Tension" in this Section. The purpose of this criterion is to ensure plane-strain conditions at the crack tip—i.e., to ensure that throughout the test the crack tip is embedded in elastically deformed material and is remote from any plastically deformed material.

A procedure is described in ASTM A 370, Section VI, for converting percentage elongation of a standard round tension test specimen to equivalent percentage elongation of a standard flat specimen (Ref 3). Such a conversion can be useful for comparing results from two different specimen shapes of the same grade of material. The empirical conversion equation for many steels is:

$$e = e_o\left(4.47\sqrt{\frac{A}{L}}\right)^a \qquad \text{(Eq 1)}$$

where e is the percent elongation after fracture on a standard flat test specimen having a gage length L and a cross-sectional area A, and e_o is the percent elongation after fracture on a standard round test specimen having a 50-mm (2-in.) gage length and a 12.5-mm (0.5-in.) diameter. The superscript a is an empirical constant (0.4 for many grades of carbon and low-alloy steel and 0.13 for annealed austenitic steels). This equation does not apply to sheet steel specimens under 0.64 mm (0.025 in.) thick, nor is it valid for materials other than steel.

A challenging testing problem arises when a longitudinal strip test specimen is taken

Fig. 2 Standard metallic tension specimens

(a) Round specimen. The gage length to diameter ratio is maintained at 4:1 in round specimens to provide a standard basis for comparing elongation values. (b) Rectangular (flat) specimen for testing material in the form of sheet, plate, flat wire, strip, band, hoop, rectangles, and shapes ranging in nominal thickness from 0.13 to 16 mm (0.005 to 0.63 in.). (c) Rectangular (flat) specimen for testing materials in the form of plate, shapes, and flat material having a nominal thickness of 4.8 mm (0.188 in.)

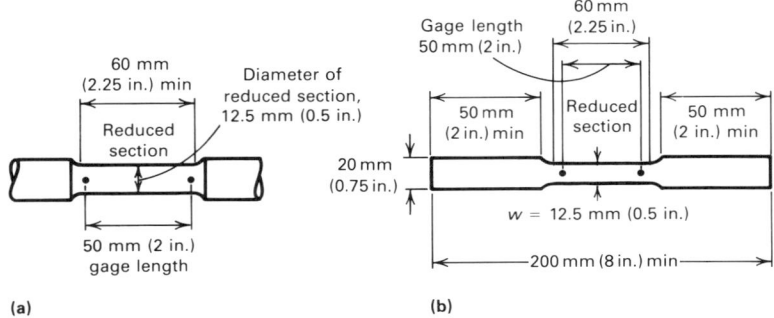

(a)

(b)

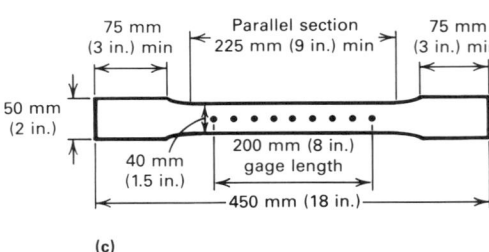

(c)

from the wall of a tubular product. Such a test specimen is curved along its entire length. Standardized test methods allow flattening of the ends to facilitate gripping. However, during flattening there is limited control over the strain hardening introduced into the specimen. Furthermore, the stress state in the region of geometric transition from flat to curved during uniaxial tensile

Fig. 3 Effect of specimen size on the ductility measured in a tension test

(a) Elongation (%) as a function of gage length for a fractured tension specimen. (b) Distribution of elongation along fractured tension specimen. Original spacing between gage marks, 12.5 mm (0.5 in.)

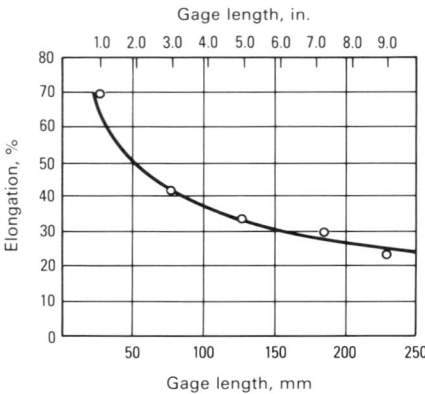

(a)

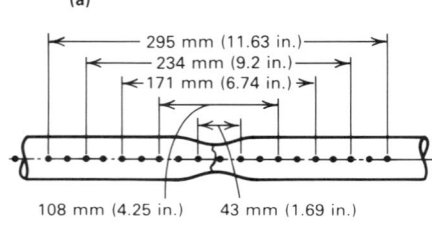

(b)

testing is usually unknown. At present, little is known about the influence of these factors on test results.

Tension test specimens are sometimes intentionally notched in the center of the gage length (Fig. 5). ASTM E 338 (Ref 8) and E 602 (Ref 9) describe procedures for testing notched specimens. Results obtained using notched specimens are useful for evaluating the response of a material to a localized stress concentration. Detailed information on the notch tensile test and a discussion of the related material characteristics (notch sensitivity and notch strength) can be found in the article "Mechanical Behavior of Materials Under Tension" in this Section.

The surface finish of test specimens received from a machine shop is not always smooth, in part because existing ASTM standard methods for tension testing usually do not specify surface finish. If surface finish in the gage length of a tensile test specimen is extremely poor (with machine tool marks deep enough to act as stress-concentrating notches, for example), test results may exhibit a tendency toward decreased and variable ductility. This is especially true for high-strength, low-ductility steels. It is good practice to examine the test specimen surface for deep scratches, gouges, or edge tears and to minimize these discontinuities by polishing or, if necessary, by further machining.

Test Setup

Most test setups for production testing of commercial materials are established according to standard practice on existing commercial equipment (see the article "Tension Testing Machines and Extensometers" in this Section for a description of test machine components and instrumentation). Components of commercial testing equipment are manufactured according to design tolerances

that can unfavorably affect the positioning of components in a test setup and introduce unwanted bending stresses. Constant-crosshead-speed testing machines used for uniaxial tensile testing may have deviations between top and bottom grip centerline positions of 0.025 to 3.2 mm (0.001 to 0.125 in.). Larger deviations may develop as applied loads cause machine frame deflection or as nonaxial crosshead separation occurs. Sideways deflection in commercial testing machines can range between 0.020 and 0.040 mm/mm (in./in.) of crosshead travel.

Components of a testing system such as pull rods, clevises, and grips are either supplied by manufacturers or custom made. Tolerances in ordinary machine shop practice usually range from ±0.025 to ±0.25 mm (±0.001 to ±0.010 in.). It is possible that when the components in a testing system are assembled, the cumulative tolerances from various components will result in displacements from the intended loading axis, causing unwanted stresses. For example, if a test specimen is fastened to wedge grips that are rigidly fastened into the testing machine, the

Fig. 4 Compact tension specimen used for plane-strain fracture toughness evaluation

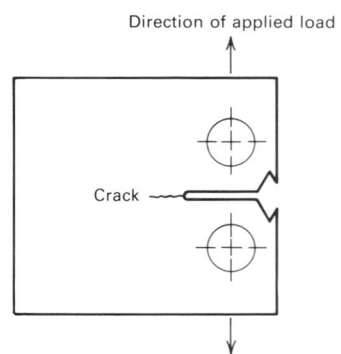

Fig. 5 Standard notched tension specimen
Source: Ref 8

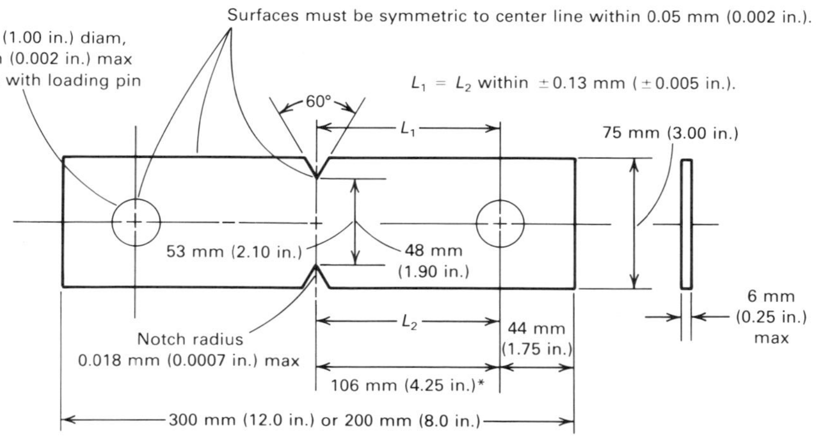

Surfaces must be symmetric to center line within 0.05 mm (0.002 in.).

25 mm (1.00 in.) diam, 0.05 mm (0.002 in.) max clearance with loading pin

60°

$L_1 = L_2$ within ± 0.13 mm (± 0.005 in.).

L_1

75 mm (3.00 in.)

53 mm (2.10 in.)

48 mm (1.90 in.)

6 mm (0.25 in.) max

Notch radius 0.018 mm (0.0007 in.) max

L_2

44 mm (1.75 in.)

106 mm (4.25 in.)*

300 mm (12.0 in.) or 200 mm (8.0 in.)

*This dimension is 56 mm (2.25 in.) for the substandard length specimen.

specimen may elastically bend when the final grip is tightened.

All testing systems can be characterized by their elastic compliance or by the inverse of compliance, elastic stiffness. The loading frame of the test machine must be extremely stiff to avoid deflections that can cause testing error. The entire testing system can be viewed as an elastic spring, with a characteristic spring constant. When a large test specimen fractures under tensile loading, the stored elastic energy in the testing system is usually released in milliseconds with a loud noise, and occasionally with a jolt to the floor on which the testing machine is standing. The influence of test machine stiffness on test results is discussed in the article "Effect of Strain Rate on Flow Properties" in this Section.

Strain-Measuring Devices. Commercial extensometers that fasten mechanically to test specimens are often used in production testing to measure elastic and plastic strain. Extensometers are usually electromechanical devices; strain is measured from an electrical output generated by a linear mechanical displacement of the test specimen. Extensometers must be calibrated to ensure reliable results, and calibration records should be retained. Furthermore, extensometers should only be used to measure displacements within their design range. When testing conditions permit, an extensometer nearly at the limit of its extension range can be lifted off the test specimen manually by the operator. ASTM E 83 deals with the verification and classification of extensometers (Ref 14).

Capacitance extensometers are sometimes used to measure extremely small displacements, e.g., 25×10^{-6} mm (1×10^{-6} in.). Both parallel-plate and concentric-cylinder capacitance extensometers are used (Ref 15).

Output of the parallel-plate capacitance extensometer is inversely proportional to displacement, whereas output of the concentric-cylinder device is linearly proportional to displacement. An electrical circuit has been designed to linearize the output of the parallel-plate capacitance extensometer (Ref 16).

Strain gages, which are also electromechanical, are sometimes used in tension testing setups to measure strain. It is important that the gages be suitably bonded to the test specimens with glue or epoxy that will remain stable at the test temperature. As with extensometers, strain gages should be properly calibrated. Additional information on the classification and calibration of strain-measuring devices can be found in the article "Calibration of Testing Equipment" in this Volume.

Load cells are electromechanical devices that convert a mechanical displacement into an electrical output signal. Most mechanical testing systems include a load cell for indicating the load magnitude. Changes in load are usually recorded during a tension test. Load cells must be calibrated periodically, and records should be kept of the calibration results. ASTM E 4 and E 74 are standard methods for verifying the load indication of testing machines (Ref 17, 18).

Temperature Control. Sometimes tension testing is performed at other than room temperature (see the article "Elevated/Low Temperature Tension Testing" in this Section). ASTM E 21 describes standard procedures for elevated temperature tension testing of metallic materials (Ref 10). There is no ASTM standard procedure for cryogenic testing.

Temperature gradients may occur in temperature-controlled systems. Gradients must be kept within tolerable limits. It is not

uncommon for more than one temperature-sensing device, e.g., thermocouples, to be used in testing systems run at other than room temperature. Besides the temperature-sensing device used in the control loop, auxiliary sensing devices may be used to measure temperature gradients along the gage length of the test specimen.

Alignment of the tension testing system is critical, especially when testing materials of low ductility or when testing in the microstrain range (Ref 16). Misalignment in the grips can cause significant elastic bending stresses that superimpose on uniaxial stresses. This causes localized microplastic flow and/or localized rupture at lower applied stresses than the true average uniaxial tensile stress typically required. ASTM E 1012 is a standard method for verification of alignment under tensile loading (Ref 19).

Test Procedures

After the test specimen has been properly prepared and measured and the test setup established, conducting the test is fairly routine. The test specimen should be installed in the grips with a minimum of distortion. Extensometers or other strain-measuring devices should be properly fastened to the specimen and calibrated; provision should be made to record extension. Load measuring and recording systems should be checked. It is sometimes useful to repetitively apply small initial loads and vibrate the load train (a metallographic engraving tool is a suitable vibrator) to overcome friction in various couplings, as shown in Fig. 6. A check to ensure that the test will run at the intended strain rate and temperature is suitable. The test is then begun by initiating load application.

Strain rate can be determined accurately from the extension per unit time indicated by extensometers or strain gages and is expressed in units of seconds^{-1}. When a strain-measuring device is not attached to the test specimen, strain rate can be estimated by:

$$\dot{\epsilon} = \frac{s}{L_o} \qquad \text{(Eq 2)}$$

where $\dot{\epsilon}$ is strain rate, s is crosshead speed, and L_o is gage length.

A basic assumption in the use of Eq 2 is that the entire crosshead displacement occurs in the gage length of the test specimen. Because the elastic compliance of the entire testing system is finite, this assumption is not completely valid during elastic loading. Consequently, calculated elastic strain rates will be slightly higher than the actual elastic strain rate. However, Eq 2 is valid for estimating the plastic strain rate once load has reached a magnitude that causes uniform plastic extension throughout the gage length

Fig. 6(a) Effectiveness of vibrating the loading train in overcoming friction in the spherical ball and seat shown in Fig. 6(b)

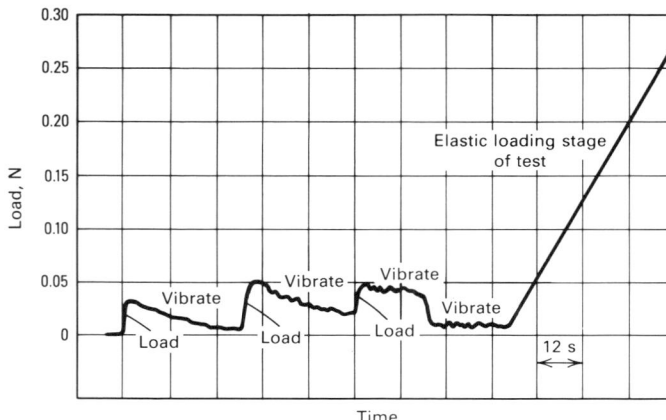

Fig. 6(b) Spherically seated gripping device for shouldered tension specimen

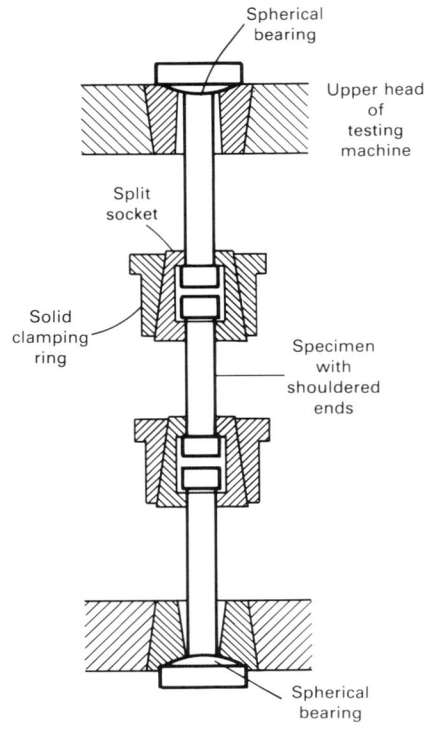

without significant further load increase. At the lower yield point in low-carbon steel, plastic strain occurs at Lüders band fronts, and the local plastic strain rate at the band fronts will be much higher than calculated by Eq 2.

If something unusual should occur during testing—for example, if fracture occurs outside the gage length or the testing equipment malfunctions—the test result should be discarded and another test conducted on an identical specimen (Ref 5).

Data Analysis and Reporting. Data on load and extension at a given strain rate and temperature are obtained during a tension test. Test specimen dimensions before and after the test are also recorded. These data are analyzed to determine yield strength, tensile strength, elongation, and reduction in area. Standard procedures to determine these values are described in ASTM E 8 (Ref 5). Procedures to evaluate plastic strain ratio and strain-hardening exponent are described in ASTM E 517 and E 646, respectively (Ref 7, 20). Young's modulus can be determined according to procedures in ASTM E 111 (Ref 21) and E 231 (Ref 22), and determination of Poisson's ratio is described in E 132 (Ref 23).

Precision and accuracy of tensile test results usually are not reported. Most ASTM methods are incomplete in this area, partly because standard reference materials with certified values of tensile properties are unavailable. Usually, the reproducibility of results from multiple specimens is reported. It is interesting to compare the coefficient of variation (standard deviation ÷ mean value) of various tensile properties with the coefficient determination for other mechanical properties. Table 1 shows that the coefficient

Table 1 Coefficients of variation for mechanical properties

Mechanical property	Coefficient of variation, C_N(a)
Elastic modulus	0.03
Ultimate tensile strength	0.05
Brinell hardness	0.05
Tensile yield strength	0.07
Fracture toughness	0.07
Fatigue strength	0.08-1.0

(a) Coefficient of variation, C_N, is the standard deviation divided by the mean value (Ref 24).

of variation for tensile properties ranges from 0.03 to 0.07 (Ref 24).

Test results should be reported in a suitable format. Laboratory data sheets may be sufficient, or circumstances may require a more detailed report. ASTM E 21 is one of the few standard methods to describe test report content in detail (Ref 10). Generally, it is the investigator's responsibility to prepare a report that facilitates evaluation of the reliability and utility of the data.

REFERENCES

1. E.R. Parker, *Brittle Behavior of Engineering Structures*, John Wiley & Sons, New York, 1957, p 304
2. R. Fields, Metallurgy Division, National Bureau of Standards, Gaithersburg, MD, private communication
3. "Standard Methods and Definitions for Mechanical Testing of Steel Products," A 370, *Annual Book of ASTM Standards*, Vol 03.01, ASTM, Philadelphia, 1984, p 1-56
4. "Standard Methods of Tension Testing of Metallic Foil," E 345, *Annual Book of ASTM Standards*, Vol 03.01, ASTM, Philadelphia, 1984, p 491-496
5. "Standard Methods of Tension Testing of Metallic Materials," E 8, *Annual Book of ASTM Standards*, Vol 03.01, ASTM, Philadelphia, 1984, p 130-150
6. "Standard Methods of Tension Testing Wrought and Cast Aluminum and Magnesium Alloys Products," B 557, *Annual Book of ASTM Standards*, Vol 03.01, ASTM, Philadelphia, 1984, p 82-99
7. "Standard Test Method of Plastic Strain Ratio *r* for Sheet Metal," E 517, *Annual Book of ASTM Standards*, Vol 03.01, ASTM, Philadelphia, 1984, p 589-596
8. "Standard Method of Sharp-Notch Tension Testing of High-Strength Sheet Materials," E 338, *Annual Book of ASTM Standards*, Vol 03.01, ASTM, Philadelphia, 1984, p 483-490
9. "Standard Method of Sharp-Notch Tension Testing with Cylindrical Specimens," E 602, *Annual Book of ASTM Standards*, Vol 03.01, ASTM, Philadelphia, 1984, p 632-640
10. "Standard Recommended Practice for Elevated Temperature Tension Tests of

Metallic Materials," E 21, *Annual Book of ASTM Standards*, Vol 03.01, ASTM, Philadelphia, 1984, p 200-209

11. C.D. Pears, "Ultimate Strength, Elastic Modulus and Poisson's Ratio to 5500 °F in Tension," Report to SRI/International from Southern Research Institute, Birmingham, AL, Jan 15, 1976

12. P.F. Foster, *The Mechanical Testing of Metals and Alloys*, Isaac Pitman & Sons, London, 1948, p 76

13. "Standard Test Method for Plane-Strain Fracture Toughness of Metallic Materials," E 399, *Annual Book of ASTM Standards*, Vol 03.01, ASTM, Philadelphia, 1984, p 519-554

14. "Standard Method of Verification and Classification of Extensometers," E 83, *Annual Book of ASTM Standards*, Vol 03.01, ASTM, Philadelphia, 1984, p 247-252

15. B.W. Christ, Effects of Misalignment on the Pre-Macroyield Region of the Uniaxial Stress-Strain Curve, *Met. Trans.*, Vol 4, April 1973, p 1961-1965

16. B.W. Christ and S.R. Swanson, *J. Test. Eval.*, Vol 4 (No. 6), Nov 1976, p 405-417

17. "Standard Practices for Load Verification of Testing Machines,"E 4, *Annual Book of ASTM Standards*, Vol 03.01, ASTM, Philadelphia, 1984, p 111-118

18. "Standard Practice of Calibration of Force-Measuring Instruments for Verifying the Load Indication of Testing Machines," E 74, *Annual Book of ASTM Standards*, Vol 03.01, ASTM, Philadelphia, 1984, p 238-246

19. "Standard Recommended Practice for Verification of Specimen Alignment Under Tensile Loading," E 1012, *Annual Book of ASTM Standards*, Vol 03.01, ASTM, Philadelphia, 1985 (to be published)

20. "Standard Test Method for Tensile Strain-Hardening Exponents (*n*-Values) of Metallic Sheet Materials," E 646, *Annual Book of ASTM Standards*, Vol 03.01, ASTM, Philadelphia, 1984, p 701-710

21. "Standard Test Methods for Young's Modulus, Tangent Modulus, and Chord Modulus," E 111, *Annual Book of ASTM Standards*, Vol 03.01, ASTM, Philadelphia, 1984, p 271-279

22. "Standard Test Methods for Static Determination of Young's Modulus of Metals at Low and Elevated Temperatures," E 231, *Annual Book of ASTM Standards*, Vol 03.01, ASTM, Philadelphia, 1984, p 387-393

23. "Standard Test Methods for Poisson's Ratio at Room Temperature," E 132, *Annual Book of ASTM Standards*, Vol 03.01, ASTM, Philadelphia, 1984, p 280-282

24. P.H. Wirsching and J.E. Kempert, A Fresh Look at Fatigue, *Machine Design*, Vol 48 (No. 12), 1976, p 120-123

Elevated/Low Temperature Tension Testing

By J. Daniel Whittenberger
Materials Engineer
NASA Lewis Research Center

and

Michael V. Nathal
Materials Engineer
NASA Lewis Research Center

TENSION TESTING at elevated or low temperatures is conducted as part of quality control measures or general characterization studies for engineering materials that are utilized at temperatures ranging from cryogenic, as in liquid-gas refrigeration systems, to very high, as in gas turbine engines and rocket nozzles.

A tension test has several advantages. A simple specimen design that can be directly cast or machined from parts is used. Measurements can be taken with equipment that is available in most metallurgical laboratories. The test can be carried out in a few minutes, exclusive of the heating and cooling periods. Although these advantages simplify quality control testing, a tension test should be used as a quality-level indicator with caution, because the significance of the test is limited by its correlation with performance. In other words, a material may pass a previously established quality control standard for tensile strength and ductility and yet be unacceptable for the desired operating conditions. For example, many different thermomechanical processing schedules can be applied to a heat-treatable steel to produce a desired combination of tensile properties, but only one of these heat treatments may be able to provide the other mechanical properties critical for its use. Thus, additional tests, such as fatigue, fracture toughness, environmental resistance, and creep, as well as microstructural examination, must be conducted for proper materials qualification.

Elevated/low temperature tension tests are conducted with basically the same specimens and procedures as room-temperature tension tests. However, the specimens must be heated or cooled in an appropriate environmental chamber (Fig. 1). Also, the test fixtures must be sufficiently strong and corrosion resistant, and the strain-measuring system must be usable at the test temperature.

Once determined, load-elongation data can be transformed into engineering stresses and strains using standard formulas. True stresses and strains, at least up to the onset of necking, can also be calculated. The onset of necking is difficult to identify, particularly at elevated temperatures.

General Characteristics

Elevated/low temperature stress-strain diagrams are similar in appearance to those determined at room temperature (Fig. 2). Relative to ambient temperature, materials become stronger, but less ductile, as temperature decreases. Conversely, materials become weaker with increasing temperature (Fig. 3). Although simple, stable alloys exhibit increased ductility with rising temperatures, the temperature-ductility behavior for most engineering materials (Fig. 3b) varies greatly. Such discontinuities in ductility with increasing temperature usually can be traced to metallurgical instabilities—carbide precipitation, for example—that affect the failure mode.

Due to the relatively high strain rates— usually 8.33×10^{-5} s^{-1} (0.5%/min) and 8.33×10^{-4} s^{-1} (5%/min)—involved in tensile testing, deformation occurs by slip (glide of dislocations along definite crystallographic planes). Thus, changes in strength and ductility with temperature generally can be related to the effect of temperature on slip.

At low temperatures (less than 0.3 homologous temperature*), the number of slip systems is restricted, and recovery processes are not possible. Therefore, strain-hardening mechanisms, such as dislocation intersections and pileups, lead to the increasingly higher forces required for continued deformation. This continues until the local stresses at pileups exceed the fracture stress, and failure occurs.

At higher temperatures (between 0.3 and 0.5 homologous temperature), thermally activated processes such as multiple slip and cross slip allow the high local stresses to be relaxed, and strength is decreased. For sufficiently high temperatures in excess of half the homologous temperature, diffusion processes become important, and mechanisms such as recovery, dislocation climb, recrystallization, and grain growth can reduce the dislocation density, prevent pileups, and further reduce strength.

Deformation under tensile conditions is governed to some extent by crystal structure. Face-centered cubic materials generally exhibit a gradual change in strength and ductility as temperature decreases. Such changes for type 304 stainless steel are illustrated in Fig. 3. Some body-centered cubic alloys, however, exhibit an abrupt change at the ductile-to-brittle transition temperature (approximately 200 °C, or 390 °F, for tungsten in Fig. 3), below which there is little plastic flow. In close-packed hexagonal and body-centered cubic materials, mechanical twin-

*The homologous temperature is the ratio of the test temperature to the melting point of the material being tested, both temperatures expressed in degrees Kelvin.

Fig. 1 Tension testing apparatus with environmental chamber for testing up to 540 °C (1000 °F)
Courtesy of Instron Corp.

ning also can occur during testing. However, twinning by itself contributes little to the overall elongation; its primary role is to reorient previously unfavorable slip systems to positions in which they can be activated.

Other factors can affect tensile behavior; however, the specific effects cannot be predicted easily. For example, re-solutioning, precipitation, and aging (diffusion-controlled particle growth) can occur in two-phase alloys during heating prior to testing and during the actual testing. These processes can produce a wide variety of responses in mechanical behavior depending on the material. Diffusion processes also are involved in yield point and strain-aging phenomena. Under certain combinations of strain rate and temperature, interstitial atoms can be dragged along with dislocations, or dislocations can alternately break away and be re-pinned, producing serrations in the stress-strain curves.

There are exceptions to the above generalizations, particularly at elevated temperatures. For example, at sufficiently high temperatures, the grain boundaries in polycrystalline materials are weaker than the grain interiors, and intergranular fracture occurs at relatively low elongation. In complex alloys, hot shortness, in which a liquid phase forms at grain boundaries, or grain boundary precipitation can lead to low strength and/or ductility.

Experimental Methods

Specific procedures and methods for conducting room-temperature and elevated-temperature tension tests have been standardized by ASTM in test methods E 8 (Ref 5) and E 21 (Ref 6), respectively. Although no standard has been adopted for low-temperature testing, the general requirements of ASTM E 8 and E 21 must also be satisfied at lower temperatures. Assuming that a test machine

with appropriate load-measuring and speed-control capabilities is used for testing, the validity of a test rests on proper concern for strain measurement, temperature control, and materials behavior.

Strain Measurement. The simplest method for strain measurement is based on crosshead displacement in the test apparatus. However, the crosshead extension includes not only the strain in the gage section, but also the deformation in the rest of the sample, load train, and testing machine. This method is suitable only for measurement of large plastic strains in which the other factors can be overlooked. For accurate determination of strain, an extensometer or strain gage must be used (see the article "Tension Testing Machines and Extensometers" in this Section for additional information). These are attached directly to the specimen and thus exclude any contributions from the testing apparatus. However, errors can still arise from bending strains and strain gradients due to nonaxial loading, specimen nonuniformities, and inherent inaccuracies in the extensometer.

Specimen strain frequently is magnified to increase the accuracy of measurement. Mechanical extensometers that magnify by using lever arms and optical extensometers that magnify by using mirrors and lenses have been employed. Electronic magnification is widely used; one common device is the linear variable differential transformer (LVDT), in which the displacement of the specimen is converted to an electrical signal by motion of a ferrite core within a coil. Electrical resistance gages that make use of the change in resistance of a material with strain are also available. These gages are generally thin and small in size and can be cemented directly on the specimen. They are useful in both elastic and plastic regimes.

Fig. 2 Stress-strain diagrams for type 304 stainless steel

(a) At low temperatures. (b) At elevated temperatures. Source: Ref 1

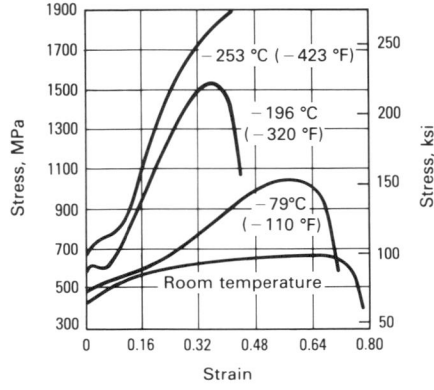

(a)

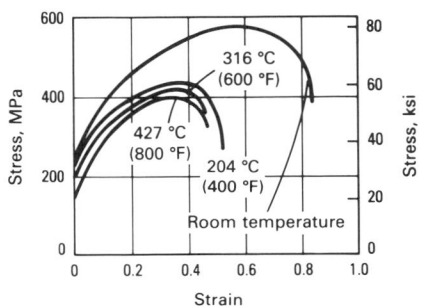

(b)

Fig. 3 Effect of temperature on strength and ductility of various materials

(a) 0.2% offset yield strength. (b) Tensile elongation. Materials tested include aluminum alloy 7075 in two heat-treated conditions (Ref 1); Ti-6Al-4V (Ref 1); AISI 1015 low-carbon steel (Ref 2); type 304 stainless steel (Ref 1); cobalt-based alloy MAR-M509 (Ref 3); directionally solidified nickel-based alloy MAR-M200 (Ref 3); and pure tungsten (Ref 4).

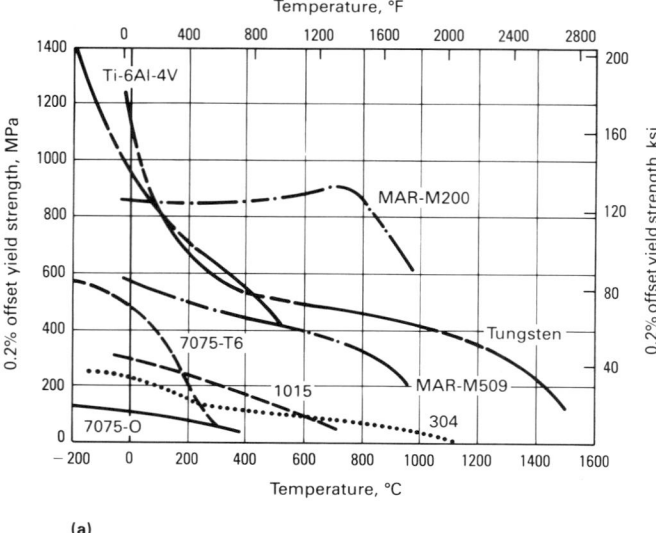

(a)

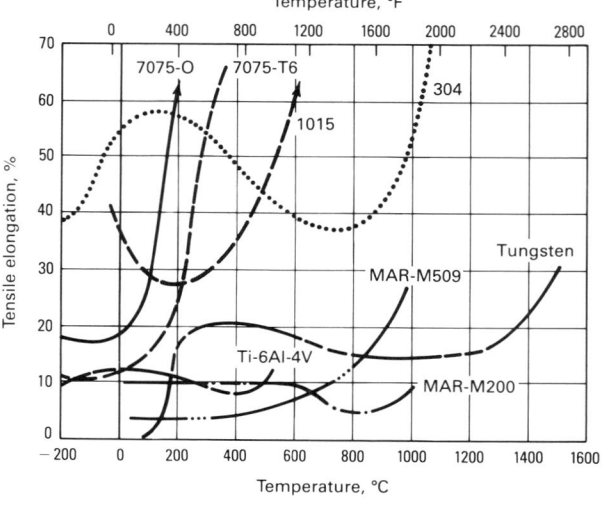

(b)

Testing at elevated or low temperatures requires strain-measuring equipment somewhat different from that used at ambient temperatures. Strain gages generally are adequate between cryogenic temperatures and approximately 600 °C (1100 °F), but at higher temperatures other devices must be used. Rod and tube extensometers are commercially available; in these, the rod and tube are attached to the top and bottom of the gage section, respectively, and they transmit the relative motion outside the test environment to a sensing device, such as an LVDT, maintained at room temperature. Frequently, an averaging extensometer is used where two or more sensors are connected to the specimen to determine nonaxial loading.

Rod and tube extensometers have general applicability over the entire range of temperatures and test environments, because they are manufactured from a variety of materials. When testing at below room temperature, Teflon is suitable. Nickel-based superalloys are adequate for testing in air to 1100 °C (2010 °F). Above 1100 °C (2010 °F), ceramics can be used in reactive atmospheres, while refractory metals are adequate for inert environments. Although ceramic extensometers can be used for essentially all test conditions, they are significantly less abuse resistant than metallic or polymeric devices; their use is thus limited to elevated-temperature testing.

Temperature Control. The actual temperature is best controlled within a few degrees of the desired temperature and

should not fluctuate with time or vary along the gage length. Temperature can be monitored by thermocouples attached near each end and at the center of the gage length.

Proper selection of thermocouples depends on the test environment and temperature. Typical examples for use in air are copper-Constantan at temperatures below 350 °C (660 °F), Chromel-Alumel at intermediate temperatures, and Pt-13Rh/Pt between 1000 and 1400 °C (1830 and 2550 °F). A detailed discussion of thermocouple materials can be found in Ref 7. The test temperature can be maintained adequately at the desired level with one of the many types of commercially available temperature controllers that reference the electromotive force (emf) output from the central thermocouple.

Minimizing the temperature gradient along the gage length depends on the heating method. With the multiple-zone resistance-wound tube furnaces typically used at temperatures greater than or equal to 300 °C (570 °F), appropriate adjustment of power to each zone lengthens the uniform temperature region. For oil bath or air ovens operating between room temperature and 300 °C (570 °F), improved flow of the heated medium and/or improved mixing may be required.

Although deep immersion into liquid gases or mixtures such as dry ice/acetone can yield uniform temperatures, refrigerant systems with a gaseous or liquid cooling fluid can be problematic and require mixing and/or increased flow of coolant to minimize gradients. Heating methods such as induction

heating with and without susceptors may require changes in the coil or susceptor design, while single-piece or wire mesh heaters typically used in vacuum testing may require material to be cut from the heating element to lessen the temperature differences. Self-resistance heating of the specimen can also be used, particularly when very rapid heat-up times are desired. However, it is difficult to produce a uniform temperature along the gage length, and special specimen designs may be required.

Proper control of the test temperature and minimization of thermal gradients along the specimen gage length are important because of the dependence of plastic flow mechanisms and material behavior on temperature. For example, specimen "hot spots" generally will be weaker than the remainder of the specimen and can become sites for localized deformation and premature failure, thus leading to inaccurate and misleading tensile data.

Material Behavior. Because alloys undergoing elevated tensile testing will, in effect, be subject to annealing prior to loading, changes in microstructure can occur and produce a material that is not characteristic of the original stock. Thus, very slow heating or prolonged holds at temperature should be avoided. Figure 4 illustrates the influence of hold time on the yield strength and ductility of a precipitation-strengthened aluminum alloy. Holding at 150 °C (300 °F) changes the amount and distribution of the reinforcing phases in such a manner that strengthening

Fig. 4 Effect of exposure time and temperature on the tensile properties of naturally aged aluminum alloy 2024-T4

(a) Yield strength. (b) Percent elongation. Source: Ref 1

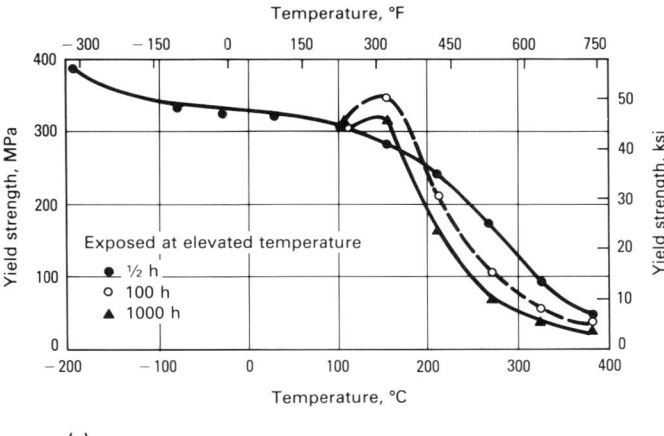

(a)

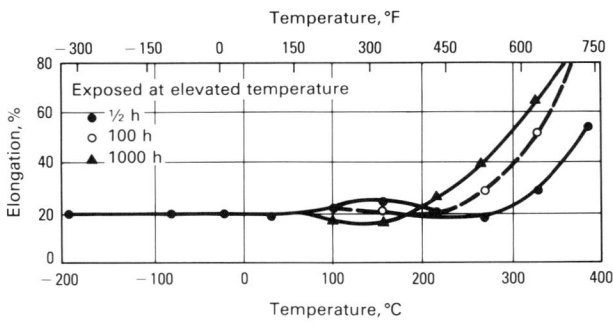

(b)

initially occurs. This is subsequently followed by weakening. Clearly, the exposed material is not the same as one that is tested rapidly.

Test environment can also affect the measured properties. Generally, the atmosphere should reflect the intended or proposed use of the material. Although the environment can rarely be a complete simulation of operating conditions, it should produce the same basic effects and should not introduce foreign attack mechanisms. For example, it would be appropriate to test oxidation-resistant alloys at elevated temperature in air; however, such conditions cannot be used for refractory metals that undergo catastrophic oxidation.

When it is desirable to examine the fracture surface, it is necessary to test in a hard vacuum or nonreactive atmosphere, even if such an environment has a detrimental effect (loss of a volatile alloying element, for instance). An inert atmosphere is required to prevent secondary oxidation/corrosion from masking details on the fracture surface.

REFERENCES

1. W.F. Brown, Jr., Ed., *Aerospace Structural Metals Handbook*, Metals and Ceramic Information Center, Columbus, OH, 1982
2. T.D. Moore, Ed., *Structural Alloys Handbook*, Metals and Ceramic Information Center, Columbus, OH, 1982
3. "High Temperature, High Strength Nickel Base Alloys," International Nickel Company, New York, 1977
4. T.E. Tietz and J.W. Wilson, *Behavior and Properties of Refractory Metals*, Stanford University Press, Stanford, CA, 1965
5. "Standard Methods of Tension Testing of Metallic Materials," E 8, *Annual Book of ASTM Standards*, ASTM, Philadelphia, 1984, p 130-150
6. "Standard Recommended Practice for Elevated Temperature Tension Tests of Metallic Materials," E 21, *Annual Book of ASTM Standards*, ASTM, Philadelphia, 1984, p 200-209
7. T.P. Wang and E.D. Zysk, Thermocouples for Industrial Applications, in *Metals Handbook*, Vol 3, 9th ed., American Society for Metals, 1980, p 696-720

SELECTED REFERENCES

- E.A. Brandes, Ed., *Smithells Metals Reference Book*, Butterworths, London, 1983
- R.F. Bunshah, Ed., *Techniques of Metals Research*, Vol V, *Measurement of Mechanical Properties*, Part 1, Interscience, New York, 1971
- G.E. Dieter, *Mechanical Metallurgy*, McGraw-Hill, New York, 1976
- *Metals Handbook*, Vol 1-3, 9th ed., American Society for Metals, 1978-80

Effect of Strain Rate on Flow Properties

By Peter P. Gillis
Professor of Materials Science
University of Kentucky

and

Todd S. Gross
Assistant Professor of Materials Science
University of Kentucky

STRAIN RATE, or the rate at which a specimen is deformed, is an important consideration in the production, fabrication, and testing of materials. This rate can have an important influence on the mechanical properties, particularly the flow stress, of a material. During a conventional (pseudostatic) tension test, ASTM prescribes an upper limit for the deformation rate; however, during actual fabrication processes, economics prescribes that the deformation be as rapid as possible. Under these different conditions, the mechanical response of a given material may vary. For most materials, strength properties tend to increase at higher rates of deformation.

Figure 1 shows stress-strain curves obtained from niobium monocrystals deformed at various rates. Clearly, the yield strength of the material increases as the deformation rate increases. This is the most consistently observed effect of strain rate. Figure 1 also shows a trend for strain hardening to decrease (a decrease in the slope of the curve) with increasing strain rate.

The strength results of Ref 1 are summarized in Fig. 2. At low deformation rates, relatively large rate changes are required to effect noticeable changes in the yield point. For example, at the lowest rates about a 1000-fold increase in strain rate is required to double the yield strength. Above some critical rate, however, strength changes become much more sensitive to the rate.

Figure 3 illustrates true yield stress at various strains for a low-carbon steel at room temperature. Between strain rates of 10^{-6} s^{-1} and 10^{-3} s^{-1} (a 1000-fold increase),

yield stress increases only by 10%. Above 1 s^{-1}, however, an equivalent rate increase doubles the yield stress. For the data in Fig. 3, at every level of strain the flow stress increases with increasing strain rate. However, a decrease in strain-hardening rate is exhibited at the higher deformation rates.

The results of the combined effects of strain rate and temperature at 200, 400, and 600 °C (390, 750, and 1110 °F) are shown in Fig. 4. At the highest temperature (600 °C, or 1110 °F), yield strength increases with increasing strain rate, as do room-temperature results, but strain hardening increases (rather than decreases) with increasing strain rate. At the intermediate temperatures shown in Fig. 4(a) and (b), however, regions of negative strain rate sensitivity are visible; that is, under certain conditions of strain, strain rate, and temperature, the flow stress of carbon steels can decrease with an increase in strain rate. This is opposite to the usual strain rate effect.

Another class of common metals is the structural aluminums, which are less strain rate sensitive than steels. Figure 5 shows data obtained for 1060-O aluminum. Between strain rates of 10^{-3} s^{-1} and 10^3 s^{-1} (a million-fold increase), the stress at 2% plastic strain increases by less than 20%. Another contrast to the behavior of steel demonstrated in Fig. 5 is that strain hardening increases with increasing deformation rate.

Reference 5 summarizes several data sets relating to the yield stress dependence on strain rate in steel and aluminum (Fig. 6). The comparison shows that in aluminum the effect of strain rate on yield stress is less sig-

nificant and occurs only at extremely high rates. (Note the difference in vertical scales in Fig. 6.)

The effect of strain rate on the ductility of metals is relatively slight (Ref 6). For moderate strain rates and temperatures, the ductility of titanium does not vary substantially, as shown in Fig. 7(a). For conditions of low temperature and high extension rates, the ductility decreases somewhat, as shown in Fig. 7(b). The ductility of two titanium alloys and Zircaloy has been shown to increase with the strain rate sensitivity parameter, m, as shown in Fig. 8. This can be rationalized by the reduced tendency to form a neck as m approaches 1. For more information on the strain rate sensitivity parameter, see the section of this article on analysis of strain rate.

The effect of strain rate on the strength of several polymeric materials is shown in Fig. 9. The explosive material LX-04-M in Fig. 9(a) has a very high strain rate sensitivity; the stress axis is logarithmic in the figure. The effect in polypropylene (Fig. 9c) is shown at six different temperatures. The effect diminishes as temperature increases.

The effect of strain rate on the ductility of polymers is much more complex than it is for metals. The ductility is at first strongly dependent on strain rate and increases with increased strain rate. At some point, the ductility peaks and then decreases with further increases in extension rate. The tensile behavior of a representative polymer is shown in Fig. 10. The strain at failure reaches a maximum at some intermediate strain rate and then decreases with further increases in

Fig. 1 Stress-strain curves for single crystals of niobium

(a) For long specimens, 6.4 mm (0.25 in.) in length and 4.8 mm (0.19 in.) in diameter. (b) For short specimens, 2 mm (0.08 in.) in length and 4.3 mm (0.17 in.) in diameter. Source: Ref 1

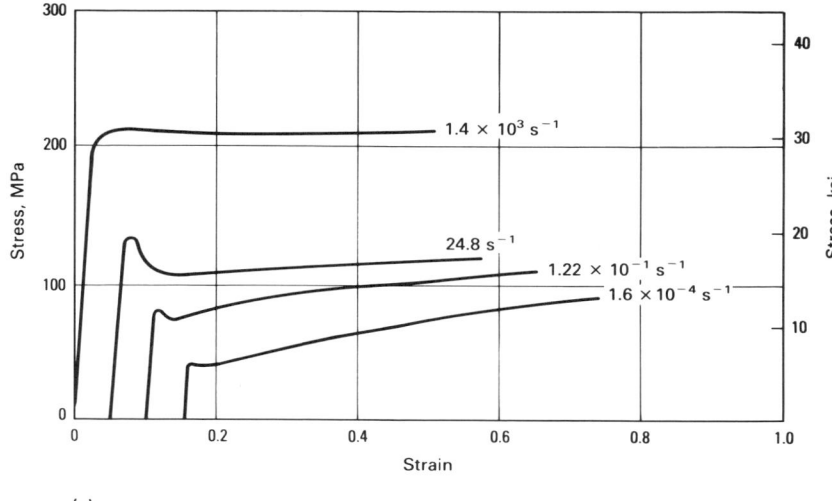

(a)

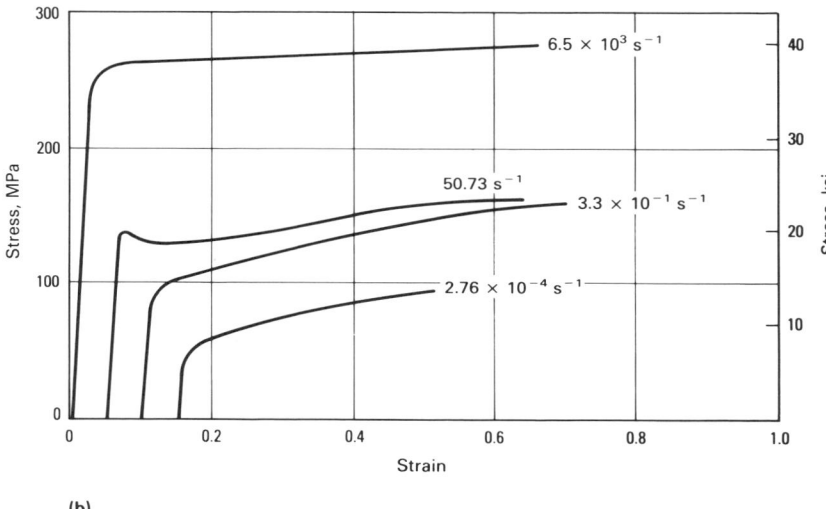

(b)

Fig. 2 Variation of yield stress versus the logarithm of the strain rate for niobium single crystals

The use of the Hopkinson bar for testing is described in the Section "High Strain Rate Testing" in this Volume. Source: Ref 1

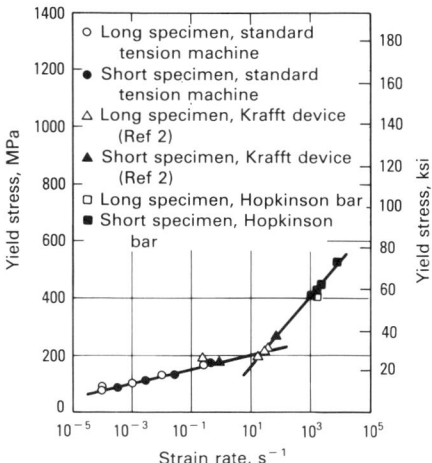

Fig. 3 True yield stresses at various strains versus strain rate for a low-carbon steel at room temperature

Source: Ref 3

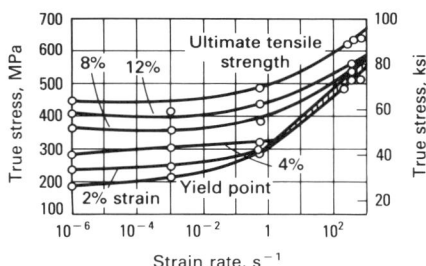

strain rate. Figure 11 shows the ductility dependence on temperature-compensated strain rate for a cross-linked rubber.

ASTM Specifications for Deformation Rates

To quantify the effect of deformation rate on strength and other properties requires specific definitions of such rates. For example, the "strain rate" is the rate at which the strain, e, accumulates ($\dot{e} = de/dt$), and it usually is specified in s^{-1} or Hz. During elastic deformation, the stress rate is proportional to the strain rate through Hooke's law. In a tension test:

$$\dot{s} = E\dot{e} \qquad (Eq\ 1)$$

where s is stress, E is the modulus of elasticity, and the superposed dots denote time derivatives. Thus, the deformation rate can also be specified by the "loading rate" units of stress per unit of time when the load is below the proportional limit.

ASTM Standard E 8, "Tension Testing of Metallic Materials," lists three quantities in addition to $\dot{e}$ and $\dot{s}$ for prescribing the speed of testing: "Speed of testing may be defined (a) in terms of free-running crosshead speed (rate of movement of the crosshead of the testing machine when not under load), (b) in terms of rate of separation of the two heads of

the testing machine during a test, (c) in terms of the elapsed time for completing part or all of the test, (d) in terms of the rate of stressing the specimen, or (e) in terms of the rate of straining the specimen. . . . For some materials the free-running crosshead speed, which is the least accurate, may be adequate, while for other materials one of the remaining methods, listed in increasing order of precision, may be necessary in order to obtain test values within acceptable limits."

ASTM E 8 specifies that the test speed must be low enough to permit accurate determination of loads and strains. When the rate of stressing is stipulated, ASTM E 8 requires that it not exceed 690 MPa/min (100 ksi/

Fig. 4 True stresses at various strains versus strain rate for a low-carbon steel

(a) At 200 °C (390 °F). Note that while the yield stress continually increases with an increase in strain rate, the ultimate tensile strength and some of the strengths at various strains decrease with increases in strain rate. This is attributed to dynamic strain aging. (b) At 400 °C (750 °F). The decrease in tensile strength with increase in strain rate is again attributed to dynamic strain aging. This temperature and the temperature in (a) are in the "blue brittle" region for steels. (c) At 600 °C (1110 °F). The abrupt change in the slope of the yield stress versus strain rate present at room temperature (Fig. 3) is not present at this temperature. Source: Ref 3

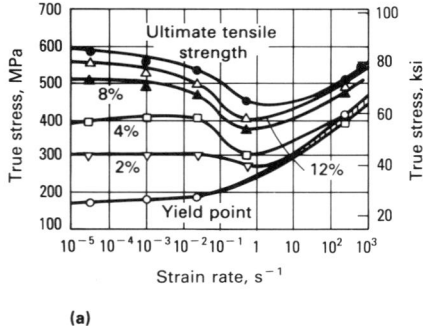

(a)

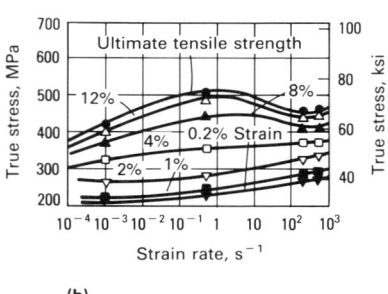

(b)

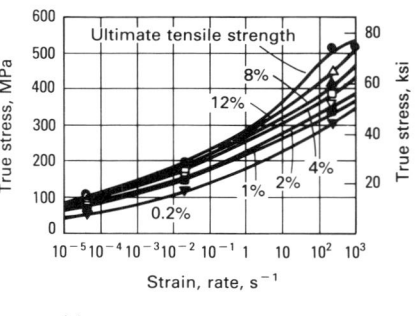

(c)

min). This corresponds to an elastic strain rate of about 5×10^{-5} s^{-1} for steel or 15×10^{-5} s^{-1} for aluminum. When the rate of straining is stipulated, ASTM E 8 prescribes that after the yield point has been passed, the rate can be increased to about 1000×10^{-5} s^{-1}; presumably, the stress rate limitation must be applied until the yield point is passed.

In ASTM Standard E 345, "Tension Testing of Metallic Foil," the same upper limit on the rate stressing is recommended. In addition, a lower limit is given, but its value is unclear. ASTM E 345 further specifies that when the yield strength is to be determined, the strain rate must range from approximately 3×10^{-5} s^{-1} to 15×10^{-5} s^{-1}.

There are a number of tests that produce strain rates higher or lower than these ranges. Creep tests yield lower rates (see the Section "Creep, Stress-Rupture, and Stress-Relaxation Testing" in this Volume); some of the highest deformation rates are produced by explosive loading (see the Section "High Strain Rate Testing" in this Volume). Table 1 lists the strain rate ranges for various tests.

Effects of Inertia

Analysis of high strain rate test results requires consideration of the effect of stress wave propagation along the length of the test specimen. For high loading rates, the strain in the specimen may not be uniform. Figure 12 illustrates an elemental length dx_o of a tension test specimen whose initial cross-sectional area is A_o and whose initial location is prescribed by the coordinate x. Neglecting gravity, no forces act on this element in its initial configuration. After the test has begun, the element is shown displaced by a distance u, deformed to new dimensions dx and A, and subjected to forces F and

$F + dF$. The difference, dF, between these end-face forces causes the motion of the element that is manifested by the displacement u. This motion is governed by Newton's second law, force equals mass times acceleration:

$$dF = \rho_o A_o dx_o \left(\frac{d^2 u}{dt^2} \right) \qquad \text{(Eq 2)}$$

where $\rho_o A_o dx_o$ is the mass of the element, $A_o dx_o$ is the volume, ρ is the density of the material, and $(d^2 u/dt^2)$ is its acceleration. Tests that are conducted very slowly involve extremely small accelerations. Thus, Eq 2 shows that the variation of force dF along the specimen length is negligible.

However, for tests of increasingly shorter durations, the acceleration term on the right side of Eq 2 becomes increasingly significant. This produces an increasing variation of axial force along the length of the specimen. As the force becomes more nonuniform, so must the stress. Consequently, the strain and strain rate will also vary with axial position in the specimen. When these effects become pronounced, the concept of average values of stress, strain, and strain rate become meaningless, and the test results must be analyzed in terms of the propagation of waves through the specimen. This is shown in Table 1 as beginning near strain rates of 10^2 s^{-1}.

In an intermediate range of strain rates (denoted as dynamic tests in Table 1), an effect known as "ringing" of the load-measuring device obscures the interpretation of test data. An example of this condition is shown in Fig. 13, which is a tracing of the oscilloscope record of load cell force versus time during a dynamic tension test of a 2024-T4 aluminum specimen. Calculation showed that the oscillations apparent in the figure are consistent with vibrations at the approximate

Table 1 Strain rate ranges for different tests

Type of test	Strain rate range, s^{-1}
Creep tests	10^{-8} to 10^{-5}
Pseudostatic tension or compression tests	10^{-5} to 10^{-1}
Dynamic tension or compression tests	10^{-1} to 10^2
Impact bar tests involving wave propagation effects	10^2 to 10^4

Source: Ref 12

natural frequency of the load cell used for this test (Ref 13, 14).

In many machines currently available for dynamic testing, electronic signal processing is used to filter out such vibrations, thus making the instrumentation records appear much smoother than the actual load cell signal. However, there is still a great deal of uncertainty in the interpretation of dynamic

Fig. 5 Analysis of the uniaxial stress/strain/strain rate data for aluminum 1060-O

Source: Ref 4

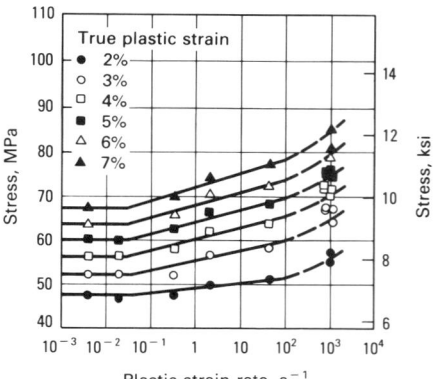

test data. Consequently, the average value of the high-frequency vibrations associated with the load cell can be expected to differ from the force in the specimen. This difference is caused by vibrations near the natural frequency of the testing machine, which are so low that the entire test can occur in less than one tenth of a cycle. Hence, these low-frequency vibrations usually are impossible to detect in a test record, but can produce significant errors in the analysis of test results. The ringing frequency for typical load cells ranges from 2400 to 3600 Hz.

Testing Machine Stiffness (Ref 15)

The most common misconception relating to strain rate effects is that the testing machine is much stiffer than the specimen. Such an assumption leads to the concept of deformation of the specimen by an essentially rigid machine. However, for most tests the opposite is true: the conventional tensile specimen is much stiffer than most testing machines. As shown in Fig. 14, for example, if crosshead displacement is defined as the relative displacement, Δ, that would occur under conditions of zero load, then with a specimen gripped in a testing machine and the driving mechanism engaged, the crosshead displacement equals the deformation in the gage length of the specimen plus elastic deflections in the machine frame, load cell, grips, specimen ends, etc. Before yielding, the gage length deformation is a small fraction of the crosshead displacement.

After the onset of gross plastic yielding of the specimen, conditions change. During this phase of deformation, the load varies slowly as the material strain hardens. Thus, the elastic deflections in the machine change slowly, and most of the relative crosshead displacement produces plastic deformation in the specimen. Qualitatively, in a test at approximately constant crosshead speed, the initial elastic strain rate in the specimen will be small, but the specimen strain rate will increase when plastic flow occurs.

Quantitatively, this effect can be estimated as follows. Consider a specimen having an initial cross-sectional area A_o and modulus of elasticity E gripped in a testing machine so that its axially stressed gage length initially is L_o. (This discussion is limited to the range of testing speeds where wave propagation effects are negligible. This restriction implies that the load is uniform throughout the gage length of the specimen.) Denote the stiffness of the machine, grips, etc., by K and the crosshead displacement rate (nominal head speed) by S. The ratio S/L_o is sometimes called the nominal rate of strain, but because

Fig. 6 Dynamic yield stress (minus the static yield stress) versus strain rate for steel and aluminum

The strain-rate-induced increase in strength for (a) steel is considerably greater than the corresponding increase in strength for (b) aluminum. For an explanation of the symbols and citations in this figure, see Ref 5.

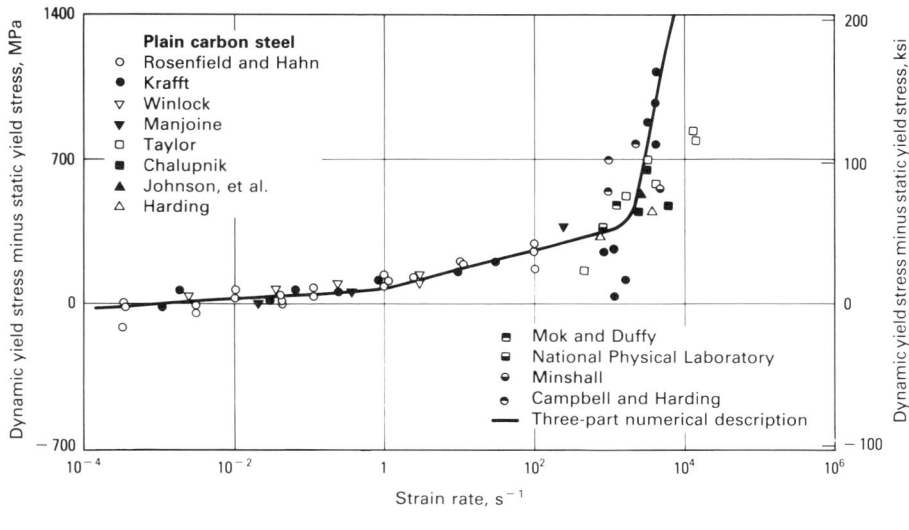

(a)

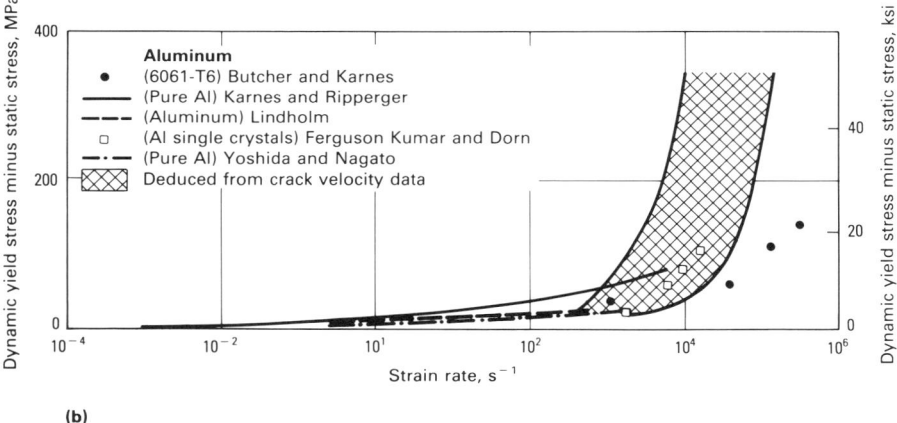

(b)

it is often substantially different from the rate of strain in the specimen, the term specific crosshead rate is preferred (Ref 16).

Let loading begin at time t equal to zero. At any moment thereafter, the displacement of the crosshead must equal the elastic deflection of the machine plus the elastic and plastic deflections of the specimen. Letting s denote the engineering stress in the specimen, the machine deflection is then sA_o/K. It is reasonable to assume that Hooke's law adequately describes the elastic deformation of the specimen at ordinary stress levels. Thus, the elastic strain e_e is s/E.

Denoting the average plastic strain in the specimen by e_p, the above displacement balance can be expressed as:

$$\int_0^t S\,dt = s\left(\frac{A_o}{K} + \frac{L_o}{E}\right) + e_p L_o \qquad \text{(Eq 3)}$$

Differentiating Eq 3 with respect to time and dividing by L_o gives:

$$\frac{S}{L_o} = \left(\frac{\dot{s}}{E}\right)\left(\frac{A_o E}{KL_o} + 1\right) + \dot{e}_p \qquad \text{(Eq 4)}$$

The strain rate in the specimen $\dot{e}$ is the sum of the elastic and plastic strain rates:

$$\dot{e} = \dot{e}_e + \dot{e}_p = \left(\frac{\dot{s}}{E}\right) + \dot{e}_p \qquad \text{(Eq 5)}$$

Using Eq 4 to eliminate the stress rate from Eq 5 yields:

Fig. 7 Effect of strain rate on the ductility of commercially pure titanium

(a) At room temperature, ductility is not affected. (b) At −195 °C (−319 °F), the ductility decreases at higher strain rates due to a change in deformation mechanism (slip versus twinning). Source: Ref 6

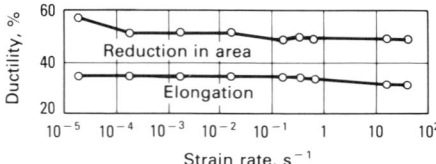

(a)

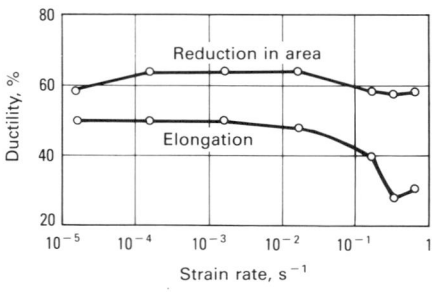

(b)

$$\dot{e} = \frac{\left(\dfrac{SK}{A_oE} + \dot{e}_p\right)}{\left(\dfrac{KL_o}{A_oE} + 1\right)} \qquad \text{(Eq 6)}$$

Thus, it is seen that the specimen strain rate usually will differ from the specific crosshead rate by an amount dependent on the rate of plastic deformation and the relative stiff-

Fig. 8 Percent elongation versus strain rate sensitivity parameter, m, for two titanium alloys and Zircaloy 4

Note that materials with a high m value possess high ductility. The phenomenon of superplasticity occurs in materials with high m values. Source: Ref 7

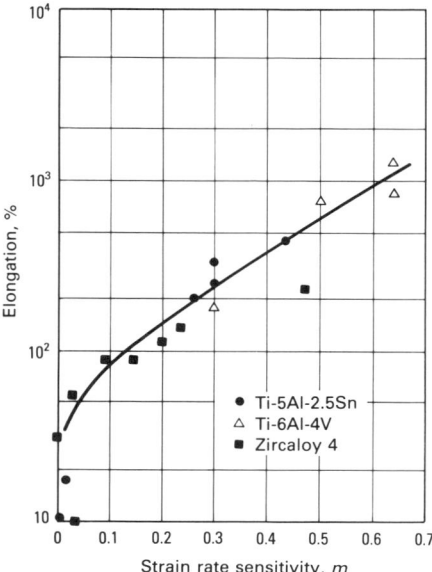

nesses of the specimen (A_oE/L_o) and the machine, K.

Tests at Constant Strain Rates

To maintain a constant average strain rate during a test, the crosshead speed must be adjusted as plastic flow occurs so that the sum ($SK/A_oE + \dot{e}_p$) remains constant. For

most metallic materials at the beginning of a test, the plastic strain rate is ostensibly zero, and from Eq 6 the initial strain rate is:

$$\dot{e}_o = \frac{\left(\dfrac{S_o}{L_o}\right)}{1 + \left(\dfrac{A_oE}{KL_o}\right)} \qquad \text{(Eq 7)}$$

where S_o is the crosshead speed at the beginning of the test. For materials that have a definite yield, $\dot{s} = 0$ at the yield point. Therefore, from Eq 4 and 5, the yield point strain rate is:

$$\dot{e}_1 = \left(\frac{S_1}{L_o}\right) \qquad \text{(Eq 8)}$$

where S_1 is the crosshead speed at the yield point. Equating these two values of strain rate shows that the crosshead speed must be reduced from its initial value to its yield point value by a factor of:

$$\frac{S_o}{S_1} = \left(1 + \frac{A_oE}{KL_o}\right) \qquad \text{(Eq 9)}$$

For particular measured values of machine stiffness given in Table 2, this factor for a standard 12.8-mm (0.505-in.) diam steel specimen is typically greater than 20 and can be as high as 100. Only for specially designed machines will the relative stiffness of the machine exceed that of the specimen. Even for wire-like specimens, the correspondingly delicate gripping arrangement will ensure that the machine stiffness is less than that of the specimen. Thus, large changes in crosshead speed usually are required to maintain a constant strain rate from the beginning of the test through the yield point.

Furthermore, for many materials, the onset of yielding is quite rapid, so that this large

Fig. 9 The effect of strain rate on the strength of polymers

(a) Stress as a function of strain rate for LX-04-M, a polymeric explosive material. Source: Ref 8. (b) Stress as a function of strain rate for two commercial polymers. Source: Ref 9. (c) Temperature-normalized stress as a function of strain rate for polypropylene at various temperatures. Source: Ref 10

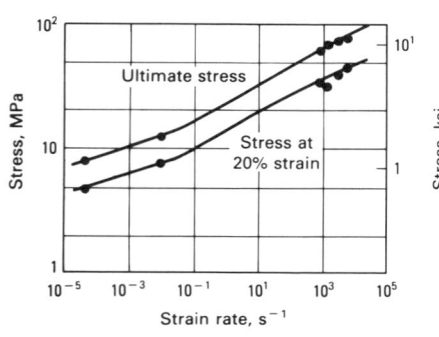

(a)

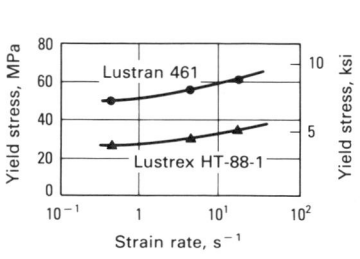

(b)

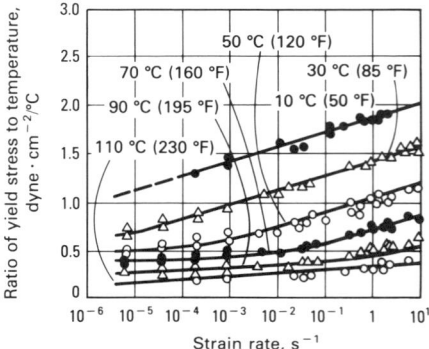

(c)

Table 2 Experimental values of testing machine stiffness

Machine stiffness		
kg/mm	lb/in.	Source
740	41 500	Ref 17
460	26 000	Ref 18
1800	100 000	Ref 19
1390-2970	77 900-166 500	Ref 20

change in speed must be accomplished quickly. Making the necessary changes in speed generally requires not only special strain-sensing equipment, but also a driving unit that is capable of extremely fast response. Thus, constant strain rate tests through yielding usually cannot be performed using screw-driven testing machines. Servohydraulic machines and electromagnetic machines may be capable of conducting tests at constant strain rate for materials with a yield point.

Equation 8 indicates the magnitude of speed changes required only for tests in which there is no yield drop. For materials having upper and lower yield points, the direction of crosshead motion may have to be reversed after initial yielding to maintain a constant strain rate. This reversal may be necessary, because plastic strains beyond the upper yield point can be imposed at a strain rate greater than the desired rate by recovery of elastic deflections of the machine as the load decreases. For a description of yield point phenomena, see the article "Mechanical Behavior of Materials Under Tension" in this Section.

Another important test feature related to the speed change capability of the testing

Fig. 10 Schematic of a hypothetical failure envelope illustrating tensile behavior of a typical polymer

Solid line represents the failure envelope. The dashed lines inside the failure envelope represent stress-strain curves at different strain rates.

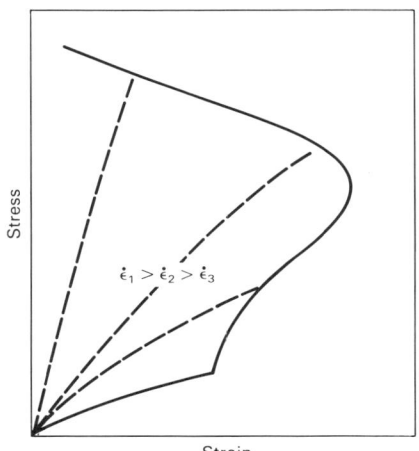

machine is the rate at which the crosshead can accelerate from zero to the prescribed test speed at the beginning of the test. For a slow test this may not be critical, but for a high-speed test the yield point could be passed before the crosshead achieves full testing speed. Thus, the crosshead may still be accelerating when it should be decelerating, and accurate information concerning the strain rate will not be obtained. With the advent of closed-loop servohydraulic machines and electromagnetic shakers, the speed at which the ram (crosshead) responds is two orders of magnitude greater than for screw-driven machines.

Tests at Constant Crosshead Speeds

Tension tests usually can be carried out at a constant crosshead speed on a conventional testing machine, provided the machine has an adequate speed controller and the driving mechanism is sufficiently powerful to be insensitive to changes in the loading rate. Because special accessory equipment is not required, such tests are relatively simple to perform. Also, constant crosshead speed tests typically provide as good a comparison among materials and as adequate a measure

of strain rate sensitivity as constant strain rate tests.

Two of the most significant test quantities—yield strength and ultimate tensile strength—frequently can be correlated with initial strain rate and specific crosshead rate, respectively. The strain rate up to the proportional limit equals the initial strain rate. Thus, for materials that yield sharply, the time-average strain rate from the beginning of the test to yield is only slightly greater than the initial strain rate:

$$\dot{e}_o = \frac{\left(\dfrac{S}{L_o}\right)}{\left(1 + \dfrac{A_o E}{K L_o}\right)} \qquad \text{(Eq 10)}$$

even though the instantaneous strain rate at yield is the specific crosshead rate:

$$\dot{e}_1 = \left(\frac{S}{L_o}\right) \qquad \text{(Eq 11)}$$

However, beyond the yield point, the stress rate is small so that the strain rate remains close to the specific crosshead rate (Eq 11). Thus, ductile materials, for which a rather long time will elapse before reaching ultimate strength, have a time-average strain

Fig. 11 Strain to failure, ϵ_b, versus a temperature-compensated strain rate, $\dot{\epsilon} a_T$, for a cross-linked rubber

This figure depicts the peak in the ductility dependence on strain rate. The temperature compensation is possible because polymers exhibit strong thermally activated deformation dependence. Source: Ref 11

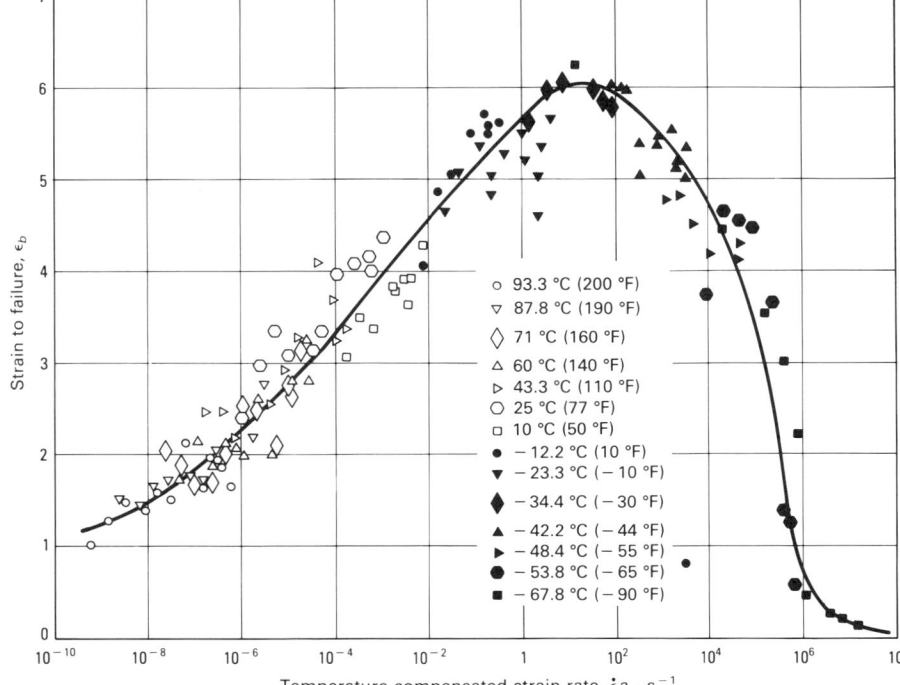

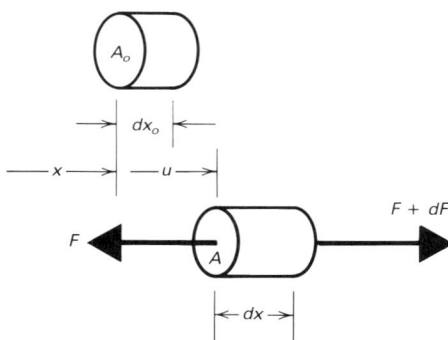

rate from the beginning of the test to ultimate that is only slightly less than the specific crosshead rate. Also, because the load rate is zero at ultimate as well as at yield, the instantaneous strain rate at ultimate equals the specific crosshead rate.

During a test at constant crosshead speed, the variation of strain rate from initial to yield point values is precisely the inverse of the crosshead speed change required to maintain a constant strain rate (Eq 9):

$$\frac{\dot{e}_1}{\dot{e}_o} = \left(1 + \frac{A_oE}{KL_o}\right) \qquad (Eq\ 12)$$

Consequently, in an ordinary tension test, the yield strength and ultimate tensile strength may be determined at two different strain rates, which can vary by a factor of 20 to 100, depending on machine stiffness. If a yield drop occurs, elastic recovery of machine deflections will impose a strain rate even greater than the specific crosshead rate given by Eq 11.

A point of interest arising from the analysis involves testing of different sized specimens at about the same initial strain rate. Assuming that these tests are to be made on one machine, under conditions for which K remains substantially constant, the crosshead speed must be adjusted to ensure that specimens of different lengths, diameters, or materials will experience the same initial strain rate. In the typical case where the specimen is much stiffer than the machine, $(1 + A_oE/KL_o)$ in Eq 10 can be approximated simply by (A_oE/KL_o), so that the initial strain rate is approximately $\dot{e}_o = SK/A_oE$. Thus, specimens of various lengths, tested at the same crosshead speed, will generally experi-

ence nearly the same initial strain rate. However, changing either the specimen cross section or material necessitates a corresponding change in crosshead speed to obtain the same initial rate.

A change in specimen length has substantially the same effect on both the specific crosshead rate (S/L_o) and the stiffness ratio of specimen to machine (A_oE/KL_o) and, therefore, has no net effect. For example, an increase in specimen length tends to decrease the strain rate by distributing the crosshead displacement over the longer length, but at the same time the increase in length reduces the stiffness of the specimen so that more of the crosshead displacement goes into deformation of the specimen and less into deflection of the machine. These two effects are almost exactly equal in magnitude. Thus, no change in initial strain rate is expected for specimens of different lengths tested at the same crosshead speed.

Experimental Determination of Machine Stiffness

For most metallic materials at low homologous temperatures, the initial strain rate will be elastic. Setting $\dot{e}_o$ in Eq 7 equal to $\dot{s}/E$ enables the machine stiffness to be written as:

$$K = \left[\frac{S}{A_o\dot{s}} - \frac{L_o}{A_oE}\right]^{-1} \qquad (Eq\ 13)$$

where the stiffness is expressed in terms of the measurable quantities crosshead speed, loading rate ($dF/dt = A_o\dot{s}$), specimen dimensions, and modulus of elasticity of the specimen material. A similar analysis of compression tests has shown how the results can be used in data reduction (Ref 21).

In Eq 13, the value for the modulus of elasticity can be obtained from textbooks when the specimen material is known. To experimentally determine this modulus, the actual strains in the specimen must be measured by strain gages or an extensometer. At the beginning of a test when the average plastic strain, e_p, is zero, if the crosshead displacement is incorrectly equated to specimen elongation in contradiction to Eq 3, the apparent modulus will be too small by the factor $(1 + A_oE/KL_o)$. Only for a perfectly rigid (infinite K) machine could the modulus be found from crosshead motion.

Analysis of Strain Rate

Although "strain rate" has been used throughout this article, the intended meaning is "deformation rate." For a specimen of initial gage length L_o and deformed length L,

the specific deformation rate is $(1/L_o)d(L - L_o)/dt$. If the deformation occurs homogeneously throughout the specimen, then the specific deformation rate corresponds everywhere to the strain rate. However, if the deformation is nonhomogeneous, then the strain (and strain rate) varies along the specimen length, and the specific deformation rate represents the spatial average strain rate. A well-known example of nonhomogeneous deformation is the propagation of deformation bands called Lüders bands.

Under relatively slow straining, most materials are assumed to transfer the heat generated by plastic deformation to their surroundings; that is, the straining is assumed to be isothermal (no change of temperature). The degree to which slow tension tests remain truly isothermal has been investigated (Ref 22). The flow stress, which is the uniaxial stress needed to continue plastic deformation of the material at a given stage of a test, is then assumed to depend only on strain and strain rate:

$$s = s(e, \dot{e}) \qquad (Eq\ 14)$$

or

$$\sigma = \sigma(\epsilon, \dot{\epsilon}) \qquad (Eq\ 15)$$

where s is the nominal or engineering stress F/A_o, e is the nominal or engineering strain $(L - L_o)/L_o$, σ is the true stress F/A, ϵ is the logarithmic strain $\ln(L/L_o)$, and superposed dots denote time derivatives. Of course, these quantities are interrelated.

For example, $\epsilon = \ln(1 + e)$ and $\dot{\epsilon} = (L_o/L)\dot{e}$. Furthermore, if the deformation is assumed to occur at constant volume (as it frequently is), $AL = A_oL_o$ so that $A = A_o \exp\{-\epsilon\} = A_o/(1 + e)$; then $\sigma = s \exp\{\epsilon\} = s(1 + e)$. Therefore, the discussion can be lim-

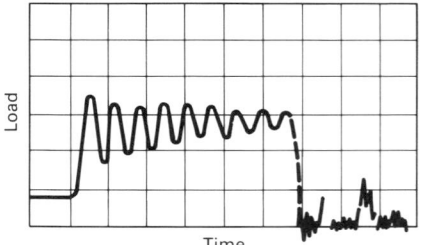

Fig. 14 Schematic illustrating crosshead displacement and elastic deflection in a tension testing machine

Δ is the displacement of the crosshead relative to the zero load displacement; L_o is the initial gage length of the specimen; K is the composite stiffness of the grips, loading frame, load cell, specimen ends, etc.; F is the force acting on the specimen. The development of Eq 3 to 13 describes the effects of testing machine stiffness on tensile properties. Source: Ref 15

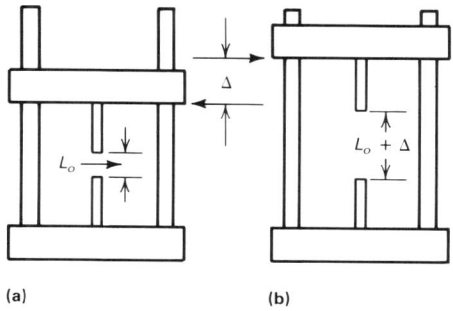 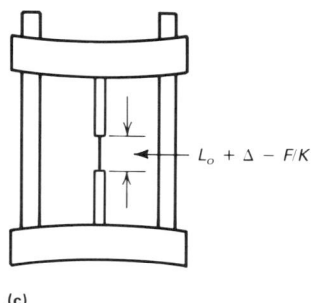

(a)   (b)   (c)

ited to either Eq 14 or 15 without loss of generality.

The strain hardening parameter n of a material can be defined as:

$$n = \left(\frac{\epsilon}{\sigma}\right)\left(\frac{\partial\sigma}{\partial\epsilon}\right) \qquad \text{(Eq 16)}$$

In an analogous manner, the strain rate sensitivity parameter m can be defined as:

$$m = \left(\frac{\dot{\epsilon}}{\sigma}\right)\left(\frac{\partial\sigma}{\partial\dot{\epsilon}}\right) \qquad \text{(Eq 17)}$$

The data given earlier in this discussion clearly indicate that both n and m are functions of strain and strain rate. Figures 4(a) and (b) show that m can be negative under some conditions. However, average values frequently are selected for these parameters, which are then treated as constants.

One analytical representation sometimes used to approximate the isothermal behavior of metals is:

$$\sigma = C\epsilon^n\left(\frac{\dot{\epsilon}}{\dot{\epsilon}_o}\right)^m \qquad \text{(Eq 18)}$$

where C is a strength parameter that has the units of stress; $\dot{\epsilon}_o$ is an arbitrary rate; and n and m are the strain hardening and strain rate sensitivity parameters, respectively, and are treated as constants. However, $\dot{\epsilon}_o$ and C are interrelated; the stress-strain curve given by Eq 18 will pass through $\sigma = C$ and $\epsilon = 1$ when the strain rate is $\dot{\epsilon} = \dot{\epsilon}_o$. Values of n usually are between 0.1 and 0.5 for metals; they are determined from, but not identical to, strain hardening rates. Values of m for metals usually are much smaller than the corresponding n values. However, fine-grained metals have relatively large rate sensitivity parameters ($m > 0.1$) under specific deformation conditions. Under such conditions, these materials can be deformed to extremely large strains and are called superplastic metals.

For extremely high rates of testing, it is commonly assumed that deformation occurs under adiabatic (no heat transfer) conditions. Plastic work is mostly (about 90%) converted to heat. The remainder is inelastically stored as changes in defect structure. In high-speed tests, this heat raises the temperature of the material. Consequently, the material properties are changed, as discussed in the article "High Strain Rate Tension Testing" in this Volume. This is another major complication in analyses of high-speed tests.

One further phenomenon that can dramatically affect high strain rate deformation is the occurrence of shear bands, which entails the accommodation of overall deformation within a small volume of material. This volume is heated rapidly so that it becomes weaker than the surrounding material and more easily deformable. Thus, deformation continues to be concentrated to the point at which the material melts locally. Detailed information on shear band formation due to high strain rate deformation can be found in Ref 23.

REFERENCES

1. J.W. Edington, Effect of Strain Rate on the Dislocation Substructure in Deformed Niobium Single Crystals, in *Mechanical Behavior of Materials under Dynamic Loads*, U.S. Lindholm, Ed., Springer-Verlag, New York, 1968, p 191-240
2. J.M. Krafft and J.C. Hahn, U.S. Patent No. 3,194,062, 1965
3. M.J. Manjoine, Influence of Rate of Strain and Temperature on Yield Stresses of Mild Steel, *J. Appl. Mech.*, Vol II, 1944, p A-211 to A-218
4. A.H. Jones, C.J. Maiden, S.J. Green, and H. Chin, Prediction of Elastic-Plastic Wave Profiles in Aluminum 1060-O Under Uniaxial Strain Loading, in *Mechanical Behavior of Materials under Dynamic Loads*, U.S. Lindholm, Ed., Springer-Verlag, New York, 1968, p 254-269
5. M.F. Kanninen, A.K. Mukherjee, A.R. Rosenfield, and G.T. Hahn, The Speed of Ductile-Crack Propagation and the Dynamics of Flow in Metals, in *Mechanical Behavior of Materials under Dynamic Loads*, U.S. Lindholm, Ed., Springer-Verlag, New York, 1968, p 96-133
6. T.S. Desisto and D.E. Driscoll, Effect of Strain Rate and Temperature on the True Stress—True Strain Properties of Commercially Pure Titanium, in *High Speed Testing*, Vol 1, A.G.H. Dietz and F.R. Eirich, Ed., Interscience, New York, 1960, p 97
7. D. Lee and W.A. Backofen, Superplasticity in Some Titanium and Zirconium Alloys, *Trans. Met. Soc. AIME*, Vol 239, 1967, p 1034
8. K.G. Hoge, The Behavior of Plastic-Bonded Explosives Under Dynamic Compressive Loads, *J. Appl. Polymer Sci.*, Applied Polymer Symposia, No. 5, 1967, p 19-40
9. J.K. Lund, Geometry Effects and High Speed Tensile Behavior in Styrene Polyblend Systems, in *High Speed Testing*, Vol 4, A.G.H. Dietz and F.R. Eirich, Ed., Interscience, New York, 1963, p 79
10. J.K. Roetling, Yield Stress Behavior of Some Thermoplastic Polymers, *J. Appl. Polymer Sci.*, Applied Polymer Symposia, No. 5, 1967, p 166
11. F. Rodriguez, *Principles of Polymer Systems*, 2nd ed., McGraw-Hill, New York, 1982, p 240
12 G.E. Dieter, *Mechanical Metallurgy*, McGraw-Hill, New York, 2nd ed., 1976, p 349
13. D.J. Shippy, P.P. Gillis, and K.G. Hoge, Computer Simulation of a High Speed Tension Test, *J. Appl. Polymer Sci.*, Applied Polymer Symposia, No. 5, 1967, p 311-325
14. P.P. Gillis and D.J. Shippy, Vibration Analysis of a High Speed Tension Test, *J. Appl. Polymer Sci.*, Applied Polymer Symposia, No. 12, 1969, p 165-179
15. M.A. Hamstad and P.P. Gillis, Effective Strain Rates in Low-Speed Uniaxial Tension Tests, *Mat. Res. Stand.*, Vol 6 (No. 11), 1966, p 569-573
16. P.P. Gillis and J.J. Gilman, Dynamical Dislocation Theories of Crystal Plasticity, *J. Appl. Phys.*, Vol 36, 1965, p 3375-3386

17. W.G. Johnston, Yield Points and Delay Times in Single Crystals, *J. Appl. Phys.*, Vol 33, 1962, p 2716
18. H.G. Baron, Stress-Strain Curves of Some Metals and Alloys at Low Temperatures and High Rates of Strain, *J. Iron Steel Inst. (Brit.)*, Vol 182, 1956, p 354
19. J. Miklowitz, The Initiation and Propagation of the Plastic Zone in a Tension Bar of Mild Steel as Influenced by the Speed of Stretching and Rigidity of the Testing Machine, *J. Appl. Mech.*, Vol 14, 1947, p A-31
20. M.A. Hamstead, ''The Effect of Strain Rate and Specimen Dimensions on the Yield Point of Mild Steel,'' Lawrence Radiation Laboratory Report UCRL-14619, April 1966
21. J.E. Hockett and P.P. Gillis, Mechanical Testing Machine Stiffness, Parts I & II, *Int. J. Mech. Sci.*, Vol 13, 1971, p 251-275
22. A.K. Sachdev and J.E. Hunter, Jr., Thermal Effects during Uniaxial Straining of Steels, *Met. Trans. A*, Vol 13, 1982, p 1063-1067
23. J. Mescall and V. Weiss, Ed., *Material Behavior Under High Stress and Ultrahigh Loading Rates*, Sagamore Army Materials Research Conference Proceedings, Vol 29, Plenum Press, New York, 1983

Tension Testing Machines and Extensometers

By James J. Martin
Manager, Technical Publications
Instron Corporation

TENSION TESTING INSTRUMENTATION has evolved from purely mechanical or electromechanical machines (circa 1900) to sophisticated instrumentation that employs advanced electronics and microcomputers. This transition has made it possible to determine rapidly and with great precision ultimate tensile strength and elongation, yield strength, modulus of elasticity, and other material properties. Test speeds typically ranging from 0.05 to 500 mm/min (0.002 to 20 in./min) can be achieved with 0.1% precision. Loads can be measured to 0.5% accuracy from 10 gf full scale to 45 000 kgf (100 000 lbf) or more.

This article reviews the current technology and examines force application systems, force measurement, strain measurement, important instrument considerations, gripping of test specimens, environmental equipment, and the use of computers for materials testing. Emphasis is placed on screw-driven tension testing apparatus. It should be noted, however that tension tests can also be performed using hydraulic and servohydraulic machines, which have higher load capacities (up to 450 000 kgf, or 1 000 000 lbf). Information on hydraulic test systems can be found in Ref 1.

Principles of Operation

Tension testing machines apply uniaxial loading in a uniform manner and generally are universal in their capabilities and applications, rather than specific to one type of test or material. Conducting load versus elongation (stress versus strain) tests involves control of forces up to 45 000 kgf (100 000 lbf) or more, holding onto a specimen that may be thin sheet metal, machined flat plate, or a cylindrical polished sample, and measuring forces (stresses) and deformations (strains) with accuracy and repeatability. The effect of the instrumentation on the test data must be minimized to avoid testing errors. Automatic recording of stress and strain is highly desirable, and modules to allow specific functions, such as automatic shutoff when the specimen breaks, simplify machine use.

A tension tester should have adequate loading capacity to cover the expected testing force range and interchangeable grips to hold various test specimens that can be easily mounted and dismounted. Typically, the test load is applied through multiple drive screws and drive nuts that move a crosshead, which is attached to a grip holding one end of the test specimen. The other end of the specimen is attached by a grip to a fixed base. By driving the screws at various rates, tension tests can be conducted over a wide range of test conditions. Components and controls for a screw-driven tension testing machine are shown in Fig. 1.

Measurement of stresses and strains in a tension tester typically is accomplished by using load- and strain-sensing transducers that create an electrical signal proportional to the applied stress or strain. The output can be seen on a recorder chart or digital meter, or can be transmitted to a computer.

Force Application Systems

As shown in Fig. 1, tensile forces are applied by a moving crosshead that is operated by two vertical lead screws. These are driven by a high-torque motor, typically powered by direct current for high starting torque, through either gears or belts. A closed-loop servodrive system ensures that the crosshead moves at a constant speed. Although the drive systems of individual units may differ, they essentially operate as follows.

The desired or user-selected speed and direction information is compared with a known reference signal. Positional control of the moving crosshead is controlled by the servomechanism to reduce any error or difference. This produces the desired crosshead speed. Direction and sequence of crosshead movement may be manually controlled using pushbuttons or automatic controls. Top crosshead speeds of 1250 mm/min (50 in./min) can be attained in screw-driven machines; servohydraulic machines can be driven at speeds of 250×10^3 mm/min (10×10^3 in./min) or more.

Because of the high forces involved, bearings and gears require particular attention to reduce friction and wear. Backlash, which is the free movement between the mechanical drive components, is particularly undesirable. Many instruments incorporate anti-backlash preloading so that forces are translated evenly through the lead screw and crosshead. However, when the testing direction is constantly in one direction, anti-backlash devices may be unnecessary.

Constant rate of extension machines are the most common type of tension testers and are characterized by a constant rate of crosshead travel regardless of applied loads. They permit testing without speed variations that might alter test results; this is particularly important when testing rate-sensitive materials such as polymers, which exhibit different ultimate strengths and elongations when tested at different speeds.

Constant Load Rate Devices. With appropriate modules, tension testing at a constant load rate can be accomplished easily. In this configuration, a load-control module allows the constant rate of extension machine to function as a constant load rate device. Usually, the servomechanism system response is particularly critical when materials are loaded through the yield point.

Constant Strain Rate Testing. Using strain signal for feedback, a constant rate of extension machine may be used for constant strain rate testing. Again, servomechanism response time is particularly critical when materials are taken through yield.

Force Measurement

The force-measurement or load-weighing systems in tension testing machines typically employ load transducers that are strain-gage-type load cells. Strain gages are devices that undergo electrical resistance changes in the presence of mechanical deformation. A typical system uses a load cell connected to a bridge circuit to measure these minute resistance changes and thus the applied loads or forces. The circuit is excited with a signal generated by the load cell amplifier, and an applied force causes the strain-gage bridge circuit to be unbalanced. The resulting signal is returned to the amplifier, where it is amplified and converted into an output signal that is proportional to the applied force. The output can drive a digital display, strip chart recorder, or computer.

A load cell amplifier enables the bridge circuit to compensate for different tare weights. It also includes a means of calibration, load range selection, and zeroing adjustments that accommodate different load cells and load ranges. Figure 2 is a block diagram of a load-weighing system along with a cross-sectional view of a load cell.

Within individual load cells, mechanical stops can be incorporated to minimize possible damage that could be caused by accidental overloads. Also, guidance and supports can be included to prevent the deleterious effects of side loading and to give desired rigidity and ruggedness. This is important in tension testing of metals because of the elastic recoil that can occur when a stiff specimen fails.

The calibration of load-weighing systems can be accomplished using precision weights, electrical calibration with factory-set standards, or proving rings. ASTM Standard E 4 (Ref 2) covers the load verification

Fig. 1 Components and controls for a screw-driven tension testing machine

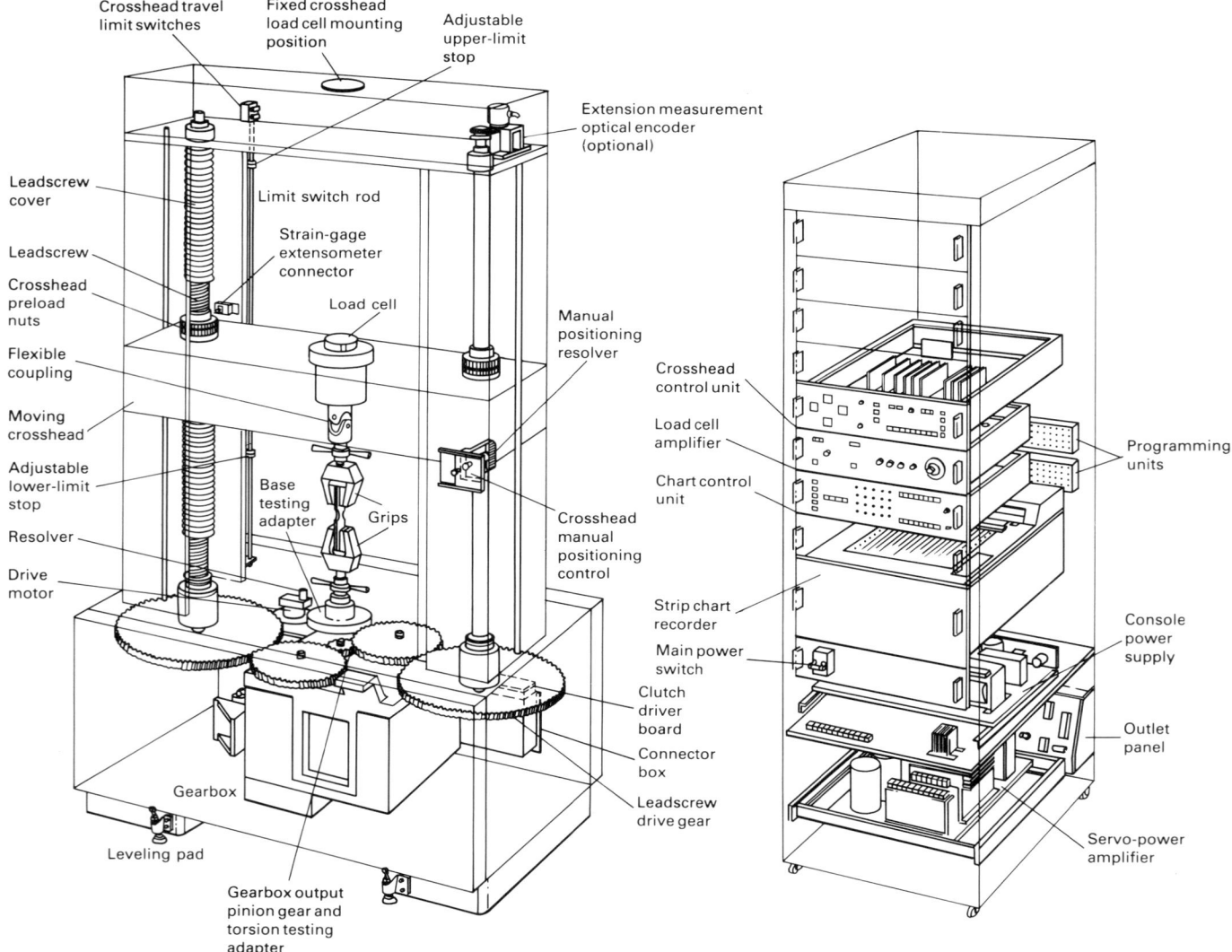

Loading frame   **Control console**

Fig. 2 (a) Typical block diagram of a load-weighing system with (b) cross-sectional view of a load cell

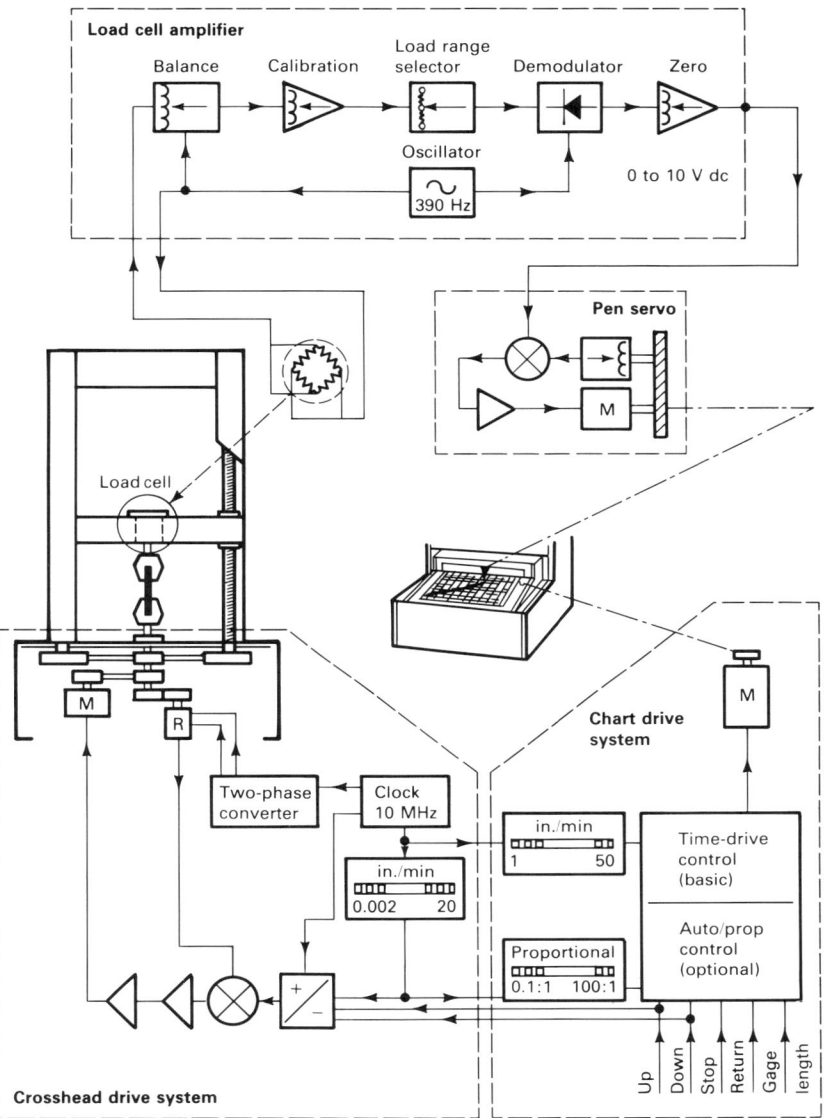

(a)

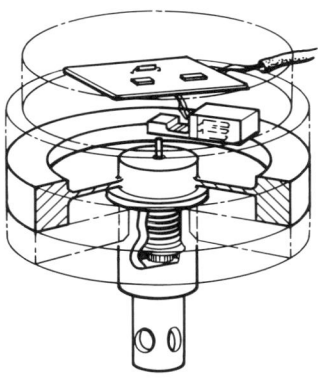

(b)

procedure for tension testing machines (see also the article "Calibration of Testing Equipment" in this Volume).

Ambient temperature changes can affect the resistance and therefore the accuracy of strain gages. Use of temperature-compensating gages eliminates this source of error.

Strain Measurement

Extensometers can be attached to a test specimen to measure elongation or strain as the load is applied. This is particularly important for metals and similar materials that exhibit high stiffness. Typical extensometers have fixed gage lengths such as 25 or 50 mm (1 or 2 in.). They are also classified by maximum percent elongation so that a typical 25-mm (1-in.) gage length unit would have different models for 10%, 50%, or 100% maximum strain. Axial and diametral extensometers attached to cylindrical and flat tension specimens are shown in Fig. 3.

There also are transverse strain-measuring devices that indicate the reduction in width or diameter as the specimen is tested. Averaging extensometers are also available for testing high modulus materials. This type utilizes dual-measuring elements that measure elongation on both sides of a sample; the measurements are then averaged to obtain a mean strain.

Extensometers can use either linear variable differential transformers (electromechanical devices that provide an output voltage proportional to displacement) or, more commonly, strain gages such as the load cells described earlier. Operation of strain-gage extensometers is based on gages that are bonded to a metallic element and connected to a bridge circuit. Deflection of the element due to specimen strain changes the gage resistance and sends an electrical output to an amplifier. The output is then relayed to a digital readout, chart recorder, or computer.

The circuitry in the strain-measuring system allows multiple ranges of sensitivity, so one transducer can be used over broad ranges. The magnification ratio, which is the ratio of output to extensometer deflection, can be as high as 10 000 to 1. Extensometers are highly sensitive, but must be calibrated with precision micrometer-type fixtures to ensure accuracy. A strain-gage extensometer attached to a precision calibration fixture is shown in Fig. 4. ASTM Standard E 83 describes the verification and classification of extensometers (Ref 3). See also the article "Calibration of Testing Equipment" in this Volume.

Materials that are more extensible can be tested using indirect or time-based strain measurements. In this manner, the y axis of a recorder can be driven on a time base, as is

Fig. 3 Typical test specimen, grips, and extensometer setups

(a) Wedge grips holding rod sample with axial extensometer. (b) Extensometer attached to a plate-type specimen held in 14 000-kgf (30 000-lbf) capacity wedge-type grips. (c) Wedge grips holding round specimen with diametral extensometer. Courtesy of Instron Corp.

(a)

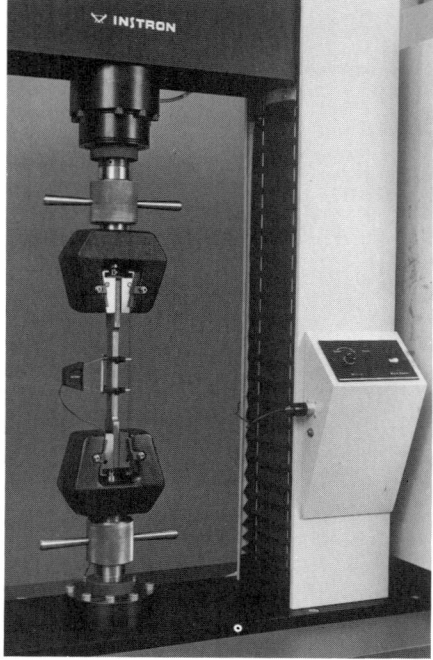

(b)

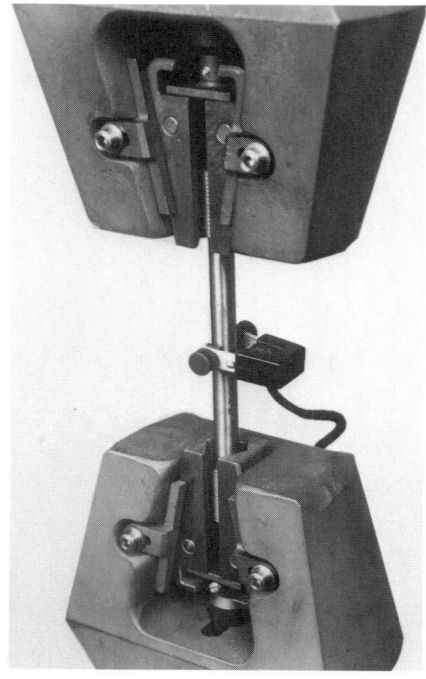

(c)

the crosshead of the testing machine. The ratio of the two speeds provides a strain axis magnification. Digital readout devices and computers that use accurate crystal clocks function similarly.

Testing Machine Considerations

Many aspects of testing, particularly the accuracy and reproducibility of test results, are dependent on the tension testing machine and its use. This section describes some factors that must be considered to avoid testing problems.

Stiffness. The loading frame of a tension testing machine must be extremely stiff to avoid deflections that can cause testing error. Often, rigid crosshead guidance systems are used in a test machine to prevent side loading of specimens. The load transducers in the test system should also have a maximum stiffness to avoid similar errors. More detailed information on testing machine stiffness can be found in the article "Effect of Strain Rate on Flow Properties" in this Section.

Precision. For reliable data, the testing machine must be accurate and provide reproducible results. Effects of mechanical inertia or offsets could be devastating to a testing program. A highly accurate, responsive

load- and strain-measuring system should be employed that maintains the same accuracy whether a given load cell is operating on a low load range or a high load range. In addition, the testing speed must be controlled within close limits, as specified by the appropriate test method.

Flexibility. A testing machine must cover a variety of uses that typically require different speeds and load ranges. Modular transducers, specimen grips, and accessories for machine control and data readout greatly extend the useful life of a testing machine.

Gripping Techniques

The use of proper grips and faces for testing materials in tension is critical in obtaining meaningful results. Trial and error often will solve a particular gripping problem. Tension testing of most flat or round specimens can be accommodated with wedge-type grips (Fig. 3). Wire and other forms may require different grips, such as capstan or snubber types. The load capacities of grips range from under 4.5 kgf (10 lbf) to 45 000 kgf (100 000 lbf) or more. ASTM Standard E 8 describes the various types of gripping devices used to transmit the measured load applied by the test machine to the tension test specimen (Ref 4).

Fig. 4 Strain-gage extensometer on precision calibration fixture

Courtesy of Instron Corp.

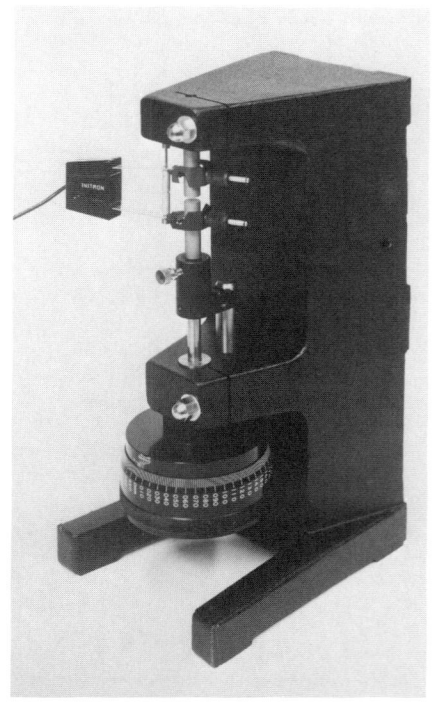

Fig. 5 Block diagram of a computerized tension testing machine

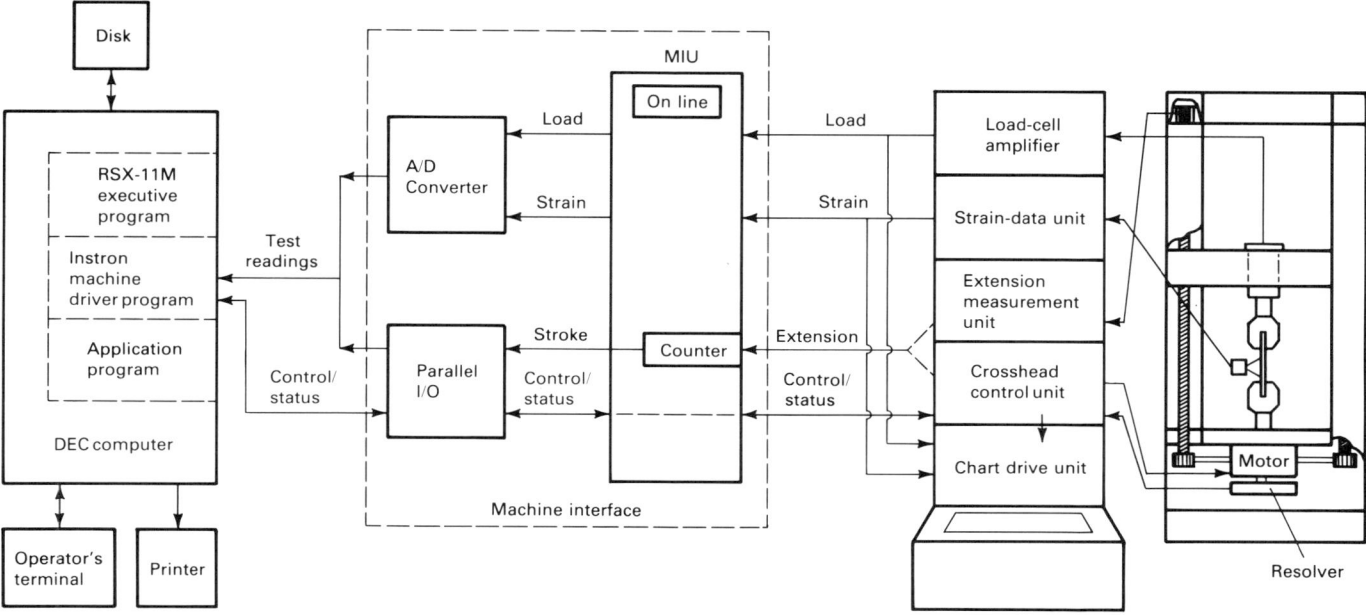

Computer system   Test instrument

Screw-action grips or mechanical grips are low in cost and are available with load capacities of up to 450 kgf (1000 lbf). This type of grip, which is normally used for testing flat specimens, can be equipped with interchangeable grip faces that have a variety of surfaces. Faces are adjustable to compensate for different specimen thicknesses.

Wedge-type grips (Fig. 3) are self-tightening and are built with capacities of up to 45 000 kgf (100 000 lbf) or more. Some units can be tightened without altering the vertical position of the faces, making it possible to preselect the exact point at which the specimen will be held. The wedge-action design works well on hard-to-hold specimens and prevents the introduction of compressive forces that cause specimen buckling.

Pneumatic-action grips are available in various designs with capacities of up to 90 kgf (200 lbf). This type of grip clamps the specimen by lever arms that are actuated by compressed air cylinders built into the grip bodies. A constant force maintained on the specimen compensates for decrease of force due to creep of the specimen in the grip. Another advantage of this design is the ability to optimize gripping force by adjusting the air pressure, which makes it possible to minimize specimen breaks at the grip faces.

Buttonhead grips enable the rapid insertion of threaded-end or mechanical-end specimens. They can be manually or pneumatically operated, as required by the type of material or test conditions.

Computerization

With current computer technology, it is possible to add a powerful data reduction capability to a tension testing machine, ranging from units that simply accept test results and perform minor calculations to large systems that control the complete testing machine sequence, calculate test results, and store millions of bits of data. Besides providing accurate control, a computer provides flexibility in generating reports and in the storage and retrieval of test results.

Many manufacturers offer different computers with standardized programs that are written to conform to ASTM or other test methods. Figure 5 illustrates how a test machine is connected to a computer through an interface that converts the analog load and strain data and presents it to the computer. Any control signals generated in the computer are transmitted to the test machine in a similar fashion.

A computerized tension testing system eliminates manual interpretation of test curves, reducing human error and improving productivity. Detailed information on computer automation of materials testing can be found in Ref 5.

REFERENCES

1. H.E. Davis, G.E. Troxell, and G.F.W. Hauck, *The Testing of Engineering Materials*, 4th ed., McGraw-Hill, New York, 1982, p 80-85
2. "Standard Practice for Load Verification of Testing Machines," E 4, *Annual Book of ASTM Standards*, Vol 03.01, ASTM, Philadelphia, 1984, p 111-118
3. "Standard Practice of Verification and Classification of Extensometers," E 83, *Annual Book of ASTM Standards*, Vol 03.01, ASTM, Philadelphia, 1984, p 247-252
4. "Standard Methods of Tension Testing of Metallic Materials," E 8, *Annual Book of ASTM Standards*, Vol 03.01, ASTM, Philadelphia, 1984, p 130-150
5. B.C. Wonsiewicz, Ed., *Computer Automation of Materials Testing*, STP 710, ASTM, Philadelphia, 1980

Compression, Bearing, and Shear Testing

Axial Compression Testing

By Ralph Papirno
Mechanical Engineer
Army Materials and Mechanics
Research Center

COMPRESSION LOADS are applied to many engineering structures that vary in dimension from massive suspension bridge piers to the thin sheet of aircraft wings. In addition, metalforming processes involve large compressive deformations. Analyses of structural behavior or metalforming require knowledge of compression stress-strain properties. This is best obtained by using the procedures described in ASTM Standard E 9 (Ref 1). This article discusses factors that contribute to obtaining valid test data, including the effects of barreling and buckling, a definition of compressive failure, and compression test procedures or techniques that differ from those of tension testing.

Several assumptions are inherent in stress-strain testing. In any test used to obtain uniaxial compression stress-strain properties, the measured quantities are generally load and strain, if a strain gage is used, or displacement, if a compressometer is used. Strain or displacement is measured at the surface of the specimen, and longitudinal and transverse strains are assumed to be uniform in each cross section along the entire gage length. The measured surface strain is assumed to be the same as the internal strain. The stress is always an inferred quantity that is calculated by dividing the applied load by the cross-sectional area of the test piece.

Additionally, it is assumed that the cross-sectional area is constant over the gage length and that the stress is uniaxial and uniform in each cross section along the gage length. Errors in stress or strain occur if the assumptions of uniformity do not exist in a test. Buckling and barreling cause nonuniform stress and strain distributions, and elimination of these phenomena in compression tests can lead to more accurate stress-strain data.

Specimen Buckling

When a specimen buckles during a compression test, the stress data calculated from the applied load will be erroneous. However, the risk of specimen buckling can be reduced by careful attention to alignment of the loading train and by careful manufacture of the specimen according to the specifications of flatness, parallelism, and perpendicularity given in Ref 1. Even with well-made specimens tested in a carefully aligned loading train, buckling may still occur. Conditions that typically induce buckling are discussed below.

Alignment. The loading train, including the loading faces, must maintain initial alignment throughout the entire loading process. Alignment, parallelism, and perpendicularity tests should be conducted at maximum load conditions of the testing apparatus.

Specimen Tolerances. The tolerances given in Ref 1 for specimen end-flatness, end-parallelism, and end-perpendicularity should be considered as upper limit values. This is also true for concentricity of outer surfaces in cylindrical specimens and uniformity of dimensions in rectangular sheet specimens. If tolerances are reduced from these values, the risk of premature buckling is also reduced.

Inelastic Buckling. Only elastic buckling is discussed in Ref 1. This may be somewhat unrealistic, because for the most slender specimen recommended, the calculated elastic buckling stresses are higher than can be achieved in a test. This specimen has a length-to-diameter ratio of $L/D = 10$. An approximate calculation using the elastic Euler equation for a steel specimen with flat ends on a flat surface (assumed value of end-fixity coefficient is 3.5) yields a buckling stress in excess of 4100 MPa (600 ksi); the comparable value for an aluminum specimen would be 1380 MPa (200 ksi). These values, however, are not realistic.

Buckling stress in the above example should not be calculated by an elastic formula but by an inelastic buckling relation. In terms of inelastic buckling it has been concluded that the following relation appropriately calculates inelastic buckling stresses (Ref 2):

$$S_{cr} = C\pi^2 \left[\frac{E_t}{\left(\frac{L}{\rho}\right)^2} \right] \qquad \text{(Eq 1)}$$

where S_{cr} is the buckling stress, MPa (ksi); C is the end-fixity coefficient; E_t is the tangent modulus of the stress-strain curve, MPa (ksi); L is the specimen length, mm (in.); and ρ is the radius of gyration of specimen cross section, mm (in.). Equation 1 reduces to the Euler equation if E, the modulus of elasticity, is substituted for E_t.

Rearranging Eq 1 to combine the stress-related factors results in:

$$\left(\frac{1}{C\pi^2}\right)\left(\frac{L}{\rho}\right)^2 = \frac{E_t}{S_{cr}} \qquad \text{(Eq 2)}$$

Note that the value of the right side of Eq 2 decreases as stress increases in a stress-strain curve. In a material whose response is elastic-pure plastic, the right side of Eq 2 vanishes, because E_t becomes zero, and buckling will always occur at the yield stress. When the material exhibits strain hardening, calculations using Eq 2 will yield the appropriate specimen dimensions to resist buckling for given values of stress.

Side Slip. One form of buckling of cylindrical specimens that may result from misalignment of the loading train under load or from loose tolerances on specimen dimensions is illustrated in Fig. 1. The ends of the specimen undergo side slip, resulting in a sigmoidal central axis. This form of buckling could be described by Eq 1 and 2, provided an appropriate value of the end-fixity coefficient can be assigned.

Thin Sheet Specimens. In testing thin sheet in a compression jig, approximately 2% of the specimen length protrudes from the jig. Buckling of this unsupported length can occur if there is misalignment of the loading train such that it does not remain coaxial with the specimen throughout the test (Ref 3). A typical compression jig and contact-point compressometer are shown in Fig. 2.

Barreling of Cylindrical Specimens

When a cylindrical specimen is compressed, Poisson expansion occurs. If this expansion is restrained by friction at the loading faces of the specimen, nonuniform states of stress and strain in the entire specimen result. The specimen acquires a barreled shape, as shown in Fig. 3. The effect on the stress and strain distributions is of consequence only when the deformations are on the order of 10% or more.

Friction at the Loading Surfaces. Friction on the loading face causes roll-over. As shown in Fig. 3, points originally on the sides of the specimen are ultimately located on the specimen end face. A computer code has been used to study the details of a compression test when friction is present (Ref 4).

Fig. 2(a) Sheet compression jig suitable for room-temperature or elevated-temperature testing

Figure 4(a) shows the computer results after a 50% height reduction in a steel specimen (loading axis is horizontal). In the computer simulation, a friction coefficient of 0.3 was assumed. The grid in the undeformed state consisted of orthogonal sets of equally spaced lines. Points A were originally on the circumference of the undeformed face (see also Fig. 3).

Figure 4(b) illustrates a polished and etched longitudinal section of a compressed AISI-SAE 4340 steel specimen. The height

Fig. 2(b) Contact-point compressometer installed on specimen removed from jig

Contact points fit in pre-drilled shallow holes in edge of specimen.

reduction was 56%. Internal line configurations are close to that of the computer simulation shown in Fig. 4(a). The lines, originally straight in the undeformed cylinder, resulted from segregation during processing and clearly reveal the internal deformations. A more detailed quantitative description of the compression of cylinders is given in Ref 4.

Reduction of Friction. Use of a high-pressure lubricant at the loading surface of the specimen reduces friction. One such material commonly used is 0.1-mm (0.004-in.) thick Teflon sheet. The action of the lubricant may be enhanced if the bearing surfaces that apply the load are hard and highly pol-

Fig. 1 Schematic diagram of side-slip buckling

The original position of the specimen center line is indicated by the dashed line.

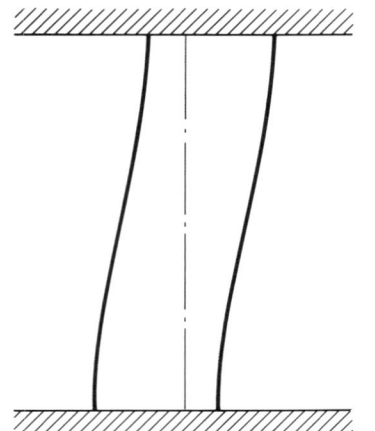

Fig. 3 Barreling during a test when the friction coefficient is 1.00 at the specimen loading face

Note that as the deformation increases, points A, B, and C originally on the specimen sides move to the loading face.

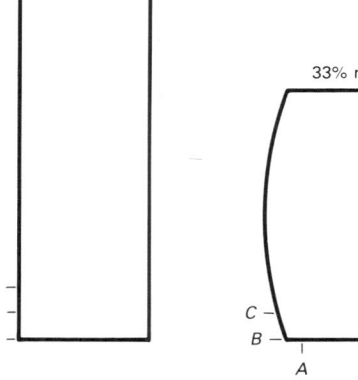
Undeformed

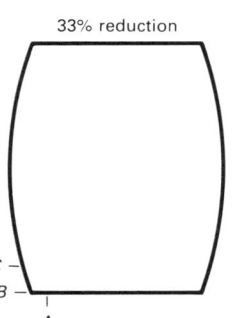
33% reduction

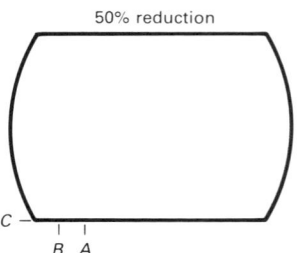
50% reduction

Fig. 4(a) Computer-code simulation for 50% compression of a steel specimen

The loading axis is horizontal. This specimen had a length-to-diameter ratio (L/D) of 2. The original grid was square. The points marked A were on the circumference of the undeformed specimen. Source: Ref 4

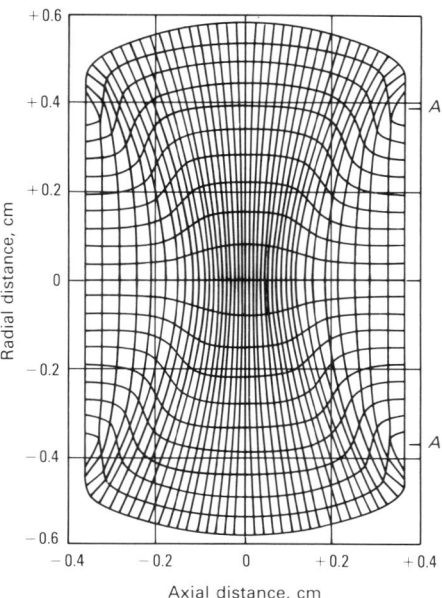

Fig. 4(b) Polished and etched section of a specimen of AISI-SAE 4340 steel compressed 56%

The lines in the specimen, which are similar to the computer code simulation in Fig. 4(a), result from segregation during processing and were originally all parallel to the horizontal axis. Source: Ref 4

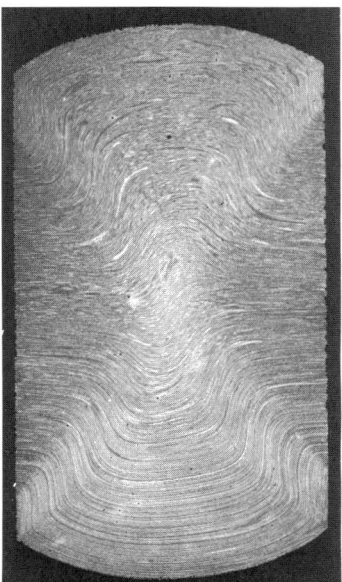

ished. The use of tungsten carbide bearing blocks is recommended for all materials undergoing compression testing. Other techniques have been used to reduce nonuniformity of stress and strain distributions along the gage length (Ref 5, 6).

Sheet Specimen Lubrication. The contact area between the specimen lateral faces and the lateral support guides of the testing jig must be well lubricated. Personnel engaged in sheet compression testing should become familiar with the literature on the subject. A selected bibliography on this subject is given in Ref 1.

Compression Fracture

A major deficiency of the compression test standard (Ref 1) is that compressive strength is uniquely defined only for catastrophic failure by crushing or fracture. However, cylindrical specimens of all but the most ductile materials will develop cracks when they are compressed. The cracks generally initiate on the outer surface of the compressed specimen. As the specimen is further deformed, the already initiated cracks propagate, and new cracks initiate. Several examples of various modes of compression cracking are given in Ref 7. In these examples, cracking occurred after substantial deformation, and it is not certain that a valid crack-initiation stress could be calculated because of barreling. Different modes of compression fracture are described below (Ref 7).

Orange Peel Cracking. In many materials, roughening or wrinkling of the surface (orange peel effect) occurs prior to compressive cracking. This effect is particularly prominent in some aluminum alloys. An extreme example is illustrated for an aluminum alloy 7075-T6 specimen in Fig. 5. The specimen is shown after 72% deformation. Wrinkling first appeared at 10 to 15% compressive deformation, and macrocracking occurred after 50 to 60% deformation. Microscopic examination revealed many microcracks in the valleys of the wrinkles, with greatest concentration in the equatorial region of the specimen. Defining a compression strength or a strain criterion of fracture would be difficult for this material.

Macrocracks in Steel. A case in which macrocracks form without apparent precursor microcracks is shown in Fig. 6. The material is AISI-SAE 4340 steel tempered at 204 °C (400 °F), yielding a hardness of 52 HRC. The cracks initiated one at a time and extended across the surface of the specimen almost instantaneously. The first crack appeared when the compressive deformation reached 30%, and other cracks continued to initiate until the test was concluded at 72% deformation, which is the condition shown in Fig. 6. The specimen was still intact, and subsequent sectioning revealed that the cracks penetrated inward a distance of ¼ diameter.

Microcrack to Macrocrack Coalescence. In some tungsten alloys, the first

Fig. 5 Two views of a 72% compressed specimen of aluminum alloy 7075-T6 displaying orange peel effect

The loading axis is vertical. Extensive macrocracking is evident in the severely wrinkled surface. Microscopic examination of the surface revealed extensive microcracking in the valleys of the wrinkles. Source: Ref 6

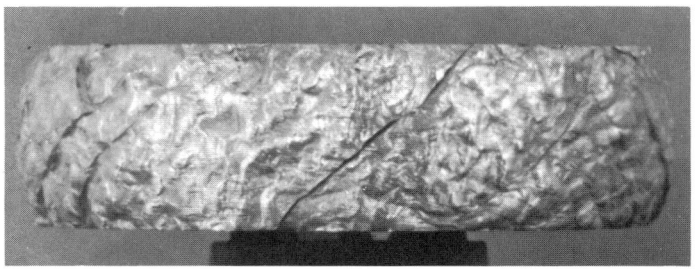

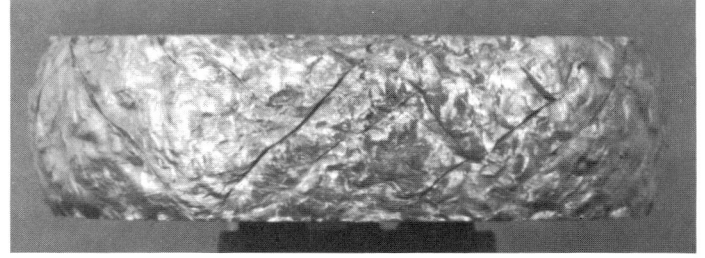

Fig. 6 Shear cracks in a 72% compressed specimen of AISI-SAE 4340 steel

The cracks initiated one at a time, starting when the deformation was 30%. Source: Ref 6

Fig. 7 Load requirements for compressing specimens of various diameters made of a material with a yield stress of 1380 MPa (200 ksi) and a strain-hardening exponent of 0.05

Diameters: A = 28.4 mm (1.12 in.); B = 25.4 mm (1.00 in.); C = 20.3 mm (0.80 in.); D = 12.7 mm (0.50 in.). Length: L/D = 3

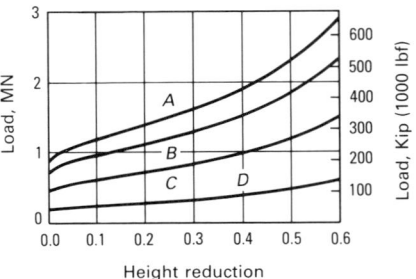

visible evidence of fracture is a shear macrocrack that appears at the equator of the specimen after 45 to 50% compressive deformation. However, using fluorescent dye penetrant methods, microcrack initiation was detected at 25% deformation (Ref 7). For this material, if crack initiation is the criterion of failure, it would be necessary to state the method of crack detection with the selected parameter for strength.

Compressive Strength or Strain to Failure. It is valid to calculate a compressive strength based on load and instantaneous area only if barreling is absent or minimal. Rather than use compressive strength as a parameter, the strains to failure have been used as a parameter. Flanged and tapered specimens on which gridlines were inscribed to measure the strain to failure have been used (Ref 8). In this case, it is also necessary to qualify the failure in terms of the method of crack detection and the size of the detected crack.

Experimental Techniques

Conducting a compression stress-strain test requires close attention to the technique described in Ref 1. Personnel that are inexperienced in the procedure should refer to the literature for assistance. Also, there are two primary areas in which compression testing differs from tension testing: strain measuring and testing machine capacity.

Strain or Deflection. Autographic recording of strain in excess of the range of resistance strain gages can become difficult in a compression test. Not only is the space for installation of a compressometer limited by the loading train, but the space decreases

as the test continues. As an alternative to recording strain, test personnel have recorded relative head motion of the testing apparatus. When this is done, the quantity measured should be clearly identified as deflection or specimen end-shortening and not as strain.

Testing Machine Capacity. In a compression test performed to large strains (to obtain fracture data, for example), large load capacity may be required. Four medium-length cylindrical specimen diameters that range from 12.7 to 28.4 mm (0.50 to 1.12 in.) with length to diameter ratios equal to 3 are suggested in Ref 1. Using these specimen sizes, consider the testing of a material with a yield stress of 1380 MPa (200 ksi) and a compression strain-hardening exponent of 0.05. Figure 7 illustrates the load-capacity requirements to reach a height reduction of 60% for each of the four cylinders recommended in Ref 1. Note that the maximum required load is approximately 3.5 times the load at yield. The required capacity for testing the same specimens to failure at 60% strain in tension would be no more than 1.5 times the yield loads.

REFERENCES

1. "Standard Methods of Compression Testing of Metallic Materials at Room Temperature," E 9, *Annual Book of ASTM Standards*, ASTM, Philadelphia, 1983, p 140-152
2. G. Gerard, *Introduction to Structural Stability Theory*, McGraw-Hill, New York, 1962, p 19-29
3. R. Papirno and G. Gerard, Compression Testing of Sheet Materials at Elevated Temperatures, in *Elevated Compression Testing of Sheet Materials*, STP 303, ASTM, Philadelphia, 1962, p 12-31
4. J.F. Mescall, R. Papirno, and J. McLaughlin, Stress and Deformation States Associated with Upset Tests in Metals, in *Compression Testing of Homogeneous Materials and Composites*, R. Chait and R. Papirno, Ed., STP 808, ASTM, Philadelphia, 1983, p 7-23
5. T.C. Hsu, A Study of the Compression Test for Ductile Materials, *Materials Research and Standards*, Materials Research and Standards, Vol 9 (No. 12), Dec 1969, p 20
6. R. Chait and C.H. Curll, Evaluating Engineering Alloys in Compression, in *Recent Developments in Mechanical Testing*, STP 608, ASTM, Philadelphia, 1976, p 3-19
7. R. Papirno, J.F. Mescall, and A.M. Hansen, Fracture in Axial Compression of Cylinders, in *Compression Testing of Homogeneous Materials and Composites*, R. Chait and R. Papirno, Ed., STP 808, ASTM, Philadelphia, 1983, p 40-63
8. E. Ermin, H.A. Kuhn, and G. Fitzsimons, Novel Test Specimens for Workability Testing, in *Compression Testing of Homogeneous Materials and Composites*, R. Chait and R. Papirno, Ed., STP 808, ASTM, Philadelphia, 1983, p 279-290

Pin Bearing Testing

By Robert E. Zinkham
Supervisor, Mechanical Metallurgy
Reynolds Metals Company

and

Cecil E. Parsons
Senior Principal Engineer
Boeing Commercial Airplane Company

PIN BEARING TESTING is conducted on metal products that must sustain loads that are applied when the material is riveted, bolted, or similarly mechanically fastened. The purpose of the test is to determine the bearing strength properties and to indicate the bearing stress versus the deformation of the hole. The bearing load is applied to the specimen through a cylindrical pin, which is fitted into a hole normal to the surface of the specimen. The data obtained by this test procedure are used to calculate minimum properties that can be utilized in the design of structural members used in the aerospace industry.

The properties of primary concern are bearing yield strength and bearing ultimate strength. The bearing yield strength is the bearing stress at which a material exhibits a specified limiting deviation from the proportionality of bearing stress to bearing strain. The bearing ultimate strength is the maximum bearing stress a material is capable of sustaining. These values are normally determined at edge distance ratios (e/D) of 1.5 and 2.0.

Edge distance is the distance from the edge of a bearing specimen to the center of the hole in the direction of the applied force (Fig. 1). The edge distance ratio is the ratio of the edge distance, e, to the pin diameter, D. Bearing area is the product of the pin diameter and specimen thickness. The bearing loads are divided by the bearing area to yield the bearing stress or strength, which is the force per unit of bearing area.

Although a standard method for the pin type bearing test is covered in ASTM E 238 (Ref 1), it is oriented to testing aluminum and magnesium. Consequently, problems with pin distortion or failure may be encountered when higher strength materials such as titanium and high-strength steel at ultimate strengths of 1860 to 2070 MPa (270 to 300 ksi) are to be tested.

Much of ASTM E 238 is aimed at obtaining consistent results among laboratories. Consequently, it must be recognized that values obtained under laboratory-type conditions will not be representative of those achieved under actual loading conditions of a part or structure.

Test Specimens

Pin bearing tests are conducted on sheet-type specimens, using the full thickness of the material when possible. A typical bearing test specimen is shown in Fig. 1. If the specimen is too thick in relation to the pin diameter, the test pin may bend or break before the bearing strength can be achieved. Conversely, buckling may occur if the specimen is too thin in relation to the pin diameter. To avoid pin deformation or failure and specimen buckling, a pin diameter to specimen thickness ratio (D/t) of 2 to 4 is recommended in ASTM E 238. To test high-strength steel, a pin material at a D/t of 2 was not strong enough; therefore, use of a D/t of 4 was necessary (Ref 2). Using a higher D/t ratio, however, increases the possibility of buckling, and most testing of aluminum is conducted using a D/t of 2.

A specimen width to hole diameter ratio (W/D) of 4 to 8 is also suggested in ASTM E 238. However, at a W/D of 4, lower bearing strengths have been obtained (Ref 3, 4). This may be due to plastic deformation in the net section; therefore, a W/D of 6 is normally used. Differences in bearing strength, however, are negligible between a W/D of 5 and 8. The diameter of the specimen hole should not exceed the pin diameter by more than 0.02 mm (0.001 in.). A loose fit between the specimen and pin tends to lower results.

When characterizing a new alloy and temper, a range of material thicknesses is usually considered. The properties can then be used in designing different types of fasteners, i.e., both rivets and bolts. Typically, D/t and W/D ratios are standardized to promote consistency over the total thickness range tested. The full thickness of the material is generally used up to a thickness of 3.2 mm (0.125 in.),

Fig. 1 Typical pin bearing test specimen

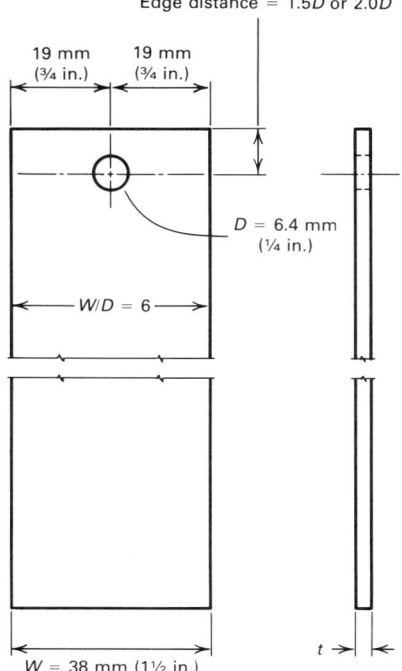

Edge distance = 1.5D or 2.0D

19 mm (¾ in.)   19 mm (¾ in.)

D = 6.4 mm (¼ in.)

W/D = 6

W = 38 mm (1½ in.)

t

although some testing laboratories use specimens up to 6.35 mm (0.250 in.) in thickness.

Above this level, the material is machined to a thickness of 3.2 mm (0.125 in.). This provides specimens of a reasonable size without having to obtain or fabricate large-diameter test pins and large loading fixtures. When machining specimens from thick material, specimens should be taken from a specified location with regard to the total thickness, *t*. For aluminum, the specified location is *t*/2 for product forms less than or equal to 38 mm (1.50 in.) and *t*/4 for plate, extrusions, and rolled and drawn rod and bar product forms greater than 38 mm (1.50 in.) thick. This follows practices specified in ASTM B 557 for tension testing (Ref 5).

Cleaning Procedure

Cleaning of the specimen and pin is necessary to provide consistent, comparable results (Ref 6). However, cleaning produces results that are higher than in actual practice, in which fasteners may have platings, sealants, or lubricants to facilitate installation. In aluminum and magnesium alloys, lubrication

can cause reductions of up to 15% in bearing yield strength values obtained in tests with clean, dry bearing surfaces. This includes the unintentional application of oil from human fingers during handling of the specimens and test fixtures. Table 1 shows the effects of lubricants and cleaners on three aluminum-based alloys.

During the mid-1960's, the Military Handbook-5 Committee could not agree on a consistent reduction factor for reducing the bearing strength from the ASTM E 238 "dry pin" values to the "dirty pin" values representative of actual installation. Caution is therefore used when applying the properties or establishing the effect of lubricants on the properties under a specific set of conditions.

Test Fixtures

ASTM E 238 recommends that the fixtures used to apply the load through the pin to the specimen should not constrain the expansion of the specimen material in front of the pin. Although this practice ensures consistency of data, it does not produce results that are representative of usage conditions. In use, any joint would have "full holes" (metal com-

pletely surrounding the hole) and some degree of clamping action that would resist expansion of the material in front of the loading pin. This would result in higher loads. Trying to simulate the amount of clamp-up for various joints would be difficult and time consuming. A typical test arrangement is shown in Fig. 2.

Determination of the deformation at the hole due to the bearing load is generally accomplished by measuring from some point on the loading straps to some point on the specimen. The primary use of the deflection measurement is to obtain the yield strength at the specified plastic deflection offset equal to 2% of the pin diameter. The plastic deformation should be at the hole bearing surface, because the loading straps and specimen generally should be in a state of elastic deformation.

The preferred method of measurement is to average the output of two measuring devices on each side of the specimen (see Fig. 2). Use of only one measuring device will result in a variety of load-deflection curves that lead to considerable variation in results. Even with two devices, care must be taken in aligning the setup, or curves will be obtained that do not have a sufficient initial straight line segment (linearity) to establish the slope.

Table 1 Effects of lubricants and cleaners on the bearing strength properties of 1.6 mm (0.063 in.) aluminum alloy sheet

Lubricant or cleaner	Average bearing ultimate strength MPa e/D = 1.5	MPa e/D = 2.0	psi e/D = 1.5	psi e/D = 2.0	Average bearing yield strength MPa e/D = 1.5	MPa e/D = 2.0	psi e/D = 1.5	psi e/D = 2.0
2024-T6								
Lubricant:								
Graphite grease	728	929	105 600	134 700	510	589	74 000	85 400
Petroleum jelly	736	938	106 800	136 100	520	597	75 400	86 600
Oil, SAE 40	756	963	109 600	139 700	532	633	77 100	91 800
Oil, SAE 10	756	965	109 600	140 000	535	629	77 600	91 200
Human Oil	742	942	107 600	136 700	520	601	75 500	87 200
Cleaner:								
Acetone, manually	767	975	111 200	141 400	549	631	79 700	91 500
Acetone, ultrasonically	769	960	111 600	139 300	556	667	80 600	96 700
6061-T6								
Lubricant:								
Emulsifying oil	518	688	75 200	99 800	408	459	59 200	66 600
Graphite grease	485	667	70 400	96 700	391	431	56 700	62 500
Petroleum jelly	497	666	72 100	96 600	395	454	57 300	65 800
Oil, SAE 40	517	669	75 000	97 000	412	463	59 800	67 100
Oil, SAE 10	524	689	76 000	100 000	411	469	59 600	68 100
Human oil	506	677	73 400	98 200	405	450	58 800	65 300
Cleaner:								
Acetone, manually	523	684	75 900	99 200	417	475	60 500	68 900
7075-T6								
Lubricant:								
Emulsifying oil	907	1160	131 600	168 200	712	836	103 300	121 200
Graphite grease	857	1071	124 300	155 400	700	791	101 600	114 800
Petroleum jelly	889	1113	128 900	161 400	724	811	105 000	117 600
Oil, SAE 40	909	1144	131 800	166 000	736	833	106 700	120 800
Oil, SAE 10	911	1111	132 200	161 200	725	852	105 100	123 600
Human Oil	878	1117	127 300	162 000	701	810	101 700	117 500
Cleaner:								
Acetone, manually	894	1154	129 600	167 400	716	855	103 800	124 000
Acetone, ultrasonically	920	1170	133 500	169 700	778	909	112 800	131 900

Note: Surface finish of the bearing pin, 0.1 to 0.2 μm (4 to 8 μin.). All pins were made of steel hardened to 61 HRC. Tests generally made in quadruplicate.
Source: Ref 6

Fig. 2 Typical pin bearing test fixture used on aluminum sheet

Courtesy of Kaiser Aluminum and Chemical Corp.

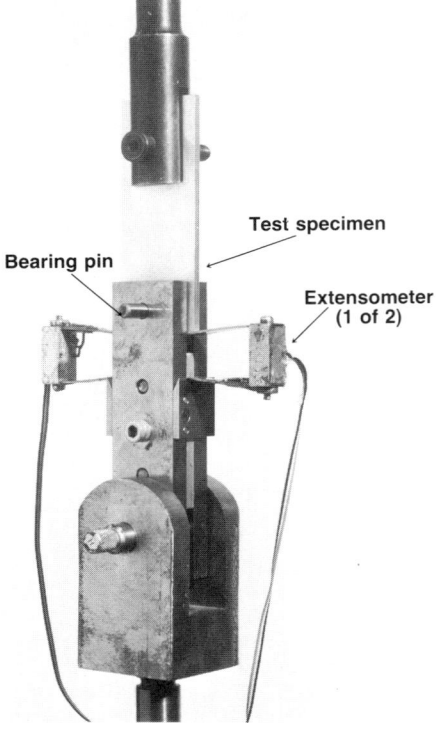

Test specimen

Bearing pin

Extensometer (1 of 2)

Fig. 3 Autographic bearing load versus bearing deformation curves for 7150-T651 aluminum alloy plate at room temperature

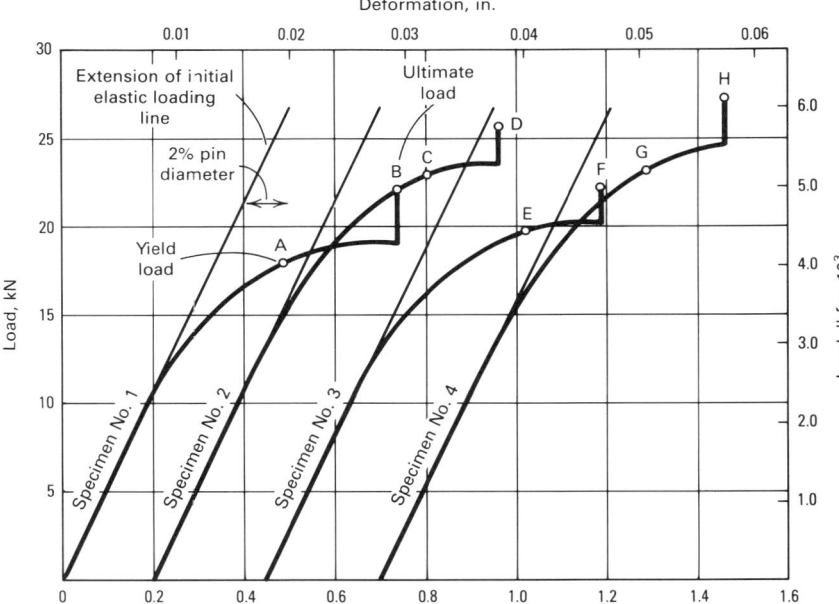

Specimen No. 1
Test direction . . . . Longitudinal
e/D ratio 1.5
Yield load (point A) . . . . 17 840 N (4010 lbf)
Ultimate load (point B) . . . . . . 21 350 N (4800 lbf)
Specimen No. 2
Test direction . . . . Longitudinal
e/D ratio 2.0
Yield load (point C) . . . . 22 700 N (5100 lbf)
Ultimate load (point D) . . . . . 26 470 N (5950 lbf)
Specimen No. 3
Test direction . . . . Long-transverse
e/D ratio 1.5
Yield load (point E) . . . . 20 000 N (4500 lbf)
Ultimate load (point F) . . . . . . 22 000 N (4900 lbf)
Specimen No. 4
Test direction . . . . Long-transverse
e/D ratio 2.0
Yield load (point G) . . . . 22 860 N (5140 lbf)
Ultimate load (point H) . . . . . 27 000 N (6100 lbf)

Fig. 4 Pin bearing test specimen orientation

Edgewise and flatwise specimens machined from thick plate. The loading conditions can also occur in the long transverse direction.

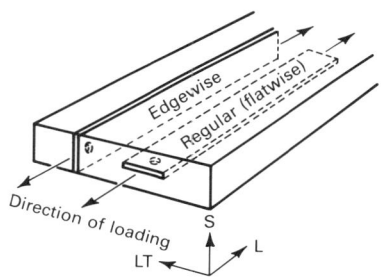

If the slope cannot be established in a reasonable manner, wide variations will exist in the yield strength values obtained by different operators from the same curves.

The determination of the bearing yield and ultimate loads is normally taken from the autographic curve. Such a curve is shown in Fig. 3. Dividing the applied loads by the projected area of the pin, which is the prod-

uct of the pin diameter and specimen thickness, gives the bearing yield and ultimate strengths. Because these strengths generally show good correlation with the corresponding tensile yield and ultimate strengths, minimum bearing properties for use in design are calculated from the product of the statistically reduced ratios (bearing/tensile) and the appropriate minimum tensile property. *The Military Standardization Handbook* (Ref 7) publishes design values for both bearing and tensile properties for most aerospace alloys.

In Ref 7, bearing strengths are given without reference to grain direction. The values shown are the lower of the longitudinal and long-transverse directions. However, a reduction factor is used for bearing strengths of edgewise specimens in some alloys, which are up to 15% lower than those taken parallel to the surface (flatwise). Test directions for both edgewise and flatwise pin bearing specimens are shown in Fig. 4.

REFERENCES

1. "Standard Method for Pin-Type Bearing Test of Metallic Materials," E 238, *Annual Book of ASTM Standards*, ASTM, Philadelphia, 1984, p 394-399

2. C. Parsons, Boeing Co., personal communication, July 18, 1984

3. R.L. Moore and C. Wescoat, "Bearing Strengths of Some Wrought-Aluminum Alloys," NACA-TN-901, Aug 1943

4. R.E. Zinkham, "Bearing Strength of X7002-T6 Alloy," Internal Reynolds Metals Company Report, MRD 38-028-500, June 1964

5. "Standard Methods of Tension Testing Wrought and Cast Aluminum and Magnesium Alloy Products, B 557, *Annual Book of ASTM Standards*, ASTM, Philadelphia, 1984

6. B.W. Stickley and A.A. Moore, Effects of Lubrication and Pin Surface on Bearing Strength of Aluminum and Magnesium Alloys, *Materials Research and Standards*, MTRSA, Vol 2 (No. 9), Sept 1962, p 747

7. Metallic Materials and Elements for Flight Vehicle Structures, in *Military Standardization Handbook*, Military Handbook-5D, Department of Defense, June 1983

Shear Testing

Shear Testing of Mill Products

By Clayton L. Harmsworth
Technical Manager
AFWAL Materials Laboratory

SHEAR DATA are primarily used in the design of mechanically fastened components, shear webs, torsion members, and other structural components subject to the application of parallel, opposing loads. Although industry-approved standards do exist for the shear testing of fasteners, adhesive joints, and composites, standards do not currently exist for the development of shear data on mill products such as forgings, extrusions, sheet, and plate.

This article discusses the most common tests used to obtain shear data on mill products, including their advantages and limitations. These tests include (1) the double-shear test, which uses solid round bar specimens, (2) the single-shear test, which uses carefully dimensioned sheet and thin plate specimens, (3) the blanking-shear (punching-shear) test, which uses thin sheet specimens, and (4) the torsion-shear test, which uses tubular specimens.

Double-Shear Tests

Shear testing for the development of engineering data on mill products is most commonly performed using the double-shear test. An advantage of the double-shear test is that it uses easily machined solid round bars as test specimens. As such, the test can be used on mill products with thicknesses of 4.8 mm (³⁄₁₆ in.) or more. For sheet less than 1.8 mm (0.071 in.) thick, the blanking-shear test, described later in this article, can be used.

The action of the shearing blades and the test procedures used in double-shear testing of mill products are similar to those described in the following section "Shear Testing of Fasteners," except that additional controls on fixture rigidity are necessary if the data are to be used for the development of design allowables.

Effect of Test Fixtures. Double-shear tests have been performed with a rigid Amsler tool as well as with a double-shear tool used primarily for testing of rivets. These setups are shown schematically in Fig. 1. In an Amsler double-shear tool similar to that illustrated in Fig. 2, specimens were sheared simultaneously on two planes 25 mm (1 in.) apart. The specimens were 76 mm (3 in.) in length and either 9.5 mm (³⁄₈ in.) in diameter for shearing on the longitudinal (L) or long-transverse (T) planes, or 4.8 mm (³⁄₁₆ in.) in diameter for shearing on the short-transverse (S) plane. Planes of shear and loading directions are defined by the TLS coordinate system shown schematically in Fig. 3. In a rivet shear tool, specimens 9.5 mm (³⁄₈ in.) in diameter and at least 19 mm (³⁄₄ in.) in length were sheared simultaneously on two planes 9.5 mm (³⁄₈ in.) apart.

The results obtained with the conventional rivet tool averaged about 10% lower than those obtained with the rigid Amsler tool. This difference was attributed to the relative length of support given a specimen in the tool, the clearance of the shearing blade edges, and the flexibility of the fixture side plates. Table 1 summarizes these results.

In tests using the rivet tool, specimens are supported over a short length, and some bending of the specimen occurs. The outside loading plates typically are not stiff enough to prevent bending; in fact, the side plates flex outward, causing even more bending of the specimen. In the more rigid Amsler tool, the specimen is supported over a relatively long length, and because the clearance between dies remains minimal (0.025 mm, or 0.001 in.), bending of the specimen is minimized.

As a result, a more realistic measure of the shear strength of the material is obtained in the more rigid test fixture. ASTM currently is developing a shear test standard for mill products based on the more rigid double-shear configuration. It should be noted, however, that this configuration is not preferred for fastener testing, because the majority of fasteners are shorter than the 76-mm (3-in.) length required using the Amsler fixture.

Effect of Specimen Orientation. As shown in Fig. 3, six different combinations of specimen grain orientations (normal to the shear plane) and loading directions are possible in double-shear testing. Test results will vary, depending on the direction of loading and plane of shear. Because of these influences, the specimen ends must be marked (scribed) with the correct orientation prior to insertion into the test fixture. The results of single- and double-shear tests in which the orientation of the plane of shear was varied are shown in Table 2. In double-shear tests of aluminum alloys, the differences ranged

Fig. 1 Schematic representation of rigid Amsler and rivet double-shear tooling

Source: Ref 1

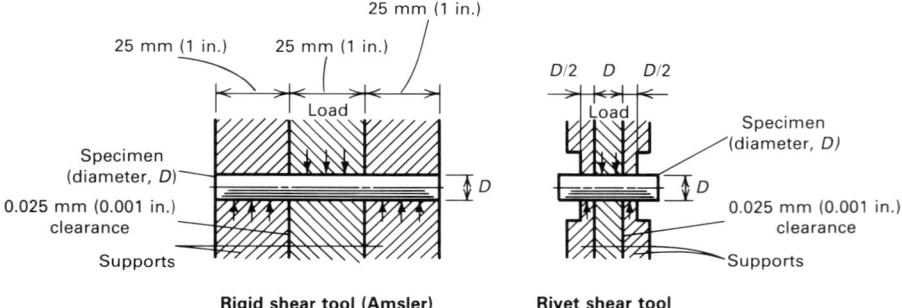

Rigid shear tool (Amsler)   **Rivet shear tool**

Fig. 2 Amsler double-shear fixture

Die clearance adjustment is shown on the right side. Courtesy of Air Force Wright Aeronautical Laboratories

Fig. 3 Planes of shear and loading directions for shear specimens

The first letter defines the grain orientation normal to the plane of shear; the second letter defines the direction of loading. L, longitudinal; T, long transverse; S, short transverse. Adapted from Ref 1

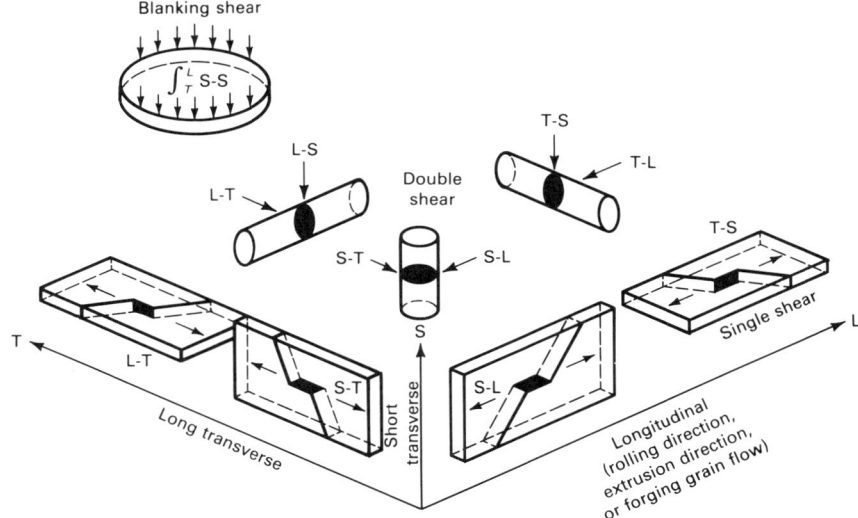

from about 3% (7079-T651) to about 10% (2014-T651). In single-shear tests of aluminum sheet (see discussion below), the differences ranged from about 5% (6060-T651) to about 24% (2219-T62).

Single-Shear Tests

The single-shear test is used to obtain ultimate shear strength values for sheet and thin plate. The test is conducted on standard tension testing equipment with machined (slotted) sheet specimens of the design shown in Fig. 4. To ensure that failure occurs along the shear path of the specimen, the geometric dimensions of the specimen are critical. In a study of the failure mechanisms of single-shear sheet specimens, three possible modes of failure were examined (Ref 2), as discussed below.

Single-Shear Failure Modes. Ideally, failure should occur in shear along the plane normal to the surface of the specimen and along the axis of loading. This is depicted in Fig. 4, with the dashed line between B and C representing the shear path. To avoid undesirable tensile failure (indicated by the dashed line CD in Fig. 4), the dimension CD must be sufficiently greater than the dimension CB so that the critical shear stress is developed before the critical tension stress. Depending on the material being tested, this ratio (CD/CB) can vary from a minimum of 2 to 14. To achieve this ratio, the dimension CB cannot be reduced to the extent that the specimen becomes fragile and easily damaged upon application of load.

A third and more complex mode of failure is caused by buckling. The critical load for buckling depends not only on the dimensions

Table 1 Results of double-shear tests of 25- to 133-mm (1- to 5¼-in.) thick aluminum alloy plate in rigid Amsler and rivet shear tools

Alloy and temper	Plane of shear(a)	T Strength, MPa (psi) Amsler	Rivet	L Strength, MPa (psi) Amsler	Rivet	S Strength, MPa (psi) Amsler	Rivet
2014-T651	T	...	...	296 (43 000)	263 (38 200)	284 (41 200)	251 (36 400)
	L	293 (42 500)	265 (38 400)	...	...	281 (40 800)	255 (37 000)
	S	268 (38 900)	248 (36 000)	263 (38 200)	236 (34 300)	...	...
2024-T351	T	...	...	297 (43 100)	262 (38 000)	283 (41 000)	256 (37 100)
	L	294 (42 600)	263 (38 200)	...	...	279 (40 300)	253 (36 700)
	S	284 (41 200)	254 (36 800)	286 (41 500)	258 (37 400)	...	...
5456-H321	T	...	...	214 (31 100)	192 (27 900)	198 (28 700)	181 (26 200)
	L	218 (31 600)	194 (28 200)	...	...	194 (28 100)	176 (25 500)
	S	209 (30 300)	186 (26 900)	201 (29 200)	186 (27 000)	...	...
6061-T651	T	...	...	209 (30 300)	185 (26 800)	202 (29 300)	179 (26 000)
	L	207 (30 100)	185 (26 800)	...	...	201 (29 100)	177 (25 600)
	S	197 (28 600)	179 (26 000)	199 (28 900)	178 (25 800)	...	...
7075-T651	T	...	...	353 (51 200)	318 (46 100)	334 (48 500)	305 (44 200)
	L	347 (50 300)	316 (45 800)	...	...	329 (47 700)	293 (42 500)
	S	325 (47 200)	292 (42 400)	325 (47 100)	290 (42 100)	...	...
7079-T651	T	...	...	342 (49 600)	311 (45 100)	333 (48 300)	303 (43 900)
	L	339 (49 200)	310 (44 900)	...	...	327 (47 400)	298 (43 200)
	S	321 (46 600)	292 (42 300)	332 (48 100)	298 (43 200)	...	...
7178-T651	T	...	...	365 (53 000)	328 (47 600)	342 (49 600)	315 (45 700)
	L	360 (52 200)	331 (48 000)	...	...	334 (48 400)	306 (44 400)
	S	330 (47 800)	290 (42 100)	328 (47 600)	297 (43 100)	...	...

(a) Plane of shear and loading direction defined by TLS coordinate system in Fig. 3
Source: Ref 1

Table 2   Influence of shear plane and loading direction on shear strength values of slotted single-shear and double-shear (Amsler tool) specimens taken from aluminum alloy plate

Plane of shear and loading direction defined by TLS coordinate system in Fig. 3

Alloy and temper	Nominal thickness mm	in.	Type of test	Shear strength L-T MPa	psi	T-L MPa	psi	S-T MPa	psi	S-L MPa	psi
2014-T651 25		1.0	Single	292	42 300	296	43 000	256	37 200	256	37 200
			Double	293	42 500	296	43 000	268	38 900	263	38 200
2024-T351 25		1.0	Single	318	46 100	327	47 400	277	40 200	278	40 400
	38	1.5	Single	309	44 800	318	46 100	277	40 200	276	40 000
			Double	294	42 600	297	43 100	284	41 200	286	41 500
2219-T62 25		1.0	Single	276	40 000	284	41 200	200	29 000	229	33 200
			Double	278	40 300	288	41 700	. . .	. . .	. . .	. . .
5456-O 64		2.5	Single	245	35 600	241	35 000	199	28 800	201	29 200
			Double	214	31 100	211	30 600	192	27 800	187	27 200
5456-H321 32		1.25	Single	236	34 300	234	34 000	197	28 600	205	29 800
			Double	218	31 600	214	31 100	209	30 300	201	29 200
6061-T651 32		1.25	Single	211	30 600	210	30 400	198	28 800	200	29 000
			Double	207	30 100	209	30 300	197	28 600	199	28 900
7075-T651 25		1.0	Single	342	49 600	357	51 800	305	44 300	300	43 500
	32	1.25	Single	367	53 300	372	54 000	303	43 900	303	44 000
			Double	347	50 300	353	51 200	325	47 200	325	47 100
7079-T651 38		1.5	Single	343	49 700	345	50 100	312	45 200	314	45 600
			Double	337	48 900	340	49 300	321	46 600	332	48 100
7178-T651 32		1.25	Single	368	53 400	381	55 200	306	44 400	294	42 700
			Double	360	52 200	365	53 000	330	47 800	333	48 300

Source: Ref 1

Fig. 4 Single-shear slotted-sheet specimen

Dashed line between B and C represents failure along the shear path. Tensile failure depicted by dashed line CD. See text for the relative dimensions of ABCD and the specimen thickness needed to ensure shear failure.

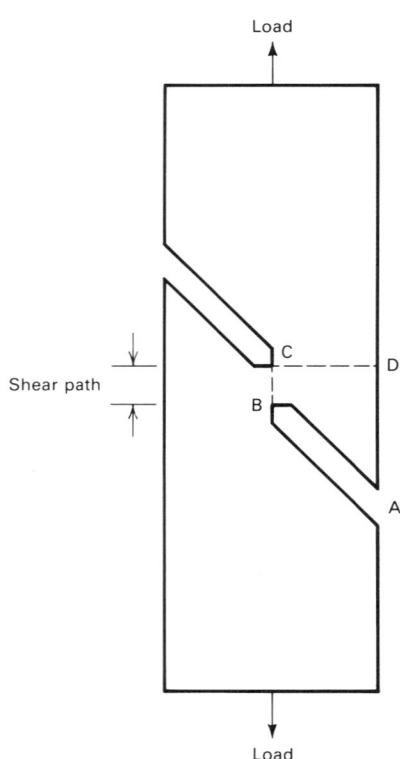

of CB and CD, but also on the thickness and work-hardening characteristics of the material. If the dimensions CB and CD are held constant but the thickness is reduced, the bending moment in the plane of the specimen will become critical before either the critical shear or tension stress value is produced, resulting in lateral buckling.

Single-Shear Versus Blanking-Shear Tests. Despite following the recommendations in Ref 2 for selection of critical specimen dimensions (thickness and width and length of shear path), results obtained using this specimen design were not consistent and resulted in shear strengths 2 to 35% higher than those obtained by blanking-shear (punching-shear) tests, the difference varying with temper (Ref 1). These results are shown in Table 3. Some differences in results, however, are to be expected, because these test procedures involve different loading directions and specimen grain orientations (see Fig. 3).

In tensile single-shear and double-shear testing of aluminum alloy plate, shear strengths differed by as much as 9%, depending on the direction of loading, and by as much as 24%, depending on the plane of shear (see Table 2).

Blanking-Shear Tests

Another method of shear testing is the simple punch-and-die method for blanking a disk out of a flat strip, which has been used primarily in the aluminum industry on material 1.8 mm (0.071 in.) thick or less. This method is shown schematically in Fig. 5. Blanking-shear tests average shear data in the longitudinal and long-transverse planes of a sheet that must be tested in the short-transverse direction (Fig. 3). A radial clearance between punch and die of approximately 12 to 14% of sheet thickness results in a clean, sheared edge.

In actual service, there is seldom a need for such data, except for applications in which a fastener or other device is loaded at right angles against sheet material. This is not, however, an ideal design configuration. Table 3 compares results of single-shear and blanking-shear tests on aluminum alloy sheet.

Torsion-Shear Tests

The major advantage of the torsion-shear test is that it can be used to obtain strain data along the plane of loading. Consequently, it can be used to obtain the shear yield strength and modulus of rigidity (G) of a material. It

Fig. 5 Schematic representation of the blanking-shear test

Radial clearance between punch and die should be approximately 12 to 14% of the sheet thickness, which is ≤1.8 mm (0.071 in.).

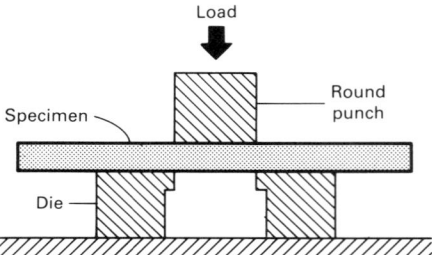

also produces a more pure state of shear stress without introducing local stress fields caused by machined slots or by an impinging knife edge. The disadvantages of the torsion-shear test are that the specimens are relatively expensive to machine, the test cannot be used on sheet or thin plate products, and a torsion testing machine may not be available at all laboratories. Additionally, the optimum specimen design for obtaining the shear yield and modulus of rigidity is not an optimum design for obtaining the ultimate shear

strength. Figure 6 illustrates a typical specimen design.

Care must be taken to ensure that the length of the reduced section, L, is not too long and that the thickness of the reduced section, t, is adequate to prevent buckling of the specimen prior to reaching shear failure. For ultimate shear strength tests, L equals one half of the diameter, D, of the reduced section ($L = 0.5D$). The diameter of the gripped section, d, is machined to properly fit the test machine grips and must be

Fig. 6 Typical torsion-shear specimen

See text for the relative dimensions of D, d, t, and L required to prevent premature buckling of the specimen.

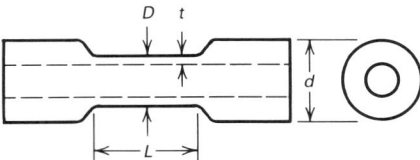

smoothly blended into the diameter of the reduced section. Some grip designs may require a reinforced (plugged) end to prevent specimen collapse (buckling) due to gripping forces.

Although direct values of shear yield strength and modulus of rigidity can be obtained from a tubular specimen, this advantage is not significant. Shear yield strength is seldom used in design. If a limiting value of elasticity is required, common practice is to use the proportional limit in shear. This property cannot be obtained directly from torsion tests. The National Bureau of Standards has shown that for common structural materials the ratio of proportional limit in shear to the proportional limit in tension is approximately 0.557 (Ref 3).

The modulus of rigidity (G) can be calculated indirectly from the modulus of elasticty (E) and Poisson's ratio (v) by:

$$G = \frac{E}{2(1 + v)}$$

This indirect method is used to compute all modulus of rigidity data given in Ref 3 and most other design handbooks. From an applications standpoint, the torsion-shear test is most useful for testing shaft and axial components in a manner similar to that in which they will be used. In service, however, these applications generally are of a rotating nature, resulting in torsion-fatigue failures rather than classical tensile-shear failures. For more detailed information on the interpretation and application of torsional loading, see the Section on "Torsion Testing" in this Volume.

Shear Testing of Fasteners

By John A. Aurentz
Executive Vice President
Briles Rivet Corporation

SHEAR TESTING is commonly performed as part of routine quality control procedures during manufacturing and as part of

Table 3   Results of single-shear and blanking-shear tests of 1.6-mm (0.063-in.) thick aluminum alloy sheet

Alloy and temper	No. of lots	Average single-shear strength(a)		Blanking-shear strength	
		MPa	psi	MPa	psi
Heat-treatable alloys					
2014-T6 2		300	43 500	284	41 200
Alclad 2014-T6 3		303	44 000	292	42 300
2024-T3 6		324	47 000	294	42 600
Alclad 2024-T3 7		312	45 300	283	41 000
2024-T4 3		330	47 900	294	42 600
Alclad 2024-T4 3		313	45 400	285	41 300
Alclad 2024-T81 2		274	39 800	267	38 800
Alclad 2024-T86 6		295	42 800	287	41 600
6061-T6 5		208	30 200	196	28 500
Alclad 6061-T6 6		196	28 400	190	27 500
7075-T6 5		361	52 400	341	49 400
Alclad 7075-T6 5		336	48 700	322	46 700
7079-T6 1		319	46 300	295	42 800
7178-T6 1		398	57 800	363	52 700
Non-heat-treatable alloys					
1100-O 1		90	13 000	66	9 600
1100-H14 1		97	14 100	81	11 800
1100-H18 1		117	17 000	107	15 500
3003-O 1		98	14 200	85	12 400
3003-H14 1		109	15 800	98	14 200
3004-O 1		141	20 400	121	17 600
3004-H34 1		161	23 300	154	22 200
3004-H38 1		190	27 600	181	26 300
5050-O 1		116	16 900	103	15 000
5050-H34 1		132	19 200	121	17 500
5050-H38 1		130	18 800	127	18 400
5086-O 1		186	27 000	171	24 800
5086-H32 1		198	28 700	184	26 700
5086-H34 1		207	30 000	196	28 400
5154-O 1		178	25 800	154	22 400
5154-H34 1		203	29 400	185	26 800
5154-H38 1		200	29 000	186	27 000
5454-O 1		186	27 000	160	23 200
5454-H32 1		187	27 200	172	24 900
5454-H34 1		205	29 800	193	28 000
5456-O 2		239	34 700	203	29 400
5456-H24 1		238	34 600	211	30 600

(a) Average values from tests taken in the L-T and T-L orientation. The plane of shear and the loading direction are defined by the TLS coordinate system shown in Fig. 3.
Source: Ref 1

the final inspection of precision fasteners. Shear testing is also a critical step in determining fastener strength during research and development of new fasteners. The degree and complexity of the shear testing usually is dictated by the fastener standard and/or the service conditions of the part. Test methods may vary significantly, and the accuracy of test results depends on adequately matching a specific fastener with the test method.

Shear testing of fasteners is accomplished by exerting pressure (shear force) in the transverse plane of the fastener until shear failure occurs. Shear force causes the two contiguous portions of the fastener to slide in opposite directions parallel to their contact plane. The force that results in shear failure is expressed as the ultimate load. The force required to cause permanent deformation is expressed as the yield load.

Double-Shear Tests

Double-shear testing is accomplished by shearing a segment of the fastener shank, which is supported on each side a distance equal to at least one half the shank diameter. Double-shear testing is illustrated schematically in Fig. 7. The most commonly used testing methods, equipment, and fixtures are those outlined in MIL-STD-1312, Test No. 13 (Ref 4), and ASTM B 565 (Ref 5).

A double-shear fixture of the type described in Ref 4 is shown in Fig. 8. It features two lower shear blades in the upright position. The thickness of these blades is equal to one half the diameter of the applicable fastener. These blades are spaced a distance equal to the specimen diameter. An upper blade (one-diameter thickness) is placed in a parallel position above and centered exactly between the lower blades. All blades have one hole, sized to the applicable fastener diameter, transverse to the blade and centrally located on a common centerline. The double shearing action occurs when the upper blade is forced downward between the lower blades, shearing a one-diameter segment of the fastener.

In conducting double-shear tests, the blades should be parallel, centered, and have the correct radius. The shearing edges should not exhibit any galling, broken edges, or uneven wear. Sheared segments that are gripped within the lower blades must be removed carefully, using a soft pry bar that will not damage the shear blades. The test is stopped after the ultimate load is attained.

Using the test setup shown in Fig. 8, fasteners with shear strength values of 586 MPa (85 ksi) or less tend to bend upward around the blade. This bending deflection results in distorted test values. This tendency can be alleviated by using the test fixture shown in

Fig. 9, which is described in ASTM B 565 (Ref 5). These double-shear fixtures have a central blade enclosed between two outer blades. The double shearing action occurs when a fastener is inserted through the holes, and the inner blade is pulled out of the two outer blades, thus shearing a segment one diameter in length from the fastener. Removal of the sheared segment requires a complete withdrawal of the central shear blade. Precautions are the same as for the test illustrated in Fig. 8.

Single-Shear Tests

A single-shear testing fixture employs two shear blades with centrally located transverse holes. The single shearing action occurs when one blade is held stationary with a fas-

tener in place, while the second blade is moved in a parallel plane, thus shearing the fastener. A schematic presentation of single-shear testing is shown in Fig. 7. Single shear is used to determine shear values of fasteners that are too short to permit double shearing. These include fasteners with shank lengths less than 2.5 times the shank diameter. Accuracy is considered less than in double shear tests, but is very close. With the wide variety of short fasteners that is available, some airframe manufacturers specify single-shear testing for fasteners, regardless of fastener length. Double-shear values can be used for comparison when the accuracy of single-shear values is in question. Fasteners can be tested in the installed or uninstalled condition, in accordance with specified requirements.

Fig. 7 Schematic of double- and single-shear testing

In both, the force is compression. Similar arrangements can be used for double- and single-shear testing in tension.

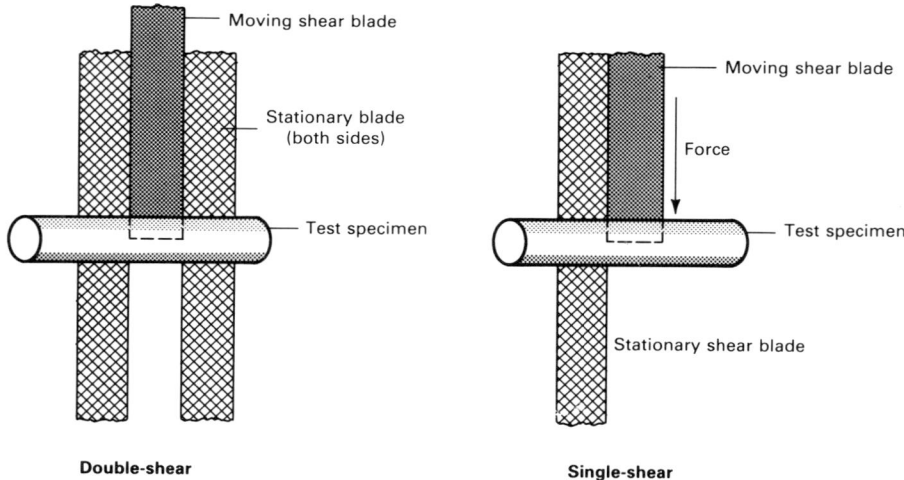

Double-shear   Single-shear

Fig. 8 Double-shear test fixture recommended in MIL-STD-1312, Test No. 13

Specimen and fixture dimensions are given in Ref 4.

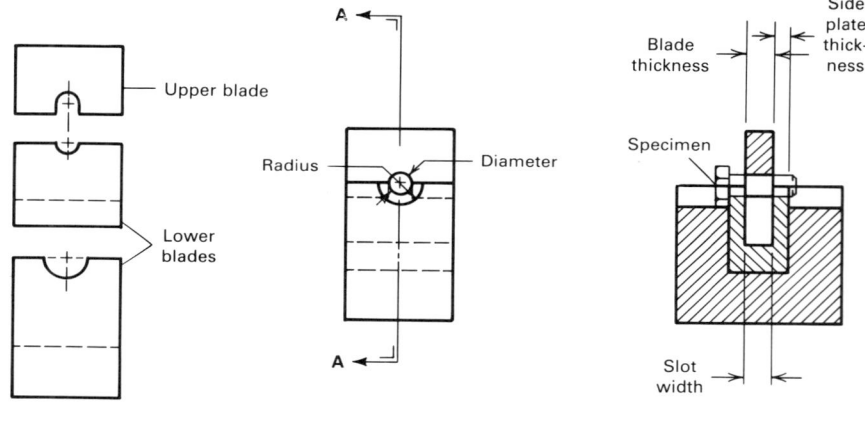

Section A-A

Fig. 9 Double-shear test fixture recommended in ASTM B 565

The tolerances of the diameter of the specimen being tested under shear must be known when drilling the hole in the jig. The recommended drilled hole size should be equal to 1.02d plus 0.12 mm (0.005 in.), less the minus tolerance applicable to the specimen. Source: Ref 5

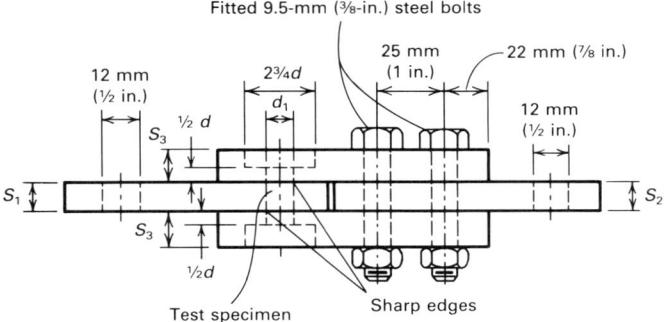

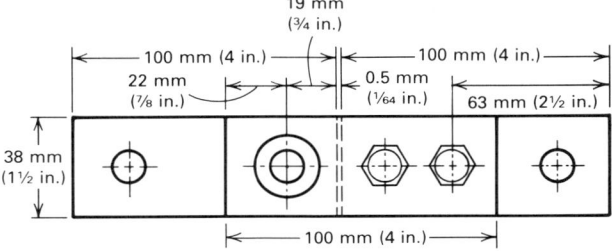

d = nominal diameter of specimen
d_1 = actual diameter of hole
d_2 = actual diameter of specimen; clearance $(d_1 - d_2)$, maximum = $0.02d + 0.12$ mm (0.005 in.)

$S_1 = d$
$S_2 = S_1 + 0.025$ mm (0.001 in.)
$S_3 = 6$ mm (¼ in.) for 1.5 to 4 mm (¹⁄₁₆ to ⁵⁄₃₂ in.) diam; 12 mm (½ in.) for 5 to 10 mm (³⁄₁₆ to ³⁄₈ in.) diam

Precautions for single-shear testing are similar to those for double shear testing. It is critical to maintain clean, properly radiused shearing edges and proper alignment of the contact faces of the blades. Testing of fasteners with ultimate tensile strengths in excess of 1517 MPa (220 ksi) results in rapid shear-blade wear. Methods, equipment, and fixtures for single-shear testing of fasteners are described in Ref 6.

Lap-Joint Shear Tests

Lap-joint shear tests, which simulate installed conditions in various structural materials, are used to determine the shear strength of fasteners installed in sheet material. Several types of lap-joint specimens are used, but single-lap specimens are the most common and are preferred. A two-fastener single-lap shear specimen described in Ref 3 is shown in Fig. 10. A similar lap-joint specimen with only one fastener is also discussed in Ref 3, but is less suitable for most uses.

Lap-joint specimen testing to develop load allowables for a fastener series and sheet material normally includes several diameters installed in various sheet thicknesses. Specimens conforming to Fig. 10 are required for load-allowable testing. The test data generated should include test curves that reveal ultimate load and yield load values. The computations from basic data are plotted and graphically analyzed to determine a single design curve representing all diameters tested. This design curve (Fig. 11) has three regions:

- Region 1, in which the joint failure mode is by sheet bearing

Fig. 10 Lap-joint shear test specimen configuration

Source: Ref 3

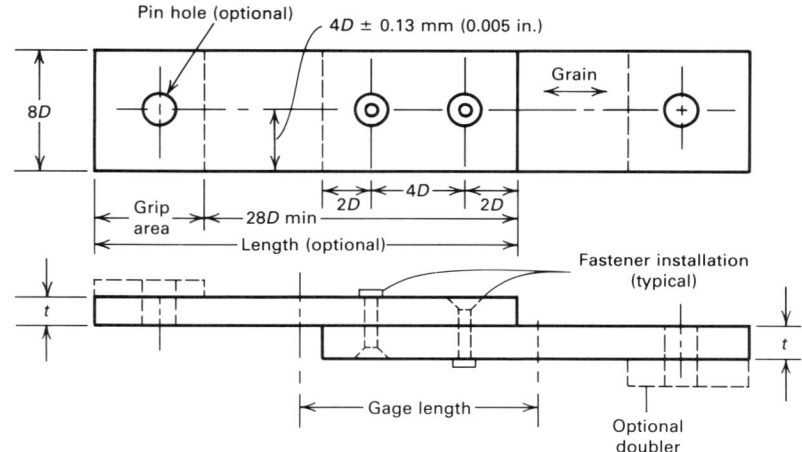

- Region 2, in which the joint failure mode is a combination of sheet bearing and fastener shear
- Region 3, in which the joint failure mode is by fastener shear

A complete and detailed description of the criteria, procedures, and presentation of lap-shear test data to develop load allowables is included in Ref 3. This material should be reviewed prior to conducting a testing program.

Testing Machines

Testing machines should be capable of applying a compressive or tensile load at a controllable rate. The machine and associ-

Fig. 11 Design curve for lap-joint shear testing

t, sheet thickness; D, fastener diameter. See text for failure region description.

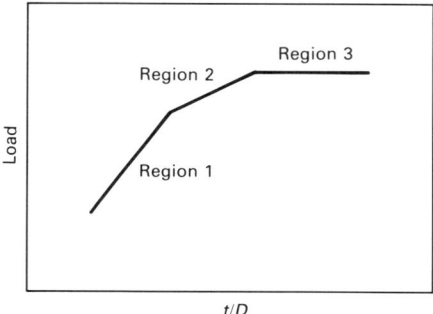

ated equipment such as extensometers and recorders should be calibrated every 6 months in compliance with ASTM E 4 (Ref 7) and ASTM E 83 (Ref 8), using a calibration system that is in conformance with MIL-STD-45662 (Ref 9).

Many types of test machines and equipment are commercially available to perform shear testing of fasteners and lap joints. Selection of the test machine should be based on the range of fastener shear loads and the degree of accuracy desired or required. When lap-joint shear tests are conducted, averaging extensometers of the appropriate gage length and test curve recording equipment are also required.

Test Fixtures and Shear Blades

The life of test fixtures and shear blades can be extended by careful selection of materials, hardness, surface finish, and cutting edge radii. For occasional or limited shear testing, fixtures made of alloy or high-carbon steels may be adequate. However, if testing is performed frequently, fixture life can be lengthened by using a hardened tool steel, such as D2 or M2, with a hardness of 60 to 64 HRC at all load-bearing surfaces and load points.

When galling and fastener wear particles are present on shear blades, the use of oil-hardened, cold worked tool steels such as AISI O6 helps alleviate the problem. Surface finishes of at least eight root mean square and polished radii at hole/shear-face intersections increase the consistency of tests and extend fixture life.

Burrs or rough (sawtooth) edges on shear blades are unacceptable. A small polished radius (0.025 to 0.05 mm, or 0.001 to 0.002 in.) increases shear blade life.

Test Procedures

The shear testing machine, test fixtures, and associated equipment must be in good working order, properly calibrated, and aligned in accordance with the designated test specification. The loads should be applied uniformly at the rate indicated in the specification. Loads that are applied suddenly and/or accelerated rapidly will result in erratic test values and possible damage to the test fixture and machine.

The test report should identify the testing method, the type of fastener tested, fastener material and dimensions, sheet material and dimensions (if a lap-shear test has been conducted), load or strain rate, ultimate and yield load values, and failure modes. The preparation of a detailed test report, analyses of data and modes of failure, and final conclusions should be in accordance with specifications and/or the requirements of the application.

Variables That Affect Fastener Shear Strength

A number of conditions may contribute to variations in the shear strength of fasteners, including surface finish, plating, coatings, hardness (both core and surface hardness), and temper. Shear strength values are also different for sleeve and multiple-piece fasteners, bimetallic fasteners, and tubular, tapered, and ellipsoidal fasteners. Variations may occur between lots, within a lot, and even within a grip length. Small-diameter fasteners with long grip lengths are particularly vulnerable to these conditions.

Careful analysis of the conditions that contribute to differences in ultimate load values can often determine the cause of test result variations. Metallographic examination can reveal plating or coating thickness variables,

internal cracks, enlarged grain size, distorted grain flow or orientation, dimensional discrepancies, and hardness variations that are not apparent by visual examination.

REFERENCES

1. J.G. Kaufman and R.E. Davies, Effects of Test Method and Specimen Orientation on Shear Strengths of Aluminum Alloys, *ASTM Proceedings*, Vol 64, ASTM, Philadelphia, 1964, p 999
2. W.W. Breindel, C.L. Seal, and R.L. Carlson, Evaluation of a Single-Shear Specimen for Sheet Material, *ASTM Proceedings*, Vol 58, ASTM, Philadelphia, 1958, p 862
3. *Metallic Materials and Elements for Aerospace Vehicle Structures MIL-HDBK-5D*, Naval Publications and Forms Center, Philadelphia, June 1983
4. MIL-STD-1312, Test No. 13, "Double Shear Test," Department of Defense, 1984
5. "Standard Method for Shear Testing of Aluminum and Aluminum Alloy Rivets and Cold Heading Wire and Rods," B 565, *Annual Book of ASTM Standards*, ASTM, Philadelphia, 1984, p 100-102
6. MIL-STD-1312, Test No. 20, "Single Shear Test," Department of Defense, 1973
7. "Standard Practices for Load Verification of Testing Machines," E 4, *Annual Book of ASTM Standards*, ASTM, Philadelphia, 1984, p 111-118
8. "Standard Method of Verification and Classification of Extensometers," E 83, *Annual Book of ASTM Standards*, ASTM, Philadelphia, 1984, p 247-252
9. MIL-STD-45662, "Calibration Systems Requirements," Department of Defense, 1980

Hardness Testing

Introduction

By Andrew R. Fee
Technical Consultant
Page-Wilson Corporation
Measurement Systems Division

Robert Segabache
Eastern Regional Sales Manager
Page-Wilson Corporation
Measurement Systems Division

and

Edward L. Tobolski
Engineering Manager
Page-Wilson Corporation
Measurement Systems Division

HARDNESS implies resistance to deformation; in the case of metals, this characteristic is a measure of their resistance to permanent or plastic deformation. The definition of hardness varies depending on the experience or background of the person conducting the test or interpreting the test data. To the metallurgist, hardness is the resistance to indentation; to the design engineer, a measure of flow stress; to the lubrication engineer, the resistance to wear; to the mineralogist, the resistance to scratching; and to the machinist, the resistance to cutting.

The types of tests used to determine hardness are:

- *Static indentation tests*, in which a ball, diamond cone, or pyramid is forced into the material being tested. The relationship of total test force to the area or depth of indentation provides the measure of hardness. The Rockwell, Brinell, Knoop, Vickers, and ultrasonic hardness tests, which will be described in articles contained in this Section, are of this type.
- *Dynamic hardness tests*, in which an object of standard mass and dimension is bounced from the workpiece; its height of rebound becomes a measure of hardness. The Scleroscope, which is described in the article "Miscellaneous Hardness Tests" in this Section, is employed in dynamic, or rebound, hardness tests.
- *Scratch tests*, in which one material is judged as capable of scratching another.

The Mohs and file hardness tests are of this type. These test procedures are also described in the article "Miscellaneous Hardness Tests."

In this Section, articles describing the procedures, equipment, applications, advantages, and limitations associated with static indentation hardness testing are presented. Although other forms of tests are important in hardness testing, indentation tests are the most widely used. Indentation testing is reproducible and can be accurately quantified. Tests to evaluate hardness by cutting, abrasion, and erosion, which generally are established on an empirical basis, will not be discussed.

Plasticity and Elasticity

Because the major portion of all hardness testing is based on some form of indentation, understanding of plasticity and elasticity theories is essential. These theories can be complex, partly because of the work hardening that occurs when an indentation is made.

The mean loading pressure, which relates to the theoretical hardness values obtained by static indentation tests, may be expressed as $C \times Y$, where C is the constraint factor for the test, and Y is the uniaxial flow (true yield) stress of the material being tested, usually given in kPa or psi. The value of the constraint factor depends primarily on the shape of the indenter used. For many of the common indenters (sphero-conical diamond, Brinell, Vickers, and Knoop), which are all relatively blunt, the constraint factor is approximately 3.

Prandtl (Ref 1) explained the constraint factor (C) by comparing the blunt hardness indenters to a flat-ended punch and calculating the stress on a two-dimensional punch—that is, one having infinite length and finite width, but no appreciable thickness—for the onset of plastic flow beneath the punch, assuming a flow pattern beneath the punch that satisfied kinematics. Thus, the material within the pattern was assumed to flow plastically in plane strain, while the material surrounding the flow pattern was considered to be rigid.

The flow pattern determined by Prandtl, which is shown in Fig. 1(a), predicts a constraint factor (C) of $1 + (\pi/2) = 2.57$. His approach was generalized by Hill (Ref 2) into what is now known as the slip-line field solution (Fig. 1b), which leads to the same value of C (2.57). According to these theories, the material displaced by the punch is accounted for by upward plastic flow. The C factor is a flow constraint from the slip-line field point of view.

The calculated value of 2.57 is reasonably close to the observed value of three, considering that the calculation is based on a two-dimensional punch, whereas the actual indenter is three dimensional. Extending the plane-strain slip-line field to a flat-ended circular punch results in a calculated value of

Fig. 1 Slip-line field solutions for a flat-ended two-dimensional punch having a width of 2a

(a) Prandtl's flow pattern. Flow in center area is downward and to left and right, as indicated by arrows in the adjoining areas. (b) Hill's flow pattern. Flow is to left and right in directions indicated by arrows in (a), but is separated.

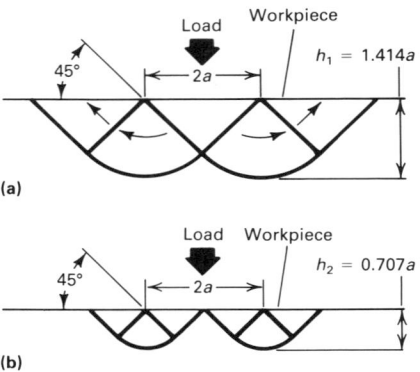

Fig. 2 Deformed grid pattern on a meridianal plane in a Brinell hardness test

(a) Modeling clay. (b) Low-carbon steel

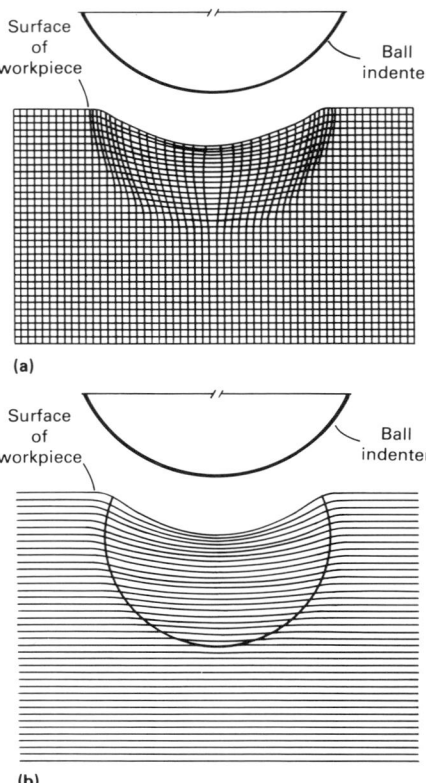

Fig. 3 Hertz lines of constant maximum shear stress on a meridianal plane below the surface of an elastic, semi-infinite body, caused by a frictionless load from a rigid sphere of very large radius

M' = maximum shear stress in the body divided by the mean stress (unit load) on the sphere.

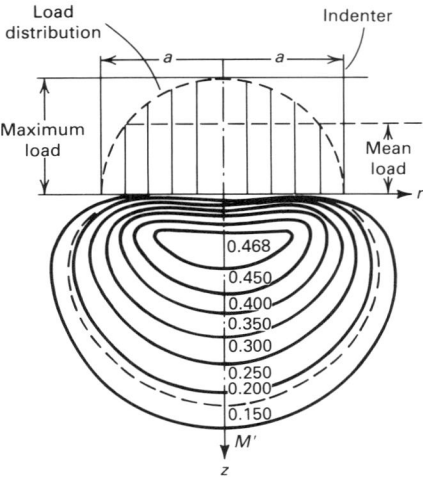

$C = 2.82$, a value closer to the observed value of three (Ref 3).

Because of these close agreements on the value of C, the slip-line field theory has been widely accepted. However, when a large block of material is loaded with a spherical indenter, flow patterns, such as those shown in Fig. 2(a) and (b), are obtained. These flow patterns do not resemble those shown in Fig. 1. The extent of the region of fully developed plastic flow may be determined by sighting along the deformed grid lines and by observing the area outlined by the circular lines in Fig. 2(b). This line separates the elastic region (outside the line) from the plastic region (inside the line).

The circular line in Fig. 2(b) resembles one of the lines of constant maximum shear stress for the elastic theory of Hertz (Ref 4) for a rigid spherical indenter of very large radius that is pressed against an elastic body. Figure 3 illustrates the spherical interface between the indenter and the body as a flat surface. The values of M' on the lines represent the ratio of constant maximum shear stress in the body to the mean stress (unit force) on the indenter.

The elastic-plastic boundary (Fig. 2b) is found to correspond closely to the dashed line of Fig. 3, for which $M' = 0.18$. If the maximum shear stress on this line is assumed to equal $(Y/2)$ by the maximum shear theory, then the mean stress on the indenter (P_m) will be:

$$P_m = \frac{Y}{2} \div M' = 2.78Y$$

This agrees with the results described in Ref 3.

Investigators (Ref 5) have shown that if the workpiece extends at least $10d$ (where d is the diameter of the indentation) in all directions from the indenter, the displaced volume may be accounted for by the elastic decrease in volume. Therefore, there is no need for the upward flow, and the elastic theory is in agreement with Fig. 2(b). The constraint factor that arises here is an elastic constraint factor, because the displaced volume is accounted for by an elastic decrease in volume instead of by upward flow, as in the slip-line field approach.

The basic difference between the slip-line field and the elastic theories of hardness lies in the assumption regarding the behavior of the material that surrounds the plastic zone. The slip-line field theory assumes the material to be rigid, whereas the elastic theory assumes it to be elastic.

Figure 2(b) supports the elastic theory of hardness for a blunt indenter. Further support is furnished by the fact that shot peening produces residual compressive stresses that improve fatigue life. This phenomenon is consistent with the elastic theory.

According to the elastic theory, when a blunt indenter is pressed into a plane surface, material beneath the indenter deforms plastically without upward flow, and the elastic stress field is the same as though there were no plastic flow. When the force is removed, there is plastic recovery—that is, plastic deformation in a direction opposite to the initial flow, but smaller in volume. Because plastic recovery is not complete, biaxial residual stresses remain in the planes parallel to the free surface after the force is removed.

Test Classification

Of all the types of indentation tests available, there are two general methods of classification. The measuring method involves forcing an indenter into the material with a specified force and measuring either the depth of penetration or the area of the unrecovered indentation to determine the hardness value. The Rockwell test, discussed in detail in the article "Rockwell Hardness Testing" in this Section, utilizes the depth of penetration method. The Brinell, Vickers, Knoop, and ultrasonic tests are examples of indentation tests that employ the area measurement method of determining hardness. See the articles "Brinell Hardness Testing" and "Microhardness Testing" in this Section for more information.

The second classification of indentation testing is governed by the load applied to the indenter. All tests can be referred to as either macrohardness or microhardness.

In all instances, a macrohardness test utilizes a force of 1 kgf or greater, usually much greater, and varies with the type of test. For the standard Vickers test, the force applied to the indenter may range from 1 to 120 kgf; 10 to 50 kgf is the most common. For the Rockwell test, this force ranges from 15 to 150 kgf. For the Brinell test, forces applied to the indenter are much higher, ranging from 500 to 3000 kgf.

Microhardness testing uses extremely light forces applied to either a Vickers or a Knoop indenter (see descriptions of these indenters in the article ''Microhardness Testing'' in this Section). The force for microhardness testing ranges from 1 gf to 1 kgf, the most common range being from 100 to 500 gf.

REFERENCES

1. L. Prandtl, Über die Haerte Plastischer Koerper, *Nachr. Akad. Wiss. Goettingen. Math-Physik. Kl.*, 1920, p 74
2. R. Hill, *The Mathematical Theory of Plasticity*, Clarendon Press, Oxford, 1950
3. R.T. Shield, On the Plastic Flow of Metals Under Conditions of Axial Symmetry, *Proc. Roy. Soc.*, Vol A233, 1955, p 267
4. H. Hertz, *Gesammelte Werke*, Leipzig, 1895
5. M.C. Shaw and G.J. DeSalvo, On the Plastic Flow Beneath a Blunt Axisymmetric Indenter, *Trans. ASME*, Vol 92, 1970, p 480

Rockwell Hardness Testing

By Andrew R. Fee
Technical Consultant
Page-Wilson Corporation
Measurement Systems Division

Robert Segabache
Eastern Regional Sales Manager
Page-Wilson Corporation
Measurement Systems Division

and

Edward L. Tobolski
Engineering Manager
Page-Wilson Corporation
Measurement Systems Division

ROCKWELL HARDNESS TESTING is the most widely used method for determining hardness, primarily because the Rockwell test is simple to perform and does not require highly skilled operators. By use of different loads (forces)* and indenters, Rockwell hardness testing can determine the hardness of most metals and alloys, ranging from the softest bearing materials to the hardest steels. Readings can be taken in a matter of seconds with conventional manual operation and in even less time with automated setups. Optical measurements are not required; all readings are direct.

Test Principle

Rockwell testing differs from Brinell testing in that the Rockwell hardness number is based on an inverse relationship to the measurement of the additional depth to which an indenter is forced by a heavy (major) load beyond the depth resulting from a previously applied (minor) load. Initially a minor load is applied, and a zero datum position is established. The major load is then applied for a specified period and removed, leaving the minor load applied. The resulting Rockwell number represents the difference in depth from the zero datum position as a result of the

---

*"Force" is the technically correct term. However, "load" is commonly used by manufacturers and users of hardness testing equipment.

application of the major load. The entire procedure requires only 5 to 10 s.

Use of a minor load greatly increases the accuracy of this type of test, because it eliminates the effect of backlash in the measuring system and causes the indenter to break through slight surface roughness. The basic principle involving minor and major loads is shown in Fig. 1. Although the principle is illustrated with a diamond indenter, the same principle applies for hardened steel ball indenters and other loads.

Indenters. The 120° sphero-conical diamond indenter (Fig. 2) is used mainly for testing hard materials such as hardened steels and cemented carbides. Its range is defined as harder than 100 HRB, and harder than 83.1 HR30T (see the section "Rockwell Scales" in this article for further explanation). Hardened steel ball indenters with diameters of $\frac{1}{16}$, $\frac{1}{8}$, $\frac{1}{4}$, and $\frac{1}{2}$ in. (1.588, 3.175, 6.35, and 12.7 mm) are used for testing softer materials such as fully annealed steels, softer grades of cast irons, and a wide variety of nonferrous metals.

Load Selection. There are two types of Rockwell tests: Rockwell and superficial Rockwell. In Rockwell testing, the minor load is 10 kgf, and the major load is 60, 100, or 150 kgf. In superficial Rockwell testing, the minor load is 3 kgf, and major loads are 15, 30, or 45 kgf. In both tests, the indenter may be either a diamond cone or a steel ball, depending principally on the characteristics of the material being tested.

Rockwell Scales

Rockwell hardness values are expressed as a combination of a hardness number and a scale symbol representing the indenter and the minor and major loads. The hardness number is expressed by the symbol HR and the scale designation. For example, 64 HRC represents the Rockwell hardness number of 64 on the Rockwell C scale; 81 HR30N represents the Rockwell superficial hardness number of 81 on the Rockwell 30N scale.

There are 30 different scales, defined by the combination of the indenter and the minor and major loads (Tables 1 and 2). In many instances, Rockwell hardness tolerances are specified or are indicated on drawings. At times, however, the Rockwell scale must be selected to suit a given set of circumstances.

The majority of applications are covered by the Rockwell C and B scales for testing steel, brass, and other metals. However, the increasing use of materials other than steel and brass as well as thin materials necessitates a basic knowledge of the factors that must be considered in choosing the correct scale to ensure an accurate Rockwell test. The choice is not only between the regular hardness test and superficial hardness test, with three different major loads for each, but also between the diamond indenter and the $\frac{1}{16}$-, $\frac{1}{8}$-, $\frac{1}{4}$-, and $\frac{1}{2}$-in. (1.588-, 3.175-, 6.35-, and 12.7-mm) diam steel ball indenters.

Fig. 1 Principle of the Rockwell test

Although a diamond indenter is illustrated, the same principle applies for steel ball indenters and other loads.

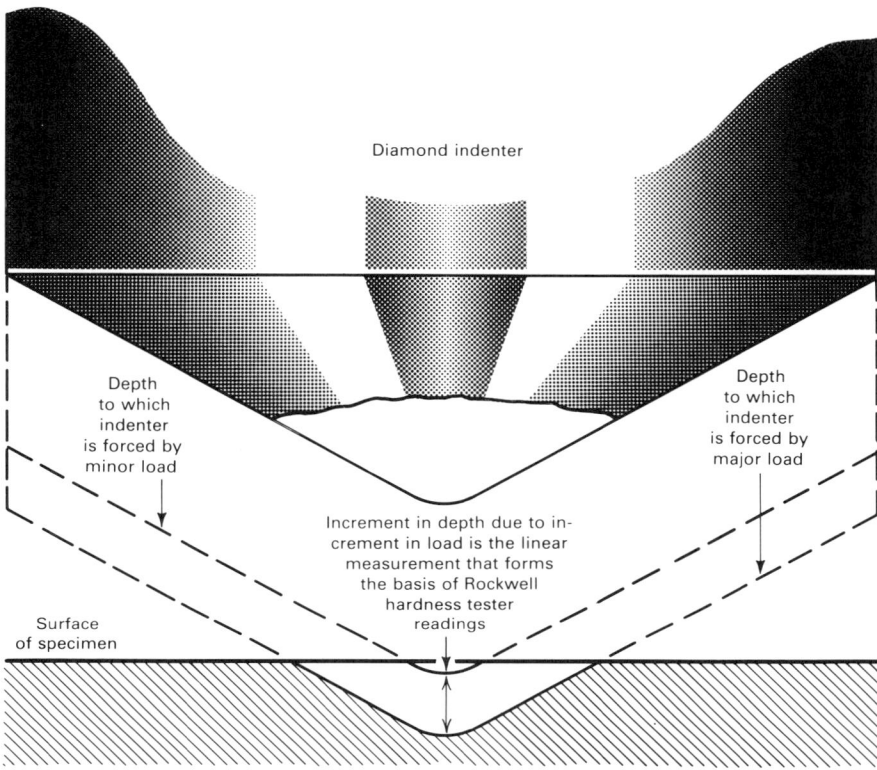

Fig. 2 120° sphero-conical diamond indenter

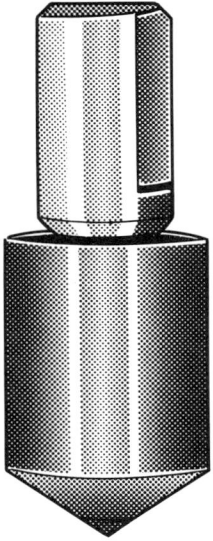

If no specification exists or there is doubt about the suitability of the specified scale, an analysis should be made of the following factors that control scale selection:

- Type of material
- Specimen thickness
- Test location
- Scale limitations

Type of Material

Standard Rockwell scales and typical materials for which these scales are applicable are listed in Table 1. This information can also be helpful if one of the superficial scales is required. For example, if a hard material such as steel or tungsten carbide is to be tested, a diamond indenter would be used. This automatically limits the choice of scale to one of six: Rockwell C, A, D, 45N, 30N, or 15N. The next step is to determine which scale will guarantee accuracy, sensitivity, and repeatability.

Specimen Thickness

The material immediately surrounding a Rockwell indentation is cold worked. The extent of the cold worked area depends on the type of material and previous work hard-

ening of the test specimen. The depth of material affected has been found by extensive experimentation to be on the order of ten times the depth of the indentation. Therefore, unless the thickness of the material being tested is at least ten times the depth of the indentation, an accurate Rockwell test cannot be ensured. This "minimum thickness" ratio of 10 to 1 should be regarded only as an approximation.

The depth of the indentation can be determined as follows. When the reading is taken with a diamond indenter, the Rockwell number is subtracted from 100, and the result is multiplied by 0.002 mm. Therefore, a reading of 60 HRC indicates an indentation depth from minor to major load of:

$$(100 - 60) \times 0.002 \text{ mm} = 0.08 \text{ mm}$$

$$\text{Depth} = 0.08 \text{ mm } (0.003 \text{ in.})$$

When a ball indenter is used, the hardness number is subtracted from 130; therefore, for a dial reading of 80 HRB, the depth is determined by:

$$(130 - 80) \times 0.002 \text{ mm} = 0.10 \text{ mm}$$

$$\text{Depth} = 0.10 \text{ mm } (0.004 \text{ in.})$$

In Rockwell superficial tests, regardless of the type of indenter used, one number represents an indentation of 0.001 mm (0.00004 in.). Therefore, a reading of 80 HR30N indicates a depth of indentation from minor to major load of:

$$(100 - 80) \times 0.001 \text{ mm} = 0.02 \text{ mm}$$

$$\text{Depth} = 0.02 \text{ mm } (0.0008 \text{ in.})$$

As indicated above, computation of the depth of penetration for any Rockwell test requires only simple arithmetic. However, in actual practice, computation is not necessary, because minimum thickness values have been established (Table 3). These minimum thickness values generally follow the 10 to 1 ratio, but they are based on experimental data accumulated for varying thicknesses of low-carbon steels and of hardened and tempered strip steel.

Consider a requirement to check the hardness of a strip of steel 0.36 mm (0.014 in.) thick with an approximate hardness of 63 HRC. According to Table 3, material in the 63 HRC range must be approximately 0.71 mm (0.028 in.) thick for an accurate Rockwell C scale test. Therefore, 63 HRC must be converted to an approximate equivalent hardness on other Rockwell scales. These values, taken from a conversion table, are 73 HRD, 82.8 HRA, 69.9 HR45N, 80.1 HR30N, and 91.4 HR15N. See the Appendix to this Section for hardness conversion tables.

Referring to Table 3, there are only three appropriate Rockwell scales—45N, 30N, and 15N—for hardened 0.356-mm (0.014-in.) thick material. The 45N scale is not suit-

Table 1 Rockwell standard hardness

Scale symbol	Indenter	Major load, kgf	Typical applications
A	Diamond (two scales—carbide and steel)	60	Cemented carbides, thin steel, and shallow case-hardened steel
B	1/16-in. (1.588-mm) ball	100	Copper alloys, soft steels, aluminum alloys, malleable iron
C	Diamond	150	Steel, hard cast irons, pearlitic malleable iron, titanium, deep case-hardened steel, and other materials harder than HRB 100
D	Diamond	100	Thin steel and medium case-hardened steel and pearlitic malleable iron
E	1/8-in. (3.175-mm) ball	100	Cast iron, aluminum and magnesium alloys, bearing metals
F	1/16-in. (1.588-mm) ball	60	Annealed copper alloys, thin soft sheet metals
G	1/16-in. (1.588-mm) ball	150	Phosphor bronze, beryllium copper, malleable irons. Upper limit HRG 92 to avoid possible flattening of ball
H	1/8-in. (3.175-mm) ball	60	Aluminum, zinc, lead
K	1/8-in. (3.175-mm) ball	150	Bearing metals and other very soft or thin materials. Use smallest ball and heaviest load that do not produce anvil effect.
L	1/4-in. (6.350-mm) ball	60	Bearing metals and other very soft or thin materials. Use smallest ball and heaviest load that do not produce anvil effect.
M	1/4-in. (6.350-mm) ball	100	Bearing metals and other very soft or thin materials. Use smallest ball and heaviest load that do not produce anvil effect.
P	1/4-in. (6.350-mm) ball	150	Bearing metals and other very soft or thin materials. Use smallest ball and heaviest load that do not produce anvil effect.
R	1/2-in. (12.70-mm) ball	60	Bearing metals and other very soft or thin materials. Use smallest ball and heaviest load that do not produce anvil effect.
S	1/2-in. (12.70-mm) ball	100	Bearing metals and other very soft or thin materials. Use smallest ball and heaviest load that do not produce anvil effect.
V	1/2-in. (12.70-mm) ball	150	Bearing metals and other very soft or thin materials. Use smallest ball and heaviest load that do not produce anvil effect.

Source: ASTM Standard E 18

Table 2 Rockwell superficial hardness scales

Scale symbol	Indenter	Major load, kgf
15N	Diamond	15
30N	Diamond	30
45N	Diamond	45
15T	1/16-in. (1.588-mm) ball	15
30T	1/16-in. (1.588-mm) ball	30
45T	1/16-in. (1.588-mm) ball	45
15W	1/8-in. (3.175-mm) ball	15
30W	1/8-in. (3.175-mm) ball	30
45W	1/8-in. (3.175-mm) ball	45
15X	1/4-in. (6.350-mm) ball	15
30X	1/4-in. (6.350-mm) ball	30
45X	1/4-in. (6.350-mm) ball	45
15Y	1/2-in. (12.70-mm) ball	15
30Y	1/2-in. (12.70-mm) ball	30
45Y	1/2-in. (12.70-mm) ball	45

Note: The Rockwell N scales of a superficial hardness tester are used for materials similar to those tested on the Rockwell C, A, and D scales, but of thinner gage or case depth. The Rockwell T scales are used for materials similar to those tested on the Rockwell B, F, and G scales, but of thinner gage. When minute indentations are required, a superficial hardness tester should be used. The Rockwell W, X, and Y scales are used for very soft materials. The letter N designates the use of the diamond indenter; the letters T, W, X, and Y designate steel ball indenters. Superficial Rockwell hardness values are always expressed by the number suffixed by a number and a letter that show the load and indenter combination. For example, 80 HR30N indicates a reading of 80 on the superficial Rockwell scale using a diamond indenter and a major load of 30 kgf.

able, because the material should be at least 74 HR45N. The 30N scale requires the material to be at least 80 HR30N; on the 15N scale, the material must be at least 76 HR15N. Therefore, either the 30N or 15N scale can be used.

If a choice remains after all criteria have been applied, then the scale applying the heavier load should be used. A heavier load produces a larger indentation covering a greater portion of the material, as well as a Rockwell hardness number more representative of the material as a whole. In addition, the heavier the load, the greater the sensitivity of the scale.

In the example under consideration, a conversion chart will indicate that, in the hard steel range, a difference in hardness of one point on the Rockwell 30N scale represents a difference of only 0.5 points on the Rockwell 15N scale. Therefore, smaller differences in hardness can be detected when using the 30N scale. This approach also applies when selecting a scale to accurately measure hardness when approximate case depth and hardness are known.

Minimum thickness charts and the 10 to 1 ratio serve only as guides. After determining which Rockwell scale should be used based on minimum thickness values, an actual test should be performed, and the side directly beneath the indentation should be examined to determine whether the material was disturbed or a bulge exists. If so, the material is not sufficiently thick for the applied load. This results in a condition known as "anvil effect." When anvil effect or flow exists, the Rockwell hardness number obtained may not be a true value. The Rockwell scale applying the next lighter load should then be used.

Use of several specimens, one on top of the other, is not recommended. Slippage between the contact surfaces of the specimens makes a true value impossible to obtain. The only exception is in the testing of plastics; use of several thicknesses for elastomeric materials when anvil effect is present is recommended in ASTM Standard D 785, "Rockwell Hardness of Plastic and Elastomeric Materials." Testing performed on soft plastics does not have an adverse effect when the test specimen is composed of a stack of several pieces of the same thickness, provided that the surfaces of the pieces are in total contact and not held apart by sink marks, burrs from saw cuts, or other protrusions.

When testing specimens for which the anvil effect results, the condition of the supporting surface of the anvil must be observed carefully. After several tests, this surface may become marred, or a small indentation may be produced. Either condition affects the Rockwell test, because under the major

Table 3   Minimum work-metal hardness values for testing various thicknesses of metals with standard and superficial Rockwell hardness testers

| Metal thickness mm | in. | Minimum hardness for standard hardness testing | | | | | | Minimum hardness for superficial hardness testing | | | | | |
| | | Diamond indenter | | | Ball indenter, 1/16 in. (1.588 mm) | | | Diamond indenter | | | Ball indenter, 1/16 in. (1.588 mm) | | |
		A (60 kgf)	D (100 kgf)	C (150 kgf)	F (60 kgf)	B (100 kgf)	G (150 kgf)	15N (15 kgf)	30N (30 kgf)	45N (45 kgf)	15T (15 kgf)	30T (30 kgf)	45T (45 kgf)
0.127	0.005	...	...	...	...	...	...	...	...	...	93	...	...
0.152	0.006	...	...	...	...	...	...	92	...	...	...	...	...
0.203	0.008	...	...	...	...	...	...	90	...	...	...	...	...
0.254	0.010	...	...	...	...	...	...	88	...	...	90	87	...
0.305	0.012	...	...	...	...	...	...	83	82	77	...	...	...
0.356	0.014	...	...	...	...	...	...	76	80	74	...	...	...
0.381	0.015	...	...	...	...	...	...	...	...	...	78	77	77
0.406	0.016	86	...	...	...	...	...	68	74	72	...	...	...
0.457	0.018	84	...	...	...	...	...	(a)	66	68	...	...	...
0.508	0.020	82	77	...	100	...	...	(a)	57	63	(a)	58	62
0.559	0.022	78	75	69	...	...	...	(a)	47	58	...	...	...
0.610	0.024	76	72	67	...	...	...	(a)	(a)	51	...	...	...
0.635	0.025	...	...	...	92	92	90	(a)	(a)	...	(a)	(a)	26
0.660	0.026	71	68	65	...	...	...	(a)	(a)	37	...	...	...
0.711	0.028	67	63	62	...	...	...	(a)	(a)	20	...	...	...
0.762	0.030	60	58	57	67	68	69	(a)	(a)	(a)	(a)	(a)	(a)
0.813	0.032	(a)	51	52	...	...	...	(a)	(a)	(a)	...	...	...
0.864	0.034	(a)	43	45	...	...	...	(a)	(a)	(a)	...	...	...
0.889	0.035	...	...	...	(a)	44	46	(a)	(a)	(a)	(a)	(a)	(a)
0.914	0.036	(a)	(a)	37	...	...	...	(a)	(a)	(a)	...	...	...
0.965	0.038	(a)	(a)	28	...	...	...	(a)	(a)	(a)	...	...	...
1.016	0.040	(a)	(a)	20	(a)	20	22	(a)	(a)	(a)	(a)	(a)	(a)

Note: These values are approximate only and are intended primarily as a guide; see text for example of use. Material thinner than shown should be tested with a microhardness tester. The thickness of the workpiece should be at least one and one half times the diagonal of the indentation when using a Vickers indenter, and at least one half times the long diagonal when using a Knoop indenter.
(a) No minimum hardness for metal of equal or greater thickness

load, the test material will sink into the indentation in the anvil and a lower reading will result. If a specimen is found to have been too thin during testing, the anvil surface should be inspected; if damaged, it should be relapped or replaced.

When using a ball indenter and a superficial scale load of 15 kgf on a specimen in which anvil effect or material flow is present, a diamond spot anvil can be used in place of the standard steel anvil. Under these conditions, the hard diamond surface is not likely to be damaged when testing thin materials. Furthermore, with materials that flow under load, the hard polished diamond provides a somewhat standardized frictional condition with the underside of the specimen, which improves repeatability of readings.

Test Location

If an indentation is placed too close to the edge of a specimen, the workpiece edge will bulge, and the Rockwell hardness number will decrease accordingly. To ensure an accurate test, the distance from the center of the indentation to the edge of the specimen must be at least two and one-half diameters. Therefore, when testing in a narrow area, the width of this area must be at least five diameters when the indentation is placed in the center. The appropriate scale must be selected for this minimum width. Although the diameter of the indentation can be calculated, for practical purposes the minimum distance can be determined visually.

An indentation hardness test cold works the surrounding material. If another indentation is placed within this cold worked area, the reading usually will be higher than that obtained had it been placed outside this area. Generally, the softer the material, the more critical the spacing of indentations. However, a distance three diameters from the center of one indentation to another is sufficient for most materials.

Scale Limitations

Because diamond indenters are not calibrated below values of 20, they should not be used when readings fall below this level. If used on softer materials, results may not agree when the indenters are replaced, and another scale—for example, the Rockwell B scale—should be used.

There is no upper limit to the hardness of a material that can be tested with a diamond indenter. However, the Rockwell C scale should not be used on tungsten carbide, because the material will fracture or the diamond life will be reduced considerably. The Rockwell A scale is the accepted scale in the carbide products industry.

Although scales that use a ball indenter (for example, the Rockwell B scale) range to 130, readings above approximately 100 should not be accepted, except under special circumstances. Between approximately 100 and 130, the tip of the ball is used. Because of the blunt shape of an indenter, the sensitivity of most scales is poor in this region.

Also, with smaller diameter indenters, flattening of the ball is possible because of the high pressure developed. However, because there is a loss of sensitivity as the size of the ball increases, the smallest possible ball should be used.

If values above 100 are obtained, the next heavier load, or next smaller indenter, should be used. If readings below 0 are obtained, the next lighter load, or next larger indenter, should be used. Readings below 0 are not recommended on any Rockwell scale, because misinterpretation may result when negative values are used. On nonhomogeneous materials, a scale should be selected that gives relatively consistent readings. If the ball indenter is too small in diameter or the load is too light, the resulting indentation will not cover an area sufficiently representative of the material to yield consistent hardness readings.

Rockwell Testing Machines

Many different types of Rockwell testers are currently produced. Test loads can be applied in a number of ways; most utilize dead weight or springs. Many testers use a dial (analog) measuring device. However, digital-readout testers are becoming more popular because of improved readability. Some testers use microprocessors to control the test process and can be used to interface with computers.

Various methods of performing the function of a Rockwell test have been developed by manufacturers. A common deadweight-type tester is shown in Fig. 3. Generally, different machines are used to make standard Rockwell and superficial Rockwell tests. However, there are combination machines available that can perform both types of tests. The principal components of a deadweight-type Rockwell tester are shown in Fig. 3.

Bench-Type Testing Machines. Routine testing is commonly performed with bench-type machines (Fig. 4), which are available with vertical capacities of up to 405 mm (16 in.). A machine of this type can accommodate a wide variety of part shapes by capitalizing on standard as well as special anvil designs. The usefulness of this standard type of machine can be greatly extended by the use of various accessories, such as:

- Outboard or counterweighted anvil adapters for testing unwieldy workpieces such as long shafts
- Clamps that apply pressure on the part, which are particularly suited for testing parts that have a large overhang or long parts such as shafts
- Gooseneck anvil adapters for testing inner and outer surfaces of cylindrical objects

Production Testing Machines. When large quantities of similar workpieces must be tested, conventional manually operated machines may not be adequate. With a motorized tester, hourly production can be increased by up to 30%.

To achieve still greater production rates, an automatic tester is required (Fig. 5). This type of tester can be fully automated to include automatic feeding, testing, and tolerance sorting. Upper and lower tolerance limits can be set from an operator control panel. This type of tester allows test loads to be applied at high speed with short dwell times. Up to 1000 parts per hour can be tested. These testers are normally dedicated to specific hardness ranges.

Portable Testing Machines. For hardness testing of large workpieces that cannot be moved, portable units are available in most regular and superficial scales and in a wide range of capacities (up to about a 355-mm, or 14-in., opening between anvil and indenter) (Fig. 6). Most portable hardness testers follow the Rockwell principle of minor and major loads, with the Rockwell hardness number indicated directly on the measuring device. Both digital and analog models are available.

The test principle is identical to that of bench-type models. The workpiece is clamped in a C-clamp arrangement, and the indenter is recessed into a ring-type holder that is part of the clamp. The workpiece is held by the clamp between what is normally the anvil and the holder (which, in effect, serves as an upper anvil). The indenter is lowered to the workpiece through the holder.

Fig. 4 Bench-type Rockwell testing machine

Fig. 3 Schematic of Rockwell testing machine

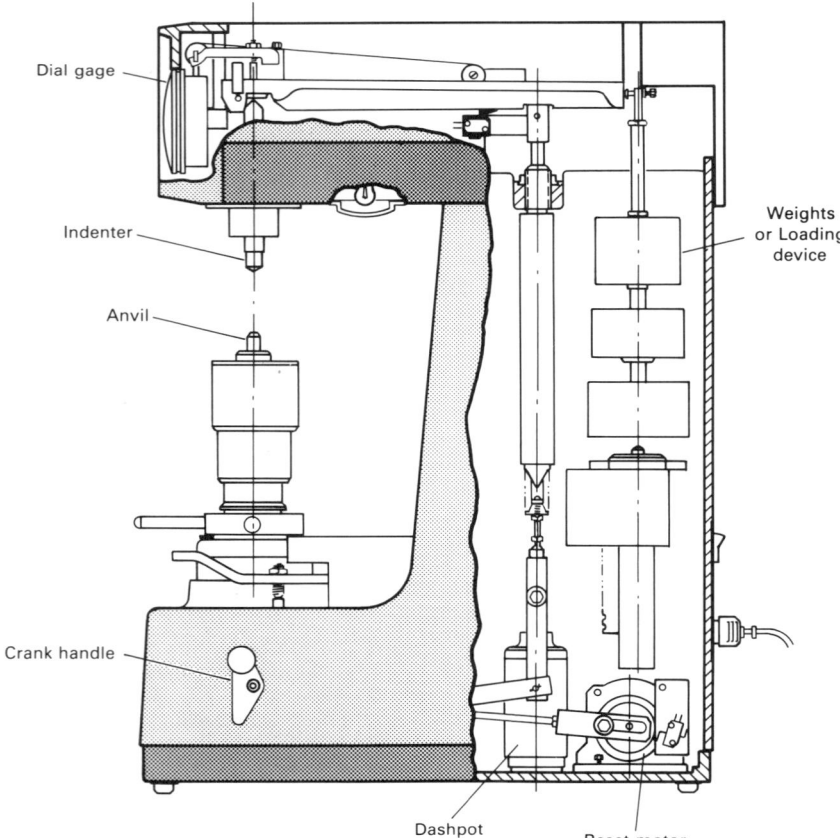

Dial gage

Indenter

Anvil

Crank handle

Weights or Loading device

Dashpot

Reset motor

Fig. 5 Automatic Rockwell testing system for high-production work

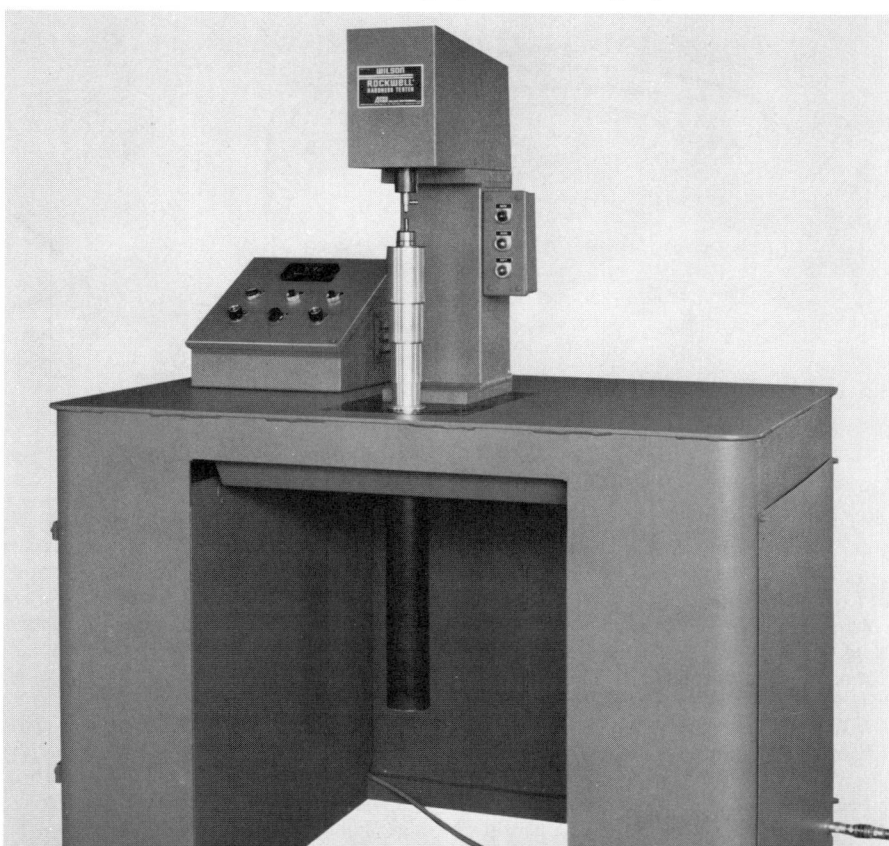

Fig. 6 Portable Rockwell hardness tester for testing large workpieces such as plates

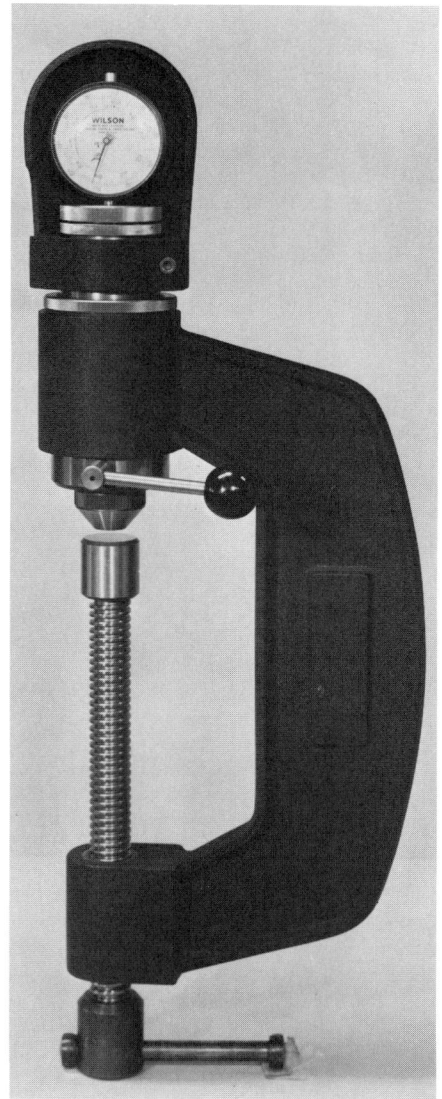

Calibration of Testing Machines

If a testing system is in constant use, a calibration check should be performed daily. Testers not used regularly should be checked before use. This check uses standardized test blocks to determine whether the tester is out of calibration and whether the indenter is damaged.

Rockwell test blocks are made from high-quality materials for uniformity of test results. To maintain the integrity of the test block, only the calibrated surface should be used. Regrinding of this surface is not recommended due to possible variations between the new and original surfaces.

If a tester is used throughout a given hardness scale, the recommended practice is to check it at the high, middle, and low ranges of the scale. For example, to check the complete Rockwell C scale, the tester should be checked at values such as 63, 45, and 25 HRC. On the other hand, if only one or two ranges are used, test blocks should be chosen that fall within ±5 hardness numbers of the testing range on any scale using a diamond indenter, and within ±10 numbers on any scale using a ball indenter.

A minimum of five tests should be made on the standardized surface of the block. The tester is in calibration if the average of these tests falls within the tolerances indicated on the side of the test block. For best results, a pedestal spot anvil should be used for all calibration work.

If the average of the five readings falls outside the Rockwell test block limits, the ball in the steel ball indenter should be inspected visually; in the case of a diamond indenter, the point should be examined using at least a 10× magnifier. If there is any indication of damage, the damaged component must be replaced.

Testing Methodology

Although the Rockwell test is simple to perform, accurate results depend greatly on proper testing methods.

Indenters. The mating surfaces of the indenter and plunger rod should be clean and free of dirt, machined chips, and oil, which prevent proper seating and can cause erroneous test results. After replacing an indenter, a ball in a steel ball indenter, or an anvil, several tests should be performed to seat these parts before a hardness reading is taken. Indenters should be visually inspected to determine whether any obvious physical damage is present that may affect results.

Anvils should be selected to minimize contact area of the workpiece while maintaining stability. Figure 7 illustrates five common types of anvils that can accommodate a broad range of workpiece shapes. An anvil with a large flat surface (Fig. 7a) should be used to support flat-bottom workpieces of thick section. Anvils with a surface diameter greater than about 75 mm (3 in.) should be attached to the elevating screw by a threaded section, rather than inserted in the anvil hole in the elevating screw (Fig. 7a).

Sheet metal and small workpieces that have flat undersurfaces are best tested on a spot anvil with a small, elevated, flat bearing surface (Fig. 7b and c). Workpieces that are not flat should have the convex side down on the bearing surface. Round workpieces should be supported in a V-slot anvil (Fig. 7d and e).

Other anvil designs are available. Special anvils to accommodate specific workpiece configurations can be fabricated. Regardless of anvil design, rigidity of the part to prevent movement during the test is absolutely essential for accurate results, as is cleanliness of the mating faces of the anvil and its supporting surface.

Specimen Surface Preparation. The degree of workpiece surface roughness that can be tolerated depends on the Rockwell scale being used. As a rule, for a load of 150 kgf on a diamond indenter, or 100 kgf on a ball indenter, a finish ground surface is sufficient to provide accurate readings. As loads become lighter, surface requirements become more rigorous. For a 15-kgf load, a polished or lapped surface usually is required.

Surfaces that are visibly ridged due to rough grinding or coarse machining offer unequal support to the indenter. Loose or flaking scale on the specimen at the point of indenter contact may chip and cause a false test. Scale should be removed by grinding or filing. Decarburized surface metal must also be removed to permit the indenter to test the true metal beneath.

Workpiece Mounting. The test specimen must be solidly supported by the anvil. The depth of indentation when the major load is applied is measured by the movement of the plunger rod holding the indenter; any slippage or movement of the workpiece will be followed by the plunger rod. The motion will be transferred to the measuring system. Errors of this type always produce softer hardness values. Because one point of hardness represents a depth of only 0.002 mm

Fig. 7 Common anvil types designed to support various shapes of workpieces during Rockwell hardness testing

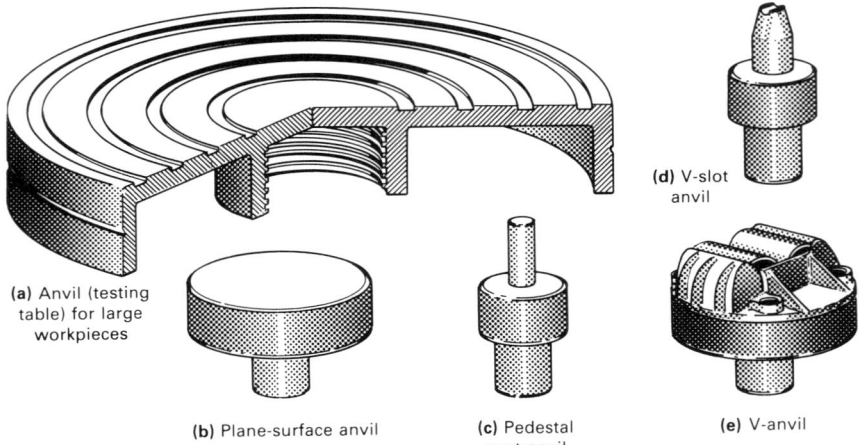

(a) Anvil (testing table) for large workpieces

(b) Plane-surface anvil

(c) Pedestal spot anvil

(d) V-slot anvil

(e) V-anvil

(0.00008 in.), a movement of only 0.025 mm (0.001 in.) could cause an error of over 10 Rockwell points.

Angle of Test Surface. The test surface should be perpendicular to the indenter axis. Extensive experimentation has found errors of 0.1 to 1.5 HRC, depending on the hardness range being tested, with an angle deviation of 3°. Such errors produce softer hardness values.

Load Application. The minor load should be applied to the test specimen in a controlled manner, without inducing impact or vibration. With manually operated testing machines, the measuring device must then be set to zero datum, or set point, position. The major load is then applied in a controlled fashion.

During the test cycle on a manually operated tester, the operator should not force the crank handle, because inaccuracies and damage to the tester may result. When the large pointer comes to rest or slows appreciably, the full major load has been applied and should dwell for up to 2 s. The load is then removed by returning the crank handle to the latched position. The hardness value can then be read directly from the measuring device. Semiautomatic digital testers perform most of these steps automatically.

Homogeneity. A Rockwell tester measures the hardness of a specimen at the point of indentation, but the reading is also influenced by the hardness of the material under and around the indentation. The effects of indentation extend about ten times the depth of the indentation. If a softer layer is located in this depth, the impression will be deeper, and the apparent hardness will be less.

This factor must be taken into account when testing material with a superficial hardness, such as case-hardened work. To obtain the average hardness of materials such as cast iron with graphite particles, or nonferrous metals with crystalline aggregates that are greater than the area of the indenter, a larger indenter must be used. In many instances, a Brinell test may be more valid for this type of material.

Spacing of indentations is very important. The distance from the center of one indentation to another must be at least three indentation diameters, and the distance to the edge should be a minimum of 2.5 diameters. Readings from any indentation spaced closer should be disregarded. These guidelines apply for all materials.

Adjustments for Specimen Size and Configuration

When performing a Rockwell test, specimen size and configuration may require that modifications in the test setup be made. For example, large specimens and thin-wall rings and tubing may need additional support equipment, and test results obtained from curved surfaces may require a correction factor.

Large Specimens. Figure 8 illustrates one of many specially designed Rockwell hardness testers that have been developed to accommodate the testing of large specimens that cannot conveniently be brought to or placed in a bench-type tester. For large and heavy workpieces, or workpieces of peculiar shape that must rest in cradles or on blocks, use of a large testing table is recommended.

Long Specimens. Work supports are available for long workpieces that cannot be firmly held on an anvil by the minor load. Because manual support is not practical, a

Fig. 8 Specially designed Rockwell hardness tester for testing large workpieces

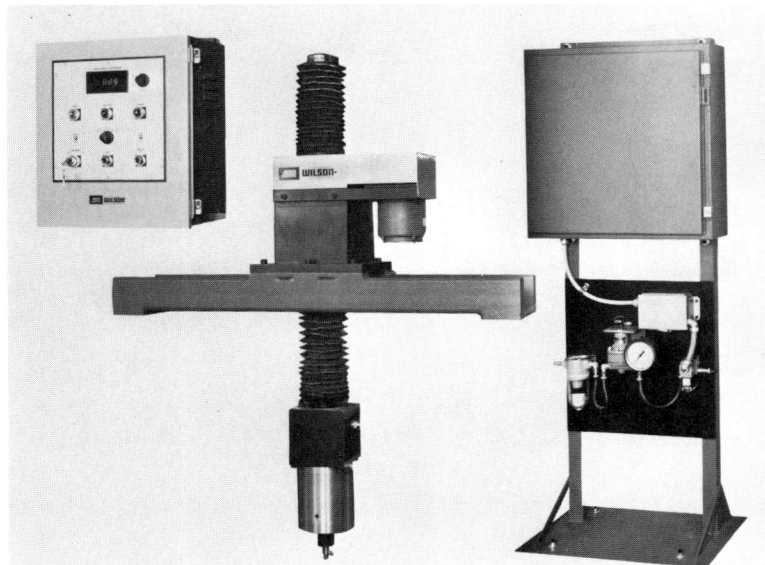

jack-rest should be provided at the overhang end to prevent pressure between the specimen and the penetrator. Figure 9 illustrates correct and incorrect methods for testing long, heavy workpieces.

Workpieces With Curved Surfaces. When an indenter is forced into a convex surface, there is less lateral support supplied for the indenting force; consequently, the indenter will sink farther into the metal than it would into a flat surface of the same hardness. Therefore, for convex surfaces, low readings will result. On the other hand, when testing a concave surface, opposite conditions prevail; that is, additional lateral support is provided, and the readings will be higher than when testing the same metal with a flat surface.

Results from tests on a curved surface may be in error and should not be reported without stating the radius of curvature. For diameters of more than 25.4 mm (1 in.), the difference is negligible. For diameters of less than 25.4 mm (1 in.), particularly for softer materials that involve larger indentation, the curvature, whether concave or convex, must be taken into account if a comparison is to be made with different diameters or with a flat surface.

Correction factors should be applied when workpieces are expected to meet a specified value. Typical correction factors for regular and superficial hardness values are presented in Table 4. The corrections are added to the hardness value when testing on convex surfaces and subtracted when testing on concave surfaces.

On cylinders with diameters as small as 6.35 mm (0.25 in.), standard Rockwell scales can be used; for the superficial Rockwell test, correction factors for diameters as small as 3.175 mm (0.125 in.) are given in Table 4. Diameters smaller than 3.175 mm (0.125 in.) should be tested by microhardness methods (see the article ''Microhardness Testing'' in this Section).

When testing cylindrical pieces such as rods, the shallow V or standard V anvil should be used, and the indenter should be applied over the axis of the rod. Care should be taken that the specimen lies flat, supported by the sides of the V. Figure 10 illustrates correct and incorrect methods of supporting cylindrical work while testing.

Inner Surfaces. The most basic approach to Rockwell hardness testing of inner surfaces is to use a gooseneck adapter for the indenter, as illustrated in Fig. 11. This

adapter can be used for testing in holes or recesses as small as 11.11 mm (0.4375 in.) in diameter or height.

Thin-Wall Rings and Tubes. When testing pieces such as thin-wall rings and tubing that may deform permanently under load, a test should be conducted in the usual manner to see if the specimen becomes permanently deformed. If it has been permanently deformed, either an internal mandrel on a gooseneck anvil or a lighter test load should be used.

Excessive deformation of tubing (either permanent or temporary) can also affect the application of the major load. If through deformation the indenter travels to its full extent, complete application of the major load will be prevented, and inaccurately high readings will result.

Gears and other complex shapes often require the use of relatively complex anvils in conjunction with holding fixtures. When hardness testing workpieces that have complex shapes—for example, the pitch lines of gear teeth—it is sometimes necessary to design and manufacture special anvils and fixtures; specially designed hardness testers may be required to accommodate these special fixtures.

Rockwell Testing at Elevated Temperatures

Of the several methods that have been devised to determine hardness at elevated temperatures, the modified Rockwell test is used most often. A system for elevated-temperature testing is shown in Fig. 12, consisting of a Rockwell tester into which is built a furnace with provisions for a controlled atmosphere—usually argon, although a vacuum furnace may be used.

As illustrated in Fig. 12, this furnace is connected to a temperature control system. This testing system also features an indexing fixture that makes it possible to bring any

Fig. 9 Method for mounting and testing long, heavy workpieces
(a) Correct method requires a support of the extended end of the piece to prevent any pressure of specimen against indenter. The jack-rest support is available as an accessory. (b) Incorrect method causes damage to indenter and, through leverage action, causes drag and jamming of plunger rod, producing inaccurate readings. When testing, the specimen must be pressed rigidly on the anvil by the pressure of the minor load. Because of this, only short or lightweight material may be permitted much overhang.

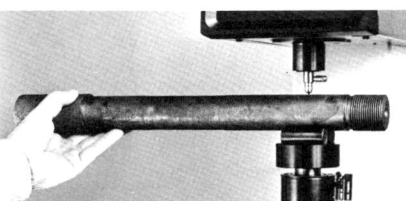

Correct method   Incorrect method

Table 4 Correction factors for cylindrical workpieces tested with standard and superficial Rockwell hardness testers

Observed reading	Correction factor for workpiece with diameter of:							
	3.175 mm (0.125 in.)	6.350 mm (0.250 in.)	9.525 mm (0.375 in.)	12.700 mm (0.500 in.)	15.875 mm (0.625 in.)	19.050 mm (0.750 in.)	22.225 mm (0.875 in.)	25.400 mm (1.000 in.)
Standard hardness testing, 1/16-in. (1.588-mm) ball indenter (Rockwell B, F, and G scales)								
100 ...	...	3.5	2.5	1.5	1.5	1.0	1.0	0.5
90 ...	...	4.0	3.0	2.0	1.5	1.5	1.5	1.0
80 ...	...	5.0	3.5	2.5	2.0	1.5	1.5	1.5
70 ...	...	6.0	4.0	3.0	2.5	2.0	2.0	1.5
60 ...	...	7.0	5.0	3.5	3.0	2.5	2.0	2.0
50 ...	...	8.0	5.5	4.0	3.5	3.0	2.5	2.0
40 ...	...	9.0	6.0	4.5	4.0	3.0	2.5	2.5
30 ...	...	10.0	6.5	5.0	4.5	3.5	3.0	2.5
20 ...	...	11.0	7.5	5.5	4.5	4.0	3.5	3.0
10 ...	...	12.0	8.0	6.0	5.0	4.0	3.5	3.0
0 ...	...	12.5	8.5	6.5	5.5	4.5	3.5	3.0
Standard hardness testing, diamond indenter (Rockwell C, D, and A scales)								
80 ...	...	0.5	0.5	0.5	...	...	...	...
70 ...	...	1.0	1.0	0.5	0.5	0.5	...	...
60 ...	...	1.5	1.0	1.0	0.5	0.5	0.5	0.5
50 ...	...	2.5	2.0	1.5	1.0	1.0	0.5	0.5
40 ...	...	3.5	2.5	2.0	1.5	1.0	1.0	1.0
30 ...	...	5.0	3.5	2.5	2.0	1.5	1.5	1.0
20 ...	...	6.0	4.5	3.5	2.5	2.0	1.5	1.5
Superficial hardness testing, 1/16-in. (1.588-mm) ball indenter (Rockwell 15T, 30T, and 45T scales)								
90 ...	1.5	1.0	1.0	0.5	0.5	0.5	...	0.5
80 ...	3.0	2.0	1.5	1.5	1.0	1.0	...	0.5
70 ...	5.0	3.5	2.5	2.0	1.5	1.0	...	1.0
60 ...	6.5	4.5	3.0	2.5	2.0	1.5	...	1.5
50 ...	8.5	5.5	4.0	3.0	2.5	2.0	...	1.5
40 ...	10.0	6.5	4.5	3.5	3.0	2.5	...	2.0
30 ...	11.5	7.5	5.0	4.0	3.5	2.5	...	2.0
20 ...	13.0	9.0	6.0	4.5	3.5	3.0	...	2.0
Superficial hardness testing, diamond indenter (Rockwell 15N, 30N, and 45N scales)								
90 ...	0.5	0.5	...	...	...	...	...	...
85 ...	0.5	0.5	0.5	...	...	...	...	...
80 ...	1.0	0.5	0.5	0.5	...	...	...	...
75 ...	1.5	1.0	0.5	0.5	0.5	0.5	...	...
70 ...	2.0	1.0	1.0	0.5	0.5	0.5	...	0.5
65 ...	2.5	1.5	1.0	0.5	0.5	0.5	...	0.5
60 ...	3.0	1.5	1.0	1.0	0.5	0.5	...	0.5
55 ...	3.5	2.0	1.5	1.0	1.0	0.5	...	0.5
50 ...	3.5	2.0	1.5	1.0	1.0	1.0	...	0.5
45 ...	4.0	2.5	2.0	1.0	1.0	1.0	...	1.0
40 ...	4.5	3.0	2.0	1.5	1.0	1.0	...	1.0

Note: These correction factors are added to the dial-gage reading when hardness testing on the outer (convex) surface and subtracted when testing on the inner (concave) surface. The values are approximate only and represent the averages, to the nearest half Rockwell number, of numerous actual observations by different investigators, as well as mathematical analyses of the same problem. The accuracy of tests on cylindrical workpieces will be seriously affected by alignment of elevating screw, V-anvil, and indenters, and by surface finish and straightness of the cylinders.

area of the specimen under the indenter without contaminating the atmosphere or disturbing the temperature equilibrium. This arrangement permits several tests to be made on a single specimen while maintaining temperature and atmosphere.

In addition to modified Rockwell testers, hot hardness testers using a Vickers sapphire indenter with provisions for testing in either vacuum or inert atmospheres have also been described (Ref 1, 2). An extensive review of hardness data at elevated temperatures is presented in Ref 3. The development and design

of hot hardness testing furnaces is described in Ref 4.

Rockwell Testing of Specific Materials

Most homogeneous metals or alloys, including steels of all product forms and heat-treatment conditions and the various wrought and cast nonferrous alloys, can be accurately tested by one or more of the 30 indenter-load combinations listed in Tables 1 and 2. How-

ever, some nonhomogeneous materials and case-hardened materials present problems and therefore require special consideration.

Cast irons, because of graphite inclusions, usually show indentation values that are below the matrix value. For small castings or restricted areas in which a Brinell test is not feasible, tests may be made with either the Rockwell B or C scale. If the hardness range permits, however, the Rockwell E or K scale is preferred, because the 1/8-in. (3.175-mm) diameter ball provides a better average reading.

Fig. 10 Anvil support for cylindrical workpieces

(a) Correct method places the specimen centrally under indenter and prevents movement of the specimen under testing loads. (b) Incorrect method of supporting cylindrical work on spot anvil. The test piece is not firmly secured, and rolling of the specimen can cause damage to the indenter or erroneous readings.

(a)

(b)

Fig. 11 Setup for Rockwell hardness testing the inner surfaces of thin-walled cylindrical workpieces using a gooseneck adapter

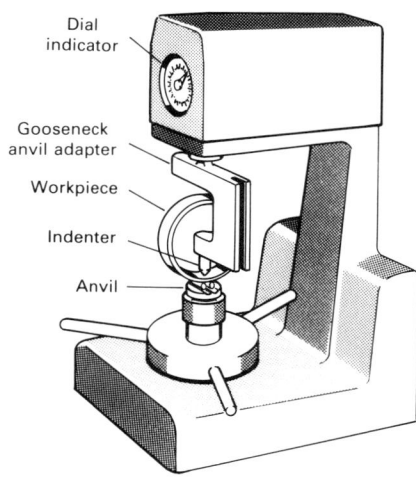

Dial indicator

Gooseneck anvil adapter

Workpiece

Indenter

Anvil

Fig. 12 Apparatus for Rockwell testing at temperatures up to 760 °C (1400 °F)

Powder metallurgy parts usually are tested on the Rockwell F, H, or B scale. Where possible, the Rockwell B scale should be used. In all instances, the result is apparent hardness, because of the voids present in the powder metallurgy parts. Therefore, indentation testing does not provide accurate results of matrix hardness, although it serves well as a quality control tool.

Cemented carbides are usually tested with the Rockwell A scale. If voids exist, the result is apparent hardness, and matrix evaluations are possible only by microhardness testing (for more information, see the article "Microhardness Testing" in this Section).

Case-Hardened Parts. For accuracy in testing case-hardened workpieces, the effective case depth should be at least ten times the indentation depth. Generally, cases are quite hard and require the use of a diamond indenter; thus, a choice of six scales exists, and the scale should be selected in accordance with the case depth.

If the case depth is not known, a skilled operator can, by using several different (sometimes only two) scales and making comparisons on a conversion table, determine certain case characteristics. For example, if a part shows a reading of 91 HR15N and 62 HRC, this indicates a case that is hard at the surface, as well as at an appreciable depth, because the equivalent of 62 HRC is 91 HR15N. However, if the reading shows 91 HR15N and only 55 HRC, this indicates that the indenter has broken through a relatively thin case. See the Appendix to this Section for hardness conversion tables.

Decarburization can be detected by the indentation hardness test, essentially by reversing the technique described above for obtaining an indication of case depth. Two indentation tests—one with the Rockwell 15N scale and another with the Rockwell C scale—should be performed. If the equivalent hardness is not obtained in converting from the Rockwell 15N to the Rockwell C scale, a decarburized layer is indicated. This technique is most effective for determining very thin layers of decarburization—0.1 mm (0.004 in.) or less. When decarburization is present, other methods such as microhardness testing should be used to determine its extent.

REFERENCES

1. F. Garofalo, P.R. Malenock, and G.V. Smith, Hardness of Various Steels at Elevated Temperatures, *Trans. ASM,* Vol 45, 1953, p 377-396
2. M. Semchyshen and C.S. Torgerson, Apparatus for Determining the Hardness of Metals at Temperatures up to 3000 °F, *Trans. ASM,* Vol 50, 1958, p 830-837
3. J.H. Westbrook, Temperature Dependence of the Hardness of Pure Metals, *Trans. ASM,* Vol 45, 1953, p 221-248
4. L. Small, "Hardness—Theory and Practice," Service Diamond Tool Co., Ferndale, MI, 1960, p 363-390

SELECTED REFERENCES

● Hardness Testing, in *Metals Handbook,* Vol 11, 8th ed., American Society for Metals, 1976, p 1-20
● V.E. Lysaght and A. DeBellis, "Hardness Testing Handbook," American Chain and Cable Co., 1969
● "Standard Test Methods for Rockwell Hardness and Rockwell Superficial Hardness of Metallic Materials," E 18, *Annual Book of ASTM Standards,* ASTM, Philadelphia, 1984

Brinell Hardness Testing

By Andrew R. Fee
Technical Consultant
Page-Wilson Corporation
Measurement Systems Division

Robert Segabache
Eastern Regional Sales Manager
Page-Wilson Corporation
Measurement Systems Division

and

Edward L. Tobolski
Engineering Manager
Page-Wilson Corporation
Measurement Systems Division

THE BRINELL TEST is a simple indentation test for determining the hardness of a wide variety of materials. The test consists of applying a constant load (force),* usually between 500 and 3000 kgf, for a specified time (10 to 30 s) using a 5- or 10-mm-diam hardened steel or tungsten carbide ball on the flat surface of a workpiece (Fig. 1a). The time period is required to ensure that plastic flow of the work metal has ceased. After removal of the load, the resultant recovered round impression is measured in millimeters using a low-power microscope (Fig. 1b).

Hardness is determined by taking the mean diameter of the indentation (two readings at right angles to each other) and calculating the Brinell hardness number (HB) by dividing the applied load by the surface area of the indentation according to the following formula:

$$HB = \frac{2P}{\pi D (D - \sqrt{D^2 - d^2})}$$

where P is load, kg; D is ball diameter, mm; and d is diameter of the indentation, mm.

It is not necessary to make the above calculation for each test. Calculations have already been made and are available in tabular form for various combinations of diameters of impressions and load. Table 1 lists Brinell hardness numbers for indentation diameters of 2.00 to 6.45 mm for 500-, 1000-, 1500-, 2000-, 2500-, and 3000-kgf loads.

Before using the Brinell test, several points must be considered. The size and shape of the workpiece must be capable of accommodating the relatively large indentation and heavy test loads. Because of this large indentation, some workpieces may not be usable after testing and others may require further machining. In addition, the maximum range of Brinell hardness values is 16 HB for very soft aluminum to 627 HB for hardened steels (approximately 60 HRC).

Indenter Selection and Geometry

The standard ball for Brinell hardness testing is 10.000 mm in diameter. ASTM E 10, "Standard Test Method for Brinell Hardness of Metallic Materials," specifies that the 10-mm ball indenter shall not deviate more than 0.005 mm in any diameter and have a hardness of at least 850 HV. When balls smaller than 10 mm in diameter are used, both the test load and ball size should be specifically stated in the test report. Balls differing in size from the standard 10-mm ball should conform to the requirements in ASTM E 10, as specified in Table 2.

Hardened steel balls can be used for testing material up to 444 HB (2.90-mm diam). Testing at higher hardness may cause appreciable error due to the possible flattening and permanent deformation of the steel ball. A special hardened and burnished steel ball called the Hultgren ball may be used up to 500 HB. Tungsten carbide ball indenters are recommended for hardnesses of 444 to 627 HB (2.90- to 2.45-mm indentation). However, slightly higher hardness values result when using carbide balls instead of steel balls because of the difference in elastic properties between these materials. Use of balls other than steel should be stated in the test report.

Load Selection

The standard loads used are 500, 1000, 1500, 2000, 2500, and 3000 kgf. The test load used is dependent mainly on size of impression, specimen thickness, and test surface. The 500-kgf load is usually used for testing relatively soft metals such as copper and aluminum alloys. The 3000-kgf load is most often used for testing harder materials such as steels and cast irons.

Size of Impression. It is recommended that the test load be of such magnitude that the diameter of the impression be in the range 2.50 to 6.00 mm (25.0 to 60.0% of ball diameter). Upper and lower limits of impression diameters are necessary, because the sensitivity of the test is reduced as impression size exceeds the above specified limits. In addition, the upper limit may be influenced by limitations of the travel of the indenter in certain types of testers. Other nonstandard lighter loads can be used as required on softer or thinner materials.

*"Force" is the technically correct term. However, "load" is commonly used by hardness tester manufacturers and users of hardness testing equipment.

Fig. 1 Brinell indentation process

(a) Schematic of the principle of the Brinell indentation process. (b) Brinell indentation with measuring scale in millimeters

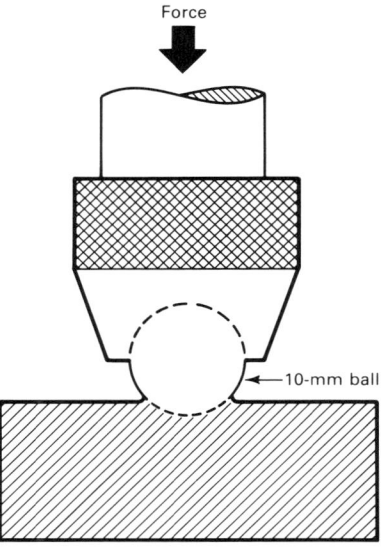

(a)

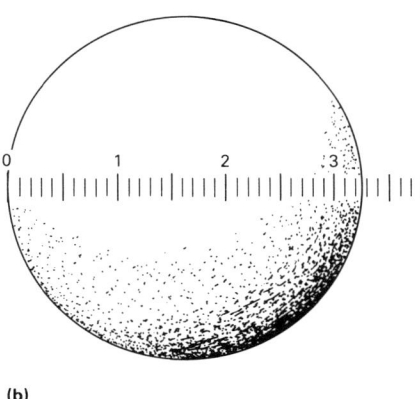

(b)

Indentation Measurement

The diameter of the indentation is measured to the nearest 0.01 mm by means of a specially designed microscope having a built-in millimeter scale. To eliminate error in the measurements due to slightly out-of-round impressions, two diameter measurements should be taken at 90° to each other. The Brinell hardness number is based on the average of these two measurements. Table 1 provides a simple way to convert the indentation diameter to the Brinell hardness number.

The indentations produced in Brinell hardness tests may exhibit different surface char-

acteristics, which have been carefully studied and analyzed. In some instances there is a ridge around the indentation that extends above the surface of the workpiece. In other instances the edge of the indentation is below the original surface. Sometimes there is no difference at all. The first phenomenon, called "ridging," is illustrated in Fig. 2(a). The second phenomenon, called "sinking," is illustrated in Fig. 2(b). An example of no difference is shown in Fig. 2(c). Cold worked metals and decarburized steels are those most likely to exhibit ridging. Fully annealed metals and light case hardened steels more often show sinking around the indentation.

The Brinell hardness number is related to the surface area of the indentation. This is obtained by measuring the diameter of the indentation, based on the assumption that it is the diameter with which the indenter was in actual contact. However, when either ridging or sinking is encountered there is always some doubt as to the exact part of the visible indentation with which the actual contact was made. When ridging is present the apparent diameter of the indentation is greater than the true value, whereas the reverse is true when sinking occurs.

Because of the above conditions, measurements of indentation diameters require experience and some judgment on the part of the operator. Experience can be gained by measuring calibration indents in the standardized test block.

Even when all precautions and limitations are observed, the Brinell indentations for some materials vary in shape. For example, materials that have been subjected to unidirectional cold working often exhibit extreme elliptical indentations. In such cases, where best possible accuracy is required, the indentation is measured in four directions approximately 45° apart, and the average of these four readings is used to determine the Brinell hardness number. Other techniques such as depth measurement are used with high-production equipment.

General Precautions and Limitations

To avoid misapplication and errors in Brinell hardness testing, the fundamentals and limitations of the test must be thoroughly understood. The following precautions should be observed before testing.

Thickness of the piece tested should be such that no bulge or other marking showing the effect of the load appears on the side of the piece opposite the impression. The thickness of the specimen should be at least ten times the depth of the indentation. Depth of

Fig. 2 Sectional views of Brinell indentations

(a) Ridging-type Brinell impression. (b) Sinking-type Brinell impression. (c) Flat-type Brinell impression

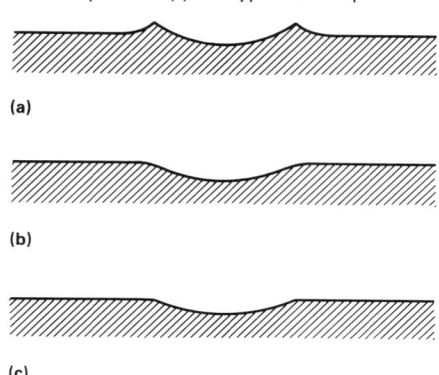

(a)

(b)

(c)

indentation may be calculated from the formula:

$$\text{Depth (mm)} = \frac{P}{\pi D \times (\text{HB})}$$

where P is load, kgf; D is ball diameter, mm; and HB is Brinell hardness number. For example, a reading of 300 HB indicates:

$$\text{Depth} = \frac{3000}{\pi D \times 300}$$

$$\text{Depth} = \frac{1}{\pi}$$

$$= 0.32 \text{ mm}$$

Minimum thickness $\cong$ 3.2 mm (0.125 in.)

Table 3 gives minimum thickness requirements.

Test surfaces that are flat give best results. Curved test surfaces of less than 25.4-mm (1-in.) radius should not be tested.

Spacing of Indentations. For accurate results, indentations must not be made near the edge of the workpiece. Lack of sufficient supporting material on one side will result in abnormally large, unsymmetrical indentations. In most instances the error in Brinell hardness number will not be significant if the distance from the center of the indentation to any edge of the workpiece is more than three times the diameter of the indentation.

Similarly, Brinell indentations must not be made too close to one another. The first indentation may cause cold working of the surrounding area that could affect the subsequent test if made within this affected region. It is generally agreed that the distance between centers of adjacent indentations should be at least three times the diameter of the indentation to eliminate significant errors.

Anviling. The part must be anviled properly to minimize workpiece movement dur-

Table 1 Brinell hardness numbers

Ball diameter 10 mm

Ball impression, diam, mm	Brinell hardness number — Load, kgf					
	500	1000	1500	2000	2500	3000
2.00	158	316	473	632	788	945
2.05	150	300	450	600	750	899
2.10	143	286	428	572	714	856
2.15	136	272	408	544	681	817
2.20	130	260	390	520	650	780
2.25	124	248	372	496	621	745
2.30	119	238	356	476	593	712
2.35	114	228	341	456	568	682
2.40	109	218	327	436	545	653
2.45	104	208	313	416	522	627
2.50	100	200	301	400	500	601
2.55	96.3	193	289	385	482	578
2.60	92.6	185	278	370	462	555
2.65	89.0	178	267	356	445	534
2.70	85.7	171	257	343	429	514
2.75	82.6	165	248	330	413	495
2.80	79.6	159	239	318	398	477
2.85	76.8	154	230	307	384	461
2.90	74.1	148	222	296	371	444
2.95	71.5	143	215	286	358	429
3.00	69.1	138	207	276	346	415
3.05	66.8	134	200	267	334	401
3.10	64.6	129	194	258	324	388
3.15	62.5	125	188	250	313	375
3.20	60.5	121	182	242	303	363
3.25	58.6	117	176	234	293	352
3.30	56.8	114	170	227	284	341
3.35	55.1	110	165	220	276	331
3.40	53.4	107	160	214	267	321
3.45	51.8	104	156	207	259	311
3.50	50.3	101	151	201	252	302
3.55	48.9	97.8	147	196	244	293
3.60	47.5	95.0	142	190	238	285
3.65	46.1	92.2	138	184	231	277
3.70	44.9	89.8	135	180	225	269
3.75	43.6	87.2	131	174	218	262
3.80	42.4	84.8	127	170	212	255
3.85	41.3	82.6	124	165	207	248
3.90	40.2	80.4	121	161	201	241
3.95	39.1	78.2	117	156	196	235
4.00	38.1	76.2	114	152	191	229
4.05	37.1	74.2	111	148	186	223
4.10	36.2	72.4	109	145	181	217
4.15	35.3	70.6	106	141	177	212
4.20	34.4	68.8	103	138	172	207
4.25	33.6	67.2	101	134	167	201
4.30	32.8	65.6	98.3	131	164	197
4.35	32.0	64.0	95.9	128	160	192
4.40	31.2	62.4	93.6	125	156	187
4.45	30.5	61.0	91.4	122	153	183
4.50	29.8	59.6	89.3	119	149	179
4.55	29.1	58.2	87.2	116	145	174
4.60	28.4	56.8	85.2	114	142	170
4.65	27.8	55.6	83.3	111	139	167
4.70	27.1	54.2	81.4	108	136	163
4.75	26.5	53.0	79.6	106	133	159
4.80	25.9	51.8	77.8	104	130	156
4.85	25.4	50.8	76.1	102	127	152
4.90	24.8	49.6	74.4	99.2	124	149
4.95	24.3	48.6	72.8	97.2	122	146
5.00	23.8	47.6	71.3	95.2	119	143
5.05	23.3	46.6	69.8	93.2	117	140
5.10	22.8	45.6	68.3	91.2	114	137
5.15	22.3	44.6	66.9	89.2	112	134
5.20	21.8	43.6	65.5	87.2	109	131
5.25	21.4	42.8	64.1	85.6	107	128
5.30	20.9	41.8	62.8	83.6	105	126
5.35	20.5	41.0	61.5	82.0	103	123
5.40	20.1	40.2	60.3	80.4	101	121
5.45	19.7	39.4	59.1	78.8	98.5	118
5.50	19.3	38.6	57.9	77.2	96.5	116
5.55	18.9	37.8	56.8	75.6	95.0	114
5.60	18.6	37.2	55.7	74.4	92.5	111
5.65	18.2	36.4	54.6	72.8	90.8	109
5.70	17.8	35.6	53.5	71.2	89.2	107
5.75	17.5	35.0	52.5	70.0	87.5	105
5.80	17.2	34.4	51.5	68.8	85.8	103
5.85	16.8	33.6	50.5	67.2	84.2	101
5.90	16.5	33.0	49.6	66.0	82.5	99.2
5.95	16.2	32.4	48.7	64.8	81.2	97.3
6.00	15.9	31.8	47.7	63.6	79.5	95.5
6.05	15.6	31.2	46.8	62.4	78.0	93.7
6.10	15.3	30.6	46.0	61.2	76.7	92.0
6.15	15.1	30.2	45.2	60.4	75.3	90.3
6.20	14.8	29.6	44.3	59.2	73.8	88.7
6.25	14.5	29.0	43.5	58.0	72.6	87.1
6.30	14.2	28.4	42.7	56.8	71.3	85.5
6.35	14.0	28.0	42.0	56.0	70.0	84.0
6.40	13.7	27.4	41.2	54.8	68.8	82.5
6.45	13.5	27.0	40.5	54.0	67.5	81.0

Table 2 Brinell hardness balls other than standard

Ball diameter, mm	Tolerance (a), mm
1 to 3, inclusive	±0.0035
More than 3 to 6, inclusive	±0.004
More than 6 to 10, inclusive	±0.0045

(a) Steel balls for ball bearings normally satisfy these tolerances.

Table 3 Minimum thickness requirements for Brinell hardness tests

Minimum thickness of specimen mm	in.	Minimum hardness for which the Brinell test may be made safely — 3000-kgf load	1500-kgf load	500-kgf load
1.6	0.0625	602	301	100
3.2	0.125	301	150	50
4.8	0.1875	201	100	33
6.4	0.250	150	75	25
8.0	0.3125	120	60	20
9.6	0.375	100	50	17

ing the test and to position the test surface perpendicular to the test force within 2°.

Surface Finish. The degree of accuracy attainable by the Brinell test can be greatly influenced by the surface finish of the workpiece. The surface of the workpiece should be milled, ground, or polished, so that the indentation is defined clearly enough to permit accurate measurement. Care should be taken to avoid overheating or cold working the surface, as that may affect the hardness of the material.

In addition, for accurate results, the workpiece surface must be representative of the material. Surface decarburization or any form of superficial hardening must be removed prior to testing.

Testing Machines

Various kinds of testers are available for laboratory, production, automatic, and portable testing. These testers commonly use deadweight, hydraulic, pneumatic, or elastic members (i.e., springs) to apply the test loads. All testers must have a rigid frame to maintain the load and a means of controlling the rate of load application to avoid errors due to impact (500 kg/s maximum). The loads must be consistently applied within ±1.0% for laboratory testers and ±2.0% for production testing as indicated in ASTM E 10. In addition, the load should be applied so that the direction of load (force) is perpendicular to the workpiece surface within 2°.

Laboratory Testers. Because of their high degree of accuracy, deadweight testers are most commonly used in laboratories and shops that do low- to medium-rate production. These units are constructed with weights connected mechanically to the Brinell ball indenter. Minimum maintenance is required because there are few moving parts. The handwheel is turned to raise the elevating screw until the specimen nearly touches the ball. Operation of the handle allows the weights to apply a test load to the penetrator. A dashpot and flow control valve are provided to control the rate of load application. A typical deadweight tester is illustrated in Fig. 3.

Machines for Production Testing. Hydraulic testers were developed to reduce testing time and operator fatigue. A typical hydraulic tester is shown in Fig. 4. Other advantages of this machine are operating economy, simplicity of controls, and dependable accuracy. The controls prevent the operator from applying the load too quickly and thus overloading. The load is applied by a hydraulic cylinder and monitored by a pres-

Fig. 3 Deadweight Brinell hardness tester

Courtesy of Page-Wilson Corporation

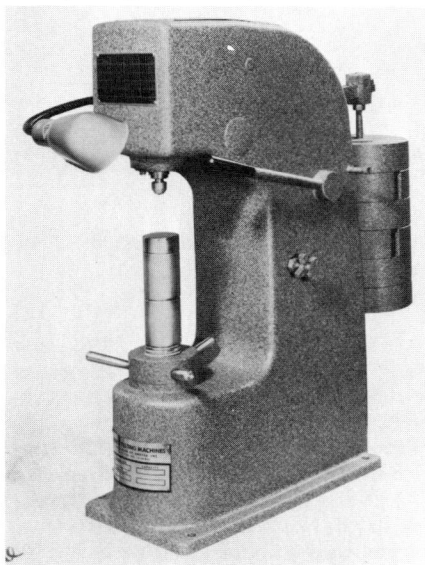

Fig. 4 Hydraulic Brinell hardness tester

Courtesy of Page-Wilson Corporation

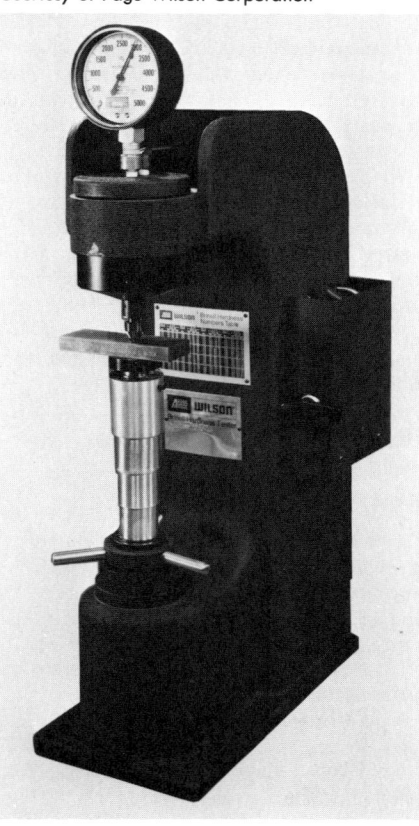

sure gage. Normally the pressure can be adjusted to apply any load between 500 and 3000 kgf.

Automatic Testers. Many types of automatic Brinell testers are currently available. Most of these testers use a depth-measurement system to eliminate the time-consuming and operator-biased measurement of the diameters. All of these testers use a preliminary load (similar to the Rockwell principle) in conjunction with the standard Brinell loads. Simple versions of this technique provide only comparative "go/no go" hardness indications; more sophisticated models offer a microprocessor-controlled digital readout to convert the depth measurement to Brinell numbers. Conversion from depth to diameter frequently varies for different materials and may require correlation studies to establish the proper relationship.

These units can be fully automated to obtain production rates up to 600 test per hour and can be incorporated into in-line production equipment. The high-speed automatic tester complies with ASTM E 103, "Standard Method of Rapid Indentation Hardness Testing of Metallic Materials."

Fig. 5 Hydraulic, manually operated portable Brinell hardness tester

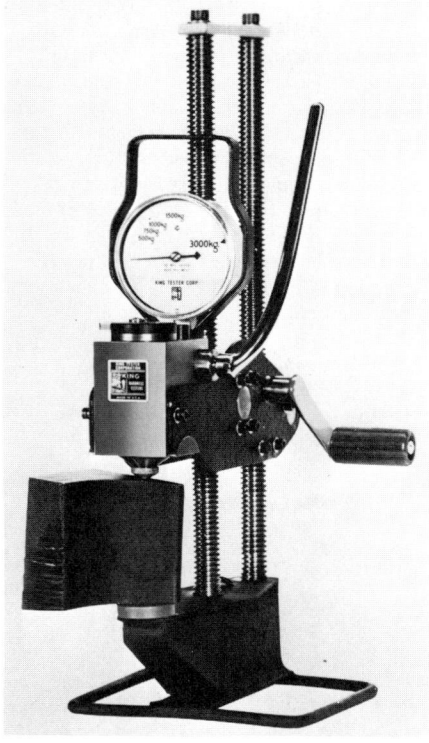

Portable Testing Machines. The use of conventional hardness testers may occasionally be limited because the work must be brought to the machine, and because the workpieces must be placed between the anvil and the indenter. Portable Brinell testers that circumvent these limitations are available. A typical portable instrument is shown in Fig. 5. This type of tester weighs only about 11.4 kg (25 lb), so it can be easily transported to the workpieces. Portable testers can accommodate a wider variety of workpieces than can the stationary types. For instance, Fig. 5 shows how a portable tester is used to test a workpiece. The tester attaches to the workpiece like a C-clamp with the anvil on one side of the workpiece and the indenter on the other. For very large parts an encircling chain is used to hold the tester in place as pressure is applied.

Portable testers generally apply the load hydraulically, employing a spring-loaded relief valve. The load is applied by operating the hydraulic pump until the relief valve opens momentarily. With this type of tester, the hydraulic pressure should be applied three times when testing steel with a 3000-kgf load. This is equivalent to a holding time of 15 s, as required by the more conventional method. For other materials and loads, comparison tests should be made to determine the number of load applications required to give results equivalent to the conventional method. A comparison-type tester that uses a calibrated shear pin is shown in Fig. 6. In this method, a small pin of a known shear load is placed in the indenter assembly against the indenter (Fig. 6b). Through hammer impact load or static clamping load the indenter is forced into the material only as far as it takes to shear the pin. Excessive force is absorbed after shear by upward movement of the indenter into an empty cavity. The resulting impression is measured by the conventional Brinell method.

Fig. 6 Pin Brinell hardness tester

(a) Clamp loading tester. (b) Schematic of pin Brinell principle

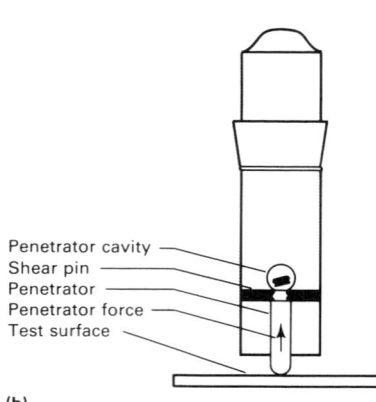

Penetrator cavity
Shear pin
Penetrator
Penetrator force
Test surface

(b)

Equipment Maintenance. To maintain accurate results from Brinell testing, equipment must be calibrated and serviced regularly, especially when machines are exposed to shop environments. The frequency of servicing depends on whether the testers are used in a production line or for making an occasional test. However, it is important that they be serviced and calibrated on a regular basis. Regular checking of the ball indenter for deformation is particularly important. Indenters are susceptible to wear as well as to damage. When an indenter becomes worn or damaged so that indentations no longer meet the standards, it must be replaced. Under no circumstance should attempts be made to compensate for a worn or damaged indenter.

Verification of Loads, Indenters, and Microscopes

As with any procedure that is dependent on several components, the accuracy of each must be verified to determine the accuracy of the result. In the case of Brinell hardness testing, load, indenter, and microscope accuracies must lie within a specified tolerance to ensure accurate results.

Load Verification. ASTM E 10 specifies that a Brinell hardness tester used to make laboratory or referee tests is acceptable for use in the loading range within which the tester error does not exceed ±1%. A tester used for routine or production testing is acceptable for use over a loading range within which the error does not exceed ±2%.

Test loads should be checked by periodic calibration with a proving ring, load cell, or a weight-and-lever system whose accuracy is traceable to the National Bureau of Standards. The proving ring (see Fig. 7) is an elastic calibration device that is placed on the anvil of the tester. The deflection of the ring under the applied load is measured either by a micrometer screw and a vibrating reed or a reading dial gage. The amount of elastic deflection is then converted into load in kilograms and compared with required accuracies.

Ball Indenter Verification. The ball indenter should be measured to an accuracy of ±0.0005 mm at not less than three positions. The mean of these readings should not differ from the nominal diameter by more than the tolerance specified in Table 2.

Microscope Verification. The measuring microscope or other device used for measuring the diameter of the impression should be verified at five intervals over the working range by the use of a scale of known accuracy such as a stage micrometer. The adjustment

Fig. 7 Proving ring used for calibrating Brinell hardness testers

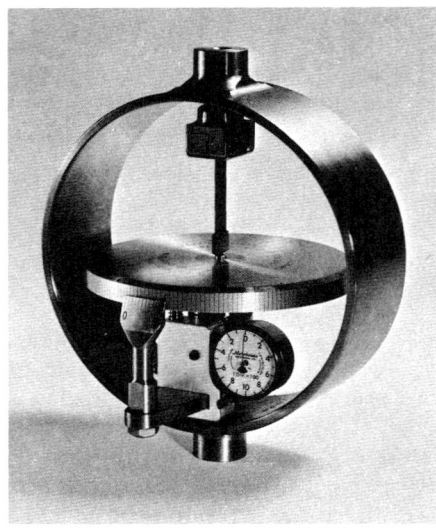

of the micrometer microscope should be such that, throughout the range covered, the difference between the scale divisions of the microscope and of the calibrating scale does not exceed 0.01 mm.

Verification by Test Block. Standardized Brinell test blocks are available so that the accuracy of the Brinell hardness tester can be verified at the hardness level of the work being tested. Commonly available ranges are:

Steel	500 HB
	400 HB
	350 HB
	300 HB
	250 HB
	200 HB
Brass	90 HB
Aluminum	140 HB

Good practice is to verify the tester throughout the hardness range encountered. This ensures that all test parameters are within tolerance.

Application of Brinell Testing to Specific Materials

As is true for other indentation methods of testing hardness, the most accurate results are obtained when testing homogeneous materials, regardless of the hardness range.

Steels. Virtually all hardened and tempered or annealed steels within the range of hardness mentioned provide accurate results with the Brinell test. However, as a rule, case hardened steels are totally unsuitable for

Brinell testing. In most instances, the surface hardness is above the practical range and is rarely thick enough to provide the required support for a Brinell test. Thus, "cave in" results, and grossly inaccurate readings are obtained.

Cast Irons. The large area of the test serves to average out the hardness difference between the iron and graphite particles present in most cast irons. This averaging effect allows the Brinell test to serve as an excellent quality-control tool.

Nonferrous metals (especially the wrought types) are generally amenable to Brinell testing, usually with the 500-kgf load, but occasionally with the 1500-kgf load. Some high-strength alloys such as titanium-based and nickel-based alloys that are phase transformation or age hardened can utilize the 3000-kgf load. In this situation, practical limits must be observed and some testing may be required to establish the optimum technique for testing a specific metal or alloy.

There are sometimes certain multiphase cast nonferrous alloys that are simply too soft for accurate Brinell testing. Microhardness testing is then employed. The lower limit of 16 HB with a 500-kgf load must always be observed.

Powder Metallurgy Parts. Testing of P/M parts with a Brinell tester (or any sort of macrohardness tester) involves the same problem as encountered with cast iron. Instead of a soft graphite phase (some P/M parts also contain free graphite), P/M parts contain voids that may vary widely in size and number. Light-load Brinell testing is sometimes used successfully for testing of P/M parts, but its only real value is as a quality-control tool.

SELECTED REFERENCES

- Hardness Testing, in *Metals Handbook*, Vol 11, 8th ed., American Society for Metals, 1976, p 1-20

- V.E. Lysaght and A. DeBellis, *Hardness Testing Handbook*, American Chain and Cable Co., Bridgeport, CT, 1969

- L. Small, *Hardness: Theory and Practice*, Service Diamond Tool Co., Ferndale, MI, 1960

- "Standard Method of Rapid Indentation Hardness Testing of Metallic Materials," E 103, *Annual Book of ASTM Standards*, Vol 03.01, ASTM, Philadelphia, 1984, p 264-267

- "Standard Test Method for Brinell Hardness of Metallic Materials," E 10, *Annual Book of ASTM Standards*, Vol 03.01, ASTM, Philadelphia, 1984, p 164-169

Microhardness Testing

MICROHARDNESS TESTING can be defined as indentation hardness testing that involves forcing a diamond indenter of specific geometry into the surface of the test material at loads (forces)* ranging from 1 to 1000 gf.

In Knoop and Vickers microhardness testing, the hardness value is determined by measuring the size of the resulting unrecovered indentation by using a microscope and established formulas or look-up tables in accordance with ASTM E 384 (Ref 1). In ultrasonic hardness testing, the measured change in the frequency of a vibrating, diamond-tipped rod applied to the specimen indicates hardness. No microscopic examination is performed. All three microhardness test methods produce an indentation depth of less than 19 μm.

Knoop and Vickers Microhardness Testing

By Andrew R. Fee
Technical Consultant
Page-Wilson Corporation
Measurement Systems Division

Robert Segabache
Eastern Regional Sales Manager
Page-Wilson Corporation
Measurement Systems Division

and

Edward L. Tobolski
Engineering Manager
Page-Wilson Corporation
Measurement Systems Division

THE DEVELOPMENT of the Knoop test by the National Bureau of Standards in 1939 and the Vickers test (also called the diamond-pyramid hardness test) in England in 1925 has made microhardness testing a routine laboratory procedure. Both of the tests use precisely shaped diamond indenters and various load ranges to determine the hardness of a variety of materials. Specific applications for microhardness testing include:

- Measuring hardness of precision workpieces that are too small to be measured by conventional macroscopic hardness testing methods
- Measuring hardness of product forms such as foil or wire that are too thin or too small in diameter to be measured by conventional macroscopic methods
- Monitoring of carburizing or nitriding operations, which is usually accomplished by hardness surveys taken on cross sections of test pieces that accompanied the workpieces through production operations
- Measuring hardness of individual microconstituents
- Measuring hardness close to edges, thus detecting undesirable surface conditions such as grinding burn and decarburization
- Measuring hardness of surface layers such as plating or bonded layers

Indenter Selection

Figure 1 compares indentations made in the same material with Knoop and Vickers indenters under loads of 3000, 1000, 500, and 100 gf. For a given load, the Vickers indenter penetrates about twice as far into the specimen as the Knoop indenter, and its diagonal is about one third the length of the Knoop indentation. Therefore, the Vickers test is less sensitive to minute differences in surface conditions than the Knoop test.

The Knoop indenter (Fig. 2) is a highly polished, rhombic-based pyramidal diamond that produces a diamond-shaped indentation with a ratio between long and short diagonals of about 7 to 1. The pyramid shape used has an included longitudinal angle of 172° 30′ and an included transverse angle of 130° 0′. The depth of the indentation is about ⅓₀th of its length.

The Knoop hardness number (HK) is the ratio of the load applied to the indenter to the unrecovered projected area:

$$HK = \frac{P}{A} = \frac{P}{CL^2}$$

where P is the applied load, kgf; A is the unrecovered projected area of indentation, mm^2; L is the measured length of long diagonal, mm; and C (0.07028) is a constant for the indenter relating projected area of the indentation to the square of the length of the long diagonal.

The geometry of the Knoop indenter also facilitates testing of hard, brittle materials, such as glass. Minimal deformation of the material occurs at the top and bottom of the short diagonal.

The Vickers indenter (Fig. 3) is a highly polished, pointed, square-based pyramidal diamond with face angles of 136°. With the Vickers indenter, the depth of indentation is about one seventh of the diagonal length.

The Vickers hardness number (HV) is the ratio of the load applied to the indenter to the surface area of the indentation:

$$HV = \frac{2P \sin\left(\frac{\theta}{2}\right)}{D^2}$$

where P is the applied load, kgf; D is the mean diagonal of the indentation, mm; and θ

Fig. 1 Indentations made by Knoop and Vickers indenters in the same work metal at the same load

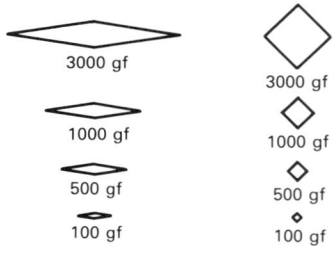

3000 gf   3000 gf

1000 gf   1000 gf

500 gf   500 gf

100 gf   100 gf

Knoop indentations   Vickers indentations

Fig. 2 Pyramidal Knoop indenter and resulting indentation in the workpiece

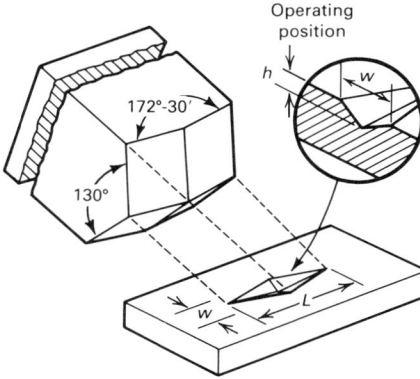

is the angle between opposite faces of the diamond, 136°.

The Vickers hardness scale is unique in that it extends beyond the microhardness range to the macrohardness range, using the same indenter. Vickers hardness tests are made at test loads up to 120 kgf, which is comparable to a Rockwell C scale test (150 kgf). The wide range of loads allows the Vickers test to be used on virtually any material.

Measuring the Indentation

A microscope, in conjunction with the hardness testing unit, is used to determine the size of the indentation. For maximum resolution, provisions should be made for the adjustment of the illumination intensity and concentration and for aligning and adjusting the aperture diaphragm and the field diaphragm of the microscope. Proper magnification is important for accurate measurement. Magnification selection for the corresponding indentation length is as follows:

Indentation length, μm	Magnification max	min
Less than 76	· · ·	400
76 to 125	800	300
Greater than 125	600	200

Whether made by the Knoop or the Vickers indenter, the ends of the indentation's diagonals must be brought into sharp focus. With the Knoop indenter, one leg of the long diagonal should never be more than 20% longer than the other. If one diagonal is abnormally long or not in focus, the surface of the workpiece should be checked to ensure that it is perpendicular to the axis of the indenter.

With the Vickers indenter, both diagonals are measured and the average is used for cal-

Fig. 3 Diamond pyramid indenter used for the Vickers test and resulting indentation in the workpiece

D is the mean diagonal of the indentation in millimeters.

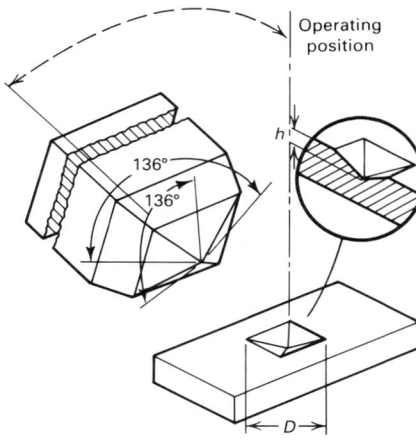

culating the HV value. With anisotropic materials, such as those that have been heavily cold worked, there may be a difference in the lengths of the two diagonals of the indentation. The test specimen should be oriented so that the diagonals of a new indentation are of approximately equal length.

Determining the Hardness Number. The indentation (Knoop or Vickers) is measured in filar units of the measuring eyepiece. Some eyepiece-objective combinations read directly in microns, while others require multiplication of filar units by an objective calibration factor to convert the reading to microns.

Conversion to hardness numbers is simple. Table 1 gives the Knoop hardness numbers for indentations of 1 to 200 μm for a load of 1 gf. To determine a hardness number, multiply the micron reading by the load in gram-force. For example, for a diagonal reading of 100 μm using a load of 500 gf, the 1-gf reading (1.423) yields:

$$HK = 500 \times 1.423 = 711$$

For Table 2 (Vickers hardness numbers), assume an average diagonal reading of 40 μm. For a 1-gf load, the HV number is 1.159. Assuming that the load is 500 gf:

$$HV = 500 \times 1.159 = 579$$

The hardness value for either Vickers or Knoop is a stress value and is expressed as kilograms per square millimeter (Kg/mm^2).

Many hardness tester manufacturers offer a microprocessor-controlled, digital read-out display that automatically converts from filar

units to Knoop or Vickers hardness values. This feature reduces testing time and eliminates operator calculation errors.

Microhardness Testing Apparatus

Several types of microhardness testers are available. Most testers can accommodate either the Knoop or Vickers indenters and operate through direct application of load by dead weight or by weights and a lever. Testers vary primarily in load range capabilities and are usually bench-mounted. Some floor-mounted models are available.

Bench-mounted testers, such as the widely used Tukon tester (Fig. 4), feature an automatic test cycle (load, application, time of application, and removal of the test load), which ensures standardization of the test. The automatic test cycle is particularly critical for test loads below 500 gf. At such low testing loads, manual removal of the load can cause vibrations in the tester that can seriously affect the accuracy and repeatability of the test.

The load of the tester shown in Fig. 4 ranges from 1 to 10 000 gf. Loads from 1 to 100 gf are applied directly to the top of the indenter; loads greater than 100 gf are ap-

Fig. 4 Bench-mounted microhardness tester with optional digital-readout package
Courtesy of Page-Wilson Corporation

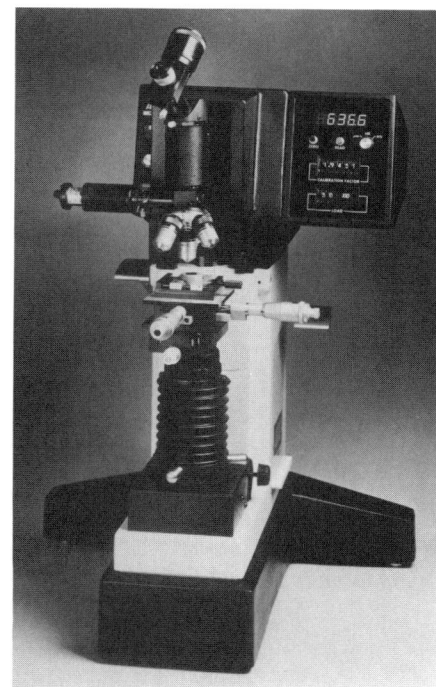

plied by a lever arrangement. Some micro-hardness testers are designed to apply loads as high as 50 000 gf. Such loads are helpful in relating microhardness to macrohardness.

An integral component of the testing apparatus is the movable stage that supports the workpiece. In many applications, the indentation must be in a select area, usually limited to a few thousandths of a square millimeter. With a bench-mounted tester, the required area is first located with the microscope. The stage is then moved under the indenter, and the indentation is made. The specimen is returned to its original location under the microscope, so that the dimensions of the indentation can be determined. The stage permits the specimen to be moved with such precision that the indentation, often invisible to the human eye, is positioned in the exact center of the field.

The stage also has two micrometer screws at right angles to each other, so that a row or even a network pattern of indentations can be made with very accurate spacing. This capability is important in applications such as case depth determinations.

Optical Equipment. Testers commonly have a microscope with three objectives to provide magnifications of 150×, 300×, and 600×. Optical equipment used in microhardness testing must focus on both ends (or on diagonally opposite corners) of the indentation at the same time and must be rigid and free from effects of vibration.

Dry objectives with high numerical aperture and resolving power generally are used. For these lenses, an accuracy of ±0.5 μm has been established. If oil-immersion lenses are used, the degree of accuracy in measuring can be increased to ±0.3 μm.

Calibration of the optical system must be precise, using a stage micrometer with a high

Table 1 Knoop hardness numbers for a test load of 1 gf

Diagonal of indentation, μm	Knoop hardness number for diagonal measured to 0.1 μm									
	.0	.1	.2	.3	.4	.5	.6	.7	.8	.9
1.	14 230.0	11 760.0	9 881.0	8 420.0	7 260.0	6 324.0	5 558.0	4 924.0	4 392.0	3 942.0
2.	3 557.0	3 227.0	2 940.0	2 690.0	2 470.0	2 277.0	2 105.0	1 952.0	1 815.0	1 692.0
3.	1 581.0	1 481.0	1 390.0	1 307.0	1 231.0	1 162.0	1 098.0	1 039.0	985.4	935.5
4.	889.3	846.5	806.6	769.5	735.0	702.7	672.4	644.1	617.6	592.6
5.	569.2	547.1	526.2	506.5	488.0	470.4	453.7	437.9	423.0	408.8
6.	395.2	382.4	370.2	358.5	347.4	336.8	326.7	317.0	307.7	298.9
7.	290.4	282.3	274.5	267.0	259.8	253.0	246.3	240.0	233.9	228.0
8.	222.3	216.9	211.6	206.5	201.7	196.9	192.4	188.0	183.7	179.6
9.	175.7	171.8	168.1	164.5	161.0	157.7	154.4	151.2	148.2	145.2
10.	142.3	139.5	136.8	134.1	131.6	129.1	126.6	124.3	122.0	119.8
11.	117.6	115.5	113.4	111.4	109.5	107.6	105.7	103.9	102.2	100.5
12.	98.81	97.19	95.60	94.05	92.54	91.07	89.63	88.22	86.85	85.51
13.	84.20	82.91	81.66	80.44	79.24	78.07	76.93	75.81	74.72	73.65
14.	72.60	71.57	70.57	69.58	68.62	67.68	66.75	65.85	64.96	64.09
15.	63.24	62.40	61.59	60.78	60.00	59.23	58.47	57.73	57.00	56.28
16.	55.58	54.89	54.22	53.55	52.90	52.26	51.64	51.02	50.41	49.82
17.	49.24	48.66	48.10	47.54	47.00	46.46	45.94	45.42	44.91	44.41
18.	43.92	43.43	42.96	42.49	42.03	41.57	41.13	40.69	40.26	39.83
19.	39.42	39.00	38.60	38.20	37.81	37.42	37.04	36.66	36.29	35.93
20.	35.57	35.22	34.87	34.53	34.19	33.86	33.53	33.21	32.89	32.57
21.	32.27	31.96	31.66	31.36	31.07	30.78	30.50	30.22	29.94	29.67
22.	29.40	29.13	28.87	28.61	28.36	28.11	27.86	27.61	27.37	27.13
23.	26.90	26.67	26.44	26.21	25.99	25.77	25.55	25.33	25.12	24.91
24.	24.70	24.50	24.30	24.10	23.90	23.71	23.51	23.32	23.14	22.95
25.	22.77	22.59	22.41	22.23	22.05	21.88	21.71	21.54	21.38	21.21
26.	21.05	20.89	20.73	20.57	20.42	20.26	20.11	19.96	19.81	19.66
27.	19.52	19.37	19.23	19.09	18.95	18.82	18.68	18.54	18.41	18.28
28.	18.15	18.02	17.89	17.77	17.64	17.52	17.40	17.27	17.15	17.04
29.	16.92	16.80	16.69	16.57	16.46	16.35	16.24	16.13	16.02	15.92
30.	15.81	15.71	15.60	15.50	15.40	15.30	15.20	15.10	15.00	14.90
31.	14.81	14.71	14.62	14.52	14.43	14.34	14.25	14.16	14.07	13.98
32.	13.90	13.81	13.72	13.64	13.55	13.47	13.39	13.31	13.23	13.15
33.	13.07	12.99	12.91	12.83	12.75	12.68	12.60	12.53	12.45	12.38
34.	12.31	12.24	12.17	12.09	12.02	11.95	11.89	11.82	11.75	11.68
35.	11.62	11.55	11.48	11.42	11.35	11.29	11.23	11.16	11.10	11.04
36.	10.98	10.92	10.86	10.80	10.74	10.68	10.62	10.56	10.51	10.45
37.	10.39	10.34	10.28	10.23	10.17	10.12	10.06	10.01	9.958	9.906
38.	9.854	9.802	9.751	9.700	9.650	9.600	9.550	9.501	9.452	9.403
39.	9.355	9.307	9.260	9.213	9.166	9.120	9.074	9.028	8.983	8.938
40.	8.893	8.849	8.805	8.761	8.718	8.675	8.632	8.590	8.548	8.506
41.	8.465	8.423	8.383	8.342	8.302	8.262	8.222	8.183	8.144	8.105
42.	8.066	8.028	7.990	7.952	7.915	7.878	7.841	7.804	7.768	7.731
43.	7.695	7.660	7.624	7.589	7.554	7.520	7.485	7.451	7.417	7.383
44.	7.350	7.316	7.283	7.250	7.218	7.185	7.153	7.121	7.090	7.058
45.	7.027	6.996	6.965	6.934	6.903	6.873	6.843	6.813	6.783	6.754
46.	6.724	6.695	6.666	6.638	6.609	6.581	6.552	6.524	6.497	6.469
47.	6.441	6.414	6.387	6.360	6.333	6.306	6.280	6.254	6.228	6.202
48.	6.176	6.150	6.125	6.099	6.074	6.049	6.024	6.000	5.975	5.951
49.	5.926	5.902	5.878	5.854	5.831	5.807	5.784	5.761	5.737	5.714
50.	5.692	5.669	5.646	5.624	5.602	5.579	5.557	5.536	5.514	5.492

(continued)

Note: To obtain the hardness number from this table, read the HK (1 gf) corresponding to the measured diagonal length in microns and multiply by the test load in gram-force.

degree of accuracy. The entire optical system, with the filar micrometer eyepiece locked in place as it would be during actual measurement, must be calibrated. The stage micrometer should be accurate to at least ±0.5 μm, or ±0.05% of any interval.

Surface Preparation

To measure accurately a Knoop or a Vickers indentation, it must be clearly defined. Therefore, requirements for surface finish are stringent. These requirements become increasingly stringent as the load decreases.

When the load is 100 gf or less, a metallographic finish is mandatory. When grinding, polishing, or both operations are necessary in specimen preparation, care should be taken to minimize heating and distortion of the specimen surface. Polishing should be performed according to the procedures outlined in ASTM E 3, "Standard Methods for Preparation of Metallographic Specimens."

When the specimen to be tested for microhardness will also be used for metallographic examination, mounting (usually in plastic) and polishing are justified. In other instances, only polishing is required.

When mounting is not necessary, fixtures may be used for holding the specimens or workpieces. Typical fixture arrangements are shown in Fig. 5. Most workpieces can be adapted to one of these fixtures. Fixtures must maintain a rigid surface perpendicular to the indenter. The holding and polishing vise shown in Fig. 5(b) can reduce preparation time, because the specimen can be polished and tested without removing it from the vise. The turntable vise fixture shown in Fig. 5(f) is convenient for holding mounted specimens.

Table 1 (continued)

Diagonal of indentation, μm	Knoop hardness number for diagonal measured to 0.1 μm									
	.0	.1	.2	.3	.4	.5	.6	.7	.8	.9
51.	5.471	5.449	5.428	5.407	5.386	5.365	5.344	5.323	5.303	5.282
52.	5.262	5.242	5.222	5.202	5.182	5.162	5.143	5.123	5.104	5.085
53.	5.065	5.046	5.027	5.009	4.990	4.971	4.953	4.934	4.916	4.898
54.	4.880	4.862	4.844	4.826	4.808	4.790	4.773	4.756	4.738	4.721
55.	4.704	4.687	4.670	4.653	4.636	4.619	4.603	4.586	4.570	4.554
56.	4.537	4.521	4.505	4.489	4.473	4.457	4.442	4.426	4.410	4.395
57.	4.379	4.364	4.349	4.334	4.319	4.304	4.289	4.274	4.259	4.244
58.	4.230	4.215	4.201	4.186	4.172	4.158	4.144	4.129	4.115	4.102
59.	4.088	4.074	4.060	4.046	4.033	4.019	4.006	3.992	3.979	3.966
60.	3.952	3.939	3.926	3.913	3.900	3.887	3.875	3.862	3.849	3.837
61.	3.824	3.811	3.799	3.787	3.774	3.762	3.750	3.738	3.726	3.714
62.	3.702	3.690	3.678	3.666	3.654	3.643	3.631	3.619	3.608	3.596
63.	3.585	3.574	3.562	3.551	3.540	3.529	3.518	3.507	3.496	3.485
64.	3.474	3.463	3.452	3.442	3.431	3.420	3.410	3.399	3.389	3.378
65.	3.368	3.357	3.347	3.337	3.327	3.317	3.306	3.296	3.286	3.276
66.	3.267	3.257	3.247	3.237	3.227	3.218	3.208	3.198	3.189	3.179
67.	3.170	3.160	3.151	3.142	3.132	3.123	3.114	3.105	3.095	3.086
68.	3.077	3.068	3.059	3.050	3.041	3.032	3.024	3.015	3.006	2.997
69.	2.989	2.980	2.971	2.963	2.954	2.946	2.937	2.929	2.921	2.912
70.	2.904	2.896	2.887	2.879	2.871	2.863	2.855	2.847	2.839	2.831
71.	2.823	2.815	2.807	2.799	2.791	2.783	2.776	2.768	2.760	2.752
72.	2.745	2.737	2.730	2.722	2.715	2.707	2.700	2.692	2.685	2.677
73.	2.670	2.663	2.656	2.648	2.641	2.634	2.627	2.620	2.613	2.605
74.	2.598	2.591	2.584	2.577	2.571	2.564	2.557	2.550	2.543	2.536
75.	2.530	2.523	2.516	2.509	2.503	2.496	2.490	2.483	2.476	2.470
76.	2.463	2.457	2.451	2.444	2.438	2.431	2.425	2.419	2.412	2.406
77.	2.400	2.394	2.387	2.381	2.375	2.369	2.363	2.357	2.351	2.345
78.	2.339	2.333	2.327	2.321	2.315	2.309	2.303	2.297	2.292	2.286
79.	2.280	2.274	2.268	2.263	2.257	2.251	2.246	2.240	2.234	2.229
80.	2.223	2.218	2.212	2.207	2.201	2.196	2.190	2.185	2.179	2.174
81.	2.169	2.163	2.158	2.153	2.147	2.142	2.137	2.132	2.127	2.121
82.	2.116	2.111	2.106	2.101	2.096	2.091	2.086	2.080	2.075	2.070
83.	2.065	2.060	2.056	2.051	2.046	2.041	2.036	2.031	2.026	2.021
84.	2.017	2.012	2.007	2.002	1.998	1.993	1.988	1.983	1.979	1.974
85.	1.969	1.965	1.960	1.956	1.951	1.946	1.942	1.937	1.933	1.928
86.	1.924	1.919	1.915	1.911	1.906	1.902	1.897	1.893	1.889	1.884
87.	1.880	1.876	1.871	1.867	1.863	1.858	1.854	1.850	1.846	1.842
88.	1.837	1.833	1.829	1.825	1.821	1.817	1.813	1.809	1.804	1.800
89.	1.796	1.792	1.788	1.784	1.780	1.776	1.772	1.768	1.765	1.761
90.	1.757	1.753	1.749	1.745	1.741	1.737	1.733	1.730	1.726	1.722
91.	1.718	1.715	1.711	1.707	1.703	1.700	1.696	1.692	1.688	1.685
92.	1.681	1.677	1.674	1.670	1.667	1.663	1.659	1.656	1.652	1.649
93.	1.645	1.642	1.638	1.635	1.631	1.628	1.624	1.621	1.617	1.614
94.	1.610	1.607	1.604	1.600	1.597	1.593	1.590	1.587	1.583	1.580
95.	1.577	1.573	1.570	1.567	1.563	1.560	1.557	1.554	1.550	1.547
96.	1.544	1.541	1.538	1.534	1.531	1.528	1.525	1.522	1.519	1.515
97.	1.512	1.509	1.506	1.503	1.500	1.497	1.494	1.491	1.488	1.485
98.	1.482	1.479	1.476	1.473	1.470	1.467	1.464	1.461	1.458	1.455
99.	1.452	1.449	1.446	1.443	1.440	1.437	1.434	1.431	1.429	1.426
100.	1.423	1.420	1.417	1.414	1.412	1.409	1.406	1.403	1.400	1.398

(continued)

Note: To obtain the hardness number from this table, read the HK (1 gf) corresponding to the measured diagonal length in microns and multiply by the test load in gram-force.

Testing Considerations

As with any other standard hardness test, care must be taken in the setup and performance of a microhardness test to ensure accurate, repeatable results. Because the margin for error is smaller in microhardness testing, the importance of accurate loads and controlled load application, precise indenter placement, and specimen surface preparation becomes increasingly critical.

Load Versus Indentation Size. As with other indentation hardness testing, judgment is required in obtaining the optimum indentation size. When the indentation is too small to obtain an accurate reading, the load should be increased; if the indentation is too large, the load should be decreased. When new materials are being tested, some experimentation with indenter loads is often required. Once an acceptable practice has been established for a given material, load selection is no longer a problem.

Spacing of Indentations. The same guidelines used for spacing of indentations for Brinell and Rockwell testing apply to Vickers and Knoop microhardness testing. The indentations should be spaced so that the distance between any two of them is greater than twice the extent of any stress deformation (cold working, butterfly fractures, etc.) that may occur when the indentation is made. This ensures that there is no overlap of the deformation between two indentations.

Hardness Number Versus Load. Before the development of the microhardness tester, the Vickers indenter (as well as other indenters that produce geometrically similar indentations) was assumed to produce a hardness number that was independent of the indenting load. Although this is generally true for loads of approximately 1000 gf and

Table 1 (continued)

Diagonal of indentation, μm	Knoop hardness number for diagonal measured to 0.1 μm									
	.0	.1	.2	.3	.4	.5	.6	.7	.8	.9
101.	1.395	1.392	1.389	1.387	1.384	1.381	1.378	1.376	1.373	1.370
102.	1.368	1.365	1.362	1.360	1.357	1.354	1.352	1.349	1.346	1.344
103.	1.341	1.339	1.336	1.333	1.331	1.328	1.326	1.323	1.321	1.318
104.	1.316	1.313	1.311	1.308	1.305	1.303	1.301	1.298	1.296	1.293
105.	1.291	1.288	1.286	1.283	1.281	1.278	1.276	1.274	1.271	1.269
106.	1.266	1.264	1.262	1.259	1.257	1.255	1.252	1.250	1.247	1.245
107.	1.243	1.240	1.238	1.236	1.234	1.231	1.229	1.227	1.224	1.222
108.	1.220	1.218	1.215	1.213	1.211	1.209	1.206	1.204	1.202	1.200
109.	1.198	1.195	1.193	1.191	1.189	1.187	1.185	1.182	1.180	1.178
110.	1.176	1.174	1.172	1.170	1.167	1.165	1.163	1.161	1.159	1.157
111.	1.155	1.153	1.151	1.149	1.147	1.145	1.142	1.140	1.138	1.136
112.	1.134	1.132	1.130	1.128	1.126	1.124	1.122	1.120	1.118	1.116
113.	1.114	1.112	1.110	1.108	1.106	1.105	1.103	1.101	1.099	1.097
114.	1.095	1.093	1.091	1.089	1.087	1.085	1.083	1.082	1.080	1.078
115.	1.076	1.074	1.072	1.070	1.068	1.067	1.065	1.063	1.061	1.059
116.	1.057	1.056	1.054	1.052	1.050	1.048	1.047	1.045	1.043	1.041
117.	1.039	1.038	1.036	1.034	1.032	1.031	1.029	1.027	1.025	1.024
118.	1.022	1.020	1.018	1.017	1.015	1.013	1.012	1.010	1.008	1.006
119.	1.005	1.003	1.001	0.9998	0.9981	0.9964	0.9947	0.9931	0.9914	0.9898
120.	0.9881	0.9865	0.9848	0.9832	0.9816	0.9799	0.9783	0.9767	0.9751	0.9735
121.	0.9719	0.9703	0.9687	0.9671	0.9655	0.9639	0.9623	0.9607	0.9591	0.9576
122.	0.9560	0.9544	0.9529	0.9513	0.9498	0.9482	0.9467	0.9451	0.9436	0.9420
123.	0.9405	0.9390	0.9375	0.9359	0.9344	0.9329	0.9314	0.9299	0.9284	0.9269
124.	0.9254	0.9239	0.9224	0.9209	0.9195	0.9180	0.9165	0.9150	0.9136	0.9121
125.	0.9107	0.9092	0.9078	0.9063	0.9049	0.9034	0.9020	0.9005	0.8991	0.8977
126.	0.8963	0.8948	0.8934	0.8920	0.8906	0.8892	0.8878	0.8864	0.8850	0.8836
127.	0.8822	0.8808	0.8794	0.8780	0.8767	0.8753	0.8739	0.8726	0.8712	0.8698
128.	0.8685	0.8671	0.8658	0.8644	0.8631	0.8617	0.8604	0.8591	0.8577	0.8564
129.	0.8551	0.8537	0.8524	0.8511	0.8498	0.8485	0.8472	0.8459	0.8446	0.8433
130.	0.8420	0.8407	0.8394	0.8381	0.8368	0.8355	0.8343	0.8330	0.8317	0.8304
131.	0.8291	0.8279	0.8266	0.8254	0.8241	0.8229	0.8216	0.8204	0.8191	0.8179
132.	0.8166	0.8154	0.8142	0.8129	0.8117	0.8105	0.8093	0.8080	0.8068	0.8056
133.	0.8044	0.8032	0.8020	0.8008	0.7996	0.7984	0.7972	0.7960	0.7948	0.7936
134.	0.7924	0.7913	0.7901	0.7889	0.7877	0.7866	0.7854	0.7842	0.7831	0.7819
135.	0.7807	0.7796	0.7784	0.7773	0.7761	0.7750	0.7738	0.7727	0.7716	0.7704
136.	0.7693	0.7682	0.7670	0.7659	0.7648	0.7637	0.7626	0.7614	0.7603	0.7592
137.	0.7581	0.7570	0.7559	0.7548	0.7537	0.7526	0.7515	0.7504	0.7493	0.7483
138.	0.7472	0.7461	0.7450	0.7439	0.7429	0.7418	0.7407	0.7396	0.7386	0.7375
139.	0.7365	0.7354	0.7343	0.7333	0.7322	0.7312	0.7301	0.7291	0.7281	0.7270
140.	0.7260	0.7249	0.7239	0.7229	0.7218	0.7208	0.7198	0.7188	0.7177	0.7167
141.	0.7157	0.7147	0.7137	0.7127	0.7117	0.7107	0.7097	0.7087	0.7077	0.7067
142.	0.7057	0.7047	0.7037	0.7027	0.7017	0.7007	0.6997	0.6988	0.6978	0.6968
143.	0.6958	0.6949	0.6939	0.6929	0.6920	0.6910	0.6900	0.6891	0.6881	0.6872
144.	0.6862	0.6852	0.6843	0.6834	0.6824	0.6815	0.6805	0.6796	0.6786	0.6777
145.	0.6768	0.6758	0.6749	0.6740	0.6731	0.6721	0.6712	0.6703	0.6694	0.6684
146.	0.6675	0.6666	0.6657	0.6648	0.6639	0.6630	0.6621	0.6612	0.6603	0.6594
147.	0.6585	0.6576	0.6567	0.6558	0.6549	0.6540	0.6531	0.6523	0.6514	0.6505
148.	0.6496	0.6487	0.6479	0.6470	0.6461	0.6452	0.6444	0.6435	0.6426	0.6418
149.	0.6409	0.6401	0.6392	0.6383	0.6375	0.6366	0.6358	0.6349	0.6341	0.6332
150.	0.6324	0.6316	0.6307	0.6299	0.6290	0.6282	0.6274	0.6265	0.6257	0.6249

(continued)

Note: To obtain the hardness number from this table, read the HK (1 gf) corresponding to the measured diagonal length in microns and multiply by the test load in gram-force.

above, microhardness, when determined using loads of less than 500 gf with the Knoop indenter and less than 100 gf with the Vickers indenter, is a function of the magnitude of the test load. In most instances, Knoop microhardness decreases with increasing loads, as shown in Fig. 6. However, an increase in Vickers values has been observed with increasing loads. This increase is followed by a range in which the hardness becomes independent of the load and approaches a constant value (Fig. 6). This effect occurs in a wide range of materials, from soft copper to fully hardened steel.

The apparent increase in hardness with decrease in load for the Knoop test in properly prepared surfaces is caused primarily by errors in measuring the indentation and by deviations in the elastic recovery of the indentation. As the size of the indentation decreases, readings are less accurate because of these factors. These inaccuracies can be attributed to the relationship between the size of the indentation and the constituents of the workpiece material.

It would seem likely that load dependence is based to some extent on elastic recovery of the indentation after the load is removed.

Although this is a factor with Knoop numbers, experimental studies have indicated that elastic recovery of Vickers indentations is too slight to explain load dependence completely. Another factor that may help explain load dependence of the Vickers hardness numbers is that the shape of the Vickers indentation often deviates from the square form. This is caused by the formation of a bulge at the sides of the indentation.

Although this phenomenon is not completely understood, some probable explanations of load dependence may be found in the design of the microhardness tester or in the

Table 1 (continued)

Diagonal of indentation, μm	Knoop hardness number for diagonal measured to 0.1 μm									
	.0	.1	.2	.3	.4	.5	.6	.7	.8	.9
151.	0.6241	0.6232	0.6224	0.6216	0.6208	0.6199	0.6191	0.6183	0.6175	0.6167
152.	0.6159	0.6151	0.6143	0.6134	0.6126	0.6118	0.6110	0.6102	0.6094	0.6086
153.	0.6078	0.6071	0.6063	0.6055	0.6047	0.6039	0.6031	0.6023	0.6015	0.6008
154.	0.6000	0.5992	0.5984	0.5976	0.5969	0.5961	0.5953	0.5946	0.5938	0.5930
155.	0.5923	0.5915	0.5907	0.5900	0.5892	0.5885	0.5877	0.5869	0.5862	0.5854
156.	0.5847	0.5839	0.5832	0.5825	0.5817	0.5810	0.5802	0.5795	0.5787	0.5780
157.	0.5773	0.5765	0.5758	0.5751	0.5743	0.5736	0.5729	0.5722	0.5714	0.5707
158.	0.5700	0.5693	0.5685	0.5678	0.5671	0.5664	0.5657	0.5650	0.5643	0.5635
159.	0.5628	0.5621	0.5614	0.5607	0.5600	0.5593	0.5586	0.5579	0.5572	0.5565
160.	0.5558	0.5551	0.5544	0.5537	0.5531	0.5524	0.5517	0.5510	0.5503	0.5496
161.	0.5489	0.5483	0.5476	0.5469	0.5462	0.5455	0.5449	0.5442	0.5435	0.5429
162.	0.5422	0.5415	0.5408	0.5402	0.5395	0.5389	0.5382	0.5375	0.5369	0.5362
163.	0.5356	0.5349	0.5342	0.5336	0.5329	0.5323	0.5316	0.5310	0.5303	0.5297
164.	0.5290	0.5284	0.5278	0.5271	0.5265	0.5258	0.5252	0.5246	0.5239	0.5233
165.	0.5226	0.5220	0.5214	0.5208	0.5201	0.5195	0.5189	0.5182	0.5176	0.5170
166.	0.5164	0.5157	0.5151	0.5145	0.5139	0.5133	0.5127	0.5120	0.5114	0.5108
167.	0.5102	0.5096	0.5090	0.5084	0.5078	0.5072	0.5066	0.5060	0.5054	0.5047
168.	0.5041	0.5035	0.5030	0.5024	0.5018	0.5012	0.5006	0.5000	0.4994	0.4988
169.	0.4982	0.4976	0.4970	0.4964	0.4959	0.4953	0.4947	0.4941	0.4935	0.4929
170.	0.4924	0.4918	0.4912	0.4906	0.4900	0.4895	0.4889	0.4883	0.4878	0.4782
171.	0.4866	0.4860	0.4855	0.4849	0.4843	0.4838	0.4832	0.4827	0.4821	0.4815
172.	0.4810	0.4804	0.4799	0.4793	0.4787	0.4782	0.4776	0.4771	0.4765	0.4760
173.	0.4754	0.4749	0.4743	0.4738	0.4732	0.4727	0.4721	0.4716	0.4711	0.4705
174.	0.4700	0.4694	0.4689	0.4684	0.4678	0.4673	0.4668	0.4662	0.4657	0.4652
175.	0.4646	0.4641	0.4636	0.4630	0.4625	0.4620	0.4615	0.4609	0.4604	0.4599
176.	0.4594	0.4588	0.4583	0.4578	0.4573	0.4568	0.4562	0.4557	0.4552	0.4547
177.	0.4542	0.4537	0.4532	0.4526	0.4521	0.4516	0.4511	0.4506	0.4501	0.4496
178.	0.4491	0.4486	0.4481	0.4476	0.4471	0.4466	0.4461	0.4456	0.4451	0.4446
179.	0.4441	0.4436	0.4431	0.4426	0.4421	0.4416	0.4411	0.4406	0.4401	0.4397
180.	0.4392	0.4387	0.4382	0.4377	0.4372	0.4367	0.4363	0.4358	0.4353	0.4348
181.	0.4343	0.4339	0.4334	0.4329	0.4324	0.4319	0.4315	0.4310	0.4305	0.4300
182.	0.4296	0.4291	0.4286	0.4282	0.4277	0.4272	0.4268	0.4263	0.4258	0.4254
183.	0.4249	0.4244	0.4240	0.4235	0.4230	0.4226	0.4221	0.4217	0.4212	0.4207
184.	0.4203	0.4198	0.4194	0.4189	0.4185	0.4180	0.4176	0.4171	0.4167	0.4162
185.	0.4158	0.4153	0.4149	0.4144	0.4140	0.4135	0.4131	0.4126	0.4122	0.4117
186.	0.4113	0.4109	0.4104	0.4100	0.4095	0.4091	0.4087	0.4082	0.4078	0.4073
187.	0.4069	0.4065	0.4060	0.4056	0.4052	0.4047	0.4043	0.4039	0.4034	0.4030
188.	0.4026	0.4022	0.4017	0.4013	0.4009	0.4005	0.4000	0.3996	0.3992	0.3988
189.	0.3983	0.3979	0.3975	0.3971	0.3967	0.3962	0.3958	0.3954	0.3950	0.3946
190.	0.3942	0.3937	0.3933	0.3929	0.3925	0.3921	0.3917	0.3913	0.3909	0.3905
191.	0.3900	0.3896	0.3892	0.3888	0.3884	0.3880	0.3876	0.3872	0.3868	0.3864
192.	0.3860	0.3856	0.3852	0.3848	0.3844	0.3840	0.3836	0.3832	0.3828	0.3824
193.	0.3820	0.3816	0.3812	0.3808	0.3804	0.3800	0.3796	0.3792	0.3789	0.3785
194.	0.3781	0.3777	0.3773	0.3769	0.3765	0.3761	0.3757	0.3754	0.3750	0.3746
195.	0.3742	0.3738	0.3734	0.3731	0.3727	0.3723	0.3719	0.3715	0.3712	0.3708
196.	0.3704	0.3700	0.3695	0.3693	0.3689	0.3685	0.3681	0.3678	0.3674	0.3670
197.	0.3666	0.3663	0.3659	0.3655	0.3652	0.3648	0.3644	0.3641	0.3637	0.3633
198.	0.3630	0.3626	0.3622	0.3619	0.3615	0.3611	0.3608	0.3604	0.3600	0.3597
199.	0.3593	0.3590	0.3586	0.3582	0.3579	0.3575	0.3572	0.3568	0.3564	0.3561
200.	0.3557	0.3554	0.3550	0.3547	0.3543	0.3540	0.3536	0.3533	0.3529	0.3525

Note: To obtain the hardness number from this table, read the HK (1 gf) corresponding to the measured diagonal length in microns and multiply by the test load in gram-force.

Fig. 5 Typical fixtures for holding and clamping workpieces for microhardness testing

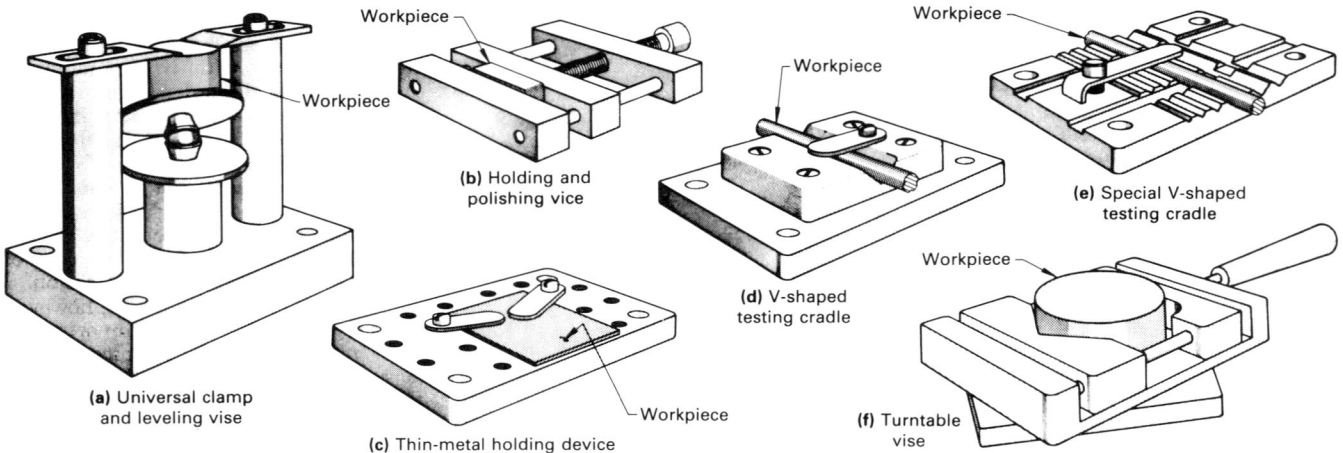

(a) Universal clamp and leveling vise

(b) Holding and polishing vice

(c) Thin-metal holding device

(d) V-shaped testing cradle

(e) Special V-shaped testing cradle

(f) Turntable vise

testing procedure. Depending on the load, microhardness readings may be affected by:

- Indenter shape
- Vibrations
- Microscope
- Friction within tester
- Surface preparation
- Cold working
- Indentation size
- Elastic recovery
- Bulge formation
- Other indentation characteristics

From a practical standpoint, as long as a single load is used throughout a series of tests, the load dependence is not significant. The choice of load depends on the size of indentation desired; usually it is made as large as practicable to obtain the greatest accuracy. Using different loads in any particular investigation alters the as-measured hardness; the lighter the load, the more significant the change. Generally, any compari-

Fig. 6 Relationship of hardness number and load for Knoop and Vickers indenters

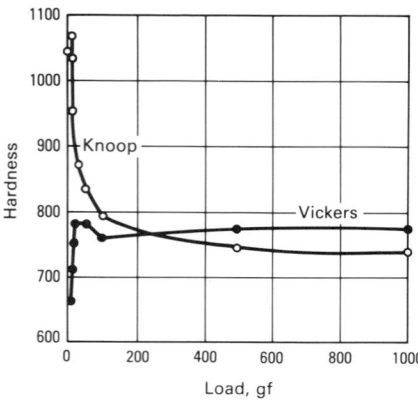

son of Knoop hardness numbers with loads of less than 500 gf and Vickers hardness numbers of less than 100 gf is invalid unless load dependence is considered. Consequently, the load should be reported when listing Knoop or Vickers numbers.

Applications

Microhardness testing is useful in research and development, materials testing, and quality control programs. Results from microhardness tests can be used to monitor hardness during design and development, fabrication, heat treatment, and performance analysis of many products and components. Typical applications are discussed below.

Small Parts. Many products or devices that are too small for conventional methods can be tested with microhardness techniques. These include small precision parts, screw machine products, stampings, metal bearings, tiny watch gears, ends of balance staff pivots, hair springs, pellets of pen points, hypodermic needles, electrical connectors, printed circuit board terminals, screws, and other fastening and retaining devices. In many instances, these parts can be tested without mounting preparations by using holding and clamping fixtures as shown in Fig. 5.

Thin Foil and Wire. Microhardness testing can also be applied to product forms such as thin foils and small-diameter wires, using fixtures such as those shown in Fig. 5(c), (d), and (e), or by using metallographic mounting techniques. The minimum thickness of sheet or foil that can be tested depends on hardness and load, as given in the Knoop minimum-thickness chart (Fig. 7). For example, a maximum load value is needed for testing a carbon steel workpiece 0.051 mm (0.002 in.) thick with an approximate hardness of 32

HRC. First, a conversion table should be consulted for the equivalent Knoop hardness number (HK) of 32 HRC, which is 326 HK (hardness conversion tables can be found in the Appendix to this Section). Next, referring to Fig. 7 at 326 HK, the intersection at the 0.051 mm (0.002 in.) line is observed, which is at about 525 gf on the load line. The next lower standard load should be used—in this case, 500 gf verifies that the results obtained are correct. This chart should be used only as a guide.

Testing Materials Using Very Light Loads. Soft or very thin materials, such as plastic sheet and rod, dental materials, epoxies, porcelain, ceramics, glass minerals, lacquers, paints, and other coatings, may require testing at loads as light as 0.5 gf. Vibration-free testing is more critical as the load decreases.

Monitoring Surface Hardening Operations. Microhardness testing is the best method for measuring depth and hardness of carburized and nitrided cases, and sometimes for studying induction-hardened zones.

In most instances, tests are made on cross sections of test coupons that have accompanied the workpieces during processing. To ensure accurate readings close to the edge, a 100-gf load is most often used, although a 500-gf load is sometimes preferred. Indentations are taken at preestablished locations and intervals, often beginning at 0.025 mm (0.001 in.), then at intervals of 0.127 mm (0.005 in.). Results usually are plotted as shown in Fig. 8, in which the hardness at the edge is near 800 HK, gradually tapering to core hardness.

Not only does microhardness testing show the hardness gradient (Fig. 8a), but it can also detect a soft skin (Fig. 8b). This condition can be caused by either retained austenite or decarburization.

Measuring Hardness of Microconstituents. Many metals consist of a mixture of microconstituents that may vary widely in hardness. Conventional macrohardness tests do not register the true hardness of such materials. Much can be learned about alloys and their potential properties (for example, their resistance to wear) by knowing the hardness of the various microconstituents.

Highly alloyed tool steels are examples of this phenomenon. Figure 9 shows a micrograph of polished and etched D2 tool steel. Knoop indentations located on the matrix (darker constituent) and on the particle of carbide (white constituent) show an obvious difference in size. Knoop hardnesses of the matrix and the carbide particle were calculated to be 801 and 1930 HK, respectively. A Rockwell test indicated a carbide particle hardness of 64 HRC, which converts to only 822 HK.

For nonhomogeneous materials such as cast irons and powder metallurgy parts, microhardness testing is the only method of making accurate hardness determinations because of graphite particles in cast irons and voids in powder metallurgy parts.

Occasionally, problems are encountered when machining parts whose macrohardness measured in Brinell or Rockwell indicates that they should be machinable. Microhardness tests often reveal that the problem is due to a mixture of both soft and very hard microconstituents. Steps can then be taken to provide a more homogeneous material.

Analyzing Metal Failure. Microhardness testing often plays an important role in determining the cause of metal failure, because it makes possible evaluations of hardness under conditions or in areas where conventional hardness testers cannot be used.

Tool failures can often be explained by microhardness testing. For example, taps

Table 2   Vickers hardness numbers for load of 1 gf

Diagonal of indentation, μm	Vickers hardness number for diagonal measured to 0.1 μm									
	.0	.1	.2	.3	.4	.5	.6	.7	.8	.9
1.	1854.0	1533.0	1288.0	1097.0	946.1	824.2	724.4	641.6	572.3	513.7
2.	463.6	420.5	383.1	350.5	321.9	296.7	274.3	254.4	236.5	220.5
3.	206.0	193.0	181.1	170.3	160.4	151.4	143.1	135.5	128.4	121.9
4.	115.9	110.3	105.1	100.3	95.78	91.57	87.64	83.95	80.48	77.23
5.	74.17	71.29	68.58	66.01	63.59	61.30	59.13	57.07	55.12	53.27
6.	51.51	49.83	48.24	46.72	45.27	43.89	42.57	41.31	40.10	38.95
7.	37.84	36.79	35.77	34.80	33.86	32.97	32.10	31.28	30.48	29.71
8.	28.97	28.26	27.58	26.92	26.28	25.67	25.07	24.50	23.95	23.41
9.	22.89	22.39	21.91	21.44	20.99	20.55	20.12	19.71	19.31	18.92
10.	18.54	18.18	17.82	17.48	17.14	16.82	16.50	16.20	15.90	15.61
11.	15.33	15.05	14.78	14.52	14.27	14.02	13.78	13.55	13.32	13.09
12.	12.88	12.67	12.46	12.26	12.06	11.87	11.68	11.50	11.32	11.14
13.	10.97	10.81	10.64	10.48	10.33	10.17	10.03	9.880	9.737	9.598
14.	9.461	9.327	9.196	9.068	8.943	8.820	8.699	8.581	8.466	8.353
15.	8.242	8.133	8.026	7.922	7.819	7.718	7.620	7.523	7.428	7.335
16.	7.244	7.154	7.066	6.979	6.895	6.811	6.729	6.649	6.570	6.493
17.	6.416	6.342	6.268	6.196	6.125	6.055	5.986	5.919	5.853	5.787
18.	5.723	5.660	5.598	5.537	5.477	5.418	5.360	5.303	5.247	5.191
19.	5.137	5.083	5.030	4.978	4.927	4.877	4.827	4.778	4.730	4.683
20.	4.636	4.590	4.545	4.500	4.456	4.413	4.370	4.328	4.286	4.245
21.	4.205	4.165	4.126	4.087	4.049	4.012	3.975	3.938	3.902	3.866
22.	3.831	3.797	3.763	3.729	3.696	3.663	3.631	3.599	3.567	3.536
23.	3.505	3.475	3.445	3.416	3.387	3.358	3.329	3.301	3.274	3.246
24.	3.219	3.193	3.166	3.140	3.115	3.089	3.064	3.039	3.015	2.991
25.	2.967	2.943	2.920	2.897	2.874	2.852	2.830	2.808	2.786	2.764
26.	2.743	2.722	2.701	2.681	2.661	2.641	2.621	2.601	2.582	2.563
27.	2.544	2.525	2.506	2.488	2.470	2.452	2.434	2.417	2.399	2.382
28.	2.365	2.348	2.332	2.315	2.299	2.283	2.267	2.251	2.236	2.220
29.	2.205	2.190	2.175	2.160	2.145	2.131	2.116	2.102	2.088	2.074
30.	2.060	2.047	2.033	2.020	2.077	1.993	1.980	1.968	1.955	1.942
31.	1.930	1.917	1.905	1.893	1.881	1.869	1.857	1.845	1.834	1.822
32.	1.811	1.800	1.788	1.777	1.766	1.756	1.745	1.734	1.724	1.713
33.	1.703	1.693	1.682	1.672	1.662	1.652	1.643	1.633	1.623	1.614
34.	1.604	1.595	1.585	1.576	1.567	1.558	1.549	1.540	1.531	1.522
35.	1.514	1.505	1.497	1.488	1.480	1.471	1.463	1.455	1.447	1.439
36.	1.431	1.423	1.415	1.407	1.400	1.392	1.384	1.377	1.369	1.362
37.	1.355	1.347	1.340	1.333	1.326	1.319	1.312	1.305	1.298	1.291
38.	1.284	1.277	1.271	1.264	1.258	1.251	1.245	1.238	1.232	1.225
39.	1.219	1.213	1.207	1.201	1.195	1.189	1.183	1.177	1.171	1.165
40.	1.159	1.153	1.147	1.142	1.136	1.131	1.125	1.119	1.114	1.109
41.	1.103	1.098	1.092	1.087	1.082	1.077	1.072	1.066	1.061	1.056
42.	1.051	1.046	1.041	1.036	1.031	1.027	1.022	1.017	1.012	1.008
43.	1.003	0.9983	0.9936	0.9891	0.9845	0.9800	0.9755	0.9710	0.9666	0.9622
44.	0.9578	0.9535	0.9492	0.9449	0.9407	0.9364	0.9322	0.9281	0.9239	0.9198
45.	0.9157	0.9117	0.9077	0.9036	0.8997	0.8957	0.8918	0.8879	0.8840	0.8802
46.	0.8764	0.8726	0.8688	0.8650	0.8613	0.8576	0.8539	0.8503	0.8467	0.8430
47.	0.8395	0.8359	0.8324	0.8288	0.8254	0.8219	0.8184	0.8150	0.8116	0.8082
48.	0.8048	0.8015	0.7982	0.7949	0.7916	0.7883	0.7851	0.7819	0.7787	0.7755
49.	0.7723	0.7692	0.7661	0.7630	0.7599	0.7568	0.7538	0.7507	0.7477	0.7447
50.	0.7417	0.7388	0.7359	0.7329	0.7300	0.7271	0.7243	0.7214	0.7186	0.7158

(continued)

Note: To obtain the hardness number from this table, read the HV (1 gf) corresponding to the measured diagonal length in microns and multiply by the test load in gram-force.

were failing prematurely from dulling of the teeth. Rockwell C hardness measurements showed 65 HRC, which was acceptable. However, when one of the taps was sectioned and teeth areas were examined, the reason for failure was evident. Figure 10 shows that the lower part of the tap tooth was 850 HK (approximately 65 HRC by conversion). At the tooth crest, however, the hardness was only 480 HK (46 HRC). The decrease was caused by overheating in grinding. The Knoop indentations shown in Fig. 10 were made with a 100-gf load on a highly polished specimen.

Ultrasonic Microhardness Testing

By Paul A. Meyer
Technical Director
Krautkrämer Branson

and

Douglas P. Lutz
Hardness Product Manager
Krautkrämer Branson

ULTRASONIC MICROHARDNESS TESTING offers an alternative to conventional tests based on the visual (microscopic) evaluation of an indentation after the load has been removed. The ultrasonic method uses a maximum indentation load of approximately 800 gf. Therefore, as in other microhardness techniques, the indentation depth is relatively small (from 4 to 18 μm). In most test situations, the resulting surface is still quite usable, classifying this as a nondestructive test. Measured values in either Vickers or Rockwell C scale are displayed on a digital-readout display directly after penetration of

Table 2 (continued)

Diagonal of indentation, μm	Vickers hardness number for diagonal measured to 0.1 μm									
	.0	.1	.2	.3	.4	.5	.6	.7	.8	.9
51.	0.7129	0.7102	0.7074	0.7046	0.7019	0.6992	0.6965	0.6938	0.6911	0.6884
52.	0.6858	0.6832	0.6805	0.6779	0.6754	0.6728	0.6702	0.6677	0.6652	0.6627
53.	0.6602	0.6577	0.6552	0.6527	0.6503	0.6479	0.6455	0.6431	0.6407	0.6383
54.	0.6359	0.6336	0.6312	0.6289	0.6266	0.6243	0.6220	0.6198	0.6175	0.6153
55.	0.6130	0.6108	0.6086	0.6064	0.6042	0.6020	0.5999	0.5977	0.5956	0.5934
56.	0.5913	0.5892	0.5871	0.5850	0.5830	0.5809	0.5788	0.5768	0.5748	0.5728
57.	0.5708	0.5688	0.5668	0.5648	0.5628	0.5609	0.5589	0.5570	0.5551	0.5531
58.	0.5512	0.5493	0.5475	0.5456	0.5437	0.5419	0.5400	0.5382	0.5363	0.5345
59.	0.5327	0.5309	0.5291	0.5273	0.5256	0.5238	0.5220	0.5203	0.5186	0.5168
60.	0.5151	0.5134	0.5117	0.5100	0.5083	0.5066	0.5050	0.5033	0.5016	0.5000
61.	0.4984	0.4967	0.4951	0.4935	0.4919	0.4903	0.4887	0.4871	0.4855	0.4840
62.	0.4824	0.4809	0.4793	0.4778	0.4762	0.4747	0.4732	0.4717	0.4702	0.4687
63.	0.4672	0.4657	0.4643	0.4628	0.4613	0.4599	0.4584	0.4570	0.4556	0.4541
64.	0.4527	0.4513	0.4499	0.4485	0.4471	0.4457	0.4444	0.4430	0.4416	0.4403
65.	0.4389	0.4376	0.4362	0.4349	0.4336	0.4322	0.4309	0.4296	0.4283	0.4270
66.	0.4257	0.4244	0.4231	0.4219	0.4206	0.4193	0.4181	0.4168	0.4156	0.4143
67.	0.4131	0.4119	0.4106	0.4094	0.4082	0.4070	0.4058	0.4046	0.4034	0.4022
68.	0.4010	0.3999	0.3987	0.3975	0.3964	0.3952	0.3941	0.3929	0.3918	0.3906
69.	0.3895	0.3884	0.3872	0.3861	0.3850	0.3839	0.3828	0.3817	0.3806	0.3795
70.	0.3884	0.3774	0.3763	0.3752	0.3742	0.3731	0.3720	0.3710	0.3699	0.3689
71.	0.3679	0.3668	0.3658	0.3648	0.3638	0.3627	0.3617	0.3607	0.3597	0.3587
72.	0.3577	0.3567	0.3557	0.3548	0.3538	0.3528	0.3518	0.3509	0.3499	0.3489
73.	0.3480	0.3470	0.3461	0.3451	0.3442	0.3433	0.3423	0.3414	0.3405	0.3396
74.	0.3386	0.3377	0.3368	0.3359	0.3350	0.3341	0.3332	0.3323	0.3314	0.3305
75.	0.3297	0.3288	0.3279	0.3270	0.3262	0.3253	0.3245	0.3236	0.3227	0.3219
76.	0.3211	0.3202	0.3194	0.3185	0.3177	0.3169	0.3160	0.3152	0.3144	0.3136
77.	0.3128	0.3120	0.3111	0.3103	0.3095	0.3087	0.3079	0.3072	0.3064	0.3056
78.	0.3048	0.3040	0.3032	0.3025	0.3017	0.3009	0.3002	0.2994	0.2986	0.2979
79.	0.2971	0.2964	0.2956	0.2949	0.2941	0.2934	0.2927	0.2919	0.2912	0.2905
80.	0.2897	0.2890	0.2883	0.2876	0.2869	0.2862	0.2855	0.2847	0.2840	0.2833
81.	0.2826	0.2819	0.2812	0.2806	0.2799	0.2792	0.2785	0.2778	0.2771	0.2765
82.	0.2758	0.2751	0.2744	0.2738	0.2731	0.2725	0.2718	0.2711	0.2705	0.2698
83.	0.2692	0.2685	0.2679	0.2672	0.2666	0.2660	0.2653	0.2647	0.2641	0.2634
84.	0.2628	0.2622	0.2616	0.2609	0.2603	0.2597	0.2591	0.2585	0.2579	0.2573
85.	0.2567	0.2561	0.2555	0.2549	0.2543	0.2537	0.2531	0.2525	0.2519	0.2513
86.	0.2507	0.2501	0.2496	0.2490	0.2484	0.2478	0.2473	0.2467	0.2461	0.2456
87.	0.2450	0.2444	0.2439	0.2433	0.2428	0.2422	0.2417	0.2411	0.2406	0.2400
88.	0.2395	0.2389	0.2384	0.2378	0.2373	0.2368	0.2362	0.2357	0.2352	0.2346
89.	0.2341	0.2336	0.2331	0.2325	0.2320	0.2315	0.2310	0.2305	0.2300	0.2294
90.	0.2289	0.2284	0.2279	0.2274	0.2269	0.2264	0.2259	0.2254	0.2249	0.2244
91.	0.2239	0.2234	0.2230	0.2225	0.2220	0.2215	0.2210	0.2205	0.2200	0.2196
92.	0.2191	0.2186	0.2181	0.2177	0.2172	0.2167	0.2163	0.2158	0.2153	0.2149
93.	0.2144	0.2139	0.2135	0.2130	0.2126	0.2121	0.2117	0.2112	0.2108	0.2103
94.	0.2099	0.2094	0.2090	0.2085	0.2081	0.2077	0.2072	0.2068	0.2063	0.2059
95.	0.2055	0.2050	0.2046	0.2042	0.2038	0.2033	0.2029	0.2025	0.2021	0.2016
96.	0.2012	0.2008	0.2004	0.2000	0.1995	0.1991	0.1987	0.1983	0.1979	0.1975
97.	0.1971	0.1967	0.1963	0.1959	0.1955	0.1951	0.1947	0.1943	0.1939	0.1935
98.	0.1931	0.1927	0.1923	0.1919	0.1915	0.1911	0.1907	0.1904	0.1900	0.1896
99.	0.1892	0.1888	0.1884	0.1881	0.1877	0.1873	0.1869	0.1866	0.1862	0.1858
100.	0.1854	0.1851	0.1847	0.1843	0.1840	0.1836	0.1832	0.1829	0.1825	0.1821

(continued)

Note: To obtain the hardness number from this table, read the HV (1 gf) corresponding to the measured diagonal length in microns and multiply by the test load in gram-force.

the test piece. This feature makes the method suitable for automated on-line testing. Up to 1200 parts per hour can be tested. Table 3 compares various aspects of indentation-type hardness testing techniques, including the ultrasonic method.

Principles of Operation

In ultrasonic microhardness testing, a Vickers diamond is attached to one end of a magnetostrictive metal rod. The diamond-tipped rod is excited to its natural frequency by a piezoelectric converter. The resonant frequency of the rod changes as the free end of the rod is brought into contact with the surface of a solid body.

Once the device is calibrated for the known modulus of elasticity of the tested material, the area of contact between the diamond tip and the tested surface can be derived from the measured resonant frequency. The area of contact is inversely proportional to the hardness of the tested material, provided the force pressing the surface is constant. Consequently, the measured frequency value can be converted into the corresponding hardness number. More detailed information on the theory and principles of ultrasonic hardness measurement can be found in Ref 2 to 4.

Components of an ultrasonic hardness tester, which automatically bring the diamond-tipped oscillating rod into contact with the test piece and electronically perform all necessary measurements and calculations, are shown schematically in Fig. 11. The hardness number is displayed on a digital readout while the oscillating rod is retracted to protect it until the next reading. The entire process generally takes less than 15 s.

Table 2 (continued)

Diagonal of indentation, μm	Vickers hardness number for diagonal measured to 0.1 μm									
	.0	.1	.2	.3	.4	.5	.6	.7	.8	.9
101.	0.1818	0.1814	0.1811	0.1807	0.1804	0.1800	0.1796	0.1793	0.1789	0.1786
102.	0.1782	0.1779	0.1775	0.1772	0.1769	0.1765	0.1762	0.1758	0.1755	0.1751
103.	0.1748	0.1745	0.1741	0.1738	0.1734	0.1731	0.1728	0.1724	0.1721	0.1718
104.	0.1715	0.1711	0.1708	0.1705	0.1701	0.1698	0.1695	0.1692	0.1688	0.1685
105.	0.1682	0.1679	0.1676	0.1672	0.1669	0.1666	0.1663	0.1660	0.1657	0.1654
106.	0.1650	0.1647	0.1644	0.1641	0.1638	0.1635	0.1632	0.1629	0.1626	0.1623
107.	0.1620	0.1617	0.1614	0.1611	0.1608	0.1605	0.1602	0.1599	0.1596	0.1593
108.	0.1590	0.1587	0.1584	0.1581	0.1578	0.1575	0.1572	0.1569	0.1567	0.1564
109.	0.1561	0.1558	0.1555	0.1552	0.1549	0.1547	0.1544	0.1541	0.1538	0.1535
110.	0.1533	0.1530	0.1527	0.1524	0.1521	0.1519	0.1516	0.1513	0.1511	0.1508
111.	0.1505	0.1502	0.1500	0.1497	0.1494	0.1492	0.1489	0.1486	0.1484	0.1481
112.	0.1478	0.1476	0.1473	0.1470	0.1468	0.1465	0.1463	0.1460	0.1457	0.1455
113.	0.1452	0.1450	0.1447	0.1445	0.1442	0.1440	0.1437	0.1434	0.1432	0.1429
114.	0.1427	0.1424	0.1422	0.1419	0.1417	0.1414	0.1412	0.1410	0.1407	0.1405
115.	0.1402	0.1400	0.1397	0.1395	0.1393	0.1390	0.1388	0.1385	0.1383	0.1381
116.	0.1378	0.1376	0.1373	0.1371	0.1369	0.1366	0.1364	0.1362	0.1359	0.1357
117.	0.1355	0.1352	0.1350	0.1348	0.1345	0.1343	0.1341	0.1339	0.1336	0.1334
118.	0.1332	0.1330	0.1327	0.1325	0.1323	0.1321	0.1318	0.1316	0.1314	0.1312
119.	0.1310	0.1307	0.1305	0.1303	0.1301	0.1299	0.1296	0.1294	0.1292	0.1290
120.	0.1288	0.1286	0.1284	0.1281	0.1279	0.1277	0.1275	0.1273	0.1271	0.1269
121.	0.1267	0.1265	0.1262	0.1260	0.1258	0.1256	0.1254	0.1252	0.1250	0.1248
122.	0.1246	0.1244	0.1242	0.1240	0.1238	0.1236	0.1234	0.1232	0.1230	0.1228
123.	0.1226	0.1224	0.1222	0.1220	0.1218	0.1216	0.1214	0.1212	0.1210	0.1208
124.	0.1206	0.1204	0.1202	0.1200	0.1198	0.1196	0.1194	0.1193	0.1191	0.1189
125.	0.1187	0.1185	0.1183	0.1181	0.1179	0.1177	0.1176	0.1174	0.1172	0.1170
126.	0.1168	0.1166	0.1164	0.1163	0.1161	0.1159	0.1157	0.1155	0.1153	0.1152
127.	0.1150	0.1148	0.1146	0.1144	0.1143	0.1141	0.1139	0.1137	0.1135	0.1134
128.	0.1132	0.1130	0.1128	0.1127	0.1125	0.1123	0.1121	0.1120	0.1118	0.1116
129.	0.1114	0.1113	0.1111	0.1109	0.1108	0.1106	0.1104	0.1102	0.1101	0.1099
130.	0.1097	0.1096	0.1094	0.1092	0.1091	0.1089	0.1087	0.1086	0.1084	0.1082
131.	0.1081	0.1079	0.1077	0.1076	0.1074	0.1072	0.1071	0.1069	0.1068	0.1066
132.	0.1064	0.1063	0.1061	0.1059	0.1058	0.1056	0.1055	0.1053	0.1052	0.1050
133.	0.1048	0.1047	0.1045	0.1044	0.1042	0.1041	0.1039	0.1037	0.1036	0.1034
134.	0.1033	0.1031	0.1030	0.1028	0.1027	0.1025	0.1024	0.1022	0.1021	0.1019
135.	0.1018	0.1016	0.1015	0.1013	0.1012	0.1010	0.1009	0.1007	0.1006	0.1004
136.	0.1003	0.1001	0.1000	0.0998	0.0997	0.0995	0.0994	0.0992	0.0991	0.0989
137.	0.0988	0.0987	0.0985	0.0984	0.0982	0.0981	0.0979	0.0978	0.0977	0.0975
138.	0.0974	0.0972	0.0971	0.0970	0.0968	0.0967	0.0965	0.0964	0.0963	0.0961
139.	0.0960	0.0958	0.0957	0.0956	0.0954	0.0953	0.0952	0.0950	0.0949	0.0948
140.	0.0946	0.0945	0.0943	0.0942	0.0941	0.0939	0.0938	0.0937	0.0935	0.0934
141.	0.0933	0.0931	0.0930	0.0929	0.0928	0.0926	0.0925	0.0924	0.0922	0.0921
142.	0.0920	0.0918	0.0917	0.0916	0.0915	0.0913	0.0912	0.0911	0.0909	0.0908
143.	0.0907	0.0906	0.0904	0.0903	0.0902	0.0901	0.0899	0.0898	0.0897	0.0896
144.	0.0894	0.0893	0.0892	0.0891	0.0889	0.0888	0.0887	0.0886	0.0884	0.0883
145.	0.0882	0.0881	0.0880	0.0878	0.0877	0.0876	0.0875	0.0874	0.0872	0.0871
146.	0.0870	0.0869	0.0868	0.0866	0.0865	0.0864	0.0863	0.0862	0.0861	0.0859
147.	0.0858	0.0857	0.0856	0.0855	0.0854	0.0852	0.0851	0.0850	0.0849	0.0848
148.	0.0847	0.0845	0.0844	0.0843	0.0842	0.0841	0.0840	0.0839	0.0838	0.0836
149.	0.0835	0.0834	0.0833	0.0832	0.0831	0.0830	0.0829	0.0828	0.0826	0.0825
150.	0.0824	0.0823	0.0822	0.0821	0.0820	0.0819	0.0818	0.0817	0.0815	0.0814

(continued)

Note: To obtain the hardness number from this table, read the HV (1 gf) corresponding to the measured diagonal length in microns and multiply by the test load in gram-force.

This type of instrument is quite small and is battery powered for portability. The automatic probe allows hardness measurements to be made in any orientation, further enhancing the usefulness of the device.

Advantages and Applications

Using an ultrasonic hardness testing system, it is possible to measure instantly the area of indentation under load. In principle, this ability is indispensible, because hardness numbers are derived by dividing the load by the surface area of the indentation. In conventional microhardness tests, this area is calculated from microscopic measurements of the lengths of the diagonals of the impression. However, using this method of indentation area measurement can lead to erroneous hardness values due to elastic recovery upon unloading.

For example, a perfect indentation made with a perfect Vickers indenter would be a square (Fig. 12a). However, anomalies are frequently observed with a pyramid indenter. The pincushion indentation in Fig. 12(b) is the result of sinking-in of the metal around the flat faces of the pyramid. This condition is observed with annealed metals and results in an overestimate of the diagonal length. The barrel-shaped indentation in Fig. 12(c) is found in cold-worked metals. It results from ridging or piling up of the metal around the faces of the indenter. The diagonal measurement in this case produces a low value of the contact area so that the hardness numbers are erroneously high. Because the area of the indentation is measured under load in ultrasonic hardness testing, elastic recovery does not affect results.

Table 2 (continued)

Diagonal of indentation, μm	Vickers hardness number for diagonal measured to 0.1 μm									
	.0	.1	.2	.3	.4	.5	.6	.7	.8	.9
151.	0.0813	0.0812	0.0811	0.0810	0.0809	0.0808	0.0807	0.0806	0.0805	0.0804
152.	0.0803	0.0802	0.0801	0.0800	0.0798	0.0797	0.0796	0.0795	0.0794	0.0793
153.	0.0792	0.0791	0.0790	0.0789	0.0788	0.0787	0.0786	0.0785	0.0784	0.0783
154.	0.0782	0.0781	0.0780	0.0779	0.0778	0.0777	0.0776	0.0775	0.0774	0.0773
155.	0.0772	0.0771	0.0770	0.0769	0.0768	0.0767	0.0766	0.0765	0.0764	0.0763
156.	0.0762	0.0761	0.0760	0.0759	0.0758	0.0757	0.0756	0.0755	0.0754	0.0753
157.	0.0752	0.0751	0.0750	0.0749	0.0749	0.0748	0.0747	0.0746	0.0745	0.0744
158.	0.0743	0.0742	0.0741	0.0740	0.0739	0.0738	0.0737	0.0736	0.0735	0.0734
159.	0.0734	0.0733	0.0732	0.0731	0.0730	0.0729	0.0728	0.0727	0.0726	0.0725
160.	0.0724	0.0724	0.0723	0.0722	0.0721	0.0720	0.0719	0.0718	0.0717	0.0716
161.	0.0715	0.0715	0.0714	0.0713	0.0712	0.0711	0.0710	0.0709	0.0708	0.0708
162.	0.0707	0.0706	0.0705	0.0704	0.0703	0.0702	0.0701	0.0701	0.0700	0.0699
163.	0.0698	0.0697	0.0696	0.0695	0.0695	0.0694	0.0693	0.0692	0.0691	0.0690
164.	0.0690	0.0689	0.0688	0.0687	0.0686	0.0685	0.0684	0.0684	0.0683	0.0682
165.	0.0681	0.0680	0.0680	0.0679	0.0678	0.0677	0.0676	0.0675	0.0675	0.0674
166.	0.0673	0.0672	0.0671	0.0671	0.0670	0.0669	0.0668	0.0667	0.0667	0.0666
167.	0.0665	0.0664	0.0663	0.0663	0.0662	0.0661	0.0660	0.0659	0.0659	0.0658
168.	0.0657	0.0656	0.0656	0.0655	0.0654	0.0653	0.0652	0.0652	0.0651	0.0650
169.	0.0649	0.0649	0.0648	0.0647	0.0646	0.0645	0.0645	0.0644	0.0643	0.0642
170.	0.0642	0.0641	0.0640	0.0639	0.0639	0.0638	0.0637	0.0636	0.0636	0.0635
171.	0.0634	0.0633	0.0633	0.0632	0.0631	0.0631	0.0630	0.0629	0.0628	0.0628
172.	0.0627	0.0626	0.0625	0.0625	0.0624	0.0623	0.0623	0.0622	0.0621	0.0620
173.	0.0620	0.0619	0.0618	0.0617	0.0617	0.0616	0.0615	0.0615	0.0614	0.0613
174.	0.0613	0.0612	0.0611	0.0610	0.0610	0.0609	0.0608	0.0608	0.0607	0.0606
175.	0.0606	0.0605	0.0604	0.0603	0.0603	0.0602	0.0601	0.0601	0.0600	0.0599
176.	0.0599	0.0598	0.0597	0.0597	0.0596	0.0595	0.0595	0.0594	0.0593	0.0593
177.	0.0592	0.0591	0.0591	0.0590	0.0589	0.0589	0.0588	0.0587	0.0587	0.0586
178.	0.0585	0.0585	0.0584	0.0583	0.0583	0.0582	0.0581	0.0581	0.0580	0.0579
179.	0.0579	0.0578	0.0578	0.0577	0.0576	0.0576	0.0575	0.0574	0.0574	0.0573
180.	0.0572	0.0572	0.0571	0.0570	0.0570	0.0569	0.0569	0.0568	0.0567	0.0567
181.	0.0566	0.0565	0.0565	0.0564	0.0564	0.0563	0.0562	0.0562	0.0561	0.0560
182.	0.0560	0.0559	0.0559	0.0558	0.0557	0.0557	0.0556	0.0556	0.0555	0.0554
183.	0.0554	0.0553	0.0553	0.0552	0.0551	0.0551	0.0550	0.0550	0.0549	0.0548
184.	0.0548	0.0547	0.0547	0.0546	0.0545	0.0545	0.0544	0.0544	0.0543	0.0542
185.	0.0542	0.0541	0.0541	0.0540	0.0540	0.0539	0.0538	0.0538	0.0537	0.0537
186.	0.0536	0.0535	0.0535	0.0534	0.0534	0.0533	0.0533	0.0532	0.0531	0.0531
187.	0.0530	0.0530	0.0529	0.0529	0.0528	0.0528	0.0527	0.0526	0.0526	0.0525
188.	0.0525	0.0524	0.0524	0.0523	0.0522	0.0522	0.0521	0.0521	0.0520	0.0520
189.	0.0519	0.0519	0.0518	0.0518	0.0517	0.0516	0.0516	0.0515	0.0515	0.0514
190.	0.0514	0.0513	0.0513	0.0512	0.0512	0.0511	0.0510	0.0510	0.0509	0.0509
191.	0.0508	0.0508	0.0507	0.0507	0.0506	0.0506	0.0505	0.0505	0.0504	0.0504
192.	0.0503	0.0503	0.0502	0.0502	0.0501	0.0500	0.0500	0.0499	0.0499	0.0498
193.	0.0498	0.0497	0.0497	0.0496	0.0496	0.0495	0.0495	0.0494	0.0494	0.0493
194.	0.0493	0.0492	0.0492	0.0491	0.0491	0.0490	0.0490	0.0489	0.0489	0.0488
195.	0.0488	0.0487	0.0487	0.0486	0.0486	0.0485	0.0485	0.0484	0.0484	0.0483
196.	0.0483	0.0482	0.0482	0.0481	0.0481	0.0480	0.0480	0.0479	0.0479	0.0478
197.	0.0478	0.0477	0.0477	0.0476	0.0476	0.0475	0.0475	0.0474	0.0474	0.0474
198.	0.0473	0.0473	0.0472	0.0472	0.0471	0.0471	0.0470	0.0470	0.0469	0.0469
199.	0.0468	0.0468	0.0467	0.0467	0.0466	0.0466	0.0465	0.0465	0.0465	0.0464
200.	0.0464	0.0463	0.0463	0.0462	0.0462	0.0461	0.0461	0.0460	0.0460	0.0459

Note: To obtain the hardness number from this table, read the HV (1 gf) corresponding to the measured diagonal length in microns and multiply by the test load in gram-force.

Fig. 7 Knoop minimum thickness chart

Indicates load and hardness combinations for determining minimum thickness of sheet or foil that can be tested

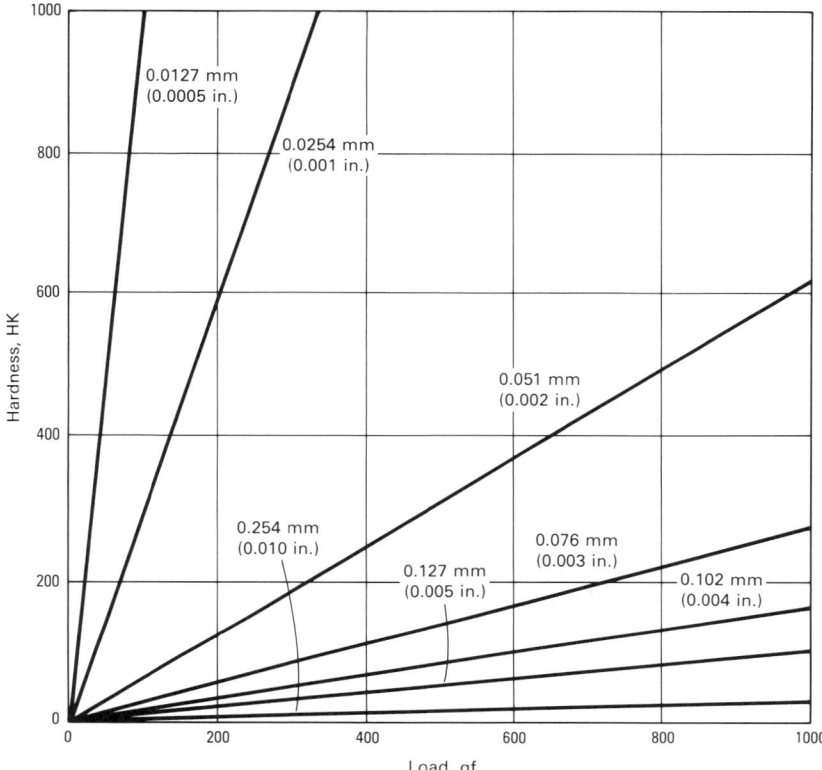

Fig. 9 Knoop indentations in two microconstituents of quenched and tempered D2 tool steel

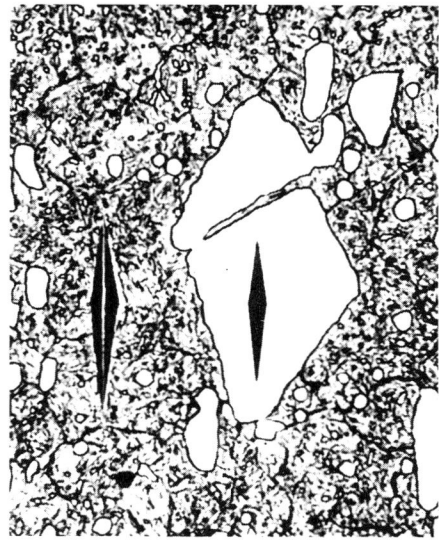

This advantage is further illustrated in Fig. 13. Whereas procedures based on the measurement of diagonals would provide the same area measurement for all three indentation geometries shown, ultrasonic measurement would result in three different values.

As in conventional Vickers and Brinell hardness testing, a single loading force is utilized. Therefore, no time is lost as a result of consecutive application of several loads as in Rockwell testing.

Because only one test load is used, no sensitive displacement-measuring instruments are necessary, and no rigid machine frames are needed. It is also possible in many cases to perform the measurement without clamping or rigidly supporting the test piece, which leads to further simplification in machine design and handling.

Because the sensitivity and the resolving power of the ultrasonic instrument can be raised to high levels, it is possible to measure

Fig. 10 Cross section of tap tooth showing hardness variations caused by overheating during grinding

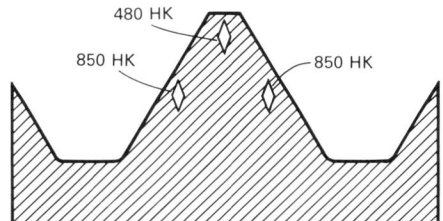

Fig. 8 Knoop hardness readings correlated with case depth

(a) Correlation of Knoop hardness readings with indentations on a cross section of a carburized case. (b) Effect of retained austenite (soft constituent) on the surface hardness is shown.

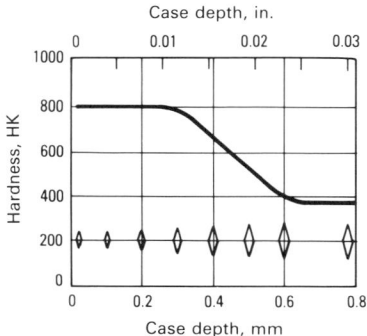
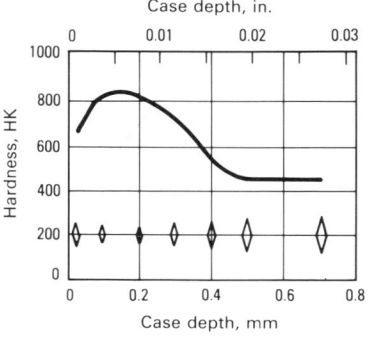

Fig. 11 Components of an ultrasonic hardness tester

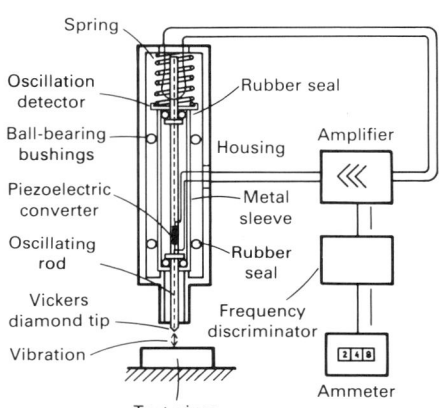

even the smallest indentations. Hardness profile curves can be obtained by untrained personnel automatically in a fraction of the time previously required. The digital display virtually eliminates operator interpretation errors. A memory feature, which will hold the last reading displayed for up to 3 min or until another reading is taken, facilitates any manual recording of data that is necessary.

A one-point calibration procedure allows the instrument to be set up quickly and easily. The few controls and adjustments that are required coupled with a motor-driven probe, facilitate repeatable test results. The portability of ultrasonic microhardness testers allows hardness evaluations to take place not only in a laboratory environment but also on-site, in the field, and in any spec-

imen orientation. Inspection of large parts and on-line in-process inspection hardness testing is possible.

Typical applications of ultrasonic microhardness testing are in the automotive, nuclear, petrochemical, aerospace, and machinery manufacturing industries, including finished goods with hardened surfaces, thin case-hardened parts, thin sheets, strips, coils, platings, and coatings. Often, 100% inspection is possible on critically stressed components. Small components and difficult-to-access parts can also be tested by the ultrasonic microhardness method, either in a hand-held or a fixtured mode (Fig. 14).

Fig. 12 Distortion of diamond pyramid indentations due to elastic effects

(a) Perfect indentation. (b) Pincushion indentation due to material sinking in around the flat faces of the pyramid. (c) Barreled indentation due to ridging of the material around the faces of the indenter

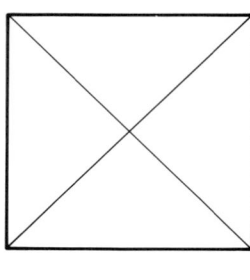 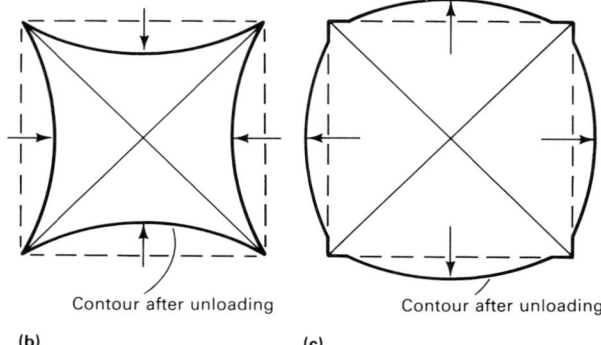

Contour after unloading   Contour after unloading

(a)   (b)   (c)

Fig. 13 Indentations with equal diameters but different areas

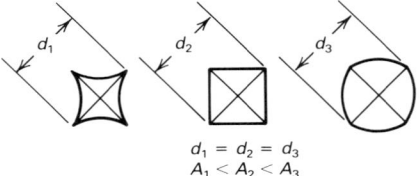

$d_1 = d_2 = d_3$
$A_1 < A_2 < A_3$

Table 3   Comparison of indentation hardness tests

The minimum material thickness for a test is usually taken to be 10 times the indentation depth.

Test	Indenter(s)	Indent — Diagonal or diameter	Depth	Load(s)	Method of measurement	Surface preparation	Tests per hour	Applications	Remarks
Brinell	Ball indenter, 10 mm or 2.5 mm in diameter	1-7 mm	Up to 0.3 mm and 1 mm, respectively, with 2.5-mm and 10-mm-diam balls	3000 kgf for ferrous materials down to 100 kgf for soft metals	Measure diameter of indentation under microscope; read hardness from tables	Specially ground area for measurements of diameter	50 with diameter measurements	Large forged and cast parts	Damage to specimen minimized by use of lightly loaded ball indenter. Indent then less than Rockwell
Rockwell	120° diamond cone, 1/16- to 1/2-in.-diam ball	0.1-1.5 mm	25-375 μm	Major 60-150 kgf Minor 10 kgf	Read hardness directly from meter or digital display	No preparation necessary on many surfaces	300 manually 900 automatically	Forgings, castings, roughly machined parts	Measure depth of penetration, not diameter
Rockwell superficial	As for Rockwell	0.1-0.7 mm	10-110 μm	Major 15-45 kgf Minor 3 kgf	As for Rockwell	Machined surface, ground	As for Rockwell	Critical surfaces of finished parts	A surface test of case hardening and annealing
Vickers	136° diamond pyramid	Measure diagonal, not diameter	0.03-0.1 mm	1-120 kgf	Measure indent with low-power microscope; read hardness from tables	Smooth clean surface, symmetrical if not flat	Up to 180	Fine finished surfaces, thin specimens	Small indent but high local stresses
Microhardness	136° diamond indenter or a Knoop indenter	40 μm	1-4 μm minimum	1 gf-1 kgf	Measure indentation with low-power microscope; read hardness from tables	Polished surface	Up to 60	Surface layers, thin stock down to 200 μm	Laboratory test used on brittle materials or microstructural constituents
Ultrasonic	136° diamond pyramid	15-50 μm	4-18 μm	800 gf	Direct readout onto meter or digital display	Surface better than 1.2 μm for accurate work. Otherwise, up to 3 μm	1200 (limited by speed at which operator can read display)	Thin stock and finished surfaces in any position	Calibration for Young's modulus necessary. 100% testing of finished parts. Completely nondestructive

Fig. 14 Ultrasonic hardness testers

(a) Hand-held instrument with modified probe shoe for testing curved surfaces. (b) Ultrasonic tester mounted in a support for testing small parts

(a)

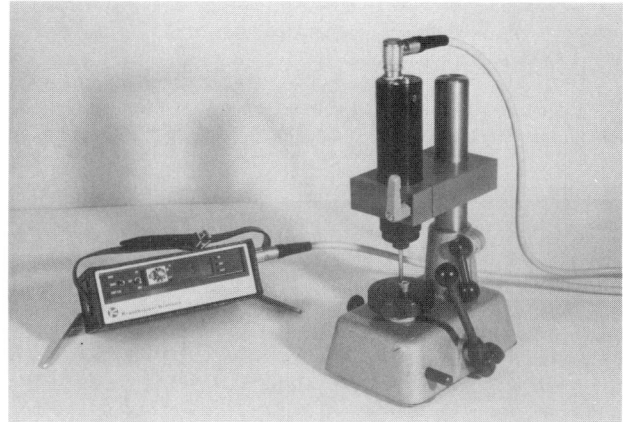

(b)

REFERENCES

1. "Standard Test Methods for Microhardness of Materials," E 384, *Annual Book of ASTM Standards*, ASTM, Philadelphia, 1984, p 497-518
2. C. Kleesattel, "Resonant Sensing Devices," U.S. Patent No. 3,153,388, 1961
3. G.M.L. Gladwell, The Calculation of Mechanical Impedances Relating to the Indenter Vibrating on the Surface of a Semi-Infinite Elastic Body, *J. Sound Vibration*, Vol 8, 1968, p 215-228
4. C. Kleesattel, The Ultrasonic Contact Impedance Testing Method, in *The Echo*, Vol 27, Krautkrämer GmbH, Cologne, West Germany

SELECTED REFERENCES

- "Hardness Testing," in *Metals Handbook*, Vol 11, 8th ed., American Society for Metals, 1976, p 1-20
- V.E. Lysaght and A. DeBellis, *Hardness Testing Handbook*, American Chain and Cable Co., 1969
- L. Small, *Hardness: Theory and Practice*, Service Diamond Tool Co., Ferndale, MI, 1960
- "Standard Test Method for Vickers Hardness of Metallic Materials," E 92, *Annual Book of ASTM Standards*, ASTM, Philadelphia, 1984, p 253-263

Miscellaneous Hardness Tests

MISCELLANEOUS HARDNESS TESTS encompass a number of test methods that have been developed for specific applications. These include dynamic, or "rebound," hardness tests using the Scleroscope, static indentation tests on rubber and plastic products using the durometer, and scratch hardness tests. This article will review the procedures, equipment, and applications associated with these alternate hardness test methods. For information on the more commonly used static indentation tests, see the articles "Rockwell Hardness Testing," "Brinell Hardness Testing," and "Microhardness Testing" in this Section.

Scleroscope Hardness Testing

The Scleroscope* dynamic hardness tester was invented by Albert F. Shore in 1907 and was the first commercially available metallurgical hardness tester produced in the United States. The Scleroscope is used most often when numerous quality checks must be made or for testing very large specimens such as forged steel or wrought alloy steel rolls. In this procedure, a diamond-tipped hammer is dropped from a fixed height onto the surface of the material being tested. The height of rebound of the hammer is a measure of the hardness of the metal.

The Scleroscope scale consists of units determined by dividing into 100 units the average rebound of a hammer from a quenched (to maximum hardness) and untempered water-hardening tool steel. The scale is continued above 100 units to permit testing of materials with hardnesses greater than that of fully hardened tool steel.

Testers. Two types of Scleroscope hardness testers are available. The Model C Scleroscope (Fig. 1) consists of a vertically disposed barrel containing a precision-bore glass tube. A scale, graduated from 0 to 140, is set behind the tube and is visible through it. Hardness is read from the vertical scale, usually with the aid of a reading glass attached to the tester. A pneumatic actuating

*Scleroscope is a registered trademark of Shore Instrument and Manufacturing Co., Inc.

head, affixed to the top of the barrel, is manually operated by use of a rubber bulb and tube. The hammer drops and rebounds within the glass tube.

The Model D Scleroscope hardness tester (Fig. 2) is available with either analog (dial-recording) or digital readouts. The tester consists of a vertically disposed barrel that contains a clutch to arrest the hammer at the maximum height of rebound. This is possible because of the short rebound height. The hammer is longer and heavier than the hammer used in the Model C Scleroscope, developing the same striking energy even though dropping a shorter distance.

Models C, D, and D digital Scleroscopes are available in two calibrations: standard and roll. Standard calibration, which conforms to ASTM E 448, "Scleroscope Hardness Testing of Metallic Materials," has a direct correlation to Rockwell C, Brinell, and Vickers hardness values (Table 1). Roll calibration conforms to ASTM A 427, "Wrought Alloy Steel Rolls for Cold and Hot Reduction," and is used to determine hardness values of homogeneous wrought hardened alloy steel rolls for use in reduction of flat rolled products. This calibration is symbolized by HFRSc or HFRSd. The Model C Scleroscope is also available cali-

Fig. 1 Model C Scleroscopes mounted in stands

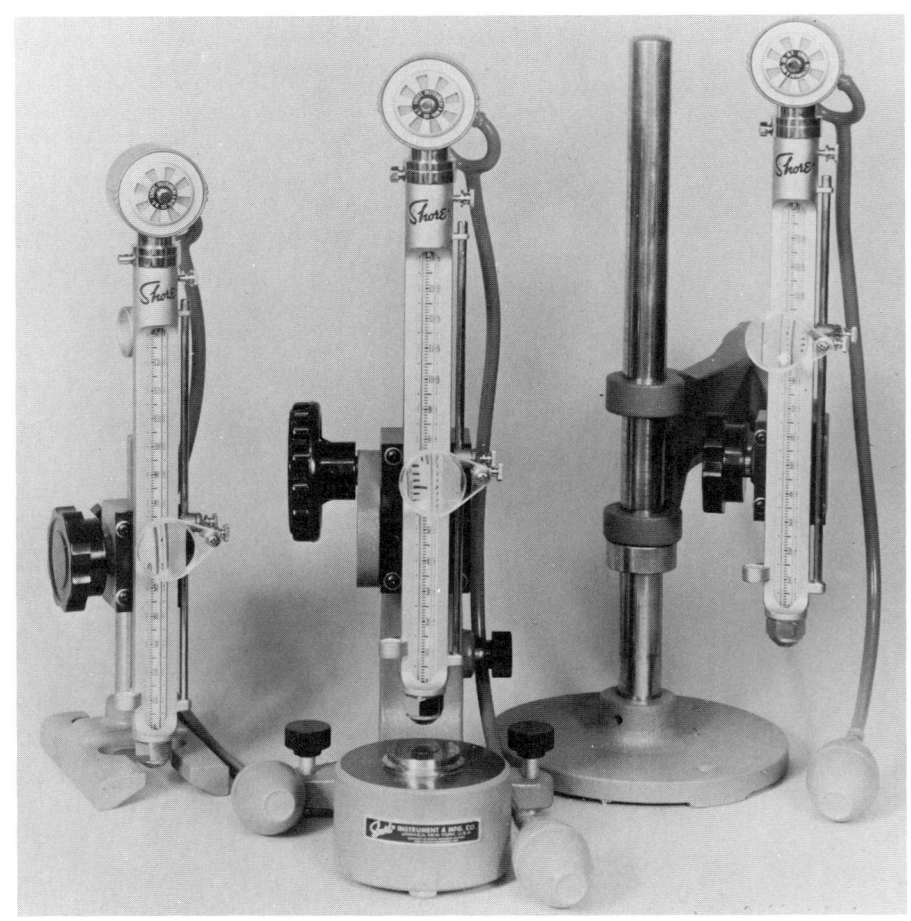

Fig. 2 Model D Scleroscopes mounted in stands
A digital model is also available.

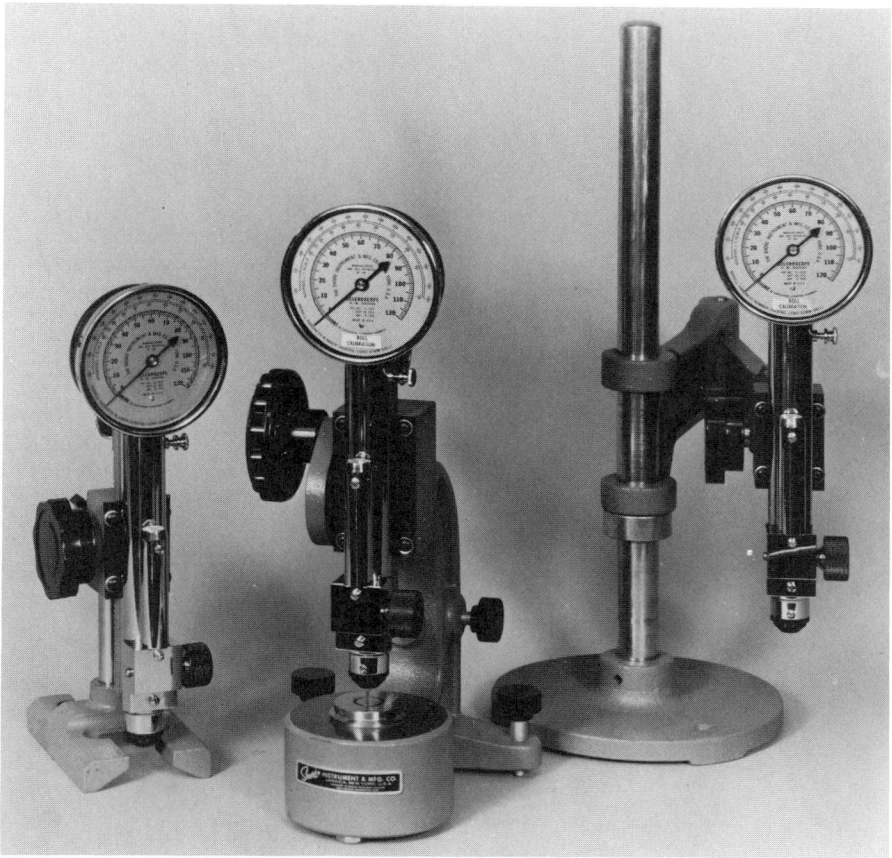

brated in accordance with ASTM C 886, "Scleroscope Testing of Fine Grained Carbon and Graphite Materials." This is referred to as Model C Carbon Calibration.

Both models of Scleroscopes can be mounted on various types of bases, although the Model C Scleroscope is commonly used unmounted when testing large workpieces with a minimum weight of 2.3 kg (5 lb). Due to its critical vertical alignment, the Model D Scleroscope should not be used unmounted, as erroneous readings may result.

Workpiece Surface Finish. In Scleroscope hardness testing, certain workpiece surface finish requirements must be met in order to obtain accurate, consistent readings. An excessively coarse surface finish will yield erratic readings; when necessary, the surface of the workpiece should be filed, machined, ground, or polished.

Care should be taken to avoid overheating or excessively cold working the surface. The surface finish required to obtain reproducible results varies with the hardness of the workpiece. In proceeding from soft metals to hardened steel, the required surface finish ranges from a minimum finish, as produced

by a No. 2 file, to a finely ground or polished finish.

Limitations on Workpiece and Case Thickness. Case-hardened steel with cases as thin as 0.25 mm (0.010 in.) can be accurately hardness tested provided the core hardness is no less than 30 Scleroscope. Softer cores require a minimum case thickness of 0.38 mm (0.015 in.) for accurate results.

Thin strip or sheet may be tested, with some limitations, but only when the Scleroscope hardness tester is mounted in a clamping stand. Ideally, the sheet should be flat and without undulation. If the sheet material is bowed, the concave side should be placed upwards to preclude any possibility of erroneous readings due to spring effect. The minimum thicknesses of sheet in various categories that may be hardness tested with a Scleroscope are as follows:

Hardened steel	0.13 mm (0.005 in.)
Cold finished steel strip	0.25 mm (0.010 in.)
Annealed brass strip	0.38 mm (0.015 in.)
Half-hard brass strip	0.25 mm (0.010 in.)

Test Procedure. To perform a hardness

test with either the Model C or the Model D Scleroscope, the tester should be held or set in a vertical position, with the bottom of the barrel in firm contact with the workpiece. The hammer is elevated and then allowed to fall and strike the surface of the workpiece. The height of rebound is then measured, which indicates the hardness.

When using the Model C Scleroscope, the hammer is elevated by squeezing a pneumatic bulb. The hammer is released by again squeezing the bulb. When using the Model D Scleroscope, the hammer is elevated by turning a knurled control knob clockwise until a definite stop is reached. The hammer strikes the workpiece when the control knob is released, and the reading is recorded on a dial. Hard steel tests about 100, medium-hard about 50, and soft metals 10 to 15.

Vertical Alignment. To minimize error, the hardness tester must be set or held in a vertical position, using the plumb rod or level on the machine to determine vertical alignment. The most accurate readings are obtained with the Scleroscope hardness tester mounted in a C-frame base that rests on three points, two of which are adjustable to facilitate leveling of the anvil and to ensure vertical alignment of the barrel. When using a mounted tester, the opposite sides of the workpiece must be parallel to each other. Vibration impedes the free fall of the hammer, thereby producing low readings, and must be avoided.

Spacing of Indentations. Indentations should be singly spaced at least 0.50 mm (0.020 in.) apart. Flat workpieces with parallel surfaces may be hardness tested within 6 mm (0.25 in.) of the edge when properly clamped.

Taking the Readings. Experience is necessary to interpret readings accurately on a Model C Scleroscope hardness tester. Thin materials, or those weighing less than 2.3 kg (5 lb), must be securely clamped to absorb the inertia of the hammer. The sound of the impact is an indication of the effectiveness of the clamping; a dull thud indicates that the workpiece has been clamped solidly, whereas a hollow ringing sound indicates that the workpiece is not tightly clamped or is warped and not properly supported. Five hardness determinations should be made, and their average taken as representative of the hardness of a particular workpiece.

Calibration. Scleroscope hardness testers are supplied with reference bars (or test blocks) of known hardness. The reference bars can be used correctly only with the Scleroscope mounted in a clamping stand, because they do not have sufficient mass to produce a full rebound of the hammer unless firmly clamped. If actual Scleroscope readings do not correspond to the values of the

Table 1 Approximate hardness conversion numbers for nonaustenitic steels

Scleroscope hardness	Rockwell C hardness number, HRC	Vickers hardness number, HV	Brinell hardness number, HB	
			10-mm standard ball, 3000-kgf load	10-mm carbide ball, 3000-kgf load
97.3 .68		940	. . .	. . .
95.0 .67		900	. . .	. . .
92.7 .66		865	. . .	. . .
90.6 .65		832	. . .	739
88.5 .64		800	. . .	722
86.5 .63		772	. . .	705
84.5 .62		746	. . .	688
82.6 .61		720	. . .	670
80.8 .60		697	. . .	654
79.0 .59		674	. . .	634
77.3 .58		653	. . .	615
75.6 .57		633	. . .	595
74.0 .56		613	. . .	577
72.4 .55		595	. . .	560
70.9 .54		577	. . .	543
69.4 .53		560	. . .	525
67.9 .52		544	500	512
66.5 .51		528	487	496
65.1 .50		513	475	481
63.7 .49		498	464	469
62.4 .48		484	451	455
61.1 .47		471	442	443
59.8 .46		458	432	432
58.5 .45		446	421	421
57.3 .44		434	409	409
56.1 .43		423	400	400
54.9 .42		412	390	390
53.7 .41		402	381	381
52.6 .40		392	371	371
51.5 .39		382	362	362
50.4 .38		372	353	353
49.3 .37		363	344	344
48.2 .36		354	336	336
47.1 .35		345	327	327
46.1 .34		336	319	319
45.1 .33		327	311	311
44.1 .32		318	301	301
43.1 .31		310	294	294
42.2 .30		302	286	286
41.3 .29		294	279	279
40.4 .28		286	271	271
39.5 .27		279	264	264
38.7 .26		272	258	258
37.8 .25		266	253	253
37.0 .24		260	247	247
36.3 .23		254	243	243
35.5 .22		248	237	237
34.8 .21		243	231	231
34.2 .20		238	226	226

Note: These Scleroscope hardness conversions are based on Vickers/Scleroscope hardness relationships developed from Vickers hardness data provided by the National Bureau of Standards for 13 steel reference blocks. Scleroscope hardness values obtained on these blocks by the Shore Instrument and Mfg. Co., Inc., the Roll Manufacturers Institute, and members of this institute, and also on hardness conversions previously published by the American Society for Metals and the Roll Manufacturers Institute.
Source: ASTM E 140, "Standard Hardness Conversion Tables for Metals"

reference bars, the instrument should be returned to the manufacturer for service.

Advantages. The Scleroscope hardness test has several advantages. Tests can be made very rapidly; over 1000 tests per hour are possible. Operation is simple and does not require highly skilled technicians.

The Model C Scleroscope is portable and can be used unmounted for testing workpieces of unlimited size (rolls and large dies). The Scleroscope hardness test is considered a non-marring test; no obvious crater is left, and only in the most unusual instances would the tiny hammer mark be objectionable on a finished workpiece. Additionally, a single scale covers the entire hardness range from the softest to the hardest metals.

Limitations of the Scleroscope hardness test include the necessity of keeping the test instrument in a vertical position so that the hammer can fall freely. Scleroscope hardness tests are more sensitive to variations in surface conditions than other hardness tests. Because readings taken with the Model C Scleroscope are indicated by the maximum rebound of the hammer on the first bounce, even the most experienced operators may disagree by one or two points. Also, the mass or configuration of the part can affect the accuracy of readings.

Hardness Testing Using the Durometer

The durometer is a pocket-sized instrument that measures the indentation hardness of rubber and plastic products. It is manually applied to the test specimen, and the reading is observed on a dial. Laboratory accuracy can also be obtained by mounting the durometer on one of several types of operating stands.

Durometer selection depends on the material being tested. Several types of durometers are available, as shown in Table 2. Of these, only two (Type A and Type D) conform to ASTM Standard D 2240, "Indentation Hardness of Rubber and Plastic by Means of a Durometer."

With the exceptions of Types 00 and 000, all other durometer types are variations of the ASTM Type A and D specifications, although they are not considered ASTM durometers. All durometers are available with either a quadrant or round face (Fig. 3), except for the Type 000, which is available only with a round face.

The scales on both styles run from 0 to 100, in increments of five for the quadrant style and in units for the round. The sweep of the dial hand is 74° for the quadrant style and 265° for the round.

Testing Procedure. Tests specimens should have a minimum thickness of at least 6 mm (0.25 in.), unless it is known that identical results are obtained on thinner specimens. However, since the thickness of the specimen is of progressively less significance on harder specimens, materials harder than 50 HFRSd may have a minimum thickness of 3 mm (0.12 in.). Thinner specimens may be stacked to obtain at least an indicative reading.

Readings should not be taken on an uneven, irregular, or coarsely grained surface. Round or cylindrical surfaces, such as rubber rollers, can be tested by "rocking" the durometer on the convex surface and observing the maximum reading that is attained when the indenter is aligned with the axis of the roller.

Application pressure should be sufficient to ensure firm contact between the flat bottom of the durometer and the test specimen; the reading should be taken within 1 s after firm contact has been established. However, after attaining an initially high reading, the dial hand may gradually recede on specimens exhibiting cold flow or creep characteristics (such as nitrile rubber stock). In such instances, both the instantaneous, or maximum, reading and the reading after a specified time interval—for example, 10 or 15 s—should be recorded.

Table 2 Specifications of durometers

Durometer type	Main spring	Indenter	For use on:
A (conforms to ASTM D 2240) ......822 g		Frustum cone	Soft vulcanized rubber and all elastomeric materials, natural rubber, GR-S, GR-I, neoprene, nitrile rubbers, Thiokol, flexible polyester cast resins, polyacrylic esters, wax, felt, leather, etc.
B ....822 g		Sharp 30° included angle	Moderately hard rubber such as typewriter rollers, platens, etc.
C ....10 lb		Frustum cone	Medium hard rubber and plastics
D (conforms to ASTM D 2240) ......10 lb		Sharp 30° included angle	Hard rubber and the harder grades of plastics such as rigid thermoplastic sheet, Plexiglas, polystyrene, vinyl sheet, cellulose acetate and thermosetting laminates such as formica, paper-filled calendar rolls, calendar bowls, etc.
D0 ....10 lb		3/32-in. sphere	Very dense textile windings, slasher beams, etc.
0 ....822 g		3/32-in. sphere	Soft printers rollers, Artgum, medium-density textile windings of rayon, orlon, nylon, etc.
00 ....4 oz		3/32-in. sphere	Sponge rubber and plastics, low-density textile windings; not for use on foamed latex
000 (available with round dial only) ......4 oz		1/2-in. diam, spherical	Ultrasoft sponge rubber and plastic
T ....822 g		3/32-in. sphere	Medium-density textile windings on spools and bobbins with a maximum diameter of 100 mm (4 in.), types T and T-2 have a concave bottom plate to facilitate centering on cylindrical specimens

Testing Results. Durometer hardness numbers, although arbitrary, have an inverse relationship to indentation by the indenter in thousandths of an inch. For example, a reading of 30 on the Type A durometer on a soft rubber roller indicates an indenter indentation of 1.8 mm (0.070 in.). Similarly, a reading of 90 on a neoprene faucet washer indicates an indenter indentation of 0.25 mm (0.010 in.). The use of the durometer at the extreme ends of the scale (below 20 and above 90) is not recommended. Materials reading above 90 on the Type A scale should be tested with the Type D durometer. Materials reading below 20 on the Type D scale should be tested with a Type A durometer.

One of the most common causes of disagreement in readings among operators is variation in the speed with which the durometer is applied to the elastomer. For example, in testing a high-creep nitrile rubber, if the durometer is applied too rapidly to the test specimen, an erroneously high reading is initially attained, with the dial hand dropping as the durometer is held in contact. At the other extreme, the durometer may be applied too slowly, causing a significant percentage of indenter penetration to occur before the presser foot of the durometer is in flush contact with the test specimen and resulting in an inaccurately low reading.

Disagreement can also occur when an insufficient number of tests have been made.

Reporting the average of five readings gives a testing error of less than one point. Tests on a particular material should all be run at the same temperature.

Durometer Calibration. All durometers are equipped with a test block, which enables the user to ascertain whether the durometer mainspring is in correct calibration at one point on the scale (usually 60 Durometer). The test block consists of a channel with an integral spring on one end of which is a blind hole. When held flush with the test block, the durometer reading should agree with the hardness number stamped on the side of the block, within plus or minus one point.

However, a correct reading on the test block does not mean that the durometer is accurate over its entire scale. Also, the test block itself may be damaged and out of calibration. The test block is not intended for use with the durometer when mounted in an operating stand. Sets of test blocks covering the range of the scale are available.

A durometer calibrating device is also available. This mechanism is recommended for end users who have several durometers to keep in calibration. The durometer may be returned to the manufacturer for periodic inspection.

Auxiliary Equipment. The operating stand is designed to enable absolutely flush application of the durometer to the test specimen, thus eliminating errors in readings due to out-of-perpendicular contact. The stand is intended primarily for testing specimens whose opposite sides are parallel.

Adjustable operating stands are also available to ensure constant velocity and constant load application. This eliminates user error due to too rapid or too slow an application, as discussed earlier. Figure 4 shows durometers mounted in operating stands. Durometers equipped with special application features that facilitate accuracy and ease of operation, including digital readout durometers, are also available.

Scratch Hardness Tests

Scratch hardness tests represent the oldest type of hardness evaluation procedures. The two most common techniques for measuring scratch hardness are the Mohs scale, which is used for testing minerals, and the file hardness test, which is used for testing steels. A third type of scratch hardness test, sometimes referred to as the "plowing test," will not be discussed in detail. This test measures the width of a scratch made by drawing a diamond indenter across the surface under a definite load. Loads on the indenter of 1, 2, 5, 10, and 25 g are commonly used. This is a useful tool for studying the relative hardness

Fig. 3 Round- and quadrant-style durometers

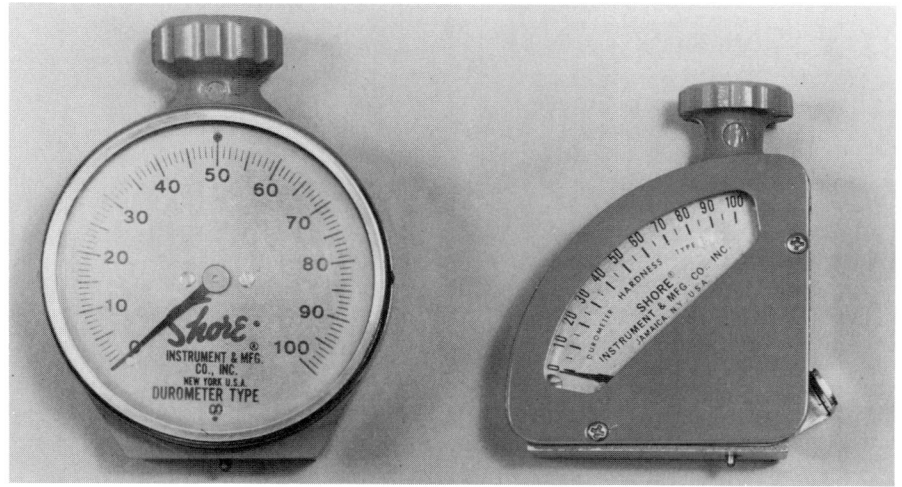

of microconstituents, but it does not lend itself to high reproducibility or extreme accuracy.

The Mohs scale of hardness was devised in 1822 by German mineralogist Friedrich Mohs. The Mohs scale consists of 10 minerals arranged in order from 1 (softest) to 10 (hardest). Each mineral in the scale will scratch all those below it.

Diamond	10
Corundum	9
Topaz	8
Quartz	7
Orthoclase (feldspar)	6
Apatite	5
Fluorite	4
Calcite	3
Gypsum	2
Talc	1

The steps between numbers on the scale are not of equal value; for example, the difference in hardness between 9 and 10 is much greater than between 1 and 2. To determine the hardness of a mineral, one must determine which of the standard materials the unknown will scratch. The hardness will lie between two points on the scale, the point between the mineral which may be scratched and the next one harder.

Materials engineers and metallurgists find little use for the Mohs scale due to its non-quantitative nature. However, the hardness of iron with 0.1% C maximum is between 3 and 4 on the Mohs scale, and copper is between 2 and 3. Fully hardened high-carbon tool steel is between 7 and 8.

The file hardness test was one of the first scratch tests used for evaluating the hardness of metallic materials. The file test is useful in estimating the hardness of steels in

Fig. 4 Durometers mounted in operating stands

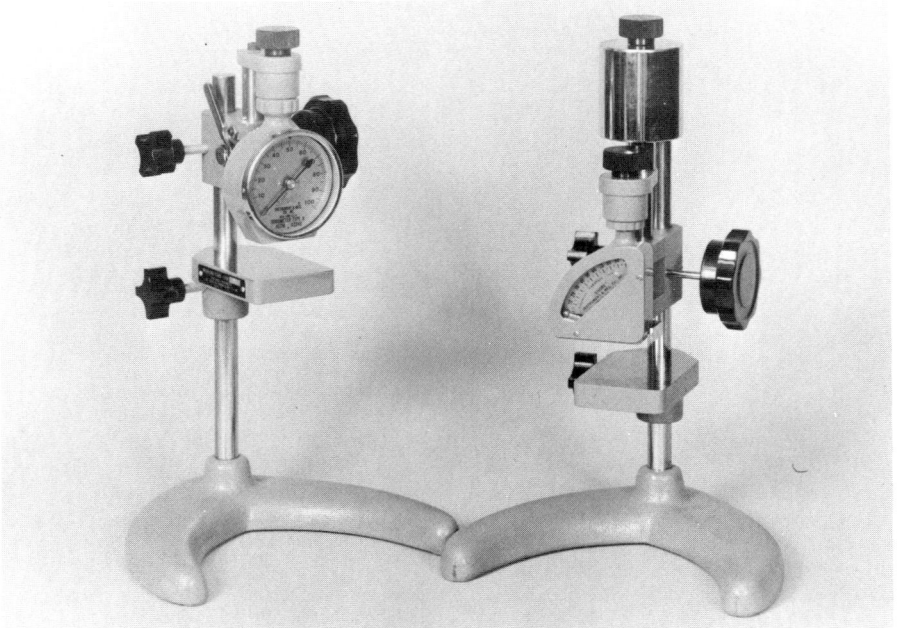

the high hardness ranges. It provides information on soft spots and decarburization quickly and easily, and is readily adaptable to odd shapes and sizes that are difficult to test by other methods.

Standard test files are heat treated to approximately 67 to 70 HRC. The flat face of the file is pressed firmly against, and slowly drawn across, the surface to be tested. If the file does not bite, the material is designated as file hard. The results of the test are influenced by a number of factors, such as pressure, speed, angle of contact, and surface roughness. Consequently, its ability to give reproducible hardness values is rather lim-

ited, and reasonable accuracy is obtained only at the highest hardness levels.

SELECTED REFERENCES

- V.E. Lysaght, *Indentation Hardness Testing*, Reinhold Publishing, New York, 1949
- V.E. Lysaght and A. DeBellis, *Hardness Testing Handbook*, American Chain and Cable Co., Bridgeport, CT, 1969
- L. Small, *Hardness—Theory and Practice*, Service Diamond Tool Co., Ferndale, MI, 1960

Appendix:
Hardness Conversion Tables

From a practical standpoint, it is important to be able to convert the results of one type of hardness test into those of a different test. Because a hardness test does not measure a well-defined property of a material and because all the tests in common use are not based on the same type of measurements, it is not surprising that universal hardness-conversion relationships have not been developed.

Hardness conversions are empirical relationships. The most reliable hardness-conversion data exist for steel, which is harder than 240 HB. The following tables for conversion among Rockwell, Brinell, and dia-

mond pyramid hardness are applicable to heat-treated carbon and alloy steels and to almost all alloy constructional steels and tool steels in the as-forged, annealed, normalized, and quenched and tempered conditions. However, different conversion tables are required for materials with greatly different elastic moduli, or with greater strain-hardening capacity. Conversion tables for nickel and high-nickel alloys, cartridge brass, austenitic stainless steel plate and sheet, and copper can be found in ASTM E 140, "Standard Hardness Conversion Tables for Metals."

The indentation hardness of soft metals

depends on the strain-hardening behavior of the material during the test, which in turn is dependent on the previous degree of strain hardening of the material before the test. The modulus of elasticity also has been shown to influence conversions at high hardness levels. At low hardness levels, conversions between hardness scales measuring depth and those measuring diameter are likewise influenced by differences in the modulus of elasticity.

The standard procedure for reporting converted hardness numbers indicates the measured hardness and test scale in parentheses—for example, 451 HB (48 HRC).

Approximate equivalent hardness numbers for Rockwell B hardness numbers for steel(a)

| Rockwell hardness No. | Vickers hardness No. | Brinell hardness No., 10-mm ball | | Rockwell hardness No. | | | Rockwell superficial hardness No., 1/16-in.-diam ball | | | Knoop hardness No., 500-g load and greater | Scleroscope hardness No. |
		500-kg load	3000-kg load	A scale, 60-kg load, diamond indenter	C scale, 150-kg load, indenter	F scale, 60-kg load, 1/16-in.-diam ball	15T scale, 15-kg load	30T scale, 30-kg load	45T scale, 45-kg load		
98	228	189	228	60.2	(19.9)	...	92.5	81.8	70.9	241	34
97	222	184	222	59.5	(18.6)	...	92.1	81.1	69.9	236	33
96	216	179	216	58.9	(17.2)	...	91.8	80.4	68.9	231	32
95	210	175	210	58.3	(15.7)	...	91.5	79.8	67.9	226	...
94	205	171	205	57.6	(14.3)	...	91.2	79.1	66.9	221	31
93	200	167	200	57.0	(13.0)	...	90.8	78.4	65.9	216	30
92	195	163	195	56.4	(11.7)	...	90.5	77.8	64.8	211	...
91	190	160	190	55.8	(10.4)	...	90.2	77.1	63.8	206	29
90	185	157	185	55.2	(9.2)	...	89.9	76.4	62.8	201	28
89	180	154	180	54.6	(8.0)	...	89.5	75.8	61.8	196	27
88	176	151	176	54.0	(6.9)	...	89.2	75.1	60.8	192	...
87	172	148	172	53.4	(5.8)	...	88.9	74.4	59.8	188	26
86	169	145	169	52.8	(4.7)	...	88.6	73.8	58.8	184	26
85	165	142	165	52.3	(3.6)	...	88.2	73.1	57.8	180	25
84	162	140	162	51.7	(2.5)	...	87.9	72.4	56.8	176	...
83	159	137	159	51.1	(1.4)	...	87.6	71.8	55.8	173	24
82	156	135	156	50.6	(0.3)	...	87.3	71.1	54.8	170	24
81	153	133	153	50.0	...	...	86.9	70.4	53.8	167	...
80	150	130	150	49.5	...	...	86.6	69.7	52.8	164	23
79	147	128	147	48.9	...	...	86.3	69.1	51.8	161	...
78	144	126	144	48.4	...	...	86.0	68.4	50.8	158	22
77	141	124	141	47.9	...	...	85.6	67.7	49.8	155	22
76	139	122	139	47.3	...	...	85.3	67.1	48.8	152	...
75	137	120	137	46.8	...	99.6	85.0	66.4	47.8	150	21
74	135	118	135	46.3	...	99.1	84.7	65.7	46.8	148	21
73	132	116	132	45.8	...	98.5	84.3	65.1	45.8	145	...
72	130	114	130	45.3	...	98.0	84.0	64.4	44.8	143	20
71	127	112	127	44.8	...	97.4	83.7	63.7	43.8	141	20
70	125	110	125	44.3	...	96.8	83.4	63.1	42.8	139	...
69	123	109	123	43.8	...	96.2	83.0	62.4	41.8	137	19
68	121	107	121	43.3	...	95.6	82.7	61.7	40.8	135	19

(continued)

Note: Values in parentheses are beyond normal range and are given for information only.
(a) For carbon and alloy steels in the annealed, normalized, and quenched-and-tempered conditions; less accurate for cold worked condition and for austenitic steels

Approximate equivalent hardness numbers for Rockwell B hardness numbers for steel(a) (continued)

Rockwell hardness No.	Vickers hardness No.	Brinell hardness No., 10-mm ball		Rockwell hardness No.			Rockwell superficial hardness No., 1/16-in.-diam ball			Knoop hardness No., 500-g load and greater	Scleroscope hardness No.
		500-kg load	3000-kg load	A scale, 60-kg load, diamond indenter	C scale, 150-kg load, indenter	F scale, 60-kg load, 1/16-in.-diam ball	15T scale, 15-kg load	30T scale, 30-kg load	45T scale, 45-kg load		
67	119	106	119	42.8	...	95.1	82.4	61.0	39.8	133	19
66	117	104	117	42.3	...	94.5	82.1	60.4	38.7	131	...
65	116	102	116	41.8	...	93.9	81.8	59.7	37.7	129	18
64	114	101	114	41.4	...	93.4	81.4	59.0	36.7	127	18
63	112	99	112	40.9	...	92.8	81.1	58.4	35.7	125	18
62	110	98	110	40.4	...	92.2	80.8	57.7	34.7	124	...
61	108	96	108	40.0	...	91.7	80.5	57.0	33.7	122	17
60	107	95	107	39.5	...	91.1	80.1	56.4	32.7	120	...
59	106	94	106	39.0	...	90.5	79.8	55.7	31.7	118	...
58	104	92	104	38.6	...	90.0	79.5	55.0	30.7	117	...
57	103	91	103	38.1	...	89.4	79.2	54.4	29.7	115	...
56	101	90	101	37.7	...	88.8	78.8	53.7	28.7	114	...
55	100	89	100	37.2	...	88.2	78.5	53.0	27.7	112	...

Note: Values in parentheses are beyond normal range and are given for information only.
(a) For carbon and alloy steels in the annealed, normalized, and quenched-and-tempered conditions; less accurate for cold worked condition and for austenitic steels

Approximate equivalent hardness numbers for Rockwell C hardness numbers for steel(a)

Rockwell hardness No.	Vickers hardness No.	Brinell hardness No., 3000-kg load, 10-mm ball		Rockwell hardness No.			Rockwell superficial hardness No., diamond indenter			Knoop hardness No., 500-g load and greater	Scleroscope hardness No.
		Standard ball	Tungsten carbide ball	A scale, 60-kg load, diamond indenter	B scale, 100-kg load, 1/16-in.-diam ball	D scale, 100-kg load, diamond indenter	15N scale, 15-kg load	30N scale, 30-kg load	45N scale, 45-kg load		
68	940	...	...	85.6	...	76.9	93.2	84.4	75.4	920	97
67	900	...	...	85.0	...	76.1	92.9	83.6	74.2	895	95
66	865	...	...	84.5	...	75.4	92.5	82.8	73.3	870	92
65	832	...	(739)	83.9	...	74.5	92.2	81.9	72.0	846	91
64	800	...	(722)	83.4	...	73.8	91.8	81.1	71.0	822	88
63	772	...	(705)	82.8	...	73.0	91.4	80.1	69.9	799	87
62	746	...	(688)	82.3	...	72.2	91.1	79.3	68.8	776	85
61	720	...	(670)	81.8	...	71.5	90.7	78.4	67.7	754	83
60	697	...	(654)	81.2	...	70.7	90.2	77.5	66.6	732	81
59	674	...	(634)	80.7	...	69.9	89.8	76.6	65.5	710	80
58	653	...	615	80.1	...	69.2	89.3	75.7	64.3	690	78
57	633	...	595	79.6	...	68.5	88.9	74.8	63.2	670	76
56	613	...	577	79.0	...	67.7	88.3	73.9	62.0	650	75
55	595	...	560	78.5	...	66.9	87.9	73.0	60.9	630	74
54	577	...	543	78.0	...	66.1	87.4	72.0	59.8	612	72
53	560	...	525	77.4	...	65.4	86.9	71.2	58.6	594	71
52	544	(500)	512	76.8	...	64.6	86.4	70.2	57.4	576	69
51	528	(487)	496	76.3	...	63.8	85.9	69.4	56.1	558	68
50	513	(475)	481	75.9	...	63.1	85.5	68.5	55.0	542	67
49	498	(464)	469	75.2	...	62.1	85.0	67.6	53.8	526	66
48	484	(451)	455	74.7	...	61.4	84.5	66.7	52.5	510	64
47	471	442	443	74.1	...	60.8	83.9	65.8	51.4	495	63
46	458	432	432	73.6	...	60.0	83.5	64.8	50.3	480	62
45	446	421	421	73.1	...	59.2	83.0	64.0	49.0	466	60
44	434	409	409	72.5	...	58.5	82.5	63.1	47.8	452	58
43	423	400	400	72.0	...	57.7	82.0	62.2	46.7	438	57
42	412	390	390	71.5	...	56.9	81.5	61.3	45.5	426	56
41	402	381	381	70.9	...	56.2	80.9	60.4	44.3	414	55
40	392	371	371	70.4	...	55.4	80.4	59.5	43.1	402	54
39	382	362	362	69.9	...	54.6	79.9	58.6	41.9	391	52
38	372	353	353	69.4	...	53.8	79.4	57.7	40.8	380	51
37	363	344	344	68.9	...	53.1	78.8	56.8	39.6	370	50
36	354	336	336	68.4	(109.0)	52.3	78.3	55.9	38.4	360	49
35	345	327	327	67.9	(108.5)	51.5	77.7	55.0	37.2	351	48
34	336	319	319	67.4	(108.0)	50.8	77.2	54.2	36.1	342	47
33	327	311	311	66.8	(107.5)	50.0	76.6	53.3	34.9	334	46
32	318	301	301	66.3	(107.0)	49.2	76.1	52.1	33.7	326	44
31	310	294	294	65.8	(106.0)	48.4	75.6	51.3	32.5	318	43
30	302	286	286	65.3	(105.5)	47.7	75.0	50.4	31.3	311	42
29	294	279	279	64.7	(104.5)	47.0	74.5	49.5	30.1	304	41
28	286	271	271	64.3	(104.0)	46.1	73.9	48.6	28.9	297	40
27	279	264	264	63.8	(103.0)	45.2	73.3	47.7	27.8	290	39
26	272	258	258	63.3	(102.5)	44.6	72.8	46.8	26.7	284	38
25	266	253	253	62.8	(101.5)	43.8	72.2	45.9	25.5	278	38
24	260	247	247	62.4	(101.0)	43.1	71.6	45.0	24.3	272	37
23	254	243	243	62.0	100.0	42.1	71.0	44.0	23.1	266	36
22	248	237	237	61.5	99.0	41.6	70.5	43.2	22.0	261	35
21	243	231	231	61.0	98.5	40.9	69.9	42.3	20.7	256	35

Note: Values in parentheses are beyond normal range and are given for information only.
(a) For carbon and alloy steels in the annealed, normalized, and quenched-and-tempered conditions; less accurate for cold worked condition and for austenitic steels

Approximate equivalent hardness numbers for Brinell hardness numbers for steel(a)

Brinell indentation diam, mm	Brinell hardness No. (b), 3000-kg load, 10-mm ball Standard ball	Tungsten carbide ball	Vickers hardness No.	A scale, 60-kg load, diamond indenter	B scale, 100-kg load, 1/16-in. diam ball	C scale, 150-kg load, diamond indenter	D scale, 100-kg load, diamond indenter	15N scale, 15-kg load	30N scale, 30-kg load	45N scale, 45-kg load	Knoop hardness No., 500-g load and greater	Sclero-scope hardness No.
2.25	···	(745)	840	84.1	···	65.3	74.8	92.3	82.2	72.2	852	91
2.30	···	(712)	783	83.1	···	63.4	73.4	91.6	80.5	70.4	808	···
2.35	···	(682)	737	82.2	···	61.7	72.0	91.0	79.0	68.5	768	84
2.40	···	(653)	697	81.2	···	60.0	70.7	90.2	77.5	66.5	732	81
2.45	···	627	667	80.5	···	58.7	69.7	89.6	76.3	65.1	703	79
2.50	···	601	640	79.8	···	57.3	68.7	89.0	75.1	63.5	677	77
2.55	···	578	615	79.1	···	56.0	67.7	88.4	73.9	62.1	652	75
2.60	···	555	591	78.4	···	54.7	66.7	87.8	72.7	60.6	626	73
2.65	···	534	569	77.8	···	53.5	65.8	87.2	71.6	59.2	604	71
2.70	···	514	547	76.9	···	52.1	64.7	86.5	70.3	57.6	579	70
2.75	(495)	···	539	76.7	···	51.6	64.3	86.3	69.9	56.9	571	···
2.75	···	495	528	76.3	···	51.0	63.8	85.9	69.4	56.1	558	68
2.80	(477)	···	516	75.9	···	50.3	63.2	85.6	68.7	55.2	545	···
2.80	···	477	508	75.6	···	49.6	62.7	85.3	68.2	54.5	537	66
2.85	(461)	···	495	75.1	···	48.8	61.9	84.9	67.4	53.5	523	···
2.85	···	461	491	74.9	···	48.5	61.7	84.7	67.2	53.2	518	65
2.90	444	···	474	74.3	···	47.2	61.0	84.1	66.0	51.7	499	···
2.90	···	444	472	74.2	···	47.1	60.8	84.0	65.8	51.5	496	63
2.95	429	429	455	73.4	···	45.7	59.7	83.4	64.6	49.9	476	61
3.00	415	415	440	72.8	···	44.5	58.8	82.8	63.5	48.4	459	59
3.05	401	401	425	72.0	···	43.1	57.8	82.0	62.3	46.9	441	58
3.10	388	388	410	71.4	···	41.8	56.8	81.4	61.1	45.3	423	56
3.15	375	375	396	70.6	···	40.4	55.7	80.6	59.9	43.6	407	54
3.20	363	363	383	70.0	···	39.1	54.6	80.0	58.7	42.0	392	52
3.25	352	352	372	69.3	(110.0)	37.9	53.8	79.3	57.6	40.5	379	51
3.30	341	341	360	68.7	(109.0)	36.6	52.8	78.6	56.4	39.1	367	50
3.35	331	331	350	68.1	(108.5)	35.5	51.9	78.0	55.4	37.8	356	48
3.40	321	321	339	67.5	(108.0)	34.3	51.0	77.3	54.3	36.4	345	47
3.45	311	311	328	66.9	(107.5)	33.1	50.0	76.7	53.3	34.4	336	46
3.50	302	302	319	66.3	(107.0)	32.1	49.3	76.1	52.2	33.8	327	45
3.55	293	293	309	65.7	(106.0)	30.9	48.3	75.5	51.2	32.4	318	43
3.60	285	285	301	65.3	(105.5)	29.9	47.6	75.0	50.3	31.2	310	42
3.65	277	277	292	64.6	(104.5)	28.8	46.7	74.4	49.3	29.9	302	41
3.70	269	269	284	64.1	(104.0)	27.6	45.9	73.7	48.3	28.5	294	40
3.75	262	262	276	63.6	(103.0)	26.6	45.0	73.1	47.3	27.3	286	39
3.80	255	255	269	63.0	(102.0)	25.4	44.2	72.5	46.2	26.0	279	38
3.85	248	248	261	62.5	(101.0)	24.2	43.2	71.7	45.1	24.5	272	37
3.90	241	241	253	61.8	100.0	22.8	42.0	70.9	43.9	22.8	265	36
3.95	235	235	247	61.4	99.0	21.7	41.4	70.3	42.9	21.5	259	35
4.00	229	229	241	60.8	98.2	20.5	40.5	69.7	41.9	20.1	253	34
4.05	223	223	234	···	97.3	(19.0)	···	···	···	···	247	···
4.10	217	217	228	···	96.4	(17.7)	···	···	···	···	242	33
4.15	212	212	222	···	95.5	(16.4)	···	···	···	···	237	32
4.20	207	207	218	···	94.6	(15.2)	···	···	···	···	232	31
4.25	201	201	212	···	93.7	(13.8)	···	···	···	···	227	···
4.30	197	197	207	···	92.8	(12.7)	···	···	···	···	222	30
4.35	192	192	202	···	91.9	(11.5)	···	···	···	···	217	29
4.40	187	187	196	···	90.9	(10.2)	···	···	···	···	212	···
4.45	183	183	192	···	90.0	(9.0)	···	···	···	···	207	28
4.50	179	179	188	···	89.0	(8.0)	···	···	···	···	202	27
4.55	174	174	182	···	88.0	(6.7)	···	···	···	···	198	···
4.60	170	170	178	···	87.0	(5.4)	···	···	···	···	194	26
4.65	167	167	175	···	86.0	(4.4)	···	···	···	···	190	···
4.70	163	163	171	···	85.0	(3.3)	···	···	···	···	186	25
4.75	159	159	167	···	83.9	(2.0)	···	···	···	···	182	···
4.80	156	156	163	···	82.9	(0.9)	···	···	···	···	178	24
4.85	152	152	159	···	81.9	···	···	···	···	···	174	···
4.90	149	149	156	···	80.8	···	···	···	···	···	170	23
4.95	146	146	153	···	79.7	···	···	···	···	···	166	···
5.00	143	143	150	···	78.6	···	···	···	···	···	163	22
5.10	137	137	143	···	76.4	···	···	···	···	···	157	21
5.20	131	131	137	···	74.2	···	···	···	···	···	151	···
5.30	126	126	132	···	72.0	···	···	···	···	···	145	20
5.40	121	121	127	···	69.8	···	···	···	···	···	140	19
5.50	116	116	122	···	67.6	···	···	···	···	···	135	18
5.60	111	111	117	···	65.4	···	···	···	···	···	131	17

Note: Values in parentheses are beyond normal range and are given for information only.

(a) For carbon and alloy steels in the annealed, normalized, and quenched-and-tempered conditions; less accurate for cold worked condition and for austenitic steels. (b) Brinell numbers are based on the diameter of impressed indentation. If the ball distorts (flattens) during test, Brinell numbers will vary in accordance with the degree of such distortion when related to hardnesses determined with a Vickers diamond pyramid, Rockwell diamond indenter, or other indenter that does not sensibly distort. At high hardnesses, therefore, the relationship between Brinell and Vickers or Rockwell scales is affected by the type of ball used. Standard steel balls tend to flatten slightly more than tungsten carbide balls, resulting in a larger indentation and a lower Brinell number than shown by a tungsten carbide ball. Thus, on a specimen of about 539 to 547 HV, a standard ball will leave a 2.75-mm indentation (495 HB), and a tungsten carbide ball a 2.70-mm indentation (514 HB). Conversely, identical indentation diameters for both types of ball will correspond to different Vickers and Rockwell values. Thus, if indentations in two different specimens both are 2.75 mm in diameter (495 HB), the specimen tested with a standard ball has a Vickers hardness of 539, whereas the specimen tested with a tungsten carbide ball has a Vickers hardness of 528.

Approximate equivalent hardness numbers for Vickers (diamond pyramid) hardness numbers for steel(a)

Vickers hardness No.	Brinell hardness No., 3000-kg load, 10-mm ball Standard ball	Tungsten carbide ball	Rockwell hardness No. A scale, 60-kg load, diamond indenter	B scale, 100-kg load, 1/16-in.-diam ball	C scale, 150-kg load, diamond indenter	D scale, 100-kg load, diamond indenter	Rockwell superficial (diamond pyramid) hardness No., diamond indenter 15N scale, 15-kg load	30N scale, 30-kg load	45N scale, 45-kg load	Knoop hardness No., 500-g load and greater	Scleroscope hardness No.
940	...	...	85.6	...	68.0	76.9	93.2	84.4	75.4	920	97
920	...	...	85.3	...	67.5	76.5	93.0	84.0	74.8	908	96
900	...	...	85.0	...	67.0	76.1	92.9	83.6	74.2	895	95
880	...	(767)	84.7	...	66.4	75.7	92.7	83.1	73.6	882	93
860	...	(757)	84.4	...	65.9	75.3	92.5	82.7	73.1	867	92
840	...	(745)	84.1	...	65.3	74.8	92.3	82.2	72.2	852	91
820	...	(733)	83.8	...	64.7	74.3	92.1	81.7	71.8	837	90
800	...	(722)	83.4	...	64.0	73.8	91.8	81.1	71.0	822	88
780	...	(710)	83.0	...	63.3	73.3	91.5	80.4	70.2	806	87
760	...	(698)	82.6	...	62.5	72.6	91.2	79.7	69.4	788	86
740	...	(684)	82.2	...	61.8	72.1	91.0	79.1	68.6	772	84
720	...	(670)	81.8	...	61.0	71.5	90.7	78.4	67.7	754	83
700	...	(656)	81.3	...	60.1	70.8	90.3	77.6	66.7	735	81
690	...	(647)	81.1	...	59.7	70.5	90.1	77.2	66.2	725	...
680	...	(638)	80.8	...	59.2	70.1	89.8	76.8	65.7	716	80
670	...	(630)	80.6	...	58.8	69.8	89.7	76.4	65.3	706	...
660	...	620	80.3	...	58.3	69.4	89.5	75.9	64.7	697	79
650	...	611	80.0	...	57.8	69.0	89.2	75.5	64.1	687	78
640	...	601	79.8	...	57.3	68.7	89.0	75.1	63.5	677	77
630	...	591	79.5	...	56.8	68.3	88.8	74.6	63.0	667	76
620	...	582	79.2	...	56.3	67.9	88.5	74.2	62.4	657	75
610	...	573	78.9	...	55.7	67.5	88.2	73.6	61.7	646	...
600	...	564	78.6	...	55.2	67.0	88.0	73.2	61.2	636	74
590	...	554	78.4	...	54.7	66.7	87.8	72.7	60.5	625	73
580	...	545	78.0	...	54.1	66.2	87.5	72.1	59.9	615	72
570	...	535	77.8	...	53.6	65.8	87.2	71.7	59.3	604	...
560	...	525	77.4	...	53.0	65.4	86.9	71.2	58.6	594	71
550	(505)	517	77.0	...	52.3	64.8	86.6	70.5	57.8	583	70
540	(496)	507	76.7	...	51.7	64.4	86.3	70.0	57.0	572	69
530	(488)	497	76.4	...	51.1	63.9	86.0	69.5	56.2	561	68
520	(480)	488	76.1	...	50.5	63.5	85.7	69.0	55.6	550	67
510	(473)	479	75.7	...	49.8	62.9	85.4	68.3	54.7	539	...
500	(465)	471	75.3	...	49.1	62.2	85.0	67.7	53.9	528	66
490	(456)	460	74.9	...	48.4	61.6	84.7	67.1	53.1	517	65
480	(448)	452	74.5	...	47.7	61.3	84.3	66.4	52.2	505	64
470	441	442	74.1	...	46.9	60.7	83.9	65.7	51.3	494	...
460	433	433	73.6	...	46.1	60.1	83.6	64.9	50.4	482	62
450	425	425	73.3	...	45.3	59.4	83.2	64.3	49.4	471	...
440	415	415	72.8	...	44.5	58.8	82.8	63.5	48.4	459	59
430	405	405	72.3	...	43.6	58.2	82.3	62.7	47.4	447	58
420	397	397	71.8	...	42.7	57.5	81.8	61.9	46.4	435	57
410	388	388	71.4	...	41.8	56.8	81.4	61.1	45.3	423	56
400	379	379	70.8	...	40.8	56.0	80.8	60.2	44.1	412	55
390	369	369	70.3	...	39.8	55.2	80.3	59.3	42.9	400	...
380	360	360	69.8	(110.0)	38.8	54.4	79.8	58.4	41.7	389	52
370	350	350	69.2	...	37.7	53.6	79.2	57.4	40.4	378	51
360	341	341	68.7	(109.0)	36.6	52.8	78.6	56.4	39.1	367	50
350	331	331	68.1	...	35.5	51.9	78.0	55.4	37.8	356	48
340	322	322	67.6	(108.0)	34.4	51.1	77.4	54.4	36.5	346	47
330	313	313	67.0	...	33.3	50.2	76.8	53.6	35.2	337	46
320	303	303	66.4	(107.0)	32.2	49.4	76.2	52.3	33.9	328	45
310	294	294	65.8	...	31.0	48.4	75.6	51.3	32.5	318	...
300	284	284	65.2	(105.5)	29.8	47.5	74.9	50.2	31.1	309	42
295	280	280	64.8	...	29.2	47.1	74.6	49.7	30.4	305	...
290	275	275	64.5	(104.5)	28.5	46.5	74.2	49.0	29.5	300	41
285	270	270	64.2	...	27.8	46.0	73.8	48.4	28.7	296	...
280	265	265	63.8	(103.5)	27.1	45.3	73.4	47.8	27.9	291	40
275	261	261	63.5	...	26.4	44.9	73.0	47.2	27.1	286	39
270	256	256	63.1	(102.0)	25.6	44.3	72.6	46.4	26.2	282	38
265	252	252	62.7	...	24.8	43.7	72.1	45.7	25.2	277	...
260	247	247	62.4	(101.0)	24.0	43.1	71.6	45.0	24.3	272	37
255	243	243	62.0	...	23.1	42.2	71.1	44.2	23.2	267	...
250	238	238	61.6	99.5	22.2	41.7	70.6	43.4	22.2	262	36
245	233	233	61.2	...	21.3	41.1	70.1	42.5	21.1	258	35
240	228	228	60.7	98.1	20.3	40.3	69.6	41.7	19.9	253	34
230	219	219	...	96.7	(18.0)	...	...	...	...	243	33

(continued)

Note: Values in parentheses are beyond normal range and are given for information only.
(a) For carbon and alloy steels in the annealed, normalized, and quenched-and-tempered conditions; less accurate for cold worked condition and for austenitic steels

Approximate equivalent hardness numbers for Vickers (diamond pyramid) hardness numbers for steel(a) (continued)

| Vickers hardness No. | Brinell hardness No., 3000-kg load, 10-mm ball | | Rockwell hardness No. | | | | Rockwell superficial (diamond pyramid) hardness No., diamond indenter | | | Knoop hardness No., 500-g load and greater | Sclero-scope hardness No. |
	Standard ball	Tungsten carbide ball	A scale, 60-kg load, diamond indenter	B scale, 100-kg load, 1/16-in.-diam ball	C scale, 150-kg load, diamond indenter	D scale, 100-kg load, diamond indenter	15N scale, 15-kg load	30N scale, 30-kg load	45N scale, 45-kg load		
220	209	209	...	95.0	(15.7)	...	...	...	...	234	32
210	200	200	...	93.4	(13.4)	...	...	...	...	226	30
200	190	190	...	91.5	(11.0)	...	...	...	...	216	29
190	181	181	...	89.5	(8.5)	...	...	...	...	206	28
180	171	171	...	87.1	(6.0)	...	...	...	...	196	26
170	162	162	...	85.0	(3.0)	...	...	...	...	185	25
160	152	152	...	81.7	(0.0)	...	...	...	...	175	23
150	143	143	...	78.7	...	...	...	...	...	164	22
140	133	133	...	75.0	...	...	...	...	...	154	21
130	124	124	...	71.2	...	...	...	...	...	143	20
120	114	114	...	66.7	...	...	...	...	...	133	18
110	105	105	...	62.3	...	...	...	...	...	123	...
100	95	95	...	56.2	...	...	...	...	...	112	...
95	90	90	...	52.0	...	...	...	...	...	107	...
90	86	86	...	48.0	...	...	...	...	...	102	...
85	81	81	...	41.0	...	...	...	...	...	97	...

Note: Values in parentheses are beyond normal range and are given for information only.
(a) For carbon and alloy steels in the annealed, normalized, and quenched-and-tempered conditions; less accurate for cold worked condition and for austenitic steels

Bend Testing

Introduction

BEND TESTS are conducted to determine the ductility or strength of a material. The tests typically used are discussed in this Section in articles that detail test methods, apparatus, procedures, specimen preparation, and interpretation and reporting of results. The article "Bending Ductility Tests" also includes representative test data for many metals.

Bend tests for ductility differ fundamentally from other mechanical tests in that most mechanical tests are designed to give a quantitative result and have an objective endpoint. In contrast, bending ductility tests give a pass/fail result with a subjective endpoint; the test operator judges whether a surface has undergone cracking.

The bending ductility test developed as a shop floor material inspection test because of its pass/fail qualities and the simplicity and low cost of the required tooling. As a consequence, the development of bending ductility test methods and apparatus has been carried out by users rather than by mechanical test equipment manufacturers.

Test procedures and specimen preparation methods have evolved without close attention to detail. Therefore, despite the value of the test and its long history of use, there has been minimal standardization. There are, however, two ASTM standards—ASTM E 190, "Guided Bend Test for Ductility of Welds" (Ref 1), and ASTM E 290, "Semi-Guided Bend Test for Ductility of Metallic Materials" (Ref 2)—which provide guidelines for testing strip, sheet, plate, and weldments.

Tests for determining the bending strength of metals have not been used widely, although the information from such tests is clearly useful. Because of the relative complexity of stress-strain relationships in bending, bending strength usually is not quantified beyond that provided by the assumption of linear elastic behavior. For this reason, a review of stress-strain relationships in bending is included at the beginning of this Section. Presentation of analytical information on stress and strain with descriptions of bend test methodology is intended to stimulate more quantitative development in this area.

ASTM E 855, "Standard Methods of Bend Testing of Metallic Flat Materials for Spring Applications" (Ref 3), discusses the techniques for determining the bending strength of metal. These testing methodologies are described extensively in this Section, with greater emphasis on interpretation, analysis, and the influence of metallurgical variables.

Bending tests that have been developed for brittle materials, coatings, construction (girder and beam) sections, and other specific product forms are not covered in this Section. However, descriptions of these test methods can be found in the ASTM standards.

REFERENCES

1. "Standard Method for Guided Bend Test for Ductility of Welds," ASTM E 190, *Annual Book of ASTM Standards*, Vol 03.01, ASTM, Philadelphia, 1984, p 336-339
2. "Standard Method for Semi-Guided Bend Test for Ductility of Metallic Materials," ASTM E 290, *Annual Book of ASTM Standards*, Vol 03.01, ASTM, Philadelphia, 1984, p 430-434
3. "Standard Methods of Bend Testing of Metallic Flat Materials for Spring Applications," ASTM E 855, *Annual Book of ASTM Standards*, Vol 03.01, ASTM, Philadelphia, 1984, p 788-804

Stress-Strain Relationships in Bending

By P. Dadras
Associate Professor
School of Engineering
Wright State University

A CHARACTERISTIC FEATURE of bending is the inhomogeneous (nonuniform) nature of the deformation. Therefore, in a bent specimen the strain and stress at a given point are dependent on the location of the point with respect to the neutral axis of the cross-sectional area of the specimen. In cases where the applied bending moment varies along the length of the specimen (as in three-point bending), the strains and stresses become dependent on axial location as well. Because of these inhomogeneities, a full appreciation of stress and strain distributions is of utmost importance in bending analyses and computations. Stress-strain relationships, strain curvature, and stress-moment equations are discussed in this article. The formulations are for elastic, elastic-plastic, and fully plastic bending conditions.

Elastic Bending

Elastic analysis of bending deformation can be performed by simple-beam theory (Ref 1), elasticity solutions (Ref 2), and numerical methods such as the finite-difference and Rayleigh-Ritz methods (Ref 3). Generally, numerical methods are suitable for bending of specimens that are subjected to complex loading patterns and that have irregular and/or varying cross-sectional areas. Elasticity solutions are useful when accuracies better than ~5% are desired. Simple-beam theory is used in most testing applications in which plates, strips, bars, and rods are bent in three- or four-point bending modes. The basic assumptions of the simple-beam theory for pure elastic bending (shear force = 0) are: (1) all sections that are initially plane and perpendicular to the axis of the beam remain plane and perpendicular to it after bending; (2) all longitudinal elements

(fibers) bend into concentric circular arcs (hence, cylindrical bending); and (3) a one-dimensional stress state is assumed, and the same stress-strain relationship is used for tension and compression.

The first assumption implies a linear distribution for fiber elongations and contractions, as shown in Fig. 1(a). The resulting strain distributions (Fig. 1b and 1c) are given by: Engineering bending strain,

$$\epsilon_x = -\frac{y}{R_n} \qquad \text{(Eq 1)}$$

True bending strain,

$$\epsilon_x = \ln\left(1 - \frac{y}{R_n}\right) \qquad \text{(Eq 2)}$$

where R_n is the radius of curvature of the neutral axis. For $\epsilon_x \leq 0.1$, the difference between these two strain definitions is $\leq 5\%$. Therefore, for most elastic bendings, the engineering strain definition is sufficiently accurate and more convenient. As shown in Fig. 1, the true strain description indicates a nonlinear strain distribution and a maximum compressive strain in the concave inner fiber that is greater than the maximum tensile strain in the outermost fiber.

For a linear elastic material:

$$\epsilon_x = \frac{1}{E}[\sigma_x - \nu(\sigma_y + \sigma_z)] \qquad \text{(Eq 3)}$$

where E is the modulus of elasticity for axial loading (Young's modulus), and ν is the Poisson's ratio. From the third assumption, $\sigma_y = \sigma_z = 0$. Therefore,

$$\sigma_x = -E\frac{y}{R_n} \qquad \text{(Eq 4)}$$

When expressions similar to Eq 3 are written for ϵ_z and ϵ_y, the following results are obtained:

$$\epsilon_z = \epsilon_y = \nu\frac{y}{R_n} \qquad \text{(Eq 5)}$$

Also:

$$\epsilon_{xy} = \epsilon_{yz} = \epsilon_{xz} = 0 \qquad \text{(Eq 6)}$$

Figure 2 illustrates grid deformations in the longitudinal and cross directions. A transverse curvature, called anticlastic curvature (Ref 4) develops with a radius of curvature equal to (R_n/ν). Experimental evidence indicates that the actual radius of anticlastic curvature depends on $(b^2/2R_nh)$, where b is the width and $2h$ is the thickness of the beam. For $(b^2/2R_nh) \leq 1$ (i.e., narrow beams), the R_n/ν estimate is sufficiently accurate. For plates and wide beams $(b^2/2R_nh > 20)$, the anticlastic deformation is primarily concentrated at the edges.

The location of the neutral axis, which is the line of zero fiber stress in any given section of a member subject to bending, is determined from the condition of zero axial forces acting on the beam. Therefore,

$$\int_A \sigma_x \, dA = -\frac{E}{R_n}\int_A y \, dA = 0 \qquad \text{(Eq 7)}$$

where A is the cross-sectional area. This equation indicates that the first moment of the cross-sectional area about neutral axis is zero, which implies that the neutral and the central (centroidal) axes are coincident.

The moment-curvature and moment-stress relationships are found by equating the externally applied bending moment to the internal bending moment at any cross section:

$$M = -\int_A (\sigma_x dA)y \qquad \text{(Eq 8)}$$

where M is the bending moment.

Fig. 1 Distribution of strain determined by the simple-beam theory

(a) Linear distribution for fiber elongations and contractions. (b) Distribution of engineering strain.
(c) Distribution of true strain. R_n = radius of neutral axis; R_i = inner radius; R_o = outer radius

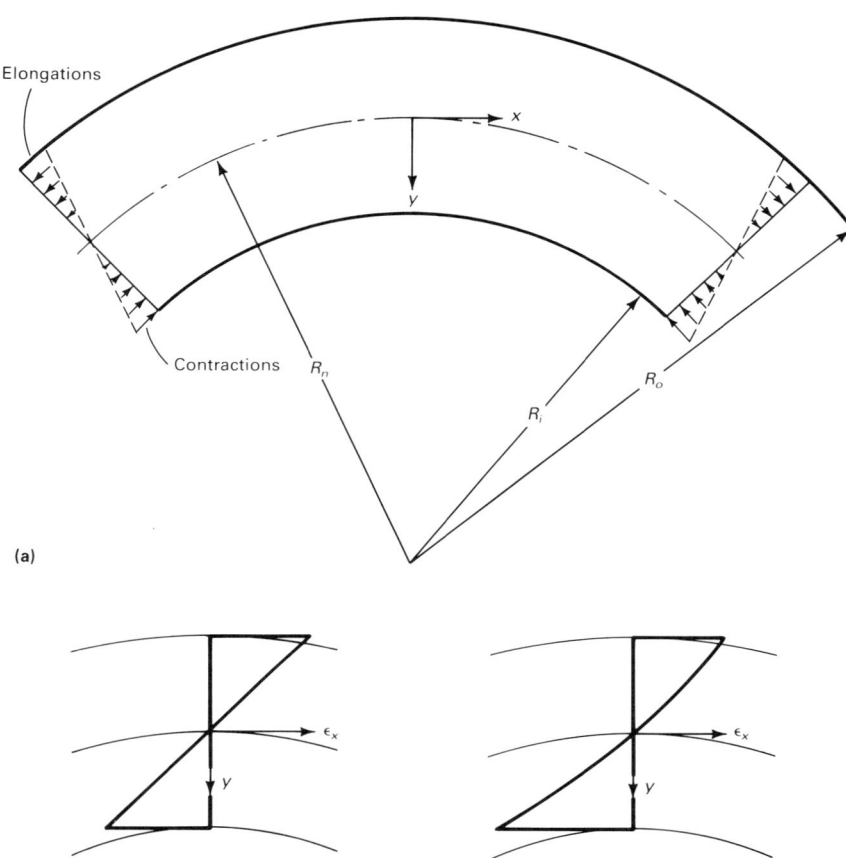

(a)

(b)   (c)

For the linear stress distribution shown in Fig. 3, the results are:

$$R_n = \frac{E}{M} I_z \qquad \text{(Eq 9)}$$

and

$$\sigma_x = -\frac{My}{I_z} \qquad \text{(Eq 10)}$$

where I_z is the area moment of inertia of the cross section about the z axis, which is coincident on the centroidal axis.

The sign (positive or negative) of the bending moment is found from the following relationship: (sign of the bending moment) = (sign of the moment vector) × (sign of the outward normal to the section). For example, the bending moment acting on section ABCD in Fig. 4 is positive, because the moment vector, $\vec{M}$, is in the positive z direction (right-hand rule) and outward normal of the plane, $\vec{n}$, is also in the positive x direction.

Noncylindrical Bending

In the previous section it was stated that the simple beam theory considers the effect of bending moments alone (shear force = 0) and assumes a bent configuration consisting of concentric circular arcs (cylindrical bending). For cylindrical bending to occur, these conditions must be met:

1. Bending must occur under the action of bending moments alone, which implies zero applied shear force.
2. The cross-sectional area of the beam must possess at least one axis of symmetry.
3. The vector of the applied bending moment must be in the direction of an axis of symmetry.

For asymmetrical beams such as Z-sections and unequal L-sections, conditions 2 and 3 for cylindrical bending are not satisfied; for unsymmetrical bending of symmetrical beams, condition 3 is not met. These

cases are significant in structural design and will not be considered here. Information on unsymmetrical loading of straight beams can be found in Ref 2.

In a majority of testing applications such as three-point bending, roll bending, and press-brake forming, the applied bending moment varies along the length of the specimen. Because shear force $V = (dM/dx)$, such variations in the bending moment imply a non-zero shear force. Therefore, condition 1 for cylindrical bending is not met. The resulting shear-stress, τ_{xy}, which is determined from the equilibrium considerations at a typical section m-n (Fig. 5a), is:

$$\tau_{xy} = \frac{VQ}{I_z b} \qquad \text{(Eq 11)}$$

where

$$Q = \int_{A_s} y \, dA \qquad \text{(Eq 12)}$$

is the first moment of the shaded area (A_s) with respect to the neutral axis (Fig. 5b). The first moment of the unshaded area with respect to the neutral axis gives the same Q. The distribution of τ_{xy} for a rectangular cross section is shown in Fig. 5c.

Elastic-Plastic Bending

The limit for elastic bending, which is the onset of elastic-plastic bending, is reached when the maximum fiber strain $(\epsilon_x)_{max} = (h/R_n)$ becomes equal to (σ_y/E), where σ_y is the yield strength and E is Young's modulus of the material. For bending beyond this limit, the beam consists of a central elastic core and two plastically deforming zones remote from the neutral axis. For accurate analysis of elastic-plastic bending, factors such as the shift of the neutral axis from the centroidal axis and the effect of radial (transverse) stresses must be considered. However, in view of analytical and computational difficulties, an approximate method, which is an extension of the simple-beam theory, is commonly employed. Therefore, the three assumptions stated for elastic bending will be enforced. The location of the neutral axis is assumed to be fixed at the centroidal axis, as it is for elastic bending.

For a beam with the stress-strain curve shown in Fig. 6(a), the development of longitudinal strain and stress at different stages of deformation are shown in Fig. 6(b) and 6(c), respectively. In Fig. 6(b), a linear strain distribution given by $\epsilon_x = (-y/R_n)$ was used. When the strain at the outermost fiber exceeds 0.1, it is suggested that the true strain description (Eq 2) be used. For bending to a radius of curvature equal to R_n, the strain distribution and the subsequent stress distribution (from the σ-ϵ curve or from a

Fig. 2 Grid deformations in the longitudinal and cross directions of a beam

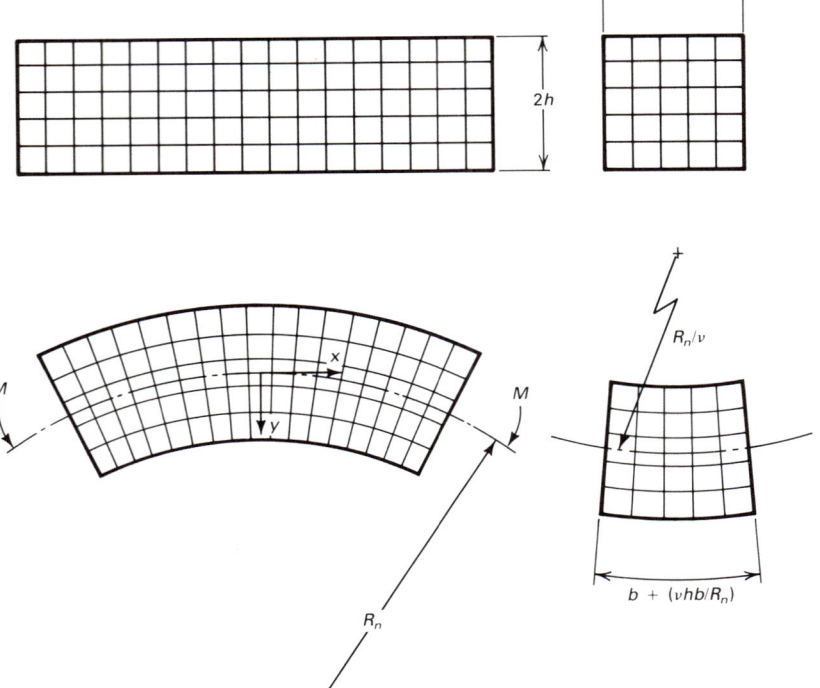

Fig. 3 Linear stress distributions in a beam

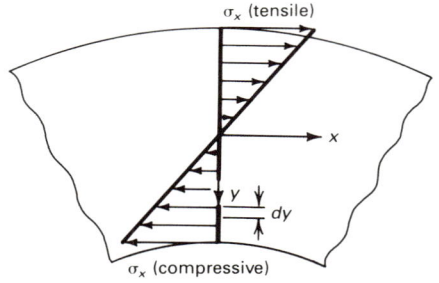

Fig. 4 Sign convention for the bending moment

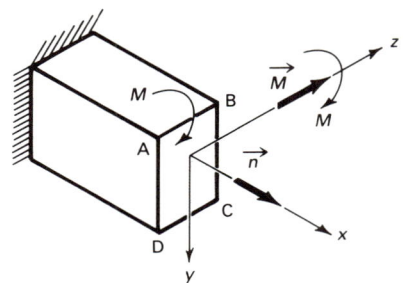

known constitutive equation for the σ-ε dependence) can be found. The moment, M, required to produce R_n is:

$$M = -\int_A (\sigma_x \, dA) y \qquad \text{(Eq 13)}$$

The thickness of the elastic core, C (Fig. 6c), is:

$$C = 2R_n\left(\frac{\sigma_y}{E}\right) \qquad \text{(Eq 14)}$$

Therefore, in bending plates of the same material to the same radius of curvature R_n, the fractional thickness of the elastic core $C/2h$ becomes smaller as the thickness increases.

For elastic-plastic bending, a general equation for the relationship between R_n and M (analogous to Eq 9 for the elastic case) does not exist. For the simple case of an elastic-perfect plastic material (Fig. 6d), the following equation is obtained:

$$R_n = \frac{E}{\sigma_y}\left[3\left(\frac{\sigma_y bh^2 - M}{b\sigma_y}\right)\right]^{1/2} \qquad \text{(Eq 15)}$$

The predictions of this equation at large plastic deformations $(C/2h \leq 0.02)$ are not reliable.

Pure Plastic Bending

The fractional thickness of the elastic core decreases as the ratio $R_n/2h$ decreases. For $R_n/2h \leq 10$, the thickness of the elastic core in a hot rolled strip of AISI 1020 steel is ≤3.2%. In such cases it is possible to ignore the elastic core altogether and use the analysis for pure plastic bending.

Most plastic bending analyses are for sheet-type specimens in which $b \gg 2h$. The width strain is small for such a geometry and plane-strain bending deformation $(\epsilon_z = 0)$ is commonly assumed. Also, assumptions 1 and 2 of the simple-beam theory are still applied. However, the radial (transverse) stresses and strains induced by higher curvatures are considered, and a plane-strain state, rather than a one-dimensional stress state (assumption 3 of the simple-beam theory), is assumed. Also, the neutral axis is not fixed at the centroid, and thickness variations due to bending sometimes are incorporated into the solution. Figure 7 shows the geometry of deformation and the strain and stress states at different zones.

At the onset of the assumed full plastic condition, $R_c = R_n = R_u$. In the current

state, the fiber with radius of curvature equal to R_n is being overtaken by the neutral axis and is experiencing unloading from a compressive tangential stress field. Accordingly, all fibers below the neutral axis $r \leq R_n$ have been progressively compressed, while those situated above R_c have been consistently stretched during deformation. All fibers in the interval $R_n \leq r < R_c$ have been overtaken by the neutral axis. Fibers between $R_u < r < R_c$ have now been stretched beyond their original length due to reverse loading, while those located in $R_n \leq r \leq R_u$ have yet to recover their original undeformed length.

As expected, a comprehensive analysis accounting for the described fiber movements is very complicated (Ref 5). A compromise solution (Ref 6), which ignores thickness variations and assumes rigid-perfect plastic material behavior, provides useful approximations for the stress-strain distributions in plastic bending.

When a rigid-perfect plastic material model is used (Fig. 7c), the same stress-strain relationship applies to all fibers with $r \geq R_n$. As a result, the distinction among R_n, R_u, and R_c becomes inconsequential. This eliminates the complicated task of describing the behavior of fibers in reversed loading and the Bauschinger effect.

The state of stress acting on a typical element is shown in Fig. 8, where σ_θ is the

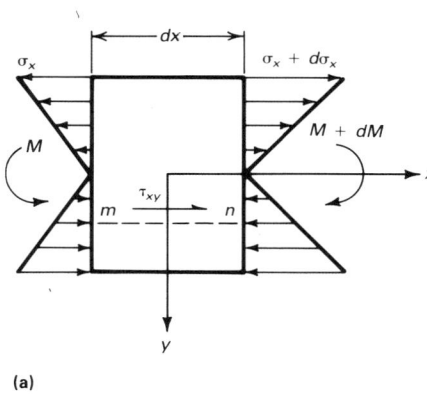

$$\sigma = \sqrt{\frac{1}{2}\left((\sigma_1-\sigma_2)^2+(\sigma_2-\sigma_3)^2+(\sigma_1-\sigma_3)^2\right)}$$

WHERE $\sigma_1 = \sigma_\theta$
$\sigma_2 = \sigma_r$

$$=\sqrt{\frac{1}{2}\left((\sigma_\theta-\sigma_r)^2+\left(\sigma_r-\frac{\sigma_\theta}{2}-\frac{\sigma_r}{2}\right)^2+\left(\sigma_\theta-\frac{\sigma_\theta}{2}-\frac{\sigma_r}{2}\right)^2\right)}$$

$\sigma_3 = \gamma(\sigma_1+\sigma_2) = .5(\sigma_\theta+\sigma_r)$ PLANE STRAIN

$$=\sqrt{\frac{1}{2}\left((\sigma_\theta-\sigma_r)^2+\frac{1}{4}(\sigma_r-\sigma_\theta)^2+\frac{1}{4}(\sigma_\theta-\sigma_r)^2\right)}=\sqrt{\frac{1}{2}(\sigma_\theta-\sigma_r)^2\frac{6}{4}}=\frac{\sqrt{3}}{2}(\sigma_\theta-\sigma_r)$$

CHECK

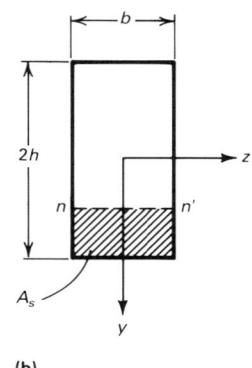

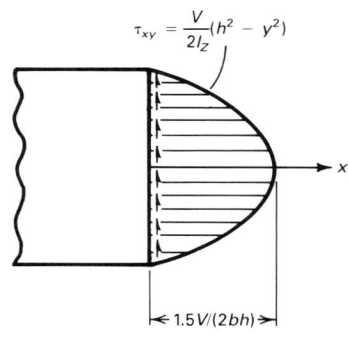

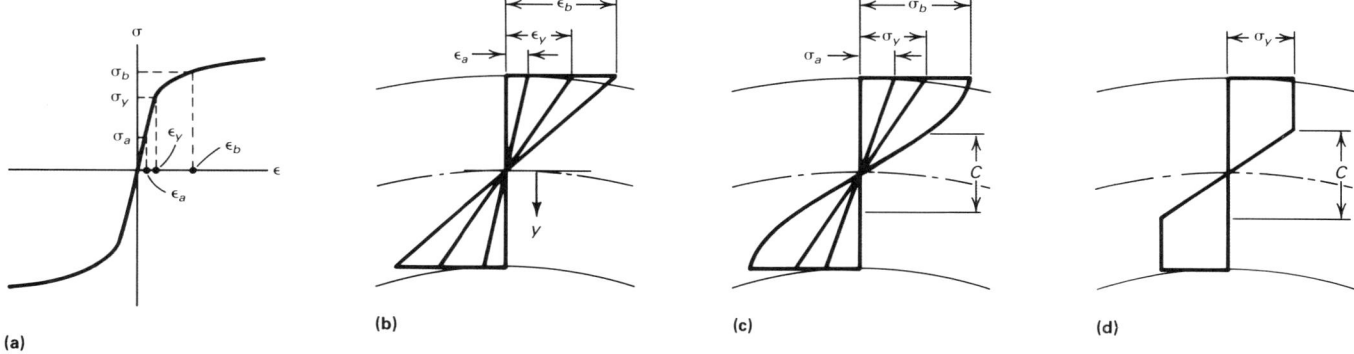

Fig. 5 Distribution of shear stress, τ_{xy}, for a rectangular specimen
See text for details.

(a)   (b)   (c)

circumferential (tangential) stress, and σ_r is the radial (transverse) stress. The equilibrium equation for plane strain deformation is:

$$\frac{d\sigma_r}{dr} = \frac{\sigma_\theta - \sigma_r}{r} \qquad \text{(Eq 16)}$$

The effective or significant stress, $\bar{\sigma}$, and strain, $\bar{\epsilon}$, for plane strain deformation, using von Mises criterion, are, respectively:

$$\bar{\sigma} = \pm\frac{\sqrt{3}}{2}(\sigma_\theta - \sigma_r) \quad \begin{array}{l}(+)\ \text{for}\ R_o \geq r \geq R_n \\ \\ (-)\ \text{for}\ R_n > r > R_i\end{array}$$

(Eq 17)

and

$$\bar{\epsilon} = \frac{2}{\sqrt{3}}\,|\epsilon_\theta| \qquad \text{(Eq 18)}$$

where $\epsilon_\theta = \ln r/R_n$. Substituting $(\sigma_\theta - \sigma_r)$ from Eq 16 into Eq 15 and putting $\sigma_r = 0$ at $r = R_o$ and $r = R_i$, these equations for the distribution of σ_r are obtained:

$$\sigma_r = -\left(\frac{2\bar{\sigma}}{\sqrt{3}}\right)\ln\left(\frac{R_o}{r}\right) \text{ for } R_n \leq r \leq R_o$$

(Eq 19)

$$\sigma_r = -\left(\frac{2\bar{\sigma}}{\sqrt{3}}\right)\ln\left(\frac{r}{R_i}\right) \text{ for } R_i \leq r \leq R_n$$

(Eq 20)

From the expressions for σ_r and Eq 16, the following expressions for σ_θ are determined:

$$\sigma_\theta = \left(\frac{2\bar{\sigma}}{\sqrt{3}}\right)\left[1 - \ln\left(\frac{R_o}{r}\right)\right]$$

for $R_n \leq r \leq R_o$ (Eq 21)

$$\sigma_\theta = -\left(\frac{2\bar{\sigma}}{\sqrt{3}}\right)\left[1 + \ln\left(\frac{r}{R_i}\right)\right]$$

for $R_i \leq r \leq R_n$ (Eq 22)

Because of equilibrium considerations, the radial stress must be continuous at $r = R_n$. Applying this condition to Eq 19 and 20, the location of the neutral axis can be found:

$$R_n = \sqrt{R_i R_o} \qquad \text{(Eq 23)}$$

Figure 9 is a schematic of the distributions of σ_r and σ_θ. In the figure, σ_r is continuous and compressive throughout the plate thickness, while σ_θ changes from tension to compression at the neutral axis. The bending moment, which according to this solution is independent of R_n, becomes:

$$M = \int_{R_i}^{R_o}(\sigma_\theta b\,dr)r = \left(\frac{2}{\sqrt{3}}\right)\bar{\sigma}bh^2$$

(Eq 24)

where h is half thickness, and b is the plate width.

The maximum radial stress occurs at the neutral axis. Its magnitude from Eq 19 and 23 is:

$$(\sigma_r)_{max} = -\left(\frac{\bar{\sigma}}{\sqrt{3}}\right)\ln\left(\frac{R_o}{R_i}\right) \qquad \text{(Eq 25)}$$

The ratio between $(\sigma_r)_{max}$ and the tangential stress at $r = R_o$ for four plates of different thicknesses, all bent to an inside radius of $R_i = 25$ mm (1 in.), is given in Table 1. This table shows that for plastic bending to $(R_n/$

Fig. 6 Stress-strain distributions in a beam
(a) Stress-strain curve. (b) Strain distribution. (c) Stress distribution. (d) Stress distribution for elastic-perfect plastic material

(a)   (b)   (c)   (d)

Fig. 7 Pure plastic bending of strip specimen

(a) Geometry of deformation. R_i = inner radius of curvature; R_o = outer radius of curvature; R_n = radius of curvature of the neutral axis; R_u = radius of currently unstretched fiber; R_c = current radius of curvature of original center fiber. (b) Strain and stress states in different zones. -----, deformed state. ϵ_θ = ln r/R_n is the circumferential strain. (c) Stress-strain curves and stress-strain states at various locations. N = stress-strain state at R_n; U = stress-strain state at R_u; C = stress-strain state at R_c.

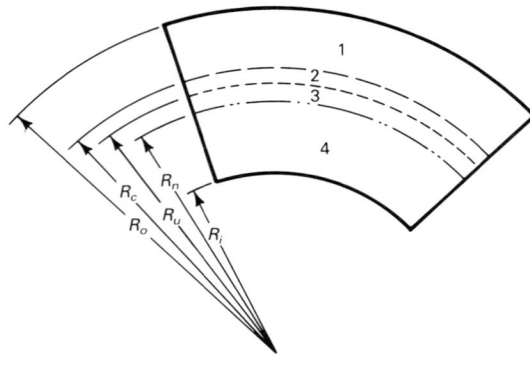

(a)

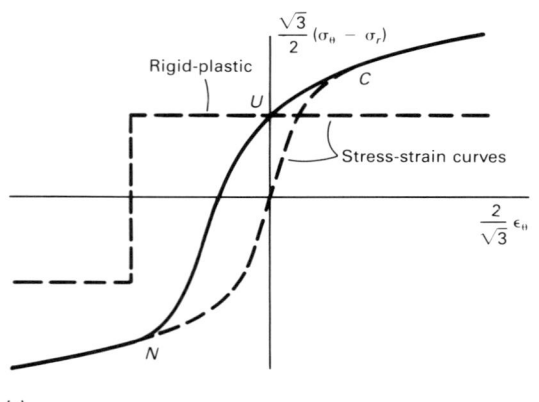

(c)

Zone	Strain state	Stress state
1	$\epsilon_\theta > 0,\ d\epsilon_\theta > 0$	
2	$\epsilon_\theta \geq 0,\ d\epsilon_\theta > 0$	
3	$\epsilon_\theta < 0,\ d\epsilon_\theta \geq 0$	
4	$\epsilon_\theta < 0,\ d\epsilon_\theta < 0$	

(b)

$2h) > 10$, the magnitude of $(\sigma_r)_{max}$ becomes very small. In such cases, the effect of σ_r can be neglected in the analysis and the elastic-plastic bending solution based on the simple-beam theory can be used.

More elaborate analyses of plastic bending (Ref 5, 7, and 8) commonly involve complicated numerical computations. Hence, no equations can be given. A comparison between the results of the analysis in Ref 6 and a solution for rigid work-hardening material behavior (Ref 5) is shown in Fig. 10. In this case, the result from Ref 5 is for a model material with a high rate of strain hardening.

$$\bar{\sigma} = 70 + 300\ \bar{\epsilon}^{0.5}\ \text{MPa} \qquad \text{(Eq 26)}$$

In using the analysis from Ref 6, $\bar{\sigma} = 120$ MPa (17.4 ksi), which is the approximate average flow stress for bending to $\bar{\epsilon} = 0.11$, and $\bar{\sigma} = 169.5$ MPa (24.6 ksi), as shown in Fig. 10(b), have been assumed. As expected, some differences in the predicted stress distributions are observed.

However, by using $\bar{\sigma} = 169.5$ MPa (24.6 ksi) in the solution from Ref 6, a close agreement (percent difference <6) between the estimates for the fiber stresses at $r = R_i$ and $r = R_o$ is obtained. Because the prediction of maximum fiber stress and strain is of special interest, the following procedure based on the solution in Ref 6 is suggested. First, find the maximum fiber strain:

$$\epsilon_o = -\epsilon_i = \ln\ \sqrt{\frac{R_o}{R_i}} \qquad \text{(Eq 27)}$$

where ϵ_o and ϵ_i represent the maximum fiber strain at the outer and inner radii, respectively. Using the stress-strain equation or stress-strain curve for the material, determine $\bar{\sigma}$ as the flow stress at ϵ_o. The maximum fiber stress is $2\bar{\sigma}/\sqrt{3}$.

Residual Stress and Springback

When a specimen that has been bent beyond the elastic limit is unloaded, the applied moment M becomes zero and the radius of

Table 1 Ratio between maximum radial stress and tangential stress for plate of various thicknesses
See text for explanation of symbols

R_i		Thickness		$R_n/2h$	(σ_r)max$/(\sigma_\theta$ at $r = R_o)$
mm	in.	mm	in.		
25	1................	1.59	0.0625	16.49	0.030
25	1................	3.17	0.125	8.48	0.059
25	1................	6.35	0.25	4.47	0.112
25	1................	12.7	0.5	2.45	0.203

Fig. 8 State of stress acting on a typical element in plane-strain bending

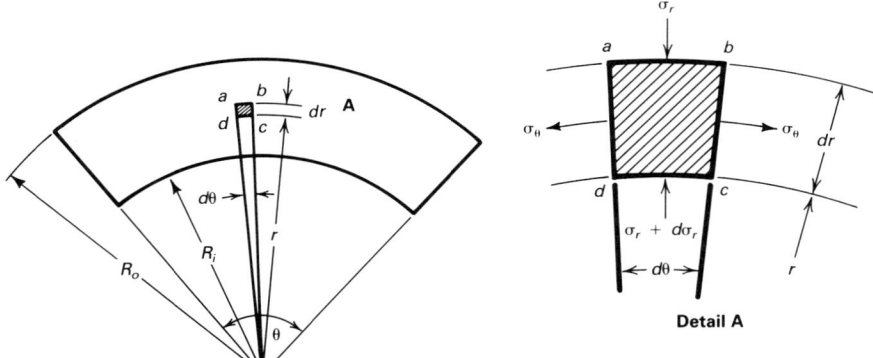

Detail A

Fig. 9 Schematic of circumferential, σ_θ, and radial, σ_r, stresses in a plate during bending

Source: Ref 6

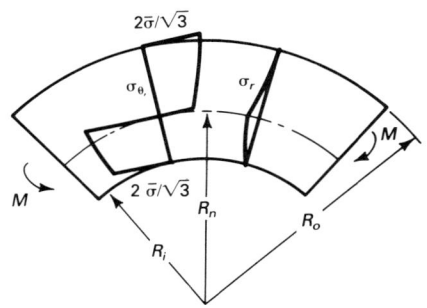

curvature increases from R_n to R'_n. For a fiber at distance y from the neutral axis, this produces a strain difference:

$$\Delta\epsilon_x = y\left(\frac{1}{R_n} - \frac{1}{R'_n}\right) \quad \text{(Eq 28)}$$

The removal of the bending moment, which is an unloading event, is assumed to be elastic. Therefore:

$$\Delta\sigma_x = E\Delta\epsilon_x = Ey\left(\frac{1}{R_n} - \frac{1}{R'_n}\right) \quad \text{(Eq 29)}$$

The change in bending moment for complete unloading is $\Delta M = M$, where M is the applied bending moment prior to unloading. Therefore,

$$\int_A\left[Ey\left(\frac{1}{R_n} - \frac{1}{R'_n}\right)dA\right]y = -\int_A(\sigma_x dA)y \quad \text{(Eq 30)}$$

which reduces to

$$\frac{1}{R_n} - \frac{1}{R'_n} = -\frac{\int_A \sigma_x y dA}{EI_z} \quad \text{(Eq 31)}$$

The distribution of the residual stresses can be found from either of the following equations:

$$\sigma'_x = \sigma_x + \Delta\sigma_x = \sigma_x - \left(\frac{y}{I_z}\right)\int_A \sigma_x y dA \quad \text{(Eq 32)}$$

$$\sigma'_x = \sigma_x + yE\left(\frac{1}{R_n} - \frac{1}{R'_n}\right) \quad \text{(Eq 32a)}$$

It is important that the correct signs for σ_x and y be used when applying these equations.

As an example, the springback and the residual stress distribution of a strip of annealed 1095 steel was examined (Ref 9). For this material, yield strength is $\sigma_y = 308$

MPa (44.7 ksi), Poisson's ratio is $\nu = 0.28$, and the approximate constitutive equation for σ in metric units of measure is:

$$\bar\sigma = (2 \times 10^5 \text{ MPa})\epsilon \quad \text{for } \epsilon \leq 0.00154 \quad \text{(Eq 33)}$$

$$\bar\sigma = (896 \text{ MPa})\epsilon^{0.16} \quad \text{for } \epsilon \geq 0.00154 \quad \text{(Eq 34)}$$

In English units of measure, σ is:

$$\bar\sigma = (29 \times 10^6 \text{ psi})\epsilon \quad \text{for } \epsilon \leq 0.00154$$

$$\bar\sigma = (126 \text{ ksi})\epsilon^{0.16} \quad \text{for } \epsilon \geq 0.00154$$

The width of the strip, b, is 50 mm (2 in.), and its thickness, $2h$, is 5 mm (0.2 in.). It is assumed that the strip is bent to $R_n = 100$ mm (4 in.). Because $(b/2h) = 10$, plane-strain deformation prevails. Because of this, the elastic modulus in plane-strain

$$E' = \frac{E}{1 - \nu^2} \quad \text{(Eq 35)}$$

is employed, and the plastic flow stresses (Eq 34) are multiplied by $(2/\sqrt{3})$. These approximate plane-strain adjustments are considered adequate when the simplified elastic-plastic analysis, which was discussed earlier in this article, is used. The thickness of the elastic core in metric units of measure is

$$C = \frac{2R_n\sigma_y}{E'} = 2(100)\left(\frac{\frac{2 \times 308}{\sqrt{3}}}{2.17 \times 10^5}\right)$$

$$C = 0.33 \text{ mm} \quad \text{(Eq 36)}$$

In English units of measure, the thickness of the elastic core is:

$$C = \frac{2R_n\sigma_y}{E'} = \frac{2(3.937)\left(2 \times \frac{44.7}{\sqrt{3}}\right)}{(31465)}$$

$$C = 0.013 \text{ in.}$$

which is 6.6% of the total plate thickness. The final radius of curvature in metric units of measure after springback is found from Eq 31:

$$\frac{1}{100} - \frac{1}{R'_n} =$$

$$-\frac{2}{EI_z}\left[\int_0^{\frac{C}{2}} - 2.17 \times 10^5\left(\frac{y}{100}\right)b\,dy +\right.$$

$$\left. \int_{\frac{C}{2}}^h -\left(\frac{2}{\sqrt{3}}\right)869\left(\frac{y}{100}\right)^{0.16}by\,dy\right] \quad \text{(Eq 37)}$$

which results in $R'_n = 116.9$ mm (4.60 in.). A more elaborate analysis of springback (Ref 9) for this case predicts $R'_n/2h = 23.41$ (or $R'_n = 117.05$ mm, or 4.608 in.). Also, for bending the same strip to $R_n = 40$ mm (1.6 in.) and $R_n = 500$ mm (19.7 in.), the final radii of curvature from Eq 31 are 42.8 mm (1.68 in.) and 1150 mm (45.3 in.), respectively. The corresponding results from Ref 9 are 42.9 mm (1.69 in.) and 1075 mm (42.3 in.).

The distribution of residual stresses after bending to $R_n = 100$ mm (3.937 in.) is obtained from Eq 32a. Therefore, in this case, in metric units:

$$\sigma'_x = \sigma_x + yE(0.0014457) \quad \text{(Eq 38)}$$

In English units:

Fig. 10 Comparison of results for determining plastic bending in a plate

(a) Distribution of tangential and radial stresses for a 25-mm (1-in.) thick plate bent to R_i = 100 mm (4 in.). (b) Stress-strain diagrams used in the analyses for (a)

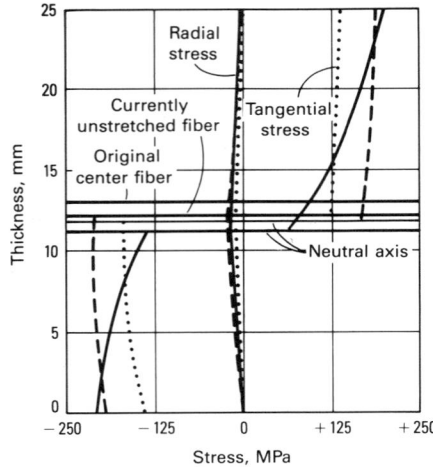

(a)

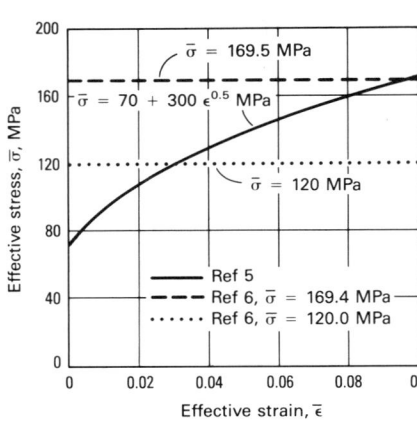

(b)

Fig. 11 Distribution of applied and residual stresses

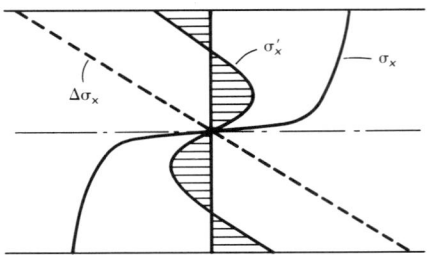

$$\sigma'_x = \sigma_x + yE(0.03672)$$

At $R = R_i$, $y = h$ = 2.5 mm (0.098 in.):

$$\epsilon_x = -0.025$$

and

$$\sigma_x = -\left(\frac{2}{\sqrt{3}}\right)(869)(0.025)^{0.16}$$

$$= -556 \text{ MPa}$$

or

$$\sigma_x = -\left(\frac{2}{\sqrt{3}}\right)(126)(0.025)^{0.16}$$

$$= -80.63 \text{ ksi}$$

For this location:

$$\sigma'_x =$$

$$-556 + 2.5(2.17 \times 10^5)$$

$$\times (0.0014457)$$

$$= +228 \text{ MPa}$$

or

$$\sigma'_x = -80.63 + 0.098(31465)(0.03672)$$

$$= +32.6 \text{ ksi.}$$

Similarly, the magnitude of σ_x for other values of y can be determined. Figure 11 shows the distribution of applied and residual stresses.

REFERENCES

1. J.M. Gere and S.P. Timoshenko, *Mechanics of Materials*, 2nd ed., Wadsworth, Belmont, CA, 1984

2. A.P. Boresi *et al.*, *Mechanics of Materials*, 3rd ed., John Wiley & Sons, New York, 1978

3. A.C. Ugural and S.K. Fenster, *Advanced Strength and Applied Elasticity*, Elsevier, New York, 1981

4. D. Horrocks and W. Johnson, On Anticlastic Curvature with Special Reference to Plastic Bending: A Literature Survey and Some Experimental Investigations, *Int. J. Mech. Sci.*, Vol 9, 1967, p 835-861

5. P. Dadras and S.A. Majlessi, Plastic Bending of Work Hardening Materials, *ASME Trans. J. Eng. Ind.*, Vol 104, 1982, p 224-230

6. R. Hill, *The Mathematical Theory of Plasticity*, Oxford University Press, London, 1950

7. H. Verguts and R. Sowerby, The Pure Plastic Bending of Laminated Sheet Metals, *Int. J. Mech. Sci.*, Vol 17, 1975, p 31

8. S.A. Majlessi and P. Dadras, Pure Plastic Bending of Sheet Laminates Under Plane Strain Conditions, *Int. J. Mech. Sci.*, Vol 25 (No. 1), 1983, p 1-14

9. O.M. Sidebottom and C.F. Gebhardt, Elastic Springback in Plates and Beams Formed by Bending, *Exp. Mech.*, Vol 19, 1979, p 371-377

Bending Ductility Tests

By Frank N. Mandigo
Research Scientist
Olin Corporation

BENDING DUCTILITY TESTS determine the smallest radius around which a specimen can be bent without cracks being observed in the outer fiber (tension) surface. This forming limit commonly is called the minimum bend radius and is expressed in multiples of specimen thickness, t. A material with a minimum bend radius of $3t$ can be bent without cracking through a radius equal to three times the specimen thickness. It thus follows that a material with a minimum bend radius of $1t$ has greater ductility than a material with a minimum bend radius of $5t$. Alternatively, the bend radius can be fixed, and the angle of bend at which fracture occurs noted. Figure 1 illustrates bend radius, angle of bend, and other concepts associated with bending tests.

This article describes apparatus, specimen preparation, and test procedures used in bend ductility testing. Bending tests are usually performed on strip, sheet, or plate; however, the same methods and apparatus can be used to evaluate the bending ductility of other product forms (drawn or extruded rounds, squares, or polygonal shapes) and of weldments.

Evaluating and reporting results of bending tests is also discussed. This discussion includes a description of failure criteria, strain distribution, directionality, and factors that affect bend ductility. Finally, a compilation of bend data for select engineering materials is presented.

Apparatus

Bending test apparatus includes wrap, wipe, V-block, and soft tooling devices that may have interchangeable die radii and are able to bend test specimens to several preset angles. The pins, mandrels, rollers, radiused flats, and clamping devices must be longer than the specimen width, and they must be strong and rigid enough to resist deformation and wear. Basic types of bending devices are described below.

Wrap bending devices (Fig. 2) grip the test specimen at one end; a mandrel, reaction pin, or block with the desired bend radius is positioned at midlength. A roller that sweeps concentrically around the bend radius applies the bending force. The distance from the mandrel to the loading roller generally is equal to the thickness or diameter of the test piece, plus clearance. The clearance is adjusted to allow the test specimen to bend to the desired radius or angle without scuffing, smearing, or galling of strip and die surfaces.

Wipe bending devices (Fig. 3) are similar to wrap bending devices, except that the bending force is applied by a mandrel or roller that moves perpendicular to the clamped specimen.

V-block bending devices consist of a mandrel and a bottom block (Fig. 4a) or specimen supports (Fig. 4b). The sample rests on supports or on the bottom block and is not clamped during the test. The distance between supports is selected to force the specimen to conform to the mandrel radius without excessive interference. This clearance is often the mandrel diameter, D, plus three times the specimen thickness, t. Bending force is applied at the center of the specimen. The bottom block normally is a V or U shape. Bends made with conforming bottom block radii are bottoming or closed-die bends (Fig. 4a); those without conforming bottom block radii are air or free bends (Fig. 4b).

Soft tool bending, or rubber pad bending, is similar to V-block bending, except

Fig. 1 Terms used in bend testing

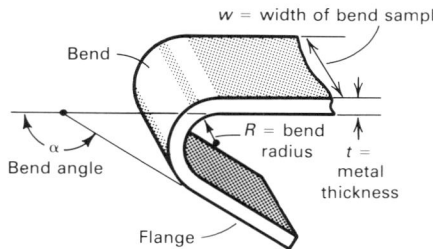

Fig. 2 Wrap bending device
Bending force is applied by a roller that sweeps concentrically around the bend radius.

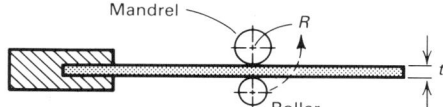

Fig. 3 Wipe bending devices
(a) Mandrel type with force applied near free end.
(b) Die type with force applied near free end

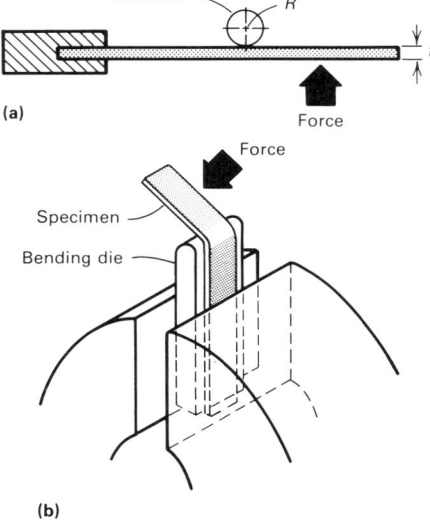

Fig. 4 V-block bending devices

(a) Closed V-block. (b) Open V-block

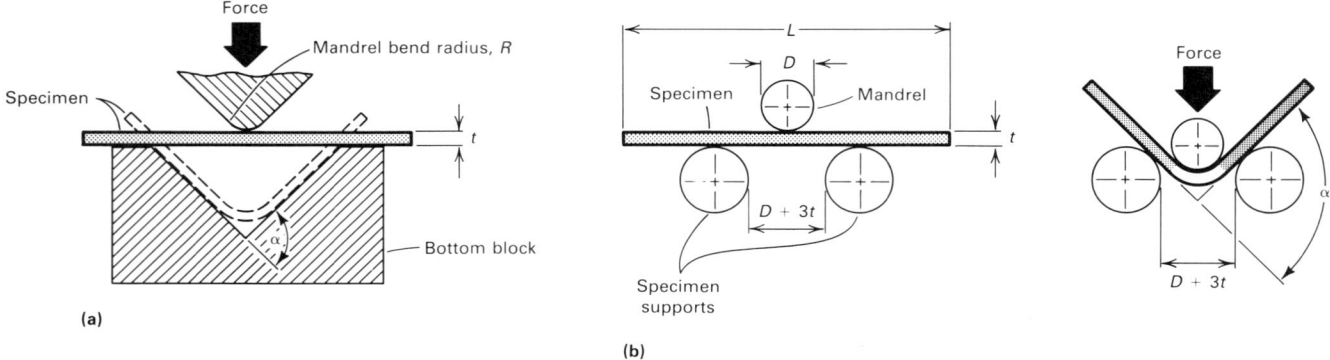

that the bottom block is replaced by highly compressible materials such as polyurethane. Figure 5 illustrates various setups for soft tool bending.

Specimen Preparation

Sheet and plate specimens normally are full thickness and are prepared by shearing or sawing. Narrow strip and bar sections can be tested with specimens cut to length. For polygonal shapes, it is sometimes necessary to machine or grind a flat, with the unmachined surface to be placed on the tension side of the bend. The rough edges of cut samples can be removed by belt sanding, filing, etc. The edges of rectangular-section test pieces can be rounded to a radius up to one tenth of the sample thickness. Edge preparation is not required for specimens with width-to-thickness ratios greater than 8 to 1, unless cracking initiates from the edge during bending.

A specimen can be machined to fit if it is too thick for the bending device or if the required bending forces exceed the capacity of the bending device. The unmachined surface is placed on the tension side. If reduced thickness samples are used, the specimen dimensions must be held constant during testing and noted in the test report.

Specimens must be long enough to be clamped securely to prevent slippage in wrap and wipe tests. Specimens for V-block and soft tool bending can be of any length greater than the distance between specimen supports (Fig. 4b).

Test specimens usually are cut parallel or perpendicular to the direction of rolling, drawing, or extrusion. However, any orientation in the plane of the product width and length can be tested. Specimen dimensions should be the same for any test orientation. Figure 6 shows the orientation of a bending test specimen with the rolling direction for a longitudinal orientation and a transverse ori-

Fig. 5 Setups for soft tool (rubber pad) bending of sheet metal

(a) Simple 90° V-bend. Air space below die pad permits deep penetration. (b) Simple U-bend or channel. Spacers enable U-bends of varying widths to be formed in the same die-pad retainer. Deflector bars help provide uniform distribution of punch pressure. (c) Modified U-bend with partial air bending. (d) Acute-angle U-bend. High side pressures are obtained by using a conforming rubber die pad and deflector bars.

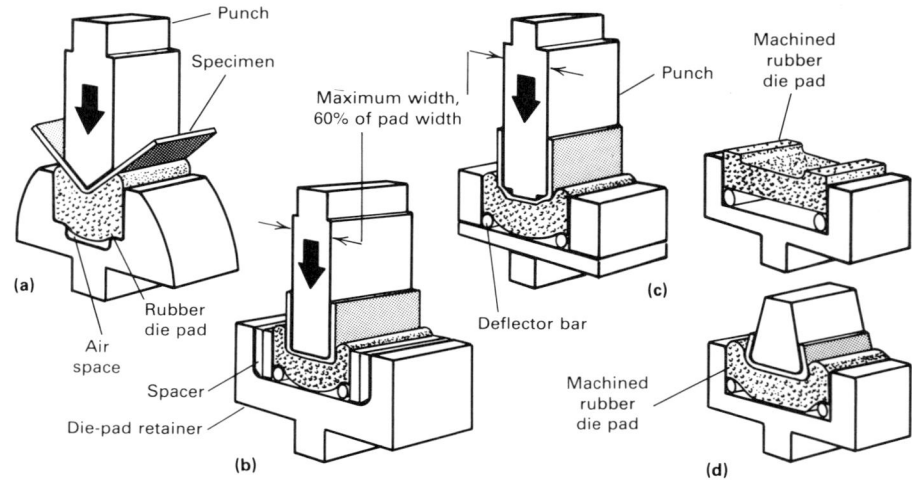

Fig. 6 Relative orientation of longitudinal and transverse bending tests

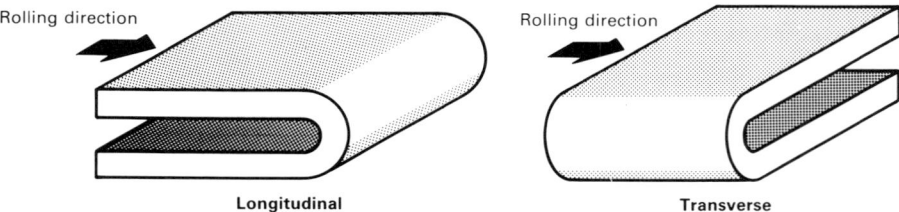

entation. The transverse orientation generally produces lower ductility, because the tensile bending stresses are oriented perpendicular to the fiber structure developed by the rolling procedure used to produce sheet and plate products.

Cold rolling oils or other protective coatings usually are left on specimens to serve as a lubricating film during bending tests. Other lubricants can be used if they reflect field conditions.

The principal stress and strain directions developed during bending are defined in Fig. 7(a). The specimen width-to-thickness ratio (w/t) determines the respective stress and strain states. At $w/t > 8$, bending occurs under plane strain conditions ($\epsilon_2 = 0$ and $\sigma_2/\sigma_1 = 0.5$) and bend ductility is independent of the exact width-to-thickness ratio (Fig. 7b). At $w/t < 8$, bending occurs under plane stress conditions ($\sigma_2/\sigma_1 < 0.5$) with plastic deformation in all principal strain directions and the measured bend ductility is strongly dependent on the width-to-thickness ratio (Fig. 7b). Therefore, bending tests are conducted at width-to-thickness ratios greater than 8 to 1 whenever possible to eliminate geometric effects on the test results. Al-

though specimens with width-to-thickness ratios less than 8 to 1 can be tested, the entire test lot must have the same width-to-thickness ratio.

Test Method

Specimens and apparatus should be carefully inspected. Specimens with scuff marks, scratches, excessive curvature, twisting, or other surface defects should be discarded. Bend radii, mandrels, and support blocks must be free of scuff marks or other visible damage.

In bending tests, specimens are bent around progressively tighter radii, or to large bend angles, until failure or cracking occurs on the convex surface. If wipe or wrap bending apparatus is used, the clearance must be adjusted for each radius. Any method can be used to force the specimen to obtain the desired radius or angle; however, it must applied slowly and steadily without significant lateral motion. Bend angles of 180° are obtained by pressing bent specimens between platens (Fig. 8), maintaining the bend radius with a spacer block twice as thick as the radius between the legs of the specimen.

During bending, the specimen can be removed to inspect the convex surface for

cracks. The test is complete when product specifications have been achieved.

Interpretation

Specimens are examined for cracking at the apex of the bend with magnifications of up to 20×. A specimen is acceptable if there are no visible cracks on its outside surface. Surface rumples and orange peeling are not considered fracture sites. If cracking occurs at the edges of the bent sample when the specimen width-to-thickness ratio is 8 to 1 or greater, the edges of the sample should be polished or ground and the specimen retested. At width-to-thickness ratios less than 8 to 1, edge cracks are expected and edge preparation may be required to obtain reproducible measurements of minimum bend radii.

The bending method can influence the strain distribution on the surface of the specimen (Fig. 9). V-block bending (Fig. 9b) develops a nonuniform strain distribution, while wrap or wipe bending produces strain that increases progressively with the bend angle until saturation. Circumferential strain becomes uniform only after the bend angle exceeds certain minimum values (Fig. 10 and

Fig. 7(a) Schematic of the bend region defining direction of principal stresses and strains

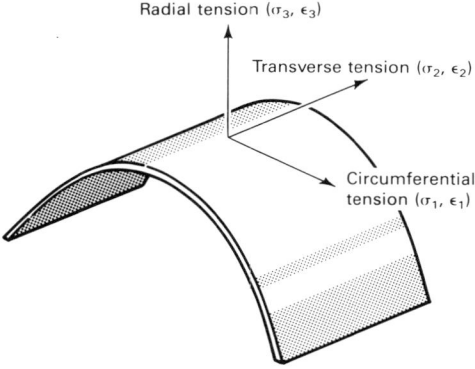

Fig. 7(b) Outer fiber strain at fracture versus width-to-thickness ratio

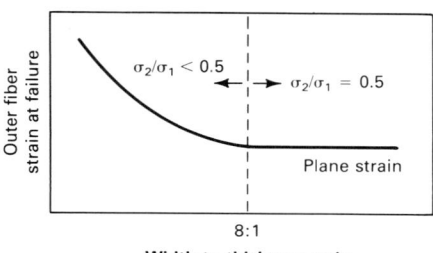

Fig. 8 Methods used to develop 180° bend angles

(a) Bend sample from wipe or V-block placed between platens. (b) Sharp (180°) bend. (c) Bend with radius equal to one half the spacer-block thickness

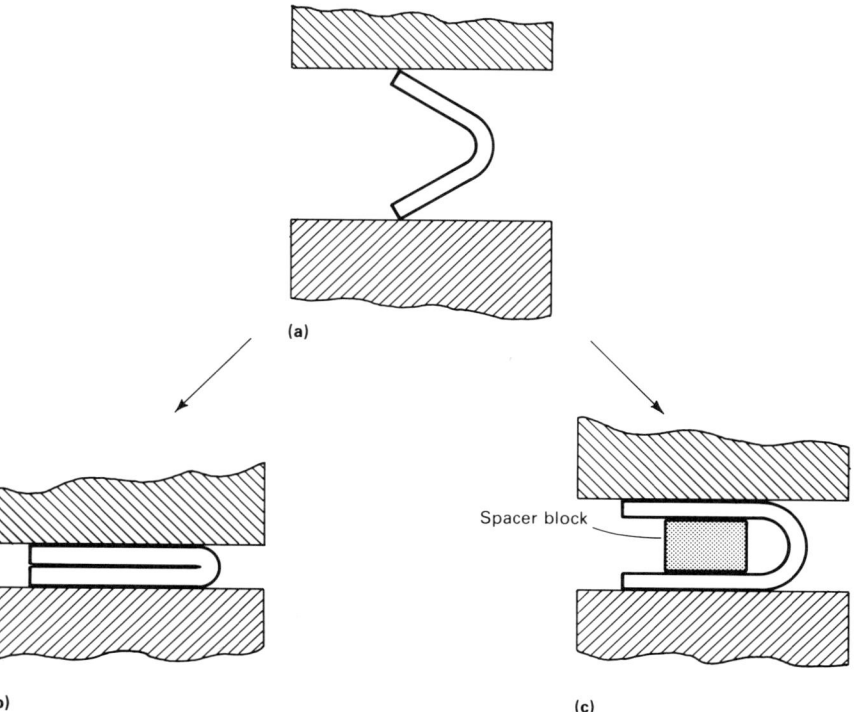

Fig. 9 Comparison of strain distributions produced by different bending methods

(a) Application of a pure bending moment (not achievable in commercial bending devices). (b) V-block bending. (c) Wipe bending

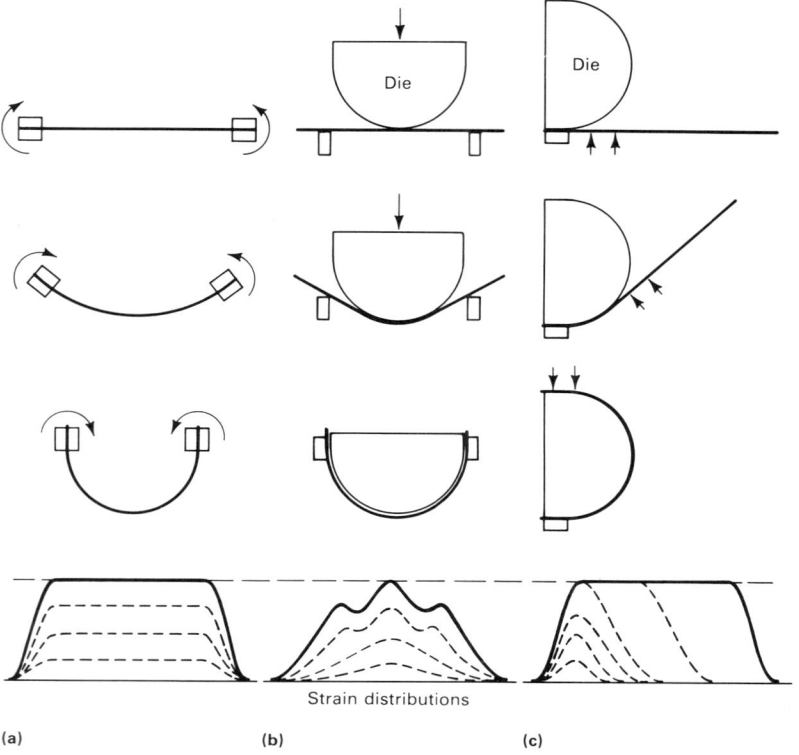

Strain distributions

(a)   (b)   (c)

Fig. 11 Dependence of measured circumferential strain on bend angle for tempered aluminum alloy 2024 sheet

Radius expressed in terms of thickness, *t*

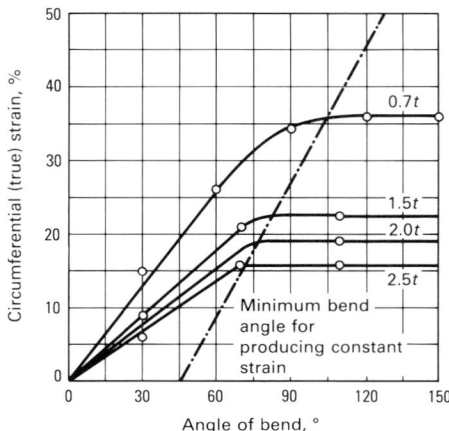

11). For wipe or wrap bending, these values are a 90° bend angle and a 1*t* radius.

Minimum bend radii reported are subjective measurements, not intrinsic material properties. This subjectivity is due to visual assessment of pass/fail criteria and the incremental steps of the available bend radii. The problems associated with visual assessment include reproducibility by one tester over time and the difficulty of two or more testers agreeing on the definition of a visible crack. The problems associated with the incremental steps of the bend radii can be minimized by utilization of closely spaced bend radii.

A number of characteristics can be expected when performing bending tests on metallic materials. The minimum bend radius is dependent on alloy composition. The minimum bend radius increases as strip or bar temper increases. Annealed strips generally have isotropic bend characteristics in the plane of the sheet (the minimum bend radii is similar both parallel or perpendicular to the rolling/drawing/extrusion direction of the strip). Strips in highly cold rolled tempers

Fig. 10 Effect of bend angle on strain distribution along the circumference of bends for tempered aluminum alloy 2024 sheet

(a) 3.2-mm (0.125-in.) thick sheet, *R/t* = 0.7. (b) 6.4-mm (0.25-in.) thick sheet, *R/t* = 2.5

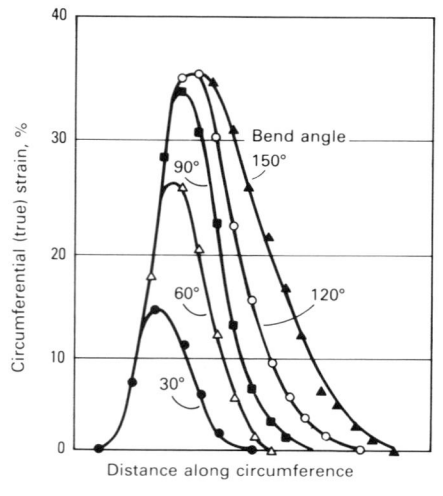

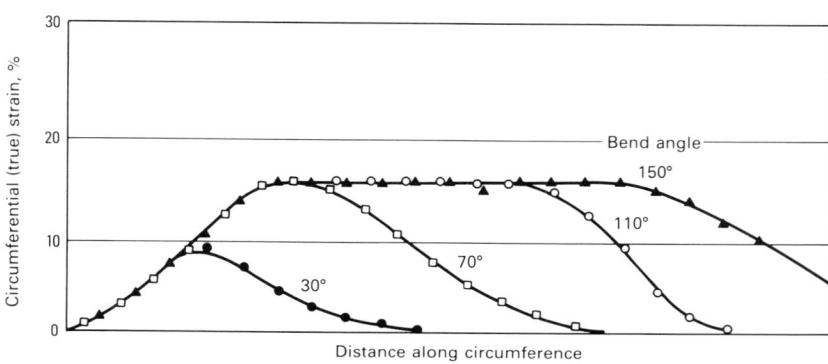

(a)   (b)

usually have better bend properties when bends are made perpendicular to the rolling direction. For a given alloy and temper, the minimum bend radius usually is directly proportional to strip thickness, as indicated in the following equation:

$$\frac{\text{Minimum bend radius}}{\text{Strip thickness}} = \text{Constant}$$

Because of these characteristics, caution should be used in choosing a bending test method or using tabulated data. The best practice is to use the same test method, specimen dimensions, bend angle, and bend radii that are used during part fabrication. Representative data for aluminum, copper, and ferrous sheet are provided in Tables 1 to 5.

Table 1 Recommended minimum bend radii for 90° cold forming of aluminum alloy sheet and plate(a)

Alloy	Temper	0.4 mm (1/64 in.)	0.8 mm (1/32 in.)	1.6 mm (1/16 in.)	3.2 mm (1/8 in.)	4.8 mm (3/16 in.)	6.4 mm (1/4 in.)	9.5 mm (3/8 in.)	12.7 mm (1/2 in.)
					Radii for various thicknesses expressed in terms of thickness, t				
1100	O	0	0	0	0	½t	1t	1t	1½t
	H12	0	0	0	½t	1t	1t	1½t	2t
	H14	0	0	0	1t	1t	1½t	2t	2½t
	H16	0	½t	1t	1½t	1½t	2½t	3t	4t
	H18	1t	1t	1½t	2½t	3t	3½t	4t	4½t
2014	O	0	0	0	½t	1t	1t	2½t	4t
	T3	1½t	2½t	3t	4t	5t	5t	6t	7t
	T4	1½t	2½t	3t	4t	5t	5t	6t	7t
	T6	3t	4t	4t	5t	6t	8t	8½t	9½t
2024	O	0	0	0	½t	1t	1t	2½t	4t
	T3	2½t	3t	4t	5t	5t	6t	7t	7½t
	T361(c)	3t	4t	5t	6t	6t	8t	8½t	9½t
	T4	2½t	3t	4t	5t	5t	6t	7t	7½t
	T81	4½t	5½t	6t	7½t	8t	9t	10t	10½t
	T861(c)	5t	6t	7t	8½t	9½t	10t	11½t	11½t
2036	T4	...	1t	1t	...	...	...	...	...
3003	O	0	0	0	0	½t	1t	1t	1½t
	H12	0	0	0	½t	1t	1t	1½t	2t
	H14	0	0	0	1t	1t	1½t	2t	2½t
	H16	½t	1t	1t	1½t	2½t	3t	3½t	4t
	H18	1t	1½t	2t	2½t	3½t	4½t	5½t	6½t
3004	O	0	0	0	½t	1t	1t	1t	1½t
	H32	0	0	½t	1t	1t	1½t	1½t	2t
	H34	0	1t	1t	1½t	1½t	2½t	2½t	3t
	H36	1t	1t	1½t	2½t	3t	3½t	4t	4½t
	H38	1t	1½t	2½t	3t	4t	5t	5½t	6½t
3105	H25	...	½t	½t	...	...	...	...	...
5005	O	0	0	0	0	½t	1t	1t	1½t
	H12	0	0	0	½t	1t	1t	1½t	2t
	H14	0	0	0	1t	1½t	1½t	2t	2½t
	H16	½t	1t	1t	1½t	2½t	3t	3½t	4t
	H18	1t	1½t	2t	2½t	3½t	4½t	5½t	6½t
	H32	0	0	0	½t	1t	1t	1½t	2t
	H34	0	0	0	1t	1½t	1½t	2t	2½t
	H36	½t	1t	1t	1½t	2½t	3t	3½t	4t
	H38	1t	1½t	2t	2½t	3½t	4½t	5½t	6½t
5050	O	0	0	0	½t	1t	1t	...	...
	H32	0	0	0	1t	1t	1½t	...	...
	H34	0	0	1t	1½t	1½t	2t	...	...
	H36	1t	1t	1½t	2t	2½t	3t	...	...
	H38	1t	1½t	2½t	3t	4t	5t	...	...
5052	O	0	0	0	½t	1t	1t	1½t	1½t
	H32	0	0	1t	1½t	1½t	1½t	1½t	2t
	H34	0	1t	1½t	2t	2t	2½t	2½t	3t
	H36	1t	1t	1½t	2½t	3t	3½t	4t	4½t
	H38	1t	1½t	2½t	3t	4t	5t	5½t	6½t
5083	O	...	...	½t	1t	1t	1t	1½t	1½t
	H321	...	...	1t	1½t	1½t	1½t	2t	2½t
5086	O	0	0	½t	1t	1t	1t	1½t	1½t
	H32	0	½t	1t	1½t	1½t	2t	2½t	3t
	H34	½t	1t	1½t	2t	2½t	3t	3½t	4t
	H36	...	...	...	3t	3½t	4t	4½t	5t

(continued)

(a) The radii listed are the minimum recommended for bending sheets and plates without fracturing in a standard press brake with air bend dies. Other bending operations may require larger radii or permit smaller radii. The minimum permissible radii also vary with the design and condition of the tooling. (b) Alclad sheet in the heat treatable alloys can be bent over slightly smaller radii than the corresponding tempers of the bare alloy. (c) Tempers T361 and T861 formerly designated T36 and T86, respectively.

Table 1 (continued)

Alloy	Temper	Radii for various thicknesses expressed in terms of thickness, t							
		0.4 mm (1/64 in.)	0.8 mm (1/32 in.)	1.6 mm (1/16 in.)	3.2 mm (1/8 in.)	4.8 mm (3/16 in.)	6.4 mm (1/4 in.)	9.5 mm (3/8 in.)	12.7 mm (1/2 in.)
5154	O	0	0	½t	1t	1t	1t	1½t	1½t
	H32	0	½t	1t	1½t	1½t	2t	2½t	3½t
	H34	½t	1t	1½t	2t	2½t	3t	3½t	4t
	H36	1t	1½t	2t	3t	3½t	4t	4½t	5t
	H38	1½t	2½t	3t	4t	5t	5t	6½t	6½t
5252	H25	0	0	1t	2t	...	...	...	...
	H28	1t	1½t	2½t	3t	...	...	...	...
5254	O	0	0	½t	1t	1t	1t	1½t	1½t
	H32	0	½t	1t	1½t	1½t	2t	2½t	3½t
	H34	½t	1t	1½t	2t	2½t	3t	3½t	4t
	H36	1t	1½t	2t	3t	3½t	4t	4½t	5t
	H38	1½t	2½t	3t	4t	5t	5t	6½t	6½t
5454	O	0	½t	1t	1t	1t	1½t	1½t	2t
	H32	½t	½t	1t	2t	2t	2½t	3t	4t
	H34	½t	1t	1½t	2t	2½t	3t	3½t	4t
5456	O	...	...	...	1t	1½t	1½t	2t	2t
	H321	...	...	...	2t	2t	2½t	3t	3½t
5457	O	0	0	0	½t	1t	1t	1t	1½t
5652	O	0	0	0	½t	1t	1t	1½t	1½t
	H32	0	0	1t	1½t	1½t	1½t	1½t	2t
	H34	0	1t	1½t	2t	2t	2½t	2½t	3t
	H36	1t	1t	1½t	2½t	3t	3½t	4t	4½t
	H38	1t	1½t	2½t	3t	4t	5t	5½t	6½t
5657	H25	0	0	0	1t	...	...	...	...
	H28	1t	1½t	2½t	3t	...	...	...	...
6061(b)	O	0	0	0	1t	1t	1t	1½t	2t
	T4	0	0	1t	1½t	2½t	3t	3½t	4t
	T6	1t	1t	1½t	2½t	3t	3½t	4½t	5t
7072	O	0	0	...	...	...	...	...	...
	H12	0	0	...	...	...	...	...	...
	H14	0	0	...	...	...	...	...	...
	H16	0	½t	...	...	...	...	...	...
	H18	1t	1t	...	...	...	...	...	...
7075(b)	O	0	0	1t	1t	1½t	2½t	3½t	4t
	T6	3t	4t	5t	6t	6t	8t	9t	9½t
7178(b)	O	0	0	1t	1½t	1½t	2½t	3½t	4t
	T6	3t	4t	5t	6t	6t	8t	9t	9½t

(a) The radii listed are the minimum recommended for bending sheets and plates without fracturing in a standard press brake with air bend dies. Other bending operations may require larger radii or permit smaller radii. The minimum permissible radii also vary with the design and condition of the tooling. (b) Alclad sheet in the heat treatable alloys can be bent over slightly smaller radii than the corresponding tempers of the bare alloy. (c) Tempers T361 and T861 formerly designated T36 and T86, respectively.

Table 2 Maximum thicknesses of aluminum alloy sheet that can be cold bent 180° over zero radius

Alloy	Temper	Sheet thickness, max mm	in.	Alloy	Temper	Sheet thickness, max mm	in.
1100	O	3.2	1/8	5005	H32	3.2	1/8
	H12	1.6	1/16		H34	0.8	1/32
	H14	1.6	1/16	5050	O	3.2	1/8
	H16	0.4	1/64		H32	1.6	1/16
Alclad					H34	0.8	1/32
2014	O	1.6	1/16	5052	O	3.2	1/8
2024	O	1.6	1/16		H32	0.8	1/32
3003	O	3.2	1/8		H34	0.4	1/64
	H12	1.6	1/16	5086	O	3.2	1/8
	H14	1.6	1/16	5154	O	1.6	1/16
3004	O	3.2	1/8		H32	0.8	1/32
	H32	0.8	1/32	5457	O	3.2	1/8
	H34	0.4	1/64		H25	0.8	1/32
5005	O	3.2	1/8	6061	O	1.6	1/16
	H12	1.6	1/16		T4	0.8	1/32
	H14	1.6	1/16	7075	O	0.8	1/32

Table 3   Minimum bend radii for 1008 or 1010 steel sheet

Quality or temper	Minimum bend radius	
	Parallel to rolling direction	Across rolling direction
Cold rolled		
Commercial	0.25 mm (0.01 in.)	0.25 mm (0.01 in.)
Drawing, rimmed	0.25 mm (0.01 in.)	0.25 mm (0.01 in.)
Drawing, killed	0.25 mm (0.01 in.)	0.25 mm (0.01 in.)
Enameling	0.25 mm (0.01 in.)	0.25 mm (0.01 in.)
Cold rolled, special properties		
Quarter hard(a)	1t	1/2t
Half hard(b)	NR	1t
Full hard(c)	NR	NR
Hot rolled		
Commercial:		
Up to 2.3 mm (0.090 in.)	3/4t	1/2t
More than 2.3 mm (0.090 in.)	1-1/2t	1t
Drawing:		
Up to 2.3 mm (0.090 in.)	1/2t	1/4t
More than 2.3 mm (0.090 in.)	3/4t	1/2t

Note: t, sheet thickness; NR, not recommended
(a) 60 to 75 HRB. (b) 70 to 85 HRB. (c) 84 HRB minimum

Table 4   Typical bending limits for six commonly formed stainless steels

	Minimum bend radius		
		Quarter hard, cold rolled	
	Annealed to 4.7 mm (0.187 in.) thick (180° bend)	To 1.27 mm (0.050 in.) thick (180° bend)	1.3 to 4.7 mm (0.051 to 0.187 in.) thick (90° bend)
Type			
301, 302, 304	1/2t	1/2t	1t
316	1/2t	1t	1t
410, 430	1t	...	...

Note: t, stock thickness

Table 5   Bend ductility data for copper-based alloys

Values shown are the maximum tensile strengths at which bending to the required bend radius can be achieved.

Alloy	Maximum tensile strength, MPa (ksi)	
	Parallel to rolling direction Gage: 0.75 mm (0.030 in.) Radius: 1.2 mm (3/64 in.)	Across rolling direction Gage: 0.75 mm (0.030 in.) Radius: 1.2 mm (3/64 in.)
C11000	352 (51)	345 (50)
C15100	400 (58)	400 (58)
C19400	496 (72)	490 (71)
C19500	572 (83)	558 (81)
C23000	593 (86)	538 (78)
C26000	662 (96)	524 (76)
C35300	572 (83)	469 (68)
C41100	496 (72)	434 (63)
C42500	621 (90)	462 (67)
C50500	469 (68)	469 (68)
C51000	648 (94)	538 (78)
C52100	731 (106)	552 (80)
C63800	793 (115)	696 (101)
C65400	731 (106)	627 (91)
C66400	641 (93)	579 (84)
C68800	745 (108)	731 (106)
C70600	496 (72)	483 (70)
C72500	517 (75)	503 (73)
C73500	579 (84)	517 (75)
C74000	586 (85)	552 (80)
C75200	579 (84)	558 (81)
C77000	717 (104)	676 (98)

Bending Strength Tests

By A. Guha
Senior Metallurgist
Brush Wellman Inc.

BENDING STRENGTH TESTS offer a means of determining the modulus of elasticity in bending and the bending strength of flat metallic materials in the form of strip, sheet, or plate. Three standard bending load-deflection tests are discussed in this article: the cantilever beam bend test, the three-point bend test, and the four-point bend test. The variables that can affect the results of these test procedures are also examined.

The cantilever beam bend test measures the angle of deflection and corresponding bending moment of metallic strip or sheet subjected to continuous loading in bending. A bending stress-strain curve analogous to the stress-strain curve in tension is developed from the data. The modulus of elasticity is determined at stresses below the proportional limit. The bending yield strength is determined by the offset method similar to that used for stress-strain curves in tension or compression.

In the three-point bend test, the test specimen is supported near each end and is loaded at one point equidistant from each support. The modulus of elasticity in bending is obtained by the measurements of load and deflection at stresses below the proportional limit. The bending proof strength is determined by increasing the load in steps and unloading until a specified permanent set is obtained.

The four-point bend test consists of a simple beam resting on two supports and loaded at two points equally spaced from each support. The modulus of elasticity in bending and the bending proof strength in this method are obtained by a procedure similar to the three-point method of bend testing.

Cantilever Beam Bend Test

In the cantilever beam bend test, the test specimen is loaded as a simple cantilever beam, in which one end of the beam is sup-ported and the free end is loaded. The test procedure involves measurements of the applied bending moment and the corresponding angular deflection of a cantilever beam in continuous loading. The bending moment-deflection data are used to determine the maximum bending stress and the bending strain occurring in the outer fibers at the fixed end of the beam. A bending stress-strain curve is constructed from the data, from which the modulus of elasticity and off-set yield strengths in bending are calculated. The offset yield strength is the stress at which a specified limiting deviation from bending stress to bending strain is exhibited.

Apparatus. A cantilever beam bend test apparatus is shown in Fig. 1. This tester consists of a specimen holder, a pendulum weighing system, a moment scale, and an angular deflection scale.

The specimen holder is a vise, V, to which an angular deflection indicator, I_1, is attached. The specimen holder is rotated counterclockwise about point 0 at a nominal rate of 60° of arc per minute.

The pendulum weighing system is composed of a set of detachable weights, an angular deflection scale with a moment pointer indicator I_2, a 6.35-mm (0.25-in.) diam loading pin that transmits the bending force of the pendulum system to the free end of the cantilever specimen, and a weight to counterbalance the loading pin. The pendulum weighing system pivots about the center of rotation designated by point 0 in Fig. 1.

The moment scale measures the applied bending moment as a function of the angle through which the pendulum system rotates. A full-scale reading of 100 corresponds to the maximum bending moment of the pendulum.

The angular deflection scale, A, is graduated in degrees of arc and indicates the angle through which the specimen vise rotates relative to the pendulum system. The difference between the angle through which the vise has been rotated and the angle through which the load pendulum has been deflected is designated as angular deflection.

Cantilever beam bend testers are available in capacities ranging from 0.1 to 5.6 N · m (1 to 50 lbf · in.). All are similar in principle and operation.

Test Specimens. As outlined in ASTM Standard E 855 (Ref 1), rectangular specimens are used for testing metallic materials in the form of strip. Before preparing test specimens, the orientation relative to the rolling direction must be identified. To ensure precision and accuracy in test results, specimens must be prepared carefully—preferably by machining, such as edge milling. Sheared or slit specimens should not be used. Shearing causes tensile residual stresses to develop in the outer fibers of a specimen, thus altering the stress distribution that existed in the as-received condition (Ref 2). Defects such as burrs that act as stress raisers at the specimen surface must be removed.

Specimen curvature due to coil set is permitted, provided the ratio of the radius of curvature to thickness of strip exceeds 500. The finished test specimens should not be twisted or wavy. No attempt should be made to correct poorly shaped test specimens by leveling or stretching prior to bend testing unless it is necessary to determine the effect of such operations on the material. Care should also be taken not to disturb the microstructure of strip prior to sampling.

The specimen size should be such that the width-to-thickness ratio is always greater than ten. The recommended minimum specimen strip thickness is 0.38 mm (0.015 in.). The recommended minimum width of the specimen is 12.7 mm (0.5 in.). The specimen width should not exceed that of the vise or the loading pin. The ratio of the span length, which is the distance between the center of rotation and the center of the loading pin, to the specimen thickness should be greater than 15. Because the span length is

Fig. 1 Schematic of a cantilever beam bend tester
Source: Ref 1

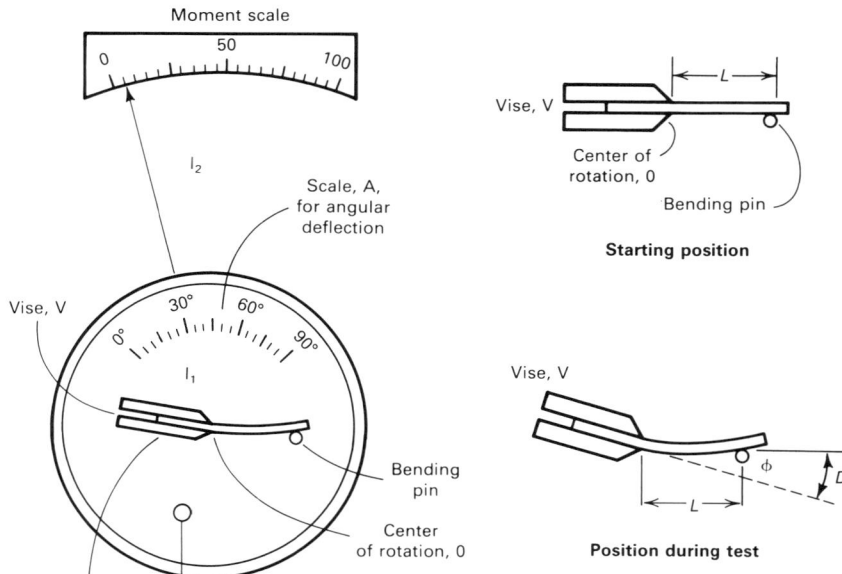

thickness, mm (in.); and L is span length, mm (in.).

In the cantilever beam test, the maximum bending moment is related to the applied load, P, by:

$$M = PL \qquad \text{(Eq 3)}$$

The maximum attainable bending moment in the pendulum system is expressed in terms of the moment scale reading, f, and the bending moment, M, measured at an angular deflection of ϕ as follows:

$$f = \frac{M}{M_m} \times 100 \qquad \text{(Eq 4)}$$

where a full-scale reading of 100 corresponds to the maximum bending moment of the pendulum. As shown in Fig. 1, the beam deflection, D, is approximated by the length of an arc having radius L and an included angle of ϕ radians. Using this approximation:

$$D = \phi L \qquad \text{(Eq 5)}$$

Combining Eq 1 through 5:

$$E_b = \frac{f M_m L}{25 \phi b h^3} \qquad \text{(Eq 6)}$$

Combining Eq 2 through 4:

$$\sigma_b = \frac{3 M_m f}{50 b h^2} \qquad \text{(Eq 7)}$$

The bending strain in the outer fibers at the clamped end corresponding to the stress given by Eq 7 is:

$$\epsilon_b = \frac{\sigma_b}{E_b} = \frac{3}{2} \frac{\phi h}{L} \qquad \text{(Eq 8)}$$

Stress-strain curves in bending are similar to those in tension or compression (see the article "Mechanical Behavior of Materials Under Tension" in this Volume). The bending stress (Eq 7) is plotted as the ordinate, and the bending strain (Eq 8) is plotted as the abscissa.

The modulus of elasticity in bending, E_b, is determined by the slope of a straight line extending from the maximum deflection data point (max) to the permanent set point (ps) obtained from the unloading curve at zero load:

$$E_b = \frac{\dfrac{M_m f}{25 b h^2}}{\left(\dfrac{\phi h}{L}\right)_{max} - \left(\dfrac{\phi b}{L}\right)_{ps}} \qquad \text{(Eq 9)}$$

To construct the bending stress-strain curve, a straight line having slope E_b is drawn so that it passes through the origin. The actual data points for elastic loading may be slightly displaced from this line due to shape deficiencies in the specimen that would prevent

typically 50.8 mm (2.0 in.) in the cantilever beam bend tester, the maximum specimen thickness is limited to 3.3 mm (0.13 in.).

Test Procedure. The thickness of the test specimen should be measured at the four corners and at the center, and the average of these five measurements reported. The width of the test specimen should be measured at both ends and at the center, and the average of these three measurements reported.

A minimum of six specimens must be tested. Each specimen should be labeled (marked) at one end on the top side. For specimens having an initial residual curvature, three specimens should be tested with the concave surface facing upward and three with the convex surface facing upward.

The specimen is firmly clamped with the long edge approximately parallel to the face of the dial plate. The angular deflection scale is set to indicate zero angle when contact is made between the specimen and the loading pin. The specimen holder is rotated to force the specimen against the loading pin. The moment scale readings are recorded at increments of 2° angular deflection until the desired deflection (not exceeding 30°) is reached. If the deflection exceeds 30°, the actual curved length of the cantilever beam exceeds 50.8 mm (2.0 in.), which is the span length in the cantilever beam bend tester. However, if the deflection is limited to 30°, the increase in length due to curvature is negated by the rotation of the contact point about the circumference of the loading pin;

therefore, the span length remains constant at 50.8 mm (2.0 in.).

The specimen is then unloaded. With the specimen in contact with the loading pin, the permanent set in degrees is recorded by reading the angular deflection scale at zero load.

Stress-Strain Relationships. To construct a stress-strain curve in a cantilever beam bend test, the bending moment-deflection data must be normalized with respect to specimen geometry. Also, a relationship must be established between the applied bending moment and the maximum bending stress occurring in the outer fibers at the clamped end of the beam. The angle of deflection must be replaced by the deflection of the loaded end of the beam to estimate the outer fiber bending strain at the clamped end. The following standard formulas for bending stress and deflection in the simple cantilever beam are used to derive these relationships (Ref 3).

The deflection, D, of the loaded end of a cantilever beam is given by:

$$D = \frac{4 P L^3}{E_b b h^3} \qquad \text{(Eq 1)}$$

The maximum bending stress, σ_b, in the outer fibers at the clamped end is given by:

$$\sigma_b = \frac{6 P L}{b h^2} \qquad \text{(Eq 2)}$$

where P is applied load, N (lbf); E_b is modulus of elasticity in bending, Pa (psi); b is specimen width, mm (in.); h is specimen

uniform distribution of load over the full width.

The nonlinear portion of the bending stress-strain curve is constructed by drawing a curve through the remaining data points and connecting it with the modulus of elasticity line. To determine the bending yield strength from the stress-strain curve, the "offset method," analogous to that used for tensile or compressive stress-strain curves, is used. The offset yield strengths in bending should be determined for strains of 0.01, 0.05, and 0.10%, provided the maximum allowable deflection angle of 30° is not exceeded.

Example Calculations. Bending moment-deflection data were obtained for spring-tempered C77000 copper alloy strip in the longitudinal orientation using the test procedure described above. The permanent set angle was measured after the load was removed. The bending moment and deflection data are shown in Table 1.

Using Eq 7 and 8, the data in Table 1 were normalized with regard to specimen geometry and plotted on the stress-strain curve shown in Fig. 2. In this diagram, the ordinate is equal to the bending stress multiplied by a constant that is equal to $50bh^2/3M_m$. Similarly, the abscissa is equal to the bending strain multiplied by a constant that is equal to $2L/3h$.

The modulus of elasticity in bending is obtained from Fig. 2 by drawing a straight line from the maximum deflection data point (max) to the permanent set point (ps). The slope of the unloading line is the modulus of elasticity in bending. To construct the linear

portion of the bending stress-strain curve, the elastic modulus line is translated along the abscissa until it passes through the origin. The nonlinear portion is constructed by drawing a curve through the remaining data points and connecting this curve with the elastic modulus line.

To determine the yield strengths in bending for strains of 0.01, 0.05, and 0.10% from Fig. 2, the offset method can be used. The results are given in Table 2.

Three-Point and Four-Point Bend Tests

Three-point and four-point bend tests consist of deflection of a simple beam subjected to either three- or four-point symmetrical loading. The modulus of elasticity is determined by load-deflection measurements at stresses below the proportional limit. The bending proof strength is determined by a load-unload sequence until a specified permanent strain is measured on unloading. The bending proof stress is the nominal stress in the outer fibers of a beam that results in permanent strain in the outer fibers upon unloading the specimen.

The apparatus for three- and four-point tests consists of two adjustable supports and a means for measuring deflection and applying load. Three- or four-point bend fixtures containing supports and load applicators are normally used, and are shown schematically in Fig. 3. The fixtures can be used with a material testing system, such as a universal testing machine, to apply the load to the test specimen. A deflection-measuring device, such as a deflectometer or cathetometer, can be used to measure deflection at midspan. The load at a given deflection is determined from the load-cell output on the machine x-y recorder, which is calibrated with dead-weights.

A spring-material test apparatus, described in German Specification DIN 50 151, also meets the requirements of loading under three-point bending. As shown schematically in Fig. 4, the specimen is supported on both sides. The spindle (4) adjusts the distance

Fig. 2 Cantilever bending normalized stress-strain curve for spring-tempered C77000 copper alloy strip in the longitudinal orientation
max, maximum deflection; ps, permanent set at zero load. Source: Olin Corp.

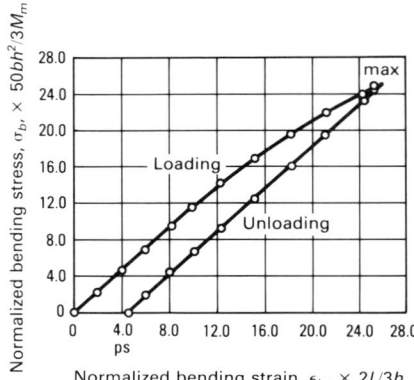

between the supports (2 and 3) along the specimen axis. The supports are symmetrically placed with respect to the load applicator (6).

To measure the modulus of elasticity, the knurled disk (9) is set at a position indicated by the letter E. A 100-gf load is applied to the specimen by pressing the activator key (7). The deflection corresponding to the load is measured from the dial (5) as the needle (10) makes contact with the specimen, and the test light (11) is observed.

The bending proof strength, which is also called the spring-bending limit, is measured as the load is increased in equal intervals by turning the knurled disk. Upon removal of the load, the residual deflection is measured from the dial. This procedure is continued until a deflection of 0.05 mm (0.002 in.) is measured. The outer fiber stress causing a permanent deflection of 0.05 mm (0.002 in.) is the spring-bending limit of the material.

Test Specimens. The recommendations with regard to specimen preparation and specimen curvature outlined earlier for cantilever beam specimens also apply for three-

Table 1 Bending moment-deflection data for spring-tempered C77000 copper alloy strip in the longitudinal orientation

Obtained by cantilever beam bend testing

Angle of deflection, degrees	Bending moment, N · m (lbf · in.) Loading	Unloading
0	0	NA(a)
2	0.2 (2.0)	NA(a)
4	0.5 (4.4)	NA(a)
4.5(b)	No data	0
6	0.8 (6.8)	0.2 (2.0)
8	1.0 (9.2)	0.5 (4.4)
10	1.3 (11.6)	0.8 (6.8)
12	1.6 (14.0)	0.9 (8.8)
15	1.9 (16.8)	1.4 (12.4)
18	2.2 (19.6)	1.8 (16.0)
21	2.5 (22.0)	2.2 (19.6)
24	2.7 (24.0)	2.6 (23.2)
25	2.8 (24.8)	2.75 (24.4)

Note: Span length, L, is 50.8 mm (2.0 in.); specimen width, b, is 25.4 mm (1.0 in.); specimen thickness, h, is 0.805 mm (0.0317 in.); maximum attainable bending moment, M_m, is 4.5 N · m (40 lbf · in.).
(a) Not applicable. (b) Permanent set angle measured during unloading at zero load is 4.5°. Source: Olin Corp.

Table 2 Modulus of elasticity and yield strengths in bending for 0.805-mm (0.0317-in.) thick spring-tempered C77000 copper alloy strip

Obtained by cantilever beam bend testing

Orientation	Modulus of elasticity GPa	psi	Bending yield strength, MPa (ksi) 0.01% strain	0.05% strain	0.10% strain
Longitudinal	118.6	17.2×10^6	451.6 (65.5)	568.8 (82.5)	653.0 (94.7)

Note: Results are based on an average of three specimens tested with the convex side facing upward and three specimens tested with the concave side facing upward.
Source: Olin Corp.

Fig. 3 Schematics of (a) three-point bend test apparatus and (b) four-point bend test apparatus

The load applicator has a 60° angle with a radius, R, of 0.127 mm (0.005 in.). In three-point bending, the load is applied at midspan using one applicator. In four-point bending, two load applicators, evenly spaced from the support, are used. The distance between the load applicators equals two thirds of the span length, L. In both three- and four-point loading, one of the load applicators has a convex (12.7-mm, or 0.50-in.) radius of curvature. P, load increment as measured from preload; L, span length between supports; R, radius; δ, deflection increment at midspan as measured from preload; a, distance from support to load applicator when specimen is straight. Source: Ref 1

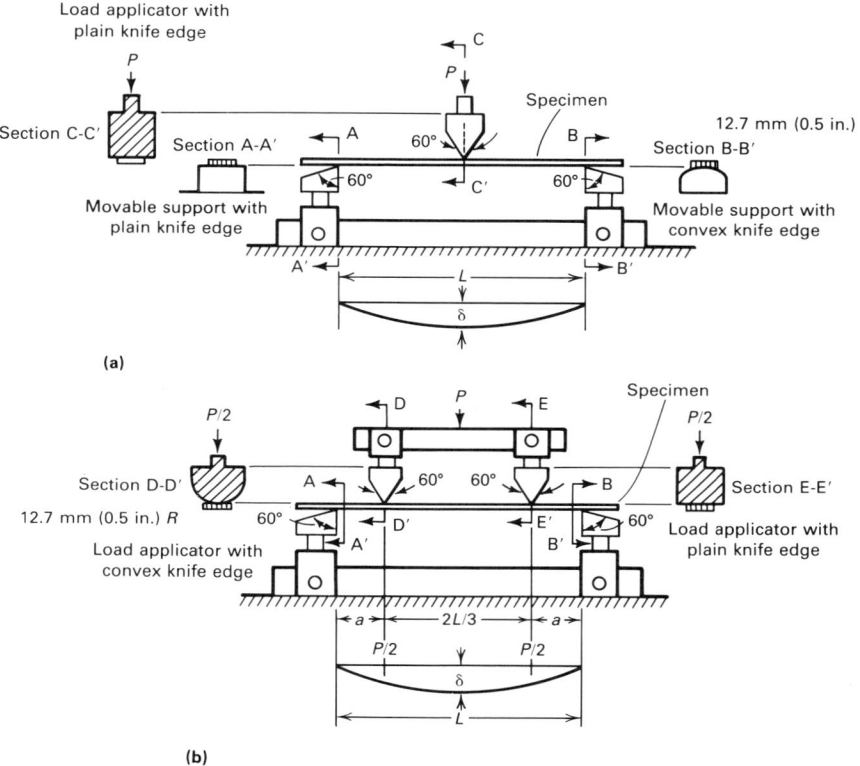

Fig. 4 Schematic of spring-material test apparatus

1, specimen; 2 and 3, specimen support blocks; 4, spindle; 5, dial; 6, load applicator; 7, activator key; 8, stroke-limiting pegs; 9, knurled disk; 10, contacting needle; 11, test light

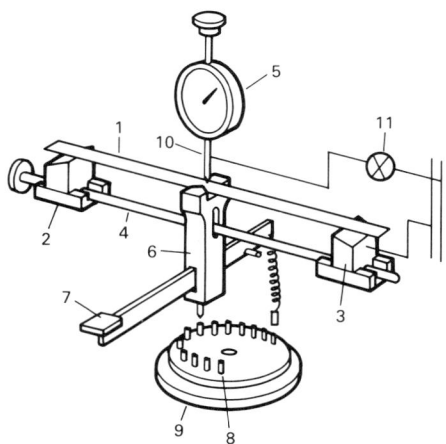

Load versus deflection data up to a maximum load corresponding to 50% of the estimated proof strength are plotted with load as the ordinate and deflection as the abscissa. The resulting straight line is similar to the linear portion of the load-elongation curves in tension or compression.

The modulus of elasticity in bending is calculated from the load increment and the corresponding deflection increment between the two points on the straight line as far apart as possible, using the formulas:

$$E_b = \frac{L^3}{4bh^3}\frac{\Delta P}{\Delta \delta} \quad \text{(three-point loading)}$$

(Eq 10)

$$E_b = \frac{a(3L^2 - 4a^2)}{4bh^3}\frac{\Delta P}{\Delta \delta} \text{ (four-point loading)}$$

(Eq 11)

where E_b is modulus of elasticity in bending, Pa (psi); L is span length, mm (in.); h is specimen thickness, mm (in.); b is specimen width, mm (in.); a is distance from the support to the load applicator (for four-point loading); ΔP is load increment as measured from preload, N (lbf); and $\Delta \delta$ is deflection increment at midspan as measured from preload, mm (in.).

The average modulus of elasticity in bending is determined for a minimum of six specimens, half of which are tested with the marked side facing upward and half with the marked side facing downward. The permanent deflection δ_p, which produces the permanent strain in the outer fiber of 0.01%, is obtained by:

and four-point bend specimens. However, size requirements differ.

The minimum specimen strip thickness for three- and four-point bend specimens (as recommended in Ref 1) is 0.25 mm (0.010 in.). The span length should be 150 times the nominal thickness in the range 0.25 to 0.51 mm (0.010 to 0.020 in.) and 100 times the nominal thickness in the range exceeding 0.51 mm (0.020 in.). The specimen width should be 3.81 mm (0.150 in.) in the thickness range 0.25 to 0.51 mm (0.010 to 0.020 in.) and 12.7 mm (0.500 in.) in the thickness range exceeding 0.51 mm (0.020 in.). Total specimen length should be 250 times the nominal thickness in the range 0.25 to 0.51 mm (0.010 to 0.020 in.) and 165 times the nominal thickness in the range exceeding 0.51 mm (0.020 in.).

Test Procedure. The thickness and width of three- and four-point bend specimens are measured as described in the section on cantilever beam testing in this article. Six specimens should be tested, and should be

marked as described earlier for cantilever beam specimens.

The modulus of elasticity is determined by applying a preload corresponding to approximately 20% of the bending proof strength, the value of which is estimated by a preliminary test. Friction at the support is minimized by gently tapping the specimen. Load and deflection are measured incrementally or continuously up to a maximum load corresponding to 50% of the estimated proof strength. At least five load and deflection measurements from the preload to the maximum load are needed to determine the modulus of elasticity in bending.

To determine bending proof strength, the specimen is loaded to within 90% of the estimated proof strength value and unloaded to the preload value. The load is progressively increased to at least 92% of the proof strength until a permanent deflection corresponding to a specified permanent strain in the outer fiber of 0.01% is obtained upon unloading.

$$\delta_p = \frac{0.0001L^2}{6h} \quad \text{(three-point loading)}$$

$$\text{(Eq 12)}$$

$$\delta_p = \frac{0.0001(3L^2 - 4a^2)}{12h}$$

$$\text{(four-point loading)} \quad \text{(Eq 13)}$$

The load, P_p, corresponding to deflection, δ_p, is determined from a linear interpolation of the data points above and below the exact value of δ_p desired.

The bending proof strength, σ_p for 0.01% strain in the outer fiber is determined by:

$$\sigma_p = \frac{3P_pL}{2bh^2} \quad \text{(three-point loading)} \quad \text{(Eq 14)}$$

$$\sigma_p = \frac{3P_pa}{bh^2} \quad \text{(four-point loading)} \quad \text{(Eq 15)}$$

The proof strengths in bending for strains of 0.05 and 0.10% can be determined similarly, provided the deflections are sufficiently small to disregard axial forces and friction at the supports. The average bending proof strength is also determined from a minimum of six specimens.

Example Calculations. Bending load-deflection data were obtained using the three-point bend test for 0.76-mm (0.030-in.) thick spring-tempered C51000 phosphor bronze strip in the longitudinal orientation. The data for one such specimen are shown in Table 3. The data are also plotted as a straight line, as shown in Fig. 5, from the five measured values of load and deflection. The modulus of elasticity in bending is calculated from Eq 10 from the load increment and the corresponding deflection increment between the two points on the straight line as far apart as possible.

To calculate the proof strength in bending, the deflection, δ_p, that produces a permanent set of 0.0001 mm/mm (in./in.) is determined as outlined in Eq 12. The load, P_p, corresponding to deflection, δ_p, is shown in Table 3 and is determined by the procedure described in the preceding section on load versus deflection data. The proof strength for 0.01% strain in the outer fiber is calculated from Eq 7; results are given in Table 4.

REFERENCES

1. "Standard Methods of Bend Testing of Metallic Flat Materials for Spring Applications," E 855, *Annual Book of ASTM Standards*, Vol 03.01, ASTM, Philadelphia, 1984, p 788-804
2. F.P.J. Pimrott and H.K. Weikinger, The Origin and Measurement of Residual Stresses, *Design Eng.*, Vol 8 (No. 4 and 5), April and May 1962, p 48-50, 60-63
3. S.P. Timoshenko and J.N. Goodier, *Theory of Elasticity*, 3rd ed., McGraw-Hill, New York, 1970

Fig. 5 Load-deflection plot for 0.76-mm (0.030-in.) thick spring-tempered C51000 phosphor bronze strip
Data obtained by three-point bend testing for determination of modulus of elasticity in bending. Source: Bell Laboratories

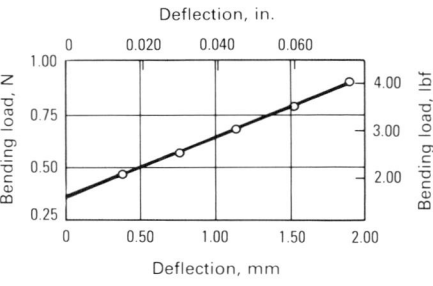

Table 3   Bending load-deflection data for spring-tempered C51000 phosphor bronze strip in the longitudinal orientation
Obtained by three-point bend testing

N	Load lbf	Deflection mm	in.
9.2	2.07	0.38	0.015
11.3	2.55	0.76	0.030
13.6	3.05	1.14	0.045
15.3	3.45	1.52	0.060
17.9	4.02	1.90	0.075

Note: Specimen thickness, h, is 0.76 mm (0.030 in.); specimen width, b, is 12.7 mm (0.500 in.); span length, $L = 100h = 76$ mm (3.0 in.). The load, P_p, corresponding to deflection, δ_p, causing a permanent strain of 0.0001 mm/mm (in./in.) is estimated to be 45.5 N (10.22 lbf).
Source: Bell Laboratories

Table 4   Modulus of elasticity and proof strength in bending at 0.01% permanent strain for 0.76-mm (0.030-in.) thick spring-tempered C51000 phosphor bronze strip
Obtained by three-point bend testing

Specimen residual curvature	Modulus of elasticity GPa	psi	Bending proof strength at 0.01% permanent strain MPa	ksi
Convex side up	100.6	14.6×10^6	688.8	99.0
Concave side up	100.6	14.6×10^6	639.1	92.7

Note: Results are based on an average of three specimens tested with the convex side facing upward and three specimens with the concave side facing upward.
Source: Bell Laboratories

Torsion Testing

Fundamental Aspects of Torsional Loading

By John A. Bailey
Professor
Department of Mechanical and Aerospace Engineering
North Carolina State University

TORSION TESTS can be carried out on most materials to determine mechanical properties such as modulus of elasticity in shear, yield shear strength, ultimate shear strength, modulus of rupture in shear, and ductility. The torsion test can also be used on full-size parts (shafts, axles, and twist drills) and structures (beams and frames) to determine their response to torsional loading. In torsion testing, unlike tension testing and compression testing, large strains can be applied before plastic instability occurs, and complications due to friction between the test specimen and dies do not arise.

Torsion tests are most frequently carried out on prismatic bars of circular cross section by applying a torsional moment about the longitudinal axis. The shear stress versus shear strain curve can be determined from simultaneous measurements of the torque and angle of twist of the test specimen over a predetermined gage length.

The following section discusses the torsional deformation of prismatic bars of circular cross section. Discussion of the torsional response of prismatic bars of noncircular section (rectangular, elliptical, or triangular) in the elastic range can be found in Ref 1.

Torsion Testing of Prismatic Bars of Circular Cross Section

In torsion testing of prismatic bars of circular cross section it is assumed that:

- Bar material is homogeneous and isotropic.
- Twist per unit length along the bar is constant.
- Sections that are originally plane to the torsional axis remain plane after deformation.

- Initially straight radii remain straight after deformation.

Figure 1 shows the torsional deformation of a long, straight, isotropic prismatic bar of circular section. Assuming the above-mentioned constraints, the displacements are given by:

$$u_r = 0$$

$$u_z = 0$$

$$u_\theta = rz\left(\frac{d\theta}{dz}\right) \qquad \text{(Eq 1)}$$

where $d\theta/dz$ is the angle of twist per unit length (θ/L), and L is the gage length of the test specimen. The strains are given by:

$$\epsilon_{zz} = 0$$

$$\epsilon_{rr} = 0$$

$$\epsilon_{\theta\theta} = 0$$

$$\gamma_{zr} = 0$$

$$\gamma_{r\theta} = 0$$

$$\gamma_{z\theta} = \frac{rd\theta}{dz} \qquad \text{(Eq 2)}$$

For an isotropic material that obeys Hooke's law, the corresponding stress state is given by:

$$\sigma_{zz} = 0$$

$$\sigma_{rr} = 0$$

$$\sigma_{\theta\theta} = 0$$

$$\tau_{zr} = 0$$

$$\tau_{r\theta} = 0$$

$$\tau_{z\theta} = G\gamma_{z\theta} \qquad \text{(Eq 3)}$$

where G is the shear modulus that is related to Young's modulus (E) and Poisson's ratio (v) by:

$$G = \frac{E}{2(1 + v)} \qquad \text{(Eq 4)}$$

The stress distribution across the prismatic bar of circular cross section is given by:

$$\tau_{z\theta} = \frac{Gr\theta}{L} \qquad \text{(Eq 5)}$$

Fig. 1 Torsion of a solid circular prismatic section

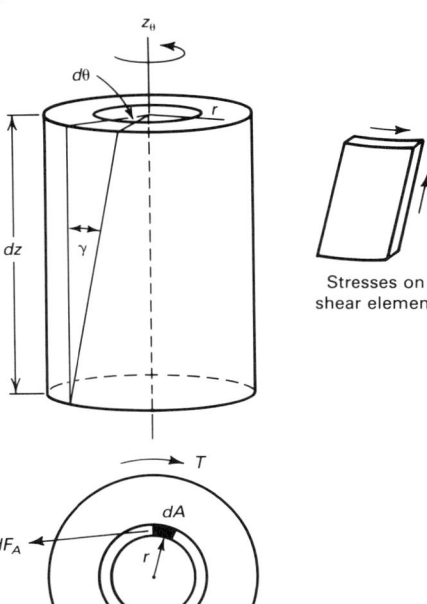

Stresses on shear element

Thus, the shear stress is zero at the center of the bar ($r = 0$) and increases linearly with radius. The maximum value of the shear stress occurs at the surface of the bar ($r = a$) and is given by:

$$\tau_{z\theta_{max}} = \frac{Ga\theta}{L} \qquad \text{(Eq 6)}$$

The torque (T) transmitted by the elemental section shown in Fig. 1 is given by:

$$dT = (\tau_{z\theta}r)dA \qquad \text{(Eq 7)}$$

or

$$T = \int_0^a \tau_{z\theta} 2\pi r^2 dr \qquad \text{(Eq 8)}$$

Combining Eq 5 and 8 gives:

$$T = \frac{G\theta}{L} \int_0^a 2\pi r^3 dr \qquad \text{(Eq 9)}$$

Integration gives:

$$T = \frac{\pi a^4 G\theta}{2L} \qquad \text{(Eq 10)}$$

or on arrangement:

$$\theta = \frac{TL}{GJ} \qquad \text{(Eq 11)}$$

where $J = \pi a^4/2$ is the polar moment of inertia of a prismatic bar of circular section about its axis of symmetry. Thus, the angle of twist can be calculated from knowledge of the applied torsional load, shear modulus, and bar geometry. Combining Eq 6 and 10 gives:

$$\tau_{z\theta_{max}} = \frac{2T}{\pi a^3} \qquad \text{(Eq 12)}$$

or

$$\tau_{z\theta_{max}} = \frac{Ta}{J} \qquad \text{(Eq 13)}$$

Thus, the maximum shear stress can be calculated from knowledge of the torsional loading and bar geometry. When the surface shear stress ($\tau_{z\theta_{max}}$) reaches the yield shear stress (k) of the test material, plastic deformation (flow) occurs. The deformation zone begins at the surface of the bar and advances inward as an annulus surrounding an elastic core. The stress distributions are shown schematically in Fig. 2 for a non-work-hardening and a work-hardening material.

For a non-work-hardening material, the total torque transmitted by the bar, according to Ref 2, is given by:

Fig. 2 Section through prismatic bar of circular section

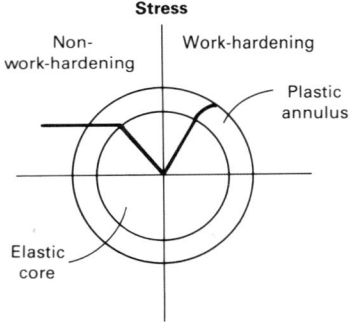

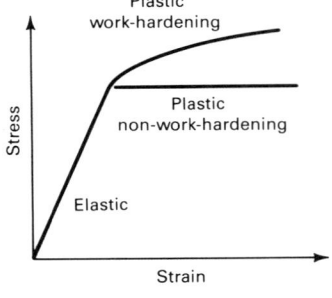

$$T = \int_{r=0}^{r=r_p} \tau_{z\theta} 2\pi r^2 dr + \int_{r=r_p}^{r=a} k\, 2\pi r^2 dr \qquad \text{(Eq 14)}$$

The first term on the right side of Eq 14 is the torque transmitted by the elastic core, where the shear stress varies linearly with r. The second term on the right side of Eq 14 is the torque transmitted by the plastic annulus, where the shear stress is constant and independent of r. The elastic-plastic boundary occurs at $r = r_p$. Integration of Eq 14 gives:

$$T = \frac{\pi r_p^4 G\theta}{2L} + \frac{2\pi a^3 k}{3} - \frac{2\pi r_p^3 k}{3} \qquad \text{(Eq 15)}$$

Compatibility at the elastic-plastic boundary requires that:

$$k = \frac{G\theta r_p}{L} \qquad \text{(Eq 16)}$$

Combining Eq 15 and 16 and rearranging gives:

$$T = \frac{2\pi a^3 k}{3} \left\{ 1 - \frac{1}{4} \left(\frac{r_p}{a} \right)^3 \right\} \qquad \text{(Eq 17)}$$

or

$$T = \frac{2\pi a^3 k}{3} \left\{ 1 - \frac{1}{4} \left(\frac{\theta_y}{\theta} \right)^3 \right\} \qquad \text{(Eq 18)}$$

where θ_y is the angle of twist at which yielding begins. When θ is very large compared to θ_y, then:

$$T_p = \frac{2\pi a^3 k}{3} \qquad \text{(Eq 19)}$$

where T_p is the torque required for plastic flow. Equation 18 can now be rewritten as:

$$T = T_p \left\{ 1 - \frac{1}{4} \left(\frac{\theta_y}{\theta} \right)^3 \right\} \qquad \text{(Eq 20)}$$

When the bar becomes fully plastic, the torque becomes independent of the angle of twist. In the fully elastic regime, the shear stress at the surface of the bar is given by Eq

6. In the elastic-plastic and fully plastic regimes, the shear stress at the surface of the bar is k. The shear strain at the surface of the bar is:

$$\gamma = \frac{a\theta}{L} \qquad \text{(Eq 21)}$$

for all regimes.

In practice, work hardening usually occurs, and the torque continues to increase up to fracture. The shear stress versus shear strain curve in the plastic range can be computed from the torque-twist curve using the procedure given below. It is important to note that the computed values of stress and strain are those that occur at the surface of the bar, and that the material is insensitive to the rate of deformation.

The torque, according to Ref 3, is given by:

$$T = \int_0^a \tau 2\pi r^2 dr \qquad \text{(Eq 22)}$$

where the subscripts on the shear stress are dropped. Changing the variable from r to γ gives:

$$T = \int_0^{\gamma_a} \frac{\tau 2\pi \gamma^2 d\gamma}{\theta_l^3} \qquad \text{(Eq 23)}$$

where θ_l is the twist per unit length. In general, the shear stress versus shear strain curve can be written as:

$$\tau = f(\gamma) \qquad \text{(Eq 24)}$$

Thus, Eq 24 becomes:

$$T = \int_0^{\gamma_a} \frac{2\pi f(\gamma)\gamma^2 d\gamma}{\theta_l^3} \qquad \text{(Eq 25)}$$

Differentiating Eq 25 with respect to θ_l gives:

$$d(T\theta_l^3) = 2\pi f(\gamma_a)\gamma_a^2 d\gamma_a \qquad \text{(Eq 26)}$$

At the specimen surface:

$$\tau_a = f(\gamma_a) \tag{Eq 27}$$

and

$$\gamma_a = a\theta_l \tag{Eq 28}$$

Substituting Eq 27 and 28 into Eq 26 gives:

$$d(T\theta_l^3) = 2\pi\tau_a a^3\theta_l^2 d\theta_l$$

or

$$\frac{d(T\theta_l^3)}{d\theta_l} = 2\pi\tau_a^3\theta_l^2 a^3 \tag{Eq 29}$$

Expanding Eq 29 gives:

$$\frac{dT}{d\theta_l}\theta_l^3 + 3T\theta_l^2 = 2\pi\tau_a a^3\theta_l^2$$

or

$$\tau_a = \frac{1}{2\pi a^3}\left[3T + \theta_l\frac{dT}{d\theta_l}\right] \tag{Eq 30}$$

The first term on the right side of Eq 30 is the torque due to the maximum yield shear stress of τ_a in a fully plastic non-strain-hardening material, whereas the second term is a correction for strain hardening. These terms can be readily derived from the torque-twist curve shown in Fig. 3, where:

$$\frac{dT}{d\theta_l} = \frac{BC}{CD}$$

$$\theta_l = CD$$

$$\theta_l\frac{dT}{d\theta_l} = BC$$

so that:

$$\tau_a = \frac{1}{2\pi a^3}\{3BA + BC\} \tag{Eq 31}$$

The shear strain at the surface is given by Eq 28. Thus, the shear stress versus shear strain curve can be deduced by drawing tangents to the torque versus the angle of twist per unit length curve.

In experimental work it has often been found that the torque (T) is related to the angle of twist per unit length by the expression:

$$T = T_0\theta_l^n \tag{Eq 32}$$

where T_0 is the torque at unit angle of twist, and n is the exponent. A graph of the logarithm of the torque (T) versus the logarithm of the angle of twist per unit length (θ_l) at constant rate of twist ($\dot\theta_l$) is linear and of slope n. Differentiating Eq 32 gives:

$$\left(\frac{dT}{d\theta_l}\right)_{\dot\theta_l} = \frac{nT}{\theta_l} \tag{Eq 33}$$

Combining Eq 30 and 33 gives:

$$\tau_a = \frac{T}{2\pi a^3}[3 + n] \tag{Eq 34}$$

This expression has been derived in Ref 4.

Shear stress versus shear strain curves may also be derived by the method of differential testing, where tests are carried out on two specimens of slightly different radii, a_1 and a_2. The shear stress and shear strain are given by:

$$\tau = \frac{3(T_1 - T_2)}{2\pi(a_1^3 - a_2^3)} \tag{Eq 35}$$

and

$$\gamma = \frac{(a_1 + a_2)\theta_l}{2} \tag{Eq 36}$$

respectively. An excellent critical review of existing methods for converting torque to shear stress is given in Ref 4.

The stress gradient across the diameter of a solid bar allows the less highly stressed inner fibers to restrain the surface fibers from yielding. Thus, the onset of yielding is generally not apparent. This effect can be minimized by the use of thin-walled tubes, in which the stress across the tube wall can be assumed to be constant. For a thin-walled tube, the shear stress and shear strain are given by:

$$\tau = \frac{T}{2\pi a^2 t} \tag{Eq 37}$$

and

$$\gamma = \frac{a\theta}{L} \tag{Eq 38}$$

respectively, where a is now the mean radius of the tube, t is the thickness of the tube wall, θ is the angle of twist, and L is the specimen gage length. Thus, from measurements of the torque (T) and angle of twist (θ), it is possible to construct the shear stress (τ) versus shear strain (γ) curve. The dimensions of the tube must be chosen carefully to avoid buckling.

Effect of Strain Rate

In the analysis presented in the previous section, it is inherently assumed that the shear stress is independent of strain rate. The assumption is approximately valid at low homologous temperatures, but is not valid at high homologous temperatures, where the strain rate sensitivity of materials is usually large. A graphical procedure based on the derivations presented in the previous section is presented in Ref 2 and 5.

If it is assumed that the torque is a function of both the angle of twist and the twisting rate, i.e., $T = f(\theta, \dot\theta)$, then the change in torque with respect to a change in the angle of twist is given by:

Fig. 3 Torque-twist curves

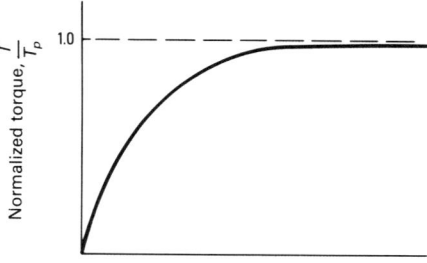

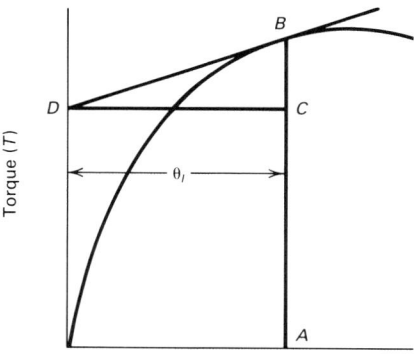

$$\frac{dT}{d\theta_l} = \left(\frac{\partial T}{\partial\theta_l}\right)_{\dot\theta_l} + \left(\frac{\partial T}{\partial\dot\theta_l}\right)_{\theta_l}\frac{\dot\theta_l}{\theta_l} \tag{Eq 39}$$

The first term on the right-hand side of this equation has been evaluated (Eq 33). The second term can be evaluated from the experimental observation that a logarithmic graph of torque (T) versus the rate of twist ($\dot\theta_l$) at a constant angle of twist per unit length (θ_l) is often linear. The slope of the graph corresponds to the twist rate sensitivity (m). Strain hardening predominates at low temperatures, whereas twist rate sensitivity predominates at elevated temperatures.

If the effect of twist rate on torque can be expressed by:

$$T = T_l\dot\theta_l^m \tag{Eq 40}$$

then, at constant strain:

$$\left(\frac{dT}{d\dot\theta_l}\right)_{\theta_l} = \frac{mT}{\dot\theta_l} \tag{Eq 41}$$

Substitution of Eq 33 and 41 into Eq 39 gives:

$$\frac{dT}{d\theta} = \frac{T}{\theta_l}(m + n) \tag{Eq 42}$$

Combining Eq 30 and 42 gives:

$$\tau_a = \frac{T}{2\pi a^3}(3 + m + n) \tag{Eq 43}$$

The shear strain is again given by:

$$\gamma = a\theta_l \tag{Eq 44}$$

Equations 43 and 44 can be used to plot graphs of shear stress versus shear strain for all temperatures and strain rates up to the point of torsional instability.

A new method of converting torque to surface shear stress (Ref 4) is based on the assumption that the shear stress at radius r is affected only by the history of this particular location. The torque is given by:

$$T = \int_0^a 2\pi\tau r^2 dr \tag{Eq 45}$$

The derivative of this integral at a given angle of twist and strain rate is:

$$\frac{dT}{dr} = 2\pi\tau_a r_a^2 \tag{Eq 46}$$

In this method, torque versus angle of twist curves are determined on specimens of increasing radii from which the torque versus radii relationship can be determined at a given strain (twist) and strain rate (twist rate). The slope of this curve at any given radius can be substituted into Eq 46 to determine the current shear stress at this radius. This method appears to reduce significantly the errors inherent in previous methods.

Relationship of Effective Stresses and Strains to Shear Stresses and Strains

It is often helpful to convert data derived under one state of stress to another state of stress. This can be accomplished by the use of so-called effective or tensile equivalent stresses and strains. The form of the relationships derived depends on the particular yield criterion used (Ref 6). For the distortional energy (von Mises) criterion, the effective stress and strain are given by:

$$\bar{\sigma} = C_1[(\sigma_1 - \sigma_2)^2 + (\sigma_2 - \sigma_3)^2 + (\sigma_3 - \sigma_1)^2]^{1/2} \tag{Eq 47}$$

and

$$d\bar{\epsilon} = C_2[(d\epsilon_1 - d\epsilon_2)^2 + (d\epsilon_2 - d\epsilon_3)^2 + (d\epsilon_3 - d\epsilon_1)^2]^{1/2} \tag{Eq 48}$$

respectively, where the variables have their usual significance. The constants C_1 and C_2 are now chosen so that the effective stresses and strains are identical to the stresses and strains in uniaxial tension (or compression). For uniaxial tension:

$$\sigma_1 = \sigma_0 \quad d\epsilon_1 = d\epsilon_1$$
$$\sigma_2 = 0 \quad d\epsilon_2 = \frac{-d\epsilon_1}{2}$$
$$\sigma_3 = 0 \quad d\epsilon_3 = \frac{-d\epsilon_1}{2} \tag{Eq 49}$$

Substituting Eq 49 into Eq 47 and 48 gives:

$$C_1 = \frac{1}{\sqrt{2}}$$

and

$$C_2 = \frac{\sqrt{2}}{3}$$

Thus, the effective stresses and strains become:

$$\bar{\sigma} = \frac{1}{\sqrt{2}}[(\sigma_1 - \sigma_2)^2 + (\sigma_2 - \sigma_3)^2 + (\sigma_3 - \sigma_1)^2]^{1/2} \tag{Eq 50}$$

and

$$d\bar{\epsilon} = \frac{\sqrt{2}}{3}[(d\epsilon_1 - d\epsilon_2)^2 + (d\epsilon_2 - d\epsilon_1)^2 + (d\epsilon_3 - d\epsilon_1)^2]^{1/2} \tag{Eq 51}$$

respectively. For the state of pure shear (torsion):

$$\sigma_1 = k \quad d\epsilon_1 = \frac{d\gamma}{2}$$
$$\sigma_2 = 0 \quad d\epsilon_2 = 0$$
$$\sigma_3 = -k \quad d\epsilon_3 = -\frac{d\gamma}{2} \tag{Eq 52}$$

Substitution of Eq 52 into 50 and 51 gives:

$$\bar{\sigma} = \sqrt{3}k$$

$$d\bar{\epsilon} = \frac{2}{\sqrt{3}}d\epsilon_1$$

or

$$d\bar{\epsilon} = \frac{1}{\sqrt{3}}d\gamma$$

Thus, the effective stresses and strains are related to the shear stresses and strains by the factors $\sqrt{3}$ and $1/\sqrt{3}$, respectively; that is, the shear stress versus shear strain curve can be converted to a true (tensile) stress versus strain curve by using:

$$\sigma = \sqrt{3}\tau$$

$$\bar{\epsilon} = \frac{\gamma}{\sqrt{3}}$$

For the Tresca (maximum shear stress) criterion, the effective stresses and strains are given by:

$$\bar{\sigma} = C_3(\sigma_1 - \sigma_3) \tag{Eq 53}$$

and

$$d\bar{\epsilon} = C_4(d\epsilon_1 - d\epsilon_3) \tag{Eq 54}$$

respectively. The constants C_3 and C_4 are again chosen so that the effective stresses and strains are identical to the stresses and strains in uniaxial tension (or compression). Using the conditions defined by Eq 49 gives:

$$C_3 = 1$$

and

$$C_4 = \frac{2}{3}$$

Thus, the effective stresses and strains become:

$$\bar{\sigma} = (\sigma_1 - \sigma_3) \tag{Eq 55}$$

and

$$d\bar{\epsilon} = \frac{2}{3}(d\epsilon_1 - d\epsilon_3) \tag{Eq 56}$$

Substitution of Eq 52 into Eq 55 and 56 gives:

$$\bar{\sigma} = 2k$$

and

$$d\bar{\epsilon} = \frac{4}{3}d\epsilon_1$$

or

$$d\bar{\epsilon} = \frac{2}{3}d\gamma$$

Thus, the effective stresses and strains are related to the shear stresses and shear strains by the factors 2 and ⅔, respectively; that is, the shear stress versus shear strain curve can be converted to a true (tensile) stress versus strain curve by using:

$$\sigma = 2\tau \tag{Eq 57}$$

and

$$\bar{\epsilon} = \frac{2}{3}\gamma \tag{Eq 58}$$

The work of deformation per unit volume in terms of the effective stresses is given by:

$$u = \int \bar{\sigma} d\bar{\epsilon} \tag{Eq 59}$$

The deformation in torsion can be calculated from the expressions:

$$\bar{\sigma} = 2\tau$$

$$d\bar{\epsilon} = \frac{2}{3}d\gamma \tag{Eq 60}$$

and

$$\bar{\sigma} = \sqrt{3}\tau$$

$$d\bar{\epsilon} = \frac{d\gamma}{\sqrt{3}} \qquad \text{(Eq 61)}$$

for the Tresca (maximum shear stress) theory and distorsional energy theory, respectively. For the Tresca theory, substitution of Eq 60 into Eq 59 gives:

$$u = \int 2\tau \cdot \frac{2}{3} d\gamma$$

$$u = \int \frac{4}{3}\tau d\gamma \qquad \text{(Eq 62)}$$

For the distorsional energy criterion, substitution of Eq 61 into Eq 59 gives:

$$u = \int \sqrt{3}\,\tau \cdot \frac{1}{\sqrt{3}} d\gamma$$

$$u = \int \tau d\gamma \qquad \text{(Eq 63)}$$

It is evident that the work obtained by the Tresca theory is too high and that the distorsional energy theory gives the correct result.

Anisotropy in Plastic Torsion

Marked dimensional changes can occur during the torsional straining of solid bars and hollow cylinders of circular cross section (Ref 6-9). These changes may produce either an increase or a decrease in the length of the test specimens. Changes in length produced in hollow cylinders are considerably greater than those produced in solid bars because of the constraining effect of the solid core with the latter geometry. If changes in length are suppressed, then large axial stresses may be produced.

These dimensional changes have been attributed, for the most part, to the development of crystallographic anisotropy that arises because of a continuous change in the orientation of individual grains. This produces preferred orientation, in which the yield stresses and macroscopic stress versus strain relationships vary with direction. The general observation is that the torsional deformation of solid bars and tubes produces axial extension at ambient temperatures and contraction at elevated temperatures. Specific results, however, depend on the initial state (anisotropy) of the test material.

A general phenomenological theory of anisotropy (Ref 7) proposes that the criterion describing the yield condition for an anisotropic material is quadratic in stress components and of the form:

$$2f(\sigma_{ij}) = F(\sigma_y - \sigma_z)^2 + G(\sigma_z - \sigma_x)^2$$
$$+ H(\sigma_x - \sigma_y)^2 + 2L\tau_{yz}^2$$
$$+ 2M\tau_{zx}^2 + 2N\tau_{xy}^2 \qquad \text{(Eq 64)}$$

where F, G, H, L, M, and N are six parameters describing the current state of anisotropy; $f(\sigma_{ij})$ is the plastic potential; and the remaining symbols have their usual significance.

The basic theory of anisotropy (Ref 7) has been applied to the torsional straining of a thin-walled cylinder in an attempt to describe the changes that occur in dimensions. For a thin-walled cylinder, the radius is large compared to the wall thickness, and thus anisotropy can be considered to be uniformly distributed throughout the volume of material deformed. It was also assumed that the axes of anisotropy along the surface of an initially isotropic cylinder were coincident with the directions of greatest accumulated tensile and compressive strain. These axes were also assumed to be mutually perpendicular and oriented at an angle ϕ to the transverse axis of the cylinder. This geometry is shown in Fig. 4. For an initially isotropic cylinder, the angle ϕ is a function of the shear strain (γ) and increases from $\pi/4$, approaching $\pi/2$ at large strains. This rotation is confined to the (x,y) plane about the z-axis that is perpendicular to the surface of the cylinder.

From an analysis of the deformation, it was shown that the change in axial strain with shear strain is given by:

$$\frac{d\epsilon_a}{d\gamma} =$$

$$\frac{(N - G - 2H)\sin\phi\,\sin 2\phi}{2N + (F + G + 4H - 2N)\sin^2 2\phi}$$
$$- \frac{(N - F - 2H)\cos^2\phi\,\sin 2\phi}{2N + (F + G + 4H - 2N)\sin^2 2\phi}$$

$$\text{(Eq 65)}$$

It is clear from Eq 65 that measurement of the change in axial strain with shear strain is insufficient to determine the anisotropic parameters and yield stresses along the anisotropic axes and thereby insufficient to describe quantitatively the state of anisotropy.

Simple expressions for the anisotropic parameters and yield stresses along the anisotropic axes in terms of the changes in axial strain, tangential strain, principal yield shear stress, and through thickness yield stress of the hollow cylinder with shear strain have been developed (Ref 8), all of which can be determined easily by experiment. It was found that the anisotropic parameters decrease and that the yield stresses along the anisotropic axes increase with an increase in

strain, eventually becoming independent of strain when the test material is fully work hardened. A direct relationship between the axial forces generated (positive, negative, zero) and the crystallographic texture developed for several materials has been proposed (Ref 9, 10). The sign and approximate magnitude of the effects can be predicted from knowledge of the ideal orientations.

The changes in length of a twisted bar during straining result in a continuous change in test specimen cross-sectional area. Thus, if the true shear stress versus shear strain curve is required, then instantaneous values of specimen dimensions must be used in computing shear stress and shear strain from the measured torque and angle of twist.

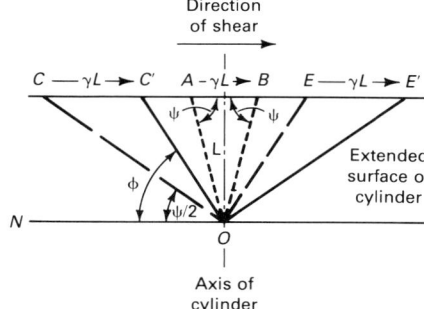

Fig. 4 Geometry of deformation for the plastic straining of a hollow cylinder

γ = shear strain; L = initial length of cylinder; OC = initial direction of greatest compression; OC' = final direction of greatest compression; OE = initial direction of greatest extension; OE' = final direction of greatest extension

REFERENCES

1. S.P. Timoshenko and J.N. Goodier, *Theory of Elasticity*, 3rd ed., McGraw-Hill, New York, 1970, p 291
2. W.J. McGregor Tegart, *Elements of Mechanical Metallurgy*, Macmillan, New York, 1967, p 64
3. A. Nadai, *Theory of Flow and Fracture of Solids*, Vol 1, McGraw-Hill, New York, 1950, p 349
4. G.R. Canova, *et al.*, in *Formability of Metallic Materials—2000 A.D.*, J.R. Newby and B.A. Niemeier, Ed., STP 753, ASTM, Philadelphia, 1982, p 189
5. D.S. Fields and W.A. Backofen, *Proceedings ASTM*, Vol 57, 1957, p 1259
6. S. Kalpakjian, *Mechanical Processing of Materials*, D. Van Nostrand, New York, 1967, p 31
7. R.R. Hill, *The Mathematical Theory of Plasticity*, Clarendon Press, Oxford, 1950, p 317
8. J.A. Bailey, S.L. Haas, and K.C. Na-

wab, *Trans. ASME, J. Basic Eng.*, March 1972, p 231

9. F. Montheillet, M. Cohen, and J.J. Jonas, Axial Stresses and Texture Development During Torsion Testing of Al, Cu and α-Fe, *Acta Met.*, Vol 32, 1984, p 2077-2089

10. F. Montheillet, P. Gilormini, and J.J. Jonas, Relation Between Axial Stresses and Texture Development During Torsion Testing: A Simplified Theory, *Acta Met.*, Vol 33, 1985, in press

Quasi-Static Torsional Testing

By James L. Oess
Materials Engineer
Dana Corporation

QUASI-STATIC TORSIONAL TEST-ING is used to determine material properties other than actual static torsional strength—usually either surface shear characteristics (strength, stress, or strain) or spring rate (modulus). The same basic test is also used to compare and select materials by testing actual parts or components. Torsional testing is also used in the design of torsionally loaded members to determine the mode of failure (buckling, localized yield, or violent fracture) when one mode is more desirable than others.

Surface shear stress affects component design when the notch sensitivity of a material requires that the surface stress remain below a predetermined maximum value. The surface stress characteristics, particularly modulus, should be known when designing and testing springs to aid in determining acceptable internal stress levels. Surface shear stresses of a material may be determined using a specimen with uniform circular cross section and by converting torsional load to surface shear stress for a solid cylinder by:

$$S_s = \frac{16\,T}{\pi\,D_o^3} \qquad \text{(Eq 1)}$$

and for a hollow (tubular) cylinder by:

$$S_s = \frac{16\,T\,D_o^3}{\pi(D_o^4 - D_i^4)} \qquad \text{(Eq 2)}$$

where T is torsional load (in. · lb); D_o and D_i are outside and inside diameters, respectively (in.); and S_s is surface shear stress (psi).

The spring rate (modulus) in torsion is the load per unit of angular deflection, such as inch-pounds per degree, or newton-meters per radian. Testing in torsion provides values for these variables that are substantially unaffected by internal discontinuities, such as porosity and inclusions not open to the surface, or by other material properties.

Test Selection

Selection of a particular test is determined by the type of data desired and the material of interest. Cast materials, bars, and wire are commonly tested "as received," as a cylindrical specimen either cut from wire or machined into a bar similar to a standard ASTM E 11 tensile test bar. The standard testing procedures for surface shear determination are given in ASTM E 143, "Standard Test Method for Shear Modulus at Room Temperature." The general methods discussed in this article conform to this standard when applicable.

Materials such as gray iron have no distinct yield point or plastic deformation in torsion. For these materials, yield and ultimate strength are considered the same, as in tension testing. Long, thin wire or tubing may withstand several full rotations of one end relative to the other before actual breakage, depending on the length of the specimen.

The expected loads can be calculated approximately from Eq 1 and 2 using established values for ultimate shear strength and solving for T in appropriate units. For example, a particular aluminum alloy has an ultimate shear strength of 276 MPa (40 000 psi). A 25-mm (1-in.) diam bar is to be tested. Using Eq 1 and solving for T gives an anticipated ultimate torque of 887.5 N · m (7854 in. · lb).

Machining operations should be chosen carefully to prevent changes in the surface structure of the material, which could affect the test results. For instance, an overly coarse machined surface will not be consistently loaded in shear, and the "burn" from roughly ground steel will exhibit decreased ductility and weaken the location with the highest stress. If practical, cast materials should be tested in the as-cast condition without surface machining, which may remove the surface layer that occurs on castings. Frequently, this skin has a substantially different structure than the underlying bulk material, which will affect the results obtained in the high-stress area. This does not apply to cast material used in a fully machined state.

Tubular material is usually tested "as received," and sheet material can be tested in tubular form if the forming process does not affect the surface stress. Uniformity in the test section should be maintained when a material rather than a component is being tested. This uniformity includes wall thickness in tubes and cross-sectional area and dimensions to within 0.1% of the nominal diameter, although this becomes increasingly impractical for larger diameters. Components that are used in a torsionally stressed mode (torsion bars, driveshafts, etc.) can be tested with different materials as complete assemblies. Gaging should be restricted to the specimen being tested.

Surface shear determination is best carried out on uniform cylindrical bars with a length of 8 to 10 diameters. Alternatively, a nonsymmetrical specimen may be strain gaged in an appropriate mode, and gage readings can be used for analysis instead of overall twist. Special circumstances may dictate deviation from ideal test conditions, or may require the use of tests such as torsional creep or impact. Users of such tests must be very careful to ensure that data obtained are characteristic of the test specimen and not the test system. The nature of torsional testing can make it difficult to isolate the specimen from the system when nonsymmetrical parts are being tested.

Equipment

Torsional testing equipment consists of a pair of facing fixturing devices (chucks, circular faceplates, or clamps) on a common axis of rotation. One end is the drive end, which is rotated by mechanical, electrical, or hydraulic means. The other end is the load end, containing torque-sensing devices (beam balance or strain gages). The load end, or both ends in some systems, is movable along the common axis to adjust for specimen length. Alignment of the two ends is critical to prevent transverse loading. The drive system must be capable of exerting a smooth twist on the specimen. Figure 1 illustrates a typical torsional testing machine.

Fixturing is a function of the test specimen geometry. Solid circular pieces similar to tensile bars with square or hexagonal expanded ends tapering to the base circle are clamped in jaws at both ends. Tubular specimens are clamped to tightly fitting plugs to eliminate distortion caused by fixturing. Thin wire specimens are turned 90° on each end and are fitted into slots on the ends of the fixtures. Provision should be made for one of the machine ends to be free to move along the common axis—in other words, a zero axial load condition. This prevents introduction of tensile loads in the specimen, as the length of the specimen changes with twisting. A slight "float" or transverse free play (usually a few millimeters) is desirable to minimize transverse loading from slight alignment errors. The primary function of fixturing is to transmit only torsional forces to the specimen. Typical examples of fixtures are shown in Fig. 2.

Suitable guards or shields must be provided around the test specimen and the fixturing, if appropriate, particularly with brittle materials that may fracture violently and suddenly. Shields or guards must not interfere with the test apparatus.

Instrumentation

Instrumentation used for torsional testing must reliably and accurately record torque and displacement angle. Occasionally, only torque is recorded, but parameters other than these are seldom used. When converted from torque and angular displacement, torsional stress-strain has the same units as engineering stress-strain, but the variance from "true" stress-strain is typically much less. On a cylindrical specimen that does not buckle, the difference is 5% or less from engineering to "true" stress-strain, even in the plastic (nonlinear) range.

Stress is recorded in torsional units of force times moment arm and is reported in newton-meters or inch-pounds. Measurement of stress is usually incorporated in the machine via a beam balance or calibrated strain-gaged element. When testing small specimens, greater accuracy is required than

for tensile testing due to the magnified effect of small variations in properties. Although a correction may be necessary for a known low-constant error (less than 1%), linearity within 0.1% and accuracy to 0.05% is not unreasonable, and commercial machines conform to these limits.

Deformation of the specimen is recorded as deflection in degrees or radians and is mathematically converted to strain as required during data reduction. It is not sufficient to record only driving head motion, because errors from the motion of the fixed head are sufficient to invalidate the results. A variety of optical, mechanical, or electrical sensors have been devised to measure and record the relative motions of the two ends of the test specimen. One of these devices is illustrated in Fig. 3.

Ideal testing allows measurement and comparison of deflections at two known points on the specimen; the area between these points becomes, by definition, the gage section. This section should be at least two diameters long and a distance equal to two diameters from each end. For thin wires or thin-walled tubing with a relatively small stiffness compared to the fixturing, the relative motion of the machine heads is recorded,

Fig. 1 Typical torsional testing machine

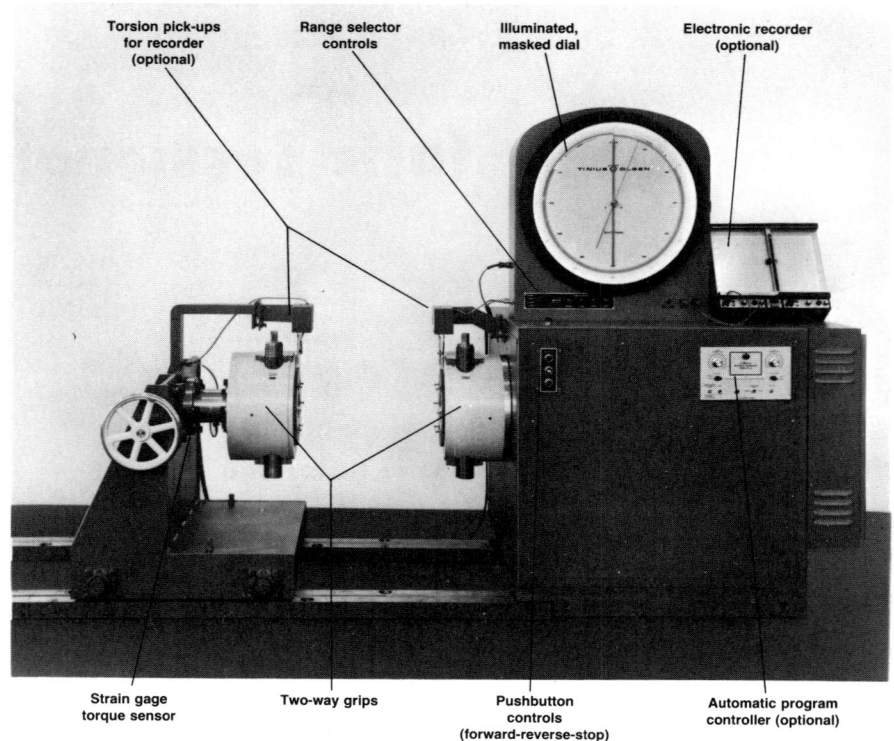

Torsion pick-ups for recorder (optional) — Range selector controls — Illuminated, masked dial — Electronic recorder (optional)

Strain gage torque sensor — Two-way grips — Pushbutton controls (forward-reverse-stop) — Automatic program controller (optional)

Fig. 2 Examples of fixturing used in torsional testing

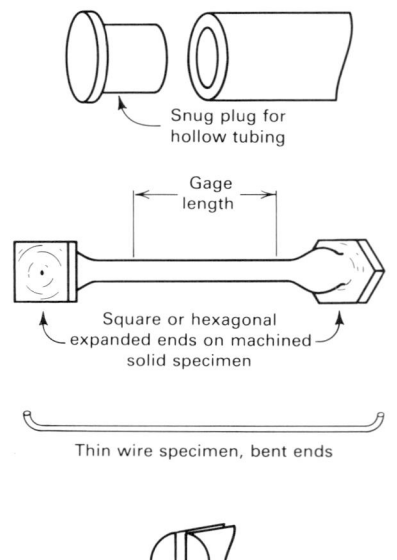

Snug plug for hollow tubing

Gage length

Square or hexagonal expanded ends on machined solid specimen

Thin wire specimen, bent ends

Chuck for wire

Fig. 3 Torsional indicator or troptometer used to measure and record test piece deflections in torsional testing

and the distance between them is considered the gage length.

Torque and angular deformation can be recorded simultaneously. These data typically are recorded on a drum recorder or *x-y* plotter, producing a ready-made data bank in graphical form. When this is not done, the stress and strain points taken must be correlated with one another, and as many points as possible should be recorded during the testing process. A curve can later be plotted from these paired points, or a real-time computer data logger can generate this plot as the test is run. A typical machine-generated plot of a torsional test of thin-walled tubing is shown in Fig. 4. Note the lack of a definite yield point.

Standard Test Methods

The simple torsional test to failure or to a predetermined load is the standard test. Other test methods are variations of this standard. When prepared equipment and specimens are available, this is a quick and simple process. The test machine is started and allowed to stabilize, fixturing and instrumentation are set up, the test is run, and the data are analyzed later.

The design of the test machine and the fixturing ensure that the specimen is mounted and aligned free of any axial or bending loads at the start of the test. For fractional errors limited to 0.1%, the rate of load application during the test must be constant within a factor of 2 at temperatures less than one half the melting temperature. The load is normally applied at a constant rate of torque increase

so that the applied strain rate is the independent variable. As in tensile testing, the rate of loading thus varies.

Data recording equipment response should be essentially linear. A small amount of free play at zero load is permissible in a unidirectional test. If the desired data are bi-directional, the free play or hysteresis at the zero-load condition should be minimized. The rate of load application is only a factor in testing ductile materials that are highly strain

Fig. 4 Schematic of a machine-generated plot of a torsional test of thin-walled tubing

Angular deformation taken directly from specimen gage length. Point V = proportional limit. Point W = offset yield strength. Offset of 0.001 radian per inch of gage length, or as assigned. Point X = ultimate torsional load

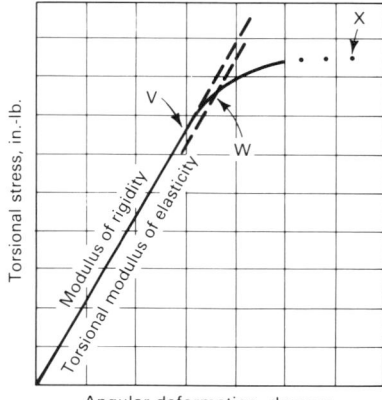

Angular deformation, degrees

hardened. A constantly applied load in these cases becomes a creep test. However, when the test time ranges from a minute or two and is constant from one test to another, rate effects are minimal. The effect of strain rate is low enough to be safely disregarded in brittle materials, provided test time exceeds a few seconds (nonshock condition). Materials that are brittle in tension usually are also brittle in torsion.

Tests in which slippage between specimen and fixturing or gaging occurs or is suspected should not be included in the results. The affected pieces should be discarded and not retested. Several applications and relaxations of about 10% of the expected elastic limit are sometimes recommended to detect slippage or machine irregularities. Hysteresis can be detected in this manner also.

Nonstandard Tests

Environmental tests consist of performing the standard test with the desired environment superimposed (heat, salt, etc.). Fixturing can become complex because of the need to restrict the exposure of a hostile environment to the test piece and not to the test machine.

High- or low-temperature tests have varied effects on results depending on the behavior of the material under specific test conditions. For example, a low-carbon steel bar tested at 500 °C (930 °F) will exhibit a much lower modulus than at room temperature. Parts tested after long exposure to corrosive conditions also exhibit lower yield points, because the corrosive conditions most strongly affect the surface of the metal, the point at which the stress level is highest.

A torsional creep test is conducted by maintaining a constant load and recording deflection versus time (or log time). Some type of feedback controller usually is required. Certain test machines come equipped with this function, or can be fitted to accomplish these requirements. A torsional stress-relaxation test is described in the article "Creep, Stress-Rupture, and Stress-Relaxation Testing" in this Volume.

Data Reporting

Material Conditions. Composition of the test piece, thermomechanical history (such as heat treatment, annealing, or cold working), and dimensions are recorded in SI units or those customary to the industry involved. Surface condition (as cast, lathe turned, ground, etc.) is an important characteristic that should be reported. The presence of any surface coating, plating, or discontinuities is another recorded parameter, as is any measured taper on a cylindrical specimen.

Test conditions, such as test speed, duration of testing (whether to failure or not), instrumentation, and environmental conditions, should be noted. The nature of fracture (brittle, spiral convolution, buckling, tearing, etc.) and the parameters defining failure (load drop, deformation, etc.) should be reported.

Interpretation of Data

Small differences in the surface conditions of specimens have significant effects on the data collected. Consequently, considerable scatter in test results on almost identical specimens is not unusual. The main tendencies of the data, if reliably and properly gathered and interpreted, provide the best indication of typical conditions for the specimens tested.

Reliable conclusions cannot be drawn from only a few widely scattered data points. If a sufficiently large number of test points are plotted on a histogram, a central grouping should emerge as the data are plotted, which is the best source for drawing conclusions. If a central grouping does not emerge, testing conditions, apparatus, and specimens should

be re-examined for anomalies. If necessary, the tests should be rerun with new parts.

Table 1 lists typical ultimate shear stress values in torsion for a variety of materials. Modulus data are subject to interpretation. Most materials do not show a completely straight plot. The modulus must therefore be derived from a predetermined set of parameters. Typically, the average slope from 10% of highest recorded load to two thirds of this load is used.

The yield point also must be arbitrarily defined. Few materials other than highly brittle ones demonstrate an obvious, sharp yield point. This is because surface strain is higher than interior strain, and surface yield occurs while the subsurface material is still behaving elastically. The yield point is most frequently obtained geometrically from the stress-strain graph, which measures secant or tangent based on initial slope to an arbitrarily defined point or slope increase.

SELECTED REFERENCES

- *ASM Metals Reference Book,* 2nd ed., American Society for Metals, 1983
- M. Bassin, S. Brodsky, and H. Wolkoff, *Statics and Strength of Materials,* McGraw-Hill, New York, 1969

Table 1 Typical values for ultimate shear stress in torsion

Material	Condition	Shear stress MPa	ksi
Wrought iron	· · ·	262	38
Gray iron	G4000	345	50
Ductile iron	D7003	414	60
SAE 1020	As hot rolled	358.5	52
SAE 1095	Normalized	724	105
SAE 4140	Normalized	772	112
AISI 302 stainless steel	Annealed	448	65
Aluminum	2024-T4	283	41
Aluminum	6061-T6	214	31
Aluminum	360-F	221	32
85% red brass	C23000	207-448	30-65
65% yellow brass . .	C27000	227-276	33-40

- *Marks' Standard Handbook for Mechanical Engineers,* 8th ed., T. Baumeister, E.A. Avallone, and T. Baumeister III, Ed., McGraw-Hill, New York, 1978
- "Standard Test for Shear Modulus," E 143, in *Annual Book of ASTM Standards,* ASTM, Philadelphia, 1983
- "Torsion Testing of Wire," E 558, in *Annual Book of ASTM Standards,* ASTM, Philadelphia, 1983
- *Universal Joint and Driveshaft Design Manual,* Society of Automotive Engineers, Warrendale, PA, 1979

Cyclic Torsional Testing

By Gail E. Leese
NASA Lewis Research Center

CYCLIC TORSIONAL TESTING generally is not part of routine material evaluation. Mechanical testing and analysis to determine the properties of materials, including monotonic and cyclic loading, have concentrated mainly on uniaxial stress states (Ref 1-4). The stress-strain response of a material due to a single monotonic load application typically is not representative of the type of response obtained when repeated cyclic loads are applied. In the latter case, progressive reversal of slip on planes of maximum shear stress introduces the cycle-dependent phenomenon of fatigue. The methodology of baseline fatigue testing centers on repeated axial loading of coupon specimens, as do most life prediction techniques (see the Section "Fatigue Testing" in this Volume).

Torsional fatigue testing sometimes may be necessary for prototypical tests of actual machine components (e.g., in the power transmission industry) that experience cyclic torsional loading in service. More frequently, such testing is included in multiaxial fatigue research programs that incorporate torsional loading as one stress state (among others) of interest. The need for multiaxial life prediction capabilities has stimulated experimental efforts in multiaxial fatigue response. Although engineering approaches duplicating specific multiaxial histories of particular components are not uncommon, fundamental research programs targeted at cyclic torsional response have been few.

Currently, there are no ASTM standards governing cyclic torsion testing. Where torsional properties are required, it is common for axial cyclic response to be extrapolated to the torsional regime. Typically, this transition is accomplished through an "effective" stress or strain parameter (often incorporating the von Mises or Tresca criteria) that can be readily determined from axial response (Ref 5-7) (see also the article "Fundamental Aspects of Torsional Loading" in this Section). The effective stress-strain versus life approach frequently is seen not only in torsional fatigue, but also in many multiaxial

loading situations in which the three-dimensional stress-strains can be resolved.

Originally conceived as yield criteria to characterize monotonic response, effective stress-strain estimations have been useful for extrapolating from one simple stress state to another (e.g., completely reversed axial fatigue to completely reversed constant amplitude torsional fatigue). However, in more complicated loading environments, this extrapolation is not straightforward. Therefore, experimental and analytical efforts in torsional fatigue as part of the general multiaxial environment are required.

Given the current immature status of torsional fatigue, it would be misleading to dictate specific testing and analysis procedures. This article will point out the various options (and associated ramifications) available to the experimentalist and will emphasize testing procedures to characterize baseline materials response in torsion rather than component history simulation.

Probably the most crucial parameter to establish in planning or evaluating a cyclic torsional testing program is the control mode. There are three basic modes: load/torsional moment, rotational angle, and strain control. The control mode governing the test may impose certain limitations on the ability to resolve stable stress-strain response and on the life regime (high or low cycle) to which the results may be applied. These limitations must be considered when planning cyclic torsional tests and when using test results for specific applications.

High-Cycle Torsional Fatigue

High-cycle fatigue refers to material response in the long-life regime. Although many investigators have included cyclic torsional response in the study of multiaxial fatigue, testing methods for high-cycle torsional fatigue have not been standardized.

The extensive work of Sines (Ref 7-9) is representative of contemporary approaches to long-life nonaxial fatigue situations.

In the high-cycle regime, stress and strain amplitudes are low, and the material response is primarily elastic. That is, of the total strain range imposed on the test specimen, the predominant portion reflects recoverable work, with shear stress and shear strain being linearly related through Hooke's law. Hence, the relationships among torsional moment, angular deflection, shear stress, and shear strain are assumed to be linear throughout most of the test. While selection of the control mode is less critical than in circumstances in which lower lives are required, torsional load or rotational angle control is common.

ASTM E 466 (Ref 10) gives the standard recommended practice for performing constant-amplitude axial fatigue tests on metallic materials in air at room temperature. Transposed into the torsional stress state, portions of this standard may be useful as guidelines for high-cycle torsional fatigue testing procedures.

For any one specimen, the torsional load is cycled around zero with a constant amplitude. This infers a zero mean shear stress and a fully reversed amplitude, $\pm\tau_a$, representing one cycle. Note that the sign of the shear stress in this case reflects only a reversal in direction of load application, whereas in axial fatigue, it corresponds to tensile or compressive loads. Controlled cycling is continued until some predetermined failure condition is observed and recorded. Each specimen tested contributes one data point relating shear stress amplitude, τ_a, to cycles to failure, N_f.

ASTM E 468 (Ref 11) establishes the desirable and minimum information for reporting results, including torsional fatigue tests in air at room temperature. Suggested data reduction varies from an empirical fit of the stress-life results on linear-log or log-log coordinates to a least squares regression, straight line fit on a log-log graph. Such a regression represents a power law relationship, such that:

$$\tau_a \propto (N_f)^b$$

where b is an exponent characteristic of a particular material (Fig. 1).

Fig. 1 Torsional stress versus life in a high-cycle regime wrought aluminum alloy

○, static compression of 150 MPa (21.8 ksi) with alternating torsion; ●, alternating torsion only. Source: Ref 8

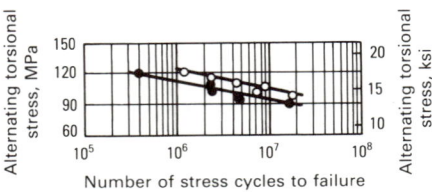

When very long lives (for example, around 10^7 cycles) are desired, the concept of an endurance limit is still popular. The idea of such a defined quantity is to indicate a stress level below which fatigue failure will never occur. However, a very low stress level that appears to represent such a limit in laboratory testing may be eradicated in actual applications by a few cycles of overstrain. Many materials never exhibit even an apparent limit. Hence, such defined quantities must be used with caution.

Low-Cycle Torsional Fatigue

Low-cycle fatigue response encompasses those cases in which cyclic stress and strain amplitudes are sufficiently high to result in relatively short lives. Plastic response dominates in this life regime; consequently, testing and analysis focus on the very local stress-strain behavior within the deforming region.

Because the relationships among torsional moment, local shear stress, and local shear strain are not necessarily linear, there are considerably different mechanical ramifications of each control mode. Controlling torsional load (moment) presents experimental and analytical difficulties in establishing a stable response. Cyclic hardening or softening of the material may be manifested in changes in local strain measurements. Any such deviations in response could cause extreme changes in the width of (and the area enclosed by) the stress-strain hysteresis loop.

Conversely, in a constant amplitude strain-controlled test, such changes in cyclic behavior result in relatively minor fluctuations. For all practical purposes, the loop encompasses a relatively constant area throughout most of the test, which is indicative of the plastic work imparted to the specimen on each cycle. In these circumstances, a material settles into a "stable" stress-strain response, making characterization of baseline material behavior possible. Hence, in shear strain control, a known strain amplitude (resolvable into elastic and plastic components) is applied, and the stress response is measured.

Rotational angle (analogous to linear stroke in axial testing) is usually measured at the end of the torsional actuator. Controlling the gross rotational angle should not be confused with shear strain control. The latter refers to control of the angle of twist within the gage length of deformation and is therefore the preferred operational mode for low-cycle torsional fatigue testing.

Local Strain Approach. The underlying purpose of most low-cycle fatigue research or applied engineering efforts is to attain or improve finite life prediction capabilities. In this regard, the analytical aspects of the local stress-strain approach and the experimental methods of axial strain-controlled low-cycle fatigue testing have been well established. References 1 to 5 and 11 to 13 cite only a few of the documents on these subjects.

Although application of this same methodology in the investigation of shear-strain-controlled torsional fatigue is not standardized, it has been demonstrated with several engineering metals (Ref 14-17). To characterize a material's baseline torsional fatigue response, each individual test specimen is cycled at a constant, fully reversed total shear strain, $\pm\gamma_T$, until a predetermined failure condition occurs, recorded as the cycles to failure, N_f, or the reversals to failure, $2N_f$. Torsional load is monitored for input (along with specimen geometry) into shear stress calculations.

Typically, cyclic changes due to hardening and softening occur early in life. With most wrought engineering metals, it is technically sound to assume that cyclic stabilization has occurred by the half-life of a test specimen. Given a stable half-life hysteresis loop, the magnitude of the crucial parameters that characterize response at that particular total strain range can be measured or calculated, including shear stress amplitude, τ_a, and the elastic and plastic shear strain amplitudes, γ_e and γ_p, respectively.

Several methods can be used to resolve the total shear strain amplitude into its elastic and plastic components. The preferred, most straightforward approach is to calculate the

elastic component using the linear Hooke's law relationship:

$$\gamma_e = \frac{\tau_a}{G} \qquad (\text{Eq } 1)$$

where G is the elastic shear modulus of the material. Assuming that the total shear strain is composed only of elastic and plastic components, the plastic shear strain magnitude using the difference of the known quantities can be determined:

$$\gamma_p = \gamma_T - \gamma_e \qquad (\text{Eq } 2)$$

Alternately, the loop width at zero stress may be measured as an indicator of the plastic strain range.

The reported test data should include the following parameters for each specimen: $\Delta\gamma$ and γ_T (or γ_a) are total shear strain range or amplitude, respectively; G is shear modulus of elasticity; $\Delta\tau$ and τ_a are shear stress range or amplitude, respectively; $\Delta\gamma_e$ and γ_e are elastic shear strain range or amplitude, respectively; $\Delta\gamma_p$ and γ_p are plastic shear strain range or amplitude, respectively; and N_f and $2N_f$ are cycles or reversals to failure, respectively.

For example, hysteresis loops from the first cycle and half-life of a 1045 hot rolled and normalized steel specimen cycled at $\gamma_a = \pm 0.025$ are shown in Fig. 2. Note the upper and lower yield point behavior on the first quarter cycle (typical of this class of

Fig. 2 Torsional hysteresis loops of 1045 hot rolled and normalized steel

(a) First cycle. (b) Stable response. Source: Ref 15

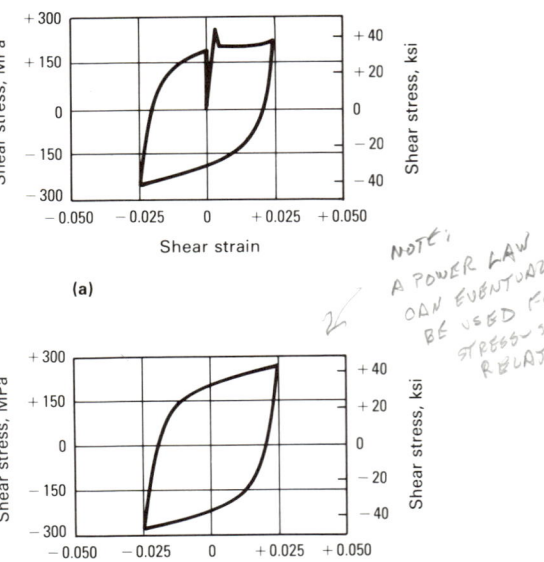

(a)

(b)

NOTE: A POWER LAW OAN EVENTUALLY BE USED FOR STRESS-STRAIN RELATION

materials). On ensuing cycles, the response generates smooth hysteresis loops. Note also the graphical representation of the parameters listed above, as labeled on the generalized hysteresis loop in Fig. 3.

Fig. 3 Stable torsional hysteresis loop

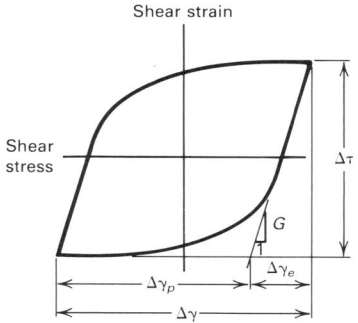

Upon completion of a series of cyclic torsional tests, data pairs relating γ_e and $2N_f$, γ_p and $2N_f$, and τ_a and γ_p for each specimen should be available. As in the axial fatigue case with analogous parameters, these can be related to the following power law relationships:

$$\gamma_e = \frac{\tau_f'}{G}(2N_f)^b \qquad \text{(Eq 3)}$$

$$\gamma_p = \gamma_f'(2N_f)^c \qquad \text{(Eq 4)}$$

$$\tau_a = Z'(\gamma_p)^{n'} \qquad \text{(Eq 5)}$$

When plotting these relationships on logarithmic coordinates, b, c, and n' are the slopes of straight lines, and τ_f'/G and γ_f' are characteristic intercepts at $2N_f = 1$. Here, $Z' = \tau_a$ at $\gamma_p = 1$.

Since the representation of such results has not been standardized, one effective way to communicate the torsional fatigue results is to modify the nomenclature of axial fatigue properties to indicate the analogous torsional fatigue quantities. For example, τ_f' is the torsional fatigue strength coefficient; b is the torsional fatigue strength exponent; γ_f' is the torsional fatigue ductility coefficient; c is the torsional fatigue ductility exponent; Z' is the cyclic torsional strength exponent; and n' is the cyclic torsional strain-hardening exponent.

These coefficients and exponents can be established by linear regression of the logarithmic values of the raw data pairs, as indicated by Eq 3 to 5. Results may be summarized through the relationships determined by total shear strain versus life and cyclic shear

stress versus shear strain. The total shear strain/life relationship is merely a summation of the elastic and plastic components of strain:

$$\gamma_T = \gamma_e + \gamma_p \qquad \text{(Eq 6)}$$

$$\frac{\Delta\gamma}{2} = \frac{\tau_f'}{G}(2N_f)^b + \gamma_f'(2N_f)^c \qquad \text{(Eq 7)}$$

Similarly with the cyclic shear stress/shear strain relationship:

$$\gamma_T = \gamma_e + \gamma_p \qquad \text{(Eq 8)}$$

$$\gamma_T = \frac{\tau_a}{G} + \left(\frac{\tau_a}{Z'}\right)^{1/n'}$$

Figures 4 and 5 illustrate these relationships using data obtained from the 1045 steel in Fig. 2.

Fig. 4 Shear strain versus life of 1045 hot rolled and normalized steel

$\bigcirc$, total strain; $\bullet$, elastic strain; $\triangle$, plastic strain. Source: Ref 15

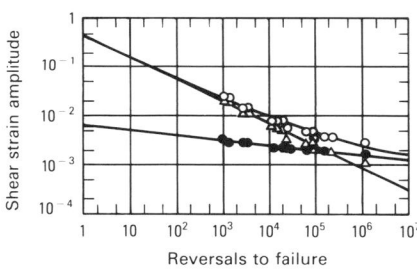

Fig. 5 Cyclic and monotonic shear stress versus shear strain of 1045 hot rolled and normalized steel

Source: Ref 15

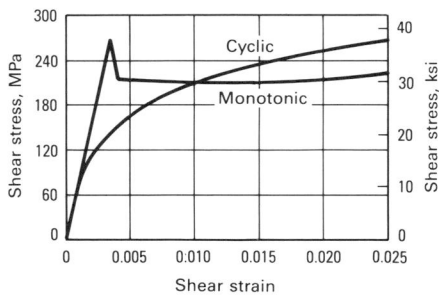

It must be emphasized that the above discussion directly translates axial low-cycle fatigue methodology to the torsional case. Due to the nature of torsional loading, there are experimental complications that have not yet been considered; these will be discussed in the following section. There are also limitations inherent in the local stress-strain low-cycle fatigue concepts, which have been

shown to be valid primarily for wrought metals at room temperature in laboratory air.

Experimental Considerations

Torsional fatigue testing requires hardware capable of imparting a known, controllable torsional load to the test specimen. Generally, this involves either offset arms carrying equal and opposite loads, thereby producing known torsional moments, or a rotary actuator and torsional load cell coupled directly in line with the specimen. In either case, the most suitable equipment for low-cycle fatigue testing is a closed-loop, servo-controlled electrohydraulic test system. Such equipment, as well as the philosophy of closed-loop testing, is described in Ref 18; this discussion will address some of the complications peculiar to the nature of torsional loading. Although these topics are not all inclusive, they encompass the most obvious issues that arise when a cyclic torsional testing program is undertaken or evaluated.

Extensometry. Commercially available axial extensometers measure and/or control axial strain within a specified gage length in low-cycle fatigue tests. Although currently available equipment is certainly capable of measuring and/or controlling torsional load and overall angular deflection, shear strain extensometers for use in torsional strain control have neither been marketed nor gained widespread acceptance.

The shear strain extensometer measures and/or controls the angle of twist (hence, the shear strain) within a gage length of a torsional fatigue specimen. Typically, such hardware consists of some type of transducer with an electrical output (reflecting the local shear strain) that can be incorporated into the closed-loop control signal of the test system. Difficulties in attachment to the specimen, signal stability, resolution, and compatibility with existing equipment have paced the development of such devices. The most desirable extensometer would be supported on the specimen without introducing geometric stress raisers such as notches and/or identations and would also expose most of the specimen surface for observation. Examples of such devices are described in Ref 14 to 17 and 19 to 21.

Specimen Design and Geometry. Just as there are no standard cyclic torsional test methods, there are no standard test specimen geometries. Of the typical specimens used, the working section is usually designed with a uniform gage length and a round cross section. The major geometric factor for consideration is whether the gage length cross section constitutes a solid round or a thick-walled cylinder or thin-walled tube.

Solid rounds and thick-walled tubes may present complications in calculating surface shear stress after the test material yields in a strain-controlled test. After the onset of plasticity, the shear stress distribution across the radius is nonlinear and cannot be calculated directly. Analytical methods for calculating surface shear stress directly from measurable test parameters are available (Ref 22-24).

Thin-walled tubular specimens are frequently used in low-cycle torsional fatigue testing. According to thin-wall theory, shear stress is uniform throughout the wall thickness in the regime of plastic response and can be calculated directly from the known geometry and torsional moment. The accuracy of this approach depends on having a sufficiently small wall thickness relative to the overall dimensions of the specimen. Typically, the ratio of wall thickness to outer diameter is less than 0.1. Prior to plastic yielding, however, the exact elasticity calculations (available in most mechanics textbooks and handbooks) are applicable to either specimen design (Ref 25-27).

Grips. Specimens must be securely mounted into the test fixture to avoid slippage and backlash during loading, unloading, and load reversals. Many gripping systems designed for axial loading applications are inadequate for torsion; for example, mated threaded ends merely tighten and release under cyclic torsional loading.

Frequently, squared specimen shoulders and matching holders are used to prevent rotation within the grips. However, these designs often prove impractical for several reasons, including the high cost of machining suitable specimens, the required degree of alignment accuracy, and excessive wear of reusable parts in gripping assemblies. Mechanical and hydraulic collet gripping systems are gaining the most acceptance due to ease of use and functionality.

Failure Criteria. This fundamental issue is perhaps the most confusing when discussing low-cycle torsional fatigue. However, fatigue data are rendered useless without a concise definition of the condition constituting failure.

Although there are many possible definitions (Ref 13), failure or crack initiation in axial-strain-controlled low-cycle fatigue tests is often defined as a predetermined drop in the tensile load required to enforce the controlled strain amplitude of the test. A drop in tensile load indicates a decreased cross-sectional area and thus the presence of a crack. The crack indicated is normal to the loading direction and hence normal to the tensile stress. From a practical viewpoint, this crite-

rion is easily incorporated into automated testing.

Such a convenient, consistent criterion is not obvious in torsional fatigue. Particularly in ductile metals, surface cracks that are quite large relative to the gage length dimensions may exist on the two planes of maximum shear stress throughout most of the full load-carrying life of the specimen. There also may be multiple crack systems throughout the gage length. Hence, the physical condition that represents failure or meaningful crack initiation in torsional fatigue must be determined.

Although it is agreed that a precise description of the condition constituting failure or crack initiation is essential in reporting torsional fatigue results, there is no widely accepted definition. Such a description should be quantitative (e.g., crack length and plane) and also should include the method of failure determination (e.g., visual inspection, surface replication, measured bulk parameters, etc.).

Special Considerations

Deviating from a room-temperature laboratory air environment is likely to affect the fatigue of a material, regardless of the cyclic stress state imposed. For example, changes in atmospheric conditions, such as testing in a vacuum or inert gas, may result in decreased chemical interaction in the surface layer of a specimen, which is believed to enhance resistance to fatigue crack initiation. Conversely, the detrimental effects of harsh conditions such as highly corrosive environments (for example, high humidity and/or temperature, or atmospheres promoting chemical oxidation of the specimen) are of continual interest for practical applications.

Mechanical response at elevated temperatures, which often involves fatigue and creep interactions, is a particularly important engineering problem. Testing and analytical methods for dealing with low-cycle fatigue/creep have been devised for the axial loading situation (Ref 28), most notably strain range partitioning (Ref 29). While theoretical extensions to multiaxial cyclic loading (e.g., torsion) have been proposed (Ref 30), this area has remained relatively untouched by experimentalists.

This article has discussed only some of the many complications arising in applying laboratory results or in designing experiments to simulate a particular prototypical environment rather than using a baseline materials response. Different complexities may be introduced when nonmetals are considered. The amount of cyclic torsional testing performed on metals is limited, but it is even

less frequent for other classes of materials, such as ceramics, polymers, and composites. Practical applications of ceramics usually are limited more by their tendency toward brittle fracture rather than by low-cycle fatigue response. Similarly, there is little information available concerning torsional fatigue testing of polymer or composite materials systems.

REFERENCES

1. J. Morrow, Cyclic Plastic Strain Energy and Fatigue of Metals, in *Internal Friction, Damping and Cyclic Plasticity*, STP 378, ASTM, Philadelphia, 1965, p 45-87
2. R.W. Landgraf, J. Morrow, and T. Endo, Determination of the Cyclic Stress-Strain Curve, *J. Mater.*, Vol 4 (No. 1), 1969, p 176-188
3. J. Morrow, Modern View of Materials Testing, in *Mechanical Behavior of Materials*, Vol 5, The Society of Materials Science, Japan, 1972, p 362-379
4. L.F. Coffin and E. Krempl, Ed., *Cyclic Stress-Strain Behavior, Analysis, Experimentation and Failure Prediction*, STP 519, ASTM, Philadelphia, 1973
5. S.S. Manson, *Thermal Stress and Low-Cycle Fatigue*, McGraw-Hill, New York, 1966, p 88, 125-192
6. E. Krempl, *The Influence of State of Stress on Low-Cycle Fatigue of Structural Materials: A Literature Survey and Interpretation Report*, STP 549, ASTM, Philadelphia, 1974
7. G. Sines, Behavior of Metals Under Complex Stresses, in *Metal Fatigue*, G. Sines and J.L. Waisman, Ed., McGraw-Hill, New York, 1959, p 145-169
8. G. Sines, "Failure of Materials under Combined Repeated Stresses with Superimposed Static Stresses," National Advisory Committee for Aeronautics Technology Note 3495, Nov 1955
9. G. Sines and G. Ohgi, Fatigue Criteria Under Combined Stresses or Strains, *J. Eng. Mater. Technol., Trans. ASME*, Vol 103, April 1981, p 82-91
10. "Standard Recommended Practice for Constant Amplitude Axial Fatigue Tests of Metallic Materials," E 466, *Annual Book of ASTM Standards*, Vol 03.01, ASTM, Philadelphia, 1984, p 567-572
11. "Standard Practice for Presentation of Constant Amplitude Fatigue Test Results for Metallic Materials," E 468, *Annual Book of ASTM Standards*, Vol 03.01, ASTM, Philadelphia, 1984, p 578-586
12. D.F. Socie, Fatigue Life Prediction Using Local Stress-Strain Concepts, *Exp.*

Mech., Vol 17 (No. 2), Feb 1977, p 50-56

13. "Standard Recommended Practice for Constant-Amplitude Low-Cycle Fatigue Testing," E 606, *Annual Book of ASTM Standards,* Vol 03.01, ASTM, Philadelphia, 1984, p 653-670

14. G.R. Halford and J. Morrow, "Low Cycle Fatigue in Torsion," Department of Theoretical and Applied Mechanics Rep. No. 203, University of Illinois, Urbana, Oct 1961

15. G.E. Leese and J. Morrow, Low Cycle Fatigue Properties of a 1045 Steel in Torsion, in *Biaxial/Multiaxial Fatigue,* STP 853, ASTM, Philadelphia, 1985

16. N.E. Dowling, "Torsional Fatigue Life of Power Plant Equipment Rotating Shafts," Department of Energy DOE/RA/29353-1, Sept 1982

17. R.A. Williams *et al.,* Torsional Fatigue of Turbine Generator Rotor Steel, in *Biaxial/Multiaxial Fatigue,* STP 853, ASTM, Philadelphia, 1985

18. *Manual on Low Cycle Fatigue Testing,* STP 465, ASTM, Philadelphia, 1969

19. J.R. Ellis, A Multiaxial Extensometer for Measuring Axial, Torsional and Diametral Strains at Elevated Temperatures, ORNL/TM-8760, Oak Ridge National Laboratory, June 1983

20. K.C. Liu, personal communication, March 1984

21. D.F. Socie, L.A. Waill, and D.F. Dittmer, Biaxial Fatigue of Inconel 718 Including Mean Stress Effects, in *Biaxial/Multiaxial Fatigue,* STP 853, ASTM, Philadelphia, 1985

22. A. Nadai, *Theory of Flow and Fracture of Solids,* 2nd ed., Vol 1, McGraw-Hill, New York, 1950, p 347-349

23. M.W. Brown, Torsional Stresses in Tubular Specimens, *J. Strain Analysis,* Vol 13 (No. 1), 1978, p 23-28

24. N.E. Dowling, Stress-Strain Analysis of Cyclic Plastic Bending and Torsion, *J. Eng. Mater. Technol.,* Vol 100, April 1978, p 157-163

25. A.P. Boresi *et al., Advanced Mechanics of Materials,* 3rd ed., John Wiley & Sons, New York, 1932, p 209-257

26. S. Timoshenko, *Strength of Materials,* Parts I and II, 3rd ed., D. Van Nostrand, Princeton, NJ, 1958

27. R.J. Roark and W.C. Young, *Formulas for Stress and Strain,* 5th ed., McGraw-Hill, New York, 1975

28. A.E. Carden, A.J. McEvily, and C.H. Wells, Ed., *Fatigue at Elevated Temperatures,* STP 520, ASTM, Philadelphia, 1973

29. S.S. Manson, G.R. Halford, and M.H. Hirschberg, Creep Fatigue Analysis by Strain Range Partitioning, *Symposium on Design for Elevated Temperature Environment,* American Society of Mechanical Engineers, New York, 1971, p 12-28 (NASA TM X-67838, 1971)

30. S.S. Manson and G.R. Halford, Multiaxial Rules for Treatment of Creep-Fatigue Problems by Strain Range Partitioning, *ASME-MPC Symposium on Creep-Fatigue Interaction,* MPC-3, American Society of Mechanical Engineers, New York, 1976, p 299-322 (NASA TM X-73488, 1976)

Application of the Torsion Test to Determine Workability

By S.L. Semiatin
Principal Research Scientist
Battelle Columbus Laboratories

G.D. Lahoti
Research Scientist
The Timken Company

and

John J. Jonas
Professor of Mechanical Metallurgy
McGill University

WORKABILITY is generally defined as the ability to impart a particular shape to a piece of metal under the load capacity of the available tooling and equipment, without the introduction of fracture or the development of undesirable microstructures. A complete description of the workability of a material is therefore specified by its flow stress dependence on processing variables (e.g., strain, strain rate, preheat temperature, and die temperature), its failure behavior, and the phase transformations that characterize the alloy system to which it belongs.

Very few mechanical tests are capable of providing information about all of these aspects, which is primarily a result of the large deformations that are common in massive forming processes such as forging, extrusion, and rolling. For the most common mechanical tests—simple tension and compression tests—the maximum uniform strains achieved are rather low because of necking and barreling, respectively. By contrast, strains in excess of 0.3 to 0.7, the levels typical of these uniaxial tests, are readily achieved in the torsion test.

Because of the large strains that can be achieved under relatively uniform deformation conditions, the torsion test is the preferred test for obtaining flow stress data and unambiguous indications of failure response and microstructural response from deformation processing. Provided failure does not intercede, it is not unusual to be able to obtain flow stress results in torsion to deformation levels equivalent to a true axial strain of 5 or more in tension, or a reduction of 90 to 95% in compression.

When failure modes are of importance, torsion is also a valuable diagnostic test. These failures are usually divided into two broad categories: fracture-controlled failures, in which deformation is relatively uniformly distributed prior to failure, and flow-localization-controlled failures, in which plastic deformation has been concentrated in a particular area of the metal specimen or workpiece prior to the actual fracturing process.

Fracture of metals occurs by a variety of mechanisms that depend largely on deformation rate and temperature. At cold working temperatures, or temperatures corresponding approximately to one fourth or less of the absolute melting or solidus temperature ($T/T_M \leq 0.25$), ductile fracture, which is characterized by void initiation at second-phase particles and inclusions, void growth, and final coalescence, is most prevalent.

At warm ($0.25 \leq T/T_M \leq 0.6$) and hot ($T/T_M \geq 0.6$) working temperatures, processes such as wedge cracking (the opening of voids and their propagation at triple junctions, or the intersection of three grain boundaries) and cavitation (the formation of voids or cavities around second-phase particles, particularly at grain boundaries) are the most important fracture mechanisms. Frequently, the deformation rate determines which of these fracture mechanisms predominates. Typically, the former mechanism occurs at high strain rates, and the latter occurs at low strain rates.

The different fracture regimes are best illustrated by deformation processing maps, as shown in Fig. 1, which are frequently determined by tension or torsion tests. Such maps may also depict regions in which large strains may be achieved without fracture, or so-called "safe" processing regimes, as well as those in which certain microstructural changes, such as dynamic recovery or recrystallization, can be expected during processing.

The torsion test can also be used to determine conditions under which flow-localization-controlled failures occur. These failures

Fig. 1 Typical deformation processing map for austenitic stainless steel

Shows regions of ductile fracture, wedge cracking, dynamic recrystallization, and "safe" forming. Note that the boundaries for safe forming vary depending on the required ductility (or fracture strain ϵ_f). —·— wedge cracking, —— ductile fracture, -------- dynamic recrystallization. Source: Ref 1

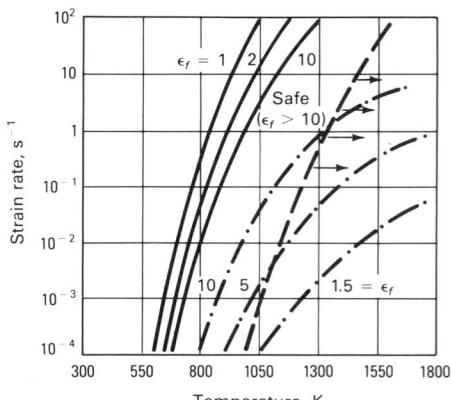

are manifested by shear bands or regions of localized shear deformation, because torsion is basically a simple shear deformation (Fig. 2). During torsion, shear bands may originate at a point of material inhomogeneity or temperature nonuniformity. Under the former conditions, localization is driven by flow softening (a decrease of flow stress with increasing strain) due to (1) dynamic recovery or dynamic recrystallization; (2) microstructural instabilities, such as the breakup and spheroidization of lamellar microstructures; (3) texture softening; or (4) deformation heating.

When deformation heating occurs, very high strain rates prevent the dissipation of heat, which in conjunction with the large negative temperature dependence of flow stress found in many metals leads to particularly strong flow localizations sometimes referred to as adiabatic shear bands. At moderate strain rates, deformation heat generation and heat conduction can introduce substantial temperature gradients during torsion testing, as in conventional metalworking processes, and can cause noticeable flow localization in materials with flow stresses that are strongly temperature dependent.

Because failure mode and microstructural development depend on process variables, torsion testing equipment and procedures to assess workability must be carefully designed. Of importance is the ability to control and measure strain and strain rate (through monitoring of twist and twist rate) as well as test temperature. Furthermore, the ability to impose arbitrary deformation or loading histories should be considered when designing torsion apparatus to be used to simulate processes such as multistage rolling or wire drawing.

This article discusses equipment design, procedures, and interpretation of torsion tests to establish workability. Experimental considerations are discussed, along with the application of torsion testing to obtain flow stress data and to gage fracture-controlled workability. Use of the torsion test to gage flow-localization-controlled failure and the effect of processing on microstructure using torsion is detailed as well. For information dealing with alternate test procedures for determining workability, see the article "Bulk Workability Testing" in this Volume.

Materials Considerations

The microstructure of the material to be tested in torsion greatly affects specimen design and interpretation of workability data. The most significant material characteristics are grain size, crystallographic texture, and mechanical texture. Because metals are composed of individual grains with specific crystallographic orientations and directional plas-

Fig. 2 Analogy between (a) torsion of a thin-walled tube, and (b) simple shear
Note the equivalence between the deformations of initially circular grid elements in the two modes.

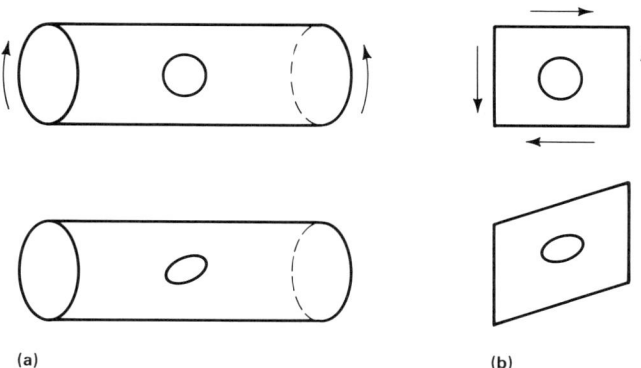

(a)   (b)

tic deformation properties, the torsion specimen gage section diameter should include at least 20 to 30 grain diameters. Torsion measurement thus reflects an average of the flow and failure response of the material. This requirement on gage section diameter may therefore result in specimens of varying sizes, depending on whether they are cut from fine-grained wrought products or large-grained cast products.

The presence of crystallographic texture must also be taken into consideration. In specimens taken from wrought bar stock, for example, wire textures (a specific crystallographic direction parallel to the bar axis) are common. The strong textures in plate materials should also be noted when torsion specimens are fabricated from these materials. Samples taken from forgings or extrusions often exhibit preferred crystallographic orientations, which also may vary from one location to another.

Mechanical texture, or the presence of inclusions, grain boundaries, or other microstructural features with a preferred direction, should also be taken into account when designing a torsion specimen. Such texture, or fibering as it is sometimes called, can greatly affect the torsional ductility, depending on the orientation relationship between the torsion axis and the mechanical "fiber." The relationship between ductility and mechanical texture is analogous to the relationship between these two properties in tensile tests performed on specimens cut from the rolling direction (higher ductility) and transverse direction (lower ductility) of rolled plate.

The mechanical behavior in torsion testing is somewhat more complex, however, because of the tendency of the fiber to rotate as twisting proceeds. Consider, for example, the torsion of a solid round bar, in which inclusions are initially in the form of stringers parallel to the torsion axis (Fig. 3). After twisting, the inclusions at the center of the

bar have not rotated, but those at the surface have rotated to form an angle, the magnitude of which depends on the gage length and diameter and the amount of twist. It is apparent that the generation of such a mechanical texture can substantially reorient the planes of weakness into the form of a "wolf's ear," as evidenced by tensile tests on bars prestrained in torsion (Fig. 4). Rotation of the mechanical texture and the general level of inclusions may have a significant effect on the torsional fracture strain. Because of this, torsion testing is frequently used to uncover lot-to-lot variations in workability for a given material.

Specimen Design

Once a rough blank has been cut, the exact torsion specimen geometry must be determined. Typical specimen designs are illustrated in Fig. 5. The most important components of design are gage section, fillet radius, shoulder length, and grip design. Gage section geometry determines the deformation level and deformation rate for a given amount of twist and a given twist rate, based on:

$$\Gamma = \frac{r\theta}{L}$$

and

$$\dot{\Gamma} = \frac{r\dot{\theta}}{L}$$

where Γ is the engineering shear strain (equal to twice the tensorial shear strain or root three times the von Mises effective strain); r is the radial position; L is the gage length; $\dot{\Gamma}$ is the engineering shear strain rate; and $\dot{\theta}$ is the twist rate (in radians per unit time).

For a given amount of twist, large values of r and small values of L promote high values of Γ. Similarly, large r and small L

Fig. 3 Torsion specimen with two typical inclusions at the center and the surface

(a) Prior to twisting. (b) After twisting

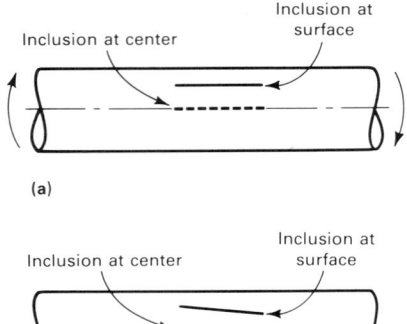

(a)

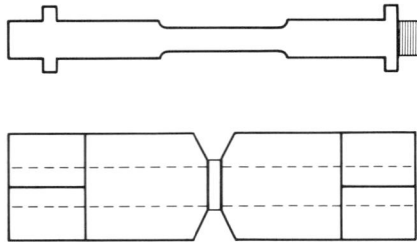

(b)

Fig. 5 Typical torsion specimen geometries used for workability testing

See text for discussion of dimensions.

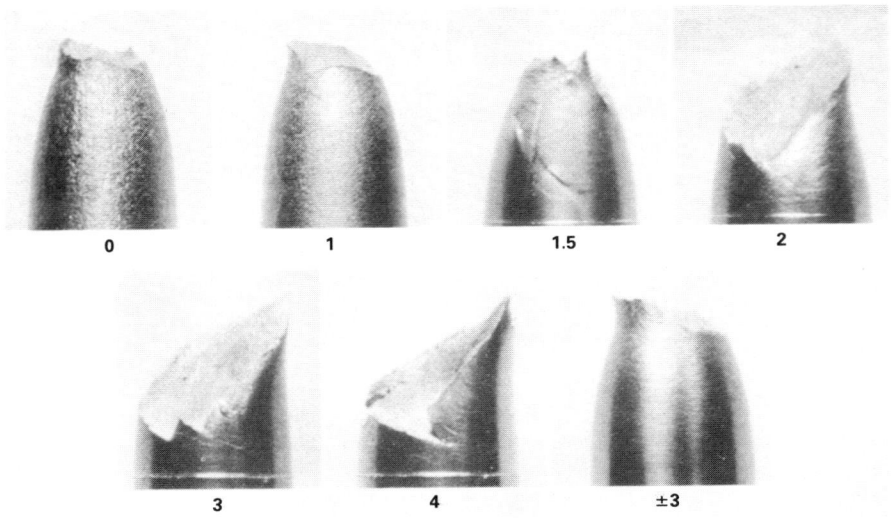

Fig. 4 Tensile fracture in torsionally prestrained copper specimens

The shear strain in the surface of each specimen is indicated below the photograph. Specimen marked ±3 shows the tensile fracture after twisting to a shear strain of 3 and then completely untwisting. Source: Ref 2

values result in large values of $\dot{\Gamma}$ for a given θ. Values for Γ and $\dot{\Gamma}$ vary with radius in a solid bar or thick-walled tube; they are greatest at the surface and zero at the center.

Strain and strain rate are typically reported as the values that pertain to the surface of the specimen. When torsion data are compared to strain and strain rate values obtained from other deformation modes, it is this strain and this strain rate that are of importance.

Although the optimum gage section geometry for a particular material is frequently determined by trial and error, the most commonly used shapes for workability assessment are a solid round bar and a thin-walled tube. With a solid bar design, problems associated with buckling of tubes are avoided. For wrought materials, typical geometry consists of a gage diameter of 6.5 mm (0.25 in.) and a length ranging from 6.5 to 50 mm (0.25 to 2.0 in.). The shorter lengths are used to obtain higher strain rates.

The gage length selected also influences the fillet radius that should be used in the torsion specimen, because some deformation usually occurs in this portion of the specimen

as well. When the gage length is short (e.g., 6.5 mm, or 0.25 in.), a sharp fillet radius (e.g., 0.25 mm, or 0.010 in.), which is blended to the shoulder area by a sharp conical taper of about 120° included angle, is often used to prevent excessive deformation outside the gage section. For longer gage length specimens, substantially larger fillet radii (e.g., 4 mm, or 0.15 in.) can be used.

Thin-walled tubular specimens provide an advantage over solid bar specimens in that shear-strain and shear-strain rate gradients are virtually eliminated in the design. This is desirable in workability testing, because ambiguities associated with definition of the failure strain and strain rate are eliminated. Moreover, the use of thin-walled tubes eliminates the variation of temperature across the section thickness that occurs in solid bars twisted at high rates. These temperature variations are a consequence of the variations in shear strain and thus deformation work converted into heat.

Thin-walled specimens must have a rather small length-to-diameter ratio to avoid buckling. Typical dimensions for such specimens are 12.5 mm (0.50 in.) outer diameter, 11 mm (0.44 in.) inner diameter, and a gage length of 2.5 mm (0.10 in.).

In both solid bars and tubular specimens, the gage length-to-diameter ratio may have a marked effect on the actual specimen temperature during moderate-speed ($\dot{\Gamma} = 10^{-2}$ to 10 s^{-1}) torsion tests due to the effects of heat conduction. Because of this, flow curves derived from data obtained at these rates tend to show a dependence on the length-to-diameter ratio (L/d). Flow curves for large L/d specimens tend to fall below those for small L/d ratios, in which most of the deformation heat

is dissipated into the shoulders (Fig. 6). Interpretation of fracture strain data from such tests should take into account not only the nominal (initial) test temperature, but also the temperature history during the test.

When workability is limited by flow localization, gage length can noticeably affect the failure process, particularly at moderate deformation rates. At these rates, heat conduction along the length of the gage section and into the shoulder can set up a substantial temperature gradient, the magnitude of which depends on the gage length, shoulder size, and deformation rate. In turn, the temperature gradient influences the deformation resistance along the length of the bar and hence the strain profile.

The design of the shoulder and grip ends of torsion specimens is determined largely by the method of heating and the type of torsion machine to be used. In all cases, the shoulder diameter should be at least one and one half times the gage section diameter and preferably two to three times as large to prevent plastic deformation as well as to minimize elastic distortion. Shoulder length can be very short for torsion tests conducted at room temperature. If testing is to be carried out at elevated temperature, the shoulder should be longer to ensure temperature uniformity, typically about 25 mm (1.0 in.) for specimens heated by induction and 25 to 50 mm (1.0 to 2.0 in.) for specimens heated in furnaces.

The grip ends of torsion specimens for workability studies are generally of two types: threaded and geometric cross sections. For threaded grip ends, a surface against which the torque can be reacted is provided by making the major diameter of the threads less than the shoulder diameter, or by includ-

Fig. 6 Stress-strain curves for solid torsion specimens of 3.3% silicon steel showing effect of gage length to diameter ratio (L/d) on flow stress at high strain rates when adiabatic heating occurs

The flow curves are in terms of von Mises effective stress-strain ($\bar{\sigma}$-$\bar{\epsilon}$), defined by $\bar{\sigma} = \sqrt{3}\tau$ and $\bar{\epsilon} = \Gamma/\sqrt{3}$, where τ-Γ is the shear stress/shear strain curve obtained in torsion testing. Source Ref 3

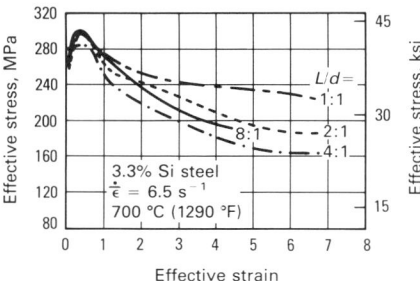

the line may still be straight (i.e., it will form a helix at a fixed angle to the torsion axis), indicating fracture-controlled failure, or it will exhibit a "kink" at a larger angle to the torsion axis than the remainder of the line, indicating flow-localization-controlled failure.

In the former case, the surface shear strain at fracture (Γ_{sf}) is given by $\Gamma_{sf} = \tan \varphi_f$, where φ_f is the angle between the scribe line and torsion axis at failure. This value of Γ_{sf} should agree with that obtained from the equation given above for shear strain as a function of r, θ, and L. When flow localization has occurred prior to fracture, the tangent of the angle between the line away from the localization region and the torsion axis provides an estimate of the workability of the material. However, the tangent of the angle between the line and the torsion axis in the localization region also yields an estimate of the fracture strain, provided the test has been carried out to this point.

Torsion Equipment

Equipment for torsion testing designed to assess workability is available as a complete system or may be specially designed to meet individual needs. Equipment usually consists of some combination of these components:

- Motor for applying torque
- Twist and torque transducers
- Heating system with temperature controller
- Water-cooled grips
- Support frame

Motors

Three types of motors are generally used for torsion testing to assess workability: electric, hydraulic, and hydraulic rotary actuator. Variable-speed electric motors are the simplest and were the first to be used for torsion tests. These motors can have speeds ranging from fractions of a revolution per minute (rpm) to several thousand rpm. However, the available speed range for a given motor may be limited. Other limitations are the difficulty of maintaining a constant rotation rate if the torque requirement changes during a test and the generally low torques that such motors generate. Electric motors may be fitted with gearboxes and energy storage devices to prevent these problems.

To overcome the drawbacks of electric motors, hydraulic motors and hydraulic rotary actuators are used in sophisticated torsion systems. Hydraulic motors are typically of the fixed displacement type, in which the rotation and displacement of the shaft is proportional to the volume of oil flowing through the motor. The torque is produced by the oil as it pushes against a set of spring-

Fig. 7 Dependence of shear stress and mean axial stress on effective strain in fixed end torsion tests at high temperatures

Source: Ref 4

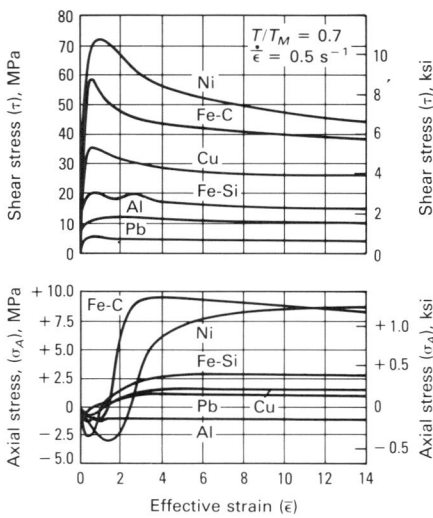

loaded vanes that rotate around the motor housing or spherical pistons that ride up and down a contoured cam. In such motors, multiple turns are achieved by supplying a continuous flow of oil. Typical speeds range from 1 to about 1000 rpm. Slower speeds are obtained by using gear reducers in series with the motor.

In hydraulic rotary actuators, the torque is also supplied by oil pressure against a set of vanes. However, the rotation is limited to only a fraction of a revolution. The specimen gage length therefore must be short or the diameter must be large to achieve large strains. Such actuators are suitable when loading histories involving rotation direction reversals are required, or when the system is to be used for torsional fatigue testing as well.

In addition, rotary actuators can produce a greater range of rotation rates than hydraulic motors, particularly at the low end of the scale where fractions of an rpm are possible. The rotation rates required to simulate various metalworking operations are given in Table 1.

The flow of oil through either hydraulic motors or rotary actuators may be at a given rate that is determined by the hydraulic pump supplying the oil (an open-loop arrangement), or may be accurately controlled by a servo-valve in a closed-loop circuit. In an open-loop system, the rotation rate of the motor can vary as the torque requirement changes. This difficulty is overcome in closed-loop, servo-controlled systems. The servo-valve is an electrically actuated valve

---

ing flanges between the threaded end and the shoulders in the specimen design. Triangular, hexagonal, or other simple geometric cross sections are also cut into ends. In these instances, the torque is reacted against the flat faces of the ends.

Selection of threaded ends versus triangular or hexagonal cross-sectional ends is based largely on whether torsion testing is conducted under load control or stroke control. In load control, axial loads are applied to prevent buckling or to enable combined tension-torsion deformation; threaded ends are required for this type of test.

Triangular or hexagonal ends are most common when tests are to be run in stroke control to prevent changes in gage section geometry that may result from the axial extension that characterizes the torsion testing of many metals. Frequently, threaded ends can also be used in such situations; however, triangular or hexagonal ends have the added advantage of allowing rapid specimen removal for purposes of quenching, which is beneficial in subsequent metallurgical analysis. The possible development of axial stresses and their effect on flow and failure response in fixed-end tests should be carefully considered, because these stresses can be a significant fraction of the torsional shear stress (Fig. 7). The subject of axial stresses is discussed in greater detail in the Appendix at the end of this article.

The final step in the preparation of torsion specimens for determination of workability is the application of a fine axial line on the gage section surface. This may be done using a felt-tip pen for room-temperature tests or a fine metal scribing instrument for elevated-temperature tests. After twisting to failure,

Table 1 Torsional rotation rates corresponding to various metalworking operations

Operation	von Mises effective strain rate ($\dot{\bar{\epsilon}}$) (a), s^{-1}	Corresponding surface shear strain rate in torsion ($\dot{\Gamma}$), s^{-1}	Rotation rate (b), rpm
Isothermal forging	10^{-3}	1.73 × 10^{-3}	0.02
Hydraulic press forging	1	1.73	16.5
Extrusion	20	34.6	330.4
Mechanical press forging	50	86.6	827.0
Sheet rolling	200	346.4	3307.9
Wire drawing	500	866.0	8269.7

(a) $\dot{\bar{\epsilon}} = \dot{\Gamma}/\sqrt{3}$. (b) Assuming specimen geometry with $r/L = 1.0$.

that forms part of a feedback circuit. A rotary transducer measures the rotation of the hydraulic motor or rotary actuator, and a function generator provides a predetermined twist rate (or twist history) in terms of an electrical voltage that varies with time. A sample circuit of this type is shown in Fig. 8.

During operation, oil is fed through the servo-valve into the motor or rotary actuator. The rotation is sensed by a transducer that sends a voltage to the servo-controller. The voltage is then compared to the predetermined voltage from the function generator for that given point in time. Depending on the difference between the two voltages, the servo-valve is either opened or closed to increase or decrease the flow of oil to the hydraulic motor or rotary actuator. This sequence of operations is performed continuously during the torsion test to ensure that the proper twist history is being applied.

Twist and Torque Transducers

Specially designed transducers are necessary to measure the twist of the motor shaft and specimen and the applied torque. As mentioned previously, a rotation transducer is a component of the feedback circuitry required to control the hydraulic motor. These transducers are of three types: optical encoders, variable resistors (rheostats), and rotary variable-differential transformers.

An optical encoder produces voltage pulses whose number is proportional to the amount of rotation. The encoder pulse count is balanced against that from a function generator in a closed-loop feedback circuit in order to control the rotation of the hydraulic motor.

Similarly, the rheostat shaft may be connected to one end of the hydraulic motor. As it turns, the rheostat resistance increases. Because it is connected across a constant current source, the rheostat produces an increasing voltage.

The rotary variable-differential transformer is perhaps the most accurate of the rotation measuring devices and is used primarily in sophisticated rotary actuator systems. Like a linear variable-differential transformer, a rotary variable-differential transformer generates a variable voltage, as an iron core moves through a current-carrying coil.

When the rotation transducer is located in the load train close to the torsion specimen, the voltage produced by it is often used to measure the actual twist applied to the specimen. However, if the load train is long and the transducer is not close to the specimen, elastic deflections of the system can introduce systematic errors in the specimen twist measurement. To overcome this difficulty, an auxiliary twist transducer is sometimes placed near the specimen. For example, a rheostat system to which a drive pulley is attached may be coupled to the specimen grip using a tight-fitting rubber O-ring or gear mechanism. This arrangement is satisfactory for rotation rates of about several hundred rpm.

Torque sensors for torsion testing are usually of the reaction-torque type and are usually placed on the side of the specimen opposite the hydraulic motor or rotary actuator. Load transducers that are capable of monitoring both axial and torsional loads are also available. In either case, the torque is measured through a structural member in the load cell, to which foil-resistance strain gages have been applied. As torque is applied, the structural member deflects elastically. The response is measured with the strain gages. In practice, the torque can be calibrated indirectly using a calibration resistance in a Wheatstone-bridge circuit, or directly by imposing fixed amounts of torque by hanging weights from a lever arm attached to the torque cell.

The capacity of the torque cell should accommodate the specimen geometry, test material, and test temperature. For typical specimen geometries, torque cells with ratings of 0 to 565 N · m (5000 lbf · in.) are sufficient. When testing is carried out on very soft metals or on most metals at hot working temperatures (at which the flow stress is very low), a torque cell of one tenth this capacity (55 N · m, or 500 lbf · in.) is required to maintain adequate signal-to-noise ratios for data acquisition. With modern electronics, most torque cells can be calibrated and used at various percentages (usually 10, 20, 50, or 100%) of their maximum rating, which allows a particular transducer to be used over a range of temperatures and strain rates.

Heating Systems and Water-Cooled Grips

Torsion tests to establish workability are frequently conducted at elevated temperatures. For this purpose, specially designed heating arrangements and specimen grips are necessary. The two most common methods of heating are furnace and induction. For typical heating times, furnaces or induction generators should have power ratings of 2 to 10

Fig. 8 Schematic of typical control and data acquisition system for torsion testing
Source: Ref 5

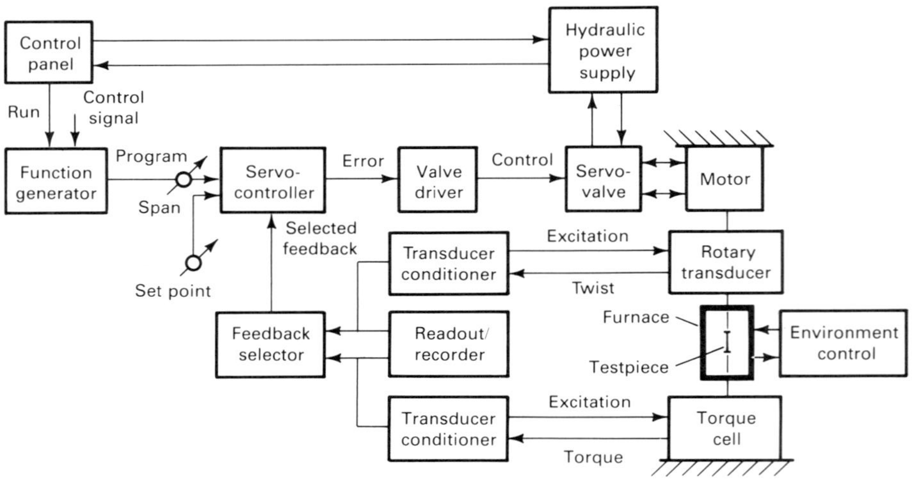

kW. Furnaces are usually electric-resistance or quartz-tube radiant heating types. The hot zone should be long enough to ensure uniform heating of the gage section. Initial temperature nonuniformities may give an incorrect picture of failure if the test material is sensitive to flow localization. This is particularly true of titanium- and nickel-based alloys, in which the flow stress is very sensitive to temperature in the hot working regime.

The induction coil for torsion specimen heating should overlap the shoulders of the test specimen to ensure temperature uniformity. In general, the most efficient application of induction heating is for ferrous alloys (carbon and ferritic alloy steels, austenitic stainless steels, etc.) and nickel- and titanium-based alloys. Generator frequencies for heating torsion specimens of these alloys are generally in the radio-frequency range (~500 kHz). Aluminum and copper alloys do not induction heat readily and thus should be heated by this method only with the use of susceptors.

To prevent contamination of torsion specimens during elevated-temperature torsion studies, specimens may be tested in specially designed chambers in which a controlled atmosphere, such as argon, is maintained. A simpler, less expensive technique is to apply protective coatings to the specimens. These coatings are usually vitreous and are supplied as frits or as commercial metalworking glass lubricants. Frits are applied at room temperature by grinding them into a fine powder, preparing a slurry with an alcohol carrier, and dipping the specimens into the coating preparation. Commercial lubricants used for specimen protection can be applied by a variety of methods, such as dipping, brushing, and spraying. For both types of glass, the coating melts and forms a viscous, protective coating as the specimen is heated.

Another problem associated with elevated-temperature testing is the necessity of preventing heat from being conducted into parts of the load train located away from the specimen grips (load cells, motors, etc.). This is accomplished by providing water cooling to the grips. This provision is relatively straightforward for the stationary grip; it is accomplished by brazing copper tubing that contains a water supply onto the grip.

Water cooling is more difficult to accomplish in grips that rotate. In this case, the grip, or an adjacent fixture comprised of a core that can rotate and an outer casing that is held stationary, must be constructed. A typical example of this type of fixture is shown in Fig. 9. Water leakage is prevented by a set of rubber O-rings. Depending on the actual test temperature, the water-cooled fixtures

should be constructed of tool steels or stainless steels. Grips or parts of tooling that are subjected to temperatures higher than approximately 540 °C (1000 °F) should be fabricated from high-temperature superalloys, such as Waspaloy, Inconel 718, or IN-100.

Design of the Load Train

Once the individual components of the torsion tooling have been assembled, the load train must be constructed. The two major types are horizontal load trains mounted on lathe beds and vertical load trains mounted in modified test machines. The use of a lathe bed, as illustrated in Fig. 10, enables accurate alignment of the tooling components. However, this type of arrangement does not allow for control of axial loads. Tests can be conducted with the tailstock (saddle) either fixed or floating.

Control of axial loads can be achieved by mounting the components in a modified test machine, such as the electrohydraulic setup shown in Fig. 11. A hydraulic motor is in series with an independently controlled, linear, hydraulic actuator that may be run either in load or stroke control. The torque is reacted against the posts of the test frame using a reaction plate to which a set of cam followers is attached, which allow the ram-hydraulic motor assembly to translate up and down.

A series of adapters and the torque and axial load cells are located at the top of the load train. The use of many adapters as well as individual load cells for axial load and torque is undesirable, because they increase

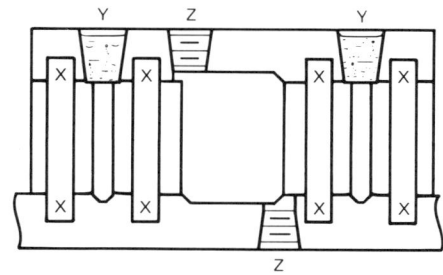

Fig. 9 Schematic of water-cooled rotating grip for high-temperature torsion testing

(a) Outer housing. (b) Inner core. Note that the outer housing, to which water lines are attached, is held stationary during testing by reaction rods against which the torque is reacted. X, O-ring grooves; Y, water inlet/outlet; Z, tapped hole for reaction rod

Fig. 10 A servo-controlled, hot torsion machine mounted on a lathe bed

Source: Ref 5

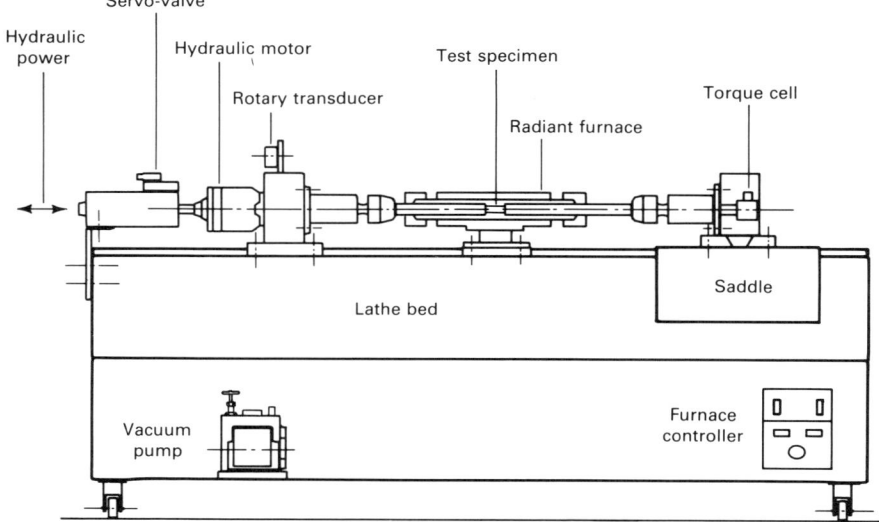

the overall length of the load train. As the length increases, the stiffness of the torsion machine decreases, thus increasing the possibility of problems associated with wave effects and "ringing" in the load cells during high strain rate experiments.

The top of the load train is attached to the upper crosshead by a set of bolts and spherical washers that provide the primary means

Fig. 11 Electrohydraulic testing machine modified for combined torsional and axial loading

1, upper crosshead; 2, tension load cell; 3, torque cell adapter No. 1; 4, torque cell; 5, torque cell adapter No. 2; 6, water-cooled grip; 7, specimen holder; 8, specimen; 9, induction coil; 10, ram; 11, lower crosshead; 12, linear (tension-compression) actuator; 13, torque reaction plate; 14, hydraulic motor; 15, incremental optical encoder; 16, electrical leads; 17, water lines; 18, hydraulic lines. Source: Ref 6

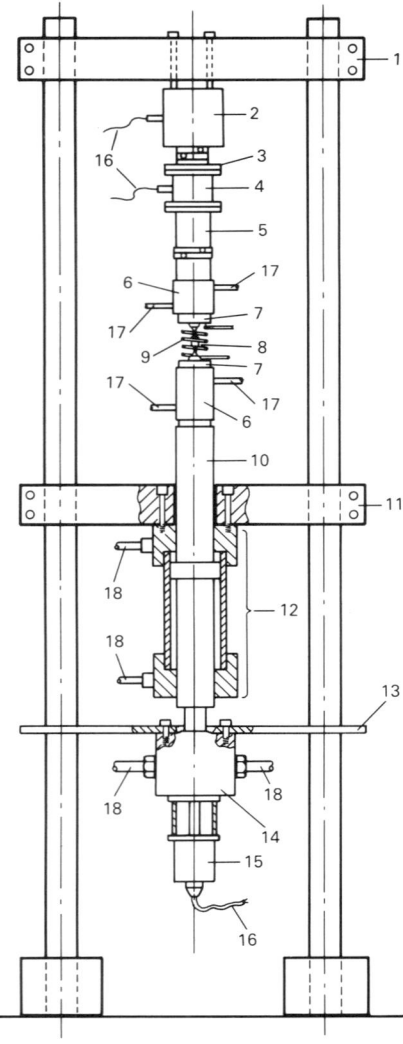

of aligning the system. If the grips and other components are accurately machined and the system is accurately aligned, the total indicator runout on the specimen rotation can be kept to several thousandths of an inch with a modified testing machine system. Such tolerances are comparable to the tolerances and deflections maintained in a linear actuator and are usually acceptable for most torsion work. These tolerances can only be improved by placing special reinforcing structures or die sets between the crossheads of the machine, thereby substantially increasing the cost of the equipment.

Flow Stress Data

When failure is controlled by fracture without prior flow localization, flow curves and fracture strains can be readily derived from measurements of torque versus twist during testing. Frequently, these data indicate increasing or nondecreasing torque with increasing rotation. However, at hot working temperatures (and sometimes even at cold working temperatures), the material may show softening due to dynamic recovery, dynamic recrystallization, microstructural softening, or deformation heating. In these cases, the only reliable method of confirming that failure is fracture-controlled is by examination of the surface scribe line or the specimen microstructure. If the deformation has been fracture-controlled, the inclination of the scribe line to the torsion axis will remain uniform, and the microstructure in the gage section will be deformed uniformly. The tangent of the angle between the scribe line and the torsion axis should be equal to the shear strain corresponding to that value of twist at which the torque-twist record shows a sudden drop, indicating fracture.

When microstructural observations are made, a particular type of cross section, which is neither axial nor transverse, should be selected. This section should be one in which the outer diameter surface, or near surface region of the round gage section, is examined. Axial-tangential sections reveal the true simple shear nature of the deformation. If other types of cross sections are selected, microstructural variations due to the variation of shear strain with radius may obscure variations due to flow localization.

Reduction of Torque-Twist Data

Methods for reducing torque-twist (M-θ) data to the shear stress/shear strain (τ-Γ) form vary with specimen geometry. Reduction of torque-twist data is easiest with thin-walled tubes. In this case, the shear strain is found from $\Gamma = r^*\theta/L$, where r^* is the mean

tube radius and L is the gage length. A relation between the torque and the (assumed) uniform shear stress is found from the torque equilibrium relation $M = 2\pi r^{*2}t\tau$, which results in $\tau = M/2\pi r^{*2}t$. Here, t is the tube wall thickness, assumed to be small in comparison to r^*.

For round solid bars, the derivation of τ-Γ from M-θ is more complex. Several methods are available; however, all overlook the occurrence of deformation heating effects at high strain rates. The simplest method was developed by Fields and Backofen (Ref 7). In this technique, the shear strain in the τ-Γ relation is taken to be that for the outer diameter surface of the solid bar ($\Gamma = r_s\theta/L$). The corresponding shear stress is obtained from the relation $\tau = (3 + n^* + m^*)M/2\pi r_s^3$. The quantity n^* is the instantaneous slope of log M versus log θ. If the material strain hardens in a power-law manner ($\tau \propto \Gamma^n$), n^* is a constant equal to the strain-hardening exponent, n. The value of m^* is found from plots of log M versus log $\dot\theta$ at fixed values of θ.

Thus, tests at a variety of strain rates are required for proper evaluation. If the material exhibits power-law strain rate hardening ($\tau \propto \dot\Gamma^m$), m^* is equal to the strain-rate sensitivity exponent, m. At cold working temperatures, m^* is usually small and can often be neglected with respect to the value of $3 + n^*$. At hot working temperatures, m^* can take values comparable to the values of m at these temperatures, which typically range from 0.1 to 0.5, or values that are not negligible in comparison to $3 + n^*$.

If the specimen design is a thick-walled tube (the wall thickness is comparable to the radius), a variation of the Fields and Backofen formula may be used to derive the τ-Γ relation. Γ is taken to be the shear strain at the outer diameter, $\Gamma = r_s\theta/L$. However, the shear stress in this case is:

$$\tau = \frac{(3 + n^* + m^*)M}{2\pi r_s^3}\left[1 - \left(\frac{r_i}{r_s}\right)^{3+n^*+m^*}\right]$$

where n^* and m^* are the slopes of the two logarithmic plots, and r_i denotes the internal specimen radius.

The equations given above for the solid bar and thick-walled tube are only rigorously valid for materials displaying power, exponential, or other simple hardening laws—i.e., when the hardening rate, at a given strain, is independent of strain rate or temperature. When this condition does not apply, or when flow softening is observed, accurate τ-Γ relations can still be determined, but this requires the use of a series of samples of increasing diameter. This method is described in the Appendix to this article.

Deformation Heating

When the surface shear strain rate in torsion is less than approximately 0.01 s^{-1}, the τ-Γ curves established from the measurement of torque and twist may be assumed to be representative of isothermal (constant temperature) behavior. Above this strain rate, thermal conduction is not sufficiently rapid to dissipate the deformation heat into the specimen shoulders (for typical metals, alloys, and specimen geometries). Because of this, the τ-Γ results are representative of flow behavior at temperatures that increase with the amount of deformation. Furthermore, because the temperature increase ΔT is a function of Γ, a radial, as well as an axial, temperature gradient can be developed in solid bar specimens. For this reason, tubular specimens, in which radial effects are negligible, are preferable for high-rate torsion tests.

The temperature increase that occurs during a torsion test is a function of the amount of heat generated and the amount of heat conducted into the shoulders. The amount of heat conducted into the shoulders depends on specimen geometry and requires numerical analysis, which is beyond the scope of this Handbook.

During an extremely rapid test ($\dot{\Gamma}$ typically greater than or equal to about 20 s^{-1}), heating is essentially adiabatic. Under these conditions, all of the deformation work converted to heat is assumed to be retained in the gage section, giving rise to a temperature increase of $\Delta T = (\eta \int \tau d\Gamma)/\rho c$. The parameter η is the fraction of the deformation work per unit volume ($\int \tau d\Gamma$) converted into heat; ρ is the material density; and c is the specific heat. The value of η usually ranges from 0.90 to 1.00; 0.95 is the value most frequently assumed. The plastic work per unit volume ($\int \tau d\Gamma$) is simply the area under the τ-Γ curve. Once the temperature increase (ΔT) has been estimated, the shear stress associated with the test strain rate and particular values of Γ and T (equal to nominal test temperature plus ΔT) can be determined.

By conducting torsion tests at several temperatures at a given adiabatic strain rate, a high-rate isothermal torsion flow curve can be determined by estimating the ΔT values and plotting τ versus the actual temperature at various levels of shear strain. This procedure is shown schematically in Fig. 12. For each level of Γ, a smooth curve is drawn through the data points. Equivalent isothermal flow curves are then obtained at various temperatures by "reading" points off the various curves at these fixed temperatures. This procedure assumes that shear strain is a "state variable," i.e., that the shear stress is uniquely determined by the value of Γ, re-

Fig. 12 Stress-temperature plots used to estimate equivalent isothermal high strain rate flow curves

The intersections of the dashed lines with the flow stress-temperature curves yield estimates of the isothermal flow behavior at temperatures T_1, T_2, and T_3.

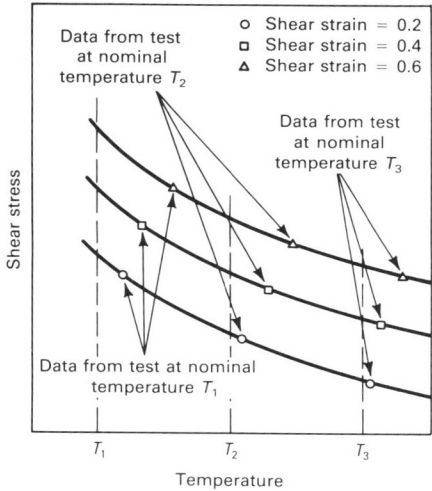

gardless of prior shear strain history. Over small intervals of Γ and ΔT, such an assumption is probably sound from an engineering viewpoint.

Reduction of τ-Γ Data to $\bar{\sigma}$-$\bar{\epsilon}$

To make use of shear stress/shear strain (τ-Γ) data for the prediction of load and the analysis of metal flow in actual metalworking operations, these quantities can be converted to effective stress ($\bar{\sigma}$) and effective strain ($\bar{\epsilon}$). These stresses and strains are the equivalent quantities that would result in an identical amount of deformation work in a state of uniaxial compression or tension. For a given deformation level, then:

$$\int \tau d\Gamma = \int \sigma d\epsilon = \int \bar{\sigma} d\bar{\epsilon}$$

where σ-ϵ is the uniaxial flow curve. In uniaxial tension of a material with a random crystallographic texture, $\bar{\sigma} = \sigma$ and $\bar{\epsilon} = \epsilon$. For uniaxial compression, $\bar{\sigma} = -\sigma$ and $\bar{\epsilon} = -\epsilon$.

In torsion, the relation between τ-Γ and $\bar{\sigma}$-$\bar{\epsilon}$ depends on the plasticity theory formulation that is used. The most common concept, first proposed by von Mises (Ref 8), leads to the relations $\bar{\epsilon} = \Gamma/\sqrt{3}$ and $\bar{\sigma} = \sqrt{3}\tau$. Note that the concepts of effective stress and effective strain are also approximations, because the $\bar{\sigma}$-$\bar{\epsilon}$ relations obtained from different mechanical tests are often different, depending on the particular material and test conditions.

Nevertheless, for load estimation, these constructs are sufficiently accurate for engineering purposes. Unless otherwise specified, the effective stress, effective strain, and effective strain rate definitions used in this article are based on the von Mises formulation.

Typical Flow Curves

Flow stress results for type 304L austenitic stainless steel at two strain rates and a variety of cold and warm working temperatures and hot working temperatures are shown in Fig. 13. Note that the shear strain levels in these curves are about 5, corresponding to von Mises effective strains in excess of 2.5 or to values much greater than those obtained in tension or compression tests. All tests shown in Fig. 13 were taken to failure. However, the flow stresses for the hot working tests are only shown for $\Gamma \leq 5$.

The flow stress behavior observed in torsion is best interpreted in terms of test temperature and strain rate. For the type 304L alloy (Fig. 13), the flow stresses generally increase with strain (or adopt stationary values) at cold and warm working temperatures. Only at 20 °C (68 °F) and at 200 °C (390 °F) and a von Mises surface effective strain rate of $\dot{\bar{\epsilon}}_s = \dot{\Gamma}_s/\sqrt{3} = 10$ s^{-1} do the flow curves exhibit softening. Under these conditions, scribe line measurements indicated that flow localization had occurred prior to fracture. All other scribe line observations indicated fracture-controlled failure.

In contrast to these flow curves are those obtained at hot working temperatures, all of which reveal maxima followed by decreasing flow stresses and eventually steady-state flow. This trend is indicative of dynamic microstructural changes characterized by recrystallization phenomena. However, scribe line measurements confirmed that, despite the flow stress decrease, failure was fracture controlled in all cases, due to the stabilizing influence of the increased strain-rate sensitivity.

The flow curves for type 304L stainless steel shown in Fig. 13 also indicate the influence of strain rate on flow behavior. At cold and warm working temperatures, strain rate has only a slight effect on flow response. In fact, the 10 s^{-1} curve at a given temperature eventually drops below the curve measured at 0.01 s^{-1}. The lower rate can be considered isothermal and the higher rate adiabatic. Thus, the crossover of flow curves at the two rates is a result of deformation heating and a relatively small strain-rate sensitivity (as shown by the initial portions of the flow curves, in which thermal effects are unimportant). Isothermal flow curves for 10 s^{-1} can be deduced by estimating the associated ΔT values and by constructing σ-T plots.

Fig. 13 Flow curves from type 304L stainless steel torsion tests
(a) Cold and warm working temperatures. (b) Hot working temperatures. Source: Ref 9

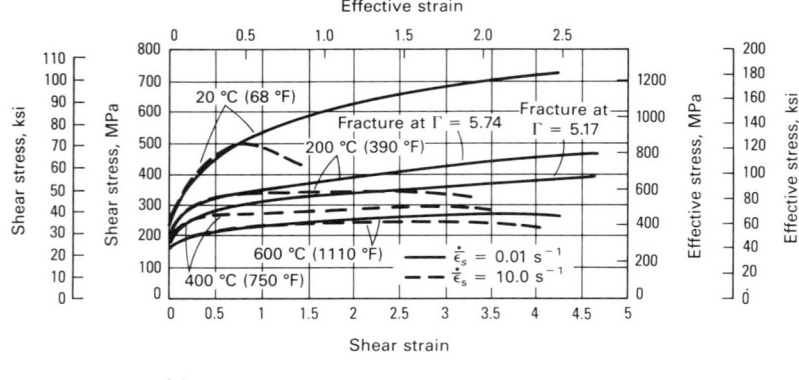

(a)

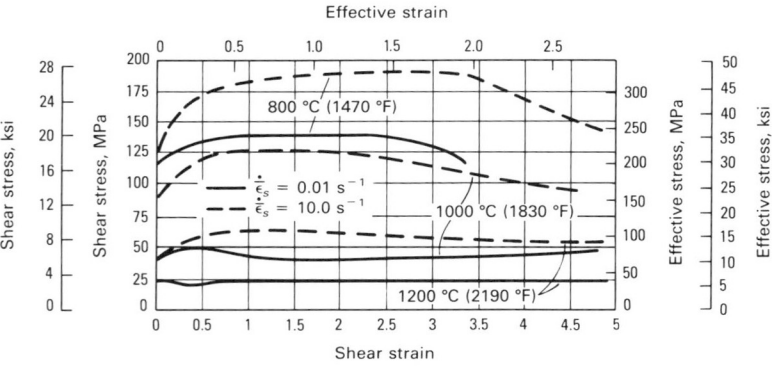

(b)

relationship between τ and Z is of the form:

$$(A \sinh \sqrt{3}\, \alpha'\, \tau)^{m'} = Z$$

in which A, α', and m' are material constants. An example of the fit for a 0.25% carbon steel is shown in Fig. 15.

Comparison of Torsion Flow Stress Data

Shear stress/shear strain data from torsion tests are often converted to "effective" flow stress data using the von Mises relations. In this form, they are useful for load prediction and metalworking analysis. However, even when converted to effective terms, flow stress data obtained from torsion tests do not always show perfect agreement with data from other mechanical tests.

The differences are usually greatest at cold working temperatures. They are a consequence of one or a combination of several factors, such as the effect of deformation path (tension, compression, etc.) on the development of crystallographic texture and on the nature of the microscopic slip or twinning processes that control the observed macroscopic strain-hardening rate. The variability of such factors is less likely at hot working temperatures because of the dynamic restorative processes that may prevent sharp textures from being formed. Also, a significant amount of deformation occurs at grain boundaries as well as in the matrix of metals.

In Fig. 16, torsion flow stress data for type 304L are compared to compression and tension data in terms of von Mises effective stress and strain. At cold and warm working temperatures, as well as at low hot working temperatures (800 °C, or 1470 °F), the flow curves from the various tests do not coincide. Generally, there is a lower level of strain hardening in torsion. Thus, although the overall stress levels are similar, the actual shapes of the curves are quite different. Even if other definitions of effective stress and strain are employed, the differences between the curves cannot be eliminated.

However, an estimate of the working loads can still be derived from torsion data plotted in von Mises terms. In contrast to the 20, 400, and 800 °C (68, 750, and 1470 °F) behavior, comparison of type 304L torsion data to tension and compression data is quite good at the hot working temperature of 1000 °C (1830 °F) (Fig. 16b). This is most likely a result of the absence of textural and strain-hardening effects.

Other metals show the same divergence between torsion flow curves and those obtained by other test techniques. Data for copper and aluminum (tested at room tem-

This leads to isothermal high strain rate flow curves that are consistently above the lower strain rate flow curves.

In contrast to the trends at cold and warm working temperatures, the type 304L flow response in the hot working regime reveals a noticeable strain-rate effect. Under these conditions, the high strain-rate curves are considerably above their low strain-rate counterparts at a given test temperature. Such a response is the result of the high strain-rate sensitivity of most metals at hot working temperatures. The strain-rate sensitivity effect offsets any possible crossover due to deformation heating at the higher strain rates. Flow stresses in this temperature regime are much lower than those at cold and warm working temperatures. Because of this, ΔT values associated with the higher strain rate, which vary with the magnitude of τ and Γ, tend to be smaller at the higher temperatures.

Flow stress data from the torsion test for other alloys (such as Waspaloy, Fig. 14)

exhibit similar variations with temperature and strain rate. At cold and warm working temperatures, strain hardening often persists to large strains, except for high-rate tests, which are characterized by deformation heating and a general decrease in flow stress with temperature. At hot working temperatures, flow curves frequently show flow stress maxima followed by flow softening and a steady-state flow stress. This behavior is associated with a variety of microstructural changes. Furthermore, the strain-rate sensitivity tends to be small ($m \leq 0.02$)—except under hot working conditions, at which the value of m is typically between 0.1 and 0.5 for many metals and alloys.

The steady-state flow stress under hot working conditions measured in torsion tests (and a variety of other mechanical tests) frequently is a function of the Zener-Hollomon parameter, $Z = \dot{\epsilon} \exp(Q/RT)$, where Q is the apparent activation energy for the flow process involved; R is the gas constant; and $\dot{\epsilon} = \Gamma/\sqrt{3}$. For many single-phase materials, the

Fig. 14 Flow curves for Waspaloy

(a) Effect of temperature at a fixed effective strain rate of 1 s⁻¹. (b) Effect of strain rate at a fixed test temperature of 1038 °C (1900 °F). Flow softening at the higher temperature is a result of dynamic recrystallization. Source: Ref 5

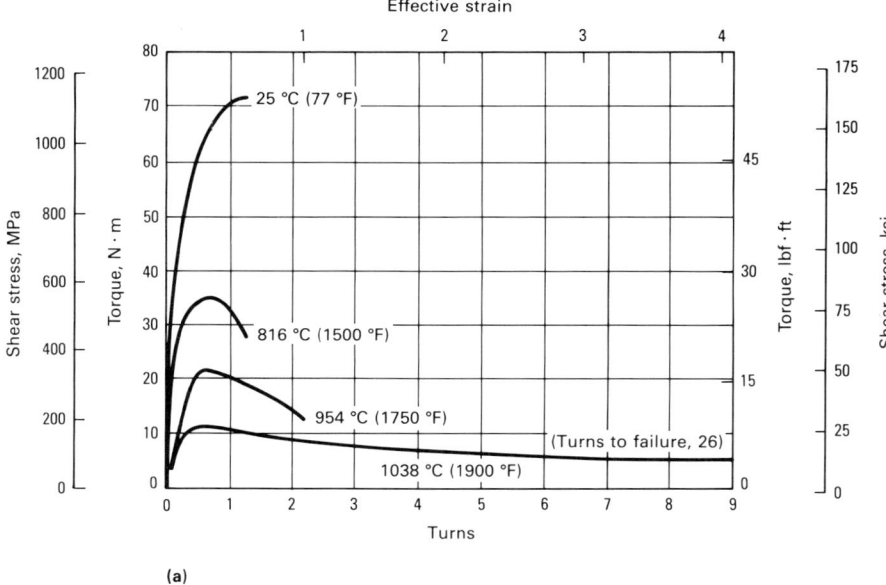

(a)

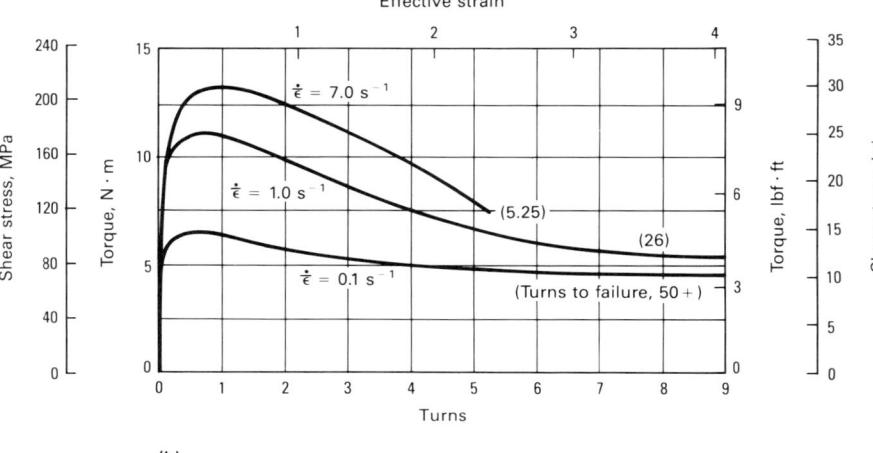

(b)

Fig. 15 Correlation of torsional flow stress data for a 0.25% carbon steel using a temperature-compensated strain rate parameter (the Zener-Hollomon parameter, Z)

Source: Ref 10

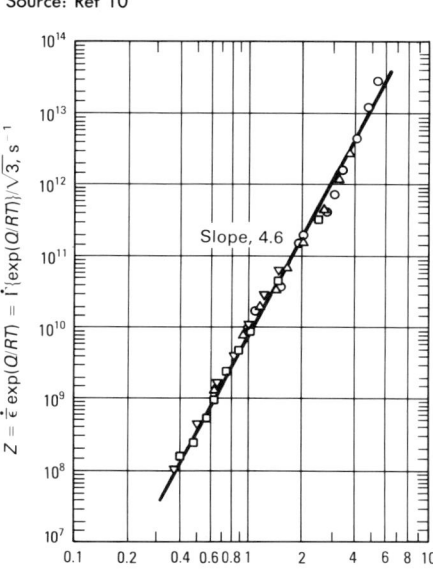

predictions than on the estimation of formability such as in sheet stretching. In these operations, workability is controlled by the onset of instability and flow localization, which are influenced strongly by the rate of strain hardening, as expressed by the strain-hardening exponent, n. Because the n values obtained from torsion tests are typically below values obtained from the other tests, they should be used with caution in formability modeling.

Interpretation of Torsional Fracture Data

Torsion testing is commonly used to estimate fracture limits. Frequently, torsion tests are conducted to determine the effects of process conditions (e.g., strain rate and temperature) and material composition on workability. In many cases, the surface strain at fracture is relatively independent of specimen design (e.g., length-to-radius ratio), as shown by the results from torsion tests at hot working temperatures for type 304 stainless steel and aluminum given in Fig. 20. Variations with specimen geometry are most likely to occur when experiments are run at moderate strain rates or at rates at which specimen geometry affects the temperature profile due to heat conduction. In these cases, the fracture strain, initial test temper-

perature) are given in Fig. 17 and 18. Comparisons with tension and plane-strain compression results are given in Fig. 17 and 18, respectively. The effective stress-strain curves from torsion show consistently lower levels of strain hardening as well as lower flow stresses.

Fig. 19 illustrates results for a material with a crystal structure different from the face-centered cubic materials in Fig. 17 and 18. These effective stress-strain curves are for body-centered cubic carbon steels that were tested in torsion and tension. To achieve the high deformation levels in ten-sion, samples were prestrained by wire drawing. The von Mises flow curves for torsion lie below those for tension for both the low-carbon steel and the pearlitic, near-eutectoid high-carbon steel.

Also shown in Fig. 19 are effective stress-strain curves from torsion tests that were calculated on the basis of the Tresca criterion, in which the effective stress is equal to 2τ and the effective strain to $\Gamma/2$. A divergence between torsion and tension is still present.

The differences between torsion flow stresses and results from other tests described above have much less of an effect on load

Fig. 16 Comparison of effective stress-strain curves determined for type 304L stainless steel in compression, tension, and torsion

(a) Cold and warm working temperatures. (b) Hot working temperatures. Source: Ref 11

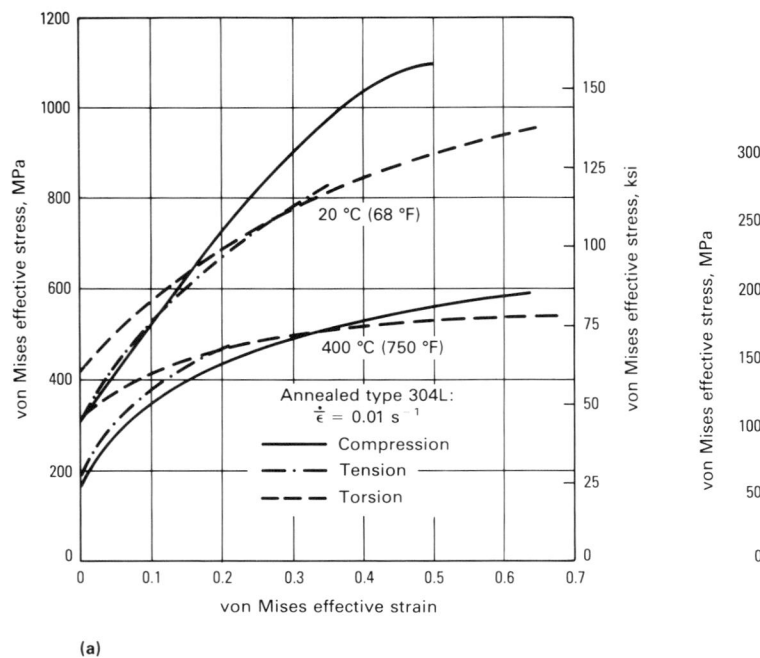

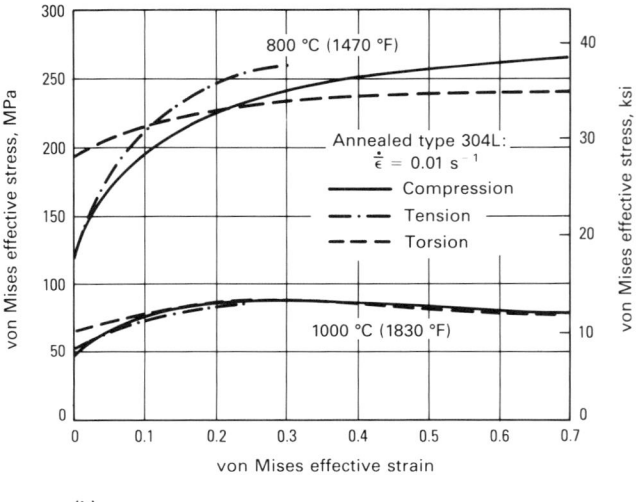

(a)

(b)

ature, and estimated temperature at fracture should be reported together.

Effect of Temperature and Alloying on Torsional Ductility

In torsion, the ductility of most metals is moderate at cold working temperatures ($\Gamma_f \leq 1$ to 5), least at warm working temperatures, and greatest at hot working temperatures, at which Γ at fracture often exceeds 10 or more. At cold working temperatures, fracture occurs by ductile fracture initiated at second-phase particles and inclusions. The decrease in ductility at warm working temperatures occurs because of the thermal activation of grain-boundary sliding, which culminates in brittle intergranular failures. At hot working temperatures, dynamic restoration processes such as recovery and recrystallization act to heal incipient voids (due to cavitation) and microcracks, thereby increasing ductility substantially.

These processes are most rapid in single-phase pure metals, which tend to have the largest ductilities at hot working temperatures. Alloys with solute elements tend to have lower ductilities because of the increased difficulty of dynamic restoration. Moreover, in such materials, the onset of true hot working conditions occurs at higher temperatures than in pure metals. This is also

Fig. 17 Flow curves determined at room temperature in tension and torsion on oxygen-free high-conductivity copper

Source: Ref 12

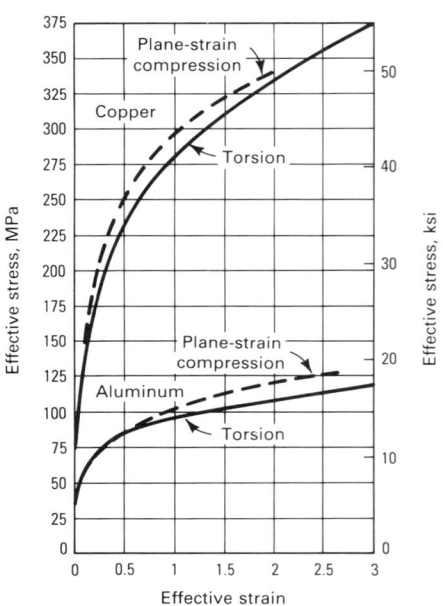

true for alloys with two or more phases in which deformation tends to be inhomogeneous because of the variation of properties between the two phases, which results in poor ductility.

In the hot working regime, ductility does not increase monotonically with temperature up to the melting or solidus temperature. Frequently, ductility passes through a maximum, which usually correlates well with the optimal temperature for forging or extrusion. Above this temperature, ductility may decrease because of effects such as grain growth or deformation heating. Increases in grain size promoted by very high tempera-

Fig. 18 Comparison of room-temperature flow curves from torsion and plane-strain compression tests on copper and aluminum

Source: Ref 13

tures increase the tendency for intergranular fracture, while retained deformation heat at moderate to high strain rates may be sufficient to increase the instantaneous local temperature above the incipient melting or solidus temperature, thus resulting in poor ductility.

Figure 21 summarizes the qualitative dependence of workability on temperature for a wide range of alloys. These trends are classified by the number of phases, the types of compounds that are present at various tem-

peratures, and so on, and are a valuable reference tool when new alloys for which quantitative data are unavailable are being formed.

Quantitative fracture data from torsion tests are usually reported in terms of twists to failure or surface fracture strain. Although using twists to failure as a measure of ductility is somewhat qualitative unless the specimen geometry is specified, it is often used to document the effect of temperature and to establish optimal working conditions. Examples of this type of torsion data obtained from hot torsion experiments are illustrated in Fig. 22 for several steels and nickel alloys.

As shown in Fig. 22, carbon and alloy steels such as 1040 and 4340 are very workable. These steels fall into group VII of Fig. 21. As the temperature is decreased, a two-

phase ferrite plus austenite structure ($\alpha + \gamma$) is formed from the single-phase austenite (γ), causing a sharp decrease in ductility. This is also shown for a high-oxygen iron alloy in Fig. 23.

Types 304 and 410 stainless steels are generally single-phase alloys at hot working temperatures and exhibit increasing ductility with temperature. However, at very high temperatures, delta ferrite is formed in type 410, causing a drop in workability. Therefore, type 304 exhibits a behavior like a group I material, and type 410 behaves like a group V alloy in Fig. 21.

The nickel-based superalloys in Fig. 22 show inferior ductility compared to types 304 and 410 steels due to the generally high alloy content of the former and the formation of hard, second-phase particles at low to mod-

Fig. 19 Flow curves determined via torsion testing and tension testing (following wire drawing)

(a) 0.06% carbon steel. (b) 0.85% carbon steel (in pearlitic condition). Note that the torsion data are expressed in terms of both the von Mises and the Tresca effective stress-strain definitions. Source: Ref 14

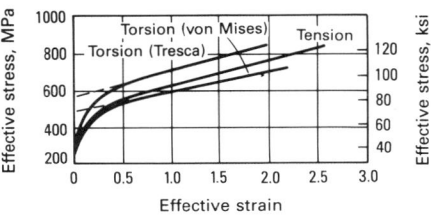

(a)

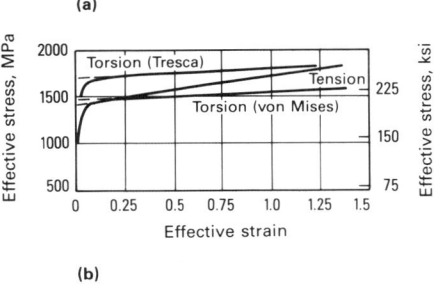

(b)

Fig. 20 Effect of the gage length-to-radius ratio on the effective strain to failure ($\bar{\epsilon}_f$) in torsion tests

Lines join results at similar strain rate and temperature. Source: Ref 3

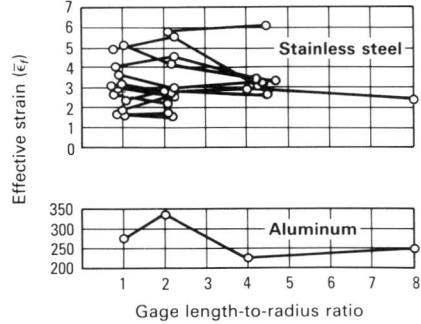

Gage length-to-radius ratio

Fig. 21 Typical workability behaviors exhibited by different alloy systems

T_M = melting temperature. Source: Ref 15

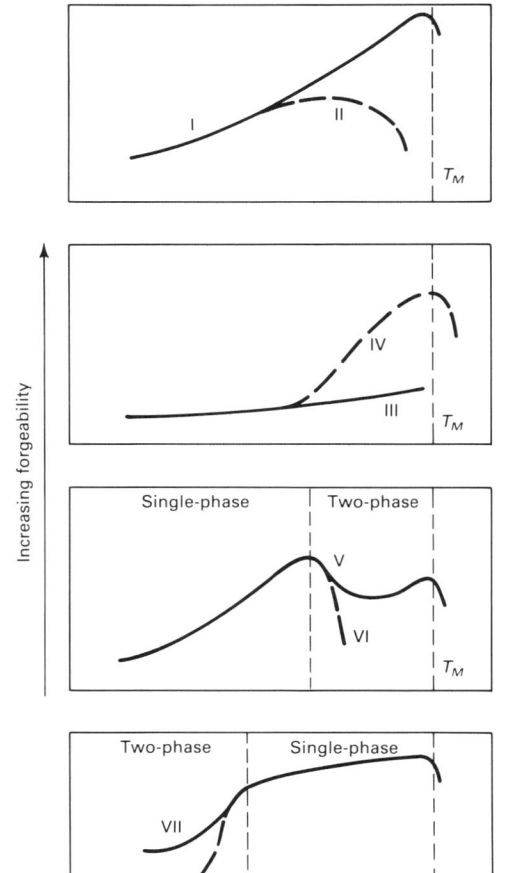

I. **Pure metals and single phase alloys**
Aluminum alloys
Tantalum alloys
Niobium alloys

II. **Pure metals and single-phase alloys exhibiting rapid grain growth**
Beryllium
Magnesium alloys
Tungsten alloys
All-beta titanium alloys

III. **Alloys containing elements that form insoluble compounds**
Resulfurized steel
Stainless steel containing selenium

IV. **Alloys containing elements that form soluble compounds**
Molybdenum alloys containing oxides
Stainless steel containing soluble carbides or nitrides

V. **Alloys forming ductile second phase on heating**
High-chromium stainless steels

VI. **Alloys forming low-melting second phase on heating**
Iron containing sulfur
Magnesium alloys containing zinc

VII. **Alloys forming ductile second phase on cooling**
Carbon and low-alloy steels
Alpha-beta and alpha-titanium alloys

VIII. **Alloys forming brittle second phase on cooling**
Superalloys
Precipitation-hardenable stainless steels.

erate hot working temperatures in these materials. These alloys exhibit group VIII behavior. Because the second phase is formed at relatively high temperatures, the superalloys must be worked close to the melting point. Hot torsion data in this range often show a decrease in ductility with temperature because of incipient melting problems.

Several other alloying effects in steels and aluminum alloys are illustrated in Fig. 24 to 26. In Fig. 24, the effect of carbon content on the torsional ductility of plain-carbon steels and electrolytic iron-carbon alloys at 0.5 T_M is shown. For both types of alloys, workability is controlled by ductile fracture and cavitation at precipitates and inclusions.

For the plain-carbon steels, the ductility increases initially with carbon content because of a decrease in the number of oxide inclusions, which serve as void nucleation sites. Similarly, with carbon contents above 0.1%, ductility decreases with carbon content as a result of the increasing volume percent of carbides at which voids may also nucleate. The variation of ductility with carbon content is similar for the electrolytic alloys, although the level is generally higher. This trend is the result of a decrease in the volume fraction of manganese sulfide and oxide inclusions.

Quantitative analysis of the effect of sulfides on the workability of steels is shown in Fig. 25, in which the ductility of several rimmed steels is presented as a function of sulfide content. As inclusion content increases, the fracture strain decreases and thus workability during an actual forming operation is affected deleteriously.

Fig. 22 Ductility determined in hot torsion tests

Source: Ref 16 and 17

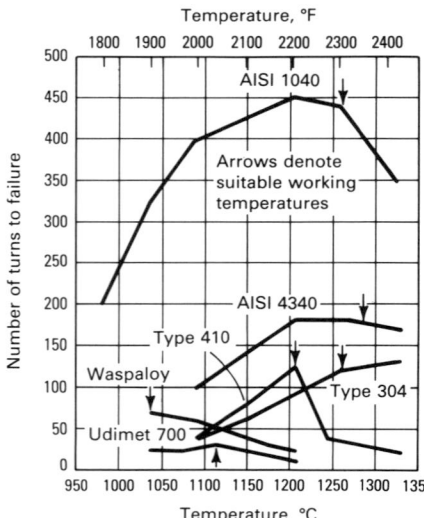

In Fig. 26, the torsional ductility of pure aluminum and a series of aluminum-magnesium alloys is given as a function of temperature. At the testing temperatures used, magnesium is totally in solution. The observed trend is caused by increased amounts of grain-boundary sliding in the magnesium alloys as well as by a decrease in the kinetics of the dynamic softening processes. The latter leads to an increase in flow stress, thereby reducing the ability of the alloy to accommodate stress concentrations at inclusions and grain boundaries.

Effect of Strain Rate on Torsional Ductility

The effect of strain rate on workability should be taken into account when attempting to apply laboratory torsion measurements to metalworking processes carried out at much higher speeds. In many cases, the general dependence of ductility on temperature will be similar at different strain rates. However, the ductility peak and thus optimal working temperature may be shifted.

An example of this effect is shown in Fig. 27 for the nickel-based superalloy U-700. For this material, the ductility maximum is shifted to higher temperatures with increases in strain rate. This behavior occurs because the alloy is single phase at the temperatures under consideration and the flow stress of single-phase alloys is a function of the Zener-Hollomon parameter ($Z = \dot{\epsilon} \exp(Q/RT)$). Thus, as the strain rate is increased, similar flow and fracture processes should occur at correspondingly higher temperatures. As shown in Fig. 27, however, the peak ductility is not as large at high strain rates. This can be ascribed to the increased amounts of deformation heating and the temperature increase at these rates, which raise the alloy to its incipient melting point.

Fig. 23 Variation of ductility with temperature for a high-oxygen Swedish iron tested in torsion at an effective strain rate of 0.5 s⁻¹

Source: Ref 18

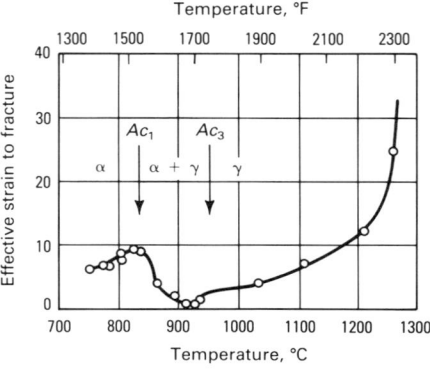

Fig. 24 Influence of carbon on the ductility of bcc iron tested in torsion at 0.03 s⁻¹ and 650 °C (1200 °F), or one half the absolute melting point

Source: Ref 19

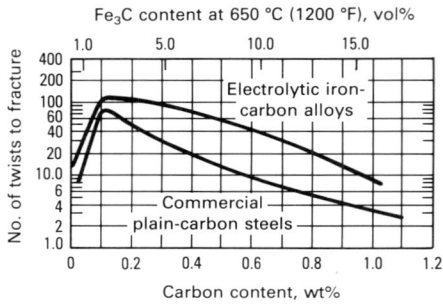

Fig. 25 Relation between manganese sulfide content and ductility in hot torsion tests on a variety of rimmed steels

Source: Ref 20

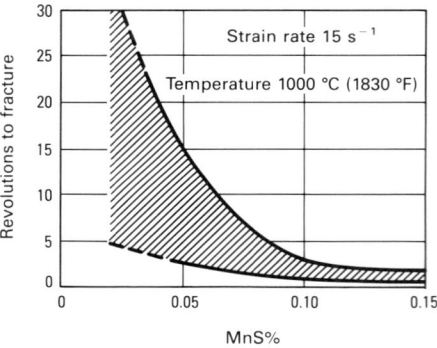

crease at these rates, which raise the alloy to its incipient melting point.

The influence of strain rate on torsional ductility is also illustrated in Fig. 28(a) for type 304L stainless steel, in which the ductilities from tests conducted at von Mises surface effective strain rates of 0.01 and 10.0 s⁻¹ are plotted. The values of Γ_{sf} from the lower rate tests are plotted versus the (constant) test temperature at which they were conducted. The high strain-rate ductilities are plotted versus the estimated temperature at fracture. These estimated temperatures are equal to the nominal test temperature plus the adiabatic ΔT from the plots shown in Fig. 28(b). Ductility data for high-strain rate tests performed at 20 °C (68 °F) and 200 °C (390 °F), in which flow localization occurred, are also shown. In these two instances, Γ_{sf} was obtained from scribe line measurements, and the ΔT versus Γ plot was extrapolated to obtain the fracture temperature.

Fig. 26 Effect of the amount of magnesium in solid solution in aluminum on torsional ductility

Tested at an effective strain rate of 2.3 s⁻¹. With 5% magnesium, ductility is severely reduced over the entire temperature range. Source: Ref 21

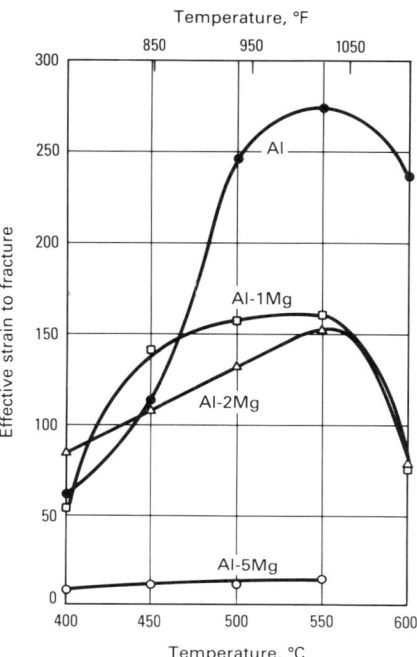

Fig. 27 Effect of test temperature on the torsional ductility of Udimet 700

Source: Ref 22

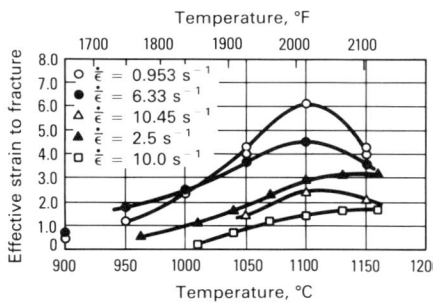

Fig. 28 (a) Fracture-strain data from type 304L austenitic stainless steel torsion tests and (b) estimated temperature changes during high rate tests

Low strain-rate ($\dot{\epsilon}$ = 0.01 s⁻¹) data are plotted versus the actual test temperature. High strain-rate (10.0 s⁻¹) data are plotted versus temperatures estimated from (b) a ΔT-Γ plot and the nominal test temperature, which is shown beside each data point. Source: Ref 9

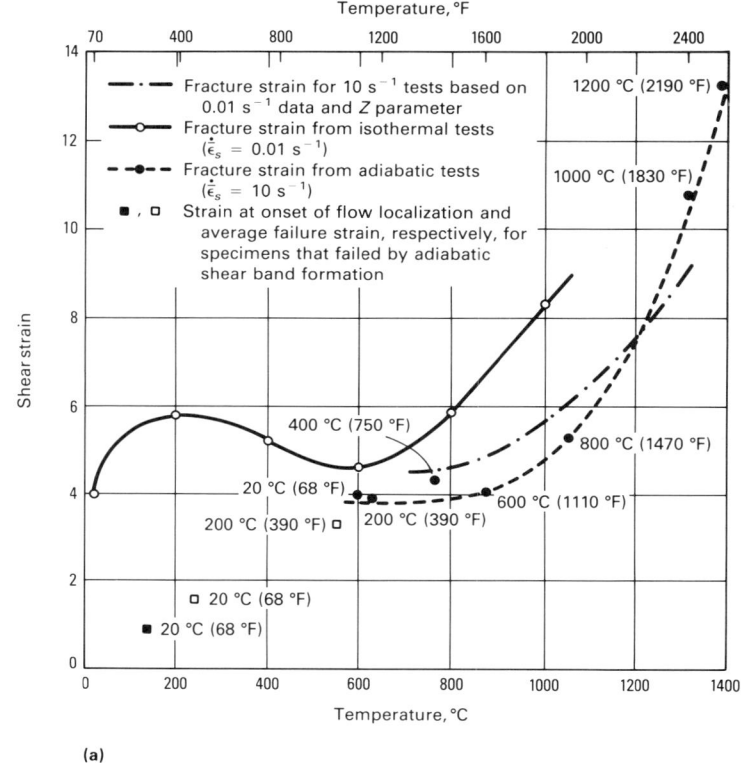

(a)

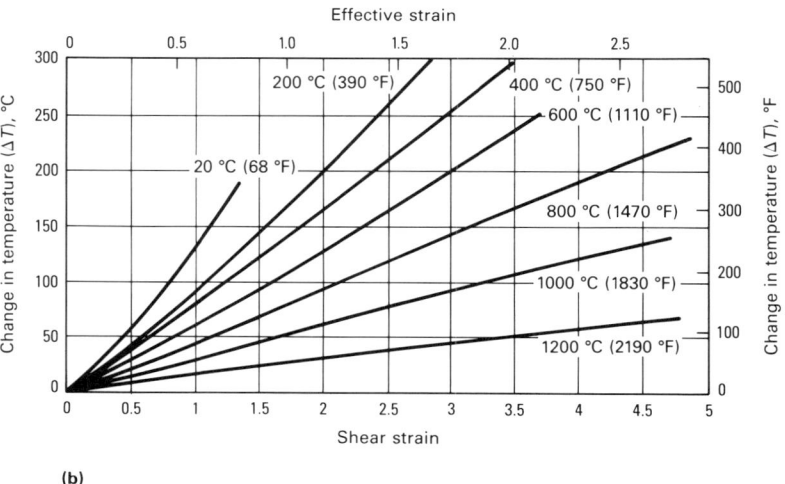

(b)

The low strain rate data for type 304L shown in Fig. 28(a) illustrate a classical dependence on temperature, i.e., a modest ductility at cold working temperatures, a ductility minimum at warm working temperatures, and large fracture strains at hot working temperatures. The major effect of the higher strain rate is a translation of the lower strain rate data to higher temperatures. For example, the ductility minimum appears to be shifted.

A more detailed interpretation of low and high strain rate data can be made through the Zener-Hollomon parameter, $Z = \dot{\epsilon} \exp (Q/RT) = \{\dot{\Gamma} \exp (Q/RT)\}/\sqrt{3}$. For this alloy, the steady-state flow stress at equal values of Z is nearly the same. Therefore, equal failure strains can be expected at fixed values of Z.

This hypothesis may be checked by "translating" the low strain-rate fracture locus $\Gamma_{f_a}(T_{f_a})$ to $\Gamma_{f_b}(T_{f_b})$ defined by $\Gamma_{f_b} = \Gamma_{f_a}$ and $\dot{\Gamma}_b \exp (Q/RT_{f_b}) = \dot{\Gamma}_a \exp(Q/RT_{f_a})$, where $\dot{\Gamma}_b/\dot{\Gamma}_a = 10/0.01 = 10^3$. For these type 304L stainless steel results, Q is taken to be the same as the activation energy from flow-stress data (98.2 kcal/mol, or 411 kJ/mol). The results of the calculation are shown in Fig. 28(a) and appear to be in fair agreement with the measured high strain rate fracture locus.

Correlation of Torsional Ductility Data

Because metalworking processes are not carried out under a state of pure torsional or shear loading, it is often necessary to convert the workability parameter measured in torsion to an index that is compatible with other deformation modes. Previously, attempts were made to correlate torsion, tension, and other types of data using the effective strain concept. Although a definite relationship exists between torsion and tension effective fracture strains (Fig. 29), such a method is incapable of explaining how fracture can be totally avoided in homogeneous compression by preventing barreling. For fracture to occur, deformation must also involve tensile stresses to promote ductile fracture, wedge cracking, or some other failure mechanism.

One of the most successful hypotheses incorporating the effects of deformation and tensile stress is that proposed by Cockcroft and Latham (Ref 23). They postulated that fracture occurs after the maximum tensile stress (σ_T) carries out a fixed amount of work through the applied effective strain, or:

$$\int_0^{\bar{\epsilon}_f} \sigma_T d\bar{\epsilon} = \text{a constant } (C)$$

The constant in the relation is a function of material, purity, and test temperature. In a torsion test, the maximum tensile stress occurs at 45° to the torsion axis and is equal to τ. As mentioned previously, $\bar{\epsilon}$ for torsion is equal to $\Gamma/\sqrt{3}$. If the material under consideration exhibits power-law strain hardening, $\tau = G\Gamma^n$, then C is equal to $(G\Gamma_{sf}^{n+1})/\sqrt{3}(n+1)$, where Γ_{sf} is the shear fracture strain. If the material does not harden in a power-law manner, the integration to evaluate C at fracture can be performed graphically.

In uniaxial tension, σ_T = the axial tensile stress, and $\bar{\epsilon}$ = the axial strain. Upon necking, σ_T is higher than the effective flow stress by a correction factor that can be estimated from the work of Bridgman (Ref 24). This factor is on the order of 0 to 33% for true axial strains of 0 to 2.5 in the necked region. For a power-hardening material, $\sigma = K\epsilon^n$. For tension, $C = (1 + \langle CF \rangle) \cdot \{K\epsilon_f^{n+1}/(n+1)\}$ where $\langle CF \rangle$ is the average correction factor, and ϵ_f is the axial fracture strain determined by reduction in area measurements and the constant volume assumption of plastic flow.

To apply the Cockcroft and Latham criterion to other deformation modes, the critical value of C must be known, and the maximum tensile stress and effective strain must be estimated. For arbitrary deformation paths, estimating these quantities can be very difficult and often requires sophisticated mathematical techniques, such as finite-element methods. Once σ_T and $\bar{\epsilon}$ are known, however, numerical techniques can be used to estimate the value of the maximum work integral, which can then be compared to the critical value of C at fracture determined by a simple workability test such as the torsion test.

The Cockcroft and Latham criterion has found its greatest success in correlating tensile and torsion fracture strains. The degree of success in these instances is judged by comparing the critical values of C established from flow and fracture results for the different tests. An example of such a comparison for type 304L stainless steel is given in Table 2. At 20 and 400 °C (68 and 750 °F), the material strain hardens in a power-law manner, enabling closed-form integration to obtain C values. At the two higher temperatures of 800 and 1000 °C (1470 and 1830 °F), a graphical integration procedure using flow curves (Fig. 16) was used. The integration procedure was straightforward for torsion, because deformation was uniform to fracture.

For tension, in which necking precludes obtaining flow stress data to fracture, the initial portion of the tensile curves was extrap-

Fig. 29 Relation between effective fracture strain from tension and from torsion tests for several alloys
Source: Ref 22

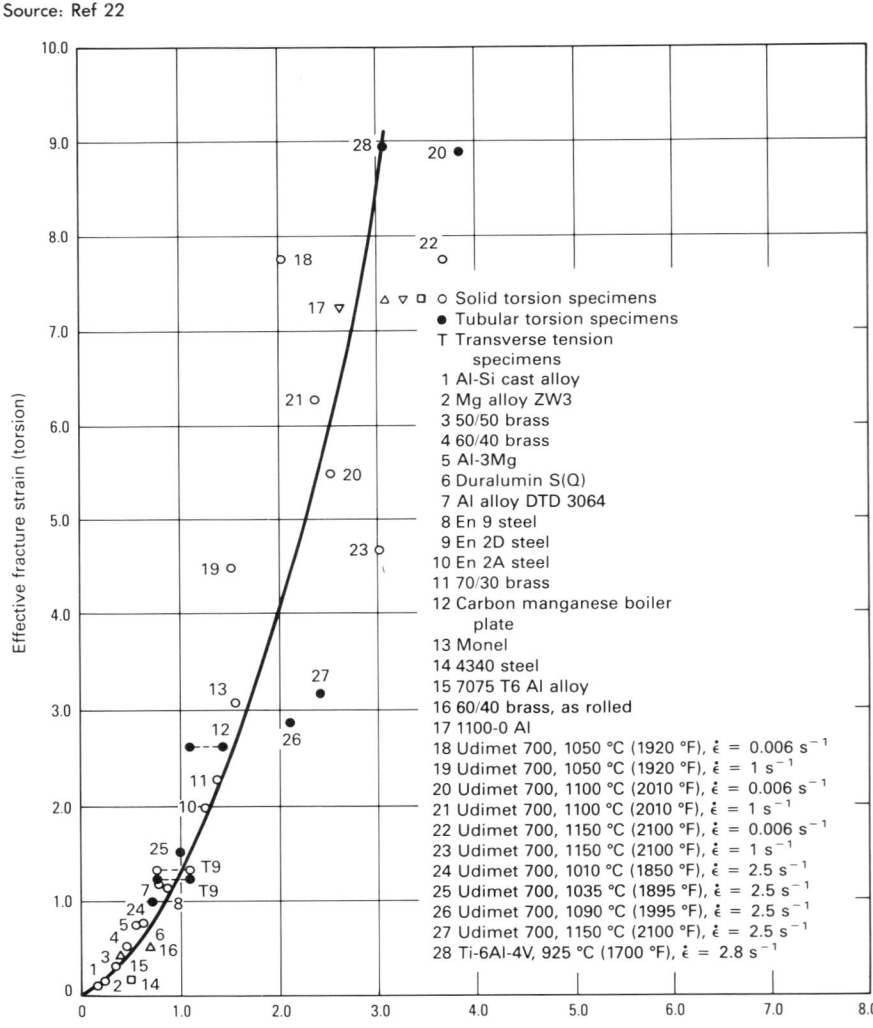

Effective fracture strain (torsion)

Effective fracture strain (tension)

△ ▽ □ ○ Solid torsion specimens
● Tubular torsion specimens
T Transverse tension specimens
1 Al-Si cast alloy
2 Mg alloy ZW3
3 50/50 brass
4 60/40 brass
5 Al-3Mg
6 Duralumin S(Q)
7 Al alloy DTD 3064
8 En 9 steel
9 En 2D steel
10 En 2A steel
11 70/30 brass
12 Carbon manganese boiler plate
13 Monel
14 4340 steel
15 7075 T6 Al alloy
16 60/40 brass, as rolled
17 1100-0 Al
18 Udimet 700, 1050 °C (1920 °F), $\dot{\epsilon} = 0.006$ s^{-1}
19 Udimet 700, 1050 °C (1920 °F), $\dot{\epsilon} = 1$ s^{-1}
20 Udimet 700, 1100 °C (2010 °F), $\dot{\epsilon} = 0.006$ s^{-1}
21 Udimet 700, 1100 °C (2010 °F), $\dot{\epsilon} = 1$ s^{-1}
22 Udimet 700, 1150 °C (2100 °F), $\dot{\epsilon} = 0.006$ s^{-1}
23 Udimet 700, 1150 °C (2100 °F), $\dot{\epsilon} = 1$ s^{-1}
24 Udimet 700, 1010 °C (1850 °F), $\dot{\epsilon} = 2.5$ s^{-1}
25 Udimet 700, 1035 °C (1895 °F), $\dot{\epsilon} = 2.5$ s^{-1}
26 Udimet 700, 1090 °C (1995 °F), $\dot{\epsilon} = 2.5$ s^{-1}
27 Udimet 700, 1150 °C (2100 °F), $\dot{\epsilon} = 2.5$ s^{-1}
28 Ti-6Al-4V, 925 °C (1700 °F), $\dot{\epsilon} = 2.8$ s^{-1}

olated to the required large strains using the compression and torsion curves as models. Results of the C parameter calculations for a given test temperature agree within 10% except at 800 °C (1470 °F), where the two values differ by approximately 15%. The magnitudes of these differences are considered typical and acceptable, particularly because a gross continuum model is being used to model a mechanism that greatly depends on the presence of microstructural features such as inclusions. The same continuum model is successful over a wide range of temperatures, within which the fracture mechanism may vary significantly.

Attempts to include microstructural features and fracture mechanisms have met with less success than the Cockcroft and Latham formulation. One such attempt is that of Hoffmanner (Ref 22), who proposed a model that incorporates a factor describing the dependence of fracture strain on the magnitude of the normal stress perpendicular to the mechanical texture. Such a concept may have application in the analysis of fracture in torsion, during which the mechanical texture rotates with respect to the stress axes.

Measuring Flow-Localization-Controlled Workability

Because torsion consists essentially of deformation in simple shear, torsion testing is frequently used to determine the material and process variables that contribute to the formation of shear bands during metalworking. Either solid or tubular specimens can be used for testing. When solid specimens are used, however, data analysis and interpretation are usually more complex than when tubular specimens are used. This difficulty arises because of the variation of strain, strain rate, and (at moderate to high strain rates) temperature across the specimen diameter.

Observation of Flow Localization in Torsion

The development of flow localization during torsion is usually detected by scribe line measurements, variations in microstructure along the gage length, or torque-twist behavior. The first two of these techniques are unequivocal. Examples of flow localization detected by these two methods for the shear bands developed in tubular type 304L stainless steel torsion specimens twisted at $\dot{\epsilon}_s = 10\ s^{-1}$ at room temperature are shown in Fig. 30 and 31.

The use of the torque-twist behavior, on the other hand, to measure flow localization may be misleading. Although a monotonically increasing torque-twist curve usually

Table 2 Cockcroft and Latham criterion for annealed type 304L fracture data

Test temperature °C	°F	Deformation mode	Strain at fracture (torsional shear or tensile), Γ_{sf} or $\bar{\epsilon}_f$	Strength coefficient, G or K MPa	ksi	Strain-hardening exponent, n	Average correction factor, $<CF>$	C parameter MPa	ksi
20	68	Torsion	4.01	517	75	0.25	· · ·	1360	197
		Tension	1.26	1172	170	0.352	0.075	1275	185
400	750	Torsion	5.17	310	45	0.137	· · ·	1020	148
		Tension	1.51	772	112	0.307	0.105	1120	162
800	1470	Torsion	5.89	· · ·	· · ·	· · ·	· · ·	445	64.5
		Tension	1.94	· · ·	· · ·	· · ·	0.12	525	76.1
1000	1830	Torsion	8.33	· · ·	· · ·	· · ·	· · ·	215	31.2
		Tension	2.75	· · ·	· · ·	· · ·	0.14	235	34.1

Note: $\dot{\epsilon} = 0.01\ s^{-1}$

indicates homogeneous or nearly homogeneous flow, a curve that exhibits a torque maximum (often referred to as torque instability) and torque softening does not necessarily signify flow localization, particularly when the test alloy possesses a significant strain-rate sensitivity index.

The difficulties associated with the interpretation of torque-twist behavior are illustrated in Fig. 32 and 33, in which experimental data are presented for hot torsion tests on thick-walled tubular specimens of Ti-6Al-2Sn-4Zr-2Mo-0.1Si (Ti-6242Si) twisted at rates sufficiently high to minimize the effects of axial heat transfer. In Fig. 32, the alloy

Fig. 30 Failed type 304L stainless steel torsion specimens

From $\dot{\epsilon} = 10\ s^{-1}$ tests showing evidence of flow localization. (a) 20 °C (68 °F), average $\Gamma_s = 1.3$. (b) 200 °C (390 °F), average $\Gamma_s = 2.9$. Magnification: 2×. Source: Ref 9

(a)

(b)

had a starting microstructure of equiaxed alpha ($\alpha + \beta$ microstructure); in Fig. 33, it had one of acicular, Widmanstätten alpha (β microstructure). The two torque-twist curves are similar in that they both exhibit torque maxima early in the deformation and subsequently decreasing torque with increasing twist.

For the β microstructure, however, the rate of softening is substantially greater. Scribe line measurements revealed that the $\alpha + \beta$ specimen deformation was uniform, whereas the β specimen had undergone flow localization. These results can be explained on the basis of a parameter that relates the degree of flow localization to the ratio of torque-softening rate to the strain rate-sensitivity index. For the $\alpha + \beta$ microstructure test, the detrimental effects of torque softening are counterbalanced by a rather high strain-rate sensitivity, and localization is prevented in much the same way that it is during the tensile testing of superplastic materials.

In contrast, for the β microstructure Ti-6242Si specimen, which also has a high strain-rate sensitivity, the larger amount of flow softening caused by the basic instability of the acicular microstructure is the main reason for the localization behavior. For either case, however, a localization analysis that deals with the problem as a process rather than as an event is required to fully understand such observations.

Flow Localization Analyses

The main variables that control flow localization during torsion are material properties (mechanical and thermal), twist rate, and the presence of material or specimen imperfections. Of the material properties, strain and strain-rate hardening rates, temperature sensitivity of the flow stress, and thermal conductivity are the most important. Twist rate affects heat conduction and heat transfer. Material or torsion specimen imperfections

Fig. 31 Micrographs from room-temperature ($\bar{\epsilon} = 10$ s^{-1}) torsion tests on type 304L stainless steel

(a) $\bar{\epsilon} < 1.16$ (outside shear band). (b) $\bar{\epsilon} > 1.16$ (inside shear band). Magnification: 310×. Source: Ref 9

(a)

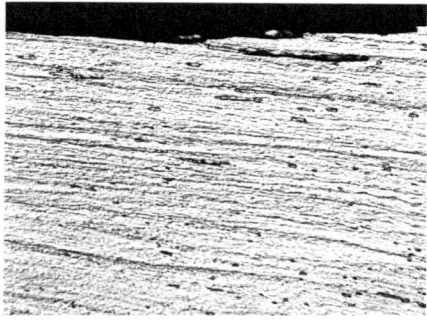

(b)

offer sites for the initiation of flow localization. Such sites are particularly important when torsion is conducted at very low or very high rates, at which the axial temperature field is relatively uniform. Under these circumstances, temperature gradients and heat transfer phenomena, which serve as a prime source of flow localization, are absent. The interpretation of torsion data under flow localization conditions is discussed below for two cases. In one, heat transfer effects are not considered, while in the other their influence is taken into account during the analysis.

Analysis of Flow Localization in the Absence of Heat Transfer

Flow Localization Parameter. When heat transfer is minimal, the study of flow localization in torsion aids understanding of the occurrence of shear bands in low strain rate isothermal metalworking operations, as well as those at very high strain rates. Examples of these processes include isothermal forging ($\dot{\bar{\epsilon}} \cong 10^{-3}$ s^{-1}) and high energy rate

Fig. 32 Comparison of experimental and theoretical torque-twist curves for $\alpha + \beta$ microstructure Ti-6242Si hot torsion specimens

Tested at $\dot{\bar{\epsilon}} \cong 0.9$ s^{-1}; $T = 913$ °C (1675 °F). Source: Ref 6

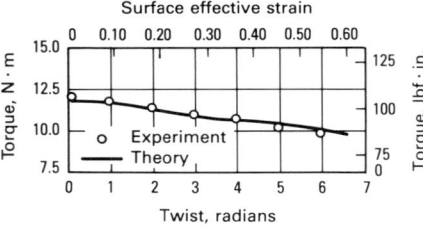

forming ($\dot{\bar{\epsilon}} \cong 10^3$ s^{-1}). In these cases, a defect is assumed to be the source of localization. For a torsion specimen, the defect may be a deficiency in radius (or wall thickness), strength coefficient, or another material property. The simplest imperfection to visualize is the geometric (or radius/wall thickness) defect, although the localization behavior is usually similar for similar sizes of geometric or material-property inhomogeneities.

When a torsion specimen with a geometric defect is twisted, variations in twist, twist rate, and temperature exist between the defect region and the nominally uniform region from the onset of deformation. Generally, all of these quantities will be higher in the defect area. These differences are generated to maintain torque equilibrium, so that the defect and uniform regions transmit an identical torque. They change as the deformation proceeds, depending on the material properties. Thus, localization, or the lack of it, must be viewed as a process.

In contrast to the localization process and analysis, shear banding is often interpreted in terms of an instability condition. Such treatments are concerned only with material properties and with determining the twist, under a nominally homogeneous deformation field, at which the torque maximum ($dM = 0$) occurs. The results for Ti-6242Si discussed previously demonstrated that a torque maximum is not a sufficient condition for localization. Rapid localization cannot occur until after the instability condition is satisfied. However, if the material shows only a small amount of flow softening or has a large positive rate sensitivity, localization and the occurrence of shear banding in torsion will be minimal. Thus, the quantitative degree of localization cannot be predicted by the instability (torque maximum) analysis alone.

As its name implies, flow localization analysis concerns variations of twist, twist rate, and temperature between a defect re-

Fig. 33 Comparison of experimental and theoretical torque-twist curves for β microstructure Ti-6242Si hot torsion specimens

Tested at $\dot{\bar{\epsilon}} \cong 0.9$ s^{-1}; $T = 816$ °C (1500 °F). Source: Ref 6

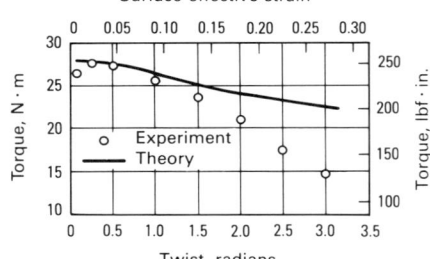

gion and a nominally uniform region. During the entire strain concentration process, the torque must be uniform along the axis, or $\delta M = 0$, where δ denotes a variation. In the present context, it is used to define variations of the various field quantities between the two regions of the torsion specimen. If $M = M(\theta, \dot{\theta}, T, R)$, where R is the specimen radius, the equilibrium condition $\delta M = 0$ may be used to obtain:

$$\frac{\delta \ln \dot{\theta}}{\delta \theta} = \frac{-\left\{ G + \left(\frac{\partial \ln M}{\partial \ln R} \right)\Big|_{\theta, \dot{\theta}, T} \frac{\delta \ln R}{\delta \theta} \right\}}{\left(\frac{\partial \ln M}{\partial \ln \dot{\theta}} \right)\Big|_{\theta, T, R}} \quad \text{(Eq 1)}$$

where G denotes the normalized torque hardening (or softening) rate at fixed $\dot{\theta}$:

$$G \equiv \left(\frac{d \ln M}{d\theta} \right)\Big|_{\dot{\theta}} = \frac{\left(\frac{\partial \ln M}{\partial \theta} \right)\Big|_{\dot{\theta}, T, R} d\theta + \left(\frac{\partial \ln M}{\partial T} \right)\Big|_{\theta, \dot{\theta}, R} dT}{d\theta} \quad \text{(Eq 2)}$$

The derivation of Eq 1 and a detailed discussion of flow localization analyses are presented in Ref 25. Once localization becomes noticeable, $\delta\theta \gg \delta \ln R$, and the second term in the braces in Eq 1 becomes negligible. Defining $(\delta \ln M / \delta \ln \dot{\theta}|_{\theta, T, R}$ as m^*, Eq 1 reduces to an expression for the torsional flow localization parameter A:

$$A \equiv \frac{\delta \ln \dot{\theta}}{\delta \theta} = \frac{-G}{m^*} \quad \text{(Eq 3)}$$

The parameter A describes the variations of twist and twist rate that can be sustained under equilibrium conditions, as a function of the material properties through their influence on G and m^*. A material whose constant strain rate flow curve shows a large amount of flow softening will exhibit large negative G values. Similarly, a high strain-rate sensitivity, $m = \partial \ln\tau/\partial \ln\dot{\Gamma}$, will result in large values of m^*.

A somewhat modified form of Eq 3 is useful in obtaining a quantitative estimate of the level of flow localization that can be expected during torsion testing. This form is obtained by substituting

$$\dot{\theta} = \frac{\sqrt{3}L\dot{\bar{\epsilon}}}{r_s} \quad \text{and} \quad \theta = \frac{\sqrt{3}\,L\bar{\epsilon}}{r_s}$$

where r_s and L are the outer radius and gage length of the specimen, respectively, into this equation to yield:

$$\left(\frac{\delta \ln\dot{\bar{\epsilon}}}{\delta \epsilon}\right) = -\frac{\sqrt{3}\,L}{r_s}\left(\frac{G}{m^*}\right) \tag{Eq 4}$$

The left side of Eq 4 is known as the alpha (α) parameter. Much experimental and theoretical work has demonstrated that when α is equal to 5 or more, noticeable flow localization should be expected, either in torsion, compression, or some other deformation mode. When $G = 0$, which corresponds to the torque instability condition, $dM = 0$, α is equal to zero, and minimal localization is expected at this point. Beyond the torque instability, sufficiently large amounts of torque softening and/or low-rate sensitivity are required to obtain values of α on the order of 5. Only when m^* ($\cong m$) is very small (as at cold working temperatures) will the onset of noticeable flow localization follow soon after the occurrence of the torque maximum.

Application of the Flow Localization Parameter. An examination of Eq 4 indicates a dependence of the flow localization parameter on specimen geometry, G, and m^*. As stated in Eq 2, G is the normalized torque softening rate under constant θ conditions. This rate varies with θ, leading to the conclusion that α, and thus the rate of localization, can vary during torsion testing.

Difficulty arises, however, when attempting to use the measured M-θ curve from a torsion test in which localization has occurred, because the strain rate will have varied along the gage length even though the overall twist rate and average strain rate may have been held constant. Thus, it is often necessary to estimate G from the measured torque-twist behavior, or to use some other mechanical test in which localization does not occur.

The values of G and m^* are readily estimated for thin-walled tube specimens. In this instance, m^* is identical to the strain-rate sensitivity index, m. G is obtained from the torque expression for this geometry:

$$M = 2\pi r_s^2 t\tau \tag{Eq 5}$$

Resulting in:

$$G = \frac{d\ln M}{d\theta} = \frac{d\ln\tau}{d\theta} = \frac{r_s}{L}\frac{d\ln\tau}{d\Gamma} \tag{Eq 6}$$

Substituting this into Eq 4 and assuming $m^* = m$ yields:

$$\alpha = \left(\frac{\delta \ln\dot{\bar{\epsilon}}}{\delta \bar{\epsilon}}\right) = -\sqrt{3}\,\frac{\left.\left(\dfrac{d\ln\tau}{d\Gamma}\right)\right|_{\dot{\Gamma}}}{m} \tag{Eq 7}$$

or

$$\alpha = -\frac{\left.\dfrac{d\ln\bar{\sigma}}{d\bar{\epsilon}}\right|_{\dot{\bar{\epsilon}}}}{m} \tag{Eq 8}$$

in terms of von Mises effective stress and strain.

Equation 8 indicates that the alpha parameter for a thin-walled torsion test specimen can be calculated directly from the normalized flow softening rate (at constant strain rate) and the strain-rate sensitivity index. Specimen geometry has no effect on the results. Furthermore, Eq 8 suggests that other mechanical tests may be useful in estimating values of $(d\ln\bar{\sigma}/d\bar{\epsilon})|_{\dot{\bar{\epsilon}}}$. In particular, tests in which flow localization can be avoided are preferable.

The mechanical test in which localization occurs the most slowly is the uniaxial compression test due to the stabilizing effect of increases in cross-sectional area. To obtain values of the alpha parameter for thin-walled torsion tests, compression data from tests conducted at constant strain rates equivalent to those in torsion can be conducted to estimate the torque-softening rate and strain-rate sensitivity parameters.

When a thick-walled tube or solid bar torsion test is to be analyzed, compression data are also useful in estimating the magnitude of the α parameter. Assuming that localization has been avoided in the compression tests, flow stresses as a function of strain, strain rate, and temperature can be converted to the equivalent shear stress τ as a function of Γ, $\dot{\Gamma}$, and T. For a given torsional $\dot{\theta}$, the shear stress can then be used via numerical integration to obtain the torque ($M = \int 2\pi r^2\tau dr$) required for the specific torsion specimen geometry. Note that this is the torque that would be required under uniform deformation conditions.

Calculations of M versus θ in such a case, which were based on the compression data,

were performed for the thick-walled torsion specimens of the Ti-6242Si alloy discussed above. The results, shown in Fig. 34, demonstrate that the absolute magnitude of $G = (d\ln M/d\theta)|_{\dot{\theta}}$ is certainly much greater for the β microstructure than the $\alpha + \beta$ microstructure. This behavior reflects the larger amount of flow softening that occurs in uniaxial compression at a variety of deformation rates.

An estimated constant strain rate M-θ curve is also plotted with each of the experimental results in Fig. 32 and 33. For the $\alpha + \beta$ microstructure, the predicted curve agrees quite well with the measured curve, confirming that the assumption of uniform deformation was valid. The maximum value of α over the twist range investigated was only 3.0. In contrast, the predicted M-θ curve for the β microstructure in Fig. 33 lies substantially above the experimental results. This divergence, or the fact that deformation proceeded at torques below those required for uniform flow, is an indication of flow localization. In addition, the α parameter in this case was found to be as high as 14.3— much greater than the critical α of 5.

Application of the Flow Localization Parameter to Other Deformation Modes. The rate of flow localization in torsion is related to the α parameter, as defined by Eq 4. Basically, this expression establishes that flow localization occurs only after a critical value of the ratio of the torque soft-

Fig. 34 Torque-twist curves for Ti-6242Si predicted from a numerical deformation/heat transfer simulation and measured compression flow stress data

Results are for testing at 913 °C (1675 °F) and various average effective strain rates. Average effective strain rate = 0.6 × surface effective strain rate for the tubular specimen design used. Source: Ref 6

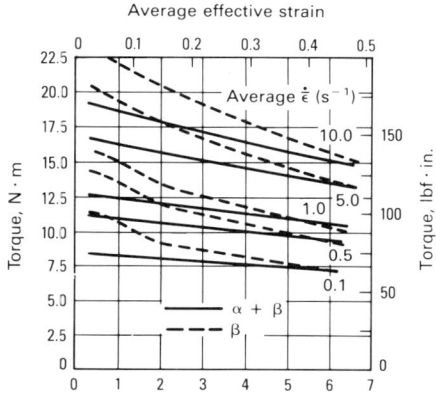

ening rate to the rate sensitivity of the torque is achieved.

Related metalforming research has demonstrated that similar α parameters can be used to gage the rate of localization in isothermal uniaxial compression and plane-strain forging operations. In compression, α is defined as

$$\alpha_c = -\frac{\gamma' - 1}{m} \qquad \text{(Eq 9)}$$

and is used to predict when unstable, localized bulging may occur. In isothermal plane-strain forging processes, an α defined by

$$\alpha_p = -\frac{\gamma'}{m} \qquad \text{(Eq 10)}$$

can be used to predict the formation of shear bands. In both cases, m is the strain-rate sensitivity and γ' is the normalized strain hardening (or softening) rate at fixed strain rate:

$$\gamma' = \left(\frac{1}{\bar{\sigma}}\frac{d\bar{\sigma}}{d\bar{\epsilon}}\right)\bigg|_{\dot{\bar{\epsilon}}} =$$

$$\frac{\left(\frac{\partial \ln\bar{\sigma}}{\partial \bar{\epsilon}}\right)\bigg|_{\dot{\bar{\epsilon}},T} d\bar{\epsilon} + \left(\frac{\partial \ln\bar{\sigma}}{\partial T}\right)\bigg|_{\bar{\epsilon},\dot{\bar{\epsilon}}} dT}{d\bar{\epsilon}}$$

$$\text{(Eq 11)}$$

and is readily determined from constant strain rate compression tests.

As in torsion, criteria based on $\alpha_c = 5$ or $\alpha_p = 5$ are useful for predicting the occurrence of flow localization. Examples are given in Fig. 35 and 36 for the isothermal hot compression and isothermal plane-strain sidepressing, respectively, of Ti-6242Si. In each case, the $\alpha + \beta$ microstructure specimens, which developed low values of the α parameter (α_c or α_p), deformed uniformly. However, the β microstructure specimens in both cases had sufficiently high degrees of flow softening to promote $\alpha \geq 5$ and thus developed regions of nonuniform flow.

Effect of Temperature and Strain Rate on the Alpha Parameter. The α parameter used to gage flow localization generally shows a sharp dependence on temperature and strain rate. These variations can be predicted, at least qualitatively, from the magnitudes of the terms that comprise it. For simplicity, only α_p, the flow localization parameter for the occurrence of shear bands in plane-strain forging operations, is discussed here.

As defined by Eq 10, α_p depends on γ' and m. In turn, γ' is specified by Eq 11. At cold working temperatures, γ' is usually positive due to strain hardening. At high strain rates, however, the effect of thermal softening in Eq 11 may outweigh the strain hardening effect (the first parenthetical term on the right

side of the equation), resulting in an overall negative γ', or flow softening. Furthermore, at cold working temperatures, m is usually small (usually between 0 and 0.02). Therefore, small amounts of flow softening coupled with low m may be sufficient to generate $\alpha \geq 5$ and hence cause noticeable flow localization. Shear bands generated at cold working temperatures at high strain rates are often called adiabatic shear bands.

At hot working temperatures, flow softening (negative γ') is prevalent not only as a result of thermal softening effects at high strain rates, but also because of microstructural softening at all deformation speeds. These microstructural effects, quantified by the first parenthetical term on the right side of Eq 11, are due to softening processes such

Fig. 35 Specimens of Ti-6242Si from isothermal hot compression tests

Tested at 913 °C (1675 °F); $\dot{\bar{\epsilon}} \simeq 2\ \text{s}^{-1}$. Starting microstructures were (a) $\alpha + \beta$ (equiaxed alpha) and (b) β (Widmanstätten alpha). Magnification: 3×. Source: Ref 26

(a)

(b)

as dynamic recovery, dynamic recrystallization, and the breakup of Widmanstätten microstructures to form equiaxed microstructures.

In addition, at hot working temperatures, m values tend to be substantially larger (usually between 0.1 and 0.5) than at cold working temperatures. Thus, substantial amounts of flow softening, much more than at lower temperatures, are required to produce marked flow localizations.

Analysis of Flow Localization in the Presence of Heat Transfer

At intermediate strain rates, heat generation and heat transfer effects must be taken into account to describe flow localization during torsion testing. Axial temperature (and hence flow stress) gradients are established as a result of uneven deformation heating caused by the presence of defects, as well as by the conduction of heat into the colder shoulders of the torsion specimen. In this case, the analysis involves the development of a torque equilibrium equation as well as a relation to describe the heat transfer aspects (Ref 25, 27, 28). In its most basic form, the heat transfer relation considers axial heat conduction and deformation heat generation. In most applications, radiation, convection, and radial heat transfer effects can be neglected.

The equilibrium and heat transfer equations are solved incrementally, subject to the imposed boundary conditions (typically, a constant overall twist rate). Calculations have been carried out to determine the effect of heat transfer on flow localization during the torsion of type 304 torsion specimens at room temperature (Ref 27) and Ti-6242Si specimens at hot working temperatures (Ref 28).

For the type 304 specimen, the material coefficients required for the analysis were determined from low-speed tests at which localization does not occur. Using a specimen with a premachined 8% defect in radius at the center of the gage section, localization occurred during tests at $\dot{\Gamma} \sim 0.05\ \text{s}^{-1}$. The localization rate (measured using scribe lines) showed good agreement with the localization simulation based on material parameters, when the additional effect of geometry changes occurring during testing because the specimen ends were not clamped was taken into account (Fig. 37).

In the Ti-6242Si testing, torsion samples without geometric defects, but along which there was an initial axial temperature gradient, were tested. Even if an initial axial temperature gradient had not existed, such a gradient would have eventually been set up due to deformation heat generation and heat

Fig. 36 Transverse metallographic sections of Ti-6242Si bars

(a) α + β (equiaxed alpha). (b) β (Widmanstätten alpha) microstructures. Isothermally sidepressed at 913 °C (1675 °F); $\dot{\epsilon} \simeq 2$ s⁻¹. Magnification: 5×. Source: Ref 6

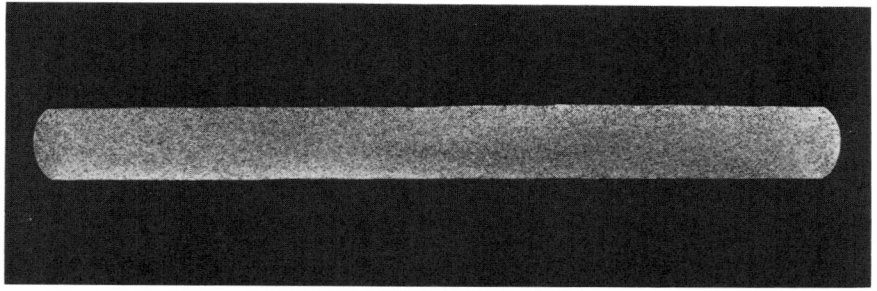

(a)

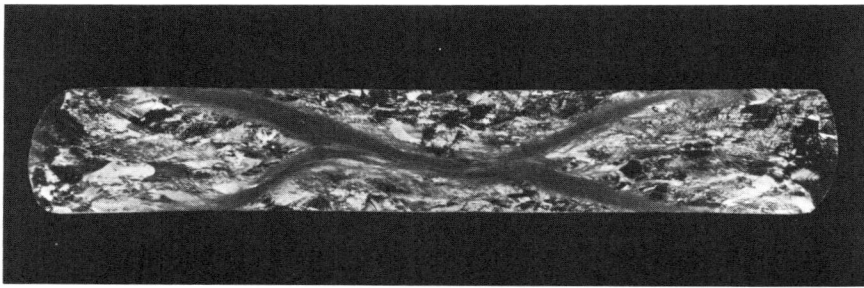

(b)

Fig. 37 Comparison of experimentally observed localization kinetics (data points) with simulation results (solid lines) in type 304 stainless steel specimens

Specimens had premachined radius defects at the center of the gage section and were tested in torsion at room temperature. The simulations were run with two rate sensitivities: $m^* = 0.01$ and $m^* = 0.005$, whose values bounded those measured in torsion tests on specimens without geometric defects. Average surface shear strain rate was approximately 0.05 s⁻¹ in both experiments and simulations. Source: Ref 27

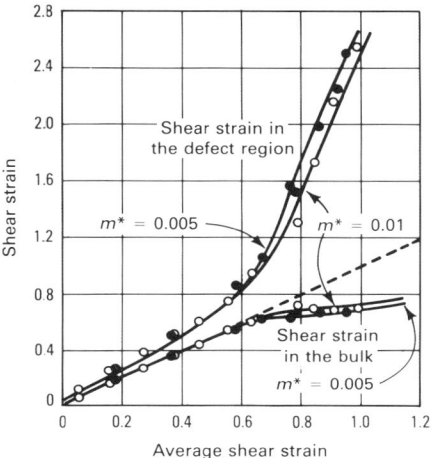

conduction. In this case, localization followed the expected trends.

Microstructure Development During Deformation Processing

In previous sections of this article, the application of torsion testing to determine workability limited by excessive loads, fracture-controlled failure, and flow-localization-controlled failure has been discussed. Often, however, the ability to form a piece of metal into a particular shape with the available equipment comprises only the basic considerations of workability. It is usually important to control the microstructure that is developed, which ultimately determines the properties of the finished product and its suitability for further deformation processing, heat treating, final machining, or service. The ability to accurately control test variables during torsion makes the technique attractive to establish the processing parameters that are required to produce the desired microstructures.

At cold and warm working temperatures, the microstructure of most metals changes only slightly. The changes are largely a distortion of the metal grains (observable at optical magnifications) and a process of dynamic recovery, which is only detectable

with the aid of transmission electron microscopy. The latter process consists of the regrouping of individual dislocations to form cells and subgrains.

However, recovery, as well as many other microstructural transformations, occurs much more readily under hot working conditions. The hot torsion test is thus frequently used to detect the effects of deformation and deformation rate on these changes. Significant microstructural phenomena that can be studied with the hot torsion test include dynamic recovery and dynamic recrystallization, which are the main mechanisms controlling microstructural development for single-phase metals and alloys, and dynamic spheroidization and precipitation in two-phase materials.

Dynamic Recovery and Recrystallization in Single-Phase Materials

In single-phase materials, the primary microstructural features are dislocations and grain boundaries. The torsion test is used to determine how these characteristics are affected by deformation temperature and deformation rate. Dynamic recovery at hot working temperatures leads to a reduction in the number of dislocations at a given strain, without noticeably affecting the gross deformation of the grains. By contrast, dynamic

recrystallization is characterized by the motion of grain boundaries and annihilation of large numbers of dislocations in a single event, thereby producing new strain-free grains.

Metals with high-stacking fault energies (e.g., pure aluminum and α-iron), in which climb of dislocations is easy because they are not dissociated, tend to soften primarily by dynamic recovery processes. On the other hand, low-stacking fault energy materials (e.g., γ-iron, copper, and nickel), in which dislocations are dissociated and thus able to climb and cross-slip only with difficulty, tend to store large reservoirs of strain energy at hot working temperatures. Dynamic recrystallization is initiated from these large stores of dislocations.

The flow curve from a torsion test on a metal in which dynamic recovery predominates at hot working temperatures has an appearance typified by iron tested in the α-regime at 700 °C (1290 °F) (Fig. 38). Initially, strain hardening occurs, during which the rate of dislocation multiplication exceeds the rate of recovery. This is followed by a plateau in flow stress and steady-state flow. At this point, well-defined, equiaxed subgrains form whose walls (subboundaries) consist of fairly regular arrays of dislocations. These subgrains remain equiaxed dur-

Fig. 38 Stress-strain curves for Armco iron

Strain-rate dependence of the flow stress at 700 °C (1290 °F), or 0.54 × the absolute melting point, is evident. Data are from compression tests; torsion results exhibit similar behavior. Source: Ref 29

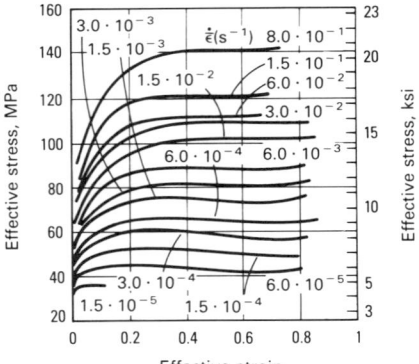

Fig. 39 Effect of twist reversal on specimen appearance and structure of super-purity aluminum after deformation at 400 °C (750 °F) and a strain rate of 2 s^{-1}

(a) 5 + 5 revolutions (2×). (b) 5 − 5 revolutions (2×). (c) 5 + 5 revolutions (25×). (d) 5 − 5 revolutions (25×). Source: Ref 30

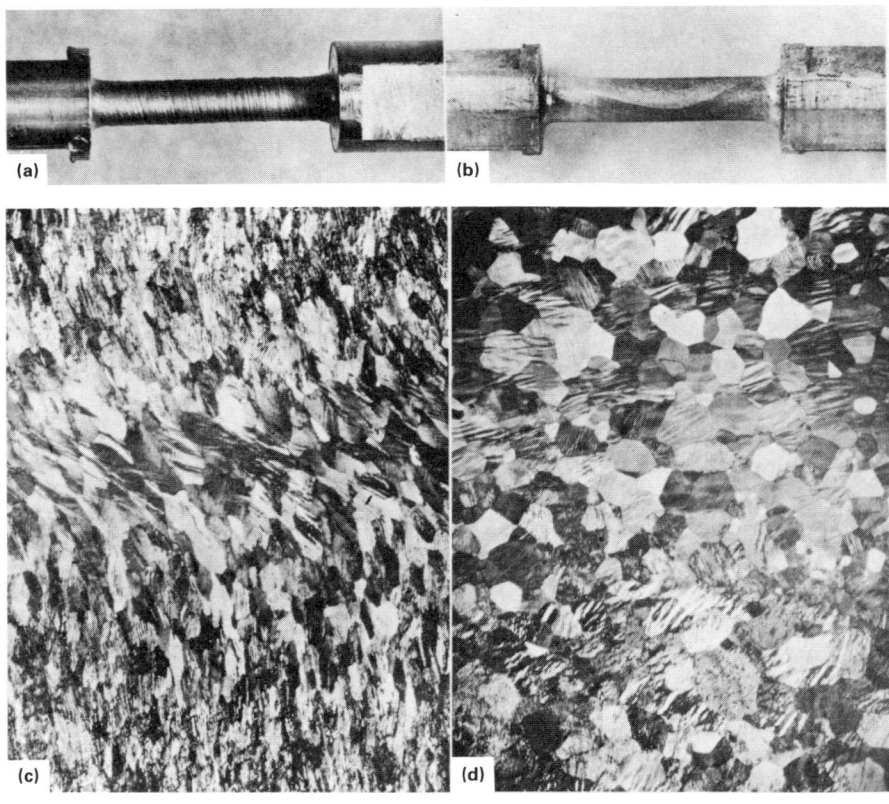

ing the remainder of the deformation process, even though the actual grain boundaries do not, as shown by the optical micrographs of Fig. 39.

Figures 39(a) and (b) are macrographs of two different hot torsion specimens twisted a total of ten revolutions. The sample in Fig. 39(a) was given ten twists in the forward direction, while that in Fig. 39(b) was twisted forward for five turns and then given five turns of reverse twist. The grains in both specimens undergo a large degree of deformation, which increases from the center to the surface because of the strain variation inherent in the torsion of solid round bars.

The large degree of deformation can be readily seen from the micrograph of the first test specimen, Fig. 39(c), in which the highly distorted nature of the twisted grains is evident at the bottom of the micrograph. The micrograph of the second test sample of identical geometry that was twisted five revolutions in one direction and then five in the reverse direction is illustrated in Fig. 39(d). It is apparent that the original equiaxed grain structure has been restored by this deformation schedule, even though these grains have received the same equivalent strain as those of Fig. 39(c).

The torsional flow curves for materials that recrystallize dynamically at hot working temperatures are often quite different from those typical of metals in which dynamic recovery predominates. These curves may be used to determine the onset of dynamic recrystallization as a function of strain rate and temperature.

Stress-strain curves for nickel and a carbon steel (in the single-phase austenite range) are shown in Fig. 40 and 41. The gross behavior consists of an initial strain hardening stage,

during which the dislocation density increases rapidly. At some point, a flow stress maximum is achieved, after which marked flow softening occurs. The strain at the peak flow stress decreases with increasing temperature and decreasing strain rate.

Such flow softening can be identified with the onset of dynamic recrystallization. At low temperatures and/or high strain rates, the flow curve eventually achieves a steady state, indicative of continuous dynamic recrystallization. At low strain rates and/or high temperatures (as under conditions approaching those typical of creep deformation), the flow curve following the peak flow stress often exhibits cyclic hardening and softening. This phenomenon has been attributed to periodic cycles of work hardening and recrystallization.

The current interpretation of torsional flow curves that indicate dynamic recrystallization is somewhat more detailed than described above. The dynamically recrystallized grain size formed under equilibrium conditions is a function of the deformation temperature and strain rate. For example, the stable grain size

Fig. 40 Stress-strain curves derived from hot torsion data for nickel at an effective strain rate of 0.016 s^{-1}

The dependence of flow behavior on test temperature in a material which undergoes dynamic recrystallization is shown. Source: Ref 31

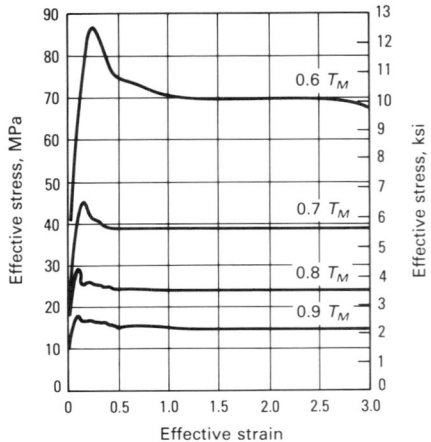

Fig. 41 Flow curves for 0.25% low-carbon steel in the austenitic state tested in torsion at 1100 °C (2010 °F)

The strong influence of strain rate and a behavior indicative of dynamic recrystallization are shown. Source: Ref 32

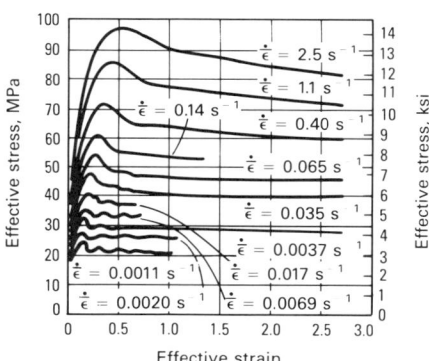

Fig. 42 Micrographs of type 304L stainless steel hot torsion specimens tested under various strain rate/temperature conditions

(a) 0.01 s^{-1}, 800 °C (1470 °F) ($\bar{\epsilon}$ = 1.99). (b) 10 s^{-1}, 800 °C (1470 °F) ($\bar{\epsilon}$ = 3.81). (c) 0.01 s^{-1}, 1000 °C (1830 °F) ($\bar{\epsilon}$ = 3.73). (d) 0.01 s^{-1}, 1200 °C (2190 °F) ($\bar{\epsilon}$ = 4.64). Magnification: 350×. Source: Ref 9

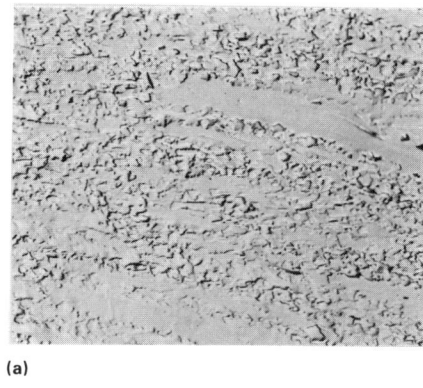

(a)

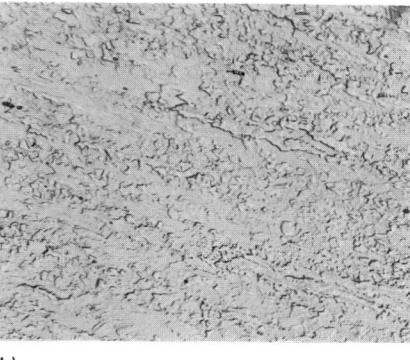

(b)

Fig. 43 Dynamically recrystallized grain sizes of copper and nickel as a function of the Zener-Hollomon parameter (Z)

Source: Ref 33

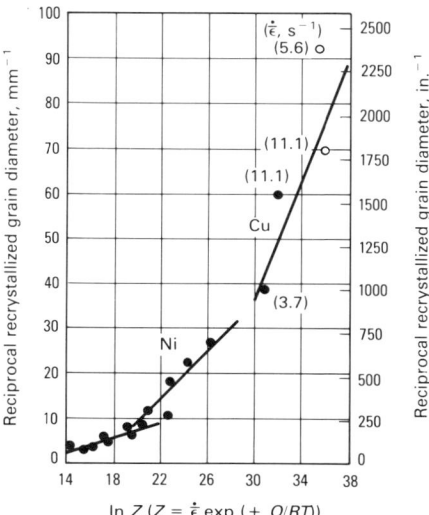

(c)

(d)

the average recrystallized grain size and Z is demonstrated for copper and nickel in Fig. 43.

A similar dependence between Z and the stable, dynamically recrystallized grain, D_s, for a 0.16% carbon steel tested in the austenite regime is shown in Fig. 44. Also illustrated are the critical Z or Z_c values as a function of starting grain size, D_o; the values of Z_c represent the conditions associated with the transition between single-peak and multiple-peak (periodic) flow curves. This curve is labeled Z_c-D_o. In addition, a curve representing Z versus $2D_s$ is shown.

The similarity between the Z_c-D_o and Z-$2D_s$ curves establish that the transition in flow behavior occurs when the equilibrium or stable, dynamically recrystallized size is equal to one half of the initial grain size. Thus, grain refinement leading to at least a halving of the starting grain size produces a single-peak flow curve, whereas grain coarsening (or refinement of less than one half) results in cyclic flow curves (Ref 34).

Fig. 44 Dependence of the critical parameter Z_c on initial austenite grain size D_o in a 0.16% carbon steel (open data points)

The solid line fitted to the filled points is the Z-D_s relationship (D_s = stable grain size achieved during dynamic recrystallization). Note that the Z-$2D_s$ (broken line) and Z_c-D_o relations are nearly coincident. Source: Ref 34

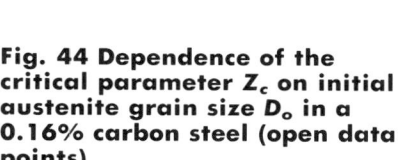

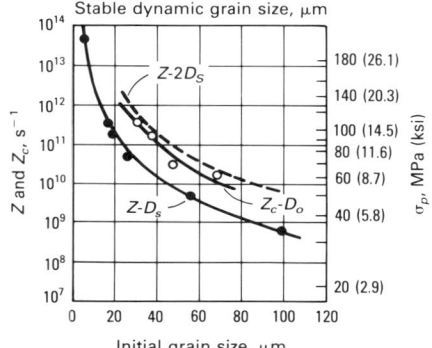

in type 304L stainless steel twisted at several temperatures and strain rates is shown in Fig. 42, which illustrates that grain size increases with increasing temperature and decreasing strain rate. As with the flow stress and fracture behavior at hot working temperatures, the dependence on these two variables is best expressed in terms of the product $\dot{\epsilon}$ exp(Q/RT), or the Zener-Hollomon parameter. The strong correlation between the reciprocal of

These observations are useful in rationalizing the dependence of the shape of the torsional flow curve on temperature and strain rate. As mentioned previously, cyclic curves are most frequently observed at low strain rates and high temperatures, or the regime in which the stable, recrystallized grain size is large. Because all but coarse grain materials undergo grain coarsening during torsion under these conditions, the flow curves are cyclic. These coarsening cy-

cles continue to be observed until the recrystallized grain size attains the equilibrium value.

Thus, the torsion test can be very useful in establishing the occurrence of recovery or recrystallization during the hot working of single-phase materials. When dynamic recrystallization occurs, the shape of the flow curve may be used to determine the temperatures and strain rates at which refinement of the grain size, a characteristic important with

Fig. 46 The mean free path (λ) between spheroidite particles in hot worked eutectoid steels as a function of the Zener-Hollomon parameter

The right ordinate is scaled to show the strain rate at 500 °C (930 °F) which would produce the indicated spacing after large plastic deformation. Data are from torsion, compression, and rolling experiments. Source: Ref 35

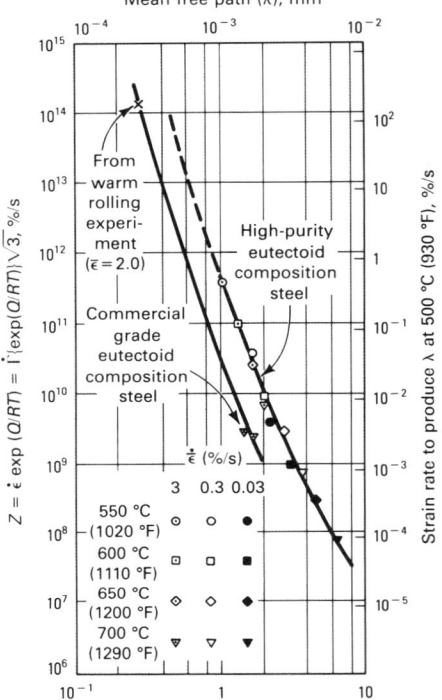

Fig. 45 Effect of test temperature on the torsional flow curve of a high-purity 0.8% carbon pearlitic Fe-C alloy

Numbers in parentheses refer to the number of twists to fracture. The flow softening at the three lower temperatures can be attributed to pearlite spheroidization. Source: Ref 19

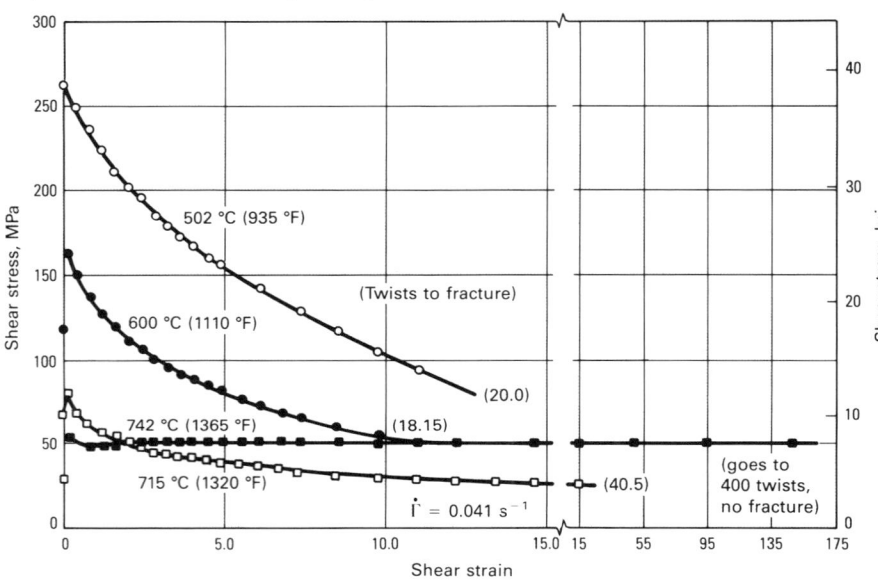

Fig. 47 Starting microstructure of Udimet 700 billet material used in torsion and extrusion studies

Magnification: 465×. Source: Ref 22

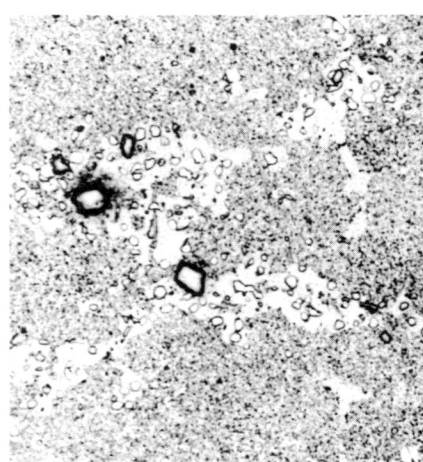

Fig. 48 Comparison of (a) torsion and (b) extrusion microstructures in Udimet 700

Deformed under nearly identical conditions of $\dot{\bar{\epsilon}} \simeq 7\ s^{-1}$, $T = 1060$ °C (1940 °F), and $\bar{\epsilon} \simeq 1.85$. Magnification 465×. Source: Ref 22

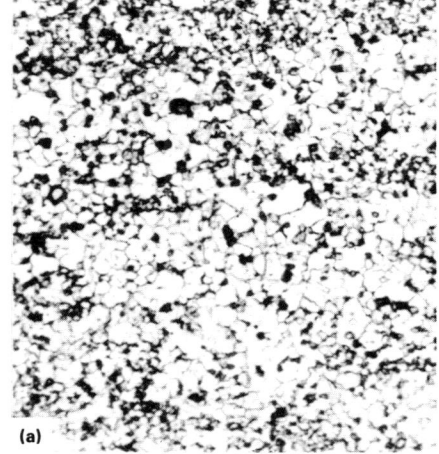
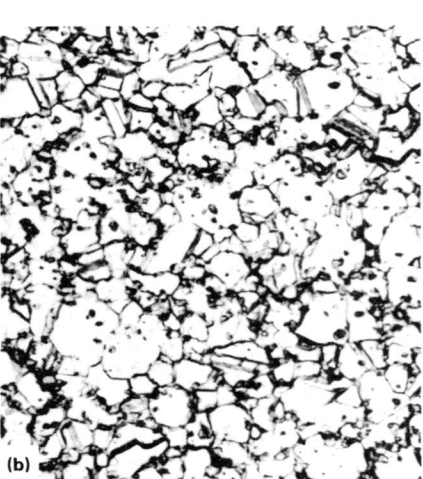

Fig. 49 Comparison of (a) torsion and (b) extrusion microstructures in Udimet 700

Deformed under nearly identical conditions of $\dot{\bar{\epsilon}} \simeq 4\ s^{-1}$, $T \simeq 1145\ °C$ (2090 °F), and $\bar{\epsilon} \simeq 2.15$. Magnification: 465×. Source: Ref 22

(a)

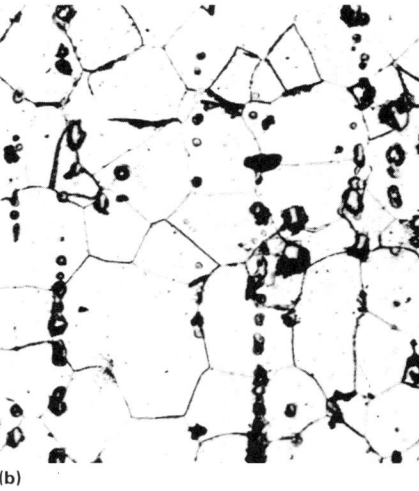

(b)

Fig. 50 Effect of continuous heating or cooling on the steady-state flow stress of vacuum-melted iron

Deformed in torsion at an effective strain rate of 1.5×10^{-3}. Source: Ref 36

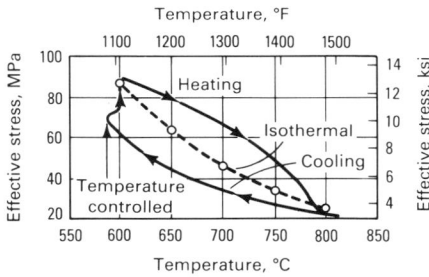

Fig. 51 Effect of increasing or decreasing strain rate on the flow stress of copper deformed in torsion at 750 °C (1380 °F)

Source: Ref 37

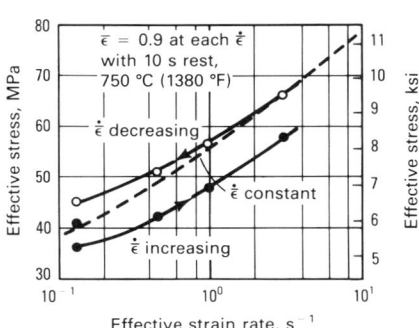

respect to service properties, occurs. Moreover, it can be used to establish equivalent combinations of temperature and strain rate at which a given grain size is produced.

Development of Microstructure in Alloys With More Than One Phase

The torsion test has also been used to determine the effects of deformation parameters on the microstructures developed in two-phase and multiphase alloys. Torsion testing has the ability to impose large strains at rates up to and including those used in commercial metalworking operations.

For example, torsion testing in two-phase alloys can be used to study flow softening and the break-up of unstable microstructures. These effects are especially strong in materials with lamellar or Widmanstätten phases, such as carbon steels and alpha-beta titanium alloys. In carbon steels, torsion may be used to determine the flow response and microstructural changes that occur in pearlitic steels subject to deformation below the lower critical temperature. For steel of a eutectoid composition, large amounts of flow softening are measured in torsion, except when tested above the critical temperature at 742 °C (1368 °F) (Fig. 45). This softening is associated with the break-up and spheroidization of the cementite (Fe_3C) lamellae.

As the temperature is lowered (or the strain rate is increased), the torsion results suggest that the rate of spheroidization increases. At the highest subcritical temperature in Fig. 45 (715 °C, or 1320 °F), the amount of softening is rather low, suggesting that microstructural changes are not as drastic as at lower temperatures, consisting primarily of the development of coarse cementite.

This conclusion is supported by microstructural examination of the hot torsion

Fig. 52 Typical stress-strain curves for tests involving instantaneous changes in strain rate for an austenitic stainless steel and a ferritic low-alloy steel

Note that the rate-change strain-rate sensitivity is either lower (stainless steel) or higher (low-alloy steel) than that based on continuous (constant strain-rate) torsion tests. Source: Ref 38

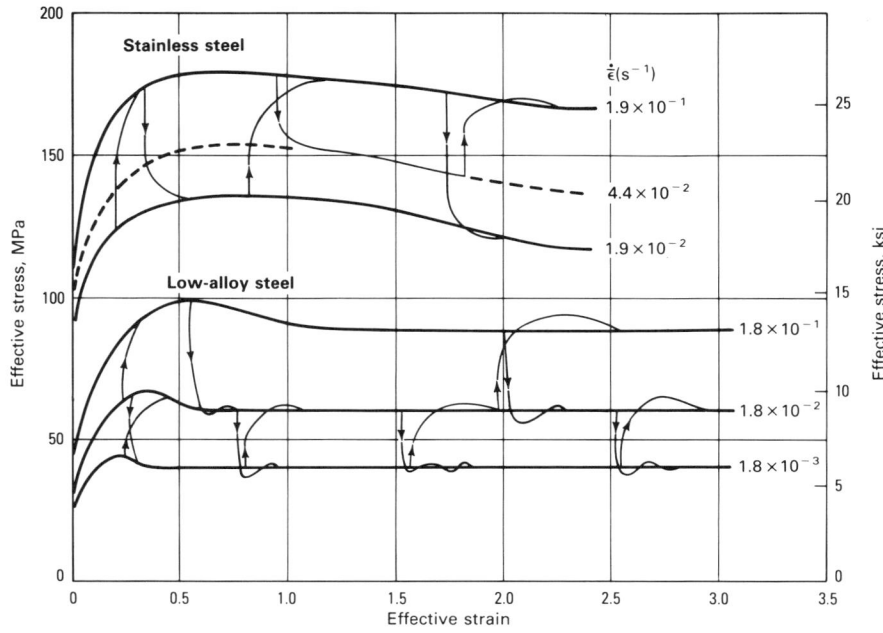

specimens, which indicates a definite relationship among strain rate, temperature, and spacing of the spheroidal particles after a fixed amount of deformation (Fig. 46). Again, the strain rate and temperature dependence is expressed through the Zener-Hollomon parameter, $\dot{\epsilon} \exp(Q/RT)$. When high strain rates or low temperatures are used, a fine, closely-spaced dispersion is produced. On the other hand, low strain rates and high temperatures, which enhance diffusion processes, result in a large interparticle spacing, indicative of a coarse dispersion of cementite.

In a similar two-phase system, torsion can be used to study the modification of the Widmanstätten alpha microstructure of alpha-beta titanium alloys subjected to deformation below the beta transus temperature. As with the eutectoid steel, tests of this type have shown that coarsening prevails at temperatures near the transus, whereas spheroidization results from deformation at high strain rates and lower subtransus temperatures.

The torsion test can also be used to study microstructural modification in multiphase alloys and to establish processing conditions based on this information. It is particularly useful for nickel-based superalloys. These materials are expensive and have a very limited working temperature range due to the presence of second phases (gamma prime, intermetallic carbides), which only go into solution at relatively high temperatures (if at all), and because of the problems of grain growth or incipient melting when deformation is performed near the melting point.

Figures 47 through 49 illustrate a typical application of the torsion test to study the development of microstructure in the nickel-based alloy Udimet 700. In the as-received condition (Fig. 47), the material had fine gamma prime, carbides, and borides dispersed through the gamma matrix, whose grain structure was not resolvable. After torsion at approximately 1060 °C (1940 °F), the structure consists of a well-defined fine gamma grain structure with carbides and borides situated at the grain boundaries (Fig. 48a).

This microstructure is very similar to that observed in Udimet 700 material extruded under almost identical conditions of strain, strain rate, and temperature. In addition, the microstructures developed in torsion and extrusion at the higher temperature of 1150 °C (2100 °F) (Fig. 49) closely resemble one another. These results and similar ones on other alloys establish the torsion test as a valuable means of determining the microstructures that can be developed during actual metalforming processes.

Documentation of Processing History

Torsion testing generally determines the patterns of deformation and failure resistance and microstructure development for a given alloy. In actual metalforming processes, however, the thermomechanical history of the workpiece is rarely so simple. Frequently, it will be preheated in a furnace and transferred to the processing equipment (forging press, rolling mill, etc). During this dwell period, it will have cooled a certain amount. Also, the workpiece will chill when it is in contact with the tooling during conventional metalworking processes. This is in contrast to normal torsion tests in which the test specimen is heated to temperature, soaked for a period of time, and then twisted. Furthermore, during deformation processing, the strain rate is rarely constant, unlike that in the conventional torsion experiment.

With proper controls, the effects of temperature and strain-rate history on workability (flow stress levels, fracture behavior, and microstructure development) can be assessed using the torsion test, provided means exist to replicate the thermal and/or deformation rate history. The thermal effect of greatest importance is cooling during processing. Cooling histories are best controlled during testing through forced air or argon convection around the specimen at rates that must be determined experimentally to obtain the desired results. Strain-rate histories are more readily controlled by interfacing closed-loop test systems with computers or function generators that provide the proper control signals representing the rotation-time dependence needed.

Figures 50 and 51 illustrate the effects of temperature and strain-rate history during torsion on the flow stress of α-iron and copper, respectively. The α-iron was tested at a constant strain rate, but the specimen temperature was increased or decreased at a continuous rate of 50 °C/min (90 °F/min). The flow stresses from such tests are compared to iso-

Fig. 53 Deformation-temperature-time schedule and resulting flow behavior of super-purity aluminum deformed in torsion at an effective strain rate of 2.3 s⁻¹

Source: Ref 30

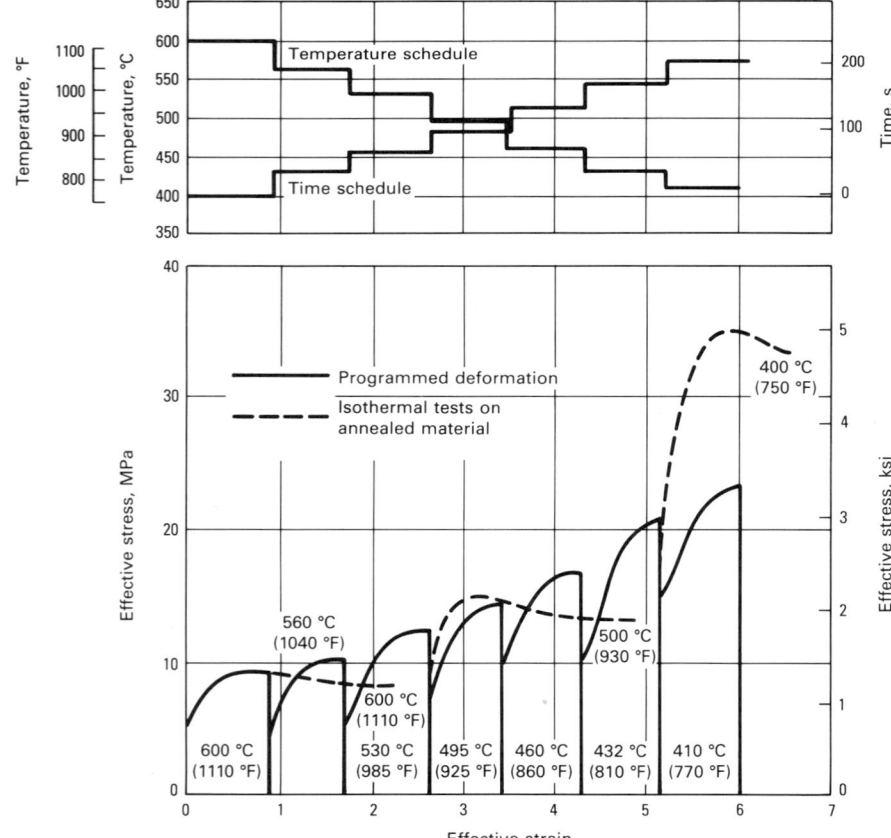

thermal stresses. If the specimen is heated during testing, the flow stress is higher than the isothermal stress. The reverse is true if the specimen is cooled.

During either heating or cooling, the dislocation substructure does not change instantaneously. During the heating experiments, a lower temperature, less highly recovered substructure is retained, giving rise to higher flow stresses than the isothermal tests. Similarly, during cooling, a softer, more highly recovered substructure leads to lower flow stresses than those observed in isothermal experiments.

The effect of strain-rate history on the flow stress of copper is shown in Fig. 51. Under constant strain-rate conditions, harder substructures are produced at higher strain rates. However, if the strain rate is increased or decreased during torsion testing, the inertia of the acquired substructure prevents changes in flow stress as high as those observed in a series of constant strain-rate tests. Thus, strain-rate sensitivities measured in rate change tests are often lower than those based on constant strain-rate or so-called continuous flow curves in materials such as copper, aluminum, and austenitic stainless steels. In materials that exhibit dynamic strain aging (e.g., carbon steels at cold working temperatures), the relationship between the two rate sensitivity parameters may be reversed depending on the strain-rate regime and the kinetics of strain aging (Fig. 52).

The use of the torsion test to study the effects of history during actual deformation processes is illustrated in Fig. 53 through 57. Figure 53 shows the type of flow stress behavior that might be expected for high-purity aluminum during processes such as rolling, in which the temperature decreases continuously. As the temperature decreases, the flow stress increases, but not as much as would be expected based on isothermal measurements. This phenomenon is particularly evident once the temperature has dropped below 450 °C (840 °F) and is a result of the retention of a soft high-temperature substructure.

Simulation of the rolling of high-strength low-alloy steels through the torsion test (Fig. 54 to 57) provides insight into how equipment requirements and final microstructures can be determined by this technique. In this instance, torsion tests were conducted in a servo-hydraulic test machine at a fixed strain rate. However, temperature was controlled to decrease continuously at a rate almost equivalent to that measured during actual production (Fig. 54). Under these conditions, the torsion flow stresses increased rapidly (Fig. 55).

Using these data, the roll-separating force and rolling torque were estimated for the rolling schedule under study using standard for-

Fig. 54 Deformation-temperature-time sequence imposed during torsion testing of microalloyed steels

The temperature-time profile followed in a production plate mill (— — —) is compared with that experienced by the sample in the torsion machine (———). Source: Ref 39

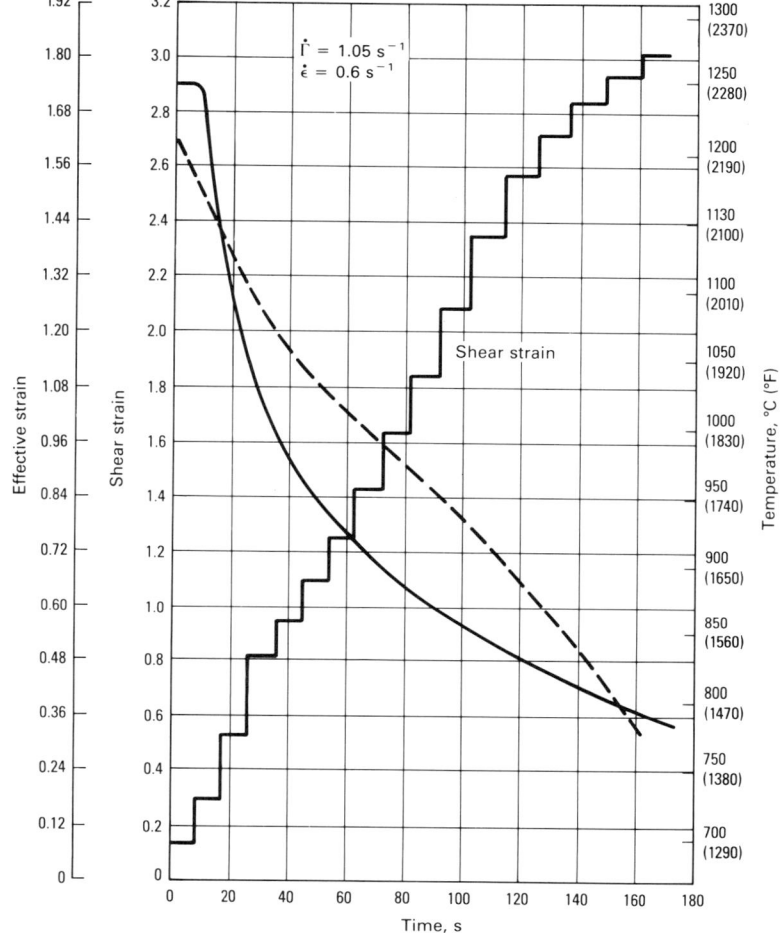

Fig. 55 Flow behavior for a Nb-V microalloyed steel deformed in 17 passes in a torsion machine

The specimen temperatures are represented by the upper bold line. Source: Ref 39

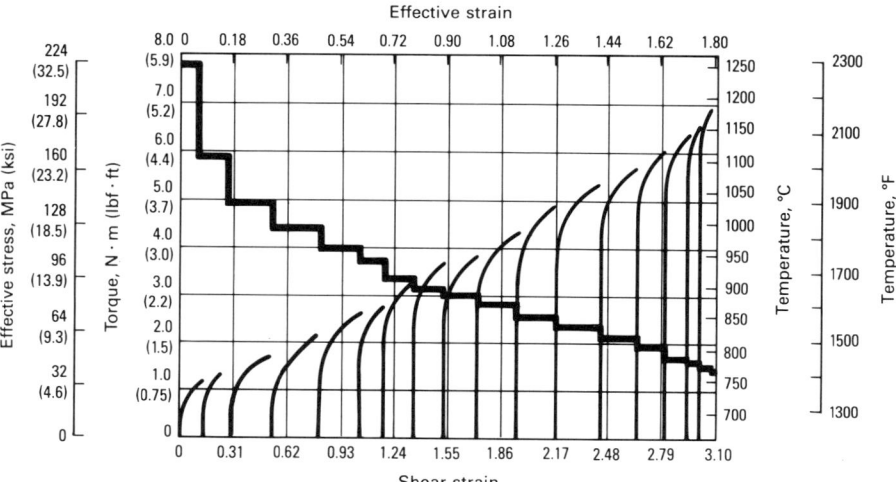

mulas. These were compared to actual measurements made at the individual stands of the rolling facility. Agreement was fairly good (Fig. 56), indicating the usefulness of the torsion test in assessing the influence of processing parameters (rolling speed, temperatures, dwell time between stands, etc.) on equipment requirements.

In addition, the final microstructure from the torsion simulation was almost identical to that from the production run (Fig. 57). Other experiments using similar computer-controlled torsion setups have provided valuable information on the required loads and micro-

structures developed during rolling of a wide range of alloy steels.

The torsion test can also be used to simulate the effects of die chilling on workability in multiphase alloys. In particular, the effects of solutioning of second phases (during preheating) and subsequent reprecipitation during working because of chilling are readily simulated by the test. This is accomplished by preheating the torsion specimen to a high temperature and testing it on cooling. The torsional ductility obtained in such tests frequently may be used to determine a potential workability problem.

Fig. 56 Comparison of the predicted and measured (a) roll separating forces and (b) roll torques associated with rolling of a Nb-V microalloyed steel

The pass number is shown beside each data point.
Source: Ref 39

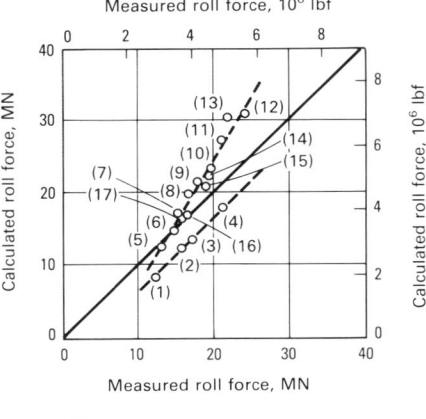

(a)

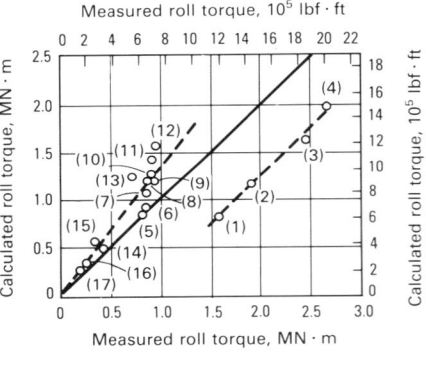

(b)

Fig. 57 Ferrite structure obtained in a Nb-V microalloyed steel
(a) After 17 passes in the torsion machine. (b) After 17 passes in a production plate mill. Source: Ref 39

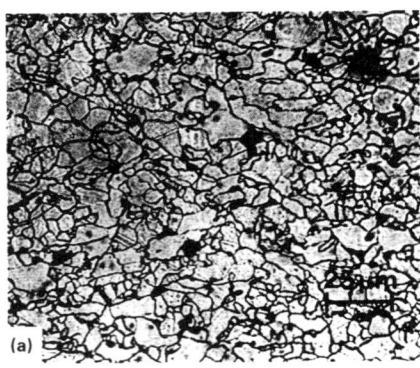

(a)

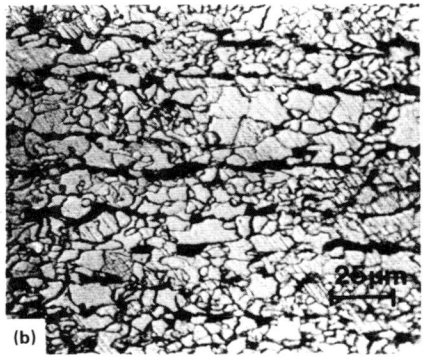

(b)

Appendix: Axial Effects and Alternative Data Analysis Methods in Torsion Testing

Axial Stresses and Texture Development During Torsion Testing

The axial stresses developed during fixed-end torsion testing are related directly to the deformation textures produced by straining (Ref 40, 41), as are the length changes observed during free-end testing. The sign and magnitude of these effects depend on the orientation of the mean value of all the burgers (i.e., slip) vectors involved in a given increment of twist (see Fig. 58). The applied shear stress causes slip (shear) to occur on a variety of crystallographic slip (shear) planes lying either in the transverse plane of the specimen or on planes that are inclined with respect to the transverse plane. Such shear

displacements are produced as multiples of the elemental unit of shear, the magnitude and direction of which are given by the burgers vector. In most metals, the latter corresponds to a single atom diameter.

When the average burgers vector lies in the transverse (shear) plane of the specimen, there is no length change or axial stress induced. By contrast, when it is inclined either away from or toward the fixed end of the specimen, lengthening or shortening of the sample, respectively, occurs (free-end conditions). Alternatively, under fixed-end conditions, compressive or tensile stresses are developed, respectively.

The inclination of the mean burgers vector depends in turn on the texture that is developed by twisting and, in particular, on the rotation of the ideal orientation away from strict coincidence with the axial (shear plane normal) and tangential (shear direction) directions of the sample. For example, when copper is deformed at 100 to 300 °C (212 to 570 °F), the $\{1\bar{1}1\}\langle110\rangle/\{\bar{1}1\bar{1}\}\langle\bar{1}\bar{1}0\rangle$ component predominates at strains of 1 to 2. This set of ideal orientations indicates that two particular (111) glide planes lie in the transverse plane of the specimen (in different grains) and that these are oriented so that certain $\langle110\rangle$ glide directions are aligned along

the shear direction. Actually, however, this set of orientations is rotated slightly about the radial direction of the sample in the sense opposite to that of the shear, so that the (111) planes do not lie *exactly* in the transverse plane, and the $\langle110\rangle$ directions are not aligned *exactly* with the shear direction. This inclination is responsible for the compressive stresses (fixed ends) and lengthening (free ends) observed in this strain range (Ref 40, 41).

At deformations of 2 to 4, the above component is replaced by the $\{001\}\langle110\rangle$ orientation, which is rotated in the same sense as the shear. Such an inclination of this component also produces a compressive force (or lengthening). Finally, at strains greater than about 5, a steady state of flow is attained, in which the $\{1\bar{1}1\}\langle110\rangle/\{\bar{1}1\bar{1}\}\langle\bar{1}\bar{1}0\rangle$ set again predominates, inclined in this case in the same sense as the shear. This inclination is in the opposite sense to the one observed at low strains

Fig. 58 Effect of the inclination of the mean burgers vector on specimen length change during simple shear deformation

(a) When the mean burgers vector **b** is inclined away from the fixed end, specimen lengthening occurs. τ_{res} is resolved shear stress; τ_{app} is applied shear stress; Δz is change in specimen length. (b) When the mean burgers vector **b** is inclined toward the fixed end, specimen shortening is observed.

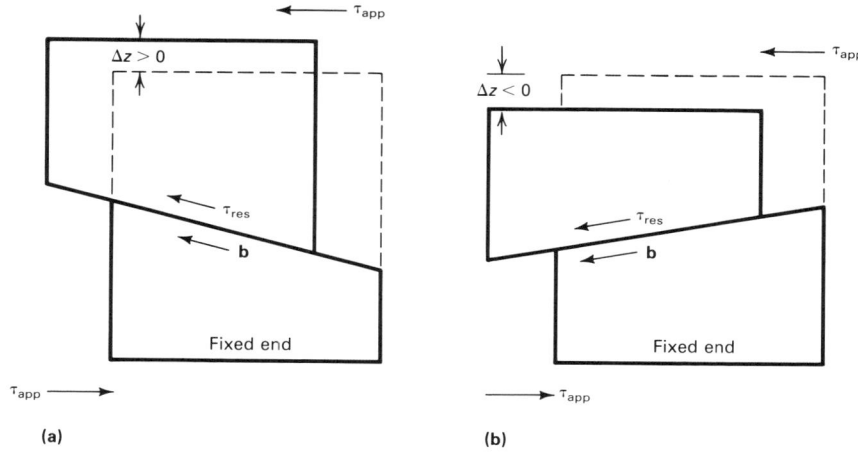

(a)   (b)

Fig. 59 Dependence of the axial force on temperature in copper

$\dot{\bar{\epsilon}} = 5 \times 10^{-3}$ s^{-1} from 20 to 200 °C (70 to 390 °F); $\dot{\bar{\epsilon}} = 5 \times 10^{-2}$ s^{-1} from 300 to 500 °C (570 to 930 °F). Source: Ref 40

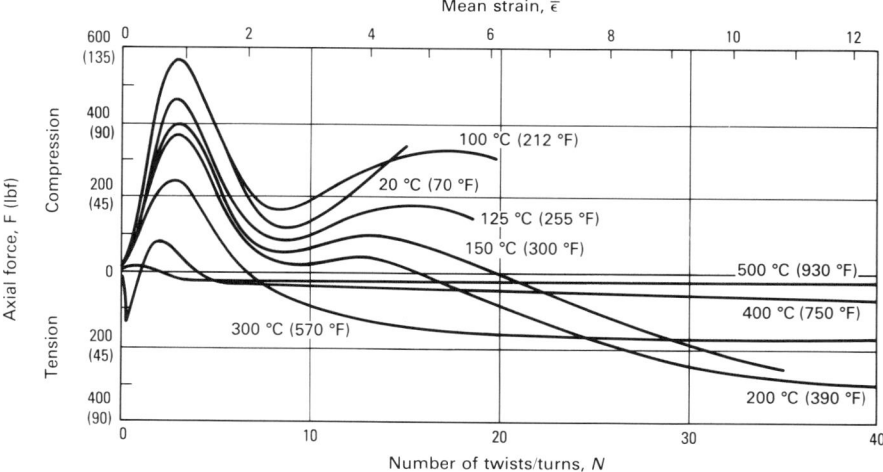

and is responsible for the tensile stress that develops at large strains (Fig. 59). In a similar manner, specimen shortening is induced when testing is carried out under free-end conditions.

The principal textures developed in face-centered cubic (fcc) metals are listed in Table 3, with their associated axial forces or length changes. Some texture components, e.g., the $\{1\bar{1}2\}\langle110\rangle/\{\bar{1}1\bar{2}\}\langle\bar{1}\bar{1}0\rangle$ set, do not lead to any axial force or length change, even when the pole figure is rotated about the radial direction away from full symmetry with respect to the shear plane normal (axial) and shear (tangential) directions. By contrast, other texture components, e.g., the $\{\bar{1}\bar{1}1\}\langle112\rangle/\{11\bar{1}\}\langle112\rangle$ set, produce axial effects even when they are in the maximum symmetry position. Some examples of the textures produced in copper that is twisted at 20 to 500 °C (68 to 930 °F) are illustrated in Fig. 60.

The textures developed in twisted body-centered cubic (bcc) metals differ somewhat from those of fcc metals and are listed in Table 4. A given bcc texture component of the form $(hkl) \langle uvw \rangle$ produces an axial effect that is qualitatively similar to that of an fcc component of the form $(uvw) \langle hkl \rangle$. Rigorously, the stress deviator tensors for the bcc and fcc cases are related through a simple

symmetry operation. The twisting of bcc iron, which induces a compressive force (lengthening) at small strains, as in the case of the fcc metals, is due to the inclination of the $\{0\bar{1}1\}\langle111\rangle/\{01\bar{1}\}\langle\bar{1}\bar{1}\bar{1}\rangle$ set in the sense opposed to the shear. In a similar manner, the tensile stresses (shortening) developed at large strains ($\epsilon > 5$) are associated with the $\{\bar{1}\bar{1}2\}\langle111\rangle$ component, which in this case is rotated in the same sense as the shear.

Free-End Versus Fixed-End Testing

In performing torsion tests, a choice must be made between the fixed- and free-end conditions of straining. Generally, fixed-end conditions are preferred. If the free-end method is chosen, the length can increase (at ambient temperatures) or decrease (at elevated temperatures) by about 10 to 40% (Ref 4). The derived shear stress τ, or $\sigma_{\theta z}$, is proportional to M/R^3 (see above), where M is the measured torque and R is the current sample radius. In most tests, the radius change is not monitored, and the initial value is used instead. Thus, a length change of 10%, which involves a radius change of about 5%, leads to an error in the derived shear stress $\sigma_{\theta z}$ of about 15%. This error increases with the $\frac{3}{2}$ power of the length change and can become quite significant.

By contrast, if the length is held constant, axial stresses, σ_{zz}, develop, which range from 2 to 20% of the developed shear stress $\sigma_{\theta z}$. The effect of these axial stresses can be estimated from the von Mises relation for the effective stress in the absence of other stress components:

$$\sigma_{vm} = \sqrt{3}\, |\sigma_{\theta z}| \qquad \text{(Eq 12)}$$

When σ_{zz} and the circumferential stress $\sigma_{\theta\theta}$ are non-zero (the radial stress σ_{rr} can be taken to be zero at the surface of a torsion specimen), the following relation applies (Ref 40):

$$\sigma_{vm} = [3\sigma_{\theta z}^2 + (\sigma_{\theta\theta}^2 + \sigma_{zz}^2 - \sigma_{\theta\theta}\sigma_{zz})]^{1/2} \qquad \text{(Eq 13)}$$

Detailed calculations (Ref 40, 41) show that the correction associated with the inner bracketed term on the right side of Eq 13 is only about 1% for fcc metals and attains a maximum of about 8% for bcc metals, which tend to develop higher ratios of axial to transverse shear stress. Because the magnitude of the correction depends on the square of the $\sigma_{zz}/\sigma_{\theta z}$ ratio, where the latter is always $\ll 1$, the error introduced by neglecting σ_{zz} in the calculation of the effective stress is generally small.

Table 3 Common texture components and axial forces observed during the torsion testing of fcc metals

		Axial force(c)		
Type(a)	Miller indices(b)	Rotated in the sense opposite to the shear	No rotation	Rotated in the same sense as the shear
A	$\{1\bar{1}1\}\ \langle110\rangle$	C	0	T
$\bar{A}$	$\{11\bar{1}\}\ \langle\bar{1}\bar{1}0\rangle$	C	0	T
A_1^*	$\{\bar{1}\bar{1}1\}\ \langle112\rangle$	T	T	T
A_2^*	$\{11\bar{1}\}\ \langle112\rangle$	C	C	C
B	$\{1\bar{1}2\}\ \langle110\rangle$	0	0	0
$\bar{B}$	$\{\bar{1}1\bar{2}\}\ \langle\bar{1}\bar{1}0\rangle$	0	0	0
C	$\{001\}\ \langle110\rangle$	T	0	C

(a) The components $A/\bar{A}$, A_1^*/A_2^*, and $B/\bar{B}$ are observed in pairs, as required by the symmetry of the torsion test. The C component is self-symmetric. (b) The plane $\{hkl\}$ is the crystallographic plane parallel to the macroscopic shear (transverse) plane; the direction $\langle uvw \rangle$ is the crystallographic direction parallel to the macroscopic shear (circumferential) direction. (c) C, compression or lengthening; T, tension or shortening. Source: Ref 41

The axial stress $\sigma_{zz}(R)$ at the surface of a solid torsion bar sample can be estimated from the mean stress $\sigma_m = F/\pi R^2$ by using the formula derived in Ref 40:

$$\sigma_{zz}(R) = \frac{F}{\pi R^2} \times$$

$$\left(1 + \frac{1}{2}\frac{\partial \ln F}{\partial \ln N} + \frac{1}{2}\frac{\partial \ln F}{\partial \ln \dot{N}}\right) \quad \text{(Eq 14)}$$

where F is the axial force, and N and $\dot{N}$ are the number of revolutions and twist rate, respectively. This relation only applies rigorously to materials displaying simple harden-ing laws, as discussed above in the section on the Fields and Backofen relation for deriving the shear stress $\sigma_{\theta z}$ (referred to as τ) from the moment M.

Shear Stress Derivations for Arbitrary Flow Laws

Methods for converting the measured torque to shear stress at the outer radius of solid bar samples and thick-walled tubes have been discussed extensively in this arti-cle. These derivations are based on the anal-ysis of Ref 7 and therefore apply only to

materials obeying simple constitutive rela-tions such as the parabolic law. When accu-rate determinations of the large strain effec-tive stress versus effective strain relations are required for materials that exhibit more com-plex behaviors, a choice must be made among the following three alternative meth-ods:

● Thin-walled tube testing
● Differential testing
● Multiple test piece method

These techniques are more accurate than the methods discussed earlier in this article for metals subject to flow softening, dynamic recrystallization, dynamic recovery, etc. Generally, they are appropriate when the flow stress at a given strain depends on the temperature and strain rate history of the test, because these techniques are based on deduc-ing the properties of a thin incremental layer of the sample without requiring any prior assumptions about the material behavior. The thin-walled tube method was described earlier in this article and will not be discussed in this Appendix.

Differential testing is a modification of the thin-walled technique that requires the use of two solid samples of slightly different radius, R_1 and R_2. In contrast to the tube

Fig. 60 Dependence of the copper textures on temperature

(a) At 20 °C (68 °F), $\bar{\varepsilon}$ = 4.7. (b) At 100 °C (212 °F), $\bar{\varepsilon}$ = 4.65. (c) At 125 °C (255 °F), $\bar{\varepsilon}$ = 5.89. (d) At 150 °C (300 °F), $\bar{\varepsilon}$ = 10.85. (e) At 200 °C (390 °F), $\bar{\varepsilon}$ = 31. (f) At 300 °C (570 °F), $\bar{\varepsilon}$ = 31. (g) At 400 °C (750 °F), $\bar{\varepsilon}$ = 31. (h) At 500 °C (930 °F), $\bar{\varepsilon}$ = 31. Same strain rates as in Fig. 59. Source: Ref 40.

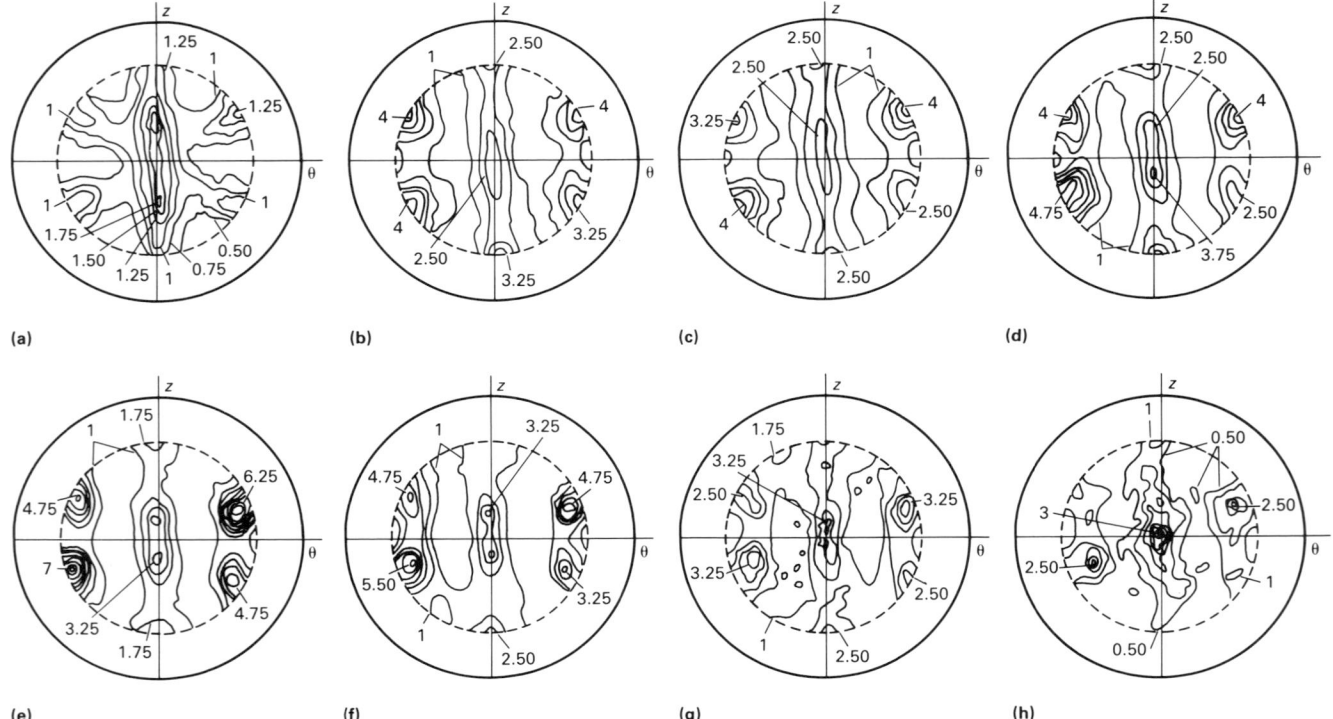

Table 4 Common texture components and axial forces observed during the torsion testing of bcc metals

Type(a)	Miller indices(b)	Axial force(c)		
		Rotated in the sense opposite to the shear	No rotation	Rotated in the same sense as the shear
D_1	$\{11\bar{2}\}\langle111\rangle$	C	C	C
D_2	$\{\bar{1}\bar{1}2\}\langle111\rangle$	T	T	T
E	$\{0\bar{1}1\}\langle111\rangle$	C	0	T
$\bar{E}$	$\{01\bar{1}\}\langle\bar{1}\bar{1}\bar{1}\rangle$	C	0	T
F^*	$\{110\}\langle001\rangle$	0	0	0

(a) The $E/\bar{E}$ orientations occur as a twin symmetric set, as required by the symmetry of the torsion test. The D_1, D_2, and F^* components are self-symmetric. (b) The plane $\{hkl\}$ is the crystallographic plane parallel to the macroscopic shear (transverse) plane; the direction $\langle uvw \rangle$ is the crystallographic direction parallel to the macroscopic shear (circumferential) direction. (c) C, compression or lengthening; T, tension or shortening. Source: Ref 41

method, samples tested according to this technique are not subject to buckling because of axial compression at ambient temperatures or to rupture because of axial tension at elevated temperatures. Consequently, large maximum strains can be attained. In a sense, the R_1 bar serves as a rotating mandrel and supports the attached $(R_2 - R_1)$ layer.

The difference between the two measured torques M_1 and M_2 determined at the same twist rate is used to deduce the properties of the incremental layer. The average shear stress $\bar{\tau}$ in the layer between R_1 and R_2 is then given by:

$$\bar{\tau} = \frac{3}{2\pi}\left(\frac{M_2 - M_1}{R_2^3 - R_1^3}\right) \tag{Eq 15}$$

This shear stress is considered to be developed at the mean radius of the layer $(R_1 + R_2)/2$, so that the associated mean shear strain is given by $\bar{\Gamma} = [(R_1 + R_2)/2] \cdot \theta/L$, where θ/L is the twist per unit length of the specimen. However, because of the nature of the torque integral and the flow-hardening characteristics of most materials, $\bar{\tau}$ generally corresponds to the stress acting at a radius greater than $(R_1 + R_2)/2$ and therefore to a shear strain greater than $\bar{\Gamma}$. The magnitude of this error, which also affects the results of the thin-walled tube tests, depends on the details of the work-hardening relation as well as on $(R_2 - R_1)$ and varies during a given test.

The differential method suffers from errors that are inherent to the definitions of the stress and the strain introduced above. This method is therefore best suited to samples with a small radius difference $(R_2 - R_1)$. However, under these conditions, a large scatter in the values of the torque difference $M_2 - M_1$ exists due to the usual sources of experimental error.

Multiple Test Piece Method. The shortcomings of the differential method can be overcome by using multiple test pieces of increasing radius (Ref 42), at the cost, however, of increasing the complexity of the test.

The multiple test piece method is based on the following derivative of the torque integral

with respect to the outer radius R at a given twist (θ), twist rate $(\dot{\theta})$, and outer radius (R_o):

$$\left.\frac{\partial M}{\partial R}\right|_{\theta,\dot{\theta}} = 2\pi\,\tau_{R_o}\,R_o^2 \tag{Eq 16}$$

Thus, the precise value of the current shear stress at R_o can be determined from a knowledge of the slope of the torque/radius curve at this radius:

$$\tau_{R_o} = \frac{\left.\dfrac{\partial M}{\partial R}\right|_{\theta,\dot{\theta}}}{2\pi\,R_o^2} \tag{Eq 17}$$

The magnitudes of the corresponding strain and strain rate are well defined at this position and are given by $\Gamma = R_o\theta/L$ and $\dot{\Gamma} = R_o\dot{\theta}/L$.

The results of a series of experiments analyzed in this way are given in Fig. 61. Ten samples of increasing diameter (5 to 10 mm, or 0.2 to 0.39 in.) were tested. These curves were used to construct a set of torque/radius curves at selected intervals of twist; the relation for $\theta = 35$ radians (≈5.6 turns) is illustrated in Fig. 62.

Each curve was fitted by means of a log M versus R polynomial, where a polynomial of degree 5 was found to produce satisfactory results. The derivative of this polynomial gives the slope $\partial M/\partial R|_{\theta,\dot{\theta}}$ corresponding to a selected R_o (e.g., 3.5 mm, or 0.14 in., in Fig. 62), from which the value of τ_{R_o} can be found from Eq 17. The complete τ versus Γ curve is obtained from a series of such determinations at increasing angles of twist θ.

Torque and Angle of Elastic Unloading. When the torsion testing of solid bars is carried out at ambient temperatures (i.e., on rate-insensitive materials and in the absence of anelastic effects), a torque cell is not absolutely indispensable for the measurement of the moment or couple. The latter can be deduced from the elastic unloading angle per

Fig. 61 Experimental torque/twist curves determined in torsion on annealed electrolytic tough pitch copper

Twisted at room temperature at 0.01 turns/s. Source: Ref 42

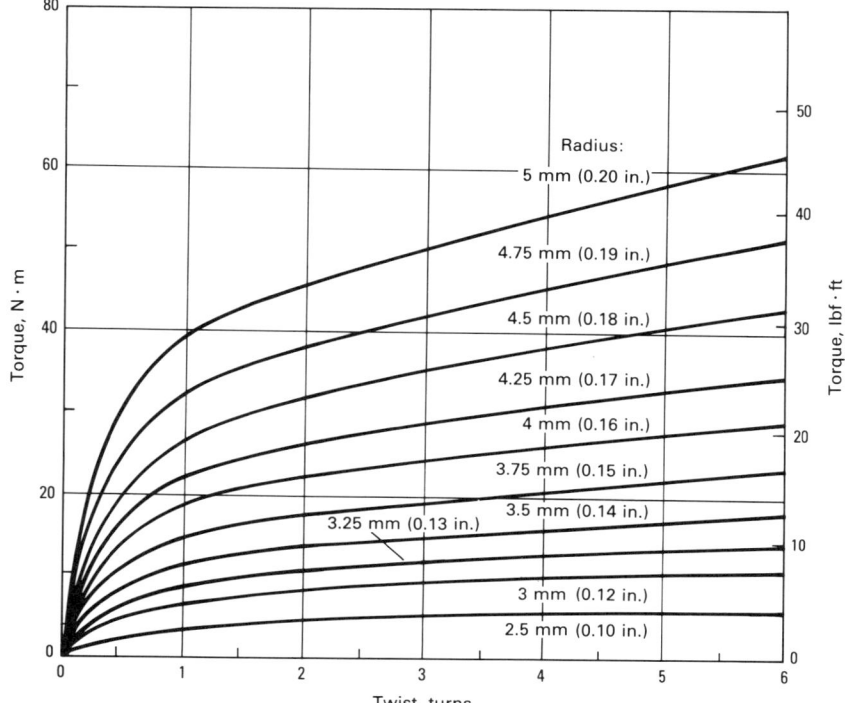

Fig. 62 Torque/radius data taken from the results of Fig. 61 at $\theta = 35$ radians

The influence of the degree of the log M versus R polynomial on the smoothing of the curve is apparent. Source: Ref 42

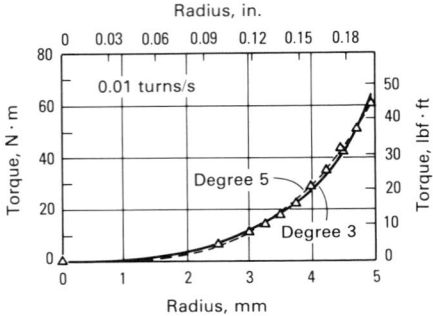

unit length θ^*/L, as determined from (Ref 43):

$$M = \left(\frac{\pi}{2}\right)GR_o^4\left(\frac{\theta^*}{L}\right) \qquad \text{(Eq 18)}$$

where G is the elastic shear modulus of the material, and R_o is the outer radius. By progressively applying measured amounts of twist $\Delta\theta$ and determining the unloading angle θ^* at each step as the experiment proceeds, the full torsion stress-strain curve can be deduced. Here, the accumulated strain is given by $\Gamma = \theta R_o/L$, where $\theta = \Sigma\,\Delta\theta$, and τ must be derived from the moment M by one of the methods described above.

REFERENCES

1. R. Raj, *Deformation Processing Maps*, American Society for Metals, to be published
2. W.A. Backofen, A.J. Shaler, and B.B. Hundy, *Trans. ASM*, Vol 46, 1954, p 655
3. D.R. Barraclough, H.J. Whittaker, K.D. Nair, and C.M. Sellars, *J. Test. Eval.*, Vol 1 (No. 3), 1973, p 220
4. D. Hardwick and W.J.McG. Tegart, *Mem. Sci. Rev. Met.*, Vol 58, 1961, p 869
5. S. Fulop, K.C. Cadien, M.J. Luton, and H.J. McQueen, *J. Test. Eval.*, Vol 5 (No. 6), 1977, p 419
6. S.L. Semiatin and G.D. Lahoti, *Met. Trans. A*, Vol 12, 1981, p 1719
7. D.S. Fields, Jr. and W.A. Backofen, *Proc. ASTM*, Vol 57, 1957, p 1259
8. R. von Mises, *Göttinger Nachrichten, Math.-Phys. K1*, 1913, p 582
9. S.L. Semiatin and J.H. Holbrook, *Met. Trans. A*, Vol 14, 1983, p 2091
10. W.J.McG. Tegart, *Ductility*, American Society for Metals, 1968, p 133
11. S.L. Semiatin, J.H. Holbrook, and M.C. Mataya, unpublished research, Battelle Columbus Laboratories, Columbus, OH, 1983
12. S.C. Shrivastava, J.J. Jonas, and G. Canova, *J. Mech. Phys. Solids*, Vol 30, 1982, p 75
13. F.A. Hodierne, *J. Inst. Metals*, Vol 91, 1962-1963, p 267
14. E. Aernoudt and J.G. Sevillano, *J. Iron Steel Inst.*, Vol 211, 1973, p 718
15. A.M. Sabroff, F.W. Boulger, and H.J. Henning, *Forging Materials and Practices*, Rheinhold, New York, 1968
16. "Evaluating the Forgeability of Steel," The Timken Company, Canton, OH, 1974
17. H.J. Henning and F.W. Boulger, *Mechanical Working of Steel I*, P.H. Smith, Ed., Gordon and Breach, New York, 1964, p 107
18. R.A. Reynolds and W.J.McG. Tegart, *J. Iron Steel Inst.*, Vol 200, 1962, p 1042
19. J.L. Robbins, O.G. Shepard, and O.D. Sherby, *Trans. ASM*, Vol 60, 1967, p 205
20. G. Mima, F. Inoko, and K. Ishikawa, *Trans. Iron Steel Inst. Jpn.*, Vol 10 (No. 3), 1970, p 216
21. J.R. Cotner and W.J.McG. Tegart, *J. Inst. Metals*, Vol 97, 1969, p 73
22. C.M. Young and O.D. Sherby, *Metal Forming—Interrelation between Theory and Practice*, A.L. Hoffmanner, Ed., Plenum Press, New York, 1971,
23. M.G. Cockcroft and D.J. Latham, National Engineering Laboratory Report No. 240, National Engineering Laboratory, East Kilbride, Glasgow, July 1966
24. P.W. Bridgman, *Studies in Large Plastic Flow and Fracture*, McGraw-Hill, New York, 1952, p 9
25. S.L. Semiatin and J.J. Jonas, *Formability and Workability of Metals: Plastic Instability and Flow Localization*, American Society for Metals, 1984
26. S.L. Semiatin, G.D. Lahoti, and T. Altan, *Process Modelling—Fundamentals and Applications to Metals*, T. Altan *et al.*, Ed., American Society for Metals, 1980, p 387
27. E. Rauch, G.R. Canova, J.J. Jonas, and S.L. Semiatin, *Acta Met.*, Vol 33, 1985, p 465
28. E. Rauch, M. Eng. thesis, McGill University, Montreal, 1983
29. J.-P.A. Immarigeon and J.J. Jonas, *Acta Met.*, Vol 22, 1974, p 1235
30. M.M. Farag, C.M. Sellars, and W.J.McG. Tegart, "Deformation under Hot Working Conditions," Special Report No. 108, Iron and Steel Institute, London, 1968, p 64
31. M.J. Luton and C.M. Sellars, *Acta Met.*, Vol 17, 1969, p 1033
32. C. Rossard, *Metaux*, Vol 35, 1960, p 102, 140, 190
33. H.J. McQueen and J.J. Jonas, *Treatise on Material Science and Technology*, Vol 6, *Plastic Deformation of Materials*, R.J. Arsenault, Ed., Academic Press, New York, 1975, p 394
34. T. Sakai and J.J. Jonas, *Acta Met.*, Vol 32, 1984, p 189
35. O.D. Sherby, M.J. Harrigan, L. Chamagne, and C. Sauve, *Trans. ASM*, Vol 62, 1969, p 575
36. G. Glover, Ph.D. thesis, University of Sheffield, 1969
37. R. Bromley, Ph.D. thesis, University of Sheffield, 1969
38. D.R. Barraclough and C.M. Sellars, *Mechanical Properties at High Rates of Strain*, Institute of Physics Conference Series No. 21, J. Harding, Ed., Institute of Physics, London, 1974, p 111
39. I. Weiss, J.J. Jonas, and G.E. Ruddle, *Process Modelling Tools*, J.F. Thomas, Jr., Ed., American Society for Metals, 1981, p 97
40. F. Montheillet, M. Cohen, and J.J. Jonas, Axial Stresses and Texture Development during the Torsion Testing of Al, Cu and α-Fe, *Acta Met.*, Vol 32, 1984, p 2077-2089
41. F. Montheillet, P. Gilormini, and J.J. Jonas, Relation between Axial Stresses and Texture Development During Torsion Testing, *Acta Met.*, Vol 33, 1985, p 705
42. G. Canova, S. Shrivastava, J.J. Jonas, and C. G'Sell, *Formability of Metallic Materials—2000 A.D.*, STP 753, J.R. Newby and B.A. Niemeier, Ed., ASTM, Philadelphia, 1981, p 189
43. J.J. Jonas, F. Montheillet, and S. Shrivastava, The Elastic Unloading of Torsion Bars Subjected to Prior Plastic Deformation, *Scripta Met.*, Vol 19, 1985, p 235-240

High Strain Rate Testing

Introduction

By M.R. Staker
Materials Research Engineer
Army Materials and Mechanics
Research Center

HIGH STRAIN RATE TESTING had a slow beginning in the early 19th century, as discussed by Rinehart in a review of high strain rate deformation and fabrication (Ref 1). A surge in the study of high strain rate deformation began around World War II. This activity has been reviewed numerous times (Ref 2-7). Thus, some of the effects of high strain rate on material properties have been well established in the high strain rate community since the early 1950's. However, the impact of these effects on the larger scientific community has been much slower. The significant changes in mechanical properties, as well as changes in failure mode, at high strain rates, demonstrated by the experimental techniques discussed in the following articles, may serve as motivation for the development of new techniques.

Characterization of the deformation, fracture, and load-carrying capability of materials subjected to high loading rates, although still evolving and problematic, is paramount for optimum materials selection for design purposes. In 1980, the National Materials Advisory Board recommended an iterative procedure in ordnance design that incorporates dynamic material property testing and computer modeling, but relies on ordnance test firings followed by a refined design that is test fired until the design objectives are obtained (Ref 8). Three reasons supported and encouraged this expensive approach: (1) dynamic material failure was inadequately understood; (2) there was a lack of accurate dynamic material response properties; and (3) it was unclear whether existing dynamic testing techniques were appropriate.

The design of transportation vehicles for impact requires optimum understanding of dynamic material response and failure and is still dependent on expensive component and full-scale testing. In the metalworking industry, the greatest need is for knowledge of dynamic material response, whereas the high-speed machining industry is interested in dynamic failure criteria. In the failure of

dynamic machine elements, the inability of failure analysis techniques to differentiate between improper design, misuse, and manufacturing defects has resulted in a growing need for the use of appropriate dynamic testing techniques to determine product liability.

This Section reviews the state of the art of high strain rate testing techniques used to characterize dynamic material properties, including failure criteria. Methods of analyzing the appropriateness of techniques to various applications are presented where possible. These techniques are listed in Table 1 with the range of strain rate (or loading rate in the case of fracture toughness) that is appropriate for each experimental technique.

Each technique has limitations that restrict the useful range of strain rate, leading to the rather lengthy list of techniques in Table 1. Although the need for many techniques may appear to be a hindrance to progress, it may have the unexpected advantage observed by Kumar (Ref 9): "Since the apparent behavior of a material is quite often influenced by these [different experimental] conditions, it is important that the true behavior be studied by different methods to isolate any excessive influence of the technique and to verify the validity of the data."

This is particularly true for those techniques in Table 1 that have ranges of strain rate that coincide or overlap. Recently, much attention has been focused on the influence of the testing technique on the apparent mate-

Table 1 Experimental methods for high strain rate testing

Mode	Applicable strain rate, s^{-1}	Testing technique
Compression	<0.1	Conventional load frames
	0.1 to 100	Special servohydraulic frames
	0.1 to 500	Cam plastometer and drop test
	200 to 10^4	Hopkinson pressure bar in compression
	10^4 to 10^5	Taylor impact test
Tension	<0.1	Conventional load frames
	0.1 to 100	Special servohydraulic frames
	100 to 10^4	Hopkinson pressure bar in tension
	10^4	Expanding ring
	>10^5	Flyer plate
Shear	<0.1	Conventional shear tests
	0.1 to 100	Special servohydraulic frames
	10 to 10^3	Torsional impact
	100 to 10^4	Hopkinson (Kolsky) bar in torsion
	10^3 to 10^4	Double-notch shear and punch
	10^4 to 10^7	Pressure-shear plate impact
Fracture toughness (rates are stress intensification rates, $\dot{K}$, MPa$\sqrt{m}$ s^{-1})	2.75 to 2×10^5	Plane-strain fracture toughness testing
	10^5	Charpy test
	10^5 to 10^6	One-point bend
	2×10^6	Dynamic notched round bar Crack arrest
Fatigue	0.1 to 100	Ultrasonic fatigue

rial behavior, thus giving rise to increased understanding via complementary analytical methods. Some of the computer and analytical methods that accompany these test methods are introduced in this Section.

Because of the limited audience and specialization required, the methodology of ballistic testing is not presented. These high strain rate tests (such as ballistic limits, merit ratings, resistance to spalling, and residual velocity) are discussed in detail by Mascianica (Ref 10).

The techniques listed in Table 1 are at various stages of standardization. For example, the Charpy test is currently a standard test method (ASTM E 23, ''Standard Methods for Notched Bar Impact Testing of Metallic Materials''), as is the high strain rate fracture toughness tests described in Annex 7 of ASTM E 399, ''Standard Test Method for Plane-Strain Fracture Toughness of Metallic Materials.'' The crack arrest procedure is a proposed test method with ASTM. Efforts are underway to provide a recommended practice for fatigue testing at ultrasonic frequencies, which could eventually become an ASTM standard test method. All other test methods listed in Table 1 currently are not standardized by ASTM.

Of the three major considerations intrinsic to high strain rate testing, wave propagation has been the most troublesome. Wave propagation becomes important in testing when the time interval of the applied force is so short and the dimensions of the specimen are such that the inertia force in the material is locally comparable to the local force for deformation. As the size of the specimen and strain rate increase, the effects of wave propagation generally become more significant, because the time available for a stress wave to propagate and reflect multiple times becomes a significant portion of the test duration time. This phenomenon is discussed in detail in the articles on compression and tension testing that follow. In addition, the stress waves may not propagate with a constant wave form, as discussed in the article on shear testing. Consequently, the intended applied stress form is modified quickly. Alterations of specimen dimensions to avert these problems eventually are ineffective at higher strain rates. Wave propagation effects must then be accounted for in an analog or digital manner. Most of the effort to develop high strain rate testing has concentrated on accounting for wave propagation or modifying specimens to eliminate its effects.

The second major consideration in high strain rate testing is the influence of strain rate on deformation and failure modes of materials. The influence of temperature in changing deformation mode from slip to twinning in body-centered cubic metals is well known. Other factors such as crystal orientation also influence this change, particularly in hexagonal metals. Equally well known is the effect of low temperature in changing failure mode from ductile to brittle in body-centered cubic metals.

An example of a deformation mode change due to strain rate is illustrated by Grant in Fig. 1 (Ref 11). As shown, there are definite breaks in the stress-rupture curves at the equicohesive temperature. The equicohesive temperature is the temperature at which resistance to deformation offered by the grain boundary is equal to the resistance offered by the grain interior. These breaks in the curves represent (for these conditions of temperature, strain rate, and grain size) a change in deformation mode from one of slip to grain boundary sliding by slipless flow. To the left of these breaks, where the strain rate is higher, the deformation proceeds by slip and is accompanied by work hardening. To the right of these break points, the strain rate is lower, and deformation lacks the characteristics of work hardening and slip.

These different deformation modes eventually culminate in different failure modes (transgranular versus intergranular), as anticipated. Even though the strain rates involved in this case are very low compared to those of high strain rate testing, this example illustrates major changes in both deformation and failure modes due to a change in strain rate.

At a high strain rate, competition between various deformation and failure modes also exists. At high strain rates, however, the changes in deformation and failure modes are not as well established. Harding (Ref 7) has reviewed the change in deformation mode from one that is thermally activated (at medium high strain rates) to a mechanism of viscous damping (at ultra-high strain rates). There is some discussion of this change in deformation mode in the article on shear testing that follows, and it is a topic of ongoing study.

At high strain rates, there is also some evidence of at least two mechanisms of failure in shear (Ref 12-14). The first, ductile shear, has all the appearances of a low or medium strain rate failure and appears to be the termination of globally uniform deformation and void coalescence. The second mechanism appears to be the result of plastic instability, with flow localization terminating in failure along adiabatic shear bands. The main distinction between these mechanisms is that, in the second, the normal low strain rate mechanism is short-circuited by the shear instability, which is believed to be a direct result of insufficient time for local temperature equilibrium by conduction and therefore only occurs at high strain rates. The article on shear testing in this Section also discusses examples of both shear deformation that remains uniform and shear deformation that localizes. The above examples illustrate the influence of strain rate on deformation and failure modes. These changes should always be accounted for when performing high strain rate tests.

The third general consideration in performing high strain rate tests is the direct effect of high strain rate on properties. Even if wave propagation is properly handled and no change in deformation or failure mode occurs over the range of strain rate of interest, an increase in strength and a decrease in mode I (opening mode) toughness usually occurs. The effect of strain rate on properties varies for each material. These general trends are well known, but because the magnitude of the change in properties with strain rate are so varied for each material, no general quantitative theory exists that satisfactorily predicts the mechanical behavior of materials over a wide range. If the capability to predict material behavior under high strain rate conditions is to develop, more effort must be devoted to high strain rate testing.

The articles in this Section are divided according to the modes of testing listed in Table 1. In each article, the development of the discussion is from strain rates just above conventional testing rates to the highest rates attainable.

Fig. 1 Log stress versus log rupture time for alloy S 590 (Fe-0.5C-20Cr-20Ni-20Co-4W-4Mo-4Nb)

The breaks in each of the curves represent the equicohesive temperature for the strain rate and grain size in the test. To the left of these breaks, the strain rate is higher than on the right. Source: Ref 11

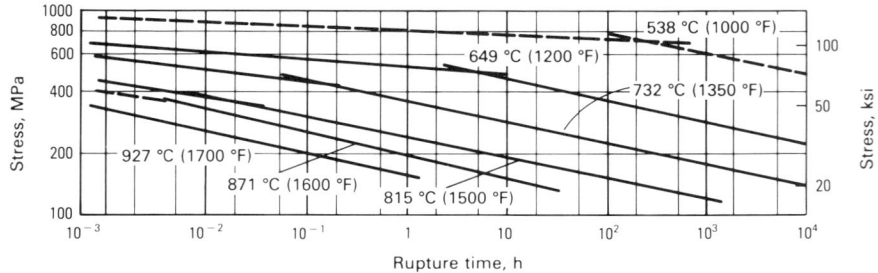

REFERENCES

1. J.S. Rinehart, Historical Perspective: Metallurgical Effects of High Strain Rate Deformation and Fabrication, in *Shock Waves and High-Strain Rate Phenomena in Metals*, M.A. Meyers and L.E. Murr, Ed., Plenum Press, New York, 1981, p 3-20
2. U.S. Lindholm, *Techniques of Metals Research*, Vol 5, Part 1, R.F. Bunshah, Ed., Interscience, New York, 1971
3. G.E. Duvall, Problems in Shock Waves Research, in *Metallurgical Effects at High Strain Rates*, R.W. Rohde, B.M. Butcher, J.R. Holland, and C.H. Karnes, Ed., Plenum Press, New York, 1973, p 1-13
4. U.S. Lindholm, Review of Dynamic Testing Techniques and Material Behavior, in *Mechanical Properties at High Rates of Strain*, J. Harding, Ed., Institute of Physics Conference Series No. 21, London, 1974, p 3-21
5. J.D. Campbell, Dynamic Plasticity: Macroscopic and Microscopic Aspects, *Mat. Sci. Eng.*, Vol 12 (No. 1), 1973, p 3-21
6. L. Davison and R.A. Graham, Shock Compression of Solids, *Phys. Rep.*, Vol 55 (No. 4), 1979, p 255
7. J. Harding, High-Rate Straining and Mechanical Properties of Materials, in *Explosive Welding, Forming and Compaction*, T.Z. Blazynski, Ed., Applied Science Publishers, New York, 1983
8. Committee on Materials Response to Ultra-High Loading Rates, in National Materials Advisory Board of the National Research Council Report NMAB-356, 1980
9. S. Kumar, Introduction, in *Mechanical Behavior of Materials Under Dynamic Loads*, U.S. Lindholm, Ed., Springer-Verlag, New York, 1968, p xiii
10. F.S. Mascianica, Ballistic Testing Methodology, in *Ballistic Materials and Penetration Mechanics*, R.C. Laible, Ed., Elsevier Scientific, Amsterdam, 1980, p 41-72
11. N.J. Grant and A.W. Mullendore, *Deformation and Fracture at Elevated Temperatures*, MIT Press, Cambridge, MA, 1965, p 16
12. M.R. Staker, On Adiabatic Shear Band Determinations by Surface Observations, *Scripta Met.*, Vol 14 (No. 6), 1980, p 677-680
13. B. Dodd and A.G. Atkins, Flow Localization in Shear Deformation of Void-Containing and Void-Free Solids, *Acta Met.*, Vol 31 (No. 1), 1983, p 9-15
14. S.L. Semiatin, M.R. Staker, and J.J. Jonas, Plastic Instability and Flow Localization in Shear at High Rates of Deformation, *Acta Met.*, Vol 32 (No. 9), 1984, p 1347-1354

High Strain Rate Compression Testing

Introduction

By Paul S. Follansbee
Staff Member
Los Alamos National Laboratory

MANY DEFORMATION PROCESSES occur at strain rates that are well above those possible with standard mechanical testing procedures. High-rate forming operations and impact events (automobile bumpers or the impact of a projectile with an armor plate) are examples of high-rate deformation processes. Because mechanical properties such as strength and ductility can vary with strain rate, it is often necessary to determine these properties under conditions that closely match the expected deformation rates in service. This article reviews experimental techniques for performing compression tests at high strain rates.

Strain Rate Regimes

Strain rate ($\dot{\epsilon}$) is the rate of change of strain (ϵ) with time (t):

$$\dot{\epsilon} = \frac{d\epsilon}{dt} \qquad \text{(Eq 1)}$$

where ϵ can be either the engineering or the true strain. Although compressive strain and strain rate are negative quantities, the negative sign is often omitted when it is understood that the test is a compression test. For a constant strain rate experiment, the strain rate is simply the total strain divided by the duration of the test:

$$\dot{\epsilon} = \frac{\epsilon}{t} \qquad \text{(Eq 2)}$$

The units of strain rate are inverse time; s^{-1} is most frequently used and will be used throughout this article. When ϵ in Eq 1 is the engineering strain, then:

$$\frac{d\epsilon}{dt} = \frac{1}{L_o}\frac{dL}{dt} = \frac{V}{L_o} \qquad \text{(Eq 3)}$$

where L is the length of the specimen of original length L_o, and V is the velocity at which the specimen is being deformed. A constant crosshead speed in a mechanical testing machine yields a constant engineering strain rate defined by Eq 3.

A typical mechanical test is performed at a strain rate of around 10^{-3} s^{-1}, which yields a strain of 0.5 in 500 s. The equipment and techniques generally can be extended to strain rates as high as 0.1 s^{-1} without difficulty. Tests at higher strain rates necessitate new experimental techniques; these tests fall into regimes roughly defined by the experimental techniques that are used to achieve the strain rates.

Medium-Rate Regime. In this regime, the experimental techniques are similar to those used at low strain rates, but special equipment is required. Servo-hydraulic load frames equipped with high-capacity valves can be used to generate strain rates as high as 200 s^{-1}. These tests are complicated by load and strain measurement and data acquisition. They also require specialized equipment and are somewhat limited in load capacity. Another experimental technique developed for compression testing in this strain rate regime is the cam plastometer, which is described later in this article. When more qualitative medium-rate test results are desired, particu-larly when high loads are required, the drop test, which is also described later, may be appropriate.

High-Rate Regime. At strain rates greater than 200 s^{-1}, the crosshead speed required exceeds that easily achieved with screw-driven or servo-hydraulic test frames. To generate high strain rates, experimental techniques that make use of projectile impacts and wave propagation phenomena have been employed. Techniques based on the Hopkinson bar have been used successfully to perform dynamic compression tests at strain rates exceeding 10^4 s^{-1} and strains to 0.30. At these strain rates, inertial constraint begins to affect the test validity. Higher strain rates and larger strains are possible with the rod impact (Taylor) test. In this experiment, the complicated stress wave propagation phenomena strongly influence the measurement and are accounted for numerically.

Very High Strain Rates. Still higher strain rates are possible, but these generally are found only in shock fronts produced by very high velocity impacts or by the detonation of explosives. These conditions apply to many deformation processes and have been the subject of recent reviews (Ref 1).

The strain rate regimes are summarized in Table 1, although the delineations are not

Table 1 Experimental techniques for various strain rate regimes in compression testing

Strain rate regime	Experimental techniques	Wave propagation
Low rate: $\dot{\epsilon} < 0.1$ s^{-1}	Standard mechanical testing procedures	Not significant
Medium rate: 0.1 $s^{-1} \le \dot{\epsilon} \le 200$ s^{-1}	Servo-hydraulic frames, cam plastometer, drop test	Influences load measurement
High rate: 200 $s^{-1} \le \dot{\epsilon} \le 10^5$ s^{-1}	Hopkinson pressure bar	Affects uniform stress approximation
	Rod impact (Taylor) test	Analysis required for interpretation of results
Very high rate: $\dot{\epsilon} > 10^5$ s^{-1}	Flyer plate impact	Critical

absolute. For instance, there are very few published experimental results using the Hopkinson bar at strain rates as high as 10^5 s^{-1}, and the drop test can be extended to higher strain rates through the use of specimens with small dimensions.

Stress Wave Propagation

The importance of stress wave propagation is also summarized in Table 1 for each experimental technique. It is evident that as the strain rate increases stress wave propagation considerations become more important. As the strain rate is increased through the medium strain rate regime, the measurement of load is the first to be affected by stress wave propagation. At these strain rates, the "ringing" of a load cell can mask the desired measurement. As the strain rate is increased even further, establishment of uniform deformation within the specimen also becomes paramount.

A common feature of all of the experimental techniques discussed in this article is that load is applied to only one surface of the test specimen. Inertia initially opposes uniform deformation. The stress wave generated at the loaded surface propagates at the speed of sound. The equilibration of stress throughout the test specimen requires a few reverberations of the stress wave. As a first approximation, Davies and Hunter (Ref 2) have estimated that three reverberations are required for stress equilibration. If the deformation is purely elastic, then the longitudinal sound velocity is simply $\sqrt{E/\rho}$, where E is the elastic modulus, and ρ is the specimen density. A one-dimensional strain-rate-independent theory predicts that the plastic wave propagates at a velocity V_p, determined by:

$$V_p = \sqrt{\frac{\dfrac{d\sigma}{d\epsilon}}{\rho}} \qquad \text{(Eq 4)}$$

where $d\sigma/d\epsilon$ is the slope of the true stress/true strain curve. For many materials, the initial work-hardening rate $d\sigma/d\epsilon$ is approximately 1% of E; thus for these materials, the plastic wave velocity is approximately 10% of the elastic wave velocity.

To illustrate the importance of stress wave propagation, consider the compression of a 10-mm (0.4-in.) specimen, in which the elastic wave velocity is 5.0×10^3 m/s (16.4×10^3 ft/s) and the initial plastic wave velocity may be approximated as 5.0×10^2 m/s (16.4×10^2 ft/s). Assuming that three reverberations of the slower moving plastic wave are required for uniform stress within the deforming specimen, the time for these

reverberations is computed to equal 60 μs. At a strain rate of 10^3 s^{-1}, the specimen will have compressed to a strain of 6% during this interval. Thus, data at strains less than this may be invalid, because it cannot be assumed that the specimen was deforming uniformly.

Experimental evidence in a variety of metals suggests that the above estimate is too conservative; stress waves generated by impacts in which the stress far exceeds the yield stress appear to travel, at least initially, at the elastic wave velocity rather than at the plastic wave velocity given by Eq 4. This implies that the material behavior is more complicated than that assumed in the derivation of Eq 4. Nevertheless, even if the elastic wave velocity is used in the above example, the critical strain at which uniform stress within the specimen is achieved is computed to be equal to 0.6%. At a strain rate of 10^4 s^{-1}, this critical strain becomes 6%.

The above example illustrates that wave propagation becomes an important consideration affecting test validity in the high strain rate regime of Table 1. It also illustrates a major advantage of compression testing over tension testing at high strain rates; a compression test specimen can be fabricated to relatively small dimensions to minimize wave propagation times. However, there are limits to the specimen dimensions, which are described in the discussion of Hopkinson bar test techniques.

Departure From Isothermal Test Conditions

It is well known that most of the work of deformation is expended as heat; only 5 to 10% of this work is actually stored in the defect structure of the deformed specimen. As the strain rate increases, there is insufficient time for the transport of this heat from the specimen to the grips (platens in the case of a compression specimen) or atmosphere. Thus, the specimen temperature can increase during deformation.

The work of deformation (W) is simply:

$$W = \int_{L_o}^{L_f} P \, dL \qquad \text{(Eq 5)}$$

where P is the applied load, and L is the length of the specimen of final length, L_f, and original length, L_o. Equation 5 can be rewritten as:

$$W = A_o L_o \int^{e_f} s(e) \, de \qquad \text{(Eq 6)}$$

where A_o and L_o are the original specimen area and length, respectively, s is the engineering stress, and e is the engineering strain.

If it is assumed that this work is transformed to heat adiabatically (i.e., without heat flow), then the specimen temperature must increase by:

$$\Delta T = \frac{1}{\rho C_p} \int^{e_f} s(e) \, de \qquad \text{(Eq 7)}$$

where ρ is the density, and C_p is the heat capacity of the specimen. For example, consider the deformation of a stainless steel specimen to a strain of 0.5 and assume, for simplicity, that the flow stress of the material is constant and equal to 750 MPa (109 ksi), the density is 8 g/cm^3 (0.29 lb/in.^{3}), and the heat capacity is 500 J/kg $\cdot$ K (0.120 cal/g $\cdot$ °C). From Eq 7, the temperature increase at a strain of 0.5 is estimated to be 94 K. This is a significant increase and demonstrates that this is not an isothermal test.

The above estimate assumed adiabatic conditions, which are only the case in the high strain rate regime. However, at least under these conditions, the temperature increase can be estimated fairly accurately, and deformation within the specimen remains uniform. The medium strain rate regime is further complicated by heat flow, which in quiescent environments is dominated by conduction through the ends of the specimen. In this regime, there is the possibility of generating a substantial axial temperature gradient from the end to the midpoint of the specimen. This gradient can result in nonuniform deformation throughout the specimen, which complicates interpretation of the experimental results. In fact, if the temperature dependence of the flow stress is high enough, deformation may become highly localized at the center of the specimen; this situation is less likely in a compression specimen than it is in a tensile specimen, which has a much higher aspect ratio.

Experimental Problems

Measurement of stress and strain and data acquisition becomes more difficult as the strain rate increases. Although specific procedures for each of the experimental techniques will be discussed in more detail in the following sections, a few general comments concerning these topics and the effect of friction are included.

Measurement of Stress and Strain. Many of the standard techniques for measuring stress and strain at low rates become unsuitable for testing at higher strain rates. The frequency response of a measurement device, such as a load cell, linear variable-differential transformer, extensometer, or strain gage, must be considered as well as that of the signal conditioning. The use of a fragile device, such as an extensometer, may not be appropriate for systems that involve large, rapid deformations. Elastic wave propagation may complicate the operation of a load cell at strain rates well below that at which stress wave propagation within the specimen is important.

All of the above factors can influence testing at medium strain rates, where the experimental techniques represent an extension of those used at low rates. Testing in the high strain rate regime involves unique experimental techniques that were developed largely to circumvent difficulties encountered in directly measuring stress and strain in a specimen deforming at high strain rates. For example, a following section describing the split Hopkinson pressure bar illustrates that stress and strain within the deforming specimen are related in a straightforward manner to strain gage measurements made on the elastic pressure bars.

Data Acquisition. Many devices typically used to record load and displacement data, such as strip chart recorders and *x-y* plotters, cannot be used at strain rates as low as 1 s^{-1} because of inadequate frequency response. The computer data acquisition systems that accompany mechanical testing systems are also frequently too slow at medium rates of strain. The use of an oscilloscope and photographic film is a widely used option, but this restricts accuracy to approximately ±5%.

Computer data acquisition systems are now available for sampling at rates as fast as 100 kHz, which is sufficient for tests at strain rates as high as 10^2 s^{-1}. For higher strain rates, analog-to-digital transient recorders are available with sampling rates inversely related to the desired accuracy of the digitized word; for a 10-bit word, state-of-the-art transient recorders sample at rates as fast as 20 MHz.

Friction. An important consideration in compression testing is the friction condition imposed at the interfaces between the ends of the specimen and the platens. If these ends are constrained due to a lack of lubricant, the stress state within the sample will not be uniform, and the measured flow stress can exceed that for uniform deformation. This increased flow stress may be incorrectly identified as a strain rate effect. Nonuniform deformation can also result from excessive lubrication or insufficient constraint at the ends of the specimen.

The design of the compression test specimen and the use of lubrication to minimize these effects are described in detail in ASTM Standard E 9, "Compression Testing of Metallic Materials at Room Temperature" and in Ref 3. One recommendation cited in these references is that the length-to-diameter ratio of the machined compression specimen be within the range of 1.5 to 2.0.

There is growing evidence that maintenance of lubrication is less of a problem at higher strain rates than it is at low strain rates. Loss of lubricant can result from its extrusion from the interface during compression. The viscous nature of lubricants introduces time as a variable for this process, and as the strain rate increases, there is less time available for loss of lubrication. Nevertheless, the use of lubrication is essential, and specific techniques for each of the experimental procedures will be discussed in the following sections of this article. For additional information on the variables and limitations of compression testing at room temperature, see the article "Axial Compression Testing" in this Volume.

Compression Testing by Conventional Load Frames at Medium Strain Rates

By Paul S. Follansbee
Staff Member
Los Alamos National Laboratory

and

Philip E. Armstrong
Staff Member
Los Alamos National Laboratory

MEDIUM STRAIN RATE COMPRESSION TESTING with conventional load frames is very similar to low strain rate compression testing, which is described in the article "Axial Compression Testing" in this Volume. For medium-rate testing, the load frames require the capability to generate higher crosshead or ram velocities. This section describes some of the techniques used to obtain medium strain rates. The measurement of load and strain at medium rates of strain poses additional experimental complexities, which are also described.

Equipment and Testing Techniques

Conventional load frames are divided into two general classes of mechanisms. Screw-driven machines generally involve the movement of a crosshead and offer excellent stability at low strain rates. Maximum crosshead velocity in commercially available machines is on the order of 17 mm/s (0.7 in./s), which yields a maximum engineering strain rate of 1 s^{-1} in a 10-mm (0.4-in.) gage length specimen. This is at the lower end of the medium strain rate regime; thus, screw-

driven machines have limited use in this regime.

Servo-Hydraulic Test Frames. Servo-hydraulic machines using oil as the working fluid have the capability of generating much higher velocities. In these machines, the crosshead is generally stationary, and a ram provides the desired motion. With state-of-the-art control valves, which can provide fluid flow rates of 1665 L/min (440 gal/min), ram velocities approaching 10 m/s (33 ft/s) and loads as high as 90 kN (10 ton) are available. This velocity yields a strain rate of 10^3 s^{-1} in a 10-mm (0.4-in.) long compression specimen. However, the difficulty of measuring load and displacements within the deforming specimen limits the strain rate capability with these machines to a strain rate of approximately 200 s^{-1}.

An important consideration when performing mechanical tests with conventional test frames is the stiffness of the machine (Ref 4). This affects the measurement of displacements within the sample and can profoundly affect the interpretation of a stress relaxation test, a test of a material exhibiting a yield drop, or of a test in which the strain rate is abruptly changed during deformation. The latter technique is typically used to evaluate the strain rate sensitivity of a material. For a test at a uniform strain rate, a high machine stiffness is desired; techniques to increase the stiffness of a hydraulic machine are described in Ref 5.

In current servo-hydraulic test frames, motion of the ram is controlled by regulating the flow of oil through the servo-valve with an error signal, which is the difference between the measured ram position and that which has been specified. Usually the position of the ram is sensed with a linear variable differential transformer and the motion of the ram is prescribed to follow a constant velocity. However, feedback control, as this type of control is termed, also provides the ability to specify a constant true strain rate, which is often a requirement when evaluating strain rate sensitivity.

Grip design for compression testing at medium strain rates requires the same considerations that apply to grip design for low strain rates. The compression specimen typically is sandwiched between two hard, polished platens that are placed in a subpress designed to maintain parallel faces during deformation. A typical grip assembly is shown in Fig. 1, in which a 5.1-mm-long by 5.1-mm-diam (0.2-in. by 0.2-in.) compression specimen is in place and ready for testing. The ram is shown in position and is separated from the subpress by approximately 20 mm (0.8 in.). This gap allows time (approximately 2 ms at the highest ram velocity) for the ram to accelerate to the speci-

Fig. 1 Subpress assembly for medium strain rate testing with conventional load frame

The specimen (5.1-mm diam by 5.1-mm long, or 0.2 in. by 0.2 in.) is sandwiched between two highly polished platens. A quartz load washer is shown positioned above the subpress assembly.

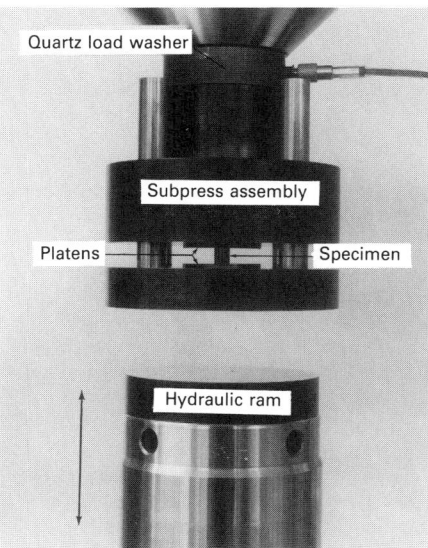

Quartz load washer

Subpress assembly

Platens — Specimen

Hydraulic ram

fied velocity. In this test, the stroke of the ram must be set accurately to ensure the desired deformation.

Instrumentation

As the strain rate increases, the measurement of load and displacement becomes increasingly more difficult. The requirement for adequate frequency response in the signal conditioners and the problems associated with load cell ringing were discussed in the introduction to this article. In this section, the measurement of load and displacement at medium strain rates is described in more detail.

Measurement of Load. A typical load cell determines load by measuring displacement in an elastic member, such as a diaphragm or cylinder. The displacements are measured with bonded strain gages; this gives the load cell sufficient intrinsic frequency response for testing at medium strain rates. However, a problem often arises due to ringing in the load cell. The load cell has a natural frequency of vibration determined by the geometry and physical properties such as density and elastic modulus. Typical load cells have a natural frequency in the range of 500 to 5000 Hz. In effect, the natural frequency of vibration sets the bandwidth of the load-measuring system.

By this criterion alone, load cells should be sufficient for compression testing at strain rates as high as 100 s^{-1}. However, the transient response of the load cell in practice limits the measurement to much lower strain rates. When a constant strain rate test is desired, deformation must be initiated by an impact because of the acceleration time required by the ram. This impact can excite the natural vibrational mode of the load cell, which will produce oscillations in the output signal that can mask the actual load measurement. Unless the impact is dampened by some means, load measurement at strain rates greater than about 1 s^{-1} can be subject to load cell ringing.

Ringing of the load cell can be minimized by selecting a load cell with a high vibrational frequency. If the natural frequency is sufficiently high, the vibrational mode may not be excited by the impact, or if excited, it may be possible to remove it from the signal with a low pass filter. Another method to reduce ringing is to dampen the impact that initiates deformation within the specimen. Often, a thin layer of deformable material placed between the impacting surfaces is sufficient to remove the higher frequencies generated by the impact that can excite the natural frequency of the load cell. For example, in the configuration shown in Fig. 1, a single loop (~50-mm, or 2-in.) of 0.51-mm (0.02-in.) diam lead-tin solder wire placed on the impacting face of the hydraulic ram was found to be effective in minimizing load cell ringing. Such layers, however, may complicate measurement of displacement within the specimen.

At strain rates close to 100 s^{-1}, the standard load cell either may not possess the necessary frequency response, or it may ring excessively. These characteristics can make the load cell inadequate for load measurement. Under these conditions, a quartz piezoelectric device such as a load washer (Fig. 1) is useful. The load washer is convenient in that it is easily adapted to a compression test; it also has excellent intrinsic frequency response and a high fundamental vibrational frequency. However, these devices require special signal conditioning and low-capacitance cables.

Measurement of Strain. The direct measurement of strain at medium strain rates presents a challenge. Many of the devices typically used for low strain rate testing are inappropriate at medium strain rates. Extensometers, for example, may have the necessary response characteristics for medium strain rate testing. However, it is difficult to ensure that the rapid and large displacement in small compression specimens will not damage the fragile extensometer.

Many hydraulic test frames use a linear variable-differential transformer (LVDT) to control the motion of the hydraulic ram. This LVDT signal is comprised of displacements within the specimen as well as elastic displacements throughout the test frame. To relate this signal to displacements within the specimen, the latter contribution must be subtracted; this problem also is encountered at low strain rates. If a deformable material is placed between the impact surfaces to dampen the impact, the displacements within this layer also must be subtracted from the LVDT signal.

A common practice is to mount the LVDT at an off-axis position adjacent to the specimen. The benefit of this configuration is that a displacement measurement is possible between two points that are quite close to the specimen; this measurement includes less of the elastic deformation in the load frame. Whenever a measurement is made at an off-axis position, it is important to verify that the measurement truly represents displacements within the sample. Often, two LVDT units are mounted at diametrically opposite positions, and their outputs are processed to eliminate the effects of nonplanar motion. The LVDT suffers from an intrinsic frequency response limitation determined by the excitation frequency. Standard excitation frequencies are in the range of 1 to 5 kHz, which limits the frequency response to around 100 to 500 Hz.

Velocity transducers, which have good intrinsic frequency response, have been used to measure the motion of the specimen and grip assembly (Ref 5). Their output can be integrated electronically or by computer to obtain the displacement. Generally, these also require mounting at off-axis locations.

The use of noncontact methods to measure strain is becoming more common. Optical extensometers that can measure the diameter of a compression specimen to within a spatial resolution of 25 μm and at a sampling rate of 1 kHz are available commercially. This is adequate for testing at strain rates approaching 10 s^{-1}. Higher sampling rates should be available soon. Laser interferometers, with much higher sampling rates, have been used to measure strain at strain rates exceeding 10^3 s^{-1} (Ref 6). However, this represents a fairly sophisticated and costly measurement technique.

The Cam Plastometer

By John E. Hockett
President
John E. Hockett, Inc.

THE CAM PLASTOMETER is designed specifically for medium strain rate compression testing. The axial load to compress the specimen is transferred from massive rotating flywheels via a cam; this provides the distinct advantage of being able to obtain a constant true strain rate experiment. This section reviews the history of the cam plastometer and discusses the basic principles involved in the test technique and the equipment used.

Historical Perspective

The cam plastometer is a relatively recent addition to mechanical testing equipment. The first cam plastometer was a constant compression rate machine built by Los at Sheffield University. It was later used by Loizou and Sims (Ref 7) in compression testing of pure lead from 1951 to 1953.

The first constant true strain rate cam plastometer was designed and built by Orowan at the British Iron and Steel Research Association (BISRA) in 1950. It was used to obtain true stress-strain curves for several steels at various true strain rates and temperatures. This facility was later used by Alder and Phillips (Ref 8) in constant true strain rate compression testing of commercial-purity aluminum at strain rates of 1 to 40 s^{-1} and temperatures from 291 to 723 K. Cook used the BISRA machine in 1956 to obtain true stress-strain curves for 12 steels at true strain rates of 0.05 to 100 s^{-1} and temperatures from 1173 to 1473 K (Ref 9).

Others have built cam plastometers to perform specific studies. Machines have been built to study the compressive deformation of aluminum and aluminum alloys (Ref 10, 11), depleted uranium (Ref 12, 13), zinc and zinc alloys (Ref 14, 15), and other ferrous and nonferrous metals (Ref 16). A modern version of the cam plastometer has also been designed and built at the Canadian Department of Mines, Energy, and Resources (Ref 17).

The cam plastometer described later in this article was originally used to study the rolling characteristics of uranium plate and sheet by the classic Orowan method (Ref 13). This machine underwent several revisions in both mechanical design and instrumentation; constant true strain rate testing is now possible at strain rates from 0.5 to ≥200 s^{-1}. The temperature range of the apparatus is 77 to 1200 K, and load-frame deflections of 0.5 mm (0.02 in.) occur at a force of 222 kN (25 ton) in compression.

Basic Principles

The cam plastometer is used to obtain the resistance to compressive deformation of materials, principally metals, at constant strain rates over a useful and significant range of strain rates and a practical range of

testing temperatures. Most plastometers have a capacity to compress cylindrical specimens homogeneously to a 50% reduction in height, assuming the material is tested at temperatures at which it is ductile enough to permit this reduction.

The cam plastometer was a significant development, because it allowed an extension of mechanical testing capability by three orders of magnitude in strain rate over that obtained in standard mechanical test machines. Furthermore, the cam plastometer mechanism lends itself to the generation of a constant true strain rate experiment, which is ideally suited for metal deformation studies.

True strain (ϵ) is defined as:

$$\epsilon = \ln \frac{L_f}{L_o} = \ln(L_f + e) \qquad \text{(Eq 8)}$$

where L_o and L_f are the original and final specimen heights, respectively, and e is the engineering strain. Both the engineering strain and true strain are negative quantities, because L_f is smaller than L_o. It can be shown that for an experiment at constant true strain rate ($\dot{\epsilon}$), the length of the sample must change with time according to:

$$L_f = L_o \exp{(\dot{\epsilon}t)} \qquad \text{(Eq 9)}$$

where t is the time measured from the instant deformation begins.

True strain rate compression tests are achieved in all cam plastometers by designing the load-applying mechanism in the form of a logarithmic cam; the radius of the rising (or lowering) part of the cam increases (or decreases) to satisfy Eq 9. The load applied by the cam to the specimen (via a platen on each end) is measured with a load cell, and load versus time is recorded. From the prescribed strain rate $\dot{\epsilon}$, the strain ϵ_i at any time t_i during the test may be obtained from:

$$\epsilon_i = \dot{\epsilon}t_i \qquad \text{(Eq 10)}$$

The load P_i at time t_i is recorded continuously, and from ϵ_i the instantaneous area A_i can be calculated. The instantaneous true stress σ_i is simply P_i/A_i, and a true stress/true strain curve may be plotted from a series of σ_i and ϵ_i points.

The true strain rate is constant only if the rotational speed of the cam is constant. This is achieved by storing large amounts of energy in flywheels and in the many rotating parts of transmissions, as well as in the cam. To maintain a constant strain rate, this stored energy must greatly exceed the energy required to compress the specimen.

As in all compression tests, deformation is homogeneous throughout the test only if frictional constraint is minimized at the interfaces between the platens and the specimen ends. That is, the specimen should start out

Fig. 2 Typical cam plastometer facility

Direct-current motor and generator are on the left; three flywheels and two transmissions are on the right; two-post loading frame is at rear. Courtesy of Los Alamos National Laboratory

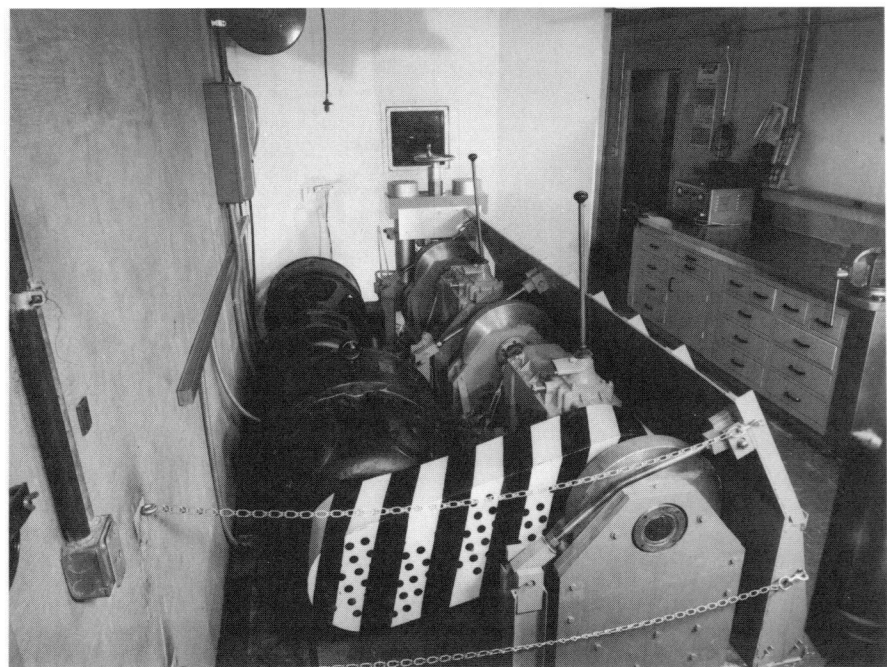

as a cylinder and maintain this cylindrical geometry throughout the test. This requires the use of a suitable lubricant at the interfaces. Techniques for lubrication are described below.

Equipment and Testing

A typical cam plastometer setup is shown in Fig. 2, in which a 37-kW direct-current motor (behind the belt guard) and the 40-kV · A generator are depicted. The motor drives three flywheels distributed between and at both ends of the drive train, which consists of two large off-highway-type transmissions. The drive train drives the cam, which is visible in Fig. 3. In this figure, the cam is stopped about midway through its loading cycle (counterclockwise rotation), but the carriage and cam follower assembly is in the rest position.

The cam follower is shown in Fig. 4 in loading position riding on the cylindrical portion of the cam; the lifting portion of the cam has just passed the riser block. The carriage has been shot into position so that in the next revolution of the cam the specimen will be compressed, and the carriage will then be retracted into its rest position. The timing of this sequence, which must occur in a fraction of a revolution at rotational speeds from 1.5 to 2000 rpm of the cam, is of critical importance and introduces one of the difficulties in operation of the cam plastometer. A cam that loads in 45° of rotation is used at high strain rates, while one that loads in 180° is used at low strain rates.

Also shown in Fig. 3 and 4 are the specimen-platen assembly and the location of the load cell for a room-temperature test. The specimens are typically 12.7 mm (0.5 in.) in diameter and are 19.8 mm (0.78 in.) long. These are rather large dimensions, which demonstrates one advantage of the high load capacity of the cam plastometer.

As shown in Fig. 2 to 4, the cam plastometer is a massive piece of equipment. Size introduces problems in dynamically balancing the cam and in providing sufficient strength and minimum deflection of the loading frame and drive train. Some cam plastometers in Japan have been equipped with four-post loading frames to circumvent these problems.

Effects of Friction. Minimizing the effects of friction in a compression test involves selection of a suitable length-to-diameter (L/D) ratio for the specimen and the selection of proper lubrication for the specimen-platen interfaces. An L/D ratio of 1.56 is generally considered optimum. Ratios much larger than this lead to buckling, whereas ratios much smaller than this yield higher stress-strain curves, indicating exces-

Fig. 3 Working area of the loading frame with the cam follower in its rest position

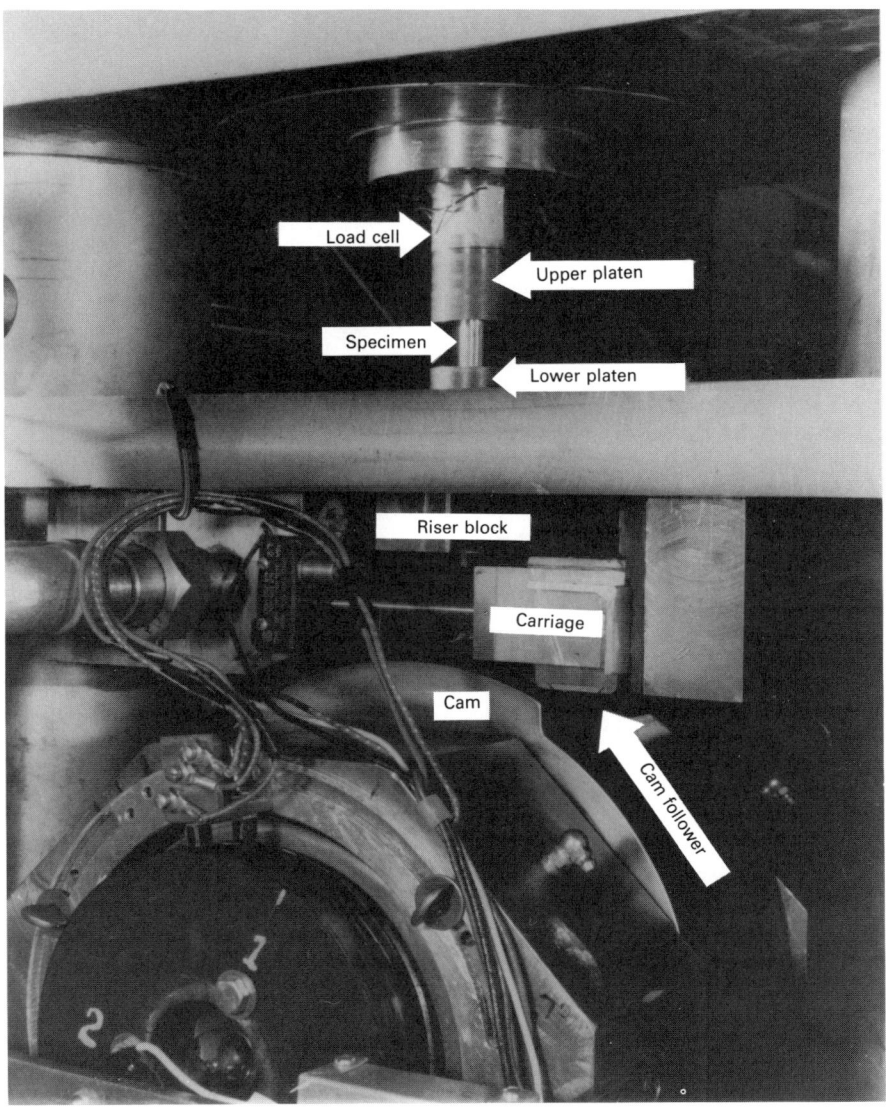

sive specimen-platen frictional constraint. An L/D ratio of 1.56 is within the range specified in ASTM Standard E 9, "Compression Testing of Metallic Materials at Room Temperature."

Many lubricants (e.g., viscous oils, powdered graphite, or powdered, liquid-based, and grease-based molybdenum disulfide) have been used for room-temperature testing. A satisfactory lubricant for room-temperature and lower temperature (to 77 K) testing is "Motor-Mica," which is a mixture of finely ground powdered mica and a highly viscous commercial automotive oil.

At slightly elevated temperatures, powdered tungsten disulfide has been used successfully. At higher temperatures, various

glass frits exhibiting viscosities in the range 4000 to 6000 Pa · s (400 to 600 poise) at temperature can be used. For a series of tests on a titanium alloy, the commercial glaze Deltaglaze 347m was used above 1283 K, and Deltaglaze 349m was used between 1173 and 1283 K.

Coefficient of friction measurements by the method proposed in Ref 18 may be helpful in the preliminary selection of lubricants. Concentric grooves in the ends of the specimen are necessary for most plastometer testing. Depth, spacing, and included angle of the grooves must be determined empirically.

The influence of lubrication on test validity is illustrated in Fig. 5, which shows a grooved specimen prior to testing, a specimen compressed at 873 K with inadequate lubrication, and a specimen compressed at the same temperature using a suitable lubricant. The center specimen represents an invalid test because of inhomogeneous deformation, commonly called ''barreling.'' See the article ''Axial Compression Testing'' in this Volume for more information on barreling of compression test specimens.

Elevated-Temperature Testing. Most cam plastometer facilities have elevated-temperature testing capabilities. Some of these are equipped for testing in vacuum or in selected atmospheres. These require a vacuum chamber around the specimen-platen assembly and the use of O-ring or bellows grips. The elevated-temperature capability for some facilities up to 1300 K has been achieved through the use of *in situ* induction heating in a reducing atmosphere and the use of aluminum oxide platens. These platens appear to dampen some load cell ringing that occurs when using much denser tungsten-carbide platens.

Drop Tower Compression Test

By Gordon B. Dudder
Senior Research Engineer
Battelle Pacific Northwest Laboratory

THE DROP TOWER COMPRESSION TEST utilizes a falling weight to provide a compressive load to the specimen. The test technique has the capability to generate high loads at medium strain rates, which cannot be readily obtained by servo-hydraulic load frames or cam plastometers. Load capacities of 900 kN (100 tons) have been demonstrated for this technique, with test durations from 0.1 to 20 ms resulting in loading rates as high as 1 GN/s (115×10^3 tons/s).

Equipment

A typical drop tower compression system is shown in Fig. 6. The test system consists of a drop tower with a massive foundation, a dynamic compression fixture, and stop blocks. The 900-kN (100-ton) capacity system utilizes a commercially available drop tower with an adjustable crosshead weight of 225 to 1000 kg (500 to 2200 lb) and a maximum drop height of 1.5 m (5 ft). A 12 000-kg (26 500-lb) steel-reinforced concrete foundation is used to provide maximum rigidity to the drop tower.

The dynamic compression fixture (Fig. 7) transfers the impact load from the crosshead

Fig. 4 Working area of the loading frame with the cam follower positioned in line with the specimen for loading upon the next revolution of the cam

to the specimen through a shaft. Proper alignment of the shaft with the specimen axis is maintained throughout the test by linear ball bearings. Small nonparallelisms in the specimen ends are accommodated by a spherical seat beneath the lower platen. The impact load is sensed by a load cell located directly below the spherical seat. Stop blocks arrest the falling crosshead to avoid continued loading of the dynamic compression fixture after the shaft has reached its maximum travel. The development of the dynamic compression fixture is described in Ref 19.

A principal consideration in the design of the dynamic compression system is that the entire load train compliance must be kept quite low to achieve high loads. Gaps, mismatches, and foreign material at any of the interfaces in the load train adversely alter the rate and form of the impact loading. A second consideration is to minimize the mass of the fixture between the specimen and the centerline of the load cell. Acceleration of this mass to the test velocity causes inertia loads (Ref 20) that can obscure the actual mechanical loading of the specimen. The control of inertial effects is described in Ref 21.

Instrumentation

Equipment designed specifically for instrumented impact testing is used to acquire the load-versus-time data for the drop tower

Fig. 5 Typical testing specimens

(a) As-machined. Note the grooves on the face of the specimen. (b) Deformed specimen illustrating the barreling that occurs when lubrication is inadequate. (c) Compression specimen that has been adequately lubricated; no barreling has occurred.

(a)   (b)   (c)

Fig. 6 Drop tower with compression fixture and stop blocks in place

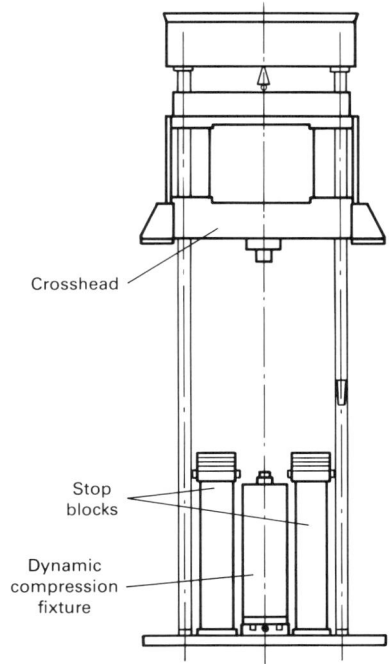

Crosshead

Stop blocks

Dynamic compression fixture

compression test. Minimum instrumentation requirements are a transducer to sense the specimen loading, the associated signal conditioning equipment, and an autographic recording instrument to provide a permanent record of the load-sensing transducer. The integrated system must be capable of sampling rates as fast as 35 kHz. The specific requirements for the load instrumentation, as well as the effect of limiting the frequency response, are described in Ref 20 and 22. The high sampling rates required for instrumented impact testing generally limit the recording equipment to oscilloscopes, transient signal recorders, and computers. Microprocessor-based data acquisition and analysis systems using high-speed analog-to-digital converters have been developed specifically for instrumented impact testing (Ref 23).

Specimen displacement or strain measurement can be added when data other than load-time history are required. Specimen displacements can be measured in the dynamic compression fixture with two parallel LVDT units symmetrically located between the upper and lower platens. The average displacements from the symmetrically located LVDT units compensate for any out-of-plane vibrations of the LVDT fixturing.

An averaging dual LVDT signal conditioner with a 20-kHz excitation frequency can be used to provide an output of the average displacement sensed by the two transducers. The resulting 2-kHz frequency response and corresponding response time of approximately 0.2 ms is an order of magnitude slower than the load-measuring capability. Accurate measurement of specimen strain can be obtained by direct strain gaging of the specimen. Signal conditioning requirements for the strain gage are analogous to those required for the load measurement.

The load output from the instrumented impact system and the averaged displacement or strain outputs can be recorded by a two-channel digital oscilloscope as a function of test time. The oscilloscope allows the recorded load and either the displacement or strain data to be displayed as a function of time, or as load versus either displacement or strain. The graphical display is output to a pen recorder to provide a permanent record of the test.

Testing Methods

The drop tower compression test has been used to measure compressive fracture strengths, to determine the compressive stress-strain behavior of material at medium strain rates, and to evaluate the dimensional stability of components subjected to impact compressive loads. Two general approaches can be used for drop tower compression testing. Fracture or compressive stress-strain tests require that the strain rate does not vary significantly during the test. This is accomplished in the drop tower test by ensuring that the available energy of the crosshead (weight times height) is no less than three times (preferably an order of magnitude) greater than the energy absorbed by the specimen and the test system.

The second test approach is used to evaluate the effect of subcritical, or low blow, loadings on the test specimen or components. In this case, the available energy of the crosshead is selected to generate a specific load. During the test, the crosshead comes to a stop and then rebounds. To avoid secondary loadings, a device for removing the specimen or catching the crosshead on the rebound must be used.

Specific specimen requirements have not been developed for the drop tower compres-

sion test. However, several specimen configurations based on right-circular cylinders and lengths greater than two times their diameter have been used. As with all compression tests, the end constraint of the specimen due to friction is a primary concern. Hardened steel and modulus-matched inserts have been used between the platens and the specimen to minimize the constraint. Other techniques for reducing the friction have been described in previous sections of this article.

Auxiliary systems can be incorporated into the design of the drop tower compression test. Possible additions include a particle collection system for post-test analysis of specimen fragments and an environmental chamber for control of temperature and atmosphere.

Limitations

The drop tower compression test is neither a constant-displacement rate nor a constant-loading rate test. The rate and form of the compressive loading depend on the specimen and test system compliances, as well as the impact velocity and available energy of the falling weight. Test conditions must be determined by trial and error or from empirically derived parameters. The test velocities and therefore the loading and strain rates are limited by both the response time of the instrumentation and the inertia loading of the

Fig. 7 Compression test fixture

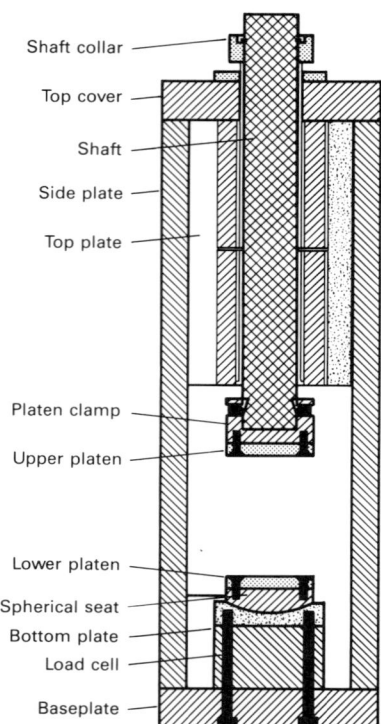

Shaft collar
Top cover
Shaft
Side plate
Top plate
Platen clamp
Upper platen
Lower platen
Spherical seat
Bottom plate
Load cell
Baseplate

system. In many cases, it is the latter that limits the test velocity. In general, it is better to drop a larger weight from a lower height than the converse when energy requirements are a consideration. The lower velocity will reduce the inertia loading and still generate comparable maximum loads.

Specimen displacement and energy calculations, which are commonly made for instrumented impact (i.e, Charpy) tests, are based on measuring the momentum impulse acting on the falling weight and calculating the resultant temporal impact velocity. These calculated values can be highly inaccurate for the drop tower compression test, because deformations in the system can be greater than the specimen deflections.

The Hopkinson Bar

By Paul S. Follansbee
Staff Member
Los Alamos National Laboratory

DEVELOPMENT OF TEST TECHNIQUES based on the Hopkinson bar has led to significant advances in high strain rate testing capabilities. These techniques yield the highest possible strain rates in a uniaxial compression test under uniform deformation conditions. In addition, the determination of stress within the deforming specimen is made without use of a load cell, and the measurement of strain is made without directly monitoring the specimen length. This section discusses the development of Hopkinson bar test techniques and provides a general description of the principles involved and of the test method, particularly the split Hopkinson pressure bar. Techniques for elevated-temperature testing are also described.

Historical Perspective

The Hopkinson bar derives its name from its developer, who in 1914 used a long elastic bar to study the pressures produced by the impact of a bullet or by the detonation of an explosive (Ref 24). In designing this experiment, Hopkinson recognized that, as long as the pressure bar remains elastic, the displacements in the pressure bar are directly related to the stresses and that the length of the wave in the bar was related to the duration of the impact through the velocity of sound in the bar, a quantity that was well known. A significant feature of this work was that the pressures were estimated by measuring the momentum (with a ballistic pendulum) acquired by a small section of the bar placed in contact with the bar at the far end.

Further developments in the experimental techniques occurred a few decades after these original experiments, when Davies (Ref 25) and Kolsky (Ref 26) designed condensers to measure displacements in the pressure bars. Kolsky also introduced the split Hopkinson pressure bar technique, in which the specimen is sandwiched between two pressure bars. He demonstrated how stress and strain within the deforming specimen are related to displacements in the pressure bars. Because of these contributions, the split Hopkinson pressure bar is often referred to as the Kolsky bar.

In the last two decades, there have been numerous advances in experimental procedures based on the Hopkinson bar. The use of strain gages to measure surface displacements on the elastic pressure bars was first reported in 1961 (Ref 27), and a technique for performing elevated-temperature tests was described in 1963 (Ref 28). The Hopkinson bar was first configured for torsional loading in 1966 (Ref 29). Lindholm introduced a unique procedure for testing in tension (Ref 30) and has provided reviews of Hopkinson bar testing techniques that have become standards for current procedures (Ref 31, 32).

Basic Principles

As identified by Hopkinson, the use of a long, elastic bar to study high-rate phenomena is feasible, because the wave propagation behavior in such a geometry is well understood and mathematically predictable. Thus, the displacements or stresses generated at any point can be deduced by measuring the elastic wave at any point as it propagates along the bar.

There are two basic configurations of the Hopkinson bar for compression testing. The most widely used is the split Hopkinson pressure bar, shown in Fig. 8(a). Several investigators have recently used the single pressure bar configuration shown in Fig. 8(b). The basic principles of the two techniques are similar. Thus, the remainder of this discussion will focus on the operation of the split Hopkinson pressure bar; differences between the two techniques will be emphasized as appropriate.

The split Hopkinson pressure bar consists of two elastic pressure bars that sandwich the specimen between them. Typically, a striker bar is propelled toward the incident bar. Upon impact, an elastic compressive wave is generated within the incident bar, and the time-dependent strain, $\epsilon_I(t)$, in the pressure bar is measured at strain gage A, located at the midpoint of the incident bar. At the incident bar/specimen interface, the wave is partially reflected and partially transmitted into the specimen. The portion that is reflected travels back along the incident bar as a tensile wave, and the strain, $\epsilon_R(t)$, is measured by strain gage A.

The strain gage on the incident bar is located at the bar midpoint so that the incident and reflected waves can be measured independently. If the strain gage were located close to the incident bar/specimen interface, the leading edge of the reflected wave could interfere with the trailing edge of the incident wave. The compressive strain, $\epsilon_T(t)$, associated with the portion of the wave that is transmitted through the sample into the output bar is measured by strain gage B, located at the midpoint of the output bar.

When the specimen is deforming uniformly, the strain rate within the specimen is directly proportional to the amplitude of the reflected wave. Likewise, the stress within the sample is directly proportional to the amplitude of the transmitted wave. These two signals can be recorded, the former integrated to yield strain, and combined to give the dynamic stress-strain curve.

Relating Strain Gage Measurements to Stress-Strain Behavior. The strain rate in the deforming specimen is:

$$\frac{d\epsilon}{dt} = \frac{V_1 - V_2}{L} \qquad \text{(Eq 11)}$$

where V_1 and V_2 are the velocities at the incident bar/specimen and specimen/output bar interfaces, respectively, and L is the length

Fig. 8 Typical configurations of the Hopkinson bar

(a) Split Hopkinson pressure bar with specimen sandwiched between two long elastic pressure bars, each of which is instrumented at its midpoint with a strain gage. (b) Single pressure bar with striker bar impacting the specimen directly. With the latter configuration, the motion of the striker bar or displacement within the specimen must be monitored independently.

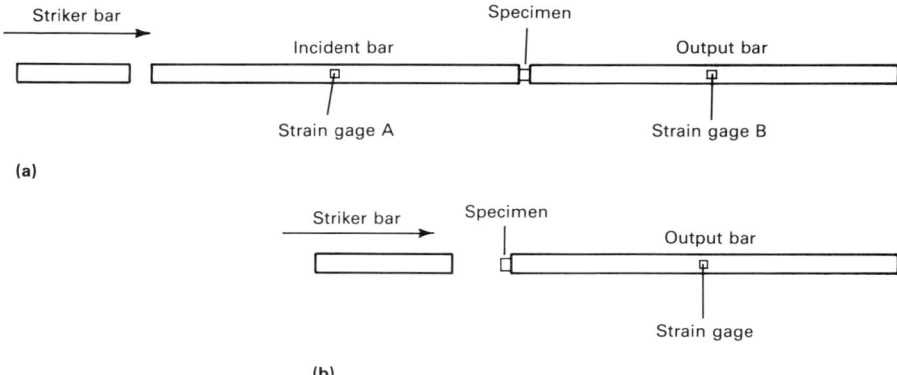

(a)

(b)

of the specimen. The velocity V_1 is the product of the longitudinal sound velocity (C_o) in the pressure bar and the total strain at the incident bar/specimen interface, which is $\epsilon_I - \epsilon_R$. Similarly, the velocity V_2 is equal to $C_o \epsilon_T$. In this development, ϵ_I, ϵ_T, and ϵ are all compressive strains, but are considered positive, whereas the quantity ϵ_R, which represents a tensile strain, is negative. For more information, see Ref 31 and 32. Replacing V_1 and V_2 in Eq 11 with these expressions yields:

$$\frac{d\epsilon(t)}{dt} = \frac{C_o}{L} [\epsilon_I(t) - \epsilon_R(t) - \epsilon_T(t)] \quad \text{(Eq 12)}$$

The average stress on the specimen is:

$$\sigma(t) = \frac{P_1(t) + P_2(t)}{2A} \quad \text{(Eq 13)}$$

where P_1 and P_2 are the forces at the incident bar/specimen and specimen/output bar interfaces, respectively, and A is the instantaneous cross-sectional area of the specimen. At the incident bar/specimen interface, the force is:

$$P_1(t) = E[\epsilon_I(t) + \epsilon_R(t)]A_o \quad \text{(Eq 14)}$$

where E is Young's modulus, and A_o is the cross-sectional area of the bar. Likewise, the force at the specimen/output bar interface is:

$$P_2(t) = E \, \epsilon_T(t)A_o \quad \text{(Eq 15)}$$

Combining Eq 14 and 15 with Eq 13 yields:

$$\sigma(t) = \frac{E}{2} \frac{A_o}{A} [\epsilon_I(t) + \epsilon_R(t) + \epsilon_T(t)] \quad \text{(Eq 16)}$$

When the specimen is deforming uniformly, the stress at the incident bar/specimen interface equals that at the specimen/output bar interface and from Eq 14 and 15:

$$\epsilon_I(t) + \epsilon_R(t) = \epsilon_T(t) \quad \text{(Eq 17)}$$

from which Eq 12 and 16 may be simplified:

$$\frac{d\epsilon(t)}{dt} = \frac{-2C_o}{L} \epsilon_R(t) \quad \text{(Eq 18)}$$

$$\sigma(t) = E \frac{A_o}{A} \epsilon_T(t) \quad \text{(Eq 19)}$$

Thus, the stress-strain behavior of the specimen is determined simply by measurements made on the elastic pressure bars in a split Hopkinson pressure bar test. The analysis applies as well to the single bar configuration shown in Fig. 8(b). However, because the striker bar impacts directly with the specimen in this test, the strain $\epsilon_R(t)$, which now represents the wave generated in the striker bar, cannot be measured with a strain gage, and another measurement technique is required. One investigator has chosen to measure $\epsilon_R(t)$ with a coaxial capacitor (Ref 33), while another has monitored the strain in the specimen with high-speed photography (Ref 34).

The above equations relating strain gage measurement to stress-strain behavior in the deforming specimen require that two important conditions be met. The first is that wave propagation within the pressure bars must be one-dimensional, if surface displacement measurements are used to represent the axial displacement over the entire cross-sectional area of the pressure bar. The second condition is that the specimen must deform uniformly; this is opposed by both radial and longitudinal inertia and by frictional constraint at the specimen/pressure bar interfaces. These two conditions are discussed in more detail below.

Wave Propagation in Elastic Pressure Bars. When the striker bar impacts the

incident bar, the wave that is generated in the latter is highly complex because of end effects due to nonuniformities at the striker bar/incident bar interface and the propagation of other types of elastic waves (e.g., spherical dilational waves). However, these end effects quickly dampen after the wave has propagated about ten bar diameters. The wave propagation behavior then becomes fully described by the equation of motion in the infinite cylindrical solid. The solution of this equation for these boundary conditions was provided by Pochhammer (Ref 35) and Chree (Ref 36), and its application to the Hopkinson bar is described in Ref 25, 37, and 38. It should be noted that the pressure bars vibrate in the fundamental mode. That is, although there are an infinite number of solutions to the equation of motion according to the vibrational mode, only one of these appears to dominate vibration in the conditions of the typical Hopkinson bar test.

There are additional consequences of the predicted wave propagation behavior. If the wave is a pure cosine wave of wavelength λ, then axial displacements and stresses are uniform over the cross section of the bar as long as $R/\lambda \ll 1$, where R is the radius of the bar. In fact, when $R/\lambda < 0.1$, the displacements at the surface of the bar differ from those along the bar axis by less than 5% (Ref 25). The wave $f(t)$ generated in the incident bar is not a pure cosine wave, but it can be represented as a Fourier series, or a summation of many cosine waves:

$$f(t) = \frac{A_o}{2} + \sum_{n=1}^{\infty} D_n \cos (n\omega_o t - \delta_n) \quad \text{(Eq 20)}$$

where ω_o is the frequency of the longest wavelength component, $A_o/2$ is the mean value of the wave, D_n is the amplitude of frequency component $n\omega_o$, and δ_n is the phase angle of this component. The majority of the energy in waves generated in the split Hopkinson pressure bar is contained in wavelengths that exceed $10R$; thus, the wave can be assumed to be one-dimensional, and surface displacement measurements are accurate indicators of axial displacements in the pressure bar (Ref 38).

Another important consequence of wave propagation in the pressure bars is that the longitudinal velocity varies with the wavelength, which leads to wave dispersion. For $R/\lambda \ll 1$, the propagation velocity is the well-known longitudinal sound velocity $C_o = \sqrt{E/\rho}$, but as λ approaches R, the velocity of these short wavelength components decreases to $0.576 \, C_o$ (for a bar material with a Poisson's ratio of 0.29). The lower velocity of the short wavelength components causes dispersion of the initial wave and leads to the appearance of oscillations, called Pochhammer-Chree oscillations, that ride on the waves and build during propagation down the bar. These oscillations are simply short wavelength frequency components that lag behind the leading edge of the wave. These oscillations have only a minor influence on the predicted stress-strain behavior (Ref 38).

Conditions for Uniform Deformation Within the Sample. Whereas wave propagation in the elastic pressure bars has been shown to be predictable mathematically, it has been far more difficult to establish conditions for uniform deformation within the sample, because this involves plastic wave propagation and the consideration of frictional constraint. These topics are discussed in the "Introduction" to this article. In terms of Hopkinson bar testing, when the stress wave enters the sample, particles of the sample are accelerated both longitudinally and radially. The first consideration is the time t required for the stress to equilibrate within the sample. It has been estimated (Ref 39) that this requires three (actually π) reverberations of the stress wave, which for a plastically deforming solid obeying the Taylor-von Karman theory yields:

$$t = \frac{\pi^2 \rho_s L^2}{\dfrac{d\sigma}{d\epsilon}} \qquad \text{(Eq 21)}$$

where ρ_s is the density of the specimen, L is the specimen length, and $d\sigma/d\epsilon$ is the slope of the true stress/true strain curve. At times less than that given by Eq 21, the sample may not be assumed to be deforming uniformly, and stress-strain data may be in error.

An important consideration in split Hopkinson pressure bar testing is the design of the experiment, such that the time t from Eq 21 is reached early in the test. One way to decrease t is to decrease the specimen length. Because other considerations, described below, limit the length-to-diameter ratio of the specimen, the specimen length may not be decreased without a concomitant decrease in the specimen and bar diameters. The use of smaller diameter bars to achieve higher strain rates is a common practice in split Hopkinson pressure bar testing. Tests in copper and aluminum at strain rates as high as 6×10^4 s^{-1} have been performed using 4.7-mm (0.19-in.) diam pressure bars (Ref 40). An additional benefit of decreasing the radius of the bar is that this also decreases the ratio R/λ, which reduces the effects caused by dispersion discussed above.

Because the value of t from Eq 21 cannot be decreased significantly, an alternate method that will yield valid data at earlier strains is to increase the rise time of the incident wave. A symmetric impact between the striker bar and the incident bar yields a short rise time pulse that approximates a square wave. The rise time of this wave is likely to be less than t in Eq 21, but if the rise time of the wave is increased to a value more comparable with t, then the data will be valid at an earlier strain than if a short rise strain pulse is used. Furthermore, because the highly dispersive short wavelength components arise from the leading and trailing edges of the waves, a long rise time pulse will contain fewer of these components than will a sharply rising pulse. Experimental techniques to shape the pulse to increase the rise time will be discussed later in this article.

The two material properties in Eq 21 are the density and the slope of the stress-strain curve. The latter is important because it can vary significantly from material to material. Thus, certain experimental conditions of strain rate and bar and specimen dimensions may satisfy Eq 21 for a strongly strain-hardening material such as copper, but these same conditions may not satisfy Eq 21 for a weakly strain-hardening material or for a material exhibiting a yield point phenomena.

Even when the specimen is determined to be deforming uniformly, longitudinal and radial inertia due to the rapid particle accelerations imposed at high strain rates can influence the predicted stress-strain behavior. The errors due to both longitudinal and radial inertia have been analyzed, and corrections have been derived for these errors (Ref 39). This approximate analysis resulted in:

$$\sigma(t) = \sigma_m(t) + \rho_s \left[\frac{L^2}{6} - v_s \frac{D^2}{8} \right] \frac{d^2\epsilon(t)}{dt^2}$$

$$\text{(Eq 22)}$$

where σ_m is the measured stress, ρ_s is the density of the specimen, v_s is the Poisson's ratio, L is the specimen length, and D is the specimen diameter. This expression has proved to be useful, because it predicts that errors are minimized if the strain rate is held constant, or if the term inside the brackets is set to zero by choosing specimen dimensions such that:

$$\frac{L}{D} = \sqrt{\frac{3v_s}{4}} \qquad \text{(Eq 23)}$$

For a Poisson's ratio of 0.33, Eq 23 suggests that the optimum L/D ratio to minimize errors due to inertia is 0.5. This L/D ratio is less than that determined to be the most favorable for the minimization of errors due to friction in ASTM Standard E 9, which specifies that $1.5 \geq L/D \leq 2.0$ (Ref 41). However, the total strain in a split Hopkinson pressure bar test is limited to approximately 25% to reduce the area mismatch between the specimen and the pressure bar. Thus, it is not expected that a L/D ratio of 0.5 will introduce serious errors, provided that the interfaces are well lubricated. The subject of friction is discussed later in this article.

Equipment and Test Method

The split Hopkinson pressure bar test apparatus consists of the pressure bars, associated mounting and alignment hardware, gas gun or alternate method producing the compressive wave, strain gage or condenser for measuring the waves, and the associated instrumentation and data acquisition system. There currently is no standard design for a split Hopkinson pressure bar test apparatus. Thus, although the following discussion is as general as possible, specific examples are based on one specific facility, as shown in Fig. 9.

Pressure bars should be constructed from a high-strength material; AISI-SAE 4340 steel or maraging steel are common choices. The yield strength of the pressure bar determines the maximum stress attainable within the deforming specimen, because the cross-sectional area of the specimen approaches that of the pressure bar during deformation. The length, l, and diameter, d, of the pressure bars are test variables. The diameter chosen depends on the strain rate desired; the highest strain rate tests require the smallest diameter bar. The length of the pressure bars is determined by conditions required for one-dimensional wave propagation, which requires approximately 10 bar diameters. Thus, each bar should exceed an l/d ratio of 20.

A second consideration affecting the bar length is the amount of strain desired in the specimen, which is related to the length of the incident wave. The pressure bar must be at least twice as long as the incident wave if the incident and reflected waves are to be measured independently. To obtain strains as high as 25%, it may be necessary to specify an l/d ratio of 100 or more.

Another important consideration for the pressure bars is the minimization of axial curvature. This can be a particular problem for materials that are hardened by heat treatment and for bars with high aspect ratios. However, optimum alignment is essential to the split Hopkinson pressure bar test technique, and the amount of curvature within the pressure bar affects the alignment.

Bar Alignment. Accurate bar alignment is required for ideal one-dimensional wave propagation within the pressure bars and for uniaxial compression within the specimen. However, alignment cannot be forced by overconstraining the pressure bars, because this violates the boundary conditions for one-

Fig. 9 Split Hopkinson pressure bar test facility at Los Alamos National Laboratory

The gas gun is at the far end; the gas-activated bar stopper and bore scope are mounted at the near end. The pressure bars are 9.2 mm (0.36 in.) in diam by 1.22 m (4 ft) long. Also shown are the bar-mover assembly and diffusion pump for elevated-temperature tests.

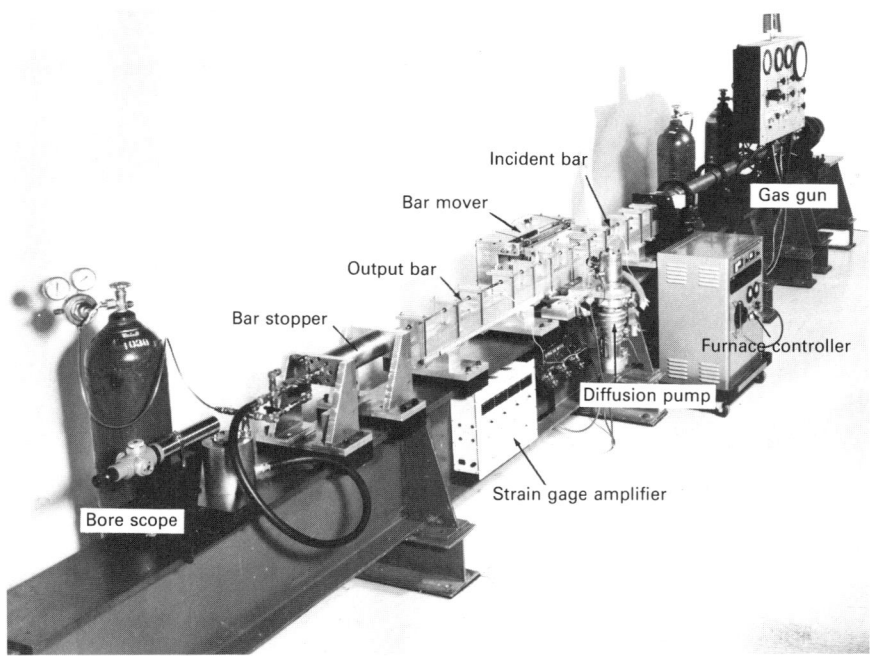

dimensional wave propagation in an infinite cylindrical solid. This is one reason for a tight specification on curvature in the pressure bars. In the device shown in Fig. 9, the pressure bars are mounted to a rigid beam. Mounting brackets with bearings, through which bars pass, are spaced every 100 to 200 mm (4 to 8 in.), depending on the bar diameter, and are designed so that each can be individually translated for alignment adjustment.

An optical bore scope at the near end of the beam is used to set and regularly verify alignment. The bearings in the mounting brackets are fabricated from oilite and are drilled to a diameter of 25 to 50 μm (0.001 to 0.002 in.) greater than the bar diameter to allow for radial expansion. A simple check of alignment is to verify that each pressure bar translates and rotates freely through the mounting brackets and that faces of the incident and output bars meet identically when brought together without the specimen in place.

Generating the Incident Wave. The most common method of generating the incident wave is to propel a striker bar toward the incident bar. Waves also can be generated by the detonation of explosives at the free end of the incident bar. However, it is more difficult to ensure a one-dimensional excitation within the incident bar using explosive techniques.

For the striker bar configuration, the projectile can be propelled using stored elastic energy of, for example, a spring, but the use of a gas gun is most common (Fig. 9). The striker bar is fabricated from the same material and is of the same diameter as the pressure bars. The length and velocity of the striker bar are chosen to yield the desired total strain and strain rate within the specimen. At a constant strain rate, the maximum strain in the specimen is directly proportional to the length of the striker bar, l:

$$\epsilon = 2\dot{\epsilon}\,\frac{l}{C_o} \qquad \text{(Eq 24)}$$

From inspection of Eq 18 and consideration of momentum conservation between the striker and incident bars, it can be shown that:

$$\dot{\epsilon} \leq \frac{V_o}{L} \qquad \text{(Eq 25)}$$

where V_o is the velocity of the striker bar, and L is the length of the specimen. The equality condition in Eq 25 is approximated for soft metals at high impact velocities. Equations 24 and 25 can be used to approximate the striker bar length and velocity required for a desired strain and strain rate.

The need for a long rise time pulse was discussed above concerning the conditions required for the equilibration of stress within

the specimen. Experimentally, the rise time of the incident wave can be increased by placing a soft, deformable disk at the striker bar/incident bar interface. The choice of material and thickness for this disk depends on the strain rate and the strength of the specimen. Typically, the disk is the same material as the specimen and is 0.1 to 2 mm (0.004 to 0.08 in.) thick. An additional benefit of this layer is that it often provides a more uniform strain rate experiment.

Friction is an important consideration in all compression testing. The optimum L/D ratio for a split Hopkinson pressure bar compression test specimen is approximately one half that determined to be most favorable for the minimization of errors due to friction. Thus, lubrication is required at the specimen/pressure bar interfaces. However, the presence of a layer of lubricant at these interfaces can affect the timing between the waves recorded on the incident and output pressure bars.

For example, consider the test of a 5-mm (0.2-in.) long specimen at a strain rate of $5 \times 10^3 \text{ s}^{-1}$. Equation 25 indicates that, for these conditions, the projectile velocity, and thus the particle velocity in the incident bar, must be at least 25 m/s (82 ft/s). For a 25-μm (0.001-in.) layer of lubricant at the specimen/incident bar interface, approximately 0.5 μs is required for the incident bar, traveling at this velocity, to contact the specimen. There will be a similar delay at the specimen/output bar interface. These delays are not negligible and can alter the timing relation between the reflected and transmitted waves used to compute stress and strain within the deforming specimen. Consequently, it is important to maintain a thin layer of lubricant.

The use of an oil-based molybdenum disulfide lubricant is successful for room-temperature testing. Typically, the lubricant is applied to the faces of the specimen, and the excess is lightly removed. For elevated-temperature tests, a thin layer of fine boron nitride powder can be used to lubricate the specimen/pressure bar interfaces.

Measurement of Displacement in the Pressure Bars. Equations 18 and 19 showed that the strain rate and stress within the samples are related to axial strains in the pressure bars, and it also was shown that, for conditions of most split Hopkinson pressure bar tests, the surface strains accurately reflect the axial strains. There are two common techniques for measuring these strains. The use of strain gages is most popular. Two gages generally are mounted at diametrically opposite positions on each bar and connected so as to average out any bending strain. Standard strain gage technology, and the use of paperback gages, has been successful. The major difficulty has been the development of techniques to attach lead wires to the strain gages. These lead wires experience large accelerations and can easily break when the stress wave passes.

An example of a strain gage installation on a 9.2-mm (0.36-in.) bar is shown in Fig. 10. The strain gage is barely visible on the left, and the 28-gage (0.3785-mm, or 0.0149-in.) insulated copper lead wires approach the strain gage from the right. There is another strain gage mounted at a diametrically opposite position on the bar. The lead wires are securely lashed to the bars with monofilament nylon fishing line. A short exposed piece of the lead wire is soldered to the internal gage lead wire. In Fig. 10, the solder joint is covered by the black vinyl electric tape. Note also that the gage lead wire is slightly arched; this added flexibility appears to protect the lead wire and solder junction when the stress wave passes.

Bonded strain gages do have a finite response capability. The rise time of the gage in response to a step change in displacement increases with the length of the strain gage (Ref 42). Typically, gages attached to pressure bars may be expected to have response times less than 1 μs, which is sufficient for conditions of the split Hopkinson pressure bar test.

Another device used to measure strain in the pressure bars is the coaxial condenser or capacitor. Although the advance in strain gage technology has all but eliminated the use of condensers in the split Hopkinson pressure bar configurations, their use continues for the single pressure bar configuration.

Instrumentation and Data Acquisition. The instrumentation required for the split Hopkinson pressure bar test includes two strain gage signal conditioners and a means of recording these signals. The strain gage signal conditioners must have a frequency response of at least 1 MHz and as high as 10 MHz. Until recently, oscilloscopes were used almost exclusively to capture and record the pulses. The reflected wave may be fed through an analog integrator to yield a signal proportional to strain. Thus, the transmitted wave and integrated reflected wave can be fed to an oscilloscope with x-y capability to directly yield the stress-strain curve.

With the advent of precise, high-speed analog-to-digital recorders, it is now possible to digitize the raw data directly and perform the integration numerically. The availability of the digitized data also facilitates adjustment of the timing between the transmitted and reflected waves to account for the transit time through the specimen; this is described in more detail in the section of this article on test limitations.

Testing at Elevated Temperatures

There are several complications encountered in split Hopkinson pressure bar testing at elevated temperatures. Because the stress and strain within the deforming specimen are determined by strain gage measurements made on the elastic pressure bars, the velocity of sound and elastic modulus of the pressure bars, both of which vary with temperature, are important parameters. The combined length of both pressure bars (which can easily exceed 2 m, or 6.5 ft) makes it difficult to heat the entire assembly. This also would require a bar material capable of withstanding the high temperature and the development of new strain gage technology.

The more common procedure for elevated-temperature testing is to heat only the sample and perhaps a short section of the pressure bars and then to allow a temperature gradient in the pressure bars. If the temperature gradient can be estimated or measured, the strain gage signals can be corrected for the temperature-dependent sound velocity and elastic modulus (Ref 29, 30). This procedure has been used at temperatures up to 613 °C (1135 °F).

Scientists at Southwest Research Institute have circumvented these problems by using short sections of aluminum oxide pressure bars in direct contact with the specimen inside a furnace. Longer steel pressure bars are placed in contact with the cold ends of the ceramic bars. Because the acoustic impedance of aluminum oxide is almost equal to that of steel, stress waves pass through these interfaces unperturbed. Tests have been performed on ceramic specimens at temperatures up to 1500 °C (2730 °F) with this configuration (Ref 43).

Another method of performing elevated-temperature split Hopkinson pressure bar tests has been demonstrated recently (Ref 44). With this procedure, the sample and pressure bars initially are separated, and the sample is heated independently with a specially designed miniature furnace. Just before the striker bar is propelled toward the incident bar, the two pressure bars are mechanically brought into contact with the hot specimen. Contact time is minimized to less than 200 ms so that the specimen temperature and temperature at the ends of the bars remain approximately constant. This technique, which does not require numerical corrections to the strain gage data, has been used successfully at temperatures up to 1000 °C (1830 °F). However, each test requires the fabrication of a furnace assembly, and the timing between the movement of the pressure bars and the firing of the striker bar is critical.

Fig. 10 Strain gage installation on a 9.2-mm (0.36-in.) diam pressure bar
Note the arch in the strain gage lead wire; this appears to provide additional protection to the lead wire and solder joint (located beneath the black tape) when the stress wave passes.

Test Limitations

Strong interest in the effect of strain rate on mechanical properties has led to the application of split Hopkinson pressure bar testing techniques at strain rates that are at the limit of, and perhaps beyond, conditions for a valid test. Some necessary conditions for a valid test are given in Eq 21 to 23. However, these expressions are only approximate and do not guarantee accurate test results. The basic limitation is the condition of uniform deformation within the sample. It has been emphasized that this problem cannot be circumvented simply by decreasing the length of the sample, because the aspect ratio of the specimen is specified by consideration of frictional constraint and the balancing of error due to radial and axial inertia.

Determination of test validity in the split Hopkinson pressure bar experiment has stimulated a number of analyses of wave propagation in the split Hopkinson pressure bar test configuration. The most extensive of these was a two-dimensional finite difference analysis of a split Hopkinson pressure bar test on aluminum (Ref 45). This analysis considered the effects of wave shape, friction, strain rate, and length-to-diameter ratio. One conclusion was that if the incident wave is a ramp wave with a rise time t_r given by

$$\frac{t_r}{d} \geq 16 \ \mu s \cdot cm^{-1} \tag{Eq 26}$$

where d is the bar diameter, then the results of experiments with aluminum are valid provided that:

$$d\dot{\epsilon} \leq 5 \times 10^3 \ cm \cdot s^{-1} \tag{Eq 27}$$

These equations quantify the combined effect of the errors discussed in this article and provide an indication of the test limitations for aluminum. Equation 26 is analogous to Eq 21, in that both emphasize the importance of increasing the rise time of the wave, while Eq 27 shows that the bar diameter strongly influences test validity.

The optimum design of a split Hopkinson pressure bar test would call for a long rise time during the elastic loading of the specimen so that the specimen would be deforming uniformly when the yield stress was reached. However, this is seldom possible to achieve, and at high strain rates it is likely that the data may not be valid until a strain of 5 or 10% is achieved. The problems associated with establishing uniform deformation within the sample make it very difficult, if not impossible, to accurately determine the dynamic yield strength using the split Hopkinson pressure bar.

The timing between the reflected and transmitted waves strongly affects interpretation of the test results at high strain rates. Typically, the strain gages are placed equidistant from the sample, and it is assumed that these waves arrive at the respective strain gages at the same instant. There are two sources of error that affect this assumption. The first is the transit time through the specimen, which is determined by the elastic sound velocity and the specimen length. This error is predictable, decreases with the bar diameter (for a constant aspect ratio specimen), and is easily incorporated into the data reduction procedure.

However, another source of error is the transit time at the imperfect specimen/pressure bar interfaces. This error is far less reproducible, does not scale with bar diameter, and may be as large as 1 or 2 μs. Uncertainty in the timing between the reflected and transmitted waves affects the predicted stress-strain curve, because the strain and stress are derived from these signals.

This error increases the uncertainty in the measurements as the strain rate increases and sets the practical limit of split Hopkinson pressure bar testing closer to a strain rate of $10^4 \ s^{-1}$ than to $10^5 \ s^{-1}$.

Rod Impact (Taylor) Test

By David C. Erlich
Physicist
SRI, International

DETERMINATION of high strain rate constitutive behavior of materials is of increasing interest to researchers in many fields. However, the availability of experimental techniques that allow such determinations at high strain rates (around 10^4 or 10^5 s^{-1}) and large plastic strains (50 to 150%) has been extremely limited, if not nonexistent. Recently, the rod impact test, based on the Taylor test for measuring the dynamic yield strength (Ref 46), has been developed. The technique has been improved so that the entire stress-strain flow curve can be determined for a material at ambient or elevated temperatures.

Historical Perspective and Basic Principles

Classic Taylor Test. In 1947, Taylor and Whiffin (Ref 46, 47) accelerated cylindrical specimen rods into a "rigid" plate (Fig. 11a). The plastic deformation at the impact end shortens the rod, and the fractional change in rod length can, by one-dimensional rigid-plastic analysis, be related to the dynamic yield strength. This relationship was shown to be independent of both the rod aspect ratio and the impact velocity for a wide variety of materials, including copper, lead, paraffin wax, and various steels.

Although appealing in its simplicity, the Taylor test received only moderate interest. In 1954, Lee and Tupper (Ref 48), using a one-dimensional characteristics code with an elastic-plastic model, attempted a theoretical determination of the strain distribution in a Taylor test specimen rod. In 1968, Hawkyard et al., (Ref 49) performed Taylor tests with copper and low-carbon steel specimens at temperatures from 20 to 700 °C (68 to 1290 °F). They investigated several one-dimensional analyses, but found that none could successfully predict the final rod deformation.

Twenty-five years after the original Taylor tests, the use of two-dimensional wave propagation codes enabled a better understanding of and renewed interest in this technique. In 1972, Wilkins and Guinan (Ref 50), using a two-dimensional finite difference code and an elastic-plastic model with work hardening, were able to correctly simulate the final shapes and final lengths of Taylor test specimens of several metallic alloys at ambient temperatures. Their results showed good correlation between the dynamic yield strength and the fractional change in rod length for a wide range of impact velocities and rod aspect ratios, thus confirming many of the Taylor/Whiffin conclusions.

Fig. 11 Two rod impact configurations
(a) Classic Taylor test. (b) Symmetric rod impact test

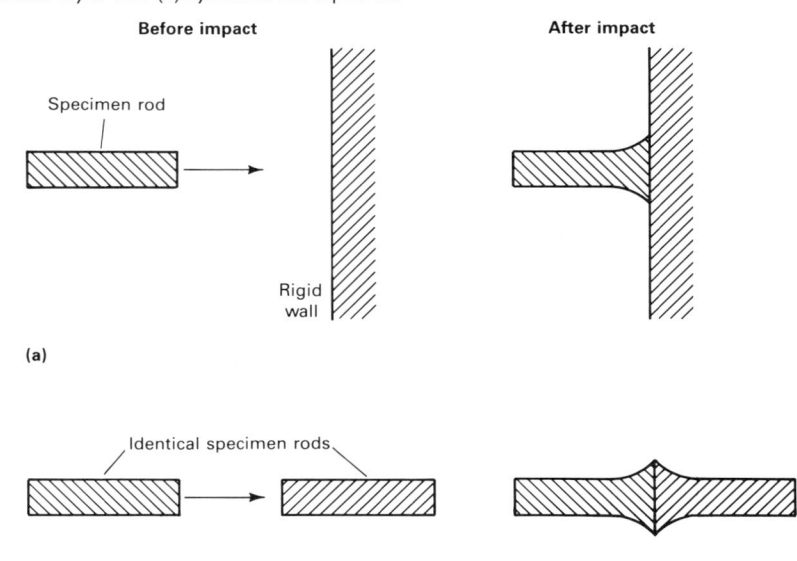

Symmetric Rod Impact Test. In the early 1980's, Erlich *et al.* (Ref 51) implemented two major modifications to the classic Taylor technique as a method for obtaining the entire stress-strain flow curve for materials undergoing high strain rate compressive and shear loading to large plastic strains. The first was to use ultrahigh-speed photography to monitor the deformation history of the specimen rod. This allows intermediate as well as final deformation states to be compared with two-dimensional computer simulations, thus improving the reliability of the flow curve determination.

The second modification was to replace the rigid plate with another rod of the same geometry and material as the impacting rod (Fig. 11b). This arrangement, referred to as the ''symmetric rod impact'' technique, allows the impacting ends of the two specimen rods to deform together symmetrically, thus eliminating boundary condition uncertainties in the analysis that arise from the unknown friction conditions at the rod/plate interface and from the deformation of the rigid plate adjacent to that interface. By using the symmetric rod impact technique, dynamic flow curves at ambient temperature were obtained for 6061-T6 aluminum (Ref 51), a titanium alloy, and 4340 steel over a wide range of initial hardnesses (Ref 52).

Asymmetric Rod Impact Test. In 1982, Gust (Ref 53) used a reverse ballistics variation of the classic Taylor test to measure fractional changes in the length of various metallic rods at initial temperatures up to about 1000 °C (1830 °F). In this variation, a rigid plate is launched into a stationary specimen rod that is preheated to the desired temperature.

Concurrently, researchers, frustrated in their attempts to use the symmetric rod impact technique to determine dynamic flow curves at elevated temperatures (because of the infeasibility of heating the moving specimen rod), were investigating several alternatives. One possibility was to combine the reverse ballistics impact of a heated rod with high-speed photographic measurements of the resulting deformation to form the asymmetric rod impact. However, the boundary condition uncertainties inherent in the asymmetric rod impact technique, mainly the effect of friction at the rod/plate interface, needed to be examined.

At room temperature, identical 4340 specimen rods were therefore impacted in two experiments—one a symmetric rod impact test and the other an asymmetric rod impact test at half the impact velocity. The same stress obtained by symmetric impact is reached by impact into a rigid target at half the velocity. The resulting specimen rod deformations were not exactly the same, and

metallographic examination of the recovered asymmetric rod impact specimen showed some frictional effects in a very narrow region at the center of the impact end.

However, by allowing the rigid impacter (an ultrahigh-strength steel) to deform according to its known material properties, instead of assuming incompressibility, and by using a frictionless rod/plate interface condition, computer simulation of both tests yielded identical flow curves to within the experimental uncertainties. Thus, it was concluded that frictional effects, although existent, were insufficient to significantly influence the analysis of the asymmetric rod impact technique, at least for this combination of specimen and impacter materials. Further tests are planned to validate this conclusion for a wider range of materials. Once accomplished, the asymmetric rod impact technique may be used to determine dynamic flow curves at ambient or elevated temperatures.

Symmetric Rod Impact Test at Ambient Temperature

A typical experimental arrangement for the symmetric rod impact test at ambient temperature is shown in Fig. 12. The specimen rods are identical right circular cylinders, 44.4 mm long by 9.5 mm (1.75 by 0.37 in.) in diameter (arbitrary dimensions), the ends of which are machined flat and parallel to within about 0.01 mm (0.0004 in.). The impacting rod is mounted on the front end of a projectile, which is accelerated by expand-

ing helium in a 63.5-mm (2.5-in.) gas gun. Projectile velocity is recorded by a series of contact pins located near the end of the gun barrel. The stationary rod is held in place by six ceramic fingers attached to a target-mounting fixture, which in turn is affixed to an alignment plate at the muzzle of the gun. The position of the latter rod can be adjusted by rotating the threaded bars, into which the ceramic fingers are inserted. Alignment of the two rods is critical to ensure that the impacting ends are parallel and coaxial.

The specimen rods are backlit by a fast rise- and fall-time, high-intensity light source (xenon flash tube or exploding bridge wires) triggered just before impact, and the silhouettes of the deforming rods are recorded by a high-speed framing camera at framing rates between one half and one million frames per second. Select frames from a typical SRI test are shown in Fig. 13.

After deformation is complete (about 30 to 40 μs after impact), the specimen rods fly into a recovery pipe filled with rags or other energy-absorbing materials that minimize additional deformation. The pipe is sufficiently narrow to prevent the projectile from entering and re-impacting the specimen. The recovered rods are then sectioned along the axis and examined metallographically to ascertain the extent of internal damage.

The impact velocity must be low enough to suppress the formation of tensile voids (which may occur at early times by the focusing of the radial release waves on the rod axis) or shear bands (which may occur at later times as a result of large plastic deformation near the impact end). Although a small amount of incipient damage can be tol-

Fig. 12 Symmetric rod impact tests at ambient temperature

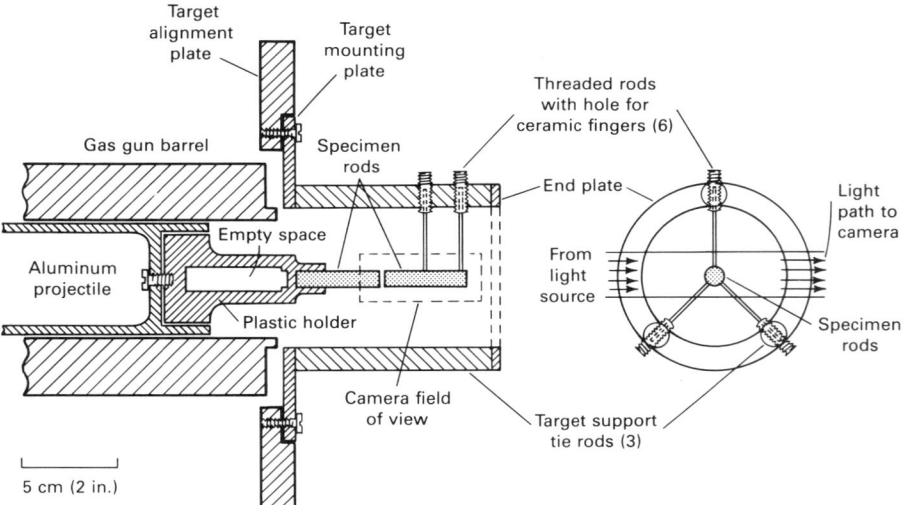

Side view   End view

Fig. 13(a) Silhouettes of 4340 steel (94 HB) rods during symmetric impact at 457 m/s (1500 ft/s)

Times shown are approximate times from impact. Original rod dimensions are 9.5 by 44.5 mm (0.37 by 1.75 in.). Moving rod impacts from the right.

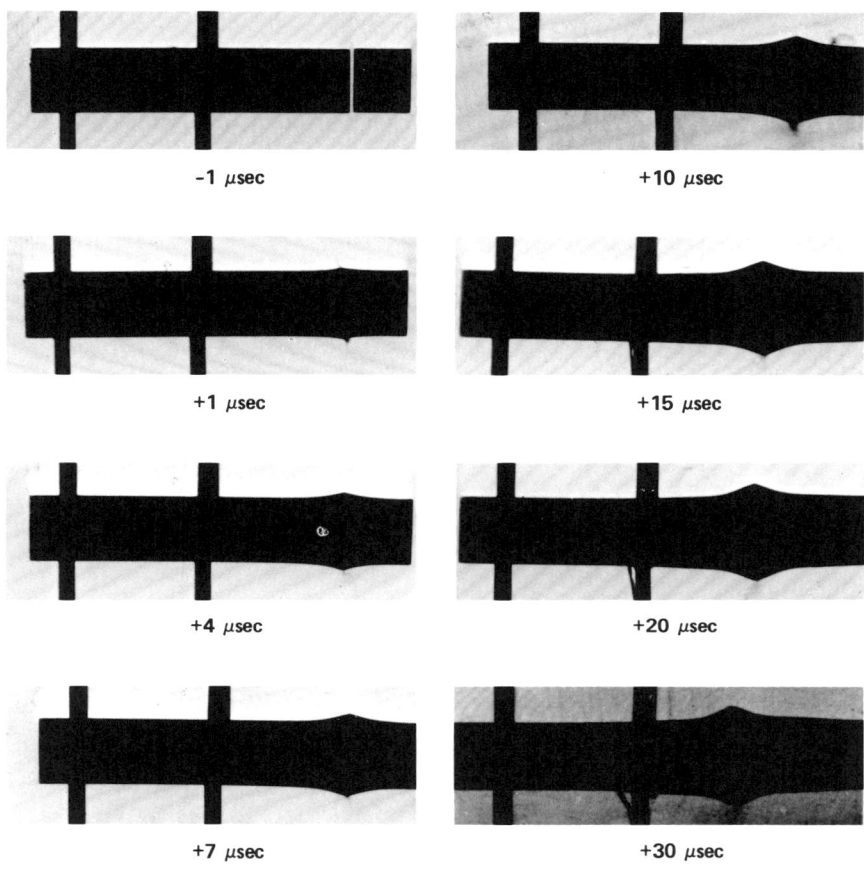

−1 μsec   +10 μsec

+1 μsec   +15 μsec

+4 μsec   +20 μsec

+7 μsec   +30 μsec

erated, any significant amount of damage may affect the shape of the deforming rod profiles.

Asymmetric Rod Impact Test at Elevated Temperatures

As mentioned previously, one advantage of the asymmetric rod impact test is that elevated temperature tests can be performed without heating a moving projectile, as required with the symmetric rod impact test. An elevated-temperature test with the asymmetric rod impact technique only requires that the target specimen be preheated to the desired test temperature. The specimen is preheated with three infrared line heaters. Radiation from each linear filament is focused onto the specimen by an elliptical reflector. A temperature of 1000 °C (1830 °F) can be obtained in about 150 s using a 280-V, 100-A power supply.

Test Procedure

In both symmetric and asymmetric rod impact tests, the heater assembly (for elevated-temperature tests) and the specimen recovery pipe are placed in position near the muzzle of the gas gun. The target alignment plate is adjusted so that it is concentric with that of the gun barrel, and on it the target assembly containing the specimen rod is mounted, positioned to be concentric with the front target support plate. The projectile is inserted at the breech of the gun, and the target chamber is closed and evacuated to about 10^{-5} Torr. The gas gun breech is then pressurized, and the specimen is preheated to the desired temperature. The firing sequence then begins.

The room is darkened, the framing camera shutter is opened, and the rotating mirror in the camera is activated. When the rotational velocity of the mirror, as measured by a frequency counter, falls within the desired range, the fire button on the gas gun control panel is pushed, launching the projectile

down the barrel. As the projectile nears the end of the barrel, it contacts a series of pins, which records its velocity and activates a delay generator. After a preset delay, at a few microseconds before impact, a signal is sent that turns on the light source and records the rotational velocity of the mirror (from which the framing rate at impact can be determined).

Impact occurs, and for the next 30 to 80 μs the camera takes pictures of the deforming silhouettes. Then the light intensity drops rapidly to avoid double exposure. The deformed specimen rods travel into the recovery pipe, where they are gradually decelerated and recovered for subsequent metallographic analysis. Profiles of the silhouetted rod from the framing camera records are digitized for subsequent comparison with computer simulations. The digitized profiles (averaged top-to-bottom) from the test depicted in Fig. 13(a) are shown at 5-μs intervals in Fig. 13(b).

Analysis of Rod Impact Tests

Determination of the dynamic flow curve of a material from a rod impact test is made by computationally simulating the experiment with a two-dimensional wave propagation computer code (such as C-HEMP, described in Ref 54). The flow parameters are varied until the computed profiles agree with those determined experimentally at various times during the deformation history.

The specimen rods are divided into a series of rectangular zones, or computational cells, each of which, in the cylindrical geometry appropriate to the rod impact test, represents an annulus of revolution about the rod axis. The corners, or nodes, of the cells are given an appropriate initial velocity. Then the subsequent node velocities and the resulting rod deformation are determined by solving the Lagrangian equations of motion for a continuous medium. Figure 14 shows the original, an intermediate, and the final cell profiles for a typical test simulation.

A rate-independent elastic-plastic model with work hardening is used to describe the plastic deformations in each cell. Various flow curves are tried. A quasi-static compressive or tensile flow curve is a good starting point, because there are few dynamic curves to be found in the literature. The computed results are then compared with the experimental results until agreement within the experimental error is obtained.

Computer simulations have confirmed that the plastic strain rate, as well as the total strain and the temperature resulting from the local plastic work, varies significantly as a function of time and axial and radial position

Fig. 13(b) Deformation contours from 4340 steel (94 HB)
Symmetric rod impact at 457 m/s (1500 ft/s)

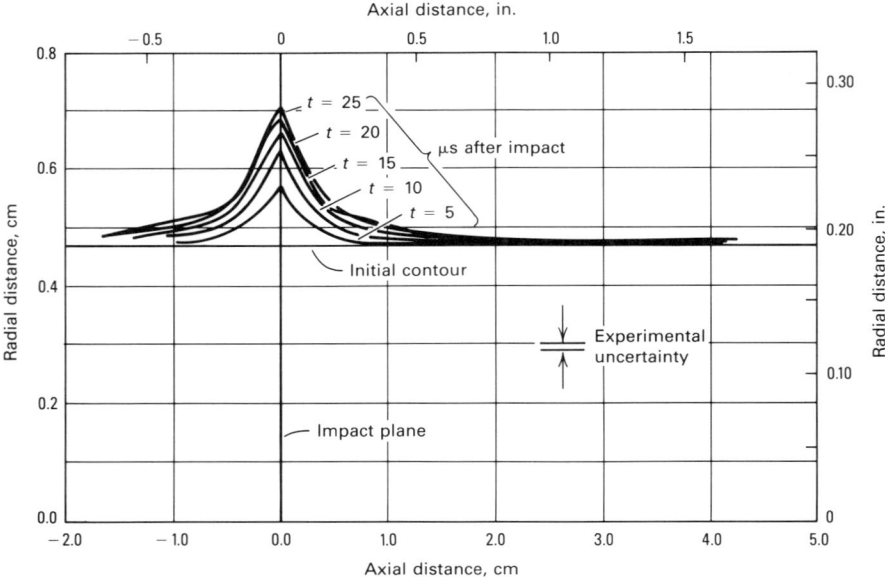

Fig. 14 Cell outlines at three times during simulation of symmetric rod impact
Horizontal and vertical axes not to scale. Cells near impact plane are originally square.

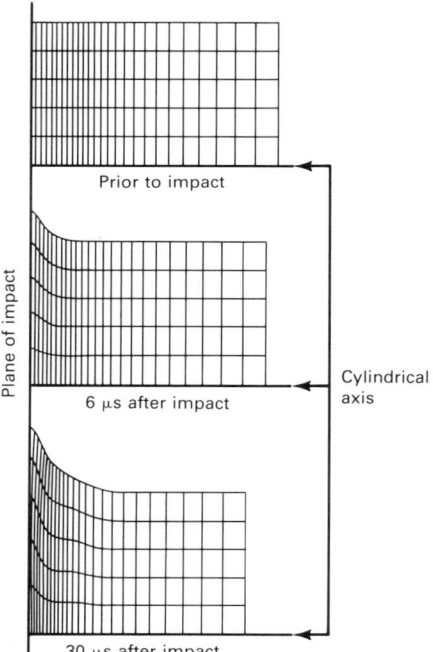

within the specimen rod. For a particular axial location, strain rates (either peak or average) along the rod axis are often three to four times higher than those near the periphery. Also, there is often an order of magnitude difference in the strain rates among locations near the impact interface and locations one rod diameter distant.

The experimentally measured rod thickness at a particular axial position is not the result of a region of material undergoing deformation at a specific strain rate, but rather the result of various regions of material deforming at a range of strain rates. Therefore, the rod impact technique should not be viewed as a method for accurately determining the strain rate sensitivity of the flow curve over the range of strain rate observed in the test, namely between about 10^4 and 10^5 s^{-1}, but rather as a means of determining the average flow curve over that range.

REFERENCES

1. L. Davison and R.A. Graham, Shock Compression of Solids, *Phys. Rep.*, Vol 55 (No. 4), 1979, p 255
2. E.D.H. Davies and S.C. Hunter, The Dynamic Compression Testing of Solids by the Method of the Split Hopkinson Pressure Bar, *J. Mech. Phys. Solids*, Vol 11, 1963, p 155
3. R. Chait and C.H. Curll, Evaluating Engineering Alloys in Compression, in *Recent Developments in Mechanical Testing*, STP 608, ASTM, Philadelphia, 1976, p 3
4. G.E. Dieter, *Mechanical Metallurgy*, 2nd ed., McGraw-Hill, New York, 1976, p 358
5. R.H. Cooper and J.D. Campbell, Testing of Materials at Medium Rates of Strain, *J. Mech. Eng. Sci.*, Vol 9, 1967, p 278
6. J.F. Bell, Diffraction Grating Strain Gauge, *Proc. Soc. Exper. Stress Anal.*, Vol 17 (No. 2), 1959, p 51
7. N. Loizou and R.B. Sims, The Yield Stress of Pure Lead in Compression, *J. Mech. Phys. Solids*, Vol 1, 1953, p 234
8. J.F. Alder and V.A. Phillips, The Effect of Strain Rate and Temperature on the Resistance of Aluminum, Copper, and Steel to Compression, *J. Inst. Metals*, Vol 83, 1954-1955, p 80
9. P.M. Cook, True Stress-Strain Curves for Steel in Compression at High Temperatures and Strain Rates, For Application to the Calculation of Load and Torque in Hot Rolling, *Conference on The Properties of Materials at High Rates of Strain*, Institution of Mechanical Engineers, 1957, p 86
10. R.R. Arnold and R.J. Parker, Resistance to Deformation of Aluminum and Some Aluminum Alloys, *J. Inst. Metals*, Vol 88, 1959-1960, p 255
11. J.A. Bailey and A.R.E. Singer, A Plane-Strain Cam Plastometer for Use in Metal-Working Studies, *J. Inst. Metals*, Vol 92, 1963-1964, p 288
12. J.E. Hockett, Compression Testing at Constant True Strain Rates, *Proc. ASTM*, Vol 59, 1959, p 1309
13. E. Orowan, The Calculation of Roll Pressure in Hot and Cold Flat Rolling, *Proc. Inst. Mech. Eng.*, Vol 150, 1943, p 143
14. J.M. Jacquerie, The Plasticity of Zinc, *C.R.M. Metallurgical Reports*, No. 9, 1966, p 51 (in French)
15. A. Hannick and J.M. Jacquerie, The Compression Test with the Cam Plastometer and its Application to the Determination of Rolling Pressures *C.R.M. Metallurgical Reports*, No. 3, 1965, p 49 (in French)
16. H. Suzuki, S. Hashizume, Y. Yabuki, Y. Ichihara, S. Nakajima, and K. Kenmochi, Studies on the Flow Stress of Metals and Alloys, *Report of the Institute of Industrial Science*, The University of Tokyo, Vol 18 (No. 3), Serial No. 117, 1968
17. M.J. Stewart, Hot Deformation of C-Mn Steels From 1000 to 2200 F (600 to 1200 C) With Constant True Strain Rates From 0.5 to 140 s^{-1}, in *The Hot Deformation of Austenite*, J.B. Ballance, Ed., The Metallurgical Society of AIME, Warrendale, PA, 1977, p 47
18. G.T. van Rooyen and W.A. Backofen, A Study of Interface Friction in Plastic Compression, *Int. J. Mech. Sci.*, Vol 1, 1960, p 1

19. W.F. Adler, T.W. James, and P.E. Kukuchek, "Development of a Dynamic Compression Fixture for Brittle Materials," Technical Report CR-81-918, Effects Technology, Inc., Santa Barbara, April 1981

20. D.R. Ireland, Procedures and Problems Associated with Reliable Control of the Instrumented Impact Test, in *Instrumented Impact Testing*, STP 563, ASTM, Philadelphia, 1974, p 3-29

21. H.J. Saxton, D.R. Ireland, and W.L. Server, Analysis and Control of Inertial Effects During Instrumented Impact Testing, in *Instrumented Impact Testing*, STP 563, ASTM, Philadelphia, 1974, p 50-73

22. D.R. Ireland, "Critical Review of Instrumented Impact Testing," Paper 5, Dynamic Fracture Toughness Conference, London, July 1976

23. S.G. Wogulis, B.W. Whitney, and D.R. Ireland, "Automated Data Acquisition and Analysis for the Instrumented Impact Test," Symposium on Computer Automation of Materials Testing, ASTM, Philadelphia, 1978

24. B. Hopkinson, A Method of Measuring the Pressure Produced in the Detonation of Explosives or by the Impact of Bullets, *Phil. Trans. A*, Vol 213, 1914, p 437

25. R.M. Davies, A Critical Study of the Hopkinson Pressure Bar, *Phil. Trans. A*, Vol 240, 1948, p 375

26. H. Kolsky, An Investigation of the Mechanical Properties of Materials at Very High Rates of Loading, *Proc. Royal Soc. B*, Vol 62, 1949, p 676

27. F.E. Hauser, J.A. Simmons, and J.E. Dorn, Strain Rate Effects in Wave Propagation, in *Response of Metals to High Velocity Deformation*, Metallurgical Society Conferences, Vol 9, P.G. Shewmon and V.F. Zackay, Ed., Interscience, New York, 1961, p 93

28. J.L. Chiddister and L.E. Malvern, Compression-Impact Tests of Aluminum at Elevated Temperature, *Exper. Mech.*, Vol 3, 1963, p 81

29. W.E. Baker and C.H. Yew, Strain-Rate Effects in the Propagation of Torsional Plastic Waves, *J. Appl. Mech.*, Vol 33, 1966, p 917

30. U.S. Lindholm and L.M. Yeakley, High Strain-Rate Testing: Tension and Compression, *Exper. Mech.*, Vol 8, 1968, p 1

31. U.S. Lindholm, Some Experiments with the Split Hopkinson Pressure Bar, *J. Mech. Phys. Solids*, Vol 12, 1964, p 317

32. U.S. Lindholm, High Strain Rate Tests, in *Measurement of Mechanical Properties*, Vol V, Part 1, R.F. Bunshah, Ed., Interscience, New York, 1971, p 199

33. G.L. Wulf, Dynamic Stress-Strain Measurement at Large Strains, in *Mechanical Properties at High Rates of Strain*, Institute of Physics Conference Series No. 21, London, 1974, p 33

34. D.A. Gorham, Measurement of Stress-Strain Properties of Strong Metals at Very High Rates of Strain, in *Mechanical Properties at High Rates of Strain*, J. Harding, Ed., Institute of Physics Conference Series No. 47, London, 1979, p 16

35. L. Pochhammer, On the Propagation Velocities of Small Oscillations in an Unlimited Isotropic Circular Cylinder, *J. Reine Angewandte Math.*, Vol 81, 1876, p 324

36. C. Chree, The Equations of an Isotropic Elastic Solid in Polar and Cylindrical Coordinates, Their Solutions and Applications, *Cambridge Phil. Soc. Trans.*, Vol 14, 1889, p 250

37. H. Kolsky, *Stress Waves in Solids*, Dover Publications, New York, 1963

38. P.S. Follansbee and C.E. Frantz, Wave Propagation in the Split Hopkinson Pressure Bar, *J. Eng. Mat. Technol.*, Vol 105, 1983, p 61

39. E.D.H. Davies and S.C. Hunter, The Dynamic Compression Testing of Solids by the Method of the Split Hopkinson Pressure Bar, *J. Mech. Phys. Solids*, Vol 11, 1963, p 155

40. U.S. Lindholm, Deformation Maps in the Region of High Dislocation Velocity, in *High Velocity Deformation of Solids*, K. Kawata and J. Shioiri, Ed., Springer-Verlag, New York, 1978, p 26

41. "Compression Testing of Metallic Materials at Room Temperature," E 9, *Annual Book of ASTM Standards*, ASTM, Philadelphia, 1984

42. K. Oi, Transient Response of Bonded Strain Gages, *Exper. Mech.*, Vol 6, 1966, p 463

43. J. Lankford, Temperature-Strain Rate Dependence of Compressive Strength and Damage Mechanisms in Aluminum Oxide, *J. Mat. Sci.*, Vol 16, 1981, p 1567

44. C.E. Frantz, P.S. Follansbee, and W.E. Wright, New Experimental Techniques with the Split Hopkinson Pressure Bar, in *High Energy Rate Forming—1984*, I. Berman and J.W. Schroeder, Ed., ASME, New York, 1984, p 229

45. L.D. Bertholf and C.H. Karnes, Two-Dimensional Analysis of the Split Hopkinson Pressure Bar System, *J. Mech. Phys. Solids*, Vol 23, 1975, p 1

46. G.I. Taylor, The Use of Flat-Ended Projectiles for Determining Dynamic Yield Strength: I. Theoretical Considerations, *Proc. Roy. Soc. A*, Vol 194, 1948, p 289-299

47. A.C. Whiffin, The Use of Flat-Ended Projectiles for Determining Dynamic Yield Strength: II. Tests on Various Metallic Materials, *Proc. Roy. Soc. A*, Vol 194, 1948, p 200-232

48. E.H. Lee and S.J. Tupper, Analysis of Plastic Deformation in a Steel Cylinder Striking a Rigid Target, *J. Appl. Mech.*, Vol 21, 1954, p 63-70

49. J.B. Hawkyard, D. Eaton, and W. Johnson, The Mean Dynamic Yield Strength of Copper and Low Carbon Steel at Elevated Temperatures from Measurements of the "Mushrooming" of Flat-Ended Projectiles, *Int. J. Mech. Sci.*, Vol 10, 1968, p 929-948

50. M.L. Wilkins and M.W. Guinan, Impact of Cylinders on a Rigid Boundary, *J. Appl. Phys.*, Vol 44, 1973, p 1200-1206

51. D.C. Erlich, D.A. Shockey, and L. Seaman, "Symmetric Rod Impact Technique for Dynamic Yield Determination," AIP Conference Proceedings, No. 78, Second Topical Conference on Shock Waves in Condensed Matter, Menlo Park, CA, 1981, p 402-406

52. D.C. Erlich and D.A. Shockey, "Dynamic Flow Curve of 4340 Steel, as Determined by the Symmetric Rod Impact Test," AIP Conference Proceedings, Third Topical Conference on Shock Waves in Condensed Matter, Sante Fe, NM, 1983

53. W.H. Gust, High Impact Deformation of Metal Cylinders at Elevated Temperatures, *J. Appl. Phys.*, Vol 53 (No. 5), 1982, p 3566-3575

54. T. Cooper, D.C. Erlich, and L. Seaman, "C-HEMP Users' Manual," SRI International Final Report for Ballistic Research Laboratory, 1984

High Strain Rate Tension Testing

By Theodore Nicholas
Materials Research Engineer
AFWAL Materials Laboratory

and

S.J. Bless
Group Leader, Impact Physics
University of Dayton Research Institute

HIGH STRAIN RATE TENSION TEST-ING is necessary to understand the response of materials to dynamic loading. Strain rates ranging from 100 s^{-1} to $>10^4 \text{ s}^{-1}$ occur in many processes or events of practical importance, such as foreign object damage, explosive forming, earthquakes, blast loading, structural impacts, terminal ballistics, and metalworking. The behavior of materials under high strain rate tensile loads may differ considerably from that observed in conventional tensile tests.

High strain rate sensitivity is primarily manifested in variations in yield and failure criteria. Yielding and failure are also affected by stress state, ratio of mean stress to deviatoric stress, stress amplitude, stress history, and temperature. Tests must be designed to simulate the most relevant load characteristics. For example, many processes involving dynamic tensile stress include compressive prestress. Strain rate sensitivity also depends on whether engineering or true strain formulations are used, because local instabilities (such as necking) are often suppressed at high rates.

Measurement of strain is a major problem in high strain rate tension testing. In quasi-static testing, the diameter of the minimum cross section in a cylindrical specimen can be measured; such measurements are virtually impossible or highly impractical in high-rate testing. Furthermore, although strains are easily measured over a uniform gage length section in quasi-static testing, the same measurements are considerably more difficult to obtain at high strain rates. Mechanical extensometers are the primary tool used in quasi-static tests, but they are of little use at high rates of strain because of the effects of inertia.

Therefore, high-rate tests use strain gages, optical extensometers, and displacement measurements between loading fixtures to determine or infer the dynamic tensile strains in a test specimen. At very high rates of strain, strains may be measured in some experimental configurations only through wave propagation analysis. This procedure generally requires that assumptions be made about the constitutive behavior, that wave propagation analysis be carried out, and that predictions and experimental observations be compared. Unique solutions cannot be guaranteed, because some other constitutive model may conceivably provide similar results in a particular wave propagation problem.

Conventional Load Frames

Strain rate effects in tension are determined by performing conventional tensile tests at varying loading rates up to approximately 100 s^{-1}. Conventional test machines are available with increased ram velocities, as are high-speed pneumatic and hydraulic machines. The speed capability of a machine may be influenced by several factors. Speed may be a function of the load that the ram is attempting to apply, and the no-load speed may be much higher than the full-load speed. The distance traveled may also affect the speed capability. A long stroke machine may attain a given speed only after a significant amount of travel. Depending on the specimen length, considerable specimen strain could occur before final maximum velocity is obtained. Finally, the ability to control speed is a function of the response capability of a servo-controlled machine working in a closed-loop mode. Open-loop machines provide speeds that may be influenced by specimen strength and cannot easily reproduce predetermined velocities or strain rates on materials with different yield strengths or strain-hardening behaviors. Additional information on the operational characteristics of tension testing machines such as open- versus closed-loop control can be found in the article "Tension Testing Machines and Extensometers" in this Volume.

Effects of Inertia and Wave Propagation. A fundamental difference between a high strain rate tension test and a quasi-static tension test is that inertia and wave propagation effects are present at high rates. It must be determined how fast a uniaxial tension test can be run to obtain valid stress-strain data. To determine this, consider a specimen of initial length L subjected to a uniform velocity v_0 at time $t = 0$, as shown in Fig. 1. This hypothesis could represent a test in a constant crosshead velocity testing machine, or a drop-weight type of test in which a large

Fig. 1 Schematic of tensile test configuration
See text for details and explanation of symbols.

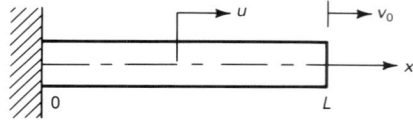

mass impacts one end of the specimen. If (x,t) denotes the displacement of any point in the x direction and assuming purely uniaxial motion—that is, neglecting radial inertia effects—the equation of motion is:

$$\frac{\partial^2 u}{\partial t^2} = c^2 \frac{\partial^2 u}{\partial x^2} \qquad \text{(Eq 1)}$$

where u is displacement and

$$c = \left(\frac{E}{\rho}\right)^{1/2} \qquad \text{(Eq 2)}$$

is the longitudinal wave velocity in a bar or rod, where E is Young's modulus, and ρ is mass density. Applying the boundary conditions of the left end fixed and the right end moving with constant velocity v_0 and assuming initial conditions of zero displacement and velocity, the solution is:

$$u(x,t) = \frac{v_0 L}{2c}\left[f\left(\tau + \frac{x}{L}\right) - f\left(\tau - \frac{x}{L}\right)\right]$$
$$\text{(Eq 3)}$$

where $\tau = tc/L$ is a dimensionless time, and $\tau = 1$ represents the time it takes a wave to propagate the length of the specimen. The function $f(\tau)$ is shown in Fig. 2. Strain can be obtained from $\epsilon = \partial u/\partial x$ and stress from $\sigma = E\epsilon$. By introducing the dimensionless variables:

$$\xi = \frac{x}{L} \qquad \text{(Eq 4a)}$$

$$v^* = \frac{v_0}{c} \qquad \text{(Eq 4b)}$$

plots of stress and strain can be constructed as a function of time. Figure 3 illustrates strain normalized with respect to v^* against dimensionless time τ at an arbitrary position ξ along the bar. The dashed line indicates the average strain in the bar, which is normally total displacement divided by bar length.

The localized strain is measured by a strain gage with a gage length that is small compared to the length of the specimen. Figure 4

Fig. 2 Graph of the function $f(\tau)$

See text for details and explanation of symbols.

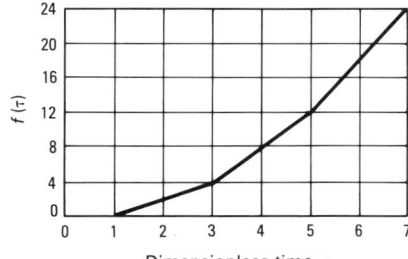

Fig. 3 Nondimensional strain profile

See text for details and explanation of symbols.

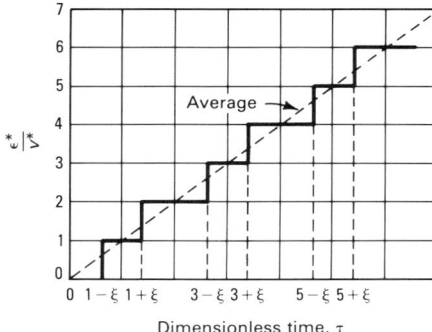

Dimensionless time, τ

illustrates the normalized stresses at both ends of the bar. The response at the fixed end is recorded by a load cell. Figures 3 and 4 illustrate that stresses and strains accumulate from numerous waves propagating back and forth in the bar. Note that the solution to the mathematical problem has assumed an instantaneous jump in velocity at $t = 0$, whereas some finite rise time usually occurs because of imperfect impact or machine response.

If many wave transits occur during a test, the use of average stresses and strains appears justified. However, if the velocity is high, then only a few wave reflections may occur before the specimen fails. In this case, individual wave propagation must be considered; average values alone cannot be considered, and the use of this test to determine dynamic stress-strain response is precluded. Note that this analysis is based on a material that is linear-elastic and assumes a zero rise time in the applied velocity. Stress waves are propagated at the elastic wave velocity. With material that has deformed into the plastic region, the plastic wave velocity is more appropriate and generally can be an order of magnitude smaller than the elastic wave velocity.

One factor in determining whether or not wave propagation effects limit the validity of a tensile test is the sample ring-up time, which is the time required for a sample to achieve a uniform state of stress. Generally, measurements are not valid for times such that $L \sim ct$. This corresponds to a situation in which strain $\epsilon \gg v/c = \dot\epsilon L/c$. Consequently, small strain measurements are difficult to obtain at very high strain rates. Another concern is that local failure may occur at the end to which the load is applied. The magnitude of the stress transient associated with the sudden application of velocity v_0 is $\sigma_m = \rho c v_0$.

The test must be designed so that $\sigma_m < Y$, the yield stress. For example, consider a bar

Fig. 4 Stress history at ends of bar

See text for details and explanation of symbols.

Fixed end
Loaded end

Dimensionless time, τ

25 mm (1 in.) in length that is accelerated at one end to 2.5 m/s (8.2 ft/s). For many engineering materials, including steels, aluminum alloys, and titanium alloys, the elastic wave velocity is about 5000 m/s (16 400 ft/s). The maximum stress generated at the accelerated end of the bar is $\rho c v$. For a steel bar, the first stress pulse is 100 MPa (14.5 ksi), and the average strain rate is 100 s^{-1}. If a steel with a strength of 1 GPa (145 ksi) is being tested, the maximum allowable driving velocity is 25 m/s (82 ft/s). At that velocity, instantaneous failure would occur at the driven end.

Assuming that wave propagation effects may be neglected in a given test, the second aspect that must be checked is the response of the load cell. Load cell ringing is frequently encountered in high-rate tensile testing. Generally, this time period (reciprocal of the natural frequency in hertz) must be small compared to the total duration of the test. For example, if a load cell has a natural frequency of 1 kHz, its period of vibration is 10^{-3} s. This load cell could then be used only for experiments that lasted over ten times that amount, or over 10 ms.

Another condition that must be satisfied is the distance of the load cell from the end of the specimen. If a sufficient distance exists between the specimen and load cell, the finite elastic wave transit time may result in load data that are not time-coincident with strain data. To prevent phase lags from obscuring the experimental data, the wave transit time from the specimen to load cell should be negligibly small compared to the test duration. Otherwise, the load data must be corrected for the delay, and such corrections seldom are precise.

Strain Measurement. The final aspect of high-speed tensile testing is determination of strain. The most direct, reliable method uses electrical resistance strain gages. The frequency response capability of strain gages is considerably greater than the mechanical response of the combination of load train, specimen, and load cell.

Another method of measuring strain involves the use of optical extensometers, in which displacement measurements across the loading fixtures are divided by an actual or effective gage length. When using crosshead displacement measurements, caution must be exercised to ensure that these represent only specimen elongation and not machine, ram, or load train elongations. The same precautions that apply in quasi-static tests also apply in dynamic tests.

If the above precautions are observed, valid stress-strain data can be obtained up to maximum strain rates in the range of 10 to 100 s^{-1}. For higher strain rates, or for cases in which the above criteria are not met, highly specialized testing techniques may have to be used, as discussed below.

Expanding Ring Test

The expanding ring test is a highly sophisticated technique for subjecting metals to tensile strain rates over 10^4 s^{-1} (Ref 1, 2). Although the testing principle is simple, its performance requires specialized equipment available in only a few laboratories. The ring test can determine the high-rate stress-strain relationships, but a simplified, more widely used version can be employed to determine ultimate strain only (Ref 3, 4).

This test involves the sudden radial acceleration of a ring due to detonation of an explosive charge or electromagnetic loading. The ring rapidly becomes a free-flying body, expanding radially, and decelerating due to its own internal circumferential stresses. A thin ring must be used for the analysis to be valid; the wall thickness should be less than one tenth the ring diameter, which is typically 25 mm (1 in.). If R is the radius of the ring, ρ the density, and σ the hoop stress:

$$\sigma = -\rho R \frac{d^2 R}{dt^2} \qquad \text{(Eq 5)}$$

To obtain stress-strain data, radial displacement as a function of time must be calculated. Strain is proportional to change in radius (just as engineering strain in tension is $\Delta L/L_0$); thus:

$$\epsilon = \ln \frac{R}{R_0} \qquad \text{(Eq 6)}$$

where R_0 is the initial radius. Stress may be computed from Eq 5 by double differentiation of radial displacement data as a function of time. Ring displacement can be obtained through the use of high-speed photography, streak cameras, displacement interferometers, or other methods for measuring radius as a function of time.

It is difficult to determine stress accurately by double differentiation of displacement data. Several laboratories have used a laser velocity interferometer to measure ring velocity directly (Ref 5, 6). Thus, only a single differentiation is necessary to calculate stress, and precision is improved considerably.

Advantages of the Ring Test. The ring test has two principal advantages. The expanding ring test subjects the material to a state of dynamic uniaxial stress without the wave propagation complications that accompany other high strain rate tests. Also, the maximum strain rate available in the ring test is higher than in any other common tension tests involving large plastic strains.

Limitations of the Ring Test. Strain rate in the expanding ring test is not usually constant. The strain rate is computed from $(dR/dt)/R$, and both of these terms vary continually. Strain rate is usually greatest at the start of ring deceleration, when strain is smallest. Values in excess of 10^4 s^{-1} are readily obtained. If the ring does not rupture, the strain rate falls to zero at the end of the test.

Ring specimens also experience a compressive preload in the radial direction that often exceeds the yield stress during the acceleration phase. Because load history is known to affect the subsequent stress-strain behavior of many materials, data obtained from expanding ring tests do not always agree with results from other tests at slightly lower strain rates.

The difficulties, expense, and limitations of the expanding ring test preclude its use as a standard test technique for generating high strain rate stress-strain data in tension. Only a few laboratories are capable of performing this test. However, if subjecting a material to high strain rates in tension without determining stress-strain data is of primary interest, the expanding ring test is much easier to conduct. A number of investigators have used this test to determine strain to failure under dynamic loading (Ref 3, 4). Here, the accurate determination of radial displacement versus time is not as critical, because stresses are not calculated. Less precise displacement data provide reasonably accurate determinations of strain rate. The ambiguity arising from possible strain rate history effects still exists when the expanding ring test is used in this simpler manner.

The expanding cylinder test, a variation of the ring test, provides a dynamic stress state equivalent to that produced in a quasi-static tensile test on a wide sheet versus a thin strip of material. A difficulty encountered in this type of test is the need for an impulse to be generated simultaneously in time along the axis of the cylinder. Because

explosive detonation along a wire, for example, propagates at a finite wave speed, uniform deformation along the length of the axis cannot be ensured. Dimensions, detonation wave speeds, and synchronization of multiple detonation all must be considered carefully to ensure that the cylinder is deformed as uniformly as possible and that axial stress waves are not generated (Ref 7).

Flyer Plate and Short Duration Pulse Loading

Traditionally, flat plate impact tests have been used to obtain high strain rate yield data, shock wave response data, and equation of state data for materials undergoing uniaxial strain. Uniaxial strain refers to a three-dimensional state of stress in which deformation or strain occurs in only one direction—the direction of loading. The uniaxial strain condition persists for only a short period of time, until stress waves originating at lateral boundaries reach the specimen interior. In a typical experiment, this time period is on the order of several to tens of microseconds. Uniaxial strain is defined mathematically as:

$$u_x \neq 0, \; u_y = u_z = 0 \qquad \text{(Eq 7)}$$

where x is the direction of loading; u_x is the displacement in that direction; and y and z are orthogonal directions in a plane normal to x. The strains are obtained from the displacement derivatives, thus:

$$\epsilon_x \neq 0, \quad \epsilon_y = \epsilon_z = 0 \qquad \text{(Eq 8)}$$

The flat plate impact test is performed by launching a flat flyer plate against a second stationary target plate. Compressed gas guns, propellant guns, magnetic accelerators, and explosives have all been used to launch the flyer plate (Ref 8). Extreme precision must be achieved to eliminate relative tilt at the instant of impact. A typical experimental setup using a gas gun is shown in Fig. 5. The flyer plate is carried in the gas gun in a plastic sabot. Velocity of the flyer is determined from the transit time between the shorting pin in the gun barrel and time-of-arrival pins in the target. The target is supported by a spall ring that suppresses late-time radial tensile waves.

The stress waves along the axis normal to the impact plane are shown in Fig. 6. A flyer plate of thickness d, moving left to right, strikes an initially stationary target of thickness T; the impact occurs at the origin, O, of the (x,t) coordinates. Elastic-plastic behavior is assumed in Fig. 6. Elastic waves propagate at approximately c_L, the longitudinal elastic sound speed. Plastic waves propagate at approximately $\sqrt{(B/\rho)}$, where B is the bulk modulus. The arrivals of the elastic and plas-

Fig. 5 Schematic of gas-gun-launched flyer plate impact test setup

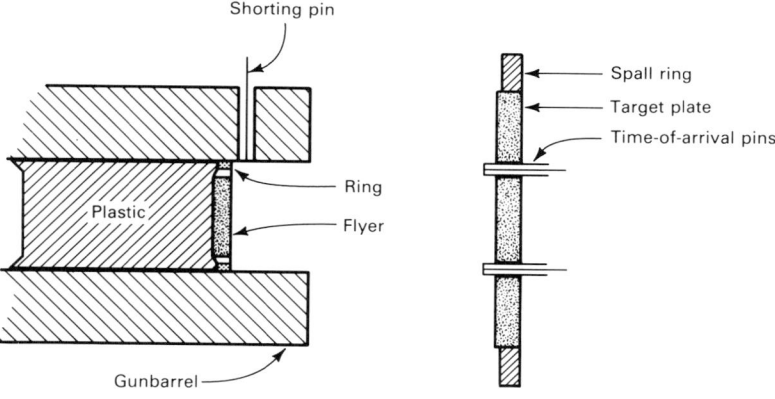

Fig. 6 Lagrangian diagram showing stress waves in flyer plate experiment

See text for details and explanation of symbols.

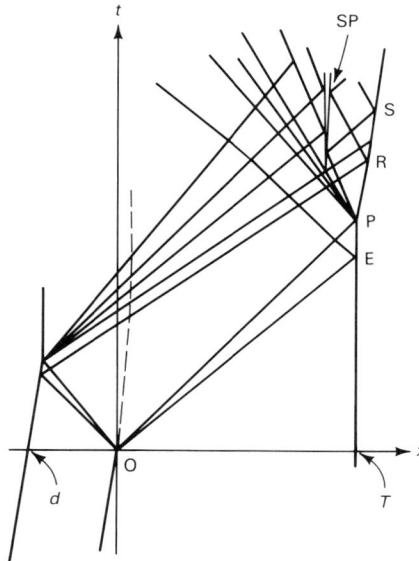

tic waves at the target rear surface are denoted as E and P.

Propagation speeds are always relative to the material into which the wave is moving. Strain occurs only at the wave fronts. The amplitude of the E wave in Fig. 6 is known as the Hugoniot elastic limit (HEL) and is simply related to the uniaxial yield stress, Y, as:

$$\sigma_{HEL} = \left[\frac{B + 4\mu/3}{2\mu}\right] Y \qquad \text{(Eq 9)}$$

where μ is the shear modulus. The final state of the shocked material is characterized by a stress and particle velocity. The functional relationship between these two variables depends on the material and is known as the Hugoniot. The state behind the P wave in Fig. 6 lies on the Hugoniot (see Ref 9 for a discussion of Hugoniots). If the flyer and target plates are composed of the same material, the particle velocity behind the P wave is one half the impact velocity.

Reference 10 discusses determination of particle velocity when the flyer and target are composed of different materials. If the Hugoniot of the target is known, then the stress can be calculated from the particle velocity. Hugoniots for most engineering metals can be found in Ref 11.

Compressive waves reflect from a free surface as tensile (rarefaction) waves, which begin to arrive at the target rear surface at point R in Fig. 6. The tensile (rarefaction) waves may interact and cause spall failure. This causes material separation in the target, which is indicated at point SP in Fig. 6. The sudden relaxation of tensile stress generates a shock wave that arrives at the free surface at point S.

Spall is a form of tensile failure under an extremely high strain rate and a nearly spherical stress tensor. Spall usually is characterized by the spall stress, σ_{spall}, defined as the highest tensile stress that exists in the material prior to rupture. When designing spall

experiments, the flyer plate diameter, a, must be large enough so that the phenomena of interest occur within a time $a/2c_L$ after the impact.

Flat plate impact tests normally are used to measure σ_{HEL} and spall strength. For example, consider the characterization of a steel by this technique. The value of σ_{HEL} for steel is usually between 5 and 15 kilobar (kbar), a useful unit for analyzing shock experiments; 1 kbar = 0.1 GPa. When density is expressed as g/cm^3 × 10 and velocity is given in km/s (or, equivalently, mm/μs), stress is given in kbar.

To measure the Hugoniot elastic limit, the impact velocity must be sufficient for the peak stress to exceed σ_{HEL}. Peak stress is given by:

$$\sigma = \rho U u \qquad \text{(Eq 10)}$$

where U is shock propagation speed, and u is particle velocity. Peak particle velocity is half the impact velocity, u_0, for a symmetric impact. For steel-on-steel impacts, Eq 10 becomes approximately $\sigma = 200\, u_0$. For $\sigma > \sigma_{HEL} = 15$ kbar, $u_0 > 75$ m/s (245 ft/s) is required. This presents no problem when a gas gun is used.

Experiments with $u_0 < 100$ m/s (<330 ft/s) are often difficult because of impact tilt, which becomes more critical at low velocities. Also, impact velocity must not be so high that the velocity of the P wave (Fig. 6) exceeds c_L. That limit for steels usually is greater than 1 km/s (0.6 mile/s). The limit for other materials can be found by consulting the tables in Ref 11.

Given an appropriate impact velocity, to determine σ_{HEL} one of the following measurements must be made. The peak particle velocity behind the E wave can be measured. This can be accomplished at the free surface with capacitor gages, sloping mirrors, or a velocity interferometer. The velocity behind the wave is half the free surface velocity. The

stress is related to the free surface velocity by Eq 10 with $u = c_L$.

Direct measurement of σ_{HEL} can be obtained by embedded piezoresistive gages. Manganin and carbon gages frequently are used for this purpose. This technique requires sectioning the target or using a backing plate and correcting for partial transmission of the wave transmitted through the interface. Magnetic particle velocity gages can be used for nonconducting targets such as plastics and rocks, but they are not suitable for metals.

Spall stress can be determined by two methods. The simplest, in terms of analysis, interpretation, and experimental technique, is to vary systematically the flyer plate thickness, d, and impact velocity, u_0, to determine the critical values at which rupture occurs. As the flyer plate thickness is increased, the duration of the compressive and tensile load increases; the load duration is approximately $2d/c_L$.

Eventually, for flyer plate thicknesses exceeding about 5 mm (0.2 in.), the spall stress reaches a load duration limit. In many metals, the limiting spall strength is several times the value of σ_{HEL}. Figure 7 illustrates typical spall stress data for low-carbon steel. The data illustrate that for pulse durations longer than a few microseconds, the greatest tensile stress that the material can sustain without rupture is 25 kbar.

Interpretation of experiments using thinner flyer plates is more complex, because a computer code must be used to calculate the

Fig. 7 Spall data for low-carbon 1020 steel

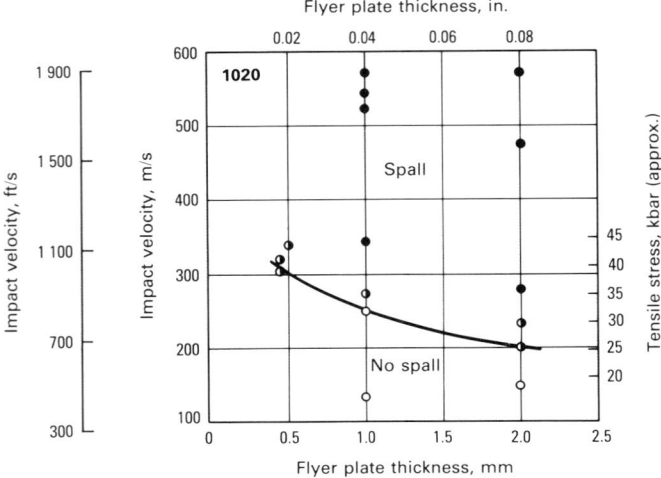

Fig. 8 Free surface velocity data when spall occurs

See text for details and explanation of symbols.

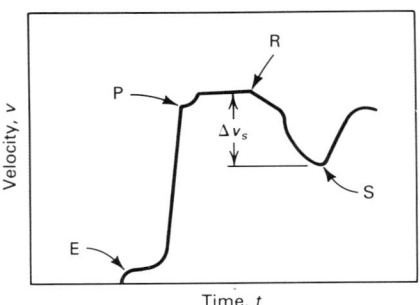

stress history on the spall plane. Finite difference codes (Ref 9) or method of characteristics codes (Ref 12) can be used. Finite difference codes are more accurate and more widely applicable than method of characteristics codes, but the user must be specially trained in this subject.

Another approach to spall characterization is to initiate impact above the spall threshold and to deduce the material behavior from the free surface velocity, Δv_s, data. Figure 8 illustrates a typical free surface velocity history with spalling. E, P, R, and S refer to the same arrivals as explained in text for Fig. 6. The spall stress is given approximately by $\rho c_L \Delta v_s/2$. However, a more exact determination requires code analysis.

Split-Hopkinson Bar in Tension

The principles of the split-Hopkinson bar in tension are similar to those in compression, as discussed in the article "High Strain Rate Compression Testing" in this Section.

The primary differences are the methods of generating a tensile loading pulse, specimen geometry, and the method of attaching the specimen to the two bars (incident and transmitter).

Basically, three types of tension split-Hopkinson bars have been developed. The first (Ref 13) involves the use of compressive pulses in the input and transmitter bars, as in the compressive test. The input bar is solid, while the transmitter or output bar is a hollow tube of the same cross-sectional area as the input bar. The specimen is a complex "top-hat" type of geometry, as shown in Fig. 9. The specimen actually is comprised of four parallel tensile bars of equal cross-sectional area. Although specimen machining is somewhat complex, the test is conducted in the same manner as compressive testing.

The second type of tension split-Hopkinson bar test involves the use of a smaller standard type of threaded tension specimen and generation of a tensile pulse directly on the end of the loading or input bar (Ref 14, 15). This can be accomplished in several

ways, as shown in Fig. 10. In Fig. 10(a), a mass is impacted directly on an anvil attached to the end of the input bar. In Fig. 10(b), an anvil is loaded by a compressive wave transmitted through another loading bar that is a hollow tube. The compressive pulse in the loading tube is generated by the same techniques as the compression split-Hopkinson bar. In Fig. 10(c), a pulse is generated by the detonation of an explosive against the anvil. In Fig. 10(a) and (c), it is difficult to generate a pulse of constant amplitude, while in 10(b) a long time duration is accomplished in the same manner as in compressive testing.

The third type of tension test also uses a threaded specimen, but uses the reflection of a compression pulse at a free end and a collar to protect the specimen from initial precompression (Ref 16). Figure 11(a) illustrates an experimental setup. Figure 11(b) is a Lagrangian x-t diagram, which illustrates the details of wave propagation in the bars and the experimental procedures. When the striker bar is accelerated against bar No. 1, the impact generates a compression pulse, the amplitude of which depends on the striker velocity and the length of which is twice the elastic wave transit time in the striker bar.

Fig. 9 Details of tension specimen

Source: Ref 13

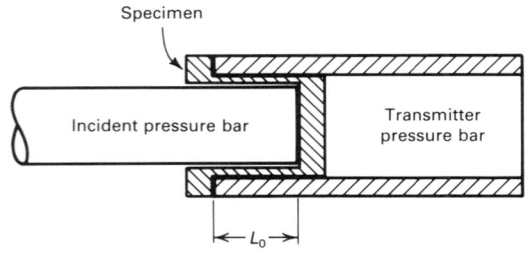

Specimen configuration

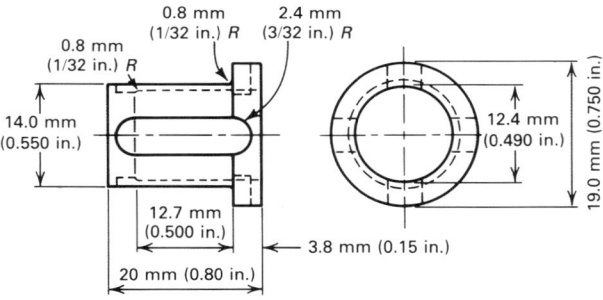

Dimensions of test specimen

Fig. 10 Schematic of three types of tension-loading techniques

See text for description.

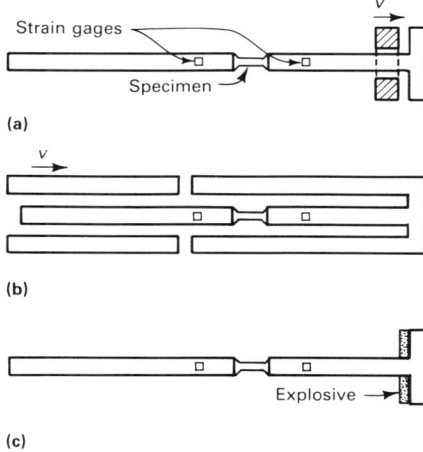

The pulse travels down the bar until it reaches the specimen. The threaded tensile specimen is attached to the two pressure bars, as shown in detail ''A'' of Fig. 11(a).

After the specimen has been screwed into the bars, a split shoulder or collar is placed over the specimen, and it is screwed in until the pressure bars fit tightly against the shoulder. The shoulder is made of the same material as the pressure bars, has the same outer diameter, and has an inner diameter that just clears the specimen. The ratio of the cross-sectional area of the shoulder to that of the pressure bars is typically 3:4, while the ratio of the area of the shoulder to the net cross-sectional area of the specimen is typically 12:1.

The compression pulse travels through the composite cross section of the shoulder and specimen in an essentially undispersed manner. Tightening of the specimen by twisting the pressure bars, the relatively loose fit of the threaded joint of the specimen into the bars, and the large area ratio of the shoulder to the specimen ensure that no compression beyond the elastic limit is transmitted to the specimen.

Ideally, the entire compression pulse passes through the supporting shoulder as if the specimen were not present, although in practice it is difficult to prevent prestraining of the specimen. The compression pulse continues to propagate until it reaches the free end of bar No. 2. There, it reflects and propagates back as a tensile pulse, ϵ_i, and passes gage No. 2. Upon reaching the specimen at point A as shown in Fig. 11(b), the tensile pulse is partially transmitted through the specimen (ϵ_t) and partially reflected back into bar No. 2 (ϵ_r). Note that the shoulder, which carried the entire compressive pulse around the specimen, is unable to support any tensile loads because it is not fastened to the bars.

Tight fitting of the shoulder against the bars is critical in transmitting the compression pulse down the bars without significant wave dispersion. Similarly, the tight fit of the threaded tensile specimen against the bars is essential to achieve smooth and rapid loading of the specimen as the tensile pulse arrives. Failure to remove all ''play'' from the threaded joint results in uneven loading of the specimen and spurious wave reflections.

Analysis of the tensile split-Hopkinson bar test is almost identical to that of the compres-

sion test. The major difference is the actual or effective gage length of the specimen. Contrary to the compression test, in which a right circular cylinder is used, the tensile test uses a cylindrical specimen with an attached shoulder and additional gripping such as threads. Because the split-Hopkinson bar test can only provide data on the relative displacement between the ends of the incident and transmitter bars, an effective gage length generally must be used. This is equivalent to determining strain in a tensile test through crosshead displacement measurement. The use of strain gages on test samples to determine an effective gage length is strongly recommended. This calibration is accomplished easily at low strain rates, preferably in a conventional test machine in which the crosshead displacement is monitored separately.

Split-Hopkinson bar data in the form of dynamic stress-strain curves can be compared with quasi-static data to assess strain-rate effects. As an example, typical stress-strain curves in tension for electrolytic tough pitch (ETP) copper at three different strain rates are shown in Fig. 12 (Ref 16). The highest strain rate was achieved using the split-Hopkinson bar apparatus; the intermediate and lowest rate data were obtained in a servohydraulic machine.

The increase in flow stress in ETP copper ranges from approximately 100% for small strains to approximately 50% for larger strains when moving from conventional quasi-static rates (10^{-3} s^{-1}) to dynamic rates (10^3 s^{-1}). Variations in flow stress over this strain rate range depend on the material being tested. For example, the increase in flow stress in mild steel specimens is similar to

Fig. 11 Split-Hopkinson bar test using threaded tension specimen

(a) Schematic of tensile loading apparatus. CRO, cathode ray oscilloscope. Source: Ref 16. (b) Lagrangian diagram for tensile loading apparatus. See text for details and explanation of symbols.

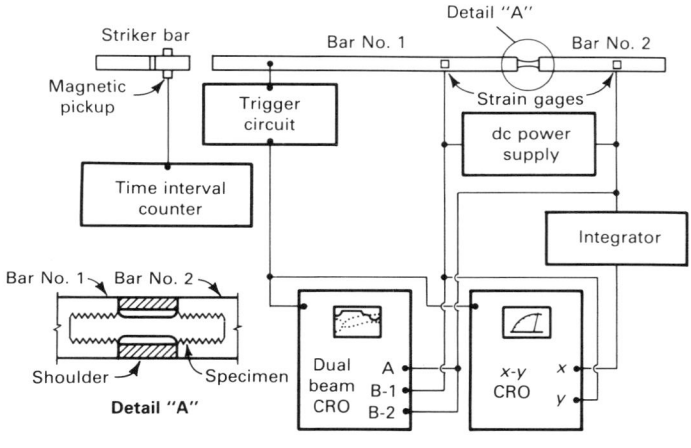

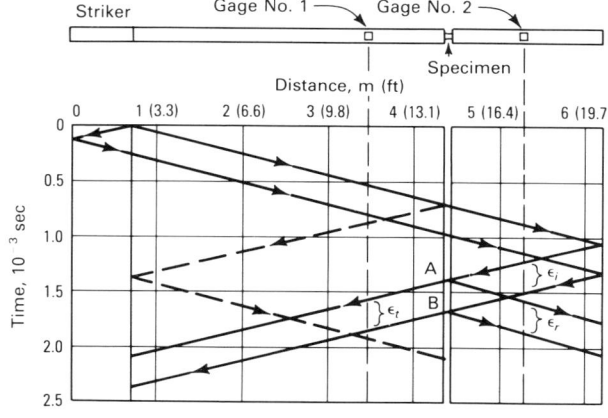

(a)

(b)

Fig. 12 Stress-strain curve for ETP copper in tension
Source: Ref 16

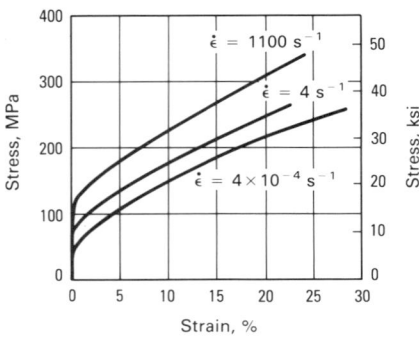

Fig. 14 Sequence of necking of specimen
Source: Ref 18

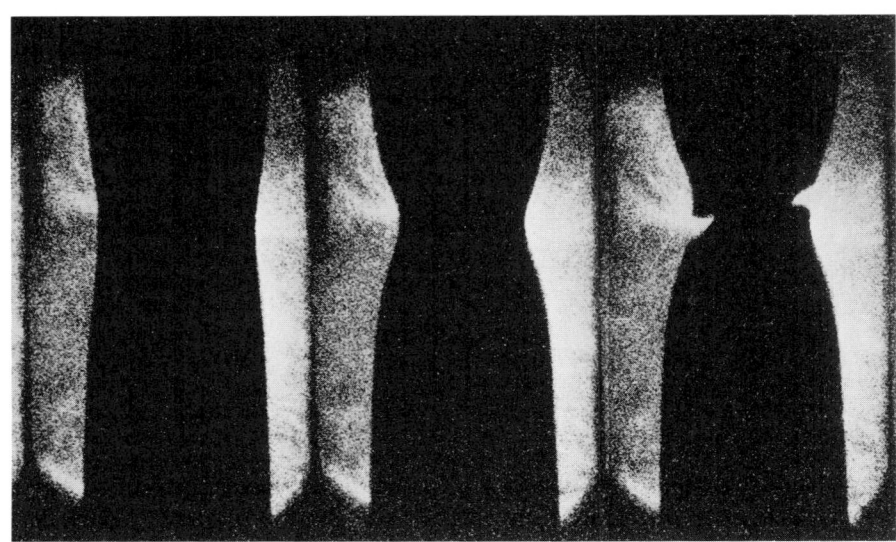

that encountered in ETP copper, while structural grades of aluminum exhibit little change in flow stress when tested under quasi-static and dynamic conditions (Ref 10).

An alternate way of presenting high strain rate data is to plot stress against log strain rate for fixed values of strain, as shown in Fig. 13 for several grades of titanium (Ref 16). Data of this type can also be plotted using nondimensional stress as the vertical axis by referencing all stress values to the quasi-static value at the lowest strain rate.

As with any uniaxial tensile test, once localized necking occurs, it is no longer possible to simply convert load-displacement data to stress-strain data. The range of application of the Hopkinson bar test can be extended by high-speed photography of necking specimens, as shown in Fig. 14. An analysis that allows estimation of effective stress and strain from the profile of the necking specimen is described in Ref 17. Photographs can be made with a suitable high-speed camera system through windows provided in the collar. The major technical difficulty is the precise synchronization of the exposures with the Hopkinson bar record (Ref 18).

Fig. 13 Stress-log strain rate data for titanium alloys
Source: Ref 16

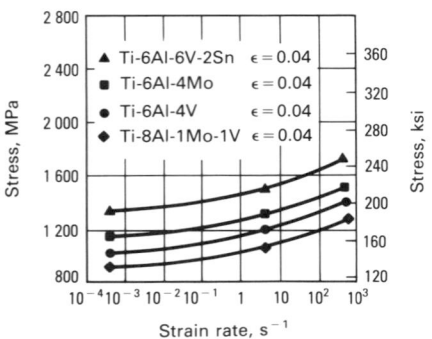

REFERENCES

1. F.I. Niordson, A Unit for Testing Materials at High Strain Rates, *Exp. Mech.*, Vol 5, 1965, p 29-32
2. C.R. Hoggatt and R.F. Recht, Stress-Strain Data Obtained at High Strain Rates Using an Expanding Ring, *Exp. Mech.*, Vol 9, 1969, p 441-448
3. D.E. Grady and D.A. Benson, Fragmentation of Metal Rings by Electromagnetic Loading, *Exp. Mech.*, Vol 28, 1983, p 393-400
4. A.M. Rajendran and I.M. Fyfe, Inertia Effects on the Ductile Failure of Thin Rings, *J. Appl. Mech.*, Vol 104, 1982, p 31-36
5. L.M. Barker and R.E. Hollenback, Laser Interferometer for Measuring High Velocities of Any Reflecting Surface, *J. Appl. Phys.*, Vol 43, 1972, p 4669-4674
6. R.H. Warnes *et al.*, An Improved Technique for Determining Dynamic Material Properties Using the Expanding Ring, in *Shock Waves and High-Strain-Rate Phenomena in Metals,* M.A. Meyers and L.E. Murr, Ed., Plenum Press, New York, 1981
7. D. Bauer and S.J. Bless, Strain Rate Effects on Ultimate Strain of Copper, in *Shock Waves in Condensed Matter,* North Holland, Amsterdam, 1983
8. G.R. Fowles, Experimental Technique and Instrumentation, in *Dynamic Response of Materials to Intense Impulse Loading,* P.C. Chou and A.K. Hopkins, Ed., Air Force Materials Laboratory, Wright-Patterson AFB, OH, 1973
9. J.A. Zukas, T. Nicholas, H.F. Swift, L.B. Greszczuk, and D.R. Curran, *Impact Dynamics*, John Wiley & Sons, New York, 1982
10. R.G. McQueen, S.P. Marsh, J.W. Taylor, J.N. Fritz, and W.J. Carter, The Equation of State of Solids from Shock Wave Studies, in *High Velocity Impact Phenomena*, R. Kinslow, Ed., Academic Press, New York, 1970
11. S.P. Marsh, *LASL Shock Hugoniot Data*, University of California Press, Berkeley, 1980
12. L.M. Barker and E.G. Young, "SWAP-9: An Improved Stress Wave Analyzing Program," Sandia National Laboratories Report No. SLA-74-0009, Albuquerque, NM, 1974
13. U.S. Lindholm and L.M. Yeakley, High Strain Rate Testing: Tension and Compression, *Exp. Mech.*, Vol 8, 1968, p 1-9
14. J. Harding, E.D. Wood, and J.D. Campbell, Tensile Testing of Material at Impact Rates of Strain, *J. Mech. Eng. Sci.*, Vol 2, 1960, p 88-96
15. C. Albertini and M. Montagnani, Testing Techniques Based on the Split Hopkinson Bar, in *Mechanical Properties at High Rates of Strain*, J. Harding, Ed., Institute of Physics, London, 1974
16. T. Nicholas, Tensile Testing of Materials at High Rates of Strain, *Exp. Mech.*, Vol 21, 1980, p 177-185
17. P.W. Bridgeman, *Studies in Large Plastic Flow and Fracture*, 1st ed., McGraw-Hill, New York, 1952, chapter 1
18. L.A. Cross, S.J. Bless, A.M. Ranjendran, E.A. Strader, and D.S. Dawicke, New Technique to Investigate Necking in a Tensile Hopkinson Bar, *Exp. Mech.*, to be published

High Strain Rate Shear Testing

Introduction

By K.A. Hartley
Research Assistant
Brown University

and

J. Duffy
Professor of Engineering
Brown University

SHEAR DEFORMATION is the primary deformation mode encountered in applications such as punching, grinding, machining, and forming operations and in events or processes that result in penetration. To accurately predict material behavior in these operations, the material should be tested in a laboratory under a large range of strain rates and temperatures; it should also be deformed directly in shear.

Shear testing has several advantages. For instance, shear eliminates the problems of necking, which occurs in tension tests, and barreling, which occurs in compression tests. Many metal specimens can be tested in tension or compression and the test results converted to shear stress and shear strain; however, this conversion seldom extends into a high strain range. The limit is frequently about 20% strain, which is far less than the strains reached in the applications mentioned above. At higher plastic strains, the deviation between results can become quite significant, with axial tests invariably providing higher values on flow stress than shear tests (after conversion by either the von Mises or Tresca flow rule).

As in axial loading, a variety of testing machines are available for shear loading. Selection usually depends on the rate of deformation required. Reference 1 provides a table for axial loading, which lists the approximate strain rate range for a variety of machines (see also Fig. 6 in this article). A similar table can be constructed for shear experiments, with rates that are not substantially different. For rates up to about 10 s^{-1}, a servohydraulic machine provides the most convenient testing method. High-speed hydraulic torsional machines are described in a subsequent section of this article. However, with servohydraulic machines, wave propagation effects within the loading column place an upper theoretical limit on the attainable strain rates. Hence, other methods of testing become necessary.

Torsional impact loading has been widely used to obtain strain rates up to about 10^2 s^{-1}. Between strain rates of 10^2 and 10^4 s^{-1}, the torsional Kolsky bar has proved to be a convenient method of testing. Other methods of high strain rate shear testing include double shear and punching, which provide somewhat higher rates than the torsional Kolsky bar.

For rates above 10^4 s^{-1} up to about 10^7 s^{-1}, a plate impact test is used to subject thin specimens to combined pressure and shear. All of the experimental techniques described in this article are capable of producing a complete stress-strain curve for a given material.

Based on a combination of results obtained from tests using the methods described above, a complete analysis of material behavior in shear over a wide range of dynamic rates can be obtained (10 to 10^7 s^{-1}). For penetration, punching, machining, etc., a strain rate range of 10^3 to 10^5 s^{-1} is of particular interest, but knowledge of material response over the entire range is very useful.

Considerable data are available in the literature on the high strain rate behavior of many materials, including steels, aluminum and copper alloys, titanium, beryllium, magnesium, and zinc. See the list of Selected References at the end of this article for more information on these subjects. However, few data are available at the higher strain rates, particularly at rates above 10^3 or 10^4 s^{-1}, at which testing becomes more difficult.

Results indicate that for many metals a linear relation exists between flow stress and the logarithm of plastic strain rate in the range from quasi-static rates to about 10^3 s^{-1}. Above this range, however, the flow stress generally rises far more rapidly with strain rate. Thus, the linear relationship is no longer valid, and extrapolation from the lower strain rate regime becomes unreliable.

In addition to immediate practical applications, high strain rate shear data are invaluable for the development of macroscopic constitutive models and for studies into possible deformation mechanisms for metals. Macroscopic results can be combined with microscopic observations to make fundamental contributions to the development of constitutive equations based on the micromechanisms that are dominant during deformation at a particular rate and temperature.

High-Speed Hydraulic Torsional Machines

By Ulric S. Lindholm
Director
Department of Materials Sciences
Southwest Research Institute

HYDRAULICALLY DRIVEN TORSIONAL SYSTEMS can be used as an alternative to mechanically driven torsional devices to provide transient loading. High-speed hydraulic torsional devices are directly analogous to similar axial tension or compression devices. A torsional actuator is used instead of a standard linear hydraulic actuator. Applications of hydraulic torsional systems are similar to those of axial machines, described in the article "High Strain Rate Compression Testing" in this Section.

A typical high-speed torsional system is illustrated schematically in Fig. 1. As in the torsional impact device discussed in the following section of this article, a stiff loading system and reaction frame are required, as well as low inertia in the moving parts. As illustrated in Fig. 1, these requirements are accomplished by enclosing the system, which features an integral actuator shaft, specimen, and load transducer, within a rigid housing. The vaned torsional actuator allows a rotation of about 3 rad.

Hydraulic loading of the actuator is maintained by a servo-valve and servo-control

Fig. 1 Dynamic hydraulic torsion test facility
Source: Ref 2

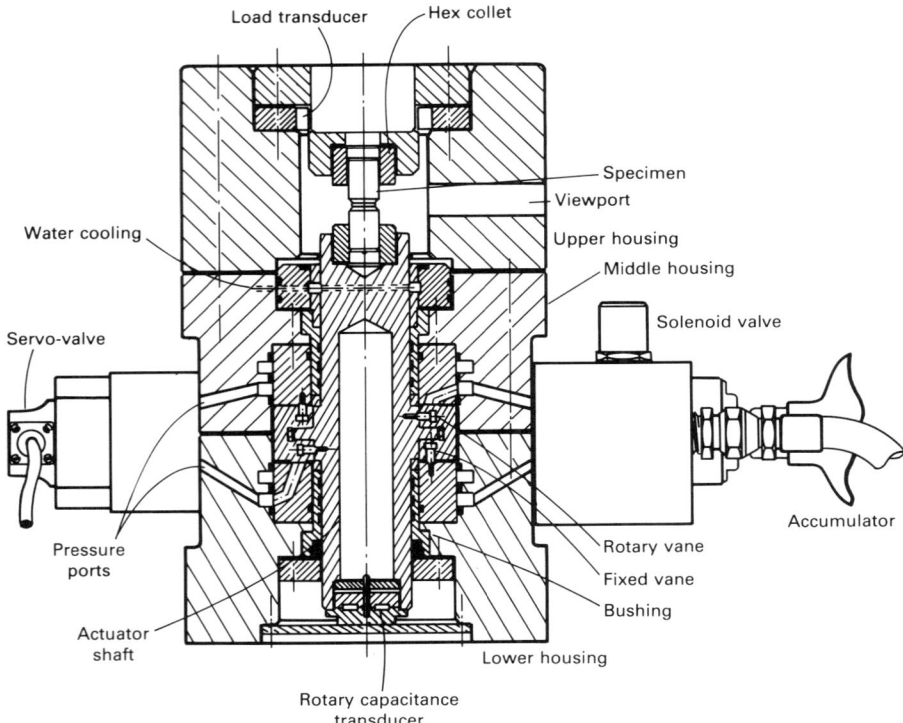

loop for operating at low rotational velocities, which provide shear strain rates to about 10 s^{-1}. For higher velocities, a solenoid-triggered quick-release valve conveys pressure from an accumulator directly to the actuator. In this open-loop mode, strain rates of up to 300 s^{-1} have been reported.

In contrast to mechanical impact devices, the rotary actuator shaft, specimen, and load transducer are initially attached (as in a standard test machine), and no transient mechanical engagement or impact is involved. This facilitates maintenance of coaxiality and alignment. However, the specimen still undergoes a rapid acceleration at the higher speeds, and inertial or resonance effects are still experienced by the load transducer.

Figure 2 illustrates a typical record of shear stress, τ, and shaft rotation, θ, as a function of time. The rotation is obtained from a rotary capacitance transducer that is attached to the bottom of the actuator shaft. The load transducer is a strain-gage element placed between the specimen and upper housing. Dynamic effects or "ringing" in the load signal are evident. A finite-element analysis of the shaft, specimen, load transducer system, as shown schematically in Fig. 3, was used to ensure that the mean signal from the load transducer was an accurate representation of the shear stress in the specimen. The computed and measured stresses

are compared in Fig. 2. Such dynamic analysis is often required for proper interpretation or verification of dynamic test procedures.

The rotation versus time record shows that the shear strain rate, $\dot{\gamma}$, is reasonably constant during the open-loop tests. The specimen geometry is similar to that used with torsional impact devices (thin-walled with a gage length of 3.2 mm, or 0.125 in.).

Figure 4 shows shear stress-strain curves for OFHC copper. Note the magnitude of the strains and the effect of strain rate on yield, strain hardening, and instability, as well as the softening at the highest rates. Localized shear bands were observed at the higher rates.

Torsional Impact Testing

By Ulric S. Lindholm
Director
Department of Material Sciences
Southwest Research Institute

TORSIONAL IMPACT LOADING METHODS have been used with a number of testing devices that were developed to test metals at strain rates up to about 10^3 s^{-1} (Ref

3-7). Early tests of this type used solid, round specimens. More recently, short, thin-walled tubular specimens have been used, in which nearly homogeneous stress and strain states can be achieved.

The devices generally consist of a stationary supported specimen, initially uncoupled from the drive mechanism. The energy to deform the specimen is stored in a rotating flywheel and/or a drive shaft. The rotating system is brought to the desired angular velocity, and a release and engagement mechanism quickly couples the torsional load to the specimen.

Torsional impact systems require:

- Alignment and concentricity of the specimen with the drive and engagement system

- Low inertia and high stiffness of the load train to minimize inertial effects and resonances during impact

- Adequate frequency response in the load- and strain-measuring devices

- Sufficient energy in the drive system to maintain nearly constant strain rate in the test specimen during deformation

A typical torsional impact system is illustrated in Fig. 5. In this device, a commercial lathe bed and drive mechanism are utilized; however, the arrangement is typical of all torsional impact systems. The specimen is rotated in the chuck of the lathe, which has speeds from 500 up to 2000 rpm. Other devices attach either the drive unit or the specimen to a flywheel, which is driven by a variable-speed motor. In either case, the drive unit is brought up to constant rotational speed before engagement of the specimen.

When the rotational speed is established, a release mechanism (the trigger rod in Fig. 5) allows a compressed drive spring to engage the drive unit with the specimen by means of mated, tapered engagement beads. In the mechanical design of this type of device, alignment, coaxiality, and details of the engagement mechanism are important. An alternative to the arrangement in Fig. 5 consists of a stationary specimen and load cell and a rotating drive unit. Such an arrangement is described in Ref 3.

Torsional load is measured with a strain-gage elastic tubular element in the load train. The stiffness of this gage section should be as high as possible so that torsional resonances of the system induced by the impact upon engagement do not produce excessive "ringing" in the load signal. These resonant frequencies are also governed by the total length of the load train between the rigid supports.

The shear strain rate, $\dot{\gamma}$, in the specimen is related to the angular velocity, $\dot{\theta}$, of the two ends of the specimen by:

Fig. 2 Comparison of computed response with experimental results for dynamic torsion test
Source: Ref 2

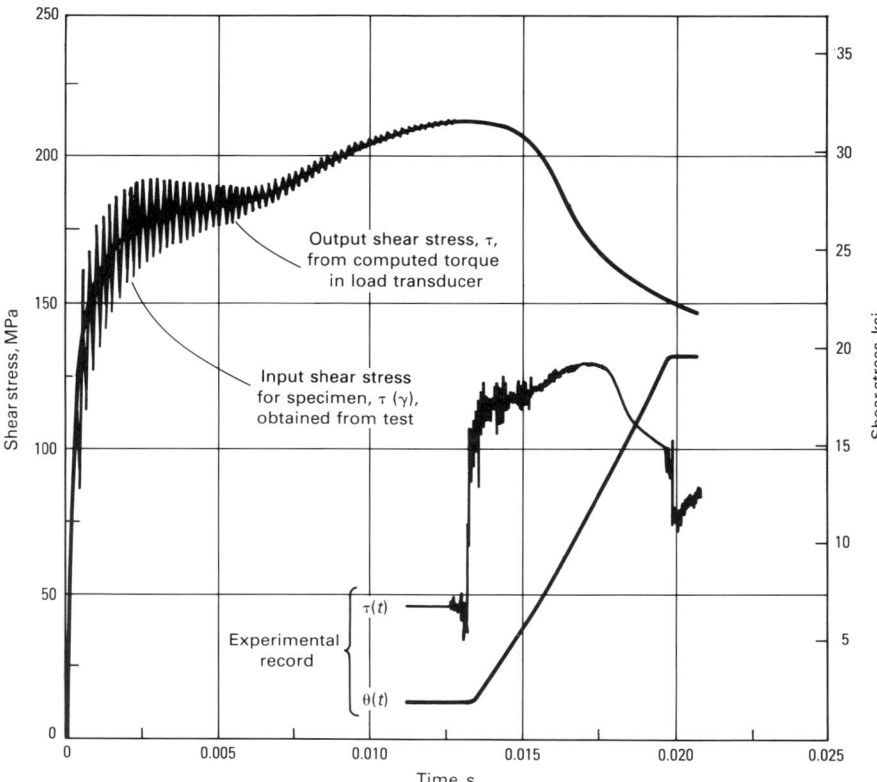

Fig. 3 Finite-element model of the torsional hydraulic actuator
Source: Ref 2

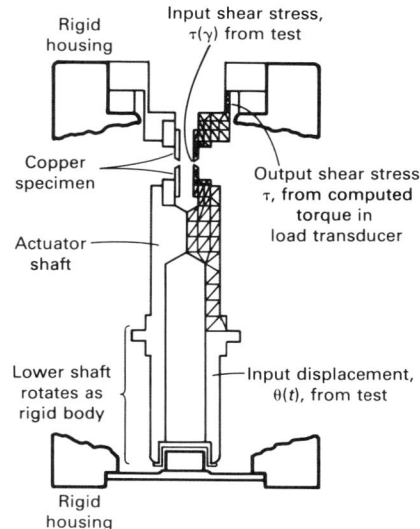

Fig. 4 Shear stress-strain curves for OFHC copper obtained with hydraulic machine
Source: Ref 2

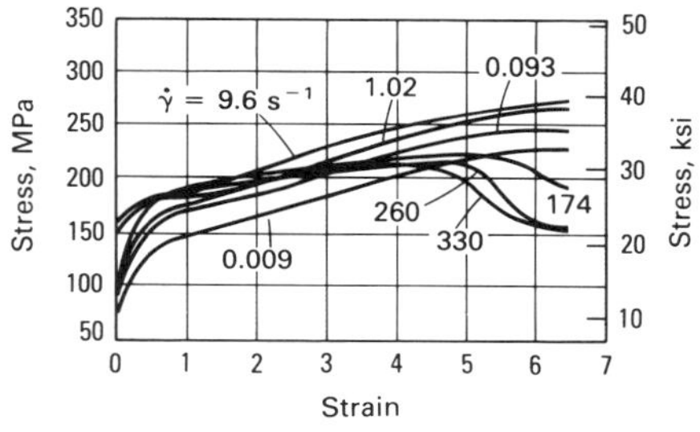

$$\dot{\gamma} = \frac{R\dot{\theta}}{L} \qquad \text{(Eq 1)}$$

where R and L are the mean radius and the length of the specimen gage section, respectively. Typical gage section dimensions are on the order of 3.2 mm (0.125 in.) in length, 6.4 mm (0.250 in.) in radius, and 0.76 mm (0.030 in.) in wall thickness. Thus, the shear strain rate is approximately twice the rotational speed of the drive mechanism.

Depending on the total angle of twist (θ) allowed, very large strains can be achieved, often in the range of $\gamma = 1$ to 5. Because such large strains generally are involved, the rotational velocity (either initial or transient) is used to determine strain rate (Eq 1) or strain. If the energy stored in the drive mechanism is large compared with the strain energy to deform the specimen, a nearly constant strain rate can be assumed to be proportional to the initial $\dot{\theta}$. Generally, direct strain measurements on the specimen have not been made. A photographic technique employing a drum camera to record the transient strain distribution in the specimen during deformation is described in Ref 3.

A significant aspect of dynamic torsion testing is the occurrence of plastic instabili-

ties due to thermal feedback when the test conditions approach adiabatic deformation (Ref 2, 7, 8). For fully adiabatic conditions, the temperature rise, ΔT_a, in the specimen is given by:

Fig. 5 Cross section of torsional impact machine
Source: Ref 3

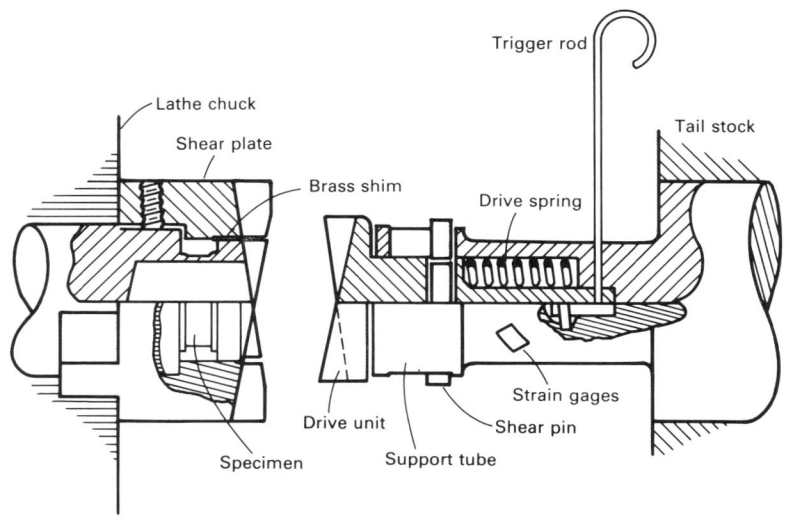

$$\Delta T_a = \frac{k}{\rho c} \int_0^{\gamma} \tau d\gamma \qquad \text{(Eq 2)}$$

where τ and γ are the shear stress and strain, respectively; k is a proportionality constant (ratio of plastic work converted to thermal energy); ρ is density; and c is specific heat. Instability occurs when the thermal softening due to the adiabatic temperature rise overcomes the intrinsic strain-hardening capacity of the test material.

Observation of adiabatic shear instabilities in torsion test specimens are reported in Ref 2, 3, and 8. Depending on material properties and specimen geometry, such instabilities occur in the strain rate range from 10 to 10^3 s^{-1}. When instability occurs, sharp temperature and strain gradients may arise in the specimen. Transient measurement of these profiles is difficult, but post-test examination of the test specimen may reveal the existence of localized straining in the form of either deformation bands or, in some cases (e.g., steels), transformation bands (Ref 7, 8).

Fig. 6 Dynamic aspects of materials testing
Source: Ref 1

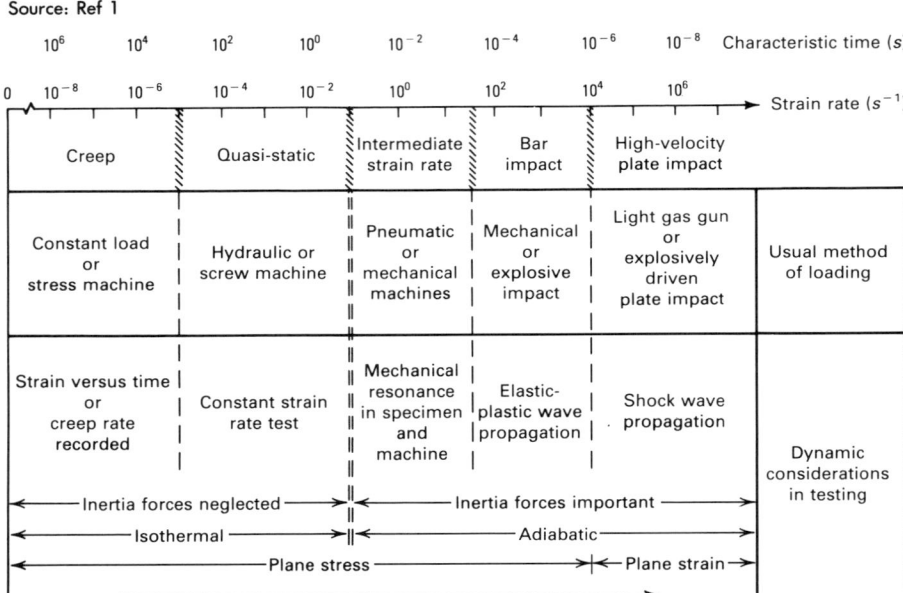

The Torsional Kolsky (Split-Hopkinson) Bar

By K.A. Hartley
Research Assistant
Brown University

J. Duffy
Professor of Engineering
Brown University

and

R.H. Hawley
Senior Research Engineer
Brown University

THE TORSIONAL KOLSKY BAR is a reliable apparatus for testing materials in the 10^2 to 10^4 s^{-1} strain rate regime (Fig. 6). The axial bar first proposed by Kolsky (and frequently referred to as the split-Hopkinson bar) is discussed in the article "High Strain Rate Compression Testing" in this Section. A description of the torsional Kolsky bar and the advantages and limitations of its use are presented in this article.

Advantages

In testing materials at high strain rates, torsional loading has some advantages over axial loading. In the original axial apparatus (Ref 9, 10), a compressional pulse propagates down a cylindrical bar to load the specimen (Fig. 7). However, due to the Poisson's ratio effect, compressional loading is accompanied by a radial expansion of the specimen. This radial expansion is opposed by a radial inertia that becomes greater with a shorter pulse and larger amplitude.

This effect results in a radial stress component in the specimen superposed on the axial stress component. Hence, in axial loading, the state of stress in the specimen is not fully uniaxial. Furthermore, the radial component of stress is difficult to evaluate, because it depends on dynamic material properties that have not as yet been determined. In addition, a radial traction is imposed at the interfaces between the specimen and the loading bars due to friction between the surfaces. Although friction can be minimized by proper lubrication, it cannot be completely eliminated. For these reasons, dynamic compression loading of a short specimen does not constitute a completely uniaxial state of stress.

In torsional testing, because of the absence of a Poisson's ratio effect, radial expansion or contraction does not occur. Hence, the inertial and frictional effects present in axial testing are absent when testing in torsion. Although this consideration originally led to the development of the torsional Kolsky bar, it has since been shown that the strain rate sensitivities determined by torsion tests on a variety of materials compare well with those determined by axial tests.

A further difference exists between an axial pulse (tension or compression) traveling down a cylindrical bar and a torsional pulse. This difference affects both the instrumenta-

tion requirements and the shape of the pulse that loads the specimen. When a compressive pulse travels down a bar, it undergoes geometric dispersion, because the different frequency components in the compressive pulse have different velocities. In particular, the higher frequency components travel at a lower velocity than the main pulse.

In contrast, there is no geometric dispersion when a torsional pulse travels along an elastic bar in its primary mode; i.e., all frequency components of the torsional pulse have the same velocity. Hence, a torsional pulse initiated either by explosive detonation or by the release of a stored torque does not change its shape as it propagates toward the specimen.

In Kolsky bar experiments, this absence of geometric dispersion in torsion means that a pulse initiated with a short rise time will maintain this rise time until it reaches the specimen, independent of the length of the Kolsky bar. Also, the torsional strain gage stations can be located as near or as far from the specimen as desired and still reveal the

Fig. 7 Schematic of Kolsky dynamic compression apparatus
Displacements at the ends of the specimen are denoted by u_1 and u_2.

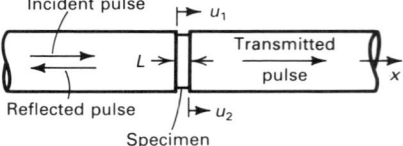

correct shape pulse incident on or reflected from the specimen. In an axial Kolsky bar, however, a gage placed too close to the specimen is subject to errors due to three-dimensional end effects, while a gage placed too far from the specimen produces unsatisfactory results because of geometric dispersion.

There is, however, a disadvantage in a nondispersive pulse. If a torsional pulse is noisy when initiated (i.e., if some high-frequency components are superposed on the main pulse), this characteristic will be maintained, regardless of the length of the Kolsky bar. As a result, the strain rate imposed on the specimen will not be constant. In an axial Kolsky bar, such high-frequency components gradually disappear. Hence, an axial pulse tends to become flat—i.e., tends to smooth out—as it travels along an elastic bar. A noisy torsional pulse, such as that associated with explosive initiation of the pulse (Ref 11), will retain this characteristic; it will never become flat and hence will not provide deformation at a constant strain rate.

As a result, for explosive initiation of a torsional pulse, a mechanical filter or pulse smoother is required to eliminate the high-frequency components. This type of device usually is not required when the torsional pulse is initiated by the stored-torque technique, because the amplitude is constant.

In general, the torsional Kolsky bar is well suited for conducting quasi-static and incremental strain rate tests, as well as dynamic strain rate tests. Thus, strain rate and strain rate history studies can be conducted using the same apparatus, thereby eliminating differences in test apparatus as a source of error in determining these effects. In the torsional Kolsky bar, a dynamic strain rate can be superposed on the quasi-static strain rate so that unloading does not occur, and the stress and strain rate increments can be measured directly for greater accuracy (Ref 12).

These tests can also be conducted with the specimen surrounded by an environmental chamber or a tube furnace to determine the effects of temperature or increments in temperature on flow stress (Ref 13). This is more difficult in an axial bar, because the strain gages must be placed close to the specimen to eliminate error due to pulse dispersion. Details of incremental strain rate and high-temperature experiments are given in subsequent sections of this article.

Torsional testing has additional advantages in determining the material behavior of single crystals due to the absence of frictional and inertial effects in torsion. In compressive loading, the added radial stress component from specimen/bar friction tends to cause barreling in the specimen, thus changing the

direction of the strain components relative to the crystallographic orientations. This can be avoided by testing in shear, where the loading direction remains well defined thoughout deformation. Hence, the torsional Kolsky bar provides a suitable technique for deforming (testing) single crystals in shear at high strain rates.

Disadvantages

Polycrystalline specimens tested in a torsional Kolsky bar experiment are shaped as short, thin-walled tubes. These specimens require substantially more material than those used in an axial bar. Generally, these specimens are more expensive to machine. Gripping the tubular specimen is difficult, and special care must be taken that the specimens are held firmly in the bar and that there are no reflections due to the gripping method. In addition, the torsional Kolsky bar system generally is more costly to build and implement than the axial testing system.

Historical Development

In 1949, Kolsky used a modified pressure bar to test thin, wafer-like specimens at high strain rates (Ref 9). The specimens were placed between two long elastic cylindrical bars that act as waveguides for the loading pulse (see Fig. 8 in the article "High Strain Rate Compression Testing" in this Section). This experiment was referred to as the split-Hopkinson bar, although the purpose of the test, the instrumentation, and the specimen were quite different from those in Hopkinson's original experiment (Ref 14).

In Kolsky's test, loading was accomplished by propagating a compressive wave down one of the bars toward the specimen. Measurements of the waves in the elastic bars were made on each side of the specimen. Kolsky showed that the portion of the incident loading wave that is transmitted through the specimen provides a measure of the axial stress in the specimen, while the magnitude of the wave that is reflected is proportional to its strain rate.

This same general analysis applies to torsional loading, with angular displacements and shear components replacing the axial components. By combining the output from the strain gages on either side of the specimen and by integration of the strain rate/time pulse to yield strain versus time, a complete record of the stress-strain curve can be obtained.

Lindholm further illustrated that by locating the gages equidistant from the specimen the reflected and transmitted record will occur at the same moment, thus facilitating data reduction and increasing accuracy (Ref 15).

This increase in accuracy is due to the precision with which a strain gage can be located, e.g., an error of 1 mm in its location corresponds to an error of only one fifth of a microsecond in the timing of a torsional pulse.

The first torsional Kolsky bar was developed by Baker and Yew (Ref 16), who used a pretwisted elastic bar to produce a step function wave. With an analysis similar to Kolsky's for axial loading, they showed that the stress-strain curves in shear for the specimen could be obtained by an analysis of the waves in the elastic bars on each side of the specimen.

Several other investigators have contributed to the development of the torsional Kolsky bar (Ref 11, 17-19). Duffy et al. originally used explosive loading to initiate the loading pulse (Ref 11). This method has the advantage of producing a shorter pulse rise time, whereas a stored-torque loading system provides potentially higher strain. Frantz and Duffy used a torsional Kolsky bar to perform tests with increments in strain rate; i.e., the specimen was first twisted statically and then a dynamic strain rate was imposed with no intermediate unloading (Ref 12). They demonstrated that the torsional Kolsky bar is a convenient method of studying strain rate history effects in metals. For high-temperature testing, various methods have been employed to account for the multiple reflections of the loading pulse due to the presence of temperature gradients along the length of the bar (Ref 20, 21).

It has been shown that the torsional Kolsky bar can be used successfully to test materials in the strain rate range of 10^2 to 10^4 s^{-1}. Selection of an axial Kolsky bar or a torsional Kolsky bar depends on the individual test requirement. Results of quasi-static, dynamic, and incremental tests over a range of temperatures for the torsional Kolsky bar and the axial Kolsky bar are given in Ref 22 to 24.

Stored-Torque Torsional Kolsky Bar

A stored-torque Kolsky bar is illustrated schematically in Fig. 8. The 25-mm (1-in.) diam bar is made of 6061-T6 aluminum alloy. Steel and titanium have also been used (Ref 18, 25). The bar must be supported along its length and aligned properly. In Fig. 8, it is supported by a series of Teflon bearings that allow the bar to rotate freely in either direction.

The loading pulse is produced by the sudden release of a stored torque. This requires a torque pulley at the end of the bar and a clamp positioned within a short distance to prevent rotation (see Fig. 8). The design of

Fig. 8 Schematic of the stored-torque torsional Kolsky bar

DCDT: direct current differential transformer

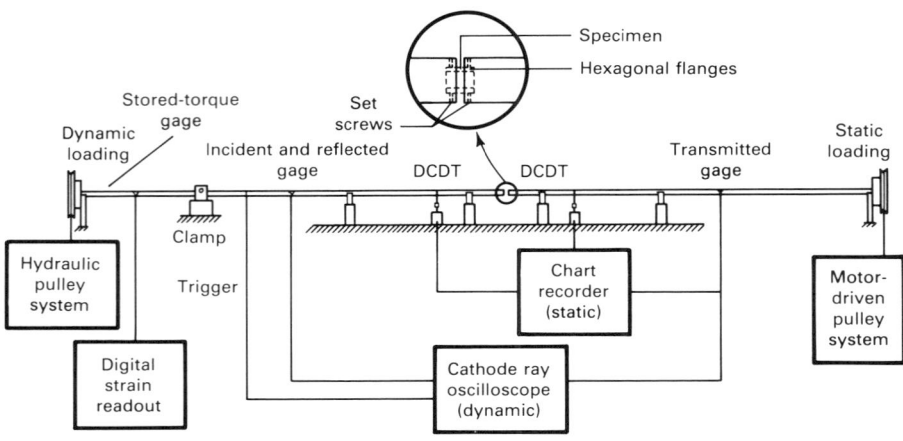

the clamp is crucial for good results. The clamp must be able to hold the desired torque without slipping and release the torque rapidly enough to produce a sharp-fronted stress pulse traveling toward the specimen.

Upon release of the clamp, the torsional pulse, with a torque of constant amplitude equal to half that of the stored torque, propagates down the bar toward the specimen. Simultaneously, an unloading pulse of equal magnitude propagates from the clamp toward the torque pulley. The torsional mechanical impedance of the pulley is sufficiently large that after reflection the unloading wave reduces the torque in the incident bar to zero as it propagates back along the bar.

The wave diagram illustrated in Fig. 9 details the positions of the wave fronts with time. The duration of the loading pulse is the time required for the pulse to travel twice the distance along the bar between the clamp and the torque pulley.

As the pulse travels down the bar, it is detected by two four-arm electric resistance strain gage bridges mounted at 45° to the axis of the bar at equal distances from the specimen and powered by 24-V direct-current supplies. The gages must be placed at a sufficient distance from the specimen so that a clear record of both the incident and the reflected pulses is obtained without overlap or interference.

For example, if the distance between the clamp and the torsional pulley is 760 mm (30 in.), the pulse length would be 490 μs for an aluminum bar with a shear wave speed of 3100 m/s (10 200 ft/s). To ensure that the reflected pulse does not overlap with the incident pulse at the gage station and thus cause interference, the gages must be placed at least 760 mm (30 in.) from the specimen. Furthermore, to ensure that the stress in the bar returns to zero after the pulse has

passed, the gages should be located at an even greater distance from the specimen. For example, for the bar shown in Fig. 9, the gages are located 915 mm (36 in.) from the specimen. A similar strain gage bridge for

monitoring the transmitted pulse is located the same distance from the specimen as the incident gages (915 mm, or 36 in.). The signals are recorded on a digital oscilloscope of sufficiently high frequency response.

The incident, reflected, and transmitted pulses should not overlap, and each should be recorded to ensure that the test is valid— i.e., that the strain rate is nearly constant, that there is a rapid rise time, and that the signals are free of noise. A schematic of the output from the incident and transmitter gages is shown in Fig. 10. Figure 11 illustrates an oscilloscope record from an actual test. The incident pulse is nearly flat, with a rise time of about 30 μs and a duration of 170 μs. These data are easily reduced to yield the stress-strain curve for the specimen; details of data reduction and calibration are presented below.

Quick-Release Clamp. The design of the quick-release clamp has a direct influence on the quality of the incident loading wave. Ideally, the incident pulse should rise instantly to a constant amplitude and then drop off immediately to zero at the end of the

Fig. 9 Wave characteristic diagram for the stored-torque Kolsky bar

Torsional wave speed is denoted by c.

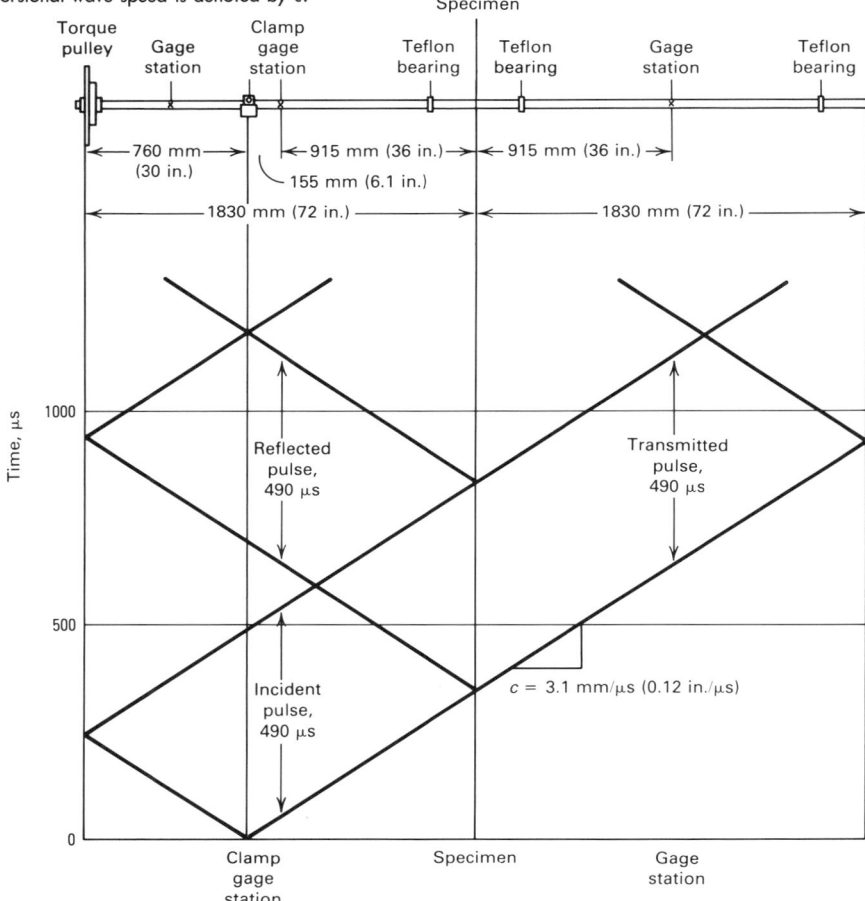

Fig. 10 Schematic of the output from the incident and transmitter gages during a Kolsky bar experiment
Transmitter gage output is enlarged relative to incident gage output.

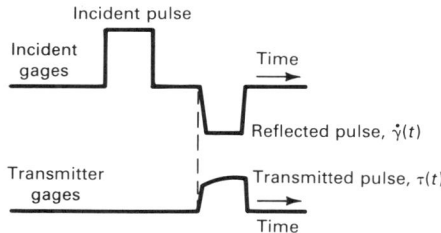

Fig. 11 Oscilloscope record from a test with the stored-torque Kolsky bar
Each horizontal division equals 200 μs. Transmitted output magnified 2×. Fracture of the specimen causes transmitted pulse magnitude to drop to zero.

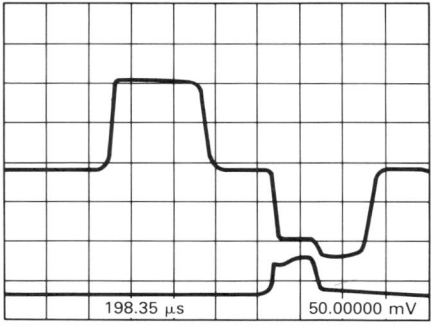

pulse, thus forming a square loading pulse and providing deformation at a constant strain rate. In practice, a finite time is required to achieve maximum torque in the loading pulse and to reduce the torque to zero at the end of the incident wave.

One clamp design (Ref 26) consists of two arms with hinged bottoms that press against the sides of the Kolsky bar under the action of a notched bolt (see Fig. 12). To initiate the pulse, the bolt is tightened until it fractures. In this design, the clamp arms are shaped to match the circumference of the loading bar. Between tests, the bare surface that was in contact with the clamp is roughened with sandpaper and cleaned with acetone to ensure good gripping.

A variation of this design is illustrated in Fig. 13. In this design, the arms of the clamp are flat and are held together at the top by a notched pin. The clamp is tightened by a hydraulic ram that pushes the lower ends of the clamp arms together. After the desired torque is loaded between the pulley and the clamp, the clamp pressure is increased until

Fig. 12 Clamp designed by Cleland and Stevenson (Ref 26) for the torsional Kolsky bar

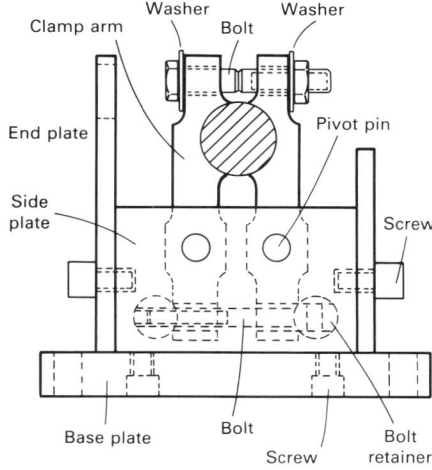

the notched pin fractures, thus releasing the stored torque.

The amount of torque required to achieve the desired strain rate is a determining factor in selecting the notched-pin material and the depth of the notch. Pulse rise time is also affected by the choice of material for the notched pin. The pin material must exhibit minimal ductility, but must not be so brittle as to fracture before the clamp is tight enough to hold the desired stored torque. Functional pin materials include 6061-T6 and 2024-T6 aluminum alloys. The desired clamping force and fracture point are achieved by varying the depth of the notch. Steel pins can also be used, but even brittle steels produce a longer rise time than aluminum alloys. It might be expected that the mass of the clamp would affect the rise time of the pulse, but this appears not to be the case. The rise time of the pulse can be decreased much more significantly by proper selection of pin material.

Misalignment of the various components of the Kolsky bar can result in a small unloading pulse immediately preceding the incident loading pulse, particularly if misalignment is caused by the action of the clamp. The entire clamp mechanism must be allowed to slide relative to the bar, via a movable carriage (Fig. 13), so that the clamp is properly aligned. Care should be taken to prevent bending of the incident bar when the clamp is tightened.

Test Specimens

Polycrystals. The specimens used in torsional Kolsky bar experiments usually are thin-walled tubes with integral flanges machined from bar stock (Fig. 14). In materials

Fig. 13 Clamp designed by Duffy for the torsional Kolsky bar

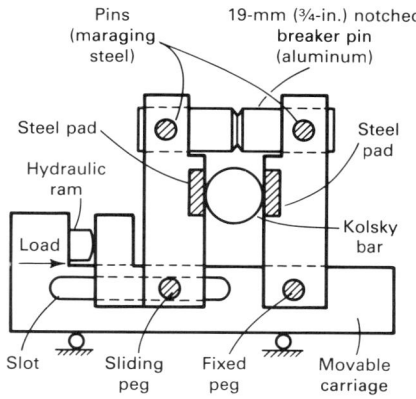

that have a relatively low flow stress (1100-O aluminum, OFHC copper, or zinc, for example), the specimen can be held in position by cementing its flanges to the Kolsky bar with an epoxy cement (Fig. 14a). This provides sufficient strength and a rigid connection.

For stronger materials, such as steel alloys, the specimen must be held in position by mechanical means. The gripping mechanism must be rigid so that it prevents lost motion between the specimen and the Kolsky bar as the stress pulses pass. Furthermore, the torsional mechanical impedance of the connection must match that of the bar so that the stress pulses will not undergo reflections before reaching the specimen.

Threaded connections do not meet these requirements. However, success has been achieved using hexagonally shaped flanges on the specimen with matching sockets in the ends of the Kolsky bar. To prevent lost motion between the socket and the flanges, small set screws are used to hold the specimen flanges against the driving faces of the hexagonal sockets (see Fig. 14b). For harder specimen materials, such as alloy steels, the set screws may not grip the specimen tightly enough to prevent rotation. To ensure rigidity, the hexagonal socket can be filled with warm glycol phthalate, the specimen inserted, and the set screws tightened.

The use of short specimens in the torsional Kolsky bar implies that a nearly homogeneous state of strain is obtained after a few reflections of the loading pulse from the ends of the specimen. To examine the development of the elastic-plastic boundary within the wall thickness of the tubular specimen, a finite-element analysis of the strain distribution has been performed (Ref 27). In this analysis, a bilinear elastic-plastic stress-strain relation was used with an elastic modulus and a plastic-hardening rate approximately equal to those of commercially pure aluminum. Figure 15 illustrates the growth of

Fig. 14 Details of the polycrystalline specimen used in the torsional Kolsky bar experiment

(a) Tubular specimen with cylindrical flanges for cementing. (b) Tubular specimen with hexagonal flanges

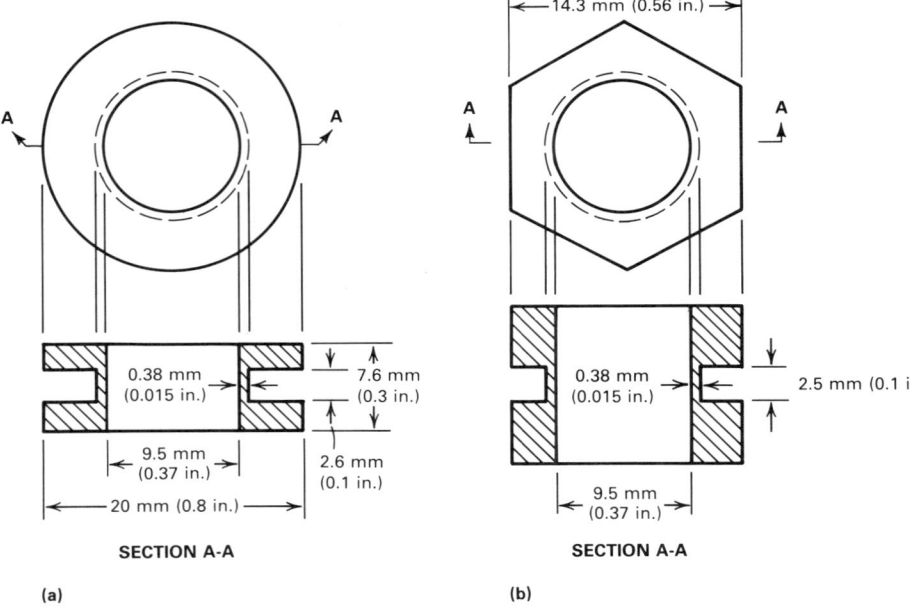

(a)   (b)

the plastic zone (the dark areas) at various stages of deformation.

The plastic zone begins at the reentrant corner between the thin-walled tube and the flange and from there spreads gradually throughout the thickness of the wall. Each diagram in Fig. 15 represents one quarter of the cross section of the tube and the adjacent part of the flange; thus, the central axis of the specimen lies below in each diagram. The values of the applied torque are given.

The plastic zone proceeds radially toward the inside wall surface and then across the gage length along the outside surface of the specimen. The plastic zone is contained until almost the entire specimen begins to flow. These results are based on a static analysis, but because the wavelength of the pulse is much longer than the specimen, these data can be applied to the dynamic deformation and appear to agree well with experimental results.

With continued loading into the plastic range, the strain distribution in the thin-walled tube may not remain homogeneous. For example, depending on the material, shear bands may form that completely encircle the thin-walled tube. Whenever the strain is inhomogeneous, it significantly alters interpretation of the strain gage records.

Although the transmitted pulse provides a measure of stress as a function of time, the reflected pulse yields only an average strain rate and hence an average strain. However, shear bands generally are not encountered

until some plastic strain has accumulated. For instance, Costin et al. (Ref 28) observed shear bands in 1018 cold rolled steel at a plastic strain of 8 to 10%, using a torsional Kolsky bar.

The presence of shear bands in a thin-walled tubular specimen is easy to detect by scribing fine axial lines on the inside wall surface of the specimen before loading. If the strain remains homogeneous throughout the deformation process, after testing, each of these lines appears tilted at the shear angle within the gage length of the specimen, but remains straight and axial in the flange area (Fig. 16b). Any departure from a straight line within the gage length is evidence of nonhomogeneous strain. When nonuniform strain is present, as in the case of a shear band, the lines depart drastically from straight lines (Fig. 16a).

Single-crystal specimens can be tested in the torsional Kolsky bar, but the specimen shape must be changed to a small rectangular parallelepiped. A matched set of four single-crystal specimens having the same crystallographic orientation are tested together. The specimens should be arranged circumferentially at 90° intervals (Fig. 17).

The specimens are secured with an epoxy cement. These cements are sufficiently strong to test pure metals, which generally have a low flow stress. However, with strong metals, the epoxy breaks before any significant amount of strain accumulates. Each test yields the average stress-strain properties of

the four specimens (Ref 29). The loading technique and instrumentation are the same for testing single-crystal or polycrystalline specimens.

Impedance Matching and High-Temperature Testing

A constant mechanical impedance must be maintained along the entire Kolsky bar to prevent undesirable wave reflections from locations other than the specimen. This condition requires that the product $\rho c J$ be held constant at each cross section along the bar. Here, ρ is the mass density, J is the polar moment of inertia, and c is the torsional wave speed equal to $\sqrt{G/\rho}$, where G is the shear modulus. For a constant impedance, any instrumentation attached to the bar must be extremely light.

Furthermore, care must be taken when there is a difference in bar material, such as at the specimen flange, or when a change occurs in the shear modulus. The latter is particularly important in high-temperature tests where the area around the specimen is enclosed in a furnace, causing thermal gradients along the incident and transmitter bars. Such gradients produce a variation in the shear modulus of the Kolsky bars and hence variations in impedance.

It has been claimed by some investigators that the gradients counter each other and hence can be neglected. However, simple analysis shows that they actually reinforce each other and thus can produce a large error. Various solutions to this problem have been proposed. For a test temperature of 100 to 150 °C (210 to 300 °F), the error can probably be ignored. Beyond that, a correction is necessary.

Chiddister and Malvern (Ref 21) corrected for this effect through a calculation of the repeated partial reflections as the pulse propagates along the bar in the heat-affected zone. Although this approach is acceptable, the calculations can be cumbersome. A different solution was proposed by Eleiche and Duffy, who tapered the Kolsky bars to counteract the effect of the thermal gradients (Ref 20). In this approach, the taper in the Kolsky bar, which provides a variable impedance, is determined by calculation to counteract the change in impedance due to the temperature gradient, thus resulting in a constant impedance bar.

An example of a tapered bar used for tests conducted at 250 °C (480 °F) is given in Fig. 18. The disadvantage of this method is that it requires a temperature survey before testing so that the Kolsky bar can be machined to the right taper. Furthermore, a bar tapered for testing at 350 °C (660 °F) cannot be used for

Fig. 15 Results of a finite-element analysis showing the growth of the plastic zone within the tubular specimen used in the torsional Kolsky bar

Each diagram shows a cross section through half the length of the tube wall, plus the adjacent flange. The plastic zone is represented by the dark area and the applied torque is T.

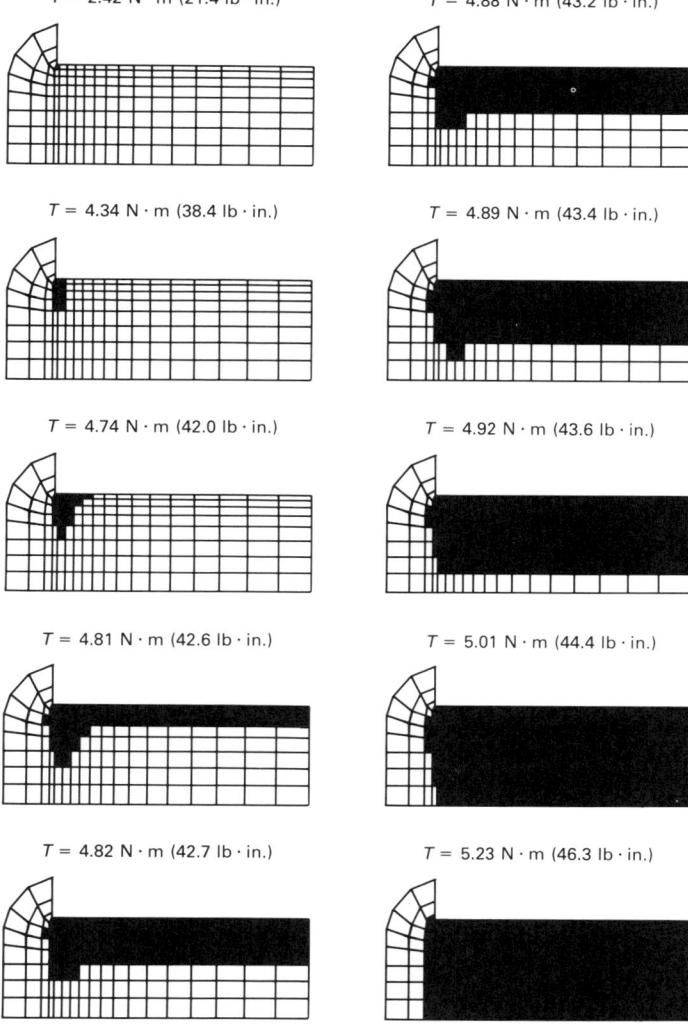

$T = 2.42$ N · m (21.4 lb · in.)

$T = 4.88$ N · m (43.2 lb · in.)

$T = 4.34$ N · m (38.4 lb · in.)

$T = 4.89$ N · m (43.4 lb · in.)

$T = 4.74$ N · m (42.0 lb · in.)

$T = 4.92$ N · m (43.6 lb · in.)

$T = 4.81$ N · m (42.6 lb · in.)

$T = 5.01$ N · m (44.4 lb · in.)

$T = 4.82$ N · m (42.7 lb · in.)

$T = 5.23$ N · m (46.3 lb · in.)

bar between the direct-current differential transformers.

The difference between the two outputs is obtained through simple electronic circuitry and fed into the x-axis of an x-y recorder. The strain gage signal proportional to the torque is fed into the y-axis to produce a record proportional to the torque-twist in the specimen. The strain gage signal is calibrated by applying a known shunt resistance in parallel with one leg of the Wheatstone bridge.

To obtain the shear strain in the specimen, the elastic rotation of the bar between the two differential transformers is subtracted from the total rotation. This elastic rotation is measured by cementing the loading bars together without a specimen and loading them quasi-statically. Typical test results obtained at a variety of temperatures using the Kolsky bar to test 1020 steel at a quasi-static strain rate of 5×10^{-4} s^{-1} are given in Fig. 19.

Incremental Strain Rate Testing

The apparent strain rate sensitivity of a material is defined by:

$$\mu_A = \frac{\partial \tau_A}{\partial \ln \dot{\gamma}} \approx \frac{\Delta \tau}{\ln \left(\dfrac{\dot{\gamma}_r}{\dot{\gamma}_i} \right)} \qquad \text{(Eq 3)}$$

where $\Delta \tau$, as shown in Fig. 20, is the difference in flow stress between two constant strain rate curves at $\dot{\gamma}_i$ and $\dot{\gamma}_r$. Experiments are performed at various constant strain rates, and the stress (at a given value of strain) is plotted as a function of strain rate on a logarithmic scale. The slope of this line provides the value of the derivative in Eq 3.

The true strain rate sensitivity is defined by:

$$\bar{\mu}_T = \frac{\partial \tau_T}{\partial \ln \dot{\gamma}} \approx \frac{\Delta \tau_s}{\ln \left(\dfrac{\dot{\gamma}_r}{\dot{\gamma}_i} \right)} \qquad \text{(Eq 4)}$$

where $\Delta \tau_s$ is the difference in flow stress due to an increment in strain rate from $\dot{\gamma}_i$ and $\dot{\gamma}_r$. To evaluate the true strain rate experimentally, the specimen is first loaded quasi-statically to a predetermined strain; without unloading, the strain rate is then suddenly increased by several orders of magnitude to a dynamic strain rate. This test is also useful in studying strain rate history effects.

Figure 20 illustrates a typical stress-strain curve developed from such an experiment. As shown, the flow stress after the strain rate increment is different from the flow stress at the same value of strain in a specimen strained entirely at the dynamic strain rate. This leads to a difference between the apparent and true strain rate sensitivities. For face-centered cubic and hexagonal close-packed

testing at room temperature. The advantage of this method is that the records of stress and strain rate as functions of time require no further interpretation beyond the usual treatment for a room-temperature test. This advantage was demonstrated by Eleiche and Duffy (Ref 20).

Quasi-static Testing

Specimens can also be tested in torsion quasi-statically using the stored-torque Kolsky bar shown in Fig. 8. In this case, the Kolsky bar is clamped against rotation using the clamp described earlier. Loading is accomplished by a pulley system at the far end

of the transmitter bar and is powered by a variable-drive low-speed electric motor.

As in dynamic testing, stress is measured by a four-arm bridge on the transmitter bar. To determine the strain in the specimen, two linear direct-current differential transformers are placed equidistant from the specimen. Extremely fine wires attached to the cores of the differential transformers are wound around the bars. As the bars rotate, the cores are pulled vertically out of the transformer coils. The difference between the output signals from the direct-current differential transformers is directly proportional to the amount of twist undergone by the specimen plus the elastic twist of the portion of the Kolsky

Fig. 16 Section of specimen showing scribe line after testing

(a) Specimen with nonhomogeneous strain distribution due to the formation of a shear band. (b) Specimen with homogeneous strain distribution

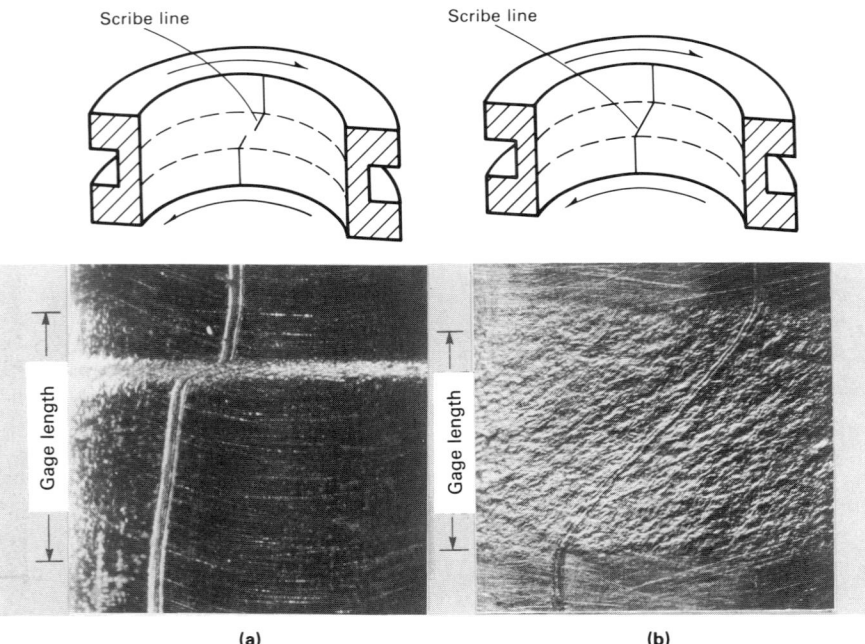

(a)   (b)

Fig. 17 Aluminum single-crystal specimens showing the arrangement relative to the Kolsky bar

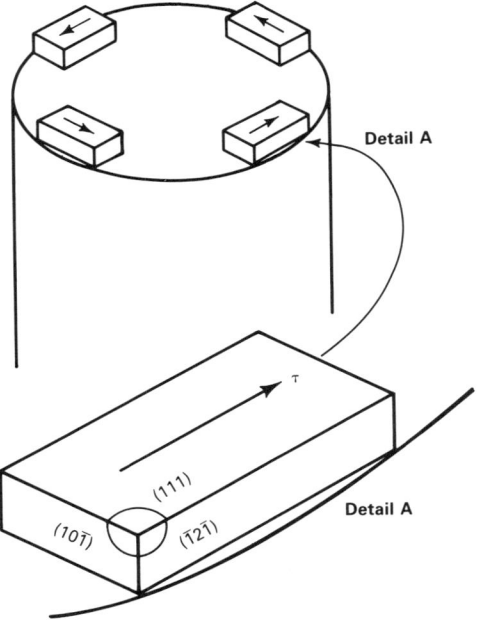

metals, the all-dynamic flow stress is greater than the flow stress immediately after the increment in strain rate. This difference is commonly referred to as the strain rate history effect, because in both cases the strain is the same, but the history of strain rate is different (Ref 12, 17, 30).

Frequently, the difference between the two values of flow stress diminishes with further deformation, and eventually the two dynamic curves merge. In this case, the material is said to have a "fading memory" (Ref 31). Body-centered cubic metals, under certain conditions and particularly at very low temperatures, may exhibit a negative strain rate history effect; i.e., the flow stress after an increment in strain rate is greater than the flow stress for the material strained entirely at the dynamic rate (Ref 13). At room temperature, body-centered cubic metals generally show only a small history effect, although the influence of strain rate on flow stress usually is considerably greater than that for face-centered cubic metals.

In incremental strain rate tests, the Kolsky bar provides an important advantage: the transmitted signal furnishes a measure not of the total stress in the specimen, but of the excess stress $\Delta\tau_s$ imposed by the stress pulse above the existing stress as a result of loading at the quasi-static strain rate (Fig. 21). Thus, rather than evaluation of a small difference between two large numbers, measurement of the stress increment $\Delta\tau_s$ can be made directly

from oscilloscope records. The results of incremental tests on aluminum are given in Fig. 22. The static portion of the curve is obtained from the output of the x-y recorder. The dynamic incremental results are obtained separately from the oscillogram.

Explosively Loaded Torsional Kolsky Bar

An alternate method of dynamic torsion testing using a Kolsky bar is to initiate the torsional pulse by the detonation of explosive charges. Generally, the advantage to explosive loading is that a shorter rise time is attained in the loading pulse; the strain rates are approximately the same as in the stored-torque bar. An explosively loaded bar made of thick-walled aluminum tubing is illustrated in Fig. 23.

The torsional pulse is generated by the simultaneous detonation of two small charges at the loading end (Fig. 24). To obtain a pure torque with no bending, each charge must provide an equal impulse, and they must be detonated at the same instant. The first requirement is met by using equal weights of explosive, and the second by using a single detonator connected to the explosive charges by means of "leaders" that are also made of the explosive. These leaders are of equal length, and tests have shown that the explosive charges are detonated less than 0.1 μs apart.

This method ensures that the stress pulse that is initiated is almost entirely torsional, although some low-amplitude axial and bending stress pulses are also initiated. The stresses must be monitored in each test, but generally are so small that they can be disregarded. However, a mechanical filter, resembling a rather stiff bellows, is used to decrease their magnitude. This filter, shown at the top of Fig. 25, is made of an aluminum alloy with a relatively high yield stress such as 6061-T6.

The pulse produced at the loading end of the Kolsky bar by the explosion is neither "square" nor "smooth." The rise time is short, as desired, but the amplitude of the impulse varies considerably with time. This variation in amplitude is undesirable, because it results in loading of the specimen at a variable strain rate. This is not a problem in the axial Kolsky bar, because the high-frequency components travel more slowly and the pulse is smoothed as it travels by geometric dispersion.

Because no geometric dispersion occurs for a torsional pulse in its first mode, a second mechanical filter, or pulse smoother, is required. The pulse smoother is a short length of tubing with a narrow neck made of 1100-O aluminum so that it deforms plastically during passage of the pulse. It reduces the magnitude of the higher frequency com-

Fig. 18 Tapered steel torsional Kolsky bar for test at 250 °C (480 °F)
Source: Ref 20

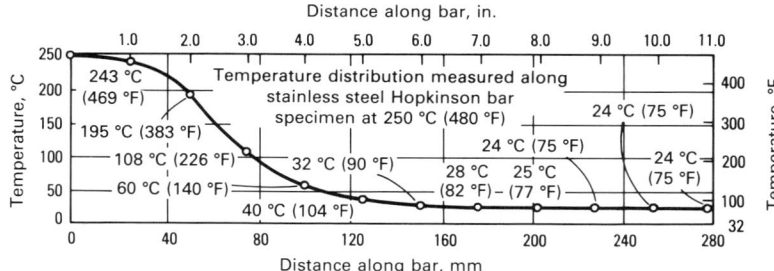

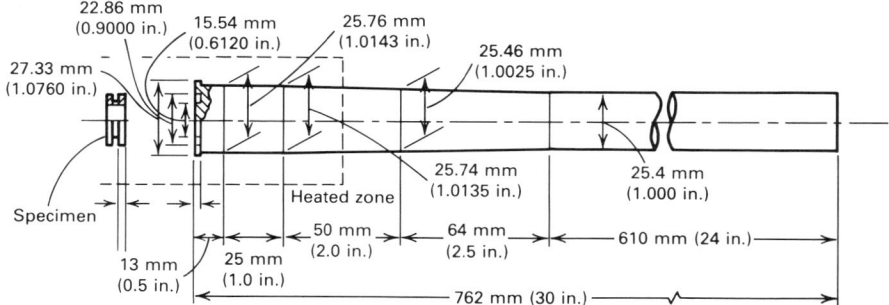

ponents, thus smoothing the pulse. As shown in Fig. 25, the two filters are placed next to each other in the Kolsky tube. The strain gages used to monitor the pulses incident on the specimen are located at stations between these filters and the specimen.

The explosively loaded Kolsky bar can be loaded quasi-statically in the same manner as a stored-torque Kolsky bar. In addition, incremental tests can be performed with equal ease. The advantage of explosive loading is that it provides rapid rise times compared to a stored-torque system. A typical rise time is only 7 to 10 μs, compared with 20 to 40 μs with the stored-torque system, depending on pulse amplitude. However, the stored-torque system requires no mechanical filtering and is inherently much safer to employ than are explosive charges.

The explosive charges produce a relatively short pulse, so that the total strain is generally small. With a stored-torque bar, the distance between pulley and clamp can be increased to produce as long a pulse as the available laboratory space will allow. Oscilloscope records obtained with an explosively loaded torsional Kolsky bar are shown in Fig. 26.

Calculating Stress, Strain, and Strain Rate

Values of stress and strain in the specimen can be inferred from the measured records of the strain in the incident and transmitter bars. These relations were first derived by Kolsky

for the axial bar; in this section they are transposed to shear values. In a later section of this article, calibration of the gage output and data reduction are discussed in detail.

In accordance with the analysis of Kolsky, the reflected pulse in a torsion bar provides a measure of the shear strain rate in the specimen $\dot{\gamma}_s(t)$ and, through a single integration, provides the shear strain in the specimen, $\gamma_s(t)$. The shear strain in the specimen is

given by the difference in rotation between its two ends divided by its length:

$$\gamma_s = \frac{D_s\phi_1 - D_s\phi_2}{2L_s} \quad \text{(Eq 5)}$$

where ϕ_1 and ϕ_2 are the angles of twist in the incident and transmitter bars, respectively; D_s is the mean diameter of the thin-walled specimen; and L_s is its length. The value of ϕ_2 can be determined from the shear strain measured at the surface of the transmitter bar through:

$$\gamma_T = \frac{D}{2}\frac{\partial\phi_2}{\partial x} = \frac{D}{2c}\frac{\partial\phi_2}{\partial t}$$

where D is the diameter of the Kolsky bar, and $c = \sqrt{G/\rho}$ is the torsional velocity in the Kolsky bar. As a result:

$$\phi_2 = \frac{2c}{D}\int_0^t \gamma_T(t)dt \quad \text{(Eq 6)}$$

Similarly, ϕ_1 can be determined from the difference in strains due to the incident and reflected pulses:

$$\phi_1 = \frac{2c}{D}\int_0^t [\gamma_I(t) - \gamma_R(t)]dt \quad \text{(Eq 7)}$$

where the negative sign is necessary because the reflected pulse travels in the $-x$ direction. If Eq 5 is differentiated and Eq 6 and 7 are utilized:

$$\dot{\gamma}_s(t) = \frac{c}{L_s}\frac{D_s}{D}[\gamma_T(t) - \{\gamma_I(t) - \gamma_R(t)\}]$$

$$\text{(Eq 8)}$$

Fig. 19 Static and dynamic shear stress/shear strain curves for 1020 hot rolled steel

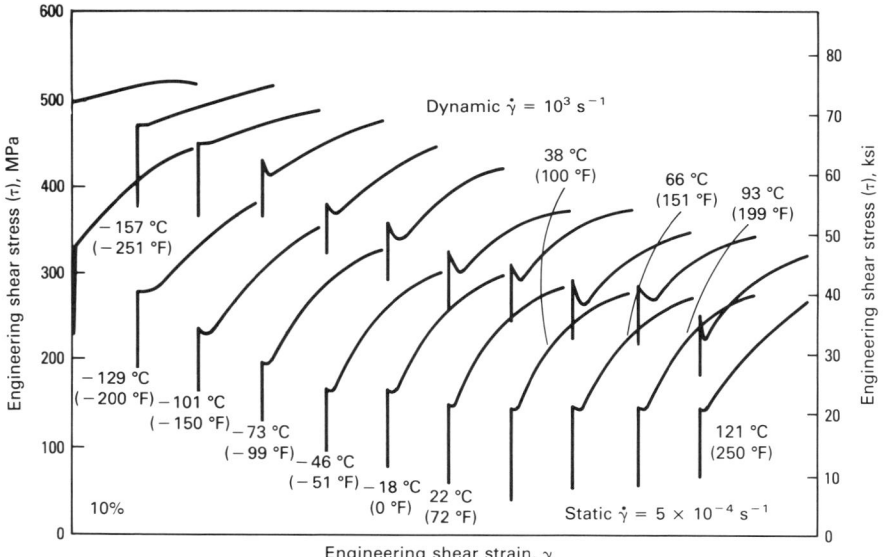

Fig. 20 Schematic representation of the effect of rapid changes in strain rate or temperature on flow stress for face-centered cubic metals

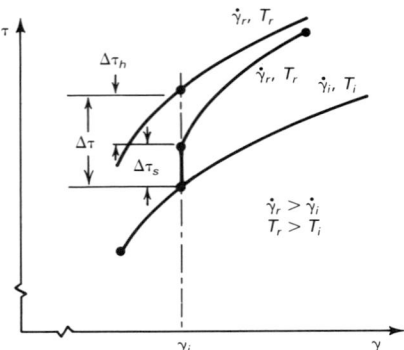

Fig. 21 Oscilloscope record of dynamic portion of incremental strain rate test

Vertical deflection represents excess stress over τ_s, the maximum static stress. Horizontal deflection represents excess strain over γ_s.

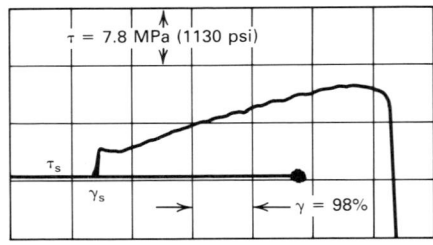

Thus, Eq 9 and 12 provide stress and strain in the specimen as functions of time. Eliminating time yields the stress-strain curve for the material at the strain rate provided through Eq 9.

Limitations on Strain Rate

Many factors must be considered in the design of a torsional Kolsky bar. However, there are practical limits on the attainable

where the strains are all functions of time. For a homogeneous state of strain in the specimen, the transmitted pulse is the difference between the incident and reflected pulses, so that $\gamma_T \approx \gamma_I - (-\gamma_R)$ and hence:

$$\dot{\gamma}_s(t) = \frac{2cD_s}{L_s D} \gamma_R(t) \qquad \text{(Eq 9)}$$

Because γ_R is determined from the output of the strain gages on the incident bar during passage of the reflected pulse and because all other quantities in Eq 9 are known, it provides a measure of the strain rate in the specimen as a function of time. The signal is integrated electronically to yield $\gamma_s(t)$, which is the strain in the specimen.

Kolsky also showed that the transmitted pulse provides a direct measure of the average shear stress in the specimen, $\gamma_s(t)$. For a thin-walled tube in torsion, the stress is given by:

$$\tau_s = \frac{2T_s}{(\pi D_s^2)t_s} \qquad \text{(Eq 10)}$$

where t_s is the wall thickness, and T_s is the average torque. The average torque in the specimen is also given by the average of the torques at each of its ends:

$$T_s = \frac{1}{2}(T_1 + T_2) \qquad \text{(Eq 11)}$$

where T_1 is the torque at the interface with the incident bar, and T_2 is the torque at the interface with the transmitter bar. The former is given in terms of the strain at the surface of the bar—i.e., by the strain in the incident and reflected pulses—by $T_1 = G(\gamma_I + \gamma_R)\pi D^3/16$. For a homogeneous state of stress in the specimen, $\gamma_I + \gamma_R \approx \gamma_T$ so that:

$$\tau_s(t) = \frac{GD^3}{8} \frac{\gamma_T(t)}{D_s^2 t_s} \qquad \text{(Eq 12)}$$

strain rate. The practical upper limit on the strain rate that can be achieved using a torsional Kolsky bar is about 10^4 s^{-1}. This can be observed by examining the torque balance equation:

$$T_I - (-T_R) = T_T \qquad \text{(Eq 13)}$$

For a solid circular cylinder, the torque is equal to $2\tau J/D$, where D is the diameter of the Kolsky bar, and J (the polar moment of inertia) is $\pi D^4/32$. Using this value in combination with Eq 9 and 12, the relation for the strain rate in terms of the stress in the incident pulse, τ_i, and the stress in the specimen, τ_s, can be obtained as:

$$\dot{\gamma}_s = \frac{1}{\rho c L_s}\left[2\tau_i\left(\frac{D_s}{D}\right)\right.$$
$$\left. - 8\left(\frac{D_s}{D}\right)^3\left(\frac{2t_s}{D}\right)\tau_s\right] \qquad \text{(Eq 14)}$$

From this equation, it is evident that the maximum strain rate is attained when the quantity on the right side is maximized. One means is to make the specimen shorter, i.e., to decrease L_s. However, with very short specimens, the end effects can become important; i.e., the strain in the specimen may never be uniform.

Fig. 22 Behavior of 1100-O aluminum under static, dynamic, and incremental strain rate loading in shear

Strain rate changes from 5×10^{-5} to 850 s^{-1} in all incremental rate tests. Source: Ref 12

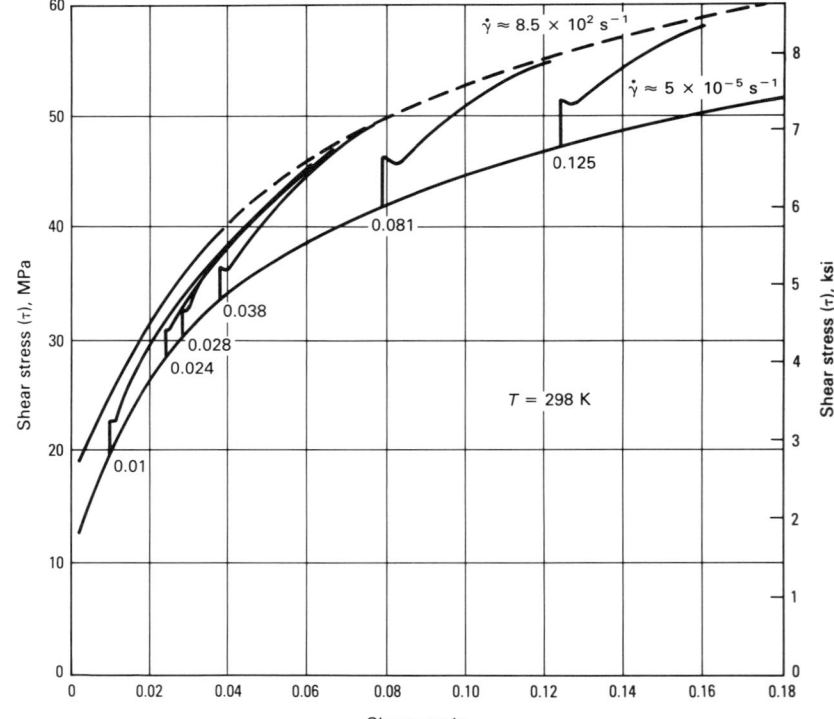

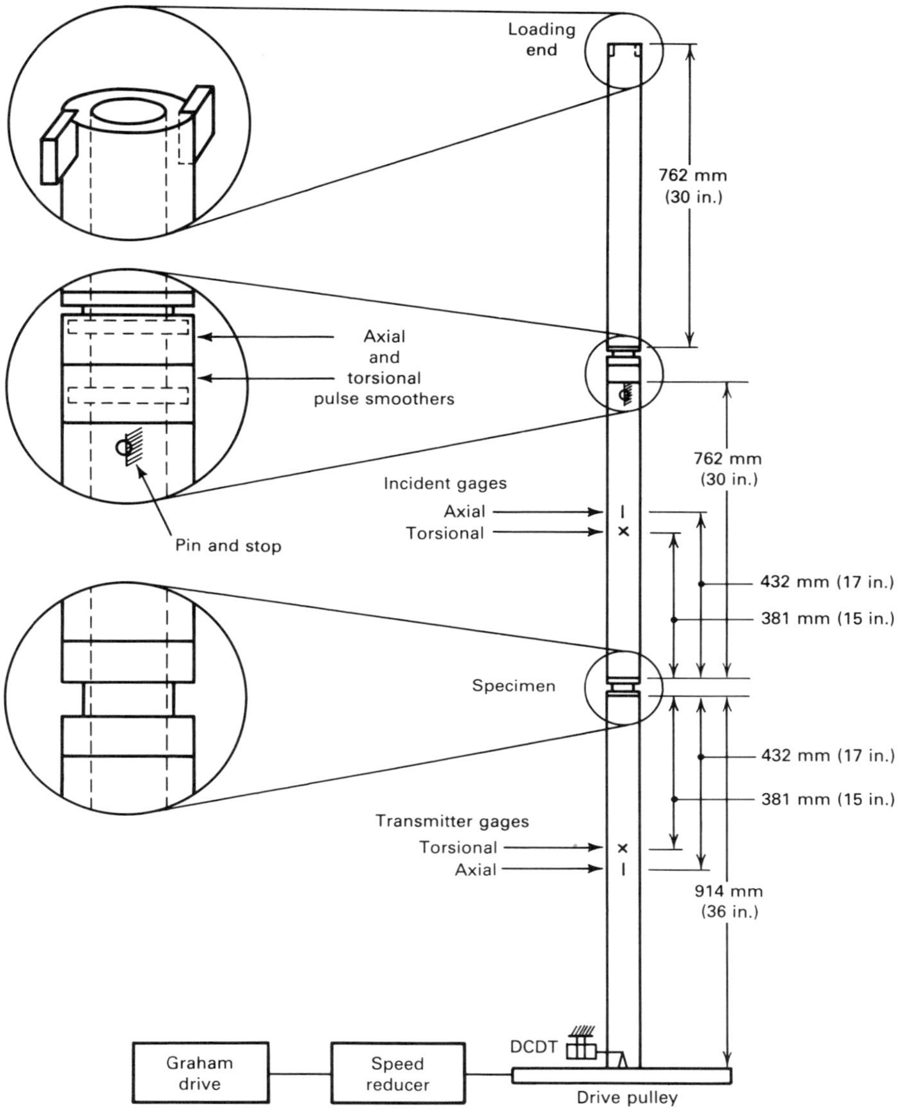

Fig. 23 Schematic of explosively loaded torsional Kolsky bar

Loading
end

762 mm
(30 in.)

Axial
and
torsional
pulse smoothers

Pin and stop

762 mm
(30 in.)

Incident gages
Axial
Torsional

432 mm (17 in.)
381 mm (15 in.)

Specimen

432 mm (17 in.)
381 mm (15 in.)

Transmitter gages
Torsional
Axial

914 mm
(36 in.)

Graham
drive

Speed
reducer

DCDT

Drive pulley

Fig. 24 Loading end of explosively loaded torsional Kolsky bar

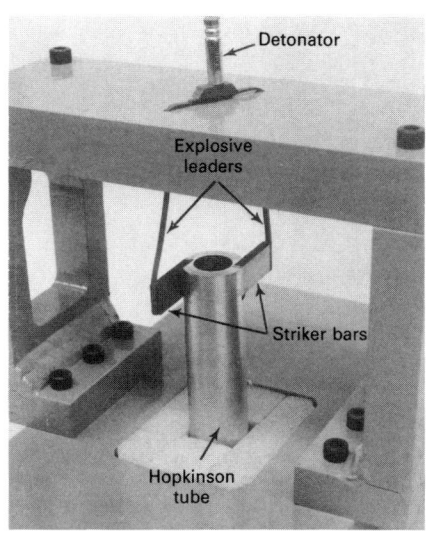

Detonator

Explosive
leaders

Striker bars

Hopkinson
tube

Fig. 25 Cross section of pulse filter showing the axial pulse attenuator (top) and the torsional pulse smoother (bottom)

Because the second term in Eq 14 generally is small, another method of increasing the strain rate is to increase the ratio of D_s/D by decreasing the diameter of the bar, or by increasing the mean diameter of the specimen. The limiting ratio of D_s/D that can be used is 1, because D_s cannot be greater than D without a significant change in impedance.

Yet another method of increasing the strain rate is to select a bar material with a high yield stress, such as an alloy steel. This would appear to be an advantage, because a much greater torque can be stored in the bar. However, two other factors affect the selection of bar material. For steel, the mass density, ρ, is about three times that of aluminum. Thus, unless the steel has a yield stress

that is more than three times that of aluminum, there will be no increase in strain rate when using a steel bar. Secondly, it is as difficult to prevent slipping when clamping a steel as it is when clamping aluminum. Hence, attaining a torque three times greater with steel requires high clamp pressures, which may produce an axial pulse upon release of the clamp.

It is evident from Eq 14 that an increase in the diameter of the Kolsky bar will not increase strain rate significantly. Increased diameter alters only the second term in Eq 14, which is considerably smaller than the first term. From Eq 14, it is evident that, although optimum bar dimensions and materials exist to achieve a maximum strain rate, there is a physical limit to the strain rate that

is attainable. To achieve rates above about 10^4 s^{-1}, a different testing system must be used, such as double shear, punching, or the plate impact test.

The practical lower limit on strain rate in a Kolsky bar is about 10^2 s^{-1}, because a lower strain rate cannot be held constant throughout the test. A constant value of strain rate requires that the magnitude of the reflected pulse be held constant, and this in turn requires that the input pulse be considerably larger than that of the transmitted pulse. Depending on the rate of work hardening of the specimen material, this imposes a lower limit on the strain rate that can be attained with the Kolsky bar.

In practice, it is easiest to design a torsional Kolsky bar to reach a strain rate of about 10^3 s^{-1}. This value is easily doubled by shortening the specimen, or can be reduced by imposing less torque. Tests outside the range of 500 to 2000 s^{-1} require special considerations.

Calibration and Data Reduction of Dynamic Tests

Data from dynamic plasticity experiments with the torsional Kolsky bar appear in the form of strain gage output voltages, which vary proportionally to the amplitude of the stress waves incident on, reflected from, and transmitted through the test specimen as a function of time. These data can be recorded using an analog oscilloscope, a digital oscilloscope, or a transient recorder with a frequency response of at least 1 MHz.

If an analog recorder is used, the output of the incident and transmitter gages must be recorded, as well as the integrated reflected signal. The stress-strain curve can be obtained by applying the integrated reflected signal to the x plate of an oscilloscope and the transmitted signal to the y plate of an oscilloscope.

With a digital oscilloscope, only the outputs of the incident and transmitter gages are required; the integration and development of a stress-strain plot can be computerized. In either case, however, calibration is required for each strain gage bridge. This is best accomplished through the use of a shunt resistance placed in parallel with one arm of each Wheatstone bridge. Results should be verified by using the usual strain gage formula with a gage factor.

Figure 26 illustrates analog records obtained in a test that was performed on 1100-O aluminum specimens. The upper trace in Fig. 26(a) represents the voltage output of the gages on the incident bar, while the lower trace shows the output of those on the transmitter bar. The oscilloscope sweeps were initiated with a 300-μs delay, and each centimeter division represents 50 μs. Because the two sets of gages are mounted at equal distances from the specimen, the reflected and transmitted pulses start at the same instant, and their curves overlap as shown in Fig. 26(a). These curves are interpreted, respectively, in terms of the strain rate $\dot{\gamma}_s$ and the stress τ_s in the specimen.

This interpretation requires that stress and strain be effectively uniform along the length of the specimen. The reflected pulse is then proportional to strain rate. From Fig. 26, it is evident that strain rate increases in about 5 to 6 μs and then remains constant to within 10% for about 50 μs. The transmitted pulse is proportional to the shear stress in the specimen, and there is a fairly well-defined yield stress.

Numerical values are assigned to the vertical scales in Fig. 26 in the following manner. It may be shown that when using a four-arm bridge:

$$\dot{\gamma}_s = \frac{2cD_s}{L_sD} \cdot \frac{2e_r}{F_1E_1} \qquad \text{(Eq 15)}$$

where e_r is the voltage output of the bridge during the passage of reflected pulse, E_1 is the bridge voltage, and F_1 is the gage factor. With a shunt resistance R_1 in parallel with one arm of the bridge, the output of the bridge will correspond to a strain rate given by:

$$\dot{\gamma}_s^* = \frac{2cD_s}{L_sD} \cdot \frac{R}{2F_1(R + R_1)} \qquad \text{(Eq 16)}$$

where R is the resistance of individual arms of this bridge. By determining the oscilloscope deflection due to the insertion of a given shunt resistance, R_1, and relating this to the strain rate using Eq 15, the calibration in this example is determined as 350 s^{-1} for 1-cm deflection.

A similar computation can be made for the torsional stress. The mean shear stress in the specimen is given by:

$$\tau_s = \frac{GD^3(1 - \lambda^4)}{8D_s^2t_s} \cdot \frac{2e_2}{F_2E_2} \qquad \text{(Eq 17)}$$

where λ is the ratio of inside to outside diameters for the transmitter tube ($\lambda = 1$ for solid bar), e_2 is the output in volts of the bridge on the transmitter bar, E_2 is the bridge voltage, and F_2 is the gage factor. Calibration is performed with a shunt resistance R_2, for which the output of the bridge is $e_2 = E_2R/4(R + R_2)$ and from Eq 17 corresponds to a specimen stress given by:

$$\tau_s^* = \frac{GD^3(1 - \lambda^4)}{8D_s^2t_s} \cdot \frac{R}{2F_2(R + R_2)} \qquad \text{(Eq 18)}$$

The calibration thus effected shows that 4.5 mm vertically in Fig. 26(a) represents a specimen stress of 17.8 MPa (2580 psi). The maximum stress is thus about 42.9 MPa (6220 psi). The same calibration holds for the vertical scale in Fig. 26(b).

The outputs representing stress and strain as functions of time are applied to the y and x plates of an oscilloscope to obtain the stress-strain curve. Figure 26(b) shows an enlargement of a portion of the record obtained in this manner. The strain scale is determined by equating the maximum horizontal travel of the sweep in the stress-strain diagram to the maximum strain found by integrating the reflected wave. The same computations are performed for the vertical scale. The apparent magnitude of the shear modulus, G, cannot be determined from the slopes of the initial loading and unloading lines. This is because the strain and strain rate are probably not homogeneous at the start and end of a loading pulse.

Double-Notch Shear Testing and Punch Loading

By John Harding
Department of Engineering Science
Oxford University

KOLSKY OR SPLIT-HOPKINSON BAR TESTING in compression, tension, or torsion is governed by an upper limit on the strain rate that can be achieved of approximately 3000 s^{-1}. However, situations arise where strain rates several orders of magnitude greater than this limit are expected, such as in high-speed metalforming operations or around rapidly propagating cracks. Attempts have thus been made to modify the Kolsky bar technique for use at strain rates in this range.

The most common method involves a drastic reduction in the effective gage length of the specimen, either by the use of a miniaturized version of the Kolsky bar, which is generally associated with an increase in the velocity of impact (Ref 32-34), or by the use of specialized specimen designs. Two examples of specialized specimen designs are the double-notch shear test and the high-speed punching test. In both, the specimen material is subjected to a high rate of shear.

Double-Notch Shear Testing

In double-notch shear testing, the output bar is replaced by a tube, into which the input bar can slide (Fig. 27). The lower end of the input bar and the upper end of the output tube are slotted to accommodate the thin plate specimen (see Fig. 28), into which two pairs of notches have been cut. With an effective gage length in this specimen of 0.84 mm (0.033 in.), a maximum shear strain rate of 40 000 s^{-1}, an order of magnitude greater than that reached in the standard Kolsky bar apparatus, has been achieved (Ref 35, 36).

The principal disadvantage of this technique is that at shear strains greater than

Fig. 26 Oscilloscope records obtained from high strain rate shear tests performed on 1100-O aluminum specimens

(a) The upper trace shows the incident and reflected pulses. The lower trace shows the transmitted pulse (each horizontal division equals 50 μs). (b) The stress-strain diagram obtained. See text for details.

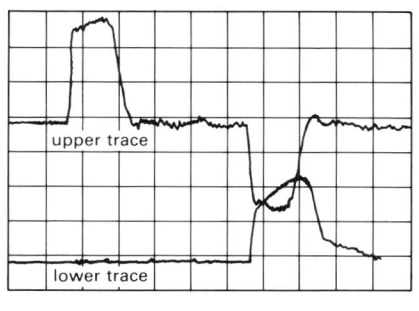
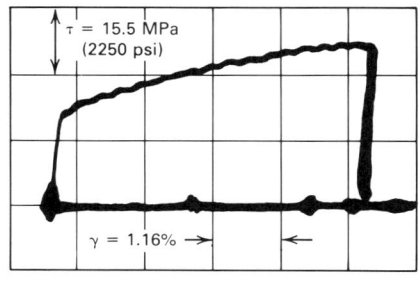

(a)

(b)

Specimen dimensions					
Gage length		Wall thickness		Outside diameter	
mm	in.	mm	in.	mm	in.
2.71	0.1066 .... 0.48	0.0188		16.6	0.6533

Fig. 27 Kolsky bar apparatus for double-notch shear testing at very high strain rates

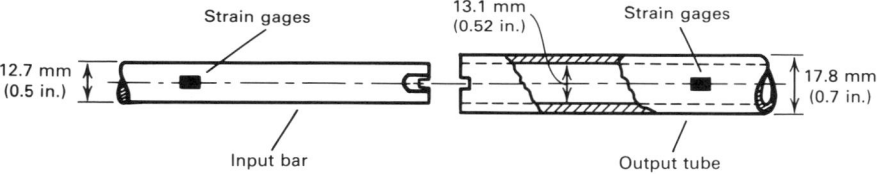

Strain gages

13.1 mm (0.52 in.)

Strain gages

12.7 mm (0.5 in.)

17.8 mm (0.7 in.)

Input bar

Output tube

Fig. 28 Double-notch shear specimen

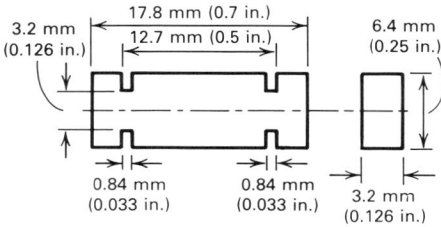

3.2 mm (0.126 in.)

17.8 mm (0.7 in.)

12.7 mm (0.5 in.)

6.4 mm (0.25 in.)

0.84 mm (0.033 in.)

0.84 mm (0.033 in.)

3.2 mm (0.126 in.)

Fig. 29 Statically deformed double-notch shear specimen
Source: Ref 37

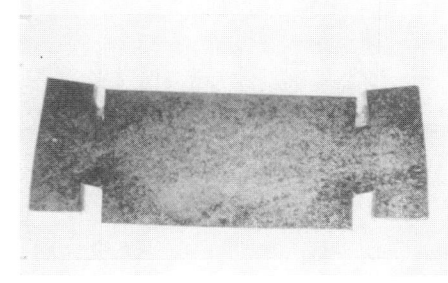

Fig. 30 Variation of lower yield stress with strain rate

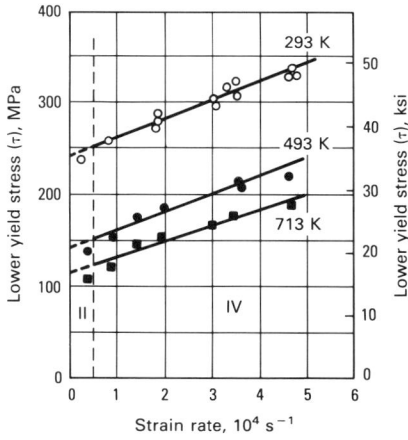

about 20% the specimen ceases to deform in pure shear. A double-notch shear specimen statically deformed to an apparent shear strain of about 60%, determined from the displacement across the loading surface, is shown in Fig. 29.

As shown, the end pieces have rotated with respect to the center and have deformed in compression. A calibration of the actual shear strain in the region under the notches yields an average value of about 20%. Reliable results can only be obtained at relatively low values of shear strain. Because about 20 μs are required for a constant strain rate to become established, results obtained at low strains can only be associated with an average value of strain rate.

Despite these problems, measurements can be taken at strain rates in excess of 10^4 s^{-1}. In this region, the lower yield stress, as shown in Fig. 30, is often found to be directly proportional to strain rate, rather than to the logarithm of strain rate. This implies a new region of mechanical response (region IV in Fig. 30), often considered to be controlled by a viscous damping mechanism, in contrast to region II, where thermally activated processes are usually assumed to control deformation.

Punch Loading

This version of the Kolsky bar is similar to the modification described above. In punch loading, however, the specimen is a flat plate in which a circular hole is punched. A typical setup is shown in Fig. 31. Although the clearance between the punch bar and the die tube is about 0.025 mm (0.001 in.) in Fig. 31, the effective gage length—i.e., the width of the shear zone in an actual test specimen—was found to be considerably greater. The width of the shear zone is not very clearly defined, so in this case the quoted shear strains and strain rates are only approximate.

A section through a low-carbon steel specimen after impact at about 15 m/s (50 ft/s) showing the zone of concentrated shear and the initiation of a shear crack is illustrated in Fig. 32. Estimates of the effective gage width for the test may be made by optical microscopy (see Fig. 32), from microhardness traverses across the shear zone (Fig. 33), or by comparing a derived shear stress/shear strain curve based on an assumed gage width with the results of a conventional torsion test at a similar rate of straining (Fig. 34). Using this technique, shear strain rates

in excess of 10^4 s^{-1} have been obtained (Ref 38).

Typical punch load/displacement curves for tests on commercial-purity aluminum specimens at mean punching speeds of 14.7 $\times$ 10^{-7} to 19.5 m/s (4.8×10^{-6} to 64 ft/s) are shown in Fig. 35. As anticipated, there is a marked increase in the punching load with increased speed of punching. For the aluminum tests shown in Fig. 35, the work done in punching—i.e., the area under the load-displacement curve—increases with punching speed. For other materials, such as a high-strength steel, the opposite effect is ob-

Fig. 31 Punch-loading Kolsky bar apparatus

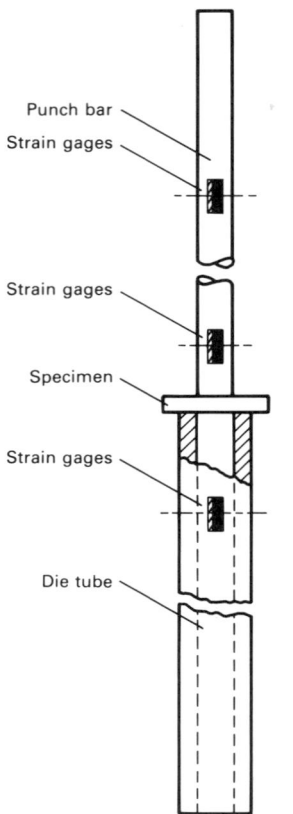

Fig. 33 Microhardness traverses

(a) Low-carbon steel. (b) Copper. ●, low-speed test; ○, high-speed test

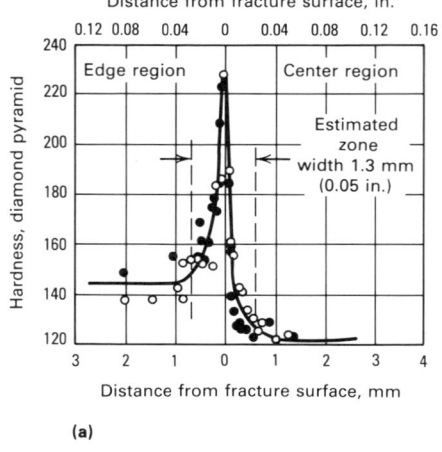

(a)

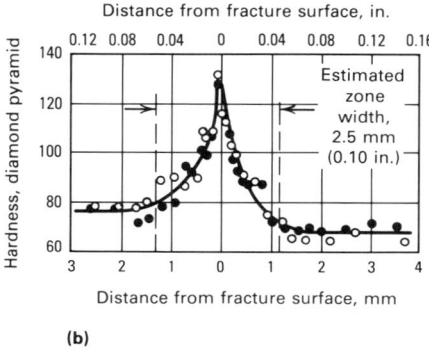

(b)

Fig. 32 Shear zone in low-carbon steel specimen impacted at ~15 m/s (~50 ft/s)

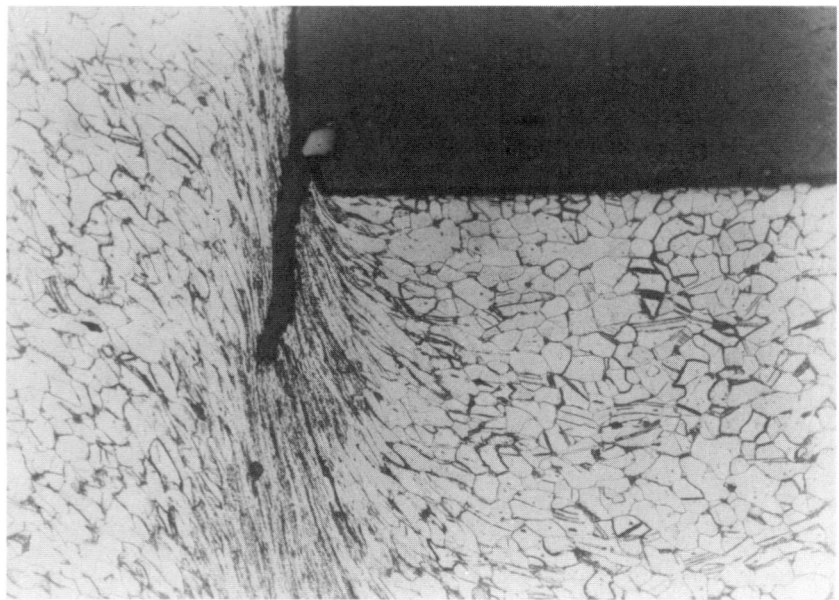

served, due to a marked reduction in punch displacement at fracture which cancels out the increased punch loads.

Although principally a technological test, this technique has been used to study the strain rate sensitivity of several materials at strain rates up to about 10^4 s^{-1}. As shown in Fig. 36, aluminum becomes strongly rate dependent at strain rates on this order. Such behavior corroborates that reported in tests using the double-notch shear technique. Nevertheless, the validity of and the reason for the apparently greatly increased rate sensitivity exhibited by many materials at these strain rates remain unclear.

Pressure-Shear Plate Impact Testing

By Rodney J. Clifton
Professor of Engineering
Brown University

and

Richard W. Klopp
Graduate Student
Brown University

PRESSURE-SHEAR PLATE IMPACT TESTING* is a relatively new procedure that is used to obtain stress-strain curves at the highest strain rates attainable under well-characterized loading conditions. Shear strain rates on the order of 10^5 s^{-1} are obtained for specimens thinned to thicknesses of about 0.2 mm (0.008 in.). Even higher strain rates, up to 10^7 s^{-1}, can be obtained for very thin specimens (with thicknesses of 2.5 μm) prepared by vapor deposition. In addition to providing high strain rates, this type of testing provides simple shearing deformation and controllable levels of nearly hydrostatic pressure.

Pressure-shear plate impact testing is limited to fine-grained materials, because the grain size must be small compared to specimen thickness to ensure that a representative average polycrystalline response is measured. The specimen material should also be soft relative to the plate materials used to impose the deformation so that these plates will remain elastic under the impact loading conditions.

*The development of the high strain rate pressure-shear test and its application to the metals reported in this section of this article was supported by the Army Research Office. Partial support for equipment and for the experiments on vapor-deposited specimens was provided by the NSF/MRL at Brown University.

Fig. 34 Shear stress-strain curves for copper

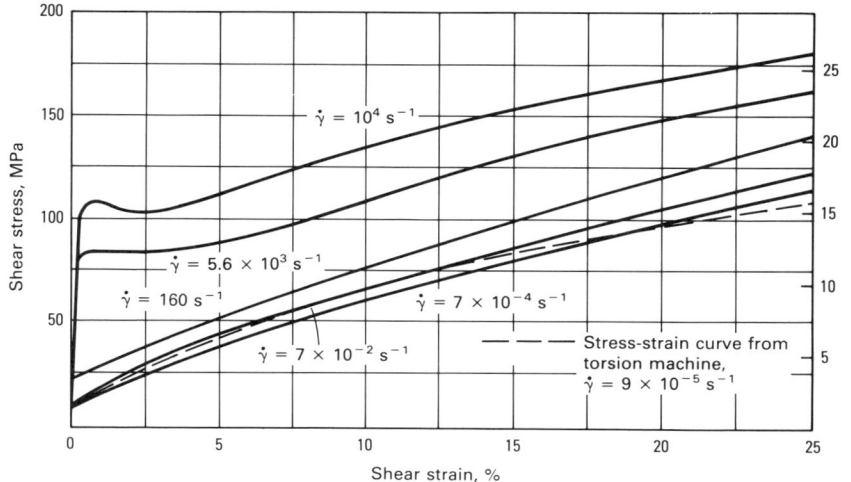

Fig. 35 Load-displacement curves for commercial-purity aluminum
Curve B corresponds to a test condition for which punching was incomplete.

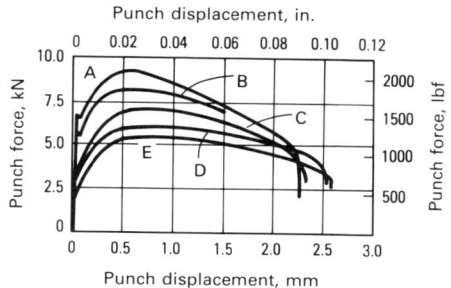

Curve	Punch speed	
	m/s	ft/s
A.............	19.5	64
B.............	12.4	41
C.............	23.8×10^{-2}	7.8×10^{-1}
D.............	14.7×10^{-5}	4.8×10^{-4}
E.............	14.7×10^{-7}	4.8×10^{-6}

Fig. 36 Strain rate sensitivity of aluminum
LYS: lower yield stress

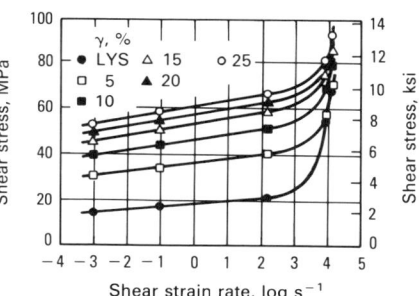

Pressure-shear experiments require a plate impact facility and instrumentation for measuring the shear waves that are generated by pressure-shear impact. Such experiments are lengthy because of the time required for specimen preparation. Although these limitations restrict the materials that can be studied and the facilities that can perform the testing, the scientific and technical importance of understanding the plastic response of metals at strain rates up to 10^5 s^{-1} and above necessitates further development of the technique. The method has been used successfully to obtain dynamic stress-strain curves of several face-centered cubic and body-centered cubic metals at strain rates of 10^5 s^{-1} and higher.

The method is based on concepts drawn from plate impact experiments and Kolsky bar experiments. From the former is taken the concept that plane waves should be used in experiments designed to measure dynamic material properties to simplify the interpretation of experimental results by making a one-dimensional wave theory applicable. Plate impact experiments also provide the methodology and instrumentation required for conducting such experiments. From Kolsky bar experiments is taken the concept of sandwiching a thin, soft specimen between two hard, elastic materials in order to sustain high strain rates in the specimen and to allow its response to be determined from measurements of wave profiles in the elastic materials. These measured profiles are related to the stresses and nominal strain rates in the specimen by one-dimensional elastic wave theory.

The pressure-shear loading configuration was introduced originally as a means of generating shear waves in symmetric plate impact experiments involving the impact of two plates of the same material. Shear waves are used, because shear wave profiles provide a more sensitive indication of the dynamic plastic response of materials than do longitudinal waves generated in conventional normal impact of plates. Shear waves have been generated by the impact of parallel plates inclined relative to their direction of approach (Ref 39, 40) and by the normal impact of an anisotropic elastic plate of y-cut quartz (Ref 41, 42). Shear wave profiles have been monitored by means of a transverse displacement interferometer (Ref 41), two velocity interferometers (Ref 43) with beams at nonnormal incidence (Ref 42), and, for nonmetallic targets, an embedded electromagnetic gage (Ref 40).

Pressure-shear waves have been used to study the plastic response of 6061-T6 aluminum (Ref 42, 44-46) and alpha-titanium (Ref 46). These studies are important because of the greater sensitivity of the shear wave profiles to the plastic flow characteristics of materials and because of the information they provide on the effects of nonproportional loading on the dynamic plastic response of materials. However, as with other wave propagation experiments, limitations include the fact that the constitutive relation between stress, strain, and strain rate is not obtained directly, but must be inferred by comparison of the recorded wave profiles with those predicted for various assumed constitutive models.

Another limitation of plastic wave propagation experiments is that, although high shear strain rates are generated near the impact face, the recorded wave profiles are determined primarily by plastic response of the material at remote positions where the wave profiles have spread and strain rates have decreased. Both of these drawbacks to pressure-shear wave propagation experiments are overcome in the high strain rate pressure-shear plate experiment.

Basic Concepts

High strain rate pressure-shear impact testing is performed by impacting a thin, soft specimen plate with a hard, elastic flyer plate inclined at an angle, θ, as shown in Fig. 37. The thin specimen plate is backed by a thick elastic anvil plate, which creates a state of high pressure and high shear strain rate in the thin specimen. Because the flyer and anvil plate remain elastic during the experiment, the stresses in the elastic plates can be inferred by measuring the projectile velocity and the particle velocity at the rear surface of the anvil plate.

During early stages of the test, before release waves reach the center of the plate from the periphery, one-dimensional wave theory applies. All states at the impact face

Fig. 37 Schematic representation of high strain rate pressure-shear impact configuration

u_o is the initial velocity of the flyer plate in the normal direction; v_o is the initial velocity of the flyer plate in the transverse direction; V_o is the projectile velocity; θ is the flyer plate angle.

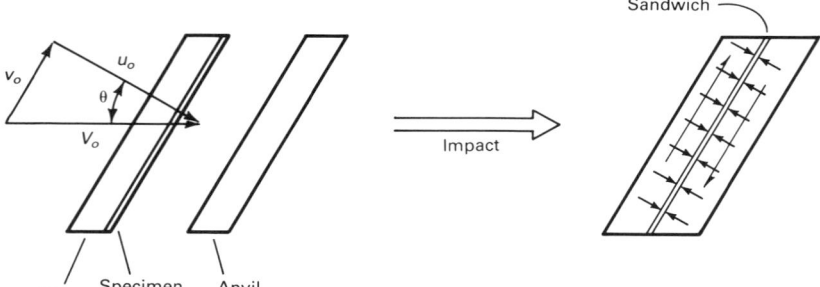

Fig. 39 Loci of shear stress/transverse particle velocity states for flyer, anvil, and specimen

v_A is the anvil particle velocity; v_{fs} is the transverse particle velocity at the free surface of the anvil; V_F is the particle velocity; v_o is the particle velocity states for the flyer and anvil plates in the transverse direction.

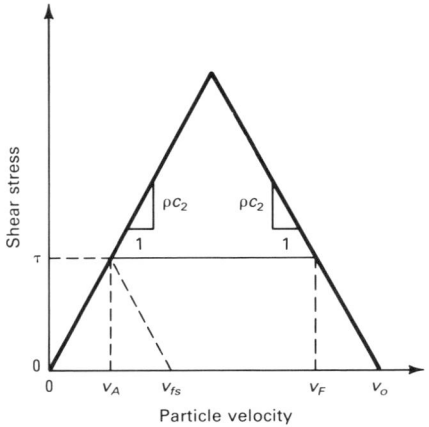

of an elastic flyer plate must satisfy the following characteristic relations (Ref 47):

$$-\sigma + \rho c_1 u = \rho c_1 u_o \quad \text{(Eq 19)}$$

$$\tau + \rho c_2 v = \rho c_2 v_o \quad \text{(Eq 20)}$$

All states at the impact face of an elastic anvil plate must satisfy:

$$-\sigma - \rho c_1 u = 0 \quad \text{(Eq 21)}$$

$$\tau - \rho c_2 v = 0 \quad \text{(Eq 22)}$$

where σ is the normal stress in the x-direction, τ is the shear traction in the transverse direction, ρc_1 is the longitudinal acoustic impedance, ρc_2 is the shear impedance, u is the particle velocity in the normal direction, and v is the particle velocity in the transverse direction. The initial normal and transverse components of the velocity of the flyer plate are u_o and v_o, respectively.

The loci of stress and particle velocity states for the flyer and anvil plates are shown in Fig. 38 for the normal components and in Fig. 39 for the transverse components. At impact, shear waves and longitudinal waves are sent forward into the anvil and backward into the specimen, as shown in the x-t diagram in Fig. 40. The flyer and anvil plates are selected to have impedances that are greater than or equal to that of the specimen so that unloading does not occur as the waves reflect back and forth.

Because the waves in the specimen are plastic, they are quickly attenuated, and the stress state in the specimen becomes nominally homogeneous. This has been verified computationally for aluminum specimens sandwiched between steel plates (Ref 47). It was also verified that after a few reflections the hydrostatic pressure in the specimen becomes nearly equal to the value of the homogeneous normal stress (Ref 47). The homogeneous normal stress attained is the value at

Fig. 38 Loci of normal stress-particle velocity states for flyer, anvil, and specimen

ρc_1 is the longitudinal acoustic impedance; u_o is the particle velocity states for the flyer and anvil plates in the normal direction.

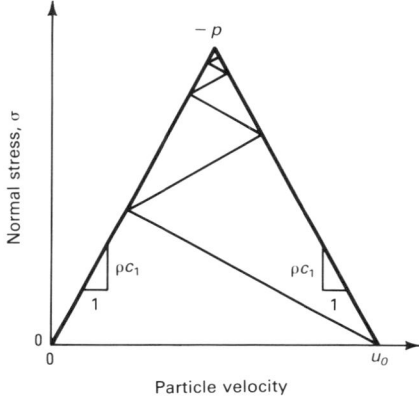

the point of intersection of the two lines given by Eq 19 and 21.

$$-p \approx \sigma = \frac{1}{2}\rho c_1 u_o \quad \text{(Eq 23)}$$

This point, which is the apex in Fig. 38, is attained because the elastic resistance of the specimen to volume change will not allow a finite normal velocity difference to be maintained across the thickness of the specimen, while flow in the radial direction is prohibited.

It is possible, however, to maintain a finite transverse velocity difference across the thin specimen for the duration of the experiment. If the specimen behaves viscoplastically and if the stress state is homogeneous, then the shear stress will equilibrate at a value τ in Fig. 39, and the flyer and anvil particle velocities will be v_F and v_A, respectively. The nominal shear strain rate is:

Fig. 40 x-t diagram illustrating the shear waves and longitudinal waves at impact

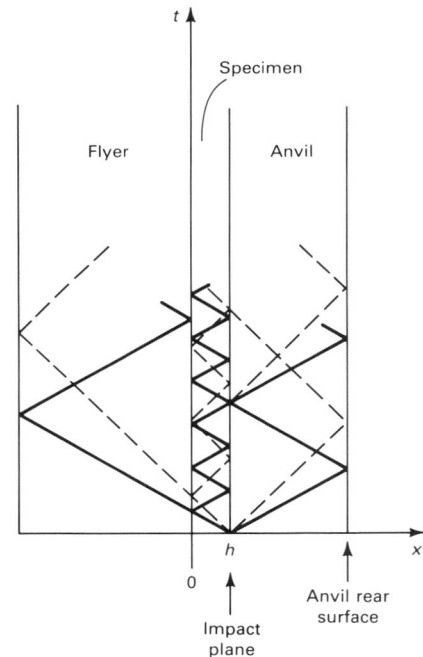

$$\dot{\gamma} = \frac{v_F - v_A}{h} \quad \text{(Eq 24)}$$

where h is the specimen thickness. Note that:

$$v_F - v_A = v_o - v_{fs} \quad \text{(Eq 25)}$$

where v_{fs} is the transverse particle velocity at the free surface of the anvil. The shear strain rate can be integrated over time to give the shear strain. From Eq 22, the shear stress is:

$$\tau = \rho c_2 v_A = \frac{1}{2}\rho c_2 v_{fs} \qquad \text{(Eq 26)}$$

Thus, it is only necessary to measure the incident projectile velocity, the skew angle, and the particle velocity at the rear surface of the anvil to construct a shear stress versus shear strain curve at high strain rates and high hydrostatic pressures.

Plate Impact Facility

The plate impact experiments discussed in this article are carried out in a 64-mm gas gun using nitrogen as the driving gas. The muzzle end of the gun is depicted schematically in Fig. 41. A keyway is broached in the barrel to prevent rotation of the projectile. The projectile, shown in Fig. 42, consists of a fiberglass-reinforced plastic tube with an aluminum cap, containing an O-ring seal and key at the back and an aluminum plate at the front.

Both end pieces are bonded to the tube by an epoxy. The front plate carries the flyer plate and the specimen. Impact occurs in a vacuum chamber at the muzzle end of the gun. This chamber is evacuated by a mechanical pump to a pressure of 120×10^{-6} torr or less to prevent the formation of a lubricating gas cushion between the impacting plates.

The velocity of the projectile, V_o, is measured just prior to impact by recording the times of contact of five pairs of wire pins spaced a measured distance apart. The pins are placed in the path of the front plate of the projectile, and the contact times are recorded with cascaded 10-ns counters. The accuracy of this system is better than 1%.

The oscilloscopes that record the motion of the free surface of the anvil plate are triggered when the first of four isolated voltage-biased pins lapped flush with the front surface of the anvil is shorted to the grounded anvil plate by the impacting specimen. The times at which these pins, which are biased at a ratio of 1:2:4:8, are grounded are recorded stairstep fashion on the first oscilloscope (see Fig. 43). This provides a redundant measure of tilt or lack of parallelism of the impacting faces. Reported values of tilt are obtained by fitting a plane through the tilt pin times by a least squares analysis and by finding the horizontal component of this tilt. The sign of the horizontal component is used to determine whether the horizontal transverse component of the normal wave that appears in the transverse motion record due to tilt is to be added

or subtracted to obtain the motion caused by the shear wave.

The motion of the stress-free rear surface of the anvil plate is measured by a combined normal velocity and transverse displacement interferometer system (Fig. 44). For the normal velocity interferometer, each "fringe" (one peak-to-peak variation in intensity) corresponds to a change, Δu, in the normal component of the free surface velocity given by:

$$\Delta u = \frac{\lambda}{2T} \qquad \text{(Eq 27)}$$

where λ is the wavelength of the laser light source, and T is the time required for light to traverse the delay leg of the normal velocity interferometer.

For the transverse displacement interferometer, each fringe corresponds to a change in the transverse displacement, ΔV, given by:

$$\Delta V = \frac{d}{2n} \qquad \text{(Eq 28)}$$

where d is the pitch of a diffraction grating on the rear surface of the anvil, and n is the order of the diffracted beams used in the transverse displacement interferometer. The light source used is a 5-W argon-ion laser with single-mode, single-frequency light at a wavelength $\lambda = 5.144 \times 10^{-4}$ mm (20.3 μin.).

A typical delay time is $T = 7.0$ ns, for which the fringe constant defined by Eq 27 is 0.0367 mm/μs (1445 μin./μs) per fringe. The commonly used diffraction grating is a 200 lines per millimeter bar and groove-type grating with a pitch of 5 μm (197 μin.).

Fig. 41 Muzzle end of the gas gun used for plate impact experiments

M, mirrors

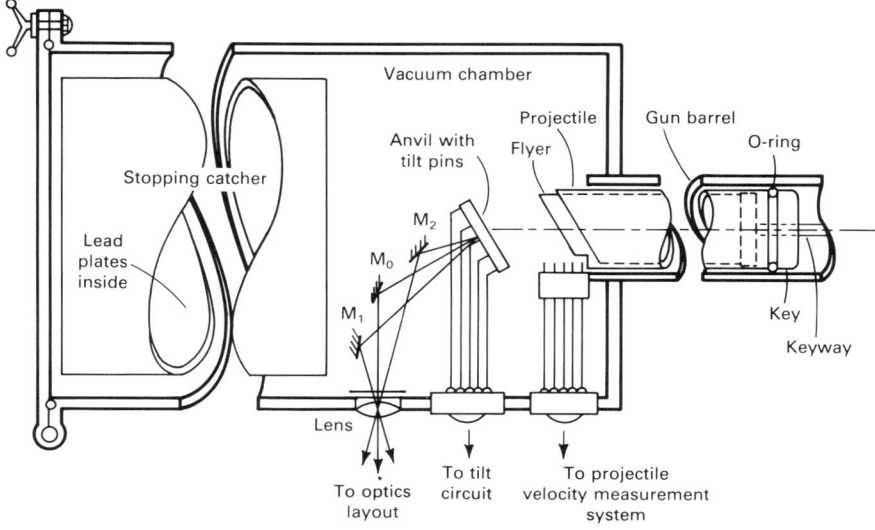

Fig. 42 Projectile used in the plate impact test

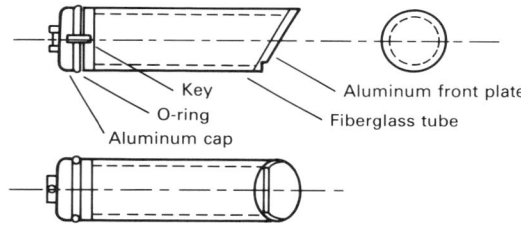

Fig. 43 Tilt pin oscilloscope record used in plate impact tests

(a) Tilt pin voltages before shot. (b) Pin contact sequence at impact

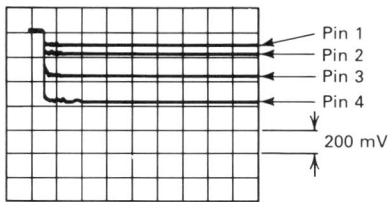

(a)

(b)

Fig. 44 Combined normal velocity and transverse displacement interferometers

M, mirrors; BS, beam splitters. Source: Ref 41

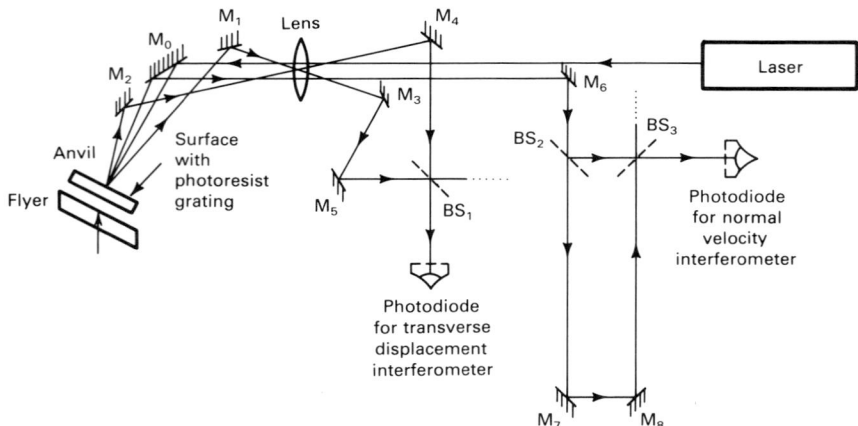

First-order fringes (i.e., $n = 1$) are normally used so that, from Eq 28, the sensitivity of the measurement of transverse displacement is 2.5 μm (98 μin.) per fringe. The interferometer optical components are mounted on a rigid magnetic optical table, isolated from vibrations by soft pneumatic supports. The interference fringes produced by the laser interferometers are detected by photodiodes, and the resulting signals are displayed on 500-MHz oscilloscopes and are recorded on high-speed film.

Preparation of Flyer, Anvil, and Specimen

The flyer and anvil plates, depicted schematically in Fig. 45, are generally made of oil-hardened tool steel in the as-quenched condition. A hardness of more than 66 HRC results, and a proportional limit of approximately 2800 MPa (405 ksi) in compression under uniaxial stress conditions is achieved. The properties of several suitable flyers and anvils are given in Table 1.

The impact faces of the plates are lapped on a cast iron lap with 15-μm (590-μin.) diamond paste to a flatness of better than 0.4 μm (15.7 μin.) over a 30-mm (1.2-in.) diameter circle. Polishing with a smaller diam-

eter diamond paste is not performed in order to preserve the surface roughness, which helps transmit the shear traction across the impact faces. Flatness is measured by placing a transparent optical flat over the surface and observing the Newton's rings that appear when the surface is illuminated by a monochromatic light source. The rear surface of the anvil, lapped to 0.5 μm (19.7 μin.) flatness, is polished on cloth with 3-μm (120-μin.) diamond paste to a mirror finish.

The plate dimensions are such that the first unloading of the center of the specimen is due to the arrival of the longitudinal wave reflected from the rear surface. The release waves from the periphery arrive slightly later, and the reflected wave from the flyer arrives much later. This provides optimum plate stiffness for lapping and also provides sharp unloading signals.

Copper tilt pins are inserted into holes through the anvil and are affixed with epoxy. The thin specimen is attached to the flyer with epoxy in the step around the periphery of the flyer. Before the epoxy is applied, the specimen and flyer are clamped together tightly to prevent epoxy from entering the interface between the specimen and flyer. The holes through the flyer allow the interface to be evacuated as the chamber containing the target assembly is evacuated.

Specimens are generally cut from bulk material in the form of 38-mm (1.5-in.) diameter discs with thicknesses of 0.4 to 0.6 mm (0.016 to 0.024 in.). Thicker discs are used when surface microstructural damage may be caused by heat treatment. In these cases, the thickness of the specimen discs is reduced to 0.40 mm (0.016 in.) by hand grinding of each face with 120-grit silicon carbide paper cooled with a soap solution. The specimens are bonded to hard-backing discs with glycol phthalate during hand grinding.

After grinding, the specimens are parted from their backing discs and lapped to the final thickness (0.15 to 0.35 mm, or 0.006 to 0.014 in.) using the modified diamond-stop lapping fixture shown in Fig. 46. The modification is the addition of the vacuum chuck to pull the warped specimens into contact with the lapped diamond-stop. With this modification, specimen parallelism can be held to 0.1 mrad. Normally, the lapping abrasive is 3-μm aluminum oxide, although 15-μm diamond has also been used.

Once the anvil has been lapped and polished, a 200 lines per millimeter diffraction grating is applied to the polished surface for use with the transverse displacement interferometer. The grating is applied by spinning on a thin film of photoresist, exposing the photoresist to an ultraviolet source through a 200 lines per millimeter ruled master grating, and then developing the photoresist. This produces a bar-space replica of the master grating.

Once the anvil has been placed in the holder and the flyer-specimen combination attached to the projectile, the impact surfaces are cleaned carefully by wiping with lens tissue soaked in trichloroethylene, then acetone, and finally methyl alcohol. Rear surface mirrors are then attached to the impact surface, as shown in Fig. 47. This setup allows alignment of the impacting faces to a parallelity of 2×10^{-5} rad, which is a closer tolerance than can be maintained during impact due to tolerance between the projectile and the barrel. Once aligned, the mirrors are removed, the projectile is retracted to the breech end of the gun, the velocity measurement system is attached, the disposable mirrors for the interferometer are aligned, and the catcher filled with lead plates is installed. The chamber is then evacuated, and the shot proceeds.

Interpretation of Experimental Records

An example of normal velocity interferometer and transverse displacement interferometer experimental records is illustrated in Fig. 48 from tests on a 0.297-mm (0.0117-

Table 1 Properties of flyers and anvils used in plate impact testing

Material	Longitudinal acoustic impedance, MPa · μs/mm	Shear impedance, MPa · μs/mm	Yield stress MPa	ksi
Carpenter Stentor steel(a)	2.49×10^4	4.55×10^4	1930	280
Tungsten carbide (3% cobalt)	5.89×10^4	9.68×10^4	3850	560
Carpenter Hampden steel(a)	2.49×10^4	4.55×10^4	2770	400

(a) Oil-hardened tool steel in the as-quenched condition

Fig. 45 Anvil and flyer plate details before lapping

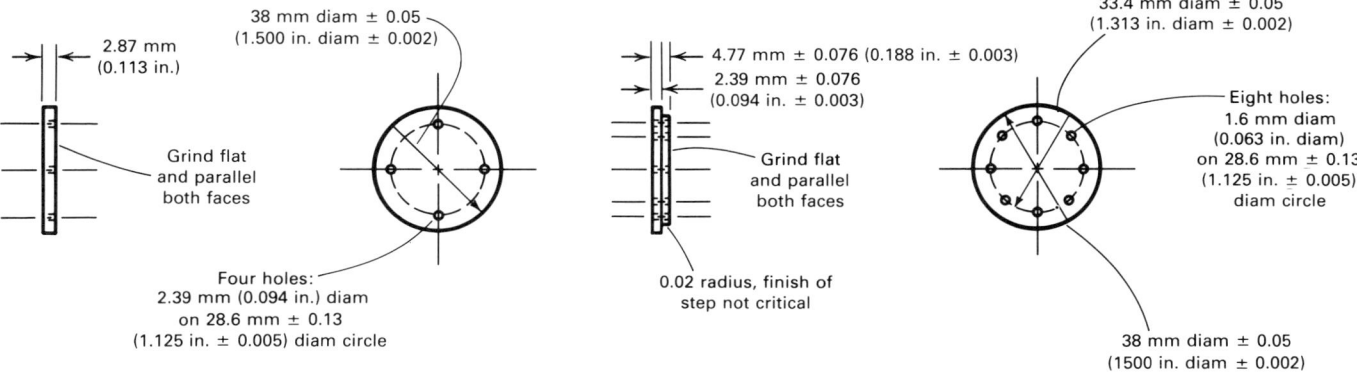

Anvil plate

Flyer plate

in.) thick specimen of vacuum arc remelted 4340 steel (600 °C, or 1110 °F, temper). The tilt is calculated by fitting the least squared error plane through the four pins (see Fig. 43), as described in Appendix B of Ref 49. The component of tilt in the horizontal direction is calculated for use in correcting the transverse displacement interferometer record.

Fig. 46 Modified diamond stop lapping fixture used to obtain final thickness of plate impact specimens

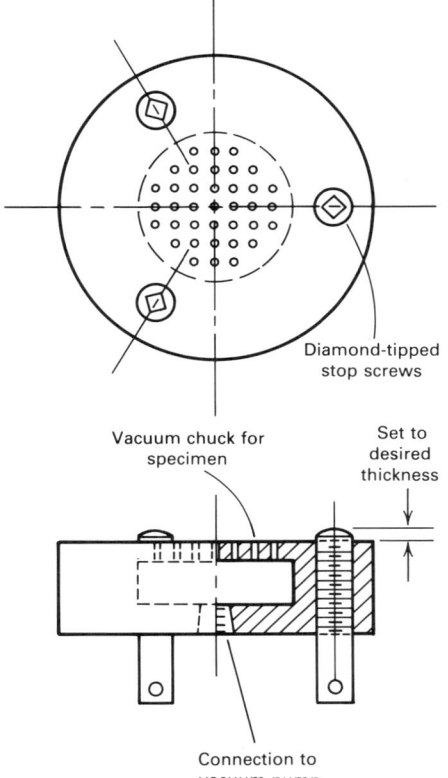

Fig. 47 Optical alignment of impact faces for plate impact testing
Source: Ref 48

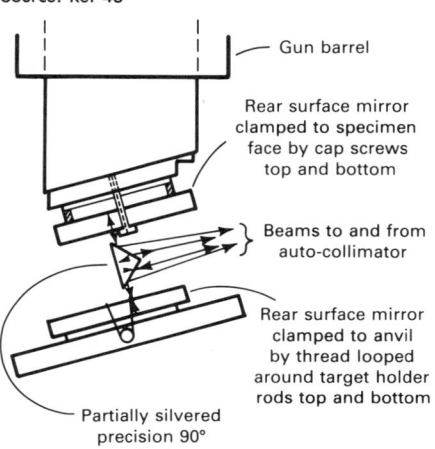

The transverse displacement interferometer picture is reduced by measuring the horizontal coordinates of the peaks and valleys of each fringe with a measuring microscope. These coordinates are converted to times by comparison with the time markings at the bottom of the photograph. Because first-order diffracted beams are used, one fringe represents 2.5 μm of motion. The free-surface transverse velocity, v_{fs}, is calculated by dividing the 2.5 μm per fringe by the time between fringes. The shear stress in the specimen is calculated from Eq 26, and the shear strain rate is calculated from Eq 24 and 25.

The velocity (v_{fs}), shear stress (τ), and strain rate ($\dot{\gamma}$) are all assigned to the mean time between fringes. The shear strain is calculated by multiplying $\dot{\gamma}$ by the time interval and adding this increment to the previous strain. The resulting v_{fs} and τ versus time profiles and stress-strain curves are shown in Fig. 49 and 50, respectively.

In Fig. 50, the strain correction $\gamma_c = 0.0165$ is an apparent strain that appears to be due to relative sliding at the interface between the flyer and the specimen. This sliding can be caused by the presence of a gap of a few microns in thickness between the flyer and the specimen. If this gap is sufficiently large so that it is not closed by the longitudinal compressive wave before the shear wave arrives, then the shear traction is not transmitted and local relative sliding occurs until the gap is closed. The correction strain is subtracted from the strain in plotting final dynamic stress-strain curves.

Fig. 48 Interferometer records from shot 82-008 of a vacuum arc remelted 4340 steel specimen
(a) Normal velocity interferometer. (b) Transverse displacement interferometer

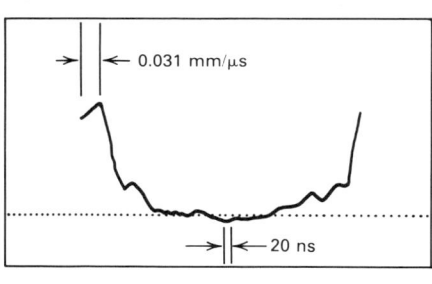

(a)

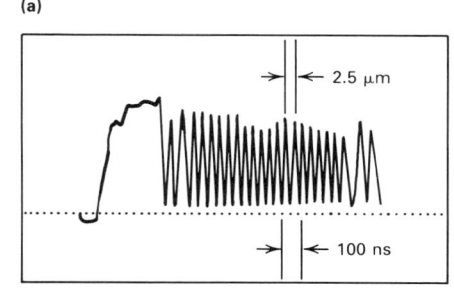

(b)

Fig. 49 Transverse velocity at the free surface of the anvil and shear stress at the specimen-anvil interface of vacuum arc remelted 4340 steel

Shot 82-008

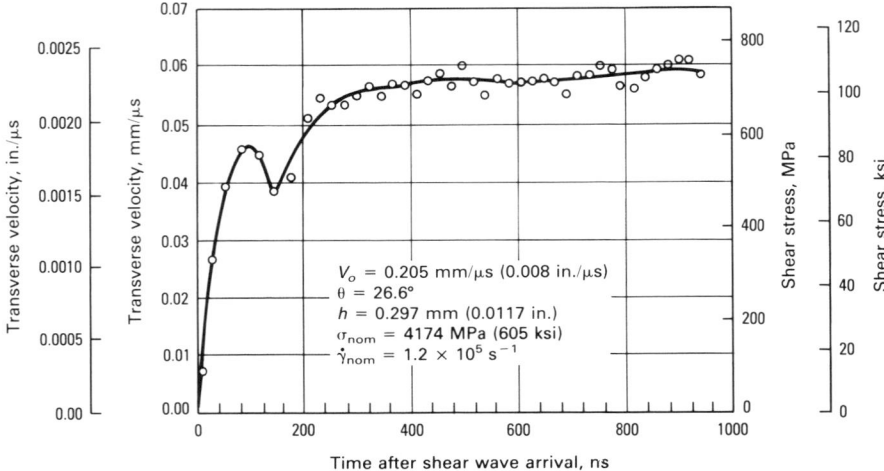

Fig. 50 Dynamic stress-strain curve of vacuum arc remelted 4340 steel in shear with superimposed pressure

Shot 81-008

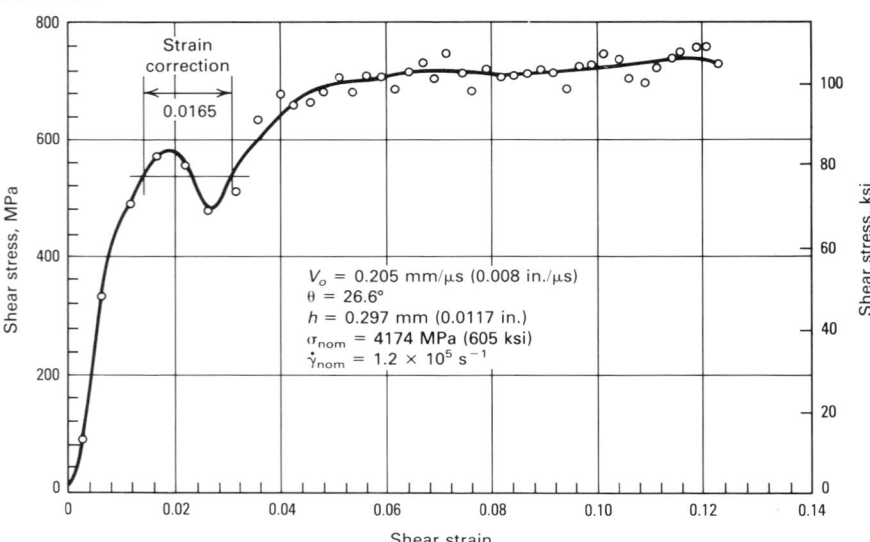

The normal velocity interferometer photograph is also reduced by measuring coordinates with the microscope. From Eq 27 with $T = 8.4$ ns, each fringe represents a change of velocity of $u = 0.031$ mm/µs (0.0012 in./µs). The normal velocity values are converted to stress by means of Eq 23, and normal strain rates are calculated from $\dot{\varepsilon} = (u_o - u_{fs})/h$. Generally, the normal velocity rises to its final values so rapidly that the fringes are close together and difficult to distinguish.

Nominal values of shear flow stress, τ, and shear strain rate, $\dot{\gamma}$, at a selected strain are calculated by fitting a straight line to the steady-state plateau region of the stress-strain curves and then interpolating or extrapolating the stress value of the selected strain. The fitted line is chosen to have the same slope in all tests. This procedure is used to account for strain hardening, which is evident in tests on some materials such as vacuum arc remelted 4340 steel. The nominal strain rate is assumed to be the strain rate at the selected strain.

The nominal hydrostatic pressure is calculated from Eq 23, where the second equality holds, because the normal stress quickly

"rings up" to the apex in Fig. 38. The first equality applies when planes parallel to the impact plane are planes of maximum shear stress. Computer simulations of similar experiments on 1100-O aluminum specimens indicate that this is approximately true (Ref 47).

Experimental Results

High strain rate pressure-shear experiments have been conducted on specimens of 1100-O aluminum, 6061-T6 aluminum, OFHC copper, vacuum arc remelted 4340 steel, and high-purity iron (Ref 47, 49, 50). Principal features of the dynamic plastic response obtained in these experiments can be illustrated by considering the experimental results for 1100-O aluminum and vacuum arc remelted 4340 steel.

Several dynamic stress-strain curves for 1100-O aluminum are shown in Fig. 51. The initial, steeply rising part of these curves should be disregarded, because this corresponds to time before nominally homogeneous states of stress have been established in the specimens. The flow stress in pressure-shear experiments at strain rates of 10^5 s^{-1} is clearly much higher than in torsional Kolsky bar experiments at strain rates of 10^3 s^{-1}. This increase in flow stress is apparently not due to the effect of hydrostatic pressure that is present under pressure-shear loading, but not under torsional loading.

This relative independence of hydrostatic pressure is demonstrated by the stress-strain curves shown in Fig. 51 for four shots conducted at the same nominal hydrostatic pressure of 2.8 GPa (406 ksi). Furthermore, systematic variation of the impact angle θ and the projectile velocity V_o in 27 tests designed to study the relative importance of strain rate and pressure on the flow stress leads to the conclusion that, in these experiments, no measurable change occurred in flow stress for changes in hydrostatic pressure up to approximately 1.0 GPa (145 ksi).

The dependence of flow stress on strain rate for 1100-O aluminum is illustrated in Fig. 52. Results of pressure-shear experiments are shown as a cluster of points at strain rates of about 10^5 s^{-1}. The stress and strain rate coordinates of these points are the values corresponding to the plateau level of stress for the type of stress-strain curves shown in Fig. 51. Points for all successful experiments are plotted, independent of the respective values of the hydrostatic pressure. Three points plotted at strain rates of 10^6 to 10^7 s^{-1} were obtained in experiments on specimens made by the vapor deposition of aluminum onto the flyer plate. Thicknesses of approximately 17 µm were obtained for the two experiments at strain rates near

Fig. 51 Dynamic stress-strain curves for 1100-O aluminum at a hydrostatic pressure of 2.8 GPa (406 ksi)
Source: Ref 47

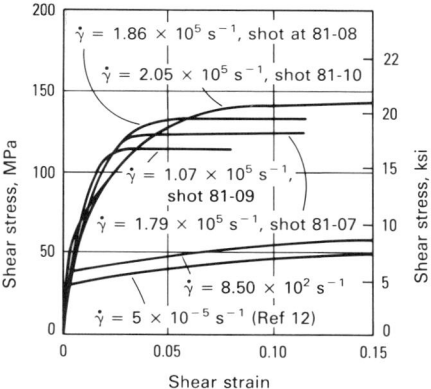

Shot No.	Impact angle, θ, degrees	Specimen thickness, h		Projectile velocity, V_o	
		mm	in.	mm/μs	in./μs
81-07	26.6	0.273	0.0107	0.0582	0.00229
81-08	26.6	0.276	0.0109	0.0624	0.00246
81-09	14	0.212	0.0083	0.0319	0.00126
81-10	26.6	0.246	0.0096	0.0623	0.00245

$5 \times 10^6 \, \mathrm{s}^{-1}$. The thickness for the strain rate of approximately $3 \times 10^7 \, \mathrm{s}^{-1}$ was 2.6 μm. From Eq 24, these small thicknesses lead to extremely high strain rates.

Interpretation of the experiments on vapor-deposited specimens and the relation of these results to those for specimens prepared from bulk aluminum samples require further research. Currently, it has been determined that the vapor-deposited specimens are crystalline and apparently fine grained. The large increase in flow stress is believed to be not an artifact of using vapor-deposited specimens, but an indication of the actual strain rate sensitivity of aluminum at high stress levels, at which the intrinsic resistance of the lattice plays an important role in determining the mobility of dislocations.

Although the dependence of flow stress on strain rate at strain rates above $10^4 \, \mathrm{s}^{-1}$ appears to be very strong on the semi-log plot in Fig. 52, the dependence is much weaker than the linear dependence often attributed to a linear viscous resistance to the motion of dislocations. This less than linear dependence of the flow stress on strain rate is viewed as an indication that the mobile dislocation density increases strongly with the stress at high stress levels (Ref 51-53).

Fig. 52 Strain rate sensitivity of polycrystalline aluminum at a plastic shear strain, γ^P, equal to 0.15 for strain rates less than $10^3 \, \mathrm{s}^{-1}$

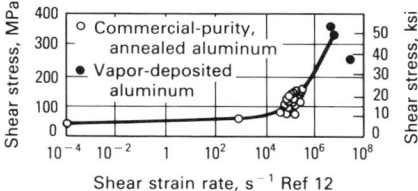

Stress-strain curves for several experiments on vacuum arc remelted 4340 steel specimens are compared in Fig. 53 with stress-strain curves obtained from torsion experiments at lower strain rates. This comparison indicates that the flow stress in the pressure-shear experiments at strain rates of $10^5 \, \mathrm{s}^{-1}$ is approximately 20% greater than in static torsion tests and less than 10% greater than in torsional Kolsky bar experiments at strain rates of $10^3 \, \mathrm{s}^{-1}$.

Systematic investigation of the effect of hydrostatic pressure indicates that there is no measurable effect for changes in pressure of as much as 1.0 GPa (145 ksi). The lack of a consistent trend in the pressure-shear experiments precludes a definitive conclusion on the strain rate sensitivity in this strain rate regime. Nevertheless, the trend is for the flow stress to increase with increasing strain rate over this regime, and the rate of increase is comparable to, and possibly slightly higher

Fig. 53 Dynamic stress-strain curves for vacuum arc remelted
4340 steel tempered at 600 °C (1110 °F)

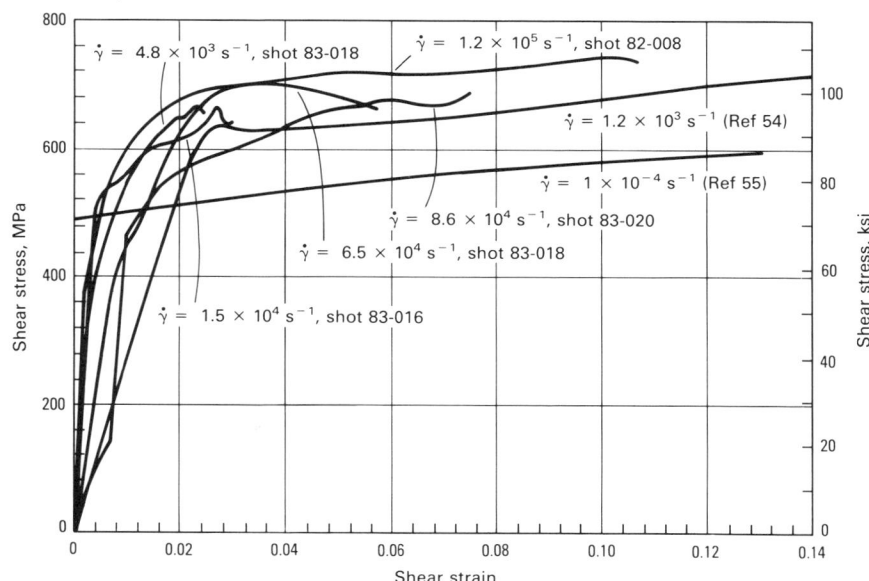

Shot No.	Impact angle, θ, degrees	Specimen thickness, h		Projectile velocity, V_o	
		mm	in.	mm/μs	in./μs
82-008	26.6	0.297	0.0117	0.205	0.0081
83-011	18.4	0.194	0.0076	0.173	0.0068
83-016	23.2	0.287	0.0113	0.145	0.0057
83-018	23.2	0.255	0.0100	0.186	0.0073
83-020	20.1	0.199	0.0078	0.201	0.0079

than, the logarithmic increase with strain rate that is observed at lower strain rates.

The weaker strain rate sensitivity of the flow stress in vacuum arc remelted 4340 steel, compared to that reported for 1100-O aluminum, is an indication that for the steel the mechanism controlling the rate of plastic flow is thermal activation of dislocations past obstacles, even at strain rates of 10^5 s^{-1}. This view is supported by the observation that the highest shear stresses obtained in the pressure-shear experiments (740 MPa, or 107 ksi) are below the values obtained in quasi-static torsion experiments at 80 K and well below the value of 900 MPa (130 ksi) obtained by extrapolation to 0 K of quasi-static flow stresses at 80 to 293 K (Ref 55).

Thus, it appears that significantly higher strain rates are required to reach flow stresses in vacuum arc remelted 4340 steel that are large enough to obtain the strong rate sensitivity associated with viscous drag mechanisms. However, a further increase in the strain rate is difficult to obtain for vacuum arc remelted 4340 steel, because slip occurs if the impact angle θ is larger than approximately 23°. Further reduction in the specimen thickness is not advisable, because there are only five to eight grains across the thickness of the specimens used in these experiments. Significantly higher impact velocities cause yielding of the flyer and anvil plates.

The strong rate sensitivity at high strain rates shown for 1100-O aluminum is representative of the response obtained for OFHC copper and high-purity iron. Furthermore, the strain rate sensitivity of a high-strength aluminum alloy (6061-T6), although stronger than that of vacuum arc remelted 4340 steel, is much weaker than that obtained for the pure metals. Based on these observations, the experimental results indicate that the flow stress increases faster than logarithmically with strain rate only when the stress levels are large enough for dislocations to overcome obstacles without thermal activation.

REFERENCES

1. U.S. Lindholm, High Strain Rate Tests, in *Techniques of Metals Research*, R.F. Bunshah, Ed., Vol 5, Part I, Measurement of Mechanical Properties, Interscience, New York, 1971, p 199-271
2. U.S. Lindholm, A. Nagy, G.R. Johnson, and J.M. Hoegfeldt, Large Strain, High Strain Rate Testing of Copper, *J. Eng. Mat. Technol.*, Trans. ASME, Vol 102, 1980, p 376-381
3. R.S. Culver, Torsional-Impact Apparatus, *Exp. Mech.*, Vol 12, 1972, p 398-405
4. O.V. Greene and R.D. Stout, The Stress-Strain Characteristics of the Torsion Impact Test, *ASM Trans.*, Vol 28, 1940, p 277-305
5. C.E. Work and T.J. Dolan, The Influence of Strain Rate and Temperature on the Strength and Ductility of Mild Steel in Torsion, *Proc. ASTM*, Vol 53, 1953, p 611-626
6. N.G. Calvert, Impact Torsion Experiments, *Inst. Mech. Eng.*, Vol 169 (No. 44), 1955, p 897
7. R.S. Culver, Thermal Instability Strain in Dynamic Plastic Deformation, in *Metallurgical Effects at High Strain Rates*, R.W. Rohde, B.M. Butcher, J.R. Holland, and C.H. Karnes, Ed., Plenum Press, New York, 1973, p 519-530
8. U.S. Lindholm and G.R. Johnson, Strain-Rate Effects in Metals at Large Shear Strains, in *Material Behavior Under High Stress and Ultrahigh Loading Rates*, J. Mescall and V. Weiss, Ed., Plenum Press, New York, 1983, p 61-79
9. H. Kolsky, An Investigation of the Mechanical Properties of Materials at Very High Rates of Loading, *Proc. Phys. Soc. London*, Vol 62-B, 1949, p 676-700
10. H. Kolsky, *Stress Waves in Solids*, Oxford University Press, London, 1953
11. J. Duffy, J.D. Campbell, and R.H. Hawley, On the Use of a Torsional Split Hopkinson Bar to Study Rate Effects in 1100-O Aluminum, *J. Appl. Mech.*, Vol 38, 1971, p 83-91
12. R.A. Frantz and J. Duffy, The Dynamic Stress-Strain Behavior in Torsion of 1100-O Aluminum Subjected to a Sharp Increase in Strain Rate, *J. Appl. Mech.*, Vol 39, 1972, p 939-945
13. K.A. Hartley and J. Duffy, Strain Rate and Temperature History Effects During Deformation of FCC and BCC Metals, in *Mechanical Properties of Materials at High Rates of Strain*, J. Harding, Ed., The Institute of Physics, London, 1984, p 21-30
14. B. Hopkinson, The Effects of Momentary Stresses in Metals, *Proc. Roy. Soc. London*, Vol 74-A, 1905, p 498-506
15. U.S. Lindholm, Some Experiments with the Split Hopkinson Pressure Bar, *J. Mech. Phys. Solids*, Vol 12, 1964, p 317-335
16. W.W. Baker and C.H. Yew, Strain Rate Effects in the Propagation of Torsional Plastic Waves, *J. Appl. Mech.*, Vol 33, 1966, p 917-923
17. T. Nicholas, Strain-Rate and Strain-Rate History Effects in Several Metals in Torsion, *Exp. Mech.*, Vol 11 (No. 8), 1971, p 370-374
18. J.L. Lewis and J.D. Campbell, The Development and Use of a Torsional Hopkinson Bar Apparatus, *Exp. Mech.*, Vol 12 (No. 11), 1972, p 520-524
19. J. Lipkin, J.D. Campbell, and J.D. Swearengen, The Effects of Strain-Rate Variations on the Flow Stress of OFHC Copper, *J. Mech. Phys. Solids*, Vol 26, 1978, p 251-268
20. A.M. Eleiche and J. Duffy, Effects of Temperature on the Static and Dynamic Stress-Strain Characteristics in Torsion of 1100-O Aluminum, *Int. J. Mech. Sci.*, Vol 17, 1975, p 85-95
21. J.L. Chiddister and L.E. Malvern, Compression Impact Testing of Aluminum at Elevated Temperatures, *Exp. Mech.*, Vol 3, 1963, p 81-90
22. J. Duffy, Testing Techniques and Material Behaviour at High Rates of Strain, in *Mechanical Properties of Materials at High Rates of Strain*, J. Harding, Ed., The Institute of Physics, London, 1984, p 1-15
23. T. Nicholas, Material Behavior at High Strain Rates, in *Impact Dynamics*, J. Zukas *et al.*, Ed., John Wiley & Sons, New York, 1981, p 277-331
24. H. Kobayashi and B. Dodd, "A Review of Dynamic Torsion Testing Techniques," Report SM 101/84, Reading University of Engineering Laboratories, Reading, England, 1984
25. A.M. Eleiche and J.D. Campbell, "The Influence of Strain Rate History on the Shear Strength of Copper and Titanium at Large Strains," Report No. 1106/74, University of Oxford, Oxford, England, 1974
26. D.L. Cleland and M.G. Stevenson, "Development of an Improved Torsional Hopkinson Bar Apparatus to Produce High Strains at Shear Strain Rates Up to 10,000/s," Report 1980/IE/2, University of New South Wales, New South Wales, Australia, 1980
27. E.K.C. Leung, An Elastic-Plastic Stress Analysis of the Specimen Used in the Torsional Kolsky Bar, *J. Appl. Mech.*, Vol 47, 1980, p 278-282
28. L.S. Costin, E.E. Crisman, R.H. Hawley, and J. Duffy, On the Localization of Plastic Flow in Mild Steel Tubes Under Dynamic Torsional Loading, in *Proc. 2nd Int. Conf. Mechanical Properties of Materials at High Rates of Strain*, J. Harding, Ed., The Institute of Physics, London, 1979, p 90-100
29. C.Y. Chiem and J. Duffy, Strain Rate History Effects and Observations of Dislocation Substructure in Aluminum Single Crystals Following Dynamic Deformation, *Mat. Sci. Eng.*, Vol 57, 1963, p 233-247
30. J. Duffy, Strain Rate History Effects and Dislocation Substructure at High Strain

Rates, in *Material Behavior Under High Stress and Ultrahigh Loading Rates*, J. Mescall and V. Weiss, Ed., Plenum Press, New York, 1983, p 21-37

31. J. Klepaczko, Strain-Rate History Effects for Polycrystalline Aluminum and the Theory of Intersections, *J. Mech. Phys. Solids*, Vol 16, 1968, p 255-266

32. U.S. Lindholm, Deformation Maps in the Region of High Dislocation Velocity, in *High Velocity Deformation of Solids*, K. Kawata and J. Shioiri, Ed., International Union of Theoretical and Applied Mechanics Symposium, Springer-Verlag, Berlin, 1978

33. G.L. Wulf, Dynamic Stress-Strain Measurements at Large Strains, in *Mechanical Properties at High Rates of Strain*, J. Harding, Ed., Institute of Physics Conference Series, London, No. 21, 1974, p 48-52

34. D.A. Gorham, Measurement of Stress-Strain Properties of Strong Metals at Very High Rates of Strain, in *Mechanical Properties at High Rates of Strain*, J. Harding, Ed., Institute of Physics Conference Series, London, No. 47, 1979, p 16

35. W.G. Ferguson, J.E. Hauser, and J.E. Dorn, The Dynamic Punching of Metals, Dislocation Damping in Zinc Single Crystals, *Brit. J. Appl. Phys.*, Vol 18, 1967, p 411-417

36. J.D. Campbell and W.G. Ferguson, The Temperature and Strain Rate Dependence of Shear Strength of Mild Steel, *Philos. Mag.*, Vol 21, 1970, p 63-82

37. J. Harding and J. Huddart, The Use of the Double-Notch Shear Test in Determining the Mechanical Properties of Uranium at Very High Rates of Strain, *Proc. 2nd Int. Conf. Mechanical Properties at High Rates of Strain*, J. Harding, Ed., The Institute of Physics, London, 1979, p 49-61

38. A.R. Dowling, J. Harding, and J.D. Campbell, The Dynamic Punching of Metals, *J. Inst. Metals*, Vol 98, 1970, p 215-224

39. A.S. Abou-Sayed, R.J. Clifton, and L. Hermann, The Oblique Plate Impact Experiment, *Exp. Mech.*, Vol 16, 1976, p 127-132

40. Y.M. Gupta, Shear Measurements in Shock Loaded Solids, *Appl. Phys. Lett.*, Vol 29, 1976, p 694-697

41. K.S. Kim, R.J. Clifton, and P. Kumar, A Combined Normal and Transverse Displacement Interferometer with an Application to Impact of Y-cut Quartz, *J. Appl. Phys.*, Vol 48, 1977, p 4132-4139

42. L.C. Chhabildas, H.J. Sutherland, and J.R. Asay, Velocity Interferometer Technique to Determine Shear-Wave Particle Velocity in Shock-Loaded Solids, *J. Appl. Phys.*, Vol 50, 1979, p 5196-5201

43. L.M. Barker and R.E. Hollenbach, Laser Interferometer for Measuring High Velocities of Any Reflecting Surface, *J. Appl. Phys.*, Vol 43, 1972, p 4669-4675

44. K.S. Kim and R.J. Clifton, Pressure-Shear Impact of 6061-T6 Aluminum and Alpha-Titanium, *J. Appl. Mech.*, Vol 47, 1980, p 11-16

45. L.C. Chhabildas and J.W. Swegle, Dynamic Pressure-Shear Loading of Materials Using Anisotropic Crystals, *J. Appl. Phys.*, Vol 51, 1980, p 4799-4807

46. A. Gilat and R.J. Clifton, Pressure-Shear Waves in 6061-T6 Aluminum and Alpha-Titanium, *J. Mech. Phys. Solids*, Vol 32, 1984 (to be published)

47. C.H. Li, A Pressure-Shear Experiment for Studying the Dynamic Plastic Response of Metals and Shear Strain Rates of 10^5 s^{-1}, Ph.D. Thesis, Brown University, Providence, RI, 1982

48. P. Kumar and R.J. Clifton, Optical Alignment of Impact Faces for Plate Impact Experiments, *J. Appl. Phys.*, Vol 48, 1977, p 1366-1367

49. R.W. Klopp, Pressure-Shear Deformation of 4340 Steel at Strain Rates of 10^5 s^{-1}, Sc.M. Thesis, Brown University, Providence, RI, 1984

50. R.W. Klopp and R.J. Clifton, Pressure-Shear Impact and the Dynamic Viscoplastic Response of Metals, *Mech. Mat.* (to be published)

51. J. Shioiri, K. Satoh, and K. Nishimura, Experimental Studies on the Behavior of Dislocations in Copper at High Rates of Strain, in *High Velocity Deformation of Solids*, K. Kawata and J. Shioiri, Ed., Springer-Verlag, New York, 1978, p 50-66

52. P.S. Follansbee and J. Weertman, On the Question of Flow Stress at High Strain Rates Controlled by Dislocation Viscous Flow, *Mech. Mat.*, Vol 1, 1982, p 345-350

53. R.J. Clifton, Dynamic Plasticity, *J. Appl. Mech.*, Vol 50, 1983, p 941-952

54. K.A. Hartley and J. Duffy, personal communication, 1984

55. S. Tanimura and J. Duffy, "Strain Rate Effects and Temperature History Effects for Three Different Tempers of 4340 VAR Steel," Technical Report No. DAAG 29-81-K-0121/4, Division of Engineering, Brown University, Providence, RI, 1984

SELECTED REFERENCES

• K. Bitans and P.W. Whitton, High-Strain-Rate Investigations, with Particular Reference to Stress-Strain Characteristics, *Int. Met. Rev.*, Vol 17, Review 161, 1972, p 66-78

• J.D. Campbell, "Dynamic Plasticity of Metals," International Centre for Mechanical Sciences, Course and Lectures, No. 46, Udine, Springer-Verlag, Vienna and New York, 1970

• R.J. Clifton, Dynamic Plasticity, *J. Appl. Mech.*, Vol 50, 1983, p 941-952

• L. Davison and R.A. Graham, Shock Compression of Solids, *Phys. Rep.*, Vol 55 (No. 4), 1979, p 255-370

• S.J. Green and S.G. Babcock, "Response of Materials to Suddenly Applied Stress Loads," Part I: "High Strain-Rate Properties of Eleven Reentry-Vehicle Materials at Elevated Temperatures," Report TR66-83, General Motors Defense Research Laboratories, Aerospace Operations Dept., Santa Barbara, CA, 1966

• A.J. Holzer and P.K. Wright, Dynamic Plasticity: A Comparison Between Results from Mechanical Testing and Machining, *Mat. Sci. Eng.*, Vol 51, 1981, p 81-92

• K. Kawata and J. Shioiri, Ed., *High Velocity Deformation of Solids*, Springer-Verlag, Berlin, 1978

• H. Kolsky, *Stress Waves in Solids*, Oxford University Press, London, 1953

• L.H.N. Lee, Dynamic Plasticity, *Nucl. Eng. Design*, Vol 27, 1974, p 386-397

• U.S. Lindholm, Ed., *Mechanical Behavior of Materials Under Dynamic Loads*, Springer-Verlag, New York, 1968

• U.S. Lindholm and R.L. Bessey, "A Survey of Rate Dependent Strength Properties of Metals," Technical Report 69-199, Air Force Materials Laboratory, Wright-Patterson Air Force Base, OH, 1969

• M.A. Meyers and L.E. Murr, Ed., *Shock Waves and High-Strain-Rate Phenomena in Metals*, Plenum Press, New York, 1980

• T. Nicholas, Material Behavior at High Strain Rates, in *Impact Dynamics*, J. Zukas *et al.*, Ed., John Wiley & Sons, New York, 1981, p 277-331

• H.C. Rogers, Adiabatic Plastic Deformation, *Ann. Rev. Mat. Sci.*, Vol 9, 1979, p 283-311

• R.W. Rohde, B.M. Butcher, J.R. Holland, and C.H. Karnes, Ed., *Metallurgical Effects at High Strain Rates*, Plenum Press, New York, 1973

• R. Sutterlin, Sur la Plasticite Dynamique. Son Application a l'Etude du Forgeage (Dynamic Plasticity. Its Application to Studies of Forging), Sci. Techn. l'Armement, Memorial de l'Artillerie Francaise, Vol 46 (No. 4), 1972, p 909-989; Vol 47 (No. 3), 1973, p 567-646

Ultrasonic Fatigue Testing

By Lewis D. Roth*
Principal Engineer
Westinghouse R & D Center

ULTRASONIC FATIGUE TESTING involves cyclic stressing of material at frequencies typically in the range of 15 to 25 kHz. The major advantage of using ultrasonic fatigue is its ability to provide fatigue-limit and near-threshold data within a reasonable length of time. High-frequency testing provides rapid evaluation of the high-cycle fatigue limit of engineering materials. Fatigue crack growth at extremely slow crack propagation rates is also possible with ultrasonic frequency testing.

Ultrasonic fatigue testing is applicable to most engineering materials, including metals, ceramics, glasses, plastics, and composites. Test data can be used for screening of high-cycle fatigue properties or extending the fatigue data already available from conventional frequency fatigue testing.

This article reviews underlying concepts and basic techniques for performing ultrasonic fatigue tests. It describes test equipment design, specimen design, and effective control over test variables. Results obtained with ultrasonic fatigue test methods are discussed with respect to strain-rate-dependent material behavior. Standardized procedures and test machinery for performing ultrasonic fatigue tests currently are not available. Efforts are in progress within ASTM to provide a recommended practice and eventually a testing standard.

Historical Perspective

Development of higher frequency testing machines began early in the 20th century. Prior to 1911, the highest fatigue testing frequency was on the order of 33 Hz, using mechanically driven systems. Electrodynamic resonance systems appeared in 1911 when Hopkinson (Ref 1) introduced a machine capable of 116 Hz. In 1925, Jenkin (Ref 2) tested wires of copper, iron, and steel at 2 kHz, using similar techniques. In 1929, Jenkin and Lehmann (Ref 3) were able to test materials up to 10 kHz using a pulsating air resonance system.

Mason (Ref 4) achieved ultrasonic frequency (20 kHz) in 1950 with the adaptation of magnetostrictive and piezoelectric-type transducers to fatigue testing. This method translated 20-kHz electrical voltage signals into 20-kHz mechanical displacements. A displacement-amplifying acoustical horn and the test specimen were driven into resonance by the transducer. This concept has remained basically unchanged and is the foundation of the practices used in modern ultrasonic fatigue test technology.

In the early 1960's, frequencies as high as 92 and 199 kHz were employed for fatigue tests using Mason's techniques (Ref 5, 6). These extremely high frequencies surpass the upper limits of practicality because of the constraints of specimen size (frequency is inversely proportional to specimen length), machining tolerances, strain amplitude measurements, and energy considerations. A review of the ultrasonic fatigue testing in the 1970's and 1980's shows that the majority of test stands operate at frequencies between 17 and 25 kHz.

This unofficial standard is primarily dictated by the availability of commercial high-power ultrasonic transducers and power supplies. These frequencies are also desirable from a safety viewpoint, because they are above the range of normal human hearing. Fatigue testing at 20 kHz proceeds quietly in comparison to testing at 1 to 10 kHz.

Strain Rates, Frequency, and Time Compression

Ultrasonic fatigue testing increases the frequency of stress cycling to reduce the time necessary to accumulate a large number of cycles. Consequently, the strain rate at these frequencies for a given strain amplitude is also increased. In Table 1, strain rate is calculated as a function of frequency and strain amplitude. For typical fatigue strain amplitudes in the range of 10^{-3} and 10^{-4}, the strain rate at 20 kHz ranges from 2 to 20 s^{-1}.

Ultrasonic fatigue techniques are particularly useful for providing fatigue data in applications where strains are being applied and removed at kilohertz frequencies; for example, high-frequency loading of turbine blades. In fact, ultrasonic fatigue testing may provide a better simulation of the higher frequency vibrations encountered in service than conventional testing does in these cases. The test method is most applicable when the test material ultimately will be applied in service at frequencies at or near the test frequency. For applications with lower frequency vibrations, the effect of frequency and strain rate on test results must be interpreted.

The time compression per cycle obtained with ultrasonic fatigue is pronounced. For

*With contributions by the participants of the First International Conference on Ultrasonic Fatigue Testing, including P. Bajons, University of Osnabrück; O. Buck, Iowa State University; L. Coffin, Jr., General Electric Corporate Research & Development; R. Ebara, Hiroshima Technical Institute; L. Fritzemeier, Columbia University; R.E. Green, Jr., Johns Hopkins University; I.L.H. Hansson, Technical University of Denmark; W. Hoffelner, Brown Boveri Research Center; K. Kromp, University of Vienna; W. Kromp, University of Vienna; C. Laird, University of Pennsylvania; R.B. Mignogna, Naval Research Laboratory; R.R. Paulson, Columbia University; A. Puskar, University of Transport and Telecommunication; K. Salama, University of Houston; C.R. Sirian, Hydronautics, Inc.; S.E. Stanzl, Massachusetts Institute of Technology; R. Stickler, University of Vienna; J.K. Tien, Columbia University; P.K. Trimmel, University of Vienna; E. Tschegg, Massachusetts Institute of Technology; B. Weiss, University of Vienna; and L.E. Willertz, Westinghouse Research & Development

Table 1 Strain rate as a function of test frequency and strain range

Frequency, Hz	Strain rate ($\dot{\epsilon}$), s^{-1}, at strain (ϵ) of:			
	$\epsilon = 10^{-5}$, m/m	$\epsilon = 10^{-4}$, m/m	$\epsilon = 10^{-3}$, m/m	$\epsilon = 10^{-2}$, m/m
10	10^{-4}	10^{-3}	10^{-2}	10^{-1}
100	10^{-3}	10^{-2}	10^{-1}	10^{0}
1 000	10^{-2}	10^{-1}	10^{0}	10^{1}
10 000	10^{-1}	10^{0}	10^{1}	10^{2}
100 000	10^{0}	10^{1}	10^{2}	10^{3}

example, a conventional fatigue test at 1 Hz would take 320 years for a 10^{10} cycle test. At 100 Hz, the test would take 3.2 years. At an ultrasonic frequency of 20 kHz, this test would be completed in less than 6 days. The time required to complete fatigue tests at different frequencies is shown in Fig. 1. This time compression is extremely attractive for situations that require high-cycle data.

In comparison to conventional frequency testing, more test conditions and/or replicate tests can be performed in a given period of time at ultrasonic frequency. This provides results and conclusions that are statistically more meaningful for planning and design. On the other hand, the minimum number of cycles that can be measured practically is limited by kHz cycling. This limit is 10^5 cycles for open-loop testing, with a testing time of 5 s. Shorter times (~1 s) are possible with closed-loop computer control of the test and data acquisition systems.

Similar time compression is possible in fatigue crack growth rate testing using ultrasonic fatigue. Figure 2 is a schematic of a typical crack growth rate, da/dN, versus stress intensity curve. The time necessary to measure a crack advance of 0.1 mm (0.004 in.) while testing at 1 Hz or 20 kHz is compared on the right side of the figure. It is obvious that ultrasonic testing is the only

practical approach to observe the extremely slow crack growth rates that are characteristic of the threshold regime. Crack growth rate measurements as low as 10^{-11} mm (4×10^{-13} in.) have been reported. Again, the practical upper bound of measurable fatigue crack growth rate at 20 kHz is on the order of 10^{-5} mm (4×10^{-7} in.) per cycle due to the rapid cycle accumulation.

The testing time compression possible with ultrasonic fatigue is an incentive for applying the technology in a more generic sense, i.e., to extend fatigue information obtained at conventional frequencies and lower numbers of cycles to higher cycle fatigue limits and threshold fatigue crack growth rates. Because this accelerated test method alters testing conditions to produce fatigue in a shorter period of time, the influence of frequency and strain rate on cyclic material behavior must be well understood.

General acceptance of ultrasonic fatigue testing also requires an understanding of how to obtain data free of testing-induced artifacts. Improper execution can have a marked effect on the property data obtained. Much of the skepticism that endures about the use of ultrasonic fatigue stems from earlier testing, where questionable techniques were used to measure cyclic strain amplitude and provide adequate cooling of the specimen. Accord-

ingly, the effects of strain rate, frequency, and test technique are the subject of most research on ultrasonic fatigue (see Ref 7, 8).

In general, testing by ultrasonic fatigue produces fatigue data that differ only slightly from those observed at more conventional frequencies. Some data reveal a shift in the ultrasonic fatigue stress-life data (S-N) for a given stress level toward increased lifetimes relative to conventional frequency results (Ref 9-11). Other reports indicate no shift in the S-N behavior (Ref 12, 13). Most reports indicate that fatigue degradation at ultrasonic frequency occurs by the same sequence of events as at conventional frequencies, namely, saturation of rapid hardening, formation of persistent slip bands, formation and growth of intrusions, and crack propagation.

Materials that exhibit clearly defined endurance limits at conventional frequencies usually exhibit endurance limits at similar cyclic stress amplitudes at ultrasonic frequencies. Similarly, materials that exhibit threshold stress intensities for fatigue crack growth at conventional frequencies also exhibit this behavior at ultrasonic frequencies. Shifts in S-N fatigue behavior to higher stress levels and longer lifetimes or da/dN behavior to slower crack growth rates do not occur for all materials tested at high frequency. Recent testing shows that the effect of frequency on S-N and da/dN performance is primarily a function of the microplasticity and slip character of the material system under test.

It might also be inferred that corrosion fatigue interactions should be negligible at ultrasonic frequency due to the short cyclic period. Again, experimental results illustrate that corrosion fatigue interactions are indeed observed at ultrasonic frequencies. Recent testing shows that ultrasonic fatigue is an effective method for the evaluation of the degradation of fatigue properties produced by environmental interactions.

Ultrasonic fatigue testing is applicable to most situations in which conventional frequency fatigue testing has been employed. Examples of a variety of results from ultrasonic fatigue are presented later in this article. As the technique continues to develop, the precise limits of applicability will become more clearly defined.

Testing Principles

The principles of ultrasonic fatigue testing are quite simple. Ultrasonic fatigue is a resonant test method, in which a large amplitude displacement wave must be established in a resonant specimen. This wave is generated by a relatively small periodic stimulus at the same frequency as the natural frequency of the test specimen. Resonance is required

Fig. 1 Testing time vs number of cycles to complete test as a function of frequency

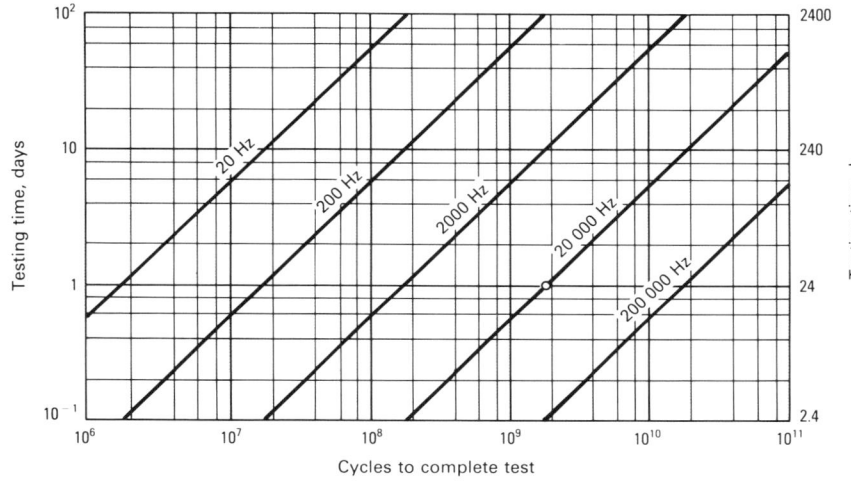

Fig. 2 Typical crack growth rate vs stress intensity curve

Difference in time to observe a finite crack growth increment at ultrasonic (20 kHz) and conventional (1 Hz) frequencies is shown.

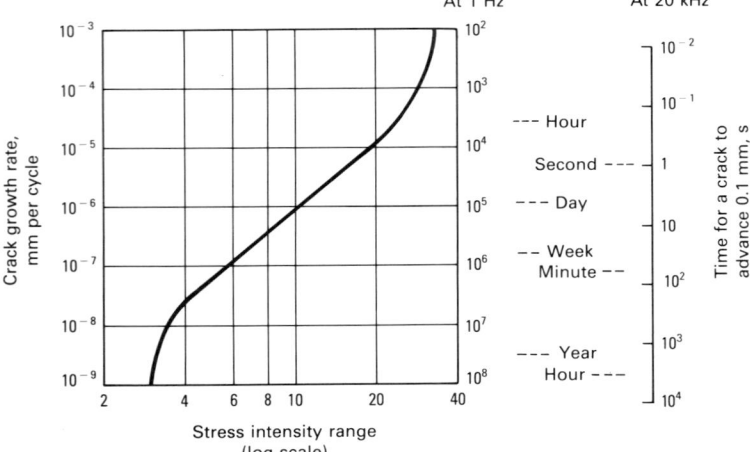

Fig. 3 Distribution of oscillatory displacement amplitude and strain amplitude over the length of a resonant bar of uniform cross section

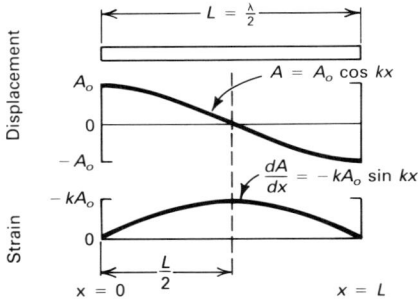

to achieve the strain amplitude needed to produce fatigue in materials.

Displacement and strain are developed in a bar of material subjected to resonant acoustic loading. Consider a straight bar of material having a uniform diameter and length L (Fig. 3). A sound wave injected longitudinally into one end of the bar travels at a certain velocity through the bar, is reflected from the opposite end, and returns to the point of entrance. The wave velocity, C, is determined by the material properties, the Young's modulus, E, and the density (mass/volume), ρ, by:

$$C = \left(\frac{E}{\rho}\right)^{1/2} \qquad \text{(Eq 1)}$$

This velocity is the speed of sound through the material. The time required to travel the length of the bar and return is $2L/C$. If this time is equal to the period of the injected sound wave, the reflected wave will be exactly in phase with the injected wave, standing wave conditions will be established, and the bar will be in resonance. The length, L, of the bar is then exactly equal to the half wavelength of the sound wave. The variation of displacement amplitude of oscillation at a point x along the length of the bar will be:

$$A(x) = A_o \cos(kx) \qquad \text{(Eq 2)}$$

where A_o is the displacement amplitude at the end of the bar; k is $2\pi/\lambda$; and λ is the wavelength of sound at the resonant frequency.

The strain distribution along the bar will be the derivative of the displacement amplitude with respect to distance or:

$$\epsilon(x) = \frac{dA(x)}{dx} = kA_o \sin(kx) \qquad \text{(Eq 3)}$$

Thus, the maximum strain occurs when $\sin(kx) = 1$, or $x = \lambda/4$. The maximum strain varies between $\pm kA_o$ during each cycle. Figure 3 shows the distribution of longitudinal displacement amplitude and strain amplitude along the length of a bar in resonance. The minimum displacement (displacement node) and maximum strain (strain antinode) occur at the center of the bar. Similarly, the maximum displacement (displacement antinode) and minimum strain (strain node) occur at the ends of the bar.

The stress distribution for each point along the bar is obtained by an elastic conversion of the strain distribution:

$$\sigma(x) = E \cdot \epsilon(x) \qquad \text{(Eq 4)}$$

where E is the dynamic Young's modulus of the material. The dynamic modulus of elasticity must be determined for the appropriate test frequency. Because of the elastic conversion, the stress maximum physically coincides with the strain maximum. Stresses cannot be obtained independent of strains in ultrasonic fatigue testing. Therefore, strict stress-controlled tests cannot be performed. Without independent per-cycle stress and strain information, plastic strain-controlled tests also are not possible at this time. For more information on plastic strain-controlled ultrasonic fatigue testing, see Ref 14.

The example of the uniform resonant bar embodies the basic concepts of ultrasonic fatigue testing. With appropriate geometric modification, these concepts can be used to design the mechanical portion of the converter, the acoustic horns, and the test specimen.

The major difference between a conventional fatigue test specimen and a high-frequency resonant specimen is that the cyclic strain amplitude varies from zero at the ends to a maximum at the center, rather than being constant over its entire length. This confines fatigue damage and hence fatigue crack initiation and propagation to the center of the specimen. Because there is minimal strain at the ends of a resonant bar, the requirements for attachment of one resonant bar to another and for gripping the specimen also are minimal.

To produce strain in a bar, only one end of a resonant bar specimen must be in acoustic contact with the source of the sound waves. This permits the testing of thin materials under reversed tension-compression loading without risk of buckling the specimen. Consequently, sheet, tubing, and wire specimens may be subjected to fully reversed loading during ultrasonic fatigue, whereas more complex gripping and alignment techniques are required to accomplish similar tests at conventional frequencies. The large and cumbersome arrangements for gripping the specimen that often are required in conventional fatigue testing are not needed in ultrasonic fatigue.

A specimen with a free end also provides the ultrasonic fatigue system with a degree of portability that is not easily obtained with conventional-frequency test methods. Fatigue testing can be performed with the specimen in an operating environment by feeding the free end of the wave train through an access port to the environment. Similarly, testing can be performed under the view of an optical or electron microscope without need of complex load-transmitting stages.

Cyclic straining can be achieved in a bar at any desired resonance frequency by appropriately choosing (tuning) the length of the bar. For a bar with a uniform cross section, the required length for fatigue testing will be $\lambda/2$ at the resonance frequency. For bars with variable cross sections or dumbbell specimen geometries, the resonant length generally is shorter than the resonant length of a uniform bar at the given test frequency. Thus, each component in a resonant testing system must be designed (tuned) to the resonance frequency to transmit the acoustic energy efficiently into the test specimen. The equations developed by Neppiras (Ref 15) are helpful in calculating the appropriate resonant lengths for variable specimen section geometries. These equations are presented later in this article in a section on specimen design.

Testing Equipment and Methods

Packaged ultrasonic fatigue test systems, with one exception, are not commercially available. However, an ultrasonic fatigue test system may be constructed easily from commercially available parts. Tien *et al.* (Ref 16) describe the construction of a test machine using ultrasonic components normally used in ultrasonic joining processes. This machine, an open-loop test stand, contains the basic equipment needed for testing. Information on test stands with additional capabilities, including double converters, mean loading, electrochemical cells, and computerized control systems, can be found in Ref 17 to 20. A portable test machine including ultrasonics, external loading frame, environmental system, and test chamber is shown in Fig. 4.

Figure 5 is a schematic of a typical ultrasonic fatigue test machine. The machine is centered around an acoustic wave train composed of a sonic energy converter, a series of acoustic amplifying horns, and the test specimen. A typical wave train is shown in Fig. 6. The acoustic energy is supplied by a high-frequency power supply. An amplitude-measuring device and a means of dissipating the heat generated by the deformation process are also necessary. This basic equipment is appropriate for stress-life (*S-N*) or fatigue crack growth rate (*da/dN*) testing. A frequency display, cycle counter, and temperature-measuring equipment are used to monitor the test. Additional monitoring equipment is necessary to measure crack length in *da/dN* testing.

Power supplies for ultrasonic fatigue testing typically range from 500 to 4000 W of electrical power. The actual output to the specimen is lower than this during normal resonant operation. Most power supplies have built-in feedback circuits, which produce a constant-amplitude oscillation in the converter. Some power supplies have circuits for automatic shutoff when the specimen or any part of the wave train goes out of resonance. This is useful for *S-N* testing. The

fatigue crack at failure will be some fraction of the cross-sectional area when the power supply shuts off. This fraction can range from a few percent to 50% of the cross-sectional area, depending on the automatic shut-off controls.

Sonic Converters. Acoustic resonance is developed in the converter by application of the electrical excitation provided by the power supply. The converter generates a standing acoustic wave that produces a cyclic displacement at the end of the converter. The acoustic wave proceeds down the rest of the resonant wave train to the specimen. Variation of the displacement and strain amplitudes along the wave train is shown in Fig. 7.

Several cycles of application of the electronic stimulus of the power supply are required to achieve the maximum resonant

amplitude in the converter and the rest of the wave train. The rise time of the converter should be known when considering a pulsed mode versus continuous cycling mode of an ultrasonic fatigue system. In a pulsed mode operation, the specimen is subjected to a series of pulses (~1 s) of high amplitude cycles followed by a cooling period without cycling. Rise time of ultrasonic equipment varies among manufacturers. If rise time is longer than pulse time, variable amplitude test conditions exist. Pulsed mode operation has been suggested by some investigators to overcome the rapid heating manifested by high damping materials upon cycling. Ultrasonic fatigue systems take several cycles for the maximum resonant amplitude to be developed. Hence, the tendency to overshoot the desired amplitude setpoint on the first cycle is small.

Fig. 4 Portable 20-kHz corrosion-fatigue machine with mean load capability

Fig. 5 Schematic of an ultrasonic fatigue test system

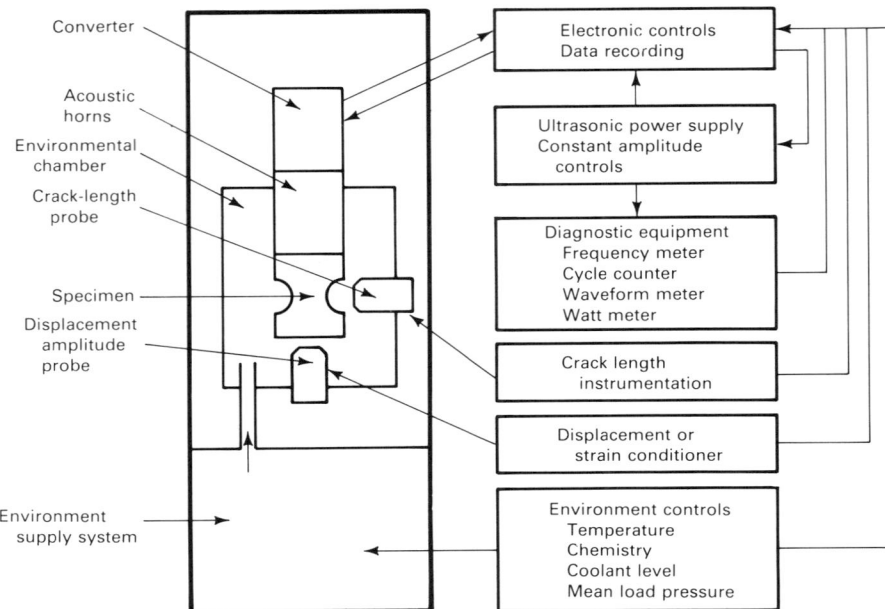

Fig. 7 Variation of the displacement and strain amplitudes along the acoustic wave train

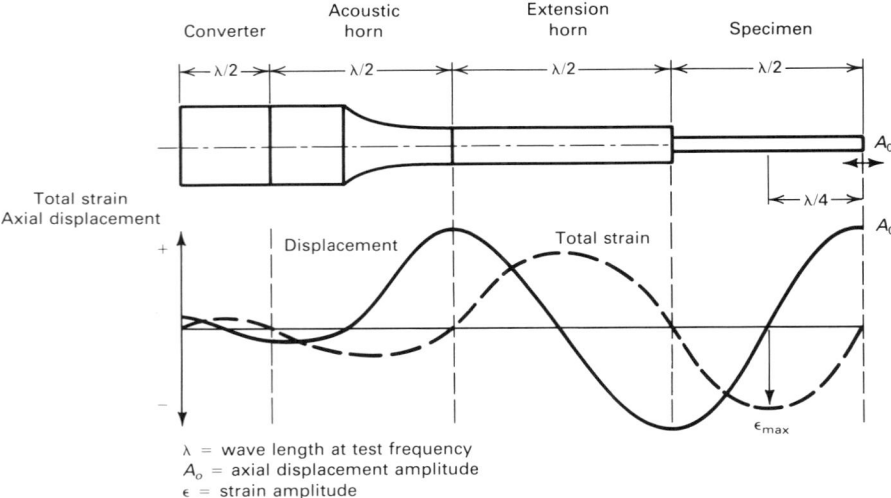

λ = wave length at test frequency
A_o = axial displacement amplitude
ϵ = strain amplitude

Fig. 6 Typical 20-kHz acoustic wave train

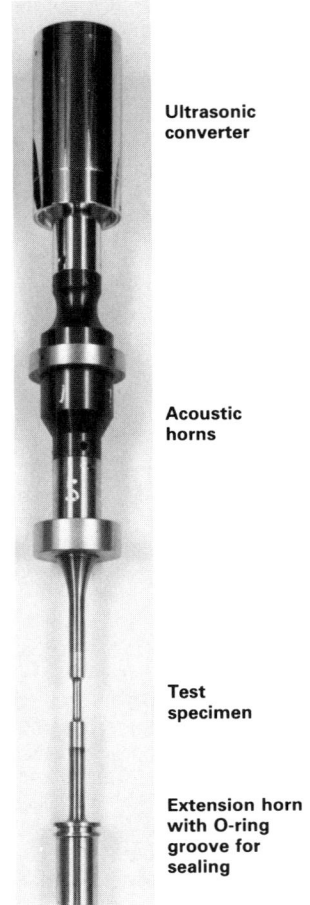

Ultrasonic converter

Acoustic horns

Test specimen

Extension horn with O-ring groove for sealing

Converters for generating ultrasonic displacement waves generally are magnetostrictive or piezoelectric devices. Most modern converters use piezoelectric materials for conversion efficiency. Magnetostrictive devices have a low (20%) conversion efficiency. Piezoelectric converters with efficiencies greater than 90% are readily available.

Converter types and designs vary among manufacturers. Some piezoelectric devices use lead-zirconium-titanate (PZT) for the converter material. The end displacement amplitude developed by a 20-kHz PZT converter ranges from 0.010 to 0.020 mm (0.0004 to 0.0008 in.). Piezoelectric plastic materials are being considered for higher amplitude ultrasonic converters.

A single- or double-converter arrangement can be used to drive the specimen into resonance. In a single-transducer system, one end of the specimen is coupled to the converter and the other end remains free. In a double-converter system, both ends of the specimen are coupled to two coaxial anti-phase-driven ultrasonic converters (Ref 17).

The advantage of a double-converter system is its symmetry.

A comparison of the displacement, strain, and specific energy parameters for a high damping perspex (Lucite) test specimen tested with a single- and double-converter system is shown in Fig. 8 (Ref 21). The symmetry of the converters is reflected in the greater symmetry of the displacement and strain distributions produced in a resonant specimen. While equivalent testing conditions can be produced with either single- or double-converter systems through precise design of the acoustic elements, the double converter is less sensitive to small differences between the resonance frequency of the specimen and the driving frequency of the converter. Data also show that the double-converter arrangement is less sensitive to detuning of the specimen due to changes in elastic properties or the growth of a fatigue crack. Fatigue crack growth testing benefits from the longer crack length attainable with a double-converter system before significant frequency degradation occurs.

Fig. 8 Comparison of single- and double-converter systems

Calculated displacement, strain, and specific energy parameters for a highly damped perspex specimen ($L = \lambda/2$) are shown. The single-converter system was excited from the left side. Assumed values for the specimen: $E = 48$ GPa (6.9 psi $\times$ 10^6); $\rho = 1.2$ g/cm^3; frequency = 20 kHz. Source: Ref 21

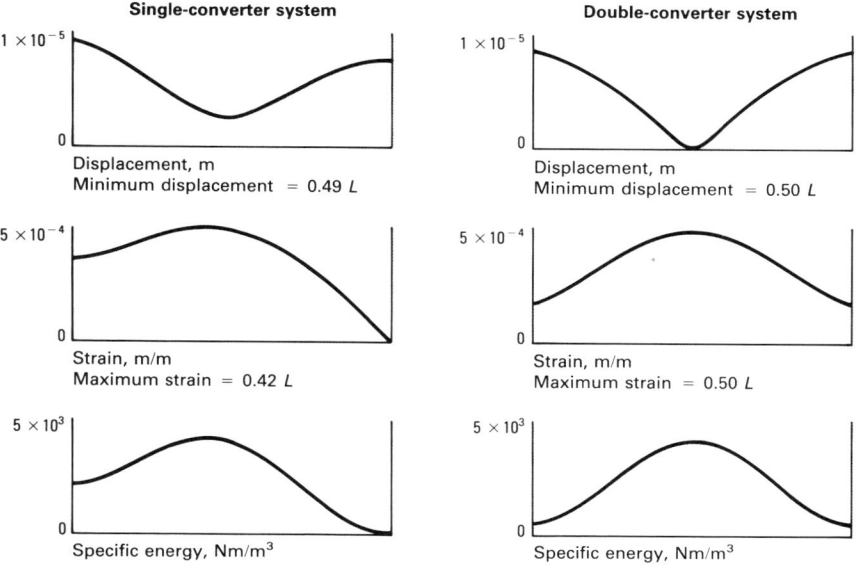

Acoustic horns transmit the resonance developed by the converter to the specimen. One or more acoustic amplifying horns generally are placed in the wave train to raise the strain amplitude in the specimen to the level required for fatigue. Design of these horns was developed by Mason (Ref 22) and Neppiras (Ref 15).

Acoustic horns are bars of resonant length with cross-sectional areas that change either continuously or discontinuously as distance from the input end varies. To maintain the requirement of continuity of particle velocity along the horn length, the vibrational amplitude must increase in areas of reduced cross-sectional area. This produces displacement and strain amplification. For the simple stepped-down horn shown in Fig. 9(a), displacement and hence strains are amplified by the ratio of the cross-sectional areas of the horn:

$$A_o = \frac{\text{Area}_{\text{input}}}{\text{Area}_{\text{output}}} \cdot A_i \qquad \text{(Eq 5)}$$

where A_o is the displacement amplitude on the output end of the horn, and A_i is the displacement amplitude on the input end of the horn. Conversely, an increase in cross-sectional area causes deamplification. A number of different horn shapes have been designed by detailed mathematical analysis (Ref 23-26); typical examples are shown in Fig. 9 along with the particle velocity and stress distribution along their length.

An obvious problem with the stepped horn is the manifestation of very high stress ampli-

tudes at the step. This eventually causes the horn to fail by fatigue at the step. The Fourier horn (Fig. 9e) is the best overall horn design for achieving the highest amplification with the greatest strength and stiffness. Ultrasonic converter suppliers typically carry a variety of acoustic horn designs with amplifications ranging from 1 to 1 to on the order of 10 to 1. Some manufacturers will design acoustic horns for custom applications including very high amplifications, special frequencies, or sealing the entrance of an environmental chamber.

Extension Acoustic Horns. The need for extension horns generally arises when the test specimen must be isolated in a controlled environmental chamber, furnace, or external load frame. Typically, a 1-to-1 extension horn is used—i.e., a uniform bar of length $\lambda/2$, having been modified with a flange for seating an environmental seal or for attaching the wave train to the external load frame.

It is important to select materials that will not affect the test. Extension horn material should be low damping whenever possible to avoid losses and excessive heating. For corrosive environments, the horn material should be such that a galvanic couple is not set up between the specimen and horn. If high temperatures are to be encountered, the horn material must possess appropriate high-temperature strength. The elastic properties of the material should be determined for the temperature of the desired test. Some properties of materials used for acoustic extension horns are presented in Table 2.

Amplitude Detection. The accuracy of determining specimen strain depends to a great extent on the accuracy of the measurement of the displacement amplitude at the end of the specimen. Much research has been directed toward the development of sensitive methods for measuring displacement amplitude at ultrasonic frequencies. The simplest, most reliable method is observation of the trajectory of a feature on the specimen surface with a dark-field optical microscope having coaxial lighting and a filar eyepiece.

At magnifications of 500× and a reticle scale in hundredths of a millimeter, displacements on the order of 2 to 3 μm can be detected. The trajectory of a point on the specimen surface appears as a bright streak whose length will be the peak-to-peak displacement amplitude. Visual observation of displacement amplitude with a microscope does not lend itself to automated recording of the displacement. However, it is the preferred method for calibrating more automated amplitude detection equipment, because secondary modes of vibration are easily detected. Secondary modes cause the normally linear trajectory of a point to skew or appear to orbit about another point.

Other amplitude detection devices more suitable for automated data acquisition or feedback to a closed-loop fatigue apparatus have also been used. These generally are displacement measuring devices, which come in many forms, including capacitance gages (Ref 16, 27), permanent magnet-coil arrangements that use eddy currents (Ref 17,

Table 2 Typical 20-kHz resonance properties for acoustic extension horn materials

See text for explanation of symbols

Material	ρ, g/cm^3	λ mm	λ in.	E MPa	E psi $\times$ 10^6
Ra-333	8.3	244	9.61	199 300	28.9
Udimet 710	8.1	264	10.40	224 090	32.5
Udimet 720	8.1	264	10.40	226 160	32.8
MP35N	8.5	246	9.68	206 160	22.9
Ti-6Al-4V	4.4	249	9.80	108 940	15.8
AISI 403	7.7	262	10.31	210 960	30.6
17-4PH	7.8	250	9.84	199 960	28.9

Fig. 9 Profiles of acoustic horns for amplifying converter output

Variations in particle velocity and stress along horns are shown below each profile. (a) Stepped. (b) Conical. (c) Exponential. (d) Catenoidal. (e) Fourier. Source: Ref 23

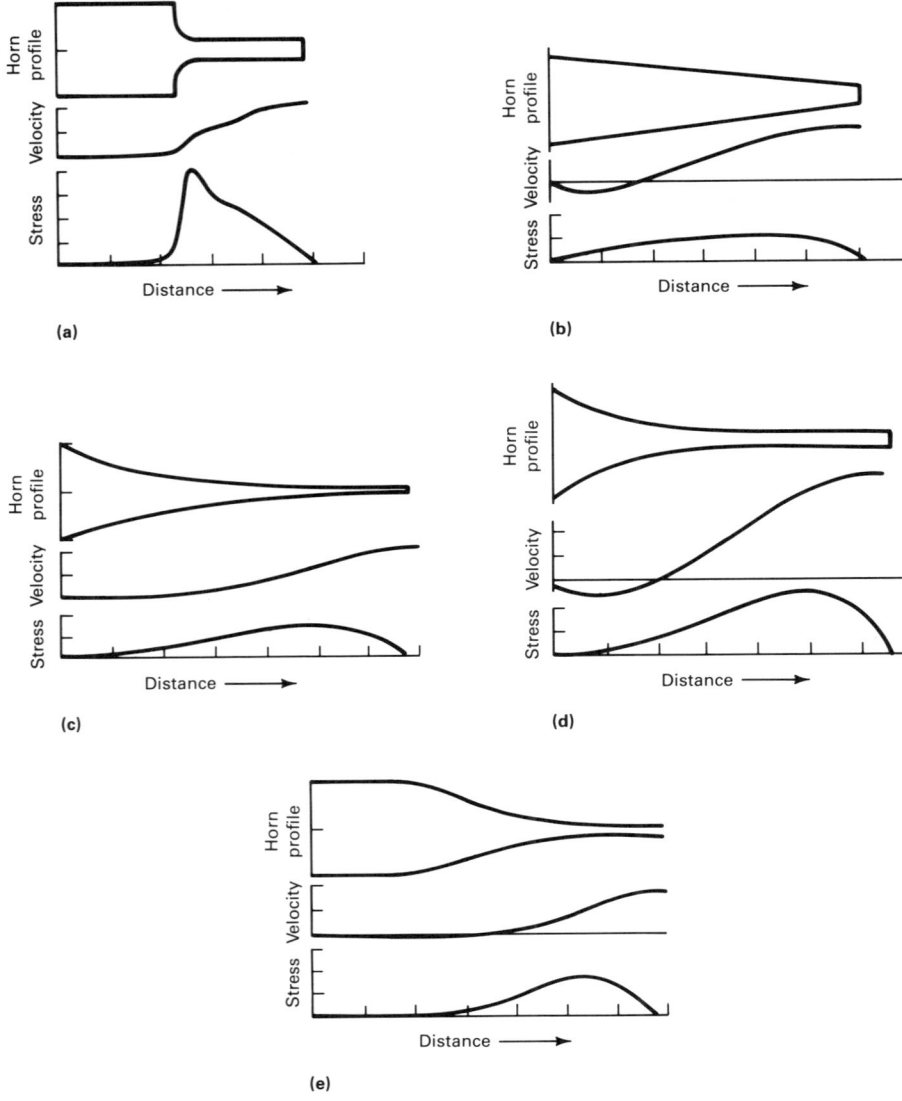

at the point of maximum strain on the specimen they generally will fail before the specimen. Attachment of the gage to the specimen may modify the specimen surface and lower its fatigue properties. The presence of a gage on the specimen may cause additional heating to occur under the gage. These problems may be avoided by placing the gage on an adjacent extension horn where the maximum strains are lower than on the specimen. The strain in the specimen then can be calibrated to the strain in the extension horn.

It is best to use amplitude detectors that can be located at or as close to the specimen as possible. Consideration should be given as to whether the environment that is used will interfere with operation of the device. Some devices give outputs that vary when they are positioned away from the specimen and require calibration for each material; electrodynamic devices operating on eddy currents induced in the vibrating specimen fall into this category. These devices are more difficult to use than capacitance gages, whose output is linear over a known range and does not depend on the properties of the test material. Capacitance gages also need to be calibrated if the working environment changes the dielectric constant.

Crack growth measuring systems for ultrasonic fatigue tests are similar to those used at conventional frequencies (see the Fatigue Testing section of this Volume). Optical methods can be used, because the fatigue crack grows in the plane of maximum strain, which is a displacement node. As a result, the crack will not appear to be vibrating. Optical methods using microscopes on traveling stages can be operated manually or automatically. Automated systems for video monitoring of the crack length are described in Ref 19. Foil-type crack growth gages can be attached to the specimen in the path of the growing crack. These gages develop a linearly varying potential as the crack tears the foil and have been shown to operate linearly under 20-kHz cycling. The usual requirements for monitoring the symmetry of fatigue crack length in double-edge-cracked or center-cracked specimens at conventional frequency also apply for ultrasonic frequency testing.

Cooling Systems. As in conventional fatigue testing, the components of fatigue deformation generate heat within the specimen in an amount equal to the area enclosed by the stress-strain hysteresis loop in each cycle. At 20 kHz, the specific energy input at high stresses can be as high as several hundred watts per cubic centimeter for high damping materials. Large temperature excursions can result if the heat is not removed. These temperature excursions not only change the properties of the test material, but

28), a microphone pickup (Ref 29), a photodiode arrangement (Ref 30), and closed-circuit television (Ref 31).

Noncontacting capacitance-type detectors to measure displacement amplitudes at ultrasonic frequency are commercially available with sensitivities on the order of 3×10^2 μm and linear frequency response up to 50 kHz. Eddy current probes also are commercially available and are frequently used to measure displacement. Because eddy currents are sensitive to composition and microstructure, eddy current probes require calibration of the output signal versus displacement for each new material that is tested. These devices generally are not useful for measuring dis-

placement of nonconducting or low-permeability materials. When the end of the specimen is inaccessible due to an environment chamber or extension horns, the displacement must be calibrated from the specimen to some other point on the wave train.

Strain can be measured by directly applying strain gages to the specimen and reading the value from a strain conditioner that has a frequency response equivalent to the test frequency. Several problems are associated with strain gages applied directly to the specimen. Strain gages, strain gage leads, and the adhesive that holds a strain gage onto the specimen are subject to fatigue loading and have finite cyclic lifetimes. If they are placed

can introduce mean tensile stresses on the surface of solid specimens because of the differential thermal expansion between the core and surface of the specimen.

In most cases, the heat generated can be dissipated by forced cooling if the heat transfer path is not too lengthy. In some cases, the heat generated cannot be eliminated by forced air cooling, and more efficient external and internal cooling may be necessary. A hollow specimen can be used to decrease the heat transfer path through the metal.

In specimens of materials with high damping and large cross sections, this heat rise can be extreme. Figure 10(a) shows the temperature contour of a 12-mm (0.48-in.) diam resonant steel bar without cooling (Ref 32). The strain antinode heated to almost 400 °C (750 °F) during ultrasonic frequency cycling at a stress level of 150 MPa (22 ksi). The highest temperatures are observed at the strain antinode. Efficient cooling of a test specimen would obviously require the greatest volume of coolant to flow over the strain antinode. Large temperature excursions can usually be minimized in practice by the choice of a hollow dumbbell specimen design.

The damping characteristic of the material and the efficiency of the coolant determine the choice of a solid or hollow specimen design for testing. In general, cooling a material with high damping (such as pure copper) with forced air heat transfer requires a thin-walled hollow specimen to ensure that the temperature rise is acceptably low. At the other end of the damping spectrum, titanium alloys generate so little heat at 20 kHz at modest stress levels that virtually any specimen geometry is compatible with the temperature rise limitations.

Martensitic steel lies somewhere between copper and titanium, and hollow or solid specimens may be selected depending on whether the coolant is forced air or water. Measurement of the temperature rise at the uncooled interior of a hollow type 403 stainless steel specimen (1.25 mm, or 0.05 in., wall thickness, with water cooling at the external surface) revealed a temperature rise of less than 10 °C (18 °F) even during fatigue at stresses high enough to cause failure in 10^7 cycles.

Monitoring the temperature rise is essential to obtaining reliable data, particularly when testing in air. Attaching a thermocouple to the specimen at the maximum strain point presents two basic problems. First, the specimen could be damaged and premature fatigue failure could occur. Second, infrared data show that attachment of the thermocouple to a nonresonant bar causes localized heating in the vicinity of the thermocouple (Fig. 10b).

Fig. 10 Infrared thermogram of specimen under 20 kHz excitation without cooling

(a) Resonant specimen exhibiting nodal heating. (b) Localized heating due to thermocouple epoxied to surface of nonresonant specimen. Source: Ref 32

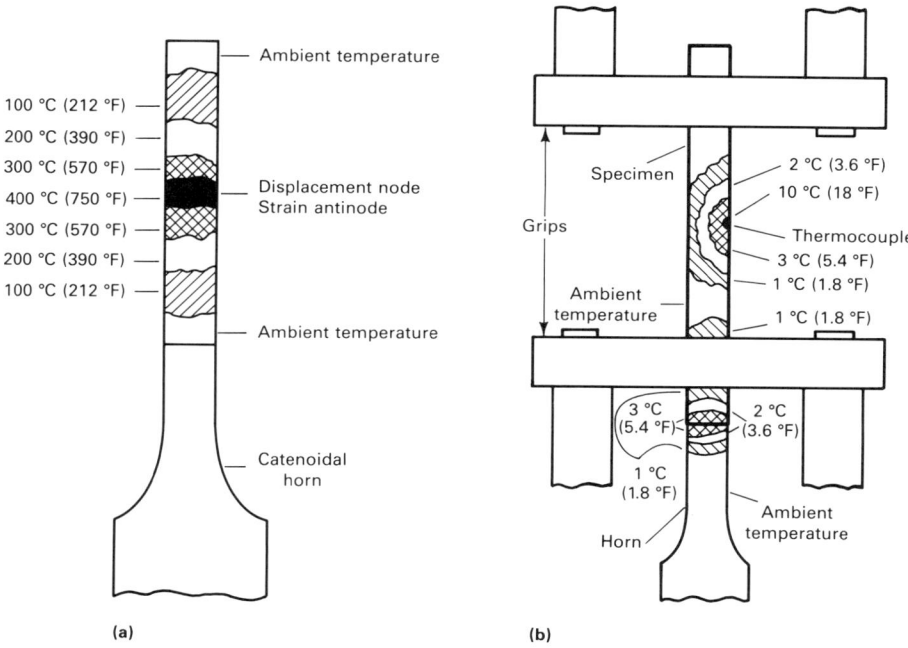

(a)   (b)

The localized heating implies that thermocouples may not be reliable for measuring the specimen temperature during ultrasonic vibration. It has been proposed that the localized heating could be caused by vibration of the leads of the thermocouple due to active displacement at all positions of a nonresonant specimen. Hence, placing the thermocouple at the displacement node of a resonant specimen might minimize this local heating. This viewpoint has not yet been verified.

Two alternate temperature-sensing methods that work well in air environments are infrared imaging of the specimen and application of temperature-sensitive paint to the specimen surface. These methods provide only surface temperature information. The subsurface temperature is hotter. In liquid environments, the alternate temperature measurements will not work, and thermocouple data may be the only possible method. In liquid environments, particularly water, temperature control is somewhat easier because the liquid acts as a coolant and moderates the temperature excursion. Generally, the liquid is controlled to some temperature below the desired temperature so that the excess heat generated from testing will be dissipated in the liquid. For elevated-temperature testing, the furnace temperature must be controlled so that the heat generated in the specimen helps achieve the desired temperature.

A forced-air cooling system with an infrared temperature monitoring system is shown in Fig. 11. Air jets are aimed at the strain antinode. With an air line pressure of 0.5 MPa (72.5 psi) and a venturi-type air cooler in the line, an air jet exit temperature of 0 °C (32 °F) can be obtained. A liquid cooling system requires that the liquid flow over the surface of the specimen.

The major difference in ultrasonic fatigue testing compared to conventional fatigue testing is that the specimen should not be totally immersed in a bath of coolant. Immersing the specimen could prevent the system from obtaining constant-amplitude resonance. If resonance can be obtained while immersed, cavitation erosion damage may occur at fillets and ends of the specimen. This can be prevented by the application of a thin film of rubber compound, such as carboline neoprene adhesive, to these areas of the specimen. A typical liquid cooling system is shown in Fig. 12. In this system, the liquid flows onto the specimen and drains off to be recirculated. This system can also be used for environmental testing.

In a liquid cooling system, an inert coolant is necessary to achieve baseline fatigue property data approaching those of air tests. The coolant must not be corrosive to the material and must have a heat capacity large enough to remove the heat from the specimen. Deionized, low-oxygen-content water is a very

Fig. 11 Test facility (20 kHz) showing positioning of forced-air cooling, infrared temperature monitor, and external load frame for mean load

1, Converter; 2, booster horn; 3, connecting horn; 4, specimen; 5, capacitance gage; 6, cooling ring; 7, four air inlets; 8, venturi air cooler; 9, air supply; 10, upper and lower support plates; 11, hydraulic pistons; 12, window; 13, infrared camera

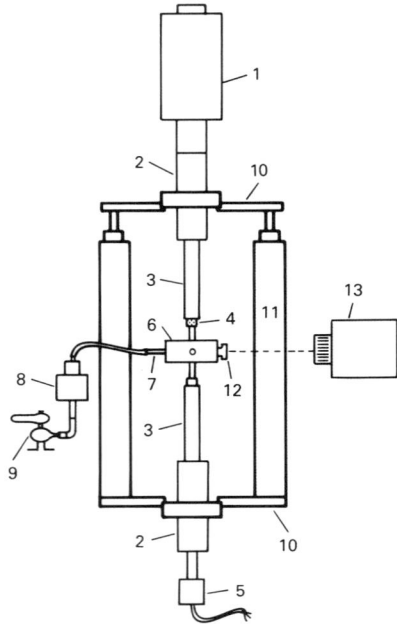

Fig. 12 Environment supply system for liquid cooling or corrosion fatigue testing

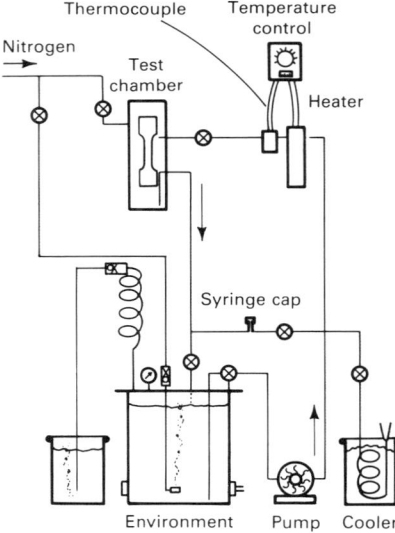

high heat capacity, minimally corrosive coolant for most materials. It is a little more difficult to use low-oxygen deionized water as a coolant, because an environmental system is necessary to control the gaseous traces in the water. Acid-free transformer oil is another coolant frequently used for ultrasonic fatigue tests. Liquid cryogen coolants, such as liquid nitrogen, generally are less effective, due to their tendency to vaporize on contact with the specimen. The vapor forms a boundary layer at the specimen surface, which reduces effective cooling instead of increasing it. Fatigue property data obtained with a nonaggressive coolant generally are slightly lower than data obtained with air cooling. Baseline data obtained with an inert coolant are usually necessary for the interpretation of corrosion fatigue data.

External Load Frame. The wave train arrangement can be used without further attachment for completely reversed tension-compression testing. The wave train can also be placed in an external load frame, such as a tensile test machine, to provide static mean loading or superposition of large-amplitude low-frequency cycling on top of the high-

frequency cycling (Ref 18, 32). The external load frame is attached to the wave train at the displacement nodes on acoustic horns on either side of the specimen, as shown in Fig. 11.

Design of specimens to be subject to superimposed external loads must take two additional factors into account. First, the elongation of the specimen due to external straining must be considered to stay within the bounds of the resonance conditions. Second, there have been observations of softening of metals during simultaneous tensile or compressive mechanical deformation and high-frequency straining, which is known as the Blaha effect (Ref 33). These mechanisms are discussed in more detail in Ref 32. Test engineers should be aware of this effect because it can result in additional plastic deformation of the material during testing.

Environmental Fatigue. Ultrasonic fatigue testing can be performed under most environmental conditions. One possible exception is vacuum, in which testing is narrowed to a few very low damping materials at low stress levels. Environmental testing requires the normal ultrasonic fatigue testing apparatus with the addition of an environmental chamber around the test specimen and an environmental supply system.

Elevated Temperatures. For high-temperature testing, the furnace serves as the environmental chamber and supply system. A high-temperature test stand with mean load capability can be constructed by placing a furnace around the specimen in the test system shown in Fig. 11. Some tuning generally is required to design extension horns that will

be resonant in the temperature gradient from the furnace midpoint to the ambient temperature outside the furnace, because the resonant frequency is temperature dependent. For high-temperature fatigue, the specimen displacement amplitude usually will have to be calibrated to a displacement antinode outside the furnace (Ref 34).

Aggressive Liquid Environments. Testing in corrosive liquid environments requires both an environmental chamber and an environmental recirculation system. This is essentially the same equipment needed for inert liquid cooling, as shown in Fig. 12. Additional features are incorporated into the environmental recirculation system to provide control of the solution composition, purity, and temperature. Ports are incorporated so that environment composition samples can be taken for documentation and solution pH can be adjusted. An inert gas overpressure is usually maintained throughout the system to control the dissolved oxygen content of the test solution. Appropriate plumbing and seals are incorporated so that the specimen chamber can be purged of air prior to circulation of the environment.

A controlled-corrosion fatigue chamber is shown in Fig. 13. This chamber exhibits features needed for electrochemical corrosion fatigue study (Ref 20). The chamber is composed of an outer chamber and an inner chamber constructed of Teflon. The inner chamber contains a finite volume of liquid around the gage section of the specimen. A platinum electrode is fitted into the inner chamber for anodic or cathodic polarization of the specimen. A window is placed at the side of the inner chamber to enable viewing of the amount of liquid in the inner chamber. A port is placed in the front of the inner

Fig. 13 Environmental fatigue test chamber for electrochemical 20-kHz testing

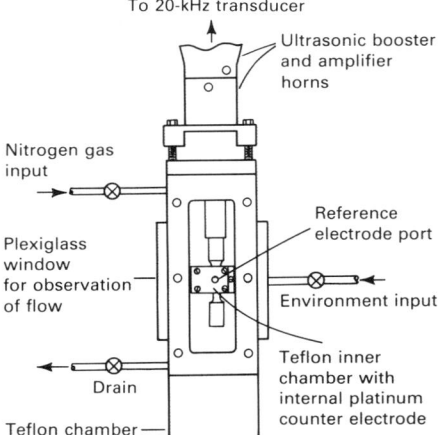

chamber so that a standard reference electrode can be inserted to measure the electrochemical potential of the specimen. The electrodes can be removed for normal corrosion fatigue testing.

Proper selection of horn material is important in corrosion fatigue testing, because a mismatched specimen and horn material combination may set up a galvanic couple when the joint is wetted with a conductive solution. Depending on the galvanic couple, the horn material may cause the gage of the specimen to be electrochemically more active or passive than normal. This could have a pronounced effect on the corrosion fatigue properties that are being determined. It is advisable to make the specimen and the extension horn out of the same material. If the horn and specimen must be made out of dissimilar metals, care must be taken to ensure that the joint is not exposed to the conductive test solution.

In fatigue crack growth testing in liquid environments, the effect of the liquid inside the growing fatigue crack also must be considered. The effect of the liquid fatigue crack growth rate is currently being investigated. Depending on the fluid properties, the fatigue crack may be wedged open, causing errors in the da/dN and threshold stress intensity range, ΔK_{th}, values. An equally important question is whether the environment ever extends to the crack tip. It has been suggested that at high crack growth rates, the environment has little influence on rapid crack growth, and the crack tip behaves as if it were in vacuum (Ref 35).

Test Specimens

Ultrasonic fatigue test specimens must be designed to resonate at the desired test frequency. The first step in designing an ultrasonic fatigue test specimen, acoustic horn, or resonant bar is to obtain the appropriate properties and constants of the materials. The material density, dynamic modulus of elasticity, and the half wavelength of sound in the material at the desired testing frequency must be determined.

Frequency, Wavelength, and Speed of Sound. The longitudinal resonance frequency of a bar test material is measured experimentally, as shown in Fig. 14. A uniform bar of test material is excited by a small converter coupled to a variable-frequency oscillator. The converter can be an electrodynamic vibrator or any other vibrator capable of ultrasonic frequencies. A piezoelectric pickup is placed against the opposite end of the bar to monitor the amplitude of vibration. For most pure metals and alloys, the bar should be about 100 to 150 mm (4 to 6 in.) long and 4 to 10 mm (0.16 to 0.4 in.) in

diameter. This diameter is comparable to the diameters of most test specimen gages.

The bar should be similar in size to the eventual specimen gage diameter, because measured resonant frequencies vary with large diameter bars. This directly affects calculation of the dynamic elastic modulus, and ultimately affects calculation of the fatigue stress amplitude. Frequencies on the order of tens of kilohertz should be measured for a bar length in this given range. Sweeping the oscillator through the frequency spectrum produces a large increase in output for some frequency; this is the resonance frequency. The length of the bar is $\lambda/2$ for the experimentally determined resonance frequency. Resonant wavelengths for several pure metals at test frequencies of 20 kHz and 2 MHz are given in Table 3. The resonant wavelength is inversely proportional to frequency. The need for a macroscopic test specimen frequently precludes high-frequency testing in the 2-MHz range. The speed of sound through the material can be calculated by relating the speed of sound, C, to frequency, f, and wavelength, λ:

$$C = \lambda_1 f_1 = \lambda_2 f_2 \qquad \text{(Eq 6)}$$

The resonance frequency of a bar of arbitrary length, as determined in the above experiment, may not be the driving fre-

quency of the converter that will be used for testing. The appropriate wavelength for testing frequency is calculated by assuming that the speed of sound through the material remains constant for a given material and temperature.

The dynamic modulus of elasticity, or dynamic Young's modulus, (E), can be determined by combining Eq 1 and 6 to obtain:

$$E = \rho(\lambda f)^2 \qquad \text{(Eq 7)}$$

The dynamic modulus differs from the static modulus (relaxed modulus) obtained from a tensile test. The static modulus is inadequate for converting the strain amplitude to stress amplitude, because it will include anelastic contributions to strain that are absent at ultrasonic frequency. Use of the static modulus gives stress estimates that are too low, because the static modulus is typically less than the dynamic modulus.

Stress-Life Specimen Design. Several specimen designs for ultrasonic fatigue stress-life testing are shown in Fig. 15. Although the uniform bar is the most easily produced specimen, it generally is not employed in testing. The stress concentration due to the screw threads used for gripping causes the local stress in the thread to exceed the maximum stress produced at the center of

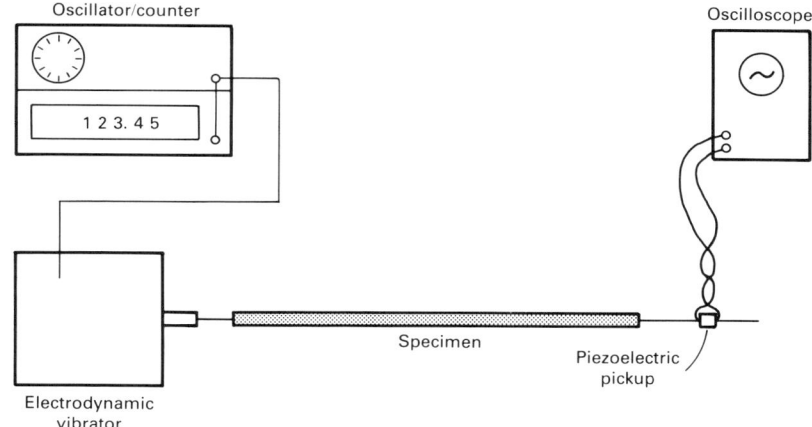

Fig. 14 Experimental measurement of the longitudinal resonance frequency of specimens

Table 3 Resonant specimen lengths for several pure metals

Metal	Young's modulus GPa	Young's modulus psi × 10⁶	Resonant specimen length, λ/2, at: 20 kHz mm	20 kHz in.	2 MHz mm	2 MHz in.
Al	62	9.0	120	4.7	1.20	0.047
Cu	110	15.9	88	3.5	0.88	0.035
Ti	116	16.8	127	5.0	1.27	0.050
Fe	196	28.4	124	4.9	1.24	0.049
Ni	207	30.0	121	4.7	1.21	0.047

Fig. 15 Profiles of specimen designs for ultrasonic fatigue

References cited provide mathematical analyses required to compute stress and strains.

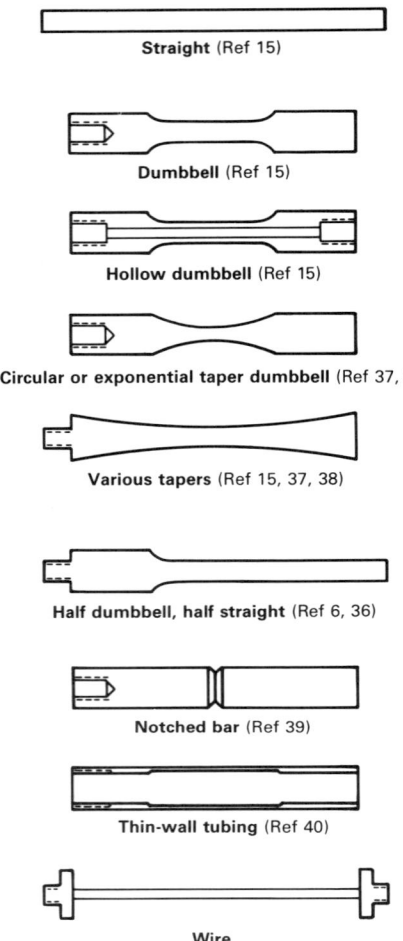

Straight (Ref 15)

Dumbbell (Ref 15)

Hollow dumbbell (Ref 15)

Circular or exponential taper dumbbell (Ref 37, 38)

Various tapers (Ref 15, 37, 38)

Half dumbbell, half straight (Ref 6, 36)

Notched bar (Ref 39)

Thin-wall tubing (Ref 40)

Wire

the specimen. Failure in the screw threads rather than in the center of the specimen would result.

Strain amplification is desirable at the center of the specimen. This is accomplished by reducing the cross-sectional area of the gage section to produce a dumbbell-shaped specimen. By reducing the cross-sectional area, the total power needed to drive the specimen into resonance is reduced. Consequently, the amount of heat produced in the specimen is reduced, which also reduces the cooling requirements.

The choice of specimen design for a particular test depends on many factors, including the amount of material available, maximum amplitude of the wave train, amount and type of cooling available or allowable, minimum diameter to ensure stiffness and eliminate flexural modes, and desired strain level in the gage section. The calculation of

the appropriate resonant specimen geometry is the primary task in designing an ultrasonic fatigue specimen. Other considerations for ultrasonic fatigue specimens are the same as those encountered in conventional fatigue testing, including surface finish, minimization of residual stresses, capabilities of the machine shop, and costs of producing the specimen.

Complex Specimen Geometries. More complicated expressions are needed to determine the dimensions and strain amplitude at the point of maximum strain for specimens with nonuniform geometries. A dumbbell-type specimen is resonant at frequency f if Eq 7 and the following equation are satisfied simultaneously (Ref 15):

$$\left(\frac{d_1}{d_2}\right)^2 = \cot\left(\frac{2\pi L_1}{\lambda}\right) \cdot \cot\left(\frac{2\pi L_2}{\lambda}\right) \quad \text{(Eq 8)}$$

The variable dimensions L_1, L_2, d_1, and d_2 are shown in Fig. 16. Equations 7 and 8 are the basic design equations for an ideal dumbbell specimen. Using this representation, any three dimensions can be selected; the fourth can be calculated for a given wavelength. The maximum elastic strain on the gage length of the dumbbell specimen shown in Fig. 16 is given by (Ref 15):

$$\epsilon_{max} = kA_o \cdot \left[\frac{\cos(kL_1)}{\sin(kL_2)}\right] \quad \text{(Eq 9)}$$

where k is $2\pi/\lambda$ and A_o is displacement amplitude at the end of the dumbbell. Comparing Eq 9 with the maximum strain obtained in a uniform bar shows that the term in square brackets is the magnitude of the amplification of strain amplitude produced by the dumbbell shape. The term in square brackets is often referred to as the strain amplification factor.

Direct measurement of the strain profile from a test specimen has been reported (Ref 41). A contacting probe aids in measuring the displacement and strain distribution for specimens with complex geometries. The displacement and strain amplitudes along the length of a circular, tapered dumbbell specimen are shown in Fig. 17. The amplification of the strain amplitude due to the dumbbell shape is clearly indicated.

If the ideal dumbbell calculations are used, small adjustments to the overall length of the specimen may be necessary during specimen design to achieve the desired test frequency. For example, if a hole is tapped in one end of the specimen to attach the specimen to the horn, the equivalent mass of material removed for the hole must be replaced in the form of extra length of that dumbbell head. If the attachment stud is the same density as the specimen, then no adjustments are required. However, if a steel stud is used to hold an

Fig. 16 Ideal dumbbell specimen dimensions

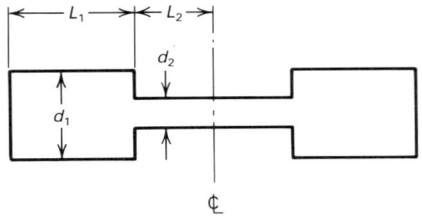

Fig. 17 Distribution of displacement and strain amplitudes obtained from a dumbbell-shaped Ti-6Al-4V specimen at 14.2 kHz

Source: Ref 41

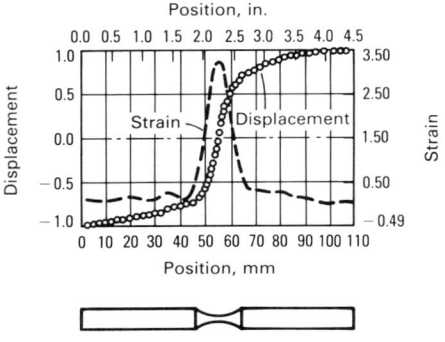

aluminum specimen, a mass adjustment to the length of the dumbbell head will be needed. Similar tuning considerations should be made when placing fillets between the dumbbell heads and the gage section.

Fillets and Radii. Efficient propagation of acoustic energy along the length of a dumbbell specimen requires that a smooth transition be provided between the heads and the gage section of the specimen. For some specimen designs such as a circular or exponential tapered dumbbell, this smooth transition is the major design element. Specimen designs that use the Neppiras formula (Eq 7 and 8) for an ideal dumbbell must provide a fillet at the transition from head to gage section. Constant-radius fillets are easily machined and can be used if other alternatives are not feasible.

Depending on the choice of radius, some heat will be generated at the fillet, because a constant radius fillet does not provide optimal transition in particle velocity. The recommended fillet design is the baud streamline fillet (Ref 42). This design has a continuously varying fillet radius. Its shape is like that of a nonturbulent stream of water as it drains out of a tank with a circular hole. It is highly efficient in providing a smooth transition of particle velocity. This is quite useful for testing high-damping or precipitation-hardened materials. One disadvantage of the baud streamline design is that it requires a tape- or computer-controlled lathe to produce the desired profile.

Notched Bar Specimens. A notched specimen is a special condition of a dumbbell specimen, where L_2 is very small and L_1 is close to $\lambda/4$ to maintain resonance. For a bar containing a narrow notch, the maximum strain in the notch (exclusive of stress concentration factors) is the product of the maximum strain at $\lambda/4$ of a uniform bar without the notch multiplied by the area ratio $(d_1/d_2)^2$ (Ref 39). The L and d dimensions are defined in the ideal dumbbell specimen shown in Fig. 16.

Finite element analysis of the notched bar specimen design shows that, despite differences in the stress distribution along the specimen length between static- and dynamic-loaded specimens, the stress distribution and configuration in the notch region are the same. The complete equation for the maximum stress, σ_{max}, in a notched resonant member can be calculated by multiplying the maximum strain, $kA_o \cdot (d_1/d_2)^2$, by the dynamic modulus, E, and the stress concentration factor, K_t', as:

$$\sigma_{max} = E \, kA_o \cdot \left(\frac{d_1}{d_2}\right)^2 \cdot K_t' \qquad \text{(Eq 10)}$$

where K_t' is the von Mises stress concentration factor. The von Mises stress concentration factor is used instead of K_t for analyzing high-cycle fatigue results (Ref 43). This is based on the findings that high-cycle fatigue failure is dictated by the alternating von Mises stress, where K_t' is approximately 10% less than K_t. Definitions of additional parameters relating to notch fatigue, including notch bar fatigue strength and notch sensitivity, are found in Ref 39.

Design of resonant fatigue crack growth rate specimens is based on physical concepts similar to stress-life specimen design. The major difference is that a sharp crack is introduced into the design at the point of maximum strain. Therefore, the relationships developed for purely elastic deformation are not exactly fulfilled when appreciable plastic deformation occurs at the crack tip, when changes occur in Young's modulus due to localized plastic deformation, or when the specimen is detuned by the growing crack. Discussion of these issues can be found in Ref 44.

The first ultrasonic fatigue crack growth test specimen was a simple resonant bar with an electrodischarge machined slot cut into one side of the bar (Ref 45). Typical geometries of fatigue crack growth specimens are shown in Fig. 18, including single-edge

cracked specimens (Ref 19), double-edge cracked specimens (Ref 46) with axial loading, center-cracked specimens with axial loading (Ref 47, 48), and single-edge crack specimens with transverse loading (Ref 49). Crack length, a, is shown.

Fatigue crack growth test specimen length is controlled by the test frequency, as in the stress-life case. However, the cross-sectional dimensions selected may vary considerably. The current trend in specimen design is to incorporate the relevant criteria of conventional frequency fatigue crack growth and fracture mechanics into the design. One factor pertains to specimen thickness, d, which should be large in comparison to the plastic zone size at the applied stress intensity range. This is consistent with the pertinent ASTM recommendation:

$$d \geq 2.5 \left(\frac{\Delta K_{max}}{\sigma_y}\right)^2 \qquad \text{(Eq 11)}$$

where ΔK_{max} is the maximum stress intensity range, and σ_y is the yield strength of the material. The exception to this rule arises when materials have high damping and produce large quantities of heat. In this case, the specimen must be thin enough to ensure adequate cooling. The specimen also must be thick enough to suppress other modes of vibration such as flexural oscillations.

The dimension of the starting notch and the permissible fractional length of fatigue crack extension also influences the design of the specimen width, b. For center-cracked specimens, the crack advance should be limited to a/b ratios of less than 0.4 (Ref 48). For specimens that are wide in comparison to the crack length, flexural contributions due to lateral displacement become significant.

Specimens tested under superimposed static loads (higher R ratios) require regions with cross sections larger than the gage section in order to transmit the required tensile load. The specimen shown in Fig. 18(e) is similar to that specified for conventional fatigue crack growth tests (Ref 50). Currently, specimen choice is equivocal. Standardization of a test specimen will be determined by the ability of a specimen to provide the necessary da/dN and ΔK data. Research is focusing on providing accurate stress intensity values under resonant conditions.

Stress Intensity Concepts. Design of fatigue crack growth rate specimens is useful only if the stress intensity at the crack tip can be calculated. In the simplest method of computing stress intensity range, the nominal stress amplitude is obtained from the displacement amplitude, and the ΔK value is calculated as:

$$\Delta K = \Delta\sigma\sqrt{a\pi} \cdot F_1\left(\frac{a}{b}\right) \qquad \text{(Eq 12)}$$

Fig. 18 Specimen geometries for crack growth measurements under high-frequency resonance excitation

(a) Center-cracked specimen. (b) Single-edge cracked specimen. (c) Double-edge cracked specimen. (d) Single-edge cracked specimen. (e) Center-cracked specimen. R, fatigue stress ratio; a, crack length; b, specimen width; t, specimen thickness. Source: Ref 44

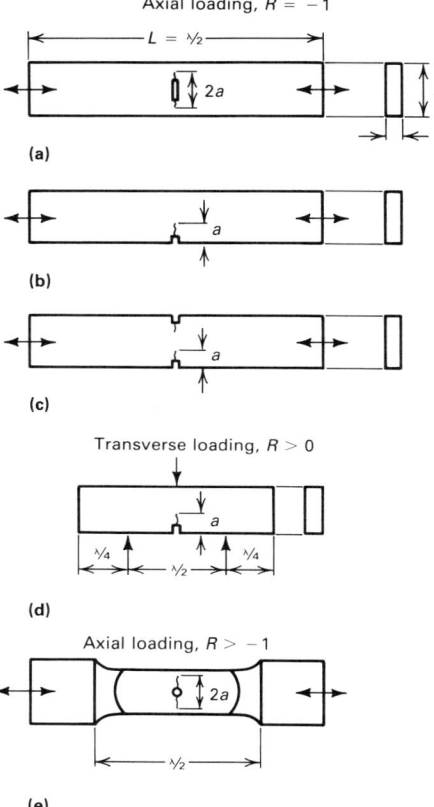

where $F_1(a/b)$ is a correction factor to account for the presence of a crack in the finite width of the specimen. This method does not account for the effects of a growing crack on the resonance behavior of the system. It assumes that the relationship between displacement and strain amplitude remains constant with detuning of the specimen due to crack growth. This problem can be avoided by measuring the strain directly with strain gages placed in the crack plane. Additional correction factors to the above stress intensity formulism (Eq 12) have been added to account for the growth of the crack by normalizing the measured effective stress to the initial cross section (Ref 51).

A dynamic correction factor for ΔK also has been determined with finite element analysis. This results in a factor that is a function of the ratio of crack length to specimen width, specimen width to length, and the instantaneous frequency.

A summary of frequently used correction formulas to calculate ΔK is given in Table 4. Further study must be undertaken to develop a standardized computation procedure for ΔK in resonant specimens.

Specimen Grips. The requirements for gripping an ultrasonic fatigue specimen are minimal. The only gripping requirements in ultrasonic fatigue are maintenance of intimate contact between the specimen and horn to allow good acoustic coupling and the absence of external forces that would disturb the resonance of the remainder of the wave train. Consequently, a fatigue specimen with a 5-mm (0.2-in.) gage can be held in place by a single 6.4-mm (0.25-in.) stud while being fatigued to failure at a stress amplitude of 485 MPa (70 ksi) or more.

Generally, gripping is accomplished by an internal thread arrangement. This arrangement is adequate, even with a specimen that is difficult to grip, such as thin-walled tubing. Gripping can be accomplished by an external screw-down collar, which grips the specimen to an extension horn. This is particularly useful for tests at high mean loads. Gripping can also be accomplished by brazing or welding the specimen to a threaded adapter or directly to an acoustic horn. A holder for wire specimens has been developed that allows multiple specimens to be tested in a batch mode (Ref 53). An interference fit of the specimen with the horn also is satisfactory, as long as good acoustic coupling with the horn is obtained. Attachment of the specimen to the acoustic horn may be accomplished using an adhesive. This is appropriate for fatigue testing of brittle materials like glass and ceramics.

Most ultrasonic converter cases are grounded. If metal-to-metal contact is maintained throughout the wave train, the specimen will be grounded also. An electrically floating specimen is needed if electrochemical potential or corrosion current is to be measured during testing. The specimen can be isolated from the horn by using an extension horn made of a nonconducting material such as Lucite to grip the specimen. Such grips limit the magnitude of the stress amplitude that can be transmitted to the specimen because of high dissipation of energy or low fatigue strength of these materials.

Applications

Ultrasonic fatigue testing has been applied successfully to many situations that require fatigue initiation and crack growth data. Testing has been performed under a variety of loading conditions, specimen geometries,

Table 4 Computation of stress intensity ranges

Specimen geometry (loading mode)	Equation	Ref
Side notch in octagonal bar (longitudinal)	$\Delta K = 1.95 \cdot \Delta\sigma \cdot f\left(\dfrac{a}{b}\right) \cdot \sqrt{\dfrac{a}{Q}}$	52
Center-notch flat bar (longitudinal) (a)	$\Delta K = \dfrac{U_o}{W} \cdot E\sqrt{a} \cdot f_3\left(\dfrac{a}{W}, \dfrac{W}{L}, v\right)$	48
Center-notch flat bar (longitudinal), $a/w < 0.5$	$\Delta K = \epsilon_t \cdot E\sqrt{a}\ \sqrt{\pi} \cdot f\left(\dfrac{a}{b}\right) \cdot f\left(\dfrac{b-a}{b}\right)$	51
Double-side-notch flat bar (longitudinal), $a/w < 0.4$	$\Delta K = \epsilon_t \cdot E\sqrt{a}\ \sqrt{\pi} \cdot f\left(\dfrac{a}{b}\right) \cdot f\left(\dfrac{b-a}{b}\right)$	46
Single-edge-notch thin, flat strip (longitudinal), $a/w < 0.6$	$\Delta K = \Delta\sigma \cdot \sqrt{a}\ \sqrt{\pi} \cdot f\left(\dfrac{a}{b}\right)$	19
Single-edge-notch rectangular bar (transverse) (a)	$\Delta K = \dfrac{U_o \cdot E}{W}\ \sqrt{a} \cdot f_2\left(\dfrac{a}{W}, \dfrac{W}{L}, v\right)$	48

Note: The definition of the symbols for each of the above expressions can be found in the original references.
(a) by finite element analysis
Source: Ref 44

and environmental constraints. With perhaps the exception of plastic strain-controlled testing and single-cycle hysteresis testing, ultrasonic fatigue techniques can be readily applied to the problems of fatigue that traditionally have been confronted at lower frequencies.

Stress-Life Results. The simplest ultrasonic fatigue test is a constant-amplitude, tension-compression test. Figure 19 shows a 20-kHz stress-life (S-N) evaluation of the fatigue limit of an aluminum alloy in two grain sizes (Ref 54). The ultrasonic fatigue test results clearly show the differences in fatigue

properties for the conditions of large and small grain size. As expected, the smaller grain size has a higher fatigue limit. The rapid accumulation of cycles allows evaluation of many specimens for routine statistical evaluation of the S-N behavior of the material. Note that the replicate data at a single strain level are spread over a decade of cycles at 10^7 cycles. This spread increases as the curve progresses to the lower strain-amplitude levels. This is related to the probability of fatigue fracture in the very-high-cycle fatigue regime, not to a problem of reproducibility with ultrasonic fatigue testing.

Fig. 19 Results of 20-kHz fatigue tests of polycrystalline aluminum with different grain sizes

Fine grain size specimen, 40 μm. Coarse grain size specimen, 250 μm. Source: Ref 54

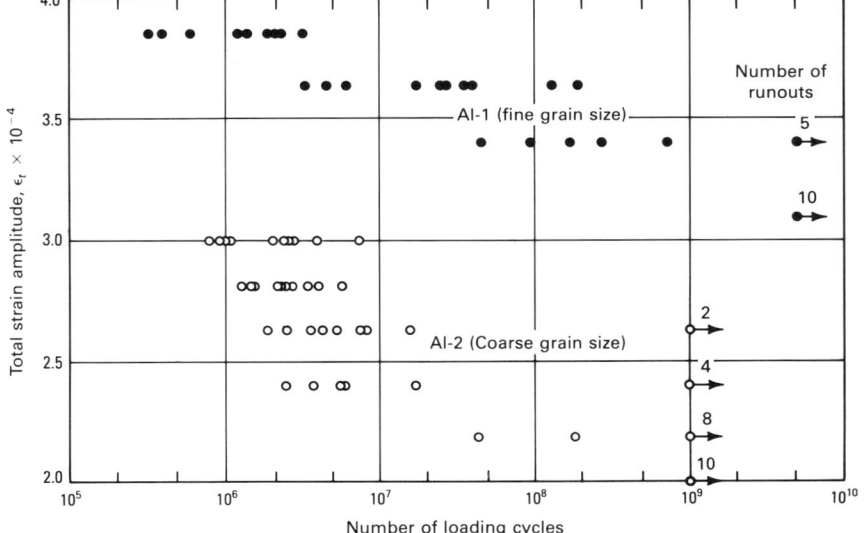

Probability of fatigue fracture can be calculated easily with data from replicate tests at a single stress or strain level (Ref 55). Fracture probability, *p*, can be computed as a function of the number of loading cycles:

$$p = \frac{i}{(n + 1)} \tag{Eq 13}$$

where *n* is the number of specimens tested, and *i* is the number order of failed specimens ranked according to increasing numbers of cycles to failure. Plotting the probability value versus the cycles to failure on probability paper for different strain levels allows the isobars for probability of fatigue fracture to be read from the graph.

The 10, 50, and 90% probability of fatigue fracture isobars are shown in Fig. 20, superimposed over a 20-kHz, *S-N* data set for a recrystallized P/M molybdenum specimen (Ref 56). Data for test temperatures of 25, 150, and 250 °C (75, 300, and 480 °F) are shown. The data show the decrease in fatigue strength of pure metals at elevated temperatures that would be expected at conventional frequencies.

Higher fatigue limits are observed for body-centered cubic (bcc) molybdenum at high frequency than at lower frequency (Ref 56). This difference is a result of bcc material deformation behavior (see the next section of this article). For applications operating at high frequency, the data can be used directly. For applications operating at lower frequency, the data should be used in conjunction with lower-frequency fatigue data. Cur-

rently, there are no empirical models or equations permitting calculation of the extent of shift of *S-N* curves resulting from testing at higher frequencies.

Ultrasonic fatigue data for face-centered cubic (fcc) materials are similar to data produced by other fatigue test methods. Figure 21 shows the results of an iterative least squares analysis of stress versus cycle to failure data for electrolytic tough pitch (ETP) copper as a function of four fatigue test methods (Ref 57). The data base included a total of 207 data points for the axial stress-control curve, 159 points for the plane-bending curve, 90 points for the rotating-bending curve, and 62 points for the ultrasonic fatigue curve. The rotating and flexure data were corrected for plasticity occurring at the outer fibers so that they could be compared to the axial data. Ultrasonic fatigue data give nominally the same results as do other test methods for high-purity copper. Also, ultrasonic fatigue testing enables testing to much higher cycles than the other three test methods illustrated in Fig. 21.

Microstructurally, fatigue deformation for copper appears to be the same at high frequency as it is at low frequency. Dislocation substructures of fatigued ETP copper are similar for constant stress amplitude at high and low frequencies (Ref 57). Formation of persistent slip bands also is observed at high frequency in copper.

The *S-N* approach in fatigue is most frequently used on an engineering basis to rate the fatigue performance of one material ver-

Fig. 21 Fatigue life as a function of stress amplitude and test method

From iterative least squares curve fit of data on annealed electrolytic tough pitch (ETP) copper. Source: Ref 57

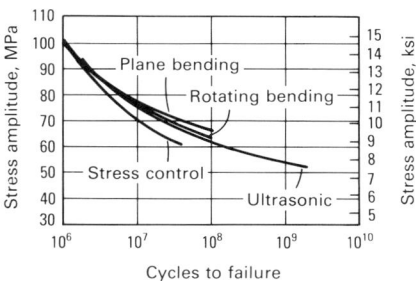

sus another to provide information for material substitution questions. Fatigue performance could be required in a nominally pure environment or a substantially corrosive one.

Figure 22 compares fatigue performance of three turbine blade alloys, AISI 403 and 17-4PH stainless steel and β-quenched Ti-6Al-4V, at ultrasonic and conventional frequencies (Ref 13). The ranking of the three materials is the same at both frequencies and remains the same whether the 10^7 or 10^9 cycle fatigue limit is used for comparison. A shift in the high-frequency fatigue limit to higher numbers of cycles to failure and higher stresses is not observed for these materials. The combined high and low frequency data appear to lie on a continuous band defining a fatigue limit over a wide range of cycles.

Corrosion Fatigue. In aggressive environments, the corrosion fatigue limit often decreases with increasing cycles, so an understanding of 10^9 and higher cycle fatigue limits is necessary for conservative design. Figure 23 compares the 10^9 cycle fatigue limit of type 403 stainless steel, type 17-4PH stainless steel, and Ti-6Al-4V in mild to very aggressive environments. More information on the environments used can be found in Ref 58. Degradation in fatigue properties due to corrosion fatigue interaction is obvious by comparing pure water with any of the environmental fatigue limits. For example, the 10^9 cycle fatigue limit of type 403 stainless steel in air-saturated 22% sodium chloride solution is one eighth of the fatigue limit in pure water. Varying degrees of degradation of the 10^9 cycle fatigue limit are observed for the other combinations of materials and environments.

As discussed above, the fatigue limit at ultrasonic frequency for some engineering alloys exhibits sensitivity to chemical and ionic species in solution similar to that ob-

Fig. 20 Fatigue fracture probability curves

Obtained from statistical evaluation of fatigue strength data from a recrystallized P/M molybdenum alloy tested at three temperatures. Source: Ref 56

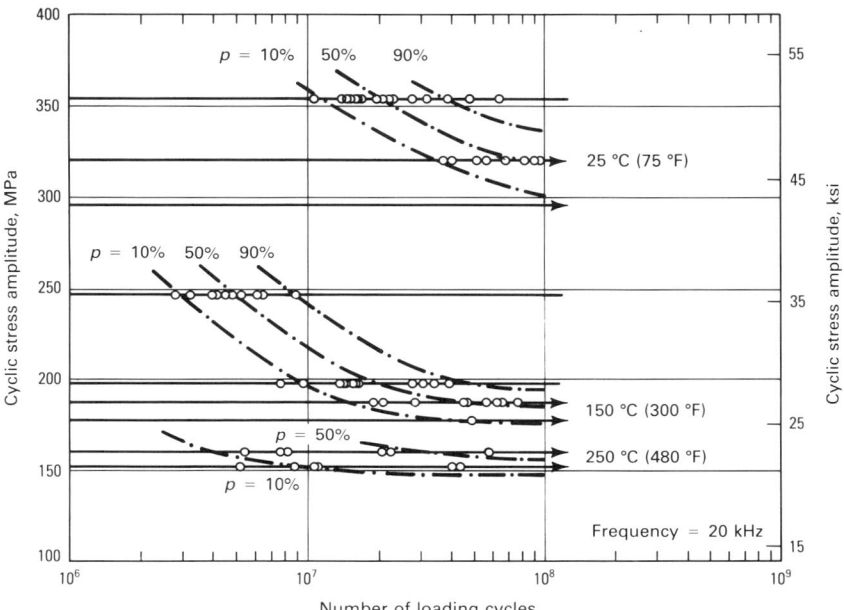

Fig. 22 Ranking of fatigue strength of three engineering alloys tested at ultrasonic and conventional frequencies

Environment: high-purity water. Temperature: 80 °C (175 °F). <20 ppb O_2; $R = -1$. Source: Ref 13

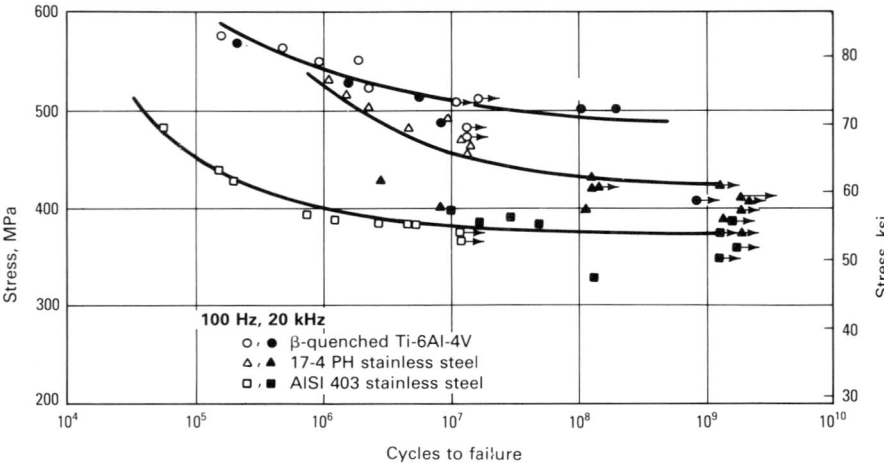

Fig. 23 Effect of various test environments on the 10^9 fatigue strength of three engineering alloys

*Extrapolation from 10^8 cycles. Fatigue strength of Ti-6Al-4V in this environment is meaningless because the material dissolves. L, low; H, high. Source: Ref 58

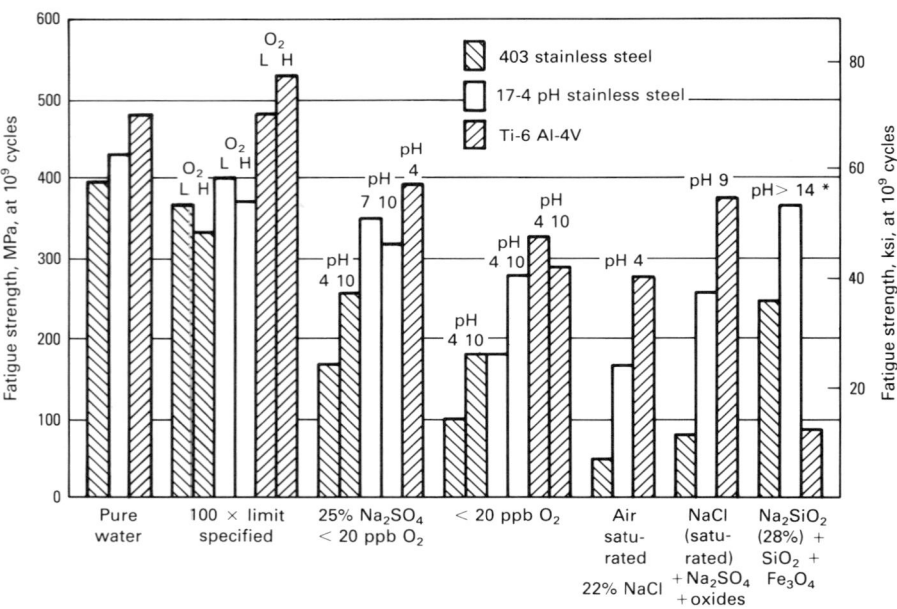

Fig. 24 Effect of cathodic polarization on ultrasonic and conventional-frequency corrosion fatigue performance of AISI 403 stainless steel in aqueous chloride solution

22% NaCl; pH, 7. $R = -1$. Open data points, 40 Hz; closed data points, 20 kHz. Source: Ref 20

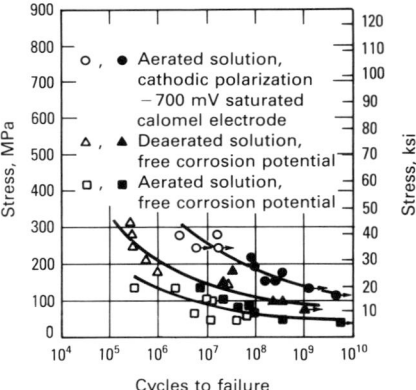

served at lower frequencies. An equivalent statement can be made concerning the imposition of electrochemical potential during fatigue testing. Figure 24 shows that cathodic polarization of a type 403 stainless steel specimen in a chloride solution has essentially the same effect on the 20-kHz corrosion fatigue data as it does on 40-Hz data (Ref 20). Pitting of the cyclically stressed gage section of type 403 stainless steel tested in chloride environments is observed at both 20 kHz and 40 Hz.

The observation of corrosion fatigue interactions at 20 kHz poses several questions about the corrosion fatigue mechanism, because the cycle period is on the order of a few tens of microseconds. A slip dissolution model that accounts for the 20-kHz corrosion fatigue behavior of type 403 stainless steel in aqueous chloride solutions is presented in Ref 20.

High Temperatures and Mean Load. Corrosion fatigue interactions at 20 kHz are not limited to aqueous environments. Figure

25 shows the high-cycle corrosion fatigue behavior of a nickel-based superalloy, Udimet 720, at 704 °C (1300 °F) at an *A* ratio ($\sigma_{alt}/\sigma_{mean}$) of 0.67 and subject to a molten potassium sulfate plus sodium chloride salt (Ref 34). The correlation between ultrasonic and conventional frequency testing is very good, because Udimet 720 deforms in a planar slip mode up to temperatures around 760 °C (1400 °F). Once diffuse slip occurs, the fatigue limits at low and high frequencies are expected to diverge due to strain rate effects.

Creep-fatigue interactions can occur under mean load and high temperatures. These will be more significant at low test frequencies. High strain rate cycling does not provide the time needed for creep to occur. A comparison of high- and low-frequency fatigue tests under creep conditions would show that fatigue test data are dependent on frequency, mean load, and temperature (Ref 59).

Notch Bar. Fatigue tests at 20 kHz using notch bar specimens of engineering materials have shown good agreement with lower frequency testing. Figure 26 shows the effect of a notch of K'_t equal to 2.41 on the smooth bar fatigue properties of 17-4PH stainless steel. K'_t was calculated by the methods given in Eq 10 (Ref 39).

Fatigue Crack Growth Results. A full compilation of data from threshold testing at ultrasonic frequencies may be found in Ref 44. Some of the values of ΔK_{th} and the corresponding minimum crack growth rate are presented in Table 5 for several pure metal and alloy systems, along with test conditions. The minimum crack growth rates obtained at ultrasonic frequency are decades

Fig. 25 Corrosion fatigue of Udimet 720 in a fused salt environment at 704 °C (1300 °F) in the presence of air
The salt contained 60 mol% sodium sulfate, 40 mol% magnesium sulfate, and 1 wt% sodium chloride. Source: Ref 34

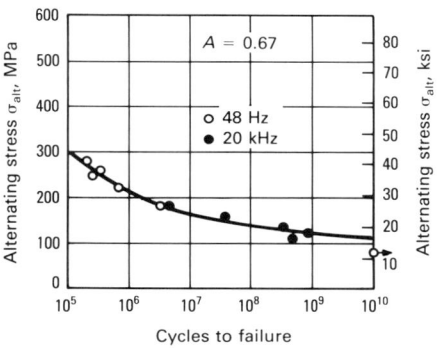

Fig. 26 Notch and plain bar fatigue properties of 17-4PH stainless steel
Tested at 100 Hz (○) and 20 kHz (●) frequencies. Source: Ref 39

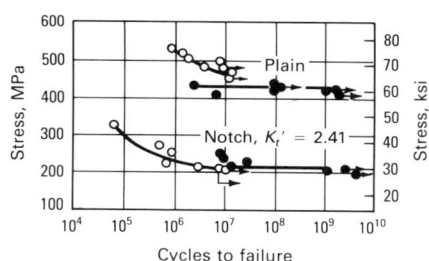

below the value of one lattice parameter per cycle. At these low crack growth rates and ΔK values, the crack tip plastic zone size is extremely small.

Figures 27 and 28 compare fatigue crack growth rate experiments at high and low frequencies. Figure 27 shows that crack growth rate data for polycrystalline copper (Ref 46) decrease almost linearly as stress intensity approaches ΔK_{th}. In this case, the crack growth rate data at high and low frequency at $R = -1$ ($\sigma_{min}/\sigma_{max}$) show similar behavior at the same ΔK levels. Figure 28 shows the fatigue crack growth rate for Hastelloy-X tested at several low frequencies and at 20 kHz (Ref 60). The 20-kHz data aid in establishing the threshold at a ΔK of 9 MPa $\sqrt{m}$ (8.2 ksi $\sqrt{in.}$). Extrapolation of lower frequency data would not indicate such a clearly defined threshold. These data were obtained at a positive R ratio. Figure 29 shows that the effect of R ratio on the threshold value of the cyclic stress intensity range varies similarly at high and low frequencies for Hastelloy-X at room temperature.

Fig. 27 Comparison of conventional and 20-kHz fatigue crack growth rate measurements in polycrystalline copper
At room temperature. Threshold stress intensity is indicated. Source: Ref 44

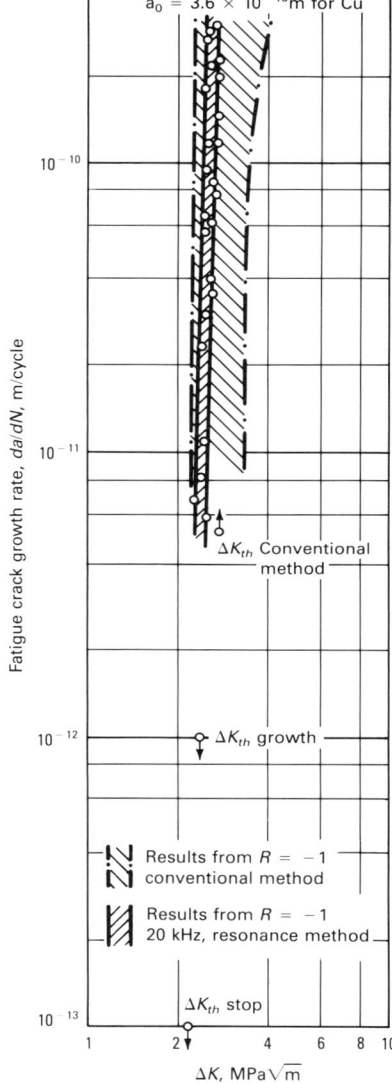

These examples show that ultrasonic fatigue testing techniques can be successfully applied to characterize fatigue properties under a wide variety of test conditions. The data show that ultrasonic fatigue data correlate well with lower frequency data in cases where the material is not particularly strain-rate sensitive. Materials that are strain-rate sensitive show a difference between high- and low-frequency tests. Environmental testing conditions may alter the strain-rate sensitivity of the material. The more that is known about the cyclic strain-rate sensitivity of a

Fig. 28 Fatigue crack growth of Hastelloy-X at room temperature as a function of cyclic stress intensity range and test frequency
Source: Ref 60

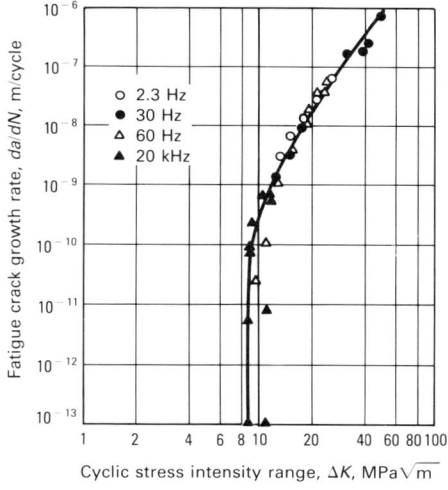

Fig. 29 Effect of mean load on the threshold value of the stress intensity range for room-temperature Hastelloy-X as a function of frequency
Threshold values were taken at a fatigue crack growth rate of 3×10^{-11} m/cycle in all cases. Source: Ref 60

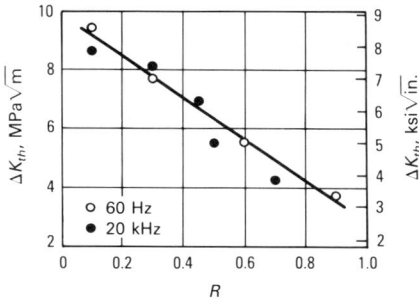

material in an environment, the easier it will be to understand the differences in high- and low-frequency fatigue properties. If proper test techniques are used and reliable data are obtained, the degree of correlation obtained between ultrasonic frequency and conventional frequency testing is a measure of the strain-rate sensitivity of the deformation process at a chosen strain amplitude and environment.

Strain-Rate-Dependent Material Behavior

The effect of frequency in ultrasonic testing can be interpreted as the effect of increas-

Table 5 Threshold stress intensity, ΔK_{th}, determined by ultrasonic resonance test methods

Material	Loading mode	R	Test conditions Environment	Temperature	Young's Modulus MPa	psi × 10⁶	ΔK_{th} at da/dN MPa $\sqrt{m}$	ksi $\sqrt{in.}$	m/cycle	ft/cycle	Remarks
Al . . . . Axial	−1	Air	20 °C (68 °F)		69 700-72 000	10.1-10.4	1.33	1.21	1×10^{-13}	3.28×10^{-13}	Grain size and work effects
Cu . . . . Axial	−1	Air	20 °C (68 °F)		122 000-126 000	17.7-18.2	1.4-2.0	1.3-1.8	1×10^{-13}	3.28×10^{-13}	Grain size effect
					126 000-126 500	17.6-18.3	1.4-2.6	1.3-2.3	1×10^{-15}	3.28×10^{-15}	Cold work effect
					138 000-184 000	20.0-26.7	1.8-2.3	1.6-2.1	1×10^{-13}	3.28×10^{-13}	Single crystals
Low-carbon steel . . . . Axial	−1	Oil	293 K		. . .		3.8	3.4	6×10^{-14}	1.97×10^{-13}	Comparison with NaCl solutions(a)
AISI 304 . . . . . Axial	−1	Oil	293 K		. . .		7.0	6.4	5×10^{-13}	1.64×10^{-12}	Comparison with NaCl solutions
X10Cr13 . . . . . Axial	−1	Oil	23 °C (73 °F)		219 600	31.8	6.7	6.1	6×10^{-14}	1.97×10^{-13}	
34CrMo4 . . . . . Axial	−1	Air	20 °C (68 °F)		196 200	28.4	2.45	2.2	1×10^{-12}	3.28×10^{-12}	
P/M Mo . . . . . . Axial	−1	Air	20 °C (68 °F)		322 000	46.7	4.8-5.2	4.4-4.7	1×10^{-13}	3.28×10^{-13}	Grain size effect
A286 . . . . Axial	−1		20 °C (68 °F)		. . .		13.0	11.8	3×10^{-12}	0.98×10^{-11}	
IN-738 . . . . Axial	−1	Air	20 °C (68 °F)		200 000	29.0	3.13	2.84	1×10^{-12}	3.28×10^{-12}	
IN-792 . . . . Axial	−1	Air	20 °C (68 °F)		206 600	30.0	4.48	4.07	1×10^{-12}	3.28×10^{-12}	
IN-600 . . . . Transverse	0.3	Air	20 °C (68 °F)		. . .		approx 5	4.5	1×10^{-11}	3.28×10^{-11}	

(a) Including comparison with low-frequency test data
Source: Ref 44

ing plastic strain rates on fracture and deformation behavior. Plastic strain rate is a function of strain amplitude and waveform as well as cyclic frequency. There is some question as to how the increased strain rate affects a particular material. Experimentation generally is required. It is clear that bcc materials exhibit a much larger strain-rate dependence than fcc materials.

The flow stress of any material can be separated into an athermal component and an effective stress, which is thermally activated or assisted. The effective stress portion of the flow stress depends on the temperature and plastic strain rate.

The effective stress for fcc materials is dependent mainly on dislocation interactions. There are three regimes associated with the cyclic stress strain curve for fcc materials (Ref 61). Different dislocation mechanisms apply to each of the three regimes (Ref 62). At low strain amplitudes, a homogeneous distribution of edge dislocation loop patches that undergo deformation during cycling exists. Screw dislocations glide between the loop patches. At intermediate strains (the plateau region of the fcc cyclic stress-strain curve), the dislocations form dipolar walls of edge dislocations arranged in persistent slip bands. The bulk of the strain is carried by screw dislocations shuttling between the walls. At higher strains, the dislocations are arranged in the well-known cell formation.

There is no evidence that a particular strain rate is uniquely applicable to any one of these three strain regimes. The reason for this is found in the large width of the dislocations in fcc metals and hence the small effect of thermal activation in the glide of the dislocations. Consequently, the strain-rate sensitivity for fcc materials is small. Because the cyclic deformation in fcc metals is weakly strain-rate dependent, it is unlikely that the mechanisms of crack initiation or propagation will be strain-rate sensitive if the environment is benign. At low crack propagation rates, however, the effect of frequency is expected to be tied strongly to the environmental effect (Ref 63). In general, fatigue data for fcc materials obtained at ultrasonic frequencies correlate with these observations. Possible exceptions to these general guidelines are discussed in Ref 62.

The cyclic deformation of bcc materials is quite different from that of fcc materials and can be quite complex. The athermal portion of the flow stress is due primarily to dislocation interactions leading to work hardening. The effective stress is dominated by the lattice friction stress. Thus, strain rate and effect of impurities are extremely important in the deformation behavior of bcc materials.

The cyclic stress-strain curve for a bcc material, e.g., iron, differs from the typical fcc curve in that there are two distinct regions to the curve with no persistent slip band plateau (Ref 64). At very low strain amplitudes, the dislocation arrangement is dominated by screw dislocations in low density. In this region, cyclic hardening and microstructural changes are negligible. At higher strain amplitudes, cyclic strain hardening is pronounced and is associated with the formation of equiaxed cells of dislocations. The slip traces are generally diffuse and, in crystals oriented for single slip, do not usually follow the trace of the primary plane (Ref 62).

At low strain rates, bcc materials behave similarly to fcc materials in their cyclic deformation. Therefore, fatigue properties will be similar at high and low frequencies. At higher strain rates, the glide of edge dislocations predominates that of the screw dislocations and produces an entirely different behavior. If screw dislocations are forced to glide as a result of cycling at high amplitudes and high strain rates, shape changes can occur in bcc crystals. These produce gross incompatibilities between adjacent grains, so that intergranular fatigue crack nucleation occurs at the expense of transgranular nucleation.

Compared to fcc materials, the flow stress for bcc materials varies more strongly with temperature. Below a transition temperature, the screw dislocations do not move freely through the lattice, which restricts the operational ability of loop-type dislocation sources. Above the transition temperature, the activation barrier for dislocation glide is overcome, allowing the sources to operate. This flow stress transition temperature for bcc materials increases with increasing frequency.

Strain aging phenomena in bcc materials produce a strong strain-rate effect, especially at temperatures above ambient. These processes cause bcc materials to be more strain-rate sensitive. This sensitivity affects fatigue properties by causing a shift to higher fatigue limits than those recorded at lower frequencies.

Because fatigue crack propagation involves cyclic deformation at the crack tip, strain-rate effects arise in fatigue crack growth of bcc material. High deformation rates especially can change the mechanism of crack propagation in bcc materials. If the local crack-tip stress increases at such a high rate that the crack cannot be blunted by thermally activated dislocation glide, cleavage results. However, if the local stress increases at a low enough rate, the crack blunts and the mechanism of crack growth remains ductile (Ref 62).

The strain-rate sensitivity of engineering alloys and materials often requires experimental verification because of the many possible combinations of alloy compositions, properties, and operating environments. Changes in alloy composition affect the activation barriers to deformation, and the high-frequency behavior might be moderated between the paradigms of bcc and fcc materials. Testing in aggressive environments can offset the expected behavior. Additional information on strain-rate-dependent fatigue behavior can be found in Ref 62, 65, and 66.

REFERENCES

1. B. Hopkinson, *Proc. R. Soc.*, Vol A86, 1911, p 101
2. C.F. Jenkin, *Proc. R. Soc.*, Vol A109, 1925, p 119
3. C.F. Jenkin and G.D. Lehmann, *Proc. R. Soc.*, Vol A125, 1929, p 83
4. W.P. Mason, *Piezoelectric Crystals and Their Application in Ultrasonics*, Van Nostrand, New York, 1950, p 161
5. F. Girard and G. Vidal, *Rev. Metall.*, Vol 56, 1959, p 25
6. M. Kikukawa, K. Ohji, and K. Ogura, *J. Basic. Eng. (Trans. ASME D)*, Vol 87, 1965, p 857
7. L.E. Willertz, *Int. Met. Rev.*, No. 2, 1980, p 65, rev. 250
8. J.M. Wells, O. Buck, L.D. Roth, and J.K. Tien, Ed., *Ultrasonic Fatigue*, TMS-AIME, Warrendale, PA, 1982
9. B.S. Hockenhull, in *Physics and Non Destructive Testing*, Gordon Breach, New York, 1967, p 195
10. H. Koganei, S. Tanaka, and T. Sakurai, *Trans. Iron Steel Inst. Jpn.*, Vol 17, 1977, p 1979
11. J. Awatani and K. Katagiri, *Bull. Jpn. Soc. Mat. Eng.*, Vol 12, 1969, p 10
12. W. Hoffelner, in *High Temperature Alloys for Gas Turbines: 1982*, R. Brunetaud, D. Coutsouradis, T.B. Gibbons, Y. Lindblum, D.B. Meadowcraft, and R. Stickler, Ed., R. Reidal Publishing, Boston, 1982, p 645
13. L.D. Roth and L.E. Willertz, in *Environment Sensitive Fracture: Evaluation and Comparison of Test Methods*, ASTM STP 821, E.N. Pugh and G.M. Ugiansky, Ed., ASTM, Philadelphia, 1984, p 497
14. P. Bajons, in *Ultrasonic Fatigue*, J.M. Wells, O. Buck, L.D. Roth, and J.K. Tien, Ed., TMS-AIME, Warrendale, PA, 1982, p 15
15. E.A. Neppiras, *Proc. ASTM*, Vol 59, 1959, p 691
16. J.K. Tien, S. Purushothoman, R.M. Arons, J.P. Wallace, O. Buck, H.L. Marcus, R.V. Inman, and G.J. Crandall, *Rev. Sci. Instrum.*, Vol 46, 1975, p 840
17. W. Kromp, K. Kromp, H. Bitt, H. Langer, and B. Weiss, *Ultrasonics International 1973 Conference Proceedings*, IPC Science and Technology Publications, Guildford, Surrey, UK, 1973, p 238
18. I. Hansson and A. Tholen, *Ultrasonics*, March 1978, p 57
19. S. Stanzl and E. Tschegg, *Metal Sci.*, April 1980, p 137
20. R.A. Yeske and L.D. Roth in *Ultrasonic Fatigue*, J.M. Wells, O. Buck, L.D. Roth, and J.K. Tien, Ed., TMS-AIME, Warrendale, PA, 1982, p 365
21. P. Trimmel and W. Kromp, in *Ultrasonic Fatigue*, J.M. Wells, O. Buck, L.D. Roth, and J.K. Tien, Ed., TMS-AIME, Warrendale, PA, 1982, p 37
22. W.P. Mason, *J. Acoust. Soc. Am.*, Vol 28, 1956, p 1207
23. J.R. Frederick, *Ultrasonic Engineering*, John Wiley & Sons, New York, 1970
24. G. Amza and D. Drimer, *Ultrasonics*, Vol 14, 1976, p 223
25. E. Eisner, *J. Acoust. Soc. Am.*, Vol 35, 1963, p 1367
26. L. Balamuth, *Trans. IRE (Ultrasonic Eng.)*, Vol 2, 1954, p 23
27. B.S. Hockenhull, C.N. Owston, and R.G. Hacking, *Ultrasonics*, Vol 9, 1971, p 26
28. A. Thiruvengadam, *J. Eng. Ind. (Trans. ASME)*, Vol 11, 1966, p 332
29. H. Konagai, S. Tanaka, and T. Sakurai, *J. Soc. Mater. Sci. Jpn.*, Vol 24, 1975, p 753
30. G.C. George, *Corrosion Fatigue: Chemistry, Mechanics and Microstructure*, O. Devereux *et al.*, Ed., National Association of Corrosion Engineers, Houston, 1972, p 459
31. V.A. Kuz'menko, G.G. Pisarenko, and A.K. Gerikhanov, *Probl. Prochn.*, Vol 4, 1977, p 120
32. R.B. Mignogna and R.E. Green, Jr., in *Ultrasonic Fatigue*, J.M. Wells, O. Buck, L.D. Roth, and J.K. Tien, Ed., TMS-AIME, Warrendale, PA, 1982, p 63
33. F. Blaha and B. Langenecker, *Die Naturwiss.*, Vol 42, 1955, p 556
34. G. Whitlow, L.E. Willertz, and J.K. Tien, in *Ultrasonic Fatigue*, J.M. Wells, O. Buck, L.D. Roth, and J.K. Tien, Ed., TMS-AIME, Warrendale, PA, 1982, p 321
35. J.K. Tien and R.P. Gamble, *Met. Trans.*, Vol 2, 1971, p 1933
36. C.H. Green and F. Guiu, *J. Phys. D. (Appl. Phys.)*, Vol 9, 1976, p 1071
37. J. Awatani, *Bull. Jpn. Soc. Mech. Eng.*, Vol 4, 1961, p 466
38. P. Bajons and W. Kromp, *Ultrasonics*, Vol 16, 1978, p 213
39. L.E. Willertz and L. Patterson, in *Ultrasonic Fatigue*, J.M. Wells, O. Buck, L.D. Roth, and J.K. Tien, Ed., TMS-AIME, Warrendale, PA, 1982, p 119
40. Westinghouse Electric Co., EPRI Tech. Rep. NP-2957, Electric Power Research Institute, Palo Alto, CA, March 1983
41. C.R. Sirian, A.F. Conn, R.B. Mignogna, and R.E. Green, Jr., in *Ultrasonic Fatigue*, J.M. Wells, O. Buck, L.D. Roth, and J.K. Tien, Ed., TMS-AIME, Warrendale, PA, 1982, p 87
42. R.E. Petersen, *Stress Concentration Factors*, John Wiley & Sons, New York, 1973
43. R.E. Petersen, *Trans. ASME (Appl. Mech. Sect.)*, Vol 58, 1936, p A-149
44. R. Stickler and B. Weiss, in *Ultrasonic Fatigue*, J.M. Wells, O. Buck, L.D. Roth, and J.K. Tien, Ed., TMS-AIME, Warrendale, PA, 1982, p 135
45. S. Purushothoman, J.P. Wallace, and J.K. Tien, *Ultrasonics International 1973 Conference Proceedings*, IPC Science and Technology Publications, Guildford, Surrey, UK, 1973, p 244
46. W. Hessler, H. Mullner, and B. Weiss, *Metal Sci.*, May 1981, p 225
47. B. Weiss, R. Stickler, J. Fembock, and K. Pffafinger, *Fatigue Eng. Mat. Struct.*, Vol 2, 1979, p 73
48. W. Hoffelner and P. Gudmundson, *Eng. Fract. Mech.*, Vol 16, 1982, p 365
49. W. Hoffelner, *J. Phys. E: Sci. Instr.*, Vol 13, 1980, p 617
50. N. Dowling, *Cyclic Stress-Strain and Plastic Deformation Aspects of Fatigue Growth*, STP 637, ASTM, Philadelphia, 1976, p 97
51. B. Weiss, *Metall.*, Vol 34, 1980, p 636

52. S. Purushothoman and J.K. Tien, *Met. Trans. A*, Vol 9, 1975, p 367

53. J. Babouk, K. Kromp, W. Kromp, and P. Bajons, in *Ultrasonic Fatigue*, J.M. Wells, O. Buck, L.D. Roth, and J.K. Tien, Ed., TMS-AIME, Warrendale, PA, 1982, p 51

54. W. Hessler, H. Mullner, B. Weiss, and H. Schmidt, in *Ultrasonic Fatigue*, J.M. Wells, O. Buck, L.D. Roth, and J.K. Tien, Ed., TMS-AIME, Warrendale, PA, 1982, p 245

55. W.W. Maennig, in *Ultrasonic Fatigue*, J.M. Wells, O. Buck, L.D. Roth, and J.K. Tien, Ed., TMS-AIME, Warrendale, PA, 1982, p 611

56. B. Weiss, R. Stickler, J. Fembock, and K. Pffafinger, in *Ultrasonic Fatigue*, J.M. Wells, O. Buck, L.D. Roth, and J.K. Tien, Ed., TMS-AIME, Warrendale, PA, 1982, p 505

57. L.D. Roth, L.E. Willertz, and T.R. Leax, in *Ultrasonic Fatigue*, J.M. Wells, O. Buck, L.D. Roth, and J.K. Tien, Ed., TMS-AIME, Warrendale, PA, 1982, p 265

58. L.E. Willertz, T.M. Rust, and V.P. Swaminathan, in *Ultrasonic Fatigue*, J.M. Wells, O. Buck, L.D. Roth, and J.K. Tien, Ed., TMS-AIME, Warrendale, PA, 1982, p 333

59. V.A. Kuz'menko, L.E. Matochnyuk, L.D. Roth, F. Cosandey, and J.K. Tien, in *Ultrasonic Fatigue*, J.M. Wells, O. Buck, L.D. Roth, and J.K. Tien, Ed., TMS-AIME, Warrendale, PA, 1982, p 229

60. W. Hoffelner, in *Ultrasonic Fatigue*, J.M. Wells, O. Buck, L.D. Roth, and J.K. Tien, Ed., TMS-AIME, Warrendale, PA, 1982, p 461

61. H. Mughrabi, *Mat. Sci Eng.*, Vol 33, 1978, p 207

62. C. Laird and P. Charlsey, in *Ultrasonic Fatigue*, J.M. Wells, O. Buck, L.D. Roth, and J.K. Tien, Ed., TMS-AIME, Warrendale, PA, 1982, p 183

63. C. Laird and G. Smith, *Phil. Mag.*, Vol 8, 1963, p 1945

64. H. Mughrabi, R. Ackermann, and K. Herz, *Fatigue Mechanisms*, STP 675, ASTM, Philadelphia, 1979, p 69

65. J.K. Tien, in *Ultrasonic Fatigue*, J.M. Wells, O. Buck, L.D. Roth, and J.K. Tien, Ed., TMS-AIME, Warrendale, PA, 1982, p 1

66. L.E. Coffin, in *Ultrasonic Fatigue*, J.M. Wells, O. Buck, L.D. Roth, and J.K. Tien, Ed., TMS-AIME, Warrendale, PA, 1982, p 423

Dynamic Fracture Testing

Introduction

By Donald A. Shockey
Director, Department of Metallurgy
and Fracture Mechanics
SRI International

DYNAMIC FRACTURE occurs under a rapidly applied load, such as that produced by impact or by explosive detonation. In contrast to quasi-static loading, dynamic conditions involve loading rates that are greater than those encountered in conventional tensile tests or fracture mechanics tests. The tests described in this article produce fracture at stress intensification rates, $\dot{K}$, that generally are greater than 10 MPa$\sqrt{\text{m}} \cdot \text{s}^{-1}$ (9.1 ksi$\sqrt{\text{in.}} \cdot \text{s}^{-1}$) and usually are greater than 10^4 MPa$\sqrt{\text{m}} \cdot \text{s}^{-1}$ (9.1 $\times$ 10^3 ksi$\sqrt{\text{in.}} \cdot \text{s}^{-1}$). A fracture toughness test per ASTM E 399, "Standard Test Method for Plane-Strain Fracture Toughness of Metallic Materials," (Ref 1) typically has a stress-intensification rate up to approximately 2.75 MPa$\sqrt{\text{m}} \cdot \text{s}^{-1}$ (2.5 ksi$\sqrt{\text{in.}} \cdot \text{s}^{-1}$).

Dynamic fracture includes the case of a stationary crack subjected to a rapidly applied load, as well as the case of a rapidly propagating crack under a quasi-stationary load. In both cases the material at the crack tip is strained rapidly and, if rate sensitive, may offer less resistance to fracture than at quasi-static strain rates. In both cases, inertial effects may influence the load history at the crack tip.

High strain rate fracture testing is of interest, because many structural components are subjected to high loading rates in service or must survive high loading rates during accident conditions. Thus, these components must be designed against crack initiation under high loading rates or designed to arrest a rapidly running crack. Furthermore, the fracture resistance of a material that is loaded rapidly is generally lower than when the load is applied slowly, consequently the dynamic fracture toughness is a more conservative value than the static value for design calculations.

Measurement and analysis of fracture behavior under high loading rates are more complex than under quasi-static conditions, and quantitative testing procedures are not as well established. Tests that are simple to perform but difficult to interpret have become standards, but often do not provide reliable quantitative fracture mechanics criteria. More recently developed tests that do provide reliable values of fracture mechanics parameters are time consuming, difficult, and expensive to perform, and hence are not suitable for standard routine testing. Nevertheless, these tests and the associated research efforts have greatly enhanced the basic understanding of dynamic fracture, which in turn should lead to simpler, more easily interpreted quantitative test procedures.

This article describes current dynamic fracture testing methodologies. The focus is on tests that produce quantitative values of fracture toughness parameters that are useful in design. Qualitative dynamic tests that measure ductile-brittle behavior such as energy absorbed in breaking a notched bar, percent of cleavage area on fracture surfaces, or temperature for nil ductility or crack arrest are not covered in this article. For information on more qualitative fracture tests such as the drop-weight test, the explosive bulge test, the Robertson test, the Esso test, and the Navy tear test, see Volume 10 of the 8th Edition of *Metals Handbook*.

Some testing procedures, such as the Charpy impact test, have been used extensively to test a wide variety of materials. Because of the simplicity of the test and the existence of a large data base, attempts have been made to modify the specimen, loading arrangement, and instrumentation to extract quantitative fracture mechanics information.

The first section of this article describes tests that use servohydraulic machines, i.e., tests that are extensions to higher loading rates of quasi-static fracture mechanics tests. This is followed by discussions of the standard Charpy V-notch impact test and subsequent modifications involving instrumentation and precracking, alternative dynamic bend tests, concepts of impact response curves, and tests using inertial loading.

Tests using explosive and impact loading techniques to achieve $\dot{K}$ values greater than 10^5 MPa$\sqrt{\text{m}} \cdot \text{s}^{-1}$ (9.1 $\times$ 10^4 ksi$\sqrt{\text{in.}} \cdot \text{s}^{-1}$) are described in a subsequent section. Following this, the effects of pulse duration on crack instability are discussed, and modifications to classical static fracture mechanics concepts are presented that allow analyses of crack instability under a short load pulse. Then the fracture toughness associated with a rapidly propagating crack is discussed, a brief history of crack arrest research is given, and a synopsis of the newly proposed ASTM standard for determining crack arrest toughness is presented.

Micromechanical aspects of dynamic fracture are discussed in the last section of this article. The phenomenology of fracture on a microscale involves nucleation, growth, and coalescence of small voids or cracks. This can be modeled, but at considerable expense. Specific problems involving many simultaneously active cracks or voids may not be treatable by continuum mechanics, but may require a micromechanical modeling approach. Plate impact experiments and a modeling methodology for micromechanics of dynamic fracture are also described.

Fracture Toughness Testing Using Servohydraulic Testing Systems

By Jacques Henri Joseph Giovanola
Research Engineer
SRI International

HIGH LOADING RATES are known to significantly decrease the value of the stress-intensity factor at which cracks become unstable in several engineering materials. However, few test procedures have been standardized to obtain dynamic fracture toughness values. This section discusses the use of ASTM Standard E 399 for fracture toughness testing of metallic alloys with rapid loading (Ref 1) and reviews the special requirements (compiled by ASTM in Annex 7 of E 399) for this type of rapid-load testing. The remainder of this section reviews test procedures that have been used successfully to measure the fracture resistance of low-strength high-toughness metallic alloys that are loaded rapidly. These procedures are modifications of the quasi-static ductile fracture procedures (Ref 2, 3) and are still being developed.

The standard concepts of stress-intensity factor K and J-integral, as well as the quasi-static test procedures used to measure the related toughness parameters K_{Ic} and J_{Ic}, are used for discussion in this article. For more information on fracture mechanics concepts, see the Section on "Fracture Mechanics" in this Volume.

Rapid-Load Plane-Strain Fracture Toughness Testing

The ASTM recommended procedure for rapid-load plane-strain fracture toughness, $K_{Ic}(t)$, testing is a modification of the quasi-static procedure of ASTM E 399 (Ref 1) to allow toughness measurements at loading rates exceeding those for conventional static testing—that is, for loading rates greater than 2.75 MPa$\sqrt{m} \cdot s^{-1}$ (2.5 ksi$\sqrt{in.} \cdot s^{-1}$). The rapid-load fracture test should be performed on servohydraulic testing machines. Requirements for performing these tests do not apply to impact or quasi-impact testing. Therefore, the ASTM rapid-load fracture toughness testing procedure covers intermediate loading rates between the quasi-static test and the impact bend tests described in subsequent sections of this article. The maximum loading rate recommended by ASTM E 399 is defined in terms of a minimum allowable test time—that is, the time elapsed between the first application of load and the moment when the crack becomes unstable. The test procedure applies only for test times greater than 1 ms.

Loading arrangement, specimen geometry, and specimen preparation used for the rapid-load tests are the same as for the standard quasi-static fracture toughness test. Reduction of the test data is also basically the same, and the rapid-load fracture toughness is related to the dynamic fracture load using static expressions for the stress-intensity factor. The main difference is that in the rapid-load test, the load is applied more rapidly to

the specimen than in the quasi-static test. The loading rate is limited by the capacity of the servohydraulic machine used for the test and more importantly by the need to limit large inertial effects and ringing of the testing system to be able to interpret the results with a static analysis. The main purpose of the requirements in Annex 7 of ASTM E 399 is therefore to specify adequate characteristics of the test instrumentation to avoid ringing problems and to define the limits for tolerable inertial effects.

To measure rapid-load fracture toughness, a record of load versus time and a record of the crack mouth opening displacement versus time must be obtained during the experiment. After the test, the load-displacement record is constructed from these two records. Most load cells are sufficiently stiff to have a high enough frequency response for the rapid-load test. However, cantilever beam clip gages commonly used in fracture experiments may cause problems. If the stiffness of the clip gage is too low, the two blades may oscillate during the test. These oscillations will cause the load-displacement record to be highly nonlinear (see Fig. 1).

The clip gage described in Annex 7 of ASTM E 399 has been suitable for tests with a loading time of 1 ms (Ref 4). The resonant frequency of this gage when mounted in a specimen is about 3300 Hz, and the free arm resonant frequency is about 750 Hz. Thus, any gage with resonant frequencies greater than these values is suitable for rapid-load tests.

The instrumentation used to amplify the transducer signals in the rapid-load test should have a frequency response of at least $20/t$ kHz, where t is the test time in milliseconds. Instruments with a frequency response equal to or greater than 25 kHz are therefore adequate for any loading rate covered by Annex 7 of ASTM E 399.

Despite the requirements imposed on the instrumentation, experience has shown that load-displacement records may still exhibit nonlinearities (Ref 4). The nonlinearities can be caused by inertial effects in the specimen grips/machine assembly and may render determination of the fracture load, P_Q, uncertain. To determine the tolerable extent of nonlinearity, a special procedure for reducing the test data has been introduced (Fig. 2). First, the straight line OA best representing the initial portion of the load deflection line (the portion that should be linear) is constructed. Then, the fracture load P_Q is determined according to the procedure used for static fracture toughness determination. A vertical line, $P_Q V_P$, is drawn through P_Q. At point X, where $P_Q V_P$ intersects OA, a horizontal line is drawn that intersects the load axis at P_V. Two points, C and E, are

Fig. 1 Load-displacement record obtained from a rapid-load fracture test showing nonlinearity in loading line

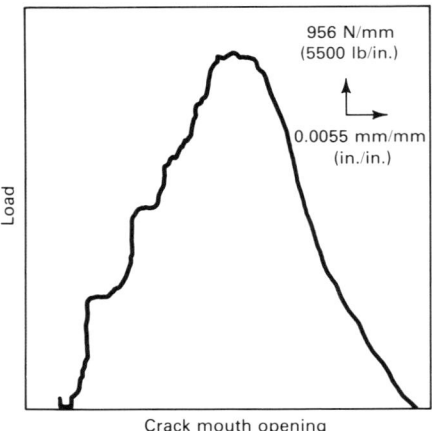

956 N/mm
(5500 lb/in.)

0.0055 mm/mm
(in./in.)

Load

Crack mouth opening

Fig. 2 Evaluation of load-displacement record from rapid-load fracture test and definition of tolerable degree of nonlinearity

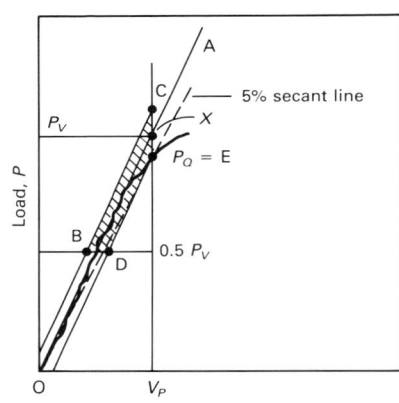

further constructed on $P_Q V_P$ at a distance $+0.05 P_V$ and $-0.05 P_V$, respectively. Note that when P_Q is obtained by the 5% secant method, $P_Q = E$.

Finally, two lines parallel to OA are drawn through C and E with a horizontal line of ordinate $0.5 P_V$, which intersects these lines in B and D, respectively. For the rapid-load test to be valid, the experimental load deflection curve up to P_Q must be within the field BCED. Nonlinearities greater than the band defined by lines BC and DE at load levels smaller than $0.5 P_V$ are considered insignificant.

The specimen size requirements for a valid rapid-load fracture test are the same as for the quasi-static fracture test, namely:

$$a, B > 2.5\left(\frac{K_Q}{\sigma_{YD}}\right)^2 \qquad \text{(Eq 1)}$$

where a is the crack length, B is the specimen thickness, K_Q is the tentative rapid-load fracture toughness, and σ_{YD} is the dynamic yield strength. When the dynamic yield strength is unknown, it can be evaluated using correlations available in the literature (Ref 5).

If the preceding requirements are met as well as the other requirements defining a valid quasi-static fracture test, the K_Q value measured in the rapid-load test can be regarded as a valid rapid-load fracture toughness value. It is denoted $K_{Ic}(t)$, where t is the test time. The loading rate relevant for the test can be obtained simply by dividing K_{Ic} by t.

The ASTM rapid-load fracture testing procedure allows measurement of dynamic fracture toughness values using a familiar test method (ASTM E 399) and readily available testing equipment. However, the loading rate covered by this procedure may be too low to fully assess the effect of loading rate on the fracture toughness. Care must therefore be exercised when using $K_{Ic}(t)$ values as lower bound toughness estimates.

Determination of Critical J-Values and J-Resistance Curves

The fracture toughness of low-strength high-toughness materials cannot be defined in terms of the parameter K_{Ic} based on linear elastic analysis. Instead, it must be defined in terms of the J-integral, a parameter characterizing the elastic-plastic stress and strain fields at the crack tip. The procedure to determine toughness values in terms of the J-integral (the J_{Ic} values) has been standardized by ASTM (Ref 2), and recommendations for determining the J-resistance curve, which characterizes the resistance of material to slow, stable crack extension, are currently being developed (Ref 3).

In the procedures described in Ref 2 and 3, only quasi-static loading is considered. To evaluate the influence of loading rate on the toughness of ductile materials, techniques to measure J_{Ic} values and J-resistance curves for rapid loading have been developed using servohydraulic machines. The major difficulties arising in determining dynamic J_{Ic} values and J-resistance curves are the occurrence of oscillations in the load and displacement records due to inertial oscillations of the loading assembly and the detection of crack initiation and subsequently the measurement of the crack growth increment.

Logsdon and Begley (Ref 6) have performed dynamic J_{Ic} and J-resistance curve measurements on A533 steel using compact tension specimens and a servohydraulic machine. The instrumentation was the same as for a quasi-static K_{Ic} or J_{Ic} test (load cell and clip gage), with the exception that the clip gage was spring loaded to increase its stiffness. When the test temperature was sufficiently low so that stable crack growth did not occur after crack initiation, the measurement of dynamic J_{Ic} values was relatively straightforward. Crack initiation was evidenced by sharp discontinuities in the load-time and load-line displacement-time records. Thus, the point of crack initiation was located easily on the load-displacement diagram. The dynamic J_{Ic} value was then calculated from the area under the load displacement curve, following the quasi-static procedure (Ref 2).

When stable crack growth occurred, however, Logsdon and Begley (Ref 6) used a multiple-specimen technique to establish the J-resistance curve. Several compact tension specimens of identical crack length were loaded dynamically to different load-line displacements to cause varying amounts of crack growth. After the test, the specimens were heat tinted and broken open. The amount of crack extension was then measured optically on the fracture surface.

For each value of the extended crack length, the corresponding value of J was calculated from the dynamic load-displacement record using the quasi-static estimation, and the J-resistance curve was obtained. Logsdon and Begley achieved equivalent stress-intensity rates up to 4.4×10^4 MPa$\sqrt{\text{m}} \cdot \text{s}^{-1}$ (4.0×10^4 ksi$\sqrt{\text{in}} \cdot \text{s}^{-1}$) without interference from inertial loading effects.

Marandet and co-workers (Ref 7, 8) have also successfully measured dynamic J_{Ic} values for steels using compact tension specimens at equivalent stress-intensity rates of 2×10^4 MPa$\sqrt{\text{m}} \cdot \text{s}^{-1}$ (1.8×10^4 ksi$\sqrt{\text{in}} \cdot \text{s}^{-1}$). To suppress inertial oscillations of the loading assembly, they introduced a variable-stiffness spring system in series with the specimen. The spring system was made of stacks of conical washers of three different stiffnesses. Marandet et al. (Ref 9) detected the onset of crack extension using a high-frequency (10-kHz) alternating-current potential drop method.

Joyce (Ref 10) has developed a single-specimen method to obtain the full dynamic J-resistance curve at the rates achievable with servohydraulic machines using simple load cell and clip gage instrumentation. The method relies on the concept of a key curve (described in Ref 11 and 12) to obtain the J-resistance curve directly from the dynamic load-displacement record. The key curve is basically a nondimensional representation of the load-displacement curve for a given spec-imen geometry and material. In his work, Joyce (Ref 10) assumed the following form for the key curve:

$$\frac{PW}{Bb^2} = F_1\left(\frac{\delta}{W}, \frac{a}{W}\right) \qquad \text{(Eq 2)}$$

where P is the load; δ is the load-line displacement; B, W, and b are the specimen thickness, width, and uncracked ligament length, respectively; and a is the crack length. F_1 is a nondimensional function characterizing the key curve for a given specimen geometry and material. It can be obtained experimentally by testing several specimens of increasing crack length into the plastic regime, but without extending the crack.

Joyce et al. (Ref 12) have further shown that the increment of the J-integral, dJ, and of the crack length, da, during a fracture test can be expressed directly in terms of the corresponding increments in load and displacement and in terms of the key curve function F_1 and its derivatives.

To measure the dynamic J-resistance curve from single-specimen tests, Joyce first determined the dynamic key curve function experimentally by testing small compact tension specimens (Ref 10). Uninterrupted dynamic fracture tests were then performed, recording load and displacement by means of load cell and clip gage. The dynamic load displacement was then evaluated on a digital minicomputer in conjunction with the dynamic key curves to yield the dynamic J-resistance curve. Tests on A533 B steel at equivalent average stress-intensity rates of about 1×10^4 MPa$\sqrt{\text{m}} \cdot \text{s}^{-1}$ (9×10^3 ksi$\sqrt{\text{in}} \cdot \text{s}^{-1}$) showed that the dynamic J-resistance curve lies significantly above the static J-resistance curve.

To date, no standard method is available for measuring the quasi-static J-resistance curve or the dynamic resistance curve. Although dynamic fracture testing in the elastic-plastic regime is feasible, it is more intricate than linear elastic fracture testing. Consequently, more work is necessary before a reliable dynamic J fracture testing procedure can be established and standardized.

Charpy Impact Testing

By William L. Server
Senior Engineering Specialist
EG&G Idaho, Inc.

BEFORE FRACTURE MECHANICS became a scientific discipline, laboratory specimens were tested under severe conditions to simulate structural failures, eliminating the need to destructively test large engineering components. Conditions chosen for simulating structural component failure were high loading rate, stress concentration, and triaxial stress state; these conditions were met by notched-bar impact tests.

The ductile-to-brittle transition temperature of low- and medium-strength ferritic steels used in structural applications such as ships, pressure vessels, tanks, pipelines, and bridges was also of interest. Ferritic steels and other body-centered cubic materials tend to go through a transition from ductile behavior (microvoid coalescence) at high temperatures to brittle fracture (cleavage) at lower temperatures. This transition dictates the need for designing structural components against brittle fracture using a service temperature above the ductile-to-brittle transition temperature.

Many notched-bar impact tests of different designs and loading have been used to predict brittle fracture of ferritic steels. In the United States, the Charpy specimen has been standardized in ASTM E 23 for impact three-point bend testing (Ref 13). The Izod specimen has also been standardized for impact cantilever bend testing (Ref 13). The Charpy three-point bend specimen is more convenient for varying test temperatures and has gained the widest acceptance for measuring notch toughness.

Measurement of notch toughness involves the fracture response of the impact specimen in terms of absorbed fracture energy. This determination usually is measured directly on a pendulum impact testing machine. The change in potential energy of the impacting head (from before impact to after fracture) is determined via a calibrated dial that measures the total energy absorbed in breaking the specimen. Other quantitative parameters such as fracture appearance (percent fibrous fracture) and degree of ductility/deformation (lateral expansion or notch root contraction) are often measured in addition to the fracture energy.

As indicated previously, these measured quantities are strongly temperature dependent for ferritic steels; this temperature dependence results in a gradual increase in toughness and ductility with increasing temperature. Because the transition is generally not sharp, it is often difficult to accurately define a ductile-to-brittle transition temperature. Thus, the transition temperature is often defined as the temperature needed to attain an arbitrary level of fracture energy (e.g., 41 J, or 30 ft · lb), a given fracture appearance (e.g., 50% fibrous fracture), or some specified level of ductility (e.g., 0.38 mm, or 15 mil, lateral expansion).

Figure 3 illustrates the changes in Charpy V-notch properties for a conventional pressure vessel steel plate tested in the transverse (open circles) and longitudinal (closed circles) orientations. Note that 50% fibrous fracture results in a higher value of transition temperature than the other conventional energy or deformation criteria.

The best criterion for selecting a particular transition temperature is to consider how it correlates with service. Few service parts undergo notched-bar impact loads, and most components are much larger than the dimensions of the Charpy specimens. Also, notch toughness energy or ductility values cannot be used directly in engineering and design applications. However, the extreme conditions of low-temperature and Charpy impact testing can be correlated with service experience through the ductile-brittle transition temperature.

The Charpy test measure of a transition temperature thus provides a quality control measure of brittle fracture for differentiating between heats or different types of ferritic steels with the advantages of small size, low cost in specimen preparation and testing, and flexibility in testing various specimen orientations.

Standard Charpy Impact Test

Notched-bar impact testing of metals has been standardized in ASTM E 23 (Ref 13). The primary specimen and test procedure involves the Charpy V-notch test. Figure 4 illustrates a standard ASTM type A Charpy impact test specimen. Other Charpy-type specimens are not used as extensively, because their degree of constraint and triaxiality is considerably less than the V-notch specimen. The V-notch specimen is discussed in this section.

ASTM E 23 provides requirements for test specimen, anvil supports and striker dimensions and tolerances, the pendulum action of the test machine, the actual testing procedure and machine verification, and the determination of fracture appearance and lateral expansion. Some of the general guidelines specified in ASTM E 23 include the following requirements. Sufficient energy must be available to completely fracture the test specimen, but the fracture energy cannot exceed 80% of the available potential energy of the machine. The machine must be securely bolted to a massive floor structure and must be sufficiently level.

Windage and frictional losses must be measured and taken into account; weekly checks are recommended. The center of strike on the specimen and the center of percussion on the striker must be measured and

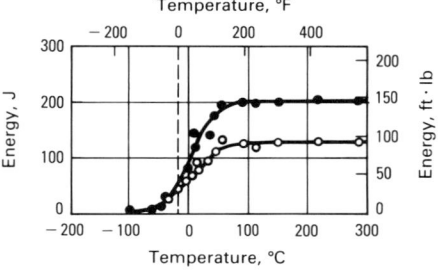

(a)

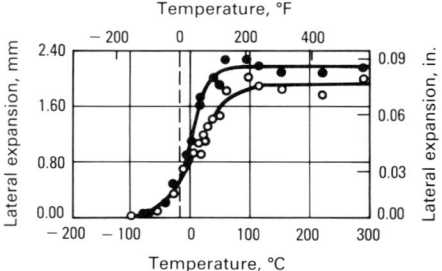

(b)

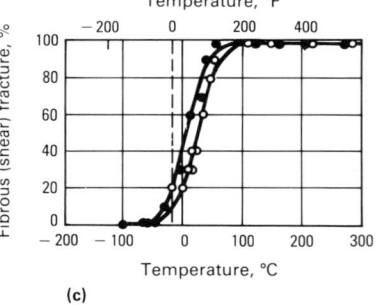

(c)

must be sufficiently close to each other. The radial play of bearings and the transverse play of the pendulum must be minimal.

The height of drop of the pendulum must be measured, and the impact velocity should be calculated to within 3 to 6 m/s (10 to 20 ft/s). The effective weight of the pendulum head must be measured so that the total potential energy of the machine can be calculated using the measured height of drop.

The release mechanism must be consistent and smooth. Test specimens must leave the impact machine freely, without jamming or rebounding into the pendulum; requirements on clearances and containment shrouds are specific to individual machine types. The test specimen must be accurately positioned on the anvil support within 5 s of removal from

Fig. 4 ASTM standard dimensions for the type A Charpy V-notch specimen and the striker-anvil arrangement
Source: Ref 13

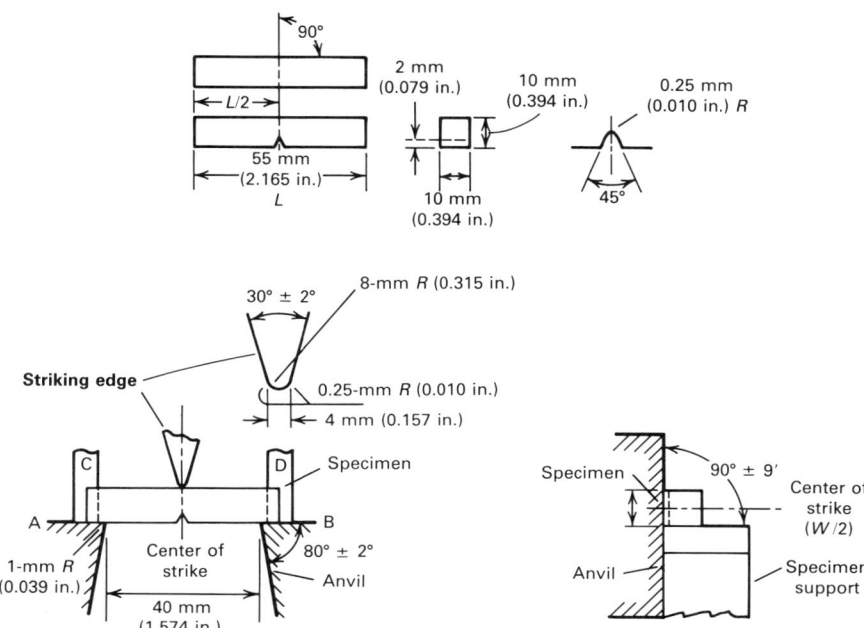

the heating (or cooling) medium; requirements for heating time depend on the heating medium. Identification marks on test specimens must not interfere with the test; also, any heat treatment of specimens should be performed prior to final machining.

A daily check procedure of the apparatus must be conducted to ensure proper performance. Verification of the testing system is required using Army Materials and Mechanics Research Center (AMMRC) standardized specimens; verification should be completed at least once a year, or after any parts are replaced or any repairs or adjustments are made to the machine. An operational testing sequence is recommended, as well as specifics on dial energy reading, lateral expansion measurement (technique and measuring fixture), and fracture appearance estimation.

Applications of the Charpy V-Notch Test

The Charpy V-notch impact test has limitations due to its blunt notch, small size, and total energy measurement (i.e., no separation of initiation and propagation components of energy). However, this test is used widely because it is inexpensive and simple to perform. Historically, extensive correlation with service performance has indicated its usefulness.

The Charpy V-notch test commonly is used as a screening test for evaluating notch toughness changes influenced by chemical

composition (alloying and impurity elements, including gases) and physical and mechanical properties of materials. Typical physical properties of interest are microstructure and grain size, which are influenced by fabrication procedures (temperature of working, cross-rolling, surface treatment, etc.). Mechanical properties generally considered are yield and flow properties and hardness.

In many structural steel applications, the Charpy V-notch test is used in procurement and quality assurance for assessing different heats of the same type of steel. Also, correlation with actual fracture toughness data is often devised for a class of steels so that fracture mechanics analyses can be applied directly. Heat-to-heat differences in Charpy properties can be quite large, and when conditions involving human safety are concerned, failure prevention is imperative. Two such examples are the nuclear pressure vessel and the steel bridge industries.

Nuclear Pressure Vessel Design Code. For nuclear pressure vessels, the American Society of Mechanical Engineers (ASME) *Boiler and Pressure Vessel Code* (Ref 14) and the *Code of Federal Regulations* (Ref 15) currently use fracture mechanics principles that dictate toughness requirements for pressure vessel steels and weldments. The specified toughness requirements are obtained using Charpy V-notch test specimens coupled with the nil-ductility transition temperature (NDTT) per ASTM Stan-

dard E 208 (Ref 16). The actual approach involves a reference temperature, designated RT_{NDT}, and the reference fracture toughness curve, K_{IR}.

The reference fracture toughness curve defined in Appendix G, Section III, of the ASME code utilizes an experimentally determined relationship between toughness and temperature that is adjusted along the temperature axis according to an index reference temperature. This procedure is shown graphically in Fig. 5(a), where T is the temperature of interest, and RT_{NDT} is the reference temperature.

The reference toughness (K_{IR}) curve is assumed to describe the minimum (lower bound) fracture toughness for all ferritic materials approved for nuclear pressure boundary applications having a minimum specified yield strength of 345 MPa (50 ksi) or less. The value of RT_{NDT} is obtained by measuring the drop weight nil-ductility transition temperature and performing standard Charpy V-notch tests. The nil-ductility transition temperature is determined initially, and then a set of three Charpy V-notch specimens is tested at a temperature that is 33 °C (60 °F) higher than the nil-ductility transition temperature to measure the temperature, T_{CV}, which ensures an increase in toughness with temperature. Charpy energies of 68 J (50 ft · lb) and lateral expansion of 0.89 mm (35 mil) are used to ensure this condition.

Fig. 5 Toughness requirements for nuclear pressure vessel steels
(a) Determined using the reference temperature, RT_{NDT}, approach. (b) Determined using the reference fracture toughness, K_{IR}, curve. See text for details. Source: Ref 14

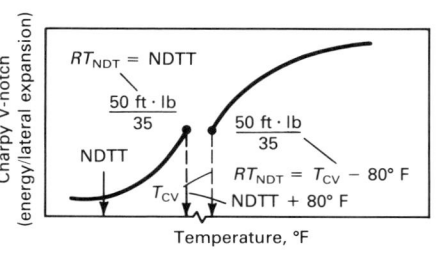

The nil-ductility transition temperature becomes the RT_{NDT} temperature if the Charpy results equal or exceed the above limits. If the Charpy values at T_{CV} or the nil-ductility transition temperature plus 33 °C (60 °F) are lower than required, additional Charpy tests should be performed at higher test temperatures, usually in increments of 5.6 °C (10 °F), until the requirements are satisfied and T_{CV} is measured. The RT_{NDT} temperature then becomes the temperature (T_{CV}) at which the criteria are met minus 33 °C (60 °F). Thus, the reference temperature is always either greater than or equal to the nil-ductility transition temperature.

The data used to construct the lower bounding function K_{IR} curve (Fig. 5b) used dynamic fracture toughness as well as crack arrest toughness results. The dynamic fracture toughness data were used to determine the lower temperature end, and the crack arrest toughness data were used to fix the higher temperature end for the lower bound envelope of the K_{IR} curve. Although a significant amount of data was used to establish the K_{IR} curve, it represented tests on only a few heats of material (Ref 17).

Additionally, the nuclear pressure vessel steels used in the reactor beltline region are susceptible to neutron radiation damage. This damage results in a degradation in material toughness, as measured by the Charpy V-notch test. Figure 6 illustrates the marked shift to higher temperatures for the transition temperature region of the Charpy curve and the decrease in the upper shelf level of toughness after neutron exposure. Reactor vessel surveillance programs have been devised to periodically monitor the radiation embrittlement of the beltline materials.

The shift in the Charpy V-notch energy curve at the 41-J (30-ft · lb) level from unirradiated to irradiated conditions is measured and then used to adjust the K_{IR} curve by an equal amount to account for the embrittling phenomenon. The shift in the K_{IR} curve cannot be excessive, or the plant cannot be operated safely. The decrease in upper shelf level is specified to not drop below 68 J (50

Fig. 6 Effect of irradiation on the Charpy impact energy for a nuclear pressure vessel steel

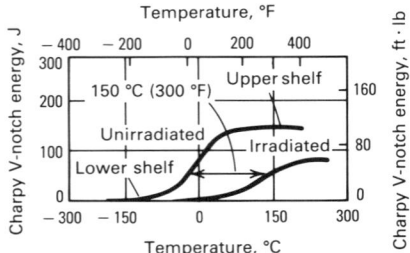

ft · lb), or further analytical studies are required to ensure plant safety. Thermal annealing of the beltline region is a possible solution to highly embrittled reactor vessels; the annealing response of irradiated materials is also assessed using the Charpy V-notch test.

Steel Bridge Toughness Criteria. The American Association of State Highway and Transportation Officials (AASHTO) has adopted Charpy impact toughness requirements for primary tension members in bridge steels based on section thickness, yield strength, and expected service temperature. These requirements are shown in Table 1. They are based on the fracture toughness corresponding to the maximum loading rate expected in service (Ref 18).

For low-strength materials with yield strengths below 448 MPa (65 ksi), brittle fracture will not occur in service at temperatures approximately 39 °C (70 °F) lower than the Charpy impact 20-J (15-ft · lb) level. When the actual (not the minimum specified) yield strength exceeds 448 MPa (65 ksi), an adjustment reduction of 8.3 °C (15 °F) for each 69 MPa (10 ksi) is required to maintain the same strain-rate-adjusted fracture toughness level. Also, for welded members greater than 50 mm (2 in.) in thickness, a higher Charpy energy level was chosen to account for the greater plane-strain constraint at a crack in the thicker welded section. For A514 steel, the same strain rate shift approach was adopted at a higher Charpy energy level to account for the greater design stresses imposed on high-strength steel tension members.

Correlations With Fracture Toughness. The critical plane-strain stress-intensity value, K_{Ic}, ahead of an atomically sharp crack at the moment of unstable crack propagation can be used directly in design applications; K_{Ic} is related to the applied stress, flaw size, and component geometry. Thus, empirical attempts have tried to correlate the Charpy impact energy with K_{Ic} to allow a quantitative assessment of critical flaw size and permissible stress levels.

Most of these correlations are dimensionally incompatible, ignore differences between the two measures of toughness (in particular, loading rate and notch acuity), and are valid only for limited types of materials and ranges of data. Additionally, these correlations can be widely scattered. However, some correlations can provide a useful guide to estimating fracture toughness; in fact, the preceding design criteria for nuclear pressure vessel and bridge steels are partially based on such correlative procedures.

Some of the more common correlations are listed in Table 2 with appropriate units. Note that some of the correlations attempt to

eliminate the effects of loading rate; the dynamic fracture toughness, K_{Id}, is correlated with Charpy energy.

Other attempts have been made to improve and explain some of the correlations (see, for example, Ref 26). A study has also been conducted using a portion of the Charpy energy to separate initiation and propagation components in the Charpy test (Ref 27). The results from this study for an upper shelf J_{Ic} correlation for pressure vessel steels were not significantly better than the Rolfe-Novak correlation listed in Table 2. A statistically based correlation for lower bound toughness has been developed for pressure vessel steels (Ref 28, 29), which may eventually replace the ASME K_{IR} curve described previously. Thus, simple and empirical correlations can be used as general guidelines for estimating K_{Ic} or K_{Id} within the limits of the specific correlation. Correlations using sharp-notch Charpy specimens have also been made and are discussed in a subsequent section of this article.

Instrumented Charpy Impact Test

The use of additional instrumentation with a standard Charpy impact machine allows augmented monitoring of the analog load/time response of Charpy V-notch specimen deformation and fracturing. The primary advantage of instrumenting the Charpy test is the additional information obtained while maintaining low cost, small specimens, and simple operation. The most commonly used approach is application of strain gages to the striker to sense the load-time behavior of the test specimen.

Figure 7 schematically illustrates the change in Charpy behavior as a function of temperature for a medium-strength steel. As shown, instrumentation clearly allows the various stages in the fracture process to be identified. The energy value, W_M, is associated with the area under the load-time (P-t) curve up to maximum load, P_M. This impulse value is converted to energy by using Newton's second law, which accounts for the pendulum velocity decrease during the deformation-fracture process. This velocity decrease is proportional to the instantaneous load on the specimen at any particular time, t_i; the actual energy absorbed (ΔE_i) simplifies to (Ref 30):

$$\Delta E = E_a \left[1 - \left(\frac{E_a}{4E_o} \right) \right] \quad \text{(Eq 3)}$$

where E_o is the total available kinetic energy of the pendulum ($\frac{1}{2}\, m \cdot V_o^2$) and:

$$E_a = V_o \int_0^{t_i} P \cdot dt \quad \text{(Eq 4)}$$

Table 1 AASHTO Charpy toughness requirements for bridge steels

ASTM designation	Thickness	Charpy V-notch impact energy(a)					
		MST ≥ −18 °C	MST ≥ 0 °F	MST from −18 to −34 °C	MST from −1 to −30 °F	MST from −34 to −51 °C	MST from −31 to −60 °F
A36 . . . . ⋯		20.3 J at 21 °C	15 ft · lb at 70 °F	20.3 J at 4.5 °C	15 ft · lb at 40 °F	20.3 J at −12 °C	15 ft · lb at 10 °F
A572 (b) . . . Up to 102 mm (4 in.), mechanically fastened		20.3 J at 21 °C	15 ft · lb at 70 °F	20.3 J at 4.5 °C	15 ft · lb at 40 °F	20.3 J at −12 °C	15 ft · lb at 10 °F
	Up to 51 mm (2 in.), welded	20.3 J at 21 °C	15 ft · lb at 70 °F	20.3 J at 4.5 °C	15 ft · lb at 40 °F	20.3 J at −12 °C	15 ft · lb at 10 °F
A440 . . . . ⋯		20.3 J at 21 °C	15 ft · lb at 70 °F	20.3 J at 4.5 °C	15 ft · lb at 40 °F	20.3 J at −12 °C	15 ft · lb at 10 °F
A441 . . . . ⋯		20.3 J at 21 °C	15 ft · lb at 70 °F	20.3 J at 4.5 °C	15 ft · lb at 40 °F	20.3 J at −12 °C	15 ft · lb at 10 °F
A242 . . . . ⋯		20.3 J at 21 °C	15 ft · lb at 70 °F	20.3 J at 4.5 °C	15 ft · lb at 40 °F	20.3 J at −12 °C	15 ft · lb at 10 °F
A588 (b) . . . Up to 102 mm (4 in.), mechanically fastened		20.3 J at 21 °C	15 ft · lb at 70 °F	20.3 J at 4.5 °C	15 ft · lb at 40 °F	20.3 J at −12 °C	15 ft · lb at 10 °F
	Up to 51 mm (2 in.), welded	20.3 J at 21 °C	15 ft · lb at 70 °F	20.3 J at 4.5 °C	15 ft · lb at 40 °F	20.3 J at −12 °C	15 ft · lb at 10 °F
	Over 51 mm (2 in.), welded	27.1 J at 21 °C	20 ft · lb at 70 °F	27.1 J at 4.5 °C	20 ft · lb at 40 °F	27.1 J at −12 °C	20 ft · lb at 10 °F
A514 . . . . . Up to 102 mm (4 in.), mechanically fastened		33.9 J at −1 °C	25 ft · lb at 30 °F	33.9 J at −18 °C	25 ft · lb at 0 °F	33.9 J at −34 °C	25 ft · lb at −30 °F
	Up to 64 mm (2.5 in.), welded	33.9 J at −1 °C	25 ft · lb at 30 °F	33.9 J at −18 °C	25 ft · lb at 0 °F	33.9 J at −34 °C	25 ft · lb at −30 °F
	Between 64 mm (2.5 in) and 102 mm (4.0 in.), welded	47.5 J at −1 °C	35 ft · lb at 30 °F	47.5 J at −18 °C	35 ft · lb at 0 °F	47.5 J at −34 °C	35 ft · lb at −30 °F

(a) Impact energy is the minimum toughness requirement at a given temperature for the indicated minimum service temperature (MST). (b) If the yield strength of the material exceeds 448 MPA (65 ksi), the temperature for the Charpy energy value for acceptability shall be reduced by 8 °C (15 °F) for each increment of 69 MPa (10 ksi) above 448 MPa (65 ksi).

Table 2 Typical Charpy/K_{Ic} correlations for steels

Correlation	Transition temperature regime	Correlation	Transition temperature regime
Barsom (Ref 18) $$\frac{K_{Id}^2}{E} = 5 \text{ (CVN)}$$ **Barsom-Rolfe (Ref 19)** $$\frac{K_{Ic}^2}{E} = 2 \text{ (CVN)}^{3/2}$$ **Sailors-Corten (Ref 20)** $$\frac{K_{Ic}^2}{E} = 8 \text{ (CVN)}$$	K_{Ic}, K_{Id} = psi$\sqrt{\text{in.}}$ E = psi CVN = ft · lb	**Marandet-Sanz—three steps (Ref 22)** $$T_{100} = 9 + 1.37\, T_{28}J$$ $$K_{Ic} = 19 \text{ (CVN)}^{1/2}$$ Shift K_{Ic} curve through T_{100} point	T_{100} = °C, for which K_{Ic} = 100 MPa$\sqrt{\text{m}}$ T_{28} = °C, for which CVN = 28 J K_{Ic} = MPa $\sqrt{\text{m}}$ CVN = J
$$K_{Id}^2 = 15.873 \text{ (CVN)}^{3/8}$$	K_{Id} = ksi$\sqrt{\text{in.}}$ CVN = ft · lb	**Wullaert-Server (Ref 23)** $$K_{Ic.d} = 2.1 \, (\sigma_y \text{ CVN})^{1/2}$$	$K_{Ic.d}$ = ksi$\sqrt{\text{in.}}$ CVN = ft · lb σ_y = ksi corresponding to approximate loading rate
Begley-Logsdon—three points (Ref 21) $(K_{Ic})_1 = 0.45\,\sigma_y$ at 0% shear fracture temperature $(K_{Ic})_2$ From Rolfe-Novak Correlation at 100% shear fracture temperature $(K_{Ic})_3 = \frac{1}{2}[(K_{Ic})_1 + (K_{Ic})_2]$ at 50% shear fracture temperature	K_{Ic} = ksi$\sqrt{\text{in.}}$ σ_y = ksi	**Upper shelf region** **Rolfe-Novak—σ_y > 100 ksi (Ref 24)** $$\left(\frac{K_{Ic}}{\sigma_y}\right)^2 = 5 \text{ (CVN/}\sigma_y - 0.05)$$ **Ault-Wald-Bertolo—ultrahigh-strength steels (Ref 25)** $$\left(\frac{K_{Ic}}{\sigma_y}\right)^2 = 1.37 \text{ (CVN/}\sigma_y) - 0.045$$	K_{Ic} = ksi$\sqrt{\text{in.}}$ CVN = ft · lb σ_y = ksi K_{Ic} = ksi$\sqrt{\text{in.}}$ CVN = ft · lb σ_y = ksi

1.0 ksi = 6.8948 MPa; 1.0 ksi$\sqrt{\text{in.}}$ = 1.099 MPa$\sqrt{\text{m}}$; 1.0 ft · lbf = 1.356 J
Note: CVN is the designation for Charpy impact energy; σ_y is the yield stress; and E is the Young's modulus.

where V_o is the initial impact velocity, and m is the effective mass of the pendulum. The ability to separate the total absorbed energy into components greatly augments the information gained by instrumentation. Load-temperature diagrams can be constructed to illustrate the various fracture process stages indicative of the fracture mode transition from brittle to ductile behavior (Ref 31).

One of the primary reasons for the development of the instrumented Charpy test was to apply existing notch bend theories (slow bend) to the dynamic three-point bend Charpy impact test. Obtaining load information during the standard Charpy V-notch impact test establishes a relationship between metallurgical fracture parameters and the transition temperature approach for assessing fracture behavior (Ref 32). Initial studies concentrated on the full range of mechanical behavior from fully elastic in the lower Charpy shelf region to elastic-plastic in the transition region to fully plastic in the upper shelf region (see Fig. 7). Almost all reported studies have been performed on structural steels, with primary emphasis on the effect of composition, strain rate, and radiation on the notch bend properties (Ref 33, 34).

Interest in instrumented impact testing has expanded to include testing of different types of specimens (precracked, large bend), variations in test techniques (low blow, full-size components), and testing of many different materials (plastics, composites, aerospace materials, ceramics, etc.). The rapid growth in the application of instrumented impact testing has produced a correspondingly large demand for standardized test methods. However, no standard currently exists for instrumented Charpy testing.

Instrumentation for a typical Charpy impact testing system includes an instrumented striker, a dynamic transducer amplifier, a signal-recording and display system, and a velocity-measuring device. The instru-

Fig. 7 Load-time response for a medium-strength steel

P_M, maximum load; P_{GY}, general yield load; P_F, fast fracture load (generally cleavage); P_A, arrest load after fast fracture propagation; t_M, time to maximum load; t_{GY}, time to general yield; W_M, energy absorbed up to maximum load

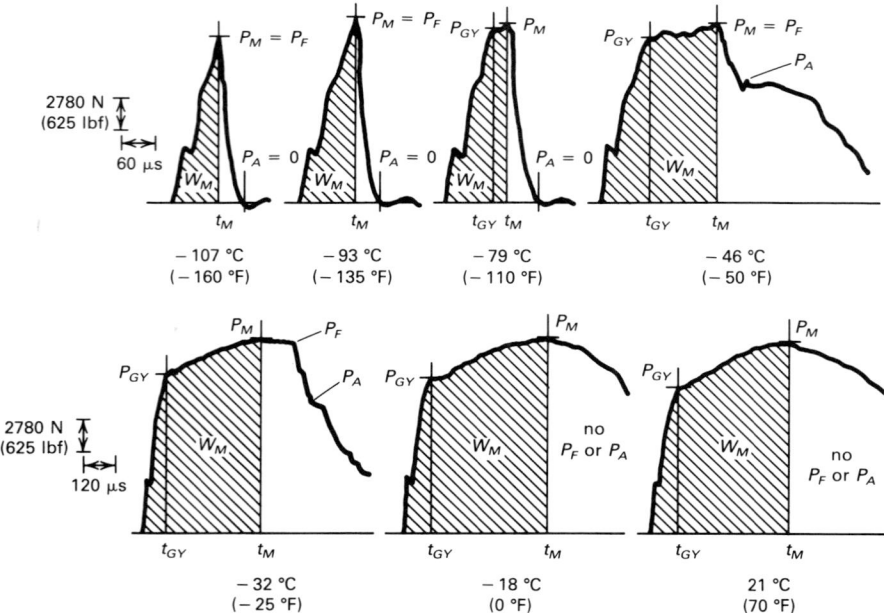

mented striker is the dynamic load cell, which is securely attached to the falling weight assembly. The striker has cemented strain gages to sense the compression loading of the tup while it is in contact with the test specimen. The dynamic transducer amplifier provides direct-current power to the strain gages and typically amplifies the strain gage output after passing through a selectable upper-frequency cutoff.

The impact signal is recorded and stored either on a storage oscilloscope or through the use of a transient signal recorder. Digital data from a transient recorder can be reconverted back to analog form and plotted on an x-y recorder, or the digital data can be transferred to a computer for direct analysis.

Triggering is best accomplished through an internal trigger that has the ability to capture the signal preceding the trigger; external triggering from the velocity-sensing device is often used instead of an appropriate internal trigger. The velocity-measuring system should be a noncontacting, optical system that clocks a flag on the impacting mass immediately before impact so that initial velocity measurements can be made. Velocities must be determined for all impact drop heights used.

The impact machine and the instrumentation package must be calibrated to ensure reliable data. Calibration of the Charpy pendulum impact machine is performed in accordance with ASTM E 23 (Ref 13), as discussed previously in this article in terms of

periodic proof testing of AMMRC calibration specimens to ensure reliable dial energy values.

Instrumentation calibration consists of a time base and load cell calibration with a system frequency response measurement. The time base calibration consists of passing a known time mark pulse through the system and calibrating accordingly. The load cell calibration is typically accomplished by testing notched specimens of 6061-T651 aluminum that are only slightly loading rate sensitive over the range used (Ref 35). The load cell is calibrated when the measured dynamic limit load is only slightly higher than the predetermined quasi-static limit load (measured using the same loading arrangement and anvil dimensions) and when the dial energy (or velocity-determined energy measurement) matches the integrated total energy. The relationship used for obtaining total absorbed energy (ΔE_o) from the area under the load-time record follows the approach in Eq 3 and 4.

The calculated ΔE_o value will match the dial energy reading when the system is calibrated (in addition to the limit load check). Because the aluminum limit load is fairly low (around 7.1 kN, or 1600 lbf), a check on load cell linearity at higher loads is also needed. To accomplish this, the integrated energy/dial energy requirement for a quenched and tempered 4340 specimen (HRC 52) that has a higher fracture load (near 27 kN, or 6000 lbf) is checked.

Low-energy AMMRC calibration specimens can be used for this procedure. If the energies match for the 4340 test at the same amplifier gain as for the aluminum calibration, the load cell calibration is usually linear throughout the usable load range. Static linearity checks can also be made if the static loading system exactly duplicates the dynamic loading conditions. Daily test checks using the aluminum calibration specimens are suggested to verify load cell calibration.

The system frequency response is determined experimentally by superimposing a constant-amplitude sine wave signal on the output of the strain gage bridge circuit (Ref 36). The peak-to-peak amplitude of the signal should be equivalent to approximately half the full-scale capacity of the load transducer at a frequency low enough to ensure no signal attenuation. The frequency of the sine wave is then increased until the amplitude is attenuated 10% (0.915 dB), and the response time, T_R, is calculated as:

$$T_R = \frac{0.35}{f_{0.915}} \quad \text{(Eq 5)}$$

where $f_{0.915}$ is the frequency at 0.915 dB (10%) attenuation.

General Test Requirements. The International Institute of Welding first attempted to standardize the instrumented Charpy test, but concluded that the test was not sufficiently documented, and the effort was discontinued (Ref 37). A few years later, two significant events prompted serious consideration of standardization. The development of the K_{IR} curve by the Pressure Vessel Research Committee and its inclusion in the ASME Code, Section III, created the need for dynamic initiation toughness, K_{Id}, data. The formation of an ASTM Task Group (E24.03.03) to standardize precracked Charpy testing also encouraged standardization.

Simultaneously, two other related groups began formulating procedures and conducting interlaboratory round robins. A Pressure Vessel Research Committee/Metals Property Council Task Group on Fracture Toughness Properties for Nuclear Components developed procedures for measuring K_{Id} values from precracked Charpy specimens (Ref 38). Also, the Electric Power Research Institute (EPRI) funded work to develop the "EPRI Procedures" (Ref 35). Currently, only subtle differences exist between the two sets of procedures (Ref 39). The following test requirements were taken from the updated EPRI procedures.

The load signal obtained from an instrumented striker during an impact test oscillates about the actual load required to deform the specimen. Therefore, the signal analysis procedure employed should minimize the deviation of the apparent load from the actual specimen deformation load. A simplistic view of the impact event allows three major areas for test specification to be identified: inertial loading, limited frequency response, and electronic curve fitting.

The impact loading of a specimen will create inertial oscillations in the contact load between striker and specimen, and a time interval between 2τ and 3τ is required for the load to be dissipated, where τ is related to the period of the apparent specimen oscillations and can be predicted empirically for a span-to-width ratio of 4 by:

$$\tau = 3.36\left(\frac{W}{S_o}\right)(EBC_s)^{1/2} \qquad \text{(Eq 6)}$$

where W is the specimen width, B is the specimen thickness, C_s is the specimen compliance, E is the Young's modulus, S_o is the speed of sound in the specimen, and τ is typically 30 μs for standard Charpy steel specimens. When any time, t, is less than 2τ, it is not possible to use the striker signal to measure the portion of the specimen load caused by inertial effects. An empirical specification for reliable load and time evaluation is:

$$t \geq 3\tau \qquad \text{(Eq 7)}$$

Control of t is obtained by control of the initial impact velocity. The constant 3 in Eq 7 may be as low as 2.3 without adversely affecting the test results, if the curve-fitting technique described below is followed. A value of 3 was chosen for the case of "unlimited" frequency response. The original EPRI procedures corresponded to the 2.3 factor and included the selective filtering for curve fitting (Ref 35). Computer simulations of the Charpy test have approximately verified the value of τ and the 3τ criterion (Ref 40).

The potential problem of limited frequency response of the transducer amplifier is avoided by specifying:

$$t \geq 1.1T_R \qquad \text{(Eq 8)}$$

where T_R is defined as the 0.915-dB response time of the instrumentation, as indicated in Eq 5. Inadequate response results in a distorted signal response. It is important to note that the electronic attenuation must be representative of a resistance-capacitance circuit for Eq 8 to apply.

The curve fitting of the oscillations is achieved by specifying a minimum T_R. The amplitude of the observed oscillations is therefore reduced such that the disparity between tup contact load and the effective deformation load is minimal. For the best

test, it has been empirically found for resistance-capacitance circuit systems that

$$T_R \geq 1.4t \qquad \text{(Eq 9)}$$

is adequate for the electronic curve fitting without altering the overall curve, when $t \geq 2.3\tau$. When $t \geq 3\tau$, it is not necessary to electronically curve fit, because the disparity between the contact load and the specimen deformation load is less than about 5%.

The requirements for obtaining acceptable load-time records (in particular, Eq 7) result in the need to control V_o. By controlling the impact velocity, a corresponding control of E_o is inherent. The reduction in striker velocity during the impact loading of the specimen should therefore be minimized. A conservative requirement is:

$$E_o \geq 3W_M \qquad \text{(Eq 10)}$$

where W_M is the system energy dissipated to maximum load P_M. This requirement ensures that the tup velocity is not reduced by more than 20% up to maximum load. This requirement is seldom a problem for full-impact Charpy V-notch tests; Eq 10 may not be met, however, when precracked Charpy tests are conducted for very tough materials.

The test requirements for reliable load measurement are summarized as follows:

Inertial effects . . . . $t \geq 3\tau$
Limited frequency response . . $t \geq 1.1\ T_R$, required only
             if $2.3\tau \leq t < 3\tau$
Electronic curve fitting . . . . . . $T_R \geq 1.4\tau$
Energy criterion . . . . $E_o \geq 3\ W_M$

The time t corresponds to the shortest time required for measurement after the specimen has been impacted; that is, t is the time to maximum load t_M for the elastic fracture, and t is the time to general yield, t_{GY}, in the post-general yield fracture (see Fig. 7). The specification for electronic curve fitting is only required if $2.3\tau \leq t < 3\tau$. Because it is often difficult to ensure that $t \geq 3\tau$ and because the filtering has no adverse effect when $t \geq 2.3\tau$, filtering at $T_R \geq 1.4\tau$ is always possible, assuming that $t \geq 1.1\ T_R$.

Limitations on Testing. Violation of any of the general test requirements presented above will invalidate the data obtained from instrumented Charpy V-notch tests. Limitations of this testing technique are the same as those for standard Charpy testing. The effects of small size relative to typical component size, the rounded machine notch, and shallow notch depth restrict general applicability and usefulness of the Charpy test. Note that the notch depth for the Charpy V-notch specimen is too shallow to prevent yielding across the gross section of the specimen (Ref 34).

Instrumentation has allowed separation of energy components and measurement of applied loads throughout the fracture event, but direct determination of the initiation component is not directly possible for ductile (microvoid coalescence) initiation from the instrumented test record. Some of these limitations have been addressed by fatigue precracking the Charpy specimen, which eliminates the notch effects and makes it a small fracture-mechanics-type specimen.

Precracked Charpy Test

By inducing a fatigue precrack in the Charpy specimen, the notch acuity and depth restrictions are eliminated. Early work concentrated on correlations with fracture toughness using only the total absorbed energy (i.e., uninstrumented testing). These energy values usually are normalized per unit area below the fatigue crack; the normalized energy values are designated as W/A.

Most of the correlations of W/A with fracture toughness have been conducted using slow-bend specimens. The basic problem in reaching an impact correlation is the difference in loading rates between the Charpy impact and the static K_{Ic} tests, particularly for loading rate sensitive materials (Ref 41). A general trend exists for a correlation between K_{Ic}^2/E and W/A, but the limited data and scatter make this difficult to utilize (Ref 25). A better correlation with K_{Id} may be possible. The reason for using K_{Ic}^2/E as the basis is the approximate proportionality between K_{Ic}^2/E and W/A, based on a presumed fracture mechanics relationship (Ref 41).

The precracked Charpy W/A values can also be used to estimate the nil-ductility transition temperature. The typical technique defines an inflection point between lower shelf and transition region behavior as the estimated nil-ductility transition temperature (Ref 42). Some exceptions have been noted to this approach (Ref 43).

Instrumented Data

The types of data and test techniques used for instrumented precracked Charpy testing are the same as those discussed earlier for instrumented Charpy impact testing. The 3τ criterion, which limits the impact velocity, becomes more important for deeply cracked, brittle materials. The greatest advantage of precracking is the transformation of the Charpy V-notch specimen into a dynamic fracture mechanics test piece. The direct calculation of fracture toughness (within certain limitations) is now possible using the instrumented load-time information. The following discussion presents the calculational aspects of these fracture toughness parameters.

If fracture is known to initiate at maximum load (as it usually does for cleavage initiation), the energy value of W_M (see Fig. 7) can be considered an initiation energy. However, W_M includes contributions other than that caused by the deflection of the specimen. Therefore, a compliance energy correction is needed to determine the true specimen energy, E_M (Ref 44). When the fracture is linear elastic (fracture before general yield; see the first two load-time records in Fig. 7), the value of E_M can be calculated directly:

$$E_M = \frac{C_{ND}(P_M^2)}{2EB} \qquad \text{(Eq 11)}$$

where C_{ND} is the nondimensional specimen compliance (Ref 45). For a fracture occurring after general yield (see Fig. 7), E_M is obtained by correcting W_M:

$$E_M =$$
$$W_M - \left\{ \left(\frac{P_M^2}{2} \right) \left[C_T - \left(\frac{C_{ND}}{EB} \right) \right] \right\} \qquad \text{(Eq 12)}$$

where C_T is the total system compliance calculated at general yield and corrected for the decrease in velocity through general yield:

$$C_T = \left(\frac{V_o t_{GY}}{P_{GY}} \right) - \left(\frac{V_o^2 t_{GY}^2}{8E_o} \right) \qquad \text{(Eq 13)}$$

This compliance correction is assumed to be linear with load, but the actual correction is not so simple. However, the error in assuming a linear relationship results in a slightly smaller (conservative) value of E_M (Ref 39).

It is often desirable to partition the total fracture energy into initiation and post-initiation (propagation) components. Assuming initiation occurs at maximum load, the propagation energy E_P is:

$$E_P = E_T - E_M \qquad \text{(Eq 14)}$$

where E_T is the total fracture energy, as determined from a dial indicator, kinetic energy change (initial and final velocity measurements), or ΔE_o.

Linear Elastic Fracture Toughness. When the fracture is elastic (fracture occurs before general yield), the stress-intensity factor, K_{Ic}, can be calculated by applying linear elastic fracture mechanics (Ref 1):

$$K_{Ic} = \frac{4P_M}{BW^{1/2}}$$
$$\times \left\{ 3 \left(\frac{a}{W} \right)^{1/2} \left[1.99 - \left(\frac{a}{W} \right) \left(1 - \frac{a}{W} \right) \right] \right.$$
$$\left. \times \left(2.15 - 3.93 \frac{a}{W} + 2.7 \frac{a^2}{W^2} \right) \right\}$$

$$\div \left\{ 2 \left(1 + \frac{2a}{W} \right) \left(1 - \frac{a}{W} \right)^{3/2} \right\}$$
$$\text{(Eq 15)}$$

where P_M is the maximum load the specimen was able to sustain, a is the crack length, W is the width of the specimen, and B is the thickness of the specimen.

The ASTM size requirements for a valid K_{Ic} are quite limiting, even if a dynamic yield strength is used. However, the general specimen size requirements of ASTM E 399 may be too conservative for dynamic testing of ferritic medium-strength steels (Ref 46). Therefore, if general yielding has not occurred, a linear elastic value of fracture toughness, K_{Ic}, generally is calculated. The stress intensification rate $\dot{K}$ is calculated as:

$$\dot{K} = \frac{K_{Ic}}{t_M} \qquad \text{(Eq 16)}$$

where t_M is the time to maximum load.

This loading rate reflects the dynamic aspect of the loading, because the lowest $\dot{K}$ for impact loading of precracked Charpy specimens is on the order of 11×10^4 MPa$\sqrt{m} \cdot s^{-1}$ (1×10^5 ksi$\sqrt{in.} \cdot s^{-1}$).

Post-General Yield Fracture Toughness. When general yielding occurs, an energy-based value of the J-integral can be used to obtain a measure of fracture toughness. The calculation of ductile fracture toughness, J_{Ic}, is contingent upon knowing the initiation point of fracture on the load-time record. For cleavage-initiated fracture, this point generally corresponds to maximum load. However, for fibrous (ductile) initiation, maximum load is generally a nonconservative assumption. When the initiation point is known or has been determined experimentally (Ref 47) and when $a/W \geq 0.5$ (Ref 48), then:

$$J_{Ic} = \frac{2E_M}{Bb} \qquad \text{(Eq 17)}$$

where b is the remaining ligament depth ($W - a$). A stress-intensity factor K_{Jc} can be obtained from the J_{Ic} value as:

$$K_{Jc} = (EJ_{Ic})^{1/2} \qquad \text{(Eq 18)}$$

An average $\dot{K}$ can also be computed, as in Eq 16, by using a K_{Jc} value. Validity criteria related to specimen dimensions appear to be (Ref 39):

$$a, b, B \geq 25 \left(\frac{J_{Ic}}{\sigma_f} \right) \text{ (fibrous initiation)}$$
$$\text{(Eq 19)}$$

and

$$a, b, B \geq 50 \left(\frac{J_{Ic}}{\sigma_f} \right) \text{ (cleavage initiation)}$$
$$\text{(Eq 20)}$$

where σ_f is the flow stress, defined as the average of the yield stress and the ultimate stress. For dynamic loading, the yield stress, σ_y, and flow stress, σ_f, of standard Charpy V-notch specimens can be estimated for post-general yield behavior as:

$$\sigma_y = 2.99 P_{GY} \frac{W}{Bb^2} \qquad \text{(Eq 21)}$$

and

$$\sigma_f = \frac{[2.99(P_{GY} + P_M)W]}{2Bb^2} \qquad \text{(Eq 22)}$$

The general form of the equation results from slip-line field solutions for blunt-notch specimens, and the constant 2.99 has been obtained from extrapolation of results from a slip-line field solution that included the tup indentation at the center loading point (Ref 49). The constant of 2.99 reduces to 2.85 for sharp-notch specimens with a fatigue precrack. The stress values obtained using this approach agree favorably with high rate tensile test results (Ref 50).

Limitations. The test requirements and data analysis procedures described in this article were developed for ferritic pressure vessel steels. Their applicability to other materials has not been evaluated fully. Work currently is underway with ASTM Committee E-24 to standardize the instrumented precracked Charpy test, but only for fracture before general yield. A review of instrumented precracked Charpy testing can be found in Ref 51, which discusses the theory and applicability of instrumented precracked Charpy testing. Not all of the relationships and approaches presented in Ref 51 are universally accepted, because standards or recommended practices do not currently exist.

Related Test Techniques

Several attempts have been made to use the precracked Charpy specimen at loading rates beyond the limits applicable to quasi-static analysis. The procedures described above assume a quasi-static situation for times greater than the limiting values near 3τ. One such attempt for larger than Charpy-size specimens is described in Ref 52, in which strain gages were mounted near the crack tip to avoid many of the spurious wave effects. Other studies have been conducted using Hopkinson bar techniques (Ref 53) and the shadow optic method of caustics (Ref 54). These studies indicate the need for dynamic analysis when using the instrumented Charpy striker approach at high loading rates.

Concept of Impact Response Curves

By Jörg F. Kalthoff
Fraunhofer-Institut für
Werkstoffmechanik

INTERPRETATION of load-time records obtained in instrumented impact tests with precracked Charpy specimens can be difficult, particularly when the early time range is considered. This difficulty and the use of a quasi-static evaluation procedure for determining the dynamic impact fracture toughness K_{Id} restricts the range of applicability of the test. To achieve sufficiently small times-to-fracture, the impact velocity must be limited. The maximum allowable velocity depends on the toughness of the material to be tested. These difficulties and restrictions are overcome by applying the concept of impact response curves for measurement of the impact fracture toughness K_{Id} (Ref 54-56).

For fixed test conditions, such as specimen geometry (particularly crack length) and impact velocity, the history of dynamic crack tip stress-intensity factor, $K_I^{dyn}(t)$, is established for the specific impact process considered. This $K_I^{dyn}(t)$ curve quantitatively relates the response of the specimen to the impact event and therefore is called the "impact response curve."

This curve depends only on the elastic reaction of the specimen-striker system and therefore is unique for the system considered and applies to all steel specimens tested under the same impact conditions. Consequently, this one relationship applies to steels of different toughnesses as long as the elastic properties of the steels, i.e., the elastic modulus and Poisson's ratio, are the same and the conditions for linear elastic fracture mechanics or small-scale yielding behavior are fulfilled.

The dynamic fracture toughness for a given structural steel is then determined by performing an impact experiment with a specimen made from the steel under study and measuring the resulting time-to-fracture. The dynamic fracture toughness value K_{Id} is obtained from the preestablished impact response curve and the measured time-to-fracture t_f (see Fig. 8) by using the relation:

$$K_{Id} = K_I^{dyn}(t = t_f) \qquad \text{(Eq 23)}$$

This fully dynamic procedure for measuring the dynamic fracture toughness does not have the previously discussed restriction that the time-to-fracture must be larger than a certain minimum value (see Eq 7). Therefore, the procedure can be applied for all experimental test conditions, particularly in the brittle fracture and high-velocity impact range, as long as the usual conditions for small-scale yielding are fulfilled.

Impact Response Curves

The impact response curves can be measured or calculated. Pure numerical calculations generally would require the consideration of both the specimen and the striking hammer. Numerical efforts are reduced considerably if experimental data, such as the load or displacement history measured at the tup of the striking hammer, are used as input data for the numerical calculation. Efforts to establish numerical relationships quantifying the specimen response during the impact process have been undertaken (Ref 57-64).

Systematic studies on the basis of semianalytical considerations discuss the specific influence of individual test parameters (Ref 65, 66). Experimental techniques based on optical methods—e.g., the photoelastic technique of isochromatic fringes and the shadow optical method of caustics (Ref 67, 68)—are applicable, but preferably are to be used with specimens that are larger than Charpy specimens, such as specimens used in drop weight arrangements (Ref 54, 59-61, 67, 68).

Strain gage instrumentation of the specimen near the crack tip provides the simplest way to establish impact response curves with precracked Charpy specimens (Ref 52, 69). In a static preexperiment, the signal obtained from a strain gage at a specific location near the crack tip is calibrated in terms of the stress-intensity factor. Because the strain gage is located near the crack tip, the strain gage signal obtained in an impact event can be assumed to represent a good measurement of the dynamic crack tip stress-intensity factor as well.

Figure 9 illustrates impact response curves for steel Charpy specimens with an initial crack length of a_o = 5 mm (0.2 in.) tested at different impact velocities of V_o = 2, 3.8, and 5 m/s (6.6, 12.5, and 16.4 ft/s). The data were obtained with a 300-J (220-ft · lb) pendulum test device (tup radius of 2 mm or 0.08 in.), with a machine compliance of $C_M = 8.1 \times 10^{-9}$ m/N (1.4×10^{-6} in./lbf). The strain gages were positioned about 2 mm (0.08 in.) to the side of the crack tip. To increase the load-carrying capacity of the specimen, blunted notches instead of fatigue-sharpened initial cracks were utilized in the experiments.

The impact response curves for different impact velocities are similar. The stress-intensity factors K_I^{dyn} depend linearly on the impact velocity V_o and are independent of the hammer mass m_o, provided the energy used for breaking the specimen is small compared

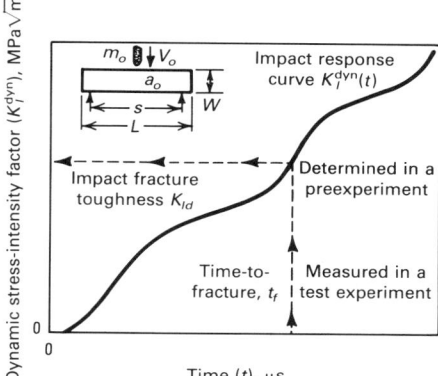

Fig. 8 Determination of the dynamic fracture toughness, K_{Id}, by impact response curves

m_o, mass of hammer; s, distance between supports

Fig. 9 Impact response curves for steel precracked Charpy specimens tested at different impact velocities

Initial crack length: 5 mm (0.2 in.) or a/W = 0.5

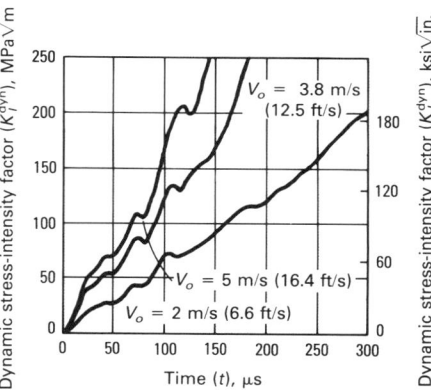

to the total impact energy (Ref 66). It can be seen from Fig. 9 that this relationship can be mathematically described by a linear dependence of K on time with superimposed dynamic corrections. This condition is usually fulfilled with conventional pendulum impact test devices, if the specimen breaks in the early time range. The impact response curves for arbitrary impact velocities can therefore be represented by a single relationship. This relationship differs with crack length, but for the small variations in crack length used in bend specimens ($0.45 < a/W < 0.55$ as recommended in ASTM E 399), the resulting differences are only modest and can be compensated for in an approximate manner.

For practical applications, it is convenient to use the expression:

$$K_I^{dyn} = R \cdot V_o \cdot t'' \qquad \text{(Eq 24)}$$

where

$$t'' = f(t') \text{ given by Table 3}$$

and

$$t' = g(t) = t\left\{1 - 0.62\left(\frac{a}{W} - 0.5\right) + 4.8\left(\frac{a}{W} - 0.5\right)^2\right\}$$

where $R = 301$ GN/m$^{5/2\dagger}$, V_o is the impact velocity, a is the crack length, W is the specimen width, t'' and t' are modified times, and t is the measured physical time. The functions of f and g account for the dynamic corrections and variations of crack length, respectively.

This approximate formula shows good agreement when compared with semianalytical results (Ref 66). Equation 24 describes the impact response curves for all practically relevant test conditions with an accuracy sufficient for engineering purposes.

Time-to-Fracture Measurements

The time-to-fracture of a precracked Charpy specimen subjected to impact loading can be obtained from signals of two uncalibrated strain gages, one of which is located on the tup of the striking hammer and the other on the specimen to the side of the crack tip. The leading edge of the signal from the hammer strain gage marks the beginning of the impact event. The onset of crack propagation, on the other hand, is indicated by the rapid drop in load registered by the crack tip strain gage. The time-to-fracture, t_f, is the interval between the two signals. Typical oscillograms of time-to-fracture measurements are shown in Fig. 10.

An indication of the time at which the crack becomes unstable can also be obtained by another procedure, which does not require specimen instrumentation. At the onset of rapid crack propagation, a magnetic signal is generated by the accelerating crack tip if the specimen has been slightly magnetized before testing—with a permanent magnet, for example (Ref 70). This signal is picked up by a magnetic sensor (e.g., a coil) located at the crack tip near, but not in contact with, the specimen surface. Very good results have been obtained with commercially available magnetic pickups that are used in tape recorders.

†This value for R applies for stiff test devices with a machine compliance $C_M = 8.1 \times 10^{-9}$ m/N. If the actual compliance of the test device should differ from this value, the resulting influence can be taken into account by multiplying the given value of R by the first-order correction factor $1.276/(1 + 0.276\ C_M/8.1 \times 10^{-9}$ m/N). Procedures for determining the machine compliance of impact test devices are described in Ref 36.

Fig. 10 Time-to-fracture measurement of a precracked Charpy specimen

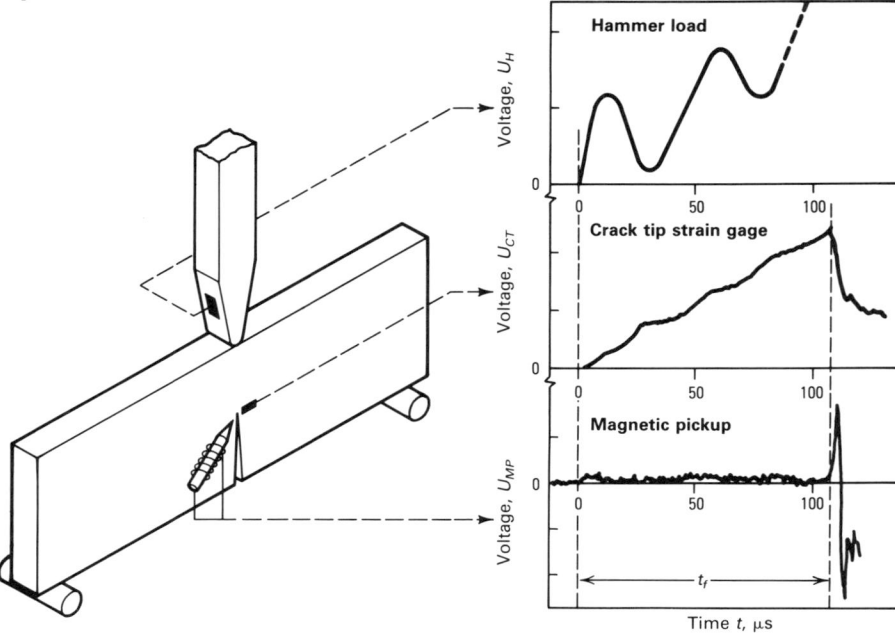

The signal thus obtained has a short rise time and gives a clear indication of the time at the moment of fracture instability. Figure 10 shows a magnetic crack initiation signal compared to a crack tip strain gage signal. This magnetic measuring procedure is inexpensive and highly advantageous for routine testing, because it does not require extensive or costly specimen instrumentation.

Examples of static and dynamic impact data are shown in Fig. 11. Specimens of normalized StE 460 (Fe-0.27C-1.52Mn-0.62Ni-0.28Si-0.18V-0.09P-0.09S) and quenched and tempered 30CrNiMo8 (Fe-0.27C-2.1Cr-1.9Ni-0.3Mo) were tested at different temperatures at an impact velocity of 5 m/s (16.4 ft/s). Despite the short times-to-fracture, $t_f < 3\tau$ (see Eq 6 for explanation of τ), the impact response curve technique yields reliable and meaningful K_{Id} data, as indicated in Fig. 11 by a comparison with equivalent static fracture toughness (K_{Ic}) data.

Advantages and Limitations

The technique of measuring impact fracture toughness, K_{Id} with impact response curves and time-to-fracture measurements has several advantages over the conventional quasi-static technique described previously in this article in the section on precracked Charpy testing. The impact response curve technique represents a fully dynamic evaluation. Kinetic effects are correctly accounted for during the entire impact event. The method can thus be applied to all experimental test conditions, particularly in the short time-to-fracture range, i.e., when high impact velocities are used or brittle materials are tested.

This method does not require a calibrated instrumentation of the hammer, which is usually a prerequisite in impact experiments designed to determine the load at crack initiation. The data-measuring procedure consists of two separate tasks: determination of the impact response curve and the measurement of the time-to-fracture. The more complicated determination of the impact response curve need only be carried out once.

An approximate formula for impact response curves has been established that applies to different experimental conditions of practical concern. The actual K_{Id} determination requires only a relatively simple and inexpensive measurement of the time-to-fracture. Time-to-fracture, in turn, can be determined from signals obtained by uncalibrated instrumentations of the hammer and specimen. Specimen instrumentation can also be avoided by sensing the magnetic signal generated by the crack at the moment of fracture instability. Thus, the dynamic fracture toughness, K_{Id}, of a given steel is determined by measuring only the time-to-fracture in an impact experiment and by utilizing the particular impact response curve that applies to the prevailing experimental conditions.

The concept of impact response curves extends the conventional quasi-static evaluation procedure into the lower time-to-fracture range. It is an appropriate measuring tool for testing specimens that fail at relatively short times after impact; i.e., the procedure is particularly suitable for testing at high velocities and for testing brittle materials. The applicability range of the procedure is limited by the usual small-scale yielding conditions, because the uniqueness of impact response curves is lost when large plastic deformations are present. Under the conditions of large plastic deformations, however, the resulting times-to-fracture generally are sufficiently long so that the quasi-static evaluation procedure can be applied successfully. Both procedures therefore have their specific ranges of applicability and complement each other.

The technique is not restricted to Charpy-size specimens. Impact response curves for large bend specimens are shown in Fig. 12. These curves were obtained by testing specimens of different sizes in a drop-weight tower with variable mass at an impact velocity of 5 m/s (16.4 ft/s) using the shadow optical method of caustics. The procedure for determining the impact fracture toughness, in particular for measuring the time-to-fracture, is analogous to the one described before for Charpy specimens. As shown in Fig. 9, these data apply for any impact velocity. The following section on the one-point bend test describes another application of the impact response curve technique.

One-Point Bend Test

By Jacques Henri Joseph Giovanola
Research Engineer
SRI International

INVESTIGATIONS of the dynamic behavior of the impacted three-point bend specimen have shown that immediately after impact by the hammer, the ends of the specimen lose contact with the supports (Ref 71). Consequently, the initial part of the response curve discussed in the preceding section is strictly the result of inertial loading, without contribution from the reactions at the supports. Therefore, if high impact velocities are applied to achieve high loading rates and short times-to-fracture, it becomes unnecessary to support the bend specimen during impact, and a so-called one-point bend test can be performed (Ref 72-74).

The one-point bend test allows reliable toughness data to be obtained at loading rates one to two orders of magnitude higher than the maximum loading rate considered in ASTM E 399 for fracture tests (Ref 1). Several tests performed with different specimen sizes and impact velocities on high-strength 4340 steel have demonstrated the usefulness of the one-point bend test and have provided dynamic fracture toughness values in good agreement with values obtained with more complicated tests.

Test Principle

The one-point bend test uses a single-edge cracked specimen and the same testing arrangement as a conventional three-point bend test, except that the end supports are removed. The procedure is illustrated in Fig. 13. When the hammer strikes the specimen, the center portion of the specimen is accelerated away from the hammer; the end portions of the specimen lag behind by an amount d^{CL} because of inertia. This causes the specimen to bend and to load the crack tip.

A typical stress-intensity history resulting from such inertial loading is shown in Fig. 14. The impact conditions for the test were such that no crack extension occurred. The stress-intensity history is approximately sinusoidal; it increases monotonically up to the maximum amplitude and then decreases monotonically. Later oscillations in the stress-intensity history are of much smaller amplitude. For comparison, the hammer load is also shown in Fig. 14. It is apparent that there is no direct relationship between the hammer load history and the crack tip stress-intensity history in the one-point bend test.

The stress-intensity history obtained in the one-point bend test depends only on the impact velocity and on the specimen geometry, size, and material. It is independent of the fracture properties of the material tested. Therefore, the concepts of response curve and time-to-fracture described earlier in this article can also be applied to the one-point bend test when used for dynamic fracture toughness measurements in the small-scale yielding regime.

As described previously, dynamic fracture measurements involve two steps: determination of the response curve, i.e., the stress-intensity history, for a given specimen geometry and class of material and measurement of the time-to-fracture with a fatigue pre-cracked specimen. The dynamic fracture toughness is then evaluated using Eq 23.

Alternatively, the stress-intensity history and the stress-intensity level at which crack

Table 3   Time-function for determination of K_I^{dyn}

t', μs	$t''=f(t')$, μs	t', μs	$t''=f(t')$, μs	t', μs	$t''=f(t')$, μs	t', μs	$t''=f(t')$, μs	t', μs	$t''=f(t')$, μs	t', μs	$t''=f(t')$, μs
0	0	50	45	100	118	150	152	200	198	250	245
2	0	52	46	102	119	152	155	202	202	252	245
4	2	54	49	104	118	154	157	204	204	254	245
6	4	56	53	106	117	156	159	206	207	256	246
8	6	58	57	108	115	158	161	208	210	258	249
10	9	60	61	110	115	160	164	210	212	260	251
12	13	62	65	112	115	162	166	212	212	262	253
14	17	64	69	114	115	164	169	214	213	264	255
16	20	66	72	116	116	166	172	216	213	266	257
18	24	68	73	118	118	168	175	218	214	268	258
20	28	70	73	120	120	170	177	220	216	270	260
22	30	72	72	122	122	172	180	222	219	272	261
24	33	74	70	124	124	174	183	224	222	274	262
26	35	76	69	126	126	176	185	226	225	276	265
28	36	78	68	128	128	178	187	228	230	278	267
30	38	80	69	130	129	180	188	230	233	280	269
32	39	82	70	132	130	182	188	232	236	282	272
34	40	84	75	134	131	184	187	234	239	284	275
36	42	86	81	136	132	186	186	236	241	286	277
38	43	88	88	138	134	188	185	238	243	288	280
40	45	90	94	140	136	190	186	240	244	290	282
42	46	92	100	142	138	192	187	242	245	292	284
44	47	94	106	144	141	194	189	244	245	294	286
46	46	96	111	146	145	196	192	246	245	296	288
48	45	98	116	148	148	198	195	248	245	298	289

Fig. 11 Measured impact fracture toughness data for low-alloy steels tested at an impact velocity of 5 m/s (16.4 ft/s)

(a) Alloy 30CrNiMo8. (b) Alloy StE 460. Data points in parentheses are invalid according to ASTM Standards.

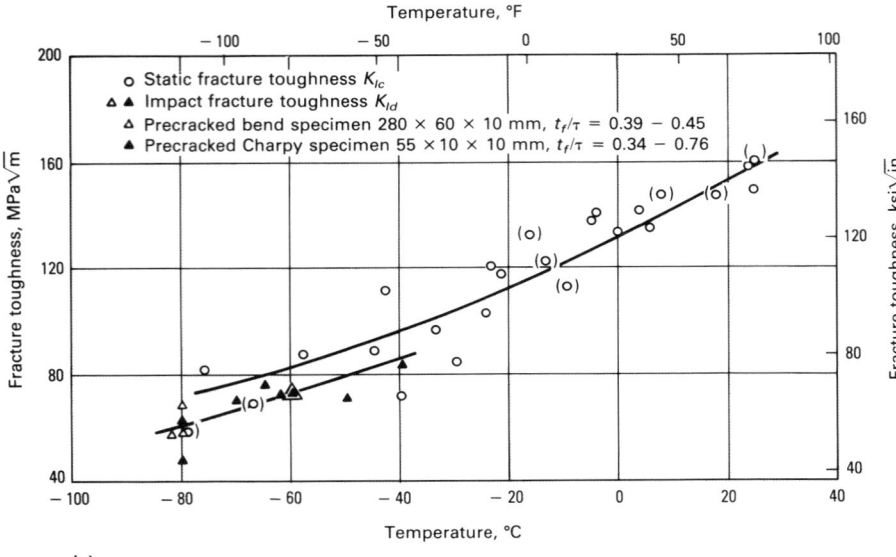

(a)

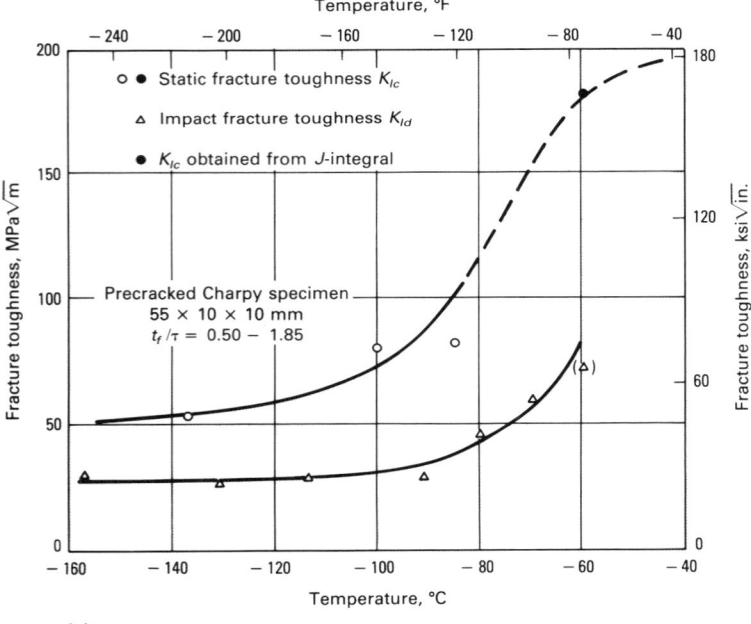

(b)

initiation occurs can be measured for each specimen by measuring the crack tip strain intensification with a calibrated strain gage placed near the crack tip. The measured strain histories can be related to the stress-intensity histories by two methods. In the first method, the strain gage signal is calibrated against a known applied stress intensity in a static three-point bend experiment. It is then assumed that the static calibration also applies in the dynamic case. In the second method, the strain history $\epsilon_y(t)$ is related directly to the stress-intensity history by means of the elastic singularity solution. The stress intensity is then given by:

$$K_I(t) = \frac{\epsilon_y(t)E\sqrt{2\pi r}}{\cos\dfrac{\theta}{2}\left[1 + \sin\dfrac{\theta}{2}\sin\dfrac{3\theta}{2} - \nu\left(1 - \sin\dfrac{\theta}{2}\sin\dfrac{3\theta}{2}\right)\right]}$$

(Eq 25)

where r and θ are the polar coordinates of the center of the strain gage with origin at the crack tip, and E and ν are Young's modulus and Poisson's ratio, respectively. Both methods work equally well if the strain gage is positioned properly. This aspect is discussed in a subsequent section.

Dependence of Stress-Intensity History on Test Parameters. As illustrated in Fig. 14, the stress-intensity history is characterized by the duration of loading, T_o, and the maximum stress-intensity amplitude, K_I^{max}. It has been shown (Ref 74) that T_o and K_I^{max} can be expressed as:

$$T_o = W\sqrt{\frac{\rho}{E}}\,h_1\left(\frac{L}{W}, \frac{a}{W}, \nu\right)$$

(Eq 26)

$$K_I^{max} = V_{imp}\sqrt{W}\sqrt{\rho E}\,h_2\left(\frac{L}{W}, \frac{a}{W}, \nu\right)$$

(Eq 27)

where W, L, and a are the width, length, and crack depth of the specimen, respectively; ρ, E, and ν are the mass density, Young's modulus, and Poisson's ratio, respectively; V_{imp} is the impact velocity; and h_1 and h_2 are nondimensional functions.

Equations 26 and 27 have been verified experimentally (Ref 74) and provide the scaling rules for the one-point bend test stress-intensity history. Given the stress-intensity history for one material, one specimen size and geometry, and one impact velocity, Eq 26 and 27 allow construction of the stress-intensity history for any material, impact velocity, and specimen size as long as geometrical similarity of the specimen is preserved. In particular, for given specimen dimensions and material, the one-point bend test stress-intensity history is proportional to the impact velocity and is independent of specimen thickness. For a specimen and impact velocity, it can further be shown that both K_I^{max} and T_o increase with an increase in crack length.

The experimental one-point bend test stress-intensity histories for several specimen geometries, normalized according to Eq 26 and 27, are shown in Fig. 15 and 16. These curves were obtained from steel specimens with blunt notches. Figure 15 illustrates the effect of changing the length-over-width ratio. Figure 16 presents the normalized stress-intensity curves for a specimen with the standard ASTM E 399 (Ref 1) proportions ($L/W = 4.2$) and for four ratios of crack length to specimen width ($a/W = 0.3, 0.4, 0.5,$ and 0.6).

Advantages and Limitations

The primary advantage of the one-point bend test is that the measured stress-intensity

Fig. 12 Impact response curves for large bend steel specimens
Impact velocity: 5 m/s (16.4 ft/s). Source: Ref 55, 56

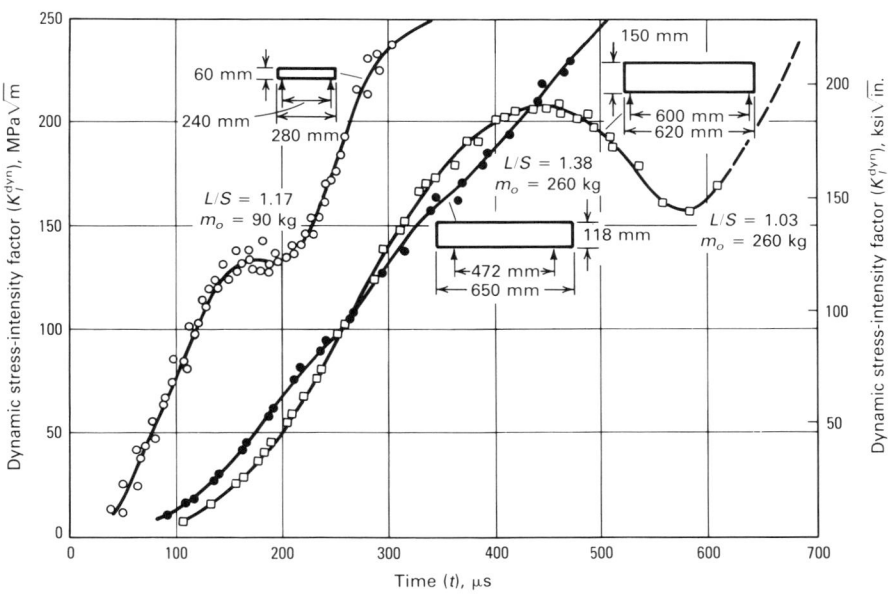

Fig. 13 Test arrangement and loading mechanism for the one-point bend test

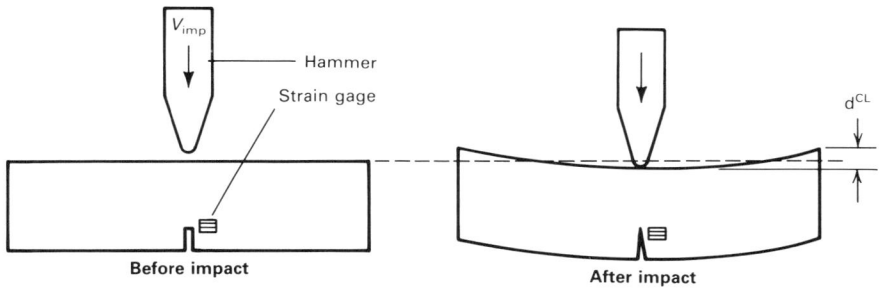

Fig. 14 Typical stress-intensity and load histories obtained in the one-point bend test

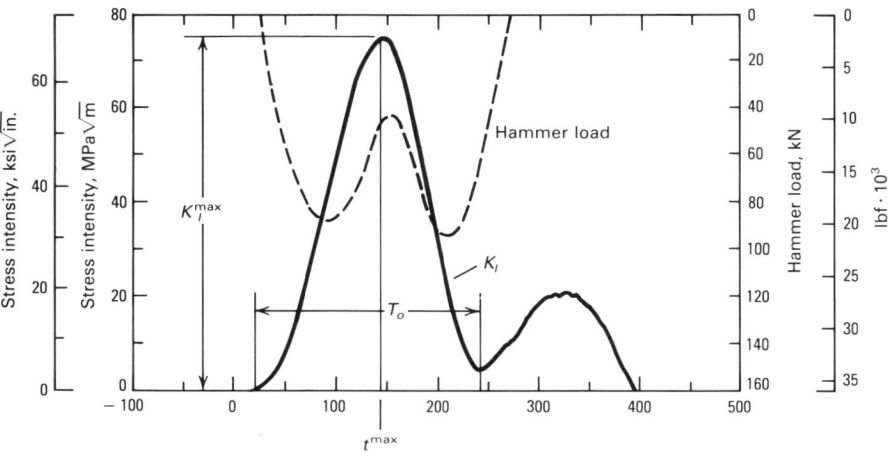

Fig. 15 Normalized stress-intensity histories for two specimens with different L/W ratios

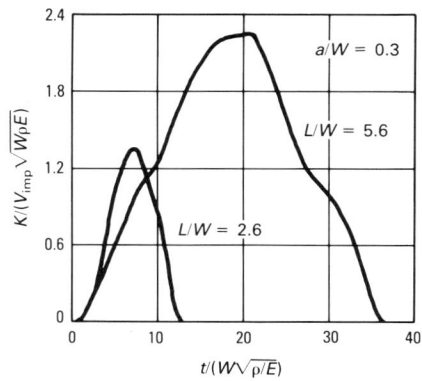

Fig. 16 Normalized stress-intensity histories for ASTM standard bend specimen with four different crack lengths

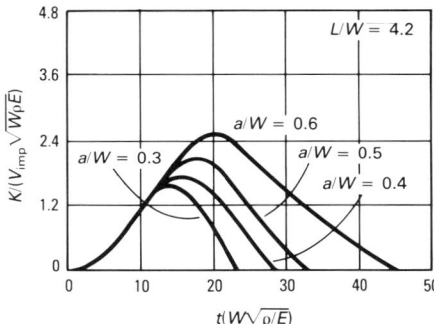

history incorporates dynamic effects completely. Therefore, no limits need to be imposed on the impact velocity and the test duration to fracture.

Combined use of the one-point bend test and the impact response curve concept produces a simple, smoothly varying response curve. For a given specimen geometry, the response curve must be measured experimentally only once. It can then be extended to other specimen sizes (of same aspect ratios), impact velocities, and materials by using the scaling rules of Eq 26 and 27.

Another advantage of the one-point bend test is that the test duration, T_o, and the stress intensification rate can be varied over an order of magnitude range (80 to 800 μs and 3×10^5 to 3×10^6 MPa$\sqrt{\text{m}} \cdot \text{s}^{-1}$, or 2.7×10^5 to 2.7×10^6 ksi$\sqrt{\text{in.}} \cdot \text{s}^{-1}$) by changing the specimen geometry and the impact velocity. Because loading occurs by inertia only, the one-point bend test response curve is insensitive to the characteristics of the loading machine, provided the hammer mass is much larger than the specimen mass.

When measurement of the stress-intensity history is direct, the test procedure to obtain dynamic fracture toughness using the one-point bend test is also simple. It requires only limited instrumentation, and interpretation of the test data is straightforward.

Use of the one-point bend test currently is restricted to small-scale yielding conditions. Furthermore, the maximum stress-intensity amplitude that can be obtained using convenient specimen sizes and a conventional pendulum impact tester is rather low, particularly for lighter alloys such as aluminum and titanium alloys. For these materials, devices capable of higher impact velocities are required. This problem can also be circumvented by attaching ballast plates to the ends of the specimen.

Test Setup and Instrumentation

Figure 17 illustrates the typical test setup for performing a one-point bend test on a pendulum impact tester. The specimen holder used in a Charpy or Izod test is replaced by a simple frame that supports the specimen, while allowing it to move freely in the horizontal plane. Depending on the design of the original pendulum and hammer, the impact tester may require retrofitting with a new hammer and striker that will not interfere with the specimen edges or the support frame. The striker should be instrumented with two foil gages—one on each lateral side—to record the impact load on the hammer.

In the response curve time-to-fracture approach, the time-to-fracture is defined as the interval between the time of impact and the time at which crack extension occurs. The time of impact can be determined conveniently from the leading edge of the hammer load signal, whereas the time of crack extension can be measured by an uncalibrated strain gage placed near the crack tip, or by a magnetic pickup (see Fig. 10). If the response curve is to be used later in conjunction with other impact velocities, it is also necessary to measure the hammer impact velocity.

When calibrated strain gages are used to perform a direct measurement of the stress-intensity history, the gage element should be approximately 2 by 2 mm (0.08 by 0.08 in.) or smaller. The center of the gage should be positioned 3 to 4 mm (0.12 to 0.16 in.) above the crack plane, on a line normal to the crack path and passing through the crack tip.

Two amplifiers with a frequency response greater than 100 kHz are required to process the hammer and specimen gage signal for recording. These signals are best recorded on a digital storage oscilloscope with a time resolution of 1 μs or better. If this recording instrument has the capability to store signals that occur before the trigger time, the hammer or specimen gage signals can be used as a trigger. Otherwise, a trigger source must be incorporated into the test setup. A laser beam interrupted by the hammer in conjunction with a photodiode is one option. Another is a simple electrical circuit closed by the contact of hammer and specimen; however, this method may cause transient noise on the strain gage signals.

Specimen Size and Geometry. To guarantee small-scale yielding conditions, specimens used in the one-point bend test are subject to the same size requirements as specimens used for static fracture measurements according to ASTM E 399 (Ref 1). In addition, restrictions on the value of the length-to-width ratio, L/W, must be imposed because of the natural modes of vibration of the specimen. It has been shown experimentally that for large values of L/W, higher mode frequencies are excited when the specimen is impacted, thus causing large oscillations in the stress-intensity history (Ref 74). To avoid this undesirable tendency, values of L/W should not exceed 7.

In principle, there is no limit to the absolute size of the specimens beyond those dictated by the development of plasticity. In practice, however, small samples are more susceptible to misalignment errors than large ones. For a given impact velocity, small samples show more test-to-test variability in the maximum amplitude of the stress-intensity history. Consequently, samples less than 100 mm (4 in.) in length are not recommended.

Determination of Dynamic Toughness From Strain Gage Records. As mentioned previously, there are two methods of determining dynamic toughness from strain gage records: the static calibration method and the direct method using Eq 25. The direct method is simpler, but requires care in positioning the strain gage to measure r and θ. The gage should be far enough from the crack tip to avoid the plastic zone, but close enough to the tip to remain in the zone dominated by the singular term of the elastic solution.

Test results show that the recommended distance of 3 to 4 mm (0.12 to 0.16 in.) meets both requirements, even for the smallest recommended specimen size. Furthermore, the crack front obtained from fatigue precracking invariably exhibits a slight curvature. Thus, when determining r and θ, it must be determined whether the origin of the coordinate system should be taken at the location of a virtual average crack front (following the averaging procedure detailed in

Fig. 17 Experimental setup used to perform the one-point bend test

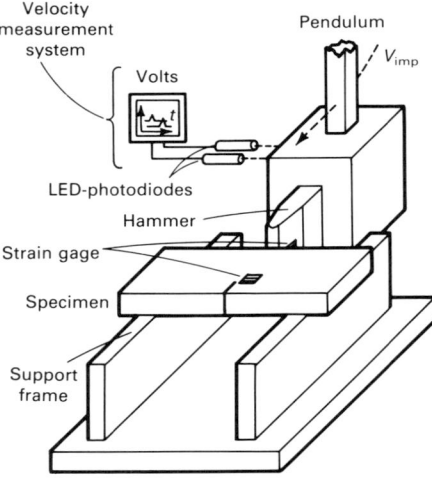

ASTM E 399), or at the actual position of the crack tip on the side of the specimen where the strain gage is mounted. Although further verification is required, experience shows that measuring r and θ from the actual crack tip on the gage side of the specimen provides more consistent results. This practice is therefore recommended.

Another concern is the determination of the point of crack initiation when the test is performed with strain gages. For experiments with long times-to-fracture (>50 μs), crack initiation causes a clear drop in the measured strain, and the point at which K_{Id} is evaluated can therefore be determined unambiguously (Fig. 18). For experiments with short times-to-fracture (<50 μs), experimental results and simulations indicate that the strain recorded by the strain gage may continue to increase once the crack has begun to extend. This may result in overestimates of the dynamic fracture toughness.

To determine the correct point of crack initiation, the procedure illustrated in Fig. 19 should be followed. The reference stress-intensity history for the specimen—i.e., the complete history untruncated by crack extension—is first scaled to the appropriate test conditions. The scaled reference stress-intensity history is then superimposed on the curve recorded during the fracture test. The point where the two curves begin separating is then selected as the point of crack initiation.

If the reference stress-intensity history is not available, a tangent straight line can be fitted to the fracture test record. As before, the point of separation of the experimental record from the tangent line determines crack initiation.

Performing Tests With Ballast at the Specimen Ends. When testing low-density

Fig. 18 Stress-intensity and hammer-load histories for fracture test with a long time-to-fracture

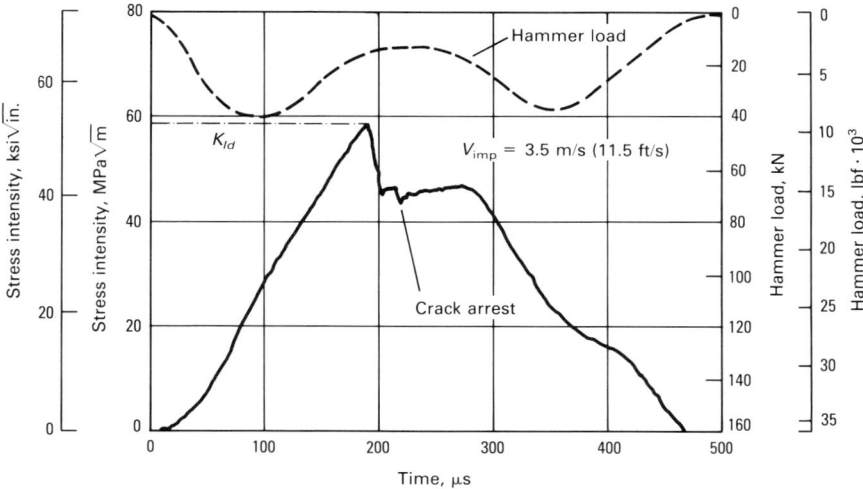

materials such as aluminum or titanium alloys, the inertial loading capacity of the one-point bend test may be insufficient to initiate crack propagation. To circumvent this limitation, ballast may be affixed to the specimen ends, as illustrated in Fig. 20. The stress-intensity history for an aluminum specimen outfitted with ballast plates is illustrated in Fig. 21 along with the stress-intensity history for the same specimen without ballast plates. The tests were performed at an impact velocity of 5.3 m/s (17.4 ft/s).

The dimensions of the specimen are $L = 241$ mm (9.5 in.), $W = 63.5$ mm (2.5 in.), $B = 12.7$ mm (0.5 in.), and $a = 19$ mm (0.75 in.). Two ballast plates of steel (89 by 25.4 by 6.4 mm, or 3.5 by 1.0 by 0.25 in.) were attached on each side of the specimen. The maximum stress-intensity amplitude is significantly higher for the specimen with ballast.

Dynamic Notched Round Bar Testing

By R.H. Hawley
Senior Research Engineer
Brown University

J. Duffy
Professor of Engineering
Brown University

and

C.F. Shih
Associate Professor of Engineering
Brown University

DYNAMIC NOTCHED ROUND BAR TESTING yields data from which a reliable value of the dynamic critical stress-intensity factor K_{Id} can be calculated easily. Hence, results are immediately related on a quantitative basis to fracture mechanics parameters. However, the test setup is rather elaborate, and more material is required for each specimen compared to Charpy testing. As a result, the technique is not suitable for routine testing. It may be used, however, when a precise evaluation of the fracture initiation properties of a particular material is required, perhaps as a function of temperature as well as of loading rate.

As shown in Fig. 22, the dynamic notched round bar specimen is a long cylindrical bar

with a fatigue precrack. During the test, the specimen is loaded in tension at one end by an impact of sufficiently large magnitude that the resulting stress pulse produces a fracture at the notch. In principle, therefore, the dynamic notched round bar test is more amenable to analysis than the Charpy test, because the fracture process is completed before the stress pulse has sufficient time to be reflected from the farthest end of the bar.

Consequently, measurements of the average stress across the fracture plane and of crack opening displacement, both as functions of time, are easily obtained. Various methods can be used to measure crack opening displacement, but the stress across the fracture plane is most easily determined by using electric resistance strain gages applied to the surface of the bar downstream from the fracture site. In this and several other respects, the dynamic notched round bar test resembles the Kolsky (or split-Hopkinson) bar used in dynamic plasticity (Ref 76). For more information on this testing procedure, see the articles "High Strain Rate Compression Testing," "High Strain Rate Tension Testing," and "High Strain Rate Shear Testing" in this Section.

An alternative geometry to the notched round bar is a long, thin strip with a perpendicular slit or slits (Ref 77). As in the notched round bar, the loading is accomplished by means of a tensile pulse. This alternative geometry is useful for simulating some practical applications more closely. However, there is no detailed analysis that relates test results directly to an evaluation of K_I, such as exists for the dynamic notched round bar test (Ref 78). In either setup, the tensile pulse can be initiated by the use of an explosive charge or by mechanical impact (Ref 75, 79).

Fig. 19 Stress-intensity history for a fracture test with a short time-to-fracture and procedure for determining K_{Id}

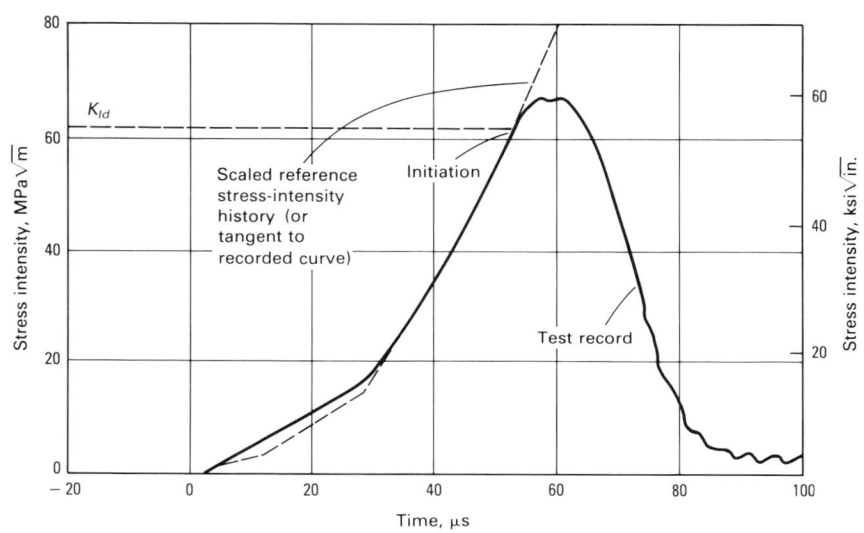

Fig. 20 Specimen and ballast plates used to obtain response curve in Fig. 21

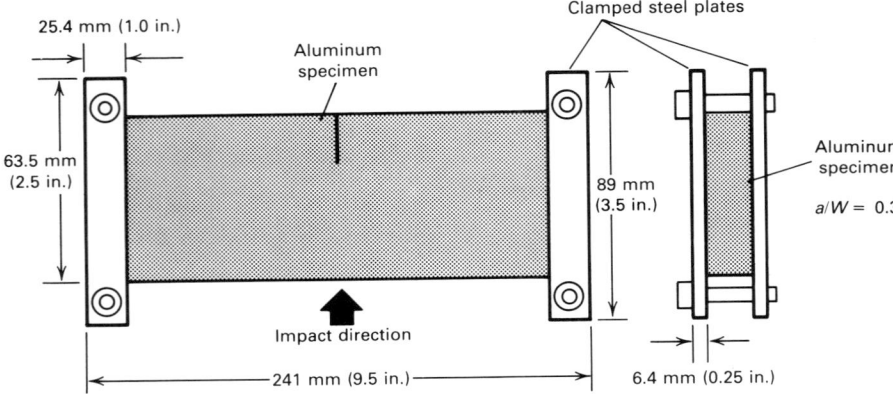

Fig. 21 Stress-intensity histories for an aluminum specimen with and without ballast plates

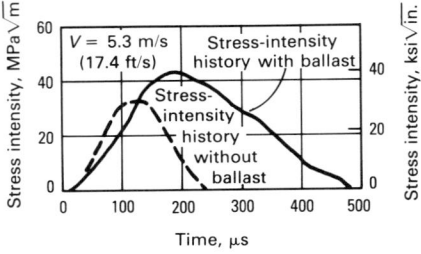

Comparison With the Charpy Test. The Charpy test is by far the most commonly used method for evaluating the fracture properties of materials, primarily due to its ease, relatively low cost, and small material requirements. The fracture energy value it provides can be compared to a wide range of accumulated data, and the test rapidly detects changes in ductility.

However, the Charpy test does have certain disadvantages. Fracture in the Charpy specimen does not occur under plane-strain conditions. Furthermore, the state of stress at the fracture site is unknown and quite complex due to multiple pulse reflections from its various surfaces. For these reasons, it is difficult to interpret Charpy results in terms of elastic or elastic-plastic fracture toughness parameters. It is thus difficult to provide reliable safeguards against fracture initiation in an engineering structure containing cracks or flaws, whether due to the manufacturing process, riveting, welding, or some other source. Detailed information on the Charpy test can be found earlier in this article.

Comparison With the Instrumented Charpy Test. Several empirical correlations have been developed to interpret the data obtained with the instrumented Charpy test in terms of fracture toughness parameters (Ref 19, 20). Although this test has some advantages compared to the Charpy test, some of the same disadvantages also apply—namely, that plane-strain conditions usually do not prevail in the specimen and the state of stress at the crack tip varies with time in a complex manner due to the numerous waves reflected from the specimen surface. Hence, it is difficult to relate test results to fracture toughness parameters. Nevertheless, as discussed earlier in this article, the test data can be used to rank the resistance of materials to fracture.

Comparison of Loading Rates. Loading rates are expressed in terms of the rate of change of the stress-intensity factor, dK_I/dt. For dynamic notched round bar experiments, the typical rate of loading is about 2×10^6 $\mathrm{MPa}\sqrt{\mathrm{m}} \cdot \mathrm{s}^{-1}$ (1.8×10^6 $\mathrm{ksi}\sqrt{\mathrm{in.}} \cdot \mathrm{s}^{-1}$) and generally is estimated at 10^5 $\mathrm{MPa}\sqrt{\mathrm{m}} \cdot \mathrm{s}^{-1}$ (9.1×10^4 $\mathrm{ksi}\sqrt{\mathrm{in.}} \cdot \mathrm{s}^{-1}$) for Charpy testing. The relatively small difference between these rates is not significant. Also, the dynamic notched round bar test specimen is fatigue precracked, whereas the notch in the Charpy specimen is as machined.

Test Setup

Typical apparatus used for dynamic notched round bar testing is illustrated in Fig. 22. The specimen consists of a solid round bar, with a circumferentially fatigued precrack. As stated previously, fracture of the specimen is produced by a tensile stress pulse initiated at one end of the bar by the detonation of an explosive charge or by some impact device (Ref 75, 79). This impact must be large enough so that when the pulse reaches the precracked section, fracture occurs while still on the rising portion of the pulse. Using explosive loading techniques, fracture is achieved within 20 to 25 μs after the arrival of the tensile pulse at the precracked section.

Fig. 22 Typical apparatus for dynamic fracture initiation experiment
Source: Ref 75

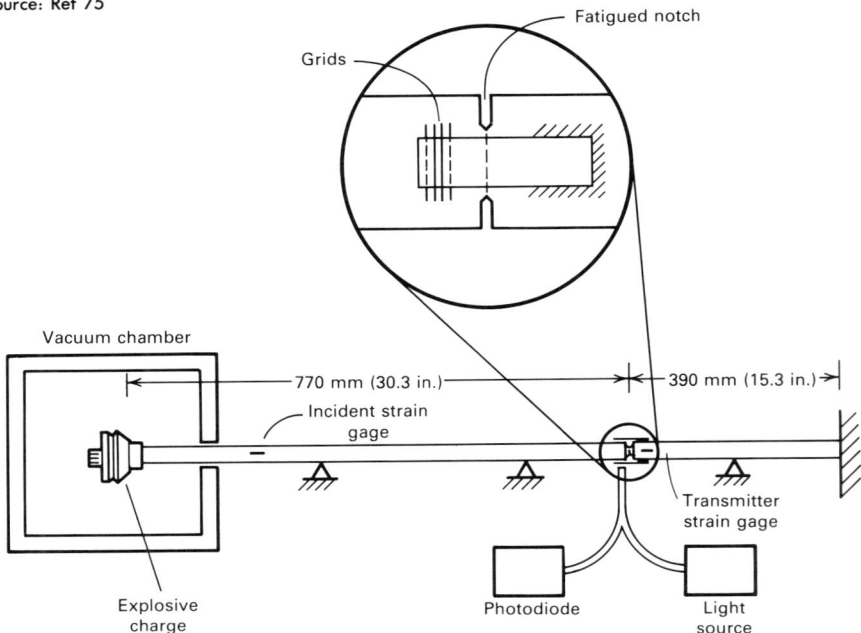

To properly conduct dynamic notched round bar testing, the specimen must be instrumented. Two sets of strain gages are required—one mounted on either side of the crack. On the upstream side, one set of gages records the pulse incident on the cracked region; on the downstream side, the gages record the pulse transmitted through the cracked region. Typical outputs of these gages are shown in Fig. 23.

The transmitted pulse provides a measure of load transmitted through the fracture site as a function of time. Crack opening displacement must be measured in a manner that does not cause interference with the fracture process. One possible method uses an optical device consisting of two overlapping photogrids of 33 lines per millimeter. A second such device may be used on the other side of the specimen to ensure that no bending occurs.

For each device, one grid is deposited photographically on a flat, polished area of the specimen surface adjacent to the crack and on its upstream side. The overlapping grid is affixed to a glass plate cemented to the specimen surface on the downstream side of the crack. This glass plate spans the crack to overlap the first grid. This arrangement produces a set of moiré fringes in the region where the two grids overlap. Because the glass plate is cemented to the specimen at a point downstream from the crack, the relative displacement that occurs between the two grids during passage of the pulse produces a shift in the position of the moiré fringes and thus provides a measure of the notch opening displacement.

With the aid of bifurcated fiber optic tubes, light is transmitted from a source and strikes the overlapping grids normal to their plane. It is then reflected by the metal surface and directed by the fiber optics to a photodiode where the intensity is measured. The motion of the moiré fringes produced by the relative displacement of the grids causes an alternating high and low intensity in the

returning light; the notch opening displacement is linearly proportional to the number of light-dark cycles recorded.

Finite element analysis of the dynamic notched round bar specimen has shown that the best measure of crack opening displacement (in conjunction with Eq 35 described below) is the notch opening displacement (Ref 78). Other possible means for measuring notch opening include interferometric techniques, capacitance gages, and photonic sensors. Regardless of the method used to monitor notch opening displacement, the instrumentation must have a sufficiently high response time to provide an accurate record of notch opening as a function of time.

As indicated in Fig. 22, the effect of a thermal environment on fracture initiation can be determined by enclosing the notched section of the specimen in an environmental chamber. The techniques described in this section have been used successfully to determine the plane-strain fracture toughness for various steels at −150 to 150 °C (−240 to 300 °F), thus offering a means for establishing values of K_{Id} on both the lower and upper shelves, as well as within the transition region between brittle and ductile fracture (Ref 80-82).

Specimen Configuration

The dynamic notched round bar specimen consists of a solid round bar 25 mm (1 in.) in diameter and 1020 mm (40 in.) in length. A circumferential notch is machined into the bar 660 mm (26 in.) from the loading end. Details of the notch are shown in Fig. 24. The notch is machined so that the faces are parallel, and the root has a sharp tip to facilitate the growth of a fatigue crack. Small flat areas about 7 mm (0.25 in.) wide are machined on opposite sides of the bar, extending approximately 75 mm (3 in.) to either side of the notch. These flats are required to accommodate the optical measuring device used to measure notch opening displacement.

Two sets of axial strain gages are attached to the specimen. The first set is placed 200 mm (8 in.) from the loading end. These gages are used to measure the incident pulse as it propagates down the bar toward the notch and are referred to as "incident gages." They are located a sufficient distance from the loading end to ensure that the transients induced by the loading are not recorded. Also, they are far enough from the notch so that the reflections returning from the notch faces do not interfere with the measurement of the incident pulse.

To determine properly the distances involved, it must be remembered that an elastic

Fig. 24 Notched region of the dynamic notched round bar specimen

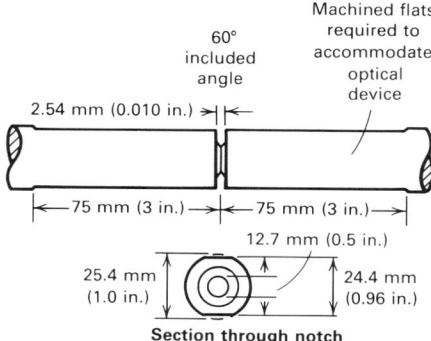

60° included angle

2.54 mm (0.010 in.)

Machined flats required to accommodate optical device

75 mm (3 in.) — 75 mm (3 in.)

12.7 mm (0.5 in.)

25.4 mm (1.0 in.)

24.4 mm (0.96 in.)

Section through notch

tensile pulse propagates along a slender bar at a speed given by $(E/\rho)^{1/2}$, where E is the Young's modulus of the bar material, and ρ is mass density. For most metals, this corresponds approximately to a speed of 5000 m/s (16 000 ft/s), so that the pulse takes about 10 μs to travel 50 mm (2 in.,). The duration of the incident pulse depends on the size of the explosive detonation with which it is initiated. For a relatively brittle steel, i.e., one whose fracture toughness is low, a smaller explosive charge is used, and the duration of the incident pulse is about 75 μs, although fracture occurs 20 or 25 μs after the arrival of the pulse front. For more ductile fractures, a large charge is required, and the pulse duration may be 100 or even 150 μs. A plastic explosive requiring an E-106 electric blasting cap or its equivalent for detonation has proved to be both reliable and safe for this application. For the experimental configuration detailed in Fig. 25, 10 to 15 g (0.4 to 0.6 oz) of explosive is required.

The second set of gages, or the transmitter gages, is located 25 mm (1 in.) beyond the notch. These gages measure the portion of the tensile pulse transmitted through the notched region. By locating the gages about one diameter away from the notch, three-dimensional wave effects are reduced suffi-

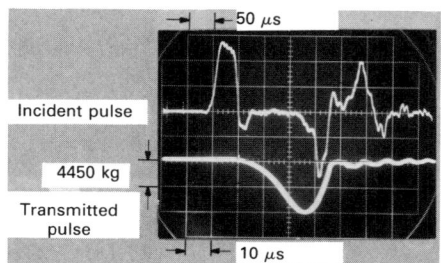

50 μs

Incident pulse

4450 kg

Transmitted pulse

10 μs

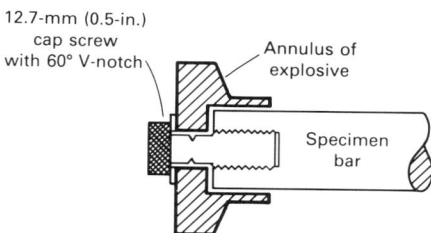

12.7-mm (0.5-in.) cap screw with 60° V-notch

Annulus of explosive

Specimen bar

ciently so that the gages provide a good measure of the total force transmitted through the notch. Hence, the magnitude of the transmitted pulse is directly proportional to the average net section stress at the fracture plane.

Furthermore, because the pulse length is relatively long compared to the notch width, material inertia has only a secondary effect. For purposes of analysis, the loading can thus be considered quasi-static. This procedure was used in Ref 75, 82, and 83 and was later confirmed by finite-element analysis that included inertia (Ref 78). An oscilloscope photograph of the outputs of the gages from a typical test is shown in Fig. 23.

Specimen Preparation

After machining the notch and the flats, a circumferentially uniform fatigue crack is grown from the root of the notch, leaving an unfractured ligament between 7 and 9 mm (0.25 and 0.36 in.) in diameter. The employed method for growing a fatigue crack at the root of circular notch is to load the bar in bending as a rotating beam. Frequently, however, the apparatus used for this purpose produces a fatigue crack of nonuniform width; it is essential for accurate application to fracture mechanics that the fatigue crack be a concentric annulus of uniform width. As shown by Mylonas and Hermann (Ref 83), this can be accomplished if the rotating beam apparatus imposes a given bending displacement and measures the force required for this displacement, whereas imposing a fixed load results in nonuniform growth of the fatigue crack.

In the fatigue apparatus shown in Fig. 26, the specimen is supported between two bearings, with the notch located midway between them. A second set of bearings, one located on either side of the notch, is mounted in a gimbal-like frame. The specimen is loaded transversely through this frame by means of the loading arm. Under these conditions, the specimen acts as a simply supported beam, with a pair of equal and opposite bending moments applied at equal distances from the notch. The notched region is thus placed in pure bending.

When the specimen is rotated, the notch root is loaded so that tension alternates continuously with compression. The imposed displacement, applied through a lever arm (see Fig. 26), is adjusted by a threaded bolt. A load cell is attached to the loading arm, and the applied load is recorded as a function of time on a strip-chart recorder as fatiguing progresses.

Because even the most carefully prepared specimen bars have a slight permanent curvature resulting from the fabrication process, two straightening collars are placed on the

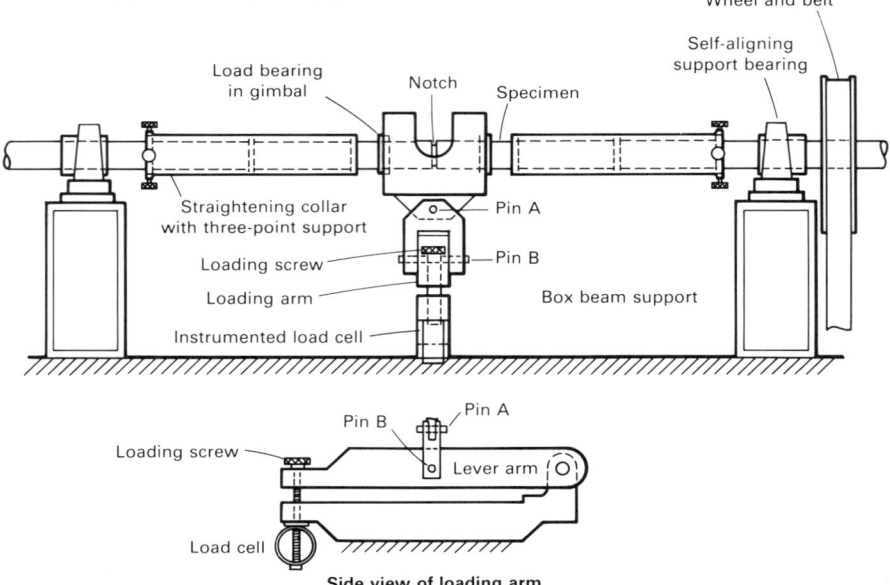

Fig. 26 Typical fatigue apparatus

specimen—one on each side of the notch between the support bearing and the load bearing. Each collar applies a straightening moment to the bar outside the notch region through three-point bending. Thus, the straightening moments do not extend across the notch.

To adjust the specimen in the apparatus, a small amount of displacement is applied through the loading arm. The specimen is then rotated slowly while the resultant load is monitored on the strip-chart. The load will vary as the bar is rotating; the magnitude of the variation depends on the degree of curvature of the bar. The adjusting screws on the straightening collars are turned until the load applied to the notch remains nearly constant as the bar rotates. Minimizing variations in load is a critical factor in growing a circumferentially uniform fatigue crack. With this apparatus, the applied load can be kept uniform to within ±3%.

After the specimen is adjusted properly, the required initial load is applied by tightening the threaded bolt, and the bar is rotated by an electric motor attached to one end of the bar through a pulley and gear box. The specimen is rotated at approximately 100 cycles/min, which is slow enough to avoid dynamic loading due to eccentricities. For 1020 hot rolled steel, a 5.5-kg (12-lb) starting load is required to grow a small fatigue crack in $5 \cdot 10^3$ to 10^4 cycles.

Because the loading is actually applied with a fixed displacement, the load does not remain constant throughout the fatiguing process. As the crack grows, the resulting increase in compliance of the specimen in bending results in a relaxation of the load. Because the load is continuously recorded on

the strip-chart, the amount of load relaxation is used as a measure of the growth of the crack. Figure 27 illustrates a plot of the load drop versus crack length for a 1020 hot rolled steel specimen. Reasonable control is obtained over the depth of the fatigue crack by making such a plot for each metal tested.

Another advantage of having the load relax as the fatigue crack grows is that the stress-intensity factor, K_I, at the crack tip during the tensile portion of the cycle remains nearly constant, because increases in K_I due to the increased crack depth are offset by the corresponding reduction in applied load. This reduces the possibility of reaching a critical condition at the crack tip during the fatiguing process and thus diminishes the likelihood of a complete fracture of the specimen during fatiguing.

Figure 28 shows the fracture surface of a specimen, with a concentric fatigue annulus that is typically obtained with rolled steel bars. Plate stock machined into a bar will not give results of this quality because of its anisotropy. This tendency thus limits the use of the dynamic notched round bar experiment to specimens of rolled round bar stock.

Measurement of Notch Opening Displacement

Experimentally, it would be difficult to measure the local crack opening displacement as a function of time during the passage of a stress wave. It is easier to measure the notch opening displacement. The notch opening displacement is also the appropriate displacement to use for calculating K_I through the load displacement formulation given in Eq 35 (Ref 78).

Fig. 27 Final root diameter of fatigued notch versus the load drop occurring during fatiguing for 1020 hot rolled steel

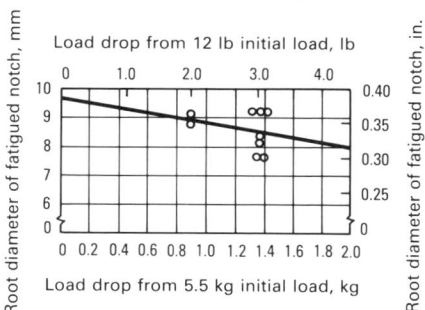

Fig. 28 Fracture surface of prenotched steel bar showing annular fatigue crack

Fig. 29 Specimen bar with grid jig attached to hold master grid glass slide

Fig. 30 Specimen with overlapping grids attached and spanning the notch
The glass grid is cemented to the right-hand portion of specimen; the grid appears as a black square to the left of the notch.

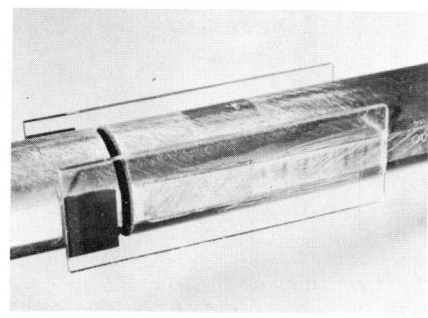

Measurement of notch opening displacement has two instrumentation requirements (Ref 75). Only 15 to 25 μs elapse from the time the pulse first arrives at the fracture site until fracture is completed; therefore, the instrument used must have a small mass and must be mounted nearly flush with the specimen so that it does not deform due to its own inertia. Furthermore, the instrumentation must have a sufficiently high frequency response. Calibration that puts the specimen under a preload, even a small one, will result in damage to the crack tip. For these reasons, optical methods are well suited to measurement of notch opening displacement.

Setup of the Optical Measuring Device. The method adopted for the first dynamic notched round bar experiment used an optical device based on the phenomenon of moiré fringes. Classically, moiré fringes are produced when light passes through two transparent gratings mounted face to face with their rulings nearly, but not quite, parallel. One such device employs matched sets of grids that are produced photographically and are affixed to the specimen so that any relative displacement between the outer edges of the notch causes an identical relative displacement between the two grids.

The first grid is produced by making a contact print of a master grid of the desired pitch (distance between rulings) on one end of a 25- by 75-mm (1- by 3-in.) high-resolution glass photographic slide. This glass slide then serves as a master for the second grid, which is reproduced directly on the specimen. In preparing the specimen, two narrow flat areas about 7 mm (0.25 in.) wide are machined on opposite sides of the bar, extending approximately 75 mm (3 in.) to either side of the notch (see Fig. 24). Each of the flat areas accommodates a set of grids—one on the glass slide and one on the specimen.

The grids on the specimen are located on the flats adjacent to the notch on the incident side of the bar. They are produced using a photoresist process in which an emulsion is spread directly upon the flat, polished metal surfaces and allowed to dry. A jig, holding the glass slides, is attached to the specimen so that the portion of each slide containing the grid is in contact with its respective emulsion-covered metal surface. Figure 29 shows the jig in place on a specimen.

The glass slides are held in the jig by leaf springs so that no relative motion between the slides and the jig is possible, even when the jig is removed from the bar. The two sides of the bar are then exposed to bright light, causing the exposed portion of the emulsion to polymerize. After exposure, the jig and attached glass slides are removed and the emulsion-covered portion of the bar is immersed in an orthoresist developer to dissolve the nonpolymerized portion of the coating, thus rendering a "print" of the grid. This grid is then dyed to enhance its light-dark contrast.

The glass grids, still held by the jig, are then repositioned over the grids on the specimen and cemented to the flat portion of the bar on the transmitter side of the notch. Once the slides are cemented, the jig is disassembled and removed. The complete assembly appears as shown in Fig. 30. It should be emphasized that the glass slides are not removed from the jig during the entire photographic process. As a result, the master grid on the slide and the duplicate on the bar are overlaid with their lines very nearly parallel.

In the cementing process, a thin shim is placed between the glass and the specimen surface, resulting in an air gap of about 0.1 mm (0.005 in.) between the two grids. This gap is required to keep the grids from rubbing when relative motion occurs during the tests. After the grids are mounted in the experimental configuration, light from a 500-W projector lamp is supplied to the grids and returned to a photodiode through a bifurcated fiber optic light tube. The output of the diode is recorded on an oscilloscope.

Moiré Fringes. To completely understand the optical measurement technique, it is first necessary to understand the use of moiré fringes. Moiré fringes are produced by optical interference when light passes through two grids of similar pitch whose lines are nearly, but not quite, parallel. If the rulings are exactly parallel, the superimposed grids will appear of uniform brightness. If the rulings are not parallel, however, the grids appear to be spanned by a number of equally spaced shadows, or fringes, that run perpendicular to the rulings. The width of these fringes depends on the rotation angle between the two grids. When relative linear motion occurs between the two grids, the moiré fringes move across the field in a direction perpendicular to themselves (parallel to the grid lines) at a speed independent of the angle between the two sets of rulings. A mechanical analog of this optical phenomenon is illustrated in Fig. 31. If the pitch of the grids is denoted by w and the small angle between the rulings is indicated by θ, it is then possible to illustrate (Ref 84, 85) that the fundamental fringe width W is given approximately by:

$$W = \frac{w}{\theta} \qquad \text{(Eq 28)}$$

The fringe width is significant, because it directly governs the amplitude of the output signal of the photodiode. The photodiode with oscilloscope readout responds only to changes in light intensity. To attain an output from the photodiode that is sufficient to record on an oscilloscope, the variations in intensity caused by the fringes passing across the field of view must be maximized. The field of view in this case is the portion of the grid covered by the end of the fiber optic tube, as illustrated below:

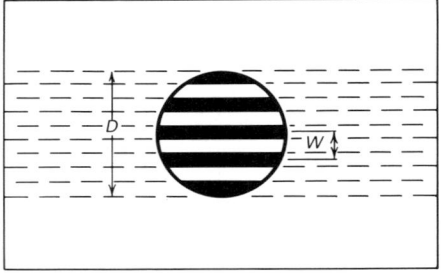

Let $\beta = D/W$, the ratio of the tube diameter to the fringe width. In the case depicted above, $\beta > 1$, so the change in intensity of transmitted light will be small as the fringes travel across the end of the tube, because the amount of light and shadow remains nearly constant. It is thus evident that the optimum value of β is approximately $\frac{1}{2}$. This case is depicted as follows:

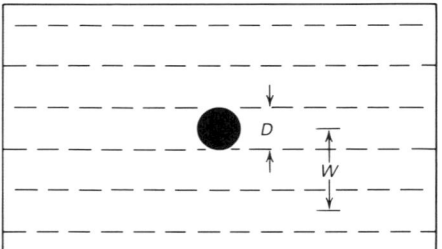

In this situation, as the fringes move across the end of the tube, the intensity change is from complete shadow to complete light, thus maximizing the photodiode output.

Because it is easier to control the fringe width than the diameter of the fiber optic tube, the grids must be aligned so that the resulting fringe width is near the optimum value for maximum signal strength. If $\beta = \frac{1}{2}$, then $W = 2D$, which combined with Eq 28 yields:

$$\theta_{optimum} = \frac{W}{2D} \qquad \text{(Eq 29)}$$

For a typical arrangement with a grid of 33 lines per millimeter, the optimum value of θ, calculated by means of Eq 29, is 0.2°. In practice, the alignment jig is used to ensure that the two grids are properly aligned when finally cemented into place. It has been determined experimentally that if $0 < \theta < 0.5°$ sufficient signal strength is achieved. Alignment of θ to within this tolerance is well within the capability of the jig employed. Note that the motion of the fringes due to a relative displacement of the grids is unaffected by the value of θ.

Calibration of the Optical Measuring Device. As the crack opens, one grid moves relative to the other, and the fringes move across the field of observation in a direction perpendicular to the grid lines. The number of fringes that pass any point of the field is equal to the number of grid lines that have moved past each other. Thus, if the relative motion is equal to one pitch of the grids in distance, one complete fringe width will be observed to move across the end of the fiber optic tube.

If the brightness distribution across one fringe width were simply light to dark, as shown below, then one cycle on an oscilloscope trace would be observed for each fringe width passing across the end of the optic tube; thus, one cycle on the trace would correspond to one pitch of relative movement of the grids. This, however, is not generally the case.

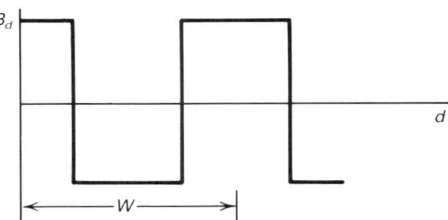

During the formation of moiré fringes, many interfering beams of light may be operative, each making various contributions to the total brightness observed across a fringe width. Guild (Ref 86) showed that a general result can be obtained by considering the brightness across the field of fringes as composed of the sum of Fourier components:

$$B_d = 1 + B_1 \cos 2\pi \left(\frac{d - d_o'}{W} \right)$$
$$+ B_2 \cos 4\pi \left(\frac{d - d_o''}{W} \right)$$
$$+ B_3 \cos 6\pi \left(\frac{d - d_o'''}{W} \right) + \cdots$$

where B_d is the brightness; B_1, B_2, ... are relative amplitudes of the various compo-

Fig. 31 Mechanical analog of moiré fringes

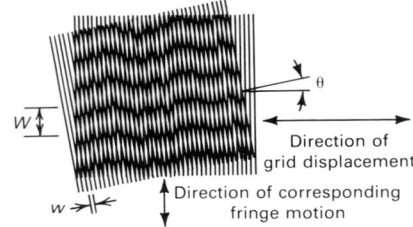

nents; W is the fundamental fringe width, given by Eq 28; and the constants d_0', d_0'', d_0''', ... indicate the positions of maximum intensity of the constituent components.

The relative amplitudes depend greatly on the circumstances of observation and must for all practical purposes be determined experimentally. If B_2, B_3, ... are all considerably less than B_1, the brightness distribution will show only one maximum and one minimum within the fundamental fringe width. However, if one or more of these coefficients is comparable to B_1, the brightness distribution may show secondary maxima or minima between those of the fundamental period, and two, three, or more cycles on the oscilloscope trace would correspond to one pitch of relative displacement. Thus, it is imperative to determine the ratios of the coefficients for the specific conditions of observation.

For the test arrangement described in this article, it has been determined experimentally that B_2 is the dominant coefficient. This results in two light-dark cycles being recorded on the oscilloscope trace when the grids are displaced a distance of one pitch relative to each other. The 33 lines per millimeter grid has a pitch of $w = 30$ μm/line (1200 μin./line). Thus, one cycle corresponds to 15 μm (600 μin.) of relative displacement. Figure 32 illustrates a typical oscillograph from a dynamic fracture test, on which the corresponding displacements are labeled.

Because it is possible to enlarge these oscillographs and plot half-cycle and quarter-cycle points, the resolution of the displacement measurement from such an oscillograph is estimated to be within 2 μm (75 μin.). Other errors due to the air gap between the two grids and small ruling errors are introduced into this arrangement. However, Guild (Ref 87) shows that errors of this nature have only a small effect.

Advantages of Optical Measurement. This measurement technique has two advantages that render it ideally suited to notch opening measurement with the dynamic notched round bar. First, the reaction time of the device is far less than the rise time of the loading pulse, resulting in accurate measurement of displacements occurring during the approximately 20 μs of loading. Second, once the grids are in place on the specimen, no calibration of the device is required; the relationship between fringe cycles and displacement, once established for the type of grids used and the experimental configuration, is fixed for all time. This is a definite advantage, because the specimen and more specifically the notch tip need not be disturbed for calibration purposes prior to performing the fracture initiation test.

Data Analysis

In order to establish valid dynamic fracture data, a load-displacement curve must first be constructed. Once this curve is established, critical fracture parameters, such as the dynamic fracture toughness, can be determined. The means of generating these curves and the resulting dynamic fracture properties are discussed below. In addition, the use and advantages of finite element analysis of the dynamic notched round bar test is described.

Load-Displacement Curve

The data recorded from each dynamic test are in the form of photographs of oscilloscope traces, as shown in Fig. 23 and 32. The load, P, transmitted through the notch is calculated as a function of time directly from the oscillogram of the transmitted pulse. This can be done by using the standard strain gage formulation for a two-arm bridge:

$$P = AE\left(\frac{2e}{ge_0}\right) \qquad \text{(Eq 30)}$$

where A is the cross-sectional area of the bar at the gage station, E is the Young's modulus of the bar material, e is the output of the strain gages in volts, g is the gage factor, and e_0 is the excitation voltage, usually a nominal 24 V.

Notch opening displacement as a function of time is calculated from the oscilloscope trace of the output from the optical extensometer, as shown in Fig. 32, in which one cycle equals 15 μm (600 μin.) of displacement. The displacement and corresponding time can be read from an enlargement of the oscilloscope trace at each quarter-cycle point. Thus, the displacement-time relationship is determined at increments of 3.75 μm (150 μin.) in displacement.

A digital oscilloscope and a small computer and plotter can greatly reduce the effort required to analyze data. To construct the complete load-displacement curve for each test, the load-time curve and displacement-time curve are plotted using the same time scale, as shown in Fig. 33.

A common zero time point is established between the two sets of data by considering the trigger time of the oscilloscopes and taking into account the time delays between the

arrival of the pulse at the various data-recording positions on the specimen. As stated previously, the pulse requires 10 μs to travel 50 mm (2 in.) in most metals. Once a common zero is established, time is eliminated from the data, resulting in a load-displacement curve for the dynamic fracture test.

Calculation of Fracture Parameters

Having determined the load-displacement curve for a test, the fracture parameters of interest can be calculated. For a nominally brittle material, the parameter of interest is K_{Id}, the critical value of the dynamic plane-strain fracture toughness. The dynamic stress-intensity factor is calculated using the static formula provided in Ref 86:

$$K_I = \left(\frac{P}{\pi R^2}\right)\sqrt{\pi R}\, F\left(\frac{2R}{D}\right) \qquad \text{(Eq 31)}$$

where R is the radius of the unfractured ligament, P is the applied load, D is the outer diameter of the bar, and $F(2R/D)$ is a size function. For the specimen geometry described in this article, the size function has a value of about 0.48. The load P_c used in the determination of K_{Id} from Eq 31 is obtained from the load-displacement curve in accordance with ASTM standards by using the 5% slope offset procedure. To apply linear elastic fracture mechanics, the size of the crack tip plastic zone must be small compared to the nominal dimensions of the specimen. A size criterion for a valid K_{Ic} test is:

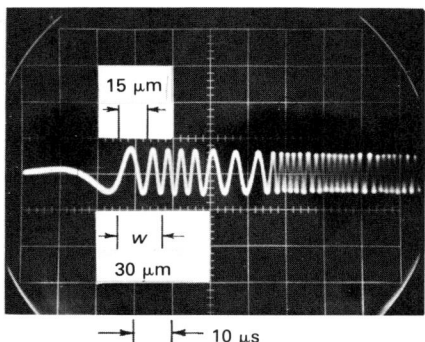

$$R \geq 2.5\left(\frac{K_{Ic}}{\sigma_y}\right)^2 \qquad \text{(Eq 32)}$$

where σ_y is a flow stress value for the material. For the dynamic notched round bar experiment, the same size criterion is employed, except that K_{Id} replaces K_{Ic} and a dynamic value of flow stress replaces the static flow stress.

For the case of small-scale yielding in which the plastic zone may be slightly larger than the given limit, a plastic zone correction can be applied to yield a more accurate value of the stress-intensity factor (Ref 87). The effect of the plastic zone can be approxi-

Fig. 33 Typical load-time and displacement-time curves for 1020 cold rolled steel

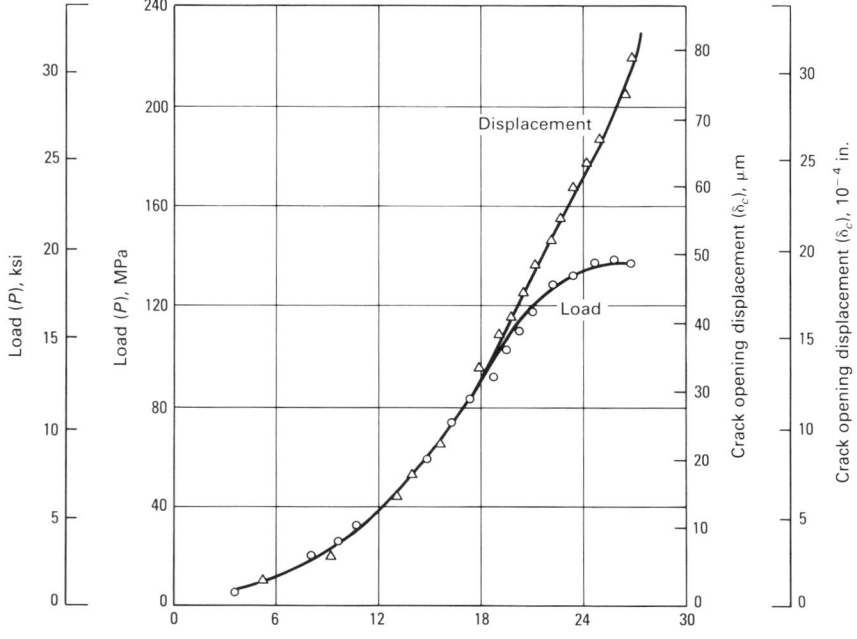

mated by taking the effective ligament radius as:

$$R_{\text{eff}} = R - \frac{1}{6\pi}\left(\frac{K_I}{\sigma_y}\right)^2 \quad \text{(Eq 33)}$$

Using R_{eff} instead of R in the calculation of K_{Id} results in a better approximation of the plane-strain fracture toughness under conditions of small-scale yielding.

When testing more ductile materials such as low-carbon steel at room or elevated temperatures, the plastic zone generally will be too large relative to the specimen size to meet the criteria given above. Such tests cannot be analyzed on the basis of linear elastic fracture mechanics; thus, a fracture mechanics approach must be used that is valid in the elastic-plastic regime. For this purpose, the J-integral proposed by Rice is used (Ref 88). Begley and Landes (Ref 89) have shown that the J-integral is the appropriate criterion for fracture under quasi-static plane-strain conditions from essentially elastic to fully plastic behavior. Details of quasi-static crack tip stress fields under large-scale yielding conditions have been provided by Tracey (Ref 90).

In applying the J-integral as a fracture criterion to the large-scale yielding problem, the region to which plane-strain conditions must apply is the region at the crack tip where the critical separation process is taking place. Thus, the specimen size requirements can be reduced considerably. Paris (Ref 91) has suggested that a possible size criterion for a valid J-integral test is:

$$R \geq \left(\frac{50J_{Ic}}{\sigma_y}\right) \quad \text{(Eq 34)}$$

Rice et al. (Ref 92) have shown that the value of J for a notched round bar may be determined from a load-displacement curve according to:

$$J = \frac{1}{2\pi R^2}\left(3\int_0^\delta Pd\delta - P\delta\right) \quad \text{(Eq 35)}$$

where δ is the notch opening displacement.

Equations 34 and 35 are intended for fractures initiated through quasi-static loading. In applying the size criterion (Eq 34) to dynamic conditions, J_{Id} replaces J_{Ic}, and σ_y is the dynamic flow stress. Furthermore, a detailed elastic-plastic dynamic analysis has shown that Eq 35 is applicable to dynamic conditions with the notched round bar if P is taken as the transmitted load and δ is the notch opening displacement (Ref 78).

The J-integral approach cannot be used for materials that exhibit significant amounts of subcritical crack growth prior to the onset of instability. However, a possible extension to materials of this type has been considered by Hutchinson and Paris (Ref 93).

A direct comparison can be made between the fracture parameters J and K by consider-

ing the specimen to be sufficiently large to be in the regime of linear elastic fracture mechanics. In this case, J and K are related by:

$$K^2 = \frac{EJ}{(1 - \nu^2)} \quad \text{(Eq 36)}$$

where E is Young's modulus, and ν is Poisson's ratio. Assuming that the value of J satisfies Eq 34, an equivalent dynamic fracture toughness K_{Id} can be calculated from Eq 36. Finally, the parameter used to specify the loading rate in the dynamic fracture test is:

$$\dot{K}_I = \frac{K_{Id}}{t_c} \quad \text{(Eq 37)}$$

where t_c is the time required to increase the equivalent stress-intensity factor from zero to the critical value K_{Id}.

In the dynamic notched round bar test, t_c is approximately 20 μs, whereas for a static test, t_c is usually on the order of 60 to 90 s or longer. It should be noted that the notched round bar described in this article is suitable for the determination of fracture toughness under quasi-static as well as under dynamic loading conditions.

Once a load-displacement curve has been plotted on the basis of a graph, such as Fig. 33, it is then possible to evaluate K_{Id}, or an equivalent K_{Id} in the case of more ductile metals. This is done by means of Eq 31 or 36. In either case, the procedure begins by an evaluation of K_{Id} through Eq 31. If the test meets the validity criterion expressed by Eq 32, then the calculation is complete.

If this criterion is not met, then J must be evaluated from the experimental load-displacement curve using Eq 35. Having evaluated J, Eq 36 can be used to determine the equivalent dynamic fracture toughness K_{Id}. Typical dynamic and static toughness values at various temperatures are shown in Fig. 34.

Finite Element Analysis

A full finite element analysis of the dynamic notched round bar experiment, including a critique of the method, has been presented by Nakamura et al. (Ref 78). In this elastic-plastic dynamic analysis, the uniaxial stress-strain curve is taken from the results of dynamic plasticity tests with specimens of an AISI 1020 hot rolled steel. A uniform rate of loading is applied at one end of the bar to generate a loading pulse similar to the pulse imposed in the experiment.

The finite element analysis was made with the mesh shown in Fig. 35, containing a zone with refined elements extending over a distance of one and one half diameters in either direction from the fracture plane. The mesh extends far enough along the bar to include the location at which the transmitter strain gages are placed during the experiment to measure the average stress at the fracture

plane. The analysis provides the complete states of stress and of strain as functions of time during passage of the stress pulse and an exact evaluation of the J-integral, including inertia effects.

It has been shown (Ref 78) that the state of strain in the neighborhood of the crack tip in the dynamic notched round bar test comes very close to that of plane-strain conditions. This analysis also shows that the stress and strain distributions are symmetric with respect to the crack plane, at least in a region large enough to enclose the fracture process zone. Hence, test results obtained with the dynamic notched round bar can be interpreted within the framework of fracture mechanics. Finally, finite element analysis clarifies the applicability of Eq 35 for the determination of dynamic fracture toughness through the dynamic notched round bar experiment. This formula is accurate if the circumferential precrack is sufficiently deep. For a precrack depth of 75% of the bar diameter, the error in J, as determined from Eq 35, is about 15% for elastic behavior, decreasing to about 5% for a fully plastic section. The corresponding error in an equivalent K_{Id} is about half as great.

The finite element analysis also shows that the location of the strain gages on the bar at a distance of one diameter from the crack plane provides an accurate measure of the transmitted load. Furthermore, the analysis reveals that the formula for the J-integral (Eq 35) must be based on the load transmitted across the ligament and on the notch opening displacement.

The finite element analysis shows that the dynamic notched round bar test can provide an accurate value of the dynamic fracture toughness if the notch in the bar is sufficiently deep. It is advantageous, therefore, to make the notch deep enough so that after fatiguing the diameter of the remaining ligament does not exceed 30% of the outside diameter of the bar. A crack of this depth provides accurate results and makes fracture easier to achieve experimentally.

Short-Pulse-Duration Tests

By Donald A. Shockey
Director, Department of Metallurgy and Fracture Mechanics
SRI International

SHORT-PULSE FRACTURE MECHANICS is a modification of classic static fracture mechanics concepts that applies to fracture under a rapidly applied, short-duration load.

Fig. 34 Static and dynamic fracture toughness of 1018 cold rolled steel and 1020 hot rolled steel versus temperature

C, cleavage (crystalline) fracture; F, fibrous (shear) fracture

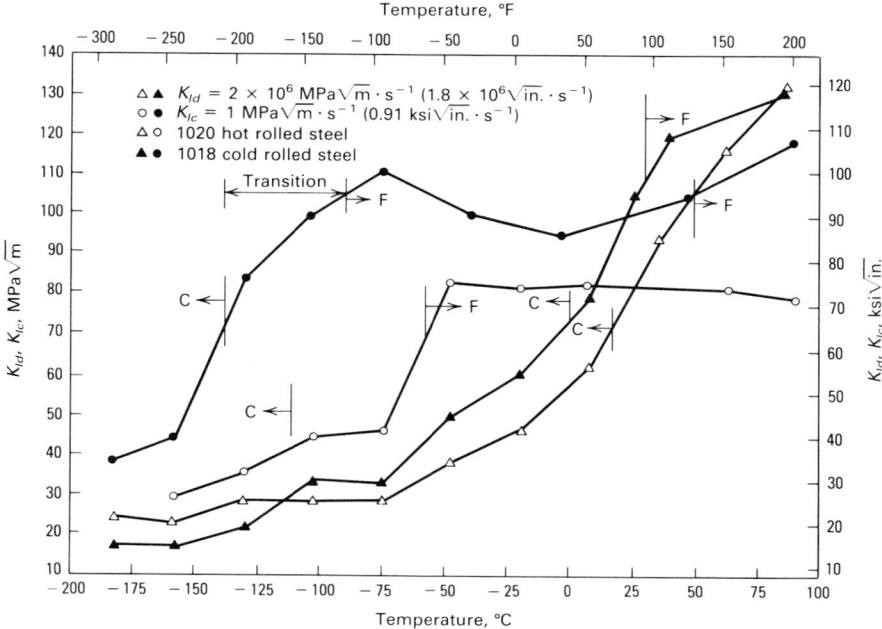

Fig. 35 Finite element elastic-plastic analysis of the dynamic notched round bar

The deformed mesh during passage of the stress pulse is shown.

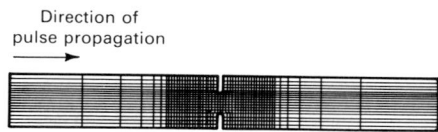

Fig. 36 Variation of dynamic crack tip stress-intensity factor/static crack tip stress-intensity factor (K_I^{dyn}/K_I^{stat}) with time for a crack loaded by a step wave

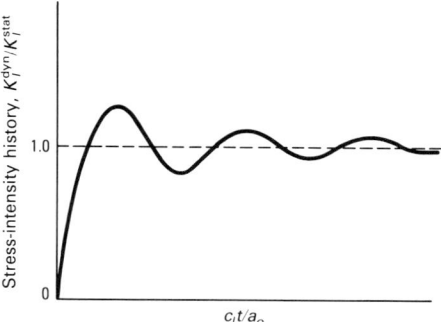

When a crack is struck by a short-lived stress pulse, classical static fracture mechanics concepts may be inadequate for predicting crack instability. When the duration of the applied load is comparable to the time required for waves to run the length of the crack, the stress-intensity history is very different from that expected from static considerations. Furthermore, if the duration of the applied load is a few microseconds, time must be included in the instability criterion.

Most dynamic load situations do not require consideration of load duration for instability prediction. However, certain situations involving impact, explosive detonation, or laser irradiation may exhibit pulse-duration effects.

According to static concepts, the critical stress for instability should continuously decrease with increasing crack length:

$$\sigma_{crit} \propto K_I \sqrt{\pi a} \qquad \text{(Eq 38)}$$

However, when a crack is loaded by a short stress pulse whose duration is comparable to the time for waves to run a distance representative of the crack length, the instability stress can be higher than that predicted by the static equations. Experiments in which controlled stress pulses were produced by plate impact or projectile impact techniques (Ref 77, 94) further showed that for cracks above a certain length (relative to the product of pulse duration and wave speed), the instabil-

ity stress is constant and independent of crack length.

These observations are understood by considering the stress-intensity history experienced by a crack subjected to a step-function load (i.e., a load that increases abruptly with time, then remains at a constant level), as described in Ref 95 and 96 and shown in Fig. 36. The stress intensity rises gradually with the square root of time, surpasses the static stress intensity considerably, then reaches the constant static value after several damped oscillations.

For a step-function load of finite duration, the stress-intensity history experienced by a crack is obtained by superimposing the behavior shown in Fig. 36 and a similar behavior of opposite sign added at the time of unloading. Figure 37 illustrates the stress-intensity histories for cracks of increasing length subjected to a rectangular stress pulse of duration T_o. For convenience, the crack length is given in units of $c_l T_o$, where c_l is the longitudinal wave speed.

For short cracks ($a_o < \frac{1}{24} c_l T_o$), the stress-intensity history is characterized by an almost rectangular shape, and the average stress intensity is identical to the equivalent static stress intensity. With increasing crack length, the stress intensification becomes larger and more triangular, but the average dynamic stress intensity is smaller than the stress-intensity factor value for the equivalent static crack. For even larger crack

lengths ($a_o > \frac{1}{3} c_l T_o$), the dynamic stress intensification does not continue to increase but remains the same, although the crack length and, accordingly, the static stress-intensity factor increase.

Assuming that the crack must experience a supercritical stress-intensity factor for at least a minimum amount of time to become unstable, a short-pulse fracture criterion has been developed (Ref 97). According to this criterion and the stress-intensity history discussed above, higher critical stresses are predicted to bring a crack above a given length to instability than in the equivalent static case.

Furthermore, the predicted instability stresses do not depend on crack length. The general instability behavior of cracks with different lengths subjected to different loading conditions (i.e., pulse amplitudes, σ_o, and pulse durations, T_o) is shown in a three-dimensional (σ_o-a_o-T_o) diagram in Fig. 38. The short-pulse fracture behavior is represented in the rear right section of the diagram; the front left regime (long pulse durations, short crack lengths) illustrates the typical static behavior.

Short-pulse fracture experiments are non-routine and require unique equipment, such as a gas gun, stress wave gages, and wave propagation computer codes. See Ref 96 for further information.

Fig. 37 Stress-intensity histories for cracks of increasing length subjected to a rectangular stress pulse of duration T_o

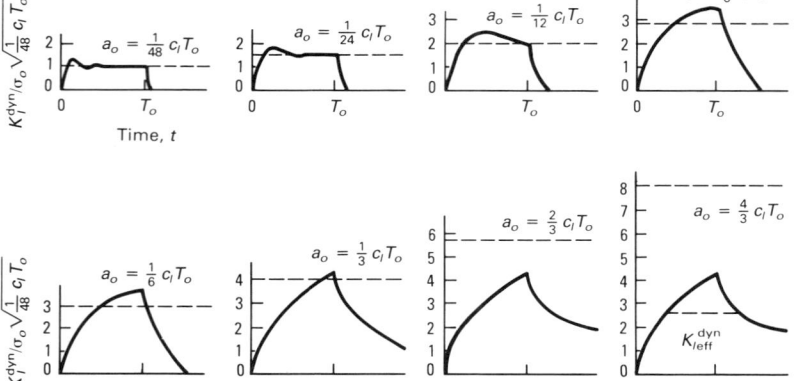

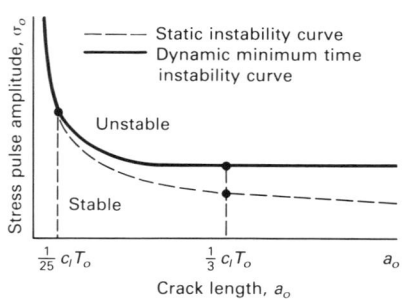

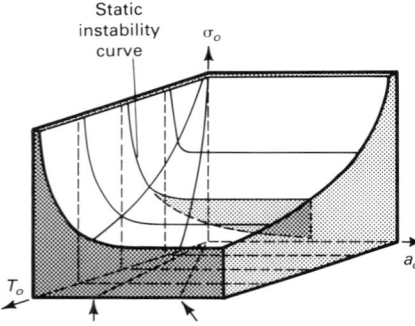

Fig. 38 Crack instability surface for short-lived stress pulses according to the minimum time criteria

Crack Propagation Tests

By Donald A. Shockey
Director, Department of Metallurgy and Fracture Mechanics
SRI International

RAPIDLY PROPAGATING CRACKS are pertinent to fracture mechanics for several reasons. In seeking lower bound toughness values, it is of interest to determine whether rate effects in the rapidly strained material at the tip of a fast running crack will result in a minimum value of fracture toughness. Those concerned with crack arrest find it necessary to study crack propagation, or the stage preceding arrest. Furthermore, the relationship between the velocity dependence of propagating crack toughness $K_{ID}(\dot{a})$

and the loading rate dependence of crack initiation toughness $K_{Id}(\dot{K})$ has been the subject of many investigations.

Standard procedures for measuring toughness associated with fast running cracks do not exist; however, photoelastic, caustic, and thermal techniques have been applied successfully and have enhanced understanding of the behavior of propagating cracks. The works of Kalthoff *et al.* (Ref 98), Kobayashi *et al.* (Ref 99) and Dally and Kobayashi (Ref 100, 101) provide details and specific results of the caustic and photoelastic techniques, respectively, applied to fast running cracks.

The toughness of steels has been determined as a function of crack velocity by an indirect method (Ref 102). Fracture specimens having a double-cantilever beam geometry were slowly wedge-loaded until a crack emerged from the notch root and propagated into the specimen. Cracks propagated at constant velocity, and crack velocity in individual specimens was varied by varying the bluntness of the starter notch. The stress intensity at crack instability was calculated from the measured critical load. Crack velocity was measured by conducting strips deposited across the expected crack path, and the crack propagation distance was measured after the experiment.

The experiment was modeled as a beam-on-elastic foundation, and a dynamic finite element simulation of the experiment was performed, using as input the load at which instability occurred and a trial toughness/crack velocity relationship. The calculated crack velocity and propagation distance were compared with the measured quantities; the toughness/crack velocity relationship was revised, and a second computational simulation was performed.

The process was repeated until a toughness/crack velocity relationship was found

that produced agreement between calculated and measured crack velocities and propagation distances. The results showed good agreement with direct measurements made by Kobayashi and Dally (Ref 101) using a bi-refringent coating method.

Many questions concerning the behavior of rapidly propagating cracks remain unanswered. Among them are the uniqueness of the stress intensity/crack velocity relationship, equivalence of dynamic crack initiation toughness and rapid crack propagation toughness, and the question of whether energy delivered to a rapidly propagating crack tip is equal to the energy absorbed at the crack tip, particularly in the early stages of the crack propagation. It is unknown at this time whether the general trend of decreasing fracture toughness with increasing loading rate continues at very high rates, or whether fracture toughness levels off or rises again at some minimum toughness.

Crack Arrest Tests

By George R. Irwin
Visiting Professor of Mechanical Engineering
University of Maryland

CRACK ARREST TOUGHNESS, K_{IA}, is a measure of the ability of a material to stop a fast running crack. Test procedures for determining crack arrest toughness are currently being examined by ASTM Task Group E24.01.06. Determination of this material property is of prime importance to the nuclear reactor pressure vessel and bridge materials industries.

At temperatures close to and below the nil-ductility temperature, test results have indicated that values of the impact fracture toughness K_{Id} obtained from tests conducted over appropriate loading times, $K_{Ic}(t)$, provide close estimates of the crack arrest toughness, K_{IA}. The required time of loading, t, to obtain good estimates of K_{IA} by K_{Id} varies from 5 ms at the nil-ductility temperature to several minutes at temperatures far below the nil-ductility temperature. At temperatures above the nil-ductility temperature, techniques for obtaining K_{IA} in this manner are difficult to implement, and the method requires additional study and development before it can be used with confidence.

For temperatures at and above the nil-ductility temperature, a variety of experimental procedures that provide a run-arrest segment of rapid fracturing have been developed and used to obtain estimates of K_{IA}. Prior to 1974, calculations of K used with these experiments assumed that K_{IA} did not differ significantly from the value of K at a short time (1 or 2 ms) after crack arrest, when the stress distribution was essentially static. For example, the method of Ripling and Crosley (Ref 103) used a pin-loaded, contoured, double-cantilever beam specimen subjected to a continuously increasing load. From static analysis, the value of K depended only on the applied load. Rapid recordings of load versus time showed that the load drop at the time of the run-arrest event was followed by load oscillations superimposed on the rising applied load. A trend line averaged through these oscillations and extrapolated back to the run-arrest time provided the load value used to calculate K at crack arrest. Side-grooved specimens and thicknesses adequate for a primarily plane-strain crack front stress pattern were used, and the result was termed K_{Ia}.

There was some discussion at that time as to the appropriate terminology, because K_{Ia} could be interpreted as the value of K closely following crack arrest, such that small high-frequency K oscillations did not cause reinitiation of the arrested crack front. The possibility that K_{Ia} might differ from the dynamic K just prior to crack arrest was discussed, but the difference was thought to be relatively small. Comparisons of K_{Ia} from this method to K_{Id} (10-ms loading time) showed agreement near the nil-ductility temperature for A302-B and A533-B steels.

From 1972 to 1974, it became apparent that the influence of dynamic effects on rapid run-arrest fracturing required additional study. Run-arrest experiments performed at Battelle Columbus Laboratories using wedge-loaded double-cantilever beam specimens showed substantial indications of dynamic effects. At the same time, computer

programs applicable to dynamic fracture problems were receiving attention. Application of computations of this type helped explain the dynamic effects observed in the run-arrest experiments with double-cantilever beam specimens.

The primary considerations related to nuclear pressure vessel safety applications were identified as the need for a K_{Ia} measurement method for which differences between K_{Ia} and K_{IA} were known to be negligible and the need for a dynamic calculation procedure that could be applied to run-arrest events in Oak Ridge National Laboratory (ORNL) experiments and to hypothesized run-arrest events in actual reactor pressure vessels.

The differences of opinion that have been expressed regarding crack arrest are related primarily to whether or not useful estimates of K_{IA} can be obtained using a static analysis test method. Model tests with wedge-loaded double-cantilever beam specimens (Ref 98, 104) showed that the post-arrest K oscillations had a decreasing trend so that the static post-arrest K value was less than K_{IA}, as shown in Fig. 39. A decrease in the initial K-value, K_o, was found to decrease both the crack jump length and the crack speed. With a sufficient reduction of dynamic effects by reduction of the initial K_o, the difference between K_{IA} and the statically determined post-arrest K_{Ia} was thought to be negligible.

Increased understanding of run-arrest cleavage fracturing was gained in 1974

through research activities supported by the U.S. Nuclear Regulatory Commission and the Electric Power Research Institute and by the formation of an ASTM Task Group on K_{Ia} measurements. Enhancement of dynamic effects in the double-cantilever beam specimen due to compliance of the specimen arms was clear, and attention was focused on the use of transverse-wedge-loaded specimens of the nearly square compact style.

A large Cooperative Test Program was conducted from 1977 to 1979 with 29 participating laboratories, a substantial number of which were outside the United States, chiefly in Europe and Japan (Ref 105). Examination of the results showed that although some features of interest could be observed using methods of dynamic analysis, the peculiarities of the test method did not provide adequate information about boundary conditions to calculate K_{IA} values. Except for an undesirably large scatter in the test results, the average post-arrest K-values computed using a static analysis agreed with expectations.

Estimated K_{Ia} values, obtained using a static analysis test method similar to that of the Cooperative Test Program, were used to assist in the planning of large-cylinder thermal shock fracture experiments at Oak Ridge National Laboratory. In addition, dynamic analysis computations indicated that for crack jumps of moderate size, the dynamic influences in the ORNL thermal shock fracture experiment cylinders would be relatively

Fig. 39 Comparison of pre- and post-arrest dynamic and static K-values observed with wedge-loaded double cantilever beam specimens for different crack jumps

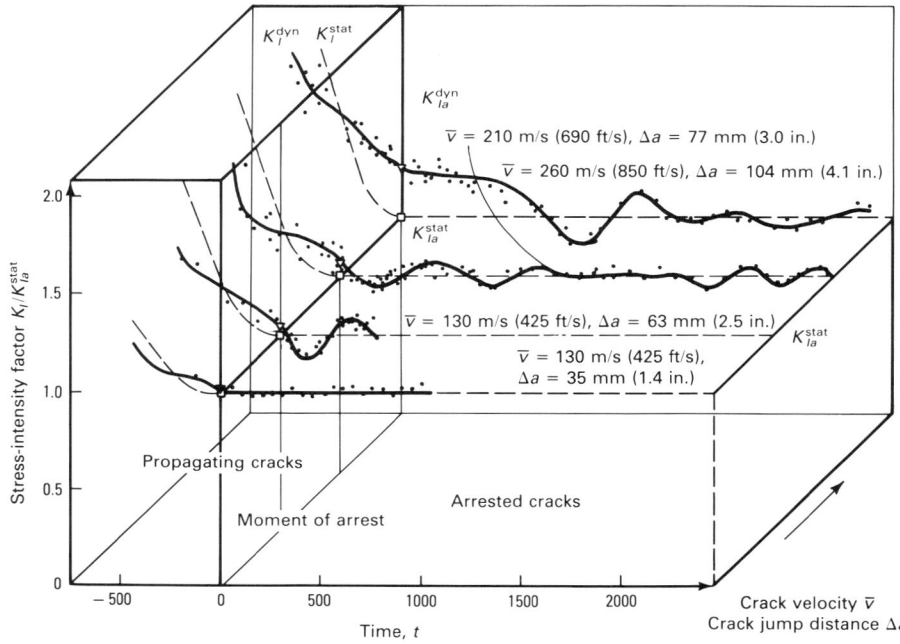

small (Ref 106). One reason for this was that the release of loading spring energy, which is the main source for dynamic effects in small-specimen tests, is quite small in the thermal-stress-loaded cylinder.

Static analysis computations of the K at crack arrest for the thermal shock fracture tests were therefore expected to provide close estimates of the dynamic K_{IA} value. Comparisons of thermal shock fracture crack arrest K_{Ia} values with those obtained at Battelle Columbus Laboratories using static analysis small-specimen tests showed good agreement (Ref 107). The close agreement of the average K_{Ia} values from the Cooperative Test Program with the K_{Ia} values from the Oak Ridge National Laboratory tests is illustrated in Fig. 40.

This information indicates that values of K_{Ia} obtained from relatively simple laboratory experiments can provide useful estimates of the actual K_{IA} pertaining to run-arrest events. On the other hand, additional information suggests that the degree of success for the small-specimen tests depends considerably on specimen geometry and the manner of load application.

For example, the Cooperative Test Program work at the University of Maryland included framing camera photographs of birefringent photoelastic coatings bonded onto the specimen surface. Although the picture quality did not permit accurate calculation of K-values for the running crack, the position of the crack front was clear. This permitted calculations of crack speed and the run-arrest time.

The run-arrest time was determined to be about 220 μs. The time required to experience significant influence from the loading spring (the local region around the specimen loading hole indented by bearing pressure from the split pins) was estimated from model tests to be about 160 μs. An approximately static stress state would be expected about 1 ms after crack arrest. Thus, a substantial influence on the run-arrest event from release of the loading spring energy was possible.

Figure 39 shows direct measurements of K versus time for a group of model experiments performed using a specimen geometry and loading arrangement similar to that of the Cooperative Test Program. The plate material used was Araldite B, and multiflash photographs of the crack tip caustic pattern permitted K determination before and after crack arrest. The K oscillations prior to and following crack arrest are small, and the average K value has no noticeable downward trend after crack arrest when the crack velocity or the crack jump distance is small.

For this group of tests, the difference between the dynamic K at crack arrest and the

value of K when the stress distribution became more nearly static was relatively small. For a test method generally similar to that of the Cooperative Test Program, close estimates of K_{IA} can be obtained using static analysis values of K at a short time after crack arrest if both the magnitude of K oscillations and the downward trend of the post-arrest average K are relatively small. The data for run-arrest events with small crack jumps shown in Fig. 39 illustrate successful fulfillment of these requirements.

Multiflash model experiments at the University of Maryland showed post-arrest K oscillations that were not relatively small. Consideration of the reasons for this difference in specimen behavior, from the results shown in Fig. 39, led to a study of transverse wedge loading methods. In the Cooperative Test Program loading technique, transverse movement of the split pins is prevented by pin flanges that bear against the front face of the specimen. In the model experiments that showed larger dynamic effects, the back face of the specimen rested on the pin flanges, which were pressed against a backup plate.

A comparison of results from the two loading configurations showed that dynamic effects were greatly reduced by the Cooperative Test Program loading method. This reduction of dynamic effects can be understood in terms of damping and a reduction in loading spring energy. The loading spring energy is reduced because friction between the pin flanges and the specimen permits load transfer with less elastic indentation of the specimen. Friction between the specimen and the backup plate adds to other sources of damping that influence the test.

An ASTM round-robin program on crack arrest testing is currently in progress using a test method similar to that of the Cooperative Test Program. There are 27 participants in this program (16 foreign laboratories and 11 American laboratories). It is hoped that the results from the round robin will facilitate ASTM standardization of crack arrest toughness measurement. Imperfections exist in the current test method, but the round-robin testing experience should lead to useful suggestions for improvement.

Two major problems remain. If the test results are to be evaluated using a static analysis procedure to determine K at a short time after crack arrest, the release of loading spring energy must be moderate, which may reduce dynamic effects, possibly by friction. Clearly, the extremes of small friction and large friction are undesirable. The degree of success that can be achieved by efforts to control the amount of friction is uncertain.

Additionally, the current method for assisting the start of cleavage fracturing uses a layer of brittle weld metal at the root of the

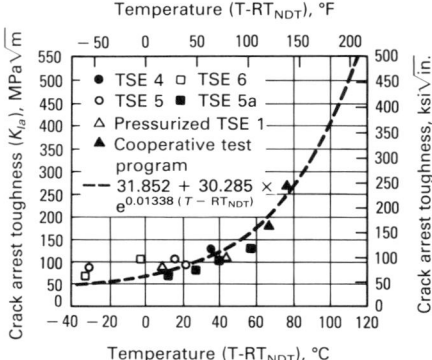

Fig. 40 Crack arrest toughness values determined from several run-arrest events in ORNL thermal shock experiments (TSE) and the mean values of the Cooperative Test Program

starter notch. The variability of successful cleavage initiation associated with this method of embrittlement has been undesirably large, particularly at temperatures that are well above the nil-ductility temperature.

General progress in dynamic fracture research will be achieved when it is recognized that the development of evaluation methods for crack arrest toughness and the study of rapid fracture with the aid of dynamic analysis computations are mutually supportive. For crack arrest toughness evaluations, emphasis on simplicity and minimum cost is desirable. This consideration conflicts directly with extensive dynamic computations, which must derive crack arrest K-values from the limited data provided by a simple small-specimen laboratory test.

Given the available information, the value of K at crack arrest can be estimated using small-specimen tests without the need for dynamic analysis computations if the dynamic influences are small. A proposed ASTM method for crack arrest testing, which is currently being used in the ASTM round-robin test program, is described in the Appendix to this article.

Micromechanics of Dynamic Fracture

By Donald A. Shockey
Director, Department of Metallurgy and Fracture Mechanics
SRI International

DYNAMIC FRACTURE occurs on the microscale by the nucleation, growth, and coalescence of roughly spherical microvoids (ductile fracture) or roughly penny-shaped planar microcracks (brittle fracture). Fracture under quasi-static loads occurs similarly. Evidence for these microfracture mechanisms can be found by examining fracture surfaces with a microscope—the hemispherical dimples and the cleavage facets are halves of voids and microcracks, respectively, that have separated during the fracture process.

In most materials, this microfracture activity occurs exclusively in the process zone at the tip of a macrocrack, where the stresses, strains, and strain rates are much larger than in the surrounding material. Examination of the material immediately below the fracture surface or near an arrested macrocrack tip usually reveals no microfractures, suggesting that nearly all microfractures that nucleate participate in the macrofracture process.

The number, size, and types of microfractures occurring in a material are strongly influenced by microstructure. Voids tend to nucleate at inclusions, second-phase particles, phase boundaries, and grain boundaries, and grow plastically under hydrostatic stress states. Coalescence may occur by impingement of these voids, but usually occurs by the nucleation, growth, and coalescence of a population of much smaller voids that nucleate at smaller heterogeneities in the ligaments of highly strained material between the larger voids.

Similarly, in more brittle materials that fracture by cleavage or by grain-boundary separation, transgranular or intergranular cracks form in favorably oriented grains or grain boundaries. Such microcracks usually arrest when they encounter adjacent grains. Coalescence of adjacent skewed microcracks occurs by rupture of the unfavorably oriented material separating the cracks. This coalescence process is often the most energy-absorbent phase of the fracture process; hence, that stage of the fracture process contributes most to material toughness.

A recently developed fractographic tool that reconstructs micromechanical details of the fracture process is the quantitative topography technique (Ref 108). The topology of conjugate fracture surfaces is determined, and the digitized data are stored in a computer. The three-dimensional conjugate surfaces are then superimposed computationally and separated by increments to show the microcrack nuclei, where they occur in the specimen, and how they grow and coalesce with each other and with the macrocrack. A particularly attractive feature of the technique is that the sequence of the various microfailure events is indicated.

In principle, macrocrack behavior can be computed by modeling the microcrack activity in advance of the macrocrack; in fact, some efforts have been made to do so (Ref 109). The micromechanics approach, however, is difficult and costly, so continuum treatments should be used when possible. However, the extra effort and expense associated with the micromechanical approach may be necessary to treat certain dynamic fracture problems, such as those involving many simultaneously active microcracks. Furthermore, the micromechanical approach holds the promise for linking fracture behavior to microstructure.

The fracture of materials under shock wave loads is an example for which continuum concepts are difficult to apply and for which the micromechanical approach has produced useful results. Dynamic fracture caused by stress waves, which is often called spalling or scabbing, occurs by near-simultaneous nucleation of microvoids or cracks over an area within a material and the subsequent growth and coalescence of the microvoids or cracks to produce a continuous macrofracture.

The process is easily studied in plate impact experiments. The arrangement shown in Fig. 41 is commonly used to produce a short-lived tensile pulse in a specimen under an easily analyzed one-dimensional strain state. Here, a flyer plate mounted on a cylindrical projectile is accelerated in a gun barrel by sudden release of pressurized gas and is made to impact a specimen plate at the gun muzzle. Care is taken to ensure simultaneous impact between the two plates at all locations so the lateral strains are zero until unloading waves from the periphery can propagate into the specimen. The specimen plate is decelerated in a chamber filled with energy-absorbing material to prevent subsequent uncontrolled impact loads.

The initial compression waves produced in the colliding plates on impact reflect at the plate boundaries to produce a tensile pulse in the specimen whose amplitude is specified and controlled by the impact velocity. The pulse duration is controlled by the plate thickness. Thus, if the flyer plate is tapered, as shown in detail A of Fig. 41, the duration of the pulse experienced by the specimen varies with location, and the dependence of ductile fracture damage on time-at-stress can be investigated in a single experiment. If a series of experiments are performed at different impact velocities, the effect of stress can also be studied.

After impact, the specimens are recovered and sectioned to reveal the internal fracture damage. Examples of ductile and brittle fracture damage observed on polished cross sections of impacted specimen plates are shown in Fig. 42. The fracture damage is analyzed quantitatively by counting and measuring the traces of the microfractures on the polished section surfaces. Measurements are made by using a large-area record reader, with which an operator positions a cross hair on one end of a void trace, records the coordinates, and repeats the process for the other end.

A simple computer program uses the data to compute the length, orientation, and position of the trace within the specimen. The size distributions on the sectioned surfaces are then converted to actual void or crack size distributions per unit volume by means of a statistical transformation implemented in a computer code. This procedure is described in detail by Seaman (Ref 110).

This quantitative damage analysis results in void or crack size distribution curves for various positions on a cross section. An example is shown in Fig. 43.

Data Analysis

Equations governing nucleation and growth of voids have been deduced from previous plate impact studies to be of the form:

$$\dot{N} = \dot{N}_o \exp \frac{\sigma - \sigma_{no}}{\sigma_1} \qquad \text{(Eq 39)}$$

Fig. 41 Plate impact arrangement for studies of shock-induced fracture in materials under uniaxial strain conditions

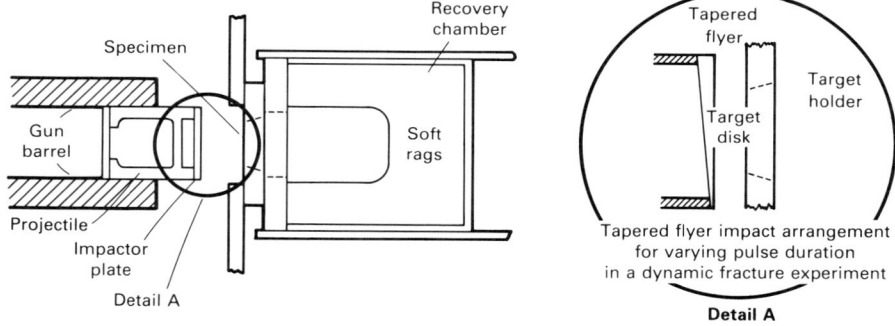

Recovery chamber

Specimen

Gun barrel

Soft rags

Projectile

Impactor plate

Detail A

Tapered flyer

Target holder

Target disk

Tapered flyer impact arrangement for varying pulse duration in a dynamic fracture experiment

Detail A

Fig. 42 Polished cross sections through plate impact specimens showing dynamic fracture damage

(a) 1145 aluminum (spherical voids). (b) Armco iron (planar cracks)

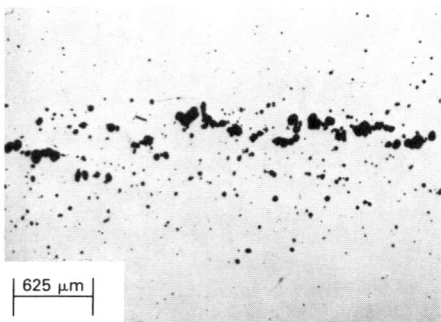

|_ 625 μm _|

(a)

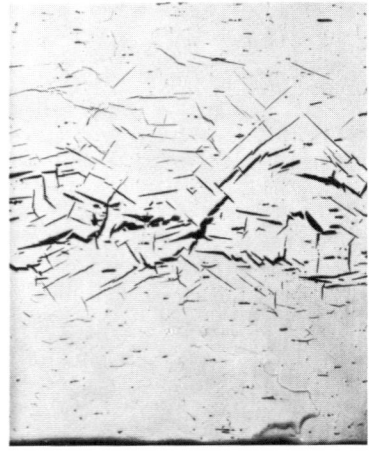

(b)

and

$$\dot{R} = \frac{\sigma - \sigma_{go}}{4\eta} R \qquad \text{(Eq 40)}$$

where $\dot{N}$ and $\dot{R}$ are time-rate-of-change of the number and size of the voids or cracks, respectively; σ is the current value of the tensile stress; and the remaining symbols are empirically determined, material-specific parameters. These equations have successfully described dynamic tensile damage in a wide variety of materials and hence are quite general.

Values for the fracture parameters are obtained from the observed damage in the following manner. The average void nucleation rate (total number of voids divided by the nominal duration of the tensile stress) is plotted versus peak stress in tension to obtain a first estimate of the nucleation threshold stress (σ_{no}) and the other nucleation parameters (σ_1 and N_o). The shape parameter R_1 of the observed distribution is plotted versus tensile impulse (peak tension times the duration of the tension) to determine the nucleation size (R_o) and the growth rate parameters (σ_{go} and η).

After the initial estimates of all five parameters are determined from plots, trial one-dimensional calculations are performed to approximate the impact conditions at several points in the target. These calculations are repeated with different fracture parameters until the computed damage and measured damage compare satisfactorily. Then a two-dimensional calculation is performed to simulate the entire impact. It is usually not necessary to modify the parameters further and repeat the two-dimensional simulation.

Appendix: Crack Arrest Fracture Toughness of Ferritic Materials*

This proposed ASTM test method provides an estimate of crack arrest fracture toughness (K_{IA}) for ferritic materials by the use of a crack-line wedge-loaded compact specimen. The estimate is denoted K_a. The test is conducted by loading a specimen to cause the run-arrest segment of crack extension; the stress-intensity factor at a very short time after arrest, K_a, is evaluated by a static analysis.

The in-plane dimensions of the specimen must be large enough to allow it to be modeled by an elastic analysis. If a condition of plane strain is required, a minimum specimen thickness is also required. In other words, the minimum specimen dimensions are a function of crack arrest toughness and yield strength, so a range of specimen sizes is possible. If the specimen does not exhibit rapid crack propagation and arrest, K_a cannot be evaluated.

Test Method

This method estimates the value of the stress-intensity factor, K, at which a fast running crack will arrest. The test is made by forcing a wedge into a split pin, which applies an opening force across the crack

*William L. Fourney, Department of Mechanical Engineering, University of Maryland; Alan R. Rosenfield, Research Leader, Battelle Columbus Laboratories; Charles W. Marschall, Principal Research Scientist, Battelle Columbus Laboratories; and Richard G. Hoagland, Associate Professor, Ohio State University

starter notch in a modified compact specimen, thus causing a run-arrest segment of crack extension. The nature of rapid crack jump suggests the need for a dynamic analysis of test results. However, experimental observations indicate that a static analysis of the test results provides a useful estimate of the actual (dynamic) value of the stress-intensity factor at crack arrest (Ref 105, 111, 112).

Calculation of the effective stress intensity at initiation, K_o, is based on measurements of the machined notch length and opening displacement at initiation. The value of K_a is based on measurements of crack length and opening displacement just after arrest.

In structures containing gradients in either toughness or stress, a crack may initiate in a region of low toughness and/or high stress and arrest in another region of higher toughness and/or lower stress. The value of K_{IA} is a measure of the ability of a material to arrest such fast running cracks. Under conditions where the dynamic effects are known to be small, static analyses using K_a values have been used successfully to estimate whether a crack will arrest in a structure (Ref 106, 113). At the instant of arrest, the state of stress in the structure may be different from that associated with static equilibrium. For this reason, the K at arrest, K_{IA}, generally will not be equal to the K after static equilibrium is reached.

For a specified material and temperature, the stress-intensity factor, K_a, obtained by this method is thought to approximate the actual stress intensity at crack arrest (Ref 105, 114). In materials research and development, crack arrest testing procedures are helpful in establishing, in quantitative terms significant to service performance, the effects of metallurgical variables such as composition or heat treatment and fabrication operations such as welding or forming on the ability of a new or existing material to arrest running cracks. In design, this method assists in the selection of material for stiffeners and arrestor plates as well as the location and sizes of such components.

Test Apparatus

Modified compact specimens that have been notched by machining are tested. Because the load required to initiate a crack from the machined notch is large compared with the load needed to maintain extension after a natural crack forms, the loading system must have a low compliance compared with the test specimen. For this reason, a wedge and split pin assembly is used to apply a load on the crack line. This loading arrangement does not permit easy measurement of opening loads. Consequently, open-

Fig. 43 Microcrack size distribution curves obtained by counting and measuring fracture damage on polished sections through plate impact specimens of Armco iron

N_g is the cumulative number of voids and/or cracks with a radius greater than a given radius R, N_t is the total number of voids and/or cracks. Dimensions given in cm^3 (1 cm^3 = 0.06 in^3) and mm (1 mm = 0.04 in.)

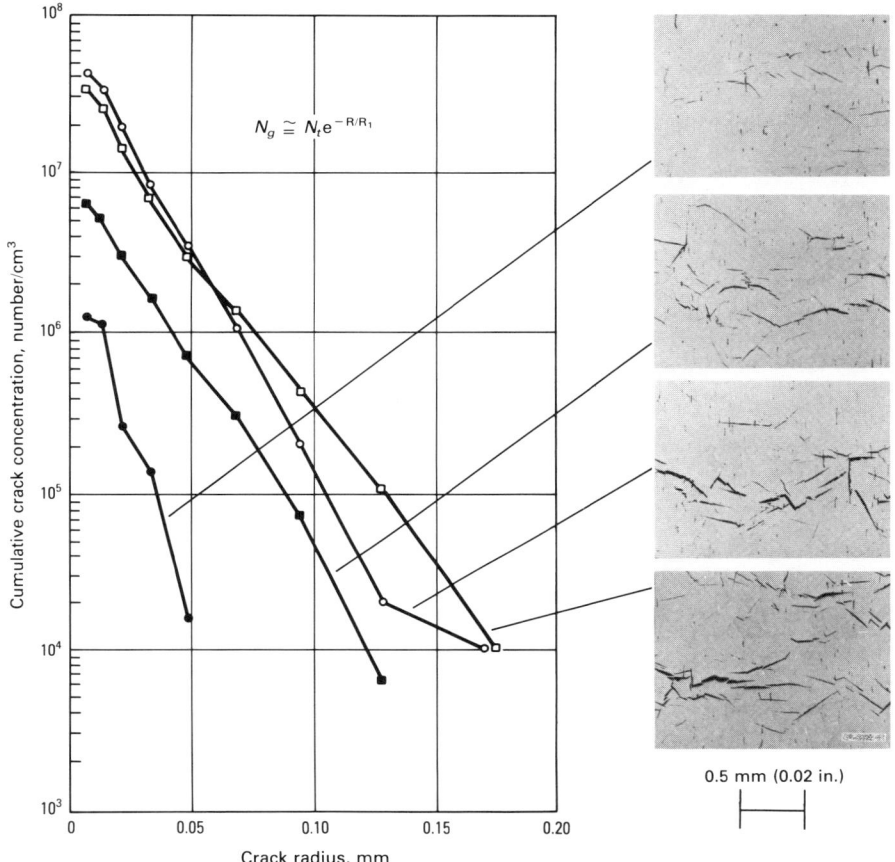

$$N_g \cong N_t e^{-R/R_1}$$

0.5 mm (0.02 in.)

Fig. 44 Cross section through the midplane of a typical load train for monotonic loading

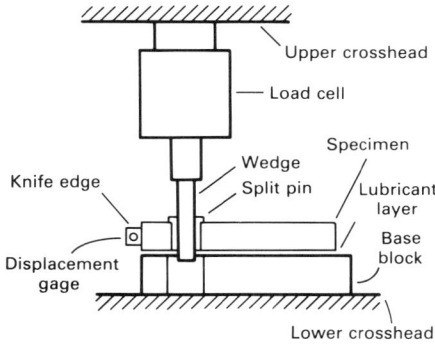

Upper crosshead
Load cell
Specimen
Wedge
Knife edge
Split pin
Lubricant layer
Base block
Displacement gage
Lower crosshead

Fig. 45 Suggested geometry and dimensions of a wedge and bushing assembly suitable for use with a 25.4-mm (1.0-in.) diam specimen hole

These dimensions should be scaled when other hole diameters are used.

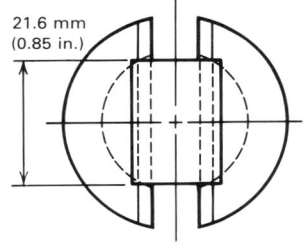

21.6 mm (0.85 in.)

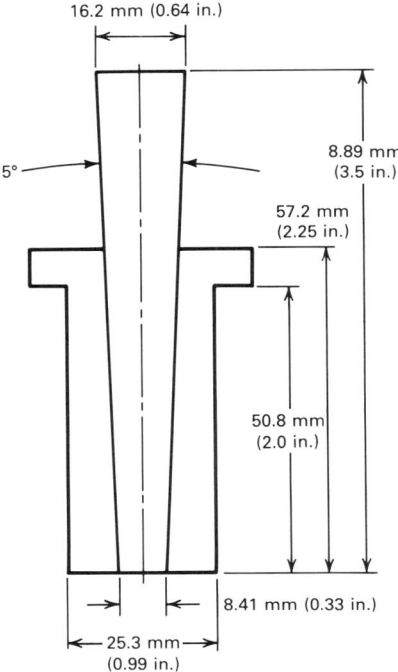

16.2 mm (0.64 in.)
5°
8.89 mm (3.5 in.)
57.2 mm (2.25 in.)
50.8 mm (2.0 in.)
8.41 mm (0.33 in.)
25.3 mm (0.99 in.)

ing displacement measurements in conjunction with crack size and compliance calibrations are used for calculating K_o and K_a.

Loading Train. A typical loading train is shown in Fig. 44. The specimen is mounted on a base block containing a hole that is aligned with the specimen hole. The diameter of the hole in the base block should be between 1.05 and 1.15 times the diameter of the hole in the specimen. The load that forces the wedge into the split pin is transmitted through a load cell.

The wedge, split pin, base plate, and specimen hole should be lubricated. Molybdenum disulfide (both dry and in a grease vehicle) has been found to be satisfactory, as have thin 0.08- to 0.13-mm (0.003- to 0.005-in.) strips of Teflon and high-temperature lubricants.

A low-taper-angle wedge and split pin arrangement is used. No special surface finish is required when Teflon strips are used. For other lubricants, a matte finish (grit blasted) on the sliding surfaces is recommended. The dimensions of a wedge and split pin assembly that has been found satisfactory for specimens having $125 < W < 170$ mm ($5 < W < 6.7$ in.) is shown in Fig. 45.

The split pin must be long enough to contact the full specimen thickness, and the radius must be large enough to avoid plastic indentation of the test specimen. In all cases, the diameter of the split pin should be 0.13 mm (0.005 in.) less than the diameter of the specimen hole. The wedge must be long enough to develop the maximum expected opening displacement. Any air- or oil-hardening tool steel is suitable for the wedge and split pins. Hardnesses of 45 to 55 HRC have been used successfully. With the recommended wedge angle and proper lubrication, a loading machine producing 1/5 to 1/10 the expected maximum opening load is adequate.

Displacement gages are used to measure the crack mouth opening displacement at $0.25W$ from the load line. Accuracy within 2% over the working range is required. The gage recommended in ASTM E 399 (Ref 1) or a similar gage modified to accommodate conical seats is satisfactory. The gage must be attached so that seating contact with the specimen is not altered by the jump of the crack. Two methods that have proved satisfactory are shown in Fig. 46. Other gages can be used if their accuracy is within 2%.

Specimen Configuration, Dimensions, and Preparation

Specimen Configuration. The configuration of a compact crack arrest specimen that is suitable for low- and intermediate-strength steels is shown in Fig. 47. The thickness, B, should be either full product plate thickness or a thickness sufficient to produce a condition of plane strain. Side grooves should be be used with a root radius of 0.25 mm (0.010 in.), a 45° included angle, and a depth of $B/8$ per side. For alloys that require notch-tip embrittlement, the side grooves should be machined after embrittlement. The specimen width, W, should be within the range of $2B \leq W \leq 8B$. The displacement gage should measure opening displacements at an offset from the load line of $0.25W$ away from the crack tip.

Specimen Dimensions. To limit the extent of plastic deformation in the specimen prior to crack initiation, certain size requirements must be met. These requirements depend on the material yield strength. They

also depend on K_a and therefore the K_o needed to achieve an appropriate run-arrest event. At this time, the procedure is incomplete with regard to the minimum and maximum allowable ratios of K_o/K_a.

The minimum in-plane size requirement is $W > 1.35(K_o/\sigma_{YS})^2$, where σ_{YS} is the static yield strength. This should limit the amount of plastic opening displacement to approximately 5% of the total opening displacement. For a test result to be termed plane-strain (K_{Ia}) by this method, the specimen thickness, B, must be equal to or exceed ($K_a/\sigma_{Yd})^2$, where σ_{Yd} is a formal dynamic yield strength estimate. This estimate is discussed below in the section on calculation and interpretation of results.

Starting Notch. The function of the starting notch is to produce crack initiation at an opening displacement (or wedging force) that will permit an appropriate length of crack extension prior to crack arrest. It is convenient to express the crack initiation condition in terms of a calculated stress-intensity factor, K_o. Different materials require different starter notch preparation procedures to produce an acceptable value of K_o.

A notched crack starter that has been used successfully with high-strength steel (Ref 115) and aluminum alloys is shown in Fig. 48(a). The basic notch parameters are the included angle, θ, and the root radius, ρ. Generally, decreasing θ and/or decreasing ρ will lead to a lower K_o.

For low- and medium-strength steels, a notched brittle weld crack starter, as shown in Fig. 48(b), can be used. It is produced by depositing a weld across the specimen thick-

ness. A welding procedure similar to that described in ASTM E 208 (Ref 16) has been found to be satisfactory for the crack starter weld. This procedure uses Fe-0.30C-1.50Mn-1.0Cr-0.5Si electrodes and requires a crack starter notch of sufficient width so that the electrode can reach the bottom of the slot.

The notch can be fabricated by drilling a hole with its center at the desired location of a_o. The sides of the notch can be produced by saw cutting to the hole. A current of 180 to 200 A, a short arc length, and a speed of about 100 mm/min (4 in./min) have been found to be satisfactory. Weld starter and runout blocks having the same shape as the specimen notch bottom are used.

The finished single-pass weld should have a relatively flat surface to facilitate subsequent notching. Embrittling the notch root by gas tungsten arc welding without filler metal has also proved to be successful. A notch is machined in the weld as shown in Fig. 48(b). A notch root diameter, ρ, of 0.25 to 0.38 mm (0.010 to 0.015 in.) or even larger can be used. Large diameters tend to produce larger K_o values. Experience has shown that weld embrittlement does not introduce serious distortion. A light surface grind operation, however, may be required.

Alternative crack starter configurations can also be used. An example is the duplex compact specimen (Ref 116), which consists of a block of high-strength low-toughness alloy containing the starter notch, electron-beam welded to the alloy being evaluated. The duplex design is particularly useful for tests carried out well above the ductile-brittle transition temperature (DBTT).

Fig. 46 Two alternative clip gage seating arrangements for modified compact specimens used in crack arrest tests

(a) Arrangement using knife edges. Dimension A is to be 0.05 to 0.25 mm (0.002 to 0.010 in.) less than the thickness of the clip gage arm. The clip gage knife edge can be attached to the specimen with mechanical fasteners or adhesives. The clip gage is installed by sliding it into the gap. (b) Arrangement using conical mounts

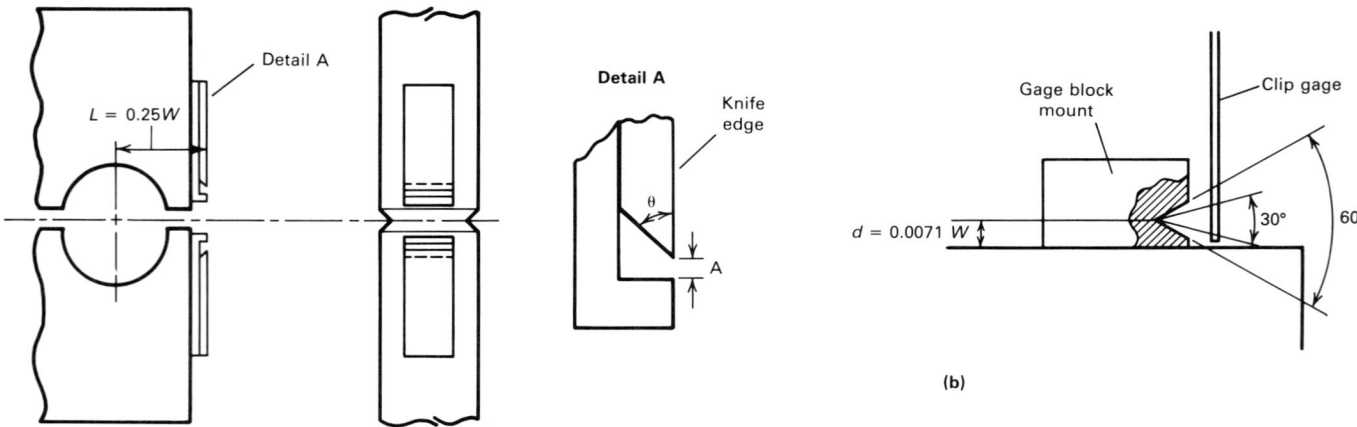

Fig. 47 Geometry of a compact crack arrest test specimen satisfactory for low- and intermediate-strength steels

$H = 0.6W \pm 0.005W$; $S = (B - B_N)/2 \pm 0.01B$; $N \leq W/10$; $0.15W \leq L \leq 0.25W$; $0.30W \leq a_o \leq 0.40W$; $0.125W \pm 0.005W \leq D \leq 0.250W \pm 0.005W$

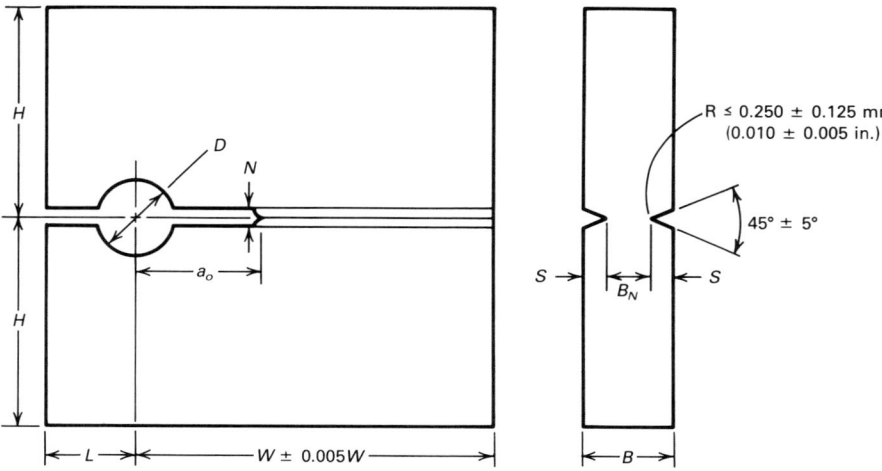

Fig. 48 Recommended starting notch configurations for crack arrest test specimens

(a) Machined notch without a brittle weld. N need not be less than 1.5 mm ($\frac{1}{16}$ in.), but must not exceed $W/10$. (b) Machined notch in a brittle weld. N must be large enough to allow entry of welding electrode. N = 13 mm (0.5 in.) has proved satisfactory.

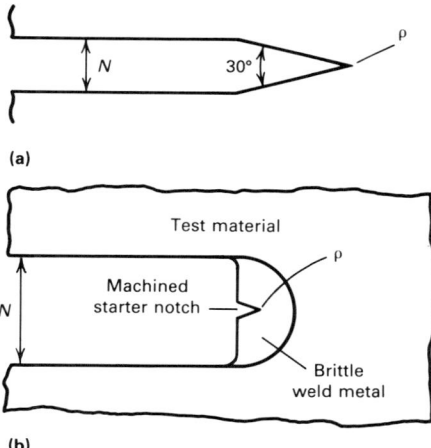

(a)

(b)

Test Procedure

Specimen and Temperature Measurement. The specimen thickness, B, and the crack plane thickness, B_N, should be measured between the end of the machined notch crack starter and the unnotched edge of the specimen. The crack plane thickness should be measured to $\pm 1\%$ of B. The specimen width, W, should be measured to $\pm 1\%$ of W. At least three replicate tests should be made at a single test temperature.

Specimens may be heated or cooled to the test temperature by any appropriate method. A method that has been used successfully for elevated-temperature tests employs electric-resistance heating tapes in combination with a variable power source. Tests at subambient temperatures have been conducted using cooling coils embedded in the specimen support base; a controlled flow of liquid nitrogen or other suitable coolant through the cooling coils permits low temperatures to be reached without difficulty.

To minimize temperature gradients through the specimen thickness, the specimen should be surrounded with a good thermal insulator. The specimen temperature should be measured with a thermocouple welded to the top surface of the specimen at a location near the side groove about 25 mm (1 in.) ahead of the starter notch.

Prior to starting the test, the specimen should be held at the test temperature for a time sufficient to eliminate significant temperature gradients. The temperature difference between any two opposing edges of the specimen should not exceed 3 °C (5.4 °F). In reporting the test results, the test temperature should be the temperature measured on the specimen at the time of the crack propagation/arrest event.

Loading Procedure. The specimen is loaded until the crack extends, or until a preselected displacement is reached. During the test, an autographic record of wedge load versus opening displacement should be obtained. It is also useful to obtain information about the final segment of the opening displacement versus time record on an oscillo-graph or other high-rate device. This provides additional information about the nature of the run-arrest event.

To measure K_a or K_{Ia}, a segment of unstable crack extension must occur. The occurrence of unstable crack extension will normally be apparent to the operator, both audibly and as an abrupt load drop on the test record. After the event, the operator should remove the load on the wedge to avoid further crack propagation.

If unstable cracking has not occurred by the time a preselected displacement, $\delta_{o(max)}$, has been attained, the load should be removed from the wedge. Any further attempts to initiate unstable cracking should use sequential load/unload cycling, as described below. The magnitude of the limiting displacement is obtained from:

$$\delta_{o(max)} = \frac{0.86 \, \sigma_{YS}W\sqrt{\dfrac{B_N}{B}}}{Ef\left(\dfrac{a_o}{W}\right)} \qquad \text{(Eq 41)}$$

where σ_{YS} is the static yield strength of the test material (or, in the case of duplex specimens, the crack-starter section material), and the other terms are defined in Eq 42 and 43. Equation 41 corresponds to $K_{o(max)} = 0.86 \, \sigma_{YS}\sqrt{W}$ and is designed to limit the departure of displacements from linear elasticity to about 5%.

Marking the Arrested Crack. The position of the arrested crack can be marked by heat tinting. For ferritic steels, heating at 260 to 370 °C (500 to 700 °F) for 10 to 90 min has proved successful. Any time and temperature combination that clearly marks the arrested crack front is acceptable. The appearance of heat tinting on freshly machined or ground and sanded surfaces may be indicative of the heat tinting progress on the fracture surfaces. If the fracture surfaces are to be examined microscopically, heat tinting temperatures at the lower end of the range are recommended.

After the crack front has been marked, the specimen is broken completely in two. This usually can be done with the wedging apparatus used in testing the specimen. The breaking open of structural steel specimens is facilitated greatly by cooling in dry ice or liquid nitrogen.

Measurement of Arrested Crack Length. The average crack front position on the heat-tinted fracture surface should be measured to within 1% at the center (mid-thickness) of the specimen and midway between the center and the bottom of the side groove on each side. The average of these three measurements defines the arrested crack length, a_f.

Crack front irregularities that may occur can make it difficult to determine the crack length. It is therefore suggested that the crack length at each of the three locations specified above be taken to be a visual average across a strip of width $B_N/4$ centered at each location. A photographic record of the heat-tinted fracture surface is recommended, particularly if there are any unusual perturbations in the crack front contours.

Calculations and Interpretation of Results

Displacement Measurement. From the autographic load-displacement record (Fig. 49), two displacement values, one measured at the onset of unstable crack growth (δ_o) and the other after arrest (δ_a), are used for calculating K_o and K_a, respectively. The δ_a point is believed to represent the displacement approximately 100 μs after crack arrest.

The preferred interpretation of δ_a is the opening displacement at about 2 ms after crack arrest. However, this measurement may not be possible with the instrumentation used. This testing practice assumes that δ_a at about 100 ms after crack arrest does not differ significantly from δ_a at 2 ms.

In the weld-embrittled specimen (Fig. 48b), a load drop of 50 to 60% has been found to indicate that a sufficient length of unstable fracture has occurred and that δ_a is a usable arrest displacement value. Limitations on the length of the run-arrest segment are outlined below in the section on size requirements.

Calculations of K_o and K_a. Calculate K_o and K_a from:

$$K = E \delta f(x) \frac{\sqrt{\dfrac{B}{B_N}}}{\sqrt{W}} \qquad \text{(Eq 42)}$$

Fig. 49 Load-displacement record for monotonic loading

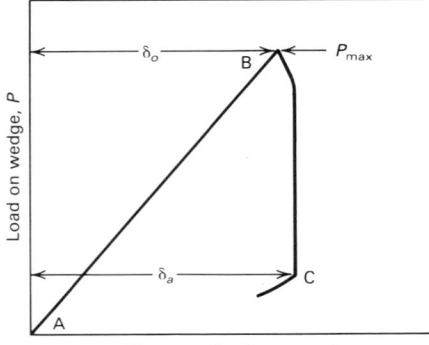

Clip gage displacement, δ

where

$$f(x) = \frac{2.24(1.72 - 0.9x + x^2)\sqrt{1-x}}{(9.85 - 0.17x + 11x^2)}$$

(Eq 43)

and where $x = a/W$; δ is the initial clip gage displacement, δ_o, or arrest clip gage displacement, δ_a; E is Young's modulus; a is the initial slot length, a_o, or final crack length, a_a; W is the specimen width; B is the specimen thickness, as shown in Fig. 47; and B_N is specimen thickness at the crack plane, as shown in Fig. 47. Values of $f(x)$ computed from Eq 43 for various values of x are given in Table 4. To calculate K_o, use $a = a_o$ and $\delta = \delta_o$. To calculate K_a, use $a = a_a$ and $\delta = \delta_a$.

Size Requirements. The value of K_a calculated from Eq 42 can be considered a linear elastic plane-strain value, K_{Ia}, provided the criteria described below and summarized in Table 5 are satisfied.

The unbroken ligament, W-a_a, must equal or exceed both $0.15W$ and $1.25(K_a/\sigma_{Yd})^2$. Here, σ_{Yd} is a formal dynamic yield strength estimate for a loading time of about 100 μs at the test temperature. For structural steels, σ_{Yd} is assumed to be 205 MPa (30 ksi) greater than the yield strength, σ_{YS}, as measured by ASTM Standard E 8 (Ref 117).

The thickness, B, must equal or exceed $(K_a/\sigma_{Yd})^2$. The minimum crack jump, $a_a - a_o$, must be at least twice the slot width, N, and greater than the plane-stress plastic zone radius associated with the initial loading, $(K_o/\sigma_{YS})^2/2\pi$. If a duplex specimen is used, the alternative requirement is that the crack penetrate a distance equal to or greater than B_N into the test section.

Sequential Load/Unload Cycling

The procedure described below can be used when the material being tested has not undergone unstable crack extension at an imposed displacement that includes about 5% plasticity. Such a failure to exhibit unstable cracking may be due to an unfavorable combination of small specimen size, low yield strength, and high toughness. Data obtained using this procedure have been shown to be equal to acceptable K_{Ia} values under some circumstances. Provided the strength level, arrest toughness level, and specimen dimensions are all comparable to those used in Ref 118 to 120, this procedure will result in useful estimates of K_{Ia}.

If, during the monotonic loading procedure specified in the main body of this proposed standard, an unstable crack is not initiated upon reaching the specified limiting displacement for the specimen being tested

Table 4 Values of f(x) for use in Eq 42

x	$f(x)$	x	$f(x)$
0.30	0.268	0.60	0.159
0.31	0.263	0.61	0.156
0.32	0.260	0.62	0.153
0.33	0.256	0.63	0.150
0.34	0.252	0.64	0.146
0.35	0.247	0.65	0.143
0.36	0.244	0.66	0.140
0.37	0.240	0.67	0.137
0.38	0.236	0.68	0.134
0.39	0.232	0.69	0.131
0.40	0.228	0.70	0.128
0.41	0.225	0.71	0.125
0.42	0.221	0.72	0.122
0.43	0.217	0.73	0.119
0.44	0.213	0.74	0.116
0.45	0.210	0.75	0.113
0.46	0.206	0.76	0.110
0.47	0.203	0.77	0.107
0.48	0.199	0.78	0.104
0.49	0.196	0.79	0.101
0.50	0.192	0.80	0.098
0.51	0.189	0.81	0.095
0.52	0.185	0.82	0.092
0.53	0.182	0.83	0.089
0.54	0.179	0.84	0.086
0.55	0.175	0.85	0.082
0.56	0.172	0.86	0.079
0.57	0.169	0.87	0.076
0.58	0.165	0.88	0.072
0.59	0.162	0.89	0.069
		0.90	0.065

Table 5 Criteria used to ensure that K_a is a linear elastic plane-strain value

Feature	Criterion
Unbroken ligament	$W - a_a \geqslant 0.15W$
Unbroken ligament	$W - a_a \geqslant 1.25 (K_a/\sigma_{Yd})^2$
Thickness	$B \geqslant 1.0 (K_a/\sigma_{Yd})^2$
Crack jump length	$a_a - a_o \geqslant 2N$
	$a_a - a_o \geqslant (K_o/\sigma_{YS})^2/2\pi$
	$a_a - 0.46W \geqslant B_N$
	(duplex specimen only)

(governed by yield strength, elastic modulus, and dimensions), the load should be removed from the wedge, and the wedge should be extracted in preparation for sequential load/unload cycling. The clip gage should remain in place during unloading and wedge removal to indicate the presence of possible residual displacement. Wedge extraction and load/unload cycling can be simplified by use of the arrangement shown in Fig. 50. Main features include a hold-down plate and a wedge that is fastened to the loading ram.

Without rezeroing the recorder, the load should be reinserted and applied to the wedge at the same displacement rate as in the initial loading. Loading should be continued until a rapid crack jump is initiated, or until the displacement measured with the clip gage reaches 105 ± 2% of that on the previous loading, whichever occurs first. If an unstable crack is not initiated upon reaching 105 ± 2% of the previous cycle displacement, the load should be removed and the wedge extracted. The load-displacement record should then be labeled with the appropriate cycle number. The load/unload cycling procedure described above can then be repeated.

Figure 51 is a schematic illustration of a load-displacement record for a specimen subjected to load/unload cycling that did not exhibit unstable cracking until the sixth cy-

cle. The initial loading and unloading path is $OA_1B_1C_1$; the second followed path is $C_1A_2B_2C_2$, etc.

If many load/unload cycles are required, it may be necessary to relubricate the wedge and split pin occasionally. Increased friction will be indicated by an increased slope in the load-displacement record, or by larger reverse loads required to extract the wedge.

If on subsequent loading cycles an unstable crack is not initiated by the time the displacement has reached a predetermined limit, the test should be discontinued. It may

be helpful at this point to retest the specimen at a lower temperature, perhaps 20 to 40 °C (36 to 72 °F) lower, in an attempt to derive useful data from the specimen. The procedure consists of reinserting the wedge and applying load until a rapid crack initiates, or until the displacement reaches the limiting value achieved on the final cycle at the higher temperature. If rapid cracking occurs, the procedures described below are followed. If not, the test temperature should be lowered again, and retesting should be performed to the same limiting displacement or until a rapid fracture is initiated. The limiting displacement can be calculated as:

$$\delta_{o\ \text{limit}} = \frac{1.35\ \sigma_{YS}\ W\ \sqrt{\dfrac{B_N}{B}}}{Ef\left(\dfrac{a_o}{W}\right)}$$

Upon initiation of an unstable crack, the subsequent procedures will be the same as if rapid fracture had occurred on the initial loading, except that the displacement recorded for initiation (δ_o) and arrest (δ_a) will be only the displacement increment applied during the particular loading cycle that led to unstable fracture. For example, in Fig. 51 the appropriate displacements are $\delta_o = \delta_D - \delta_{C_5}$ and $\delta_a = \delta_E - \delta_{C_5}$. In addition to recording δ_o and δ_a, the total residual displacement that preceded the rapid-fracture cycle, $\Sigma\delta_p$ (δ_{C_5} in Fig. 51), should be recorded.

Fig. 50 Hold-down plate that facilitates wedge extraction during sequential load/unload cycling

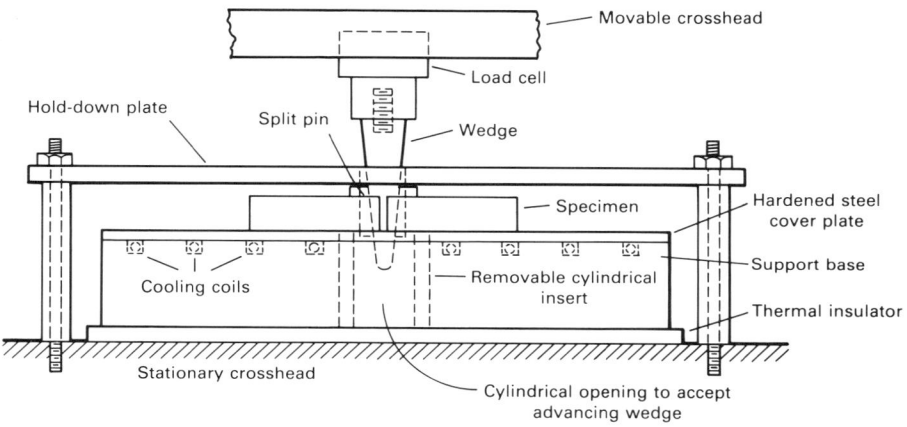

Fig. 51 Load-displacement record for a specimen subjected to sequential load/unload cycling

Rapid fracture was initiated on cycle No. 6. The curves illustrated are idealized; the actual curves may cross over each other because of changes in the friction between the wedge and split pin. Also, the point labeled B may lie at negative loads if the friction is high, and some or all of the points labeled C may be indistinguishable from zero displacement.

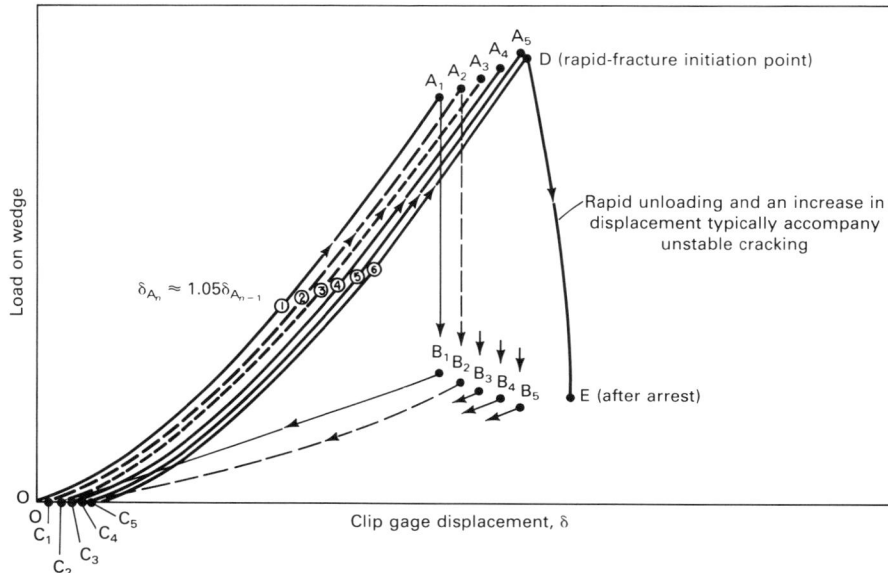

REFERENCES

1. "Standard Test Method for Plane-Strain Fracture Toughness of Metallic Materials," E 399, *Annual Book of ASTM Standards*, Vol 03.01, ASTM, Philadelphia, 1984, p 519-554
2. "Standard Test Method for J_{Ic}, A Measure of Fracture Toughness," E 813, *Annual Book of ASTM Standards*, Vol 03.01, ASTM, Philadelphia, 1984, p 763-781
3. P. Albrecht, W.R. Andrews, J.P. Gudas, J.A. Joyce, F.J. Loss, D.E. McCabe, D.W. Schmidt, and W.A. Van der Sluys, Tentative Test Procedure for Determining the Plane Strain J_I-R Curve, *J. Test. Eval.*, Vol 10, 1982, p 245-251
4. A.K. Shoemaker and R.R. Seeley, Summary Report of Round-Robin Testing by the ASTM Task Group E24.01.06 on Rapid Loading Plane-Strain Fracture Toughness K_{Ic} Testing, *J. Test. Eval.*, Vol 11, 1983, p 261-272
5. K.D. Boyer, J.N. Fisher, G.R. Irwin,

G.V. Krishna, U. Morf, R. Roberts, and R.E. Slockblower, "Determination of Tolerable Flaw Sizes in Full-Size Welded Bridge Details," Fritz Engineering Laboratory Report No. 399-3 (76), Lehigh University, Bethlehem, PA, 1976

6. W.A. Logsdon and J.A. Begley, Dynamic Fracture Toughness of SA533 Grade A Class 2 Base Plate and Weldments, in *Flaw Growth and Fracture*, STP 631, ASTM, Philadelphia, 1977, p 477-492

7. B. Marandet, G. Phelippeau, and G. Sanz, Experimental Determination of Dynamic Fracture Toughness by *J*-Integral Method, in *Advances in Fracture Research*, Vol 1, D. Francis, Ed., Pergamon Press, Oxford, 1981, p 375-383

8. B. Marandet, G. Phelippeau, and G. Sanz, Influence of Loading Rate on the Fracture Toughness of Some Structural Steels in the Transition Regime, in *Fracture Mechanics: Fifteenth Symposium*, STP 833, R.J. Sanford, Ed., ASTM, Philadelphia, 1984, p 622-647

9. B. Marandet, G. Labbe, *et al.*, "Detection de l'Amorçage et Suivi de la Propagation d'une Fissure par Variation du Potentiel Electrique en Régime Alternatif," External Report IRSID, RE 549, Institut de Recherches de la Siderurgie Francaise, Saint Germain-en Laye, France, 1978

10. J.A. Joyce, Static and Dynamic *J-R* Curve Testing of A533B Steel Using the Key Curve Analysis Technique, in *Fracture Mechanics Fourteenth Symposium, Vol I: Theory and Analysis*, STP 791, J.C. Lewis and G. Sines, Ed., ASTM, Philadelphia, 1983, p I-543 to I-560

11. H. Ernst, P.C. Paris, M. Rossow, and J.W. Hutchinson, Analysis of Load Displacement Relationship to Determine *J-R* Curve and Tearing Instability Material Properties, in *Fracture Mechanics*, STP 677, C.W. Smith, Ed., ASTM, Philadelphia, 1979, p 581-599

12. J.A. Joyce, H. Ernst, and P.C. Paris, Direct Evaluation of *J*-Resistance Curves from Load Displacement Records, in *Fracture Mechanics: Twelfth Conference*, STP 700, ASTM, Philadelphia, 1980, p 222-236

13. "Standard Methods for Notched Bar Impact Testing of Metallic Materials," E 23, *Annual Book of ASTM Standards*, Vol 03.01, ASTM, Philadelphia, 1984, p 210-233

14. "Rules for Construction of Nuclear Power Plant Components," *ASME Boiler and Pressure Vessel Code*, Section III, Division 1—Appendices, Nonmandatory Appendix G, American Society for Mechanical Engineers, New York, 1983

15. *Code of Federal Regulations*, Title 10 ("Energy"), Part 50 ("Domestic Licensing of Production and Utilization Facilities"), U.S. Government Printing Office, Washington, DC, 1981

16. "Standard Method for Conducting Drop-Weight Test to Determine Nil-Ductility Transition Temperature of Ferritic Steels, E 208, *Annual Book of ASTM Standards*, Vol 03.01, ASTM, Philadelphia, 1984, p 346-365

17. "Pressure Vessel Research Committee Recommendation on Toughness Requirements for Ferritic Materials," Welding Research Council Bulletin 175, New York, Aug 1972

18. J.M. Barsom, The Development of AASHTO Fracture Toughness Requirements for Bridge Steels, *Eng. Frac. Mech.*, Vol 7 (No. 3), Sept 1975, p 605-618

19. J.M. Barsom and S.T. Rolfe, Correlations Between K_{Ic} and Charpy V-Notch Test Results in the Transition Temperature Range, in *Impact Testing of Materials*, STP 466, ASTM, Philadelphia, 1979, p 281-302

20. R.H. Sailors and H.T. Corten, Relationship Between Material Fracture Toughness Using Fracture Mechanics and Transition Temperature Tests, in *Fracture Toughness, Proceedings of the 1971 National Symposium on Fracture Mechanics*, STP 514, Part II, ASTM, Philadelphia, 1972, p 164-191

21. J.A. Begley and W.A. Logsdon, "Correlation of Fracture Toughness and Charpy Properties for Rotor Steels," WRL Scientific Paper 71-1E7-MSLRF-P1, Westinghouse Research Laboratory, Pittsburgh, July 1971

22. B. Marandet and G. Sanz, Evaluation of the Toughness of Thick Medium-Strength Steels by Using Linear Elastic Fracture Mechanics and Correlations Between K_{Ic} and Charpy V-Notch, in *Flaw Growth and Fracture*, STP 631, ASTM, Philadelphia, 1977, p 72-95

23. R.A. Wullaert, Fracture Toughness Predictions from Charpy V-Notch Data, in *What Does the Charpy Test Really Tell Us?*, Proceedings of the American Institute of Mining, Metallurgical and Petroleum Engineers, Denver, American Society for Metals, 1978

24. S.T. Rolfe and S.R. Novak, Slow-Bend K_{Ic} Testing of Medium-Strength High-Toughness Steels, in *Review of Developments in Plane-Strain Fracture Toughness Testing*, STP 463, ASTM, Philadelphia, 1970, p 124-159

25. "Rapid Inexpensive Tests for Determining Fracture Toughness," National Materials Advisory Board, National Academy of Sciences, Washington, DC, 1976

26. *What Does the Charpy Test Really Tell Us?*, Proceedings of the American Institute of Mining, Metallurgical and Petroleum Engineers, Denver, American Society for Metals, 1978

27. D.M. Norris, J.E. Reaugh, and W.L. Server, A Fracture-Toughness Correlation Based on Charpy Initiation Energy, in *Fracture Mechanics: Thirteenth Conference*, STP 743, ASTM, Philadelphia, 1981, p 207-217

28. W.L. Server *et al.*, "Analysis of Radiation Embrittlement Reference Toughness Curves," EPRI NP-1661, Electric Power Research Institute, Palo Alto, CA, Jan 1981

29. *Reference Fracture Toughness Procedures Applied to Pressure Vessel Materials*, Metal Properties Council MPC-24, Proceedings of the Winter Annual Meeting of the American Society for Mechanical Engineers, New Orleans, ASME, New York, 1984

30. B. Augland, Fracture Toughness and the Charpy V-notch Test, *Brit. Weld. J.*, Vol 9 (No. 7), 1962, p 434

31. G.D. Fearnehough and C.J. Hoy, Mechanism of Deformation and Fracture in the Charpy Test as Revealed by Dynamic Recording of Impact Loads, *J. Iron Steel Inst.*, Vol 202, 1964, p 912

32. R.A. Wullaert, Application of the Instrumented Charpy Impact Test, in *Impact Testing of Metals*, STP 466, ASTM, Philadelphia, 1970, p 148-164

33. A.S. Tetelman and A.J. McEvily, Jr., *Fracture of Structural Materials*, John Wiley & Sons, New York, 1967

34. T.R. Wilshaw and P.O. Pratt, The Effect of Temperature and Strain Rate on the Deformation and Fracture of Mild Steel Charpy Specimens, in *Proceedings of the First International Conference on Fracture*, Vol 2, Sendai, Japan, Sept 1965, p 973-991

35. D.R. Ireland, W.L. Server, and R.A. Wullaert, "Procedures for Testing and Data Analysis," ETI Report TR-75-43, Effects Technology, Inc., Santa Barbara, CA, Oct 1975

36. D.R. Ireland, Procedures and Problems Associated with Reliable Control of the Instrumented Impact Test, in *Instrumented Impact Testing*, STP 563, ASTM, Philadelphia, 1974, p 3-29

37. E.C.J. Buys and A. Cowan, Interpretation of the Instrumented Impact Test, *Welding in The World*, Vol 8 (No. 1), 1970, p 70-76

38. "Instrumented Precracked Charpy Testing, Report I—Recommended Testing Procedure," and "Report II—Associated Test Program," Pressure Vessel Research Committee/Metal Properties Council Working Group on Instrumented Precracked Charpy Testing, Westinghouse Research Laboratory, Pittsburgh, 1974

39. W.L. Server, Impact Three-Point Bend Testing for Notches and Precracked Specimens, *J. Test. Eval.*, Vol 6 (No. 1), 1978, p 29-34

40. D.M. Norris, D. Quiñones, and B. Moran, Computer Simulation of Plastic Deformation in the Charpy V-notch Impact Test, in *What Does the Charpy Test Really Tell Us?*, American Society for Metals, 1978, p 22-32

41. T.M.F. Ronald, J.A. Hall, and C.M. Pierce, Usefulness of Precracked Charpy Specimens for Fracture Toughness Screening Tests of Titanium Alloys, *Met. Trans.*, Vol 3, April 1972, p 813-818

42. G.M. Orner and C.E. Hartbower, Transition-Temperature Correlations in Construction Alloy Steels, *Weld. J.*, Vol 40 (No. 9), Oct 1961, p 459s

43. J.H. Gross, The Effect of Strength and Thickness on Notch Ductility, *Weld. J.*, Vol 48 (No. 10), Oct 1969, p 441s

44. W.L. Server, D.R. Ireland, and R.A. Wullaert, "Strength and Toughness Evaluations from an Instrumented Impact Test," ETI TR-74-29R, Effects Technology, Inc., Santa Barbara, CA, Nov 1974

45. H.J. Saxton et al., Load Point Compliance of the Charpy Impact Specimen, in *Instrumented Impact Testing*, STP 563, ASTM, Philadelphia, 1974, p 30-49

46. W.L. Server, R.A. Wullaert, and J.W. Sheckherd, "Verification of the EPRI Dynamic Fracture Toughness Testing Procedures," ETI TR 75-42, Effects Technology, Inc. Santa Barbara, CA, Oct 1975

47. W.L. Server, W. Oldfield, and R.A. Wullaert, "Experimental and Statistical Requirements for Developing a Well-Defined K_{IR} Curve," EPRI NP-372, Electric Power Research Institute, Palo Alto, CA, May 1977

48. J.D.G. Sumpter and C.E. Turner, Method for Laboratory Determination of J_c, in *Cracks and Fracture*, STP 601, ASTM, Philadelphia, 1976, p 3-18

49. D.J.F. Ewing, Calculations on the Bending of Rigid/Plastic Notched Bars, *J. Mech. Phys. Solids*, Vol 16, 1968, p 205-213

50. W.L. Server, General Yielding of Charpy V-notch and Precracked Charpy Specimens, *J. Eng. Mat. Technol.*, Vol 100, April 1978, p 183-188

51. Committee on Safety of Nuclear Installations Specialist Meeting on Instrumented Precracked Charpy Testing, EPRI NP-2102-LD, Electric Power Research Institute, Palo Alto, CA, Nov 1981

52. F.J. Loss, Ed., "Structural Integrity of Water Reactor Pressure Boundary Components," Progress Report Ending Feb 1976, NRL Report 8006, Naval Research Laboratory, Washington, DC, Aug 1976

53. T. Nicholas, "Instrumented Impact Testing Using a Hopkinson Bar Apparatus," AFML-TR-75-54, Air Force Materials Laboratory, Wright-Patterson Air Force Base, OH, July 1975

54. J.F. Kalthoff et al., "Measurements of Dynamic Stress Intensity Factors in Impacted Bend Specimens," Committee on Safety of Nuclear Installations Specialist Meeting on Instrumented Precracked Charpy Testing, EPRI NP-2102-LD, Electric Power Research Institute, Palo Alto, CA, Nov 1981

55. J.F. Kalthoff, S. Winkler, W. Böhme, and W. Klemm, Determination of the Dynamic Fracture Toughness K_{Id} in Impact Tests by Means of Impact Response Curves, in *Advances in Fracture Research*, D. Francois et al., Ed., Pergamon Press, New York, 1980, p 363-373

56. J.F. Kalthoff, "Time Effects and Their Influences on Test Procedures for Measuring Dynamic Material Strength Values," in *Proceedings of the International Conference on the Application of Fracture Mechanics to Materials and Structures*, G.C Sih, E. Sommer, and W. Dahl, Ed., Freiburg, West Germany, June 20-24, Martinus Nijhoff Publishers, The Hague, 1983, p 107-136

57. D.J. Ayres, Dynamic Plastic Analysis of Ductile Fracture—the Charpy Specimen, *Int. J. Frac.*, Vol 12, 1976, p 567-578

58. R. Arone, On the Dynamic Fracture Toughness Determination by Instrumented Impact Tests, in *Proceedings*

of the 4th International Conference on Fracture, Vol 3, ICF 4, Waterloo, Canada, June 19-24, 1977, p 549-553

59. S. Mall, A.S. Kobayashi, and Y. Urabe, Dynamic Photoelastic and Dynamic Finite-element Analyses of Dynamic-tear-test Specimens, *Exp. Mech.*, Vol 18, 1978, p 449-510

60. S. Mall, A.S. Kobayashi, and F.J. Loss, Dynamic Fracture Analysis of Notched Bend Specimens, in *Crack Arrest Methodology and Applications*, STP 711, G.T Hahn and M.F. Kanninen, Ed., ASTM, Philadelphia, 1980, p 70-85

61. A.S. Kobayashi, A.F. Emery, and B.M. Liaw, "Dynamic Analyses of Notched Bend Specimens," Committee on Safety of Nuclear Installations Specialist Meeting on Instrumented Precracked Charpy Testing, Palo Alto, CA, Dec 1-3, 1980, EPRI NP-2102-LD, Electric Power Research Institute, Nov 1981

62. S. Mall, A Finite Element Analysis of Transient Crack Problems with a Path-Independent Integral, in *Advances in Fracture Research*, D. Francois et al., Ed., Pergamon Press, New York, 1980, p 2171-2178

63. D.M. Norris, Computer Simulation of the Charpy V-Notch Toughness Test, *Eng. Fract. Mech.*, Vol 11, 1979, p 261-274

64. T. Peuser, Dynamic Analysis of Impact Test Specimens, in *Proceedings of the International Conference on Application of Fracture Mechanics to Materials and Structures*, G.C. Sih, E. Sommer, and W. Dahl, Ed., Freiburg, West Germany, June 20-24, 1983, Martinus Nijhoff Publishers, The Hague, p 455-465

65. C.E. Turner, "Dynamic Fracture Toughness Measurements by Instrumented Impact Testing," ISPRA Courses, Advanced Seminar on Fracture Mechanics, Commission of the European Communities, Joint Research Center, Ispra, Italy, 20-24 Oct 1975

66. W. Böhme, "Eine einfache Methode zur Bestimmung der dynamischen Ri β-spitzenbeanspruchung bei schlagbelasteten Dreipunktbiegeproben," IWM-Report Z 2/84, Fraunhofer-Institut für Werkstoffmechanik, Freiburg, West Germany, March 1984

67. J.F. Kalthoff, Stress Intensity Factor Determination by Caustics, in *Proceedings of International Conference on Experimental Mechanics*, Society for Experimental Stress Analysis and Japan Society of Mechanical Engi-

neers, Honolulu, May 23-28, 1982, p 1119-1126

68. J.F. Kalthoff, W. Böhme, and S. Winkler, Analysis of Impact Fracture Phenomena by Means of the Shadow Optical Method of Caustics, in *Proceedings of the VIIth International Conference on Experimental Stress Analysis*, Haifa, Israel, Aug 23-27, 1982, p 148-160

69. F.J. Loss, "Dynamic Toughness Analysis of Pressure Vessel Steels," Third Water Reactor Safety Research Information Meeting, Oct 1975

70. J.F. Kalthoff and S. Winkler, Vorrichtung zur Erfassung des Rißstarts bei einer Bruchmechanikprobe, Patentanmeldung P 33 34 570.8, Deutsches Patentamt, Munich, Sept 24, 1983

71. W. Böhme and J.F. Kalthoff, The Behavior of Notched Bend Specimens in Impact Testing, *Int. J. Fract.*, Vol 20 (No. 1), 1982, p 139-143

72. J.F. Kalthoff, S. Winkler, W. Böhme, and D.A. Shockey, Mechanical Response of Cracks to Impact Loading, in *Proceedings of the International Conference on the Dynamical Mechanical Properties and Fracture Dynamics of Engineering Materials*, Czechoslovak Academy of Sciences, Institute of Physical Metallurgy, Brno, Czechoslovakia, June 1983

73. J.F. Kalthoff, W. Böhme, and S. Winkler, "Verfahren zum Bestimmen mechanischer Werkstoff-Kennwerte und Vorrichtung zur Durchführung des Verfahrens," patent disclosure, Deutsches Patentamt, Munich, June 1983

74. J.H. Giovanola, "Investigation and Application of the One-Point-Bend Impact Test," ASTM 17th National Symposium on Fracture Mechanics, Albany, NY, Aug 1984

75. L.S. Costin, J. Duffy, and L.B. Freund, Fracture Initiation in Metals Under Stress Wave Loading Conditions, in *Fast Fracture and Crack Arrest*, STP 627, G.T. Hahn, and M.F. Kanninen, Ed., ASTM, Philadelphia, 1977, p 301-318

76. H. Kolsky, An Investigation of the Mechanical Properties of Materials at Very High Rates of Loading, *Proc. Phys. Soc.*, Vol B62, 1949, p 676-700

77. H. Homma, D.A. Shockey, and Y. Murayama, Response of Cracks in Structural Materials to Short Pulse Loads, *J. Mech. Phys. Solids*, Vol 31, 1983, p 261-279

78. T. Nakamura, C.F. Shih, and L.B. Freund, "Elastic-Plastic Analysis of a Dynamically Loaded Circumferentially Notched Round Bar," Brown

University Technical Report, NB82RA20004/1, Providence, RI, March 1984 (to be published in *Eng. Frac. Mech.*)

79. R. Dormeval, J.M. Chevallier, and M. Stelly, Fracture Initiation of Metals of High Loading Rates, in *Proceedings of Fifth International Conference on Fracture*, D. Francois, Ed., Cannes, France, 1981, p 355-362

80. L.S. Costin and J. Duffy, The Effect of Loading Rate and Temperature on the Initiation of Fracture in a Mild, Rate-Sensitive Steel, *J. Eng. Mat. Technol.*, Vol 101, July 1979, p 258-263

81. M.L. Wilson, R.H. Hawley, and J. Duffy, The Effect of Loading Rate and Temperature on Fracture Initiation in 1020 Hot-Rolled Steel, *Eng. Frac. Mech.*, Vol 13 (No. 2), 1980, p 371-385

82. H. Couque, J. Duffy, and R.J. Asaro, "Correlation of Microstructure with Dynamic and Quasi-Static Fracture of a Plain Carbon Steel," Brown University Technical Report, No. DAAG 29-81-K-0121/7, Providence, RI, July 1984

83. C. Mylonas and L. Hermann, Brown University Technical Report, No. 40-002-080/16, Providence, RI, Aug 1976

84. J. Guild, *The Interference System of Crossed Diffraction Gratings—Theory of Moiré Fringes*, Oxford University Press, London, 1960

85. J. Guild, *Diffraction Gratings as Measuring Scales*, Oxford University Press, London, 1960

86. H. Tada, P.C. Paris, and G.R. Irwin, "The Stress Analysis of Cracks Handbook," Del Research Corp., Hellertown, PA, 1973

87. G.R. Irwin, Crack Toughness Testing of Strain Rate Sensitive Materials, *J. Eng. Power*, Oct 1964, p 444-450

88. J.R. Rice, A Path-Independent Integral in the Approximate Analysis of Strain Concentration by Notches and Cracks, *J. Appl. Mech.*, Vol 35, 1968, p 379-386

89. J.A. Begley and J.D. Landes, The *J*-integral as a Fracture Criterion, in *Fracture Toughness*, STP 514, ASTM, Philadelphia, 1972, p 1-23

90. D.M. Tracey, "On the Fracture Mechanics Analysis of Elastic-Plastic Materials Using the Finite Element Method," Brown University Technical Report, NASA NGL 40-002-080/11, Providence, RI, 1973

91. P.C. Paris, in written discussion to Begley and Landes (see Ref 89)

92. J.R. Rice, P.C. Paris, and J.C. Mer-

kle, Some Further Results of *J* Integral Analysis and Estimates, in *Progress in Flaw Growth and Fracture Toughness Testing*, STP 536, ASTM, Philadelphia, 1973, p 231-245

93. J.W. Hutchinson and P.C. Paris, Stability Analysis of *J*-Controlled Crack Growth, in *Elastic-Plastic Fracture*, STP 668, J.D. Landes, J.A. Begley, and G.A. Clarke, Ed., ASTM, Philadelphia, 1979, p 37-64

94. D.A. Shockey, J.F. Kalthoff, and D.C. Erlich, Evaluation of Dynamic Crack Instability Criteria, *Int. J. Frac.*, Vol 22, 1983, p 217-229

95. J.D. Achenbach, Brittle and Ductile Extension of a Finite Crack by a Horizontally Polarized Shear Wave, *Int. J. Eng. Sci.*, Vol 8, 1970, p 947-966

96. G.C. Sih, G.T. Embley, and R.J. Ravera, Impact Response of a Plane Crack in Extension, *Int. J. Solids Struct.*, Vol 8, 1972, p 977-993

97. J.F. Kalthoff and D.A. Shockey, Instability of Cracks under Impulse Loads, *J. Appl. Phys.*, Vol 48 (No. 3), 1977, p 986-993

98. J.F. Kalthoff, J. Beinert, and S. Winkler, Measurements of Dynamic Stress Intensity Factors for Fast Running and Arresting Cracks in Double-Cantilever-Beam Specimens, in *Fast Fracture and Crack Arrest*, STP 627, G.T. Hahn and M.F. Kanninen, Ed., ASTM, Philadelphia, 1977, p 161-176

99. W.B. Bradley and A.S. Kobayashi, An Investigation of Propagating Cracks by Dynamic Photoelasticity, *Exp. Mech.*, Vol 10, 1970, p 106-113

100. T. Kobayashi and J.W. Dally, Relation Between Crack Velocity and the Stress Intensity Factor in Birefringent Polymers, in *Fast Fracture and Crack Arrest*, STP 627, G.T. Hahn and M.F. Kanninen, Ed., ASTM, Philadelphia, p 257-273

101. T. Kobayashi and J.W. Dally, Dynamic Photoelastic Determination of the *a-K* Relation for 4340 Alloy Steel, in *Crack Arrest Methodology and Applications*, STP 711, G.T. Hahn and M.F. Kanninen, Ed., ASTM, Philadelphia, 1980, p 189-210

102. G.T. Hahn *et al.*, "Critical Experiments, Measurements and Analyses to Establish a Crack Arrest Methodology for Nuclear Pressure Vessel Steels," Annual Report to the U.S. Nuclear Regulatory Commission on Contract NUREG/CR-0057, by Battelle Memorial Institute, Columbus, OH, 1975-77

103. P.B. Crosley, and E.J. Ripling, Crack Arrest Toughness of Pressure Vessel

Steels, *J. Nucl. Eng. Design*, Vol 17, 1971, p 32-45

104. R.G. Hoagland, A.R. Rosenfield, P.C. Gehlen, and G.T. Hahn, A Crack Arrest Measuring Procedure for K_{Im}, K_{Id} and K_{Ia} Properties, in *Fast Fracture and Crack Arrest*, STP 627, G.T. Hahn and M.F. Kanninen, Ed., ASTM, Philadelphia, 1977, p 177-202

105. P.B. Crosley *et al.*, "Final Report on Cooperative Test Program on Crack Arrest Toughness Measurements," NUREG/CR-3261, U.S. Nuclear Regulatory Commission, Washington, DC, April 1983

106. R.D. Cheverton, P.C. Gehlen, G.T. Hahn, and S.K. Iskander, Application of Crack Arrest Theory to a Thermal Shock Experiment, in *Crack Arrest Methodology and Applications*, STP 711, ASTM, Phildelphia, 1980, p 392-421

107. G.D. Whitman and R.H. Bryan, "Heavy-Section Steel Technology Program Quarterly Progress Report for July-September 1982," NUREG/CR-2751/Volume 3 (ORNL/TM-8369/v3), U.S. Nuclear Regulatory Commission, Washington, DC, Jan 1983, p 38

108. T. Kobayashi and D.A. Shockey, "Environmentally Accelerated Cyclic Crack Growth Mechanisms in Reactor Steel Determined from Fracture Surface Topography," in *Fracture: Interactions of Microstructure, Mechanisms and Mechanics*, J.M. Wells and J.D. Landes, Ed., Metallurgical Society of AIME, New York, 1984, p 447-462

109. D.A. Shockey, L. Seaman, and D.R. Curran, The Microstatistical Fracture Mechanics Approach to Dynamic Fracture Problems, *Inst. J. Fract.* (to be published)

110. L. Seaman, D.R. Curran, and D.A. Shockey, Computational Models for Ductile and Brittle Fracture, *J. Appl. Phys.*, Vol 47, 1976, p 4814-4826

111. G.R. Irwin, "Notes on Crack Arrest," Appendix 1 in Quarterly Progress Report to ORNL/HSST Program, University of Maryland, Sept 1982

112. J.F. Kalthoff, J. Beinert, S. Winkler, and W. Klemm, Experimental Analysis of Dynamic Effects in Different Crack Arrest Test Specimens, in *Crack Arrest Methodology and Applications*, STP 711, G.T. Hahn and M.F. Kanninen, Ed., ASTM, Philadelphia, 1980, p 109-127

113. R.D. Cheverton *et al.*, Fracture Mechanics Data Deduced From Thermal Shock and Related Experiments with LWR Pressure Vessel Material, in *Aspects and Fracture Mechanics in Pressure Vessels and Piping*, American Society of Mechanical Engineers, New York, 1982, p 1-15

114. A.R. Rosenfield, "Validation of Compact-Specimen Crack-Arrest Data," ASME Paper No. JEMT-36-82, American Society of Mechanical Engineers, New York, 1982

115. P.B. Crosley and E.J. Ripling, "Plane Strain Crack Arrest Characterization of Steels," *J. Pressure Vessel Technol.*, Vol 97 (No. 4), Nov 1975, p 291-298

116. G.T. Hahn *et al.*, Fast Fracture Toughness and Crack Arrest Toughness of Reactor Pressure Vessel Steel, in *Crack Arrest Methodology and Applications*, STP 711, ASTM, Philadelphia, 1980, p 289-320

117. "Standard Methods of Tension Testing of Metallic Materials," E 8, *Annual Book of ASTM Standards*, Vol 03.01, ASTM, Phildelphia, 1984, p 130-150

118. C.W. Marschall, P.N. Mincer, and A.R. Rosenfield, Subsize Specimens for Crack-Arrest Testing, in *Fracture Mechanics: Fourteenth Symposium, Volume II: Testing and Applications*, STP 791, J.C. Lewis and G. Sines, Ed., ASTM, Phildelphia, 1983, p II-295 to II-319

119. A.R. Rosenfield *et al.*, "Size Requirements for Crack-Arrest Testing," Battelle Handout, Meeting of ASTM Task Group on Crack Arrest Testing, Jan 1982

120. A.R. Rosenfield *et al.*, Recent Advances in Crack-Arrest Technology, in *Fracture Mechanics: 15th Symposium*, STP 833, R.J. Sanford, Ed., ASTM, Philadelphia, 1984, p 149-164

Creep, Stress-Rupture, and Stress-Relaxation Testing

Introduction

By J. Daniel Whittenberger
Materials Engineer
NASA Lewis Research Center

CREEP is the slow deformation of a material under a stress that results in a permanent change in shape. Generally, creep pertains to rates of deformation less than 1.0%/min; faster rates are usually associated with mechanical working (processes such as forging and rolling). Shape changes arising from creep generally are undesirable and can be the limiting factor in the life of a part. For example, blades on the spinning rotors in turbine engines slowly grow in length during operation and must be replaced before they touch the housing.

Although creep can occur at any temperature, only at temperatures exceeding about 0.4 of the melting point of the material are the full range of effects visible ($T \geq 0.4\ T_M$, where T is temperature and T_M is the melting point of the material). At lower temperatures, creep is generally characterized by an ever-decreasing strain rate, while at elevated temperature, creep usually proceeds through three distinct stages and ultimately results in failure.

A schematic representation of creep in both temperature regimes is shown in Fig. 1. At time = 0, the load is applied, which produces an immediate elastic extension that is greater for high-temperature tests due to the lower modulus. Once loaded, the material initially deforms at a very rapid rate, but as time proceeds, the rate of deformation progressively decreases. For low temperatures, this type of behavior can continue indefinitely. At high temperatures, however, the regime of constantly decreasing strain rate (primary or first-stage creep) leads to conditions where the rate of deformation becomes independent of time and strain. When this occurs, creep is in its second stage or steady-state regime.

Although considerable deformation can occur under these steady-state conditions, eventually the strain rate begins to accelerate with time, and the material enters tertiary or third-stage creep. Deformation then proceeds at an ever-faster rate until the material can no longer support the applied stress and fracture occurs. With ϵ, t, and $\dot{\epsilon}$ representing strain, time, and strain rate, respectively, creep consists of the following components:

General behavior during creep

Stage	Temperature	Characteristic
First (primary)	$T > 0.4\ T_M$ or $T \leq 0.4\ T_M$	$\dot{\epsilon}$ decreases as t and ϵ increase
Secondary (steady state)	$T \geq 0.4\ T_M$	$\dot{\epsilon}$ is constant ($\dot{\epsilon}_{ss}$)
Third (tertiary)	$T \geq 0.4\ T_M$	$\dot{\epsilon}$ increases as t and ϵ increase

In addition to temperature, stress also affects creep, as shown in Fig. 2. In both temperature regimes, the elastic strain on loading increases with increasing applied

Fig. 1 Low-temperature and high-temperature creep of a material under a constant engineering stress

A and B denote the elastic strain on loading; C denotes transition from primary (first-stage) to steady-state (second-stage) creep; D denotes transition from steady-state to tertiary (third-stage) creep.

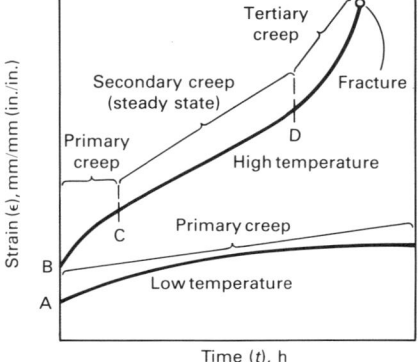

stress. At low temperatures (Fig. 2a), very high stresses (σ_4) near or above the ultimate tensile stress result in rapid deformation and fracture at time t_4. A somewhat lesser stress (σ_3) can result in a long period of constantly decreasing strain rate, followed by a short transition to an accelerating rate and failure at t_3. Finally, lowered stresses (σ_2 and σ_1) exhibit ever-decreasing creep rates, where σ_2 produces more elastic and plastic strain than σ_1 in the same period. The stress range over which behavior changes from that of σ_4 to that of σ_2 is small, and fracture under stress σ_3 is likely to be the result of microstructural and/or mechanical instabilities.

At elevated temperatures (Fig. 2b), increasing the initial stress usually shortens the period of time spent in each stage of creep. Hence, the time-to-rupture (t_6, t_7, and t_8) decreases as stress is increased. Additionally, the steady-state creep rate decreases as the applied stress is decreased. The stress range over which behavior changes from that exhibited by stresses σ_8 and σ_5 (Fig. 2b) is much broader than the range necessary to yield similar behavior at low temperatures (Fig. 2a).

Most of the behavior shown in Fig. 1 and 2 can be understood in terms of the Bailey-Orowan model (Ref 1), which views creep as the result of competition between recovery and work-hardening processes. Recovery is the mechanism(s) through which a material becomes softer and regains its ability to undergo additional deformation. In general, exposure to high temperature (stress relieving after cold working, for example) is necessary for recovery processes to be activated. Work-hardening processes make a material increasingly more difficult to deform as it is strained. The increasing load required to continue deformation between the yield stress and the ultimate tensile stress during a short-term tensile test is an example of work hardening.

After the load is applied, fast deformation begins, but this is not maintained as the

Fig. 2 Elevated-temperature creep in a material as a function of stress where the time-to-rupture is t_i for stress σ_i

(a) Low-temperature creep, where $\sigma_4 > \sigma_3 > \sigma_2 > \sigma_1$. (b) High-temperature creep, where $\sigma_8 > \sigma_7 > \sigma_6 > \sigma_5$

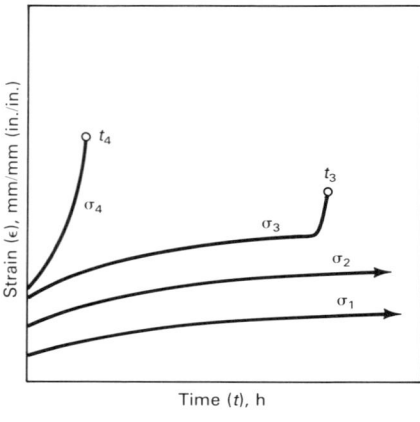

(a)

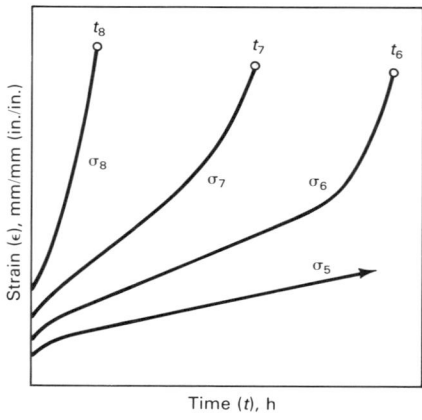

(b)

material work hardens and becomes increasingly more resistant to further deformation. At low temperatures, recovery cannot occur; hence, the creep rate is always decreasing. However, at elevated temperatures, softening can occur, which leads to the steady state, in which recovery and hardening processes balance one another. As the temperature increases, recovery becomes easier to activate and overcomes hardening. Thus, the transition from primary to secondary creep generally occurs at lower strains as temperature increases.

Third-stage creep cannot be rationalized in terms of the Bailey-Orowan model. Instead, tertiary creep is the result of microstructural and/or mechanical instabilities. For instance, defects in the microstructure, such as cavities, grain-boundary separations, and cracks, develop. These result in a local decrease in cross-sectional area that corresponds to a slightly higher stress in this region.

Because creep rate is dependent on stress, the strain and strain rate in the vicinity of a defect will increase. This then leads to an increase in the number and size of microstructural faults, which in turn further decreases the local cross-sectional area and increases the strain rate. Additionally the microstructural defects, as well as other heterogeneities, can act as sites for necking. Once formed, deformation tends to increase in this region, because the local stress is higher than in other parts of the specimen. The neck continues to grow, because more local deformation yields higher stresses.

Creep Experiments

The creep behavior of a material is generally determined by uniaxial loading of test specimens heated to temperature in some environment. Creep-rupture experiments measure the deformation as a function of time to failure. If strain-time behavior is measured, but the test is stopped before failure, this is termed an interrupted creep experiment. Finally, if an inadequate strain-measuring system or no attempt to determine length is employed, and the test is run to fracture, a stress-rupture experiment results.

In terms of data that characterize creep, the stress-rupture test provides the least amount, because only the time-to-rupture and strain-at-rupture data are available for correlation with temperature and stress. These data and other information, however, can be obtained from creep-rupture experiments. Such additional measurements can include elastic strain on loading, amount of primary creep strain, time to onset of secondary creep, steady-state creep rate, amount of secondary creep, time to onset of tertiary creep, time to 0.5% strain, time to 1.0% strain, etc. All of these data can be fitted to equations involving temperature and stress. An interrupted creep test provides much the same data as a creep-rupture experiment within the imposed strain-time limitations.

Direction of Loading. Most creep-rupture tests of metallic materials are conducted in uniaxial tension. Although this method is suitable for ductile metals, compressive testing is more appropriate for brittle, flaw-sensitive materials. In compression, cracks perpendicular to the applied stress do not propagate as they would in tension; thus, a better measure of the inherent plastic properties of a brittle material can be obtained.

In general, loading direction has little influence on many creep properties—for example, steady-state creep rate in ductile

materials (Ref 2). However, even in these materials, the onset of third-stage creep and fracture is usually delayed in compression compared to tension. This delay is due to the minimized effect of microstructural flaws and the inability to form a "neck-like" mechanical instability. For brittle materials, the difference in behavior between tension and compression can be extreme, primarily due to the response to flaws. Consequently, care must be exercised when using compressive creep properties of a brittle material to estimate tensile behavior.

Test specimens for uniaxial tensile creep-rupture tests are the same as those used in short-term tensile tests. Solid round bars with threaded or tapered grip ends or thin sheet specimens with pin and clevis grip ends (Ref 3) are typical. However, many other types and sizes of specimens have been used successfully where the choice of geometry was dictated by the available materials. For example, small threaded round bars with a 12-mm (0.47-in.) overall length and a 1.52-mm (0.06-in.) diam by 5-mm (0.2-in.) long reduced section have been used to measure transverse stress-rupture properties of a 13-mm (0.51-in.) diam directionally solidified eutectic alloy bar (Ref 4).

In the case of uniaxial compression testing, specimen design can be simple small-diameter right cylinders or parallelepipeds with length-to-diameter ratios ranging from approximately 2 to 4. Larger ratios tend to enhance elastic buckling, and smaller ratios magnify the effects of friction between the test specimen and the load-transmitting member. These specimen geometries are well suited for creep testing when only a small amount of material is available, or when the material is difficult to machine.

Environment. The optimum conditions for a creep-rupture test are those in which the specimen is influenced only by the applied stress and temperature. This rarely occurs, particularly at elevated temperatures, and these conditions do not exist for real structures and equipment operating under creep conditions. For example, turbine blades are continuously exposed to hot, reactive gases that cause corrosion and oxidation.

Reactions between the test environment and material vary greatly, ranging from no visible effect to large-scale attack. For example, creep-rupture testing of aluminum, iron-chromium-aluminum, nickel-chromium, and nickel-based superalloys at elevated temperatures in air can generally be accomplished without problems, because these materials form thin, stable, protective oxide films. This is not the case for refractory metals (molybdenum, niobium, tantalum, and tungsten) and their alloys, due to their strong reaction with oxygen, which leads to the formation of porous, nonprotective, and in some cases volatile oxides. Environmental effects such as oxidation and corrosion reduce the load-bearing cross-sectional area and can also facilitate the formation and growth of cracks.

Reactions are also possible in inert atmospheres (such as vacuum) and in reducing gas environments. Elevated-temperature testing in vacuum can result in the loss of volatile alloying elements and subsequent loss of strength. Exposure to reducing gases can result in the absorption of interstitial atoms (carbon, hydrogen, and nitrogen), which may increase strength, but also induce brittleness.

A "perfect" environment does not exist for all creep-rupture testing. The appropriate choice depends on the material, its intended use, and the available environmental protection methods. If creep mechanisms are being determined, then the atmosphere should be as inert, or nonreactive, as possible. However, if the material is to be used in an unprotected state in a reactive atmosphere, then creep-rupture testing should reflect these conditions.

Creep-rupture data from inert atmosphere tests cannot be used for design purposes when the material will be exposed to conditions of severe oxidation. However, if environmental protection methods, such as oxidation- or corrosion-resistant coatings, are available, then testing in inert gas is acceptable, and the resulting data can be used for design.

If reactions occur between the test environment and the specimen, the resultant creep-rupture data will not reflect the true creep properties of the material. Rather, the measured data are indicative of a complex interaction between creep and environmental attack, where the effects of environmental attack become more important in long-term exposure.

Strain Measurement. Care must be taken to ensure that the measured deformation occurs only in the gage section. Thus, measurements based on the relative motion of parts of the gripping system above and below the test specimen are generally inaccurate, because the site of deformation is unknown. Extensometry systems are currently available that attach directly to the specimen (shoulders, special ridges machined on the reduced section, or the gage section itself) and transmit the relative motion of the top and bottom of the gage section via tubes and rods to a sensing device such as a linear variable differential transformer (LVDT). Figure 3 illustrates such a system. These systems are quite accurate and stable over long periods of time.

Other methods of direct strain measurement exist and, under certain circumstances,

Fig. 3 Typical rod-and-tube-type extensometer for elevated-temperature creep testing

Extensometer is clamped to grooves machined in the shoulders of the test specimen.

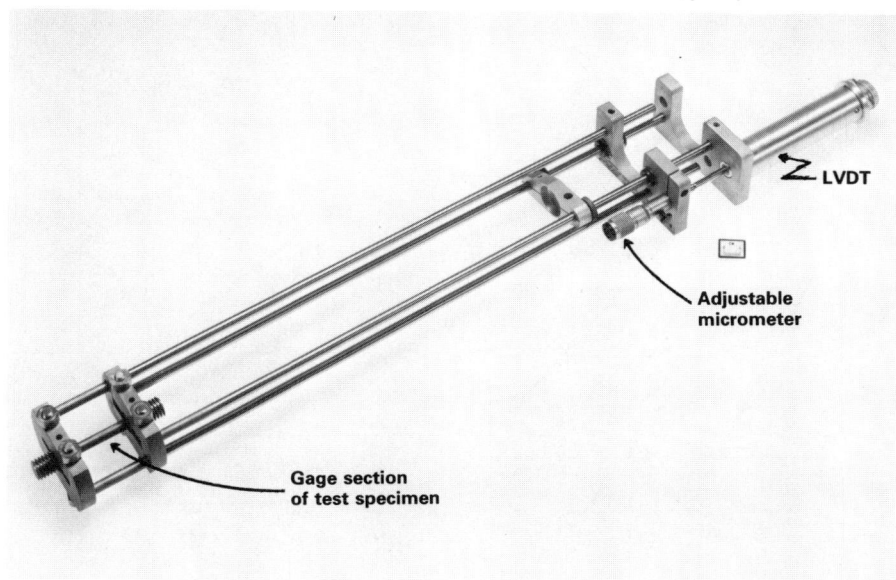

are suitable. At low temperatures, strain gages can be directly bonded to the gage section and can be used to follow deformation over the range of extension for which the strain gage is valid. For specimens that will undergo reasonable deformations ($\epsilon > 1.0\%$), the distance between two gage marks can be optically tracked with a cathetometer as a function of time. While the location of strain is known, use of this technique is operator dependent and is generally limited to tests of less than 8 h or greater than 100 h in duration in order to permit sufficient readings to properly define the creep curve.

Data Presentations

Generally, all creep- and stress-rupture-related data are analyzed in terms of three variables: time, stress, and temperature. Other factors are also important, particularly when an understanding of the process(es) in control of deformation is desired. However, for a straightforward presentation and representation of most experimental data, these three variables are sufficient.

The time-to-rupture (t_r) from either isothermal stress-rupture or creep-rupture testing is presented as a function of stress σ as:

$$t_r = K_1 \exp(a\sigma) \qquad \text{(Eq 1)}$$

and

$$t_r = K_2 \sigma^m \qquad \text{(Eq 2)}$$

where K_1, K_2, and a are constants, and m is the stress exponent for rupture. An example of time-to-rupture results and the use of these equations to describe the data at several temperatures are shown in Fig. 4(a) and (b). Generally, there is little difference between the exponential (Eq 1, Fig. 4a), or power law (Eq 2, Fig. 4b) descriptions of the times-to-rupture. In both cases, the data lie on straight lines, and coefficients of determination (R^2) for the linear regression fits of the data have high values.

When temperature effects as well as stress effects on time-to-rupture are to be considered, one common presentation is:

$$t_r = K_3 \sigma^m \exp\left(\frac{Q_r}{RT}\right) \qquad \text{(Eq 3)}$$

where K_3 is a constant; Q_r is the activation energy for rupture; R is the universal gas constant ($8.314 \text{ kJ/mol} \cdot K$); and T is the absolute temperature in Kelvin. Figure 4(c) describes time-to-rupture data for several temperatures as a single line. The Larson-Miller parameter (LMP) (Ref 6) represents another approach using a single curve to represent data gathered under a variety of conditions, where:

$$\text{LMP} = T(C + \log t_r) \qquad \text{(Eq 4)}$$

and

$$\text{LMP} = K_4 \log \sigma + K_5 \qquad \text{(Eq 5)}$$

where C, K_4, and K_5 are constants. Originally, C was set equal to 20; however, C is

Fig. 4 Time to rupture (t_r) as a function of stress (σ) for [100] oriented Ni-5.8Al-14.6Mo-6.2Ta (wt%) single crystals tested in tension at several temperatures in air

□, 927 °C (1700 °F); ○, 982 °C (1800 °F); ●, 1038 °C (1900 °F). (a) Exponential form (Eq 1). (b) Power law form (Eq 2). (c) Temperature-compensated power law form (Eq 3). (d) Larson-Miller parameter (LMP) form (Eq 5). Source: Ref 5

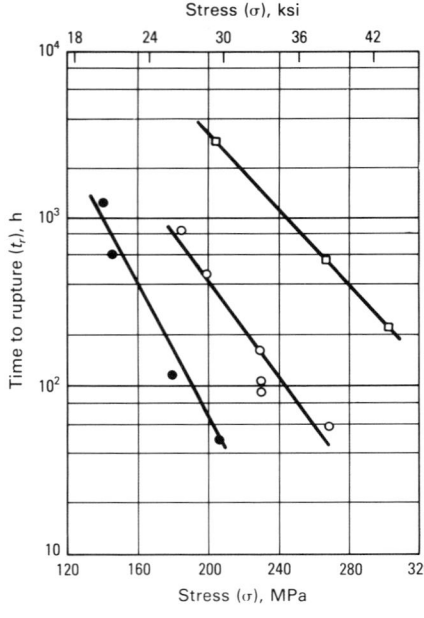

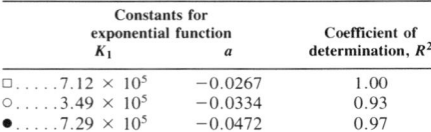

Constants for exponential function		Coefficient of determination, R^2
K_1	a	
□.....7.12×10^5	-0.0267	1.00
○.....3.49×10^5	-0.0334	0.93
●.....7.29×10^5	-0.0472	0.97

(a)

Constants for power law function		Coefficient of determination, R^2
K_2	m	
□.....7.96×10^{18}	-6.67	1.00
○.....9.61×10^{19}	-7.54	0.95
●.....2.79×10^{20}	-8.12	0.98

(b)

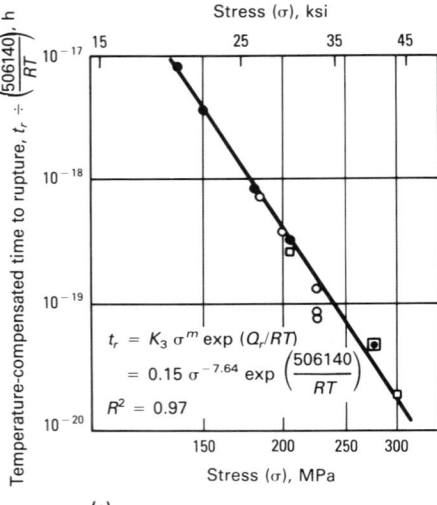

$$t_r = K_3 \sigma^m \exp(Q_r/RT)$$
$$= 0.15 \sigma^{-7.64} \exp\left(\frac{506140}{RT}\right)$$
$$R^2 = 0.97$$

(c)

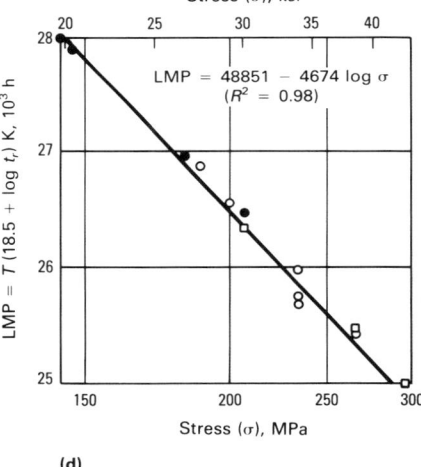

$$LMP = 48851 - 4674 \log \sigma$$
$$(R^2 = 0.98)$$

(d)

currently permitted to assume a value that best describes the data. Figure 4(d) illustrates the use of the Larson-Miller parameter to combine time-to-rupture data from several temperatures and stresses into one curve. For more information on time-temperature parameters, see the article "Assessment and Use of Creep-Rupture Properties" in this Section.

Creep curves generated by either creep-rupture or interrupted creep testing are usually presented as strain versus time or log strain versus log time. Examples of both forms are given in Fig. 5. The linear presentation (Fig. 5a) gives an accurate representation of the three stages of creep. However, the strain incurred in first- and/or second-stage creep is de-emphasized if considerable strain occurs in third-stage creep. Also, the time spent in primary and/or tertiary creep is de-emphasized if steady-state creep exists over a large fraction of the time-to-rupture.

The logarithm format, on the other hand, emphasizes the time and strain during first-stage creep and, to some extent, the strain during second-stage creep at the expense of time during second-stage creep and the strain and time of third-stage creep. De-emphasis of tertiary creep in the logarithm presentation allows the low strain behavior to be highlighted, which is perhaps of most interest for design purposes.

Steady-State Creep Rate. The most important creep parameter in terms of theoretical analysis is the steady-state creep rate $\dot{\epsilon}_{ss}$. Its dependence on stress is generally expressed as:

$$\dot{\epsilon}_{ss} = K_6 \sigma^n \qquad \text{(Eq 6)}$$

and for temperature and stress:

$$\dot{\epsilon}_{ss} = K_7 \sigma^n \exp\left(\frac{Q_c}{RT}\right) \qquad \text{(Eq 7)}$$

where K_6 and K_7 are constants, n is the stress exponent for creep, and Q_c is the activation energy for creep. Because of theoretical development, certain values of the stress exponent for creep have been correlated with deformation mechanisms, and the activation energy for creep has been correlated to the activation energy for diffusion.

Examples of steady-state creep rates as functions of stress and temperature are shown in Fig. 6, along with the results of linear regression fits to power law creep (Eq 6) and temperature-compensated power law creep (Eq 7). Note the general agreement of the compressive test results with the tensile data in Fig. 6 and the reproducibility of $\dot{\epsilon}_{ss}$ versus σ behavior from two castings of the same material (Fig. 6a).

Stress to Produce 1.0% Strain. In many cases, the objective of testing is to determine the total amount of creep strain that can be expected during stress/temperature exposures. Of particular interest are the stresses required to produce 0.5, 1.0, and 2.0% strain in a certain period of time as a function of temperature. A typical example of such a presentation is shown in Fig. 7 for several refractory alloys. Also illustrated is a case in which a metallurgical variable (grain size) was factored into the analysis to account for the difference in behavior between the two lots of the tantalum-based alloy Astar 811C.

Monkman-Grant Relationship. For elevated-temperature tensile creep-rupture experiments, the product of the time-to-rup-

Fig. 5 Creep curves for [100] oriented NASAIR 100 (Ni-5.5Al-8.5Cr-0.7Mo-3Ta-1Ti-10W (wt%)) single crystals tested in tension to rupture at 1000 °C (1830 °F) in air

(a) Strain versus time. (b) Log strain versus log time. Source: Ref 7

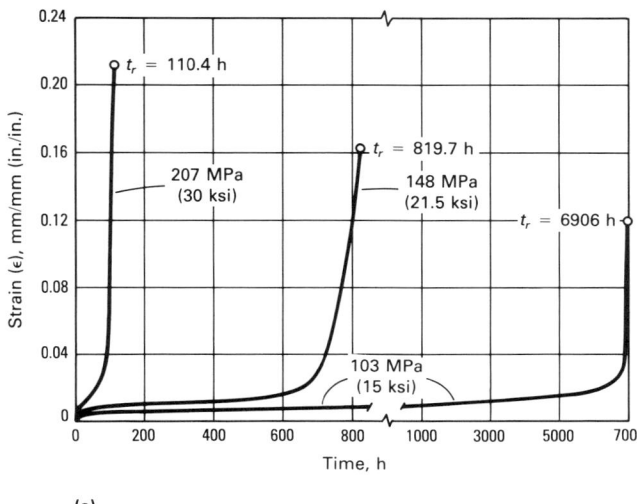

(a)

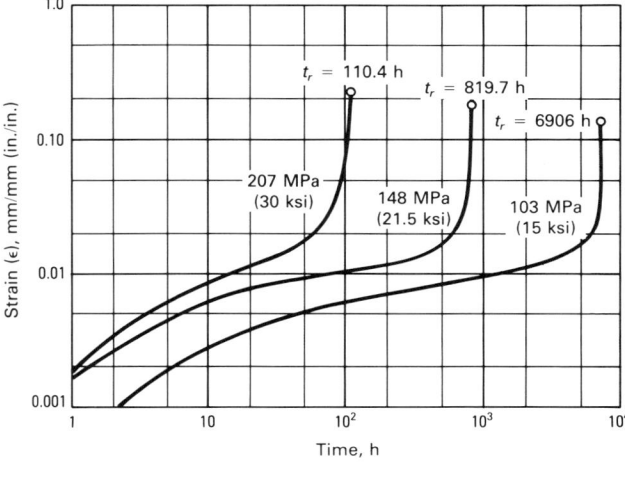

(b)

ture and steady-state creep rate raised to the power M is approximately a constant for many materials (Ref 9):

$$t_r \dot{\epsilon}_{ss}^M = K_8 \qquad \text{(Eq 8)}$$

where M and K_8 are constants with values roughly equal to 1. An example of this relationship is shown in Fig. 8 using a rearranged form of Eq 8:

$$\log t_r = \log K_8 - M(\log \dot{\epsilon}_{ss}) \qquad \text{(Eq 9)}$$

Once M and K_8 are known, reasonable predictions of either quantity can be predicted from knowledge of the other.

Other Testing Considerations

Constant Load Versus Constant Stress Testing. Most uniaxial creep and stress-rupture tests are conducted under constant-load conditions. Although the method is simple, the stress in the gage section varies with strain (time). This can be seen by considering a bar of length L_0 and cross-sectional area A_0 subjected to a tensile load P. At time $t = 0$, the initial engineering stress on the bar is:

$$s_0 = \frac{P}{A_0} \qquad \text{(Eq 10)}$$

With the assumption of uniform deformation during creep, the bar lengthens to L and the cross-sectional area decreases to A, because volume must be conserved:

$$LA = L_0 A_0 \qquad \text{(Eq 11)}$$

Therefore,

$$A = A_0 \left(\frac{L_0}{L}\right) \qquad \text{(Eq 12)}$$

and the true stress on the bar is:

$$\sigma = \left(\frac{P}{A}\right) = s_0 \left(\frac{L}{L_0}\right) \qquad \text{(Eq 13)}$$

Methods have been devised to account for the change in cross-sectional area during creep. These are based on a rearranged form of Eq 13, where:

$$\sigma = \frac{P}{A} = \frac{PL}{L_0 A_0} \qquad \text{(Eq 14)}$$

By maintaining PL at a fixed value, a constant stress test can be conducted. In general, the form of strain-time behavior under constant stress conditions is the same as those shown in Fig. 1 and 2. However, the period of time spent in primary and secondary creep under constant stress can be much longer than under an identical engineering stress (constant load). Hence, rupture life is longer under constant stress conditions.

Failure under constant stress conditions eventually occurs due to some microstructural and/or mechanical instability in the same manner as a constant load experiment. Once a local variation in cross-sectional area is formed, the actual stress is higher than the imposed constant stress and further deformation concentrates at this location.

In reality, the basic assumptions of conservation of volume and/or uniform deformation have been violated; therefore, Eq 14 is no longer valid. When nonuniform deformation starts in either a constant load or constant stress test, the local strain and strain rate vary along the gage section in an unknown manner.

Engineering Strain Versus True Strain. In creep experiments, there is little difference between strains calculated by the engineering strain or true strain definitions when the length change is approximately 10% or less. For greater length changes, the calculated values of strain deviate greatly. Although this is of no consequence for tension, the limit of a maximum engineering strain of -1.00 in compression places an artificial barrier on the description of compressive creep. Hence, true strain is a much better indicator of compressive ductility and creep characteristics. In particular, creep behavior measured in tension and compression can be compared only when both are expressed as true strain due to the limiting engineering strain in compression.

Microstructure. During creep, significant microstructural changes occur on all levels. On the atomic scale, dislocations are created and forced to move through the material. This leads to work hardening as the dislocation density increases and the dislocations encounter barriers to their motion. At low temperatures, an ever-diminishing creep rate results; however, if the temperature is sufficiently high, dislocations rearrange and annihilate through recovery events.

The combined action of hardening and recovery processes during primary creep can lead to the formation of a stable distribution of subgrains or loose three-dimensional dislocation networks in some materials, or an approximately uniform dislocation distribution without subgrains in other materials. These stable dislocation configurations are maintained and are characteristic of second-stage creep.

Creep deformation also produces change in the light optical macro- and microstructures. Such changes include slip bands,

Fig. 6 Steady-state creep rate ($\dot{\epsilon}_{ss}$) as a function of applied stress (σ) and temperature (T) for two heats of [100] oriented NASAIR 100 single crystals tested in tension at several temperatures in air

(a) Power law form (Eq 6). (b) Temperature-compensated power law form (Eq 7). Source: Ref 7

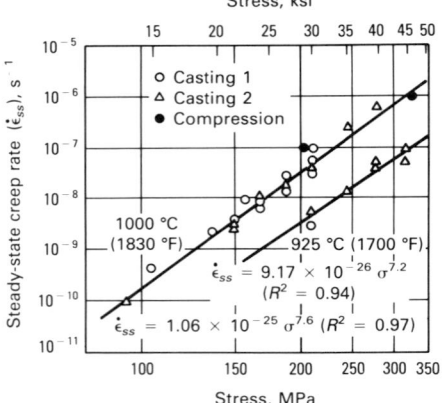

(a)

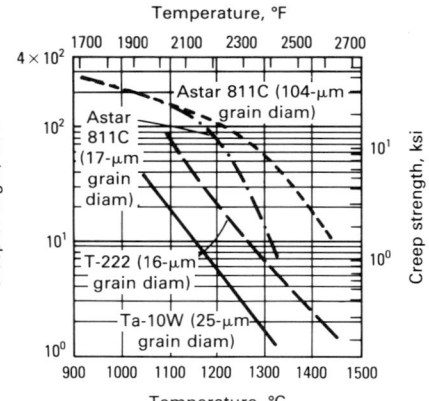

(b)

grain-boundary sliding, cavity formation and growth, and cracking (grain-boundary, interphase boundary, and transgranular). The extent of these microstructural changes is generally increased near fracture sites compared to other regions.

The microstructure of an elevated-temperature creep or stress-rupture test specimen rarely resembles the initial microstructure. Most materials are not thermodynamically stable; hence, prolonged exposure under creep conditions can result in the precipitation of new phases, dissolution or growth of desired phases, grain growth, etc. Although many of the structural changes can be duplicated through simple heat treatment, some changes will only occur under the combined influence of stress and temperature.

Figure 9 is an example of a stress/temperature-dependent microstructure. Under normal isothermal annealing, the cube-shaped γ' (Ni_3Al) strengthening phase (Fig. 9a) in the nickel-based superalloy NASAIR 100 undergoes Ostwald ripening (Fig. 9b), where ripening is characterized by an increase in particle size without any shape change. However, during creep testing, the individual precipitates grow together rapidly and form thin γ' plates where the long dimensions of each plate are perpendicular to the stress in tensile creep (Fig. 9c) and parallel to the applied stress in compressive creep (Fig. 9d).

The changes in microstructure that occur during testing affect creep properties. Although such changes may be unavoidable, in many cases thermomechanical processing schedules can be established to influence the

Fig. 7 Stress necessary for 1.0% strain in 10 000 h as a function of temperature for several tantalum alloys tested in vacuum

Materials include Ta-10W, T-222 (Ta-2.4Hf-9.6W-0.01C), and Astar 811C (Ta-1Hf-1Re-7.5W-0.02C) (all materials in wt%). Source: Ref 8

changes so that they tend to strengthen the material or minimize the overall effect. For example, if a heterogeneous precipitate is formed during creep, a simple heat treatment or cold work followed by annealing prior to testing should give a homogeneous distribution of precipitates.

Microstructural changes due to the combined influence of temperature and stress are the most difficult to control. These changes enhance creep and therefore contribute to the observed strain. Even if the changes are essentially complete after primary creep and the resultant microstructure is more creep resistant than the original structure, the creep strain from such changes may be so great that the material cannot be used. To circumvent these changes, simulation of creep exposure prior to actual use may be necessary.

Complete microstructural examination of tested and untested materials should be an essential part of any creep experiment. As a minimum, the as-received microstructure should be compared to those at and away from the fracture site for the shortest lived and longest lived test specimens at each temperature. Comparison of these aids identification of the relevant deformation mechanism, indicates whether environment is affecting creep, and reveals any significant microstructural changes. Such information is vital for interpreting and understanding creep behavior.

Stress Relaxation

Traditional creep testing to develop descriptions of strain rate, stress, and temperature behavior can be time-intensive and

Fig. 8 Time-to-rupture (t_r) as a function of steady-state creep rate ($\dot{\epsilon}_{ss}$) for [100] oriented NASAIR 100 single crystals tested in tension at several temperatures in air

Source: Ref 7

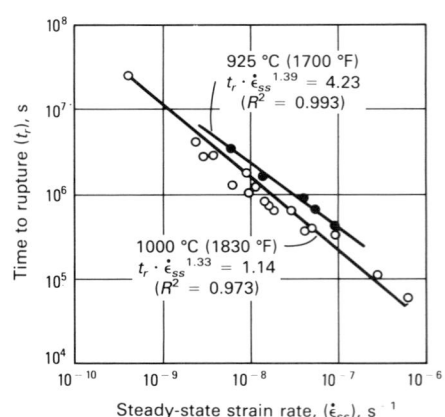

Fig. 9 Comparison of microstructural changes in a γ'-strengthened nickel-based superalloy

(a) Cube-shaped γ' strengthening phase resulting from isothermal annealing (Ref 10). (b) Ostwald ripening of strengthening phase due to isothermal annealing (Ref 7). (c) and (d) Microstructural changes due to tensile creep and compressive creep, respectively (Ref 10). See text for details.

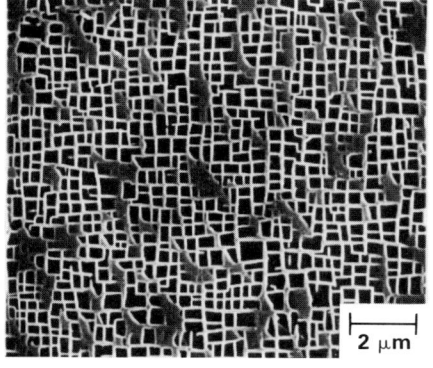

(a)

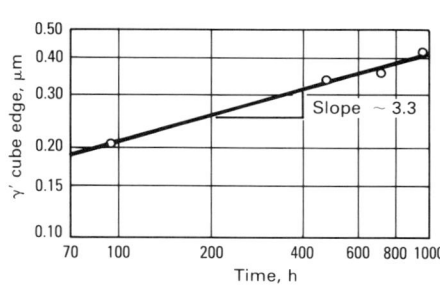

(b)

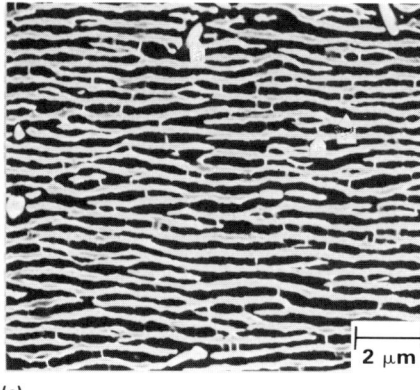

(c)

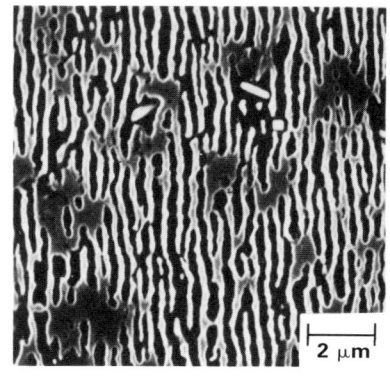

(d)

controlled throughout the experiment. This is critical, because even small fluctuations in temperature will produce thermal expansion effects that can mask changes due to relaxation. Also, the calculated stress exponents and activation energies may not be the same as those determined from creep testing. The processes that produce plastic flow could be different for these two situations. Only comparison of the results from both types of testing can detect equivalent behavior.

REFERENCES

1. S.K. Mitra and D. McLean, Work Hardening and Recovery in Creep, *Proc. Roy. Soc.*, Vol 295, 1966, p 288-299
2. G.P. Tilly and G.F. Harrison, Interpretation of Tensile and Compressive Creep Behavior of Two Nickel Alloys, *J. Strain Anal.*, Vol 8, 1973, p 124-131
3. "Standard Methods of Tension Testing of Metallic Materials," E 8, *Annual Book of ASTM Standards*, Vol 03.01, ASTM, Philadelphia, 1984
4. H.H. Gray, "Transverse Tensile and Stress Rupture Properties of $\gamma/\gamma' - \delta$ Directionally Solidified Eutectic," NASA TMX-73451, 1979
5. R.A. MacKay, "Morphological Changes of Gamma Prime Precipitates in Nickel-Base Superalloy Single Crystals," NASA TM-83698, 1984
6. F.R. Larson and J. Miller, A Time-Temperature Relationship for Rupture and Creep Stresses, *Trans. ASME*, Vol 74, 1952, p 765-775
7. M.V. Nathal, "Influence of Cobalt, Tantalum, and Tungsten on the High Temperature Mechanical Properties of Single Crystal Nickel-Base Superalloys," NASA TM-83479, 1984
8. W.D. Klopp, R.H. Titran, and K.D. Sheffler, "Long-Time Creep Behavior of the Tantalum Alloy Astar 811C," NASA TP-1691, 1980
9. F.C. Monkman and N.J. Grant, "An Empirical Relationship Between Rupture Life and Minimum Creep Rate," in *Deformation and Fracture at Elevated Temperatures*, N.J. Grant and A.W. Mullendore, Ed., MIT Press, Cambridge, MA, 1965, p 91-103
10. M.V. Nathal and L.J. Ebert, Gamma Prime Shape Changes during Creep of a Nickel-Base Superalloy, *Scripta Met.*, Vol 17, 1983, p 1151-1154

expensive, involving many creep test stands, specimens, and thousands of hours of testing. Stress relaxation offers the potential to eliminate this difficulty by producing strain rate/stress data over a wide range of rates from a single specimen. This information is developed when the elastic strain of a specimen extended (or compressed) to a certain, constant length is converted to plastic strain. As the specimen slowly deforms, the load required to maintain the constant length is reduced, hence the term "stress relaxation" test. In addition, this type of experiment simulates the real engineering problem of the long-term loosening of tightened bolts and other fasteners.

In its simplest form, a stress relaxation test involves loading (straining) a specimen to some predetermined load (strain), fixing the

position of the specimen (halting the crosshead motion in a universal test machine, for example), and measuring the load as a function of time. With knowledge of the elastic modulus of the specimen and the stiffness of the testing machine, the load-time data can be converted to stress/strain-rate data. This information can then be used to determine the stress exponents and activation energies for deformation via Eq 2 and 3.

One major drawback to stress relaxation testing is the demands placed on the experimental equipment. The load-measuring system must be capable of making accurate measurements of very small changes as a function of time. The effect of the loading rate on the relaxation rate should be evident. In addition, room temperature, as well as specimen temperature, must be precisely

Theory of Creep Deformation

By J. Wayne Jones
Associate Professor
Materials and Metallurgical Engineering
University of Michigan

THE DEPENDENCE OF CREEP RATE on stress and temperature has been well characterized. Recently, however, insight on the atomic processes that produce creep deformation and are responsible for the observed stress and temperature dependence of creep deformation has been gained. Knowledge of the important creep mechanisms for a particular alloy or alloy system aids the metallurgist in varying composition and microstructure to improve creep resistance. Thus, creep life prediction methods, particularly for alloys that undergo microstructural changes during creep, can be improved significantly.

This article summarizes the mechanisms related to creep deformation. Theories that describe the observed stress and temperature dependence of steady-state creep and that account for the generally observed characteristics of primary (transient) and secondary (steady-state) creep are discussed. A review of these theories can also be found in Ref 1.

Creep Curves

If a stress is suddenly applied to pure metals, some solid solutions, and most engineering alloys at a temperature near or greater than $0.5T_M$ (where T_M is the absolute melting point of the metal or alloy), deformation proceeds as shown in Fig. 1. The initial application of stress causes an instantaneous elastic strain, ϵ_e, to occur. If the stress is sufficiently high, an initial plastic deformation, ϵ_p, also occurs. At low temperatures, significant deformation ceases after the initial application of stress, and an increase in stress is required to cause further deformation.

At elevated temperatures ($T \geq 0.5T_M$), deformation under a constant applied load continues with time. The early stage of such deformation, called primary creep, is characterized by an initially high creep rate, $d\epsilon/dt$, which gradually decreases with time. Even-tually, a linear variation of creep strain accumulation with time is observed.

This steady-state creep region is characterized by a constant, minimum creep rate. Steady-state creep rates depend significantly on stress and temperature and are used frequently to compare the creep resistance among alloys.

After significant deformation in steady-state creep, necking occurs, or sufficient internal damage in the form of voids or cavities accumulates to reduce the cross-sectional area, resulting in an increase in stress and creep rate. The process accelerates rapidly, and failure occurs. This region of the creep curve is called tertiary creep.

Figure 1 also shows the derivative of the creep curve, or the creep rate curve. Although Fig. 1 represents the most common creep strain/time behavior, other behavior modes also occur (Ref 1, 2). Figure 2 illustrates the strain rate/time behavior at temperatures below approximately $0.3T_M$. In Fig. 2, only transient creep is observed, which is characterized by a continuously decreasing creep rate that approaches zero as the inverse of time. Such low-temperature creep behavior is called logarithmic creep.

Figure 2 also illustrates the creep rate/time behavior of alloys that exhibit a continuously increasing creep rate in the early stages of creep. Such alloys have been designated Class I alloys (Ref 3). In Class I alloys—for example, Al-3% Mg—creep may occur by viscous glide of dislocations. In pure metals, dislocation climb is thought to be the dominant creep mechanism.

Creep by viscous dislocation glide results when dislocations glide, or move, in slip planes under the action of an applied stress. These dislocations drag along solute atoms attracted to the strain fields of the dislocations. In order for the dislocations to move, the solute atmospheres must diffuse in the direction of dislocation motion. As a result, viscous dislocation glide is quite slow and is the rate-controlling creep process in Class I alloys.

In pure metals where no such atmospheres surround dislocations, dislocation glide can occur very rapidly until dislocation motion is stopped by barriers to motion, such as other

Fig. 1 Variation of creep and creep rate with time

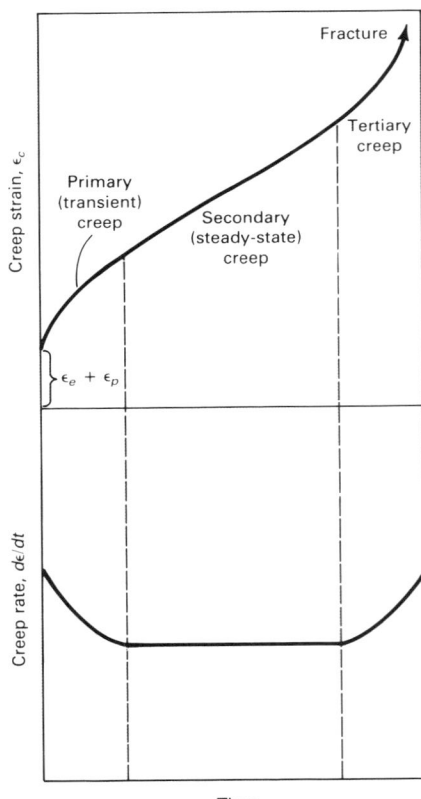

Fig. 2 Creep rate versus time for logarithmic creep and creep in Class I solid solutions

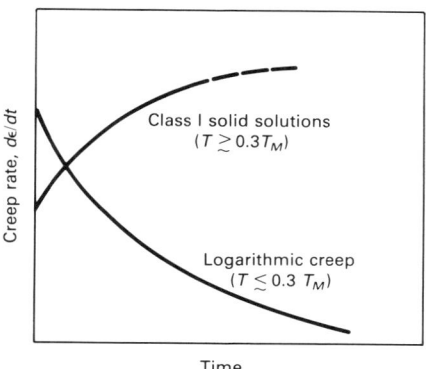

Fig. 3 Activation energy of creep for pure aluminum as a function of absolute temperature

The activation energy for creep equals the activation energy for self-diffusion only at temperatures above $0.5T_M$. Source: Ref 3

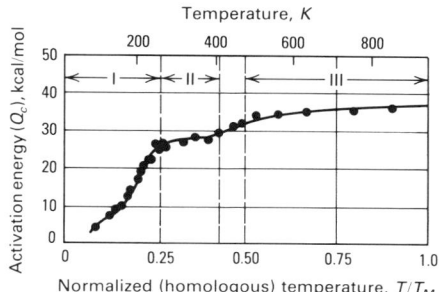

dislocations or obstacles. The dislocations may surmount the barriers by a process called climb. In dislocation climb, vacancies diffuse to or away from the dislocation and the dislocation moves perpendicular to its glide plane. When sufficient clearance of the obstacle is attained, the dislocation may combine with another dislocation or may glide to the next barrier.

At lower temperatures ($\sim 0.3T_M$) or at higher stress levels, creep occurs by thermally activated dislocation glide. Under these conditions, dislocations can overcome barriers to motion without dislocation climb. The localized motion of small dislocations is important in overcoming barriers. Because of the larger contribution of stress in thermally activated dislocation glide, a temperature and stress dependence for creep different from that for dislocation climb or viscous glide will be observed.

Stress Dependence of Steady-State Creep

At intermediate stress levels and at temperatures above $0.5T_M$, the steady-state creep rate for most metals and alloys varies with stress according to:

$$\dot{\epsilon}_s = A\sigma^n$$

where A and n are constants. For pure metals, n generally varies from 4 to 5, and for solid-solution alloys, n has a value of approximately 3. For precipitation- or dispersion-strengthened alloys, the reported values of n can range as high as 30 to 40. Such high values for precipitation- and dispersion-strengthened alloys can be explained in terms of the interaction stresses between dislocations and the barriers to dislocation motion during creep (Ref 4).

At stress levels higher than approximately $5 \times 10^{-4}G$, where G is the shear modulus,

the use of a power-law creep equation underestimates the creep rate (Ref 2, 3). At the higher stresses, creep rate varies exponentially with applied stress. The breakdown in power-law creep behavior apparently results from a transition from diffusion-controlled dislocation climb creep to thermally activated dislocation glide processes similar to deformation found at lower temperatures (Ref 5).

Temperature Dependence of Steady-State Creep

Although creep is known to be a thermally activated process, it is less generally understood that the measured activation energy varies significantly with temperature at low temperatures and becomes independent of temperature only above approximately $0.6T_M$. Figure 3 illustrates the activation energy for creep in pure aluminum as a function of temperature (Ref 3). At low temperatures, the activation energy is low and varies almost linearly with temperature.

The low activation energy for creep at low temperatures indicates that the deformation processes are only weakly temperature dependent at low temperatures, which is anticipated if dislocation glide processes are the dominant deformation mechanism in this temperature regime. Above $0.6T_M$, the activation energy for creep deformation is independent of temperature and is equal to the activation energy for self-diffusion. For many pure metals, an excellent correlation exists between the activation energy for self-diffusion and the activation energy for creep at high temperatures (Ref 3). Such a correlation is shown in Fig. 4.

The Dorn Equation

In the temperature/stress regime in which power-law stress dependence exists and in which dislocation climb determines the temperature dependence of creep, the temperature/stress dependence of steady-state creep may be expressed by the Dorn equation (Ref 6):

$$\dot{\epsilon}_s = AD_L \frac{Gb}{kT}\left(\frac{\sigma}{G}\right)^n$$

where A and n are constants; b is the burgers vector, representative of the discontinuity in the crystal caused by the dislocation; D_L is the self-diffusion coefficient; k is Boltzman's constant; T is the absolute temperature; σ is the applied stress; and G is the shear modulus. The constant A incorporates material properties that cannot be explicitly determined and is determined empirically.

The equivalence of the activation energy for self-diffusion and creep indicates that dislocation climb is an integral part of creep at high temperatures. However, several different dislocation theories of power-law creep involve dislocation climb and therefore identical activation energies. The stress exponent, n, can be determined by the dominant dislocation mechanisms involved in the creep process.

Dislocation Creep Mechanisms

Creep has been considered to result from the competing processes of work hardening and thermal recovery. The Bailey-Orowan equation defines this concept mathematically:

$$d\sigma = \left(\frac{d\sigma}{d\epsilon}\right)d\epsilon + \left(\frac{d\sigma}{dt}\right)dt$$

Fig. 4 Relationship between the activation energy for creep and the activation energy for self-diffusion for several metals and compounds

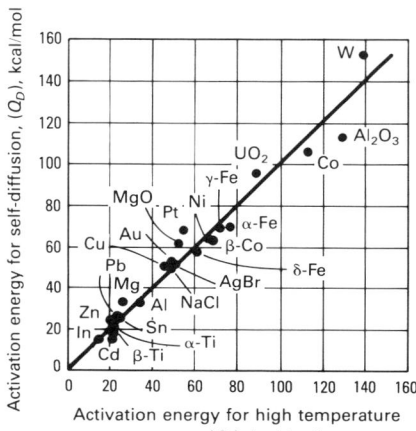

where $d\sigma$ represents the change in flow stress, $d\sigma/d\epsilon$ represents the hardening that results from an increment of plastic strain $d\epsilon$, and $d\sigma/dt$ represents the softening due to recovery in a time increment dt. For steady-state creep, the rate of work hardening equals the rate of recovery and $d\sigma = 0$. Although this approach is simplistic, the concept of work hardening balanced by recovery forms the basis for most dislocation creep theories.

The rate of creep deformation initiated by dislocation motion can be expressed as:

$$\dot{\gamma} = \rho v \mathbf{b}$$

where $\dot{\gamma}$ is the shear strain rate, ρ is the mobile dislocation density and defines the number of dislocations free to move under an applied stress with a mean velocity v, and $\mathbf{b}$ is the burgers vector. If the stress dependence of ρ and v can be determined, the stress dependence of the steady-state creep rate can also be determined. As described earlier, the dislocation processes of thermally activated dislocation glide, viscous dislocation glide, and dislocation climb contribute to creep strain.

Most current dislocation creep models (Ref 1-3) assume that immediately after application of stress all dislocations are mobile, and dislocation multiplication and glide can occur readily. This accounts for the initially high creep rate in primary creep. With continued straining, however, the dislocation configuration changes from a random distribution to a more orderly distribution, in which three-dimensional cellular networks of dislocations are formed.

Cell formation results in a decrease in the mobile dislocation density and a corresponding decrease in creep rate. The long-range stresses existing within the dislocation cells cause dislocation climb recovery. When the dislocation density in the cell is reduced by such recovery, new mobile dislocations are created, and an increment of strain is produced.

A recent model (Ref 1) illustrates that power-law creep occurs by dislocation glide in the cell interiors and dislocation climb in the cell boundaries. Such a model, for which the steady-state creep rate is the sum of the creep rates produced by each mechanism, accounts for power-law breakdown by presuming that, as stress level increases, the contribution to creep from dislocation glide is much greater than the contribution from dislocation climb.

Deformation Mechanism Maps

In addition to thermally activated glide, viscous glide, and dislocation climb, creep may also occur by strictly diffusional mass transport through grain interiors and along grain boundaries. The relative contribution of a particular mechanism depends on its stress and temperature sensitivity, as illustrated by the development of deformation mechanism maps (Ref 7).

Figure 5 is a deformation mechanism map of aluminum. The coordinates of the map are stress, normalized by shear modulus, and temperature, normalized by absolute melting temperature. Each field on the map represents the ranges of stresses and temperature at which a particular mechanism dominates the creep behavior. Thus, at low stresses and high temperatures, diffusional processes dominate. At lower temperatures and higher stresses, dislocation creep dominates. Specific knowledge of stress and temperature dependence of creep mechanisms thus allow the metallurgist to determine more efficiently alloy modifications for improved creep resistance.

Fig. 5 Deformation mechanism map for aluminum

Creep strain rate, $\dot{\epsilon}_c = 10^{-8}$ s^{-1}. Source: Ref 7

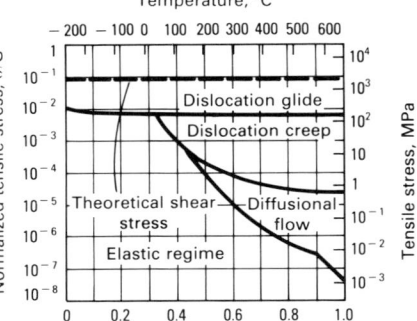

REFERENCES

1. W.D. Nix and B. Ilschner, Mechanisms Controlling Creep of Single Phase Metals and Alloys, in *Strength of Metals and Alloys*, Vol 3, P. Haasen *et al.*, Ed., Pergamon Press, New York, 1980, p 1503-1530

2. J. Weertman and J.R. Weertman, Mechanical Properties, Strongly Temperature Dependent, in *Physical Metallurgy*, R.W. Cahn, Ed., North-Holland, Amsterdam, 1970, p 983-1010
3. O.D. Sherby and P.M. Burke, Mechanical Behavior of Crystalline Solids at Elevated Temperature, *Prog. Mat. Sci.*, Vol 13 (No. 7), 1967, p 325-390
4. K.R. Williams and B. Wilshire, Effects of Microstructural Instability on the Creep and Fracture Behavior of Ferritic Steels, *Mat. Sci. Eng.*, Vol 28, 1977, p 289
5. A. Arielli and A.K. Mukherjee, On the Power-Law Breakdown During High Temperature Creep of FCC Metals, in *Creep and Fracture of Engineering Materials and Structures*, B. Wilshire and D.R.J. Owen, Ed., Pineridge Press, Swansea, UK, 1981, p 97
6. A.K. Mukherjee, J.E. Bird, and J.E. Dorn, Experimental Correlations for High Temperature Creep, *ASM Trans. Quart.*, Vol 62, 1969, p 155-179
7. M.F. Ashby, A First Report on Deformation Mechanism Maps, *Acta. Met.*, Vol 20, 1972, p 887-898

Creep, Stress-Rupture, and Stress-Relaxation Testing

USE OF TESTS that measure creep, stress-rupture, and stress-relaxation properties has grown due to the design and application of metal parts that must withstand high loads at high temperatures for long periods of time. Many parts are designed for a given expected life span. To determine that life span, accurate data are needed to predict expected deformation under the conditions of stress and temperature to be encountered in service. These data can be obtained under tension, compression, combined tension and compression (bending), or torsion.

This article discusses testing equipment and procedures for constant-load testing, constant-stress testing, and stress-relaxation testing. For more information on creep and stress-rupture data presentation, assessment, and analysis, see the articles "Assessment and Use of Creep-Rupture Properties" and "Analysis of Creep and Creep-Rupture Data" in this Volume.

The creep test measures the deformation of a metal as a function of time at constant temperature. In an engineering creep test, the load is usually maintained constant throughout the test. Thus, as the specimen elongates and decreases in cross-sectional area, the axial stress increases. The initial stress that was applied to the specimen is usually the reported value of stress. Curve A in Fig. 1 illustrates a typical creep curve for a constant-load test.

Methods of compensating for the change in dimensions of the specimen so as to carry out the creep test under constant-stress conditions have been developed. When constant-stress tests are made, no region of accelerated creep rate occurs (region III, Fig. 1), and a creep curve similar to B in Fig. 1 frequently is obtained.

The stress-rupture test determines the tendencies of materials that may break under an overload. It is used widely in the selection of materials for applications in which dimensioned tolerances are not critical, but in which rupture cannot be tolerated. The stress-rupture test is similar to a constant-load creep test, using the same type of specimen and apparatus. However, no strain measurements are made during the test. The specimen is stressed under a constant load at constant temperature, as in the creep test, and the time to fracture is measured. If, however an adequate strain-measuring system is employed, and the test is run to fracture, a creep-rupture experiment results. A comparison between stress-rupture and creep-rupture experiments is given in the Introduction to this Section. Because of their similarities with regard to test specimens and test apparatus, constant-load creep and stress-rupture test methods will be discussed together in this article.

The stress-relaxation test is somewhat similar to the creep test, but the load continually decreases instead of remaining constant. In a stress-relaxation test, the load is reduced at intervals to maintain a constant strain. The y axis in a stress-relaxation curve is stress or load rather than strain (elongation) as in the creep curve.

Creep and Stress-Rupture Testing Equipment

By Robert McDemus
Assistant Supervisor
Production Testing Laboratory
Carpenter Technology Corporation

DETERMINATION OF CREEP CHARACTERISTICS of metals at high temperatures requires the use of a loading device or test stand, an electric furnace with suitable temperature control, and an extensometer. Equipment discussed in this section is for uniaxially loaded specimens in tension. More information on creep and stress-rupture testing equipment can be found in Ref 1.

Test Stands

The test stand is designed to apply static stress to a test specimen for an extended period of time at a constant elevated temperature. Typical test stands have a balance beam that connects the test specimen to a weight pan, as shown in Fig. 2. Ratios of 3 to 1 up to 20 to 1 are commonly used between the weight pan and the specimen. The lower ratios are used to provide optimum accuracy at lower loads. The weight pan is part of the overall weights and frequently is suspended with a chain to prevent bending moments on the load train.

On the specimen side of the machine, a balance beam leveling motor is recommended to compensate for elongation of the test specimen. If this is not available, the balance beam may become unlevel, thus changing the calibration of the weight system. However, properly designed creep testing machines will maintain load accuracy well within ASTM requirements, even when out of level by as much as ±10°.

The procedure for calibration of weights is given in Ref 2. The weights should be verified within a limit of 1% at least every 5

Fig. 1 Typical creep curve showing the three stages of creep

Curve A, constant-load test; curve B, constant-stress test

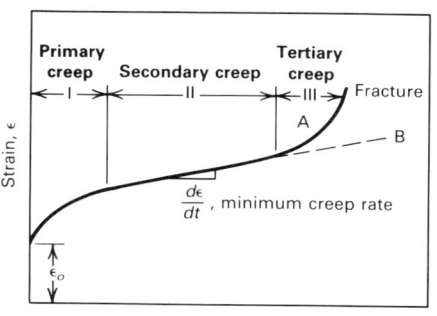

Fig. 2 Schematic of a test stand used for creep and stress-rupture testing

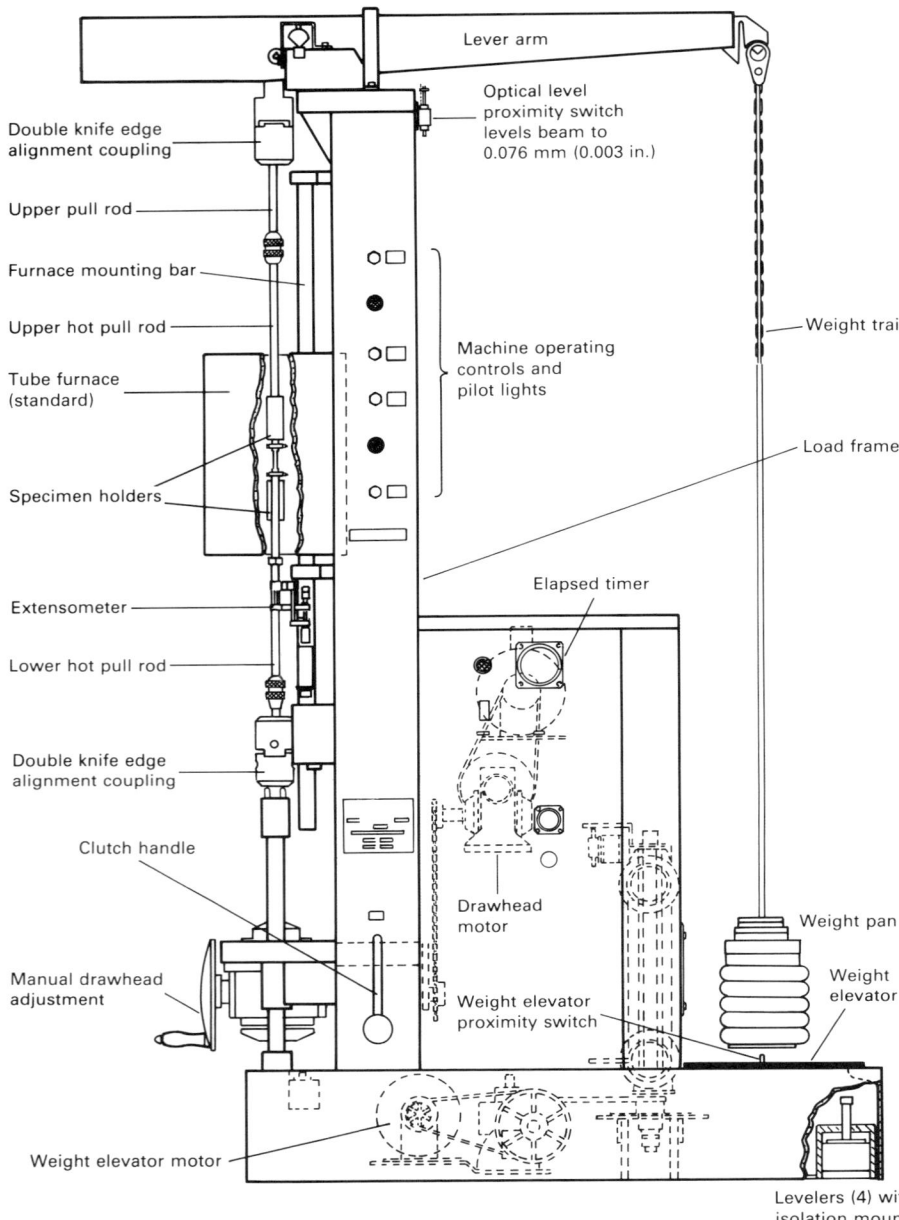

or shock is minimal. Additionally, the test stand should be isolated from the floor with a vibration-damping material such as cork or rubber. The leveling motor can introduce vibration that may affect long-term creep tests, or shorter tests if the vibration is significant.

A timer is also included on most test stands that automatically records the time to rupture. During creep tests, the time must be recorded with the creep values.

Furnaces

Furnaces used in creep and stress-rupture testing generally are tubular, with an electrical-resistance winding that heats the test specimen through radiation in an air atmosphere (Fig. 3). These furnaces can have single or multiple heating zones. The tube is located in a vertical position, with the pull rods connected to the specimen. Care must be taken to seal the opening of the furnace without interfering with the alignment of the load train or the action of the linkage for creep tests.

Temperature Control and Measurement. Material properties frequently are affected by temperature. The requirement for temperature control of creep and stress-rupture tests is ±1.7 °C (±3 °F) when testing at 982 °C (1800 °F) and below, and ±2.8 °C (±5 °F) above that value. Maintaining control requires practice.

Temperature measurement systems require a transducer to convert a temperature differential to an electrical signal. The transducer typically is a thermocouple that is attached directly to the specimen (Fig. 4). Specimens with a gage length 25.4 mm (1.0 in.) or greater require two thermocouples; specimens with gage lengths 50.8 mm (2.0 in.) or greater require a third thermocouple.

The thermocouple should maintain intimate contact with the test specimen during the entire test. Inherent errors are associated with thermocouples, including calibration error, drift due to metallurgical changes to the thermocouple junction during the test, lead wire error, attachment gap error, radiation error, and conduction error.

The present trend in temperature control incorporates computers with two basic control schemes. Stand-alone control systems are available that control only one test stand. This type of control was used for many years with analog controllers and is now used with digital controllers. One advantage of a stand-alone controller is that if it fails only one test is lost. One disadvantage is the cost of dedicating a computer to each test stand.

The second control scheme utilizes a micro- or minicomputer to control multiple test stands. The loss of multiple tests through computer failure can be overcome by using a

years. Additionally, the weight of the overall load train system should be verified within a limit of 1% at least once a year, except for test stands for long-term tests that last more than 1 year.

The test specimen is connected to the balance beam through the load train, a system of pull rods and couplings manufactured from high-temperature alloys that are capable of maintaining strength and corrosion resistance at the test temperatures encountered. The load train should be machined and assembled such that minimum bending moments are imposed on the test specimen. ASTM recommends a maximum of 10% bending strain, compared to the axial strain due to misalignment of the load train. To overcome this problem, alignment couplings (such as ball and socket or knife-edge systems) are used in the load train to facilitate self-alignment.

Vibration and shock loads can have a significant effect on the end results in creep and stress-rupture testing. Care must be taken in selecting the test site to ensure that vibration

Fig. 3 Split furnace mounted on a lever-arm testing machine

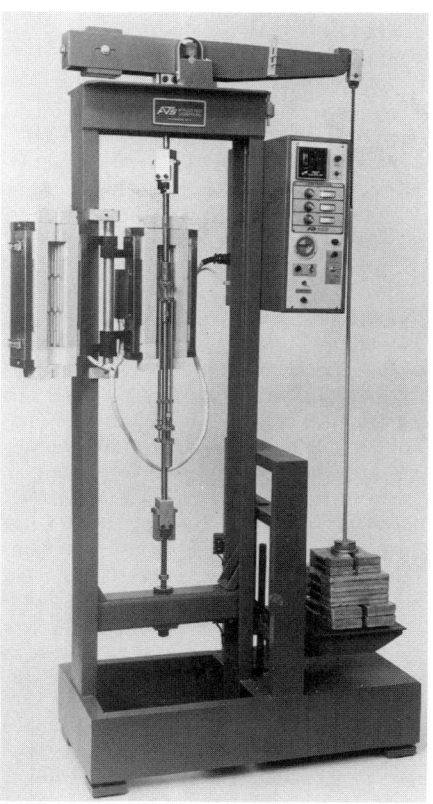

Fig. 4 Thermocouple attached to a test specimen

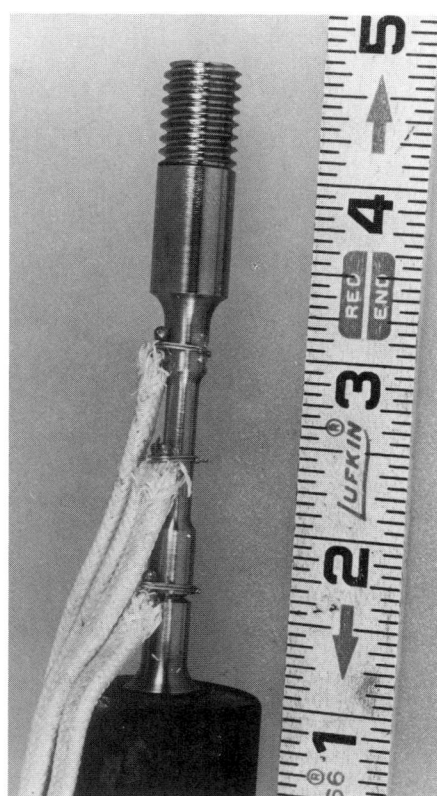

Fig. 5 Linear variable differential transformer

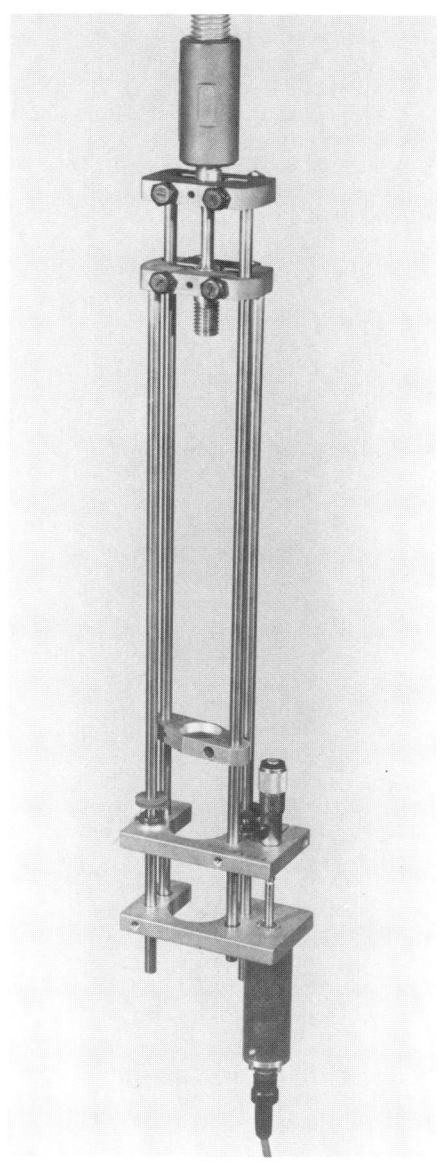

"host" computer that provides automatic backup. Control systems currently exist that have a host computer backing up multiple front-end control computers.

Extensometers

When creep data are required, the specimen strain must be measured as a function of time. With most metals this is difficult, because use of strain-measuring transducers generally is not practical at the test temperatures encountered. A mechanical linkage must then be attached to the specimen to transmit the strain to the strain-measuring equipment outside the high-temperature environment.

The most commonly used strain transducer is the linear variable differential transformer, which consists of a movable metal core that changes the electrical characteristics with the small motion associated with strain measurements (Fig. 5). Linkages that attach to the specimen typically are made of alloys that can withstand the test temperatures encountered. The linkage can be attached either to the gage length or on the shoulders outside the gage length.

Problems encountered with the use of extensometers include:

- Error due to strain in the fillet of the test specimen when the extensometer linkage is attached to the shoulder (see Ref 3 for strain corrections)
- Fracture where the extensometer linkage attaches to the specimen gage length
- Error in strain measurement when the extensometer linkage is attached such that it measures the strain only on one side of the specimen
- Error due to slippage of extensometer linkage on the gage section of ductile specimens
- Damage to linkages and extensometers when the specimen fails
- Bending strain introduced to smaller test specimens, particularly strip specimens, due to the weight of the linkage and extensometer

Constant-Load Testing

MOST UNIAXIAL creep and stress-rupture tests are conducted under constant-load conditions. Typical procedures for constant-load tests and methods of data presentation are discussed below. In addition, procedures for determining the rupture life of notched specimens are presented.

Specimen Preparation

Test specimens for constant-load tests are prepared to meet material specifications and end-use design parameters; thus, a wide variety of configurations are available.

Test specimens for creep and stress-rupture tests in tension are similar to those used for short-time tensile tests. More information on smooth, round, or flat specimens can be found in the article "Effect of Specimen Preparation, Setup, and Test Procedures on Test Results" in this Volume and in Ref 1 and 4. Notched specimens are discussed later in this article.

Specimens with round cross sections have threaded ends, except those used at very high temperatures or those of materials that are difficult to machine or display high sensitivity to notches. Specimens with shouldered ends (no threads) are called "buttonhead" specimens (Fig. 6) and are used for tests conducted at very high temperatures, particularly those in vacuum. At high temperatures, threaded specimens tend to seize in the adapters after testing due to oxidation or diffusion. The method of gripping buttonhead specimens provides self-alignment, which is advantageous with brittle materials.

For sheet or plate specimens, the load is often applied via pins placed in through holes in the shoulder ends of the specimens. If the specimen is sufficiently large, it may have elongated shoulder ends extending outside the furnace. Extension tabs sometimes are welded or brazed to the specimen shoulder and extend outside the furnace. Specially designed grips that fit the fillets at the end of the gage length are also used to apply load.

The gage length of a specimen should be uniform. The diameter or width at the ends of the reduced section should not be less than the diameter or width at the center. It is sometimes desirable to have the diameter or width of the reduced section slightly smaller at its center. The difference should range from maximums of 0.5% for a 2.54-mm (0.100-in.) diam or width specimen to 0.2% for a 12.8-mm (0.505-in.) diam or width specimen. Specimens should be smooth and free from scratches or other stress raisers and

Fig. 6 Buttonhead specimen used for tests at elevated temperatures

should be machined to minimize cold working or surface distortion. In computing the stress or the load required to provide a certain stress, the smallest original cross-sectional area should be used.

Misalignment can cause high local stresses and premature failure. If threaded specimens are used, the threaded adapters that the specimens fit into should be inspected frequently to ensure proper alignment. These devices may creep during a test and after several tests may undergo appreciable misalignment.

Buttonhead specimens and adapters tend to be self-aligning and pose fewer alignment problems. With sheet specimens, it is important that any brazing or welding of extension tabs be done in an alignment fixture. For sheet specimens using a pin in each tab to apply the load, the pin holes must be centered on a line running through the center of the reduced section rather than centered on the tabs. Brittle materials are more sensitive to misalignment than ductile materials.

Specimen Loading

Care is required to avoid straining the specimen when mounting it in the adapters and load train, particularly if the specimen is small or brittle. With the specimen in place, the load train (specimen adapters or grips, pull rods, etc.) should be examined carefully for any misalignment that may cause bending of the specimen under load.

The upper load train should be suspended from the lever arm, and the compensating weight adjusted so that the lever arm balances. Strain-measuring clamps and an extensometer or platinum strips are attached to the specimen, and the load train is inserted into the furnace with the specimen centered. The specimen must be stabilized at temperature before loading. Also, the extensometer should be adjusted and zeroed.

Loading the weight pan should be done smoothly and without excessive shock. If the specimen is to be step-loaded, the weight is placed on the weight pan in measured increments, and the strain corresponding to each step of loading is recorded. The loading curve thus obtained is used in determining the elastic modulus and plastic strain from load application. If step-loading is not used, a method of smoothly applying the load must be used. This can be done by placing a support such as a scissors jack under the load pan during loading. When all weights are in place, the supporting jack is lowered smoothly from under the weight pan.

Temperature Control

The specimen should not be overheated while brought to temperature. A common

practice is first to bring the specimen to about 10 °C (18 °F) below the desired temperature in about 1 to 4 h. Then, over a longer period, the specimen is brought to the desired temperature.

A period of time above the desired temperature is not "cancelled out" by an equal period at a temperature the same amount below the desired temperature. If the temperature rises above the desired temperature by more than a small amount, the test should be rejected. Specified limits are ±1.7 °C (±3 °F) up to 982 °C (1800 °F) and ±2.8 °C (±5 °F) above 982 °C (1800 °F). At temperatures significantly above 1093 °C (2000 °F), the limits are broadened. Variations of temperature along the specimen from the nominal test temperature should vary no more than these limits at these temperatures. These limits refer to indicated variations in temperature according to the temperature recorder.

The indicated temperature must be as close to the true temperature as possible to prevent thermocouple error or instrument error. Thermocouples, particularly base-metal thermocouples, drift in calibration with use or when contaminated. Other sources of error are incorrect lead wires, lead wires that are connected incorrectly, and direct radiation on the thermocouple bead.

Representative thermocouples should be calibrated from each lot of wires used for base-metal thermocouples. At high temperatures, base-metal thermocouples should not be reused without first removing the wire exposed to high temperature and rewelding. Noble-metal thermocouples generally are more stable. However, they are also subject to error due to contamination and must be annealed periodically by connecting a variable transformer and passing sufficient current through the wires to make them incandescent.

When attaching the thermocouple to the specimen, the junction must be kept in intimate contact with the specimen. The bead at the junction should be as small as possible, and there must be no twisting of the thermocouple that could cause shorting. Any other metal contact across the two wires will cause shorting and erroneous readings. Shielding of the thermocouple junction from radiant heating is also recommended.

Temperature measuring, controlling, and recording instruments must be calibrated periodically against a standard. This usually is done by connecting a precision potentiometer to the thermocouple terminals on the instrument and applying potential corresponding to the output of the thermocouple at each of several temperatures. Tables of millivolt output as a function of temperature for various types of thermocouples are available in ASTM Standard E 230, "Temperature

Electromotive Force (EMF) Tables for Standardized Thermocouples,'' and from manufacturers of precision potentiometers.

Most creep and stress-rupture machines are equipped with a switch that automatically shuts off a timer when the specimen breaks. In creep tests, the load usually is low enough that rupture does not occur. The microswitch that shuts off the timer often also shuts off or lowers the temperature of the furnace. In some furnaces, the life of the heating element is reduced significantly if the furnace is shut off after each test; instead, the temperature is merely lowered.

Interrupted Tests

Power failure or some other problem may make it necessary to interrupt a test, during which time the specimen cools and must then be reheated. For many materials, this appears to have little effect on either creep properties or time to rupture if cooling and heating times are not too long. However, such treatment may affect the test material. Any interruption of a test should be reported.

Data Presentation

Readings of strain should be made frequently enough to produce a well-defined curve. This necessitates more frequent readings during the early part of the test than during later stages. The elastic portion of the stress-strain curve can be obtained from the step-loading curve or estimated by measurement of the instantaneous contraction when the load is removed at the end of the test, if the specimen has not broken.

Creep data usually are presented in the form of a curve showing percent creep strain as the vertical axis and time as the horizontal axis. Time usually is plotted on a log scale to illustrate the early part of the curve in detail. Sometimes, a family of curves is plotted on the same coordinates to show the effect of different temperatures or different stresses on one material.

Other methods for plotting data include time to reach a given percent of creep versus load at a constant temperature, and time to reach a given percent of creep versus temperature at constant load. The loading curve, showing the strain versus load as the specimen is loaded, is plotted separately and is used in computing the elastic modulus of the material at temperature.

Rupture data are presented in several types of graphs. One common format has stress as the ordinate versus log of time to rupture (at constant temperature) on the abscissa. Usually, stress-rupture data are presented by means of a parameter plot; i.e., stress is plotted against a parameter value that relates it to both time and temperature. Several different parameters have been used. More information on these time-temperature parameters can be found in the article ''Assessment and Use of Creep-Rupture Properties'' in this Section.

Notched-Specimen Testing

Notched specimens are used principally as a qualitative alloy selection tool for comparing the suitability of materials for components that may contain deliberate or accidental stress concentrations. The rupture life of notched specimens is an indication of the ability of a material to deform locally without cracking under multiaxial stresses. Because this behavior is typical of superalloys, the majority of notched-specimen testing is performed on superalloys.

The most common practice is to use a circumferential 60° V-notch in round specimens, with a cross-sectional area at the base of the notch one half that of the unnotched section. However, size and shape of test specimens should be based on requirements necessary for obtaining representative samples of the material being investigated.

In a notch test, the material being tested most severely is the small volume at the root of the notch. Therefore, surface effects and residual stresses can be very influential. The notch radius must be carefully machined or ground, because it can have a pronounced effect on test results. The root radius is generally 0.13 mm (0.005 in.) or less and should be measured using an optical comparator or other equally accurate means. Size effects, stress-concentration factors introduced by notches, notch preparation, grain size, and hardness are all known to affect notch-rupture life.

Notch-rupture properties can be obtained by using individual notched and unnotched specimens, or by using a specimen with a combined notched and unnotched test section. The ratio of rupture strength of notched specimens to that of unnotched specimens varies with (1) notch shape and acuity, (2) specimen size, (3) rupture life (and therefore stress level), (4) testing temperature, and (5) heat treatment and processing history.

To avoid introducing large experimental errors, notched and unnotched specimens must be machined from adjacent sections of the same piece of material, and the gage sections must be machined to very accurate dimensions. For the combination specimen, the diameter of the unnotched section and the diameter at the root of the notch should be the same within ±0.025 mm (±0.001 in.).

Notch sensitivity in creep rupture is influenced by various factors, including material and test conditions. The presence of a notch may increase life, decrease life, or have no effect. When the presence of a notch increases life over the entire range of rupture time, as shown in Fig. 7(a), the alloy is said to be notch strengthened; that is, the notched specimen can withstand higher nominal stresses than the unnotched specimen. Conversely, when the notch-rupture strength is consistently below the unnotched-rupture strength, as in Fig. 7(c), the alloy is said to be notch sensitive, or notch weakened. Many investigators have defined a notch-sensitive condition as one for which the notch strength ratio is below unity. However, this ratio is unreliable and can vary according to class of alloy and rupture time.

Certain alloys and test conditions show notch strengthening at high nominal stresses

Fig. 7 Three general types of notch effects in stress-rupture tests

(a) Notch strengthening in 19-9 DL heat treated 50 h at 650 °C (1200 °F) and air cooled. (b) Mixed behavior in Haynes 88 heat treated 1 h at 1150 °C (2100 °F), air cooled, and worked 40% at 760 °C (1400 °F). (c) Notch weakening in K-42-B heat treated 1 h at 955 °C (1750 °F), water quenched, reheated 24 h at 650 °C, and air cooled

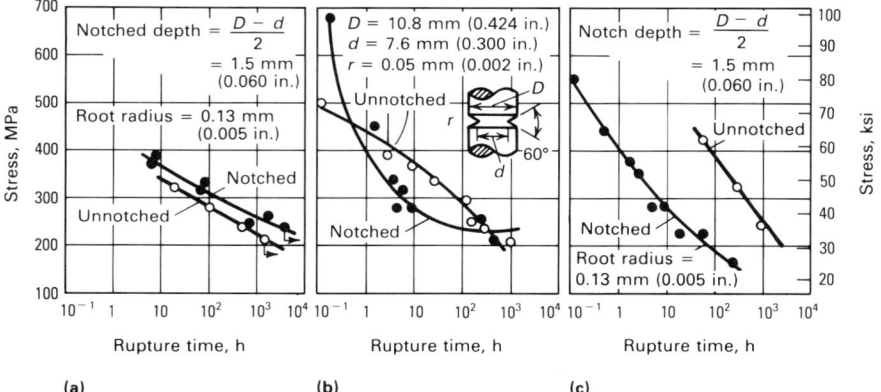

Fig. 8 Notch-rupture strength ratio versus temperature at four different rupture times for Inconel X-750

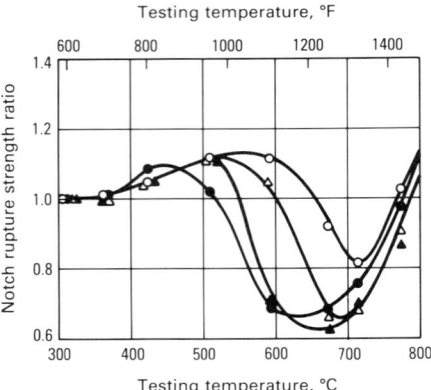

(short rupture times) and notch weakening at lower nominal stresses (longer rupture times), with the result that the stress-rupture curve for notched specimens crosses the curve for unnotched specimens as nominal stress is reduced. Figure 7(b) shows that Haynes 88 becomes notch sensitive under high nominal stresses in a rupture time of about 2 h and that the material becomes notch strengthened again at lower nominal stresses at a rupture time of approximately 400 h. This same phenomenon has been observed in many superalloys and is illustrated in a different manner in Fig. 8. The "notch ductility trough" varies with alloy composition. For example, A-286 is notch sensitive at 540 °C (1000 °F), whereas Inconel X-750 is notch sensitive at 650 °C (1200 °F). A given alloy may show notch weakening at some temperatures and notch strengthening at others. Generally, notch sensitivity appears to increase as temperature is reduced.

Changes in heat treatment of some alloys may alter notch sensitivity significantly. For example, single low-temperature aging of some alloys may produce very low rupture ductilities, because the structure is not sufficiently stabilized. Consequently, exposure of such materials for prolonged rupture times will further reduce rupture ductility because of continued precipitation of particles that enhance notch sensitivity. On the other hand, multiple aging usually stabilizes the structure and thus reduces notch sensitivity.

Notch configuration can have a profound effect on test results, particularly in notch-sensitive alloys. Most studies on notch configuration present results in terms of the elastic stress-concentration factor. The design criterion for the weakening effect of notches at normal and low temperatures is that of complete elasticity. The design stress

is the yield stress divided by the elastic stress-concentration factor K_t (Fig. 9a). The value of the peak axial (design) stress depends on the configuration of the notch.

There is no simple relationship for the effect of notches at elevated temperatures. The metallurgical effects that influence the behavior of notched material are complex and include composition, fabrication history, and heat treatment. Effects of several heat treatments on rupture time of Waspaloy are shown in Table 1.

For ductile metals, the ratio of rupture strength of notched specimens to that of unnotched specimens usually increases to some maximum as the stress-concentration factor is increased. For very insensitive alloys, there may be little further change. Metals that are more notch sensitive may undergo a reduction in ratio as the notch sharpness (stress-concentration factor) is increased beyond the maximum and may show notch weakening for even sharper notches. Very notch-sensitive alloys may undergo little or no notch strengthening, even for very blunt notches (low stress-concentration factor) and may undergo progressive weakening as notch sharpness increases.

Relationships between notch configuration and the ratio of rupture strengths of notched and unnotched specimens are shown in Fig. 9(b). In curve 1, for an alloy with an unnotched rupture ductility of 40%, the notch-strengthening factor decreases as the notch is decreased in sharpness (increase in ratio r/d). In curve 2, for an alloy with unnotched rupture ductility of 7%, the notch-strength factor increases with increasing notch sharpness, reaches a peak, and then drops to a notch-strength reduction factor of less than unity. For an alloy with a still lower unnotched rupture ductility of 3% (curve 4), the notch-strength factor is only slightly greater than unity for large radii of curvature and becomes less than unity. It continues to decrease as the notches become sharper.

Effect of Grain Size and Other Variables. The effects of grain size on notched and unnotched rupture strength are shown in Fig. 10. The coarse grain sizes (ASTM −1 to +2) were obtained by reheating bars in which small strains had been introduced by cold reducing them 1 to 1.25%. Notches had a strengthening effect on both S-816 and Waspaloy when tested at 815 °C (1500 °F). There was no measured effect of grain size on either the notched or unnotched specimens of S-816. On the other hand, the coarse-grained Waspaloy specimens showed a longer rupture time at the same rupture stress for both notched and unnotched specimens.

The rupture time for Discaloy at 650 °C (1200 °F) increases with increasing hardness up to about 290 HV for notched specimens ($K_t = 3.9$) and up to 330 HV for unnotched specimens, as shown in Fig. 11. Ductility, as measured by elongation values for unnotched bars, decreases with increasing hardness.

The peak in rupture time at 650 °C (1200 °F) corresponds to a rupture elongation of 1.5%. The continual reduction in rupture elongation with increasing hardness indicates that the alloy exhibits time-dependent notch sensitivity. Notched bars exhibit a strengthening effect at lower hardnesses and higher ductilities; for specimens of higher hardness and lower ductility, rapid notch weakening is apparent.

For this particular alloy at this temperature, 5% rupture elongation indicates the point at which no notch strengthening or weakening occurs; this point is also indicated by the crossover of the two curves in Fig. 11 at about 318 HV. For other alloys, this crossover may occur at rupture ductilities as low as 3% or as high as 25%. Alloys with lower rupture ductilities are more notch sensitive.

The effects of notches on rupture life of three superalloys are shown in Fig. 12. Nimonic 80A, S-816, Inconel 751 (formerly Inconel X-550), and Waspaloy show various

Table 1 Effect of a notch on rupture time of Waspaloy

Condition	Hours to failure at 730 °C (1350 °F) and 360 MPa (52 ksi):	
Condition	Unnotched bar	Notched bar
Solution heat treated 4 h at 1080 °C (1975 °F), air cooled, aged 16 h at 760 °C (1400 °F), air cooled	76.0	1.5
Solution heat treated 4 h at 1080 °C (1975 °F), air cooled, stabilized 4 h at 845 °C (1550 °F), air cooled, aged 16 h at 760 °C (1400 °F), air cooled	82.8	150(a)
Solution heat treated 4 h at 1080 °C (1975 °F), air cooled, stabilized 4 h at 870 °C (1600 °F), air cooled, aged 16 h at 760 °C (1400 °F), air cooled	87.4	150(a)
Solution heat treated 4 h at 1080 °C (1975 °F), air cooled, stabilized 1 h at 980 °C (1800 °F), air cooled, aged 16 h at 760 °C (1400 °F)	1.9	46.6

(a) No failure; test discontinued after 150 h.

Fig. 9 Effect of notch dimensions on stress concentration and notch-rupture strength ratio

(a) Variation of stress-concentration factor with ratio of minor to major diameter and with ratio of root radius to major diameter for notched bar stressed in tension within the elastic range. (b) Variation of notch-rupture strength ratio for 1000 h life with ratio of root radius to minor diameter. Curve 1 is for 12Cr-3W steel heated 3 h at 900 °C (1650 °F) and air cooled. Grain size, ASTM No. 12; hardness, 215 HV; unnotched rupture ductility, 40%; test temperature, 540 °C (1000 °F). Curve 2 is for Refractaloy 26 oil quenched from 1010 °C (1850 °F); reheated 20 h at 815 °C (1500 °F) and air cooled; reheated 20 h at 650 °C (1200 °F) and air cooled; reheated 20 h at 815 °C and air cooled; and finally reheated 20 h at 650 °C and air cooled. Grain size, ASTM No. 7 to 8; hardness, 330 HV; unnotched rupture ductility, 7%; test temperature 650 °C. Curve 3 is for Refractaloy 26 oil quenched from 1175 °C (2150 °F); reheated 20 h at 815 °C and air cooled; reheated 20 h at 730 °C (1350 °F) and air cooled; and finally reheated 20 h at 650 °C and air cooled. Grain size, ASTM No. 2 to 3; hardness, 325 HV; unnotched rupture ductility, 10%; test temperature, 650 °C. Curve 4 is for Refractaloy 26 oil quenched from 980 °C (1800 °F); reheated 44 h at 730 °C and air cooled; and finally reheated 20 h at 650 °C and air cooled. Grain size, ASTM 7 to 8; hardness, 375 HV; unnotched rupture ductility, 3%; test temperature, 650 °C

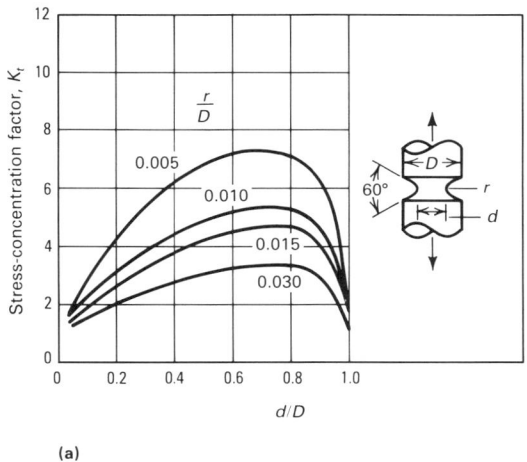

(a)

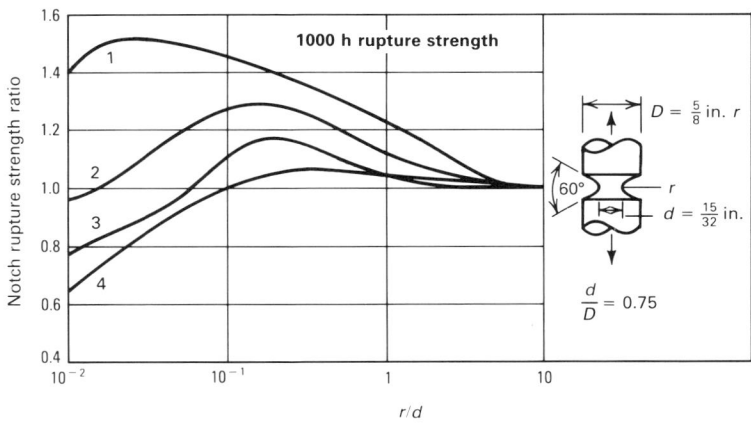

(b)

Fig. 10 Rupture strength as a function of time for notched and unnotched bars of different grain size

S-816 was heated to 1175 °C (2150 °F) and water quenched, reheated to 760 °C (1400 °F), held 12 h, and air cooled. Waspaloy was heated to 1080 °C (1975 °F), held 4 h, and air cooled; reheated to 840 °C (1550 °F), held 4 h, and air cooled; and finally reheated to 760 °C (1400 °F), held 16 h, and air cooled. Smaller grain sizes were produced by cold reducing the S-816 1%, and the Waspaloy 1.25%, by cold rolling at 24 °C (75 °F), and then heat treating. Diameter of specimens was 12.7 mm (0.5 in.), diameter at base of notch was 8.9 mm (0.35 in.), root radius was 0.1 mm (0.004 in.), and notch angle was 60°. Data are a composite of results from two laboratories.

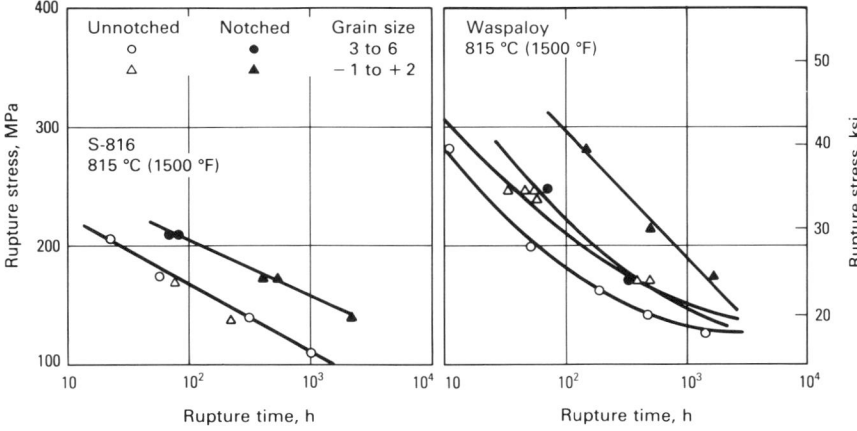

Fig. 11 Variation of rupture time at 650 °C (1200 °F) with initial hardness for Discaloy

Open symbols indicate notched-bar tests (K_t = 3.9); solid symbols indicate smooth-bar tests. Numbers adjacent to points are total elongations for these tests.

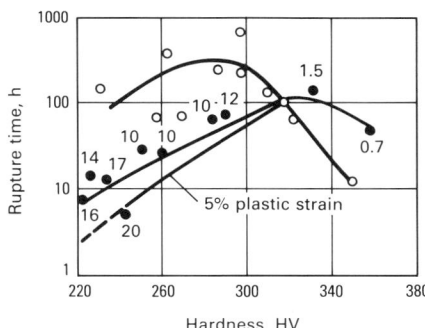

Similar tests on Inconel X-750 with rupture ductilities of 10% reduction in area at 730 °C (1350 °F) and 24% at 815 °C (1500 °F) indicated notch strengthening at both temperatures. These results support the observation that materials with high rupture ductilities under the initial test conditions will be less notch sensitive in long exposure times than materials with low initial rupture ductilities. Alloy S-816 further illustrates this theory in that the alloy has very high rupture ductilities at each test temperature and does not show signs of notch weakening at any test temperature (Table 2).

degrees of notch sensitivity. Nimonic 80A shows a notch-strengthening effect at 650 °C (1200 °F), but is notch weakened in about 100 h at 705 °C (1300 °F) and in about 40 h at 760 °C (1400 °F).

Nimonic 80A becomes notch strengthened at 815 °C (1500 °F), which illustrates that the alloy exhibits a notch ductility trough be-

tween 705 and 760 °C (1300 and 1400 °F). Using 0.13 mm (0.005 in.) as the standard notch radius, Waspaloy and Inconel 751 both exhibited notch weakening at the lower test temperatures 650 and 730 °C (1200 and 1350 °F) and notch strengthening at the higher test temperatures 815 and 870 °C (1500 and 1600 °F).

Fig. 12 Effects of notches on rupture life of three superalloys

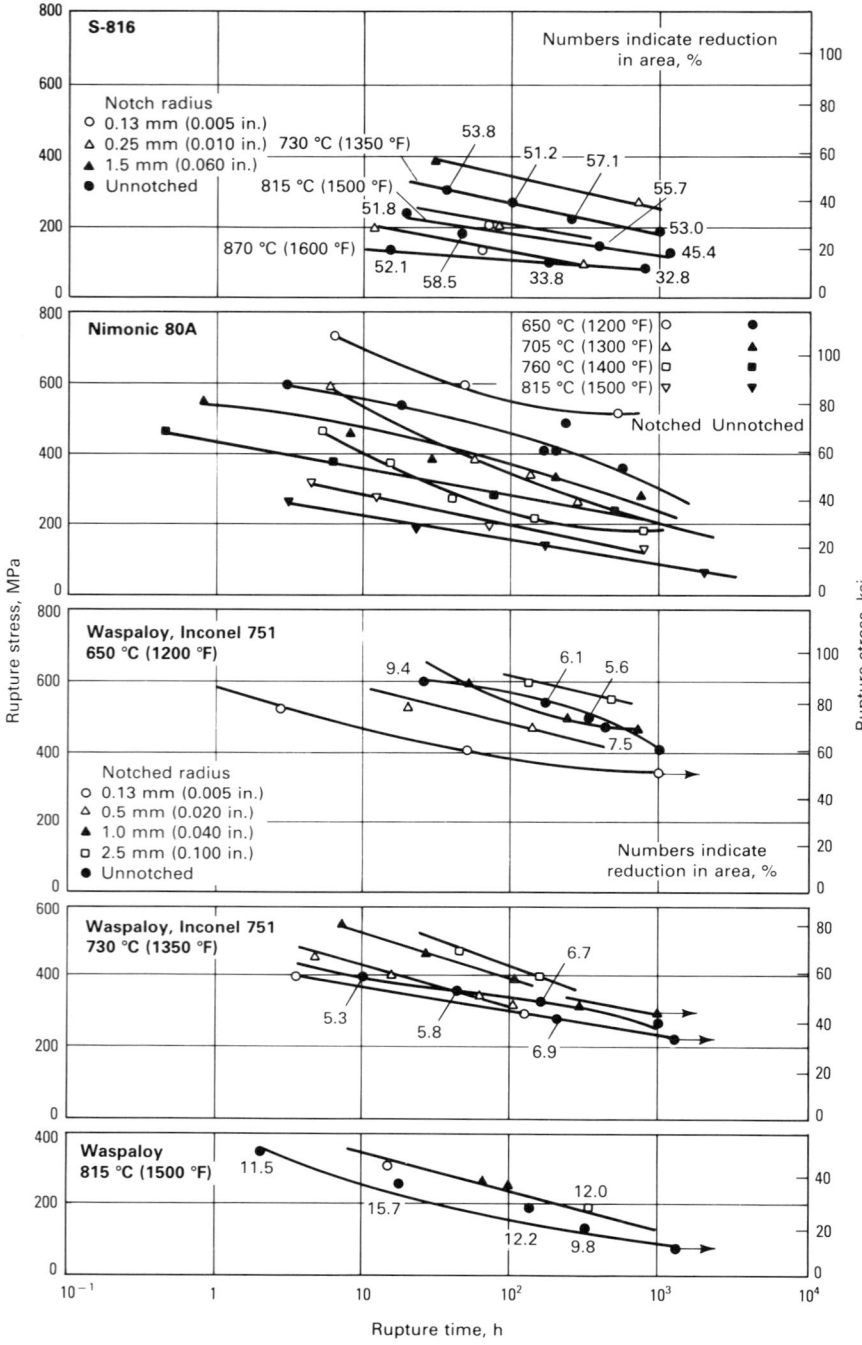

show slight notch strengthening, but only with the larger radii and very short rupture times. Thus, the larger radii compensate for notch sensitivity in Waspaloy, but not in Inconel 751 at 730 °C (1350 °F).

At the highest test temperature—815 °C (1500 °F)—Waspaloy was notch strengthened under all conditions of notch severity. The notch radius did not have any effect on alloy S-816, as evidenced by notch strengthening at all notch severities and test temperatures. However, high rupture ductilities enhanced notch strengthening.

Constant-Stress Testing

By Norman L. Carroll
President
Applied Test Systems, Inc.

CREEP AND STRESS-RUPTURE DATA usually are obtained under constant-load test conditions. However, it is sometimes desirable or necessary to obtain test data under constant-stress conditions. In this case, the applied load is adjusted as the length of the specimen changes to maintain constant stress on the specimen. Constant-stress testing is necessary to accurately determine differences between the temperature dependence and the stress dependence of a material. Early quantitative work on creep of materials demonstrated that the tensile creep of several pure metals and a selection of alloys under constant stress could best be represented by (Ref 5):

$$L = L_o (1 + \beta t^{1/3})e^{Kt}$$

where L is length of specimen after time t, L_o is the length immediately after loading, and β and K are material parameters. Typical creep curves and the effect of temperature and stress on the constants β and K in the Andrade creep equation (Ref 6) are shown in Fig. 13. The sensitivity to stress should be noted.

Testing Methods and Equipment

Hyperbolic-Weight Constant-Stress Apparatus. An early application of constant-stress methods in creep testing was the use of a hyperbolic weight by Andrade (Ref 6) in which load was reduced with extension of the specimen as the weight was lowered

The curves for the alloys with various notch radii show that, in general, notch sensitivity increases with increased notch severity at the lower test temperatures. This effect is particularly evidenced in Waspaloy. At 650 and 730 °C (1200 and 1350 °F) and for a notch radius of 0.13 mm (0.005 in.), the alloy is highly notch sensitive, but shows

notch strengthening at these same test temperatures at a blunter notch radius of 2.5 mm (0.10 in.).

The effect of the radius at the root of the notch is minimal for Inconel 751 at 730 °C (1350 °F), in that the material is notch sensitive under all conditions of notch severity at prolonged rupture times. The material does

Table 2 Comparison of creep-rupture strengths for notched and unnotched specimens of S-816

Temperature		Unnotched	Static rupture strength	Notched(a)	
°C	°F	MPa	ksi	MPa	ksi
24	75	1010	147(b)	...	...
595	1100	625	91(c)	...	...
650	1200	450	65(c)	...	...
730	1350	290	42	405	59
815	1500	170	25	255	37
900	1650	97	14	...	...

(a) Circular, 60° V-notch, D = 9.5 mm (0.375 in.), d = 6.4 mm (0.25 in.), r = 0.25 mm (0.010 in.), K_t = 3.4.
(b) Tensile strength. (c) Typical values; all other test specimens from same heat.

into a liquid (Fig. 14). The required shape of the weight is given by an equation of a hyperbola:

$$y = \sqrt{\frac{ML_o}{\rho\pi}} \cdot \frac{1}{L + x}$$

where M is the mass of the load, L_o the initial length of the wire, and ρ is the density of the liquid, usually water. The coordinates of the hyperbola, x and y, are shown in Fig. 14. After selecting a particular M and L_o, the exact size of the weight is given. A weight, once constructed, is exact for one particular initial load, but with reasonable approximation the same weight can be used over a limited range of loads.

Andrade's tests on lead wire showed a significant difference between results of constant-load and constant-stress tests for the same initial length and the same initial load (Fig. 15). Under constant stress, the rate per unit length, once past the initial effect, is constant up to breaking.

Balanced Beam With Cams. A more convenient and useful method uses a balanced beam with shaped cams (Ref 7). In Fig. 16, a beam PH is supported by a knife edge B and carries two plates F and C, one at each end. The plate C has a groove along its outer edge HK; the profile HK is an arc of a circle with center B. D is a thin steel wire resting in the groove that is fixed to the adjusting screw E. The lower end of D is attached to the upper end of the wire to be stretched. F is the second plate, in the groove of which lies a thin steel wire supporting a weight W. The profile of the bottom of the groove PQR is made such that the moment of the weight W about the axis through B is inversely proportional to the length of the wire undergoing stretch, which (assuming that any change in density of the metal that may occur during stretch is negligible) will

make the stretching force proportional to the cross section of the wire.

A compact cam-lever apparatus for application of either constant stress or constant load in tension for large uniform deformations has been developed (Ref 8). The profile of the constant-stress cam lever is similar to that of the balanced beam with cams (Ref 7), except that exact parametric equations for the cam profile have been determined. The load to the specimen is applied through a circular disk of radius R (Fig. 17). The initial mechanical advantage is r_o/R, but the ratio is reduced as the specimen elongates.

To maintain constant stress, the load P on the specimen must be reduced as the specimen elongates to compensate for the reduction in area A. Thus, the instantaneous stress P/A must remain constant. By assuming constant specimen volume and uniform strain, LA (where L is the specimen length) must remain constant. Therefore, it follows that PL also remains constant. In Fig. 17, under equilibrium conditions, $P = Wr/R$, where W is the applied weight and r is the instantaneous moment arm of the applied weight. Thus, to maintain a constant stress, the following condition must be satisfied:

$$rL = \text{constant} - r_oL_o \qquad \text{(Eq 1)}$$

Fig. 14 Hyperbolic-weight constant-stress apparatus

Source: Ref 6

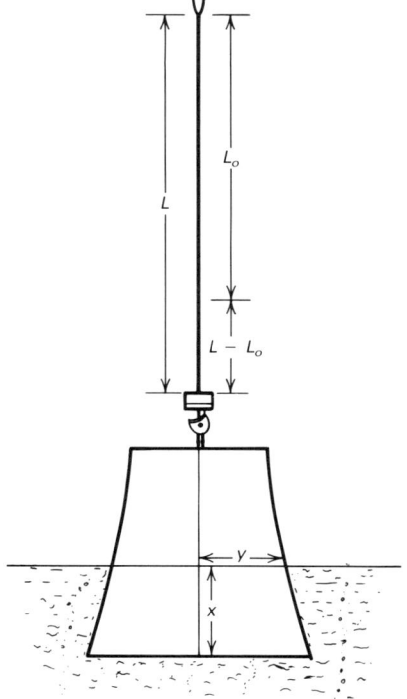

Fig. 13 Creep expression constants β and K/σ plotted as a function of stress at two temperatures

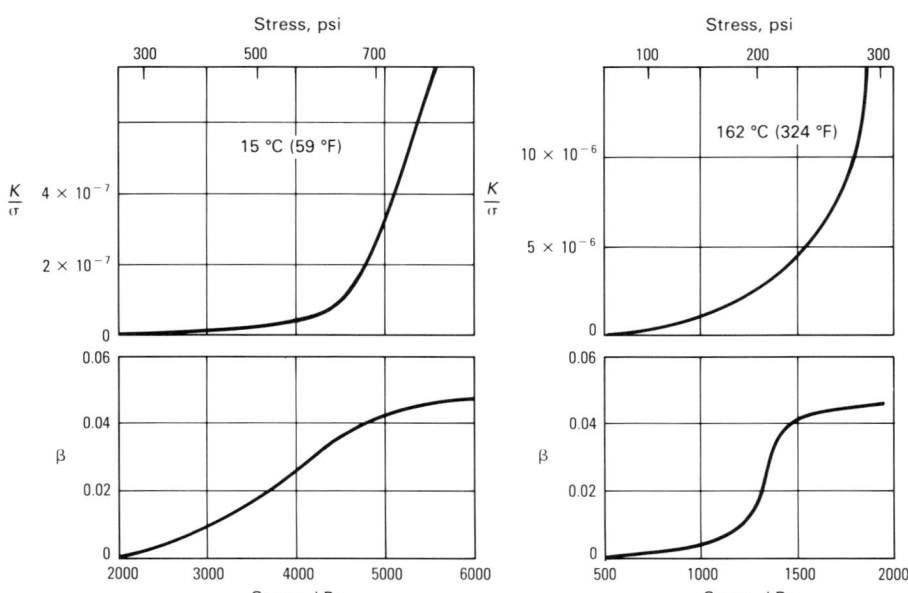

Fig. 15 Results of tests on lead wire under constant-load and constant-stress conditions

Source: Ref 6

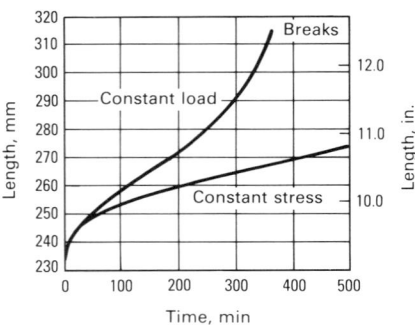

Fig. 16 Balanced beam with cams

See text for explanation of symbols. Source: Ref 7

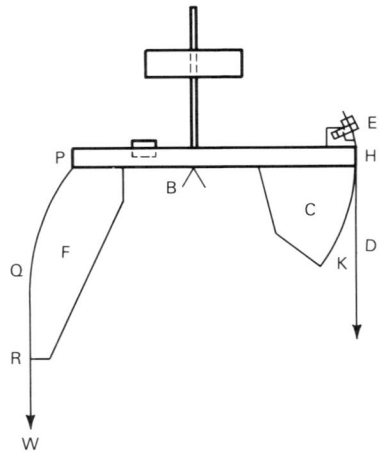

Fig. 17 Constant-stress cam-lever apparatus

Source: Ref 8

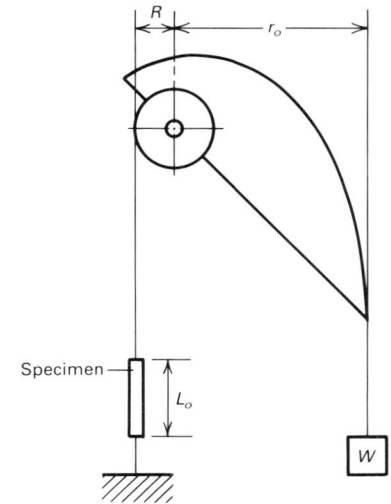

Fig. 18 Constant-stress testing system with balancing cam and constant-load wheel

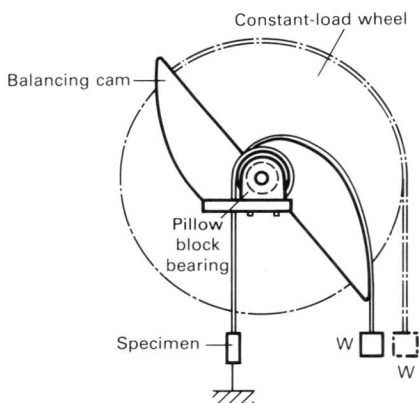

where L_o is the initial specimen length, and r_o is the initial value of r.

As the length of the specimen increases, the arc of contact on the small wheel increases by an equivalent amount. Thus, the instantaneous specimen length may be written as:

$$L = L_o + (\theta - \theta_0)R \qquad \text{(Eq 2)}$$

where θ_0 is the angle for the initial positioning of the constant-stress cam. Eliminating L from Eq 1 and 2 leads to:

$$r = \frac{r_o L_o}{L_o + R(\theta - \theta_0)} \qquad \text{(Eq 3)}$$

This is the equation of the profile of a constant-stress cam. To design an actual cam profile, it is desirable to transform Eq 3 into a fixed Cartesian coordinate system (x, y). These design considerations are discussed Ref 7 and 8.

To provide balance and proper control of loading on the specimen, a balancing cam is added to the loading cam. As shown in Fig. 18, a circular disk or constant-load wheel also may be added to the cam for the purpose of performing constant-load testing. The load can be transmitted by a number of methods; steel bands or roller chain are used most frequently. The shaft of the cam assembly usually is supported by pillow block bearings. Another method of support is the use of hardened and ground knife edges and V-blocks (Fig. 19), although this method results in reduced cam rotation and specimen elongation.

Typical constant-stress creep curves obtained for type 316 austenitic stainless steel at 705 °C (1300 °F) are shown in Fig. 20. In each specimen, the test was ended soon after tertiary creep began. The maximum strain reached in these tests is slightly less than 60%, although the apparatus was designed to accommodate uniform strains of at least 100%.

A constant-stress apparatus for use with low forces has been developed to eliminate the frictional effects of hinges or bearings (Ref 9). Reference 10 describes a method of cam assembly support by means of steel tapes. Modifications of that method were designed to achieve the accurate maintenance of constant stress when forces as low as 0.1 N (0.022 lb) are involved (Fig. 21).

To balance the mass of the cam about the point of suspension and thereby ensure that the only force on the specimen is the applied force due to the mass M, the countermass C is necessary. With the modified suspension, the cam profile equation becomes:

$$r = \frac{(r_o - R)L_o}{L_o + 2R(\theta - \theta_0)} + R \qquad \text{(Eq 4)}$$

For actually determining the cam profile, Eq 4 usually is transformed into two profile equations that express the Cartesian coordinates x and y in terms of a single parameter.

In the absence of frictional forces, the lower limit on the applied force P is determined by the sensitivity with which the suspension system can be balanced. For the present apparatus, the lower limit was 0.1 N (0.022 lb), which corresponds to 5.0 kPa (0.72 psi) if the specimen diameter is 5 mm (0.2 in.). Thus, the apparatus is suitable for use at small stresses, such as may be encountered in studies of plastic flow at temperatures close to the melting point. The system demonstrated accuracy within ±0.50%.

High-Temperature Constant-Stress Compression Creep Apparatus. A constant-stress test apparatus has been developed for high-temperature compression testing of ceramic materials in controlled atmospheres (Ref 11). The creep frame is based on the fundamental concepts developed in Ref 12 for the application of constant com-

pressive stress loading using the Andrade fulcrum principle via a lever-arm mechanism from a weight pan loaded system. The following equation ensures maintenance of constant stress:

$$L_3 = \frac{L_1 L_2}{L}$$

where L_1, L_2, and L_3 are the distances between the various knife edges (Fig. 22 and 23), and L is the sample length.

Two knife edges provide the fulcrum for the lever arm, another supports the weight pan, and a fourth applies the force to the lower sample push rod. When the load pan support columns are vertical, the frame acts as a simple fulcrum to apply a force on the lower push rod that is greater than the force on the weight pan by a factor of L_2/L_1. When

Fig. 19 Constant-stress testing system using V-block and knife-edge support

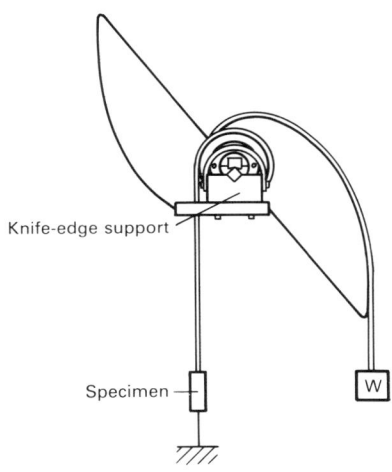

Knife-edge support

Specimen

W

Fig. 20 Typical constant-stress creep curves obtained at 705 °C (1300 °F) for type 316 austenitic stainless steel

Source: Ref 8

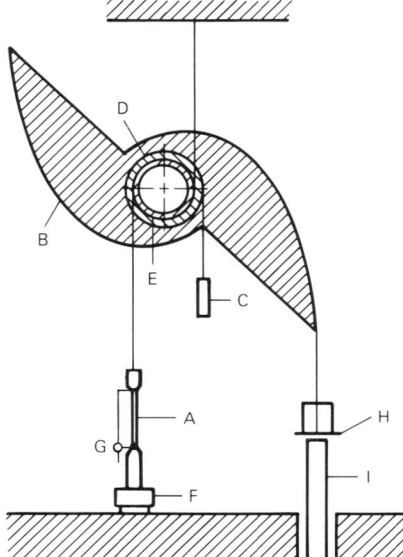

	Stress	
Specimen	MPa	ksi
A	160	23.2
B	147.5	21.4
C	128.2	18.6
D	106.9	15.5
E	91.0	13.2
	(continued to 2100 h)	
F..........	77.2	11.2
	(continued to 2900 h)	

the sample deforms, the frame tilts backward, and the columns are no longer vertical.

This movement, which is proportional to the preset L_3 and the amount of deformation, increases the force applied to the push rod and also to the sample, The increase in force compensates for the increase in area of the sample as it is compressed, thus maintaining a constant stress. Shorter samples require a greater L_3, because the area of the sample becomes greater for the same amount of deformation than for a sample of greater height. Sample area is not important in making L_3 calculations, because changes in area for a given amount of creep are of the same proportion for samples having either a large or a small area if both are of the same height. The system described also allows L_3 to be adjusted.

By increasing or decreasing the range of L_3, the range of sample lengths can be changed. This is done by using longer or shorter weight pan support columns, or by using an adjusting system that simultaneously changes L_3 and L_2 by moving the weight pan knife edge, as shown in Fig. 23. Two bars with weights are used to balance the frame.

The unit maintains constant stress to within 1%, as long as the total strain does not exceed 10% and the test is initiated with the weight pan support columns in a vertical position. However, if an approximation of the total strain during the run can be predicted and if the unit is set such that it passes through the vertical position when approximately one half of strain has occurred, the stress can be held to within 0.5% up to a total strain of 10%.

Applications of Constant-Stress Testing Apparatus. A typical constant-

stress system of 10-kN (2250-lbf) capacity similar to the design developed in Ref 8 is shown in Fig. 24. This system includes a counterbalance cam, pillow block bearing support, and three-zone furnace for elevated-temperature testing. A motorized automatic weight elevator with a platform usually provides gradual load application, reduces labor in handling load weights, and minimizes shock upon specimen failure. The cams for these systems are programmed and machined on a modern numerical control milling machine to provide optimum accuracy and minimum cost. Systems of this type have been used for constant-stress testing of a wide variety of materials at loads up to 45 kN (10 000 lbf).

Fig. 21 Constant-stress testing system for use with low forces

Designed by Hopkin (Ref 10) and modified by Holmes and Wray (Ref 9). A, specimen; B, double cam; C, countermass; D, pulley for countermass; E, hollow cylinder from which specimen is suspended; F, force cell; G, tape displacement gage; H, loaded mass pan; I, hydraulic ram

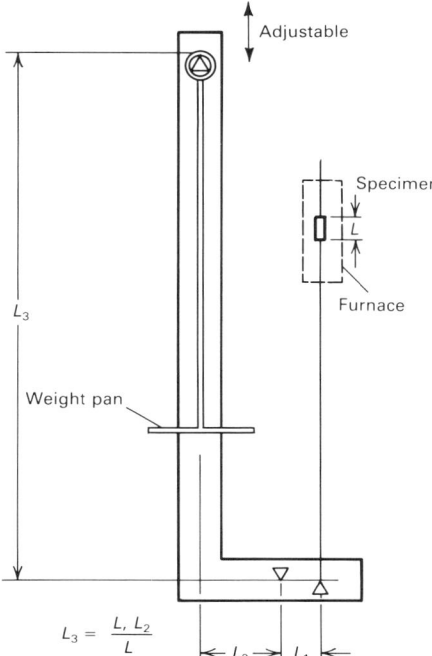

A constant-stress system designed for testing at low load forces down to 1-kN (225-lbf) load capacity is shown in Fig. 25. This system was constructed to the design principles developed in Ref 9 and 10 using a unique cam suspension system. This system, equipped with a weight elevator, furnace, and controls, is used to test wire and thin sheet materials at temperatures up to 870 °C (1600 °F).

Significant differences can exist between constant-load and constant-stress tests. Figure 26 shows creep-rate curves for silver-lead alloy cable sheath under both constant-load and constant-stress conditions (Ref 5).

Fig. 22 Knife-edge configuration for constant-stress compression testing

Source: Ref 11

Adjustable

Specimen

Furnace

L_3

Weight pan

$L_3 = \dfrac{L, L_2}{L}$

L_2 L_1

Fig. 23 Alternate knife-edge construction providing a greater range of sample lengths
Source: Ref 11

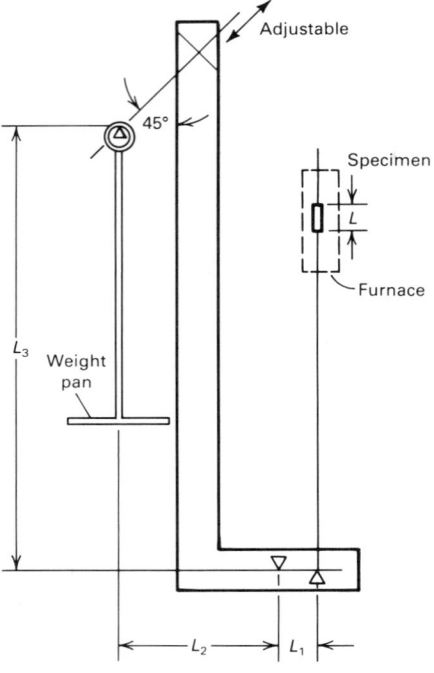

Stress-Relaxation Testing

By Alfred Fox
Member of Technical Staff
Metallurgical and Engineering
Science Department
AT&T Bell Laboratories

MAXIMUM UTILIZATION of engineering materials demands the measurement of mechanical properties that previously either were overlooked because of difficulties in determining them quantitatively, or were estimated from other more readily measured properties. Stress relaxation, the time-dependent decrease in stress in a constrained specimen, is one of these properties. Stress relaxation causes the drop in pitch frequently observed in stringed instruments. It is also the source of thermal and mechanical processes needed for the relief of residual stresses.

Relaxation also tends to reduce stress gradients resulting from geometrical factors and nonuniform loads in working parts; hence, stress-relaxation data can be used not only to

Fig. 24 Typical constant-stress testing system and furnace with 10-kN (2250-lbf) load capacity

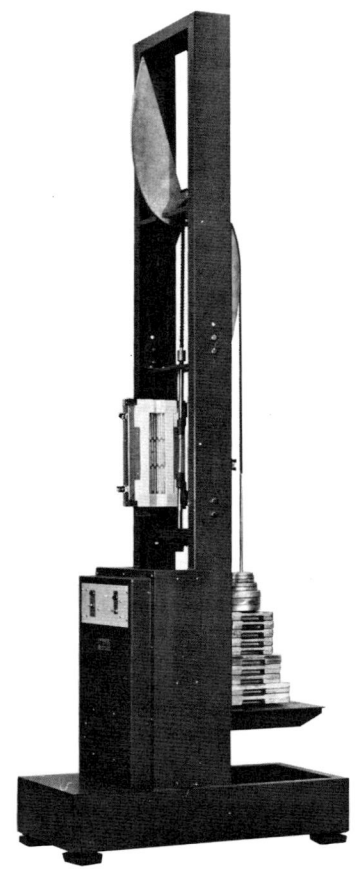

develop stress-relief treatments for reducing residual stress in castings, forgings, welded assemblies, cold worked metals, and machined surfaces, but also to obtain information on the permanent tightness of bolted joints, riveted assemblies, shrink fits, gaskets, solderless-wrapped connections, and similar devices. Stress relaxation is also of interest to designers of turbines and nuclear reactors, as in the redistribution of internal stress during reactor shutdown (Ref 13).

The effect of stress relaxation on residual stress of a wide variety of materials is discussed in Ref 14 and 15. In a conventional stress-relaxation test, whether it be in tension, compression, bending, or torsion (multiaxial stress situations are not covered in this article), the test specimen generally is held at constant total strain and temperature. The change in load, bending moment, or torque associated with the time- and temperature-dependent conversion of elastic to plastic strain is then determined.

Fig. 25 Constant-stress testing system for testing at low load forces (down to 1 kN, or 225 lbf)

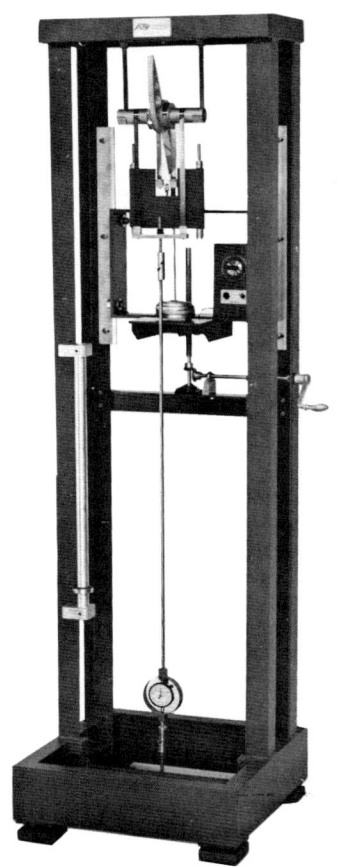

Testing Considerations (Ref 16)

When the material data are to be applied to the design of a particular class of component, the stress during the relaxation test should be similar to that imposed on the component. For example, tension tests are suitable for bolting applications and bending tests for leaf springs. Tension and compression tests have the advantage that the stress can be reported simply and unequivocally. During bending tests, the state of stress is complex, but can be accurately determined when the stresses on initial loading are elastic. If plastic strain occurs on loading, stresses usually can be determined within a bounded range only. Tension tests, when compared to those of compression, have a further advantage: it is unnecessary to guard against buckling. Therefore, when the test method is not restricted by the type of stress in the component, tension testing is recommended.

Fig. 26 Creep-rate/time curves for 0.01% silver-lead alloy cable sheath (B-33 sheath)

(a) At 5520 kPa (800 psi). (b) At 4140 kPa (600 psi). Source: Ref 5

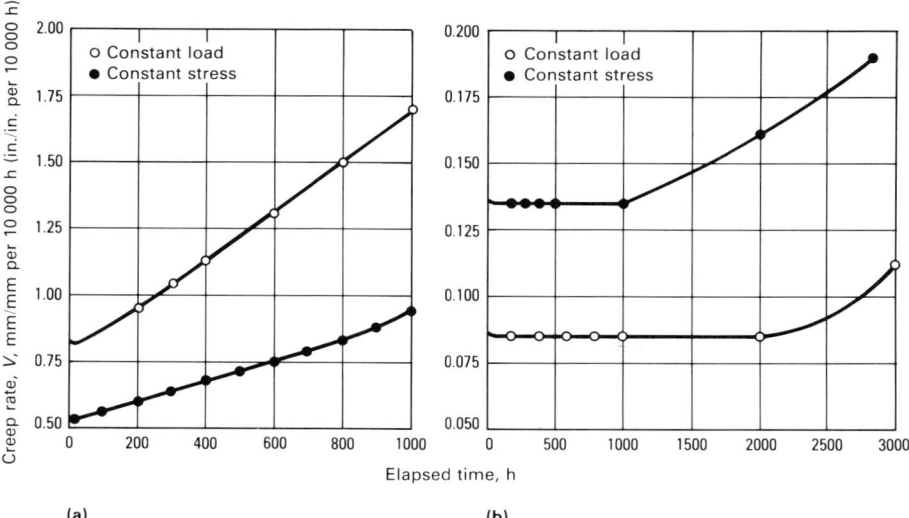

(a)

(b)

Fig. 27 Ideal stress-relaxation test in tension

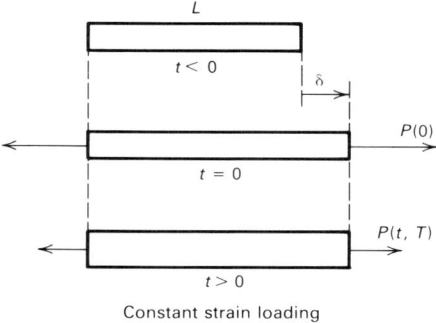

Constant strain loading

Fig. 28 Ideal stress-relaxation curve in tension

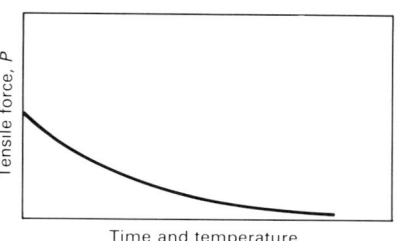

Fig. 29 Stress-strain diagram for determining relaxation in stress

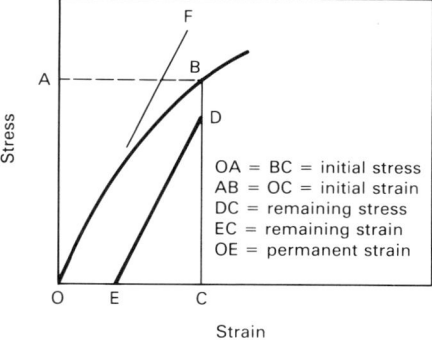

OA = BC = initial stress
AB = OC = initial strain
DC = remaining stress
EC = remaining strain
OE = permanent strain

Bending tests for relaxation, when compared to tension, compression, and torsion tests, have the advantage of using a lighter and simpler apparatus for specimens of the same cross-sectional area. Furthermore, strains usually are calculated from deflection or curvature measurements. Because the specimens usually can be designed so that these quantities are much greater than the axial deformation in a direct stress test, strain is measured more easily and is more readily used for machine control in the bending tests. Due to the small forces normally required and the simplicity of the apparatus when static fixtures are sufficient, many specimens can be placed in a single oven or furnace when tests are made in bending at elevated temperatures.

All stress-relaxation testing techniques described in Ref 16 are based on the same basic principles—namely, that force, moment, or torque required to maintain the constraint constant can be monitored either continuously or periodically. Alternately, the permanent strain present on the test specimen can be measured after completely removing the force, moment, or torque, and the stress remaining can be calculated from this value.

Stress-Relaxation Tension Testing

An ideal stress-relaxation tension test is shown in Fig. 27. A specimen of length L has a load P applied at time $t = 0$, which elongates the specimen by a change in length δ. Subsequently, the specimen is held at the same total length, and the load P decreases as

a function of time and temperature (Fig. 28).

Most reported data concern stress relaxation in materials initially stressed below their yield strength. On the other hand, if initial yielding occurs, an initial permanent or plastic strain develops (Fig. 29). The elastic strain remaining in Fig. 29, EC, is directly proportional to the remaining stress, CD, so that in effect, after the initial yielding, stress relaxation is a time-dependent conversion of elastic to plastic irrecoverable strain. As time progresses, line ED in Fig. 29 shifts to the right.

Thus, after initial yielding, an elastic strain remains, which for linear elastic materials is proportional to the applied load. This may be expressed as:

$$\epsilon_i = \frac{\dfrac{P_o}{A_o}}{E} \quad \text{or} \quad \frac{\sigma_i}{E} \qquad \text{(Eq 5)}$$

where ϵ_1 is the elastic strain remaining after initial loading, P_o is the applied load, A_o is the original cross-sectional area of the test specimen, σ_i is the initial stress, and E is the modulus of elasticity.

Because the total applied strain is kept constant, the residual stress can change only if the elastic strain is converted into time-dependent plastic strain. Thus:

$$\epsilon_i = \epsilon \qquad \text{(Eq 6)}$$

where ϵ is the total strain at time t. If stress relaxation occurs, then at any time t greater than zero, the total strain is composed of two

components—elastic or recoverable and plastic or irrecoverable. Therefore:

$$\epsilon_i = \epsilon = \epsilon_e + \epsilon_p \qquad \text{(Eq 7)}$$

where ϵ_e is the residual elastic strain at any time t, and ϵ_p is the plastic strain at time t. ϵ_p is equal to the residual strain after removal of the constraint and can be measured.

From this value, the loss in stress, $\Delta\sigma$, which cannot be measured, can be calculated as equivalent to the stress required to produce an equal amount of elastic deformation. Thus, assuming that the modulus of elasticity is constant and that the change in cross-sectional area is negligible during the test:

$$\Delta\sigma = \sigma_i - \sigma_t = E\epsilon_p \qquad \text{(Eq 8)}$$

where σ_t is the stress at time t. The constraint can first be removed and the specimen reloaded until the constrained length is again $L_o(1 + \epsilon_i)$, where L_o is the original gage length of the test specimen at zero load. Because some residual plastic strain, ϵ_p, now exists, the elastic strain, ϵ_e, required to restore the constrained length to the value $L_o(1 + \epsilon_i)$ is less than ϵ_i. From this it follows that the applied load required to deform the specimen by the reduced amount will be less than the initial load. In this case, the residual stress can be calculated either from the relation:

$$\sigma_t = \frac{P_t}{A_o} \qquad \text{(Eq 9)}$$

where P_t is the new load required to deform the specimen to the original constrained length $L_o(1 + \epsilon_i)$, or from the relation:

$$\sigma_t = E\epsilon_0 \qquad \text{(Eq 10)}$$

Repetition of either of these load/unload/reload procedures yields data such as those reported in Ref 16, from which the stress-relaxation curve can be developed. However, unless a new specimen is used for each successive loading test, the prior stress history will affect the rate at which elastic strain is converted into plastic strain. Hence, the stress-relaxation curve obtained from the repeated loading and unloading of a single specimen will differ somewhat from the true relation.

Although either of the above procedures can be used to measure stress relaxation, neither readily lends itself to the design of a machine for automatically recording the required values. Consequently, much of the early stress-relaxation data have been derived from "step-down," or interrupted, creep tests.

In the step-down creep test, a specimen is subjected to a load that causes an initial extension such as OA in Fig. 30(a). The load is applied as rapidly as possible without impact, and the maximum deformation is measured, after which an increment of load ΔP is removed. Upon removal of the load, elastic recovery corresponding to AB occurs, but under the remaining load, $P_o - \Delta P$, creep is initiated and follows the path BC. When the total length is again equal to the original value of $L_o(1 + \epsilon_i)$, a second increment of load ΔP is removed.

Under this reduced load, a somewhat longer time is required to deform the specimen to the original value $L_o(1 + \epsilon_i)$; that is, creep may now be assumed to follow the curve DE of Fig. 30(a) rather than the curve BC. By repetition of this procedure, data are obtained

which, when plotted in the form of Fig. 30(b), will approximate the stress-relaxation curve.

If in the step-down creep test the load is reduced each time the limiting strain ϵ_i is attained, the average strain will be less than this limiting value. In other tests in which the load is not reduced until after a limiting amount of creep occurs, the average strain will exceed the limiting value. Although the mechanism of these two tests is the same, the creep rates obtained will differ, because the two loads, or their equivalent stresses, are slightly different. This may produce some differences in the approximations to the derived stress-relaxation curve.

The step-down procedure can be mechanized, and if the load increments are sufficiently small, agreement between the curves derived from the data obtained in this type of test and from a stress-relaxation test should be very close. Any equipment that uses a load-reducing mechanism to maintain the specimen strain constant will produce a step-down creep curve, as shown in Fig. 30.

A typical stress-relaxation test system equipped for step-down tension testing (Fig. 31) includes a three-zone furnace, temperature control, a temperature-compensated extensometer system, electronic control module, and digital printer. It is constructed with a precision-balanced lever arm supported on knife edges. This system was designed primarily for testing to the rigid requirements of ASTM Standard E 328, "Stress-Relaxation Tests for Materials and Structures," which specifies control of total strain on the test specimen within ±0.000025 mm/mm (in./in.). This system has also been used for stress-relaxation testing of plastics, composites, and ceramics. Such a system normally has a load capacity of 45 or 90 kN (10 000 or 20 000 lbf).

The use of a lever arm (Fig. 31) has a number of advantages:

- The force (torque) required on the motor drive is reduced by a factor equal to the ratio of the lever arm, which is normally 20 to 1. This greatly reduces the power requirements and consequently the inertia of the motor drive system, resulting in finer and smoother control.
- The strain amplitude is increased by the lever arm ratio, resulting in increased sensitivity and control.
- The lever system provides a softer mechanical action.

Thus, smoother control action is possible, particularly when reducing load increments while maintaining constant strain as the specimen relaxes. Load steps are reduced to a minimum without impact or vibration. A lever system can be readily converted to con-

stant-load creep testing by disconnecting the motor drive and installing a weight pan and calibrated load weights on the weight train.

A schematic of a test setup that will produce a continuous tensile stress-relaxation curve by periodically measuring the

Fig. 30 Derivation of stress-relaxation curve for step-down creep test
(a) Constant extension approximated by a step-down creep test. (b) Stress-time relation

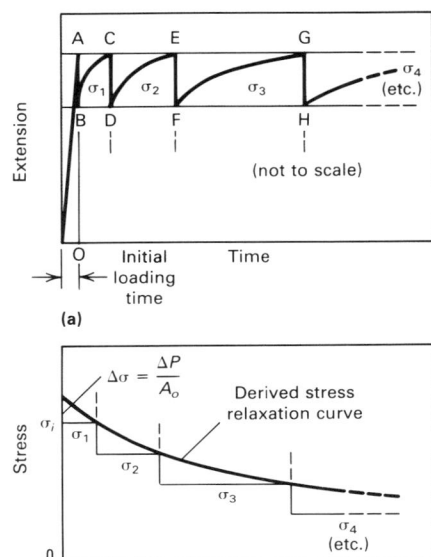

Fig. 31 Schematic of stress-relaxation test system equipped for step-down tension testing

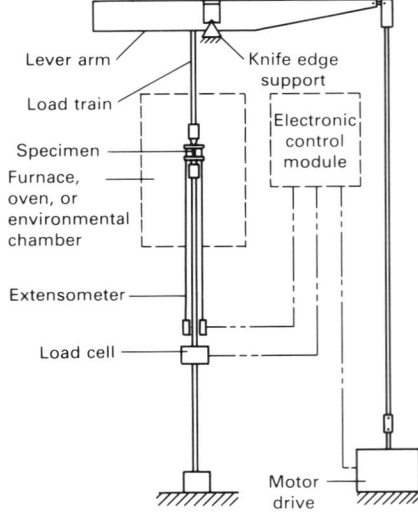

load on the test specimen is shown in Fig. 32. A method for continuously monitoring stress relaxation in wires or thin strips has been reported in Ref 17 to 19. This technique is based on the principle that the frequency of vibration of a string fixed at both ends is a function of the tension in the string.

$$f = \frac{n}{2L} \sqrt{\frac{T}{\mu}} \qquad \text{(Eq 11)}$$

where f is the frequency, n is the number of the harmonic (the fundamental harmonic $n = 1$), L is the nodal length of the string, T is the tensile force in the string, and μ is the line density (the weight per unit of length).

The principle of the equipment used in Ref 19 is illustrated in Fig. 33. The basic apparatus consists of two rigid panels: in this case, two 2.5-cm by 25-cm (1-in. by 10-in.) aluminum alloy plates mounted on vertical aluminum alloy H-beams. Two clamps, one mounted on the upper edge and one on the lower edge of the plates, fasten the wire test specimen to cylindrical stainless steel arbors, through which the wire is threaded and around which it is wrapped to apply a tensile force. One of the clamps is grounded to the frame, and the other is insulated by a piece of anodized aluminum.

In use, a portable U-shaped permanent magnet is mounted on the aluminum alloy plate so that the test specimen being measured is parallel to the planes of the pole faces (soft iron) and centrally located in the air gap of the magnet. An alternating electromotive force from a precision oscillator is then applied to the test specimen through a common ground (the grounded test panel), and the end of the specimen is attached to the insulated clamp. The rms voltage of this alternating electromotive force should be held constant for precise results.

In the circuit used, the alternating current varies with the magnetic field in accordance with the fundamental physical relation:

$$F = \frac{IL \times B}{C} \qquad \text{(Eq 12)}$$

where F is the force, I is the current, L is the length of the specimen, B is the magnetic field intensity, and C is a conversion constant.

Driven by this force, the specimen vibrates in the plane of the air gap. When the frequency of the driving force corresponds to the resonant frequency of the stressed wire, the latter vibrates at maximum amplitude. This is determined optically by means of a microscope equipped with a filar eyepiece. The precision oscillator is used not only to drive the test specimen, but also to measure the resonant frequency from which the value of the remaining stress can be calculated in accordance with Eq 11.

Periodic vibration of the wires and measurement of the resonant frequencies yield data from which the stress-relaxation curve can be developed. The amplitude of the vibrating specimen is kept to a minimum during measurement of the resonant frequency to minimize the superposition of bending and tensile stress due to the lateral vibration. This is done by keeping the driving voltage as low as possible to detect resonance.

This equipment can be used at normal temperatures and also at elevated temperatures within the limitations of the permanent magnet, provided the entire frame is at temperature. Typical stress-relaxation curves obtained by this method as a function of time and temperature for tinned, annealed 30 AWG (0.25-mm, or 0.01-in., diam) electrolytic tough pitch copper wire (C11000) are shown in Fig. 34.

Stress-Relaxation Compression Testing

A typical apparatus for conducting stress-relaxation compression tests is shown in Fig. 35. In this apparatus, a subpress is used to apply the initial compressive load to the test specimen by means of a hydraulic ram. The nut on the load strut is adjusted to maintain a constant compressive strain on the test specimen. As the force required to maintain a constant compressive strain drops as a function of time and temperature, it is recorded by a pressure transducer, shown schematically as a "pressure gage" in Fig. 35.

Fig. 32 Test setup to produce a continuous tensile stress-relaxation curve using periodic load measurement

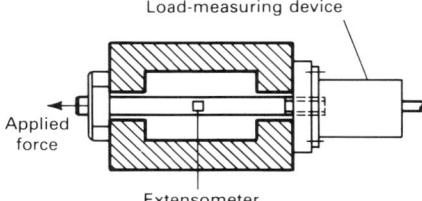

Fig. 33 Stress-relaxation tension test using the vibrating string method

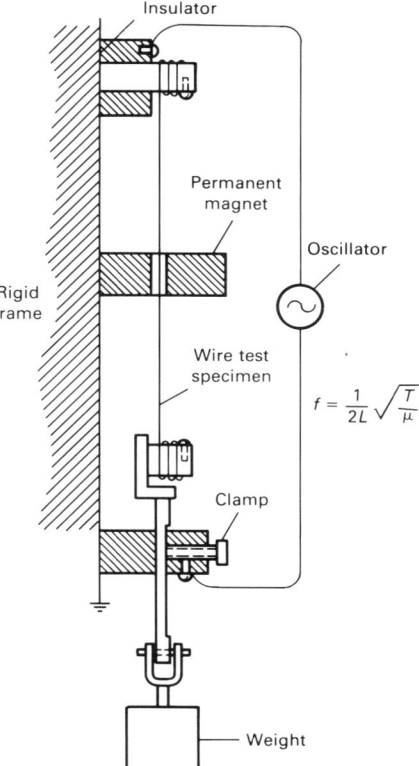

$$f = \frac{1}{2L} \sqrt{\frac{T}{\mu}}$$

Fig. 34 Tensile stress-relaxation data for C11000 copper wire obtained using the vibrating string technique

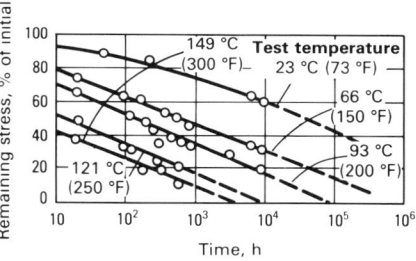

Fig. 35 Apparatus for conducting a stress-relaxation compression test

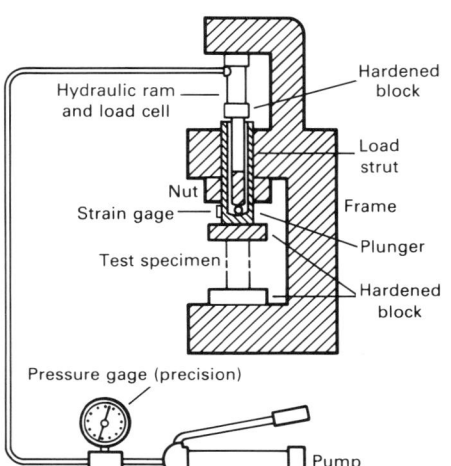

In compression testing, the possibility of buckling and barreling must be minimized (Ref 20). Thus, cylindrical test specimens are best suited for this type of testing. Specimen surfaces must be flat and parallel, and the end surfaces must be properly lubricated.

Stress-Relaxation Bend Testing

Stress relaxation in bending is important to designers of flat and wire springs—such as automotive leaf springs, electronic connectors, and relay springs. In most stress-relaxation bending tests, a bending moment is applied to the test specimen. The latter is held under constant constraint conditions, which involves either constant displacement or curvature. The change in bending moment is determined either by monitoring the applied force continuously using a load cell or force gage in conjunction with a spherometer (Fig. 36) or by intermittently lifting the specimen just off its reaction point and measuring the force (or moment) required to accomplish this (Fig. 37). Alternatively, the loss in stress (moment) may be determined by using either a mandrel (Fig. 38) or a tapered, constant-curvature specimen (Fig. 39) and measuring elastic springback upon unloading. An alternate procedure to accomplish the springback measurement involves holding the ends of a wire or strip coaxial with a clamp to form a circle (Ref 21).

Reference 16 details procedures for evaluating stress relaxation in bending and recommends that test conditions be specified in terms of strain rather than an initial stress. Tests usually are made under conditions of constant radius of curvature over the gage length of the specimen. These conditions are achieved by using either a symmetrically four-point loaded beam with uniform cross section (Fig. 36), a tapered, constant-curvature cantilever specimen (Fig. 37 and 39), or a cylindrical mandrel (Fig. 38). The specimen dimensions are chosen to provide adequate measuring sensitivity. For strip specimens, for example, the bending strain may be taken as:

$$\epsilon_b = \frac{h}{(2R_i + h)} = \frac{h}{(2R_o - h)} \qquad \text{(Eq 13)}$$

where h is the specimen thickness, R_i is the radius of the concave surface of the specimen, R_o is the radius of the convex surface of the specimen, and ϵ_b is the maximum bending strain.

In round specimens such as wire, the thickness of the beam, h, may be replaced by the diameter. However, the volume of material subjected to increasing strain will be proportionally less in round sections as the distance from the neutral axis increases compared to rectangular cross sections.

It has been shown that the stress at certain interior points in a beam is approximately the same for a given bending moment, independent of whether the stress-strain relationship is linear (Ref 22). Such interior points are called skeletal points. At these points, the elastic flexure formula can be used to estimate stress even when plastic flow, creep, or relaxation occurs. For a beam of rectangular cross section, the skeletal point is two thirds of the distance from the neutral axis of the surface. Therefore, the stress at the skeletal point is two thirds of the nominal bending stress, and the strain at the skeletal point is two thirds of the outer fiber strain. When conspicuous plastic strain occurs on loading, the stress-relaxation curve at the skeletal point is the only curve that can be plotted directly and confidently.

Fig. 38 Mandrel-type stress-relaxation bending test

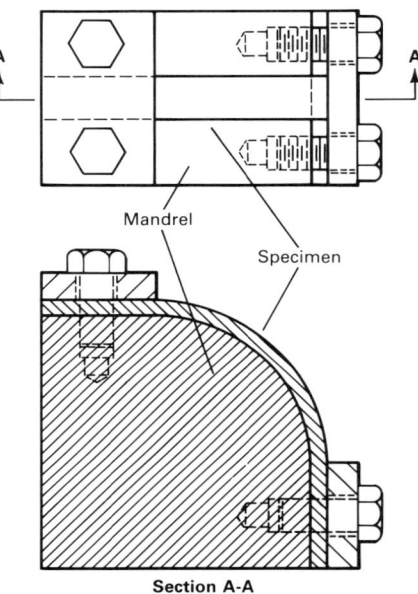

Section A-A

Fig. 36 Stress-relaxation bend test specimen and spherometer in four-point loaded beam with uniform cross section

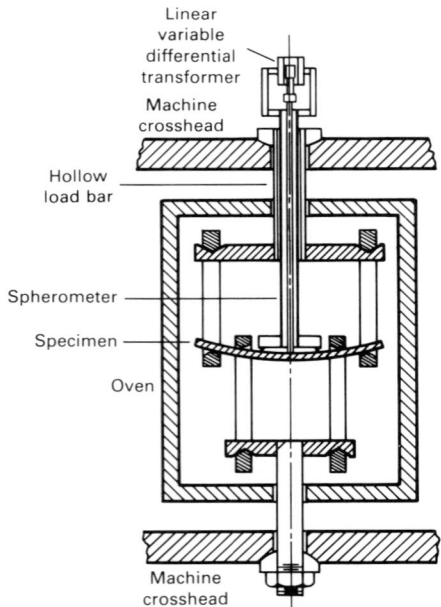

Fig. 37 Stress-relaxation bend test specimen in static fixture for lift-off measurements

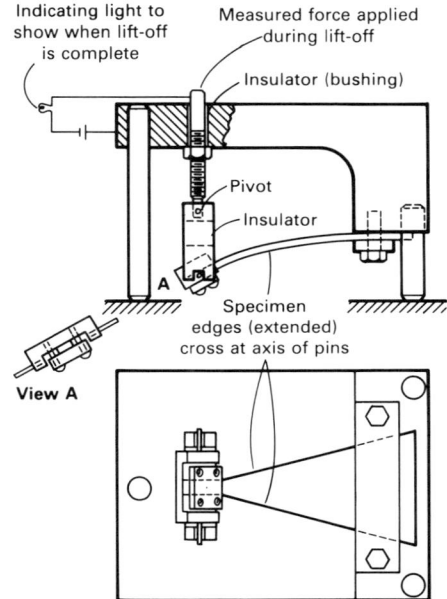

Fig. 39 Tapered, constant-curvature cantilever beam strip specimen bending test
Springback-type test for determining stress relaxation

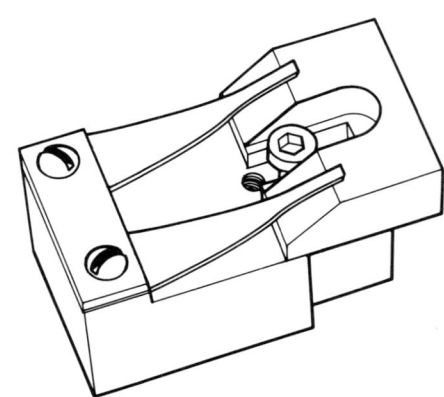

The stress at the skeletal point can be determined using the elastic flexure equation:

$$\sigma = \frac{My}{I} \qquad \text{(Eq 14)}$$

where σ is the nominal flexure stress at the outer fiber, M is the bending moment at the most highly stressed section, y is the distance from the skeletal plane to the centroidal axis of the cross section (the axis being normal to the plane of the loading forces), and I is the moment of inertia about the centroidal axis of the cross section.

For rectangular cross sections:

$$\frac{y}{I} = \frac{6}{bh^2} \qquad \text{(Eq 15)}$$

where b is the width of the beam (that is, the dimension perpendicular to the plane of the loading forces), and h is the depth of the beam (that is, the dimension in the plane of the loading forces). Consequently:

$$\sigma = \frac{6M}{bh^2} \qquad \text{(Eq 16)}$$

When using the elastic springback method on unloading to determine stress relaxation in bending, Eq 13 can be used to determine the change in curvature versus bending strain relationship.

Stress-Relaxation Torsion Testing

The techniques used for torsion testing are the same as those used for bend testing, except that a cylindrical or tubular specimen is subjected to a constant torsional strain of long duration. The change in restraining torque associated with torsional stress relaxation can be determined by continuously monitoring the latter with a torque transducer or by periodically applying a slight torque to transfer the restraining torque from the end grips to the transducer. Alternatively, the twist springback on unloading can be used to determine how much remanent twist angle remains, or how much torque would be required to twist the specimen back to its original angle of twist prior to relaxation. Because of the complexity of the stresses and strains in noncircular or noncylindrical specimens, these are not discussed in this article.

Figure 40 illustrates a cylinder subjected to torsion, including the relationship between angular displacement θ and angle of twist ϕ. For a cylindrical specimen such as that shown, if the torque versus angle of twist curve is linear, the initial maximum torsional stress is:

$$\tau_i = \frac{16T_i}{T_i d^3} \qquad \text{(Eq 17)}$$

where τ_i is the initial maximum torsional stress, T_i is the initial torque, and d is the specimen diameter. For a tubular specimen:

$$\tau_i = \frac{16T_i d_o}{\pi(d_o^4 - d_i^4)} \qquad \text{(Eq 18)}$$

where d_o is the specimen outside diameter, and d_i is the specimen inside diameter.

If the torque-twist curve is nonlinear, the initial maximum torsional stress can only be estimated for cylindrical or thin-walled tubular specimens. For a cylindrical specimen:

$$\tau_i = \frac{4(3T_i + \theta\alpha)}{\pi d^3} \qquad \text{(Eq 19)}$$

where θ is the angle of twist per unit length at torque T_i, and α is the slope of the torque-twist curve at torque T_i. For a thin-walled tubular specimen, the initial maximum torsional stress can be approximated by Eq 18. The remaining stress $\tau_{(t)}$ after time t can similarly be calculated by Eq 17 or 18, depending on whether the specimen is cylindrical or tubular, by substituting the remaining torque $T_{(i)}$ after time t.

When the elastic springback method is used, it is necessary to determine the difference in the angle of twist ϕ from the loaded to the unloaded condition, to find the torque corresponding to this differential angle from the torque versus angle of twist curve, and then to substitute this value for torque in either Eq 17 or 18.

Tests on Springs

Generally, stresses in springs are quite complex. Designers need to know how much force is left in a spring after a given service life. This information can be obtained readily by one of the following methods, provided the spring constant (relating the spring force F to its displacement δ) is known:

1. Deflecting the spring to a constant displacement and monitoring the decrease in spring force under conditions of constant displacement and temperature

Fig. 40 Shear strain resulting from torsion

Illustrated by a shaft of length L_o subjected to torque T

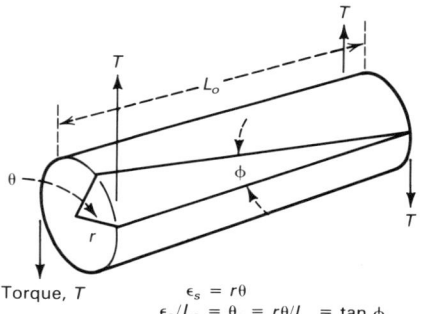

Torque, T

$$\epsilon_s = r\theta$$
$$\epsilon_s/L_o = \theta_s = r\theta/L_o = \tan\phi$$

2. Transferring the load from the deflecting member to a load-measuring device intermittently and determining the remaining spring load
3. Using the elastic springback on unloading to determine the spring force remaining immediately prior to unloading

Methods 2 and 3 are illustrated in Fig. 41 and 42 (Ref 14).

Figure 41 illustrates the apparatus used to determine the force remaining as a function of time and temperature in a complex relay pileup clamp spring. To measure the force with which the spring acts on the pileup, the spring is clamped into a rigid steel fixture that is placed on a compression load cell. Force is applied by means of a plunger, which is depressed and then twisted so that its top shoulder rests against the top shoulder of the fixture. At this time, electrical contact is made between the plunger and a spring-loaded contact. Force is applied by means of a universal testing machine. The force required to break the electrical contact may be measured as a function of time. Results of this type of test have been correlated with springback measurements in Ref 23.

Fig. 41 Setup for stress-relaxation testing of a clamp spring

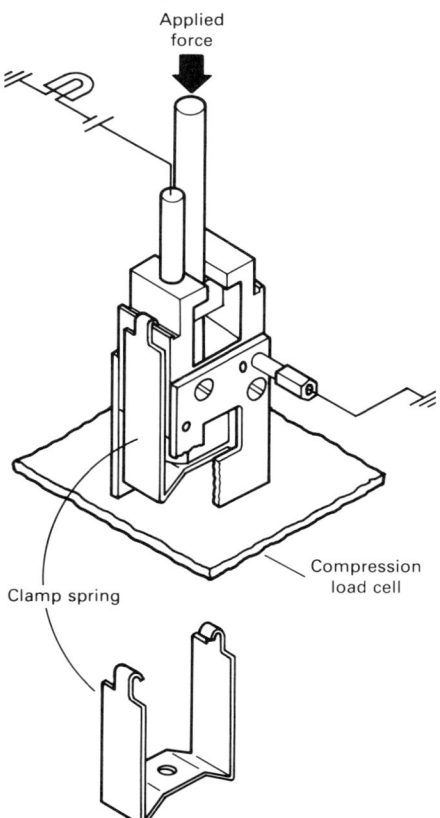

Applied force

Clamp spring

Compression load cell

Fig. 42 Application of load during a stress-relaxation test of a helical compression spring at room temperature

L_o, original free length; L_1, compressed length; L_2, free length upon release from load at temperature; ΔL, permanent set

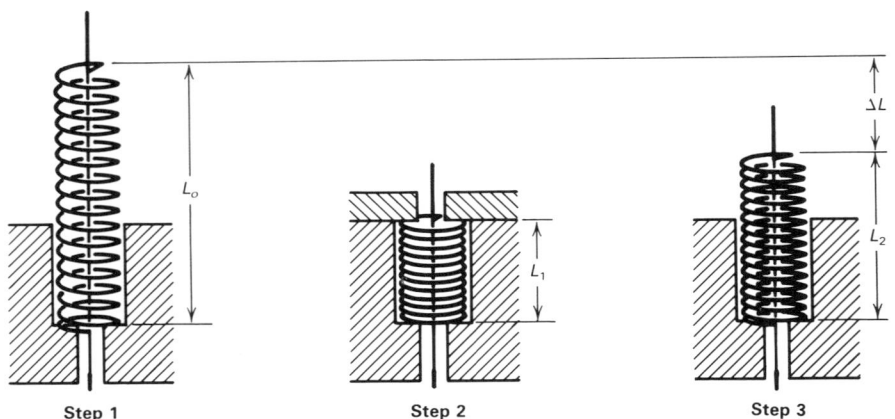

Step 1   Step 2   Step 3

In Fig. 42, a helical compression spring is deflected and held at the test temperature. The force remaining after time, t, is determined by measuring the permanent set ΔL on unloading by the relationship:

$$\frac{\Delta P}{P_o} = \frac{\Delta L}{L_o}$$

where P_o is the initial load, ΔP is the change in load, L_o is the initial length, and L is the permanent set.

REFERENCES

1. "Standard Practice for Conducting Creep, Creep-Rupture, and Stress-Rupture Tests of Metallic Materials," E 139, *Annual Book of ASTM Standards*, ASTM, Philadelphia, 1984, p 283-297
2. "Standard Practices for Load Verification of Testing Machines," E 4, *Annual Book of ASTM Standards*, ASTM, Philadelphia, 1984, p 111-118
3. J.M. Thomas and J.F. Carlson, Errors in Deformation Measurements for Elevated Temperature Tension Tests, *ASTM Bull.*, May 1955, p 47-55
4. "Standard Methods of Tension Testing of Metallic Materials," E 8, *Annual Book of ASTM Standards*, ASTM, Philadelphia, 1984, p 130-150
5. A. Fox, Lecture Notes, ASM Continuing Education Seminar, Creep and Stress Relaxation Testing, March 1980
6. E.N. Da C. Andrade, On the Viscuous Flow in Metals and Allied Phenomena, *Proc. Royal Soc. London*, Vol 84, Series A, 1910-11, p 1
7. E.N. Da C. Andrade and G. Chalmers, The Resistivity of Polycrystalline Wires in Relation to Plastic Deformation and the Mechanism of Plastic Flow, *Proc. Royal Soc. London*, Vol 138, Series A, 1932, p 348
8. F. Garofalo, O. Richmond, and W.F. Domis, Design of Apparatus for Constant-Stress or Constant-Load, *J. Basic Eng.*, Vol 84, 1962, p 287-293
9. M.F. Holmes and P.J. Wray, A Constant Stress Apparatus Suitable for Use with Small Forces, *J. Phys. E: Sci. Instr.*, Vol 3, 1970
10. L.M.T. Hopkin, A Simple Constant Stress Apparatus for Creep Testing, *Proc. Phys. Soc. B*, Vol 63, 1950, p 346-349
11. C.H. Carter, Jr., C.A. Stone, and R.F. Davis, High-temperature Multi-atmosphere Constant Stress Compression Creep Apparatus, *Rev. Sci. Instr.*, Vol 51 (No. 10), Oct 1980, p 1352-1357
12. R.L. Fullman, R.P. Carreker, and J.C. Fisher, Simple Devices for Approximating Constant Stress During Tensile Creep Tests, *J. Metals*, Vol 5, 1953, p 657
13. J.M. Beeston and T.K. Burr, In Reactor Stress Relaxation of 384 Stainless Steel In Pile Tube, in *Stress Relaxation Testing*, STP 676, A. Fox, Ed., ASTM, Philadelphia, 1979, p 155-170
14. A. Fox, Effect of Temperature on Stress Relaxation of Several Metallic Materials, in *Residual Stress and Stress Relaxation*, E. Kula and V. Weiss, Ed., Proceedings of the 28th Sagamore Army Materials Research Conference, Plenum Press, New York, 1982, p 181-203
15. "Standard Recommended Practices for Stress Relaxation Tests for Materials and Structures," E 328, *Annual Book of ASTM Standards*, ASTM, Philadelphia, 1984
16. R.E. Davis, Relaxation of Stress in Heat-Exchanger Tube of Ideal Material, *Trans. ASME*, Vol 74, 1952, p 381
17. M.G. Dawance, A New Method of Studying Relaxation Phenomena in Steel Wires, *Annales de l'Institut Technique du Bâtiment et des Travaux Publics*, No. 9, Feb 1948
18. G.R. Gohn and A. Fox, New Methods for Determining Stress Relaxation, *Mat. Res. Stand.*, Vol 1 (No. 12), Dec 1961, p 957-967
19. A. Fox, Ed., *Stress Relaxation Testing*, STP 676, ASTM, Philadelphia, 1979
20. "Standard Methods for Compression Testing of Metallic Materials at Room Temperature," E 9, *Annual Book of ASTM Standards*, ASTM, Philadelphia, 1984
21. A. Fox, A Simple Test for Evaluating Stress Relaxation in Bending, *Mat. Res. Stand.*, Vol 4 (No. 9), Sept 1964, p 480-481
22. D.L. Marriott and F.A. Leckie, "Some Observations on the Deflections of Structures During Creep," *Conference on Thermal Loading and Creep in Structures and Components*, Institute of Mechanical Engineers, London, 1964, p 4-5 to 4-15
23. A. Fox, Stress Relaxation in Bending of AISI 30, Type Corrosion Resistant Steel Strip, in *Stress Relaxation Testing*, STP 676, A. Fox, Ed., ASTM, Philadelphia, 1979, p 78-88

Assessment and Use of Creep-Rupture Properties

By Howard R. Voorhees
Technical Director
Materials Technology Corporation

USE OF CREEP-RUPTURE PROPER-TIES to determine allowable stresses for service parts has evolved with experience, although guidelines for use differ among specifications. For temperatures in the creep range, the *ASME Boiler and Pressure Vessel Code* (Ref 1) uses the following criteria as a lower limit in determining creep values: 100% of the stress to produce a creep rate of 0.01%/1000 h, which is based on a conservative average of reported tests, as evaluated by an ASME committee (in assessing data, greater weight is given to those tests run for longer times); 67% of the average stress, or 80% of the minimum stress, required to produce rupture at the end of 100 000 h, as determined from available extrapolated data.

For most commercial steels and alloys, the available raw data are obtained from many tests of durations ranging from a few hundred to a few thousand hours. Tests seldom run longer than 10 000 h, and durations of 100 000 h are rare. Measured secondary creep rates as low as 0.01%/1000 h are also unusual. Allowable stresses recommended in existing specifications usually are derived from extrapolations. Considerable scatter in test results may be observed even for a given heat of an alloy, so interpolated creep and rupture strengths are not precisely known.

Although measurement and application of creep-rupture properties is often imprecise, general trends are evident. Methods for assessing creep-rupture properties (including nonclassical creep behavior), common interpolation and extrapolation procedures, and properties estimation based on insufficient data are discussed in this article.

Scatter of Creep-Rupture Test Data

Reliable creep-rupture property measurements require that the test specimens be representative of the material to be used in service, preferably in the product form and condition of intended service. Testing results may vary significantly with sampling procedures. For example, heavy sections may exhibit variations in strength level with depth after normalizing or quenching and tempering treatments. A common practice is to take samples midway between the center and the surface of the specimen.

Similar variations in strength can occur for fully annealed plate or bar that has been cold straightened. Re-annealing may be required before valid testing results can be obtained. For some materials (e.g., stainless steel castings several inches thick, or alloys subjected to elevated temperature after critical prior plastic deformation), the grains may become so coarse that a specimen cross section contains only one or several grains. Local creep rates may then vary considerably along the specimen gage length according to the orientation of the individual large grains with respect to the loading direction.

Such effects caused cross sections of specimens from 127-mm (5-in.) thick castings of ACI CF-8M to distort from circular to oval, or even to nearly flat, as creep progressed in tests whose rupture life at 595 to 870 °C (1100 to 1600 °F) ranged between a few hours and 40 000 h (Ref 2). In this example, time to rupture was not markedly affected by gage section diameter of 8.9 versus 12.7 mm (0.35 versus 0.5 in.) or by the nonuniform deformation of the gage section. Measured creep tended to be more variable with the smaller specimens.

For materials in which grain-boundary material and that in the body of the grains differ markedly in either composition or strength, the effect of number and orientation of grains in the cross section would be expected to result in greater scatter than was noted in the above example.

Sampling direction seldom affects creep and rupture properties for materials with uniform single-phase structures of small equiaxed grains. For such materials, specimen size also exhibits minimal influence, except for the greater relative importance of surface oxidation and similar effects in small specimens. Due to solidification and processing conditions, preferred orientation of secondary phases or grains can alter test results. In critical situations, the loading direction in the test specimen should parallel the major stress direction expected under service loading. For rolled materials, ASTM Standard E 139 (Ref 3) recommends that specimens be taken in the direction of rolling, unless otherwise specified.

The presence of large discrete particles of lower ductility and strength (e.g., graphite flakes in gray cast iron, or large glassy inclusions in steel) may significantly lower the sound cross section of a small specimen, but have a lesser influence in a larger specimen. Specimen size may have the opposite effect if only a few scattered low-strength particles are present; consequently, a specimen small enough could be free of these large-scale areas of weakness. More subtle variations in local creep and rupture strengths arise at grain boundaries, precipitates, voids, composition gradients, and other regions of microscale nonuniformity.

Specimen Size. For many steels, the influence of specimen size is minimal compared to other variables. During testing of a low-alloy steel (ASTM A193, grade B16) and a high-alloy steel (ASTM A453, grade 660) using five different specimen diameters ranging from 5 to 84 mm (0.2 to 3.3 in.), rupture strength appeared to be independent of size for unnotched specimens (Ref 4). Variation in rupture time with size was erratic for notched specimens, but the largest size had about three quarters of the strength of the smallest geometrically similar specimen.

Scatter From Heterogeneities. Rupture properties were not uniform with position in the original bar, as well as among bars from the same heat that received the same treatment (Ref 4). Fracture in notched specimens sampled from the mid-radius location of the A193 alloy originated consistently in the area of the cross section nearest the outside of the original bar.

Three tests at 227.5 MPa/538 °C (33 ksi/1000 °F) produced rupture times from 10 330 to 12 000+ h (test incomplete) for specimens from one bar, whereas three specimens from the other two bars of the A193 steel lasted only 7198 to 7820 h under similar conditions. These six tests included four different specimen diameters, but scatter among bars significantly exceeded that due to test specimen size.

Heterogeneities in a low-alloy steel of commercial quality have been found to cause much of the typical 30% scatter encountered in determining creep rates under strict testing conditions (Ref 5). Scatter for rupture times is approximately half that magnitude with materials that are specially prepared to be uniform.

Temperature and Other Testing Variables. Even with the use of precise temperature controllers and high-quality pyrometric practice, care is required to hold the average indicated specimen temperature within usual specifications of ±2 °C (±3.6 °F) (Ref 3) with time. Variation from one location to another along the specimen gage length typically approaches this magnitude, even in furnaces with independent adjustment of power input within zones.

These variations in indicated temperature do not include errors in initial thermocouple and pyrometer calibrations, leadwire mismatch, or drift in thermocouple output with time. Actual temperature can differ from the reported value by at least 5 °C (9 °F) during some portion of a representative creep or rupture test.

For steel at 450 °C (842 °F), a temperature disparity of 5 °C (9 °F) corresponds approximately to 20% variation in rupture time (Ref 5). A 10% change in creep rate results from a load change of 1.5%, or a temperature change of 2.5 °C (4.5 °F) (Ref 4). An error of more than 1% in the applied load is unlikely under typical conditions.

Equipment for creep-rupture testing commonly includes features to promote uniform axial loading. However, bending loads may still occur. The increase in secondary creep rates and reduction in rupture life with eccentricity of loading is analyzed in Ref 6, which reports that extreme cases of bending introduced by threaded ends on cylindrical specimens reduce rupture life by as much as 60%. The largest effects occur for short or notched specimens and for materials with low ductility to rupture.

Eccentric loading exerts maximum influence on creep measurements in the early stages of a test. With initially bowed specimens, which are common when specimens are as-cast or are strip-type machined from thin material, early creep indications may be erroneous if the extensometer fails to accurately average the strains at opposite sides of the specimen.

Typical Data Scatter. Rolled bars from a single heat of 2.25Cr-1Mo steel were tested by 21 laboratories in eight countries (Ref 7). The largest deviations in average results for one laboratory from the arithmetic mean of values from all laboratories for the stress to cause rupture in 1000, 3000, and 10 000 h were:

Temperature	Deviation
550 °C (1022 °F)	+14.3 and −16.6%
600 °C (1112 °F)	+21.2 and −12.1%

Another group of 18 laboratories in seven countries cooperated in tests of Nimonic 105 alloy bars at 900 °C (1652 °F) (Ref 8). Mean rupture times adjusted to a common stress for individual laboratories (four tests each) ranged from a high of 1491.4 h to a low of 1090.2 h (15.5% deviation). Time to 0.5% total deformation varied from 75.4 to 182.3 h for 16 of these laboratories; deviation from the mean value thus increased to 41.5%.

Temperature control was found to be the most serious source of variation in rupture measurement. In particular, calibration drift of Chromel-Alumel thermocouples at this high test temperature was responsible for the long mean rupture times found by five of the laboratories. Not more than about 15% scatter generally can be attributed to testing variables, if a laboratory has followed standard procedures.

Multiple Heats and Product Forms. Scatter bands become much broader when data originate from tests on numerous heats, particularly when data include a broad range of product forms and sizes. Elevated-temperature properties for steel and superalloys are available (see, for example, Ref 9-19). Typically, data for each grade of steel include plots of the stress for rupture in 1000, 10 000, or 100 000 h versus the test temperature. Where appropriate, distinctive shapes of data points identify different product forms or heat treatments.

When data at a given temperature are extensive, the reported range of derived stresses to produce rupture in a given time typically consists of a two-fold ratio of the highest to the lowest stress level. The corresponding spread of rupture times for a given test stress is on the order of 100-fold.

If a new set of test results does not fall within the broad scatter bands, careful re-

view and confirmation is required before the new data are accepted as valid. Bias of new data tends toward the upper half of the scatter bands compiled for older data, because the increase of residual alloying elements from scrap recycling, higher nitrogen content, and improved alignment and temperature control in modern testers tend to raise indicated creep and rupture properties.

At a test temperature of 593 °C (1100 °F) and at a stress of 207 MPa (30 ksi), rupture life ranged from 84 to 2580 h and secondary creep rate ranged from 0.16 to 0.00077%/h for 20 heats of type 304 stainless steel in the re-annealed condition (Ref 20). Corresponding large variations were observed at all test temperatures and for tests on seven heats of type 316 stainless steel. Re-annealing of as-received material lowered time to rupture in some cases, but the degree of variation persisted in properties among heats.

Good linear correlation was obtained when the logarithm of rupture strength was plotted against ultimate tensile strength at the same temperature for various types of austenitic stainless steels in the annealed and cold worked conditions at temperatures ranging from 538 to 816 °C (1000 to 1500 °F) and for test times approaching 10 000 h. Tensile strength, in turn, was reported to be essentially proportional to $(C + N)d^{-1/2}$, where C and N represent the weight percentage of carbon and nitrogen content, respectively, and d is average grain diameter.

Although long-time performance generally cannot be accurately predicted from short-time data, the location of a particular set of rupture data within a published scatter band should agree with the relative tensile strength of the material being tested in the range of tensile strengths spanned by all heats and product forms. Tests to measure low rates of secondary creep frequently are terminated after the creep rate appears to have become reasonably constant. If the test duration had been prolonged substantially, a continued slow decline in creep rate may have been observed.

A "false" minimum rate may occur in some tests before the classical secondary creep period has been reached. Observations of many comparative creep and rupture strengths for steels has established the following general relationships:

$$\frac{0.0001\%/h \text{ creep strength}}{10\,000\text{ h rupture strength}} = 0.7 \text{ to } 0.8$$

$$\frac{0.00001\%/h \text{ creep strength}}{100\,000\text{ h rupture strength}} = 0.5 \text{ to } 0.6$$

If a new set of test results differs from these patterns, verification is suggested before the new results are accepted.

Nonclassical Creep Behavior

The curve of creep deformation versus time traditionally displays three consecutive stages (Fig. 1). The longest period of substantially constant creep rate is preceded by a primary stage during which the rate declines from an initial high value and is followed by a tertiary stage of rising creep rate as rupture is approached (Fig. 2).

Although this classical pattern can be made to fit many materials and test conditions, the relative duration of the three periods differs widely with materials and conditions. For example, in many superalloys and other materials in which a strengthening precipitate continues to age at creep temperatures, brief primary creep often shows transition to a long upward sweep of creep rate, with only a point of inflection for the secondary period.

Aging of normalized and tempered 0.5Cr-0.5Mo-0.25V steel during creep under 80 MPa (11.6 ksi) stress at 565 °C (1050 °F) has been reported to cause the creep curve to effectively exhibit only a continuously increasing creep rate to fracture (Ref 21). For twice the amount of stress, the creep curve in this case followed the classical trends. In other alloys, such as titanium alloys, with limited elongation before fracture, the tertiary stage may be brief and may show little increase in creep rate before rupture occurs.

A more obvious departure from classical behavior develops during the early portion of many tests when precise creep measurements are taken. When 34 ferritic steels were studied for as long as 100 000 h at temperatures ranging from 450 to 600 °C (842 to 1112 °F), step-form irregularities were observed, with an extended period of secondary creep preceded by a lower creep rate of shorter duration during primary creep (Ref 22).

Negative Creep. Because a variety of metallurgical processes can be involved and because the rates and direction of these processes can vary with time and temperature, departures from classical creep curves can take many forms and can be overlooked, unless accurate creep readings are taken at sufficiently close intervals, particularly during early stages of the test. For 2.25Cr-1Mo, Cr-Ni-Mo, and Cr-Mo-V steels, some tests have demonstrated an abrupt drop to negative creep rate (contraction) after a brief beginning period of positive primary creep. Once this contraction ceased, the remaining portion of the test displayed the classical succession of declining, steady, and then rising creep rates. Figure 3 gives an example for normalized and tempered 2.25Cr-1Mo steel tested at 275.8 MPa (40 ksi) at 482 °C (900 °F) (Ref 2).

Fig. 1 Schematic tension-creep curve showing the three stages of creep

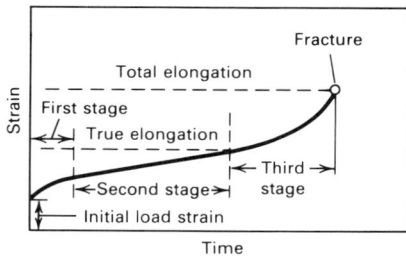

Definite negative creep was noted in at least one test each at 482, 704, 816, and 871 °C (900, 1300, 1500, and 1600 °F) for cast CF8 austenitic stainless steel in tests of boiler and pressure vessel materials (Ref 2). Rupture times for these tests ranged from 1000 h to longer than 30 000 h. For some combinations of material lots and test temperatures, nearly all creep curves of these tests showed an early "false" minimum rate during part of the primary stage. Structural changes responsible for the measured contraction were undetermined.

Short-term negative creep also was observed in tests on quenched and tempered 2.25Cr-1Mo steel at 482 °C (900 °F) and at 482 and 538 °C (900 and 1000 °F) for the same steel in the normalized and tempered condition. Two steady-state creep stages for annealed 2.25Cr-1Mo steel have been reported (Ref 23), which were due to the interaction of molybdenum and carbon atoms

Fig. 2 Relationship of strain rate, or creep rate, and time during a constant-load creep test

The minimum creep rate is attained during second-stage creep.

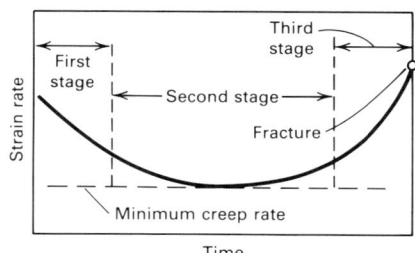

with dislocations and the subsequent decrease in the number of these atoms as Mo_2C precipitated. A volume decrease associated with the precipitation process also could account for the observed creep curve trends.

Interstitial diffusion of carbon and hydrogen into dislocations has been observed, and alloy strain-aging effects have been found to cause creep rate transitions noted for carbon steels and normalized 0.5% Mo steel (Ref 24). Negative creep in Nimonic 80A appears related to an ordering reaction in the Ni-Cr matrix and possible formation of Ni_3Cr (Ref 25).

Oxide and Nitride Strengthening. An entirely different source of variation from classical patterns occurs in creep tests at high temperature due to reaction with the air that forms the environment. Tests longer than 50 h with 80Ni-20Cr alloys at 816 and 982 °C

Fig. 3 Creep curve of 2.25Cr-1Mo steel with nonclassical early stage

Normalized and tempered to 607 MPa (88 ksi) tensile strength at room temperature. Tested at 482 °C (900 °F) at 275.8 MPa (40 ksi).

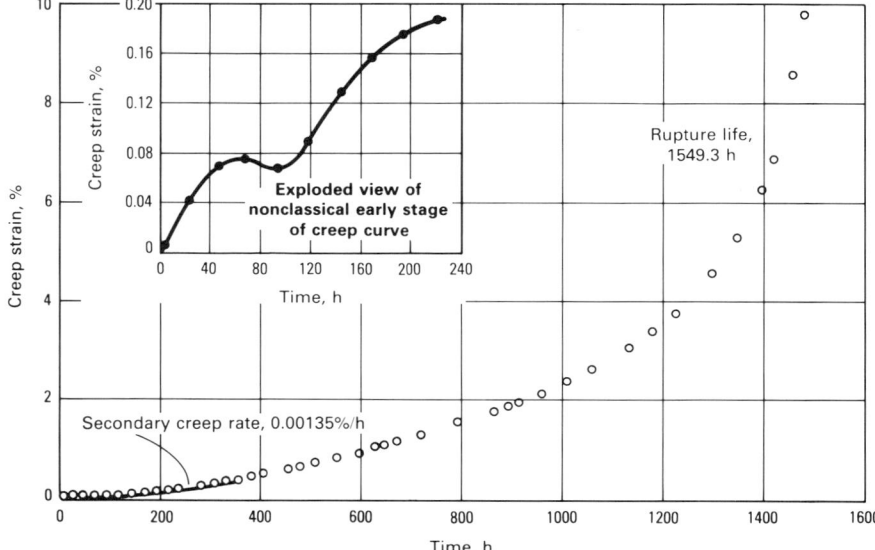

(1500 and 1800 °F) showed a deceleration of creep after the normal tertiary stage was reached, resulting in a second period of steady-state creep and later another period of last-stage creep (Ref 26). Figure 4 illustrates a creep curve showing the effect of oxide strengthening.

This behavior, which prolonged rupture life and caused a slope decrease in curves of log stress versus log rupture life, was due to oxide and nitride formation on surfaces of the intercrystalline cracks that form extensively during tertiary creep. Observed interconnection of the bulk of these cracks added substantially to strengthening against creep deformation in the late stages of the tests.

This effect also was observed in 99.8% Ni tested at 816 °C (1500 °F) under 20.7 MPa (3000 psi) stress (Ref 27). Fracture after a prolonged time occurred in a lower stressed section of the fillet. Fewer intergranular cracks in this region resulted in less oxidation strengthening than in the gage section.

Extrapolation and Interpolation Procedures

The determination of creep-rupture behavior under the conditions of intended service requires extrapolation and/or interpolation of raw data. No single method for determination of properties exists; however, a variety of techniques have evolved for data handling of most materials and applications of engineering interest. These techniques include graphical methods, time-temperature parameters, and methods used for estimations when data are sparse or hard to obtain.

Graphical Methods

Test results frequently are displayed as plots of log stress versus log rupture time and log stress versus log secondary creep rate, with a separate curve (isotherm) for each test temperature. For limited ranges of test variables, test points frequently fall in a straight line for each temperature. Nonlinearity of isotherms with broader ranges of test parameters has been treated variously, but common practice is to represent the data by two or more intersecting straight line segments. Figure 5 illustrates such treatment for an aluminum alloy.

Isotherms for lower temperatures characteristically display a flatter slope than those for higher temperatures. At a given temperature, when the test stresses drop below a given level that varies with alloy composition and metallurgical condition, the slope of the isotherm usually steepens. This steeper slope often approximates the slope for early times at the next higher test temperature.

Early investigators of engineering creep behavior introduced a "conservative" prac-

Fig. 4 Creep curves of alloy 2V tested at 980 °C (1800 °F) and 17.2 MPa (2500 psi)

(a) Tests in argon and air for same duration. (b) Entire curve of specimen tested in air is shown. ■ on both graphs represents same point.

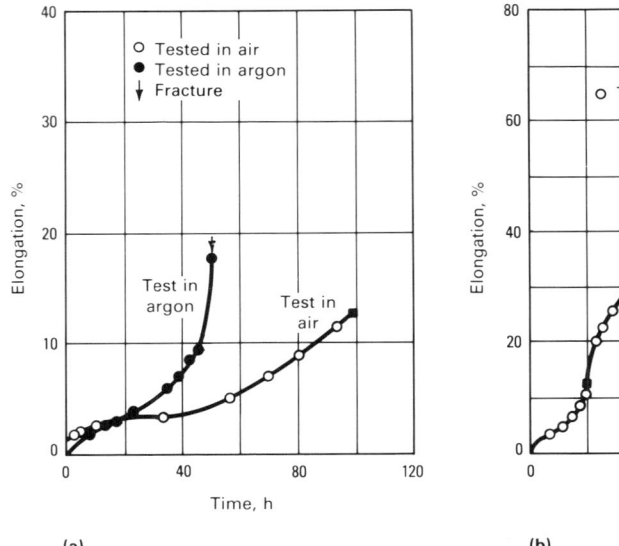

(a)   (b)

Fig. 5 Logarithmic plot of stress versus rupture life for aluminum alloy 6061-T651

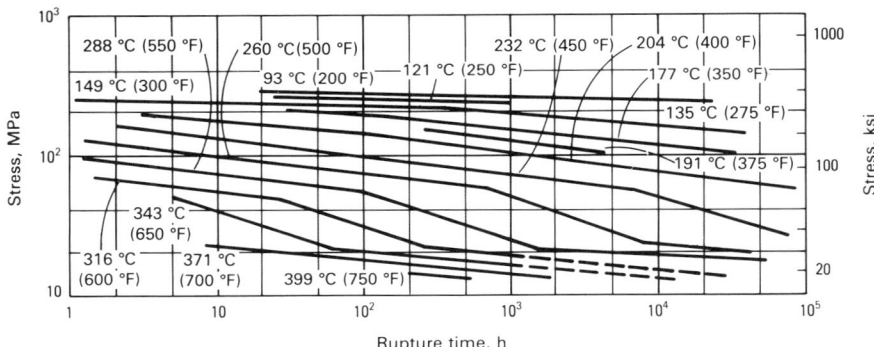

tice of using the slope from the next higher temperature when an isotherm had to be extended to longer times. Use of this method is limited to the specific temperatures of the test runs. Even under these conditions, extrapolations should be only in the direction of longer times for the lower range of test temperatures.

Because the change in slope of log stress versus log time isotherms historically appeared to be associated with a gradual change in fracture mode from transgranular at lower temperatures and higher stress to intergranular at relatively high temperatures and low stress, the belief developed that once the slope of the longer time portion became established, further slope change would not occur. Experimental data available at that time provided no indication that these linear

plots could not be extrapolated to long times with confidence. Subsequent long-time data demonstrate that such extrapolations may lead to erroneous results.

Upward Inflection of Log-Log Rupture Plots at Long Times. Review of 52 heats from 31 wrought and cast steels, each with test times longer than 50 000 h, indicated that some portion of the log stress versus log rupture time curves for all ferritic steels showed an increase in slope when tests were of sufficiently long durations (Ref 28). This upward inflection was pronounced depending on composition, heat treatment, and particularly test temperature.

A sharp inflection at one temperature (e.g., 500 °C, or 930 °F) was usually accompanied by a less distinct inflection covering a broader time range at a higher test temperature (e.g., 550 °C, or 1022 °F). Generally, these inflections shifted to shorter times and lower stresses with increasing test temperature. Existence of inflections appeared to be related to precipitation phenomena.

For the heat-treatable aluminum alloy 6061-T651, test stresses between about 20 and 50 MPa (2900 and 7250 psi) for temperatures ranging from 260 to 343 °C (500 to 650 °F) exhibited nearly the same slope on a plot of log stress versus log rupture time, which was steeper than for either higher or lower stress levels (Ref 29). The long-time rupture results obtained had been predicted (Ref 30) by separate graphical extrapolation of each of three regimes of rather constant slope (see Fig. 5).

In this instance, the curves that were actually extended were lines for fixed stress levels (isostress lines) on plots of log rupture time versus temperature, or the reciprocal of absolute temperature. However, extrapolation could have been carried out on the usual log stress versus log rupture time plot by treating the data as a family of curves, with different portions of each curve falling into different slope regimes. Direct graphical extension of isostress lines appeared to provide better extrapolation of rupture data than other common methods (Ref 31).

Curves of log stress versus log rupture life for two chromium-molybdenum steels (ASTM A387, grades 22 and 11) typically show an increase followed by a decrease in steepness for tests at 538 to 566 °C (1000 to 1050 °F). Consequently, correct prediction of 100 000-h strengths requires that these changes in slope be incorporated into the analysis (Ref 32). This requirement applies to all evaluation methods. Unless the input data include results that encompass structural changes of the type expected under intended service conditions, accurate extrapolation cannot be expected.

Some metallurgists prefer a semilogarithmic plot of stress versus log rupture time. The sigmoidal shape of isotherms is thus more evident, but extrapolation difficulties remain. The double inflections (or sigmoidal shape) for rupture curves can be greatly accentuated when notched specimens are tested. In the intermediate stress regime, rupture life can actually decrease as the level of test stress is lowered.

Time-Temperature Parameters

Temperatures that are higher than those encountered in service have traditionally been used to shorten the time required to obtain creep-rupture results. One such approach incorporates time and temperature into an expression or parameter, such that a single master curve of stress or log stress can

represent all data obtained for a given lot of material over a wide range of test conditions.

When the parameter calculated for a desired service time and temperature falls within the range of the master curve, the corresponding stress can be read directly from that curve. More than 30 parameters have been proposed; although not always developed that way, several can be derived from the following:

$$P = \frac{\dfrac{\log t}{\sigma^Q} - \log t_A}{(T - T_A)^R} \qquad \text{(Eq 1)}$$

where t is rupture time in hours; σ is stress in psi; T is test temperature in °F; and T_A, $\log t_A$, Q, and R are constants determined from the experimental test data. Figure 6 illustrates geometric requirements for lines of constant stress for several parameter models on a plot of logarithm of time versus either temperature or its reciprocal. Of these, the Larson-Miller and Manson-Haferd parameters represent early developments in time-temperature parameters that retain considerable application.

The Larson-Miller Parameter. In 1953, Larson and Miller introduced the concept of time-temperature parameters to correlate and extrapolate creep-rupture data (Ref 33). For the Larson-Miller parameter, constant stress lines on a graph of log t versus $1/T_A$ converge to a point at $1/T_A = 0$ (Fig. 6a). At that point, $\log t = C$ defines the optimum value of C for the data involved.

In Fig. 7(a), an actual set of stress-rupture data (log stress versus log time to rupture) is shown for the nickel-based alloy Inconel 718. The data are then replotted in the form shown in Fig. 6(a) for the Larson-Miller parameter. This is accomplished by using constant-time intercepts, shown in Fig. 7(b) by the dashed lines, to arrive at the constant-time curves shown in Fig. 7(b).

Constant-stress curves are then plotted as shown in Fig. 7(c), using the intercepts shown by the dashed lines in Fig. 7(b). By extending the data in Fig. 7(c), a plausible set of convergent isostress lines meeting on the ordinate at a value of log $t = -25$ can be obtained. As illustrated in Fig. 6(a), the Larson-Miller equation for this set of data is:

$$P = f(\sigma) = T_A (\log t_R + 25) \qquad \text{(Eq 2)}$$

where T is temperature in °R, and t_R is time to rupture in hours.

For each data point in the original set of stress, time, and temperature data, the proper value can be substituted in Eq 2 and plotted as shown in Fig. 7(d). This is the compact

parameter form of graphical representation known as the "master curve." A common practice when input data are limited is to assume that $C = 20$, which has been found to be reasonably true for many materials.

Other Parameters. The Manson-Haferd parameter predicts that a constant stress plot of log t versus T yields a family of straight lines converging to a point $t(T_A, t_A)$, which defines the optimum constants for that particular data set (Fig. 6b). On these same coordinates, the Manson-Succop parameter requires that isostress lines be straight and parallel (Fig. 6d). These conflicting patterns and still different patterns for the additional parameters cannot occur simultaneously over the entire range of data from a given set.

A frequent finding is that different parameters provide best fit to different portions of the same data. For example, using data obtained from tests on a 1Cr-1Mo-0.25V steel, the Larson-Miller parameter gave the best extrapolation at 482 °C (900 °F), and the Manson-Haferd parameter was preferable at 538 °F (1000 °F). However, the Orr-Sherby-Dorn parameter gave the best fit at 593 °C (1100 °F) (Ref 34, 35).

Although numerous studies have considered the relative merits of these and other proposed parameters, no one parameter has emerged as universally superior to all others. Five representative parameters were compared in terms of correlating and extrapolating extensive sets of data on the creep and rupture properties of seven steels and superalloys (Cr-Mo, Cr-Mo-V, 12%Cr, A-286, Astroloy, René 41, and Inconel 718) (Ref 36). The difference in fit among parameter methods was found to be relatively small and inconsistent from one alloy to another. The largest source of variation in the fitted values for time to rupture, time to 1% creep, and minimum creep rate was the difference between alloys, regardless of the parameter used.

Results were marginal to poor when extrapolation was beyond the range of the fitted master curve. When prediction of long-time data was confined to the master curve derived using only short-time data, no one parametric method gave consistently superior results.

For critical evaluation of the comparative ability to predict known long-time rupture lives (11 000 to 64 000 h) for the ferritic steels, data up to 10 000 h were applied to establish fit to the parametric model. For the superalloys, only data up to 1000 h were used to predict known rupture times between 1200 and 33 000 h. Table 1 lists ranges for the ratio (predicted life/actual life) extracted from Ref 36 for 46 extrapolations, including one in which slight extension was required beyond the fit of each master curve.

Fig. 6 Schematic representation of several parameter models

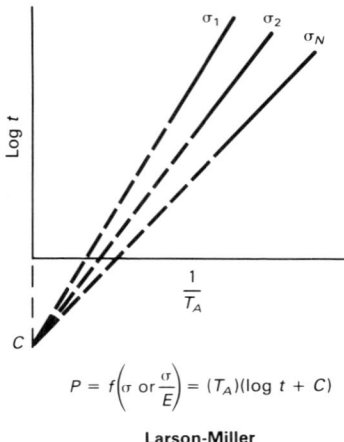

$$P = f\left(\sigma \text{ or } \frac{\sigma}{E}\right) = (T_A)(\log t + C)$$

Larson-Miller

(a)

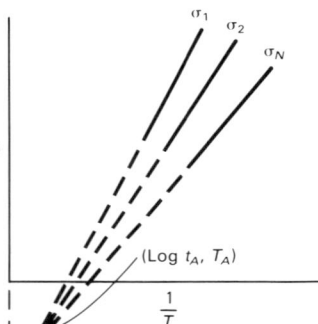

$$P = f\left(\sigma \text{ or } \frac{\sigma}{E}\right) = \frac{\log t - \log t_A}{1/T - 1/T_A}$$

Goldhoff-Sherby

(c)

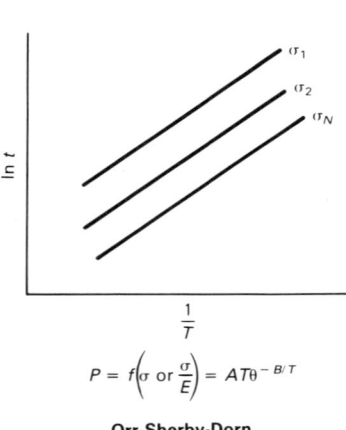

$$P = f\left(\sigma \text{ or } \frac{\sigma}{E}\right) = AT\theta^{-B/T}$$

Orr-Sherby-Dorn

(e)

$$P = f\left(\sigma \text{ or } \frac{\sigma}{E}\right) = \frac{\log t - \log t_A}{T - T_A}$$

Manson-Haferd

(b)

$$P = f\left(\sigma \text{ or } \frac{\sigma}{E}\right) = \log t - BT$$

Manson-Succop

(d)

Studies on 0.5Cr-0.5Mo-0.25V steel pipe (Ref 37) found the Manson-Haferd parameter significantly superior to predict known rupture times (8712 to 20 664 h) from data points of less than 6000 h duration than either the Larson-Miller or Orr-Sherby-Dorn parameters. The latter two parameters generally provide optimistic prediction of behavior. According to Ref 37, very short-time data should be eliminated from the analysis if predictions beyond 10 000 h are desired, because their inclusion distorts the correlation at long times. As with graphical methods, the accuracy of parametric extrapolations is related to the interval of test variables on which the prediction is based.

Minimum Commitment Method. Experimental data may deviate from the re-

quirement imposed by the form of each parameter for linearity of isostress curves or for parallelism or convergence of families of such curves. The minimum commitment method starts with a time-temperature-stress relationship sufficiently general to include all commonly used parameters. The pattern of the data is not forced in advance; instead, the actual experimental data naturally lead to the most appropriate functional relationship for the particular material.

Manson applied a "station-function" approach to $f(\log t) + p(T) = g(\sigma)$; all parametric formulations can reduce to this equation. Each of the functions f, p, and g were represented by their discrete numerical magnitudes at specific values of the corresponding independent variable. Figure 8 illustrates treatment of Astroloy data given in Ref 36.

Temperature stations were arbitrarily chosen at $T = 760$, 816, 871, 927, and 982 °C (1400, 1500, 1600, 1700, and 1800 °F); the values of the p functions at these respective temperatures are designated p_1, p_2, p_3, p_4, and p_5. For times such that $\log t = 1.0$, 1.25, 1.5, ..., 3.0, the respective corresponding f values are $f_1, f_2, f_3, \ldots, f_9$. The g values are designated $g_1, g_2, g_3, \ldots, g_{11}$ for levels of $\log \sigma = 1.0$, 1.1, 1.2, ..., 2.0, with σ given in ksi.

Consider the experimental point A ($T = 927$ °C, or 1700 °F); $\log t = 2.873$, $\log \sigma = 1.322$). At this point, p is directly p_4, but values of f and g must be interpolated. Higher order interpolation can be easily accomplished, but simple linear interpolation was chosen for this illustration. For A between $\log t$ of $f_8 = 2.75$ and $f_9 = 3.00$:

$$f(2.873) = \frac{3.0 - 2.873}{3.0 - 2.75} f_8$$

$$+ \frac{2.873 - 2.75}{3.0 - 2.75} f_9$$

$$= 0.508 f_8 + 0.492 f_9 \qquad \text{(Eq 3)}$$

In a similar manner, the relation for stress at this point becomes:

$$g(1.322) = \frac{1.4 - 1.322}{1.4 - 1.3} g_4 + \frac{1.322 - 1.3}{1.4 - 1.3} g_5$$

$$= 0.78 g_4 + 0.22 g_5 \qquad \text{(Eq 4)}$$

Introducing these results into the original general equation yields:

$$0.508 f_8 + 0.492 f_9 + p_4$$

$$= 0.78 g_4 + 0.22 g_5 \qquad \text{(Eq 5)}$$

A similar equation can be written for each experimental point. By choosing a sufficient

Fig. 7 Method of creating master Larson-Miller curve for Inconel 718 from experimental stress-rupture curves
See text for discussion of (a) through (d).

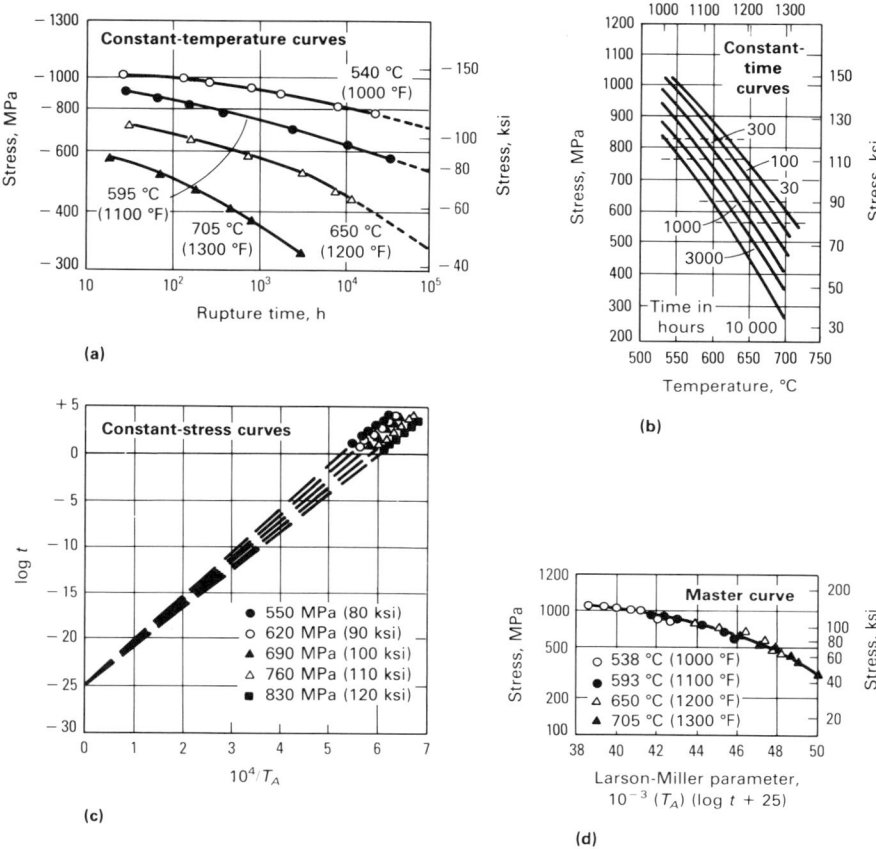

(a)

(b)

(c)

(d)

Table 1 Comparative extrapolation abilities

Parameter or method	Ratio: predicted life/actual life		
	Minimum	Average	Maximum
Larson-Miller	0.34	1.57	5.64
Manson-Haferd	0.44	1.51	6.30
Goldhoff-Sherby	0.36	1.64	8.85
Manson-Succop	0.39	1.53	4.96
Orr-Sherby-Dorn	0.11	1.09	4.01
Monkman-Grant	0.33	0.93	1.82

required to fit the data. On a plot of log stress versus log time, all isothermals converge at the extrapolated point log $t = -(1/A)$ (Ref 40). For the parallel isotherms of the Orr-Sherby-Dorn parameter, $A = 0$; Miller-Larson requires $A = +0.5$. With data fitting the Manson-Haferd parameter, negative A values result; the Astroloy data discussed above converge at about $t = 6.36$, resulting in $A = -0.157$.

Establishing an accurate individualized value of A for a specific material is difficult from a short-life data base. One acceptable approach is to use universalized values of $A = 0$ for aluminum alloys and pure metals, and $A = -0.05$ for most steels and superalloys (Ref 40). Highly unstable materials like Astroloy, as well as carbon steels, require higher negative values of A; for these materials, $A = -0.15$ was found to be adequate in most instances.

The minimum commitment method was initially developed for use with single-heat data. Application to multiple-heat analyses is discussed in Ref 41, including information on suitable computer programs.

Estimation of Required Properties Based on Insufficient Data

Complete independent evaluation of creep-rupture properties for a new lot of material, whether by graphical, parametric, or minimum commitment methods, requires numerous test points covering an extensive range of test variables. Frequently, the amount of data available is too limited for full treatment by usual procedures. Experimental difficulties often limit obtaining accurate test results at conditions of interest, such as evaluation of creep-rupture properties near the low end of the temperature range in which time-dependent effects are significant.

Tests of short or moderate duration at these temperatures frequently require use of such high stress levels that the immediate high plastic strains at load application alter the nature of the material from that which exists during service under lower stresses. Testing at or near a stress of intended application often requires more time and/or expense than is feasible before the material is to be put into use. Approximate methods permit such difficulties to be treated in a generally satisfactory manner. Established correlations also permit estimation of some unmeasured properties from other types of available results.

The Monkman-Grant Relationship. Analysis of data for a variety of aluminum-,

number of stations, the number of equations available will exceed the number of unknowns, so that a least squares solution can be used to determine the unknowns. Once the equations have been solved, the function $f(\log t)$ may be extrapolated using graphical, polynomial, or recurrence relations.

Although a better fit to the data should result by the minimum commitment method than by forcing fit to an arbitrary parameter, extrapolations can still be imprecise. As with other methods, the degree of accuracy of predictions relies on having accurate and representative raw data from tests that reflect any structural changes expected to occur in the regime of the extrapolated conditions.

A specialized form of the general equation developed in 1971 (Ref 39) contains a characteristic material constant A:

$$\log t + AP(T) \log t + P(T)$$
$$= G (\log \sigma) \qquad (Eq 6)$$

The parameter A is a measure of structural stability, because the more unstable the material, the higher the negative value of A

Fig. 8 Application of station function approach to Astroloy data

(a) Station-function representation of $p(T)$, $g(\sigma)$, and $f(\log t)$ at specific values of T, σ, and $\log t$. (b) Net point selections for solution. Source: Ref 38

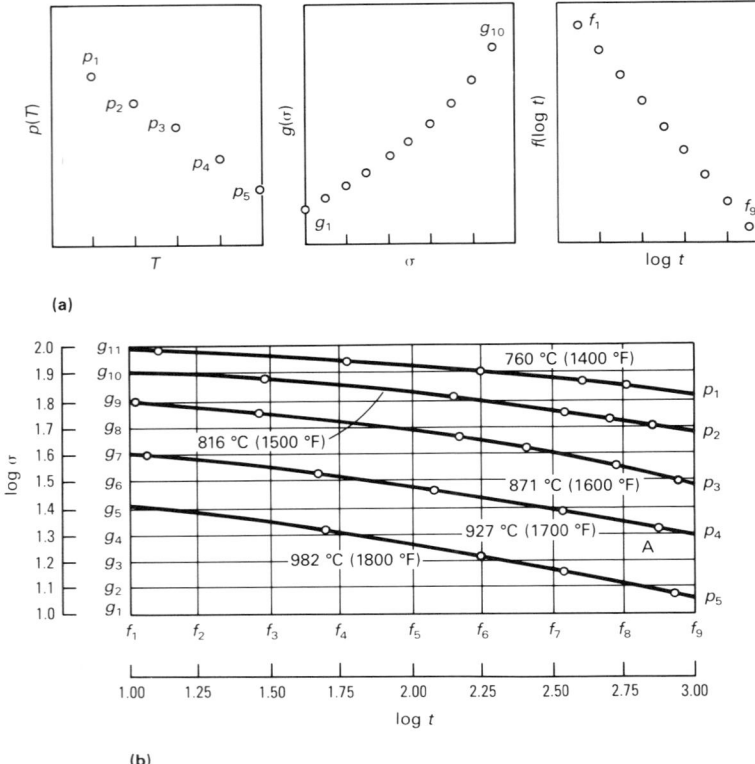

(a)

(b)

Reduced scatter was noted (Ref 45) for eight nonferrous alloys and two superalloys when the term $\log t_r$ in Eq 7 was replaced by $\log (t_r/\epsilon_c)$, where ϵ_c is the total creep deformation at fracture. This trend was confirmed by Ref 46 in tests on a 2.25Cr-1Mo steel.

Although deformation-modified rupture time may improve correlation in some instances, other cases exist where use of the original relationship is sufficient or better. Data for 17 test points for 4% cold worked type 304 stainless steel (Ref 47) exhibited a spread in creep elongation from 1.5 to 24%. Goodness of fit was identical (coefficient of determination $r_2 = 0.86$) for linear regression of the data treated by the original versus the modified log-log relationships.

Extrapolation is fast and direct when using the Monkman-Grant coordinates, but with the modified relationship, creep elongation at the given temperature and corresponding to the rupture time sought must first be estimated. This usually requires subjective extrapolation of only a few elongation values displaying wide scatter and with no evident single trend. Introduction of a creep elongation factor may have value when only correlation or interpolation of test results is desired, but it is not recommended for extrapolation.

One occasional problem in estimation of rupture life from creep data is uncertainty whether secondary creep has truly been established. Changes in creep rate with continuing test time are often sufficiently gradual and so close to the sensitivity of measurement that what appears to be a steady-rate condition may in fact still be a late portion of primary creep. Reference 48 illustrates successive apparent minimum creep rates of 2.05, 1.7, and 1.40%/10 000 h for respective test durations of 1000, 2000, and 5000 h.

A distinctive slope change in a plot of log creep rate versus time or log time often provides better assurance that the secondary creep period has been entered than study of the deformation-time curve itself. Although an equation expressing true strain in terms of elapsed time, secondary creep rate, and three constants deviates markedly from actual behavior during the early portion of primary creep, a statistical analysis such as that detailed in Ref 49 may predict acceptable values of secondary creep rate from transient data.

For type 316 stainless steel tested at 704 to 830 °C (1300 to 1525 °F), the initial transient rate at $t = 0$ was found to be almost equal to 3.3 times the secondary creep rate in the same test (Ref 50). A significantly different magnitude (near 1000) for this ratio of initial and secondary creep rates was found in Ref 51 for a high-temperature alloy. A simple

iron-, nickel-, titanium-, cobalt-, and copper-based alloys led Monkman and Grant to the following empirical relationship (Ref 42):

$$\log t_r + m \log (mcr) = C \qquad (Eq\ 7)$$

where t_r is time to rupture; mcr is minimum creep rate; and m and C are constants that differ significantly among alloy groups, but exhibit nearly fixed values for a given heat of material, or for different lots within the same alloy group.

Equation 7 enables assessment of the reliability of each individual test by examining its fit within the scatter band for all tests. Once a minimum creep rate has been determined in a long test, rupture life can be estimated without running the test to failure. Although Monkman and Grant stated that this relationship was not intended for extrapolation, it can be used for that purpose, particularly when only low-stress tests are acceptable to prevent large initial plastic strains.

Table 1 includes the results obtained when the Monkman-Grant relationship is applied to data obtained on seven materials (Ref 36). The prediction of rupture life for these 46

extrapolations using this technique was more accurate than that provided by any of the five time-temperature parameters.

For additional materials (Ref 43) where good fit is obtained to a single linear plot on the coordinates of log time versus log secondary creep rate, extrapolation of a known secondary creep rate to the corresponding rupture life appears to be as good or better than by other extrapolation methods. One advantage of Eq 7 is that it can be applied successfully to as few as five or six data points, in contrast to the approximately 30 tests needed to establish the entire Manson-Haferd master curve (Ref 44). For the minimum commitment method, even more data points are usually required.

One advantage of this correlation, particularly with materials that exhibit structural instability under testing, is that the specimens used to determine the input data for secondary creep rates experience the same history of structural change that exists during the corresponding period of a test carried to rupture. Best predictions result by concentrating on tests encompassing a limited range of stresses and temperatures, thus discounting results obtained at considerably higher combined temperatures and stresses.

proportionality of this type and the more general analysis cited above are tempting alternatives for shortened test durations, but both suffer from the need for creep measurements that are more precise than those commonly obtained. Currently, neither method is capable of replacing long-time testing.

The Gill-Goldhoff Method. Many designs for elevated-temperature service require that deformation not exceed some maximum value; in these cases, creep strain rather than rupture life becomes the focus. Although published compilations and computer banks of data include rupture properties for most materials of engineering interest, corresponding information on the time-dependency of strain is frequently sparse or nonexistent. Many early studies did not include strain measurement during rupture tests. When creep data were obtained, accuracy was sometimes questionable due to inadequate control of temperature or low precision of strain measurements. Typically, the only listed creep data are minimum creep rates. Most of these results were obtained from tests that were terminated after a few thousand hours, or even less, and true secondary creep rate may not have been established.

Studies by the Metal Properties Council and similar groups attempt to report both the total strain on loading and the times to various levels of creep strain. Until such results are more universally available, estimates may still be required of the creep strain to be expected in given design situations.

Gill and Goldhoff (Ref 48, 52) found a log-log correlation between stress to cause rupture and stress for a given creep strain for the same time and temperature. Figure 9 shows their composite plot for aluminum-based alloys and stainless steels, including several superalloys.

To obtain this correlation, tests in which 0.1% creep occurred in less than 100 h were rejected to prevent intolerable data scatter. Despite this, the "universal" curves of Fig. 9 can be associated with fairly wide data scatter, particularly at low creep strains. Some deviations from the correlation were related to microstructural instabilities, which produce differing proportions of primary, secondary, and tertiary creep between alloys and for varying test conditions.

Despite occasional anomalies, the Gill-Goldhoff correlation meets some preliminary design needs, particularly if the technique is tailored to grades of alloys similar to those of immediate concern. In principle, this technique can also be used to predict rupture properties from early creep measurements from tests that are not continued to rupture. This use is limited by the short rupture times that are derived from tests terminated at

Fig. 9 Composite graph for the Gill-Goldhoff correlation
Source: Ref 52

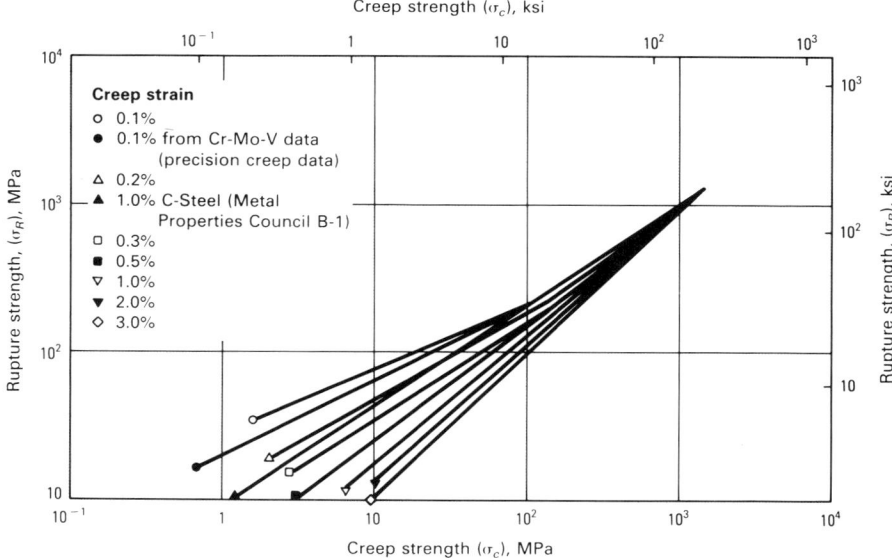

creep strains of 1% or less. If these tests were continued to higher levels of creep, improved predictions of rupture could be obtained by determining the secondary creep rate and then applying the Monkman-Grant relationship.

Treatment of Isolated Test Points. Particularly at the start of a testing program, the need may arise to extract information from a single available test. The form of most parameters limits their use to situations in which multiple test results are available. The Larson-Miller parameter is an exception if the generalized constant $C = 20$ is used.

For the stress of the test, longer rupture times within a factor of ten from the test duration frequently can be estimated satisfactorily for temperatures below that of the test. If a master curve or graph of isothermals is available for another lot of like or similar alloy, a parallel curve passed through the coordinates of the test point for the new lot can serve as an approximate representation of expected behavior for limited ranges of variables from the test conditions.

Evaluating Creep Damage and Remaining Service Life

Specimens from service occasionally are tested to compare residual creep and rupture properties with the same or similar material that has not been used in service. Diverse evaluations may result, depending on the conditions selected for the tests and on the criteria used to define damage. As shown below, a false prediction of drastic drop in

rupture strength can result if tests are of rather short duration and if a simple direct ratio is taken of rupture life after service versus before service for similar test conditions.

Discrete Changes in Test Temperature or Load. When a creep-rupture test is interrupted by cooling and reheating at a moderate rate at constant load, and if the time under changing temperature is brief compared to the original test duration, the effect of the interruption on either the creep curve or rupture life usually cannot be readily detected, unless thermal gradients cause gross plastic deformation or spalling of surface layers.

Similar results can be expected when temperature and stress rise and fall in unison, as during start-up and shutdown of a steam boiler. For the alternate situation in which unloading occurs while the creep temperature is maintained, significant recovery of primary creep can ensue; reapplying the load results in a period of primary creep.

Under step-wise alteration of load, temperature, or both during a test or service, performance frequently follows the "life-fraction rule" or "linear cumulative damage rule" (Ref 53), in which the percentage of total life consumed for each period of fixed temperature and stress is represented as:

$$\% \text{ total life} = \frac{\text{actual time at the given conditions}}{\text{rupture life at those conditions without alteration}} \times 100$$

Accuracy of this rule ranges from excellent to rather poor, with best results for multiple small excursions (Ref 54). Although solid-state reactions, which can reverse at different exposure temperatures, may introduce complications under some conditions, investigators have found the life-fraction rule more appropriate for steels under temperature changes than under stress changes (Ref 55).

Life-fraction summations at failure as low as 0.36 and as high as 2.43 have been reported (Ref 56). The spread was only from 0.75 to 1.50 for the same tests using damage fractions defined by $K(t/t_r) + (1 - K)(\epsilon/\epsilon_r)$, where t and ϵ are the time and strain under a period of fixed conditions, for which the rupture time and fracture strain are t_r and ϵ_r. K is a material constant ranging from zero to 1; the zero limit applies to materials that develop cracks early in life, and K approaches 1 for materials that exhibit no cracking until rupture is imminent. Typical values of K (Ref 56) are as follows:

Material and test temperature	K
Copper at 250 °C (482 °F)	0.3
A-286 alloy at 649 °C (1200 °F), solution treated at 1204 °C (2200 °F)	0.47
A-286 alloy at 649 °C (1200 °F), solution treated at 982 °C (1800 °F)	0.43
Inconel X-550 at 732 °C (1350 °F)	0.625

When data are insufficient for determination of K, acceptable results frequently can be obtained with an empirical rule (Ref 57), by which the life-fractions added are defined by $\sqrt{(t/t_r)(\epsilon/\epsilon_r)}$. With any of these cumulative damage rules, comparison usually is against rupture life from constant-load tests—i.e., with actual stress rising as creep reduces the cross section. When the same specimen undergoes load changes for different periods, respective stress levels have been based on the initial cross section. This corresponds to using the same original load if an interrupted test must be restarted. The actual stress at the time of test restart is $(\sigma_n)(A_o/A_c)$, where σ_n is the present nominal stress, A_o is the initial specimen cross section, and A_c is the specimen cross section after creep deformation up to the time of the test interruption.

When the specimen for a test in a later portion of creep has been machined from a part that has already undergone considerable reduction in cross section by prior creep, a corresponding area-modified stress must be employed for consistent interpretation of the results (Ref 58). The load applied to produce the desired nominal stress σ_n related to the virgin material is calculated to make the actual stress $\sigma_a = \sigma_n(A_o/A_c)$, where the latter term relates cross-sectional areas of the orig-

inal part before and after creep. Use of such an area-modified stress has been reported to improve prediction of remaining life from post-service rupture tests (Ref 58, 59).

An approximation of remaining rupture life for components that have undergone prolonged service can be calculated by introducing best estimates for operating temperatures and stresses into the above damage rules. More exact evaluation, however, can be obtained by testing representative samples removed from service.

Measurement of Rupture Properties After Service. Direct measurement of remaining life of a part at conditions near the service stress and temperature generally requires impractically lengthy testing times. Studies on material after creep service have involved increased temperature, stress, or both, with subsequent extrapolation of results to nominal service conditions.

Assessments made in this manner generally conclude that carbon, carbon-molybdenum, and chromium-molybdenum steels operated at or below allowable stresses (recommended by the *ASME Boiler and Pressure Vessel Code*) experienced negligible creep damage from service exposures up to 200 000 h (Ref 60). Significantly different, and presumably erroneous, conclusions can result from cursory studies or from poor selection of post-service tests.

Operating conditions for 1Cr-0.5Mo steel tubing removed from a superheater after 33 000 h of service (Ref 59) were sufficiently well known to permit true residual life to be determined by continuing samples to rupture under the same stress and temperature (66.26 MPa/557 °C, or 9610 psi/1035 °F). These tests indicated service to have accounted for 78% of the total life of the tubes.

In that study a series of accelerated tests at 66.26 MPa (9610 psi) stress, but with test temperatures ranging from 561 to 610 °C (1042 to 1130 °F), yielded service lives ranging from 6535 to 432 h. A plot of these results on coordinates of test temperature versus log test life was linear and extrapolated rather well to the total life established by the samples continued to rupture at service stress and temperature. The simple life-fraction rule yielded values ranging from 0.93 to 1.05 for the sum of test and service fractions when area-modified stresses were used and 0.96 to 1.11 without the area correction.

Acceleration of tests by using stress levels ranging from 77.15 to 141.75 MPa (11.2 to 20.6 ksi) resulted in test durations between 2057 and 61 h at 557 °C (1035 °F). Departure from linearity of the plot of test stress versus log test life as the service stress was approached made extrapolation to expected total life at service conditions difficult. Devi-

ation trends between mean ISO data and properties of this tubing before service may bias the life-fraction comparisons for increased temperature versus increased stress tests after service. However, the sum of test and service fractions were found to be only 0.83 for the highest test stress and were found to increase only marginally to 0.89 for the lowest accelerated stress (77.85 MPa, or 11.3 ksi) (Ref 60).

Life assessment based on extrapolation of temperature at the service stress is generally preferred to life assessment based on extrapolation of stress at the service temperature (Ref 61). However, many apparently successful evaluations of materials after creep service (Ref 62, 63) have used isothermal tests similar to those used to establish the original allowable stresses of the *ASME Boiler and Pressure Vessel Code* for creep temperatures.

A common finding for steels used to construct boilers and heated vessels has been that materials taken from service show lower rupture strengths than the same original material or other typical new material made to the same specifications. However, the slope of the log stress versus log rupture time curve is flatter than for virgin samples, and each isotherm frequently approximates a single straight line. Extrapolation of that line characteristically predicts negligible change from original long-time strength values for specimens whose actual service stress was low enough to preclude extensive creep damage after prolonged operation.

Subjecting such essentially undamaged service specimens to a heat treatment similar to that originally given the component restores short-time rupture curves to approximately the level and form reported for the steel before service. Note that sigmoidal isotherms have been reported for these steels, with a final lower slope linear portion after time-dependent structural alterations produced a slope increase encompassing tests with intermediate durations to rupture (Ref 28).

Dependable application of these techniques to estimate remaining life requires that the loading direction for the final test corresponds to the largest principal stress of service and that the specimen is representative of surface deterioration or other damage present in the part. Possible temperature and loading gradients in service must be kept in mind when selecting a sampling site and when applying test results to predictions of further serviceability. Despite these possible additional variables, published assessments of post-service rupture properties require only about the same number and duration of tests as conventional evaluation of any new lot of familiar material.

Creep-Rate Measurements After Service. An alternative testing approach has been to perform a creep test on the specimen from service and to compare the measured rate against creep for the same or like material without service. An ideal test would be conducted at the operating temperature and stress, but more extreme test conditions are common. Although simple in principle, this test can entail several difficulties.

If the specimen of previously crept material is heated to the test temperature before stressing, recovery effects are likely to produce a higher creep rate at the start of the test than the rate that existed when service was interrupted. Even if the test is brought to temperature with the load applied, an imprecise match of test and service stresses commonly leads to early nonsteady creep rates, even when a steady condition was previously established in service.

Reference creep rates, whether from tests on identical new material or not, usually are limited to reported extrapolated stresses for 0.0001%/h or 0.00001%/h secondary creep rate at several temperatures. To fix the creep rate to be expected for virgin material at the service stress and temperature requires nonlinear interpolation of a few tabulated data points.

The considerable scatter typical in creep measurement poses a particular problem, because under the specified stresses of most elevated-temperature designs, creep deformation obtained in a reasonable time can approach the sensitivity limit of most extensometer systems. A large increase in that sensitivity limit is impractical, because movements caused by thermal expansion due to unavoidable temperature fluctuations necessitate averaging of the resulting unsteady strain indications.

Even determination of test duration can be an important test decision. If the component in service was well within an extended period of secondary creep, a test conducted for about a thousand hours at near-service conditions should display a creep rate that is close to the latest period of service during the entire test. However, if secondary creep is not distinct and prolonged for the alloy under study, or if service exposure has reached a stage significantly earlier or later than the minimum rate, extended testing deviates increasingly from the last rate of service. Greatest concern arises when post-service test conditions are made more severe than those of service, either to shorten the study or to yield rates that are more convenient to measure.

Procedure modifications minimize these difficulties, but their use has not been documented. The "ideal" test does not require a virgin specimen or knowledge of its initial properties; it can be used with a single representative specimen and permits evaluation in no longer than a few months at conditions reasonably near those of service.

Assuming that prior operation typifies specified design, so that rupture life at service temperature T_s is 100 000 h or more, a suitable test procedure may be:

1. Load a cold post-service specimen to the estimated service stress (or, if this is unknown, to the applicable specified allowable stress).
2. Increase the specimen temperature to T_s + 50 °C (T_s + 90 °F) within 3 h. Immediately on reaching that temperature, gradually reduce furnace power to obtain a uniform specimen temperature of T_s + 20 °C (T_s + 36 °F) at the end of one additional hour.
3. Take creep measurements until a rate is established with a precision of two significant digits.
4. Increase the specimen temperature to T_s + 50 °C (T_s + 90 °F) (with the load maintained) and hold until about 0.5% additional creep strain occurs at that temperature.
5. Reduce the temperature to T_s + 20 °C (T_s + 36 °F) as in Step 2 and obtain a second creep rate as in Step 3.

This procedure minimizes anomalous recovery effects at the start of each creep-rate determination, permitting valid direct comparison of rates for the material as removed from service and after an additional exposure, which should account for another significant fraction of total rupture life.

A measured creep rate from Step 5 that is lower than the rate determined in Step 3 under identical conditions indicates that service has been so mild that steady secondary creep has not yet become fully established. The more common situation, in which the two rates are approximately the same or the second is faster than the first, would indicate that the component is within or past the secondary creep stage.

Creep and rupture tests appear to offer confident determination of remaining life of elevated-temperature materials; however, more rapid and less expensive methods of assessing creep damage continue to be sought.

Microstructural Evaluation of Creep Damage. Numerous studies beyond the realm of creep-rupture testing have sought immediate warning of creep damage, while monitoring its development, without the need for destructive sampling. Pending improvement and acceptance of such methods, it has been reported that remaining lifetime can be estimated from quantitative measurements of the microstructure (Ref 64). The procedure involves comparisons between a specimen taken from service and other specimens from accelerated tests at the service stress, but at increased temperatures.

Planning a Test Program

There are a number of helpful procedures to assist in the selection of appropriate stresses for the first tests on a new lot of material. Generally, data from other lots with the same or similar composition and heat treatment, which are available from compilations in the ASTM Data Series or from computer storage banks, are consulted. Approximate positions within the typical broad scatter bands frequently can be estimated by using the general observation that rupture strength increases with room-temperature tensile strength, particularly for creep rupture near the lower end of typical creep service temperatures.

For example, consider the situation in which rupture strength is to be determined for a time near 10 000 h at a single temperature. The test stress to produce a useful life near this goal value could be determined by extrapolating a series of tests covering a range of higher stresses at this same temperature, but the total testing time required by preliminary tests may well exceed that for the final test.

Several relatively short tests at each of several higher temperatures often permit equally good estimation of the required long-time stress by use of parameter methods. As discussed earlier, a favored mode is to use the same stress at different levels of temperature. Under most circumstances, these tests need not extend more than about 100 °C (180 °F) above the temperature of the desired final test. When a planned program involves a range of desired rupture lives over several temperature levels, no extra conditions outside the desired matrix should be required. A representative procedure in such a case is provided by the following example.

Selection of Rupture Stresses for Extensive Study

Current efforts by the American Petroleum Institute and the Metal Properties Council involve testing of heavy sections (>300 mm, or 12 in., thick) of 2.25Cr-1Mo steel melted to restricted chemistry. The specified test matrix calls for rupture times of 100, 300, 1000, 3000, and 12 000 h each at temperatures of 454, 482, and 510 °C (850, 900, and 950 °F).

Conditions for Initial Tests. Room-temperature tensile tests for samples from four different thick plates averaged near 638 MPa (92.5 ksi). Published data for normalized and tempered 2.25Cr-1Mo steel with tensile strengths ranging from 620 to 655 MPa (90 to 95 ksi) suggested rupture strengths near the following levels:

Temperature, °C (°F)	Typical rupture strength from published data, MPa (ksi)		
	100 h	300 h	1000 h
510	317	290	262
(950)	(46)	(42)	(38)
480	358	331	303
(900)	(52)	(48)	(44)

The first test for each heat of steel was started at the highest temperature (510 °C, or 950 °F), with the 290 MPa (42 ksi) stress chosen to give a rupture time of around 300 h. If the composition and treatment for that heat result in higher or lower rupture strength than anticipated, the results are still acceptable as an approximation for the 1000-h or 100-h point.

Heat C produced 114.0 h rupture life at the 290 MPa (42 ksi) stress. Assuming the trend of the above table to be valid, the second specimen at 510 °C (950 °F) for this lot was started at 262 MPa (38 ksi), with goal life near 300 h, and a third specimen was started at the next lower temperature (480 °C, or 900 °F) at 331 MPa (48 ksi) to yield a goal life of 100 h. The latter test lasted 154.5 h.

For lower strength Heat L, the initial test at 510 °C (950 °F) and 290 MPa (42 ksi) ruptured after only 40.4 h. The stress for the next test was therefore lowered to 248 MPa (36 ksi), and the first test at 480 °C (900 °F) used 290 MPa (42 ksi) for a goal life of 300 h.

Results from these six preliminary tests provided sufficient data for independent treatment by the Larson-Miller parameter with universalized $C = 20$ to set stresses for tests at all three temperatures to 1000 h goal life. As additional results accumulated, required stresses for the longest tests could be closely fixed by application of either refined parameter techniques or by graphical methods.

Allowance for Data Scatter. In an effort to minimize uncertainties from extrapolation procedures, undue emphasis may be placed on a test with the longest affordable duration. Normal data scatter still limits evaluation of expected performance from isolated points. Indeed, a more reliable prediction may be obtained when the same total test time is devoted to performing a greater number of tests with intermediate duration rather than only a few with extended life. An exception is when instabilities occur only with very long testing.

Some researchers study scatter by running multiple tests at one or more fixed combinations of stress and temperature. Again, a series of tests each at different conditions permits equally good statistical treatment to evaluate the scatter, while providing a broader indication of the material characteristics. Although emphasis has been placed on time to rupture, the extra expense of obtaining complete creep deformation data in all tests is often justified by the insight into the effects of structural changes.

REFERENCES

1. *ASME Boiler and Pressure Vessel Code*, Section I, Rules for Construction of Power Boilers, Paragraph A-150, Basis for Estimating Stress Values, American Society of Mechanical Engineers, New York, 1983
2. Private business communication, Materials Technology Corporation data sheets submitted to the Metal Properties Council for tests under MPC Contract No. 174-1
3. "Standard Recommended Practice for Conducting Creep, Creep-Rupture, and Stress-Rupture Tests of Metallic Materials," E 139, *Annual Book of ASTM Standards*, Vol 03.01, ASTM, Philadelphia, 1984, p 283-297
4. A.K. Schmieder, Size Effect During Rupture Tests of Unnotched and Notched Specimens of Cr-Mo-V and Cr-Ni Steels, in *Ductility and Toughness Considerations in Elevated Temperature Service*, American Society of Mechanical Engineers, New York, 1978, p 31-48
5. B. Aronsson and A. Hede, Some Observations on the Reproducibility of Creep Rate Determinations, in *High-Temperature Properties of Steel*, Iron and Steel Institute, London, 1967, p 41-45
6. D.R. Hayhurst, The Effects of Test Variables on Scatter in High-Temperature Tensile Creep-Rupture Data, *Int. J. Mech. Sci.*, Vol 16, 1974, p 829-841
7. W. Ruttman, M. Krause, and K.J. Kremer, International Community Tests on Long-Term Behavior of 2-¼%Cr-1%Mo Steel, in *High-Temperature Properties of Steel*, Iron and Steel Institute, London, 1967, p 23-29
8. D. Coutsouradis and D.K. Faurschou, Cooperative Creep Testing Program, AGARD Report No. 581, North Atlantic Treaty Organization, Advisory Group for Aerospace Research & Development, Neuilly-sur-Seine, France, 1971
9. *Report on the Elevated-Temperature Properties of Stainless Steels*, DS 5-S1, ASTM, Philadelphia, 1965
10. *An Evaluation of the Yield, Tensile, Creep, and Rupture Strengths of Wrought 304, 316, 321, and 347 Stainless Steels at Elevated Temperatures*, DS 5-S2, ASTM, Philadelphia, 1969
11. *Supplemental Report on the Elevated-Temperature Properties of Chromium-Molybdenum Steels*, DS 6-S1, ASTM, Philadelphia, 1966
12. *Supplemental Report on the Elevated-Temperature Properties of Chromium-Molybdenum Steels*, DS 6-S2, ASTM, Philadelphia, 1971
13. *Report on the Elevated-Temperature Properties of Selected Super-alloys*, DS 7-S1, ASTM, Philadelphia, 1968
14. *An Evaluation of the Elevated-Temperature Tensile and Creep-Rupture Properties of Wrought Carbon Steel*, DS 11-S1, ASTM, Philadelphia, 1969
15. *Evaluation of the Elevated-Temperature Tensile and Creep-Rupture Properties of C-Mo, Mn-Mo and Mn-Mo-Ni Steels*, DS 47, ASTM, Philadelphia, 1971
16. *Evaluation of the Elevated-Temperature Tensile and Creep-Rupture Properties of Steel*, DS 50, ASTM, Philadelphia, 1973
17. *Evaluation of the Elevated-Temperature Tensile and Creep-Rupture Properties of 3 to 9 Percent Cr-Mo Steels*, DS 58, ASTM, Philadelphia, 1975
18. *Evaluations of the Elevated-Temperature Tensile and Creep-Rupture Properties of 12 to 27 Percent Chromium Steels*, DS 59, ASTM, Philadelphia, 1980
19. *Compilation of Stress-Relaxation Data for Engineering Alloys*, DS 60, ASTM, Philadelphia, 1982
20. V.K. Sikka, H.E. McCoy, Jr., M.K. Booker, and C.R. Brinkman, "Heat-to-Heat Variation in Creep Properties of Types 304 and 316 Stainless Steels," ASME Paper No. 75-PVP-26, American Society of Mechanical Engineers, New York, 1975
21. K.R. Williams and B. Wilshire, Effects of Microstructural Instability on the Creep and Fracture Behavior of Ferritic Steels, *Mat. Sci. Eng.*, Vol 28, 1977, p 289-296
22. M. Wild, Analyse des Zeitstandverhaltens warmfester ferritischer Stähle bei Temperaturen von 450 bis 600 C, *Archiv. Eisenhüttenwes.*, Vol 34, Dec 1963, p 935-950
23. R.L. Klueh, Interaction Solid Solution Hardening in 2.25 Cr-1 Mo Steel, *Mat. Sci. Eng.*, Vol 35, 1978, p 239-253
24. J. Glen, The Shape of Creep Curves, *Trans. ASME J. Basic Eng.*, Vol 85, Series D, No. 4, Dec 1963, p 595-600

25. J.P. Milan *et al.*, "Negative Creep in Ni-Cr-Ti Alloys," Fourth Interamerican Conference on Materials Technology, Caracas, June-July 1975, p 102-106

26. R. Widmer and N. J. Grant, The Role of Atmosphere in the Creep-Rupture Behavior of 80Ni-20Cr Alloys, *Trans. ASME J. Basic Eng.*, Vol 82, Series D, No. 4, Dec 1960, p 882-886

27. P. Shahinian and M.R. Achter, "Creep-Rupture of Nickel of Two Purities in Controlled Environments," Joint International Conference on Creep, Institute of Mechanical Engineers, London, 1963, p 7-49 to 7-57

28. J.H. Bennewitz, "On the Shape of the Log Stress-Log Time Curve of Long Time Creep-Rupture Tests," Joint International Conference on Creep, Institute of Mechanical Engineers, London, 1963, p 5-81 to 5-92

29. W.C. Leslie, J.W. Jones, and H.R. Voorhees, Long-Time Creep-Rupture Tests of Aluminum Alloys, *J. Test. Eval.*, Vol 8 (No. 1), 1980, p 32-41

30. D.J. Wilson and H.R. Voorhees, Creep Rupture Testing of Aluminum Alloys to 100,000 Hours, *J. Mat.*, Vol 7 (No. 4), 1972, p 501-509

31. S.P. Agrawal, L.E. Byrnes, J.A. Yaker, and W.C. Leslie, Creep Rupture Testing of Aluminum Alloys: Metallographic Studies of Fractured Test Specimens, *J. Test. Eval.*, Vol 5 (No. 3), May 1977, p 161-173

32. D.J. Wilson, Extrapolation of Rupture Data for Type 304 (18Cr-10Ni), Grade 22 (2-¼Cr-1 Mo), and Grade 11 (1-¼Cr-½Mo-¾Si) Steels," *Trans. ASME J. Eng. Mat. Technol.*, Jan 1974, p 22-33

33. F.R. Larson and J. Miller, A Time-Temperature Relationship for Rupture and Creep Stresses, *Trans. ASME*, Vol 74, 1952, p 765

34. R.M. Goldhoff, Comparison of Parameter Methods for Extrapolating High-Temperature Data, *Trans. ASME J. Basic Eng.*, Vol 81, Series D, No. 4, Dec 1959, p 629-644

35. R.L. Orr, O.D. Sherby, and J.E. Dorn, Correlations of Rupture Data for Metals at Elevated Temperatures, *Trans. ASM*, Vol 46, 1954, p 113-118

36. R.M. Goldhoff and G.J. Hahn, Correlation and Extrapolation of Creep-Rupture Data of Several Steels and Superalloys Using Time-Temperature Parameters, in *Time-Temperature Parameters for Creep-Rupture Analysis*, American Society for Metals, 1968, p 199-245

37. W.M. Cummings and R.H. King, Extrapolation of Creep Strain and Rupture Properties of ½Cr-½Mo-¼V Pipe Steel, *Proc. Inst. Mech. Eng.*, Vol 185, 1970-71, p 285-299

38. S.S. Manson, Time-Temperature Parameters—A Re-evaluation and Some New Approaches, in *Time-Temperature Parameters for Creep-Rupture Analysis*, American Society for Metals, 1968, p 1-113

39. S.S. Manson and C.R. Ensign, "A Specialized Model for Analysis of Creep Rupture Data by the Minimum Commitment Station-Function Approach," NASA TM X-52999, 1971, p 1-14

40. S.S. Manson and C.R. Ensign, Interpolation and Extrapolation of Creep Rupture Data by the Minimum Commitment Method. Part I. Focal-Point Convergence, in *Characterization of Materials for Service at Elevated Temperature*, American Society of Mechanical Engineers, New York, 1978, p 299-398

41. S.S. Manson and A. Muralidharen, Analysis of Creep Rupture Data for Five Multiheat Alloys by the Minimum Commitment Method Using Double Heat Centering Technique, in *Progress in Analysis of Fatigue and Stress Rupture*, American Society of Mechanical Engineers, New York, 1984, p 1-46

42. F.C. Monkman and N.J. Grant, An Empirical Relationship Between Rupture Life and Minimum Creep Rate in Creep-Rupture Tests, *Proc. ASTM*, Vol 56, 1956, p 593-605

43. H.R. Voorhees, "Determination of Rupture Strength at Temperatures Near the Lower End of the Time-Dependent Range," findings reported to the Metal Properties Council, Inc., New York, 1984

44. S.S. Manson and W.F. Brown, Jr., Discussion to a paper by F. Garofalo *et al.*, *Trans. ASME*, Vol 78 (No. 7), Oct 1956, p 143

45. F. Dobeš and K. Milička, The Relation Between Minimum Creep Rate and Time to Fracture, *Met. Sci.*, Vol 10, Nov 1976, p 382-384

46. D. Lonsdale and P.E.J. Flewitt, Relationship Between Minimum Creep Rate and Time to Fracture for 2-½%Cr-1%Mo Steel," *Met. Sci.*, May 1978, p 264-265

47. M. Gold, W.E. Leyda, and R.H. Zeisloft, The Effect of Varying Degree of Cold Work on the Stress-Rupture Properties of Type 304 Stainless Steel, *Trans. ASME J. Eng. Mat. Technol.*, Vol 97, Series H, No. 4, Oct 1975, p 305-312

48. R.F. Gill and R.M. Goldhoff, Analysis of Long-Time Creep Data for Determining Long-Term Strength, *Met. Eng. Quart.*, Vol 10 (No. 3), Aug 1970, p 30-39

49. P.L. Threadgill and B.L. Mordike, The Prediction of Creep Life From Transient Creep Data, *Z. Metallkd.*, Vol 68 (No. 4), 1977, p 266-269

50. F. Garofalo, C. Richmond, W.F. Domis, and F. von Gemmingen, "Strain-Time, Rate-Stress and Rate-Temperature Relations During Large Deformations in Creep," Joint International Conference on Creep, Institution of Mechanical Engineers, London, 1963, p 1-31 to 1-39

51. P.L. Threadgill and B. Wilshire, Mechanisms of Transient and Steady-State Creep in a γ'-Hardened Austenitic Steel, in *Creep Strength in Steel and High-Temperature Alloys*, The Metals Society, London, 1974, p 8-14

52. R.M. Goldhoff and R.F. Gill, A Method for Predicting Creep Data for Commercial Alloys on a Correlation Between Creep Strength and Rupture Strength, *Trans. ASME J. Basic Eng.*, Vol 94, Series D, No. 1, March 1972, p 1-6

53. E.L. Robinson, Effect of Temperature Variation on the Creep Strength of Steels, *Trans. ASME*, Vol 60, 1938, p 253-259

54. P.N. Randall, Cumulative Damage in Creep-Rupture Tests of a Carbon Steel, *Trans. ASME J. Basic Eng.*, Vol 84, Series D, No. 2, June 1962, p 239-242

55. D.A. Woodford, Creep Damage and Remaining Life Concept, *Trans. ASME J. Eng. Mat. Technol.*, Vol 101 (No. 4), Dec 1979, p 311-316

56. M.M. Abo El Ata and I. Finnie, A Study of Creep Damage Rules, *Trans. ASME J. Basic Eng.*, Vol 94, Series D, No. 3, Sept 1972, p 533-543

57. J.W. Freeman and H.R. Voorhees, "Notch Sensitivity of Aircraft Structural and Engine Alloys," Wright Air Development Center, Part II, Further Studies with A-286 Alloys, Technical Report 57-58, ASTIA Document No. 207,850, Jan 1959

58. R.V. Hart, Concept of Area-Modified Stress for Life-Fraction Summation During Creep," *Met. Technol.*, Sept 1977, p 447-448

59. R.V. Hart, Assessment of Remaining Creep Life Using Accelerated Stress-Rupture Tests, *Met. Technol.*, Jan 1976, p 1-7

60. J.W. Freeman and H.R. Voorhees, *Literature Survey on Creep Damage in Metals*, STP 391, ASTM, Philadelphia, 1965

61. B.J. Cane and K.R. Williams, Creep Damage Accumulation and Life Assessment of a ½Cr ½Mo ¼V Steel, in *Mechanical Behavior of Materials*, Vol

2, Pergamon Press, Oxford, 1979, p 255-264

62. J.J. Bodzin, J.W. Freeman, and I.A. Rohrig, Carbon and Carbon-Moly Steam Pipe After Long-Time Service, *Trans. ASME J. Basic Eng.*, Vol 88, Series D, No. 1, March 1966, p 14-20

63. T.M. Cullen, I.A. Rohrig, and J.W. Freeman, Creep-Rupture Properties of 1.25Cr-0.5Mo Steel After Service at 1000 Deg. F, *Trans. ASME J. Basic Eng.*, Vol 88, Series D, No. 3, Sept 1966, p 669-674

64. K.F. Hale, Creep-Failure Prediction from Observation of Microstructures in 2-¼% Chromium-1% Molybdenum Steel, in *Physical Metallurgy of Reactor Fuel Elements*, Central Electricity Generating Board, United Kingdom, 1975, p 193-201

SELECTED REFERENCES

● F.R. Larson and J. Miller, Time-Temperature Relationship for Rupture and Creep Stress, *Trans. ASME*, Vol 74, 1952, p 765-771

● S.S. Manson and A.M Haferd, "A Linear Time-Temperature Relation for Extrapolation of Creep and Stress-Rupture Data," NACA TN 1890, March 1953

Influence of Multiaxial Stressing on Creep and Creep Rupture

By Blaine W. Roberts
Manager, Material Properties Evaluation
Combustion Engineering

COMPLEX STRESS STATES can be reduced to three orthogonal normal stresses, or principal stresses. If only one normal stress exists, the stress state is uniaxial. If two or three normal stresses exist, the stress state is multiaxial. Because creep life and ductility can be reduced under multiaxial stresses, the treatment of multiaxial stresses is important.

Initial study of the influence of multiaxial stressing on deformation and failure concentrated on elasticity and plasticity. The extension to creep has evolved with only minor adaptations to the classical approach. This article discusses the practical implications of multiaxial creep, multiaxial creep theories, experimental methods, and ductility.

Most engineering structures are influenced by complex multiaxial stresses that arise from loading, geometry, and/or material inhomogeneity. However, laboratory testing in the creep regime predominantly uses simple specimens that are subjected to uniaxial states of stress. Concurrently, most design codes develop allowable stresses from uniaxial test data. Because of the contrast between design-based uniaxial test data and the functional operation of components with multiaxial stresses, much research has been conducted to ensure that the relationship between the two is adequately understood (Ref 1, 2). Experimental observations of engineering alloys at high temperature have demonstrated that no single, simple theory relates uniaxial to multiaxial material response. Nevertheless, simplifications of theories can be used to ensure successful component performance.

Multiaxial Creep Theories

From a continuum mechanics viewpoint, an equivalent stress and strain are necessary to relate multiaxial and uniaxial stress states. For materials with adequate ductility in which the initial inelastic deformation involves a shear process, the Tresca (maximum shear) or von Mises (octahedral shear) definitions generally have been used. Although the Tresca criterion is generally more conservative, the von Mises relations given in Eq 1 through 3 are more suitable from a continuum and calculational standpoint (Ref 3):

$$\bar{\sigma} = \frac{1}{\sqrt{2}} [(\sigma_1 - \sigma_2)^2 + (\sigma_2 - \sigma_3)^2$$
$$+ (\sigma_3 - \sigma_1)^2]^{1/2} \qquad \text{(Eq 1)}$$

$$\bar{\epsilon} = \frac{\sqrt{2}}{3} [(\epsilon_1 - \epsilon_2)^2 + (\epsilon_2 - \epsilon_3)^2$$
$$+ (\epsilon_3 - \epsilon_1)^2]^{1/2} \qquad \text{(Eq 2)}$$

$$\dot{\bar{\epsilon}} = \frac{\sqrt{2}}{3} [(\dot{\epsilon}_1 - \dot{\epsilon}_2)^2 + (\dot{\epsilon}_2 - \dot{\epsilon}_3)^2$$
$$+ (\dot{\epsilon}_3 - \dot{\epsilon}_1)^2]^{1/2} \qquad \text{(Eq 3)}$$

where $\sigma_1 > \sigma_2 > \sigma_3$ are the principal stresses; ϵ_1, ϵ_2, and ϵ_3 are the principal strains; $\dot{\epsilon}_1$, $\dot{\epsilon}_2$, and $\dot{\epsilon}_3$ are the principal strain rates; and $\bar{\sigma}$, $\bar{\epsilon}$, and $\dot{\bar{\epsilon}}$ are the effective stress, strain, and strain rate, respectively. The relation between stress and strain rate is commonly expressed as:

$$\frac{\dot{\epsilon}_1 - \dot{\epsilon}_2}{\sigma_1 - \sigma_2} = \frac{\dot{\epsilon}_2 - \dot{\epsilon}_3}{\sigma_2 - \sigma_3} = \frac{\dot{\epsilon}_3 - \dot{\epsilon}_1}{\sigma_3 - \sigma_1}$$
$$= \frac{3}{2} \frac{\dot{\epsilon}}{\bar{\sigma}} \qquad \text{(Eq 4)}$$

Typical assumptions for material behavior are constant volume, isotropy, no influence of hydrostatic pressure, equivalent creep rates in the tensile and compressive directions, and coincidence of the principal axes of stress and strain. The isothermal uniaxial tensile data normally are used to infer the functional form of the stress-strain equation, and a wide range of mathematical models incorporating primary, secondary, and tertiary behavior have been used.

When stress or temperature changes as a function of time, a hardening law must be chosen. If the changes are small, the hardening law does not critically affect predicted results. However, for large changes, different hardening laws have a significant influence on the results. Because of mathematical simplicity, time hardening and strain hardening have been used most widely. Of these, strain hardening results generally show closer agreement with experimental values.

A multiaxial stress state complicates the prediction of component performance under creep-rupture loading. Several theories enable failure prediction. Although such theories may be stress or strain based, theories with a stress basis tend to be used more frequently. To express the failure theories, it is necessary to define the following parameters:

$$J_1 = \sigma_1 + \sigma_2 + \sigma_3 \qquad \text{(Eq 5)}$$

$$\sigma_H = \frac{J_1}{3} \qquad \text{(Eq 6)}$$

$$J_2' = \frac{1}{6} [(\sigma_1 - \sigma_2)^2 + (\sigma_2 - \sigma_3)^2$$
$$+ (\sigma_3 - \sigma_1)^2] \qquad \text{(Eq 7)}$$

$$S_1 = \sigma_1 - \sigma_H \qquad \text{(Eq 8)}$$

$$S_s = \sigma_1^2 + \sigma_2^2 + \sigma_3^2 \qquad \text{(Eq 9)}$$

These parameters are the first invariant, J_1, of the stress tensor; σ_H, the mean or hydrostatic stress; the second invariant, J_2' of the deviatoric stress tensor; S_1, the maximum deviatoric stress; and S_s a stress parameter that is invariant under coordinate rotation (Ref 4).

Creep life depends on the synergism of deformation and fracture (Ref 5). Generally, the creep-rupture process can be separated into the phases of crack initiation and subsequent crack growth. When failure is by general and gradual crack propagation on an order that is microscopically visible from the outset of testing, the maximum principal stress is the stress criterion for rupture (Ref 2). This class of materials generally is described as creep-brittle.

In contrast, when fracture is not accompanied by microscopically visible cracking until the deformation becomes localized near the end of life, the von Mises stress is the appropriate criterion. This class of materials generally is described as creep-ductile. The amount of life spent in these phases dictates which of the stress criteria are best suited to correlate the multiaxial creep rupture data.

Stress criteria using only one of the previously defined stress components are called "unmixed," and those using more than one are termed "mixed." In terms of the above parameters, some of the more common equivalent stresses, σ_e, corresponding to unmixed criteria, are:

Maximum principal stress,

$$\sigma_e = \sigma_1 \qquad \text{(Eq 10)}$$

von Mises effective stress, $\sigma_e = \bar{\sigma}$ (Eq 11)

Mixed rules that have been suggested and used on a limited basis include:

$$\sigma_e = \lambda\sigma_1 + (1 - \lambda)\sigma \quad \text{(Ref 6)} \qquad \text{(Eq 12)}$$

where λ is a constant,

$$\sigma_e = \alpha\sigma_1 + \beta J_1 + \delta J_2'^{1/2} \quad \text{(Ref 7)} \text{ (Eq 13)}$$

where α, β, and δ are constants, and

$$\sigma_e = \frac{3}{2}S_1\left(\frac{2\bar{\sigma}}{3S_1}\right)^a \times$$

$$\exp\left[b\left(\frac{J_1}{S_s} - 1\right)\right] \quad \text{(Ref 4)} \quad \text{(Eq 14)}$$

where a and b are material constants. In addition, several other unmixed creep-rupture criteria (Ref 8) are:

- Maximum strain energy (Beltrami)

- Maximum shear stress (Tresca)
- Maximum pressure reduced shear stress (Coulomb-Mohr)
- Maximum reduced stress (Hill)
- Maximum normal strain (St. Venant)

As an example of the considerations in developing a multiaxial theory, the model discussed in Ref 4 will be considered in more detail below. Although this formulation has not been verified against a sufficient amount of materials data to make a final judgment, it appears to offer great potential by incorporating the following features. Parameters S_1, $\bar{\sigma}$, and J_1 have been shown to be the key continuum stress parameters controlling the mechanistic process of creep damage, which encompasses cavity initiation, growth, and linkage leading to failure (Ref 9). Stress state and material dependency are incorporated to obtain real material behavior. Parameter J_1 allows the model to distinguish between life under multiaxial tensile ($J_1 > 0$) versus compressive ($J_1 < 0$) stress states, thus reflecting real material behavior. Parameters S_1 and $\exp(J_1/S_s)$ are positive definite and thus introduce no numerical problems. This makes the model fully three-dimensional and applicable to all regions of design space. The plane of S_1 (or σ_1) tends to be the plane of macrocrack growth, at least for biaxial tension-tension stress states.

Testing Methods

Although there are no standard specimens and procedures for multiaxial testing, use of varied methods has been reported (Ref 5, 10-12). The diversity of test methods arises from an effort to simulate the actual stress pattern found in a particular component. For example, the behavior of gas turbine disks is frequently examined by spin tests on laboratory models, whereas in the pressure vessel field, internal pressurization of cylinders is used.

In simple multiaxial testing, only the time to failure is measured. In more complex cases, instrumentation is incorporated to measure strain or deformation that occurs during testing. This instrumentation may include strain gages, extensometers, or periodic dimensional measurements between reference marks on the specimen to relate the strain history and failure to a strength theory.

One of the inherent problems is the definition of failure in multiaxial tests, compared to the complete severing that constitutes failure in a uniaxial test. Environmental and size effects also make comparison difficult. For example, the influence of the surface-to-volume ratio on oxidation may extend the life of a large multiaxial specimen compared to a smaller uniaxial specimen.

Nearly all multiaxial tests use proportional loading; i.e., the ratio of loads is independent of time. In the ideal specimen, all three components of principal stress can be applied independently and remain invariant with time, if deformations remain small. The thin-walled circular cylinder is closest to this ideal by combining axial, torsional, and pressure loadings. Other common configurations, such as a pressurized thick-walled cylinder and an axially loaded notched circular bar, experience stress redistribution with time. This redistribution often attains a stationary state early in life, such that the majority of time is spent in a stress-invariant state.

Due to the large number of test methods and configurations, a comprehensive assessment of the advantages and disadvantages of all methods is not attempted in this article. However, potential multiaxial tests include:

- Circular bars under combinations of torsion, bending, axial load, and external pressure
- Thin- and thick-walled cylinders under combinations of torsion, bending, axial load, and external pressure loads
- Notched circular bars under axial load
- Plate specimens (generally circular) with various loadings and edge conditions
- Cruciform loading of plate specimens
- Rotating disks
- Bars loaded in shear

Ductility

Ductility refers to the ability of a material to deform extensively when stressed uniaxially above its yield stress (Ref 13). This definition of ductility does not refer to a fundamental property of a material in a given metallurgical state, because the plastic strain before fracture is a function of the stress state, strain rate, temperature, and environment. Although many potential ductility indices at elevated temperature have been enumerated, a triaxiality factor (TF) to account for the influence of stress state on ductility has been of great value (Ref 14):

$$TF = \frac{J_1}{\bar{\sigma}} \qquad \text{(Eq 15)}$$

The triaxiality factor is the ratio of the normal stress to the octahedral shear stress, normalized to unity for simple tension. Ductility is an inverse function of the triaxiality factor. For uniaxial stresses, the triaxiality factor is unity, and ductility is conventionally defined. As the triaxiality factor becomes increasingly larger, approaching a state of equal triaxial tension, ductility approaches zero. Figure 1 illustrates that the ratio of the von Mises effective strain to the tensile elongation decreases with increasing triaxiality

Fig. 1 Influence of triaxiality factor on stress-rupture ductility

(1) 0.35% carbon steel. (2) 0.25% carbon steel. (3) Annealed brass. (4) 2024-T3 aluminum. (5) 7075-T651 aluminum. Source: Ref 15

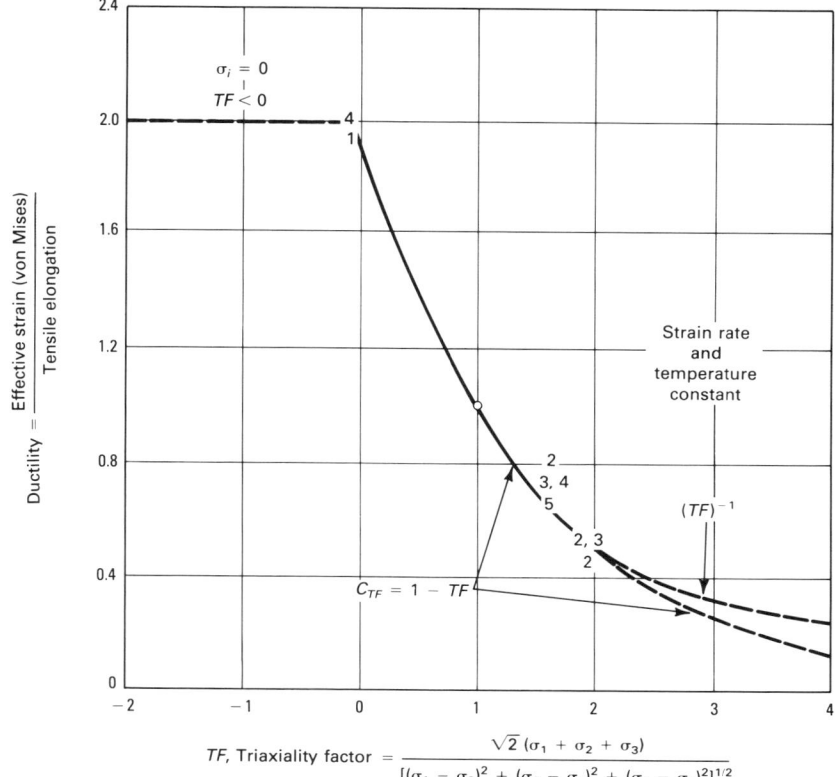

$$TF, \text{ Triaxiality factor} = \frac{\sqrt{2}\,(\sigma_1 + \sigma_2 + \sigma_3)}{[(\sigma_1 - \sigma_2)^2 + (\sigma_2 - \sigma_3)^2 + (\sigma_3 - \sigma_1)^2]^{1/2}}$$

factor for several materials (Ref 15). Hence, with increasing triaxiality, ductility can be expected to be less than that associated with the uniaxial tension or stress-rupture test.

REFERENCES

1. A.E. Johnson, J. Henderson, and B. Khan, *Complex-Stress Creep Relaxation and Fracture of Metallic Alloys*, Her Majesty's Stationary Office, Edinburgh, 1962
2. A.E. Johnson, Complex Stress Creep of Metals, *Metall. Rev.*, Vol 5, 1960, p 447-506
3. I. Finnie and W.R. Heller, *Creep of Engineering Materials*, McGraw-Hill, New York, 1959
4. R.L. Huddleston, "An Improved Multiaxial Creep-Rupture Strength Criterion," ASME Paper 84-PVP-106, American Society of Mechanical Engineers, New York, 1984
5. R.J. Browne, D. Lonsdale, and P.E.J. Flewitt, Multiaxial Stress Rupture Testing and Compendium of Data for Creep Resisting Steels, *J. Eng. Mat. Technol.*, Vol 104, 1984, p 291-296
6. M.M. Abo El Ata and I. Finnie, On the Prediction of Creep Rupture Life of Components Under Multiaxial Stress, in *Creep in Structures*, International Union of Theoretical and Applied Mechanics Symposium, Göteborg, 1970
7. D.R. Hayhurst, Creep Rupture under Multi-Axial States of Stress, *J. Mech. Phys. Solids*, Vol 20, 1972, p 381-390
8. R.J. Fields and J.H. Smith, Mechanical Testing in the 80's, *Met. Prog.*, Aug 1980, p 39-45
9. B.J. Cane, Creep Damage Accumulation and Fracture Under Multiaxial Stresses, in *Advances in Fracture Research, Fifth International Conference on Fracture*, Vol 4, Pergamon Press, Oxford, 1981
10. J. Henderson, An Investigation of Multi-Axial Creep Characteristics of Metals, *J. Eng. Mat. Technol.*, Vol 101, 1979, p 356-364
11. W.N. Findley, U.W. Cho, and J.L. Ding, Creep of Metals and Plastics Under Combined Stresses, a Review, *J. Eng. Mat. Technol.*, Vol 101, 1979, p 365-368
12. W.J. Evans, *Deformation and Failure Under Multiaxial Stresses—A Survey of Laboratory Techniques and Experimental Data*, C.P. No. 1306, Her Majesty's Stationary Office, London, 1974
13. M.J. Manjoine, Multiaxial Stress and Fracture, in *Fracture—An Advanced Treatise*, H. Liebewicz, Ed., Vol III, Academic Press, New York, 1971, p 263-309
14. M.J. Manjoine, Ductility Indices at Elevated Temperature, *J. Eng. Mat. Technol.*, April 1975, p 156-161
15. M.J. Manjoine, Creep-Rupture Behavior of Weldments, *Welding Res. Suppl.*, Feb 1982, p 505-575

SELECTED REFERENCES

- W.N. Findley, J.S. Lai, and K. Onaran, *Nonlinear Creep and Relaxation of Viscoelastic Materials*, North-Holland, Amsterdam, 1976
- J.H. Gittus, *Creep, Viscoelasticity and Creep Fracture in Solids*, John Wiley & Sons, New York, 1975
- *J. Eng. Mat. Technol.*, Vol 101, Oct 1979
- H. Kraus, *Creep Analysis*, John Wiley & Sons, New York, 1980
- J. Marin, *Mechanical Behavior of Engineering Materials*, Prentice-Hall, Englewood Cliffs, NJ, 1962
- B. Paul, A Modification of the Coulomb Mohr Theory of Fracture, *J. Appl. Mech.*, Vol 28 (No. 2), June 1961, p 259-268
- Yu. N. Robotnov, *Creep Problems in Structural Materials*, North-Holland, Amsterdam, 1969
- A.I. Smith and A.M. Nicolson, Ed., *Advances in Creep Design*, Applied Science Publishers, London, 1971

Creep-Fatigue Interaction

By Joseph B. Conway
President
Mar-Test Inc.

CREEP-FATIGUE INTERACTION is a special phenomenon that can have a detrimental effect on the performance of metal parts or components operating at elevated temperatures. When temperatures are high enough, both creep strains and cyclic (i.e., fatigue) strains can be present; interpretation of the effect that one has on the other becomes extremely important. For example, it has been found that creep strains can seriously reduce fatigue life and/or that fatigue strains can seriously reduce creep life. It is the quantification of these effects and the application of this information to life prediction procedures that constitutes the primary objective in creep-fatigue interaction studies. Creep-fatigue interaction is a problem that designers of elevated-temperature components must deal with to provide reliable service within the creep range.

Historical Perspective

Creep-fatigue interaction has been investigated for almost 50 years. The earliest attempts (Ref 1-4) to evaluate combined creep and fatigue properties were made in Germany between 1936 and 1942. In this same time period, work was underway to develop methods for predicting combined creep and fatigue behavior (Ref 5, 6). Attempts were also made to apply basic structural theories to the problem of combined creep and fatigue (Ref 7-9).

Although creep-fatigue interfaces, environments, and properties had been discussed (Ref 10, 11), the first use of the term "creep-fatigue interaction" appears to have been in Ref 12. Furthermore, in an early review of the subject (Ref 13), it was noted that higher operating temperatures and stresses accentuated the problems of creep and fatigue.

According to Ref 12, "It is not satisfactory to assess the safety of the structure in terms of fatigue or creep alone since the presence of one phenomenon may lower the limit of the other." Several questions arise when the specifics of this statement are studied. The suggestion that the presence of one phenomenon *may* lower the limit implies that the limit may or *may not* be lowered. What determines how much the limit is lowered? What is meant by the limit and how is it defined quantitatively?

Most experimental studies of creep-fatigue behavior attempt to answer these questions and to develop a basic understanding of this behavior. Such understanding is closely linked to the development of life-prediction methods. These methods outline the principles to be applied in predicting the expected operating life when a component is subjected to a creep influence of a certain magnitude in the presence of a fatigue influence of a certain magnitude.

In experiments dealing with lead tested at 32 °C (90 °F), a mean stress of 5980 kPa (867 psi) was applied and a fatigue stress equivalent to 25% of the static value was superimposed (Fig. 1). In the range of frequency of the fatigue stress (from 200 to 10 Hz), the minimum creep rate increased as the frequency decreased. Frequency had little effect on the minimum creep rate from 200 to 600 Hz. This example illustrates the presence of a fatigue or cyclic stress influence leading to a reduction of creep resistance measured in terms of an increase in the steady-state creep rate.

Another typical example of a combined effect is discussed in Ref 15, which presents an evaluation of the effect of frequency on the fatigue behavior of lead tested at 43 °C (110 °F) (Ref 16) and at 29 °C (84 °F) (Ref 17). This effect is shown in Fig. 2, which plots plastic strain range ($\Delta\epsilon_p$) versus cycles to failure (N_f). The test temperature is well into the creep range (above one half the absolute melting temperature) for lead, and the presence of the creep influence is quite prominent (Ref 15). At a plastic strain range of 0.002, a decrease in frequency from 2.38×10^6 cycles/day to 6.6 cycles/day led to a decrease in fatigue life amounting to a factor of about 100. The exponent k on the plastic strain range expression in Fig. 2 increased from 0.58 to 4.0 over this range of frequencies. In this example, the presence of a creep influence led to a significant lowering of the limit for fatigue based on the fatigue life corresponding to a frequency of 2.38×10^6 cycles/day.

The primary emphasis of this article is placed on results obtained in strain-controlled testing. This type of test yields extensive information about material response characteristics and provides the necessary data for developing the existing mathematical treatments of creep-fatigue interaction. Strain-controlled testing is discussed in Ref 18 and 19, and its development is detailed in Ref 20. Its use in providing a closer simulation of actual service conditions is described in Ref 19. Servo-controlled equipment capable of performing strain-controlled experiments became available in the 1960's. See the article "Creep, Stress-Rupture, and Stress-Relaxation Testing" in this Section for more information.

Creep-Fatigue Effects

The introduction of hold periods into the fatigue cycle of high-temperature strain-controlled fatigue tests provided a new approach to the evaluation of creep-fatigue interaction. It also simulated many service conditions more closely and yielded material information that was more realistic for design purposes. Prior to 1964, the results of elevated-temperature, strain-controlled low-cycle fatigue tests without hold periods were considered to be relevant for design (Ref 18, 19, 21). Later, it was found that simple, uninter-

Fig. 1 Effect of frequency on minimum creep rate

Source: Ref 14

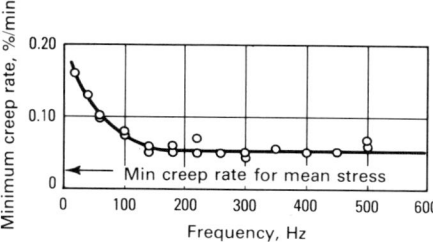

rupted low-cycle fatigue is seldom limiting in the design of pressure vessels operating at elevated temperatures (Ref 22, 23) and that the influence of repeated creep relaxation was an important consideration for hot pressure vessels (Ref 23).

Creep relaxation during hold periods in strain-controlled fatigue tests is illustrated in Fig. 3 for completely reversed strain cycling. In this case, the ratio of strain amplitude to mean strain, the A ratio, is equal to ∞; and the R ratio, or ratio of minimum strain to maximum strain, is equal to −1. The material is strained at a constant strain rate along 0A (Fig. 3a); when point A is reached, the control system functions to maintain the strain value constant for the hold period of duration AB. At point B, the material is unloaded along BC, and a hold period of duration CD is introduced at the peak compressive strain level. Hold periods can be introduced at any strain level, but these usually occur at the peak strain levels in the cycle.

Stress relaxations during AB and CD are shown in Fig. 3(b). Creep occurs during the hold periods at constant strain. To maintain the total strain constant, the stress is reduced to exchange elastic strain for creep or inelastic strain.

Studies of several stainless steels using the waveforms shown in Fig. 4 have been reported (Ref 24). In tests of type 304 stainless steel at 650 °C (1200 °F), hold times in the tension portion of the cycle only (waveform 4 in Fig. 4) were found to be particularly detrimental and resulted in reductions in fatigue life much greater than those noted with waveforms 2 and 3. Figure 5 illustrates this behavior for a strain rate of $4 \times 10^{-3} \, s^{-1}$. On the basis of this total strain-range graph, a 30-min hold period in only the compression portion of the cycle leads to a slight reduction in fatigue resistance. A slightly greater reduction is observed when 30-min hold periods are used in both the tension and compression portions of the cycle. A significant reduction is observed for a 30-min hold period in only the tension portion of the cycle.

Fig. 2 Effect of frequency of cycling on low-cycle fatigue behavior of lead

Source: Ref 15

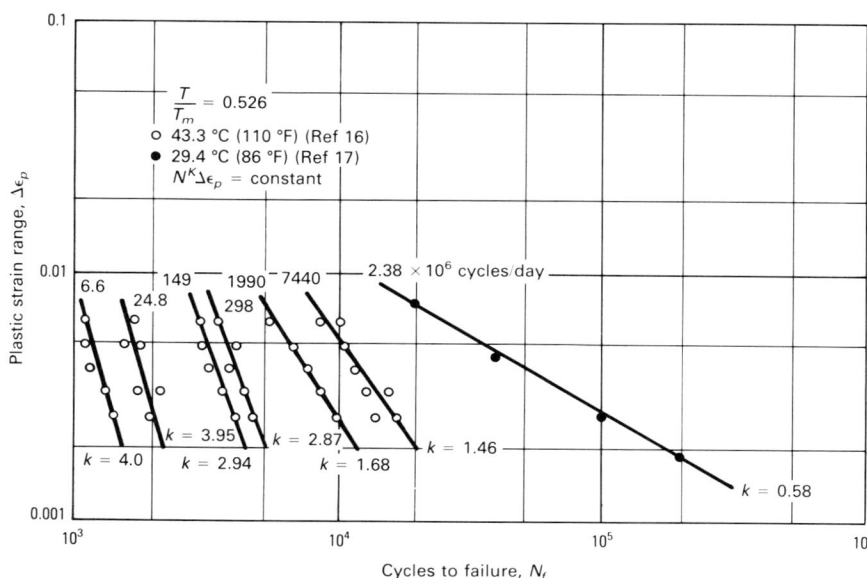

The data obtained in symmetrical-hold- and compression-hold-only tests indicated a correlation with inelastic strain range. These results, presented in Fig. 6, indicate that on the basis of inelastic strain range the data obtained using waveforms 2 and 3 are essentially identical with those obtained in no-hold tests (waveform 1). In these hold-time tests, the inelastic strain range was based on the relaxed stress range, $\Delta\sigma_r$ at $N_f/2$.

These results suggest that the decreases in the low-cycle fatigue resistance noted in Fig. 5 (except for the tension-hold-only tests) are due to increases in inelastic strain. Even though Fig. 5 is based on data obtained at

certain constant values of total strain range, the inelastic strain ranges associated with all the data points at a given total strain range are not the same. The relaxation effects are somewhat different for each waveform, thus leading to variations in the inelastic strain-range component at a given value of the total strain range. When the data (except for tension-hold-only data) are compared at the same value of the inelastic strain range, a definite consistency is noted.

Data obtained in the tension-hold-only tests are inconsistent with the inelastic strain-range correlation. As shown in Fig. 6, these results describe a fatigue life that is much

Fig. 3 Strain-controlled fatigue cycle with hold periods

(a) Effect on strain over time. (b) Effect on stress over time

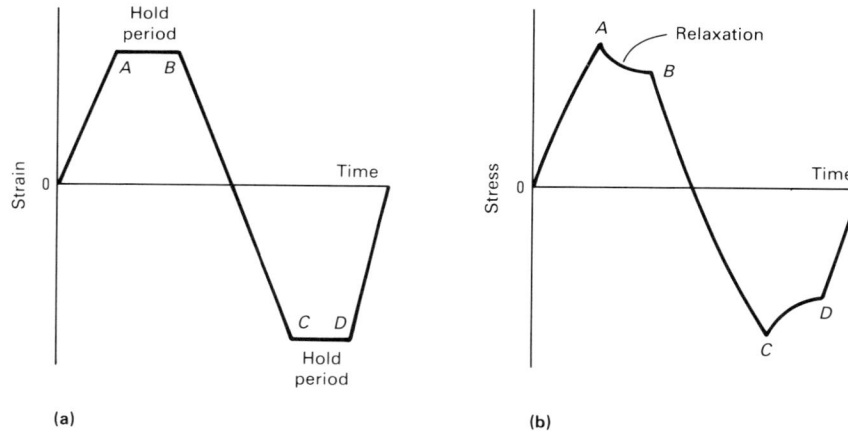

(a)   (b)

Fig. 4 Strain waveforms used in evaluating effects of hold time on low-cycle fatigue resistance

T, tension; C, compression. See text for an explanation of waveforms.

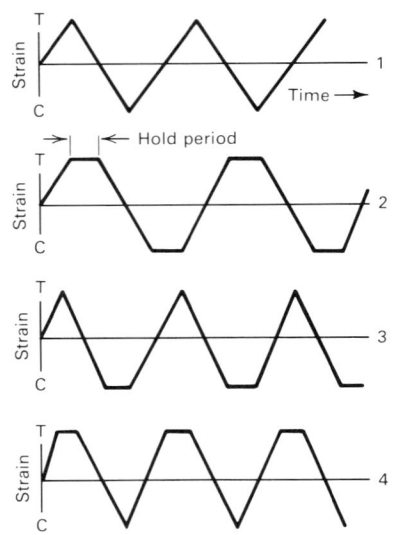

Fig. 5 Effect of hold period and strain waveform on the low-cycle fatigue resistance of AISI type 304 stainless steel

Tested in air at 650 °C (1200 °F) and at a strain rate of 4×10^{-3} s^{-1}

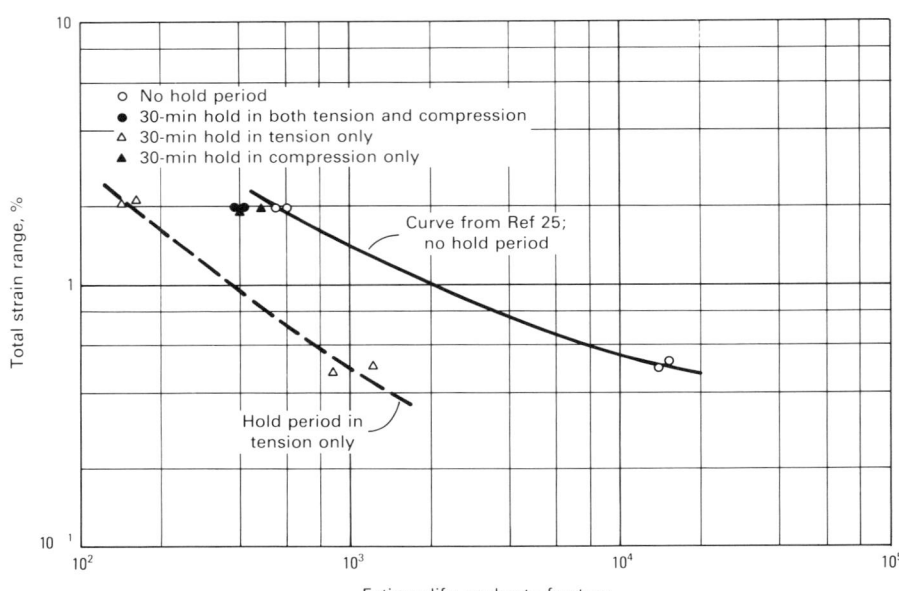

Fig. 6 Effect of hold period and strain waveform on the plastic-strain fatigue resistance of AISI type 304 stainless steel

Tested in air at 650 °C (1200 °F) and at a strain rate of 4×10^{-3} s^{-1}

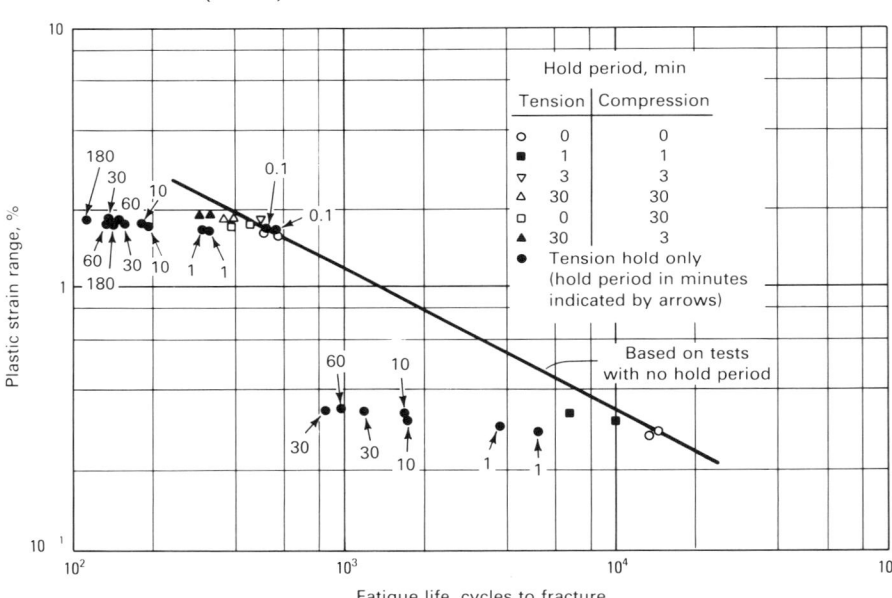

lower than would be predicted by the inelastic strain-range correlation. Apparently, holding only in the tension portion of the cycle leads to extensive material damage and drastically reduced fatigue life.

Some success has been achieved by comparing tension-hold-only data with stress-rupture behavior. This comparison is shown in Fig. 7, where the fatigue data were plotted as $\sigma_{t\,avg}$ versus the total time in the fatigue test during which the specimen is exposed to a tensile stress. Here $\sigma_{t\,avg}$ was taken as the arithmetic mean of $\sigma_{t\,max}$ and $\sigma_{t\,min}$, where $\sigma_{t\,max}$ is the maximum tensile stress imposed to attain the desired tensile-strain amplitude and $\sigma_{t\,min}$ is the tensile stress after relaxation during the hold period at a constant value of total strain. Fairly good agreement with the published (Ref 26) stress-rupture data is indicated. This correlation should be viewed with caution, because it has been tested only once. This correlation can also be criticized, because the average tensile stress used makes conversion to the corresponding stress components difficult without some knowledge of the relaxation behavior of the material.

Tension-hold-only tests have indicated that serious damage is encountered even when the hold period is on the order of 1 min. Table 1 summarizes the data obtained in the evaluation of this effect at 2.0% strain range. A saturation effect appears to be observed when the hold period approaches 30 min. These data also are presented in Fig. 8, in which the data obtained in the no-hold test

have been arbitrarily plotted at a hold period of 10^{-2} min for reference.

For no-hold, compression-hold-only, and symmetrical-hold testing, a consistent behavior is indicated, as shown in Fig. 8. Data for tension-hold-only testing indicate deviations from this graph, and the direction of the

deviation is toward reduced fatigue life. The hold period corresponding to the point of this deviation is about 10^{-1} min for both the 0.5 and 2.0% total strain range.

Some saturation is indicated at the 0.5% strain range to yield a behavior similar to that observed at the higher strain range. Only a

Fig. 7 Correlation of fatigue data involving hold periods in tension only with typical stress-rupture data for AISI 304 stainless steel

Tested in air at 650 °C (1200 °F) and at a strain rate of 4×10^{-3} s^{-1}. Source: Ref 26

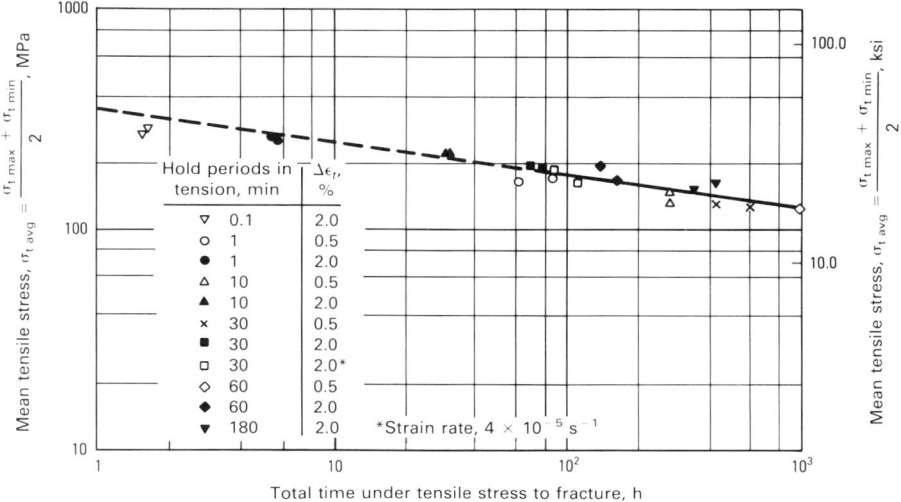

Fig. 8 Fatigue life versus hold-period time for AISI type 304 stainless steel tested in air at 650 °C (1200 °F) and at strain rates of 4×10^{-3} s^{-1} (open symbols) and 4×10^{-5} s^{-1} (closed symbols) using various strain waveforms

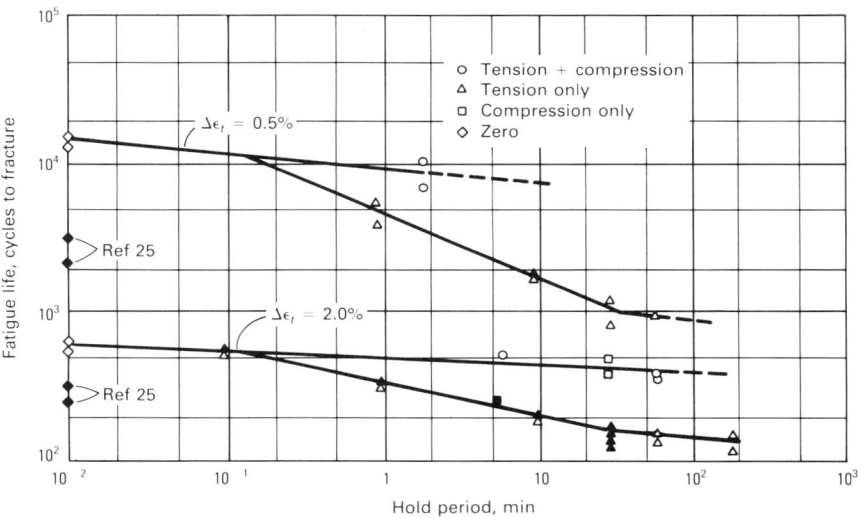

Table 1 Effect of hold-period length in tension-hold-only testing on fatigue resistance of AISI type 304 stainless steel

Tested in air at 650 °C (1200 °F) and a strain rate of 4×10^{-3} s^{-1} at a strain range of about 2.0%

Hold period, min	Cycles to failure, N_f	
	Test 1	Test 2
0	592	546
0.1	570	545
1.0	329	331
10.0	193	201
30.0	146	165
60.0	144	158
180.0	150	120

dency for internal void formation. These results involving unsymmetrical holding appear to be the first data on the subject reported in the literature.

Logarithmic graphs (Ref 27) of stress range or stress amplitude versus time to fracture are of interest because of their similarity to stress-rupture graphs. Figure 9 reveals this similarity and also illustrates that the data for type 316 stainless steel are clearly higher than those for type 304 stainless steel. For the stress amplitudes involved, it appears that linear relations are indicated, although this linearity is known to exist only over a limited range of stress. As the stress amplitude is reduced, the data define a definite curvature that is upwardly concave.

The time-to-fracture data in Fig. 9 show that decreasing the strain rate increases the time to fracture. This effect is opposite of that noted for the analyses (Ref 25) in terms of cycles to fracture; in these cases, the cycles to fracture decreased as the strain rate decreased.

A special correlation has evolved from an application of the above analysis to the hold-time data (Ref 28) for type 304 stainless steel. All the tension-hold-only data were considered in terms of time to fracture, and the correlation shown in Fig. 10 was obtained. The no-hold-time results, corresponding to a strain rate of 4×10^{-3} s^{-1} as given in Fig. 9, have been plotted along with additional no-hold-time data points. Particularly significant in this analysis is the fact that the tension-hold-only data obtained at both the 2.0 and 0.5% strain range (strain rate of 4×10^{-3} s^{-1}) yield a linear relation on this logarithmic graph. The lines described for the two strain ranges are essentially parallel (slope of about -0.05) and appear to intersect the no-hold-time line at the associated strain-range value.

If this type of behavior is typical, a method of predicting the effect of tension hold periods would be available. For example, if the stress-amplitude versus time-to-fracture graph were available from tests involving no

few tests were performed at a strain rate of 4×10^{-5} s^{-1}. In tests at the 2.0% strain range, the tension-hold-only data at the slow strain rate were identical to the data in the saturation region of the higher strain rate. Because of the limited data available at the lower strain rate, no definite conclusions can be made regarding these observations.

Tests involving a 30-min hold period in tension plus a shorter hold period in compression (unsymmetrical holding) have shown that the detrimental effect of a hold period in tension can be significantly reduced

by a short hold period in the compression portion of the cycle (Table 2). When the tension hold period is 30 min and a 3-min compression hold period is introduced, the fatigue life is within 80% of the fatigue life observed in the 30-min symmetrical-holding tests. Without this short hold period in compression, the fatigue life is reduced to about 40% of the 30-min symmetrical-holding fatigue life.

In this type of testing, the hold period in compression exerts a "healing" effect, or provides a mechanism that reduces the ten-

Table 2 Test results of AISI type 304 stainless steel obtained using a 30-min hold period in tension plus a short hold period in compression

Tested in air at 650 °C (1200 °F) and a strain rate of 4×10^{-3} s^{-1}

Hold period tension, min	Hold period compression, min	Total strain range, %	Cycles to failure, N_f Test 1	Test 2
0	0	1.98	592	546
30	30	1.98	380	416
30	0	2.08	146	⋯
30	0	2.02	⋯	165
30	3	1.98	308	⋯
30	3	2.00	⋯	336

hold periods, then hold-time effects could be estimated for a given strain range by locating this point on the no-hold-time line and drawing a line with a slope of −0.05 through this point. Estimates at other strain ranges could be made using similar constructions. It must be emphasized that these constructions lead only to qualitative predictions of hold-time effects. Any point on these construction lines (slope of −0.05) defines an operating life in a hold-time test that is larger than that corresponding to a no-hold-time test at the same strain range, but it is not possible to assign a specific value of the hold period to this particular point. Other analyses must supply this information.

Another correlation involving the tension-hold-only data for annealed AISI type 304 stainless steel tested in air at 650 °C (1200 °F) and at a strain rate of 4×10^{-3} s^{-1} is presented in Fig. 11. A logarithmic graph of time to fracture versus the tension-hold period in minutes yields a curve that is upwardly concave. However, if the value for the time to fracture with no hold period is subtracted from each time-to-fracture value, a definite linearity is obtained. This leads to:

$$t_f - t_{fo} = AH_T^{0.81} \quad \text{(Eq 1)}$$

where t_f is time to fracture (hour); t_{fo} is time to fracture using no hold periods; A is a constant; and H_T is hold period in tension, min.

A similar relation was found to apply to the 0.5% strain-range data. These relations can be used to calculate (for type 304 stainless steel and the condition involved) the time to fracture for any tension-hold-only period, which can then be used in conjunction with Fig. 10 to yield the stress amplitude for any hold period. A value for the cycles to fracture, N_f, will also follow from:

$$\text{Time to fracture} = \frac{N_f}{f} + N_f H_T \quad \text{(Eq 2)}$$

where H_T is the hold time in tension, and f is

Fig. 9 Stress-amplitude versus time-to-fracture data obtained in low-cycle fatigue tests of annealed AISI types 304 and 316 stainless steels in air at 650 °C (1200 °F)

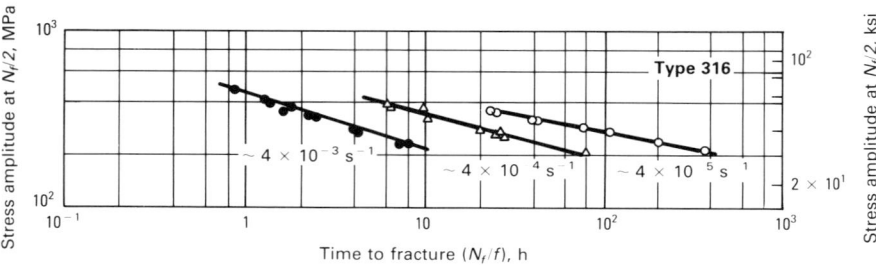

(a)

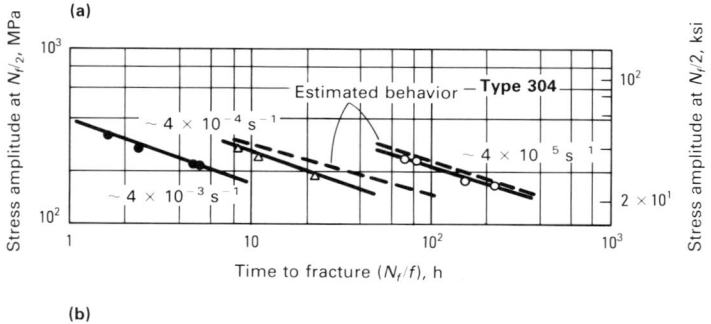

(b)

Fig. 10 Stress-amplitude vs time-to-fracture correlation of hold-time data involving hold periods in tension only for AISI type 304 stainless steel

Tested in air at 650 °C (1200 °F) and a strain rate of 4×10^{-3} s^{-1}. ○ and ●, hold-time data; numbers in parentheses are tension-hold periods in minutes. △, no-hold-time data; numbers represent total strain range in percent.

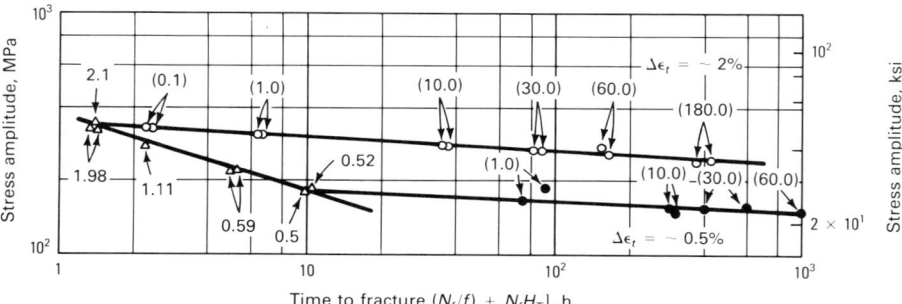

the cycling frequency. For the triangular waveform:

$$f = \frac{\dot{\epsilon}_t}{2\Delta\epsilon_t} \quad \text{(Eq 3)}$$

and

$$\text{Time to fracture} = \frac{2N_f\Delta\epsilon_t}{\dot{\epsilon}_t} + N_f H_T \quad \text{(Eq 4)}$$

Thus, it is simple to convert the data plotted in Fig. 10 to data involving cycles to fracture. This conversion leads to the results shown in Fig. 6 and 8 and indicates that the fatigue life measured in cycles to fracture decreases as the length of the hold period

increases. This effect might appear to be in contradiction to Fig. 10, which indicates an increase in the time to fracture as the length of the hold period increases. No contradiction exists, because both effects are correct and mutually consistent.

Introducing a hold period into the strain cycle leads to a decrease in N_f, but this hold period increases the cycle time by a factor that is greater than that corresponding to the reduction in N_f. As a result, even though N_f is decreased by the introduction of the hold period, the time of the test or time to fracture is increased. Consequently, the failure time of the specimen has been increased owing to the effect of the hold period.

Fig. 11 Low-cycle fatigue data for annealed AISI type 304 stainless steel obtained in tension-hold-only tests in air at 650 °C (1200 °F)

Total strain range, 2.0%; strain rate, 4×10^{-3} s^{-1}

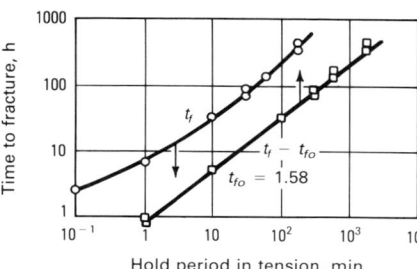

Fig. 12 Plot of N_f versus cycle time for AISI type 304 stainless steel tested in air at 650 °C (1200 °F)

Two primed points refer to 30-min hold-period test using strain rate of 4×10^{-5} s^{-1}

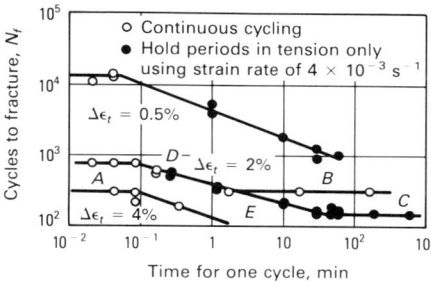

Fig. 13 Time to fracture versus length of hold period in tension for cast 1Mo steel

Tested at 510 °C (950 °F) at a total strain range of 2.0%

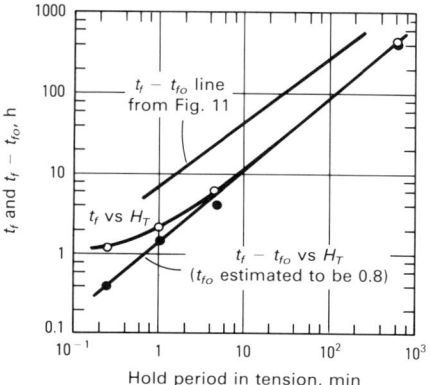

Fig. 14 Graph of $t_f - t_{fo}$ versus hold time in tension only for Cr-Mo-V steel

Data obtained at 565 °C (1050 °F) and 10 cpm for a total strain range of 1.0%

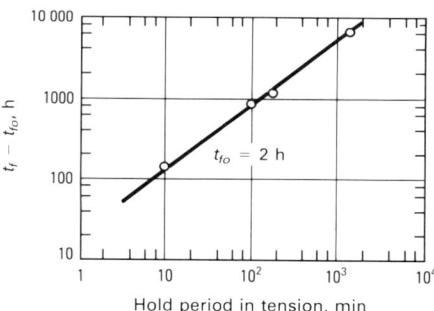

Figure 12 illustrates fatigue behavior in terms of a plot of cycles to fracture versus the cycle period (Ref 29). Various regions are:

- *Region A*: N_f is independent of strain rate above about 8×10^{-3} s^{-1} (cycle time < 0.083 min).
- *Region B*: N_f is independent of strain rate below about 4×10^{-4} s^{-1} (cycle time > 1.67 min).
- *Region C*: N_f is independent of hold period duration (for tension-hold period only) above about 50 min.
- *Region D*: N_f decreases with decreasing strain rate in the range from 8×10^{-3} to about 4×10^{-4} s^{-1}.
- *Region E*: N_f decreases with increasing hold period length (tension-hold period only) in the range to about 50 min.

Region B is viewed as a saturation in frequency degradation for continuous cycling, while region C is a tension hold-time saturation.

Confirmation of the trends noted in Fig. 10 and 11 has been reported (Ref 30, 31). In one study, hold times were introduced in only the tension portion of the cycle (push-pull loading) in tests of the following cast steels: 1Mo, 1Cr-1Mo, 12Cr-0.5Mo, and an austenitic type 316 containing niobium (Ref 31). All tests were performed in air at 510 °C (950 °F), and hold periods ranged from 15 s to 12 h.

Data for 1Mo steel are plotted in Fig. 13. As in Fig. 11, a logarithmic graph of time to fracture versus the length of the hold period (tension-hold only) yields a curve that is upwardly concave. However, when the graph is modified to represent $t_f - t_{fo}$, approximate linearity is observed (time to fracture with no hold time was not given in Ref 31, but appears to be about 0.8 h for these data). The slope of this line is essentially identical to that of the line in Fig. 11.

Some additional low-cycle fatigue data involving hold periods in only the tension portion of the strain cycle have been reported (Ref 32, 33). In these studies of Cr-Mo-V steel at 565 °C (1050 °F), reversed bending was employed at 10 cpm. Hold times in tension ranged from 6 to 1440 min, and these hold periods led to serious reductions in the cyclic fatigue life. These results for Cr-Mo-V steel also exhibit the type of behavior shown in Fig. 11. As shown in Fig. 14, a fairly definite linearity exists; the linearity extends to a hold period of 1440 min (24 h), which represents the longest hold period reported to date.

A summary of fatigue data reported in Ref 34 agrees with the $t_f - t_{fo}$ concept of Eq 1. The comparison graph derived from this study (Fig. 15) indicates a definite similarity. The positions of the lines for a given material are affected by strain range. Lower strain-range data are positioned above the data corresponding to the higher strain-range results. For a given hold time, the $t_f - t_{fo}$ value is greater for the lower strain range. Consequently, for a given hold time, the actual increase in the t_f value above the corresponding t_{fo} value is larger at the lower strain range.

This behavior appears to be related to the N_f behavior. At the 0.5% strain range, the value of N_{fo} is much greater (about 14 000 compared to about 550 at the 4×10^{-3} s^{-1} strain rate) than the N_{fo} value at the 2.0% strain range. When a 10-min hold period is introduced, the ratio of N_f/N_{fo} is about 0.12 at the 0.5% strain range and about 0.36 at the 2.0% strain range. This indicates a greater reduction in the cyclic life at the lower strain range, but the higher value of N_f at this lower strain range leads to a higher value for the time to fracture.

The effect of hold time has been found to become more and more detrimental as the

hold-period duration is increased (Ref 11). One study (Ref 36) correlated axial results with some reverse-bend hold-time data to support this proposition. However, this hypothesis cannot be applied generally, because other results (Ref 28, 30, 31) taken from axial tests provide some disagreement with this trend. In several cases, the cyclic life approaches a lower limit as the hold time is increased (Fig. 16). In general, it appears that the effects of long hold time are more detrimental for reverse bend than for axial results.

The effect of a hold period in the compressive part of the cycle is unclear. For axial tests on type 304 stainless steel at 650 °C (1200 °F), it was found that if the cycle contained a 30-min tensile hold and no compressive hold, there was a serious reduction in cyclic life and the mode of crack initiation was intergranular (Ref 28). If a small (3-min) compressive hold was added to the cycle, a

Fig. 15 Comparison graph of $t_f - t_{fo}$ data obtained in tension-hold-only low-cycle fatigue tests

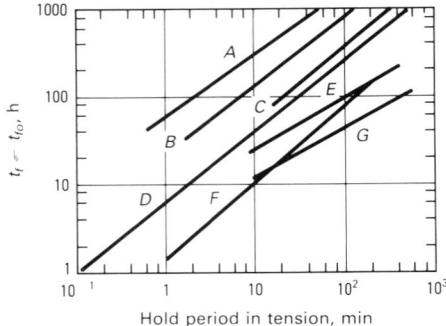

Material, type steel	Temperature °C	°F	t_{fo}, h	$\dot{\epsilon}_t$, s^{-1}	f, cpm	$\Delta\epsilon_t$, %	Ref
A 304 stainless	650	1200	10	4×10^{-3}	. . .	0.5	24
B Cr-Mo-V	565	1050	2	. . .	10	1	32, 33
C 2.25Cr-1Mo	600	1110	6.67	. . .	1	2	35
D 304 stainless	650	1200	1.6	4×10^{-3}	. . .	2	24
E 316 stainless (A7)	600	1110	2	. . .	2	3	30
F 1Mo	650	1200	0.8	. . .	~3	2	31
G 316 stainless (A8)	600	1110	2	. . .	2	3	30

healing effect was obtained, so that the loss in cyclic life was minimal, and the mode of initiation was transgranular. However, from axial tests on a nickel-based alloy at 760 °C (1400 °F), it was found that holding compressive strain was more damaging than holding tensile strain (Ref 37).

When the test temperature is high enough to produce large amounts of time-dependent strain and the hold period is in the tension portion of the cycle, a reduced life generally results (Ref 38). This is typical in most creep-fatigue interaction studies. Often, hold periods in compression at these temperatures are not damaging and can even reduce the effect of hold periods in tension when both are present in the same cycle.

When small amounts of inelastic strains are produced and the material is highly creep resistant at the test temperature, opposite effects are often observed. Hold periods in tension produce longer lives, and hold periods in compression are most detrimental. The hold periods in these tests, although producing small time-dependent strains, can cause the mean stress to shift away from zero. If the hold period is in compression, the mean stress can shift in a tension direction (and vice versa). Because the material is highly creep resistant, this mean stress can be sustained. The reduced life of such tests is attributed to the higher tensile stresses and not to the time-dependent strains.

A study of the effect of hold periods at peak strain on the fatigue life of René 95 indicated some interesting patterns (Ref 39). Waveforms with strain hold periods in tension, compression, and both tension and compression (balanced hold) were utilized. Although this testing effort focused primarily on hold period durations of 1 min, some 10-min hold periods were used to identify the different effects of the two hold-period durations and the effect of the location of the hold period within the cycle.

The effect of a 1-min hold period in tension is shown in Fig. 17. Abbreviations of

the form (1/10) are used to define the waveform imposed. The numerator represents the tension hold period time in minutes, and the denominator represents compression hold time. Generally, the effect is negligible over the strain-range regime from 0.9 to 2.0%. A few data points at a strain range of 1.0% indicate a beneficial effect of the 1-min tension hold period. The two 10-min hold-period

tests shown in Fig. 17 indicate essentially the same effect observed for the 1-min hold period. Consequently, for René 95 at 650 °C (1200 °F) ($A = \infty$) and at strain ranges of 1.2 and 1.4%, hold-period durations of 1 and 10 min in tension have no detrimental effect on fatigue life.

Similar tests at 650 °C (1200 °F) with hold periods in compression led to results that

Fig. 16 Cycles to fracture versus cycle time for reversed bending and axial push-pull tests for various steels

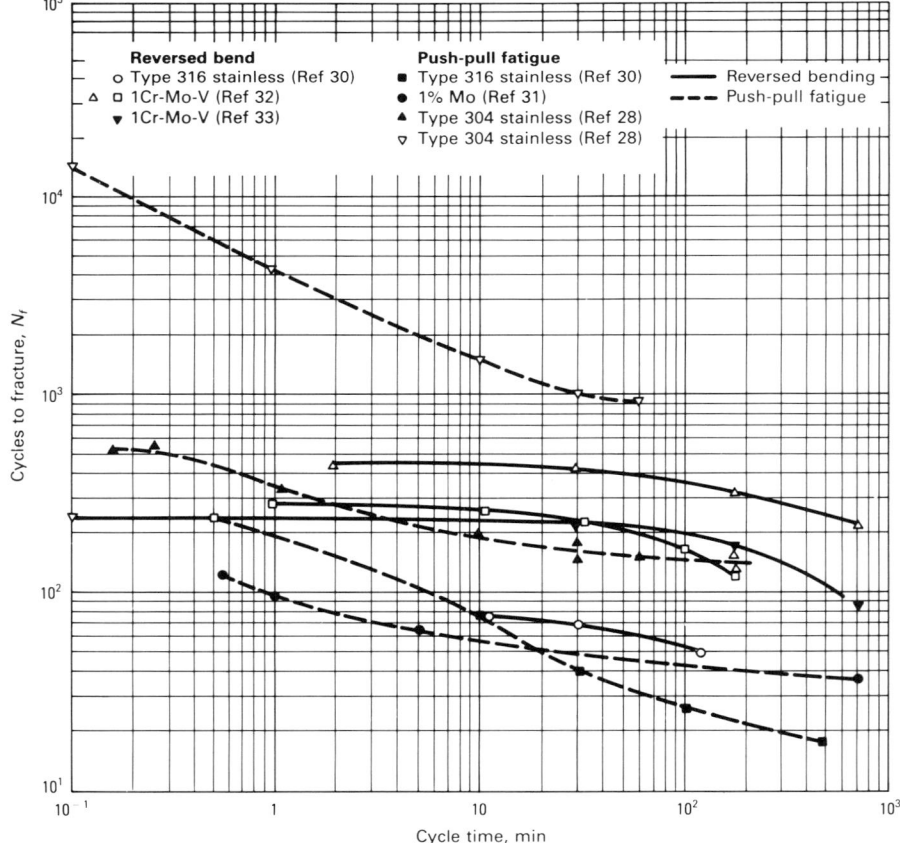

Fig. 17 Strain range versus N_f; hold periods in tension for René 95

Temperature, 650 °C (1200 °F); A = ∞

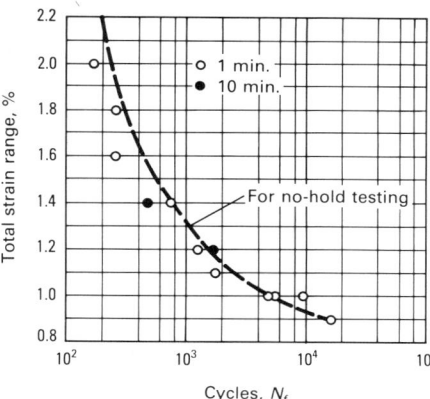

Fig. 18 Strain range versus N_f; hold period results for René 95

Temperature, 650 °C (1200 °F); A = ∞

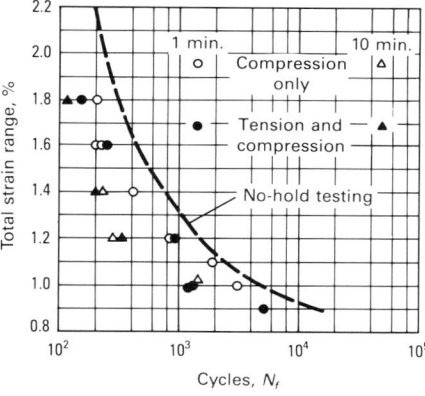

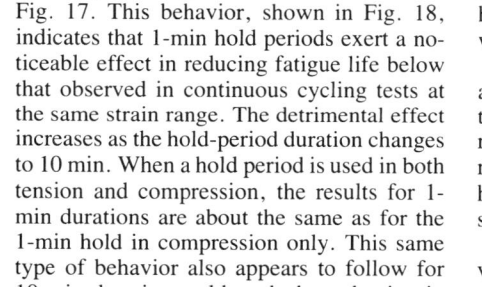

Fig. 19 Fatigue-life reduction factor versus strain range for René 95 for (1/1) and (10/10) tension-compression hold periods

Temperature, 650 °C (1200 °F); A = ∞

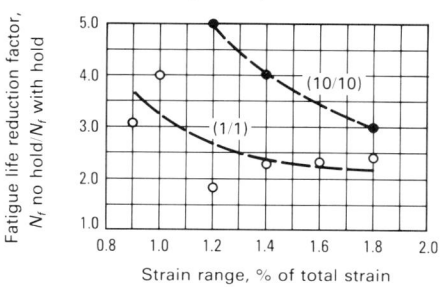

Fig. 20 Fatigue-life reduction factor versus strain range for René 95 for a range of tension-compression hold periods

Temperature, 650 °C (1200 °F); A = ∞

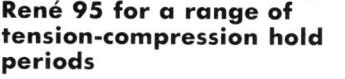

were significantly different from those in Fig. 17. This behavior, shown in Fig. 18, indicates that 1-min hold periods exert a noticeable effect in reducing fatigue life below that observed in continuous cycling tests at the same strain range. The detrimental effect increases as the hold-period duration changes to 10 min. When a hold period is used in both tension and compression, the results for 1-min durations are about the same as for the 1-min hold in compression only. This same type of behavior also appears to follow for 10-min durations, although the reduction in fatigue life is somewhat greater than that observed with 1-min durations.

A more graphic summary of these hold-time effects is illustrated by plotting the fatigue-life reduction factor. This factor is simply the ratio of the fatigue life without a hold period to the fatigue life observed when a hold period is used at a given strain range. For certain waveforms, the fatigue-life reduction factor exhibited a definite trend toward higher values as the strain range was decreased. An example of this is shown in Fig. 19 for (1/1) and (10/10) types of tests.

A composite of all the hold-time results is shown in Fig. 20. The following trends are identifiable within reasonable scatter. One-minute hold periods in tension had a negligible effect on the fatigue life over the strain-range regime from 0.9 to 2.0%. One-minute hold periods in compression and in both tension and compression define a fatigue-life reduction factor of 2.0 for strain ranges from 1.8 to about 1.0%. Below 1.0%, there appears to be a sharp increase in the slope of this trend behavior, suggesting more detrimental effects in the lower strain-range regime. Because this lower strain-range regime is in the area of special design interest (i.e., fatigue-life values greater than 10 000

cycles), a more comprehensive study of hold-time effects in this regime appears warranted.

Ten-minute hold periods in compression and in both tension and compression define a trend behavior that begins at a fatigue-life reduction factor of about 5.0 in the strain-range regime near 1.0%. Further study of hold-time effect also is needed in the lower strain-range region.

The development of different mean stress values in the hold-time tests is an important consideration, and a comparison of such behavior is shown in Fig. 21. In continuous cycling tests at 20 cpm, the absolute value of the compressive stress component is always just slightly larger than the tensile stress component. This behavior is shown in Fig. 21; over much of the strain-range regime studied, a mean compressive stress of about 34.5 MPa (5 ksi) was exhibited.

In the region near a strain range of 1.0%, the compressive mean stress increased to a value near 70 MPa (10 ksi). This result can be compared with the behavior observed for various hold-period combinations, as shown in Fig. 21. For the 1/0 combination (1-min hold in tension only), the absolute magnitude of the mean compressive stress increases above the continuous cycling value and shows an increase to 276 MPa (40 ksi) compression in the low-strain-range regime. This magnitude of the mean compressive stress is increased even more in the 10/0 combination. For the 0/1 combination (1-min hold period in compression only) the mean stress is close to zero in the high-strain-range regime and gradually increases to about 69 MPa (10 ksi) tension at the lower strain ranges.

For 0/10, the mean tensile stress is even higher and appears to approach 276 MPa (40

ksi) in the strain-range regime near 1.0%. Data for the 1/1 and 10/10 combinations are not plotted, but are not too different from the continuous cycling behavior. In the case of the 10/1 combination, the mean stress behavior is similar to that exhibited by the 1/0 combination.

Results of studies of Cr-Mo-V steel tested at 565 °C (1050 °F) using static creep, repeated tension, and reversed cyclic creep are shown in Fig. 22 (Ref 40). In all these tests, a transition from ductile to brittle fracture occurred at a stress level near 260 MPa (38 ksi). Tensile strain measurements that were made during load-controlled continuous cycling tests at zero mean load revealed an accumulation of strain (Fig. 23) and creep-type failures, except at a stress level of 325 MPa (47.1 ksi), where no strain accumulation was noted and a fatigue failure was obtained.

Tests using prior fatigue exposure (10 and 50% of N_f) followed by a static creep test led to the results shown in Fig. 24. The creep-rupture life values were consistently shorter than the equivalent tests of virgin material. The cyclic straining in the fatigue cycles led to cyclic strain softening of the material and a

Fig. 21 Mean stress versus strain range for René 95
Temperature, 650 °C (1200 °F); A = ∞

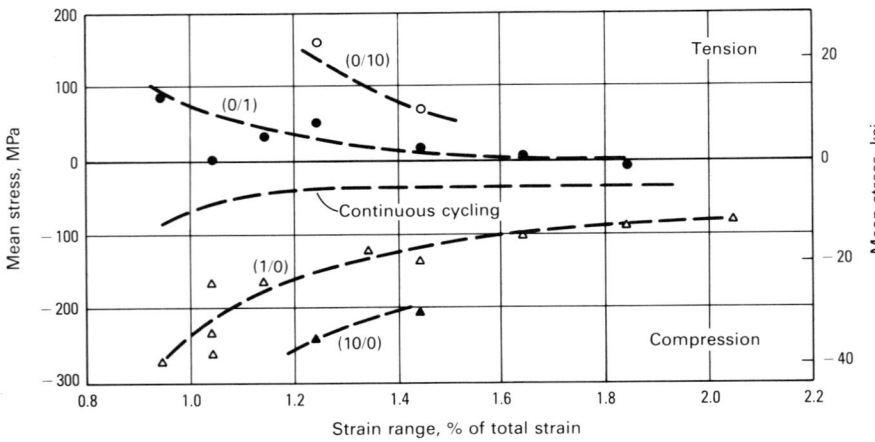

Another part of this assumption was to choose the effective time period within which this peak stress is applied as being equal to k/F, where k is a fraction of the cycle time $1/F$ (F is frequency). In this approach, the "fatigue damage" effect was taken as the ratio of the number of cycles actually applied, N'_f, to the number that would be sustained in the absence of a creep effect (the latter being equal to the Universal Slopes value for N_f). In addition, the "creep damage" effect was taken to be the ratio of the time actually spent at stress, σ_a, to the time required to cause rupture at this stress value. Then, assuming that failure occurs when:

Creep damage + fatigue damage = 1

$$(Eq\ 5)$$

$$\frac{t'}{t_R} + \frac{N'_f}{N_f} = 1 \qquad (Eq\ 6)$$

where:

$$t' = \frac{k}{F}\,(N'_f) \qquad (Eq\ 7)$$

This leads to:

$$N'_f = \frac{N_f}{1 + \dfrac{k}{AF}\,(NF)^{(m+0.12)/m}} \qquad (Eq\ 8)$$

where A is obtained as a time intercept on the stress-rupture isotherm, and m is the slope of the stress-rupture isotherm.

It was suggested (Ref 10) that the life be estimated by the 10% rule and by the application of Eq 8, and to then use the lower of the two values as a lower bound on fatigue life and a value of twice the lower-bound life as the average fatigue life.

substantial reduction in the subsequent creep-rupture life.

Most of the data reported for fatigue and creep-fatigue evaluations apply to testing in an air environment. However, it has been shown that the test environment can have a significant effect on fatigue and creep-fatigue results. Although this subject is beyond the scope of this article, environmental effects should be given proper consideration in any evaluation of creep-fatigue interactions. An interesting introduction to environmental effects is available in Ref 47.

Creep-Fatigue Prediction Techniques

Development of a mathematical formulation for life prediction is one of the most challenging aspects of creep-fatigue interaction. It is complicated by the fact that any proposed formulation must account for strain rate, relaxation at constant strain, creep at constant load, the difference between tension and compression creep and/or relaxation, or combinations of all these.

10% Rule. In an early attempt to provide estimates for fatigue life under combined creep and fatigue conditions (Ref 10), modifications to the Universal Slopes equation were considered (Ref 41). It was reasoned that the complexities introduced by high-temperature operation in the presence of a creep influence could not be accommodated in the Universal Slopes equation. Even when tensile properties determined at the temperature and strain rate of the fatigue test were used, the Universal Slopes equation led to fatigue-life estimates that were greater than those actually observed. It was concluded that the creep influence was introducing some intercrystalline cracking, which caused a portion of the crack initiation phase of the

fatigue process to be by-passed. As a result, the "10% rule" was introduced; it proposed to take 10% of the Universal Slopes value and use it as the estimate for a creep-fatigue life. Generally, this approach yields conservative or lower fatigue-life estimates. The 10% rule inherently excludes the possible effects of frequency hold-time, mean load, and other effects.

In an attempt to devise an alternate approach, the addition of a term to define "time at temperature" was considered (Ref 10). Using the diagram in Fig. 25 to illustrate the stress pattern for completely reversed strain cycling (A ratio = ∞; R = −1.0), it was assumed that the effect of the compression stress could be omitted and that the entire stress pattern could be replaced by one in which the stress is constant and equal to the maximum stress of the pattern it replaces.

Fig. 22 Static, repeated tension, and reversed cyclic creep results for Cr-Mo-V steel
Source: Ref 40

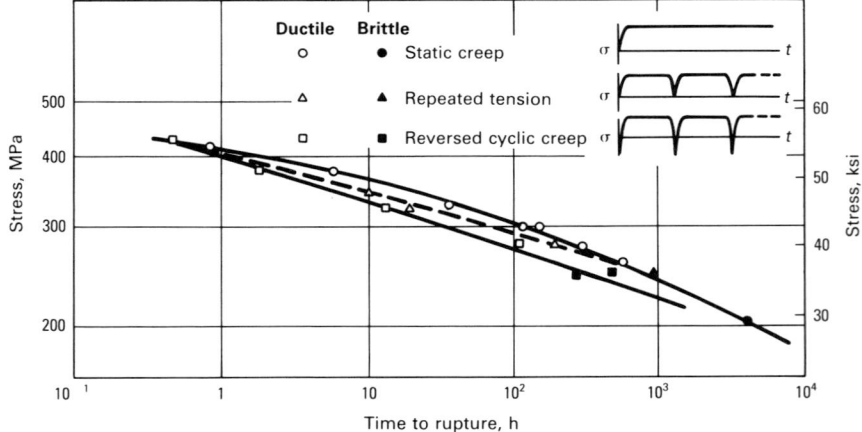

Fig. 23 Tensile strain accumulation in load-controlled continuous-cycling tests with zero mean load for Cr-Mo-V steel
Source: Ref 40

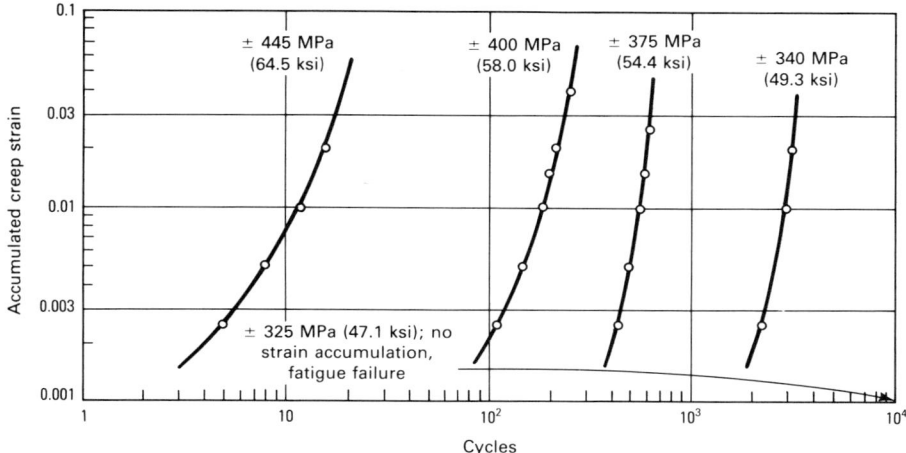

Fig. 24 Effect of prior fatigue on subsequent static creep rupture life for Cr-Mo-V steel
Source: Ref 40

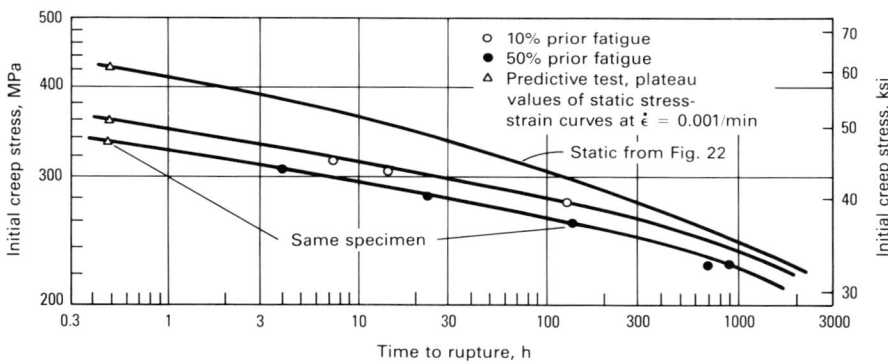

Fig. 25 Equivalent creep-rupture damage done by cyclic stress
k = effective fraction of each cycle spent at maximum tensile stress

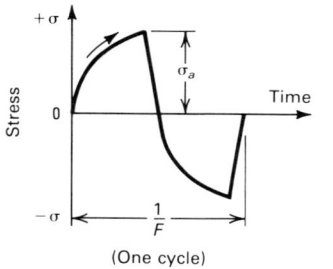 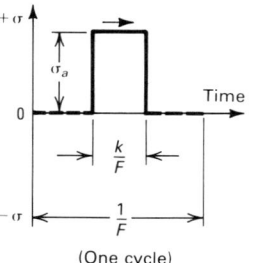

The linear damage rule has been used extensively in the evaluation of creep-fatigue interaction. It is based on the simple relationship (similar to Eq 5) that fatigue damage can be expressed as a cycle fraction and that creep damage can be expressed as a time fraction. It is also assumed that these quanti- ties can be added linearly to represent damage accumulation. Failure occurs when this summation reaches a certain value. Expressed mathematically, this damage rule is:

$$\sum_{n=1}^{f} \frac{N}{N_f} + \sum_{n=0}^{f} \frac{t}{t_R} = K \qquad \text{(Eq 9)}$$

where N is the number of cycles of exposure at a given strain range; N_f is the cycles to failure at a given strain range; t is the time of exposure to a given stress-temperature combination; t_R is the time to rupture at a given stress-temperature combination; and f indicates that summation occurs over the entire exposure life, at which point the total damage accumulation is equal to K.

Usually, K is assumed to be unity, because the cycle ratio summation should be unity when no creep damage is present. This assumption has been proven to be false for multiple exposures, although K must be unity for a single strain-range exposure. Similarly, when no fatigue is present, the time ratio summation should be unity; this also has been proven false for multiple exposures.

Equation 9 differs from Eq 6 because t' is not used and N_f is a measured quantity rather than a calculated value based on the Universal Slopes equation. Furthermore, Eq 9 applies to the complete cycle because t' of Eq 6 is automatically accounted for due to the use of a measured N_f value. The time ratio term in Eq 9 applies when constant-stress or constant-strain hold periods are introduced. With constant-strain hold periods and for a hold-period duration of t_H, when relaxation occurs this creep damage would be calculated in integral form:

$$\sum_{n=1}^{N} \int_{0}^{t_H} \frac{dt}{t_R} \qquad \text{(Eq 10)}$$

Creep-damage fractions (ratios) and fatigue-damage fractions (ratios) are usually presented graphically to form a creep-fatigue interaction diagram. In Fig. 26, the 45° diagonal has been drawn for $K = 1$ in Eq 9. The combinations of creep- and fatigue-damage fractions that define points in the region below the diagonal (for $K = 1$) represent safe operation. Points that fall on or above the diagonal correspond to failure conditions.

Generally, the creep-fatigue interaction diagram is not as shown in Fig. 26, because K usually is not equal to unity and the creep-fatigue interaction behavior usually is not linear. Several examples of creep-fatigue interaction diagrams are presented in Fig. 27 and 28. In Fig. 28, the fatigue-damage fraction is given in terms of the cycles to a 5.0% reduction in the peak cyclic stress. An example of a creep-fatigue damage summation for K greater than unity has been reported (Ref 44, 45) in creep-interspersion tests of 2.25Cr-1Mo steel at 538 °C (1000 °F).

Fractional Damage Equation. Results of some high-strain fatigue tests at room temperature and at 350 °C (660 °F) on a low-carbon, high-manganese pressure vessel steel have been reported (Ref 12). Values for the fatigue damage sustained were calculated using N_c/N_f, where N_f is the cycles to failure

Fig. 26 Creep-fatigue interaction diagram and linear creep-fatigue interaction

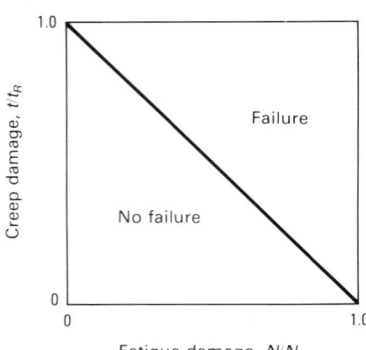

Fig. 27 Combined creep-fatigue data for austenitic stainless steel sheet tested in reversed bending at 700 °C (1290 °F)
Source: Ref 42

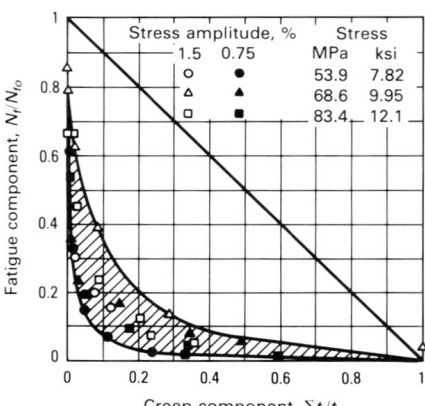

from continuous-cycle, room-temperature fatigue data for strain range ($\Delta\epsilon$), and N_c is the cycles to failure for a particular combination of hold period (t_h) and strain range ($\Delta\epsilon$). The creep-rupture damage was calculated as t_c/t_f, where t_f is the rupture time corresponding to the stress and temperature combination used during hold periods, obtained from monotonic creep data; and $t_c = N_c t_h$ is the total time spent at creep.

The computed values for N_c/N_f and t_c/t_f are plotted on a creep-fatigue interaction diagram in Fig. 29 and are represented by (Ref 12):

$$\log \frac{t_c}{t_f} + 4.25 \left[\left(\frac{N_c}{N_f}\right)^2 + 0.257 \left(\frac{N_c}{N_f}\right) \right] = 0$$

This result represents a significant deviation from the linear damage rule (Eq 5) represented by the 45° diagonal in Fig. 29.

The frequency-modified fatigue equation has been proposed (Ref 46) to introduce a time-dependency term to properly account for behavior observed at high temperature. Based on Ref 16 and 32, fatigue resistance may be represented by:

$$\nu^k t_f = \text{constant} \qquad (\text{Eq 11})$$

where ν is the frequency, t_f is the time to failure, and k is a constant that depends only on temperature. The right side of Eq 11 has been expressed (Ref 46) in terms of the plastic strain range, $\Delta\epsilon_p$, to yield:

$$\nu^k t_f = f(\Delta\epsilon_p) \qquad (\text{Eq 12})$$

and

$$\nu^k t_f = \nu^k \left(\frac{N_f}{\nu}\right) = N_f \nu^{k-1} \qquad (\text{Eq 13})$$

and a definition of the frequency-modified fatigue life. Plotting this quantity versus $\Delta\epsilon_p$ yields (Ref 46):

$$(N_f \nu^{k-1})^\beta \, \Delta\epsilon_p = C_2 \qquad (\text{Eq 14})$$

where β and C_2 are constants. The modified Basquin relationship was then employed:

$$\Delta\epsilon_e = \frac{\Delta\sigma}{E} = \frac{A'}{E} N_f^{-\beta'} \nu^{k'_1} \qquad (\text{Eq 15})$$

where $\Delta\epsilon_e$ is the elastic strain range; $\Delta\sigma$ is the stress range; E is the elastic modulus; and A', β', and k'_1 are constants.

The total strain range is then obtained by adding the elastic and plastic strain-range terms of Eq 14 and 15 and then eliminating $\Delta\epsilon_p$ to yield:

$$\Delta\epsilon = \frac{AC_2^{n'}}{E} N_f^{-\beta n'} \nu^{k+(1-k)\beta n'}$$

$$+ C_2 N_f^{-\beta} \nu^{(1-k)\beta} \qquad (\text{Eq 16})$$

Once the material constants are determined in supporting tests, Eq 16 can be used to estimate N_f values for given values of $\Delta\epsilon$ and frequency. For hold-time tests, the frequency term is taken as the reciprocal of the cycle period (i.e., hold period plus ramp time).

Equation 16 does not account for wave shape and thus does not distinguish between different strain rates existing in loading and unloading or between compression holds, tension holds, and hold periods in tension and compression. Therefore, the concept of frequency separation was introduced (Ref 47). This involves a concept for separating the hysteresis loop for a cycle into two parts, with a time dependency associated with each part. Tension-going and compression-going frequencies are defined as ν_t and ν_c, respectively, where tension-going refers to the part of the loop in which the plastic strain rate is positive. Correspondingly, the compression-going part of the loop involves a negative plastic strain rate. Stress-range and plastic strain-range terms are also defined for use in the type of expression shown in Eq 16. These expressions involve the use of tension-going and compression-going strain ratios that can be converted into frequencies.

Fig. 28 Creep-fatigue interaction plot for hold-time data obtained in tests of type 304 stainless steel at 650 °C (1200 °F)
Source: Ref 43

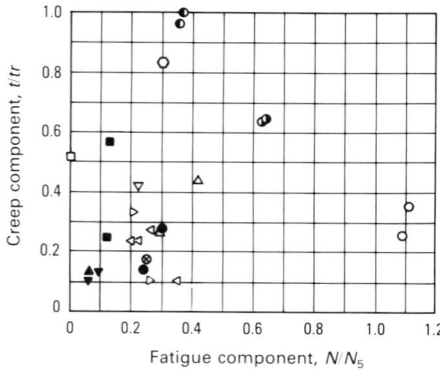

	Total strain range, %	Hold time, min		Total strain range, %	Hold time, min
□	0.25	10	◑	2.0	1
△	0.5	1	◐	2.0	10
■	0.5	10	◁	2.0	30
▼	0.5	30	▷	2.0	60
▲	0.5	60	●	2.0	180
▽	1.0	10	⊗	2.0	600
○	2.0	0.1	◯	4.0	10

Fig. 29 Creep-fatigue interaction for a low-carbon high-manganese steel at 350 °C (660 °F)

t_c and N_c refer to time and cycles associated with failure. Source: Ref 12

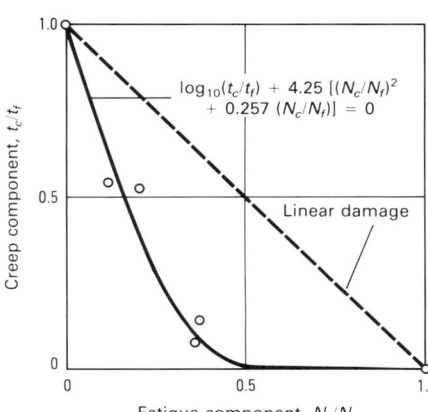

This approach has been applied (Ref 47) to data for many different alloys and has shown agreement between predicted and measured N_f values to within a factor of ± 2 (i.e., the range of measured values is between the predicted value multiplied by 2 and the predicted value divided by 2).

Strain-range partitioning focuses on the inelastic strain present in a cycle and the two directions of straining (Ref 48). These two directions are tension (positive inelastic strain rate) and compression (negative inelastic strain rate). In addition, two types of inelastic strain are considered: time-dependent (creep) and time-independent (plasticity).

By combining the two directions with the two types of strain, it is possible to achieve four types of strain ranges that might be encountered in any conceivable hysteresis loop. These define the manner in which a tensile component of strain is balanced by a compressive component to close a hysteresis loop:

- Tensile plasticity reversed by compressive plasticity is designated a *PP* strain range and is represented by $\Delta \epsilon_{PP}$.
- Tensile creep reversed by compression plasticity is designated a *CP* strain range and is represented by $\Delta \epsilon_{CP}$.
- Tensile plasticity reversed by compressive creep is designated a *PC* strain range and is represented by $\Delta \epsilon_{PC}$.
- Tensile creep reversed by compressive creep is designated a *CC* strain range and is represented by $\Delta \epsilon_{CC}$.

The subscript notation for the strain ranges indicates the type of tensile strain first, followed by the type of compressive strain.

Strain-range partitioning represents the premise that, in order to handle a complex high-temperature low-cycle fatigue problem, the inelastic strain range must first be divided or partitioned into its components.

It was then proposed that each of the inelastic strain-range types be treated on a plot of inelastic strain range versus N_f to yield four straight lines, as shown in Fig. 30. These lines must be established from results obtained in special tests; once available, they can be used in a strain-range partitioning analysis of any type of cycle. Figure 31 presents four types of idealized hysteresis loops as examples of loops that could be used to generate the life relationships in Fig. 30.

Figure 32 is a complex hysteresis loop composed of a series of loading sequences. Starting at point 1, load is applied rapidly to point 2, and then a constant stress is held until point 3 is reached. Elastic unloading then takes place to point 4, and compressive loading continues rapidly to point 5. The compressive stress is then held constant until point 6 is reached, when the strain is held constant and the stress is relaxed to point 7. The cycle is completed by unloading elastically to point 1, where the cycle started.

In this loop, the inelastic strain range is defined by the width of the loop AC. In going from A to C in the tension direction, a tensile plastic strain, AB, was accumulated along with a tensile creep strain, BC. In reversing the cycle, a compressive plastic strain, CD, was accumulated along with a compressive creep strain, DA. In this example, the tensile plastic strain AB was reversed by a portion of the compressive plastic strain CD; likewise,

the entire compressive creep strain DA was reversed by only a portion of the tensile creep strain BC. Therefore, a *PP* strain range of magnitude AB was present, along with a *CC* strain range of magnitude DA. From the excess tensile creep strain and excess compressive plastic strain, a *CP* strain range of magnitude BC-DA or CD-AB (because CD-$AB = BC$-DA) is present.

In any hysteresis loop, it is possible to have a maximum of only three of the four types of strain ranges. It is not possible for the *PC* and *CP* strain ranges to be components of the same hysteresis loop.

For this model, the inelastic strain range, $\Delta \epsilon_{IN} = AC$ is made up of three components: $\Delta \epsilon_{PP} = AB$, $\Delta \epsilon_{CC} = DA$, and $\Delta \epsilon_{CP} = (BC$-$DA)$. In equation form, $\Delta \epsilon_{IN} = \Delta \epsilon_{PP} + \Delta \epsilon_{CC} + \Delta \epsilon_{CP}$.

Fig. 30 Typical partitioned strain range/life relationships used to characterize material behavior in the creep-fatigue range

See text for explanation of symbols. Source: Ref 48

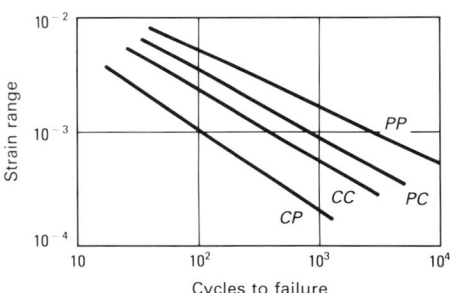

Fig. 31 Idealized hysteresis loops used in defining individual partitioned strain range/life relationships

(a) PP-type cycle. (b) CP-type cycle. (c) PC-type cycle. (d) CC-type cycle. See text for explanation of symbols. Source: Ref 48

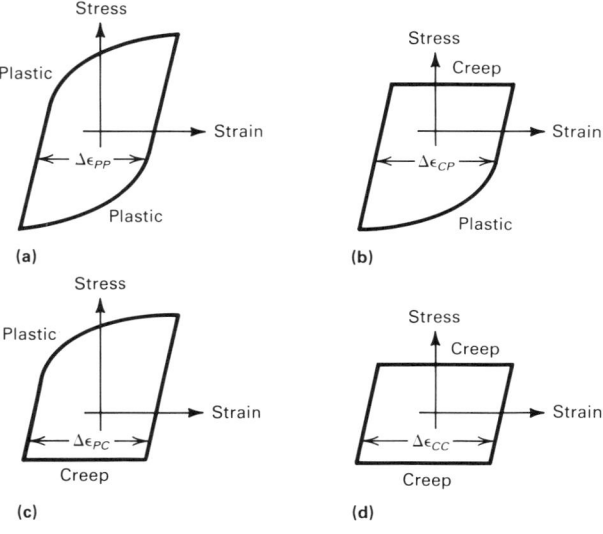

Fig. 32 Defining partitioned strain-range components of complex hysteresis loop
Source: Ref 48

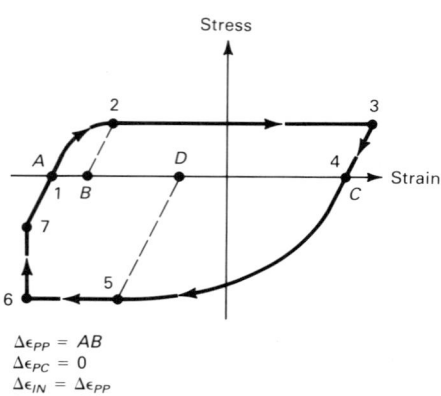

$\Delta\epsilon_{PP} = AB$
$\Delta\epsilon_{PC} = 0$
$\Delta\epsilon_{IN} = \Delta\epsilon_{PP}$
$\Delta\epsilon_{CC} = DA$
$\Delta\epsilon_{CP} = (BC - DA)$
$\Delta\epsilon_{IN} = \Delta\epsilon_{PP} + \Delta\epsilon_{CC} + \Delta\epsilon_{CP} = AC$

Prediction of cyclic life is performed by using a modification of the linear damage rule. This has been termed the interaction damage rule and is given by:

$$\frac{F_{PP}}{N_{PP}} + \frac{F_{CC}}{N_{CC}} + \frac{F_{CP}}{N_{CP}} + \frac{F_{PC}}{N_{PC}} = \frac{1}{N_f}$$

where N_f is the predicted life, and the F terms are as defined in Fig. 33. For the inelastic strain range (in this case, $\Delta\epsilon_{IN} = AC$), the values of cyclic lives N_{PP}, N_{CC}, and N_{CP} can be read. It is not necessary to obtain a value of N_{PC} for this example, because it was determined that no such strain-range component exists ($\Delta\epsilon_{PC} = 0$). As the next step, the fractions for each of the partitioned strain ranges, as shown in Fig. 33, should be calculated. For this example, $F_{PP} = \Delta\epsilon_{PP}/\Delta\epsilon_{IN} = AB/AC$, $F_{CC} = \Delta\epsilon_{CC}/\Delta\epsilon_{IN} = DA/AC$, and $F_{CP} = \Delta\epsilon_{CP}/\Delta\epsilon_{IN} = (BC - DA)/AC$. Note that $F_{PP} + F_{CC} + F_{CP} = 1$. The damage per cycle due to each of the components can be represented by F_{PP}/N_{PP}, F_{CC}/N_{CC}, and F_{CP}/N_{CP}. The total damage per cycle ($1/N_f$) is equal to the sum of the individual damage contributions. Hence:

$$\frac{1}{N_f} = \frac{F_{PP}}{N_{PP}} + \frac{F_{CC}}{N_{CC}} + \frac{F_{CP}}{N_{CP}}$$

This approach has been evaluated using data for numerous alloys. As in the case of the frequency separation approach, the agreement has been within a factor of ±2.

Other approaches to creep-fatigue interaction analysis have been proposed and have been partially successful. Among these are the tensile hysteresis energy approach (Ref 49), the *t-n* diagram approach (Ref 50), and the damage rate approach (Ref 51). Although each approach has yielded satisfac-

Fig. 33 Interaction damage rule
See text for explanation of symbols. Source: Ref 48

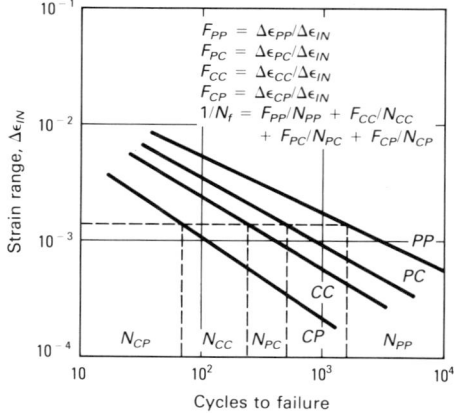

Cycles to failure

tory results in particular instances, they have not been accepted as generally as strain-range partitioning and frequency separation. The damage rate approach has some unique features and with experience will probably be viewed as an effective method.

REFERENCES

1. A.H. Meleka, *Met. Rev.*, Vol 7 (No. 25), 1962, p 43
2. M. Hempel and H.E. Tillmanns, *Mitt. K.W. Inst. Eisenforsch.*, Vol 18, 1936, p 163
3. M. Hempel and F. Ardelt, *Mitt. K.W. Inst. Eisenforsch.*, Vol 21, 1939, p 115
4. M. Hempel and H. Krug, *Mitt. K.W. Inst. Eisenforsch.*, Vol 24, 1942, p 77
5. H.J. Tapsell, P.G. Forrest, and G.R. Tremain, *Engineering*, Vol 170, 1950, p 189
6. H.J. Tapsell, *Symp. High-Temperature Steels and Alloys for Gas Turbines*, Special Rep. No. 43, Iron and Steel Institute, London, 1952, p 169
7. A.J. Kennedy, *Proc. Int. Conf. Fatigue of Metals*, Institution of Mechanical Engineers, London, 1956, p 401
8. A.H. Meleka and A.V. Evershed, *J. Inst. Met.*, Vol 88, 1959-60, p 411
9. A.H. Meleka and G.B. Dunn, *Nature*, Vol 184, 1959, p 896
10. S.S. Manson, Interfaces Between Fatigue, Creep, and Fracture, *Int. J. Fracture Mech.*, Vol 2 (No. 1), 1966, p 327
11. E.G. Ellison and E.M. Smith, *Fatigue at Elevated Temperatures*, STP 520, ASTM, Philadelphia, 1972, p 575
12. D.S. Wood, The Effect of Creep on the High-Strain Fatigue Behavior of a Pressure Vessel Steel, *Weld. J. Res. Suppl.*, Vol 45, 1966, p 92-S/96-S
13. E.G. Ellison, *J. Mech. Eng. Sci.*, Vol 11 (No. 3), 1969, p 318
14. A.V. Evershed and G.B. Dunn, *Br. Iron Steel Res. Assoc. Rep.*, 1958, p 16
15. L.F. Coffin, Jr., Introduction to High-Temperature Low-Cycle Fatigue, *Exp. Mech.*, Vol 8, May 1968, p 218
16. J.F. Eckel, The Influence of Frequency on the Repeated Bending Life of Acid Lead, *Proc. ASTM*, Vol 51, 1951, p 745
17. G.R. Gohn and W.C. Ellis, The Fatigue Test as Applied to Lead Cable Sheath, *Proc. ASTM*, Vol 51, 1951, p 721
18. R.E. Peterson, Design Approaches for Low-Cycle Fatigue Problems in Power Apparatus, *Tenth Sagamore Conference*, 1963, Syracuse University Research Institute, Syracuse, NY
19. E. Krempl and B.M. Wundt, *Hold-Time Effects in High-Temperature, Low-Cycle Fatigue: A Literature Survey and Interpretive Report*, STP 489, ASTM, Philadelphia, 1971
20. A.E. Carden, Fatigue at Elevated Temperatures: A Review of Test Methods, in *Fatigue at Elevated Temperatures*, STP 520, ASTM, Philadelphia, 1973, p 195-223
21. W.R. Berry and I. Johnson, Prevention of Cyclic Thermal-Stress Cracking in Steam Turbine Rotors, *J. Eng. Power*, Vol 86A, 1964, p 361-367
22. R.P. Kent, Some Aspects of Metallurgical Research and Development Applied to Large Steam Turbines, *Parsons J.*, Dec 1964
23. H.G. Edmunds, Repeated Cyclic Strains, *Proc. Inst. Mech. Eng.*, Vol 180 (Part 31), 1965/1966, p 373-379
24. J.B. Conway, R.H. Stentz, and J.T. Berling, *Fatigue, Tensile, and Relaxation Behavior of Stainless Steels*, TID-26135, U.S. Atomic Energy Commission, 1975
25. J.T. Berling and T. Slot, Effect of Temperature and Strain Rate on Low-Cycle Fatigue Resistance of AISI 304, 316 and 348 Stainless Steels, in *Fatigue at High Temperature*, STP 459, ASTM, Philadelphia, 1969, p 3-30; also USAEC Report GEMP-642, General Electric Co., Cincinnati, 1968
26. W.F. Simmons and H.C. Cross, *Report on the Elevated Temperature Properties of Stainless Steels*, STP 124, ASTM, Philadelphia, 1952
27. J.B. Conway, J.T. Berling, and R.H. Stentz, Correlating the Effects of Hold Time and Strain Rate on Low-Cycle Fatigue Behavior, *Proc. Int. Conf. Thermal Stresses and Thermal Fatigue*, Berkeley, England, 1969, D.J. Littler, Ed., Butterworths, London, 1971
28. J.T. Berling and J.B. Conway, Effect of Hold Time on the Low-Cycle Fatigue

Resistance of 304 Stainless Steel at 1200 °F, *Proc. 1st Int. Conf. Pressure Vessel Technology*, Part 2, *Materials and Fabrication*, American Society of Mechanical Engineers, New York, 1970, p 1233

29. J.B. Conway, J.T. Berling, and R.H. Stentz, Strain Rate and Holdtime Saturation in Low-Cycle Fatigue: Design-Parameter Plots, in *Fatigue at Elevated Temperatures*, STP 520, ASTM, Philadelphia, 1972, p 637-647

30. R.A.T. Dawson *et al.*, High Strain Fatigue of Austenitic Steels, *Proc. Int. Conf. Thermal and High-Strain Fatigue*, London, 1967, Monograph and Report Series No. 32, Metals and Metallurgy Trust, London, 1967

31. C.D. Walker, Strain Fatigue Properties of Some Steels at 950 °F (510 °C) with a Hold in the Tension Part of the Cycle, *Joint Conf. Creep*, Institution of Mechanical Engineers, London, 1963, Session 3, Paper 24, p 49

32. A. Coles *et al.*, The High Strain Fatigue Properties of Low-Alloy Creep Resisting Steels, *Proc. Int. Conf. Thermal and High-Strain Fatigue*, Monograph and Report Series No. 32, Metals and Metallurgy Trust, London, 1967

33. A. Coles and D. Skinner, Assessment of Thermal-Fatigue Resistance of High Temperature Alloys, *J. Roy. Aeronaut. Soc.*, Vol 69, 1965, p 53-55

34. J.B. Conway, J.T. Berling, and R.H. Stentz, New Correlations Involving the Low-Cycle Fatigue and Short-Time Tensile Behavior of Irradiated and Unirradiated 304 and 316 Stainless Steel, *Nucl. Appl. Tech.*, Vol 9, 1970, p 31

35. H.G. Edmunds and D.J. White, Observations of the Effect of Creep Relaxation on High-Strain Fatigue, *J. Mech. Eng. Soc.*, Vol 8 (No. 3), 1966, p 310

36. E. Krempl and C.D. Walker, *The Effect of Rupture Ductility and Hold-Time on the 1000 F Strain-Fatigue Behavior of 1Cr-1Mo-0.25V Steel*, STP 459, ASTM, Philadelphia, 1969, p 75

37. C.H. Wells and C.P. Sullivan, Low-Cycle Fatigue Damage of Udimet 700 at 1400 °F, *ASM Trans. Quart.*, Vol 58, 1965, p 391

38. R.H. Stentz, private communication, 1984

39. J.B. Conway and R.H. Stentz, "High Temperature, Low Cycle Fatigue Data for Three High Strength Nickel-Base Superalloys," Tech. Rep. AFWAL-TR-9-4077, Wright-Patterson Air Force Base, OH, June 1980

40. E.G. Ellison and A.J.F. Paterson, *Creep Fatigue Interactions in a 1 Cr-Mo-V Steel*, *Proc. Inst. Mech. Eng.*, Vol 190, 1976, p 321

41. S.S. Manson, Fatigue: A Complex Subject—Some Simple Approximations, *Exp. Mech.*, Vol 5 (No. 7), 1965, p 193

42. R. Lagneborg and R. Attermo, The Effect of Combined Low-Cycle Fatigue and Creep on the Life of Austenitic Stainless Steels, *Met. Trans.*, Vol 2, 1971, p 1821

43. R.D. Campbell, Creep/Fatigue Interaction Correlation for 304 Stainless Steel Subjected to Strain-Controlled Cycling with Hold Times at Peak Strain, *Trans. ASME, Ser. B., J. Eng. Ind.*, Vol 93 (No. 4), 1971, p 88

44. R.M. Curran and B.M. Wundt, A Study of Low-Cycle Fatigue and Creep Interactions in Steels at Elevated Temperatures, in *Reports of Current Work in Behavior of Materials at Elevated Temperatures*, A.O. Schaefer, Ed., ASME Publ. G87, American Society of Mechanical Engineers, New York, 1974, p 1-104

45. R.M. Curran and B.M. Wundt, Continuation of a Study of Low-Cycle Fatigue and Creep Interaction in Steels at Elevated Temperatures, in *1976 ASME-MPC Symp. on Creep-Fatigue Interaction*, R.M. Curran, Ed., ASME Publ. G00112 (MPC-3), American Society of Mechanical Engineers, New York, 1976, p 203-282

46. L.F. Coffin, Jr., The Effect of Frequency on High Temperature, Low-Cycle Fatigue, *Proc. Air Force Conf. Fracture and Fatigue of Aircraft Structures*, AFDL-TR-70-144, 1970, p 301-312

47. L.F. Coffin, Jr. *et al.*, *Time-Dependent Fatigue of Structural Alloys, A General Assessment*, ORNL-5073, Oak Ridge National Laboratory, Oak Ridge, TN, 1975

48. S.S. Manson, G.R. Halford, and M.H. Hirschberg, Creep-Fatigue Analysis by Strain-Range Partitioning, *First Symp. Design for Elevated Temperature Environment*, American Society of Mechanical Engineers, New York, 1971, p 12-28

49. W.J. Ostergren, Correlation of Hold Time Effects in Elevated Temperature Low-Cycle Fatigue Using a Frequency Modified Damage Function, *ASME-MPC Symp. Creep-Fatigue Interaction*, MPC-3, American Society of Mechanical Engineers, New York, Dec 1976, p 179-202

50. J.R. Ellis and E.P. Esztergar, Consideration of Creep-Fatigue Interaction in Design Analysis, *Symp. Design for Elevated Temperature Environment*, American Society of Mechanical Engineers, New York, 1971, p 29-43

51. S. Majumdar and P.S. Maiya, A Unified and Mechanistic Approach to Creep-Fatigue Damage, *Proc. 2nd Int. Conf. Mechanical Behavior of Materials*, ICM-II, American Society for Metals, 1976, p 924-928 (see also ANL-76-58, Argonne National Laboratory, Argonne, IL, Jan 1976)

SELECTED REFERENCES

- C. Amzallag *et al.*, *Low Cycle Fatigue and Life Prediction*, STP 770, ASTM, Philadelphia, 1982

- J.T. Berling and J.B. Conway, A Proposed Method for Predicting the Low-Cycle Fatigue Behavior of 304 and 316 Stainless Steel, *Trans. Met. Soc. AIME*, Vol 245, 1969, p 1137-1140

- J.T. Berling and J.B. Conway, New Approach to the Prediction of Low-Cycle Fatigue Data, *Met. Trans.*, Vol 1, 1970, p 805-809

- L.F. Coffin, Jr., "A Generalized Equation for Predicting High-Temperature Low-Cycle Fatigue, Including Hold Times," Rep. 69-C-401, General Electric Co., Research and Development Center, Schenectady, Dec 1969

- J.B. Conway, "Evaluation of Plastic Fatigue Properties of Heat-Resistant Alloys," USAEC Rep. GEMP-740, General Electric Co., Cincinnati, 1969

- J.B. Conway, "Short-Term Tensile and Low-Cycle Fatigue Studies of Incoloy 800," USAEC Rep. GEMP-732, General Electric Co., Cincinnati, 1969

- J.B. Conway and J.T. Berling, A New Correlation of Low-Cycle Fatigue Data Involving Hold Periods, *Met. Trans.*, Vol 1 (No. 1), 1970, p 324

- J.B. Conway, J.T. Berling, and R.H. Stentz, A Brief Study of Cumulative Damage in Low-Cycle Fatigue Testing of AISI 304 Stainless Steel at 650 °C, *Met. Trans.*, Vol 1, 1970, p 2034

- J.B. Conway, J.T. Berling, and R.H. Stentz, A Temperature Correlation of the Low-Cycle Fatigue Data for 304 Stainless Steel, *Met. Trans.*, Vol 2 (No. 11), 1971, p 3247

- R.M. Curran and B.M. Wundt, A Program to Study Low-Cycle Fatigue and Creep Interaction in Steel at Elevated Temperatures, *2¼ Chrome 1 Molybdenum Steel in Pressure Vessels and Piping*, American Society of Mechanical Engineers, New York, 1972, p 49

- J. Dubuc *et al.*, Unified Theory of Cumulative Damage in Metal Fatigue, *Weld. Res. Council Bull.*, No 162, 1971

- E.P. Esztergar and J.R. Ellis, Cumulative Damage Concepts in Creep-Fatigue Life Prediction, in *Proc. Int. Conf. Thermal*

Stress and Thermal Fatigue, Berkeley, England, 1969, D.J. Littler, Ed., Butterworths, London, 1971

- P.G. Forrest, The Fatigue Behavior of Mild Steels at Temperatures up to 500 °C, *J. Iron Steel Inst.*, Vol 200, 1962, p 452
- P.H. Frith, Fatigue Tests at Elevated Temperatures, *Symp. High Temperature Steels and Alloys for Gas Turbines,* Special Rep. No. 43, Iron and Steel Institute, London, 1951
- G.R. Halford, Cyclic Creep-Rupture Behavior of Three High-Temperature Alloys, *Met. Trans.*, Vol 3 (No. 8), 1972, p 2247
- A.J. Kennedy, *Processes of Creep and Fatigue in Metals*, John Wiley & Sons, New York, 1962, p 155

- A.J. Kennedy, Interactions Between Creep and Fatigue in Aluminum and in Certain of its Alloys, *Int. Conf. Creep*, Institution of Mechanical Engineers, London, 1963, Session 3, p 17
- M.M. Leven, The Interaction of Creep and Fatigue for a Rotor Steel, *Exp. Mech.*, Vol 13 (No. 9), 1973, p 353-372
- S.S. Manson and G. Halford, A Method of Estimating High Temperature Low Cycle Fatigue Behavior of Materials, in *Proc. Int. Conf. Thermal and High-Strain Fatigue*, Monograph and Rep. Series No. 32, Metals and Metallurgy Trust, London, 1967
- S.S. Manson, G.R. Halford, and D.A. Spera, The Role of Creep in High Temperature Low Cycle Fatigue, *A.E. Johnson Memorial Volume*, A.I. Smith, Ed., American Elsevier, New York, 1971
- R.W. Swindeman, *The Strain Fatigue Properties of Inconel*, Part II, ORNL-3250, Oak Ridge National Laboratory, Oak Ridge, TN, 1962
- R.W. Swindeman, The Inter-Relation of Cyclic and Monotonic Creep Rupture, *Int. Conf. Creep*, Institution of Mechanical Engineers, London, 1963, Session 3, p 71
- S. Taira, *Creep in Structures*, Academic Press, New York, 1962, p 96
- D.J. White, Effect of Environment and Hold Time on the High Strain Fatigue Endurance of ½ Percent Molybdenum Steel, *Proc. I Mechanical Engineers 184*, Part I, 1969-70, p 223

Fatigue Testing

Introduction

FATIGUE is the progressive, localized, permanent structural change that occurs in materials subjected to fluctuating stresses and strains that may result in cracks or fracture after a sufficient number of fluctuations. Fatigue fractures are caused by the simultaneous action of cyclic stress, tensile stress and plastic strain. If any one of these three is not present, fatigue cracking will not initiate and propagate. The cyclic stress starts the crack; the tensile stress produces crack growth (propagation). Although compressive stress will not cause fatigue, compression load may do so.

The process of fatigue consists of three stages:

- Initial fatigue damage leading to crack nucleation and crack initiation
- Progressive cyclic growth of a crack (crack propagation) until the remaining uncracked cross section of a part becomes too weak to sustain the loads imposed
- Final, sudden fracture of the remaining cross section

Fatigue cracking normally results from cyclic stresses that are well below the static yield strength of the material. (In low-cycle fatigue, however, or if the material has an appreciable work-hardening rate, the stresses also may be above the static yield strength.)

Fatigue cracks initiate and propagate in regions where the strain is most severe. Because most engineering materials contain defects and thus regions of stress concentration that intensify strain, most fatigue cracks initiate and grow from structural defects. Under the action of cyclic loading, a plastic zone (or region of deformation) develops at the defect tip. This zone of high deformation becomes an initiation site for a fatigue crack. The crack propagates under the applied stress through the material until complete fracture results. On the microscopic scale, the most important feature of the fatigue process is nucleation of one or more cracks under the influence of reversed stresses that exceed the flow stress, followed by development of cracks at persistent slip bands or at grain boundaries.

Prediction of Fatigue Life

The fatigue life of any specimen or structure is the number of stress (strain) cycles required to cause failure. This number is a function of many variables, including stress level, stress state, cyclic wave form, fatigue environment, and the metallurgical condition of the material. Small changes in the specimen or test conditions can significantly affect fatigue behavior, making analytical prediction of fatigue life difficult. Therefore, the designer may rely on experience with similar components in service rather than on laboratory evaluation of mechanical test specimens. Laboratory tests, however, are essential in understanding fatigue behavior, and current studies with fracture mechanics test specimens are beginning to provide satisfactory design criteria.

Laboratory fatigue tests can be classified as crack initiation or crack propagation. In crack initiation testing, specimens or parts are subjected to the number of stress cycles required for a fatigue crack to initiate and to subsequently grow large enough to produce failure.

In crack propagation testing, fracture mechanics methods are used to determine the crack growth rates of preexisting cracks under cyclic loading. Fatigue crack propagation may be caused by cyclic stresses in a benign environment, or by the combined effects of cyclic stresses and an aggressive environment (corrosion fatigue).

Following a review of the basic concepts related to fatigue testing and data generation, the above test methodologies will be described in articles on "Fatigue Crack Initiation," "Fatigue Crack Propagation," and "Environmental Effects on Fatigue Crack Propagation."

Fatigue Crack Initiation

Most laboratory fatigue testing is done either with axial loading, or in bending, thus producing only tensile and compressive stresses. The stress usually is cycled either between a maximum and a minimum tensile stress, or between a maximum tensile stress and a maximum compressive stress. The latter is considered a negative tensile stress, is given an algebraic minus sign, and therefore is known as the minimum stress.

The stress ratio is the algebraic ratio of two specified stress values in a stress cycle. Two commonly used stress ratios are the ratio, A, of the alternating stress amplitude to the mean stress ($A = S_a/S_m$) and the ratio, R, of the minimum stress to the maximum stress ($R = S_{min}/S_{max}$).

If the stresses are fully reversed, the stress ratio R becomes -1; if the stresses are partially reversed, R becomes a negative number less than 1. If the stress is cycled between a maximum stress and no load, the stress ratio R becomes zero. If the stress is cycled between two tensile stresses, the stress ratio R becomes a positive number less than 1. A stress ratio R of 1 indicates no variation in stress, making the test a sustained-load creep test rather than a fatigue test.

Applied stresses are described by three parameters. The mean stress, S_m, is the algebraic average of the maximum and minimum stresses in one cycle, $S_m = (S_{max} + S_{min})/2$. In the completely reversed cycle test, the mean stress is zero. The range of stress, S_r, is the algebraic difference between the maximum and minimum stresses in one cycle, $S_r = S_{max} - S_{min}$. The stress amplitude, S_a, is one half the range of stress, $S_a = S_r/2 = (S_{max} - S_{min})/2$.

During a fatigue test, the stress cycle usually is maintained constant so that the applied stress conditions can be written $S_m \pm S_a$, where S_m is the static or mean stress, and S_a is the alternating stress, which is equal to half the stress range. Nomenclature to describe test parameters involved in cyclic stress testing are shown in Fig. 1.

S-N Curves. The results of fatigue crack initiation tests usually are plotted as maximum stress, minimum stress, or stress amplitude to number of cycles, N, to failure using a logarithmic scale for the number of cycles. Stress is plotted on either a linear or a logarithmic scale. The resulting plot of the data is an S-N curve. Three typical S-N curves are shown in Fig. 2.

The number of cycles of stress that a metal can endure before failure increases with decreasing stress. For some engineering materials such as steel (see Fig. 2) and titanium, the S-N curve becomes horizontal at a certain limiting stress. Below this limiting stress, known as the fatigue limit or endurance limit, the material can endure an infinite number of cycles without failure. Statistical characterization of the S-N curve and the techniques for defining a mean fatigue curve and evaluating scatter or variability about that mean are discussed in the article "Fatigue Data Analysis" in this Volume.

Fatigue Limit and Fatigue Strength. The horizontal portion of an S-N curve represents the maximum stress that the metal can withstand for an infinitely large number of cycles with 50% probability of failure and is known as the fatigue (endurance) limit, S_f. Most nonferrous metals do not exhibit a fatigue limit. Instead, their S-N curves continue to drop at a slow rate at high numbers of cycles, as shown by the curve for aluminum alloy 7075-T6 in Fig. 2.

For these types of metals, fatigue strength rather than fatigue limit is reported, which is the stress to which the metal can be subjected for a specified number of cycles. Because there is no standard number of cycles, each table of fatigue strengths must specify the number of cycles for which the strengths are

reported. The fatigue strength of nonferrous metals at 100 million (10^8) or 500 million (5×10^8) cycles is erroneously called the fatigue limit.

Low-Cycle Fatigue. For the low-cycle fatigue region ($N < 10^4$ cycles) tests are conducted with controlled cycles of elastic plus plastic strain, rather than with controlled load or stress cycles. Under controlled strain testing, fatigue life behavior is represented by a log-log plot of the total strain range, $\Delta\epsilon_t$, versus the number of cycles to failure (Fig. 3).

The total strain range is separated into elastic and plastic components. For many metals and alloys, the elastic strain range, $\Delta\epsilon_e$, is equal to the stress range divided by the modulus of elasticity. The plastic strain range, $\Delta\epsilon_p$, is the difference between the total strain range and the elastic strain range. For more information on low-cycle fatigue, see the article "Fatigue Crack Initiation" in this Section.

Stress-Concentration Factor. Stress is concentrated in a metal by structural discontinuities, such as notches, holes, or scratches, which act as stress raisers. The stress-concentration factor, K_t, is the ratio of the area test stress in the region of the notch (or other stress concentrators) to the corresponding nominal stress. For determination of K_t, the greatest stress in the region of the notch is calculated from the theory of elasticity, or equivalent values are derived experimentally.

The fatigue notch factor, K_f, is the ratio of the fatigue strength of a smooth (unnotched) specimen to the fatigue strength of a notched specimen at the same number of cycles.

Fatigue notch sensitivity, q, for a material is determined by comparing the fatigue notch factor, K_f, and the stress-concentration factor, K_t, for a specimen of a given size containing a stress concentrator of a given shape and size. A common definition of fatigue notch sensitivity is $q = (K_f - 1)/(K_t - 1)$, in which q may vary between zero (where $K_f = 1$) and 1 (where $K_f = K_t$). This value may be stated as percentage.

Fatigue Crack Propagation

In large structural components, the existence of a crack does not necessarily imply imminent failure of the part. Significant structural life may remain in the cyclic growth of the crack to a size at which a critical failure occurs. The objective of fatigue crack propagation testing is to determine the rates at which subcritical cracks grow under cyclic loadings prior to reaching a size criti-

Fig. 2 Typical S-N curves for constant amplitude and sinusoidal loading

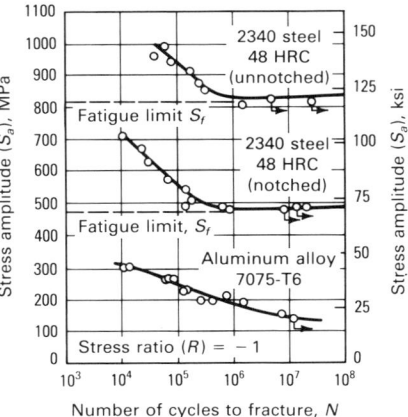

Fig. 3 Typical plot of strain range versus cycles-to-failure for low-cycle fatigue

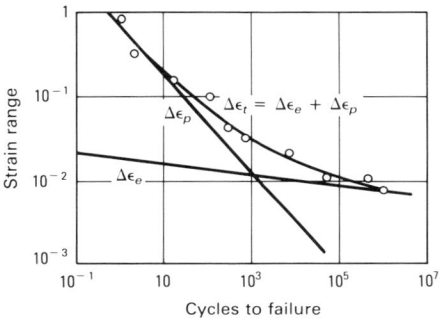

Fig. 4 Fatigue crack propagation rate data in 7075-T6 aluminum alloy

$R < 0$

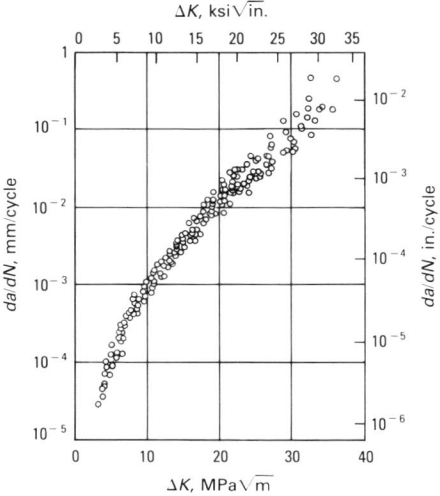

Fig. 1 Nomenclature to describe test parameters involved in cyclic stress testing

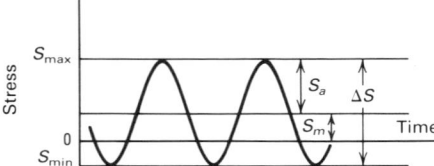

cal for fracture. For an in-depth discussion, see the article "Fatigue Crack Propagation" in this Section.

The growth or extension of a fatigue crack under cyclic loading is principally controlled by maximum load and stress ratio. However, as in crack initiation, there are a number of additional factors that may exert a strong influence, including environment, frequency, temperature, and grain direction. Fatigue crack propagation testing usually involves constant-load-amplitude cycling of notched specimens that have been pre-cracked in fatigue. Crack length is measured as a function of elapsed cycles, and these data are subjected to numerical analysis to establish the rate of crack growth, da/dN. Methods for numerically determining the crack growth rate can be found in the article "Fatigue Data Analysis" in this Volume.

Crack growth rates are expressed as a function of the crack tip stress-intensity factor range, ΔK. The stress-intensity factor is calculated from expressions based on linear elastic stress analysis and is a function of crack size, load range, and cracked specimen geometry. Detailed information on the expressions used to define stress intensity from stress field analysis can be found in the article "Fracture Mechanics" in this Volume. Fatigue crack growth data is typically presented in a log-log plot of da/dN versus ΔK (Fig. 4).

Fatigue Crack Initiation

By Ernest J. Czyryca
Metallurgist
David Taylor Naval Ship Research
and Development Center

CRACK INITIATION TESTS are procedures in which a specimen or part is subjected to cyclic loading to failure. A large portion of the total number of cycles in these tests is spent initiating the crack. Although crack initiation tests conducted on small specimens do not precisely establish the fatigue life of a large part, such tests do provide data on the intrinsic fatigue crack initiation behavior of a metal or alloy. As a result, such data can be utilized to develop criteria to prevent fatigue failures in engineering design. Examples of the use of small-specimen fatigue test data can be found in the basis of the fatigue design codes for boilers and pressure vessels, complex welded, riveted, or bolted structures, and automotive and aerospace components.

Following a description of the phenomena of crack nucleation and crack initiation, this article examines specimen design and preparation, as well as the apparatus used in crack initiation testing. In addition, variables that influence fatigue properties, such as the effect of mean stress, stress concentrations, stress amplitude, residual stresses, and surface properties, are reviewed.

For information on the planning and design of fatigue tests and the statistical analysis of the test results, see the article "Fatigue Data Analysis" in this Volume. For an explanation of the symbols, terminology, and basic concepts used in this article, see the Introduction to this Section.

Fatigue Cracking

Fatigue cracks normally result from cyclic stresses that are below the yield strength of the metal. In low-cycle fatigue, however, the cyclic stress may be above the static yield strength, especially in a material with an appreciable work-hardening rate. Generally, a fatigue crack is initiated at a highly stressed region of a component subjected to cyclic loading of sufficient magnitude. The crack then propagates in progressive cyclic growth through the cross section of the part until the maximum load cannot be carried, and complete fracture results.

On a microscopic scale, the most important feature of the fatigue cracking process is the nucleation of one or more crack sites subjected to stress reversals exceeding the local flow stress. These sites may be at persistent slip bands (those deeper than several microns below the free surface), within grains, or at grain boundaries. The development of the microcrack at the nucleation site to a detectable fatigue crack is called crack initiation and may comprise a large portion of the total fatigue life of a component.

Crack Nucleation. A variety of crystallographic features have been observed to nucleate fatigue cracks. In pure metals, tubular holes that develop in persistent slip bands, slip band extrusion-intrusion pairs at free surfaces, and twin boundaries are common nucleation sites. Grain boundaries in polycrystalline metals, even in the absence of inherent grain boundary weakness, are crack nucleation sites. At high strain rates, this appears to be the preferred site. Nucleation at grain boundaries appears to be a geometrical effect, whereas nucleation at twin boundaries is associated with active slip on crystallographic planes immediately adjacent and parallel to the twin boundary.

The foregoing processes also occur in alloys and heterogeneous materials. However, alloying and commercial production practices introduce segregation, inclusions, second-phase particles, and other features that disturb the structure. All of these phenomena have a significant influence on the crack nucleation process. In general, alloying that (1) enhances cross slip, (2) enhances twinning, or (3) increases the rate of work hardening will stimulate crack nucleation. On the other hand, alloying usually raises the flow stress of a metal, thus offsetting its potentially detrimental effect on fatigue crack nucleation.

Crack Initiation. Fatigue cracks initiate at points of maximum local stress and minimum local strength. The local stress pattern is determined by the shape of the part and by the type and magnitude of the loading. In addition to the geometric features of a part, features such as surface and metallurgical imperfections can act to concentrate stress locally. Surface imperfections such as scratches, dents, burrs, cuts, and other manufacturing flaws are the most obvious sites at which fatigue cracks initiate. Except for instances where internal defects or special surface-hardening treatments are involved, fatigue cracks initiate at the surface.

In welded structures, the site of fatigue crack initiation is most commonly the toe of the weld reinforcement or fillet. However, defects such as crater cracks at stops and starts, lack of penetration, lack of fusion, and excessive porosity may act as initiation sites. In addition, the metallurgical changes and residual stresses caused by the heat of welding influence crack initiation in weldments.

Relation to Environment. Corrosion fatigue describes the degradation of the fatigue strength of a metal by the initiation and growth of cracks under the combined action of cyclic loading and a corrosive environment. Because it is a synergistic effect of fatigue and corrosion, corrosion fatigue can produce a far greater degradation in strength than either effect acting alone or by superposition of the singular effects. An unlimited number of gaseous and liquid mediums may affect fatigue crack initiation in a given material. Fretting corrosion, which occurs from relative motion between joints, may also accelerate fatigue crack initiation. Environmental effects are discussed in more detail later in this article and in the article "Environmental Effects on Fatigue Crack Propagation" in this Section.

Fatigue Testing Regimes

The magnitude of the nominal stress on a cyclically loaded component frequently is measured by the amount of overstress—that is, the amount by which the nominal stress exceeds the fatigue limit or the long-life fatigue strength of the material used in the component. The number of load cycles that a component under low overstress can endure is high; thus, the term high-cycle fatigue is often applied.

As the magnitude of the nominal stress increases, initiation of multiple cracks is more likely. Also, spacing between fatigue striations, which indicate the progressive growth of the crack front, is increased, and the region of final fast fracture is increased in size.

Low-cycle fatigue is the regime characterized by high overstress. The arbitrary, but commonly accepted, dividing line between high-cycle and low-cycle fatigue is considered to be about 10^4 to 10^5 cycles. In practice, this distinction is made by determining whether the dominant component of the strain imposed during cyclic loading is elastic (high cycle) or plastic (low cycle), which in turn depends on the properties of the metal as well as the magnitude of the nominal stress.

Special test techniques are required for control and monitoring of low-cycle fatigue tests. Typically, strain-controlled tests (constant strain amplitude) are used. Because high-cycle fatigue tests require uninterrupted operation for long periods of time, simple, reliable test machines are used. Constant-load amplitude or constant-deflection tests commonly are conducted.

In routine low- and high-cycle fatigue crack initiation testing, complete fracture of a small specimen is the failure criteria. Approximately 30 to 40% of the low-cycle fatigue life and about 80 to 90% of the high-cycle fatigue life measured by cycles-to-failure involves nucleation of the fatigue microcrack.

Presentation of Fatigue Data. High-cycle fatigue data are presented graphically as stress (S) versus cycles-to-failure (N) in S-N diagrams or S-N curves. These are described in the Introduction to this Section along with the symbols and nomenclature commonly applied in fatigue testing. Because the stress in high-cycle fatigue tests is usually within the elastic range, the calculation of stress amplitude, stress range, or maximum stress on the S-axis is made using simple equations from mechanics of materials, i.e., stress calculated using the specimen dimensions and the controlled load or deflection applied axially, in flexure, or in torsion.

Until World War II, little attention was paid to the low-cycle range, and most of the existing fatigue results were for high cycles only. It was then realized that for some pressure vessels, pressurized fuselages, mechanisms for extending landing gears and controlling wing flaps, missiles, spaceship launching equipment, and so forth, only a short fatigue life was required. Consequently, interest in low-cycle fatigue testing developed.

Figure 1 illustrates a stress-strain loop under controlled constant-strain cycling in a low-cycle fatigue test. During initial loading, the stress-strain curve is O-A-B. Upon unloading, yielding begins in compression at a lower stress C due to the Bauschinger effect. In reloading in tension, a hysteresis loop develops. The dimensions of this loop are described by its width $\Delta\epsilon$ (the total strain range) and its height $\Delta\sigma$ (the stress range). The total strain range $\Delta\epsilon$ consists of an elastic strain component $\Delta\epsilon_e = \Delta\sigma/E$ and a plastic strain component $\Delta\epsilon_p$.

The width of the hysteresis loop depends on the level of cyclic strain. When the level of cyclic strain is small, the hysteresis loop becomes very narrow. For tests conducted under constant $\Delta\epsilon$, the stress range $\Delta\sigma$ usually changes with an increasing number of cycles. Annealed materials undergo cyclic strain hardening so that $\Delta\sigma$ increases with cycles and then levels off after about 100 strain cycles. The larger the value of $\Delta\epsilon$, the greater the increase in stress range. Materials that are initially cold worked undergo cyclic strain softening so that $\Delta\sigma$ decreases with increasing number of strain cycles.

The common method of presenting low-cycle fatigue data is to plot either the plastic strain range, $\Delta\epsilon_p$, or the total strain range, $\Delta\epsilon$, versus N. When plotted using log-log coordinates, a straight line can be fit to the $\Delta\epsilon_p$-N plot. The slope of this line in the region where plastic strain dominates has

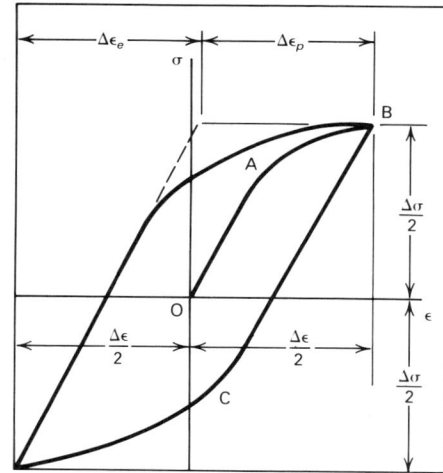

Fig. 1 Stress-strain loop for constant-strain cycling

shown little variation for the large number of metals and alloys tested in low-cycle fatigue, the average value being ½. This power-law relationship between $\Delta\epsilon_p$ and N is known as the Coffin-Manson relationship (Ref 1-6). Figure 2 is an example of the typical presentation of low-cycle fatigue test results.

When lines of curves are presented with fatigue data, the equation and the method of the fit should be indicated. Any presentation of fatigue data should include the following pertinent information, when applicable, regarding the material and the test:

- Material identification (product form)
- Tensile strength
- Orientation of the specimen
- Surface condition
- Notch description (stress-concentration factor)

Fig. 2 Low-cycle fatigue curve ($\Delta\epsilon_p$ versus N) for type 347 stainless steel

Source: Ref 1

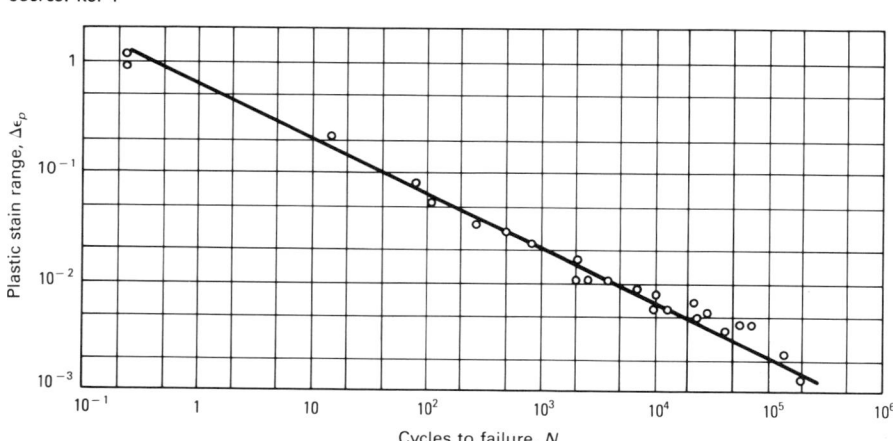

- Type of fatigue test (mode of loading)
- Controlled test parameters
- Stress ratio
- Test frequency
- Test temperature and environment

Classification of Fatigue Testing Machines

Unlike some areas of mechanical testing, many testing devices and specimen designs have been developed for fatigue testing. Many test devices are marketed commercially to provide laboratories with the capability to conduct fatigue testing. Fatigue test specimens are primarily described by the mode of loading:

- Direct (axial) stress
- Plane bending
- Rotating beam
- Alternating torsion
- Combined stress

Testing machines, however, may be universal-type machines that are capable of conducting all of the above modes of loading, depending on the fixturing used. Machines are defined by several classifications: (1) the controlled test parameter (load, deflection, strain, twist, torque, etc.), (2) the design characteristics of the machine (direct stress, plane bending, rotating beam, etc.) used to conduct the specimen test, or (3) the operating characteristics of the machine (electromechanical, servohydraulic, electromagnetic, etc.). Machines range from simple devices that consist of a cam run against a plane cantilever beam specimen in constant-deflection bending to complex servohydraulic machines that conduct computer-controlled spectrum load tests. Additional information on the design and applicability of servohydraulic and electromechanical fatigue testing systems can be found in the article "Fatigue Crack Propagation" in this Section.

Fatigue Testing Machine Components

Whether simple or complex, all fatigue testing machines consist of the same basic components: a load train, controllers, and monitors. The load train consists of the load frame, gripping devices, test specimen, and drive (loading) system. Typical load train components in an electrohydraulic axial fatigue machine are shown schematically in Fig. 3.

The load frame is the structure of the machine that reacts to the forces applied to the specimen by the drive system. Load frames vary in capacity (both in load and space for access to specimen and grips), rigidity (resonance must be avoided), convenience in use and space, and alignment sen-

sitivity. Typically, the load frame must support the drive system, grips, specimen, and all required sensors for control and readout.

The drive system is the most significant feature of a fatigue testing system and usually is electrically powered. The simplest systems use electric motors to act on test specimens via cams, levers, or rotating grips. In electrohydraulic machines, the motors drive hydraulic pumps to provide service pressure for control of the motion and force of a hydraulic piston actuator. Electromagnetic excitation can be used to excite a mass or inertia system to load a specimen.

Control Systems. The controls and controllers manually or automatically initiate power and test, adjust, and maintain the controlled test parameter(s). Controllers also terminate the test at a predefined status (failure, load drop, extension, or deflection limit). The control of time-varying deflection or displacement can be obtained in mechanical systems by cam-operated deflection levels, a rotating eccentric mass, or hydraulically through a piston limited by stops. Time-varying force can be produced by a rotating eccentric mass, electromagnetic force, dead-weight loading of a rotating beam, or pressure controlled by hydraulic systems. Programming capability frequently can be obtained in a drive system that superposes a static load by spring force or static pressure. Currently, sophisticated programming capabilities are available in servohydraulic closed-loop systems incorporating function generators, analog devices, and computer (digital) programming. The main components and programming of closed-loop servohydraulic systems are discussed in detail in the article "Fatigue Crack Propagation" in this Section.

Control in most simple machines and drive systems is obtained via the open-loop mode. In such systems, the magnitude of force and displacement initially set by the control system remains constant throughout the test. Closed-loop control of fatigue testing systems requires continual sensing of controlled and uncontrolled parameters during the test and establishment of a feedback loop to the controls, where the feedback signal is compared to the desired control valve. The controller automatically adjusts the drive system to correct and minimize the difference.

Sensors are required to measure the load, strain, displacement, deflection, and cycle count. Some devices provide an output signal to the controller, or to a readout device in the case of uncontrolled parameters. Common sensors are load cells (resistance strain gage bridges calibrated to load) inserted in the load train. Pressure transducers are used in hydraulic or pneumatic actuator devices.

Linear voltage differential transformer devices are used as extensometers, deflectom-

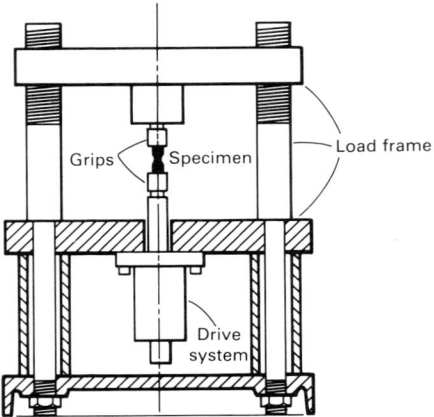

Fig. 3 Schematic of the load train in an electrohydraulic axial fatigue machine

eters, or spring force indicators. Strain gages frequently are applied directly to test specimens or to the arms of extensometers. Electronic-readout devices such as oscilloscopes, digital voltmeters (peak reading), x-y plotters, strip chart recorders, and timers are used as parameter monitors. A variety of mechanical or electrical counters or tachometers are used in cycle counting, and most systems use electrical limit switches to effect system shutdown at specimen failure or during emergency system failure.

Loading fixtures to alter the mode of loading provide versatility. Fixtures can be designed to convert the axial force provided by a hydraulic actuator to perform four-point bending or torsion testing. Similarly, fixtures attached to an oscillating platen of a rotating-eccentric-mass-type machine can facilitate axial, bending, and torsion fatigue testing of specimens. Components of a rotating-eccentric-mass-type fatigue system are illustrated in the article "Fatigue Crack Propagation" in this Section.

Grips. Proper gripping is not simply the attachment of the test specimen in the load train. Grip failure sometimes occurs prior to specimen failure. Frequently, satisfactory gripping evolves after specimen design development. Care must be taken in grip design and specimen installation in the grips to prevent misalignment.

Collet, or lathe, grips are commonly used for rotating beam specimens and must be designed so that fretting does not occur in the grip and so that the specimen does not seize in the grip and can be removed readily without grip damage. For tension-compression testing in direct stress, the specimen and grip design must prevent backlash while passing through zero load and prevent buckling of the test specimen at the maximum compression load (the grips may be part of the column).

The grips shown in Fig. 4 are typical of those used for axial fatigue tests.

Axial (Direct-Stress) Fatigue Testing Machines

The direct-stress fatigue testing machine subjects a test specimen to a uniform stress or strain through its cross section. For the same cross section, an axial fatigue testing machine must be able to apply a greater force than a static bending machine to achieve the same stress. Axial machines are used to obtain fatigue data for most applications and offer the best method of establishing, by closed-loop control, a controlled strain range in the plastic strain regime (low-cycle fatigue).

Electromechanical systems have been developed for axial fatigue studies. Generally, these are open-loop systems, but often have partial closed-loop features to continuously correct mean load. The rotating eccentric mass machine with a direct-stress fixture (illustrated in the article ''Fatigue Crack Propagation'') and the crank and lever machine (Fig. 5) are typical testing systems using this drive mechanism.

In crank and lever machines, a cyclic load is applied to one end of the test specimen through a deflection-calibrated lever that is driven by a variable-throw crank. The load is transmitted to the specimen through a flexure system, which provides straight-line motion to the specimen. The other end of the specimen is connected to a hydraulic piston that is part of an electrohydraulically controlled load-maintaining system that senses specimen yielding. This system automatically and steplessly restores the preset load through the hydraulic piston. Thus, the static and dynamic loads are applied to opposite ends of the specimen, making it possible to maintain a constant load on the specimen regardless of dimensional changes caused by specimen fatigue.

Servohydraulic closed-loop systems offer optimum control, monitoring, and versatility in fatigue testing systems. These can be obtained as component systems and can be upgraded as required. A hydraulic actuator typically is used to apply the load in axial fatigue testing. Because of the versatility the servohydraulic system provides in control modes (load, strain, or actuator displacement), the same machine can be used for both high-cycle fatigue and low-cycle (strain-controlled) fatigue testing. A wide variety of grips, particularly self-aligning types, are available for these machines.

Electromagnetic or magnetostrictive excitation is used for axial fatigue testing machine drive systems, particularly when low-load amplitudes and high-cycle fatigue lives are desired in short test durations. The high cyclic frequency of operation of these types of machines enables testing to long fatigue lives ($>10^8$ cycles) within weeks. Examples of these types of machines are illustrated in the article ''Fatigue Crack Propagation.''

Bending Fatigue Machines

The most common types of fatigue machines are small bending fatigue machines. In general, these simple, inexpensive systems allow laboratories to conduct extensive test programs with a low equipment investment. The common general-purpose fatigue machines are cantilever beam plane bending (repeated flexure) and rotating beam.

Cantilever beam machines, in which the test specimen has a tapered width, thickness, or diameter, result in a portion of the test area having uniform stress with smaller load requirements than required for uniform bending or axial fatigue of the same section size. In bending, the surface stress is highest (maximum outer fiber stress), and this test mode is preferred to study the effect of surface treatments or coatings.

The cam or eccentric principles used in the plane bending machines typically have a limited cyclic frequency compared to rotating beam types. Both deflection-controlled and load-controlled types are in common use.

Rotating Beam Machines. The earliest type of fatigue testing machine and the most commonly used is the rotating beam machine, in which a specimen with a round cross section is subjected to a dead-weight load while bearings permit rotation. A given point on the circular test section surface is subjected to a sinusoidal stress amplitude from tension on the top to compression on the bottom with each rotation.

Typical rotating beam machine types are shown in Fig. 6. The R.R. Moore-type machines (Fig. 6a) can operate up to 10 000 rpm. In all bending-type tests, only the material near the surface is subjected to the maximum stress; therefore, in a small-diameter specimen, only a very small volume of material is under test. Thus, the fatigue properties obtained from small rotating beam fatigue tests typically are higher than those obtained from axial fatigue tests on the same cross section.

Torsional Fatigue Testing Machines

Torsional fatigue tests can be performed on axial-type machines using the proper fixtures if the maximum twist required is small. Specially designed torsional fatigue testing machines consist of electromechanical machines, in which linear motion is changed to rotational motion by the use of cranks, and servohydraulic machines, in which rotary actuators are incorporated in a closed-loop testing system (Fig. 7).

Special-Purpose Fatigue Testing Machines

To perform fatigue testing of components that are prone to fatigue failure (gears, bear-

Fig. 4 Grip designs used for axial fatigue testing

(a) Standard grip body for wedge-type grips. (b) V-grips for rounds for use in standard grip body. (c) Flat grips for specimens for use in standard grip body. (d) Universal open-front holders. (e) Adapters for special samples (screws, bolts, studs, etc.) for use with universal open-front holders. (f) Holders for threaded samples. (g) Snubber-type wire grips for flexible wire or cable

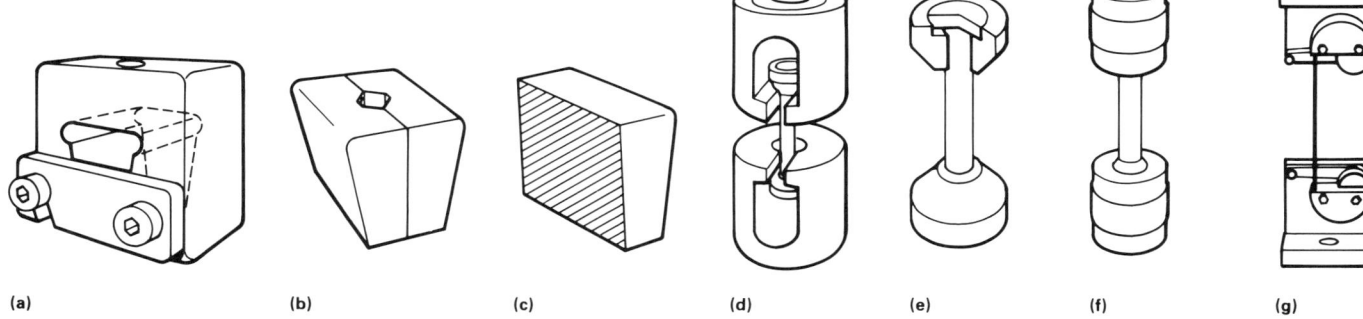

(a)　　　　(b)　　　　(c)　　　　(d)　　　　(e)　　　　(f)　　　　(g)

Fig. 5 Schematic of a crank and lever testing machine

See text for a discussion of the principle of operation.

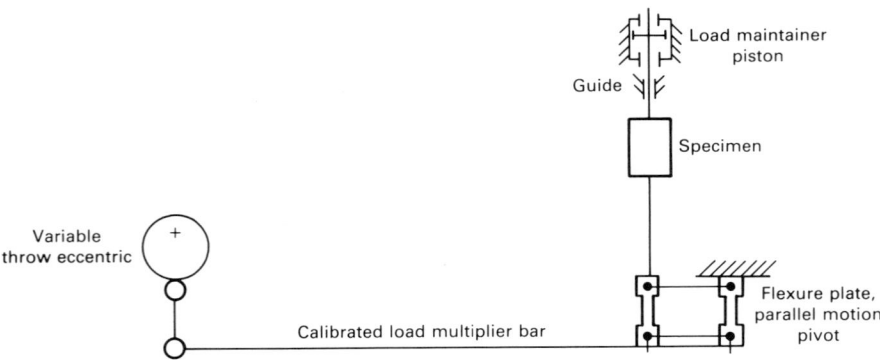

Fig. 6 Schematic of rotating beam fatigue testing machines

(a) Four-point loading R.R. Moore testing machine. (b) Single-end rotating cantilever testing machine

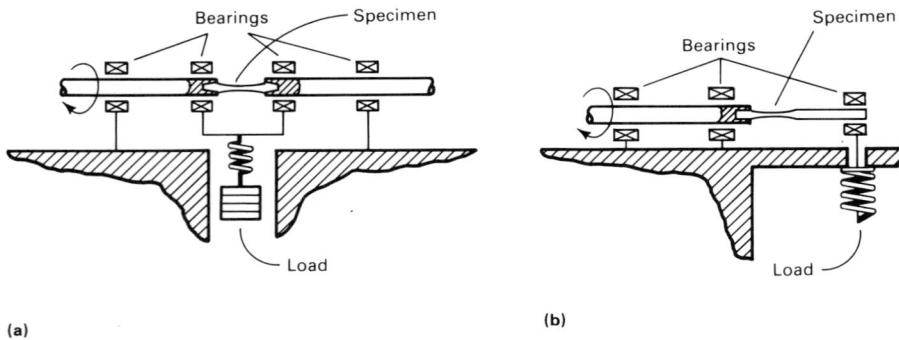

(a)   (b)

ings, wire, etc.), special devices have been used, sometimes as modifications to an existing fatigue machine. Wire testers are a modification of rotating beam machines, in which a length of the test wire is used as the beam and is deflected (buckled) a known amount and rotated.

Rolling contact fatigue testers usually are constant-load machines in which a Hertzian contact stress between two rotating bearings is applied until occurrence of fatigue failure by pitting or spalling is indicated by a vibration or noise level in the system. Rolling contact fatigue of ball and roller bearings under controlled lubrication conditions is a specialized field of fatigue testing. The objective is to produce reliable, long-life, quiet bearings that meet a rated load requirement.

Rotating and fixed-position testers are used for fatigue testing of gears. The rotating devices mechanically or hydraulically induce a static load (locked torque) between mated gear sets. In rotation, the mated teeth are subjected to the load at each contact. The nonrotating type of gear fatigue apparatus tests a single pair of teeth on the mated gears, one of which is held in a fixed position. Fatigue tests have been conducted

on simulated gear teeth specimens (cantilever beam bending fatigue specimens) to evaluate gear hardening treatments under bending fatigue or impact fatigue.

Multiaxial Fatigue Testing Machines

Many special fatigue testing machines have been designed to apply two or more

modes of loading, in or out of phase, to specimens to determine the properties of metals under biaxial or triaxial stresses. Such tests are of particular significance, because most actual service conditions involve complex stress systems, and fatigue failure criteria based on simpler modes must be verified in controlled tests in which multiaxial stress states are imposed.

Biaxial tension is a common service stress state often requiring evaluation in pressurized systems. A common biaxial fatigue test specimen is a tubular specimen subjected to simultaneous internal pressure and axial loading. In tension-torsion systems, push-pull axial loading and torsional loading are applied simultaneously. Bending-torsion machines are similar. By offsetting the applied load with respect to the specimen centerline, simple machines have been designed to apply simultaneous bending and torsion.

Fatigue Test Specimens

A specimen is a representative sample of a metal or alloy used to determine and compare basic fatigue characteristics of materials. In addition to this primary use, specimens can be designed to study the effects of notches, fillets, surface finish, environment, and various surface treatments on the material.

A typical fatigue test specimen has three areas: the test section and the two grip ends. The grip ends are designed to transfer load from the test machine grips to the test section and may be identical, particularly for axial fatigue tests. The transition from the grip ends to the test area is designed with large, smoothly blended radii to eliminate any stress concentrations in the transition.

The design and type of specimen used depend on the fatigue testing machine used and the objective of the fatigue study. The test section in the specimen is reduced in

Fig. 7 Schematic of a servohydraulic torsional fatigue testing machine

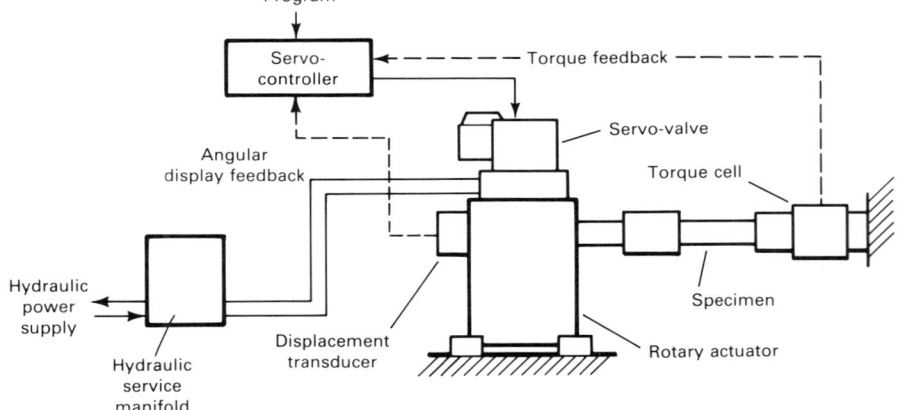

cross section to prevent failure in the grip ends and should be proportioned to use the upper ranges of the load capacity of the fatigue machine, i.e., avoiding very low load amplitudes where sensitivity and response of the system are decreased. Several types of fatigue test specimens are illustrated in Fig. 8.

Specimen Preparation. Because fatigue crack initiation typically is surface dependent, proper machining and surface preparation of test specimens is critical. Unless care is taken, scatter caused by variable surface conditions will mask the scatter that is typical for the material being studied.

Fig. 8 Typical fatigue test specimens

(a) Torsional specimen. (b) Rotating cantilever beam specimen. (c) Rotating beam specimen. (d) Plate specimen for cantilever reverse bending. (e) Axial loading specimen

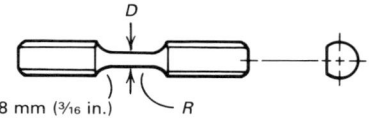

D, selected on basis of ultimate strength of material R, 12.7 mm (0.50 in.)

(a)

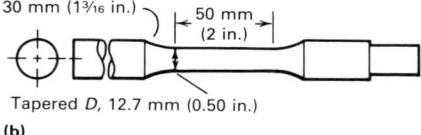

Tapered D, 12.7 mm (0.50 in.)

(b)

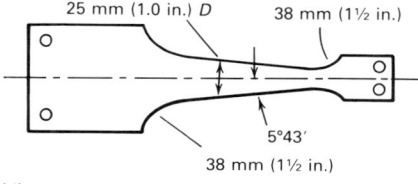

D, 5 to 10 mm (0.20 to 0.40 in.) selected on basis of ultimate strength of material R, 90 to 250 mm (3.5 to 10 in.)

(c)

25 mm (1.0 in.) D 38 mm (1½ in.)

5°43'

38 mm (1½ in.)

(d)

D

R

D, selected on basis of ultimate strength of material R, 75 to 250 mm (3 to 10 in.)

(e)

One of the primary aims of fatigue testing is comparison of materials. Therefore, uniform preparation procedures must be established. Machining operations must not alter the surface structure of the metal; thus, heat generation, heavy cutting, and severe grinding are prohibited. For as-machined surfaces, the final machining should be transverse to the plane of stress. Transition fillets must be blended into the test area without steps or undercutting. Surface polishing using metallographic techniques is preferred for smooth specimens, where machining marks are removed by a sequence of grinding steps.

The final polishing is not a buffing operation, but a cutting operation that uses lapping compounds or aluminum oxide powder in a liquid medium to remove grinding scratches. For flat sheet or plate specimens, edges should be slightly rounded and ground to eliminate nicks, dents, cuts, and sharp edges, which cause early crack initiation. The relationship between surface characteristics and fatigue properties is discussed later in this article in the section on surface effects and fatigue.

Cylindrical Specimens. Three types of specimens with circular cross sections are commonly used:

- Specimens with a continuous radius between the grip ends with the minimum diameter at the center (Fig. 8c, 8e, and 9a). These are referred to as hourglass specimens.
- Specimens with tangentially blending fillets between the test section and the grip ends (Fig. 8a and 9b)
- Specimens for use in cantilever beam loading with tapered diameters proportioned to produce nominally constant stress along the test section (Fig. 8b)

The design of the grip ends depends on the machine design and the gripping devices used. Round specimens for axial fatigue machines using grips like those shown in Fig. 4 may be threaded, buttonhead, or constant-diameter types for clamping in V-wedge pressure grips.

For rotating beam machines, short, tapered grip ends with internal threads are used, and the specimen is pulled into the grip by a draw bar. A long, constant-diameter grip end shank is used on machines with lathe-type collet chucks as grips. Torsional fatigue specimens generally are cylindrical; however, there is usually a flat or keyway in the grip ends to transmit torque from the machine into the specimen (see Fig. 8a).

Flat Sheet and Plate Specimens. Generally, flat specimens for either axial or bending fatigue tests are reduced in width in the test section, but may have thickness

reductions. The most commonly used types include:

- Specimens with tangentially blending fillets between the test section and the grip ends (Fig. 9c). These specimens are used in both axial and bending fatigue tests.
- Specimens with a continuous radius between the grip ends (Fig. 9d). These are also used in both axial and bending fatigue tests.
- Specimens for use in cantilever reverse bending tests with tapered widths (Fig. 8d)

Flat specimens generally are clamped in flat wedge-type grips, or may be held with a stiff bolted clamp/joint friction grip for reversed axial loading. Pin loading can be used when compression loads are not encountered. When pin loading is utilized, the holes drilled in the grip end must be designed to avoid shear or bearing failures at the holes, tensile failure between the holes at maximum load, and fatigue cracking at the holes in the grip end. In axial fatigue testing of flat sheet specimens, the test length must be designed to prevent premature buckling of the specimen.

Effect of Stress Concentration

Fatigue strength is reduced significantly by the introduction of a stress raiser such as a notch or hole. Because actual machine elements invariably contain stress raisers such as fillets, keyways, screw threads, press fits, and holes, fatigue cracks in structural parts usually initiate at such geometrical irregularities.

An optimum way of minimizing fatigue failure is the reduction of avoidable stress raisers through careful design and the prevention of accidental stress raisers by careful machining and fabrication. Stress concentration can also arise from surface roughness and metallurgical stress raisers such as porosity, inclusions, local overheating in grinding, and decarburization. Some of these conditions are discussed later in this article in the section on surface effects and fatigue.

The effect of stress raisers on fatigue is generally studied by testing specimens containing a notch, usually a V-notch or a U-notch. The presence of a notch in a specimen under uniaxial load introduces three effects: (1) there is an increase or concentration of stress at the root of the notch, (2) a stress gradient is set up from the root of the notch toward the center of the specimen, and (3) a triaxial state of stress is produced at the notch root.

The ratio of the maximum stress in the region of the notch (or other stress concentra-

Fig. 9 Typical round and flat fatigue test specimen configurations

(a) Hourglass specimen with continuous radius between the grip ends. (b) Round specimen with tangentially blended fillets between the test section and the grip ends. (c) Flat specimen with tangentially blended fillets between the test section and the grip ends. (d) Flat specimen with continuous radius between the grip ends

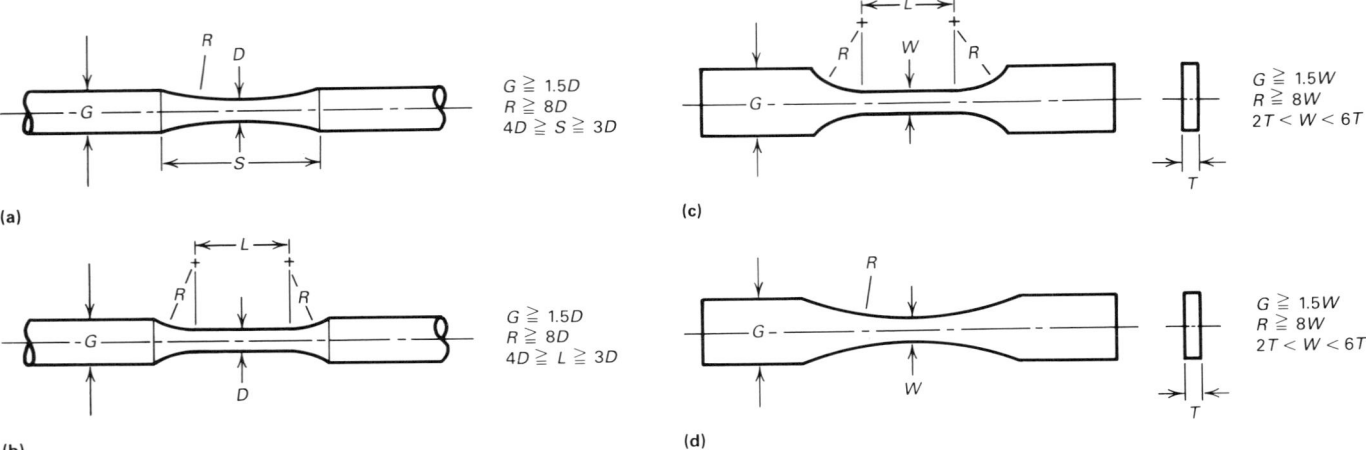

(a)

(b)

(c)

(d)

$G \geqq 1.5D$
$R \geqq 8D$
$4D \geqq S \geqq 3D$

$G \geqq 1.5D$
$R \geqq 8D$
$4D \geqq L \geqq 3D$

$G \geqq 1.5W$
$R \geqq 8W$
$2T < W < 6T$

$G \geqq 1.5W$
$R \geqq 8W$
$2T < W < 6T$

tion) to the corresponding nominal stress is the stress-concentration factor, K_t (see the Introduction to this Section). In some situations, values of K_t can be calculated using the theory of elasticity, or can be measured using photoelastic plastic models. Values for K_t are reported in Ref 7 to 10.

The effect of notches on fatigue strength is determined by comparing the *S-N* curves of notched and unnotched specimens. The data for notched specimens usually are plotted in terms of nominal stress based on the net cross section of the specimen. The effectiveness of the notch in decreasing the fatigue limit is expressed by the fatigue-notch factor, K_f. This factor is the ratio of the fatigue limit of unnotched specimens to the fatigue limit of notched specimens.

For materials that do not exhibit a fatigue limit, the fatigue-notch factor is based on the fatigue strength at a specified number of cycles. Values of K_f have been found to vary with (1) severity of the notch, (2) type of notch, (3) material, (4) type of loading, and (5) stress level. The published values of K_f are subject to considerable scatter and should be examined carefully for their limitations and restrictions. However, two general trends usually are observed for test conditions of completely reversed loading. First, K_f is usually less than K_t, and the ratio of K_f/K_t decreases as K_t increases.

The notch sensitivity of a material in fatigue is expressed by a notch-sensitivity factor, q:

$$q = \frac{K_f - 1}{K_t - 1} \qquad \text{(Eq 1)}$$

A material that experiences no reduction in fatigue due to a notch ($K_f = 1$) has a factor of $q = 0$, whereas a material in which the notch

has its full theoretical effect ($K_f = K_t$) has a factor of $q = 1$. However, q is not a true material constant, because it varies with the severity and type of notch, size of specimen, and type of loading. As shown in Fig. 10, notch sensitivity increases with tensile strength. Thus, it is sometimes possible to decrease fatigue performance by increasing the hardness or tensile strength of a material.

Effect of Test Specimen Size (Ref 12)

It is not possible to predict directly the fatigue performance of large machine members from the results of laboratory tests on small specimens. In most cases, a size effect exists; i.e., the fatigue strength of large members is lower than that of small specimens. Precise determination of this phenomenon is difficult. It is extremely difficult to prepare geometrically similar specimens of increasing diameter that have the same metallurgical structure and residual stress distribution throughout the cross section. The problems in fatigue testing of large specimens are considerable, and few fatigue machines can accommodate specimens with a wide range of cross sections.

Changing the size of a fatigue specimen usually results in variations of two factors. First, increasing the diameter increases the volume or surface area of the specimen. The change in amount of surface is significant, because fatigue failures usually initiate at the surface. Secondly, for plain or notched specimens loaded in bending or torsion, an increase in diameter usually decreases the stress gradient across the diameter and in-

creases the volume of material that is highly stressed.

Experimental data on the size effect in fatigue typically show that the fatigue limit decreases with increasing specimen diameter. Horger's data for steel shafts tested in reversed bending (Table 1) show that the fatigue limit can be appreciably reduced in large section sizes (Ref 13).

No size effect was found for smooth fatigue specimens of plain carbon steel with diameters ranging from 5 to 35 mm (0.2 to 1.4 in.) when tested in axial tension-compression loading (Ref 14). However, when a notch was introduced into the specimen, so that a stress gradient was produced, a definite size effect was observed. Thus, it may be concluded that a size effect in fatigue is due primarily to the existence of a stress gradient.

The fact that large specimens with shallow stress gradients have lower fatigue limits supports the concept that a critical value of stress must be exceeded over a given finite depth of material for failure to occur. This appears to be a more realistic criterion of size effect than the ratio of the change in surface area to the change in specimen diameter. The importance of stress gradients in size effect

Table 1 Effect of specimen size on the fatigue limit of normalized plain carbon steel in reversed bending

Specimen diameter		Fatigue limit	
mm	in.	MPa	ksi
7.6	0.30	248	36
38	1.50	200	29
152	6.00	144	21

Source: Ref 13

Fig. 10 Variation of notch-sensitivity index with notch radius for steels tested in bending or axial fatigue loading
Source: Ref 11

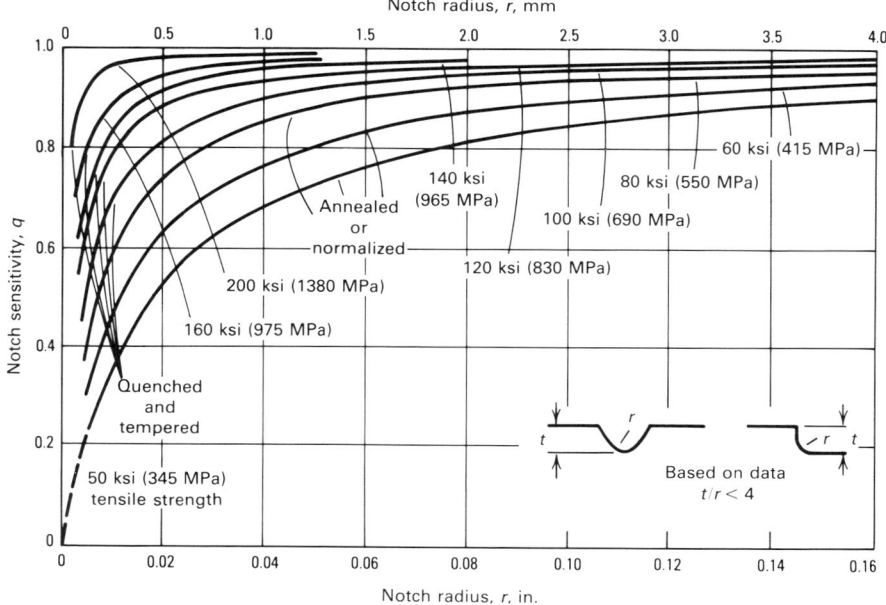

Marked improvements in fatigue properties can result from the formation of harder and stronger surfaces on steel parts by carburizing and nitriding (Ref 15). However, because favorable compressive residual stresses are produced in the surface by these processes, the higher fatigue properties are not due exclusively to the formation of higher strength material on the surface. The effectiveness of carburizing and nitriding in improving fatigue performance is greater when a high stress gradient exists, as in bending or torsion, than in an axial fatigue test.

The greatest increase in fatigue performance occurs in notched fatigue specimens that are nitrided. The amount of strengthening depends on the diameter of the part and the depth of surface hardening. Improvements in fatigue properties similar to those caused by carburizing and nitriding may also be produced by flame hardening and induction hardening. In surface-hardened parts, failure initiates at the interface between the hard case and the softer case, rather than at the surface.

Electroplating of the surface generally decreases the fatigue limit of steel, because the platings inherently contain cracks. The particular plating conditions used to produce an electroplated surface can have a significant effect on fatigue properties, because large changes in the residual stress, adhesion, porosity, and hardness of the plate can be produced. Anodized coatings on aluminum alloys can be similarly detrimental to fatigue strength.

Surface Residual Stresses. Processing, such as grinding, polishing, and machining, that work hardens or increases residual stress on the surface can influence the fatigue

explains why correlation between laboratory results and service failure is often poor. Actual failures in large parts usually are directly related to stress concentrations, either intentional or accidental, and it is usually impossible to duplicate the same stress concentration and stress gradient in a small laboratory specimen.

Surface Effects and Fatigue

Generally, fatigue properties are very sensitive to surface conditions. Except in special cases where internal defects or case hardening is involved, all fatigue cracks initiate at the surface. Factors that affect the surface of a fatigue specimen can be divided into three categories: (1) surface roughness or stress raisers at the surface, (2) changes in the properties of the surface metal, and (3) changes in the residual stress condition of the surface. Additionally, the surface may be subjected to oxidation and corrosion.

Surface Roughness. In general, fatigue life increases as the magnitude of surface roughness decreases. Decreasing surface roughness minimizes local stress raisers. Therefore, special attention must be given to the surface preparation of fatigue test specimens. Typically, a metallographic finish, free of machining grooves and grinding scratches, is necessary. Figure 11 illustrates the effects that various surface conditions have on the fatigue properties of steel.

Changes in Surface Properties. Because fatigue failure is dependent on surface condition, any phenomenon that changes the fatigue strength of the surface material will greatly alter fatigue properties. Decarburization of the surface of heat treated steel is particularly detrimental to fatigue performance. Similarly, the fatigue strength of aluminum alloy sheet is reduced when a soft aluminum coating (cladding) is applied to the stronger age-hardenable aluminum alloy sheet.

Fig. 11 Effect of surface conditions on the fatigue properties of steel (302 to 321 HB)

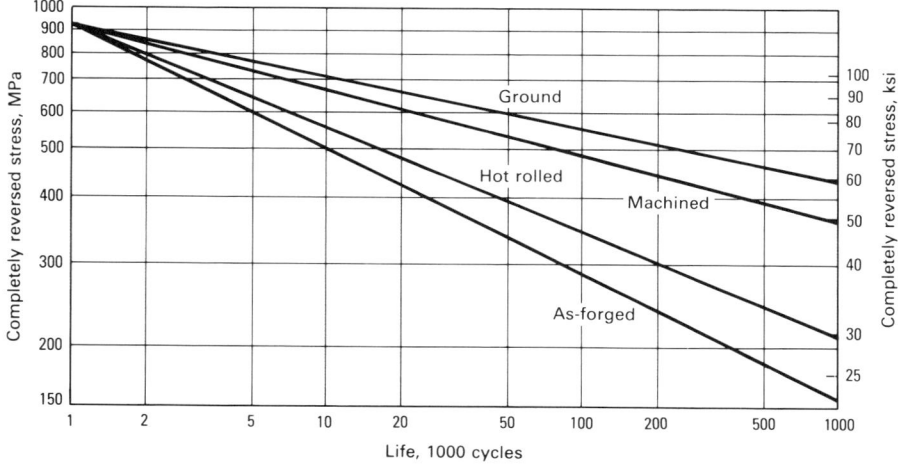

strength, although there is no generalization that predicts the extent of improved fatigue strength that can be derived from work hardening and residual stress. Compressive residual surface stresses generally increase the fatigue strength, but tensile residual surface stresses do not. There may be a gradual decrease in residual stress if the cyclic stresses cause some plastic deformation. Compressive residual surface stress provides greater improvement in the fatigue strength of harder materials (such as alloy spring steel). In softer materials (such as low-carbon steel), work hardening effectively improves fatigue strength.

In a notched high-strength steel, the beneficial effect of prestretching and the detrimental effect of precompression are much greater than in a plain carbon steel because of the type of residual stress present at the notch. A compressive residual stress introduced during quenching from a tempering temperature will increase the fatigue strength, particularly in notched specimens.

In general, residual stresses are introduced by misfit of structural parts, a change in the specific volume of a metal accompanying phase changes, a change in shape following plastic deformation, or thermal stresses resulting from rapid temperature changes such as occur in quenching.

The influence of residual stress on fatigue strength is similar to that of an externally applied static stress. A static compressive surface stress increases the fatigue strength, and static tensile surface stress reduces it.

Effect of Mean Stress

A series of fatigue tests can be conducted at various mean stresses, and the results can be plotted as a series of S-N curves. A description of applied stresses and S-N curves can be found in the Introduction to this Section. For design purposes, it is more useful to know how the mean stress affects the permissible alternating stress amplitude for a given life (number of cycles). This usually is accomplished by plotting the allowable stress amplitude for a specific number of cycles as a function of the associated mean stress.

At zero mean stress, the allowable stress amplitude is the effective fatigue limit for a specified number of cycles. As the mean stress increases, the permissible amplitudes steadily decrease. At a mean stress equal to the ultimate tensile strength of the material, the permissible amplitude is zero.

The two straight lines and the curve shown in Fig. 12 represent the three most widely used empirical relationships for describing the effect of mean stress on fatigue strength.

The straight line joining the alternating fatigue strength to the tensile strength is the modified Goodman law. Goodman's original law included the assumption that the fatigue limit was equal to one third of the tensile strength; this has since been generalized to the relation shown in Fig. 12, using the fatigue strength as determined experimentally.

Gerber found that the early experiments of Wöhler fitted closely to a parabolic relation, and this is known as Gerber's parabola (curve in Fig. 12). The third relation, known as Soderberg's law, is a conservative design rule for working stresses shown in Fig. 12 by the straight line from the fatigue strength to the static yield strength. In most designs, it is required that the static yield strength not be exceeded, and this relation is intended to fulfill the conditions that neither fatigue failure nor yielding occurs. The relations may be written as:

Modified Goodman law:

$$S_a = S\left[1 - \left(\frac{S_m}{S_u}\right)\right]$$

Gerber's law:

$$S_a = S\left[1 - \left(\frac{S_m}{S_u}\right)^2\right]$$

Soderberg's law:

$$S_a = S\left[1 - \left(\frac{S_m}{S_y}\right)\right]$$

where S_a is the alternating stress associated with a mean stress S_m, S is the alternating fatigue strength, S_u is the tensile strength, and S_y is the yield strength. More detailed information on the Goodman, or constant-life, diagram can be found in the article "Fatigue Data Analysis" in this Volume.

Stress Amplitude. Because stress amplitude varies widely under actual loading conditions, it is necessary to predict fatigue life under various stress amplitudes. The most widely used method of estimating fatigue under complex loading is provided by the linear damage law. This is a hypothesis first suggested by Palmgren and restated by Miner, and is sometimes known as Miner's rule (Ref 16).

The assumption is made that the application of n_i cycles at a stress amplitude S_i, for which the average number of cycles to failure is N_i, causes an amount of fatigue damage that is measured by the cumulative cycles ratio n_i/N_i, and that failure will occur when $\Sigma(n_i/N_i) = 1$.

This method is not applicable in all cases, and numerous alternative theories of cumula-

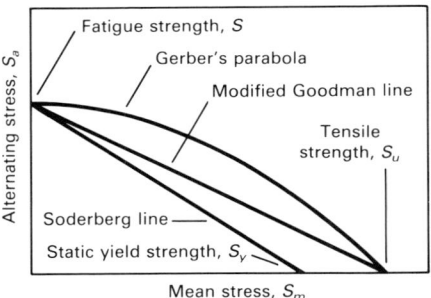

Fig. 12 Effect of mean stress on the alternating stress amplitude

As shown by the modified Goodman line, Gerber's parabola, and Soderberg line. See text for discussion.

tive linear damage have been suggested (Ref 17). Some considerations of redistribution of stresses have been clarified, but there is as yet no satisfactory approach for all situations.

The effect of varying the stress amplitude (linear damage) can be evaluated experimentally by means of a test in which a given number of stress cycles are applied to a test piece at one stress amplitude. The test is then continued to fracture at a different amplitude. Alternatively, the stress can be changed from one stress amplitude to another at regular intervals; such tests are known as block, or interval, tests. These tests do not simulate service conditions, but may serve a useful purpose for assessing the linear damage law and indicating its limitations. More detailed information on the statistical prediction of fatigue life can be found in the article "Fatigue Data Analysis" in this Volume.

Corrosion Fatigue

Corrosion fatigue is the combined action of repeated or fluctuating stress and a corrosive environment to produce progressive cracking. Usually, environmental effects are deleterious to fatigue life, producing cracks in fewer cycles than would be required in a more inert environment. Once fatigue cracks have formed, the corrosive aspect also may accelerate the rate of crack growth.

In corrosion fatigue, the magnitude of cyclic stress and the number of times it is applied are not the only critical loading parameters. Time-dependent environmental effects also are of prime importance. When failure occurs by corrosion fatigue, stress-cycle frequency, stress-wave shape, and stress ratio all affect the cracking processes.

Effect of Frequency. In nonaggressive environments, cyclic frequency generally has little effect on fatigue behavior. On the other hand, in aggressive environments fatigue strength is strongly dependent on frequency. Corrosion-fatigue strength (endurance limit at a prescribed number of cycles) will generally decrease as the cyclic frequency is decreased. This effect is most important at frequencies of less than 10 Hz.

The frequency dependence of corrosion fatigue is thought to result from the fact that interaction of a material and its environment is essentially a rate-controlled process. Low frequencies, especially at low strain amplitudes or when there is substantial elapsed time between changes in stress levels, allow time for interaction between material and environment; high frequencies do not, particularly when high strain amplitude is also involved. At very high frequencies, or in the plastic-strain range, localized heating may seriously affect the properties of the part. Such effects normally are not considered to be related to a corrosion-fatigue phenomenon.

When environments have a deleterious effect on fatigue behavior, a critical range of frequencies of loading may exist in which the mechanical/environmental interaction is significant. Above this range the effect usually disappears, while below this range the effect may diminish.

Effect of Stress Amplitude. In general, a low amplitude of cyclic stress favors relatively long fatigue life, permitting greater opportunity for involvement of the environment in the fatigue process. Where stresses are sufficiently high to cause significant macroscopic plastic deformation, environmental interaction may be insignificant unless the strain rate is in a critical range for stress-corrosion cracking in certain alloy/environment systems.

Stress amplitude must be considered together with mean stress and frequency. Low stress levels may allow more time for environmental interaction, but if the frequency is high, the crack tip may not be exposed to the environment for a time sufficient for the corrosion processes to do significant damage.

Environmental Effects on Fatigue Strength. For any given material, the fatigue strength, or fatigue life at a given value of maximum stress, generally decreases in the presence of an aggressive environment. This effect varies widely, depending primarily on the characteristics of the material/environment combination. The environment affects crack growth rate or the probability of fatigue crack initiation, or both. For many materials, the stress range required to cause fatigue failure diminishes progressively with time and with number of cycles.

Environmental Effects on Crack Initiation. Surface features at origins of corrosion-fatigue cracks vary with the alloy and with specific environmental conditions. In carbon steels, cracks often nucleate at corrosion pits and often contain significant amounts of corrosion products. The cracks are predominantly transgranular and may exhibit a slight amount of branching. More detailed information on corrosion fatigue can be found in Ref 18 to 20.

Standards and Recommended Practices for Fatigue Testing

Only a few standardized procedures for fatigue testing are available, because many of the testing machines are custom built. However, several types of fatigue testing machines, testing methods, and test specimens are accepted as standard. In the United States, ASTM issues voluntary standards and recommended practices. Those related to fatigue testing are:

- ASTM E 206, "Standard Definitions of Terms Relating to Fatigue Testing and the Statistical Analysis of Fatigue Data"
- ASTM E 466, "Standard Practice for Conducting Constant Amplitude Axial Fatigue Tests of Metallic Materials"
- ASTM E 467, "Standard Practice for Verification of Constant Amplitude Dynamic Loads in an Axial Load Fatigue Testing Machine"
- ASTM E 468, "Standard Practice for Presentation of Constant Amplitude Fatigue Test Results for Metallic Materials"
- ASTM E 513, "Standard Definitions of Terms Relating to Constant Amplitude Low-Cycle Fatigue Testing"
- ASTM E 606, "Standard Practices for Constant Amplitude Low-Cycle Fatigue Testing"
- ASTM E 739, "Standard Practice for Statistical Analysis of Linear or Linearized Stress-Life (S-N) and Strain-Life (ϵ-N) Fatigue Data"
- ASTM E 742, "Standard Definitions of Terms Relating to Fluid Aqueous and Chemical Environmentally Affected Fatigue Testing"

REFERENCES

1. L.F. Coffin, Jr., *Met. Eng. Quart.*, Vol 3, 1963, p 22
2. R.W. Smith, M.H. Hirschberg, and S.S. Manson, NASA Report TN D-1574, NASA, April 1963
3. S.S. Manson and M.H. Hirschberg, *Fatigue: An Interdisciplinary Approach*, Syracuse University Press, Syracuse NY, 1964, p 133
4. L.F. Coffin, Jr., *Trans. ASME*, Vol 76, 1954, p 931
5. J.F. Tavernelli and L.F. Coffin, Jr., *Trans. ASM*, Vol 51, 1959, p 438
6. R.W. Hertzberg, *Deformation and Fracture Mechanics of Engineering Materials*, 2nd ed., John Wiley & Sons, New York, 1983, p 503-507
7. R.E. Peterson, *Stress Concentration Design Factors*, John Wiley & Sons, New York, 1974
8. H. Neuber, "Theory of Notch Stresses: Principles for Exact Calculation of Strength With Reference to Structural Form and Material," Springer, 1958, AEC-Tr-4547; available through NTIS, U.S. Dept. of Commerce
9. R.J. Roark, *Formulas for Stress and Strain*, 4th ed., McGraw-Hill, New York, 1965
10. T. Topper, R. Wetzel, and J. Morrow, Neuber's Rule Applied to Fatigue of Notched Specimens, *J. Mat.*, Vol 4 (No. 1), 1969
11. G. Sines and J.R. Weisman, Ed., *Metal Fatigue*, McGraw-Hill, New York, 1959
12. G.E. Dieter, *Mechanical Metallurgy*, McGraw-Hill, New York, 1976, p 426-427
13. O.J. Horger, Fatigue Characteristics of Large Sections, in *Fatigue*, American Society for Metals, 1953
14. C.E. Phillips and R.B. Heywood, *Proc. Inst. Mech. Eng. (London)*, Vol 165, 1951, p 113-124
15. *Fatigue Durability of Carburized Steel*, American Society for Metals, 1957
16. M.A. Miner, Cumulative Damage in Fatigue, *Trans. ASME*, Vol 67, 1945, p A159
17. H.J. Grover, "Fatigue of Aircraft Structures," NAVAIR 01-1A-13, Naval Air Systems Command, U.S. Department of the Navy, 1966
18. O. Devereux, A.J. McEvily, and R.W. Staehle, Ed., *Corrosion Fatigue: Chemistry, Mechanics, and Microstructure*, conference proceedings, University of Connecticut, June 14-18, 1971, National Association of Corrosion Engineers, Houston
19. H.L. Craig, T.W. Crooker, and P.W. Hoeppner, Ed., *Corrosion-Fatigue Technology*, symposium proceedings, Denver, Nov 14-19, 1976, ASTM, Philadelphia
20. Corrosion-Fatigue Failures, in *Metals Handbook*, Vol 10, 8th ed., American Society for Metals, 1975, p 240-249

Fatigue Crack Propagation

By the ASM Committee on Fatigue Crack Propagation*

FATIGUE FAILURE of structural and equipment components due to cyclic loading has long been a major design problem and the subject of numerous investigations. Although considerable fatigue data are available, the majority has been concerned with the nominal stress required to cause failure in a given number of cycles—namely, *S-N* curves (see the Introduction to this Section and the preceding article, "Fatigue Crack Initiation"). Usually, such data are obtained by testing smooth or notched specimens. With this type of testing, however, it is difficult to distinguish between fatigue crack initiation life and fatigue crack propagation life.

Preexisting flaws or crack-like defects within a material reduce or may eliminate the crack initiation portion of the fatigue life of the component. Fracture mechanics methodology enhances the understanding of the initiation and propagation of fatigue cracks and assists in solving the problem of designing to prevent fatigue failures.

Linear elastic fracture mechanics technology is based on an analytical procedure that relates the stress-field magnitude and distribution in the vicinity of a crack tip to the nominal stress applied to the structure, to the size, shape, and orientation of the crack or crack-like imperfection, and to the material properties. The stress-field equations show that the magnitude of the elastic stress field can be described by a single parameter, K, which is designated the stress-intensity factor.

The relationships between the stress-intensity factor and various body configurations, crack sizes, shapes, and orientations, and loading conditions are discussed in Ref 1 and 2. For additional information on the concepts related to fracture mechanics, see the article "Fracture Mechanics" in this Volume.

This article reviews the specimen configurations, crack-measuring techniques, and apparatus used in fatigue crack propagation testing. Information on the combined effects of cyclic stresses and an aggressive environment (corrosion fatigue) can be found in the article "Environmental Effects on Fatigue Crack Propagation" in this Section. Analyses performed on the numerical output of fatigue crack propagation tests, including the analysis framework for modeling fatigue crack growth rate data, numerical methods for calculating fatigue crack growth as a function of the stress-intensity factor, and principles of fatigue crack growth damage analysis are discussed in the article "Fatigue Crack Growth Data Analysis" in this Volume.

Fatigue Crack Propagation Test Methods

The general nature of fatigue crack propagation using fracture mechanics techniques is summarized in Fig. 1. A logarithmic plot of the crack growth per cycle, da/dN, versus the stress-intensity factor range, ΔK, corresponding to the load cycle applied to a specimen is illustrated (Ref 3). The da/dN versus ΔK plot was constructed of data on five specimens of ASTM A533 B1 steel tested at 24 °C (75 °F). A plot of similar shape is anticipated with most structural alloys; the absolute values of da/dN and ΔK, however, are dependent on the material.

Results of fatigue crack growth rate tests for nearly all metallic structural materials have shown that the da/dN versus ΔK curves have three distinct regions. The behavior in Region I (Fig. 1) exhibits a fatigue crack growth threshold, ΔK_{th}, which corresponds to the stress-intensity factor range below which cracks do not propagate. Additional information on cyclic crack growth rate testing in the threshold (low-growth) regime can be found later in this article.

At intermediate values of ΔK (Region II in Fig. 1), a straight line usually is obtained on a log-log plot of ΔK versus da/dN. This is described by the power-law relationship (Ref 4):

$$\frac{da}{dN} = C(\Delta K)^n \qquad \text{(Eq 1)}$$

where C and n are constants for a given material and stress ratio.

Fatigue crack growth rate data for some steels show that the primary parameter affecting growth rate in Region II is the stress-intensity factor range and that the mechanical and metallurgical properties of these steels have negligible effects on the fatigue crack growth rate in a room-temperature air environment. Data for four martensitic steels fall within a single band, as shown in Fig. 2. The upper bound of scatter can be obtained from:

$$\frac{da}{dN} = 0.66 \times 10^{-8}(\Delta K)^{2.25} \qquad \text{(Eq 2)}$$

where a is given in inches, and ΔK is given in ksi$\sqrt{\text{in}}$.

For some steels, the stress ratio and mean stress have negligible effects on the rate of crack growth in Region II. Also, the frequency of cyclic loading and the waveform (sinusoidal, triangular, square, trapezoidal) do not affect the rate of crack propagation per cycle of load for some steels in benign environments (Ref 5).

At high ΔK values (Region III in Fig. 1), unstable behavior occurs, resulting in a rapid increase in the crack growth rate just prior to complete failure of the specimens. There are two possible causes of this behavior (Ref 6-8). First, the increasing crack length during constant load testing causes the peak stress intensity to reach the fracture toughness, K_{Ic}, of the material, and the unstable behavior is related to the early stages of brittle fracture. Second, the growing crack reduces the uncracked area of the specimen sufficiently for

*David A. Utah, *Chairman*, Manager, Materials Testing, Aircraft Engine Group, General Electric Company; William H Cullen, Senior Scientist, Materials Engineering Associates, Inc.; Lorenzo C. Majno, Assistant Product Manager, Servohydraulic Systems, Instron Corporation; Richard A. Meyer, Manager, Transducer Products, MTS Systems Corporation; Robert O. Ritchie, Professor of Metallurgy, Department of Materials Science and Mineral Engineering, University of California—Berkeley; R.H. Stentz, Vice President, Mar-Test Inc.; Roy Williams, Senior Engineer—Mechanics of Materials, General Electric Company

Fig. 1 Fatigue crack growth behavior of ASTM A533 B1 steel

Yield strength of 470 MPa (70 ksi). Test conditions: $R = 0.10$; ambient room air, 24 °C (75 °F). Source: Ref 3

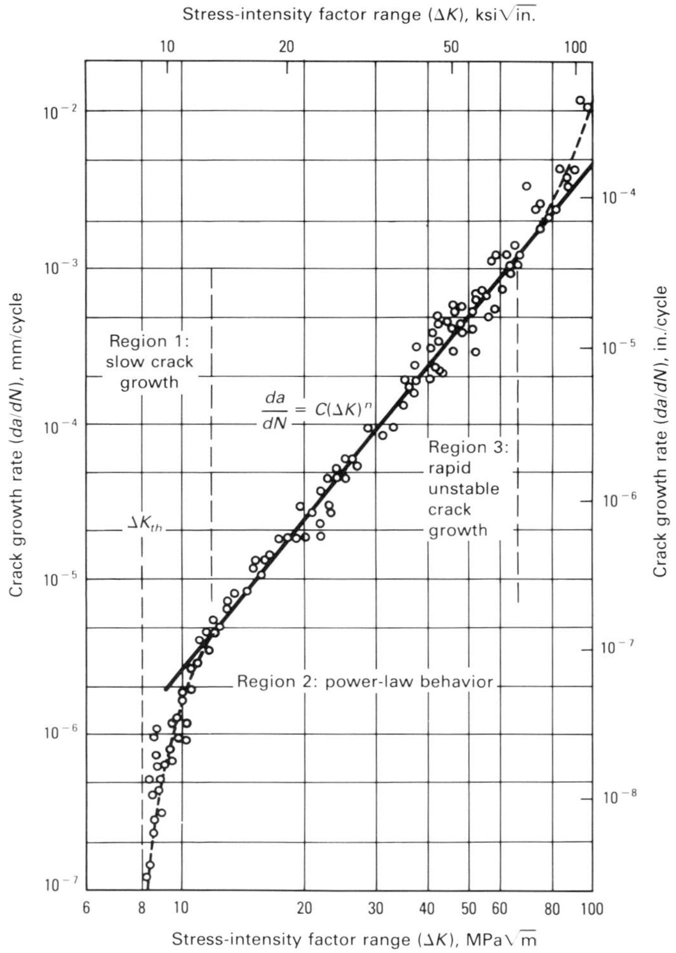

Stress-intensity factor range (ΔK), ksi$\sqrt{\text{in.}}$

Crack growth rate (da/dN), mm/cycle

$$\frac{da}{dN} = C(\Delta K)^n$$

Region 1: slow crack growth

ΔK_{th}

Region 3: rapid unstable crack growth

Region 2: power-law behavior

Crack growth rate (da/dN), in./cycle

Stress-intensity factor range (ΔK), MPa$\sqrt{\text{m}}$

Fig. 2 Summary of fatigue crack growth data for martensitic steels

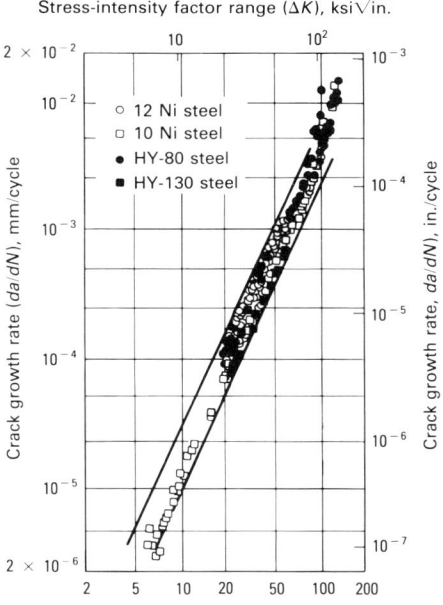

Stress-intensity factor range (ΔK), ksi$\sqrt{\text{in.}}$

○ 12 Ni steel
□ 10 Ni steel
● HY-80 steel
■ HY-130 steel

Crack growth rate (da/dN), mm/cycle

Crack growth rate, da/dN, in./cycle

Stress-intensity factor range (ΔK), MPa$\sqrt{\text{m}}$

the peak load to cause fully plastic limit load behavior. The first possibility is operative for high-strength, low-toughness metals, in which specimen sizes normally used for fatigue crack growth rate testing behave in a linear elastic manner at K levels equal to K_{Ic}. The second possibility, plastic limit load behavior, is common for ductile metals, particularly if K_{Ic} is high.

When plastic limit load behavior causes unstable crack growth, ΔK values have no meaning, because the limitations of linear elastic fracture mechanics have been exceeded. Here, the use of the *J*-integral concept, crack-opening displacement, or some other elastic-plastic fracture mechanics approach is more appropriate than ΔK for correlating the data. These concepts are discussed in the article "Fracture Mechanics" in this Volume.

Standarized testing procedures for measuring fatigue crack growth rates are de-scribed in ASTM Standard E 647 (Ref 9). This method applies to medium to high crack growth rates—that is, above 10^{-8} m/cycle (3.9×10^{-7} in./cycle). Procedures for growth rates below 10^{-8} m/cycle are under consideration by ASTM. For applications involving fatigue lives of up to about 10^6 load cycles, the procedures recommended in ASTM E 647 can be used. Fatigue lives greater than about 10^6 cycles correspond to growth rates below 10^{-8} m/cycle, and these require special testing procedures, which are related to the threshold of fatigue crack growth illustrated in Fig. 1 and discussed later in this article.

ASTM E 647 describes the use of center-cracked specimens and compact specimens (Fig. 3 and 4). The specimen thickness-to-width ratio, B/W, is smaller than the 0.5 value for K_{Ic} tests; the maximum B/W values for center-cracked and compact specimens are 0.125 and 0.25, respectively. With the thinner specimens, crack length measurements on the sides of the specimens can be used as representations of through-thickness crack growth behavior.

For tension-tension fatigue loading, the K_{Ic} loading fixtures frequently can be used. For this type of loading, both the maximum and minimum loads are tensile, and the load ratio, $R = P_{min}/P_{max}$, is in the range $0 < R < 1$. A ratio of $R = 0.1$ is commonly used for developing data for comparative purposes.

Testing often is performed in laboratory air at room temperature; however, any gaseous or liquid environment and temperature of interest may be used to determine the effect of temperature, corrosion, or other chemical reaction on cyclic loading (see the article "Environmental Effects on Fatigue Crack Propagation" in this Section). Cyclic loading may involve various waveforms for constant-amplitude loading, spectrum loading, or random loading.

Data Analysis. For constant-amplitude loading, a set of crack-length versus elapsed-cycle data (a versus N) is generated, with the specimen loading, P_{max} and P_{min}, generally held constant. Figure 5 illustrates a typical a versus N plot. The minimum crack-length interval, Δa, between data points (see Fig. 5) should be 0.25 mm (0.01 in.) or ten times the crack-length measurement precision (Ref 9), which is defined as the standard deviation on the mean value of crack length determined for a set of replicate measurements. This pre-

Fig. 3 Standard center-cracked tension specimen for fatigue crack propagation testing when the width (W) of the specimen ≤75 mm (3 in.)

$2a_n$ is the machined notch; a is the crack length; B is the specimen thickness. Source: Ref 9

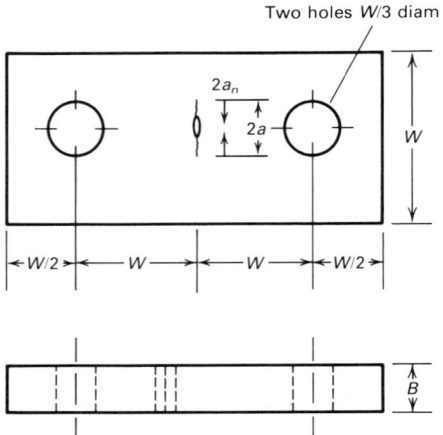

Fig. 4 Standard compact-type specimen for fatigue crack propagation testing

See Fig. 3 for explanation of symbols. Source: Ref 9

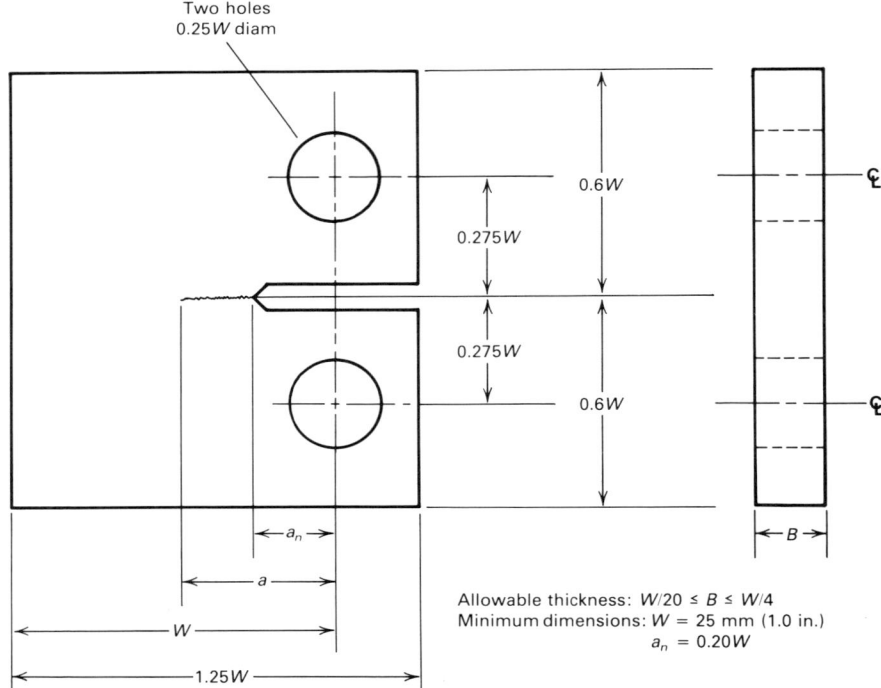

Allowable thickness: $W/20 \leq B \leq W/4$
Minimum dimensions: W = 25 mm (1.0 in.)
a_n = 0.20W

vents the measurement of erroneous growth rates from a group of data points that are spaced too closely relative to the precision of data measurement and relative to the scatter of data.

Crack measurement intervals are recommended in ASTM E 647 according to specimen type. For compact-type specimens:

$$\Delta a \leq 0.04W \text{ for } 0.25 \leq \frac{a}{W} \leq 0.40$$

$$\Delta a \leq 0.02W \text{ for } 0.40 \leq \frac{a}{W} \leq 0.60$$

$$\Delta a \leq 0.01W \text{ for } \frac{a}{W} \geq 0.60$$

For center-cracked tension specimens:

$$\Delta a \leq 0.03W \text{ for } \frac{2a}{W} < 0.60$$

$$\Delta a \leq 0.02W \text{ for } \frac{2a}{W} > 0.60$$

Fatigue crack growth rate data can be calculated by several methods. These are outlined in the article "Fatigue Crack Growth Data Analysis" in this Volume. The most commonly used methods, however, are the secant and incremental polynomial methods. The secant method consists of the slope of the straight line connecting two adjacent data points. This method, although simpler, results in more scatter in measured crack growth rate.

The incremental polynomial method fits a second-order polynomial expression (parab-

ola) to typically five to seven adjacent data points, and the slope of this expression is the growth rate. The incremental polynomial method eliminates some of the scatter in growth rate that is inherent in fatigue testing. Additional methods for calculating fatigue crack growth rate data are discussed in Ref 10.

Numerous relationships have been generated to correlate crack growth rate and stress-intensity data. The most widely accepted relationship is that proposed by Paris (see Eq 1). This is a linear relationship when plotted on log-log coordinates and generally yields a reasonable fit to the data in Region II (see Fig. 1) of the crack growth regime.

Other relationships based on the Paris equation, such as the commonly used Forman equation (Ref 11), are used to represent the variation of da/dN with other key variables, including load ratio, R, and the critical K value, K_c, at which rapid fracture of the specimen occurs (Region III in Fig. 1). The Forman equation is:

$$\frac{da}{dN} = \frac{C(\Delta K)^n}{(1 - R)(K_c - \Delta K)} \qquad \text{(Eq 3)}$$

where C and n are material constants of the same types as those in the Paris equation, but of different values. An advantage of the Forman equation is that it describes the type of

accelerated da/dN behavior that is often observed at high values of ΔK, which is not described by the Paris equation.

Additionally, the Forman equation describes the frequently observed increase in da/dN associated with an increase in R from 0 toward 1. When it is necessary to describe the effect of K approaching K_c, or the effect of R on da/dN, the Forman equation can be used to represent the da/dN behavior. When only ΔK in Region II is involved, the less complex Paris equation may be used. Additional formulas that include factors to account for Region I and III crack growth regimes and the effect of the load ratio on fatigue crack propagation characteristics can be found in the article "Fatigue Crack Growth Data Analysis" in this Volume and in Ref 12 to 18.

Cyclic Crack Growth Rate Testing in the Threshold Regime

Cyclic crack growth rate testing in the low-growth regime (Region I in Fig. 1) complicates acquisition of valid and consistent data, because the crack growth behavior becomes more sensitive to the material, environment, and testing procedures under this regime. Within this regime, the fatigue

Fig. 5 Crack growth versus constant-amplitude stress cycles for a Fe-10Ni-8Co-1Mo high-strength steel

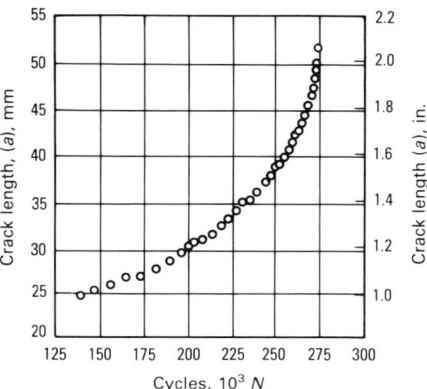

mechanisms of the material that slow the crack growth rates are more significant.

The precise definition of the cyclic crack growth rate threshold, ΔK_{th}, varies significantly. The most accurate definition would be the stress-intensity value below which fatigue crack growth will not occur. It is extremely expensive to obtain a true definition of ΔK_{th}, and in some materials a true threshold may be nonexistent. Generally, designers are more interested in the near-threshold regime, such as the ΔK that corresponds to a fatigue crack growth rate of 10^{-8} to 10^{-10} m/cycle (3.9×10^{-7} to 10^{-9} in./cycle). Because the duration of the tests increases greatly for each additional decade of near-threshold data (10^{-8} to 10^{-9} to 10^{-10}, etc., m/cycle), the precise design requirements should be determined in advance of the test.

Although the methods of conducting fatigue crack threshold testing may differ while achieving the same results, a proposed ASTM standard test method (Ref 19) provides a good starting point. As shown by Saxena *et al.* (Ref 20), the rate of change in the monotonic plastic zone size can be approximated with increasing crack size by:

$$K_{max} = K_{max_0} \exp[C(a - a_o)] \qquad \text{(Eq 4)}$$

where K_{max_0} refers to the maximum stress intensity at the initial crack length a_o, and a is the instantaneous crack length. The constant C describes the rate of change in stress intensity as a function of crack length.

More negative values of C result in sharper declines of stress intensity (the dimension C is in 1/length, or mm^{-1}, in.$^{-1}$, etc.). When C is held constant, the percentage change in K is constant for equal increments of crack length. Equation 4 can be rewritten as:

$$C = \left(\frac{1}{K}\right)\left(\frac{dK}{da}\right) \qquad \text{(Eq 5)}$$

It has been demonstrated (Ref 19) that for negative values of C greater than -0.08/mm (-2.0/in.), good agreement is present between the K-decreasing and K-increasing portions of tests on 2219-T851 aluminum and 10Ni steel at various stress ratios (see Fig. 6a and 6b). Note that the greatest amount of difference was found with tests conducted at the low stress ratio of 0.1. However, higher negative values of C have been used successfully in nickel-based superalloys when using an automated crack-measuring device with a resolution on the order of 0.002 mm (0.0001 in.). When a new crack-length measuring device is introduced, a new type of material is used, or any other factor is different from that used in previous testing, the K-decreasing portion of the test should be followed with a constant load (K-increase) to provide a comparison between the two methods. Once a consistency is demonstrated, constant-load testing in the low crack growth rate regime is not necessary under similar conditions.

In all areas of crack growth rate testing, the resolution capability of the crack-measuring technique should be known; however, this becomes considerably more important in the threshold regime. The smallest amount of crack-length resolution as possible is desired, because the rate of decreasing applied loads (load shedding) is dependent on how easily the crack length can be measured. The minimum amount of change in crack growth that is measured should be ten times the crack-length measurement precision. It is also recommended that for noncontinuous load shedding testing, where $[(P_{max_1} - P_{max_2})/P_{max_1}] > 0.02$, the reduction in the maximum load should not exceed 10% of the previous maximum load, and the minimum crack extension between load sheds should be at least 0.50 mm (0.02 in.).

In selecting a specimen, the resolution capability of the crack-measuring device and the K-gradient (the rate at which K is increased or decreased) in the specimen should be known to ensure that the test can be conducted appropriately. If the measuring device is not sufficient, the threshold crack growth rate may not be achieved before the specimen is separated in two. To avoid such problems, a plot of the control of the stress intensity (K versus a) should be generated before selection of the specimen.

Behavior of Short Cracks

Recently, it has been well documented that short cracks may behave differently from large cracks when plotted in the standard form of cyclic crack growth rate versus stress intensity (Ref 21-23). Pearson (Ref 22) has shown that in precipitation-hardened alloys cracks of sizes comparable to the grain size grew faster than larger cracks at similar stress-intensity values. Kitagawa and Takahashi (Ref 23) used arc strikes to produce small surface flaws and showed that below a critical crack size, the threshold stress intensity for short cracks decreased with decreasing crack length when the threshold stress approached that of the fatigue limit at very short crack lengths. Dowling demonstrated that the propagation of cracks in low-cycle fatigue specimens could be predicted using the J-integral concept, except for very short cracks (Ref 21).

A short crack is difficult to define. It may be small compared to the microstructure of the material to be studied (1 to 50 μm) when the concepts of continuum mechanics are of interest. It can also be small compared to the plastic zone size (10 to 1000 μm). In this situation, linear elastic fracture mechanics might be replaced with elastic-plastic fracture mechanics. The crack may also be physically small (500 to 1000 μm) when crack closure, crack tip shape, environment, and growth mechanisms are of concern. Figure 7 schematically illustrates the possible behavior of short cracks. Additional information on the behavior of short cracks can be found in Ref 24.

Fatigue Test Specimens

Selection of a fatigue crack growth test specimen is usually based on the availability of the material and the types of test systems and crack-monitoring devices to be used. The two most widely used types of specimens are the center-cracked tension specimen and the compact-type specimen (see Fig. 3 and 4). However, any specimen configuration with a known stress-intensity factor solution can be used in fatigue crack growth testing, assuming that the appropriate equipment is available for controlling the test and measuring the crack dimensions. Alternative specimen geometries are shown in Fig. 8. Stress-intensity factor solutions for center-cracked tension and compact-type specimens are given in Table 1.

Consideration of the range of application of the stress-intensity solution of a specimen configuration is very important. Many stress-intensity expressions are valid only over a range of the ratio of crack length to specimen width (a/W). For example, the expression given in Table 1 for the compact-type specimen is valid for $a/W > 0.2$; the expression for the center-cracked tension specimen is valid for $2a/W < 0.95$. The use of stress-intensity expressions outside their applicable

Fig. 6 Effect of normalized *K*-gradient on near-threshold fatigue crack growth rates established by *K*-decreasing method

The *K*-gradient is the rate at which the stress intensity is increased or decreased. (a) Aluminum alloy 2219-T851. (b) 10Ni steel. Source: Ref 19

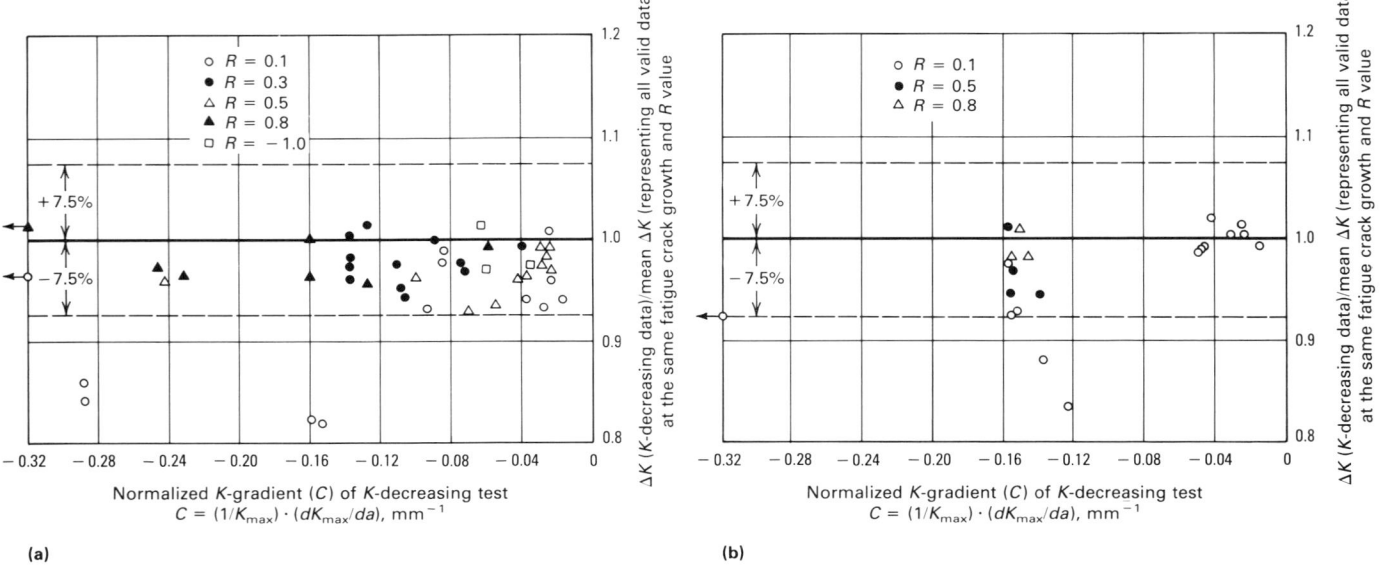

(a)   (b)

Fig. 7 Typical short crack behavior

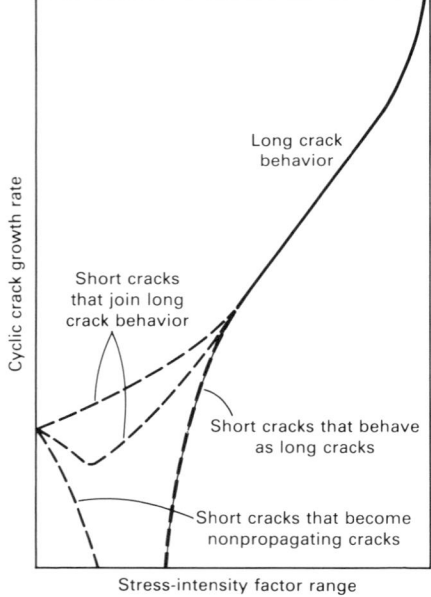

Table 1   Stress-intensity factor solutions for standardized (ASTM E 647) fatigue crack growth specimen geometries

Center-cracked tension specimens (Fig. 3)

$$\Delta K = \frac{\Delta P}{B} \sqrt{\frac{\pi\alpha}{2W} \sec \frac{\pi\alpha}{2}}$$

where $\alpha = \dfrac{2a}{W}$; expression valid for $\dfrac{2a}{W} < 0.95$

Compact-type specimens (Fig. 4)

$$\Delta K = \frac{\Delta P(2 + \alpha)}{B\sqrt{W}\,(1 - \alpha)^{3/2}}(0.886 + 4.64\alpha - 13.32\alpha^2 + 14.72\alpha^3 - 5.6\alpha^4)$$

where $\alpha = \dfrac{a}{W}$; expression valid for $\dfrac{a}{W} \geq 0.2$

crack-length region can produce significant errors in data.

The size of the specimen must also be appropriate. To follow the rules of linear elastic fracture mechanics, the specimen must be predominantly elastic. However, unlike the requirements for plane-strain fracture toughness testing, the stresses at the crack tip do not have to be maintained in a plane-strain state. The stress state is considered to be a controlled test variable. The material characteristics, specimen size, crack length, and applied load will dictate whether the specimen is predominantly elastic. Because the loading mode of different specimens varies significantly, each specimen geometry must be considered separately.

For the center-cracked tension specimen, the following is required:

$$W - 2a \geq \frac{1.25P_{max}}{B\sigma_{YS}} \qquad \text{(Eq 6)}$$

where $W - 2a$ is the uncracked ligament of the specimen (see Fig. 3), and σ_{YS} is the 0.2% offset yield strength at the temperature corresponding to the fatigue crack growth rate data.

For the compact-type specimen, the following is required:

$$W - a \geq \frac{\frac{4}{\pi}}{\left(\dfrac{K_{max}}{\sigma_{YS}}\right)^2} \qquad \text{(Eq 7)}$$

where $W - a$ is the uncracked ligament (see Fig. 4). For the compact-type specimen, the size requirement in Eq 7 limits the monotonic plastic zone in a plane-stress state to approximately 25% of the uncracked ligament. For both Eq 6 and 7, Ref 9 recommends the use of the monotonic yield strength. The size requirements in Eq 6 and 7 are appropriate for low-strain hardening materials ($\sigma_{ULT}/\sigma_{YS} \leq 1.3$), where σ_{ULT} is the ultimate tensile strength of the material.

An alternative size requirement can be used for high-strain hardening materials ($\sigma_{ULT}/\sigma_{YS} \geq 1.3$). In this case, the monotonic yield strength, σ_{YS}, is replaced with a higher effective yield strength that takes into account the strain hardening capacity of the material. This effective yield strength, termed the flow strength, is defined as:

$$\sigma_{FS} = \frac{1}{2}(\sigma_{YS} + \sigma_{ULT}) \qquad \text{(Eq 8)}$$

Use of Eq 8, however, allows mean plastic deformation to occur in the specimen. If flow strength rather than monotonic yield strength is used, results should be compared to other data or conditions, because crack growth may have been accelerated due to plastic deformation in the specimen.

Specimen size generally is governed by the material application and the loading capacity of the test system. Because most stress-intensity solutions are based on the ratio of crack length to specimen width (a/W), almost any combination of a and W can be used, as long as other criteria are satisfied. For tension loading specimens, the stresses in the plane of the crack must be uniformly distributed. Therefore, grips must be attached to the specimen at a sufficient distance from the crack. For a center-cracked tension specimen, this distance should generally be equal to the width of the specimen.

The size of compact-type specimens can vary greatly; however, the height to width ratio must remain constant ($h/W = 0.600$). These specimens are commonly referred to as $1T$, $2T$, $3T$, etc., specimens, where the specimen thicknesses are in 25.4-mm (1-in.) increments and the widths are in 50.8-mm (2-in.) increments. For example, a $4T$ compact-type specimen would have a width of 203 mm (8.0 in.) and a thickness of 101.6 mm (4 in.).

Specimen thickness can vary significantly. However, as the thickness increases, the amount of crack curvature in the specimen will increase. Because stress-intensity solutions are based on a straight through-crack, a significant amount of curvature, if not properly accounted for, can lead to an error in the data. Crack-curvature correction calculations are detailed in Ref 9.

For compact-type specimens, it is recommended that the thickness range between 5 and 25% of the width (Ref 9). For center-cracked tension specimens, the thickness should not exceed 25% of the width. The minimum allowable thickness depends on the gripping method used; however, the bending strains should not exceed 5% of the nominal strain in the specimen.

Two additional considerations regarding crack shape are the amount of crack variation from the front and back sides of the specimen and the amount of out-of-plane cracking. Due to microstructural changes through the specimen thickness, residual stresses (particularly in weldments), or misalignment of the

Fig. 8 Alternative crack growth specimen geometries

(a) Single-edge-crack bending specimen. (b) Double-edge-crack tension specimen. (c) Single-edge-crack tension specimen. (d) Surface-crack tension specimen. (e) Disc-shaped compact-type specimen

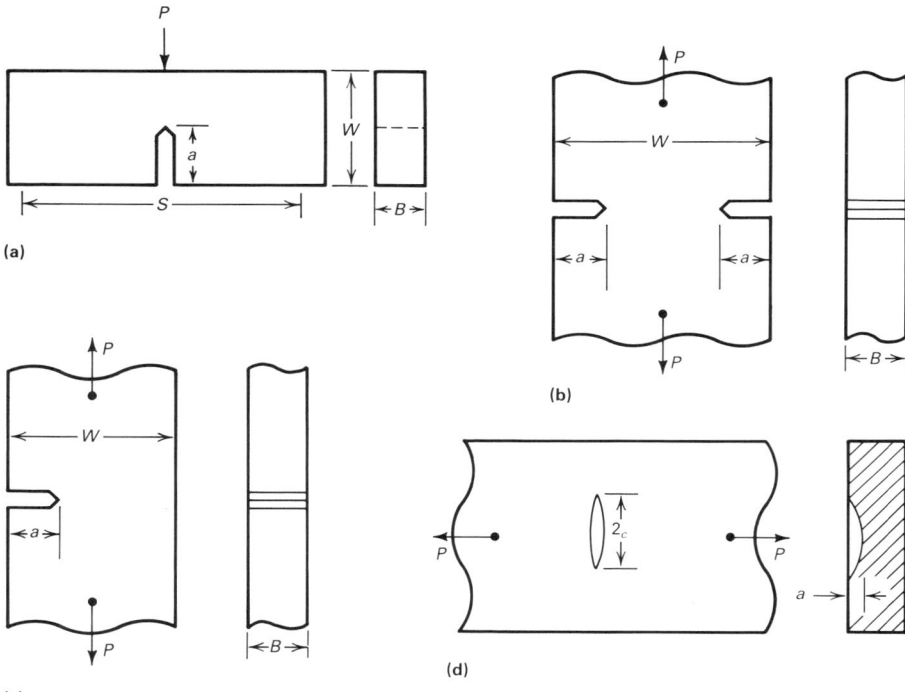

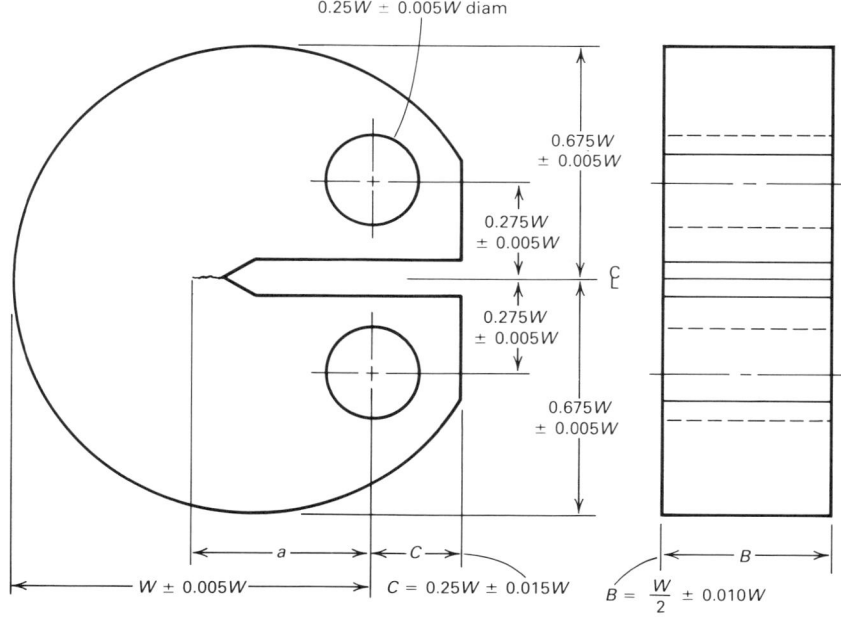

specimen in the grips, the crack may grow unevenly on the two surfaces. If any two crack-length measurements vary by more than $0.025W$ or by more than $0.25B$ (whichever is less), the precracking operation was not suitable and test results will not be valid. If a fatigue precrack departs more than $\pm5°$ from the plane of symmetry, the specimen is not suitable for subsequent testing. Additional information on precracking of the specimen can be found later in this article.

Notch Preparation. The method by which a notch is machined depends on the specimen material and the desired notch root radius (ρ). Sawcutting is the easiest method, but is generally acceptable only for aluminum alloys. For a notch root radius of $\rho \leq 0.25$ mm (0.010 in.) in aluminum alloys, milling or broaching is required. A similar notch root radius in low- and medium-strength steels can be produced by grinding. For high-strength steel alloys, nickel-base superalloys, and titanium alloys, electrical discharge machining may be necessary to produce a notch root radius of $\rho \leq 0.25$ mm (0.010 in.).

Precracking of a specimen prior to testing is conducted at stress intensities sufficient to cause a crack to initiate from the starter notch and propagate to a length that will eliminate the effect of the notch. To decrease the amount of time needed for precracking to occur, common practice is to initiate the precracking at a load above that which will be used during testing and to subsequently reduce the load.

Load generally is reduced uniformly to avoid transient effects. Crack growth can be arrested above the threshold stress-intensity value due to formation of the increased plastic zone ahead of the tip of the advancing crack. Therefore, the step size of the load during precracking should be minimized. Reduction in the maximum load should not be greater than 20% of the previous load condition. As the crack approaches the final desired size, this percentage may be decreased.

The amount of crack extension between each load decrease must also be controlled. If the step is too small, the influence of the plastic zone ahead of the crack may still be present. To avoid transient effects in the test data, the load range in each step should be applied over a crack-length increment of at least (3π) $(K'_{max}/\sigma_{YS})^2$, where K'_{max} is the terminal value of K_{max} from the previous load step. This requirement ensures that the crack extension between load sheds is at least three plastic zone diameters.

The influence of the machined starter notch must be eliminated so that the crack tip conditions are stable. For compact-type and center-cracked tension specimens, this re-

quires that the final precrack be at least 10% of the thickness of the specimen or equivalent to the height of the starter notch, whichever is greater (Ref 9).

Gripping of the specimen must be done in a manner that does not violate the stress-intensity solution requirements. For example, in a single-edge notched specimen, it is possible to produce a grip that permits rotation in the loading of the specimen, or it is possible to produce a rigid grip. Each of these requires a different stress-intensity solution. In grips that are permitted to rotate, such as the compact-type specimen grip, the pin and hole clearances must be designed to minimize friction. It is also advisable to consider lateral movement above and below the grips.

When appropriate, the use of a lubricant is recommended to reduce friction. In thick samples, the amount of bending in the pins should be minimized. Finally, the alignment of the system should be checked carefully to avoid undesirable bending stresses, which generally cause uneven cracking. Alignment can be easily checked using a strain gage specimen of a geometry similar to that used in the test program. Generally, bending strains should not exceed 5% of the nominal strain to be used in the test program.

Gripping arrangements for compact-type and center-cracked tension specimens are described in ASTM E 647 (Ref 9). For a center-cracked tension specimen less than 75 mm (3 in.) in width, a single pin grip is generally suitable. Wider specimens generally require additional pins, friction gripping, or some other method to provide sufficient strength in the specimen and grip to prohibit failure at undesirable locations, such as in the grips. Grips designed for compact-type specimens are illustrated in Fig. 9.

Crack-Length Measurement Techniques

The development of fracture mechanics as a means of predicting the mechanical behavior of materials in service has directed attention to testing techniques that clearly define the influence of sharp notches or cracks on the resistance of a material to failure. Analysis by fracture mechanics of the failure of precracked specimens in such tests requires knowledge of the critical crack length as well as the load at fracture, particularly when slow growth precedes final catastrophic failure. In studies of time-dependent fracture and subcritical crack growth, such as fatigue, creep, and stress corrosion, the need to determine crack propagation rate also requires accurate measurement of crack length at any stage during the test.

Fig. 9 Grips designed for fatigue crack propagation testing of compact-type specimens
Courtesy of MTS Systems Corp.

Several different techniques have been developed to monitor the initiation, growth, and instability of cracks, including optical (visual and photographic), electrical (eddy current and resistance), compliance, ultrasonic, and acoustic emission monitoring techniques. This section discusses the most commonly used laboratory procedures for determining crack length in fatigue crack growth testing: optical, compliance, and electrical potential techniques. Additional information on these and other techniques can be found in Ref 25 and 26.

Optical Crack Measurement Techniques

Monitoring of fatigue crack length as a function of cycles is most commonly conducted visually by observing the crack at the specimen surfaces with a traveling low-power microscope at a magnification of 20 to 50×. Crack-length measurements are made at intervals such that a nearly even distribution of da/dN versus ΔK is achieved. The minimum amount of extension between readings is commonly about 0.25 mm (0.10 in.). Information pertaining to crack measurement intervals for compact-type and center-cracked tension specimens can be found in the introduction to this article.

For planar specimens, the crack length is measured on one or both surfaces, depending on the section thickness. For example, ASTM E 647 (Ref 9) specifies a B/W value of 0.15 as the limit; measurements on only one side are sufficient if $B/W < 0.15$.

Through-thickness variations in crack length must be considered and corrected for if too severe. Typical behavior is for the crack length to lead at the midplane (crack tunneling). Because this cannot be observed *in situ* by visual monitoring, post-test observations must be made.

To account for through-thickness crack-length variation, ASTM E 647 recommends measuring the crack length at five points along the crack front contour and averaging the five readings. If the average of the five points exceeds the surface length by more than 5%, the average length is used in computing the growth rate and K.

The optical technique is straightforward and, if the specimen is carefully polished and does not oxidize during the test, produces accurate results. However, the process is time consuming, subjective, and can be automated only with complicated and expensive video-digitizing equipment. In addition, many fatigue crack growth rate tests are conducted in simulated-service environments that obscure direct observation of the crack. The trend toward laboratory automation has resulted in the development of indirect methods of determining crack extension, such as specimen compliance and electric potential monitoring.

Compliance Method of Crack Extension Measurement

The compliance of an elastically strained specimen containing a crack of length a measured from the load line to the crack tip is usually expressed as the quotient of the displacement, δ, and the tensile load, P, with the displacement measured along, or parallel to, the load line. Figure 10 illustrates that the more deeply a specimen is cracked, the greater the amount of δ measured for a specific value of tensile load. In this section, discussion will be limited to tensile loads (Mode I, in fracture mechanics nomenclature). Compliance can also be defined for shear and torsional loads applied to cracked specimens, and crack extension under these loading modes can be similarly determined. Additional information on the compliance method can be found in Ref 25 and 26.

Instrumentation. The displacement usually is measured across the crack mouth opening using either a cantilever beam clip gage (Ref 27), a linear variable differential transformer (LVDT), or an eddy current probe. In each case, the device is connected

to a conditioner/amplifier unit that is capable of producing a voltage in the ±10 V dc range, which can be used as input to *x-y* recorders, data loggers, or other computer-interfaced processors. Specimen load is simultaneously measured by an electronic load cell and conditioner/amplifier system, and the output is directed to the same data-acquisition system. A generalized schematic of the circuits involved is shown in Fig. 11.

The required sensitivity of the systems depends on specimen geometry and size; in general, noise-free, amplified output on the order of 1 V dc per 1 mm (0.04 in.) of deflection is satisfactory. Similarly, for the load range applied to the specimen, an approximately 1 V dc change in signal from the load cell is required for accurate calculation of the compliance. More specific information on calculation of compliance can be found in Ref 28 and 29.

Basic Laboratory Practice. Consider a constant-load amplitude fatigue crack growth rate test with a positive load ratio ($R = P_{min}/P_{max}$). As the test proceeds and the crack extends, the change in crack mouth opening displacement increases. This is illustrated schematically by the *x-y* recorder traces in Fig. 12. At the beginning of each trace, the pen of the recorder has been offset to the right to provide visual separation of the traces. This offset, which is not fully the result of a change in specimen response (although such a change is measurable), is in the same direction as the offset manually applied to the recorder controls. There is little or no hysteresis in the full cyclic trace, implying that there is no crack growth occurring during the cycle used to measure the compliance. This is generally true for tests conducted well within the elastic response range of the specimen. Special testing proce-

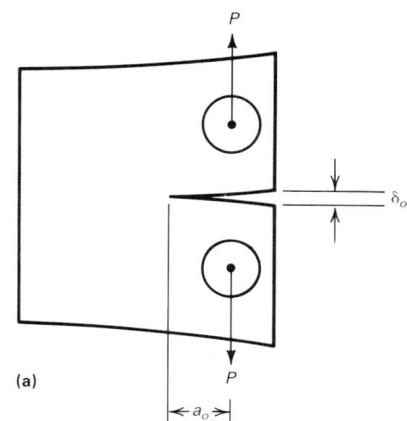

(a)

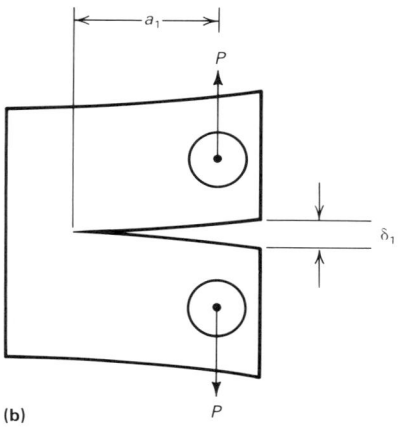

(b)

Fig. 11 Components of a compliance measurement system

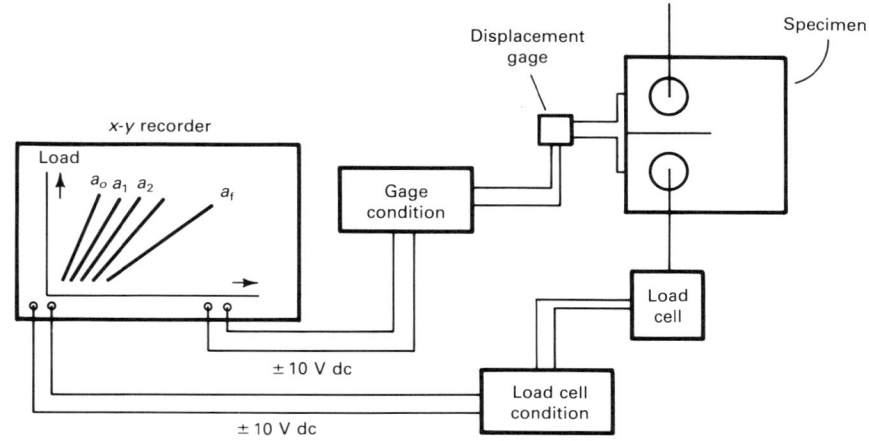

Fig. 12 Schematic of the series of x-y recorder traces that might be generated during a fatigue crack growth test

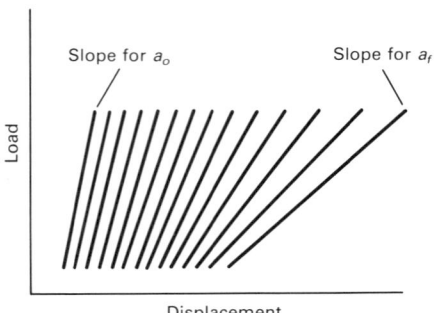

Fig. 13 Bolt-on attachment of crack-opening displacement transducer to fatigue test specimen

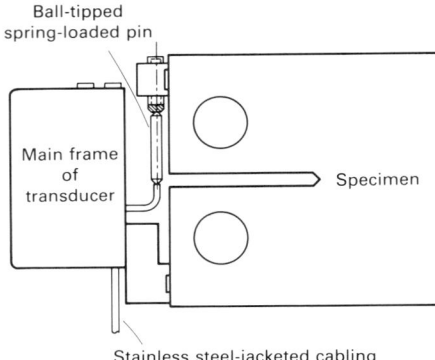

dures are required when a specimen exhibits plastic deformation; these are defined in ASTM E 647 (Ref 9).

Users of less sophisticated compliance measurement systems often find that the hysteresis loops are somewhat open and that just after load reversal at the minimum and maximum load points, the load will change while the displacement appears to be constant. This indicates a mechanical problem, such as excess backlash in the gage mounting hardware or friction in the clevis pins. Once the problem is resolved, there is no difference between using the load-increasing slope or the load-decreasing slope to make the actual compliance measurement.

Attachment of Displacement Measurement Hardware. One of the most important factors affecting the accuracy of crack-opening displacement measurements is the manner in which the displacement transducer is attached to the test sample. Transducers for measuring the crack-opening displacement commonly consist of cantilevered arms affixed across the crack. When the crack is opened, deflections either in the arms, or in a flexure attached to the arms, induces measurable strains, which are ultimately converted to displacements.

The transducer can be bolted across the crack opening at the point of testing (Fig. 13), or it can be attached to the specimen through hardened knife-edge pivots that are mechanically or adhesively affixed to the specimen (Fig. 14). The transducer can also be affixed via knife-edge contacts that are machined into the test sample (Fig. 15). For elevated-temperature testing, feed-rod systems are frequently used (Fig. 16).

The bolt-on system of attaching the transducer to the test specimen (Fig. 13) is capable of reacting to high acceleration loads, which result from higher frequency dynamic testing when the rocking moment generated

by the mass of the transducer is carried to the bolt-on attachment through the transducer frame. This attachment system is preferred when a transducer has high mass or an effective mass center that is located a great distance from the specimen contact pads.

This system also provides accurate crack-opening displacement measurements on specimens tested under environments that are not conducive to the use of knife edges—such as elevated-temperature or corrosive environments. In addition, the bolt-on attachment system allows the use of stiffer cabling without disturbing the measurement; for example, a displacement transducer with relatively rigid stainless steel-jacketed cabling can be used to make measurements in pressurized high-temperature water/steam environments (see Fig. 15).

The frame or case of this unit is fastened to the test specimen via tapped holes in the specimen. Measurements are taken through a ball-tipped, spring-loaded pin to a second attachment block that is bolted to the opposite side of the specimen notch. This allows accurate measurements to be made in difficult environments. After the test sample fails, the ball-tipped push pin simply drops free, with no damage to the transducer.

Hardened knife-edge pivot contacts (Fig. 14) provide a measurement system with minimal sliding action; the knife edge rocks in a hardened seat in the transducer arm. This allows measurements to be made with very low hysteresis levels. Contact and seat ramp angles can be designed for optimal trade-offs between static and dynamic measurement accuracy, dynamic stability, and contact durability. Male knife-edge contact replacements are relatively low in cost, and various configurations are available, such as three-point contact, line contact, large radius, and small radius.

Knife-edge contacts that are machined into the test sample (Fig. 15) eliminate the possibility of knife-edge screws loosening, which results in slippage and hysteresis. The compressed initial gage length can be machined to the required tolerance.

Fig. 14 Bolt-on hardened knife-edge attachment of crack-opening displacement transducer to fatigue test specimen

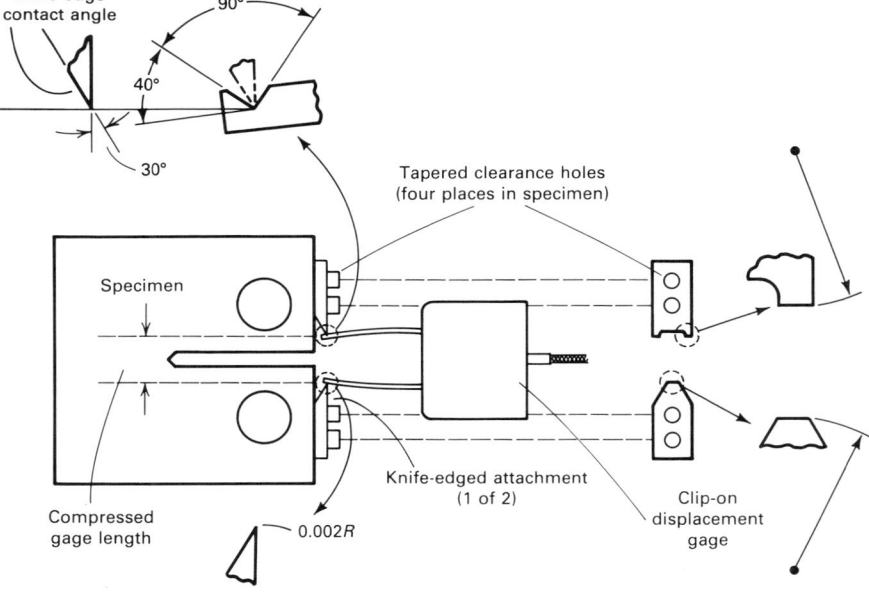

Fig. 15 Attachment of crack-opening displacement transducer to specimen by machined knife-edge contacts

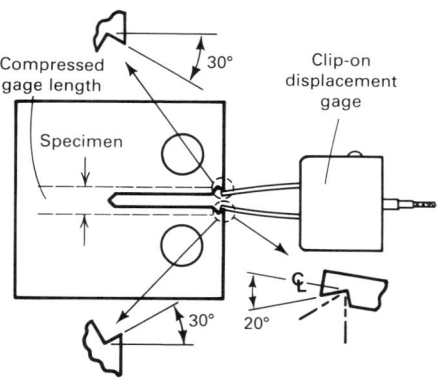

Fig. 16 Feed-rod attachment of crack-opening displacement transducer to specimen for high-temperature testing

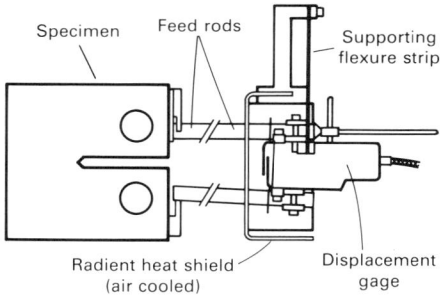

Fig. 17 Comparison of predicted and experimental compliance for a compact-type fatigue specimen

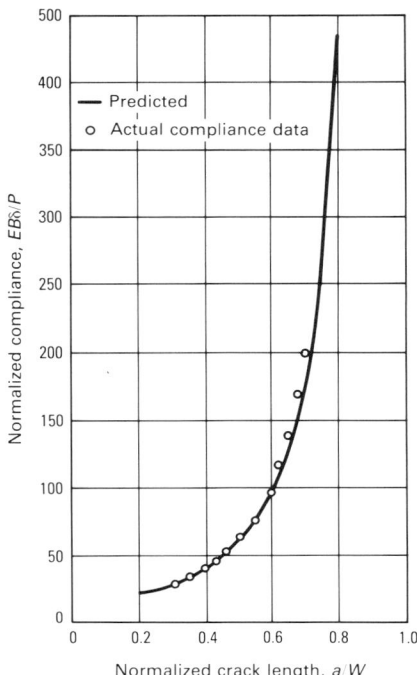

Feed-rod attachment of the transducer to the test sample (Fig. 16) makes testing of samples at elevated temperatures possible, while allowing the transducer to be operated at a much lower temperature. Feed-rods generally are made from quartz or ceramic materials. Quartz generally is used at temperatures up to 1000 °C (1830 °F), and various ceramics are used at higher temperatures. Feed-rods generally contact the specimen through bolt-on or weld-on attachment blocks. The transducer body is shielded from radiant heating and is air or water cooled.

Computing Normalized Compliance. When measuring the compliance of a fatigue specimen, the usual practice is to compute a normalized compliance, $EB\delta/P$, where E is the elastic (Young's) modulus of the material, and B is the specimen thickness. This normalized compliance is plotted against a normalized crack length, a/W, where W is the specimen width. For standard geometries, such as a compact-type specimen, this relationship has the form shown in Fig. 17. Note that when the crack is short ($a/W \sim 0.2$ to 0.4), the compliance is less sensitive to changes in crack length than when the crack is long ($a/W > 0.5$). Thus, the sensitivity of the compliance method is significantly improved for the longer crack lengths, both because of this relationship and because the amount of crack mouth opening and the resulting displacement gage signal are larger. The amount of displacement or crack mouth opening that is measured is a strong function of the location of the line of measurement of the gage with respect to the load line, which is the reference point for crack extension. The farther away from the crack tip the measurement can be made, the more displacement will be incurred, and the sensitivity of the method will be improved proportionately.

Data Acquisition and Processing. The signals from the load cell and displacement gage must be obtained simultaneously in order for this method to work to its best advantage. In the most direct case, the two signals can be fed to an x-y recorder, with the load applied to the y-axis and displacement to the x-axis; at various intervals during the test, a trace of the two signals can be made. If the test is being conducted at a reasonably high frequency (>1 Hz), then the frequency will have to be diminished so that the slow rate of the recorder can keep up with the changing voltage. This is not a problem if a transient recorder is used and the results from the two channels (load and displacement) are co-plotted. The slopes of the recorder traces can be measured, multiplied by suitable calibration factors, and used in the compliance to crack length relationship. Reviews of the application of computerized data acquisition and processing are available in Ref 30 and 31.

A more sophisticated method is to use a computerized data-acquisition system to obtain load-displacement data. These systems are usually faster, and thus can accept data from rather high-frequency waveforms. In addition, software can be developed to perform the calculations involved in processing the compliance data to crack length. Software to perform fatigue crack growth rate measurements is generally available from manufacturers, but most researchers write their own data-acquisition packages, perhaps using some of the manufacturer-supplied subroutines that are specific to the hardware involved.

It is good practice to set up a computer program so that data are acquired on either the load-increasing or load-decreasing segment of the full load-displacement cycle, but not on both, in case there is measurable hysteresis due to backlash in the mounting fixture, which is inevitable in any LVDT fixture that allows for rotation of the internal magnetic core. Additionally, data should be taken between about 10 and 90% of the load range. Eliminating the top and bottom fractions of the load range avoids problems of crack closure (at loads approaching zero) or incipient plasticity (near the load maximum, at longer crack lengths). The sets of load-displacement pairs are fitted to a straight line, the slope of which is used in the compliance expression.

Mathematics of Compliance Measurement. There are two basic ways to produce the mathematical expression of the normalized compliance versus normalized crack length curve shown in Fig. 17: (1) locate an appropriate expression in the literature or (2) create an expression appropriate to the particular specimen, gage, and hardware combination involved by deriving a table of measured compliance values versus crack lengths and curve fitting the results to obtain a useful function. Recent research has yielded a number of carefully constructed formulas for the compliance to crack length relationships for most of the common fatigue crack growth specimens (Ref 32, 33). In many cases, if the displacement measurement centerline upon which the formula was based is not the one used by the experimentalist, the formulas can be reconstructed easily using equations for the center of rotation (which are also functions of the crack length), generating a new table of location-specific compliance values versus normalized crack length, and curve fitting this tabular data. The curve-fitting routine used should not be extended beyond its capabilities. Most of the published formulas were developed with the aid of very sophisticated computer programs, which may not be available to the average laboratory. Therefore, only the range of crack length of interest should be incorporated into the curve fit, along with the fewest number of parameters or degrees of freedom that are necessary to obtain a goodness-of-fit to less than about 1%. An alternative procedure is to construct a rather extensive table of normalized compliance versus normalized crack length, and look up, manually or by computer, the various values of compliance as they are measured.

Electric Potential Crack Monitoring Technique

The electrical potential, or potential drop, technique has gained increasingly wide acceptance in fracture research as one of the most accurate and efficient methods for monitoring the initiation and propagation of cracks. This method relies on the fact that there will be a disturbance in the electrical potential field about any discontinuity in a current-carrying body, the magnitude of the disturbance depending directly on the size and shape of the discontinuity.

For the application of crack growth monitoring, the electric potential method entails passing a constant current (maintained constant by external means) through a cracked test specimen and measuring the change in electrical potential across the crack as it propagates. With increasing crack length, the uncracked cross-sectional area of the test piece decreases, its electrical resistance increases, and thus the potential difference between two points spanning the crack rises. By monitoring this potential increase, V_a, and comparing it with some reference potential, V_o, the crack length to width ratio, a/W, can be determined through the use of the relevant calibration curve for the particular test piece geometry concerned.

Accuracy of electrical potential measurements of crack length may be limited by a number of factors, including the electrical stability and resolution of the potential measurement system, electrical contact between crack surfaces where the fracture morphology is rough or where significant crack closure effects are present, and changes in electrical resistivity with plastic deformation (Ref 34-36). Another key factor is the determination of calibration curves relating changes in potential across the crack

(V_a) to crack length (a). In most instances, experimental calibration curves have been obtained by measuring the electrical potential difference (1) across machined slots of increasing length in a single test piece (Ref 36), (2) across a growing fatigue crack, where the length of the crack at each point of measurement is marked on the fracture surface by a single overload cycle or by a change in mean stress, or (3) across a growing fatigue crack in thin specimens where the length of the crack is measured by surface observation (Ref 37, 38).

Other experimental calibrations have been achieved using an electrical analog of the test piece, where the specimen design is duplicated, usually with increased dimensions for better accuracy, using graphitized analog paper (Ref 39) or thin aluminum foil (Ref 40), and where the crack length can be increased simply by cutting with a razor blade. Such calibration procedures, however, are relatively inaccurate, particularly at short crack lengths, and are tedious to perform. Furthermore, where measurements of crack initiation and early growth are required ahead of short cracks or notches of varying acuity, such procedures demand a new experimental calibration to be obtained for each notch geometry (Ref 41).

Attempts at theoretical calibrations have involved finding solutions to Laplace's equation within the boundary conditions of a particular test piece geometry, where for a strip of metal of constant thickness and width containing a transverse crack, the steady electrical potential V at a point (x, y) is given by

$$\nabla^2 (V) = 0 \qquad \text{(Eq 9)}$$

assuming that the current flows only in the plane of the strip (i.e., in the x-y plane). Analytical solutions to Eq 9 using conformal mapping techniques have been obtained for several simple geometries, including center- and edge-cracked plates with razor slit starter notches (Ref 37, 42), single-edge notched specimens with semielliptical, semicircular, and V-notches (Fig. 18a), and cylindrical geometries with cord cracks (Ref 43).

For more complex geometries, such as the compact-type specimen (Fig. 18b), conformal mapping procedures are not as easily applicable. Here, successful theoretical calibrations have been obtained numerically by utilizing finite element procedures for several types of test samples (Ref 35, 44, 45). A comparison of such calibration curves, given in the form of V_a/V_{a_o} versus a/W, where V_a is the potential drop across the crack length a, and V_{a_o} is the reference potential drop across the initial starter notch (or crack) of length a_o, is shown for the single-edge notched specimen in Fig. 19, derived from indepen-

dent experimental, electrical analog, conformal mapping, and finite element calibrations. A similar comparison of experimental and finite element calibrations for the compact-type specimen are shown in Fig. 20. Through the use of such nondimensionalized calibrations, calibration curves become independent of material properties, test piece thickness, and magnitude of input current (provided it remains constant) and are primarily a function of specimen and crack ge-

Fig. 18 Test specimen geometries and typical locations of current input and potential measurement probes

(a) Single-edge notched specimen (uniform current configuration). Initial crack length, a_o = 5 mm (0.2 in.); width, W = 20 mm (0.8 in). (b) Compact-type specimen (point application of current). Initial crack length, a_o = 17.5 mm (0.7 in.); width, W = 50 mm (2 in.); $2H$ = 60 mm (2.4 in.)

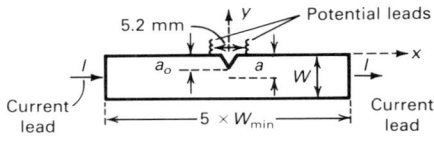

(a)

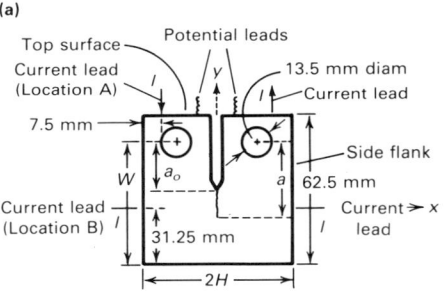

(b)

Fig. 19 Calibration curve of a single-edge notched specimen for $0.25 < a/W < 0.75$ for configuration shown in Fig. 18(a)

——, theoretical (conformal mapping) (Ref 41); ●, experimental (Ref 34); ○, electrical analog (Ref 46); ■, finite element (Ref 44)

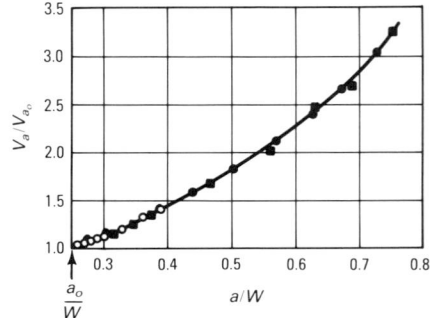

Fig. 20 Calibration curve for compact-type specimen for 0.35 < a/W < 0.8 for configuration shown in Fig. 18(b) at current lead location A

●, experimental (Ref 34); ■, experimental (Ref 47); ○, finite element (coarse mesh) (Ref 44); □, finite element (fine mesh) (Ref 45)

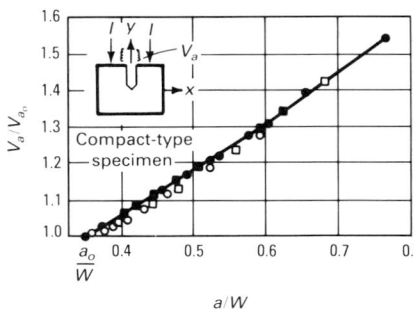

Table 2   Summary of electric potential optimization parameters

Parameter	Limiting consideration	Mathematical statement
Accuracy	Accuracy of calibration curve	...
Sensitivity	Slope of calibration curve	Maximize dV/da
Reproducibility	Ability to locate potential leads	Minimize dV/dx, dV/dy
Measurability	Magnitude of output voltage signal	Maximize $|V_a|$

ometry and the locations of current input and potential measurement leads.

Optimization Parameters. In any specimen geometry, there are numerous locations for both the current input leads and the potential measurement probes. Optimization of the technique involves finding the best locations, considering accuracy, sensitivity, reproducibility, and magnitude of output (measurability), as summarized in Table 2.

Accuracy is defined in terms of the degree to which the calibration curve approximates the real relationship between potential change and crack length, and may be assessed by comparison of experimental and theoretical calibrations. However, in practice, the accuracy of the electrical potential technique may be limited by several other factors, such as the electrical stability and resolution of the potential measurement system, crack front curvature, electrical contact between crack surfaces where the fracture morphology is particularly rough or where significant crack closure effects are present, and changes in electrical resistivity with plastic deformation, temperature variations, or both (Ref 34).

Sensitivity is defined as the ability of the method to discriminate between small differences in crack length, as illustrated by the slope of the calibration curve V_a/V_{a_o} versus a/W. Increased resolution may be achieved by maximizing the calibration curve slope, and for a particular geometry this is a function of lead placement.

Reproducibility refers to inaccuracies produced by small errors in positioning the potential measurement leads. Such leads are generally fine wires that are spot welded or screwed to the specimen, and accurate positioning is typically no better than within 0.5

mm (0.02 in.). To maximize reproducibility, these leads should be placed in an area where the calibration curve is relatively insensitive to small changes in position—that is, where dV/dx and dV/dy are small, where x and y are position coordinates—with the origin at the midpoint of the specimen, as defined in Fig. 18. This consideration is often at variance with sensitivity considerations.

Measurability is defined as the ability of the output voltage signal to be measured over background noise, such as thermal electromotive force, instrument drift, and white noise. To optimize measurability, current input and potential measurement lead locations are chosen to maximize the absolute magnitude of the output voltage signal $|V_a|$. As output voltages are generally at the microvolt level and because of the high electrical conductivity of metals, a practical means of achieving measurability is simply to increase the input current. However, there is a limit to

this increase, because when the current is too large (typically exceeding 30 A in a 12.7-mm, or 0.5-in., thick $1T$ steel compact-type specimen), appreciable specimen heating can result from contact resistance at current input positions.

Optimizations for specific test piece geometries can be easily performed using electrical analog patterns, where the distribution of equipotential lines yield information on sensitivity and reproducibility with probe position (Fig. 21). More detailed optimizations require numerical procedures (Ref 35, 45). When applied to the compact-type specimen, such procedures have shown that, whereas greatest sensitivity is obtained by placing the potential measurement probes at the notch tip, the potential gradients are so steep there that reproducibility, from minor variations in the probe positioning, is poor (Fig. 21). Conversely, by placing these probes on the top surface, as in Fig. 18, as close to the mouth

Fig. 21 Equipotential distribution from electrical analog patterns for compact-type specimens (uncracked and cracked) with different current positions

(a) "Area-contact" along top face. (b) "Area-contact" along side flanks. Source: Ref 46

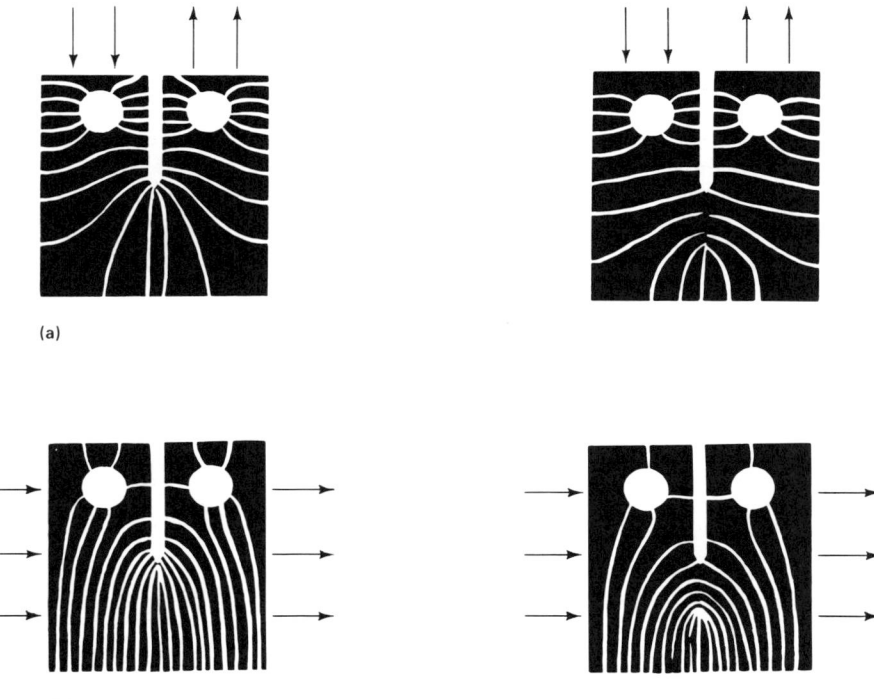

of the notch as possible, a better combination of sensitivity and reproducibility is achieved (Ref 45).

This can be seen in Fig. 22, where the non-dimensional voltage V_a/V_{a_o} is plotted as a function of potential lead position x on the top surface for five different crack lengths ($0.4 \leq a/W \leq 0.68$). Note that the curves are flattest (that is, dV/dx is a minimum) when potential leads are attached close to the notch ($x < 4$ mm, or 0.16 in.), confirming that this location yields excellent reproducibility of measurement. For example, for the case of a 2.5-mm (0.1-in.) crack emanating from the notch ($a/W = 0.4$), an error in lead placement of 5 mm (0.2 in.) results in a voltage error of less than 5%. As the potential leads are moved farther apart (x increasing), the potential error from lead misplacement substantially increases. Similarly, maximum apparent sensitivity (that is, maximum increase in V_a/V_{a_o} with increasing a/W) is achieved when the potential leads are close to the notch; V_a/V_{a_o} decreases markedly, particularly at longer crack lengths, as they are moved farther apart. Thus, from both reproducibility and sensitivity locations, potential leads are best placed on the specimen top surface close to the notch mouth.

Current input locations can similarly be optimized. Two commonly used locations on the compact-type specimen are shown in Fig. 18(b)—on the top surface or on the side flank. Although sensitivity apparently is increased by using the side flank input location (i.e., the calibration curve of V_a/V_{a_o} versus a/W becomes distinctly steeper for this configuration), the absolute magnitude of the voltage output signal (V_a) is substantially lower (~60%). In low-resistivity metals, such as aluminum, this reduced output clearly can become a problem with respect to the signal to noise ratio; for this reason, a configuration of current input probes on the top surface (i.e., location A in Fig. 18b) is preferred (Ref 45).

Apparatus. A schematic diagram of a typical experimental setup for electrical potential crack monitoring measurements is shown in Fig. 23. The technique can be used with either direct or alternating current. Although the alternating current method does offer certain advantages, particularly with lower power requirements for monitoring in large structures and the lack of errors from thermally induced electromotive forces (Ref 48-50), it has been found to be more difficult to utilize and accordingly the direct current method is more widely used.

The power supply generally must be capable of producing a large, stabilized, constant current. This can be achieved with commercial constant-current power supplies or by modification of a constant-voltage power supply through the use of a simple feedback circuit that senses the voltage across a shunt in series with the specimen. Because the output current stability is dependent on the constancy of resistance of this shunt, a large air-cooled ballast resistor with an extremely low coefficient of thermal conductivity generally is used. The resistance of the shunt (typically ~20 mΩ) is large compared to that of the specimen (typically ~1μΩ) so that changes in the resistance of the test piece due to cracking are negligible with respect to overall external resistance of the circuit.

For most regular fracture mechanics test geometries (e.g., a 25-mm, or 1-in., thick compact-type specimen) in metallic materials, stabilized direct currents of between 5 and 50 A, operating at voltages of a few volts with a current stability of 1 part in 10^4, generally are sufficient to obtain measurable

Fig. 22 Variation of V_a/V_{a_o} with potential lead position x on top surface of compact-type specimen for current (I) input location A on top surface of test piece
Source: Ref 45

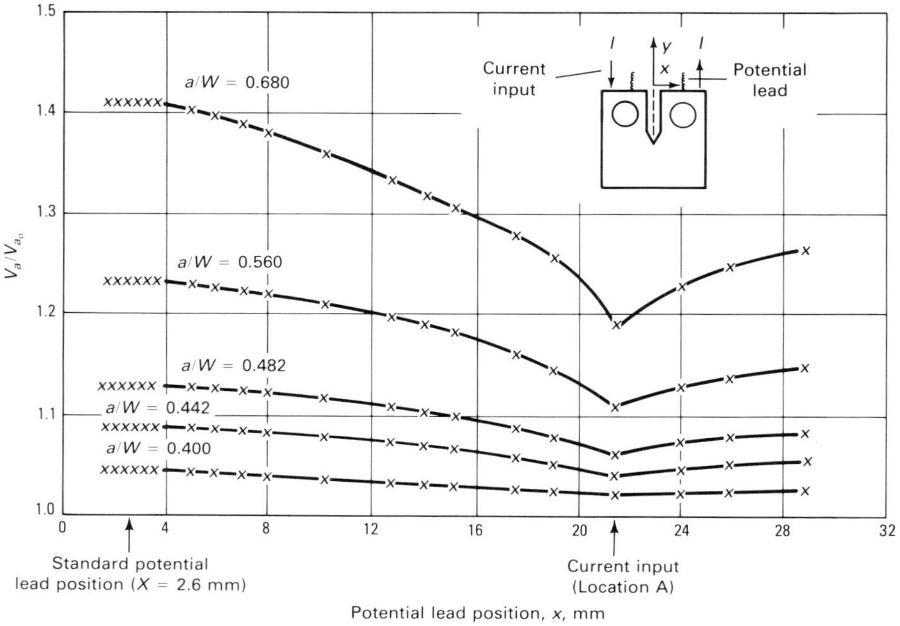

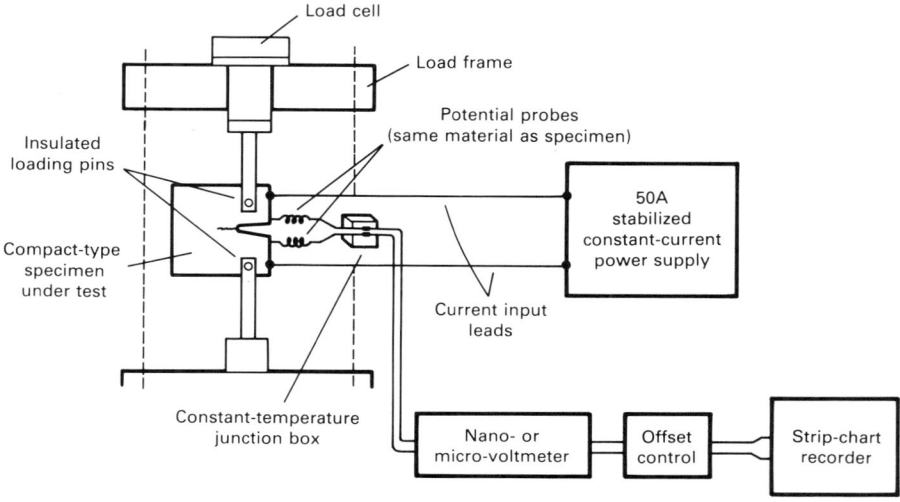

Fig. 23 Schematic of the direct current electrical potential crack monitoring system

voltage readings. In order to minimize uncontrolled heating of the sample from contact resistances, high-current (welding) cable is ideal for current input leads, provided it is tightly connected to the sample. A compact specimen with screw-threaded copper current input connections and spot-welded potential measurement probes is shown in Fig. 24.

To prevent short-circuiting of the current, insulation of the sample from the testing machine generally is necessary. This can be achieved readily by nonconducting inserts in the loading train, or in pin-loaded samples through the use of ceramic collars (e.g., alumina) inside the loading holes. For potential measurements, a low-noise, low-drift voltmeter/amplification system with a gain in excess of 10^3 is essential. Although not requiring the advanced electronics necessary in the use of the alternating current method, direct current measurement systems must be capable of detecting fractions of microvolts without noticeable drift over the duration of the test (i.e., typically from a few hours to a week). With certain precautions, this is possible with commercial nanovoltmeters or by constructing a circuit consisting of a low-drift (<0.2 μV/°C) differential amplifier (Ref 34). However, to maintain these characteristics of the amplifier, all external resistors on the input side should be of high quality with less than 1 ppm/°C temperature stabil-

ity, and the amplifier itself should be encased within a constant-temperature enclosure.

Furthermore, to minimize thermal electromotive forces from soldered joints (the thermal electromotive force for copper/solder is approximately 3 μV/°C at ambient temperatures), low thermal electromotive force cadmium solder must be used for all input connections with the tinning scraped off the resistors prior to soldering. Also, thermal electromotive forces at the test piece can be avoided by constructing the potential measurement probes from wire of the same material as the specimen. For steels, where spot-welded 0.2-mm (0.008-in.) enamel-coated iron wires are ideal (the iron/copper thermal electromotive force is approximately 12 μV/°C at ambient temperatures), connection of these wires to the voltmeter/amplification system is best achieved within a constant-temperature junction box. With the alternating current systems, thermal electromotive forces and drift are not a problem, and noise can be minimized through the use of lock-in amplifiers as both the reference source and measurement device. However, these advantages can be offset in the alternating current systems by problems of potential lead interaction and electronic instability.

All external signal leads should be twisted and held fairly rigid to prevent induced voltage pickup, and electrically and magnetically screened to ground. The provision of a single good ground is vitally important in this type of application to prevent large errors from noise. It is also useful to screen the specimen from drafts and maintain a constant specimen temperature. Moreover, under certain conditions where thermal-induced voltages cannot be avoided, periodic potential V_a readings with the current switched off can provide a means of monitoring any changes in the thermal electromotive forces.

A listing of ideal procedures for successful operation of the direct current electrical potential method is given in Table 3. With such procedures, accuracies of better than 0.1 mm (0.004 in.) are readily obtainable for measurements of absolute crack length, and the technique can resolve changes in crack length down to the order of 10 μm.

Crack Initiation Studies. There have been numerous attempts over the years to apply the electric potential technique to detect crack initiation in fatigue, creep, and fracture toughness testing (Ref 51-59), particularly under elastic-plastic conditions—for example, for J_{Ic} and crack-opening dis-

Fig. 24 Compact-type specimen showing screw-threaded copper current input connections and spot-welded potential measurement probes

The shielding on the potential probes is grounded to the negative side of the input current.

Table 3   Procedures for successful operation of the direct current electrical potential method

Current stability
Use of a constant-current power supply with feedback stabilization across a large constant-resistance shunt

Drift of amplifying system
Use of a stabilized low-noise, low-drift preamplifier
Constant-temperature enclosures for all amplifiers
Long warm-up periods

Error from thermal electromotive forces
Probe wires used for potential measurements of same material as test piece
Temperature-stabilized dissimilar metal junction at input to preamplifier
Low thermal electromotive force cadmium solder used with untinned resistors for all input connections
Periodic voltage readings with zero current input

Noise
All leads twisted and held fairly rigid
Input leads to preamplifier made as short as possible
All leads electrically and magnetically screened to a single good ground
Capacitor circuits across each amplifier to damp out alternating current pickup
Constant-voltage (isolation) transformer to prevent power surges

Reproducibility
Potential probes located on the top face of the specimen to prevent large errors from positional variations
Uniform current configuration used for single-edge notched specimens to prevent errors from slight variations in current input position
Introduction of current on the top surface in compact-type specimens

Variations in crack profile
Potential probes offset diagonally across the notch to obtain a more effective value of mean crack length

Heating effects
Use of current densities not exceeding 10 A/cm^2
High-current welding cable and large copper terminals used as current input connections
Tests conducted in a constant-temperature enclosure

Calibration errors
Measurement of a reference potential largely unaffected by crack growth
Low or constant degree of plasticity in specimen with crack extension

Short-circuiting
Fatigue tests conducted with a small minimum tensile load
Use of noninert test environments

placement determination (Fig. 25). The results have been mixed. Problems appear to arise, at least in ductile materials, by the inherent "diffuse" nature of the initiation event. This involves, in the case of ductile fracture, for example, the gradual creation of voids and their subsequent coalescence to form a major crack (Fig. 25). Also, the delineation of the initiation event can be masked by the effect on the potential of the accompanying crack tip plasticity and large geometry changes in the test piece.

The effect of plasticity on the potential variation is especially troublesome, as it can vary distinctly from material to material. An initial decrease in potential at the start of ductile fracture toughness testing has been noted in certain alloys (Fig. 25c) and has been attributed to magnetostrictive effects (Ref 57) or crack surface contact at the tip of the fatigue precrack at ductile crack initiation (Ref 58). However, these explanations are questionable because this "anomalous" potential drop is extremely material-specific,

can be observed with imposed direct currents, and is apparent in specimens containing a machined slot where no crack closure can occur (Ref 59). Instead, the effect appears to result from a change in resistivity with deformation, which can be negative in certain alloys.

The technique gives the best indication of initiation of high-strength lower toughness alloys (Ref 54-56), where the higher strength and accompanying lower strain hardening causes cleavage fracture or promotes easier strain localization, which results in a more distinct linkage of voids (i.e., direct necking) and thus a clearer point of initiation. However, in lower strength ductile alloys where void coalescence is more diffuse and accompanied by extensive inelasticity, it is extremely difficult to discern a point of crack initiation from potential/displacement traces without prior knowledge from another crack detection method.

Crack Growth Studies. By far the most useful application of the electrical potential method has been in measurements of crack length during crack propagation, where it has been utilized to monitor almost all mechanisms of subcritical crack growth and most notably to follow fatigue crack growth. Typical crack propagation rates derived from direct current potential measurements are shown in Fig. 26 for tests on a 2.25Cr-1Mo steel in air, gaseous hydrogen, and hydrogen sulfide environments.

Aside from the usual problems of mini-

mizing noise and particularly drift in the electrical system over extended time periods, one problem that can arise in crack growth studies is electrical shorting between crack surfaces, leading to underestimates of the true crack length. This is particularly evident where fracture surfaces are rough or in inert environments where insulating oxide films are less readily formed (e.g., in coarse-grained titanium alloys *in vacuo*), where shorting occurs between the highly faceted crystallographic crack surfaces, and with intergranular stress-corrosion cracking in certain reducing environments. Where the effect has not been too extensive, corrections have been applied by comparing the final crack length deduced from potential measurements with that determined optically on the broken fracture surface to derive a percentage underestimate. This assumes that the degree of electrical shorting is uniform and thus the error is proportional to crack surface area.

The technique also has been used successfully at elevated temperatures by careful consideration of thermal-induced electromotive forces and by avoiding temperature gradients in the specimen. However, care must be taken to account for the presence of time-dependent creep deformation and for an increased propensity for crack surface bridging from corrosion deposits within the crack (Ref 52).

Small Crack Behavior. One area of fatigue research that has received much attention is the behavior of cracks that are small

Fig. 25 Typical test records for load and electrical potential versus load point displacement for detecting initiation in elastic-plastic fracture toughness tests

(a) Load-displacement trace. (b) Corresponding variation in potential showing only gradual change at initiation. (c) "Anomalous" potential drop effect seen in some alloys, such as SA533-B steel. The initial rise in potential at the start of the test is due to the precracked surfaces parting and is reversible.

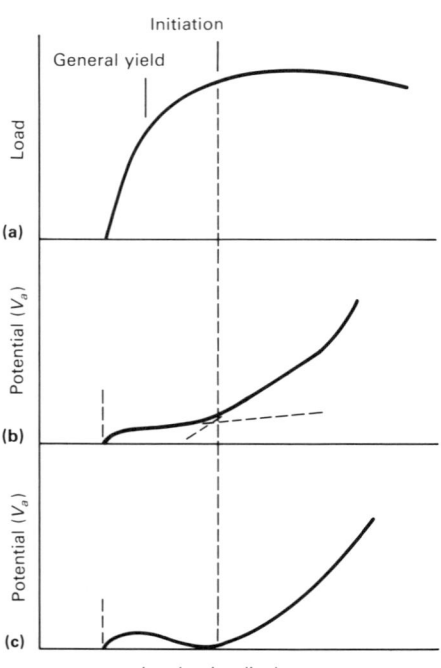

Fig. 26 Fatigue crack propagation data over a wide spectrum of growth rates

Data derived from direct current potential measurements in martensitic 2.25Cr-1Mo steel (SA542-C12) at $R = 0.05$ to 0.75 in air, hydrogen, and hydrogen sulfide at ambient temperature. Source: Ref 60

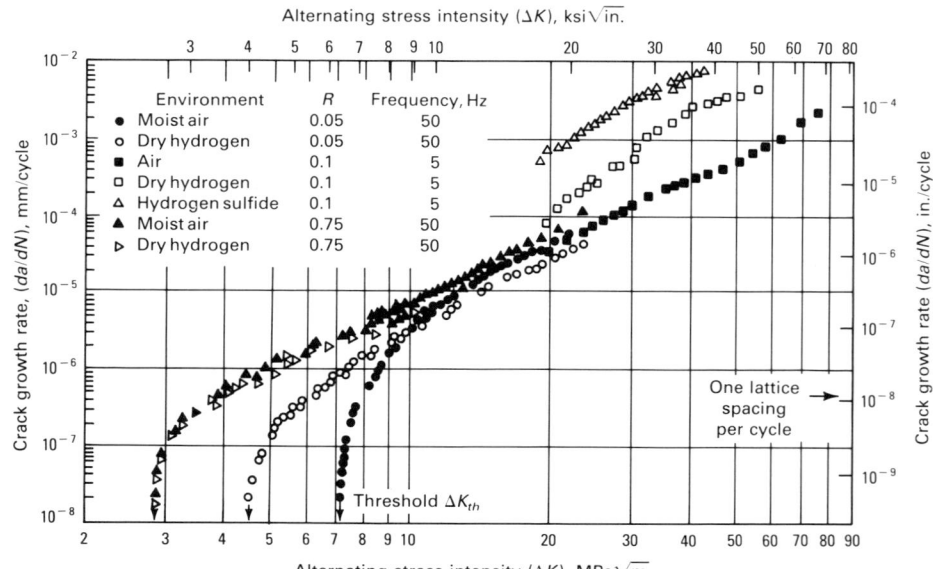

compared to either the scale of microstructure or local plasticity (Ref 61). For the measurement of such small cracks, where crack sizes can be much less than 1 mm (0.04 in.) and are often part-through surface flaws, the use of electrical potential methods poses several problems. First, the extent of plasticity relative to crack size is much larger, leading to enhanced errors from changes in resistivity (Ref 62). Second, the use of experimental calibration equations, where best-fit curves have applied to experimental points, often leads to zero-error discrepancies at small crack lengths (Ref 41). Finally, for part-through cracks, some estimate of both size and shape of the flaw is required for crack monitoring.

The latter two problems have been minimized through the use of closed-form theoretical calibration equations that account for the spread of crack advance in both length and depth dimensions. Full details of these calibrations and the use of direct current potential methods for monitoring small surface defects can be found in Ref 43. Such three-dimensional calibrations for part-through defects can be readily verified experimentally using an electrolytic task simulation where appropriately shaped insulating inserts, representing the crack, are immersed in a (conducting) aqueous solution of potassium chloride electrolyte, representing the specimen (Ref 63, 64). Calibration obtained in such a manner have been found to agree well with those obtained by experimental and numerical methods (Ref 64).

Fatigue Crack Closure. Electrical potential methods also have been utilized to assess the extent of crack closure during fatigue crack growth by detecting electrical contact between the crack surfaces during the loading cycle (Ref 36, 65-67). Due, in part, to the presence of insulating oxide films on the fracture surface, in all but inert environments this procedure has not always proved to be entirely satisfactory, and in several instances conflicting results have been obtained when compared with simultaneous compliance measurements (Ref 36, 66). An additional problem appears to be that contact between the crack surfaces at the crack mouth yields a far greater reduction in potential than similar contact at the crack tip, although methods have been devised to differentiate between such closures (Ref 67). In general, however, the use of electrical potential methods is not recommended for the quantitative evaluation of crack closure; unloading compliance techniques, particularly using back-face strain gages, appear to be much more reliable (Ref 68).

Use With Nonconducting Materials. The electrical potential method can be adapted for electrically nonconducting materials by affixing metal foils to the test piece, which crack along with the specimen. One commercially available system employs such gages, termed Krak-Gages, which with appropriate circuitry can yield a direct output of crack length. Such approaches, which are in widespread use for crack growth studies in polymers, ceramics, composites, and biomaterials, offer the advantage that they are independent of specimen geometry. However, they can provide information only on the surface lengths of the crack, and thus are inappropriate for situations where the crack front is highly nonlinear.

Electromechanical Fatigue Testing Systems

The primary function of electromechanical fatigue testers is to apply millions of cycles to a test piece at oscillating loads up to 220 kN (50 000 lbf) to investigate fatigue life, or the number of cycles to failure under controlled cyclic loading conditions. Variables associated with fatigue-life tests are frequency of loading and unloading, amplitude of loading (maximum loads and minimum loads), and control capabilities. The fundamental data output requirement is the number of cycles to failure, as defined by the application.

A variety of electromechanical fatigue testers have been developed for different applications. Forced-displacement, forced-vibration, rotational bending, resonance, and servomechanical systems are discussed in this article and are compared in Table 4. Other specialized electromechanical systems are available to perform specific tasks.

Forced-Displacement Systems

Forced-displacement motor-driven systems are the simplest type of electromechanical fatigue testers. They effectively reproduce service environments that impart fixed, reciprocating displacements to a component or test piece. An electric motor-driven flywheel is used to carry a loading arm at a variable distance from the center of rotation, much in the same manner as a connecting rod in an automotive engine. This rotational displacement is transformed into a guided, vertical displacement and is used to fatigue the specimen.

Although load can be monitored in such systems, the fixed displacement precludes the ability to control load, which is a function of specimen characteristics. Therefore, the load generally drops as failure progresses. These systems typically are custom-built, inexpensive fatigue machines, used primarily for bend tests on soft samples in which load control, high frequencies, and large loads are not required.

Forced-Vibration Systems

Forced-vibration motor-driven systems were the first production fatigue testers in commercial use. The centrifugal forces of an imbalanced rotor is used to impart a cyclic load to the test piece. The major components of the forced-vibration operating mechanism are illustrated in Fig. 27.

In operation, an electric motor is used to rotate an eccentric mass via flexible couplings. The rotating mass is mounted in a frame that is guided by flexure plates to restrict movement to vertical motion only. The centrifugal force produced by the rotating eccentric mass (m) is transmitted through the vertically guided frame to the test piece. The horizontal component of the centrifugal force is absorbed by the restraining flexure plates.

Because the centrifugal force usually is totally absorbed by the mounting frame (of mass M), the inertial reaction is separated from the centrifugal force in such a way as to transmit only the centrifugal forces to the specimen. This technique involves the use of frame-support compensator springs; the natural frequency of the spring (K)/mass (M) system is tuned to the revolutions per minute of the motor. Thus, neither the specimen nor the rotating eccentric mass (m) "sees" an inertial reaction from the frame, because the inertial effects of the frame are totally compensated for by the frame support springs (not the specimen).

This technique has two requirements: the rotating frequency (ω) must be kept constant and the mass of mounting frame (M) must be kept constant. Consequently, the loading frequency of the device is fixed at 1800 rpm, and masses must be added or removed from the frame to compensate for fixturing to keep M constant.

The magnitude of the dynamic load is determined by placing the rotating mass at a known distance from the axis of rotation (r). Because ω, m, M, and K are known, the force on the specimen, F, is calibrated directly as a function of r as follows:

$$F = M\omega^2 r \text{ (centrifugal)} - Ma_z \text{ (inertial)}$$
$$+ Kz \text{ (spring compensated)}$$

where a_z is the acceleration of the frame in the z direction, and Kz is the spring-compensated displacement in the z direction. Because Ma_z is tuned to equal Kz, $F = M\omega^2 r$. This principle is illustrated schematically in Fig. 28.

Thus, the forced-vibration rotating eccentric mass system is an open-loop, load-controlled system with the ability to accommodate up to 25 mm (1.0 in.) of total sample deflection at loads up to 220 kN (50 000 lbf) using special fixtures. The mean or static

Table 4 Comparison of electromechanical fatigue systems

Parameter	Forced displacement	Forced vibration	Rotational bending	Resonance	Servomechanical
Tension	Yes	Yes	No	Yes	Yes
Compression	Yes	Yes	No	Yes	Yes
Reverse stress	Yes	Yes	Yes	Yes	Yes
Bending	Yes	Yes	Yes	Yes	Yes
Frequency range	Fixed	Fixed, 1800 rpm	0–10 000 rpm	40–300 Hz	0–1 Hz
Load range	Typically < 450 N (<100 lbf)	Up to 220 kN (50 000 lbf)	. . .	Up to 180 kN (40 000 lbf)	Up to 90 kN (20 000 lbf)
Type:					
Control	Open-loop	Open-loop	Open-loop	Closed-loop	Closed-loop
Mode	Displacement	Load	Rotation/bending	Load	Load, displacement, strain
Maximum deflection	. . .	25.4 mm (1.00 in.)	. . .	1.0 mm (0.040 in.)	100 mm (4 in.)
Advantages	Simple, straight-forward	Versatile, efficient, durable	Efficient, durable, simple	Fully closed-loop, extremely efficient	Fully closed-loop, high precision
Disadvantages	No load control, very limited applications (soft samples)	Fixed frequency, limited control (open-loop)	Rotational bending only, limited applications	Operating frequency directly proportional to sample stiffness	Low frequency only

Fig. 27 Major components of the forced-vibration operating mechanism

(1) Stationary frame; large top provides ample work space. (2) Reciprocating platen. (3) Rotating eccentric mass is source of dynamic force, which is varied by screwing threaded rod in or out. (4) Thread screw locks threaded rod in position. (5) Scale reads in pounds of vibratory force. (6) Flexure plates absorb horizontal centrifugal force so that only vertical force is transmitted to platen. (7) Synchronous motor drives eccentric mass at constant 1800 cycles/min. (8) Springs provide seismic mounting so that no vibration is transmitted to or from surroundings. (9) Dial indicates preload. (10) Compensator springs absorb all inertia forces produced by reciprocating masses, preventing transmission to the specimen. (11) Plate holds one end of compensator springs firmly to stationary frame. (12) Preload mechanism

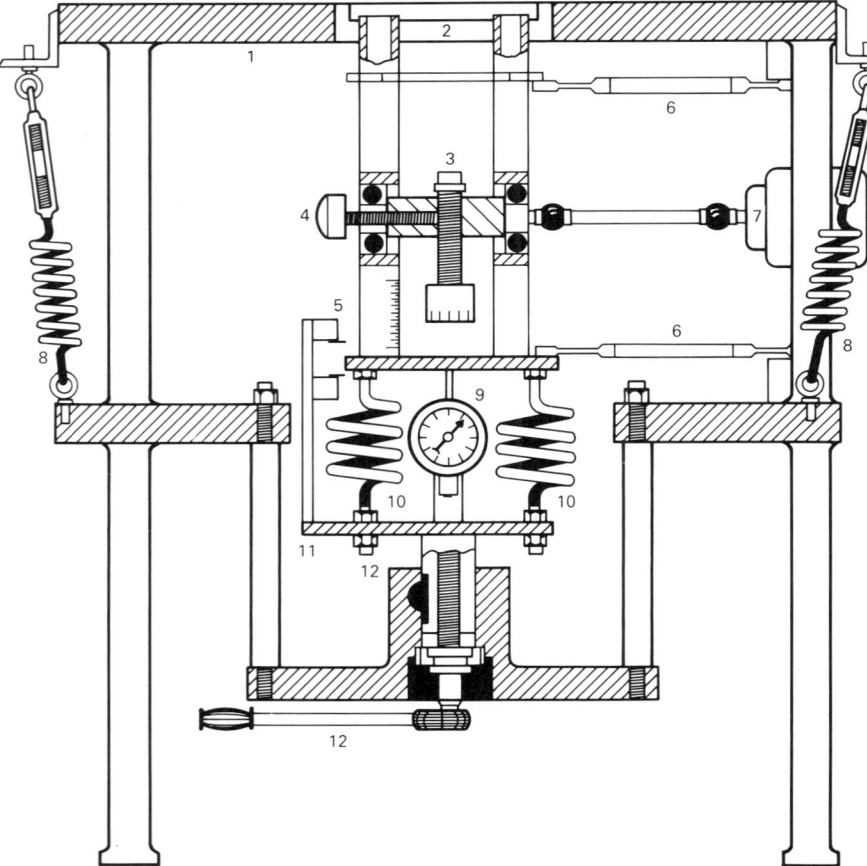

load, onto which the dynamic load is superimposed, is achieved by preloading the inertia compensator spring, K.

Through special fixturing, forced-vibration devices are capable of testing in tension, compression, bending, torsion, or reverse stress. Although servo-controlled, mean-load-maintenance systems are available, the open-loop nature of the system prevents direct load measurement or control, which is characteristic of closed-loop systems. The load applied to the specimen is assumed to be a function of r, and a graduated scale is provided to permit reasonably accurate setup.

Rotational Bending Systems

Rotational bending systems effectively apply reversed loading to the outer surface of rods or shafts. The basic operating principle of the rotating beam consists of the use of a motor to rotate a shaft of known dimensions around its longitudinal axis. By applying a known static force at the end of the shaft, a bending moment can be applied to the test section, the outer surface of which oscillates between tension and compression during each rotation.

The cantilevered specimen, however, is subjected to a nonuniform bending moment, which is large at the supported end of the specimen and zero at the free end. To produce a more meaningful, uniform bending moment throughout the test piece, a specially designed tapered specimen should be used or bending moments should be applied to each end of the specimen. Figure 29 illustrates the rotating-beam operating mechanism and the resulting stress distribution in the specimen.

Rotating-beam systems are restricted to reverse-stress, outer-fiber tests at frequencies determined by the drive motor and gearing used. These tests are useful in evaluating surface finish and residual stress effects on the fatigue lives of rotating shafts.

Resonance Systems

After World War II, a high-speed fatigue testing system was developed by Amsler that operated at 40 to 300 Hz, achieved high loads (up to 90 kN, or 20 000 lbf), and consumed minimal energy. It is based on a resonant spring/mass system, in which the specimen is used, like a spring, as an integral part of the oscillating mechanism.

The fatigue load, in the form of a sinewave, is achieved by preloading the sample in the frame via a complex optomechanical procedure and dynamically loading the sample at the natural oscillating frequency of the spring/mass system. The preload is maintained automatically during the test. The dynamic load is achieved by pulsing an electromagnet at the natural frequency of the spring/mass system. During resonance, the electromagnet restores any hysteresis energy lost during the previous cycle, thereby maintaining a constant, controllable dynamic load. Capable of tension, compression,

bending, torsional, and reverse-stress fatigue tests, the Amsler resonant fatigue testers were instrumental in obtaining the vast amount of fatigue data currently available.

The resonant system illustrated in Fig. 30 is based on a similar principle, but incorporates solid-state technology to achieve fully closed-loop control of mean and dynamic loads. This system uses dual opposing masses (unlike the single oscillating mass/seismic base of earlier systems), linked by the specimen to achieve vibration-free resonance. A strain gage load cell, in series with the specimen, senses the load and automatically triggers the electromagnet to achieve self-tuning capability.

The mean load is achieved by physically moving the upper mass up or down to

achieve tension or compression, respectively; the dynamic load is achieved by varying the width of the pulse to the magnet beneath the lower mass. The dynamic load, like the mean load, is electronically maintained at a preset command level through solid-state closed-loop circuitry. The remainder of the controls and mechanisms associated with the resonant fatigue system maintain a preset air gap between the magnet and the oscillating lower mass, maintain preset loading conditions (shutting down at preset load levels or frequencies), and power the electromagnet.

The frequency of operation of these resonant systems is highly dependent on sample stiffness; the stiffer the sample, the higher the operating frequency. The operating fre-

Fig. 28 Schematic of the forced-vibration operating principle

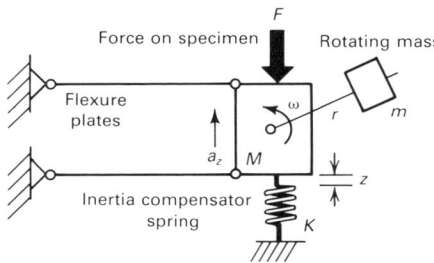

Fig. 29 Schematic of the rotating-beam operating mechanism and the resulting stress distribution in the specimen

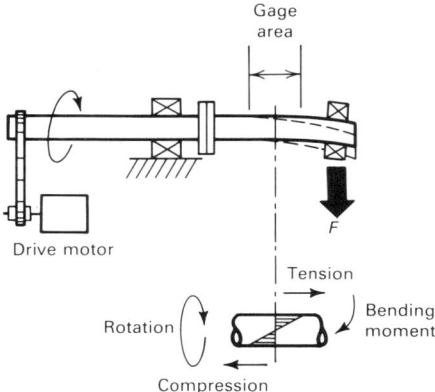

Fig. 30 Major components of a closed-loop resonant fatigue testing system

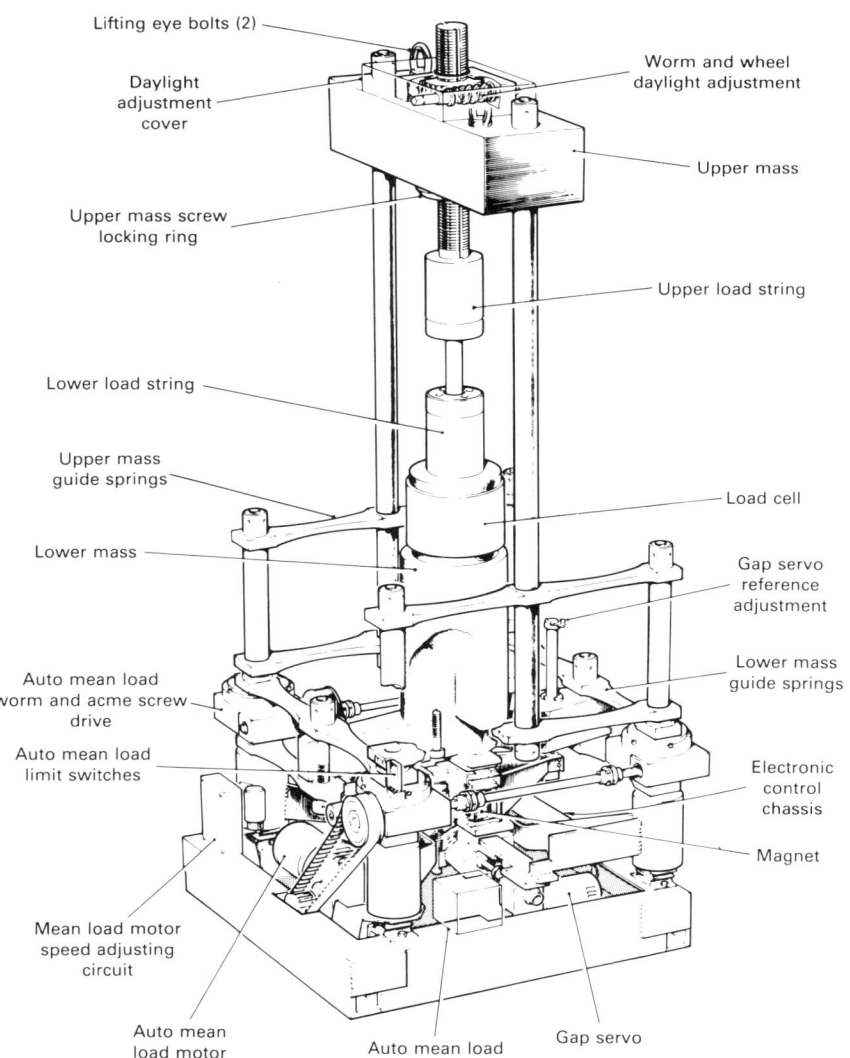

quency is not the resonant frequency of the sample (which for metallic samples is in the kilohertz range), but rather is the resonant frequency of the entire spring/mass system. Consequently, the operating frequency is limited to frequencies that can be achieved by adding or removing heavy masses from the Amsler system, or to the frequency determined by the sample on the closed-loop system (Fig. 30).

The high efficiency of resonant systems makes them well suited to high-cycle fatigue tests, in which closed-loop load control, high loads (up to 180 kN, or 40 000 lbf), low power consumption (around 750 W maxi-

mum for closed-loop systems), and high throughput are required. These systems tolerate minimal hysteresis and produce optimum testing results when used with stiff metallic samples. A computer-controlled resonant fatigue tester is illustrated in Fig. 31.

Closed-Loop Servomechanical Systems

The most recent development in electromechanical fatigue testers is based on an electric actuator/load frame assembly. The system closely resembles its servohydraulic counterpart in that it consists of an actuator (Fig. 32), load frame, load cell, power sup-

ply, and a solid-state closed-loop electronic control console. Closed-loop systems compare live feedback signals to an input command signal to maintain accurate control of preset conditions. Figure 33 illustrates the major operating elements of a closed-loop servomechanical system.

The closed-loop servomechanical system is, by virtue of its design, primarily intended for low-cycle and creep-fatigue studies. The ball-screw actuator is belt driven by an electric motor and has the ability to reach extremely low test speeds (1 μm/h) and a maximum speed of 350 mm/min (14 in./min). The stiffness of the actuator rod, which is restricted from lateral motion by cam rollers, and its quiet motion (i.e., no random motion due to servo-valves), makes closed-loop servomechanical systems ideal for testing ceramics and other stiff, displacement-sensitive materials.

Closed-loop electronics allow the system to operate in load, stroke, or strain control within the speed limitation of the actuator, which is typically less than 1 Hz. Frequently, this type of system is equipped with a high-temperature furnace (1000 °C, or 1830 °F), axial and/or diametral extensometry, or vacuum equipment. An 18-bit digital function generator allows precise command waveforms to be programmed with servohy-

Fig. 31 Computer-controlled resonant fatigue tester
Courtesy of Instron Corp.

Fig. 32 Electric actuator assembly used in closed-loop servomechanical fatigue testing systems
Courtesy of Instron Corp.

Fig. 33 Major components of a closed-loop servomechanical testing system

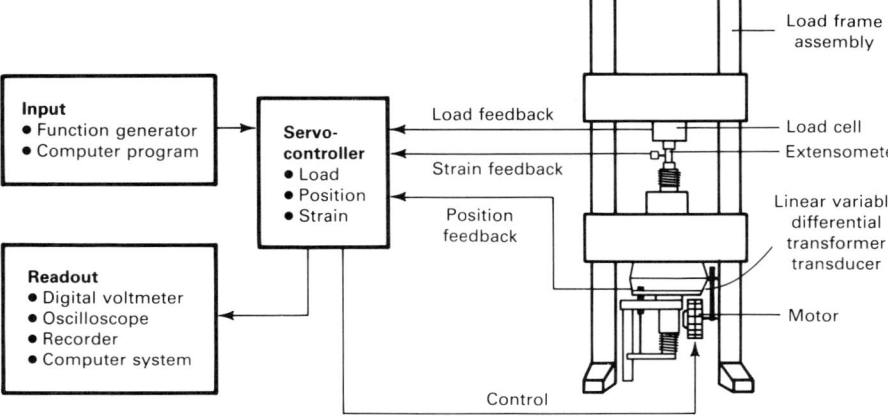

draulic-type controls achieving the closed-loop control of the electric motor and power amplifier. A closed-loop fatigue/creep servomechanical system equipped with a furnace chamber and electronic controls is shown in Fig. 34.

Other Electromechanical Fatigue Testers

In the pursuit of an "ideal" fatigue testing system, several hybrid machines have been developed. The two systems described below represent viable attempts at evolving fatigue testing technology. Other types of systems are available, with modifications to suit individual testing needs.

A rotating eccentric mass resonance system, which is similar to a resonant forced-vibration system, is shown in Fig. 35. An electric motor is used to rotate an eccentric mass mounted on a cantilevered beam. This beam, pivoted at one end, carries a mass on its other end, which essentially determines the resonant frequency of the system.

Located between the rotating mass and the pivot of the cantilevered arm, a vertically guided loading head is used to transmit the load to the specimen. The amplitude of the dynamic load is determined by varying the rotating frequency of the eccentric mass and by varying the mass supported at the end of the cantilevered beam. A torque system is used to load the cantilevered arm and apply the static load to the specimen. Superimposed on this static load is the oscillating load created by the rotating eccentric mass. The limitations of this device include its frequency range (10 to 100 Hz), load capability (180 kN, or 40 000 lbf, maximum), and the requirement that the system operate at fre-

quencies determined by the specimen, the rotating mass, and the cantilevered masses.

Dynamic Cycler. Another technologically advanced electromechanical fatigue testing system is the dynamic cycler shown in Fig. 36. Although no longer in production, the dynamic cycler used an electric motor to drive a flywheel, which in turn drove a master piston back and forth through a connecting rod attached to the flywheel at a variable distance from the center. The combined electromechanical and hydraulic systems allowed the dynamic cycler to operate at any desired frequency within the 0- to 100-Hz frequency range and to achieve fully closed-loop control of the mean and dynamic loads. This mechanically driven hydraulic system was replaced by more versatile servohydraulic systems, which do not require complicated flywheel and pulley arrangements.

Servohydraulic Fatigue Testing Systems

Servohydraulic testing machines are particularly well suited for providing the control capabilities required for fatigue testing. Extreme demands for sensitivity, resolution, stability, and reliability are imposed by fatigue evaluations. Displacements may have to be controlled (often for many days) to within a few microns, and forces can range from 100 kN to just a few newtons. This wide range of performance can be obtained with servomechanisms in general and, in particular, with the modular concept of servohydraulic systems.

Usually, the problem of selecting the appropriate system is simply a matter of optimizing the various components to form a

system best suited to the given testing application. In this section, the principles underlying closed-loop servo systems are discussed briefly. In addition, the interaction between system components is illustrated, and a brief description of their operating principles and characteristics is provided.

With any type of control system, the objective is to obtain an output that relates as closely as possible to the programmed input. In a fatigue testing system, it may be desired to vary the force on a specimen in a sinusoidal manner, at a frequency of 1 Hz over a force range of 0 to 100 kN (0 to 22 000 lbf). The only practical means to accomplish this with precision is through the use of a negative-feedback closed-loop system. An overview of the basic principles of operation of negative-feedback systems is provided in Fig. 37. The blocks shown in Fig. 37(a) represent a group of typical components of a testing machine. The transfer functions of each of these blocks can be combined to produce the more simplified diagram shown in Fig. 37(b).

Placement of the switch, S_1, has been added to the diagram to permit analysis of the system when it is open (no feedback, or an open-loop condition) and when it is closed (providing feedback to the system). The equation governing this simplified open-loop system is:

$$C = K_o D \qquad \text{(Eq 10)}$$

where C represents the controlled output, K_o represents the open-loop transfer function, and D represents the electronic demand signal. Therefore, the output is simply proportional to the system demand if K_o is a constant. Unfortunately, K_o is seldom a constant, because it can be influenced by several common system variations. The electronic components may drift slightly, or their gain may vary. The behavior of the hydraulic components may change with temperature, contamination, or wear, and the mechanical components may vary because of thermal effects or friction.

The most significant change, however, will probably be in the specimen itself. Because of exposure to the test conditions, it may harden or soften, and thus its compliance will change. Cracks are likely to develop and grow, further changing its response to loading. This behavior will result in pronounced changes in K_o and consequently in the system output.

If the switch is closed, however, the system equation is:

$$C = \left[\frac{K_o}{1 + K_o K_f}\right] D \qquad \text{(Eq 11)}$$

which can be rewritten as:

$$C = \left[\dfrac{\dfrac{1}{K_f}}{\dfrac{1}{K_o K_f} + 1} \right] D \qquad \text{(Eq 12)}$$

where K_f represents the feedback transfer function. If $K_o K_f$ is large with respect to 1, the denominator will approach unity, and the output C will be approximately equal to $(1/K_f)D$.

In effect, this indicates that the output of a high-gain (i.e., $K_o K_f$ is large) closed-loop servo system can be made insensitive to many of the variations discussed above. The system output depends primarily on the system demand and the feedback components. Of primary importance is the fact that the anticipated specimen changes in a typical fatigue test will not adversely affect system control.

Servohydraulic System Components

Many commercially manufactured units are available for each component in a typical servohydraulic testing system. In addition, complete systems in which all of the components are properly integrated and specially designed to meet particular testing specifications are available. It is thus beneficial for those involved with the selection and use of these systems to understand the basic functions and primary features of each component. The components that constitute a typical servohydraulic fatigue testing system, such as that shown in Fig. 38, are discussed below.

Figure 39 illustrates the interconnections between the various hydraulic components in a system. Several variations of this system are possible. The central hydraulic power supply in Fig. 39 provides flow and pressure to more than one testing system. With this technique, independent control of pressure is possible, and each machine can be isolated completely from the others for maintenance. Most of the components are standard and require no special discussion. Proper operation of the servo-valve requires that the hydraulic fluid be extremely clean. The filter shown in Fig. 39 should be of very high quality, and additional filters are also desirable on the main power supply.

The programmer supplies the command signal to the system, which is generally an analog of the desired behavior of the controlled parameter. For example, assume the same test conditions as previously discussed (control the force on the specimen in a sinusoidal manner at a frequency of 1 Hz and a force range of 0 to 100 kN). In this instance, the programmer might be set to produce an electronic signal with a sinusoidal waveform

Fig. 34 Closed-loop servomechanical system equipped with a 1000 °C (1830 °F) furnace chamber and associated electronic controls used for fatigue and creep testing
Courtesy of Instron Corp.

Fig. 35 Schematic of a rotating eccentric mass resonance system

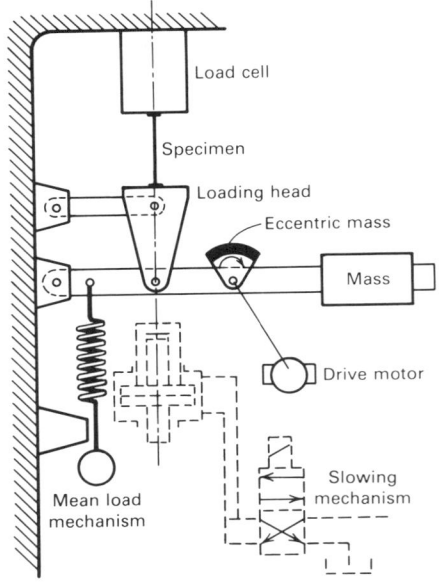

Fig. 36 Dynamic cycler fatigue testing system
Courtesy of Instron Corp.

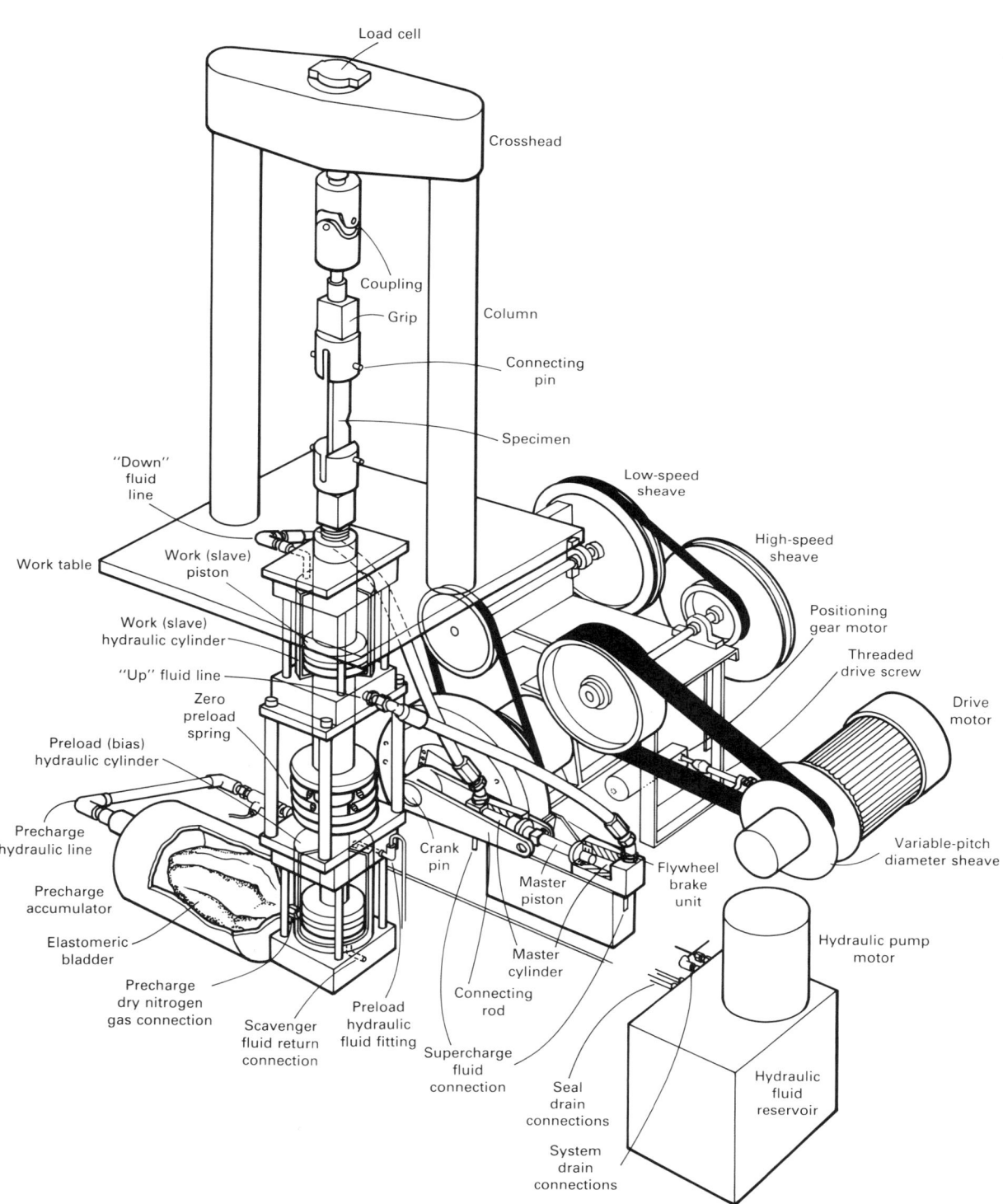

Fig. 37 Simplified block diagram for a negative-feedback closed-loop testing machine

(a) Typical components. (b) Transfer functions. See text for details and explanation of symbols.

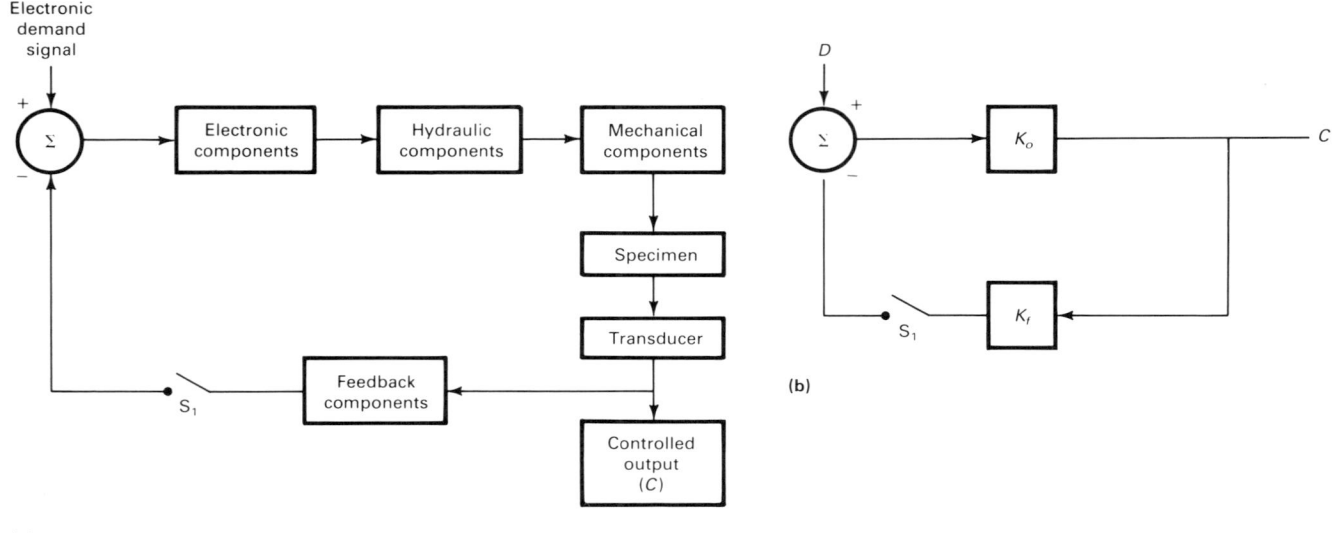

(a)

(b)

Fig. 38 Block diagram of components in a typical servohydraulic fatigue testing system

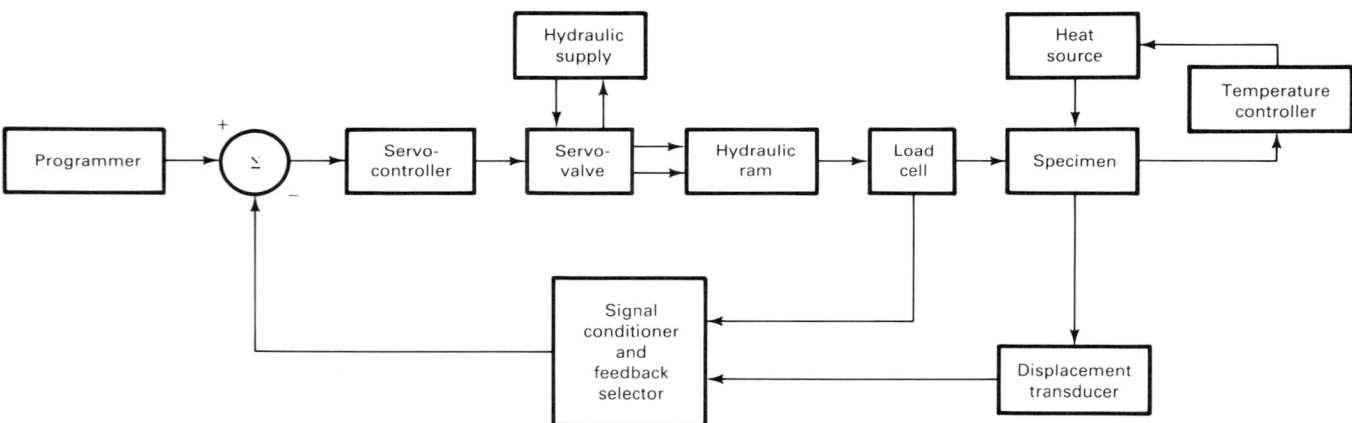

that has a frequency of 1 Hz and a voltage output of 0 to 10 V. The analog is: 1 V represents 1000 N. The system can then be adjusted to produce the correct output. Any change in the programmer signal will result in a corresponding change in the controlled parameter.

The programmer must be able to satisfy the objectives of the test program. It must have the capability to produce the frequency range and waveforms required. The mean level of the output signal should be adjustable, and the cyclic signal should be able to be started and stopped as required. The programmer must be able to provide a very stable, drift-free signal that is compatible with the input characteristics of the servo-controller.

Programmers are also available that permit an arbitrary output waveform to be created by the operator. With these devices, simulated service conditions can be readily imposed on the specimen. The programmer can be used to "play back" a prerecorded signal in the form of a time-varying output voltage. With this technique, a recorded signal taken from an in-service component can be used as a demand signal in a fatigue test.

As shown in Fig. 38, most programmers are outside the servo-loop and merely serve to "command" the loop. It is possible, however, to dynamically modify the output of the programmer during the test as a result of changes that might occur in the controlled variable of the system. Frequently, this technique uses a computer to monitor a specimen parameter, which then instructs the programmer to change its output as required. A system that monitors the crack length in a specimen and then automatically adjusts the force applied to the specimen so that the range of the stress-intensity factor remains constant is an example of this type of arrangement. In this instance, the programmer is really within the servo-loop.

The servo-controller makes most of the adjustments necessary to optimize system performance. For example, it compares the

Fig. 39 Interconnections between components in a servohydraulic system

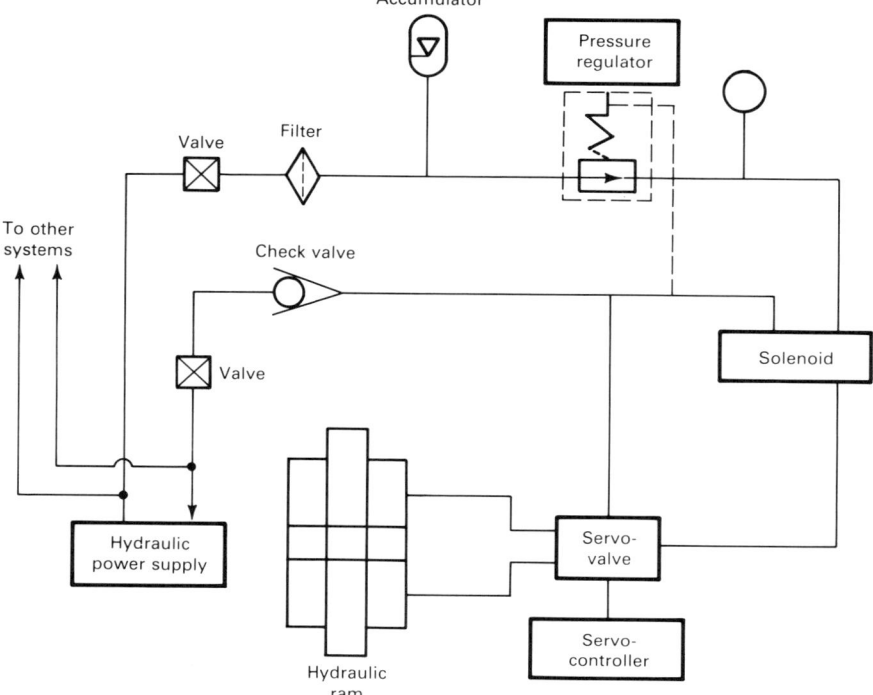

command signal with a signal produced by the controlled parameter (stress or strain, for example) and relays a correction signal, if needed, to the control device in the system (usually a flow-control servo-valve). A servo-controller incorporates numerous other compensatory features, such as:

- Means to adjust the gain or proportional band of the system
- Controls to modify the feedback or correction signals for improved stability
- Controls to adjust the mean level and amplitude of the command signal(s)
- Controls to enhance and adjust servo-valve response
- Means to monitor the system error signal (a measure of how well the command and feedback signals agree)
- Capability to select various command and feedback signals
- Auxiliary functions such as recorder signal conditioning, calibration, and system startup and shutdown

Some of these functions may be located elsewhere in the system. Of all the system components, the specimen requires the most adjustments of the servo-controller. Size, geometry, and mechanical property variations can be significant, and the servo-controller provides the means to compensate for these variations.

The servo-valve controls the volume and direction of flow of hydraulic fluid between the hydraulic power supply and the hydraulic ram. Within the control loop, it is the intermediary between the low-power servo-controller and the hydraulic ram, which can supply large forces and displacements to the specimen. Characteristics of the device are such that the output flow is approximately proportional to the input current when the output pressure is constant. Also, the output pressure is approximately proportional to the square of the input current when the flow is constant.

Various techniques are used to produce the flow, pressure, and frequency response required for fatigue testing. These performance curves are readily available from manufacturers. The valve must be compatible with the hydraulic ram (compatibility between the servo-controller and the servo-valve generally is not a problem). Actually, the hydraulic ram usually is selected first, and a servo-valve is chosen that supplies sufficient flow to produce the ram speed required.

The maximum ram speed for a triangular waveform can be estimated by multiplying the expected peak-to-peak displacement of the ram by two times the frequency; if the waveform is sinusoidal, the displacement is multiplied by 3.2 times the frequency. If the displacement is in millimeters and the fre-

quency is in hertz, the ram speed will be in millimeters per second (or $mm \cdot s^{-1}$). The flow required will then be equal to the ram speed times the effective area of the ram piston. The ram must be sized so that it can produce the maximum force required, and the servo-valve must be able to move the ram at the required rate. It is advisable, however, not to oversize the servo-valve.

Generally, the frequency response decreases as the flow rating of the valve increases. Also, the low flow characteristics of a large valve are not as beneficial as those of a smaller one. Servo-valves should be changed if necessary to meet different flow requirements, rather than using one valve for all testing. Some systems use multiple valves, operating in parallel to supply sufficient flow. This technique produces better frequency response than one large valve. Additional flexibility is also provided, because extra valves can be disconnected for low flow applications.

Hydraulic rams, or actuators or cylinders, furnish the forces and displacements required by the testing system. These rams usually are double ended to provide the greatest lateral rigidity and to produce the balanced flow and force characteristics desirable for push-pull testing. The effective area of the piston is therefore equal to the cross-sectional area of the piston minus the cross-sectional area of the piston rod. Under static conditions (very little flow), the maximum force capability of the ram will approach the hydraulic supply pressures multiplied by the effective area.

The force available during dynamic operation depends on the pressure drop and flow characteristics of the servo-valve. Reference should be made to the load/flow/pressure characteristics supplied by the servo-valve manufacturer.

Compromises are often necessary in the selection of the force rating of the hydraulic ram (as well as related components such as load frame, load cell, and fixturing). There may be a tendency to select a system that will supply sufficient force for any specimen to be tested. It should be noted, however, that the flow requirements of the system will be proportional to the effective area of the piston at a given ram speed. Thus, a large ram will require a larger servo-valve to provide the same ram speed. The frequency response and resolution of a test machine for normal testing can be degraded by selecting an oversize ram having extra capacity that may seldom be needed.

The length of the piston rod provides another degree of flexibility that is somewhat unique to servohydraulic systems. The stroke required by most fatigue tests is usually quite small (1 mm, or 0.04 in.). The ram can be designed to supply this type of displacement with extreme precision and resolution. However, the same ram can also provide a larger stroke of many millimeters to accommodate specimens of various size and to facilitate test setup. This is a significant advantage, even in test machines with movable crossheads or other mechanical adjustments. However, space limitations or response considerations may influence the stroke selected.

Hydraulic rams for materials testing may be conventional off-the-shelf cylinders, and for many applications these are adequate. Frequently, however, the ram may be designed specifically for fatigue testing. For example, it may contain special low-leakage piston seals and low-friction end bearings, thus making it better suited than a standard cylinder for smooth, long-life operation under a fatigue environment.

Load Cells. The strain gage load cell is the most widely used force-measuring and feedback device in closed-loop fatigue machines. An external applied force causes the elastic deformation of an internal member to which a strain gage bridge has been attached. An electronic signal that is proportional to the resistance change in the bridge and to the applied force can thus be produced. Some load cells are designed specifically for fatigue evaluations. Variable features include sensitivity, natural resonant frequency, temperature stability, fatigue rating, linearity, hysteresis, deflection constant, load capacity, overload rating, resistance to extraneous loading, and compatibility with the testing machine and fixtures. Most commercially available cells are very competitive with respect to these features.

An important consideration in the selection of a load cell is its capacity. As with the selection of the servo-valve and hydraulic ram, care should be taken that the load cell is not too large. Although the cell should be able to withstand the maximum force produced by the machine (or it must be protected from this force), too large a cell capacity can cause measurement problems when small forces must be monitored. Because of limitations in the output of strain gages, the maximum signal that can be obtained from a load cell is about the same regardless of the force rating of the cell. A larger signal can therefore be obtained when a small cell is used to measure a small force than when a large cell is used to measure that same force. For example, a 100-kN (22 000-lbf) load cell and a 10-kN (2200-lbf) load cell may both produce 50 mV when loaded to their maximum rating. When measuring a small force (1 kN, or 220 lbf), however, the small cell would produce 5 mV and the larger cell only 0.5 mV. Additionally, because hysteresis,

nonlinearity, and thermal change usually are proportional to maximum cell capacity, the signal obtained from the small cell will be more precise.

Load Frames. In a fatigue machine, the reaction forces to the specimen and to the housing of the ram are supplied by the load frame. Many styles of load frames are available, but for fatigue purposes the frames should be customized. The requirements of good high-frequency response demand high axial stiffness in the frame. When a deflection occurs in the load frame, additional flow is required from the servo-valve. Therefore, this deflection should be minimal compared to that imparted to the specimen.

In addition, because fatigue specimens must be subjected to fully reversed loading (i.e., compressive as well as tensile forces), lateral rigidity must be increased to resist bending. This is generally considered necessary in the design of fatigue machines. The extra rigidity can be obtained by increasing the diameter of the support columns or by utilizing three- or four-column configurations.

Exceptional alignment is required of load frames used in fatigue evaluations to minimize undesirable bending forces. In addition, some means is usually provided to refine the alignment with manual adjustments when necessary. A strain-gaged specimen can be used to make this evaluation.

Specimens. Although the actual design of fatigue specimens is not discussed in this section, the interaction between the specimen and the remainder of the system must be emphasized. The specimen is part of the servo-loop, and its requirements of force and deflection affect total system performance. Therefore, its design should be such that all unnecessary elastic deflections are eliminated; machining tolerances should also be close enough to preserve total alignment. There must be compatibility between the size and compliance of the specimens and the load- and flow-sensitive components. The servo-controller can make adjustments to accommodate different specimens, but care should be taken not to exceed the normal range of these adjustments.

REFERENCES

1. H. Tada, P.C. Paris, and G.R. Irwin, Ed., *Stress Analysis of Cracks Handbook*, Del Research Corp., Hellertown, PA, 1973
2. G.C. Sih, *Handbook of Stress-Intensity Factors for Researchers and Engineers*, Institute of Fracture and Solid Mechanics, Lehigh University, Bethlehem, PA, 1973
3. P.C. Paris, R.J. Bucci, E.J. Wessel, W.R. Clark, and T.R. Mager, in *Stress Analysis and Growth of Cracks*, STP 513, ASTM, Philadelphia, 1972, p 141-176
4. P.C. Paris and F. Erdogan, A Critical Analysis of Crack Propagation Laws, *J. Basic Eng. Trans. ASME*, Vol 85, Dec 1963, p 528-534
5. J.M. Barsom and S.T. Rolfe, *Fracture and Fatigue Control in Structures—Applications of Fracture Mechanics*, Prentice-Hall, Englewood Cliffs, NJ, 1977
6. N.E. Dowling and J.A. Begley, Fatigue Crack Growth During Gross Plasticity and the *J*-integral, in *Mechanics of Crack Growth*, STP 590, ASTM, Philadelphia, 1976, p 82-103
7. C.M. Carmen and J.M. Katlin, *J. Basic Eng. Trans. ASME*, 1966, p 792-800
8. J.M. Barsom and E.J. Imhof, in *Progress in Flaw Growth and Fracture Toughness Testing*, STP 536, ASTM, Philadelphia, 1973, p 182-205
9. "Standard Test Method for Constant-Load-Amplitude Fatigue Crack Growth Rates Above 10^{-8} m/Cycle," E 647, *Annual Book of ASTM Standards*, ASTM, Philadelphia, 1984, p 711-731
10. D.A. Virkler, B.M. Hillberry, and P.K. Goel, "The Statistical Nature of Fatigue Crack Propagation," AFFDL-TR-78-43, Wright-Patterson Air Force Base, OH, April 1978
11. R.G. Forman, V.E. Kearney, and R.M. Engle, Numerical Analysis of Crack Propagation in Cyclic Loaded Structures, *J. Basic Eng. Trans. ASME*, Vol 89, Series D, Sept 1967
12. J.E. Callipriest, An Experimentalist's View of the Surface Flaw Problem, in *The Surface Crack: Physical Problems and Computational Solutions*, American Society of Mechanical Engineers, New York, 1972, p 43-62
13. E.K. Priddle, High Cycle Fatigue Crack Propagation Under Random and Constant Amplitude Loading, *Int. J. Pressure Vessel Piping*, Vol 4, 1976, p 89
14. C.M. Branco, J.C. Radon, and L.E. Culver, Growth of Fatigue Cracks in Steel, *Met. Sci.*, Vol 10, 1976, p 149
15. C.G. Annis, Jr., R.M. Wallace, and D.L. Sins, "An Interpolative Model for Elevated Temperature Fatigue Crack Propagation," AFML-TR-76-176, Wright-Patterson Air Force Base, OH, 1976
16. D.A. Utah, "Crack Growth Modeling in an Advanced Powder Metallurgy Alloy," AFWAL-TR-80-4098, Wright-Patterson Air Force Base, OH, July 1980

17. K. Walker, *The Effect of Stress Ratio During Crack Propagation and Fatigue for 2024-T3 and 7075-T6 Aluminum*, STP 426, ASTM, Philadelphia, 1970

18. A. Saxena, S.J. Hudak, Jr., and G.M. Jouris, A Three Component Model for Representing Wide Range Fatigue Crack Growth Data, *Eng. Fract. Mech.*, Vol 12, 1979, p 103-115

19. R.J. Bucci, Development of a Proposed ASTM Standard Test Method for Near-Threshold Fatigue Crack Growth Rate Measurement, in *Fatigue Crack Growth Measurement and Data Analysis*, S.J. Hudak, Jr. and R.J. Bucci, Ed., STP 738, ASTM, Philadelphia, 1981, p 5-28

20. A. Saxena, S.J. Hudak, Jr., J.K. Donald, and D.W. Schmidt, *J. Test. Eval.*, Vol 6 (No. 3), May 1978, p 167-174

21. N.E. Dowling, Crack Growth During Low-Cycle Fatigue of Smooth Axial Specimens, in *Cyclic Stress-Strain and Plastic Deformation Aspects of Fatigue Crack Growth*, STP 637, ASTM, Philadelphia, 1977, p 97-121

22. S. Pearson, Initiation of Fatigue Cracks in Commercial Aluminum Alloys and the Subsequent Propagation of Very Short Cracks, *Eng. Fract. Mech.*, Vol 7, 1975, p 235-247

23. H. Kitagawa and S. Takahashi, Applicability of Fracture Mechanics to Very Small Cracks of the Cracks in the Early Stage, in *Proc. 2nd Int. Conf. Mechanical Behavior of Materials*, 1979, p 627-631

24. R.O. Ritchie and S. Suresh, "Mechanics and Physics of the Growth of Small Cracks," Materials and Molecular Research Division, Lawrence Berkeley Laboratory, and Department of Materials Science and Mineral Engineering, University of California, Berkeley

25. C.J. Beevers, *The Measurement of Crack Length and Shape During Fracture and Fatigue*, Engineering Materials Advisory Services Ltd., Warley, UK, 1980

26. C.J. Beevers, *Advances in Crack Length Measurement*, Engineering Materials Advisory Services Ltd., Warley, UK, 1983

27. "Standard Test Method for Plane-Strain Fracture Toughness of Metallic Materials," E 399, *Annual Book of ASTM Standards*, ASTM, Philadelphia, 1984, p 519-554

28. R.L. Hewitt, Accuracy and Precision of Crack Length Measurement Using a Compliance Technique, *J. Test. Eval.*, Vol 11, 1983, p 150-155

29. D.C. Maxwell, J.P. Gallagher, and N.E. Ashbaugh, "Evaluation of COD Compliance Determined Crack Growth Rates," AFWAL-TR-84-4062, Wright-Patterson Air Force Base, OH

30. W.H. Cullen *et al.*, A Computerized Data Acquisition System for High-Temperature, Pressurized Water, Fatigue Test Facility, in *Computer Automation of Materials Testing*, STP 710, ASTM, Philadelphia, 1980, p 127-140

31. T.A. Prater, W.R. Catlin, and L.F. Coffin, "Environmental Crack Growth Measurement Techniques," EPRI Report NP-2641, Electric Power Research Institute, Palo Alto, CA, Nov 1982

32. A. Saxena and S.J. Hudak, Jr., Review and Extension of Compliance Information for Common Crack Growth Specimens, *Int. J. Fract.*, Vol 14 (No. 5), Oct 1978, p 453-468

33. J.A. Kapp, G.S. Leger, and B. Gross, "Wide Range Displacement Expressions for Standard Fracture Mechanics Specimens," Report ARLCB-TR-84025, Army Armament Research and Development Center, 1984

34. R.O. Ritchie, "Crack Growth Monitoring: Some Considerations on the Electrical Potential Method," Departmental Report, Department of Metallurgy and Materials Science, University of Cambridge, England, Jan 1972

35. M.A. Ritter and R.O. Ritchie, On the Calibration, Optimization and Use of d.c. Electrical Potential Methods for Monitoring Mode III Crack Growth in Torsionally-Loaded Samples, *Fatigue Eng. Mat. Struct.*, Vol 5, 1982, p 91-99

36. K.D. Unangst, T.T. Shih, and R.P. Wei, Crack Closure in 2219-T851 Aluminum Alloy, *Eng. Fract. Mech.*, Vol 9, 1977, p 725-734

37. H.H. Johnson, Calibrating the Electrical Potential Method for Studying Slow Crack Growth, *Mat. Res. Stand*, Vol 5, 1965, p 442-445

38. C.Y. Li and R.P. Wei, Calibrating the Electrical Potential Method for Studying Slow Crack Growth, *Mat. Res. Stand.*, Vol 6, 1966, p 392-394

39. R.O. Ritchie and J.F. Knott, Mechanisms of Fatigue Crack Growth in Low Alloy Steel, *Acta Metall.*, Vol 21, 1973, p 639-648

40. A.A. Anctil, E.B. Kula, and E. DiCesare, Electrical Potential Technique for Determining Slow Crack Growth, in *Proceedings*, Vol 63, ASTM, Philadelphia, 1963, p 799-808

41. G. Clark and J.F. Knott, Measurement of Fatigue Cracks in Notched Specimens by Means of Theoretical Electrical Potential Calibrations, *J. Mech. Phys. Solids*, Vol 23, 1975, p 265-276

42. D.M. Gilbey and S. Pearson, "Measurement of the Length of a Central Crack or Edge Crack in a Sheet of Metal by an Electrical Resistance Method," Technical Report 66402, Royal Aircraft Establishment, Farnborough, UK, Dec 1966

43. R.P. Gangloff, "Electrical Potential Monitoring of Fatigue Crack Formation and Growth from Small Defects," General Electric Report No. 79CRD267, Schenectady, NY, Jan 1980

44. R.O. Ritchie and K.J. Bathe, On the Calibration of the Electrical Potential Technique for Monitoring Crack Growth Using Finite Element Methods, *Int. J. Fract.*, Vol 15, 1979, p 47-55

45. G.H. Aronson and R.O. Ritchie, Optimization of the Electrical Potential Technique for Crack Growth Monitoring in Compact Test Pieces Using Finite Element Analysis, *J. Test. Eval.*, Vol 7, 1979, p 208-215

46. R.O. Ritchie, G.G. Garrett, and J.F. Knott, Crack Growth Monitoring: Optimisation of the Electrical Potential Method Using an Analogue Method, *Int. J. Fract. Mech.*, Vol 7, 1971, p 462-467

47. G.G. Garrett, "Toughness and Cyclic Crack Growth Studies in Aluminum Alloys," Ph.D. thesis, University of Cambridge, England, July 1973

48. R.P. Wei and R.L. Brazill, An a.c. Potential System for Crack Length Measurement, in *The Measurement of Crack Length and Shape During Fracture and Fatigue*, C.J. Beevers, Ed., Engineering Materials Advisory Services Ltd., Warley, UK, 1980, p 190-201

49. W.D. Dover, F.D.W. Charlesworth, K.A. Taylor, R. Collins, and D.H. Michael, a.c. Field Measurement—Theory and Practice, in *The Measurement of Crack Length and Shape During Fracture and Fatigue*, C.J. Beevers, Ed., Engineering Materials Advisory Services Ltd., Warley, UK, 1980, p 222-260

50. K.R. Walt, A Consideration for an a.c. Potential Drop Method for Crack Length Measurement, in *The Measurement of Crack Length and Shape During Fracture and Fatigue*, C.J. Beevers, Ed., Engineering Materials Advisory Services Ltd., Warley, UK, 1980, p 202-221

51. A.R. Jack and A.T. Price, The Initiation of Fatigue Cracks from Notches in Mild Steel Plates, *Int. J. Fract.*, Vol 6, 1970, p 401-409

52. B.L. Freeman and G.J. Neate, The Measurement of Crack Length During Fracture at Elevated Temperatures Us-

ing d.c. Potential Drop Technique, in *The Measurement of Crack Length and Shape During Fracture and Fatigue*, C.J. Beevers, Ed., Engineering Materials Advisory Services Ltd., Warley, UK, 1980, p 435-459

53. J.E. Srawley and W.F. Brown, Fracture Toughness Testing Methods, in *Fracture Toughness Testing and Its Applications*, STP 381, ASTM, Philadelphia, 1965, p 133-196

54. M.G. Vassilaros and E.M. Hackett, *J*-Integral *R*-Curve Testing of High Strength Steels Utilizing the Direct-Current Potential Drop Method, in STP 833, R.J. Sanford, Ed., ASTM, Philadelphia, 1984

55. C.G. Chipperfield, Detection and Toughness Characterization of Ductile Crack Initiation in 316 Stainless Steels, *Int. J. Fract.*, Vol 12, 1976, p 873-886

56. D.A. Curry and I. Milne, The Detection and Measurement of Crack Growth During Ductile Fracture, in *The Measurement of Crack Length and Shape During Fracture and Fatigue*, C.J. Beevers, Ed., Engineering Materials Advisory Services Ltd., Warley, UK, 1980, p 401-434

57. B. Marandet and D. Sanz, Characterization of the Fracture Toughness of Steels by the Measurement with a Single Specimen of J_{Ic} and the Parameter K^*_{Bd} in *Fracture Mechanics: 10th Symposium*, STP 631, ASTM, Philadelphia, 1977

58. V. Bachmann and D. Munz, Unusual Potential Drop during the Application of the Electrical Potential Method in a Fracture Mechanics Test, *J. Test. Eval.*, Vol 4, 1976, p 252-260

59. R.O. Ritchie, unpublished work, Lawrence Berkeley Laboratory, University of California, 1976

60. H. Nayeb-Hashemi, F.A. McClintock, and R.O. Ritchie, Effects of Friction and High Torque on Fatigue Crack Propagation in Mode III, *Met. Trans. A*, Vol 13, 1982, p 2197-2204

61. S. Suresh and R.O. Ritchie, The Propagation of Short Fatigue Cracks, *Int. Met. Rev.*, Vol 29, 1984, p 445-476

62. W. Plumbridge, Problems Associated with Early Stage Fatigue Crack Growth, *Met. Sci*, Vol 12, 1978, p 251-256

63. J.F. Knott, The Use of Analogue and Mapping Techniques with Particular Reference to Detection of Short Cracks, in *The Measurement of Crack Length and Shape During Fracture and Fatigue*, C.J. Beevers, Ed., Engineering Materials Advisory Services Ltd., Warley, UK, 1980, p 113-135

64. C.P. You and J.F. Knott, Electrolytic Task Simulation of the Potential Drop Technique for Crack Length Determinations, *Int. J. Fract.*, Vol 23, 1983, p R139-R141

65. P.E. Irving, J. Robinson, and C.J. Beevers, A Study on the Effects of Mechanical and Environmental Variables on Fatigue Crack Closure, *Eng. Fract. Mech.*, Vol 7, 1975, p 619-630

66. C.K. Clarke and G.C. Cassatt, A Study of Fatigue Strength Crack Closure Using Electric Potential and Compliance Techniques, *Eng. Fract. Mech.*, Vol 9, 1977, p 675-688

67. C.M. Ward-Close, A New Potential-Drop Method for the Measurement of Crack Closure, *Int. J. Fract.*, Vol 16, 1980, p R211-R213

68. V.B. Dutta, S. Suresh, and R.O. Ritchie, Fatigue Crack Propagation in Dual-Phase Steels: Effects of Ferritic-Martensitic Microstructures on Crack Path Morphology, *Met. Trans. A*, Vol 15, 1984, p 1193-1207

Environmental Effects on Fatigue Crack Propagation

Introduction

By Richard P. Gangloff
Senior Staff Metallurgist
Exxon Research and Engineering
Company

CORROSION FATIGUE CRACK PROP-AGATION is produced by the interaction between cyclic mechanical loading and chemical embrittlement. This phenomenon represents an important fracture mode for engineered components. Failure analyses, life prediction methods, and development of high-performance alloys require careful consideration of environmental factors, particularly during testing. This introduction provides an overview of environmentally enhanced fatigue crack propagation. Important variables and established testing procedures are emphasized for a range of alloy-environment systems. Discussion is limited to the crack propagation component of fatigue life.

The concepts developed in this overview are amplified in the following sections for eight specific environments:

- Vacuum and gaseous environments at ambient temperatures
- Vacuum and oxidizing gases at elevated temperatures
- Aqueous solutions at ambient temperatures
- Acidified chloride environments at ambient and elevated temperatures
- High-temperature pure water under aerated conditions
- High-temperature pure water under deaerated conditions
- Liquid metal environments
- Steam or boiling water with contaminants

Detailed information on the specimen types and crack-measuring techniques discussed in the following sections can be found in the article "Fatigue Crack Propagation." Other related articles in this Volume include "Fatigue Data Analysis," "Fatigue Crack Growth Data Analysis," and "Fracture Mechanics."

Environment

The surrounding environment is the most important factor influencing the propagation kinetics of a fatigue crack. This is illustrated by the crack growth rate data for a high-strength quenched and tempered steel shown in Fig. 1. Cracking in moist air progresses at rates up to four times those in vacuum. Crack growth is increased by two orders of magnitude for samples tested in aqueous sodium chloride and by up to three orders of magnitude for a purified gaseous hydrogen environment.

In sharp contrast is the band of data about the moist air line. Crack growth rates for 13 steels stressed in moist air vary by less than a factor of three, despite large variations in yield strength (300 to 2100 MPa, or 45 to 305 ksi) and microstructure (pearlitic, martensitic, and bainitic). Corrosion fatigue is clearly dominated by chemical reactions, in this case attributable to hydrogen embrittlement.

Large differences in crack growth rate are produced by environment at constant mechanical driving force—the stress-intensity range (ΔK). Furthermore, cracking in aggressive environments proceeds by brittle, intergranular separation and transgranular cleavage micromechanisms, as compared to ductile, plasticity-based transgranular fatigue cracking in vacuum or benign environments. Thus, environmentally induced crack path transitions provide an important fingerprint of corrosion fatigue embrittlement. References 3 to 11 provide a broad overview of fatigue crack propagation.

From a practical standpoint, all environments must be assumed to enhance fatigue relative to vacuum, or the universal standard of "laboratory air," unless proven contrary by experiment. Predicting corrosion fatigue is complicated further because variables such as stress ratio, loading frequency, temperature, alloy structure, yield strength, and environment composition influence cracking in different ways for different environments.

Fracture Mechanics Approach to Corrosion Fatigue

Corrosion fatigue crack growth rates are determined by the applied stress-intensity range, ΔK, which is defined as the difference between maximum and minimum stress intensities for any time-dependent loading. As a result, fatigue crack growth rates are independent of specimen geometry, crack size, and the applied load. The basics of the fracture mechanics characterization of environmentally sensitive fatigue are discussed in Ref 12, and evidence for the dominance of ΔK is given in Fig. 2.

Fatigue crack growth in a high-strength aluminum alloy is enhanced by up to an order of magnitude through exposure to aqueous 3.5% sodium chloride compared to dry air. For each environment, crack growth rate depends on a unique function of ΔK, not applied stress. Specifically, ΔK increases with increasing crack length for the "remote" specimen and decreases with crack extension in the "wedge force" geometry. Net section stress increases as cracking reduces the remaining ligament for each configuration. Crack speed accelerates with increasing crack depth for the remote case and decreases with crack size for the wedge case.

Quantitative crack growth rate versus stress intensity data (e.g., Fig. 1 and 2) have been widely produced (Ref 3-11) and are rel-

Fig. 1 Effect of environment on fatigue crack propagation in type 4130 steel with a yield strength of 1330 MPa (195 ksi)

The band of data about the moist air line represents cracking in 13 steels with varying microstructures and yield strengths ranging from 300 to 2100 MPa (45 to 305 ksi). Temperature: 23 °C (73 °F). Frequency: 0.1 Hz. Stress ratio: $R = 0.1$. Source: Ref 1, 2

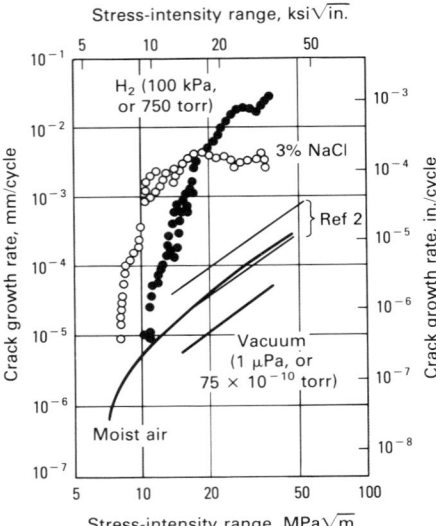

Fig. 2 Crack tip stress-intensity control of fatigue crack propagation in 7075-T6 aluminum alloy

(a) Remote and wedge force methods of loading specimens. (b) Aqueous 3.5% sodium chloride environment. (c) Benign dry air environment. Source: Ref 13

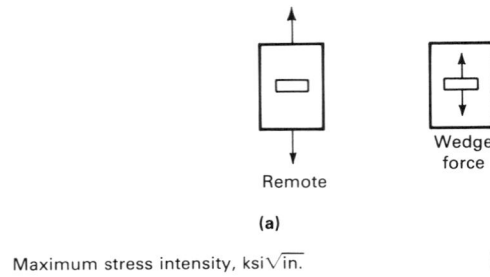

(a)

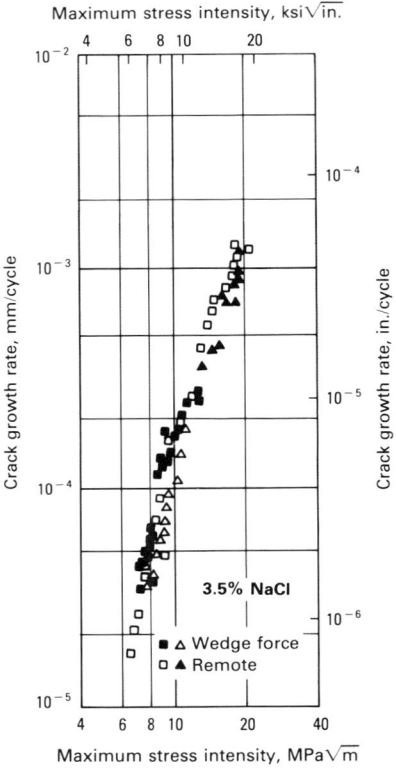

(b)

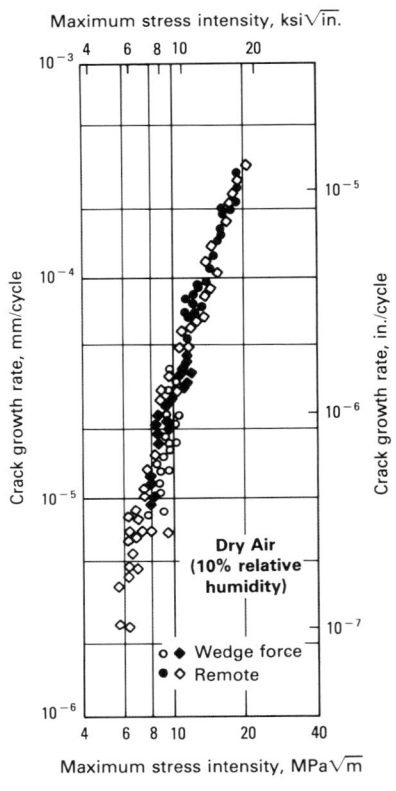

(c)

evant to mitigating corrosion fatigue from several perspectives. Critically, laboratory crack growth rate data are scalable quantitatively to predict component performance. The crack growth rate (da/dN) versus stress-intensity range (ΔK) law for a given material and environment is integrated in conjunction with the stress-intensity solution for a component to predict life (Ref 12). Specific examples pertaining to corrosion fatigue are reported in Ref 14 and 15 and are illustrated in Fig. 3. Note that the predicted 85-year life of a welded pipe is based on week-long laboratory measurements of da/dN versus ΔK for steel in an oil environment.

Alloy development for optimized corrosion fatigue resistance, defined in terms of da/dN versus ΔK for different metallurgical conditions, is directly related to component life goals based on fracture mechanics. Similarly, effects of mechanical and chemical variables on corrosion fatigue and component performance are defined through measured changes in growth rate at different applied stress-intensity levels.

Crack growth rate data are important to fundamental studies of corrosion fatigue mechanisms. The fracture mechanics approach isolates crack propagation from initiation and in terms of a precise near-tip mechanical driving force, ΔK. Crack growth rates are related directly to the kinetics of

mass transport and chemical reaction that constitute embrittlement. As shown in Fig. 4, prediction of the effect of loading frequency on crack growth rate in salt water (normalized to vacuum) identifies important rate-limiting crack tip electrochemical reactions. Modeling and measurements in Fig. 4 provide a sound basis for extrapolating short-term laboratory data to predict long-term component cracking.

The fracture mechanics approach to corrosion fatigue may be compromised by several factors. Discussion in this article is limited to crack growth in conjunction with net section elastic loading and small-scale plasticity confined to the crack tip. Studies of crack growth under large-scale cyclic plasticity

(Ref 17, 18) have not been extended to consider environmental effects. Also, stress intensity provides a mechanical description of similitude, which may not completely describe crack growth due to interacting chemical and mechanical driving forces.* Investigations of this phenomenon are currently underway. Results indicate that simple K-based approaches can be compromised for small corrosion fatigue cracks below 5 mm (0.2 in.), cases in which varying crack shape and load transients alter crack chemistry and embrittlement, and situations in which sur-

*Fracture mechanics is based on the concept of similitude, wherein cracks in different geometries grow at equal rates when subjected to equal near-tip driving forces, usually ΔK.

Fig. 3 Predicted fatigue crack extension from a weld toe crack in an API 5LX52 carbon steel pipeline carrying hydrogen-sulfide-contaminated oil

Temperature: 23 °C (73 °F). Source: Ref 14

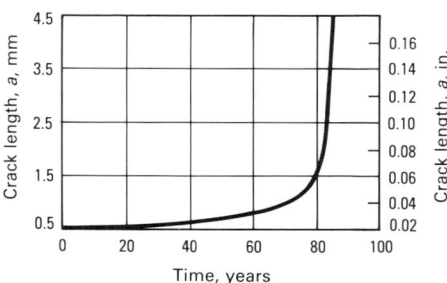

Fig. 4 Modeled effect of loading frequency on corrosion fatigue crack growth in alloy steels in an aqueous chloride solution

The determination of the normalized crack growth rate and the time constants, τ_o, from the model can be found in Ref 16.

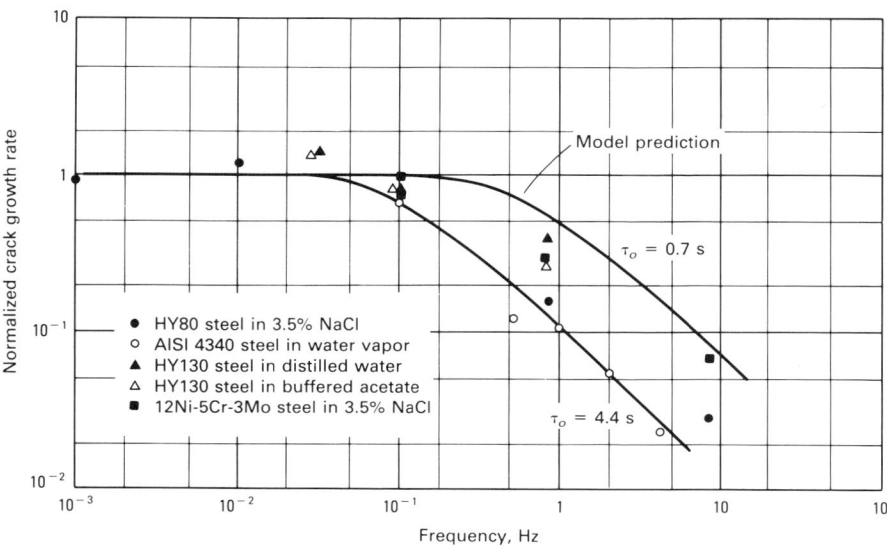

face roughness or corrosion products impede crack displacement (Ref 19-21). References 17 to 21 should be consulted if an application includes plasticity, small crack, load transient, or crack closure effects.

Variables Influencing Corrosion Fatigue

Although corrosion fatigue phenomena are diverse and specific to the environment, several variables are known to repeatedly influence crack growth rate. The following factors must be considered in any study of corrosion fatigue:

- Stress-intensity range
- Load frequency
- Stress ratio
- Aqueous environment electrode potential
- Environment contamination
- Alloy microstructure and yield strength

Effects of variables such as temperature, load history and waveform, stress state, and environment composition are unique to specific materials and environments (Ref 3-11).

Stress-Intensity Range. For embrittling environments, crack growth rate generally increases with increasing stress intensity; however, the precise dependence varies markedly. It is incorrect to assume that the three regimes (near threshold, power law, and fast fracture) of fatigue cracking observed for benign environments simply shift to higher crack speeds at all ΔK levels. Although such behavior may actually occur, as illustrated in Fig. 2 for the aluminum/dry air system, data contained in Fig. 5 and 6 are more typical of complex stress-intensity dependencies for aggressive environments.

Materials that are extremely environment sensitive, such as ultrahigh-strength steel in distilled water (see Fig. 5), are characterized by high growth rates that depend on ΔK to a

reduced power. Time-dependent corrosion fatigue crack growth occurs predominately above the threshold stress intensity for static load cracking and is modeled through linear superposition of stress-corrosion cracking and inert environment fatigue rates (Ref 12, 16).

Cycle-dependent corrosion fatigue crack propagation often occurs below the threshold for time-dependent stress corrosion (Ref 24). Typical ΔK dependencies for this mode of cracking are complex, as illustrated in Fig. 6 for Ti-6Al-4V exposed to aqueous sodium chloride. Mechanistic implications of the various stress-intensity dependencies are detailed in Ref 12, 23, and 24. The influences of variables such as frequency, stress ratio, and metallurgical factors depend on the proportions of time- and cycle-dependent corrosion fatigue.

Frequency. Cyclic load frequency is the most important variable that influences corrosion fatigue for most material, environment, and stress-intensity conditions. The rate of brittle cracking, above that produced in vacuum, generally decreases with increasing frequency. Frequencies exist above which corrosion fatigue is eliminated. The dominance of frequency is related directly to the time dependence of the mass transport and chemical reaction steps required for brittle cracking. Basically, insufficient time is available for chemical embrittlement at rapid loading rates; fatigue damage is only mechanical, equivalent to crack growth in vacuum. It is impossible to predict the frequency range at which corrosion fatigue is severe, due to the numerous chemical processes. It is

also difficult to extrapolate short-time (high-frequency) laboratory crack growth rate data to predict long-term component performance.

The literature provides qualitative guidance in terms of frequency effects on corrosion fatigue. Crack growth in environmentally sensitive materials stressed above the stress-corrosion threshold proceeds at rapidly increased rates with decreasing frequency, as illustrated in Fig. 5. The frequency effect is predicted through integration of static load data throughout each fatigue cycle (Ref 16). For such systems, the chemical contribution to fatigue cracking is suppressed above 0.5 to 5 Hz, with higher critical frequencies associated with more aggressive environments and sensitive microstructures.

Frequency effects on cycle-dependent cracking are complex and unpredictable. Data contained in Fig. 4 and 6 provide typical examples. For steels in salt water (Fig. 4), corrosion fatigue is suppressed at about 10 Hz. Crack speed increases with decreasing frequency and reaches a plateau between 0.5 and 0.1 Hz. In contrast, corrosion fatigue in Ti-6Al-4V occurs at loading frequencies at least as high as 10 Hz (Fig. 6).

Elevated-temperature corrosion fatigue is enhanced by decreased cyclic frequency and by prolonged periods of constant high-stress intensity. This time dependence is attributable to chemical damage by the oxidizing environment and to creep plasticity mechanisms. An example of the influence of an elevated-temperature environment is shown in Fig. 7 for NASA IIB-7, a high-strength nickel-based superalloy with a nominal com-

Fig. 5 Effect of stress-intensity range and loading frequency on corrosion fatigue crack growth in ultrahigh-strength 4340 steel exposed to distilled water

Temperature: 23 °C (73 °F). Source: Ref 22

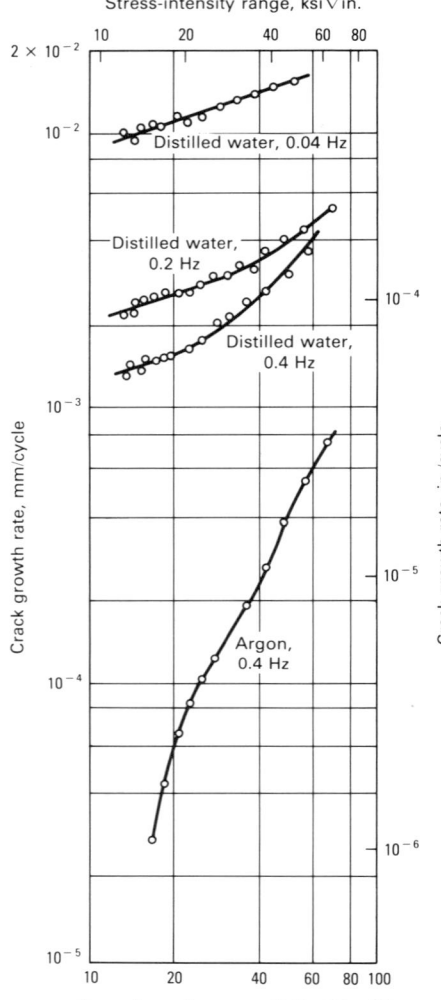

Fig. 6 Effects of stress intensity and frequency on corrosion fatigue in Ti-6Al-4V in aqueous sodium chloride

Temperature: 23 °C (73 °F). Stress ratio: $R = 0.1$. Source: Ref 23

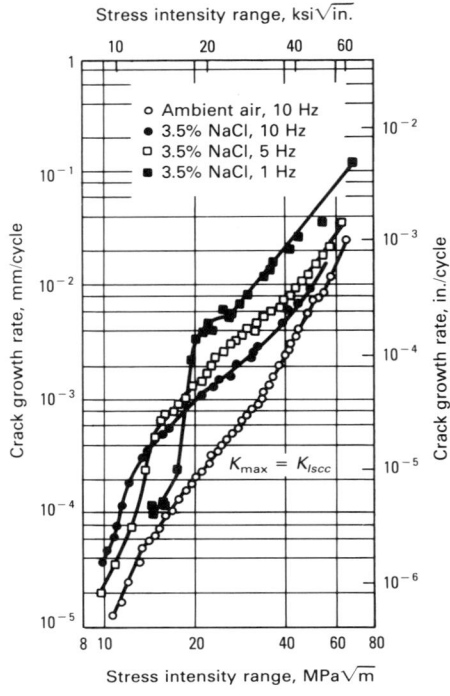

Fig. 7 Elevated-temperature fatigue crack propagation in a high-strength nickel-based superalloy (NASA IIB-7) exposed to an oxidizing environment

Alloy yield strength: 1200 MPa (175 ksi). Temperature: 650 °C (1200 °F). Stress ratio: $R = 0.05$. Source: Ref 25

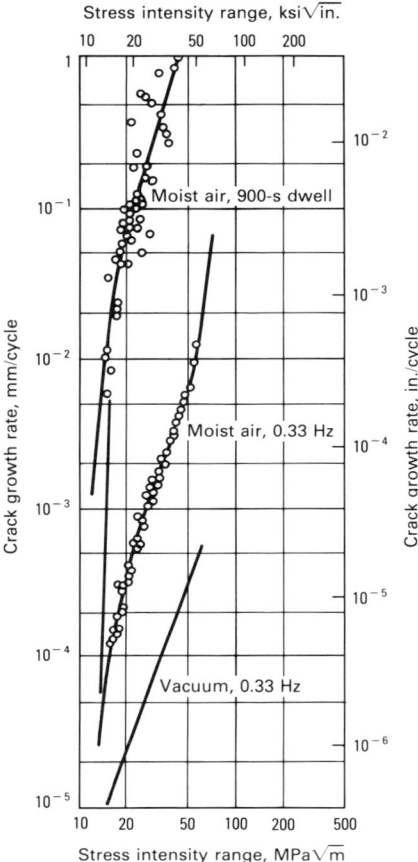

position of Ni-14Cr-8Co-3Mo-3W-3Ti-3Al-3Nb-0.05C (Ref 25). Crack growth rate is increased by one order of magnitude for continuous (0.33 Hz) cycling in moist air compared to vacuum. Cracking further accelerates by over two orders of magnitude due to prolonged constant loading (900-s dwell) at maximum K. The strength of the alloy and the air-vacuum comparison suggest that the time dependence of cracking illustrated in Fig. 7 is largely due to environmental effects.

Stress Ratio. Rates of corrosion fatigue crack propagation generally are enhanced by increased stress ratio, R, which is the ratio of the minimum stress to the maximum stress, $R = \sigma_{min}/\sigma_{max}$. As discussed in the article

"Fatigue Crack Growth Data Analysis" in this Volume, stress ratio only slightly influences fatigue crack growth rate for benign environments.

The deleterious effect of increased stress ratio is illustrated in Fig. 8 and 9 for two markedly different environments. Figure 8 demonstrates that for carbon steel stressed cyclically in pressurized nuclear reactor water at 288 °C (550 °F), the extent of the corrosion fatigue effect relative to dry air increases from about a factor of four at low R (0.11 to 0.24) to as much as 20- to 30-fold at high R (0.61 to 0.71) (Ref 26). Corrosion fatigue in this system probably proceeds by repeated passive film formation and rupture. Increased stress ratio at constant ΔK results in increased crack tip strain and strain rate, enhanced film rupture, and hence increased corrosion fatigue crack propagation (Ref 27).

The data in Fig. 9 further demonstrate the damaging effect of increased stress ratio for structural steel in seawater at 23 °C (73 °F). Crack growth rates in seawater are actually less than or just above reference data for moist air at low R values below 0.1. In contrast, crack growth is enhanced with respect to air for high-stress-ratio loading (0.5 to 0.85) for a range of ΔK values. Corrosion

fatigue in this system probably proceeds by hydrogen embrittlement, particularly at cathodic potentials that favor electrochemical hydrogen production (Ref 27). Hydrogen embrittlement is intensified by increased maximum stress, thus explaining the deleterious effect of increased R in Fig. 9. Furthermore, cracking at a low stress ratio is retarded by premature crack surface contact, due to formation of calcareous deposits from seawater at cathodic potentials. High mean stress may eliminate the potentially beneficial effects of corrosion-product-induced crack closure for a wide range of materials and environments.

Electrode Potential. Like loading frequency, electrode potential strongly influences rates of corrosion fatigue crack propagation for alloys in aqueous environments. Controlled changes in the potential of a specimen can result in either the complete elimi-

Fig. 8 Effect of stress ratio on corrosion fatigue crack propagation in ASTM A533 B and A508 carbon steels exposed to pressurized high-purity water

Temperature: 288 °C (550 °F). Frequency: 0.017 Hz. Average behavior in air is represented by the dashed line labeled "Dry." Source: Ref 26

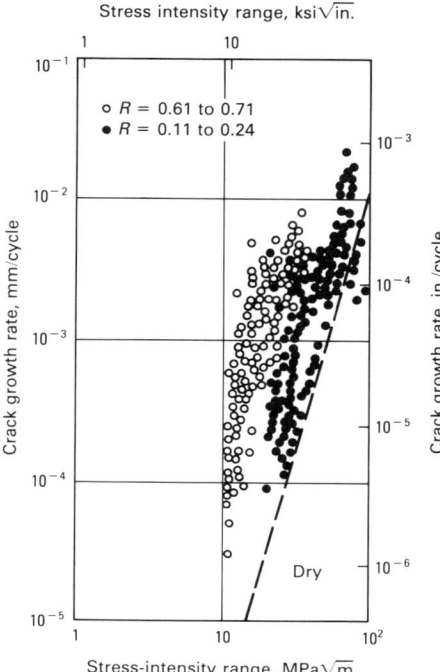

example, corrosion fatigue of austenitic stainless steel in chloride solutions at elevated temperatures is probably suppressed by polarization active to the critical potential for stress-corrosion cracking.

Another example is illustrated in Fig. 11. Aluminum alloy 7070-T651 is degraded by corrosion fatigue in several aqueous halide solutions at the free corrosion potential and at 23 °C (73 °F). Compared to dry argon, cracking in potassium iodide is enhanced by more than an order of magnitude. Cracking is further enhanced by anodic polarization above about −0.6V (standard hydrogen electrode), but is suppressed by cathodic polarization 400 mV active to the corrosion potential, E_{corr}.

Electrode potential should be monitored and, if appropriate, maintained constant during corrosion fatigue experimentation. Often, apparent effects of variables such as solution dissolved oxygen (O_2) content, flow rate, ion concentration, and alloy composition on corrosion fatigue are traceable to changing electrode potential.

Environmental Contamination. Environment composition and purity can influence corrosion fatigue crack propagation. Specific examples for aqueous environments are discussed in Ref 4, 27, 31, and 33. A striking example of an environmental-purity effect is illustrated in Fig. 12 for gaseous hydrogen embrittlement of a low-strength

Fig. 9 Effect of stress ratio on corrosion fatigue in BS4360:50D structural steel exposed to seawater

Constant cathodic potential: −1.10 V. Temperature: 5 to 10 °C (40 to 50 °F). Frequency: 0.1 Hz (sine). Source: Ref 28

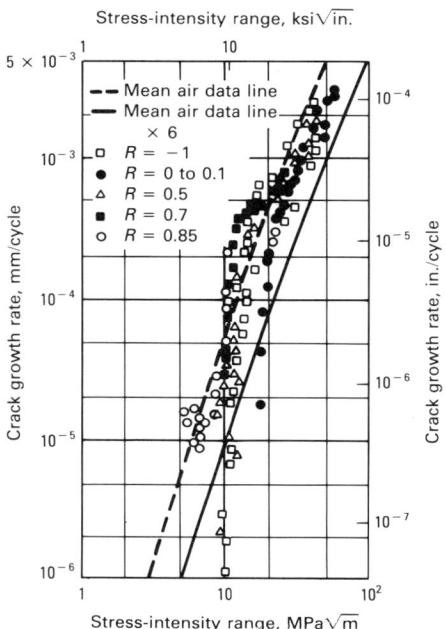

nation or the dramatic enhancement of brittle fatigue cracking. The precise influence depends on the mechanism of the environmental effect and on the anodic or cathodic magnitude of the applied potential.

For ferrous alloys that crack by hydrogen embrittlement when stressed in aqueous solutions, corrosion fatigue is enhanced by high cathodic polarization from the free corrosion potential. Specific data are contained in Fig. 10 for three steels stressed at constant ΔK in aqueous chloride (Ref 1, 17, 29). For each steel, the crack growth rate in salt water is about three times faster than that reported for air at the free corrosion potential and is enhanced by a factor of five at a very high cathodic potential. For the low-strength BS4360:50D grade, intermediate potentials appear to be mildly beneficial. Corrosion fatigue parallels electrode-potential-induced changes in the amount of cathodically evolved hydrogen in the crack tip. Precise modeling of this reaction sequence is complex and only partially completed (Ref 30).

Electrode potential control can suppress corrosion fatigue for alloys that crack through anodic dissolution/film rupture or anion adsorption mechanisms (Ref 31). For

Fig. 10 Effect of applied electrode potential on corrosion fatigue crack propagation in several steels exposed to seawater or 3% sodium chloride at constant ΔK between 20 and 40 MPa$\sqrt{m}$ (18 and 36 ksi$\sqrt{in.}$)

Temperature: 23 °C (73 °F). Frequency: 0.1 Hz. Stress ratio: $R = 0.1$.

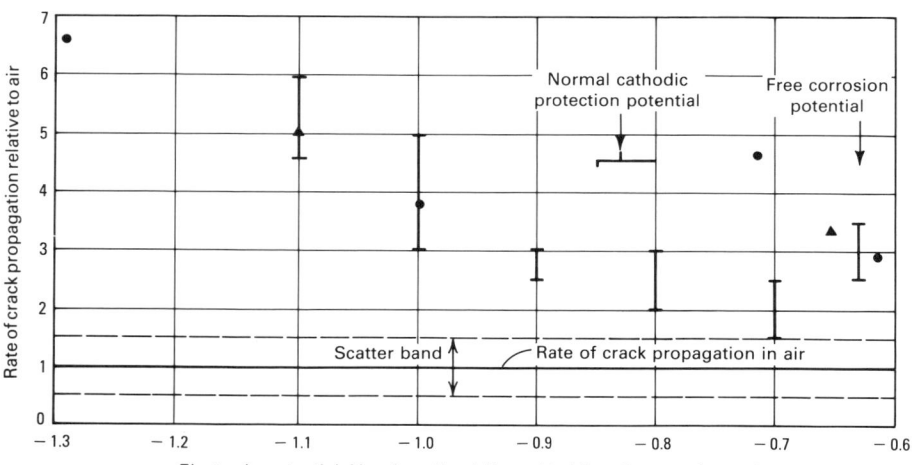

Material tested	Symbol	Yield strength		Environment	Ref
		MPa	ksi		
BS4360:50D	I	450	65	Seawater	27
HY80	●	600	85	3% NaCl	29
AISI 4130	▲	750	110	3% NaCl	1

carbon steel. Note that, relative to vacuum, crack growth is accelerated by factors of 3 and 25 for moist air and highly purified low-pressure hydrogen gas, respectively. Small additions of oxygen to the hydrogen (H_2) environment essentially eliminate the brittle corrosion fatigue component to crack growth, consistent with a trend first reported by Johnson (Ref 35). Similar effects have been reported for carbon monoxide and unsaturated hydrocarbon contamination of oth-

Fig. 11 Corrosion fatigue behavior of aluminum alloy 7079-T651

Temperature: 23 °C (73 °F). Frequency: 4 cycles/s. Stress ratio: R = 0. (a) Effect of stress-intensity range on crack growth rate. (b) Effect of electrode potential at ΔK of 6.7 MPa$\sqrt{m}$ (6 ksi$\sqrt{in.}$) in 25% potassium iodide solution. Source: Ref 32

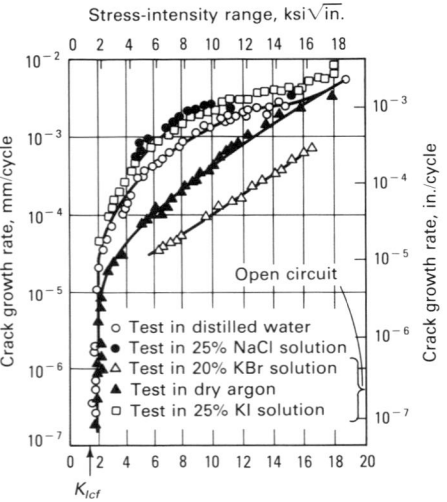

(a)

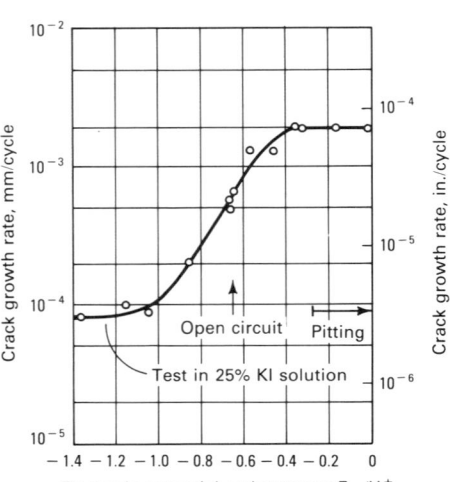

(b)

erwise pure hydrogen environments (Ref 33, 34). The composition and purity of the corrosion fatigue test environment must be carefully controlled, documented, and, when possible, equivalent to the intended application.

Metallurgical Variables. Microstructure and alloy strength influence fatigue crack propagation in embrittling gases and liquids. In Fig. 1, 13 steels of varying strengths and microstructures exhibited similar crack growth rates when tested in relatively benign moist air at high frequency and at 23 °C (73 °F). In sharp contrast, cracking in such steels progresses at rates that differ by three to four orders of magnitude, depending on the material, when each is loaded cyclically in hydrogen-producing environments such as hydrogen or salt water (Ref 3).

Data in Fig. 13 demonstrate the importance of microstructure (and strength) for superalloys stressed in high-pressure hydrogen gas. A 9-min loading cycle was used to simulate a Space Shuttle main engine mission (Ref 36). Crack growth in Inconel 718 varies by orders of magnitude, depending on thermomechanical treatment. The slowest rates of corrosion fatigue are associated with transgranular cracking in the cast product, while the presence of δ-Ni_3Nb phase (STA 2) dramatically enhances embrittlement.

Elimination of the δ phase (STA 1) does not fully mitigate cracking, because the equiaxed austenite grain boundaries in the wrought product are hydrogen sensitive. Such boundaries are not present in the casting. The influence of Inconel 718 microstructures occurs at constant strength, produced by γ"-phase precipitation hardening. For the remaining alloys shown in Fig. 13, both strength and associated microstructural changes influence corrosion fatigue. Ideally, the effects of these interacting variables should be isolated experimentally.

In general, brittle corrosion fatigue cracking is accentuated by:

- Impurity (phosphorus or sulfur, for example) segregation at grain boundaries
- Solute depletion or sensitization (chromium, for example) about grain boundaries
- Planar deformation associated with ordering or peak aged coherent precipitates
- Increased yield strength or hardness
- Large inclusions (manganese sulfide, for example)

The effects of alloy composition, grain size, and microstructure (e.g., bainitic versus martensitic steel) vary with environment and brittle cracking mechanism, as discussed in detail in Ref 37. Laboratory experiments are necessary to establish specific trends.

Fig. 12 Effect of oxygen (O_2) contamination on gaseous hydrogen embrittlement of a low-strength AISI/SAE 1020 carbon steel

Frequency: 1 Hz. Source: Ref 34

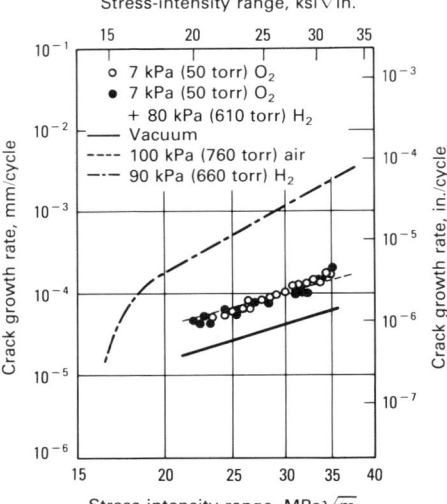

Crack Closure Effects. Premature crack surface contact during unloading, or "crack closure," can greatly reduce rates of fatigue crack propagation. The true (or effective) crack tip driving force is reduced below the applied ΔK because of the reduced crack tip displacement range. Closure phenomena are produced by a variety of mechanisms and are particularly relevant to fatigue crack propagation in the near-threshold regime, after large load excursions, or for embrittling corrosive environments, as reviewed in Ref 19, 38, and 39.

Two mechanisms of crack closure are relevant to corrosion fatigue. Rough intergranular crack surfaces, typical of environmental embrittlement, promote crack closure, because uniaxially loaded cracks open in a complex three-dimensional mode, thus allowing for surface interactions and load transfer. Roughness-induced closure is most relevant to corrosion fatigue at low ΔK and at stress-ratio levels where absolute crack opening displacements (0.5 to 3 μm) are less than fractured grain heights (5 to 50 μm).

Alternately, crack closing is impeded by dense corrosion products within the pulsating fatigue crack. For mildly oxidizing environments, such as moist air, this closure mechanism is relevant at low stress-intensity levels and contributes to the formation of a "threshold," as described in Ref 39.

For corrosive bulk environments or localized crack solutions, cracking at high ΔK values may be retarded below growth rates observed for air or vacuum due to corrosion product formation within the crack. The en-

Fig. 13 High-pressure (38 MPa, or 2.9 × 10⁵ torr) hydrogen (H₂) assisted fatigue crack propagation in several high-strength superalloys
Temperature: 23 °C (73 °F). Source: Ref 36

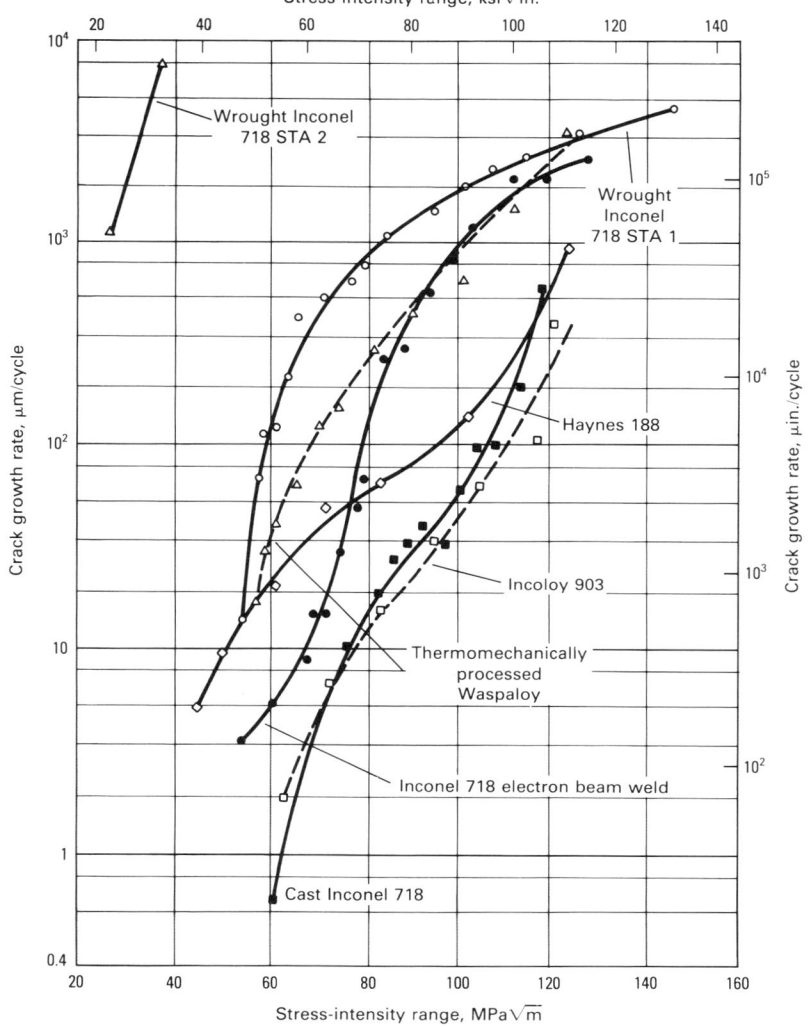

ing is clear, as emphasized by data contained in Fig. 15 for a 2.25Cr-1Mo steel (ASTM A542, class 3) stressed in hydrogen at 23 °C (73 °F). At high ΔK levels and low loading frequencies, cracking is accelerated in hydrogen compared to moist air, due to classical hydrogen embrittlement. For low stress intensities, cracking is also enhanced by hydrogen exposure; however, the effect is not due to chemical embrittlement because of the rapid loading frequencies. Oxides form on crack surfaces through a fretting mechanism during cycling in air. Crack growth rates are reduced by oxide-induced closure. Oxide formation is precluded for cycling in pure hydrogen, crack closure is absent, and growth rates are enhanced relative to moist air. Equally rapid rates of fatigue cracking have been reported for low ΔK cycling in hydrogen and in helium. Equal rates of cracking are observed for hydrogen, helium, and moist air at high R levels where closure is absent (Ref 38, 39).

Corrosion Fatigue Experimentation

Experimental characterizations of corrosion fatigue crack propagation are complicated by the numerous variables that influence the failure process. Both the mechanics

Fig. 14 Corrosion fatigue crack propagation in ASTM A471 steel exposed to moist air and steam
Temperature: 100 °C (212 °F). Frequency: 100 Hz. Stress ratio: $R = 0.35$. Source: Ref 40

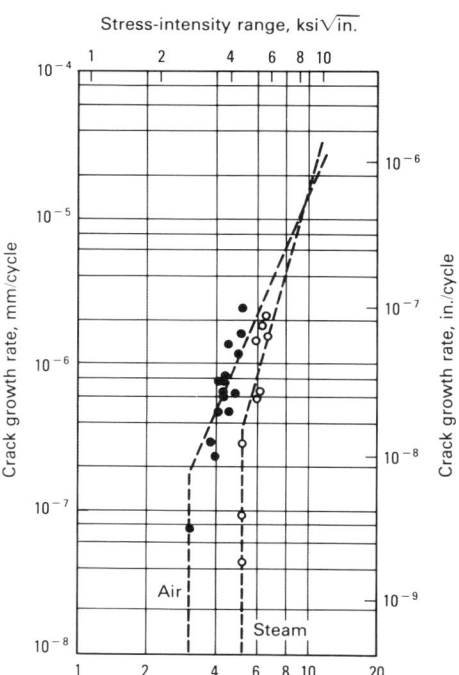

gineering significance of beneficial crack closure influences depends on the stability of the corrosion product during complex tension-compression loading and fluid conditions. The beneficial effects of environmentally induced crack closure are reported in the literature for several material environment systems.

Data contained in Fig. 9 show that corrosion fatigue in steel exposed to seawater is retarded at cathodic electrode potentials below about −850 mV (saturated calomel electrode) for low-stress-ratio cycling (Ref 27, 28, 30, 33). Calcareous deposits precipitate within the crack for these electrochemical conditions and restrict crack displacement during unloading at low R.

Another example of crack closure is presented in Fig. 14 for ASTM A471 steel exposed to either moist air or steam at 100 °C (212 °F). Crack growth is reduced for the latter environment due to crack surface oxidation and enhanced closure. Note that oxide-induced closure influences cracking even at a relatively high stress ratio of 0.35. Data presented in Fig. 14 were obtained at a high loading frequency of 100 Hz. Chemically enhanced crack growth is essentially precluded.

Crack growth at lower frequencies could be influenced deleteriously by metal embrittlement and beneficially by increased oxide formation and closure. The complexity of predicting environmentally enhanced crack-

Fig. 15 Corrosion fatigue in 2.25Cr-1Mo pressure-vessel steel in hydrogen (H₂) due to hydrogen embrittlement at high ΔK and to reduced oxide-induced closure at low ΔK

Stress ratio: $R = 0.05$. Source: Ref 38

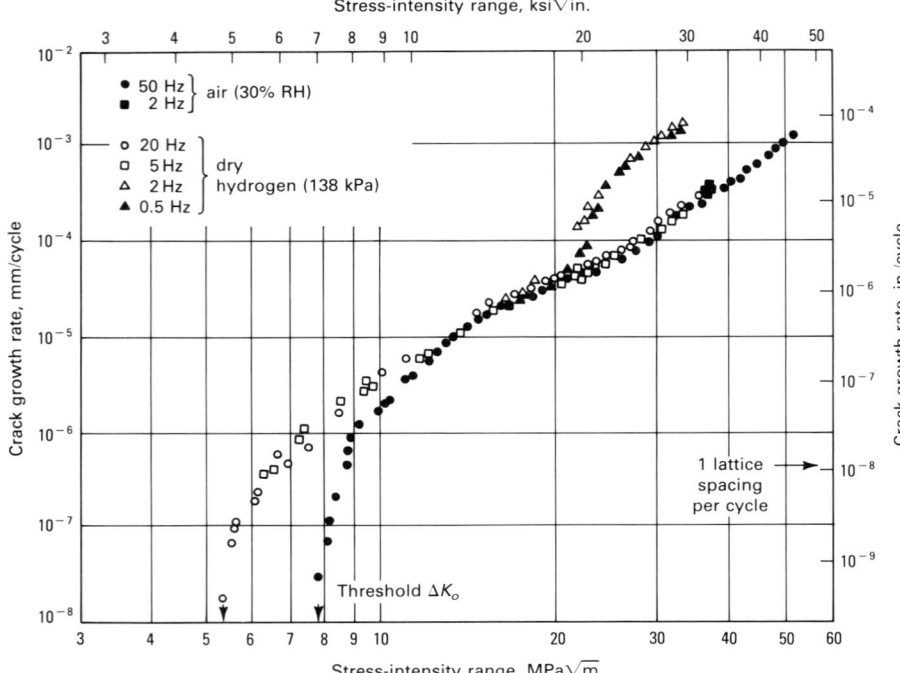

Vacuum and Gaseous Environments at Ambient Temperature

By R. P. Wei
Professor of Mechanics
Lehigh University

Gary W. Simmons
Professor of Chemistry
Lehigh University

and

Peter S. Pao
Scientist
McDonnell Douglas Research
Laboratories

FATIGUE TESTS in vacuum and gaseous environments at ambient temperatures are performed to assess the fatigue crack growth resistance of materials under simulated service conditions in relation to design, or to obtain critical fatigue data for scientific study. In both cases, one of the most critical considerations for fatigue tests in vacuum and gaseous environments is the maintenance of the purity of (and the reduction and measurement of the impurity level in) the test environment. Small amounts of contaminants (impurities) in the test environment can lead to fatigue crack growth rates that are not representative of the resistance of the material to fatigue crack growth in that environment.

If the effects of impurities are not properly considered and if such anomalous fatigue rates are used in design, an overdesigned or a nonconservatively designed structure may result, depending on whether the impurities in the test environment reduce or accelerate the fatigue crack growth rate. Also, incorrect scientific deductions are obtained. A clean environmental test chamber that provides a very low background pressure and quantifiable impurity levels (below 10^{-7} to 10^{-6} Pa, or 7.5×10^{-10} to 7.5×10^{-9} torr) is essential, even if the tests are to be carried out in gaseous environments at relatively high pressures (i.e., above the background).

The actual load and thus the real crack driving force must be determined precisely, and proper compensation for the load result from the pressure differential between the test environment and the external (atmospheric) pressure must be made. If the test environment consists of toxic or combustible gases (such as hydrogen sulfide or molecular hydrogen), special safety precautions must be incorporated into the test system design and test procedures. The reader is encouraged to follow the fatigue test procedures and

of loading and the composition of the environment must be controlled. Standard methods of measurement and fracture mechanics analysis of fatigue crack propagation in benign environments can be found in ASTM Standard E 647 (Ref 41).

ASTM Committee E24.04 is currently developing an Annex to this standard to describe testing in ambient temperature aqueous solutions (Ref 42). This Annex will be based largely on a Navy procedure developed for marine environments (Ref 43). Recent ASTM Special Technical Publications address experimental methods for fatigue crack propagation rate measurements in benign (Ref 44) and embrittling (Ref 5) environments.

Four problem areas are relevant to corrosion fatigue experimentation. The environment must be contained about the cracked specimen without affecting loading, crack monitoring, or specimen-environment composition. Parameters such as environmental purity, composition, temperature, and electrode potential must be monitored and controlled frequently.

Secondly, the deleterious effect of low cyclic frequency dictates that crack growth rates must be measured at low (often <0.2 Hz) frequencies, which lead to long test times, often from several days to weeks.

Load-control and crack-monitoring electronics and environment composition must be stable throughout long-term testing.

Thirdly, crack length must also be measured for calculations of stress intensity and crack growth rate. Optical methods are often precluded by the environment and test chamber. Indirect methods, based on specimen compliance (Ref 44) or electrical potential difference (Ref 45), have been applied successfully to monitor crack growth in a wide variety of hostile environments. Experimental and analytical requirements, however, are complex for indirect crack monitoring.

Finally, specimen geometry and size requirements for ΔK-based crack propagation data, which are scalable to components through similitude, have not been established completely for subcritical crack growth. In-plane yielding must be limited to the crack tip by guaranteeing that net section stress is below yield and that the maximum plastic zone size, defined as $\sim 0.2 \ (K_{max}/\sigma_{YS})^2$, is much less (e.g., 10- to 50-fold) than the uncracked ligament. Specimen thickness, as it influences the degree of plane-strain constraint, and crack size, as it influences the chemical driving force, may affect corrosion fatigue crack speeds. Currently, such effects are unpredictable; specimen thickness and crack geometry must be treated as variables.

data reduction methods outlined in ASTM E 647 (Ref 41).

Methods and Materials for Environment Containment

An all-metal environmental test chamber with mechanical-force feedthroughs is preferred for the study of environmentally assisted fatigue crack growth in vacuum and gaseous environments. A typical system is shown schematically in Fig. 16. Stainless steels are suitable materials for the environmental test chamber, with copper used as the gasketing material. The test chamber usually is equipped with a glass viewport that enables the operator to visually monitor the progress of the experiment. The viewport should be designed to safely withstand the maximum testing pressure.

With adequate pumping, the background pressure in the clean test chamber is usually below 10^{-6} Pa (7.5×10^{-9} torr). Maintaining an ultraclean test system is important, because a small amount of impurities can either significantly reduce or accelerate the fatigue crack growth rate, depending on the material and the types of impurities (Ref 34, 46, 47). The effects of impurities on crack growth rates is shown in Fig. 12.

To achieve a low background pressure, the test chamber frequently is baked out (with the test specimen in place) at a temperature above ambient (60 to 400 °C, or 140 to 750 °F) to remove adsorbed and absorbed gases on the chamber wall. The bake-out temperature should be considerably below the tempering or aging temperature of the test material to ensure that the microstructure and the mechanical properties of the test material are not altered by the bake-out process. For example, the first-step artificial aging temperature for high-strength 7050-T7451 aluminum alloy is 121 °C (250 °F). The bake-out temperature for the test chamber is thus normally kept below 80 °C (175 °F).

Environment

Only high-purity, laboratory-grade gases should be used. Additional purification and dehumidification of the gas is recommended by passing it through a molecular-sieve purifier and a cold trap (−196 °C, or −321 °F) before allowing the gas to enter the test chamber. Gas pressure in the environmental test chamber is usually controlled by admitting the gas through a variable-leak valve.

If the test environment contains a toxic gas (such as hydrogen sulfide) or a combustible gas (such as hydrogen and methane), a protective hood with negative suction pressure should be used to enclose the test chamber or

the entire test system. The test chamber should be purged thoroughly with an inert gas, such as argon or nitrogen, before it is reopened to the atmosphere.

If water vapor is used as the test environment, it can be drawn through the variable-leak valve from a high-purity reservoir that is attached to the test chamber. Deionized distilled water in the reservoir should be purified further by subjecting it to repeated freezing/pumping/thawing cycles to remove residual dissolved gases in the water (Ref 48).

Certain gases can decompose or react with containment vessels over time. For example, hydrogen sulfide can react with a stainless steel container to produce hydrogen. Provision must be made to remove the product gases before the test gas is admitted into the test chamber.

Finally, if the environment consists of mixed gases, the gas at the lowest partial pressure is admitted first. If premixed gases are used, they must be thoroughly mixed in the supply reservoir to minimize stratification.

If test conditions such as gas pressure, test frequency, or applied load are changed during fatigue testing, a transient period may occur before the material assumes the steady-state fatigue crack growth rate that corresponds to the new test condition. The duration of this transient period depends on several variables, including the type of material, the test environment, and the magnitude of the change in test conditions.

Measuring and Controlling Variables

Pressure. Corrosion fatigue crack growth rate frequently is related to the pressure of the deleterious gas. Gas pressure in the test environment therefore must be maintained and monitored. Corrosion-resistant pressure-sensing devices such as capacitance manometers are useful in monitoring the gas pressure. Because hot gage filaments can decompose or react with gases, gages with exposed hot filaments (such as ionization gages) should not be used.

Temperature. Because most of the rate-controlling processes for fatigue crack growth are thermally activated, test temperature can significantly affect the rate of fatigue crack growth. If the electrical potential technique is used to monitor crack length, variations in test temperature can also induce variations in the potential drop across the crack and hence affect the accuracy of crack length measurements. See the article "Fatigue Crack Propagation" in this Section for a description of the electric potential technique for fatigue crack growth measurement.

Fig. 16 Schematic of environmental chamber and associated subsystems

(1) Mechanical feedthroughs. (2) Capacitance manometer. (3) Bakeable ultrahigh vacuum valve. (4) Sorption pump. (5) Foreline valve. (6) Carbon vane pump. (7) Leak valve. (8) Variable-leak valve. (9) Residual gas analyzer. (10) Cryogenic trap. (11) Liquid nitrogen feedthrough. (12) Potential and thermocuple feedthroughs. (13) Power feedthrough (for heating). (14) Ultrahigh vacuum chamber with integral ion pump, titanium sublimation pump, and cryopanels

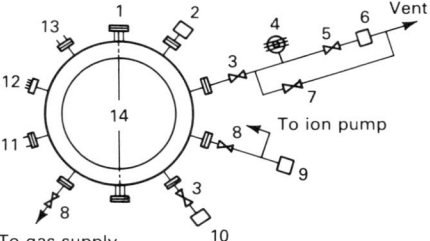

The temperature of the test system must also be monitored and maintained constant (preferably to within ± 1 °C, or ±2 °F) throughout fatigue testing.

Purity. Residual contaminants such as water vapor and oxygen can influence the rate of fatigue crack growth. The level of impurities in the background should be checked with the aid of a quadrupole residual gas analyzer.

Specimen Configuration

Compact tension and center-cracked tension specimens, as recommended in ASTM E 647 (Ref 41) and illustrated in the article "Fatigue Crack Propagation," are often used for fatigue crack growth tests, as are wedge-opening load specimens. Other specimen configurations can also be used if well-established stress-intensity factor calibrations are available and if specimens are sufficiently large to remain predominantly elastic during testing. Face or side grooves are often effective in preventing the crack from propagating out of the plane of symmetry.

Application and Measurement of Load

Load generally is applied through sealed mechanical-force feedthroughs, such as bellows, that are attached to the test chamber and is measured with an externally mounted load cell. If the test chamber pressure is different from the external (or atmospheric) pressure, the pressure differential must be compensated for to accurately determine the load actually applied to the test specimen. Performing fatigue tests without proper load compensation can lead to substantial overestimation or underestimation of the crack driving force and can introduce significant errors into the fatigue crack growth rate data.

The magnitude of load compensation that is required depends on the pressure differential between the system and the surrounding atmosphere and also on the load train design and configuration. Load compensation can be accomplished empirically by first establishing the "zero-load state" for the test specimen and then electronically compensating the load cell output.

Monitoring Crack Length

The electrical potential technique and the compliance method are frequently used to monitor fatigue crack growth in vacuum and in gaseous environments (Ref 49). Visual methods generally are not practical; often, the crack and the test specimen are obscured by the test chamber, or a microscope with a long focal length is needed.

The electrical potential technique is preferred over the compliance method for use inside an environmental test chamber, because the compliance gage may outgas and is a potential source of test environment contamination. Its use in a corrosive environment is also unsuitable. The electric potential technique, however, is noncontaminating and can be used in most environments.

Use of the compliance method is generally limited to compact tension and wedge-opening load specimens. It is not used for center-cracked tension specimens because of limitations in sensitivity and accuracy. The electrical potential technique can be readily applied to all three specimen types. The principal drawback of the electrical potential technique is that the specimen must be electrically conductive; thus, it cannot be applied directly to specimens made of nonelectrically conducting materials, such as polymer-based composites and ceramics. In addition, electrical shorting across the crack surfaces may affect its measurement accuracy, particularly for tests in vacuum. Both the electrical potential and compliance techniques can be readily interfaced with a digital computer for real-time control of the experiment and for on-line data acquisition and reduction.

Post-Test Analysis

Post-fatigue test analyses generally include fatigue data reduction and fractographic examination of fracture surfaces. For fatigue data analyses, the procedures outlined in ASTM E 647 should be followed (Ref 41). The crack length measurement interval should be at least 0.25 mm (0.01 in.) to ten times the crack length measurement precision, whichever is greater. A much smaller measurement interval relative to the measurement precision would introduce large scatter into the resultant crack growth rate (da/dN) versus stress-intensity range (ΔK) data. Fatigue crack growth rates can be determined from the crack length versus elasped cycles data by following one of the recommended procedures given in Appendix 1 of ASTM E 647.

Fracture surfaces of fatigue-fractured test specimens usually are examined by scanning electron microscopy to determine the fracture path and the fracture mode of the test material in relation to its microstructure. Such information may be valuable in identifying the fracture mechanism in certain environment and material combinations and may be used to assess the severity of the deleterious environment and to aid in analyzing service failures. Fatigue in inert (including vacuum) and in deleterious environments generally produce different fracture modes or fracture paths. Good qualitative agreement between the observed fracture modes and fatigue crack growth kinetics can be expected. In some cases, excellent quantitative correlation between fracture surface morphology and fatigue crack growth kinetics has been reported for environmentally assisted fatigue cracking of aluminum alloys and high-strength steels (Ref 50-52).

Vacuum and Oxidizing Gases at Elevated Temperatures

By David Jablonski
Manager, Research Laboratory
Instron Corporation

FATIGUE TESTING in elevated-temperature vacuum and oxidizing environments creates unique experimental problems. A carefully designed vacuum test chamber is a necessity for this type of testing. The chamber must keep the test specimen in a vacuum or oxidizing gas environment, allowing forces to be applied to the specimen, a means to measure crack length, and a method of applying and controlling the specimen temperature.

Variables that can affect fatigue crack growth rate in this environment are time and rate dependent or structure dependent. Examples of time- and rate-dependent variables are oxidation and creep. Structure-dependent variables include phase transformations, nucleation and growth of new and existing phases, and grain growth. When fatigue crack growth rate test data are reported for these environments, test temperature, vacuum pressure, partial pressure of oxidizing gas, waveform type, waveform frequency, and stress ratio must be reported. All of these test variables may affect the fatigue crack growth rate; thus, a proper definition of test conditions is required to allow comparison of test results. Discussion of the effects of these variables on fatigue crack growth rate can be found in Ref 53 to 58.

An example of the effect of air versus vacuum environment at elevated temperature is shown in Fig. 17. The crack growth rates in vacuum are substantially lower than in air. In vacuum, the test frequency still affects crack growth rate, which indicates that creep interactions are still present. Figure 18 schematically compares crack growth rates in air and vacuum. In vacuum at high frequencies (>10 cpm), crack growth rate is frequency independent, with a transgranular fracture path. At lower frequencies ($<10^{-1}$ cpm), crack growth rate is time dependent, with an intergranular fracture path. Crack growth rates in air are substantially higher than those in vacuum, indicating a strong environmental interaction.

Fatigue Test Chambers

References 59 and 60 provide complete discussions of test chamber design. Materials used in the test chamber should be selected to minimize outgassing in vacuum. For example, many plastic materials contain plasticizers, which slowly outgas in vacuum. These types of materials limit the ultimate vacuum obtainable. Stainless steel is suitable for the manufacture of the main test chamber. In test chamber design, components in the chamber should be designed for fast outgassing. When threaded components are used in the test chamber, channels should be machined in the threads to allow paths for fast outgassing.

For vacuum levels of 6.5×10^{-5} Pa (5×10^{-7} torr), O-rings provide sufficient sealing; for higher vacuum levels, copper gaskets should be used. Electricity, water, radiofrequency, and the thermocouple can be input into the chamber using standard vacuum feedthroughs.

A typical vacuum test system is shown in Fig. 19. Figure 20 shows a schematic cross section of a vacuum test chamber used for fatigue crack propagation testing. In these test chambers, metal bellows allow motion of the loading rods and provide a vacuum-tight seal. The metal bellows must possess an infinite fatigue life at the displacements required to test the specimen. This generally requires the use of relatively long bellows (>100 mm, or 4 in.), which minimizes the stresses in the bellows.

Test chambers for very high vacuums should be designed to be bakable. Test cham-

Fig. 17 Comparison of fatigue crack growth rates of Hastelloy-X in air and in a high vacuum

Temperature: 760 °C (1400 °F). Stress ratio: R = 0.05. Source: Ref 58

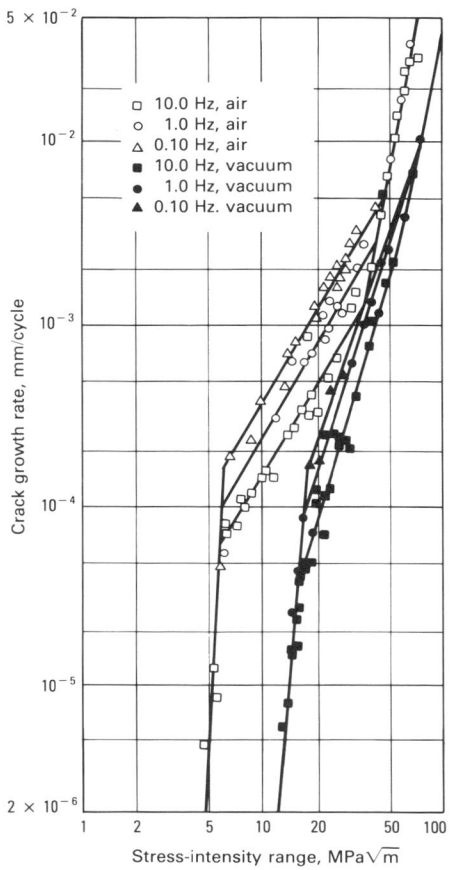

Fig. 18 Comparison of the fatigue crack growth rates of nickel-base superalloy A-286 in air and in vacuum

Temperature: 593 °C (1100 °F). Plastic strain range: 0.002 mm/mm (in./in.). Source: Ref 53

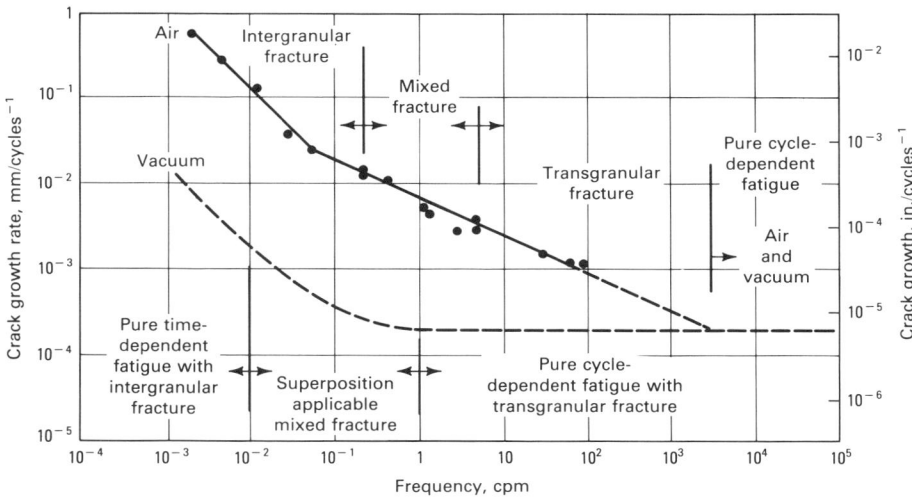

Fig. 19 Typical vacuum fatigue test system

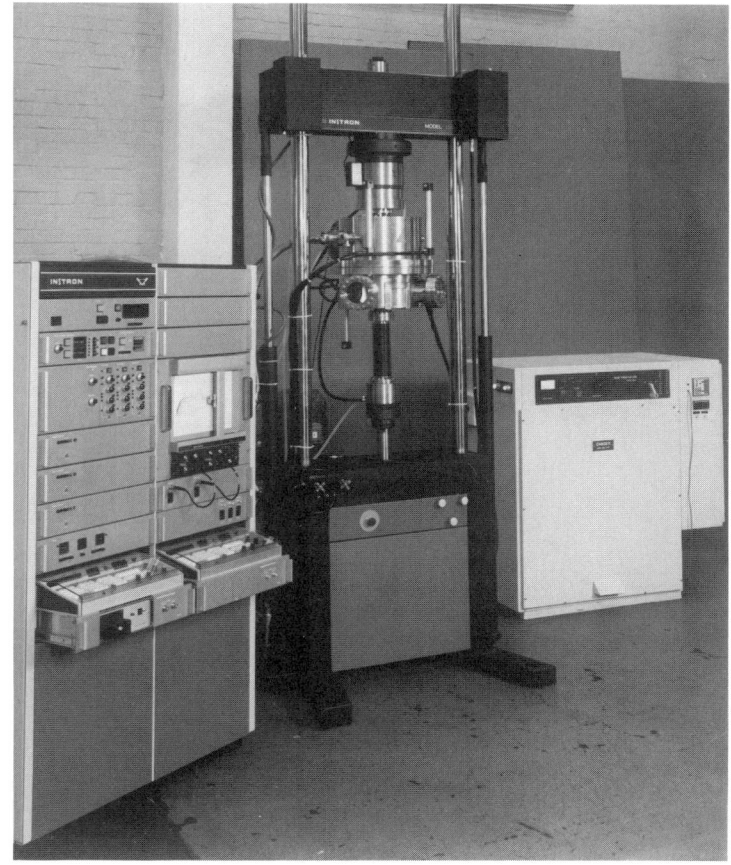

bers are baked out by heating all of the components to 300 to 450 °C (570 to 840 °F). This can be easily accomplished using heating tape wrapped around the various components.

Vacuum System

The type of vacuum system used depends on the desired vacuum level. Equipment manufacturers can assist in the choice of a proper vacuum system. The various types of vacuum pumps available are described in Ref 59 and 60.

A typical vacuum pumping system consists of a roughing mechanical vacuum pump (1.33×10^{-1} Pa, or 10^{-3} torr) and a diffusion-type vacuum pump (1.33×10^{-4} Pa, or 10^{-6} torr). The roughing pump is used for initial evacuation of the test chamber and to maintain a low-force pressure at the diffusion pump. For higher vacuum levels, a liquid-

Fig. 20 Schematic cross section of vacuum fatigue test chamber

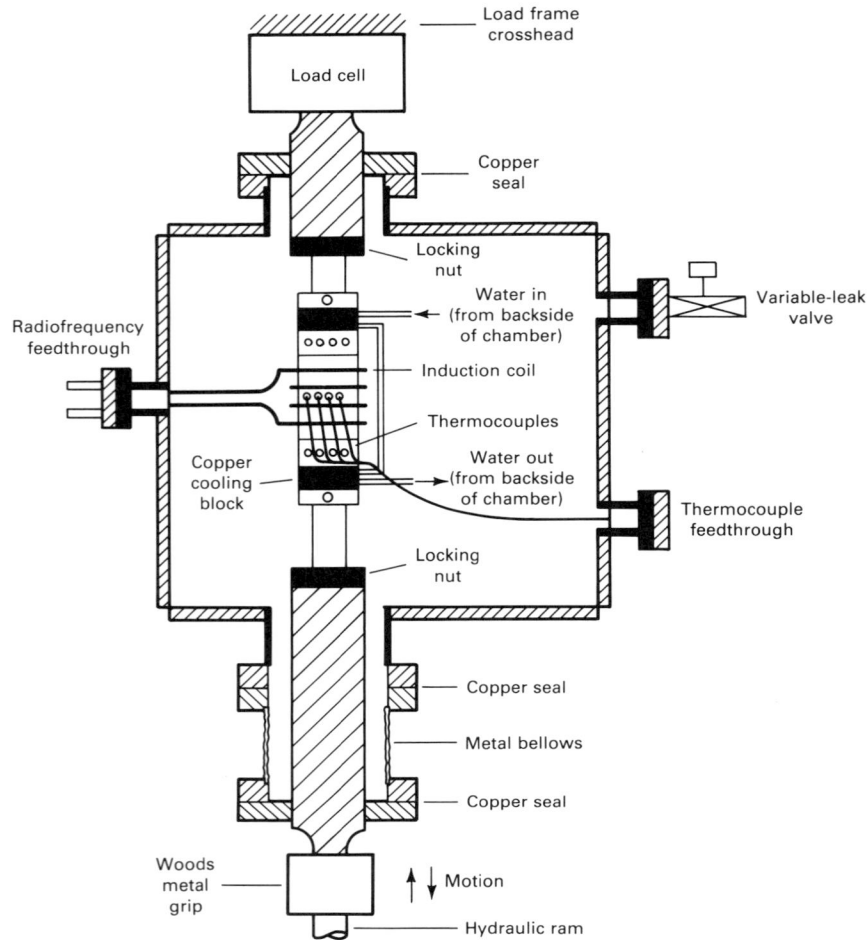

nitrogen-cooled baffle can be added to bring the vacuum level to 1.33×10^{-5} Pa (10^{-7} torr). Even higher vacuum levels are obtained by using an ion-type vacuum pump or a cryopump.

Low vacuum levels (1.33×10^{-1} Pa, or 10^{-3} torr) generally are measured with a thermocouple-type vacuum gage. High vacuum levels ($<1.33 \times 10^{-1}$ Pa, or 10^{-3} torr) are measured using either an ionization gage tube or a cold cathode-type gage.

Control of Oxidizing Gas

Oxidizing gases are introduced into the test chamber at a controlled flow rate by use of a variable-leak valve. The minimum flow rate is generally 10^{-10} L/s. The oxidizing gases should be high purity and should be free of hydrogen and water vapor. Low levels of hydrogen and water vapor can affect the fatigue crack growth rate in many materials (Ref 61). Stainless steel tubing should be used to connect the gas bottle to the leak

valve. The tubing should be flushed with the oxidizing gas and baked out before use.

The gas partial pressure can be measured by extracting a test sample from the chamber and analyzing it with a mass spectrometer. The gas partial pressure can be estimated by subtracting the steady-state vacuum pressure after the introduction of the oxidizing gas from the steady-state vacuum pressure before the introduction of the gas.

Specimen Heating and Temperature Control

Induction heating is the only suitable method to heat test specimens in vacuum and oxidizing environments. Radiofrequency generators with frequencies of 200 to 500 kHz are used for induction heating of test specimens. The induction coils should be made of copper and have no insulating coating. When oxidizing gases are introduced into the test chamber, a certain pressure range exists at which the gases will be ion-

ized between the specimen and induction coils. In this pressure range, it is impossible to heat the specimen, because the radiofrequency field arcs and shuts off the radiofrequency generator. To continue testing, the gas pressure must be either increased or decreased.

Two types of temperature controllers that are suitable for induction heating are thermocouple and infrared-type controllers. Each controller type has advantages and disadvantages. Infrared temperature controllers measure and control temperature from the spectral energy density emitted from the test specimen over a certain wavelength range. These measurements are noncontacting, but require a clear optical path from the sensor head to the test specimen. The test chamber must contain a window if infrared temperature measurements are used. The specimen emissivity must be corrected for transmission loss through the glass window. Infrared temperature controllers have a minimum temperature measurement capability of approximately 350 °C (660 °F). Two color infrared controllers eliminate errors due to transmission loss and emissivity changes, but have a minimum temperature-measuring capability of 700 °C (1290 °F).

Thermocouple-type temperature controllers are also used in vacuum test chambers. A variety of thermocouple types can be used with these controllers. Thermocouple type is determined by the temperature range and the required durability of the thermocouple. For example, American National Standards Institute type S and type K thermocouples temperature ranges overlap, but for long-term tests of more than one week, type S thermocouples are preferred, because they are more oxidation resistant. This would not be a consideration in a high-vacuum environment. Thin thermocouple wire less than 0.25 mm (0.01 in.) in diameter must be used to eliminate inductive heating of the thermocouple wire. With some temperature controllers, it is necessary to filter out radiofrequency noise in the thermocouple with a passive inductor/capacitor-type filter. Thermocouple classifications and comparisons can be found in the article "Temperature Control" in Volume 4 of the 9th Edition of *Metals Handbook*.

Test Specimens

In fatigue crack propagation tests, the specimen must be heated by induction. Therefore, many of the standard fracture mechanics test specimens cannot be used. Center-cracked tension (Ref 41) and single-edge notched (Ref 62) specimens are commonly used, because it is relatively easy to maintain the specimen gage section at uniform temperature with induction heating. When tests are conducted at high vacuum levels or low oxidizing gas partial pressures, specimen thickness may affect crack growth rate, because transport of the oxidizing gas to the crack tip may be the rate-limiting factor.

From a fracture mechanics viewpoint, the critical factors controlling crack growth rate are the stress-intensity factor and the stress ratio. The stress-intensity factor can be calculated using the standard formulas in Ref 41 and 62. All of the commonly used methods for measuring crack length (Ref 49, 63, 64)—unloading elastic compliance, direct-current potential, alternating-current potential, and optical—are applicable to high-temperature vacuum fatigue crack growth rate. Choice of method depends on the equipment available. See the article "Fatigue Crack Propagation" in this Section for a description of fatigue crack-measuring techniques.

A load cell located outside the vacuum chamber is used to measure the load applied to the specimen. The load measured by the load cell must be compensated for by the vacuum load if the chamber is attached as shown in Fig. 20. The vacuum load is caused by the imbalance of pressure from the inside to the outside of the test chamber. It is compensated for by removing the specimen, evacuating the test chamber, and rebalancing the load cell.

Calculation of Crack Growth Rate

Crack growth rate is calculated from crack length versus cycle number (da/dN) data. The data for crack growth rate determination should be chosen to eliminate transient effects, which may be caused by changes in stress-intensity range level, test frequency or waveform, or gas partial pressure or temperature. The crack length increment for transient behavior can be estimated from the plastic zone size, R_p:

$$R_p = \frac{1}{3\pi} \left(\frac{\Delta K}{\sigma_y} \right)^2$$

where ΔK is the stress-intensity factor range, and σ_y is the yield strength.

Crack length data should not be used for crack growth rate determination until the crack has propagated a distance of at least twice the plastic zone size. Crack growth rate can be calculated by the secant and incremental methods, which are described in the article "Fatigue Crack Growth Data Analysis" in this Volume and in Ref 41.

Post-Test Analysis

The fracture surface of the specimen should be examined, and the extent of crack tunneling should be determined. The final crack length should be measured on the fracture surface and compared to the final value determined during the test. If the two crack lengths differ by more than 4%, the test data should be corrected to account for this disparity. If substantial crack tunneling occurs, the crack length should be corrected to give an average crack length. The average crack length should be based on a five-point average crack length described in ASTM Standard E 399 (Ref 64). The stress-intensity factor should also be corrected for the curvature as described in ASTM E 399.

The test specimen should be examined metallographically to determine if the microstructure has changed during testing. A change in microstructure can cause a change in the crack growth rate. The microstructural effects usually appear at the near-threshold region or at high crack growth rates near the static fracture toughness. The fracture surface of the test specimen can be examined by scanning electron microscopy.

The test data should be reported in accordance with ASTM E 647 (Ref 41). The experimenter should also report specimen type, test frequency, test waveform, temperature, vacuum level, gas partial pressure, and an analysis of the gas residual components.

Aqueous Solutions at Ambient Temperature

By Peter M. Scott
U.K. Atomic Energy Authority
Atomic Energy Research
Establishment

and

A. Turnbull
Principal Scientific Officer
National Physical Laboratory

FATIGUE STUDIES in aqueous solutions at ambient temperatures present fewer problems experimentally than many of the other environments considered in this article. Nevertheless, simulation of practical conditions in relation to a specific engineering application requires careful consideration if the data are to be relevant. Similarly in an academic investigation, comparison of the results of different pieces of work is possible only if the mechanical, metallurgical, and chemical variables are properly characterized and controlled. It is often the case, however, that the most frequent problem in determining the validity of corrosion fatigue data lies with the control and monitoring of the bulk water chemistry and the monitoring and recording of the electrochemical potential.

Methods and Materials for Environment Containment

Glass and plastics are suitable materials for environmental test chambers and ancillary pipework for aqueous solutions at ambient temperatures. At elevated temperatures (>60 °C, or 140 °F), however, dissolution of silicates from glassware can inhibit corrosion (Ref 65). Dissolution of plasticizers from certain plastics (e.g., polypropylene) is also a concern. Flexible plastics, such as twin-pack casting silicone rubber, have proved to be useful in the vicinity of the fatigue specimen.

A corrosion fatigue test cell that avoids the need for a water-tight seal at the specimen is shown in Fig. 21. Normal specimen movement and any sudden fracture event can be accommodated without catastrophic consequences. Highly effective seals between plastics and metal surfaces can be made with silicone rubber caulking compounds, if necessary, although sufficient time must be allowed for escape of the acetic acid solvent base.

Fatigue specimens of passive metals such as aluminum, titanium, and stainless steel may be subject to crevice corrosion under the caulking compound unless the primer and epoxy paint coat are applied initially to the metal surface. Gasket seals using O-rings, for example, can also form a satisfactory seal, but generally are more expensive to engineer and can also be subject to crevice corrosion in some configurations.

The decision to circulate the environment depends on the application and the extent of any problems in controlling water chemistry. This is discussed in more detail in the following section. Standard glassware tubing usually is satisfactory for piping within the

Fig. 21 Typical corrosion fatigue test cell

Maintenance of the equilibrium oxygen concentration is ensured by cascading the solution in the circulation rig.

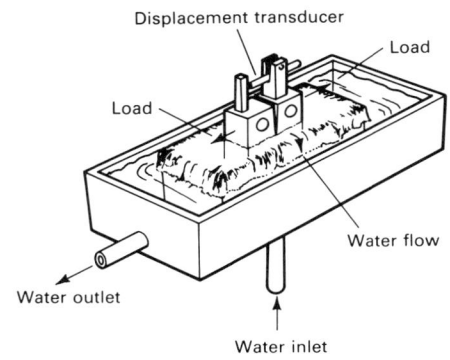

temperature constraint indicated above. A constant-head arrangement to supply several test stations is often used.

Water Chemistry

The prevailing water chemistry and the electrode potential of the material in its environment in the field are essential factors in any simulation experiment. Accelerated fatigue cracking can occur in a number of environments, including seawater, salt water/salt spray, and body fluids. These must be reproduced as closely as possible in the laboratory, although limitations are necessarily imposed in simulating aspects of complex environments, such as the biological activity of seawater.

The importance of reproducing the service environment as closely as possible is illustrated by comparing the behaviors of metals in sodium chloride and in seawater. The buffering action of seawater associated with dissolved bicarbonate/carbonate can result in the formation of calcareous scale under cathodic protection, which can precipitate in cracks and influence the cyclic crack opening and closing, thus affecting crack growth rates (Ref 66). Substitute ocean water, as described in ASTM Standard D 1141 (Ref 67), usually is a satisfactory substitute for seawater, but some differences have been observed in relation to the rate of calcareous scale formation (Ref 68) and the rate of corrosion fatigue growth (Ref 69).

Environment Synthesis

Laboratory solutions should be prepared using the purest chemicals available in distilled or de-ionized water. Concentrations at the level of parts per million can have profound effects on electrochemistry and corrosion.

Several variables must be measured and controlled when simulating an aqueous environment. These include solution purity, composition, temperature, pH, dissolved oxygen content, and the flow (circulation) rate of the solution.

Purity can be ensured by careful preparation. It can also be checked by using ion chromatography and other methods of chemical analysis.

Solution composition can vary with time, but this will usually be indicated by a change in pH or other variables. Monitoring of particular ions *in situ* can be conducted with ion-selective electrodes. The generation of chlorine from the counterelectrode during cathodic protection using potentiostatic control in chloride solutions can affect the electrochemical behavior. Accumulation of chlo-

rine should be avoided by using a large counterelectrode (e.g., platinum, graphite, or magnetite) and by bubbling air through the solution in the vicinity of the counterelectrode to remove chlorine.

Solution temperature can influence the rates of electrode reactions and diffusion and can affect the solubility of dissolved chemical species and gases. The electrode potential consequently changes with temperature unless controlled potentiostatically. In seawater or sodium chloride solution, the corrosion potential usually moves to more positive values with decreasing temperature. The effect of temperature on crack growth rates for steels in seawater is about a factor of two per 15 °C (27 °F) rise in temperature (Ref 28). The temperature of the solution should therefore be monitored and maintained to ±2 °C (±3.5 °F).

The pH of the laboratory solution should correlate to that in service. A change in bulk solution pH may influence the pH within the crack enclave, the kinetics of reaction at the crack tip, and thus crack growth rates. In some cases, if the pH varies outside the range 4 to 10, it will also affect the corrosion potential (Ref 70).

The effect of pH on fatigue crack growth rates is not well characterized; however, changes over a small range (6 to 8) may not be too significant (Ref 71). In unbuffered solutions (3.5% sodium chloride, for example), the pH can change during a test as a consequence of corrosion reactions or the accumulation of reaction products in cathodically protected systems. Fatigue data obtained during this transient period should be treated with caution. The pH can be maintained constant by using an automatic pH control system. In buffered solutions, a change in pH is an indication of exhaustion or depletion of the buffering species and the need for replenishment. Hence, monitoring of the pH either continuously or by occasional sampling is essential for any aqueous solution.

Dissolved oxygen control is vital in many applications, because it usually dictates the corrosion potential, unless this is imposed externally by means of a potentiostat. It may also influence the environment and reaction products within the crack through mixing of the bulk solution and the crack enclave. For systems open to the atmosphere, adequate maintenance of the equilibrium oxygen concentration can be ensured by bubbling air through the solution or by cascading the solution in a circulation rig (Fig. 21). In closed systems, oxygen monitoring is important, and standard meters using electrochemical probes are available. More sophisticated meters are used where parts per billion resolution is required.

Flow Rate. Ideally, the flow or circulation rate of solution past the fatigue specimen should reflect the conditions (or range of conditions) in service. If this is not possible, flow rate should be considered a variable. In relation to corrosion fatigue crack growth, the flow rate is most significant under freely corroding conditions through its effect on the corrosion potential. Higher flow rate provides a more positive potential, usually because of the enhanced mass transport of oxygen to the metal surface.

There is insufficient data to assess flow rate effects at controlled potentials. It is possible that it may influence the interchange of solution in the crack enclave with the bulk solution, although it is likely to be a second-order effect. Nevertheless, the nominal flow rate past the specimen should be given if possible as a linear flow rate.

The circulation rate may be similar to the flow rate and is primarily to ensure replenishment of the solution in the test cell. Flow rate, however, can be varied independently by a stirring arrangement. Even though the solution in service may be stagnant, some circulation in the laboratory may be necessary to ensure a large ratio of volume of solution to metal area without constructing a huge test cell. The circulation rate in relation to the number of cell volume changes per unit time should be given, and the total solution volume and regularity of replenishment should be indicated.

Electrode Potential

The rates of electrochemical reactions and the thermodynamic stability of dissolved chemical species depend on the solution composition and the electrode potential of the metal in that solution. Monitoring and reporting the electrode potential during corrosion fatigue experiments is important. The effect of electrode potential on corrosion fatigue crack growth is illustrated in Fig. 22.

The potential should be measured using a reference electrode located in the bulk solution adjacent to the specimen. When impressed currents are applied to the specimen, measurement should be made adjacent to the surface using a Luggin capillary (Ref 73) to eliminate a potential drop between the reference electrode and the metal surface, the magnitude of which will depend on the solution conductivity and flow of current.

Selection of a reference electrode depends on the particular application, but those most commonly used in laboratory tests are the saturated calomel electrode and the silver/silver chloride electrode. For some solutions in which contamination with chloride is undesirable, use of a mercury/mercurous sulfate reference electrode is an option. Con-

Fig. 22 Comparison of cathodic potential and free corrosion fatigue crack growth rates for X65 linepipe steel
Environment: 3.5% sodium chloride solution.
Source: Ref 72

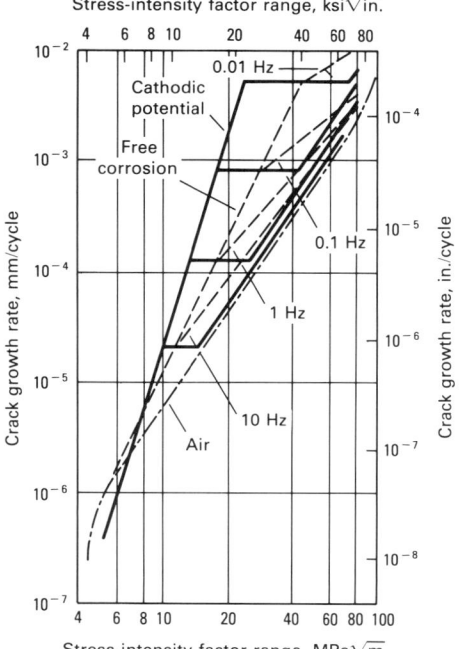

tamination can be reduced by using commercially available double-junction electrodes, in which the outer jacket is filled with test solution.

In quoting measured potentials, the potential should be referred to a standard reference electrode such as the standard hydrogen electrode (SHE) or the saturated calomel electrode (SCE) at 25 °C (75 °F). In tests remote from 25 °C (75 °F), allowance must be made for the fact that the measured potential of the reference electrode varies with temperature and should be checked with respect to the ambient temperature value using a salt bridge between solutions at the two different temperatures (Ref 74). A high-impedance meter ($>10^{12}$ Ω), such as an electrometer or a pH meter, should be used for monitoring potential.

Test Specimens

Corrosion fatigue testing often is performed at low cyclic frequencies. Consequently, multiple test stations are desirable. For this reason and for general economy, compact tension specimens as described in ASTM E 647 (Ref 41) are frequently used. Such specimens minimize the applied load required to achieve a given crack tip stress intensity, thus permitting the use of low load capacity and less expensive test machines.

Fatigue test specimen dimensions and surface finish are specified in ASTM E 647. Side grooves usually are not necessary to control crack bowing and tunneling, because interest generally focuses on the lower range of stress-intensity amplitudes where most fatigue life is spent. Side grooving is unlikely to be effective where a tendency toward crack branching exists.

From a fracture mechanics standpoint, the critical factor in relation to crack growth is the stress-intensity factor range, ΔK which depends on the applied stress, crack length (a), and compliance of the specimen. In corrosion fatigue, the electrochemistry within the crack is mass transport dependent and can vary with crack depth (Ref 65) and possibly with specimen geometry and with accessibility of solution in the through-thickness direction via the crack sides (Ref 75, 76).

Consequently, in principle these factors can influence crack growth rates despite constancy of the range of the stress-intensity factor (Ref 77, 78). The application of fracture mechanics to corrosion fatigue must therefore be considered carefully, because in some circumstances the basic concepts may be invalid. In reports of test data, information regarding crack depth should be quoted or be deducible from the data. An alternative is to determine crack depth and ΔK as independent variables to verify that ΔK is an adequate characterizing parameter for the rate of crack growth.

In applying load to specimens in a test cell, cell friction must not affect load in sealed systems. This is generally not a significant factor in most ambient-temperature applications, however. Insulation between specimens and grips, pin assemblies, and so forth is essential to avoid galvanic effects, but greases should not be used.

Monitoring Crack Length

Crack growth in aqueous solutions can rarely be observed optically. Indirect compliance or electric potential methods should be used; the former is useful only with compact tension specimens. These methods are described in the article "Fatigue Crack Propagation" in this Section and in Ref 49. Galvanic effects in relation to transducer mounting and attachments must be avoided. Most linear displacement transducers cannot be submerged.

The use of alternating-current and direct-current electric potential monitoring methods is increasing for crack length measurement. Comparative tests with and without large direct currents suggest no unwanted electrochemical side effects (Ref 71, 79). Closure

effects at low R values are more apparent in aqueous environments than in air for structural steels because of reduction of the insulating air-formed oxide film. The value of potential difference at the highest tensile peak of the cycle should be used to assess crack depth.

Difficulties can arise in using a direct-current electric potential monitoring system in conjunction with potentiostatic control of the specimen potential if the same wire connections to the specimen are used. Separate connections to the specimen for both systems should be used. One significant advantage of using compliance and electric potential methods simultaneously is that corrosion products causing crack closure can be detected early. The effect of crack branching is less certain, but some discrepancy in the apparent crack length from the two readings can be anticipated.

Post-Test Analysis

Analysis is made primarily of the record of crack length versus elapsed cycles, although some metallography or fractography of the specimens may be carried out. The latter two procedures are unlikely to yield useful information if the specimen was actively corroding in its environment, because the extent of corrosion will be greater than the scale of the original fracture surface topography. When cathodic protection is applied, or when passive materials that are not subject to crevice corrosion down the crack have been used, post-test fractographic examination of the failed specimen may yield useful information.

For analysis of the crack length versus cyclic data, the methods described in ASTM Standard E 647 may be used (Ref 41). This standard discusses the minimum crack increment that is statistically valid, taking into account the resolution of the crack measurement apparatus. Indirect methods of crack measurement are often claimed to have much higher resolution than is justified over long-term experiments.

Failure to make a true estimate of crack measurement scatter will result in variations in this measurement appearing in the reduced da/dN versus ΔK data. Not all corrosion fatigue data obeys the Paris equation; consequently, any data reduction technique that assumes this is automatically invalid. For example, some computer programs find the coefficients of a Paris equation fitted through the data regardless of whether this is a sensible physical representation of the crack advance mechanism. An example of data in which this tendency is likely to occur is shown in Fig. 22.

Acidified Chloride at Ambient and Elevated Temperatures

By Peter L. Andresen
Scientist
General Electric Company

INVESTIGATIONS performed in acidified chloride, particularly at high temperature, pose unique problems. These include not only experimental barriers, such as suitable containment and seal materials, and sensitivity to low-level oxidizing species, but also interpretational complexities, such as the effects of pitting and crevice processes on enhancement or retardation (by blunting) of crack initiation and growth. Care must be exercised in designing and conducting experiments to ensure personnel and equipment safety and to ensure proper simulation, control, and monitoring of environmental parameters.

Solution Containment

Below 100 °C (212 °F). Materials and techniques for solution containment depend on the test temperature regime. Below the boiling point in solutions containing dissolved oxygen, a primary design concern is to prevent leaks that can damage equipment. A horizontal loading frame helps ensure that sensitive components are not readily damaged by leaks. Additionally, some specimen configurations (such as compact tension) permit the loading linkage to be placed above the solution, simplifying the choice of materials and seal designs.

Testing in deaerated solutions may require careful selection of materials, depending on the sensitivity of the test to low oxygen concentration. For example, clear, flexible tubing often used in laboratories is very permeable to oxygen. Additionally, some plastics degrade in acidic environments. Guidelines for proper selection of containment materials can be found in Ref 73 and 80 to 83.

Above 100 °C (212 °F), the propensity for pitting and crevice attack increases, the internal pressure rises, the design strength of some materials (e.g, titanium) begins to decrease, and good seal design (particularly for sliding seals) is crucial. Pitting and crevice potential studies show that the resistance of iron- and nickel-base alloys in environments containing chloride decrease from room temperature to about 200 °C (390 °F) (Ref 84, 85). The potential at which pitting initiates is lowered by about 600 mV over this range for type 304 stainless steel and Hastelloy C-276 (Fig. 23). Weld metal may also exhibit different susceptibility to pitting and cracking than base metal.

The best approach for selecting pressure boundary materials is to combine published data with recommendations from autoclave manufacturers and metals producers. No assumptions should be made regarding the performance of materials with varying environment. For example, commercial-purity titanium, which is often used in neutral and acidified chloride environments, performs very poorly in acidified chloride under reducing conditions, in acidified environments containing sulfate, and in caustic environments at high temperature. Addition of a small amount (0.2%) of palladium (grade 7) greatly improves resistance in acidified environments that contain sulfate.

Above 200 °C (390 °F), materials selection is particularly difficult. In general, for acidified chlorides, commercial-purity titanium is favored under oxidizing conditions (containing oxygen, iron ion, or copper ion), while zirconium (for example, UNS R60702) is favored for reducing environments. Zirconium alloys are highly intolerant of fluoride. In some cases, high-strength materials, such as Ti-6Al-4V or the Hastelloy C series alloys, are required, although there is generally a loss in corrosion resistance. Liners of Teflon or tantalum are options in some instances.

Because of its effect on the autoclave and test results, control of the oxidizing nature of the environment is often critical. In addition to oxidizing species, such as oxygen, iron ions, and copper ions, care in the use of externally applied potential is required. If ground loops exist, or if used as the counterelectrode, the autoclave may be polarized into a harmful regime. A similar result can occur if the autoclave contacts a dissimilar metal.

Because of the rate and extent of expansion on leakage, hot pressurized water poses a serious safety hazard. Each autoclave must have a pressure-relief device attached to it, preferably in a fashion that does not permit bypassing or isolation. Selection of the pressure-relief device must account for the pressure, environment (often gold-coated elements are used in rupture disks), and temperature at which the device actually operates. Additionally, autoclaves, particularly when used in aggressive environments, must be examined regularly for damage resulting from pitting, crevice attack, general corrosion, hydriding, and so forth.

Pressure testing coupled with dimensional checks must also be performed. Manufacturers offer this service and will usually provide

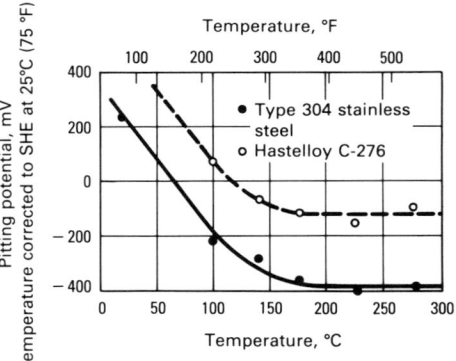

Fig. 23 Pitting potentials of type 304 stainless steel and Hastelloy C-276 on unstrained specimens as a function of temperature

Because the temperature coefficient of the standard hydrogen reaction contributes to the potential as a function of temperature, all potentials were converted to a reference value of SHE at 25 °C (75 °F). Environment: 3N (normal solution) sodium chloride. Scan rate: 100 mV/h

the test details. Test pressure and dimensional tolerances are a function of autoclave design, material, and temperature of use. Leaks may also occur in tubing, and valves, which are often difficult to inspect or test. Leaks almost always develop slowly. Nevertheless, a relatively rapid, controlled method for depressurizing the system should be included in the system design.

For some applications, inexpensive miniature autoclaves can be custom fabricated. In a design for cylindrical tensile specimens (Ref 85, 86), an autoclave was fabricated from 50-mm (2-in.) diam bar stock that incorporated integral sliding seals, influent and effluent, thermocouple, and an external reference electrode. The small internal volume of these devices is an advantage if a leak occurs in the system.

Connections, feedthroughs, and seals are available commercially. Fittings using a metal ferrule seal rather than a sliding seal and single and multiple feedthrough fittings are available in a variety of materials. Single feedthrough fittings with a Teflon sealant are often used as a sliding seal. However, these seals can exhibit inconsistent frictional loads and inadequate sealing on large-diameter rods (above approximately 15 mm at 250 °C, or 0.6 in. at 480 °F), or in thermal cycling applications. Options include more elaborate seal designs (Ref 85, 86), the use of bellows, or the use of seal rings (Ref 85). The use of Teflon is limited to about 300 to 330 °C (570 to 625 °F) by chemical breakdown, although it flows readily at much lower temperatures. Release of fluorine at lower temperatures may be a serious problem in nonrecirculating zirconium autoclaves. These shortcomings can be managed by cooling the seal (water cooling or extending the seal away from the autoclave) and by trapping the Teflon between pieces of high-temperature or reinforced plastic so that flow is minimized.

High-pressure pumps are available in a wide range of pressure, flow rate, and materials. Pressure regulation is commonly performed with a backpressure regulator. Because high-pressure pumps usually deliver flow in spurts, pressure pulse dampening using accumulators or hydropads may be desirable. Solution preheating should be performed using a regenerative heat exchanger, which can be easily fabricated from tubing of two different diameters and appropriate fittings.

In some instances (small autoclaves), the solution should be fully preheated to the test temperature. In most autoclaves, there is a large temperature gradient. A difference of 40 °C (70 °F) at a test temperature of 300 °C (570 °F) is not uncommon. Temperature uniformity can be improved greatly by rapid recirculation, involving a second high-temperature recirculation pump, or by using a sleeve or tube inside the autoclave, which creates a thermal siphon in the 10- to 20-mm (0.4- to 0.8-in.) gap between the autoclave wall and the sleeve.

Environment

Inherent to the study of environmental effects on fatigue crack growth is the likelihood of undesirable effects due to inadequate definition, simulation, and control of the solution composition. In chloride environments, particular attention should be given to factors that influence pH and electrochemical potential. Parameters affecting potential include very low concentrations of dissolved oxygen (above about 10 ppb at temperatures above 200 °C, or 390 °F, as described in Ref 87), low levels of iron and copper ion (generally above several ppm), and galvanic coupling of the specimen to more noble materials (e.g., a titanium autoclave).

Thus, unless the application of interest indicates otherwise, it is generally advisable to use a flowing test solution. For similar reasons, the design of the solution reservoir and solution recycling must be planned carefully. Optimal flow rates are dependent on the experiment and should also account for any velocity effects in service applications. In deaerated solutions with resistant materials, a low refresh rate of about one autoclave volume per hour is satisfactory. Similarly, a low flow rate may be satisfactory in air- or oxygen-saturated solutions with resistant materials.

In experiments with low, but purposeful, additions of oxygen (or if materials are more reactive), much higher refresh rates, ranging from 10 to 100 times the autoclave volume per hour, may be necessary, and oxygen should be monitored in the effluent. In open test or reservoir designs, continuous bubbling is generally necessary to ensure the proper oxygen concentration. During bubbling at atmospheric pressure, solution temperature affects the dissolved gas concentration via gas solubility, which changes rapidly below 20 °C (70 °F), and the contribution of the vapor pressure of water, which changes rapidly between 70 and 100 °C (160 and 212 °F). Pure oxygen bubbled in water at atmospheric pressure yields 69 ppm by weight at 0 °C (32 °F), 40 ppm at 25 °C (75 °F), 26 ppm at 50 °C (120 °F), 16 ppm at 75 °C (165 °F), and 4.2 ppm at 95 °C (205 °F).

Assumptions of an adequate level of deaeration can be erroneous due to purge-gas purity, permeability to oxygen of plastic tubing and equipment, inadequate time for initial deaeration, poor seals around the reservoir or pump, etc. Thus, the influent water should be monitored regularly or continuously. In addition to standard analytical techniques, dissolved oxygen concentration can be measured colorimetrically.

In some cases, other dissolved species, such as hydrogen sulfide, are of interest. Whether present in major or minor concentrations, good control over both hydrogen sulfide and species that affect electrochemical potential is necessary. High-temperature pH-potential diagrams illustrate, for example, that the stability of the sulfate or bisulfate ion is increased at the expense of hydrogen sulfide as the potential is raised (Ref 88). In other cases, species such as cathodically generated chlorine may be formed. This reinforces the need for adequate flow rate and continuous bubbling.

Electrochemical Potential

The electrochemical potential developed at the surface of a metal in solution generally is a critical parameter to monitor and control. It directly influences corrosion fatigue crack growth rates and determines the likelihood of pitting and crevice attack. Near room temperature, it generally is possible to use commercial reference electrodes such as calomel and silver chloride electrodes; some electrode designs permit use near boiling. Designs that place the reference electrode in a separate chamber at a different temperature than the test solution are complicated by formation of a thermal junction potential in the electrolyte, the magnitude of which may be large (over 0.1 V).

At temperatures over boiling, a custom reference electrode generally is necessary. Most investigators use internal or external silver/silver chloride reference electrodes (Ref 85, 89, 90). The silver chloride reaction occurs at the test temperature for internal electrodes. For external electrodes, the silver chloride reaction occurs at room temperature, with a temperature gradient occurring in the potassium chloride electrolyte as it enters the autoclave. A porous junction in the autoclave isolates the potassium chloride electrolyte from the autoclave solution. This thermal junction potential has been well characterized over a range of temperatures and potassium chloride concentrations (Ref 91).

Potentials should be reported on the standard hydrogen electrode (SHE) scale, particularly for elevated-temperature tests, for which the conversion factors to $V_{(SHE)}$ are not widely known. However, when comparing results as a function of temperature, it may be helpful to eliminate the contribution of the standard hydrogen cell, because like other reactions, it has a potential that varies with temperature. It is by convention that the standard hydrogen cell is 0 V at any temperature; relative to the standard hydrogen reaction at 25 °C (75 °F), the potential of the standard hydrogen reaction at 50 °C (120 °F) is 0.021 V, 0.057 V at 100 °C (212 °F), 0.086 V at 150 °C (300 °F), and about 0.105 V between 210 and 300 °C (410 and 570 °F), as described in Ref 92.

Application of imposed potential using a potentiostat requires electrical isolation of the specimen. In some cases, it may be difficult to insulate the specimen from the loading linkage; instead, the linkage must be insulated from the autoclave, and the current flow cannot be attributed only to reactions at the specimen. Ground loops present perpetual problems, because most potentiostats are designed to hold the specimen at ground (or virtual ground). With necessary mechanical and plumbing connections, the autoclave is connected to ground; thus, the specimen is effectively connected to the autoclave. The problem is compounded if the autoclave is used as the counterelectrode, because the ground loop shorts out the potentiostat. Options include thorough electrical isolation of the autoclave from ground and use of a fully floating potentiostat.

Attachment of a lead to the specimen to permit measurement or application of potential can be a challenge in aggressive environments. Recommendations include use of wire that is either identical to the specimen or a very noble metal, such as platinum. Attachment using a weld bead (e.g., by gas tungsten arc welding) usually is superior to spot welding. Covering the lead wire with heat-shrink Teflon and, at low test temperatures, covering the weld with an organic "stop-off" coating helps maintain a good connection and minimizes the effects of the wire (via galvanic coupling or its contribution to the measured current). Another technique involves the use of a commercially available plasma-sprayed insulating coating, which can be used in high-temperature water.

Errors in the potential applied by a potentiostat can occur in solutions of low conductivity. These current resistance drops, which result from current flow between the counter and working electrodes in the high-resistance solution, are detected by the reference electrode and summed with the electrode potential of the specimen. Electronic compensation is possible, but not straightforward, in high-temperature water. Partial compensation is possible by placing the reference electrode near the specimen, although for small specimens the measured potential becomes very sensitive to electrode positioning. Most chloride solutions possess sufficient conductivity; a rough estimate of the possible error can be made by multiplying the resistivity of the solution (preferably determined by measuring the alternating current flowing between the counter and working electrodes when a 1000-Hz alternating-current voltage is applied) by the potentiostat current that flows during a test.

Test Specimens

The compact tension specimen is widely used because of its prevalance, level of standardization, well-defined fracture mechanics relationships, and availability of high-resolution crack-monitoring techniques. It is not, however, an ideal specimen for studying crack initiation or short crack behavior, and its open sides (in contrast to specimens with thumbnail-type cracks) may poorly simulate in-service cracks in some instances. In high-temperature tests, the size and bulk of the specimen and associated linkage make autoclave design more complex, particularly in highly corrosive environments.

Alternative designs have drawbacks as well. Cylindrical tensile specimens provide only simple load versus cycles-to-failure data, which may prove adequate or necessary under some circumstances. Cylindrical or plate specimens with notches or small, semicircular defects can be used to study initiation and short crack behavior, particularly when configured with a high-resolution crack growth monitor (Ref 93).

Monitoring Crack Length

Automated high-resolution techniques for determining crack length have made obsolete most optical determinations, which are im-practical in high-temperature water. Compliance techniques (Ref 94) require relatively bulky attachments to the specimen, complicate electrical isolation and ground loop problems, and can be difficult to successfully implement in a corrosive, high-temperature environment. The required unloading cycle can affect constant-load and high-R tests.

Electric potential techniques have flourished, with the direct-current approaches generally favored for overall simplicity. A reversed direct-current technique has been developed that essentially eliminates errors due to thermocouple effects, amplifier offset, and temperature changes. The current requirements have decreased (only a few amps are generally required) as a result of improved accuracy of low-level voltage measurement. The potential drop between crack sides is typically 10 to 100 μV, and studies in several systems have found no effect of this technique. Attachments to the specimen are limited to two current-carrying leads and two to four potential-measuring leads. Techniques for measuring crack growth rate are discussed in the article "Fatigue Crack Propagation" in this Section.

Post-Test Analysis

Analysis of the fracture surface is crucial to ensure accuracy of the crack-monitoring technique, to identify branching and out-of-plane cracking, and to determine crack morphology. Because pits can act as nucleation sites for cracking in solutions containing chloride, possible interaction between pitting and fatigue behavior should also be examined (Ref 95). Accurate determination of crack growth on a cycle or time basis requires an understanding of the resolution of the monitoring technique under the actual test conditions. Use of standards such as ASTM E 647 (Ref 41) helps ensure a uniform approach to data analysis and statistical validity.

High-Temperature Pure Water: Aerated Conditions

By F. Peter Ford
Manager,
Surfaces and Reactions Unit
General Electric Company

INTERGRANULAR STRESS-CORROSION CRACKING in weld-sensitized or furnace-sensitized austenitic stainless steel has long been recognized in high-purity oxygen-ated water at elevated temperatures (Ref 96, 97). Subsequently, laboratory investigations have been extended to cover corrosion-fatigue cracking over a variety of stress/time conditions, and other similar environmentally controlled cracking studies have been made in other structural alloys, such as carbon steel (Ref 98) and low-alloy steels (Ref 99).

The environmental conditions of high-purity oxygenated water present experimental problems, particularly when small changes in impurity and oxygen content can have significant effects on cracking susceptibility (Ref 85, 100-102). For example, the degree of water purity required to simulate conditions in boiling-water reactors is:

Conductivity	0.2-0.5 μS · cm^{-1}
Chloride content	0.2-0.5 ppm
pH	6.3-7.0
Oxygen content	0.2-0.3 ppm
Hydrogen content	0.025-0.035 ppm
Temperature	288 °C (550 °F)

These data span typical reactor water chemistry conditions at 288 °C (550 °F). The experimental actions necessary to achieve such water quality control at 288 °C (550 °F) are discussed below.

Environmental Containment System

Because of the specific requirement of high-temperature (e.g, 250 to 300 °C, or 480 to 570 °F) high-purity water conditions, the containment autoclaves must be manufactured from corrosion-resistant alloys. Austenitic stainless steel and titanium- or nickel-base alloys have been used most commonly and are manufactured to the ASME VIII Boiler Code specifications. The use of low-alloy steel (e.g., A533B) autoclaves is usually avoided, because the general corrosion of such steels leads to an uncontrolled decrease in the dissolved oxygen content.

Even when constructed of a more corrosion-resistant material, however, an autoclave must be conditioned for several weeks in oxygenated water at 250 to 300 °C (480 to 570 °F) to form a stable surface oxide. This minimizes subsequent loss of oxygen from the water.

Additional steps to prevent oxygen loss involve introducing the oxygenated water directly onto the specimen surface rather than allowing a long residence time in the bulk of the autoclave (Ref 85). Use of miniature autoclaves, in which only the gage section of the specimen is exposed to the water, also minimizes the autoclave surface area and permits tighter control of the environment (Ref 103).

Sliding seals are needed to introduce the loading train to the specimen. In addition to the potential mechanical problems these devices may introduce, such as sticking or slipping, the possibility of environmental contamination exists. Seals generally are designed around modified high-pressure fittings with chemically inert Teflon inserts to produce a low-friction seal (Ref 103). These seals must be cooled, however, and inserts that contain a combination of Teflon and graphite produce superior frictional and mechanical properties at temperatures up to 300 °C (570 °F). Such seals also serve to electrically isolate the specimen and loading train from the autoclaves, which is an important consideration in galvanic coupling for dissimilar metals in the system. Alternatively, the specimen may be insulated with ceramic zirconium oxide inserts at the loading clevis.

Figure 24 illustrates the pressurization loop that is required to maintain single-phase conditions within the autoclave. To achieve this at 288 °C (550 °F), pressures greater than 7200 kPa (1047 psi) are required. These may be accomplished via a high-pressure positive-displacement pump, whose liquid-contacting surface is made from stainless steel, titanium, or Teflon. The pump capacity per minute should be at least 1% of the volume of the autoclave (Ref 104, 105), because the flow rate through the autoclave should be sufficient to counteract any loss of oxygen in the water due to corrosion of the autoclave. The precise critical flow rate to achieve this depends on the autoclave system (its volume/surface area ratio, materials, etc.) and must be checked by monitoring the inlet and outlet oxygen contents.

The piping in the system illustrated in Fig. 24 may be either stainless steel or titanium to maintain water purity, with regenerative or coaxial heat exchanges being placed immediately prior to the autoclave. This latter action minimizes the heating time (and thereby the time available for oxygen to be removed from the water by corrosion of the piping) before the water enters the electrically heated autoclave. To monitor water chemistry purity, a "once through" system is preferred—the water is discarded once it passes through the autoclave and analytical equipment (Fig. 24).

Water Purification and Treatment

Distilled water, which may have a conductivity in the range 1 to 10 $\mu S \cdot cm^{-1}$ depending on the supply, is further purified to ~0.1 to 0.2 $\mu S \cdot cm^{-1}$ by passing through ion-exchange columns that remove soluble impurities. For the neutral conditions required, the cation resin is in the H^+ (hydrogen) form and anion resin is in the OH^- (hydroxyl) form. Purified water is stored in a large stainless steel or plastic holding tank. This stock solution can then be deaerated by purging with argon gas, for example. Alternatively, this deoxygenating action can be achieved by an anion-exchange resin bed (e.g., sulfite) prior to processing through the purification beds.

Oxygenation to the desired level can be accomplished by equilibrating the deaerated solution with an appropriate inert gas/oxygen mixture. The precise mixture depends on the oxygen solubility/temperature relationship; for example, an argon/0.5% oxygen mixture at 20 °C (68 °F) yields 200 ppb of dissolved oxygen in the high-purity water. Oxygen is preferred over air in such an equilibrating gas mixture because of the possible deleterious role of dissolved carbon dioxide from this source on the cracking resistance of stainless steels (Ref 106, 107) and low-alloy steels (Ref 108) in high-temperature water. Controlled amounts of impurities can be added to the system, if required, either by direct addition to the feedwater reservoir or at a controlled rate to the feedwater entering the autoclave. By using the latter technique, the effect of transients in water purity on cracking susceptibility can be measured *in situ*, as shown in Fig. 25.

Monitoring Procedures

Dissolved oxygen content (Ref 96, 98-100), conductivity (Ref 85, 102, 106), temperature (Ref 98, 100), and stress state (Ref 101) are prime variables that can affect the cracking susceptibility of various structural materials in water at elevated temperatures. Consequently, all must be monitored, preferably continuously.

Conductivity at ambient temperature must be monitored to ensure correct solution purity. Such measurements should be made not only on the inlet water, but also on the exit water, because it has been observed that significant amounts of impurities can be generated by corrosion of the autoclave material. Analytical techniques such as atomic absorption, ion chromatography, or ion-specific electrodes can determine the species giving rise to the conductivity changes, should that information be required. However, the oxidation state of these species can change between room-temperature conditions (where the measurement is made) and the elevated-temperature conditions in the autoclave.

Stress. Small vibratory stresses on top of the main stress/time waveform may have an

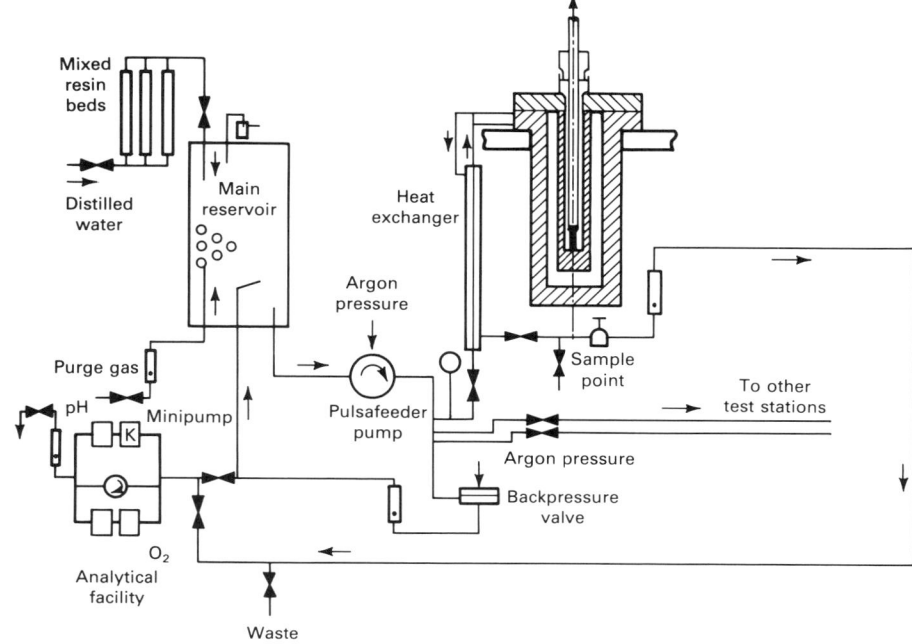

Fig. 24 Typical pressurized loop for supplying high-purity, high-pressure water to autoclaves for environmentally controlled cracking studies

Fig. 25 Variation in crack growth rate in stainless steel/water system due to transients in conductivity

Compact tension specimen of sensitized 304 stainless steel. Dissolved oxygen content: 0.2 ppm. Temperature: 288 °C (550 °F). Frequency: 0.01 Hz. Stress ratio: R = 0.9. Source: Ref 85

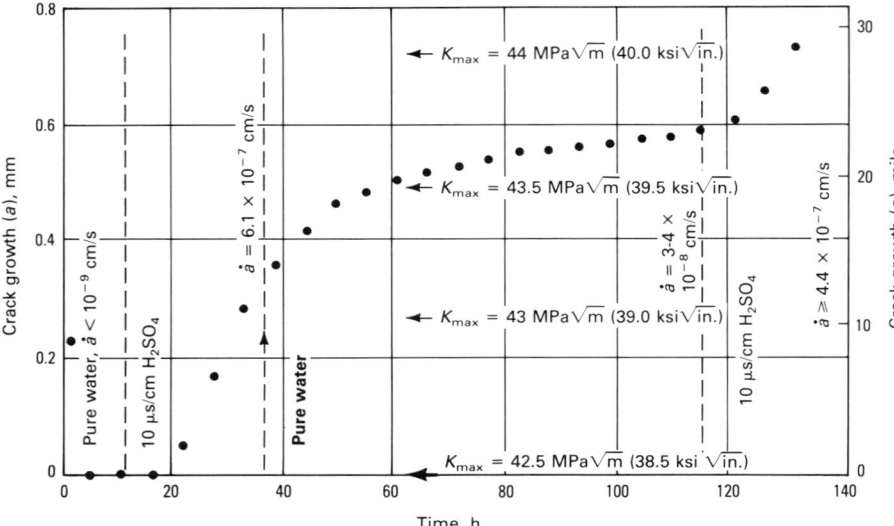

effect on the cracking response (Ref 109). In autoclave systems, such vibratory stresses can arise due to pressure oscillations in an inadequately dampened pumping system.

Flow Rate. An adequate flow through the autoclave is required to maintain the dissolved oxygen content in the autoclave water. However, it has been found that flow rate in itself can affect cracking susceptibility in high-temperature water (Ref 110). It is currently uncertain whether this is due to the effect of laminar/turbulent transitions on the mass transport of species in cracks, or whether it is due to changes in the diffusion-limited reduction rate of dissolved oxygen.

Dissolved oxygen content should be measured at both inlet and outlet lines to the autoclave to ensure that a minimal amount of dissolved oxygen is being removed by general corrosion. The dissolved oxygen content, in conjunction with the anionic impurity level, can significantly accelerate environmentally controlled crack growth rates in ductile alloy/aqueous environment systems at 288 °C (550 °F) (Ref 85, 96-103).

Corrosion Potential. Other oxidizing species, in addition to dissolved oxygen, may affect the cracking susceptibility in high-temperature solutions of a given conductivity. A normalized measurement of such oxidizing power of the solution is the corrosion potential of the specimen and the autoclave (Ref 111, 112). Reference electrodes for use in water at 250 to 300 °C (480 to 570 °F) have been developed and may be manufactured in the laboratory.

The electrodes most commonly used are based on the silver/silver chloride system (Ref 91, 103, 111, 113-114) and the stabilized zirconia system (Ref 115). It is strongly recommended that the corrosion potential be monitored. In many cases, this clarifies the unexplained scatter in crack growth rate data.

Other Parameters. Although crack propagation rates at a given corrosion potential in relatively high-purity water can be correlated with solution conductivity, the initiation process seems to correlate with the solution pH (Ref 85, 116). Thus, the use of commercially available pH electrodes on the inlet water at room temperature may be desirable. Alternatively, stabilized zirconia electrodes can be used to monitor the pH at 288 °C (550 °F), as described in Ref 117.

The cracking susceptibility of many structural alloys can be decreased markedly by decreasing oxygen content to a level that is dependent on the solution conductivity. In practice, this can be accomplished by introducing hydrogen into the feedwater. Thus, from an experimental (and practical) standpoint, the dissolved hydrogen content must be measured (Ref 118).

Test Specimens and Crack Measurement Techniques

Many standard specimen configurations, such as smooth tensile and compact tension specimens, have been used successfully for high-temperature aerated water testing. Side grooves on compact tension specimens are desirable, particularly under high-R low-frequency conditions, where crack branching may be favored along with the transition from transgranular to intergranular cracking. Different crack propagation rates can be obtained on specimens with short crack lengths compared with conventional fracture mechanics specimens at the same nominal stress-intensity value (Ref 93). It is unclear whether this specific phenomenon is confined to high-temperature aerated water environments, however.

Standard techniques of crack measurement using, for example, linear variable differential transformer compliance techniques have been used successfully. In these cases, the linear variable differential transformer has been sheathed in a corrosion-resistant alloy. Other techniques involving electric potential methods have been developed that have superior crack advance resolution (i.e., $\Delta a \sim 2.5$ μm, or 0.1 mil) in high-temperature aqueous environment systems.

Finally, direct optical observation of crack propagation is possible by use of sapphire windows in the side of the autoclave (Ref 119). This is not a commonly used technique, however, and it does not offer the accuracy of electric potential techniques, which are described in the article "Fatigue Crack Propagation" in this Section.

High-Temperature Pure Water: Deaerated Conditions

By W.A. Van Der Sluys
Technical Advisor
Babcock and Wilcox

CYCLIC CRACK GROWTH studies in simulated pressurized-water reactor environment conditions require technical expertise in several areas. The three major technical areas in which the investigator must be knowledgeable are fracture mechanics, water chemistry, and corrosion technology. If experiments are to be conducted under computer control and computer data acquisition, some knowledge is also needed in these areas.

Inadequate knowledge in one or more of these technical areas can result in erroneous data generated in the experiments. Of the three areas, it is easiest to acquire adequate knowledge in fracture mechanics from

ASTM E 647, Ref 41, which can be used as a guide for conducting the fatigue experiments. Similar sources of information for the water chemistry and corrosion areas do not exist. For this reason, more detailed discussion of the water chemistry and corrosion is included in this article.

Pressurized Water Reactor Water Chemistry

Because of the manner in which a pressurized water reactor is operated, no single typical water chemistry exists. Boric acid is added to the primary coolant in a pressurized water reactor as a chemical shim to aid in the control of the reactivity of the core. A large concentration is required early in the fuel cycle, with the concentration gradually decreasing with time. Figure 26 illustrates the relationship of typical boric acid concentration versus time for a pressurized water reactor. Because the pH of the reactor coolant would be very low at the higher boric acid concentrations, lithium hydroxide is added to the water for pH control. Figure 27 illustrates the relationship of pH versus boric acid concentration for pressurized water reactor water at 25 °C (75 °F) and 315 °C (600 °F). The pH at 315 °C (600 °F) is close to neutral due to the dissociation of the boric acid at 315 °C (600 °F).

Table 1 gives the typical water chemistry specifications for pressurized water reactor water, as well as the specifications for the water chemistry in a typical corrosion-fatigue project. The water temperature in the pressurized water reactor system depends on location within the system and particular reactor type. The temperature can vary from 288 to 330 °C (550 to 625 °F). The water pressure varies between 13.8 and 17.2 MPa (2000 and 2500 psi).

Simulating Pressurized Water Reactor Water

The steps taken in producing simulated pressurized water reactor water in the laboratory are shown schematically in Fig. 28. Initially, the water is obtained from a deep well. This water is treated with a sodium zeolite softener to remove "hardness" ions (calcium and magnesium). The water is then filtered using 0.13- to 0.019-μm filters before further removal of particulates and dissolved solids with a reverse osmosis unit.

The water from this unit is then stored until needed. Upon demand, the stored water is processed through separate anion- and cation-exchange columns, followed by two mixed bed ion-exchange columns. The ion-

Fig. 26 Boric acid concentration versus time for a pressurized water reactor coolant

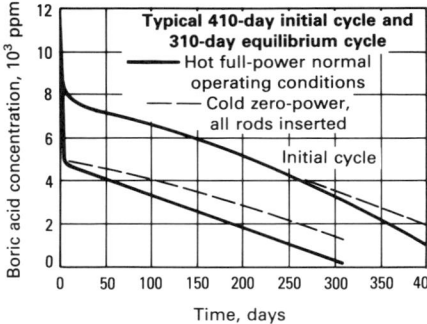

exchange columns remove ionic dissolved solids, and the electric conductivity of the purified water is less than or equal to 0.10 μS cm⁻¹.

Water used for corrosion-fatigue tests is further purified and deoxygenated by passage through a mixed bed ion-exchange column and a strong base resin deoxygenating column and a final mixed bed ion-exchange column. The sulfite or resin in the strong base resin column removes dissolved oxygen by chemical conversion of the resin from the sulfite form ($R\text{-}SO_3$) to the sulfate form ($R\text{-}SO_4$). Any leakage of sulfite or sulfate ions from the strong base resin columns is removed by the final mixed bed ion-exchange column. Water is recirculated continuously from the outlet of the final ion-exchange column to the inlet of the strong base resin column.

Fig. 27 pH versus boric acid concentration for a pressurized water reactor coolant

(a) At 25 °C (75 °F). (b) At 315 °C (600 °F)

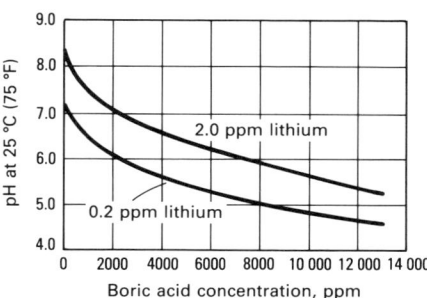

(a)

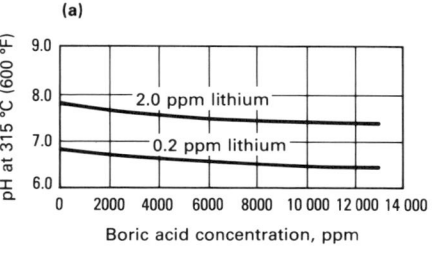

(b)

The pressurized water reactor water is prepared in a feedwater tank. The tank is purged with nitrogen and filled under nitrogen pressure with the high-purity deoxygenated water. Chemicals, boric acid, lithium hydroxide, and hydrogen are mixed with the deaerated, high-purity water in the feedwater

Table 1 Corrosion fatigue crack growth testing in a pressurized-water reactor water chemistry environment

	Typical pressurized water reactor specification	EPRI corrosion fatigue test specification	Desired concentration
Boric acid, ppm	100-13 000	2000-8000(a)	5700
Lithium, ppm	0.2-2.0	0.2-2.0(b)	2.0
pH	4.6-8.5	4.6-8.5(c, d)	...
Specific conductivity	...	<20 μS · cm⁻¹(e)	...
Chloride, ppm	<0.10	<0.10(g)	...
Fluoride, ppm	<0.10	<0.10(g)	...
Dissolved oxygen, ppm	<0.1(f)	<0.01(g, h)	...
Dissolved hydrogen std cm³/kg H₂O	15-40	15-50	...
Total dissolved gas, std cm³/kg H₂O	<100	...	<100
Silica, ppm	...	(g)	<0.020
Sodium, ppm	...	(g)	<0.010
Iron, ppm	...	(j)	<0.010
Copper, ppm	...	(j)	<0.010
Total organic carbon	...	(g)	...

(a) Control at 3000 ppm H₃BO₃ for specimen numbers 2T03, 2T05, 2T06, IHT4, and IHT80 and control at 5700 ppm H₃BO₃ for the remainder of the test specimens. (b) Control at 2.0 ppm lithium using lithium hydroxide. (c) Within the specified range, pH will vary with boric acid and lithium hydroxide concentrations. (d) Materials selected for testing should be compatible with this pH range. (e) Boric acid and lithium hydroxide concentrations specified yield conductivity of <20 S · cm⁻¹. (f) O₂ must be <0.1 ppm before heating reactor coolant above 120 °C (250 °F). (g) Harmful contaminant. (h) Oxygen concentration should be reduced to <0.01 ppm before startup of the autoclave. The addition of hydrazine (N₂H₄) to the autoclave during startup is an acceptable method for oxygen reduction. (j) Indicates corrosion of wetted surfaces, including dissolved iron from corrosion of carbon steel specimens

tanks. Water is continuously recirculated from the bottom of each tank and sprayed through the gas at the top of the tank to facilitate mixing of the chemicals with the water.

Hydrogen is dissolved in the water to the desired concentration by controlling the temperature and partial pressure of the hydrogen in the water in each tank. Saturated boric acid solution (48 000 ppm) and saturated lithium hydroxide solution (38 000 ppm) are added to the feedwater tanks in appropriate quantities to achieve the desired boric acid and lithium concentrations. Saturated lithium hydroxide solution is added to the feedwater tank through a septum with a syringe. Saturated boric acid solution is added to the feedwater tank from a boric acid feeder tank.

This feeder tank is used to prepare high-purity deoxygenated water saturated with boric acid. A saturated solution of boric acid is prepared by passing high-purity deoxygenated water through an excess of crystalline boric acid. This boric acid feeder tank is charged with crystalline boric acid, purged with nitrogen, and filled with high-purity deoxygenated water.

Care must be taken to obtain lithium hydroxide and boric acid of high enough purity so as not to contaminate the high-purity water with some undesirable chemical species. These chemicals are often contaminated with undesirable sulfur compounds. The water

from the feedwater tank then flows through the autoclave. The rate of flow depends on the conditions desired in the fatigue experiment. The chemical composition of the effluent water should be monitored periodically. Table 2 gives a typical schedule and chemical species analysis.

In the facility described above, the effluent from the autoclaves is not reused, in order to prevent buildup of contaminants in the test solutions. Some investigators reuse the effluent water. Mixed bed demineralizers are available, which can be used to clean the effluent water. Care must be exercised when using these solutions so that the water is not contaminated with the decomposition products from the demineralizers. The demineralizers must be operated at the correct cross-sectional flow rate to avoid contaminating the water. Furthermore, after a period of stagnation, the demineralizers must be purged to remove contaminants that have sloughed from the resins into the nonflowing water within the demineralizers.

Autoclave Construction

An autoclave is necessary to contain the simulated pressurized water reactor water environment. The pressurized water reactor system typically is operated at approximately 15.2 MPa (2200 psi), and temperatures vary

from 288 to 340 °C (550 to 645 °F), depending on the location in the system. A variety of materials can be used for the construction of the autoclave. Typical materials include types 316 and 347 stainless steels, ASTM A508 and A533 carbon steels, and titanium. The use of an adequate safety code such as Section VIII of the *ASME Boiler and Pressure Vessel Code*, the corrosion resistance of the materials, and the method of transmitting the load through the autoclave must be considered in constructing an autoclave.

Autoclave designs vary widely from conventional types with round cross sections to those with square or rectangular cross sections. Loading rod seal assemblies are located on either the top or bottom of the autoclave and usually in the closure head. The bottom seal location offers an advantage. High-temperature seals are hard to obtain, and the lower the temperature of the seal assembly, the longer the seal life. The bottom of the autoclave usually is cooler than the top, and forced cooling of the seal is accomplished more easily on the bottom.

In the design of the autoclave and specimen loading system, the specimen should be isolated electrically from the autoclave. The electrode potential of the specimens in the environment is an important consideration in any corrosion fatigue testing.

Fig. 28 Water purification and chemical addition systems for producing a simulated pressurized water reactor environment

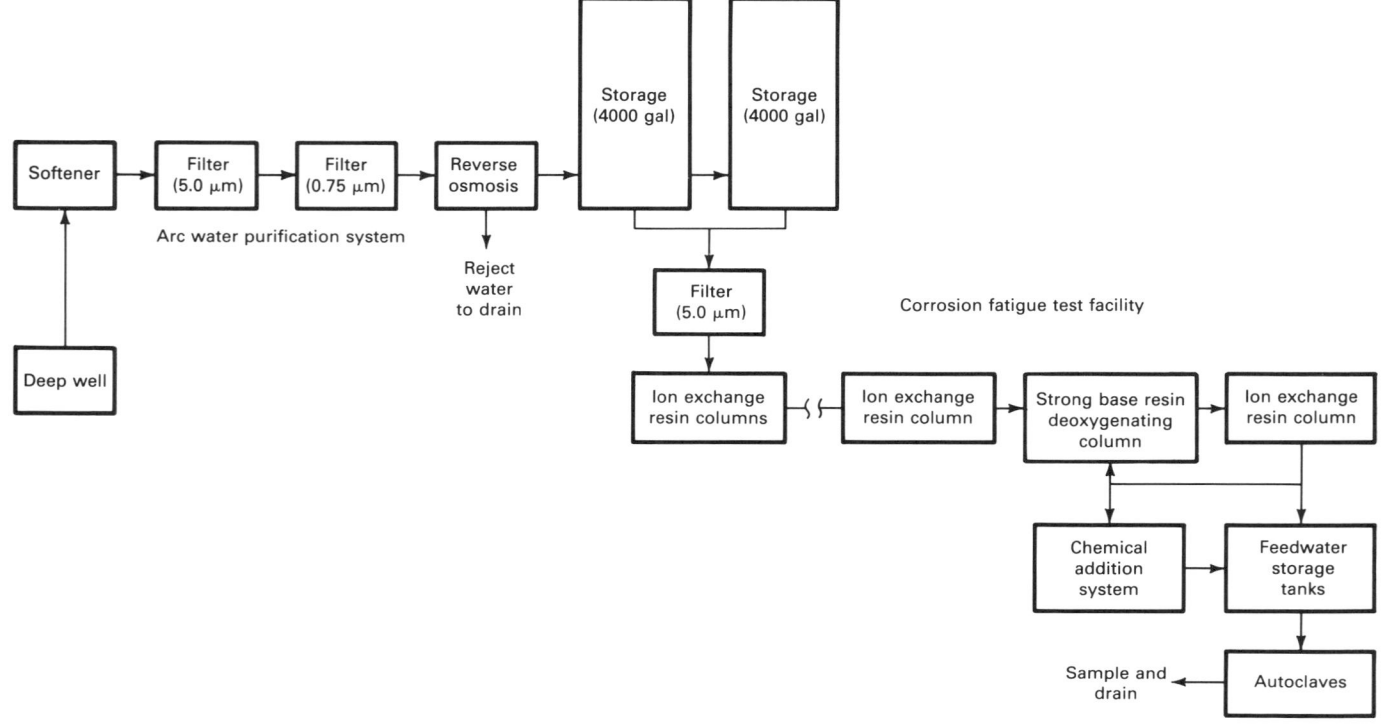

Table 2 Corrosion fatigue crack growth testing: initial analysis frequency for single autoclave operation(a)

	Feedwater tank	Autoclave effluent
Boric acid(b)	P/F	AD
Lithium(b)	P/F	AD
pH	P/F/W	D
Specific conductivity	P/F/W	D
Chloride	P/F	AD
Fluoride	P/F	AD
Dissolved oxygen	P/F/W	P/AD
Dissolved hydrogen(b)	P/F/W	W
Suspended solids	F/W	D
Silica	W	W
Sodium	W	W
Iron	W	W
Copper	W	W
Total organic carbon	...	...

Note: P, analysis should be completed prior to placing the feedwater tank in service or heating the autoclave; F, analysis should be conducted each time tank is filled; W, weekly analysis required during test operation; D, daily analysis required during test operation; AD, analysis required every 48 to 72 h during test operation.
(a) After the characteristics of this system have been established, the analysis frequency will be revised. (b) To ensure proper mixing of the chemical additives (boric acid, lithium hydroxide, and hydrogen gas) with the water in the feedwater tank, deaerated high-purity water should be added slowly, and water should be recirculated from the bottom to the top of the feedwater tank. The recirculating pump should remain on during operation of the autoclave system.

Measurement of the corrosion potential in the high-temperature, high-pressure pressurized water reactor is difficult. Several techniques have been applied successfully in a pressurized water reactor environment, but they are extremely technique dependent. External silver/silver chloride reference electrodes are used most often (Ref 86).

Testing Procedures

The procedures for obtaining the fatigue crack growth rates in a pressurized water reactor do not differ significantly from those recommended in ASTM E 647 (Ref 41). The remote measurement of crack length rather than the use of optical methods is the most difficult experimental technique to be developed. Most investigators use the compact fracture type of specimen, as described in ASTM E 647 and ASTM E 399 (Ref 64). The compliance method is the most widely used technique of remote crack length measurement. The electric potential technique is also widely used. References 120 and 121 discuss the compliance method of crack length measurement and the anticipated accuracies. A description of one approach to the use of electrical potential monitoring for crack length measurement can be found in Ref 86.

Most fatigue crack growth experiments are conducted using the constant-load amplitude approach described in ASTM E 647. An alternate approach, in which the stress-intensity factor amplitude is held constant, can be found in Ref 121. This approach is useful when the effect of a large number of variables is being investigated. A single specimen can be used to scan several variables and assess the influence of individual variables on the crack growth rate.

Liquid Metal Environments

By Lee A. James
Fellow Engineer
Westinghouse Hanford Company

LIQUID METALS (sodium, potassium, and lithium, for example) are frequently used in heat-transport applications at elevated temperatures. Such applications include liquid-metal-cooled nuclear reactors, first-wall coolant for fusion devices, and heat-transport systems in solar collectors. These applications often involve cyclic temperature and/or pressure fluctuations, as well as other sources of cyclic stresses. For this reason, knowledge of the fatigue crack propagation behavior of structural alloys in the liquid metal environments is sometimes necessary. The environments discussed in this article are relatively benign, and test techniques for the classic liquid-metal embrittlement phenomenon (e.g., brass or aluminum alloys in mercury) are not discussed.

System Design

Generally, liquid metals react (in some cases, quite violently) with air and/or water vapor; therefore, testing systems must be designed to exclude both air and water. Three basic designs have been developed to expose the specimen (or crack region of a specimen) to the liquid metal environment, while excluding air, water, and other contaminants.

The simplest method uses a sealed environmental chamber attached to the specimen that completely surrounds the notch and crack extension plane in a compact-type specimen. For a description of compact-type specimens, see ASTM E 647 (Ref 41) and the article "Fatigue Crack Propagation" in this Section. The small environmental chamber contains the liquid metal, but does not extend to the region of the loading holes; hence, the loading pins, clevis grips, and remainder of the load train are not subjected to the liquid metal environment.

Relative motion across the notch and crack area is accommodated by bellows. This type of system, which is described in Ref 122 and 123, has the advantages of simplicity and low cost. The main disadvantage is that the liquid metal is static; hence, the characteristics of large heat-transport systems (e.g., mass transport due to nonisothermal operation) cannot be studied.

The second type of system, a circulating loop, is much more costly to build and operate, but can be used to study potential effects on fatigue crack propagation such as mass transport, which occurs during carburizing, decarburizing, and dissolution of alloying elements. A typical loop system and the environmental test chamber installed in a loop system are shown in Fig. 29. Note that the system in Fig. 29(b) resembles conventional aqueous environment autoclaves. Both compact-type specimens (Ref 124) and three-point bend specimens (Ref 125) have been used in circulating loop systems.

The third type of system consists of an open crucible (containing the test specimen immersed in static liquid metal) that is located within an inert gas cell or glovebox. This type of system, which is described in Ref 126 and 127, is also relatively inexpensive to build and operate, but it has the greatest potential for exposure to air and other contaminants.

Austenitic stainless steels generally have been used in the construction of current systems, and their use has proved satisfactory. System designers should consider, however, that under some conditions mechanical properties (tensile, stress rupture, etc.) can be influenced by long-term exposure to liquid metals (Ref 128, 129).

Characterization and Control of Environment

As mentioned above, the liquid metal purity in a sealed static system can only be adjusted at the start of a test; no control is possible during the test. By contrast, in loop systems, liquid metal purity can be characterized and controlled during a test. One of the most effective means of controlling the liquid metal purity is with the use of a cold trap

Fig. 29(a) Schematic of typical circulating loop for fatigue crack growth testing in a liquid metal environment

Trace heaters, thermocouples, etc., installed on loop as required to achieve desired temperature distributions. Source: Ref 124

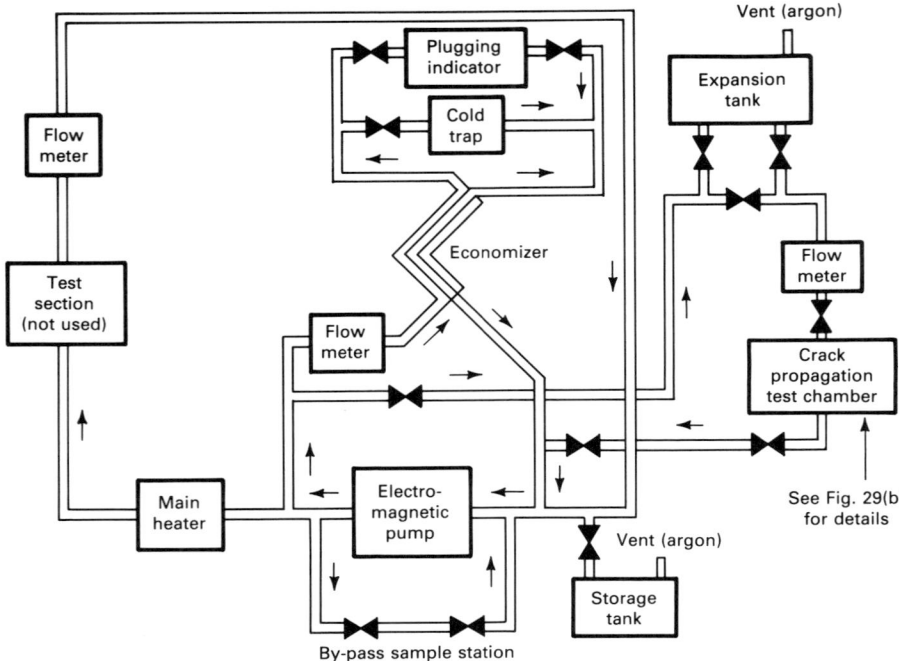

(a)

(Fig. 29a), whereby impurities are precipitated out upon the cold trap surfaces.

Several methods are available whereby impurities in liquid metals can either be measured directly or inferred. One of the simplest devices is the plugging indicator (Fig. 29a), which allows the level of impurities (primarily oxygen) to be inferred by successively lowering the temperature in the vicinity of the plugging meter orifice until the liquid metal freezes. The temperature at which flow decreases or stops (i.e., the freezing temperature) can, through proper calibration, be related to the impurity content (Ref 130, 131).

Another relatively simple method of analyzing liquid metal purity is through the use of a by-pass sample station (Fig. 29a). A representative sample of the liquid metal can be isolated using valves, frozen, and then removed and sent to the laboratory for characterization.

Electrochemical methods can also be used for on-line characterization of oxygen (Ref 132, 133), hydrogen (Ref 133, 134), and carbon (Ref 133). Reference 135 also discusses electrochemical methods.

Loading Systems

Feedback-controlled servohydraulic loading systems have been used on all current systems. In the loop-type systems (Fig. 29), bellows have been used to assist load train penetration into the chamber. The load cells have been located external to the chamber; hence, it is necessary to account for the bellows spring rate as well as the pressure differential between the inside and outside acting upon the projected area of the bellows. Almost any specimen design can be used, but currently only compact-type and three-point bend designs have been used. Additional information on servohydraulic fatigue test systems and specimen designs can be found in the article "Fatigue Crack Propagation" in this Section.

Crack Length Measurement

Visual observation of the crack for purposes of length measurement is clearly impossible. In addition, the excellent electrical conduction properties of liquid metals make it impossible to use electric potential techniques for crack length measurements. Three methods have been used successfully.

The simplest method is to grow the crack a relatively short distance in a single increment and remove the specimen for direct observation (Ref 125). This is relatively inefficient, because the specimens should not be returned to the liquid metal for further testing.

A second method grows relatively short increments of crack extension at differing stress ratios, thereby producing "beach marks" that can be examined at the conclusion of testing (Ref 122, 124, 136, 137). Finally, measurements of specimen elastic compliance can be made periodically throughout each test, and the crack length can be inferred from such measurements (Ref 123, 138).

Fatigue Crack Growth Behavior

Several variables influence fatigue crack propagation behavior in liquid metal environments. One of the most significant parameters is the medium itself; fatigue crack propagation behavior in liquid sodium (Ref 124) appears to vary from that in liquid lithium (Ref 127). Other possibilities include impurities such as oxygen, hydrogen, hydroxides, and halides. Some of these impurities have been the subject of research efforts (Ref 124, 136, 138). In addition, it has been shown that, under some conditions, fatigue crack propagation in liquid sodium varies under carburizing and decarburizing conditions (Ref 125, 136, 138).

Other potential variables that could influence crack growth include temperature, cyclic frequency, loading waveform, and stress ratio. Typical results for fatigue crack propagation behavior of austenitic stainless steels in a liquid sodium environment are shown in Fig. 30. In most cases, fatigue crack propagation rates are lower in sodium environments than in elevated-temperature air environments. The relatively benign nature of sodium environments also leaves the fracture faces in excellent condition for viewing with optical microscopes, scanning electron microscopes, or transmission electron microscopes (Ref 137).

Steam or Boiling Water With Contaminants

By Peter K. Liaw
Senior Engineer
Westinghouse R&D Center

CORROSIVE ENVIRONMENTS, such as steam or boiling water with contaminants, come in contact with many structural components. To assess the structural integrity of machine hardware, development of fatigue crack propagation rate results in the environ-

Fig. 29(b) Schematic of a typical liquid metal environmental chamber installed on the circulating loop system shown in Fig. 29(a)
Source: Ref 124

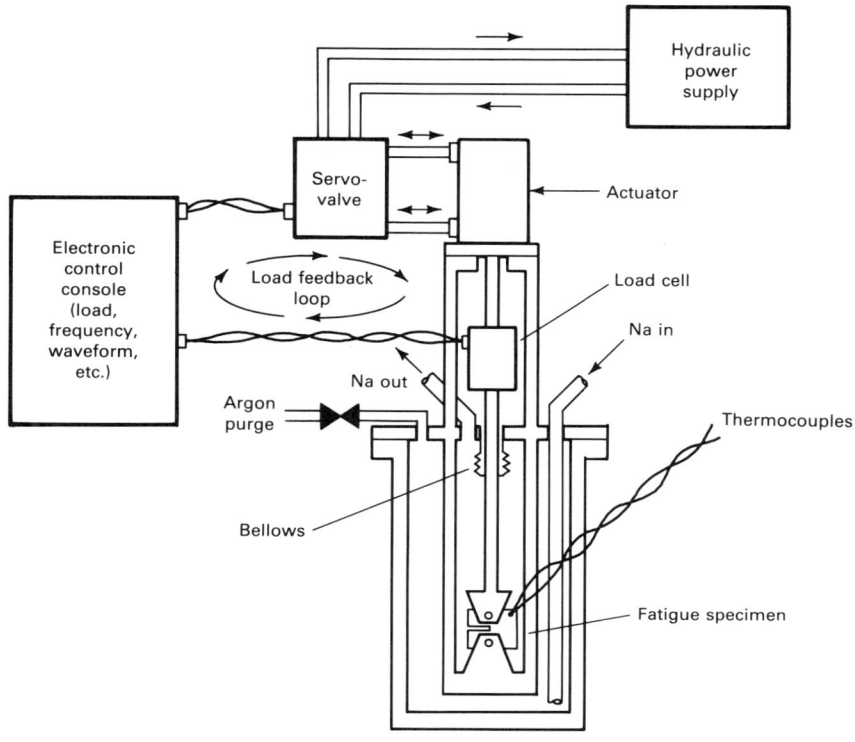

(b)

ments of concern is essential. Fatigue crack growth testing in corrosive environments requires special care because of the presence of corrosive mediums and testing complexity.

Test Methods and Materials for Environmental Containment

Special designs are required to accommodate fatigue crack growth testing in steam or boiling water with contaminants. If the environmental pressure and temperature are moderate, e.g., at a pressure of 500 kPa (72.5 psi) and a temperature of 100 °C (212 °F), simple stainless steel O-ring sealed chambers can be clamped to each side of the specimen in which cracking will occur. If necessary, the test environment can be circulated through the chamber at a controlled flow rate (Ref 139-143).

If the environmental pressure and temperature are high, e.g., in steam at a pressure of 7.2 MPa (1040 psi) and a temperature of 288 °C (550 °F), a chamber that encloses the test specimens must be constructed (Fig. 31). The test chamber in Fig. 31 can easily

accommodate fatigue crack propagation testing in steam at 7.2 MPa (1040 psi) and 288 °C (550 °F), or in pressurized water at 13.8 MPa (2000 psi) and 288 °C (550 °F) (Ref 144, 145). The chamber was fabricated from an ASTM A533 pressure vessel steel. The interior of the chamber was chrome plated to prevent corrosion. A chamber made from Inconel alloys is also suitable for use in chloride-containing steam or water environments.

Control of Variables

Composition of the test environment must be carefully analyzed before and after the experiment, because the composition can alter fatigue crack growth behavior (Ref 139, 140). For example, the value of threshold stress-intensity range (ΔK_{th}), which is typically defined as the stress-intensity range (ΔK) corresponding to a growth rate of 10^{-10} m/cycle 4×10^{-9} in./cycle) (Ref 146), varies with the concentrations of sodium chloride and sodium sulfate in boiling water for type 403 stainless steel and Ti-6Al-4V alloy (Fig. 32). In particular, the value of ΔK_{th} tends to increase with increasing sodium chloride in boiling water for both alloys.

Moreover, increasing the hydrazine (N_2H_4) content increases the resistance to near-threshold fatigue crack propagation in type 403 stainless steel (Fig. 33a). However, the effect of hydrazine on crack growth rate properties of Ti-6Al-4V alloy is minimal, as shown in Fig. 33(b).

Temperature should be monitored continuously and maintained at ±2 °C (±3.5 °F). Thermocouples should be attached to the test specimen near the crack plane to provide accurate temperature readings.

pH. The effect of pH on fatigue crack growth behavior has not been well characterized. Nevertheless, some results indicate that the rates of crack propagation are not influenced significantly by changing pH from 5 to 10 (Ref 71, 139, 140). Figures 34(a) and (b) illustrate the negligible effect of pH on fatigue crack propagation rates in type 403 stainless steel and Ti-6Al-4V alloy, respectively.

Dissolved Oxygen. Control and measurement of dissolved oxygen levels in the steam environment are of prime importance, because oxygen can affect fatigue crack propagation rate properties (Ref 140). Figure 35 demonstrates that increasing the oxygen content in steam increases crack growth rates in Ti-6Al-4V alloy.

Oxygen content can be controlled by bubbling argon or nitrogen through the water reservoir, or by maintaining a hydrogen overpressure. Oxygen content can be measured by using a colorimetric technique (Ref 139-141). Furthermore, oxygen levels can be monitored continuously by using oxygen analyzers that can achieve the resolution in the parts per billion range.

Test Specimens

Compact-type specimens usually are used to develop fatigue crack growth rate results (Ref 41). Three-point bend bars (Ref 147) and side-notched or center-notched specimens (Ref 41) can also be used, depending on the availability of test materials and the environmental setup, e.g., the design of the test chamber or the loading frame.

To maximize the effects of corrosive environments on crack propagation rates and to simulate operating conditions of machine components, test frequencies are often slow, such as 1 cycle/min (Ref 144, 145). At these slow frequencies, fatigue crack propagation testing may require 1 to 2 months. To save time and reduce cost, several fatigue crack growth experiments usually are conducted simultaneously by loading test specimens in series. An example is illustrated in Fig. 31, which shows two compact-type specimens loaded in series for steam or pressurized-water environment fatigue crack propagation testing.

Fig. 30 Typical fatigue crack propagation results for austenitic stainless steels tested in a liquid sodium environment at elevated temperatures

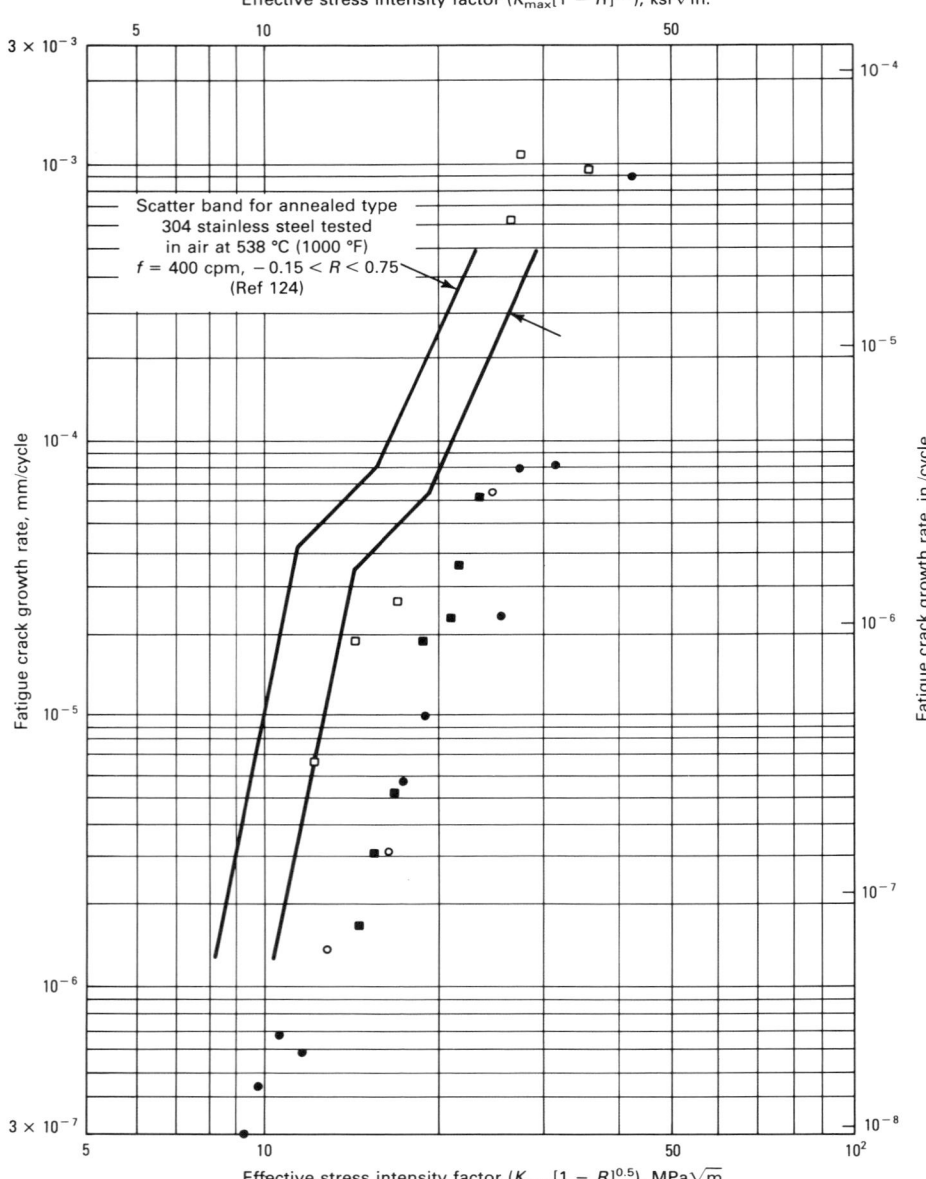

Stainless steel tested	Symbol	Environment	Temperature °C (°F)	Frequency, cpm	Stress ratio, R	Source
Annealed type 304	■	Sodium	538 (1000)	300	0.05 < R < 0.50	124
Annealed type 304	○	Vacuum	538 (1000)	400	0.086 < R < 0.50	124
Annealed type 316	●	Sodium	600 (1112)	600	0	122
Annealed type 316	□	High-carbon sodium	510 (950)	0.5	0.05	125

Monitoring Crack Length

In steam or boiling water with contaminants, it is difficult to conduct visual measurements of crack length. However, various techniques, such as Krak-Gages (Ref 148-152), compliance methods (Ref 94) using clip gages, extensometers (Ref 71, 139-152), linear variable differential transformers or eddy current gages (Ref 153-155), and electric potential (Ref 93, 156-159) can be used to determine crack extension in corrosion fatigue testing. Krak-Gages have been used successfully to measure crack length in corrosive environments (Ref 149, 150). Nevertheless, special Krak-Gages, such as sputtered Krak-Gages (Ref 150) must be developed for high-temperature (≥288 °C, or 550 °F) applications.

An extensometer can be mounted on the front face of a compact-type specimen to monitor crack opening displacements that are converted to crack lengths by using the compliance technique. Similarly, a linear variable differential transformer can be mounted on a compact-type specimen to monitor crack extension by using the compliance method (Fig. 31). The linear variable differential transformer is extended outside the chamber. Leakage around the seal between the linear variable differential transformer and the chamber must be prevented, because the moving rod in the linear variable differential transformer may cause difficulty in sealing the chamber.

An eddy current gage can also be mounted on the front face of the specimen to measure crack opening displacements (Ref 154, 155). The gage can be immersed directly in high-temperature steam or water environments. The lead wire for the gage can be sealed easily through the test chamber, because the wire is not a moving component. An experimental setup for fatigue crack growth testing using an eddy current gage is shown in Fig. 36.

Electric potential has also been used to monitor crack extension in corrosive environments (Ref 93, 157). In the presence of oxide deposits on fracture surfaces, crack closure often occurs, especially at low ΔK levels, because of oxide-wedging at the crack tip (Ref 160-163). The compliance technique provides a more effective means of monitoring crack closure than the electric potential method.

In long-term steam environment fatigue testing, data acquisition and analysis can be automated to provide a 24-h operation (Ref 139-144, 164, 165) without operator attendance. The 24-h operation in corrosion fatigue crack growth testing is particularly important, because any interruption during testing may cause unusual transient effects in the da/dN versus ΔK curve.

Application of Load

In high-pressure steam environments, test specimens frequently are enclosed in a chamber (Fig. 31). The pressure load exerted on the loading shaft should be included for calculating stress-intensity range. The applied loads during near-threshold fatigue crack propagation testing are small, particularly at threshold levels, and the pressure load may be a significant portion of the overall loading applied to the test specimen.

Fig. 31 Experimental setup for fatigue crack growth testing in high-pressure steam or pressurized-water environments
Source: Ref 144, 145

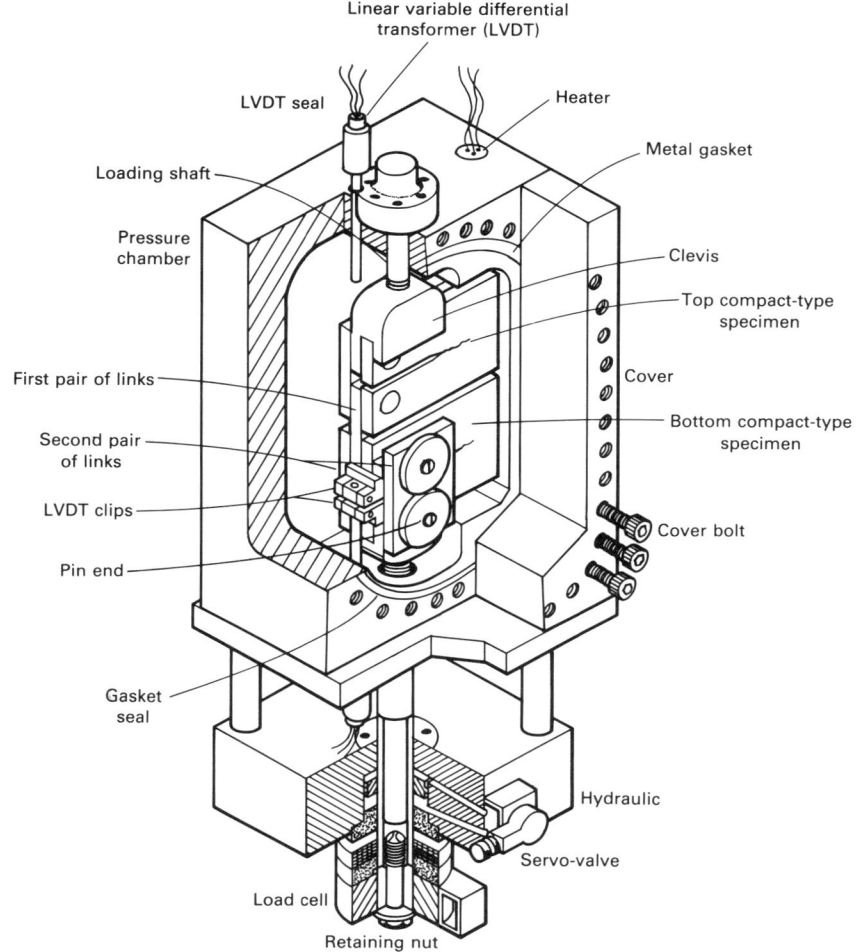

REFERENCES

1. R.P. Gangloff, unpublished research, Exxon Research and Engineering Company, Clinton, NJ, 1984
2. J.M. Barsom, E.J. Imhoff, and S.T. Rolfe, Fatigue Crack Propagation in High Yield Strength Steels, *Eng. Fract. Mech.*, Vol 2, 1971, p 301-324
3. I.M. Bernstein and A.W. Thompson, Ed., *Hydrogen Effects in Metals*, The Metals Society—American Institute of Mining, Metallurgical, and Petroleum Engineers, Warrendale, PA, 1981

Fig. 33(a) Effect of hydrazine on near-threshold fatigue crack growth rates in type 403 stainless steel
Environment: 0.1 g NaCl + 1.0 g Na_2SO_4 (g/100 mL H_2O) in boiling water (100 °C, or 212 °F). Stress ratio: R = 0.8. Source: Ref 139

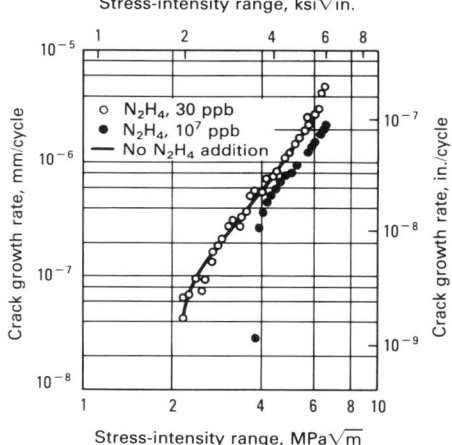

Fig. 32(a) Sodium chloride content versus the threshold stress-intensity range average regression curve for type 403 stainless steel
○: Amount of sodium sulfate, g/100 mL H_2O. Stress ratio: R = 0.8. Source: Ref 139

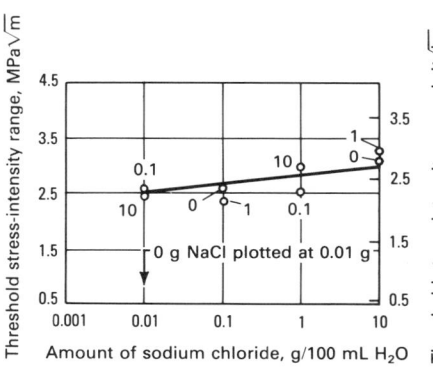

Fig. 32(b) Sodium chloride content versus the threshold stress-intensity range average regression curve for Ti-6Al-4V alloy
○: Amount of sodium sulfate, g/100 mL H_2O. Stress ratio: R = 0.8. Source: Ref 140

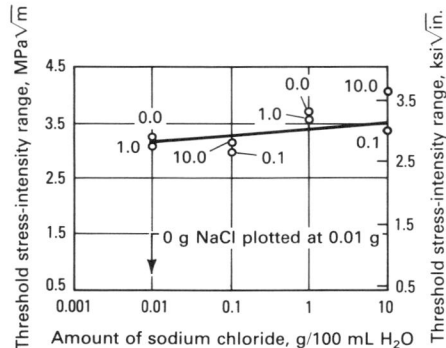

Fig. 33(b) Effect of hydrazine on near-threshold fatigue crack growth rates in Ti-6Al-4V alloy
Environment: 0.1 g NaCl + 0.1 g Na_2SO_4 (g/100 mL H_2O) in boiling water (100 °C, or 212 °F). Stress ratio: R = 0.8. Source: Ref 140

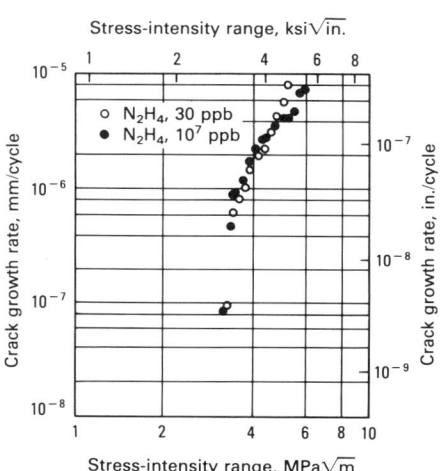

4. C.E. Jaske, J.H. Payer, and V.S. Balint, "Corrosion Fatigue of Metals in Marine Environments," MCIC Report 81-42, Battelle, Columbus, OH, 1981

5. T. Dean, E.N. Pugh, and G.M. Ugiansky, Ed., *Environment-Sensitive Fracture: Evaluation and Comparison of Test Methods*, STP 821, ASTM, Philadelphia, 1984

6. Z.A. Foroulis, Ed., *Environment Sensitive Fracture of Engineering Materials*, The Metals Society—American Institute of Mining, Metallurgical, and Petroleum Engineers, Warrendale, PA, 1978

7. J. Hochmann, J. Slater, R.D. McCright, and R.W. Staehle, Ed., *Stress Corrosion Cracking and Hydrogen Embrittlement of Iron Base Alloys*, National Association of Corrosion Engineers, Houston, 1976

8. T.W. Crooker and B.N. Leis, Ed., *Corrosion Fatigue: Mechanics, Metallurgy, Electrochemistry and Engineering*, STP 801, ASTM, Philadelphia, 1984

9. O. Deveraux, A.J. McEvily and R.W. Staehle, Ed., *Corrosion Fatigue: Chemistry, Mechanics and Microstructure*, National Association of Corrosion Engineers, Houston, 1973

10. H.L. Craig, Jr., T.W. Crooker, and D.W. Hoeppner, Ed., *Corrosion Fatigue Technology*, STP 642, ASTM, Philadelphia, 1978

11. M.H. Kamdar, Ed., *Embrittlement by Liquid and Solid Metals*, The Metals Society—American Institute of Mining, Metallurgical, and Petroleum Engineers, Warrendale, PA, 1984

12. A.J. McEvily and R.P. Wei, Fracture Mechanics and Corrosion Fatigue, in *Corrosion Fatigue: Chemistry, Mechanics and Microstructure*, O. Deveraux, A.J. McEvily and R.W. Staehle, Ed., National Association of Corrosion Engineers, Houston, 1973, p 381-395

13. J.A. Feeney, J.C. McMillan, and R.P. Wei, Environmental Fatigue Crack Propagation of Aluminum Alloys at Low Stress Intensity Levels, *Met. Trans.*, Vol 1, 1970, p 1741-1757

14. O. Vosikovsky and R.J. Cooke, An Analysis of Crack Extension by Corrosion Fatigue in a Crude Oil Pipeline, *Int. J. Pressure Vessel Piping*, Vol 6, 1978, p 113-129

15. S.J. Hudak, O.H. Burnside, and K.S. Chan, "Analysis of Corrosion Fatigue Crack Growth in Welded Tubular Joints," Offshore Technology Conference Paper OTC 4771, Houston, 1984

16. R.P. Wei and G. Shim, Fracture Mechanics and Corrosion Fatigue, in *Corrosion Fatigue: Mechanics, Metallurgy, Electrochemistry and Engineering*, STP 801, T.W. Crooker and B.N. Leis, Ed., ASTM, Philadelphia, 1984, p 5-25

17. M.H. El Haddad, T.H. Topper, and B. Mukherjee, Review of New Developments in Crack Propagation Studies, *J. Test Eval.*, Vol 9, 1981, p 65-81

18. N.E. Dowling, Crack Growth During Low Cycle Fatigue of Smooth Axial Specimens, in *Cyclic Stress-Strain and Plastic Deformation Aspects of Fatigue Crack Growth*, STP 637, ASTM, Philadelphia, 1977, p 97-121

19. R.P. Gangloff and R.O. Ritchie, Environmental Effects Novel to the Propagation of Short Fatigue Cracks, in *Fundamentals of Deformation and Fracture*, K.J. Miller, Ed., Cambridge University Press, Cambridge (in press)

20. S.J. Hudak, Jr. and R.P. Wei, Consideration of Nonsteady State Crack Growth in Material Evaluation and Design, *Int. J. Pressure Vessel Piping*, Vol 9, 1981, p 63-74

21. R.P. Gangloff, Crack Size Effects on the Chemical Driving Force for Aqueous Corrosion Fatigue, *Met. Trans. A* (in press)

22. C.S. Kortovich, Corrosion Fatigue of

Fig. 34(a) Effect of pH on near-threshold fatigue crack growth rates in type 403 stainless steel

Environment: 0.1 g NaCl + 1.0 g Na$_2$SO$_4$ (g/100 mL H$_2$O) in boiling water (100 °C, or 212 °F). Stress ratio: R = 0.8. Source: Ref 139

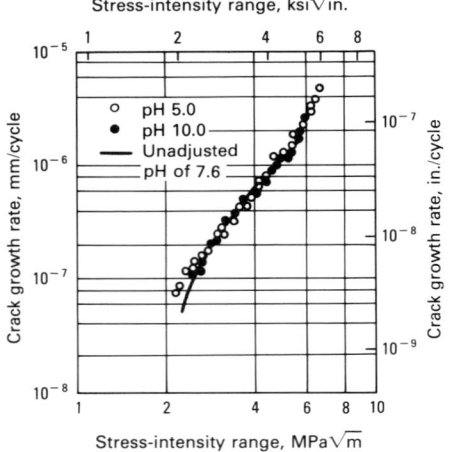

Fig. 34(b) Effect of pH on near-threshold fatigue crack growth rates in Ti-6Al-4V alloy

Environment: 0.1 g NaCl + 0.1 g Na$_2$SO$_4$ (g/100 mL H$_2$O) in boiling water (100 °C, or 212 °F). Stress ratio: R = 0.8. Source: Ref 140

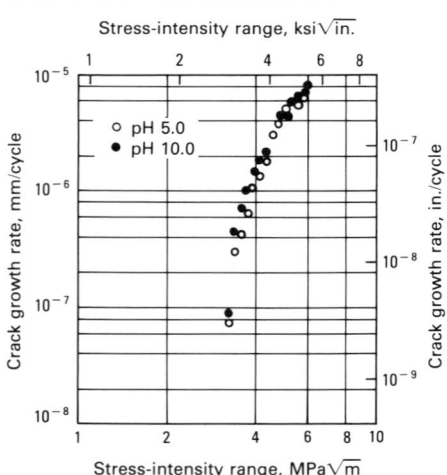

Fig. 35 Effect of oxygen content on near-threshold fatigue crack growth rates in Ti-6Al-4V alloy

Temperature: 104 °C (219 °F). Stress ratio: R = 0.5. Source: Ref 140

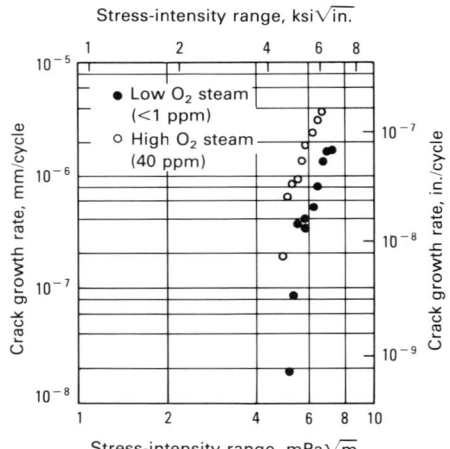

Fig. 36 Experimental setup for fatigue crack growth testing using an eddy-current gage

Source: Ref 155

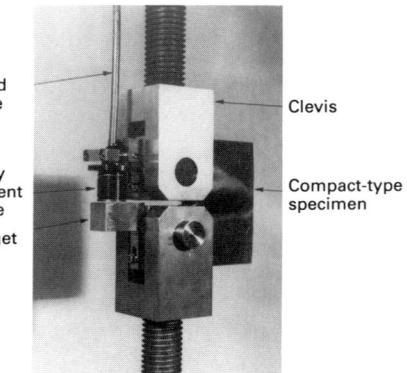

4340 and D6AC Steels Below K_{Iscc}, "Proceedings of the 1974 Triservice Conference on Corrosion of Military Equipment," AFML-TR-75-42, Air Force Materials Laboratory, Wright-Patterson Air Force Base, 1975

23. D.B. Dawson and R.M. Pelloux, Corrosion Fatigue Crack Growth of Titanium Alloys in Aqueous Environments, *Met. Trans. A*, Vol 5, 1974, p 723-731

24. J.M. Barsom, Corrosion Fatigue Crack Propagation Below K_{Iscc}, *Eng. Fract. Mech.*, Vol 3, 1971, p 15-25

25. B.A. Cowles, D.L. Sims, and J.R. Warren, "Evaluation of the Cyclic Behavior of Aircraft Turbine Disc Alloys," NASA Technical Report CR-159409, 1979

26. B. Tompkins and P.M. Scott, Environment Sensitive Fracture: Design Considerations, *Met. Tech.*, Vol 9, 1982, p 240-248

27. P.M. Scott, Effects of Environment on Crack Propagation, in *Developments in Fracture Mechanics—II*, G.G. Shell, Ed., Applied Science Publishers, London, 1979, p 221-257

28. T.W. Thorpe, P.M. Scott, A. Rance, and D. Silvester, Corrosion Fatigue of BS4360:50D Structural Steel in Seawater, *Int. J. Fatigue*, Vol 5, 1983, p 123-133

29. J.P. Gallagher, "Corrosion Fatigue Crack Growth Rate Behavior Above and Below K_{Iscc}," Naval Research Laboratory, Report NRL-7064, 1970

30. A. Turnbull, Progress in the Understanding of the Electrochemistry in Cracks, in *Embrittlement by the Localized Crack Environment*, R.P. Gangloff, Ed., The Metals Society—American Institute of Mining, Metallurgical, and Petroleum Engineers, Warrendale, PA, 1984, p 3-31

31. R.N. Parkins, Prevention of Environment Sensitive Cracking by Inhibition, in *Embrittlement by the Localized Crack Environment*, R.P. Gangloff, Ed., The Metals Society—American Institute of Mining, Metallurgical, and Petroleum Engineers, Warrendale, PA, 1984, p 385-404

32. M.O. Speidel, M.J. Blackburn, T.R. Beck, and J.A. Feeney, Corrosion Fatigue and Stress Corrosion Crack Growth in High Strength Aluminum Alloys, Magnesium Alloys and Titanium Alloys Exposed to Aqueous Solutions, in *Corrosion Fatigue: Chemistry, Mechanics and Microstructure*, O. Deveraux, A.J. McEvily, and R.W. Staehle, Ed., National Association of Corrosion Engineers, Houston, 1973, p 324-345

33. R.P. Gangloff, "Inhibition of the Chemical Driving Force for Corrosion Fatigue in High Strength Steel," Exxon Research and Engineering Company, Report CR.xxBV.85, 1984

34. H.G. Nelson, Hydrogen Induced Slow Crack Growth of a Plain Carbon Pipeline Steel Under Conditions of Cyclic Loading, in *Effect of Hydrogen on the Behavior of Materials*, A.W. Thompson and I.M. Bernstein, Ed., The Metals Society—American Institute of Mining, Metallurgical, and Petroleum Engineers, Warrendale, PA, 1976, p 602-611

35. H.H. Johnson, Hydrogen Brittleness in Hydrogen and Hydrogen-Oxygen Gas Mixtures, in *Stress Corrosion Cracking and Hydrogen Embrittlement of Iron Based Alloys*, J. Hochmann, J. Slater, R.D. McCright, and R.W. Staehle, Ed., National Association of Corrosion Engineers, Houston, 1976, p 382-389

36. R.J. Walter, J.D. Frandsen, and R.P. Jewitt, Fractography of Alloys Tested in High Pressure Hydrogen, in *Hydrogen Effects in Metals*, I.M. Bernstein and A.W. Thompson, Ed., The Metals Society—American Institute of Mining, Metallurgical, and Petroleum Engineers, Warrendale, PA, 1981, p 819-827

37. A.W. Thompson and I.M. Bernstein, The Role of Metallurgical Variables in Hydrogen-Assisted Environmental Fracture, in *Advances in Corrosion Science and Technology*, Vol 7, M.G. Fontana and R.W. Staehle, Ed., Plenum Publishing, New York, 1980, p 53-175

38. R.O. Ritchie, Application of Fracture Mechanics to Fatigue, Corrosion Fatigue and Hydrogen Embrittlement, in *Analytical and Experimental Fracture Mechanics*, G.C. Sih, Ed., Sithoff and Noorohoff, Holland, 1981, p 81-108

39. S. Suresh and R.O. Ritchie, The Propagation of Short Fatigue Cracks, *Int. Met. Rev.*, (in press)

40. L.K.L. Tu and B.B. Seth, Threshold Corrosion Fatigue Crack Growth in Steels, *J. Test. Eval.*, Vol 6, 1978, p 66-74

41. "Standard Test Method for Constant Load-Amplitude Fatigue Crack Growth Rates Above 10^{-8} m/cycle," E 647, *Annual Book of ASTM Standards*, ASTM, Philadelphia, 1984, p 711-731

42. R.J. Bucci, private communication, Alcoa Research Laboratories, Alcoa Center, PA, 1984

43. T.W. Crooker, F.D. Bogar, and G.R. Yoder, "Standard Method of Test for Constant Load Amplitude Fatigue Crack Growth Rates in Marine Environments," Naval Research Laboratory Report 4594, 1981

44. S.J. Hudak, Jr. and R.J. Bucci, Ed., *Fatigue Crack Growth Measurement and Data Analysis*, STP 738, ASTM, Philadelphia, 1981

45. R.P. Gangloff, Electrical Potential Monitoring of the Formation and Growth of Small Fatigue Cracks in Embrittling Environments, in *Advances in Crack Length Measurement*, C.J. Beevers, Ed., Engineering Materials Advisory Services, London, 1982, p 175-230

46. P.S. Pao and J.E. O'Neal, Hydrogen-Enhanced Fatigue Crack Growth in Ti-6242S, *J. Nuc. Mat.*, Vol 122 and 123, 1984, p 1587

47. R.P. Wei, P.S. Pao, R.G. Hart, T.W. Weir, and G.W. Simmons, Fracture Mechanics and Surface Chemistry Studies of Fatigue Crack Growth in an Aluminum Alloy, *Met. Trans. A*, Vol 11, 1980, p 151

48. D.J. Dwyer, G.W. Simmons, and R.P. Wei, A Study of the Initial Reaction of Water Vapor with Fe(001) Surface, *Surf. Sci.*, Vol 64, 1977, p 617

49. C.J. Beevers, Ed., *Advances in Crack Length Assessment*, Engineering Materials Advisory Service, London, 1982

50. M. Gao, P.S. Pao, and R.P. Wei, Role of Micromechanisms in Corrosion Fatigue Crack Growth in a 7075-T651 Aluminum Alloy, in *AIME Symposium on Synergism of Microstructure, Mechanism, and Mechanics in Fracture*, The Metals Society—American Institute of Mining, Metallurgical, and Petroleum Engineers, Warrendale, PA (in press)

51. M. Gao and R.P. Wei, A Hydrogen Partitioning Model for Hydrogen Assisted Crack Growth, *Met. Trans. A* (submitted for publication)

52. D.M. Ressler, "An Examination of Fatigue Crack Growth in AISI 4340 Steel in Respect to Two Corrosion Fatigue Models," M.S. thesis, Lehigh University, Bethlehem, PA, 1984

53. H.D. Solomon and L.F. Coffin, Effects of Frequency and Environment on Fatigue Crack Growth in A-286 at 1100 °F, in STP 520, ASTM, Philadelphia, 1973, p 112-122

54. M.O. Speidel, *Fatigue Crack Growth at High Temperatures*, P.R. Sahm and M.O. Speidel, Ed., Elsevier, Amsterdam, 1974, p 207-251

55. L.A. James, *Met. Trans. A*, Vol 6A,

1975, p 109-116

56. K. Snowden, *Acta Met.*, Vol 12, 1964, p 295-303

57. L.F. Coffin, S.S. Manson, A.E. Carden, L.K. Severud, and W.L. Greenstreet, "Time Dependent Fatigue of Structural Alloys," Oak Ridge National Laboratory Report, ORNL-5073, 1977

58. D.A Jablonski, "Fatigue Behavior of Hastelloy-X at Elevated Temperatures in Air, Vacuum and Oxygen Environments," Ph.D. thesis, Massachusetts Institute of Technology, Cambridge, 1978

59. L.G. Carpenter, *Vacuum Technology*, American Elsevier, New York, 1970

60. D.S. Dushman, *Scientific Foundations of Vacuum Technique*, J.M. Lafferty, Ed., John Wiley & Sons, New York, 1962

61. D.S. Transden, W.L. Morris, and H.L. Marcus, in *Hydrogen in Metals*, I.M. Bernstein and A.W. Thompson, Ed., American Society for Metals, 1974

62. H. Tada, P. Paris, and G. Irwin, *The Stress Analysis of Cracks Handbook*, Section 2.10, Del Research Corp., Hellerton, PA, 1973

63. H. Schwalbe, Test Techniques, *Proceedings of the Internal Conference on Fracture, No. 5*, Vol 4, Cannes, France, 1981

64. "Standard Test Method for Plane-Strain Fracture Toughness of Metallic Materials," E 399, *Annual Book of ASTM Standards*, ASTM, Philadelphia, 1984, p 519-554

65. A.D. Mercer and G.M. Brook, The Effect of Temperature Between 25 and 90 °C on the Corrosion of Mild Steel in Distilled Water, *La Tribune de Cebedeau*, Vol 31, 1978, p 299-306

66. W.H. Hartt and S.S. Roypathak, "Formation of Calcareous Deposits Within Simulated Fatigue Cracks in Seawater," Paper 62, Corrosion '83, Anaheim, CA, National Association of Corrosion Engineers

67. "Standard Specification for Substitute Ocean Water," D 1141, *Annual Book of ASTM Standards*, Vol 11.02, ASTM, Philadelphia, 1984, p 514-515

68. G.P. Rothwell, P.E. Francis, and K.F. Hale, "The Effects of Heat Transfer on the External Corrosion of Submarine Pipelines and Rivers," Report No. UDS 620.193.27 (OT-R-8292), Society for Underwater Technology, London, 1982

69. F.D. Bogar and T.W. Crooker, The Influence of Bulk-Solution Chemistry Conditions on Marine Corrosion Fatigue Crack Growth Rate, *J. Test. Eval.*, Vol 7 (No. 3), 1979, p 155-159

70. V. Rollins, B. Arnold, and E. Lardner, *Brit. Corr. J.*, Vol 5, 1970, p 33

71. P.M. Scott, T.W. Thorpe, and D.R.V. Silvester, Rate Determining Processes for Corrosion Fatigue Crack Growth in Ferritic Steels in Seawater, *Corr. Sci.*, Vol 23, 1983, p 559-575

72. O. Vosikovsky, Fatigue Crack Growth in an X65 Line Pipe Steel at Low Cyclic Frequencies in Aqueous Environments, *Trans. ASME J. Eng. Mat. Tech.*, Vol 97, 1975, p 298-304

73. H.H. Uhlig, *Corrosion and Corrosion Control*, 2nd ed., John Wiley & Sons, New York, 1971

74. D.J.G. Ives and G.J. Janz, Ed., *Reference Electrodes*, Academic Press, New York, 1961

75. A. Turnbull, Review of the Electrochemical Condition in Cracks with Particular Reference to Corrosion Fatigue of Structural Steels in Sea Water, *Rev. Coat. Corr.*, Vol 5 (No. 1-4), 1982, p 44-171

76. A. Turnbull, A Theoretical Evaluation of the Influence of Mechanical Variables on the Concentration of Oxygen in a Corrosion Fatigue Crack, *Corr. Sci.*, Vol 22 (No. 9), 1982, p 877-893

77. B.F. Jones, The Influence of Crack Depth on the Fatigue Crack Propagation Rate for a Marine Steel in Sea Water, *J. Mat. Sci.*, Vol 17, 1982, p 499-507

78. R.P. Gangloff, The Criticality in Crack Size in Aqueous Corrosion Fatigue, *Res. Mechanica Lett.*, Vol 1, 1981, p 299-306

79. J.A. Atkinson and T.C. Lindley, "The Effect of Frequency and Temperature on Environmentally Assisted Fatigue Crack Growth below K_{Iscc} in Steels," presented at meeting on the Influence of Environment on Fatigue, Institute of Mechanical Engineers, London, 1977

80. N.E. Hammer, "Nonmetals Section—Corrosion Data Survey," National Association of Corrosion Engineers, Houston, 1981

81. N.E. Hammer "Metals Section—Corrosion Data Survey," National Association of Corrosion Engineers, Houston, 1974

82. "Managing Corrosion With Plastics," National Association of Corrosion Engineers, Houston, 1977

83. M.G. Fontana, *Corrosion Engineering*, McGraw-Hill, New York, 1978

84. P.E. Manning and D.J. Duquette, The Effect of Temperature (25-289 C) on Pit Initiation in Single Phase and Duplex 304L Stainless Steels in 100 ppm Chloride Solutions, *Corr. Sci.*, Vol 20, 1980, p 598

85. P.L. Andresen, "The Effects of Aqueous Impurities on Intergranular Stress Corrosion Cracking of Sensitized Type 304 Stainless Steel," Final Report NP-3384, Electric Power Research Institute, Nov 1983

86. P.L. Andresen, Effects of Material and Environmental Variables on SCC Initiation in Slow Strain Rate Tests on Type 304 Stainless Steel, in *Environment-Sensitive Fracture: Evaluation and Comparison of Test Methods*, STP 821, T. Dean, E.N. Pugh, and G.M. Ugiansky, Ed., ASTM, Philadelphia, 1984

87. M.E. Indig and A.R. McIlree, High Temperature Electrochemical Studies of the Stress Corrosion Type 304 Stainless Steel, *Corrosion*, Vol 35, 1979, p 288

88. C.M. Chen, K. Aral, and G.J. Theus, "Computer Calculated Potential pH Diagrams to 300 °C. Volume 2: Handbook of Diagrams," Final Report NP3137, Electric Power Research Institute, June 1983

89. A.K. Agrawal and R.W. Staehle, A Silver-Silver Chloride Reference Electrode for High Temperature and High Pressure Electrochemistry, *Corrosion*, Vol 33, 1977, p 418

90. M.J. Danielson, The Construction and Thermodynamic Performance of an Ag-AgCl Reference Electrode for Use in High Temperature Aqueous Environments Containing H_2 and H_2S, *Corrosion*, Vol 35, 1979, p 201

91. D.D. MacDonald, A.C. Scott, and P. Wentrcek, Silver-Silver Chloride Thermocells and Thermal Liquid Junction Potentials for Potassium Chloride Solutions at Elevated Temperatures, *J. Electrochem. Soc.*, Vol 126, 1979, p 908

92. D.F. Taylor, Thermodynamic Properties of Metal-Water Systems at Elevated Temperatures, *J. Electrochem. Soc.*, Vol 125, 1978, p 808

93. T.A. Prater, W.R. Carlin, and L.F. Coffin, "Environmental Crack Growth Measurement Techniques," Final Report NP-2641, Electric Power Research Institute, Nov 1982

94. A. Saxena and S.J. Hudak, Jr., Review and Extension of Compliance Information for Common Crack Growth Specimens, *Int. J. Frac.*, Vol 14, 1978, p 453

95. P.L. Andresen and D.J. Duquette, The Effects of Dissolved Oxygen, Chloride Ion and Applied Potential on the SCC

Behavior of Type 304 Stainless Steel in 290 °C Water, *Corrosion*, Vol 36, 1980, p 409

96. S. Szklarska-Smialowska and G. Cragnolino, Stress Corrosion Cracking of Sensitized Type 304 Stainless Steel in Oxygenated Pure Water at Elevated Temperatures, *Corrosion*, Vol 36, 1980, p 653

97. R.W. Weeks, Stress Corrosion Cracking in BWR and PWR Piping, *Proceedings of Symposium on Environmental Degradation of Materials in Nuclear Power Systems*, Myrtle Beach, SC, Aug 22-25, 1983, National Association of Corrosion Engineers, Houston, p 69-86

98. D. Weinstein, "Environmental Cracking Margins for Carbon Steel Piping," Final Report NP-2406, Electric Power Research Institute, May 1982

99. B. Tomkins and J.A. Hudson, Environmental Factors Influencing the Failure Properties of Pressure Vessel Materials, *Proceedings of Symposium on Environmental Degradation of Materials in Nuclear Power Systems*, Myrtle Beach, SC, Aug 22-25, 1983, National Association of Corrosion Engineers, Houston, p 25-52

100. F.P. Ford and M.J. Povich, The Effect of Oxygen Temperature Combinations on the Stress-Corrosion Susceptibility of Sensitized Type 304 Stainless Steel in High Purity Water, *Corrosion*, Vol 35, 1979, p 569

101. R. Heut *et al.*, "Stress Corrosion Cracking of Type 304 Stainless Steel in High Purity Water: A Compilation of Crack Growth Rates," Interim Report NP-2423-LD, Electric Power Research Institute, June 1982

102. L. Ljunberg, "BWR Water Chemistry Impurity Studies: Literature Review on Stress Corrosion Cracking," Interim Report NP-3663, Electric Power Research Institute, Sept 1984

103. P.L. Andresen, "Innovations in Experimental Techniques for Testing in High Temperature Aqueous Environments," Report 81CRD088, General Electric Corporate Research and Development Center, Schenectady, NY, May 1981

104. "Autoclave Corrosion Testing of Metals in High Temperature Water," NACE Standard TM-01-71, National Association of Corrosion Engineers, Houston, approved April 1971

105. "Dynamic Corrosion Testing of Metals in High-Temperature Water," NACE Standard TM-02-74, National Association of Corrosion Engineers, Houston, approved April 1974

106. R.B. Davis and M.E. Indig, "The Effect of Aqueous Impurities on the Stress Corrosion Cracking of Austenitic Stainless Steel in High-Temperature Water," Final Report NP-3174-LD, Electric Power Research Institute, July 1983

107. N. Ohnaka *et al.*, Effects of Environmental Factors on IGSCC Susceptibility of Sensitized 304 Stainless Steel in High-Temperature Water, *Boshoku Gijutsu*, Vol 32, 1983, p 214

108. B.G. Ackland and B.W. Chewy, Stress-Corrosion Cracking of Welded Line-Pipe Steel, *Corrosion Australasia*, Vol 6, 1981, p 8

109. F.P. Ford and M. Silverman, Effect of Loading Rate on Environmentally Controlled Cracking of Sensitized 304 Stainless Steel in High Purity Water, *Corrosion*, Vol 36, 1980, p 597

110. H. Choi *et al.*, The Effect of Fluid Flow on the Stress Corrosion Cracking of ASTM A508 Cl 2 Steel and A151 Type 304 Stainless Steel in High Temperature Water, *Corrosion*, Vol 38, 1982, p 76

111. E.L. Burley, "Oxygen Suppression in Boiling Water Reactors," General Electric Report, NEDC 23856-7, San Jose, CA, Oct 1982

112. J. Leibovitz, "The Effects of Variations in BWR Water Chemistry on the Electrochemical Potential of Stainless Steel and Other Metals," Report NP-3517, Electric Power Research Institute, May 1984

113. J. Leibovitz, "Improved Electrodes for BWR In-Plant ECT Monitoring," Report NP-2524, Electric Power Research Institute, July 1982

114. D. MacDonald, Reference Electrodes for High Temperature Aqueous Systems—A Review and Assessment, *Corrosion*, Vol 34, 1978, p 75

115. L.W. Niedrach, The Use of High Temperature pH Sensor as a "Pseudo-reference Electrode" in Monitoring of Corrosion and Redox Potentials at 285 °C, *J. Electrochem. Soc.*, Vol 129, 1982, p 1445

116. W.E. Ruther *et al.*, "Effect of Sulphuric Acid, Oxygen and Hydrogen in High Temperature Water on Stress-Corrosion Cracking of Sensitized Type 304 Stainless Steel," Paper 125, Corrosion '83, Anaheim, CA, April 18-22, 1983, National Association of Corrosion Engineers, Houston

117. L.W. Niedrach, A New Membrane-Type pH Sensor for Use in High-Temperature/High Pressure Water, *J. Electrochem. Soc.*, Vol 127, 1980, p 2122

118. L.W. Niedrach and W.H. Stoddard,

Continuous Voltametric Monitoring of Hydrogen and Oxygen in Water, *Anal. Chem.*, Vol 54, 1982, p 1651

119. M. Hishida, Toshiba Corporation, Japan, private communication, May 1, 1983

120. W.A. Van Der Sluys and R.J. Futato, Computer-Controlled Single Specimen J-Test, in *Elastic Plastic Fracture: Second Symposium, Volume II, Fracture Resistance Curves and Engineering Applications*, STP 803, C.F. Shih and J.P. Gudas, Ed., ASTM, Philadelphia, 1983, p II-464, II-482

121. W.A. Van Der Sluys and D.S. DeMiglio, The Use of a Constant ΔK Test Method in the Investigation of Fatigue Crack Growth in 288 °C Water Environments, in *Environment-Sensitive Fracture: Evaluations and Comparison of Test Methods*, STP 821, T. Dean, E.N. Pugh, and G.M. Ugiansky, Ed., ASTM, Philadelphia, 1984

122. P. Marshall, "The Fatigue Behaviour of Annealed AISI 316 Stainless Steel in Air and High Temperature Sodium: Review and Preliminary Results," Report RD/B/N3236, Central Electricity Generating Board, Berkeley Nuclear Laboratories, Nov 1974

123. E.K. Priddle and C. Wiltshire, The Measurement of Fatigue Crack Propagation in Specimens Immersed in Liquid Sodium at Elevated Temperatures: Technique and Preliminary Results, *Int. J. Fract.*, Vol 11 (No. 4), 1975, p 697-700

124. L.A. James and R.L. Knecht, Fatigue-Crack Propagation Behavior of Type 304 Stainless Steel in a Liquid Sodium Environment, *Met. Trans. A*, Vol 6 (No. 1), 1975, p 109-116

125. J.L. Yuen and J.F. Copeland, Fatigue Crack Growth Behavior of Stainless Steel Type 316 Plate and 16-8-2 Weldments in Air and High-Carbon Liquid Sodium, *J. Eng. Mat. Technol.*, Vol 101 (No. 3), 1979, p 214-223

126. D.L. Hammon, S.K. DeWeese, D.K. Matlock, and D.L. Olson, Evaluation of Mechanical Properties of Materials in Liquid Lithium, *Closed Loop*, Vol 9, Nov 1979, p 3-11 (MTS Systems Corp.)

127. D.L. Hammon, D.K. Matlock, and D.L. Olson, The Effect of Liquid Lithium on Fatigue Crack Propagation in 304L Stainless Steel at Elevated Temperatures, *J. Mat. Energy Syst.*, Vol 2 (No. 4), 1981, p 26-33

128. J.L. Krankota, The Effect of Carburization in Sodium on the Mechanical Properties of Austenitic Stainless

Steels, *J. Eng. Mat. Technol.*, Vol 98 (No. 1), 1976, p 10-17

129. G.J. Lloyd, Mechanical Properties of Austenitic Stainless Steels in Sodium, *Atomic Energy Rev.*, Vol 16 (No. 2), 1978, p 155-208

130. C.C. McPheeters and J.C. Biery, The Dynamic Characteristics of a Plugging Indicator for Sodium, *Nucl. Appl.*, Vol 6 (No. 6), 1969, p 573-581

131. H. Yamamoto, M. Murase, I. Sumida, K. Kotani, and S. Shimoyashiki, Measurement of Impurity Concentration in Sodium by Automatic Plugging Indicator, *J. Nucl. Sci. Technol.*, Vol 14 (No. 6), 1977, p 452-456

132. J.M. McKee, D.R. Vissers, P.A. Nelson, B.R. Grundy, E. Berkey, and G.R. Taylor, Calibration Stability of Oxygen Meters for LMFBR Sodium Systems, *Nucl. Technol.*, Vol 21 (No. 3), 1974, p 217-227

133. M.R. Hobdell and C.A. Smith, Electrochemical Techniques for Monitoring Dissolved Carbon, Hydrogen and Oxygen in Liquid Sodium, *J. Nucl. Mat.*, Vol 110 (No. 2 and 3), 1982, p 125-139

134. D.R. Vissers, J.T. Holmes, L.G. Bartholme, and P.A. Nelson, A Hydrogen-Activity Meter for Liquid Sodium and its Application to Hydrogen Solubility Measurements, *Nucl. Technol.*, Vol 21 (No. 3), 1974, p 235-244

135. J.M. Dahlke, Ed., *Second International Conference on Liquid Metal Technology in Energy Production Proceedings*, CONF-800401-P2, U.S. Department of Energy, Aug 1980

136. P. Marshall, The Influence of Low Oxygen and Contaminated Sodium Environments on the Fatigue Behavior of Solution Treated AISI 316 Stainless Steel, in *Proceedings, Conference on the Influence of Environment on Fatigue*, Paper C101/77, The Institution of Mechanical Engineers, London, 1977, p 27-36

137. W.J. Mills and L.A. James, The Fatigue-Crack Propagation Response of Two Nickel-Base Alloys in a Liquid Sodium Environment, *J. Eng. Mat. Technol.*, Vol 101 (No. 2), 1979, p 205-213

138. E.K. Priddle, F. Walker, and C. Wiltshire, "The Effects of Helium, Sodium and Other Environments on the Fatigue Crack Propagation Characteristics of a Stainless Steel," Report RD/B/N3802, Central Electricity Generating Board, Berkeley Nuclear Laboratories, Aug 1976

139. P.K. Liaw, J. Anello, and J.K. Donald, Influence of Corrosive Environ-
ments on Near-Threshold Fatigue Crack Growth in 403 Stainless Steel, *Met. Trans. A.*, Vol 13, 1982, p 2177-2189

140. P.K. Liaw, J. Anello, and J.K. Donald, Effects of Corrosive Environments on Near-Threshold Fatigue Crack Growth Behavior of Ti-6Al-4V, *Eng. Fract. Mech.*, Vol 19 (No. 6), 1984, p 1047-1056

141. P.K. Liaw, J. Anello, N.S. Cheruvu, and J.K. Donald, Near-Threshold Corrosion Fatigue Crack Growth Behavior of Type 422 Stainless Steel at Controlled Maximum Stress Intensities, *Met. Trans. A*, Vol 15, 1984, p 693-699

142. P.K. Liaw, S.J. Hudak, Jr., and J.K. Donald, Influence of Gaseous Environments on Rates of Near-Threshold Fatigue Crack Propagation in NiCrMoV Steel, *Met. Trans. A*, Vol 13, 1982, p 1633-1645

143. P.K. Liaw, S.J. Hudak, Jr., and J.K. Donald, Near-Threshold Fatigue Crack Growth Investigation of NiMoV Steel in Hydrogen Environment, in STP 791, ASTM, 1983, p II-370 to II-388

144. L.J. Ceschini, P.K. Liaw, G.E. Rudd, and W.A. Logsdon, Automated Corrosion Fatigue Crack Growth Testing in Pressurized Water Environments, in *Environment-Sensitive Fracture: Evaluation and Comparison of Test Methods*, STP 821, T. Dean, E.N. Pugh, and G.M. Ugiansky, Ed., ASTM, Philadelphia, 1984, p 426-442

145. W.A. Logsdon, P.K. Liaw, and J.A. Begley, Fatigue Crack Growth Rate Properties of SA508 Cl 2a and SA533 Gr A Cl 2 Pressure Vessel Steels and Submerged Arc Weldments in a Steam (Secondary Side) Environment, *Corrosion* (submitted for publication)

146. J.K. Donald, "Preliminary Results of the ASTM E24.04.03 Round Robin Test Program on Low Delta-K Fatigue Crack Growth Rates," ASTM Task Group Document, ASTM, Philadelphia, 1982

147. T.R. Fabis, P.K. Liaw, L.J. Ceschini, T.R. Leax, and J.D. Landes, Computer-Controlled Fatigue Crack Growth Rate Testing on Bend Bars in a Corrosive Environment, in *Environment-Sensitive Fracture: Evaluation and Comparison of Test Methods*, STP 821, T. Dean, E.N. Pugh, and G.M. Ugiansky, Ed., ASTM, Philadelphia, 1984, p 470-483

148. P.C. Paris and B.R. Hayden, "A New System for Fatigue Crack Growth Measurement and Control," presented at ASTM Symposium on Fatigue
Crack Growth, Oct 1979

149. P.K. Liaw, H.R. Hartmann, and E.J. Helm, Corrosion Fatigue Crack Propagation Testing with the KRAK-GAGE in Salt Water, *Eng. Fract. Mech.*, Vol 18 (No. 1), 1983, p 121-131

150. P.K. Liaw, H.R. Hartmann, and W.A. Logsdon, A New Transducer to Monitor Fatigue Crack Propagation, *J. Test. Eval.*, Vol 11 (No. 3), 1983, p 202-207

151. P.K. Liaw, W.A. Logsdon, L.D. Roth, and H.R. Hartmann, *KRAK-GAGES for Automated Fatigue Crack Growth Rate Testing: A Review*, STP 877, ASTM, Philadelphia, 1985

152. A.S. Kobayashi, A.F. Emery, and B.M. Liaw, Dynamic Fracture Toughness of Reaction-Bonded Silicon Nitride, *J. Am. Ceramic Soc.*, Vol 66 (No. 2), 1983, p 151-155

153. "KD-1901 High Temperature Non-Conducting Displacement System with KD-2018 Laboratory Electronics," Operation Manual, Kamman Instrumentation Corp., Colorado Springs, 1976

154. L.A. James and L.J. Ceschini, "A New Method for In-Situ Compliance Measurements for Fatigue-Crack Growth Testing in Autoclaves," presented at the International Cyclic Crack Growth Rate (ICCGR) Committee Meeting, 1984

155. P.K. Liaw, L.J. Ceschini, W.A. Logsdon, J.A. Begley, E.J. Helm, and R.C. Brown, Automated Fatigue Crack Growth Testing Using an Eddy Current Gage, *J. Test. Eval.* (submitted for publication)

156. A. Saxena, Electric Potential Technique for Monitoring Subcritical Crack Growth at Elevated Temperatures, *Eng. Fract. Mech.*, Vol 13 (No. 4), 1980, p 741-750

157. R.P. Gangloff, Electric Potential Monitoring of Crack Formation and Subcritical Growth from Small Defects, *Fatigue Eng. Mat. Struct.*, Vol 4, 1981, p 15-33

158. R.P. Wei and R.L. Brazill, An Assessment of a.c. and d.c. Potential Systems for Monitoring Fatigue Crack Growth, in *Fatigue Crack Growth Measurement and Data Analysis*, STP 738, S.J. Hudak, Jr. and R.J. Bucci, Ed., ASTM, Philadelphia, 1981, p 103-119

159. R.O. Ritchie and K.J. Bathe, On the Calibration of the Electric Potential Technique for Monitoring Crack Growth Using Finite Element Methods, *Int. J. Fract.*, Vol 15, 1979, p 47-55

160. K. Endo, K. Komai, and Y. Matsuda,

Memo. Fac. Eng. Kyoto Univer., Vol 31, 1969, p 25

161. R.O. Ritchie, S. Suresh, and C.M. Moss, Near-Threshold Fatigue Crack Growth in 2¼Cr-1Mo Pressure Vessel Steel in Air and Hydrogen, *J. Eng. Mat. Technol.*, Vol 102, 1980, p 293-299

162. P.K. Liaw, T.R. Leax, R.S. Williams, and M.G. Peck, Influence of Oxide-Induced Crack Closure on Near-Threshold Fatigue Crack Growth Behavior, *Acta Met.*, Vol 30, 1982, p 2071-2078

163. P.K. Liaw, T.R. Leax, R.S. Williams, and M.G. Peck, Near-Threshold Fatigue Crack Growth Behavior in Copper, *Met. Trans. A*, Vol 13, 1982, p 1607-1618

164. A. Saxena, S.J. Hudak, Jr., J.K. Donald, and D.W. Schmidt, Computer-Controlled Decreasing Stress Intensity Technique for Low Rate Fatigue Crack Growth Testing, *J. Test. Eval.*, Vol 6 (No. 3), 1978, p 167-174

165. R.S. Williams, P.K. Liaw, M.G. Peck, and T.R. Leax, Computer Controlled Decreasing ΔK Fatigue Threshold Test, *Eng. Fract. Mech.*, Vol 18 (No. 5), 1983, p 953-964

Fracture Mechanics

Fracture Mechanics

By George R. Irwin
Visiting Professor of Mechanical Engineering
University of Maryland

FRACTURE MECHANICS depends on experiments and observations to suggest useful representations of the forces that cause the development and extension of cracks. The progressive fracturing concept was developed from observations (Ref 1) and is relatively simple and macroscopic. This concept dominates current fracture mechanics technology. Analysis models appropriate to fracture behaviors of interest and their applications are discussed in this article.

Materials engineers frequently need to understand macroscopic fracture behavior in terms of composition, microstructure, and other fine-scale features. The mechanical aspects of fine-scale details of progressive fracturing, such as void growth, cleavage, and grain-boundary separation, are of interest. Extension of fracture mechanics into these areas by development of appropriate analysis models has been helpful. However, research has been primarily exploratory. A basic understanding of the fracture mechanics of progressive fracturing in macroscopic terms is also required.

This article emphasizes fracture concepts, analysis, and toughness testing. Related articles of interest in this Volume include "Fatigue Crack Propagation," "Environmental Effects on Fatigue Crack Propagation," "Dynamic Fracture Testing," "Tests for Stress-Corrosion Cracking," "Tests for Hydrogen Embrittlement," and "Fatigue Crack Growth Data Analysis." Additional discussions of related fracture mechanics topics are given in Ref 2 to 5.

Progressive Fracturing

Solid material fractures occur by progressive fracturing—that is, by the spreading of a crack through the material. For crystalline solids, stationary (sessile) dislocations do not contribute to the plastic deformation mode of stress relaxation. Similarly, stationary cracks, unless large enough to cause overstress of the remaining section, do not contribute to the separational mode of stress relaxation. To understand progressive fracturing, attention must be focused on the leading edge of the crack.

The tendency of the stress-strain field to drive the crack can be represented in a relatively simple way, usually with a single stress field parameter. This permits development of a fracture control plan. Crack extension behavior, as governed by the stress field driving force, is studied using specimens of convenient size made of material similar to the service component. Estimates are made of the relationship between applied loads and the crack driving force for cracks in the service component. The relationship between laboratory test data and expected crack extension behavior in service permits estimates of the toughness, quality control parameters, and service loadings that are appropriate with regard to fracture safety.

Methods that represent the crack driving tendency of a stress field are based on simplifications. The crack itself, or the region where separation has occurred, is regarded as a smooth, continuous surface. The boundary between this surface and regions where separation has not occurred, called the crack front, is regarded as a continuous line. This degree of simplification permits mathematical description of the stress field properties at the crack front in general terms. However, the crack problem solutions available in useful form are more limited. For these, the crack front is a segment of a straight line, ellipse, or circle, and the crack is positioned on a flat plane. The neglect of irregularities, commonly observed for real cracks in structural materials, has not been a serious handicap for purposes of general macroscopic analysis and testing.

Two viewpoints exist concerning representation of the crack driving tendency by stress-strain field parameters. One involves the stress-intensity factor, K, and the other the crack-extension force, $\mathcal{G}$. These viewpoints will be discussed and shown to be equivalent.

For the condition of customary interest, in which the crack plane is normal to the direction of largest tension, a linear elastic treatment of the stress field is often used, which employs a stress-intensity (or stress-amplitude) factor, K, applicable to a region surrounding the crack front (Ref 6, 7). K represents the influences of the stress field toward drive and control of the enclosed crack-extension behavior. The assumed extreme sharpness of the crack opening at the crack front, along with its line-segment character, forces the stresses near the crack front into a characteristic annular and two-dimensional pattern, within which stress values are proportional to a single stress-intensity factor, K.

In application, it is assumed that the region within which this stress pattern dominates surrounds, but does not intrude upon, the crack front region containing nonelastic strains and advance separations. The extension of this annular region outward from the crack front is limited to a fraction of the crack size. For visualization, $K/\sqrt{2\pi r}$ may be thought of as the tensile stress acting normal to the crack plane at a small distance, r, ahead of the crack front. In the same manner, $(4K/\pi E)\sqrt{2\pi r}$ can be regarded as the crack opening displacement at a small distance, r, behind the crack front.

Another viewpoint, which was prominent during the developmental stage of fracture mechanics, assumes that the crack-extension force, $\mathcal{G}$, represents the tendency of the stress field to drive and control progressive fracturing. $\mathcal{G}$ is a generalized force concept (Ref 8, 9). From theoretical mechanics, if it is assumed that the only energy reservoir of interest is the stress field, then $\mathcal{G}$ is the system-isolated rate of disappearance of stress field energy per increment of crack extension. Because $\mathcal{G}$ is always computed per unit of crack front length, the dimensions of $\mathcal{G}$ are force per length, or energy per area, which are the same as for surface tension.

The generalized force concept used in dislocation mechanics to represent forces on dislocation lines has a similar basis. These ideas support the use of the term "crack-extension force." Also, $\mathcal{G}$ can be correlated directly to the crack theory of Griffith (Ref 10). For the ideal crack model, energy loss from the stress field can be computed for a small increment of crack extension using products of stress times crack opening near the crack front. Because these "stress × opening" quantities must be proportional to K^2, $\mathcal{G}$ is proportional to K^2. Therefore, the two viewpoints are equivalent.

Despite the complexities of plastic and nonlinear deformation, several available methods are applicable when the crack front plastic zone is not relatively small. To understand these methods, it is helpful to recognize that during loading and with negligible crack extension, nonlinear deformations can be treated analytically as reversible and elastic. Thus, the influence of the nonelastic material property has a negligible effect in a local region, until the stresses on that region begin to decrease. If the deformations are elastic (not strain path dependent), then the crack-extension force can be computed using an appropriate form of the Rice J-integral (Ref 11). Values of J derived in this manner are important to current elastic-plastic fracture technology.

Given the above nonlinear elastic behavior assumption, the nature of the crack front again forces the stress-strain field local to that region into a characteristic annular and two-dimensional pattern that can be regarded as governed by a single parameter, the value of J (Ref 12, 13). When details of the nonlinear behavior (work hardening, for example) are known, strain-amplitude and stress-amplitude crack front parameters can be developed. However, these have not been used in practical application.

An alternative method of elastic-plastic characterization uses an estimate of the crack opening, δ, at the crack front (Ref 14). This concept preceded the use of J-values and resulted from modifications of linear elastic crack stress fields, which provided approximate corrections for a plastic zone at the crack tip. The tensile plastic flow stress, σ_Y, multiplied by δ furnished an estimate of $\mathcal{G}$. Direct measurement of δ for actual cracks seemed impractical. However, calibrations could be done using saw-cut crack models. In addition, estimates of δ could be achieved

using available methods of plasticity mechanics. Currently, it is recognized that an equivalence exists between J and the product of δ times an appropriate estimate of the tensile plastic flow stress.

The relationships between K, $\mathcal{G}$, J, and δ are:

$$K^2 = \mathcal{G} \cdot \begin{cases} E \text{ for plane stress} \\ E/(1 - \nu^2) \text{ for plane strain} \end{cases}$$

$$J \simeq \mathcal{G}$$

$$J = \sigma_Y \delta$$

where E is Young's modulus; ν is Poisson's ratio; and σ_Y is the tensile plastic flow stress adjusted for strain rate, temperature, and degree of plane-strain constraint.

Fracture Behavior

For structural metals and many other solids, the mechanism of progressive fracturing consists of the opening and joining of advance separations. The advance separations are rarely coplanar with the neighboring advance openings or with the crack plane. Typically, increments of growth and joining occur abruptly. The size of the irregularities on the fracture surface serves as an indication of the crack front region, within which the stress-strain state differs substantially from predictions based on the analytical treatments of macroscopic fracture mechanics. This region is commonly known as the fracture process zone.

Nonlinear or plastic strains occur in a crack front region that is usually much larger than the fracture process zone. For metals, this region is known as the crack front plastic zone. For loading conditions well below general yielding, these two crack front zones can be illustrated schematically, as shown in Fig. 1. The compact nature of the plastic zone changes during crack extension and during loading, particularly near and beyond general yielding of the net section. These changes affect measurements of J as a function of crack extension and are discussed later in this article.

In the absence of pronounced directional weaknesses, the crack plane tends to remain normal to the direction of largest tension across the crack front region. A straight line of cracking occurs when the direction of largest tension does not change during crack extension. The development of oblique-shear separations in fracture testing of plates is not an exception to this tendency. The oblique-shear separation is a planar hole-joining behavior within the crack front plastic zone. It can be viewed as a free-surface-induced enlargement of the small oblique-shear separations that are observed in the flat, tensile region of a thick-plate fracture.

Fig. 1 Schematic view of a crack front

Nominal plastic zone (2r_Y), fracture process zone, crack tip opening displacement (δ), polar coordinates (r, θ), and coordinates for linear analysis (x, y) are depicted.

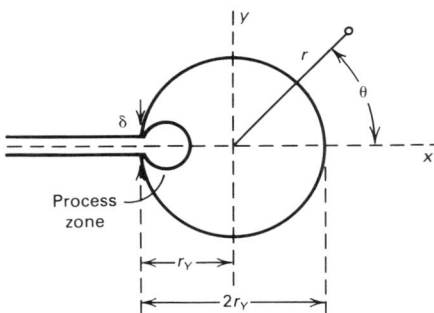

The orientation of an oblique-shear region of fracturing does not influence applications of a linear elastic analysis. However, when the flat, tensile central region of a plate fracture is bordered by shear lips, the tendency is for the crack front to lag behind at the free surfaces, which complicates the task of determining the effective crack size.

Slow, stable cracking plays an important role in the enlargement of cracks toward a size that may be detrimental to structural integrity. The crack advances at a relatively small speed, paced by weakening of the crack front region due to strain reversals, chemical reactions, hydrogen, and creep. Static methods of analysis can be used for the parameters that describe the influence of mechanical loading.

At crack speeds that are fast enough so that the appropriate stress analysis methods are dynamic (inertia dependent), crack extension can still be regarded as stable, in that the stress field driving force is balanced by the resistance to crack extension. An understanding of the rapid fracture regime assists understanding of initiation and arrest toughness measurements.

The basic information required for analysis of rapid fracturing in a material is the relationship between crack speed and the control tendency of the stress field, as represented by the value of K. This relationship has been observed for various transparent plastics, in which transparency and photoelastic sensitivity assisted in direct observation of crack speed and K. Results are shown schematically in Fig. 2.

Data on wide-plate tests of ship steel (Ref 15) and recent studies of rapid fracture in 4340 steel indicate that the principal concepts illustrated in Fig. 2 are also applicable to metals. There is a low K range, in which crack speed increases rapidly with increase in K, and a higher K range, in which the crack

Fig. 2 Relationship of crack speed to K for a semi-brittle material

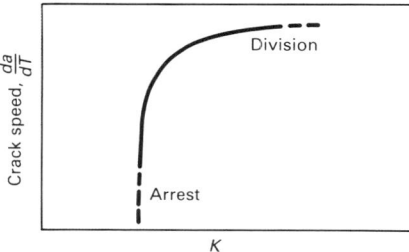

speed is approaching a maximum value and is relatively insensitive to the value of K. In the initial trials of fracture mechanics, the toughness measurement value was often referred to as the onset of rapid fracturing. The tendency of the initiation of rapid fracturing to be abrupt can be understood as a reflection of the expected crack speed behavior in the low K range. A similar degree of abruptness is expected for crack arrest when the K value drops through the low K range.

$\mathcal{G}$ and Compliance Calibration

For a fracture test specimen, the calculation of $\mathcal{G}$ can be assisted by compliance calibration. In the single-edge notched specimen shown in Fig. 3, the applied load, P, is assumed proportional to the load displacement Δ_P:

$$\Delta_P = CP \qquad \text{(Eq 1)}$$

where C is the compliance.

Experimentally, Δ_P can be measured as the increase, due to P, of vertical separation of the reinforced specimen ends. At any value of P, during loading the total stress field energy in the specimen is given by:

$$U_T = \int_0^P P \, d\Delta_P = \frac{1}{2}CP^2 \qquad \text{(Eq 2)}$$

With Δ_P fixed during an increment of crack extension, da, the loss of stress field energy would be $\mathcal{G}B \, da$. At the same time, energy, $P \, d\Delta_P$, enters the specimen if Δ_P does not remain fixed. Thus:

$$dU_T = P \, d\Delta_P - \mathcal{G}B \, da \qquad \text{(Eq 3)}$$

Use of derivatives from Eq 1 and 2 yields:

$$\mathcal{G} = \frac{P^2}{2B} \cdot \frac{dC}{da} \qquad \text{(Eq 4)}$$

Assuming careful measurement of C for a series of lengths (a) of crack-simulating notches in the specimen, methods of trend analysis can be used to determine C and dC/da as a function of a. For a test specimen calibrated in this manner, values of $\mathcal{G}$ during fracture testing can be calculated using the most suitable pair from the three variables P, Δ_P, and a.

This method has been applied to various two-dimensional fracture specimens. In concept, the same compliance calibration concept can be used for circumferential cracks in a round bar or cylinder if the stress pattern and crack exhibit axial symmetry. For use with the J concept, note that Eq 3 can be generalized to apply when the specimen is loaded at several loading positions. Assuming N loading positions:

$$\mathcal{G}B \, da = \sum_{i=1}^{N} P_i \, d\Delta_{P_i} - dU_T \qquad \text{(Eq 5)}$$

where each displacement increment, $d\Delta_{P_i}$, is parallel to the force, P_i.

Stress-Intensity Factors and Crack Tip Stress Fields

Figure 4 illustrates a segment of the crack front used for stress field analysis. For small enough values of the distance, r, from the leading edge of the crack, the stress field is dominated by singularity stress patterns associated with σ_y, τ_{xy}, and τ_{yz}. These are the three stresses that must be finite (usually zero) on the surfaces of the crack.

Figure 5 illustrates the three modes of crack deformation. The stress pattern associated with σ_y is termed Mode 1 and results in direct opening of the crack. For the stress pattern associated with τ_{xy} (Mode 2), the upper crack surface tends to slide over the lower surface in the x direction. The τ_{yz} pattern (Mode 3) is also a sliding mode. The upper crack surface moves over the lower surface in the z direction. The opening mode (Mode 1) is used more frequently in practical applications than Modes 2 and 3.

The Mode 1 crack tip stress pattern is given by:

$$\sigma_y = \frac{K}{\sqrt{2\pi r}} \cos\frac{\theta}{2} \left(1 + \sin\frac{\theta}{2}\sin\frac{3\theta}{2}\right) \quad \text{(Eq 6)}$$

$$\sigma_x = \frac{K}{\sqrt{2\pi r}} \cos\frac{\theta}{2} \left(1 - \sin\frac{\theta}{2}\sin\frac{3\theta}{2}\right) \quad \text{(Eq 7)}$$

$$\tau_{xy} = \frac{K}{\sqrt{2\pi r}} \sin\frac{\theta}{2}\cos\frac{\theta}{2}\cos\frac{3\theta}{2} \qquad \text{(Eq 8)}$$

$$\tau_{yz} = \tau_{xz} = 0 \qquad \text{(Eq 9)}$$

$$\sigma_z = \begin{cases} 0 \text{ for plane stress} \\ \nu(\sigma_x + \sigma_y) \text{ for plane strain} \end{cases} \quad \text{(Eq 10)}$$

Fig. 3 Single-edge notched fracture test specimen

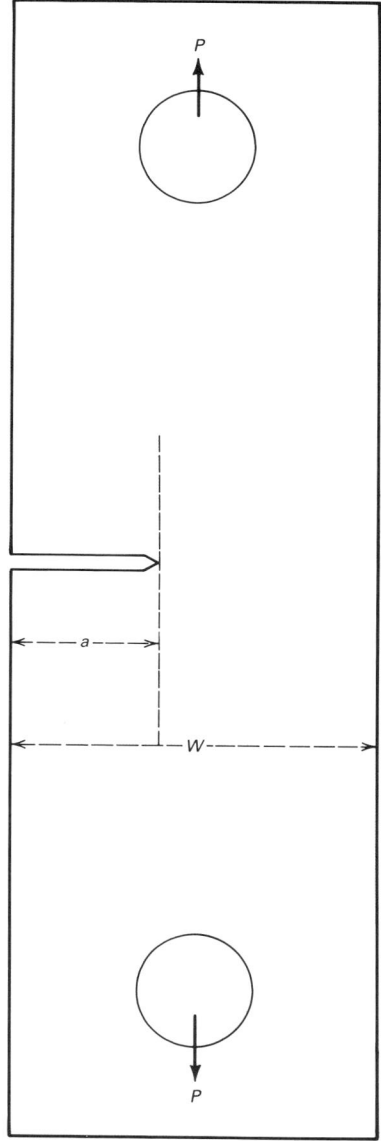

Each of the stresses is proportional to a single parameter K, the stress-intensity factor. Assuming a stress state of plane strain, the x-direction displacement, u, and the y-direction displacement, v, are given by:

$$2\pi\mu u = K_I \sqrt{2\pi r}$$

$$\times \cos\frac{\theta}{2}\left\{1 - 2\nu + \left(\sin\frac{\theta}{2}\right)^2\right\} \text{(Eq 11)}$$

$$2\pi\mu v = K_I \sqrt{2\pi r}$$

$$\times \sin\frac{\theta}{2}\left\{2(1 - \nu) - \left(\cos\frac{\theta}{2}\right)^2\right\} \text{(Eq 12)}$$

Fig. 4 Schematic crack front segment as simplified for stress field analysis

Cartesian coordinates (x, y, z), polar coordinates (r, θ), the crack plane (dashed line), and the crack front are depicted.

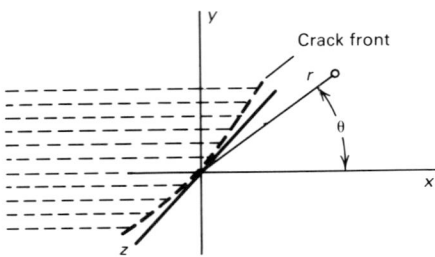

Crack front

Fig. 5 Modes of crack deformation

Mode 1 (opening mode): tension stress in y direction (perpendicular to crack surfaces). Mode 2 (edge-sliding mode): shear stress in x direction (perpendicular to crack tip). Mode 3 (screw-sliding mode): shear stress in z direction (parallel to crack tip)

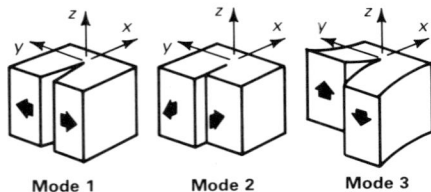

Mode 1   Mode 2   Mode 3

where μ is the modulus of rigidity, and ν is Poisson's ratio. Displacement equations for plane stress are obtained if ν is replaced by $\nu/(1 + \nu)$. For use of Young's modulus, E, in place of μ, note that $E = 2\mu(1 + \nu)$.

The formal definition of K for Mode 1 is:

$$K = \lim_{r\to 0 \text{ on } \theta=0} \sigma_y\sqrt{2\pi r} \qquad \text{(Eq 13)}$$

The stresses in a large area surrounding the crack front can be expressed in the form of a series of terms proportional to increasing integral powers of $\sqrt{r}$, beginning with -1. Each term has an amplitude parameter and a known function of θ. Multiplication of the power series for a stress (for example, σ_y) by $\sqrt{2\pi r}$ and reduction of r toward zero eliminates all power series terms except the first.

Thus, the equality between K and $\sigma_y\sqrt{2\pi r}$ on $\theta = 0$, indicated by Eq 6, is exact only for a region of infinitesimal smallness. However, it is useful to review the nature of the stresses near the crack tip in a region much smaller than the crack size and component dimensions, but substantially larger than the region containing nonlinear strains and the fracture process zone.

Because the stress terms that are inversely proportional to $\sqrt{r}$ often enclose and domi-

nate a region of that nature, Eq 6 to 10 are applicable. In linear elastic applications of fracture mechanics, this singularity (or crack tip) stress field is assumed to dominate. The use of K to characterize the tendency of the stress field to extend the crack depends on this assumption.

Mode 2. The crack tip stress equations for Mode 2 are:

$$\sigma_x = -\frac{K_2}{\sqrt{2\pi r}}\sin\frac{\theta}{2}\left\{2 + \cos\frac{\theta}{2}\cos\frac{3\theta}{2}\right\}$$
(Eq 14)

$$\sigma_y = \frac{K_2}{\sqrt{2\pi r}}\sin\frac{\theta}{2}\cos\frac{\theta}{2}\cos\frac{3\theta}{2} \qquad \text{(Eq 15)}$$

$$\tau_{xy} = \frac{K_2}{\sqrt{2\pi r}}\cos\frac{\theta}{2}\left\{1 - \sin\frac{\theta}{2}\sin\frac{3\theta}{2}\right\}$$
(Eq 16)

The values for τ_{yz}, τ_{xz}, and σ_z are the same as for Mode 1. The corresponding displacements, assuming plane strain, are:

$$2\pi\mu u =$$
$$K_{II}\sqrt{2\pi r}\sin\frac{\theta}{2}\left\{2(1 - \nu) + \left(\cos\frac{\theta}{2}\right)^2\right\}$$
(Eq 17)

$$2\pi\mu v =$$
$$-K_{II}\sqrt{2\pi r}\cos\frac{\theta}{2}\left\{(1 - 2\nu) - \left(\sin\frac{\theta}{2}\right)^2\right\}$$
(Eq 18)

The formal definition of K_2 is given by:

$$K_2 = \lim_{r\to 0 \text{ on } \theta=0}\tau_{xy}\sqrt{2\pi r} \qquad \text{(Eq 19)}$$

The above equations use the ASTM convention for mode subscripts on K. Roman numeral subscripts are used only when a stress state of plane strain is to be indicated. In addition, K, with no subscript for Mode 1, and with Arabic subscripts for Modes 2 and 3, is appropriate when specialization to a stress state of plane strain is not desirable.

The relationship of $\mathscr{G}$ to K for Mode 1 can be obtained using Eq 6 and 12. The procedure consists in putting $\theta = 0$ into the equation for σ_y and putting $\theta = 180°$ into equation for v. The energy restored to the stress field by closure of a small increment of crack length can then be calculated. Because the product of stress times displacement is proportional to K^2, $\mathscr{G}$ must be proportional to K^2. The computation thus yields:

$$K_I^2 = \frac{2\mu}{1 - \nu}\mathscr{G}_1 \qquad \text{(Eq 20)}$$

where

$$\frac{2\mu}{1 - \nu} = \frac{E}{1 - \nu^2}$$

Replacing ν in Eq 20 with $\nu/(1 + \nu)$ provides the plane-stress relationship:

$$K^2 = E\mathscr{G}_1 \qquad \text{(Eq 21)}$$

The corresponding proportionality factors for Mode 2, using Eq 16 and 17, are the same as those for Mode 1.

Crack extension behavior can be related to the crack tip stress equations for Modes 1 and 2. For example, assume an ongoing process of crack extension by opening and joining of advance separations and assume that the influences of nonelastic strains can be neglected. From the Mode 1 crack tip stress equations, $\tau_{xy} = 0$ for $\theta = \pm 60°$. The largest extensional stress (principal stress) is given by:

$$\sigma_1 = \frac{K}{\sqrt{2\pi r}}\left(\cos\frac{\theta}{2} + \frac{1}{2}|\sin\theta|\right) \quad \text{(Eq 22)}$$

For any fixed value of r, σ_1 is largest for $\theta = \pm 60°$, and the direction of the tension σ_1 (at $\theta = 60°$) is normal to the crack line. For this simplistic, two-dimensional model of crack extension, out-of-plane advance separations are favored, and this conclusion corresponds well to the irregularities commonly observed on flat, tensile fractures of structural metals.

A second behavior characteristic is indicated if one considers the stress pattern that would result from joining of the crack tip with an out-of-plane separation. For a deviation of the crack path such as that shown schematically in Fig. 6, the crack tip stress pattern is modified by addition of a small Mode 2 stress pattern, with K_2 positive. The small arrows in Fig. 6 indicate the direction of the Mode 2 crack surface displacements.

The Mode 2 addition causes a small clockwise rotation of the "far-field" tension acting across the new crack tip position, as indicated by the larger arrows in Fig. 6. For the new crack tip position, the principal tension, σ_1, is largest (for a fixed r) at the angles $(60° - \epsilon)$ and $-(60° + \epsilon)$, where ϵ is a small angle proportional to K_2/K_1. In addition, σ_1 is larger (for a fixed r) at $\theta = -(60° + \epsilon)$ than at $\theta = (60° - \epsilon)$.

Thus, the stress pattern is modified with out-of-plane deviations of the crack path to favor an average situation of Mode 1. Departures from this expectation have not been observed in materials that are isotropic with regard to separational strength.

The simplicity of the crack extension model used for the above example is rather

Fig. 6 Schematic of crack path after joining of an out-of-plane advance separation

Large arrows show tilt of the average tension across the new crack tip region. Small arrows indicate Mode 2 displacements.

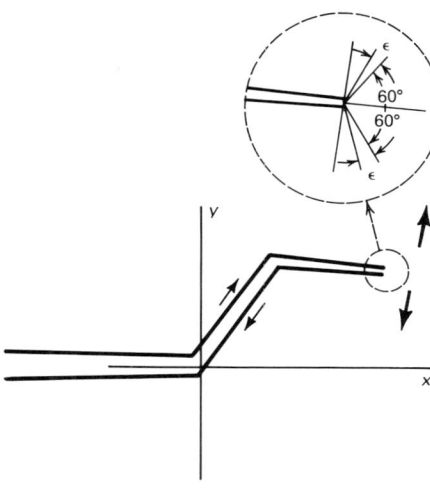

extreme. Out-of-plane advance separations would be expected, independent of the stress pattern, due to chance location of weak regions near the crack front. Furthermore, rapid extension of an advance separation parallel to the crack front is required to create an approximate two-dimensional condition.

Out-of-plane irregularities generally pertain to the crack front as well as to the advance separations. Nevertheless, observations of progressive fracturing indicate strong tendencies toward formation of a crack front that is as linear as the separational details permit. Several fine-scale mechanisms contribute to this behavior, which must be substantially successful to permit large crack front stress elevations and a continuing array of advance separations as the crack front moves forward.

Mode 3. For Mode 3 stress fields, simplification of the analysis is possible, because only z-direction displacements, w, occur and all extensional stresses are absent. Mode 3 is rarely of practical importance. However, analytical studies of Mode 3, particularly those that provide a plastic zone enclosed by a linear elastic stress field, have been helpful concerning Mode 1 behavior. The linear elastic crack tip stress pattern is given by:

$$\tau_{yz} = \frac{K_3}{\sqrt{2\pi r}} \cos\frac{\theta}{2} \qquad \text{(Eq 23)}$$

$$\tau_{xz} = \frac{-K_3}{\sqrt{2\pi r}} \sin\frac{\theta}{2} \qquad \text{(Eq 24)}$$

The shear stress, τ_{xy}, and the extensional stresses are zero. The z-direction displacement, w, is given by:

$$\pi\mu w = K_3\sqrt{2\pi r}\, \sin\frac{\theta}{2} \qquad \text{(Eq 25)}$$

By definition:

$$K_3 = \lim_{r\to 0 \text{ on } \theta=0} \tau_{yz}\sqrt{2\pi r} \qquad \text{(Eq 26)}$$

The relationship between K_3 and $\mathcal{G}_3$ is:

$$K_3 = 2\mu\mathcal{G}_3 \qquad \text{(Eq 27)}$$

From Eq 23 and 24, the maximum shear stress is given by:

$$\tau_M^2 = (\tau_{xz})^2 + (\tau_{yz})^2 = \frac{K_3^2}{2\pi r} \qquad \text{(Eq 28)}$$

The locus line for a fixed value of τ_M is a circle centered on the crack tip.

The equations presented in this section assume linear elasticity. For situations in which the crack front stress pattern is influenced by more than one mode, the stresses, strains, and displacements can be computed separately for each mode and summed. However, the K-values are in the nature of amplitudes, associated with different stress patterns for each mode, and are not individually additive. However, the energy loss rates, $\mathcal{G}$, $\mathcal{G}_2$, and $\mathcal{G}_3$, can be computed independently for each mode and then added if computation of total energy loss rate is desired.

Computations and Estimates Related to K

Several function theory methods have been applied to crack stress field problems. These methods vary in degree of complexity. Experience suggests that, for purposes of K-value determination, methods of numerical analysis usually are preferred, unless the function theory approach is relatively simple. A relatively simple stress function method (Ref 16) has been useful for various infinite plate crack stress field problems. The stress field for a central crack of length 2a in an infinite or very large plate, with remote tension acting normal to the crack, provides a useful illustration.

A solution for the above central crack problem was derived by Griffith (Ref 10) from a study (Ref 17) concerning stresses near an elliptical opening in a plate. The procedure in Ref 10 can be greatly simplified using the stress function method suggested in Ref 16. Using this method, the stress equations are given by:

$$\sigma_x = \text{Re}\, Z - y\,\text{Im}\, Z' \qquad \text{(Eq 29)}$$

$$\sigma_y = \text{Re}\, Z + y\,\text{Im}\, Z' \qquad \text{(Eq 30)}$$

$$\tau_{xy} = -y\,\text{Re}\, Z' \qquad \text{(Eq 31)}$$

In the above equations, Z is a function of the variable $z = x + iy$. Z' is the derivative of the function Z. A function, $Z(z)$, can be expressed as:

$$Z(z) = \text{Re}\, Z + i\,\text{Im}\, Z \qquad \text{(Eq 32)}$$

$\text{Re}\, Z$ (the real portion of Z) and $\text{Im}\, Z$ (the imaginary portion of Z) are functions of x and y; both functions are entirely real. The central crack problem of interest can be solved by assuming:

$$Z = \frac{\sigma z}{\sqrt{z^2 - a^2}} \qquad \text{(Eq 33)}$$

Forming the derivative of Z yields:

$$Z' = \frac{-\sigma a^2}{(z^2 - a^2)^{3/2}} \qquad \text{(Eq 34)}$$

The complexities of finding Re (real) portions and Im (imaginary) portions of the indicated functions are reduced if the vector relations are used:

$$z = re^{i\theta}$$
$$z - a = r_1 e^{i\theta_1}$$
$$z + a = r_2 e^{i\theta_2} \qquad \text{(Eq 35)}$$

These vectors are shown in Fig. 7. Restriction of the range of the three vector angles to $\pm 180°$ is a convenient choice. Using the relationships given above, the stress equations are:

$$\sigma_x = \frac{\sigma r}{\sqrt{r_1 r_2}}$$
$$\times \left\{ \cos(\theta - \phi) - \frac{a^2}{r_1 r_2}\sin\theta\sin 3\phi \right\} - \sigma$$
$$\text{(Eq 36)}$$

$$\sigma_y = \frac{\sigma r}{\sqrt{r_1 r_2}}$$
$$\times \left\{ \cos(\theta - \phi) + \frac{a^2}{r_1 r_2}\sin\theta\sin 3\phi \right\}$$
$$\text{(Eq 37)}$$

$$\tau_{xy} = \frac{\sigma r}{\sqrt{r_1 r_2}}\left\{ \frac{a^2}{r_1 r_2}\sin\theta\cos 3\phi \right\} \qquad \text{(Eq 38)}$$

where $\phi = \frac{1}{2}(\theta_1 + r_2)$.

Both $\cos(\theta - \phi)$ and the ratio $r/\sqrt{r_1 r_2}$ approach unity with an increase in distance from the crack. Thus, the remote value of σ_y is σ. Because of the added term, $-\sigma$, in Eq 36, the remote value of σ_x is zero. This added term, representing a uniform compressive stress, did not originate from the stress function solution for the problem.

However, a uniform extensional stress parallel to the crack can be added to a stress function solution when needed to represent boundary conditions. Along the line segment occupied by the crack, $y = 0$ and $|x| \le a$. For these conditions, $\phi = 90°$ and θ is $0°$ or $180°$. Either value of θ provides $\cos(\theta - \phi) = 0$ and $\sigma_y = 0$. Noting that τ_{xy} is zero along the entire x-axis, the above solution satisfies boundary conditions along the crack and remote from the crack.

To determine the K-value, attention should be focused on the crack tip at $x = a$. Note that for $\theta_1 = 0$, $\theta = \phi = 0$, $r = a + r_1$, and $r_2 = 2a + r_1$. The product, $\sigma_y\sqrt{2\pi r_1}$, then becomes:

$$\sigma_y\sqrt{2\pi r_1} = \frac{\sigma(a + r_1)\sqrt{2\pi r_1}}{\sqrt{r_1(2a + r_1)}} \qquad \text{(Eq 39)}$$

Taking the limit of the right side of this equation as r_1 approaches zero provides:

$$K = \sigma\sqrt{\pi a} \qquad \text{(Eq 40)}$$

The stress function method illustrated above has provided crack stress fields and K-values for several infinite plate problems. In practical fracture problems, usually one free boundary is adjacent to the real (or imagined) crack of interest. In such cases, useful K estimates can sometimes be made using the K-value for a selected infinite plate crack problem as a first approximation. For example, if normal stresses on the y-axis shown in Fig. 7 are ignored, the K-value for a crack of depth a at the free edge of a semi-infinite (very large) plate would be $\sigma\sqrt{\pi a}$. Because $\phi = \pm90°$ along the y-axis, Eq 38 provides $\tau_{xy} = 0$. However, from Eq 36, σ_x is not zero along this line. A region close to the crack exists where σ_x is compressive.

With increase in y, σ_x achieves moderate tensile values before diminishing to zero remote from the crack. Consideration of the effect of these σ_x stresses suggests that they exert a small closing tendency on the opening of the crack and that the K-value, when the y-axis is a free edge, should be moderately larger than $\sigma\sqrt{\pi a}$. The known K-value for this crack problem is given by:

$$K = 1.12\sigma\sqrt{\pi a} \qquad \text{(Eq 41)}$$

where σ is the remote value of σ_y.

A numerical method known as successive stress removal was used to obtain Eq 41. Boundary colocation can provide K-values for some specimens used in fracture testing. Currently, numerical calculations using the finite element approach are preferred. The advantages of this method are enhanced by its compatibility with the J-integral concept.

K for Three-Dimensional Crack Problems

Function theory methods have been used to determine crack stress fields and K-values for a limited number of three-dimensional crack problems. However, these methods are complex. The most widely used result is the distribution of K around the border of a flat, elliptical crack when uniform remote tension is applied normal to the crack. It has been shown that the displacement pattern of the crack surfaces open the crack into an ellipsoidal shape (Ref 18). For linear elastic crack stress fields, $\mathcal{G}$ is proportional to the radius of curvature of the crack opening at the crack front. Assuming that the shape of the flat, elliptical crack is specified in terms of a minor semiaxis, a, and a major semiaxis, c, visualization of the ellipsoidal crack opening shape then shows that the crack front radius of curvature and the value of $\mathcal{G}$ will be largest at the two minor axis crack front points.

Assuming that the flat, elliptical crack lies on the xz plane where $y = 0$ and that the smallest dimension, $2a$, coincides with the x-axis, and assuming that the largest dimension, $2c$, coincides with the z axis, pairs of x and z values on the elliptical crack front can be represented by the parametric equations:

$$x = a\sin\phi, \; z = c\cos\phi \qquad \text{(Eq 42)}$$

Values of K around the crack front are given by (Ref 19):

$$K^2 = \frac{\sigma^2\pi a}{E_K^2}\sqrt{1 - K^2\cos^2\phi} \qquad \text{(Eq 43)}$$

where $K^2 = 1 - (a/c)^2$, σ is the remote y-direction tension, and:

$$E_K = \int_0^{\pi/2}\sqrt{1 - K^2\cos^2 u}\;du \qquad \text{(Eq 44)}$$

E_K is a well-known elliptical shape parameter. $\mathcal{G}$ and K are largest at $z = 0$, where ϕ is $90°$.

Estimates of K for part-through surface cracks are sometimes made using Eq 43. Figure 8 illustrates such a crack in the xz plane. Usually, the surface length is more than

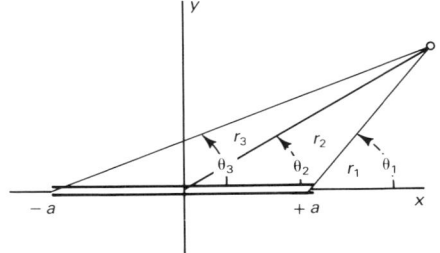

twice the midpoint crack depth, a. If the surface length is more than $4a$, E_K is nearly unity, and ϕ, for the crack front region of interest, is approximately $90°$. Taking into account some elevation of K, because the $x = 0$ plane is a free surface, leads to an estimate of the significant K-value similar to that given by Eq 41.

Three-dimensional finite element computations have been used to develop improved estimates of K for part-through surface cracks and corner cracks at rivet holes in plates (Ref 19). The validity of calculations of this type becomes uncertain as the crack front approaches a free surface. At that limit point, the basic definition of K does not apply. In addition, regions of that nature are the first to exhibit yielding with application of loads.

Crack Front Stresses and Dynamic Characterization

During rapid progressive fracturing, the resistance to crack extension is influenced by small-scale behavior in the fracture process zone that is irregular in time (for example, abrupt) as well as irregular in space. However, for purposes of continuum mechanics analysis, crack speed is thought of as a

Fig. 8 Part-through surface crack with half-elliptical shape

The phase angle ϕ for an indicated point on the crack front is shown.

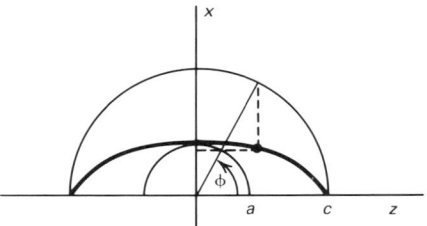

smoothly varying average across extensions much larger than the fracture process zone.

Furthermore, the fracture process zone is assumed to be small enough so that the inertia of its parts has no influence on the surrounding stress-strain field. Using these assumptions and linear elastic analysis, the analytical, singularity-dominated crack tip stress field has a simple r,θ pattern that is similar to the static analysis stress pattern (Ref 20-22).

The fact that crack speed is regarded as constant during an extension larger than the local (crack tip) stress field permits a substantial analysis simplification. Major features of the analysis results are discussed in this section. Stress analysis can be conducted by using a moving system of Cartesian and polar coordinates that remain centered on the crack front, as shown in Fig. 1. Stresses and strains are inversely proportional to $\sqrt{r}$, as in static problems. The variation of the stress-strain pattern with the angle θ changes with crack speed in a manner such that, for a given K, the opening displacements on $\theta = 180°$ increase with an increase in crack speed. It is convenient to use the same definition for K (Eq 13) as used for static crack stress field analysis. Thus, within the local crack tip stress field and for $\theta = 0$:

$$\sigma_y = \frac{K}{\sqrt{2\pi r}} \qquad \text{(Eq 45)}$$

The crack opening displacements, $2v$, directly behind the crack tip ($\theta = 180°$) are given by:

$$2\pi\mu v = \frac{2\lambda_1(1 - \lambda_2^2)}{4\lambda_1\lambda_2 - (1 + \lambda_2^2)^2} K\sqrt{2\pi r}$$

$$\text{(Eq 46)}$$

The parameters λ_1 and λ_2, as well as their functional arrangement in Eq 46, express the influence of the crack speed, c:

$$\lambda_1^2 = 1 - \left(\frac{c}{c_1}\right)^2, \lambda_2^2 = 1 - \left(\frac{c}{c_2}\right)^2 \text{ (Eq 47)}$$

where c_1 is the elastic wave speed associated with the dilatation, Δ, and c_2 is the shear wave speed associated with the rotation, ω. The value of c_2 is given by $\rho c_2^2 = \mu$. The value of c_1 is given by:

$$\rho c_1^2 = \frac{2\mu}{1 - \nu} \text{ (plane stress)}$$

$$\rho c_1^2 = \frac{2\mu(1 - \nu)}{1 - 2\nu} \text{ (plane strain)} \qquad \text{(Eq 48)}$$

where ρ is the density of the material.

The Raleigh wave speed, c_R, is the value of c for which $4\lambda_1\lambda_2$ is equal to $(1 + \lambda_2^2)^2$. This is the propagation velocity for a surface

wave and is clearly the elastic analysis limit for the crack speed. For typical values of Poisson's ratio, ν, c_R is close to $0.92\ c_2$.

Actual observations of maximum crack speed prior to crack division have ranged from $0.2\ c_2$ to $0.6\ c_2$ (Ref 23). Evidently, the succession of local deformations associated with forming and joining of advance openings represents a behavior for which the propagation speed can be much less than c_R.

Using the same method as used to obtain Eq 20, calculation of the energy loss rate, $\mathcal{G}_1$, can be made using Eq 45 and 46, resulting in:

$$K^2 = 2\mu\mathcal{G}_1 \frac{4\lambda_1\lambda_2 - (1 + \lambda_2^2)^2}{\lambda_1(1 - \lambda_2^2)} \qquad \text{(Eq 49)}$$

The coefficient of $2\mu\mathcal{G}_1$ on the right side of Eq 49 is shown as $f(c)$ in Fig. 9. As c approaches zero and the λ terms approach unity, $f(c)$ approaches $(1 + \nu)$ for plane stress and approaches $1/(1 - \nu)$ for plane strain. When determining K-values for running cracks using optical mechanics methods, increased accuracy can be obtained by assuming a dynamic crack tip stress field, as described above, along with several terms with higher powers of $\sqrt{r}$ to represent boundary influences.

Crack Speed. Propagation velocities for nonlinear and plastic deformations are substantially less than those for linear elastic deformations. Assuming that a certain non-elastic or plastic strain is required to produce advance separations, the propagation velocity for that strain would provide a crack speed upper bound. The tendency for crack extension to occur in irregular small-scale segments can be expected to establish the average crack speed at a value somewhat lower than this upper bound.

Generally, the relationship of the crack speed to K (or $\mathcal{G}$) is best obtained by direct measurements. Figure 10 illustrates measurements for various plate specimens of Homolite 100, a transparent material frequently used for optical mechanics experiments that employ photoelasticity. Curves of similar form have been observed using plates of Araldite B.

From a variety of observations with glass and with metallic and polymeric solids of relatively low toughness, the expected relationship is one in which the sensitivity of crack speed to the K-value is quite large near crack arrest and becomes quite small in the K-value region prior to crack division. In the case of strain-rate-sensitive materials, the reduction of crack speed with reduction of K becomes discontinuous toward crack arrest at a crack speed of substantial magnitude (Ref 24). An abrupt jump in crack speed at the initiation of

Calculations assume plane strain and $\nu = 0.3$.

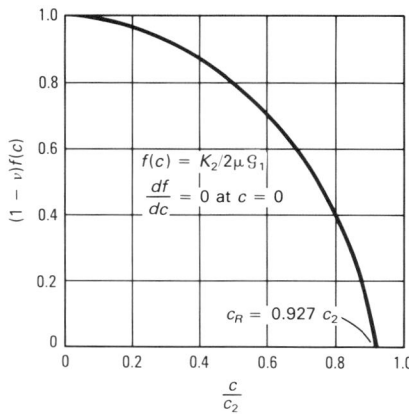

a rapid segment of fracturing is also commonly observed.

If the relationship of crack speed to K is regarded as a steady-state condition of a running crack at various crack speeds and if the stress field around the zone of nonlinear strains is well represented by the linear elastic singularity field, this crack speed versus K relationship provides a unique materials characterization. Actual observations have suggested that crack speed in a given plate material is not always governed entirely by the value of K. Plausible reasons for unusual behaviors exist. For example, at crack speeds near the maximum velocity, attempts at branching couples with restriction of the singularity-dominated zone size may cause conditions such that the singularity field does not completely dominate. For some complex polymeric solids, deformation effects related to molecular structure may interfere with the graduality of the crack speed versus K relationship.

Numerical Dynamic Analysis

Dynamic analysis computations applicable to fracture problems are primarily of two types. In computations of the first type, the crack is stationary, but loads are applied so rapidly that a stress analysis for K determination must account for material inertia. Usually, the goal of the calculation is to relate measurements at some distance from the crack to the value of K at initiation of crack propagation.

In computations of the second type, the dynamic effects are caused by rapid crack propagation. Usually, the stress field prior to initiation of rapid fracture can be established

Fig. 10 Relationship between crack speed and crack tip stress-intensity factor for several test specimen geometries of Homolite 100
Source: Ref 24

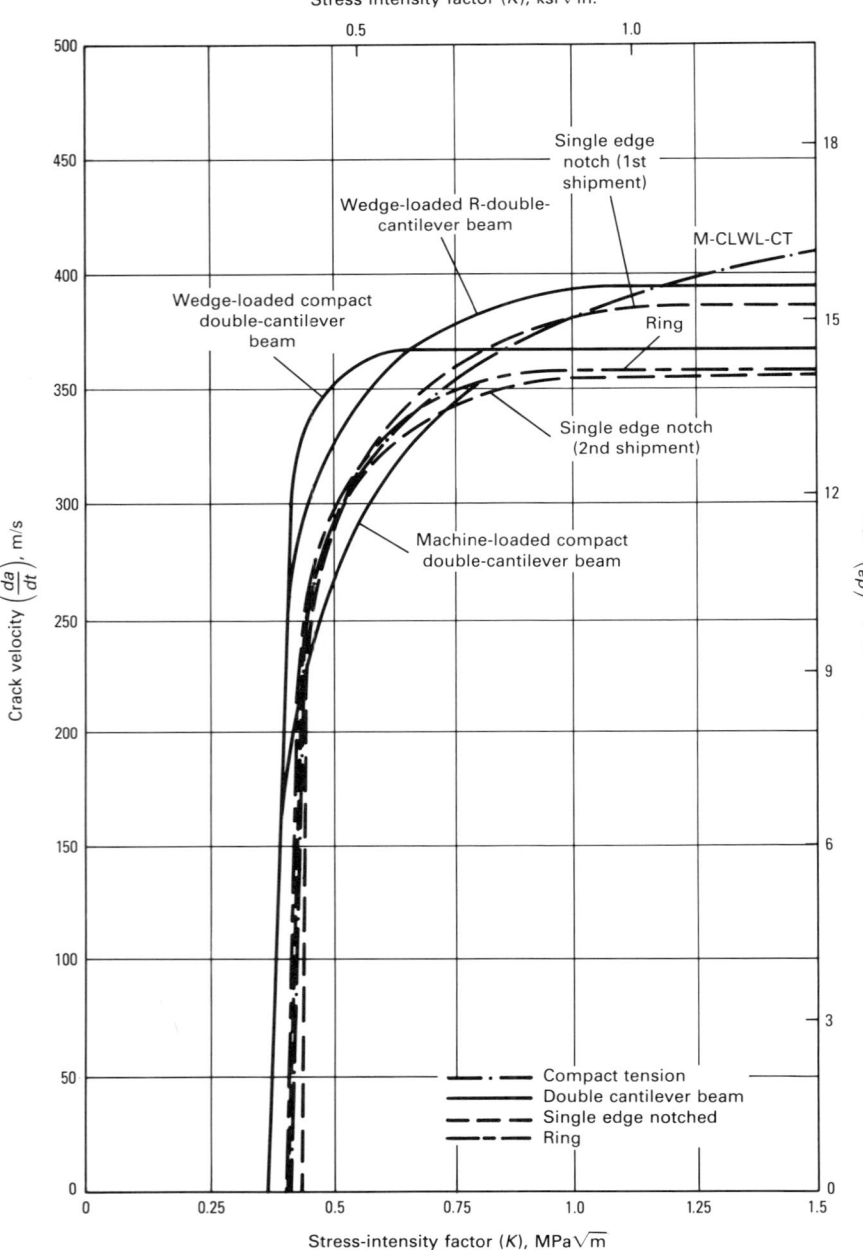

be settled by plausible assumptions, dynamic analysis calculations for a running crack are useful. They assist planning of experiments and the interpretation of observations.

Elastic-Plastic Analysis

The first elastic-plastic analysis model concerned specifically with extending cracks was developed by McClintock (Ref 25). Although restricted to Mode 3 type behavior, this and subsequent Mode 3 analyses have been helpful in terms of understanding stress-strain fields in Mode 1. The addition of a crack growth criterion, based on maintaining a fixed critical strain at a small fixed distance ahead of the crack tip, permitted illustration of the "tearing instability" concept (Ref 26).

At the same time, two other methods of analysis applicable to loading conditions prior to general yielding were presented. The first of these was the "effective crack size" concept, which positioned the crack tip (for calculation purposes) ahead of the physical crack tip by a distance, r_Y, termed the plasticity adjustment factor (Ref 7). Comparisons of the nominal plastic zone size, $2r_Y$, to the plate thickness assisted estimates of the degree of plane-strain constraint. The second method introduced the "plastic strip zone" concept, in which the displacement due to nonlinear (plastic) yielding is concentrated within a strip extending directly ahead of the crack tip (Ref 27).

All of these methods yielded similar estimation equations for the size of the crack opening near the physical crack tip, and interest developed in the use of nominal crack tip opening displacement as a measure of the condition tending to cause crack extension (Ref 28). Use of this method of crack tip characterization beyond general yielding was anticipated. It was soon apparent that the objectives of this analysis method could be achieved alternatively by the J analysis concept based on the Rice J-integral (Ref 11) and on the plastic zone singularity fields (Hutchinson-Rice-Rosengren, or H-R-R, fields) described in Ref 12 and 13.

The J analysis extends the $\mathcal{G}$ concept to provide a crack tip stress-strain field characterization factor that is applicable to nonlinear plastic as well as linear elastic behavior and that is applicable to plastic behavior when deviations from proportional loading are relatively unimportant. For the large plastic zone situations of prime interest, J is calculated as the magnitude of the energy loss rate from the stress field, using the stress-strain relationships as if the material behavior were nonlinear elastic. Actually, the occurrence of energy loss is distributed in the field of plastic strains and is not entirely at the crack tip. However, if deviations from proportional loading are small, J continues to

by static analysis computations. For an anticipated fracture event in a service component, the objective of the calculation is to ascertain whether crack arrest will occur and, if so, at what crack size. Both of these types of dynamic computations have been of special help in demonstrating that static analysis computations are often useful over a larger range than expected.

When full dynamic computations for a rapid run-arrest event are essential, the lack of crack speed versus K information for particular materials of interest is a serious handicap, as such information is essential to the dynamic calculation. Other uncertainties exist related to loading compliance and to stretch and impingement of late breaking connections. When input uncertainties can serve as an amplitude factor characterizing the intensity of the stress-strain field around the crack tip. As the distance, r, from the crack tip becomes sufficiently small, each stress-strain product becomes proportional (or nearly proportional) to J/r.

The term "proportional loading" applies to a field of plastic strain when the direction of the principal strains in a local region does not change. When this condition holds for the region of interest, the stress-strain field of that region does not differ, when computed using nonlinear elasticity (deformation plasticity), from that which would be obtained using stress-strain relations corresponding to the more realistic incremental plasticity. Actually, some nonproportional straining occurs in the Mode 1 elastic-plastic field near a crack. However, calculation comparisons indicate that in the absence of appreciable crack extension and within deformation limits which permit a substantial amount of yielding, the influence of nonproportional loading is relatively small.

Interest in elastic-plastic fracture mechanics developed rapidly following demonstration of the value of the J concept in the interpretation of fracture tests using small specimens (Ref 29, 30). Subsequently, it was learned that the elevation of J during crack extension is primarily due to crack closure influences from unloading in portions (primarily in by-passed regions) of the plastic field. In fact, for crack extension by ductile hole joining, a crack growth criterion equivalent to the critical strain concept discussed in Ref 25 appears applicable. In a related manner, the elevation of J for a substantial amount of crack extension in specimens scaled in size should decrease with a decrease in specimen size (Ref 31). The interpretation complexities introduced by the influences of nonproportional straining that accompanies appreciable amounts of crack extension may be undesirable. However, they are necessary to make the analysis framework consistent with the basic strain-governed nature of ductile crack extension.

The Rice J-Integral

Development of the J-integral concept is discussed in detail in Ref 32. The basic concept of this energy-loss-rate integral is closely related to Eq 5, in which the specimen is visualized as a two-dimensional body loaded at a number of discrete locations. The correction for non-zero load displacements during an increment of crack extension is represented as a summation of load times load displacement products. To represent the energy loss rate, J, in contour integral form, attention should be restricted to the material within a contour that surrounds the crack tip, as shown in Fig. 11. Note that a line segment

of the contour can be regarded as a discrete loading point composed of dx and dy components. An example of an increment of energy transfer due to load displacement during crack extension is illustrated as:

$$(\sigma_x dy)\frac{\partial u}{\partial a}\, da = -(\sigma_x dy)\frac{\partial u}{\partial x}\, da \quad \text{(Eq 50)}$$

which represents the product of a unit thickness load, $\sigma_x dy$, times the parallel increment of load displacement (due to crack extension), $(\partial u/\partial a)da$. Replacement of $\partial u/\partial a$ by $-(\partial u/\partial x)$ is permitted, because the difference between these gradients involves only infinitesimals of a higher order than da. Following this concept with the contour boundary stresses and displacements provides:

$$\Sigma P_i d\Delta_{P_i} = da \oint \left[dx\left(\sigma_y \frac{\partial v}{\partial x} + \tau_{xy} \frac{\partial u}{\partial x}\right) \right.$$
$$\left. - dy\left(\sigma_x \frac{\partial u}{\partial x} + \tau_{xy} \frac{\partial v}{\partial x}\right) \right] \quad \text{(Eq 51)}$$

With regard to the second component of Eq 5 on a unit thickness basis:

$$-dU_T = -da \iint \frac{\partial U}{\partial a}\, dx\, dy$$

$$-dU_T = da \iint \frac{\partial U}{\partial x}\, dx\, dy = da \oint U\, dy$$
$$\text{(Eq 52)}$$

where U is the stress field energy density, and the area integrals are within the contour. Summing the boundary integrals of Eq 51 and 52 provides the energy loss, Jda. Thus, the J-integral has the form:

$$J = \oint \left[dy\left(U - \sigma_x \frac{\partial u}{\partial x} - \tau_{xy} \frac{\partial v}{\partial x}\right) \right.$$
$$\left. + dx\left(\sigma_y \frac{\partial v}{\partial x} + \tau_{xy} \frac{\partial u}{\partial x}\right) \right] \quad \text{(Eq 53)}$$

Thus, it can be shown that the condition for this contour integral to be path independent is:

$$\frac{\partial U}{\partial x} = \sigma_x \frac{\partial \epsilon_x}{\partial x} + \sigma_y \frac{\partial \epsilon_y}{\partial x} + \gamma_{xy} \frac{\partial \gamma_{xy}}{\partial x} \quad \text{(Eq 54)}$$

The computed value of J must be path independent if the behavior is elastic, because energy loss can occur only at the crack tip. However, Eq 54 is of interest in terms of proportional loading and use of the deformation plasticity concept. If the proportionalities between the three strains in Eq 54 remain fixed during loading, then Eq 54 is satisfied and Eq 53 is path independent. There is no similar requirement for proportional loading for regions within the contour where the stress field is elastic (linear or nonlinear),

because U is always strain path independent in such regions.

The potential usefulness of characterization with the J-concept depends on the degree of enclosure of the fracture process zone by an H-R-R singularity field in which stress-strain products are proportional to J/r (Ref 12, 13). After initiation of crack extension, forward movement of the crack front causes unloading in certain regions of the plastic flow pattern, particularly in regions close to the crack front. This stress-strain field modification, introduced by crack extension, has a crack closure influence that adds to the load (and J-value) needed to continue crack extension. The positive value of dJ/da decreases with continued crack extension in agreement with the expectation that the closure influence tends toward a maximum value. This aspect affects interpretations of J-controlled crack extension, but does not interfere extensively with use of J in fracture testing.

A second feature of importance concerns the size of the H-R-R singularity field. General yielding implies intersection of the plastic field with a free edge of the specimen. Continued loading beyond this point substantially increases the influences of the boundary on the plastic field. As a result, for a given specimen material and J-value, the size of the H-R-R singularity field tends to reduce with reduction of specimen size.

The singularity field size requirement tends to increase with crack extension because of the increase of plastic-wake effects. Thus, the size requirements for J-controlled fracture testing are strongly influenced by the need to provide an H-R-R singularity field of adequate size for the anticipated amounts of crack extension.

Fig. 11 J-integral contour around a crack tip shown as a circle for simplicity

Representation of a contour segment (upper right) and the stress forces acting against it. Because the contour integration moves counterclockwise, dx is negative.

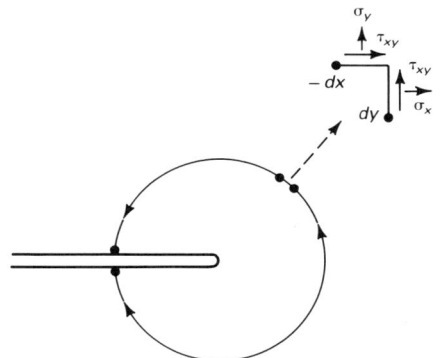

J-values for a plate-type specimen can be calculated by numerical methods. However, descriptive assumptions of stress-strain behavior in the plastic range are necessary. The usual uncertainty of plastic behavior information as well as computation complexities limit the value of stress-strain field J calculations in J-controlled fracture testing. As a result, methods of J-determination that rely on experimental observations are of special value. The basic concept underlying the experimental methods relies on the anticipated similarity of deformation behavior occurring during loading to that of a nonlinear elastic material.

The procedure described in Ref 30 is basically one of compliance calibration. However, the load versus load displacement curves must be carried into the range of large plastic strain and will be nonlinear. The calibration plan requires loading trials of several specimens of the test material. The specimens are identical, except for the size, a, of the crack-simulating notch.

Figure 12 schematically illustrates load versus load displacement curves for three crack sizes. The notch roots are prepared so that substantial loads can be applied without causing crack extension. The shaded area is the nonlinear elastic stress field energy loss which would occur during the crack extension, $a_2 - a_1$, if Δ_P remained fixed. Assuming unit thickness and with adjustment for the finite size of $a_2 - a_1$, the shaded area represents $J \times (a_2 - a_1)$ for an intermediate value of a. An appropriate J-calibration for the material and the specimen can be obtained in this manner.

J-controlled fracture tests frequently use a specimen for which the stress pattern between the crack front and a free edge corresponds primarily to bending. For a deeply

Fig. 12 Schematic P versus Δ_P curves are shown for three specimens with the fixed crack sizes a_1, a_2, and a_3

Equating the shaded area to $J \times (a_2 - a_1)$ provides J for an intermediate crack size. P is the applied load per unit plate thickness.

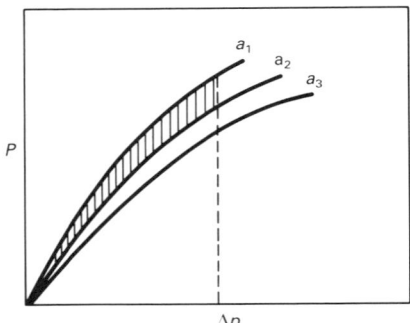

notched bend specimen, for example, it is helpful to assume that the bending moment (or load) depends on crack size only through proportionality to the square of the net section size, b (Ref 33). Using this assumption and reducing $a_2 - a_1$ in Fig. 12 so that $a_2 - a_1 = da$, the vertical separation between the curves for crack sizes a_2 and a_1 becomes:

$$\frac{\partial P}{\partial a} da = -da \frac{2P}{b} \qquad \text{(Eq 55)}$$

In other words, the decrease in P is proportional to the decrease in b^2. Thus, on a unit thickness basis, the shaded area is represented by:

$$J\, da = da\, \frac{2}{b} \int_0^P P\, d\Delta_P \qquad \text{(Eq 56)}$$

Using Eq 56, J can be calculated from the crack size and the area (to the selected measurement point) under the load versus load displacement record. The proportionality of bending moment to b^2 depends only on similitude of the net ligament strain pattern for different sizes of the net ligament at the same angle of bending.

When estimates of J are required for a situation in which the simplicities and experimental observations noted above are not available, numerical calculations may be necessary. The nature of the J-integral is well suited to the finite element numerical method. Care is required so that the energy loss calculation compensates for residual stress and body forces arising from rapid rotation. An acceptable representation of strain hardening is also required. Numerical J-computations are discussed in Ref 32.

The primary application of elastic-plastic fracture mechanics has been toward use of small-specimen tests to predict significant characteristics of crack extension in components that are large enough to ensure that the degree of general yielding is relatively small. Several simplifications can then be used to estimate J for cracks in the service component. The J characterization is two dimensional, as are $\mathcal{G}$ and K.

The concept that the two-dimensional nature of the crack front causes development of a plastic stress-strain singularity field that can be given a one-parameter characterization, J, is attractive for situations in which estimates of J are feasible using available information. When the stress-strain field that surrounds and controls the J singularity field is three dimensional as well as nonlinear, J calculations may be feasible in concept, but impractical for the given situation. For problems of that nature, useful estimation methods may be found in ASTM STP 668 (Ref 34) and in ASTM STP 803 (Ref 35).

Enclosed Plastic Zones. Models that represent the influence of a crack tip plastic zone are useful for conditions that are between general yielding and the extreme smallness of the plastic zone. The r_Y method recognizes that the ability of the linear elastic singularity field to enclose and govern a small plastic zone at the crack tip can be enhanced by positioning the tip of the analysis model crack centrally within the plastic zone. This is done by adding r_Y to the physical crack size:

$$r_Y = \frac{1}{2\pi}\left(\frac{K}{\sigma_Y}\right)^2 \qquad \text{(Eq 57)}$$

In this equation, σ_Y is a judgment estimate of the average tension, normal to the crack, acting across the plastic zone region. In practice, σ_Y frequently is estimated from tensile bar tests as the average of the yield strength and ultimate strength. For a large degree of plane-strain constraint, σ_Y usually has been estimated as the yield strength times $\sqrt{3}$ or 2. The effects of temperature and strain rate on yield properties should also be taken into account.

As an example of the r_Y method, assume that the test specimen is a long plate in tension containing a central crack. A close approximation for K for this specimen style is:

$$K^2 = \sigma^2 \pi a \sec\left(\frac{\pi a}{W}\right) \qquad \text{(Eq 58)}$$

where σ is the average tensile stress near the specimen grips, W is the specimen width, and a is the half-length of the central crack. For K calculation purposes, a_e, the effective value of a, is assumed to be the physical crack size, a_p, plus r_Y. For example:

$$\frac{\pi a}{W} = \frac{\pi a_e}{W} = \frac{\pi a_p}{W} + \frac{\pi r_Y}{W} \qquad \text{(Eq 59)}$$

Dividing Eq 58 by $2W\sigma_Y^2$ and using Eq 57 results in:

$$\frac{\pi r_Y}{W} = \frac{1}{2}\left(\frac{\sigma}{\sigma_Y}\right)^2\left(\frac{\pi a_p}{W} + \frac{\pi r_Y}{W}\right)$$
$$\times \sec\left\{\frac{\pi a_p}{W} + \frac{\pi r_Y}{W}\right\} \qquad \text{(Eq 60)}$$

When the values of σ and a_p are known for the test point of interest, graphical or iteration methods can be used with Eq 60 to obtain values for r_Y and K (Ref 7). Figure 13 shows the ratio of r_Y to the net section, b, as a function of the average (normalized) net section stress, σ_N/σ_Y, for $a_p/W = 0.2$. When $\sigma_N/\sigma_Y = 0.5$, use of the r_Y method increases K by 4.5% and r_Y is only $0.04b$.

However, calculations of σ_Y at the nominal plastic zone boundary already differ from σ_Y values of the singularity field by about 5%. The singularity field that provides

Fig. 13 Ratio of r_Y to the net ligament b as a function of σ_N/σ_Y using Eq 60 and $a_p/W = 0.2$

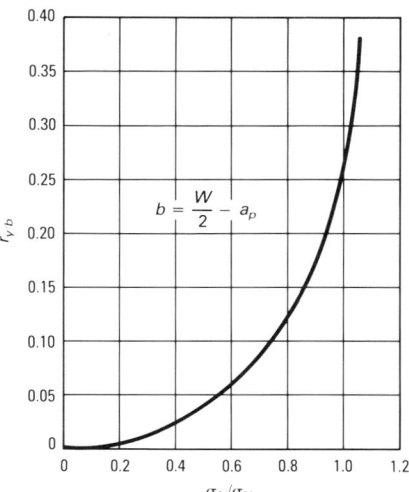

boundary values for an enclosed plastic zone is marginally accurate at $\sigma_N/\sigma_Y = 0.5$ and becomes increasingly inaccurate as σ_N/σ_Y increases above 0.5. Early experiments (Ref 7) showed that use of the r_Y method eliminated specimen size effects in plane-stress (K_c) testing for values of σ_N/σ_Y as large as 0.8.

Use of elastic-plastic analysis clarified this experimental result. For example, values of J were estimated using the strip plastic zone model to simplify calculations (Ref 36). Comparison of values of J to values of $\mathcal{G}$ (r_Y corrected) showed a close, near-unity proportionality to about $\sigma_N/\sigma_Y = 0.9$.

Reference 37 describes experiments that illustrate that careful determinations of r_Y permits $\mathcal{G}$-type estimates of J, even for a moderate degree of general yielding. From the behavior of r_Y near $\sigma_N/\sigma_Y = 1$, as shown in Fig. 13, a large sensitivity to the judgment choice of σ_Y can be anticipated near general yielding. Apparently, the r_Y method permits useful estimates of J for loadings near and below general yielding. In practical application, conservative estimates are advised in the vicinity of general yielding, where the sensitivity to σ_Y is large.

Crack Tip Opening Displacement.

An approximate estimate of the crack tip opening displacement, δ, can be obtained from Eq 12 by using $r = r_Y$, $\theta = 180°$, and by solving the equation for $2v$. The result is:

$$2v = \delta = \frac{4}{\pi} \frac{\mathcal{G}}{\sigma_Y} \qquad \text{(Eq 61)}$$

Equation 12 is in plane-strain form. However, changing the equation to plane stress yields the same result. The strip model of a plastic zone (Ref 27, 38) provides a well-defined value of the crack tip opening displacement. Figure 14 shows the strip plastic zone model for a central crack in an infinite plate. This problem can be solved by treating it as a linear elastic crack stress field problem and using a stress function from Ref 16. From the stress field solution, the value of crack tip opening displacement (plane stress) is given by:

$$\delta = \frac{8\sigma_Y a}{\pi E} \ln \left(\frac{c}{a}\right) \qquad \text{(Eq 62)}$$

where

$$\frac{c}{a} = \sec \left(\frac{\pi\sigma}{2\sigma_Y}\right) \qquad \text{(Eq 63)}$$

The value of c/a shown in Fig. 14 corresponds to $\sigma = 0.67 \, \sigma_Y$. Thus, the plastic zone shown in Fig. 14 cannot contain an H-R-R singularity field. However, the J-integral can be applied to obtain:

$$J = \sigma_Y \delta \qquad \text{(Eq 64)}$$

Comparison of J-values, computed in this manner for various values of σ_N/σ_Y, to values of $\mathcal{G}$ from the r_Y method show agreement within 2% until σ_N/σ_Y becomes larger than 0.8. As for the r_Y method, J-estimates become sensitive to the selection of σ_Y when general yielding is closely approached.

The comparisons discussed above, as well as Eq 61 and 64, suggest that a nominal value of δ may serve as a useful characterization parameter after general yielding. In fact, development of crack tip opening displacement test methods began in England prior to introduction of the J-concept. Currently, methods of analysis associated with the J-concept are used to support and improve crack tip opening displacement fracture testing. The equivalence shown in Eq 64, with allowance for a suitable σ_Y, has been verified by experimental comparisons for loading conditions beyond general yielding (Ref 39).

The r_Y method and the strip plastic zone model have been used primarily for fracture tests in which the crack front stress state is one of plane stress. Useful fracture toughness information is obtained when R-curve methods are used to study slow, stable crack extension, which often follows crack initiation. Calculated values of K (r_Y corrected) continue to increase after crack initiation and achieve nearly a plateau value after about $2r_Y$ extension of the physical crack size.

Some graduality was observed in shear lip development and thickness reduction, but it was not clear that these observations explained the R-curve increase in K. The advent of J-controlled R-curve testing provided numerous specimens with substantial segments of slow, stable cracking. Shear lips were relatively small or, for side-grooved specimens, absent.

It was then evident that soon after initiation of crack extension the fracture surface achieved a fixed appearance that did not change with a further increase in crack size or J. Parallel numerical calculation studies (Ref 40) have shown that the influence of unloading in the by-passed plastic zone was substantial and that the stress-strain conditions in the small crack tip region, severed by each small increment of crack extension, were nearly constant during stable crack growth.

This finding led to several new, but essentially equivalent, characterization parameters. The concept of a fixed critical strain, ϵ_Y, at a fixed small distance ahead of the crack front (Ref 26) is appropriate. Equivalent concepts, such as a fixed crack tip opening angle, may be more convenient for finite element computations.

The crack closure influence due to unloading in the plastic zone wake of the extending crack can be regarded as fine-scale behavior that causes an increase in J until the closure influence saturates. Use of J as the main characterization parameter linking the crack growth to boundary loads and displacements would not be affected in specimens of adequate size. However, the strength of the closure influence depends on the clamping strength of the surrounding residual elastic field.

Reduction of specimen size not only reduces the portion of the plastic field that resembles the H-R-R field, but also reduces the area of the residual elastic field in the specimen. Although the situation is complex, the basic reasons for the size effects in J-R testing reported in Ref 31 are useful and clear. The suggestion of a corrected J parameter, J_M (Ref 41), is useful and applies well to the compact tension tests of Ref 31.

Fig. 14 Crack tip strip zone model for a central crack of length $2a$ in an infinite plate

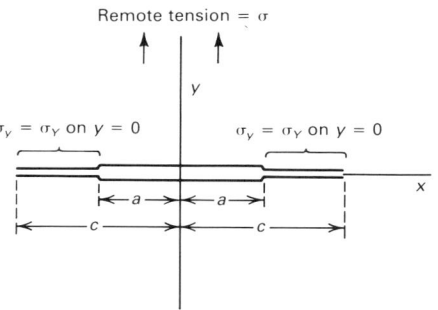

Fracture Toughness Evaluation

From the typical response of crack speed to K (Fig. 2), the onset and arrest of rapid fracturing are abrupt and, for this reason, appear to be natural measurement points for toughness evaluation. However, the degree of abruptness and the value of K pertaining to onset of rapid fracturing usually depend on the history of the prior crack and the loading arrangement. For example, if the final crack extension during formation of a prior crack in a test specimen produces a crack front with a high degree of sharpness severity (low-amplitude fatigue), the initial crack extension during the loading test may be slow and stable so that increments of additional crack extension require additional load.

Measurements of K and crack extension in tests of this type produce resistance curves, or R-curves, as shown in Fig. 15. Crack extension increases the compliance of the specimen. If the loading compliance is sufficiently large, the increase in K due to crack extension eventually will become larger than the increase of crack extension resistance so that rapid separation occurs. However, the indication of a definite and typical K for onset of rapid fracture is uncertain.

With an alternative loading arrangement—for example, crack-line wedge loading as discussed in ASTM Standard E 561 (Ref 42)—relatively long segments of slow, stable crack extension have been obtained using high-strength metallic sheet materials. R-curves determined in this manner can be used to predict loads for onset of rapid fracture in other loading arrangements. Measurements of fracture toughness using the R-curve method are more basic than toughness measurements that focus attention on initiation of crack extension.

The value of K required to cause extension of a prior crack in a service component may differ from the K required for initiation of crack extension measured for the same material under a fixed set of test conditions. Reasons for this may include the influence of unloading in regions of the by-passed plastic zone near the crack front, crack formation by large amplitude fatigue, warm prestressing in structural steels, and crack front irregularities.

Laboratory toughness measurements should duplicate as nearly as possible the stress state surrounding a crack front of typical shape in a service component. Usually, this can be accomplished by using fracture mechanics toughness measurements. However, various influences as noted above can modify the enclosed fracture process zone region and may have a substantial effect on the actual behavior. Consequently, standard

test method development has been conservative.

Special complexities are encountered in toughness measurements of structural steels due to the increased probability of the cleavage separation mode with decrease in test temperature or increase in loading rate. Dominantly cleavage fracturing is a self-stimulating process that occurs rapidly, for which the initiation phase is important. After initiation, the rapid (average) forward motion of the crack front provides the high local strain rates necessary for continuation.

The patterns of small cleavage cracks that lead the crack front are rarely coplanar. However, they are sufficient to cause rapid load transfer of a major portion of the normal tension acting across the pattern. The resistance to crack opening provided by the late-breaking connections introduces the plastic deformation and strips of hole joining fracture commonly observed on the fracture surface. Successful initiation requires spreading of an initial cleavage pattern across a region of sufficient size to trigger adjacent patterns and continuation of this mode of separation.

Uncertainties in the value of K necessary to initiate cleavage fracturing are introduced by inhomogeneity of the material and by irregularities of initial separation events along the front of the prior crack. Even in the ductile hole joining fracture mode, separation tends to occur with considerable local irregularity. Some of these local events are sufficiently abrupt so as to introduce rapid stress elevations by load transfer to adjacent regions. Such influences may introduce undesirable amounts of scatter into K_{Ic} tests for steels, even when the specimen satisfies the size requirements of ASTM Standard E 399 (Ref 43). Because of the probabilistic nature of cleavage initiations, specimen size effects pertaining to the average initiation K value are expected.

Plane-Strain Fracture Toughness

BASICALLY, FOR CRACK INITIATION ONLY

The first standard method of toughness evaluation to use the K parameter (ASTM E 399) was based on initiation of crack extension for plane-strain crack front conditions. Surface cracks are the most frequent type of prior cracking. Crack extension typically begins by deepening of the crack into the section thickness, with the crack front in a stress state of plane strain.

Thus, values of plane-strain fracture toughness, K_{Ic}, determined using ASTM E 399 can be of value in fracture control plans, particularly for high-strength metals of limited ductility. Departures from a large degree of crack front plane strain toward a plane-stress condition consistently result in larger

Fig. 15 Plane-strain initiation fracture toughness as a function of test temperature

Loading times of approximately 150 s (slow load), 0.5 s (intermediate load), and 0.5 ms (rapid load); 25-mm (1-in.) thick specimens from plate of ABS-C steel. NDTT, Nil-ductility transition temperature

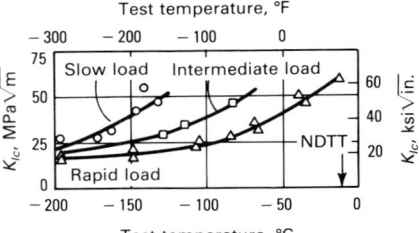

fracture toughness. Thus, values of K_{Ic} are minimum toughness values for a given material and a fixed set of testing conditions.

When the test material is structural steel and the onset of cleavage occurs at or near the measurement point, caution is advised concerning the effects discussed above. ASTM E 399 has been used to study alloy composition, cleanliness, temperature, and loading rate (Ref 2).

Figure 16 illustrates the four plate-type test specimens currently described in ASTM E 399 for use in measurement of plane-strain fracture toughness, K_{Ic}. The compact specimen is used most commonly to save material. The other pin-loaded, circular specimens are taken primarily from thick-walled cylinders or from round bars. For the bend specimen, a ratio of $S/W = 4$ is recommended. This specimen can be used with either three-point or four-point loading.

ASTM E 399 specifies that during the crack extension used to establish the size of the initial crack prior to testing, the last segment of extension should use low-amplitude fatigue. This produces a high degree of notch sharpness severity. As indicated in Fig. 16, the tip of the initial crack is positioned near the midpoint of the specimen.

The K-dominated singularity zone size was investigated for a compact specimen and for a double-cantilever beam specimen (Ref 44). Five percent error of the extensional stress normal to the crack was used as a singularity zone boundary criterion. The crack tip to boundary distance typically was several times larger at $\theta = 90°$ than at $\theta = 0°$ (directly ahead). Using the smaller of these size estimates, two conclusions were reached: (1) the size requirements of ASTM E 399 ensure satisfactory enclosure of the crack tip plastic zone by the singularity zone, and (2) the size of the singularity zone ($\theta = 0°$) is nearly a fixed fraction, $\frac{1}{70}$, of the distance from the crack tip to the nearest free edge of the specimen.

Fig. 16 Plate-type specimen shapes used for K_{Ic} measurement
Source: Ref 43

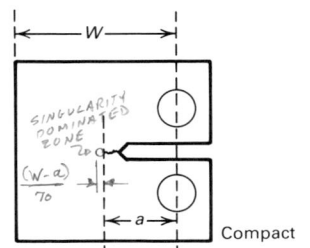

Compact

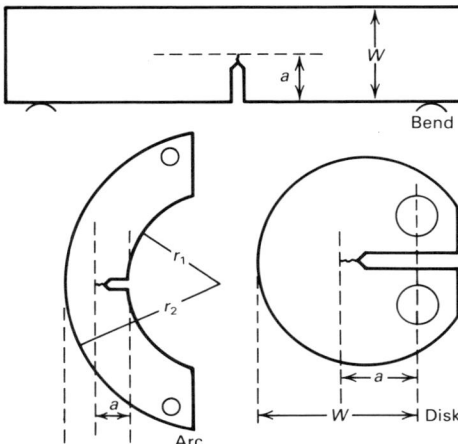

Bend

Arc

Disk

Fig. 17 Schematic P versus crack mouth opening displacement records for three types of specimen behavior
A 5% slope reduction is shown as dashed lines.

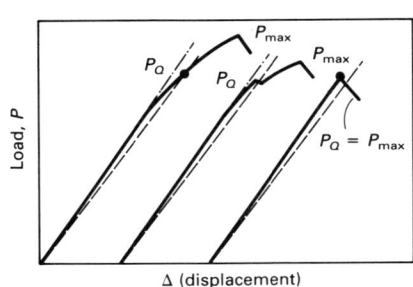

In the K_{Ic} test method, crack opening displacement measurements are made at the specimen edge behind the crack tip (crack mouth opening displacement). Figure 17 graphically records load as a function of that opening displacement for three types of specimen behavior. To establish a definite measurement point on the load versus opening displacement graph, two lines are drawn. One line continues the initial linear portion of the graph. The second line, a 5% secant, is drawn from the origin with a 5% slope reduction.

From knowledge of specimen compliance as a function of crack size, intersection of the secant line with the graph corresponds to about 2% increase in crack size. If the specimen breaks abruptly (or reaches maximum load) before the graph intersects the 5% secant line, maximum load establishes the measurement point. Otherwise, the secant line intersection is the measurement point used in a provisional calculation of K_{Ic}.

The prior crack sharpness severity permits development of slow, stable crack extension within the toughness limitations of the material. Some nonlinearity of the graph is also expected due to formation of a crack tip plastic zone. The contribution of plastic strain to

effective crack size is limited by specimen size requirements of ASTM E 399. These requirements are:

$$a, B, W - a \geq 2.5\left(\frac{K_Q}{\sigma_{YS}}\right)^2 \qquad \text{(Eq 65)}$$

where B is specimen thickness, and σ_{YS} is the material yield strength under uniaxial loading. The subscript Q is used to designate a provisional value of P or K at the K_{Ic} measurement point.

A plane-strain estimate of r_Y using Eq 65 requires an estimate of the ratio σ_Y/σ_{YS}. Commonly used values for this ratio range from $\sqrt{3}$ to 2. The resulting r_Y estimates are close to $0.02a$. However, the plane-strain crack tip plastic zone does not extend forward from the crack tip in the manner of the plane-stress plastic zone. Thus, the test method ensures values of real crack extension at the measurement point that range downward from 2% of the crack size.

An important purpose of the specimen size requirement is to establish an adequate degree of plane strain at the crack front. Using the above r_Y estimate, the nominal plastic zone size, $2r_Y$, is required to be less than 4% of the specimen thickness. Experiments with specimens of high-strength metals of various thicknesses (Ref 45) supported the concept that a ratio of B to $(K/\sigma_{YS})^2 \geq 2.5$ provides a degree of plane-strain constraint that is satisfactory for purposes of K_{Ic} testing. This guideline has been used in discussions of plane-strain constraint.

The amount of real crack extension at the measurement point may be quite small. For situations of this kind, the test method does not allow use of the term K_{Ic} for a test result unless the graph of load versus opening displacement, continued beyond the secant line, shows a maximum load no larger than 1.1 P_Q. Because the allowable nominal plastic zone size is a small fraction of the net section, $W - a$, a maximum value of P no larger than 1.1 P_Q indicates that substantial

real crack extension occurred very close to the load value, P_Q. This ensures that the reported value of K_{Ic} is not overly conservative relative to the real toughness of the material.

ASTM E 399 encourages compliance with the maximum P limitation by requiring $W = 2B$ for all test specimens. In addition, when the test results appear invalid because P becomes larger than 1.1 P_Q prior to fracture, valid results may be obtained by using larger specimens. When P_Q falls on the secant offset line, increase in specimen size increases the amount of crack growth prior to P_Q. For this reason, use of larger specimens may yield larger indications of the K_{Ic} value. Reference 46 provides illustrations of this for high-strength aluminum alloys.

For materials with strength properties that are strain rate dependent, toughness evaluations with short loading times may be of interest. For specimens of moderate size, the loading time can be reduced to about 5 ms without introducing dynamic analysis complexities. Rapid, high-quality recording methods are necessary to continue measurement into the millisecond range. Comparisons between slow-load and rapid-load K_{Ic} values for structural steels can be used to study the shift of the cleavage-fibrous transition to higher temperatures with an increase in strain rate.

The K_{Ic} value obtained from ASTM E 399 pertains only to a small initial amount of stable crack extension. Application of large values of K may then be necessary to determine the onset of rapid fracturing. However, testing modifications to yield more complete toughness information involve some form of R-curve testing and a corresponding increase in complexity both in testing and in applications.

R-Curve Measurements

When forward movement of a crack causes an elevation of K, the resistance value, K_R, after substantial amounts of crack extension may be of interest. R-curve measurements graphically represent K_R as a function of crack extension beyond the initiation phase. Primary applications include pressure vessels and pressurized airplane fuselages. In these applications, proper allowance for the influence of outward bulging on the crack driving force is a complicating factor.

R-curve measurements currently used in laboratory testing restrict out-of-plane deformation of the test specimen. Nevertheless, these tests are of considerable value. They allow direct measurement and comparison under similar test conditions of the resistance to crack extension beyond the starting phase of this separational process.

ASTM E 561 (Ref 42) is suitable only for slow crack extension measurements. However, for structural aluminum alloys and, to a lesser degree, for high-strength titanium and steel alloys, the influences of strain rate on resistance to crack extension are relatively small. For steels, *R*-curve testing is used primarily for testing conditions such that the onset of cleavage is unlikely.

An increase in the effective crack extension resistance, K_R, is expected as the separation moves forward from a prior crack of severe sharpness. The principal reason is a crack closure effect due to stress relaxation in the plastic zone near the crack tip. This influence is enhanced by development of shear lips and crack front curvature. The values of *K* used in ASTM E 561 are computed using the effective crack size. For this reason, values of K_R are nearly equal to the square root of *EJ*, and plastic zones almost large enough to initiate general yielding are acceptable. The close relationship of *R*-curve testing to basic concepts of elastic-plastic fracture mechanics is noteworthy.

The fracture process zone is embedded in a fine-scale stress-strain environment influenced by unloading in nearby regions of the by-passed plastic zone. This fine-scale region is surrounded by a stress-strain field that can be approximated as an H-R-R singularity field governed by a value of *J*. Outward from this region, the stress-strain field blends into the residual linear elastic field. For common *R*-curve testing conditions, a *K*-dominated linear elastic singularity field surrounding the field of nonlinear strains does not exist. Use of *K* as a characterization parameter in ASTM E 561 is permissible, because use of the r_Y correction to crack size (or the compliance indication of effective crack size) in calculation of *K* provides values of K^2 that are equal to *EJ*.

A second basic consideration concerns the influence of specimen geometry on the relationship of K_R plotted as a function of crack extension. *R*-curves measured using different specimen geometries and different lateral

size can be superimposed only when a crack size plasticity adjustment is used in calculations of the K_R value (Ref 47-49).

This result supports the concept that the resistance to crack extension can be visualized in terms of a *J*-controlled plastic field near the crack front (Ref 36). In addition, the influences that cause an increase in *J* and K_R with crack extension appear to operate at a sufficiently small scale so that their approximate enclosure by the *J* field is a useful assumption.

Figure 18 presents plan views of three specimens used with the *R*-curve method. The compact specimen, wedge-loaded near the crack line, has the largest *R*-curve measurement capacity and is preferred. With the center-cracked panel, the tendency toward fixed-load elevation of *K* with crack extension is greatest. This limits the amount of slow, stable crack extension that can be observed prior to the onset of rapid fracturing.

To avoid complexities from out-of-plane deformation and to permit accurate measurement of opening displacement, Ref 42 suggests use of devices that hold sheet-type specimens in a flat condition by application of moderate pressure against the specimen surfaces. The sharpness severity of the initial crack is established by use of low-amplitude fatigue. In the center-cracked panel, the recommended initial crack tip position is at about 0.2× specimen width from the center line. In compact specimens, the initial crack tip location is approximately 0.4*W*.

The specimen size requirements essentially limit the applied load to a value that is moderately less than would be required for general yielding. The ASTM E 561 size requirements allow much larger plastic zones than do those in ASTM E 399. For example, with compact specimens, requirements are defined by:

$$(W - a_p) \geq \frac{4}{\pi} \left(\frac{K_{max}}{\sigma_{YS}} \right)^2 \qquad \text{(Eq 66)}$$

where a_p is the physical crack size. When the difference between σ_Y for plane strain and σ_Y

for plane stress is taken into account, Eq 66 allows a ratio of $2r_Y$ to net section size about seven times larger than allowed in ASTM E 399.

The *R*-curve measurement method using the wedge-loaded compact specimen requires crack opening measurements at two positions. These positions, marked V_1 and V_2, are shown in Fig. 18. From compliance calibration of the specimen, the ratio of the opening displacement at these two positions indicates the effective crack size. Using this value of crack size, the load applied to the specimen by wedge action can be calculated using either one of the two opening displacements.

For the other two specimens in Fig. 18, the load is measured directly, and only one crack opening measurement is required. The observed loading and opening displacement provide a compliance value that can be used to estimate the effective crack size. Visual observations of the physical crack size are desirable. Comparisons can then be made between r_Y, estimated using a judgment value of σ_Y, and the difference between the effective crack size (from apparent compliance) and the physical crack size. An agreement should be found if the accuracy of opening displacement measurements is good.

Comparisons of this nature are useful in developing improved judgment estimates of σ_Y for materials and plate thicknesses. The application of a measured *R*-curve may require determining the critical load necessary for tearing instability and the onset of rapid fracture in a service component or model containing a crack of given size.

The conditions for tearing instability are shown in Fig. 19. The dashed line segment, tangent to the *R*-curve, shows the magnitude and slope of *K* versus *a*, calculated for a structural application at a loading condition where the application values of *K* and *dK/da* match the values of K_R and dK_R/da. As with *R*-curve *K*-values, an r_Y-type plastic zone adjustment is used for calculation of applica-

Fig. 18 Plan views of two compact-type specimens and the notch portion of a center-cracked panel
Source: Ref 42

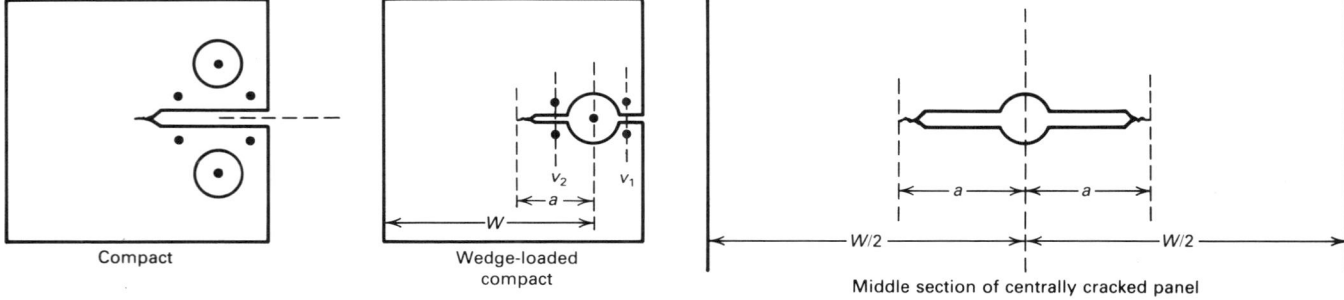

Fig. 19 Schematic R-curve

Dashed line shows a segment of the driving K-value, for which dK/da and K match the slope of the K-value of the R curve.

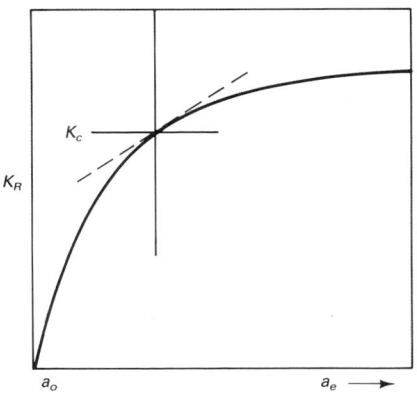

tion K values. The values of K_R and of dK_R/da must be calculated for the applications crack using r_Y-corrected estimates of K_R.

The close relationship between basic concepts used in R-curve testing and in J-R testing has been noted. However, with R-curve testing, the main emphasis has been on the crack front conditions of plane stress; with J-R testing, emphasis is on crack front conditions of plane strain.

Impact K_{cd} Testing of Structural Steels

When a fractured structural steel exhibits only a small amount of hole joining (fibrous) separation prior to the onset of cleavage and when fracture occurs prior to general yielding, a single K-value characterization of the crack front conditions for the onset of rapid cleavage is useful. When substantial fibrous separation occurs prior to cleavage, the increase in toughness is likely to cause general yielding prior to the measurement point of interest. Application of fracture mechanics methods of analysis then becomes difficult.

Restricting the testing temperature to low values so that general yielding does not occur usually provides an adequately small region of initial fibrous separation. This restriction does not prevent extension of the test results into a transition-temperature range. However, except for rapid-load testing, the toughness transition range occurs at temperatures well below the nil-ductility transition temperature.

In this discussion, to avoid confusion with K_c-values that pertain to R-curve tearing instability, the K-value for onset of cleavage obtained with rapid-load testing is termed

K_{cd}. Although several techniques have been used to obtain K_{cd}-values, no accepted testing standards exist for impact fracture toughness testing. The following method (Ref 50) serves to illustrate this concept. Specimens were of the notched-bend type with three-point loading. Width, W, was 76 mm (3 in.), the span, S, was 3.33W, and specimen thicknesses ranged from 12.7 to 76 mm (0.5 to 3 in.). Bridge steel plate materials (A36, A441, A588, and A514) were tested, and the full plate thickness was used. Rapid loading was applied by dropping a 91-kg (200-lb) weight.

Impact effects that might cause dynamic analysis complexities were avoided by using an aluminum alloy cushion between the tup of the drop-weight and the specimen. The resulting loading time (about 1 ms) was long enough so that instrumentation on the tup could be used to measure the maximum load on the specimen. K_{cd} was calculated using this maximum load and the crack size exhibited by fracture examination. The removal of dynamic effects in the specimen was marginal.

In a subsequent study, the drop-weight was increased to 182 kg (400 lb), and the cushion material was changed to half-rounds of unhardened drill rod. This change doubled the loading time. In addition, calibration determined that maximum loads could be inferred from the sizes of the cushion indents.

The K_{cd} results obtained using the above method were similar to those reported in Ref 51, which presents K_{cd}-values (termed K_{Ic}) at various temperatures for 38-mm (1.5-in.) thick A36 steel at three loading rates. Comparisons to the specimen thickness requirement of ASTM E 399 indicated that the governing stress-strain state for these results was essentially plane strain.

The disappearance of loading rate influence at low temperatures was found consistently for other steel materials. A36 had the lowest yield strength among the steels tested and showed the largest influence of loading rate on the transition temperature position. For steels with a yield strength larger than 965 MPa (140 ksi), no significant influence of loading rate was found, although a substantial increase in toughness with temperature was evident.

A summary of strain rate and temperature effects as related to fracture toughness in structural steels can be found in Ref 3. Correlations between toughness in terms of K and results obtained with tests used for toughness quality control are emphasized. All three of the toughness control tests considered involved rapid loading: V-notch Charpy impact, dynamic tear, and nil-ductility transition temperature.

Correlations between K-value tests and no-K-value tests have been examined at similar loading speeds. Thus, several slow-bend tests supplemented the impact tests with Charpy specimens. Special attention was given to the degree of superposition from shifting the temperature for results in the transition range by amounts dependent on yield strength of the steel. A correlation plan based entirely on Charpy V-notch impact testing as the toughness control method was then developed and has been used in specifications for bridge steels.

Crack Arrest Testing

Measurement of crack arrest toughness is of interest for cleavage fracturing of structural steels. Strain rates in the fracture process zone are high for a running crack and support the continuing cleavage in a structural steel. Materials with relatively small sensitivity to strain rate require values of K as large as the K_c-values observed in an R-curve test to ensure a continuing separation at any crack speed.

The practical importance of crack arrest toughness is associated primarily with run-arrest cleavage behavior, because the crack arrest value of K (K_A) may be less than the K-value for initiation of rapid fracturing (K_c). The testing methods described below pertain to crack arrest toughness of structural steels under nearly plane-strain conditions at the crack front.

Interest in measurement of crack arrest toughness began prior to the development of numerical methods of analysis that are generally applicable to a running crack. Attention therefore centered on estimation methods that used only a static method of analysis. For example, from fractographic studies of cleavage fracturing, separational irregularities occur that can be regarded as local arrest followed by reinitiation at fine scale. This suggested that a reduction in the K-value such that reinitiations along the crack front fail to cause continuing cleavage might correspond to a close equivalence between K_{Ia} and the rapid-load K_{Ic}-value.

From available data, the loading time that is short enough for this equivalence varies from several milliseconds at the nil-ductility transition temperature to several minutes at sufficiently lower temperatures. For test temperatures above the nil-ductility transition temperature, a loading time small enough for the above equivalence is likely to introduce dynamic (stress wave) loading complexities. Thus, the temperature range of principal interest for crack toughness evaluations that employ run-arrest segments of cleavage fracturing is above the nil-ductility transition temperature.

Measuring the end-point K-value for a run-arrest segment of fast fracture in a test specimen of moderate size has several advantages. The observed event appears realistic in its nature. In addition, the large elevation of plastic flow resistance, due to very high strain rates near the moving crack front, limits the plastic zone size and assists application of linear elastic analysis to the computation of test results.

However, no simple observation methods exist to permit direct measurement of K during crack arrest. A practical test method for evaluation of crack arrest toughness assumes that a static analysis estimate of K from final crack size and an opening displacement measurement at a short time (about 1 ms) after crack arrest do not differ substantially from the value of K during crack arrest. The results of static analysis tests of this nature are represented by K_{Ia}.

Figure 20 illustrates the specimen used for the first well-known K_{Ia} tests (Ref 52). The transverse wedge-loaded compact specimen shown in Fig. 20(b) was used in a cooperative test program (Ref 53) and is of current interest for measurements of K_{Ia}. The specimen in Fig. 20(b) is a modification of the wedge-loaded specimen in Fig. 18 used for R-curve measurements.

The wedge-loading arrangement is modified to provide increased loading stiffness. Only one opening displacement measurement is used, and devices to reduce out-of-plane deformation are unnecessary. The specimen usually is horizontal, and downward motion of the small included angle wedge moves the loading pins outward, applying an opening load to the notched specimen. Downward motion of the pins usually is prevented by pin flanges bearing on the top surface of the specimen.

An alternative loading arrangement provides pin flanges against the bottom surface of the specimen. A base plate is needed for either arrangement to react against the downward force acting to move the wedge. Initiation of a rapid cleavage fracture is encouraged by placing a deposit of brittle weld metal across the root of the initial notch. The development of shear lips and a large crack front curvature lead to uncertainties of the crack size at arrest. For this reason, side grooves of moderate depth are used.

For a successful test, the onset of rapid fracture must occur before the initial plastic zone size becomes excessive relative to the specimen dimensions. The desired crack arrest position depends on specimen size, but is typically in the range of $a/W = 0.55$ to 0.85. After crack arrest, the specimen is heated to tint the fracture surface and is then broken at low temperature.

Fig. 20 Typical crack arrest specimens

(a) Contoured specimen used in early crack arrest experiments. The contour provides $K = MP$, where M is a constant. Source: Ref 52. (b) Wedge-loaded compact specimen used in cooperative test program. Source: Ref 53

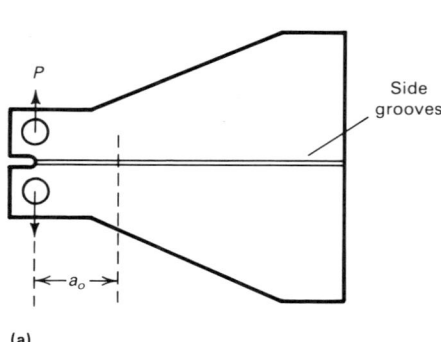

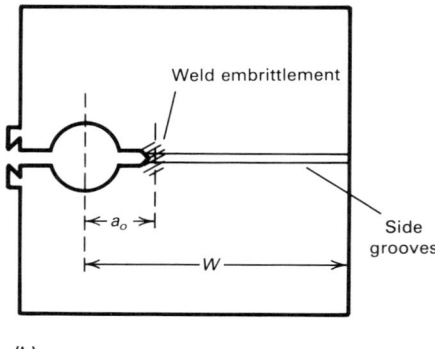

The crack opening, at the position indicated in Fig. 20(b), is measured with either a clip gage or an eddy current displacement gage. The opening displacement record determines a final value in two increments. First, there is the relatively slow opening due to wedge movement. At the time of rapid fracture, with the wedge essentially stationary, a quick additional opening occurs. This quick opening, usually 5 to 20% of the slow opening, is due to the release of compressive strains near the contacting surfaces of the specimen and the loading pins. The entire displacement, including the quick opening, is used in computations of the test result, K_{Ia}.

Based on the available experience with K_{Ia} measurements, a K_{Ia} test description (Ref 54) has been proposed with the following requirements:

$$K_o \leq 0.86\sigma_{YS}\sqrt{W}, \ B > \left(\frac{K_{Ia}}{\sigma_{YD}}\right)^2 \quad \text{(Eq 67)}$$

$$0.15 < (W - a_a) > 1.25\left(\frac{K_{Ia}}{\sigma_{YD}}\right)^2 \quad \text{(Eq 68)}$$

$$a_a - a_o > \left(\frac{1}{2\pi}\right)\left(\frac{K_o}{\sigma_{YS}}\right)^2 \quad \text{(Eq 69)}$$

where K_o is the K-estimate at the start of run-arrest, a_a is the crack size after run-arrest, and $\sigma_{YD} \approx \sigma_{YS} + 30$ ksi. Useful results have been obtained using methods similar to the one above. However, additional efforts to reduce imperfections are desirable. These include the uncertain effectiveness of the weld-embrittled notch in initiating the cleavage fracture, the tendency of test results to decrease with crack jump size, and indications that the scatter of test results is partly due to the method rather than to inequalities in the test material.

For the static-condition K-value after arrest to equal the dynamic-condition K-value during arrest, two requirements must be met: (1) oscillations of the K-value during and after arrest should be relatively small, and (2) the decay of dynamic effects after arrest should not introduce a significant increasing or decreasing trend in the K-value. The main obstacle to consistent achievement of these requirements is uncertain size and the influence of the quick release of loading-spring energy.

The influence of this dynamic impulse on crack extension occurs late during the run-arrest time, but soon enough to affect the position of crack arrest. The magnitude of this dynamic disturbance can be reduced by use of the pressure of the pin flanges against the top of the specimen. However, the degree of reduction depends on friction and is difficult to control.

Despite the testing problems, K_{Ia} values obtained in the manner indicated above have shown good correlation with crack arrest K-values for thermal shock tests of thick-walled cylinders in which K at crack arrest was computed using static analysis (Ref 55). However, trials of numerical dynamic analysis programs for selected test conditions indicated that dynamic influences on the K-value for crack arrest would be small for these tests. This is due primarily to the loading stiffness that can be obtained when loading is applied by use of a temperature gradient.

Moderate-size crack jumps were also helpful. The comparison values of K_{Ia} obtained from small-specimen tests were obtained from a laboratory experienced in crack arrest toughness testing. The average K_{Ia} values for A533B steel specimens tested at two temperatures from 28 laboratories are compared and evaluated in Ref 53. Extension of these types of crack arrest toughness measurements to temperatures near and above the start of the "shelf" region of Charpy impact test results does not currently appear feasible with compact specimens of moderate size ($W < 200$ mm, or 8 in.).

Dynamic fracture mechanics is an active field of research. Given better estimates of crack speed versus K, test refinements from use of dynamic analysis computations may become useful for small-specimen testing. As a method for minimizing testing complexities, the use of correlations between K_{Ia} and rapid-load K_{Ic}-values should not be overlooked.

Elastic-Plastic Fracture Toughness Tests

Demonstration of near equivalence between small-specimen J_{Ic}-values and large-specimen K_{Ic}-values in terms of the relationship, $EJ_I = (1 - \nu^2)K_I^2$ (Ref 30), fostered the initial interest in elastic-plastic fracture mechanics. Later it was shown that use of the J-concept for measurement of the increase of J during substantial increments of crack growth (J-R testing) was closely related and helpful. However, testing simplicity favored the development of a standard method for J_{Ic} testing as a separate task. The result is ASTM E 813, "Standard Test Method for J_{Ic}, a Measure of Fracture Toughness" (Ref 56). Guidelines for J-R testing are also included.

The objective of the J_{Ic} test is to establish the applied J-value needed to cause a relatively small amount of initial crack extension. As for the K_{Ic} method (ASTM E 399), crack growth starts from an initial crack front produced by low-amplitude fatigue. ASTM E 813 is not appropriate for conditions in which rapid cleavage starts directly from the initial crack front. The start of crack extension is pictured as a slow, stable separation. An initial increase of the physical crack size, a_p, is anticipated, due to the stretching and blunting of the initial crack tip. With a further increase in loading, the blunting behavior is replaced by crack front advancement due to formation and joining of small advance openings.

ASTM E 813 focuses on the value of J applicable to the change from blunting to hole joining behavior as a suitable J_{Ic} measurement point. Careful investigations have shown that this behavior change has substantial graduality and extends across a rather large fractional change in the J-value. However, separation by advance hole joining is clearly dominant at very small values of crack extension. This permits the use of arbitrary rules to provide a satisfactory initiation-type measurement point.

The determination of J_{Ic} in ASTM E 813 is based on a graph of J as a function of crack extension, Δa_p. The measured pairs of J and Δa_p values are expected to show a rising trend of J with an increase in Δa_p. A "blunting line" is constructed on the graph using the equation, $J = (2\sigma_Y)\Delta a_p$, where σ_Y is the average of the yield and ultimate tensile bar strengths. Observations should lie on or near this line prior to the "start" of crack extension.

Observations that deviate significantly from the blunting line (to larger values of Δa_p), but by no more than 1.5 mm (0.06 in.), are then used to determine a best-fit straight line that can be extended to intersect the blunting line. Figure 21 illustrates the concept of this measurement. The intersection of the R-curve segment line with the blunting line is the measurement point and provides the J_{Ic}-value. The restriction of ASTM E 813 to conditions of plane strain at the crack front and to a very small amount of initial crack extension eliminates several complexities that must be considered when measuring J as a function of Δa_p using the R-curve method.

ASTM E 813 specifies two specimen types (compact and bend) and two acceptable measurement techniques for obtaining the paired determinations of J and Δa_p. Calculations of J capitalize on the close relationship of J to loading work. The observations must permit determination of load versus load displacement. For a compact specimen, this is assisted by shifting the opening displacement measurement position to a line through the loading hole centers. For the bend specimen, three-point loading is specified. The load displacement measurement pertains to the region of the specimen adjacent to the central loading point, and allowance for specimen indentation must be considered.

Measurement techniques are either multi-specimen or single-specimen methods. For the multi-specimen method, five or more identical specimens are used to obtain a J_{Ic}-value. Under typical testing conditions, general yielding permits development of a maximum load during the test at a relatively small value of Δa_p. Thus, displacement control loading is recommended.

Loading of each specimen is discontinued at a displacement value that is estimated to provide a pair of J- and Δa_p-values within the region offset from the blunting line, as indicated in Fig. 21. At least four determinations are necessary. Values of Δa_p are determined after the loading test by examination of the fracture surface. Prior to breaking the specimen completely, the fracture is heat tinted.

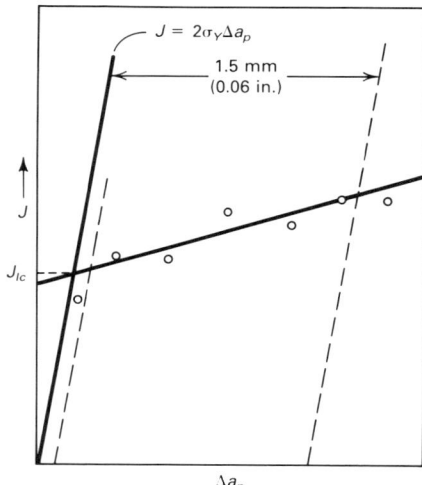

Fig. 21 Graph of J and Δa_p data points showing the line segment used to determine J_{Ic}

Data points to the far left and far right are not used in the determination.

Alternatively, a segment of fatigue cracking can be used to mark the end-point of crack extension.

Values of J, calculated without a correction for crack extension, are satisfactory for purposes of J_{Ic} determination. Omitting the correction for crack extension, J is given by a relation based on loading work, and equivalent to Eq 56, as follows:

$$J = \frac{A}{Bb} f\left(\frac{a_o}{W}\right) \qquad \text{(Eq 70)}$$

where A is the area under the load versus load displacement curve, B is specimen thickness (the net thickness for side-grooved specimens), and b is $(W - a_o)$. The function $f(a_o/W)$ is 2 for the three-point bend specimen. For the compact specimen, an adjustment to this value of f is necessary because of the tension component on the net ligament; appropriate values for f are tabulated in ASTM E 813.

The single-specimen method uses measurements of linear elastic specimen compliance at spaced intervals during the test to provide a series of Δa_p-values. The load versus load displacement curve is not significantly altered by the insertion of these unload-reload cycles, which drop the load by about 10%.

Because the linear elastic specimen compliance is known as a function of crack size, a, values of a can be determined using the compliance indicated by the linear portion of each unload-reload cycle. Satisfactory crack size determinations by this method require special care, as discussed in ASTM E 813.

Specimen size requirements are placed on the values of J used to determine J_{Ic}, as well as on the value of J_{Ic}. Each acceptable value of J must be small enough so that $25(J/\sigma_Y)$ is less than either $(W - a_o)$ or B (or B_{net} for side-grooved specimens). When values of J_{Ic} are converted to values of K_{Ic} for comparison purposes, use of the following equation is suggested in ASTM E 813:

$$K_{Ic}^2 = EJ_{Ic} \qquad \text{(Eq 71)}$$

Although Eq 71 is not the expected relationship between K and J for plane strain, it is consistent with the K_{Ic} determination used in ASTM E 399.

ASTM E 813 is designed only for tests at a relatively slow rate of loading, typically 30 s to 5 min. The method requires that the slope of the extended R-curve segment, which intersects the blunting line at J_{Ic}, must be less than one half of the blunting line slope. This condition, along with the size requirements, may render the J_{Ic} test improper for high-toughness materials. For structural steels, the J_{Ic} test is not appropriate if onset of rapid cleavage can occur during loading.

Exploratory research with rapid-load J testing of steels has indicated that the values of J required to extend a crack in the hole joining mode of separation increase with time-rate of crack extension. Thus, the limitation of the J_{Ic} test to slow loading is not a serious drawback in practical applications. Values of K_{Ic} derived from the J_{Ic} test usually are moderately lower than directly measured values of K_{Ic} due to the smaller amount of crack extension at the J_{Ic} measurement point. The initiation behavior characterized by J_{Ic} is a conservative indication of initiation of crack extension from a prior crack in a service component.

J-R Measurements

Although the J_{Ic} measurement adequately characterizes the initiation phase of ductile crack growth, the resistance to crack extension that develops after initiation is important both for practical applications and for improved understanding of the fracture toughness of materials. Measurements that provide graphical representations of J as a function of Δa_p (J-R curves) can be made using specimens of moderate size and have several advantages. The J method focuses attention on the region of plastic strain directly adjacent to the fracture process zone. Careful study of this region during crack growth is a necessary component of research efforts toward improved understanding of fracture toughness in terms of behaviors within the fracture process zone.

At present, a testing standard for J-R measurements has not been completed. The proposed ASTM J-R curve method is limited to conditions of plane strain at the crack front. Specimens and testing methods similar to those described in ASTM E 813 can be used. Computations of J using Eq 70 include the influence of additional crack size on the estimate of J. A correction for rotation is sometimes used for improved accuracy.

When the objective is a J-R curve instead of a single J-value, several measurement considerations are changed. Although the multi-specimen technique is potentially applicable, the data point scatter typical for this technique is undesirable. Use of compliance indications of crack size (single-specimen technique) has the advantage of providing reduced scatter, as well as more data points. Most J-R curve investigations use the single-specimen technique and compact specimens (Ref 31, 57, 58).

ASTM E 813 allows use of a net ligament size, $W - a_p$, substantially larger than specimen thickness, B, if B is larger than $25(J/\sigma_Y)$. However, the plane-stress influences that develop during crack extension should be restricted for plane-strain J-R curves. This is accomplished by use of values of specimen thickness at least as large as the net ligament size. The size requirements that currently appear acceptable for J-R measurements are:

$$B \text{ and } (W - a_o) > 20\left(\frac{J}{\sigma_Y}\right) \qquad \text{(Eq 72)}$$

$$\Delta a_p < 0.1(W - a_o) \qquad \text{(Eq 73)}$$

J-calculations are complex due to use of the incremental equation, the Δa_p determinations from unload-reload compliance, and (often) a rotation correction for shift of load line with large opening displacements. It has been found that these calculations, as well as testing efficiency, can be aided by computer-interactive testing methods (Ref 59, 60). Interest in J-R testing using the three-point-loaded bend specimen has been small compared to J-R testing with the compact specimen. Problems encountered with use of the bend specimen for J-R testing (Ref 61) usually disappear with additional testing experience.

During the development of crack size determinations from unload-reload compliance, the compliance indication of crack size was usually less than the nine-point average crack size obtained from the fracture surface. The error increased with curvature of the crack front and could essentially be eliminated by use of side grooves of moderate depth, which produced a nearly straight

crack front between the side grooves. The error can also be eliminated by incorporating a correction.

In regions of Δa_p larger than 1 mm (0.4 in.), J-R curves obtained using side-grooved specimens yield smaller J and dJ/da values compared to specimens that are not side grooved. One likely cause of the difference is the presence of smaller volumes of plane-stress yielding near the specimen surfaces than are present when side grooving is not used. Use of a side groove depth no larger than $0.1B$ ($B_N = 0.8B$) usually straightens the crack front to a satisfactory degree. Side grooving does not alter J_{Ic} values.

References 31 and 41 discuss J-R curve measurements with specimens ranging in W from 25 to 510 mm (1 to 20 in.). All specimens had moderate-depth side grooves. Comparison of the results for the geometrically similar specimens ($W = 2B$) showed an increase of J and dJ/da average values with an increase in specimen size. The difference was negligible if results for Δa_p larger than $0.06(W - a_o)$ were ignored. The possible usefulness of J-R results for Δa_p larger than $0.06(W - a_o)$ was studied in Ref 41, and a method of correcting results was suggested. The modified J value, J_M, is:

$$J_M = J - \int_{a_o}^{a} \left(\frac{\partial J_{pl}}{\partial a}\right)_{\Delta_p} da \qquad \text{(Eq 74)}$$

where the derivative in the integrand is taken holding the plastic component of the load displacement to a fixed value. Additionally, $J_{pl} = J - \mathcal{G}$, where $\mathcal{G}$ is the linear elastic crack extension force calculated using load and crack size.

From Eq 74, in the limit of zero crack extension, $J_M = J$, as indicated by size effect experiments. In the limit of a specimen size large enough so that the plastic zone is enclosed by the linear elastic field, J_{pl} is a negligible component of J, and Eq 74 provides the necessary result, $J_M = J$.

For intermediate specimen sizes and substantial values of crack growth, the integral in Eq 74 is always negative so that J_M is larger than J. Use of the J_M modification with the test results of Ref 31 indicated that the J_M-R curves were similar for all specimen sizes, even when the values of Δa_p were rather large fractions of $(W - a_o)$.

The results of these specimen size experiments are as would be expected, if it is recognized that the crack closure influence of the by-passed plastic zone must decrease with a decrease in the elastic clamping force capability of the specimen. The correspondence of J_M to pertinent mechanisms in detail is not currently apparent. However, J_M has appropriate end-point features and should have a substantial range of usefulness.

Figure 22 illustrates J and J_M values measured for A508 steel in three specimen sizes. Tests were conducted at 205 °C (400 °F). In this comparison, use of the modified J has a large effect only for the smallest specimen size, $W = 25.4$ mm (1.0 in.). The large increase in J with crack growth for these results indicated that substantial amounts of crack extension (i.e., 6.35 mm, or 0.25 in.) will require application of J values several times larger than the value of J_{Ic}.

Tearing Instability Analysis. The usefulness of the J-R curve for various structural conditions has been presented using tearing instability analysis (Ref 62, 63). Methods of tearing instability analysis generalize the instability concept associated with small plastic-zone R-curves (ASTM E 561). The extension made possible by J-characterization methods permits estimates of conditions for onset of rapid fracture both before and after general yielding of the net section ahead of the crack front.

Typical experimental trials of tearing instability analysis are reported in Ref 64 and 65. In its present form, the analysis is restricted to slow loading. For example, consider a fracture test specimen in which crack growth occurs due to a gradual increase in the crack opening. Although a graph of data points would show that the J-R curve slope remains positive, the decrease in net ligament size will, after some crack growth, cause a maximum load point, after which decreasing loads are sufficient to continue crack extension.

Clearly, the onset of rapid fracture will occur when the rate of decrease of the applied load permitted by the loading system becomes less than the rate of decrease of load required for slow, stable continuation of crack extension. The onset of rapid fracture in a tensile bar test is governed by similar principles.

In tearing instability concepts, certain conventions are commonly used. The increment, da, in the expression, dJ/da, always refers to the change in physical crack size, a. $(dJ/da)_{mat}$ is the slope of the measured J-R curve for the material. $(dJ/da)_{appl}$ is an estimate of the increase rate of J using load, loading compliance, specimen compliance, and yield properties of the material. $(dJ/da)_{appl} > (dJ/da)_{mat}$ is required for instability. The value of $(dJ/da)_{appl}$ is a structural rather than a toughness property. The following dimensionless forms are also used:

$$T_{mat} = \left(\frac{E}{\sigma_Y^2}\right)\left(\frac{dJ}{da}\right)_{mat}$$

$$T_{appl} = \left(\frac{E}{\sigma_Y^2}\right)\left(\frac{dJ}{da}\right)_{appl}$$

The condition for tearing instability can be expressed as $T_{appl} > T_{mat}$. Relatively simple methods for estimating T_{appl} may be possible when the plastic deformation region is isolated and clearly associated with the crack. For example, in a large component, under loading conditions prior to general yielding, an estimate of T_{appl} can be made by assuming J is the r_Y-corrected value of $\mathcal{G}$.

In large component applications, use of T_{mat} estimates from small-specimen J-R curves that are not modified for specimen size effects will be conservative. Reference 66 discusses applications of the J-concept and tearing instability to predicted cracks in large nuclear reactor vessels.

When general yielding between the crack front and a free edge is anticipated, essentially kinematic estimates of T_{appl}, using plasticity mechanics, may provide useful approximations of conditions that might cause rapid crack extension. For example, if the material has a low work-hardening rate, methods for estimating T_{appl} may be useful, assuming that the deformation resistance of the net section remains at limit load conditions during crack extension. Examples are given in Ref 62, 64, and 65. Reference 67 provides a review of instability analysis with extension to include work hardening.

Tearing instability is a basic and valuable concept. However, in practical structural applications, complex conditions requiring additional study may occur. For example, uncertainties of instability predictions due to substantial work hardening have been noted. Application difficulties have been experienced when the two-dimensional plastic zone adjacent to the crack front is surrounded by a stress-strain field that is plastic and three-dimensional. In steel structures, the possibility of the onset of cleavage must be considered.

Additional Elastic-Plastic Test Methods

Use of the J-concept for toughness evaluations under crack front conditions of plane stress have received less attention than crack front conditions of plane strain. Reference 49 describes experiments that extend ASTM E 561 into the range of net section yielding, using wedge-loaded compact specimens. In this study, the steel and aluminum alloy plate materials were thin enough so that shear lips and extensive crack front curvature did not interfere with crack size determination. The results showed that the R-curves for a given material were independent of specimen size if data points for $(W - a_p)$ less than about 35 J/J_Y were ignored.

Reference 68 describes a technique for toughness evaluation using an estimate of the

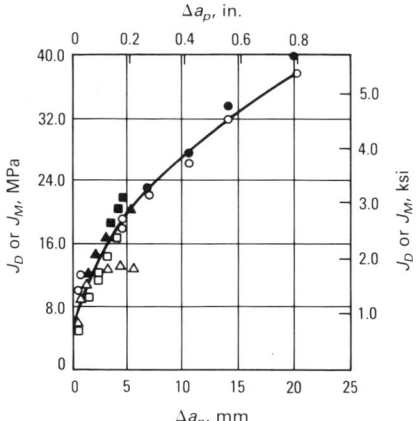

Fig. 22 Pairs of J and Δa_p observations for three compact (20% side-grooved) specimens of different size

The normal (deformation theory) J-values are termed J_D. J-results using Eq 74 are termed J_M. Source: Ref 31

Size	Deformation theory, J, J_D	Modified J, J_M
10T	○	●
1T	□	■
0.5T	△	▲

crack tip opening displacement, δ, as the characterization parameter. Bend specimens with three-point loading are precracked by fatigue. The crack opening is measured at the specimen edge as in ASTM E 399. The δ parameter is estimated as the sum of elastic and plastic contributions. The plastic part assumes rigid rotation about a hinge point located $0.4(W - a_o)$ beyond the initial crack front.

Use of the full plate thickness of specimens cut from plate materials is recommended (Ref 68). The standard measurement point is the initiation of crack extension (i.e., from heat testing with multiple specimens). However, other measurement points are suggested. Results using the initiation measurement point usually are comparable to J_{Ic} values in terms of the equation, $J_{Ic} = 2\sigma_Y \delta_{Ic}$. The close relationship of the δ parameter to J is discussed in Ref 39 and 69.

Use of δ (or crack tip opening displacement) measurements in an R-curve format has received only limited use. Reference 68 allows various degrees of plane-strain constraint. Development and use of this method have been closely associated with fracture problems in welded structures. Although the δ parameter is somewhat less basic than J from an analysis viewpoint, crack tip opening displacement testing is useful, and the δ

concept is frequently helpful in constructing estimates of J.

In fracture testing of steel plate materials in the temperature range of the cleavage fibrous transition, the influence of constraint on the onset of rapid cleavage is important. One method for determining a temperature high enough to prevent development of cleavage is to conduct three-point-loaded bend specimen tests of the material using a spring plate in the loading train. At a temperature below, but near the critical temperature for loss of cleavage, onset of cleavage occurs due to the increase in strain rate when the hole joining fracture is driven rapidly (Ref 70).

In trials of this kind, the observed loss-of-cleavage temperature varied consistently with the degree of plane-strain constraint. Relative to the temperature for the start of the shelf region of Charpy V-notch impact tests, the loss-of-cleavage temperature was moderately lower or substantially higher, depending on the degree of plane-strain constraint.

The largest difference (elevation) of the cleavage temperature was 50 °C (90 °F), observed for an A508 steel specimen 61 mm (2.4 in.) thick heat treated to a yield strength of 620 MPa (90 ksi). Moderate-depth side grooves were used to remove shear lips and to add some bias toward conservative results. In practical applications, the development of shear lips and crack front curvature may have a significant effect, separate from crack speed, on the development of cleavage.

In fracture testing of structural steels, if the measurement point of the toughness test is strongly influenced by cleavage, test result scatter can be expected, particularly in the transition temperature range. This behavior is observed in all of the small-specimen tests—e.g., Charpy impact tests (standard or precracked), impact K_{cd} tests, and slow-load "K from J" tests. The magnitude of the scatter is often objectionable. However, the observed variations in replicate tests are rarely due to the test method.

From fractography and section view studies of cleavage fractures, the variations in carbide density, along with variations of microstructure at finer scale, appear to provide adequate explanation for the observed scatter of test results (Ref 71). Under similar conditions of constraint, loading rate, and temperature in the test and in the service component, the low values of toughness observed in testing are likely to control a service fracture failure.

Fracture Control. The execution of fracture control tasks can be assisted by a sound understanding of fracture and fracture mechanics. Prior to 1960, fracture problems in structures were often ascribed improperly to inadequate fracture toughness. To some extent, a tendency to overemphasize toughness in relation to fracture control may still exist. In addition to composition, heat treatment, and fabrication of metallic components, fracture control involves design, inspection, proof testing, service loadings, and consideration of aggressive service environments. Structures often contain design details that contribute to development and growth of fatigue cracks.

In practice, inspection is rarely performed with sufficient efficiency to eliminate prior crack possibilities. Anticipation of service loading and of possible aggressive environments is sometimes difficult, but deserves attention. Because costs must be limited and perfection in the various fracture control tasks is unlikely, cost-benefit considerations play a significant role in the development of an appropriate fracture control plan. Fracture behavior information, expressed in terms of fracture mechanics, assists determination of the relative importance and necessary quality of the fracture control tasks.

Appendix: Fracture Toughness Testing of Aluminum Alloys

By Robert E. Zinkham
Supervisor, Mechanical Metallurgy
Reynolds Metals Company

FRACTURE TOUGHNESS TESTING of aluminum alloys, although conducted in essentially the same manner as fracture toughness testing of other metals, is unique because of its use in aerospace applications. Many aerospace specifications for premium toughness alloys focus on the use of aluminum. Because of the expense and extensive testing required for aluminum alloys, less expensive screening tests have been developed, and "meaningful" data have been defined.

The concept of fracture toughness in relation to the aluminum industry was expanded in the early 1970's with the advent of the B-1 bomber, the first aircraft designed on the basis of "damage tolerant" principles. The specifications for this aircraft included material quality assurance that encompassed not only aluminum, but also titanium and steel.

The immediate impact of the specification upon the aluminum industry resulted in the formation of a task group by the Aluminum Association. In 1974, this group formulated "The Aluminum Association Position on Fracture Toughness Requirements and Quality Control Testing" (Ref 72), which was revised in 1978 and updated in 1985.

This document presents the aluminum industry's position on matters related to fracture toughness testing, quality control, and material procurement, including classification of alloys, statistically reliable minimum fracture toughness indices, fracture toughness test procedures, screening tests, and a fracture toughness data base. Reference 73 provides a brief review of this document, along with a discussion of its impact in the area of applied testing.

Classification of Alloys for Fracture Toughness Testing

Fracture toughness is rarely, if ever, a design consideration in the 1000, 3000, 4000, 5000, and 6000 series alloys. The fracture toughness of these alloys is sufficiently high that thicknesses beyond those commonly produced would be required to obtain a valid test. Therefore, these alloys are excluded from further consideration in this article. Among the alloys for which fracture toughness is a meaningful design-related parameter, controlled-toughness high-strength alloys and conventional high-strength alloys merit discussion.

Controlled-toughness high-strength alloys were developed for their high fracture toughness and range in measured K_{Ic} values from about 19.8 MPa$\sqrt{m}$ (18 ksi$\sqrt{in.}$) upward. The alloys and tempers currently identified as controlled-toughness high-strength products include:

Alloy	Condition	Product form
2048	T8	Sheet and plate
2124	T3, T8	Sheet and plate
2419	T8	Sheet, plate, extrusions, and forgings
7049	T7	Plate, forgings, and extrusions
7050	T7	Sheet, plate, forgings, and extrusions
7150	T6	Sheet and plate
7175	T6, T7	Sheet, plate, forgings, and extrusions
7475	T6, T7	Sheet and plate

Typical applications include 2419-T851 used in the lower wing skins of the B-1 bomber and 7475-T7351 and 2124-T851 in the F-16 aircraft.

Conventional High-Strength Alloys. Although these alloys, tempers, and products are not used for fracture-critical components, fracture toughness can be a meaningful design parameter. Conventional aerospace alloys for which fracture toughness minimums may be useful in design include 2014, 2024, 2219, 7075, and 7079. These alloys have toughness levels that are inferior to those of their controlled-toughness counterparts. Consequently, toughness is not guaranteed.

Controlled-toughness alloys are often derivatives of conventional alloys. For example, 7475 alloy is a derivative of 7075 with maximum compositional limits on some elements that were found to decrease toughness.

Fracture toughness quality control and material procurement minimums are appropriate for controlled-toughness high-strength alloys, tempers, and products, because checks on composition and tensile properties are inadequate assurances that the proper levels of toughness have been achieved. If the minimum specified fracture toughness value is not attained, the material is not acceptable.

Development of Fracture Toughness Indices

Fracture toughness can be measured by a variety of indices, most notably K_{Ic} and K_c. When choosing the fracture toughness index for a given application, statistical reliability is an important consideration. Several methods for determining and maintaining this reliability that are in use or being developed are described below.

Fracture Toughness Minimums. For alloys, tempers, and products for which ASTM E 399 (Ref 43) is applicable, the principal index of fracture toughness is plane-strain fracture toughness (K_{Ic}). This test method is the most widely used procedure.

For sheet thicknesses less than 6.35 mm (0.250 in.), there are no standard test procedures for the generation of design-related fracture indices such as K_c. Toughness indices for these products must be agreed upon by the producer and customer. However, ASTM E 338 (Ref 74), which detail sharp-notch tension testing of sheet materials, is appropriate and has been used to establish quality control indices for high-toughness sheet by correlating data from center-notched panel tests. Similarly, correlations are being developed using K_R values based on a 25% secant offset of maximum load incorporating the load displacement curve from a compact specimen according to ASTM E 561 (Ref 42).

Variations in specifications exist sometimes in areas where standards organizations have not as yet established a consensus, or where the results of a standardized test may be subject to a variety of interpretations. Some of these areas encompass elastic-plastic fracture mechanics technology and are the subject of current research and standards development. Aircraft engineers, supported by large individual groups of empirical data, are ahead of scientists and standards writers.

Another area of divergence results from the use of inexpensive screening tests to reduce the amount of testing by a costly standard method. To obtain sufficient statistical confidence, large data bases are required to establish the correlation between the two tests. The limits imposed in the user's specification may vary from supplier to supplier and may depend on the size, and hence the statistical accuracy, of each supplier's data base.

Statistical Bases. Specific statistical bases for minimum values of K_{Ic} are being established and used by all producers in developing minimum fracture toughness indices. The 95% confidence levels or A and B values, as defined in Ref 75, usually incorporate more than a hundred test values when normally distributed. The lower confidence values of C and D only require about 50 and 20 values, respectively, at a 75% confidence level. See the article "Design Allowables for Static Metallic Material Properties" in this Volume for more information on A and B values.

Fracture Toughness Test Procedures

The aerospace industry is concerned with the prevention of both plane-strain fractures that involve minimal plastic deformation near the crack front and plane-stress fractures that are associated with extensive plasticity. Plane-strain fracture testing generally is required on materials used for the thicker sections of an aircraft, such as rolled plate wing boxes and bulkheads, forged undercarriage legs, and heavy longeron extrusions, in which plane-strain conditions can occur at a crack front. Plane-stress testing is usually specified for thinner sections, such as rolled sheet or plate wing skins or flanges and webs on some extrusions.

Individual test methods usually are divided into those relating to plane strain, mixed mode, and plane stress. The division is arbitrary, however, because there is a continuum in the fracture behavior in this range. Consequently, considerable overlap exists in the range of suitability of some test methods.

Plane-Strain Tests. Plane-strain fractures are those involving minimal or no macroscopic yielding around the crack and are characterized by a flat fracture face normal to the applied stress. They are promoted by relatively high strength and relatively low toughness and are more readily established in thick sections. In most heat-treated aluminum alloys used in aircraft applications, plane-strain tests usually are relevant only to section thicknesses greater than about 25 mm (1 in.); however, in some T6 temper alloys,

thicknesses as low as about 5 mm (0.20 in.) may produce relevant results.

All major aircraft specifications that require plane-strain testing use the ASTM E 399 test method or closely related foreign standards. This test method is also used for metals other than aluminum and measures the plane-strain fracture toughness, K_{Ic}. Within the aluminum industry, plane-strain fracture toughness usually is determined with a compact tension test specimen (Fig. 23). A fatigue crack is made in the sample, and a rising tensile load is applied. The record of load and crack mouth opening displacement (CMOD) is analyzed to obtain the crack tip stress intensity after a 2% or less increase in crack length. This critical stress intensity is termed K_{Ic}.

The K_{Ic} test results can be used in studying the fracture mechanics of aircraft structures, if plane-strain conditions prevail in the structure. However, aircraft manufacturers frequently accept materials based on a plane-strain test, although the material will be subsequently machined to a thickness in which fracture would occur by plane-stress modes. For example, heavy-section plate (100 mm, or 4 in.) is frequently machined into waffle structures, the thicknesses of which do not exceed 10 mm (0.4 in.). Wing extrusions and skins frequently are machined to taper from the fuselage to the wing tip, so that the potential fracture mode may vary continuously along the wing.

The ASTM E 399 test method incorporates a number of after-the-fact checks on the test data to determine if the measured fracture toughness is valid. If it is not, the test result is termed K_Q. These checks ensure that the specimen size was sufficiently large so that the test fracture occurred under essentially plane-strain conditions.

Many test results fail the validity checks; however, invalidity rarely implies an inferior product or an improper test procedure. Most aluminum alloy products are not ordered, or cannot be made, in a sufficiently thick section size to allow plane-strain crack growth in a test. Consequently, it is often not possible to measure a valid K_{Ic} to predict the sample size necessary to obtain a valid test.

Another cause of invalidity is uneven penetration by the fatigue starter crack, which is caused directly by residual stress in the sample. Heat-treated material that has not been stretched or compressed before aging usually contains significant residual stresses. Several test attempts may be necessary before a valid result is obtained, or it may be necessary to negotiate customer acceptance of an invalid result.

Validity checks require tensile yield strength data for the directions tested for fracture toughness, e.g., longitudinal (L)

Fig. 23 Compact tension fracture toughness specimen

The fatigue starter crack can be seen on the fracture faces of the broken specimen. ASTM E 399 test method. Courtesy of Kaiser Aluminum and Chemical Co.

Fig. 24 Round notched tensile specimen

Standard sizes are 27 mm (1.0625 in.) and 12.7 mm (0.5 in.) diam. ASTM E 602 test method. Courtesy of Kaiser Aluminum and Chemical Co.

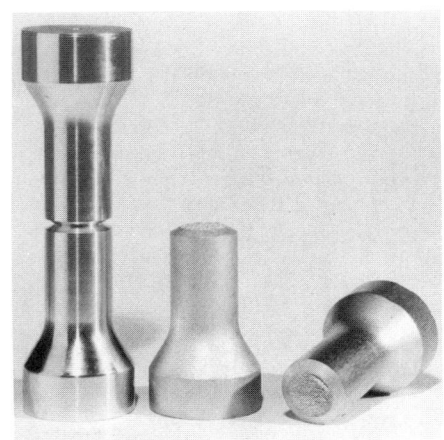

tensile data for a longitudinal-transverse (L-T) toughness test. It is usually economically advantageous for manufacturers to supply these data for the fracture testing laboratory.

Test results that are marginally invalid frequently are meaningful, and they may extend the useful range of applicability of the ASTM E 399 test result. Some validity checks in ASTM E 399 are appropriate for some metals, but are unnecessarily restrictive for aluminum alloys. Furthermore, where experience has shown that a particular test invalidity causes the measured toughness (K_Q) to be lower than K_{Ic}, the material can be considered acceptable if the conservative invalid result exceeds the specified minimum toughness.

For these reasons, ASTM B 645 (Ref 76) has been developed for aluminum alloys. This document specifies which invalidities are acceptable and which invalid ASTM E 399 test results can be considered meaningful. The terms ''valid'' and ''meaningful'' have strict legal definitions in these ASTM documents.

A growing number of specifications permit the use of ASTM B 645 to extend the range of applicability of the ASTM E 399 test method into the mixed-mode regime. Some specifications have not been revised since ASTM B 645 was written; others continue to accept specific invalidities without reference to this specification.

Plane-Strain Screening Tests. Because the test specimen recommended by ASTM E 399 is expensive and time consuming to machine and to fatigue precrack, aluminum producers and users have cooperated to reduce costs and expedite the shipment of material by using a relatively inexpensive notched round tensile test to screen material. Material that fails is retested to determine K_{Ic} or K_Q.

The test specimen (Fig. 24) and method are described in ASTM E 602 (Ref 77); however, this standard does not cover interpretation of test results. The aluminum industry generally has adopted the interpretation described in ASTM B 646 (Ref 78), which refers to the ratio of the notched tensile strength (NTS) to the tensile yield strength (TYS) of a conventional smooth tensile specimen. This is termed the notched yield ratio (NYR), or NTS/TYS ratio, which has been shown to correlate with K_{Ic} or K_Q.

The cost of this test can be decreased significantly if the required tensile yield strength data, which usually are available, are provided by the manufacturer to the testing laboratory. However, manufacturer's tensile tests must be taken adjacent to locations that will be used for notched tensile testing.

When the ASTM E 602 specimen is used as a screening test, the minimum acceptable NYR values are selected by the customer to correspond to a particular statistical confidence that the specified minimum K_{Ic} will be met (Ref 72). The probability of not passing the limit thus depends on the number of K_{Ic}-NYR correlations in the data base of the supplier for that particular product. When a lower NYR is obtained, a K_{Ic} or K_Q retest must be made in accordance with ASTM E 399.

The frequency of retesting also depends on the size of the data base. When only a small data base has been established, a high retest rate may make the use of the screening test uneconomical or too slow. Conversely, the larger the data base, the greater the savings achieved by the use of the screening test, but this must be balanced against the cost of establishing the data base and the expectations of the manufacturer regarding future orders for the product. In many cases, differ-

ent producers must pass different NYR minimums for the same product because of differences in the sizes of their data bases. In the United States, the need for standardization has prompted the Metal Properties Council, the Aluminum Association, and committees for the *Military Standardization Handbook* (Ref 75) and *Damage Tolerant Design Handbook* (Ref 79) to jointly develop industry-wide data bases for NYR/K_{Ic} correlations, which will eventually result in revised procurement specifications.

The chevron-notched short rod and short bar specimens are used in another plane-strain fracture test that is currently under development. This specimen requires no fatigue precrack, and the stress intensity (K_{SR} or K_{SB}) is measured at the front of a crack that has grown from the tip of a chevron in a heavily side-grooved rod or bar (Fig. 25). Tests on aluminum alloys have shown that these parameters correlate closely with K_{Ic} (Ref 80) and that useful plane-strain fracture toughnesses can be measured with a specimen that is smaller than that used in the ASTM E 399 test.

Although a Society of Automotive Engineers (SAE) recommended test practice (Ref 81) exists for these specimen geometries, and testing costs are considerably less than those of the ASTM E 399 test, the SAE test has not been included in any aircraft company specifications to date. One major advantage of this specimen is that it uses normal machine shop tolerances and is easy to test.

Mixed-Mode Tests. A number of specifications extend into the mixed-mode regime, evaluating the application of plane-strain fracture mechanics for material testing by using R-curve techniques. An R curve for a material describes the increase in fracture

Fig. 25 Chevron-notched short bar toughness specimens

B = 25 mm (1.0 in.) (left) and 12.5 mm (0.5 in.) (right). A broken half of a compact tension sample is shown for comparison (center). Courtesy of Kaiser Aluminum and Chemical Co.

Table 1 Typical fracture toughnesses measured on the same lot of a high-toughness 7475-T7351 aluminum alloy plate

Specified test method	Typical L-T toughness(a) MPa√m	ksi√in.
ASTM E 399, K_{Ic}	55	50
ASTM E 561, R curve		
5% secant	45-65	40-60
25% secant	75-110	70-100
Center-cracked panel 150 mm (6 in.) wide	130	120
Center-cracked panel 400 mm (16 in.) wide	200	180

Note: Test sample thicknesses are those typically specified. Data are approximate and are presented to contrast different test results, not for use for design purposes. R-curve testing also requires tensile data for test validity checks, and these should be provided to the testing laboratory.
(a) Test sample orientation code is described in ASTM E 399 (Ref 43). The first letter represents the direction of applied tensile stress; the second letter is the direction of crack growth. L, longitudinal; T, transverse; S, thickness direction

toughness, K_R, as a crack grows and is usually given as a plot of K_R versus crack extension, a.

Crack growth in relatively high-toughness materials is associated with increasing plasticity, decreasing crack sharpness, and the loss of plane-strain conditions at the crack tip. This in turn increases the measured toughness. R curves are used for a wide range of mixed-mode and plane-stress fractures, and their use for relatively thin sections is described under the section in this article on plane-stress testing.

Several standard methods for R-curve determinations are described in ASTM E 561 (Ref 42); however, interpretation of the R curve is not. Additionally, interpretation may vary from one specification to another and from one test method to another.

In relatively heavy sections, R curves may be determined using a compact tension specimen and techniques similar to those used for plane-strain ASTM E 399 tests (Fig. 23). However, the ASTM E 561 specimen has a relatively shorter initial crack length, minor geometric differences, and a different set of validity criteria. When an ASTM E 399 test is found to be invalid, some specifications require that the ASTM E 399 test record be analyzed by the R-curve methods of ASTM E 561. However, strict compliance with ASTM E 561 precludes this. The approach offers ease of testing and lower testing costs; however, its application requires data to ensure that the departures from the ASTM E 561 geometry produce conservative results.

The R curve indicates the fracture property of the material in the thickness tested, but it is inconvenient to use this curve in its entirety in a specification. It is standard practice to measure the toughness, K_c, after a particular increment of crack growth or, in some plane-stress tests, at maximum load.

The crack tip stress intensity after a particular crack growth increment, Δa, is determined by constructing secants on the load/CMOD record (Fig. 26). These slopes are, by convention, either 5 or 25% lower than the elastic portion of the record. The 5% slope is the same as that used in the ASTM E 399 plane-strain test, which corresponds to a 2% increase in crack length, whereas the 25% secant approximates a 30% crack growth. The particular value of K_c frequently is identified by using the subscripts $K_{5\%}$ or $K_{25\%}$.

Typically, different toughness values can be quoted for the same material and even for the same test, depending on the convention adopted by the specification. In the one material, the 5% secant toughness determined from the R curve, $K_{5\%}$, may be higher or lower than K_{Ic}. Additionally, it usually will be much lower than the 25% secant value, $K_{25\%}$. Relative differences for one lot of a high-strength aluminum alloy tested with typical specimen geometries are shown in Table 1.

Mixed-Mode Screening Tests. There are no extensive data bases for screening tests in this fracture regime. Consequently, relatively little use is made of screening tests.

In thicker sections, round notched tensile testing is sometimes used.

In the future, NYR values may be correlated with a selected R-curve parameter (i.e., $K_{25\%}$), and the notched yield ratio will be used to reduce the amount of R-curve testing. Some specifications permit machining of some intermediate thickness materials to thicknesses suitable for plane-stress testing, but this has little impact on testing costs.

Plane-Stress Tests. Plane-stress fractures involve extensive plasticity that does not permit sufficient restraint to develop high lateral stresses in addition to the applied stress. In common usage, this includes flat fractures that are accompanied by significant plasticity and shear-mode fractures.

The fracture in many plane-stress tests may be of a mixed mode, typically a flat fracture with marginal shear lips. The relative proportion of each cracking mode may

Fig. 26 Typical load versus crack mouth opening displacement curve for a compact tension specimen used for generating an R curve

5% and 25% secant offsets are indicated. ASTM E 561 test method. Courtesy of Kaiser Aluminum and Chemical Co.

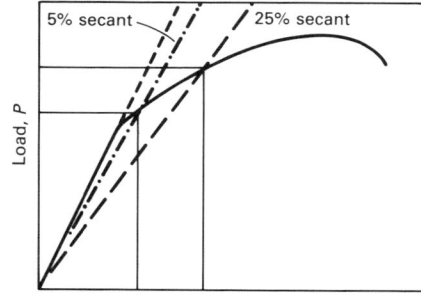

change as the crack propagates. Commonly, the fracture appearance changes from a flat fracture, with increasing shear lips, to a fully shear failure mode. The occurrence of these failure modes depends on material properties and sample thickness.

Common plane-stress tests are based on ASTM E 561 *R*-curve techniques, as described earlier for mixed-mode tests, but the center-cracked panel specimen that is tested in tension (Fig. 27) is more commonly used. A 400-mm (16-in.) wide by 1000-mm (40-in.) long panel is the industry standard plane-stress test for high-strength aluminum alloy sheet and plate less than about 25 mm (1 in.) thick.

The test record of load versus crack opening may be interpreted by 5 or 25% secants (Fig. 26) as for the compact tension geometry specimen described above, but in the center-cracked panel test, by convention, the K_c value usually is calculated at the maximum load during the test. This toughness measure also differs from K_{Ic}, $K_{5\%}$, and $K_{25\%}$ (Table 1).

To simplify testing, most specifications allow several departures from the ASTM E 561 test method. The use of sawn slots instead of a fatigue crack is permitted. Also, the net-section stress in the panel at maximum load may exceed the yield stress. Machining of the sample thickness to 6.3 mm (0.25 in.), which is generally expensive, allows the use of testing machines of normal capacity. These departures greatly reduce the difficulty of testing for material acceptance, but severely limit the usefulness of K_c data to the aircraft designer and reduce the center-cracked panel test to an empirical screening test.

Plane-Stress Screening Tests. The center-cracked panel described in ASTM E 561 is an expensive test specimen to prepare, and the removal of one L-T specimen from a production plate produces a piece of scrap about 1 m (3.28 ft) long by the plate width. Consequently, less expensive screening tests that use smaller specimens have been developed.

The most common screening test is the flat, edge-notched tensile specimen described in ASTM E 338 (Ref 74), which is shown in Fig. 28. It is used in a similar manner to the notched round specimen (Ref 42) and also necessitates the establishment of a data base for empirical correlation of edge-notched tensile data and center-cracked panel data.

Some specifications permit the use of a smaller 150-mm (6-in.) wide and 400-mm (16-in.) long center-cracked panel as a screening test for the similar 400-mm (16-in.) wide test. In these cases, conservative results are ensured by maintaining the speci-

Fig. 27 400-mm (16-in.) wide center-notched panel test specimen used for plane-stress fracture toughness testing

ASTM E 561 test method. Courtesy of Kaiser Aluminum and Chemical Co.

fied lower limit of toughness K_c at the same level as for the larger sample, even though the test usually measures a lower K_c in the smaller panel (Table 1).

Fracture Toughness Data Base

The Aluminum Association Task Group on Fracture Toughness has established a fracture toughness data base in cooperation with the Metal Properties Council. As more data are obtained from various producers in this cooperative effort, statistically reliable fracture toughness minimum values will be developed (see Ref 82).

Fig. 28 Double-edged notched tension specimen

ASTM E 338 test method. Courtesy of Kaiser Aluminum and Chemical Co.

REFERENCES

1. G.R. Irwin *et al.*, Interpretation of Fracture Marking, *J. Appl. Phys.*, Vol 21, 1950, p 716-720
2. J.E. Campbell, W.W. Gerberich, and J.H. Underwood, Ed., *Application of Fracture Mechanics for Selection of Metallic Structural Materials*, American Society for Metals, 1982
3. S.T. Rolfe and J.M. Barsom, *Fracture and Fatigue Control in Structures: Applications of Fracture Mechanics*, Prentice-Hall, Englewood Cliffs, NJ, 1977
4. R.W. Hertzberg, *Deformation and Fracture Mechanics of Engineering Materials*, John Wiley & Sons, New York, 1976
5. J.W. Fisher, *Fatigue and Fracture in Steel Bridges*, John Wiley & Sons, New York, 1984
6. G.R. Irwin, Analysis of Stresses and Strains Near the End of a Crack Transversing a Plate, *J. Appl. Mech.*, Vol 24, 1957, p 361-364
7. Report of a Special ASTM Committee, "Fracture Testing of High-Strength Sheet Materials," *ASTM Bull.*, Jan 1960, p 29-51
8. G.R. Irwin, Fracture Dynamics, in *Fracturing of Metals*, American Society for Metals, 1948, p 147-166
9. G.R. Irwin, Fracture I, in *Handbuch der Physik VI*, Springer-Verlag, Berlin, 1958, p 558-590
10. A.A. Griffith, The Phenomena of Rupture and Flow in Solids, *Phil. Trans.*, Vol 221, 1920, p 163-198
11. J.R. Rice, A Path Independent Integral and the Approximate Analysis of Strain Concentrations by Notches and Cracks, *J. Appl. Mech.*, June 1968, p 379-386
12. J.R. Rice and G.F. Rosengren, Plane-Strain Deformation near a Crack Tip in a Power-Law Hardening Material, *J. Mech. Phys. Solids*, Vol 16, 1968, p 1-12
13. J.W. Hutchinson, Singular Behavior at the End of a Tensile Crack in a Hardening Material, *J. Mech. Phys. Solids*, Vol 16, 1968, p 13-31
14. A.A. Wells, Notched Bar Tests, Fracture Mechanics, and the Brittle Strengths of Welded Structures, *Brit. Weld. J.*, Jan 1965, p 2-13
15. G.R. Irwin and A.A. Wells, A Continuum Mechanics View of Crack Propagation, *Met. Rev.*, Vol 10 (No. 38), 1965, p 223-270
16. H.M. Westergaard, Bearing Pressures and Cracks, *J. Appl. Mech.*, Vol 6 (No. 2), 1939, p A-49

17. C.E. Inglis, Stresses in a Plate Due to the Presence of Cracks and Sharp Corners, *Proc. Inst. Naval Arch.*, Vol 55, 1913, p 219

18. A.E. Green and I.N. Sneddon, *Proc. Cambridge Phil. Soc.*, Vol 46, 1950, p 159

19. G.R. Irwin, Crack Extension Force for a Part-Through Crack in a Plate, *Trans. ASME*, Vol E29, 1962, p 651

20. E.H. Yoffe, *Phil. Mag.*, Vol 42, 1951, p 739

21. G.R. Irwin, "Relatively Unexplored Aspects of Fracture Mechanics," Report TAM 240, University of Illinois, Feb 1963

22. L.B. Freund, Crack Propagation in an Elastic Solid Subjected to General Loading, *J. Mech. Phys. Solids*, Vol 20 (No. 3), 1972, p 129-152

23. A.B.J. Clark and G.R. Irwin, *Exp. Mech.*, 1966

24. J.W. Dally, Dynamic Photoelastic Studies of Fracture, *Exp. Mech.*, Vol 19 (No. 10), 1979, p 349-361

25. F.A. McClintock, Ductile Fracture Instability in Shear, *J. Appl. Mech.*, Vol 25, 1958, p 582-588

26. F.A. McClintock and G.R. Irwin, Plasticity Aspects of Fracture Mechanics, in *Fracture Toughness Testing and its Applications*, STP 381, W.F. Brown, Ed., ASTM, Philadelphia, 1965, p 84-113

27. D.S. Dugdale, Yielding of Steel Sheets Containing Slits, *J. Mech. Phys. Solids*, Vol 8, 1960, p 100-104

28. A.A. Wells, Unstable Crack Propagation in Metals—Cleavage and Fast Fracture, in *Proceedings of Crack Propagation Symposium*, Vol 1, paper B4, Cranfield, England, 1961, p 210-230

29. J.A. Begley and J.D. Landes, The *J* Integral as a Fracture Criterion, in STP 514, H.T. Corten, Ed., ASTM, Philadelphia, 1972, p 1-23

30. J.A. Begley and J.D. Landes, Effect of Specimen Geometry of J_{Ic}, in STP 514, ASTM, Philadelphia, 1972, p 24-39

31. D.E. McCabe and J.D. Landes, J_R-Curve Testing of Large Compact Specimens, in *Elastic-Plastic Fracture, Second Symposium*, STP 803, Vol 2, C.F. Shih and J.P. Gudas, Ed., ASTM, Philadelphia, 1983, p 353-371

32. J.R. Rice, Mathematical Analysis in the Mechanics of Fracture, in *Fracture*, Vol II, H. Leibowitz, Ed., Academic Press, New York, 1968, p 191-309

33. J.R. Rice, P.C. Paris, and J.G. Merkle, Some Further Results of J-Integral Analysis and Estimates, in *Progress in Flaw Growth and Fracture Toughness Testing*, STP 536, ASTM, Philadelphia, 1973, p 231-245

34. J.D. Landes, J.A. Begley, and G.A. Clarke, Ed., *Elastic-Plastic Fracture*, STP 668, ASTM, Philadelphia, 1979

35. C.F. Shih and J.P. Gudas, Ed., *Elastic-Plastic Fracture, Second Symposium*, Vol 1 and 2, STP 803, ASTM, Philadelphia, 1983

36. G.R. Irwin and P.C. Paris, Elastic-Plastic Crack Tip Characterization in Relation to *R*-Curves, in *Fracture 1977: Proceedings of the International Congress on Fracture*, Vol 1 (ICF-4), University of Waterloo Press, Waterloo, Canada, 1977, p 93-100

37. D.E. McCabe and J.D. Landes, An Evaluation of Elastic-Plastic Methods Applied to Crack Growth Resistance Measurements, in *Elastic-Plastic Fracture*, STP 668, J.D. Landes, J.A. Begley, and G.A. Clarke, Ed., ASTM, Philadelphia, 1979, p 288-305

38. F.M. Burdekin and D.E. Stone, *J. Strain Anal.*, Vol 1, 1966, p 144

39. M.G. Dawes, Elastic-Plastic Fracture Toughness Based on the COD and *J*-Contour Integral Concepts, in *Elastic-Plastic Fracture*, STP 668, J.D. Landes, J.A. Begley, and G.A. Clarke, Ed., ASTM, Philadelphia, 1979, p 307-333

40. M.F. Kanninen *et al.*, Elastic-Plastic Fracture Mechanics for Two-Dimensional Stable Crack Growth and Instability Problems, in *Elastic-Plastic Fracture*, STP 668, J.D. Landes, J.A. Begley, and G.A. Clarke, Ed., ASTM, Philadelphia, 1979, p 121-150

41. H.A. Ernst, Material Resistance and Instability Beyond *J*-Controlled Crack Growth, in *Elastic-Plastic Fracture, Second Symposium*, STP 803, Vol 1, C.F. Shih and J.P. Gudas, Ed., ASTM, Philadelphia, 1983, p 191-213

42. "Standard Practice for *R*-Curve Determination," E 561, *Annual Book of ASTM Standards*, Vol 03.01, ASTM, Philadelphia, 1984, p 612-631

43. "Standard Test Method for Plane-Strain Fracture Toughness of Metallic Materials," E 399, *Annual Book of ASTM Standards*, Vol 03.01, ASTM, Philadelphia, 1984, p 519-554

44. R. Chona, G.R. Irwin, and R.J. Stanford, Influence of Specimen Size and Shape on the Singularity Dominated Zone, in *Fracture Mechanics: Fourteenth Symposium*, STP 791, Vol I: *Theory and Analysis*, J.C. Lewis and G. Synes, Ed., ASTM, Philadelphia, 1983, p 3-23

45. M.H. Jones and W.F. Brown, Jr., Influence of Crack Length and Thickness in Plane-Strain Fracture Toughness Tests, in *Review of Developments in Plane Strain Fracture Toughness*, STP 463, ASTM, Philadelphia, 1970, p 63-101

46. J.G. Kaufman and F.G. Nelson, More on Specimen Effects in Fracture Toughness Testing, in *Fracture Toughness and Slow-Stable Cracking*, STP 559, ASTM, Philadelphia, 1974, p 74-85

47. R.H. Heyer and D.E. McCabe, Plane Stress Fracture Toughness Testing Using a Crack-Line-Loaded Specimen, *Eng. Frac. Mech.*, Vol 4, 1972,

48. *Fracture Toughness Evaluation by R-Curve Methods*, STP 527, D.E. McCabe, Ed., ASTM, Philadelphia, 1973

49. D.E. McCabe, Determination of *R*-Curves for Structural Materials by Using Nonlinear Mechanics Methods, in *Flaw Growth and Fracture*, STP 631, ASTM, Philadelphia, 1976, p 245-266

50. R.B. Madison and G.R. Irwin, Dynamic K_c Testing of Structural Steel, *J. Struct. Div. ASCE*, 1974, p 1331-1349

51. A.K. Shoemaker and S.T. Rolfe, Static and Dynamic Low-Temperature Crack Toughness Performance of Seven Structural Steels, *J. Eng. Fract. Mech.*, Vol 2, 1971, p 319-339

52. P.B. Crosley and E.J. Ripling, Crack Arrest in an Increasing *K*-Field, *Nucl. Eng. Design*, Vol 17, 1973, p 32-45

53. P.B. Crosley *et al.*, "Final Report on Cooperative Test Program on Crack Arrest," NUREG/CR-3261, U.S. Nuclear Regulatory Commission, April 1983

54. K_{Ia} test description proposed by K_{Ia} Task Group of ASTM Committee E 24 (unpublished)

55. A.R. Rosenfield *et al.*, "BCL HSST Support Program," Heavy-Section Steel Technology Program Quarterly Report (July-Sept 1981), NUREG/CR-2141, Vol 3, U.S. Nuclear Regulatory Commission, Feb 1982, p 10-43

56. "Standard Test Method for J_{Ic}, a Measure of Fracture Toughness," E 813, *Annual Book of ASTM Standards*, Vol 03.01, ASTM, Philadelphia, 1984, p 763-781

57. D.A. Davis, M.G. Vassilaros, and J.P. Gudas, Specimen Geometry and Extended Crack Growth Effects on J_I-R Curve Characteristics for HY-130 and ASTM A533B Steels, in *Elastic-Plastic Fracture*, STP 803, Vol 2, C.F. Shih and J.P. Gudas, Ed., ASTM, Philadelphia, 1983, p 582-610

58. D.E. McCabe, J.D. Landes, and H.A. Ernst, An Evaluation of the J_R-Curve Method for Fracture Toughness Evaluation, in *Elastic-Plastic Fracture*, STP 803, Vol 2, C.F. Shih and J.P. Gudas, Ed., ASTM, Philadelphia, 1983, p 562-581

59. J.A. Joyce and J.P. Gudas, Computer Interactive J_{Ic} Testing of Naval Alloys, in *Elastic-Plastic Fracture*, STP 668, J.D. Landes, J.A. Begley, and G.A. Clarke, Ed., ASTM, Philadelphia, 1969, p 451-468

60. W.A. Van Der Sluys and R.J. Futato, Computer Controlled Single-Specimen *J*-Test, in *Elastic-Plastic Fracture*, STP 803, Vol 2, C.F. Shih and J.P. Gudas, Ed., ASTM, Philadelphia, 1983, p 464-482

61. A.A. Willoughby and S.J. Garwood, On the Unloading Compliance Method of Deriving Single-Specimen *R*-Curves in Three-Point Bending, in *Elastic-Plastic Fracture*, STP 803, Vol 2, C.F. Shih and J.P. Gudas, Ed., ASTM, Philadelphia, 1983, p 372-397

62. P.C. Paris, H. Tada, Z. Zahoor, and H. Ernst, Instability of the Tearing Mode of Elastic-Plastic Crack Growth, in *Elastic-Plastic Fracture*, STP 668, J.D. Landes, J.A. Begley, and G.A. Clarke, Ed., ASTM, Philadelphia, 1979, p 5-36

63. J.W. Hutchinson and P.C. Paris, The Theory of Stability Analysis of *J*-Controlled Crack Growth, in *Elastic-Plastic Fracture*, STP 668, J.D. Landes, J.A. Begley, and G.A. Clarke, Ed., ASTM, Philadelphia, 1979, p 37-64

64. P.C. Paris, H. Tada, Z. Zahoor, and H. Ernst, An Initial Experimental Investigation of the Tearing Instability Theory, in *Elastic-Plastic Fracture*, STP 668, J.D. Landes, J.A. Begley, and G.A. Clarke, Ed., ASTM, Philadelphia, 1979, p 251-265

65. J.A. Joyce, Instability Testing of Compact and Pipe Specimens Utilizing a Test System Made Compliant by Computer Control, in *Elastic-Plastic Fracture*, STP 803, Vol 2, C.F. Shih and J.P. Gudas, Ed., ASTM, Philadelphia, 1983, p 439-463

66. P.C. Paris and R.E. Johnson, A Method of Application of Elastic-Plastic Fracture Mechanics to Nuclear Vessel Analysis, in *Elastic-Plastic Fracture*, STP 803, Vol 2, C.F. Shih and J.P. Gudas, Ed., ASTM, Philadelphia, 1983, p 5-40

67. V. Kumar, M.D. German, and C.F. Shih, "An Engineering Approach for Elastic-Plastic Fracture Analysis," EPRI Report NP-1931, Electric Power Research Institute, Palo Alto, CA, July 1981

68. "Method for Crack Opening Displacement (COD) Testing," BS-5762, British Standards Institution, London, 1979

69. J.D.G. Sumpter and C.E. Turner, Methods for Laboratory Determination of J_c, in *Cracks and Fracture*, STP 601, J.L. Swedlow and M.L. Williams, Ed., ASTM, Philadelphia, 1976, p 3-18

70. W.L. Fourney *et al.*, "Investigations of Damping and Cleavage-Fibrous Transition in Reactor Grade Steels," Heavy-Section Steel Technology Program Semiannual Progress Report (April-Sept 1984), NUREG/CR-3744, C.E. Pugh, Ed., U.S. Nuclear Regulatory Commission, Dec 1984, p 32-35

71. K. Ogawa *et al.*, Microstructural Aspects of the Fracture Toughness Transition for Reactor Grade Steel, in *Fracture Mechanics: Fifteenth Symposium*, STP 833, ASTM, Philadelphia, 1984, p 394-411

72. "The Aluminum Association Position on Fracture Toughness Requirements and Quality Control Testing," Bulletin T-5, Aluminum Association, Washington, DC, Sept 1978

73. K.R. Brown, Fracture Toughness Acceptance Testing of Aluminum Alloys for Aircraft Applications, in *Proceedings of the International Conference on Fracture Mechanics Technology Applied to Material Evaluation and Structural Design*, G.C. Sih, N.E. Ryan, and R. Jones, Ed., Martinus Nijhoff, Melbourne, Australia, 1982, p 257-270

74. "Standard Method of Sharp-Notch Tension Testing of High Strength Sheet Materials," E 338, *Annual Book of ASTM Standards*, Vol 03.01, ASTM, Philadelphia, 1984, p 483-490

75. *Metallic Materials and Elements for Aerospace Vehicle Structures*, MIL-HDBK-5D, Department of Defense, Washington, DC, June 1983, revised Jan 1984

76. "Standard Practice for Plane Strain Fracture Toughness Testing of Aluminum Alloys," B 645, *Annual Book of ASTM Standards*, Vol 03.01, ASTM, Philadelphia, 1984, p 103-105

77. "Sharp-Notch Tension Testing with Cylindrical Specimens," E 602, *Annual Book of ASTM Standards*, Vol 03.01, ASTM, Philadelphia, 1984, p 632-640

78. "Standard Practice for Fracture Toughness Testing of Aluminum Alloys, B 646, *Annual Book of ASTM Standards*, Vol 03.01, ASTM, Philadelphia, 1984, p 106-110

79. *Damage Tolerant Design Handbook*, MCIC-HB-01, Metals and Ceramics Information Center, Battelle, Columbus, OH, Jan 1975 and current updates

80. K.R. Brown, "An Evaluation of the Chevron-Notched Short-Bar Specimen for Fracture Toughness Testing in Aluminum Alloys," Kaiser Aluminum & Chemical Co., Oct 1, 1981

81. "Determination of Short Bar Fracture Toughness of Metallic Materials," Aerospace Recommended Practice, ARP 1704, Society of Automotive Engineers, Warrendale, PA, July 1, 1981

82. J.G. Kaufman and S.F. Collis, A Fracture Toughness Data Bank, *J. Test. Eval.*, Vol 9 (No. 2), March 1981, p 121-126

Micro-Fracture Mechanics

By Robert O. Ritchie
Professor of Metallurgy
Department of Materials Science and Mineral Engineering
University of California—Berkeley

FRACTURE MECHANICS is an analytical tool that permits prediction of the structural behavior of cracked or fractured components under stress. Criteria for the onset of fracture in the macroscopic sense, i.e., based on plane strain fracture toughness ($K = K_{Ic}$ in the linear-elastic range, or $J = J_{Ic}$ in the nonlinear-elastic range) (Ref 1, 2), relies on characterization of crack tip stress and deformation fields. This characterization uses asymptotic continuum mechanics; that is, the analysis becomes increasingly more accurate as the crack tip is approached.

One of the main advantages of fracture mechanics analysis is that it effectively correlates the macroscopic aspects of crack initiation and growth without developing microscopic models for the local fracture processes that depend on the nature of the microstructure and the local crack tip stress and deformation histories (Ref 3). However, for a complete understanding of fracture, such microstructural initiation and growth criteria must be defined and related to macroscopic continuum analyses. This is achieved through construction of microscopic models for specific fracture mechanisms, which are called "micromechanisms."

Unlike the continuum approach, microfracture mechanics requires a microscopic model for a given fracture mode that designates a local failure criterion and significant microstructural characteristics, as well as detailed knowledge of both the asymptotic and very-near-tip stress and deformation fields. Physical fracture processes, and consequently the local failure criterion and characteristic microstructural dimensions, vary substantially with fracture mode. Figure 1 illustrates four classical fracture morphologies: microvoid coalescence (ductile), quasi-cleavage (brittle), intergranular cracking (brittle with aspects of ductile fracture), and transgranular cleavage (brittle).

Fig. 1 Classical fracture morphologies

(a) Microvoid coalescence. (b) Quasi-cleavage. (c) Intergranular cracking. (d) Transgranular cleavage. Fractographs (a) and (c) were obtained using scanning electron microscopy. (b) and (d) are transmission electron microscopy replicas. Courtesy of A.W. Thompson

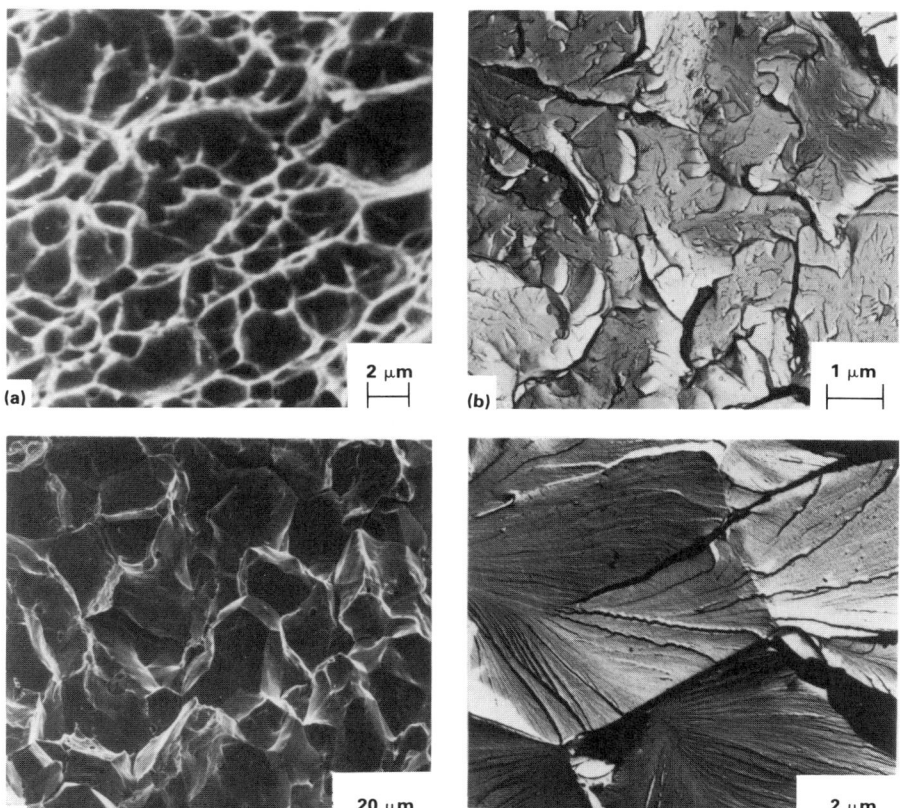

Brittle Fracture

Because of the specificity of micro-fracture models to particular fracture mechanisms for particular microstructures, a complete microscopic/macroscopic characterization of toughness has only been achieved in a few simplified cases. For example, for slip-initiated transgranular cleavage fracture (Fig. 1d) in ferritic steels, the onset of brittle crack extension at $K = K_{Ic}$ is consistent with a critical stress model in which the local tensile opening stress (σ_{yy})

directly ahead of the crack must exceed a local fracture stress (σ_f^*) over a microstructurally significant characteristic distance $(x = L_o^*)$, as depicted in Fig. 2(a) (Ref 4).

Fig. 2 Microscopic fracture criteria

(a) Critical stress-controlled model for cleavage fracture (RKR). (b) Critical stress-modified, critical strain-controlled model for microvoid coalescence. Source: Ref 9

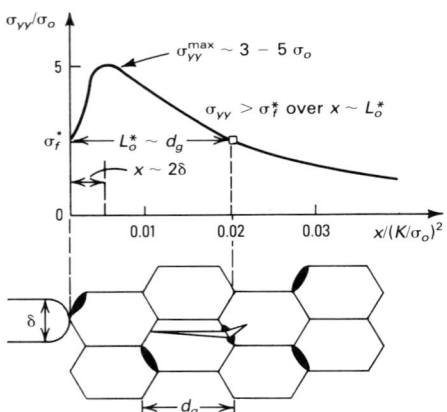

(a)

(b)

Using the stress distribution for a nonlinear elastic power-hardening solid (Ref 5, 6) to define crack tip stresses, this model (the Ritchie-Knott-Rice, or RKR, model) for cleavage fracture toughness implies that K_{Ic} will vary (Ref 4, 7, 8):

$$K_{Ic} \propto \left[\frac{(\sigma_f^*)^{(1+n)/2}}{(\sigma_o)^{(1-n)/2}} \right] (L_o^*)^{1/2} \qquad \text{(Eq 1)}$$

where n is the strain-hardening exponent, and σ_o is the flow stress.

In low-carbon steels with ferrite/carbide microstructures, the characteristic distance is on the order of the spacing of the void-initiating grain boundary carbides, i.e., typically a few grain diameters (d_g) (Ref 4, 9). However, different size scales have been found when the analysis is applied to other materials. Extensive studies on cleavage fracture in low-carbon steels indicate that σ_f^* is essentially independent of temperature below the ductile/brittle transition. In addition, recent modeling studies of cleavage in low-carbon steel, using weakest link statistical considerations of the size distribution of cracked carbides, have interpreted the characteristic distance as the carbide location with the highest elemental failure probability pertinent to crack advancement (Ref 10).

The model has been found to be particularly successful in quantitatively predicting cleavage fracture toughness for a wide range of microstructures and in rationalizing the effect of temperature (Ref 4, 8, 11), strain rate (Ref 8, 11, 12), neutron irradiation (Ref 8, 12), etc. on K_{Ic}. Similar microscopic models involving a critical stress criterion have been suggested for other fracture modes, including intergranular cracking (Fig. 1c) in temper-embrittled steels (Ref 13, 14) and hydrogen-assisted fracture (Ref 15).

Ductile Fracture

For initiation of ductile fracture by microvoid coalescence (Fig. 1a), one criterion requires that the critical crack tip opening displacement (δ) exceed half the mean void-initiating particle spacing $(2\delta \approx L_o^* \sim d_p)$, based on the notion that, in nonhardening materials, this phenomenon would occur when the void sites are first enveloped by the intense strain region at the crack tip at distance $x \sim 2\delta_i$ from the tip (Ref 16-18). This model implies:

$$\delta_i = \delta_{Ic} \approx (0.5 \text{ to } 2)d_p \qquad \text{(Eq 2a)}$$

or

$$J_{Ic} \sim \sigma_o L_o^* \qquad \text{(Eq 2b)}$$

although it is unusual to find a direct increase in fracture toughness with an increase in strength.

This problem is overcome by using a stress-modified critical strain criterion (Ref 8, 19-21). At $J = J_{Ic}$, the local equivalent plastic strain $\bar{\epsilon}_p$ must exceed a critical fracture strain or ductility $\bar{\epsilon}_f^* (\sigma_m/\bar{\sigma})$, specific to the relevant stress state, over a characteristic distance L_o^* comparable with the mean spacing (d_p) of the void-initiating particles, as shown in Fig. 2(b). The ratio $\sigma_m/\bar{\sigma}$, which defines the stress state, is the ratio of the hydrostatic to equivalent stress. Following the approach detailed in Ref 8, the near-tip strain distribution $\bar{\epsilon}_p$ is considered in terms of distance $(r = x)$ directly ahead of the crack, normalized with respect to the crack tip opening displacement δ:

$$\bar{\epsilon}_p \propto \left(\frac{J}{\sigma_o r} \right)^{1/(n+1)} \sim c_1 \left(\frac{\delta}{x} \right) \qquad \text{(Eq 3)}$$

where c_1 is of order unity. The crack initiation criterion of $\bar{\epsilon}_p$ exceeding $\bar{\epsilon}_f^* (\sigma_m/\bar{\sigma})$ over $x = L_o^* \sim d_p$ at $J = J_{Ic}$ now implies a ductile fracture toughness of (Ref 8):

$$\delta_i = \delta_{Ic} \sim \bar{\epsilon}_f^* L_o^* \qquad \text{(Eq 4a)}$$

or

$$J_{Ic} \sim \sigma_o \bar{\epsilon}_f^* L_o^* \qquad \text{(Eq 4b)}$$

or

$$K_{Ic} \equiv \sqrt{J_{Ic} E'} \sim \sqrt{E' \sigma_o \bar{\epsilon}_f^* L_o^*} \qquad \text{(Eq 4c)}$$

Unlike the critical crack tip opening displacement (CTOD) criterion (Eq 2a), the stress-modified critical strain criterion (Eq 4) now implies that J_{Ic} for ductile fracture is proportional to strength times ductility. This concept is more physically realistic and permits rationalization of the toughness-strength relation for cases in which microstructural changes that increase strength also cause a more rapid reduction in the critical fracture strain.

Microstructure and Toughness

Using microscopic models to predict, or at least to rationalize, the toughness of materials also involves consideration of the local fracture stress, σ_f^*, and fracture ductility $\bar{\epsilon}_f^*$. Values of σ_f^* are measured in V-notch bend bars, as described in detail in Ref 22. There is no conceptual difficulty with the term $\bar{\epsilon}_f^*$, but defining it as a material constant creates some practical difficulties. For example, it cannot necessarily be equated to either tensile or plane-strain ductilities as conventionally measured. Analysis for the rate of void expansion in the triaxial stress field ahead of a crack tip in a nonhardening material, in terms of the void radius (R_p), suggests (Ref 18):

$$\frac{dR_p}{R_p} = 0.28 \, d\bar{\epsilon}_p \exp \left(\frac{1.5\sigma_m}{\bar{\sigma}} \right) \qquad \text{(Eq 5)}$$

For an array of void-initiating particles of diameter D_p and mean spacing d_p, setting the initial void radius to $D_p/2$ and integrating Eq 5 to the point of ductile fracture initiation gives an expression for the fracture strain, $\bar{\epsilon}_f^*$, as:

$$\bar{\epsilon}_f^* \approx \frac{\ln \left(\dfrac{d_p}{D_p} \right)}{0.28 \exp \left(\dfrac{1.5\sigma_m}{\bar{\sigma}} \right)} \qquad \text{(Eq 6)}$$

An earlier analysis for a strain-hardening material (of exponent n) containing cylindrical holes similarly suggests:

$$\bar{\epsilon}_f^* \approx \frac{\ln\left(\dfrac{d_p}{D_p}\right)(1 - n)}{\sinh\left[\dfrac{(1 - n)(\sigma_a^\infty + \sigma_b^\infty)}{\dfrac{2\bar{\sigma}}{\sqrt{3}}}\right]} \quad \text{(Eq 7)}$$

where σ_a^∞ and σ_b^∞ are the transverse stress components (Ref 19).

Although both analyses consider the fracture strain to be limited by the impingement of the growing voids and thus tend to overestimate $\bar{\epsilon}_f^*$ by ignoring prior coalescence due to shear banding by strain localization, they correctly suggest a dependence of $\bar{\epsilon}_f^*$ on stress state ($\sigma_m/\bar{\sigma}$), strain hardening (n), and purity (d_p/D_p). For example, a significant effect of stress state (i.e., triaxiality) on fracture strain is predicted such that from Eq 7 $\bar{\epsilon}_f^*$ would be expected to be reduced by an order of magnitude by going from an unnotched plane-strain condition to that ahead of a sharp crack. Increased strain hardening, however, can enhance $\bar{\epsilon}_f^*$, particularly at high triaxiality. However, the benefits of increased purity (i.e., increased hole spacing d_p) are only pronounced at low D_p/d_p ratios due to the logarithmic terms in Eq 6 and 7. For example, reducing the volume fraction f_p of inclusions from 0.001 to 0.000001 would only increase $\bar{\epsilon}_f^*$ by a factor of 2 (Ref 23).

The success of these microscopic models for crack initiation toughness is illustrated in Fig. 3, in which the RKR critical stress model for cleavage (Eq 1) and stress-modified critical strain model for ductile fracture (Eq 4) are used to predict the respective lower and upper shelf toughness in ASME SA533B-1 nuclear pressure vessel steel (Ref 8). To achieve a correspondence between macroscopic and microscopic results, however, the characteristic distance (L_o^*) for cleavage fracture was found to scale with approximately two to four times the grain size (essentially the bainite packet size), whereas for ductile fracture L_o^* was found to be approximately five to six times the mean spacing of the major inclusions (d_p).

Such analyses, together with more recent models for the toughness of slowly growing cracks (Ref 9), serve to integrate the microscopic aspects of the fracture processes to macroscopic toughness behavior and to highlight the fundamental role of microstructure. However, their predictive capacity, in terms of engineering and structural design, must be treated carefully. Due to imprecise knowledge of the precise nature of the local mechanisms of fracture in many materials, *a priori* determination of the characteristic distance

Fig. 3 Comparison of experimentally measured fracture toughness K_{Ic} data for crack initiation in high-strength ASME SA533B-1 class 02 nuclear pressure vessel steel in the unirradiated condition ($\sigma_o \sim 500$ MPa, or 72.5 ksi)

Predicted values based on RKR critical stress model for cleavage on the lower shelf (Eq 1) and on the stress-modified critical strain model for microvoid coalescence on the upper shelf (Eq 4). Source: Ref 8

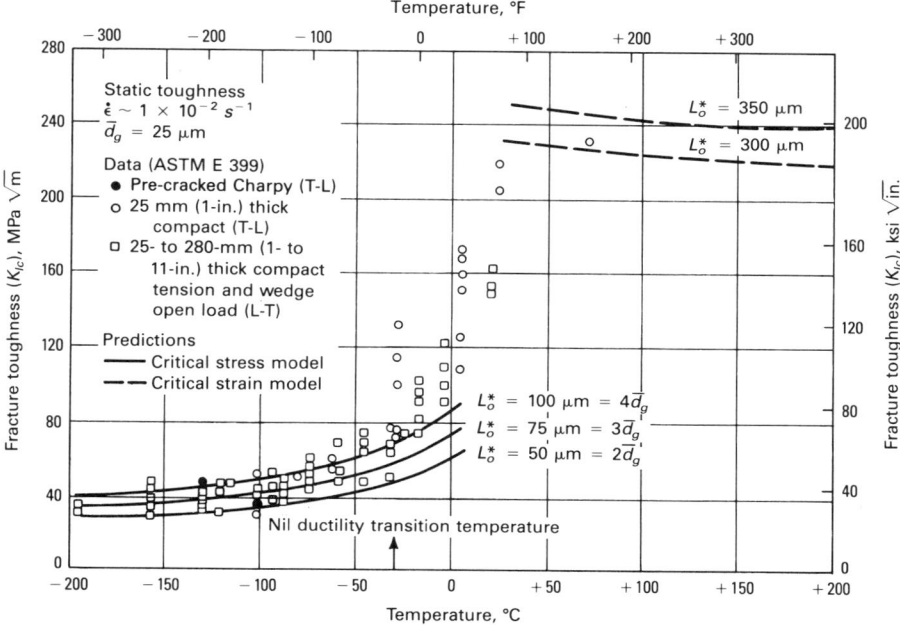

L_o^* is often simply not possible. At the present level of understanding, it appears more appropriate to regard L_o^* as essentially a fitting parameter that depends on the particular fracture micromechanism, but that must also exhibit a single scale consistent with the micromechanism and the relevant microstructural features.

REFERENCES

1. "Standard Test Method for Plane Strain Fracture Toughness of Metallic Materials," E 399, *Annual Book of ASTM Standards*, Vol 03.01, ASTM, Philadelphia, 1984, p 519-554

2. "Standard Test Method for J_{Ic}, a Measure of Fracture Toughness," E 813, *Annual Book of ASTM Standards*, Vol 03.01, ASTM, Philadelphia, 1984, p 763-781

3. G.R. Irwin, Fracture, in *Handbuch de Physik*, Vol 6, Springer, Berlin, 1958, p 551-590

4. R.O. Ritchie, J.F. Knott, and J.R. Rice, On the Relationship between Critical Tensile Stress and Fracture Toughness in Mild Steel, *J. Mech. Phys. Solids*, Vol 21, 1973, p 395-410

5. J.W. Hutchinson, Singular Behavior at the End of a Tensile Crack in a Harden-ing Material, *J. Mech. Phys. Solids*, Vol 16, 1968, p 13-31

6. J.R. Rice and G.R. Rosengren, Plane Strain Deformation near a Crack Tip in a Power-Law Hardening Material, *J. Mech. Phys. Solids*, Vol 16, 1968, p 1-12

7. D.A. Curry, Comparison between Two Models of Cleavage Fracture, *Met. Sci.*, Vol 14, 1980, p 319

8. R.O. Ritchie, W.L. Server, and R.A. Wullaert, Critical Fracture Stress and Fracture Strain Models for the Prediction of Lower and Upper Shelf Toughness in Nuclear Pressure Vessel Steels, *Met. Trans. A*, Vol 10, 1979, p 1557-1570

9. R.O. Ritchie and A.W. Thompson, On Macroscopic and Microscopic Analyses for Crack Initiation and Growth Toughness in Ductile Alloys, *Met. Trans. A*, Vol 15, 1984 (in press)

10. A.G. Evans, Statistical Aspects of Cleavage Fracture in Steel, *Met. Trans. A*, Vol 14, 1983, p 1349-1355

11. D.A. Curry, Predicting the Temperature and Strain Rate Dependencies of the Cleavage Fracture Toughness of Ferritic Steels, *Mater. Sci. Eng.*, Vol 43, 1980, p 135-144

12. D.M. Parks, Interpretation of Irradiation Effects on the Fracture Toughness of

Pressure Vessel Steels in Terms of Crack Tip Stress Analysis, *J. Eng. Mater. Technol.* Trans. ASME Series H, Vol 98, 1976, p 30-35

13. R.O. Ritchie, L.C.E. Geniets, and J.F. Knott, Effects of Grain-Boundary Embrittlement and Fracture Crack Propagation in Low Alloy Steel, in *The Microstructure and Design of Alloys*, Proc. 3rd Int. Conf. Strength Metals and Alloys, Vol 1, Institute of Metals/Iron and Steel, London, 1973, p 124-128

14. J. Kameda and C.J. McMahon, Solute Segregation and Brittle Fracture in an Alloy Steel, *Met. Trans. A*, Vol 11, 1980, p 91-101

15. K.N. Akhurst and T.J. Baker, The Threshold Stress Intensity for Hydrogen-Induced Crack Growth, *Met. Trans. A*, Vol 12, 1981, p 1059-1070

16. F.A. McClintock, Crack Growth in Fully Plastic Grooved Tensile Specimens, in *Physics of Strength and Plasticity*, A.S. Argon, Ed., MIT Press, Cambridge, MA, 1969, p 307-326

17. J.R. Rice and M.A. Johnson, The Rate of Large Crack Tip Geometry Changes in Plane Strain Fracture, in *Inelastic Behavior of Solids*, M.F. Kanninen *et al.*, Ed., McGraw-Hill, New York, 1970, p 641-672

18. J.R. Rice and D.M. Tracey, On the Ductile Enlargement of Voids in Triaxial Stress Fields, *J. Mech. Phys. Solids*, Vol 17, 1969, p 201-217

19. F.A. McClintock, A Criterion for Ductile Fracture by the Growth of Holes, *J. Appl. Mech.*, Trans. ASME Series H, Vol 25, 1958, p 363-371

20. A.C. Mackenzie, J.W. Hancock, and D.K. Brown, On the Influence of State of Stress on Ductile Failure Initiation in High Strength Steels, *Eng. Fract. Mech.*, Vol 9, 1977, p 167-188

21. R.C. Bates, Mechanics and Mechanisms of Failure, in *Metallurgical Treatises*, J.K. Tien and J.F. Elliott, Ed., TMS-AIME, Warrendale, PA, 1982, p 551-570

22. J.F. Knott, *Fundamentals of Fracture Mechanics*, Butterworths, London, 1973

23. F.A. McClintock, Mechanics in Alloy Design, in *Fundamental Aspects of Structural Alloy Design*, R.I. Jaffee and B.A. Wilcox, Ed., Plenum Press, New York, 1977, p 147-172

Fracture Toughness Testing Using Chevron-Notched Specimens

By Donald H. Sherman
Senior Research Engineer
Caterpillar Tractor Company

CHEVRON-NOTCHED SPECIMENS have gained acceptance due to their significant economic advantages over other specimen configurations for determining fracture toughness of metallic engineering alloys and other engineered materials, such as ceramics and glass. This specimen configuration is used routinely at several industrial and government laboratories. The Society of Automotive Engineers has adopted aerospace recommended procedure ARP 1704 for determination of fracture toughness using chevron-notched specimens, and the ASTM committee on chevron-notched test methods is developing a standard test procedure for the specimen.

Chevron-notched specimens were originally used for determining the fracture toughness of brittle materials that were difficult to fatigue precrack. These materials exhibited nearly ideal linear-elastic behavior and required only maximum load to failure for calculating fracture toughness. Use of chevron-notched specimens was extended to ductile materials that required impractically large specimens for determination of fracture toughness by other methods. The thin chevron slots provide good plane-strain constraint at the crack ends, resulting in smaller specimen sizes for valid fracture toughness values from a given material. Detailed data analysis methods were developed for nonideal behavior so that the significant advantages of the specimen, test procedure, and data analysis could be used on most engineering alloys.

Advantages

Chevron-notched specimens can be used to determine the fracture toughness of materials that are difficult to precrack, materials unavailable in large section sizes, and materials that are economically prohibitive to test using other specimen configurations. A natural crack initiates at the notch tip and extends in a stable manner as the load increases during testing. This eliminates fatigue precracking and simplifies interpretation of results. A chevron-notched specimen is only about 40% of the thickness and 2% of the weight of specimens required for other fracture toughness tests. For example, a 12-mm by 12-mm by 19-mm (0.5-in. by 0.5-in. by 0.75-in.) chevron-notched specimen of SAE 15B35 steel, quenched and tempered to 50 HRC with a yield strength of 1120 MPa (162 ksi) and $K \cong 80 \, \mathrm{MPa}\sqrt{m}$ (72.8 $\mathrm{ksi}\sqrt{in.}$), would weigh about 0.02 kg (0.05 lb). An ASTM E 399 compact tension specimen of the same material would be 25 mm by 64 mm by 61 mm (1.0 in. by 2.5 in. by 2.4 in.) and weigh 0.73 kg (1.62 lb).

Specimen design and preparation is simple. The test can be run on tension testing equipment, on universal testing equipment (using an attachment), or on equipment specifically designed for the test. Test times run from 5 to 20 min, depending on the degree of automation. These times are short compared to those of other fracture toughness testing methods.

The calculation of fracture toughness from test results is straightforward when using chevron-notched specimens. Other advantages include good repeatability and accuracy. In comparison testing (Ref 1), the standard deviations for results from tests using chevron-notched specimens were smaller and fell within the standard deviations for compact tension specimens, as specified in ASTM E 399, "Standard Test Method For Plane-Strain Fracture Toughness of Metallic Materials." ASTM E 399 centers attention on the start of crack extension from a fatigue precrack, whereas the chevron-notched specimen method makes use of a steady-state propagating crack or a crack at the initiation of a crack jump. Although both methods are based on the principles of linear elastic fracture mechanics, this distinction, coupled with differences in test procedures, causes fracture toughness values measured by the chevron-notched specimen method to be somewhat higher than those obtained using ASTM E 399.

Specimens and Test Equipment

Several chevron-notched specimen geometries and test equipment configurations are available for determining fracture toughness, all of which are capable of producing valid results.

Fig. 1 Single chevron-notched specimen

The shaded area denotes the crack; F denotes the opening load applied at the mouth of the specimen.

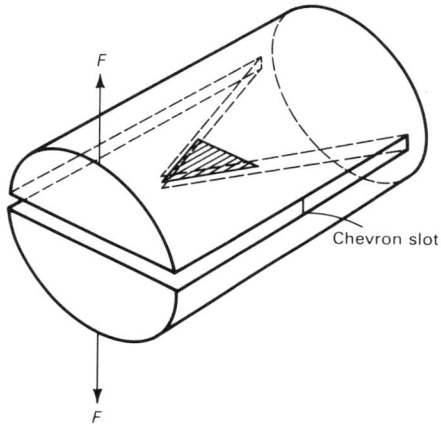

Specimen Geometry. A simple chevron-notched specimen is illustrated in Fig. 1. The very thin chevron slot significantly reduces the size of the nonplane-strain zone at the flank ends of the crack. This increased triaxial constraint allows valid fracture toughness determinations for a given metallic alloy to be made using much smaller specimens than required by ASTM E 399.

The chevron slot serves three additional critical functions. First, the slot defines the crack plane and propagation direction, providing a simple method for orienting the fracture with respect to the bulk material. Second, the slot forces crack initiation at the point of the "V" when the load is applied. Finally, the chevron slot configuration produces stable crack initiation and growth, because the crack front widens as it advances. This is advantageous when testing brittle alloys that are difficult to fatigue precrack.

Common specimen geometries and sizes are shown in Fig. 2 to 4 and in Table 1 (Ref 2-11). The indicated specimen length, width, and height ratios and chevron-notched geometries are based on extensive testing and analysis. These ratios and geometries are optimized for accuracy of fracture toughness determination and testing economy (Ref 11). Figure 5 correlates chevron slot angle θ with initial crack length a_o for curved chevron slots.

Specimen geometries can be divided into two broad categories: specimens with rectangular cross sections (short bar) and those with round cross sections (short rod). Both straight and curved chevron slots are used with either specimen cross section, depending on the side-slot machining equipment available. A curved slot is made by plunge cutting with a circular blade, whereas a straight slot is cut by using an electrical discharge wire or by feeding a circular cutter through the specimen.

Test Equipment. The maximum load to failure is the minimum data required to calculate fracture toughness from chevron-notched specimens. However, an x-y plot of load versus specimen mouth opening displacement is desirable for making a complete evaluation of material behavior. Loading devices commonly used for generating the required data include tension testing machines (Fig. 6), the Fractometer I* test system for hard, brittle materials (Fig. 7), the Fractometer II* test system for most materials (Fig. 8 to 10), and the Fracjack* test system, which attaches to the universal test machine (Fig. 11).

*Fractometer and Fracjack are trade names of Terra Tek, Inc., Salt Lake City, Utah.

Fig. 2 Typical specimen geometries with straight chevron slots

The load line is the line along which the opening load is applied in the mouth of the specimen. Common sizes have B equal to 12.7 mm (0.5 in.), 19.08 mm (0.75 in.), and 25.4 mm (1.0 in.).

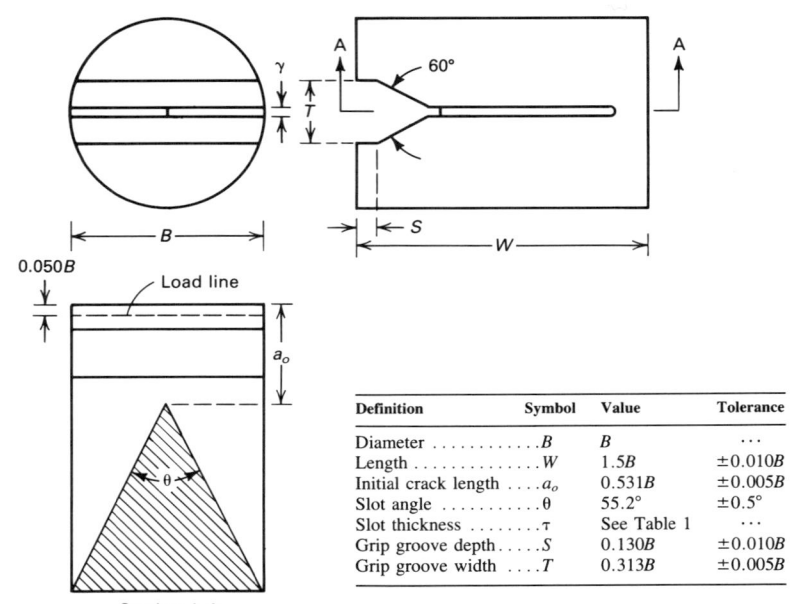

Definition	Symbol	Value	Tolerance
Diameter . . . . B	B	1.5B	· · ·
Length . . . . W	W	1.5B	±0.010B
Initial crack length . . . . a_o	a_o	0.531B	±0.005B
Slot angle . . . . θ	θ	55.2°	±0.5°
Slot thickness . . . . τ	τ	See Table 1	· · ·
Grip groove depth . . . . . S	S	0.130B	±0.010B
Grip groove width . . . . T	T	0.313B	±0.005B

Short rod specimen

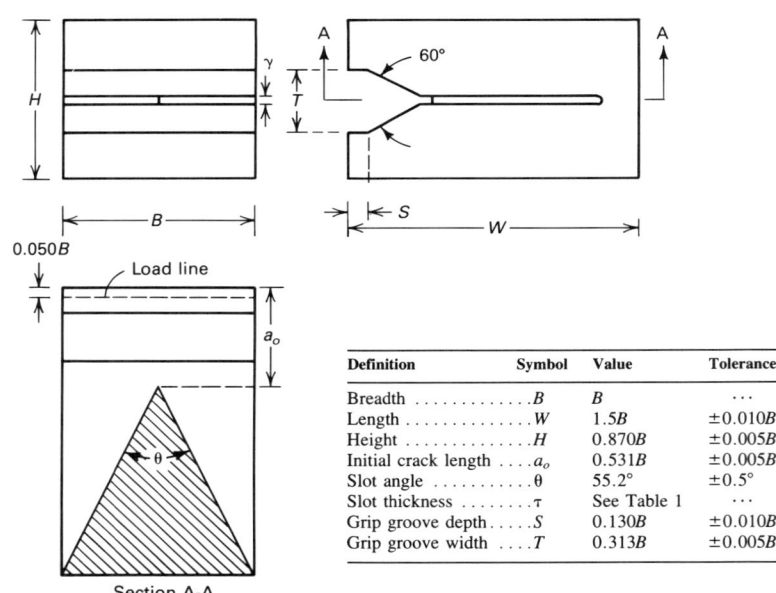

Definition	Symbol	Value	Tolerance
Breadth . . . . B	B		· · ·
Length . . . . W	W	1.5B	±0.010B
Height . . . . H	H	0.870B	±0.005B
Initial crack length . . . . a_o	a_o	0.531B	±0.005B
Slot angle . . . . θ	θ	55.2°	±0.5°
Slot thickness . . . . τ	τ	See Table 1	· · ·
Grip groove depth . . . . . S	S	0.130B	±0.010B
Grip groove width . . . . T	T	0.313B	±0.005B

Short bar specimen

The Fractometer test configurations are complete, stand-alone systems. The degree of automation and instrumentation for tension testing machines and the Fracjack system depends on the capabilities of each test laboratory.

Linear Elastic Fracture Mechanics Test

Chevron-notched specimens made from very hard, brittle materials, such as glass, ceramics, and carbides, exhibit nearly ideal

Fig. 3 Typical specimen geometries with curved chevron slots

The load line is the line along which the opening load is applied in the mouth of the specimen. Common sizes have B equal to 12.7 mm (0.5 in.), 19.08 mm (0.75 in.), and 25.4 mm (1.0 in.).

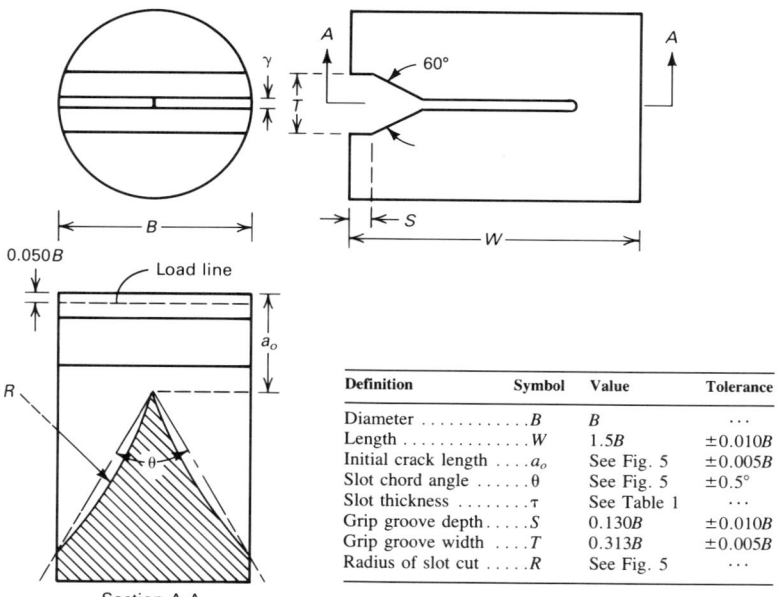

Definition	Symbol	Value	Tolerance
Diameter . . . . .B	B	. . .
Length . . . . .W	1.5B	±0.010B
Initial crack length . . . .a_o	See Fig. 5	±0.005B
Slot chord angle . . . . . .θ	See Fig. 5	±0.5°
Slot thickness . . . . .τ	See Table 1	. . .
Grip groove depth. . . . .S	0.130B	±0.010B
Grip groove width . . . .T	0.313B	±0.005B
Radius of slot cut . . . . .R	See Fig. 5	. . .

Short rod specimen

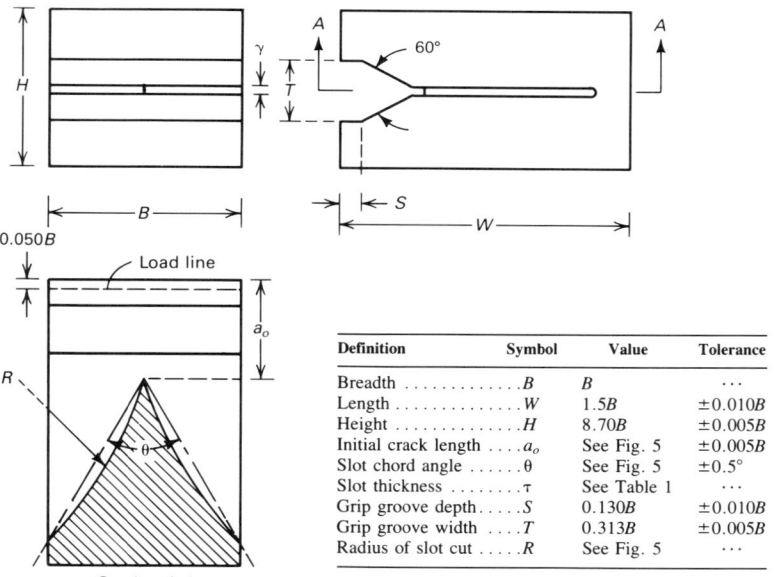

Definition	Symbol	Value	Tolerance
Breadth . . . . .B	B	. . .
Length . . . . .W	1.5B	±0.010B
Height . . . . .H	8.70B	±0.005B
Initial crack length . . . .a_o	See Fig. 5	±0.005B
Slot chord angle . . . . . .θ	See Fig. 5	±0.5°
Slot thickness . . . . .τ	See Table 1	. . .
Grip groove depth. . . . .S	0.130B	±0.010B
Grip groove width . . . .T	0.313B	±0.005B
Radius of slot cut . . . . .R	See Fig. 5	. . .

Short bar specimen

linear elastic behavior. A plot of load versus mouth opening displacement for an ideal elastic material is shown in Fig. 12. Crack initiation occurs at *I*, where the loading first deviates from linearity. The crack then extends in a stable manner throughout the test.

The load required to advance the crack increases to a smooth maximum at the critical crack length. This maximum load is used to calculate fracture toughness. The critical crack length is dependent only on specimen geometry and is independent of material.

Therefore, there is no need to measure crack length when determining fracture toughness.

The rationale for using linear elastic fracture mechanics principles for calculating fracture toughness of an ideal elastic material is illustrated in Fig. 12. Unloading from either *R* or *S* is linear to 0. Therefore, the shaded area represents the irrecoverable work to form the fracture surface (i.e., the fracture toughness of the material). The following equation is used to calculate the fracture toughness of the specimens shown in Fig. 2 and 3 (Ref 6, 8-12):

$$K = \frac{(AF)}{(B)^{3/2}}$$

where *K* is the fracture toughness obtained from chevron-notched specimens, *F* is maximum load, *B* is specimen diameter for short rods or thickness for short bars, and *A* is a dimensionless calibration constant dependent only on specimen geometry.

The currently recommended calibration constant for the specimens shown in Fig. 2 and 3 is 22.0. The equations and calibration constants for the specimens shown in Fig. 4 are given in Ref 2 to 5 and Ref 7.

Elastic-Plastic Behavior

Most chevron-notched specimens made from practical engineering alloys exhibit elastic-plastic behavior when tested. A plot of load versus mouth opening displacement for ideal elastic-plastic behavior is shown in Fig. 13. Unloading from *R* and *S* is offset from 0 to *C* and *D*, respectively. This offset is the result of yielding in two plastic zones in the specimen: the plane-stress plastic zone and the plane-strain plastic zone.

The plane-stress plastic zone occurs at the crack tip where the crack intersects the bottom of the chevron notch. The size of this zone is dependent on chevron-notch acuity. Sufficient constraint usually is generated to adequately minimize the plane-stress plastic zone size when one of the following criteria is met:

- The chevron slot thickness is less than 1.5% of the short rod diameter or short bar thickness.
- The chevron slot bottoms are sharp pointed, as with a 60° included angle at the bottom of the slots.
- The plastic zone size is less than 0.2% of the specimen diameter or thickness.

However, the plane-strain plastic zone is located at the crack tip within the bulk of the specimen. Yielding in the plane-strain plastic zone partially relieves the stress at the crack tip. Therefore, the crack is apparently longer

Fig. 4 Alternate geometries for chevron-notched specimens

(a) Short rod. (b) Short bar. These geometries are recommended for linear elastic testing of brittle materials. Note that the grip grooves are different from the grip grooves shown in Fig. 2 and 3.

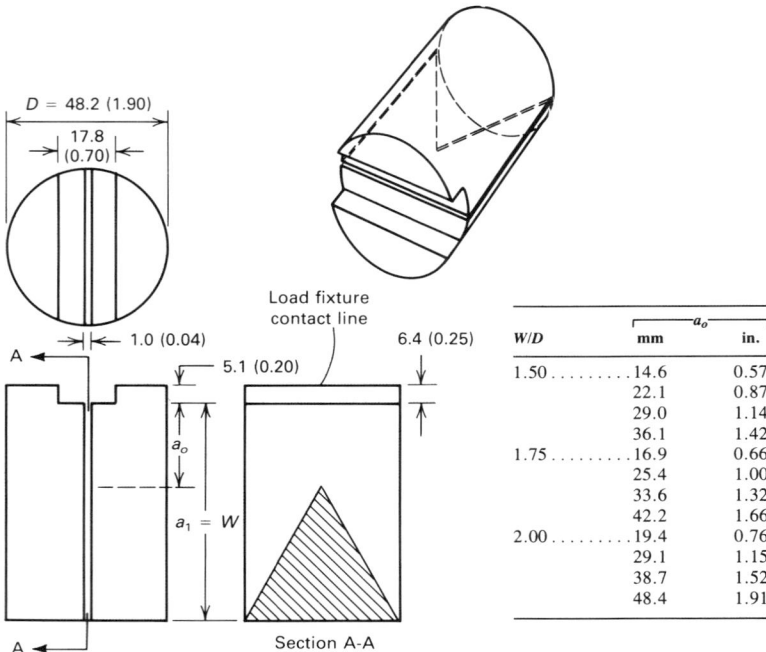

W/D	a_o	
	mm	in.
1.50	14.6	0.57
	22.1	0.87
	29.0	1.14
	36.1	1.42
1.75	16.9	0.66
	25.4	1.00
	33.6	1.32
	42.2	1.66
2.00	19.4	0.76
	29.1	1.15
	38.7	1.52
	48.4	1.91

Dimensions, mm (in.)

(a)

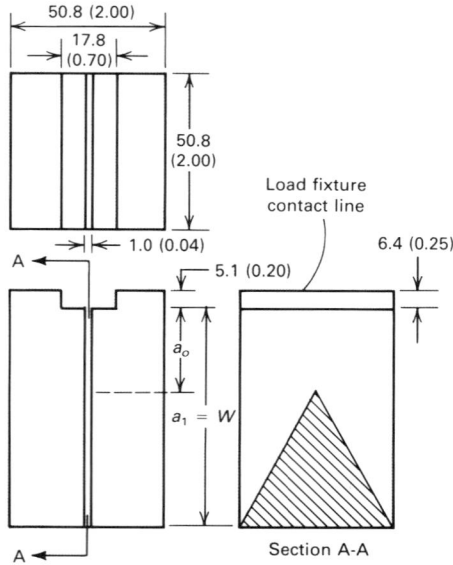

W/H	a_o/W
4	0.201
	0.345
	0.495
3	0.196
	0.342
	0.481

Dimensions, mm (in.)

(b)

Fig. 5 Chevron slot angle, θ, and initial crack length, a_o, for curved chevron slots

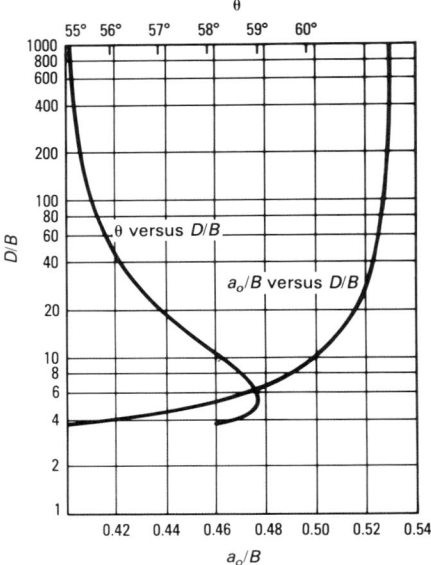

Fig. 6 Setup for testing chevron-notched specimens on a tension testing machine

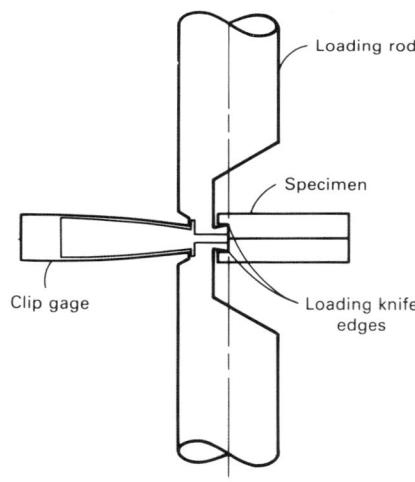

than is the actual case and thus the measured fracture toughness values are too low.

A plasticity factor, P, is used to adjust the fracture toughness to the proper value. The plasticity factor is measured experimentally from the load versus mouth opening displacement plot in Fig. 13 as follows:

$$P = \frac{(B' - A')}{(D - C)}$$

Note that the plasticity determination re-

Table 1 Results of slot geometry study

Slot thickness mm	in.	Effect on specimen calibration, %	Plane-strain constraint(a)	Slot configuration
0.38	0.015 0		Excellent	
0.8	0.031 . . . . −1		Excellent	
1.6	0.063 . . . . −3		Excellent	
0.38	0.015 0		Excellent	
0.8	0.031 . . . . −1		Good	
1.6	0.063 . . . . −3		Poor	
0.38	0.015 0		Good	
0.8	0.031 . . . . −1		Poor	
1.6	0.063 . . . . −3		Poor	

(a) Excellent, less than +2% effect on the measurement; good, less than +5% effect on the measurement; poor, more than +5% effect on the measurement.

Fig. 7 Fractometer I method for loading hard, brittle chevron-notched specimens

(a) The specimen is seated with the flatjack in the specimen slot. (b) Fluid pressure in the flatjack produces fracture of the specimen. The specimen deflection is greatly exaggerated.

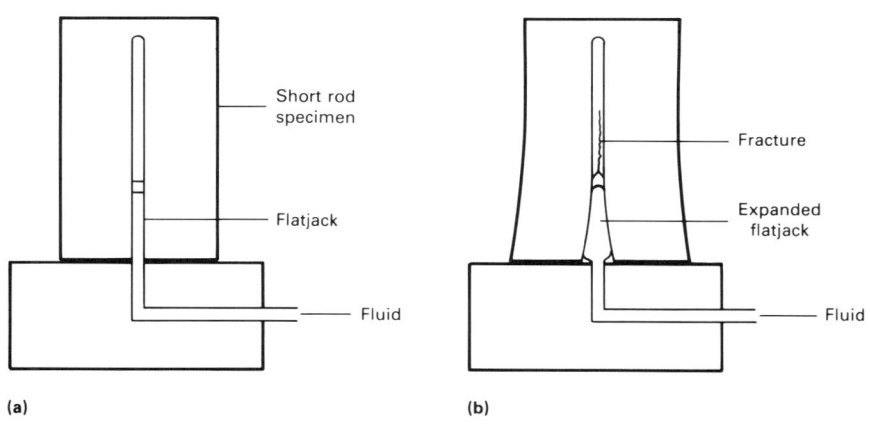

(a)   (b)

quires a plot of load versus displacement and two specimen unloadings and reloadings. Useful fracture toughness values are obtained when $-0.05 < P < 0.1$. When too large, P is reduced for a given material with sharper chevron slot acuity and/or larger specimen size.

There is no assurance that the peak load will occur at the critical crack length in elastic-plastic tests. However, for both linear elastic and elastic-plastic tests, the critical crack length occurs at a critical unloading slope ratio, r_c, which is a constant for any particular specimen configuration. The unloading slope ratio is defined as the unloading slope divided by the slope of the initial elastic loading path. Therefore, the correct load, F_c, to use for calculating fracture toughness is located where an unloading line having the critical unloading slope intersects the load displacement curve. This line is found from vertical interpolation between two actual unloading lines.

This interpolation is done either graphically on the load versus displacement plot or automatically with a computer (Ref 12). The critical slope ratios for the specimens shown in Fig. 2 and 3 are presented in Fig. 14. The critical slope ratios are not available for the specimens shown in Fig. 4 which are therefore not recommended for elastic-plastic testing.

Fracture toughness for elastic-plastic behavior for the specimens shown in Fig. 2 and 3 is then calculated using the following equation:

$$K = \left[\frac{(AF_c)}{B^{3/2}} \right] \left[\frac{(1 + P)}{(1 - P)} \right]$$

where A is a dimensionless calibration constant dependent only on specimen geometry (the currently recommended calibration constant for the specimens shown in Fig. 2 and 3 is 22.0), F_c is the load at the critical unloading slope ratio, B is specimen diameter for a short rod or thickness for a short bar, and P is the plasticity factor.

Data Analysis

Actual test records of load versus mouth opening displacement for chevron-notched specimens show marked departures from ideal behavior depending on the type of material and specimen size. Although methods for analysis are still evolving, four basic behavior types have been identified, and nearly all test records can be classified as either one or a combination of two or more of these types (Ref 12). The data analysis method depends on the type of test behavior. For the

Fig. 8 Fractometer II test system

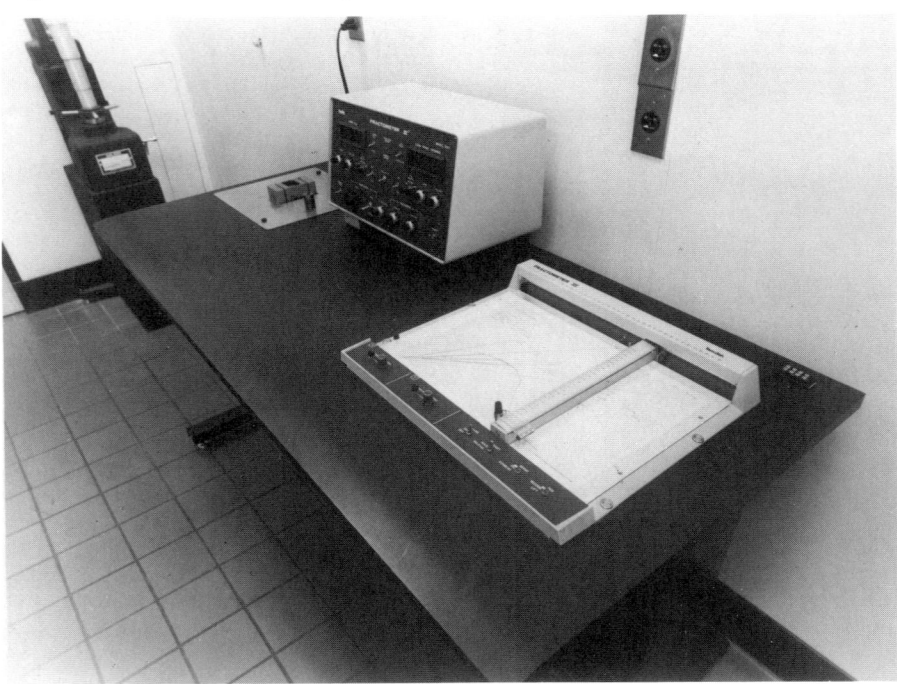

most accurate data analysis, the test should include two or more unloading and reloading cycles. One of these cycles should be somewhat before and another somewhat after the maximum load.

The four behavior types are shown in Fig. 15 and are designated as follows:

- Ideal linear elastic (Fig. 15a)
- Hysteresis in unloading-reloading paths (Fig. 15b)
- Crack jumps (Fig. 15c)
- Elastic-plastic and/or residual stress effect (Fig. 15d)

Detailed data analysis methods for each of these four basic behavior types have been developed (Ref 12).

REFERENCES

1. L.M. Barker, "Preliminary Analyses of Some Results of the Chevron Notched Short Rod/Short Bar Cooperative Test Program," presented at ASTM Subcommittee Meeting on Chevron-Notched Test Methods, Jacksonville, FL, April 10, 1984
2. D. Munz, R.T. Bubsey, and J.L. Shannon, Jr., Performance of Chevron-Notch Short Bar Specimen in Determining the Fracture Toughness of Silicon Nitride and Aluminum Oxide, *J. Test. Eval.*, Vol 8 (No. 3), 1980, p 103-107
3. D. Munz, R.T. Bubsey, and J.E. Srawley, Compliance and Stress Intensity Coefficients for Short Bar Specimens

Fig. 9 Schematic of the Fractometer II specimen loading mechanism

The strain gaged members serve as the load transducer.

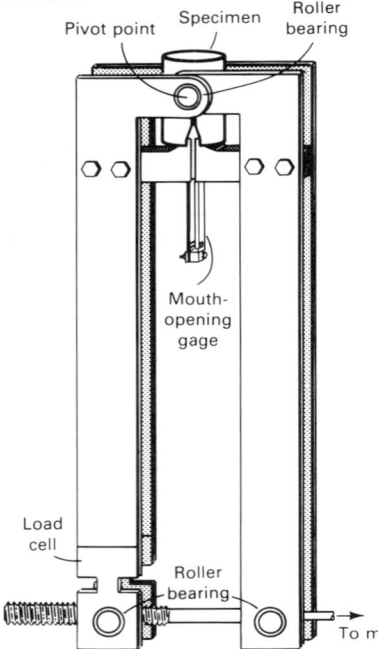

Fig. 10 Chevron-notched specimen about to be installed on the Fractometer II grips, showing the mouth opening gage configuration

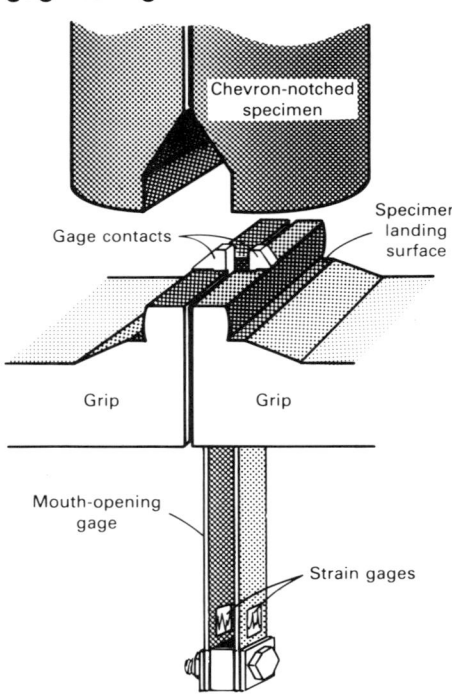

Fig. 11 Fracjack test system

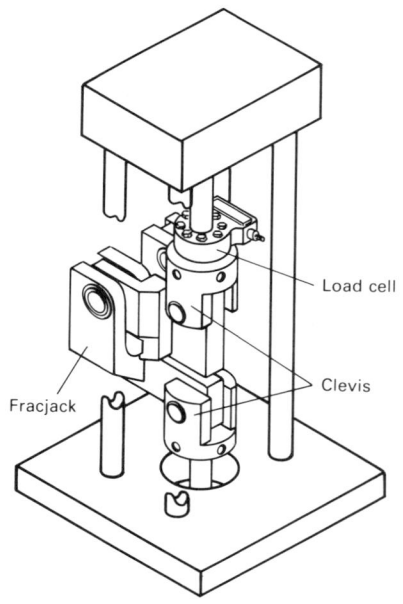

Fig. 12 Schematic of elastic specimen behavior

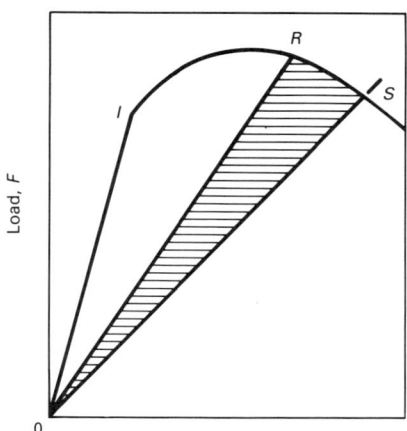

Fig. 13 Schematic of elastic-plastic specimen behavior

$\bar{F}$ is the average load between points R and S on the curve and locates both A' and B' on the unloading lines.

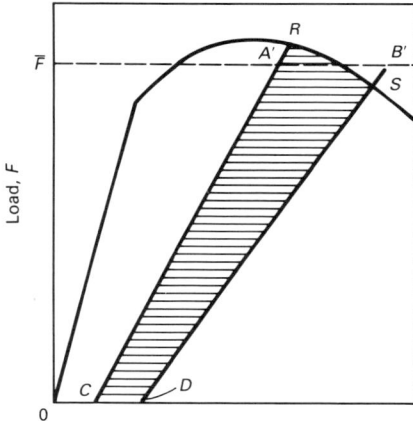

Fig. 14 Variation of the critical slope ratio, r_c, with the slot curvature parameter, D/B

Where B is the specimen diameter for a short rod or thickness for a short bar. D is twice the chevron slot radius of curvature.

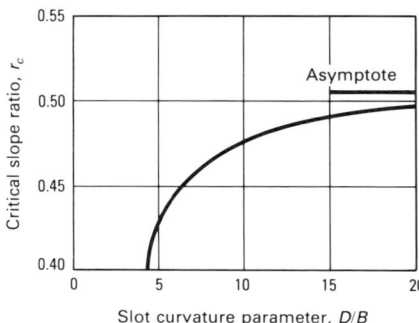

Fig. 15 Types of load versus mouth opening curves

(a) Ideal linear elastic fracture mechanics curve. (b) Hysteresis in unloading-reloading paths. (c) Crack jumps. (d) Specimen response resulting from elastic-plastic and/or residual stress effects

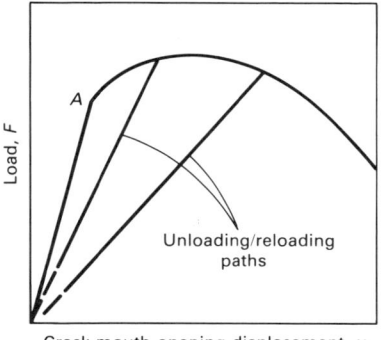

(a)

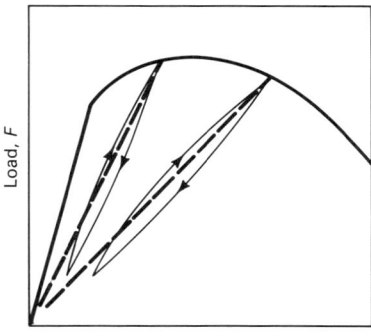

(b)

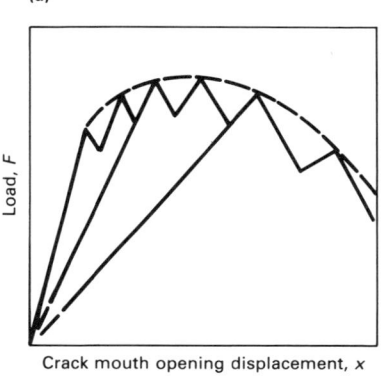

(c)

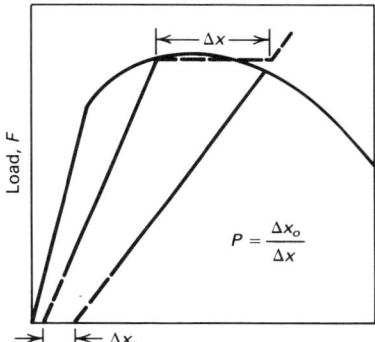

(d)

with Chevron Notches, *Int. J. Frac.*, Vol 16 (No. 4), 1980, p 359-394

4. R.T. Bubsey, D. Munz, W.S. Pierce, and J.L. Shannon, Jr., Compliance Calibration of the Short Rod Chevron-Notch Specimen for Fracture Toughness Testing of Brittle Materials, *Int. J. Frac.*, Vol 18 (No. 2), 1982, p 125-133

5. J.L. Shannon, Jr., R.T. Bubsey, W.S. Pierce, and D. Munz, Extended Range Stress Intensity Factor Expressions for Chevron-Notched Short Bar and Short Rod Fracture Toughness Specimens, *Int. J. Frac.*, Vol 19, 1982

6. J.F. Beech and A.R. Ingraffea, Three-Dimensional Finite Element Calibration of the Short Rod Specimens, *Int. J. Frac.*, Vol 18, 1982, p 217-229

7. A. Mendelson and L.J. Ghosen, Three-Dimensional Analysis of Short-Bar Chevron-Notched Specimens by Boundary Integral Method, in *Chevron-Notched Specimens: Testing and Stress Analysis*, STP 855, ASTM, Philadelphia, 1984

8. A.R. Ingraffea, R. Perucchio, T-Y Han, W.H. Gersthe, and Y.P. Huang, Three-Dimensional Finite and Boundary Element Calibration of the Short Rod Specimen, in *Chevron-Notched Specimens: Testing and Stress Analysis*, STP 855, ASTM, Philadelphia, 1984

9. L.M. Barker and R.V. Guest, "Compliance Calibration of the Short Rod Fracture Toughness Specimen," Report TR-78-20, Terra Tek, Inc., Salt Lake City, 1978

10. L.M. Barker, "Compliance Calibration of a Family of Short Rod and Short Bar Fracture Toughness Specimens," Report TR-81-07, Terra Tek, Inc., Salt Lake City, 1981

11. L.M. Barker, "Short Rod and Short Bar Fracture Toughness Specimen Geometries and Test Methods for Metallic Materials," Report TR-80-11, Terra Tek, Inc., Salt Lake City, 1980

12. L.M. Barker, "Data Analysis Methods for Short Rod and Short Bar Fracture Toughness Tests of Metallic Materials," Report TR-80-12, Terra Tek, Inc., Salt Lake City, 1980

Microstructure and Fracture

By William W. Gerberich
Professor and Associate Head
Department of Chemical Engineering
and Materials Science
University of Minnesota

FRACTOGRAPHY provides a basis for reconstructing the failure mechanism for a material of a given microstructure. Fractography and metallography are complementary to fracture mechanics testing, which determines under what conditions a component may fail. These methods are used in upgrading materials specifications, improving product design, and analyzing failures for improved product reliability. This article concentrates on how microstructure may affect fractographic observations.

Over the last two decades, four major types of failure modes have been discussed in the literature:

Dimpled rupture (microvoid coalescence)

- Ductile fracture
- Rapid overload fracture

Cleavage or quasi-cleavage

- Brittle fracture
- Premature or overload failure by catastrophic rapid fracture

Intergranular

- Stress-corrosion cracking or hydrogen embrittlement
- Subcritical growth under sustained load

Ductile striations

- Fatigue cracking
- Subcritical growth under cyclic load

The first three types are microstructurally dependent. Complications arise in determining failure type when complex microstructures are involved. For example, aluminum alloys containing extensive intermetallic compounds may fail by a combination of ductile rupture of the matrix and brittle cleavage of particles of the compound. In a similar manner, mixtures of ferrite, austenite, and martensite in dual-phase steels or delta ferrite and austenite in precipitation-hardened semi-austenitic stainless steels may produce mixed fracture modes, depending on test temperature, loading mode, and environment. Thus, a metallographic investigation is required to properly characterize the nature of the microstructure, particularly when more than one phase may exist.

Even with single-phase microstructures, the above categorization should be applied conservatively. For example, dimpled rupture is the least "controversial," because particle fracture and/or particle-matrix decohesion leading to void growth usually require the presence of locally large stresses and strains. Nevertheless, the effect of these stresses and strains can be changed by the environment, e.g., test temperature (Ref 1) or high-pressure hydrogen (Ref 2). Consequently, the dimple size may change without altering the general fractographic category.

A variety of microscopic failure modes may accompany each type of macroscopic failure. In this article, macroscopic failures are generalized into three categories: failure due to overload, fatigue failure, and environmental (sustained-load) failure. Some of the microscopic modes that are associated with these failure mechanisms are listed in Fig. 1. The four failure modes listed above represent the first horizontal row of microstructural failure modes in Fig. 1, with the exception of cleavage. Figure 1 shows that, in addition to cleavage, overload brittle failures may occur by intergranular segregation. This occurs because of segregation of "tramp elements" (phosphorus, tin, tellurium, sulfur, antimony, and indium) to grain boundaries (Ref 3, 4). Figure 1 further shows that for fatigue loading, the typical ductile striation usually results at an intermediate stress-intensity range, ΔK.

Because of the extensive evaluation of near-threshold regime fractures in the past decade, more microscopic fracture modes have become apparent. For example, Fig. 1 indicates that at a low stress-intensity factor range, ΔK, a cyclic ductile process occurs along crystallographic planes, while at low temperature, T, a brittle process occurs along crystallographic planes. In addition, microvoid separation may occur at very high ΔK values. In a similar manner, environmentally induced fracture may lead to several microscopic fracture modes. This depends on the environment, material, and microstructure combination. As indicated in Fig. 1, the number of observations decreases as alternative types of fracture modes increase. Nevertheless, improvement in materials selection depends on the effect of microstructural features on the failure mechanism. This also applies to alloy design and frequently to a determination of product liability.

This article provides a pictorial overview of possible microscopic fracture modes. Multiple microstructural phases lead to numerous combinations. When identifying failure modes, the analyst should obtain the most complete record possible of the product service history. The combination of detailed light microscopy and scanning electron or transmission electron fractography is the best method of categorizing the failure mode. It must be emphasized that the conditions in Fig. 1 are only a portion of the possible fracture-microstructure interactions.

Figure 2 illustrates typical microstructural effects with micrographs of commercially pure titanium (Fig. 2a) and Ti-6Al-4V, the most commonly used structural titanium alloy (Fig. 2b and c). The microstructure in Fig. 2(a) is recrystallized and annealed with an approximately 20-μm grain size. A coarse Widmanstätten structure is depicted in Fig. 2(b), and a fine Widmanstätten structure interspersed with coarse α platelets is shown in Fig. 2(c). The α phase is hexagonal close-packed (hcp), and the β phase is body-centered cubic (bcc). All of these microstructures have a different size, shape, and mixture of phases. The extent to which these affect crack path (e.g., crack path tortuosity) will affect both fracture toughness and subcritical crack growth. A description of the types of failure modes and the effect of microstructure on each is given below.

Failure Due to Overload

Single-phase microstructures, microstructures with different precipitate distributions and shapes, and either pretransformed or strain-induced *in situ* transformed microstructures may produce a variety of microscopic fracture processes. Classification of the fracture mechanism may be straightforward or may require high-quality transmission electron microscopy (TEM) replica studies. Because many subcritical crack growth processes are slow, the details may be even more obscure. Complications may arise because of reverse slip, asperity deformation, fretting corrosion, or dissolution.

This section will review overload fractures of metals and alloys that occur by ductile fracture, cleavage (brittle fracture), and mixed ductile/cleavage modes. Additional information on these fracture behaviors can be found in Volume 10 of the 8th Edition of *Metals Handbook*.

Ductile Fracture. Two types of ductile fracture (or rupture) (Ref 8) are illustrated in Fig. 3. Figure 3(a) shows the fracture surface of an ultrahigh-strength steel containing sulfide stringers. The stringers are typically 1 to 3 μm in diameter (Fig. 3b). The material is a

Fig. 1 Observed microscopic fracture modes under different loading conditions and environments

DBTT is the ductile-brittle transition temperature. K_{Iscc} and K_{IH} are threshold values for stress-corrosion cracking and hydrogen embrittlement. Insufficient fracture toughness may produce "premature" brittle fracture at notches, keyways, and inherent defects.

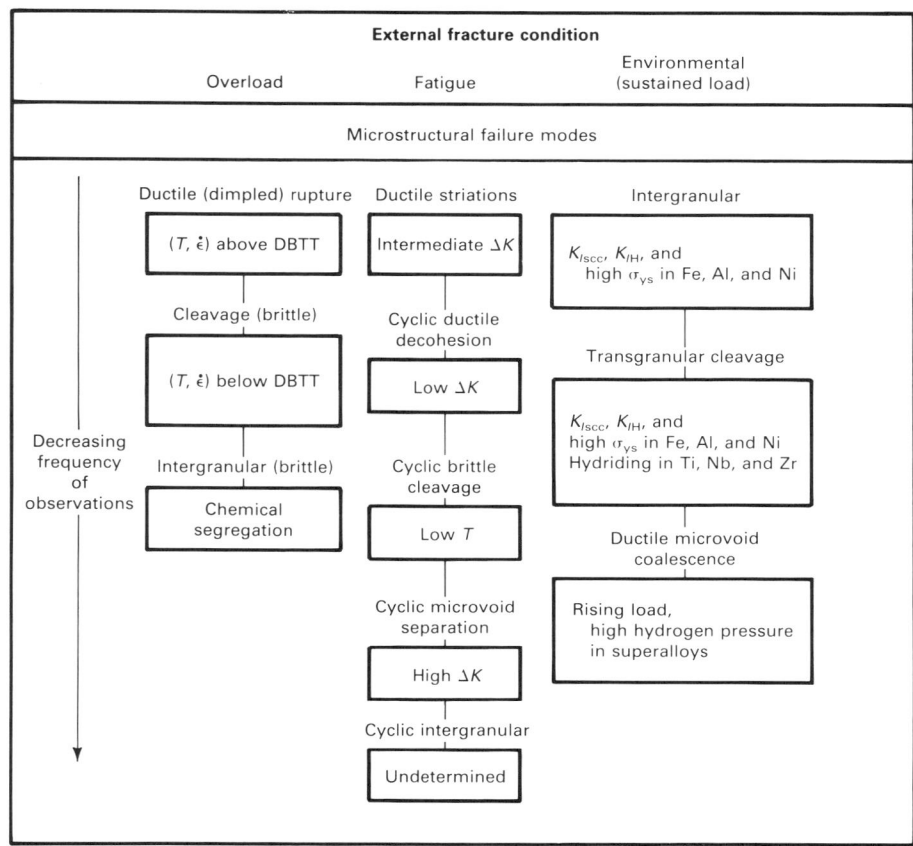

Fig. 2 Optical micrographs of titanium and titanium alloys

(a) Commercially pure annealed titanium, recrystallized and annealed. (b) Ti-6Al-4V with coarse α-β Widmanstätten structure. (c) Ti-6Al-4V with fine α-β Widmanstätten structure interspersed with coarse α platelets. Source: Ref 5-7

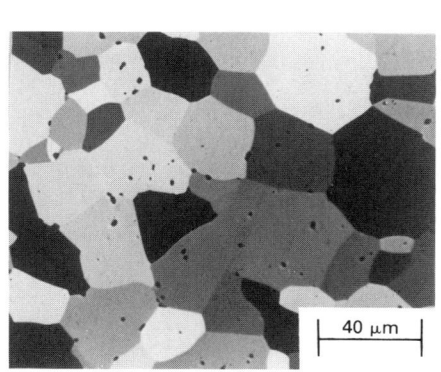

(a)

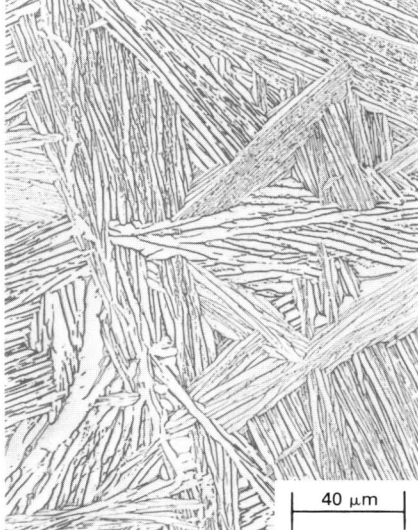

(b)

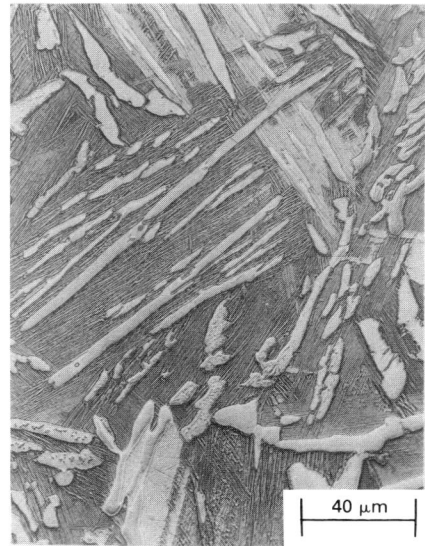

(c)

Fig. 3 Scanning electron micrographs of AISI 4130 steel

(a) and (b) Resulfurized AISI 4130 steel quenched and tempered to 1400 MPa (203 ksi) strength level. (c) Low-sulfur AISI 4130 steel spheroidize-annealed to 600 MPa (87 ksi) strength. Source: Ref 8

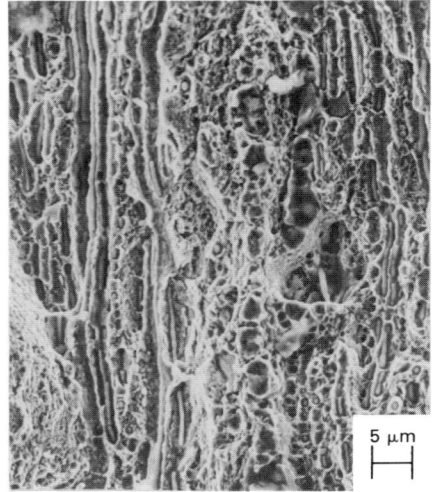
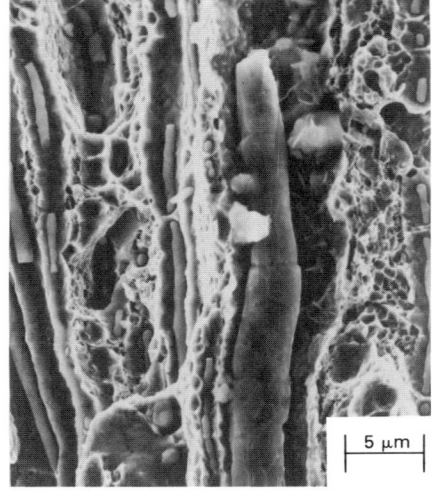
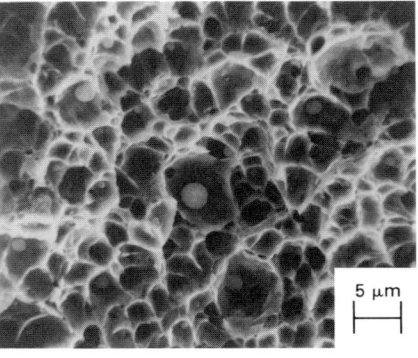

(a)
(b)
(c)

resulfurized AISI 4130 steel quenched and tempered to a 1400 MPa (203 ksi) strength level. The combination of high strength and high volume fraction of sulfide stringers led to low ductility, decohesion along the sulfide stringers, and very shallow microvoids that absorbed minimal energy in between the sulfide stringers. The plane-strain fracture toughness, K_{Ic}, was on the order of 30 MPa$\sqrt{m}$ (27.3 ksi$\sqrt{in.}$).

Figure 3(c) shows the fracture surface of a similar AISI 4130 steel, but with a low sulfur content. In addition, the microstructure was changed to a spheroidize-annealed structure

with a yield strength of 600 MPa (87 ksi). The combination of lower strength, the absence of sulfide, and the presence of spheroidal carbide particles led to larger, deeper microvoids and greater toughness. The K_{Ic} value for this specimen exceeded 100 MPa$\sqrt{m}$ (91 ksi$\sqrt{in.}$). The larger microvoids in Fig. 3(c) are associated with large carbide particles.

Large angular alloy carbides in superalloys (Ref 9) and large intermetallic compounds or constituent particles (>1 μm) in aluminum alloys (Ref 10, 11) also reduce toughness (Fig. 4a). Such constituent parti-

cles form by eutectic decomposition during solidification. These particles include relatively insoluble $AlCu_2Fe$, Mg_2Si, and $(Fe,Mn)Al_6$ to relatively soluble $CuAl_2$ and $CuAl_2Mg$. They are infrequently dissolved during processing or solution annealing. The particles fracture first on stressing, and void growth follows in the manner described in Ref 10 and 11.

The cleavage of such particles produces a large, flat area in the center of large microvoids, as shown for 7075-T651 in Fig. 4(b). Early cleavage and void growth around such particles in connection with relatively shal-

Fig. 4 Particle effects in aluminum fracture

(a) Cracked particles in alloy 2024-T851 strained 6.1% illustrating start of void formation. Source: Ref 12. (b) Similar constituent particle fracture as viewed by TEM replica techniques in 7075-T651. (c) Small microvoids associated with submicron particles producing void sheet instabilities in 7075-T651 (replica)

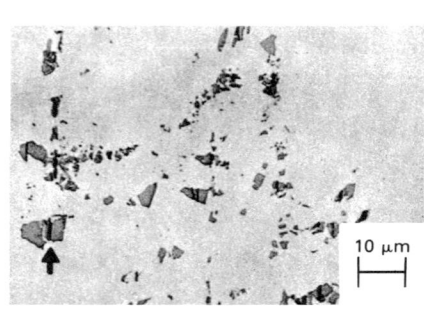
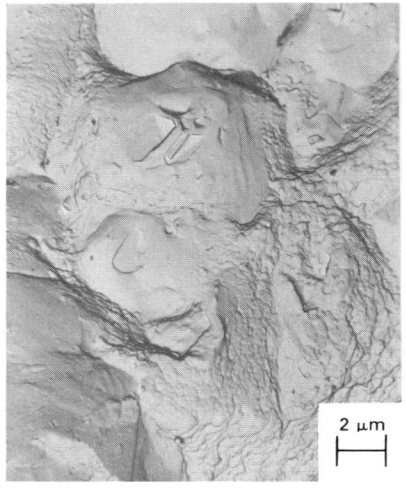
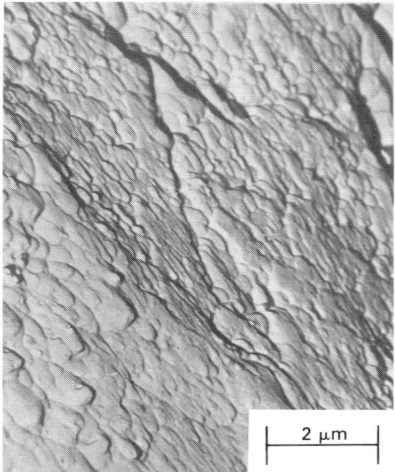

(a)
(b)
(c)

low microvoid coalescence leads to low-energy shear. The fracture toughness, K_{Ic}, was only 25 MPa$\sqrt{m}$ (22.7 ksi$\sqrt{in}$.). This material had a peak-aged microstructure that resulted in very fine, shallow dimples in the fracture. Initial separation at the dimples produced void sheets that caused local plastic instability. These dimples are usually associated with dispersoids such as $Al_{12}Mg_2Cr$ or $Al_{20}Cu_2Mn$, which form by solid-state precipitation. As shown in the high-magnification TEM replica of Fig. 4(c), even particles less than 1000 Å in diameter may nucleate microvoids. The control of such small particles is essential and is discussed in more detail in Ref 12. Control of dispersoids in aluminum alloys, carbide distributions in steels, and Laves phase formation in superalloys to suppress void sheet instability has been discussed extensively in the literature (Ref 9-11, 13).

Figure 5(a) shows a metallographic section of an ultrahigh-strength specimen of AISI 4340 steel strained just short of fracture, exhibiting a void sheet between large voids that nucleated at lower strains at manganese sulfides. In this case, the smaller voids nucleated at the cementite particles that formed during tempering. The close spacing of void initiation sites minimizes further void growth and tends to reduce fracture toughness.

According to Ref 13, void sheet formation is easier in materials with larger size strengthening precipitates. This is supported by the absence of void sheets in maraging steels with yield strengths of 1380 MPa (200 ksi) that contain precipitates a few hundred angstroms in length and by the presence of void sheets in AISI 4340 steel containing cementite particles 0.17 μm in length. Thus, maraging steels are tougher than quenched and tempered steels at similar strength levels.

The key to dimpled rupture is the void nucleation event. Void nucleation occurs more readily at larger inclusion sites, and void growth also proceeds more rapidly from larger inclusions. Toughness can be improved by increasing the spacing between inclusions. Reduction of impurities that occur as inclusions will reduce the inclusion volume fraction and increase the inclusion spacing, thereby improving K_{Ic}.

The effect of particle size has been discussed extensively in the literature, but one of the simplest quantitative relationships is that of fracture toughness as a function of inclusion spacing. Reference 1 suggests that the critical crack tip opening displacement, δ_c, should be proportional to the distance between inclusions, λ, as shown schematically in Fig. 5(b). This would result in K_{Ic} being proportional to the square root of λ

Fig. 5 Effect of particles on fracture toughness

(a) Void sheet linking three inclusions in a sectioned tensile sample of AISI 4340 steel. (b) Schematic relationship between crack tip opening displacement and inclusion spacing. (c) K_{Ic} for a martensitic 0.45C-Ni-Cr-Mo-V steel as a function of inclusion spacing and yield strength. Source: Ref 13, 14

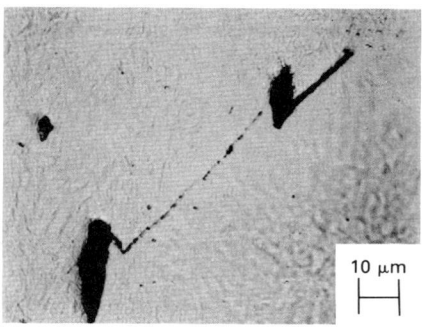

10 μm

(a)

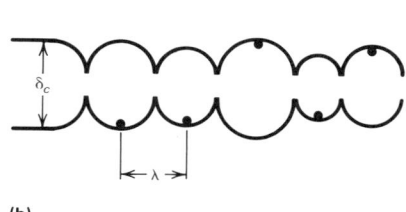

(b)

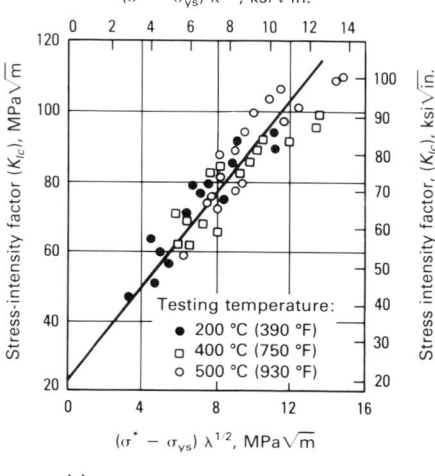

(c)

(see the article "Micro-Fracture Mechanics" in this Section for more information on this subject).

Such a semiempirical relationship has been found for a 0.45C-Ni-Cr-Mo-V steel similar to AISI 4340, but with somewhat higher chromium, molybdenum, and vanadium contents (Ref 14). This relationship is given by:

$$K_{Ic} = 23 \text{ MPa}\sqrt{m} + 7(\sigma^* - \sigma_{ys})(\lambda)^{1/2}$$

(Eq 1)

with σ^* equal to 2000 MPa (290 ksi). In Fig. 5(c), this relationship fits the data for a large variation in λ and for three different test temperatures where dimpled rupture is the microstructural fracture mode.

Cleavage. The second type of microstructural overload failure listed in Fig. 1 is cleavage. Generally a very low-energy, brittle type of fracture, cleavage in lower strength materials may occur with some plasticity. Figure 6(a) illustrates a fracture surface of an Fe-4Si (at.%) single crystal cleaved at 77 K under the rapid loading (1 m/s, or 3.3 ft/s) of a hydraulic testing machine. The typical river pattern of cleavage is apparent, but there are also ledges along the rivers that have been deformed upward, resulting in a white appearance. Some of these ledges have been separated from the surface by additional cleavage to form ligaments.

Two such ligaments are shown to be substantially deformed (Fig. 6b). Such fracture-plane demarcations can provide a significant change in fracture toughness (Ref 8). On a smaller scale, such effects are also seen in polycrystalline Fe-4Si (at.%) that was fractured at 233 K (Fig. 6c).

If more frequent out-of-plane deviations are provided by the microstructure, an improvement in toughness can be expected. One such method has been to utilize second-phase particles (Ref 16); another is to refine the grain structure. For example, ferritic steels with a grain size of about 40 μm and a yield strength of 320 MPa (46.4 ksi) at −120 °C (−184 °F) have a K_{Ic} approximately equal to 25 MPa$\sqrt{m}$ (22.7 ksi$\sqrt{in}$.) (Ref 17).

The fine lath microstructure of a modified type 403 martensitic stainless steel provides not only a higher toughness of about 40 MPa$\sqrt{m}$ (36.4 ksi$\sqrt{in}$.), but also more than doubles the strength to 700 MPa (101.5 ksi). Figure 7 shows data for very thick compact tension samples. The effects of minor alloying additions are shown by three heats of the type 403 stainless steel in Fig. 8. At about −50 °C (−58 °F), heat 933 varies from heat 484 by nearly a factor of two. As suggested by their compositions, this is due to additions of nickel and molybdenum. It was not reported whether these additions improve toughness through microstructural modification of the carbides or of the lath martensite,

Fig. 6 Scanning electron micrographs of Fe-4Si (at.%)

(a) and (b) Single crystal tested at 77 K showing ledges and ligaments. (c) Similar effect on impact testing polycrystalline Fe-4Si (at.%) with a yield strength of 302 MPa (43.8 ksi) at 233 K. Source: Ref 8, 15

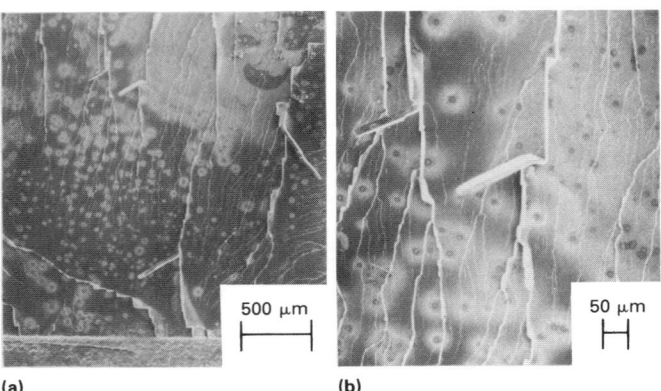

500 μm

(a)

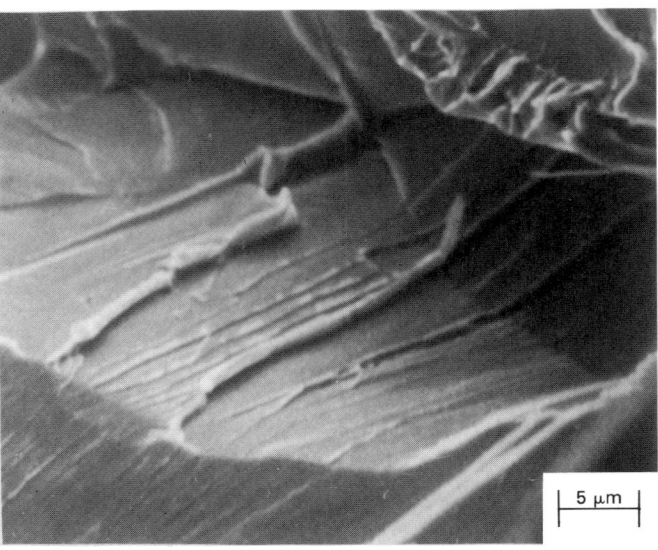

50 μm

(b)

5 μm

(c)

or through inherent toughness of the martensite. Careful fractography, metallography, and transmission electron microscopy may be required to evaluate such effects.

Composition and microstructural effects on toughness are interrelated, as illustrated by Ti-6Al-4V. If relatively coarse α platelets are incorporated into an α-β Widmanstätten matrix, they may be either detrimental or

beneficial, depending on the oxygen content. An example of a detrimental effect is a case in which high oxygen was purposely allowed to diffuse into the surface (Ref 7). This stabilized the α phase and made it brittle, resulting in an easy crack path through a high volume fraction of brittle α particles (Fig. 9a). Oxygen embrittles the α phase (Ref 20) of recrystallized and annealed microstruc-

tures (Fig. 9b). However, if the oxygen content is kept low, the transformed structure may produce increased toughness.

Transformed structures appear to be tough, primarily because fractures in such structures must proceed along tortuous, many-faceted crack paths. According to Ref 21, K_{Ic} is proportional to the fraction of transformed microstructure in Ti-6Al-4V (Table 1). It has been suggested that the α platelet size and the efficient dispersion of the β phase enhance toughness (Ref 22). Reference 23 provides direct evidence that crack tortuosity is an important factor in determining the form of fracture topography/microstructure correlations for the same sample.

A proper balance between platelet thickness and spacing in the transformed microstructure is required to achieve the highest toughness in Ti-6Al-4V. Figure 10(a) illustrates that platelets must be thick enough to turn a crack, while being spaced such that turns are frequent. This microstructure is basically the same as that shown in Fig. 2(c). The effect of this crack path tortuosity on fractography is shown by a high-magnification replica in Fig. 10(b). Although the α platelet appears to cleave, the ductile fracture around it and the accompanying crack path deviations increase toughness.

A complex experimental α-β alloy has been studied in which strength was held constant in both the equiaxed α and transformed microstructural conditions (Ref 24). In the equiaxed α condition, toughness increased

Fig. 7 Fracture toughness data obtained over ranges of temperature and specimen thickness for modified type 403 stainless steel

Source: Ref 18

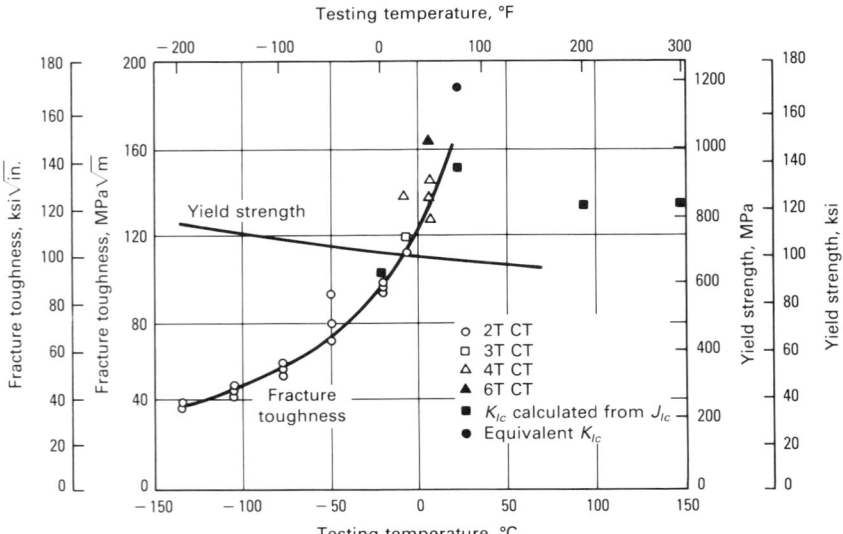

Fig. 8 Scatter bands of fracture toughness (K_{Ic}) data for three heats of type 403 stainless steel in the heat treated condition
Source: Ref 19

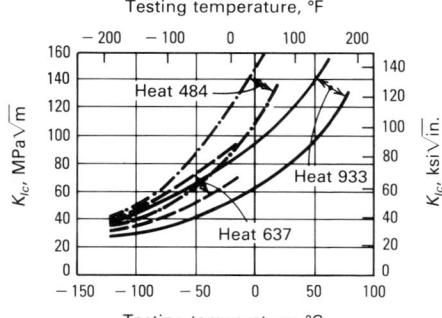

Heat No.	Composition, wt%								
	C	Mn	P	S	Si	Ni	Cr	Mo	Cu
637	0.13	0.46	0.013	0.009	0.20	0.32	12.18	0.43	0.12
933	0.15	0.54	0.010	0.008	0.32	0.38	12.37	0.08	0.10
484	0.13	0.57	0.009	0.006	0.33	1.60	12.32	0.55	. . .

with β grain boundary area per unit volume. In the transformed condition, toughness increased with the percentage of primary α, which is always elongated whether transformed or originating in the grain boundary.

Ductile/Cleavage Mixed Modes. As discussed previously, pretransformed structures lead to crack path deviation and increased toughness. Strain-induced *in situ* transformations may play a similar role. Although the actual toughening mechanics are still unclear, failure may change from ductile to ductile/cleavage mixed modes, depending on composition and/or test temperature. An example is TRIP (transformation-induced plasticity) steels, in which the transformation from metastable austenite (γ) to martensite (α') may produce unusual fractographs (Ref 25). Medium- to low-carbon 9Cr-8Ni-4Mo steels produce a strain-induced α' morphology along oblique shear bands in tensile tests (Fig. 11).

A similar α' formation along shear bands in the plastic zone below the fracture surface is shown in Fig. 12. For carbon contents less than 0.24 wt% C, even testing at −196 °C (−320 °F) produced only ductile rupture. Evidence of ductile rupture was not clear with use of scanning electron microscopy (SEM) alone, because even the high magnification used in Fig. 13 could resolve the microvoids only in the austenite but not in the delamination trenches that contained the martensite fracture.

A TEM replica did reveal 1- to 5-μm size dimples in this specimen (Fig. 14). Note that the transformed α' plates produced an unusual macroscopic fracture mode. In the γ austenite at the top, relatively equiaxed dimples were produced, while the γ at the right of the smooth area in the center consisted of elongated shear dimples. The *in situ* transformation changed the local state of stress and hence the crack path direction during propa-

gation in this region. Finally, the martensite laths produced extremely fine microvoids of about an order of magnitude smaller.

A distinctly different result is obtained if the carbon content is greater than 0.28 wt% C. In Fig. 15, a similar α' area fails by cleavage. This combined ductile/brittle fracture mode reduces toughness considerably at −196 °C (−320 °F). As shown in Fig. 16, the apparent K_{Ic} value dropped rapidly with increasing carbon content, even though the same type of TRIP process occurred.

Fatigue Failure

Fatigue is the progressive, localized, permanent structural change that occurs in a material subjected to repeated or fluctuating strains at stresses having a maximum value less than the tensile strength of the material. Fatigue may culminate in cracks or fracture after a sufficient number of fluctuations.

This section will review the microstructural failure modes listed under fatigue in Fig. 1, including complex, mixed fracture modes. See the Section on "Fatigue Testing" in this Volume for additional information on test procedures and environmental effects on fatigue crack propagation.

Ductile striations are the most commonly observed fatigue features that occur in the microstructurally insensitive intermediate ΔK regime (see Fig. 1). The undulating or ripple-like fatigue striations appear to be correlated to a normalized stress-intensity range, ΔK. An example of such striations is shown for 2024-T3 aluminum alloy in Fig. 17. Although there are exceptions, such regular striations have been found to correlate to ΔK/E (stress-intensity range over Young's modulus), as shown in Fig. 18. A good correlation is:

$$\text{Striation spacing} \approx 6\left(\frac{\Delta K}{E}\right)^2 \qquad \text{(Eq 2)}$$

Fig. 9 Compositional and microstructural effects on toughness of Ti-6Al-4V

(a) Brittle α phase (lightly etched) with a high oxygen content, resulting in an easy crack path through a high volume fraction of α particles. (b) Influence of oxygen content on the fracture toughness of Ti-6Al-4V in the recrystallized-annealed condition. Source: Ref 20

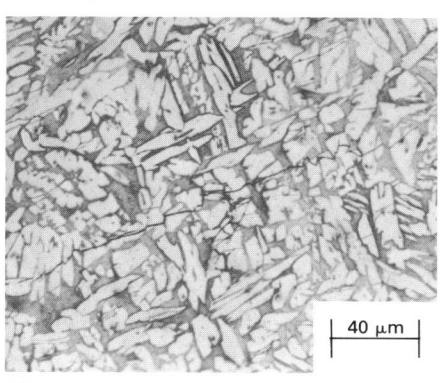

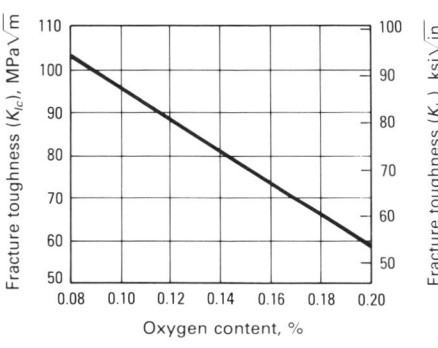

(a)

(b)

Fig. 10 Microstructural effects on crack path and the resulting fracture surface for Ti-6Al-4V

(a) Effect of higher toughness α phase changing the crack path. (b) TEM replica showing local ductile deviations and step-wise void growth on the fracture surface resulting from the crack path shown in (a). Source: Ref 7

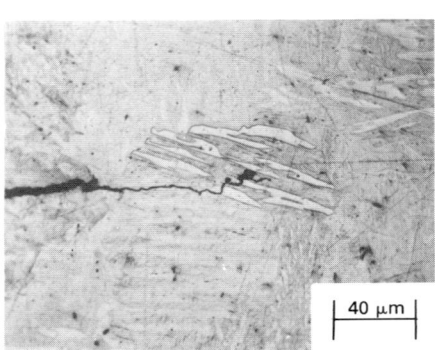

(a)

(b)

Fig. 12 Strain-induced transformation product on intersecting slip bands below a plane-strain fracture surface

Low-carbon metastable austenite fractured at −196 °C (−320 °F). Magnification: 55×. Source: Ref 27

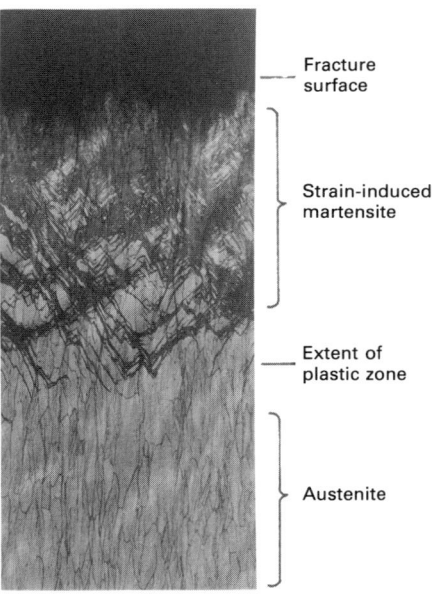

Fig. 11 Martensitic strain-induced transformation product resulting from deformation of a metastable austenitic tensile bar

The angle represented is the theoretical oblique shear slip band. Source: Ref 26

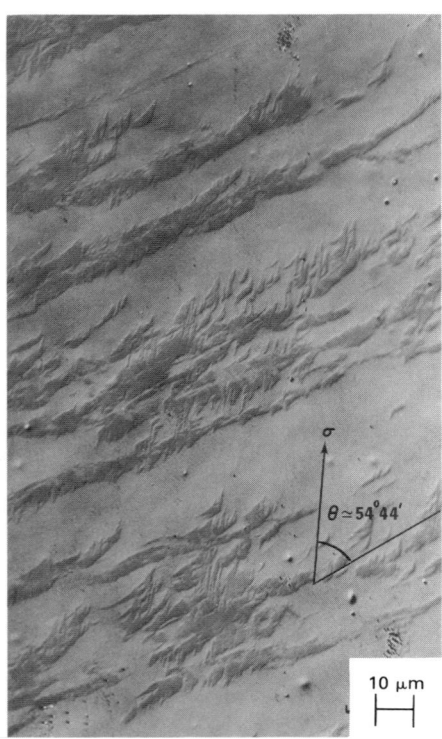

Figure 18 illustrates the application of this equation to many material types. More quantitative relationships have interpreted this growth mechanism to be proportional to the crack tip displacement (Ref 30, 31). Such displacements may be changed by microstructural variations of either the flow stress or crack path. Thus, within the scatter band of Fig. 18, and in some cases outside it, microstructure may play a role.

Figure 17 illustrates the change in crack growth in the vicinity of a cleaved intermetallic compound. Such cleaved particles are analogous to those in Fig. 4. As the crack approached the preexisting void, the growth rate decreased. The average growth rate in the out-of-plane region above the fractured particle is 3×10^{-8} m/cycle (1.2×10^{-6} in./cycle), while in the adjacent matrix it is 4×10^{-8} m/cycle (1.6×10^{-6} in./cycle). Note that the overall growth rate away from this particle is 5×10^{-8} m/cycle (1.9×10^{-6} in./cycle), so that there appears to be a minor effect of microstructure even in this

insensitive regime. A higher volume fraction of such out-of-plane deviations might produce a larger effect.

Cyclic ductile decohesion is the second most commonly observed fatigue process. It is usually observed at low ΔK values near threshold. At intermediate ΔK values for Inconel 718, a face-centered cubic nickel-based superalloy, the striations in Fig. 19 bear a nearly one-to-one relation to macroscopic measurements of da/dN. Whereas the growth rate above 22 MPa$\sqrt{\text{m}}$ (20 ksi$\sqrt{\text{in.}}$) occurs predominantly by the ductile striation process (Fig. 19a), it is crystallographic at lower ΔK values (Fig. 19b). The fracture surface contains a high density of crystallographic facets on {111} planes (Ref 32).

Similar observations have been made for a large number of face-centered cubic aluminum and nickel-base alloys (Ref 33), and it is

Table 1 Relationship between K_{Ic} and fraction of transformed structure in alloy Ti-6A1-4V

Heat treating temperature(a)		Fraction of transformed structure, %	K_{Ic}(b)	
°C	°F		MPa$\sqrt{\text{m}}$	ksi$\sqrt{\text{in.}}$
1050	1920	100	69.0 (69.9)	63 (63.6)
950	1740	70	61.5 (60.4)	56 (54.9)
850	1560	20	46.5 (44.6)	42 (40.6)
750	1380	10	39.5 (41.5)	36 (37.8)

(a) Heated for 1 h at indicated temperature and then air cooled.
(b) Values in parentheses calculated from linear least-squares expression relating percentage transformation to K_{Ic}.

Fig. 13 Scanning electron micrographs of the fracture surface shown in Fig. 12
Source: Ref 25

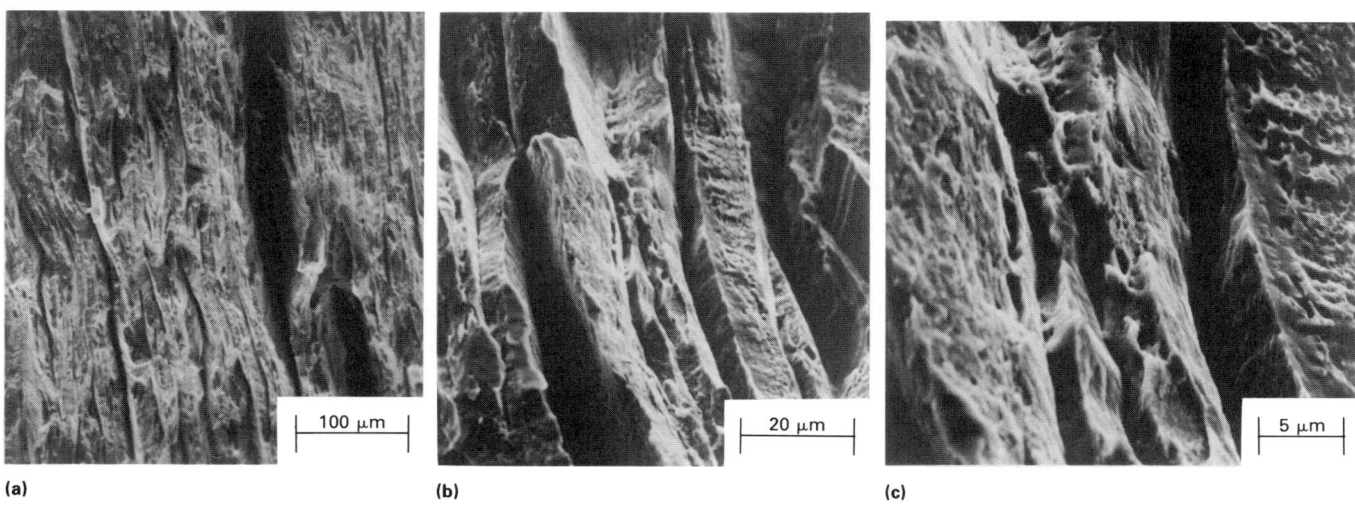

100 µm

20 µm

5 µm

(a)   (b)   (c)

Fig. 14 Transmission electron fractographs (replicas) of the fracture surface shown in Fig. 12
Source: Ref 25

Fig. 15 Transmission electron fractograph (replica) of a higher carbon metastable austenite fracture at −196 °C (−320 °F)
Source: Ref 25

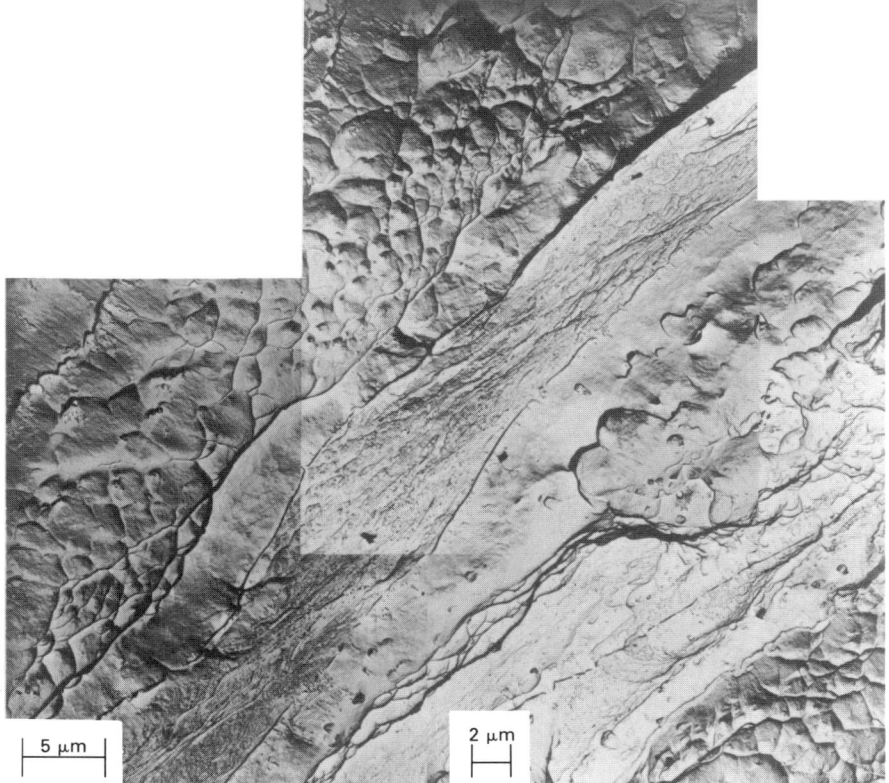

5 µm

2 µm

5 µm

probable that similar effects occur in some hexagonal close-packed and body-centered cubic systems. Noting that $\{111\}_{fcc}$ are slip planes, there is some evidence that this is an inhomogeneous cyclic-slip type of decohesion process; other evidence suggests that

microstructural transformations may also be involved (Ref 32). Such rock-candy types of fracture surface morphologies may play a significant role by affecting growth rate through either crack path deviations or through closure.

Because mode II crack tip sliding can occur along inhomogeneous shear bands, the local slip character of the microstructure may affect asperity contact during the unloading portion of the fatigue cycle. This depends on asperity size, degree of sliding, and the mini-

Fig. 16 Effect of carbon content on the apparent fracture toughness of metastable austenites at −196 °C (−320 °F)
Source: Ref 25

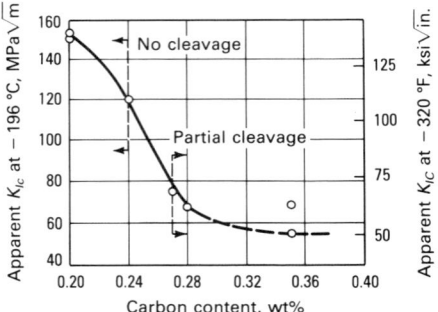

mum load. Thus, a combination of both microstructure and type of load cycle may affect the effective stress intensity near threshold. Further discussion of such effects can be found in Ref 34 to 36. It is clear, however, that the load ratio of the fatigue cycle affects the resulting microscopic morphology.

Cyclic brittle cleavage is another important microscopic fracture mode that has only recently received attention (Ref 37). For an Fe-4Ni (at.%) alloy tested at 233 K at a $\Delta K = 17$ MPa$\sqrt{\text{m}}$ (15.5 ksi$\sqrt{\text{in.}}$), step-

Fig. 17 Transmission electron fractograph (replica) of ductile fatigue striations in 2024-T3 aluminum
Source: Ref 28

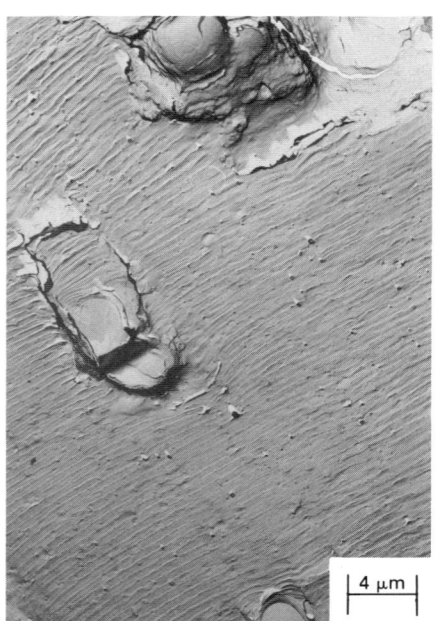

wise growth occurred on {100} cleavage planes (Fig. 20). Similar features have been observed for high-strength low-alloy steels and medium-strength steels (Ref 38). Such features have been observed both slightly above and below room temperature.

Cyclic cleavage in Fe-4Si (at.%) is shown in Fig. 21. At 75× magnification (Fig. 21a), this specimen appears to illustrate a brittle overload fracture. The predominant cleavage mode with patches of intergranular fracture suggests a catastrophic rapid fracture. However, this surface was produced under fatigue loading at 233 K (−40 °C, or −40 °F) at a ΔK of 18.4 MPa$\sqrt{\text{m}}$ (16.7 ksi$\sqrt{\text{in.}}$). The macroscopic growth rate was only 5×10^{-9} m/cycle (1.9×10^{-7} in./cycle). At higher magnifications in Fig. 21(b) to (d), clear evidence of brittle fatigue striations on one of the major cleavage steps is visible.

Cyclic microvoid and cyclic intergranular processes have also been observed. The former is shown in Fig. 22 for an ultrahigh-strength D-6 AC steel tested at a ΔK greater than 60 MPa$\sqrt{\text{m}}$ (54.6 ksi$\sqrt{\text{in.}}$). The crack advance per cycle is very rapid (Fig. 22a). With TEM replicas, the microscopic growth process in Fig. 22(b) was shown to be elongated shear dimples. Apparently few of these are nucleated by second-phase particles compared to the much larger equiaxed dimples of Fig. 3. It should be noted that this specimen is relatively thin and that, for the high ΔK involved, fatigue growth is under plane-stress conditions. This contributes to the shear dimples noted in Fig. 22(b). At much lower ΔK values at both low temperature (Ref 36) and room temperature

Fig. 18 Correlation of fatigue striation spacing with ΔK normalized with respect to elastic modulus
Source: Ref 29

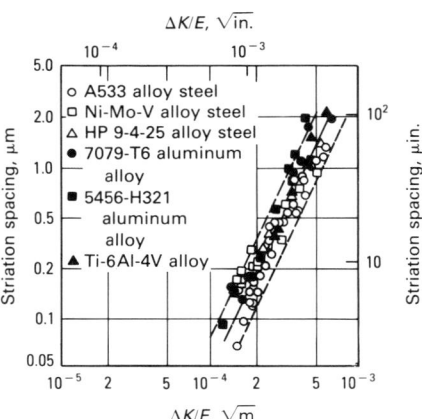

(Ref 38), intergranular ductile fatigue has been noted. For the Fe-4Ni (at.%) in Fig. 23(a), the type of loading that produced this fracture surface is not obvious. It could be brittle intergranular or stress-corrosion cracking. In fact, however, the higher magnification in Fig. 23(b) shows this to be a ductile intergranular fatigue process. This was determined not to be a moisture-related phenomenon, because the partial pressure of water vapor at 123 K is negligible.

Mixed Fracture Modes. From the above discussion, it is clear that fatigue growth may occur by means of all microscopic fracture modes. A summary of these fracture modes

Fig. 19 Scanning electron micrographs of the fracture surfaces of Inconel 718 specimens tested at room temperature

(a) $\Delta K = 30$ MPa$\sqrt{\text{m}}$ (27 ksi$\sqrt{\text{in.}}$), striation spacing (l) $\simeq 0.2$ μm, and $da/dN \simeq 0.1$ μm/cycle. Arrow indicates direction of crack propagation. (b) $\Delta K = 14$ MPa$\sqrt{\text{m}}$ (13 ksi$\sqrt{\text{in.}}$) and $da/dN = 2 \times 10^{-3}$ μm/cycle. Source: Ref 32

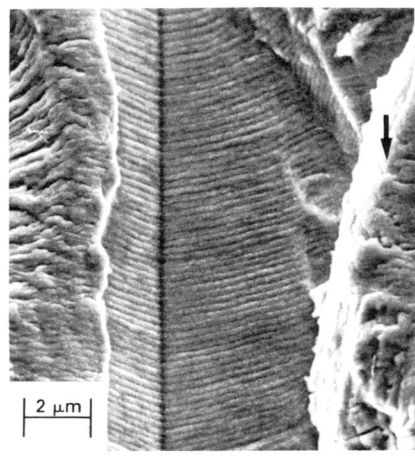

(a)

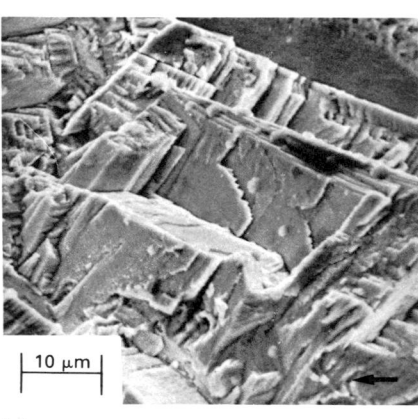

(b)

Fig. 20 Cyclic cleavage observations in Fe-4Ni (at.%) at 233 K, $\Delta K = 17$ MPa$\sqrt{\text{m}}$ (15.5 ksi$\sqrt{\text{in.}}$)

(a) Jogs formed during cyclic stepping across grain. (b) Large number of cleavage rivers formed to accommodate the twist angle misorientation between two grains. (c) Cleavage striations superimposed on rivers during cyclic crack growth. Source: Ref 37

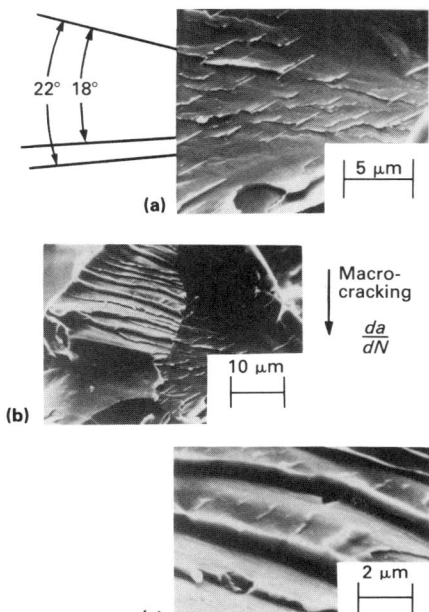

Fig. 21 Scanning electron micrographs of cyclic cleavage of Fe-4Si (at.%) at 233 K, $\Delta K = 18.4$ MPa$\sqrt{\text{m}}$ (16.7 ksi$\sqrt{\text{in.}}$)

(a) Overload cleavage appearance at 75×. (b), (c), and (d) Increasing magnifications at 750, 1500, and 3800× showing brittle striations on large cleavage river

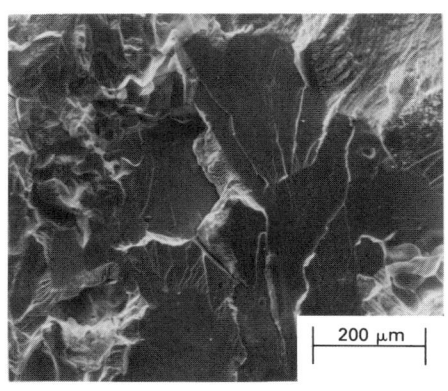
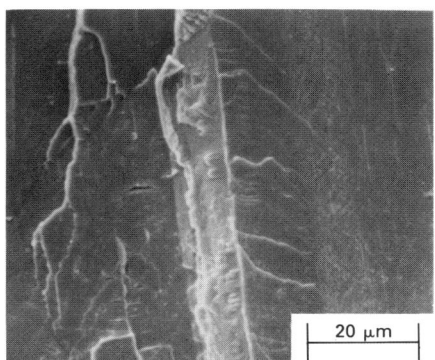
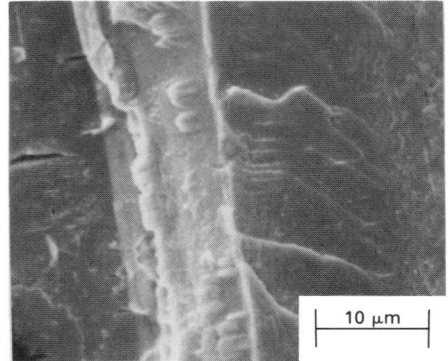
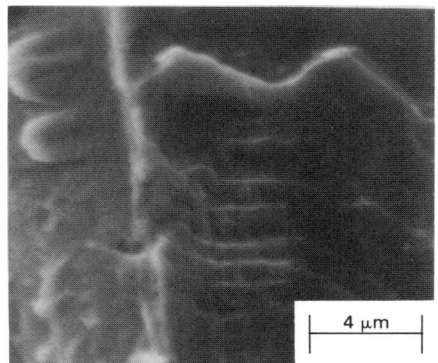

for three classes of materials is given in Fig. 24. In all three, there is a tendency for some alternative fracture mode to exist at either very low or very high ΔK values. At intermediate ΔK values, the tendency for striations to dominate is observed. The one exception is illustrated in Fig. 24(c), in which cyclic microvoid growth occurs at high R values (load ratio) and thus with K_{max} values approaching K_{Ic}. In all of these cases, microstructure may or may not have played a major role.

Some of the possible growth mechanisms are shown in Fig. 25. For single-phase materials, the transgranular and intergranular modes shown in Fig. 25(a), (b), (e), (f), (g), and (i) are possible, if the type of behavior observed in Fig. 24 is present. It is also possible that second phases with different properties and distributions may promote the existence of dual fracture modes. An example is given in Fig. 26, which is a microstructure of Ti-6Al-4V similar to that shown in Fig. 2(c). The optical micrograph in Fig. 26(a) and the fractograph in Fig. 26(b) clearly show the effect of microstructure. The coarse α platelets tend to promote cyclic

Fig. 22 High-cycle fatigue in D-6 AC steel at $\Delta K > 60$ MPa$\sqrt{\text{m}}$ (54.6 ksi$\sqrt{\text{in.}}$)

(a) Macroscopic view of rapid fatigue growth under plane-stress conditions. (b) High-quality TEM replica showing elongated shear dimples from fatigue region. Source: Ref 39

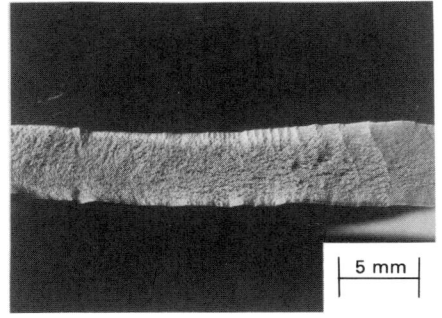

cleavage, while a more ductile fatigue process occurs in the Widmanstätten α-β structure in between.

Another effect of microstructure is suggested in Fig. 27. The substructure is shown for thin films taken from the sample shown in

Fig. 19(b). It is clear that the fracture surface is located along deformation twinning planes in the substructure. Thus, inhomogeneous shear with twinning as a prominent deformation mode led to the ductile decohesion process indicated in Fig. 19(b).

Fig. 23 Fracture surface topography of Fe-4Ni (at.%) fatigue cracked at 123 K with $\Delta K < 10$ MPa$\sqrt{m}$ (9.1 ksi$\sqrt{in.}$)

(a) Overall intergranular appearance at low magnification. (b) Higher magnification SEM resolution of ductile intergranular fatigue striations. Source: Ref 36

(a)

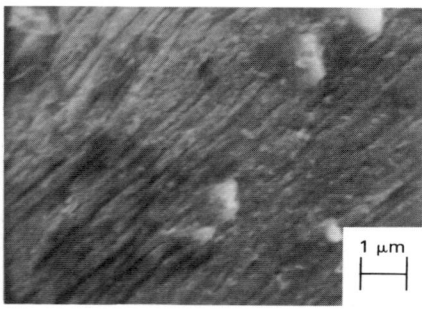

(b)

Fig. 24 Effect of ΔK on mixed fracture modes

(a) α-β titanium alloy. (b) EN-24 and 300M high-strength steels. (c) Precipitation-hardened stainless steel. Source: Ref 34, 40-42

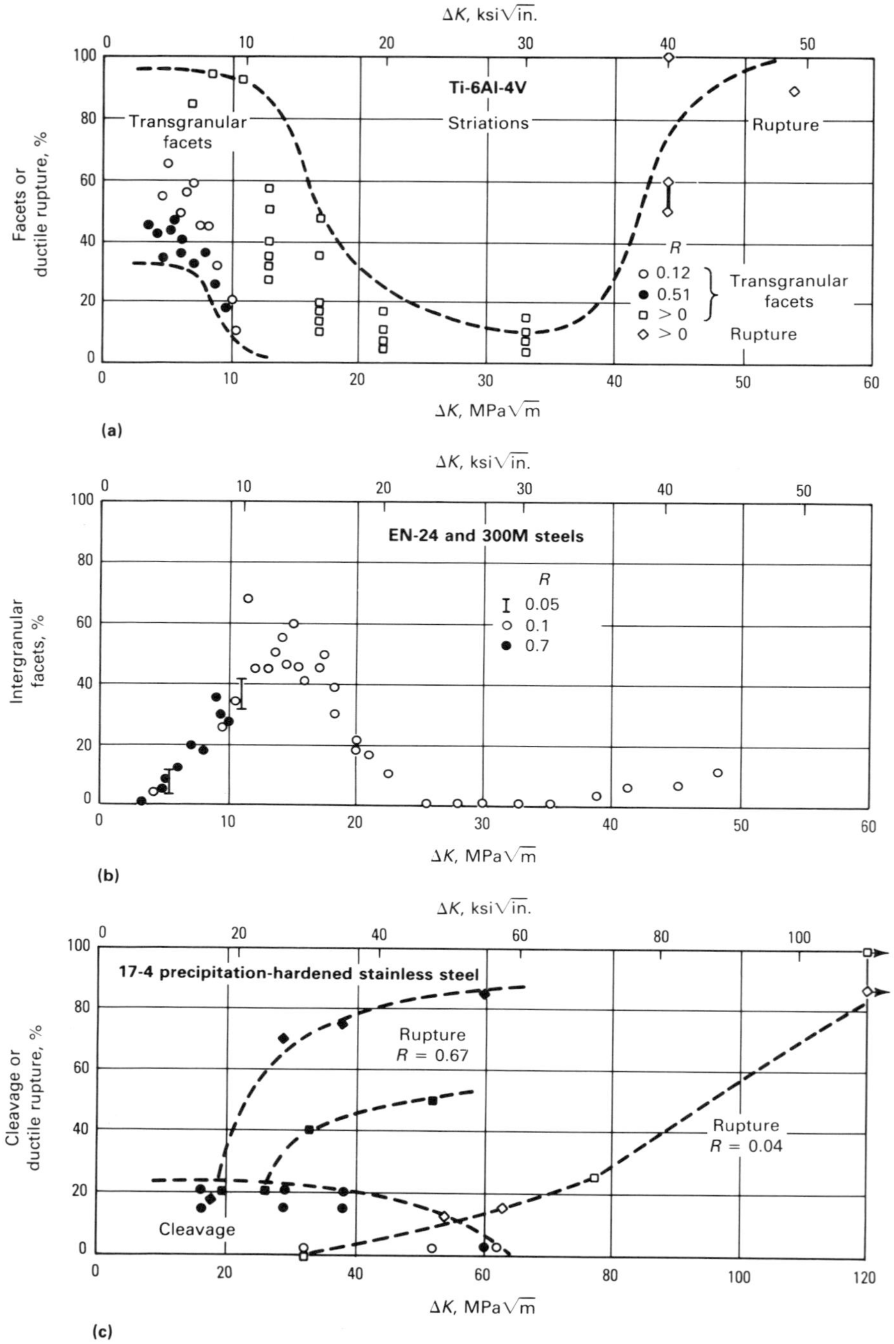

(a)

(b)

(c)

Environmental (Sustained-Load) Failure

The third type of external fracture condition listed in Fig. 1 combines the effects of environment and stress. This group includes stress-corrosion cracking, liquid-metal embrittlement, corrosion fatigue, and damage due to hydrogen.

This section will review intergranular and transgranular cleavage microscopic failure modes listed under environment in Fig. 1. A discussion of environmentally enhanced fatigue is also included. For more information see the articles "Tests for Stress-Corrosion Cracking," "Tests for Hydrogen Embrittlement," and "Environmental Effects on Fatigue Crack Propagation" in this Volume.

Intergranular fracture processes, whether hydrogen or stress-corrosion enhanced, are frequently observed in high-strength steels and aluminum alloys. The

manner in which composition or microstructure affects the fracture mode is still under investigation.

The resistance to stress-corrosion cracking or hydrogen embrittlement of single-phase

materials or materials with dispersoids or precipitate phases for hardening is affected by heat treatment. In these materials, microstructure seems to play a secondary role in terms of how well hardening precipitates,

Fig. 25 Types of alternate microscopic fracture modes in fatigue

(a) Ductile striations triggering cleavage. (b) Cyclic cleavage. (c) α-β interface fracture. (d) Cleavage of α in an α-β phase field. (e) Forked intergranular cracks in a hard matrix. (f) Forked intergranular cracks in a soft matrix. (g) Ductile intergranular striations. (h) Particle nucleated ductile intergranular voids. (i) Discontinuous intergranular facets. Source: Ref 36

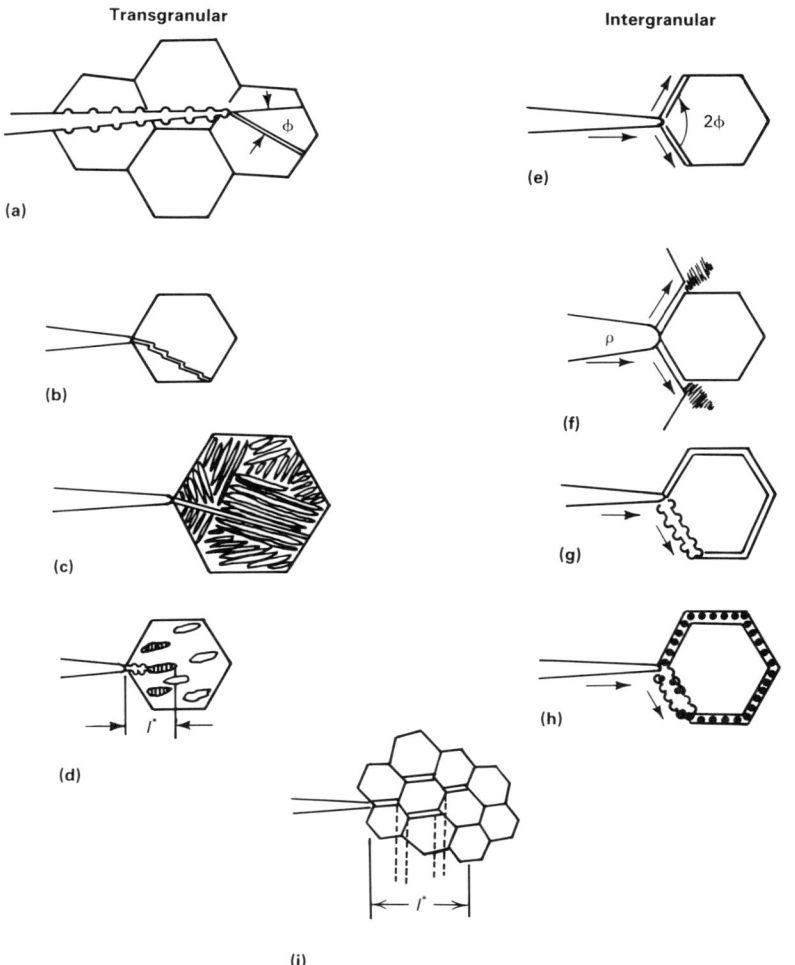

dispersoids, or other second-phase interfaces "tie up" segregants. For example, for ultrapure high-strength steels, refining the grains (increasing the grain-boundary area) decreases the tramp element concentration in a given boundary. This refinement increases the threshold stress intensity in AISI 4340 steel (Ref 4).

Addition of some alloying elements may either enhance or diminish the susceptibility of a boundary to hydrogen embrittlement or stress-corrosion cracking. For example, manganese added to AISI 4340 steel decreases resistance to hydrogen embrittlement. In a series of special heats, it was demonstrated that manganese additions reduced thresholds by nearly 100% (Ref 43). As shown in Fig. 28(a), crack velocity increased by an order of magnitude with additions of 2 wt% Mn. Increasing the manga-

nese content also increases the frequency of intergranular fracture as opposed to ductile microvoid coalescence, because manganese promotes segregation of phosphorus at prior austenite grain boundaries. Figure 28(b) compares the fracture surfaces of steel specimens containing various manganese contents that were tested in hydrogen gas.

It is well known that the combination of tramp elements and hydrogen lowers grain boundary fracture resistance (Ref 3, 4). "Tying up" similar deleterious species at other interfaces may change the microstructural fracture mode. For example, 7075-T651, an Al-Mg-Zn-Cu alloy, fails by intergranular fracture. Evidence indicates that this is a hydrogen embrittlement process along grain boundaries with excess magnesium at the primary boundaries (Ref 44). If this excess magnesium can be removed either by adding

excessive zinc to promote matrix-hardening precipitates or by tying it up in dispersoids such as $Al_{12}Mg_2Cr$, then the stress-corrosion cracking resistance may be improved. There is evidence that additions of dispersoid-forming elements such as chromium, manganese, zirconium, titanium, and vanadium tend to reduce susceptibility to slow crack growth (Ref 12). It is unclear whether this slow crack growth is due to changing grain boundary composition through the dispersoids or through grain refinement.

Direct microstructural influences on sustained-load slow crack growth have been observed in hydriding materials. Both zirconium and α titanium alloys fail along hydride phases. This has been well demonstrated in niobium, also a strong hydride former (Ref 45). In α-β titanium microstructures, it is apparent that either hydrides in the α phase or a hydrogen-rich interface phase in Widmanstätten microstructures tends to fail. An example of hydride-induced fracture in a Ti-6Al-6V-2Sn microstructure is shown in Fig. 29.

Transgranular Cleavage. As indicated in Fig. 1, transgranular cracking has been observed in iron-, aluminum-, and nickel-base alloys. It has also been found in brass, which is known to proceed to failure by a series of crystallographic cleavage events. It has been reported that copper in 1 M sodium nitrite solution may also fail by a series of microcleavage events (Ref 47). Typical fracture surfaces are shown in Fig. 30. Even these very ductile single crystals may fail by a "brittle" process induced by dynamic straining and oxide formation. This brittle process leaves striations on the fracture surface (Fig. 30b).

Environmentally Enhanced Fatigue. Although not indicated in Fig. 1, interactions occur between the environment and fatigue. This may be fatigue caused by corrosion or by gaseous environments such as hydrogen or hydrogen sulfide, or internal (microstructural) related fatigue resulting from hydrogen or carbon in solid solution. All that is required is a temperature for diffusion, adsorption, and/or segregation at internal sites.

Figure 31 illustrates a low-strength (σ_{ys} = 195 MPa, or 28.3 ksi) Fe-2.5Si (at.%) alloy that was cycled at a ΔK near threshold. For the noncharged material, the ΔK_{th} was 8 MPa$\sqrt{m}$ (7.3 ksi$\sqrt{in}$.). In humidified room-temperature air, the fracture surface was nearly 50% intergranular (Fig. 31a). The same material cathodically precharged with 3.5 wt ppm hydrogen resulted in nearly 100% intergranular fracture (Fig. 31b), a failure mode not normally associated with fatigue. Similar effects have been observed in high-strength low-alloy steels and

Fig. 26 Second-phase cleavage fracture in Ti-6Al-4V

(a) Light micrograph of polished and etched surface. (b) Scanning electron micrograph of fracture surface. Source: Ref 6

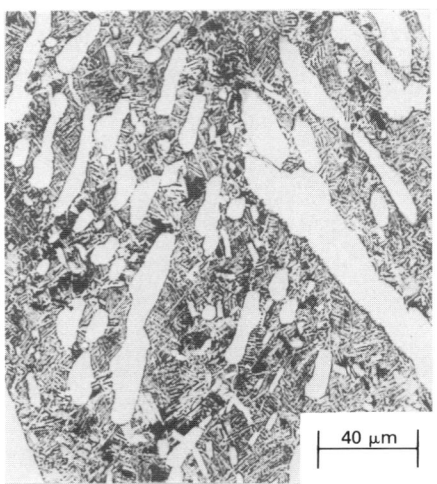

(a)

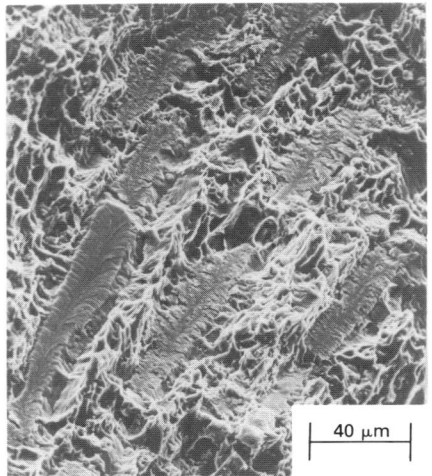

(b)

Fig. 27 Identification of deformation twins in the threshold regime in Inconel 718 specimens

Room temperature; $\Delta K = 14$ MPa$\sqrt{m}$ (13 ksi$\sqrt{in}$.); $da/dN = 0.002$ μm/cycle. (a) Dark field and (b) diffraction pattern. Twins are marked by arrows. Source: Ref 32

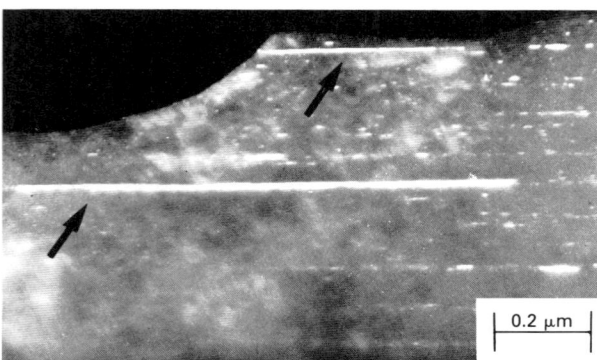

(b)

AISI 4340 steel with strengths ranging from 300 to 1500 MPa (43.5 to 217.6 ksi) (Ref 49, 50).

REFERENCES

1. K.H. Schwalbe, On the Influence of Microstructure on Crack Propagation Mechanisms and Fracture Toughness of Metallic Materials, *Eng. Frac. Mech.*, Vol 9, 1977, p 795-832
2. A.W. Thompson, Microstructure and Hydrogen Embrittlement, in *Hydrogen Effects in Metals*, I.M. Bernstein and A.W. Thompson, Ed., TMS-AIME, Warrendale, PA, 1981, p 291-308
3. W.C. Johnson and D.F. Stein, A Study of Grain Boundary Segregants in Thermally Embrittled Maraging Steel, *Met. Trans.*, Vol 5, 1974, p 549-554
4. C.J. McMahon, Jr., Intergranular Fracture in Steels, *Mat. Sci. Eng.*, Vol 25, 1976, p 233-239
5. Y.A. Katz, Ph.D. thesis, University of California—Berkeley, 1969
6. J.C. Chesnutt, C.G. Rhodes, and J.C. Williams, in *Fractography—Microscopic Cracking Processes*, STP 600, ASTM, Philadelphia, 1976, p 99
7. W.W. Gerberich and G.S. Baker, On the Toughness of Two-Phase 6Al-4V Titanium Microstructures, in *Applications Related Phenomena in Titanium Alloys*, STP 432, ASTM, Philadelphia, 1967, p 80-99
8. W.W. Gerberich, K.A. Esaklul, and E. Kurman, unpublished data, 1984
9. M.G. Stout and W.W. Gerberich, Structure/Property/Continuum Synthesis of Ductile Fracture in an IN 718 Superalloy, *Met. Trans. A*, Vol 9, 1978, p 649-658
10. R.H. Van Stone, J.R. Low, Jr., and R.H. Merchant, *Investigation of the Plastic Fracture of High Strength Aluminum Alloys*, STP 556, ASTM, Philadelphia, 1974, p 93-124
11. J.T. Staley, "Fracture Toughness of Microstructure of High Strength Alloys," Aluminum Company of America, Alcoa Center, PA, paper presented at the Spring Meeting of the American Institute of Metallurgical Engineers, Pittsburgh, May 23, 1974
12. J.G. Kaufman and J.S. Santner, Fracture Properties of Aluminum Alloys, in *Application of Fracture Mechanics for Selection of Metallic Structural Materials*, J.E. Campbell, W.W. Gerberich, and J.A. Underwood, Ed., American Society for Metals, 1982, p 169-211
13. T.B. Cox and J.R. Low, Jr., An Investigation of the Plastic Fracture, AISI

Fig. 28(a) Dependence of the crack propagation rate, *da/dt*, on the applied *K*

Hydrogen pressure P_{H_2} = 98 kPa (14.2 psi).
Source: Ref 43

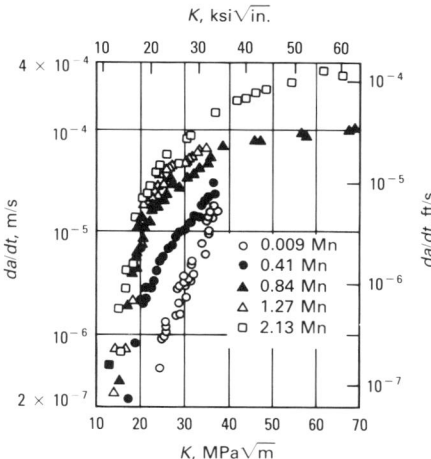

4340 and 18 Nickel-200 Grade Maraging Steels, *Met. Trans. A*, Vol 5, 1974, p 1457-1470

14. A.H. Priest, in *Effect of Second-Phase Particles on the Mechanical Properties of Steel*, The Iron and Steel Institute, London, 1971

15. W.W. Gerberich, Y.T. Chen, D.G. Atteridge, and T. Johnson, Dislocation Dynamics of Fe-Binary Alloys. II. Application to the Ductile-Brittle Transition, *Acta Met.*, Vol 29, 1981, p 1187-1201

16. G.T. Hahn, The Influence of Microstructure on Brittle Fracture Toughness, *Met. Trans. A*, Vol 15, 1984, p 947-959

17. D.A. Curry and J.F. Knott, The Relationship Between Fracture Toughness and Microstructure in the Cleavage Fracture of Mild Steel, *Met. Sci.*, Vol 10, 1976, p 1-6

18. W.A. Logsdon, *Elastic-Plastic (J_{Ic}) Fracture Toughness Values: Their Experimental Determination and Comparison with Conventional Linear Elastic (K_{Ic}) Fracture Toughness Values for Five Materials*, STP 590, ASTM, Philadelphia, 1976, p 43-60

19. W.A. Logsdon, An Evaluation of the Crack Growth and Fracture Properties of AISI 403 Modified 12Cr Stainless Steel, *Eng. Frac. Mech.*, Vol 1, 1975, p 23-40

20. R.R. Ferguson and R.G. Berryman, "Fracture Mechanics Evaluation of B-1 Materials," Report AFML-TR-76-137, Vol 1, Rockwell International, Los Angeles, 1976

21. T.W. Hall and C. Hammond, The Relation Between Crack Propagation Characteristics and Fracture Toughness in Alpha + Beta Titanium Alloys, in *Titanium Science and Technology*, Vol 2, Plenum Press, New York, 1973, p 1365-1376

22. R.E. Curtis and W.F. Spurr, Effect of Microstructure on Fracture Properties of Titanium Alloys in Air and Salt Solution, *ASTM Trans. Quart.*, Vol 61 (No. 1), March 1968, p 115-127

23. J.C. Chesnutt and R.A. Spurling, Fracture Topography—Microstructure Correlations in SEM, *Met. Trans. A*, Vol 8 (No. 1), 1977, p 216-218

24. M.A. Greenfield and H. Margolin, Interrelationship of Fracture Toughness and Microstructure in a Ti-5.25Al-5.5V-0.9Fe-0.5Cu Alloy, *Met. Trans. A*, Vol 2 (No. 3), March 1971, p 841-847

25. W.W. Gerberich, P.L. Hemmings, and V.F. Zackay, Fracture and Fractography of Metastable Austenites, *Met. Trans. A*, Vol 2, 1971, p 2243

26. J.A. Hall, V.F. Zackay, and E.R. Parker, *Trans. ASM*, Vol 62, 1969, p 965

27. W.W. Gerberich, "Metastable Austenite Steels with Ultra-High Strength and Toughness," Society of Automotive Engineers, Detroit, paper 69062, Jan 1969, p 1154-1159

28. C. Beachem, private communication, Naval Research Laboratories, Washington, DC, 1970

29. R.C. Bates and W.G. Clark, Jr., *Trans. Quart. ASM*, Vol 62 (No. 2), 1969, p 380

30. B. Tomkins, *Philos. Mag.*, Vol 18, 1968, p 1041

31. J.R. Rice, in *Fatigue Crack Propagation*, STP 415, ASTM, Philadelphia, 1967, p 247

32. M. Clavel and A. Pineau, Frequency

Fig. 28(b) Fracture surfaces of specimens containing varying manganese contents tested in hydrogen gas

Source: Ref 43

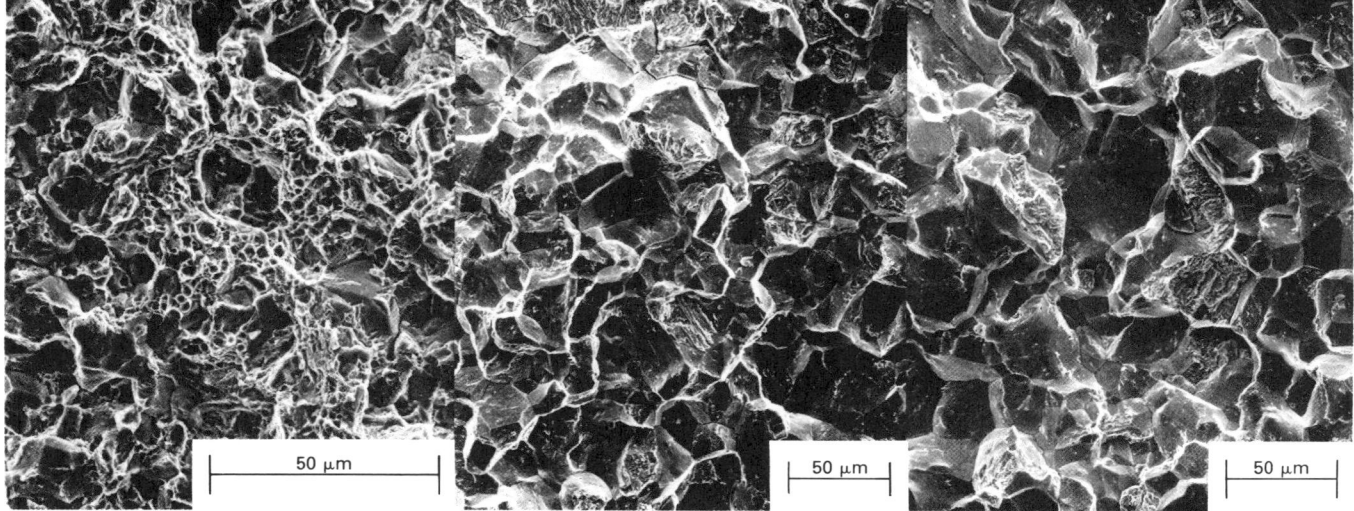

0.009Mn-0.41C   0.84Mn-0.42C   2.13Mn-0.43C

Fig. 29 Cleaved α grains in a hydrogen-charged Ti-6Al-6V-2Sn α-β microstructure

(a) TEM. (b) SEM. Source: Ref 46

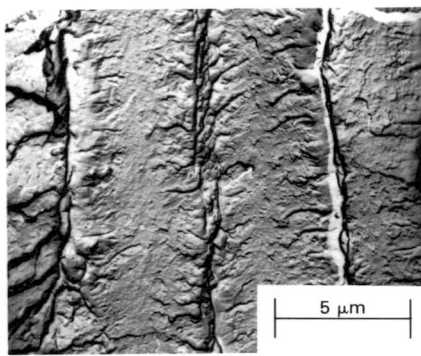

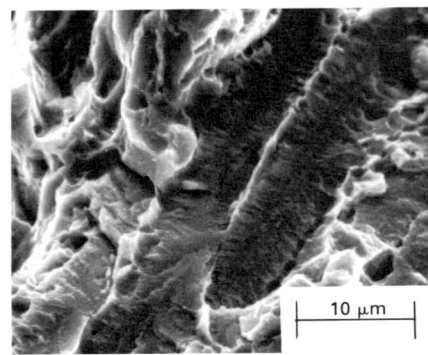

(a) (b)

Fig. 30 Typical fracture surfaces for a smooth copper tensile specimen in 1 M sodium nitrite solution

(a) Surface cracking approximately 300 μm deep in smooth tensile specimen tested at 25 °C (77 °F) and 0 V (saturated calomel electrode). The tensile axis is vertical. The nominal strain rate was 6 × 10⁻⁷/s, and the test was interrupted after 24 h. Note the rotation and necking of ligaments between adjacent crack segments. (b) Crack front striations (stereo pair). Crack growth top to bottom. The surface rumpling is considered to result from slip after the passage of the crack. Source: Ref 47

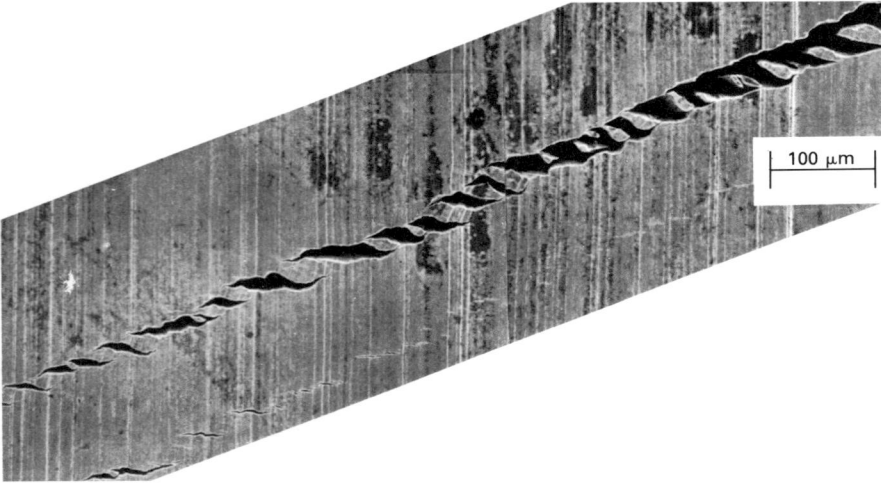

(a)

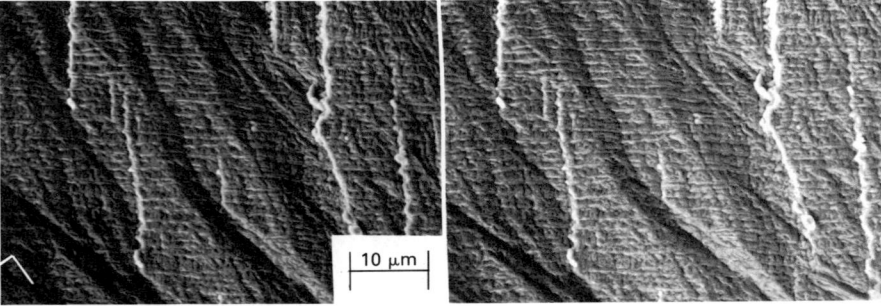

(b)

and Waveform Effects on the Fatigue Crack Growth Behavior of Alloy 718 at 298 °K and 823 °K, *Met. Trans. A*, Vol 9 (No. 4), April 1978, p 471-480

33. S.D. Antolovich and J.E. Campbell, Fracture Properties of Superalloys, in *Application of Fracture Mechanics for Selection of Metallic Structural Materials*, J.E. Campbell, W.W. Gerberich, and J.H. Underwood, Ed., American Society for Metals, 1982, p 253-310

34. T.W. Crooker, D.F. Hasson, and G.R. Yoder, in *Fractography—Microscopic Cracking Processes*, STP 600, ASTM, Philadelphia, 1976, p 220

35. R.O. Ritchie, Near-Threshold Fatigue-Crack Propagations in Steels, *Int. Met. Rev.*, Vol 24 (No. 5 and 6), 1979, p 205-230

36. W.W. Gerberich and N.R. Moody, A Review of Fatigue Fracture Topology Effects on Threshold and Growth Mechanisms, in *Fatigue Mechanisms*, STP 675, J.T. Fong, Ed., ASTM, Philadelphia, 1979, p 292-346

37. W.W. Gerberich and K. Jatavallabhula, Quantitative Fractography and Dislocation Interpretations of the Cyclic Cleavage Crack Growth Process, *Acta Met.*, Vol 31 (No. 2), 1983, p 241-255

38. W. Yu, K. Esaklul, and W.W. Gerberich, Fatigue Threshold Studies in Fe, Fe-Si and HSLA Steel: Part II. Thermally Activated Behavior of the Effective Stress Intensity at Threshold, *Met. Trans. A*, Vol 15, 1984, p 889-900

39. W.W. Gerberich and C.E. Hartbower, Some Observations on Dimple Size in Cyclic-Load-Induced Plane-Stress Fracture, *ASM Trans. Quart.*, Vol 61 (No. 1), 1968, p 184

40. C.J. Beevers, *Met. Sci.*, Vol 11, 1977, p 362

41. A. Yuen, S.W. Hopkins, G.R. Leverant, and C.A. Rau, *Met. Trans.*, Vol 5, 1974, p 1833

42. R.O. Ritchie, *J. Eng. Mat. Technol.*, Vol 99, 1977, p 195

43. M. Nakamura and E.-I. Furubayashi, Crack Propagation of High Strength Steels in a Gaseous Hydrogen Atmosphere, *Met. Trans. A*, Vol 14, 1983, p 717-726

44. E.C. Pow, W.W. Gerberich, and L.E. Toth, Analysis of Stress Corrosion Crack-Tip Surface Chemistries in 7075 Type Aluminum Alloys, *Scripta Met.*, Vol 15, 1981, p 55

45. M.L. Grossbeck, P. Williams, C. A. Evans, Jr., and H.K. Birnbaum, *Phys. Stat. Solidi*, Vol 34, 1976

46. K. Peterson, J.C. Schwanebeck, and W.W. Gerberich, In-situ Scanning Auger Analysis of Hydrogen-Induced Frac-

ture in Ti-6Al-6V-2Sn, *Met. Trans. A*, Vol 9, 1978, p 1169

47. K. Sieradzki, R. L. Sabatini, and R.C. Newman, Stress-Corrosion Cracking of Copper Single Crystals, *Met. Trans. A*, Vol 15, 1984, p 1941-1946
48. K.A. Esaklul, Ph.D. thesis, University of Minnesota, 1984
49. K.A. Esaklul, A.G. Wright, and W.W. Gerberich, On the Effect of Hydrogen Induced Surface Asperities on Fatigue Crack Closure in Ultrahigh Strength Steel, *Scripta Met.*, Vol 17, 1983, p 1073-1078
50. K.A. Esaklul and W.W. Gerberich, On the Influence of Internal Hydrogen on Fatigue Thresholds of HSLA Steel, *Scripta Met.*, Vol 17, 1983, p 1079-1082

Fig. 31 Scanning electron micrographs of Fe-2.5Si (at.%) cycled near the ΔK threshold at room temperature

(a) Nonhydrogen-charged specimen. (b) Hydrogen-charged specimen. Source: Ref 48

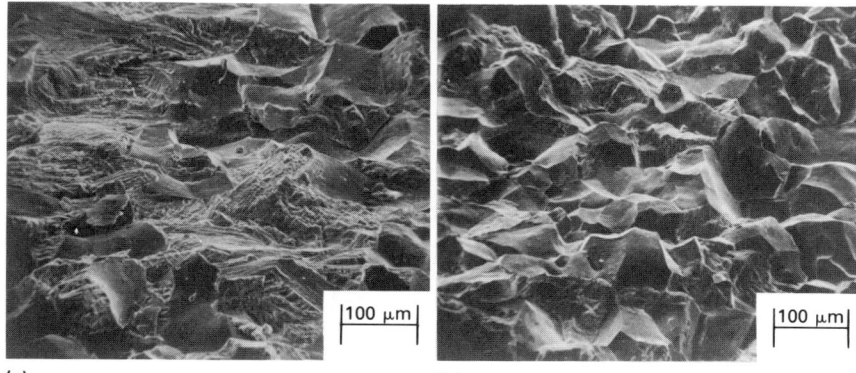

(a)　　　　　　　　(b)

Corrosion Testing

Tests for Stress-Corrosion Cracking

By Donald O. Sprowls
Consultant
Alcoa Technical Center (retired)

STRESS-CORROSION CRACKING (SCC) is a time-dependent process in which a metallurgically susceptible material fractures prematurely due to the synergistic interaction of a corrosive environment and sustained tensile stress at the metal surface. The tensile stress may be residual stress resulting from heat treatment or fabrication of the metal, may be developed by external loading, or may be a combination of these conditions. This article describes the principal stress-corrosion test methods and discusses their relative advantages and limitations.

Stress-corrosion cracking failures resemble brittle fracture; typically there is little, if any, indication of metal ductility at the origin of fracture. The cracking actually is a form of subcritical flaw growth, either intergranular or transgranular, depending on the particular combination of microstructure, environment, and strain rate. However, it is more difficult to design against environmentally assisted cracking than against fracture because of the compositional, mechanical, and metallurgical synergism and the consequent need to consider a range of environmental variables, as well as their variations with time and their interactions with loading and metallurgical variables (Ref 1). The necessary conditions required for stress-corrosion cracking are depicted in Fig. 1.

Although stress-corrosion cracking is a potential problem for some alloys and tempers of all commonly used structural metals, it is but one of a number of causes of premature fracture influenced by corrosion of a structural component (Ref 2). The most common causes of metal failure are compared in Fig. 2.

Unlike cracking due to corrosion fatigue, which propagates only under cyclic operating loads, stress-corrosion cracking and the other processes shown in Fig. 2 typically occur under the mechanical driving force of sustained residual and applied tensile stress (intensity), aggravated in some cases by su-

perimposed dynamic loading. Moreover, stress-corrosion cracking appears to occur only in the presence of certain electrochemical conditions related to the composition and microstructure of the metal, the environment, and film rupture due to dynamic straining (creep).

Hydrogen embrittlement cracking may have a major role in the stress-corrosion cracking of many alloy systems; however, the actual mechanism of embrittlement caused by hydrogen has yet to be established, even for alloys in which hydrogen-assisted cracking has been recognized (for more information on hydrogen embrittlement, see the article "Tests for Hydrogen Embrittlement" in this Section). The role of hydrogen in the stress-corrosion cracking of various materials is reviewed in Ref 3, and stress-corrosion cracking and the causes of premature (brittle) fracture are discussed in Ref 2. Thus, in stress-corrosion cracking tests, the results must be derived from true stress-corrosion cracking processes.

Verification of stress-corrosion cracking in a specimen corroded under stress is determined primarily by use of the optical metallograph and the electron microscope. Direct microscopic examination of fracture surfaces with the scanning electron microscope or by transmission electron microscopy using replicas of the fracture surface can be helpful in determining the cause of fracture. Successful use of these tools relies on recognition of modes of stress-corrosion cracking characteristic for the alloy system under study.

Valid environmental characterization of materials is vital to engineers who are responsible for materials development and selection and component structural integrity. Hence, accelerated stress-corrosion cracking tests are used extensively. The purposes of such tests fall into two broad categories: commercial (industrial) and academic. The primary objective of commercial testing is the estimation of risks or the prediction of

serviceability of an alloy or mill product. The objective of academic testing is to develop an understanding of stress-corrosion mechanisms, which in turn enables the selection of more appropriate testing methods for commercial purposes.

Although the long-term goal of all stress-corrosion testing is to predict the stress-corrosion performance of an alloy in a given service environment, additional objectives include:

Screening tests for alloy development

- Compare experimental compositions and tempers
- Compare variations in environment
- Develop quality control tests

Predicting stress-corrosion performance in service

- Determine the resistance to stress-corrosion cracking of an actual or simulated structure in anticipated service environment
- Determine optimum material selection through accelerated tests
- Determine the effectiveness of protective treatments
- Duplicate response of a material that failed in service

Fig. 1 Conditions required for stress-corrosion cracking

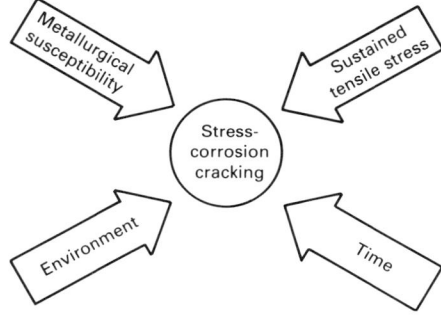

Fig. 2 Causes of premature fracture influenced by the corrosion of a structural component

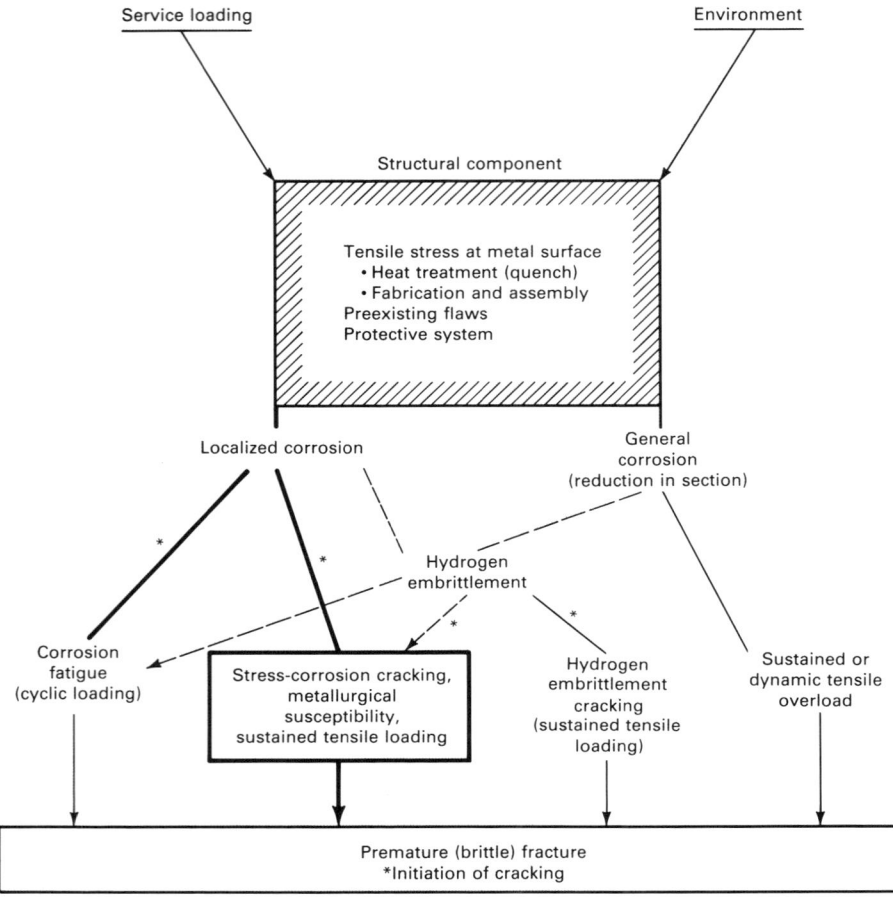

Determining if a product meets specifications

Investigating mechanisms of stress corrosion

Because of the diversity of objectives and the varied product forms that must be tested, a single all-purpose method for evaluation of stress-corrosion cracking susceptibility does not exist. Hence, the objective of a stress-corrosion study must be defined to ensure selection of the most appropriate test procedure and the most useful type of test results (Ref 4).

Accelerated Stress-Corrosion Cracking Testing

Stress-corrosion cracking has been of interest for the last 40 years (Ref 5, 6), and there have been significant advancements in the understanding of stress-corrosion mechanisms and in state-of-the-art stress-corrosion cracking testing in the last 20 years. As industrial stress-corrosion cracking problems increased during the 1950's and 60's, testing methods proliferated as investigators studied problems and metal producers developed improved alloys.

Before 1965, constant-load or constant-strain tests of smooth and notched test specimens of various configurations were used to assess stress-corrosion cracking. During the 1960's, two accelerated test techniques based on different mechanical approaches emerged. One technique tests and analyzes statically loaded mechanically precracked test specimens using linear elastic fracture mechanics concepts. The second technique consists of constant (slow) strain rate tests on smooth or precracked specimens. Laboratory testing with these techniques frequently has produced stress-corrosion cracking, whereas older traditional tests have not. Valid testing must take into consideration the performance requirements of the intended engineering structure.

Stress-corrosion cracking is frequently discussed in terms of initiation (incubation and nucleation) and propagation. A gradual transition exists between these micropro-

cesses, with no distinct separation of stages, as suggested in Fig. 3. However, a precise model has not been established. Propagation may be viewed as a repetitive series of reinitiation and propagation microscopic events. In the design and selection of testing methods, attention should be paid to tests that examine initiation as well as propagation (Ref 7).

Static Loading

Historically, service failures due to stress-corrosion cracking have been identified with sustained tensile stress; thus, stress-corrosion cracking testing has developed around the use of static loading. Engineering conditions exist in which the entire component appears to be the optimum choice for a test specimen. However, this usually is not practical; it is often necessary to select smaller specimens that afford the required predictive capability. Tests are conducted by exposing specimens to appropriate environments while stressed by application of a constant load or constant strain.

Smooth Specimens. As shown in Fig. 3, when the stress-corrosion crack has attained a sufficient length, the critical combination of stress and crack length is met for unstable (purely mechanical) fracture. The crack length reached before the onset of the terminal fracture depends on the magnitude of the exposure stress and the fracture toughness of the alloy. The chemical environment has no effect on this terminal, fast-fracture process.

The kinetics of the various processes shown in Fig. 3 are illustrated schematically in Fig. 4 for alloys with varying fracture toughnesses. Curves A, B, and C represent materials with decreasing toughness, with the latter showing fast fracture initiated by corrosion pits or fissures without exhibiting stress-corrosion cracking. The behavior of a

Fig. 3 Initiation and propagation of stress-corrosion cracking in an SCC-sensitive material

A, localized breakdown of oxide film; B, formation of corrosion fissures, localized concentration of stress, and nucleation of stress-corrosion cracking; C, propagation of stress-corrosion cracking in two or three stages with changing dependency on the stress-intensity factor (see also Fig. 5)

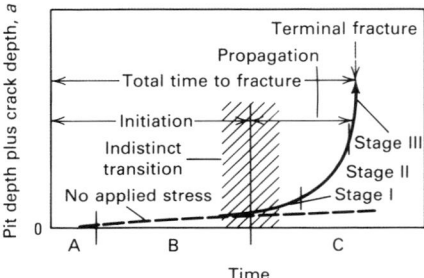

Fig. 4 Various processes in stress-corrosion cracking as influenced by the fracture toughness of the metal

Kinetics for pitting (or, in material D, nonpitting), stress-corrosion cracking (materials A and B only), and fast fracture (lower plot). Line at top illustrates how time-to-failure data can be misleading.
Source: Ref 8

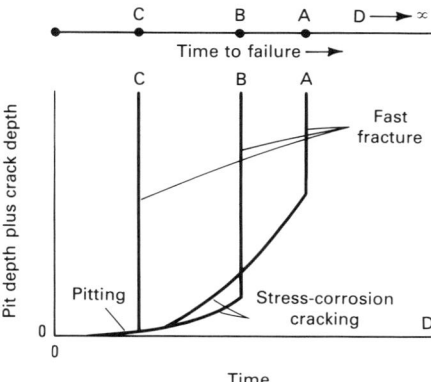

material that does not develop localized pitting or intergranular attack is represented by a line coincident with the abscissa, designated D in Fig. 4. The time-to-failure ranking above the graph indicates D as best and A, B, and C as poorest. Actually, the stress-corrosion cracking responses of C and D were not measured, and the true stress-corrosion cracking ranking of A and B (indicated by depth of stress-corrosion cracking at the time of fracture) exhibits an opposing trend to that inferred from the time-to-failure data above.

Thus, undifferentiated time-to-failure data can be highly misleading, particularly when comparing high-strength alloys, in which brittle fracture occurs before stress-corrosion cracking can cause significant differences in total time-to-fracture. Further difficulties may arise, because the total time-to-fracture is also influenced by non-stress-corrosion cracking factors such as specimen type and size, method of loading, initial stress level, and initiation behavior of the alloy. Consequently, the stress-corrosion cracking ranking of materials may vary among investigators using different testing techniques.

Although stress-corrosion cracking comparisons generally are conducted in terms of threshold stresses and percent of specimens failing, these parameters are also based on arbitrary exposure times. Test results are significantly influenced by the time required to initiate stress-corrosion cracking, which can vary widely. Additionally, under some conditions, random corrosion effects may override mild to moderate stress-corrosion cracking response. Hence, the test results for alloys with improved stress-corrosion crack-

ing resistance require time-consuming exposures with attendant scatter in data. Careful interpretation is required to prevent non-stress-corrosion cracking factors from erroneously affecting test results.

Precracked (Fracture Mechanics) Specimens. One method of distinguishing stress-corrosion cracking characteristics from the effects of both long incubation time and purely mechanical fracturing uses precracked test specimens loaded with the corrodent surrounding the region of the notch. The use of a precracked specimen is based on the concept that large structures with thick components are apt to contain crack-like defects. After a stress-corrosion crack commences to grow, or if the specimen is provided with a mechanical precrack, classical stress analysis is inadequate to determine the response of the material subjected to stress in the presence of a corrodent.

The mechanical driving force for cracks can be measured with linear elastic fracture mechanics theory in terms of the crack-tip stress-intensity factor, K, which is expressed in terms of the remotely applied loads, crack size, and test specimen geometry. At or above a certain level of K, stress-corrosion cracking in a susceptible material will initiate and grow, whereas below that level no measurable propagation is observed (Ref 9).

The apparent threshold for the propagation of stress-corrosion cracking (assuming that crack nuclei form in a manner that cannot be described by fracture mechanics such as localized corrosion) is designated K_{Iscc} (or K_{th}). This value defines the largest crack that can exist in a structure in the chosen environment without propagation in the form of stress-corrosion cracking. Thus, in terms of linear elastic fracture mechanics theory, for a surface crack in a large plate remotely loaded in tension, the shallowest crack (of a shape that is long compared to its depth) that will propagate as a stress-corrosion crack is $a_{cr} = 0.2(K_{Iscc}/\text{TYS})^2$, where TYS is tensile yield strength. Thus, a crack that is shallower than this value will not propagate.

The value of a_{cr} incorporates the stress-corrosion cracking resistance, K_{Iscc}, and the contribution of stress levels (of the order of the yield strength) to stress-corrosion cracking due to residual or assembly stresses in thick component sections (Ref 2). Thus, the application of fracture mechanics does not provide independent information about stress-corrosion cracking; it simply provides a usable method for treating the stress factor in the presence of a crack.

When stress-corrosion cracking propagation rate is determined and plotted as a function of K, the test results for a highly susceptible alloy will exhibit the general trend shown in Fig. 5. Actual curves vary, depending upon the stress-corrosion cracking

resistance and fracture toughness of the alloy. Development of such a curve (or a band if several production lots are tested) for a material provides two types of stress-corrosion cracking data that are potentially useful for predicting the behavior of large structural components for which fracture is a critical design factor. In addition to determining a K_{th} value for stress-corrosion cracking propagation, indications of a "plateau velocity" can be observed. This may be regarded as an estimate of the maximum propagation rate that can be sustained in a particular material environment system. These values can be used for ranking materials that are sensitive to stress-corrosion cracking or stress-corrosion cracking environments that promote the phenomenon.

The use of precracked specimens in crack-tip stress analysis by fracture mechanics permits prediction of the stress-corrosion response in terms of flaw size and stress from laboratory tests on specimens with simple geometries. Hence, laboratory stress-corrosion cracking test results can be used in quantitative failure analysis studies and can be applied to damage-tolerant design of structural parts. Although precracking may shorten or modify the initiation period, it does not circumvent it. Thus, this method of testing also requires arbitrary and sometimes long exposure periods.

The evolution of elastic-plastic fracture mechanics has extended the validity of fracture mechanics to more extensive ranges of plasticity. Local crack-tip stress and strain field equations (Ref 10 and 11) provide an

Fig. 5 Generalized kinetics of stress-corrosion cracking

Fast fracture at one of the points marked K_{Ic} may preclude development of Region III, or of Regions II and III. There may be a true threshold K_{Iscc} or the kinetics may simply decrease continuously to smaller but finite values as K decreases, denoted by the dashed line prolongation of Region I.
Source: Ref 8

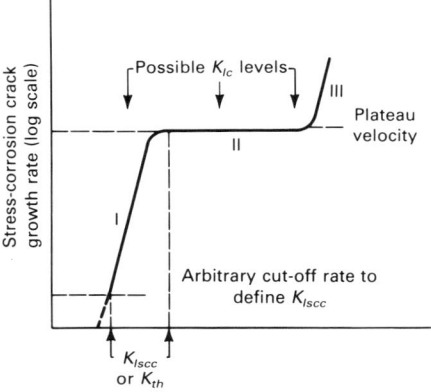

elastic-plastic fracture mechanics driving force parameter, J, that is more appropriate for small cracks in small specimens. It can also be related directly to the linear elastic fracture mechanics stress-intensity factor, K, in cases of more limited plasticity.

Correlation of Smooth and Precracked Specimens. Due to the fracture mechanics relationship of crack depth, gross section stress, and crack-tip stress intensity, the results of tests on crack-free and fracture mechanics specimens can be correlated. Support of this hypothesis was obtained by experimental testing with a magnesium-based alloy (Mg-7Al) (Ref 12), which demonstrated the type of relationship to be anticipated if fracture mechanics were relevant to the cracking of small-sized smooth specimens (Fig. 6).

The concept of combining stress-corrosion cracking thresholds obtained with smooth and with linear elastic fracture mechanics specimens from a sample of aluminum alloy plate to yield a conservative assessment of materials for design (Ref 13) is shown schematically in Fig. 7. The exclusive use of either one of the test methods may yield nonconservative conclusions. Establishment of a "safe zone" requires results from both types of tests, but careful interpretation of the specific test condition is required.

As shown in Fig. 7, the K_{th} analysis breaks down in the small flaw region (ABE) when the smooth specimen threshold stress is exceeded. Application and further development of elastic-plastic fracture mechanics theory should lead to improved estimates of critical stress/flaw size combinations for the onset of stress-corrosion cracking and mechanical fracture, as suggested in Fig. 8.

Dynamic Loading: Slow Strain Rate Testing

The most recent method used for accelerating the stress-corrosion cracking process in laboratory testing involves relatively slow strain rate tension testing of a specimen during exposure to appropriate environmental conditions. The application of slow dynamic strain above the elastic limit assists in the initiation of stress-corrosion cracking.

This accelerating technique, known as slow strain rate testing, is consistent with the proposed general mechanisms of the stress-corrosion cracking process, most of which involve plastic microstrain and film rupture. Based on experimental evidence, it has been suggested that the most important role of stress is the dynamic strain that it produces (Ref 12, 15).

Slow strain rate testing is not terminated after an arbitrary period of time. Testing always ends in specimen fracture, and the mode of fracture is then compared with the

Fig. 6 Correlation of stress-corrosion thresholds obtained with smooth and precracked specimens

Threshold stress values and maximum size for nonpropagating cracks calculated from K_{th} values for Mg-7Al alloy after various heat treatments. Threshold stresses based on complete fracture of 2.3-mm (0.090-in.) diam cylindrical tension specimens stressed by constant total strain in a sodium chloride and potassium chloride solution. Source: Ref 12

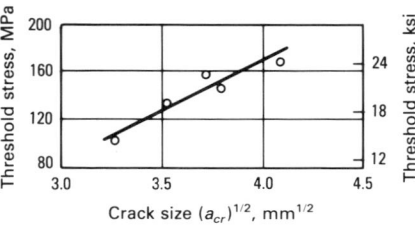

criteria of stress-corrosion cracking susceptibility. In addition to its time-saving benefits, less scatter occurs in the test results.

The most significant variable in slow strain rate testing is the magnitude of strain rate. If too high, ductile fracture will occur

Fig. 7 Concept for combining stress-corrosion thresholds obtained on smooth and linear elastic fracture mechanics specimens to yield a conservative assessment of materials

(1) Minimum stress at which small tensile specimens fail by stress-corrosion cracking when stressed in environment of interest. (2) Minimum stress intensity at which significant stress-corrosion crack growth occurs in environment of interest. Source: Ref 13

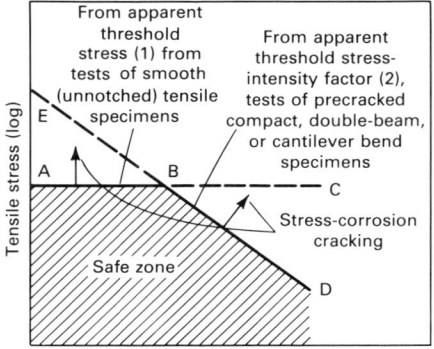

Fig. 8 Proposed linear elastic and elastic-plastic models for describing critical combinations of stress and flaw size at stress-corrosion cracking thresholds and at the onset of rapid tensile fracture

Source: Ref 14

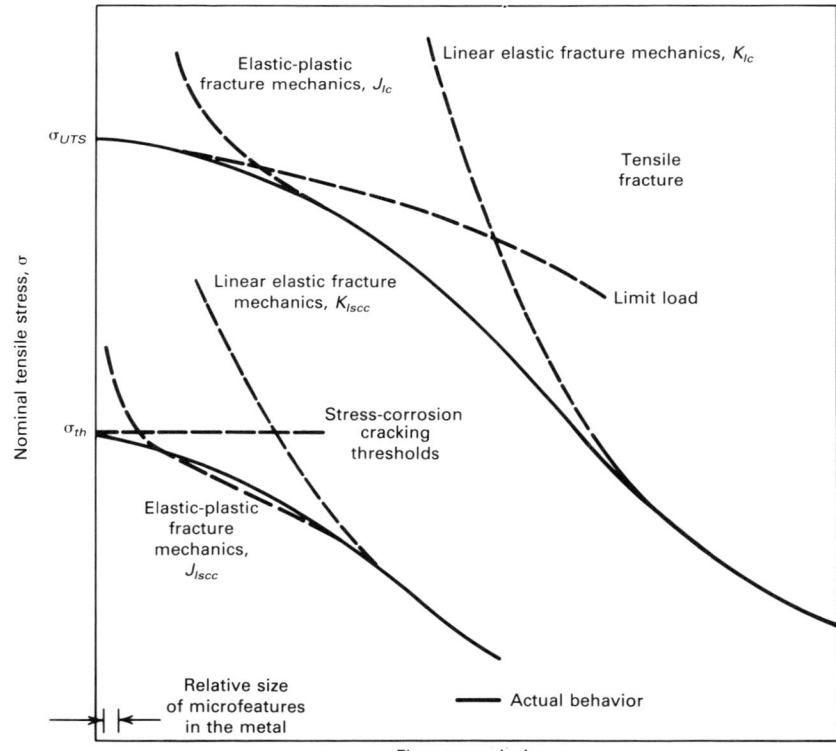

Fig. 9 Plot of typical results for constant-strain tests on one material in two environments

One environment was inert and the other caused stress-corrosion cracking. Source: Ref 2

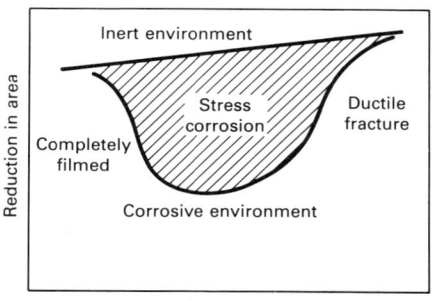

Fig. 10 Nominal stress-elongation curves for carbon-manganese steel in constant-strain tests in boiling 4N sodium nitrate and in oil at the same temperature

Source: Ref 15

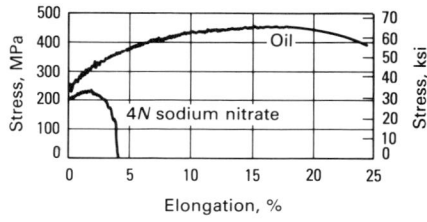

Fig. 11 Comparison of threshold stresses indicated by constant-load and constant-strain tests

Stress-corrosion crack velocities observed in a carbon-manganese steel immersed in carbonate-bicarbonate solution at 75 °C (167 °F) and −650 mV for different initial net section stresses in constant-load and constant-deflection tests. Source: Ref 15

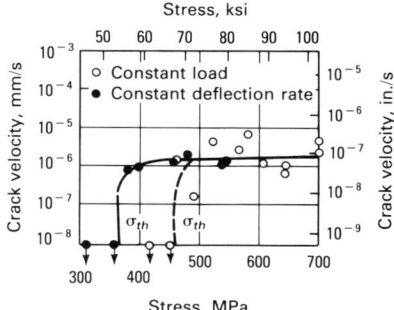

before the necessary corrosion reactions can occur. Thus, relatively low strain rates should be used. However, at too low a strain rate, corrosion may be prevented due to film repair so that the necessary reactions of bare metal cannot be sustained, and stress-corrosion cracking may not occur (Fig. 9). Although typical critical strain rates range from 10^{-8} to $10^{-4}/s^{-1}$ depending on the alloy and environment system, the most severe strain rate must be determined in each instance.

The effects of stress-corrosion cracking are reflected in the load-deflection curve recorded during slow strain rate testing. Because stress-corrosion failure usually is associated with relatively little macroscopic plastic deformation during crack propagation, a comparison of the load-deflection curves for specimens with and without stress-corrosion cracking usually reveals significant differences. Figure 10 depicts stress-elongation curves that illustrate how stress-corrosion cracks influence the elongation to fracture, as well as the maximum load. Therefore, crack susceptibility can be expressed by maximum load, as well as by the usual measures of ductility, elongation, and reduction in area.

Time-to-fracture, or fracture energy (area under the load extension curve), can also be used to indicate crack susceptibility. The results are normalized in terms of a ratio of the parameter measured in the corrodent divided by measurements taken in an inert environment under the same conditions. Thus, increasing susceptibility is marked by an increasing departure from unity.

Slow strain rate testing as generally used does not provide data that can be used for design purposes. However, recent work has shown that average stress-corrosion cracking velocities, threshold stresses, and threshold strain rates can be obtained with modified techniques for some alloy and environment systems (Ref 12, 15, 16). Additional devel-

opment is needed to ensure applicability to various design areas.

Stress Corrosion and Corrosion Fatigue. The threshold stress for stress-corrosion cracking generally should be lower in applied strain rate tests than in constant-load or constant-strain tests, because the nonpropagating cracks that form in constant-load strain tests below the threshold stress (when determination of the threshold is based on fracture) will propagate if an appropriate strain rate is maintained. Tests on precracked cantilever beam specimens of a carbon-manganese steel alloy (Fig. 11) have demonstrated this phenomenon.

The threshold stress is reduced due to applied strain rates, as opposed to static loads, and can be reduced even further with cyclic loading. With appropriate frequency, load change, and temperature, average creep rates can be sustained over extended periods, whereas with static loading the creep rate falls below the level needed to promote stress-corrosion cracking.

The interface between stress-corrosion cracking and corrosion fatigue has been studied in carbon-manganese steels in a carbonate-bicarbonate solution (Ref 15). It was shown that corrosion fatigue under certain conditions can be indistinguishable from stress-corrosion cracking under constant or monotonic loading for a given alloy and environment combination that is sensitive to stress-corrosion cracking.

Environmental Factors

In addition to an alloy of appropriate composition and structure and a stress, stress-intensity factor, or strain rate of suitable magnitude, certain environmental requirements are necessary to promote stress-corrosion cracking. Stress-corrosion cracking will not occur in a vacuum, but traces of water in vapor or liquid form are sufficient to promote stress-corrosion cracking in some aluminum and ferrous alloys. Typical contaminants in seacoast and inland industrial (urban) atmospheres increase the range of environments

for common structural metals that are sensitive to stress-corrosion cracking. Figure 12 illustrates this for 7079-T651, a highly susceptible aluminum alloy (Ref 17).

The primary environmental factors are the nature and concentration of anions and cations in aqueous solutions, electrochemical potential, solution pH, the partial pressure and nature of species in gaseous mixtures, and temperature. Separately or in combination, environmental variables can affect the thermodynamics and kinetics of the electrochemical processes that control environmentally assisted fracture. This is achieved, for example, in a corrosion fissure or at a crack tip by bringing the local environment into a regime where metal dissolution occurs, the production of embrittling hydrogen atoms occurs, or passive film formation becomes impossible and crack blunting results. The kinetics of the cracking process are controlled by environmental variables through their effect on the critical balance between active and passive behavior related to the formation of protective films at the crack tip. The extent to which metal surfaces exhibit active/passive transitions at rates that sustain stress-corrosion cracking depends not only on the nature of the environment, but also on the composition and structure of the metal (Ref 1).

Thus, the environmental requirements for stress-corrosion cracking vary with different alloys. Although a mechanical precrack or a critical strain rate provides a "worst case" for stress-corrosion cracking from a mechanical standpoint, there does not appear to be a generally accepted "worst case" from an environmental standpoint. However, because the presence of moisture and salt water is universal, the stress-corrosion cracking characteristics of alloys in these environ-

Fig. 12 Effect of corrosive environment on stress-corrosion cracking propagation rate in 64-mm (2.5-in.) thick 7079-T651 plate stressed in the short-transverse orientation

Stress-corrosion cracking did not occur during 3 years of exposure to dry air in a desiccator. Plateau velocity (horizontal part of each curve) and the apparent threshold stress intensity (K_{th}) varied with the environment. Source: Ref 17

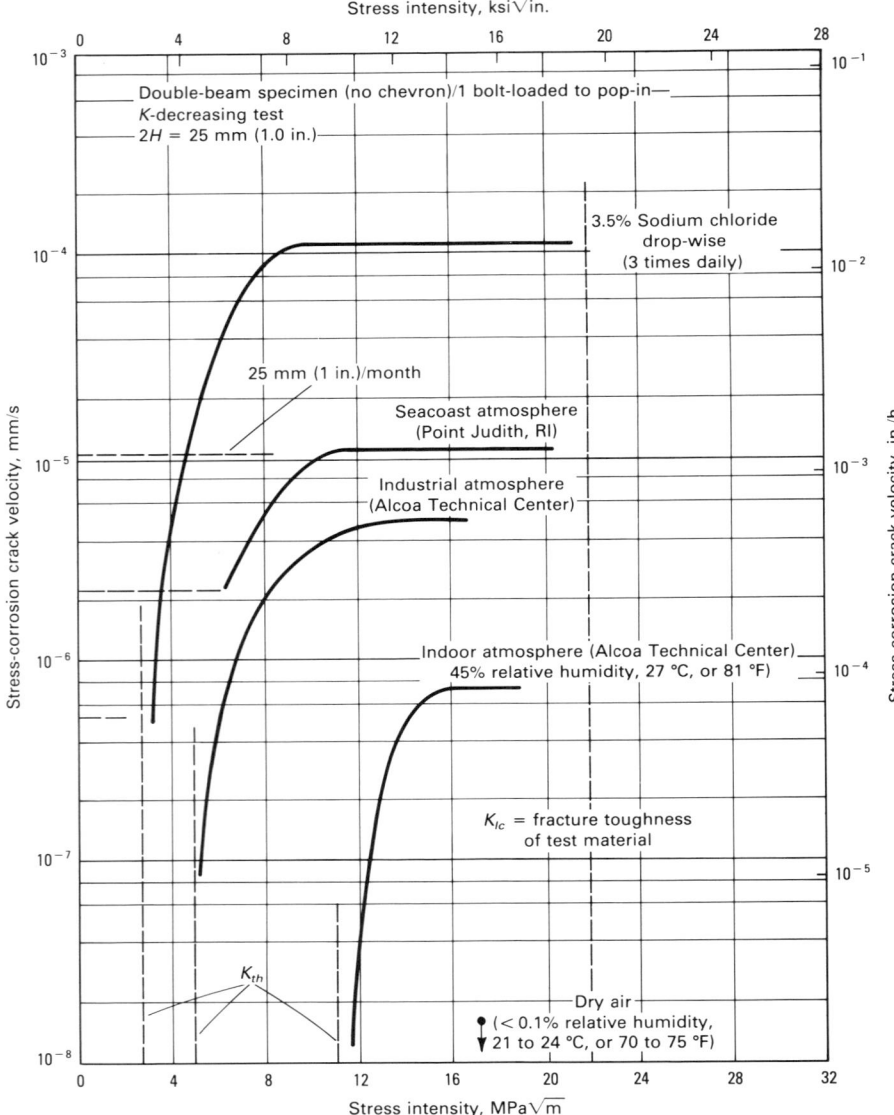

Fig. 13 Cantilever beam specimens of precipitation-hardened 13-8 Mo stainless steel after testing

Experiments demonstrate that electrochemical factors can override mechanical factors in selecting site of initiation of stress-corrosion cracks. Source: Ref 18

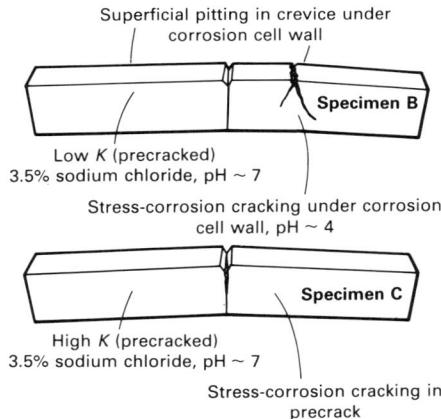

sharp as a fatigue crack, the K level would have been much lower than at the machined and fatigued notch. In the stagnant situation under the cell wall, the stainless steel reacted with the salt water to form hydrochloric acid and other corrosion products from the metal. Thus, the low pH in a crevice, due to the hydrolysis of chromium corrosion product, overcame the mechanical disadvantage of the lack of a precrack. Specimen A served to test the effectiveness of electrochemical conditions in crack initiation. Saturated ferric chloride was selected to lower the pH to the range inside an active corrosion pit in the stainless steel, and on application of the solution to the unnotched beam, many cracks initiated immediately in the smooth surface. Hydrochloric acid was found to be equally effective.

Information about the nature of the corrodent within the chemical process zone of stress-corrosion cracks is highly incomplete and almost without independent verification. If more data were available about the local corrodent at initiation as well as during propagation, the holding time required for crack initiation could possibly be shortened. However, models for estimating the risks of serviceability of alloys and tempers in terms of stress-corrosion cracking do not presently exist.

ments are always of interest, as well as in any special environment a given engineering structure may experience (Ref 8).

The complete effects of local chemistry on the control of stress-corrosion cracking have not been established. Figure 13 illustrates that electrochemical factors can override mechanical factors in selecting a site for initiation of stress-corrosion cracks (Ref 18). Three cantilever beam specimens of precipitation-hardened 13-8 Mo stainless steel were tested in salt water. Specimen C was tested at a high K level. With the participation of the chloride ions, the protective oxide film rup-

tured at the bottom of the precrack and initiated stress-corrosion cracking, which was halted before the beam fractured completely. Specimen B was loaded at a lower K level. After 1300 h, a stress-corrosion crack initiated, but not in the precrack. Crack initiation occurred under the wall of the cell that surrounded the central portion of the specimen and contained the salt water.

Careful examination of this specimen and replicate specimens revealed small crevice corrosion pits under the wall, which initiated stress-corrosion cracks in an almost smooth surface. Even if these small pits had been as

Interpretation of Results

Experimental difficulties are inherent to all accelerated stress-corrosion cracking testing methods because of the complex compositional, mechanical, and metallurgical interactions. Unlike standardized mechanical property tests, results from stress-corrosion cracking tests require interpretation, as only few have been standardized. Interpretation problems consist of:

• Quantitative measures of the degree of susceptibility to stress-corrosion cracking; the need for stress-corrosion cracking parameters that are suitable to statistical analyses
• Correlation of test results; laboratory tests versus actual structures and relationships among stress-corrosion cracking parameters obtained from different testing methods
• Definition of service requirements to assess serviceability of materials in specific applications, or to define safe environments
• Prediction of service life of a structural component

Standardization

Standardization of stress-corrosion testing methods in the United States was initiated in 1964 by ASTM. Currently, 12 ASTM standards concern statically loaded stress-corrosion tests using smooth test specimens; five pertain to the preparation and use of various types of specimens, and six apply to the corrodents that are applicable to specific alloy systems (Ref 19). Although tests using precracked specimens and slow strain rate tests are widely used, no standards pertaining to these tests have been issued by ASTM or the International Standards Organization (ISO) to date.

The recognition of various procedures for conducting stress-corrosion cracking testing does not imply that all methods are adequate and valid for all situations. Selection of the most appropriate test conditions depends on the test objective, existing knowledge of the metallurgy of the alloy, the mechanical properties of the metal product, the composition of the test medium, and the intended service application.

Sampling of Test Materials

The resistance to stress-corrosion cracking of many materials, notably high-strength aluminum alloys, is markedly influenced by microstructure, which usually varies with different mill product forms such as sheet, plate, and forgings. The microstructure can be directional, depending on the type of mill product, thickness, and metallurgical history.

Such directionality is illustrated in Fig. 14, which presents results of a stress-corrosion cracking test performed on specimens of 7075-T651 aluminum alloy plate that demonstrate the pronounced effect of test specimen orientation. Tests on specimens with the loading stress acting in the longitudinal and long-transverse orientations produced greater load-carrying ability than tests on specimens stressed in the short-transverse direction. Service experience with high-strength aluminum alloys has shown, in fact, that stress-corrosion cracking problems usually resulted from stresses oriented in the short-transverse direction of the grain structure (Ref 17, 21).

Directional grain structures cannot be predicted reliably from the geometry of the product without knowledge of the metallurgy of the particular alloy and the fabrication history of the product. Therefore, the macrostructure and microstructure of the product to be sampled should be checked by visual examination of macroetched sections or by metallographic examination.

The effect of a directional grain structure on the stress-corrosion cracking performance of a material is greatest in materials that develop stress-corrosion cracking along grain boundaries or along preferred metallurgical features such as phase boundaries or crystallographic planes. In this discussion, emphasis is placed on the orientation with respect to metallurgical structure, rather than to the geometry of the mill product, which is typical of most mechanical property tests.

Static Loading

Tests to predict the stress-corrosion performance of an alloy in a particular service application should be conducted with a stress system similar to that anticipated in service (Ref 22). Table 1 lists the numerous sources of sustained tension that are known to have initiated stress-corrosion cracking in service and the applicable methods of stressing. Most of the stress-corrosion cracking service problems involve tensile stresses of unknown magnitude that are usually very high. As such, tests that incorporate a high total strain usually are the most realistic in terms of duplicating service environment.

Fig. 14 Effects of the magnitude of sustained tensile stress and its orientation relative to the microstructure on the stress-corrosion cracking resistance of a metallurgically susceptible material

Tests were performed on 3.18-mm (0.125-in.) diam tension specimens machined from the midplane of 7075-T651 plates of various thicknesses. The solid line, lower bound, defines the stress-corrosion cracking performance of test specimens with different orientations to the grain structure. Note the relatively low stress level at which short-transverse specimens failed compared to the long-transverse and longitudinal specimens. Source: Ref 20

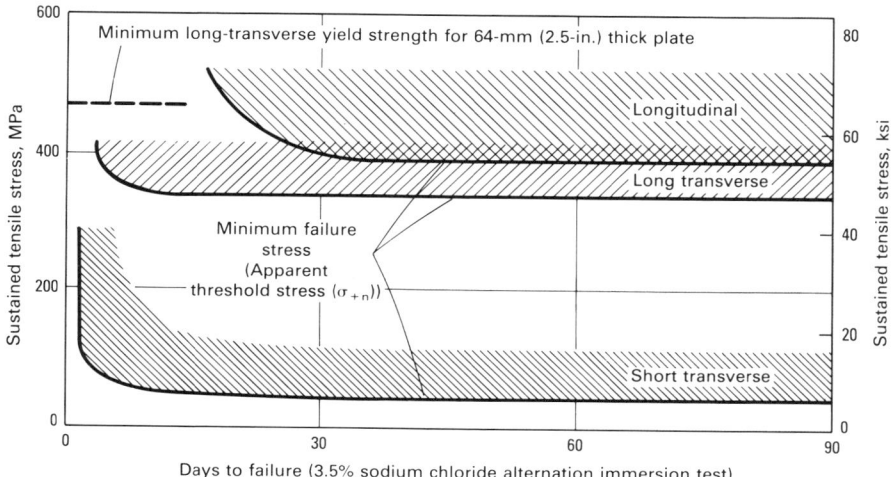
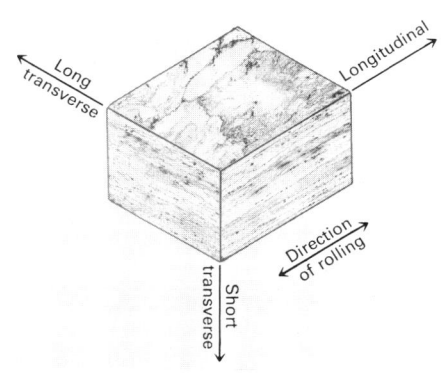

Table 1 Methods of stressing applicable to various sources of sustained tension in service

Source of sustained tension in service	Constant strain	Constant load
Residual stress:		
Quenching after heat treatment . . .	X	. . .
Forming.	X	. . .
Welding.	X	. . .
Misalignment (fit-up stresses)	X	. . .
Interference fasteners	X	. . .
Interference bushings:		
Rigid	X	. . .
Flexible	. . .	X
Flareless fittings	X	. . .
Clamps	X	. . .
Hydraulic pressure	X	X
Dead weight	. . .	X
Faying surface corrosion	X	X

Note: The greatest hazard arises when residual, assembly, and operating stresses are combined.

Constant-Strain Versus Constant-Load Stressing

Constant-strain (fixed-displacement) tests are widely used, primarily because a variety of simple and inexpensive stressing jigs can be used. Problems associated with such techniques usually are related to poor reproducibility of stress level. Sophisticated procedures have thus been developed to improve this facet of testing.

Constant-strain tests are sometimes called decreasing-load tests, because after the onset of stress-corrosion cracking, the gross section exposure stress and the average stress on the remaining cross section begin to decrease in small test specimens. This results from the opening of the crack (or cracks) under the high stress (intensity) concentrated at the crack tips, thus causing some of the applied elastic strain to change to plastic strain, with an attendant reduction in the initial load (Ref 7, 23). Such trends in changing stress during crack growth are illustrated in Fig. 15.

Consequently, stress relaxation in SCC-sensitive ductile alloy specimens stressed in very stiff, rigid frames can result in incomplete failure (fracture) of specimens. The stiffer the frame and the smaller the cross section of the test specimens, the less the elastic strain is likely to remain in the net section after opening of the crack. When complete fracture does not occur, auxiliary mechanical property tests or metallographic examination of longitudinal sections of the unbroken specimens is necessary to determine whether stress-corrosion cracking has occurred.

When comparing the stress trends for a constant-strain test (Fig. 15a) with those for a constant-load test (Fig. 15b), it is evident that neither method of loading provides a constant-stress test after growth of macrocracks has occurred. True constant-load (dead-load) tests are more likely to lead to

earlier failure with complete fracture and lower estimates of a threshold stress than constant-strain tests. Figures 15(a) and 15(b) illustrate basic trends that may be applied to all types of test specimens, including precracked specimens. However, specific curves will differ depending on other test conditions.

The stiffness of the combined stressing frame/test specimen system can have a significant effect on the reproducibility of test results if identical test procedures are not used (Ref 7). Many so-called constant-strain tests, particularly if a spring is included in the stressing system, are not actually constant-strain tests, because a significant amount of elastic strain energy may be contained in the stressing system. Depending on the "softness" of the spring or the elasticity of the stressing jig, the stiffness (compliance) of the stressing system can be varied greatly between zero stiffness (dead load) and infinite stiffness (true constant total strain). An example of the typical change in net section stress with the onset of stress-corrosion cracking in an intermediate stiffness stressing frame is shown in Fig. 16.

The corrosion pattern on the test specimen, particularly the number and distribution of cracks, can also contribute to the lack of precision in results obtained by either constant-strain or constant-load tests. When isolated stress-corrosion cracks are initiated in a specimen stressed by either method, the average tensile stress on the net section increases rapidly until the notch fracture strength is reached and the specimen breaks. Less penetration is required for fracture of specimens under dead load, thus indicating that specimen life is shorter with frames that are less stiff. When microcracks initiate close

Fig. 15 Schematic comparison of changing stress during initiation and growth of isolated stress-corrosion cracking in constant-strain and constant-load tests of a uniaxially loaded tension specimen

(a) Constant-strain test. (b) Constant-load test. σ_M is the maximum stress at crack tip, σ_N is the average stress in the net section, and σ_G is the applied stress to the gross section. Source: Ref 23

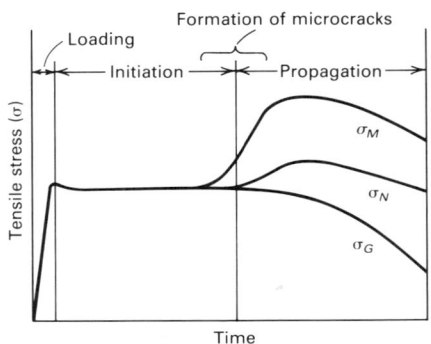

(a)

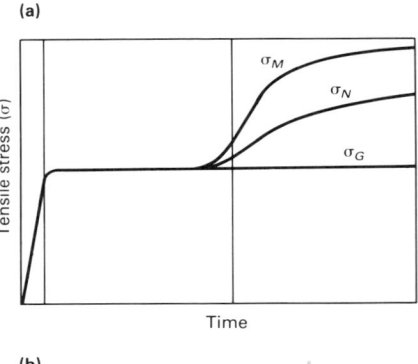

(b)

Fig. 16 Effect of loading method and extent of cracking or corrosion pattern on average net section stress in a uniaxially loaded tension specimen

Behavior is generally representative, but curves will vary with specific alloys and tempers. Source: Ref 19

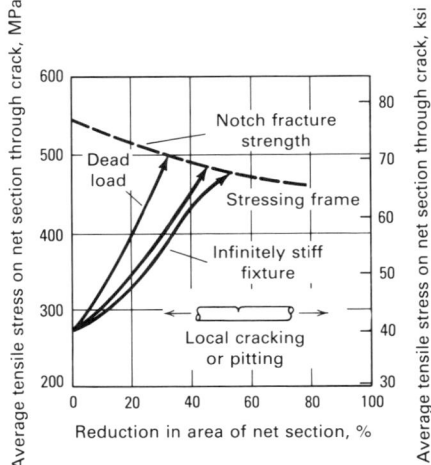

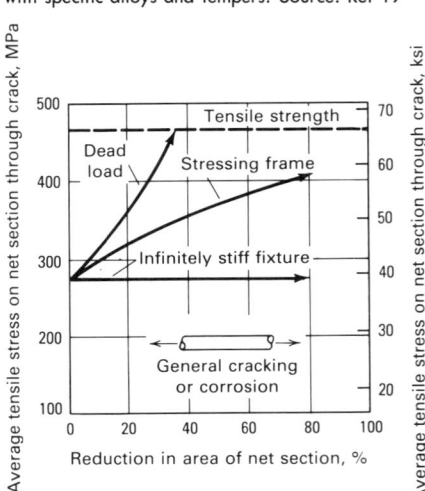

to one another, their individual stress concentrations interact and are relaxed. Consequently, there may not be a sufficient stress concentration in the true constant-strain test to propagate further stress-corrosion cracking, and the specimen will not break (Fig. 16).

Under a constant load, however, growth of many cracks continues, and the specimen ultimately breaks. The difference in stress-corrosion cracking response can be strongly influenced by frame stiffness under this type of cracking pattern, which is typical of cracking behavior at exposure stress levels near the threshold stress for a given set of test conditions. Thus, the stress-corrosion cracking comparison of specimens tested at stress levels just above their thresholds, as in materials with low susceptibility to stress-corrosion cracking, is complicated by variations in the cracking pattern, particularly when tested with relatively stiff stressing systems.

Although constant-load stressing is preferred for testing materials relatively resistant to stress-corrosion cracking, difficulties arise when small-diameter specimens are used, which are generally preferred to avoid the use of massive loads or lever systems. In some stress-corrosion cracking environments, highly stressed specimens may fail from simple pitting and an attendant increase in the effective stress. Such non-stress-corrosion cracking failures complicate interpretation of test results, unless failure by stress-corrosion cracking is confirmed by metallographic examination. Such extraneous failures are less likely to occur with specimens loaded under constant strain.

Thus, small-sized test specimens, which generally are preferred for laboratory screening tests and research studies, must be used with caution, particularly when estimates of serviceability are required. To properly determine serviceability, larger specimens should be used and a stressing system should be selected that best duplicates the anticipated service conditions. Because of the inability of classical stress analyses to cope with the conditions that exist with a growing crack, fracture mechanics are used to bridge the gap between conveniently sized laboratory test coupons and large structural components.

Bending Versus Uniaxial Tension

The most extensively used stressing systems historically have incorporated constant-deformation-loaded bend specimens. This method is versatile because of the variety of simple techniques that can be used to test most metal products in all types of corrosive environments. The state of stress in a bend specimen, however, is much more complex than in a tension specimen. The tensile stress

theoretically is uniform through and around the cross section in the tension specimen, except at corners in rectangular sections, whereas the tensile stress in bend specimens varies at the metal surface and through the thickness.

Tensile stress is maximum on the convex surface and decreases steeply to zero at the neutral axis. It then changes to a compressive stress, which reaches a maximum on the concave surface. Thus, only about 50% of the metal surface is under tension and can vary from maximum to zero, depending on the stressing system. The stress gradient through the section thickness produces changes in stresses and strains as stress-corrosion cracking penetrates the metal that are different from those in a uniaxial tension specimen. This tendency yields a significantly different stress-corrosion cracking response for the two types of stressing, as shown in Fig. 17.

Bending stress specimens experience other sources of variability in stress that are not present with direct tension stressing. Variations occur in the principal longitudinal stress across the width of the specimen as well as biaxial stresses, both of which are influenced by the design of the specimen. Thus, like constant-load stressing, optimum control of stress and more severe testing conditions are provided by uniaxial tension stressing.

Smooth Test Specimens

Statically loaded smooth test specimens for stress-corrosion cracking tests can be divided into three general categories: elastic strain specimens, plastic strain specimens, and residual stress specimens. Commonly used test specimen geometries for each of these categories are discussed below. Following these discussions, information on smooth specimen preparation is provided.

Elastic Strain Specimens

To control the surface tensile stress applied by deformation loading, strain usually is restricted to the elastic range for the test material. The magnitude of the applied stress can then be calculated from the measured strain and modulus of elasticity. In constant-load stressing, the load typically is measured directly, and the stress is calculated using the appropriate formula for the specimen configuration and the method of loading. Load cells or calibrated springs may be useful for applying and monitoring possible changes in load during the test. Commonly used types of specimens for tests under measured elastic-range stress are described below.

Bent-beam specimens can be used to test a variety of product forms. Although used principally for sheet, plate, or flat extruded sections, which conveniently provide flat specimens of rectangular cross section,

Fig. 17 Comparison of the stress-corrosion cracking response with bending versus direct tension stressing under constant load for Al-5.3Zn-3.7Mg-0.3Mn-0.1Cr T6 temper alloy sheet Source: Ref 24

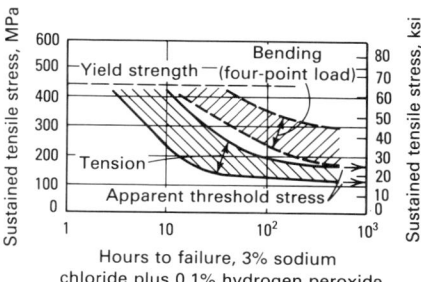

Hours to failure, 3% sodium chloride plus 0.1% hydrogen peroxide

they are also used for cast materials, rod, or for machined specimens of circular cross section. This method is applicable to specimens of any metal that are stressed to levels less than the elastic limit of the material; therefore, the applied stress can be calculated or measured accurately.

Stress calculations by this method are not applicable to plastically stressed specimens. Bent-beam specimens usually are tested under nominally constant-strain conditions, but nominally constant-load conditions can also be used. In either case, local changes in the curvature of the specimen when cracking occurs result in changes in stress and strain during crack propagation. The "test stress" is taken as the highest surface tensile stress existing at the start of the test, i.e., before the initiation of stress-corrosion cracking (Fig. 17).

Several configurations of bent-beam specimens and stressing systems are illustrated in Fig. 18 and are described in detail in ASTM Standard G 39, "Standard Practice for Preparation and Use of Bent-Beam Stress-Corrosion Test Specimens." The following stress calculations are used to determine the maximum longitudinal stress in the outer fibers of the convex surface of a specimen. When specimens are tested at elevated temperatures, the possibility of stress relaxation should be investigated. For more information on stress relaxation, see the article "Creep, Stress-Rupture, and Stress-Relaxation Testing" in this Volume. Relaxation can be estimated from creep data for the specimen holder and insulating materials.

Two-point loaded specimens can be used for materials that do not deform plastically when bent to $(L - H)/H = 0.01$. The specimens should be approximately 25- by 250-mm (1- by 10-in.) flat strips cut to appropriate lengths to produce the desired stress after bending, as shown in Fig. 18(a).

Fig. 18 Schematic specimen and holder configurations for bent-beam stressing

(a) Two-point loaded specimen. (b) Three-point loaded specimen. (c) Four-point loaded specimen. (d) Double-beam specimen. (e) Modified double-beam specimen. Formula for stressing: $\Delta d = 2fa/3Et(3L - 4a)$, where Δd is deflection (in.), f is nominal stress (psi), and E is modulus of elasticity (psi). Source: Ref 26

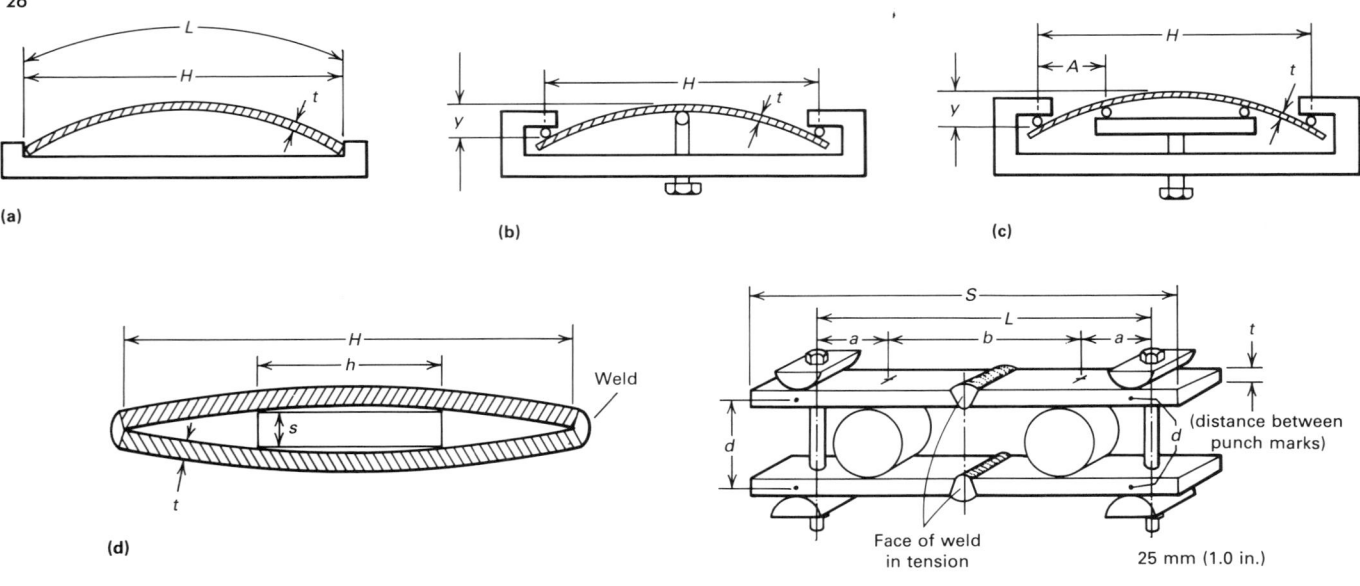

Modified double-beam specimen dimensions for various plate thicknesses

t		a		b		L		S	
mm	in.	mm	in.	mm	in.	mm	in.	mm	in.
3.2	0.125	100	4.0	50	2.0	250	10.0	305	12.0
6.4	0.25	100	4.0	50	2.0	250	10.0	305	12.0
9.5	0.375	120	4.75	90	3.5	330	13.0	380	15.0
13	0.5	120	4.75	90	3.5	330	13.0	380	15.0
19	0.75	140	5.5	150	6.0	430	17.0	480	19.0
25	1.0	150	6.0	200	8.0	510	20.0	560	22.0
38	1.5	165	6.5	305	12.0	635	25.0	685	27.0

The elastic stress in the outer fibers at the midlength of a two-point loaded specimen can be calculated from relationships derived from theoretically exact large-deflection analysis as follows:

$$\epsilon = 4(2E - K) \left[\frac{k}{2} - \frac{2E - K}{12} \left(\frac{t}{H} \right) \right] \frac{t}{H}$$

(Eq 1)

and

$$\frac{(L - H)}{H} = \left[\frac{K}{(2E - K)} \right] - 1 \qquad \text{(Eq 2)}$$

where L is the length of the specimen; H is the distance between supports (holder span); t is the thickness of the specimen; ϵ is the maximum tensile strain; $K = \int_0^{\pi/2} (1 - k^2 \sin^2 z)^{-1/2} dz$ (complete elliptic integral of the first kind); $E = \int_0^{\pi/2} (1 - k^2 \sin^2 z)^{1/2} dz$ (complete elliptic integral of the second kind); $k = \sin \theta/2$, where θ is the

maximum slope of the specimen (that is, at the end of the specimen); and z is the integration parameter (Ref 25).

Mathematical analysis establishes that Eq 1 and 2 define the relationship between the strain ϵ and $(L - H)/H$ in parameter form. The common parameter in these equations is the modulus k of the elliptic integrals. Thus, the following procedure can be used to determine the specimen length, L, that is required to produce a given maximum stress σ. To determine the strain, ϵ, divide the stress, σ, by the modulus of elasticity, E_m:

$$\epsilon = \frac{\sigma}{E_m} \qquad \text{(Eq 3)}$$

From Eq 1, the value of k corresponding to the required value of ϵ can be determined. By using appropriate values of k, Eq 2 can be evaluated for L. The deflection of the specimen can be calculated as follows:

$$\frac{y}{H} = \frac{k}{(2E - K)} \qquad \text{(Eq 4)}$$

where y is the maximum deflection. This relationship can be used to ensure that the maximum stress does not exceed the proportional limit. If it should exceed the proportional limit, the measured deflection will be greater than that calculated from Eq 4.

Specimen thickness and length and holder span should be selected to obtain a value for $(L - H)/H$ of between 0.01 and 0.50, thus keeping the error of stress within acceptable limits. A specimen thickness of about 0.8 to 1.8 mm (0.03 to 0.07 in.) and a holder span of 178 to 216 mm (7.00 to 8.50 in.) has been acceptable for high-strength steels and aluminum alloys with applied stresses ranging from about 205 MPa (30 ksi) for aluminum to 1380 MPa (200 ksi) for steel. Specimen dimensions for all bent-beam specimens can be modified to suit specific needs, but approximate dimensional proportions should be preserved.

In two-point loaded specimens, the maximum stress occurs at the midlength of the specimen and decreases to zero at specimen ends. The two-point loaded specimen is preferred over the three-point loaded specimen, because in many instances crevice corrosion of the specimen occurs at the central support of the three-point loaded specimen. Because this corrosion site is very close to the point of highest tensile stress, it may cathodically protect the specimen and prevent possible crack formation, or it may cause hydrogen embrittlement. Furthermore, the pressure of the central support at the point of highest load introduces biaxial stresses at the area of contact and can introduce tension stresses where compression stresses normally are present.

Three-point loaded specimens typically are flat strips 25 to 51 mm (1 to 2 in.) wide and 127 to 254 mm (5 to 10 in.) long. The thickness of a specimen is usually dictated by the mechanical properties of the material and the available product form. The specimen should be supported at the ends and bent by forcing a screw (equipped with a ball or knife-edge tip) against it at a point halfway between the end supports, as shown in Fig. 18(b).

The elastic stress at midspan in the outer fibers of a three-point loaded specimen can be calculated as:

$$\sigma = \frac{6Ety}{H^2} \qquad \text{(Eq 5)}$$

where σ is the maximum tensile stress, E is the modulus of elasticity, t is the thickness of the specimen, y is the maximum deflection, and H is the distance between the outer supports.

Equation 5 is based on small deflections ($y/H < 0.1$). In sheet-gage bent-beam specimens, the deflections usually are large, and thus the relationship is only approximate. To obtain more accurate stress values, a prototype specimen equipped with strain gages for calibration should be used. The prototype should have the same dimensions as the test specimens and should be stressed in the same manner. In a three-point loaded specimen, the maximum stress occurs at the midlength of the specimen and decreases linearly to zero at the outer supports.

Four-point loaded specimens are flat strips typically 25 to 51 mm (1 to 2 in.) wide and 127 to 254 mm (5 to 10 in.) long. The thickness of a specimen usually is dictated by the mechanical properties of the material and the available product form. The specimen is supported at the ends and is bent by forcing two inner supports against it, as shown in Fig. 18(c). The two inner supports are located symmetrically around the midpoint between the outer supports.

The elastic stress can be calculated for the midportion of the specimen (between the contact points of the inner support) in the outer fibers of a four-point loaded specimen as:

$$\sigma = \frac{12Ety}{(3H^2 - 4A^2)} \qquad \text{(Eq 6)}$$

where σ is the maximum tensile stress, E is the modulus of elasticity, t is the thickness of the specimen, y is the maximum deflection (between the outer supports), H is the distance between the outer supports, and A is the distance between the inner and outer supports. The dimensions frequently are chosen

so that $A = H/4$. The above equations are based on small deflections ($y/H < 0.1$).

In a four-point loaded specimen, the maximum stress occurs between the contact points of the inner supports; in this area the stress is constant. From the inner supports, the stress decreases linearly toward zero at the outer supports. The four-point loaded specimen is preferred over the three-point and two-point loaded specimens, because it provides a large area of uniform stress.

Double-beam specimens consist of two flat strips 25 to 51 mm (1 to 2 in.) wide and 127 to 254 mm (5 to 10 in.) long. The strips are bent against each other over a centrally located spacer until both ends touch. They are held in position by welding the ends together, as shown in Fig. 18(d).

The elastic stress for the midportion of the specimen (between the contact points of the spacer) in the outer fibers of a double-beam specimen can be calculated from the following:

$$\sigma = \frac{3Ets}{H^2\left[1 - \left(\dfrac{h}{H}\right)\right]\left[1 + \left(\dfrac{2h}{H}\right)\right]} \qquad \text{(Eq 7)}$$

where σ is the maximum tensile stress, E is the modulus of elasticity, t is the thickness of the specimen strips, s is the thickness of the spacer (H), and h is the length of the spacer, as shown in Fig. 18(d).

When h is chosen so that $H = 2h$, Eq 7 is simplified to:

$$\sigma = \frac{3Ets}{H^2} \qquad \text{(Eq 8)}$$

The above relationships are based on small deflections ($s/H < 0.2$).

In a double-beam specimen, the maximum stress occurs between the contact points of the spacer; in this area, the stress is constant. From contact with the spacer, the stress decreases linearly toward zero at the ends of the specimen.

A modified double-beam specimen is shown in Fig. 18(e), with suggested specimen dimensions for various thicknesses of plate and the formula for stressing specimens (Ref 26). Beam deflections required to develop the intended tension stress are calculated with the formula and are then applied by bolting the ends of the beams together. The deflections are measured with a dial gage to within ±0.0127 mm (±0.0005 in.). Thus, the error in stress application, if the beams are of homogeneous material and the cross sections are uniform, is within 2%. The precision of the deflection measurement is within 0.5%, and the error in determining the modulus of elasticity, E, is within 1%.

Fully supported specimens consist of specially bent beams (Fig. 19) designed so

that a constant moment exists from one end to the other, which produces equal stress along the length of the specimen (Ref 27, 28). The width-to-thickness ratio is less than 4 so that biaxial stresses are eliminated.

This type of specimen offers the advantage of a relatively large area of material under a uniform stress. A fully supported specimen can be used when the dimensions of the specimen are too small for other bent-beam specimens—for example, when specimens are taken in the short-transverse direction in plate. The elastic stress in the convex surface is calculated as:

$$\sigma = \frac{4Ety}{h^2} \qquad \text{(Eq 9)}$$

where h is the distance between the edges of the inner supports, and y is the deflection between the edges of the inner supports. This is a special case of Eq 6 when $A = 0$.

C-Ring Specimens. As defined by ASTM G 38, "Standard Recommended Practice for Making and Using C-Ring Stress-Corrosion Test Specimens," the C-ring is a versatile, economical specimen for quantitatively determining the susceptibility to stress-corrosion cracking of all types of alloys in a wide variety of product forms. It is particularly well suited for testing tubing and for making short-transverse tests on various product forms, as shown in Fig. 20. Sizes of C-rings may be varied over a wide range, but rings with outer diameters less than about 16 mm (⅝ in.) are not recommended because of increased difficulties in machining and decreased precision in stressing.

Circumferential stress is of principal interest in the C-ring specimen. This stress is not uniform (Ref 29), as discussed previously in the section on bent-beam specimens. It varies around the circumference of the C-ring from

Fig. 19 Bent beam designed to produce pure bending
Source: Ref 27

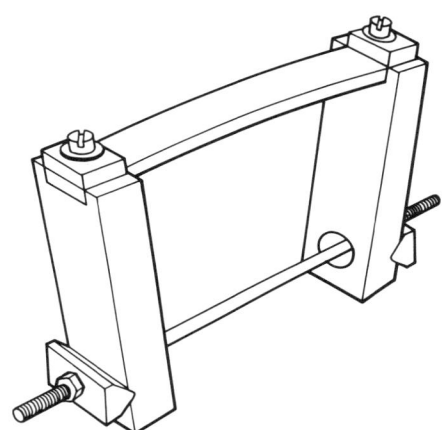

Fig. 20 Sampling procedure for testing various products with C-rings

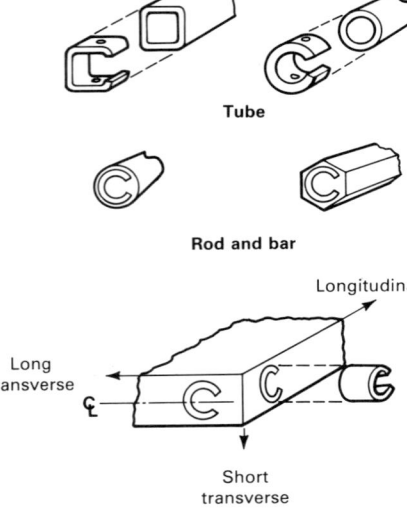

Tube

Rod and bar

Plate

Fig. 21 Methods of stressing C-rings

(a) Constant strain. (b) Constant load. (c) Constant strain. (d) Notched C-ring; a similar notch could be used on the side of (a), (b), or (c).

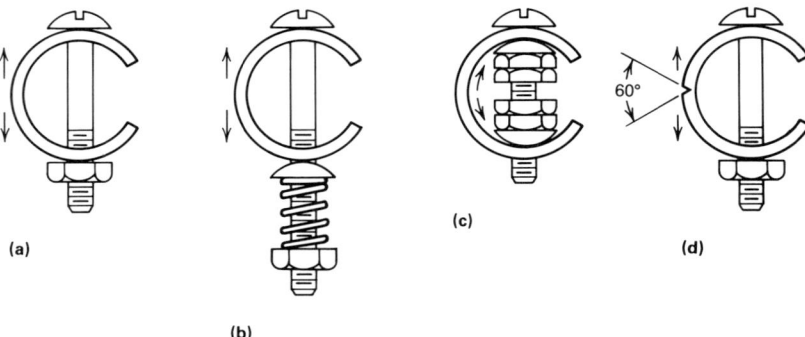

(a)

(b)

(c)

(d)

zero at each bolt hole to a maximum at the middle to the arc opposite the stressing bolt. In a notched C-ring, a triaxial stress state is present adjacent to the root of the notch (Ref 30). For all notches, the circumferential stress at the root of the notch is greater than the nominal stress and generally may be expected to be in the plastic range.

The C-ring typically is a constant-strain specimen, with tensile stress produced on the exterior of the ring by the tightening of a bolt centered on the diameter of the ring. However, an almost constant load can be developed by placing a calibrated spring on the loading bolt. C-rings can also be stressed in the reverse direction by spreading the ring and creating a tensile stress on the inside surface. These methods of stressing are illustrated in Fig. 21.

Generally, the C-ring can be stressed with high precision. The most accurate stressing procedure consists of attaching circumferential and transverse electrical strain gages to the surface stressed in tension, followed by tightening the bolt until the strain measurements indicate the desired circumferential stress. If the stresses and strains do not exceed the proportional limit, the circumferential (σ_c) and transverse (σ_t) stresses are calculated as follows:

$$\sigma_c = \left(\frac{E}{1 - \mu^2}\right)(\epsilon_c + \mu\epsilon_t) \qquad \text{(Eq 10)}$$

$$\sigma_t = \left(\frac{E}{1 - \mu^2}\right)(\epsilon_t + \mu\epsilon_t) \qquad \text{(Eq 11)}$$

where E is modulus of elasticity, μ is Poisson's ratio, ϵ_c is circumferential strain, and ϵ_t is transverse strain.

The amount of compression required on the C-ring to produce elastic straining and the degree of elastic strain can be predicted theoretically. Therefore, C-rings can be stressed by calculating the deflection required to develop a desired elastic stress. In notched specimens, a nominal stress is assumed, using a ring outer diameter measured at the root of the notch and taking into consideration the stress-concentration factor (K_t) for the specific notch.

O-ring specimens are used to develop a hoop stress in a particular part (Fig. 22)—for example, a cylindrical die forging in which a critical "end-grain" structure associated with the parting plane of the forging exists only at the surface of the forging. A relatively large surface area of metal is placed under a uniform tensile stress, and the O-ring stressing plug assembly simulates service conditions in structures containing interference-fit components. Stressed O-rings have also been used to evaluate protective treatments for the prevention of stress-corrosion cracking (Ref 31).

An O-ring is stressed by pressing it onto an oversized plug that is machined to a predetermined diameter to develop the desired stress at the outside surface of the ring. The nominal dimensions of this specimen can be varied to suit the part being tested, but certain characteristics should be observed to achieve adequate control of the stresses. The ring width should be not more than four times the wall thickness to ensure maximum uniformity of the hoop stress from the center line to the edges of the ring. The tensile stress varies through the thickness of the ring and is highest at the inside surface. The following equation can be used to calculate the interference required for stressing an O-ring:

$$I = \frac{F\,(OD)^2}{E\,(ID)} \qquad \text{(Eq 12)}$$

where I is the interference (on the diameter) between the O-ring and the plug, E is the modulus of elasticity, ID is the inside diameter, OD is the outside diameter, and F is the circumferential stress desired on the outside surface. Additional information regarding the design and stressing of O-rings is given in Ref 32.

Tension Specimens. Specimens used to determine tensile properties in air are well suited and easily adapted to stress-corrosion testing, as discussed in ASTM G 49, "Standard Practice for Preparation and Use of Direct Tension Stress-Corrosion Test Specimens." When uniaxially loaded in tension, the stress pattern is simple and uniform, and the magnitude of the applied stress can be determined accurately. Specimens can be stressed quantitatively with equipment for application of either a constant load, a constant strain, or an increasing load or strain.

This type of test is one of the most versatile methods of stress-corrosion testing because of the flexibility permitted in the type and size of test specimen, stressing procedures, and range of stress levels. It allows the simultaneous exposure of unstressed specimens (no applied load) with stressed specimens and subsequent tension testing to distinguish between the effects of true stress-corrosion cracking and mechanical overload.

A wide range of test specimen sizes can be used, depending primarily on the dimensions of the product to be tested. Stress-corrosion test results can be influenced significantly by the cross section of the test specimen. Although large specimens may be more representative of most structures, they often cannot be machined from the product forms being evaluated. They also present more difficulties in stressing and handling in laboratory testing.

Smaller cross-sectional specimens are widely used. They have a greater sensitivity to the initiation of stress-corrosion cracking, usually yield test results rapidly, and permit

Fig. 22 O-ring stress-corrosion cracking test specimens and stressing plug

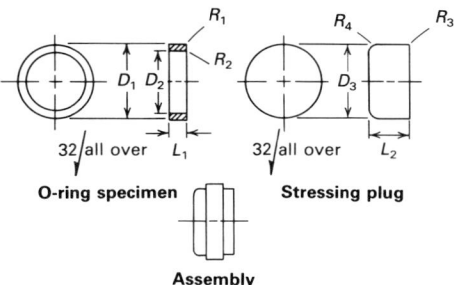

Fig. 23 Spring-loaded fixture used to stress 3.2-mm (0.125-in.) thick sheet tensile specimens in direct tension
Source: Ref 26

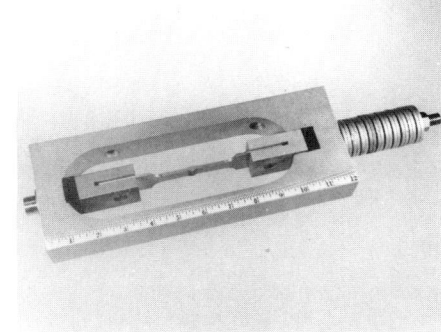

Fig. 24 Ring-stressed tension specimen for field testing
Source: Ref 33

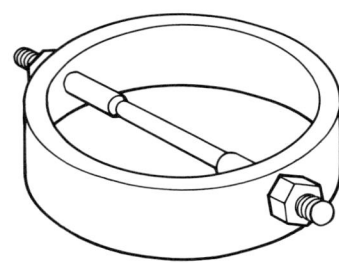

Fig. 25(a) Constant-strain stress-corrosion testing frame and specimen loading device
Exploded view (left) showing the 3.2-mm (0.125-in.) diam tension specimen and various parts of the stressing frame. Final stressed assembly (right). Source: Ref 34

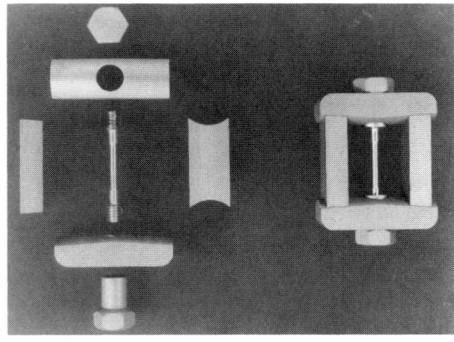

Fig. 25(b) Synchronous loading device used to stress specimens
The specimen is loaded to a prescribed strain value determined from a clip-on gage. The applied stress is given by the product of the strain and the material elastic modulus. A stressed assembly and one assembled "finger-tight" ready for stressing are shown.

greater convenience in testing. However, the smaller specimens are more difficult to machine, and test results are more likely to be influenced by extraneous stress concentrations resulting from nonaxial loading, corrosion pits, etc. Therefore, use of specimens less than about 10 mm (0.4 in.) in gage length or 3 mm (0.12 in.) in diameter is not recommended.

Tension specimens containing machined notches can be used to study stress-corrosion cracking and hydrogen embrittlement. The presence of a notch induces a triaxial stress state at the root of the notch, in which the actual stress will be greater by a concentration factor that is dependent on the notch geometry. Advantages of such specimens include the localization of cracking to the notch region and acceleration of failure. However, unless directly related to practical service conditions, results may not be relevant.

Tension specimens can be subjected to a wide range of stress levels associated with either elastic or plastic strain. Because the stress system is intended to be essentially uniaxial (except in the case of notched specimens), great care must be exercised in the construction of stressing frames to prevent or minimize bending stresses.

The simplest method of providing a constant load consists of a dead weight hung on one end of the specimen. This method is particularly useful for wire specimens. For specimens of larger cross section, however, lever systems such as those used in creep testing machines are more practical. The primary advantage of any dead-weight loading device is the constancy of the applied load.

A constant-load system can be modified by the use of a calibrated spring, such as that shown in Fig. 23 (Ref 26). The proving ring, as used in the calibration of tension testing machines, has also been adapted to stress-corrosion testing to provide a simple, compact, easily operated device to apply axial load (Fig. 24). The load is applied by tightening a nut on one of the bolts and is determined by carefully measuring the change in ring diameter.

Constant-strain stress-corrosion tests are performed in low-compliance tension testing machines. The specimen is loaded to the required stress level, and the moving beam is then locked in position. Other laboratory stressing frames have been used, generally for testing specimens of smaller cross section. Figure 25(a) shows an exploded view of such a stressing frame, and Fig. 25(b) illustrates a special loading device developed to ensure axial loading with minimal torsion and bending of the specimen.

For stressing frames that do not contain any mechanism for the measurement of load, the stress level can be determined from measurement of the strain. However, only when the intended stress is below the elastic limit of the test material is the average linear stress (σ) proportional to the average linear strain (e), $\sigma/e = E$, where E is the modulus of elasticity.

When tests are conducted at elevated temperatures with constant-strain-loaded specimens, consideration should be given to the possibility of stress relaxation. When stress relaxation or creep occurs in the test specimen, some of the elastic strain is converted to plastic strain and the nominal applied test stress is reduced.

Even though eccentricity in loading can be minimized to levels acceptable for tension testing machines, tensile stress around the circumference of the test specimen varies to some extent. Several factors may introduce bending moments on specimens, such as longitudinal curvature, misalignment of threads on threaded-end round specimens, and the corners of sheet-type specimens. These factors have a greater effect on specimens with smaller cross sections.

When stress-corrosion cracking occurs, it generally results in complete fracture of the specimen, which is easy to detect. However, when testing relatively ductile materials at stress levels close to the threshold of susceptibility, fracture may not occur during the period of exposure. The presence of stress-corrosion cracking in such instances must be determined by mechanical tests or by metallographic examination, as discussed previously.

To study trends in stress-corrosion cracking susceptibility, such as in alloy development research, it is often necessary to detect small differences in susceptibility. For this purpose, it is advantageous to employ replicate sets of specimens stressed at several levels, including zero applied stress. The sets are then removed for tension tests after appropriate periods of exposure.

Use of this procedure with samples of 7075 aluminum alloy given thermal treatments to decrease susceptibility to stress-corrosion cracking is illustrated in Fig. 26. These data are better suited to statistical comparisons than the typical pass-fail data (Ref 14).

The tuning fork specimen is a special-purpose specimen with numerous modifications (Fig. 27). In Europe, the metal is strained into the plastic range, and stresses and strains usually are not measured (Ref 28, 35). In the United States, however, these specimens have been used with measured strains in the elastic range. Specimens of the type shown in Fig. 27(b) are convenient when a small self-contained specimen is required that will afford some insight into the applied stresses. Such a specimen is particularly well suited for testing thin plate material in the longitudinal or long-transverse direction while keeping the original mill-finished surface intact.

Tuning fork specimens are stressed by closing the specimen tines and restraining them in the closed position with a bolt placed at the tine ends. The amount of closure is determined from the following empirical relationship derived from the data obtained with strain gages placed at the base of the tines on calibration specimens (Ref 33):

$$S = A\Delta t \qquad \text{(Eq 13)}$$

where S is the maximum tension stress in the outer fiber of either tine, A is the calibration constant, Δ is the total amount of closure at the tine ends, and t is the thickness of the tines.

The stress on tuning forks with straight tines is maximum in a small area at the base of the tines. In tuning forks with tapered tines, the maximum stress extends uniformly along the tapered section. Tuning forks must be given the same consideration with regard to biaxial stresses as other flexurally loaded specimens.

The "mini" tuning fork shown in Fig. 27(c) was devised (Ref 36) to conduct short-transverse tests on sections that are too thin for tensile specimens or C-rings to be obtained. As with tuning fork specimens, the relationship between strain on the grooved surface and the deflection at the ends of the legs can be determined through the use of strain gages.

Plastic Strain Specimens

Much accelerated stress-corrosion cracking testing is performed with plastically deformed specimens, because they are simple and economical to manufacture and use.

These specimens are convenient for multiple replication tests of self-stressed (fixed-deflection) specimens in all environments. Because they usually contain large amounts of elastic and plastic strain, they provide one of the most severe tests available for smooth stress-corrosion test specimens.

Generally, the stress conditions are not known precisely. However, the anticipated high level of stress can be obtained consistently only if precautions dictated by the particular specimen are observed. During forming of specimens, varying amounts of cold work may be introduced; this deformation may influence the stress-corrosion cracking differently compared to material behavior in the original condition.

Tests of this type are primarily used as screening tests to detect large differences between the stress-corrosion cracking resistance of (1) one alloy in several environments, (2) one alloy in several metallurgical conditions in a given environment, and (3) different alloys in the same environment. Although these tests are sometimes claimed to be too severe and therefore unsuitable for many applications, the stress conditions are nevertheless representative of the high locked-in fabrication and assembly stresses frequently responsible for stress-corrosion cracking in service.

U-bend specimens are rectangular strip specimens that are bent approximately 180° around a predetermined radius and maintained in this plastically (and elastically) deformed condition during the stress-corrosion test. Standardized test methods for this type of specimen are discussed in ASTM G 30, "Standard Recommended Practice for

Fig. 26 Mean breaking stress versus exposure times for short-transverse 3.2-mm (0.125-in.) diam tension specimens exposed to 3.5% sodium chloride solution by alternate immersion test (ASTM G 44) at various exposure stress levels
Each point represents an average of five specimens. Source: Ref 14

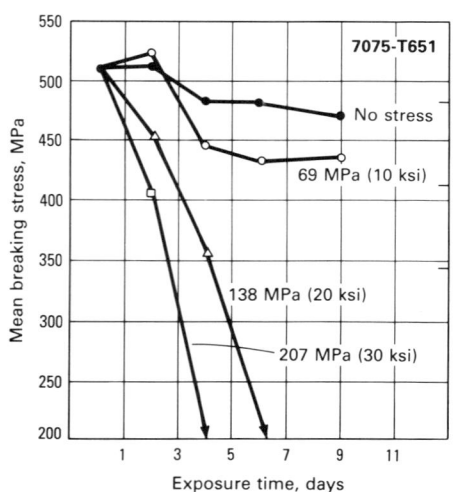
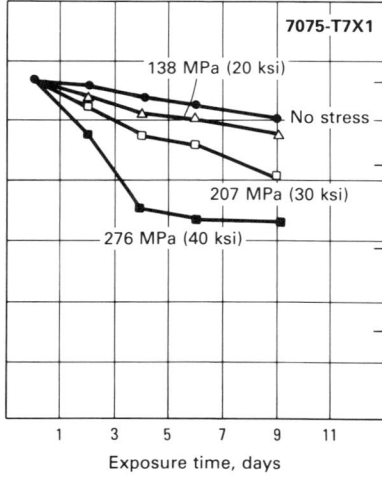
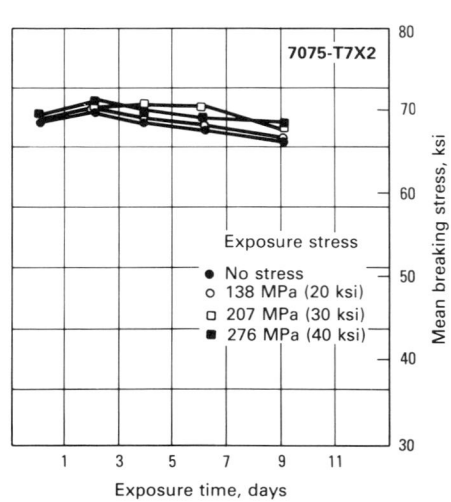

Fig. 27 Typical tuning fork specimens

(a) Source: Ref 28. (b) Source: Ref 33. (c) Source: Ref 36

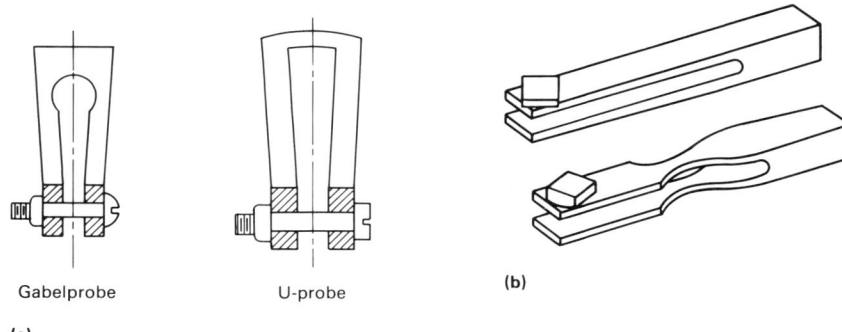

Gabelprobe   U-probe   **(b)**

(a)

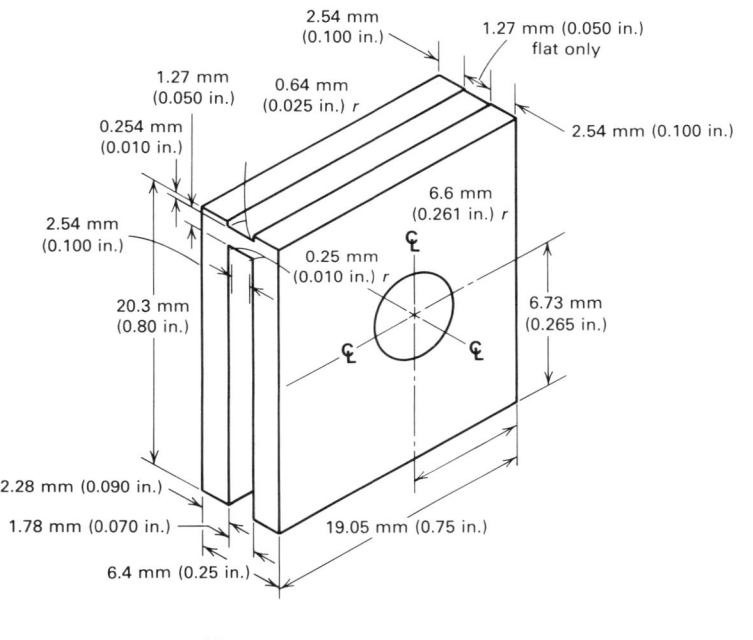

(c)

Making and Using U-Bend Stress-Corrosion Test Specimens.'' Bends slightly less than or greater than 180° are also used, but the term ''U-bend'' generally is applied to test specimens bent past their elastic limit. Typical U-bend configurations showing several different methods of maintaining the applied stress are shown in Fig. 28. Also shown are typical dimensions that have been used successfully for a wide range of materials.

U-bend specimens can be used for all materials sufficiently ductile to be formed into a U-configuration without cracking. A U-bend specimen is most easily made from strip or sheet, but specimens can be machined from plate, bar, wire, castings, and weldments. Of primary interest in U-bend specimens is circumferential stress, which is not uniform, as discussed previously in the section on bent-beam specimens. Stress distribution in the U-

bend specimen is discussed in detail in Ref 37.

A good approximation of applied strain can be obtained by:

$$\epsilon = \frac{T}{2R} \text{ when } T < R \qquad \text{(Eq 14)}$$

where T is the specimen thickness, and R is the radius of curvature at the point of interest. To determine the stress, knowledge of the stress-strain curve is necessary. When a U-bend specimen is stressed, the material in the outer fibers of the bend is strained into the plastic portion of the true stress/true strain curve, such as in section AB in Fig. 29(a). Several other stress-strain relationships that can exist in the outer fibers of a U-bend test specimen are shown in Fig. 29(b) through (e). The actual relationship obtained depends on the method of stressing used.

Stressing usually is achieved by a one- or two-stage operation. Single-stage stressing is accomplished by bending the specimen into shape and maintaining it in that shape without allowing relaxation of the tensile elastic strain. Typical stressing sequences are shown in Fig. 30. The method illustrated in Fig. 30(a) generally is performed in a tension testing machine and is often the most suitable method for stressing U-bends that are difficult to form manually because of large-thickness and/or high-strength specimen material.

The methods illustrated in Fig. 30(b) and (c) generally are suitable for thin and/or low-strength material, but generally are inferior to the method illustrated in Fig. 30(a). The method illustrated in Fig. 30(b) results in a more complex strain system in the outer surface and can cause scratching. The technique illustrated in Fig. 30(c) does not enable control of the bend radius.

The two types of stress conditions that can be obtained by single-stage stressing are defined by point X in Fig. 29(b) and (c). In Fig. 29(c), some elastic strain relaxation has occurred by allowing the U-bend legs to spring back slightly at the end of the stressing sequence.

Two-stage stressing involves first forming the approximate U-shape and allowing the elastic strain to completely relax before the second stage of applying the test stress (Fig. 31a-d). The type of equipment shown in Fig. 30(a) and (b) can also be used to preform the U-shape. The applied test strain can be a percentage (from 0 to 100%) of the tensile elastic strain that occurred during preforming (Fig. 29d) or can involve additional plastic strain (Fig. 29e).

The slope MN of the curve shown in Fig. 29(d) is steep. Therefore, it is often difficult to apply reproducibly a constant percentage of the total elastic prestrain, and the specimen surface may remain under compressive stress. Thus, because it results in a more severe test (i.e., higher applied stress), the stress conditions in Fig. 29(b) and (e) are recommended. Hence, the final applied strain prior to testing consists of plastic and elastic strain. To achieve the conditions illustrated in Fig. 29(b) and (e), prestraining should be less than the final test strain, and ''springback'' of the U-bend legs after achieving the final plastic strain should be avoided.

Residual Stress Specimens

Most industrial stress-corrosion cracking problems are associated with residual stresses developed in the metal during processes such as heat treatment, fabrication, and welding. Thus, residual stress specimens simulating anticipated residual stress conditions in service are useful for predicting the

Fig. 28 Typical U-bend stress-corrosion specimens

(a) Various methods of stressing U-bends. (b) Typical U-bend specimen dimensions

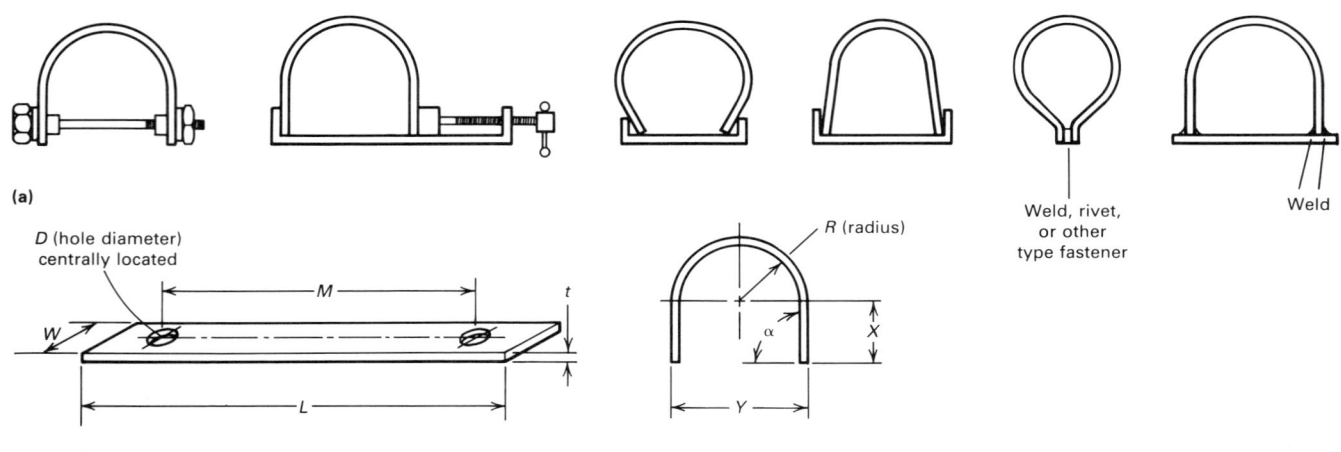

Alternative size	L		M		W		t		D		X		Y		R	
	mm	in.	mm	in.	mm	in.	mm	in.	mm	in.	mm	in.	mm	in.	mm	in.
A	80	3.2	50	2.0	20	0.8	2.5	0.098	10	0.4	32	1.26	14	0.55	5	0.2
B	100	4.0	90	3.5	9	0.35	3.0	0.12	7	0.28	25	0.98	38	1.50	16	0.6
C	120	4.7	90	3.5	20	0.8	1.5	0.06	8	0.31	35	1.4	35	1.4	16	0.6
D	130	5.1	100	4.0	15	0.6	3.0	0.12	6	0.24	45	1.77	32	1.26	13	0.51
E	150	5.9	140	5.5	15	0.6	0.8	0.03	3	0.12	61	2.40	20	0.8	9	0.35
F	310	12.2	250	9.8	25	0.98	13.0	0.51	13	0.51	105	4.13	90	3.5	32	1.26
G	510	20.1	460	18.1	25	0.98	6.5	0.26	13	0.51	136	5.35	165	6.5	76	3.0

Note: $\alpha = 1.57$ rad

stress-corrosion performance of some alloys and tempers in particular structures and in specific environments.

Plastic Deformation Specimens. Residual stresses resulting from fabricating operations such as forming, straightening, and swaging that involve localized plastic deformation at room temperature can exceed the elastic limit of the material. Examples of specimens of this type that have been used are shown in Fig. 32 and 33. Other specimen types used include panels with sheared edges, punched holes, or stamped identification numbers and specimens that show evidence of other practical fabricating operations.

Weld Specimens. Residual stresses developed in and adjacent to butt welds are frequently a source of stress-corrosion cracking in service. Longitudinal stresses in the vicinity of a single weld are unlikely to be as large as stresses developed in plastically deformed components, because stress in the weldment is limited by the lower yield strength of the hot metal that shrinks as it cools. High stresses can be built up, however, when two or more weldments are welded together into a more complex structure.

Test specimens containing residual welding stresses are shown in Fig. 34. In fillet welds, residual tensile stress transverse to the weld can be critical, as indicated in Fig. 34(a) for a situation in which the tension stress acts in the short-transverse direction in an aluminum-zinc-magnesium alloy plate.

Surface Preparation of Smooth Specimens

The pronounced effect of surface conditions on the time required to initiate stress-corrosion cracking in test specimens is well known (Ref 22). Unless the as-fabricated surface is being studied, the final surface preparation generally preferred is a mechanical process followed by degreasing. However, to remove heat treating films or thin layers of surface metal that may have become distorted during machining, chemical etches or electrochemical polishes may be used.

Care should be exercised to select an etchant that will not selectively attack constituents or phases in the metal and that will not deposit undesirable residues on the surface. Etching or pickling should not be used with alloys that are susceptible to hydrogen embrittlement.

When machining specimens, precautions should be taken to avoid overheating, plastic deformation, or the development of residual stress in the metal surface. Machining should be performed in stages so that the final cut leaves the principal surface with a clean finish of 0.7 μm (30 μin.) rms or more. The required machining sequences, types of tools, and feed rate depend on the alloy and temper of the test piece. Lapping, mechanical polishing, and similar operations that produce flow of the metal should be avoided.

Precracked Test Specimens

Precracked test specimens and related fracture mechanics analyses have been used extensively to study stress-corrosion cracking (Ref 38). Almost all standard plane-strain fracture toughness test specimens can be adapted to stress-corrosion cracking testing. These standard configurations should be used to ensure valid fracture analyses. Comprehensive discussions on stress-corrosion testing with precracked specimens can be found in Ref 1, 39, and 40.

Typical precracked specimens are illustrated schematically in Fig. 35 and are classified with respect to loading methods and the relationship with the stress-intensity factor as stress-corrosion cracking propagates. Proportional dimensions and tolerances per ASTM E 399 (Ref 41) for commonly used specimens are given in Fig. 36. Minor mod-

Fig. 29 True stress/true strain relationships for stressed U-bends
See text for discussion of (a) to (e).

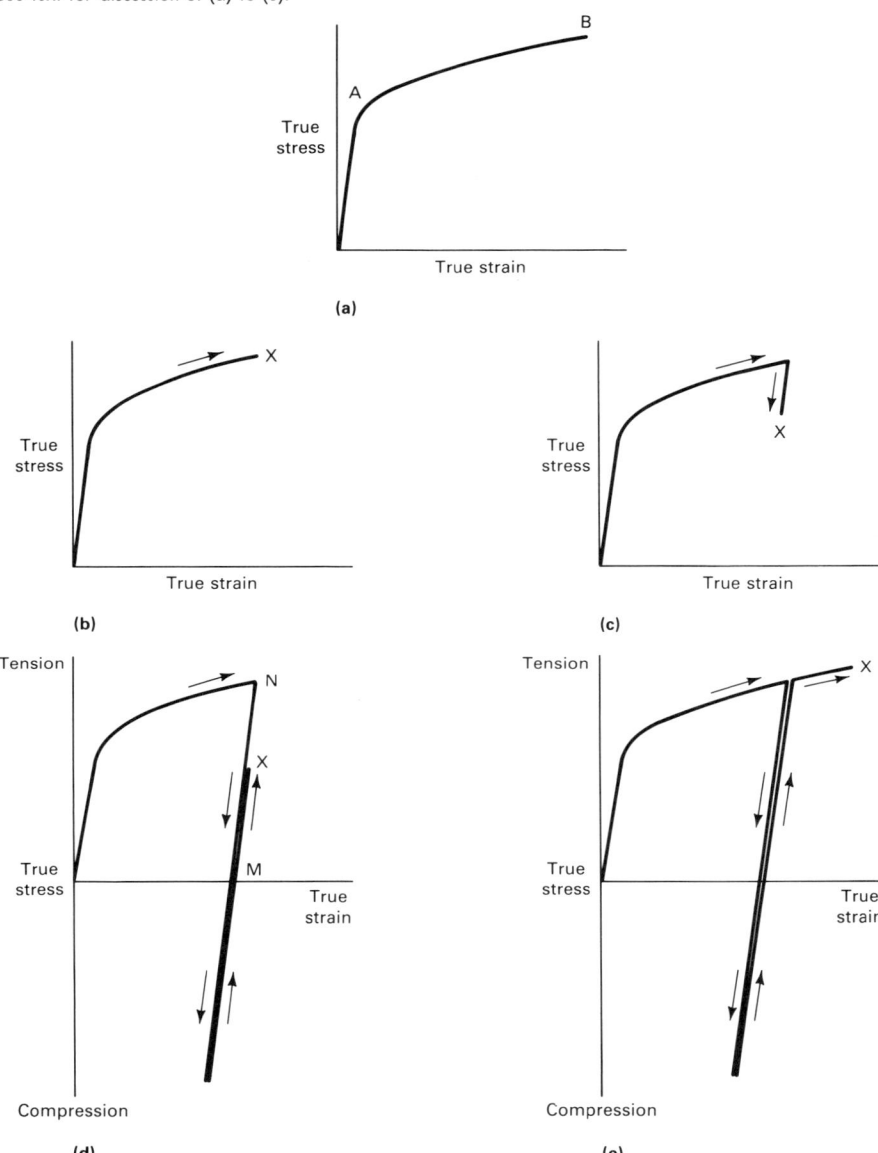

$$K = \frac{6M}{BW^2} \left(\frac{2W}{\pi a} \tan \frac{\pi a}{2W} \right)^{1/2}$$

$$\times \left[\frac{0.923 + 0.199 \left(1 - \sin \dfrac{\pi a}{2W} \right)^4}{\cos \dfrac{\pi a}{2W}} \right]$$

(Eq 15)

$$\frac{EBW(2V_o)}{M} = e^x$$

(Eq 16)

where e is the base of natural logarithm (2.7188) and $x = [0.1426 + 11.92(a/W) - 17.42(a/W)^2 + 15.84(a/W)^3 - 2.235(a/W)^4]$,

$$\frac{EBW\, V_{LL}}{M} = e^y$$

(Eq 17)

where $y = [6.188 + 12.98(a/W) - 41.19(a/W)^2 + 54.98(a/W)^3 - 22.28(a/W)^4]$. In Eq 15 to 17, M is the applied bending moment, B is the thickness of the specimen (face grooves when present may be accounted for by replacing B with $\sqrt{BB_n}$, where B_n is the net thickness at the base of the face grooves), W is the depth of the specimen, a is the depth of the notch plus crack, E is the modulus of elasticity, $2V_o$ is the total crack mouth opening displacement at the top face of the specimen, and V_{LL} is the total crack mouth opening displacement measured at the point of load application, which will vary depending on the load arm length.

Equation 15 is an expression for the stress intensity of a rectangular beam in pure bending and is valid over a wide range of a/W values. It applies to Mode I loading only, however, and the usual tests include a Mode II component from resulting shear stresses.

Equations 16 and 17 were determined by fitting experimental compliance data for cantilever bend specimens with a polynomial equation expressing the natural log of the normalized compliance as a function of a/W. These experimental values are in excellent agreement with those determined from Eq 15 for pure bending, even though the stress state at the crack tip will differ for cantilever bending. It has been suggested that analyses using pure bending expressions related to compliance measurement are suitable for testing with the cantilever bend configuration (Ref 43).

Crack growth measurements can be made with clip gage readings in conjunction with the crack opening displacement calibrations given above or by any other method that can be verified within ±0.127 mm (±0.005 in.). Examples of various methods are given in Ref 43 and 45.

ifications to accommodate different loading arrangements and to facilitate mechanical precracking can be made to these configurations without invalidating the plane-strain constraints on the specimens. Figure 37 illustrates alternate chevron-notch and face-groove designs.

Standards for stress-corrosion cracking tests using precracked specimens have not yet been developed, although recommended test procedures have been published for certain uses (Ref 42). Typical stress-intensity and compliance calibration relationships and guidelines for testing the specimens illustrated in Fig. 36 are discussed below. Stan-

dard names for these specimens and methods of loading per ASTM E 616, "Standard Terminology Relating to Fracture Testing," are used in Fig. 36 and in paragraph headings for the following discussion. Reference is also made to familiar names used in the literature, which may appear elsewhere in this article.

Cantilever bend specimens (Fig. 36a), which are also referred to as single-edge-notched cantilever bend specimens, have been used in constant-load tests (K-increasing) for characterizing high-strength steels and titanium alloys (Ref 42). The following equations are recommended (Ref 43, 44):

Fig. 30 Single-stage method of stressing U-bend specimens

See text for discussion of (a) to (c).

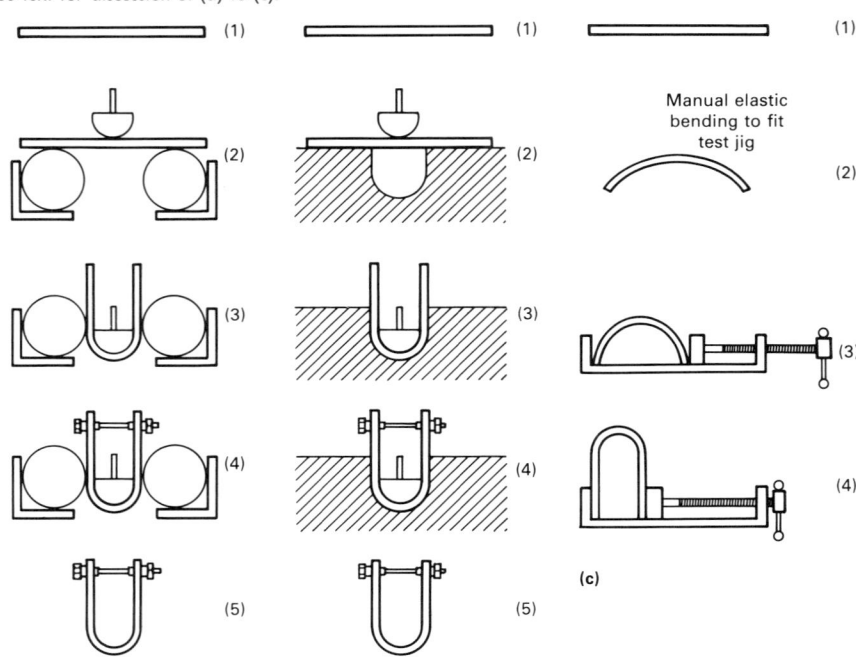

$$K_{Io} = \frac{P\left(2 + \dfrac{a_o}{W}\right)}{B\sqrt{W}\left(1 - \dfrac{a_o}{W}\right)^{3/2}}$$

$$\times \left[1.308 + 5.278\left(\frac{a_o}{W}\right) - 19.67\left(\frac{a_o}{W}\right)^2 \right.$$

$$\left. + 24.57\left(\frac{a_o}{W}\right)^3 - 10.27\left(\frac{a_o}{W}\right)^4 \right]$$

$$(Eq\ 18)$$

$$2V_o = \frac{P}{EB} \times e^x \qquad (Eq\ 19)$$

where $x = [1.830 + 4.307(a/W) + 5.871 (a_o/W)^2 - 17.53(a_o/W)^3 + 14.57(a_o/W)^4]$

$$2V_{LL} = 2V_o \times \frac{e^y}{e^x} \qquad (Eq\ 20)$$

where $y = [1.623 + 3.352(a_o/W) + 8.205 (a_o/W)^2 - 19.59(a_o/W)^3 + 15.23(a_o/W)^4]$

Modified compact specimens (K-decreasing or K-increasing), as shown in Fig. 36(b), are frequently referred to as 1T-WOL (wedge-opening loaded) or modified WOL specimens. Although most frequently used with constant-displacement (bolt) loading (Ref 42, 46), these specimens have also been used with constant load (Ref 14, 47, 48). The specimen configuration shown in Fig. 36(b) is similar to that adopted by the Navy (Ref 42), except that it does not incorporate face grooves.

The following equations can be used to calculate stress-intensity levels and normalized crack opening displacements for fatigue precracking, for initiation of stress-corrosion testing, and for subsequent intervals during the test. These equations are based on boundary colocation values determined for this type of specimen configuration with face grooves and bolt loading (threaded bolt against a rigid loading tup) (Ref 43). The polynomial regression equation agrees with experimentally determined colocation values within 1% for $0.2 \leq a/W \leq 0.95$.

$$K_{Ii} = \frac{E(2V_{LL})\left(2 + \dfrac{a_i}{W}\right)}{\sqrt{W}\left(1 - \dfrac{a_i}{W}\right)^{3/2}}$$

$$\times \left\{ \left[1.308 + 5.278\left(\frac{a_i}{W}\right) \right.\right.$$

$$\left. - 19.67\left(\frac{a_i}{W}\right)^2 + 24.57\left(\frac{a_i}{W}\right)^3 \right.$$

$$\left.\left. - 10.27\left(\frac{a_i}{W}\right)^4 \right] \div e^z \right\} \qquad (Eq\ 21)$$

Fig. 31 Two-stage method of stressing U-bend specimens

See text for discussion of (a) to (d).

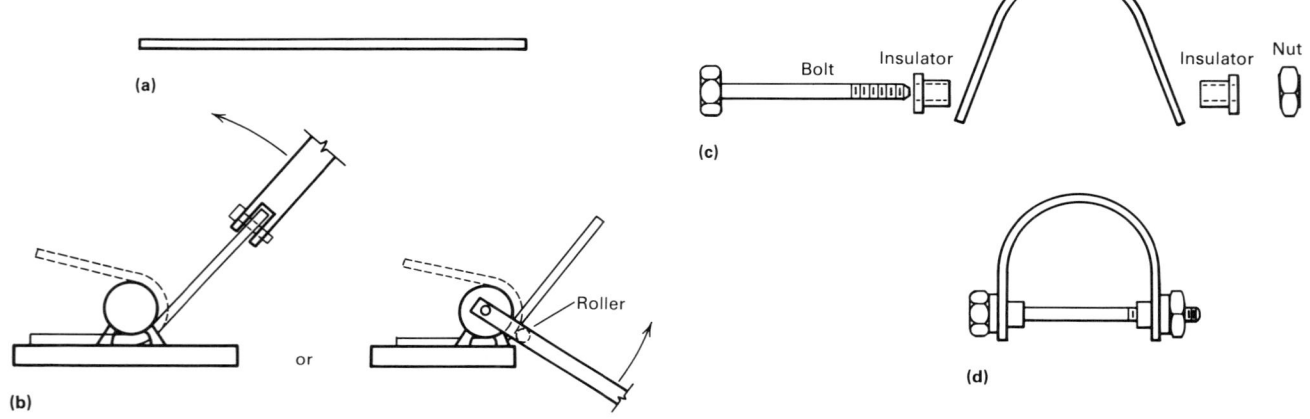

Fig. 32 Stress-corrosion test specimens containing residual stresses from cold working

Depicted are 12.7-mm (0.5-in.) diam stainless steel tubular specimens after stress-corrosion testing. (a) and (b) Annealed tubing that was cold worked before testing. (c) Cold worked tubing tested in the as-received condition. Source: Ref 33

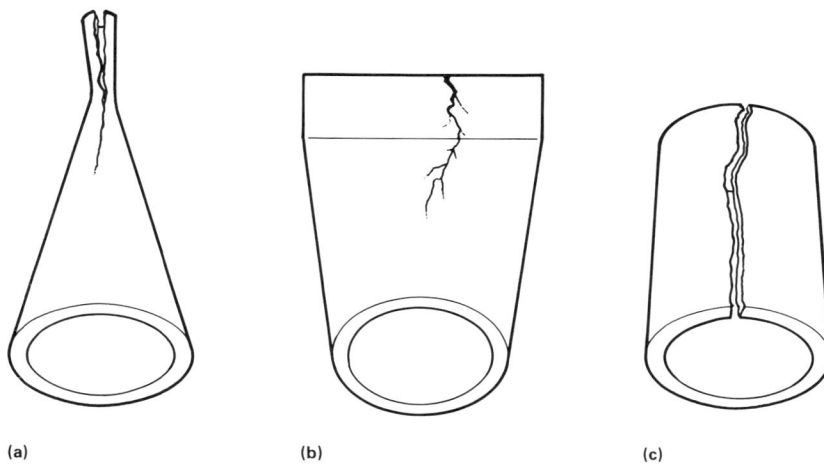

(a)   (b)   (c)

Fig. 34 Stress-corrosion test specimen containing residual stresses from welding

(a) Sandwich specimen simulating rigid structure. Note stress-corrosion cracking in edge of center plate. Source: Ref 26. (b) Cracked ring-welded specimen. Source: Ref 33

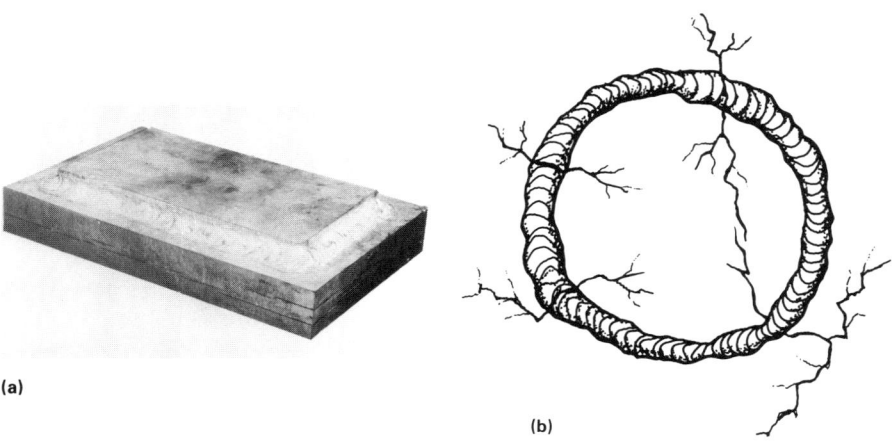

(a)

(b)

Fig. 33 Stress-corrosion test specimens containing residual stresses from plastic deformation

(a) Cracked cup specimen (Ericksen impression). Source: Ref 33. (b) Joggled extrusion containing stress-corrosion cracking in the plastically deformed region. Source: Ref 24

(a)

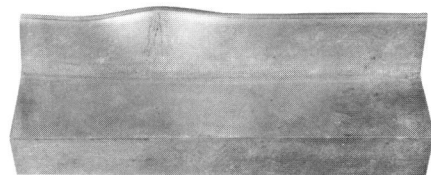

(b)

where $z = [1.623 + 3.352(a_i/W) + 8.205(a_i/W)^2 - 19.59(a_i/W)^3 + 15.23(a_i/W)^4]$. In Eq 18 to 21, K_{Io} is the desired starting stress intensity, a_o is the starting crack length, P is the load calculated to develop K_{Io} with measured a_o, W is the net width of the specimen measured from the load line, K_{Ii} is the stress intensity after time interval i, and a_i is the crack length after time interval i, and $2V_{LL}$ is the total crack mouth opening displacement at the load line. All other quantities are as defined previously.

Double-beam specimens (K-decreasing or K-increasing), which are also referred to as double-cantilever beam specimens, are similar to modified compact specimens, but because of their greater width or length, they are well suited for studying stress-corrosion cracking growth rates over a greater range of K_I values. The smaller height of these specimens (Fig. 36c) allows more versatility in performing short-transverse tests from moderate thicknesses of material. Like compact specimens, double-beam specimens generally are used with constant-displacement (bolt) loading for convenience, but also may be used with constant loading.

Bolt-loaded specimens used with a test procedure similar to that described in Ref 49 have been used extensively for short-transverse tests of aluminum alloy products (Ref 48, 50, 51).

The following equations are recommended for general use with double-beam specimens:

$$K_{Io} = \frac{P\sqrt{12}}{B\sqrt{H}}$$

$$\times \left[\frac{a_o}{H} + 0.6728 + 0.0377\left(\frac{H}{a_o}\right)^2\right]$$

(Eq 22)

$$2V_o = \frac{4K_{Io}\sqrt{H}}{\sqrt{3}\,E}\left(\frac{a_o}{H} + 0.6728\right)^2$$

$$\times \left[1 + 1.5\left(\frac{C_o}{a_o}\right) - 1.15\left(\frac{C_o}{a_o}\right)^2\right]$$

(Eq 23)

$$2V_{LL} = \frac{2V_o}{1 + 1.5\left(\frac{C_o}{a_o}\right) - 1.15\left(\frac{C_o}{a_o}\right)^2}$$

(Eq 24)

$$K_i = \frac{\sqrt{3}\, E\, (2V_{LL})}{4\sqrt{H} \left(\dfrac{a_i}{H} + 0.6728 \right)^2} \qquad \text{(Eq 25)}$$

Equation 22 is an expression reported in Ref 52. The simplified Eq 25 provides more versatility with high accuracy for a wider range of specimen configurations and K values (crack growth) than equations previously published (Ref 49).

Two early K_I calibrations based on stress analysis (Ref 53) and compliance (Ref 54) are illustrated in Fig. 38 and are in excellent agreement. The shape of these curves can also be used as a design guide for preparing specimens. If the test must be completed in

Fig. 35 Classification of precracked specimens for stress-corrosion testing
Asterisks denote common configurations. Source: Ref 40

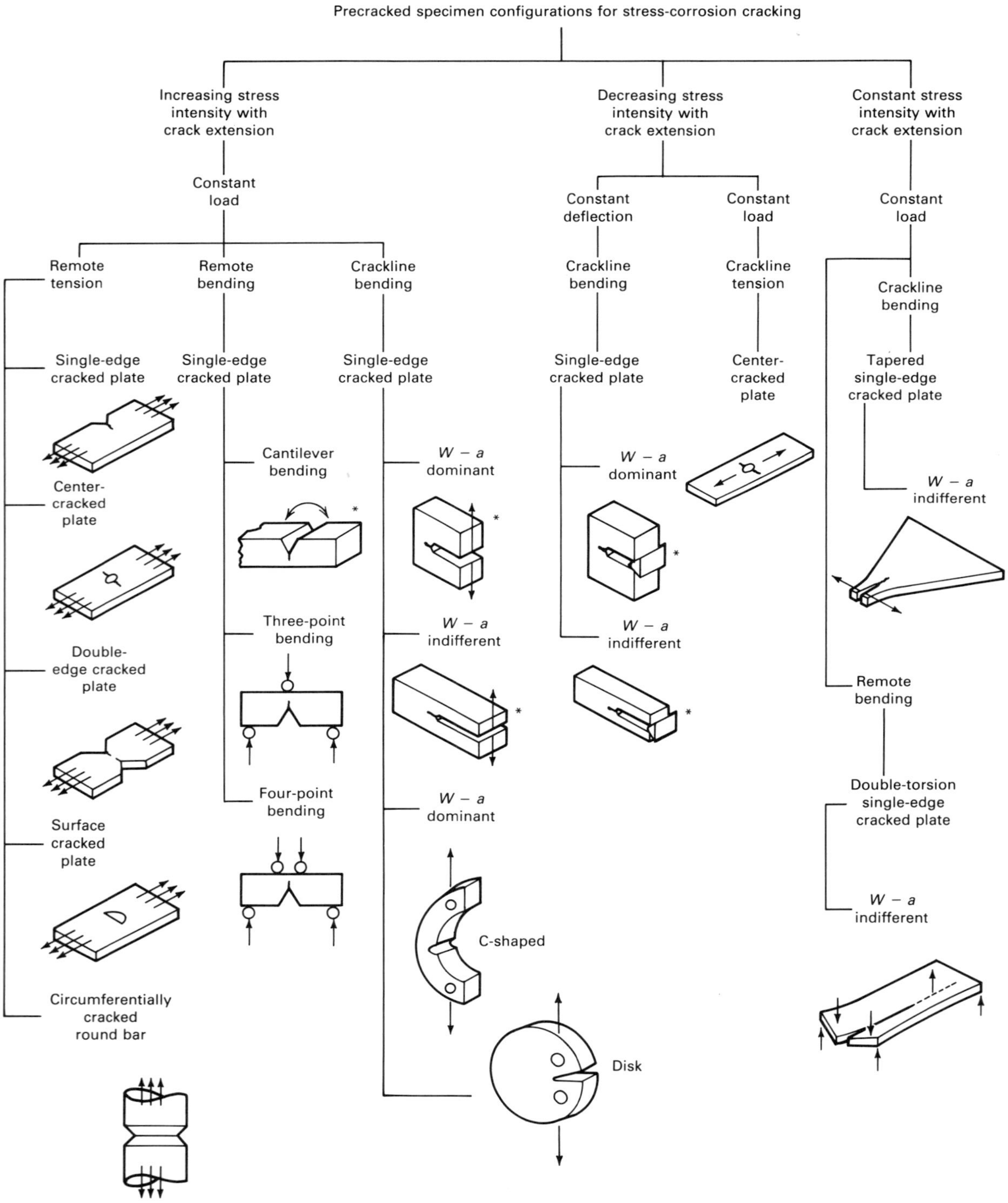

the shortest possible time, a_o should be short to capitalize on the fact that the rate of decrease of K_I with crack extension is maximum for short cracks. However, if maximum accuracy is desired, a longer crack (effective notch length M in Fig. 36c) should be chosen so that errors in crack length measurement do not cause significant errors in K_I.

Although in early work with aluminum alloys (Ref 49, 50) a relatively short effective notch length was used ($a_o/H \simeq 0.9$), deeper notches have been used recently ($a_o/H \simeq 1.2$ to 2.2), all with a $2H$ value of 25.4 mm (1.0 in.) (Ref 14, 48, 50, 51). The recommended starting a/H value shown in Fig. 36(c) is about 2 to 2.2, depending on the length of the precrack. Limited tests of a smaller beam height of $2H = 12.7$ mm (0.5 in.) have shown little effect on the amount and rate of crack growth in plate of 7075 aluminum alloy (Ref 55); however, additional study is needed in this area.

Constant K_I specimens are well suited for studying the mechanisms of stress-corrosion cracking, because the stress intensity, K_I, is not dependent on crack length and can be neglected in kinetic studies. Other attractive features are the relatively simple expressions for stress intensity and compliance and the apparent retention of plane-strain conditions in thin plate and sheet specimens. The cost of specimen preparation and instrumentation, however, prohibits its use for extensive stress-corrosion cracking characterizations.

Reference 40 provides equations for analysis of two types of constant K_I specimens—the tapered double-beam specimen and the double-torsion-loaded single-edge-cracked specimen. A recent evaluation of the double-torsion method (Ref 56) used aluminum-zinc-magnesium alloy sheet 3.2 mm (0.125 in.) thick. Using the double-torsion specimen, $V\text{-}K$ curves were produced for 7075-T651 sheet with conventional two-stage growth and plateau velocities that were slightly higher than those for conventional double-cantilever beam tests of plate.

Other precracked specimen configurations, such as those shown in Fig. 35, can be used for special testing conditions. Information on the preparation and use of these specimens and the related fracture mechanics equations are given in Ref 40 and 57 to 59.

Preparation of Precracked Specimens

When using precracked stress-corrosion test specimens, the experimentalist must consider the dimensional (size) requirements of the specimen, its crack configuration and orientation, and machining and precracking of the specimen. These considerations are discussed below. Additional guidelines and recommendations on specimen preparation in conjunction with fracture toughness testing are given in Ref 40, 41, and 57 to 59.

Dimensional Requirements. A basic requirement of all precracked specimen configurations is that the dimensions be sufficient to maintain predominantly triaxial stress (plane-strain) conditions, in which plastic deformation is limited to a very small region in the vicinity of the crack tip. Experience with fracture toughness testing has

Fig. 36(a) Proportional dimensions and tolerances for cantilever bend test specimens

Width = W; thickness (B) = 0.5W; half loading span (L) = 2W; notch width (N) = 0.065W maximum if $W > 25$ mm (1.0 in.); $N = 1.5$ mm (0.06 in.) maximum if $W \leq 25$ mm (1.0 in.); effective notch length (M) = 0.25 to 0.45W; effective crack length (a) = 0.45 to 0.55W.

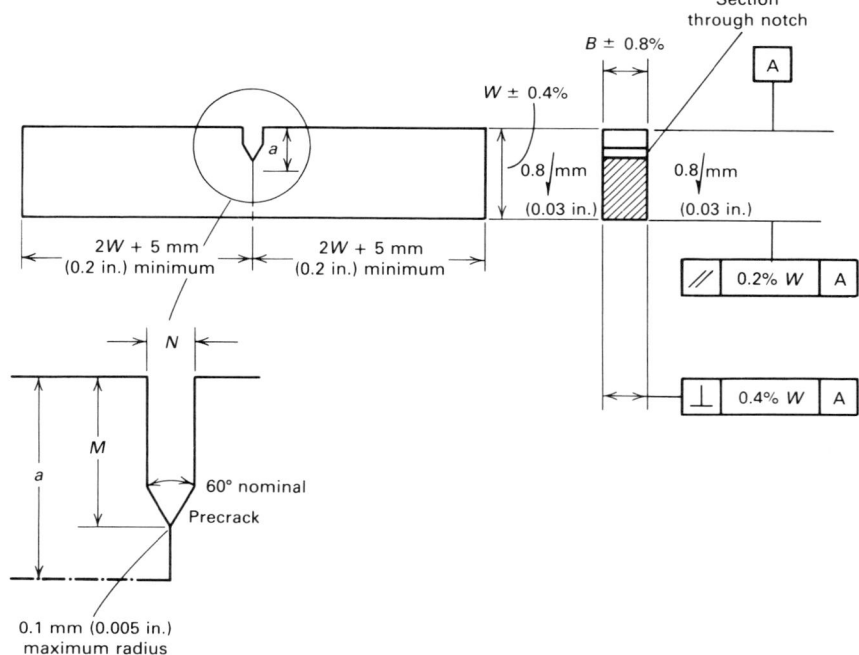

Fig. 36(b) Proportional dimensions and tolerances for modified compact specimens

Surfaces should be perpendicular and parallel as applicable to within 0.002H total indicator readout. The bolt centerline should be perpendicular to the specimen centerline within 1°. Bolt of material similar to specimen: fine threaded, square or Allen head. Thickness = B; net width (W) = 2.55B; total width (C) = 3.20B; half height (H) = 1.24B; hole diameter (D) = 0.718B + 0.003B; effective notch length (M) = 0.77B; notch width (N) = 0.06B; thread diameter (T) = 0.625B

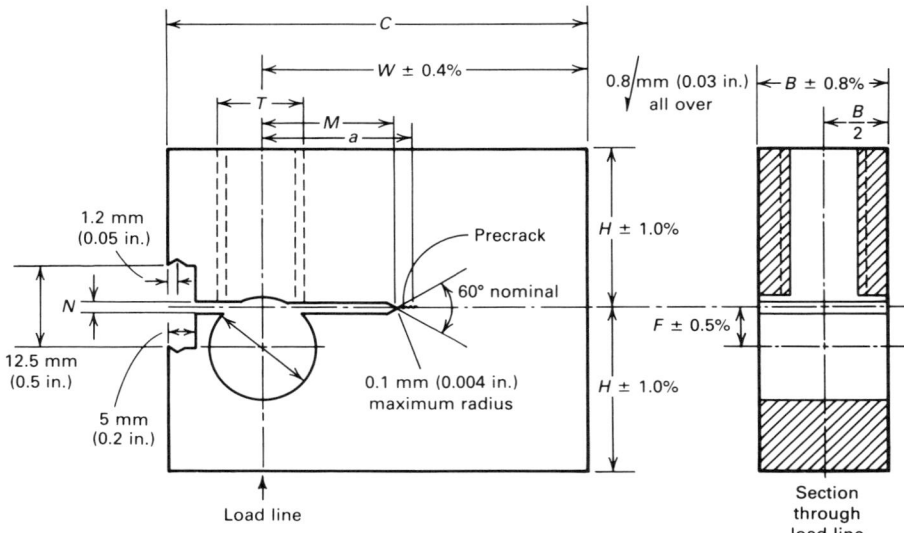

Fig. 36(c) Proportional dimensions and tolerances for double-beam specimens

"A" surfaces should be perpendicular and parallel as applicable to within 0.002H total indicator readout. At each side, the point "B" should be equidistant from the top and bottom surfaces to within 0.001H. The bolt centerline (load line) should be perpendicular to the specimen centerline to within 1°. Bolt of material similar to specimen: fine threaded, square or Allen head. Half height = H; thickness (B) = 2H; net width (W) = 10H minimum; total width (C) = W + T; thread diameter (T) = 0.75H minimum; notch width (N) = 0.14H maximum; effective notch length (M) = 2H

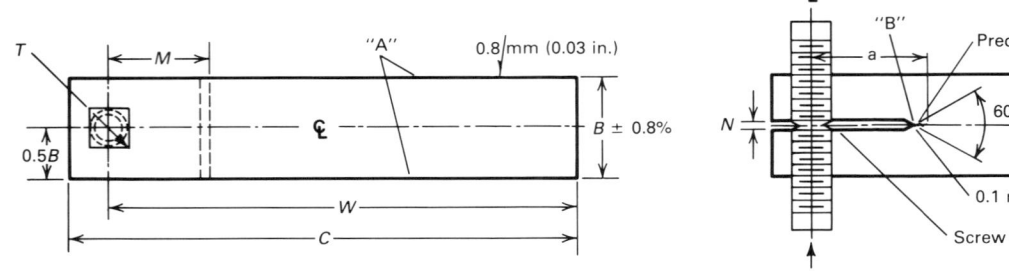

shown that for a valid K_{Ic} measurement both the crack length, a, and the thickness, B, should not be less than $2.5(K_{Ic}/YS)^2$, where YS is the yield strength of the material (Ref 42). Because of the present uncertainty regarding a minimum thickness for which an invariant value of K_{Iscc} can be obtained, the guidelines for designing the fracture mechanics test specimen should also be followed for stress-corrosion cracking test specimens. The threshold stress-intensity value should

be substituted for K_{Ic} in the above expression as a test of its validity.

If specimens are to be used for determination of K_{Iscc}, the initial specimen size should be based on an estimate of the K_{Iscc} of the material. Overestimation of the K_{Iscc} value is recommended; therefore, a larger specimen should be used than may eventually be necessary. When determining stress-corrosion crack growth behavior as a function of stress intensity, specimen size should be based on the highest stress intensity at which crack growth rates are to be measured (substitute K_{Io} in the $2.5(K_{Io}/YS)^2$ expression).

Crack Configuration and Orientation. For stress-corrosion testing, the length of the initial crack-starter notch—i.e., the machined slot with a fatigue or mechanical pop-in crack at its apex—can be as short as 0.2W. Guidelines for the length (depth) of the notch depend on the limits of accurate K_I calibration with respect to the range of a/W or a/H and the considerations discussed previously for double-beam specimens.

Several designs of crack-starter notches are available for most plate specimens. The machined slot is used to simulate a crack, because it is impractical to produce plane cracks of sufficient size and accuracy in plate specimens. ASTM E 399 (Ref 41) recommends that the notch root radius should not be greater than 0.127 mm (0.005 in.), unless the chevron form is used, in which case it may be 0.3 mm (0.01 in.) or less (Fig. 37). This tolerance can be achieved easily with conventional milling and grinding equipment.

A significant factor in stress-corrosion cracking testing of thick sections of some metals, such as aluminum and titanium, is the direction of applied stress relative to the grain structure. A standardized plan for the identification of the loading direction, the fracture plane, and the direction of crack propagation is shown in Fig. 39.

Machining. Specimens of the required orientation should be machined from prod-

ucts in the fully heat treated and stress-relieved condition to avoid complications due to residual stresses in the finished specimens. For specimens of material that cannot easily be completely machined in the fully heat treated condition, the final thermal treatment can be given prior to the notching and finishing operations. However, heat treatment of fully machined specimens should be performed only in cases in which the heat treatment will not result in distortion, residual stress, quench cracking, or detrimental surface conditions.

Fig. 37 Alternate chevron notch and face grooves for single-edge cracked specimens

(a) Chevron notch. (b) Face grooves

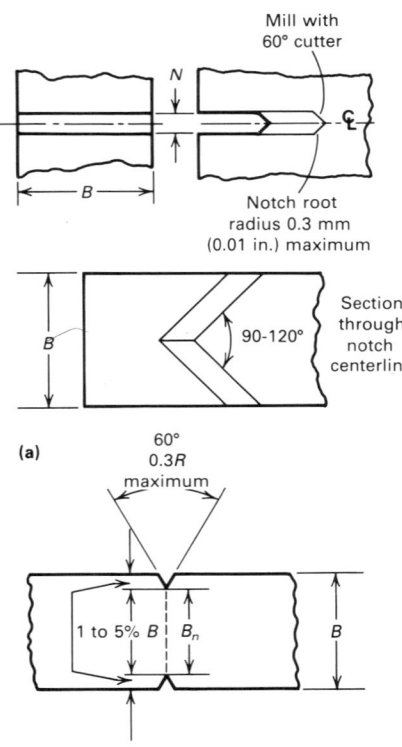

Fig. 38 Configuration and K_I calibration of a double-beam plate specimen

Normalized stress intensity, K_I (y-axis), plotted against a/H ratio (x-axis). (W − a) indifferent, crackline-loaded, single-edge cracked specimen. Source: Ref 40

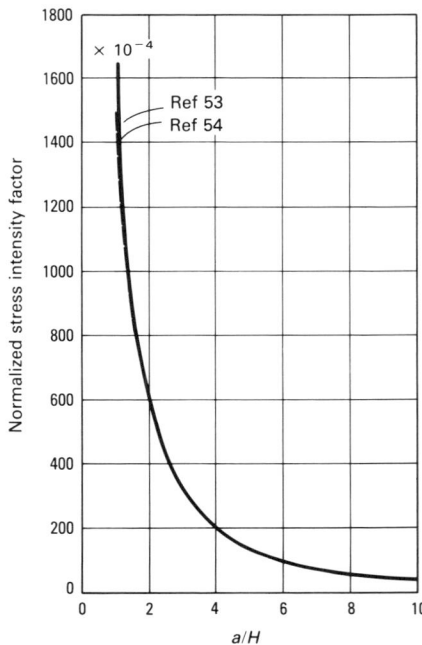

Fig. 39 Specimen orientation and fracture plane identification

L, length, longitudinal, principal direction of metal working (rolling, extrusion, axis of forging); T, width, long-transverse grain direction; S, thickness, short-transverse grain direction; C, chord of cylindrical cross section; R, radius of cylindrical cross section. First letter: normal to the fracture plane (loading direction); second letter: direction of crack propagation in fracture plane. Source: Ref 41

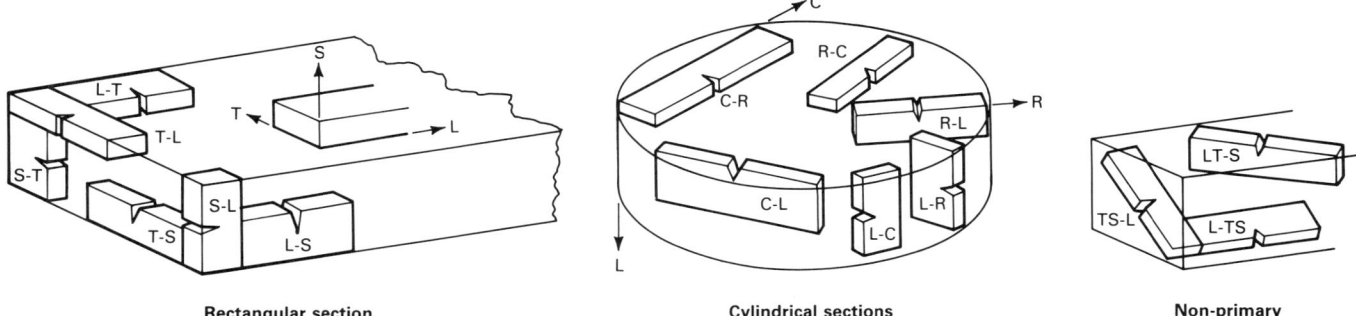

Rectangular section   Cylindrical sections   Non-primary

Precracking. Fatigue precracking should be done in accordance with ASTM E 399. The K level used for precracking each specimen should not exceed about two thirds of the intended starting K-value for the environmental exposure. This prevents fatigue damage or residual compressive stress at the crack tip, which may alter the stress-corrosion behavior, particularly when testing at a K-level near the threshold stress intensity for the specimen.

Aluminum alloy specimens also may be precracked by "pop-in" methods (wedge-opening loaded to the point of tensile overload), but steel and titanium alloys are usually too strong and tough to pop in without breaking off one of the specimen arms. Chevron notches usually are used to facilitate starting such mechanical precracks, and face grooves are sometimes necessary to produce straight precracks in tougher alloys (Fig. 37). These modifications also may be necessary to control fatigue precracking of some materials.

When a specimen is mechanically precracked by pop-in, the load should be maintained and should not be reduced for testing at a lower initial K-value. Reducing the load (crack mouth opening displacement) required for pop-in will result in residual compressive stress at the crack tip, which could interfere with the initiation of stress-corrosion cracking. When testing specimens at a relatively low fraction of K_{Ic}, fatigue precracking is recommended.

Exposure to Environment. When practical for laboratory accelerated testing, the test environment should be brought into contact with the specimen before it is stressed; this enhances access of the corrodent to the crack tip to promote earlier initiation of stress-corrosion cracking and to decrease variability in test results. Similarly, it is also beneficial to introduce the corrodent even earlier, i.e., during precracking. However, unless facilities are available to commence

environmental exposure immediately following the precracking operation, corrodent remaining at the crack tip may promote blunting due to corrosive attack.

Testing Procedure

For all methods using precracked specimens, the primary objective usually is to determine K_{Iscc}, a threshold stress intensity for stress-corrosion cracking for the alloy and environment combination. One procedure, similar to that used with smooth specimens, depends on initiation of stress-corrosion cracking. Both constant-load (K-increasing) and constant-displacement (K-decreasing) tests can be used. The other procedure, which is unique to precracked specimens, involves crack arrest. This technique requires a K-decreasing constant-displacement test. These methods are compared in Fig. 40, which illustrates the shift in the stress-intensity factor as stress-corrosion cracking growth occurs.

K-Increasing Versus K-Decreasing Tests. In constant-load specimens (K-increasing tests), stress parameters can be quantified with confidence. Because crack growth results in an increasing crack opening, there is less likelihood that oxide films will block the crack or wedge it open. Crack-length measurements can be made readily with several continuous-monitoring methods.

A wide selection of constant-load specimen geometries are available to suit the test material, experimental facilities, and test objective. Thus, crack growth can be studied under either bend or tension loading conditions. Specimens can be used to determine K_{Iscc} by the initiation of a stress-corrosion crack from a preexisting fatigue crack using a series of specimens or to measure crack growth rates.

The principal disadvantages of constant-load specimens are the expense and bulk associated with the need for an external load-

ing system. Bend specimens can be tested in relatively simple cantilever beam equipment, but specimens subjected to tension loading require constant-load creep-rupture equipment or similar testing machines. In this case, expense can be minimized by testing chains of specimens connected by loading links that are designed to prevent unloading upon failure of individual specimens. Because of the size of these loading systems, it is difficult to test constant-load specimens under operating conditions, but they can be tested in environments obtained from operating systems.

Constant-displacement specimens (K-decreasing tests) are self-loaded; thus, external stressing equipment is not required. Their compact dimensions also facilitate exposure to operating service environments. They can

Fig. 40 Schematic comparison of determination of K_{Iscc} by crack initiation versus crack arrest

(a) Constant-load test. (b) Constant crack opening displacement test

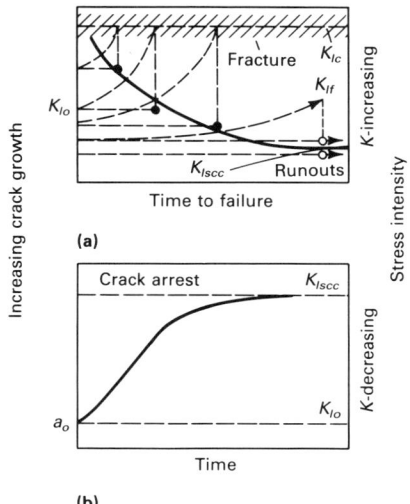

be used to determine K_{Iscc} by the initiation of stress-corrosion cracks from the fatigue pre-crack, in which case a series of specimens must be used to pinpoint the threshold value. This can also be achieved by the arrest of a propagating crack, because under constant-displacement testing conditions, the stress intensity decreases progressively as crack propagation occurs. In this case, a single specimen suffices in principle; in practice, the use of several replicates is recommended to assess variability in test results.

Constant-displacement specimens are subject to several inherent disadvantages. Oxide formation or corrosion products can wedge the crack surfaces open, thus changing the applied displacement and load. Oxide formation can also block the crack mouth, thus preventing the ingress of corrodent, and can impair the accuracy of crack-length measurements by electrical resistance methods. Applied loads can only be measured indirectly by displacement changes or by other sophisticated instrumentation. Crack arrest must be defined by an arbitrary crack growth rate, below which it is impracticable to measure cracks accurately.

Loading Arrangements and Crack Measurement. To monitor crack propagation rate as a function of decreasing stress intensity when testing constant-displacement loaded specimens, two of the three testing variables must be measured—crack length (a_i), load (P_i), and crack opening displacement at the load line (V_{LL}). Although crack initiation and growth can be detected from change in either load or crack length, load change usually is more sensitive to these conditions. Thus, crack advance is easier to detect in specimens loaded in a testing machine, an elastic loading ring, or an instrumented bolt than in specimens loaded with a bolt or wedge. Figure 41 illustrates typical loading arrangements for which load changes can be monitored automatically (Ref 14, 60).

Figure 42 illustrates an ultrasonic method of measuring crack length at the interior (midwidth and quarter widths) of a bolt-loaded double-beam specimen, which provides a more accurate measure of crack length than visual measurements made on the specimen surfaces. Various other techniques have been used, such as measurement of beam deflection for the cantilever beams (Ref 43) and changes in electrical resistance. Such arrangements, however, require calibration.

Calculation of Crack Growth Rates. There is no generally accepted procedure for calculating crack growth rate (da/dt) as a function of stress intensity from crack growth curves. Various approaches exist; the simplest is a graphical $\Delta a/\Delta t$ technique that may

incorporate smoothing of the a versus t curve (Ref 49-51). Another widely used approach is smoothing of the crack growth curve by computer techniques for curve fitting the entire a versus t curve by a multiple-term polynomial function (Ref 43).

Other techniques include a secant method and an incremental polynomial method, in which derivatives of the smoothed crack growth curve are calculated at various points to determine instantaneous crack growth rates. Instantaneous growth rates are then plotted against the instantaneous stress intensities, K_{Ii}, at corresponding time intervals to obtain graphs similar to that shown in Fig. 5. Additional information on the secant and incremental methods, which are often used in fatigue studies, can be found in the article "Fatigue Crack Growth Data Analysis" in this Volume.

A limited study of the above four methods of treating crack growth data is discussed for a high-strength aluminum alloy in Ref 14. All of the methods used to calculate crack growth rates produced the same general results, which were difficult to interpret because of large amounts of scatter resulting from the use of small crack-growth increments. Moreover, the significance of such graphs is dubious when the corrosivity of the environment and the length of exposure can invalidate the estimate of K by causing gross corrosion product wedging effects and/or crack branching.

Reduction of crack length data becomes useless without prior subjective interpretation of crack length versus time curves. Allowances should be made for extraneous effects caused by erratic or "apparent" initiation of stress-corrosion crack growth, scatter in the measurement data due to excessive crack front curvature, multiple crack planes,

crack-tip branching, and gross wedging caused by corrosion products.

A simple method of comparing materials using crack growth curves is based on average growth rates taken from an exposure time of zero to an arbitrary time that is sufficient to achieve significant crack extension in the most SCC-susceptible materials being compared (Ref 55). This method not only rapidly identifies materials with relatively low resistance to stress-corrosion cracking, but also provides numerical test results for highly resistant materials that may not develop a K_I

Fig. 41(a) Wedge-opening load specimen loaded with instrumented bolt
Source: Ref 60

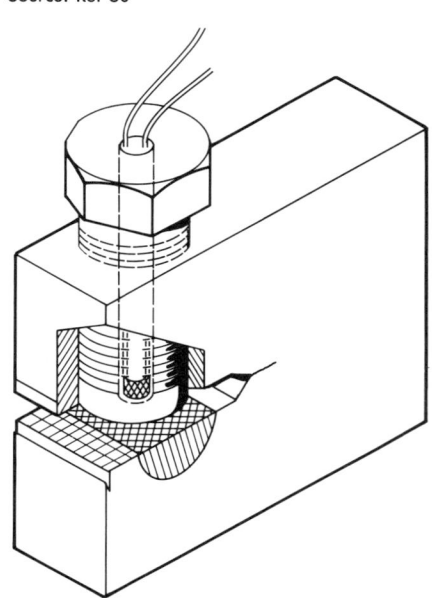

Fig. 41(b) Ring-loaded wedge-opening load specimen test setup
Box to the left of loading rings contains analog signal conditioning for load and displacement signals. The digital data-acquisition system consists of a scanner connected to the analog load and displacement signals, a digital voltmeter, and a portable computer used to read and store data and to control the other instruments. Source: Ref 14

Fig. 42 Ultrasonic crack measurement system for double-beam specimens

Bolt-loaded specimen is mounted on translation stage at center. Ultrasonic transducer is located above specimen, and the oscilloscope at left shows peaks indicating the top of specimen (left peak), the crack plane (center peak), and bottom face reflection (third peak). Digital readouts of stage position and peak height for the crack front measurement used to make consistent positioning measurements are shown (right). This system has a crack growth resolution of approximately 0.127 mm (0.005 in.) and an absolute crack length measurement error less than 2.54 mm (0.1 in.).

Fig. 43 Strain rate regimes for studying stress-corrosion cracking of various aluminum alloys

Corrodent: 3% sodium chloride plus 0.3% hydrogen peroxide. Source: Ref 61

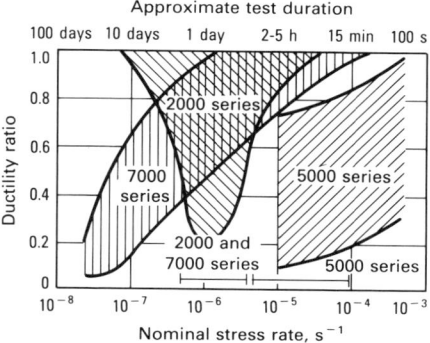

versus da/dt curve with a definite plateau. Comparison must be made with test results obtained under identical test conditions.

Dynamic Loading: Slow Strain Rate Testing

Slow strain rate tests can be used to test a wide variety of product forms, including parts joined by welding. Tests can be conducted in tension, in bending, or with plain, notched, or precracked specimens. The principal advantage of this test is the rapidity with which the stress-corrosion cracking susceptibility of a particular alloy and environment can be assessed.

Critical Strain Rate. The most important characteristic of the test is the relatively slow strain rate generated at the region of crack initiation and growth in the metal. For many systems, a tensile strain rate of 10^{-5} to 10^{-6} s^{-1} promotes stress-corrosion cracking, but the absence of cracking in tests conducted at such rates should not be construed as indications of immunity to cracking for a given system until tests have been performed at faster and slower strain rates. Once necking begins in a ductile metal stressed in tension, the effective strain rate in the necked region may increase by as much as an order of magnitude, which can cause the strain rate to move in or out of the critical range.

The fastest strain rate that will promote stress-corrosion cracking in a given system depends on crack velocity. Generally, the lower the stress-corrosion cracking velocity, the slower the strain rate required. Applied strain rates known to have promoted stress-corrosion cracking in metal/environment systems are listed in Table 2.

Table 2 Critical strain rate regimes promoting stress-corrosion cracking in various metal/environment systems

System	Applied strain rate, s^{-1}
Aluminum alloys in chloride solutions	10^{-4} to 10^{-7}
Copper alloys in ammoniacal and nitrite solutions	10^{-6}
Ferritic steels in carbonate, hydroxide, or nitrate solutions	10^{-6}
Magnesium alloys in chromate/chloride solutions	10^{-5}
Stainless steels in chloride solutions	10^{-6}
Stainless steels in high-temperature solutions	10^{-7}
Titanium alloys in chloride solutions	10^{-5}

The most relevant strain rates for various aluminum alloys are illustrated in Fig. 43. These trends illustrate that slow strain rate tests should be performed in a strain rate regime that is appropriate for the given alloy and environment system.

Test Specimen Selection. Standard tension specimens (ASTM E 8, "Standard Methods of Tension Testing of Metallic Materials") generally are recommended for use with the specified conditions of gage lengths, radii, etc., unless specialized studies are being conducted. For initially smooth specimens, the strain rate at the onset of the test is clearly defined; however, once cracks have initiated and grown, straining is likely to concentrate in the vicinity of the crack tip, and the effective strain rate is unknown. Rigorous solutions for determining the strain rate at crack tips or notches are not available, but effective strain rates are likely to be higher

than for the same deflection rates applied to plain specimens.

Notched or precracked specimens can be used to restrict cracking to a given location—for example, when testing the heat-affected zone associated with a weld. Notched or precracked specimens can also be used to restrict load requirements where bending, as opposed to tensile loading, may offer an added benefit. The section thickness or diameter of such specimens usually is relatively small, so the testing duration is short.

Testing Equipment. Constant strain rate apparatus requirements include sufficient stiffness to resist significant deformation under loads required to fracture the test specimens, a system to provide reproducible, constant strain rates over the range of 10^{-4} to 10^{-8} s^{-1}, and a cell to contain the test solution. Auxiliary equipment is used to control environmental conditions and to record test data. A typical constant strain rate unit is illustrated schematically in Fig. 44. Various types of corrosion cells may be required to control the test conditions for specific studies (Fig. 45).

In addition to uniaxial tensile units, cantilever constant strain rate apparatus also has been used, in which an extension arm attached to a cantilever beam specimen is lowered at a constant rate. This technique has been used successfully to study stress-corrosion cracking of low-carbon steel in carbonate-bicarbonate environments to determine crack velocity, critical strain rates, and inhibitor effectiveness (Ref 16). Additional information on slow strain rate testing equipment and procedures is available in Ref 58 and 64.

Assessment of Results. Historically, the principal methods of assessment of stress-corrosion cracking were based on pa-

Fig. 44 Typical slow strain rate test apparatus
Source: Ref 62

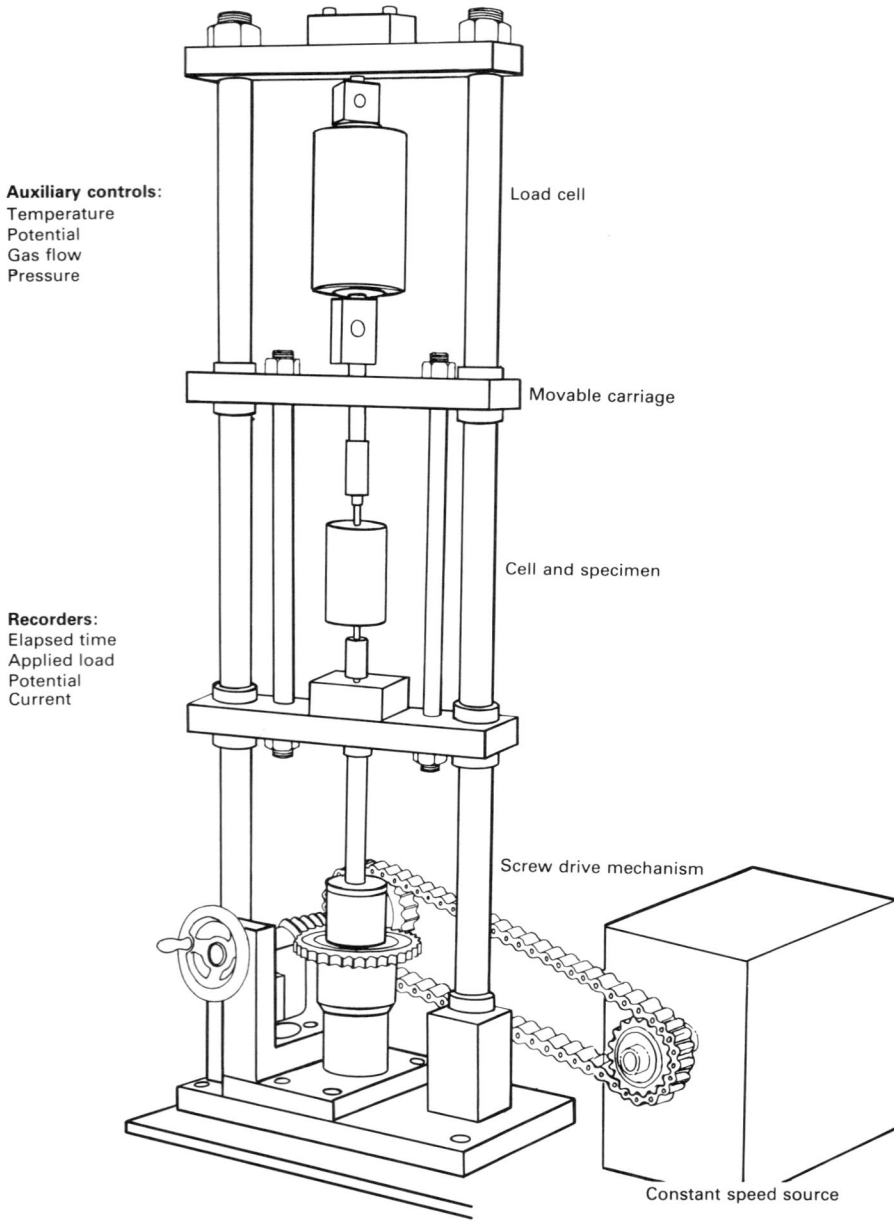

Auxiliary controls:
Temperature
Potential
Gas flow
Pressure

Load cell

Movable carriage

Cell and specimen

Recorders:
Elapsed time
Applied load
Potential
Current

Screw drive mechanism

Constant speed source

rameters derived from slow strain rate tension testing, including time-to-failure, maximum gross section stress developed during the tension test, percent elongation, area bounded by the load-deflection curve, and reduction in area. To eliminate non-stress-corrosion cracking effects, parallel tests are conducted in an inert environment, and a ratio of the result obtained in the corrodent divided by the result obtained in the inert environment is commonly used as an index of stress-corrosion cracking susceptibility. For example, in Fig. 43 higher resistance to

stress-corrosion cracking is denoted by higher ductility ratios. The visual appearance of a stress-corroded specimen containing many secondary stress-corrosion cracks and reduced ductility at fracture is depicted in Fig. 46.

Some alloys experience rapid deterioration of mechanical properties on contact with certain corrosive environments; any additional effect of applied straining can best be assessed by comparison with the behavior of unstrained specimens. Thus, it is essential that the cause of environmental degradation

be verified as the result of stress-corrosion cracking.

Recent studies using modified slow strain rate testing techniques combined with microscopy have demonstrated that average stress-corrosion cracking velocities can be determined from the depth of the largest crack measured on the fracture surfaces of specimens that have failed completely, or in longitudinal sections on the diameter of specimens that have not experienced total failure, divided by the time of testing. With this parameter, stress-corrosion cracking is assumed to initiate at the start of the test, which is not always true.

With precracked specimens, other methods can be used to monitor crack growth, whereby crack velocities can be determined. The stress-corrosion cracking behavior of a pipeline steel (Fig. 47) has been studied using a precracked cantilever bend specimen in terms of threshold strain rate for crack growth and also in terms of crack growth rates analogous to the Stage II plateau velocities illustrated in Fig. 5 and 12. Material properties, such as strength and toughness, that influence the stress-corrosion cracking, performance when measured by tension testing are eliminated as factors, thus enabling valid comparisons of alloys with widely different structures and mechanical properties.

Further modification of the slow strain rate testing technique combined with metallographic evaluation of cracking tendencies has led to determination of stress-corrosion cracking threshold stresses. Additional information on this method of assessment and the effects of strain rate can be found in Ref 65 to 67.

Testing of Weldments

ASTM G 58, "Standard Practice for the Preparation of Stress-Corrosion Test Specimens for Weldments," covers test specimens in which stresses are developed by (1) the welding process only, i.e., residual stress (Fig. 34), (2) an externally applied load in addition to the stresses due to welding (Fig. 18e), and (3) an externally applied load only, with residual welding stresses removed by annealing.

The National Materials Advisory Board Committee on Environmentally Assisted Cracking Test Methods for High-Strength Weldments recently published the following guidelines on stress-corrosion cracking testing of weldments (Ref 1). Fracture mechanics of cracked bodies was found to be a valid and useful approach for designing against environmentally assisted cracking, although several limitations and difficulties must be taken into consideration. For static loading, K_{Iscc} and da/dt versus K_I are useful parame-

Fig. 45 Typical slow strain rate cell for testing the effectiveness of inhibitors in coatings to prevent stress-corrosion cracking
Source: Ref 63

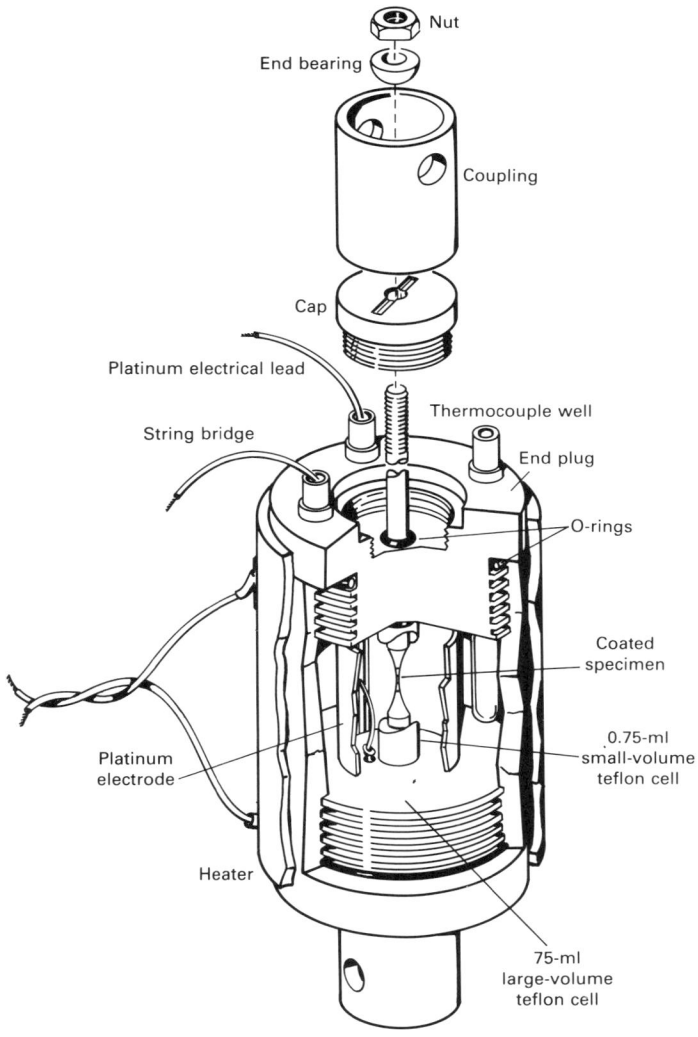

Fig. 46 Photomacrographs of two carbon steel specimens after slow strain rate tests conducted at a strain rate of 2.5×10^{-6} s^{-1} and 80 °C (180 °F)
The ductility ratio in this example was 0.74 (original diameter: 2.54 mm, or 0.100 in.). (left) Ductile fracture in oil. (right) Stress-corrosion cracking in carbonate solution

Fig. 47 Effects of beam deflection rate on stress-corrosion cracking velocity in precracked cantilever bend specimens of a carbon-manganese steel
In a carbonate-bicarbonate solution at 75 °C (165 °F) and at a potential of −650 mV (saturated calomel electrode). Source: Ref 15

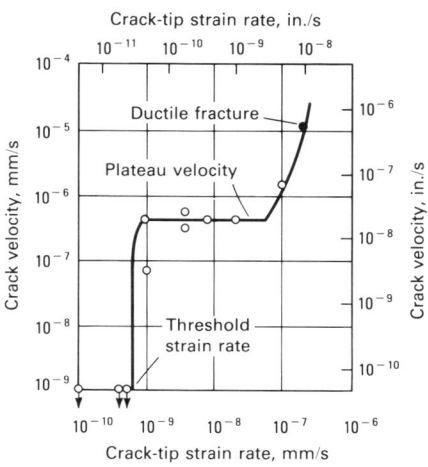

ters. They are specific to a material, temperature, and metal/environment system and also are functions of local chemical composition, microstructure, etc.

Superimposed minor load fluctuations and infrequent changes in load can alter environmental cracking response. This effect, which cannot be predicted from K_{Iscc} and da/dt values, may be significant and detrimental. Reexamination of static loading as a design premise may be required. Existing test methodology for environmentally assisted cracking tendency is applicable to the evaluation of weldments. As in other structural components, residual stress must be treated in a quantitative and realistic manner.

The National Materials Advisory Board report supports current design emphasis based on the presumption of preexisting crack-like flaws in the structure and covers testing with precracked (fracture mechanics) specimens only. It contains a critical assessment of the problems associated with environmentally assisted cracking in high-strength alloys and of state-of-the-art design and test methodology.

Selection of Test Environment

Accelerated corrosion tests are needed to aid in the development of alloys and processes and in evaluation of new products. However, in no area is the hazard of accelerated testing greater and the selection of appropriate tests more fallible. Stress-corrosion tests can be divided into two broad classes: those conducted in service environments and those conducted under laboratory conditions.

Field Testing and Service Environments. Field testing consists of placing a metal specimen in an environment in which conditions simulate those anticipated in service. Typical examples include immersion in seawater, exposure to the atmosphere at marine or industrial sites, chemical plant streams, etc. Field tests can be performed with test coupons or with actual or simulated structural components.

The following example illustrates the value, and in some cases the necessity, of exposure tests performed in actual service environments as an adjunct to laboratory evaluation. The standard 3.5% sodium chloride alternate immersion test data for 2024 and 7075 alloys proved useless in predicting the serviceability of these alloys for handling rocket propellant oxidizers such as nitrogen tetroxide and inhibited red fuming nitric acid (Ref 68). The alternate immersion test showed 2024-T351 and 7075-T651 to be susceptible to stress-corrosion cracking at low short-transverse stresses, whereas 2024-T851 and 7075-T351 were quite resistant. These data were supported by outdoor field tests in seacoast and industrial atmospheres.

However, in proof tests consisting of exposure to the actual service environment—inhibited red fuming nitric acid at 74 °C (165 °F)—stress-corrosion cracking occurred in both tempers of 7075 alloy and did not occur in either temper of 2024 alloy (Fig. 48). There were no unexpected failures with the 2219-T87 and 6061-T651 materials, however.

Laboratory Environments. Simulated service tests should be conducted under conditions duplicating exactly the service environment, as illustrated by the following example with Ti-6Al-4V alloy pressure tanks for propellant-grade nitrogen tetroxide (<0.20 wt% moisture) (Ref 69). Preliminary laboratory stress-corrosion cracking tests using specimens of Ti-6Al-4V demonstrated satisfactory compatibility with the nitrogen tetroxide and gave no indication of stress-corrosion cracking.

In subsequent tests of actual pressurized tanks, however, stress-corrosion cracking occurred rapidly, and the tanks failed. It was subsequently shown that the nitrogen tetroxide used in the tanks was of a higher purity grade than that used in the laboratory tests. When test specimens were exposed to this grade of nitrogen tetroxide, they also failed. The small quantities of water and nitric oxide (impurities) present in the nitrogen tetroxide used in previous studies were sufficient to inhibit stress-corrosion cracking of Ti-6Al-4V in nitrogen tetroxide.

A characteristic of stress-corrosion phenomena is that a given alloy may be sensitive to a specific chemical factor in the environment. Therefore, tests should be carried out under conditions simulating the total corrosive environment. In addition to the presence or absence of chemical agents, cracking frequently is controlled critically by physical parameters such as temperature, humidity, or the rate of diffusion or convection of solution. Chemical environments are sensitive to changes in temperature and pH and to dissolved gases, such as oxygen, hydrogen, hydrogen sulfide and other sulfur compounds, carbon dioxide, and ammonia and other nitrogen compounds.

Mercury, other liquid metal systems, and molten salts can also cause failure by cracking of metal under stress. In most such instances, oxidation-reduction reactions (corrosion) are not involved with the alloy, and the mechanism of the cracking under such conditions is quite different from the mechanism of stress-corrosion cracking.

Cracking caused by hydrogen embrittlement has been found to occur in titanium-based alloys and many high-strength ferrous alloys in various service environments. The similarities between stress-corrosion cracking and hydrogen embrittlement cracking of alloys under tensile stress, although still controversial, will be discussed below in connection with these alloys.

Hydrogen, under some conditions, enters the stressed metal without the presence of a corrosion reaction, and cracking ensues. As a corrosion reaction product, hydrogen enters stressed metal, and cracking is stimulated or will ensue. Thus, to characterize adequately the environmental cracking of such alloys under sustained tensile stress, testing should be performed both with and without conditions that could involve hydrogen damage. For more information, see the article "Tests for Hydrogen Embrittlement" in this Section.

Accelerated Testing Mediums for Various Alloy Systems

Ideally, the results of a short exposure in an accelerated test will enable a reliable prediction of the stress-corrosion cracking performance of an alloy over a long period of service under particular environmental conditions. To fulfill this function, test conditions should be selected with regard to the anticipated service conditions. However, accelerated test mediums that yield reliable results for one alloy may not yield dependable results for another alloy, even though both alloys are of the same basic metal (Ref 24, 66). Thus, caution must be exercised in the development and use of standardized test mediums.

Correlation of Accelerated Testing Mediums. The soundness of an accelerated test medium requires correlation with practical service experience or with the results of appropriate field tests. Some methods used to correlate stress-corrosion test environments are inadequate and can be misleading.

For example, Fig. 49 illustrates an example in which the resistance to stress-corrosion cracking is plotted as a function of one of several variables that influence stress-corrosion performance. Tests conducted at extreme values, W and Z, of the independent variable resulted in a high degree of survival and a high degree of failure, respectively, corresponding to the service environment. Although it may appear that a reliable test can be provided by any one of the three accelerated test mediums, other tests at inter-

Fig. 48 Stress-corrosion cracking resistance of various aluminum alloys in inhibited red fuming nitric acid versus alternate immersion in 3.5% sodium chloride solution

Each bar graph represents an individual short-transverse C-ring test specimen machined from rolled plate and stressed at the indicated level.

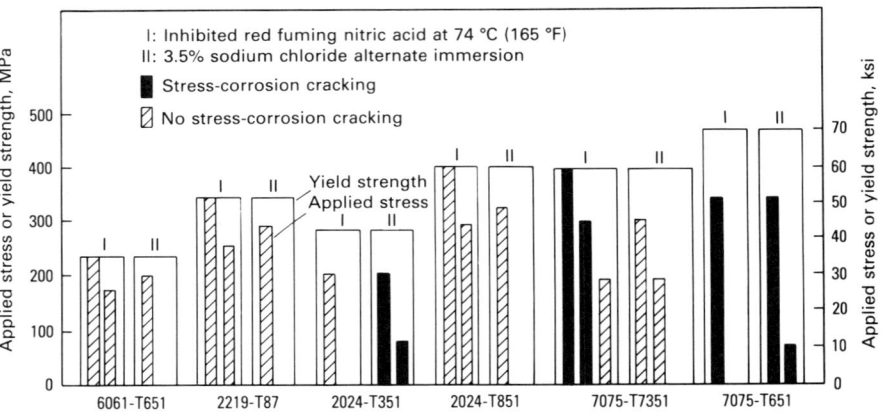

Fig. 49 Correlation of accelerated stress-corrosion test mediums with service environment

Examples of the independent variable are percentage of alloy element, duration of precipitation treatment, yield strength, applied stress, etc. Source: Ref 24

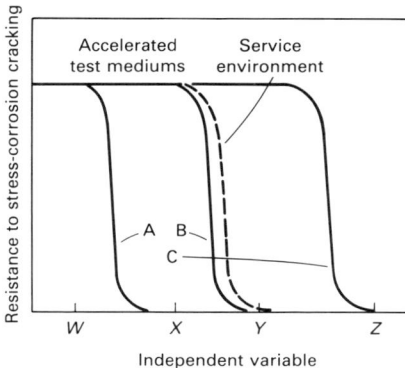

Fig. 50 Correlation of accelerated test mediums with service environment

Combined data for five lots of rolled plate of aluminum alloy 7039-T64 (4.0Zn-2.8Mg-0.3Mn-0.2Cr). Source: Ref 24

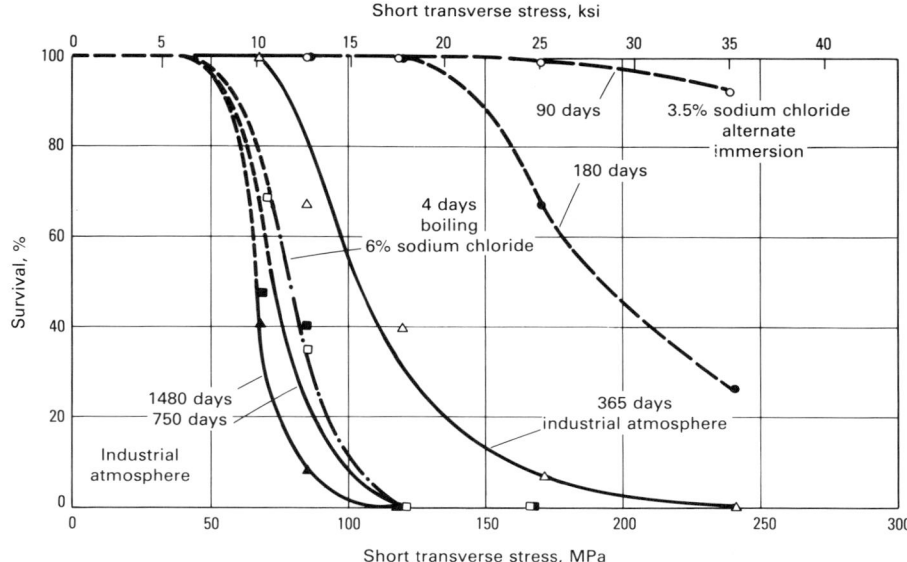

mediate values, X and Y, showed correlation to be poor for medium A and C.

Although none of the mediums can be expected to predict service performance exactly, test medium B most accurately correlated to the service environment and hence should be the most useful for predicting service performance. Thus, investigations must be conducted at several values of the independent variable to establish an adequate calibration of test environments.

Correlation of two accelerated test mediums with a service environment for an aluminum alloy is illustrated in Fig. 50. These data demonstrate that the 4-day test in boiling 6% sodium chloride solution correlates better with the industrial atmosphere exposure than either a 90-day or a 180-day exposure to the 3.5% sodium chloride alternate immersion test. Additionally, a 180-day test period cannot be considered an accelerated test situation. The data also illustrate that in a service environment the length of exposure required to determine the stress-corrosion performance of an alloy can require several years.

Testing of Aluminum Alloys

The alternate immersion test in 3.5% sodium chloride is described in ASTM G 44, "Standard Recommended Practice for Alternate Immersion Stress Corrosion Testing in 3.5% Sodium Chloride Solution," and in MIL-STD 1568, "Materials and Processes for Corrosion Prevention and Control in Aerospace Weapons Systems." It consists of a 1-h cycle that includes a 10-min soak in an aqueous solution of 3.5% sodium chloride followed by a 50-min period out of the solution, during which specimens are allowed to dry.

This 1-h cycle is repeated continuously for the total number of days recommended for the particular alloy being tested. Typically, aluminum and steel alloys are exposed from 10 to 90 days or longer, depending on the resistance of the alloy to corrosion by salt water. This test method is used widely for most types of aluminum alloys and some ferrous materials and is well suited for testing all types of smooth specimens.

The alternate immersion test is used primarily for alloy development studies and for quality control of alloys with improved resistance to stress-corrosion cracking (Ref 20, 70). This test method is specified in ASTM G 47, "Standard Recommended Practice for Determining Susceptibility to Stress-Corrosion Cracking of High-Strength Aluminum Alloy Products," which covers the method of sampling, type of specimen, specimen preparation, test environment, and method of exposure for determining the susceptibility to stress-corrosion cracking of high-strength 2000 series alloys (1.8 to 7.0% Cu) and 7000 series alloys (0.4 to 2.8% Cu).

The alternate immersion test provides an environment for detecting materials that are likely to be susceptible to stress-corrosion cracking in outdoor environments, particularly those with marine effects. It is not suitable for predicting serviceability in unusual chemical environments, such as inhibited red fuming nitric acid. Although the test is applicable to all types of aluminum alloys, it is not equally discriminative of all types.

Table 3 compares the relative stress-corrosion cracking performances of several high-strength aluminum alloys used in aircraft construction. Although the 3.5% sodium chloride alternate immersion test enabled good prediction of the performance observed in service environments for four of the five alloys, it did not indicate the poor performance observed for the two lower stress level tests of alloy 7079-T651. However, when this alloy was tested with bolt-loaded pop-in double-beam specimens, the stress-corrosion cracking propagation rate in a simulated 3.5% sodium chloride alternate immersion test was actually higher than in the atmospheric environments (see Fig. 12). A poor service record of stress-corrosion cracking problems with 7079 alloy products (Ref 21) confirmed the field testing of smooth specimens in atmospheric environments and the stress-corrosion cracking propagation rate tests using 3.5% sodium chloride in laboratory tests.

Continuous Immersion in Boiling 6% Sodium Chloride. This rapid (4-day) test is widely used by American aluminum producers to evaluate the stress-corrosion cracking behavior of 7000 series alloys containing less than 0.25% Cu (Ref 24, 71). Standards are currently being developed for testing smooth test specimens in this test medium. The continuous immersion test is not applicable to the 2000 series (aluminum-copper) or the 5000 series (aluminum-magnesium) alloys.

Table 3 Comparison of relative stress-corrosion cracking performances of several high-strength aluminum alloys

20-year atmospheric exposure in different environments(a)

Alloy and temper	Composition, %										Exposure stress(b), Yield strength, %	MPa (ksi)	Seacoast at Point Judith, RI	
	Cu	Fe	Si	Mn	Mg	Zn	Cr	Ti	Zr	V			F/N(d)	Days to failure
Susceptible alloys														
7079-T651	0.64	0.23	0.10	0.20	3.48	4.59	0.15	0.03	⋯	⋯	75	345 (50)	3/3	27, 27, 56
											50	228 (33)	4/4	27, 27, 27, 49
											25	117 (17)	3/3	653, 1284, 3605
7178-T651	1.97	0.16	0.10	0.02	2.60	6.72	0.21	0.01	⋯	⋯	75	380 (55)	3/3	27, 27, 49
											50	255 (37)	4/4	49, 76, 90, 208
											25	124 (18)	2/3	6122, 6331, OK 7510
2014-T651	4.41	0.26	0.88	0.82	0.57	0.04	0.01	0.02	⋯	⋯	75	310 (45)	3/3	49, 56, 56
											50	207 (30)	4/4	56, 161, 252, 548
											25	103 (15)	0/3	3 OK 7510
Resistant alloys														
7075-T7351	1.54	0.21	0.10	0.02	2.35	5.68	0.19	0.04	⋯	⋯	75	303 (44)	0/5	5 OK 7510
2219-T87	6.28	0.18	0.11	0.29	0.01	0.03	0.01	0.06	0.17	0.10	75	269 (39)	0/5	5 OK 7510

Alloy and temper	Seacoast at Point Comfort, TX		Inland industrial at New Kensington, PA		3.5% NaCl alternate immersion(c)	
	F/N(d)	Days to failure	F/N(d)	Days to failure	F/N(d)	Days to failure
Susceptible alloys						
7079-T651	3/3	59, 79, 88	3/3	98, 98, 116	2/2	15, 21
	4/4	137, 137, 137, 165	4/4	126, 172, 192, 197	0/2	2 OK 84
	3/3	1054, 1054, 2529	3/3	640, 673, 976	0/2	2 OK 84
7178-T651	3/3	40, 79, 79	3/3	192, 192, 198	2/2	5, 5
	4/4	88, 96, 104, 165	4/4	219, 258, 258, 258	2/2	14, 14
	3/3	1090, 4446, 5997	3/3	2376, 2639, 4689	0/2	2 OK 84
2014-T651	3/3	40, 40, 45	3/3	205, 623, 1629	2/2	5, 5
	4/4	96, 96, 137, 508	0/4	4 OK 7540	1/4	19, 3 OK 84
	1/3	2529, 2 OK 7360	0/3	3 OK 7540	0/4	4 OK 84
Resistant alloys						
7075-T7351	0/5	5 OK 7360	0/4	4 OK 7540	0/5	5 Ok 84
2219-T87	0/5	5 OK 7360	0/4	4 OK 7540	0/5	5 OK 84

(a) Transverse 3.175-mm (0.125-in.) diam tension specimens machined from 63.5-mm (2.5-in.) diam rolled rod. (b) Stressed in constant-strain wedge-type stressing frames (Fig. 25). (c) Similar to ASTM G 44, except salt solution was made with commercial-grade sodium chloride and New Kensington tap water. (d) F/N denotes number of stress-corrosion cracking failures/number of replicate specimens exposed.

Other Testing Mediums. Additional procedures have been found effective in accelerating the stress-corrosion cracking of aluminum alloys in aqueous chloride solutions. These include:

- Increased chloride concentration
- Increased acidity (lower pH)
- Increased temperature, particularly effective for aluminum-zinc-magnesium alloys
- Addition of oxidizers; aeration of the solution or exposure by intermittent immersion or spraying, or the addition of oxidants such as hydrogen peroxide, ni-

trates, or chromates, particularly effective for aluminum-copper, aluminum-magnesium, and aluminum-zinc-magnesium-copper alloys
- Anodically driving the stress-corrosion cracking process by impressing an appropriate potential or electrical current density

These techniques have been used, usually in combination, to create special test conditions. Use of these techniques has given rise to:

- Continuous immersion test for 7000 series (aluminum-zinc-magnesium) alloys (Ref

72); aqueous solution containing 3% sodium chloride, 0.5% hydrogen peroxide (30%), 100 mL/L 1N sodium hydroxide, and 20 mL/L acetic acid (100%), pH 4.0
- Impressed current test for 7000 series (aluminum-zinc-magnesium) alloys (Ref 73); aqueous solution containing 2% sodium chloride plus 0.5% sodium chromate, pH 8.1, current density 4.65×10^{-4} mA/mm^2 (0.3 mA/in.^{2}), 30-day maximum exposure time
- Substitute ocean water as a less corrosive substitute for 3.5% sodium chloride in the alternate immersion test (Ref 19). Aqueous solution containing 2.86% sodium chloride plus 0.52% magnesium chloride (total chloride equal to that in ocean water) (Ref 74)
- Continuous immersion test for 2000 series (aluminum-copper) and 7000 series (aluminum-zinc-magnesium-copper) alloys (Ref 75); aqueous 1% sodium chloride plus 2% potassium dichromate at 60 °C (140 °F), 168-h maximum exposure time
- Continuous immersion test for 2000 series (aluminum-copper) and 7000 series (aluminum-zinc-magnesium-copper) alloys (Ref 50, 51); aqueous solution containing 0.6 M sodium chloride, 0.02 M sodium dichromate, 0.07 M sodium acetate, plus acetic acid to pH 4, used principally for tests with precracked specimens
- Impressed current of 6.2×10^{-2} mA/mm^2 (40 mA/in.^{2}) of specimen surface in aqueous solution of 3.5% sodium chloride solution; test designed for 5000 series (aluminum-magnesium) alloys (Ref 76)

Testing With Precracked Specimens. Currently, there are no standard test mediums for specific use with precracked specimens. A periodic moistening procedure (drop-wise application of 3.5% sodium chloride solution three times per day) as a substitute for the alternate immersion procedure for smooth specimen testing has been used widely (Ref 48-51). This technique produces more rapid stress-corrosion cracking growth in 2000 series (aluminum-copper) and 7000 series (aluminum-zinc-magnesium-copper) alloys than continuous immersion in 3.5% sodium chloride solution (Fig. 51). However, it also causes corrosion-product wedging effects that interfere with stress-corrosion cracking arrests.

Exposure to substitute ocean water by alternate immersion produced alloy rankings similar to those in atmospheric exposure, with significantly less evidence of corrosion-product wedging (Ref 51, 77). Some tests have been made with 7000 series alloys in distilled water (Ref 78) or in water vapor at 40 °C (104 °F) (Ref 79).

Slow strain rate testing currently is not governed by any standards. Various aqueous solutions have been used in addition to plain 3.5% sodium chloride. Because the 3.5% sodium chloride solution did not appear aggressive enough for slow strain rate testing, more corrosive test mediums have been used, including oxidant additions to the sodium chloride solution or more acidic solutions such as aluminum chloride (Ref 61, 80).

A round-robin testing program using several aluminum alloy types and several corrodents to determine that a solution containing 3% sodium chloride plus 0.3% hydrogen peroxide was the most promising candidate for possible standardization. Another promising candidate was a solution of 2% sodium chloride plus 0.5% sodium chromate, with a pH of 3 (Ref 61, 81). Additional study is needed to determine the optimum composition of the sodium chloride plus hydrogen peroxide solution.

Testing of Copper Alloys

Testing in Mattsson's Solution. ASTM G 37, "Standard Recommended Practice for Use of Mattsson's Solution of pH 7.2 to Evaluate the Stress-Corrosion Cracking Susceptibility of Copper Zinc Alloys," requires that a stressed test specimen be completely and continuously immersed in an aqueous solution containing 0.05 g-atom/L of Cu^{++} and 1 g-mol/L of ammonium with a pH of 7.2. The copper is added as hydrated copper sulfate and the ammonium is a mixture of ammonium hydroxide and ammonium sulfate. The ratio of the latter two compounds is adjusted to achieve the desired pH.

Mattsson's pH 7.2 solution is recommended only for brasses (copper-zinc based alloys). The use of this test environment is not recommended for other copper alloys, because the results may be erroneous. This is particularly true for alloys containing aluminum or nickel.

The test environment is believed to provide an accelerated ranking of the relative or absolute degree of stress-corrosion cracking susceptibility for different brasses. Test environment correlates well with the corresponding service ranking in environments that cause stress-corrosion cracking, which may be due to the combined presence of traces of moisture and ammonia vapor. The extent to which the accelerated ranking correlates with the ranking obtained after long-term exposure to environments containing corrodents other than ammonia is not known. Such environments may be severe marine atmospheres (chloride), severe industrial atmospheres (predominantly sulfur dioxide), or superheated ammonia-free steam.

It is currently not possible to specify a time-to-failure in Mattsson's pH 7.2 solution that corresponds to a distinction between ac-

Fig. 51 Environmental crack growth in S-L double-beam specimens wedge-loaded with a bolt to pop-in and exposed to various corrodents

Crack growth values represent average of two specimens. See Fig. 39 for an explanation of specimen orientation and fracture plane identification.

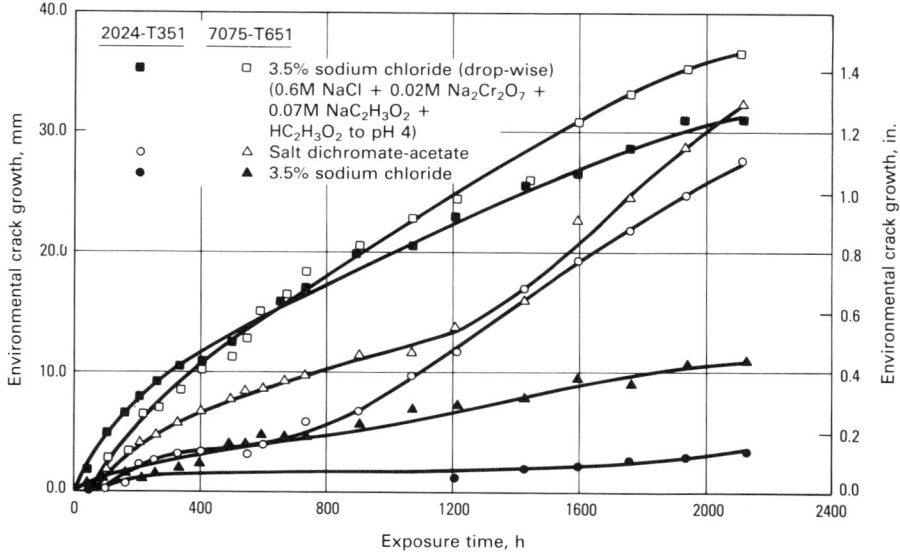

ceptable and unacceptable stress-corrosion behavior in brass alloys. Such correlations must be determined on an individual basis.

Mattsson's pH 7.2 solution may also cause some stress-independent general and intergranular corrosion of brasses. Thus, stress-corrosion failure may possibly be confused with mechanical failure induced by corrosion-reduced net cross section. This is most likely with small cross-sectional specimens, high applied stress levels, long exposure times, and stress-corrosion-resistant alloys. Careful metallographic examination is recommended to accurately determine the cause of failure. Alternatively, unstressed control specimens can be exposed to corrosive environments to determine the extent to which stress-independent corrosion degrades mechanical properties.

Other Testing Mediums. The most widely used stress-corrosion cracking agent for copper and copper alloys is ammonia (NH_3) (Ref 82). The ammonium ion does not appear to cause cracking in a stable salt such as ammonium sulfate. In a salt in which it dissociates (such as in ammonium carbonate) to form ammonia, cracking will occur.

The $Cu (NH_3)_x^{++}$ ion is thought to be necessary to induce stress-corrosion cracking in copper metals (Ref 83), where x usually is 4 or 5. NH_2 groups also cause cracking, or are easily converted to ammonia. Amines and sulfuric acid also cause cracking. Dry ammonia does not cause stress-corrosion cracking of brass, as demonstrated by the successful use of brass valves and gages on tanks of anhydrous ammonia.

Stress-corrosion cracking of copper metals will not occur in the absence of oxygen or an oxidizing agent. Carbon dioxide is also a requisite (Ref 84). Thus, air rather than pure oxygen is necessary. As a practical matter, moisture is essential. When other factors are favorable, a very small amount of NH_3 is sufficient to cause cracking. Hence, the controlling factor may be moisture, as cracking may appear to be caused by the presence of a condensed moisture film.

Other than ammonia, the most effective agents for causing cracking are the fumes from nitic acid or moist nitrogen dioxide. Sulfur dioxide will also crack brass, but both maximum and minimum concentration limits exist and the reaction is slow (Ref 82). Alloy development studies have been conducted using a test environment of moist ammoniacal atmosphere containing 80% air, 16% ammonia, and 4% water vapor at 35 °C (95 °F) (Ref 85). However, none of these corrodents has received the attention that ammonia has garnered.

Historically, immersion of a copper alloy product in a mercurous nitrate solution has been used to test for residual stresses (Ref 86, 87). Because these residual stresses are possible sources of failure by stress-corrosion cracking in other environments, some have regarded this test as a stress-corrosion test. However, it is only an indirect method of identifying stress-corrosion cracking tendencies and does not correlate to its presence as well as test methods based on specific attack by ammonia (Ref 82). It does indicate, however, that mercury and other low-melting-point liquid metals can cause embrittlement and failure due to cracking.

Testing of Low-Carbon Steels

Generally, steels with lower strengths are susceptible to stress-corrosion cracking only on exposure to a small number of specific environments, such as the hot caustic solutions encountered in steam boilers, hot nitrate solutions, anhydrous ammonia, and hot carbonate-bicarbonate solutions (Ref 88, 89).

Boiler Water Embrittlement Detector Testing. Caustic cracking failures frequently originate in riveted and welded structures in the vicinity of faying surfaces, where small leaks cause soluble salts to accumulate in high local concentrations of caustic soda and silica. This type of intergranular cracking failure has been produced with concentrations of sodium hydroxide as low as 5%, but usually a concentration of 15 to 30% is required at 200 to 250 °C (390 to 480 °F) to produce this phenomenon. ASTM D 807, "Standard Method of Assessing the Tendency of Industrial Boiler Waters to Cause Embrittlement (USBM Embrittlement Detector Method)," details the apparatus and procedures used to determine the embrittling or nonembrittling characteristics of the water in an operating boiler.

Other Testing Mediums. Caustic cracking occurs in digester vessels used in chemical process industries, and laboratory studies have been conducted using sodium hydroxide concentrations of about 30 to 35% (Ref 90). Tests in boiling nitrate solutions frequently have been used to study the effects of composition and metallurgical variables (Ref 89). In studies of low-carbon steel in boiling nitrate solutions with different cations, solutions containing the more acidic cations in greater concentrations were found to be the most potent. This tendency is illustrated by the apparent threshold stresses for failure of a 0.05% carbon steel in nitrate solutions with a range of concentrations, as shown in Table 4.

Cracking can be accelerated by additions of small amounts of acid or oxidants such as potassium permanganate, manganese sulfate, sodium nitrite, and potassium dichromate, whereas hydroxides and other salts, particularly those forming insoluble iron products, such as sodium carbonate or sodium hydrogen phosphate, retard or prevent failure. A standard test environment has not been established, and conditions should be tailored to individual testing requirements.

The ranking of a given series of alloys may vary with exposure conditions (Ref 66). Consequently, selection of a particular alloy for use in an environment that varies from that used in laboratory ranking tests may result in

Table 4 Apparent threshold stress values for 0.05% carbon steel in nitrate solutions of varying concentrations

Nitrate	Apparent threshold stress values, MPa (ksi), at a solution concentration of:							
	8N		4N		2.5N		N	
Ammonium nitrate......14	(2)	21	(3)	48	(7)	83	(12)
Calcium nitrate 34	(5)	48	(7)	83	(12)	159	(23)
Lithium nitrate.........34	(5)	55	(8)	131	(19) at 2N	159	(23)
Potassium nitrate.......41	(6)	62	(9)	97	(14)	165	(24)
Sodium nitrate 55	(8)	131	(19)	152	(22)	179	(26)

unexpected service failure. This tendency is illustrated by the effects of alloying additions in ferritic steels on cracking in two different environments (Ref 91). Figure 52(a) illustrates that each of the alloying additions is beneficial in the carbonate-bicarbonate solution, with molybdenum having the greatest effect. However, the molybdenum addition has an adverse effect in the 35% sodium hydroxide solution, although the beneficial effects of nickel and chromium additions remain the same (Fig. 52b). Although nickel additions are beneficial in the above example, a similar addition of nickel to a carbon-manganese steel produced susceptibility to stress-corrosion cracking in boiling magnesium chloride, which did not occur in the steel without the addition of nickel (Ref 92).

The use of laboratory testing mediums that duplicate service conditions is equally important when accelerated tests are used for quality control via the acceptance or rejection of production lots of a particular alloy. Reference 93 discusses tests of prestressing steels for use in concrete (rebar) in which an ammonium thiocyanate solution was used to discriminate between heats of steel.

When using carbonate-bicarbonate solutions for testing pipeline steels by the slow strain rate method, the susceptibility to stress-corrosion cracking was found to be dependent on the electrochemical potential of the specimen surface in the test environment, as shown in Fig. 52. A critical range in which stress-corrosion cracking occurred was established. The critical range varies with the test environment and alloy composition. Several tests at various carbonate-bicarbonate concentrations, temperatures, pH levels, and corrosion potentials indicated that test conditions using an impressed potential of −650 mV saturated calomel electrode and a temperature of 75 °C (167 °F) were optimal for various studies (Fig. 11, 47).

Testing of High-Strength Steels (Ref 2, 94)

For steels with yield strengths greater than about 690 MPa (100 ksi)—low-alloy and alloy steels, hot work die steels, maraging steels, and martensitic and precipitation-

hardenable stainless steels—the environments that cause stress-corrosion cracking are not specific. In many alloy systems, the phenomena of stress-corrosion cracking and hydrogen embrittlement cracking are indistinguishable (Fig. 2). This is particularly the case in environments that contain sulfides or other promoters of hydrogen entry.

Environments of major concern are natural waters—rainwater, seawater, and atmosphere moisture. Any of these environments

Fig. 52 Comparison of stress-corrosion cracking behavior of a low-alloy ferritic steel with various alloying additions in two different corrosive environments

Behavior indicated by time-to-failure ratios in a slow strain rate test. (a) Immersed in 1 N sodium carbonate plus 1 N sodium bicarbonate at 75 °C (165 °F). (b) Immersed in boiling 35% sodium hydroxide. Source: Ref 91

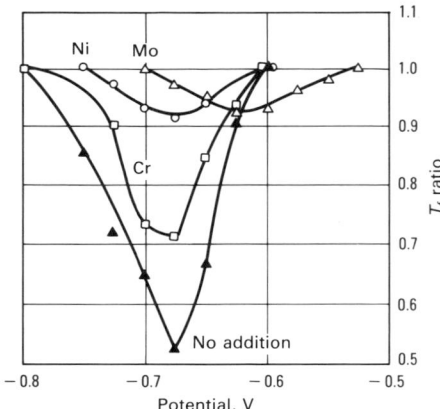

(a)

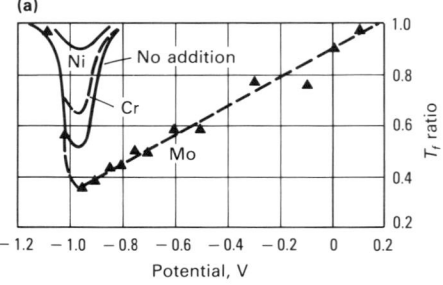

(b)

may become contaminated, which significantly increases the likelihood of stress-corrosion cracking. Contamination with hydrogen sulfide is particularly serious; consequently, the presence of hydrogen sulfide in high concentrations in salt water associated with certain deep oil wells (termed "sour crude" wells) places an upper limit of approximately 620 MPa (90 ksi) on the yield strength that can be tolerated in stressed steel in such environments without cracking.

Sulfide Stress Cracking. Determination of sulfide stress cracking is covered in the National Association of Corrosion Engineers (NACE) Standard TM-01-77 (Ref 95). Stressed specimens are immersed in acidified 5% sodium chloride solution saturated with hydrogen sulfide at ambient pressure and temperature. The solution is acidified with the addition of 0.5% acetic acid, yielding an initial pH of approximately 3. Applied stress at convenient increments of the yield strength is used to obtain cracking data that are plotted as shown in Fig. 53. A 30-day test period is considered sufficient to reveal failure of susceptible material in most cases.

The purpose of this test method is to facilitate conformity in testing. Evaluation of data requires individual judgment on several points based on the specific requirements of the end use. Consequently, the test should not be used as a single criterion for evaluating an alloy for use in environments containing hydrogen sulfide or other hydrogen charging elements. Attention should be paid to other factors that may affect stress-corrosion cracking, such as pH, temperature, hydrogen sulfide concentration, corrosion potential, stress level, etc., when determining the suitability of a metal for use.

The NACE test method recommends the use of smooth, small-diameter tension specimens stressed with constant-load or sustained-load devices (Ref 95). However, different types of beam and fracture mechanics specimens may be included in the testing standard in the future.

Testing in sodium chloride solution constitutes a "worst case" determination for high-strength steels; as such, it is generally considered unrealistically aggressive for useful ranking of steels in service environments that do not contain hydrogen sulfide or other conditions favoring entry of hydrogen. Tests usually are performed in water containing about 3.5% sodium chloride, artificial seawater, natural seawater (rarely), or a marine atmosphere (Ref 2), unless specific environmental conditions are under study. ASTM G 44 is used where applicable.

In salt water and fresh water, a true threshold K_{Iscc} exists for high-strength steels, which is useful for characterizing resistance to stress-corrosion cracking (Fig. 40). Ideally, K_{Iscc} defines the combination of applied stress and defect size, below which stress-corrosion cracking will not occur under static loading conditions in a given alloy and environment system. However, the reported value of K_{Iscc} for a given system often reflects the initial K_I level and the exposure time associated with the testing. Table 5 illustrates the risk of overestimating K_{Iscc} by terminating the exposure test too soon when using the stress-corrosion cracking initiation method (Ref 1, 9). A similar risk exists in tests conducted with the arrest method. Table 6 shows that K_{Iscc} values determined by the initiation and arrest methods may be the same when the testing times are sufficiently long and when compatible criteria are used for establishing the threshold (Ref 1).

Figure 54 illustrates a method used to compare various high-strength steels (Ref 2, 96). Data were obtained in salt water or seawater, and K_{Iscc} values are plotted versus yield strength. Envelopes are used to enclose all known valid data for the various steels. The cross-hatched envelopes or individual data points represent the featured steels, which allows comparison with characteristics of the other steels. Straight lines in Fig. 54 illustrate how K_{Iscc} values relate to the maximum depth of long surface flaws that can be tolerated without stress-corrosion crack growth.

Electrochemical Polarization. Although the mechanism of cracking in hydrogen sulfide environments is predominantly one of hydrogen embrittlement, the mechanism of environmentally induced failures in environments not containing sulfides or other promoters of hydrogen entry is not clearly agreed upon (Ref 94). Time-to-failure in a sodium chloride solution depends on the corrosion potential (Ref 2, 97), which also determines whether failure results from active path corrosion or hydrogen embrittlement cracking.

Figure 55 compares the various types of cracking behavior that can be expected from electrochemical polarization (Ref 98). All of the curves were obtained experimentally, except for curve G. Curve A represents the case where only hydrogen embrittlement is obtained, whereas curve B shows only active path corrosion. Both processes are shown in curves C and D.

When both anodic and cathodic polarization shorten the cracking time, as in curve E, it is not possible to determine which mechanism prevails without applied current. Curves F and G can be expected in acid solutions when the corrosion potential is anodic to the reversible hydrogen potential.

In curve H, neither anodic nor cathodic polarization has any effect on cracking time. Thus, it is possible that a hydrogen embrittlement mechanism is involved. However, the mechanism by which hydrogen enters the steel is not electrochemical. To perform realistic accelerated tests, the end use of the material and the environmental conditions involved should be considered so that the test procedure represents the appropriate cracking mechanism.

Testing of Non-Heat-Treatable Stainless Steels

The corrosive environments causing stress-corrosion cracking that are encountered in the chemical industry are specific and are limited primarily to chloride and caustic solutions at elevated temperatures and sulfide environments at ambient temperatures. In seawater at or near room temperature, austenitic steels (iron-chromium-nickel alloys) and ferritic (iron-chromium) steels do

Table 5   Influence of cut-off time on apparent K_{Iscc} using the stress-corrosion cracking initiation method

Exposure time, h	Apparent K_{Iscc}	
	MPa$\sqrt{m}$	ksi$\sqrt{in.}$
100	187	170
1 000	127	110
10 000	28	25

Note: The initiation method was used on a constant-load cantilever bend specimen (K-increasing) of high-alloy steel with a yield strength of 1240 MPa (180 ksi). Test environment: synthetic seawater at room temperature

Table 6   Comparison of K_{Iscc} values determined by initiation and arrest methods

Steel alloy	K_{Iscc}, MPa$\sqrt{m}$ (ksi$\sqrt{in.}$)	
	Initiation	Arrest
10Ni, normal purity	24 (22)	26 (24)
10Ni, high purity	59 (54)	57 (52)
18Ni (250), normal purity	22-33 (20-30)	28 (25)
18Ni (250), high purity	<33 (<30)	<33 (<30)

Note: Based on a crack growth rate of 2.5×10^{-4} mm/h (10^{-5} in./h). Modified compact specimens: constant load for initiation and wedge-loaded with a bolt for arrest. Test environment: salt water at room temperature

Fig. 53 Method of plotting results of sulfide stress cracking tests

Open symbols indicate failure. Closed symbols indicate runouts. Source: Ref 95

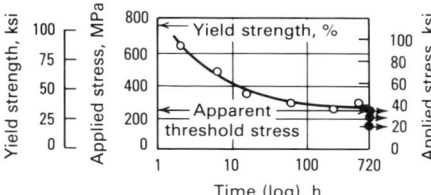

Fig. 54 Comparison of stress-corrosion cracking behavior of several high-strength steel alloys based on threshold stress-intensity values in salt water
Source: Ref 96

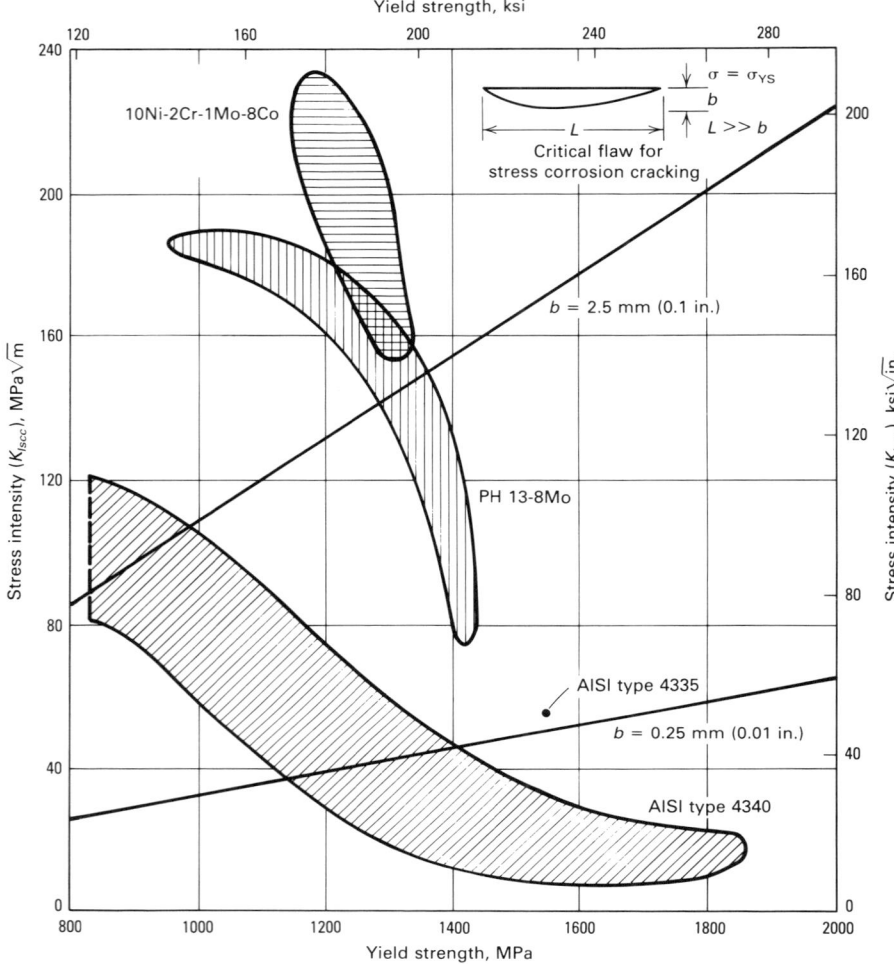

not experience stress-corrosion cracking. Fully ferritic stainless steels are highly resistant to stress-corrosion cracking in chloride and caustic environments that cause austenitic stainless steels to crack. However, laboratory studies have shown that additions of small amounts of nickel or copper to ferritic steels may render them susceptible to stress-corrosion cracking in severe environments (Ref 2, 85).

Testing in Boiling Magnesium Chloride Solution. ASTM G 36, "Standard Practice for Performing Stress-Corrosion Cracking Tests in a Boiling Magnesium Chloride Solution," is applicable to wrought, cast, and welded stainless steels and related nickel-based alloys. This method determines the effects of composition, heat treatment, surface finish, microstructure, and stress on the susceptibility of these mate-

rials to chloride stress-corrosion cracking. Although this test can be performed using various concentrations of magnesium chloride, ASTM G 36 recommends a test solution maintained at a constant boiling temperature of 155.0 ± 1.0 °C (311.0 ± 1.8 °F), i.e., approximately 45% magnesium chloride. Test apparatus capable of maintaining solution concentration and temperature within the recommended limits for extended periods of time is also described. Typical exposure times are up to 1000 h.

Most chloride cracking testing has been carried out in accelerated test mediums such as boiling magnesium chloride (Ref 2, 99, 100). All austenitic stainless steels are susceptible to chloride cracking, with no significant differences exhibited among them, as shown in Fig. 56. Although this solution causes rapid cracking, it does not necessarily

simulate the cracking observed in field applications.

Other ions in addition to chloride can cause cracking. Thus, in diagnosing service failures, it is necessary to establish which ions (and other environmental and stress conditions as well) have caused failure. In this manner, an appropriate test procedure can be designed for the evaluation of alternative materials.

Reference 102 discusses laboratory reproduction of an environment that caused stress-corrosion cracking at the top of a distillation tower in a crude oil refinery. The service environment consisted of a very dilute hydrochloric acid solution (36 ppm chloride) with a pH of 3.0 saturated with hydrogen sulfide gas at 80 °C (175 °F). In this test environment, austenitic stainless steels such as type 304 or type 316 failed, whereas type 430 and type 434 did not.

Testing in Polythionic Acids. Petrochemical refinery equipment is subject to polythionic acid cracking, which may occur after shutdown. Polythionic acid forms by decomposition of sulfides on metal walls in the presence of oxygen and water. ASTM G 35, "Standard Practice for Determining the Susceptibility of Stainless Steels and Related Nickel-Chromium-Iron Alloys to Stress-Corrosion Cracking in Polythionic Acids," describes procedures for preparing and conducting exposures to polythionic acids ($H_2S_nO_6$, where n is usually 2 to 5) at room temperature (22 to 25 °C, or 72 to 77 °F) to determine the relative susceptibility of sensitized stainless steels or other related materials (high nickel-chromium-iron alloys) to intergranular stress-corrosion cracking.

This test method can be used to evaluate stainless steels or other material in the as-received condition, or after high-temperature service (480 to 815 °C, or 900 to 1500 °F) for prolonged periods of time. Wrought products, castings, and weldments of stainless steels or other related materials used in environments containing sulfur or sulfides can also be evaluated. Other materials that are capable of being sensitized can also be tested.

A variety of smooth stress-corrosion test specimens, surface finishes, and methods of applying stress can be used. Stressed specimens are immersed in the polythionic acid solution, which can be prepared by passing a slow current of hydrogen sulfide gas for 1 h through a fritted glass tube into a flask containing chilled (0 °C, or 32 °F) 6% sulfurous acid, after which the liquid is kept in a stoppered flask for 48 h at room temperature. Solutions can also be prepared by passing a slow current of sulfur dioxide gas through a fritted glass bubbler submerged in a container of distilled water at room temperature.

Fig. 55 Use of electrochemical polarization to distinguish between stress-corrosion and hydrogen embrittlement mechanisms in a high-strength steel immersed in sodium chloride solution

See text for explanation of curves A through H. Source: Ref 98

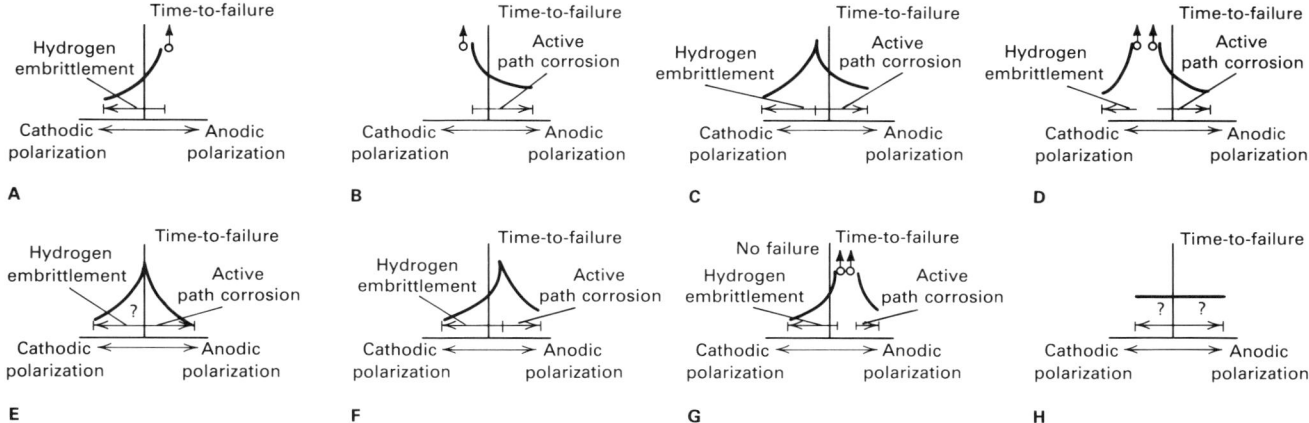

This is continued until the solution becomes saturated, then the hydrogen sulfide gas is slowly bubbled into the sulfurous acid solution.

Prior to use, the polythionic acid solution should be filtered to remove elemental sulfur and then tested for acid content. This can be done by analytical tests or by using a control test specimen of sensitized type 302 stainless steel. The control should fail by cracking in less than 1 h.

The wick test can be used to evaluate the chloride cracking characteristics of thermal insulation for applications in the chemical process industry. ASTM C 692, "Standard Method of Evaluating the Influence of Wicking Type Thermal Insulations on the Stress Corrosion Cracking Tendency of Austenitic Stainless Steel," covers the methodology and apparatus used to conduct this procedure. When a dilute aqueous solution is transmitted to a metal surface by capillary action through an absorbent fibrous material, the process is called wicking. Cracking occurs at much lower temperatures when alternate wetting and drying is used than when the specimens are kept wet continuously.

Other Testing Mediums. Hot concentrated caustic solutions are another type of environment encountered in chemical industries that causes stress-corrosion cracking of stainless steels. However, the conditions leading to caustic cracking are more restrictive than those leading to chloride cracking, and caustic environments have not received the attention that chlorides have. There is little difference in the susceptibilities among types 304, 304L, 316, 316L, 347, and USS 18-18-2 austenitic steels. All of these alloys crack rapidly in solutions of 10 to 50% sodium hydroxide at 150 to 370 °C (300 to 700 °F) (Ref 2, 100, 103).

Certain strong acid solutions containing chlorides, such as 5 N sulfuric acid plus 0.5 N sodium chloride, 3 N perchloric acid plus 0.5 N sodium chloride, and 0.5 N to 1.0 N hydrochloric acid are capable of causing stress-corrosion cracking in austenitic stainless steels at room temperature (Ref 2). Cracking in these environments is similar to the type of cracking that occurs in hot chloride environments.

Electrochemical Polarization. Stress-corrosion cracking in austenitic and ferritic stainless steels can be delayed or prevented by the application of cathodic current; however, if ferritic steels are overprotected by relatively large cathodic current, they are apt to blister or crack due to the hydrogen discharged by the cathodic protection action. Anodic polarization significantly accelerates the initiation of stress-corrosion cracking, but appears to have a smaller accelerating effect on crack propagation (Ref 104).

Testing of Magnesium Alloys

There is no standard accelerated test environment recommended for assessing the susceptibility of magnesium-based alloys to stress-corrosion cracking. Exposure of stressed specimens to the atmosphere generally has been used to determine the stress-

Fig. 56 Relative stress-corrosion cracking behavior of austenitic stainless steels in boiling magnesium chloride

Source: Ref 101

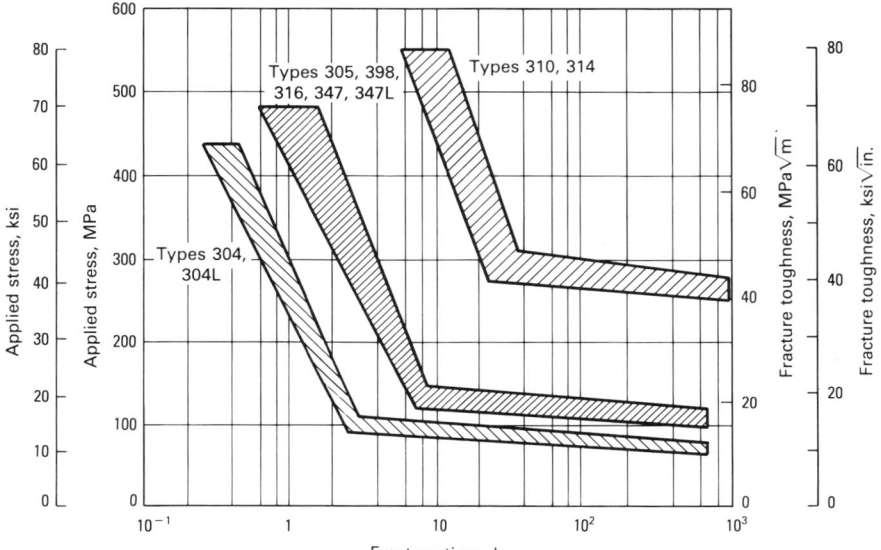

corrosion susceptibility of specific products.

The chloride-containing solutions typically used in accelerated tests for aluminum alloys are unsatisfactory for stress-corrosion cracking tests of magnesium alloys due to excessive general corrosion. A chromate-inhibited chloride solution (35 g/L sodium chloride plus 20 g/L potassium chromate with a pH of 8) was found suitable for testing magnesium alloys (Ref 105). Good correlation was found between the stress-corrosion cracking behavior of magnesium-aluminum-zinc alloys exposed by total immersion in this solution and the behavior of the same alloy exposed to an industrial atmosphere. Cracking of highly stressed susceptible alloys occurs within a few hours, but exposures can be continued up to 1000 h without incurring excessive pitting. Laboratory tests also have been conducted using potassium hydrogen fluoride and a dilute solution of sodium chloride plus sodium bicarbonate as the test medium (Ref 71).

Testing of Nickel Alloys

Nickel-based alloys are highly resistant to the chloride stress-corrosion cracking that affects stainless steels. Iron-chromium-nickel alloys with nickel contents greater than 50% are immune to cracking in boiling 42% magnesium chloride (Fig. 57). However, stress-corrosion cracking of nickel and high-nickel alloys has been experienced in high-temperature caustic soda and caustic potash solutions and in molten caustic.

Cracking of some nickel-based alloys also has occurred under special conditions in fluorosilic acid, hydrofluoric acid, mercuric salt solutions, and high-temperature water and steam that are contaminated with trace amounts of oxygen, lead, fluorides, or chlorides (Ref 99, 106, 107). Sensitized alloys are susceptible to stress-corrosion cracking in sulfur compounds such as sodium sulfite, sodium thiosulfate, and polythionic acids.

The standard test environments that are most frequently used for high-nickel alloys are the same as for stainless steels. In a study of sulfur-induced stress-corrosion cracking of sensitized Inconel 600 steam generator tubing in water contaminated by air and sodium thiosulfate at temperatures ranging from 22 to 95 °C (72 to 203 °F), a solution of 0.1 M sodium tetrathionate with a pH of 3.5 to 4.0 at 22 °C (72 °F) appeared to be an excellent test medium for sensitization in nickel alloys and stainless steels. Slow strain rate testing was also found to be more effective than tests with statically loaded U-bend specimens (Ref 108).

Slow strain rate testing was also effective for evaluating several nickel- and cobalt-based alloys in hot chloride and hot caustic solutions. The average length of secondary

stress-corrosion cracking, as determined by metallographic examination, appeared to be a more appropriate parameter for quantifying the severity of stress-corrosion behavior than loss in ductility or loss in fracture strength parameters (Table 7). However, it was noted that when using slow strain rate testing methods, care must be taken not to confuse stress-assisted localized corrosion with stress-corrosion cracking (Ref 109).

Testing of Titanium Alloys

Although titanium alloys are not susceptible to stress-corrosion cracking in either boiling 42% magnesium chloride or boiling 10% sodium hydroxide solutions, which are commonly used to study stress-corrosion cracking in stainless steels, the susceptibility of titanium and its alloys to stress-corrosion cracking has been demonstrated in several environments, as summarized in Table 8.

Testing in a Hot Salt Environment. The hot salt test consists of exposing a stressed salt-coated test specimen to an elevated temperature for various predetermined lengths of time. The exposure periods are determined by the alloy, stress level, temperature, and selected damage criterion (that is, embrittlement, cracking, or rupture, or a combination of these phenomena). Exposures typically are carried out in laboratory ovens or furnaces equipped with loading equipment for stressing specimens. Environmental conditions, the degree of control required, and the means for obtaining control are described in ASTM G 41, "Standard Practice for Determining Crack Susceptibility of Metals Exposed Under Stress to a Hot Salt Environment."

This recommended practice can be used to test all metals if service conditions warrant. The test limits maximum operating temperatures and stress levels, or categorizes different alloys as to their susceptibility, if hot salt

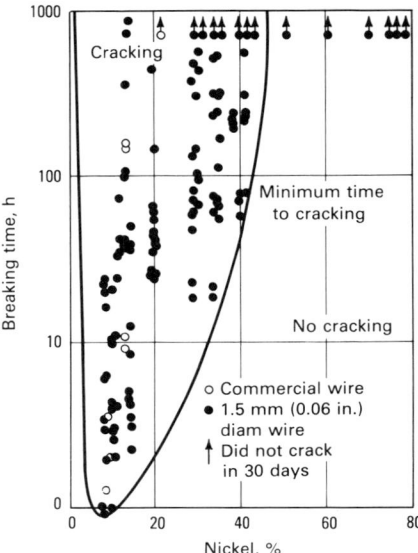

Fig. 57 Effect of nickel additions to a 17 to 24% chromium steel on the resistance to stress-corrosion cracking in boiling 42% magnesium chloride

1.5-mm (0.06-in.) diam wire specimens dead-weight loaded to 228 or 310 MPa (33 or 45 ksi). Source: Ref 106

damage has been found to accelerate failure by creep, fatigue, or rupture. Although limited evidence relates this phenomenon to actual service failures, cracking under stress in a hot salt environment is a potential design-controlling factor.

The hot salt test should not be misconstrued as being related to the stress-corrosion cracking of materials in other environments. It should be used only in an environment that may be encountered in service.

Table 7 Results of slow strain rate tests on nickel-based alloy C-276

Alloy condition	Strain rate, s^{-1}	Environment	$\Delta A/A$, %	Ultimate tensile strength MPa	ksi	Time-to-failure, h	Average length of secondary stress-corrosion cracks
Mill annealed	3.4×10^{-6}	Air	71	745	108	60	0
	3.4×10^{-6}	50% sodium hydroxide, 147 °C (297 °F)	61	593	86	52	13×10^{-5} m (5 mils)
50% cold swaged	3.4×10^{-6}	Air	49	1524	221	17	0
	3.4×10^{-6}	50% sodium hydroxide, 147 °C (297 °F)	51	1503	218	18	$<1 \times 10^{-5}$ m (<0.1 mils), no obvious stress-corrosion cracking
	9×10^{-7}	Air	53	1558	226	29	0
	9×10^{-7}	50% sodium hydroxide, 147 °C (297 °F)	47	1524	221	30	2.5×10 m (1 mil)
	5.3×10^{-7}	Air	51	1593	231	51	0
	5.3×10^{-7}	50% sodium hydroxide, 147 °C (297 °F)	47	1565	227	60	4.8×10^{-5} m (1.9 mils)

Table 8 Environments and temperatures conducive to stress-corrosion cracking of titanium alloys

Environment	Temperature
Hot dry chloride salts	260-480 °C (500-900 °F)
Seawater, distilled water, aqueous solutions	Ambient
Nitric acid, red fuming	Ambient
Nitrogen tetroxide	Ambient to 75 °C (165 °F)
Methyl and ethyl alcohols	Ambient
Chlorine	Elevated
Hydrogen chloride	Elevated
Hydrochloric acid, 10%	Ambient to 40 °C (105 °F)
Trichloroethylene	Elevated
Trichlorofluoroethane	Elevated
Chlorinated diphenyl	Elevated

Source: Ref 85

Hot salt testing can be used for alloy screening to determine the relative susceptibility of metals to embrittlement and cracking and to determine the time-temperature-stress threshold levels for the onset of embrittlement and cracking. However, certain types of specimens are more suitable for each of these types of characterizations. For titanium alloys, precracked specimens are unsuitable for testing, because cracking reinitiates at salt-metal-air interfaces, thus resulting in many small cracks that extend independently. For this reason, use of smooth specimens is recommended.

Testing in Water and Aqueous Solutions. Water, seawater, and almost any neutral aqueous solution (except atmospheric water vapor) may cause stress-corrosion cracking in many titanium alloys in the presence of preexisting crack-like flaws, although susceptibility in these environments cannot be detected by smooth specimens. Therefore, fracture-mechanics-type characterizations are necessary. For titanium alloys, the extremely rapid growth of stress-corrosion cracks in salt water and the dependency on specimen geometry precludes the possibility of design use based on crack growth rate data.

Therefore, ranking of materials must be based on K_{Iscc} values, and a true threshold stress intensity for stress-corrosion cracking apparently does exist. Titanium alloys do not exhibit region I type crack growth kinetics (Fig. 5) in neutral aqueous solutions. Tests have been performed for sufficient periods of time to allow detection rates of 10^{-9} m/s (1.4×10^{-4} in./h), but stress-corrosion cracking was not observed. The slowest crack velocity that has been detected is 10^{-8} m/s (1.4×10^{-3} in./h). Thus, in neutral aqueous solutions, a threshold K_{Iscc} exists at which stress-corrosion cracking will not propagate (Ref 2, 110). The above rates, however, are not as slow as those observed in

high-susceptibility aluminum alloys (Fig. 12). Tests are commonly performed in water containing about 3.5% sodium chloride, artificial seawater, or natural seawater, unless specific environments are being tested.

Electrochemical Polarization. The halide ions (chloride, bromide, and iodide) are stress-corrosion cracking agents unique to titanium alloys in aqueous solutions at room temperature. The crack initiation load and velocity are controlled by the applied potential, as illustrated for the crack initiation load in Fig. 58 (Ref 110, 111). At potentials more negative than about −700 to −1400 mV, depending on the solution, specimens were cathodically protected. The sodium fluoride solution and solutions of the other anions that do not produce stress-corrosion cracking (hydroxide, sulfide, sulfate, nitrite, nitrate, perchlorate, cyanide, and thiocyanate) yielded results at all potentials in the same scatterband as the air values.

At potentials more positive than the above values, susceptibility in varying degrees occurred in the chloride, bromide, and iodide solutions. The width of the critical potential range and the potential for maximum susceptibility varies with the anion. A region of anodic protection occurred in the chloride and bromide solutions, but not in the iodide solution.

Crack propagation can be stopped by switching the potential to either the anodic or

cathodic protection zone. The corrosion potential of titanium alloys in 3.5% sodium chloride and seawater, about −800 mV saturated calomel electrode, is similar (slightly more negative) to the potential at which stress-corrosion cracking susceptibility reaches a maximum (Ref 110).

Testing in Organic Fluids. A wide variety of organic fluids may cause stress-corrosion cracking in some alloys under specific test conditions (Table 8). Most of these fluids attack the passive surface film that is characteristic of titanium alloy products. Consequently, precracked specimens do not have to be used to accelerate the initiation of stress-corrosion cracking. A standard environment does not exist. As such, test conditions must be selected with appropriate consideration given to the type of environmental service required.

Sustained-Load Cracking in Inert Environments. High-strength titanium alloys for use in highly stressed components for military aircraft and other similar applications may be susceptible to sustained-load cracking in inert environments (including dry air). Sustained-load cracking is similar to stress-corrosion cracking except that it is much slower and occurs in the total absence of a reactive environment. Sustained-load cracking is caused by, or is greatly aggravated by, hydrogen dissolved in the titanium during processing. Vacuum annealing can

Fig. 58 Variation of crack initiation load with potential in 0.6 M halide solutions for Ti-8Al-1Mo-1V

Specimen: single-edge-cracked sheet tension-loaded by constant displacement

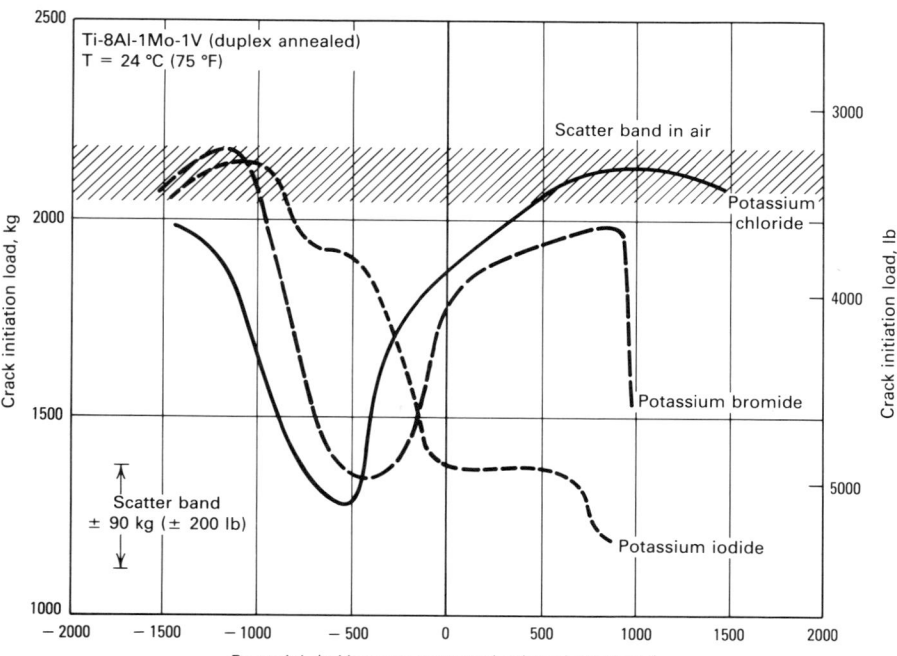

reduce the hydrogen level to less than 10 ppm, at which concentration the tendency toward sustained-load cracking is greatly reduced (Ref 2, 112).

An example of the occurrence of sustained-load cracking in mill-annealed plate of Ti-8Al-1V containing 48 ppm hydrogen is illustrated in Fig. 59. As shown in Fig. 59, the threshold stress-intensity factor for sustained-load cracking in dry air is designated K_{IH} because it is attributed to hydrogen in the metal. When the hydrogen concentration was reduced to 2 ppm by vacuum annealing, the K_{IH} value was increased to equal the inherent plane-strain fracture toughness, K_{Ix}. However, the K_{Iscc} value was not affected (Ref 113). Thus, in addition to the practical importance of sustained-load cracking, its potential contribution to cracking should be taken into account when evaluating environmental effects, particularly in mechanistic studies.

Electrochemical Tests

Recognition of the importance of potential as one of the controlling parameters in stress-corrosion cracking has resulted in increasing use of tests with potentiostatic control or impressed electrical current. No standards exist for such electrochemical tests, although several methods are used routinely. These tests involve specific conditions that are applicable only to given alloy and environment systems under consideration. These types of tests offer greater rapidity and precision than free corrosion tests.

Sophisticated approaches for studying systems involving active path corrosion use potentiodynamic methods. Anodic polarization curves can provide reasonably accurate predictions of the critical potential ranges and kinetic factors controlling stress-corrosion cracking in a given system. One procedure described by Parkins (Ref 114) involves first scanning a range of potentials in the anodic direction at a relatively high scan rate (~1000 mV/min) to determine regions with high current density, in which intense anodic activity is likely. This is followed by a relatively low scan rate (~20 mV/min), which indicates regions in which relative inactivity is likely.

Scans should be started at a potential at which the surface is film free. The rapid scan minimizes film formation so that the observed currents relate to relatively film-free conditions. The slow scan allows time for filming to occur. Comparison of the two curves reveals any ranges of potential within which high anodic activity in the film-free region reduces to insignificant activity when the time requirements for film formation are met, thus indicating the critical potential range in which stress-corrosion cracking is likely.

Figure 60 illustrates this type of experiment for a low-carbon steel in a carbonate-bicarbonate solution with the predicted domains of behavior. Establishing a precise range of critical potentials for stress-corrosion cracking requires subjective determination of boundary conditions relating to the current densities.

A different approach with the use of anodic polarization curves has been used to characterize the stress-corrosion cracking behavior of aluminum alloys (Ref 115). Corrosion of a susceptible microstructure in aqueous chloride solutions was exclusively intergranular only for a limited range of potentials between the first and second breakdown potentials (E_{BR}) determined from the anodic polarization curve, where E_{BR_1} approximates the critical pitting potential of the active corrosion path at the grain boundaries and E_{BR_2} approximates the critical pitting potential of the grain bodies. Figure 61 depicts a schematic anodic polarization curve for aluminum alloy 7075-T651 that illustrates the predicted domains of behavior.

ASTM G 5, "Standard Practice for a Standard Reference Method for Making Potentiostatic and Potentiodynamic Anodic Polarization Measurements," covers the use of anodic polarization curves and related measurements. Information obtained from the anodic polarization curve can be useful in developing optimum environmental test mediums to reduce the number of actual stress-corrosion tests required. However, such electrochemical tests are concerned with testing a bulk environment in contact with an exposed surface, and it should not be assumed that such measurements represent the crack tip under conditions of crack growth. However, even when an environment is identified as innocuous, a potent local environment may exist, such as in a pit when the pitting potential is exceeded or in a crevice or precrack.

Moreover, when laboratory stress-corrosion tests involve cracks or crevices, the potential at the crack tip may differ appreciably from that measured at the surface where the crack or crevice emerges. In laboratory tests, it must be considered whether or not environmental changes in composition or potential are involved and how they may relate to service conditions (Ref 66).

REFERENCES

1. "Characterization of Environmentally Assisted Cracking for Design—State of the Art," National Materials Advi-

Fig. 59 Effect of sustained-load cracking compared to stress-corrosion cracking in Ti-8Al-1Mo-1V mill-annealed sheet

Hydrogen concentration, 48 ppm; yield strength, 850 MPa (123 ksi); cantilever bend specimen (T-S); B = 6.35 mm (0.25 in.). See Fig. 39 for an explanation of specimen orientation and fracture plane identification. Source: Ref 113

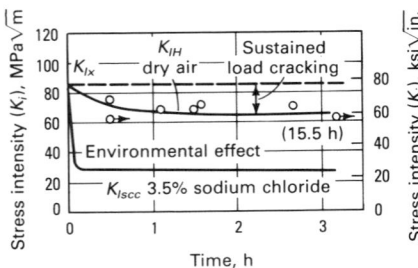

Fig. 60 Potentiodynamic polarization curves for carbon-manganese steel in 1 N sodium carbonate plus 1 N sodium bicarbonate at 90 °C (195 °F) showing the domains of behavior predicted from the curves

Source: Ref 114

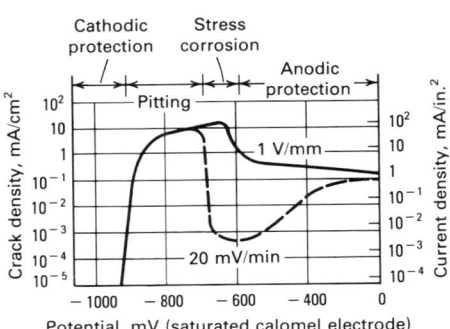

Fig. 61 Anodic polarization curves for aluminum alloy 7075-T651 in deaerated 3.5% sodium chloride solution showing the domains of behavior predicted from the curve

Source: Ref 116

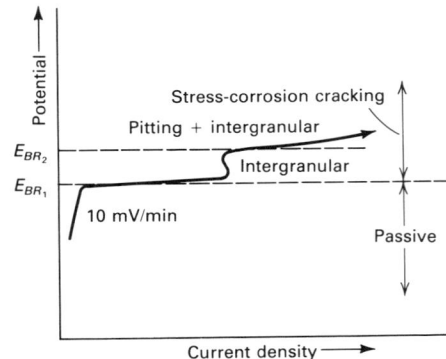

sory Board Report No. NMAB-386, National Academy of Sciences, Washington, DC, 1982

2. B.F. Brown, *Stress Corrosion Cracking Control Measures,* National Bureau of Standards Monograph 156, U.S. Department of Commerce, June 1977

3. A.W. Thompson, Current Status of the Role of Hydrogen in Stress-Corrosion Cracking, *Mat. Sci. Eng.,* Vol 43, 1980, p 41-46

4. L.C. Wasson, Designing a Corrosion Experiment, *Mat. Protect.,* Vol 9 (No. 2), 1970, p 31-33

5. E.H. Dix, Jr., "Symposium on Stress-Corrosion Cracking, Introduction," *Symposium on Stress-Corrosion Cracking of Metals,* ASTM and American Institute of Mechanical Engineers, New York, 1945, p 1-5

6. E.H. Dix, Jr., Aluminum-Zinc-Magnesium Alloys, Their Development and Commercial Production, *Trans. ASM,* Vol 42, 1950, p 1057-1127

7. R.N. Parkins, Stress Corrosion Test Methods—Physical Aspects, in *The Theory of Stress Corrosion Cracking in Alloys,* J.C. Scully, Ed., NATO Scientific Affairs Division, Brussels, 1971, p 449-468

8. B.F. Brown, in *Stress Corrosion Cracking in High Strength Steels and in Titanium and Aluminum Alloys,* Chap. I, B.F. Brown, Ed., Naval Research Laboratory, Washington, DC, 1972, p 13

9. R.P. Wei, S.R. Novak, and D.P. Williams, Some Important Considerations in the Development of Stress Corrosion Cracking Test Methods, *Mat. Res. Stand.,* Vol 12, 1972, p 25

10. J.W. Hutchinson, Singular Behavior at the End of a Tensile Crack in a Hardening Material, *J. Mech. Phys. Solids,* Vol 16, 1968, p 1-12

11. J.R. Rice and G.R. Rosengren, Plane Strain Deformation Near a Crack Tip in a Power Hardening Material, *J. Mech. Phys. Solids,* Vol 16, 1968, p 13-31

12. W.R. Wearmouth, G.P. Dean, and R.N. Parkins, Role of Stress in the Stress Corrosion Cracking of a Mg-Al Alloy, *Corrosion,* Vol 29 (No. 6), 1973, p 251-258

13. J.G. Kaufman, Stress Corrosion—Traditional Versus Fractured Mechanisms, *Corrosion,* Vol 35 (No. 4), April 1979, p i

14. D.O. Sprowls *et al.,* "A Study of Environmental Characterization of Conventional and Advanced Aluminum Alloys for Selection and Design: Phase II—The Breaking Load Test Method," Contract NASI-16424, NASA Contractor Report 172387, Aug 1984

15. R.N. Parkins, Development of Strain-Rate Testing and Its Implications, in *Stress-Corrosion Cracking—The Slow Strain-Rate Technique,* STP 665, G.M. Ugiansky and J.H. Payer, Ed., ASTM, Philadelphia, 1979, p 5-25

16. R.N. Parkins, 5th Symposium on Line Pipe Research, American Gas Association, Catalog No. L30174, 1974, U1-40

17. D.O. Sprowls and E.H. Spuhler, "Avoiding Stress-Corrosion Cracking in High Strength Aluminum Alloy Structures," Alcoa Green Letter 188 (Rev. 1982-01), Aluminum Company of America, Alcoa Center, PA

18. B.F. Brown, The Contributions of Physical Metallurgy and of Fracture Mechanics to Containing the Problem of Stress Corrosion Cracking, *Philos. Trans. Royal Soc. London,* Vol 282 (No. 1307), Series A, 1976, p 235-245

19. *Annual Book of ASTM Standards,* Section 3, Vol 03.02, ASTM, Philadelphia, 1984

20. D.O. Sprowls and R.H. Brown, What Every Engineer Should Know About the Stress Corrosion of Aluminum, *Met. Prog.,* Vol 81 (No. 4), April 1962, p 79-85; Vol 81 (No. 5), May 1962, p 77-83

21. M.O. Speidel, Stress Corrosion Cracking of Aluminum Alloys, *Met. Trans. A,* Vol 6, April 1975, p 631-651

22. R.N. Parkins *et al.,* Report Prepared for the European Federation of Corrosion Working Party on Stress Corrosion Test Methods, *Brit. Corrosion J.,* Vol 7, July 1982, p 154-167

23. G. Vogt, Comparative Survey of Type of Loading and Specimen Shape for Stress Corrosion Tests, *Werkstoffe Korrosion,* Vol 29, 1978, p 721-725

24. H.L. Craig, Jr., D.O. Sprowls, and D.E. Piper, Stress-Corrosion Cracking, in *Handbook on Corrosion Testing and Evaluation,* W.H. Ailor, Ed., John Wiley & Sons, New York, 1971, p 231-290

25. G. Haaijer and A.W. Loginow, Stress Analysis of Bent-Beam Stress Corrosion Specimens, *Corrosion,* Vol 21 (No. 4), 1965, p 105-112

26. M.B. Shumaker *et al.,* Evaluation of Various Techniques for Stress Corrosion Testing Welded Aluminum Alloys, in *Stress Corrosion Testing,* STP 425, ASTM, Philadelphia, 1967, p 317-341

27. R.A. Davis, Stress Corrosion Cracking Investigation of Two Low Alloy, High Strength Steels, *Corrosion,* Vol 19 (No. 2), 1963, p 45t-55t

28. German Standard DIN 50908, "Testing of Light Metals, Stress Corrosion Test," 1964

29. S.O. Fernandez and G.F. Tisinai, Stress Analysis of Un-notched C-rings Used for Stress Cracking Studies, *J. Eng. Ind.,* Feb 1968, p 147-152

30. F.S. Williams, W. Beck, and E.J. Jankowsky, A Notched Ring Specimen for Hydrogen Embrittlement Studies, *Proc. ASTM,* Vol 60, 1960, p 1192

31. D.O. Sprowls *et al.,* "Investigation of the Stress Corrosion Cracking of High Strength Aluminum Alloys," Final Technical Report for U.S. Government NASA Contract NAS 8-5340, Control No. 1-4-50-01167-01(lf), CPB-02-1215-64, 1967

32. Report of Task Group 1, of ASTM Subcommittee B-3/X, Stress Corrosion Testing Methods, *Stress Corrosion Testing,* STP 425, ASTM, Philadelphia, 1967, p 3-20; *Proc. ASTM,* Vol 65, 1965, p 182-197

33. A.W. Loginow, Stress Corrosion Testing of Alloys, *Mat. Protect.,* Vol 5 (No. 5), 1966, p 33-39

34. B.W. Lifka and D.O. Sprowls, Stress Corrosion Testing of 7079-T6 Aluminum Alloy in Various Environments, in *Stress Corrosion Testing,* STP 425, ASTM, Philadelphia, 1967, p 342-362

35. P. Brenner, Realistic Stress Corrosion Testing, *Metallurgy,* Vol 23 (No. 9), 1969, p 879-886

36. F.H. Haynie *et al.,* "A Fundamental Investigation of the Nature of Stress Corrosion Cracking in Aluminum Alloys," Technical Report AFML 66-267, June 1966, under USAF Contract No. AF 33(615)-1710

37. H. Nathorst, Stress Corrosion Cracking in Stainless Steels, Part II—An Investigation of the Suitability of the U-Bend Specimen, *Welding Res. Council Bull.,* Series No. 6, Oct 1950

38. B.F. Brown, A New Stress Corrosion Cracking Test for High Strength Alloys, *Mat. Res. Stand.,* Vol 6 (No. 3), 1966, p 129-133

39. B.F. Brown, The Application of Fracture Mechanics to Stress Corrosion Cracking, *Met. Rev.,* Vol 13, 1968, p 171-183

40. H.R. Smith and D.E. Piper, Stress Corrosion Testing with Precracked Specimens, in *Stress Corrosion Cracking in High Strength Steels and in Titanium and Aluminum Alloys,* B.F. Brown, Ed., Naval Research Laboratory, Washington, DC, 1972, p 17-78

41. "Standard Test Method for Plane-Strain Fracture Toughness of Metallic Materials," E 399, *Annual Book of ASTM Standards*, ASTM, Philadelphia, 1984, p 710-730

42. J.A. Hauser, II, R.W. Judy, Jr., and T.W. Crooker, "Draft Standard Method of Test for Plane-Strain Stress-Corrosion-Cracking Resistance of Metallic Materials in Marine Environments," NRL Memorandum Report 5295, Naval Research Laboratory, Washington, DC, March 1984

43. W.B. Lisagor, Influence of Precracked Specimen Configuration and Starting Stress Intensity on the Stress Corrosion Cracking of 4340 Steel, in *Environment-Sensitive Fracture: Evaluation and Comparison of Test Methods*, STP 821, S.W. Dean, E.N. Pugh, and G.M. Ugiansky, Ed., ASTM, Philadelphia, 1984, p 80-97

44. H. Tada, P. Paris, and G. Irwin, *The Stress Analysis of Cracks Handbook*, Del Research Corp., Hellertown, PA, 1973

45. J.A. Joyce, D.F. Hasson, and C.R. Crowe, Computer Data Acquisition Monitoring of the Stress Corrosion Cracking of Deplated Uranium Cantilever Beam Specimens, *J. Test. Eval.*, Vol 8 (No. 6), 1980, p 293-300

46. S.R. Novak and S.T. Rolfe, Modified WOL Specimen for K_{Iscc} Environmental Testing, *J. Met.*, Vol 4 (No. 3), 1969, p 701-728

47. J.G. Kaufman, J.W. Coursen, and D.O. Sprowls, An Automated Method for Evaluating Resistance to Stress-Corrosion Cracking with Ring-Loaded Precracked Specimens, in *Stress-Corrosion—New Approaches*, STP 610, H.L. Craig, Jr., Ed., ASTM, Philadelphia, 1976, p 94-107

48. C. Micheletti and M. Buratti, New Testing Methods for the Evaluation of the Stress-Corrosion Behavior of High-Strength Aluminum Alloys by the Use of Precracked Specimens, in *Symposium Proceedings, Aluminum Alloys in the Aircraft Industry*, Turin, Italy, 1-2 Oct 1976, Technicopy Ltd., England, 1978, p 149-159

49. M.V. Hyatt, Use of Precracked Specimens in Stress Corrosion Testing of High Strength Aluminum Alloys, *Corrosion*, Vol 26 (No. 11), 1970, p 487-503

50. D.O. Sprowls *et al.*, "Evaluation of Stress Corrosion Cracking Susceptibility Using Fracture Mechanics Techniques," Contract NAS 8-21487, Contractor Report NASA CR-124469, May 1973

51. R.C. Dorward and K.R. Hasse, "Flow Growth of 7075, 7475, 7050, and 7049 Aluminum Plate in Stress Corrosion Environments," Final Technical Report for U.S. Government Contract NAS 8-30890, Oct 1976 (also published in *Corrosion*, Vol 34 (No. 11), 1978, p 386-395)

52. W.B. Fichter, The Stress Intensity Factor for the Double Cantilever Beam, *Int. J. Fract.*, Vol 22, 1983, p 133-143

53. J.E. Srawley and B. Gross, Stress Intensity Factors for Crackline-Loaded Edge-Crack Specimens, *Mat. Res. Stand.*, Vol 7 (No. 4), 1967, p 155-162

54. S. Mostovoy, R.B. Crosley, and E.J. Ripling, Use of Crackline Loaded Specimens for Measuring Plane Strain Fracture Toughness, *J. Mat.*, Vol 2 (No. 3), 1967, p 661-681

55. D.O. Sprowls and J.D. Walsh, Evaluating Stress-Corrosion Crack Propagation Rates in High Strength Aluminum Alloys with Bolt Loaded Precracked Double Cantilever Beam Specimens, in *Stress Corrosion—New Approaches*, STP 610, H.L. Craig, Jr., Ed., ASTM, Philadelphia, 1976, p 143-156

56. T.L. Bond, R.A. Yeske, and E.N. Pugh, Studies of Stress Corrosion Crack Growth in Al-Zn-Mg Alloys by the Double Torsion Method, in *Environment-Sensitive Fracture: Evaluation and Comparison of Test Methods*, STP 821, S.W. Dean, E.N. Pugh, and G.M. Ugiansky, Ed., ASTM, Philadelphia, 1984, p 128-149

57. J.C. Lewis and G. Sines, Ed., *Fracture Mechanics: Fourteenth Symposium*, Vol II: Testing and Application, STP 791, ASTM, Philadelphia, 1983

58. S.W. Dean, E.N. Pugh, and G.M. Ugiansky, Ed., *Environment-Sensitive Fracture: Evaluation and Comparison of Test Methods*, STP 821, ASTM, Philadelphia, 1984

59. J.H. Underwood, S.W. Freiman, and F.I. Baratta, Ed., *Chevron-Notched Specimens: Testing and Stress Analysis*, STP 855, ASTM, Philadelphia, 1984

60. W.B. Gilbreath and M.J. Adamson, Aqueous Stress Corrosion Cracking of High Toughness D6AC Steel, in *Stress-Corrosion—New Approaches*, STP 610, H.L. Craig, Jr., Ed., ASTM, Philadelphia, 1976, p 176-187

61. N.J.H. Holroyd and G.M. Scamans, Slow-Strain-Rate Stress Corrosion Testing of Aluminum Alloys, in *Environment-Sensitive Fracture: Evalua-*

tion and Comparison of Test Methods, STP 821, S.W. Dean, E.N. Pugh and G.M. Ugiansky, Ed., ASTM, Philadelphia, 1984, p 202-241

62. J.H. Payer, W.E. Berry, and W.K. Boyd, Constant Strain Rate Technique for Assessing Stress-Corrosion Susceptibility, in *Stress-Corrosion—New Approaches*, STP 610, H.L. Craig, Jr., Ed., ASTM, Philadelphia, 1976, p 82-93

63. J.H. Payer, W.E. Berry, and R.N. Parkins, Application of Slow Strain-Rate Technique to Stress Corrosion Cracking of Pipeline Steel, in *Stress-Corrosion Cracking—The Slow Strain-Rate Technique*, STP 665, G.M. Ugiansky and J.H. Payer, Ed., ASTM, Philadelphia, 1979, p 222-234

64. G.M. Ugiansky and J.H. Payer, Ed., *Stress Corrosion Cracking—The Slow Strain-Rate Technique*, STP 665, ASTM, Philadelphia, 1979

65. R.N. Parkins and Y. Suzuki, Environment Sensitive Cracking of a Nickel-Aluminum Bronze Under Monotonic and Cyclic Loading Conditions, *Corrosion Sci.*, Vol 23 (No. 6), 1983, p 577-599

66. R.N. Parkins, A Critical Evaluation of Current Environment-Sensitive Fracture Test Methods, in *Environment-Sensitive Fracture: Evaluation and Comparison of Test Methods*, STP 821, S.W. Dean, E.N. Pugh, and G.M. Ugiansky, Ed., ASTM, Philadelphia, 1984, p 5-31

67. J. Yu, N.J.H. Holroyd, and R.N. Parkins, Application of Slow-Strain-Rate Tests to Defining the Stress for Stress Corrosion Crack Initiation in 70/30 Brass, in *Environment-Sensitive Fracture: Evaluation and Comparison of Tests Methods*, STP 821, S.W. Dean, E.N. Pugh, and G.M. Ugiansky, Ed., ASTM, Philadelphia, 1984, p 288-309

68. D.O. Sprowls and R.H. Brown, Stress Corrosion Mechanisms for Aluminum Alloys, in *Fundamental Aspects of Stress-Corrosion Cracking*, R.W. Staehle, A.J. Forty, and D. van Rooyen, Ed., National Association of Corrosion Engineers, Houston, 1969, p 466-512

69. J.D. Jackson, W.K. Boyd, and R.W. Staehle, "Corrosion of Titanium," DMIC Memorandum 218, Battelle Memorial Institute, Columbus, OH, Sept 1966

70. D.O. Sprowls *et al.*, Evaluation of a Proposed Standard Method of Testing for Susceptibility to SCC of High Strength 7XXX Series Aluminum Al-

loy Products, in *Stress-Corrosion—New Approaches*, STP 610, H.L. Craig, Jr., Ed., ASTM, 1976, p 3-31

71. H.B. Romans, Stress Corrosion Test Environments and Test Durations, in *Stress Corrosion Testing*, STP 425, ASTM, Philadelphia, 1967, p 182-208

72. W. Pistulka and G. Lang, Accelerated Stress Corrosion Test Methods for Al-Zn-Mg Type Alloys, *Aluminum*, Vol 53 (No. 6), 1977, p 366-371

73. P.W. Jeffrey, T.E. Wright, and H.P. Godard, An Accelerated Laboratory Stress Corrosion Test for Al-Zn-Mg Alloys, in *Proceedings of the Fourth International Congress on Corrosion*, National Association of Corrosion Engineers, Houston, 1969, p 133-139

74. T.S. Humphries and J.E. Coston, "An Improved Stress Corrosion Test Medium for Aluminum Alloys," NASA Technical Memorandum NASA TM-82452, George C. Marshall Space Flight Center, Alabama, Nov 1981

75. W.J. Helfrich, "Development of a Rapid Stress Corrosion Test for Aluminum Alloys," Final Summary Report, Contract No. NAS 8-20285, George C. Marshall Space Flight Center, Alabama, May 1968

76. F.F. Booth and H.P. Godard, An Anodic Stress-Corrosion Test for Aluminum-Magnesium Alloys, in *First International Congress on Metallic Corrosion*, Butterworths, London, 1962, p 703-712

77. L. Schra and J. Faber, "Influence of Environments on Constant Displacement Stress-Corrosion Crack Growth in High Strength Aluminum Alloys," National Aerospace Laboratory NLR, The Netherlands, NLR TR 81138 U, 1981

78. M.V. Hyatt and M.O. Speidel, High Strength Aluminum Alloys, in *Stress Corrosion Cracking in High Strength Steels and in Titanium and in Aluminum Alloys*, B.F. Brown, Ed., Naval Research Laboratory, Washington, DC, 1972, p 148-244

79. G.M. Scamans, Discontinuous Propagation of Stress Corrosion Cracks in Al-Zn-Mg Alloys, *Scripta Met.*, Vol 13, 1979, p 245-250

80. S. Maitra, Determinaton of SCC Resistance of Al-Cu-Mg Alloys by Slow Strain Rate and Alternate Immersion Testing, *Corrosion*, Vol 37 (No. 2), 1981, p 98-103

81. "Stress-Corrosion Cracking Testing of Aluminum Alloys for Aircraft Parts," (in German) German Aircraft Standard LN 65666, July 1974

82. D.H. Thompson, Stress Corrosion Cracking of Copper Metals, in *Stress Corrosion Cracking of Metals—A State of the Art*, STP 518, ASTM, Philadelphia, 1972, p 39-57

83. E.N. Pugh, J.V. Craig, and A.J. Sedricks, The Stress-Corrosion Cracking of Copper, Silver, and Gold Alloys, in *Proceedings of Conference on Fundamental Aspects of Stress-Corrosion Cracking*, Ohio State University, 1967, National Association of Corrosion Engineers, Houston, 1969, p 118-158

84. G. Edmunds, E.A. Anderson, and R.K. Waring, Ammonia and Mercury Stress-Corrosion Cracking Tests for Brass, in *Symposium on Stress-Corrosion Cracking of Metals*, ASTM and American Institute of Mining, Metallurgical, and Petroleum Engineers, New York, 1944, p 7-18

85. T. Lyman, Ed., *Metals Handbook*, Vol 10, 8th ed., *Failure Analysis and Prevention*, American Society for Metals, 1975

86. "Standard Method of Mercurous Nitrate Test for Copper and Copper Alloys," B 154, *Annual Book of ASTM Standards*, ASTM, Philadelphia, 1984, p 352-354

87. "Tubing of Copper and Copper Alloys for Condensers and Heat Exchangers," German Standard DIN-1785, 1983 (available from American National Standards Institute, New York)

88. H.L. Logan, *The Stress Corrosion of Metals*, John Wiley & Sons, New York, 1966

89. R.N. Parkins, Stress Corrosion Cracking of Low Carbon Steels, in *Proceedings of Conference on Fundamental Aspects of Stress Corrosion Cracking*, Ohio State University, 1967, National Association of Corrosion Engineers, Houston, 1964, p 361-373

90. M.J. Humphries and R.N Parkins, Stress Corrosion Cracking of Mild Steels in Sodium Hydroxide Solutions Containing Various Additional Substances, *Corrosion Sci.*, Vol 7, 1967, p 747-761

91. R.N. Parkins, P.W. Slattery, and B.S. Poulson, The Effects of Alloying Additions to Ferritic Steels Upon Stress-Corrosion Cracking Resistance, *Corrosion*, Vol 37 (No. 11), 1981, p 650-664

92. B.S. Poulson and R.N. Parkins, Effect of Nickel Additions Upon the Stress Corrosion of Ferritic Steels in a Chloride Environment, *Corrosion*, Vol 29, Nov 1973, p 414-422

93. "Stress Corrosion Cracking Resistance Test for Prestressing Tendons," Technical Report 5, Proceedings of the Federation Internationale de la Precontrainte, 8th Congress, London, Cement and Concrete Association, Wexham Springs, Slough, UK, 1978

94. E.H. Phelps, A Review of the Stress Corrosion Behavior of Steels with High Yield Strength, in *Proceedings of Conference on Fundamental Aspects of Stress Corrosion Cracking*, Ohio State University, 1967, National Association of Corrosion Engineers, Houston, 1969, p 398-410

95. "Test Method for Testing of Metals for Resistance to Sulfide Stress Cracking at Ambient Temperatures," National Association of Corrosion Engineers Standard TM-01-77, National Association of Corrosion Engineers, Houston, 1977, p 77-84

96. G. Sandoz, High Strength Steels, in *Stress Corrosion Cracking in High Strength Steels and in Titanium and in Aluminum Alloys*, B.F. Brown, Ed., Naval Research Laboratory, Washington, DC, 1972, p 79-145

97. H.J. Bhatt and E.H. Phelps, The Effect of Electrochemical Polarization on the Stress Corrosion Behavior of Steels with High Yield Strength, in *Proceedings of the Third International Congress on Metallic Corrosion*, Moscow, May 16-26, 1966

98. H.J. Bhatt and E.H. Phelps, Effect of Solution pH on the Mechanism of Stress Corrosion Cracking of a Martensitic Stainless Steel, *Corrosion*, Vol 17, 1961, p 430t-434t

99. R.M. Latanision and R.W. Staehle, Stress Corrosion Cracking of Iron-Nickel-Chromium Alloys, in *Proceedings of Conference on Fundamental Aspects of Stress Corrosion Cracking*, Ohio State University, 1967, National Association of Corrosion Engineers, Houston, 1969, p 214-307

100. S.W. Dean, Review of Recent Studies on the Mechanism of Stress Corrosion Cracking in Austenitic Stainless Steels, in *Stress Corrosion—New Approaches*, STP 610, H.L. Craig, Jr., Ed., ASTM, Philadelphia, 1976, p 308-337

101. E. Denhard, Effect of Composition and Heat Treatment on the Stress Corrosion Cracking of Austenitic Stainless Steels, *Corrosion*, Vol 16 (No. 7), 1960, p 131-141

102. S. Takemura, M. Onoyama, and T. Ooka, Stress Corrosion Cracking of Stainless Steels in Hydrogen Sulfide Solutions, *Corrosion*, Vol 16 (No. 7), 1960, p 338-348

103. J.L. Wilson, F.W. Pement, and R.G. Aspden, Effect of Alloy Structure, Hydroxide Concentration, and Temperature on the Caustic Stress-Corrosion Cracking of Austenitic Stainless Steel, *Corrosion*, Vol 30 (No. 4), 1974, p 139-149

104. H. Kohl, A Contribution to the Examination of Stress-Corrosion Cracking of Austenitic Stainless Steel in Magnesium Chloride Solution, *Corrosion*, Vol 23 (No. 12), 1967, p 39-49

105. G.F. Sager, R.H. Brown, and R.B. Mears, Tests for Determining Susceptibility to Stress Corrosion Cracking, *Symposium on Stress Corrosion Cracking of Metals*, ASTM and American Institute of Mechanical Engineers, New York, 1945, p 255-272

106. H.R. Copson, Effect of Composition on Stress Corrosion Cracking of Some Alloys Containing Nickel, in *Physical Metallurgy of Stress Corrosion Fracture*, T.N. Rhodin, Ed., Interscience, New York, 1959, p 247-272

107. W.K. Boyd and W.E. Berry, Stress Corrosion Cracking of Nickel and Nickel Alloys, in *Stress Corrosion Cracking of Metals—A State of the Art*, STP 518, ASTM, Philadelphia, 1972, p 58-78

108. R.C. Newman, R. Roberge, and R. Bandy, Evaluation of SCC Test Methods for Inconel 600 in Low-Temperature Aqueous Solutions, in *Environment-Sensitive Fracture: Evaluation and Comparison of Test Methods*, STP 821, S.W. Dean, E.N. Pugh, and G.M. Ugiansky, Ed., ASTM, Philadelphia, 1984, p 310-322

109. A.I. Asphahani, Slow Strain-Rate Technique and Its Applications to the Environmental Stress Cracking of Nickel-Base and Cobalt-Base Alloys, in *Stress Corrosion Cracking—The Slow Strain Rate Technique*, STP 665, G.M. Ugiansky and J.H. Payer, Ed., ASTM, Philadelphia, 1979, p 279-293

110. M.J. Blackburn, W.H. Smyrl, and J.A. Sweeny, Titanium Alloys, in *Stress Corrosion Cracking in High Strength Steels and in Titanium and in Aluminum Alloys*, B.F. Brown, Ed., Naval Research Laboratory, Washington, DC, 1972, p 246-363

111. T.R. Beck, Electrochemical Aspects of Titanium Stress Corrosion, in *Proceedings of Conference on Fundamental Aspects of Stress Corrosion Cracking*, Ohio State University, 1967, National Association of Corrosion Engineers, Houston, 1969, p 605-619

112. D.A. Meyn, Effect of Hydrogen on Fracture and Inert Environment Sustained Load Cracking Resistance of Alpha-Beta Titanium Alloys, *Met. Trans.*, Vol 5 (No. 11), 1974, p 2405-2414

113. G. Sandoz, Subcritical Crack Propagation in Ti-8Al-1Mo-1V Alloy in Organic Environments, Salt Water and Inert Environments, in *Proceedings of Conference on Fundamental Aspects of Stress Corrosion Cracking*, Ohio State University, 1967, National Association of Corrosion Engineers, Houston, 1969, p 684-690

114. R.N. Parkins, Predictive Approaches to Stress Corrosion Cracking Failure, *Corrosion Sci.*, Vol 20, 1980, p 147-166

115. J.R. Galvele and S.M. DeMichelli, Mechanism of Intergranular Corrosion of Al-Cu Alloys, *Corrosion Sci.*, Vol 10, 1970, p 795-807

116. S. Maitra and G.C. English, Mechanism of Localized Corrosion of 7075 Alloy Plate, *Met. Trans. A*, Vol 12, March 1981, p 535-541

Tests for Hydrogen Embrittlement

By L. Raymond
Independent Consulting Engineer
METTEK Laboratories

HYDROGEN EMBRITTLEMENT is a time-dependent fracture process caused by the absorption and diffusion of atomic hydrogen into a metal, which results in a loss in ductility and tensile strength. Hydrogen embrittlement is distinguished from stress-corrosion cracking (see the article "Tests for Stress-Corrosion Cracking" in this Section) generally by the interactions of the specimens with applied currents (Ref 1). Cases where the applied current makes the specimen more anodic and accelerates cracking are considered to be stress-corrosion cracking, with the anodic-dissolution process contributing to the progress of cracking. On the other hand, cases where cracking is accentuated by current in the opposite direction, which accelerates the hydrogen-evolution reaction, are considered to be hydrogen embrittlement.

Hydrogen-induced failures have been reported for parts that have been forged, heat treated, welded, chemically milled, pickled, or exposed to paint removers. More recently, hydrogen cracking has been a serious concern when hydrogen is used as the liquid fuel in engines for spacecraft and long-range missiles.

Hydrogen embrittlement has been detected on parts installed in boilers, pressurized-water reactors, high-pressure hydrogenation units, and parts protected cathodically. Hydrogen sulfide stress cracking has been a difficult problem for the petroleum industry. Hydrogen embrittlement presents problems in welding because of the inherent complex nature of a weld and the various sources of hydrogen.

Tests for hydrogen embrittlement are performed to determine the effect of hydrogen damage in combination with residual or applied stresses. In the past decade, conventional testing methods have been modified to incorporate fracture mechanics, and the various types of hydrogen damage have been classified further in terms of crack nucleation, crack growth rates, and threshold stress-intensity measurements.

This article discusses current methods of hydrogen embrittlement testing, and focuses on accelerated small-specimen testing methods for failure analysis and production control of hydrogen embrittlement. Additional information on hydrogen damage in metals and test methods for hydrogen embrittlement can be found in Ref 2 to 9.

Testing Methods

As described in the article "Tests for Stress-Corrosion Cracking" in this Section, the cantilever beam test and the wedge-opening load test result in a parameter called K_{Iscc}, which is the threshold stress intensity for stress-corrosion cracking. Many different designations, such as K_{Ith}, K_{Ihem}, and K_{Ish}, denote this parameter for steels that undergo a similar phenomenon in which the mechanism is internal hydrogen embrittlement.

In this article, the threshold stress intensity for hydrogen stress cracking is designated by K_{Ihem}, and K_{Iscc} is used for stress-corrosion cracking. The mechanisms are different in that stress-corrosion cracking occurs under anodic polarization conditions, whereas hydrogen embrittlement and hydrogen stress cracking occur under cathodic polarization conditions, which normally are generated to protect steels from corrosion. Such is the case when a sacrificial anode is galvanically coupled to the steel hull of a ship to prevent the hull from corroding. In such a couple, the steel is the cathode and hydrogen is produced at the cathode in an electrochemical reaction. This results in a steel structure, apparently free of corrosion (with a clean, metallic luster), that fails by intergranular cracking due to internal diffusion of hydrogen generated at the surface. This type of hydrogen embrittlement is found in types 410 and 17-4PH stainless steel and AISI type 4340 steel.

The cantilever beam test is a constant-load test in which a V-notched specimen is inserted along a portion of the beam and enclosed by an environmental chamber (Fig. 1). A crack at the root of the V-notch is initiated and extended by fatigue before testing. Notch-root thickness is prescribed by ASTM, although the requirement often is excessive for high-toughness steels. The specimen is subjected to a constant load over a preset time period. As the crack grows, the stress intensity increases. Time-to-failure is plotted versus applied stress intensity. The lower limit of the resultant curve is a threshold for hydrogen embrittlement (Fig. 2).

The K_{Ihem} results of a cantilever beam test depend on how much time elapses before the test is terminated. Recommended testing periods to establish the true stress-intensity threshold vary (Ref 11), ranging from 200 h, which is typical for hydrogen embrittlement testing, to as long as 5000 h. Another limitation of this testing method is that it can be expensive in terms of materials and machining. As many as 12 specimens, placed under

Fig. 1 Fatigue-cracked cantilever beam test specimen and fixtures
Source: Ref 10

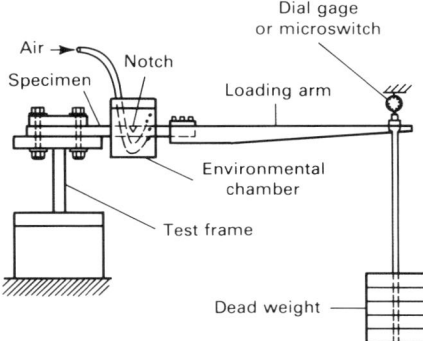

Fig. 2 Procedure to obtain K_{Ihem} with precracked cantilever beam test specimen
Source: Ref 10

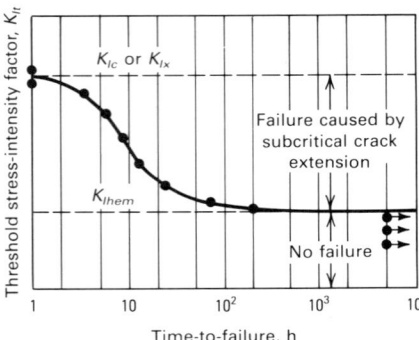

different loads in separate test machines, are needed per test to obtain valid K_{Ihem} values.

The wedge-opening load test applies a constant wedge or crack opening displacement; as the crack extends, stress intensity decreases until crack arrest occurs (Fig. 3). The initial load is assumed to be slightly above K_{Ihem}. The specimen is maintained under these conditions for about 5000 h to establish the threshold. The crack grows to a point after which further growth is not measured (K_{Ihem}). However, it is difficult to determine precisely when the "no-growth" criterion is met. Crack tip opening displacement should also be monitored. Corrosion reactions accompanied by expansion in volume may occur at the crack tip. This changes the opening displacement and increases the load, thus altering desired testing conditions.

As subcritical crack extension occurs, stress intensity increases in the cantilever beam test and decreases in the wedge-opening load test (Fig. 4). Generally, the threshold stress intensity measured with the wedge-opening load test is lower than with the cantilever beam test (Ref 12). The advantage of the wedge-opening load test is that only a single specimen is required to measure K_{Ihem}.

In long-term tests, it is essential to ensure that the environmental solution does not change in concentration or composition over time. For example, evaporation may cause the solution level to drop below the crack line of the specimen, rendering the test invalid.

Results of wedge-opening load and cantilever beam tests on 25.4-mm (1-in.) thick Fe-Ni-Co steel specimens are shown in Fig. 5. The data imply that the wedge-opening load K_{Ihem} crack-arrest result is the lower limit for the cantilever beam K_{Ihem} threshold.

The open circles in Fig. 5 represent "no fracture" at various exposure times for an overaged Fe-10Ni-8Co alloy in a cantilever beam test. The open squares indicate failure at increased stress intensities, following step-loading at various exposure times at the lower stress intensity. The results suggest that increasing the load during the test produces more aggressive hydrogen embrittlement conditions than a constant load.

This may be due to the possible formation of an oxide film on the surface. When the load is increased, the oxide film is broken, exposing fresh metal, and more hydrogen is produced at the crack tip. For this reason, the test should use a rising load, because the constant-load cantilever beam test does not provide "worst-case" (fresh metal exposed) loading conditions.

The data also suggest that the cantilever beam test can generate an artificially high K_{Ihem} threshold, depending on the time limit selected. If the test had been terminated at 200 h rather than at 5000 h, the reported K_{Ihem} values from the cantilever beam test (Fig. 5) would have been four times higher than those measured after the longer time period. Similarly, if insufficient time is allowed in the wedge-opening load test, the incubation period may not be exceeded, and no crack growth will result.

Both testing methods require costly and time-consuming steps and result in a parameter, K_{Ihem}, whose design significance is questionable. However, the parameter does provide a relative ranking of susceptibility to hydrogen embrittlement or, more generally, stress-corrosion cracking.

Fig. 3 Schematic showing basic principle of modified wedge-opening load test specimen

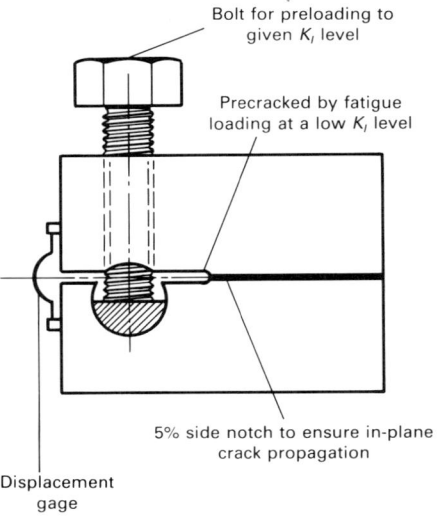

The contoured double-cantilever beam test is used to measure crack growth rate at a constant stress-intensity factor. This test simplifies the calculation of stress intensity by using a contoured specimen so that stress intensity is proportional to the applied load and is independent of the crack length. Under a constant load, stress intensity also remains constant with crack extension. For the test geometry shown in Fig. 6, the stress-

Fig. 4 Influence of time, crack extension, and load on stress-intensity behavior of modified wedge-opening load, cantilever beam, and contoured double-cantilever beam test specimens
Source: Ref 10

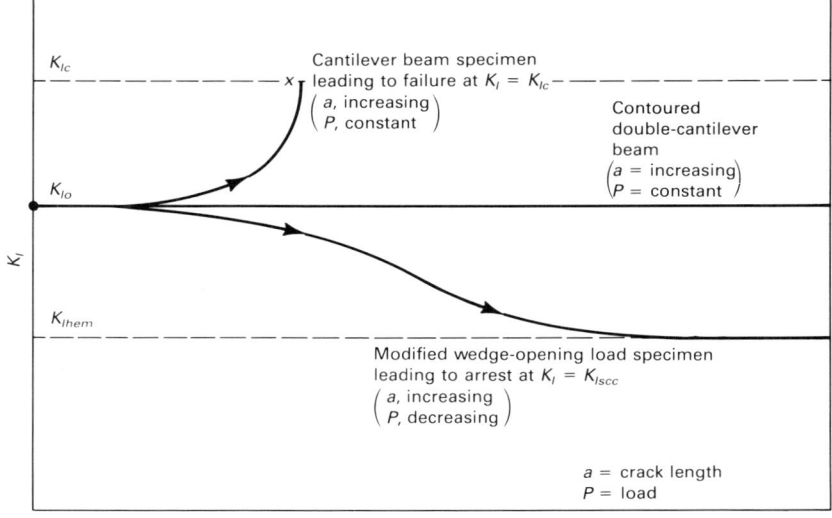

Fig. 5 Comparison of single-edge-notched cantilever beam and wedge-opening load test results for hydrogen embrittlement cracking of Fe-Ni-Co steels

Source: Ref 13

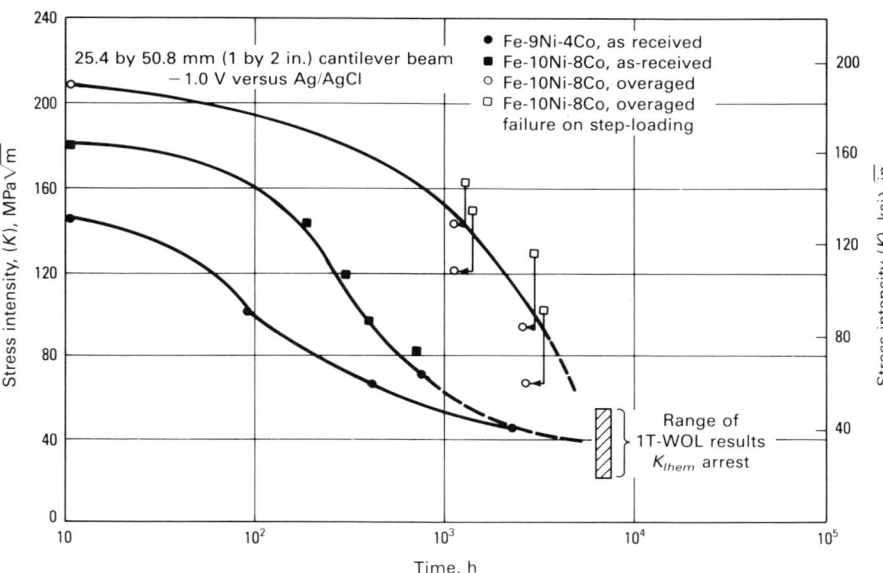

intensity factor equals 20 times the load ($K = 20P$).

Data on hydrogen embrittlement can be obtained with subthickness specimens (Ref 14), even in excess of the ASTM requirement of $K_{max}^2 < 0.4\ B/YS^2$ (where B is thickness and YS is yield strength of the specimen), by using side grooves, which provide additional constraint on the material being tested. Side grooves enable the main-

tenance of a plane-strain condition in a thin specimen by enhancing stress triaxiality. This method has been used extensively to study the effect of heat treatment (hardness) and environment on hydrogen stress cracking of AISI type 4340 steels (Fig. 7).

The contoured double-cantilever beam test has also been used to study the stress-history effect that produces an incubation time before hydrogen stress cracking. Figure 8

shows that incubation time is dependent on the type of steel. A decrease in the stress-intensity factor from 44 to 22 MPa$\sqrt{m}$ (40 to 20 ksi$\sqrt{in.}$) may change the incubation time from less than 1 h for AISI type 4340 steel to about 1 year for type D-6AC steel.

Three-Point and Four-Point Bend Tests. The contoured double-cantilever beam test uses a constant load to maintain a constant stress-intensity factor with crack extension. The same effect can be produced by using a three- or four-point bend test under displacement control. These tests use heavily side-grooved Charpy V-notch specimens (Fig. 9). Because crack opening displacement is constant as the crack extends, the load decreases, so that there is a slight initial increase in stress intensity to a maximum value that drops slightly as the ratio of crack depth to specimen width exceeds 0.5. Typically, stress intensity is constant, within a small range. Figure 10 compares the change in stress-intensity factor with crack extension as a function of load control to that of displacement control for a three-point bend specimen.

The rising step-load test provides a stress intensity that is different at each load, but that remains constant with crack extension as each load level is sustained. Crack initiation is signaled by a drop in load (Fig. 11). The rising step-load test was developed as an accelerated low-cost test to measure resistance of steels (particularly weldments) to hydrogen embrittlement (Ref 13, 17). The threshold obtained by this method will be somewhat high, as test duration at each load is short.

To index susceptibility to hydrogen-assisted cracking, the test should last no longer than 24 h, and the hydrogen source

Fig. 6 Dimensions and configuration for double-cantilever beam test specimen

Specimen contoured to $3a^2/h^3 + 1/h = C$, where C is a constant. All values given in inches (1.0 in. = 25.4 mm).

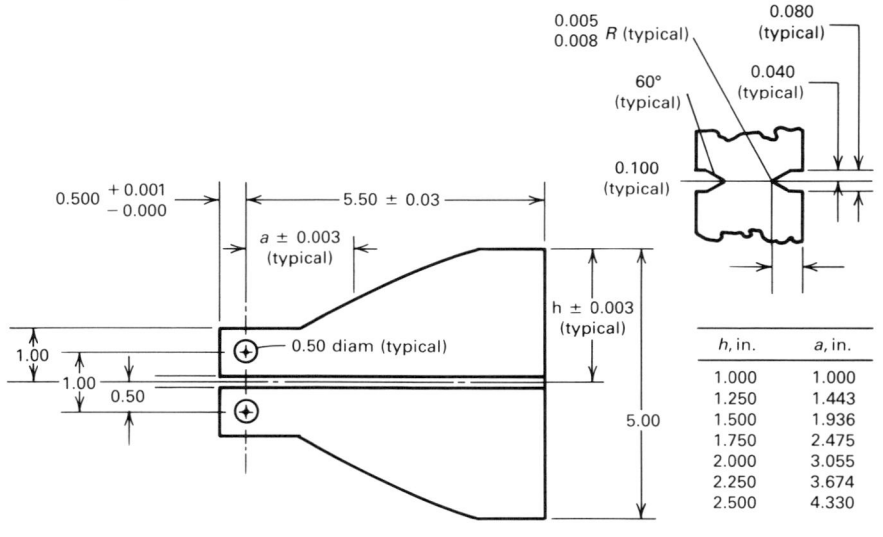

Fig. 7 Hydrogen embrittlement crack growth rate as a function of applied stress intensity for two different hardnesses and environments for an AISI 4340 steel contoured double-cantilever beam test specimen

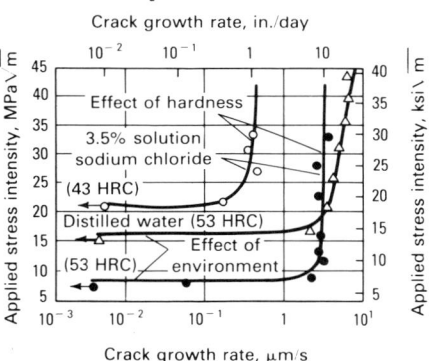

Fig. 8 Incubation time prior to hydrogen stress cracking for AISI type 4340 and type D-6AC steel contoured double-cantilever beam test specimens as a function of decrease in stress intensity
Source: Ref 15

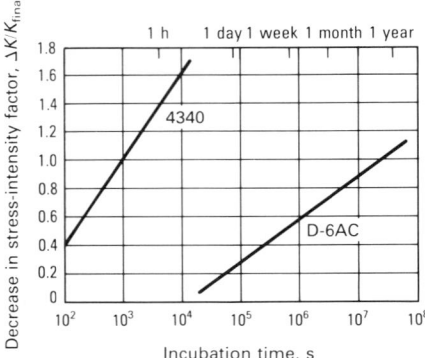

Fig. 9 Standard side-grooved Charpy V-notch test specimen used for three- and four-point bend tests

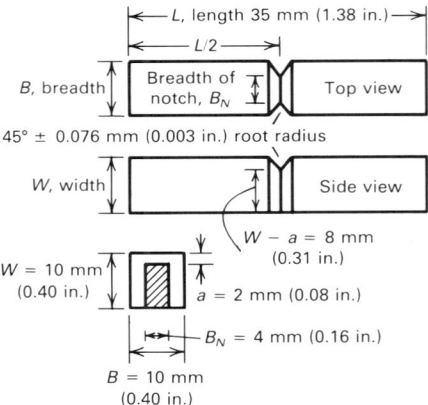

Fig. 10 Use of three-point bend displacement control as constant-K specimen

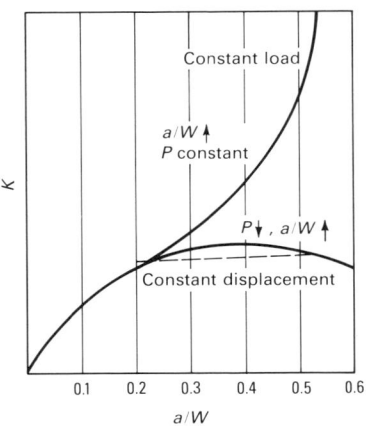

should reflect the most aggressive environment. In one experiment, a 3.5% sodium chloride solution was selected to simulate seawater, and a cathodic potential of -1.2 V (saturated calomel electrode) was used to generate hydrogen to reproduce the extreme conditions of sacrificial anodic protection generally found on a ship hull.

A Charpy specimen was chosen, because such specimens are small and easy to machine and handle. In this test, however, the specimen was modified. Instead of using a fatigue precrack, the notch-root radius was machined to less than 7.6 μm (3 mil). This was done to lower the cost and give less ambiguous environmental conditions at the crack tip. Also, hydrogen cracks nucleate below the surface.

The specimen was deeply side grooved, a common practice used in hydrogen stress cracking tests to prevent the crack from branching. Side grooves are also used in crack opening displacement or J-integral testing to cause load-displacement curves to increase monotonically to fracture by inducing a highly triaxial stress field at the crack tip. Because a Charpy specimen is small, deep side grooves produce a triaxial stress field at the notch to promote hydrogen stress cracking. The extent of the side grooving is such that the remaining ligament is only 40% of the original thickness. The modified Charpy specimen dimensions are shown in Fig. 9.

The specimen was loaded by means of beams and an instrumented bolt (Fig. 12). Four-point bending under constant displacement control and stress intensity produced crack growth. Once cracking initiated at the notch ($a/W = 0.2$, where a is crack length

and W is width of the specimen), arrest did not occur until the crack was nearly through the specimen. The load was increased manually at 1-h intervals. An environmental chamber encompassed the specimen and included a potentiostat to produce hydrogen while under stress.

The rising step-load test was used to evaluate high-strength HY ship steels and welyments in an environment simulating seawater under conditions of cathodic protection commonly used to protect ship hulls (Ref 17). Samples from the heat-affected zone and other locations in the weld metal were tested. Interlayer gas tungsten arc heating was evaluated as a means of providing a refined, homogeneous, tempered microstructure with improved resistance to hydrogen stress cracking. As a baseline, comparison was made between HY-130 and HY-180 steels.

Figure 13 plots rising step-load test results for HY-130 and HY-180 base metals, in addition to combinations of modified HY steel compositions and programmed-cooling-rate thermal cycles for the base metal and weld wire. The vertical axis is a plot of a parameter derived from the specimen strength ratio in ASTM E 399, "Test Method for Plane-Strain Fracture Toughness of Metallic Materials"—i.e., $6\ P_{max}/B(W - a)^2 YS$, where P_{max} is the maximum load that the specimen is able to sustain, B is the specimen thickness, W is the specimen width, a is the crack length, and YS is the yield strength in tension. For the data shown in Fig. 13, P_{max} was replaced by the crack initiation load. The horizontal axis is a ratio of K_{Ihem}/YS, measured in a separate test program with cantilever beam and wedge-opening load specimens.

The resistance to hydrogen embrittlement of the two base metals and six locations in

Fig. 11 Typical load-time record for four-point rising step-load test
Source: Ref 16

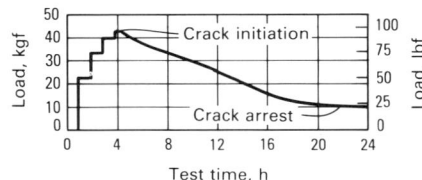

HY-130 weldments was ranked using this testing method. Test results showed that HY-180 is more susceptible to hydrogen stress cracking than HY-130 and that the resistance to hydrogen embrittlement of specimens taken from the heat-affected zone and fusion line is consistently higher than that of weld-metal specimens. The resistance of the weld metal is affected by grain structure; interlayer gas tungsten arc reheating homogenized the weld structure, but did not temper the weld metal. Specimens from the gas tungsten arc reheated weldment consistently exhibited higher hardness and lower resistance to hydrogen embrittlement than similar specimens from the standard HY-130 weld metal (Ref 12, 17).

The disk-pressure testing method measures susceptibility to hydrogen embrittlement of metallic materials under a high-pressure gaseous environment (Ref 18). The test is used for the selection and quality control of materials, protective coatings, surface finishes, and other processing variables.

A thin disk of the metallic material to be tested is placed as a membrane in a test cell and subjected to helium pressure until it bursts. Because helium is inert, the fracture is caused by mechanical overload; no sec-

Fig. 12 Loading frame used for rising step-load test

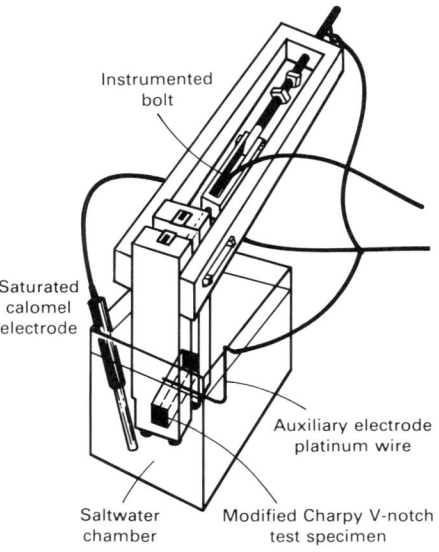

ondary physical or chemical action is involved. An identical disk is placed in the same test cell and subjected to hydrogen pressure until it bursts. Metallic materials that are susceptible to environmental hydrogen embrittlement fracture under a pressure lower than the helium-burst pressure; materials that are not susceptible fracture under the same pressure for both hydrogen and helium. The ratio (S_{H_2}) between the helium-burst pressure (P_{He}) and the hydrogen-burst pressure (P_{H_2}) indicates the susceptibility of the material to environmental hydrogen embrittlement:

$$S_{H_2} = \frac{P_{He}}{P_{H_2}}$$

If S_{H_2} is equal to or less than 1, the material is not susceptible to environmental hydrogen embrittlement. When S_{H_2} is greater than 2, the material is considered to be highly susceptible. At values between 1 and 2, the material is moderately susceptible, with failure expected after long exposure to hydrogen; therefore, the material must be protected against exposure.

A compilation of test results is shown in Fig. 14. Alloys with little or no sensitivity are 7075-T6 aluminum, Haynes 188 (cobalt base), beryllium copper (copper base), type 304, 316, and 310 austenitic stainless steels, type 430 ferritic steel, and age-hardened austenitic A286 steel. Titanium-base alloy Ti-6Al-4V exhibits moderate sensitivity. Alloys with high sensitivity are Haynes 25 (cobalt base) and iron-base alloys, including medium- and high-strength steels. Conventional testing methods, such as the cantilever beam,

wedge-opening load, and contoured double-cantilever beam tests, have also been adapted for testing in high-pressure gaseous hydrogen environments.

Slow strain rate tensile tests can be used to evaluate many product forms, including plate, rod, wire, sheet, and tubing, as well as welded parts. Smooth, notched, or precracked specimens can be used. The principal advantage of this standardized test (Ref 19) is that the susceptibility to hydrogen stress cracking for a particular metal-environment combination can be assessed rapidly.

A variety of specimen shapes and sizes can be used; the most common is a smooth bar tensile coupon, as described in ASTM E 8, "Methods of Tension Testing of Metallic Materials." The specimen is exposed to the environment and is stressed under displacement control. For stainless steel in chloride solution, the strain rate is 10^{-6}/s. One or more of the following parameters are applied to the tensile test at the same initial strain rate: time-to-failure; ductility, as assessed by reduction in area or elongation to fracture, for example; maximum load achieved; and area bounded by a nominal stress-elongation curve or a true stress/true strain curve.

Potentiostatic Slow Strain Rate Tensile Testing. The use of dissociated water under potentiostatic conditions that produce hydrogen on the surface of the tensile test specimen while under slow strain rate displacement control has been studied. Results suggest that hydrogen is the most significant parameter in stress cracking under conditions of hydrogen sulfide stress-corrosion cracking found in oil fields (Ref 20).

Hydrogen Embrittlement in Processing

Two sources of hydrogen embrittlement in processing are the plating bath and various cleaners and paint strippers (see Ref 5 for additional discussion). AISI 4340 steel heat treated to 52 to 54 HRC usually is used in testing, because this hardness is a sensitized temper that requires a prescribed minimum of hydrogen to cause hydrogen embrittlement cracking. If the plating and baking process does not generate sufficient hydrogen to cause AISI 4340 steel to crack, other steels considered more resistant will not be susceptible.

Plating Baths. The test sequence for evaluating a plating bath is to immerse prepared test specimens in the plating bath along with production parts, and then heat treat the specimens and parts (Ref 5). After heat treatment, the specimens are exposed to a stress within 1 h after removal from the baking furnaces. If failure does not occur in 200 h, the

Fig. 13 Analytical correlation of strength ratio with threshold stress-intensity data

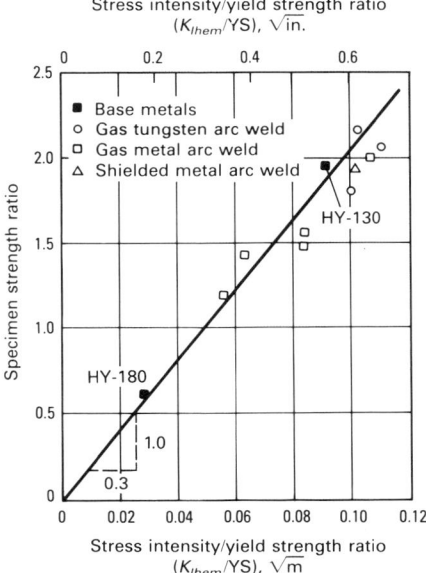

parts may not have sufficient residual hydrogen to cause hydrogen embrittlement cracking.

Exposing a prestressed part to hydrogen is more damaging than introducing the stress after exposure to hydrogen. High residual tensile stresses prior to plating actual components introduce the possibility that the parts may crack during the plating operation. For this reason, the test specimens are not stressed during the plating operation, because they would crack in the plating bath. Unstressed specimens may not be representative of the actual components. Two frequently used test specimen configurations are the notched round tensile bar, which requires a sustained-loading frame with a capacity often in excess of 53.4 kN (12 kip), and the notched round bending bar, which is an integral part of a self-loading frame.

Cleaning Solutions, Paint Strippers, and Maintenance Chemicals. The testing sequence for these products varies somewhat from that used to evaluate plating baths, although the specimens used are the same. The specimens are under stress during exposure to the chemicals.

Evaluation of Manufactured Components

Fasteners. Testing for hydrogen embrittlement of steel self-drilling tapping screws consists of inserting the self-drilling tapping screw against a Type B standard plane

Fig. 14 Relative hydrogen susceptibility of various metals and alloys tested at a rate of 65 bars/min at room temperature

c.w., cold worked; ann., annealed. Source: Ref 18

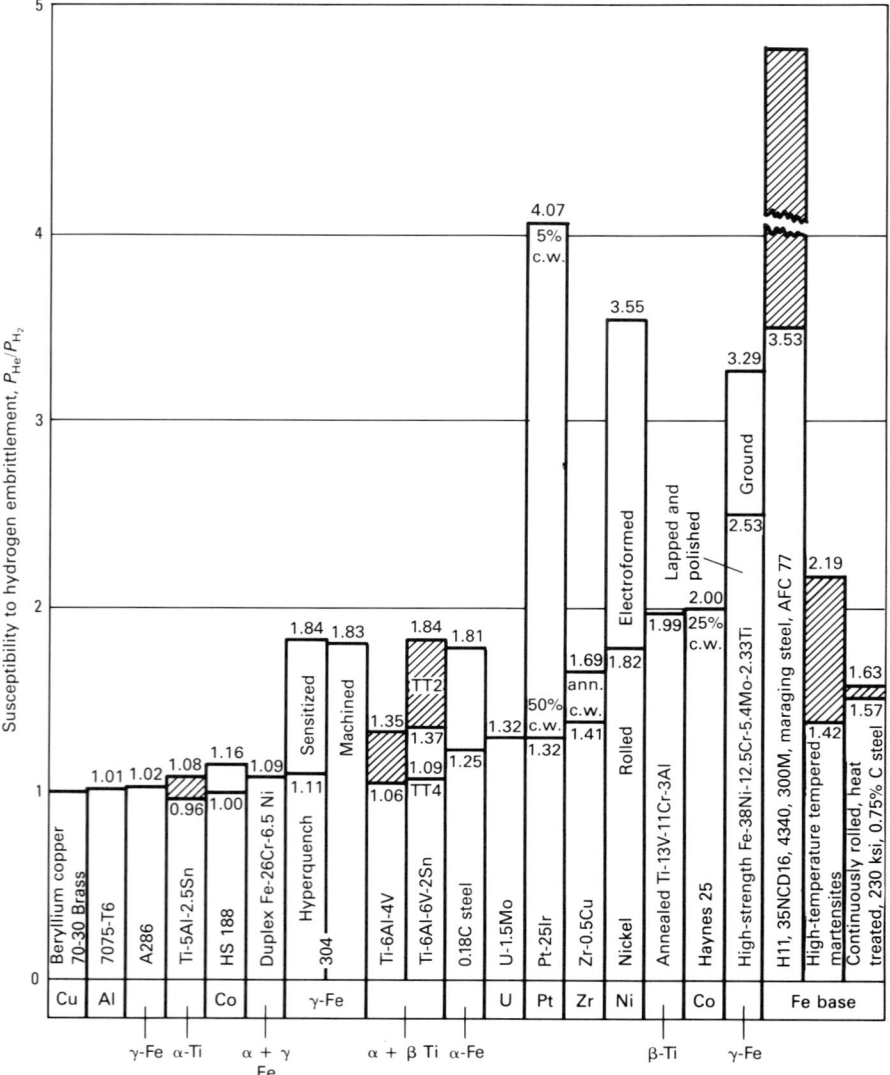

Fig. 15 Notched four-point bend beam test

All values given in inches (1.0 in. = 25.4 mm). Source: Ref 5

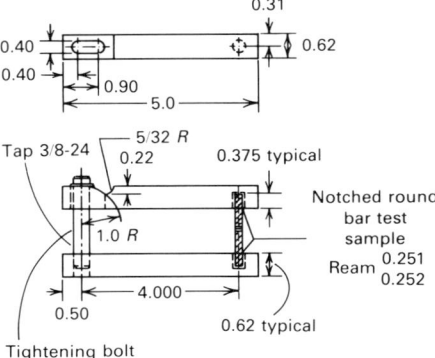

men. The assembly is inserted in a four-point bend test fixture (Fig. 15). The rising step-load test technique is used, and the specimen is loaded directly to failure. Overload fracture is measured and observed with scanning electron microscopy, and the maximum fracture load is thus established.

A second sample is then step-loaded at 1-h intervals at each step until crack initiation begins, usually within 8 h. Crack extension continues, and the decreasing load is recorded. Scanning electron microscopy can be used to verify the presence of brittle cracking typical of hydrogen embrittlement-type failures (i.e., flat facets instead of dimpled rupture).

Service Performance Evaluation. In actual use, a manufactured component often is painted or coated with cadmium or zinc. The coating should be scratched or scraped and exposed to a potentially damaging environment. A 3.5% salt water solution typically is used.

The environment should be specific to the anticipated service environment. The relationship of the plating to the exposed base metal is important in establishing the corrosion current and thus the amount of hydrogen generated at the metal surface. Therefore, the entire system should be examined after the individual constituents have been monitored.

washer and tightening to a prescribed torque. This stress is maintained for 24 h; the prescribed torque is then reapplied, and the screw is removed by the application of a removal torque. The fastener has failed the test if the reapplication of the torque cannot be obtained, or if the screw cannot be removed without shearing the fastener.

Reference 21 describes a hydrogen embrittlement test for a metallic-coated externally threaded fastener. This test requires the use of a wedge to produce a sustained combined tension and bending load. The tension load is specified as 75% of the minimum ultimate tensile strength. The time specified is 48 h, after which the test fastener is visually and

microscopically examined for hydrogen-embrittlement-induced failure. The torque is then reapplied to attain at least 90% of the initial tightening torque. The fastener should show no evidence of hydrogen-induced cracking when visually examined, and the retightening torque should not be less than 90% of the initial tightening torque to indicate successful testing.

Modified Charpy V-Notch Testing of Bolts. Plated fasteners have been evaluated using a modified Charpy specimen configuration. Fixture adapters are screwed onto the shank of the fastener and tightened to a total length of 55 mm (2.16 in.), which is the length of a standard Charpy V-notch speci-

REFERENCES

1. M.G. Fontana and N.D. Greene, *Corrosion Engineering*, McGraw-Hill, New York, 1978, p 113

2. A.R. Troiano, The Role of Hydrogen and Other Interstitials on the Mechanical Behavior of Metals, *Trans. ASM*, Vol 52, 1960, p 54-80

3. W. Beck, E.J. Jankowsky, and P. Fisch-

er, "Hydrogen Stress Cracking of High Strength Steels," Report No. NADC-MA-7140, Naval Air Development Center, Warminster, PA, Dec 1971

4. L. Raymond, Ed., *Hydrogen Embrittlement Testing*, STP 543, ASTM, Philadelphia, 1972

5. "Standard Method for Mechanical Hydrogen Embrittlement Testing of Plating Processes and Aircraft Maintenance Chemicals," F 519, *Annual Book of ASTM Standards*, Vol 02.05, ASTM, Philadelphia, 1984, p 692-705

6. "Standard Method for Electronic Hydrogen Embrittlement Test for Cadmium-Electroplating Processes," F 326, *Annual Book of ASTM Standards*, Vol 02.05, ASTM, Philadelphia, 1984, p 683-691

7. J.M. Bernstein and A.W. Thompson, Ed., *Hydrogen in Metals*, American Society for Metals, 1973

8. C.D. Beachem, Ed., *Hydrogen Damage*, American Society for Metals, 1977

9. C.G. Interrante and G.M. Pressouyre, Ed., *Current Solutions to Hydrogen Problems in Steels*, American Society for Metals, 1983

10. S.T. Rolfe and I.M. Barsom, *Fracture and Fatigue Control in Structures*, Prentice-Hall, Englewood Cliffs, NJ, 1977

11. R.P. Wei, S.R. Novak, and D.R. Williams, Some Important Considerations in the Development of Stress Corrosion Cracking Test Methods, *Mater. Res. Stand.*, Vol 12 (No. 9), Sept 1972, p 25

12. C.A. Zanis, Subcritical Cracking in High Strength Steel Weldments—A Materials Approach, *SAMPE Quart.*, Jan 1978, p 8-12

13. C.A. Zanis, P.W. Holsberg, and E.C. Dunn, Jr., Seawater Subcritical Cracking of HY-Steel Weldments, *Welding J. Res. Suppl.*, Vol 59, Dec 1980

14. National Material Advisory Board, "Rapid Inexpensive Tests for Determining Fracture Toughness," NMAB-328, National Academy of Sciences, Washington, DC, 1976

15. D.L. Dull and L. Raymond, Stress History Effect on Incubation Time for Stress Corrosion Crack Growth in AISI 4340 Steel, *Met. Trans.*, Vol 3, Nov 1972, p 2943-2947

16. D.L. Dull and L. Raymond, Electrochemical Techniques, in *Hydrogen Embrittlement and Testing*, STP 543, ASTM, Philadelphia, 1974, p 20-33

17. P.J. Fast, C.S. Susskind, and L. Raymond, "Charpy V-Notched Specimens for Indexing Stress Corrosion Cracking in HY Ship Steels and Weld Metals," Final report under contract F04701-78-C-0079, The Aerospace Corp., El Segundo, CA, Aug 1979

18. J.P. Fidelle, R. Bernardi, R. Broudeur, C. Roux, and M. Rapin, *Disk Pressure Testing of Hydrogen Environment Embrittlement*, STP 543, ASTM, Philadelphia, 1974, p 221-253

19. "Recommendations for Conducting Slow Strain Rate Stress Corrosion Tests," Document No. ISO/TC 156/WG 2, International Standards Organization, British Standards Institute

20. D.L. Dull and L. Raymond, Surface Cracking of Inconel 718 During Cathodic Charging, *Met. Trans.*, Vol 4, June 1973, p 1635-1638

21. "Standard Method for Conducting Tests to Determine the Mechanical Properties of Externally and Internally Threaded Fasteners, Washers and Rivets," F 606, *Annual Book of ASTM Standards*, Vol 15.08, ASTM, Philadelphia, 1984, p 414-426

Formability Testing

Sheet Formability Testing

By Brian Taylor
Staff Research Scientist
General Motors Corporation

SHEET METAL FORMING is the process of converting a flat sheet of metal into a part of desired shape without fracture or excessive localized thinning. The process may be simple, such as a bending operation, or may be a sequence of very complex operations such as those performed in high-volume stamping plants. In the manufacture of most large stampings, a sheet metal blank is held on its edges by a blankholder ring and is deformed by means of a punch and die. The movement of the blank into the die cavity is controlled by pressure between the upper and lower parts of the blankholder ring.

This control usually is increased by means of one or more sets of drawbeads. These consist of an almost semicylindrical ridge on the upper part of the blankholder and a corresponding groove in the lower part (the positions are sometimes reversed). The drawbeads force the perimeter of the blank to bend and unbend as it is pulled into the die, which increases the restraining force considerably. Presses with capacities of up to 17.8 MN (2000 tons) are commonly used for the manufacture of large stampings, and presses up to 26.7 MN (3000 tons) are used for heavy-gage parts.

Sheet metal forming operations are so diverse in type, extent, and rate that no single test provides an accurate indication of the formability of a material in all situations. However, knowledge of material properties and careful analysis of the various types of forming involved in making a particular part are indispensable in determining the probability of successful parts production and in developing the most efficient process.

Types of Forming

Many forming operations are complex, but all consist of combinations or sequences of the basic forming operations—bending, stretching, drawing, and coining.

Bending is the most common type of deformation and occurs in almost all forming operations. Bending around small radii can lead to splitting in the early stages of a forming process, because it localizes strain and prevents its distribution throughout the part. Ideally, strain should be distributed as uniformly as possible to maximize the amount of deformation that can be obtained. Sometimes, even a slight increase in the radius in a given location can significantly improve strain distribution. Frequently, designs specify smaller radii than are necessary, which results in manufacturing problems and increases costs.

Lubrication is not recommended when bending over a sharp radius, because die friction reduces strain localization by restricting metal movement away from the radius. The orientation of the sheet in relation to the rolling direction can also be important in a bending operation. During rolling, inclusions and other defects become elongated in the rolling direction, producing lines of weakness. When the axis of bending is in this direction, a tendency to split along the lines of weakness exists. This lowers resistance to fracture compared to when the axis is inclined to the rolling direction.

Outer and inner panels of a part frequently are assembled by bending (hemming) the edges of the outer panel around the inner panel. This requires material that can be easily bent over very small radii. In the absence of other types of deformation, bending produces tensile stresses on the outside surface. These decrease to zero at an interior level known as the neutral axis. These stresses then become compressive on the inside of the bend. They can cause springback (shape distortion) upon removal of the applied forces. If tensile deformation is also present, the compressive stresses may be reversed; generally, however, through-thickness stress and strain gradients will still exist.

Many forming operations involve pulling metal over a die radius so that it is initially bent and subsequently straightened. The net strain resulting from this process may be quite small, depending on the size of the die radius and the tensile forces involved. However, the bending and straightening process cold works the metal, particularly at the surfaces, and reduces its subsequent formability.

Stretching is caused by tensile stresses in excess of the yield stress. When they are applied in perpendicular directions in the plane of the sheet, these forces produce biaxial stretching. When the perpendicular forces are equal, balanced biaxial stretching occurs. Much higher levels of deformation, as measured by an increase in area, can be reached in balanced biaxial stretching than in any other forming mode.

Many forming operations involve stretching of some areas within the stamping. Automotive outer body panels are typical examples of parts formed primarily by stretching. Parts with regions containing domes, ribs, and embossments also involve stretching.

Plane-strain stretching results in elongation in one direction and no dimensional change in the perpendicular direction. It frequently occurs when a wide, flat area of sheet metal is stretched longitudinally, e.g., in the side wall of a stamping. In this case, strain in the transverse direction is prevented by the adjacent metal. Plane-strain stretching is an important type of deformation, because most materials fracture at a lower level of strain in plane strain than in any other condition. Many of the fractures occurring in stamping operations are in the plane-strain region.

Drawing of sheet metal causes elongation in one direction and compression in the perpendicular direction. The simplest example is the drawing of a flat-bottomed cylindrical cup. In this process, a circular disk is held between two flat annular dies and impacted in the center by a flat-bottomed punch. This draws (pulls) the edges of the disk inward to form the wall of the cup. The metal is stretched radially by the tensile forces produced by the punch, but it is compressed circumferentially as its diameter decreases. Many other forming operations involve substantial drawing.

Coining occurs when metal is compressed between two die surfaces. It is used extensively for making coins and parts with similar surface features, for flattening, and for reducing springback (shape distortion) upon removal of parts from a die. In many stretching and drawing operations, coining is undesirable, because it restricts metal movement, localizes strain, and produces surface damage. Much of the die preparation for these operations concentrates on locating and eliminating coining.

Combinations of Types of Forming. Most forming operations involve combinations of different types of forming. Figure 1 illustrates a part design that requires drawing in the corners, biaxial stretching in the dome, bending, straightening, and plane-strain stretching in the walls, and bending and plane-strain stretching at the tops and bottoms of the walls.

Formability Problems

The major problems encountered in sheet metal forming are fracturing, buckling and wrinkling, shape distortion, loose metal, and undesirable surface textures. The occurrence of any one or a combination of these conditions can render the sheet metal part unusable. The effects of these problems are discussed below.

Fracturing occurs when a sheet metal blank is subjected to stretching or shearing (drawing) forces that exceed the failure limits of the material for a given strain history, strain state, strain rate, and temperature. In stretching, the sheet initially thins uniformly, at least in a local area. Eventually, a point is reached at which deformation concentrates and causes a band of localized thinning, known as a neck, which ultimately fractures. Generally, the formation of a neck is regarded as failure, because it produces a visible defect and a structural weakness. Most

Fig. 1 Part design requiring a combination of types of forming
B, bending; BS, biaxial stretching; D, drawing; P, plane-strain stretching; U, unbending (straightening). Source: Ref 1

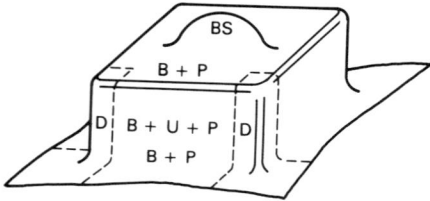

current formability tests are concerned with fracture occurring in stretching operations.

In shearing, fracture can occur without prior thinning. The most common examples of this type of fracture occur in slitting, blanking, and trimming. In these operations, sheets are sheared by knife edges that apply forces normal to the plane of the sheet. Shearing failures sometimes are produced in stamping operations by shearing forces in the plane of the sheet, but they are much less common than stretching failures.

Buckling and Wrinkling. In a typical stamping operation, the punch contacts the blank, stretches it, and starts to pull it through the blankholder ring. The edges of the blank are pulled into regions with progressively smaller perimeters. This produces compressive stresses in the circumferential direction. If these stresses reach a critical level characteristic of the material and its thickness, they cause slight undulations, known as buckles. Buckles may develop into more pronounced undulations or waves, known as wrinkles, if the blankholder pressure is not sufficiently high.

This effect can also cause wrinkles in other locations, particularly in regions with abrupt changes in section and in regions where the metal is unsupported or contacted on one side only. In extreme cases, folds and double or triple metal may develop. These in turn may lead to splitting in another location by preventing metal flow, or by "locking the metal out." For this reason, increasing the blankholder pressure often corrects a splitting problem.

Shape Distortion. In forming operations, metal is deformed elastically and plastically by applied forces. Upon removal of the external forces, the internal elastic stresses relax. In some locations, they can relax completely, with only a very slight change in the dimensions of the part. However, in areas subjected to bending, through-thickness gradients in the elastic stresses will occur; i.e., the stresses on the outer surfaces will be different from those on the inner surfaces.

If these stresses are not constrained or "locked in" by the geometry of the part, relaxation will cause a change in the part shape, known as shape distortion or springback. Springback can be compensated for in die design for a specific set of material properties, but may still be a problem if there are large material property or process variations from blank to blank.

Loose metal occurs in undeformed regions and is undesirable, because it can be easily deflected. A phenomenon usually referred to as "oil canning," in which a local area can be either concave or convex, can also be encountered. In stampings with two

or more sharp bends of the same sign in roughly the same direction, such as a pair of feature lines, a tendency exists for the metal between them to be loose, because it is difficult to pull metal across a sharp radius.

It is sometimes possible to circumvent the problem by ensuring that the metal is not contacted by both lines at the same time; thus, some stretching can occur before the second line is contacted. There is a tendency for loose metal to occur toward the center of large, flat, or slightly curved parts. Increasing the restraining forces on the blank edges usually improves this condition.

Undesirable Surface Textures. Heavily deformed sheet metal, particularly if it is coarse grained, often develops a rough surface texture, commonly known as "orange peel." This usually is unacceptable in parts that are visible in service. Another source of surface problems occurs in metals that have a pronounced yield point elongation, i.e., materials that stretch several percent without an increase in load after yielding. In these metals, deformation at low strain levels is concentrated in irregular bands known as Lüders lines or bands, or stretcher strains.

These defects disappear at moderate and high strain levels. However, almost all parts have some low-strain regions. These defects are unsightly and are not concealed by painting. Aged rimmed steels and some aluminum-magnesium alloys develop severe Lüders lines.

In some cases, zinc-coated steels exhibit surface defects known as "spangles." This phenomenon occurs only in hot dipped products and is caused by the development of a coarse grain size in the galvanic coating, which makes the individual grains clearly visible. This problem can be corrected in the coating process. In addition to the above occurrences, handling damage, dents caused by dirt or slivers in the die, and scoring or galling caused by a rough die surface or inadequate lubrication sometimes produce unacceptable surfaces.

Measurement of Deformation

The principal methods for measuring deformation are gage marks, strain gages, optical extensometers, and thickness and shape measurements.

Gage Marks. The most widely used method for measuring deformation is to mark the sheet by etching or scribing, or by means of ink, dye, or paint, and measuring the changes in the separations of the marks caused by the deformation. Rectangular grid markings and arrays of small-diameter (e.g.,

2.5-mm, or 0.1-in.) circles frequently are used.

In most production forming operations and in the later stages of tensile testing, deformation varies rapidly with location, which can lead to large differences in strain measurements made over different gage lengths. For this reason, small gage lengths, such as the diameters of small, closely spaced circles, are commonly used. Circular markings provide an additional advantage in that it is easy to identify the directions of the maximum (or major) and minimum (or minor) strains and to thus measure their values. On deforming, the circles change into ellipses with their major axes in the direction of the maximum strain and minor axes in the direction of the minimum strain.

This information is essential in determining how close the local strain state is to the maximum the material can withstand without fracturing, which depends on the ratio of the strains. It is also useful in determining how the geometry of a die must be modified when the formability limits of the work material are exceeded.

Strain Gages and Extensometers. In some cases, a strain gage or a strain gage extensometer is attached to the sheet or test sample. Accurate strain measurements are thus obtained continuously during a forming operation or test. Optical extensometers, which are particularly effective at high strain rates, can also be used.

Thickness and Shape Measurements. In some cases, thickness measurements, which can be made rapidly by ultrasonic methods, can be used to determine strains. In practice, this method is limited to situations in which the ratio of the major and minor strains is known from previous measurements, as many different combinations of strains can lead to the same change in thickness.

Part shape is measured by means of templates, checking fixtures, or shadowgraphs, or by means of a profile meter that uses a stylus to contact the surface. Profile meters may give two- or three-dimensional digital representations of the part. Noncontacting surface digitizers and systems for measuring deformation by locating grid markings in three dimensions are also used.

Representation of Strain

The most common method of representing strain defines the engineering strain, e, as the ratio of the change in length, ΔL, to the original length, L_o:

$$e = \frac{\Delta L}{L_o} = \frac{L - L_o}{L_o} = \frac{L}{L_o} - 1 \qquad \text{(Eq 1)}$$

The second method defines the true strain, ϵ, in the region of uniform elongation, as the integral of the incremental change in length, dL, divided by the actual (instantaneous) length, L:

$$\epsilon = \int_{L_o}^{L} \frac{dL}{L} = \ln\left(\frac{L}{L_o}\right)$$

$$= \ln(1 + e) \qquad \text{(Eq 2)}$$

The engineering strain is easier to calculate and is satisfactory for many applications. The true strain is used in theoretical analysis of formability and has the advantage that successive strains can be added to give the cumulative strain.

The strain state of a deformed sheet is frequently represented graphically by plotting the maximum or major strain, e_1, on the vertical axis and the minimum or minor strain, e_2, which can be positive or negative, on the horizontal axis. This is illustrated in Fig. 2, in which five strain paths, each leading to the same major strain of 40%, but with minor strains ranging from -40 to $+40\%$, are represented. The ellipses shown were originally circles (shown dashed) in the undeformed sheet.

On the right side of Fig. 2, the circles have transformed into ellipses that are larger in all directions than the original circles. This is the region of biaxial stretching and, in the diagonal (45°) direction, of balanced biaxial stretching. In this region, the circles have expanded without changing shape.

On the left side of Fig. 2, the circles have transformed into ellipses, which are larger in one direction, but smaller in the perpendicular direction, than the original circles. This is the region of drawing and is the strain state

developed in the tensile test. On the vertical axis, the ellipses are larger in one direction, but unchanged dimensionally from the original circles in the perpendicular direction. This is the region of plane strain.

Effect of Material Properties on Formability

The properties of sheet metals vary considerably, depending on the base metal (steel, aluminum, copper, etc.), alloying elements present, processing, heat treatment, gage, and level of cold work. In selecting material for a particular application, a compromise usually must be made between the functional properties required in the part and the forming properties of the available materials. For optimum formability in a wide range of applications, the work material should:

- Distribute strain uniformly
- Reach high strain levels without necking or fracturing
- Withstand in-plane compressive stresses without wrinkling
- Withstand in-plane shear stresses without fracturing
- Retain part shape on removal from the die
- Retain a smooth surface and resist surface damage

Some production processes can be operated successfully only when the forming properties of the work material are within a narrow range. More frequently, the process can be adjusted to accommodate shifts in work material properties from one range to another, although sometimes at the cost of lower production and higher material waste. Some processes can be operated successfully

Fig. 2 Schematic of several major strain/minor strain combinations

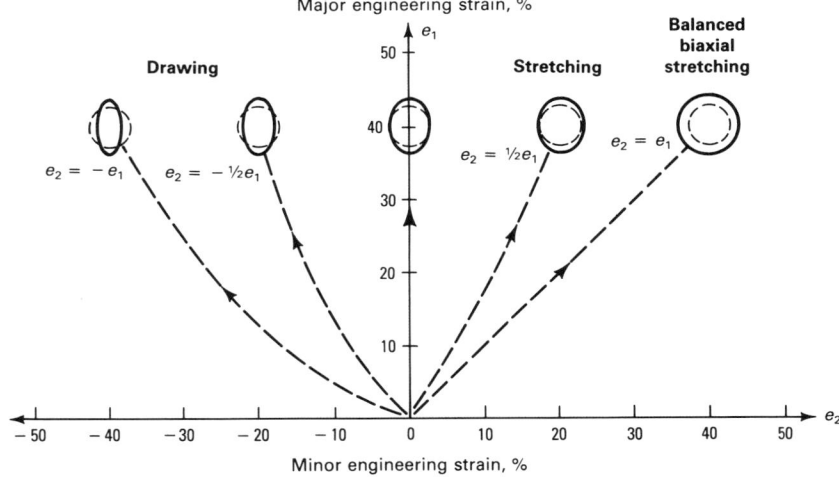

using work material that has a wide range of properties. In general, consistency in the forming properties of the work material is an important factor in producing a high output of dimensionally accurate parts.

Strain Distribution

Three material properties determine the strain distribution in a forming operation: the strain-hardening coefficient (also known as the work-hardening coefficient or exponent) or n value, the strain-rate sensitivity or m value, and the plastic strain ratio (anistropy factor) or r value. The ability to distribute strain evenly depends on the n value and the m value. The ability to reach high overall strain levels depends on many factors, such as the base material, alloying elements, temper, n value, m value, r value, thickness, uniformity, and freedom from defects and inclusions.

The n value, or strain-hardening coefficient, is determined by the dependence of the flow (yield) stress on the level of strain. In materials with a high n value, the flow stress increases rapidly with strain. This tends to distribute further strain to regions of lower strain and flow stress. A high n value is also an indication of good formability in a stretching operation.

In the region of uniform elongation, the n value is defined as:

$$n = \frac{d \ln \sigma_T}{d \ln \epsilon} \qquad \text{(Eq 3)}$$

where σ_T is the true stress (load/instantaneous area). This relationship implies that the true stress/strain curve of the material can be approximated by a power law constitutive equation proposed in Ref 2:

$$\sigma_T = k\epsilon^n \qquad \text{(Eq 4)}$$

where k is a constant known as the strength coefficient.

This equation provides a good approximation for most steels, but is not very accurate for dual-phase steels and some aluminum alloys. For these materials, two or three n values may need to be calculated for the low, intermediate, and high strain regions.

When Eq 4 is an accurate representation of material behavior, n is equal to $\ln(1 + e_u)$, where e_u is the uniform elongation, or elongation at maximum load in a tensile test. By definition, $\ln(1 + e_u)$ is identical to ϵ_u, which is the true strain at uniform elongation.

Most steels with yield strengths below 345 MPa (50 ksi) and many aluminum alloys have n values ranging from 0.2 to 0.3. For many higher yield strength steels, n is given by the relationship (Ref 3):

$$n \simeq \frac{70}{\text{(yield strength in MPa)}} \qquad \text{(Eq 5)}$$

A high n value leads to a large difference between yield strength and ultimate tensile strength (engineering stress at maximum load in a tensile test). The ratio of these properties therefore provides another measure of formability.

The m value, or strain-rate sensitivity, is defined by:

$$m = \frac{d \ln \sigma_T}{d \ln \dot{\epsilon}} \qquad \text{(Eq 6)}$$

where $\dot{\epsilon}$ is the strain rate, $d\epsilon/dt$. This implies a relationship of the form:

$$\sigma_T = f(\epsilon) \cdot \dot{\epsilon}^m$$

or

$$\sigma_T = k \, \epsilon^n \cdot \dot{\epsilon}^m \qquad \text{(Eq 7)}$$

where the second equation incorporates Eq 4 between stress and strain.

A positive strain-rate sensitivity indicates that the flow stress increases as the rate of deformation increases. This has two consequences. Higher stresses are required to form parts at higher rates. Also, at a given forming rate, the material resists further deformation in regions that are being strained more rapidly than adjacent regions by increasing the flow stress in these regions. This helps to distribute the strain more uniformly.

The need for higher stresses in a forming operation is usually not a major consideration, but the ability to distribute strains can be crucial. This becomes particularly important in the post-uniform elongation region, where necking and high strain concentrations occur. An approximately linear relationship has been reported between the m value and the post-uniform elongation for a variety of steels and nonferrous alloys (Ref 4). As m increases from -0.01 to $+0.06$, the post-uniform elongation increases from 2 to 40%.

Metals in the superplastic range have high m values of 0.2 to 0.7, which is one to two orders of magnitude higher than typical values for steel. At ambient temperatures, some metals, such as aluminum alloys and brass, have low or slightly negative m values, which explains their low post-uniform elongation.

High n and m values lead to good formability in stretching operations, but have little effect on drawability. In a drawing operation, metal in the flange must be drawn in without causing fracture in the wall. In this instance, high n and m values strengthen the wall, which is beneficial, but they also strengthen the flange and make it harder to draw in, which is detrimental.

The r value, or plastic strain ratio, relates to drawability and is known as the anisotropy factor. This is defined as the ratio of the true width strain to the true thickness strain in the uniform elongation region of a tensile test:

$$r = \frac{\epsilon_w}{\epsilon_t} = \frac{\ln\left(\dfrac{w}{w_o}\right)}{\ln\left(\dfrac{t}{t_o}\right)} \qquad \text{(Eq 8)}$$

The r value is a measure of the ability of a material to resist thinning. In drawing, material in the flange is stretched in one direction (radially) and compressed in the perpendicular direction (circumferentially). A high r value indicates a material with good drawing properties.

The r value frequently changes with direction in the sheet. In a cylindrical cup drawing operation, this variation leads to a cup with a wall that varies in height, which is known as earing (Fig. 3). It is therefore common to measure the average r value, or average normal anisotropy, r_m, and the planar anisotropy, Δr.

The property r_m is defined as $(r_0 + 2r_{45} + r_{90})/4$, where the subscripts refer to the angle between the tensile specimen axis and the rolling direction. Δr is defined as $(r_0 - 2r_{45} + r_{90})/2$. It is a measure of the variation of r with direction in the plane of a sheet. r_m determines the average depth (i.e., the wall height) of the deepest draw possible. Δr determines the extent of earing. A combination of a high r_m value and a low Δr value provides optimum drawability.

Hot rolled low-carbon steels have r_m values ranging from 0.8 to 1.0; cold rolled rimmed steels range from 1.0 to 1.4, and cold rolled aluminum-killed (deoxidized) steels range from 1.4 to 2.0. Interstitial-free steels have values ranging from 1.8 to 2.5, and aluminum alloys range from 0.6 to 0.8. The theoretical maximum r_m value for a ferritic steel is 3.0; a measured value of 2.8 has been reported (Ref 5).

Fig. 3 Drawn cup with ears in the directions of high r value

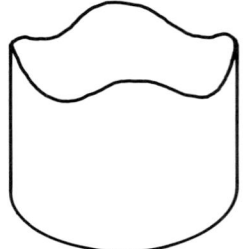

Maximum Strain Levels: The Forming Limit Diagram

Each type of steel, aluminum, brass, or other sheet metal can be deformed only to a certain level before local thinning (necking) and fracture occur. This level depends principally on the combination of strains imposed, i.e., the ratio of major and minor strains. The lowest level occurs at or near plane strain, i.e., when the minor strain is zero.

This information was first represented graphically (Ref 6 and 7) as the forming limit diagram, which is a graph of the major strain at the onset of necking for all values of the minor strain that can be realized. Figure 4 shows a typical forming limit diagram for steel. The diagram is used in combination with strain measurements, usually obtained from circle grids, to determine how close to failure (necking) a forming operation is, or whether a particular failure is due to inferior work material or to a poor die condition (Ref 8).

For most low-carbon steels, the forming limit diagram has the same shape as the one shown in Fig. 4, but the vertical position of the curve depends on the sheet thickness and the n value. The intercept of the curve with the vertical axis, which represents plane strain and is also the minimum point on the curve, has a value equal to n in the (extrapolated) zero thickness limit. The intercept increases linearly with thickness to a thickness of about 3 mm (0.12 in.).

The rate of increase is proportional to the n value up to $n = 0.2$, as shown in Fig. 5 (Ref 9). Beyond these limits, further increases in thickness and n value have little effect on the position of the curve. The level of the form-ing limits also increases as the m value increases (Ref 4).

The shape of the curve for aluminum alloys, brass, and other materials differs from that in Fig. 4 and varies from alloy to alloy within a system. The position of the curve also varies and rises with an increase in the thickness, n value, or m value, but at rates that generally are not the same as for low-carbon steel.

The forming limit diagram is also dependent on the strain path. The standard diagram is based on an approximately uniform strain path. Diagrams generated by uniaxial straining followed by biaxial straining, or the reverse, differ considerably from the standard diagram. For this reason, the effect of the strain path must be taken into account when using the diagram to analyze a forming problem.

Material Properties and Wrinkling

The effect of material properties on the formation of buckles or wrinkles is the subject of extensive research. In drawing operations, there is general agreement, based primarily on experiments with conical and cylindrical cups, that a high r_m value and low Δr value reduce buckling in both flanges and walls (Ref 10-12). In addition to the above correlations, a low flow stress to elastic modulus ratio (σ_F/E) decreases wall wrinkling (Ref 13). The n value has an indirect effect. When the binder force is kept constant, the n value has no effect. However, high n values enable higher binder forces to be used, which reduces buckling.

In stretching operations, the situation appears to be different. A close correlation between the formation of buckles at low strain levels and the yield strength to tensile strength ratio (YS/TS) has been reported, as well as an inverse correlation with the low strain n value and an absence of correlation with the r_m value and uniform elongation (Ref 14). Some of the differences between these results may be attributed to the fact that the experiments with cups involved high strains and high compressive stresses, whereas the stretching experiments were conducted at low strain and low compressive stress levels. In both situations, the problem becomes significantly more severe as the sheet thickness decreases.

Material Properties and Shear Fracture

Shear fractures due to in-plane shear stresses are more prevalent in high-strength cold worked materials, particularly when internal defects such as inclusions are present. Typical strain combinations that cause shear fracture are shown on the forming limit diagram in Fig. 6. For this material, Fig. 6 shows that at high strain levels in the regions close to $\epsilon_2 = \pm\epsilon_1$, failure occurs by shearing before the initiation of necking.

The position and shape of the shear fracture curve depends on the material, its temper, and the type and degree of prestrain or cold work (Ref 15-17). The data available on shear fracture are currently limited.

Fig. 4 Typical forming limit diagram for steel

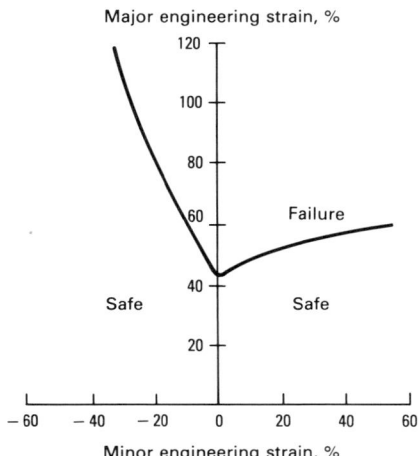

Major engineering strain, %

Fig. 5 Effect of thickness and n value on the plane-strain intercept of forming limit diagram
Source: Ref 9

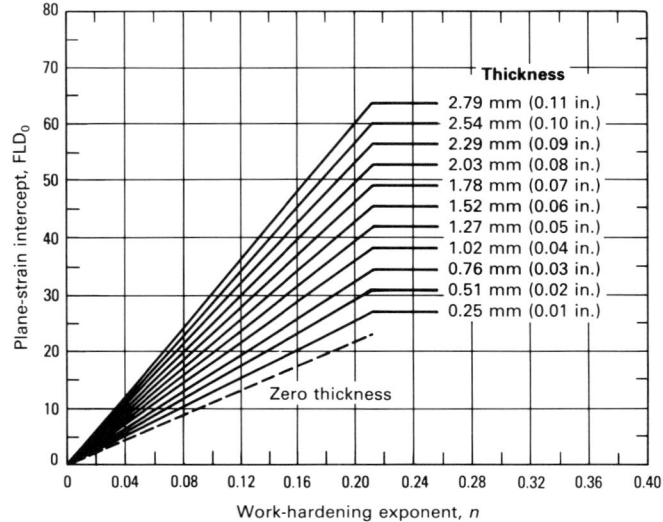

Thickness
2.79 mm (0.11 in.)
2.54 mm (0.10 in.)
2.29 mm (0.09 in.)
2.03 mm (0.08 in.)
1.78 mm (0.07 in.)
1.52 mm (0.06 in.)
1.27 mm (0.05 in.)
1.02 mm (0.04 in.)
0.76 mm (0.03 in.)
0.51 mm (0.02 in.)
0.25 mm (0.01 in.)

Zero thickness

Plane-strain intercept, FLD_0

Work-hardening exponent, n

Fig. 6 Forming limit diagram including shear fracture
Source: Ref 15

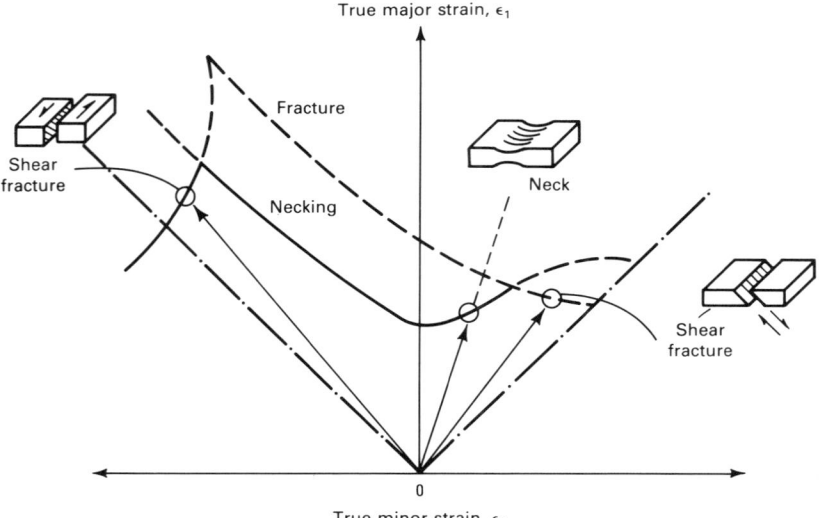

Material Properties and Springback

Material properties that control the amount of springback that occurs after a forming operation are:

- Elastic modulus, E
- Yield strength, σ_y
- Slope of the true stress/strain curve, or tangent modulus, $d\sigma_T/d\epsilon$

Springback is best described by means of three examples involving a rectangular beam: elastic bending below the yield stress, simple bending with the yield stress exceeded in the outer layers of the beam, and combined stretching and bending. In an actual part, springback is determined by the complex interaction of the residual internal elastic stresses, subject to the constraints of the part geometry.

Elastic Bending Below the Yield Stress. Tensile elastic stresses are generated on the outside of the bend. These stresses decrease linearly from a maximum at the surface to zero at the center (neutral axis). They then become compressive and increase linearly to a maximum at the inner surface. Upon removal of the externally applied bending forces, the internal elastic forces cause the beam to unbend as they decrease to zero throughout the cross section (Fig. 7a).

The maximum amount of elastic deflection that can be produced without entering the plastic range is proportional to the yield stress divided by the elastic modulus. The strain at the yield point is equal to σ_y/E ($E = \sigma/\epsilon$). The springback moment for a given deflection is thus proportional to the elastic modulus ($\sigma = E\epsilon$).

Simple Bending. In this example, the yield stress is exceeded in the outer layers of the beam. The outer layers deform plastically, and their stored elastic stresses continue to increase, but at a much lower rate that is proportional to the slope of the true stress/strain curve, or tangent modulus, $d\sigma_T/d\epsilon$, instead of the elastic modulus. Figure 7(b) illustrates this condition for a beam bent so that 50% of its volume is in the plastic range.

On removal of the externally applied bending forces, the stored elastic stresses cause the beam to unbend until their combined bending moment is zero. This produces compressive stresses at the outer surface and tensile stresses at the inner surface.

The springback in this case is less than for a material whose yield strength is not exceeded at the same strain level. This can result from either a higher yield stress or a lower elastic modulus. It is also apparent that higher values of the tangent modulus cause greater springback when the yield strength is exceeded.

In actual conditions, the neutral axis moves inward on bending, because the outer part of the beam is stretched and becomes thinner, and the inner part is compressed and becomes thicker. This effect is analyzed in detail in Ref 18.

Combined Stretching and Bending. In this case, the entire beam can be plastically deformed in tension by as little as 0.5% stretching. However, a stress gradient still exists from the outer to the inner surface (Fig. 7c). On removing the external forces, the internal elastic stresses recover.

This causes unbending, but to a lesser extent than in the previous cases. As the level of stretching is increased, the amount of springback decreases, because the tangent modulus and therefore the stress gradient through the beam decrease at higher strains. The yield strength ceases to be a factor in springback once all regions are plastically deformed in tension.

In the bending of wide sheets, the metal is deformed in plane strain and the plane-strain properties (elastic modulus, yield stress, and tangent modulus) should be used. The effects of a low elastic modulus and a high yield stress and tangent modulus in increasing springback have been experienced in forming operations. Springback is more severe with aluminum alloys than with low-carbon steel (1-to-3 modulus ratio). High-strength steels exhibit more springback than low-carbon steels ($\sim$2-to-1 yield strength ratio), and

Fig. 7 Springback of a beam in simple bending
(a) Elastic bending. (b) Elastic and plastic bending. (c) Bending and stretching

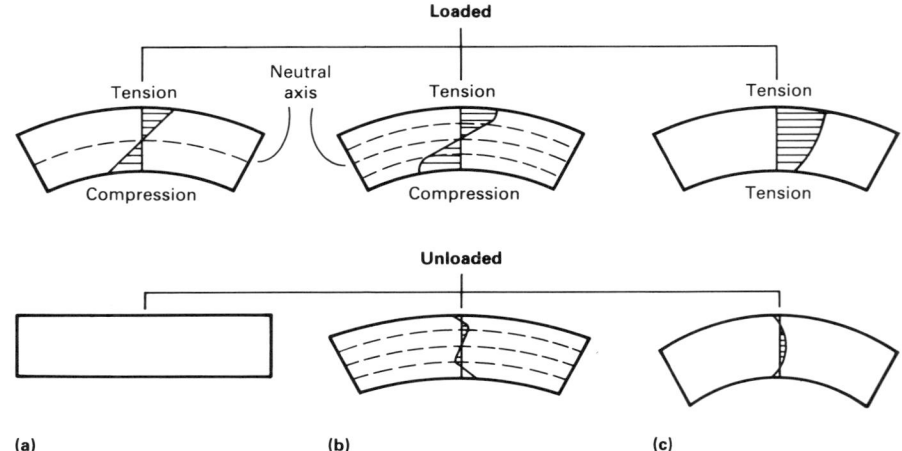

dual-phase steels spring back more than high-strength steels of the same yield strength (higher tangent modulus).

The effect of stretching in reducing springback to very low levels has also been reported (Ref 19). Springback is also greatly influenced by geometrical factors and increases as the bend angle and ratio of bend radius to sheet thickness increase.

Surface Quality

The previously mentioned conditions that lead to undesirable surface textures can be minimized or prevented. The formation of orange peel in heavily deformed regions can be minimized by using a fine-grained material. The development of Lüders lines in rimmed steels can be prevented by temper rolling to 0.25 to 1.25% extension, or by flex rolling, which produces mobile dislocations for a limited period of time, until they are trapped by nitrogen atoms. This also reduces elongation slightly. This problem is becoming less common with the increased use of continuous casting, which requires killed steels. These steels have less free nitrogen to interact with the dislocations and do not develop Lüders lines. Similar treatments can be applied to aluminum-magnesium alloys to prevent this defect.

Effect of Temperature on Formability

A change in the overall temperature changes the properties of the material, which thus affects formability. Also, local temperature differences within a deforming blank lead to local differences in properties that affect formability.

At high temperatures, above one half of the melting point on the absolute temperature scale, extremely fine-grained aluminum, copper, magnesium, nickel, stainless steel, steel, titanium, zinc, and other alloys become superplastic. Superplasticity is characterized by extremely high elongation, ranging from several hundred to more than 1000%, but only at low strain rates, usually below about 10^{-2}/s, at high temperatures.

The requirements of high temperatures and low forming rates have limited superplastic forming to low-volume production. In the aerospace industry, titanium is formed in this manner. The process is particularly attractive for zinc alloys, because they require comparatively low temperatures (~270 °C, or ~520 °F).

At intermediate elevated temperatures, steels and many other alloys have less ductility than at room temperature (Ref 20, 21). Aluminum and magnesium alloys are exceptions and have minimum ductility near room temperature. Alloys of these metals have been formed commercially at slightly ele-

vated temperatures (~250 °C, or ~480 °F). It has been found (Ref 22) that the strain-rate sensitivity (*m* value) and post-uniform elongation for aluminum-magnesium alloys increase significantly in this temperature range.

Low-temperature forming has potential advantages for some materials, based on their tensile properties, but practical problems have limited application. Local increases in temperature occur during forming due to surface friction and internal heating produced by the deformation. Generally, this is detrimental, because it lowers the flow stress in the area of greatest strain and tends to localize deformation.

A method of improving drawability by creating local temperature differences has been developed (Ref 23) and is in use commercially. It entails water cooling the punch in a deep drawing operation. This lowers the temperature of the blank where it contacts the punch, which is the principal failure zone, and increases the local flow stress. Heating the die to lower the flow stress in the deformation zone at the top of the draw wall has also been found to be beneficial. The combination of these procedures has produced an increase of over 20% in the drawability of an austenitic stainless steel.

Types of Formability Tests

Formability tests are of two basic types: intrinsic tests, which measure basic characteristic properties of materials that can be related to their formability, and simulative tests, which subject the material to deformation that closely resembles the deformation that occurs in a particular forming operation.

Intrinsic tests provide comprehensive information that is insensitive to the thickness and surface condition of the material. The most important and extensively used intrinsic test is the uniaxial tensile test, which provides the values of many material properties for a wide range of forming operations. Other commercially important intrinsic tests are the plane-strain tensile test, the Marciniak stretching and sheet torsion tests, the hydraulic bulge test, the Miyauchi shear test, and hardness tests.

Simulative tests provide limited and specific information that is usually sensitive to thickness, surface condition, lubrication, and geometry and type of tooling. This information usually relates to only one type of forming operation. Many simulative tests, such as the Olsen and Swift cup test, have been used extensively for many years, with good correlation to production in specific cases. Several simulative tests are described later in this article.

Uniaxial Tensile Testing

The most widely used intrinsic test of sheet metal formability is the uniaxial tensile test. A "dog bone"-shaped specimen with sides that are accurately parallel over the gage length, which is usually 50.8 mm (2.00 in.) long and 12.7 mm (0.50 in.) wide, is used (Fig. 8). The specimen is gripped at each end and stretched at a constant rate in a tensile machine until it fractures, as described in ASTM Standard E 8, "Methods of Tension Testing of Metallic Materials." The applied load and extension are measured by means of a load cell and strain gage extensometer.

The load-extension data can be plotted directly. However, data are usually converted to engineering (conventional) stress, σ_E (load/original cross section), and engineering strain, e (elongation/original length), or to true stress, σ_T (load/instantaneous cross section), and true strain, ϵ (natural logarithm of (strained length/original length)).

In addition, for formability testing it is common practice to measure the width of the specimen during the test. This is done either intermittently by interrupting the test at preselected elongations to make measurements manually, or continuously by means of width extensometers. From these measurements, the plastic strain ratio (anisotropy factor), or *r* value, can be determined.

During the rolling process used to produce metals in sheet form and subsequent annealing, the grains and any inclusions present become elongated in the rolling direction, and a preferred crystallographic orientation develops. This causes a variation of properties with direction. Thus, it is common practice to test specimens cut parallel to the rolling direction and at 45° and 90° to this direction. These are known as longitudinal, diagonal, and transverse specimens, respectively. This also enables the values of r_m and Δr to be calculated. Because the mechanical properties and elongation tend to be lower in the transverse direction, tests in this direction are often used as the basis for specifications.

The rate at which the test is performed can have a significant effect on the end results. Two methods are commonly used to deter-

Fig. 8 Sheet tensile test specimen

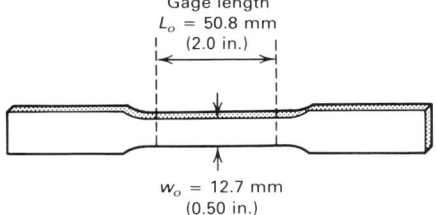

Gage length
L_o = 50.8 mm
(2.0 in.)

w_o = 12.7 mm
(0.50 in.)

mine this effect. In the first, replicate samples are tested at different rates, and variations between samples influence the results. In the second method, the test rate is alternated between two levels. This avoids the problem of variation between samples, but cannot be used at very high rates and is complicated by transients, which occur each time the rate is changed. From these tests, the strain-rate sensitivity, or *m* value, can be calculated.

Figure 9 shows a typical engineering stress/strain curve and the corresponding true stress/strain curve for a material that has a smooth transition between the very low strain (elastic) and the higher strain (plastic) regions of the curve. When the load is removed in the elastic region, the sample returns to its original dimensions. When this is done in the plastic region, the sample retains permanent deformation.

In the tensile test, the load rises to a maximum value and then decreases prior to fracture. The decrease is due to localization of the deformation, which causes a reduction in cross section. This reduction has a greater effect than the opposing increase in flow stress due to strain hardening.

Some materials such as aged rimmed steels do not have a smooth transition between the elastic and plastic regions of the stress-strain curve. The load they can support decreases at the beginning of the plastic region and remains approximately constant for up to about 7% elongation. Subsequently, the load increases to a maximum and then decreases again at high elongations. This type of stress-strain curve is shown in Fig. 10. With the increasing use of continuous casting, which requires killed steels (steels deoxidized by small additions of aluminum, for example), rimmed steels are becoming less common.

Test Procedure

For accurate and reproducible results, uniaxial tensile testing must be performed in a carefully controlled manner. The main steps in the procedure are discussed in detail in the Section on "Tension Testing" in this Volume. These procedures are summarized below.

Specimen Preparation. The surfaces of the specimen should be free from scratches or other damage that can act as stress raisers and cause early failure. The edges should be smooth and free from irregularities. Care should be taken not to cold work the edges, or to ensure that any cold work introduced is removed in a subsequent operation, because this changes mechanical properties and lowers ductility.

It is a common practice to mill and grind the edges, but other procedures such as fine milling, nibbling, and laser cutting are also used. When a new method is used, initial tests should be performed to compare the results with those obtained by conventional methods.

The width of a nominally 12.7-mm (0.50-in.) wide specimen should be measured to the nearest 0.025 mm (0.001 in.), and the thickness for specimens in the range 0.5 to 2.5 mm (0.02 to 0.1 in.), to the nearest 0.0025 mm (0.0001 in.). If this is impractical due to surface roughness, the thickness should be measured to the nearest 0.025 mm (0.001 in.).

The tensile test is sensitive to variations in the width of the specimen, which should be accurately controlled. For a specimen 12.7 mm (0.50 in.) wide, the width of the reduced section should not deviate by more than ±0.25 mm (±0.01 in.) from the nominal value and should not differ by more than ±0.05 mm (±0.002 in.) from end to end.

Some investigators intentionally taper the reduced section slightly toward the center to increase the probability that fracture will occur within the gage length. In this case, the center should not be narrower than the ends by more than 0.10 mm (0.004 in.).

Alignment of Specimens. The specimen should be accurately aligned with the

Fig. 10 Engineering stress/strain curve for rimmed steel

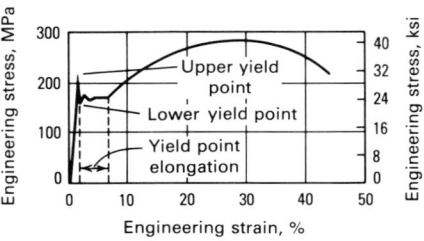

centerline of the grips. The effect of small displacements (10% of the specimen width) of one or both ends from the centerline has been calculated (Ref 24). It was determined that the latter case is the more serious, but both strongly affect the strain in the outermost fibers. It was also concluded that the calculated stress-strain curve is not significantly affected at strains above 0.3%.

Measurement of Load and Elongation. The applied load is measured by means of a load cell in the test machine, for which the usual calibration procedures must be followed (ASTM E 4, "Practices for Load Verification of Testing Machines"). Elongation usually is determined by means of a clip-on strain gage extensometer (ASTM E 83, "Method of Verification and Classification of Extensometers"). In addition, small scratches are often scribed across the specimen at the ends of the gage length so that the total elongation can be determined from the broken specimen.

Circle grids are sometimes etched or printed on the specimen. These can be used to measure the strain distribution and width strain as well as the overall strain. This can be done continuously by means of a television camera and data processing system if required. For some applications, particularly high-speed testing, optical extensometers are used. These require well-illuminated boundaries that are clearly delineated by means of high-contrast coatings, such as black and white paint.

An approximate measure of elongation can be obtained from the crosshead travel. This involves errors due to elongation of the specimen outside the gage length and elastic strain in the grips, which can be compensated for to some extent. This method is used when the specimen is inaccessible, such as in nonambient testing. The signals from the load cell and extensometer can be plotted on a chart recorder or processed by a data processing system to the required form, such as plots of stress versus strain, or tables of mechanical and forming properties.

Measurement of Width and Thickness. In addition to the initial measurements of specimen width and thickness, which are

Fig. 9 Typical engineering and true stress/strain curves

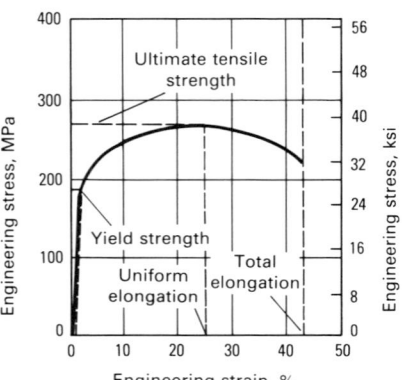
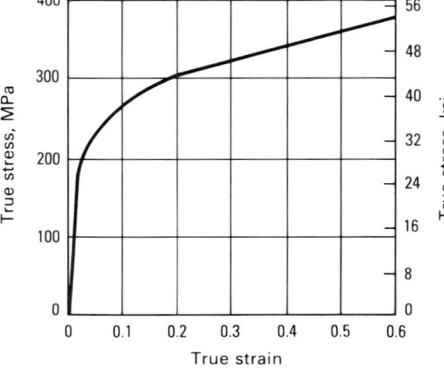

required to calculate the stress, measurements may be made at intervals during the test to determine the r value (ASTM E 517, "Test Method for Plastic Strain Ratio r for Sheet Metal") and to determine the reduction in area and true strain. The r value is measured at a specified strain level between the yield point and the uniform elongation (e.g., at 15% elongation). It can be measured by stopping the test at this strain level and measuring the width accurately ($\pm$0.013 mm, or $\pm$0.0005 in.) at a minimum of three equally spaced points in the gage length (for a 50.8-mm, or 2.0-in., gage length). In practice, the thickness is calculated from the specimen width and length, assuming no change in volume.

Alternatively, width measurements can be made during the test using width extensometers, which complicates the procedure. Currently, attempts are underway to develop combined width and length extensometers to simplify this method.

Reduction in area is the ratio $(A_o - A)/A_o$, where A is the instantaneous cross-sectional area and A_o is the original cross-sectional area. It is used to calculate the true strain in the region of post-uniform elongation. A large reduction in area at fracture correlates with a small minimum bend radius, a high m value, and high energy absorption. To calculate the reduction in area, the width and thickness must be measured in the narrowest part of the necked region.

Effect of Gage Length on Elongation.

In post-uniform elongation, part of the specimen is elongated uniformly, and the remainder is narrowed into a necked region of higher strain level. A change in the gage length alters the ratio of these two regions and has a significant effect on the total elongation measurement. This phenomenon is discussed in detail in Ref 25.

To obtain results that are comparable for different gage lengths, the ratio of the square root of the cross-sectional area to the length, $\sqrt{A}/L$, should be the same. When comparing samples of different thickness, this implies that the gage length or the width should be adjusted to maintain this ratio.

Rate of Testing.

Most tensile tests are performed on screw-driven or hydraulic testing machines at strain rates of 10^{-5} to 10^{-2}/s. The strain rate is defined as the increase in length per unit length per second. These tests are known as low strain rate or static tests.

Most high-volume production forming operations are performed at considerably higher strain rates, in the range of 1 to 10^2/s. To determine the tensile properties in this range, dynamic test machines, which operate at rates of 10^{-1} to 10^2/s, are used (Ref 25). As mentioned previously, steels have higher tensile properties and lower elongations at high strain rates. The properties of aluminum alloys exhibit little sensitivity to the strain rate.

Material Properties

The stress-strain curve determined by uniaxial tension testing provides values of many formability-related material properties. Several of these properties and methods for measurement are discussed below. Table 1 lists typical values of properties measured in tensile tests on thin (0.5 to 1.0 mm, or 0.02 to 0.04 in.) sheet materials.

Young's Modulus.

The initial slope of the stress-strain curve—i.e., the ratio of the stress to the strain in the elastic region before any plastic deformation has occurred—is the Young's modulus (E) of the material. This property affects springback and shape distortion at low strains. For accurate measurement of Young's modulus, a low strain rate and a high data acquisition rate should be used in the elastic region (below about 0.5% elongation), and a very stiff tension testing machine should be used if strain is inferred from crosshead displacement.

Yield Strength.

The stress at which the stress-strain curve deviates in elongation from the initial elastic slope by a specified amount, commonly 0.2%, is known as the yield strength (YS). The yield strength determines the load necessary to initiate deformation in a forming operation, which is usually a high percentage (40 to 90%) of the maximum load required.

For accurate measurement of yield strength, a rate of loading of less than 690 MPa/min (100 ksi/min) is specified. Beyond this point, the strain rate should not exceed 0.08/s. Some materials elongate without an increase in load, or at a decreased load, at the transition between the elastic and plastic regions. The point at which this initiates is known as the yield point.

With a decrease in load, the material has an upper yield point and a lower yield point. The upper yield point is difficult to measure reproducibly. The lower yield stress usually fluctuates, and the minimum value is used.

The elongation that occurs after yielding before the load starts to increase monotonically is known as the yield point elongation. Yield point elongation leads to nonuniform deformation at low strains in forming operations. If it exceeds about 1.5%, irregular surface markings known as Lüders lines or stretcher strains may occur to an extent that is unacceptable in visible parts.

Tensile Strength.

The maximum stress observed in the test is known as the tensile strength (TS), or ultimate tensile strength. Tensile strength determines the maximum load that can be usefully applied in a forming operation.

Uniform Elongation.

The engineering strain at the maximum engineering stress is known as the uniform elongation (e_u). Prior to this point, the sample deforms uniformly. Subsequently, deformation concentrates—initially in a fairly large region known as a diffuse neck, and ultimately in a localized

Table 1 Typical tensile properties of select sheet metals

Material	Young's modulus, E		Yield strength		Tensile strength		Uniform elongation, %	Total elongation, %	Work-hardening factor, n	Average normal anisotropy, r_m	Planar anisotropy, Δr	Strain-rate sensitivity, m
	GPa	10^6 psi	MPa	ksi	MPa	ksi						
Aluminum-killed draw quality steel	207	30	193	28	296	43	24	43	0.22	1.8	0.7	0.013
Interstitial-free steel	207	30	165	24	317	46	25	45	0.23	1.9	0.5	0.015
Rimmed steel	207	30	214	31	303	44	22	42	0.20	1.1	0.4	0.012
High-strength low-alloy steel	207	30	345	50	448	65	20	31	0.18	1.2	0.2	0.007
Dual-phase steel	207	30	414	60	621	90	14	20	0.16	1.0	0.1	0.008
301 stainless steel	193	28	276	40	690	100	58	60	0.48	1.0	0.0	0.012
409 stainless steel	207	30	262	38	469	68	23	30	0.20	1.2	0.1	0.012
3003-O aluminum	69	10	48	7	110	16	23	33	0.24	0.6	0.2	0.005
6009-T4 aluminum	69	10	131	19	234	34	21	26	0.23	0.6	0.1	−0.002
70-30 brass	110	16	110	16	331	48	54	61	0.56	0.9	0.2	0.001

region of sharply reduced cross section known as a local neck. Deformation continues to concentrate in this region until fracture occurs.

Total Elongation. Elongation at the point of fracture is known as total elongation (e_T). It has been used extensively as an approximate indication of sheet metal formability. However, no single property is a reliable indicator of formability under all conditions.

Reduction in area, $(A_o - A)/A_o$, is calculated from measurements of actual specimen width and thickness in the narrowest part of the necked region. The true strain, which cannot be determined from length measurements in the post-uniform elongation region, is also calculated from these values.

The true strain in the necked region is equal to $\ln (dL/dL_o)$, where dL is a small element of length in this region, whose original length was dL_o. Equating the original and final volumes of this element of length gives:

$$V_o = A_o \, dL_o = V = A \, dL$$

or

$$\epsilon = \ln \left(\frac{dL}{dL_o}\right) = \ln \left(\frac{A_o}{A}\right) \qquad \text{(Eq 9)}$$

The relation between the reduction in area at fracture and the minimum bend radius is given in Ref 25 as follows. For values of reduction in area at fracture (q) below 0.2, the ratio of the minimum bend radius (R_m) to sheet thickness (t) is given by:

$$\frac{R_m}{t} = \frac{1}{2q} - 1 \qquad \text{(Eq 10)}$$

For values of q greater than 0.2:

$$\frac{R_m}{t} = \frac{(1 - q)^2}{(2q - q^2)} \qquad \text{(Eq 11)}$$

Strain-Hardening Exponent. The n value, $d \ln \sigma_T/d \ln\epsilon$, is given by the slope of a graph of the logarithm of the true stress versus the logarithm of the true strain in the region of uniform elongation. For materials that closely follow the Holloman constitutive equation (Eq 4), an approximate n value can be obtained from two points on the stress-strain curve by the Nelson-Winlock procedure (Ref 26). The two points commonly used are at 10% strain and at the maximum load. The ratio of the loads or stresses at these two points is calculated, and the n value and uniform elongation can then be determined from a table or graph. The accuracy of the n value determined in this way is ±0.02.

The n value can be determined more accurately by linear regression analysis, as in ASTM E 646, "Standard Test Method for Tensile Strain-Hardening Exponents (n-values) of Metallic Sheet Materials." For some materials, n is not constant, and initial (low strain), terminal (high strain), and sometimes intermediate n values are determined. The initial n value relates to the low deformation region, in which springback is often a problem. The terminal n value relates to the high deformation region in which fracture may occur.

Plastic Strain Ratio. The r value, or anisotropy factor, is defined as the ratio of the true width strain to the true thickness strain in a tensile test. Generally, its value depends on the elongation at which it is measured. It usually is measured at 10, 15, or 20% elongation.

The r value is calculated from the measured width and length as:

$$\epsilon_w = \ln \left(\frac{w}{w_o}\right)$$

$$\epsilon_t = \ln \left(\frac{t}{t_o}\right) = \ln \left(\frac{L_o w_o}{Lw}\right) \qquad \text{(Eq 12)}$$

where constancy of volume ($Lwt = L_o w_o t_o$) has been used and:

$$r = \frac{\epsilon_w}{\epsilon_t} = \frac{\ln \left(\dfrac{w}{w_o}\right)}{\ln \left(\dfrac{L_o w_o}{Lw}\right)} \qquad \text{(Eq 13)}$$

The average r value, or normal anisotropy (r_m), and the planar anisotropy, or Δr value, can be calculated from the values of r in different directions using the formulas given previously.

Strain-Rate Sensitivity. The m value, $d \ln \sigma_T/d \ln\dot\epsilon$, is determined either from duplicate tensile tests performed at different strain rates or from a single test in which the rate is alternated between two levels during the test. These methods are shown schematically in Fig. 11. The m value can be determined at various strain levels in the region of uniform elongation:

$$m = \frac{\ln \left(\dfrac{\sigma_1}{\sigma_2}\right)}{\ln \left(\dfrac{\dot\epsilon_1}{\dot\epsilon_2}\right)} \qquad \text{(Eq 14)}$$

In some materials, m is insensitive to strain (Ref 4, 27). In other materials, however, m is sensitive to strain and strain rate (Ref 28). In many materials, m increases and n decreases with an increase in temperature (Ref 29), sometimes to the extent that superplastic properties develop.

Determining n and r Values

The time and facilities required for sample preparation and for performing the uniaxial tensile test make it difficult to use for on-line process control. The following simplified tests for determining n and r are more suitable for this purpose. The circle arc elongation test and the rapid-n test utilize tensile specimens with two sections that differ in width by about 5% to determine n values. Fracture almost always occurs in the narrow section, but the final measurements are made on the wide section, which elongates uniformly. The r value can be obtained from these tests, but for ferritic steels, the Modul-r test is faster and easier to perform. This test actually measures the elastic modulus of the specimen and uses an empirically determined correlation between the modulus and the r value.

The circle arc elongation test does not require measurement of the applied loads (Ref 30). It uses a rectangular tensile specimen with a reduced width section produced by milling a pair of small circular arc notches on opposite sides. The gage length is marked in the full-width section, and the specimen is pulled to fracture, which usually occurs in the narrow section. The uniform elongation is measured in the full-width section. The value is slightly lower than that obtained in the conventional tensile test and gives a slightly lower n value. However, it is suitable for production control. The r value can be determined by the additional measurement of the change in width in the full-width section.

The rapid-n test (Ref 31) provides rapid and fairly accurate measurements of yield and tensile strengths, elongation, and n and r values. It requires relatively simple equipment and can be performed in less than 5 min, including specimen preparation. The test is suitable for sheet metals whose properties are represented accurately by the Holloman equation, $\sigma_T = k\epsilon^n$. It has been used successfully on low-carbon and stainless steels and on a variety of nonferrous alloys.

The test specimen, which is punched directly from the sheet sample, has the dimensions shown in Fig. 12. Generally, 25-mm (1.0-in.) gage lengths are marked on both the wide and narrow sections, and the specimen is strained to fracture in a manual or motorized load frame, or in a tension testing machine. The yield load is measured if there is discontinuous yielding, and the maximum load is measured. Yield and tensile strengths are calculated from the measured loads and the initial dimensions of the narrow section. If the yielding is continuous, the yield strength can be calculated from the tensile strength and n value, as indicated below.

Fig. 11 Methods for determining strain-rate sensitivity (*m* value)

(a) Duplicate test method. (b) Changing rate method

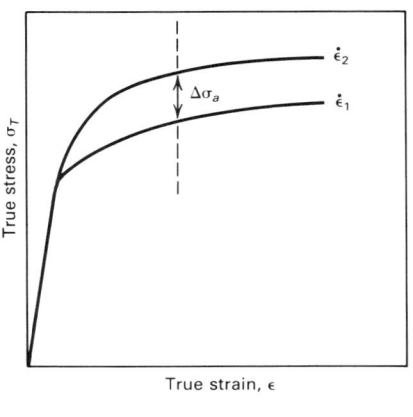

(a)

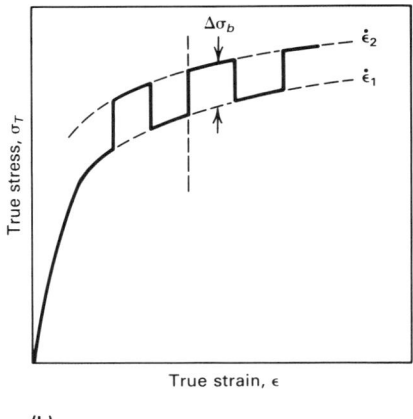

(b)

Fig. 12 Rapid-*n* test specimen

Source: Ref 31

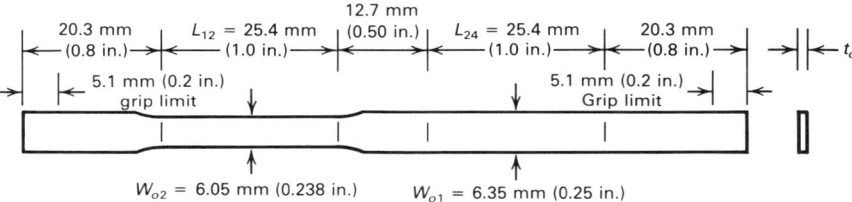

An empirically determined correction is applied to compensate for the effect of the sheared edges of the specimens. For steel, this correction reduces the measured yield and tensile strengths by 13.6 MPa/mm (50.0 ksi/in.) of initial sample thickness. The *n* value is calculated from the following equation:

$$n = \frac{\ln\left(\dfrac{w_{o2}t_o}{w_{f1}t_f}\right) - n}{\ln\left[\dfrac{\ln\left(\dfrac{w_{o1}t_o}{w_{f1}t_f}\right)}{n}\right]} \qquad \text{(Eq 15)}$$

where w_{o1} and w_{o2} are the initial widths of the wide and narrow sections, respectively; w_f is the final width of the wide section; and t_o and t_f are the initial and final thicknesses of the wide section, respectively. The equation can be solved iteratively in four steps, or by means of a simple computer program, beginning with a trial value of 0.24 for *n*.

For materials that do not have a discontinuous yield point, the yield strength can be calculated as:

$$YS = TS\left(\frac{C}{n}\right)^n \qquad \text{(Eq 16)}$$

where TS is the tensile strength and C is a constant. For low-carbon steels, $C \approx 0.02$. For other materials, C must be determined empirically. The *r* value can be computed from the initial and final dimensions of the wide section as described previously.

The Modul-*r* test (Ref 32) measures the elastic modulus (Young's modulus) of low-carbon steel samples by determining their resonant frequencies by exciting them using an oscillating magnetic field. The elastic modulus is directly proportional to the square of the resonant frequency, and simple empirical relationships exist between the directionally averaged elastic modulus and the r_m value and between the planar variation of the modulus and the Δr value.

This test uses a flat 102- by 6.35-mm (4.0- by 0.25-in.) punched specimen and a specially designed, commercially available magnetostrictive oscillator. The specimen is placed inside drive and pickup coils in the oscillator. An alternating current passed through the drive coil produces an alternating magnetic field, which causes magnetostrictive oscillations in the sample. The oscillations induce an alternating current in the pickup coil. This current is used to change the frequency of the current in the drive coil to maximize the amplitude of the oscillations, i.e., to obtain the resonant frequency, which is displayed digitally.

The relationships used to determine r_m and Δr from the resonant frequency, f, are:

$$E = 4\rho L^2 f^2 \qquad \text{(Eq 17)}$$

$$r = \frac{4822.6}{(E_m - 267.7)^2} - 0.564 \qquad \text{(Eq 18)}$$

$$\Delta r = 0.031 - 0.0468 \, \Delta E \qquad \text{(Eq 19)}$$

where E is the elastic modulus in GPa; ρ and L are the density and length of the specimen, respectively; and E_m and ΔE are defined analogously to r_m and Δr.

This test provides a more reproducible measure of r_m and Δr than the conventional tensile testing method and is less sensitive to differences between operators. It can be performed in 5 min, including specimen preparation. When alloy steels or ferritic stainless steels are tested, a different correlation between the modulus and *r* value must be used. This must be determined experimentally. When the test is used on coated products, the coating must be removed chemically prior to testing.

Plane-Strain Tensile Testing

In conventional uniaxial tensile testing, the sample is strained in the region of drawing; i.e., the minor or width strain is negative. The test does not provide information on the response of sheet materials in the plane-strain state, in which the minor strain is zero. However, it can be modified to produce this strain state in part of the sample. This modification involves the use of a very wide, short sample or the use of knife edges to prevent transverse (width) strain in part of the sample.

Wide Sample Methods. Increasing the width of the sample and decreasing the gage length changes the strain state from one with a large negative minor strain component toward the plane-strain state, in which the minor strain component is zero. In the rectangular sheet tension test (Ref 33), samples with length-to-width ratios of 1 to 1, 1 to 2, and 1 to 4 are used to approach the plane-strain conditions. Gage lengths are constrained further by reinforcements welded onto each side of the sample at both ends, thus making the samples three layers thick except in the gage length.

The minimum minor strain obtained with the 1-to-4 length-to-width ratio is -0.05 times the major strain, which is close to the plane-strain condition of zero minor strain. The in-plane strains are measured by means of grid markings on the samples, and through-thickness deformations can be observed by holographic interferometry.

A similar approach was used in testing many wide specimen designs to determine the effect of edge profile and length-to-width ratio on strain state (Ref 34-36). The specimen geometry that yielded the highest center strain at failure with a large region of plane strain is shown in Fig. 13. The plane-strain region, which is arbitrarily taken as the region where $|e_2/e_1|$ is less than 0.2, occupies about 80% of the specimen width. The outer part of the specimen deforms in a similar manner to a standard tensile test specimen.

Special grips were developed that exert a high clamping force at the inner contact lines. This minimizes distortion and slippage in these regions, giving the test well-defined boundary conditions. The results of both types of wide specimen tensile tests described above correlated well with stress-strain predictions obtained by finite-element modeling using material properties obtained in the standard tensile test (Ref 34, 37).

Width Constraint Method. In the width constraint method (Ref 38), a rectangular sample is used that has a central gage section reduced in width by circular notches. The gage section is clamped between two pairs of opposing parallel knife edges (stingers) aligned with the sample axis. The knife edges prevent transverse (width) strain in this region. The sample is pulled to fracture in a tension testing machine, and the plane-strain limit (necking) and fracture strains are determined from thickness measurements made on the fractured sample. This procedure is described in detail in Ref 38. The use of a spring-loaded clamp around the knife edges makes adjustment of the clamp during testing unnecessary.

Biaxial Stretch Testing

Two tests that determine the properties of sheet metals in biaxial stretching without involving surface friction effects are the Marciniak biaxial stretching test (Ref 39) and the hydraulic bulge test (Ref 40). The Marciniak

test subjects the sample to in-plane biaxial stretching, but does not determine the stresses. In the hydraulic bulge test, the stresses can be determined, but the sample is deformed into a dome, which involves out-of-plane stresses and strains.

Marciniak Biaxial Stretching Test. A disk of the test material is stretched over a flat-bottomed punch of cylindrical or elliptical cross section. This creates uniform in-plane biaxial strain in the center of the sample, with a strain ratio that is determined by the ratio of the major and minor diameters of the punch. Most testing has been performed with a cylindrical punch, which produces balanced biaxial stretching.

The center of the punch is hollowed out to eliminate friction in this area, and a spacer is placed between the sample and the punch. The spacer is a disk of similar material to that under test, with the same diameter, but with a hole at the center. The experimental arrangement is shown in Fig. 14. As the disk and spacer are stretched over the punch, the hole in the spacer enlarges, and the central part of the test sample is deformed in uniform in-plane biaxial stretching.

The function of the spacer is to reverse the direction of the surface friction experienced by the sample. In the absence of the spacer, the surface friction opposes the movement of the sample over the punch and reduces the maximum strain level attainable. The spacer deforms more easily than the test sample, because of the hole in the center, and exerts a frictional force on the sample directed outward over the punch radius.

For the material to stretch freely, the punch and die radii must be adequate for the thickness of the material under test. A spacer hole diameter to punch diameter ratio of 1 to 3 has been used successfully. The strains can be measured by using grid circles, squares, or other suitable markings. The test has the following applications:

- Determination of the limiting strains of materials in uniform in-plane biaxial stretching without surface friction

- Application of a carefully controlled level of uniform in-plane biaxial strain to samples with large areas to be used in other tests, e.g., tests to determine the effect of different strain paths on the limiting strain levels
- Detection of defects, such as inclusions, by straining a sample of large area to a uniformly high level. Defects will cause early localized fracture, usually parallel to the rolling direction.

Hydraulic Bulge Test. The periphery of a sheet metal sample is clamped between circular or elliptical die rings, and hydraulic pressure is applied on one side of the sample to deform it into a dome, as shown in Fig. 15. The flange of the sample is prevented from slipping by a lock bead placed in the die rings. This consists of a ridge with small radii on one ring and a matching groove on the other.

With circular die rings, the center of the dome has been found to be nearly spherical (Ref 40). The stress and strain states in this region can be determined from the curvature and extension and the fluid pressure. A biaxial test extensometer has been developed that measures the extension and curvature by means of a spherometer and an extensometer that are in direct contact with the dome (Ref 41).

More recently, a system for controlling the strain rate in this test has been developed (Ref 42), because significantly different test results have been obtained (Ref 43) under conditions of constant strain rate and constant fluid flow. The system uses feedback from the extensometer signal to operate a servo-valve, which controls the flow of hydraulic oil to the bulge. This system was used to determine the strain-rate sensitivity of aluminum alloys to a much higher strain level than is possible in the tensile test.

A computerized system is available that uses electronic vision to measure the principal strains and closed-loop feedback to control the strain rate (Ref 44). This system monitors the relative positions of the centers of three closely spaced white dots painted on a black background at the center of the sam-

Fig. 13 Plane-strain tensile test specimen

Source: Ref 36

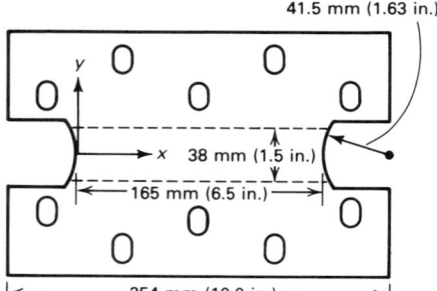

Fig. 14 Marciniak biaxial stretching test

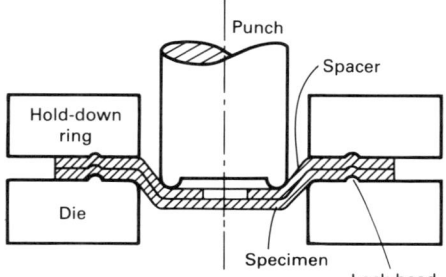

Fig. 15 Schematic of hydraulic bulge test

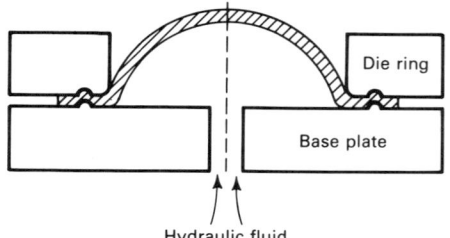

ple. Initially, the dots form a right-angled isoceles triangle. The principal strains are computed from the change in the spacings of the dots and changes in the angle they subtend. This information is recorded and also used to maintain a constant strain rate by controlling the hydraulic pressure. Strains can thus be computed once per second, which limits the maximum controllable strain rate.

A camera is mounted so that it maintains a constant distance from the top of the dome and the dots, except for the effect of the curvature of the dome, which is negligible. The stress state is determined by measuring the curvature of the dome with a contacting spherometer and by measuring the hydraulic pressure with a strain gage pressure transducer.

For thin samples, bending stresses can be neglected, and the radial (meridional) stress, σ_r, is given by:

$$\sigma_r = \frac{pR}{2t} \qquad \text{(Eq 20)}$$

where p is the hydraulic pressure, R is the radius of curvature, and t is the instantaneous thickness. The thickness is calculated from the measured strains using constancy of volume. At the top of the dome, the sample is in balanced biaxial stretching, and the circumferential stress, σ_c, is equal to the radial stress.

For convenience, it is customary to express the results of hydraulic bulge tests in terms of the true thickness stress and strain. This is done by theoretically superimposing a hydrostatic compressive stress that does not influence deformation and that has in-plane components equal to the actual radial and circumferential tensile stresses. This converts the stress state to a simple uniaxial thickness compressive stress, as shown in Fig. 16.

The true thickness strain, ϵ_t, can be obtained from the radial strain, ϵ_r, and circumferential strain, ϵ_c, by using the constancy of volume condition:

$$\epsilon_t = -\epsilon_r - \epsilon_c \qquad \text{(Eq 21)}$$

This enables the results to be represented by a true compressive stress-strain curve in the thickness direction. The hydraulic bulge test has the following applications:

- Intrinsic material characterization in biaxial stretching, which is a very common strain state in production stampings
- Testing to much higher strain levels than those achievable in tensile testing (in some cases by as much as a factor of ten), particularly for heavily cold worked materials
- Checking the validity of plasticity theories that attempt to predict the yielding behav-

Fig. 16 Addition of compressive hydrostatic stress to biaxial tensile stress

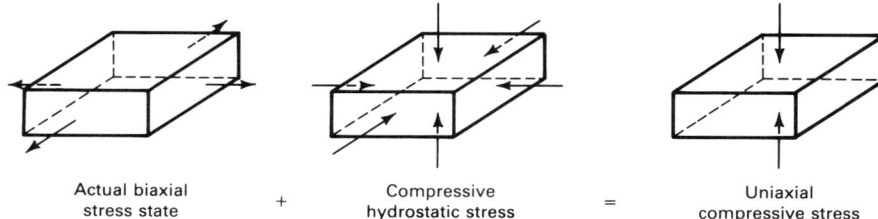

Actual biaxial stress state $+$ Compressive hydrostatic stress $=$ Uniaxial compressive stress

ior of metals in all stress states from properties measured in uniaxial and plane-strain tensile testing

Shear Testing

Two tests have been developed to determine the properties of sheet metals subjected to planar shear deformation: the Marciniak in-plane sheet torsion test (Ref 45) and the Miyauchi shear test (Ref 46).

Marciniak In-Plane Sheet Torsion Test. A flat 50-mm (1.97-in.) square sample is effectively divided into three zones: an inner circular zone, which is clamped; a ring-shaped middle zone, which surrounds the inner zone and is free to deform; and an outer ring-shaped zone, which is clamped. The inner zone is rotated in its plane relative to the outer zone, which deforms the middle zone in shear. The sample is deformed to fracture, and the angular rotations at two radii in the middle zone are measured by means of calibrated drums that rotate with the sheet. This is shown schematically in Fig. 17.

For materials that follow the power law shear strain-hardening relationship:

$$\tau = C\gamma^n \qquad \text{(Eq 22)}$$

where τ is the shear stress, C is a constant, and γ is the shear strain, it can be shown (Ref 45) that:

$$n = \frac{2\log\left(\dfrac{R_a}{R_b}\right)}{\log\left(\dfrac{\alpha_a}{\alpha_b}\right)} \qquad \text{(Eq 23)}$$

where α_a and α_b are the angular displacements at radii R_a and R_b.

Shear fracture occurs where the shear stress is greatest—at the inner radius, R_o, of the deforming ring. The shear fracture strain, γ_f, is given by:

$$\gamma_f = \frac{2\alpha_a\left(\dfrac{R_a}{R_o}\right)^{2/n}}{n} \qquad \text{(Eq 24)}$$

Fig. 17 Deformed Marciniak in-plane torsion test specimen

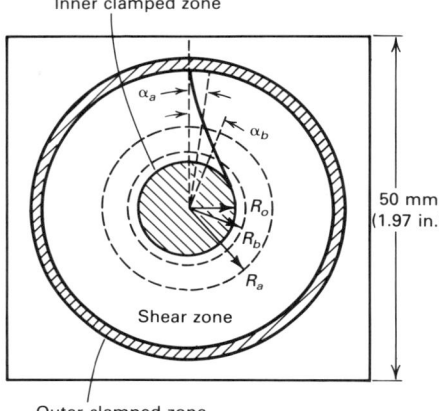

Inner clamped zone

Shear zone

Outer clamped zone

50 mm (1.97 in.)

This test enables the forming properties of sheet metals to be determined at much higher strain levels than is possible in the uniaxial tensile test and in a different strain state, i.e., in shear.

The Miyauchi shear test determines the properties of sheet metals in planar shear deformation by means of a modified tensile technique (Ref 46). The test uses flat, rectangular specimens whose ends are divided into three equal sections by parallel longitudinal slits, as shown in Fig. 18(a).

The specimen is clamped in a fixture that prevents out-of-plane deformation. The inner and outer sections are then pulled in opposite directions in a tensile machine. This produces a shear stress in the regions between the inner and outer sections and deforms the specimen, as shown in Fig. 18(b). The deformation in these regions is uniform, except at the ends.

The shear strain, γ, is the tangent of angle θ, which is the change in direction of lines scribed across the specimen as they pass through the shear zone, as shown in Fig. 18(b). The strain can also be determined from the displacement of the inner section, once a relationship has been established between the displacement and θ, which must be

Fig. 18 Miyauchi shear test specimen
(a) Undeformed. (b) Deformed. Source: Ref 46

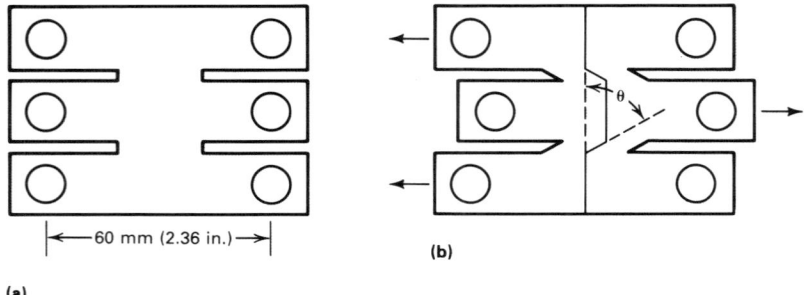

Fig. 19 Simple bending test
Source: ASTM E 290

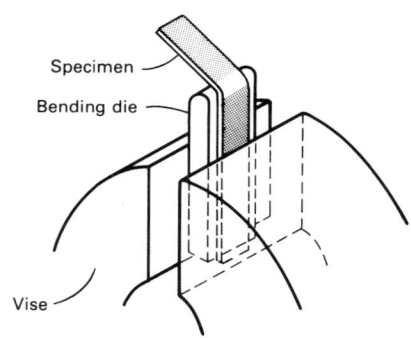

done for each type of sheet metal tested. Shear stress-strain curves for three different steels which show differences in the strain dependence of the work-hardening coefficient from that in the tensile test have been published (Ref 46).

Hardness Testing

Hardness, or the resistance to indentation by a concentrated load applied by a suitable indenter, has been used in many stamping plants as a measure of formability. Generally, formability decreases with increasing hardness, but the fine-scale correlation between these properties has not been reliable. The test can be used effectively to monitor changes in a particular grade of material caused by changes in processing that may affect formability.

For steels, hardness measurements correlate well with yield strength values (Ref 47). Hardness testing is thus useful in quality control to ensure that the material in use is the specified grade and has the required strength level.

The Rockwell hardness test (ASTM E 18, "Standard Test Methods for Rockwell Hardness and Rockwell Superficial Hardness of Metallic Materials"), which is described in detail in the article "Rockwell Hardness Testing" in this Volume, is typically used to determine the hardness of such materials. The load and indenter must be selected for the gage and hardness range of the material according to the test specification to ensure that the indentation is the appropriate size. If the indentation is too deep in a sheet sample, the reading will be artificially high because of the influence of the supporting anvil. This becomes a more serious consideration with the use of thinner gage sheet metal in many industries. For thicker and harder materials the Rockwell B scale is commonly used, and for thinner or softer materials the Rockwell 30T superficial scale is used.

Hardness readings are influenced by the degree of flatness and surface conditions.

Also, the presence of cold worked surface layers can cause unrepresentatively high hardness readings that suggest a lower formability level than actually exists.

Simulative Tests

For many forming operations, tests that simulate the operation are more useful and relevant than fundamental intrinsic property measurement tests. These tests subject the work material to deformation that closely approximates the production operation, including the effects of factors not present in the intrinsic tests, such as bending and unbending and friction between the work materials and die surfaces. Because these additional factors are present, simulative tests tend to be less reproducible than intrinsic tests and must be performed under carefully controlled conditions to minimize variability in the results.

Simulative tests can be classified on the basis of the predominant forming operation involved: bending, stretching, drawing, and stretch-drawing. In addition, tests to measure wrinkling and the springback occurring after bending or another forming operation have been developed.

Bending Tests

Two types of bending tests relate to sheet metal forming: simple bending and stretch-bending tests. Simple bending tests are useful in predicting how the sheet metal will perform when bent without tension, as in a hemming operation. Stretch-bending tests relate to the response to combined bending and stretching, as when sheet metal is pulled over a punch or die radius, for example.

Simple bending tests can be performed in various ways (ASTM E 290, "Standard Method for Semi-Guided Bend Test for Ductility of Metallic Materials"). The simplest method for thin sheet material is to clamp a specimen and a bending die in a vise, as shown in Fig. 19, and to bend the specimen

over the die manually or with a nonmetallic mallet.

If the specimen bends through 180° without fracturing or cracking, the experiment is repeated using a bending die of smaller radius. In the case of highly ductile metals that have extremely small bend radii, a modified test is performed. The specimen is bent initially at its midpoint, through less than 90°, over a small radius. The test is then completed by pressing the ends of the specimen together between flat platens without a bending die placed between the platens.

The specimen width-to-thickness ratio should be greater than 8 to 1, and sheared edges should be machined, filed, or sanded to remove the heavily cold worked metal present. The orientation of the specimen with respect to the rolling direction may be important, because it affects the resistance of the specimen to fracture. Specimens cut perpendicular to the rolling direction usually require a larger bend radius and therefore provide a more conservative measure of this property.

For low-carbon sheet steels, the minimum bend radius is usually not a limiting factor. For high-strength steels and aluminum alloys it sometimes is, and methods such as "rope hemming," which increase the bend radius, have been developed to prevent cracking during hemming of these materials.

Stretch-Bending Tests. A rectangular strip of sheet metal is clamped at its ends in lock beads and deformed in the center by a punch, as shown in Fig. 20. There are two types of stretch-bending tests: the hemispherical test, in which a hemispherical-tipped punch and a concentric circular lock bead are used, and the angular test, in which a wedge-shaped punch and straight parallel lock beads are used. The hemispherical test involves a range of strain states. The angular test produces the plane-strain state.

The punch travel between initial contact and specimen fracture is measured. The conditions are chosen so that fracture occurs in the region of punch contact. When fracture

Fig. 20 Stretch-bending test

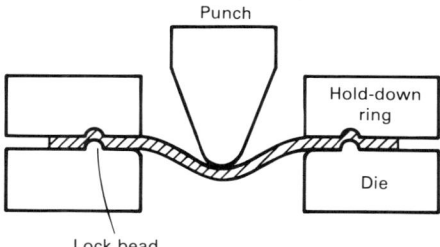

occurs in the unsupported region, which tends to happen with narrow, thin-gage specimens and large punch radii, the test effectively becomes a tensile test.

The results of several hemispherical and angular stretch-bending tests on three types of steels and an aluminum alloy have been reported (Ref 48). For the hemispherical test, the effects of variations in punch tip radii ranging from 3.2 to 51 mm (0.13 to 2.0 in.), in sheet thicknesses ranging from 0.5 to 3.3 mm (0.02 to 0.13 in.), and in specimen widths ranging from 25 to 203 mm (1.0 to 8.0 in.) were investigated in the dry and lubricated conditions. The tests showed that the height at fracture increased with increasing punch radius and sheet thickness and with the use of lubricants. It decreased with increasing specimen width in the range of 102 to 203 mm (4.0 to 8.0 in.), in which fracture occurred in the region of punch contact. The ranking of two of the steels was found to be dependent on specimen thickness.

Fewer conditions were investigated in the angular test. The results for a 76-mm (3.0-in.) wide specimen and punch radii ranging from 1.6 to 6.4 mm (0.06 to 0.25 in.) showed much greater heights than for the same conditions in the hemispherical test. Increases in height with increasing punch radius were also evident, but in contrast to the hemispherical case, a decrease with increasing thickness was observed. Preliminary correlation between the results of these tests and production experience is reported to be fairly good.

Data from the angular stretch-bending test have been analyzed (Ref 49) and indicate that fracture occurs at a constant limit strain that is independent of sheet thickness and punch radius. Stretch-bending tests are useful for material selection and for predicting the effects of material substitution and gage reduction in many forming operations.

Stretching Tests

Historically, ball punch tests, such as the Olsen cup test and Erichsen cup test, have been used to determine the properties of sheet metals in stretching. These tests stretch a specimen over a hardened steel ball and measure the height of the cup produced. More recently, tests that stretch the specimen over a much larger hemispherical dome have been developed, including the limiting dome height test, which uses specimens of different widths to control the strain ratio at fracture.

Many forming operations involve stretching an edge of a part or of a cutout (hole) in a part. For example, when a concavely contoured edge is flanged, the metal is stretched. The ability of the material to undergo this type of forming operation can be measured by the hole expansion test. In this test, a cylindrical, hemispherical, or conical punch is pushed through a circular hole of smaller diameter in the specimen. This initially increases the diameter of the hole and then forms a rim of stretched metal. The edge ductility of the material is indicated by the amount of hole expansion that occurs without edge cracking.

Ball Punch Tests. The Olsen and Erichsen cup tests are similar, differing principally in the dimensions of the tooling used. The Olsen test (ASTM E 643, "Standard Test Method for Ball Punch Deformation of Metallic Sheet Material") uses a 22.2-mm (0.875-in.) diam hardened steel ball and a die with a 25.4-mm (1.0-in.) internal diameter (28.6 mm, or 1.125 in., for gages over 1.5 mm, or 0.06 in.) and a 0.81-mm (0.032-in.) die profile radius, as shown in Fig. 21. The Erichsen test, which is used extensively in Europe, uses a 20-mm (0.79-in.) diam ball and a die with a 27-mm (1.06-in.) internal diameter and a 0.75-mm (0.03-in.) die profile radius.

In both tests, the cup height at fracture is used as the measure of stretchability. The preferred criterion for determining this point is the maximum load. When this cannot be determined, the onset of a visible neck or fracture can be used, but this yields a slightly different value. The cup height measured by means of a visible fracture is 0.3 to 0.5 mm (0.012 to 0.020 in.) greater than the height measured at the maximum load.

These tests, as indicators of stretchability, should correlate with the n value, but the correlation is not satisfactory. Improved correlations with the total elongation (Ref 50) and reduction in area (Ref 51) have been reported. Some investigators (Ref 52, 53) have reported poor reproducibility of test results in the Olsen and Erichsen tests and poor correlation with production experience. Satisfactory reproducibility and correlation in specific cases have been reported when experimental conditions were carefully controlled (Ref 50).

The variability in tests has been attributed (Ref 52, 53) to the small size of the pene-

Fig. 21 Olsen cup test
Source: ASTM E 643

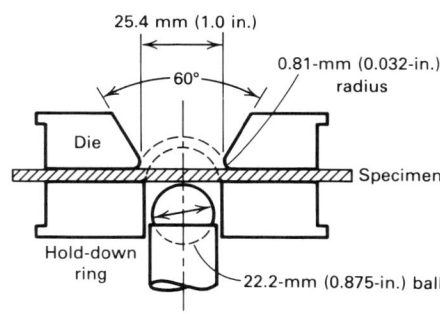

trator, uncontrolled drawing-in of the flange, and inconsistent lubrication. The small size of the penetrator leads to excessive bending, particularly in thicker sheet, and is generally unrepresentative of production conditions. Drawing-in can be controlled somewhat by standardizing the specimen size and by using a high (~71-kN, or ~8-ton) clamping force. Even greater control can be achieved by the use of lock beads or serrated dies (dies with concentric circular ridges of triangular cross section that dig into the specimen and prevent slippage).

Consistent lubrication can be achieved by the use of oiled polyethylene between the specimen and penetrator. The problems with the Olsen and Erichsen tests have led to the development of stretching tests that use a much larger diameter punch and a lock bead to prevent drawing-in.

Hemispherical dome tests using 50.8-, 76.2-, and 101.6-mm (2.0-, 3.0-, and 4.0-in.) punches have been reported (Ref 52, 53). A 100-mm (3.94-in.) test is now the most widely used. Typical tooling designed for this test is shown in Fig. 22. The lock bead, in combination with a hold-down force of about 222 kN (25 tons), completely prevents drawing-in of the flanges.

Fig. 22 Tooling for the 101.6-mm (4.0-in.) hemispherical dome test
Source: Ref 52

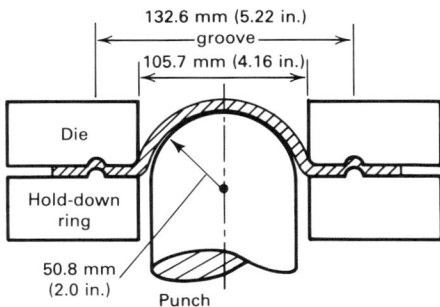

The specimens fracture circumferentially at a distance (for lightly lubricated low-carbon steel) of 35 to 40 mm (1.38 to 1.57 in.) from the pole, at which point the radial strain peaks sharply. The circumferential strain varies gradually from a maximum of 10 to 20% at the pole to zero at the lock bead.

The hemispherical dome test yields more reproducible results than the Olsen and Erichsen cup tests. For low-carbon steels, the dome height, which is measured at the point of maximum load, increases linearly with the *n* value. For a wide range of material (including brasses, aluminum alloys, and zinc), optimum correlation is found between the dome height and the total elongation, which incorporates the effects of strain-rate hardening and limiting strains.

Overall, the use of lubrication in hemispherical dome tests is beneficial. A thin layer of a standard lubricant, applied in a consistent manner, reduces scatter in test results, simulates production conditions more closely, reduces damage to the tooling, and simplifies specimen preparation. The improved sensitivity obtained in the dry condition is negated by the increased scatter in the results.

The use of lubrication makes the strain ratio at fracture more biaxial. This is undesirable for production simulation, because most production failures occur in the region of plane strain, i.e., in a less biaxial manner. To control the strain ratio at fracture, specimens of different widths have been used (Ref 54). This technique has been developed further (Ref 55, 56) into the limiting dome height test.

Limiting Dome Height (LDH) Test. Specimens of various widths are held in a circular lock bead and stretched over a 100-mm (3.94-in.) dome using tooling of the type shown in Fig. 22. In principle, this test can be used to duplicate a large range of production failure strain states and to select the most suitable material for each particular operation. In practice, most production failures occur close to plane strain, which is generally the strain state at the minimum on a plot of dome height versus specimen width. Consequently, attention has concentrated on this minimum value.

When testing a new material, initial tests should be performed to determine the specimen width that yields the minimum dome height, or LDH value, and the corresponding minor strain. Once this has been established, tests can be run at this width only. For low-carbon steels, the minimum dome height occurs at a width of approximately 124 mm (4.9 in.). This can also be used as an approximation for other materials. Increments in test specimen width of ±3 mm (±0.12 in.) are sufficiently close.

It has been found that for specimens lubricated lightly with a wash oil, the dome height increases with decreasing hold-down force below about 250 kN (28 tons). This is attributed to drawing-in of the flange. A hold-down force of at least 250 kN (28 tons) should therefore be used. The limiting dome height is taken as the height at which the maximum load occurs.

Preliminary tests (Ref 57) have shown a correlation between the limiting dome height test and production stamping performance. Some problems with test reproducibility over a period of time and among different test facilities have been encountered. Numerous attempts have been made to determine a correlation between the limiting dome height test and mechanical and forming property measurements. The dome height depends on the ability of the material to distribute strain and the limiting strain level and would therefore be expected to correlate with the total elongation. Correlation for a range of different materials has been reported (Ref 58).

The specimens used in the limiting dome height test can be sheared or blanked from the sheet sample, and the test can be performed rapidly on equipment that automatically measures the dome height at the maximum punch load. The test has considerable potential for production control and research applications.

Hole Expansion Test. A flat sheet specimen with a circular hole in the center is clamped between annular die plates and deformed by a punch, which expands and ultimately cracks the edge of the hole. Flat-bottomed, hemispherical, and conical punches have been used, and in some cases die plates have been equipped with lock beads to prevent drawing-in of the flange. The punch should be well lubricated and should have a large profile radius. A spacer may be used between the punch and the sample, as in the Marciniak test. Figure 23 illustrates the hole expansion test using a flat-bottomed punch.

The test is terminated when a visible crack is observed, and the hole expansion is expressed as the percentage increase in hole diameter:

$$\text{Hole expansion (\%)} = \frac{100(D_f - D_o)}{D_o} \quad \text{(Eq 25)}$$

where D_o and D_f are the initial and final hole diameters, respectively.

The results of several hole expansion tests on eight different types of steel are reported in Ref 59. Square specimens 203 mm (8.0 in.) on each side with a 25-mm (1.0-in.) diam punched hole, a 101.6-mm (4.0-in.) diam hemispherical punch, and die plates with a 2-mm (0.08-in.) radius lock bead were

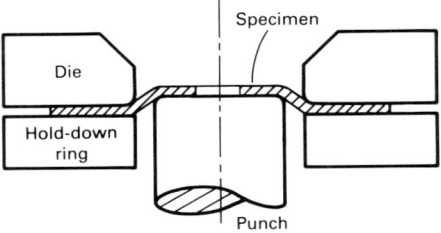

Fig. 23 Hole expansion test with a flat-bottomed punch

used. The measured hole expansion ranged from 24 to 82% for steels with yield strengths ranging from 253 to 537 MPa (36.7 to 77.9 ksi).

In most cases, removing the burr and cold worked metal from the edge of the punched hole increased the hole expansion considerably. The hole expansion also increased with increasing total elongation and r_m value and decreased with increasing tensile strength (which was anticipated, because total elongation decreases with increasing tensile strength). Inclusions were observed in crack locations, and inclusion shape control improved hole expansion performance.

Drawing Test

Swift Cup Test. The most commonly used test for deep drawability is the Swift cup test. Circular blanks of various diameters are clamped in a die ring and deep drawn into cups by a flat-bottomed cylindrical punch. The standard tooling for this test is shown in Fig. 24. Drawability is expressed as either the limiting draw ratio (LDR) or the percent reduction. The limiting draw ratio is the ratio of the diameter, D, of the largest blank that can be drawn successfully to the diameter, d, of the punch:

$$\text{Limiting draw ratio} = \frac{\text{maximum blank diameter}}{\text{punch diameter}} = \frac{D}{d} \quad \text{(Eq 26)}$$

Percent reduction is defined as:

$$\text{Percent reduction} = \frac{100(D - d)}{D} \quad (\%) \quad \text{(Eq 27)}$$

Cup height, h, is approximately (Ref 60):

$$h = \frac{(D^2 - d^2)}{4d} \quad \text{(Eq 28)}$$

An alternative method for determining the limiting draw ratio (Ref 61) uses blanks of a single diameter, which is less than the critical

Fig. 24 Standard tooling for the Swift flat-bottomed cup test

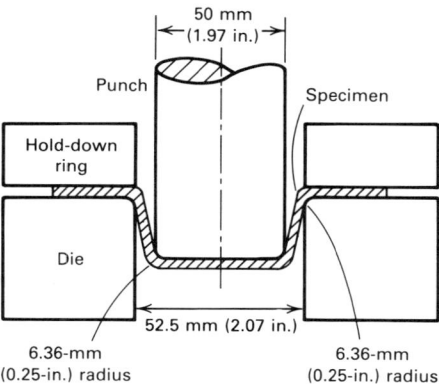

diameter in the standard test. The blanks are drawn to the maximum load, which usually occurs before 50% of the draw has occurred. The clamping force is then increased to prevent further drawing-in of the flange, and the load is increased to the point of fracture. The limiting blank diameter (LBD) is defined by:

$$LBD = \left[\frac{\text{Fracture load} \times (\text{blank diameter} - \text{die diameter})}{\text{Maximum drawing load}} \right] + \text{Die diameter} \qquad \text{(Eq 29)}$$

The limiting draw ratio (LDR) is given by:

$$LDR = \frac{LBD}{\text{Punch diameter}} \qquad \text{(Eq 30)}$$

This method has been shown to correlate well with the standard test for a range of materials of widely different drawability (Ref 61).

The limiting draw ratio increases with increasing normal anisotropy (r_m) and thickness, particularly at the low ends of the ranges for these variables, but is not sensitive to the n value (Ref 62). The limiting draw ratio also increases as the punch profile radius increases up to about eight times the sheet metal thickness, as the die profile radius increases up to about 12 times the metal thickness, and as the punch speed increases. The height of the ears formed in this test is proportional to the Δr value.

Too low a blankholder force may cause wrinkling, and too high a blankholder force may cause fracture at the punch profile radius. The die rings should be well lubricated, but the punch should not be lubricated. By not lubricating the punch, the amount of stretching that occurs over the punch profile radius and the tendency for splitting to occur at this location are reduced.

Stretch-Drawing Tests

Many forming operations involve stretching and drawing; for example, square cups have drawn corners and stretched sides. The ratio of stretching to drawing in an actual part can be measured by a shape analysis technique (Ref 63). A line is drawn from a reference point, e.g., the center of the blank, to the edge of the blank, through the critical forming area. After forming, the ratio of the increases in length of this line inside and outside the initial die contact line is taken as the ratio of stretching to drawing. Two tests are commonly used for stretch-drawing: the Swift round-bottomed cup test and the Fukui conical cup test.

The Swift round-bottomed cup test resembles the Swift flat-bottomed cup test described above. However, the top of the punch is hemispherical, which causes stretching in the center of the specimen in addition to drawing-in of the flange to produce the wall of the cup.

This test was used (Ref 64) to evaluate 50 different steels with a 50-mm (1.97-in.) diam punch and 127-mm (5.0-in.) diam specimens and with a 65-mm (2.56-in.) diam punch and 165-mm (6.5-in.) diam specimens. Hold-down forces of 490 and 981N (110 and 221 lb), respectively, were used at a test speed of 1 mm/s (0.04 in./s). Both sides of the specimens were lubricated with thin polyethylene sheet.

The end point of the test is determined by observing fracture visually or by detecting a drop in the punch load. Multiple regression analysis of the test results showed that the cup height at fracture increased linearly with increases in the r_m value, n value, and metal thickness.

To determine the correlation between performance of the steels in the stretch-drawing test and in actual parts production, four automotive stampings were made, using 12 different steels for each. The stampings had stretch-to-draw ratios ranging from approximately 1-to-5 to 2-to-1, and minor-to-major strain ratios in critical areas ranging from −0.3 to +0.45. The correlation coefficients between the test and stamping results had an average value of 0.92 and ranged from 0.89 to 0.94 (a value of 1.00 indicates perfect correlation). In another trial on a stamping with a stretch-to-draw ratio of 4.5 to 1, the test results did not correlate. These tests indicate that for parts that involve both stretching and drawing, without excessive stretching, the Swift flat-bottomed cup test is useful as a quality control tool.

Fukui Conical Cup Test. In the Fukui conical cup test (Ref 62 and 65, JIS Z 2249), circular specimens punched from a sample of sheet metal are deformed into conical cups

by means of a 12.5- to 27-mm (0.5- to 1.1-in.) diam ball and tooling of the type shown in Fig. 25. The ball size depends on the sheet thickness. The specimens are lubricated on the die side only. Lubrication on the punch side leads to tilting of the specimens. Specimens are centered and held in place by the hold-down ring and deformed to fracture by the punch.

The diameter of the base of the conical cup formed is measured and divided by the diameter of the original specimen to give the Fukui conical cup value. The end point of the test is not critical, because the diameter of the cone does not change after fracture. Usually, a constant punch travel is used. When the test material has a high level of planar anisotropy (a high Δr value), the conical cup is asymmetric, and an average diameter must be determined. A high correlation between the Fukui conical cup value and the product of the average n value and the average r value has been reported for low-carbon steels (Ref 62).

An alternative method has been developed for performing this test (Ref 50). The punch travel between the initial contact with the specimen and the onset of a drop in the punch load, which coincides with the formation of a visible neck, is measured and used instead of the ratio of the diameters. This value, which is known as the formability index, correlates with the uniform elongation and therefore with the n value for low-carbon steels.

Wrinkling and Buckling Tests

Two principal types of tests are used for wrinkling and buckling, the conical cup wrinkling test and the Yoshida buckling test. The conical cup wrinkling test is similar to the Swift flat-bottomed cup test, but uses a punch that is much smaller than the die opening. Consequently, the cup wall is conical and is not in contact with the punch. Under some conditions, wrinkles form in the cup

Fig. 25 Fukui conical cup test
Source: Ref 62

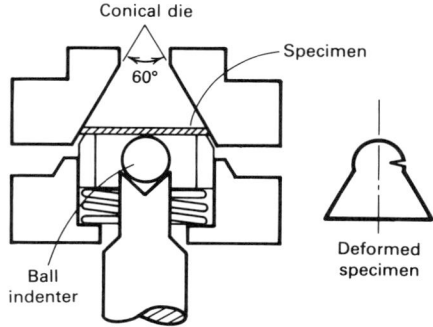

wall. In the Yoshida buckling test (Ref 66), a flat, square specimen is stretched slightly in the diagonal direction, and the height of the buckle that is formed is measured.

Conical Cup Wrinkling Test. A circular blank is clamped between annular dies and deformed by a flat-bottomed punch with a diameter that is typically about 75% of the internal diameter of the die. This procedure is illustrated in Fig. 26. At very low levels of hold-down force, wrinkling occurs in the flange. At higher levels, flange wrinkling is suppressed, but wrinkling occurs in the unsupported wall. This is caused by compressive stresses in the circumferential direction (hoop stresses), due to the local reduction in diameter as drawing progresses. For example, with a 75-mm (2.96-in.) diam punch and a 100-mm (3.94-in.) diam die, the top of the wall has a diameter of 100 mm (3.94 in.). If the cup depth is doubled, the original top of the wall becomes the new midpoint and must decrease in diameter to 87.5 mm (3.44 in.).

At high levels of blankholder force, the tensile stresses in the radial direction in the wall prevent the formation of wrinkles, and fracture at the punch or die radius becomes the limiting factor. The maximum cup height occurs at the intersection of the wall wrinkling and fracture limits, as shown in Fig. 27.

The results of experiments on several types of steel with different thicknesses and tooling of various dimensions have been reported in Ref 13 and 68. Wrinkling occurred in the unsupported wall when the true compressive hoop strain exceeded a certain value for each level of the tensile radial strain for all tooling geometries and forming conditions. The critical wrinkling strains were plotted on the forming limit diagram, as shown in Fig. 28.

Attainment of the critical wrinkling strain is strongly influenced by the dimensions of the specimen and tooling, lubrication, and the hold-down force. Changes in these variables that reduce the radial stress (i.e., an increase in the die radius, improved lubrica-

Fig. 26 Conical cup wrinkling test

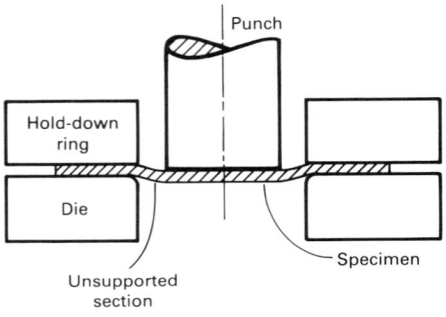

Fig. 27 Wrinkling and fracture limits in conical cup drawing
Source: Ref 67

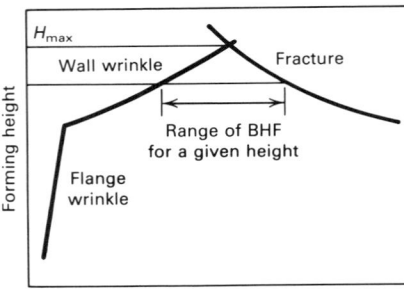

tion, or a reduction in the blank diameter or the hold-down force) increase the tendency to form wrinkles.

Material properties that affect wrinkling in the conical cup test are the r_m, Δr, and n values and the ratio of the flow stress to the elastic modulus. A high r_m value and low Δr value reduce wrinkling, which initiates in the directions of lowest r value. A high n value enables the hold-down force to be increased, which increases the radial force and reduces wrinkling. A low flow stress-to-elastic modulus ratio also reduces wrinkling.

Yoshida Buckling Test. A flat, square specimen is gripped at opposite corners and pulled in tension in the diagonal direction, as shown in Fig. 29 (Ref 66, 67). The standard specimen is 100 mm (3.94 in.) square with 41-mm (1.6-in.) wide grips and a gage length of 75 mm (2.95 in.). The buckle height is measured over a 25.4-mm (1.0-in.) width at the center of the specimen.

Nonuniform stresses are generated in the specimen, which cause a buckle to form in the center along the direction of loading. The height of the buckle at a given elongation, e.g., 2%, is used as the measure of buckling.

Several investigations have been conducted on the correlation between buckle height and test material properties. The Yoshida buckling test and a conical cone wrinkling test (using a hemispherical punch) were performed on several ferrous and nonferrous materials in different tempers (Ref 69). A direct correlation for both tests between the buckling or wrinkling height and the yield strength, an inverse correlation with the work-hardening exponent, and lack of correlation with the normal anisotropy were reported. The Yoshida test was not successful for aluminum, because the specimens fractured before buckling.

The Yoshida test was performed on 31 steels of different types and thicknesses (Ref 14), and correlations between the slope of the buckle height versus elongation curve, which

Fig. 28 Combined forming and wrinkling limit diagram
Source: Ref 68

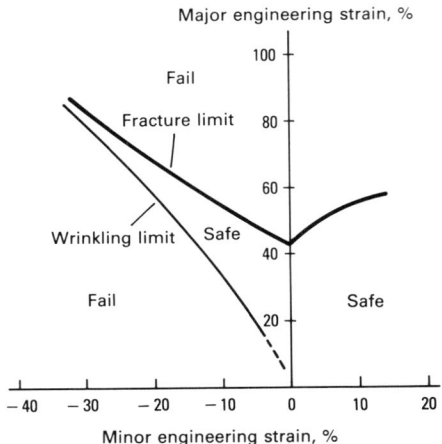

Fig. 29 Yoshida buckling test
Source: Ref 67

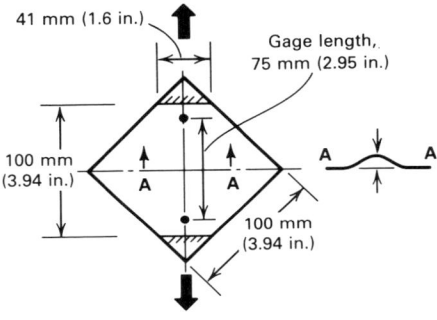

is easier to determine than the height at a particular elongation, and the yield strength and the ratio of yield strength to tensile strength were obtained. An inverse correlation with the instantaneous (2%) strain-hardening exponent and a lack of correlation with the uniform elongation and normal anisotropy were also noted.

Springback Tests

Springback tests that bend a specimen about a mandrel and determine the change in the angle of bending on removal of the bending load have been used as indicators of yield strength. This test was developed (Ref 70) with a 12.5-mm (0.5-in.) radius mandrel, as shown in Fig. 30, for use as a quality control tool with sheet materials with thicknesses ranging from 0.15 to 0.38 mm (0.006 to 0.015 in.). Previously, hardness measurements had been used for this purpose, but they were found to be insufficiently accurate for hard, thin-gage steels and aluminum alloys, and did not provide any information on anisotropy.

Fig. 30 Springback tester for determining yield strength
Source: Ref 70

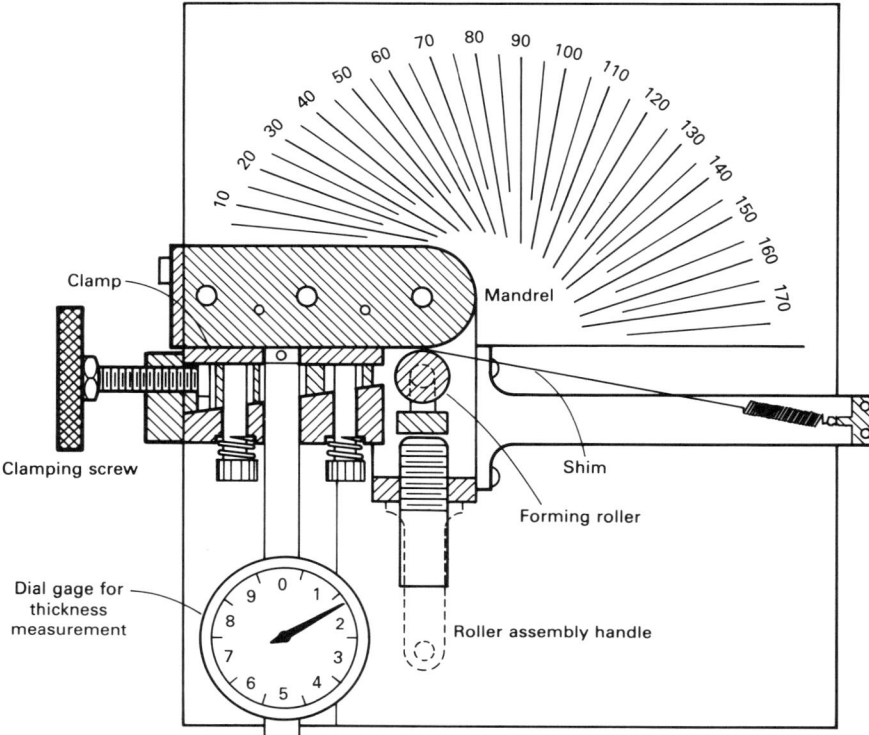

In the test, specimens are bent through 180° and released, and the angle of springback is read on the scale. The yield stress can then be determined from the springback angle and material thickness by means of a previously determined nomograph. Springback depends on the elastic modulus, which necessitates different nomographs for materials with different moduli. The test is most accurate in the range of springback angles of 60° to 120° and should be modified by changing the mandrel radius if the angle is less than 30° or greater than 150°.

The nomograph was calculated assuming an elastic/perfectly plastic stress-strain relationship, which is the same in tension and compression (i.e., zero strain hardening and no Bauschinger effect). This calculation has been refined (Ref 71) by using an average of the experimentally determined tensile and compressive stress-strain curves, including strain hardening. This improves the average ratio of the yield strength predicted using springback measurements to the tensile test yield strength from 0.80 to 0.91.

More recently, a similar test was developed (Ref 72) that uses a larger radius mandrel (19 mm, or 0.75 in.). Twenty steels with thicknesses ranging from 0.56 to 2.36 mm (0.022 to 0.093 in.) and with tensile strengths ranging from 285 to 714 MPa (41.6 to 103.5 ksi) were tested. The measured springback correlated better with the forming strength, which is the average of the yield and tensile strengths, than with the individual strength values. For the same forming strength, steels with high (>1.5%) levels of yield point elongation develop less springback than those without. In these cases, the tangent modulus is almost zero in the region of yield point elongation.

Springback after a 90° flanging operation has been measured (Ref 73) as a function of material flow stress, thickness, degree of cold work, and die radius and clearance for a low-carbon steel, two high-strength low-alloy steels, and a dual-phase steel. Springback increased with increases in flow stress and die radius and clearance, and with a decrease in material thickness.

For minimum springback, the ratio of the thickness to the die radius should be greater than 0.4, beyond which a further increase has little effect. The springback developed by the dual-phase steel increased more rapidly with the level of cold work than that developed by the high-strength steels, as anticipated from its higher strain-hardening exponent.

Springback after the combined effects of bending and stretching has been investigated (Ref 19) using a tensile machine and a three-point bending fixture. Tests were performed on four thin-gage (~0.4 mm, or ~0.016 in.) low-carbon steels in various tempers with yield strengths ranging from 155 to 670 MPa (22.5 to 97.0 ksi). For applied tensile stresses that are below the yield strength of the material, the springback decreased linearly with the tensile stress to the same extent whether the stress was applied during or after bending.

For tensile stresses above the yield strength, tension during bending decreased the springback to a level that was independent of the initial yield strength and thickness. Stretching after bending, which deforms the entire cross section plastically in tension, decreases the springback progressively to extremely low levels.

Correlation Between Simulative Tests and Material Properties

Quantitative correlations between the results of simulative tests and select tensile properties have been determined for the Olsen cup, Swift flat-bottomed cup, and Fukui conical cup tests (Ref 50) and for the Swift round-bottomed cup test (Ref 64). In the first correlation, 48 materials were tested, including aluminum-killed, rimmed, and stainless steels and aluminum alloys in various tempers (Ref 50). Tensile properties included the directionally averaged percentage of total elongation, $\bar{e}_T$, to indicate the stretchability, and the normal anisotropy, r_m, to indicate drawability. The following relationships and correlation coefficients were obtained:

Test parameter	Relationship	Correlation coefficient
Olsen cup height/ punch diameter	$0.217 + 0.00474\bar{e}_T + 0.00392r_m$	0.925
Limiting draw ratio (Swift)	$1.93 + 0.00216\bar{e}_T + 0.226r_m$	0.835
Formability index (Fukui)	$0.525 + 0.0134\bar{e}_T + 0.207r_m$	0.757

The Olsen test involves a much greater ratio of stretching to drawing than the Fukui test, which in turn involves a much greater ratio than the Swift flat-bottomed cup test.

In the second correlation, 50 different steels were tested (Ref 64), and the results were correlated with the average n and r values and thickness, t, as follows:

$$\frac{\text{Swift round-bottomed cup height}}{\text{Blank diameter}} = 0.0830t$$

$$+ 0.679\bar{n} + 0.0594r_m - 0.036$$

Determination of Forming Limit Diagrams

Forming limit diagrams indicate the limiting strains that sheet metals can sustain over a range of major-to-minor strain ratios. Two main types of laboratory tests are used to determine these limiting strains. The first type of test involves stretching test specimens over a punch or by means of hydraulic pressure—for example, the hemispherical punch method. This produces some out-of-plane deformation and, when a punch is used, surface friction effects. The second test produces only in-plane deformation and does not involve any contact with the sample within the gage length.

The first type of test has been used much more extensively (Ref 6, 74-76) than the second and provides slightly different results (Ref 33, 77). Good correlation has been obtained between forming limit diagrams determined in the laboratory and production experience.

The hemispherical punch method for determining forming limit diagrams uses circle-gridded strips of the test material ranging in width from 25.4 to 203 mm (1.0 to 8.0 in.) that are clamped in a die ring and stretched to incipient fracture by a 102-mm (4.0-in.) diam steel punch (Ref 74, 76). The narrowest strip fractures at a minor-to-major strain ratio of about −0.5, which is comparable to that obtained in a tensile test. As the strip width is increased, the strain ratio increases to a slightly positive value for a full-width specimen. Further increases in the ratio to a maximum value of +1.0 (balanced biaxial stretching) are achieved by using progressively improved punch lubrication (oiled polyethylene, oiled neoprene) and by increasing thicknesses of polyurethane rubber.

The strains are measured in and around regions of visible necking and fracture. The forming limit curve is drawn above the strains measured outside the necked regions and below those measured in the necked and fractured regions, as shown in Fig. 31.

In-plane determination of the forming limit diagram can be achieved by using the uniaxial tensile test, rectangular sheet tension test, and Marciniak biaxial stretching test with elliptical and circular punches, as described earlier in this article. The forming limit curve can be determined over the full range of strain ratios, without introducing any out-of-plane deformation. A comparison of the in-plane and punch methods showed close agreement for negative strain ratios and slightly higher values in the punch test at plane strain and for positive strain ratios (Ref 33).

Fig. 31 Strain measurements and forming limit diagram for aluminum-killed steel

Source: Ref 76

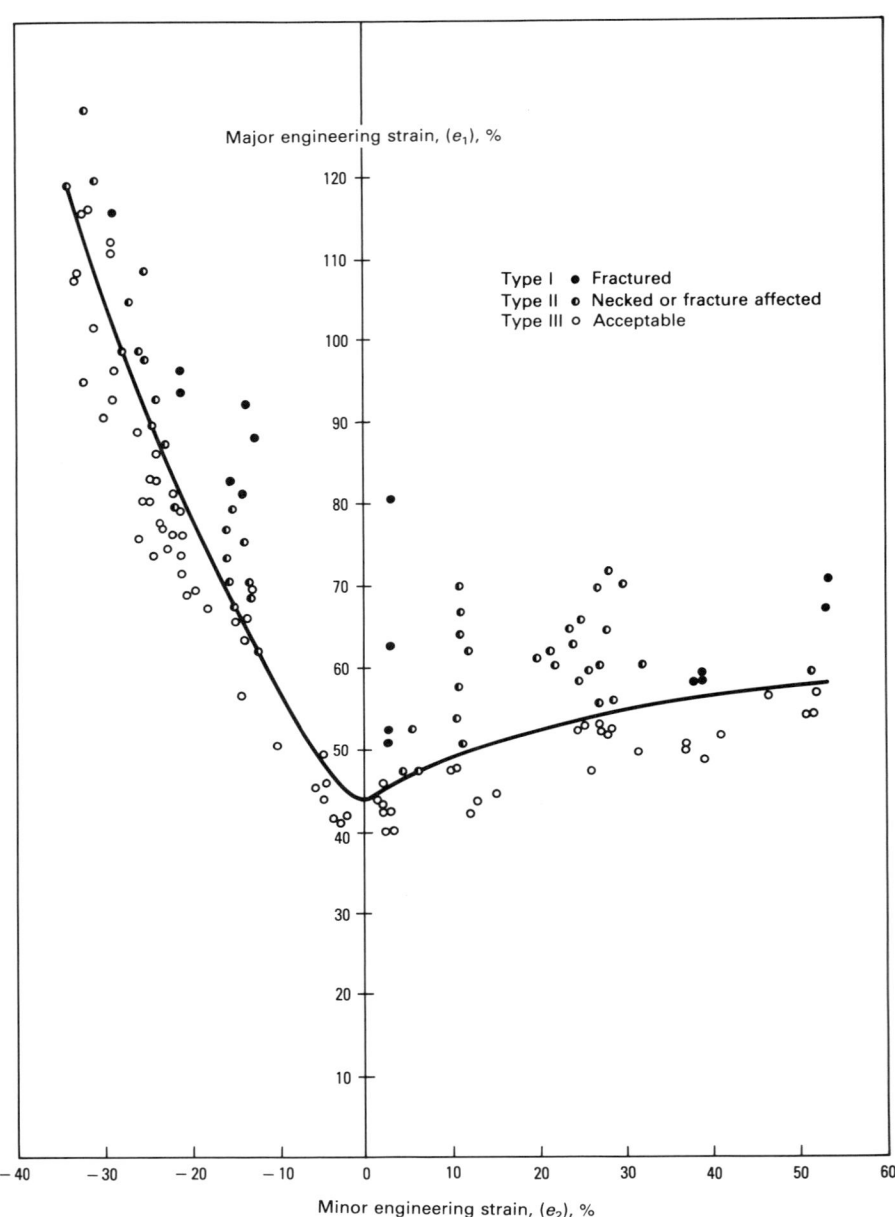

Circle Grid Analysis

Circle grid analysis is a useful technique for ensuring that a die is adequately prepared for production and for diagnosing the causes of necking and splitting failures in production (Ref 78). The forming limit diagram for the type and gage of work material selected must first be obtained. Arrays of small (2.5-mm, or 0.1-in.) diam evenly spaced circles are printed or etched on several blanks in the critical strain regions, preferably in the same location on each blank. Some of the blanks are formed into parts, and the major and minor axes of the deformed circles are measured in the critical locations. The critical strain regions of the part are identified by visual observation of necking or splitting, or from previous experience with similar parts. The local strains are then calculated from the measured dimensions and plotted on the forming limit diagram.

If the maximum strains measured are close to or above the forming limit strains, problems with the tooling, lubrication, blank size or positioning, or press variables are indicated, whether or not necking or splitting actually occurs. Fluctuations that occur in operating conditions and in the properties of the work material over a production run will eventually cause failure if the material is strained to its full capacity. In this case, some of these variables should be changed until the circle grid strain measurements fall below the material limits by a given safety factor, such as 10%.

If the maximum strains measured are significantly below the limit strains and necking or splitting occurs, the batch of work material is substandard. The material used in die tryout must have typical, or slightly lower, forming properties than the production material. The use of superior material may have an adequate forming safety margin that will disappear when a more typical or lower formability material is used. It is good practice to periodically form a few gridded blanks of a standard (nonaging) reference material during a production run to determine the trends in the maximum strains. If the strains are approaching the forming limits, corrective measures can be taken before any actual failures occur.

Circle Grids. Many types of circle grid patterns have been used, such as square arrays of contacting or closely spaced non-contacting circles and arrays of overlapping circles. The contacting and overlapping circles provide improved coverage, but are more difficult to measure manually and cannot at this time be measured automatically.

With small, closely spaced circles, it is possible to determine strain gradients accurately, provided the circles are not too small for accurate measurement. Circles with 2.5-mm (0.1-in.) diameters have been found to be a good size. Both open and solid circles have been used successfully, and automatic systems have been developed for measuring both types.

Applying Circle Grids to the Blanks. The circle grids can be applied to the blanks by a printing or photographic technique or by electrochemical etching. Printed and photographically applied circles are easily damaged and tend to rub off in areas contacted by the dies. This has led to general acceptance of etched circles.

In the electrochemical etching process, an electric stencil with the required grid pattern is placed on the blank and covered with a felt pad soaked in an etching solution. An electrode is placed on the pad, and a low-voltage (up to 14 V) current is passed between the electrode and the blank for a short time, usu-

ally less than 1 min. This produces a lightly etched and oxidized pattern on the surface of the blank. The stencils, etching solutions, and power supplies for this process are commercially available. Different metals require different solutions, levels and types of voltage, and etching times.

Measuring Strains From Deformed Circles. Deformed circles can be measured manually by means of dividers and a ruler, graduated transparent tapes, or a low-power microscope with a graduated stage. Automatic systems, known as grid circle analyzers, have also been developed (Ref 79, 80) for measuring the dimensions of the circles and calculating and displaying the major and minor strains. These systems are now commercially available.

In regions of high curvature, the most accurate method of measurement is the transparent tape, because it follows the contour of the part and measures the arc length, whereas the other methods measure the chord length. The tapes have a pair of diverging lines graduated to give direct readings of the strain, as shown in Fig. 32.

Grid circle analyzers use a solid-state digital array camera with a built-in light source, a minicomputer, keyboard, cathode ray tube (CRT) display, and printer. An image of a given deformed circle is displayed on the CRT, and a least squares curve fitting program selects the most suitable ellipse, which is displayed simultaneously. The major and minor strains, computed from the equation for the ellipse and the diameter of the original circle, are displayed on the screen and printed. A typical layout for the equipment is shown in Fig. 33.

Drawbead Forces

It is common practice in production stamping operations to control the movement of the edges of the blank into the die cavity by means of drawbeads placed in the blankholder. These consist of a semicylindrical ridge in the upper part of the blankholder and a corresponding groove with rounded shoulders in the lower part, or a similar but opposite configuration. The drawbeads cause the

perimeter of the blank to bend and straighten three times as it passes through each bead, as shown in Fig. 34.

The repeated bending and straightening produces a restraining force in addition to that caused by surface friction. A method has been devised for measuring the restraining force due to deformation independently of the effects of friction, using a drawbead simulator with low-friction rollers instead of a fixed bead and groove (Ref 81, 82). A second drawbead simulator with nonrotating parts can be used to measure the combined effects of friction and deformation. Figure 35 shows both types of simulators.

Strips of 0.75- to 1.00-mm (0.03- to 0.04-in.) thick and 50-mm (1.97-in.) wide rimmed and aluminum-killed steels and two aluminum alloys were tested using simulators and a universal testing machine. The contribution from deformation to the total restraining force depended on the lubricant used and ranged from an average of 60% with poor lubrication to 85% with very good lubrication. The required clamping forces, surface strains in the workpiece at various locations in the draw bead simulators, effect of drawbead radius, and effect of rate of testing were also investigated.

Lubricants

The use of lubricants is essential in most forming operations. An effective lubricant provides the following advantages:

- Reduction or elimination of direct sheet metal to die contact and the associated wear and galling
- Control of friction
- More uniform distribution of strain and therefore an increase in the overall level of deformation
- Reduction of heating

Lubricants must meet many requirements to be used in a production operation, such as:

Fig. 33 Layout of a grid circle analyzer
Source: Ref 79

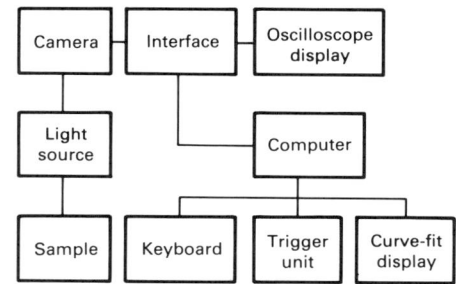

Fig. 32 Transparent tape measurement of deformed circles
Source: Ref 78

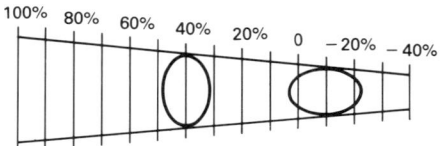

Fig. 34 Repeated bending and straightening of a blank edge in a drawbead

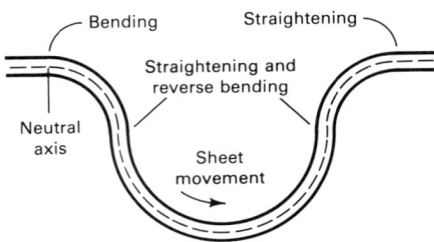

Fig. 35 Drawbead simulators

(a) Frictionless simulator. (b) Standard simulator. Source: Ref 82

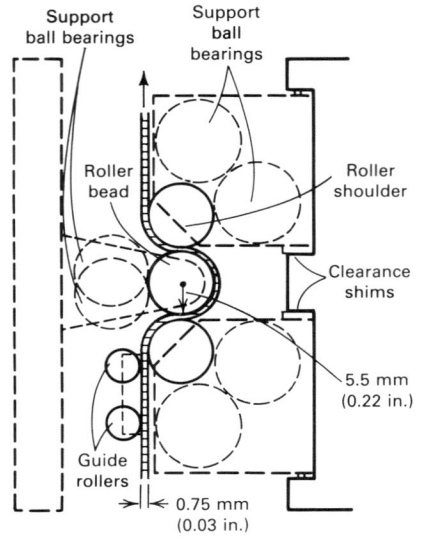

(a)

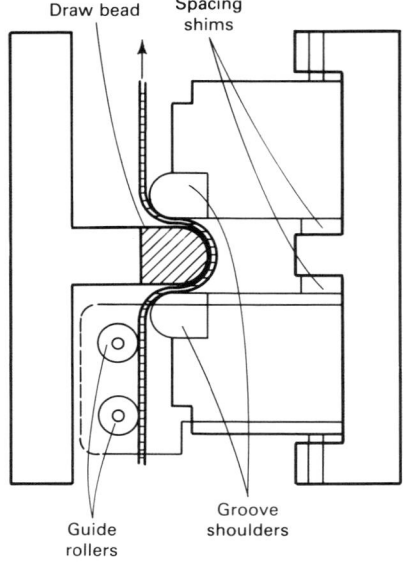

(b)

- Suitable viscosity over the ranges of temperatures and pressures encountered
- Chemical and physical compatibility with work and die materials
- Ease of application, removal, and disposal
- Compatibility with welding operations, sealants, and paint systems

No single lubricant is optimum for all types and rates of forming and all combinations of work and die materials. Rankings of lubricants change considerably for different types of operations and material combinations, which necessitates evaluation on an individual basis. Differences in the performance of lubricants are to be anticipated in view of the differences in surface composition, roughness, and texture of work and die materials and the different strain paths and rates of different forming operations. For example, some stretching operations involve local increases in area in excess of 100%, whereas in many drawing operations a negligible or negative change in area occurs. In addition, the rate at which the work material slides over the dies varies widely.

A simple test to evaluate lubricants measures the frictional force exerted on a lubricated strip of sheet metal when it is pulled between two rectangular blocks of die material. The force used between the blocks and the test rate can be varied. The number of strips that can be tested before the onset of galling provides an additional measure of the effectiveness of the lubricant. However, this test does not include some of the important aspects of actual forming operations, such as plastic deformation of the work material, the ranges of sliding speeds and rates of straining involved, die geometry, which influences the amount of residual lubricant in various locations as the operation progresses, and the heating that occurs in high-volume production operations.

The drawbead simulator described earlier is more realistic and has been used to measure friction forces with various lubricants (Ref 81, 82). Under most conditions tested, friction was described by Coulomb's law,

which states that the friction force, F, is directly proportional to the normal force, N, between the contacting surfaces:

$$F = \mu N \qquad \text{(Eq 31)}$$

where μ is the coefficient of friction. Coulomb's law did not apply at the highest contact pressure tested. The value of the coefficient of friction for mill oil, a poor lubricant, was found to be 0.17; for the best lubricant tested—a soap-based lubricant—it was 0.06.

In addition, simulative tests can be used as lubricant evaluation tests. The production operation should be characterized in terms of the principal types of forming operations involved and their relative severities, and the appropriate simulative test selected. The Swift flat-bottomed cup test has been used extensively for the evaluation of lubricants for deep drawing, and the Swift round-bottomed and Fukui conical cup tests have been used for stretch-drawing operations. The 100-mm (3.94-in.) hemispherical dome test can be used for stretching operations.

The various regimes of lubrication—i.e., thick film, thin film, mixed, and boundary lubrication—and their characteristics are reviewed in Ref 83, which also discusses the limited validity of Coulomb's law, methods for lubricant evaluation, and some of the current limitations in this area.

REFERENCES

1. S.P. Keeler, "Understanding Sheet Metal Formability, Vol 2," Paper No. 350A, Metal Fabricating Institute, Rockford, IL, 1970
2. J.H. Holloman, Tensile Deformation, Trans. AIME, Vol 162, 1945, p 268-290
3. W.A. Backofen, Massachusetts Institute of Technology Industrial Liaison Symposium, Chicago, March 1974
4. A.K. Ghosh, The Influence of Strain Hardening and Strain-Rate Sensitivity on Sheet Metal Forming, Trans. ASME, Vol 99, July 1977, p 264-274
5. I.S. Brammar and D.A. Harris, Production and Properties of Sheet Steel and Aluminum Alloys for Forming Applications, J. Austral. Inst. Met., Vol 20 (No. 2), 1975, p 85-100
6. S.P. Keeler and W.A. Backofen, Plastic Instability and Fracture in Sheets Stretched over Rigid Punches, Trans. ASM, Vol 56 (No. 1), 1963, p 25-48
7. G.M. Goodwin, "Application of Strain Analysis to Sheet Metal Forming Problems in the Press Shop," Paper No. 680093, Society of Automotive Engineers, Warrendale, PA, 1968
8. S.P. Keeler, Determination of Forming Limits in Automotive Stampings, Sheet Metal Ind., Vol 42, Sept 1965, p 683-691
9. S.P. Keeler and W.G. Brazier, "Relationship Between Laboratory Material Characterization and Press-Shop Formability," Microalloying 75 Proceedings, Union Carbide Corp., New York, 1977, p 517-530

10. H. Naziri and R. Pearce, The Effect of Plastic Anisotropy on Flange-Wrinkling Behavior During Sheet Metal Forming, *Int. J. Mech. Sci.*, Vol 10, 1968, p 681-694

11. K. Yoshida and K. Miyauchi, Experimental Studies of Material Behavior as Related to Sheet Metal Forming, in *Mechanics of Sheet Metal Forming*, Plenum Press, New York, 1978, p 19-49

12. W.F. Hosford and R.M. Caddell, in *Metal Forming, Mechanics and Metallurgy*, Prentice-Hall, Englewood Cliffs, NJ, 1983, p 273, 309

13. J. Havranek, "The Effect of Mechanical Properties of Sheet Steels on the Wrinkling Behavior During Deep Drawing of Conical Shells," 9th Congress of the International Deep Drawing Research Group, Ann Arbor, MI, American Society for Metals, p 245-263

14. J.S.H. Lake, "The Yoshida Test—A Critical Evaluation and Correlation with Low-Strain Tensile Parameters," 13th Congress of the International Deep Drawing Research Group, Melbourne, Australia, Feb 1984, p 555-564

15. J.L. Duncan, R. Sowerby, and M.P. Sklad, "Failure Modes in Aluminum Sheet in Deep Drawing Square Cups," Conference on Sheet Forming, University of Aston, Birmingham, England, Sept 1981

16. G. Glover, J.L. Duncan, and J.D. Embury, Failure Maps for Sheet Metal, *Met. Technol.*, March 1977, p 153-159

17. J.D. Embury and J.L. Duncan, Formability Maps, *Ann. Rev. Mat. Sci.*, Vol 11, 1981, p 505-521

18. J. Datsko, *Materials in Design and Manufacturing*, J. Datsko Consultants, Ann Arbor, MI, 1977, p 7-16

19. N. Kuhn, On the Springback Behavior of Low-Carbon Steel Sheet after Stretch Bending, *J. Australian Inst. Met.*, Vol 12 (No. 1), Feb 1967, p 71-76

20. G.V. Smith, *Elevated Temperature Static Properties of Wrought Carbon Steel*, STP 503, ASTM, Philadelphia, 1972

21. F.N. Rhines and P.J. Wray, Investigation of the Intermediate Temperature Ductility Minimum in Metals, *Trans. ASM*, Vol 54, 1961, p 117-128

22. B. Taylor, R.A. Heimbuch and S.G. Babcock, "Warm Forming of Aluminum," 2nd International Conference on Mechanical Behavior of Materials, Boston, American Society for Metals, Aug 1976, p 2004-2008

23. W.G. Granzow, The Influence of Tooling Temperature on the Formability of Stainless Sheet Steel, in *Formability of Metallic Materials—2000 A.D.*, STP 753, J.R. Newby and B.A. Niemeier, Ed, ASTM, Philadelphia, 1981, p 137-146

24. H.C. Wu and D.R. Rummler, Analysis of Misalignment in the Tension Test, *Trans. ASME*, Vol 101, Jan 1979, p 68-74

25. G.E. Dieter, *Mechanical Metallurgy*, 2nd ed., McGraw-Hill, New York, 1976, p 347, 349, 681

26. R.L. Whitely, *Correlation of Deep Drawing Press Performance with Tensile Properties*, STP 390, ASTM, Philadelphia, 1965

27. S.J. Green, J.J. Langan, J.D. Leasia, and W.H. Yang, Material Properties, Including Strain-Rate Effects, as Related to Sheet Metal Forming, *Met. Trans. A*, Vol 2, 1971, p 1813-1820

28. G. Rai and N.J. Grant, On the Measurements of Superplasticity in an Al-Cu Alloy, *Met. Trans A.*, Vol 6, 1975, p 385-390

29. W.J. McGregor Tegart, in *Elements of Mechanical Metallurgy*, MacMillan, New York, 1966, p 29-38

30. R.H. Heyer and J.R. Newby, Measurement of Strain Hardening and Plastic Strain Ratio Using the Circle-Arc Specimen, *Sheet Metal Ind.*, Vol 43, Dec 1966, p 910-914

31. D.C. Ludwigson, A Rapid Test for the Assessment of Steel Sheet and Tinplate Properties, *J. Test. Eval.*, Vol 7 (No. 6), 1979, p 301-109

32. P.R. Mould and T.E. Johnson, Rapid Assessment of Drawability of Cold-Rolled Low-Carbon Steel Sheets, *Sheet Metal Ind.*, Vol 50, June 1973, p 328-348

33. M.L. Devenpeck and O. Richmond, Limiting Strain Tests for In-Plane Sheet Stretching, in *Novel Techniques in Metal Deformation Testing*, The Metallurgical Society of AIME, Warrendale, PA, 1983, p 79-88

34. R.H. Wagoner and N.M. Wang, An Experimental and Analytical Investigation of In-Plane Deformation of 2036-T4 Aluminum, *Int. J. Mech. Sci.*, Vol 21, 1979, p 255-264

35. R.H. Wagoner, Measurement and Analysis of Plane-Strain Work Hardening, *Met. Trans. A*, Vol 11, Jan 1980, p 165-175

36. R.H. Wagoner, "Plane-Strain and Tensile Hardening Behavior of Three Automotive Sheet Alloys," Experimental Verification of Process Models, ASM Symposium, Cincinnati, OH, Sept 1981

37. E.J. Appleby, M.L. Devenpeck, L.M. O'Hara, and O. Richmond, Finite Element Analysis and Experimental Examination of the Rectangular-Sheet Tension Test, in *Applications of Numerical Methods to Forming Processes*, ASME Winter Annual Meeting, San Francisco, Applied Mechanics Division, Vol 28, Dec 1978, p 95-105

38. H. Sang and Y. Nishikawa, A Plane Strain Tensile Apparatus, *J. Metals*, Feb 1983, p 30-33

39. Z. Marciniak and K. Kuczynski, Limit Strains in the Processes of Stretch-Forming Sheet Metal, *Int. J. Mech. Sci.*, Vol 9, 1967, p 609-620

40. J.L. Duncan, J. Kolodziejski, and G. Glover, "Bulge Testing as an Aid to Formability Assessment," 9th Congress of the International Deep Drawing Research Group, Ann Arbor, MI, American Society for Metals, Oct 1976, p 131-150

41. W. Johnson and J.L. Duncan, The Use of the Biaxial Test Extensometer, *Sheet Metal Ind.*, Vol 42, April 1965, p 271-275

42. J.E. Bird, Hydraulic Bulge Testing at Controlled Strain Rate, "*Novel Techniques in Metal Deformation Testing*, The Metallurgical Society of AIME, Warrendale, PA, 1983, p 403-416

43. A.J. Ranta-Eskola, Use of the Hydraulic Bulge Test in Biaxial Tensile Testing, *Int. J. Mech. Sci.*, Vol 21, 1979, p 457-465

44. D.N. Harvey, "Electronic Vision as Input to Calculated Variable Control of the Hydraulic Bulge Test," Proceedings of the International Symposium on Automotive Technology and Automation, Wolfsburg, West Germany, Sept 1982

45. Z. Marciniak, "Aspects of Material Formability," Metalworking Research Group, McMaster University, Ontario, Canada, 1973, p 84-91

46. K. Miyauchi, "Stress-Strain Relationship in Simple Shear of In-Plane Deformation for Various Steel Sheets," 13th Congress of the International Deep Drawing Research Group, Melbourne, Australia, Feb 1984, p 360-371

47. R.A. George, S. Dinda, and A.S. Kasper, Estimating Yield Strength From Hardness Data, *Met. Prog.*, May 1976, p 30-35

48. M.Y. Demeri, The Stretch-Bend Forming of Sheet Metal, *J. Appl. Metalwork.*, Vol 2 (No. 1), 1981, p 3-10

49. O.S. Narayanaswamy and M.Y. Demeri, Analysis of the Angular Stretch Bend Test, in *Novel Techniques in Metal Deformation Testing*, The Metallurgical Society of AIME, Warrendale, PA, 1983, p 99-112

50. A.S. Kasper, "Forming Sheet Metal Parts," Paper MF 69-516, Society of

Manufacturing Engineers, Dearborn, MI, 1969

51. T.R. Thompson, G. Glover, and R. Jackson, "The Formability of Sheet Steels," Broken Hill Proprietary Co., Australia, Tech. Bull. 18 (No. 2), 1974, p 15-19

52. S.S. Hecker, A Cup Test for Assessing Stretchability, *Met. Eng. Quart.*, Nov 1974, p 30-36

53. J.R. Newby, A Practical Look at Biaxial Stretch Cupping Tests, *Sheet Metal Ind.*, Vol 54, March 1977, p 240-252

54. K. Nakazima, T. Kikuma, and K. Hasuka, "Study on the Formability of Sheet Steels," Yawata Tech. Rept. No. 264, Yawata Iron & Steel Co., Tokyo, Sept 1968, p 141

55. A.K. Ghosh, How to Rate Stretch Formability of Sheet Metals, *Met. Prog.*, May 1975, p 52-54

56. A.K. Ghosh, The Effect of Lateral Drawing-In on Stretch Formability, *Met. Eng. Quart.*, Aug 1975, p 53-64

57. R.A. Ayres, W.G. Brazier, and V.F. Sajewski, Evaluating the Limiting Dome Height Test as a New Measure of Press Formability, *J. Appl. Metalwork.*, Vol 1 (No. 1), 1979, p 41-49

58. R. Stevenson, Correlation of Tensile Properties with Plane-Strain Limiting Dome Height, *J. Appl. Metalwork.*, Vol 3 (No. 3), July 1984, p 272-280

59. R.D. Adamczyk, D.W. Dickinson, and R.P. Krupitzer, "The Edge Formability of High-Strength Cold-Rolled Steel," Paper 830237, Society of Automotive Engineers, Warrendale, PA, 1983

60. D.F. Eary and E.A. Reed, *Techniques of Pressworking Sheet Metal*, 2nd ed., Prentice-Hall, Englewood Cliffs, NJ, 1974, p 136-172

61. D.V. Wilson, B.J. Sunter, and D.F. Martin, A Single-Blank Test for Drawability, *Sheet Metal Ind.*, Vol 43, June 1966, p 465-476

62. K.M Frommann, "The Prediction of Metal Stamping Behavior," American Metal Stamping Association Conference, Detroit, April 1968, p 1-55

63. A.S. Kasper and P.J. VanderVeen, "A New Method of Predicting the Formability of Materials," Paper 720019, Society of Automotive Engineers, Warrendale, PA, 1972

64. M.W. Boyles and H.S. Chilcott, Recent Developments in the Use of the Stretch-Draw Test, *Sheet Metal Ind.*, Vol 59, Feb 1982, p 149-156

65. G.M. Goodwin, Formability Index, Paper MF 71-165, Society of Manufacturing Engineers, Dearborn, MI, 1971

66. K. Yoshida, H. Hayahsi, K. Miyauchi, M. Hirata, T. Hira, and S. Ujihara, "Yoshida Buckling Test," International Symposium on New Aspects Sheet Metal Forming, Iron and Steel Institute of Japan, Tokyo, 1981, p 125-148

67. H. Abe, K. Nakagawa, and S. Sato, Proposal of Conical Cup Buckling Test," Congress of the International Deep Drawing Research Group, Kyoto, Japan, May 1981, p 1-19

68. J. Havranek, Wrinkling Limit of Tapered Pressings, *J. Austral. Inst. Met.*, Vol 20, 1975, p 114-119

69. A.M. Szacinski and P.F. Thompson, The Effect of Mechanical Properties on the Wrinkling Behavior of Sheet Material in the Yoshida Test and in a Cone Forming Test, 13th Congress of the International Deep Drawing Research Group, Melbourne, Australia, Feb 1984, p 532-542

70. J.E. O'Donnell, E.J. Ripling, and R.M. Brick, A New Tool for Testing Thin Sheet, *Met. Prog.*, May 1962, p 67-71

71. J.E. Bower, Use of Springback Tester with Steels Having Nonlinear Stress-Strain Curves, *Mat. Res. Stand.*, Dec 1965, p 607-610

72. W.G. Granzow, "A Portable Springback Tester for In-Plant Determination of the Strength of Sheet Steels," Paper 830238, Society of Automotive Engineers, Warrendale, PA, 1983

73. R.G. Davies, Springback in High-Strength Steels, *J. Appl. Metalwork.*, Vol 1 (No. 4), 1981, p 45-52

74. C.C. Veerman, L. Hartman, J.J. Peels, and P.F. Neve, Determination of Appearing and Admissible Strain in Cold-Reduced Sheets, *Sheet Metal Ind.*, Vol 48, Sept 1971, p 678-694

75. A.B. Haberfield and M.W. Boyles, Laboratory Determined Forming Limit Diagrams, *Sheet Metal Ind.*, Vol 50, July 1973, p 400-411

76. S.S. Hecker, Simple Technique for Determining Forming Limit Curves, *Sheet Metal Ind.*, Vol 52, Nov 1975, p 671-676

77. A.K. Ghosh and S.S. Hecker, Stretching Limits in Sheet Metals: In-Plane Versus Out-of-Plane Deformation, *Met. Trans.*, Vol 5, Oct 1974, p 2161-2164

78. S. Dinda, K.F. James, S.P. Keeler, and P.A. Stine, *How to Use Circle Grid Analysis for Die Tryout*, American Society for Metals, 1981

79. R.A. Ayres, E.G. Brewer, and S.W. Holland, Grid Circle Analyzer: Computer Aided Measurement of Deformation, *SAE Trans.*, Vol 88 (No. 3), 1979, p 2630-2634 (SAE Paper 790741)

80. D.N. Harvey, "Optimising Patterns and Computational Algorithms for Automated, Optical Strain Measurement in Sheet Metal," 13th Congress of the International Deep Drawing Research Group, Melbourne, Australia, Feb 1984, p 403-414

81. H.D. Nine, Drawbead Forces in Sheet Metal Forming, in *Mechanics of Sheet Metal Forming*, Plenum Press, New York, 1978, p 179-211

82. H.D. Nine, The Applicability of Coulomb's Friction Law to Drawbeads in Sheet Metal Forming, *J. Appl. Metalwork.*, Vol 2 (No. 3), July 1982, p 200-210

83. W.R.D. Wilson, Friction and Lubrication in Sheet Metal Forming, in *Mechanics of Sheet Metal Forming*, Plenum Press, New York, 1978, p 157-177

SELECTED REFERENCES

● A.K. Ghosh, S.S. Hecker, and S.P. Keeler, Sheet Metal Forming and Testing, in *Workability Testing Techniques*, American Society for Metals, 1984, p 135-195

● S.S. Hecker and A.K. Ghosh, The Forming of Sheet Metal, *Sci. American*, Vol 235, Nov 1976, p 100-108

● R.M. Hobbs and J.L. Duncan, *Material Selection—Tests for Formability*, Metals Engineering Institute, American Society for Metals, April 1979

● S.P. Keeler, Understanding Sheet Metal Formability, Vol 1, *Machinery*, Feb-July 1968

● S.P. Keeler, Sheet Metal Forming in the 80's, *Met. Prog.*, July 1980, p 25-29

● J.R. Newby, Formability Testing and Deformation Analysis of Sheet Steel, in *Formability Topics—Metallic Materials*, STP 647, ASTM, Philadelphia, 1978, p 4-38

● R. Pearce, "Sheet Metal Testing—From the 19th Century Until Now," 10th Congress of the International Deep Drawing Research Group, University of Warwick, Warwick, England, April 1978, Portcullis Press Ltd, Redhill, Surrey, UK 1978, p 355-362

● "Sheet Steel Formability," AISI Committee of Sheet Steel Producers, American Iron and Steel Institute, Washington, DC, Aug 1984

Bulk Workability Testing

By George E. Dieter
Dean of Engineering
University of Maryland

WORKABILITY refers to the relative ease with which a metal can be shaped through plastic deformation. The term "workability" often is used interchangeably with "formability," which is the preferred term when referring to the shaping of sheet metal parts (see the article "Sheet Formability Testing" in this Section). However, workability is usually used to refer to the shaping of materials by bulk deformation processes such as forging, extrusion, and rolling.

The characterization of the mechanical behavior of a material by tension testing measures two different types of mechanical properties: strength properties (e.g., yield strength and ultimate strength) and ductility properties (e.g., percent elongation and reduction in area). Similarly, evaluation of workability involves both measurement of the resistance of deformation (strength) and determination of the extent of possible plastic deformation before fracture (ductility).

However, the major emphasis in workability is on measurement and prediction of the limits of deformation before fracture. Therefore, the emphasis in this article is on test methods for determining the extent of deformation a metal can withstand before cracking or fracture occurs. It is important, however, to allow for a more general definition in which workability is defined as the degree of deformation that can be achieved in a particular metalworking process without creating an undesirable condition. The undesirable condition generally is cracking or fracture, but may be another condition such as poor surface finish, buckling, or the formation of laps, which are defects created when metal folds over itself during forging.

Evaluating Workability

Generally, workability depends on the local conditions of stress, strain, strain rate, and temperature in combination with mate-

rial factors, such as the resistance of a metal to ductile fracture. In addition to a review of the many process variables that influence the degree of workability, the mathematical relationships that describe the occurrence of room-temperature ductile fracture under workability conditions are summarized. The most common testing techniques for workability prediction are also discussed. These test methods are described in more detail in subsequent sections of this article.

Material Factors Affecting Workability

Fracture Mechanisms. Fracture in bulk deformation processing usually occurs as ductile fracture, rarely as brittle fracture. However, depending on temperature and strain rate, the details of the ductile fracture mechanism vary. Figure 1 schematically illustrates the different modes of ductile fracture obtained in a tension test over a wide range of strain rates and temperatures. At temperatures below about half of the melting point of a given material (below the hot working region), a typical dimpled rupture type of ductile fracture usually occurs.

The three stages of ductile fracture are shown schematically in Fig. 2. The first stage is void initiation, usually at second-phase particles or inclusions. Voids are initiated because particles do not deform, which

Fig. 1 Tensile fracture modes as a function of temperature (measured on the absolute temperature scale) and strain rate

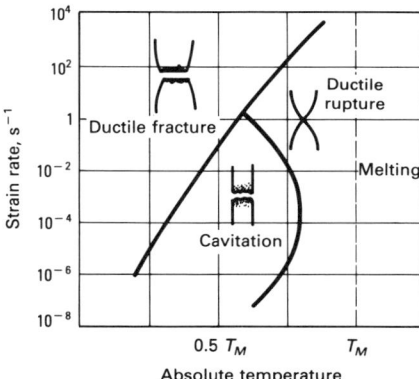

forces the ductile matrix around the particle to deform more than normal. This in turn produces more strain hardening, thus creating a higher stress in the matrix near the particles. When the stress becomes sufficiently large, the interface may separate or the particle may crack. As a result, ductility is strongly dependent on the size and density of the second-phase particles, as shown in Fig. 3 (Ref 1).

Fig. 2 Stages in the dimpled-rupture mode of ductile fracture
(a) Void initiation. (b) Void growth. (c) Void linking

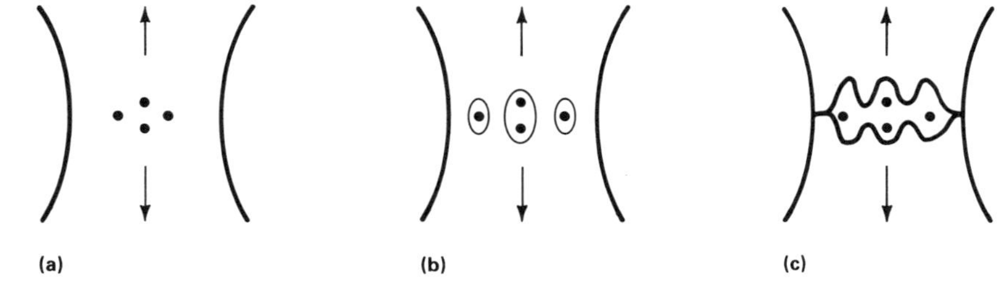

| (a) | (b) | (c) |

Fig. 3 Effect of volume fraction of second-phase particles on tensile ductility of steel

Source: Ref 1

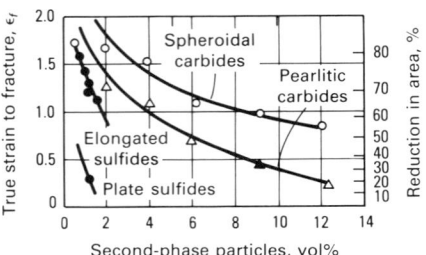

Fig. 4 Formation of grain-boundary voids (cavitation) and triple-point cracks at warm and hot working temperatures

(a) Schematic showing how grain-boundary voids are formed under the action of matrix deformation and how grain-boundary sliding in the absence of grain-boundary migration and recrystallization may cause cracks to open at triple points. (b) Examples of grain-boundary voids and triple-point cracking at the prior beta grain boundaries in hot forged Ti-6Al-2Sn-4Zr-2Mo-0.1Si with a Widmanstätten alpha starting microstructure. Source: Ref 2, 3

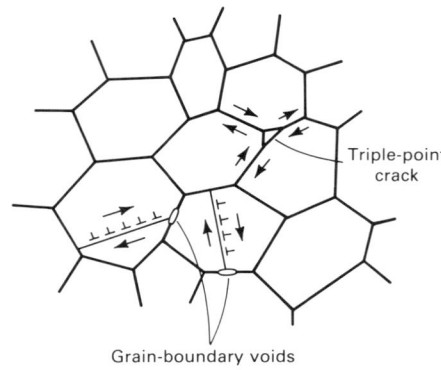

(a)

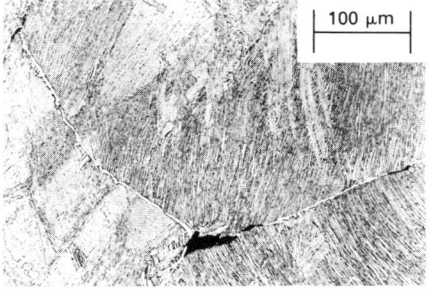

(b)

The second stage of ductile fracture is void growth, which is a strain-controlled process. As voids grow they elongate, and the ligaments of matrix material between the voids thin. Thus, the final stage of ductile fracture is hole coalescence via the separation of the ligaments, which links the growing voids.

Ductile fracture by void growth and coalescence can occur by two modes. Fibrous tearing (mode I) occurs by void growth in the crack plane that is essentially normal to the tensile axis. In mode II void growth, voids grow in sheets at an oblique angle to the crack plane under the influence of shear strains. This type of shear band tearing is found on the surface of the cone in a ductile cup-and-cone tensile fracture. It commonly occurs in deformation processing, in which friction and/or geometric conditions produce inhomogeneous deformation, leading to local shear bands. Localization of deformation in these shear bands leads to adiabatic temperature increases that produce local softening.

Increasing the temperature of deformation leads to significant changes in deformation behavior and fracture mode. At temperatures above one half the melting point, particularly at low strain rates, grain-boundary sliding becomes prominent. This leads to wedge-shaped cracks that propagate along grain boundaries and result in low ductility (Fig. 4).

The probability of wedge cracking varies with the applied strain rate. If the strain rate is so high that the matrix deforms at a faster rate than the boundaries can slide, then grain-boundary sliding effects will be negligible, and wedge cracking will not occur. If the strain rate is very low, sufficient time is available to relax the high stresses at the triple points, and a different fracture mechanism comes into play.

Round or elliptical cavities form in grain boundaries at high temperatures, but at lower stresses than those that produce wedge-shaped cracks. Nevertheless, like wedge cracks, these cavities are formed by grain-boundary shearing. With low strain rates (below 1 s^{-1}), initiation of voids by grain-boundary sliding and the reduced rate of dynamic recrystallization can result in extensive internal void formation, or cavitation. Although cavitation is germane with respect to ductility in creep, it is not generally a factor in fracture in hot working.

For high-temperature fracture initiated by grain-boundary sliding, the processes of void growth and coalescence, rather than void initiation, are the prime factors that control ductility. When voids initiated at the original grain boundaries experience difficulty in linking because boundary migration is high as a result of dynamic recrystallization, hot ductility is high. In extreme cases, this can lead to highly ductile rupture, as shown in Fig. 1.

Compressive stresses superimposed on tensile or shear stresses by the deformation process can have a significant influence on closing small cavities or limiting their growth and thus enhancing workability. Because of this important role of the stress state, it is not possible to express workability in absolute terms. Workability depends not only on material characteristics, but also on process variables such as strain, strain rate, temperature, and stress state.

By considering all the failure mechanisms that can operate in a material over a range of strain rates and temperatures, a processing map can be developed. The processing map is a useful concept that should gain wider acceptance in understanding material response in deformation processing.

Figure 5 illustrates a processing map for aluminum that is based on theoretical models; however, it agrees well with experimental work. A "safe" region is indicated in which the material should be free from cavitation damage or flow localization. The processing map predicts that, at constant temperature, there should be a maximum in ductility with respect to strain rate. For example, at 500 K, ductility should be at a maximum at a strain rate of 10^{-3} to 10^{-1} s^{-1}. Below the lower value, wedge cracking will occur, whereas above this level, ductile fracture would reduce ductility.

The safe region shown in Fig. 5 is sensitive to the microstructure of the metal. Decreasing the size and volume fraction of

Fig. 5 A composite processing map delineating the safe region for forming

The boundaries will shift with microstructure. Instabilities due to purely continuum effects, such as shear localization in sheet metal forming, have not been considered. Source: Ref 4

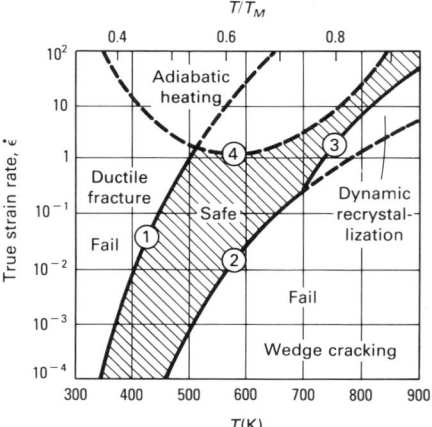

hard particles would move boundary 1 to the left, while increasing the size or fraction of hard particles in the grain boundary would make sliding more difficult and would move boundary 2 to the right.

Flow Localization. Workability problems can arise when metal deformation is localized to a narrow zone. This results in a region of different structures and properties that can be the site of failure in service. Localization of deformation can also be so severe that it leads to failure in the deformation process. In either mode, the presence of flow localization needs to be recognized.

Flow localization is commonly caused by the formation of a dead-metal zone between the workpiece and the tooling. This can arise from poor lubrication at the workpiece-tool interface. Figure 6 illustrates upsetting of a cylinder with poorly lubricated platens. When the workpiece is constrained from sliding at the interface, it barrels, and a friction-hill pressure distribution is created over the interface. The inhomogeneity of deformation throughout the cross section leads to a dead zone at the tool interface and a region of intense shear deformation.

A similar situation can arise when the processing tools are cooler than the workpiece, so that heat is extracted at the tools. Consequently, the flow stress of the metal near the interface is higher because of the lower temperature.

However, flow localization may occur during hot working in the absence of frictional or chilling effects. In this case, localization results from flow softening (negative strain hardening). Flow softening, which arises during hot working as a result of instabilities in structure such as adiabatic heating, generation of a softer texture during deformation, grain coarsening, or spheroidization, has been correlated with materials properties (Ref 5) by the parameter:

$$\alpha = \frac{(\gamma - 1)}{m}$$

where $\gamma = (1/\sigma)d\sigma/d\epsilon$ is the normalized flow softening rate, and $m = d \log \sigma/d \log \dot{\epsilon} = (\dot{\epsilon}/\sigma)d\sigma/d\dot{\epsilon}$ is the strain rate sensitivity. In $\alpha + \beta$ titanium alloys and other materials that exhibit a strong tendency toward flow localization, this phenomenon is likely to occur when the parameter α is greater than 5. Figure 7 shows a crack that initiated in a shear band during high-energy-rate forging of a complex austenitic stainless steel.

Metallurgical Considerations. The common failure modes occurring in deformation processing are summarized in Table 1. Three temperature regions are common: cold working, warm working, and hot working. Two types of ductile fracture are distinguished in cold working, depending on the

Fig. 6 Consequences of friction illustrated in the upsetting of a cylinder

(a) Direction of shear stresses. (b) Consequent rise in interface pressure. (c) Inhomogeneity of deformation. τ_i, average frictional shear stress; p, normal pressure; σ_f, flow stress; p_a, average die pressure

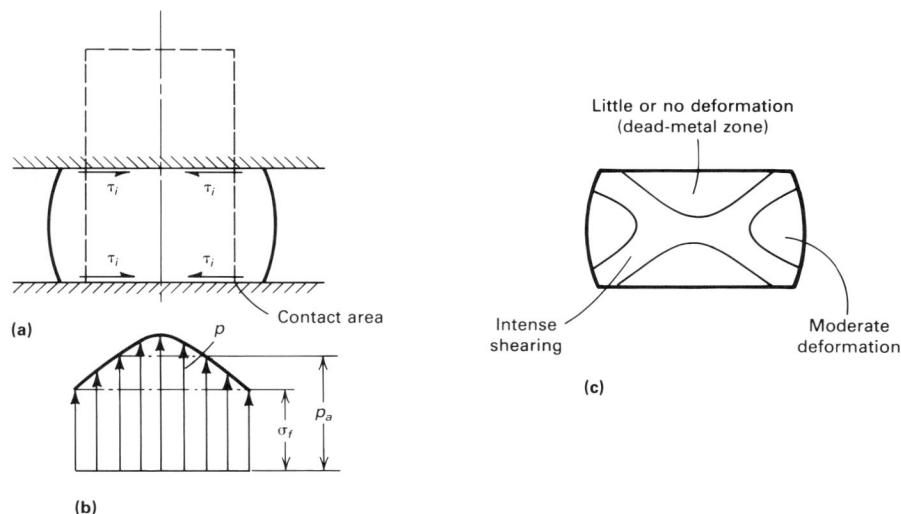

Fig. 7 Austenitic stainless steel high-energy-rate-forged extrusion

Forging temperature, 815 °C (1500 °F); 65% reduction in area; $\dot{\epsilon} = 1.4 \times 10^3$ s^{-1}. (a) View of extrusion showing spiral cracks. (b) Optical micrograph showing the microstructure at the tip of one of the cracks in the extrusion (area A). Note that the crack initiated in a macroscopic shear band that formed first at the lead end of the extrusion. Etchant: Oxalic acid. Source: Ref 6

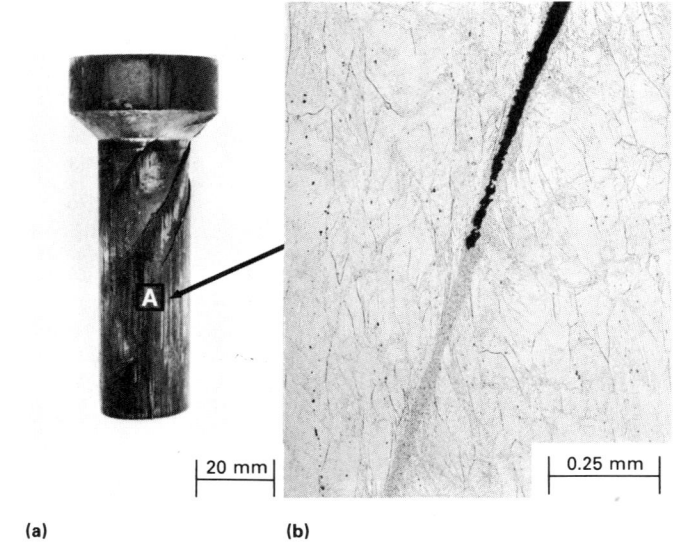

stress state. Free surface fracture occurs at a free surface when there is no stress normal to the surface, while centerburst-type failure occurs at the center of the bar or forging due to high hydrostatic tension.

Workability problems depend greatly on grain size and grain structure. When the grain size is large relative to the overall size of the workpiece, as in conventionally cast ingot structures, workability is lower, be-

cause cracks may initiate and propagate easily along the grain boundaries. Moreover, with cast structures, impurities frequently are segregated to the center and top or to the surface of the ingot, which creates regions of low workability. Because chemical elements are not distributed uniformly on either a microscopic or a macroscopic scale, the temperature range over which an ingot structure can be worked is rather limited.

Typically, cast structures must be hot worked. The melting point of an alloy in the as-cast condition is usually lower than that of the same alloy in the fine-grained, recrystallized condition because of chemical inhomogeneities and the presence of low-melting-point compounds that frequently occur at grain boundaries. Deformation at temperatures too close to the melting point of these compounds may lead to grain-boundary cracking when the heat developed by deformation increases the workpiece temperature and produces local melting.

This fracture mode is called hot shortness. This type of fracture can be prevented by using a sufficiently low deformation rate that allows the heat developed by deformation to be dissipated by the tooling, by using lower working temperatures, or by subjecting the workpiece to a homogenization heat treatment prior to hot working. The relationship between the workability of cast and wrought structures and temperature is shown in Fig. 8.

The intermediate temperature region of low ductility shown in Fig. 8 is found in many metallurgical systems (Ref 7). This occurs at a temperature that is sufficiently high for grain-boundary sliding to initiate grain-boundary cracking, but not so high that the cracks are sealed off from propagation by a dynamic recrystallization process.

The relationship of workability and temperature for various metallurgical systems is summarized in Fig. 9. Generally, pure metals and single-phase alloys exhibit the best workability, except when grain growth occurs at high temperatures. Alloys that contain low-melting-point phases (e.g., iron alloys with sulfur) or brittle phases (e.g., γ'-strengthened nickel-base superalloys) tend to be difficult to deform and have a limited range of working temperature.

Process Variables Controlling Workability

Strain. The prime objective of plastic deformation processes is to change the shape of the deformed product. A secondary objective is to improve or control the properties of the deformed product. In dense metals, unlike porous powder compacts, the volume of the workpiece remains constant as it decreases in cross-sectional area, A, and increases in length, L:

$$A_0 L_0 = A_1 L_1$$

The strain produced in a deformation process is described by the engineering strain:

$$e = \frac{L_1 - L_0}{L_0} = \frac{A_0 - A_1}{A_1}$$

or alternatively by the true strain:

Table 1 Common failure modes in deformation processing

Temperature regime	Metallurgical defects in: Cast grain structure	Wrought (recrystallized) grain structure
Cold working	(a)	Free surface fracture, dead-metal zones (shear bands, shear cracks), centerbursts, galling
Warm working	(b)	Triple-point cracks/fractures, grain-boundary cavitation/fracture
Hot working	Hot shortness, centerbursts, triple-point cracks/fractures, grain-boundary cavitation/fracture, shear bands/fractures	Shear bands/fractures, triple-point cracks/fractures, grain-boundary cavitation/fracture, hot shortness

(a) Cold working of cast structures is typically performed only for very ductile metals (e.g., dental alloys) and usually involves many stages of working with intermediate recrystallization anneals. (b) Warm working of cast structures is rare.

Fig. 8 Relative workabilities of cast metals and wrought recrystallized metals at cold, warm, and hot working temperatures

The melting point (or solidus temperature) is denoted as MP_c (cast metals) or MP_w (wrought and recrystallized metals).

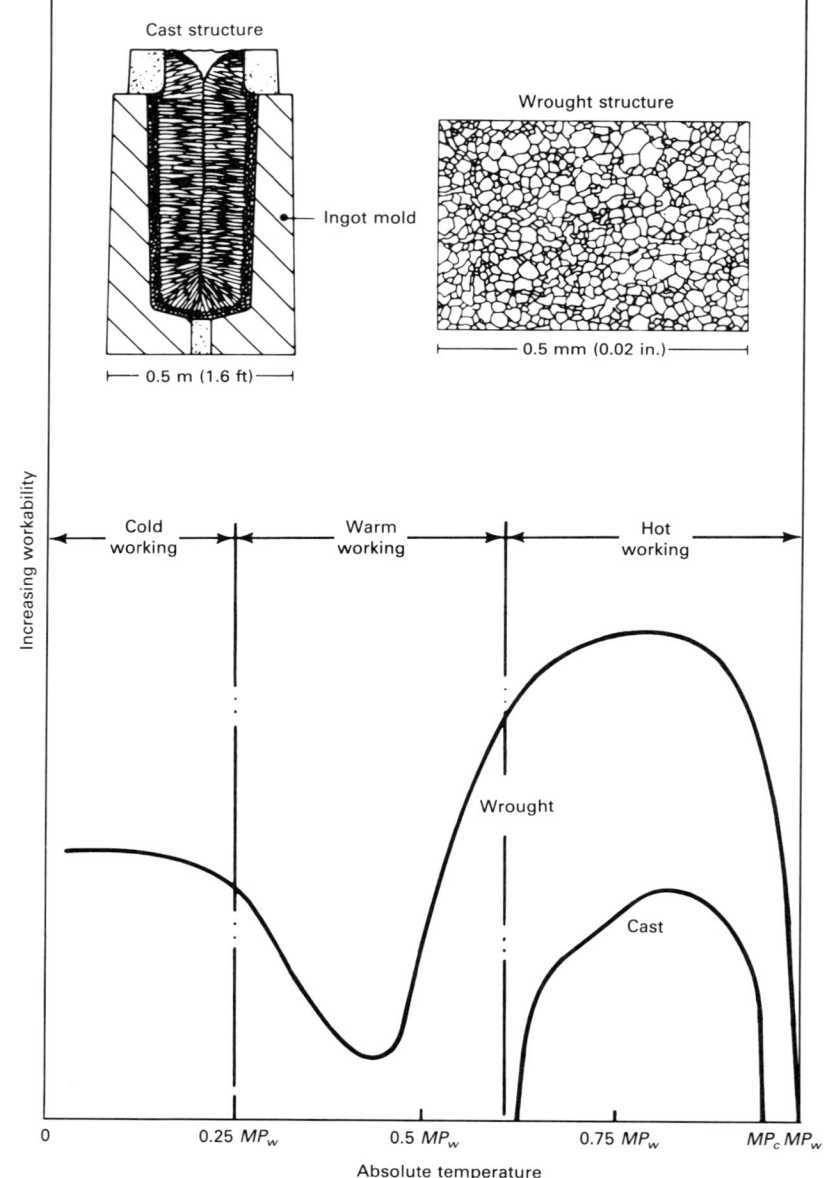

$$\epsilon = \ln\frac{L_1}{L_0} = \ln\frac{A_0}{A_1} = \ln(e + 1)$$

Deformation in metalworking frequently is expressed by the cross-sectional area reduction, R:

$$R = \frac{A_0 - A_1}{A_0}$$

but from constancy of volume, $A_0L_0 = A_1L_1$, and:

$$\epsilon = \ln\frac{L_1}{L_0} = \ln\frac{A_0}{A_1} = \ln\frac{1}{1 - R}$$

The relationship between area reduction and the various measures of strain is given in Table 2.

Strain rate is the time rate of change of strain, i.e., the rate at which deformation proceeds. It can be an important variable in workability experiments and may be difficult to control. The true strain rate for a cylinder of height h upset in compression at a deformation velocity v is:

$$\dot{\epsilon} = \frac{d\epsilon}{dt} = \frac{1}{h}\frac{dh}{dt} = \frac{v}{h}$$

Temperature. Metalworking processes commonly are classified as hot working or cold working operations. Hot working refers to deformation under conditions of temperature and deformation velocity such that recovery processes occur simultaneously with deformation. Cold working refers to deformation carried out under conditions for which recovery processes are not effective during the process. In most hot working, the strain hardening and distorted grain structure produced by deformation are eliminated rapidly by the formation of new strain-free grains as a result of recrystallization during or immediately after deformation.

Very large deformations are possible in hot working, because the recovery processes keep pace with the deformation. Hot working occurs at essentially constant flow stress. Flow stress decreases with the increasing temperature of deformation. In cold working, strain hardening is not relieved, and the flow stress increases continuously with deformation. Therefore, the total deformation that is possible prior to fracture is less for cold working than for hot working, unless the effects of strain hardening are relieved by annealing.

About 95% of the mechanical work expended in deformation is converted into heat. Some of this heat is conducted away by the tools or lost to the atmosphere. However, a portion remains to increase the temperature of the workpiece. The faster the deformation process, the greater the percentage of heat energy that goes to increase the temperature of the workpiece.

Friction. An important concern in all practical metalworking processes is the friction between the deforming workpiece and the tools and/or dies that apply the force and constrain the shape change. Friction occurs because metal surfaces, at least on a microscale, are never perfectly smooth and flat. Relative motion between such surfaces is impeded by contact under pressure.

The existence of friction increases the value of the deformation force and makes deformation more inhomogeneous (Fig. 6c), which in turn increases the propensity for fracture. If friction is very high, seizing and

Fig. 9 Typical workability behavior exhibited by different alloy systems

T_M, absolute melting temperature. Source: Ref 8

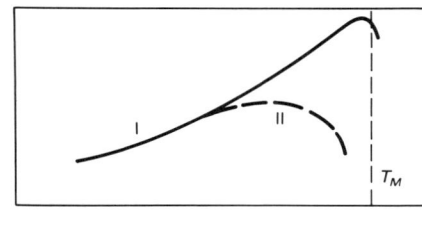
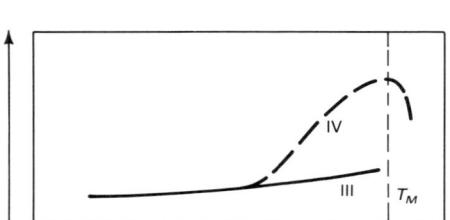
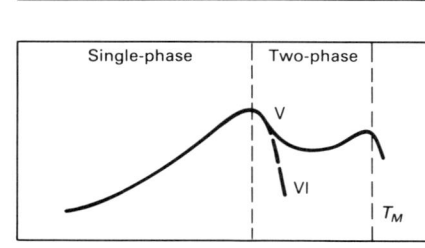
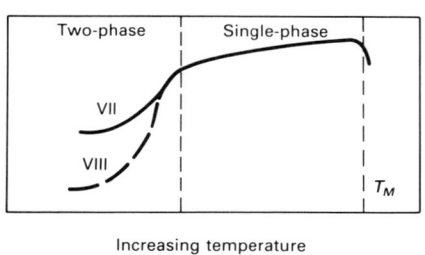

Increasing forgeability (vertical axis)

Increasing temperature (horizontal axis)

I. Pure metals and single phase alloys
Aluminum alloys
Tantalum alloys
Niobium alloys

II. Pure metals and single-phase alloys exhibiting rapid grain growth
Beryllium
Magnesium alloys
Tungsten alloys
All-beta titanium alloys

III. Alloys containing elements that form insoluble compounds
Resulfurized steel
Stainless steel containing selenium

IV. Alloys containing elements that form soluble compounds
Molybdenum alloys containing oxides
Stainless steel containing soluble carbides or nitrides

V. Alloys forming ductile second phase on heating
High-chromium stainless steels

VI. Alloys forming low-melting second phase on heating
Iron containing sulfur
Magnesium alloys containing zinc

VII. Alloys forming ductile second phase on cooling
Carbon and low-alloy steels
Alpha-beta and alpha-titanium alloys

VIII. Alloys forming brittle second phase on cooling
Superalloys
Precipitation-hardenable stainless steels

Table 2 Relationships among area reduction (R), engineering strain (e), and true strain (ε)

| | Area reduction (%) | | | | | | | | | | | |
	5	10	15	20	25	30	40	50	60	70	80	90
e	0.052	0.111	0.176	0.250	0.333	0.428	0.667	1.000	1.500	2.333	4.000	9.000
ϵ	0.051	0.105	0.162	0.223	0.287	0.356	0.511	0.693	0.916	1.204	1.609	2.303

Note: $e = R/(1 - R)$; $\epsilon = \ln(e + 1)$; $\epsilon = \ln(1/1 - R)$

galling of the workpiece surface occur and surface damage results.

The mechanics of friction at the tool-work-piece interface is very complex; therefore, simplifying assumptions usually are used. One such assumption is that friction can be described by Coulomb's law of friction:

$$\mu = \frac{\tau_i}{p}$$

where μ is the Coulomb coefficient of friction, τ_i is the shear stress at the interface, and p is the stress (pressure) normal to the interface.

Another simplification of friction is to assume that the shear stress at the interface is directly proportional to the flow stress, σ_0, of the material:

$$\tau_i = m \frac{\sigma_0}{\sqrt{3}}$$

where m, the constant of proportionality, is the interface friction factor. For given conditions of lubrication and temperature and for given die and workpiece materials, m is usually considered to have a constant value independent of the pressure at the interface. Values of m vary from 0 (perfect sliding) to 1 (no sliding). In the Coulomb model of friction, τ_i increases with p up to a limit where the interface shear stress equals the shear yield stress of the workpiece material.

Control of friction through lubrication is an important aspect of metalworking. High friction leads to various defects that limit workability. However, for most workability tests, conditions under which friction are either absent or easily controlled are selected. Most workability tests make no provision for reproducing the frictional conditions existing in the production process; consequently, serious problems can result in correlating test results with actual production conditions.

Stress State. Because of the different geometries of the tools and workpiece and the different manners in which the forces of deformation are applied, different metalworking processes produce different stress states. A common system of classifying the stress state found in metalworking processes is:

Tensile-compressive systems

- Biaxial tension-uniaxial compression, such as under the roll of a two-roll rotary piercer
- Uniaxial tension-uniaxial compression, such as in the flange of a cup in deep drawing
- Uniaxial tension-biaxial compression, such as in the deformation zone in wire and rod drawing

Compressive stress systems

- Uniaxial stress, such as in forging and upsetting in closed dies
- Biaxial stress, such as between the rolls of a rolling mill when operating without front or back tension
- Triaxial stress, such as in extrusion and in certain parts of a forging deformed in a die

Tensile stress systems

- Biaxial stress, such as stretch forming

Two states that occur frequently in either structural or metalworking applications—plane stress and plane strain—deserve special mention. The use of the adjective plane implies that the condition is confined to a two-dimensional situation. Plane stress occurs when the stress state lies in the plane of the member. Typically, this occurs when one dimension of the member is very small compared to the other two, and the member is loaded by a force lying in the plane of symmetry of the body. Examples are thin, plate-type structures such as thin-walled pressure vessels.

Plane strain occurs when the strain in one of the three principal directions is zero (e.g., $e_3 = 0$), such as an extremely long member subjected to lateral loading or a thick plate with a notch loaded in tension. A common plane-strain condition in metalworking is the rolling of a wide sheet. In this case, there is no strain in the width direction. Although $e_3 = 0$ for plane strain, there is a stress acting in that direction. For plastic deformation, the stress in the principal direction for which strain is zero is the average of the other two principal stresses, i.e., $\sigma_3 = (\sigma_1 + \sigma_2)/2$.

Stress systems in metalworking processes usually are complex. In general, the severity of the stress system with regard to workability increases with the level of tensile stress. For a given material, temperature, and strain rate of deformation, workability is much greater for a more compressive state of stress. This can be described by the mean state of stress, also known as the hydrostatic component of the stress state:

$$\sigma_m = \frac{1}{3} (\sigma_1 + \sigma_2 + \sigma_3)$$

Figure 10 shows that when σ_m is predominantly compressive (a negative value), workability is enhanced.

Fracture is a local process that depends on the local state of stress. Even in a deformation process that is predominantly compressive in nature, local tensile stresses can arise because of friction or other contributors to nonuniform deformation. These stresses, often called secondary tensile stresses, become

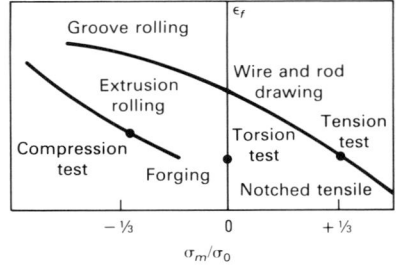

Fig. 10 Influence of stress state σ_m/σ_0 on strain to fracture (ϵ_f)

primary stresses in determining the workability of a material.

Yielding Criteria. The ease with which a metal yields plastically or flows is an important factor in workability. If a metal can be deformed at low stresses, as in superplastic deformation, then the stress levels throughout the deforming workpiece are low and fracture is less likely. The dominant metallurgical conditions and temperature are important variables, as is the stress state. Plastic flow is produced by slip within the individual grains, and slip is induced by a high resolved shear stress. Thus, the beginning of plastic flow can be predicted by a maximum shear-stress, or Tresca, criterion:

$$\tau_{max} = \frac{1}{2} (\sigma_1 - \sigma_3) = \frac{\sigma_0}{2}$$

where τ_{max} is the maximum shearing stress, and σ_0 is the yield (flow) stress measured in either a uniaxial tension or uniaxial compression test. Although this yield criterion is adequate, it neglects the intermediate principal stress σ_2. A more complete and generally applicable yielding criteria is that proposed by von Mises:

$$2\sigma_0^2 = (\sigma_1 - \sigma_2)^2 + (\sigma_2 - \sigma_3)^2 + (\sigma_3 - \sigma_1)^2$$

where $\sigma_1 > \sigma_2 > \sigma_3$ are the three principal stresses, and σ_0 is the uniaxial flow stress of the material.

The significance of yield criteria is best illustrated by examining a simplified stress state, in which $\sigma_3 = 0$ (plane stress). Then, the Tresca yield criterion defines a hexagon and the von Mises criterion an ellipse (Fig. 11).

Yielding (plastic flow) can be initiated in several modes. In pure tension, flow occurs at the flow stress σ_0 (points 1 in Fig. 11, corresponding to two directions in the plane of a sheet). In pure compression, the material yields at the compressive flow stress, which in ductile materials is usually equal to the tensile flow stress σ_0 (points 2 in Fig. 11). When a sheet is bulged by a punch or a pressurized medium, the two principal stresses in the surface of the sheet are equal (balanced biaxial tension) and must reach σ_0 (point 3 in Fig. 11).

A technically important condition is reached when deformation of the workpiece is prevented in one of the principal directions (plane strain). This occurs because a die element keeps one dimension constant, or only one part of the workpiece is deformed and adjacent nondeforming portions exert a restraining influence. In either case, the restraint creates a stress in that principal direction; the stress is the average of the two other principal stresses (corresponding to points 4 in Fig. 11). The stress required for deformation is still σ_0 according to Tresca, but is $1.15\sigma_0$ according to von Mises. The latter is usually regarded as the plane-strain flow stress of the material, $2k$. It is sometimes called the constrained flow stress.

Another important stress state is pure shear, in which the two principal stresses are of equal magnitude, but of the opposite sign (points 5 in Fig. 11). Flow now occurs at the shear flow stress τ_0, which is equal to $0.5\sigma_0$ according to Tresca and $0.577\sigma_0$ according to von Mises. The shear flow stress according to von Mises is often denoted as k. Figure 11 illustrates how the stress required to produce plastic deformation varies significantly with the stress state and how it can be related to the basic uniaxial flow stress of the material through a yield criterion.

Workability Fracture Criteria

Workability is not a unique property of a given material. It depends on process variables such as strain, strain rate, temperature, friction conditions, and the stress system imposed by the process. For example, metals can be deformed to a greater extent by extrusion than by drawing, due to the compressive nature of the stresses in the extrusion process that makes fracture more difficult. Thus, workability can be expressed as:

Workability = f_1(material) $\times$ f_2(process)

where f_1 is a function of the basic ductility of the material, and f_2 is a function of the stress and strain imposed by the process. Because f_1 depends on the material condition and the fracture mechanism, it is a function of temperature and strain rate. Likewise, f_2 depends on process conditions such as lubrication (friction) and die geometry.

Therefore, to describe workability in a fundamental sense, a fracture criterion must be established that defines the limit of strain as a function of strain rate and temperature. Also required is a description of stress,

Fig. 11 (a) Directions of principal stresses and (b) yield criteria with some typical stress states

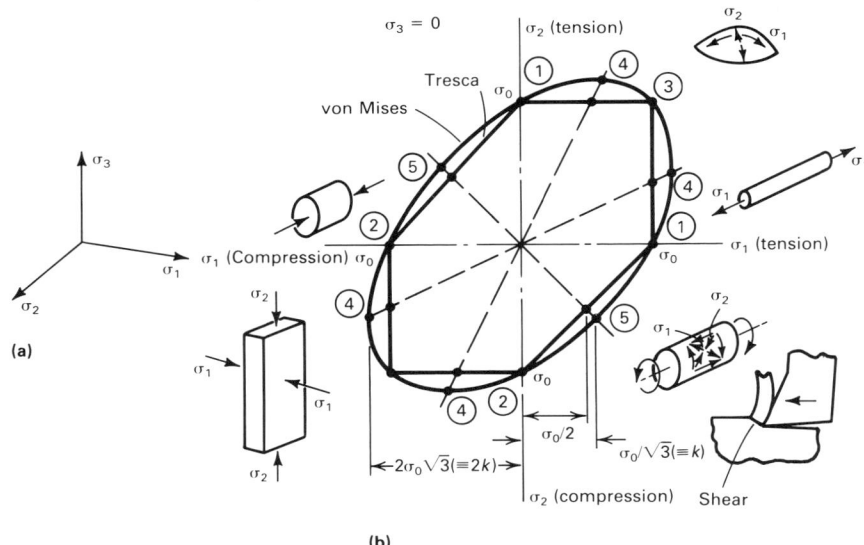

strain, strain rate, and temperature history at potential fracture sites. The application of computer-aided finite element analysis of plastic deformation and heat flow has facilitated this goal. A completely experimental approach that uses this concept of workability is the forming-limit diagram discussed below.

The simplest and most widely used fracture criterion was proposed by Cockcroft and Latham (Ref 9). This fracture criterion is not based on a micromechanical model of fracture, but simply recognizes the joint roles of tensile stress and plastic strain in producing fracture:

$$\int_0^{\epsilon_f} \bar{\sigma}\left(\frac{\sigma^*}{\bar{\sigma}}\right)d\bar{\epsilon} = C$$

where $\bar{\sigma}$ is the effective stress, which is equal to $(\sqrt{2}/2)[(\sigma_1 - \sigma_2)^2 + (\sigma_2 - \sigma_3)^2 + (\sigma_3 - \sigma_1)^2]^{1/2}$; σ^* is the maximum tensile stress; $d\bar{\epsilon}$ is the effective strain, which is equal to $(\sqrt{2}/3)[(d\epsilon_1 - d\epsilon_2)^2 + (d\epsilon_2 - d\epsilon_3)^2 + (d\epsilon_3 - d\epsilon_1)^2]^{1/2}$; and C is a material constant evaluated from the compression test. This fracture criterion indicates that fracture occurs when the tensile strain energy per unit volume reaches a critical value.

Use of this fracture criterion is shown in Fig. 12. The values of the reduction ratio at which centerburst fracture occurs in the cold extrusion of two aluminum alloys is illustrated. The energy conditions for different die angles are given by the three curves that reach a maxima. The fracture curves for the two materials slope down to the right. Centerburst occurs in the regions of reduction, for which the process energies exceed the

material fracture curve. No centerburst occurs at small or large reduction ratios.

Assessment of Bulk Workability

Development of a general approach to workability prediction has been slow because of the inherent complexity of plastic deformation processes. Such an approach requires both a measure of material ductility and the ability to describe the stress, strain, and temperature state in a deforming material. The advent of powerful finite-element methods of analysis and the widespread availability of computers have enhanced this. However, because ready access to computer-based plasticity analysis has become available only recently, specialized tests have been developed for each class of deformation process. The primary testing techniques used to assess bulk workability are the compression test, tension test, torsion test, and bend test.

The compression test, in which a cylindrical specimen is upset into a flat "pancake," is usually considered a standard bulk workability test. The average stress state during testing is similar to that in many bulk deformation processes, without introducing the problems of necking or material reorientation. Therefore, a large amount of deformation can be achieved before fracture occurs. By controlling the barreling of the specimen through variations in geometry and reducing friction between the specimen ends and the anvil with lubricants, the stress state can be varied over wide limits.

Compression testing has developed into a highly sophisticated test of workability in cold upset forging and is a common quality control test in hot forging operations. Compression forging is a useful method for assessing the frictional conditions in hot working. The main disadvantage of the compression test is that tests at a constant true strain rate require special equipment.

Descriptions of both cold (upset testing) and hot compression testing for determining workability can be found later in this article. Additional information can be found in the articles "Axial Compression Testing" and "High Strain Rate Compression Testing" in this Volume.

The tension test is widely used to determine the mechanical properties of a material. Uniform elongation, total elongation, and reduction in area at fracture frequently are used as indices of ductility. However, the extent of deformation possible in a tension test is limited by the formation of a necked region in the tension specimen. This introduces a triaxial tensile stress state and leads to fracture.

For most metals, the uniform strain that precedes necking rarely exceeds a true strain of 0.5. For hot working temperatures, this uniform strain frequently is less than 0.1. Although tension tests are easily performed, necking makes control of strain rate difficult and leads to uncertainties about the value of strain at fracture because of the complex stresses arising from necking. Therefore, the usefulness of the tension test is limited in workability testing. This test is used primarily under special high strain rate hot tension test conditions to establish the range of hot working temperatures. A description of this test method can be found later in this article. Additional information on tension testing can be found in the Section on Tension Testing and the articles "Sheet Formability Testing" and "High Strain Rate Tension Testing" in this Volume.

Torsion Test. In this test, deformation is caused by pure shear, and large strains can be achieved without the limitations imposed by necking. Because the strain rate is proportional to rotational speed, high strain rates are readily obtained. Moreover, friction has no effect on the test, as it does in compression testing. Although the stress state in torsion may represent the typical stress in metalworking processes, deformation in the torsion test is not an accurate simulation of metalworking processes because of excessive material reorientation at large strains. The use of the torsion test in workability studies is discussed in the article "Application of the Torsion Test to Determine Workability" in this Volume.

The bend test is useful for assessing the workability of thick sheet and plate. Generally, this is most applicable to cold working

operations. The stress and strain states in bending and the use of the bend test for assessing workability are discussed in the Section on Bend Testing and in the article "Sheet Formability Testing" in this Section.

Cold Upset Testing

The upsetting of a small cylinder in room-temperature deformation is one of the most widely used workability tests (Ref 11). As a cylinder is compressed, it usually tends to barrel, and a biaxial stress state develops at the equator of the upset cylinder (Fig. 13). The stress state usually consists of a circumferential tension stress and an axial compression stress, although an axial tension stress may develop when barreling is severe. The surface strains measured at fracture for a wide range of test conditions lead to the construction of a fracture-limit line for the material. Comparison of the strains in a material during an actual deformation process with the fracture-limit line for the material predicts the possibility of fracture.

Specimen Design and Friction Conditions. The aspect (height-to-diameter) ratio affects the strain occurring at the bulging free surface. The upper limit on this ratio is 2.0, because at a higher h/D there is a strong tendency toward buckling. The lower limit is based on a convenient height for application of the grid marks. Normally, specimen aspect ratios range between 0.75 and 1.75.

In addition to right circular cylinders, tapered and flanged test specimens may be

used to increase circumferential tension (Fig. 14). These specimens are used most frequently with ductile materials that do not fracture when deformed as cylindrical specimens. The height of the cylindrical surface at midheight ranges from 0.2 to 0.75 times the specimen height. The reduced diameter of the flanged compression test specimens is 0.8 times the original cylinder diameter, and the angle of the tapered compression speci-

Fig. 12 Workability criteria for centerbursting in aluminum alloy 2024

Based on a maximum tensile stress-strain energy criterion. Source: Ref 10

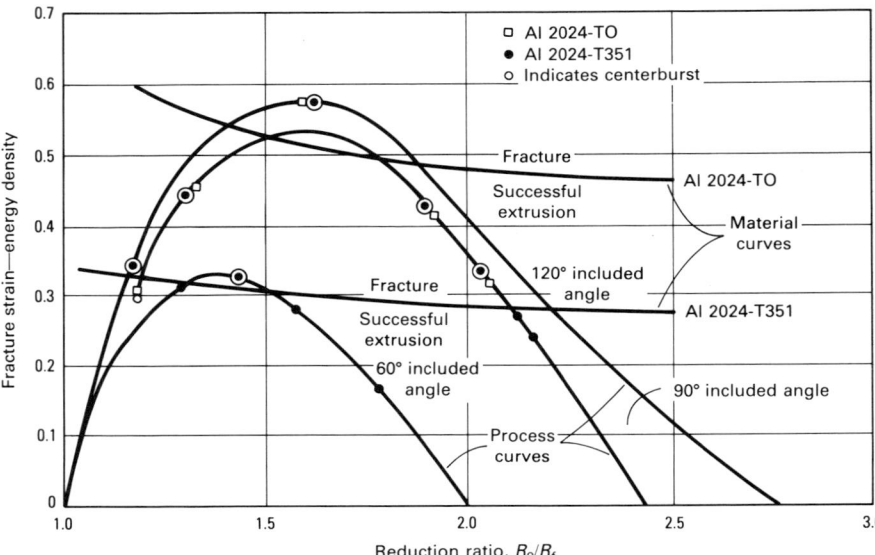

Fig. 13 Circumferential tensile and axial compressive stresses at the equator of an upset cylinder

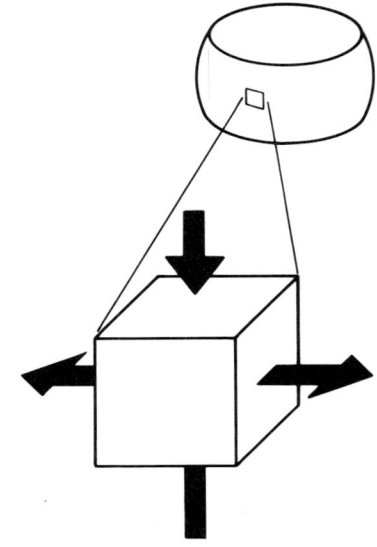

Fig. 14 (a) Cylindrical, (b) tapered, and (c) flanged compression test specimens

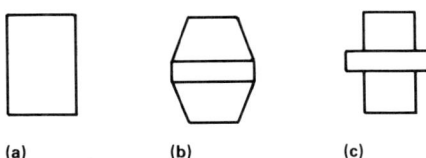

(a)   (b)   (c)

Fig. 15 Grids for strain measurement on upset cylinders
Axial strain, $\epsilon_z = \ln(h/h_0)$. Hoop strain, $\epsilon_\theta = \ln(W/W_0)$ or $\ln(D/D_0)$

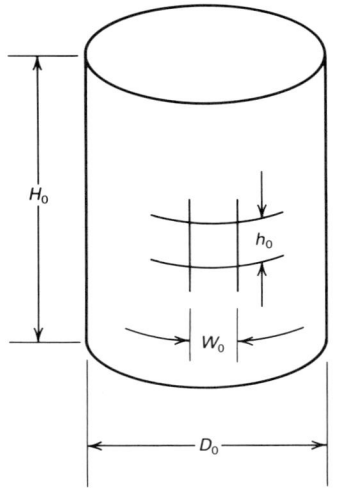 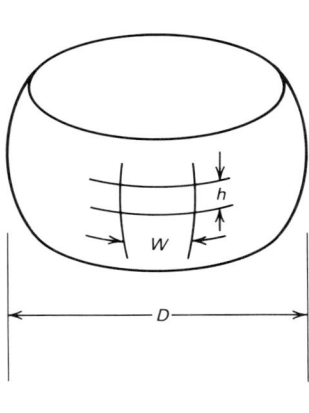

mens is 20°. The end-face diameter should not be less than one half of the overall specimen height.

The friction at the interface between the specimen and the flat die contact surface has an important influence on the strains developed at the free surface. Usually, the end faces of the specimen are polished to prevent lubricant entrapment. Three common die conditions are steel dies with knurled surfaces ($m = 0.45$), unlubricated steel dies with 0.05-μm (2-μin.) polished surface ($m = 0.21$), and polished steel dies with molybdenum disulfide solid lubricant ($m = 0.08$). Teflon film is used when friction must be minimized to prevent barreling.

Strain measurements consist of measuring the axial strain, ϵ_z, and the hoop or circumferential strain ϵ_θ by the distortion of a grid placed on the surface of the cylinder at the equatorial diameter (Fig. 15). One of the following methods can be used for applying the gage marks.

- Two strips of cellophane tape are wrapped around the cylindrical surface so that there is a thin band of bare metal around the circumference at midheight. The specimens are then sprayed with blue steel dye, and the strips of tape are removed when the dye has dried. The gage band is about 3 mm (0.12 in.) wide and has sharply defined edges that deform with the specimen surface without cracking as the test proceeds.
- The gage band is produced by electrochemical etching with the electromark system or with the photoresist method.
- The gage band is marked by scribing fine lines about 3 mm (0.12 in.) apart with a carbide-tipped height gage.
- Four gage points are indented at midheight on the cylindrical specimen with a Vickers or diamond brale hardness indenter.

Scribe lines or hardness indentations should be applied only to very ductile materials, because these gage marks could be potential stress concentrations, leading to premature fracture during upset testing.

Test specimens complete with gage bands are compressed incrementally between flat dies under one of the friction conditions until cracks at the bulged surface are visible. For each incremental deformation, displacement of the gage band is measured with a toolmaker's microscope to ±0.013 mm (±0.0005 in.), and the diameter is measured with a micrometer to the same degree of accuracy. Average values are calculated from three measurements to obtain more accurate data.

The axial and circumferential true strains on the barreled surface are calculated by:

Axial strain:

$$\epsilon_z = \ln\left(\frac{h}{h_0}\right)$$

Circumferential strain:

$$\epsilon_\theta = \ln\left(\frac{w}{w_0}\right) \quad \text{or} \quad \ln\left(\frac{D}{D_0}\right)$$

where h_0 and h are initial and final gage heights, respectively; W_0 and W are initial and final gage widths, respectively; and D_0 and D are initial and final diameters, respectively (Fig. 15).

Crack Detection. Most materials exhibit some orange peel effect or surface roughening shortly after plastic flow begins. Fine networks of microcracks may also appear if the surface is observed at a magnification of about 30×. These cracks are stable and do not grow in size as deformation progresses. Consequently, they should be ignored, because they do not represent a limitation to useful deformation of the material. After further deformation, one or more large cracks will form and grow rapidly; these cracks are easily visible.

The fracture strain is measured, in accordance with the above equations, when the large crack is observed. There is always some uncertainty regarding the initiation of a large crack, which affects the accuracy of the results. Because large cracks grow rapidly, however, the strains measured at the first sight of a crack closely approximate the strains that existed at crack initiation.

Material Considerations. Because cracks are initiated at the bulged surface, the ductility of the specimen during upsetting is greatly influenced by surface conditions, such as surface defects, decarburization, and residual stress. Material inhomogeneities have adverse effects on workability. Localized variations that can reduce the workability of a steel include ferrite grain size, nonmetallic inclusion content, banding, center segregation, and decarburization. Therefore, the free surfaces of the test specimen should contain the same surface structural features as the actual material to be used in the deformation process of interest.

The effect of mechanical anisotropy on ductile fracture should be considered. Generally, the direction of inclusion alignment relative to the secondary tensile hoop stress generated by friction during the uniaxial compression test strongly influences the strain to fracture. It is therefore necessary that the test specimen be prepared so that the direction of tensile stresses relative to the direction of inclusion alignment during testing is the same as that in the actual process.

Deformation Characteristics. In a cylindrical compression test carried out under frictionless conditions, deformation of the test specimen is uniform, and no bulging is produced (Fig. 16). Almost frictionless conditions can be achieved through the use of Teflon film under laboratory conditions. In this manner, the material flow-stress characteristics can be determined without the

Fig. 16 Cylindrical compression test specimens

(a) Undeformed specimen. (b) Specimen compressed with friction (note crack). (c) Specimen compressed without friction

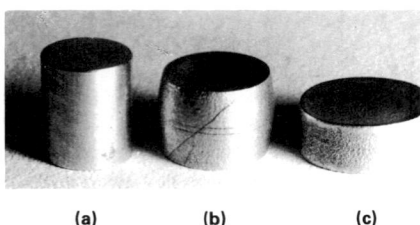

(a)   (b)   (c)

Fig. 17 Strain paths in upset test specimens

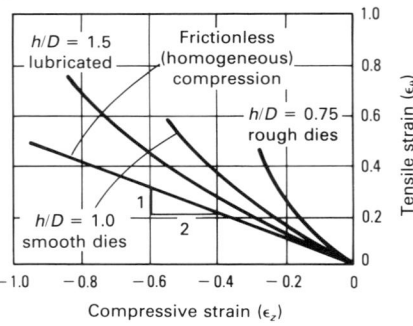

Fig. 18 Stresses at the equatorial surface of an upset test specimen

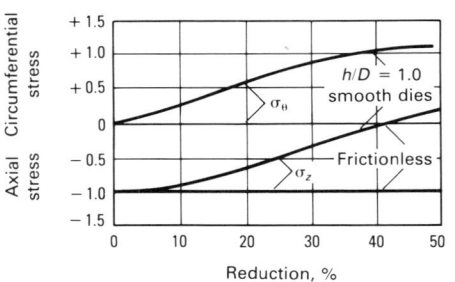

effects of friction. Frictional constraint at the end faces of the specimen under conventional compression prevents uniform deformation and leads to a bulged surface at the mid-height of the specimen.

Free-Surface Strains. At the free surfaces of the compressed cylinders, the strains consist of circumferential tension and axial compression. For frictionless (homogeneous) compression of cylindrical specimens, the tensile strain is equal to one half of the compressive strain. With increasing frictional constraint, bulge severity increases, the tensile strain becomes larger, and the compressive strain decreases. Effects of the bulged profile, which develops naturally in cylindrical compression specimens, are imposed artificially through tapered and flanged compression tests.

Figure 17 summarizes the effects of friction, aspect ratio, and specimen profile on the measured free-surface strains at midheight. Measured strain paths are shown in terms of circumferential (tensile) strain versus axial (compressive) strain. Beginning with the strain ratio of one half for homogeneous deformation, the strain-path slope increases with increasing friction. For a given value of friction, a decreasing aspect ratio increases the strain-path slope slightly. Tapered compression specimens further increase the strain-path slope, and flanged compression specimens result in strain paths that lie nearly along the circumferential tensile strain axis (Ref 12, 13).

Stress States. Stresses at the free surface of compressed specimens can be calculated from measured strains and plasticity theory (Ref 14):

$$\sigma_z = \frac{\sigma_0}{\sqrt{3}} \left(\frac{2\alpha + 1}{\sqrt{\alpha^2 + \alpha + 1}} \right)$$

where $\alpha = d\epsilon_z / d\epsilon_\theta$, which is determined graphically from the measured strain path. The flow stress, σ_0, is taken from the stress-strain curve of the material obtained by a homogeneous (frictionless) compression test. Figure 18 illustrates typical changes in

σ_θ and σ_z in a cylindrical compression test at progressively larger reductions in height. At large deformations, as the barreling becomes severe, the stress in the z direction, σ_z, becomes tensile.

Upset Test Fracture Limits

A wide range of stress and strain states can be produced at the free surfaces of cylindrical, tapered, or flanged test specimens, which permits evaluation of the effects of variations in stress and strain states with fracture. The most convenient representation of the limits of fracture is a plot of the circumferential and axial strains that existed on the specimen surface at fracture. This plot, as shown in Fig. 19, is a fracture-limit line. At strain combinations below the line, the material has not fractured. For strains above the line, the material has fractured. The fracture limit line is parallel to the line for homogeneous compression of a cylinder that has a slope of $-\frac{1}{2}$.

This relationship is appropriate for a wide range of materials (Ref 15, 16). The intersection of ϵ_θ (tensile strain axis) corresponds to the fracture strain in plane-strain tensile testing.

The use of flanged and tapered specimens expands the range of strains that can be obtained in the upset test. As shown in Fig. 20, tests with tapered and flanged specimens on aluminum alloy 2024-T351 have permitted the accumulation of data close to the $\epsilon_\theta = 0$ position. In this case, the data fit a straight line of constant slope. For a 1045 steel, however, the fracture-limit line shows a bilinear behavior, with a larger slope in the small strain region (Fig. 21). A discussion on the use of the flanged (collar) specimen and other specimen designs can be found in Ref 13.

The anisotropy inherent in wrought products such as plate can be examined with the upset test (Ref 17). Figure 22 shows the orientation of the compression test specimens cut from hot rolled 1045 steel plate. The arrows indicate the direction of the critical

Fig. 19 Fracture loci in cylindrical upset test specimens of two materials deformed at room temperature

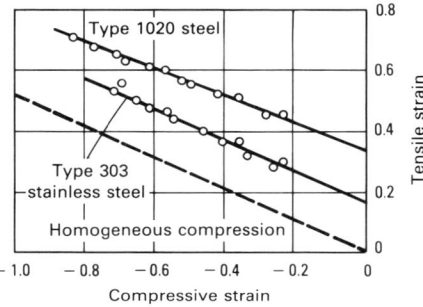

Fig. 20 Fracture loci in cylindrical, tapered, and flanged upset test specimens of aluminum alloy 2024-T351 deformed at room temperature

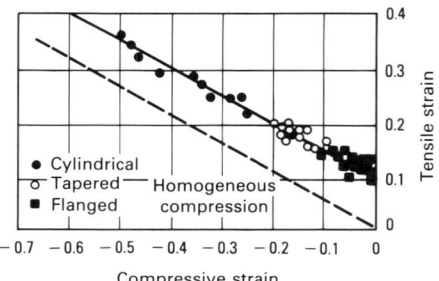

tensile hoop stress for the compression specimens and the direction of the tensile stress in tension specimens cut from the plate. The tensile stress direction in the long transverse tension test is identical to the critical tensile direction in the short transverse compression tests. The fracture-limit lines for the upset tests taken from the three orientations in the plate are shown in Fig. 23.

Specimens taken in the short transverse direction show a higher fracture-limit line, while longitudinal and long transverse upset

Fig. 21 Fracture loci in cylindrical, tapered, and flanged upset test specimens of type 1045 cold finished steel deformed at room temperature

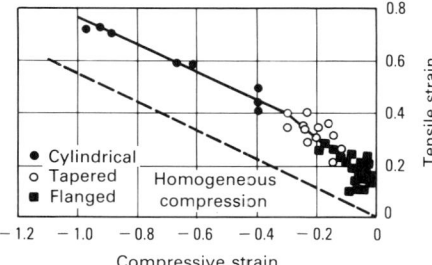

Fig. 22 Orientation of the upset (compression) test and tension test specimens cut from hot rolled plates

Arrows indicate directions of critical secondary tensile stress in upset tests and applied tensile stress in tension tests. L (longitudinal), T (long transverse), and S (short transverse) refer to orientation of specimen axes.

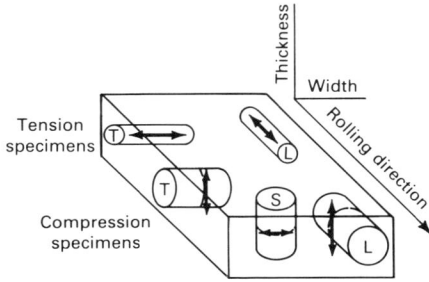

Fig. 23 Comparison of the fracture strains from tension tests with those from upset tests

Data points are tension test results for zero-gage-length true strain at fracture. Source: Ref 17

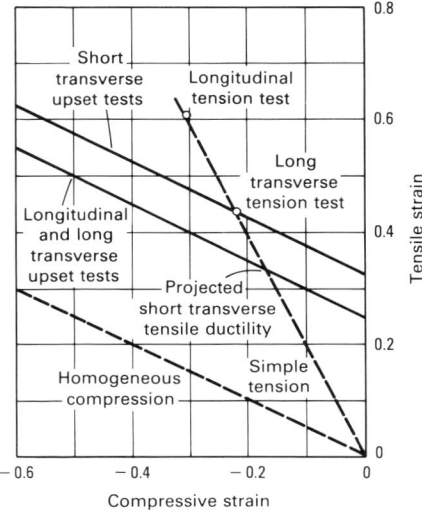

tests fall on the same curve. These results are explained by the orientation of the tensile stress imposed by the test relative to the direction of the sulfide inclusions, which are the main structural feature causing fracture in this steel plate. Figure 22 shows that in short transverse specimens the tensile strain is parallel to the plane of the inclusions, while in longitudinal and long transverse specimens, the critical strain is perpendicular to the plane of the inclusions. The latter condition is a more critical fracture condition and results in a lower fracture-limit line.

Ductility measurements from the tension test can also be correlated with the fracture limit line, if ductility is measured by the zero-gage-length fracture strain ϵ_f:

$$\epsilon_f = 2 \ln \left(\frac{D_0}{D_f} \right)$$

where D_0 is the original diameter, and D_f is the minimum diameter at fracture. Tension testing subjects the material to a strain ratio (tension/compression) of 2. Figure 22 shows that the long transverse tension test has the same orientation of tensile stress as the short transverse upset test. When the value of ϵ_f is measured for the long transverse tension test, it plots on Fig. 23 at the point where a line with slope 2 intersects the fracture-limit line for the short transverse upset test orientation. This leads to the prediction that the short transverse zero-gage-length fracture strain for the plate, which is difficult to measure because of thickness limitations, should be where the simple tension line intersects the fracture-limit line for longitudinal and long transverse upset tests.

Forming-Limit Diagram. Figures 19 to 21 are examples of forming-limit diagrams for free-surface fracture in bulk deformation processes. A fracture-limit diagram can be used as a forming-limit diagram. For example, consider a simple bolt-heading process such as that illustrated in Fig. 24. If material A is used for the product, strain path a will

cross the fracture line on its way to its position in the final deformed geometry, and cracking is likely to occur. Defects can be avoided in two ways: material B, which has a higher forming-limit line, can be used, or the strain path can be altered to follow b, which in this case is accomplished through improved lubrication. In other deformation processes, the strain path can be altered by changing the preform design.

Hot Compression Testing

Processes for hot working of metals, such as hot rolling, hot forging, and isothermal forging of high-performance aerospace materials, are of great importance industrially because of the generally low resistance to deformation and high ductility exhibited by most metals at temperatures greater than one half of the melting point on the absolute temperature scale. High rates of working, which are desirable for economic reasons and to

minimize heat transfer, generally have the effect of increasing resistance to deformation and, in some instances, of decreasing ductility.

The sensitivity of the material to the rate of straining and to the deformation temperature should be considered when conducting hot compression workability testing, particularly when test data are to be used to define parameters for industrial processing.

Many of the features discussed earlier in this article in relation to cold upset testing are equally pertinent to hot compression testing. However, much closer attention must be paid to temperature and strain-rate effects. In addition, because lubrication of the workpiece and dies is more difficult at high tempera-

Fig. 24 Comparison of strain paths and fracture locus lines

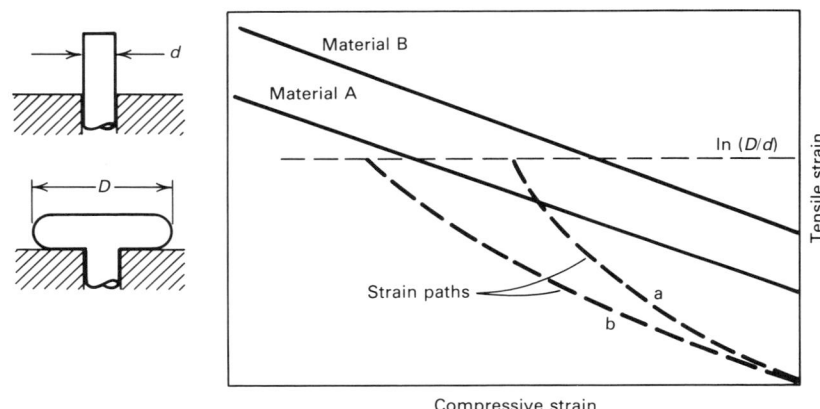

tures, frictional effects must be given greater attention.

Test Conditions. Unless lubrication at the ends of the specimen is very good, frictional restraint retards the outward motion of the end face, and part of the end face is actually formed by folding the sides of the original cylinder onto the end face in contact with the platens. The barreling that results introduces a complex stress state, which is beneficial in fracture testing, but which is detrimental when the compression test is used to measure flow stress. The frictional restraint also causes internal inhomogeneity. Slightly deforming zones develop adjacent to the platens, while severe deformation is concentrated in zones that occupy roughly diagonal positions between opposing edges of the specimen.

Hot upsetting of a cylinder under conditions of poor lubrication where the platens are cooler than the specimen is shown schematically in Fig. 25. The cooling at the ends restricts the flow, so the deformation is concentrated in a central zone, with dead-metal zones forming adjacent to the platen surfaces (Fig. 25a).

As deformation proceeds, severe inhomogeneity develops, and growth of the end faces is attributed entirely to folding over of the sides (Fig. 25b). When the value of D/h exceeds about 3, expansion of the end faces occurs (Fig. 25c). The conditions described here are extreme and should not be allowed to occur in hot compression testing unless the objective is to simulate cracking under forging conditions. Although adequate lubrication cannot improve the situation so that homogeneous deformation occurs, with glass lubricants and isothermal conditions it is possible to conduct hot compression testing without appreciable barreling (Ref 19). Isothermal test conditions can be achieved by using a heated subassembly, such as that shown in Fig. 26, or special heated dies that provide isothermal conditions (Ref 21).

The true strain rate in a compression test is:

$$\dot{\epsilon} = \frac{d\epsilon}{dt} = \frac{\dfrac{-dh}{h}}{dt} = -\frac{1}{h}\frac{dh}{dt} = -\frac{v}{h}$$

where v is the velocity of the platen, and h is the height of the specimen at time t. Because h decreases continuously with time, the velocity must decrease in proportion to $(-h)$ if $\dot{\epsilon}$ is to be held constant. In a normal test, if v is held constant, the engineering strain rate $\dot{e}$ will remain constant:

$$\dot{e} = \frac{de}{dt} = \frac{\dfrac{-dh}{h_0}}{dt} = -\frac{1}{h_0}\frac{dh}{dt} = \frac{-v}{h_0}$$

Fig. 25 Deformation patterns in nonlubricated, nonisothermal hot forging

(a) Initial barreling. (b) Barreling and folding over. (c) Beginning of end-face expansion. Source: Ref 18

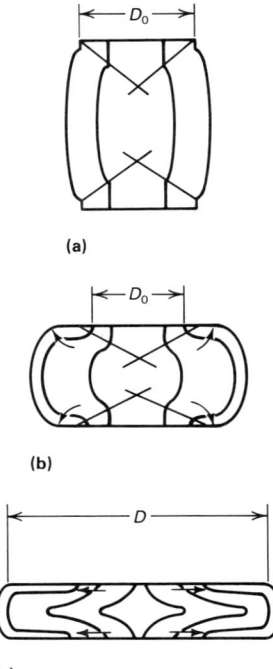

(a)

(b)

(c)

The true strain rate, however, will not be constant. A special machine called a cam plastometer, which is described in the article "High Strain Rate Compression Testing" in this Volume, causes the bottom platen to compress the specimen through cam action at a constant true strain rate to a strain limit of $\epsilon = 0.7$. Use of cam plastometers is limited; there probably are not more than ten in existence. However, an essentially constant true strain rate can be achieved on a standard closed-loop servo-controlled testing machine. Strain rates up to 20 s^{-1} have been achieved (Ref 19).

When a constant true strain rate cannot be obtained, the mean strain rate may be adequate. The mean true strain rate, $\langle\dot{\epsilon}\rangle$, for constant velocity v_0, when the specimen is reduced in height from h_0 to h, is given by:

$$\langle\dot{\epsilon}\rangle = \frac{v_0}{2}\frac{\ln\left(\dfrac{h_0}{h}\right)}{(h_0 - h)}$$

Flow Stress in Compression. Optimally, the determination of flow stress in hot compression must be carried out under isothermal conditions (no die chilling) at a constant strain rate and with a minimum of

Fig. 26 Heated subassembly with specimen in position used to achieve isothermal test conditions

Thermocouple is removed prior to compression. Source: Ref 20

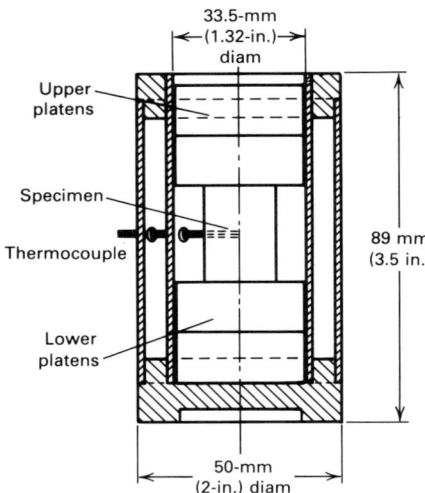

friction to minimize barreling. With conventional servohydraulic testing machines these conditions can be met. For an essentially homogeneous upsetting test, a cylinder of diameter D_0 and initial height h_0 will be compressed to height h and spread out to diameter D, according to the law of constancy of volume:

$$D_0^2 h_0 = D^2 h$$

If friction can be neglected, the uniaxial compressive stress (flow stress) corresponding to a deformation force P is:

$$\sigma_0 = \frac{P}{A} = \frac{4P}{\pi D^2} = \frac{4Ph}{\pi D_0^2 h_0}$$

If substantial friction is present, the average pressure, $\bar{p}$, required to deform the cylinder is greater than the flow stress of the material, σ_0:

$$\frac{\bar{p}}{\sigma_0} = \left(\frac{h}{4\mu a}\right)^2 \left(e^{2\mu a/h} - \frac{2\mu a}{h} - 1\right)$$

where a is the radius of the cylinder, and μ is the Coulomb coefficient of friction. The true compressive strain is given by:

$$\epsilon = \ln\left(\frac{h_0}{h}\right)$$

The effects of friction and die chilling can be minimized by using a long, thin specimen. Thus, most of the specimen volume is unaffected by the dead-metal zones at the platens. However, this approach is limited, because buckling of the specimen will occur if h/D exceeds about 2.

An extrapolation method involves testing cylinders of equal diameters but varying heights so that the D_0/h_0 ratio ranges from about 0.5 to 3.0 (Ref 22). A specific load is applied to the specimen, the load is removed, and the new height is determined to calculate a true strain. Upon relubrication, the specimen is subjected to an increased load, unloaded, measured, and the cycle is repeated.

The same test procedure is followed with each specimen so that the particular load levels are duplicated exactly. The results are plotted as shown in Fig. 27. For the same load, the actual strain (due to height reduction) is plotted against the D_0/h_0 ratio for each test cylinder. A line drawn through the points is extrapolated to a value of $D_0/h_0 = 0$. This would be the anticipated ratio for a specimen of infinite initial height for which the end effects would be restricted to a small region of the full test height. The true stress corresponding to each of these true strains is given by:

$$\sigma_0 = \frac{4Ph}{\pi D_0^2 h_0}$$

The fracture-locus line, which was discussed earlier in this article (see Fig. 19 to 21 and corresponding text), may be determined for hot working conditions using the appropriate isothermal fixtures. For some materials, the fracture locus lines are straight and parallel with a slope of $-\frac{1}{2}$ (Fig. 28), while for other materials the fracture line is observed to have two slopes (Fig. 29). In this case, a slope of unity is found at small strains using the tapered and flanged compression specimen. This is similar to the situation shown previously in Fig. 20 and 21 for cold compression tests.

Ductility Testing. The basic hot ductility test consists of compressing a series of cylindrical or square specimens to various thicknesses, or to the same thickness with varying

specimen length-to-diameter (length-to-width) ratios. The limit for compression without failure by radial or peripheral cracking is considered a measure of workability. This type of test has been widely used by the forging industry. Longitudinal notches are sometimes machined into the specimens prior to compression, because the notches apparently cause more severe stress concentrations, thus providing a more reliable index of the workability to be expected in a complex forging operation.

Plastic Instability in Compression. Several types of plastic instabilities can be developed in the compression test. The first type is associated with a maximum in the true stress-strain curve. The second type concerns inhomogeneous deformation and shear band formation. Figure 30 shows the type of plastic instability that occurs in some materials in hot compression testing. At certain temperatures and strain rates some of the typical strengthening mechanisms become unstable. Because the rate of flow softening exceeds the rate of area increase as the specimen is compressed, a maximum results in the flow stress curve.

Analysis of the process indicates that the plastic deformation is stable (no maximum in the flow curve) as long as $(\gamma + m) \leq 1$, where γ is the dimensionless work-hardening coefficient, and m is strain-rate sensitivity. Both of these material parameters are defined below (Ref 24). A material with a high strain rate sensitivity is more resistant to flow localization in the tension test (necking), whereas in compression testing, a higher rate sensitivity leads to earlier flow localization.

Flow softening or negative strain hardening can also produce flow localization effects in compression independently of the effects of die chilling or high friction. The constant-strain-rate isothermal hot compression test is

useful for detecting and predicting flow localization. Nonuniform flow in compression is likely if:

$$\alpha_c = \frac{\gamma - 1}{m} \geq 5$$

where

$$\gamma = \left(\frac{1}{\sigma}\right)\left(\frac{d\sigma}{d\epsilon}\right)$$

and

$$m = \frac{\partial \ln \sigma}{\partial \ln \dot{\epsilon}}\bigg|_{T,\epsilon} \simeq \frac{\Delta \log \sigma}{\Delta \log \dot{\epsilon}}\bigg|_{T,\epsilon}$$

Figure 31 illustrates the differences in deformation of titanium alloy samples. Specimens in Fig. 31(a) to (c) were deformed at a temperature where α_c was high. In Fig. 31(a), $\dot{\epsilon} = 10^{-3}$ s^{-1} and $\alpha_c = 2$; in Fig. 31(b), $\dot{\epsilon} = 10^{-1}$ s^{-1} and $\alpha_c = 5$; and in Fig. 31(c), $\dot{\epsilon} = 10$ s^{-1} and $\alpha_c = 5$, while

Fig. 29 Fracture loci for 2024 aluminum alloy tested at room temperature and 300 °C (570 °F at $\dot{\epsilon} = 0.1$ s^{-1}

Source: Ref 12

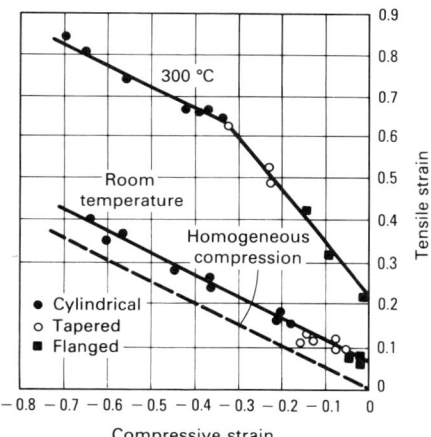

Fig. 27 Extrapolation method for correcting for end effects in compressive loading

Source: Ref 22

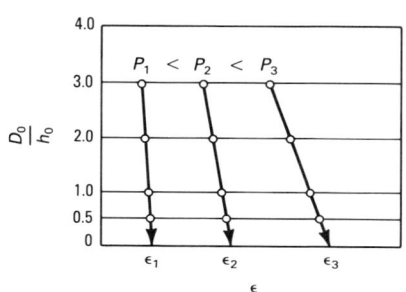

Fig. 28 Fracture loci for elevated-temperature tests on sintered 4620 steel (80% of theoretical density) tested at $\dot{\epsilon} = 10$ s^{-1}

Source: Ref 23

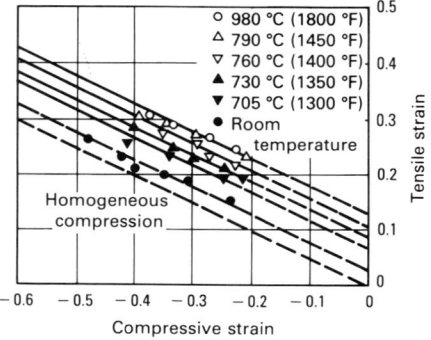

Fig. 30 Example of compressive flow stress curve showing strain softening

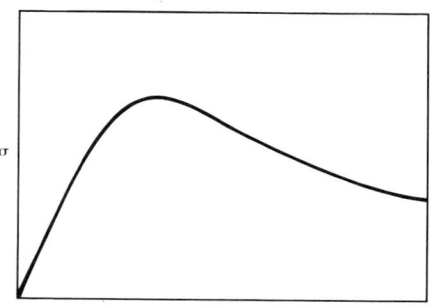

(a), (b), and (c) were tested at 704 °C (1300 °F). (d), (e), and (f) were tested at 816 °C (1500 °F). Strain rates were (a, d) 10^{-3} s^{-1}, (b, e) 10^{-1} s^{-1}, and (c, f) 10 s^{-1}. Prior to testing, alloy had been beta annealed to yield an equiaxed beta starting microstructure.

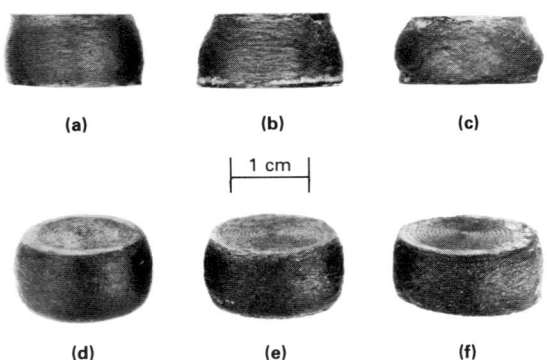

(a) (b) (c)

| 1 cm |

(d) (e) (f)

Fig. 32 Plane-strain compression test

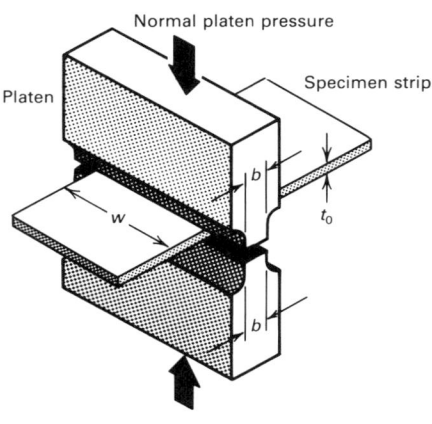

Plane-Strain Compression Test

In the plane-strain compression test, the difficulties encountered with bulging and high friction at the platens in compression of cylinders can be minimized (Ref 22). As shown in Fig. 32, the specimen is a thin plate or sheet that is compressed across the width of the strip by narrow platens that are wider than the strip. The elastic constraints of the undeformed shoulders of material on each side of the platens prevent extension of the strip in the width dimension—hence the name plane strain.

Deformation occurs in the direction of platen motion and in the direction normal to the length of the platen. To ensure that lateral spread is negligible, the width of the strip should be at least six to ten times the breadth of the platens. To ensure that deformation beneath the platens is essentially homogeneous, the ratio of platen breadth to strip thickness (b/t) at all times should be between 2 and 4. It may be necessary to change the platens during testing to maintain this condition. By carrying out the test in increments, so as to provide good lubrication and maintain the proper b/t ratio, true strains of 2 can be achieved. Although the plane-strain compression test is used primarily for measurement of flow properties at room temperature, it can be used for elevated-temperature tests (Ref 25, 26).

The true stress and true strain determined from the plane-strain compression test can be expressed as:

$$p = \frac{p}{wb}$$

the specimens in Fig. 31(d) to (f) were deformed at a temperature where α_c was less than zero.

and

$$\epsilon_{pc} = \ln \frac{t_0}{t}$$

Because of the stress state associated with plane-strain deformation, the mean pressure on the platens is 15.5% higher than in uniaxial compression testing. The true stress-strain curve in uniaxial compression (σ_0 versus ϵ) may be obtained from the corresponding plane-strain compression curve (p versus ϵ_{pc}) by the following:

$$\sigma_0 = \frac{2}{\sqrt{3}} p = \frac{p}{1.155}$$

and

$$\epsilon = \frac{\sqrt{3}}{2} \epsilon_{pc} = 1.155 \, \epsilon_{pc}$$

Plane-strain compression testing is useful for determining the flow properties of sheet metal. Because of the geometry of the test specimen, this test does not cause cracking in the material.

Partial-Width Indentation Test

The partial width indentation test is a new test for evaluating the workability of metals. It is similar to the plane-strain compression test, but does not subject the test specimen to true plane-strain conditions (Ref 27). A simple slab-shaped specimen is deformed over part of its width by two opposing rectangular anvils with widths that are less than that of the specimen. On penetrating the workpiece, the anvils longitudinally displace metal from the center. This creates overhangs (ribs) that are subjected to secondary, nearly uniaxial, tensile straining. The ductility of the material under these conditions is indicated as the reduction in the rib height at fracture. The

Fig. 33 Partial-width indentation test

$L \approx h$; $b = h/2$; $w_a = 2L$; $l = 4L$

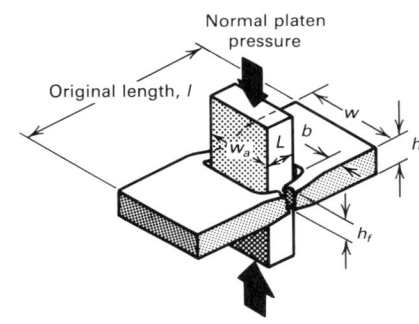

test geometry has been standardized, as shown in Fig. 33.

One advantage of this test is that it uses a specimen of simple shape. Also, as-cast materials can be readily tested. One edge of the specimen can contain original surface defects. The test can be conducted hot or cold. Thus, the partial-width indentation test is suitable not only for determining the intrinsic ductilities of materials, but also for evaluating the inhomogeneous aspects of workability. This test has been used to establish the fracture-limit loci for ductile metals (Ref 28).

The secondary tension test is a modification of the partial-width indentation test that imposes more severe strain in the rib for testing highly ductile materials. In this test, a hole or a slot is machined in the slab-type specimen adjacent to where the anvils indent the specimen. Preferred dimensions of the hole and slot are given in Fig. 34. With this design, the ribs are sufficiently stretched to ensure fracture in even the most ductile materials. The fracture strain is based on reduc-

Fig. 34 Secondary-tension test showing the geometries of holes and slots

$L \simeq h; w_a \geq 2h; b = h/4; D = h/2$

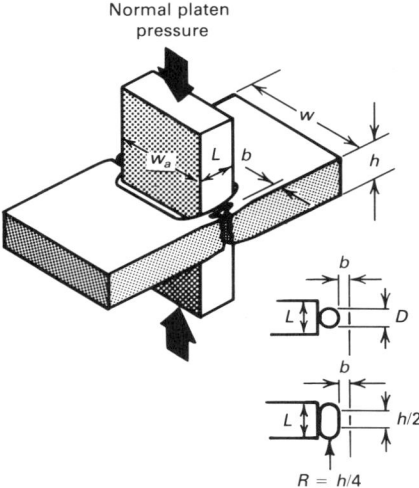

Normal platen pressure

Fig. 35 Variation in shape of ring test specimens deformed the same amount under different frictional conditions

Left to right: undeformed specimen; deformed 50%, low friction; deformed 50%, medium friction; deformed 50%, high friction

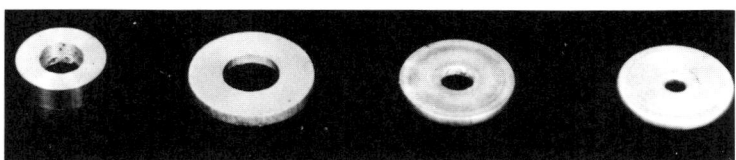

tion in area where the rib is cut out so that the fracture area can be photographed or traced on an optical comparator.

Ring Compression Test

When a flat, ring-shaped specimen is upset in the axial direction, the resulting shape change depends only on the amount of compression in the thickness direction and the frictional conditions at the die-ring interfaces. If the interfacial friction were zero, the ring would deform in the same manner as a solid disk, with each element flowing outward radially at a rate proportional to its distance from the center.

In the case of small, but finite, interfacial friction, the outside diameter is smaller than in the zero-friction case. If the friction exceeds a critical value, frictional resistance to outward flow becomes so high that some of the ring material flows inward to the center. Measurements of the inside diameters of compressed rings provide a particularly sensitive means of studying interfacial friction, because the inside diameter increases if the friction is low and decreases if the friction is higher, as shown in Fig. 35.

The ring test is therefore a compression test with a built-in frictional measurement. Thus, it is possible to measure the ring dimensions and compute both the friction value and the basic flow stress of the ring material at the strain under the given deformation conditions.

Analysis of Ring Compression. The mechanics of the compression of flat ring-shaped specimens between flat dies have

been analyzed using an upper bound plasticity technique (Ref 29, 30). Values of p/σ_0 (where p is the average forging pressure on the ring and σ_0 is the flow stress of the ring material) can be calculated in terms of ring geometry and the interfacial shear factor, m. In these equations, neither σ_0 nor the interfacial shear stress, τ, appears in terms of independent absolute values, but only as the ratio m.

The analysis assumes that this ratio remains constant for a given material and deformation conditions. If the analysis is carried out for a small increment of deformation, σ_0 and τ can be assumed to be approximately constant for this increment, and the solution is valid. Thus, if the shear factor m is constant for the entire operation, the mathematical analysis can be continued in a series of small deformation increments using the final ring geometry from one increment as the initial geometry for the subsequent increment. As long as the ratio of the interfacial shear stress, τ, to the material flow stress, σ_0, remains constant, strain hardening of the ring material during deformation has no effect, provided the increase in work hardening in any single deformation increment can be neglected.

The progressive increase in interfacial shear stress accompanying strain hardening is also immaterial, provided it can be assumed to be constant over the entire die-ring interface during any one deformation increment. Thus, the analysis can be justifiably applied to real materials even though it was initially assumed that the material would behave according to the von Mises stress-strain rate laws, provided the assumption of a constant interfacial shear factor, m, is correct. However, it has been shown that a highly strain-rate-sensitive material requires a different analysis (Ref 31).

Based on these assumptions, the plasticity equations have been solved for several different ring geometries over a complete range of m values from zero to unity (Ref 32), as shown in Fig. 36. By measuring the change in internal diameter of the ring, the friction factor can be determined.

The ring thickness is usually expressed in relation to the inside and outside diameters.

The maximum thickness that can be used while still satisfying the mathematical assumption of "thin specimen" conditions varies, depending on the actual friction conditions. Under conditions of maximum friction, the largest usable specimen height is obtained with rings of dimensions in the ratio of 6:3:1. Under conditions of low friction, thicker specimens can be used while still satisfying the above assumption. For normal lubricated conditions, a geometry of 6:3:2 can be used to obtain results of sufficient accuracy for most applications.

For experimental conditions in which specimen thicknesses are greater that those permitted by a geometry of 6:3:1 and/or the interface friction is relatively high, the resulting side barreling or bulging must be considered. Analytical treatment of this more complex situation is available (Ref 33).

The ring compression test can be used to measure the flow stress under high-strain practical forming conditions. The only instrumentation required is for the measure-

Fig. 36 Theoretical calibration curve for standard ring with an OD:ID:thickness ratio of 6:3:2

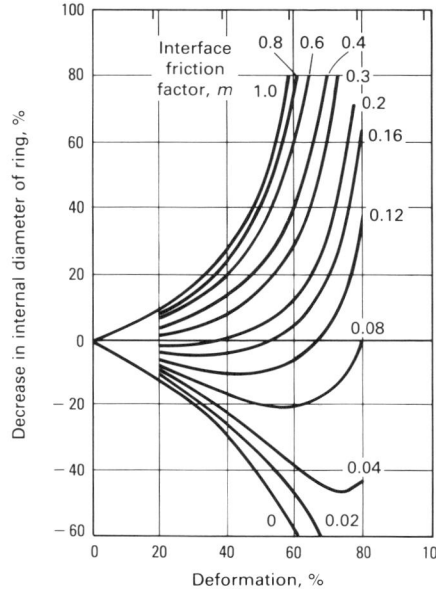

ment of the force needed to produce the reduction in height. Measurement of the change in diameter of a 6:3:1 ring is used to obtain a value of the ratio p/σ_0 by solving the analytical expression for the deformation of the ring or by using computer solutions for the ring (Ref 34). Measurement of the area of the ring surface formerly in contact with the die and knowledge of the deformation load facilitate calculation of p and hence the value of the material flow stress, σ_0, for a given amount of deformation. Repeating this process with other ring specimens over a range of deformation allows the generation of a complete flow stress-strain curve for a given material under particular temperature and strain rate deformation conditions.

Hot Tension Testing

Although necking in tension testing is a fundamental limitation, the tension test is nevertheless practically useful for establishing temperature limits for hot working. The major advantage of this test for industrial applications is that it clearly establishes maximum and minimum hot working temperatures (Ref 35).

Most commercial hot tensile testing is done with a Gleeble unit, which is a high strain rate, high-temperature testing machine (Ref 36). A solid buttonhead specimen that has a reduced diameter of 6.35 mm (0.250 in.) and an overall length of 88.9 mm (3.5 in.) is held horizontally by water-cooled copper jaws (grips), through which electric power is introduced to resistance heat the test specimen (Fig. 37). Specimen temperature is monitored by a thermocouple welded to the specimen surface at its midlength. The thermocouple, via a function generator, controls the heat fed into the specimen according to a programmed cycle. Thus, a specimen may be tested under time-temperature conditions that simulate hot working sequences.

The specimen is loaded by a pneumatic-hydraulic system. The load may be applied at any desired time in the thermal cycle. Temperature, load, and crosshead displacement are measured as a function of time. In the Gleeble test, the crosshead speed can be maintained constant throughout the test, but the true strain rate decreases until necking occurs, according to the relationship:

$$\dot{\epsilon} = \frac{d\epsilon}{dt} = \frac{1}{L}\frac{dL}{dt}$$

When the specimen necks, the strain rate increases suddenly in the deforming region, because deformation is concentrated in a narrow zone. Although this variable strain-rate history introduces some uncertainty into the determination of strength and ductility val-

ues, it does not negate the practical usefulness of the hot tension test.

The percent reduction in area is the primary result obtained from the hot tension test. This measure of ductility assesses the ability of the material to withstand crack propagation. Reduction in area adequately detects small ductility variations in materials caused by composition or processing when the material is of low to moderate ductility. It does not reveal small ductility variations in materials of very high ductility.

A general qualitative rating scale between reduction in area and workability is given in Table 3. This correlation was originally based on superalloys. In addition to ductility measurement, the ultimate tensile strength

can be determined with the Gleeble test. This gives a measure of the force required to deform the material.

Test Procedure Variations. Two variations of the hot tension test can be used to establish the temperature limits of hot working: "on heating" tests and "on cooling" tests. The "on heating" test method is used for a material for which little or no hot working information is available. Specimens are resistance heated to the test temperature, held for 1 to 10 min, and pulled to fracture at a crosshead rate approximating the strain rate of plant practice.

Reduction in area versus test temperature obtained by "on heating" testing of a heat-resistant alloy is shown in Fig. 38. The opti-

Fig. 37 Gleeble test unit
Showing specimen held between grips and thermocouple welded to center of specimen

Table 3   Qualitative hot workability ratings for specialty steels and superalloys

Hot tensile reduction in area(a), %	Expected alloy behavior under normal hot reductions in open-die forging or rolling	Remarks regarding alloy hot working practice
<30	Poor hot workability, abundant cracks	Preferably not rolled or open-die forged; extrusion may be feasible; rolling or forging should be attempted only with light reductions, low strain rates, and an insulating coating
30 to 40	Marginal hot workability, numerous cracks	This ductility range usually signals the minimum hot working temperature; rolled or forged with light reductions and lower-than-usual strain rates
40 to 50	Acceptable hot workability, few cracks	Rolled or press forged with moderate reductions and strain rates
50 to 60	Good hot workability, very few cracks	Rolled or press forged with normal reductions and strain rates
60 to 70	Excellent hot workability, occasional cracks	Rolled or press forged with heavier reductions and higher strain rates than normal if desired
>70	Superior hot workability, rare cracks, ductile ruptures can occur if strength is too low	Rolled or press forged with heavier reductions and higher strain rates than normal provided that alloy strength is sufficiently high to prevent ductile ruptures

(a) Ratings apply for Gleeble tension testing of 6.35-mm (0.250-in.) diam specimens with 25.4-mm (1-in.) head separation.
Source: Ref 35

Fig. 38 Reduction in area versus test temperature obtained by hot tension testing "on heating"

Specimens were heated to the test temperature, held 5 min, and pulled to fracture.

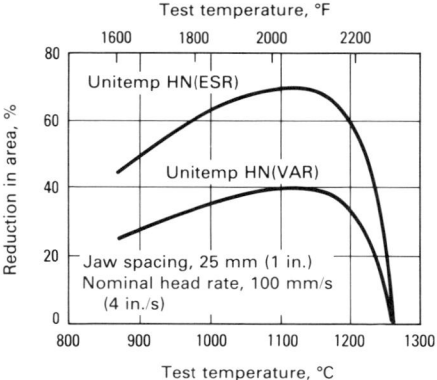

mum reheat temperature for working lies between the peak ductility temperature (PDT) and the zero ductility temperature (ZDT). The test clearly distinguishes between ingots prepared by electroslag (ESR) and vacuum arc (VAR) melting practices.

The "on cooling" test procedure is used to establish the optimum preheat temperature in this range. The objective is to determine which hot working temperature provides the highest ductility over the broadest temperature range without risking permanent damage to the material from overheating. Unmachined specimen blanks are heat treated in a furnace at a given preheat temperature and time to duplicate a furnace soak commensurate with the workpiece size and the hot working operation. Samples are water quenched from the soak temperature to retain the high-temperature structure.

After machining, tensile specimens are heated to the preheat temperature in the Gleeble unit and are held for 1 to 10 min to dissolve any phases that may have precipitated during cooling. Specimens are then cooled to a series of temperatures below the preheat temperatures at 28 to 55 °C (50 to 100 °F) intervals, held 5 s at the test temperature, and pulled to fracture at the appropriate head speed.

Data obtained from "on cooling" tests conducted on three test specimens that were subjected to varying preheat temperatures are shown in Fig. 39. A preheat temperature of 1205 °C (2200 °F) was selected as optimum in this example, because it produced a slightly higher ductility and a rather broad band of high ductility. The minimum hot working temperature was established as the temperature at which the reduction in area decreases to the 50% level for typical work-

piece reductions and 30 to 40% for smaller reductions.

"On cooling" hot tension testing is useful, because the short hold times for "on heating" tests may not develop a grain size representative of that temperature, or they may be insufficient to dissolve or precipitate a phase that will occur during an actual furnace soak prior to hot working. Also, most industrial hot working operations are performed while the workpiece temperature cools slowly. "On cooling" tests also indicate how closely the ZDT can be approached before hot ductility is severely reduced.

Workability Tests for Forging

Basically, all forging processes consist of compressive deformation of a metal workpiece between a pair of dies (Ref 37). The two broad categories of forging processes are open-die and closed-die modes. The simplest open-die forging operation is the upsetting of a cylindrical billet between two flat dies (Ref 38). The compression test is a small-scale prototype of this process. As the metal flows laterally between the advancing die surfaces, there is less deformation at the die interfaces (because of the friction forces) than at the midheight plane. Thus, the sides of the upset cylinder barrel. Generally, metal flows most easily toward the nearest free surface, because this represents the lowest frictional path.

Closed-die forging is done in closed or impression dies that impart a well-defined shape to the workpiece (Ref 37). The degree of lateral constraint varies with the shape of the dies and the design of the peripheral areas where flash is formed, as well as with the same factors that influence metal flow in open-die forging (amount of reduction, frictional boundary conditions, and heat transfer between the dies and the workpiece).

Because forging is a complex process, a single workability test cannot be relied on to determine forgeability. However, several testing techniques have been developed for predicting forgeability, depending on alloy type, microstructure, die geometry, and process variables. This section summarizes some of the common tests for determining workability in both open-die and closed-die forging.

Wedge-Forging Test. In this test, a wedge-shaped piece of metal is machined from a cast ingot or wrought billet and forged between flat, parallel dies (Fig. 40). The dimensions of the wedge must be selected so that a representative structure of the ingot is tested. Larger grained materials require

Fig. 39 Reduction in area versus testing temperature for Unitemp HN (ESR) generated by testing "on cooling"

Specimen blanks were furnace soaked 2 h at the preheat temperatures. Then, specimens were heated to the preheat temperatures in the Gleeble unit, held 5 min, cooled to the test temperature, held 5 s, and pulled to fracture.

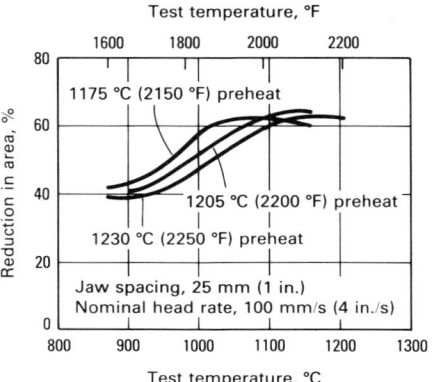

Fig. 40 Specimens for the wedge test

(a) As-machined specimen. (b) Specimen after forging

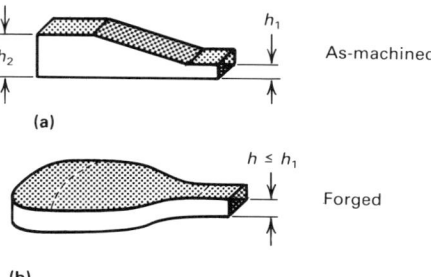

larger specimens than fine-grained materials. The wedge-forging test is a gradient test in which the degree of deformation varies from a large amount at the thick end (h_2) to a small amount, or no deformation, at the thin end (h_1). The specimen should be used on the actual forging equipment in which production will occur to allow for the effects of deformation velocity and die chill on workability.

Tests can be made at a series of preheat temperatures, beginning at about nine tenths of the solidus temperature or the incipient melting temperature. After testing at each temperature, the vertical deformation that causes cracking can be established. Also, by performing metallographic examination in the direction of the strain gradient, the extent of recrystallization as a function of strain and temperature can be determined.

The sidepressing test consists of compressing a cylindrical bar between flat, parallel dies where the axis of the cylinder is parallel to the surfaces of the dies. Because the cylinder is compressed on its side, this testing procedure is called sidepressing. This test is sensitive to both surface-related cracking and to general unsoundness of the bar, because high tensile stresses are created at the center of the cylinder (Fig. 41).

For a cylindrical bar deformed against flat dies, the tensile stress is greatest at the start of deformation and decreases as the bar assumes more of a rectangular cross section. As shown in Fig. 41, the degree of tensile stress can be reduced at the outset of the tests by changing from flat dies to curved dies that support the bar around part of its circumference.

The typical sidepressing test is conducted with unconstrained ends. In this case, failure occurs by ductile fracture on the expanding end faces. If the bar is constrained to deform in plane strain by preventing the ends from expanding, deformation will be in pure shear, and cracking will be less likely. Plane-strain conditions can be achieved if the ends are blocked from longitudinal expansion by machining a channel or cavity into the lower die block.

The notched-bar upset test is similar to the conventional upset test, except that axial notches are machined into the test specimens (Ref 8, 40). It is used with materials of marginable forgeability in which the standard upset test may indicate an erroneously high degree of workability. The introduction of notches produces high local stresses that induce fracture. The high levels of tensile stress in the test are believed to be more typical of those occurring in actual forging operations.

To prepare test specimens, a forging billet is quartered longitudinally, exposing center material along one corner of each test specimen (Fig. 42). Notches with either 1.0-mm (0.04-in.) or 0.25-mm (0.01-in.) radii are machined into the faces as shown. A weld button is frequently placed on one corner to identify the center and surface material of alloys that experience difficulties in forgeability due to segregation.

Specimens are heated to predetermined temperatures and upset about 75%. Because of the stress concentration effect, ruptures are most likely to occur in the notched areas. These ruptures may be classified by the rating system shown in Fig. 43. A rating of 0 indicates that no ruptures are observed, and higher numbers indicate an increasing frequency and depth of rupture.

Tests for Flow Localization Controlled Fracture. Complex forgings frequently develop regions of highly localized

Fig. 41 Effects of billet shape and degree of enclosure on stress state in forging with good lubrication and no chilling
Source: Ref 39

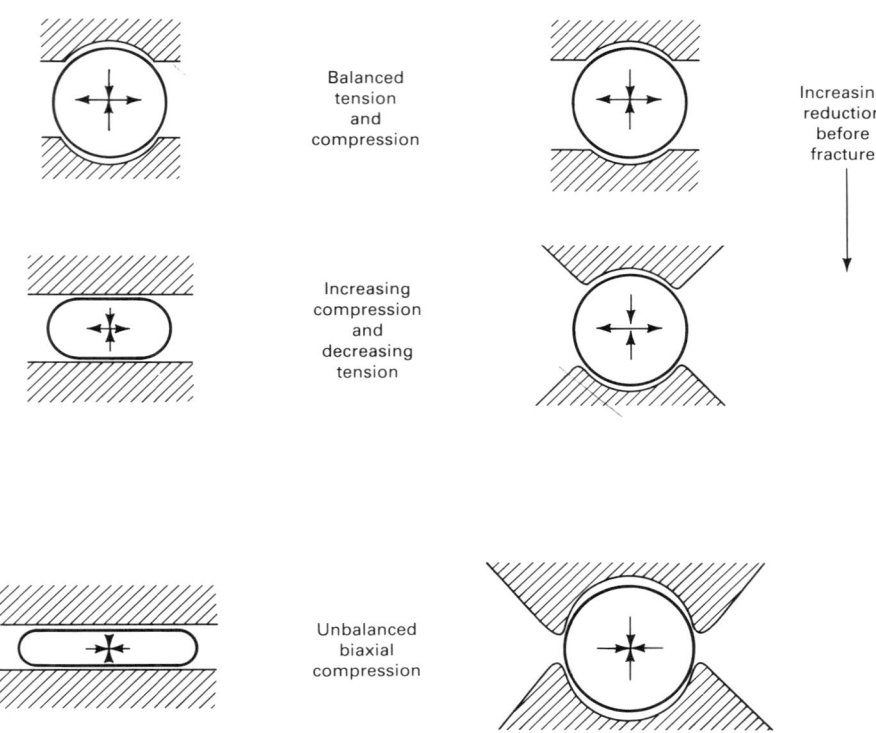

Fig. 42 Method of preparing specimens for notched bar upset forgeability test
Source: Ref 8, 40

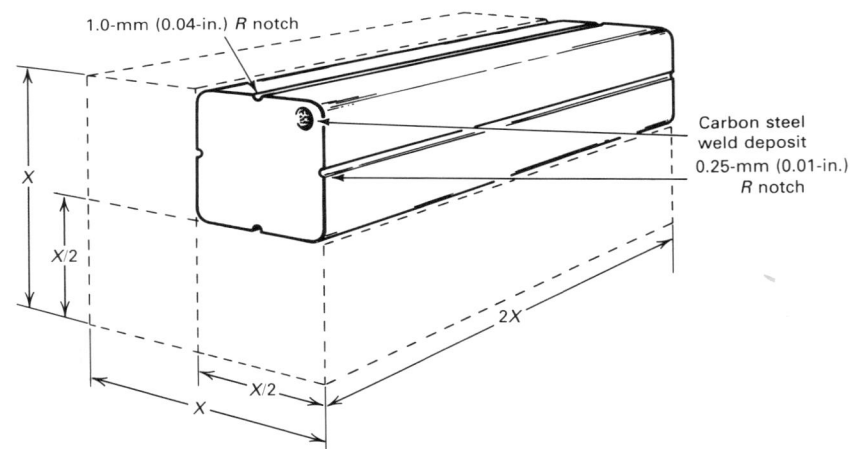

deformation. Shear bands may span the entire cross section of a forging and, in extreme cases, produce shear cracking. Flow localization can arise from constrained deformation due to die chill or high friction. However, it can also occur in the absence of these effects if the metal undergoes flow softening or negative strain hardening.

The simplest workability test for detecting the influence of heat transfer (die chilling) on flow localization is the nonisothermal upset test, in which the dies are much colder than the workpiece. Figure 44 illustrates zones of flow localization made visible by sectioning and metallographic preparation.

The sidepressing test conducted in a non-

Fig. 43 Suggested rating system for notched bar upset test specimens that exhibit progressively poorer forgeability

A rating of 0 would indicate freedom from ruptures in the notched area. Source: Ref 8, 40

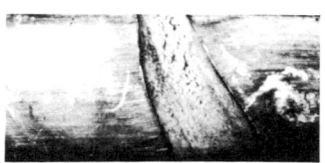

Rating 1

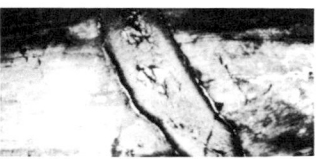

Rating 2

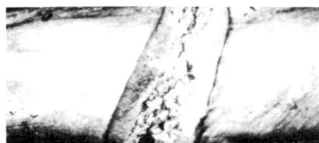

Rating 3

Rating 4

isothermal manner can also be used to detect flow localization. Several test specimens are sidepressed between flat dies at several workpiece temperatures, die temperatures, and working speeds. The formation of shear bands is determined by metallography (Fig. 45). Flow localization by shear band formation is more likely in the sidepressing test than in the upset test. This is due to the absence of a well-defined die chilling region and the fact that sidepressing is basically a plane-strain operation with surfaces of zero extension, along which shearing can initiate and propagate.

Testing to evaluate material susceptibility to localized deformation may also make use of a cylindrical upset specimen with a reduced gage section, as shown in Fig. 46 (Ref 37). The ability of the material to distribute deformation (Fig. 47) is measured by an empirical parameter, percent distributed gage volume (DGV). The larger the DGV percentage, the greater the penetration of the deformation into the heavy ends of the specimen and the greater the ability of the material to distribute deformation. Figure 48 shows the metallographic appearance of the condition with distributed flow (Fig. 48a)

Fig. 46 Shape and dimensions of cylindrical compression specimen with a reduced gage section

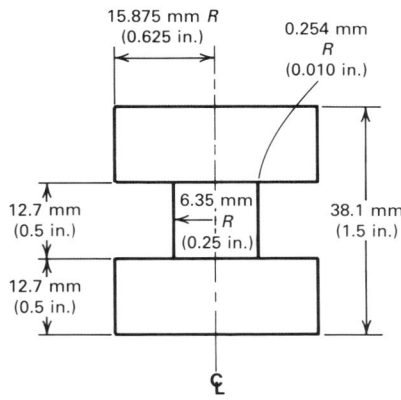

and concentrated deformation (Fig. 48b). Figure 49 illustrates that the DGV percentage is a sensitive parameter for detecting flow localization.

Workability Tests for Cold Forging. The types of cracks that develop in cold forging by upsetting-type processes are discussed

Fig. 45 Transverse metallographic sections of specimens of Ti-6Al-2Sn-4Zr-2Mo-0.1Si with an equiaxed alpha structure

Specimens were nonisothermally sidepressed with zero dwell time in a mechanical press ($\dot{\varepsilon} \approx 30 \text{ s}^{-1}$) between dies heated to 191 °C (376 °F). Specimen preheat temperatures (T_s) and percent reductions (R), relative to the initial specimen diameter, were as follows: (a) T_s, 913 °C (1675 °F); R, 14%. (b) T_s, 913 °C (1675 °F); R, 54%. (c) T_s, 913 °C (1675 °F); R, 77%. (d) T_s, 982 °C (1800 °F); R, 21%. (e) T_s, 982 °C (1800 °F); R, 57%. (f) T_s, 982 °C (1800 °F); R, 79%. Source: Ref 41

Fig. 44 Axial cross sections of specimens of Ti-6Al-2Sn-4Zr-2Mo-0.1Si with an equiaxed alpha starting microstructure

Specimens were nonisothermally upset at 954 °C (1749 °F) to 50% reduction in a mechanical press ($\dot{\varepsilon} \approx 30 \text{ s}^{-1}$) between dies at 191 °C (376 °F). Dwell times on the dies prior to deformation were (a) 0 s and (b) 5 s. Source: Ref 41

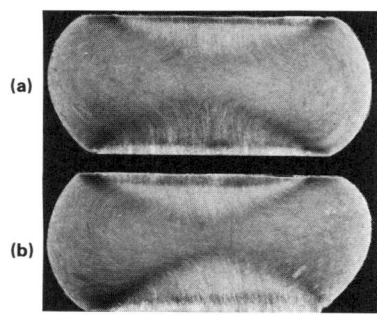

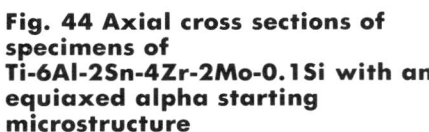

Fig. 47 Schematic of specimen cross sections showing the relative amount of gage volume penetration (DGV) into the specimen ends for two different deformation behaviors

(a) Distributed deformation. (b) Concentrated deformation. Source: Ref 6

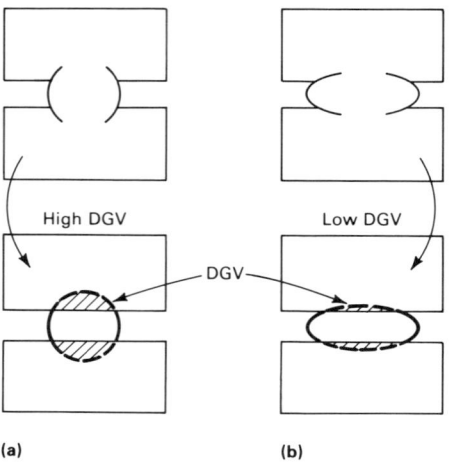

(a)   (b)

in Ref 42. The various geometric forms of cold forging are shown in Fig. 50, and crack classification is shown in Fig. 51. Table 4 provides a detailed description of cracking.

The upsetting of a right cylinder may not be a suitable workability test for material with high workability, because cracks are not produced when the material is upset with flat, smooth dies. The use of a tapered or flanged specimen (Fig. 14) may provide conditions that produce fracture. The use of a V-notch on the surface of the cylinder usually is ineffective in very ductile materials, because it disappears during upsetting due to swelling of the groove. The use of dies with a V protrusion that indents the specimen on its end surfaces has been proposed (Ref 13).

Workability in Closed-Die Forging. The defects discussed above may also occur in closed-die forging. However, other defects in addition to fracture and localized flow occur in closed-die forging. These defects often result from factors such as improper selection of the starting or preform shape of the workpiece, poor die design, improper selection of lubricant, temperature, or working speed.

The principal defects in closed-die forging are laps, flow-through defects, extrusion defects, and cold shuts. Laps are defects that form when metal folds back over itself during forging. For example, in finish forging of a webbed forging in which the preform web is too thin, the web may buckle and fold back onto itself. Also, in forging of a web, the metal may flow nonuniformly and cause a lap (Fig. 52). Frequently, a lap results from an excessively sharp radius in the forging die.

Flow-through defects are flaws that form when metal is forced to flow past a recess after the recess has filled or material in the recess has ceased to deform because of chilling (Fig. 53). Similar to laps in appearance, flow-through defects may be shallow, but are indicative of an undesirable grain-flow pattern or shear band that extends much deeper into the forging. Flow-through defects may also occur when trapped lubricant forces metal to flow past an impression.

Extrusion-type defects are formed when centrally located ribs formed by extrusion-type flow draw too much metal from the main body or web of the forging. A defect similar to a pipe cavity is thus formed (Fig. 54). Methods of minimizing the occurrence of these defects include increasing the thickness of the web or designing the forging with a small rib opposite the larger rib, as shown in Fig. 54.

Most of the defects summarized above occur in hot forging, which is most common for

Fig. 49 Variation in DGV percentage for pressed specimens of Ti-6Al-4V as a function of upset temperature

Closed circles indicate starting microstructures of globular alpha (α). Open circles indicate starting microstructure of acicular α. Source: Ref 6

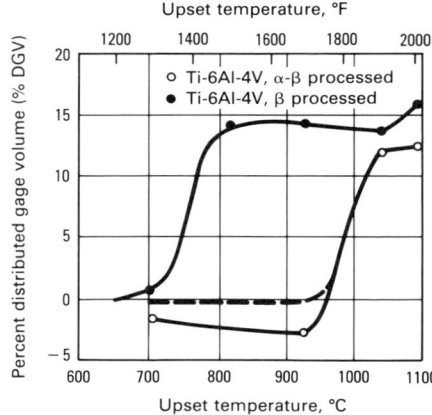

impression-die forging. Thus, defect formation may also involve entrapment of oxides and lubricant. When this occurs, the metal is incapable of "rewelding" under the high forging pressures; the term "cold shut" is frequently applied in conjunction with laps, flow-through defects, etc., to describe the flaws generated.

These defects all are related to the velocity field distribution of the deforming metal. They can be avoided by proper die design, preform design, and choice of lubrication system. Strictly speaking, these defects are not fundamental to the workability of the material. However, these common forging defects, and other defects for other metalworking processes discussed in the remainder of this article, are mentioned briefly to allow the reader to relate to workability on a practical level. These are the defects that

Fig. 48 Light micrographs showing variations in flow-line contours and gage penetration into the specimen ends

After press forging at (a) 650 °C (1200 °F), (b) 815 °C (1500 °F), and (c) 870 °C (1600 °F). Etched in oxalic acid

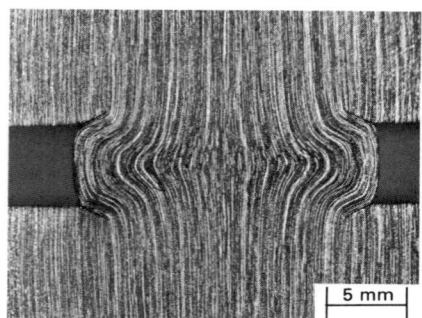

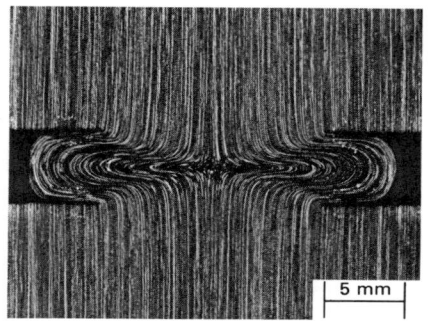

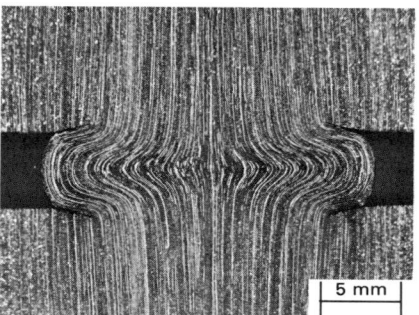

(a)   (b)   (c)

Fig. 50 Working methods in cold working
Hatching portions show the tool shapes. See Table 4 for a description of the working method numbers. Source: Ref 42

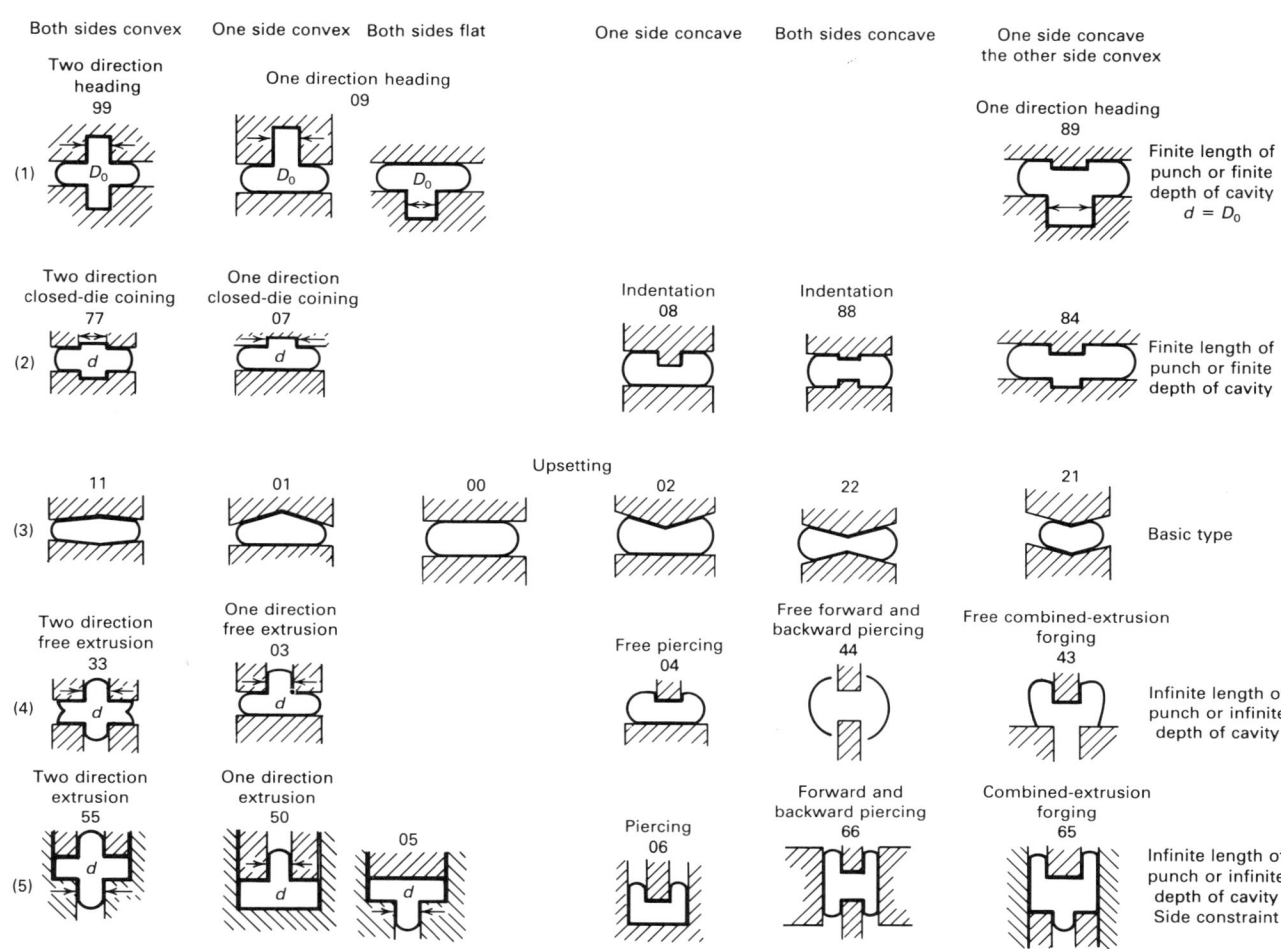

commonly limit deformation in the basic metalworking processes.

Workability in Extrusion and Drawing

Extrusion and drawing are similar in the manner in which the metal flows through the die. However, in extrusion the workpiece is pushed through the die, while in drawing it is pulled through the die. This has a great influence on workability, because the high degree of hydrostatic compression developed in extrusion leads to enhanced workability. Conversely, the tensile conditions in wiredrawing can impair workability.

Another significant difference between the two processes is that much extrusion is carried out hot, while drawing usually is performed cold. In addition, extrusion generally is a one-pass process, while the typical wiredrawing process requires multiple passes.

Therefore, in wiredrawing, multipass workability is most important (Ref 45).

The limits of workability in extrusion and drawing may be associated with the development of localized surface cracks, centerbursts, or gross fracture. Although surface cracks and centerbursts are subcritical flaws that may not necessarily produce fracture, they are serious defects in that they may go unnoticed, thus creating inferior mechanical properties in shipped material or leading to gross fracture in subsequent processing or service conditions. Defects can also be produced in high-quality material by improper processing conditions.

Surface cracking occurs in the circumferential direction (fir tree cracking), at an angle roughly 45° to the circumferential direction (crows feet cracking), or in the longitudinal direction (splitting), as shown in Fig. 55. Surface cracking, whether circumferential or 45°, results from poor lubrication,

which causes sticking and slipping that create intense shearing and tearing of local regions. In hot extrusion, surface cracking is related to surface heating. Surface frictional heating, intensified by high ram speed, may increase the surface temperature into the region of hot shortness, and surface fracture ensues.

When a hot billet is extruded from a colder container, chilling contributes to the inhomogeneity of deformation. Deformation is very complex, and toward the later stages of extrusion a dead-metal zone spreads throughout the entire length of the billet. Thus, material near the punch face moves toward the center and carries surface oxide films with it. This leads to the development of the extrusion defect, a ring of oxide inclusions in the form of a cone of oxide (pipe) protruding into the center of the extrusion.

In cold extrusion, the surface finish is strongly influenced by the lubrication conditions (Fig. 56). Pickup from the die or die

Fig. 51 Classification of cracks

Greek letters indicate cracking types. See also Table 4. Source: Ref 42

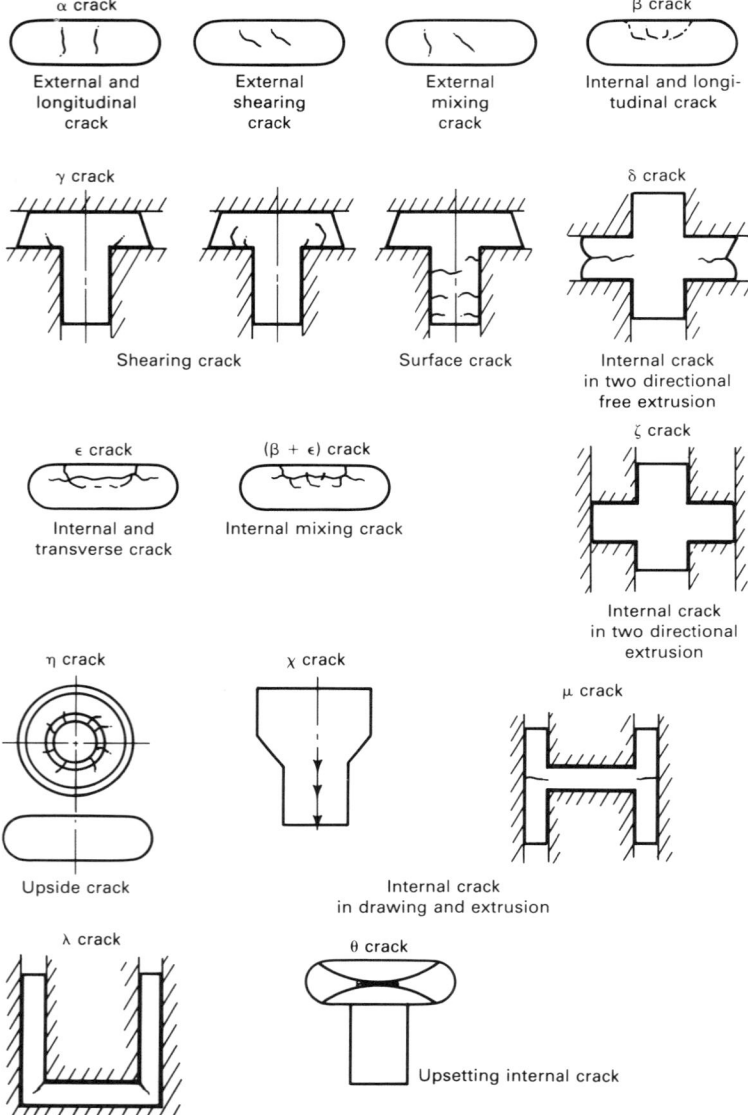

Fig. 52 Lap formation in the rib of a rib-web part due to improper preform geometry

Source: Ref 43

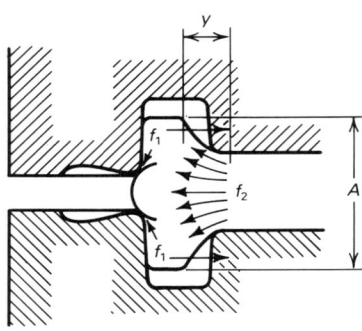

Formation of flash

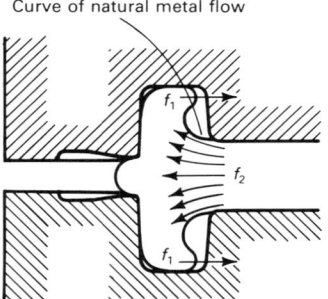

Reverse flow forming a fold

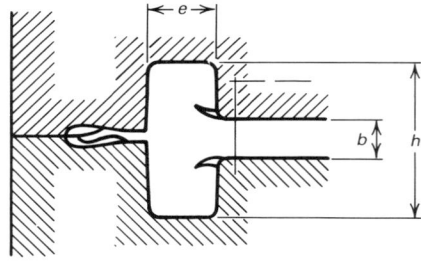

Formed forging defect

land leads to deep scoring of the extrusion surface (Fig. 56a). A thick lubricant film results in surface roughness (Fig. 56b), while an excessively thick, unstable film leads to a bamboo surface (Fig. 56c). In less ductile materials, this can result in shear cracks (Fig. 56d). If the billet has been turned in preparation for extrusion, the machining marks are stabilized by the lubricant (Fig. 56e).

Centerbursts are chevron-like shear fractures that form at the center of extruded or drawn products (Fig. 57). In the early stages of development, they may simply be pores along the centerline. Centerbursts arise from nonuniform flow through the die and

the associated development of tensile stresses at the center of the workpiece. This can be rationalized via the deformation zone parameter, Δ:

$$\Delta = \frac{\alpha}{R}\left(1 + \sqrt{1 - R}\right)^2$$

where α is the semi-die angle, and R is the reduction per pass, which is equal to $(A_0 - A_1)/A_0 = 1 - (D_1/D_0)^2$. Values of Δ less than 1 are associated with uniform metal flow. Centerbursting is associated with nonuniform flow, usually when Δ is 2.5 or more.

Chevron cracking may be slow to develop

and may require considerable plastic deformation before becoming visible. For example, light extrusion passes ($R < 0.2$) frequently do not result in centerburst, despite high Δ values (low R produces high Δ). In multipass wiredrawing, centerburst may develop gradually and eventually lead to failure. These failures are called "cuppy-core" fractures.

Workability in Drawing. Good estimates of workability in drawing can be made with a laboratory-scale drawbench or a tensile testing machine outfitted wth a die holder. The following testing procedure is recommended (Ref 45).

Table 4 Characteristics of cracks and crack growth mechanism

Cracking type	No. of working method in the chart (see Fig. 50)	Characteristics of cracking and estimation of crack growth mechanism
α	00, 01, 11, 02, 21, 22, 08, 87, 07, 77, 09, 03, 33, 04, 44, 43	External cracking that appears at midheight of side surface of the specimen in the upsetting; two types of cracks, longitudinal and oblique, occur according to the degree of end constraint of specimen; they are all shear cracks
β	08, 88, 87, 89, 04, 44, 43	Longitudinal cracking that occurs at the bottom of the concave part of the specimen (in upsetting by a circular truncated cone punch); it is a shear crack based on the section of the specimen and is caused by the expansion of material under the cone punch due to circumferential flow; prevention requires selection of a suitable punch shape
γ	03, 33, 05, 55, 07, 77, 87, 43, 06, 66, 65	Shear cracking that appears at the corner of extruded material in free extrusion (coining); cracking occurs at the boundary between dead-metal and plastic zones; surface cracking is caused by cracks that occur at the corners; prevention requires selection of a suitable diameter for the die
δ	33	Cracking that occurs at midheight of material in flange in two direction free extrusion without side constraint; the material is extruded forward and backward; therefore, cracking is caused by the depression of material in flange; prevention requires selection of a suitable diameter die
ε	08, 87, 88, 89	Cracking that occurs at midheight on inside surface of the concave part and is advanced in a circumferential direction in upsetting by the circular truncated cone punch; crack starts at the point where the material around the concave portion bends toward the inside; prevention requires selection of a suitable punch shape
α + β	08, 87, 88, 89	Cracking in which α and β cracks coexist
ζ	55	Cracking that occurs at the center of material with the sides constrained and forward and backward extrusion; the cavity occurs at the center of material, as material is extruded forward and backward
η	08, 87, 88, 89	Cracking that appears to advance from upper and side surface of concave portion to the top of specimen (in upsetting by a circular truncated cone punch); in cross section, the cracks are distributed radially; in the case of a large tapered punch, cracks are caused by the expansion of upper part of specimen by the punch; prevention requires selection of a suitable punch shape
κ	05, 55	Cracking that occurs at the center of specimen when excessive reduction is imposed on the specimen in extrusion and drawing
μ	66	Peripheral cracking that occurs at midheight on the side surface of the specimen in forward and backward piercing and 45° shearing crack with respect to longitudinal axis
λ	06, 65, 66	Cracking that occurs at the bottom of the concave portion in piercing
θ	07, 99	Microscopic cracks that occur at the boundary between top and bottom dead metal and at the point of inflection of the metal flow in case of excessive upsetting of a bolt head in bolt forging; in practical use, the splitting off of the head is caused by these microscopic cracks

Source: Ref 42

A series of specimens should be drawn once each through dies with different die angles, using the same reduction for each specimen. The value of Δ increases with die angle, and the die angle at which centerbursting is first noticed can be used as the index of centerburst resistance for the material. A 10% reduction with α values of 4 to 16° will provide a range of Δ values from 2.7 to 10.6.

For materials with too much ductility for failure to occur in a single pass, a sequence of dies should be set up to allow successive reductions at relatively high values of Δ (Δ ≃ 5). The reductions and die angles can be the same for each pass. The total reduction at which centerbursting is first observed is used as the index of centerburst resistance for the material. A series of 15% reductions at α = 12° (Δ = 5.2) is recommended.

As an alternative to waiting for gross fracture by centerbursting to occur, the ductility damage at the centerline can be evaluated by metallography or precision density measure-ments. Pores on the order of 1-μm diam can be observed easily.

Some measure of resistance to surface cracking can be obtained by drawing under conditions of no lubrication and by examining the surface with a scanning electron microscope. Three or four passes of 20% each may be necessary to establish a consistent pattern of surface damage.

Workability in Rolling

Although rolling of flat products appears to be a simple, well-defined process, the actual deformation that a workpiece experiences in rolling can be rather complex (Ref 46). The primary variable controlling deformation is the deformation zone parameter, Δ, which for rolling is defined as:

$$\Delta = \frac{h}{L} \approx \sqrt{\frac{h_0}{2DR}}\,(2 - R)$$

where h is the average workpiece height; L is the projected length of contact with the rolls; D is roll diameter; and R is reduction, which is equal to $(h_0 - h_f)/h_0$.

At large h/L values (in excess of 3), rigid metal zones develop in the deforming metal under the rolls, and tensile stresses are generated in the center plane of the slab. This could lead to the type of centerburst defects found in extrusion and wiredrawing, although in practice centerburst defects are not common. With decreasing h/L ratios, the stress state becomes fully compressive at the center of the workpiece, but longitudinal tensile stresses are developed on the surface of the workpiece. At $h/L < 0.3$, deformation is much more homogeneous, and the strain differential between the surface and the center layer is smaller.

Inhomogeneous Deformation and Rolling Defects. The ends and sides of the workpiece will deform in response to the inhomogeneity in strain in the through-thickness direction. With a high h/L ratio, the side will form a double-barreled (fish tail) shape (Fig. 58a). However, at a lower h/L ratio, the edge will form a single bulge (Fig. 58b). The material in the bulge is no longer directly compressed and is forced to elongate by the laterally and longitudinally adjacent material.

As a first approximation, the strain state in the middle of a barreled edge is close to uniaxial tension, and the secondary tensile stress generated is close to the fracture stress. Such a stress state induces cracking in a material of limited ductility at much lighter reductions than in a square-edge profile. A double-barreled edge can lead to fracture even in highly ductile materials such as pure aluminum.

A common method of producing a double-barreled edge is to roll materials with light

Fig. 53 Flow-through defect in Ti-6Al-4V rib-web structural part
Source: Ref 44

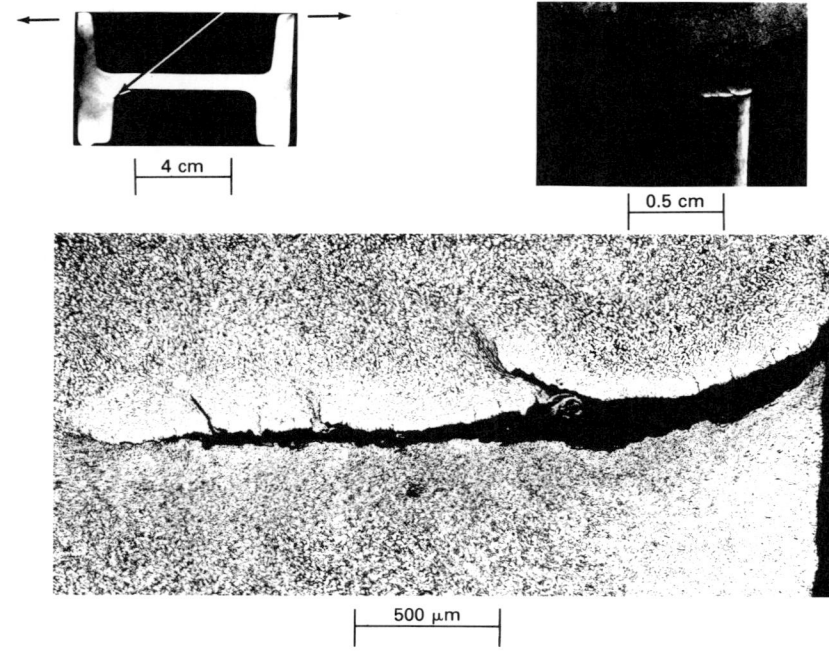

Fig. 54 (a) Extrusion-type defect in centrally located rib and (b) die-design modification used to avoid defect
Source: Ref 8

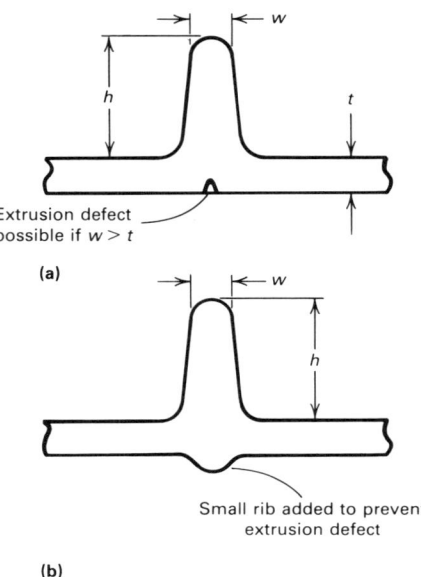

Extrusion defect possible if $w > t$

(a)

Small rib added to prevent extrusion defect

(b)

Table 5   Workability limits in cold rolling

Edge	1.3% C steel (2.5 mm, or 0.1 in.)	1.5% W steel (2 mm, or 0.08 in.)	Al-7Mg (4.75 mm, or 0.19 in.)	Al-7Mg (9.5 mm, or 0.38 in.)
Square	59 to 62	64	82 to 88	86 to 90 (69)
Fully rounded	18 to 21	18	65 (51)	63 (47)
Chamfered:				
30°	...	...	35	34
60°	...	...	36	34
90°	...	...	42	43
120°	...	...	50	53
150°	...	...	53	62
160°	...	...	62	63
170°	...	...	74	73
180°	...	...	88	86

(a) Roll diameter, 250 mm (10 in.). Numbers in parentheses refer to rolling with light (2 to 5%) reduction per pass.
Source: Ref 46

Fig. 55 Schematic of three types of surface cracks that occur in extruded and drawn products
Source: Ref 45

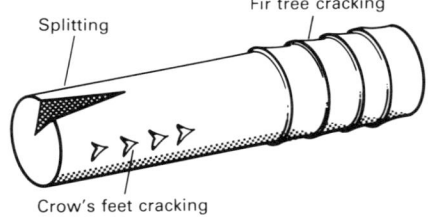

Splitting

Fir tree cracking

Crow's feet cracking

reductions and thus relatively large h/L ratios. When the h/L ratio is below unity, the side surfaces develop a single-barreled edge (Fig. 59). Once double barreling develops in the early stages of rolling, it usually persists even when the h/L ratio diminishes as rolling proceeds. Frequently, the side surfaces close down to form a lamination that may be quite deep on both sides of the slab. Table 5 shows the correlations between the limit in cold rolling reduction at which edge cracking occurs and the shape (profile) of the edge.

Edge cracking can be minimized by either improving the workability of the material or by changing the process. The greatest im-

provements in workability have been achieved by advances in solidification processing that minimize segregation and large columnar grain structures. Continuous casting, which results in thinner slabs, allows rolling with more favorable h/L ratios. One approach to minimize double barreling in thick slabs is to cast the slabs with V-shaped edges so that double barreling superimposed on the starting shape will result in a straight edge. The worst conditions for edge cracking is an as-cast structure with concave edges.

If barreling can be removed during processing, edge cracking is greatly reduced. In cold rolling, the edges can be machined. In

hot rolling, a much more practical approach is to use a universal mill with vertical-axis rolls on both sides of the main horizontal-axis work rolls. The edging rolls not only improve workability by limiting nonhomogeneous edge deformation, but they also ensure better control of the width of the rolled strip.

Spread in the roll gap can be prevented with a special edge-restraint device (Ref 47). Because spread is not allowed, secondary tensile stresses cannot develop, and materials of poor workability can be successfully rolled. Another method of handling difficult-to-roll materials is to contain them in a heavy frame of high-strength, high-ductility metal (a picture frame). A somewhat similar effect can be produced by rolling a less ductile

Fig. 56 Typical defects observed in cold extrusion of aluminum alloys
See text for descriptions of defects.

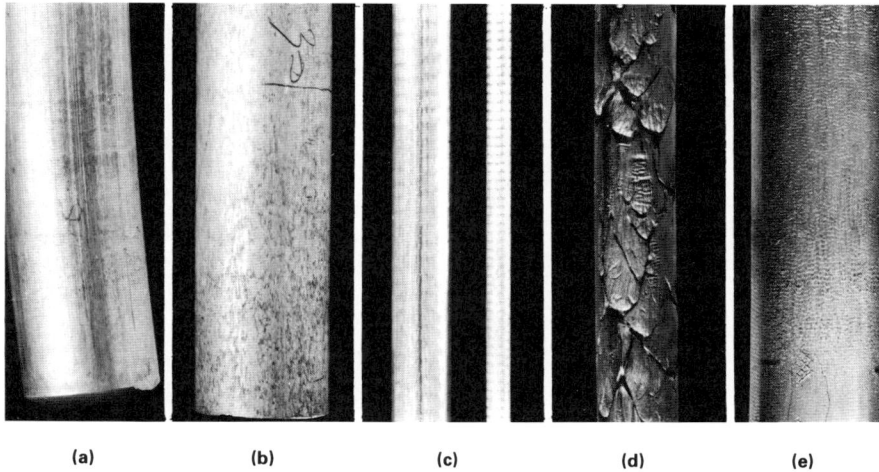

(a)   (b)   (c)   (d)   (e)

sheet between more ductile cladding sheets that are sufficiently wide to close around the center sheet.

Several other defects commonly occur in rolling. Differential elongation between the center and edge of the workpiece due to mismatch between the strip cross section and the roll profile can result in nose splitting (Fig. 60). Another defect, which is related to inhomogeneity in plane-strain deformation, is alligatoring. With this type of defect, the slab fractures down the center plane. Although the causes of this defect are uncertain, alligatoring tends to occur most frequently at low h/L ratios, when deformation is most severe in the billet and when the material contains structural weakness at the center of the billet.

Surface cracking is encountered in some materials with very narrow hot working ranges. Chilling can bring the surface layers into a low-ductility region and can produce cracks parallel to the roll axis.

Currently, the onset of edge cracking, alligatoring, or surface cracking cannot be predicted theoretically. However, correlations have been made between the fracture strain in tension and the thickness strain at which edge cracks occur in cold rolling (Fig. 61).

Simulation experiments on a laboratory rolling mill are reliable for cold rolling, if the

ratios of roll diameter to slab thickness and slab width to slab thickness are scaled proportionately. However, simulation is much more difficult in hot rolling, because smaller laboratory workpieces cool much more rapidly. Simulation should be conducted at strain rates typical to production rolling to ensure that similar metallurgical responses will occur. Because more severe cooling is inevitable, barreling is more pronounced and edge conditions are more severe than in production.

A wedge-shaped workpiece that can be rolled to produce reductions from 20 to 80% front to back is a convenient method of determining workability in rolling. The critical reduction at which edge cracking or surface cracking will occur can be approximated. A disadvantage of this technique is the variation in h/L throughout the specimen, which results in a continuously varying deformation pattern. Moreover, each section being rolled is also affected by the strain histories of the adjacent portions. When this history is continually variable, the strain history of the critical section is also uncertain.

REFERENCES

1. T. Gladman, B. Holmes, and L.D. McIvor, *Effect of Second-Phase Particles on the Mechanical Properties of Steel*, Iron and Steel Institute, London, 1971, p 78
2. S.L. Semiatin, unpublished research,

Fig. 57 Centerbursts in sectioned cold extruded steel bars
Courtesy of Bethlehem Steel Corp.

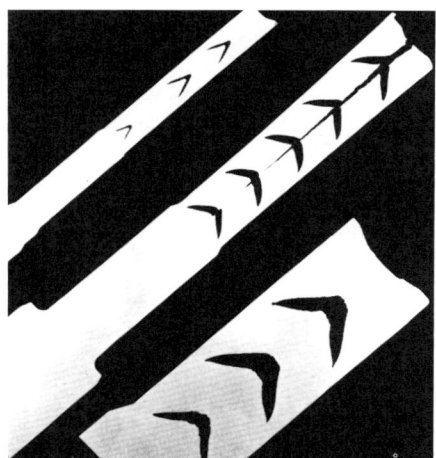

Fig. 58 Schematic of end and side deformations in rolling with (a) high and (b) low h/L ratios
Source: Ref 46

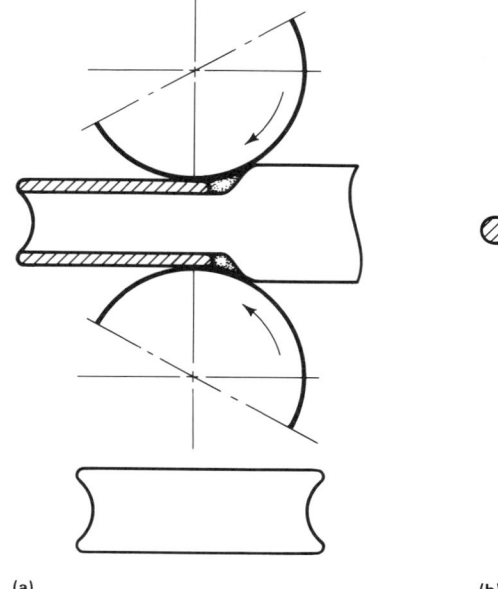

(a)

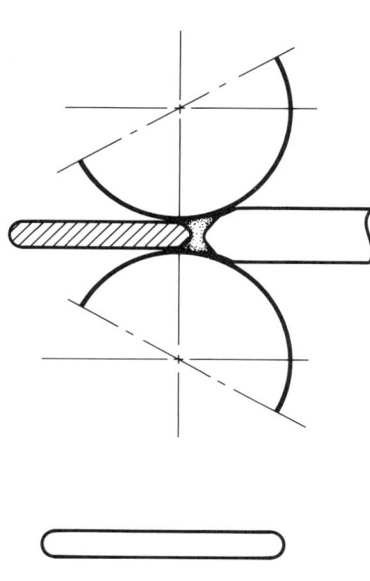

(b)

Fig. 59 Edge cracking in cast Al-8Mg alloy workpiece after single-edge barreling
Source: Ref 46

Fig. 60 Nose-splitting in cast Al-8Mg alloy
Source: Ref 46

Fig. 61 Correlation of maximum cold rolling reduction with reduction in area in tension testing
Source: Ref 9

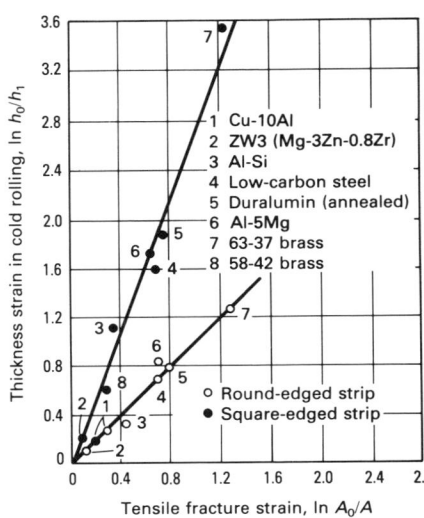

Legend:
1 Cu-10Al
2 ZW3 (Mg-3Zn-0.8Zr)
3 Al-Si
4 Low-carbon steel
5 Duralumin (annealed)
6 Al-5Mg
7 63-37 brass
8 58-42 brass

○ Round-edged strip
● Square-edged strip

Y-axis: Thickness strain in cold rolling, ln h_0/h_1

X-axis: Tensile fracture strain, ln A_0/A

Battelle Columbus Laboratories, Columbus, OH, March 1979

3. H. Gegel, S. Nadiv, and R. Raj, Dynamic Effects on Flow and Fracture During Isothermal Forging of a Titanium Alloy, *Scripta Met.*, Vol 14, 1980, p 241

4. R. Raj, Development of a Processing Map for Use in Warm-Forming and Hot-Forming Processes, *Met. Trans. A*, Vol 12, 1981, p 1089-1097

5. S.L. Semiatin and G.D. Lahoti, The Occurrence of Shear Bands in Isothermal Hot Forging, *Met. Trans. A*, Vol 13, 1982, p 275-288

6. M.C. Mataya and G. Krauss, A Test to Evaluate Flow Localization During Forging, *J. Appl. Metalworking*, Vol 2, 1981, p 28-37

7. F.N. Rhines and P.J. Wray, Investigation of the Intermediate Temperature Ductility Minimum in Metals, *Trans. ASM*, Vol 54, 1961, p 117

8. A.M. Sabroff, F.W. Boulger, and H.J. Henning, *Forging Materials and Practices*, Rheinhold, New York, 1968

9. M.G. Cockcroft and D.J. Latham, Ductility and the Workability of Metals, *J. Inst. Metals*, Vol 96, 1968, p 33-39

10. A.L. Hoffmanner, U.S. Air Force Materials Lab. Tech. Report, AFML-TR-69-174, June 1969

11. P.W. Lee and H.A. Kuhn, Cold Upset Testing, in *Workability Testing Techniques*, G.E. Dieter, Ed., American Society for Metals, 1984

12. E. Erman and H.A. Kuhn, Novel Test Specimens for Workability Measurement, in *Compression Testing of Homogeneous Materials and Composites*, STP 808, ASTM, Philadelphia, 1983, p 279-290

13. R. Sowerby *et al.*, Materials Testing for Cold Forging, *Trans. ASME J. Eng. Mat. Technol.*, Vol 106, 1984, p 101-106

14. N. Chandrasekaran *et al.*, The Assessment of the Free Surface Strains and Stresses During the Upsetting of Circular Cylinders, *Scripta Met.*, Vol 16, 1982, p 697-701

15. P.W. Lee and H.A. Kuhn, Fracture in Cold Upset Forging—A Criterion and Model, *Met. Trans. A*, Vol 4, 1973, p 969-974

16. H.A. Kuhn, P.W. Lee and T. Erturk, A Fracture Criterion for Cold Forging, *Trans. ASME J. Eng. Mat. Technol.*, Vol 95, 1973, p 213-218

17. T. Erturk, W.L. Otto, and H.A. Kuhn, Anisotropy of Ductile Fracture in Hot-Rolled Steel Plates—An Application of the Upset Test, *Met. Trans. A*, Vol 5, 1974, p 1883-1886

18. J.A. Schey, T.R. Venner, and S.L. Takomana, Shape Changes in the Upsetting of Slender Cylinders, *Trans. ASME J. Eng. Ind.*, Vol 104, 1982, p 79

19. G. Fitzsimmons, H.A. Kuhn, and R. Venkateshwar, Deformation and Fracture Testing for Hot Working Processes, *J. Met.*, May 1981, p 11-17

20. J.F. Alder and V.A. Phillips, The Effect of Strain Rate and Temperature on the Resistance of Aluminum, Copper and Steel to Compression, *J. Inst. Met.*, Vol 83, 1954-55, p 80-86

21. F.J. Gurney and D.J. Abson, Heated Dies for Forging and Friction Studies on a Modified Hydraulic Forge Press, *Met. Mat.*, Vol 7, 1973, p 535

22. A.B. Watts and H. Ford, On the Basic Yield Stress Curve for a Metal, *Proc. Inst. Mech. Eng.*, Vol 169, 1955, p 1141-1149

23. H.A. Kuhn, unpublished research, University of Pittsburgh, Oct 1984

24. J.J. Jonas, R.A. Holt, and C.E. Coleman, Plastic Stability in Tension and Compression, *Acta Met.*, Vol 24, 1976, p 911

25. J.A. Bailey, The Plane Strain Forging of Aluminum at Low Strain Rates and Elevated Temperatures, *Int. J. Mech. Sci.*, Vol 11, 1969, p 491

26. O. Pawelski, U. Rüdiger, and R. Kaspar, The Hot Deformation Simulator, *Stahl und Eisen*, Vol 98, 1978, p 181-189

27. S.M. Woodall and J.A. Schey, Development of New Workability Test Techniques, *J. Mech. Work. Technol.*, Vol 2, 1979, p 367-384

28. S.M. Woodall and J.A. Schey, Determination of Ductility for Bulk Deformation, in *Formability Topics—Metallic Materials*, STP 647, ASTM, Philadelphia, 1978, p 191-205

29. B. Avitzur, *Metal Forming: Processes and Analysis*, McGraw-Hill, New York, 1968

30. B. Avitzur and C.J. Van Tyne, Ring Forming: An Upper Bound Approach, *Trans. ASME J. Eng. Ind.*, Vol 104, 1982, p 231-252

31. G. Garmong, N.E. Paton, J.C. Chesnutt, and L.F. Nevarez, An Evaluation of the Ring Test for Strain-Rate Sensitive Materials, *Met. Trans. A*, Vol 8, 1977, p 2026, 2027

32. A.T. Male and V. DePierre, The Validity of Mathematical Solutions for Determining Friction from the Ring Compression Test, *Trans. ASME J. Lub. Technol.*, Vol 92, 1970, p 389-397

33. V. DePierre, F.J. Gurney, and A.T. Male, "Mathematical Calibration of the Ring Test with Bulge Formation," U.S. Air Force Materials Lab Tech. Report No. AFML-TR-37, March 1972

34. G. Saul, A.T. Male, and V. DePierre, "A New Method for the Determination of Material Flow Stress Values under Metalworking Conditions," U.S. Air Force Materials Lab Tech. Report No. AFML-TR-70-19, Jan 1970

35. R.E. Bailey, R.R. Shiring, and H.L. Black, Hot Tension Testing, in *Workability Testing Techniques*, G.E. Dieter, Ed., American Society for Metals, 1984

36. E.F. Nippes, W.F. Savage, B.J. Bastian, H.F. Mason, and R.M. Curran, An Investigation of the Hot Ductility of High Temperature Alloys, *Welding J.*, Vol 34, April 1955, p 183-196s

37. S.L. Semiatin, Workability in Forging, in *Workability Testing Techniques*, G.E. Dieter, Ed., American Society for Metals, 1984

38. G.E. Dieter, *Mechanical Metallurgy*, 2nd ed., McGraw-Hill, New York, 1976, p 585

39. A.L. Hoffmanner, "Plasticity Theory as Applied to Forging of Titanium Alloys," Symposium on the Thermal-Mechanical Treatment of Metals, London, May 1970

40. R.P. Daykin, unpublished research, Ladish Co., Cudahy, WI, 1951

41. S.L. Semiatin and G.D. Lahoti, The Occurrence of Shear Bands in Nonisothermal, Hot Forging of Ti-6Al-2Sn-4Zr-2Mo-0.1Si, *Met. Trans. A*, Vol 14, 1983, p 105

42. T. Okamoto, T. Fukuda, and H. Hagita, Material Fracture in Cold Forging—Systematic Classification of Working Methods and Types of Cracking in Cold Forging, in *Source Book on Cold Forming*, American Society for Metals, 1975, p 216-226

43. A. Chamrouard, *Closed Die Forging*, Part I, Dunod, Paris, 1964 (in French)

44. F.N. Lake and D.J. Moracz, "Comparison of Major Forging Systems," Tech. Report AFML-TR-71-112, TRW, Inc., Cleveland, May 1971

45. R.N. Wright, Workability in Extrusion and Wire Drawing, in *Workability Testing Techniques*, G.E. Dieter, Ed., American Society for Metals, 1984

46. J.A. Schey, Workability in Rolling, in *Workability Testing Techniques*, G.E. Dieter, Ed., American Society for Metals, 1984

47. J.A. Schey, Prevention of Edge Cracking in Rolling by Means of Edge Restraint, *J. Inst. Met.*, Vol 94, 1966, p 193-200

Wear Testing

Wear Testing

By Raymond G. Bayer
Senior Engineer—Manager
IBM Corporation

WEAR, generally defined as a progressive loss or displacement of material from a surface as a result of relative motion between that surface and another, has long been a subject of practical interest, yet until recently has not received a great deal of theoretical attention. The belief that it is easier to replace a part rather than to design a part with adequate life may have been true at one time. Today, however, this can be a costly practice.

Wear testing traditionally has been used by materials engineers and scientists to rank wear resistance of materials for the purpose of optimizing material selection or development for a given application. Standardization, repeatability, convenience, short testing time, and simple measuring and ranking techniques are desirable in these tests. Currently, more demanding and complex methods of wear testing are being used by mechanical and reliability engineers to determine wear parameters that can project performance and to establish the influence of various factors on these parameters.

Wear is closely related to friction and lubrication; the study of these three subjects is known as tribology (Ref 1-7). Although the apparatus used for one tribological test frequently can be used for another, friction, wear, and lubrication are distinct phenomena. Consequently, test procedures and interpretations vary (Ref 8-18). For example, a lubricant test evaluates the ability of a lubricant to withstand temperature, speed, or load and still provide protection against wear; the degree or amount of wear is a measure of lubricant response. In a wear test, which can be conducted lubricated or dry, the area of interest is the wear response of the material. A friction test measures friction force or the coefficient of friction. In addition, friction tests can be used to evaluate lubricant performance, and wear tests often monitor friction. This article centers on wear testing, with other tribological factors discussed only in reference to their effects on such testing.

A single general-purpose wear test that establishes a unique wear parameter or rating of a material does not exist. Consequently, a general discussion of wear testing must encompass overall methodology. This article presents a brief review of testing equipment and an overview of significant wear mechanisms and phenomena. Elements of wear testing are described, including measurement and reporting of data. Finally, two case histories are presented.

Testing Equipment

In the 1960's the American Society for Lubrication Engineers (ASLE) listed more than 200 types of wear tests and equipment in use (Ref 19), and the list has since grown. This wide variety is the result of a desire to ensure appropriate controls and test convenience and a need to simulate the wear conditions of the intended application.

The equipment described in Ref 19 includes apparatus specially designed for laboratory use. Although many of these test configurations are one-of-a-kind machines, others are available as commercial units. Typical configurations are illustrated in Fig. 1. However, wear testing is not limited to such equipment. Tests often are performed with replicas or facsimiles of actual devices. This automated equipment provides more control and instrumentation and offers a high degree of simulation to actual wear conditions. Tests with such apparatus generally are complex and time consuming, but simplify extrapolation and prediction.

Wear Mechanisms and Phenomena

Wear is not a basic material property, but a system response of the material. Any material can wear by a variety of mechanisms—e.g., adhesion, abrasion, fatigue erosion, and oxidation—influenced by such factors as ambient temperature, loading conditions, and counterface conditions.

One of the most important aspects of wear testing is simulation of actual wear conditions. This requires an understanding not only of the intended application, but also of the various mechanisms that result in wear. Several wear classifications have been developed. The two most common are based on (1) physical mechanisms of material removal or displacement and (2) operational mechanisms such as mechanical action (rolling wear, sliding wear, etc.) and materials interaction (metal versus metal, metal versus polymer, etc.). Both of these classifications are considered here, because both illustrate the complexity of wear and the need for accurate simulation in wear testing.

Physical Mechanisms

There are many physical wear mechanisms; this discussion will be limited to four basic types—adhesion, abrasion, fatigue, and oxidation (Ref 21, 22). These representative mechanisms serve to illustrate two points: (1) there is more than one distinct mechanism for wear, and (2) sensitivity to parameters such as load, speed, etc. can be different for different wear situations.

Adhesive wear occurs when one surface bonds to another, and with subsequent motion, rupture occurs in one of the materials (Fig. 2). This type of sliding wear is described as:

$$V = \frac{K(PS)}{H_m} \qquad \text{(Eq 1)}$$

where V is the volume of wear in mm^3 or $in.^3$ produced in a sliding distance (S, in mm or in.) under load (P, in kg or lb); H_m is the Knoop hardness of the softer material; and K, the wear coefficient, is the probability that the rupture of a junction (localized contact spot between the two surfaces) will result in adhesive wear. K can vary over several orders of magnitude (Table 1), depending on such factors as the similarity of the materials,

Fig. 1 Typical wear testing configurations

Specimen sizes range from less than 25 mm (1.0 in.) (pins and blocks) to more than 75 mm (3.0 in.) (rings and disks). Loads range from fractions to thousands of kilograms or pounds. Some apparatus provide for a single load, and others are adjustable; some increase load automatically in preset increments. Speeds are available in a wide range, depending on the apparatus (2.5 mm/s to 2540 mm/s, or 0.1 in./s to 100 in./s). Source: Ref 20, adapted from Ref 19

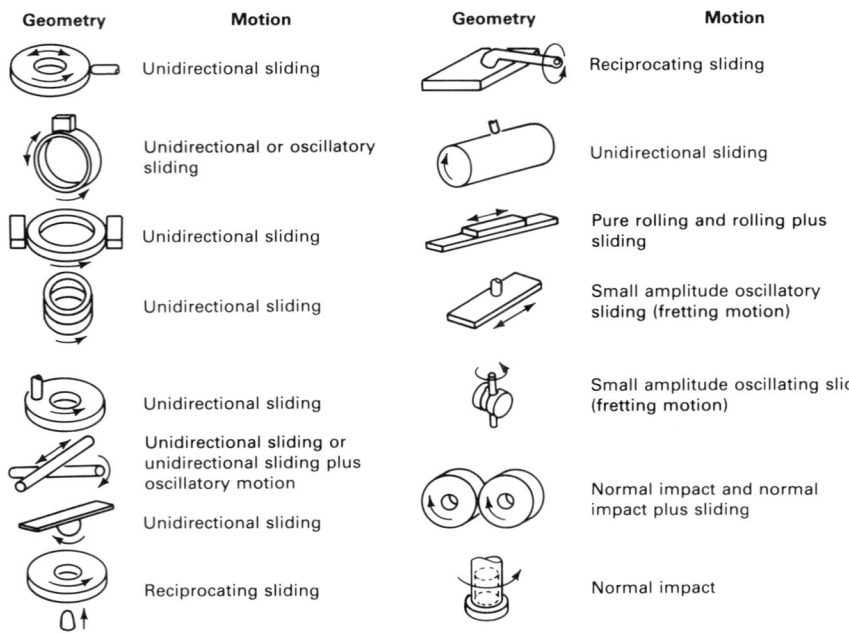

Geometry	Motion
	Unidirectional sliding
	Unidirectional or oscillatory sliding
	Unidirectional sliding
	Unidirectional sliding
	Unidirectional sliding
	Unidirectional sliding or unidirectional sliding plus oscillatory motion
	Unidirectional sliding
	Reciprocating sliding

Geometry	Motion
	Reciprocating sliding
	Unidirectional sliding
	Pure rolling and rolling plus sliding
	Small amplitude oscillatory sliding (fretting motion)
	Small amplitude oscillating sliding (fretting motion)
	Normal impact and normal impact plus sliding
	Normal impact

Fig. 2 Schematic depiction of adhesive wear

(a) Contact between two surfaces is characterized by A_a, apparent area of contact, and A_R, real area of contact and sum of all junction areas. (b) Adhesive junction wears if surface movement causes failure along path 1. No wear results if the failure is along path 2. Source: Ref 22

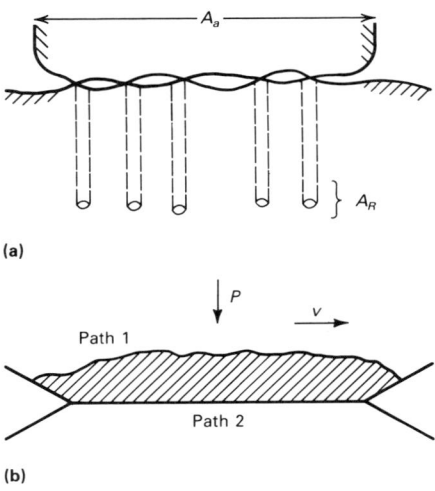

(a)

(b)

Fig. 3 Schematic of abrasive wear, depicting an abrasive particle plowing out a groove in a softer material.
Source: Ref 22

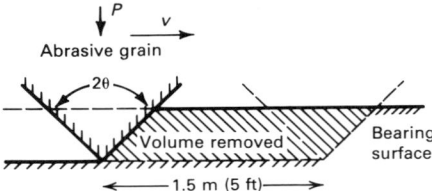

Table 1 Range of wear coefficients for sliding wear

$K = (V \cdot H_m)/PS$

Condition	Wear coefficient, K
Metal/metal sliding	
Clean:	
like	5×10^{-3}
unlike	2×10^{-4}
Poorly lubricated:	
like	2×10^{-4}
unlike	2×10^{-4}
Average lubrication	
like	2×10^{-5}
unlike	2×10^{-5}
Excellent lubrication	
like	2×10^{-6} to 10^{-7}
unlike	2×10^{-6} to 10^{-7}
Non-metal/metal sliding	
Clean	5×10^{-6}
Poorly lubricated	5×10^{-6}
Average lubrication	5×10^{-6}
Excellent lubrication	2×10^{-6}
Abrasion	
2-body, 0(100 μm) particular	1×10^{-1}
3-body, 0(100 μm) particular	5×10^{-3}
3-body, 0(≤1 μm) particular	10^{-7} to 10^{-4}

Note: See text for identification of variables.
Source: Ref 21

the smoothness of the surfaces, the normal load, and the lubrication or cleanliness of the surfaces.

In general, the more similar the materials and the cleaner the surfaces, the higher the value of K. Too smooth or too rough a surface results in K values higher than the optimum. Because high loads may result in oxide layer fracture and lubricant penetration, K generally increases with load. Lubrication decreases K; the extent of the decrease depends on the ability of the lubricant to separate the surfaces. Poor lubrication results in a relatively large amount of contact between surfaces, while excellent lubrication allows virtually none.

Abrasive wear occurs when a hard protuberance (asperity) on the surface of a material or a hard, loose particle trapped between surfaces plastically deforms or cuts a surface as a result of motion. The basic process is illustrated in Fig. 3. This mode can be further subdivided in terms of the exact nature of the deformation or cutting process involved. This type of wear is described by an equation similar to Eq 1, in which K is related to the sharpness, or geometry, of the asperity or particle causing the wear. As with adhesive wear, K can vary over several orders of magnitude. One factor that can influence this is the degree of freedom of the abrasive particle or asperity. For example, a sand particle

produces significantly higher wear when bound to a surface (two-body abrasion), as in sandpaper, than when it is free to move (three-body abrasion), as in a slurry. In the former it is dragged across the surface, while in the latter it may roll. The resistance of a material to abrasive wear is fairly constant when the abrasive is much harder than the material (Fig. 4). As the hardnesses of both become similar, and the abrasive action approaches polishing, wear resistance generally improves by one to two orders of magnitude. Then, a simple description like Eq 1 is no longer applicable.

Fatigue. Unlike abrasive and adhesive wear, which explain loss or displacement of material resulting from a single interaction, fatigue requires multiple interactions. In fatigue, a surface experiences repeated stress cycling, leading to cracking. The cracks link and result in the formation of a loose wear

Fig. 4 Comparison of wear produced by SiO₂ and SiC abrasives

Ratio drops to zero when surfaces much harder than SiO₂ are abraded. Source: Ref 21, 23

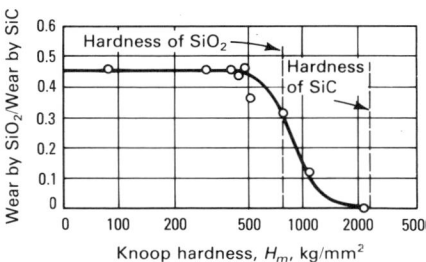

Fig. 5(a) Mechanism of fatigue wear

At time t_1, a crack forms. With subsequent engagement, crack propagates, t_2. At t_3, after sufficient cycles, crack propagates to surface, and a loose particle is formed. Source: Ref 22

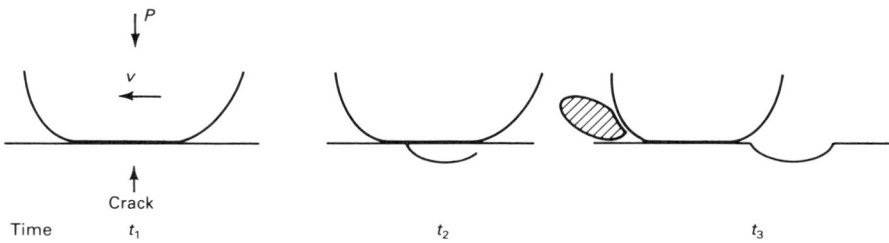

particle (Fig. 5). Fatigue is strongly stress dependent; many studies indicate that the wear rate is proportional to (stress)^n, where n is on the order of 8 or 9 (Ref 24). Generally, fatigue results in lower wear rates than abrasion or adhesion.

Oxidation. In oxidative wear, the surface of the material is changed chemically by environmental factors. This results in the formation of a surface layer with properties that differ from those of the parent material (Ref 25); consequently, this layer may wear differently. This in turn may induce fracture at the interface of the surface layer and parent material, resulting in removal of the entire surface layer. In such a case, the effective wear behavior of the system is a function of the time for surface layer growth and the wear contacts (Fig. 6). A mathematical description of simple oxidative wear is:

$$\dot{V} = K_F \dot{T} \left[\frac{AP}{H_m} \right]^{1/2}$$ (Eq 2)

where $\dot{V}$ is the volume wear rate, mm³/s or in.³/s; K_F is the probability of failure; $\dot{T}$ is the rate of growth of the oxide layer, mm/s or in./s; A is the apparent area of contact, mm² or in.²; P is the load, kg or lb; and H_m is the Knoop hardness. It is assumed that some oxide layer forms between contacts and that it fractures between the interface and the substrate with a probability of K_F.

Comparison of Eq 1 and 2, a linear versus a square root relationship, shows that in oxidative wear the chemical parameters are more significant than the mechanical. With oxidative wear, sharp transitions in wear behavior are expected as a function of parameters such as temperature and speed. On one side of the transition, the material shows little or no layer formation, while on the other side significant layer formation is evident.

A sliding contact illustrates this point. At a high sliding speed, the surface may not have time to oxidize between contacts; at a lower speed, it can. Layer formation does not necessarily mean increased wear. In fact, certain

Fig. 5(b) Subsurface cracks formed in a copper crystal due to low-stress sliding

These cracks result in surface fatigue wear. Source: Ref 22

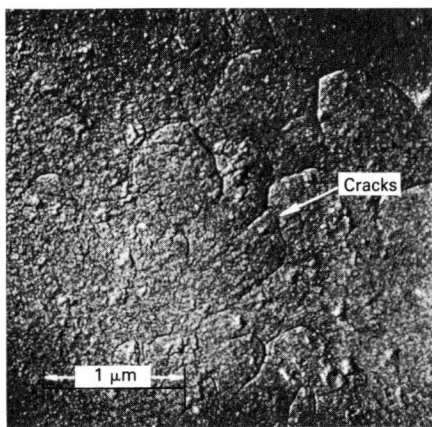

layers reduce wear. For example, some oxides of copper and steel have excellent wear resistance.

Mechanism Interactions. The interactions, competition, and simultaneous occurrence of physical mechanisms in actual wear situations can be demonstrated. For example, an adhesive wear fragment forms because the surface is weakened by fatigue. The fragment abrades the surface and, through several deformation cycles, produces a larger fatigue wear particle, or it is swept from the contact and causes no further wear.

The sequence depends on such factors as the motions and geometry involved. Alternatively, adhesive wear can have a removal rate so high that significant fatigue cracking does not develop before the material is removed.

Operational Mechanisms

The second method of wear classification is based on the materials, environment, and motions involved:

- Dry or lubricated sliding wear
- Rolling wear

Fig. 5(c) Plotted combination of stress and strokes that produces surface fatigue damage in copper in a reciprocating ball-plane wear test

τ_{max} is the maximum shear stress. See Fig. 5(b) for resulting microstructure. Source: Ref 22

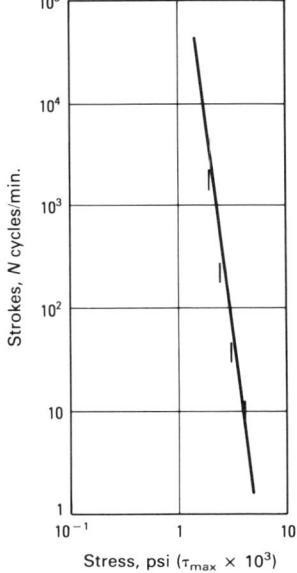

- Impact wear
- Filtering
- Metal/metal wear
- Polymer/metal wear
- Mild wear
- Severe wear
- High-temperature wear

Each of these categories involves different combinations of basic wear mechanisms, thus illustrating the influence of the conditions of and around the wear contact.

Control and simulation are critical in wear testing, because transitions in wear behavior frequently occur. Figure 7 shows two transitions in the wear of a steel, related to load and moisture. The wear phenomena of these regions are different; therefore, a wear test valid for one region may not be valid for an

Fig. 6 Schematic depiction of oxidative wear

At t_1, particle cleared of oxide; t_2, oxide re-forms; t_3, re-engagement results in layer removal.

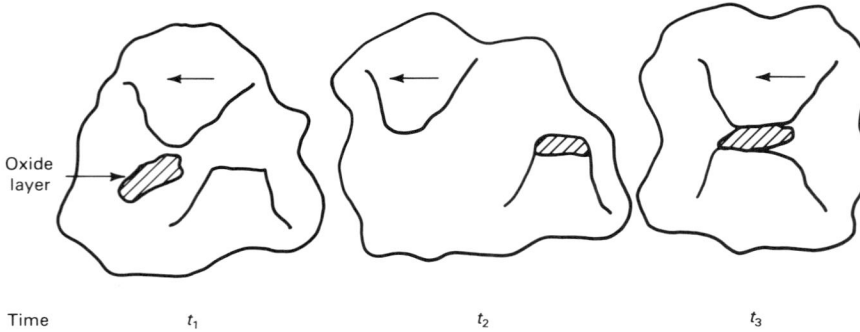

Oxide layer

Time t_1 t_2 t_3

application in another. The appearance of wear scars and debris generally is different for each region, or on either side of a transition.

Transitions from region to region often indicate a shift from mild to severe wear, with which adhesive and abrasive wear generally are associated. Coarseness of wear scars and degree of surface damage are different for mild and severe wear, and the presence, absence, or rupture of surface or lubricant layers is often related to a transition from mild to severe wear or from one wear mechanism to another.

Parameters Influencing Wear

A variety of parameters influence all wear mechanisms:

- *Material parameters*: Composition, grain size, modulus, thermal conductivity, degree of work hardening, hardness, etc.
- *Design parameters*: Shape, loading, type of motion, roughness, vibration, cycle time, etc.
- *Environmental parameters*: Temperature, humidity, atmosphere, contamination, etc.
- *Lubrication parameters*: Type of lubricant, lubricant stability, type of fluid lubrication, etc.
- Presence or absence of wear-in

These parameters apply to both members of the wearing contact.

Transfer films are layers generated in a sliding contact by the deposition of wear debris on the surface; frequently they reduce wear. The type of motion, relative area of one surface to the other, and environmental conditions influence transfer film formation. The presence or absence of these films can result in different wear behavior for the same material. A wear test in which a film does not form is not a valid simulation for an application in which it does.

Lubrication, particularly with greases and fluids, influences surface energies, causes oxide or other layers to form, or affects the degree of contact between the two surfaces, depending on the thickness of the lubricant layer (Table 1). With a grease or solid lubricant, the latter effect may depend simply on the amount and strength or hardness of the lubricant. However, with a fluid, the relative speeds, contact geometry, and loading conditions also influence separation of the surfaces through hydrodynamic lubrication (Ref 27, 28). With sufficient lubrication, light loads, and high speed, hydroplaning—the complete separation of two surfaces—can occur. The extent and type of lubrication must be controlled to simulate application if wear testing is to be valid. Figure 8 summarizes various lubrication regimes.

Variations in lubrication can cause nonlinear behavior and transitions in wear behavior. Figure 9 shows the nonlinear sensitivity of wear to the amount of lubricant supplied to a device. Frequently, transitions in wear behavior measure lubricant performance.

Wear-In and Friction. Long-term wear behavior is influenced by the initial experience of the system. Sometimes a wear-in (or break-in) cycle is required, during which the surfaces change, resulting in better wear. Changes in roughness and alignment and formation of surface layers or transfer films are factors that influence wear-in and therefore must be simulated in wear testing.

Although different phenomena, friction and wear react to many of the same factors and are therefore related. High friction does not always imply high wear; however, stable wear systems usually have characteristic values for the coefficient of friction. Changes in these values frequently reflect changes in system wear. Consequently, monitoring friction behavior in a wear test is useful. One way in which friction relates to wear is through heat dissipated in the wearing contact. Surface temperature severely affects the wear behavior of a material through its impact on mechanical properties and chemical reaction rates. Also, transfer film or surface layer formation can be monitored through friction, because their coefficients of friction generally differ from those of parent surfaces.

Elements of a Wear Test

General elements of a wear test are simulation, acceleration, specimen preparation, control, measurement, and reporting. Simulation is the most critical; however, no element of the application should be overlooked. As stated above, wear and wear phenomena can be influenced by load, environment, geometry, motion, the wearing mediums and counterface, and other factors.

Simulation ensures that the behavior experienced in the test is the same as in the application. Given the complexity of wear

Fig. 7 Changes in wear behavior

(a) Wear rate of AISI 1052 steel pin at various applied loads from a pin-on-ring apparatus. Sliding speed, 1.0 m/s (39.3 in./s). Sharp transitions are a function of load. (b) Accumulated wear after various times of sliding of 1.5 Mn steel using a cross-cylinder apparatus. Changes in behavior are a function of relative humidity. Source: Ref 20, 26

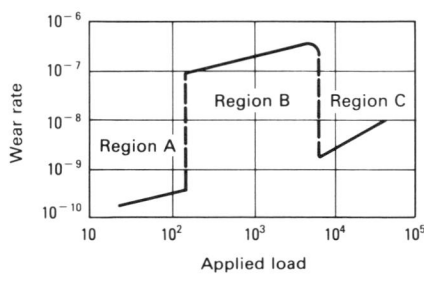

(a)

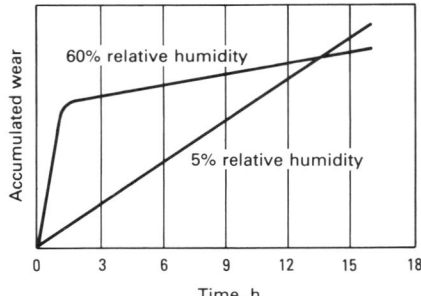

(b)

Fig. 8 Friction-velocity curve for surfaces capable of hydrodynamic lubrication

Particle contact and wear decrease from boundary to full-fluid lubrication. Wear occurs in the full-fluid region by fatigue, cavitation, and erosion, generally in orders of magnitude smaller than with boundary lubrication. Source: Ref 21

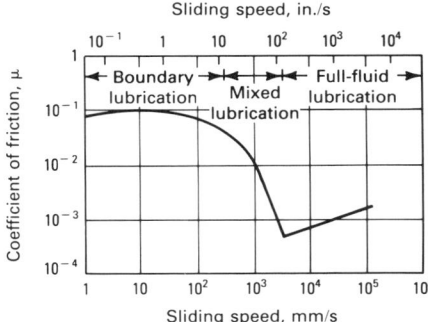

and the current incomplete understanding of wear and its phenomena, test development is subject to trial and error and is dependent on the capability of the developer. The ideal test exactly duplicates a wear situation. Generally, this is not practical, but any modifications in the test procedure should be evaluated carefully to obtain the most useful wear data.

The literature, prior data, and results of auxiliary or preliminary tests are useful in assessing the possible effects of proposed modifications. Published wear data should be reviewed carefully to compare test conditions for the data with those of the intended application. Key differences imply that the data are not relevant.

The engineer concerned with reliability and life generally requires precise simulation. However, the material developer interested in a convenient test to rank the wear resistance of materials usually requires only that the test fall within the general area of application.

Wear test simulation does not require that an application be replicated to provide valid data. For example, a sliding wear test is used to evaluate the wear resistance of material used for print elements in mechanical printers (Ref 30, 31). In this application, the apparent key element is impact. Print element wear, however, is caused by abrasive sliding that occurs during impact. As another example, the dry-sand rubber wheel test, useful in ranking material wear situations involving dry, low-stress scratching abrasion (Fig. 10), is not typical of the situations to which the results are applied. In the test, a rotating rubber wheel presses and rubs sand across the face of a specimen (Ref 32). This test is frequently used to select materials for farm tools operating in sandy soils and for

other machinery operating in abrasive environments. In both the test and these applications, the wear may be associated with low-stress scratching and abrasion—for example, fine scratches.

Although general knowledge and experience can aid in assessing the differences between test and application, correlations between test and application should also be studied. The most helpful correlation in developing a test is comparison of the worn surface and wear debris produced in the test to those produced in the application. The morphology of the scar, the presence or absence of oxidized or other surface layers, changes in the microstructure of the material, and wear debris size, shape, and composition can be compared. If major features of the wear scar and debris are different, valid simulation is unlikely. Wear mechanisms frequently result in characteristic wear particles. Consequently, comparing wear debris can be useful (Ref 20).

Selection of test geometry is another factor that must be considered when simulating wear conditions. For example, laboratory sliding contact wear tests employ three general types of contact—point contacts (such as a sphere on a plane), line contacts (such as a cylinder on a flat), and conforming contacts (such as a flat on a flat). Each of these geometries has advantages and disadvantages. Point-contact geometry eliminates alignment problems and allows wear to be studied from the start of the test. However, stress levels change as wear progresses, requiring more complex data analysis and comparison techniques.

Conforming-contact tests generally allow the parts to "wear-in" to establish uniform

Fig. 9 Variation in wear rate with change in oil supply rate in a printer

Source: Ref 29

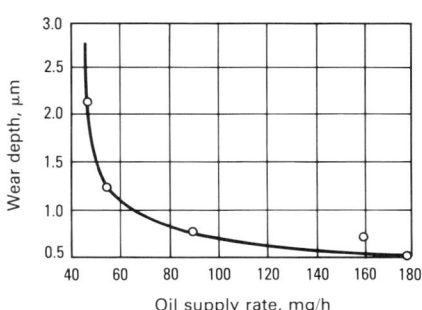

and stable contact geometry before taking data. As a result, it is difficult to identify wear-in phenomena, because there is no continuous observation of wear behavior. Consequently, it is difficult to differentiate surface modifications from simple alignment improvements. In addition, for applications in which allowed wear is small, the wear-in period may generate key information. However, conforming contact provides constant load and stress conditions once the parts are worn-in.

Because of the differences in stress behavior, a point or line contact is more sensitive to stress-dependent wear mechanisms than a conforming contact. For example, a point or line contact results in a different exponent in a log-log plot when the wear is a function of stress, compared with when it is not (Ref 33), because the stress level changes as wear progresses. A conforming contact with constant stress does not show this response.

Fig. 10 Dry sand low-stress rubber wheel abrasive wear tester

Source: Ref 20

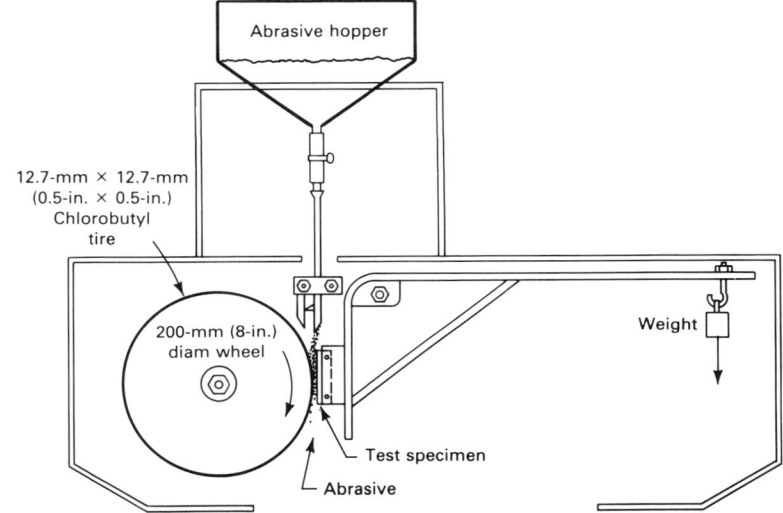

Stress dependency of the line contact lies between the point and conforming contact. The differences in these geometries must be recognized to obtain the required simulation.

Acceleration is desirable; unaccelerated tests frequently are costly and time consuming. However, acceleration may threaten simulation by introducing different phenomena. Wear mechanisms generally have threshold acceleration values for transition from mild to severe wear behavior. In addition, acceleration of such parameters as load or speed can emphasize one wear mechanism over another, thus causing different wear behavior. Nevertheless, most wear tests incorporate some element of acceleration—continuous operation, measurement of smaller quantities of wear, or higher loads, speeds, or temperatures.

Specimen preparation and test control are important for precision and repeatability. Lack of attention to these areas can cause scatter in wear tests; however, when properly carried out, scatter can be reduced to a level typical of other engineering tests (e.g., 20% or less).

To reduce scatter and to monitor control in wear testing, a test standard should be built around uniform, consistent, and readily obtainable reference material. Periodic standard tests should monitor the condition of the test rig, skill of the operator, and such factors as the influence of environmental effects.

Generally, close simulation or replication exists in tests that show good correlation to practice, and tight controls are evident in tests that provide good repeatability and low scatter. The ASTM procedures described in Ref 18 and 34 to 36 are examples of the detail and care that are necessary to obtain good repeatability and minimum scatter. The precision of the apparatus, specimen preparation, conditions of the counterface and the abrasive, and details of wear measurement and reporting are discussed in each procedure.

Specimen preparation and the details of test control vary with the test and materials involved. For metals, surface roughness, geometry of the specimens, microstructure, homogeneity, hardness, and presence of surface layers usually must be controlled. Similar controls are also necessary for the counterface and the wear-producing mediums. For example, in a test using sand as an abrasive, the purity, particle shape and size, and moisture content of the sand must be controlled. In wear tests involving fluids (e.g., as an erosive medium or lubricant), the properties of these fluids must be controlled.

In addition, parameters such as load, speed, instrument construction, ambient environment, location and alignment, and sup-

ply of abrasive or fluid require adequate control. In test development, investigation is necessary to assess the degree of control required and to establish repeatability.

Wear Measurement. Common wear measures are mass or weight loss, volume loss or displacement, scar width or depth or other geometrical measures, and indirect measures, such as the time required to wear through a coating or the load required to cause severe wear or a change in surface reflectance. The selection of variables to measure wear is often based on convenience, the nature of the wear specimens, and available techniques.

For large amounts of wear, weight-loss measurement is suitable, because it is simple and scales usually are available. However, weight-loss measurement has two major limitations. First, wear is related primarily to volume of material removed or displaced. If the tested materials differ in density, weight loss does not provide a true ranking. Second, this measure does not account for wear by material displacement; a specimen may gain weight by transfer. Therefore, weight-loss measurement is valid only when material densities are the same and when displacement wear and transfer do not occur.

Volume loss or displacement, although directly attributable to wear, frequently is difficult to measure. Except for simple wear scar geometries, determination of volume loss is complex and time consuming. A linear dimension, such as the depth or width of the scar, often is measured, because it is related to volume through the test geometry. However, the applicability of this type of measurement is limited to one test geometry and test. For example, wear scars frequently are not uniform in depth or width.

Wear measurement by indirect techniques is viable in some cases. For example, when comparing the wear resistance of very thin coatings, the time required to wear through may be the only convenient way of measuring performance. However, indirect techniques generally are limited in scope and applicability and do not easily provide or establish fundamental wear parameters.

In wear tests used to rank materials, the wear data are used directly. The wear measurement also establishes parameters that both rank material performance and project behavior in an application. In the zero and measurable wear model proposed in the 1960's (Ref 33, 37), wear test results establish a zero wear factor, γ_R, and determine whether the system exhibits variable- or constant-energy wear behavior. These behaviors influence wear measurement selection and data development.

Material wear behavior can be compared by determining a wear curve, or by measur-

ing wear at a single point in the test. Because wear behavior frequently is nonlinear, a wear curve provides more information and allows evaluation of more complex behavior than single-point measurement.

For example, the wear behaviors of two materials in the same test are plotted as functions of the number of sliding cycles (Fig. 11). Both materials demonstrate nonlinear behavior. However, single measurements taken at different points in the test result in different rankings. Therefore, the potential for nonlinear and transitional behavior should be considered when a wear test is developed.

In engineering applications for which material life and reliability are concerns, the wear curve provides more complete information about material behavior and aids in data extrapolation, particularly if data analysis includes a log-log plot. Appropriate exponential relationships between the wear measure and key parameters are easily developed. However, single-point measurement frequently is selected when quick evaluation and simple ranking of materials are desired.

Reporting Testing Data. Wear is a system response; when reporting wear data, a description of the wearing system must be supplied, including:

- Apparatus
- Geometry of contact
- Type of motion
- Load
- Speed
- Environmental condition
- Condition of wearing mediums
- Description of materials
- Description of lubricant and lubrication used
- Description of wear-in period, if appropriate
- Unusual observation, e.g., evidence of transfer

Fig. 11 Wear behavior of two materials in the same test

A single measurement below 5×10^6 cycles would rank these materials differently from a single measurement above 5×10^6 cycles.

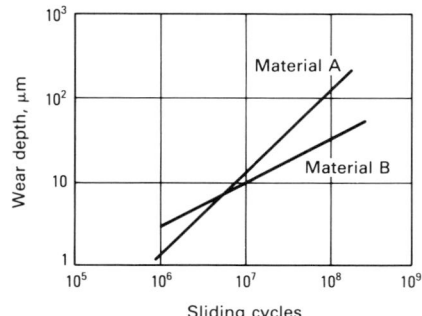

- Surface and material preparations
- Roughness

The report should describe the material tested, general nature of the test, conditions of the counterface, testing environment, and any other significant features. For example, in metal/metal wear, transfer film formation should be recorded and reported. Frequently, such observations lead to a greater understanding of the wear situation and material response and to improved test development.

As discussed previously, wear data scatter can be reduced by careful testing methods. Nevertheless, several tests, rather than one, should form the basis for conclusions. Two tests are good practice; a minimum of three is preferred. The need for replication beyond three runs depends on the testing purpose, degree of control, and scatter.

Wear Test Case Histories

The general methodology for wear testing (e.g., simulation and control requirements) is the same for all materials and all types of wear situations. Only the parameters that must be controlled, the degree of control required, and the features that must be simulated vary with materials and applications. The large number of wear tests and test configurations in existence attests to this. Some of the problems associated with inadequate simulation and poor test techniques are described in the following case histories.

Oil Pump Wear. An oil pump driven by a cam follower system that operated in an oil bath was developed. The cam speed was slow—in the range of a few revolutions per week. During the development stage of the pump, it was determined that wear tests should be performed to determine the service

Fig. 12 Result of a metal band/metal platen wear test for a high-speed printer

Solid line is observed behavior. Dashed line is behavior projected from low-speed sliding tests. After 10 min, severe adhesive wear is observed in the band/platen test, changing the wear behavior. Similar behavior is not observed in the low-speed sliding test.

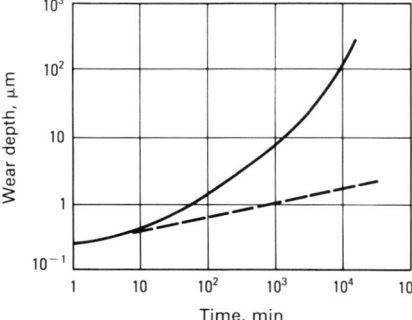

life of the device. Wear tests using the actual device were conducted. Acceleration was necessary, because the operations per unit of time was low, but the required life was long, about 5 years. In the test, the pump ran approximately 100 times faster than in the field. Several materials were tested, and no wear problems were found.

Several months after release to the field, pumps were failing; the cam followers were severely worn. Failure analysis was initiated to determine whether thrust parts were out of spec (too rough, too soft, wrong dimension, etc.). No out-of-spec conditions were found. Reviewing the earlier tests led to speculation that acceleration in the tests caused fluid lubrication that did not occur in field operation. The original test specimens revealed virtually no evidence of contact, while specimens from short-term tests run at normal operating speed showed considerable contact. New tests, performed at a speed that did not result in fluid lubrication but that did provide some acceleration, enabled selection of material and shapes that resolved the failure problem.

This case history illustrates that possible effects of acceleration on a simulation must be considered. Also, if more care had been taken in performing the test and evaluating the worn specimens, the test problems would have been identified earlier. The lack of contact marks and the similar results for several materials should have been indicative of the problem.

Band Printer Wear. A print band in a printer slides across the surface of platen under negligible load. To print, the band is pressed against the platen for a short time (0.1 μs) as it moves across the platen. No lubrication is used. To rank materials for this application, a reciprocating ball-plane test was used. The relative sliding speed for the tests was much lower than for the application, and the stresses were significantly higher, although not in the plastic range. Several metallic combinations were tested and ranked. Analytical models, based on the test data, projected that several of the material pairs provided adequate performance. However, validation tests in a printing robot resulted in failure.

Detailed studies showed that early-life wear in the robot agreed with the projections and that wear scar features in the tests and robot generally were the same. However, as use continued in the printing robot, adhesive wear and galling occurred with an avalanching effect (Fig. 12). It was concluded that the sliding speed in the machine encouraged adhesive wear, probably as a result of localized high temperature, which did not occur in the test. Subsequently, several metal/high-temperature polymer combinations and thinly

lubricated metal/metal cases were tested. These combinations provided good correlation. The wear mode did not change with continued robot testing, and wear scars were similar in the laboratory and robot tests.

Although the differences between the test and the application resulted in poor simulation for the dry metal/metal systems, it did not for other combinations. This case history demonstrates that simulation is needed, but that the degree required varies. Also, careful examination of the wear produced in the test and in the application is required. Additionally, wear curves generally are superior to single-point measurements.

REFERENCES

1. D. Rigney and W. Glaeser, Ed., *Source Book on Wear Control Technology*, American Society for Metals, 1978
2. D. Scott, Ed., *Treatise on Materials Science and Technology*, Vol 13, Academic Press, New York, 1979
3. N.P. Suh and N. Saka, Ed., *Fundamentals of Tribology*, MIT Press, Cambridge, MA, 1980
4. M.B. Peterson and W.O. Winer, Ed., *Wear Control Handbook*, American Society of Mechanical Engineers, New York, 1980
5. H. Czichos, *Tribology*, Elsevier, Amsterdam, 1978
6. M.J. Neale, Ed., *Tribology Handbook*, John Wiley & Sons, New York, 1973
7. J.A. Schey, *Tribology in Metalworking: Friction, Lubrication and Wear*, American Society for Metals, 1983
8. "Standard Method for Calibration and Operation of The Alpha Model LFW-1 Friction and Wear Testing Machine," D 2714, *Annual Book of ASTM Standards*, ASTM, Philadelphia, 1984, p 658-662
9. "Standard Test Method for Wear Life of Solid Film Lubricants in Oscillating Motion," D 2981, *Annual Book of ASTM Standards*, ASTM, Philadelphia, 1984, p 1-5
10. "Standard Test Method for Wear Preventive Properties of Lubricating Greases Using the (Falex) Ring and Block Test Machine in Oscillating Motion," D 3704, *Annual Book of ASTM Standards*, ASTM, Philadelphia, 1984, p 427-429
11. "Standard Method for Measurement of Extreme Pressure Properties of Lubricating Grease (Timken Method)," D 2509, *Annual Book of ASTM Standards*, ASTM, Philadelphia, 1984, p 446-455
12. "Standard Method for Measurement of Extreme Pressure Properties of Lubri-

cating Fluids (Timken Method)," D 2782, *Annual Book of ASTM Standards*, ASTM, Philadelphia, 1984, p 700-709

13. "Standard Method for Measuring Wear Properties of Fluid Lubricants (Falex Pin and Vee-Block Method)," D 2670, *Annual Book of ASTM Standards*, ASTM, Philadelphia, 1984, p 619-626

14. "Standard Test Method for Endurance (Wear) Life and Load-Carrying Capacity of Solid Film Lubricants (Falex Method)," D 2625, *Annual Book of ASTM Standards*, ASTM, Philadelphia, 1984, p 593-599

15. "Standard Methods for Measurement of Extreme Pressure Properties of Fluid Lubricants (Falex Methods)," D 3233, *Annual Book of ASTM Standards*, ASTM, Philadelphia, 1984, p 88-98

16. "Standard Test Method for Wear Preventive Characteristics of Lubricating Grease (Four-Ball Method)," D 2266, *Annual Book of ASTM Standards*, ASTM, Philadelphia, 1984, p 249-252

17. "Standard Test Method for Wear Preventive Characteristics of Lubricating Fluid (Four-Ball Method)," D 4172, *Annual Book of ASTM Standards*, ASTM, Philadelphia, 1984, p 697-702

18. "Standard Practice for Ranking Resistance of Materials to Sliding Wear using Block-on-Ring Wear Test," G 77, *Annual Book of ASTM Standards*, ASTM, Philadelphia, 1984, p 446-462

19. R. Benzing *et al.*, *Friction and Wear Devices*, American Society for Lubrication Engineers, Park Ridge, IL, 1976

20. R. Bayer, Ed., *Selection and Use of Wear Tests for Metals*, STP 615, ASTM, Philadelphia, 1976

21. E. Rabinowicz, *Friction and Wear of Materials*, John Wiley & Sons, New York, 1966

22. R.G. Bayer, Understanding the Fundamentals of Wear, *Mach. Design*, Dec 1972, p 73-76

23. F.K. Aleinikov, *Soviet Physics—Technical Physics*, Vol 2, 1957, p 505-511

24. R.G. Bayer and R.A. Schumacher, On the Significance of Surface Fatigue in Sliding Wear, *Wear*, Vol 12, 1968, p 173-183

25. G. Yoshimoto and G. Tsukozol, On the Mechanism of Wear Between Metal Surfaces, *Wear*, Vol 1, 1957-1958

26. R. Bayer, Ed., *Wear Tests for Plastics: Selection and Use*, STP 701, ASTM, Philadelphia, 1979

27. E.R. Booser, Ed., *Handbook of Lubrication, Theory & Design*, Vol II, Chemical Rubber Co., 1984

28. F.F. Ling, E.E. Klaus, and R.S. Fein, Ed., *Boundary Lubrication*, American Society for Mechanical Engineers, New York, 1969

29. R.G. Bayer, The Influence of Lubrication Rate on Wear Behavior, *Wear*, Vol 35, 1975, p 35-40

30. P.A. Engel and R.G. Bayer, Abrasive Impact Wear of Type, *J. Lubricat. Technol.*, Vol 98, 1976, p 330-334

31. R.G. Bayer, Wear by Ribbon and Paper, *Wear*, Vol 49, 1978, p 147-168

32. F. Borik, Testing for Abrasive Wear, in *Selection and Use of Wear Tests for Metals*, R.G. Bayer, Ed., STP 615, ASTM, Philadelphia, 1976, p 30-44

33. R.G. Bayer and T.C. Ku, *Handbook of Analytical Design Procedures for Wear*, Plenum Press, New York, 1964

34. "Standard Practice for Conducting Dry Sand/Rubber Wheel Abrasion Tests," G 65, *Annual Book of ASTM Standards*, ASTM, Philadelphia, 1984, p 351-368

35. "Standard Test Method for Wear Testing with a Crossed-Cylinder Apparatus," G 83, *Annual Book of ASTM Standards*, ASTM, Philadelphia, 1984, p 495-502

36. "Standard Practice for Jaw Crusher Gouging Abrasion Test," G 81, *Annual Book of ASTM Standards*, ASTM, Philadelphia, 1984, p 476-484

37. R.G. Bayer, Prediction of Wear in a Sliding System, *Wear*, Vol 11, 1968, p 319-332

Calibration of Testing Equipment

Calibration of Testing Equipment

By Robert S. Strimel
Technical Director
Tinius Olsen Testing
Machine Company, Inc.

VERIFICATION AND CALIBRATION of testing equipment are critical to ensure accurate determination of material strength characteristics. In the United States, the procedures for test system verification and calibration are outlined in the *Annual Book of ASTM Standards*. Section 3 of the 1984 edition contains six volumes that pertain to test methods and analytical procedures for metals. Of these, Volume 03.01 relates specifically to mechanical testing of metals. Each of the standards contained in Volume 03.01 that concern strength properties references ASTM E 4, "Standard Practices for Load Verification of Testing Machines" (Ref 1). Many of the standards listed in this volume also refer to ASTM E 83, "Standard Method of Verification and Classification of Extensometers" (Ref 2). In some cases, these standards include recommended methods to verify specialized testing equipment, whereas others simply stipulate the standard practices outlined in ASTM E 4.

Basically, ASTM E 4 details procedures for the load (force) verification of testing machine systems, and ASTM E 83 covers procedures for the verification and classification of extensometers and associated strain measurement systems. ASTM E 4 was first issued in 1923, and ASTM E 83 was issued in 1950. Both standards have been revised and updated several times to encompass the technological advances incorporated in testing machines and related instrumentation, as well as modern microprocessor-based data acquisition systems. The most recent update of these two standards includes provisions for verifying the load (force) and strain (extension) aspects of testing machine data systems. In addition, the 1983 revision of ASTM E 4 was approved for use by agencies of the Department of Defense and is listed in the "DOD Index of Specifications and Standards."

Accordingly, if testing machine users ensure that the verification procedures of ASTM E 4 and ASTM E 83 are followed exactly, and if the verification results are within the stipulated requirements, the test results obtained from the equipment should be accurate to within existing tolerances for all production and research testing programs. Values obtained will also be defensible in a court of law, if necessary. To further strengthen this legal aspect, procedures have been incorporated in these standards to ensure the accuracy of the testing systems between annual verifications. Although all details of compliance to ASTM E 4 and ASTM E 83 are not included in this article, the most important aspects are discussed.

Verification of Test Equipment

The terms "calibration" and "verification" are not synonymous. Calibration of testing machines refers to the procedure of determining the magnitude of error in the indicated loads. Verification is a calibration to ascertain whether the errors are within a predetermined range. Verification also implies certification that a machine meets stated accuracy requirements.

To ensure valid verification, calibration procedures should be performed by skilled personnel who are knowledgeable about testing machines and related instruments and the proper use of calibration standards. Only load-indicating mechanisms that comply with ASTM E 74, "Standard Practice of Calibration of Force-Measuring Instruments for Verifying the Load Indication of Testing Machines" (Ref 3), and displacement devices that comply with ASTM E 83 should be used.

Users of testing equipment should designate to the calibrator or calibration agency the various systems of the testing machines such as dials, scales, marked or unmarked recorder charts, digital displays, extensometers, and data acquisition systems, that are to be verified. After verification is performed, the calibrator or agency must issue reports and certificates attesting to compliance of the equipment with the verification requirements, including the loading range(s) for which the system may be used.

Load Verification. For the load verification to be valid, the weighing system(s) and associated instrumentation and data systems must be verified annually. In no case should the time interval between verifications exceed 18 months. Testing systems and their loading range(s) should be verified immediately after relocation of equipment, after repairs or parts replacement (mechanical or electric/electronic) that could affect the accuracy of the load-measuring system(s), or whenever the accuracy of indicated loads is suspect, regardless of when the last verification was made.

Additionally, to comply with ASTM E 4, one or a combination of the three allowable verification methods must be used in the determination of the loading range or multiple loading ranges of the testing system. These methods are based on the use of:

- Standard weights
- Standard weights and lever balances
- Elastic calibration devices

For each loading range, at least five (preferably more) verification load levels must be selected. The difference between any two successive test loads must not be larger than one third of the difference between the maximum and minimum test loads. Although the maximum can be the full capacity of an individual range, the minimum can be no less than 10% of the range capacity. For example, acceptable test load levels could be 10,

25, 50, 75, and 100%, or 10, 20, 40, 70, and 100%, of the stated machine range.

Regardless of the load verification method used at each of the test levels, the values indicated by the load-measuring system(s) of the testing machine must be accurate to within ±1% of the loads indicated by the calibration standard. If all five or more of the successive test load deviations are within the ±1% required in ASTM E 4, the loading range(s) may be established and reported to include all of the values. If any deviations are larger than ±1%, the system should be corrected or repaired immediately. For determining accuracy of values at various test loads (or the deviation from the indicated load of the standard), ASTM E 74 specifies the required calibration accuracy tolerances of the three allowable types of verification methods.

For determining material properties, the testing machine loads should be as accurate as possible. In addition, deformations resulting from load applications should be measured as precisely as possible. This is particularly important because the relationship of load to deformation, which may be extension, compression, etc., is the main factor in determining material properties.

As described previously, load accuracy may be ensured by following the ASTM E 4 procedure. In a similar manner, the methods contained in ASTM E 83, if followed precisely, will ensure that the devices or instruments used for deformation (strain) measurements will operate satisfactorily.

Extensometer Verification. Procedures for the verification and classification of extensometers can be found in ASTM E 83. The six classes of extensometers, which are based on allowable error deviations, are discussed later in this article. This standard also establishes a verification procedure to ascertain compliance of an instrument to a particular classification. In addition, it stipulates that a certified calibration apparatus must be used for all applied displacements and that the accuracy of the apparatus must be five times more precise than allowable classification errors. Ten displacement readings are required for verification of a classification.

Universal Testing Machines

Although there are many types of test systems in current use, the most commonly used are universal testing machines, which are designed to test specimens in tension, compression, or bending. Consequently, the following discussion of load application and load-measuring systems and their conformance to ASTM E 4 and ASTM E 83 is limited to universal testing machines.

Load Application Systems

Universal testing machines are designed to apply a load to a material to determine its strength and resistance to deformation. The load-applying mechanism may be a hydraulic piston and cylinder with an associated hydraulic power supply, or the load may be administered via precision-cut machine screws driven by the necessary gears, reducers, and motor to provide a variable travel speed. In some light-capacity machines (only a few hundred pounds maximum), the load is applied by an air piston and cylinder.

Regardless of the method of load application, testing machines are designed to drive a crosshead or platen at a controlled rate, thus applying a tensile or compressive load to a specimen at a uniform rate. Such testing machines measure and indicate the applied load in pound-force (lbf), kilogram-force (kgf), or Newtons (N). These customary force units are related by the following:

$$1 \text{ lbf} = 4.448222 \text{ N}$$

$$1 \text{ kgf} = 9.80665 \text{ N}$$

All current testing machines are capable of indicating the applied load in either lbf or N (the use of kgf is not recommended). This feature applies equally to machines that incorporate a load dial indicator (see Fig. 1) and to modern machines that utilize digital display of the applied load. Direct conversions of testing machine load levels or capacities, however, rarely are used. Consequently, only English units of measure have been used to indicate load levels and capacities in this article. Dual units of measure are used in all other cases. It should be noted that the same standardized procedures must be followed when calibrating a machine in metric units of measure.

Hydraulic Testing Machines. The major components of a typical hydraulic testing machine are shown in Fig. 1. The space between the table and the lower head is the compression space, and the area between the lower and upper head is the tension space. In operation, hydraulic fluid from the motor-driven pump is directed into the cylinder, forcing the piston upward, which moves the table and the upper head.

When a specimen is placed between the table and the lower head, it is in compression, whereas a specimen placed between the two heads is in tension. If the piston and cyl-

Fig. 1 Components of a typical hydraulic testing machine

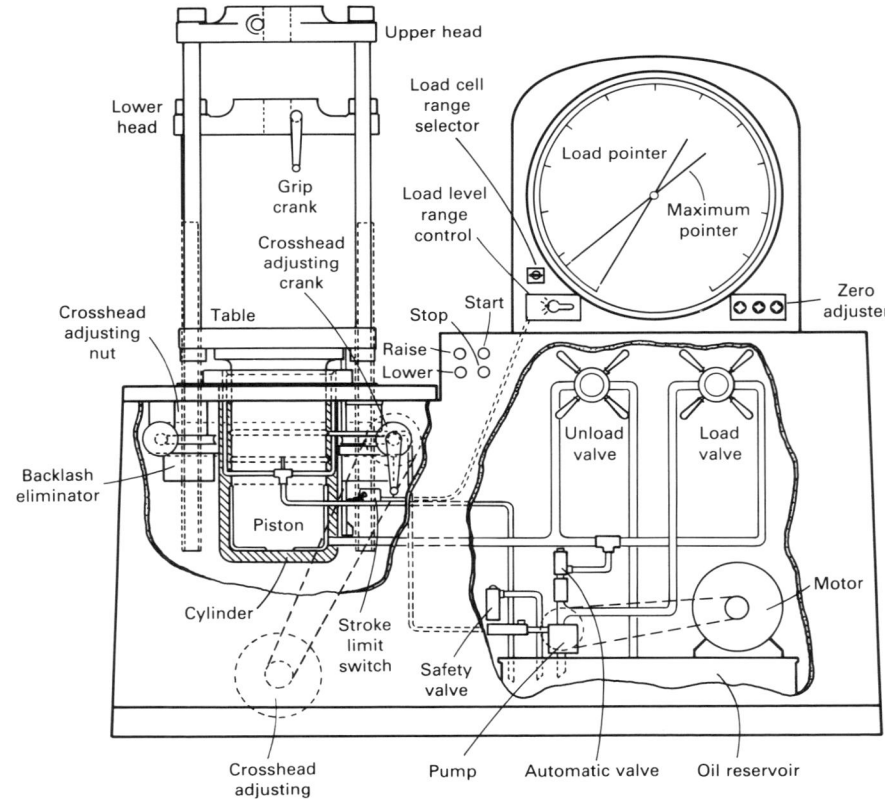

inder are of the "packless type" (no piston seals used), which minimizes friction, the force resulting from pressure in the cylinder acting on the piston face is nearly the same as the force on the specimen.

However, if the piston and cylinder are of the "packed type," then friction, which is not constant, is present, and the force on the specimen is less than the force resulting from pressure in the cylinder acting on the piston face. Load cells, mounted between the piston and the table or in the specimen load train, are then used to determine the force on the specimen.

Some machines built in the United States and many machines built in Europe have the piston and cylinder mounted overhead on top or above the upper head. The upper head and table remain stationary, and the lower or intermediate head is movable so that the lower space is used for tension testing and the upper space for compression, bending, or shear testing.

Screw-driven testing machines currently used are of either one-, two-, or four-screw design. In this article, only the two-screw design is described. Single- and multiple-screw-driven machines operate in the same fashion. The screws are rotated by a variable-control motor and drive the lower (intermediate) head downward. This action produces compression between the moving head and the table and tension in the upper space (between the heads).

To eliminate twist in the specimen from the rotation of the screws, one screw has a right-hand thread, and the other has a left-hand thread. For alignment and lateral stability, the screws are supported in bearings on each end. In some machines, loading crossheads are guided by columns or guideways to achieve alignment. Components and controls of screw-driven testing machines are illustrated in Fig. 1 of the article "Tension Testing Machines and Extensometers" in this Volume.

Load-Measuring Systems

Several types of measuring and indicating systems are incorporated in currently used machines. Early systems, some of which are still in use, employed a graduated balanced beam similar to platform scale weighing systems (Fig. 2). Subsequent systems included Bourdon tube hydraulic test gages, Bourdon tubes with various support and assist devices, and load cells of several types.

One of the most common load-measuring systems used by testing machine manufacturers prior to the development of load cells was the displacement pendulum (Fig. 3). The load applied to the specimen (whether transmitted hydraulically or mechanically) was

Fig. 2 Screw-driven balanced beam universal testing machine
1890 model

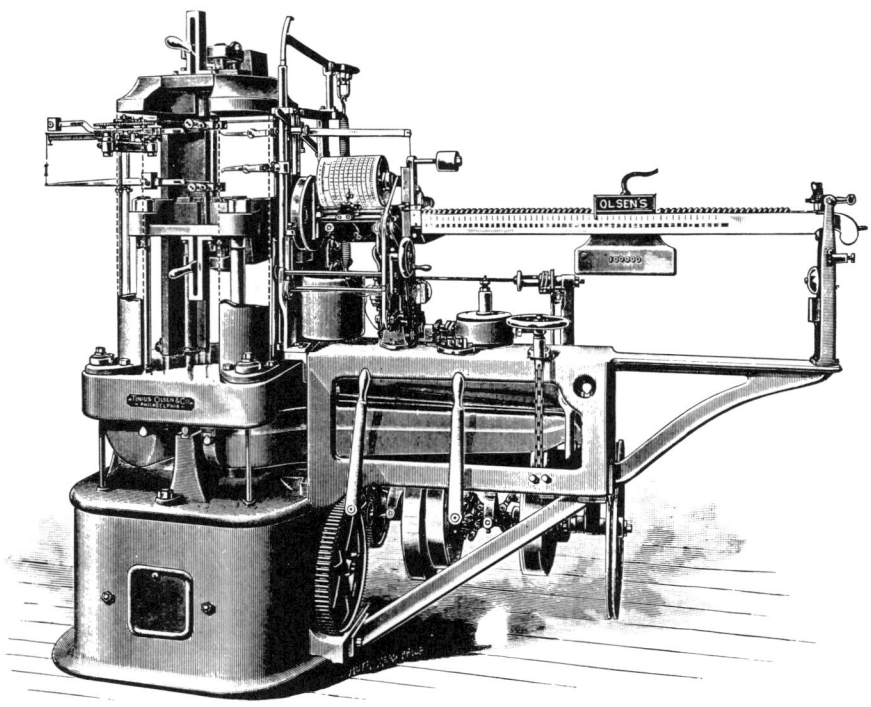

measured by the movement of the balance displacement pendulum.

The pendulum measuring system was used widely, because it is applicable to both hydraulic and screw-driven machines and has a high degree of reliability and stability. Many machines of this design are still in use, and they are still manufactured in Europe, India, South America, and Asia.

Another widely used testing system was the Emery-Tate oil/pneumatic system (Fig. 4a), which accurately sensed the hydraulic pressure in a closed, flat capsule. Many of these machines are also still in operation; however, lack of replacement parts for their precision components (Fig. 4b) limits usage.

For load measuring in current testing machines, strain-gage load cells and pressure transducers are used. In a load cell, strain gages are mounted on precision-machined alloy steel elements, hermetically sealed in a case with the necessary electrical outlets and arranged for tension and/or compressive loading. The load cell can be mounted so that the specimen is in direct contact, or the cell can be indirectly loaded through the machine crosshead, table, or columns of the load frame.

In pressure transducers, which are variations of strain-gage load cells, the strain-gaged member is activated by the hydraulic

Fig. 3 Displacement pendulum weighing system

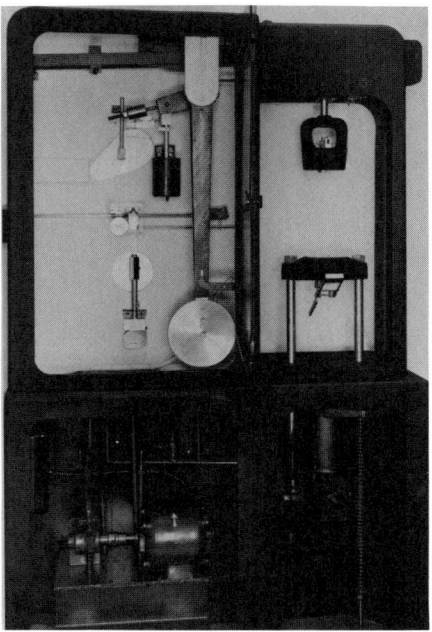

pressure of the system. The electrical circuits are the same in both cases.

Strain gages, strain-gage load cells, and pressure transducers are manufactured to

Fig. 4(a) 120 000-lbf hydraulic testing machine utilizing the Emery-Tate oil/pneumatic system

Fig. 4(b) Rear view of the load dial indicator system shown in Fig. 4(a)

several degrees of accuracy; however, when used as the load-measuring mechanism of a testing machine, the mechanisms must conform to ASTM E 4, as well as to the manufacturer's quality standards. Load cells are also discussed in the following section.

Elastic Calibration Devices

Several types of designs of elastic devices are available, but two designs are universally accepted in initial calibration and for subsequent in-field recalibration and certification. These designs are the elastic proving ring and strain-gage load cells.

The elastic proving ring (Fig. 5a and 5b), is a flawless forged steel ring that is precisely machined to a fine finish and closely maintained tolerances. This device has a uniform and repeatable deflection throughout its loaded range. Elastic proving rings usually are designed to be used only in compression, but special rings are designed to be used in tension or compression.

As the term "elastic device" implies, the ring is used well within its elastic range, and the deflection is read by a precise micrometer. Proving rings are available with capacities ranging from 1000 to 1 200 000 lbf. Their usable range is from 10 to 100% of load capacity, based on compliance with the ASTM E 74 verification procedure.

Proving rings vary in weight from 4 to 5 lb to several hundred pounds. They are portable and easy to use. After initial certification, they should be recalibrated and recertified at intervals not exceeding 2 years.

Proving rings are not load rings. Although the two devices are of similar design and construction, only proving rings that use a precise micrometer for measuring deflection can be used for calibration. Load rings employ a dial indicator to measure deflection and usually do not comply with the requirements of ASTM E 74.

Strain-gage load cells are precisely machined high-alloy steel elements designed to have a positive and predetermined uniform deflection under load. The steel load cell element contains one or more reduced sections, onto which wire or foil strain gages are attached to form a balanced circuit containing a temperature-compensating element.

Strain-gage load cells used for calibration purposes are either compression or tension/ compression types and have built-in capacities ranging from about 100 to 1 000 000 lbf. Their usable range is from 5 to 100% of

capacity load, and their accuracy is ±0.05%, based on compliance with procedures described in ASTM E 74. Figure 6 illustrates a load cell system used to calibrate a universal testing machine. This particular system incorporates a digital load indicator unit.

Comparison of Elastic Calibration Devices. The deflection of a proving ring is measured in divisions that are assigned a value in lbf, kgf, or N. The force is then calculated in the desired units. Although the deflection of a load cell is given numerically and a force value can be assigned with a load cell reading, electric circuits can provide direct readout in lbf, kgf, or N. Thus, certified load cells are more practical and convenient to use and minimize errors in calculation.

In small capacities (1000 to 5000 lbf), proving rings and load cells are of similar size and weight (4 to 10 lb). In large capacities (400 000 to 600 000 lbf), load cells are about one half the size and weight of proving rings. Proving rings are a single-piece, self-contained unit. A load cell calibration kit consists of two parts: the load cell and the display indicator (see Fig. 6). Although the

Fig. 5(a) Elastic proving ring with precision micrometer for deflection/load readout

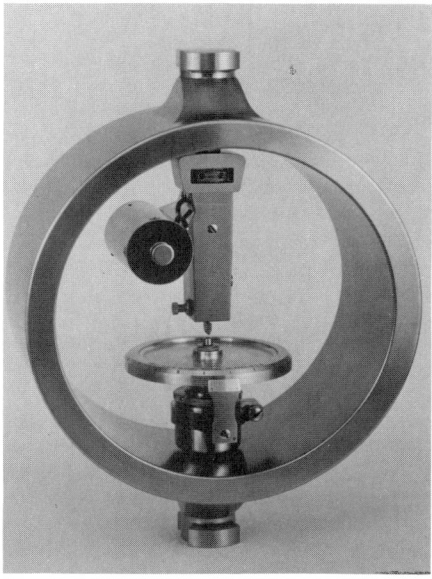

Fig. 5(b) 120 000-lbf screw-driven testing machine being calibrated with a proving ring

display indicator is designed to be used with a load cell of any capacity, it can only be used with load cells that have been verified with it as a system.

Although both proving rings and load cells are portable, the lighter weight and smaller size of high-capacity load cells enhance their suitability for general use. Load cells and

their display indicators require a longer setup time; however, their direct readout feature reduces the overall calibration and reporting time. After initial certification, the load cell should be recalibrated after 1 year and thereafter at intervals not exceeding 2 years.

Both types of calibration devices are certified in accordance with the provisions of ASTM E 74 and the verification values determined by the National Bureau of Standards. The National Bureau of Standards maintains a 1 000 000-lbf deadweight calibrator that is kept in a temperature- and humidity-controlled environment. This force-calibrating machine incorporates twenty 50 000-lb stainless steel weights, each accurate to within ±0.25 lb. This machine, and six others of smaller capacities, are used to calibrate elastic calibrating devices, which in turn are employed to accurately calibrate other testing equipment.

Elastic calibrating devices for verification of testing machines are calibrated to primary standards, which are weights. The masses of the weights used are determined to 0.005% of their values.

Calibration of Universal Testing Machines

In the United States and Canada, the manufacturer calibrates the testing machine before shipping and certifies that it conforms to ASTM E 4 and to the manufacturer's guarantee of accuracy. Subsequent calibrations can be made by the manufacturer or another organization with recognized equipment that is properly maintained and recertified periodically.

Except in Mexico and other Central and South American countries, recertification must be performed by a governmental standard bureau, or by an organization or department under governmental jurisdiction. It cannot be performed by the manufacturer or another organization.

Example: Calibrating a 60 000-lbf-Capacity Testing Machine. A 60 000-lbf-capacity dial-type universal testing machine of either hydraulic or screw-driven design will have the following typical scale ranges:

- 0 to 60 000 lbf reading by 50-lbf divisions
- 0 to 30 000 lbf reading by 25-lbf divisions
- 0 to 12 000 lbf reading by 10-lbf divisions
- 0 to 1200 lbf reading by 1-lbf divisions

As discussed previously, the ASTM required accuracy is ±1% of the indicated load above 10% of each scale range. Most manufacturers produce equipment to an accuracy of ±0.5% of the indicated load or ± one division, whichever is greater.

According to ASTM specifications, the 60 000-lbf scale range must be within 1% at

Fig. 6 Load cell and digital load indicator used to calibrate a 200 000-lbf hydraulic testing machine

60 000 lbf (±600 lbf) and at 6000 lbf (±60 lbf). In both cases, the increment division is 50 lbf. Although the initial calibration by the manufacturer is to closer tolerance than ASTM E 4, subsequent recalibrations are usually to the ±1% requirement. In the low range, the machine must be accurate (±1%) from 120 to 1200 lbf. Thus, the machine must be verified from 120 to 60 000 lbf.

If proving rings are used in calibration, a 60 000-lbf-capacity proving ring is usable down to a 6000-lbf load level. A 6000-lbf-capacity proving ring is usable down to a 600-lbf load level, and a 1000-lbf-capacity proving ring is usable down to a 100-lbf load level.

If calibrating load cells are used, a 60 000-lbf-capacity load cell is usable down to a 3000-lbf load level, a 6000-lbf-capacity load cell is usable to a 300-lbf load level, and a 600-lbf-capacity load cell is usable down to a 120-lbf load level.

Before use, proving rings and load cells must be removed from their cases and allowed to stabilize to ambient (surrounding) temperature. Upon stabilization, either type of unit is placed on the table of the testing machine. At this stage, proving rings are ready to operate, but load cells must be connected to an appropriate power source and again be allowed to stabilize, generally for 5 to 15 min.

Each system is set to zero, loaded to the full capacity of the machine or elastic device, then unloaded to zero for checking. Loading to full capacity and unloading must be repeated until a stable zero is obtained, after which the load verification readings are made at the selected test load levels.

For the highest load range of 60 000 lbf, loads are applied to the calibrating device from its minimum lower limit (6000 lbf for proving rings and 3000 lbf for load cells) to its maximum 60 000 lbf in a minimum of five steps, or test load levels, as discussed earlier in this article in the section on load verification. In the verification loading procedure for proving rings, a "set-the-load" method usually is used. The test load is determined, and the nominal load is preset on the proving ring. The machine load readout is read when the nominal load on the proving ring is achieved. For load cells, a "follow-the-load" method can be used, wherein the load on the display indicator is followed until the load reaches the nominal load, which is the preselected load level on the readout of the testing machine.

In both methods, the load of the testing machine and the load of the calibration device are recorded. The error E and the percent error, E_p, can be calculated as:

$$E = A - B$$

$$E_p = \frac{(A - B)}{B} \times 100$$

where A is the load indicated by the machine being verified in lbf, kgf, or N, and B is the correct value of the applied load (lbf, kgf, or N), as determined by the calibration device.

This procedure is repeated until each scale range of the testing machine has been calibrated from minimum to maximum capacity. The necessary reports and certificates are then prepared, with the loading range(s) indicated clearly as required by ASTM E 4. Figures 5(b) and 6 illustrate universal testing machines being calibrated with elastic proving rings and calibration load cells.

Extensometers

Extensometers are instruments for measuring the elongation of a specimen as a result of the applied load. They are used extensively and can provide a high degree of deformation (strain) measurement accuracy. Early extensometers were held to the specimen with center points matching the specimen gage length punch marks, and elongation was indicated between the points by a dial indicator.

Because of mechanical problems associated with these early devices, most dial extensometers in current applications use knife edges and leaf-spring pressure for specimen attachment. An extensometer using a dial indicator to measure elongation is shown in Fig. 7. The dial indicator usually is

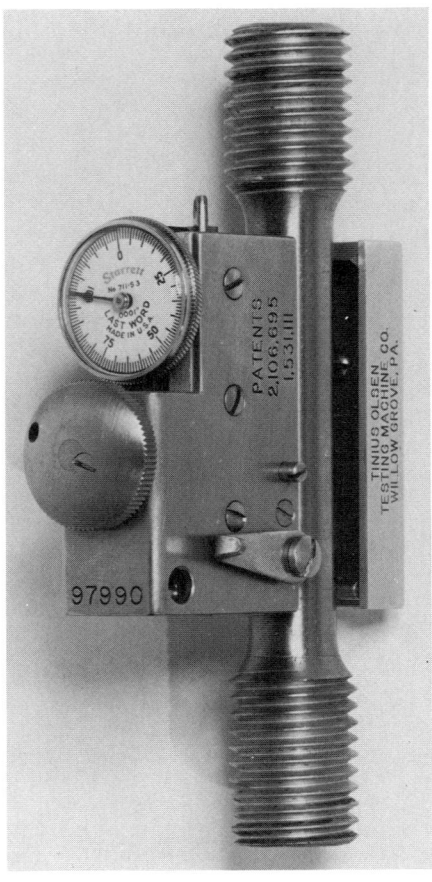

Fig. 7 2-in. (50-mm) gage length averaging dial-type extensometer

marked off in 0.0001-in. (0.0025-mm) increments and measures the total extension between the gage points. This value divided by the gage length gives strain in inches per inch or millimeters per millimeter.

Extensometers typically are either nonaveraging or averaging types. A nonaveraging extensometer has one fixed nonmovable knife edge or center point and one movable knife edge or center point on the same side of the specimen. This arrangement results in extension measurements that are taken on one side of the specimen only; such measurements do not take into account that elongation may be slightly different on the other side.

For most specimens, notably those with machined rounds or reduced gage length flats, there is no significant difference in elongation between the two sides. However, for as-cast specimens, some forged parts, and specimens made from tubing, a difference in elongation sometimes exists on opposite sides of the specimen when subjected to a tensile load. This is due to part configura-

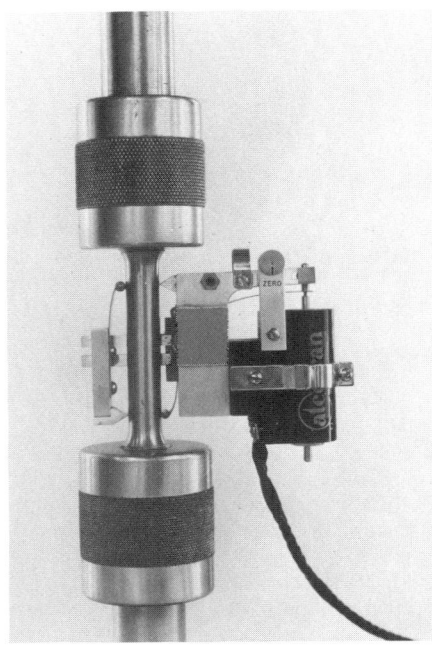

Fig. 8 2-in. (50-mm) gage length averaging LVDT extensometer mounted on a threaded tension specimen

tion and/or internal stress. Misalignment of grips also contributes to elongation measurement variations in the specimen.

When precision must be ensured and test results are to be used in a courtroom or in authoritative reports, averaging-type instruments are preferred and sometimes specified. An averaging dial-type extensometer (Fig. 7) has one fixed knife edge or center point on one side and one movable knife edge or center point on the opposite side. Thus, elongation on each side of the specimen is measured and averaged.

When a graphical record of load versus elongation is required, electrical-readout extensometers are used rather than dial-type extensometers. Readout-type extensometers, in conjunction with the test machine recorder, automatically provide a load-extension curve, thus eliminating the time-consuming task of constructing a curve from dial readings.

The most commonly used extensometers, which are similar to the device shown in Fig. 8, employ a linear variable differential transformer (LVDT). Deformation of the specimen causes the core of the LVDT to move proportionately. In universal testing machines with built-in drum-type x-y recorders, this core movement produces an electrical signal. After amplification, this signal activates a servomotor that rotates the recorder drum and a cam (Fig. 9). As the cam turns, it

Fig. 9 Schematic of the load-strain recorder system

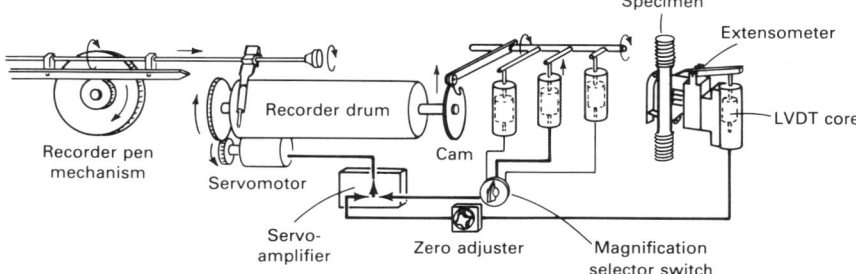

moves the cam follower and the attached duplicate LVDT core(s) in the recorder. This core motion produces a signal that is opposite and equal to the strain signal produced by the extensometer.

The pen line drawn by the rotation of the recorder drum indicates the specimen strain. While the recorder drum is being rotated at a rate proportional to the specimen strain rate, the load-activated pen mechanism moves the pen point across the drum in direct proportion to the increasing or decreasing stress (load) on the specimen. A typical drum *x-y* recorder is shown in Fig. 10. Flatbed *x-y* recorders are also frequently used.

LVDT extensometers are small, lightweight, and easy to use. Knife edges provide an exact point of contact and are mechanically set to the exact gage length. Unless the test report specifies total elongation, center punch marks or scribed lines are not required to define the gage length. They are available with gage lengths ranging from 0.4 to 100 in. (10 to 2500 mm) and can be fitted with breakaway features (Fig. 11), sheet metal clamps, low-pressure clamping arrangements (film clamps, as shown in Fig. 12), and other devices. Thus, they can be used on small specimens, such as thread, yarn, and foil, and on large test specimens, such as reinforcing bars, heavy steel plate, and tubing up to 3 in. (75 mm) in diameter.

Modifications of the LVDT extensometer described above permit linear measurements at temperatures ranging from −100 to 2200 °F (−75 to 1205 °C). Accurate measurements can also be made in a vacuum. For standard instruments, the working tempera-

ture range is approximately −100 to 250 °F (−75 to 120 °C). However, by substituting an elevated-temperature transformer coil, the usable range of the instrument can be extended to −200 to 500 °F (−130 to 260 °C).

Extensometers using strain gages rather than linear variable differential transformers are also available and are lighter in weight and smaller in size, although strain gages are somewhat more fragile than linear variable differential transformers. The strain gage usually is mounted on a pivoting beam, which is an integral part of the extensometer. The beam is deflected by the movement of the extensometer knife edge when the specimen is stressed. The strain gage attached to the beam is an electrically conductive smallsized grid that changes its resistance when deformed in tension, compression, bending, or torsion. Thus, strain gages can be used to supply the information necessary to calculate strain, stress, angular torsion, and pressure.

Fig. 11 2-in. (50-mm) gage length breakaway-type LVDT extensometer that can remain on the specimen through rupture

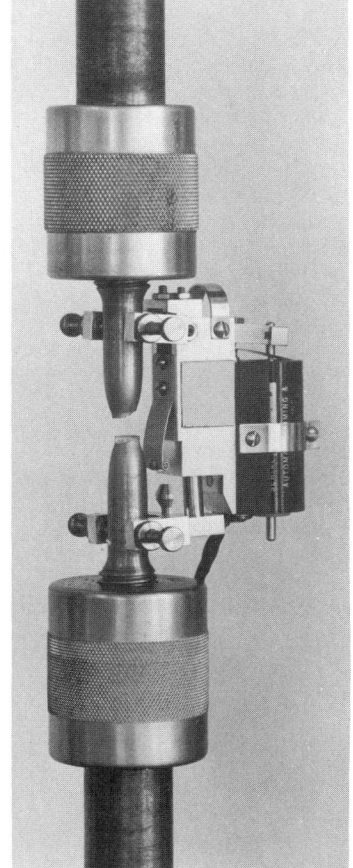

Fig. 10 Drum-type x-y recorder

Fig. 12 2-in. (50-mm) averaging LVDT extensometer mounted on a 0.005-in. (0.127-mm) wire specimen

The extensometer is fitted with a low-pressure clamping arrangement (film clamps) and is supported by a counterbalance device.

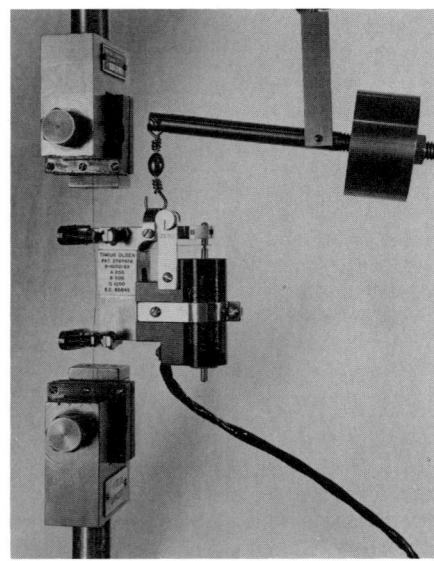

Fig. 13 Fatigue test specimen with bonded resistance strain gages and a 1-in. (25-mm) gage length extensometer mounted on the reduced section

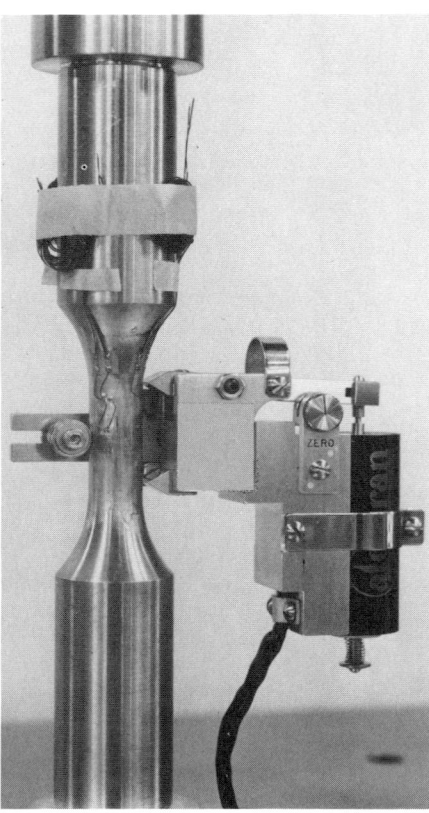

Fig. 14 Calibrator used for mechanical and electrical extensometers and load-elongation recorders

Strain gages have been improved and refined, and their use has become widespread. Basic types include wire gages, foil gages, and capacitive gages. Wire and foil bonded resistance strain gages are used for measuring stress and strain and for calibration of load cells, pressure transducers, and extensometers. These gages typically measure ⅜ to ½ in. (9.5 to 13 mm) in width and ½ to ¾ in. (13 to 19 mm) in length and are adhesively bonded to a metal element (Fig. 13).

For some strain measurements, strain gages are mounted on the part being tested. When used in this manner, they differ from extensometers in that they measure average unit elongation over nominal gage length rather than total elongation between definite gage points. Thus, strain gages are seldom used in production testing of standard tension specimens.

In conventional use, wire or foil strain gages, when mounted on structures and parts for stress analysis, are discarded with the tested item. Foil strain gages currently are the most widely used, due to the ease of their attachment.

Optical systems, lasers, and other systems can also be used to obtain linear strain measurements. Optical extensometers are particularly useful with materials such as rubber, thin films, plastics, and other materials where the weight of a conventional extensometer would distort the workpiece and affect the readings obtained. Such strain-

measuring systems are expensive, however, and their principal use is in research and development work.

Extensometer Classification. All types of extensometers must be verified, classified, and calibrated in accordance with ASTM E 83 to be acceptable for materials testing. This designation establishes six classifications of instruments based on the maximum error of indicated strain (see Table 1). To determine accuracy or error deviations, the calibration displacement apparatus must

be certified to be five times more accurate than the allowable classification error.

Class A extensometers, if available, would be used for determining precise values of the modulus of elasticity and for precise measurements of permanent set or very slight deviations from Hooke's law. Currently, however, there are no commercially available extensometers manufactured that are certified to comply with class A requirements.

Class B-1 extensometers are frequently

Table 1   Classification of extensometer systems

| Classification | Error of strain(a) not to exceed the greater of: | | Error of gage length not to exceed the greater of: | |
	Fixed error, in./in.	Variable error, % of strain	Fixed error, in.	Variable error, % of gage length
Class A	0.00002	±0.1	±0.001	±0.1
Class B-1	0.0001	±0.5	±0.0025	±0.25
Class B-2	0.0002	±0.5	±0.005	±0.5
Class C	0.001	±1	±0.01	±1
Class D	0.01	±1	±0.01	±1
Class E	0.1	±1	±0.01	±1

Note: This classification system has been proposed by ASTM Subcommittee E 28.01, Calibration of Mechanical Testing Machines and Apparatus, for inclusion in ASTM E 83 in 1985.
(a) Strain of extensometer system—ratio of applied extension to the gage length

used to determine values of the modulus of elasticity and to measure permanent set or deviations from Hooke's law. They are also used for determining values such as the yield strength of metallic materials.

Class B-2 extensometers are used for determining the yield strength of metallic materials. This classification applies to extensometers with gage lengths of less than 2 in. (50 mm), which do not meet class B-1 accuracy requirements.

All LVDT and strain-gage extensometers comply with class B-1 or class B-2 requirements if their measuring ranges do not exceed 0.02 in. (0.5 mm). Instruments with measuring ranges of over 0.02 in. (0.5 mm) are class C instruments.

Most electrical differential transformer extensometers of 500 strain magnification and higher conform to class B-1 requirements throughout their measuring range. Extensometers of less than 500 strain magnification comply only with class B-1 requirements in their lower (40%) measuring range and are basically class B-2 instruments.

Although all dial instruments usually are considered class C instruments, the majority (up to a gage length of 8 in., or 200 mm) are class B-1 and class B-2 in their initial 40% measuring range and class C throughout the remainder of the range. Dial instruments are used universally for determining yield strength by the extension-under-load method and yield strength of 0.1% offset and greater.

Extensometers with a gage length of 24 in. (610 mm) begin in class C, although their overall measuring range must be considered as class D. For gage lengths of 24 to 100 in. (610 to 2500 mm), these instruments are class D. Verification and classification of extensometers are applicable to instruments of both the averaging and nonaveraging type.

Extensometer Calibration. Calibration of extensometers and extensometer systems must be very precise. Several devices can be used, including an interferometer, calibrated standard gage blocks and an indicator, and a micrometer screw.

An extensometer and recorder calibrator such as that shown in Fig. 14 is widely used. This device contains a precision micrometer screw and etched dial that are certified by the National Bureau of Standards. The scribed increments are 0.00005 in. (0.00127 mm) and are readable to 0.00001 in. (0.000254 mm). The calibrator, when fitted with proper precise spindles and adapters, is capable of calibrating extensometers with gage lengths of up to 24 in. (610 mm).

REFERENCES

1. "Standard Practices for Load Verification of Testing Machines," ASTM E 4, *Annual Book of ASTM Standards*, Vol 03.01, ASTM, Philadelphia, 1984, p 111-118

2. "Standard Method of Verification and Classification of Extensometers," ASTM E 83, *Annual Book of ASTM Standards*, Vol 03.01, ASTM, Philadelphia, 1984, p 247-252

3. "Standard Practice of Calibration of Force-Measuring Instruments for Verifying the Load Indication of Testing Machines," ASTM E 74, *Annual Book of ASTM Standards*, Vol 03.01, ASTM, Philadelphia, 1984, p 238-246

Statistics and Data Analysis

Introduction

By Richard C. Rice
Manager, Structural Integrity Projects Office
Mechanics Section
Battelle Columbus Laboratories

APPLIED STATISTICS AND DATA ANALYSIS are important tools for the materials engineer involved in mechanical testing who must consider the issues of how much data is enough, how each variable is factored into an experimental study, and how data should be analyzed and interpreted. Because mechanical property data are subject to uncertainty and variability, these factors must be addressed directly if the results are to have significance.

Statistical concepts that form the basis of most of the following articles on specific areas of applied statistics and data analysis in this Section are reviewed here. Readers with prior statistical background may find it sufficient to scan this Introduction before proceeding to the detailed discussions that follow. In addition, several basic statistical references are cited for further information.

In this Section, fundamental statistical concepts are discussed first to provide a common base of understanding. Three articles on applied statistics that present practical examples for illustration follow. The first of these, "Statistical Distributions," deals with six statistical distribution functions used in engineering applications. This article describes the characteristics of these distributions and identifies methods for computing the significant parameters associated with each of them.

The next article, "Planning of Comparative Experiments," addresses the important problem of constructing well-designed experiments. It describes several specific types of experimental designs, ranging from very simple two-level factorial designs to more complex experimental designs involving multiple levels, randomized blocks and partial factorial procedures.

The third article, "Analysis of Designed Experiments," addresses some of the basic statistical procedures that can be applied to mechanical property data. Correlation and regression analysis is dealt with briefly, followed by a discussion of the analysis of designed experiments. In this article in particular, sources of further information are given.

Data-specific analysis methods are covered in the last four articles in this Section. Types of data analyses perceived to be especially complicated, important, and/or controversial are addressed. Specifically, the types of data analyses that are covered include "Design Allowables for Static Metallic Material Properties," "Fatigue Crack Growth Data Analysis," "Analysis of Creep and Creep-Rupture Data," and "Fatigue Data Analysis." Each article offers detailed commentary on the recommended ways to evaluate statistically and interpret these important mechanical property data. Limits on the applicability of different procedures are generally identified, and sample cases illustrate the practical application of these methods.

Statistical Concepts

No test method, however well specified and controlled, will produce exactly the same results every time it is applied, and no material property is absolute. There is some inherent variability in even the most basic material properties, such as Young's modulus, coefficient of thermal expansion, and density.

The variability or uncertainty in a measured property is as important as the absolute value of the initial measurement. Variability, or scatter, must be dealt with realistically if the material property values are to be useful.

For example, consider a situation in which a material supplier produces new heats (batches or lots) of material and wishes to monitor the "quality" of the product, which in this case will be described simply by the ultimate tensile strength. If one coupon is cut from a particular heat and tested, a measure of the tensile strength of that material is obtained. If several more coupons are cut from the same heat and tested in exactly the same manner, additional measures of the tensile strength almost certainly will not be identical.

If another heat of material is tested in the same manner, variability in ultimate tensile strength will also be observed. Table 1 illustrates a possible outcome of the efforts to characterize these two heats of material. If only one specimen had been taken from each heat, the probable conclusion would have been that the two heats were comparable in tensile strength. If two specimens had been taken from each heat, a possible conclusion would have been that Heat 2 was inferior in strength. However, subsequent pairs of tests, taken separately, could lead to the opposite conclusion. After five tests of each heat, Heat 1 shows a higher average strength, but a lower minimum strength and much higher scatter.

Which heat is better? Material properties are not exact quantities, and variability among specimens and products occurs. A better estimate of the value of a particular material property can be obtained if repeat measurements are made and a sample average is obtained. If accuracy were the only concern, then many samples would be tested for each property to be determined.

Table 1 Results of a hypothetical test series to determine the ultimate tensile strength of two heats of material

Sample No.	Tensile strength			
	Heat 1		Heat 2	
	MPa	ksi	MPa	ksi
1	522	75.7	517	75.0
2	511	74.1	490	71.1
3	489	70.9	499	72.4
4	554	80.3	514	74.6
5	500	72.5	503	72.9
Average	515	74.7	505	73.2

Economic constraints generally prevent a large number of repeat, or replicate, tests. The trade-off of accuracy and validity of results versus testing time and cost is often complicated by a variety of factors, some controllable and others not. The discussion below reviews some of the basic concepts that must be understood to successfully apply the statistical procedures that are reviewed in subsequent articles in this Section.

Probability

Although probability and statistics are often discussed in conjunction with one another, it is not always understood where one concept begins and ends and in what areas the two disciplines complement each other. In fact, statistics and probability are interrelated; the theory of statistics provides much of the structure for the theory of probability.

Statistics is commonly defined as the mathematical process of collecting, organizing, and interpreting numerical data, particularly the assessment of population characteristics by inference from sampling. A probability is a number that expresses the likelihood of occurrence of a specific event. This concept can be expressed as:

$$\text{Probability of occurrence} = \frac{\text{Number of desired occurrences}}{\text{Total number of trials}}$$

$$= \lim_{n \to \infty} \frac{m}{n}$$

$= p(x)$ for a continuous random variable

$= p(x_i)$ for discrete random variable

Basic properties of such a probability can be defined. First, a probability is a nonnegative quantity that must lie in the range between 0.0 and 1.0. An event that is almost certain to occur has a probability close to unity. An event that is very unlikely to occur has a probability near zero.

In practical terms, and in the context of this article, the idea of probability becomes important when an engineer changes his question from "Is it strong enough?" or "Will it fail?" to "What is the likelihood that it is strong enough?" or "What is the chance that it will fail?" The latter questions are perhaps more difficult to answer than the former, but they are also more realistic. The underlying truth is that uncertainty arises from a variety of sources, including random measurement error, bias in the test procedure, and the intrinsic material property. In the context of a random variable, the probability that such a variable will take on any

one of its possible values is known as its density function. There are many well-known theoretically established probability density functions that display a wide variety of shapes for both discrete and continuous random variables.

Statistical distributions that may be used to describe samples and populations of data are derived mathematically from the basic probability density functions. Confidence limits and intervals are constructed to contain the population parameters at a specific probability or confidence level. Null and alternate hypotheses are tested for significance at specific probability levels. The basis for reliability analyses centers around the joint probabilities of low strength and high stress. Several practical texts on the subject of probability theory and its application are available (Ref 1-4).

Random Variables

Table 1 illustrates that even apparently similar samples tested under identical conditions generally do not give the same result, producing data scatter or variability. Each observation was based on randomly selected specimens from each heat of material. Even if it were possible to measure with absolute precision the strength of a large number of randomly selected specimens, the resultant strength values would be distributed over a wide stress range. In this case, ultimate tensile strength is considered a continuous random variable because it can take on an infinite number of values between zero and infinity. Most mechanical properties are considered to be continuous random variables, including properties such as tensile strength, hardness fracture toughness, fatigue resistance, and creep resistance.

By comparison, consider a component with five possible and equally probable failure locations, identified as 1, 2, 3, 4, and 5. The number of the failure location in this case is a discrete random variable, because the variable can only take on one of the five values. Discrete random variables become part of engineering applications in counting and partitioning. The number of defective samples drawn from a large lot of samples or the number of specimens tested in an 8-h shift would be discrete random variables.

Randomness of the observations of strength, hardness, or toughness must not be significantly violated, or the commonly used statistical procedures to analyze those data will be invalid. Randomness can be disturbed by a variety of factors, including variations in specimen fabrication procedures, test methods, system calibration, operator bias, and data analyst bias. In most cases, the goal of the experimenter/analyst is to minimize

unnecessary sources of variability, carefully control the desired variability factors, and then randomize the experiments within the entire matrix of conditions to be examined (Ref 5, 6).

Descriptive Statistics

Descriptive statistics define the characteristics of samples of data. A sample is defined as a group of items under test, selected randomly from a lot of similar items of "infinite" size. This infinitely large lot is the population. In most engineering situations the problem is one of predicting the characteristics of the population based on test results that comprise a small sample of that population. Clearly, a sample represents an approximation of the real population. Generally, if a sample is randomly selected from the population, its statistical characteristics converge toward the statistical characteristics of the population as the sample size increases. Note, however, that the observed characteristics of the population under study depend on the measurement process used.

Measures of Central Tendency

Whether dealing with a discrete or continuous random variable, it often becomes important to identify a "typical" value. Most engineers think in terms of typical or average values when they consider the mechanical properties of a particular material. Even though most mechanical properties can vary over a significant range, some values are more typical than others. These typical values are described statistically as the mean or expected value, median, and mode.

Mean or Expected Value. The population mean or the expected value of a continuous random variable may be expressed as:

$$\mu = E(x) = \int_{-\infty}^{+\infty} x f(x) dx \qquad \text{(Eq 1)}$$

where x is the continuous random variable, and $f(x)$ is the distribution function.

Six commonly used statistical distributions are reviewed in the article "Statistical Distributions" in this Section. For a discrete random variable, the population mean can be computed as:

$$\mu = E(x) = \sum_{i=1}^{n} x_i P(x_i) \qquad \text{(Eq 2)}$$

where x_i is the value of the discrete random variable, and $P(x_i)$ is the probability of occurrence associated with x_i. The mean of a sample or sample average is obtained by dividing the sum of all observations by their total number. In direct mathematical terms, the mean of a sample of size n is computed as:

$$\bar{x} = \frac{\sum\limits_{i=1}^{n} x_i}{n} \qquad \text{(Eq 3)}$$

Median. The median of a population is that value of the random variable at which the cumulative distribution function, $F(x)$, is 0.50. In other words, it is the value that divides the population into equal parts. For symmetrical distributions such as the normal distribution, the median and the mean are equal. For nonsymmetrical distributions such as the Weibull distribution, the median and mean generally are not equal (see the article "Statistical Distributions" in this Section for more information on symmetrical and nonsymmetrical distributions).

The median of a sample is the number in the "middle" when all the observations are ranked in their order of magnitude. For example, the median observation in Table 1 for Heat 1 is 511 MPa (74.1 ksi), because the observations in rank order are 489 (70.9), 500 (72.5), 511 (74.1), 522 (75.7), and 554 (80.3). If there is an even number of observations, the estimator of the median value is the average of the two middlemost rank observations. For example, for the rank values 489, 500, 511, 522, 554, and 560, the sample median is 516.5 [(511 + 522)/2].

Mode. The mode of a population, whether it is a continuous or discrete random variable, is the value of the variable corresponding to the maximum probability of occurrences. It is defined mathematically for a continuous random variable as that value of x at which:

$$\frac{dp(x)}{dx} = \frac{df(x)}{dx} = 0 \qquad \text{(Eq 4)}$$

The mode of a sample is the value that occurs most frequently. If a frequency of occurrence diagram is constructed, as in Fig. 1, the mode represents the "peak" in the histogram. In this example, the mode is the fracture-toughness interval ranging from 37 to 39 MPa$\sqrt{\text{m}}$ (33.7 to 35.5 ksi$\sqrt{\text{in.}}$).

Measures of Variability

The next logical step in the statistical characterization of a sample of data obtained from a material is to quantify the amount of dispersion or scatter. The simplest way to specify the variability of a sample is to define the range from the lowest to highest observations. In Fig. 1, the observations clearly range from 29 to 49 MPa$\sqrt{\text{m}}$ (26.4 to 44.6 ksi$\sqrt{\text{in.}}$). In cases such as this, however, which show clear central tendency, it is possible to compute a more meaningful measure

of variability. This measure of variability is known as the standard deviation. The standard deviation of a sample, σ, is normally computed by taking the square root of the variance, σ^2, which is the second moment about the sample average, $\bar{x}$:

$$\sigma = \sqrt{\frac{\sum\limits_{i=1}^{n} (x_i - \bar{x})^2}{n-1}} \qquad \text{(Eq 5)}$$

Because $\bar{x}$ represents the average of the sample and x_i represents individual observations, it is apparent that high scatter leads to high values of standard deviation. Two equivalent formulations for computing sample standard deviations are defined as:

$$\sigma = \sqrt{\frac{n\sum\limits_{i=1}^{n} x_i^2 - \left(\sum\limits_{i=1}^{n} x_i\right)^2}{n(n-1)}} \qquad \text{(Eq 6)}$$

$$\sigma = \sqrt{\frac{\sum\limits_{i=1}^{n} x_i^2 - n\bar{x}^2}{n-1}} \qquad \text{(Eq 7)}$$

Equations 5 to 7 give unbiased estimates of the sample standard deviation. In some cases, a biased estimate of the standard deviation is computed in which the divisor used is n instead of $n - 1$. This represents the root mean square (rms) value of the sample.

A useful measure of relative variability can be obtained by dividing the sample standard deviation by the sample mean. This quantity is known as the coefficient of variation. Mechanical properties displaying low variability exhibit coefficients of variation below about 0.02 or 2%, while more highly scattered mechanical properties may display coefficients of variation of 10% or more.

If a fixed quantity is added or subtracted from each of the sample observations, the mean will be increased or decreased, respectively, by that amount, but the standard deviation and variance will not be affected. For two populations with the same standard deviation, the population with the smaller mean will have the larger coefficient of variation. If each sample observation is multiplied by a fixed quantity, m, the new mean is m times the old mean, the new variance is m^2 times the old variance, the new standard deviation is m times the old standard deviation, and the coefficient of variation is unaffected.

Degrees of Freedom

The degrees of freedom of a sample are important to consider when computing descriptive statistics (i.e., the standard devia-

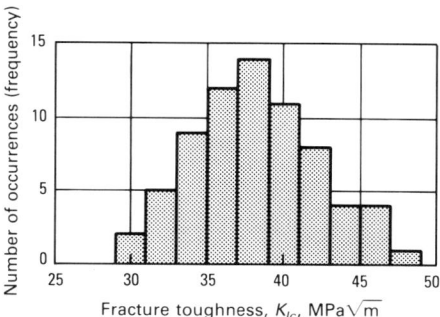

Fig. 1 Example of a distribution of fracture toughness values for a material
1.0 MPa$\sqrt{\text{m}}$ = 0.91 ksi$\sqrt{\text{in.}}$

Fracture toughness, K_{Ic}		Observations
MPa$\sqrt{\text{m}}$	ksi$\sqrt{\text{in.}}$	
29-31	26.4-28.2	2
31-33	28.2-30.0	5
33-35	30.0-31.8	9
35-37	31.8-33.7	12
37-39	33.7-35.5	14
39-41	35.5-37.3	11
41-43	37.3-39.1	8
43-45	39.1-40.9	4
45-47	40.9-42.8	4
47-49	42.8-44.8	1
Total		70

tion) or performing other more complex statistical analyses. Degrees of freedom, ν, can be defined as the number of observations made in excess of the minimum needed to estimate a statistical parameter or quantity.

For example, only one measurement is theoretically required to identify the length of a fatigue crack in a specimen. If the measurement were repeated six times, then the sample variance of crack length measurement has five degrees of freedom, because five more measurements were made than were required. If only one measurement were made, the variance would be indeterminate. If two measurements were made, the variance could be estimated as the range, or difference, of the two observations.

When a regression analysis that involves estimation of multiple parameters is performed, such as:

$$y = a_1 + a_2 x_1 + a_3 x_2 + a_4 x_1 x_2 + a_5 x_3$$
$$\text{(Eq 8)}$$

the degrees of freedom to establish the correlation are $n - m$, where m is the number of regression coefficients, a_i. In Eq 8, if the regression analysis were completed on 15 observations, the degrees of freedom for the residuals would be $n - m$, or $15 - 5$, which equals 10.

Degrees of freedom are important in a number of statistical calculations. When confidence intervals on regression coefficients are calculated, the degrees of freedom indicate whether the coefficients are significantly different from zero. In testing the correlation coefficient, r, the degrees of freedom indicate whether the correlation is significant.

More information on basic descriptive statistics and the concept of degrees of freedom can be found in elementary and intermediate-level statistical analysis texts (Ref 7-10).

Confidence Limits and Intervals

In the design and analysis of engineering experiments, the concept of confidence limits and confidence intervals is useful. When dealing with test data, computed sample statistics are only estimates of the true population statistics. Therefore, it must be determined how close these estimates are likely to be to the true values for the population.

A confidence interval defines a range within which, at some specified confidence level, the population parameter can be expected to lie. A confidence limit or bound defines a value above or below which, at some specified confidence level, the population parameter can be expected to lie. Both one- and two-sided confidence limits can be constructed.

The degree of confidence associated with a confidence interval or limit is known as its confidence level. Confidence levels of 90, 95, and 99% are commonly used. For example, a 95% lower confidence limit for the unknown population mean formed by use of the sample average and sample standard deviation provides a value above which the unknown population mean is expected to lie with 95% confidence. A 95% confidence interval on the standard deviation could be similarly established.

The confidence interval can be given physical meaning if random samples of size n drawn from a normal population are considered. For example, interval estimates of the population mean, μ, and standard deviation, σ, can be constructed for each sample by use of $\bar{x}$ and s for a 90% confidence level. The intervals defined for each sample either contain the true parameter value (μ or σ, whichever is being considered) or do not. The confidence level of 90% simply implies that the confidence interval established for 90% of all repeated samples of size n would bracket the true parameter value. Larger samples tend to give narrower confidence intervals for the same level of confidence. Also, the size of a computed interval is larger when a higher confidence level is desired. This illustrates a basic trade-off between level of confidence, size of a confidence interval, and sample size. If the size of a confidence interval is to remain fixed, sample sizes should be increased in situations where higher confidence levels are required.

The difference between confidence intervals, prediction intervals, and tolerance intervals must also be noted. As already discussed, a confidence interval can be expected to contain a population parameter or characteristic, such as the population average, at some confidence level. A prediction interval can be expected to contain a single future value at some confidence level. A lower confidence limit on the average of a collection of tensile test results will always be greater than a lower prediction limit on a single future test result if both limits are established at the same confidence level.

A tolerance interval can be expected to contain a particular percentage of future observations with a specified level of confidence. A tolerance limit is constructed in the same manner, but it defines only a single bound, above or below which a certain percentage of future observations can be expected to fall with a specified level of confidence. An example of a tolerance limit commonly used for structural materials is the "A" value (Ref 1), which is a statistically determined quantity that represents the stress level above which 99% of all future observations of tensile strength are expected to fall at a 95% confidence level. The determination of tolerance limits on mechanical property data is discussed in the article "Design Allowables for Static Metallic Material Properties" in this Section. Other discussions of confidence limits and intervals can be found in Ref 8 to 13.

Testing Hypotheses

A hypothesis is an assertion that is subject to verification or proof. In statistical terms, a hypothesis is a statement about the relationship between a number of random variables, i.e., mechanical property data.

A statistical test always involves a null hypothesis—the hypothesis under test. One or more alternative hypotheses derive from the null hypothesis. The null hypothesis might involve the absolute comparison of two means [H_0: ($\mu = \mu_0$)] or two variances [H_0: ($\sigma^2 = \sigma_0^2$)]. Alternatively, the null hypothesis might involve the relative comparison of two means [H_0: ($\mu_x - \mu_y$) = 0] or two variances [H_0: (σ_x^2/σ_y^2) = 1].

Alternative hypotheses may be one sided or two sided. One-sided alternative hypotheses for an absolute comparison of two means are [H_A: ($\mu < \mu_0$)] and [H_A: ($\mu > \mu_0$)], while the two-sided alternative hypothesis for an absolute comparison of two variances is [H_A: $\sigma^2 \neq \sigma_0^2$)]. Similarly, the two-sided alternative hypothesis for a relative comparison of two means could be [H_A: ($\mu_x - \mu_y$) $\neq$ 0] if the two variances are unknown and assumed unequal, and the one-sided hypothesis for a relative comparison of two variances is [H_A: ($\sigma_x^2/\sigma_y^2 > 1$)]. The null hypothesis is usually the base or beginning hypothesis.

When testing statistical hypotheses, there is always a risk of error. Two types of errors can occur. If the null hypothesis is rejected when it is actually true—for example, if two means are found to be different when they really are equal—then an error of the first kind, or Type I error, has been made. If the null hypothesis is not rejected when it actually is false, then an error of the second kind, or Type II error, has been made. In a specific instance it is never known whether a Type I or Type II error has been made. However, the probability of making either type of error can be defined.

For example, it may be necessary to judge whether the mean strength value of a new heat of material conforms with the properties known to be typical of that product. In this case one might assume that the samples tested came from a heat with the same strength properties as the "parent" heat. This represents the null hypothesis [H_0: ($\mu = \mu_0$)]. One could also assume that the null hypothesis is not true. If this second assumption is made, one of several alternative hypotheses could be formulated: [H_A: ($\mu \neq \mu_0$)], [H_A: ($\mu > \mu_0$)], or [H_A: ($\mu < \mu_0$)]. If the null hypothesis chosen is accepted, the alternative hypothesis would automatically be rejected.

In any decision regarding the equality or inequality of the properties for the two materials, there is some probability of making a mistake. Therefore, it is important to determine the probability of making a Type I error by rejecting the null hypothesis when it is true. The probability of this occurrence is normally denoted by α. The quantity $(1 - \alpha)$ is often called the confidence level, but it is really the probability of making a correct decision when the null hypothesis is true. It is generally desirable for α to be small, because this would imply that the result was statistically significant and did not occur by chance.

Often there is some value of mean strength, μ, which is undesirable. In most cases too low a strength is critical, but in other cases too high a strength may also be considered undesirable. In this case, let us consider a strength level below μ to be undesirable. It is important to construct a hypothesis that addresses whether the sample mean is below μ_1, that is [H_A: ($\mu < \mu_1$)].

The next step is to find the probability of accepting the null hypothesis as true when the alternate hypothesis is actually true; that is, what is the probability of stating that

$\mu = \mu_0$ when actually $\mu < \mu_1$? This is a Type II error. Its probability of occurrence is commonly denoted by β.

For a particular case, the critical values of α and β should be defined before the statistical analysis is complete. Then, calculations can be made based on either the null or alternate hypothesis, and the calculated probabilities of α or β can be compared with the critical values. If the actual value of α or β exceeds the critical value, then the corresponding hypothesis should be rejected. If the computed values of α and β are both small relative to the critical values, there is a good statistical basis for deciding whether the mean strength of the new heat is different from that of the standard or other heat. If α and β are not small enough, then there will be insufficient evidence to draw a statistically significant conclusion. Generation of additional data may be required to draw a statistically significant conclusion.

The significance level represents the level of reluctance to reject the null hypothesis. A high significance level of 0.01 or even 0.001 implies great reluctance to state that the populations are different when they really are the same. As a consequence of this conservative decision, the probability of not rejecting the null hypothesis when it is really false will be large, unless the actual difference in the means is very large. This may be an acceptable situation if standard material properties are considered most desirable.

However, if the intent is to identify a heat that is better than standard, it may be preferable to adopt a higher risk level or lower significance level of perhaps $\alpha = 0.05$ or 0.10. Such a decision would increase the chance of detecting a small but important improvement in properties, without the expense of a large testing effort. Obviously, the choice of a significance level, α, for a statistical test is a trade-off proposition that depends on the circumstances and objectives of the test program. Significance levels of 0.05 and 0.01 are commonly used and are presented in most statistical tables that are used to complete a statistical test. Details on how such tests are performed are included in the article "Analysis of Designed Experiments" in this Section. Several other sources may also be useful if an in-depth review of testing hypotheses is desired (Ref 8-10, 14).

Reliability

Reliability is a measure of the adequacy of a design in its intended service environment for a specified period of time. Reliability applies to a part, system, or assembly rather than to the materials that make up the design. Failure probability and reliability are in-

versely related. If the failure probability is near zero, then the reliability level is near unity because:

Reliability = 1 − failure probability

Numerically, the reliability level relates to the percentage of survivors from a large number of units.

Reliability levels and confidence levels are related, but they are not the same and are generally not equal in a given situation. A particular part may have 95% reliability at 80% confidence. This means that 1 out of 20 parts tested under the conditions of interest would be expected to fail. If such a test series were repeated ten times, it would be expected that no more than one failure would occur eight out of ten times.

Because reliability and failure probability are related, it is sufficient to identify the probability of failure and then infer the level of reliability. The concept of failure probabilities is tied to the idea that failure occurs when a stress on a part exceeds the part strength. It is also based on the idea that a given type of part has a distribution of strengths from one part to the next, and that these parts, when placed in service, will experience different stress levels from one usage to another. The overlap of these two distributions of stress and strength represents the probability of failure (Fig. 2).

In Fig. 2, as the overlap increases, the shaded area increases, and the failure probability increases. This illustrates that two different components designed to the same factor of safety (the ratio of $\bar{S}/\bar{s}$ in Fig. 2) may not have the same reliability. If one component is made of material that is too variable in its strength properties (displaying a high coefficient of variation), it is likely to be less reliable than a more controlled product. Also, if one component design experiences different stresses in different usages, it is likely to be less reliable than a part design subjected to the same stresses from one usage to another. This argument addresses only statically loaded components, although the same concepts could be applied for a fatigue critical application. Accurate reliability estimates depend on the proper statistical characterization of mechanical property data generated on the structural materials involved. Detailed presentations on reliability analysis can be found in Ref 4 and 15 to 17.

REFERENCES

1. W. Feller, *An Introduction to Probability Theory and Its Applications*, Vol 1, 3rd ed., John Wiley & Sons, New York, 1968
2. E. Parzen, *Modern Probability Theory*

and Its Applications, John Wiley & Sons, New York, 1960
3. P.G. Hoel, *Introduction to Probability Theory*, Houghton-Mifflin, Boston, 1971
4. E.B. Haugen, *Probabilistic Approaches to Design*, John Wiley & Sons, New York, 1968
5. G.J. Hahn, Must I Randomize?, *Chemtech*, Vol 7, Oct 1977, p 630-632
6. G.J. Hahn, More On Randomization, *Chemtech*, Vol 8, March 1978, p 164-168
7. J. Neter, W. Wasserman, and G.A. Whitmore, *Applied Statistics*, 2nd ed., Allyn and Bacon, Boston, 1982
8. G.W. Snedecor and W.G. Cochran, *Statistical Methods*, 7th ed., Iowa State University Press, Ames, IA, 1980
9. M.G. Natrella, *Experimental Statistics*, National Bureau of Standards Handbook No. 91, U.S. Government Printing Office, Washington, DC, 1963 (also available from John Wiley & Sons, New York)
10. C. Lipson and N.J. Sheth, *Statistical Design and Analysis of Engineering Experiments*, McGraw-Hill, New York, 1973
11. G.J. Hahn, Putting Bounds on a Prediction, *Chemtech*, Vol 4, June 1974, p 381-383
12. A.E. Mace, *Sample Size Determination*, Reinhold, New York, 1964
13. G.J. Hahn, What Confidence Level Should I Select?, *Chemtech*, Vol 5, March 1975, p 186-187
14. A.H. Bowker and G.J. Lieberman, *Engineering Statistics*, Prentice-Hall, Englewood Cliffs, NJ, 1959
15. *Reliability Handbook*, W.G. Ireson, Ed., McGraw-Hill, New York, 1966
16. D.K. Lloyd and M. Lipon, *Reliability: Management, Methods, and Mathematics*, 2nd ed., Redondo Beach, CA, 1977
17. M.L. Shooman, *Probabilistic Reliability: An Engineering Approach*, McGraw-Hill, New York, 1968

Fig. 2 Stress (s) and strength (S) distributions

Including probability of failure (shaded area); commonly called the "Warner diagram"

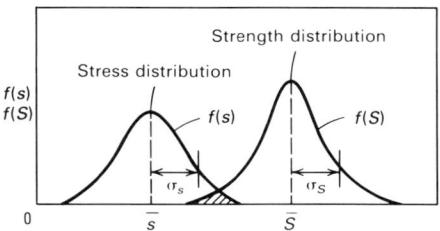

Statistical Distributions

By Steven W. Rust
Principal Research Statistician
Battelle Columbus Laboratories

and

Richard C. Rice
Manager, Structural Integrity Projects Office
Mechanics Section
Battelle Columbus Laboratories

STATISTICAL DISTRIBUTIONS model various physical, mechanical, electrical, and chemical properties, as well as the number of occurrences of events. There are six well-known and widely used statistical distributions: normal, log normal, Weibull, and exponential, which are continuous in nature, and binomial and Poisson, which are discrete functions. This article discusses the applicability of each distribution, including information on the probability density function, cumulative distribution function, population mean and variance, and parameter and percentile estimation. Table 1 summarizes this information. The final section of this article also addresses the problem of selecting statistical distributions.

The probability density function of a continuous population is the limiting function approached by histograms of samples drawn from the population as the sample size grows large. Thus, a probability density function is essentially a smooth histogram. For discrete data, the probability density function assigns probability to each of the possible values of the variable.

Figure 1 illustrates a histogram that describes the distribution of tensile yield strengths for a particular material. The number of observations in each interval is recorded. Continuous probability density functions are superimposed over the histogram to indicate the conformity of these data to a normal distribution as compared with either a two-parameter or a three-parameter Weibull distribution.

In most practical engineering situations, a limited number of observations, called a sample, are available to describe a particular property. Statistical distributions often represent apparent trends of samples.

The cumulative distribution function of a population (both discrete and continuous) indicates the proportion of the population that lies below each possible value in the population. The mean value of a population is the arithmetic average of all the values in the population. The population variance is the average squared deviation of the population items from the population mean; thus, the variance is a measure of population spread.

The population mean and variance are sometimes referred to as the first and second moments of a population. Common symbols for these population parameters are mean = μ and variance = σ^2 (where σ = population standard deviation).

All of the distributions discussed in this article depend on parameters that are generally unknown in specific applications, where only limited samples of data are available. Methods for estimating these unknown parameters (primarily maximum likelihood estimation procedures) are provided for each distribution. It is assumed either that the ana-

Fig. 1 Histogram of tensile yield strength values with superimposed probability density functions

Note: 1 ksi = 6.8948 MPa.

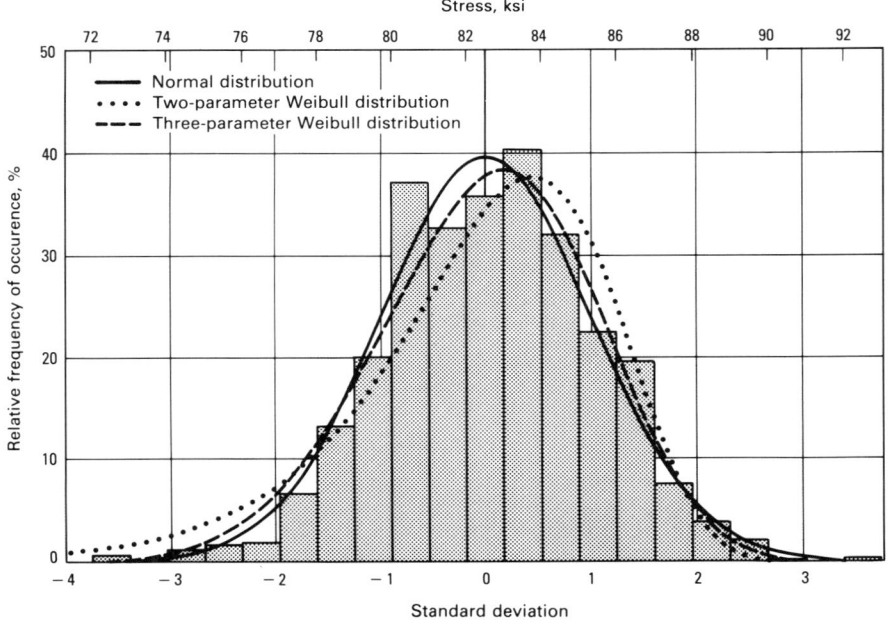

lyst has prior information concerning the most appropriate distribution or that one of the standard goodness-of-fit criteria will be applied in order to select the distribution that provides the minimum goodness-of-fit statistic.

Population percentiles are also considered for the continuous distributions. Formulas for calculating population percentiles and estimating unknown population percentiles are provided. A population percentile is a value below which a specified percentage of the population lies. For example, assuming a normal distribution, a value 2.32 standard deviations below the mean represents approximately the first percentile; that is, only 1% of the population would be expected to fall below this value.

Distributions that model data directly are discussed in this article. Common distributions that are derived from the normal distribution (chi-square, t, F) are discussed in Ref 1 to 6. The estimators presented here are based on the assumption that the measurements taken have not been censored in any way. For excellent treatments of censored data problems, see Ref 4 and 6. References 2 and 3 provide a comprehensive collection of information about numerous statistical distributions used in the development of statistical theory and practice. Reference 5 is a useful layman's source on statistical distributions and their practical application.

Normal Distribution

The normal distribution is the most widely used and best understood statistical distribution. In addition to its widespread use as an approximation to the distributions of many computed statistics, the normal distribution is frequently used to model various physical, mechanical, electrical, and chemical properties—for example, tensile strength, hardness, conductivity, and elastic modulus. Many properties that scatter randomly about a well-defined mean value, without either positive or negative bias, also tend to follow a normal distribution.

The primary limitation associated with the modeling of physical measurements with the normal distribution is its symmetry about an average value. Thus, skewed or asymmetric populations are not modeled well by the normal distribution. The main advantage of the normal distribution is the large body of literature available describing procedures for the analysis of normally distributed data.

Probability Density Function. The normal probability distribution has two parameters: the population mean and the population standard deviation. The population mean is denoted by μ, and the population

Table 1 Statistical distribution information

Statistical distribution	Probability density function	Cumulative distribution function(a)	Mean
Normal ..... $f(x) = \dfrac{1}{\sqrt{2\pi}\,\sigma} \exp\left[-\dfrac{1}{2}\left(\dfrac{x-\mu}{\sigma}\right)^2\right]$		$F(x) = F_o\left(\dfrac{x-\mu}{\sigma}\right)$	μ
Log normal . $f(x) = \dfrac{1}{\sqrt{2\pi}\,\sigma x} \exp\left[-\dfrac{1}{2}\left(\dfrac{\ln(x)-\mu}{\sigma}\right)^2\right]$, $x > 0$		$F(x) = F_o\left(\dfrac{\ln(x)-\mu}{\sigma}\right)$	$\exp\left(\mu + \dfrac{\sigma^2}{2}\right)$
Weibull ..... $f(x) = \dfrac{\beta}{\alpha}\left(\dfrac{x}{\alpha}\right)^{\beta-1}\exp\left[-\left(\dfrac{x}{\alpha}\right)^\beta\right]$, $x > 0$		$F(x) = 1 - \exp\left[-\left(\dfrac{x}{\alpha}\right)^\beta\right]$	$\alpha\Gamma\left(1 + \dfrac{1}{\beta}\right)$
Exponential . $f(x) = \dfrac{1}{\alpha}\exp\left(\dfrac{-x}{\alpha}\right)$, $x > 0$		$F(x) = 1 - \exp\left(\dfrac{-x}{\alpha}\right)$	α
Binomial .... $f(x) = \binom{n}{x} p^x (1-p)^{n-x}$, $x = 0, 1, \ldots, n$		$F(x) = \sum_{y=0}^{x} \binom{n}{y} p^y (1-p)^{n-y}$	np
Poisson ..... $f(x) = \dfrac{\mu^x e^{-\mu}}{x!}$, $x = 0, 1, \ldots, n$		$F(x) = \sum_{y=0}^{x} \dfrac{\mu^y e^{-\mu}}{y!}$	μ

Statistical distribution	Variance	pth percentile(b)	Parameter estimates	Percentile estimates
Normal	σ^2	$Q_p = \mu + z_p\sigma$	$\hat{\mu} = \bar{x}$ $\hat{\sigma} = s$	$\hat{Q}_p = \bar{x} + z_p s$
Log normal	$(\exp(\sigma^2) - 1)\exp(2\mu + \sigma^2)$	$Q_p = \exp(\mu + z_p\,\sigma)$	$\hat{\mu} = \bar{x}_L$ $\hat{\sigma} = s_L$	$\hat{Q}_p = \exp(\bar{x}_L + z_p\,s_L)$
Weibull ...	$\alpha^2\left[\Gamma\left(1 + \dfrac{2}{\beta}\right) - \Gamma^2\left(1 + \dfrac{1}{\beta}\right)\right]$	$Q_p = \alpha\left[-\ln\left(1 - \dfrac{p}{100}\right)\right]^{1/\beta}$	See text	See text
Exponential	α^2	$Q_p = \alpha\left[-\ln\left(1 - \dfrac{p}{100}\right)\right]$	$\hat{\alpha} = \bar{x}$	$\hat{Q}_p = \hat{\alpha}\left[-\ln\left(1 - \dfrac{p}{100}\right)\right]$
Binomial	$np(1-p)$		$\hat{p} = \dfrac{x}{n}$	
Poisson	μ		$\hat{\mu} = \bar{x}$	

(a) F_o is the standard normal cumulative distribution function. (b) z_p is the pth percentile of the standard normal distribution.

standard deviation is denoted by σ. The normal probability density function is:

$$f(x) = \frac{1}{\sqrt{2\pi}\,\sigma}\exp\left[-\frac{1}{2}\left(\frac{x-\mu}{\sigma}\right)^2\right]$$

$$\text{(Eq 1)}$$

A graph of the normal density function is illustrated in Fig. 2. The numbers under the curve represent the areas under the curve for intervals of length σ. In Fig. 2, the normal distribution is symmetric about its mean μ. Furthermore, virtually all of the area (99.74%) under the normal density is contained within the interval $(\mu - 3\sigma, \mu + 3\sigma)$.

Consequently, almost all of the probability associated with a normal distribution falls within three standard deviations of the population mean μ. Also, 95.44% of the area falls within two standard deviations of μ and 68.26% within one standard deviation.

Cumulative Distribution Function. The normal cumulative distribution function is obtained by integrating the normal density function (Eq 1) from $-\infty$ to x to obtain the

Fig. 2 Normal probability density function

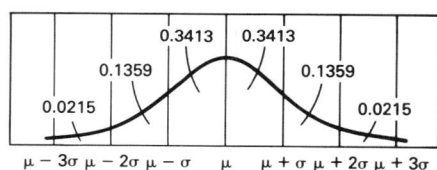

0.3413 0.3413
0.1359 0.1359
0.0215 0.0215

$\mu - 3\sigma \quad \mu - 2\sigma \quad \mu - \sigma \quad \mu \quad \mu + \sigma \quad \mu + 2\sigma \quad \mu + 3\sigma$

area under the density function to the left of the arbitrary point x. Figure 3 illustrates the cumulative distribution function, $F(x)$, for the standard normal distribution. $F(x)$ has a minimum value of zero for very small values of x relative to μ and a maximum value of unity for very large values of x relative to μ. The cumulative distribution function identifies the percentage of a population that lies below the value x. The normal cumulative distribution function is:

$$F(x) = \int_{-\infty}^{x} \frac{1}{\sqrt{2\pi}\,\sigma}$$
$$\exp\left[-\frac{1}{2}\left(\frac{y-\mu}{\sigma}\right)^2\right] dy \qquad \text{(Eq 2)}$$

There is no closed form for $F(x)$ that does not require integration. The formula for $F(x)$ can be changed by a linear transformation to a simpler form in which the integrand does not depend on any unknown parameters. The simplified form for $F(x)$ is:

$$F(x) = \int_{-\infty}^{(x-\mu)/\sigma} \frac{1}{\sqrt{2\pi}}$$
$$\exp\left[-\frac{1}{2}z^2\right] dz \qquad \text{(Eq 3)}$$

If $F_o(x)$ is used to denote the cumulative distribution function (Eq 2) of the standard normal distribution ($\mu = 0$, $\sigma = 1$), then Eq 3 can be simplified further to:

$$F(x) = F_o\left(\frac{x-\mu}{\sigma}\right) \qquad \text{(Eq 4)}$$

Thus, the area under a general normal density to the left of x is equal to the area under the standard normal curve to the left of $(x - \mu)/\sigma$. Values of the standard normal cumulative distribution function $F_o(x)$ are presented in Table 2.

Population Mean and Variance. The two parameters of the normal distribution—

Fig. 3 Cumulative distribution function for the standard normal distribution

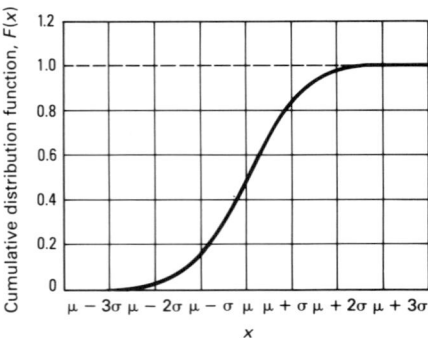

μ and σ—represent the mean and standard deviation of the population described by the distribution. The variance of a normal population is σ^2. As illustrated in Fig. 2, the normal distribution is symmetric about its mean μ.

Population Percentiles. The pth percentile of a normal distribution, Q_p, is the solution to the equation $F(Q_p) = p/100$, where the function $F(\cdot)$ is the cumulative distribution function of the normal distribution (Eq 2 to 4). Assuming that z_p denotes the pth percentile of the standard normal distribution, from Eq 4 it follows:

$$F(\mu + \sigma z_p) = F_o(z_p) = \frac{p}{100} \qquad \text{(Eq 5)}$$

Thus, Q_p is:

$$Q_p = \mu + \sigma z_p \qquad \text{(Eq 6)}$$

Table 2 may be used to determine z_p for use in Eq 6. Many statistical computer packages also contain programs that calculate values of z_p.

Estimation of Parameters and Percentiles. Given a set of n measurements x_1, $x_2, \ldots, x_n$ to be modeled by a normal distribution, population parameters and percentiles are estimated as follows. Calculate the sample mean, $\bar{x}$, and the sample standard deviation, s, by the following formulas:

$$\bar{x} = \frac{1}{n}\sum_{i=1}^{n} x_i \qquad \text{(Eq 7)}$$

$$s^2 = \frac{1}{n-1}\sum_{i=1}^{n}(x_i - \bar{x})^2$$
$$= \frac{1}{n-1}\left[\sum_{i=1}^{n} x_i^2 - \frac{1}{n}\left(\sum_{i=1}^{n} x_i\right)^2\right]$$
$$\text{(Eq 8)}$$

Then $\bar{x}$ is the maximum likelihood estimate of the population mean ($\hat{\mu} = \bar{x}$), and s is an estimate of the population standard deviation ($\hat{\sigma} = s$).

Estimates of population percentiles can be obtained by replacing μ and σ with their estimates in Eq 6. Thus, an estimate of the pth percentile of the normal population from which the sample was drawn is:

$$\hat{Q}_p = \bar{x} + s z_p \qquad \text{(Eq 9)}$$

The sample mean and standard deviation of the tensile yield strengths in Fig. 1 are 82.53 and 2.805, respectively. Assuming that this material property is distributed normally, 82.53 is the estimated population mean and 2.805 is the estimated population standard deviation. The fitted normal distri-

bution is simply the normal probability density function with this mean and standard deviation. When examining the tensile yield strength of a particular material, values of the lower percentiles of the strength distribution are often the quantities of interest. For example, in Fig. 1, estimates of the first and tenth percentiles are, respectively, $75.99 = 82.53 - 2.33(2.805)$ and $78.94 = 82.53 - 1.28(2.805)$.

Log Normal Distribution

The log normal distribution frequently is used as a model for lifetime and durability phenomena. In particular, fatigue data developed on simple specimens and components have been found to be distributed approximately in accordance with the log normal distribution. For example, a series of fatigue test data was generated on a rail steel material at a single stress value. When the logarithms of the fatigue lives were computed and the computed values were ranked and plotted on normal probability paper, a nearly straight line resulted (Fig. 4). The data trends deviate from linearity at the very high and low lives, but the trends are reasonably linear on the log normal scale for observations between the 10th and 90th percentiles.

The log normal distribution assigns zero probability to negative values of the measured property, and it always represents an asymmetric distribution that is positively skewed (long right tail). Thus, the log normal distribution is often used as an alternative to the normal distribution for positively valued, positively skewed data.

If the values of a measured property follow a log normal distribution, then the logarithms of the measured values follow a normal distribution. Thus, procedures for the analysis of normally distributed data are applicable to the log normal distribution through transformation.

Probability Density Function. The log normal probability distribution has two parameters: μ and σ. Assume that a measured property x follows a log normal distribution with parameters μ and σ. The log normal distribution is so named because the natural logarithm of x, $\ln(x)$, follows a normal distribution with mean μ and standard deviation σ. The log normal probability density function is:

$$f(x) = \frac{1}{\sqrt{2\pi}\,\sigma x}$$
$$\exp\left[-\frac{1}{2}\left(\frac{\ln(x)-\mu}{\sigma}\right)^2\right], x > 0$$
$$\text{(Eq 10)}$$

Table 2 Cumulative normal distribution

$$F(x)_{x=x} \frac{1}{\sqrt{2\pi}} \int_{-\infty}^{x} e^{-u^2/2}\, du$$

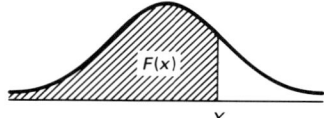

X	0.00	0.01	0.02	0.03	0.04	0.05	0.06	0.07	0.08	0.09
0.0	0.50000	0.50399	0.50798	0.51197	0.51595	0.51994	0.52392	0.52790	0.53188	0.53586
0.1	0.53983	0.54380	0.54776	0.55172	0.55567	0.55962	0.56356	0.56749	0.57142	0.57535
0.2	0.57926	0.58317	0.58706	0.59095	0.59483	0.59871	0.60257	0.60642	0.61026	0.61409
0.3	0.61791	0.62172	0.62552	0.62930	0.63307	0.63683	0.64058	0.64431	0.64803	0.65173
0.4	0.65542	0.65910	0.66276	0.66640	0.67003	0.67364	0.67724	0.68082	0.68439	0.68793
0.5	0.69146	0.69497	0.69847	0.70194	0.70540	0.70884	0.71226	0.71566	0.71904	0.72240
0.6	0.72575	0.72907	0.73237	0.73565	0.73891	0.74215	0.74537	0.74857	0.75175	0.75490
0.7	0.75804	0.76115	0.76424	0.76730	0.77035	0.77337	0.77637	0.77935	0.78230	0.78524
0.8	0.78814	0.79103	0.79389	0.79673	0.79955	0.80234	0.80511	0.80785	0.81057	0.81327
0.9	0.81594	0.81859	0.82121	0.82381	0.82639	0.82894	0.83147	0.83398	0.83646	0.83891
1.0	0.84134	0.84375	0.84614	0.84850	0.85083	0.85314	0.85543	0.85769	0.85993	0.86214
1.1	0.86433	0.86650	0.86864	0.87076	0.87286	0.87493	0.87698	0.87900	0.88100	0.88298
1.2	0.88493	0.88686	0.88877	0.89065	0.89251	0.89435	0.89617	0.89796	0.89973	0.90147
1.3	0.90320	0.90490	0.90658	0.90824	0.90988	0.91149	0.91309	0.91466	0.91621	0.91774
1.4	0.91924	0.92073	0.92220	0.92364	0.92507	0.92647	0.92786	0.92922	0.93056	0.93189
1.5	0.93319	0.93448	0.93574	0.93699	0.93822	0.93943	0.94062	0.94179	0.94295	0.94408
1.6	0.94520	0.94630	0.94738	0.94845	0.94950	0.95053	0.95154	0.95254	0.95352	0.95449
1.7	0.95543	0.95637	0.95728	0.95818	0.95907	0.95994	0.96080	0.96164	0.96246	0.96327
1.8	0.96407	0.96485	0.96562	0.96638	0.96712	0.96784	0.96856	0.96926	0.96995	0.97062
1.9	0.97128	0.97193	0.97257	0.97320	0.97381	0.97441	0.97500	0.97558	0.97615	0.97670
2.0	0.97725	0.97778	0.97831	0.97882	0.97932	0.97982	0.98030	0.98077	0.98124	0.98169
2.1	0.98214	0.98257	0.98300	0.98341	0.98382	0.98422	0.98461	0.98500	0.98537	0.98574
2.2	0.98610	0.98645	0.98679	0.98713	0.98745	0.98778	0.98809	0.98840	0.98870	0.98899
2.3	0.98928	0.98956	0.98983	0.99010	0.99036	0.99061	0.99086	0.99111	0.99134	0.99158
2.4	0.99180	0.99202	0.99224	0.99245	0.99266	0.99286	0.99305	0.99324	0.99343	0.99361
2.5	0.99379	0.99396	0.99413	0.99430	0.99446	0.99461	0.99477	0.99492	0.99506	0.99520
2.6	0.99534	0.99547	0.99560	0.99573	0.99585	0.99598	0.99609	0.99621	0.99632	0.99643
2.7	0.99653	0.99664	0.99674	0.99683	0.99693	0.99702	0.99711	0.99720	0.99728	0.99736
2.8	0.99744	0.99752	0.99760	0.99767	0.99774	0.99781	0.99788	0.99795	0.99801	0.99807
2.9	0.99813	0.99819	0.99825	0.99831	0.99836	0.99841	0.99846	0.99851	0.99856	0.99861
3.0	0.99865	0.99869	0.99874	0.99878	0.99882	0.99886	0.99889	0.99893	0.99897	0.99900
3.1	0.99903	0.99906	0.99910	0.99913	0.99916	0.99918	0.99921	0.99924	0.99926	0.99929
3.2	0.99931	0.99934	0.99936	0.99938	0.99940	0.99942	0.99944	0.99946	0.99948	0.99950
3.3	0.99952	0.99953	0.99957	0.99957	0.99958	0.99960	0.99961	0.99962	0.99964	0.99965
3.4	0.99966	0.99968	0.99969	0.99970	0.99971	0.99972	0.99973	0.99974	0.99975	0.99976
3.5	0.99977	0.99978	0.99978	0.99979	0.99980	0.99981	0.99981	0.99982	0.99983	0.99983
3.6	0.99984	0.99985	0.99985	0.99986	0.99986	0.99987	0.99987	0.99988	0.99988	0.99989
3.7	0.99989	0.99990	0.99990	0.99990	0.99991	0.99991	0.99992	0.99992	0.99992	0.99992
3.8	0.99993	0.99993	0.99993	0.99994	0.99994	0.99994	0.99994	0.99995	0.99995	0.99995
3.9	0.99995	0.99995	0.99996	0.99996	0.99996	0.99996	0.99996	0.99996	0.99997	0.99997

Source: Ref 1

Graphs of the log normal density function for $\mu = 0$ and $\sigma = 0.25, 0.5, 1.5$, and 3 are given in Fig. 5. As illustrated in Fig. 5, the log normal distribution is positively skewed (long right tail). The degree of positive skewness decreases as σ approaches zero, and log normal distributions with $\sigma \leq 0.1$ are similar to normal distributions.

The cumulative distribution function of the log normal distribution is:

$$F(x) = F_o\left(\frac{\ln(x) - \mu}{\sigma}\right), \quad x > 0 \quad \text{(Eq 11)}$$

where $F_o(\cdot)$ is the standard normal cumulative distribution function. See the section of this article on the cumulative distribution function of the normal distribution for more information on $F_o(\cdot)$.

Population Mean and Variance. The mean of the log normal population is:

$$\text{Mean} = \exp\left(\mu + \frac{\sigma^2}{2}\right) \quad \text{(Eq 12)}$$

and the variance is:

$$\text{Variance} = [\exp(\sigma^2) - 1]$$
$$\exp(2\mu + \sigma^2) \quad \text{(Eq 13)}$$

Note that $\exp(\mu)$ is the population median—i.e., the point that divides the population into two groups of equal size.

Population Percentiles. The pth percentile of a log normal distribution, Q_p, is the solution to the equation $F(Q_p) = p/100$, where the function $F(\cdot)$ is the cumulative distribution function of the log normal distribution (Eq 2). Assuming that z_p denotes the pth percentile of the standard normal distribution, it follows:

$$F[\exp(\mu + \sigma z_p)] = F_o(z_p) = \frac{p}{100} \quad \text{(Eq 14)}$$

Thus, Q_p is given by:

$$Q_p = \exp(\mu + \sigma z_p) \quad \text{(Eq 15)}$$

For more information on z_p, see the section of this article on population percentiles of normal distributions.

Estimation of Parameters and Percentiles. Given a set of n measurements $x_1, x_2, \ldots, x_n$ to be modeled by a log normal distribution, population parameters and percentiles are estimated as follows. Calculate the sample mean, $\bar{x}_L$, and the sample standard deviation, s_L, of the natural logarithms of the observations according to:

$$\bar{x}_L = \frac{1}{n}\sum_{i=1}^{n}\ln(x_i) \quad \text{(Eq 16)}$$

$$s_L^2 = \frac{1}{n-1}\sum_{i=1}^{n}[\ln(x_i) - \bar{x}_L]^2$$

$$= \frac{1}{n-1}\left[\sum_{i=1}^{n}(\ln(x_i))^2 - \frac{1}{n}\left(\sum_{i=1}^{n}\ln(x_i)\right)^2\right] \quad \text{(Eq 17)}$$

Fig. 4 Schematic distribution of fatigue lives for 64 steel rail samples at a single stress level

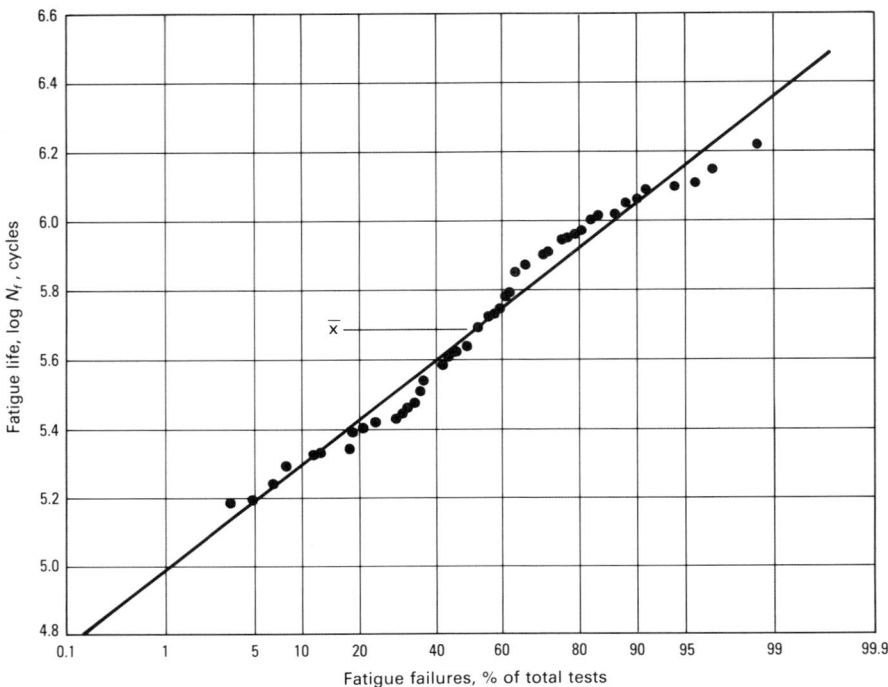

Fig. 5 Log normal probability density function
$\mu = 0$, and $\sigma = 0.25, 0.5, 1.5,$ and 3.0.

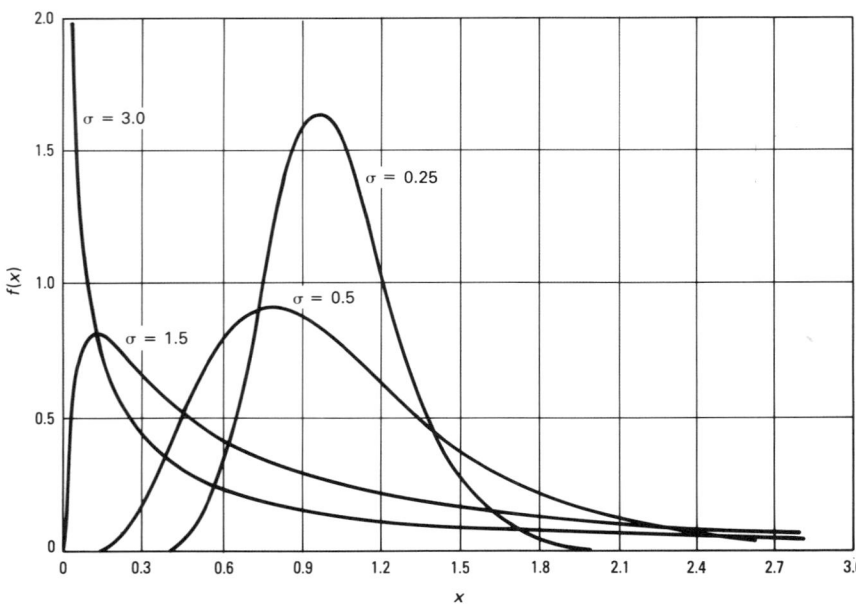

Then, $\bar{x}_L$ is the maximum likelihood estimate of $\mu(\hat{\mu} = \bar{x}_L)$ and s_L is an estimate of $\sigma(\hat{\sigma} = s_L)$. The quantity $\exp(\bar{x}_L)$ is the maximum likelihood estimate of the population median.

Estimates of population percentiles are obtained by replacing μ and σ with their estimates in Eq 15. Thus, an estimate of the pth percentile of the log normal population from which the sample was drawn is:

$$\hat{Q}_p = \exp(\bar{x}_L + s_L z_p) \qquad \text{(Eq 18)}$$

Weibull Distribution

The Weibull distribution frequently is used as a model for lifetime and durability phenomena. This distribution also assigns zero probability to negative values of the measured property. However, the Weibull distribution can be positively skewed, symmetric, or negatively skewed. This flexibility has led to its widespread use.

Figure 6 illustrates a data set that conforms to the two-parameter Weibull distribution. The ranked values of fatigue life fall on a straight line on the Weibull probability paper.

The three-parameter Weibull distribution is simply an extension of the two-parameter Weibull case that includes a nonzero threshold value. Figure 7 shows scattered strength values that range from only 80 up to 290. These data plotted in terms of the two-parameter Weibull distribution do not conform to a straight line. However, the data plot as a straight line if the correct value is subtracted from each observation. The correct value (70, in this case) represents the threshold value for the population. Threshold estimates that are only modestly high or low still produce significantly nonlinear trends for the data.

Probability Density Function. The Weibull probability distribution has two parameters: a population scale parameter α and a population shape parameter β. The three-parameter form of the Weibull distribution, which uses a threshold value, is also discussed in Ref 4. The Weibull probability density function is:

$$f(x) = \frac{\beta}{\alpha}\left(\frac{x}{\alpha}\right)^{\beta-1}$$

$$\exp\left[-\left(\frac{x}{\alpha}\right)^{\beta}\right], x > 0 \qquad \text{(Eq 19)}$$

Graphs of the Weibull density function for $\alpha = 1$ and $\beta = 0.5, 1.0, 1.5,$ and 3.0 are given in Fig. 8.

The shape of the Weibull distribution depends only on the shape parameter β and not on the scale parameter α. For values of β less than approximately 3.6, the distribution is positively skewed (long right tail). For values of β greater than approximately 3.6, the distribution is negatively skewed (long left tail). Weibull distributions with shape parameters of approximately 3.6 are similar to normal distributions.

Cumulative Distribution Function. The Weibull cumulative distribution function is obtained by integrating the Weibull density function (Eq 19) from $-\infty$ to x to obtain the area under the density function to the left of the arbitrary point x. This integral has a closed-form solution, and the Weibull cumulative distribution function is:

$$F(x) = 1 - \exp\left[-\left(\frac{x}{\alpha}\right)^{\beta}\right], x > 0$$

(Eq 20)

Population Mean and Variance. The mean of a Weibull population is:

$$\text{Mean} = \alpha\Gamma\left(1 + \frac{1}{\beta}\right)$$

(Eq 21)

where $\Gamma(\cdot)$ is the standard gamma function given by:

$$\Gamma(\theta) = \int_0^{\infty} x^{\theta-1}\, e^{-x}\, dx$$

(Eq 22)

See Ref 1 for tabular values of the gamma function. The variance of a Weibull population is:

$$\text{Variance} = \alpha^2\left[\Gamma\left(1 + \frac{2}{\beta}\right)\right.$$

$$\left. - \Gamma^2\left(1 + \frac{1}{\beta}\right)\right]$$

(Eq 23)

Population Percentiles. The pth percentile of a Weibull distribution, Q_p, is the solution to the equation $F(Q_p) = p/100$, where the function of $F(\cdot)$ is the cumulative distribution function of the Weibull distribution (Eq 20). Solving this equation yields:

$$Q_p = \alpha\left[-\ln\left(1 - \frac{p}{100}\right)\right]^{1/\beta}$$

(Eq 24)

as the pth percentile of a Weibull distribution.

Estimation of Parameters and Percentiles. Given a set of n measurements x_1, $x_2, \ldots, x_n$ to be modeled by a Weibull distribution, maximum likelihood estimates of the population shape and scale parameters, β and α, respectively, are obtained as follows. Given the maximum likelihood estimate of β, the maximum likelihood estimate of α is calculated as:

$$\alpha = \bar{x}_G\left[\frac{1}{n}\sum_{i=1}^{n}\left(\frac{x_i}{\bar{x}_G}\right)^{\beta}\right]^{1/\beta}$$

(Eq 25)

where

$$\bar{x}_G = \exp\left[\frac{1}{n}\sum_{i=1}^{n}\ln(x_i)\right]$$

(Eq 26)

is the geometric mean of the x values. Define the function $D(\beta)$ as:

$$D(\beta) = \frac{1}{n}\sum_{i=1}^{n}\ln(x_i)\left[\left(\frac{x_i}{\alpha}\right)^{\beta} - 1\right] - \frac{1}{\beta}$$

(Eq 27)

where α is calculated according to Eq 25.

Fig. 6 Schematic of fatigue life data generated at a single stress level that conform to the two-parameter Weibull distribution

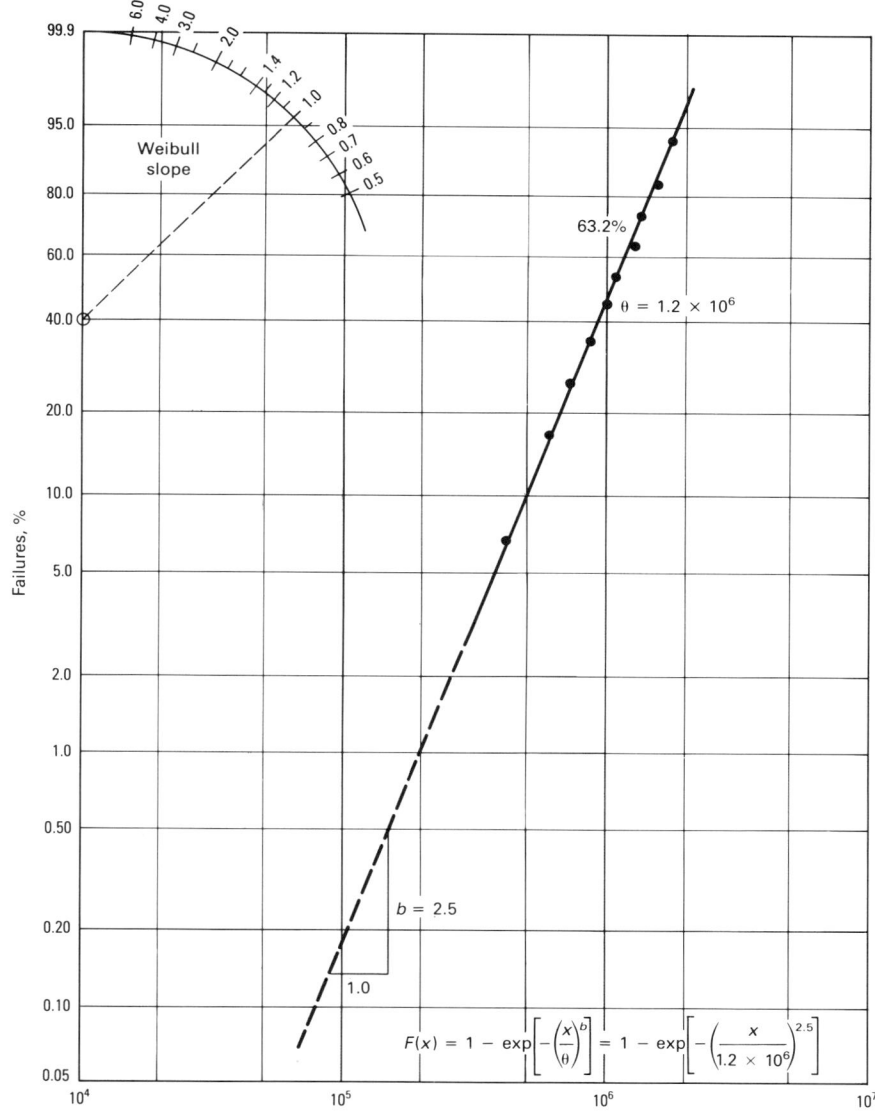

The maximum likelihood estimate of β is then the solution to the equation $D(\beta) = 0$. The function $D(\beta)$ is a monotonically increasing continuous function of β. A simple method for finding the solution is as follows. Calculate $I = 1.28/s_L$ as an initial approximation of the solution. (See Eq 29 for the definition of s_L.) If $D(I) > 0$, then find the smallest positive integer k such that $D(I/2^k) < 0$, and let $L = I/2^k$ and $H = I/2^{k-1}$. If $D(I) < 0$, then find the smallest positive integer k such that $D(2^k I) > 0$, and let $L = 2^{k-1}I$ and $H = 2^k I$. In either case, the interval (L, H) contains the solution to $D(\beta) = 0$.

Now calculate $D(M)$, where $M = (L + H)/2$. If $D(M) = 0$, then the solution is $\beta = M$. If $D(M) > 0$, then let $H = M$. If $D(M) < 0$, then let $L = M$. The new interval (L, H) still contains the solution to $D(\beta) = 0$, but is only half as long as the old interval. Calculate a new M value and begin the process of interval halving again. The process is repeated until $H - L < 2I/10^6$. The solution to $D(\beta) = 0$ is then taken to be $M = (L + H)/2$. The solution is in error by at most $I/10^6$. Once the maximum likelihood estimate of β is determined by solving this equation, the maximum likelihood estimate of α is calculated according to Eq 25.

An alternative to the maximum likelihood method described above is to obtain a method of moments estimator of β and then use Eq 25 to obtain an estimate of α. Calculate the sample mean, $\bar{x}_L$, and the sample standard deviations, s_L, of the natural logarithms of the observations according to:

$$\bar{x}_L = \frac{1}{n} \sum_{i=1}^{n} \ln(x_i) \qquad \text{(Eq 28)}$$

$$s_L^2 = \frac{1}{n-1} \sum_{i=1}^{n} [\ln(x_i) - \bar{x}_L]^2$$

$$= \frac{1}{n-1} \left[\sum_{i=1}^{n} (\ln(x_i))^2 \right.$$

$$\left. - \frac{1}{n} \left(\sum_{i=1}^{n} \ln(x_i) \right)^2 \right] \qquad \text{(Eq 29)}$$

A method of moments estimator of β is $\hat{\beta} = 1.28/s_L$. An estimate of α is then obtained by substituting $\hat{\beta}$ for β in Eq 25.

Let $\hat{\alpha}$ and $\hat{\beta}$ denote estimates of the population scale and shape parameters obtained by one of the methods described above. Estimates of population percentiles are obtained by replacing α and β with their estimates in Eq 24. Thus, an estimate of the pth percentile of the Weibull population from which the sample was drawn is:

$$Q_p = \hat{\alpha} \left[-\ln\left(1 - \frac{p}{100}\right) \right]^{1/\hat{\beta}} \qquad \text{(Eq 30)}$$

Exponential Distribution

The exponential distribution is a special case of the Weibull distribution ($\beta = 1$) and is a very positively skewed distribution. Historically, this distribution has been applied to a wide variety of lifetime, durability, and reliability phenomena because of the ease of developing analytical procedures for exponentially distributed data.

Realization that many inferences are sensitive to departures from the exponential model has led to greater caution in the use of the exponential distribution. For example, when used to model component lifetimes, it requires that failures occur by chance alone with no dependence on the time that an item has been in service, which is often an unrealistic assumption. This implies that the probability an item will fail in the next hour is the same as the probability of failure in any given hour in the past or future. If this assumption is invalid, then very unrealistic life predictions can be made.

The exponential distribution can be used in some cases to identify failure rates in complete systems or in large collections of items. For example, Fig. 9 illustrates failures as a function of time in service. The percentage of failures with time follows an exponential distribution or a Weibull distribution with a slope (shape) parameter of unity.

Probability Density Function. The exponential probability distribution has one parameter, a population scale parameter denoted by α. An exponential distribution is simply a Weibull distribution with shape parameter $\beta = 1$. The exponential probability density function is:

$$f(x) = \frac{1}{\alpha} \exp\left(\frac{-x}{\alpha}\right), \, x > 0 \qquad \text{(Eq 31)}$$

A graph of the exponential density function with $\alpha = 1$ is shown in Fig. 10. The cumu-

Fig. 7 Strength data that conform to the three-parameter Weibull distribution

Threshold value (x_o) = 70

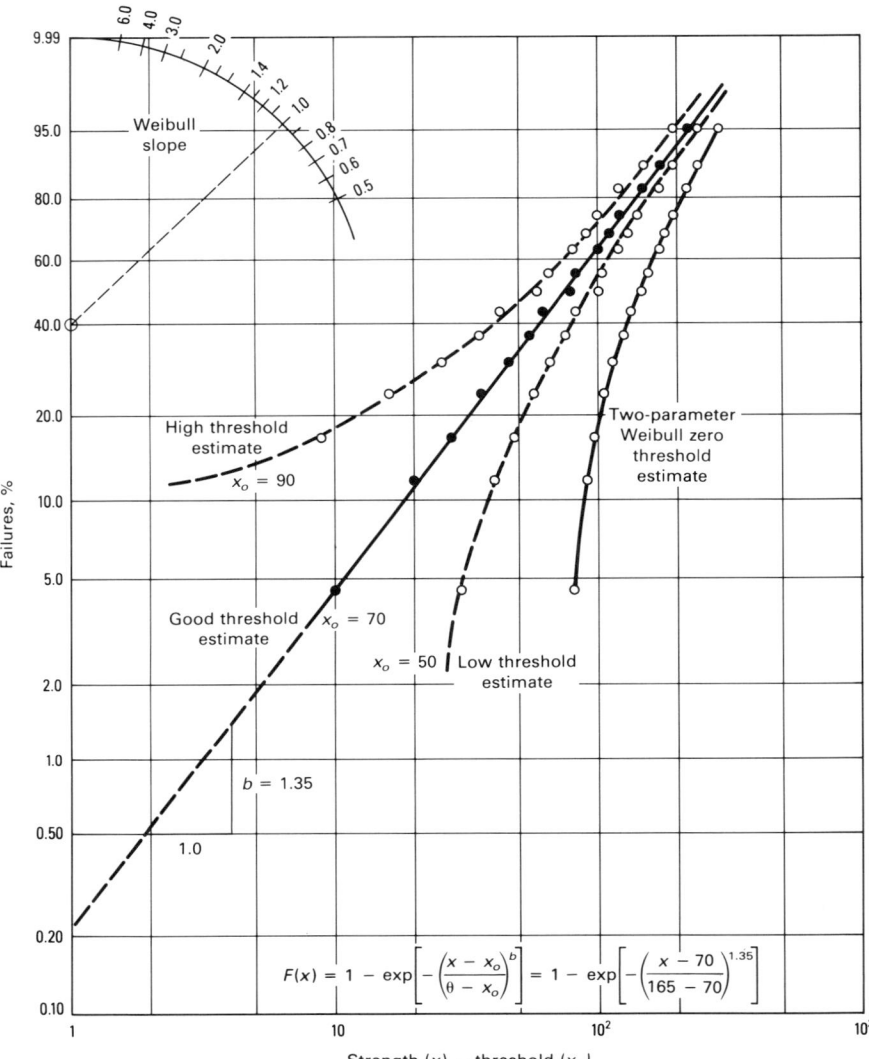

$$F(x) = 1 - \exp\left[-\left(\frac{x - x_o}{\theta - x_o}\right)^b\right] = 1 - \exp\left[-\left(\frac{x - 70}{165 - 70}\right)^{1.35}\right]$$

Fig. 8 Weibull probability density function

$\alpha = 1$, and $\beta = 0.5, 1.0, 1.5,$ and 3.0.

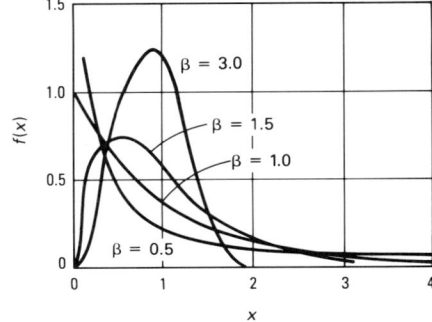

Fig. 9 Failures as a function of time

The failure percentages follow an exponential distribution with a Weibull slope of unity.

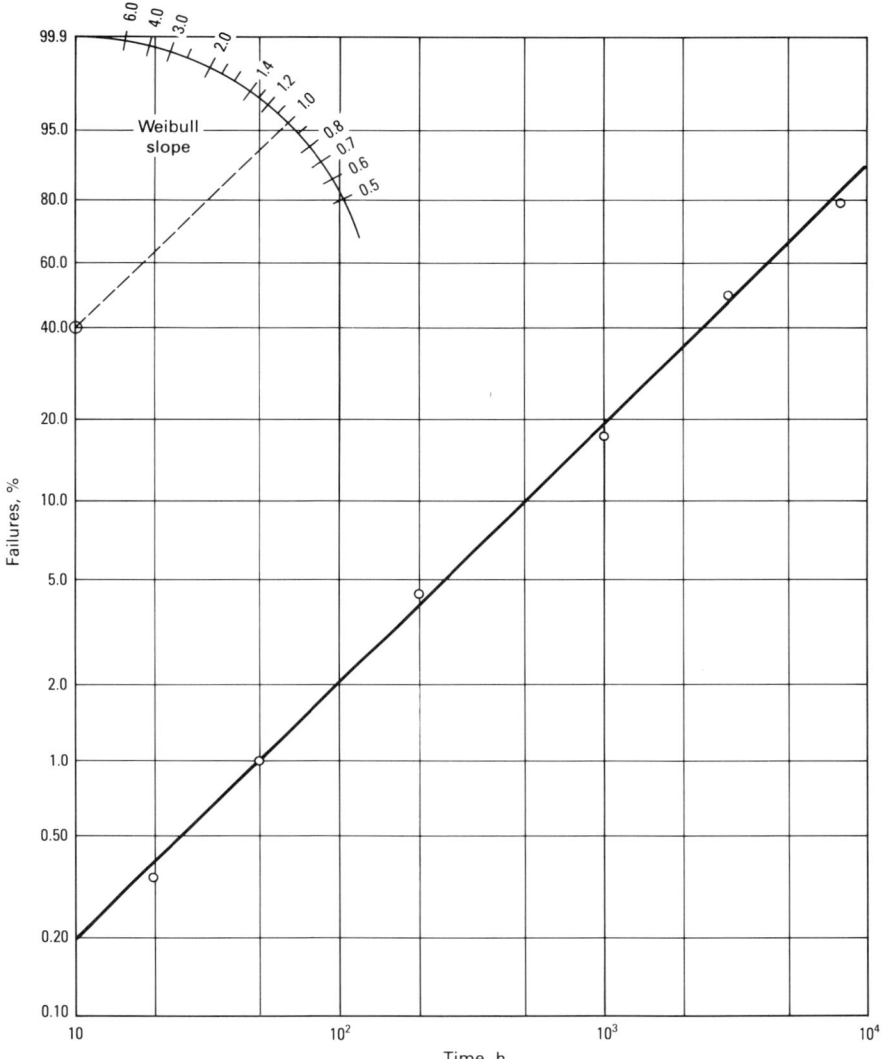

Fig. 10 Exponential probability density function

$\alpha = 1$.

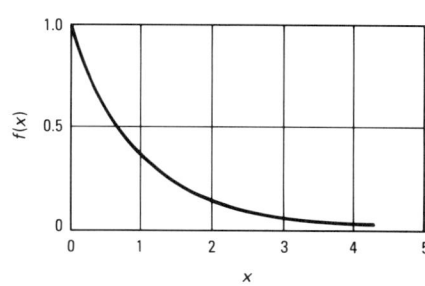

$$\hat{Q}_p = \bar{x}\left[-\ln\left(1 - \frac{p}{100}\right)\right] \qquad (\text{Eq } 36)$$

as the maximum likelihood estimate of the pth percentile of an exponential population.

Binomial Distribution

The binomial distribution frequently is used to model the number of defective items in samples drawn from large lots of items (such as mechanical and electrical parts) submitted to inspection. The binomial theory is correct only when the lots are sampled with replacement; that is, each sampled item must be returned to the lot after inspection and before the next item is selected. However, in many applications, the size of the sample drawn is small in comparison to the lot size, which allows the binomial distribution to provide an excellent approximation even if lots are sampled without replacement.

Most applications involve the estimation of the proportion of defective items in a lot through sampling of a small proportion of the lot. In the following, n is the number of items in the sample from the lot selected for inspection, x is the number of defective items in the sample of size n, and p is the proportion of defective items in the entire lot.

Probability Mass Function. The binomial probability distribution is a discrete probability distribution with two parameters: the sample size, n, and the proportion defective, p. The binomial probability mass function is:

$$f(x) = \binom{n}{x} p^x (1 - p)^{n-x},$$

$$x = 0, 1, \ldots, n \qquad (\text{Eq } 37)$$

Example. Lots of 200 parts are shipped with the assertion that 90% of the parts will exceed a certain mechanical property. For every lot, 205 parts are made and five are selected at random for destructive testing. If more than one of the five samples fail the criteria, the lot is considered suspect.

lative distribution function of the exponential distribution is:

$$F(x) = 1 - \exp\left(\frac{-x}{\alpha}\right), \; x > 0 \qquad (\text{Eq } 32)$$

Population Mean and Variance. The mean and variance of an exponential population are:

Mean $= \alpha$

Variance $= \alpha^2 \qquad (\text{Eq } 33)$

Population Percentiles. The pth percentile of an exponential population, Q_p, is the solution to the equation $F(Q_p) = p/100$, where the function $F(\cdot)$ is the cumulative distribution function of the exponential distribution (Eq 30). Solving this equation yields:

$$Q_p = \alpha\left[-\ln\left(1 - \frac{p}{100}\right)\right] \qquad (\text{Eq } 34)$$

as the pth percentile of an exponential distribution.

Estimation of Parameters and Percentiles. Given a set of n measurements x_1, $x_2, \ldots, x_n$ to be modeled by an exponential distribution, the population scale parameter and percentiles are estimated as follows. Calculate the sample mean:

$$\bar{x} = \frac{1}{n} \sum_{i=1}^{n} x_i \qquad (\text{Eq } 35)$$

of the observations. Then $\bar{x}$ is the maximum likelihood estimate of $\alpha(\hat{\alpha} = \bar{x})$. Replacing α with $\bar{x}$ in Eq 34 yields:

The probability of selecting one below-strength part on the first trial is 10%, or $0.10p$. The probability of selecting one good part on the first trial is 90%, or $q = 1 - p$. The probability of selecting two below-strength parts out of two trials is p^2 or 0.01, while the probability of selecting one below-strength part out of two trials is $2pq$ or 0.18. Using Eq 37, the probability of selecting two below-strength parts out of five trials is $(5!/2!3!)p^2q^3$ or 0.0729. The probability of selecting x number of substandard specimens out of five from a population containing 90% good parts is plotted in Fig. 11. The same kinds of trends are shown for lower percentages of good parts in the tested population. As the percentage of good parts in the tested population decreases, the likelihood of seeing a high percentage of substandard parts increases. Note that for a p value of 0.50 the probability plot is symmetric. For high p values, the greatest probability of occurrence tends toward a low number of substandard parts.

Cumulative Distribution Function. The binomial cumulative distribution function is:

$$F(x) = \sum_{y=0}^{x} \binom{n}{y} p^y (1 - p)^{n-y},$$

$$x = 0, 1, \ldots, n \qquad \text{(Eq 38)}$$

Fig. 11 Probability of obtaining x substandard parts out of five from a population containing p% defective parts

Curved lines represent trends and do not imply a continuous function.

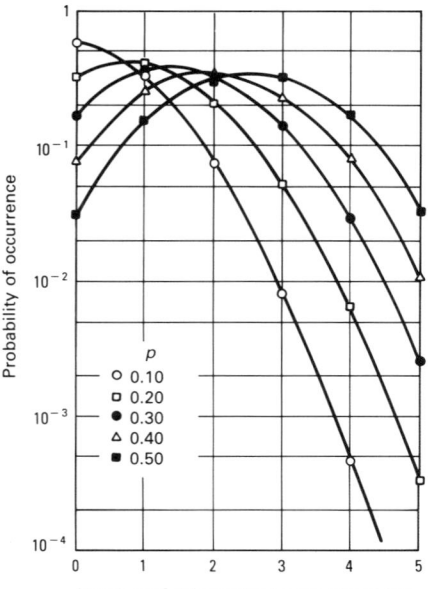

p
- ○ 0.10
- □ 0.20
- ● 0.30
- △ 0.40
- ■ 0.50

Number of substandard parts out of five

Probability of occurrence

Tables of this distribution function for various values of p and small values of n may be found in Ref 1, as well as in many introductory statistical textbooks. Alternatively, many statistical computer packages contain programs to calculate values of $F(x)$ and $f(x)$ for any value of p and n.

For values of n and p for which $np \geq 5$ and $n(1 - p) \geq 5$, the normal approximation to the binomial distribution may be used to obtain approximate values of $F(x)$. Using this approximation yields:

$$F(x) \approx F_o \left(\frac{x + 0.5 - np}{\sqrt{np\,(1 - p)}} \right) \qquad \text{(Eq 39)}$$

as an approximation to $F(x)$, where $F_o(\cdot)$ is the standard normal cumulative distribution function. For more information on $F_o(\cdot)$, see the section of this article on the cumulative distribution function of the normal distribution.

Example. For each lot, 2050 parts are made and 50 are selected at random for destructive testing. If more than 8 of the 50 sampled parts fail the test criterion, the entire lot is considered suspect. The probability that a lot that contains 10% below-strength parts will pass inspection may be estimated using the normal approximation (Eq 39):

$$F(8) = \text{the probability that the number of}$$
$$\text{defective parts in the sample is}$$
$$\text{less than or equal to } 8$$
$$\approx F_o \left(\frac{8 + 0.5 + 50(0.1)}{\sqrt{50(0.1)\,(0.9)}} \right)$$
$$\approx F_o\,(1.65) = 0.95$$

Thus, approximately 95% of the lots containing 10% below-strength parts will pass inspection.

Population Mean and Variance. The mean and variance of a binomial distribution are:

$$\text{Mean} = np$$

$$\text{Variance} = np(1 - p) \qquad \text{(Eq 40)}$$

Parameter Estimation. The sample size, n, is usually a known parameter in applications. Given an observed number of defective items, in a sample of n items, the maximum likelihood estimate of the unknown proportion defective can be estimated by:

$$\hat{p} = \frac{x}{n} \qquad \text{(Eq 41)}$$

This is the sample proportion of defective items.

Poisson Distribution

The Poisson distribution is used to model the occurrence of isolated events in a continuum; that is, the distribution is used to model the number of times that a rare event occurs in a given period of time, an area, or a volume. The primary assumptions of the Poisson distribution are that the period of time, area, or volume can be divided into a large number of elements such that: (1) the probability of the occurrence of the rare event in a given element is proportional to the element size; (2) the probability of two occurrences within a single element is infinitesimally small; and (3) the occurrence of the event in a given element is independent of the occurrence of the event in all other elements. These assumptions are often approximately valid, and the Poisson model provides a reasonable model for x, where x is the number of flaws or defects in materials produced in large quantities and the number of failures occurring during the performance of a repetitive task in a specified period of time, area, or volume.

Probability Mass Function. The Poisson probability distribution is a discrete probability distribution with one parameter, the population mean denoted by μ. The Poisson probability mass function is:

$$f(x) = \frac{\mu^x e^{-\mu}}{x!}, x = 0, 1, 2, \ldots \qquad \text{(Eq 42)}$$

Example. An ultrasonic system is used to detect certain types of defects in railroad rails. It has been found that sections of rail containing clusters of defects are most troublesome in terms of eventual rail failures. The output from a survey covering 540 miles of track is as follows:

Number of defects per mile	Number of miles where this number was observed	Total number of defects
0	136	0
1	170	170
2	113	208
3	67	201
4	40	180
5	10	50
6	3	36
7	1	21
Total	540	866

To determine the probabilities of detecting a certain number of defects in a mile using Poisson statistics, the average number of defects per mile should be computed: in this case, 1.604.

From Eq 42 the predicted frequencies of occurrence are computed and then compared with the actual frequencies of occurrence (Fig. 12). For example, the predicted frequency of occurrence for three defects per mile is:

Fig. 12 Predicted frequency of occurrence as compared with actual frequency of occurrence of event

Event is defects per mile; see text for explanation. Curved line represents a trend and does not imply a continuous function.

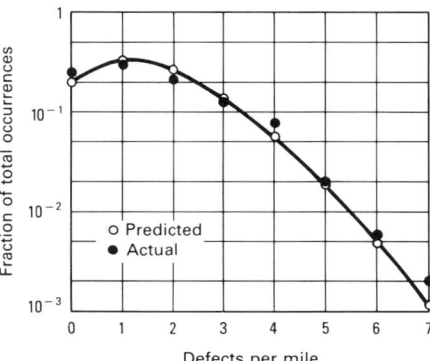

$$p(x = 3) = 0.2 \left(\frac{1.60^3}{3!} \right) = 0.136$$

The actual frequency of occurrence was 0.124 (67/540). In general, the agreement between predicted and actual occurrence rates is good, which suggests the applicability of the Poisson distribution.

Poisson probability paper can be used to evaluate the validity of the assumed Poisson distribution or to rapidly estimate the expected occurrence rates for different numbers of occurrences per unit interval (Fig. 13).

If the actual frequencies of occurrence are known, then the expected frequencies can be computed. If the data are distributed according to the Poisson distribution, the values defined by the difference between the actual cumulative failure percentages and unity

(plotted as cumulative probability values for each value of x) should fall on a nearly straight vertical line about the expectation. The cumulative probability values for the rail survey data fall on a nearly straight vertical line about the expectation value of 1.604 (Fig. 13). If a Poisson distribution is known to hold, one occurrence rate at a single condition is sufficient to estimate the occurrence rates at other conditions.

Cumulative Distribution Function. The Poisson cumulative distribution function is:

$$F(x) = \sum_{y=0}^{x} \frac{\mu^y e^{-\mu}}{y!}, \; x = 0, 1, 2, \ldots$$

$$(Eq \; 43)$$

Tables of this distribution function for various values of μ and small values of x may be found in Ref 1, as well as in many introductory statistical textbooks. Alternatively, many of the statistical computer packages contain programs that calculate values of $F(x)$ for any values of μ and x.

For $\mu \geq 10$, the normal approximation to the Poisson distribution may be used to obtain approximate values of $F(x)$. Using this approximation yields:

$$F(x) \approx F_o \left(\frac{x + 0.5 - \mu}{\sqrt{\mu}} \right) \qquad (Eq \; 44)$$

as an approximation to $F(x)$, where $F_o(\cdot)$ is the standard normal cumulative distribution function. For more information on $F_o(x)$, see the section of this article on the cumulative distribution function of the normal distribution.

Population Mean and Variance. The mean and variance of a Poisson distribution are:

Mean $= \mu$
Variance $= \mu$ $\qquad (Eq \; 45)$

Parameter Estimation. Given an observed value, x, from a Poisson distribution, the unknown population mean is estimated by the observed value ($\hat{\mu} = x$). Given a set of n measurements $x_1, x_2, \ldots, x_n$ to be modeled by a Poisson distribution, the maximum likelihood estimate of the unknown population mean is the sample mean:

$$\hat{\mu} = \bar{x} = \frac{1}{n} \sum_{i=1}^{n} x_i \qquad (Eq \; 46)$$

Selecting Statistical Distributions

Before a set of data can be analyzed, the statistical distribution that will be used to model the data must be selected. The choice of a statistical distribution is often difficult, depending on engineering judgment and prior experience. This process is often facilitated by the construction of probability plots on various probability papers (normal, log normal, Weibull, etc.), so that experimental data will plot as approximately a straight line if the data follow the appropriate underlying distribution.

Once a statistical distribution has been chosen, a statistical test should be performed to determine whether the chosen distribution provides an adequate fit to the data. Common statistical tests for carrying out this determination fall into two categories: chi-square goodness-of-fit tests and empirical distribution function (EDF) goodness-of-fit tests. The EDF tests are often preferred because they do not require the large samples of test data required by the chi-square tests.

References 4 to 6 contain discussions of probability plotting and goodness-of-fit tests.

Fig. 13 Poisson probability paper illustrating the cumulative probability p that an event will occur at least r_i times when the expected number of occurrences has the value of y

Data points are based on Eq 43. See text for explanation.

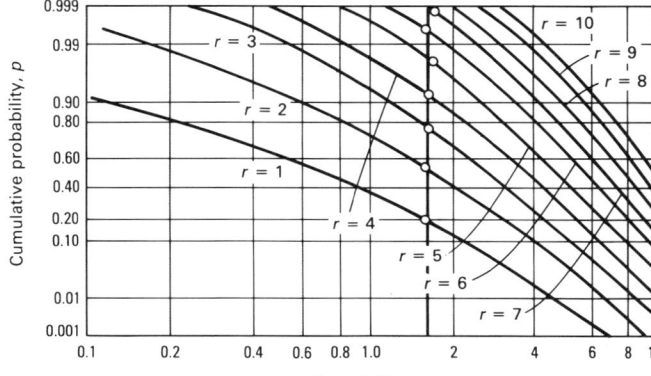

Event No., r	No. of defects, x	Probability mass function, $f(x)$	Cumulative distribution function, $F(x)$, or probability, P
1 0		0.201	0.201
2 1		0.322	0.523
3 2		0.259	0.782
4 3		0.138	0.920
5 4		0.055	0.975
6 5		0.018	0.993
7 6		0.005	0.998
8 7		0.001	0.999

Once a statistical distribution has been verified by a goodness-of-fit test procedure, the set of measurements of interest can be analyzed.

REFERENCES

1. W.H. Beyer, *Handbook of Tables for Probability and Statistics*, 2nd ed., Chemical Rubber Company, Cleveland, 1968
2. N.L. Johnson and S. Kotz, *Discrete Distributions*, Houghton Mifflin, New York, 1969
3. N.L. Johnson and S. Kotz, *Continuous Univariate Distributions—I & II*, Houghton Mifflin, New York, 1970
4. J.F. Lawless, *Statistical Models and Methods for Lifetime Data*, John Wiley & Sons, New York, 1982
5. C. Lipson and N.J. Sheth, *Statistical Design and Analysis of Engineering Experiments*, McGraw-Hill, New York, 1973
6. W. Nelson, *Applied Life Data Analysis*, John Wiley & Sons, New York, 1982

Planning of Comparative Experiments

By Mary G. Natrella
Mathematical Statistician
Statistical Engineering Division
National Bureau of Standards

PLANNING OF EXPERIMENTS does not consist merely of identifying a few key parameters and then selecting a specific plan. Selection of the proper experimental plan depends on the purpose of the experiment, physical restrictions on the taking of measurements, and other restrictions imposed by time, economic considerations, and materials and personnel availability. The novice experimenter should consult a competent statistician and be prepared to provide all available information about the experiment being contemplated.

It has been stated (Ref 1) that "participation in the initial stages of experiments in different areas of research leads to a strong conviction that too little time and effort is put into the planning of experiments. The statistician who expects that his contribution to the planning will involve some technical matter in statistical theory finds repeatedly that he makes a much more valuable contribution simply by getting the investigator to explain clearly why he is doing the experiment, to justify the experimental treatments whose effects he proposes to compare, and to defend his claim that the completed experiment will enable its objectives to be realized." It is good practice to draw up a written proposal for any experiment, including a statement of the objectives, a description of the experiment (covering such matters as experimental treatments, the size of the experiment, and the experimental material), and an outline of the method of analysis of the results.

The following recommendations should be followed in outlining the methods of conducting and analyzing an experiment to ensure successful results (Ref 2). The experimenter should clearly establish the experiment objectives before proceeding with the experiment. Is this a preliminary experiment to determine the future course of experimentation, or is it intended to furnish answers to immediate questions? Are the results to be carried into practical use at once, or are they to be used to explain aspects of theory not adequately understood before? Are estimates or determination of significance desired? Over what range of experimental conditions are the results to be extended?

The experiment should be described in detail. Treatments should be clearly defined. Is it necessary to use a control treatment to make comparisons with past results? Also, the size of the experiment should be determined. If insufficient funds are available to conduct an experiment from which useful results can be obtained, the experiment should not be initiated. Most importantly, material necessary to conduct the experiment should be available.

Methods of analysis of data obtained from the designed experiments described in this article can be found in Ref 3 to 5 and the Selected References. See also the article "Analysis of Designed Experiments" in this Section. Computer packages are also available.

The Nature of Experimentation

An experiment has been broadly defined as a considered course of action aimed at answering one or more carefully framed questions. Observational programs in the natural sciences and sample surveys in the social sciences are clearly included in this general definition. This article discusses a more restricted type of experiment, in which the experimenter varies the parameters, or factors, under study and then observes the effects of this action.

These factors may be quantitative (such as temperature, which can be varied along a continuous scale), or they may be qualitative (such as different machines, operators, composition of charge, etc.). The use of the proper experimental pattern aids in evaluation of factors.

In addition to factors that are varied in a controlled fashion, the experimenter should be aware of background variables that may affect the outcome of the experiment. Sometimes these variables cannot be included as factors in an experiment. However, it is frequently possible to plan an experiment so that background variables do not affect information obtained about the factors of primary interest. Some information about the effects of the background variables can also be obtained.

Additionally, variables may exist of which the experimenter is unaware. These may also have an effect on the outcome of an experiment. The effects of these variables may be "balanced out" by the introduction of randomization into the experimental pattern; Ref 6 provides in-depth discussion of the general principles of experimentation.

Certain characteristics, or requisites, of an experiment, must be met to achieve satisfactory results. There must be a clearly defined objective. To the greatest extent possible, the effects of factors should not be obscured by other variables. Also, the results should not be influenced by conscious or unconscious bias in the experiment or on the part of the experimenter.

An experiment should provide a measure of precision. However, this requisite can be relaxed in some situations, i.e., when there is a well-known history of the measurement process and consequently good estimates of precision. An experiment must have sufficient precision to accomplish its purpose.

To aid in achieving these requisites, statistical design of experiments can provide some tools for sound experimentation. These tools include experimental pattern, planned grouping, randomization, and replication. Their functions in experimentation are listed in Table 1 and are discussed below.

Experimental Design Terminology. The "experimental area" is the scope of the planned experiment. A "block" is a group of results from a particular operator, from a particular machine, or obtained on a particular day. It consists of any planned or natural grouping where it is expected that results from one block will be more alike than results from different blocks. A "treatment" is the factor being investigated (material, environmental condition, etc.) in a single-factor experiment. In factorial experiments (where several variables are studied at the same time), a treatment combination consists of the prescribed levels of the factors to be applied to an experimental unit. A "response" or "yield" is a measured result.

Experimental pattern refers to the planned schedule for taking measurements. A particular pattern may include planned grouping, randomization, and replication. Each of these tools can improve the experimental pattern in given situations. The proper pattern for an experiment aids in control of bias and in measurement of precision, simplifies the requisite calculations of the analysis, and permits clear estimation of the effects of factors.

A common experimental pattern is the so-called factorial design experiment, in which several factors are controlled and their effects are investigated at two or more levels. If two levels of each factor are involved, the experimental plan consists of taking an observation at each of the 2^n possible combinations. Factorial design is discussed in greater detail later in this article.

Planned Grouping. Block designs, an important class of experimental patterns, are characterized by planned grouping. The use of planned grouping (blocking) arose in comparative experiments in agricultural research, which recognized that plots that are close together in a field are usually more alike than plots that are far apart. In industrial and engineering research, planned grouping can be used to capitalize on naturally homogeneous groupings in materials, machines, time, etc. This technique takes into account background variables that are not direct factors in the experiment.

Suppose the experimenter compares the effect of five different treatments of a plastic

Table 1 Requisites and tools for sound experimentation

Requisites	Tools
1. The experiment should have carefully defined objectives	The definition of objectives requires all of the specialized subject matter knowledge of the experimenter and results in (1) choice of factors, including their range; (2) choice of experimental materials, procedure, and equipment; (3) knowledge of to what the results pertain
2. As far as possible, effects of factors should not be obscured by other variables	The use of an appropriate experimental pattern helps to free the comparisons of interest from the effects of uncontrolled variables and simplifies the analysis of the results
3. As far as possible, the experiment should be free from bias (conscious or unconscious)	Some variables may be taken into account by planned grouping; for other variables, use randomization; the use of replication aids randomization to do a better job
4. Experiment should provide a measure of precision (experimental error)(a)	Replication provides the measure of precision; randomization ensures validity of the measure of precision
5. Precision of experiment should be sufficient to meet objectives set forth in requisite 1	Greater precision may be achieved by refinements of technique, experimental pattern (including planned grouping), and replication

(a) Except where there is a well-known history of the measurement process

material. Plastic properties vary considerably within a given sheet. To obtain adequate comparison of the five treatment effects, the plastic sheet should be divided into homogeneous areas, and each area subdivided into five parts. The five treatments could then be allocated to the five parts of a given area. Each set of five parts may be termed a block. In some cases, the naturally homogeneous area (block) may not be large enough to accommodate all the treatments of interest.

If the wear resistance of automobile tires is the property of interest, the natural block is a block of four—the four wheels of an automobile. Each automobile may travel over different terrain or have different drivers. However, the four tires on any given automobile will undergo much the same conditions, particularly if they are rotated frequently. In testing different types of plastic soles for shoes, for example, the natural block consists of two units—the two feet of an individual. The block may consist of observations taken at nearly the same time or place. If a machine can test four items at one time, then each run may be regarded as a block of four units, each item being a unit.

Statisticians have developed a variety of advantageous configurations of block designs, which are classified by their structures into randomized blocks, Latin squares, incomplete blocks, lattices, etc., with subcategories for each. Several block designs are discussed later in this article.

Randomization. To eliminate bias from an experiment, experimental variables (often unknown or unidentified) that are not specifically controlled as factors, or "blocked out" by planned grouping are "controlled" by

randomization. This control is achieved in that any treatment or method has the same chance as any other to be affected by the unknown variables of the experiment. Randomization also ensures valid estimates of experimental error and enables the application of statistical tests of significance and the construction of confidence intervals. Failure to randomize at a crucial stage of an experimental plan can lead to misleading results. The beneficial effects of randomization are obtained in long-term testing and not in a single isolated experiment.

In general, the experimenter should try to identify all variables that could possibly affect the results, select as factors as many variables as can reasonably be studied, and use planned groupings wherever possible. Ideally, then, the test is randomized in terms of all other variables. However, the ideal cannot always be realized in practice.

The term "randomization" has been used rather than randomness to emphasize the fact that experimental material rarely, if ever, has a random distribution in itself. It is unwise to assume randomness; consequently, randomness has to be ensured by formal or mechanical randomization.

Replication. To evaluate the effects of factors, a measure of precision (experimental error) must be available. In some experiments, notably biological or agricultural research, this measure must be obtained from the experiment itself, because no other source provides an appropriate measure. In some industrial and engineering experimentation, however, records may be available on a relatively stable measurement process, and these data may provide an appropriate measure. When the measure of precision must be obtained from the experiment, replication allows the effects of uncontrolled factors to balance each other and thus makes randomization a bias-decreasing tool. In successive replications, the randomization features must be independent. Replication also helps detect gross errors in measurements.

Factorial Experiments

A factorial experiment is an experiment in which several factors are controlled and their effects at each of two or more levels are investigated. An experimental plan consists of taking an observation at each one of all possible combinations that can be formed for the different levels of the factors. Each different combination is called a treatment combination.

Suppose that the effect of pressure and temperature on the yield of some chemical process is the subject under investigation. Pressure and temperature are the factors in the experiment. Each specific value of pressure to be included is a level of the pressure factor, and each specific value of temperature to be included is a level of the temperature factor.

In the past, one common experimental approach has been the so-called "one-at-a-time" approach. This type of experiment studies the effect of varying pressure at a constant temperature, and then studies the effect of varying temperature at a constant pressure. Factors were varied "one at a time." The results of such an experiment are fragmentary in the sense that information is obtained about the effect of different pressures at one temperature only (and the effect of different temperatures at one pressure only). The reaction of the process to different pressures may depend on the temperature used; if a different temperature is selected, the observed response to varying pressure may be quite different. In statistical language, there may be an interaction effect between the two factors within the range of interest, which the "one-at-a-time" procedure does not detect.

In a factorial experiment, the levels of each factor to be studied are selected, and a measurement is made for each possible combination of levels. Suppose that two levels for both pressure and temperature have been chosen: 7 and 14 cm (2.76 and 5.51 in.) and 21 and 30 °C (70 and 100 °F). Four possible combinations of pressure and temperature exist, and the factorial experiment would consist of four trials. In this example, the term "level" is used in connection with quantitative factors, but the same term is also used when the factors are qualitative.

Analysis of factorial experiments encompasses main effects and interaction effects, known simply as interactions. Main effects of a given factor are always functions of the average response or yield at the various levels of the factor. When a factor has two levels, the main effect is the difference between the responses at the two levels averaged over all levels of the other factors.

When the factor has more than two levels, there are several independent components of the main effect, the number of components being one less than the number of levels. If the difference in the response between two levels of factor A is the same regardless of the level of factor B (except for experimental error), there is no interaction between A and B, or the AB interaction is zero.

Figure 1 illustrates two response or yield curves; one example shows the presence of an interaction, and the other exhibits no interaction. If there are two levels of each of the factors A and B, then the AB interaction (neglecting experimental error) is the difference in the yields of A at the second level of B minus the difference in the yields of A at the first level of B. If there are more than two levels of either or both A and B, then the AB interaction is composed of more than one component. If a levels of factor A exist and b levels of factor B exist, then the AB interaction has $(a - 1)(b - 1)$ independent components.

For factorial experiments with three or more factors, interactions can be defined similarly. For example, the ABC interaction is the interaction between the factor C and the AB interaction (or equivalently between the factor B and the AC interaction, or A and the BC interaction).

Factorial Experiments With Each Factor at Two Levels

A factorial experiment with n factors, each at two levels, is known as a 2^n factorial experiment. The experiment consists of 2^n trials, one at each combination of levels of the factors. To identify each of the trials, a conventional notation is adopted. A factor is identified by a capital letter, and the two levels of a factor by the subscripts zero and one. For three factors A, B, and C, the corresponding levels of the factors are A_0, A_1, B_0, B_1, C_0, and C_1. By convention, the zero subscript refers to the lower level, to the normal condition, or to the absence of a condition. A trial is represented by a combination of lower case letters denoting the levels of the factors in the trial. The presence of a lower case letter indicates that the factor is at the level denoted by the subscript 1 (the higher level for quantitative factors); the absence of a lower case letter indicates that the factor is at the level denoted by the subscript zero (the

lower level for quantitative factors). Thus, the symbol a represents the treatment combination where A is at the level A_1, B is at B_0, and C is at C_0. The symbol bc represents the treatment combination where A is at the level A_0, B is at B_1, and C is at C_1. Conventionally, the symbol (1) represents the treatment combination with each factor at its zero subscript level. In an experiment with three factors, each at two levels, the $2^3 = 8$ combinations, and thus the eight trials, are represented by (1), a, b, ab, c, ac, bc, and abc.

Blocked Factorial Experiments

When there are several factors to be studied, the required number of trials 2^n may be too large to be carried out under reasonably uniform conditions (e.g., on one batch of raw material or on one piece of equipment). In such cases, the design can be arranged in groups or blocks so that conditions affecting each block can be made as uniform as possible.

The use of planned grouping within a factorial design (i.e., a blocked factorial) improves the precision of estimation of experimental error and enables estimation of the main effects free of block differences. However, the structure of the designs is such that certain interaction effects will be inextricable from block effects. In most designs, however, only three-factor and higher order interactions will be confounded with blocks.

Blocked factorial designs have not been widely used in experimentation in the physical sciences; they may not be well suited for the type of nonhomogeneity that occurs in these applications.

Fractional Factorial Experiments

With many factors, a complete factorial experiment requiring all possible combinations of levels of the factors involves a large number of tests. This is the case even when only two levels of each factor are being investigated. In such cases, the complete factorial experiment may overtax the available facilities.

In other situations, it may not be practical to plan the entire experimental program in advance, and a few smaller experiments may be conducted to serve as a guide to future work. The complete set of experiments may furnish more information or precision than is needed for the purpose at hand.

In these cases, it is useful to have a plan that requires fewer tests than the complete factorial experiment. Recent developments in statistics have considered the problem of planning multifactor experiments that require measuring only a fraction of the total number

Fig. 1 Examples of response curves showing presence or absence of interaction

(a) Effect of different levels of A on the response for three different levels of B (interaction present). (b) Effect of different levels of A on the response for two different levels of C (no interaction present)

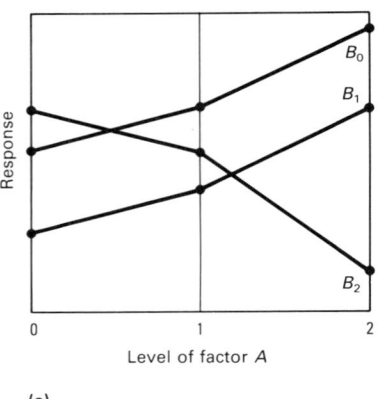

(a)

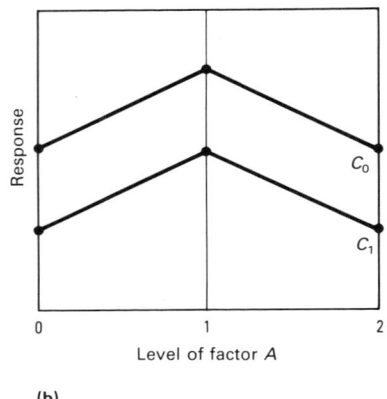

(b)

of possible combinations. The fraction is a carefully prescribed subset of all possible combinations. Its analysis is relatively straightforward, and the use of a fractional factorial does not preclude the possibility of completion of the full factorial experiment at a later time.

In Fig. 2 to 4, let the letters A, B, C, D, E, F, and G represent seven factors to be investigated, and let the subscripts zero and one denote two alternative levels of each factor. The 128 (2^7) possible experimental conditions are represented by the 128 cells in Fig. 2. The shaded squares represent those experimental combinations to be investigated if the experimenter wishes to measure only half of the 128 possible combinations. In the same manner, the shaded cells in Fig. 3 and 4 illustrate plans requiring only 32 and 16 measurements, respectively.

Fractional factorial experiments cannot produce as much information as the full factorial. Economy is achieved at the expense of assuming that certain of the interactions between factors are negligible. Some of the large fractions (e.g., the half-replicate shown in Fig. 2) require only that third-order (and higher) interactions be assumed negligible, and this assumption is not uncommon. However, the plan calling for one eighth of the possible combinations, as shown in Fig. 4, can only be used for evaluating the main effects of each of the seven factors and does not allow the evaluation of any two-factor interactions.

A complete factorial experiment consists of 2^n tests. Analysis of a complete factorial contains n main effects, $2^n - n - 1$ interaction effects, and an overall average effect. The 2^n tests can be used to obtain independent estimates of the 2^n effects. In a frac-

tional factorial (for example, the fraction $\frac{1}{2}^b$), only 2^{n-b} tests are made; therefore, 2^{n-b} independent estimates are possible. In designing the fractional plans (i.e., in selecting an optimum subset of the 2^n total combinations), the goal is to keep each of the 2^{n-b} estimates as "clean" as possible (i.e., to keep the estimates of main effects and, if possible, second-order interactions free of confusion with each other).

To test whether certain effects are significant, an estimate of the variation due to experimental error is needed, which must be independent of the estimates of the effects. Table 2 presents several useful two-level fractional factorial plans, as well as the effects that can be estimated (assuming three-factor and higher order interaction terms are

negligible). The treatment combinations should be randomly allocated to the experimental material. More two-level plans may be found in Ref 7, and fractional factorial plans for factors at three levels may be found in Ref 8.

Estimates of Experimental Error

Internal Estimates of Error. As in any experiment, a measure of experimental error is required to judge the significance of the observed differences in treatments. In large factorial designs, estimates of higher order interactions are available. The usual assumption is that higher order interactions are physically impossible and that these estimates are actually estimates of experimental error.

As a working rule, third-order and higher order interactions may be used for error. This does not imply that third-order interactions are always nonexistent. The judgment of the experimenter will determine which interactions may reasonably be assumed meaningful and which may be assumed to be error. These latter interactions may be combined to provide an internal estimate of error for a factorial experiment of reasonable size. For very small factorials (e.g., 2^3 or smaller), there are no estimates of higher order interactions, and the experiment must be replicated (repeated) to obtain an estimate of error from the experiment.

In blocked factorial designs, some of the higher order interactions will be confounded with blocks and will not be available as estimates of error. Table 3 presents some blocked factorial plans. The plan for a 2^3 factorial is arranged in two blocks of four observations. The single third-order interaction

Fig. 2 A one half replicate of a 2^7 factorial

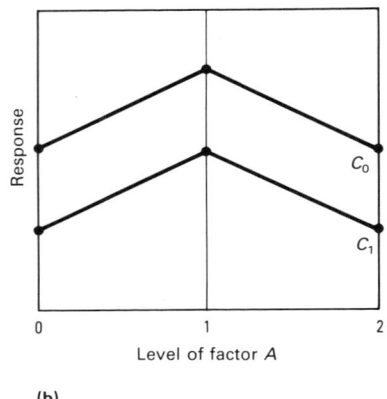

Fig. 3 A one quarter replicate of a 2^7 factorial

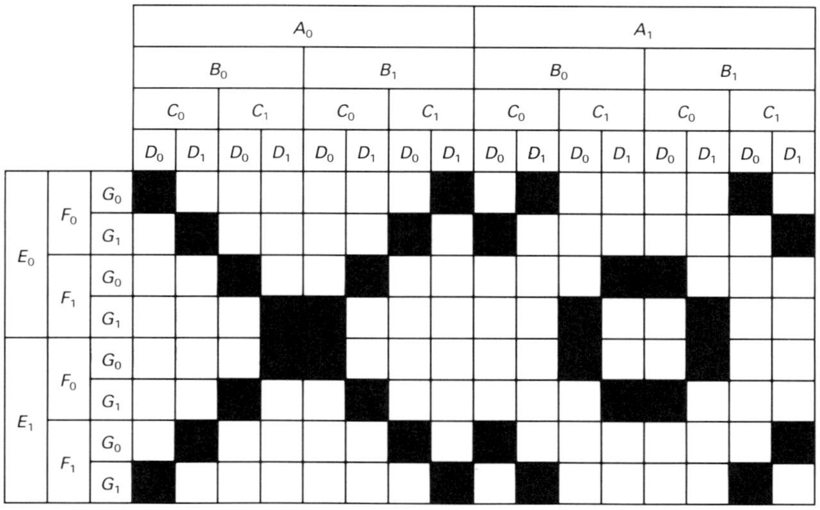

provides the blocking, i.e., the means of subdividing the experiment into homogeneous groups, and therefore will estimate block effects, not error. Replication of the experiment may be necessary to estimate experimental error.

With fractional factorials, replication of the experiment is not desirable; further experimentation would probably be aimed at completing the full factorial or obtaining a larger fraction of the full factorial. The smaller fractional factorial designs do not contain higher order interactions that can be assumed to be error. In fact, none of the plans given in Table 2 provides a suitable internal estimate of error. Accordingly, an independent estimate of error is required

when using a small fractional factorial. Occasionally, and cautiously, second-order interaction effects may be used to test main effects, if the purpose of the experiment is to determine very large main effects (much larger than second-order effects). In using interactions as estimates of error, however, the experimenter must decide before conducting the experiment (or at least before having a knowledge of the responses or yields) which of the effects may be assumed to be zero, so that they may be used in the estimate of the variation due to experimental error.

Estimates of Error Obtained from Previous Runs. For experiments that do not provide adequate estimates of error, an esti-

mate based on past experience with the measurement process may be adequate. In laboratory and industrial situations, this information is often readily available or can be obtained by simple analysis of previously recorded data.

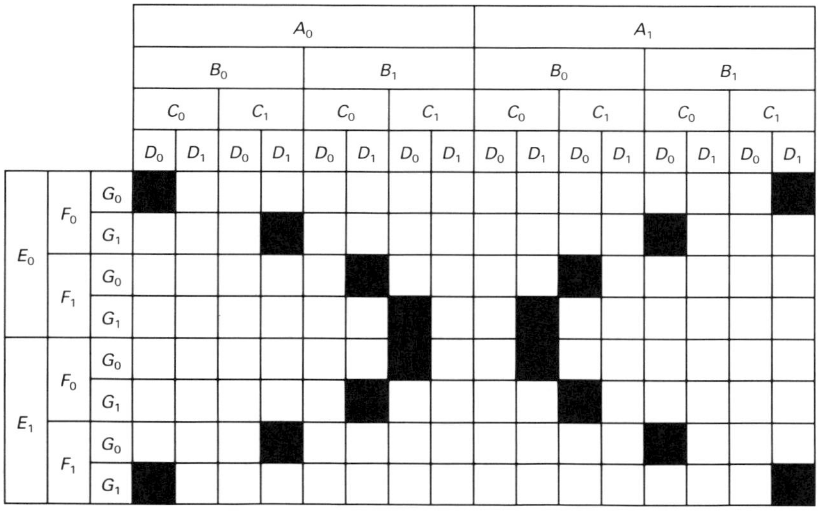

Randomized Designs and Block Designs

The experimental designs discussed in this section use planned grouping, except for the completely randomized design. The simplest type of blocked design is randomized blocks, in which blocking is made with respect to one source of inhomogeneity and the block is large enough to accommodate all treatments to be tested.

For incomplete block designs, blocking is one-way, but the block size is not large enough for all treatments to be tested in every block. One design example is the balanced incomplete block plan, in which restrictions on the assignment of treatments to blocks lead to equal precision in the estimation of differences between treatments. The chain block design, a special type of incomplete block design without this balance in the precision of the estimates, is also discussed.

When the experimental plan is designed to eliminate two sources of inhomogeneity, two-way blocking is used. The Latin squares and Youden squares are examples of such designs.

Completely randomized plans are simple and are most appropriate when the experimental material is homogeneous and background conditions can be controlled during the experiment. If there are a total of N available experimental units, and $n_1, n_2, \ldots, n_t$ experimental units are assigned respectively to each of the t treatments or products, then the experimental units are assigned to the treatments at random.

For example, suppose three types of ammunition of a given size and caliber are to be tested to determine which type has the highest velocity. There are n_1, n_2, n_3 shells, respectively, of the three types. If the conditions under which the shells are fired (i.e., temperature, barrel conditions, etc.) are assumed to be the same for each shell, then the simplest plan is to select the shells at random and to fire them in that order. If all the shells of one type are fired first and then all the shells of the succeeding types are fired, there is no assurance against influences on velocity such as the wearing of the gun barrel or changes in atmospheric conditions such as temperature.

Randomization affords assurance against uncontrollable disturbances in the sense that such disturbances have the same chance of

Fig. 4 A one eighth replicate of a 2^7 factorial

Table 2 Select fractional factorial plans

Plans	Treatment combinations(a)	Estimated effects(b)	Plans	Treatment combinations(a)	Estimated effects(b)
Plan 1	(1)	T	**Plan 5 (continued)**	bde	$BD + CF$
Three factors ($n = 3$)	ac	$A - BC$		abd	(c)
½ replication ($b = 1$)	bc	$B - AC$		cde	$CD + BF$
Four observations	ab	$-C + AB$		acd	(c)
				bcdf	F
Plan 2	(1)	T		abcdef	$AF + DE$
Four factors ($n = 4$)	ad	A			
½ replication ($b = 1$)	bd	B	**Plan 6**	(1)	T
Eight observations	ab	$AB + CD$	Six factors ($n = 6$)	adf	$A - DE - CF$
	cd	C	⅛ replication ($b = 3$)	bde	$B - CE - DF$
	ac	$AC + BD$	Eight observations	abef	$AB + CD + EF$
	bc	$BC + AD$		cdef	$C - AF - BE$
	abcd	D		ace	$-F + AC + BD$
				bcf	$-E + AD + BC$
Plan 3	(1)	T		abcd	$D - AE - BE$
Five factors ($n = 5$)	ae	A			
½ replication ($b = 1$)	be	B	**Plan 7**	(1)	T
Sixteen observations	ab	AB	Seven factors ($n = 7$)	aeg	A
	ce	C	⅛ replication ($b = 3$)	befg	B
	ac	AC	Sixteen observations	abf	$AB + CE + DG$
	bc	BC		cef	C
	abce	$-DE$		acfg	$AC + BE + FG$
	de	D		bcg	$BC + AE + DF$
	ad	AD		abce	E
	bd	BD		dfg	D
	abde	$-CE$		adef	$AD + EF + BG$
	cd	CD		bde	$BD + CF + AG$
	acde	$-BE$		abdg	G
	bcde	$-AE$		cdeg	$CD + BF + EG$
	abcd	$-E$		acd	(c)
				bcdf	F
Plan 4	(1)	T		abcdefg	$AF + DE + CG$
Five factors ($n = 5$)	ad	$A - DE$			
¼ replication ($b = 2$)	bde	$B - CE$	**Plan 8**	(1)	T
Eight observations	abe	$AB + CD$	Eight factors ($n = 8$)	aegh	A
	cde	$C - BE$	1/16 replication ($b = 4$)	befg	B
	ace	$AC + BD$	Sixteen observations	abfh	$AB + CE + DG + FH$
	bc	$-E + BC + AD$		cefh	C
	abcd	$D - AE$		acfg	$AC + BE + FG + DH$
				bcgh	$BC + AE + DF + GH$
Plan 5	(1)	T		abce	E
Six factors ($n = 6$)	ae	A		dfgh	D
¼ replication ($b = 2$)	bef	B		adef	$AD + EF + BG + CH$
Sixteen observations	abf	$AB + CE$		bdeh	$BD + AG + CF + EH$
	cef	C		abdg	G
	acf	$AC + BE$		cdeg	$CD + AH + BF + EG$
	bc	$BC + AE + DF$		acdh	H
	abce	E		bcdf	F
	df	D		abcdefgh	$AF + DE + CG + BH$
	adef	$AD + EF$			

(a) The order given is the order in which the data are to be listed in the first column of the Yates method of analysis. (b) The order given is the order in which estimated effects come out in the last column of the Yates method of analysis. (c) To be used in the estimate of the variation due to experimental error.

affecting each of the factors under study and will balance out in long-term testing. Results of a completely randomized plan are displayed as in Fig. 5.

Randomized Block Plans. To compare several treatments, all other conditions must be maintained as nearly constant as possible. Frequently, the required number of tests is too large to be carried out under similar conditions. In such cases, the experiment is divided into blocks, or planned homogeneous groups. When each such group in the experiment contains exactly one observation on every treatment, the experimental plan is called a randomized block plan.

There are many situations in which a randomized block plan is suitable. For example, a testing scheme may take several days to complete. If some systematic differences between days is expected, each item can be observed on each day, or one test per day can be conducted on each item. One day would then represent a block.

In another situation, several persons may be conducting the tests or making the observations, and differences between operators are expected. The tests or observations made by a given operator can be considered a block. The size of a block may also be restricted by physical considerations. In general, a randomized block plan is one in which each of the treatments appears only once in each block. The treatments are allocated to experimental units at random within a given block.

The results of a randomized block experiment can be displayed in a two-way table such as Fig. 6, which assumes that b blocks and t treatments have been used. Because each treatment occurs only once in each block, the treatment totals, or means, are directly comparable without adjustment.

Incomplete block plans are similar to randomized block plans in that planned grouping is used. In incomplete block plans,

Table 3   Select blocked factorial plans
For use when factorial experiment must be subdivided into homogeneous groups

Plans for three factors: $2^3 = 8$ observations
  (i) Four observations per block (*ABC* confounded with block effects)
      Block 1   (1), *ab, ac, bc*
      Block 2   *a, b, c, abc*

Plans for four factors: $2^4 = 16$ observations
  (i) Eight observations per block (*ABCD* interaction confounded with block effects)
      Block 1   (1), *ab, ac, bc, ad, bd, cd, abcd*
      Block 2   *a, b, c, abc, d, abd, acd, bcd*
  (ii) Four observations per block (*AD, ABC, BCD,* confounded with block effects)
      Block 1   (1), *bc, abd, acd*
      Block 2   *a, abc, bd, cd*
      Block 3   *b, c, ad, abcd*
      Block 4   *d, bcd, ab, ac*

Plans for five factors: $2^5 = 32$ observations
  (i) Sixteen observations per block (*ABCDE* interaction confounded with block effects)
      Block 1   (1), *ab, ac, bc, ad, bd, cd, abcd, ae, be, ce, abce, de, abde, acde, bcde*
      Block 2   *a, b, c, abc, d, abd, acd, bcd, e, abe, ace, bce, ade, bde, cde, abcde*
  (ii) Eight observations per block (*BCE, ADE, ABCD,* confounded with block effects)
      Block 1   (1), *ad, bc, abcd, abe, bde, ace, cde*
      Block 2   *a, d, abc, bcd, be, abde, ce, acde*
      Block 3   *b, abd, c, acd, ae, de, abce, bcde*
      Block 4   *e, ade, bce, abcde, ab, bd, ac, cd*
  (iii) Four observations per block (*AD, BE, ABC, BCD, CDE, ACE, ABDE,* confounded with block effects)
      Block 1   (1), *bce, acd, abde*
      Block 2   *a, abce, cd, bde*
      Block 3   *b, ce, abcd, ade*
      Block 4   *c, be, ad, abcde*
      Block 5   *d, bcde, ac, abe*
      Block 6   *e, bc, acde, abd*
      Block 7   *ab, ace, bcd, de*
      Block 8   *ae, abc, cde, bd*

Plans for six factors: $2^6 = 64$ observations
  (i) Thirty-two observations per block (*ABCDEF* confounded with block effects)
      Block 1   (1), *abcdef,* plus all treatment combinations represented by two letters (e.g., *ab, ac,* etc.) and by four letters (e.g., *abcd, bcde,* etc.)
      Block 2   All treatment combinations represented by a single letter, by three letters, and by five letters
  (ii) Sixteen observations per block (*ABCD, BCEF, ADEF,* confounded with block effects)
      Block 1   (1), *bc, ad, abcd, ef, bcef, adef, abcdef, bde, cde, abe, ace, bdf, cdf, abf, acf*
      Block 2   *a, abc, d, bcd, aef, abcef, def, bcdef, abde, acde, be, ce, abdf, acdf, bf, cf*
      Block 3   *b, c, abd, acd, bef, cef, abdef, acdef, de, bcde, ae, abce, df, bcdf, af, abcf*
      Block 4   *e, bce, ade, abcde, f, bcf, adf, abcdf, bd, cd, ab, ac, bdef, cdef, abef, acef*
  (iii) Eight observations per block (*ADE, BCE, ACF, BDF, ABCD, ABEF, CDEF,* confounded with block effects)
      Block 1   (1), *ace, bde, abcd, adf, cdef, abef, bcf*
      Block 2   *a, ce, abde, bcd, df, acdef, bef, abcf*
      Block 3   *b, abce, de, acd, abdf, bcdef, aef, cf*
      Block 4   *c, ae, bcde, abd, acdf, def, abcef, bf*
      Block 5   *d, acde, be, abc, af, cef, abdef, bcdf*
      Block 6   *e, ac, bd, abcde, adef, cdf, abf, bcef*
      Block 7   *f, acef, bdef, abcdf, ad, cde, abe, bc*
      Block 8   *ab, bce, ade, cd, bdf, abcdef, ef, acf*
  (iv) Four observations per block (*AD, BE, CF, ABC, BCD, CDE, DEF, ACE, AEF, ABF, BDF, ABDE, BCEF, ACDF, ABCDEF,* confounded with block effects)
      Block 1   (1), *bcef, acdf, abde*
      Block 2   *a, abcef, cdf, bde*

Plans for six factors: $2^6 = 64$ observations (continued)
      Block 3   *b, cef, abcdf, ade*
      Block 4   *c, bef, adf, abcde*
      Block 5   *d, bcdef, acf, abe*
      Block 6   *e, bcf, acdef, abd*
      Block 7   *f, bce, acd, abdef*
      Block 8   *ab, acef, bcdf, de*
      Block 9   *ac, abef, df, bcde*
      Block 10   *ad, abcdef, cf, be*
      Block 11   *ae, abcf, cdef, bd*
      Block 12   *af, abce, cd, bdef*
      Block 13   *bc, ef, abdf, acde*
      Block 14   *bf, ce, abcd, adef*
      Block 15   *abc, aef, bdf, cde*
      Block 16   *abf, ace, bcd, def*

Plans for seven factors: $2^7 = 128$ observations
  (i) Sixty-four observations per block (*ABCDEFG* confounded with block effects)
      Block 1   (1), and all treatment combinations represented by two letters, four letters, or six letters (e.g., *ab, abcd,* etc.)
      Block 2   All treatment combinations represented by a single letter, by three letters, and by five letters, plus *abcdefg*
  (ii) Thirty-two observations per block (*ABCD, ABEFG, CDEFG,* confounded with block effects)
      Block 1   (1), *ab, abcd, ace, acf, acg, ade, adf, adg, bce, bcf, cdef, cdeg, cdfg, abcdef, abcdeg, abcdfg, abef, bcg, bde, bdf, bdg, abeg, abfg, cd, ef, eg, fg, acefg, adefg, bcefg, bdefg*
      Block 2   *a, b, bcd, ce, cf, cg, de, df, dg, abce, abcf, acdef, acdeg, acdfg, bcdef, bcdeg, bcdfg, bef, abcg, abde, abdf, abdg, beg, bfg, acd, aef, aeg, afg, cefg, defg, abcefg, abdefg*
      Block 3   *c, abc, abd, ae, af, ag, acde, acdf, acdg, be, bf, def, deg, dfg, abdef, abdeg, abdfg, abcef, bg, bcde, bcdf, bcdg, abceg, abcfg, d, cef, cfg, aefg, acdefg, befg, bcdefg, ceg*
      Block 4   *e, abe, abcde, ac, acef, aceg, ad, adef, adeg, bc, bcef, cdf, cdg, cdefg, abcdf, abcdg, abcdefg, abf, bceg, bd, bdef, bdeg, abg, abefg, cde, f, g, efg, acfg, adfg, bcfg, bdfg*
  (iii) Sixteen observations per block (*ABCD, BCEF, ADEF, ACFG, BDFG, ABEG, CDEG,* confounded with block effects)
      Block 1   (1), *bde, adg, abeg, bcg, cdeg, abcd, ace, efg, bdfg, adef, abf, bcef, cdf, abcdefg, acfg*
      Block 2   *a, abde, dg, beg, abcg, acdeg, bcd, ce, aefg, abdfg, def, bf, abcef, acdf, bcdefg, cfg*
      Block 3   *b, de, abdg, aeg, cg, bcdeg, acd, abce, befg, dfg, abdef, af, cef, bcdf, acdefg, abcfg*
      Block 4   *c, bcde, ac, abceg, bg, deg, abd, ae, cefg, bcdfg, acdef, abcf, bef, df, abdefg, afg*
      Block 5   *d, be, ag, abdeg, bcdg, ceg, abc, acde, defg, bfg, aef, abdf, bcdef, cf, abcefg, acdfg*
      Block 6   *e, bd, adeg, abg, bceg, cdg, abcde, ac, fg, bdefg, adf, abef, bcf, cdef, abcdfg, acefg*
      Block 7   *f, bdef, adfg, abefg, bcfg, cdefg, abcdf, acef, eg, bdg, ade, ab, bce, cd, abcdeg, acg*
      Block 8   *g, bdeg, ad, abe, bc, cde, abcdg, aceg, ef, bdf, adefg, abfg, bcefg, cdfg, abcdef, acf*
  (iv) Eight observations per block (*ACF, ADE, BCE, BDF, CDG, ABG, EFG, ABEF, CDEF, ABCD, BDEG, ACEG, ADFG, BCFG, ABCDEFG,* confounded with block effects)
      Block 1   (1), *aceg, bdeg, abcd, adfg, cdef, abef, bcfg*
      Block 2   *a, ceg, abdeg, bcd, dfg, acdef, bef, abcfg*
      Block 3   *b, abceg, deg, acd, abdfg, bcdef, aef, cfg*
      Block 4   *c, aeg, bcdeg, abd, acdfg, def, abcef, bfg*
      Block 5   *d, acdeg, beg, abc, afg, cef, abdef, bcdfg*
      Block 6   *e, acg, bdg, abcde, adefg, cdf, abf, bcefg*
      Block 7   *f, acefg, bdefg, abcdf, adg, cde, abe, bcg*
      Block 8   *g, ace, bde, abcdg, adf, cdefg, abefg, bcf*
      Block 9   *ab, bceg, adeg, cd, bdfg, abcdef, ef, acfg*
      Block 10   *ac, eg, abcdeg, bd, cdfg, adef, bcef, abfg*
      Block 11   *ad, cdeg, abeg, bc, fg, acef, bdef, abcg*
      Block 12   *ae, cg, abdg, bcde, defg, acdf, bf, abcefg*
      Block 13   *af, cefg, abdefg, bcdf, dg, acde, be, abcg*
      Block 14   *ag, ce, abde, bcdg, df, acdefg, befg, abcf*
      Block 15   *bg, abce, de, acdg, abdf, bcdefg, aefg, cf*
      Block 16   *abg, bce, ade, cdg, bdf, abcdefg, efg, acf*

Fig. 5 Schematic presentation of results for completely randomized plans

Observation	Treatment			
	1	2	. . .	t
1				
2				
3				
.				
.				
.				
Total				
Mean				

Fig. 6 Schematic presentation of results for randomized block plans

Block	Treatment				Total	Block mean = B/t
	1	2	. . .	t		
1					B_1	
2					B_2	
.					.	
.					.	
.					.	
b					B_b	
Total	T_1	T_2	. . .	T_t	G	
Treatment Mean = T/b						

the block size is not large enough to accommodate all treatments in one block. For example, suppose that a block is one day, but that the time required for each test is so long that all experimental treatments cannot be run in one day. The limitation might also be due to a lack of space, such as in spectrographic analysis in which a block may be one photographic plate and the number of specimens to be compared may exceed the capacity of the plate.

Two types of randomized incomplete block plans are discussed in this article: balanced incomplete block plans and chain block plans. Balanced incomplete block plans offer easy analysis. Additionally, all differences between treatment effects are estimated with the same degree of precision. Chain block plans limit the number of duplicate observations on treatments and are useful when the difference in treatments is large in comparison to the amount of experimental error. Experimental error may be thought of as the difference between an observed treatment and the average of a large number of similar observations under similar conditions. Other incomplete block designs are

also available that offer flexibility in the number of blocks, size of blocks, and number of treatments—e.g., partially balanced incomplete block designs (Ref 9).

Balanced Incomplete Block Plans. The description of these plans involves the use of the following symbols: r is the number of replications (number of times each treatment appears in the plan); b is the number of blocks in the plan; t is the number of treatments; k is the number of treatments that appear in each block; λ is number of blocks in which a given treatment pair appears, $\lambda = r(k - 1)/(t - 1)$; E is a constant used in the analysis, $E = t\lambda/rk$; and N is the total number of observations, $N = tr = bk$. Balanced incomplete block plans are listed in Table 4 for $4 \leq t \leq 10$, $r \leq 10$. A schematic representation of the results for a balanced incomplete block plan is shown in Fig. 7, which shows the arrangement for Plan 7 of Table 4. For other balanced incomplete block plans, see Ref 1.

In order to estimate and to test block effects as well as treatment effects, plans where $b = t$ (i.e., the number of blocks equals the number of treatments) should be

used. In such plans, known as symmetrical balanced incomplete block designs, differences between block effects are estimated with equal precision for all pairs of blocks.

To use a given plan from Table 4:

- Rearrange the blocks at random. In a number of the plans in Table 4, the blocks are arranged in groups. In these plans, rearrange the blocks at random within their respective groups.
- Randomize the positions of the treatment numbers within each block.
- Assign the treatments at random to the treatment numbers in the plan.

Chain block plans are useful when observations are expensive and the experimental error is small. Such a plan can accommodate a large number of treatments relative to the total number of observations. The required number of observations is only slightly larger than the number of treatments. A chain block plan, however, requires differences in treatment effects to be substantially larger than experimental error.

In a chain block design, some treatments are observed once, and some treatments are observed twice. A typical chain block plan is illustrated in Fig. 8, in which A'_1 represents either a treatment or a group of treatments, and A''_1 represents the same treatment or group of treatments. x represents treatments for which there is only one observation, and the same number of such treatments is not required in each block.

When the experimental conditions are appropriate for their use, chain blocks offer a flexible and efficient design. They are easy to construct. Following the example below, the user should be able to produce a chain block plan suited to his own needs. For additional information, see Ref 3.

For a given number of blocks b and a given number of treatments t, various plans can be constructed. Analysis is not difficult, but is not as straightforward as the analysis of some simpler designs. Two examples of chain block designs are given below. The numbers in each block represent treatments. Plan 1 consists of four blocks ($b = 4$) and 13 treatments ($t = 13$). It is represented as:

Block			
1	2	3	4
$\begin{Bmatrix}1\\2\end{Bmatrix}$	$\begin{Bmatrix}3\\4\end{Bmatrix}$	$\begin{Bmatrix}5\\6\end{Bmatrix}$	$\begin{Bmatrix}7\\8\end{Bmatrix}$
$\begin{Bmatrix}3\\4\end{Bmatrix}$	$\begin{Bmatrix}5\\6\end{Bmatrix}$	$\begin{Bmatrix}7\\8\end{Bmatrix}$	$\begin{Bmatrix}1\\2\end{Bmatrix}$
9	10	11	12
13			

Schematically, Plan 1 may be written as:

	─────Block─────		
1	**2**	**3**	**4**
A'_1	A'_2	A'_3	A'_4
A''_2	A''_3	A''_4	A''_1
x	x	x	x
x			

In Plan 1, treatments 1 and 2 constitute the group A_1, which appears in blocks 1 and 4; treatments 3 and 4 constitute group A_2 (in blocks 1 and 2); treatments 5 and 6 constitute group A_3 (in blocks 2 and 3); and treatments 7 and 8 constitute group A_4 (in blocks 3 and 4). The remaining treatments (9 through 13) are distributed among the blocks to make the number of treatments per block as equal as possible.

Treatments 1 through 8 appear twice each; treatments 9 through 13 appear only once. Treatment 1 never occurs without treatment 2; treatment 3 never occurs without treatment 4, etc. Thus, the treatments that are replicated twice fall into four groups (schematically A_1, A_2, A_3, and A_4), and these groups are the links in the chain of blocks. Treatments 3 and 4 link blocks 1 and 2; treatments 5 and 6 link blocks 2 and 3; treatments 7 and 8 link blocks 3 and 4; and treatments 1 and 2 complete the chain by linking blocks 4 and 1.

Plan 2 consists of three blocks ($b = 3$) and 11 treatments ($t = 11$). It is represented as:

	─────Block─────	
1	**2**	**3**
$\begin{Bmatrix}1\\2\\3\end{Bmatrix}$	$\begin{Bmatrix}4\\5\\6\end{Bmatrix}$	$\begin{Bmatrix}7\\8\\9\end{Bmatrix}$
$\begin{Bmatrix}4\\5\\6\end{Bmatrix}$	$\begin{Bmatrix}7\\8\\9\end{Bmatrix}$	$\begin{Bmatrix}1\\2\\3\end{Bmatrix}$
10	11	

Schematically, Plan 2 may be written as:

	─────Block─────	
1	**2**	**3**
A'_1	A'_2	A'_3
A''_2	A''_3	A''_1
x	x	

In Plan 2, treatments 1, 2, and 3 constitute group A_1; treatments 4, 5, and 6 constitute group A_2; and treatments 7, 8, and 9 constitute group A_3. The remaining two treatments (10 and 11) are assigned to blocks 1 and 2. Treatments 1 through 9 appear twice each, and treatments 10 and 11 appear once each. Treatments 1, 2, and 3 always occur together as a group; treatments 4, 5, and 6 always occur together; and treatments 7, 8, and 9 always occur together. Thus, the treatments that are replicated twice fall into three groups

Fig. 7 Schematic representation of results for a balanced incomplete block plan

Block	Treatment							Total
	A	B	C	D	E	F	G	
1	X	X		X				B_1
2		X	X		X			B_2
3			X	X		X		B_3
4				X	X		X	B_4
5	X				X	X		B_5
6		X				X	X	B_6
7	X		X				X	B_7
Total	T_A	T_B	T_C	T_D	T_E	T_F	T_G	G

Fig. 8 Schematic representation of a chain block plan

Blocks				
1	2	. . .	$b-1$	b
A'_1	A'_2	. . .	A'_{b-1}	A'_b
A''_2	A''_3	. . .	A''_b	A''_1
x	x			x
x	x			x
.	.			.
.	.			.
.	.			.
x	x			x
Total: B_1	B_2	. . .	B_{b-1}	B_b G (= grand total)

(schematically A_1, A_2, and A_3). Group A_2 links blocks 1 and 2, group A_3 links blocks 2 and 3, and group A_1 completes the chain by linking blocks 3 and 1. To use a given chain block plan, the numbers should be allocated to the treatments at random.

Latin square plans (or Youden square plans) are useful when it is necessary to allow for two specific sources of nonhomogeneity in the conditions affecting test results. Such designs were originally applied in agricultural experimentation, in which the two-directional sources of nonhomogeneity were simply the two directions on the field, and the "square" was literally a square plot of ground.

Its usage has been extended to many other applications, in which there are two sources of nonhomogeneity that may affect experimental results, such as machines, positions, operators, runs, days, etc. A third variable, the experimental treatment, is then associated with the two source variables in a prescribed fashion. The use of Latin squares is restricted by two conditions: (1) the number of rows, columns, and treatments must all be the same; and (2) there must be no interactions between row and column factors.

To illustrate a Latin square, suppose it is necessary to compare the wear resistance of four materials. A wear testing machine is to be used that can accommodate four samples simultaneously. Two sources of inhomogeneity might be the variations from run to run and the variation among the four positions on the wear machine. In this situation, a 4×4 Latin square allows for both sources of inhomogeneity if four runs are made. The Latin square plan for this example is given below. The four materials are designated as A, B, C, D:

Run	─────Position No.─────			
	1	**2**	**3**	**4**
1	A	B	C	D
2	B	C	D	A
3	C	D	A	B
4	D	A	B	C

Examples of Latin squares from sizes 3×3 to 12×12 are given in Fig. 9. For a 4×4 Latin square, four examples are given. When a 4×4 Latin square is needed, one of the four examples should be selected at random. The procedure for using a given Latin square is:

Table 4   Balanced incomplete block plans (4 ≤ t ≤ 10, r ≤ 10)

t	k	r	b(b)	λ	E(a)	Plan No.
4	2	3	6	1	2/3	1
	3	3	4	2	8/9	(c)
5	2	4	10	1	5/8	2
	3	6	10	3	5/6	(c)
	4	4	5	3	15/16	(c)
6	2	5	15	1	3/5	3
	3	5	10	2	4/5	4
	3	10	20	4	4/5	5
	4	10	15	6	9/10	6
	5	5	6	4	24/25	(c)
7	2	6	21	1	7/12	(c)
	3	3	7	1	7/9	7
	4	4	7	2	7/8	8
	6	6	7	5	35/36	(c)
8	2	7	28	1	4/7	9
	4	7	14	3	6/7	10
	7	7	8	6	48/49	(c)
9	2	8	36	1	9/16	(c)
	3	4	12	1	3/4	11
	4	8	18	3	27/32	12
	5	10	18	5	9/10	13
	6	8	12	5	15/16	14
	8	8	9	7	63/64	(c)
10	2	9	45	1	5/9	15
	3	9	30	2	20/27	16
	4	6	15	2	5/6	17
	5	9	18	4	8/9	18
	6	9	15	5	25/27	19
	9	9	10	8	80/81	(c)

(a) The constant $E = t\lambda/rk$ is used in the analysis. (b) The number of blocks b serves as a check that no block has been missed. (c) Plans that may be constructed by forming all possible combinations of the t treatments in blocks of size k.

Plan 1: t = 4, k = 2, r = 3, b = 6, λ = 1, E = 2/3(d)

Group I	Group II	Group III
(1) 1, 2	(3) 1, 3	(5) 1, 4
(2) 3, 4	(4) 2, 4	(6) 2, 3

Plan 2: t = 5, k = 2, r = 4, b = 10, λ = 1, E = 5/8(d)

Group I	Group II
(1) 1, 2	(6) 1, 3
(2) 2, 5	(7) 2, 4
(3) 3, 4	(8) 3, 2
(4) 4, 1	(9) 4, 5
(5) 5, 3	(10) 5, 1

Plan 3: t = 6, k = 2, r = 5, b = 15, λ = 1, E = 3/5(d)

Group I	Group II	Group III	Group IV	Group V
(1) 1, 2	(4) 1, 3	(7) 1, 4	(10) 1, 5	(13) 1, 6
(2) 3, 4	(5) 2, 5	(8) 2, 6	(11) 2, 4	(14) 2, 3
(3) 5, 6	(6) 4, 6	(9) 3, 5	(12) 3, 6	(15) 4, 5

Plan 4: t = 6, k = 3, r = 5, b = 10, λ = 2, E = 4/5(d)

(1) 1, 2, 5	(5) 1, 4, 5	(8) 2, 4, 6
(2) 1, 2, 6	(6) 2, 3, 4	(9) 3, 5, 6
(3) 1, 3, 4	(7) 2, 3, 5	(10) 4, 5, 6
(4) 1, 3, 6		

Plan 5: t = 6, k = 3, r = 10, b = 20, λ = 4, E = 4/5(d)

Group I	Group II	Group III	Group IV
(1) 1, 2, 3	(3) 1, 2, 4	(5) 1, 2, 5	(7) 1, 2, 6
(2) 4, 5, 6	(4) 3, 5, 6	(6) 3, 4, 6	(8) 3, 4, 5

Group V	Group VI	Group VII	Group VIII
(9) 1, 3, 4	(11) 1, 3, 5	(13) 1, 3, 6	(15) 1, 4, 5
(10) 2, 5, 6	(12) 2, 4, 6	(14) 2, 4, 5	(16) 2, 3, 6

Group IX	Group X
(17) 1, 4, 6	(19) 1, 5, 6
(18) 2, 3, 5	(20) 2, 3, 4

Plan 6: t = 6, k = 4, r = 10, b = 15, λ = 6, E = 9/10(d)

Group I	Group II	Group III
(1) 1, 2, 3, 4	(4) 1, 2, 3, 5	(7) 1, 2, 3, 6
(2) 1, 4, 5, 6	(5) 1, 2, 4, 6	(8) 1, 3, 4, 5
(3) 2, 3, 5, 6	(6) 3, 4, 5, 6	(9) 2, 4, 5, 6

Group IV	Group V
(10) 1, 2, 4, 5	(13) 1, 2, 5, 6
(11) 1, 3, 5, 6	(14) 1, 3, 4, 6
(12) 2, 3, 4, 6	(15) 2, 3, 4, 5

(d) Block numbers are in parentheses followed by numbers which indicate treatments. In a number of the plans given, the blocks are arranged in groups. In setting up the experiment, make the groups as homogeneous as possible, i.e., if possible, there should be more difference between blocks in different groups than between blocks in the same group.

Plan 7: t = 7, k = 3, r = 3, b = 7, λ = 1, E = 7/9

(1) 1, 2, 4	(3) 3, 4, 6	(5) 5, 6, 1	(7) 7, 1, 3
(2) 2, 3, 5	(4) 4, 5, 7	(6) 6, 7, 2	

Plan 8: t = 7, k = 4, r = 4, b = 7, λ = 2, E = 7/8

(1) 1, 2, 3, 6	(3) 3, 4, 5, 1	(5) 5, 6, 7, 3	(7) 7, 1, 2, 5
(2) 2, 3, 4, 7	(4) 4, 5, 6, 2	(6) 6, 7, 1, 4	

Plan 9: t = 8, k = 2, r = 7, b = 28, λ = 1, E = 4/7

Group I	Group II	Group III	Group IV
(1) 1, 2	(5) 1, 3	(9) 1, 4	(13) 1, 5
(2) 3, 4	(6) 2, 8	(10) 2, 7	(14) 2, 3
(3) 5, 6	(7) 4, 5	(11) 3, 6	(15) 4, 7
(4) 7, 8	(8) 6, 7	(12) 5, 8	(16) 6, 8

Group V	Group VI	Group VII
(17) 1, 6	(21) 1, 7	(25) 1, 8
(18) 2, 4	(22) 2, 6	(26) 2, 5
(19) 3, 8	(23) 3, 5	(27) 3, 7
(20) 5, 7	(24) 4, 8	(28) 4, 6

Plan 10: t = 8, k = 4, r = 7, b = 14, λ = 3, E = 6/7

Group I	Group II	Group III	Group IV
(1) 1, 2, 3, 4	(3) 1, 2, 7, 8	(5) 1, 3, 6, 8	(7) 1, 4, 6, 7
(2) 5, 6, 7, 8	(4) 3, 4, 5, 6	(6) 2, 4, 5, 7	(8) 2, 3, 5, 8

Group V	Group VI	Group VII
(9) 1, 2, 5, 6	(11) 1, 3, 5, 7	(13) 1, 4, 5, 8
(10) 3, 4, 7, 8	(12) 2, 4, 6, 8	(14) 2, 3, 6, 7

Plan 11: t = 9, k = 3, r = 4, b = 12, λ = 1, E = 3/4

Group I	Group II	Group III	Group IV
(1) 1, 2, 3	(4) 1, 4, 7	(7) 1, 5, 9	(10) 1, 8, 6
(2) 4, 5, 6	(5) 2, 5, 8	(8) 7, 2, 6	(11) 4, 2, 9
(3) 7, 8, 9	(6) 3, 6, 9	(9) 4, 8, 3	(12) 7, 5, 3

Plan 12: t = 9, k = 4, r = 8, b = 18, λ = 3, E = 27/32

Group I	Group II
(1) 1, 4, 6, 7	(10) 1, 2, 5, 7
(2) 2, 6, 8, 9	(11) 2, 3, 6, 5
(3) 3, 8, 9, 1	(12) 3, 4, 7, 9
(4) 4, 1, 3, 2	(13) 4, 9, 2, 1
(5) 5, 7, 1, 8	(14) 5, 1, 9, 6
(6) 6, 9, 4, 5	(15) 6, 8, 1, 3
(7) 7, 3, 2, 6	(16) 7, 6, 4, 8
(8) 8, 2, 5, 4	(17) 8, 5, 3, 4
(9) 9, 5, 7, 3	(18) 9, 7, 8, 2

Table 4   (continued)

Plan 13: $t = 9$, $k = 5$, $r = 10$, $b = 18$, $\lambda = 5$, $E = 9/10$

Group I	Group II
(1) 1, 2, 3, 7, 8	(10) 1, 2, 3, 5, 9
(2) 2, 6, 8, 4, 1	(11) 2, 6, 5, 1, 8
(3) 3, 8, 5, 9, 2	(12) 3, 5, 1, 4, 6
(4) 4, 3, 9, 2, 6	(13) 4, 3, 2, 8, 7
(5) 5, 1, 7, 3, 4	(14) 5, 7, 9, 2, 4
(6) 6, 4, 2, 5, 7	(15) 6, 8, 7, 3, 5
(7) 7, 9, 1, 6, 3	(16) 7, 4, 8, 9, 1
(8) 8, 5, 4, 1, 9	(17) 8, 9, 4, 6, 3
(9) 9, 7, 6, 8, 5	(18) 9, 1, 6, 7, 2

Plan 14: $t = 9$, $k = 6$, $r = 8$, $b = 12$, $\lambda = 5$, $E = 15/16$

Group I	Group II
(1) 1, 2, 4, 5, 7, 8	(4) 1, 2, 5, 6, 7, 9
(2) 2, 3, 5, 6, 8, 9	(5) 1, 3, 4, 5, 8, 9
(3) 1, 3, 4, 6, 7, 9	(6) 2, 3, 4, 6, 7, 8

Group III	Group IV
(7) 1, 3, 5, 6, 7, 8	(10) 4, 5, 6, 7, 8, 9
(8) 1, 2, 4, 6, 8, 9	(11) 1, 2, 3, 4, 5, 6
(9) 2, 3, 4, 5, 7, 9	(12) 1, 2, 3, 7, 8, 9

Plan 15: $t = 10$, $k = 2$, $r = 9$, $b = 45$, $\lambda = 1$, $E = 5/9$

Group I	Group II	Group III	Group IV	Group V
(1) 1, 2	(6) 1, 3	(11) 1, 4	(16) 1, 5	(21) 1, 6
(2) 3, 4	(7) 2, 7	(12) 2, 10	(17) 2, 8	(22) 2, 9
(3) 5, 6	(8) 4, 8	(13) 3, 7	(18) 3, 10	(23) 3, 8
(4) 7, 8	(9) 5, 9	(14) 5, 8	(19) 4, 9	(24) 4, 10
(5) 9, 10	(10) 6, 10	(15) 6, 9	(20) 6, 7	(25) 5, 7

Group VI	Group VII	Group VIII	Group IX
(26) 1, 7	(31) 1, 8	(36) 1, 9	(41) 1, 10
(27) 2, 6	(32) 2, 3	(37) 2, 4	(42) 2, 5
(28) 3, 9	(33) 4, 6	(38) 3, 5	(43) 3, 6
(29) 4, 5	(34) 5, 10	(39) 6, 8	(44) 4, 7
(30) 8, 10	(35) 7, 9	(40) 7, 10	(45) 8, 9

Plan 16: $t = 10$, $k = 3$, $r = 9$, $b = 30$, $\lambda = 2$, $E = 20/27$

(1) 1, 2, 3	(11) 1, 2, 4	(21) 1, 3, 5
(2) 2, 5, 8	(12) 2, 3, 6	(22) 2, 7, 6
(3) 3, 7, 4	(13) 3, 4, 8	(23) 3, 8, 9
(4) 4, 1, 6	(14) 4, 9, 5	(24) 4, 2, 10
(5) 5, 8, 7	(15) 5, 7, 1	(25) 5, 6, 3
(6) 6, 4, 9	(16) 6, 8, 9	(26) 6, 1, 8
(7) 7, 9, 1	(17) 7, 10, 3	(27) 7, 9, 2
(8) 8, 10, 2	(18) 8, 1, 10	(28) 8, 4, 7
(9) 9, 3, 10	(19) 9, 5, 2	(29) 9, 10, 1
(10) 10, 6, 5	(20) 10, 6, 7	(30) 10, 5, 4

Plan 17: $t = 10$, $k = 4$, $r = 6$, $b = 15$, $\lambda = 2$, $E = 5/6$

(1) 1, 2, 3, 4	(6) 1, 6, 8, 10	(11) 3, 5, 9, 10
(2) 1, 2, 5, 6	(7) 2, 3, 6, 9	(12) 3, 6, 7, 10
(3) 1, 3, 7, 8	(8) 2, 4, 7, 10	(13) 3, 4, 5, 8
(4) 1, 4, 9, 10	(9) 2, 5, 8, 10	(14) 4, 5, 6, 7
(5) 1, 5, 7, 9	(10) 2, 7, 8, 9	(15) 4, 6, 8, 9

Plan 18: $t = 10$, $k = 5$, $r = 9$, $b = 18$, $\lambda = 4$, $E = 8/9$

(1) 1, 2, 3, 4, 5	(7) 1, 4, 5, 6, 10	(13) 2, 5, 6, 8, 10
(2) 1, 2, 3, 6, 7	(8) 1, 4, 8, 9, 10	(14) 2, 6, 7, 9, 10
(3) 1, 2, 4, 6, 9	(9) 1, 5, 7, 9, 10	(15) 3, 4, 6, 7, 10
(4) 1, 2, 5, 7, 8	(10) 2, 3, 4, 8, 10	(16) 3, 4, 5, 7, 9
(5) 1, 3, 6, 8, 9	(11) 2, 3, 5, 9, 10	(17) 3, 5, 6, 8, 9
(6) 1, 3, 7, 8, 10	(12) 2, 4, 7, 8, 9	(18) 4, 5, 6, 7, 8

Plan 19: $t = 10$, $k = 6$, $r = 9$, $b = 15$, $\lambda = 5$, $E = 25/27$

(1) 1, 2, 4, 5, 8, 9	(6) 2, 3, 4, 6, 8, 10	(11) 1, 4, 5, 7, 8, 10
(2) 5, 6, 7, 8, 9, 10	(7) 1, 2, 6, 7, 9, 10	(12) 1, 2, 3, 5, 7, 10
(3) 2, 4, 5, 6, 9, 10	(8) 1, 3, 5, 6, 8, 9	(13) 2, 3, 5, 6, 7, 8
(4) 1, 2, 4, 6, 7, 8	(9) 1, 2, 3, 8, 9, 10	(14) 1, 3, 4, 5, 6, 10
(5) 3, 4, 7, 8, 9, 10	(10) 2, 3, 4, 5, 7, 9	(15) 1, 3, 4, 6, 7, 9

Fig. 9 Select Latin squares

3 × 3

```
A B C
B C A
C A B
```

4 × 4

1	2	3	4
A B C D	A B C D	A B C D	A B C D
B A D C	B C D A	B D A C	B A D C
C D B A	C D A B	C A D B	C D A B
D C A B	D A B C	D C B A	D C B A

5 × 5

```
A B C D E
B A E C D
C D A E B
D E B A C
E C D B A
```

6 × 6

```
A B C D E F
B F D C A E
C D E F B A
D A F E C B
E C A B F D
F E B A D C
```

7 × 7

```
A B C D E F G
B C D E F G A
C D E F G A B
D E F G A B C
E F G A B C D
F G A B C D E
G A B C D E F
```

8 × 8

```
A B C D E F G H
B C D E F G H A
C D E F G H A B
D E F G H A B C
E F G H A B C D
F G H A B C D E
G H A B C D E F
H A B C D E F G
```

9 × 9

```
A B C D E F G H I
B C D E F G H I A
C D E F G H I A B
D E F G H I A B C
E F G H I A B C D
F G H I A B C D E
G H I A B C D E F
H I A B C D E F G
I A B C D E F G H
```

10 × 10

```
A B C D E F G H I J
B C D E F G H I J A
C D E F G H I J A B
D E F G H I J A B C
E F G H I J A B C D
F G H I J A B C D E
G H I J A B C D E F
H I J A B C D E F G
I J A B C D E F G H
J A B C D E F G H I
```

11 × 11

```
A B C D E F G H I J K
B C D E F G H I J K A
C D E F G H I J K A B
D E F G H I J K A B C
E F G H I J K A B C D
F G H I J K A B C D E
G H I J K A B C D E F
H I J K A B C D E F G
I J K A B C D E F G H
J K A B C D E F G H I
K A B C D E F G H I J
```

12 × 12

```
A B C D E F G H I J K L
B C D E F G H I J K L A
C D E F G H I J K L A B
D E F G H I J K L A B C
E F G H I J K L A B C D
F G H I J K L A B C D E
G H I J K L A B C D E F
H I J K L A B C D E F G
I J K L A B C D E F G H
J K L A B C D E F G H I
K L A B C D E F G H I J
L A B C D E F G H I J K
```

- Permute the columns at random.
- Permute the rows at random.
- Assign letters randomly to the treatments.

If squares 5×5 and higher are used frequently, each time it is used a square should be chosen at random from the set of all possible squares. Reference 10 provides complete representation of the squares from 4×4 to 6×6 and sample squares up to 12×12.

The results of a Latin square experiment are recorded in a two-way table that is similar to the plan itself. The treatment totals and the row and column totals of the Latin square plan are each directly comparable without adjustment.

Youden square plans, like Latin square plans, are used when two kinds of inhomogeneity are involved. The conditions for the use of a Youden square, however, are less restrictive than for the Latin square. The use of Latin square plans is restricted by the fact that the number of rows, columns, and treatments must all be the same. Youden squares have the same number of rows and treatments, but fairly wide flexibility in the number of columns is possible.

The following notation is typically used: t is the number of treatments to be compared; b is the number of levels of one source of inhomogeneity (rows); k is the number of levels of the other source of inhomogeneity (columns); and r is the number of replications of each treatment. In a Youden square, $t = b$ and $k = r$. The procedure for using a given Youden square is:

- Permute the rows at random.
- Permute the columns at random.
- Assign letters at random to the treatments.

The results of an experiment using a Youden square plan are recorded in a two-way table that is similar to the plan itself. Table 5 provides examples of Youden square arrangements. In some cases where there are two sources of inhomogeneity, a suitable Latin or Youden square may not exist.

Determining Optimum Conditions or Levels

In many industrial-type processes, there is a measurable end-property whose value is of primary interest, which the experimenter would like to have attain some optimum value. This end-property is termed "yield" or "response" in experimental design. For example, the end-property might be the actual yield of the process to be maximized, a strength property to be maximized, cost to be minimized, or some chemical or physical

characteristic that would be most desirable at a maximum or at a minimum, as specified.

The value of this primary end-property depends on the values of settings of a number of factors in the process that affect the end-property. The goal of experimentation is to determine which settings will result in an optimum response. Often, it is necessary to determine not only the values of the variables that result in optimum response, but also how much change in response results from small deviations from the optimum settings (i.e., the nature of the response function in the vicinity of this optimum).

The Response Function. In a factorial experiment where the levels of all factors are quantitative (e.g., time, temperature, pressure, amount of catalyst, purity of ingredients, etc.), the response y can represent a function of the levels of the experimental factors. For an n-factor experiment, true yield is represented by:

$$y = \Phi \, (x_1, x_2, \ldots, x_n) \qquad \text{(Eq 1)}$$

where x_1 is the level of factor 1, x_2 is the level of factor 2, etc.

For observed values of y,

$$Y_u = \Phi \, (x_{1u}, x_{2u}, \ldots, x_{nu}) + e_u \qquad \text{(Eq 2)}$$

where Y_u is the uth observation of y, where $u = 1, 2, \ldots, N$ represent the N observations in the factorial experiment; x_{1u} is level of factor 1 for the uth observation; x_{2u} is the level of factor 2 for the uth observation, etc.; and e_u is experimental error of the uth observation.

The function Φ represents the response function. If the function Φ is determined, the results of the experiment can be described completely and y can be predicted for values of the factors that were not included in the experiment (but the function should not be used for prediction outside the range of experiment).

Usually, the mathematical form of the function is completely unknown. Frequently, however, it can be satisfactorily approximated within a limited region by a polynomial in x_{iu}. As the relation $y = \Phi(x)$ can be represented by a curve, the relation between y and two factors, x_1 and x_2 (i.e., $y = \Phi(x_1, x_2)$, can be represented by the response surface, as shown in Fig. 10. Alternatively, it can be represented by a contour diagram, which traces contours of equal response, as shown in Fig. 11.

Experimental designs and methods of analysis have been developed for fitting polynomials of the first and second degree; these designs are called first- and second-order designs, respectively. These designs may be described, for example, as a first-

order design in two dimensions, or a second-order design in four dimensions. In general, these designs are referred to as kth-order designs in n dimensions. The dimension n refers to the number of independent variables (x_i) in the response function, and the order k refers to the degree of the fitted polynomial function.

A design in which only one variable is controlled is a one-dimensional design, and y is observed as a function of the single variable x, i.e., $y = \Phi(x)$. The first approach in describing such a relationship may be that of fitting a first-order equation; i.e., a straight line $y = \beta_0 + \beta_1 x$. If it has been determined that the relationship cannot be adequately represented by a straight line, a second-degree (or higher degree) polynomial may be fitted. A one-dimensional design, however, is not usual in this kind of experimentation; typically, more variables are involved.

If the response y as a function of two variables (x_1 and x_2) is the subject of interest, it can be represented as:

$$y = \Phi \, (x_1, x_2) \qquad \text{(Eq 3)}$$

Fig. 10 A response surface

Source: Ref 11

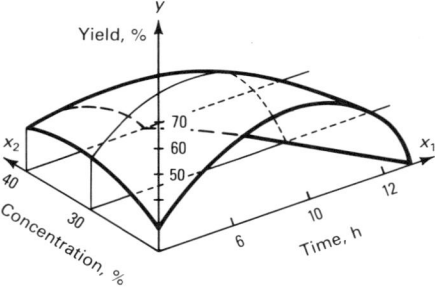

Fig. 11 Yield contours for the surface of Fig. 10 with 2^2 factorial design

Source: Ref 11

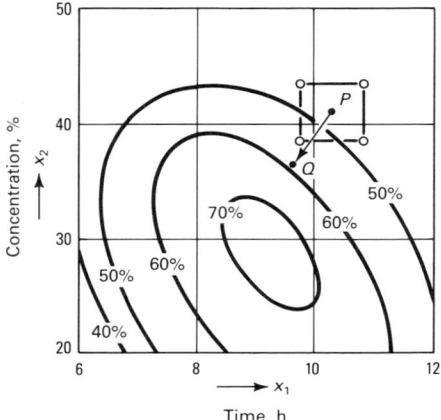

Table 5 Youden square arrangements (r ≤ 10)

Plan No.	t = b	r = k	Index λ	Index E = tλ/rk	Remarks
1	3	2	1	3/4	(a)
2	4	3	2	8/9	(a)
3	5	4	3	15/16	(a)
4	6	5	4	24/25	(a)
5	7	3	1	7/9	
6	7	4	2	7/8	Complement of Plan 5(b)
7	7	6	5	35/36	(a)
8	8	7	6	48/49	(a)
9	9	8	7	63/64	(a)
10	10	9	8	80/81	(a)
11	11	5	2	22/25	
12	11	6	3	11/12	Complement of Plan 11(b)
13	11	10	9	99/100	(a)
14	13	4	1	13/16	
15	13	9	6	26/27	Complement of Plan 14(b)
16	15	7	3	45/49	
17	15	8	4	15/16	Complement of Plan 16(b)
18	16	6	2	8/9	
19	16	10	6	24/25	
20	19	9	4	76/81	
21	19	10	5	19/20	Complement of Plan 20(b)
22	21	5	1	21/25	
23	25	9	3	25/27	
24	31	6	1	31/36	
25	31	10	3	93/100	
26	37	9	2	74/81	
27	57	8	1	57/64	
28	73	9	1	73/81	
29	91	10	1	91/100	See Ref 1

(a) Blocks in these plans are columns of Latin squares with one row deleted. (b) The complement of a plan is developed as follows: Construct the first block (column) by writing all treatments that did not appear in the first block of the original plan. With these letters as starting points, complete each row by writing in alphabetical order all remaining treatment letters followed by A, B, C, etc., until every treatment letter appears once in each row. For example, Plan 6 is developed from Plan 5 as follows: The first block of Plan 5 is ABD; its complement and therefore the first block of Plan 6 is CEFG. The complete layout for Plan 6 is:

Row	Block 1	2	3	4	5	6	7
1	C	D	E	F	G	A	B
2	E	F	G	A	B	C	D
3	F	G	A	B	C	D	E
4	G	A	B	C	D	E	F

Note: The detailed plans given are only those that are not easily derivable from other designs. (see Index columns above).

Plan 5: t = b = 7, r = k = 3

Row	Block 1	2	3	4	5	6	7
1	A	B	C	D	E	F	G
2	B	C	D	E	F	G	A
3	D	E	F	G	A	B	C

Plan 11: t = b = 11, r = k = 5

Row	Block 1	2	3	4	5	6	7	8	9	10	11
1	A	B	C	D	E	F	G	H	I	J	K
2	E	F	G	H	I	J	K	A	B	C	D
3	F	G	H	I	J	K	A	B	C	D	E
4	G	H	I	J	K	A	B	C	D	E	F
5	I	J	K	A	B	C	D	E	F	G	H

Plan 14: t = b = 13, r = k = 4

Row	Block 1	2	3	4	5	6	7	8	9	10	11	12	13
1	A	B	C	D	E	F	G	H	I	J	K	L	M
2	B	C	D	E	F	G	H	I	J	K	L	M	A
3	D	E	F	G	H	I	J	K	L	M	A	B	C
4	J	K	L	M	A	B	C	D	E	F	G	H	I

Plan 16: t = b = 15, r = k = 7

Row	Block 1	2	3	4	5	6	7	8	9	10	11	12	13	14	15
1	A	B	C	D	E	F	G	H	I	J	K	L	M	N	O
2	B	C	D	E	F	G	H	I	J	K	L	M	N	O	A
3	C	D	E	F	G	H	I	J	K	L	M	N	O	A	B
4	E	F	G	H	I	J	K	L	M	N	O	A	B	C	D
5	F	G	H	I	J	K	L	M	N	O	A	B	C	D	E
6	I	J	K	L	M	N	O	A	B	C	D	E	F	G	H
7	K	L	M	N	O	A	B	C	D	E	F	G	H	I	J

Plan 18: t = b = 16, r = k = 6

Row	Block 1	2	3	4	5	6	7	8	9	10	11	12	13	14	15	16
1	A	B	C	D	E	F	G	H	I	J	K	L	M	N	O	P
2	B	C	D	A	F	G	H	E	J	K	L	I	N	O	P	M
3	C	D	A	B	G	H	E	F	K	L	I	J	O	P	M	N
4	E	F	G	H	I	J	K	L	M	N	O	P	A	B	C	D
5	I	J	K	L	M	N	O	P	A	B	C	D	E	F	G	H
6	M	N	O	P	A	B	C	D	E	F	G	H	I	J	K	L

Plan 19: t = b = 16, r = k = 10

Row	Block 1	2	3	4	5	6	7	8	9	10	11	12	13	14	15	16
1	A	B	C	D	E	F	G	H	I	J	K	L	M	N	O	P
2	C	A	B	D	F	G	H	E	J	K	L	I	N	O	P	M
3	D	C	A	B	G	H	E	F	K	L	I	J	O	P	M	N
4	N	D	E	K	H	D	L	I	M	A	O	C	H	E	D	I
5	M	N	J	H	B	A	O	M	N	B	C	H	E	F	C	H
6	B	J	K	I	A	L	P	J	O	O	B	N	C	O	B	O
7	J	K	I	F	L	O	A	K	A	P	H	O	B	N	E	B
8	I	H	F	L	O	M	K	J	P	H	A	A	I	I	D	G
9	H	I	L	O	M	K	J	A	E	I	P	P	K	K	F	C
10	O	M	K	J	P	I	I	O	H	D	E	B	D	B	I	H

Plan 20: t = b = 19, r = k = 9

Row	Block 1	2	3	4	5	6	7	8	9	10	11	12	13	14	15	16	17	18	19
1	A	B	C	D	E	F	G	H	I	J	K	L	M	N	O	P	Q	R	S
2	C	D	E	F	G	H	I	J	K	L	M	N	O	P	Q	R	S	A	B
3	D	E	F	G	H	I	J	K	L	M	N	O	P	Q	R	S	A	B	C
4	E	F	G	H	I	J	K	L	M	N	O	P	Q	R	S	A	B	C	D
5	G	H	I	J	K	L	M	N	O	P	Q	R	S	A	B	C	D	E	F
6	H	I	J	K	L	M	N	O	P	Q	R	S	A	B	C	D	E	F	G
7	K	L	M	N	O	P	Q	R	S	A	B	C	D	E	F	G	H	I	J
8	N	O	P	Q	R	S	A	B	C	D	E	F	G	H	I	J	K	L	M
9	O	P	Q	R	S	A	B	C	D	E	F	G	H	I	J	K	L	M	N

Plan 22: t = b = 21, r = k = 5

Row	Block 1	2	3	4	5	6	7	8	9	10	11	12	13	14	15	16	17	18	19	20	21
1	A	B	C	D	E	F	G	H	I	J	K	L	M	N	O	P	Q	R	S	T	U
2	B	C	D	E	F	G	H	I	J	K	L	M	N	O	P	Q	R	S	T	U	A
3	G	H	I	J	K	L	M	N	O	P	Q	R	S	T	U	A	B	C	D	E	F
4	I	J	K	L	M	N	O	P	Q	R	S	T	U	A	B	C	D	E	F	G	H
5	S	T	U	A	B	C	D	E	F	G	H	I	J	K	L	M	N	O	P	Q	R

Again, as a first step, a first-order model (now the equation of a plane) could be fit:

$$y = \beta_0 + \beta_1 x_1 + \beta_2 x_2 \qquad \text{(Eq 4)}$$

Where three or more variables are controlled, there is a function of the type:

$$y = \Phi(x_1, x_2, \ldots, x_n) \qquad \text{(Eq 5)}$$

A general goal in selecting and constructing experimental designs when observing a function of several quantitative variables is that the selected design should permit relatively simple and straightforward estimation of the coefficients of the fitted equation. Two-level factorial designs are used for fitting first-order models, particularly in the two-dimensional case. New designs have been developed (Ref 7, 4, 5).

Finding the Optimum. In general, experimentation proceeds sequentially. Initial levels of the variables are selected so that the levels are near either present operating conditions or the optimum response. A design is chosen and experimental observations are made at values of the variables that are specified by the design. Typically, first-order designs provide information on the adequacy of the first-order model, indicate whether the response is near the optimum, and indicate the direction to move to approach the optimum. Another first-order design may then be run at a new position, or a second-order design may be run at the original position. The methods are extremely flexible and useful.

REFERENCES

1. W.G. Cochran and G.M. Cox, *Experimental Designs*, 2nd ed., John Wiley & Sons, New York, 1957
2. R.L. Anderson and T.A. Bancroft, *Statistical Theory in Research*, McGraw-Hill, New York, 1952
3. M.G. Natrella, *Experimental Statistics*, NBS Handbook 91, U.S. Government Printing Office, Washington, DC, 1966 (reprinted by John Wiley & Sons, New York, 1983)
4. G.E.P. Box, W.G. Hunter, and J.S. Hunter, *Statistics for Experimenters*, John Wiley & Sons, New York, 1978
5. V.L. Anderson and R.A. McLean, *Design of Experiments*, 2nd ed., Marcel Dekker, New York, 1983
6. E.B. Wilson, Jr., *An Introduction to Scientific Research*, McGraw-Hill, New York, 1952
7. "Fractional Factorial Experiment Designs For Factors at Two Levels," National Bureau of Standards, Applied Mathematics Series, No. 48, U.S. Government Printing Office, Washington, DC, 1957 (available from the National Technical Information Service, Springfield, VA 22161, order No. PB-176119)
8. "Fractional Factorial Experiment Designs For Factors at Three Levels," National Bureau of Standards, Applied Mathematics Series, No. 54, U.S. Government Printing Office, Washington, DC, 1959 (available from the National Technical Information Service, Springfield, VA 22161, order No. COM73-11111)
9. W.H. Clatworthy, "Tables of Two Associate Class Partially Balanced Designs, National Bureau of Standards," Applied Mathematics Series, No. 63, U.S. Government Printing Office, Washington, DC, 1973
10. R.A. Fisher and F. Yates, *Statistical Tables for Biological, Agricultural and Medical Research*, 5th ed., Oliver and Boyd, Ltd., Edinburgh, and Hafner Publishing, New York, 1957
11. O.L. Davies, Ed., *The Design and Analysis of Industrial Experiments*, Hafner Publishing, New York, 1956 (reprinted by Longman, New York, 1978)

SELECTED REFERENCES

- D.R. Cox, *Planning of Experiments*, John Wiley & Sons, New York, 1958
- C. Daniel, *Applications of Statistics to Industrial Experimentation*, John Wiley & Sons, New York, 1976
- W.J. Diamond, *Practical Experimental Designs*, Lifetime Learning Publications, Belmont, CA, 1981
- C.R. Hicks, *Fundamental Concepts in the Design of Experiments*, 3rd ed., Holt, Rinehart & Winston, New York, 1982
- N.L. Johnson and F.C. Leone, *Statistics and Experimental Design in Engineering and Physical Sciences*, Vol 2, 2nd ed., John Wiley & Sons, New York, 1977
- W. Mendenhall, *Introduction to Linear Models and the Design and Analysis of Experiments*, Wadsworth Publishing, Belmont, CA, 1968
- D.C. Montgomery, *Design and Analysis of Experiments*, 2nd ed., John Wiley & Sons, New York, 1984

Analysis of Designed Experiments*

ANALYSIS OF DESIGNED EXPERIMENTS is a broad topic that has been addressed in many books and articles, ranging from the very elementary to the most complex and rigorous. The methods described in this article are oriented toward the analysis of planned cooperative experiments that involve multiple factors, blocks, and/or treatment levels, as discussed in the article "Planning of Comparative Experiments" in this Section.

The parametric and nonparametric statistical procedures that are commonly used for evaluating single and multiple samples of data are not addressed in this article. Numerous elementary statistics textbooks cover this material quite adequately (see Ref 1 to 9). Other more advanced sources may also be useful when considering multivariate analysis methods (Ref 10, 11), nonparametric methods (Ref 12, 13), and regression analysis problems (Ref 14–16). Many statistical procedures can also be easily applied using computer software (Ref 17, 18).

Four types of designed experiments are analyzed in this article: factorial experiments, fractional factorial experiments, randomized block experiments, and incomplete block experiments. Analysis of other types of designed experiments can be found in Ref 19 to 21. It should be noted that the analysis methods described in the following paragraphs are "longhand" procedures that can be applied in a systematic fashion without the aid of a computer or special statistical software routines. Other methods are also available for analyzing the results of designed experiments (Ref 22). For additional information, see the article "Planning of Comparative Experiments" in this Section.

Factorial Experiments

Assume that tests were performed to determine the importance of different processing variables on the elongation properties of a cast aluminum. The experiment has four factors, each at two levels; therefore, it is a 2^4

factorial experiment. Note that most of the factors are qualitative in this experiment. The experimental factors and levels are:

Symbol	Factors	Levels
A	Mold type	A_0: composite
		A_1: sand
B	Type of specimen	B_0: cut from casting
		B_1: cut from appendage
C	Cooling conditions	C_0: at a riser
		C_1: at a chill
D	Thickness	D_0: > 12.7 mm (0.500 in.)
		D_1: < 12.7 mm (0.500 in.)

The observations reported in Table 1 are percent elongation values, measured on 25.4-mm (1.0-in.) gage length samples with special extensometers that allow accurate readings to 0.1% elongation. The conventional symbol representing the treatment combination appears beside the resulting observation.

Estimation of Main Effects and Interactions. A systematic method for estimation of main effects and interactions for two-level factorials was originally described by Yates (Ref 19) and may be found in Ref 20 and 21. The method, as discussed in this article, applies to factorials, blocked factorials, and fractional factorials for which there are 2^n observations.†

The systematic procedure for the Yates method of analysis is as follows: A table is constructed with $n + 2$ columns. The data summarized in Table 1 is a 2^4 factorial ($n = 4$) experiment. Therefore, Table 2, which was constructed from these data, has six columns. In the first column, the treatment combinations are listed.

†In a $\frac{1}{2}b$ fraction of a 2^n factorial, there are $2n'$ observations, where $n' = n - b$.

For factorials or blocked factorials, the treatment combinations should be listed in "standard order" in the first column—i.e., for two factors: (1), a, b, ab; for three factors: (1), a, b, ab, c, ac, bc, abc; for four factors: (1) a, b, ab, c, ac, bc, abc, d, ad, bd, abd, cd, acd, bcd, $abcd$, . . ., etc. Standard order for five factors is obtained by listing all the treatment combinations given for four factors, followed by e, ae, be, abe, . . ., $abcde$ (i.e., the new element multiplied by all previous treatment combinations). Standard order for a higher number of factors is obtained in similar fashion, beginning with the series for the next smaller number of factors and continuing by multiplying that series by the new element.

In column 2, the observed yield or response is entered that corresponds to each of the treatment combinations listed in column 1. In the top half of column 3, the sums of consecutive pairs of entries in column 2 are entered in order. In the bottom half of the column, the differences between the same consecutive pairs of entries are entered in order—the second entry minus the first entry, the fourth entry minus the third entry, etc. For example:

$$4.2 + 3.1 = 7.3$$
$$4.5 + 2.9 = 7.4$$
$$3.9 + 2.8 = 6.7, \text{ etc., and}$$
$$3.1 - 4.2 = -1.1$$
$$2.9 - 4.5 = -1.6$$
$$2.8 - 3.9 = -1.1, \text{ etc.}$$

Columns 4, 5, and 6 are obtained in the same manner as column 3—by obtaining in each

Table 1 Example of test data from a factorial experiment

Results of elongation measurements on cast aluminum tensile specimens. Values reported in percent elongation.

		A_0				A_1			
		B_0		B_1		B_0		B_1	
C_0	D_0	4.2	(1)	4.5	b	3.1	a	2.9	ab
	D_1	4.0	d	5.0	bd	3.0	ad	2.5	abd
C_1	D_0	3.9	c	4.6	bc	2.8	ac	3.2	abc
	D_1	4.0	cd	5.0	bcd	2.5	acd	2.3	$abcd$

*Based on Ref 1

Table 2 Yates method of analysis using data from Table 1

1 Treatment combination	2 Response (yield)	3	4	5	6 g
(1)	4.2	7.3	14.7	29.2	$57.5 = g_T$
a	3.1	7.4	14.5	28.3	$-12.9 = g_A$
b	4.5	6.7	14.5	-5.2	$2.5 = g_B$
ab	2.9	7.8	13.8	-7.7	$-3.5 = g_{AB}$
c	3.9	7.0	-2.7	1.2	$-0.9 = g_C$
ac	2.8	7.5	-2.5	1.3	$-0.5 = g_{AC}$
bc	4.6	6.5	-3.5	-0.8	$1.3 = g_{BC}$
abc	3.2	7.3	-4.2	-2.7	$0.5 = g_{ABC}$
d	4.0	-1.1	0.1	-0.2	$-0.9 = g_D$
ad	3.0	-1.6	1.1	-0.7	$-2.5 = g_{AD}$
bd	5.0	-1.1	0.5	0.2	$0.1 = g_{BD}$
abd	2.5	-1.4	0.8	-0.7	$-1.9 = g_{ABD}$
cd	4.0	-1.0	-0.5	1.0	$-0.5 = g_{CD}$
acd	2.5	-2.5	-0.3	0.3	$-0.9 = g_{ACD}$
bcd	5.0	-1.5	-1.5	0.2	$-0.7 = g_{BCD}$
abcd	2.3	-2.7	-1.2	0.3	$0.1 = g_{ABCD}$
Total	57.5				
Sum of squares	219.15				3506.40

case the sums and differences of the pairs in the preceding column in the manner described above.

The estimated main effects and interactions in the last column (column $n + 2$) also appear in a standard order—for two factors: T, A, B, AB; for three factors: T, A, B, AB, C, AC, BC, ABC, . . ., etc.; where T corresponds to the overall average effect; A to the main effect of factor A, AB to the interaction of factors A and B, etc. The entries in column 6 are called g_T, g_A, g_B, g_{AB}, etc., corresponding to the ordered effects T, A, B, AB, etc. Estimates of main effects and interactions are obtained by dividing the appropriate g by 2^{n-1}. The overall mean is equal to g_T divided by 2^n. In Table 2, $g_A = -12.9$, and the estimated main effect of A is $-12.9/8 = -1.6$. Given that $g_{AD} = -2.5$, the estimated effect of AD interaction is $-2.5/8 = -0.3$, etc.

Several calculations can be made to check the computations just described. The sum of all the 2^n individual responses (column 2, Table 2) should equal the total given in the first entry of column 6 (column $n + 2$): $57.5 = 57.5$. The sum of the squares of the individual responses (column 2) should equal the sum of the squares of the entries in column 6 (column $n + 2$) divided by 2^n:

$$219.15 = \frac{3506.40}{16}$$

$$= 219.15$$

For any main effect, the entry in the last column (column $n + 2$) should equal the sum of the responses in which that factor is at its higher level minus the sum of the responses in which that factor is at its lower level. In this example, for g_A:

$$g_A = (a + ab + ac + abc + ad + abd$$
$$+ acd + abcd) - [(1) + b + c + bc$$
$$+ d + bd + cd + bcd]$$
$$= (22.3) - (35.2)$$
$$= -12.9$$

Testing for Significance of Main Effects and Interactions. The significance of the main effects and interactions identified by the methods described above can be evaluated in the following manner. Let α be the level of significance. In this case, $\alpha = 0.05$. If there is no available estimate of the variation due to experimental error, the sum of squares of the g's corresponding to interactions of three or more factors in Table 2 are calculated as:

$$g_{ABC}^2 + g_{ABD}^2 + g_{ACD}^2 + g_{BCD}^2 + g_{ABCD}^2$$
$$= 5.17$$

To obtain s^2, the sum of squares is divided by $2^n\nu$, where ν is the number of interactions included. In a 2^n factorial, the number of third and higher interactions will be $2^n - (n^2 + n + 2)/2$. If an independent estimate of the variation due to experimental error is available, use that s^2 value:

$$n = 4$$
$$\nu = 5$$
$$2^n\nu = 16(5) = 80$$
$$s^2 = \frac{5.17}{80}$$
$$= 0.0646$$
$$= 0.254$$

Consult Table 3 for the ν degrees of freedom for $t_{(1-\alpha)/2}$. If higher order interactions are used to obtain s^2, ν is the number of interactions included. If an independent estimate of s^2 is used, ν is the degrees of freedom associated with this estimate. In this case, $t_{0.975}$ for 5 d.f. = 2.571. Then, compute:

$$w = (2^n)^{1/2} t_{(1-\alpha)/2}s$$
$$= 4(2.571)(0.254)$$
$$= 2.61$$

For any main effect or interaction X, if the absolute value of g_X is greater than w, it can be concluded that X is different from zero; e.g., if $|g_A| > w$, it can be concluded that the A effect is different from zero. Otherwise, there is no basis for assuming that X is different from zero. From Table 2, $|g_A| = 12.9$, and $|g_{AB}| = 3.5$, which are greater than w; therefore, the main effect of A and the interaction AB are believed to be significant. In this example, the effects of mold type, and mold type combined with specimen type, are believed to be significant factors in maximizing elongation properties.

Fractional Factorial Experiments

In some cases, it is impossible to investigate all elements of a factorial design. The use of fractional factorials is a commonly used "shortcut" to the full factorial experimental procedure. The analysis of data generated from a fractional factorial design is discussed in the following example.

Assume that it is possible to investigate only 8 of the 16 possible combinations of processing variables given in Table 1. The conditions selected and the data that resulted are given in Table 4.

Estimation of Main Effects and Interactions. The Yates procedure can be applied by replacing n with n', where $n' = n - b$ for the particular fractional factorial used (see Table 5). For fractional factorials, the treatment combinations in column 1 should be listed in the order given in Table 5. The preferred order of the estimated effects is also given in Table 5. For fractional factorial plans other than those given in Table 5, see Ref 21 for the necessary ordering for the Yates method of analysis.

The last column of the Yates table (column $n' + 2$) provides the g values corresponding to the effects in the order listed in the "estimated effects" column of Table 5. To obtain the estimates of main effects and interaction, each g is divided by $2^{n'-1}$. This procedure is illustrated in Table 6 for the data given in Table 4.

Table 3   Percentiles of the *t* distribution

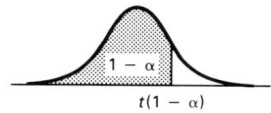

Entry is $t(1 - \alpha, \nu)$, where $P\{t(\nu) \leq t(1 - \alpha, \nu)\} = 1 - \alpha$.

ν	0.55	0.60	0.65	0.70	0.75	0.80	0.85	0.90	0.95	0.975	0.99	0.995	0.9995
1	0.158	0.325	0.510	0.727	1.000	1.376	1.963	3.078	6.314	12.706	31.821	63.657	636.619
2	0.142	0.289	0.445	0.617	0.816	1.061	1.386	1.886	2.920	4.303	6.965	9.925	31.598
3	0.137	0.277	0.424	0.584	0.765	0.978	1.250	1.638	2.353	3.182	4.541	5.841	12.924
4	0.134	0.271	0.414	0.569	0.741	0.941	1.190	1.533	2.132	2.776	3.747	4.604	8.610
5	0.132	0.267	0.408	0.559	0.727	0.920	1.156	1.476	2.015	2.571	3.365	4.032	6.869
6	0.131	0.265	0.404	0.553	0.718	0.906	1.134	1.440	1.943	2.447	3.143	3.707	5.959
7	0.130	0.263	0.402	0.549	0.711	0.896	1.119	1.415	1.895	2.365	2.998	3.499	5.408
8	0.130	0.262	0.399	0.546	0.706	0.889	1.108	1.397	1.860	2.306	2.896	3.355	5.041
9	0.129	0.261	0.398	0.543	0.703	0.883	1.100	1.383	1.833	2.262	2.821	3.250	4.781
10	0.129	0.260	0.397	0.542	0.700	0.879	1.093	1.372	1.812	2.228	2.764	3.169	4.587
11	0.129	0.260	0.396	0.540	0.697	0.876	1.088	1.363	1.796	2.201	2.718	3.106	4.437
12	0.128	0.259	0.395	0.539	0.695	0.873	1.083	1.356	1.782	2.179	2.681	3.055	4.318
13	0.128	0.259	0.394	0.538	0.694	0.870	1.079	1.350	1.771	2.160	2.650	3.012	4.221
14	0.128	0.258	0.393	0.537	0.692	0.868	1.076	1.345	1.761	2.145	2.624	2.977	4.140
15	0.128	0.258	0.393	0.536	0.691	0.866	1.074	1.341	1.753	2.131	2.602	2.947	4.073
16	0.128	0.258	0.392	0.535	0.690	0.865	1.071	1.337	1.746	2.120	2.583	2.921	4.015
17	0.128	0.257	0.392	0.534	0.689	0.863	1.069	1.333	1.740	2.110	2.567	2.898	3.965
18	0.127	0.257	0.392	0.534	0.688	0.862	1.067	1.330	1.734	2.101	2.552	2.878	3.922
19	0.127	0.257	0.391	0.533	0.688	0.861	1.066	1.328	1.729	2.093	2.539	2.861	3.883
20	0.127	0.257	0.391	0.533	0.687	0.860	1.064	1.325	1.725	2.086	2.528	2.845	3.850
21	0.127	0.257	0.391	0.532	0.686	0.859	1.063	1.323	1.721	2.080	2.518	2.831	3.819
22	0.127	0.256	0.390	0.532	0.686	0.858	1.061	1.321	1.717	2.074	2.508	2.819	3.792
23	0.127	0.256	0.390	0.532	0.685	0.858	1.060	1.319	1.714	2.069	2.500	2.807	3.767
24	0.127	0.256	0.390	0.531	0.685	0.857	1.059	1.318	1.711	2.064	2.492	2.797	3.745
25	0.127	0.256	0.390	0.531	0.684	0.856	1.058	1.316	1.708	2.060	2.485	2.787	3.725
26	0.127	0.256	0.390	0.531	0.684	0.856	1.058	1.315	1.706	2.056	2.479	2.779	3.707
27	0.127	0.256	0.389	0.531	0.684	0.855	1.057	1.314	1.703	2.052	2.473	2.771	3.690
28	0.127	0.256	0.389	0.530	0.683	0.855	1.056	1.313	1.701	2.048	2.467	2.763	3.674
29	0.127	0.256	0.389	0.530	0.683	0.854	1.055	1.311	1.699	2.045	2.462	2.756	3.659
30	0.127	0.256	0.389	0.530	0.683	0.854	1.055	1.310	1.697	2.042	2.457	2.750	3.646
40	0.126	0.255	0.388	0.529	0.681	0.851	1.050	1.303	1.684	2.021	2.423	2.704	3.551
60	0.126	0.254	0.387	0.527	0.679	0.848	1.046	1.296	1.671	2.000	2.390	2.660	3.460
120	0.126	0.254	0.386	0.526	0.677	0.845	1.041	1.289	1.658	1.980	2.358	2.617	3.373
∞	0.126	0.253	0.385	0.524	0.674	0.842	1.036	1.282	1.645	1.960	2.326	2.576	3.291

Source: Ref 2

Testing for Significance of Main Effects and Interactions.

The significance of the main effects and interactions identified by the above method can be evaluated through the following procedure.

Let α be the level of significance. In this example, let $\alpha = 0.5$. If no external estimate of the variation due to experimental error is available, check the lines in the Yates table that correspond to the estimated effects that are expected to be zero.

The sum of squares of the *g*'s for the lines checked are then computed. To obtain s^2, the sum of squares obtained is divided by $2^{n'}\nu$, where ν is the number of interactions included. If an independent estimate of the variation due to experimental error is available, use this s^2 value. In this analysis, an independent estimate of s^2, from 24 pairs of duplicate measurements obtained in another part of the larger program, is used:

$$s^2 = 0.0408$$

$$s = 0.202$$

$$\nu = 24$$

In this example, obtain $t_{(1 - \alpha)/2}$ from Table 3. For ν degrees of freedom, $t_{0.975}$ for 24 d.f. = 2.064. Then, compute:

$$w = (2^{n'})^{1/2} t_{(1 - \alpha)/2} s$$
$$= \sqrt{8}\,(2.064)\,(0.202)$$
$$= 1.18$$

For any main effect or interaction X, if the absolute value of g_X is greater than w, it can

Table 4   Results of elongation measurements on cast aluminum tensile specimen

Fractional factorial experiment of data given in Table 1

		A_0				A_1	
		B_0	B_1	B_0		B_1	
C_0 ....D_0	4.2	(1)				2.9	ab
D_1		5.0	bd	3.0	ad		
C_1 ....D_0		4.6	bc	2.8	ac		
D_1	4.0	cd				2.3	abcd

be concluded that X is different from zero. For example, if $|g_A| > w$, it can be concluded that the A effect is different from zero. Otherwise, there is no basis to assume that X is different from zero. According to

Table 6, $|g_A| = 6.8$, $|g_C| = 1.4$, and $|g_{AB + CD}| = 2.0$, which are all greater than w. Therefore, the main effect of A, the main effect of C, and the mixed interaction $AB + CD$ are assumed to be significant.

Table 5   Selected fractional factorial plans

Plan	Treatment combinations(a)	Estimated effects(b)
Plan 1:	(1)	T
Three factors ($n = 3$)	ac	$A - BC$
½ replication ($b = 1$)	bc	$B - AC$
4 observations	ab	$-C + AB$
Plan 2:	(1)	T
Four factors ($n = 4$)	ad	A
½ replication ($b = 1$)	bd	B
8 observations	ab	$AB + CD$
	cd	C
	ac	$AC + BD$
	bc	$BC + AD$
	abcd	D
Plan 3:	(1)	T
Five factors ($n = 5$)	ae	A
½ replication ($b = 1$)	be	B
16 observations	ab	AB
	ce	C
	ac	AC
	bc	BC
	abce	$-DE$
	de	D
	ad	AD
	bd	BD
	abde	$-CE$
	cd	CD
	acde	$-BE$
	bcde	$-AE$
	abcd	$-E$
Plan 4:	(1)	T
Five factors ($n = 5$)	ad	$A - DE$
¼ replication ($b = 2$)	bde	$B - CE$
8 observations	abe	$AB + CD$
	cde	$C - BE$
	ace	$AC + BD$
	bc	$-E + BC + AD$
	abcd	$D - AE$
Plan 5:	(1)	T
Six factors ($n = 6$)	ae	A
¼ replication ($b = 2$)	bef	B
16 observations	abf	$AB + CE$
	cef	C
	acf	$AC + BE$
	bc	$BC + AE + DF$
	abce	E
	df	D
	adef	$AD + EF$
	bde	$BD + CF$
	abd	(c)
	cde	$CD + BF$
	acd	(c)
	bcdf	F
	abcdef	$AF + DE$
Plan 6:	(1)	T
Six factors ($n = 6$)	adf	$A - DE - CF$
⅛ replication ($b = 3$)	bde	$B - CE - DF$
8 observations	abef	$AB + CD + EF$
	cdef	$C - AF - BE$
	ace	$-F + AC + BD$
	bcf	$-E + AD + BC$
	abcd	$D - AE - BF$

(a) The order given is the order in which the data are to be listed in the first column of the Yates method of analysis.
(b) The order given is the order in which estimated effects come out in the last column of the Yates method of analysis.
(c) To be used in the estimate of the variation due to experimental error.

Randomized Block Experiments

The analysis of a randomized block experiment is based on several assumptions. It is normally assumed that each of the observations is the sum of three components. If y_{ij} is defined as the observation on the ith treatment in the jth block, then:

$$y_{ij} = \phi_i + \beta_j + e_{ij}$$

where β_j is the amount by which the response of a given treatment in the jth block differs from the response of the same treatment averaged over all blocks, assuming no experimental error. ϕ_i is a term pertaining only to the ith treatment and is constant for all blocks, regardless of the block in which the treatment occurs. It may be regarded as the average value of the ith treatment averaged over all blocks in the experiment, assuming no experimental error (e_{ij}) associated with the measurement y_{ij}.

To make interval estimates for, or to make tests on, the ϕ_i or the β_j values, it is generally assumed that the experimental errors, e_{ij}, are independent and normally distributed. However, if the experiment was randomized properly, failure of this assumption generally will not cause serious difficulty.

For analysis purposes, the results of a randomized block experiment can be exhibited in a two-way table (Table 7), which assumes that there are b blocks and t treatments. The following analysis evaluates the possible differences in tensile strength in ksi of a 4140 steel plate as a function of location through the thickness and position along its length. The data and block and treatment statistics are given in Table 8.

Estimation of Treatment Effects. A treatment effect ϕ_i is estimated by the mean of the observations on the ith treatment. That is, the estimate of ϕ_i is $t_i = T_i/b$. For example, in Table 8, the estimate of the effect of heat treatment on tensile properties at the beginning of the plate is $t_1 = T_1/4 = 585.2/4 = 146.30$.

Testing and Estimation of Differences in Treatment Effects. To test the significance of differences in treatment effects, the following procedure can be used. Let α be the significance level of the test. For this example, let $\alpha = 0.05$. Obtain $q_{1 - \alpha}(t, \nu)$ in Table 9, where $\nu = (b - 1)(t - 1)$. From the data given in Table 8, $q_{0.95} (6, 15) = 4.59$. Then compute:

$$S_t = \frac{T_1 + T_2 + \ldots + T_t}{b} - \frac{G^2}{tb}$$

$$S_t = 517\,181.998 - 517\,176.400$$

$$= 5.598$$

Table 6 Yates method of analysis using data given in Table 4

1 Treatment combination	2 Response (yield)	3	4	5 g	Estimated effect
(1)........................	4.2	7.2	15.1	28.8	T
ad	3.0	7.9	13.7	−6.8	A
bd	5.0	6.8	−3.3	0.8	B
ab	2.9	6.9	−3.5	−2.0	$AB + CD$
cd	4.0	−1.2	0.7	−1.4	C
ac	2.8	−2.1	0.1	−0.2	$AC + BD$
bc	4.6	−1.2	−0.9	−0.6	$BC + AD$
abcd	2.3	−2.3	−1.1	−0.2	D
Total	28.8				
Sum of squares	110.34			882.72	

Checks: The sum of column 2 should equal g_T, the first entry in column 5. The sum of squares of entries in column 2 should equal the sum of squares of the g's divided by $2^{n'} = 2^3 = 8$ (110.34 = 882.72/8 = 110.34). g_A is equal to the sum of all yields in which A is at its higher level minus sum of all yields in which A is at its lower level. $g_A = 11.0 − 17.8 = −6.8$. Similarly, $g_B = 14.8 − 14.0 = 0.8$. $g_C = 13.7 − 15.1 = −1.4$.

Compute:

$$S_b = \frac{B_1 + B_2 + \ldots + B_b}{t} - \frac{G^2}{tb}$$

$$S_b = 518\,104.065 - 517\,176.400$$

$$= 927.665$$

Compute:

$$S = \sum_{i=1}^{t} \sum_{j=1}^{b} y_{ij}^2 - \frac{G^2}{tb}$$

That is, compute the sum of the squares of all the observations and subtract G^2/tb.

$$S = 518\,123.13 - 517\,176.40$$

$$= 946.73$$

Compute:

$$s^2 = \frac{S - S_b - S_t}{(b - 1)(t - 1)}$$

$$s^2 = \frac{13.467}{15}$$

$$= 0.8978$$

$$s = 0.9475$$

Compute:

$$w = \frac{q_{1-\alpha}s}{\sqrt{b}}$$

$$w = \frac{(4.59)(0.9475)}{\sqrt{4}}$$

$$= 2.175$$

If the absolute difference between any two estimated treatment effects exceeds w, the treatment effects differ; otherwise, there is no basis to assume that the treatment effects differ. Because there is no pair of treatment means whose difference exceeds 2.175, there is no basis to conclude that nonuniform

properties exist along the length of the plate. For all possible pairs of treatments i and j, the statements

$$[(t_i - t_j) - w] \le (\phi_i - \phi_j)$$

$$\le [(t_i - t_j) + w]$$

can be made with $(1 - \alpha)100\%$ confidence that all the statements are simultaneously true.

Estimation of Block Effects. The block effect, β_j, is estimated by the mean of the observations in the jth block minus the grand mean. That is, the estimate of β_j, the jth block effect, is $b_j = (B_j/t) - (G/bt)$.

For example, using the data sample given in Table 8, the grand average equals $G/bt = 3523.1/24 = 146.80$.

$$b_1 = 139.52 - 146.80$$
$$= -7.28$$
$$b_2 = 152.25 - 146.80$$
$$= 5.45$$
$$b_3 = 153.63 - 146.80$$
$$= 6.83$$
$$b_4 = 141.78 - 146.80$$
$$= -5.02$$

Testing and Estimating Differences in Block Effects. To test the significance of differences in block effects, a procedure very similar to that described for treatments can be used. It is not described here, but is shown in detail in Ref 1.

Incomplete Block Experiments

The analysis of balanced, incomplete block plans is similar to the analysis of randomized block plans, except that each block does not contain all of the treatments. The following example illustrates such an analysis.

Table 7 Schematic presentation of results for randomized block plans

Block	Treatment 1	2	...	t	Total	Block mean = B/t
1					B_1	
2					B_2	
.					.	
.					.	
.					.	
b					B_b	
Total	T_1	T_2	...	T_t	G	
Treatment mean = T/b						

Table 8 Typical randomized block test data

Tensile strength values (ksi) as a function of specimen location

Depth and lateral position (blocks)	Position along length, ft (treatments)						Total	Mean
	0.0	2.0	4.0	6.0	8.0	10.0		
Mid-thickness, Side A	138.0	141.6	137.5	141.8	138.6	139.6	$B_1=837.1$	$b_1=139.52$
Surface, Side A	152.2	152.2	152.1	152.2	152.0	152.8	$B_2=913.5$	$b_2=152.25$
Surface, Side B	153.6	154.0	153.8	153.6	153.2	153.6	$B_3=921.8$	$b_3=153.63$
Mid-thickness, Side B	141.4	141.5	142.6	142.2	141.1	141.9	$B_4=850.7$	$b_4=141.78$
Total	$T_1=$ 585.2	$T_2=$ 589.3	$T_3=$ 586.0	$T_4=$ 589.8	$T_5=$ 584.9	$T_6=$ 587.9	$G=$ 3523.1	
Mean	$t_1=$ 146.30	$t_2=$ 147.32	$t_3=$ 146.50	$t_4=$ 147.45	$t_5=$ 146.22	$t_6=$ 146.98		

A heat-resistant alloy is tested, with emphasis on the effect of several alloying constituents on low-cycle fatigue properties. Previous experience indicates the fatigue properties should be log-normally distributed. Four different levels of two different alloying constituents are identified in 12 samples. The combination of the different levels of the two constituents produced a partially complete matrix of alloying levels, as shown in Table 10. In the plan, different levels of alloying constituent A were considered treatment levels, and different levels of alloying constituent B were considered block levels. All tests were run in strain control at 540 °C (1000 °F) and a strain amplitude of 0.02. All values given in Table 10 are logarithmic values of fatigue life. Note that for this example, it is also possible to identify the main effects and interactions through a two-way analysis of variance (ANOVA) procedure, such as described in Ref 2, 5, and 10. In fact, an ANOVA approach would be preferred over the incomplete block procedure in situations where the treatment and block effects are both likely to be significant and interdependent. The "longhand" procedures described below are effective, however, and can be used when the necessary computer facilities and/or software are not available.

Estimation of Treatment Effects. Because the block plan is incomplete, the treatment effects cannot be estimated directly from the treatment averages. Any apparent effects of treatment must be adjusted for possible block effects. The estimate of ϕ_i, the effect of the ith treatment, is:

$$t_i = \frac{Q_i}{Er} + \frac{G}{rt}$$

where $Q_i = T_i - [$(sum of totals of all blocks containing treatment i)$/k]$. The notation used here is the same as that defined in the article "Planning of Comparative Experiments" in this Section. For example, using the data given in Table 10:

$$Q_1 = T_1 - \frac{B_1 + B_2 + B_3}{3}$$

$$Q_1 = 4.41 - \frac{11.00}{3}$$

$$= 4.41 - 3.6667$$

$$= 0.7433$$

Similarly:

$$Q_2 = 3.55 - \frac{12.06}{3}$$

$$= 3.55 - 4.0200$$

$$= -0.4700$$

Table 9 Percentiles of the studentized range, q

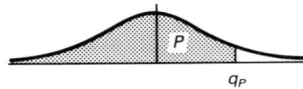

$q = w/s$, where w is the range of t observations, and ν is the number of degrees of freedom associated with the standard deviation s.

$q_{0.90}$

ν \ t	2	3	4	5	6	7	8	9	10
1	8.93	13.44	16.36	18.49	20.15	21.51	22.64	23.62	24.48
2	4.13	5.73	6.77	7.54	8.14	8.63	9.05	9.41	9.72
3	3.33	4.47	5.20	5.74	6.16	6.51	6.81	7.06	7.29
4	3.01	3.98	4.59	5.03	5.39	5.68	5.93	6.14	6.33
5	2.85	3.72	4.26	4.66	4.98	5.24	5.46	5.65	5.82
6	2.75	3.56	4.07	4.44	4.73	4.97	5.17	5.34	5.50
7	2.68	3.45	3.93	4.28	4.55	4.78	4.97	5.14	5.28
8	2.63	3.37	3.83	4.17	4.43	4.65	4.83	4.99	5.13
9	2.59	3.32	3.76	4.08	4.34	4.54	4.72	4.87	5.01
10	2.56	3.27	3.70	4.02	4.26	4.47	4.64	4.78	4.91
11	2.54	3.23	3.66	3.96	4.20	4.40	4.57	4.71	4.84
12	2.52	3.20	3.62	3.92	4.16	4.35	4.51	4.65	4.78
13	2.50	3.18	3.59	3.88	4.12	4.30	4.46	4.60	4.72
14	2.49	3.16	3.56	3.85	4.08	4.27	4.42	4.56	4.68
15	2.48	3.14	3.54	3.83	4.05	4.23	4.39	4.52	4.64
16	2.47	3.12	3.52	3.80	4.03	4.21	4.36	4.49	4.61
17	2.46	3.11	3.50	3.78	4.00	4.18	4.33	4.46	4.58
18	2.45	3.10	3.49	3.77	3.98	4.16	4.31	4.44	4.55
19	2.45	3.09	3.47	3.75	3.97	4.14	4.29	4.42	4.53
20	2.44	3.08	3.46	3.74	3.95	4.12	4.27	4.40	4.51
24	2.42	3.05	3.42	3.69	3.90	4.07	4.21	4.34	4.44
30	2.40	3.02	3.39	3.65	3.85	4.02	4.16	4.28	4.38
40	2.38	2.99	3.35	3.60	3.80	3.96	4.10	4.21	4.32
60	2.36	2.96	3.31	3.56	3.75	3.91	4.04	4.16	4.25
120	2.34	2.93	3.28	3.52	3.71	3.86	3.99	4.10	4.19
∞	2.33	2.90	3.24	3.48	3.66	3.81	3.93	4.04	4.13

$q_{0.90}$

ν \ t	11	12	13	14	15	16	17	18	19	20
1	25.24	25.92	26.54	27.10	27.62	28.10	28.54	28.96	29.35	29.71
2	10.01	10.26	10.49	10.70	10.89	11.07	11.24	11.39	11.54	11.68
3	7.49	7.67	7.83	7.98	8.12	8.25	8.37	8.48	8.58	8.68
4	6.49	6.65	6.78	6.91	7.02	7.13	7.23	7.33	7.41	7.50
5	5.97	6.10	6.22	6.34	6.44	6.54	6.63	6.71	6.79	6.86
6	5.64	5.76	5.87	5.98	6.07	6.16	6.25	6.32	6.40	6.47
7	5.41	5.53	5.64	5.74	5.83	5.91	5.99	6.06	6.13	6.19
8	5.25	5.36	5.46	5.56	5.64	5.72	5.80	5.87	5.93	6.00
9	5.13	5.23	5.33	5.42	5.51	5.58	5.66	5.72	5.79	5.85
10	5.03	5.13	5.23	5.32	5.40	5.47	5.54	5.61	5.67	5.73
11	4.95	5.05	5.15	5.23	5.31	5.38	5.45	5.51	5.57	5.63
12	4.89	4.99	5.08	5.16	5.24	5.31	5.37	5.44	5.49	5.55
13	4.83	4.93	5.02	5.10	5.18	5.25	5.31	5.37	5.43	5.48
14	4.79	4.88	4.97	5.05	5.12	5.19	5.26	5.32	5.37	5.43
15	4.75	4.84	4.93	5.01	5.08	5.15	5.21	5.27	5.32	5.38
16	4.71	4.81	4.89	4.97	5.04	5.11	5.17	5.23	5.28	5.33
17	4.68	4.77	4.86	4.93	5.01	5.07	5.13	5.19	5.24	5.30
18	4.65	4.75	4.83	4.90	4.98	5.04	5.10	5.16	5.21	5.26
19	4.63	4.72	4.80	4.88	4.95	5.01	5.07	5.13	5.18	5.23
20	4.61	4.70	4.78	4.85	4.92	4.99	5.05	5.10	5.16	5.20
24	4.54	4.63	4.71	4.78	4.85	4.91	4.97	5.02	5.07	5.12
30	4.47	4.56	4.64	4.71	4.77	4.83	4.89	4.94	4.99	5.03
40	4.41	4.49	4.56	4.63	4.69	4.75	4.81	4.86	4.90	4.95
60	4.34	4.42	4.49	4.56	4.62	4.67	4.73	4.78	4.82	4.86
120	4.28	4.35	4.42	4.48	4.54	4.60	4.65	4.69	4.74	4.78
∞	4.21	4.28	4.35	4.41	4.47	4.52	4.57	4.61	4.65	4.69

(continued)

Table 9   (continued)

$q_{0.95}$

v \ t	2	3	4	5	6	7	8	9	10
1	17.97	26.98	32.82	37.08	40.41	43.12	45.40	47.36	49.07
2	6.08	8.33	9.80	10.88	11.74	12.44	13.03	13.54	13.99
3	4.50	5.91	6.82	7.50	8.04	8.48	8.85	9.18	9.46
4	3.93	5.04	5.76	6.29	6.71	7.05	7.35	7.60	7.83
5	3.64	4.60	5.22	5.67	6.03	6.33	6.58	6.80	6.99
6	3.46	4.34	4.90	5.30	5.63	5.90	6.12	6.32	6.49
7	3.34	4.16	4.68	5.06	5.36	5.61	5.82	6.00	6.16
8	3.26	4.04	4.53	4.89	5.17	5.40	5.60	5.77	5.92
9	3.20	3.95	4.41	4.76	5.02	5.24	5.43	5.59	5.74
10	3.15	3.88	4.33	4.65	4.91	5.12	5.30	5.46	5.60
11	3.11	3.82	4.26	4.57	4.82	5.03	5.20	5.35	5.49
12	3.08	3.77	4.20	4.51	4.75	4.95	5.12	5.27	5.39
13	3.06	3.73	4.15	4.45	4.69	4.88	5.05	5.19	5.32
14	3.03	3.70	4.11	4.41	4.64	4.83	4.99	5.13	5.25
15	3.01	3.67	4.08	4.37	4.59	4.78	4.94	5.08	5.20
16	3.00	3.65	4.05	4.33	4.56	4.74	4.90	5.03	5.15
17	2.98	3.63	4.02	4.30	4.52	4.70	4.86	4.99	5.11
18	2.97	3.61	4.00	4.28	4.49	4.67	4.82	4.96	5.07
19	2.96	3.59	3.98	4.25	4.47	4.65	4.79	4.92	5.04
20	2.95	3.58	3.96	4.23	4.45	4.62	4.77	4.90	5.01
24	2.92	3.53	3.90	4.17	4.37	4.54	4.68	4.81	4.92
30	2.89	3.49	3.85	4.10	4.30	4.46	4.60	4.72	4.82
40	2.86	3.44	3.79	4.04	4.23	4.39	4.52	4.63	4.73
60	2.83	3.40	3.74	3.98	4.16	4.31	4.44	4.55	4.65
120	2.80	3.36	3.68	3.92	4.10	4.24	4.36	4.47	4.56
∞	2.77	3.31	3.63	3.86	4.03	4.17	4.29	4.39	4.47

$q_{0.95}$

v \ t	11	12	13	14	15	16	17	18	19	20
1	50.59	51.96	53.20	54.33	55.36	56.32	57.22	58.04	58.83	59.56
2	14.39	14.75	15.08	15.38	15.65	15.91	16.14	16.37	16.57	16.77
3	9.72	9.95	10.15	10.35	10.52	10.69	10.84	10.98	11.11	11.24
4	8.03	8.21	8.37	8.52	8.66	8.79	8.91	9.03	9.13	9.23
5	7.17	7.32	7.47	7.60	7.72	7.83	7.93	8.03	8.12	8.21
6	6.65	6.79	6.92	7.03	7.14	7.24	7.34	7.43	7.51	7.59
7	6.30	6.43	6.55	6.66	6.76	6.85	6.94	7.02	7.10	7.17
8	6.05	6.18	6.29	6.39	6.48	6.57	6.65	6.73	6.80	6.87
9	5.87	5.98	6.09	6.19	6.28	6.36	6.44	6.51	6.58	6.64
10	5.72	5.83	5.93	6.03	6.11	6.19	6.27	6.34	6.40	6.47
11	5.61	5.71	5.81	5.90	5.98	6.06	6.13	6.20	6.27	6.33
12	5.51	5.61	5.71	5.80	5.88	5.95	6.02	6.09	6.15	6.21
13	5.43	5.53	5.63	5.71	5.79	5.86	5.93	5.99	6.05	6.11
14	5.36	5.46	5.55	5.64	5.71	5.79	5.85	5.91	5.97	6.03
15	5.31	5.40	5.49	5.57	5.65	5.72	5.78	5.85	5.90	5.96
16	5.26	5.35	5.44	5.52	5.59	5.66	5.73	5.79	5.84	5.90
17	5.21	5.31	5.39	5.47	5.54	5.61	5.67	5.73	5.79	5.84
18	5.17	5.27	5.35	5.43	5.50	5.57	5.63	5.69	5.74	5.79
19	5.14	5.23	5.31	5.39	5.46	5.53	5.59	5.65	5.70	5.75
20	5.11	5.20	5.28	5.36	5.43	5.49	5.55	5.61	5.66	5.71
24	5.01	5.10	5.18	5.25	5.32	5.38	5.44	5.49	5.55	5.59
30	4.92	5.00	5.08	5.15	5.21	5.27	5.33	5.38	5.43	5.47
40	4.82	4.90	4.98	5.04	5.11	5.16	5.22	5.27	5.31	5.36
60	4.73	4.81	4.88	4.94	5.00	5.06	5.11	5.15	5.20	5.24
120	4.64	4.71	4.78	4.84	4.90	4.95	5.00	5.04	5.09	5.13
∞	4.55	4.62	4.68	4.74	4.80	4.85	4.89	4.93	4.97	5.01

(continued)

$$Q_3 = 4.01 - \frac{11.05}{3}$$

$$= 4.01 - 3.6833$$

$$= 0.3267$$

$$Q_4 = 2.97 - \frac{10.71}{3}$$

$$= 2.97 - 3.5700$$

$$= -0.6000$$

$$E = \frac{8}{9}, \; r = 3, \; Er = 2.6667,$$

$$t = 4, \; rt = 12$$

$$\frac{G}{rt} = \frac{14.94}{12}$$

$$= 1.2450$$

$$t_1 = \frac{Q_1}{Er} + \frac{G}{rt}$$

$$= \frac{0.7433}{2.6667} + 1.2450$$

$$= 1.5237$$

$$t_2 = \frac{-0.4700}{2.6667} + 1.2450$$

$$= 1.0688$$

$$t_3 = \frac{0.3267}{2.6667} + 1.2450$$

$$= 1.3675$$

$$t_4 = \frac{-0.6000}{2.6667} + 1.2450$$

$$= 1.0200$$

Testing and Estimation of Differences in Treatment Effects. To test the significance of differences in treatment effects, the following procedure can be used. Let α be the significance level of the test. As before, $\alpha = 0.05$. Obtain $q_{1-\alpha}(t, v)$ in Table 9, where $v = tr - t - b + 1$. From the data given in Table 10, compute:

$$t = 4$$

$$v = 5$$

$$q_{0.95}(4, 5) = 5.22$$

Compute Q_i and t_i for each treatment. The sum of the Q_i should equal zero. Thus compute:

$$S_t = \frac{Q_1^2 + Q_2^2 + \ldots + Q_t^2}{Er}$$

Table 9 (continued)

$q_{0.99}$

v	2	3	4	5	6	7	8	9	10
1	90.03	135.0	164.3	185.6	202.2	215.8	227.2	237.0	245.6
2	14.04	19.02	22.29	24.72	26.63	28.20	29.53	30.68	31.69
3	8.26	10.62	12.17	13.33	14.24	15.00	15.64	16.20	16.69
4	6.51	8.12	9.17	9.96	10.58	11.10	11.55	11.93	12.27
5	5.70	6.98	7.80	8.42	8.91	9.32	9.67	9.97	10.24
6	5.24	6.33	7.03	7.56	7.97	8.32	8.61	8.87	9.10
7	4.95	5.92	6.54	7.01	7.37	7.68	7.94	8.17	8.37
8	4.75	5.64	6.20	6.62	6.96	7.24	7.47	7.68	7.86
9	4.60	5.43	5.96	6.35	6.66	6.91	7.13	7.33	7.49
10	4.48	5.27	5.77	6.14	6.43	6.67	6.87	7.05	7.21
11	4.39	5.15	5.62	5.97	6.25	6.48	6.67	6.84	6.99
12	4.32	5.05	5.50	5.84	6.10	6.32	6.51	6.67	6.81
13	4.26	4.96	5.40	5.73	5.98	6.19	6.37	6.53	6.67
14	4.21	4.89	5.32	5.63	5.88	6.08	6.26	6.41	6.54
15	4.17	4.84	5.25	5.56	5.80	5.99	6.16	6.31	6.44
16	4.13	4.79	5.19	5.49	5.72	5.92	6.08	6.22	6.35
17	4.10	4.74	5.14	5.43	5.66	5.85	6.01	6.15	6.27
18	4.07	4.70	5.09	5.38	5.60	5.79	5.94	6.08	6.20
19	4.05	4.67	5.05	5.33	5.55	5.73	5.89	6.02	6.14
20	4.02	4.64	5.02	5.29	5.51	5.69	5.84	5.97	6.09
24	3.96	4.55	4.91	5.17	5.37	5.54	5.69	5.81	5.92
30	3.89	4.45	4.80	5.05	5.24	5.40	5.54	5.65	5.76
40	3.82	4.37	4.70	4.93	5.11	5.26	5.39	5.50	5.60
60	3.76	4.28	4.59	4.82	4.99	5.13	5.25	5.36	5.45
120	3.70	4.20	4.50	4.71	4.87	5.01	5.12	5.21	5.30
∞	3.64	4.12	4.40	4.60	4.76	4.88	4.99	5.08	5.16

$q_{0.99}$

v	11	12	13	14	15	16	17	18	19	20
1	253.2	260.0	266.2	271.8	277.0	281.8	286.3	290.4	294.3	298.0
2	32.59	33.40	34.13	34.81	35.43	36.00	36.53	37.03	37.50	37.95
3	17.13	17.53	17.89	18.22	18.52	18.81	19.07	19.32	19.55	19.77
4	12.57	12.84	13.09	13.32	13.53	13.73	13.91	14.08	14.24	14.40
5	10.48	10.70	10.89	11.08	11.24	11.40	11.55	11.68	11.81	11.93
6	9.30	9.48	9.65	9.81	9.95	10.08	10.21	10.32	10.43	10.54
7	8.55	8.71	8.86	9.00	9.12	9.24	9.35	9.46	9.55	9.65
8	8.03	8.18	8.31	8.44	8.55	8.66	8.76	8.85	8.94	9.03
9	7.65	7.78	7.91	8.03	8.13	8.23	8.33	8.41	8.49	8.57
10	7.36	7.49	7.60	7.71	7.81	7.91	7.99	8.08	8.15	8.23
11	7.13	7.25	7.36	7.46	7.56	7.65	7.73	7.81	7.88	7.95
12	6.94	7.06	7.17	7.26	7.36	7.44	7.52	7.59	7.66	7.73
13	6.79	6.90	7.01	7.10	7.19	7.27	7.35	7.42	7.48	7.55
14	6.66	6.77	6.87	6.96	7.05	7.13	7.20	7.27	7.33	7.39
15	6.55	6.66	6.76	6.84	6.93	7.00	7.07	7.14	7.20	7.26
16	6.46	6.56	6.66	6.74	6.82	6.90	6.97	7.03	7.09	7.15
17	6.38	6.48	6.57	6.66	6.73	6.81	6.87	6.94	7.00	7.05
18	6.31	6.41	6.50	6.58	6.65	6.73	6.79	6.85	6.91	6.97
19	6.25	6.34	6.43	6.51	6.58	6.65	6.72	6.78	6.84	6.89
20	6.19	6.28	6.37	6.45	6.52	6.59	6.65	6.71	6.77	6.82
24	6.02	6.11	6.19	6.26	6.33	6.39	6.45	6.51	6.56	6.61
30	5.85	5.93	6.01	6.08	6.14	6.20	6.26	6.31	6.36	6.41
40	5.69	5.76	5.83	5.90	5.96	6.02	6.07	6.12	6.16	6.21
60	5.53	5.60	5.67	5.73	5.78	5.84	5.89	5.93	5.97	6.01
120	5.37	5.44	5.50	5.56	5.61	5.66	5.71	5.75	5.79	5.83
∞	5.23	5.29	5.35	5.40	5.45	5.49	5.54	5.57	5.61	5.65

Source: Ref 1

$$S_t = \frac{1.24012778}{2.6667}$$

$$= 0.46504$$

Compute:

$$S_b = \frac{B_1^2 + B_2^2 + \ldots + B_b^2}{K} - \frac{G^2}{rt}$$

$$S_b = \frac{56.8430}{3} - 18.60030$$

$$= 0.34737$$

Compute:

$$S = \Sigma y_{ij}^2 - \frac{G^2}{rt}$$

That is, compute the sum of the squares of all the observations and subtract G^2/rt:

$$S = 19.4812 - 18.6003$$

$$= 0.88090$$

Compute:

$$s^2 = \frac{S - S_t - S_b}{tr - t - b + 1}$$

$$s^2 = \frac{0.06849}{5}$$

$$= 0.0137$$

$$s = 0.117$$

Compute:

$$w = \frac{q_{1-\alpha}s}{\sqrt{Er}}$$

$$w = \frac{(5.22)(0.117)}{1.63}$$

$$= 0.375$$

If the absolute difference between two estimated treatment effects exceeds w, then the treatment effects differ. Otherwise, there is no basis to conclude that treatment effects differ. Because differences exist between pairs of treatment effects that do exceed 0.375 in this example, significant differences are concluded to exist in fatigue resistance for different levels of alloying constituent A. Thus, simultaneous confidence interval statements can be made about the differences between pairs of treatments i and j, with $(1 - \alpha)100\%$ confidence that all statements are simultaneously true. The statements are, for all i and j:

$$[(t_i - t_j) - w] \leq (\phi_i - \phi_j)$$
$$\leq [(t_i - t_j) + w]$$

Table 10   Results from an incomplete block plan

Low-cycle fatigue tests for different levels of two alloying constituents

Alloy B (blocks)	Level 1	Alloy A (treatments) Level 2	Level 3	Level 4	Total
Level 1	1.11		0.95	0.82	$B_1 = 2.88$
Level 2	1.70	1.22		0.97	$B_2 = 3.89$
Level 3	1.60	1.11	1.52		$B_3 = 4.23$
Level 4		1.22	1.54	1.18	$B_4 = 3.94$
Total	$T_1 = 4.41$	$T_2 = 3.55$	$T_3 = 4.01$	$T_4 = 2.97$	$G = 14.94$

Note: $t = 4$, $k = 3$, $b = 4$, $r = 3$, $\lambda = 2$, $E = {}^8\!/_9$, $N = 12$.

Estimation of Block Effects. Like treatment effects, block effects cannot be estimated directly from block averages, but must be adjusted according to which treatments occur. In the following discussion, only the estimation of block effects for symmetrical plans are discussed, i.e., where $b = t$, or the number of blocks equals the number of treatments. To estimate or test block effects in a balanced incomplete block plan that is not symmetric, see Ref 20.

For symmetric plans, the estimate of β_j, the jth block effect, is:

$$b_j = \frac{Q_j'}{Er}$$

where $Q_j' = B_j -$ [(sum of totals of all treatments occurring in the jth block)/r]. For example, using the data given in Table 10:

$$Q_1' = B_1 - \frac{T_1 + T_3 + T_4}{3}$$

$$Q_1' = 2.88 - \frac{11.39}{3}$$

$$= 2.88 - 3.7967$$

$$= -0.9167$$

Similarly,

$$Q_2' = 3.89 - \frac{10.93}{3}$$

$$= 3.89 - 3.6433$$

$$= 0.2467$$

$$Q_3' = 4.23 - \frac{11.97}{3}$$

$$= 4.23 - 3.9900$$

$$= 0.2400$$

$$Q_4' = 3.94 - \frac{10.53}{3}$$

$$= 3.94 - 3.5100$$

$$= 0.4300$$

$$Er = 2.6667$$

$$b_1 = \frac{-0.9167}{2.6667}$$

$$= -0.34376$$

$$b_2 = \frac{0.2467}{2.6667}$$

$$= 0.09251$$

$$b_3 = \frac{0.2400}{2.6667}$$

$$= 0.09000$$

$$b_4 = \frac{0.4300}{2.6667}$$

$$= 0.16125$$

Testing and Estimation of Differences in Block Effects. To test the significance of differences in block effects, similar procedures to those used to identify treatment effects can be used. They are not described here, but are shown in detail in Ref 1.

REFERENCES

1. M. Natrella, *Experimental Statistics*, U.S. Government Printing Office, Washington, DC, 1966 (also reprinted by John Wiley & Sons, NY, 1983)
2. J. Neter and W. Wasserman, *Applied Linear Statistical Models*, Richard D. Irwin, Inc., Homewood, IL, 1974
3. M. Hollander and D.A. Wolfe, *Nonparametric Statistical Methods*, John Wiley & Sons, New York, 1973
4. C. Lipson and N.J. Sheth, *Statistical Design and Analysis of Engineering Experiments*, 1st ed., McGraw-Hill, New York, 1973
5. J. Neter, W. Wasserman, and G.A. Whitmore, *Applied Statistics*, 2nd ed., Allyn and Bacon, Boston, 1982
6. G.W. Snedecor and W.G. Cochran, *Statistical Methods*, 7th ed., Iowa State University Press, Ames, IA, 1980
7. A.H. Bowker and G.J. Lieberman, *Engineering Statistics*, 2nd ed., Prentice-Hall, Englewood Cliffs, NJ, 1972
8. R.G.D. Steel and J.H. Torrie, *Principles and Procedures of Statistics, A Biometrical Approach*, 2nd ed., McGraw-Hill, New York, 1980
9. R.E. Walpole and R.H. Meyers, *Probability and Statistics for Engineers and Scientists*, 2nd ed., Macmillan, New York, 1978
10. R.A. Johnson and D.W. Wichern, *Applied Multivariate Analysis*, Prentice-Hall, Englewood Cliffs, NJ, 1982
11. D. Morrison, *Multivariate Statistical Methods*, 2nd ed., McGraw-Hill, New York, 1976
12. W.J. Conover, *Practical Nonparametric Statistics*, 2nd ed., John Wiley & Sons, New York, 1980
13. J.D. Gibbons, *Nonparametric Methods for Quantitative Analysis*, Holt, Rinehart, and Winston, New York, 1976
14. J. Neter, W. Wasserman, and M.H. Kutner, *Applied Linear Regression Models*, Richard D. Irwin, Inc., Homewood, IL, 1983
15. N.R. Draper and H. Smith, *Applied Regression Analysis*, 2nd ed., John Wiley & Sons, New York, 1981
16. S. Chatterjee and B. Price, *Regression Analysis By Example*, John Wiley & Sons, New York, 1977
17. I. Francis, *Statistical Software, A Comparative Review*, Elsevier North-Holland, New York, 1981
18. W.J. Kennedy and J.E. Gentle, *Statistical Computing*, Marcel Dekker, New York, 1980
19. F. Yates, "The Design and Analysis of Factorial Experiments," Technical Communication No. 35, Imperial Bureau of Soil Science, Harpenden, England, 1937
20. W.G. Cochran and G.M. Cox, *Experimental Designs*, 2nd ed., John Wiley & Sons, New York, 1957
21. O.L. Davies, Ed., *The Design and Analysis of Industrial Experiments*, Hafner Publishing, New York, 1954

Design Allowables for Static Metallic Material Properties

By Paul E. Ruff
Principal Research Scientist
Battelle Columbus Laboratories

STATISTICAL ANALYSIS of mechanical property data is the most reliable method for determination of minimum design allowables. In the design of minimum-weight, highly reliable structural members (where failures cannot be tolerated), a safety factor is usually applied to the minimum design strength (stress) values for the critical mechanical property. Therefore, minimum design values for these critical properties must be available.

Typically, only tensile tests are routinely conducted for metallic materials, especially for quality assurance testing. The purchaser may specify that tests for other mechanical strength properties, e.g., elevated temperature tensile, shear ultimate, compressive yield, and fracture toughness, be conducted by the supplier in addition to the room-temperature tensile test.

Determination of Minimum Design Values

Minimum design values usually are determined first for tensile yield and ultimate strengths. Often, the minimum mechanical property value specified in the governing material specification (federal, military, or industrial) is used as the design value. However, the statistical assurance associated with the specification value is not known, because information regarding the quantity and nature of data and the statistical procedures used to establish the specification minimum values is not available.

In 1950, Alcoa adopted a statistical technique to predict guaranteed tensile properties that afforded a 0.95 confidence that at least 0.99 of the distribution would fall above the predicted minimum value. This statistical basis was adopted by the aluminum industry and has subsequently been accepted by suppliers in other metal industries. As a result, design values have been determined on the A, B, S, and typical bases (Ref 1).

For the A basis, at least 0.99 of the population of values is expected to equal or exceed the mechanical property allowable, with a confidence of 0.95. At least 0.90 of the population of values is expected to equal or exceed the B basis mechanical property allowable, with a confidence of 0.95. The B basis provides design values that are higher than A values and that sometimes are used for noncritical parts or for structures with redundant load paths.

The S value is the minimum value specified by the governing federal, military, industry, or company specifications. The statistical assurance associated with this value is not known. The S value normally is adequate for elongation and reduction of area. The typical property value is an average value and has no statistical assurance associated with it. Typical values are utilized for elastic modulus, Poisson's ratio, and physical properties.

Mechanical Property Symbols. The mechanical properties for commonly determined minimum design values are shown in Table 1. To simplify the description of analytical procedures, the symbols shown in Table 1 are used.

General Procedures

The procedures used to determine design allowables vary somewhat from sample to sample. All involve a number of steps illustrated by the flowcharts in Fig. 1 and 2. These steps include specifying the population to which the allowable applies, selecting the procedure for computing the allowable, and computing the allowable.

Specifying the Population. For computational purposes, the definition of a population must be sufficiently restrictive to ensure that the computed design allowables are realistic and useful. This is done by establishing the range of products and test conditions for which a given mechanical property can be characterized by a single statistical distribution. A homogeneous population should not include more than one alloy, heat treated condition, test temperature, and measured test parameter.

Table 1 Mechanical property terms and symbols

Property	Unit	Symbol Room-temperature minimum value	Symbol Individual or typical value
Tensile ultimate strength	MPa (ksi)	F_{tu}	TUS
Tensile yield strength	MPa (ksi)	F_{ty}	TYS
Compressive yield strength	MPa (ksi)	F_{cy}	CYS
Shear ultimate strength	MPa (ksi)	F_{su}	SUS
Bearing ultimate strength	MPa (ksi)	F_{bru}	BUS
Bearing yield strength	MPa (ksi)	F_{bry}	BYS
Elongation	%	e	Elongation
Reduction in area	%	RA	Reduction in area

Note: The listed mechanical property symbols should be followed by one of the following additional symbols: L—longitudinal direction; parallel to the principal direction of flow in a worked metal. T—Transverse direction; perpendicular to the principal direction of flow in a worked metal; may be further defined as LT or ST. LT—long-transverse direction; the transverse direction having the largest dimension, often called the "width" direction. ST—short-transverse direction; the transverse direction having the smallest dimension, often called the "thickness" direction.

Fig. 1 General procedures for determining design allowables

See Table 1 for explanation of mechanical property abbreviations and symbols.

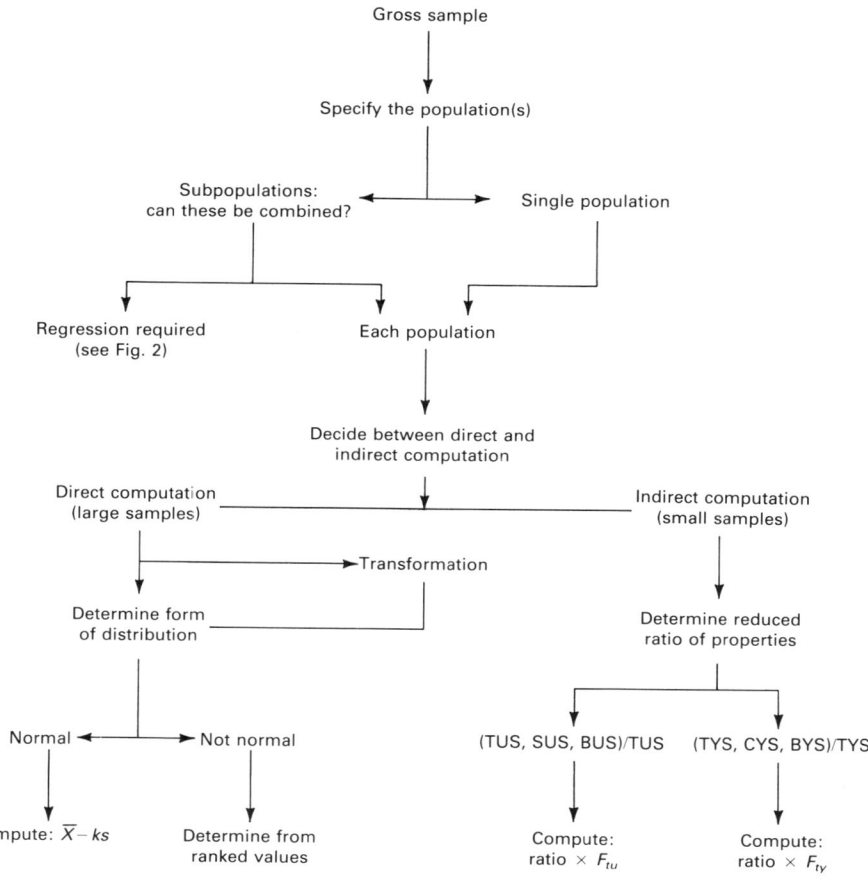

It is not always obvious whether a population may include more than one product form, size, processing history, etc. Sheet, strip, plate, bars, and forgings of one alloy may have essentially the same tensile yield strength (TYS), while the TYS may differ greatly among these product forms when made of another material. To investigate differences in populations, appropriate statistical tests of significance should be applied to the respective groups of data. These relatively simple tests are described in detail later in this article.

In most cases, these tests provide a clear division between one population and another. However, properties may sometimes vary continuously with some dimensional characteristic such as thickness. In this case, it may be necessary to analyze the data for the relationship between the property and the material dimension. Regression analysis, a useful technique for this purpose, is described later in this article.

When data for the determination of A and B values are supplied by several producers,

the observations from each producer should be analyzed separately. Statistical tests of significance should be applied to determine if all the data can be combined to constitute one homogeneous population. If the data from the various producers constitute different populations, an effort should be made to discover the reason for the difference (material processing differences, test conditions or methods, data errors) and thereby categorize the data. If no reason can be found, design values should be based on that producer's data that yield the lowest A and B values.

Direct and Indirect Computation.
The only room-temperature design allowables that are regularly determined by direct computation are F_{tu} and F_{ty}. This procedure usually is limited to a specified or usual testing direction, because there are seldom enough data available to determine the properties in other test directions.

The choice between direct and indirect computation is governed by two rules. First, F_{tu} and F_{ty} in the specified or usual testing

direction may be determined by direct computation only. Second, F_{tu} and F_{ty} in other testing directions (as well as F_{cy}, F_{su}, F_{bru}, and F_{bry} in all directions) may be determined by direct computation only if (1) data are adequate to determine the distribution form and reliable estimates of the population mean and standard deviation, or (2) the sample includes 300 or more individual, representative observations of the property to be determined.

For example, suppose that the available data for a relatively new alloy consist of 50 observations of tensile ultimate strength (TUS) in the specified testing direction. This sample is not considered large enough to determine the distribution form and reliable estimates of the population mean and standard deviation; therefore, the determination of A and B values must be postponed until a larger sample is available. The S basis is used for design values when the quantity of data is insufficient to compute A and B values.

If the number of observations increases to 100, this may be adequate to allow the determination of A and B values, provided the data are near normally distributed. If the distribution is not normal, at least 300 observations are required so that computation can proceed without knowledge of the distributional form.

If the above example involved observations of shear ultimate strength (SUS) instead of TUS, the same criteria would apply for direct computation. However, F_{su} could be determined by indirect computation with as few as ten paired observations of SUS and TUS (representing at least ten lots and two heats), provided F_{tu} has been established.

Determining Distribution Form

Experience with many alloy systems indicates that the distributional form of room-temperature strength observations usually is normal (or Gaussian). In this discussion, references to normal distribution apply to the original data or an appropriate transformation of the data that results in a normal distribution. In some instances, another parametric form may be found that more nearly approximates the population distribution; however, statistical techniques for determining A and B values for other parametric distributions either have not been determined or would prove to be exceedingly difficult to manage.

The decision regarding whether a distribution is normal can be made on a statistical basis by employing a "chi-squared" test or other "goodness-of-fit" tests. In using these tests, the risk that one may conclude errone-

Fig. 2 General procedures for determining design allowables when regression is required

See Table 1 for explanation of mechanical property abbreviations and symbols.

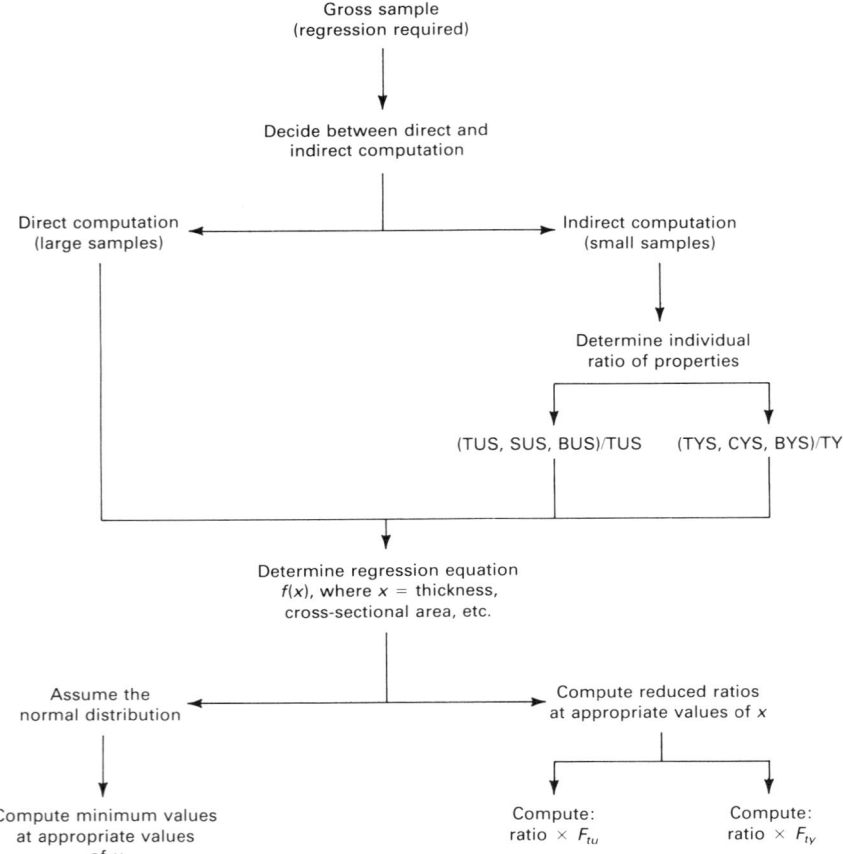

ously that a normal distribution is not normal is set at 5%.

If the distribution is found to be normal and the sample is adequate in other respects, the computational procedures for the normal distribution should be followed. If the distribution is not normal, the procedures for an unknown distribution should be employed.

Direct Computation for Normal Distribution

This procedure should be used when a mechanical property value is to be computed directly (not paired with another property for computational purposes) and the population may be interpreted to signify either the property measured, e.g., TUS, or some transformation of the measured value that is normally distributed. This procedure is applicable to F_{tu} and F_{ty}. It may also be used for F_{cy}, F_{su}, F_{bru}, and F_{bry} if sufficient data are available.

Data Requirements. Direct calculation of A and B allowables requires adequate data to determine the form of distribution and reliable estimates of the population mean and standard deviation. Prior experience with the material under consideration aids in determining sample size requirements. For a material, each population should be represented by a sample containing at least 100 observations that are normally distributed. The sample should include multiple lots, representing at least ten production heats, from a majority of the suppliers. The sample should be distributed somewhat evenly over the size range applicable to the allowable.

To avoid an undesirable biasing of the sample in favor of lots represented by more observations than other lots, the number of observations from each lot must be nearly equal. Considerations regarding the employment of averaged data are discussed later in this article.

Computational Procedure. To compute design allowables for a normally distributed population, it is necessary to have

available estimates of the mean and standard deviation of this population and tables of one-sided tolerance-limit factors for the normal distribution. A one-sided factor is used to establish a limit, above or below which it is predicted that at least some percentage of the values lie. A two-sided factor is used to establish an interval or range within which it is predicted that at least some percentage of the values lie. In both instances, a confidence is specified for the prediction. Thus, in determining an A value, one predicts, with 0.95 confidence in the prediction, that at least 0.99 of the values in the population lie above the A value, and a one-sided tolerance factor is used in the prediction.

The mean and standard deviations are computed from random samples drawn from the population. One-sided tolerance-limit factors for the normal distribution are tabulated as functions of sample size, population proportion covered by the tolerance interval, and confidence coefficients. Tolerance-limit factors are presented in Table 2.

With a suitable sample of size n, the computation of design allowables is carried out by use of the following formulas:

$$A = \bar{X} - k_A s \qquad \text{(Eq 1)}$$

$$B = \bar{X} - k_B s \qquad \text{(Eq 2)}$$

where $\bar{X}$ is the sample mean based on n observations; s is the standard deviation; k_A is the one-sided tolerance-limit factor corresponding to a proportion at least 0.99 of a normal distribution and a confidence coefficient of 0.95; and k_B is the one-sided tolerance-limit factor corresponding to a proportion at least 0.90 of a normal distribution and a confidence coefficient of 0.95.

If the variable that is normally distributed represents a transformation of the mechanical property for which design allowables are required, the A and B values computed by the above formulas must be transformed back to the original units in which the mechanical property is conventionally reported.

Use of Lot Averages. Occasionally, individual measurements are not available for computation, but averages of several tests per lot are available. This presents a problem, because the lot averages have a smaller variance than do the individual data. However, it can be demonstrated that the variance of the lot averages is inversely proportional to the number of observations comprising each lot average. To convert the standard deviation for a sample of lot averages to the equivalent standard deviation for individual

Table 2 One-sided tolerance-limit factors, k, for normal distribution, 0.95 confidence, and n − 1 degrees of freedom

Use values for P = 0.99 to determine A allowables and P = 0.90 to determine B allowables

n	0.90000	0.99000
2	20.581	37.094
3	6.155	10.553
4	4.162	7.042
5	3.407	5.741
6	3.006	5.062
7	2.755	4.642
8	2.582	4.354
9	2.454	4.143
10	2.355	3.981
11	2.275	3.852
12	2.210	3.747
13	2.155	3.659
14	2.109	3.585
15	2.068	3.520
16	2.033	3.464
17	2.002	3.414
18	1.974	3.370
19	1.949	3.331
20	1.926	3.295
21	1.905	3.263
22	1.886	3.233
23	1.869	3.206
24	1.853	3.181
25	1.838	3.158
26	1.824	3.136
27	1.811	3.116
28	1.799	3.098
29	1.788	3.080
30	1.777	3.064
31	1.767	3.048
32	1.758	3.034
33	1.749	3.020
34	1.740	3.007
35	1.732	2.995
36	1.725	2.983
37	1.717	2.972
38	1.710	2.961
39	1.704	2.951
40	1.697	2.941
41	1.691	2.932
42	1.685	2.923
43	1.680	2.914
44	1.674	2.906
45	1.669	2.898
46	1.664	2.890
47	1.659	2.883
48	1.654	2.876
49	1.650	2.869
50	1.646	2.862
51	1.641	2.856
52	1.637	2.850
53	1.633	2.844
54	1.630	2.838
55	1.626	2.833

n	0.90000	0.99000
56	1.622	2.827
57	1.619	2.822
58	1.615	2.817
59	1.612	2.812
60	1.609	2.807
61	1.606	2.802
62	1.603	2.798
63	1.600	2.793
64	1.597	2.789
65	1.594	2.785
66	1.591	2.781
67	1.589	2.777
68	1.586	2.773
69	1.584	2.769
70	1.581	2.765
71	1.579	2.762
72	1.576	2.758
73	1.574	2.755
74	1.572	2.751
75	1.570	2.748
76	1.568	2.745
77	1.565	2.742
78	1.563	2.739
79	1.561	2.736
80	1.559	2.733
81	1.557	2.730
82	1.556	2.727
83	1.554	2.724
84	1.552	2.721
85	1.550	2.719
86	1.548	2.716
87	1.547	2.714
88	1.545	2.711
89	1.543	2.709
90	1.542	2.706
91	1.540	2.704
92	1.538	2.701
93	1.537	2.699
94	1.535	2.697
95	1.534	2.695
96	1.532	2.692
97	1.531	2.690
98	1.530	2.688
99	1.528	2.686
100	1.527	2.684
101	1.525	2.682
102	1.524	2.680
103	1.523	2.678
104	1.521	2.676
105	1.520	2.674
106	1.519	2.672
107	1.518	2.671
108	1.517	2.669
109	1.515	2.667
110	1.514	2.665

n	0.90000	0.99000
111	1.513	2.663
112	1.512	2.662
113	1.511	2.660
114	1.510	2.658
115	1.508	2.657
116	1.507	2.655
117	1.506	2.654
118	1.505	2.652
119	1.504	2.651
120	1.503	2.649
121	1.502	2.648
122	1.501	2.646
123	1.500	3.645
124	1.499	2.643
125	1.498	2.642
126	1.497	2.640
127	1.496	2.639
128	1.496	2.638
129	1.495	2.636
130	1.494	2.635
131	1.493	2.634
132	1.492	2.632
133	1.491	2.631
134	1.490	2.630
135	1.489	2.628
136	1.489	2.627
137	1.488	2.626
138	1.487	2.625
139	1.486	2.624
140	1.485	2.622
141	1.485	2.621
142	1.484	2.620
143	1.483	2.619
144	1.482	2.618
145	1.481	2.617
146	1.481	2.616
147	1.480	2.615
148	1.479	2.613
149	1.479	2.612
150	1.478	2.611
151	1.477	2.610
152	1.476	2.609
153	1.476	2.608
154	1.475	2.607
155	1.474	2.606
156	1.474	2.605
157	1.473	2.604
158	1.472	2.603
159	1.472	2.602
160	1.471	2.601
161	1.470	2.600
162	1.470	2.600
163	1.469	2.599
164	1.469	2.598
165	1.468	2.597

n	0.90000	0.99000
166	1.467	2.596
167	1.467	2.595
168	1.466	2.594
169	1.466	2.593
170	1.465	2.592
171	1.464	2.592
172	1.464	2.591
173	1.463	2.590
174	1.463	2.589
175	1.462	2.588
176	1.462	2.587
177	1.461	2.587
178	1.460	2.586
179	1.460	2.585
180	1.459	2.584
181	1.459	2.583
182	1.458	2.583
183	1.458	2.582
184	1.457	2.581
185	1.457	2.580
186	1.456	2.580
187	1.456	2.579
188	1.455	2.578
189	1.455	2.577
190	1.454	2.577
191	1.454	2.576
192	1.453	2.575
193	1.453	2.575
194	1.452	2.574
195	1.452	2.573
196	1.451	2.572
197	1.451	2.572
198	1.450	2.571
199	1.450	2.570
200	1.450	2.570
205	1.447	2.566
210	1.445	2.563
215	1.443	2.560
220	1.441	2.557
225	1.439	2.555
230	1.437	2.552
235	1.436	2.549
240	1.434	2.547
245	1.432	2.544
250	1.431	2.542
255	1.429	2.540
260	1.428	2.537
265	1.426	2.535
270	1.425	2.533
275	1.423	2.531
280	1.422	2.529
285	1.421	2.527
290	1.419	2.525
295	1.418	2.524
300	1.417	2.522

n	0.90000	0.99000
305	1.416	2.520
310	1.415	2.518
315	1.413	2.517
320	1.412	2.515
325	1.411	2.514
330	1.410	2.512
335	1.409	2.511
340	1.408	2.509
345	1.407	2.508
350	1.406	2.506
355	1.405	2.505
360	1.404	2.504
365	1.404	2.502
370	1.403	2.501
375	1.402	2.500
380	1.401	2.499
385	1.400	2.498
390	1.399	2.496
395	1.399	2.495
400	1.398	2.494
425	1.394	2.489
450	1.391	2.484
475	1.388	2.480
500	1.385	2.475
525	1.382	2.472
550	1.380	2.468
575	1.378	2.465
600	1.376	2.462
625	1.374	2.459
650	1.372	2.456
675	1.370	2.454
700	1.368	2.451
725	1.367	2.449
750	1.365	2.447
775	1.364	2.445
800	1.363	2.443
825	1.361	2.441
850	1.360	2.439
875	1.359	2.438
900	1.358	2.436
925	1.357	2.434
950	1.356	2.433
975	1.355	2.432
1000	1.354	2.430
1500	1.340	2.411
2000	1.332	2.399
3000	1.323	2.385
5000	1.313	2.372
10 000	1.304	2.358
∞	1.282	2.326

[handwritten note: NORMAL DISTRIBUTION VALUES]

measurements, the standard deviation of the lot average, $s_{\bar{X}}$, should be multiplied by $\sqrt{m}$, where m is the number of measurements per lot average value.

For example, assume that averages for 30 lots are reported and that four measurements were averaged for each lot. Statistics for the lot averages were as follows: number of lots, 30; number of measurements per lot, (m), 4; mean of lot averages, 690 MPa (100 ksi); standard deviation of lot averages, 17.2 MPa (2.5 ksi). The following statistics should be used to compute A and B values, as described previously: number of measurements (*n*), $30 \times 4 = 120$; mean ($\bar{X}$), 690 MPa (100 ksi); standard deviation (*s*), $17.2 \times \sqrt{4} = 34.4$ MPa ($2.5 \times \sqrt{4} = 5.0$ ksi).

Direct Computation for an Unknown Distribution

This procedure should be used when a mechanical property value is to be computed directly (not paired with another property for computational purposes) and the form of the distribution of the population is unknown (not normal). The distribution should not be considered unknown if tests show it to be nearly normal, if it can be transformed to a nearly normal distribution, or if it can be separated into nearly normal subpopulations. This procedure is applicable to F_{tu} and F_{ty}. It may also be used for F_{cy}, F_{su}, F_{bru}, and F_{bry} if a sufficient quantity of data is available.

Data Requirements. Data must be adequate to ensure that the sample is representative of the population. Although censoring is highly undesirable, parametric techniques will tolerate a limited degree of censoring. In contrast, nonparametric techniques will not tolerate censoring.

The determination of an A value requires at least 300 individual observations; the inclusion of additional data is very desirable. The selection of the number 300 is not arbitrary. Rather, 300 (or, more precisely, 299) represents the smallest sample for which the lowest observation is an A value. For smaller samples, the A value is below the lowest observation and thus cannot be determined without knowledge of the form of the population distribution.

The lowest of 30 observations is a B value; in practice, however, 100 observations is the smallest sample normally employed to compute a B value. For 100 observations, a B value would be the 5th ranked value from the lowest observation according to Table 3. The requirement for number of heats and distribution of lots for the sample is comparable to that required for a parametric analysis.

Computational Procedure. Nonparametric (or distribution-free) data analysis assumes a random selection of test points and uses only the ranks of the individual test points and the total number of test points. If an unknown number of test points have been deleted from a sample, the random basis is violated; consequently, the procedure described must not be used when there is reason to suspect that the sample may have been censored.

As an example, assume that a sample consists of 300 test points selected randomly. The test point having the lowest value has the rank 1, the test point having the next lowest

value has the rank 2, etc. Thus, an array of ranked test points might appear as follows:

Rank of test point	Value of test point, MPa (ksi)
1	505.4 (73.3)
2	510.9 (74.1)
3	518.5 (75.2)
4	519.2 (75.3)
5	521.2 (75.6)
.	.
.	.
.	.
299	590.9 (85.7)
300	592.9 (86.0)

For each rank from a sample of size *n*, it is possible to predict with 0.95 confidence the least fraction of the population that exceeds the value of the test point having rank, *r*. Because only two fractions, or probabilities, are of interest in the determination of A and B values, only the ranks of test points having the probability and confidence of A and B values are presented in Table 3.

For example, to use Table 3 with a sample size of 300, one would designate the value for the lowest ($r = 1$) test measurement as an A value and the 22nd lowest ($r = 22$) test

Table 3 Ranks, *r*, of observations, *n*, for an unknown distribution having the probability and confidence of A and B values

A basis n	r	A basis n	r	B basis n	r	B basis n	r
≤298	(a)	6 322	51	≤28	(b)	1 802	160
299	1	6 433	52	29	1	1 910	170
473	2	6 545	53	46	2	2 015	180
628	3	6 656	54	61	3	2 121	190
773	4	6 766	55	76	4	2 228	200
913	5	6 878	56	89	5	2 330	210
1 049	6	6 990	57	103	6	2 437	220
1 182	7	7 100	58	116	7	2 542	230
1 312	8	7 211	59	129	8	2 647	240
1 441	9	7 322	60	142	9	2 752	250
1 568	10	7 432	61	154	10	2 858	260
1 693	11	7 543	62	167	11	2 962	270
1 817	12	7 652	63	179	12	3 066	280
1 940	13	7 761	64	191	13	3 171	290
2 063	14	7 875	65	203	14	3 276	300
2 185	15	7 984	66	215	15	3 379	310
2 308	16	8 092	67	227	16	3 484	320
2 426	17	8 205	68	239	17	3 589	330
2 547	18	3 313	69	251	18	3 693	340
2 665	19	8 426	70	263	19	3 798	350
2 785	20	8 533	71	275	20	3 902	360
2 902	21	8 645	72	298	22	4 006	370
3 021	22	8 752	73	321	24	4 110	380
3 137	23	8 864	74	345	26	4 214	390
3 254	24	8 970	75	368	28	4 317	400
3 371	25	9 081	76	391	30	4 421	410
3 486	26	9 193	77	413	32	4 524	420
3 604	27	9 298	78	436	34	4 629	430
3 719	28	9 409	79	459	36	4 733	440
3 835	29	9 520	80	481	38	4 837	450
3 948	30	9 624	81	504	40	4 940	460
4 063	31	9 735	82	526	42	5 044	470
4 180	32	9 845	83	550	44	5 148	480
4 293	33	9 955	84	571	46	5 250	490
4 408	34	10 065	85	592	48	5 354	500
4 520	35	10 175	86	615	50	5 613	525
4 634	36	10 278	87	640	52	5 870	550
4 749	37	10 387	88	661	54	6 130	575
4 862	38	10 496	89	683	56	6 389	600
4 975	39	10 606	90	704	58	6 645	625
5 088	40	10 714	91	726	60	6 903	650
5 122	41	10 823	92	781	65	7 161	675
5 247	42	10 932	93	837	70	7 418	700
5 372	43	11 040	94	890	75	7 727	730
5 497	44	11 148	95	945	80	8 036	760
5 622	45	11 256	96	1 000	85	8 344	790
5 747	46	11 364	97	1 054	90	8 651	820
5 872	47	11 472	98	1 109	95	8 961	850
5 987	48	11 579	99	1 162	100	9 268	880
6 102	49	11 685	100(c)	1 269	110	9 576	910
6 208	50			1 376	120	9 884	940
				1 483	130	10 191	970
				1 590	140	10 498	1 000(d)
				1 696	150		

(a) A value is lower than value of lowest observation. (b) B value is lower than value of lowest observation. (c) For *n* > 11 685, A basis; $r \sim 100 + (n - 11\,685)/107$. (d) For *n* > 10 498, B basis; $r \sim 1000 + (n - 10\,498)/10.25$.

measurement as a B value. For sample sizes lying between tabulated values, interpolation is permissible. For sample sizes smaller than 300, the A value is smaller than the value of the lowest test point and cannot be determined in this manner.

Computation of Derived Properties

Ideally, it is desirable to determine F_{cy}, F_{su}, F_{bru}, and F_{bry} as well as F_{tu} and F_{ty} in other than the specified test direction by direct computation, as described previously. The direct computation procedures should be used if sufficient data are available.

Compression, shear, and bearing tests are complicated, time consuming, and expensive; consequently, statistical quantities of these data are not normally available. For these properties, it is more practical to determine the ratios of these properties to the corresponding tensile yield or ultimate strengths with an acceptable degree of certainty, and to derive the minimum design values for these properties based on the product of these lower confidence limits (reduced ratios) and the minimum tensile yield or ultimate strengths (S, A, or B values) that have already been established. It has been shown that these properties can be related to tensile properties, and reliable minimum design values can be derived in this manner.

A derived property is a mechanical property value that is determined by its relationship to an established tensile property (F_{tu} or F_{ty}, A, B, or S basis). This indirect method of computation is applicable to F_{tu} and F_{ty} in grain directions other than the specified testing direction as set forth in the applicable material specification, and for all grain directions for F_{cy}, F_{su}, F_{bru}, and F_{bry}.

The procedure involves the pairing of individual TUS, SUS, or bearing ultimate strength (BUS) measurements with TUS measurements for which F_{tu} has been established. It also involves the pairing of individual TYS, compressive yield strength (CYS), and bearing yield strength (BYS) measurements with TYS measurements for which F_{ty} has been established.

This technique is based on the premise that the mean ratio of paired observations representing two related properties provides an estimate of the ratio of the corresponding population means. The ratio consists of the individual measurements of the property to be derived as the numerator and the measurement of the established tensile property as the denominator. Thus, TUS or TYS in the specified testing direction always appears in the denominator of the ratio of observed values.

The grain direction to be used for the denominator is the specified test direction as delineated in the applicable material specification. For most materials, routine quality control (certification) tests are usually conducted only in one grain direction even though the specification may contain mechanical property requirements for two or three grain directions. For guidance, the specified or primary test direction for the different product forms of each alloy system is shown in Table 4.

Data Requirements. The computation of a derived value for each significant test direction requires at least ten paired measurements from ten lots of material obtained from at least two production heats for each product form and heat treat condition. For a single form and thickness, data from no more than one heat treat lot per heat may be used to meet the ten-lot requirement. The thicknesses of the ten lots should span the thickness ranges of the product form covered by the material specification.

Test specimens for the paired ratios should be located in close proximity and should be taken from the same sheet, plate, bar, extrusion, forging, or casting. If coupons or specimens are machined prior to heat treatment, the coupons or specimens representing paired ratios should be heat treated simultaneously in the same heat treat load through all heat treating operations.

In those cases where multiple observations are available from a single lot, the average of those observations should be treated as an individual observation. Because some variation in strength may be expected from one specimen location to another, the use of lot averages minimizes the effect of this variable.

Computational Procedure. Four basic steps are involved in determining design allowable properties by indirect (ratioing) computation:

- Determine the ratios of paired observations for each lot of material.
- Compute the statistics, $\bar{r}$ and s, for the ratios of paired observations.
- Determine the lower confidence interval estimate (reduced ratio) for the mean ratio.

- Use the reduced ratio as the ratio of the derived to the establishing design allowable.

The ratio of two paired observations is obtained by dividing the measurement of the property to be derived, e.g., CYS (LT) for heat treatable aluminum sheet, by the measurement for the established tensile property, e.g., TYS (LT), in the specified testing direction.

The ratio of the two population means for CYS (LT) and TYS (LT), respectively, is expected to exceed the lower confidence limit defined as:

$$\bar{r} - \frac{t_{1-\alpha}s}{\sqrt{n}} \qquad \text{(Eq 3)}$$

where n is number of ratios, r is the average of n ratios, s is the standard deviation of the ratios, and $t_{1-\alpha}$ is the $1 - \alpha$ fractile of the t distribution for $n - 1$ degrees of freedom. At a risk level of $\alpha = 0.05$, the appropriate t value is $t_{0.95}$.

Because the lower confidence interval estimate is used as the ratio between the design allowable properties, the reduced ratio, R, may be defined as:

$$R = \bar{r} - \frac{t_{0.95}s}{\sqrt{n}} \qquad \text{(Eq 4)}$$

Values of $t_{0.95}$ for various degrees of freedom, $n - 1$, are given in Table 5.

The reduced ratio may now be used to establish the design allowable for the property to be derived using the example of aluminum sheet: $R = F_{cy}(LT)/F_{ty}(LT) = $ allowable to be derived/established allowable in specified test direction. The derived allowable property is computed by cross multiplying:

$$F_{cy}(LT) = RF_{ty}(LT)$$

The basis (A, B, or S) for the computed, or derived, allowable is assumed to be the same as the basis for the F_{ty} or F_{tu} tensile allowable in the right side of the equation.

Table 4 Primary testing direction for various alloy systems(a)

Product form	Steel alloys	Non-heat-treatable Al alloys	Heat treatable Al alloys	Mg alloys	Ti alloys	Heat-resistant alloys
Sheet and plate	LT	L	LT	L	(b)	LT
Bar	L	L	L	L	L	L
Tubing	L	L	L	L	L	L
Extrusion	L	L	L	L	L	L
Die forging	⋯	L	L(c)	L	⋯	L
Hand forging	⋯	LT	LT	LT	(b)	L

(a) Although material specifications may contain mechanical-property requirements for two or three grain directions, the primary test direction indicates the grain direction that is tested regularly. (b) Either L or LT may be used. (c) Although material specifications require testing in both L and T directions, the T direction by definition includes all orientations not within $\pm 15°$ of parallel to grain flow. Hence, the L direction is preferred for analytical purposes.

Table 5 The 0.95 and 0.975 fractiles of the *t* distribution associated with degrees of freedom (d.f.)

d.f.	$t_{0.95}$	$t_{0.975}$	d.f.	$t_{0.95}$	$t_{0.975}$
1	6.314	12.706	21	1.721	2.080
2	2.920	4.303	22	1.717	2.074
3	2.353	3.182	23	1.714	2.069
4	2.132	2.776	24	1.711	2.064
5	2.015	2.571	25	1.708	2.060
6	1.943	2.447	26	1.706	2.056
7	1.895	2.365	27	1.703	2.052
8	1.860	2.306	28	1.701	2.048
9	1.833	2.262	29	1.699	2.045
10	1.812	2.228	30	1.697	2.042
11	1.796	2.201	40	1.684	2.021
12	1.782	2.179	50	1.676	2.009
13	1.771	2.160	60	1.671	2.000
14	1.761	2.145	80	1.664	1.990
15	1.753	2.131	100	1.660	1.984
16	1.746	2.120	120	1.658	1.980
17	1.740	2.110	200	1.653	1.972
18	1.734	2.101	500	1.648	1.965
19	1.725	2.093	∞	1.645	1.960
20	1.725	2.086			

In a sample of ratios for a given product, the effect of thickness on the ratio should be examined. If no effect of thickness is noted, the ratios for the various thicknesses can be pooled to compute the average and reduced ratio. If a thickness effect exists, a regression with thickness should be computed and the average and reduced ratios determined from the regression.

When the computed design allowable results in a fractional number, the actual design allowable value is determined in the following manner. Fractions greater than 0.75 usually are raised to the next larger integer while lesser decimal fractions are disregarded.

Treatment of Grain Direction. Tensile allowables are usually listed according to grain direction in material specifications. Some specifications do not indicate a grain direction, which implies isotropy. It is recommended that tension allowables be determined for each grain direction. When the material is shown to be isotropic, then the same properties should be used for each direction.

Compression allowables are determined by grain direction similar to tension allowables. An example of computing compression allowables for heat treatable plate is given below. The reduced ratio, R, for the longitudinal grain direction is determined from ratios, r, formed from paired observations for each lot of material, CYS(L)/TYS(LT). Although a longitudinal ratio is being obtained, the divisor is long transverse because this is the specified testing direction (refer to Table 4). The reduced ratio, R, for the long transverse grain direction is determined from ratios, r, formed from paired observations for each lot of material, CYS(LT)/TYS(LT).

Similarly, the reduced ratios, R, for the short transverse grain direction are determined from ratios, r, formed from paired observations for each lot of material, CYS(ST)/TYS(LT). The ratios, r, determined in the above manner are used in conjunction with Eq 4 to obtain a reduced ratio, R, for each grain direction. Equating the reduced ratios, design allowable values are determined from the resulting relationships:

$$R = \frac{F_{cy}(L)}{F_{ty}(LT)}$$

or

$$F_{cy}(L) = RF_{ty}(LT)$$

Similarly:

$$F_{cy}(LT) = RF_{ty}(LT)$$

and

$$F_{cy}(ST) = RF_{ty}(LT)$$

Shear and bearing properties should be analyzed according to grain direction, and the design allowables should be based on the lowest reduced ratio obtained for the longitudinal, long transverse, and short transverse (when applicable) directions. An exception is aluminum hand forgings, for which shear values should be established according to grain direction.

In the computation of derived properties, paired ratios representing different grain directions should not be combined in the determination of a reduced ratio. If the ratio for two paired measurements is to provide an estimate of the population mean ratio, then the paired measurements must represent the same grain direction as that of the corresponding population means.

For aluminum die forgings, the transverse, T, grain direction by definition includes all orientations not within $\pm 15°$ of parallel to grain flow; consequently, only longitudinal tensile data should be used as the bases to establish compression, shear, and bearing allowables for aluminum die forgings.

Determining Design Allowables by Regression Analysis

Before employing regression analysis in the determination of design allowables, it must be ascertained that the property or ratio to be regressed varies continuously and linearly with some dimensional parameter, such as some function of thickness (for example, t, $\sqrt{t}$, $i/\sqrt{t}$, etc., where t is thickness). If the variation is attributable to other causes, e.g., differences in processing operations between sheet products and plate products, then the regression should not be used.

The significance and linearity of a regression equation of the form:

$$TUS' = a + bx$$

or

$$\left(\frac{SUS}{TUS}\right)' = a + bx$$

can be evaluated through an analysis of variance, as depicted in Fig. 3 through 5. Through this procedure, one F statistic is computed (F_1) that defines the significance of the regression, and another statistic is developed (F_2) that verifies the linearity of the regression equation. If F_2 indicates nonlinearity in the data, a transformation on thickness may account for the nonlinearity. If F_1 indicates an insignificant regression, one of the normal analysis techniques (without regression) should be used, as described in the preceding section of this article.

If there are too few data to establish linearity with 0.95 confidence (as typically happens with ratios) and the data appear to follow a linear pattern with thickness, linearity may be assumed and a linear regression completed on that assumption. The significance of that linear regression equation can be evaluated through a check of the confidence limits on the slope (b) of the regression equation.

The upper and lower confidence limits can be expressed as:

$$b_{upper} = b + t_{0.95}s'\sqrt{\frac{1}{\Sigma(X_i - \bar{X})^2}}$$

Fig. 3 Analysis of variance procedure of linear regression

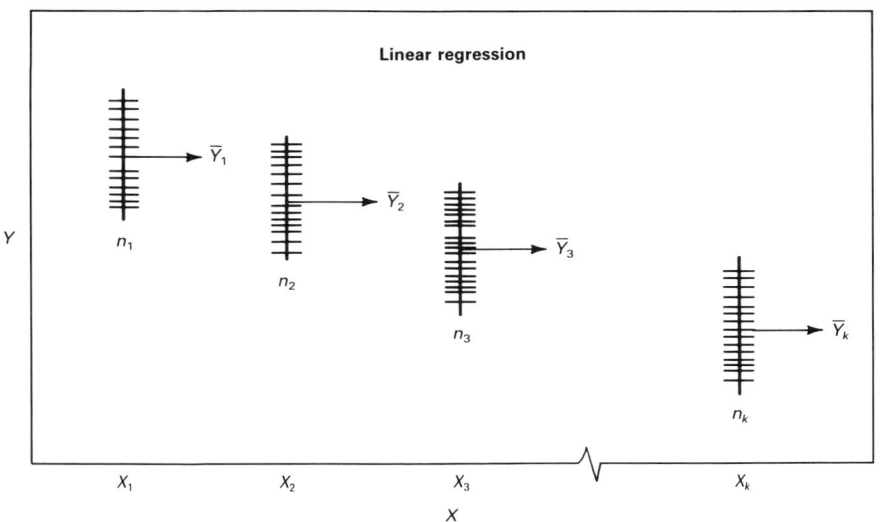

Fig. 4 Graphical display of *F*-test for significance of a regression

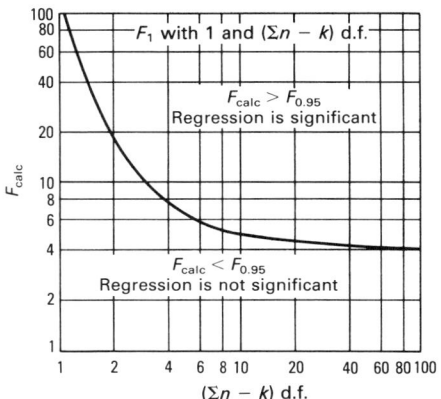

Preliminary calculations for Fig. 3

	X				
	X_1	X_2	X_3	X_k	
Observations, Y	Y_1 data	Y_2 data	Y_3 data	Y_k data	$\rightarrow$ $\Sigma Y = \sum\limits_{i=1}^{k} Y_i$ $\Sigma Y^2 = \sum\limits_{i=1}^{k} Y_i^2$
Treatment totals, T	$T_1 = \Sigma Y_1$	$T_2 = \Sigma Y_2$	$T_3 = \Sigma Y_3$	$T_k = \Sigma Y_k$	$\rightarrow$ $G = \sum\limits_{i=1}^{k} T_i$
Treatment sample sizes, n	n_1	n_2	n_3	n_k	$\rightarrow$ $n = \sum\limits_{i=1}^{k} n_i$
$\bar{Y}$	$\bar{Y}_1 = \Sigma Y_1/n_1$	$\bar{Y}_2 = \Sigma Y_2/n_2$	$\bar{Y}_3 = \Sigma Y_3/n_3$	$\bar{Y}_k = \Sigma Y_k/n_k$	$\bar{Y} = G/n$
T^2/n	T_1^2/n_1	T_2^2/n_2	T_3^2/n_3	T_k^2/n_k	$\rightarrow$ $\Sigma(T^2/n) = \sum\limits_{i=1}^{k} (T_i^2/n_i)$

Analysis of variance for Fig. 3

Source of variation	Sum of squares, ss	Degrees of freedom, d.f.	Mean squares, ms	F_{calc}
Treatments...............I. $\Sigma(T^2/n) - G^2/n$		$k - 1$		
Linear regression(a) ......II. sp^2/ss_x		1	A. II	$F_1 = A/C$
Deviation from linearity ...III. $= I - II$		$(k - 1) - 1$	B. III/$(k - 2)$	$F_2 = B/C$
Error ....IV. $\Sigma Y^2 - \Sigma(T^2/n)$		$\Sigma n - k$	C. IV/$(\Sigma n - k)$	See attached curves
Total ....V. $\Sigma Y^2 - G^2/n$		$\Sigma n - 1$		

(a) $sp^2 = (\Sigma XY - \Sigma X\Sigma Y/n)^2$; $ss_x = \Sigma X^2 - (\Sigma X)^2/n$

$$b_{lower} = b - t_{0.95}s' \sqrt{\frac{1}{\Sigma(X_i - \bar{X})^2}}$$

where $t_{0.95}$ is Students' t factor for $n - 2$ degrees of freedom, X_i is individual thick-

ness (or other dimensional parameter) values, $\bar{X}$ is the average thickness (or other dimensional parameter), b is the slope of the regression line, and s' is the standard error of estimate.

After computing b_{upper} and b_{lower}, it must be determined whether the interval between the two limits contains zero. If it does, the regression should be considered insignificant, and tolerance limits or confidence limits should be established without a regression analysis. If the confidence interval on b does not contain zero, a significant regression is indicated. When both the upper and lower limits on b are positive, it indicates increasing property values with thickness (or another dimensional parameter). When both limits on b are negative, it indicates decreasing property values with thickness (or another dimensional parameter).

The steps involved in the use of regression techniques may be summarized as follows:

Express the property or ratio as a function of the dimensional parameter: e.g., TUS' $= a + bx$, and (SUS/TUS)' $= a + bx$, where x is thickness, area, and so on; a and b are constants from the a and b least squares equation; and prime denotes computed values.

Determine the standard deviation of the observed points from the computed values at corresponding values of x. Thus, the deviation of an individual observation is TUS observed $-$ TUS computed, and SUS/TUS observed $-$ (SUS/TUS) computed. The standard deviation computed by regression analysis has $n - 2$ degrees of freedom, because the regression line contains two constants (the intercept a and slope b).

At selected values of x, determine either the lower tolerance limit for the regression line for the property, or the lower confidence interval estimate for the regression line for the ratio.

The procedure takes into account errors in the estimates of the true intercept and slope of the regression line. Thus, using the examples above, to compute an A value for F_{tu} at $x = x_0$:

$$A = TUS'$$

$$- k_A s' \sqrt{1 + \frac{1}{n} + \frac{\left(x_0 - \frac{\Sigma x}{n}\right)^2}{(\Sigma x^2) - \frac{(\Sigma x)^2}{n}}} \quad (Eq\ 5)$$

where k_A is selected corresponding to a sample size of $n - 1$.

The equation for computing a B value is similar, with k_B being used in place of k_A. To compute the reduced ratio for F_{su}/F_{tu} at $x = x_0$:

$$\text{Reduced ratio} = \frac{SUS}{TUS}$$

$$- t_{0.95} s' \sqrt{\frac{1}{n} + \frac{\left(x_0 - \frac{\Sigma x}{n}\right)^2}{(\Sigma x^2) - \frac{(\Sigma x)^2}{n}}} \quad (Eq\ 6)$$

where s' is the standard error of estimate, and $t_{0.95}$ is selected corresponding to $n - 2$ degrees of freedom. This equation is identical to Eq 4, except that $1/\sqrt{n}$ is replaced by a term containing the correction factor.

The procedures described in this section permit determination of design allowables only for specific values of x. When it is desirable to present a single allowable covering a range of products, e.g., 2.542- to 5.080-cm (1.001- to 2.000-in.) plate, the lowest allowable for the range is used. Thus, if TUS(LT) decreases continuously with increasing thickness for the material in the above example, the TUS(LT) corresponding to $x = 5.080$ cm (2.000 in.) would be used. If the decrease is large, a decrease in the product size interval can be made, e.g., by splitting the 2.542- to 5.080-cm (1.001- to 2.000-in.) interval into two intervals of 2.542 to 3.810 cm (1.001 to 1.500 in.) and 3.811 to 5.080 cm (1.501 to 2.000 in.), respectively. The Appendix at the end of this article gives examples of computational procedures.

Low- and Elevated-Temperature Design Properties

Minimum design values for various mechanical properties are frequently required at temperatures other than room temperature.

Fig. 5 Graphical display of F-test for significant deviation from linearity

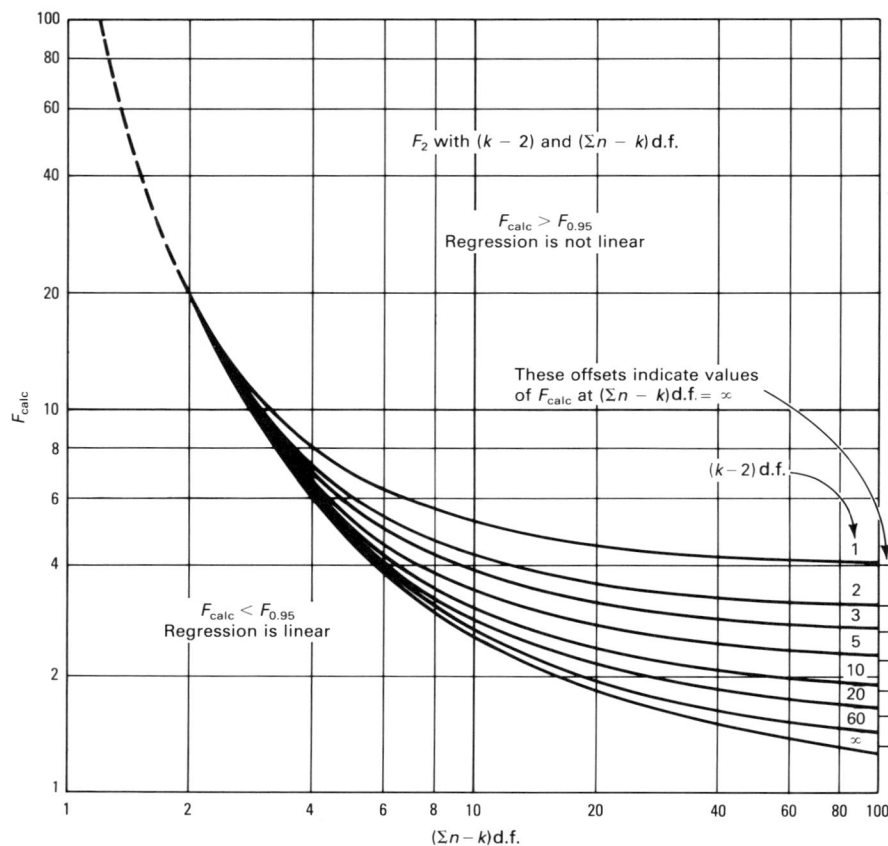

With metallic materials, strength usually decreases with increasing temperature and increases with decreasing temperature. Testing at low and elevated temperatures to determine various mechanical properties is time consuming and expensive; consequently, statistical quantities of data at various low or elevated temperatures are seldom available.

However, a ratio technique can be employed to derive a percentage value of room temperature strength at various temperatures. Ratios of the strength at each temperature to the strength at room temperature are determined, and from these ratios a lower confidence limit for each temperature is derived. The minimum design value at temperature can be obtained by multiplying the lower confidence limit (percentage) at temperature by the room-temperature design value that has already been established.

Strength Properties. Tensile ultimate and yield strengths, compressive yield strength, shear ultimate strength, and bearing ultimate and yield strengths at temperatures other than room temperature (27 °C, or 80 °F) should be established as percentages of the room-temperature value of each property. The use of percentage curves allows a

single curve to be used in place of multiple curves when more than one room-temperature value is presented for a property, e.g., for differing A and B design values for each of several thickness ranges.

No significance level is attached to these curves. For practical purposes, however, the product of a room-temperature A design value and an appropriate percentage value from the curve may be regarded as an A design value at the indicated temperature.

Data Requirements. To establish the shape of an "effect-of-temperature" curve, the sample should include observations from at least five lots of material composed of at least two heats at each of several temperatures. For a single form and thickness, data from no more than one heat treat lot per heat may be used to meet the five-lot requirement.

The choice of temperature should be guided by the probable range of service temperatures anticipated for the material, as well as by its metallurgical characteristics. For materials used at cryogenic temperatures, testing is normally conducted at -79, -196, and -253 °C (-110, -320, and -423 °F); however, no attempt should be made to extrapolate the curve below the lowest temperature for which adequate data are available.

For elevated-temperature applications, data should normally be available at temperature intervals of from 110 to 167 °C (200 to 300 °F), except in the regions of time- and temperature-dependent metallurgical change, where temperature intervals of perhaps 55 to 83 °C (100 to 150 °F) are appropriate. Again, extrapolation beyond the range of temperatures covered by adequate data is not advisable.

For a number of alloys, particularly the heat-resistant alloys, procurement specifications may designate minimum property values at temperatures other than room temperature, and either A- or S-basis values may be available at the elevated testing temperature. In this case, the "effect-of-temperature" curve may be scaled by using this elevated temperature value.

Determination of Working Curves. Working curves for each product form, heat treat condition, property, and grain direction should be constructed. The separate curves should be examined to determine if some data can be combined. For example, it may be possible to combine data for sheet and plate, T73 and T7351 tempers, tensile and compressive yield strengths, or longitudinal and long transverse grain directions.

The dimensional units of these working curves should be in terms of percentages of the corresponding room-temperature value for the property. A percentage should be determined for each lot by dividing the average value of the individual measurements (other than at room temperature) by the room-temperature average value for the same lot of material in the same testing direction, then multiplying by 100 to convert from a fraction to a percentage.

For each temperature, the lower 0.95 confidence interval estimate (reduced ratio) of the mean percentage should be determined from the individual percentage values at that temperature. If r equals the individual percentage values, $\bar{r}$ the average of these values,

and n the number of such percentages, the estimated standard deviation (s) and the reduced ratio (R) can be determined:

$$s^2 = \frac{\Sigma(r - \bar{r})^2}{(n - 1)} \qquad \text{(Eq 7)}$$

or

$$s^2 = \frac{\Sigma(r^2) - \dfrac{(\Sigma \bar{r})^2}{n}}{(n - 1)} \qquad \text{(Eq 8)}$$

and

$$R = \bar{r} - t\frac{s}{\sqrt{n}} \qquad \text{(Eq 9)}$$

where t is the 0.95 fractile of the t distribution corresponding to $n - 1$ degrees of freedom (see Table 5).

The working curve should be a smooth curve drawn through 100% at room temperature and not higher than the computed values of R at each temperature. When only room-temperature minima are applicable, no further adjustment of the working curve is required.

However, when a secondary testing temperature is specified for the property, the working curve should be lowered, if necessary, so that the product of the percentage from this curve and a room-temperature S value do not exceed the S value at the secondary testing temperature. Also, if A-basis values have been established for this temperature, the working curve should be lowered, if necessary, so that the product of the percentage from this curve and the room-temperature A value do not exceed the A value at the secondary testing temperature.

Because strength at temperature is a function of time as well as temperature, minimum design values must be established for each exposure time (0.5, 1, 10, 100+ h exposure). Figure 6 illustrates an example of a working curve.

Preparation of Final Curves. When two or more working curves can be com-

bined into a single curve, the data for these working curves should be combined and new reduced ratios computed. These new computed ratios should be used to construct a final curve.

Final curves should not exhibit "humps," such as might appear with a temperature range in which aging takes place. When humps appear in the working curves, they should be leveled by means of horizontal line segments. An example of a final percentage curve is shown in Fig. 7.

REFERENCE

1. *Metallic Materials and Elements for Aerospace Vehicle Structures*, Military Standardization Handbook MIL-HDBK-D, revised Jan 1984

Fig. 6 Working curve drawn through reduced ratios converted to percentages

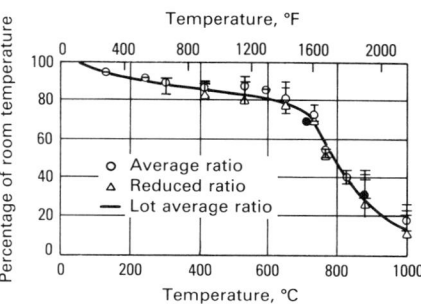

Fig. 7 Percentage curve representing two elastic moduli

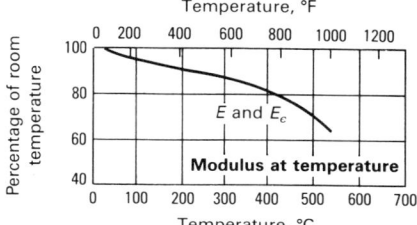

Appendix: Examples of Computational Procedures

To review the computational procedures described in this article, a hypothetical set of input data has been invented that represents a composite of many actual cases. In this hypothetical case, the flow-charts of Fig. 1 and 2 have been followed.

Assume that a quantity of test data has been amassed, representing one long transverse tensile test per lot, as well as other tests from a portion of the lots, at a frequency of one test per lot. Progressive steps of the computations are illustrated in Problems 1 through 7. The input data are:

- *Material identification:* Alloy X sheet, annealed
- *Specified testing direction:* Long transverse (LT)
- *Specified properties:*
  ≤ 0.125 in.—$F_{tu}(LT) = 140$ ksi; $F_{ty}(LT) = 115$ ksi
  0.126 to 0.249 in.—$F_{tu}(LT) = 135$ ksi; $F_{ty}(LT) = 110$ ksi
- *Available test results:*
  Group 1. 300 observations of TUS(LT) for thickness range of 0.020 to 0.125 in. from Supplier A; no variation with thickness. Proceed to Problems 1 and 3.
  Group 2. 30 observations of TUS(LT) for thickness range of 0.020 to 0.125 in. from Supplier B; no variation with thickness. Observations may be paired with TUS(L) if desired. Proceed to Problem 1.
  Group 3. 300 observations of TYS(LT) for thickness range of 0.020 to 0.125 in. from Supplier A; no variation with thickness. Proceed to Problem 2.
  Group 4. 30 observations of TYS(LT) for thickness range of 0.020 to 0.125 in. from Supplier B; no variation with thickness. Proceed to Problem 2.
  Group 5. 30 observations of SUS(LT) for thickness range of 0.020 to 0.249 in. apparent decrease in SUS(LT) with increasing thickness. Observations may be paired with TUS(LT) if desired. Proceed to Problem 7.
  Group 6. 100 observations of TUS(LT) for thickness range of 0.126 to 0.249 in.; no variation with thickness. Proceed to Problem 3.

Problem 1

Should the Data in Groups 1 and 2 Be Combined?

Neither property varies with thickness. Sample statistics are:

Subpopulation	n	$\bar{X}$, ksi	s, ksi
Group 1 TUS (LT), 0.020 to 0.249 300		150.0	4.00
Group 2 TUS (L), 0.020 to 0.249 30		151.0	5.00

Step 1. Test to determine whether the variances differ significantly.

$$F = (s_1)^2/(s_2)^2 = (4.00)^2/(5.00)^2 = 0.64$$

Degrees of freedom, numerator $= n_1 - 1 = 300 - 1 = 299$

Degrees of freedom, denominator $= n_2 - 1 = 30 - 1 = 29$

$F_{0.975}$ (299,29 d.f.) from Table 6 $= 1.87$ (approximately)

$1/F_{0.975}$ (29,299 d.f.) $= 1/1.69 = 0.59$

Because the computed value of F (0.64) lies within the 0.95 confidence interval (0.59 to 1.87), conclude the variances do not differ significantly.

Step 2. Test to determine whether the averages differ significantly.

Difference between averages, $D_{\bar{X}} = 150.0 - 151.0 = 1.0$ ksi

$$u = t_{0.975} S_p \sqrt{\frac{n_1 + n_2}{n_1 n_2}}$$

Degrees of freedom $= n_1 + n_2 - 2 = 300 + 30 - 2 = 328$

$t_{0.975}$ (328 d.f.) from Table 5 $= 1.969$

$$s_p = \sqrt{\frac{(n - 1)s_1^2 + (n_2 - 1)s_2^2}{n_1 + n_2 - 2}}$$

$$= \sqrt{\frac{(300 - 1)(4.00)^2 + (30 - 1)(5.00)^2}{300 + 30 - 2}} = 4.10 \text{ ksi}$$

$$u = 1.969 \times 4.10 \times \sqrt{\frac{n_1 + n_2}{n_1 n_2}}$$

$$= 1.969 \times 4.10 \times \sqrt{\frac{300 + 30}{300 \times 30}} = 1.54 \text{ ksi}$$

Because the observed difference between the averages, $D_{\bar{X}}$(1.0 ksi), is less than u (1.54 ksi), conclude the averages do not differ significantly.

Step 3. Because there is no reason to conclude that the subpopulations represented by Groups 1 and 2 do not belong to the same population, combine the sample statistics for these groups.

$$n = n_1 + n_2 = 300 + 30 = 330$$

$$\bar{X} = \frac{n_1 \bar{X}_1 + n_2 \bar{X}_2}{n_1 + n_2} = \frac{300 \times 150.0 + 30 \times 151.0}{300 + 30} = 150.1 \text{ ksi}$$

$$s = s_p = 4.10 \text{ ksi}$$

The sample statistics for the combined groups are:

Subpopulation	n	$\bar{X}$, ksi	s, ksi
Groups 1 and 2 TUS(LT), 0.020 to 0.125, Suppliers A and B 330		150.1	4.10

Proceed to Problem 4.

Problem 2

Should the Data in Groups 3 and 4 Be Combined?

Neither property varies with thickness. Sample statistics are:

Subpopulation	n	$\bar{X}$, ksi	s, ksi
Group 3 TYS (LT), 0.020 to 0.125, Supplier A 300		130.0	4.00
Group 4 TYS (LT), 0.020 to 0.125, Supplier B 30		131.0	5.00

The steps involved in this problem are identical to those in Problem 1, and similar conclusions were obtained from the input; namely, that Groups (3) and (4) should be combined. The sample statistics for the combined groups are:

Subpopulation	n	$\bar{X}$, ksi	s, ksi
Groups 3 and 4 TYS (LT), 0.020 to 0.125, Suppliers A and B 330		130.1	4.10

Proceed to Problem 5.

Problem 3

Should the Data in Groups 1 and 6 Be Combined?

Neither property varies with thickness. Sample statistics are:

Subpopulation	n	$\bar{X}$, ksi	s, ksi
Group 1 TUS (LT), 0.020 to 0.125 300		150.0	4.00
Group 6 TUS (LT), 0.126 to 0.249 100		145.0	4.47

Step 1. Test to determine whether the variances differ significantly.

$$F = (s_1)^2/(s_2)^2 = (4.00)^2/(4.47)^2 = 0.80$$

Degrees of freedom, numerator $= n_1 - 1 = 300 - 1 = 299$

Degrees of freedom, denominator $= n_6 - 1 = 100 - 1 = 99$

$F_{0.975}$ (299,99 d.f.) from Table 6 $= 1.46$ (approximately)

$1/F_{0.975}$ (99,299 d.f.) $= 1/1.43 = 0.700$

Because the computed value of F (0.80) lies within the 0.95 confidence interval (0.700 to 1.46), conclude that the variances do not differ significantly.

Step 2. Test to determine whether the averages differ significantly.

Difference between averages, $D_{\bar{X}} = (150.0 - 145.0) = 5.0$ ksi

$$u = t_{0.975}\, s_p \sqrt{\frac{n_1 + n_6}{n_1 n_6}}$$

Degrees of freedom $= n_1 + n_6 - 2 = 300 + 100 - 2 = 398$

$t_{0.975}$ (398 d.f.) from Table 5 $= 1.968$

$$s_p = \sqrt{\frac{(n_1 - 1)s_1^2 + (n_6 - 1)s_6^2}{n_1 + n_6 - 2}}$$

$$= \sqrt{\frac{(300 - 1)(4.00)^2 + (100 - 1)(4.47)^2}{300 + 100 - 2}} = 4.20 \text{ ksi}$$

$$u = 1.968 \times 4.20 \times \sqrt{\frac{n_1 + n_6}{n_1 n_6}}$$

$$= 1.968 \times 4.20 \times \sqrt{\frac{300 + 100}{300 \times 100}} = 0.95 \text{ ksi}$$

Because the observed difference between the averages $D_{\bar{X}}$ (5.0 ksi) is greater than u (0.95 ksi), conclude that the averages do differ significantly and that the subpopulations represented by Groups 1 and 6 do not belong to the same population.

Step 3. Do not combine the sample statistics for these groups. Proceed to Problem 6.

Problem 4

What Computational Method Should Be Used for the Combined Observations of Groups 1 and 2?

This property does not vary with thickness. Sample statistics for the combined observations are:

Population	n	$\bar{X}$, ksi	s, ksi
Groups 1 and 2 TUS (LT), 0.020 to 0.125 330		150.1	4.10

Form of the distribution has not been determined. The sample is large enough to permit direct computation of A and B values. Consequently, the computational method will be determined by whether or not the observations are normally distributed.

Step 1. Test to determine whether the distribution is normal. A chi-square test will be employed in this example. Because the sample statistics have already been computed for the observations in Groups 1 and 2, it will be convenient to establish intervals based on X and s, and ten intervals having equal areas under the normal curve would appear to be an appropriate choice.

The intervals can be determined from Table 7, which lists the area under the normal curve from $-\infty$ to mean $+z$ standard deviations. For example, to find the 0.9 to 1.0 interval, look for the z value corresponding to the 0.9000 area:

z	Area
1.29 .... 0.9015	
1.28 .... 0.8997	

By interpolation, 1.282 corresponds to 0.9000. If the mean $\bar{X} = 150.1$ ksi and $s = 4.10$ ksi, then:

$$\bar{X} + zs = 150.1 + 1.282(4.10)$$
$$= 150.1 + 5.26$$
$$= 155.36 \text{ ksi}$$

The 0.9 to 1.0 interval is 155.36 ksi and over. The 0.8 to 0.9 interval is found similarly:

z	Area
0.85 .... 0.8023	
0.84 .... 0.7995 and by interpolation	
0.842 .... 0.8000	

$$\bar{X} + zs = 150.1 + 0.842(4.10)$$
$$= 153.55$$

To find the intervals for areas less than 0.5000 (or less than that contained from $-\infty$ to mean), refer to the z for 1-area and subtract zs from the mean. For the 0.4 to 0.5 interval:

z	1-area
0.26 .... 0.6026	
0.25 .... 0.5987 and by interpolation	
0.253 .... 0.6000	

The area from $-\infty$ to 0.4 is:

$$\bar{X} - zs = 150.1 - 0.253(4.10)$$
$$= 150.1 - 1.04$$
$$= 149.06$$

The 0.4 to 0.5 interval is 149.06 to 150.1.

The following table shows the 10 intervals selected in this manner. In the second column, the intervals are presented in ksi. The next two columns show the number of observations observed (f_o) and expected (f_e) within each interval, and the last column is the fraction $(f_o - f_e)^2/f_e$ computed for this interval.

Table 6 The 0.975 fractiles of the F distribution associated with n_1 and n_2 degrees of freedom

$F_{0.975}(n_1, n_2)$

n_1 = degrees of freedom for numerator; n_2 = degrees of freedom for denominator

n_2 \ n_1	1	2	3	4	5	6	7	8	9	10	12	15	20	24	30	40	60	120	∞
1	647.8	799.5	864.2	899.6	921.8	937.1	948.2	956.7	963.3	968.6	976.7	984.9	993.1	997.2	1001	1006	1010	1014	1018
2	38.51	39.00	39.17	39.25	39.30	39.33	39.36	39.37	39.39	39.40	39.41	39.43	39.45	39.46	39.46	39.47	39.48	39.49	39.50
3	17.44	16.04	15.44	15.10	14.88	14.73	14.62	14.54	14.47	14.42	14.34	14.25	14.17	14.12	14.08	14.04	13.99	13.95	13.90
4	12.22	10.65	9.98	9.60	9.36	9.20	9.07	8.98	8.90	8.84	8.75	8.66	8.56	8.51	8.46	8.41	8.36	8.31	8.26
5	10.01	8.43	7.76	7.39	7.15	6.98	6.85	6.76	6.68	6.62	6.52	6.43	6.33	6.28	6.23	6.18	6.12	6.07	6.02
6	8.81	7.26	6.60	6.23	5.99	5.82	5.70	5.60	5.52	5.46	5.37	5.27	5.17	5.12	5.07	5.01	4.96	4.90	4.85
7	8.07	6.54	5.89	5.52	5.29	5.12	4.99	4.90	4.82	4.76	4.67	4.57	4.47	4.42	4.36	4.31	4.25	4.20	4.14
8	7.57	6.06	5.42	5.05	4.82	4.65	4.53	4.43	4.36	4.30	4.20	4.10	4.00	3.95	3.89	3.84	3.78	3.73	3.67
9	7.21	5.71	5.08	4.72	4.48	4.32	4.20	4.10	4.03	3.96	3.87	3.77	3.67	3.61	3.56	3.51	3.45	3.39	3.33
10	6.94	5.46	4.83	4.47	4.24	4.07	3.95	3.85	3.78	3.72	3.62	3.52	3.42	3.37	3.31	3.26	3.20	3.14	3.08
11	6.72	5.26	4.63	4.28	4.04	3.88	3.76	3.66	3.59	3.53	3.43	3.33	3.23	3.17	3.12	3.06	3.00	2.94	2.88
12	6.55	5.10	4.47	4.12	3.89	3.73	3.61	3.51	3.44	3.37	3.28	3.18	3.07	3.02	2.96	2.91	2.85	2.79	2.72
13	6.41	4.97	4.35	4.00	3.77	3.60	3.48	3.39	3.31	3.25	3.15	3.05	2.95	2.89	2.84	2.78	2.72	2.66	2.60
14	6.30	4.86	4.24	3.89	3.66	3.50	3.38	3.29	3.21	3.15	3.05	2.95	2.84	2.79	2.73	2.67	2.61	2.55	2.49
15	6.20	4.77	4.15	3.80	3.58	3.41	3.29	3.20	3.12	3.06	2.96	2.86	2.76	2.70	2.64	2.59	2.52	2.46	2.40
16	6.12	4.69	4.08	3.73	3.50	3.34	3.22	3.12	3.05	2.99	2.89	2.79	2.68	2.63	2.57	2.51	2.45	2.38	2.32
17	6.04	4.62	4.01	3.66	3.44	3.28	3.16	3.06	2.98	2.92	2.82	2.72	2.62	2.56	2.50	2.44	2.38	2.32	2.25
18	5.98	4.56	3.95	3.61	3.38	3.22	3.10	3.01	2.93	2.87	2.77	2.67	2.56	2.50	2.44	2.38	2.32	2.26	2.19
19	5.92	4.51	3.90	3.56	3.33	3.17	3.05	2.96	2.88	2.82	2.72	2.62	2.51	2.45	2.39	2.33	2.27	2.20	2.13
20	5.87	4.46	3.86	3.51	3.29	3.13	3.01	2.91	2.84	2.77	2.68	2.57	2.46	2.41	2.35	2.29	2.22	2.16	2.09
21	5.83	4.42	3.82	3.48	3.25	3.09	2.97	2.87	2.80	2.73	2.64	2.53	2.42	2.37	2.31	2.25	2.18	2.11	2.04
22	5.79	4.38	3.78	3.44	3.22	3.05	2.93	2.84	2.76	2.70	2.60	2.50	2.39	2.33	2.27	2.21	2.14	2.08	2.00
23	5.75	4.35	3.75	3.41	3.18	3.02	2.90	2.81	2.73	2.67	2.57	2.47	2.36	2.30	2.24	2.18	2.11	2.04	1.97
24	5.72	4.32	3.72	3.38	3.15	2.99	2.87	2.78	2.70	2.64	2.54	2.44	2.33	2.27	2.21	2.15	2.08	2.01	1.94
25	5.69	4.29	3.69	3.35	3.13	2.97	2.85	2.75	2.68	2.61	2.51	2.41	2.30	2.24	2.18	2.12	2.05	1.98	1.91
26	5.66	4.27	3.67	3.33	3.10	2.94	2.82	2.73	2.65	2.59	2.49	2.39	2.28	2.22	2.16	2.09	2.03	1.95	1.88
27	5.63	4.24	3.65	3.31	3.08	2.92	2.80	2.71	2.63	2.57	2.47	2.36	2.25	2.19	2.13	2.07	2.00	1.93	1.85
28	5.61	4.22	3.63	3.29	3.06	2.90	2.78	2.69	2.61	2.55	2.45	2.34	2.23	2.17	2.11	2.05	1.98	1.91	1.83
29	5.59	4.20	3.61	3.27	3.04	2.88	2.76	2.67	2.59	2.53	2.43	2.32	2.21	2.15	2.09	2.03	1.96	1.89	1.81
30	5.57	4.18	3.59	3.25	3.03	2.87	2.75	2.65	2.57	2.51	2.41	2.31	2.20	2.14	2.07	2.01	1.94	1.87	1.79
40	5.42	4.05	3.46	3.13	2.90	2.74	2.62	2.53	2.45	2.39	2.29	2.18	2.07	2.01	1.94	1.88	1.80	1.72	1.64
60	5.29	3.93	3.34	3.01	2.79	2.63	2.51	2.41	2.33	2.27	2.17	2.06	1.94	1.88	1.82	1.74	1.67	1.58	1.48
120	5.15	3.80	3.23	2.89	2.67	2.52	2.39	2.30	2.22	2.16	2.05	1.94	1.82	1.76	1.69	1.61	1.53	1.43	1.31
∞	5.02	3.69	3.12	2.79	2.57	2.41	2.29	2.19	2.11	2.05	1.94	1.83	1.71	1.64	1.57	1.48	1.29	1.27	1.00

Table 7 Area under the normal curve from $-\infty$ to mean $+z$ standard deviations

z_p	0.00	0.01	0.02	0.03	0.04	0.05	0.06	0.07	0.08	0.09
0.0	0.5000	0.5040	0.5080	0.5120	0.5160	0.5199	0.5239	0.5279	0.5319	0.5359
0.1	0.5398	0.5438	0.5478	0.5517	0.5557	0.5596	0.5636	0.5675	0.5714	0.5753
0.2	0.5793	0.5832	0.5871	0.5910	0.5948	0.5987	0.6026	0.6064	0.6103	0.6141
0.3	0.6179	0.6217	0.6255	0.6293	0.6331	0.6398	0.6406	0.6443	0.6480	0.6517
0.4	0.6554	0.6591	0.6628	0.6664	0.6700	0.6736	0.6772	0.6808	0.6844	0.6879
0.5	0.6915	0.6950	0.6985	0.7019	0.7054	0.7088	0.7123	0.7157	0.7190	0.7224
0.6	0.7257	0.7291	0.7324	0.7357	0.7389	0.7422	0.7454	0.7486	0.7517	0.7549
0.7	0.7580	0.7611	0.7642	0.7673	0.7704	0.7734	0.7764	0.7794	0.7823	0.7852
0.8	0.7881	0.7910	0.7939	0.7967	0.7995	0.8023	0.8051	0.8078	0.8106	0.8133
0.9	0.8159	0.8186	0.8212	0.8238	0.8264	0.8289	0.8315	0.8340	0.8365	0.8389
1.0	0.8413	0.8438	0.8461	0.8485	0.8508	0.8531	0.8554	0.8577	0.8599	0.8621
1.1	0.8643	0.8665	0.8686	0.8708	0.8729	0.8749	0.8770	0.8790	0.8810	0.8820
1.2	0.8849	0.8869	0.8888	0.8907	0.8925	0.8944	0.8962	0.8980	0.8997	0.9015
1.3	0.9032	0.9049	0.9066	0.9082	0.9099	0.9115	0.9131	0.9147	0.9162	0.9177
1.4	0.9192	0.9207	0.9222	0.9236	0.9251	0.9265	0.9279	0.9292	0.9306	0.9319
1.5	0.9332	0.9345	0.9257	0.9270	0.9382	0.9394	0.9406	0.9418	0.9429	0.9441
1.6	0.9452	0.9463	0.9474	0.9484	0.9495	0.9505	0.9515	0.9525	0.9535	0.9545
1.7	0.9554	0.9564	0.9573	0.9582	0.9591	0.9599	0.9608	0.9616	0.9625	0.9633
1.8	0.9641	0.9649	0.9656	0.9664	0.9671	0.9678	0.9686	0.9693	0.9699	0.9706
1.9	0.9713	0.9719	0.9726	0.9732	0.9738	0.9744	0.9750	0.9756	0.9761	0.9767
2.0	0.9772	0.9778	0.9783	0.9788	0.9793	0.9798	0.9803	0.9808	0.9812	0.9817
2.1	0.9821	0.9826	0.9830	0.9834	0.9838	0.9842	0.9846	0.9850	0.9854	0.9857
2.2	0.9861	0.9864	0.9868	0.9871	0.9875	0.9878	0.9881	0.9884	0.9887	0.9890
2.3	0.9893	0.9896	0.9898	0.9901	0.9904	0.9906	0.9909	0.9911	0.9913	0.9916
2.4	0.9918	0.9920	0.9922	0.9925	0.9927	0.9929	0.9931	0.9932	0.9934	0.9936
2.5	0.9938	0.9940	0.9941	0.9943	0.9945	0.9946	0.9948	0.9949	0.9951	0.9952
2.6	0.9953	0.9955	0.9956	0.9957	0.9959	0.9960	0.9961	0.9962	0.9963	0.9964
2.7	0.9965	0.9966	0.9967	0.9968	0.9969	0.9970	0.9971	0.9972	0.9973	0.9974
2.8	0.9974	0.9975	0.9976	0.9977	0.9977	0.9978	0.9979	0.9979	0.9980	0.9981
2.9	0.9981	0.9982	0.9982	0.9983	0.9984	0.9984	0.9985	0.9985	0.9986	0.9986
3.0	0.9987	0.9987	0.9987	0.9988	0.9988	0.9989	0.9989	0.9989	0.9990	0.9990
3.1	0.9990	0.9991	0.9991	0.9991	0.9992	0.9992	0.9992	0.9992	0.9993	0.9993
3.2	0.9993	0.9993	0.9994	0.9994	0.9994	0.9994	0.9994	0.9995	0.9995	0.9995
3.3	0.9995	0.9995	0.9995	0.9996	0.9996	0.9996	0.9996	0.9996	0.9996	0.9997
3.4	0.9997	0.9997	0.9997	0.9997	0.9997	0.9997	0.9997	0.9997	0.9997	0.9998

Note: For negative values of z, subtract the tabular value of area from unity.

Interval	(ksi)	f_o	f_e	$(f_o - f_e)^2/f_e$
0 to 0.1 $(-\infty$ to $\bar{X} - 1.282s)$	Up to 144.84	37	33	0.48
0.1 to 0.2 $(\bar{X} - 1.282s$ to $\bar{X} - 0.842s)$	144.84 to 146.65	26	33	1.49
0.2 to 0.3 $(\bar{X} - 0.842s$ to $\bar{X} - 0.524s)$	146.65 to 147.95	31	33	0.12
0.3 to 0.4 $(\bar{X} - 0.524s$ to $\bar{X} - 0.253s)$	147.95 to 149.06	39	33	1.09
0.4 to 0.5 $(\bar{X} - 0.253s$ to $\bar{X})$	149.06 to 150.10	29	33	0.48
0.5 to 0.6 $(\bar{X}$ to $x + 0.253s)$	150.10 to 151.14	33	33	0.00
0.6 to 0.7 $(\bar{X} + 0.253s$ to $\bar{X} + 0.524s)$	151.14 to 152.25	38	33	0.76
0.7 to 0.8 $(\bar{X} + 0.524s$ to $\bar{X} + 0.842s)$	152.25 to 153.55	27	33	1.09
0.8 to 0.9 $(\bar{X} + 0.842s$ to $\bar{X} + 1.282s)$	153.55 to 155.36	36	33	0.27
0.9 to 1.0 $(\bar{X} + 1.282s$ to $\infty)$	155.36 and over	34	33	0.03
	Total	330	330	5.81

The total of $(f_o - f_e)^2/f_e$ for the ten intervals is the chi-squared value for this test and is equal to 5.81. Table 8 shows that the 0.95 fractile of the chi-squared distribution for $10 - 3 = 7$ degrees of freedom is 14.07. Because the computed value of 5.81 is less than the tabulated value of 14.07, it is concluded that the distribution is normal.

Step 2. Compute F_{tu} (LT), 0.020 to 0.125, for Alloy X, using procedures for the normal distribution.

Population	n	$\bar{X}$, ksi	s, ksi
Groups 1 and 2 TUS (LT), 0.020 to 0.125	330	150.1	4.10

$k_A = 2.512$
$k_B = 1.410$
F_{tu} (LT), A basis $= \bar{X} - k_A s = 150.1 - 2.512 \times 4.10$
$\quad = 139.8$ or 140 ksi
F_{tu} (LT), B basis $= \bar{X} - k_B s = 150.1 - 1.410 \times 4.10$
$\quad = 144.3$ or 144 ksi

Problem 5

What Computational Method Should Be Used for the Combined Observations of Groups 3 and 4?

This property does not vary with thickness. Sample statistics for the combined observations are:

Population	n	$\bar{X}$, ksi	s, ksi
Groups 3 and 4 TYS (LT), 0.020 to 0.125	330	130.1	4.10

Table 8 The 0.95 fractiles of the chi-squared distribution associated with d.f.

d.f.	$\chi^2_{0.95}$	d.f.	$\chi^2_{0.95}$
1	3.84	16	26.30
2	5.99	17	27.59
3	7.81	18	28.87
4	9.49	19	30.14
5	11.07	20	31.41
6	12.59	21	32.67
7	14.07	22	33.92
8	15.51	23	35.17
9	16.92	24	36.42
10	18.31	25	37.65
11	19.68	26	38.88
12	21.03	27	40.11
13	22.36	28	41.34
14	23.68	29	42.56
15	25.00	30	43.77

Form of the distribution has not been determined. The sample is large enough to permit direct computation of A and B values. Consequently, the computational method to be used will be determined by whether or not the observations are normally distributed.

Step 1. Test to determine whether or not the distribution is normal. In this example, it will be assumed that the distribution was lopsided or "skewed" and that the chi-squared value determined as in the preceding example is 30.00, indicating that the distribution cannot be assumed to be normal. For TUS or TYS, if the test for normality fails, one for log-normality will probably fail also; thus, this will be treated as an unknown distributional form.

Step 2. Compute F_{ty} (LT), 0.020 to 0.125, using procedures for the unknown distribution. This procedure requires the ranking of observations from lowest to highest. Referring to Table 3, it is found that for a sample size of 330, the lowest observation (rank = 1) is an A value, and the 24th lowest (rank = 24) is a B value. The 24 lowest observations are:

Rank	TYS, ksi	Rank	TYS, ksi	Rank	TYS, ksi
1	115.6	9	119.6	17	121.0
2	116.4	10	119.8	18	121.1
3	116.6	11	119.9	19	121.2
4	116.9	12	120.2	20	121.5
5	117.9	13	120.4	21	121.5
6	118.4	14	120.5	22	121.6
7	118.5	15	120.9	23	121.9
8	119.2	16	121.0	24	122.0

Consequently, from these data, the following allowables have been computed for Alloy X:

$$F_{ty} \text{ (LT), A basis} = 115.6 \text{ ksi}$$
$$F_{ty} \text{ (LT), B basis} = 122.0 \text{ ksi}$$

Problem 6

What Computational Procedure Should Be Used for the Observations in Group 6?

The data in Group 6 represent a borderline situation. They cannot be combined with data for lesser thicknesses, because there is a significant difference between the TUS(LT) averages for the two thickness ranges, as shown in Problem 3. The sample size is just barely adequate for direct computation if the distribution is found to be normal. If the distribution is not normal, the properties for this product would be presented on an S basis, pending the accumulation of more data. The test for normality would be conducted as described in Problem 4 and will not be illustrated here.

Problem 7

What Computational Procedure Should Be Used for the Observations in Group 5?

SUS(LT) decreases with increasing thickness, while TUS(LT) does not vary with thickness. Sample statistics are:

Population	n	$\bar{X}$, ksi	s, ksi
Group 5 SUS (LT), 0.020 to 0.249	30	Not determined	

The sample size for these data is too small to permit direct computation. Thus, the procedure that should be used is indirect computation by pairing observations of SUS(LT) with observations of TUS(LT). Also, because a thickness effect was suspected in the original data, a regression against thickness should be made and checked for significance.

Step 1. Pair SUS(LT) with TUS(LT). Ratios of SUS(LT)/TUS(LT) are plotted against thickness in Fig. 8.

Step 2. Determine regression equation in the form $(SUS/TUS)' = r' = a + bx$, where $x =$ thickness, using least squares techniques. The following sums were obtained from analysis of the ratios plotted in Fig. 8:

Number of ratios, $n = 30$			
$\Sigma(x)$ = 2.94		$\Sigma(xr)$ = 1.8723	
$\Sigma(x^2)$ = 0.4260		$(\Sigma x)^2$ = 8.6436	
$\Sigma(r)$ = 19.53		$(\Sigma r)^2$ = 381.4209	
$\Sigma(r^2)$ = 12.7319		$(\Sigma x)(\Sigma r)$ = 57.4182	

$$\text{Slope, } b = \frac{(\Sigma \times r) - \dfrac{(\Sigma x)(\Sigma r)}{n}}{(\Sigma x^2) - \dfrac{(\Sigma x)^2}{n}}$$

$$= \frac{1.8723 - 57.4182/30}{0.4260 - 8.6436/30} = -0.302$$

$$\text{Intercept, } a = \frac{\Sigma r - b(\Sigma x)}{n} = \frac{19.53 - (-0.302)(2.94)}{30}$$

$$= 0.6806$$

Fig. 8 Ratios of input data for Problem 7

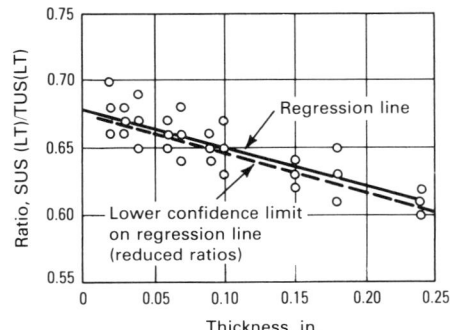

Standard error of estimate:

$$s_r' = \sqrt{\frac{\Sigma(r^2) - (\Sigma r)^2/n - \dfrac{[\Sigma(xr) - (\Sigma x)(\Sigma r)/n]^2}{[\Sigma(x^2) - (\Sigma x)^2/n]}}{(n-2)}}$$

$$= \sqrt{\frac{[12.7319 - 381.4209/30] - \dfrac{[1.8723 - 57.4182/30]^2}{[0.4260 - 8.6436/30]}}{(30-2)}}$$

$$= 0.014$$

The equation of the regression line is $r' = 0.6806 - 0.302x$. The regression line is shown in Fig. 8.

Step 3. Perform an analysis of variance to check the significance and linearity of the regression: The only information missing from Step 2 required for the analysis of variance is the values of T, or the summed values of r for each x. They are:

x_i	T_i	x_i	T_i
0.02	2.04	0.09	1.95
0.03	2.01	0.10	1.95
0.04	2.01	0.15	1.89
0.06	1.98	0.18	1.89
0.07	1.98	0.24	1.83

Using these values, the analysis of variance can be completed as illustrated in Fig. 3 as follows:

Source of variation	Sum of squares	Degrees of freedom	Mean squares	F_{calc}
Treatments	0.0130	9	...	...
Linear regression	0.0126	1	0.0126	51.4
Deviation from linearity	0.0004	8	0.00005	0.208
Error	0.0049	20	0.00024	...
Total	0.0179	29	...	...

The first calculated F statistic of 51.4 exceeds the value required by Fig. 4 for a significant regression with 95% confidence, and the second F statistic lies below the value required by Fig. 5 for a linear regression. Consequently, both conditions are satisfied; thus, the linear relationship and regression are significant.

Step 4. Compute the ratio F_{su}/F_{tu}. In performing this step, the reduced ratio will be computed at each of four thicknesses (0.020, 0.062, 0.125, and 0.249 in.). This is done by determining the lower confidence limit for the regression line at the desired thicknesses. The computation will be worked in detail for $x_o = 0.020$ in.:

$$\text{Reduced ratio} = (\text{SUS/TUS}) - t_{0.95}\, s_r' \sqrt{\frac{1}{n} + \frac{(x_o - \Sigma x/n)^2}{(\Sigma x^2) - (\Sigma x)^2/n}}$$

$$(\text{SUS/TUS})' = r' = 0.681 - 0.302x_o \text{ (from Step 2, Problem 7)}$$
$$= 0.681 - 0.302 \times 0.020 = 0.6746$$

$t_{0.95}$ (for $n - 2$) $= 30 - 2 = 28$ degrees of freedom
$$= 1.701 \text{ (from Table 5)}$$

$s_r' = 0.014$ (from Step 2)

$$\sqrt{\frac{1}{n} + \frac{(x_o - \Sigma x/n)^2}{(\Sigma x^2) - (\Sigma x)^2/n}}$$

$$= \sqrt{\frac{1}{30} + \frac{(0.020 - 2.94/30)^2}{0.4260 - 8.6436/30}} = 0.2783$$

Reduced ratio $= 0.6746 - 1.701 \times 0.014 \times 0.2783 = 0.668$

The corresponding ratios for the other thicknesses are tabulated in Step 5.

Step 5. Compute F_{su}. This computation will be illustrated for a thickness of 0.020 in., using the reduced ratio from Step 4. From Problem 4:

$$F_{tu} \text{ (LT)} = 140 \text{ ksi (A basis)}$$
$$F_{tu} \text{ (LT)} = 144 \text{ ksi (B basis)}$$
$$F_{su} \text{ (LT)} = \text{reduced ratio} \times F_{tu} \text{ (LT)}$$
$$F_{su} \text{ (LT) (A basis)} = 0.668 \times 140 = 93.5 \text{ ksi}$$
$$F_{su} \text{ (LT) (B basis)} = 0.668 \times 144 = 96.2 \text{ ksi}$$

For the four thicknesses listed:

t, in.	Reduced ratio	F_{su} (LT), ksi — A basis	B basis	S basis
0.020	0.668	93.5	96.2	...
0.062	0.657	92.0	94.6	...
0.125	0.638	89.3	91.9	...
0.249	0.595	...	...	80.3

Because F_{su} is shown to decrease with increasing thickness, only the lowest value applicable to the range should be used. By dividing the 0.020 to 0.125 in. thickness range into two ranges, a somewhat higher F_{su} (LT) value may be presented for thinner material as shown below.

The results of the computations in Problems 1 through 7 have produced the following results (fractions greater than 0.75 usually are raised to the next higher ksi, while lesser fractions are dropped):

Mechanical property	<0.020 S basis	0.020 to 0.062 A basis	B basis	0.063 to 0.125 A basis	B basis	0.126 to 0.249 S basis
F_{tu} (LT), ksi	140	140	144	140	144	135
F_{ty} (LT), ksi	115	115	122	115	122	110
F_{su}, ksi	...	92	94	89	92	80

Fatigue Crack Growth Data Analysis

By Alan P. Berens
Senior Research Statistician
University of Dayton Research Institute

FATIGUE CRACK GROWTH RATE TESTING AND DATA ANALYSIS are performed to characterize the crack propagation resistance of material-environment combinations in order to predict crack growth life under anticipated stress histories. When a structure is subjected to cyclic loads that are greater than a threshold, but not large enough to cause fracture, fatigue cracks initiate, propagate, and eventually lead to failure. In cases involving limited plasticity, it has been shown that the crack propagation rate, da/dN, during the subcritical crack growth phase can be correlated with a stress intensity factor, K. This correlation permits transferring crack growth rates from laboratory specimens to estimates of crack growth life of real structures subjected to variable amplitude loads.

Analyses performed on the numerical output of crack growth rate tests, including the analysis framework for modeling fatigue crack growth rate data, numerical methods for calculating da/dN as a function of stress intensity factor, and principles in fatigue crack growth damage analysis are discussed in this article. For a discussion of fatigue crack growth rate test methods, see the article on "Fatigue Crack Propagation" in this Volume.

Analysis

The basic linear elastic fracture mechanics equation for correlating crack propagation with the stress intensity factor is (Ref 1):

$$\frac{da}{dN} = f(\Delta K) \qquad \text{(Eq 1)}$$

where da/dN is crack extension during the nth stress cycle. The stress intensity factor, K, is a measure of the magnitude of the stress field in the vicinity of the crack tip. The stress intensity factor is a function of the remote stress, σ, the geometry, and the crack length, a. The general form of the function is:

$$K = \sigma \cdot \beta(a) \cdot \sqrt{\pi a}$$

where $\beta(a)$ is a geometry factor. The fatigue crack growth rate correlates with the range of the stress intensity ΔK:

$$\Delta K = K_{max} - K_{min}$$

This gives

$$\Delta K = \Delta\sigma \cdot \beta(a) \cdot \sqrt{\pi a} \qquad \text{(Eq 2)}$$

where

$$\Delta\sigma = \sigma_{max} - \sigma_{min}$$

Characterizing the fatigue crack growth rate for a material is a three-step process. Constant-amplitude fatigue tests are performed on specimens with known crack length/stress intensity relationships. Typically, these are compact-type or center-cracked-tension specimens (Fig. 1). Crack length measurements are generally made at known numbers of total stress cycles. The results of fatigue crack growth rate tests comprise a pair of data points, (a_i, N_i), which represent crack length as a function of the number of stress cycles when plotted. Figure 2 is an example of an a versus N plot, as obtained from a 10Ni-8Co-1Mo steel compact-type specimen.

Because the fracture mechanics characterization is expressed in terms of da/dN versus ΔK, the a versus N data must be reduced to this format. The most commonly used numerical techniques for accomplishing this are discussed in the next section of this article. Figure 3 presents da/dN versus ΔK for the data of Fig. 2, using the numerical method known as the incremental polynomial method.

The $f(\Delta K)$ function of Eq 1 is then obtained from the discrete da/dN versus ΔK data points. Typically, all data points from comparable tests are combined, and the parameters of a specific crack growth model are estimated or a smoothed empirical model is

Fig. 1 Fatigue crack growth rate test specimens

a, crack length; a_n, notch length; B, thickness; D, diameter; W, width; W_1, length; W_2, notch-to-diameter width; W_3, notch-to-outside width. (a) Compact-type specimen. (b) Center-cracked tension specimen

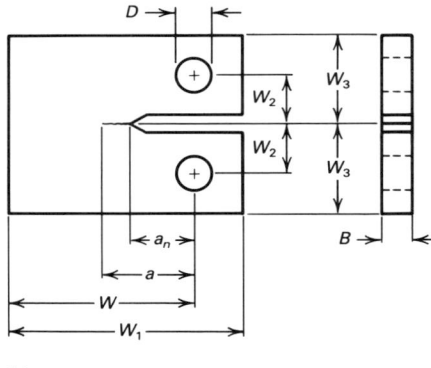

(a)

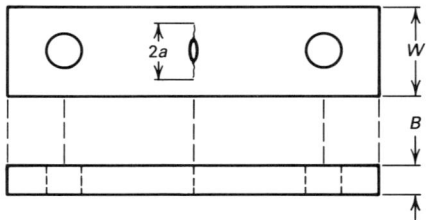

(b)

Fig. 2 Crack growth versus constant-amplitude stress cycles

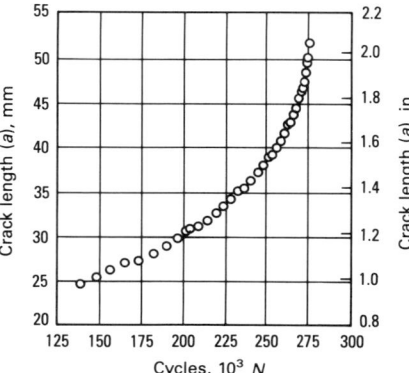

fit to the data. The resulting $f(\Delta K)$ value is considered representative of the crack propagation rate. Models and methods for estimating $f(\Delta K)$ are presented later in this article.

The a versus N behavior exhibited in Fig. 2 is specific to the material, specimen geometry, stress levels, and experimental conditions of the test. In addition, the experimental data reflect error associated with crack length measurements, variability of crack growth within a specimen, and the variability associated with the particular specimen under test. An "identical" test of a second specimen would yield somewhat different results, the difference depending on the homogeneity of the material and the care taken in testing. Although such variability will always exist, the testing and analysis procedures must be directed at reducing variability that is not related to the material and measuring the variability that is important to a comprehensive characterization of the material's resistance to crack growth.

Random errors associated with a particular specimen and with the uncontrolled sources of variation can be modeled:

$$\frac{da}{dN} = f[\Delta K, \epsilon_j(\Delta K), \delta_j(\Delta K)] \qquad \text{(Eq 3)}$$

where $\epsilon_j(\Delta K)$ is the random effect due to specimen j at a stress intensity of ΔK, and $\delta_j(\Delta K)$ is the random effect at a stress intensity of ΔK due to all sources other than specimen j during the test of specimen j. In other words, ϵ_j effects are those caused primarily by material differences between specimens, and δ_j effects are caused by experimental errors during a single test.

Separate terms are introduced for the between-specimen and within-specimen random effects to demonstrate the data processing effects in reducing the within-experiment random errors and to emphasize the need for multiple test specimens in the characterization of the variability of crack growth rates of

a material. Because crack growth rate data are usually plotted on a log-log scale, one form of Eq 3 that has been used (Ref 2, 3) is given by:

$$\log \frac{da}{dN} =$$

$$f(\Delta K) + \epsilon_j(\Delta K) + \delta_j(\Delta K) \qquad \text{(Eq 4)}$$

The normal distribution has been used as a model for describing the scatter about $f(\Delta K)$ in this equation; i.e., $\log da/dN$ will have a normal distribution with a median value of $f(\Delta K)$.

The magnitude of the within-specimen random component, δ, is determined not only by variations in experimental conditions, but also by the random errors in crack length measurement, the increment between crack length measurements, and the data processing methods. The use of rigid controls during any experiment is an established method of reducing test-to-test variability. In fatigue crack growth rate testing, special emphasis must be placed on controlling stress levels. The actual applied stress affects the rate of crack growth, while the perceived crack length affects the stress intensity factor. Therefore, errors in applied stress can produce significant bias in the crack growth rate characterization. Fluctuations about preset norms in other experimental conditions (e.g., temperature, humidity, and frequency) generally produce only second-order effects in crack growth rate characterization.

The effects of random errors associated with crack length measurement cannot be isolated from those of the increment between measurements and the method of data processing. Because the data are eventually reduced to derivatives of the a versus N data, additive biases cancel each other. Furthermore, the numerical methods used to generate da/dN versus ΔK data points filter the random errors depending on the numerical technique used. Finally, the da/dN versus ΔK data points are further smoothed in the generation of the $f(\Delta K)$ function. This last step frequently combines data from multiple specimens. Therefore, the processing methods reduce the effect of the random errors in the a versus N data. Generally, the crack growth increments between readings, Δa, should exceed ten times the estimate of the measurement precision (Ref 4).

When multiple specimens are fatigued under identical conditions, variations in da/dN values at fixed ΔK are observed (Ref 5). Although one component of this variation, δ, can be correlated to the variability within the experiment, a second component, ϵ, can be attributed to the effects of differences in material composition and physical fabrication of the specimens. Therefore, a distribution of

Fig. 3 Crack growth rate as a function of stress intensity factor

a versus N data from Fig. 2 reduced using the incremental polynomial method

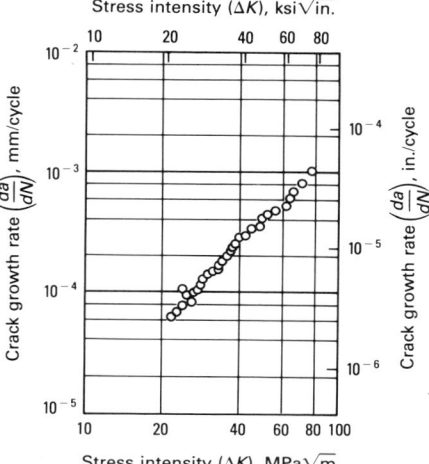

da/dN values at each ΔK can be postulated for an experimental condition.

The analysis methods currently used to characterize crack growth rates have estimated the mean da/dN as a function of ΔK (i.e., $f(\Delta K)$ of Eq 4). Methods for modeling the variability about this mean function are not well established. In particular, statistical methods are needed for measuring the precision of the estimate of the da/dN versus ΔK as a function of sample size (number of test specimens) and for correlating variations in crack growth life predictions with variations in crack growth rates. Considering the current state of analysis methods, caution should be exercised in interpreting scatter in da/dN values.

Crack Growth Rate Calculation

A number of different numerical techniques have been used to calculate crack growth rates from the set of (a_i, N_i) data points of a given crack growth rate test. The secant and incremental polynomial methods (see discussion below) are the most widely used. When the data are processed to a final smoothed $da/dN = f(\Delta K)$ format, these methods provide approximately equivalent results. However, the scatter of individual da/dN values about the average depends greatly on the data reduction method.

Analysis of the analytical techniques discussed below for calculating crack growth rates (Ref 6) indicates that the apparent variability in the resulting da/dN value depends on the method. However, none of the methods introduces a significant bias to an overall mean trend curve. Figure 4 plots da/dN versus ΔK, as calculated from the data of Fig. 2, using the secant, incremental polynomial, and five-point modified difference methods. Because methods of analyzing and interpreting the scatter in da/dN are not currently available, the simpler techniques—i.e., the secant and incremental polynomial methods—are often chosen.

Secant Method. In the secant method for differentiating the a versus N data, the average crack extension per cycle is calculated for each pair of data points, and ΔK is calculated at the midpoint of the crack lengths:

$$\left.\frac{da}{dN}\right|_{\hat{a}_i} = \frac{a_{i+1} - a_i}{N_{i+1} - N_i} \qquad \text{(Eq 5)}$$

$$\hat{a}_i = \frac{a_{i+1} + a_i}{2} \qquad \text{(Eq 6)}$$

This method is simple, but exhibits the most scatter in da/dN values.

The incremental polynomial method in practice is actually the incremental second-order polynomial method. A least squares, second-order polynomial is obtained for successive sets of $(2k + 1)$ data points:

$$\hat{a}_i = b_0 + b_1\left(\frac{N_i - C_1}{C_2}\right) + b_2\left(\frac{N_i - C_1}{C_2}\right)^2 \qquad \text{(Eq 7)}$$

where $C_1 = (N_{i+k} + N_{i-k})/2$ and $C_2 = (N_{i+k} - N_{i-k})/2$ are centering and scaling constants, respectively, that are introduced to prevent numerical problems in obtaining the least squares fit. The crack growth rate at $\hat{a}_i$, the predicted central crack length at N_i, is given by the derivative of Eq 7:

$$\left.\frac{da}{dN}\right|_{\hat{a}_i} = \frac{b_1}{C_2} + 2b_2\frac{(N_i - C_1)^2}{C_2} \qquad \text{(Eq 8)}$$

Typical values of k are 1, 2, or 3, resulting in the second-order polynomial being estimated on the basis of 3, 5, or 7 successive data points. The estimated crack growth rate function loses k data points at each end. Less apparent scatter is obtained for larger k values.

In the graphical method, a smooth, subjectively determined curve is drawn through the (a_i, N_i) data points, and the slope of the tangent line at the desired $\hat{a}_i$ values is calculated.

The modified difference methods are finite difference techniques for estimating the derivative at the midpoint of a data set. These methods use numerical derivatives (di-

Fig. 4 Comparison of da/dN calculation methods

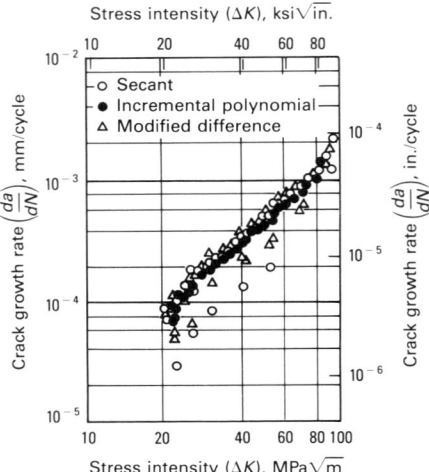

vided differences) to fit a polynomial to the data. Alternative methods exist for combining the numerical derivatives. The formula for estimating the derivative at the midpoint, a_i, of three successive data points (Ref 7) is given by:

$$\left.\frac{da}{dN}\right|_{a_i} = \frac{a_i - a_{i-1}}{N_i - N_{i-1}} + \left[\frac{N_i - N_{i-1}}{N_{i+1} - N_{i-1}}\right]$$

$$\left[\frac{a_{i+1} - a_i}{N_{i+1} - N_i} - \frac{a_i - a_{i-1}}{N_i - N_{i-1}}\right] \qquad \text{(Eq 9)}$$

The derivative would be estimated at each successive set of three points. The formula for estimating the derivative at the center of five data points (Ref 7) is given by:

$$\left.\frac{da}{dN}\right|_{a_i} = \frac{1}{hA_2}\left[-a_{i+2}\frac{A_3}{B_0}\right.$$

$$+ a_{i+1}\left(\frac{1}{A_0} + \frac{A_3B_2}{B_0}\right)$$

$$+ a_i\left(\frac{A_3B_1}{B_0} - \frac{A_1}{C_0}\right)$$

$$+ a_{i-1}\left(\frac{A_1C_2}{C_0} - \frac{1}{A_0}\right)$$

$$\left. + a_{i-2}\left(\frac{A_1C_1}{C_0}\right)\right] \qquad \text{(Eq 10)}$$

where $h = N_i - N_{i-1}$; $a = (N_{i+1} - N_i)/h$; $\beta = (N_{i+2} - N_{i+1})/\alpha h$; $\gamma = h/(N_{i-1} - N_{i-2})$; $A_0 = (\alpha + 1)/6\alpha$; $A_1 = \alpha(2 - \alpha)$; $A_2 = (\alpha + 1)^2$; $A_3 = 2\alpha - 1$; $B_0 = \alpha\beta(1 + \beta)$; $B_1 = \beta^2$; $B_2 = 1 - \beta^2$; $C_0 = (1 + \gamma)$; $C_1 = \gamma^2$; and $C_2 = 1 - \gamma^2$. The derivative is estimated at the first and each succeeding set of five consecutive data points. The method

cannot provide an estimate of da/dN at the first two or last two data points.

The total a versus N curve method assumes a single model, $a = g(N)$, for the entire set of (a_i, N_i) data points rather than segmenting the a versus N curve, as was done in the secant, incremental polynomial, and modified difference methods. The mathematical techniques used to estimate the parameters of the model depend on the particular form of the function (Ref 8). A variation of this method is to assume a particular model for the da/dN versus ΔK relationship and to estimate the parameters by a finite integral optimization calculation (Ref 9).

This method can be illustrated by integrating the equation:

$$\frac{da}{dN} = C(\Delta K)^b$$

where $\Delta K = f(\Delta\sigma, \beta(a), a)$, and making a least squares fit to the original a versus N data. In many cases the equation cannot be integrated in closed form, and a numerical method is required.

Stress Intensity Factor Calculation

Crack growth rates are correlated with the stress intensity factor, ΔK, which correlates the local stresses in the region of the crack tip with the remote stress, crack geometry, and structural geometry (Eq 2). The geometry factor, $\beta(a)$, depends on the crack length and the geometric configuration and dimensions of the test specimens. Figure 1 illustrates the compact-type and center-cracked tension specimens, which are most often used in fatigue crack growth rate testing. The specific relation between ΔK and a for the compact-type specimen depends on the values of the dimensions indicated in the figure. The equation for ΔK for the center-cracked tension specimen is given as:

$$\Delta K = \Delta\sigma\left[\pi a \sec\left(\frac{\pi a}{W}\right)\right]^{1/2} \qquad \text{(Eq 11)}$$

This equation is valid for $2a/W < 0.95$ and $R = \sigma_{min}/\sigma_{max} \geq 0$.

Crack Growth Rate Modeling

A log-log plot of the da/dN versus ΔK data generally has a sigmoidal shape, as illustrated in Fig. 5. The effect of R, the ratio of the minimum to the maximum stress, is also indicated, where $R = 0$ is commonly assigned to negative stress ratios. The earliest model for the crack growth rates was the Paris power law (Ref 1), expressed by the equation:

Fig. 5 Crack growth rate as a function of stress intensity factor and stress ratio

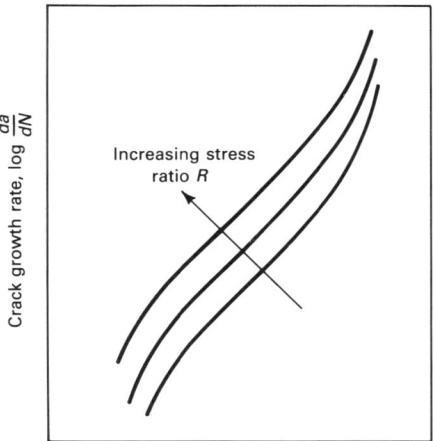

$$\frac{da}{dN} = C(\Delta K)^n \tag{Eq 12}$$

This model assumes a straight-line relation between log da/dN and log ΔK, which generally is a reasonable assumption in the mid-ΔK ranges.

To account for the stress ratio effect and to fit the sigmoidal shape at the high and low stress intensity factor levels, many models have been proposed. For example, to modify the Paris equation to account for the stress ratios, Walker (Ref 10) proposed:

$$\frac{da}{dN} = C\left[\frac{\Delta K}{(1-R)^{1-m}}\right]^n \tag{Eq 13}$$

The Walker model yields a series of parallel lines on a log-log plot, whose separation is a function of the stress ratio. The parameters m, n, and c are determined empirically from data. To partially account for the sigmoidal shape in the high ΔK region, Forman (Ref 11) proposed:

$$\frac{da}{dN} = \frac{C(\Delta K)^n}{[(1-R)K_c - \Delta K]} \tag{Eq 14}$$

where K_c is the critical stress intensity factor at which fracture occurs. To better model the stress ratio effect, the Forman equation was modified (Ref 12):

$$\frac{da}{dN} =$$

$$\frac{C[(1-R)^{1+m}\Delta K]^n}{[(1-R)^m K_c - (1-R)^{1+m}\Delta K]^L} \tag{Eq 15}$$

where m, L, n, and C are empirical constants.

Although this equation permits empirical modeling of the effect of R and fits the general shape of the da/dN versus ΔK values, it overpredicts crack growth rates at low ΔK values. Use of the above equations to predict crack growth for small cracks or at low stress levels produces conservative results.

To model crack growth rates over the entire stress intensity range, many mathematical functions have been proposed (Ref 13). These include four- and five-parameter inverse hyperbolic tangent models (Ref 12, 14), four-parameter hyperbolic sine models (Ref 15), five-parameter mixtures of exponentials (Ref 16), etc. Some of the models are expressed in terms of material properties (K_c and K threshold) and other parameters to be estimated from crack growth rate test data. Other models are expressed only in parameters to be estimated from the data. No model has universally fit all data or gained universal acceptance. However, some organizations have adopted particular equations for their specific applications (Ref 16).

Because crack growth rate characterizations are used to compare materials and to predict crack growth by means of computer integration, there is no need for an equation to describe the da/dN versus ΔK relation. Comparisons are made on the basis of plots, not parameter values of a model, and numerical integration can be efficiently and accurately performed with the da/dN versus ΔK data for each stress ratio stored in a tabular format. The incremented crack growth for a stress cycle is obtained by interpolating or extrapolating between stress ratios given in tabular form.

A round-robin comparison of the effect of nine crack growth rate models on crack growth prediction was performed (Ref 16). No analytical model performed better than the table look-up procedure, and only one of the analytical models (with five parameters estimated for each stress ratio) did as well. Therefore, there is a definite trend against the use of a mathematical function to represent the crack growth rate data.

Methods for estimating the parameters of a crack growth rate model depend on the specific equation of the model. The parameters of the simple Paris model (Eq 12) can be estimated by simple least squares methods, while those of the more complex models require nonlinear regression methods. All valid da/dN versus ΔK data points for comparable tests are combined in defining the relationship.

A mean trend fit of da/dN versus ΔK for use in the table look-up method, i.e., the "no-model" approach, can be obtained from a simple visual fit to a plot of the data points. The subjectivity of a visual fit can be removed by fitting a least squares spline function to the data (Ref 17). The spline function fit is obtained by segmenting the data points into ΔK ranges and fitting a polynomial to each segment. The algorithm is written to ensure a smooth transition between segments by constraining the polynomials to have equal second derivatives of the common abscissas (knots). The spline function fit is often considered to be an analytical french curve fit. For fitting crack growth rate data, cubic polynomials with five or fewer knots have proven sufficient.

Figure 6 illustrates a spline function fit to crack growth data from type 2219-T851 aluminum at $R = 0.3$. The values for the table look-up procedure can be determined by evaluating the spline function at the desired set of ΔK values. The tables can be found in Ref 17.

Principles of Damage Analysis

The representation of fatigue crack growth rate as a function of stress intensity factor can be used to calculate the growth of real or potential damage in a structure. In addition to the da/dN versus ΔK representation for the environment of interest, the process requires (1) a representation of the predicted or actual stress cycles, (2) the crack size/stress intensity factor relationship for the structural detail of interest, (3) a model that takes into account load interactions (if necessary), and (4) an integration routine. Although the following paragraphs present a brief overview of these topics and the use of the crack growth life curve, a detailed discussion can be found in Ref 18. Whenever possible, crack growth predictions should be verified by tests.

Stress Sequences. The service stress histories experienced by a structural detail will, in general, result from variable ampli-

Fig. 6 Cubic spline curve fit to fatigue crack growth rate data for 2219-T851 aluminum at a stress ratio of $R = 0.3$

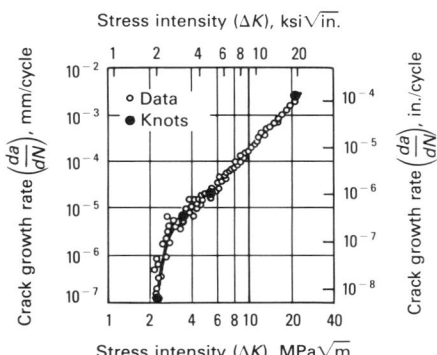

tude loading rather than constant-amplitude loading. Before a crack growth calculation can be made, a representation of the variable-amplitude stress history that summarizes the history in terms of a sequence of complete stress cycles in proper order is necessary. Two distinct approaches to this problem have been used: the equivalent constant-amplitude stress approach and the cycle-by-cycle approach.

If the stress history can be represented as a stationary random process about a relatively constant mean stress, a root-mean-square type of characterization of stress history provides reasonable crack growth predictions (Ref 19, 20). In this approach, ΔK is replaced by ΔK_{rms}, where ΔK_{rms} is the root-mean-square type of stress intensity for the load history.

In many applications, service stress histories are highly irregular, and sequence effects can significantly affect the crack growth calculations. To account for sequence effects, the history is represented by a sequence of stress cycles stored in a computer, and the crack growth calculation is made on a cycle-by-cycle basis. However, resolving a complex stress history to a sequence of cycles can be performed in several ways (Ref 21-23).

The rainflow method (Ref 23) for associating minimum and maximum values is generally considered the most representative for effects on crack growth. This method describes an imaginary "rainflow" down a vertically plotted stress history. The most positive maximums are associated with the most negative minimums for each segment of the history. It assumes that crack growth occurs during the increasing portions of the cycle and that ordering is determined by the order of the observed maxima. Other constraints defined by operational use patterns are typically imposed in the generation of the stress sequences. For example, the load history of an airplane structure must first reflect the load sequence within a flight and then the flight-to-flight variation of different missions.

Stress intensity factors in the structure depend not only on crack length and stress range, but also on the geometry of the crack as well as the structural detail being analyzed. Closed form solutions for stress intensity factors have been obtained for many typical crack geometries (Ref 24, 25). However, when existing solutions do not exist because of complex geometry, boundary conditions, or load transfer, the stress intensity factor can be obtained from finite-element analysis (Ref 26-28).

Load Interactions. When a high-load cycle occurs in the middle of a sequence of smaller load cycles, the effect on crack growth of the cycles immediately following

the large cycle is often reduced. This crack retardation effect is due to compressive residual stresses in front of the crack tip, or to crack closure. Similarly, a compressive overload can eliminate the beneficial effects of the tensile overload, or can cause crack growth acceleration. These effects have been observed in variable-amplitude load histories (Ref 29, 30) and should be accounted for in crack growth predictions. A number of models have been proposed for load interaction effects (Ref 31-37), and most computation routines incorporate at least one such model.

Integration Routine. The crack growth life curve is obtained by integrating over the desired number of cycles, or until a predetermined crack size is reached:

$$a = a_o + \sum_{j=1}^{N} \Delta a_j \qquad \text{(Eq 16)}$$

and

$$\Delta a_j = f(\Delta K_j, R_j) \qquad \text{(Eq 17)}$$

where ΔK_j is the incremental stress intensity factor during the jth cycle, and R_j is the stress ratio during the jth cycle.

Numerous computer programs are available to perform this integration with select or alternative choices for stress cycle representation, stress intensity factor solution format, and load interaction effect models. See Ref 38 to 42 for brief discussions of several routines.

Although the cycle-by-cycle integration of Eq 16 and 17 represents the most accurate prediction of crack growth life, extensive computer time may be required to complete the task. An alternative approach has been devised (Ref 43) in which crack growth over a miniblock (e.g., a flight) of stress sequences is correlated with a statistically defined stress intensity factor. This calculation method permits multiple analyses to be made on the basis of one crack-by-crack analysis.

Application. The crack growth rate, as a function of stress intensity factor and stress ratio, can be used to calculate (predict) the growth of real or potential damage in a structural detail. Figure 7 illustrates this process for three severities of load sequences. The initial damage size, a_o, is the largest equivalent crack length that could be (or is) present at the critical location at the beginning of a usage period. As loading cycles are encountered, the crack will grow, and the residual strength of the structure will decrease. When the crack reaches the size a_f, for which an applied load produces the critical stress intensity factor, K_c, fracture is predicted.

The predicted crack growth curves, as a function of time in the usage environment, provide the basis for a number of potential

Fig. 7 Predicted crack growth

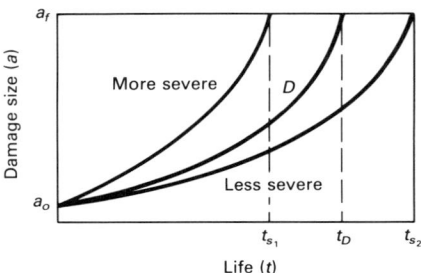

applications. Among these are (1) comparing materials for specific applications, (2) evaluating design configurations, (3) evaluating damage tolerance (the ability of a structure to perform its function in the presence of damage for a predefined operational period), (4) demonstrating compliance with warranty requirements or specifications, (5) calculating life estimates, and (6) scheduling inspection, repair, or replacement intervals.

Crack growth predictions discussed in this article are based on deterministic methods. In reality, crack growth lives are statistical in nature and the variability of life predictions should also be evaluated. Probabilistic fracture mechanics (Ref 44-46) can be used to predict distributions of crack sizes at fixed intervals or distributions of usage times to reach a known crack size. These distributions, in turn, provide the basis for evaluating structural maintenance costs (Ref 44, 46-48) and the risks of structural failure (Ref 44, 49).

REFERENCES

1. P. Paris and F. Erdogan, A Critical Analysis of Crack Propagation Laws, *Trans. ASME, J. Basic Eng.*, Vol 85, 1963, p 528-534
2. A.P. Berens, J.P. Gallagher, and P.W. Hovey, Statistics of Crack Growth, *Proceedings of the Ninth U.S. National Congress of Applied Mechanics*, Cornell University, Ithaca, NY, June 1982
3. P.W. Hovey, A.P. Berens, and J.P. Gallagher, Estimating the Statistical Properties of Crack Growth for Small Cracks, *Eng. Fract. Mech.*, Vol 18 (No. 2), 1983, p 285-294
4. R.P. Wei, W. Wei, and G.A. Miller, Effect of Measurement Precision and Data Processing Procedures on Variability in Fatigue-Crack Growth-Rate Data, *J. Test. Eval.*, submitted for publication
5. D.A. Virkler, B.M. Hilberry, and P.K. Goel, "The Statistical Nature of Fatigue Crack Propagation," AFFDL-TR-78-

43, Air Force Wright Aeronautical Laboratories, Wright-Patterson Air Force Base, OH, April 1978

6. W.G. Clark, Jr. and S.J. Hudak, Jr., Variability in Fatigue Crack Growth Rate Testing, *J. Test. Eval.*, Vol 3 (No. 6), 1975, p 454-476

7. M.G. Salvadori and M.L. Baron, *Numerical Methods in Engineering*, Prentice-Hall, New York, 1952

8. C. Daniel and F.S. Wood, *Fitting Equations to Data*, John Wiley & Sons, New York, 1971

9. D.F. Ostergaard and B.M. Hillberry, Characterization of the Variability in Fatigue Crack Propagation Data, in *Probabilistic Fracture Mechanics and Fatigue Methods: Applications for Structural Design and Maintenance*, STP 798, ASTM, Philadelphia, 1983

10. E.K. Walker, The Effect of Stress Ratio During Crack Propagation and Fatigue for 2024-T3 and 7075-T6, in *The Effects of Environment and Complex Load History on Fatigue Life*, STP 462, ASTM, Philadelphia, 1970

11. R.G. Forman, V.E. Kearney, and R.M. Engle, Numerical Analysis of Crack Propagation in Cyclic Loaded Structures, *J. Basic Eng., ASME*, Vol 89, Series D, 1967, p 459-464

12. L.E. Jaske *et al.*, "Analysis of Fatigue, Fatigue Crack Propagation, and Fracture Data," NASA CR-132332, National Aeronautics and Space Administration, Nov 1973

13. D.W. Hoeppner and W.E. Krupp, Prediction of Component Life by Application of Fatigue Crack Growth Knowledge, *Eng. Frac. Mech.*, Vol 6, 1974, p 47-70

14. J.E. Collipriest and R.M. Ehret, "A Generalized Relationship Representing the Sigmoidal Distribution of Fatigue Crack Growth Rates," Rockwell International Report SD74-CE-0001, March 1974

15. J.N. Yang, G.C. Salivar, and C.G. Annis, Jr., "Statistics of Crack Growth in Engine Materials—Volume 1: Constant Amplitude Fatigue Crack Growth at Elevated Temperatures," AFWAL-TR-82-4040, Air Force Wright Aeronautical Laboratories, Wright-Patterson Air Force Base, OH, July 1982

16. M.S. Miller and J.P. Gallagher, An Analysis of Several Fatigue Crack Growth Rate (FCGR) Descriptions, in *Fatigue Crack Growth Measurement and Data Analysis*, STP 738, ASTM, Philadelphia, 1981

17. J.P. Gallagher *et al.*, "Damage Tolerant Design Handbook—A Compilation of Fracture and Crack Growth Data for High Strength Alloys," MCIC-HB-01R, Metals and Ceramics Information Center, Battelle Columbus Laboratories, Columbus, OH, Dec 1983

18. J.P. Gallagher *et al.*, "USAF Damage Tolerant Design Handbook: Guidelines for the Analysis and Design of Damage Tolerant Aircraft Structures," AFWAL-TR-82-3073, Air Force Wright Aeronautical Laboratories, Wright-Patterson Air Force Base, OH, Oct 1982

19. J.M. Barsom, Fatigue Crack Growth Under Variable Amplitude Loading in ASTM A514-B Steel, in *Progress in Flaw Growth and Fracture Toughness Testing*, STP 536, ASTM, Philadelphia, 1973

20. J.B. Chang, R.M. Hiyama, and M. Szamossi, "Improved Methods for Predicting Spectrum Loading Effects, Volume 1—Technical Summary," AFWAL-TR-81-3092, Air Force Wright Aeronautical Laboratories, Wright-Patterson Air Force Base, OH, Nov 1981

21. J. Schijve, The Analysis of Random Load Time Histories with Relation to Fatigue Tests and Life Calculations, in *Fatigue of Aircraft Structures*, W. Barrois and E.L. Ripley, Ed., Macmillan, New York, 1963, p 115-148

22. G.M. Van Dÿk, "Statistical Load Data Processing," *Advanced Approaches to Fatigue Evaluation*, NASA SP-309, National Aeronautics and Space Administration, 1972

23. N.E. Dowling, Fatigue Failure Predictions for Complicated Stress-Strain Histories, *J. Mat.*, Vol 7 (No. 1), March 1972, p 71-87

24. H. Tada, H.C. Paris, and G.R. Irwin, *The Stress Analysis of Cracks Handbook*, Del Research Corp., Hellertown, PA, 1973

25. G.C. Sih, *Handbook of Stress Intensity Factors*, Institute of Fracture and Solid Mechanics, Lehigh University, Bethlehem, PA, 1973

26. S.K. Chan, I.S. Tuba, and W.K. Wilson, On the Finite-Element Method in Linear Fracture Mechanics, *Eng. Frac. Mech.*, Vol 2, 1970, p 1-12

27. E. Byskov, The Calculation of Stress-Intensity Factors Using the Finite Element Method with Cracked Elements, *Int. J. Frac. Mech.*, Vol 6, 1970, p 159-167

28. D.M. Tracey, Finite Elements for Determination of Crack-Tip Elastic Stress-Intensity Factors, *Eng. Frac. Mech.*, Vol 3, 1971, p 255-265

29. J. Schijve, Cumulative Damage Problems in Aircraft Structures and Materials, *Aeronaut. J.*, Vol 74, 1970, p 517-532

30. J.P. Gallagher, H.D. Stalnaker, and J.L. Rudd, "A Spectrum Truncation and Damage Tolerance Study Associated with the C-5A Outboard Pylon Aft Truss Lug," AFFDL-TR-74-5, Air Force Flight Dynamics Laboratory, Wright-Patterson Air Force Base, OH, 1974

31. O.E. Wheeler, Spectrum Loading and Crack Growth, *J. Basic Eng.*, Vol 94D, 1972, p 181-186

32. J.D. Willenborg, R.M. Engle, and H.A. Wood, "A Crack Growth Retardation Model Using an Effective Stress Concept," AFFDL-TM-71-1 FBR, Air Force Flight Dynamics Laboratory, Wright-Patterson Air Force Base, OH, 1971

33. P.D. Belland and M. Creager, "Crack-Growth Analysis for Arbitrary Spectrum Loading," AFFDL-TR-74-129, Air Force Flight Dynamics Laboratory, Wright-Patterson Air Force Base, OH, 1974

34. J.P. Gallagher, "A Generalized Development of Yield Zone Models," AFFDL-TM-74-28 FBR, Air Force Flight Dynamics Laboratory, Wright-Patterson Air Force Base, OH, 1974

35. J.C. Newman, Jr., A Crack Closure Model for Predicting Fatigue Crack Growth under Aircraft Spectrum Loading, in *Methods and Models for Predicting Fatigue Crack Growth Under Random Loading*, STP 748, ASTM, Philadelphia, 1981

36. W.S. Johnson, Multiparameter Yield Zone Model for Predicting Spectrum Crack Growth, in *Methods and Models for Predicting Fatigue Crack Growth Under Random Loading*, STP 748, ASTM, Philadelphia, 1981

37. J.B. Chang, M. Szamossi, and K-W Liu, Random Spectrum Fatigue Crack Life Predictions with or without Considering Load Interactions, in *Methods and Models for Predicting Fatigue Crack Growth Under Random Loading*, STP 748, ASTM, Philadelphia, 1981

38. R.M. Engle, "CRACKS, A Fortran IV Digital Computer Program for Crack Propagation Analysis," AFFDL-TR-70-107, Air Force Flight Dynamics Laboratory, Wright-Patterson Air Force Base, OH, 1970

39. J.B. Chang, M. Szamossi, and K-W Liu, "A User's Guide Manual for a Detailed Level Fatigue Crack Growth Analysis Computer Code, Volume I," AFWAL-TR-81-3093, Air Force Wright Aeronautical Laboratory, Wright-Patterson Air Force Base, OH, 1980

40. D.S. Dawicke and D.A. Skinn,

"MODCRKS: Volume 1, User's Guide," UDR-TR-84-40, University of Dayton Research Institute, Dayton, OH, May 1984

41. G.G. Chell, "FATPAC: A Computer Program for Calculating Fatigue Crack Growth," RD/L/N205/78, Central Electricity Research Laboratories, Manchester, U.K., 1979

42. J.C. Newman, "Instructions for Use of FAST-2, A Program for the Fatigue-Crack Growth Analysis of Various Crack Configurations," submitted to COSMIC (Computer Software Management and Information Center), University of Georgia, Athens, Nov 1982

43. J.P. Gallagher and H.D. Stalnaker, Predicting Flight-by-Flight Crack Growth Rates, *J. Aircraft*, Vol 12, 1975, p 699-705

44. P.M. Besuner and A.S. Tetelman, "Probabilistic Fracture Mechanics," EPRI 217-1 Technical Report No. 4, Electric Power Research Institute, Palo Alto, CA, 1975

45. J.M. Bloom and J.C. Ekvall, Ed., *Probabilistic Fracture Mechanics and Fatigue Methods: Application of Structural Design and Maintenance*, STP 798, ASTM, Philadelphia, 1983

46. S.D. Manning and J.N. Yang, "USAF Durability Design Handbook: Guidelines for the Analysis and Design of Durable Aircraft Structures," AFWAL-TR-83-3027, Air Force Wright Aeronautical Laboratories, Wright-Patterson Air Force Base, OH, 1984

47. J.N. Yang, Statistical Estimation of Economic Life for Aircraft Structures, *J. Aircraft*, Vol 17 (No. 7), 1980, p 528-535

48. A.P. Berens, "Predicted Crack Repair Costs for Aircraft Structures," ASD-TR-78-39, Aeronautical Systems Division, Wright-Patterson Air Force Base, OH, 1978

49. J.W. Lincoln, "Method for Computation of Structural Failure Probability for an Aircraft," ASD-TR-80-5035, Aeronautical Systems Division, Wright-Patterson Air Force Base, OH, 1980

Analysis of Creep and Creep-Rupture Data*

By M.K. Booker
Research Staff Member
Oak Ridge National Laboratory

ANALYSIS of creep and creep-rupture data has many inherent problems. Often, experimentalists are not aware of the types of information needed for analysis; therefore, analysts are unsure of the adequacy of data available to them. An understanding of the information needed for analysis can facilitate generation of data.

Additionally, an understanding of the factors that can affect testing results, such as the design of experimental testing programs, can help ensure that analysts have sufficient data in terms of accuracy and adequacy. Typical problems encountered in the analysis of experimental creep and creep-rupture data and possible solutions to these drawbacks are presented in this article.

Test Planning

The most important prerequisite of a reliable data analysis is the availability of "good" data. However, test data generally are not obtained with the specific goal of analysis. To ensure sound analysis of data, the design of experimental programs must be considered. Variables that significantly affect the data obtained are discussed below.

A simple method for planning a test program for creep and creep-rupture data is detailed in Ref 1. This method involves developing a test layout in matrix form by listing the test temperatures in rows and tabulating the target test times in columns. Before the test matrix can be formed, however, the required temperature and life intervals must be determined.

Temperature. A range of temperature is usually required. For example, if information on material behavior is required over the range 482 to 649 °C (900 to 1200 °F), the intermediate temperatures must be chosen. This can be influenced by such factors as the

expected magnitude of the temperature effect on properties, the likelihood that the different stress-property isothermals will be similar in shape and slope, and the anticipated occurrence of temperature-dependent instabilities or transformations.

In general, the test temperatures should be spaced so that at least one test stress level is common to each pair of adjacent isotherms. Temperature spacings of 55 °C (100 °F) are generally adequate for most applications. Rounded temperatures in degrees Fahrenheit should be used rather than Celsius, because existing data, if generated in the United States, are usually obtained at such temperatures. Thus, for direct comparison with previous data, Fahrenheit temperatures are suggested—for example, testing at 900, 1000, 1100, and 1200 °F as opposed to 500, 550, 600, and 650 °C.

Life. Three tests, spaced about equally in log time, should be sufficient to characterize one log life cycle. Thus, between 100 and 1000 h, the target rupture lives would be 180, 320, and 560 h. Generally, times may be extrapolated by a factor of 2^η beyond the range of available data, where η is the number of log cycles spanned by the data.

Thus, if available data span times from 10 to 1000 h, they may be extrapolated to 4000 h with reasonable accuracy. If tests range from 10 to 30 000 h in duration, they may be extrapolated to approximately 300 000 h, which is a typical power plant lifetime. Tests with durations less than 10 h should not be factored into the calculation of η, however.

Stress. The above life estimates must, of course, be converted to stress before development of an actual test matrix. This requires some prior knowledge of the behavior of the material being tested. If such knowledge is not available, a series of scoping tests must be conducted to obtain stress estimates.

Lot-to-Lot Variations. Different lots of the same material, or batches of material from the same heat having a common processing and heat treatment history, can have significantly different creep properties. Tests of several lots are required to yield the information needed for analysis of such data. Generally, data from at least eight lots, ideally from several different vendors, should be available to support analyses.

If possible, tests on the different lots should be conducted at common stresses and temperatures. At least one lot should be tested at least three times at a common stress and temperature for at least two temperatures to yield estimates of scatter. These repeated test conditions should result in test times of at least 1000 h.

A full test matrix need not be completed for every lot of material tested. However, at least one lot should be tested at all conditions in the matrix, and each lot should be tested at least once within each log cycle in time at each temperature.

To characterize completely a material over the range of 482 to 649 °C (900 to 1200 °F), for example, the following minimum test matrix should be completed:

- Test one lot about 11 times (for times ranging from 10 to 30 000 h) at four temperatures (482, 538, 593, and 649 °C; or 900, 1000, 1100, and 1200 °F) (44 tests).
- Repeat one of the above tests twice at 482 and 649 °C (900 and 1200 °F) (4 tests).
- Test seven additional lots four times at each temperature (112 tests).

Thus, a minimum of about 160 tests is suggested to fully characterize a material. However, this characterization will apply only to lots representative of the population. This matrix is shown schematically in Fig. 1.

*Research sponsored by the Office of Converter Reactor Deployment, U.S. Department of Energy, under contract DE-AC05-840R21400 with Martin Marietta Energy Systems, Inc.

Fig. 1 Test matrix for creep testing

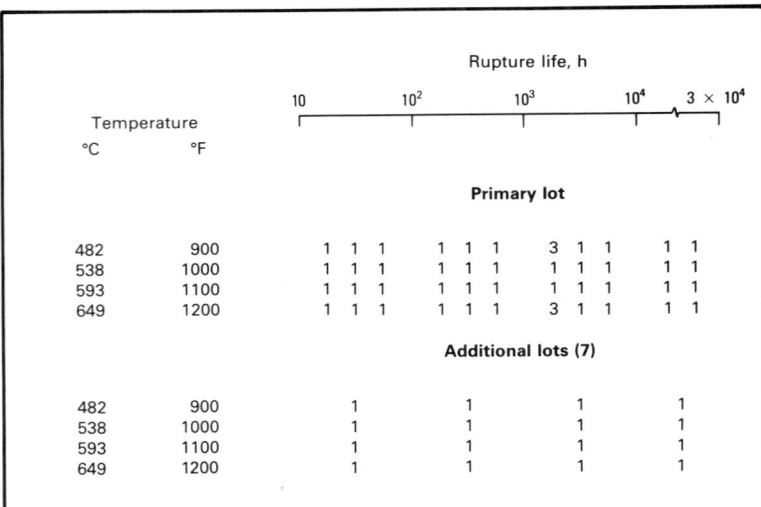

Creep Strain/Time Relationships

Elevated-temperature inelastic design requires an expression for creep strain as a function of time (t), stress (σ), and temperature (T). There are several methods for relating creep strain to time under constant load (or stress) in isothermal uniaxial tests.

The time (t_x) to accumulation of given amounts of creep strain can be analyzed by techniques similar to those commonly used for rupture life data. This method was used to describe data for Incoloy* alloy 800 (grade 2), Inconel* alloy 625 (grade 2), and 2.25Cr-1Mo steel in Ref 2. Using a series of strain levels and interpolations, a creep curve can be reconstructed.

This method fails to yield an expression for the creep strain rate, which is required in design calculations. However, it can be used to generate isochronous stress-strain curves such as those given in ASME Code Case N-47 (Ref 3). The method used in Ref 4 for alloy 800 and 2.25Cr-1Mo steel assumed an equation relating strain level to time and thus analytically "tied together" the various strain levels. This equation can yield an expression for creep rate as well as for creep strain.

Empirical relationships between creep strain/time properties and established properties such as rupture life, $\dot{\epsilon}_m$, or minimum creep rate, t_r, can also be used. These relationships may involve equations relating time (t_x) to $x\%$ creep strain, to $\dot{\epsilon}_m$, or to t_r. Stress-based correlations have also been proposed that relate the stress causing $x\%$ strain

*Trade names of the International Nickel Co., Inc.

in a given time to the stress causing rupture in that time (Ref 5-7). Such relationships, however, do not yield estimates of creep rate.

The parametric approach of Ref 8 has also found practical application. At a given total strain, ϵ (loading plus creep), this parameter is given by:

$$\phi(\epsilon) = \sigma(1 + \alpha t^n) \qquad \text{(Eq 1)}$$

where σ and t are stress and time, respectively; α is a constant; and n is a constant whose value is about one third. Monotonic stress-strain curves ($t = 0$) can be used to determine the value of ϕ at various strains, from which α can be determined from creep curves by assuming $n = \frac{1}{3}$ and solving Eq 1 to yield:

$$\alpha = \left(\frac{\phi}{\sigma} - 1\right) t^{-\frac{1}{3}} \qquad \text{(Eq 2)}$$

Because α is independent of σ and t, only one point from one creep curve should be required to determine α at each temperature. Alternatively, the values of ϕ, α, and n can be determined by least squares fits to creep data (Ref 9).

Another method involves the use of a creep equation (i.e., an analytical expression for creep strain as a function of time, stress, and temperature). Because these equations can generally be differentiated with respect to time to yield expressions for creep rate, they are more useful for design purposes than the other methods mentioned above. In addition, a creep equation can be more easily related to the actual microstructural processes occurring during creep. This method is adopted here because it is useful not only in practical design applications but also in fundamental studies of material behavior.

Developing Creep Equations

In general, the creep strain (ϵ_c) is given by ϵ_c (σ, t, T). Two basic methods for evaluation of this function have been proposed. The method described in Ref 10 assumes that the functional dependence of ϵ_c on σ, t, and T is separable, such that:

$$\epsilon_c = f_1(\sigma)f_2(t)f_3(T) \qquad \text{(Eq 3)}$$

Such a separation simplifies analysis procedures, but it may restrict unnecessarily the analytic description of material behavior.

A method that has recently been applied with some success involves first describing individual creep curves by a strain/time equation (Ref 11-13). The parameters in this equation are then expressed as functions of stress and temperature. Such an expression, however, may be indirect and may involve correlations among various properties such as rupture life, minimum creep rate, and the other creep equation parameters. Examples of such expressions may be found in Ref 11 to 14.

Another approach that has found some use involves the construction of a "master creep equation," in which creep curves derived from various loading conditions are normalized onto a single curve.

Selection of Strain/Time Equation Form

For purposes of a design equation, the creep equation must be valid at least through primary and secondary creep, although some design rules specifically forbid the onset of tertiary creep in service (Ref 3). Many materials exhibit very nearly classical creep curves, as shown in Fig. 2, further simplifying the choice of a strain/time equation form.

Many equation forms have been proposed (Ref 15-18). Of these, the most widely used are listed in Table 1 (Ref 16). The most commonly used forms for austenitic stainless

Fig. 2 Quantities used to characterize the creep behavior of type 304 stainless steel

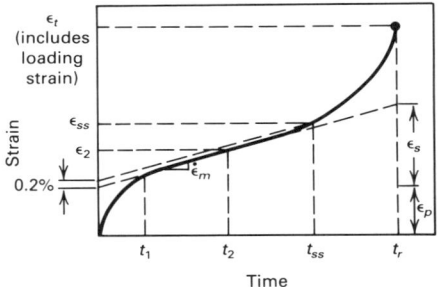

Table 1 Typical creep strain/time relations

Equation	Equation form	T/T_m	Materials
Logarithmic	$\epsilon = \alpha \ln t + c$	0.05-0.3	Al, Ag, Au, Cd, Cu, Mg, Ti-steel, Al-10%Cu, and Cu-3%Ag
Power law........	$\epsilon = \beta t^m$ or $\epsilon = \beta t^m + kt$	0.2-0.7	Al, Ag, brass, Cd, Cu, iron and steel, Mg, Mg-2%Al, Ni, Pb, Pt, Sn, and Zn
Exponential.......	$\epsilon = \epsilon_t(1 - \epsilon^{-rt})$ $+ \dot{\epsilon}m^t$ or $\epsilon = \epsilon t(1 - \epsilon^{-rt})$ $+ \epsilon_x(1 - \epsilon^{-st}) + \epsilon_m t$	0.4-0.6	Ferritic and stainless steels
Linear	$\epsilon = \dot{\epsilon}_s t$	0.96-0.99	Al, Au, Cu, and δ-Fe

Source: Ref 16

steels have been the so-called single and double exponential equations:

$$\epsilon_c = \epsilon_t[1 - \exp(-rt)] + \dot{\epsilon}_m t \qquad \text{(Eq 4)}$$

where ϵ_c is creep strain, %; ϵ_t is primary creep strain, %; r is creep rate parameter, 1/h; t is time, h; and $\dot{\epsilon}_m$ is minimum creep rate, %/h, and

$$\epsilon_c = \epsilon_x[1 - \exp(-st)]$$
$$+ \epsilon_t[1 - \exp(-rt)] + \dot{\epsilon}_m t \qquad \text{(Eq 5)}$$

in which the primary strain component is divided into two separate terms with amplitudes ϵ_x and ϵ_t (%), and rate parameters s and r (1/h).

Equation 4, first introduced in Ref 19, gained acceptance for type 316 stainless steel in Ref 20. Subsequent investigators have applied Eq 4 to a variety of materials, and considerable attention has been given to the meanings of and interrelationships among the parameters ϵ_t, r, and ϵ_m (Ref 21-31). Some investigators have extended the validity of the equation through tertiary creep by the addition of a positive exponential term (Ref 26, 28-31).

Although Eq 4 has been widely applied, it often underestimates the creep rate in the initial part of the creep curve (Ref 11, 25-31). Blackburn used Eq 5, in which the second exponential term, the so-called "fast transient" term, was added to increase the accuracy of the equation in the initial stages (Ref 11). However, Wilshire and co-workers fit Eq 4 to experimental data in the form:

$$\epsilon = \epsilon_L + \epsilon_t[1 - \exp(-rt)] + \dot{\epsilon}_m t \qquad \text{(Eq 6)}$$

where ϵ_L is the instantaneous strain incurred on loading, which thus corresponds to the strain at zero time (Ref 25-30). In fitting this equation, ϵ_L was generally overestimated, resulting in the situation shown schematically in Fig. 3. The curve-fitting procedure used by Blackburn suffers from a similar shortcoming (Ref 11). However, if one subtracts the loading strain and fits only creep strain using Eq 5, the equation forces a better fit in the initial stages, because ϵ_c is always zero at $t = 0$. Thus, the problems incurred with fitting the single-exponential equation

Fig. 3 Results of fitting Eq 6 directly to an experimental creep curve

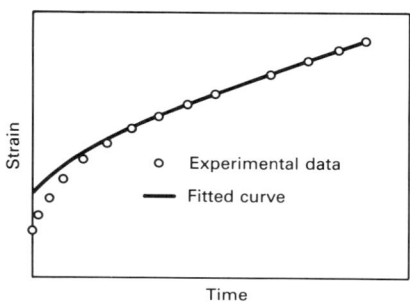

were probably due as much to the particular curve-fitting procedures used as to shortcomings in the equation form.

For design applications, any additional precision obtained by adding the second exponential term is probably not justified by the increase in the complexity of the model. The double-exponential equations developed by Blackburn (Ref 11) for types 304 and 316 stainless steel were used to develop the isochronous stress-strain curves in ASME Code Case N-47 (Ref 3). These equations have several shortcomings, which have been eliminated by more recent results, as discussed below.

Many recent results describe applications of a rational polynomial equation (Ref 11-14):

$$\epsilon_c = \frac{Cpt}{1 + pt} + \dot{\epsilon}_m t \qquad \text{(Eq 7)}$$

where ϵ_c is creep strain, %; t is time, h; $\dot{\epsilon}_m$ is minimum creep rate, %/h; and C is the limiting value of the transient primary creep strain, %. The parameter p is related to the sharpness of the curvature of the primary creep region.

Finally, an equation of this form has recently been used with success:

$$\epsilon_c^* = \exp[a(t^* - 1)] \, (t^*)^b \qquad \text{(Eq 8)}$$

where a and b are empirical constants; t^* is t/t_{ss}; and ϵ_c^* is ϵ_c/ϵ_{ss}. The parameters t_{ss} and

ϵ_{ss} are the time and strain at the onset of tertiary creep as defined by the 0.2% strain offset method. This equation has a concave downward initial primary region, followed by a nearly linear region, and finally a concave upward tertiary creep region.

Exponential Creep Equation

The exponential creep equation form was first introduced on an empirical basis, because its mathematical properties resemble those of primary and secondary creep curves (Ref 19-20, 25-28). By differentiating Eq 4, an estimate of the creep rate ($\dot{\epsilon}_c$) can be obtained:

$$\dot{\epsilon}_c = \epsilon_t r \epsilon^{-rt} + \dot{\epsilon}_m \qquad \text{(Eq 9)}$$

Thus, the initial creep rate $\dot{\epsilon}_0$ is given by:

$$\dot{\epsilon}_0 = \epsilon_t r + \dot{\epsilon}_m \qquad \text{(Eq 10)}$$

and $\dot{\epsilon}_c$ decreases exponentially and asymptotically approaches the minimum creep rate. The contribution of the primary term approaches a maximum strain value of ϵ_t. The parameter r, often called the time constant, controls the rate of decay of the transient contribution and thus is related to the sharpness of curvature of the primary region of the creep curve. The same basic comments apply to the double-exponential form of the equation.

Recently, this same equation has been derived by several investigators on a phenomenological basis. The work in Ref 31 assumes that primary and secondary creep obey first-order kinetics:

$$\frac{d\dot{\epsilon}_c}{dt} = -(\dot{\epsilon}_c - \dot{\epsilon}_m)r \qquad \text{(Eq 11)}$$

Integrating this expression twice yields Eq 4. Equation 5 can be derived from the assumption that two separate mechanisms operate independently, each obeying its own first-order kinetics. References 23, 24, and 32 also present phenomenological derivations.

Rational Polynomial Creep Equation

The properties of the rational polynomial creep equation given in Eq 7 are reviewed extensively in Ref 33 and will be described only briefly in this article. Although not as widely used as the exponential equations, the rational polynomial equation is not new. In fact, an equation equivalent to Eq 7, without the linear secondary term, was used more than 40 years ago (Ref 34). The equation was introduced in its present form more than 30 years ago (Ref 35).

Clearly, the parameter C in the rational polynomial equation is analogous to ϵ_t in the single-exponential equation, and $\dot{\epsilon}_m$ is the minimum creep rate in both equations. Differentiating Eq 7, creep rate $\dot{\epsilon}_c$ is expressed as:

$$\dot{\epsilon}_c = \frac{Cp}{(1 + pt)^2} + \dot{\epsilon}_m \qquad \text{(Eq 12)}$$

while the initial creep rate is given by:

$$\dot{\epsilon}_0 = Cp + \dot{\epsilon}_m \qquad \text{(Eq 13)}$$

The parameter p is related to the rate of approach of the creep rate from $\dot{\epsilon}_0$ to $\dot{\epsilon}_m$. Figure 4 summarizes the properties of the rational polynomial equation.

Equations 4 and 7 cannot be made equal, but they can be made to approximate one another within the accuracy of engineering creep data. To accomplish this, $\dot{\epsilon}_m$ represents the same parameter in each equation, and $C = \epsilon_t$. Then the two equations can be set equal at any point along the curve to determine the relationship between p and r. A suitable choice is the point at which half of the transient strain has been exhausted. Defining this time to be t^*, from Eq 7 $t^* = 1/p$. Equating Eq 4 and 7 at $t = t^*$ yields:

$$p = 1.44r \qquad \text{(Eq 14)}$$

Thus, the single-exponential and rational polynomial equations can be used almost interchangeably. Equations 13 and 14 imply that the rational polynomial equation yields a slightly higher initial creep rate than the exponential equation.

The rational polynomial creep equation is best viewed as a simple, flexible means of empirically representing experimental creep data. However, it has also been derived from various formulations of creep behavior. Oding derived the equation from a self-diffusion theory of creep (Ref 17, 35). The equation has also been derived on a phenomenological basis (Ref 17). The similarity of the rational polynomial and exponential equations also indicates that interpretations involving the parameters in the exponential equations (e.g., Ref 23, 24) can be applied to the rational polynomial equation.

Equation 7 can be solved explicitly for time, yielding:

$$t = \frac{\epsilon_c p - Cp - \dot{\epsilon}_m}{2p\dot{\epsilon}_m}$$
$$+ \frac{\sqrt{(Cp + \epsilon_m - \epsilon_c p)^2 + 4\epsilon_c p\epsilon_m}}{2p\dot{\epsilon}_m} \qquad \text{(Eq 15)}$$

Thus, the equation can be easily applied to variable stress situations involving strain hardening analyses.

Fits to Experimental Curves

The rational polynomial equation can be fitted to experimental curves by various least squares and graphical techniques, as described in Ref 33. In general, the least squares techniques yield the most objective fit. However, many low-stress and low-temperature tests are available in which the amount of strain incurred is so small compared with the accuracy of the strain-measuring equipment used that significant scatter occurs in the data. To avoid wrong conclusions based on least squares fits due to this scatter, the semigraphical approach described in Ref 17 and 36 may be preferred.

Reference 36 showed that if three points— (t_1, ϵ_1), (t_2, ϵ_2), and (t_3, ϵ_3)—are chosen from an experimental creep curve, the values of p and $\dot{\epsilon}_m$ can be found by:

$$p = \frac{\dfrac{t_1 - t_2}{t_3 - t_2}\left(\dfrac{\epsilon_3}{t_3} - \dfrac{\epsilon_2}{t_2}\right) + \dfrac{\epsilon_2}{t_2} - \dfrac{\epsilon_1}{t_1}}{\dfrac{t_1 - t_2}{t_3 - t_2}(\epsilon_2 - \epsilon_3) + \epsilon_1 - \epsilon_2} \qquad \text{(Eq 16)}$$

The value of C can then be determined by setting the prediction of Eq 7 equal to the experimental strain at any point on the curve. Theoretically, any three points on the creep curve can be used to evaluate these expressions. In practice, however, it is generally best to choose one point midway through the primary creep region, one near the end of primary creep, and one well into secondary creep. In an analysis of data for type 304 stainless steel, this method was used to determine C at each of the above points; these results were then averaged, although they did not vary significantly (Ref 37). Figures 5 to 9 illustrate fits to experimental curves obtained by this method.

Finally, in many cases, the accuracy of experimental data is such that it is acceptable to use the geometric interpretations in Fig. 4 to directly "pick" the values of the equation parameters from experimental curves.

Fig. 4 Properties of the rational polynomial creep equation

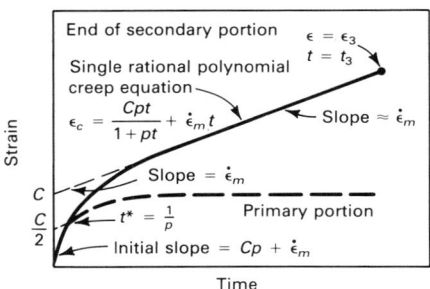

Stress and Temperature Dependence of Equation Parameters

The stress and temperature dependence of the creep equation parameters (Ref 11-13) are often evaluated as follows. First, available experimental creep curves are individually fitted by use of the chosen model to yield a set of parameter values for each test. Then, the individual equation parameters are directly expressed as functions of stress and temperature, with techniques similar to those used to express rupture life as a function of stress and temperature.

Although this method has met with some success, it has several limitations. The num-

Fig. 5 Fit of the rational polynomial creep equation to an experimental curve at 427 °C (800 °F) and 97 MPa (14 ksi)

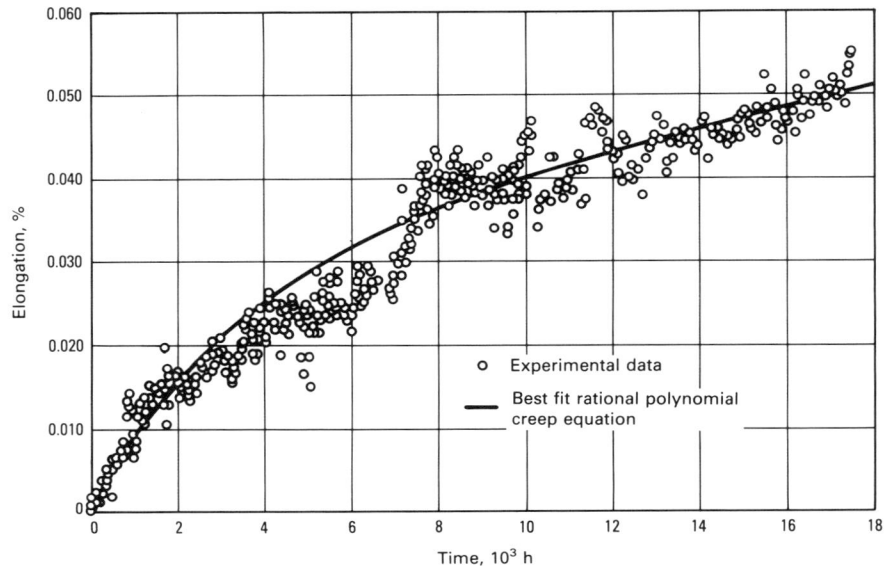

Fig. 6 Fit of the rational polynomial creep equation to an experimental curve at 482 °C (900 °F) and 138 MPa (20 ksi)

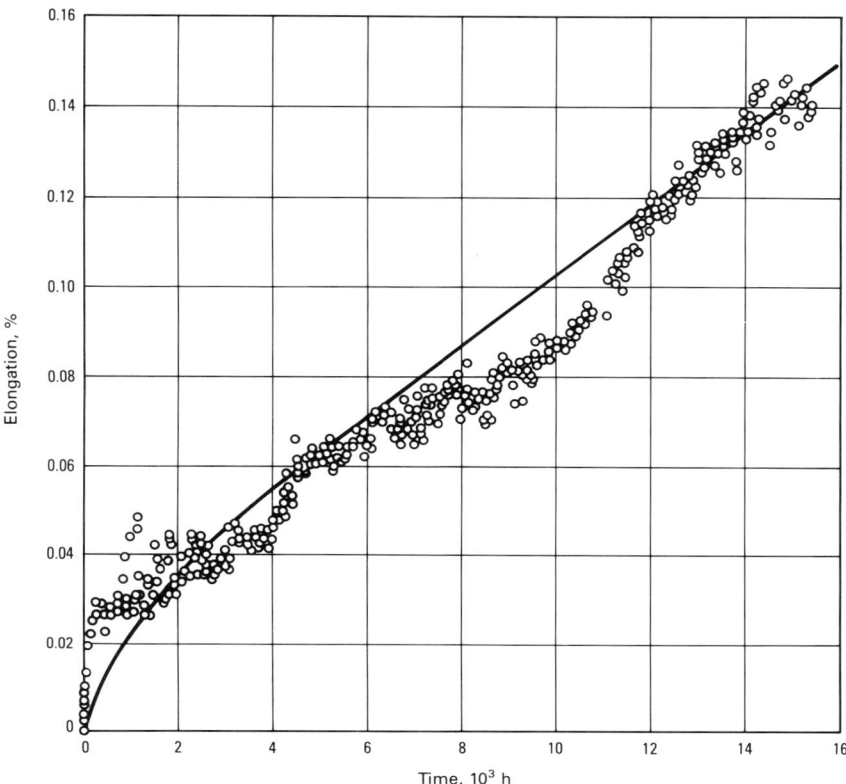

type 304 stainless steel, the following relationships were used (see Fig. 10 and 11):

$$Cp = 2.77\dot{\epsilon}_m^{0.78} \qquad \text{(Eq 17)}$$

$$p = 1.77\dot{\epsilon}_m^{0.77}\left(\dot{\epsilon}_m > 4 \times \frac{10^{-3}\%}{h}\right) \quad \text{(Eq 18)}$$

and

$$p = 0.5\dot{\epsilon}_m^{0.54}\left(\dot{\epsilon}_m < 4 \times \frac{10^{-3}\%}{h}\right) \quad \text{(Eq 19)}$$

Note the use of Cp—which has units of creep rate, as does p—rather than C in Eq 17.

One of the advantages of the above approach is that lot-to-lot variations frequently can be easily factored into the analysis for $\dot{\epsilon}_m$, but equations such as Eq 17 to 19 often minimize such variations through normalization. Thus, lot-to-lot variations in the entire creep equation can be handled by the variations in $\dot{\epsilon}_m$ alone. Figures 12 to 17 illustrate the accuracy that can be expected in making strain-time predictions with the above approach compared with the accuracy of the "Blackburn equation" (of the form of Eq 5), which did not include lot-to-lot variability. Note the variations in load and temperature for Fig. 12, 16, and 17. These figures illustrate that it is possible to predict material behavior under different loads at different temperatures.

Methods of Analysis

The various techniques that are available to express rupture life as a function of stress and temperature as well as of lot-to-lot variability in strength are discussed below. Although the discussion concerns rupture life, the same techniques may be applicable to similar expressions for other creep param-

ber of available creep curves is often quite small in comparison with rupture and other data. Additionally, measured creep strain values can be too small to adequately identify the dependence of the equation parameters on stress and temperature. Experimental variations and complex interrelationships among equation parameters can cause considerable scatter in the values of these parameters. As a result, the correlations among these parameters and the test conditions (σ, T) can be very poor. This problem becomes

particularly significant when extrapolation in one or more of the independent variables is necessary.

Separate analysis of the equation parameters as functions of stress and temperature can cause problems when the parameters are recombined into a unified equation, because they are actually strongly interrelated. Finally, it is difficult to account for heat-to-heat variations with this method.

An alternative method helps to overcome these limitations. One of the advantages of forms such as the rational polynomial equation is the explicit minimum creep rate term. Although analytical fits of Eq 7 to experimental curves will, in general, yield slightly different values of $\dot{\epsilon}_m$ than graphical measurement, this difference is insignificant compared to the overall uncertainty in creep strain/time predictions. Techniques for analysis of minimum creep rate data are similar to those used for rupture life (see Eq 17-19), and it is common to express $\dot{\epsilon}_m$ as a function of stress and temperature.

The parameters C and p can then be expressed as functions of minimum creep rate. Thus, in a recent analysis of data for

Fig. 7 Fit of the rational polynomial creep equation to an experimental curve at 538 °C (1000 °F) and 55 MPa (8 ksi)

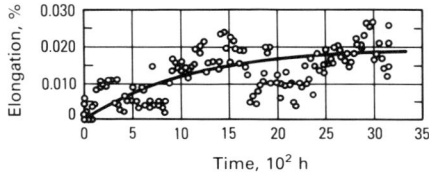

Fig. 8 Fit of the rational polynomial creep equation to an experimental curve at 593 °C (1100 °F) and 83 MPa (12 ksi)

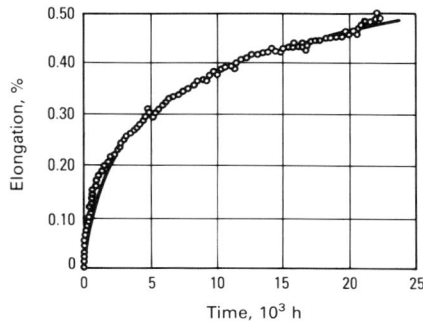

Fig. 9 Fit of the rational polynomial creep equation to an experimental curve at 649 °C (1200 °F) and 21 MPa (3 ksi)

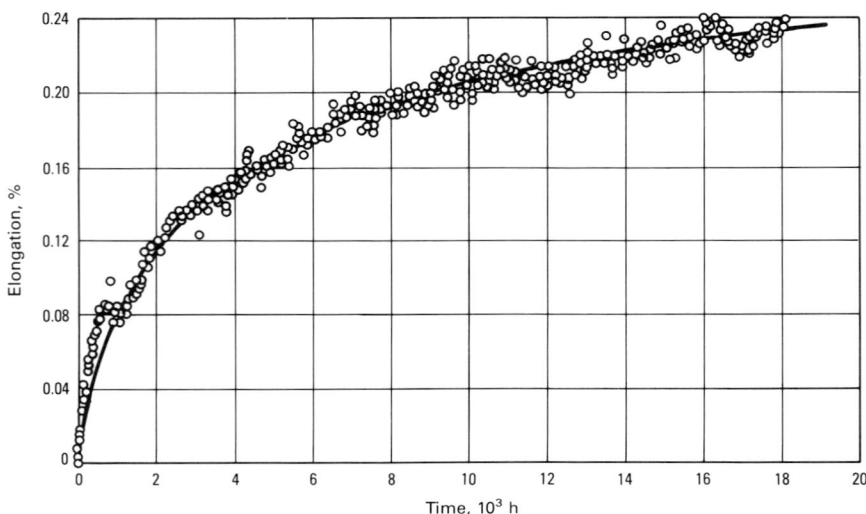

Time, 10^3 h

eters. For example, quantities such as minimum creep rate and time to tertiary creep may be expressed as functions of stress and temperature by techniques similar to those used for rupture life.

Some common methods used in the past for analysis of creep-rupture data are described in Ref 37. These methods consist of two basic categories: (1) direct extrapolation of isothermal log σ-log t_r curves and (2) analysis by standard time-temperature parameters. Direct extrapolation is usually accomplished on an individual lot basis, with the 10^5-h rupture strength values from the indi-

Fig. 10 Relationship between minimum creep rate and primary creep parameter Cp

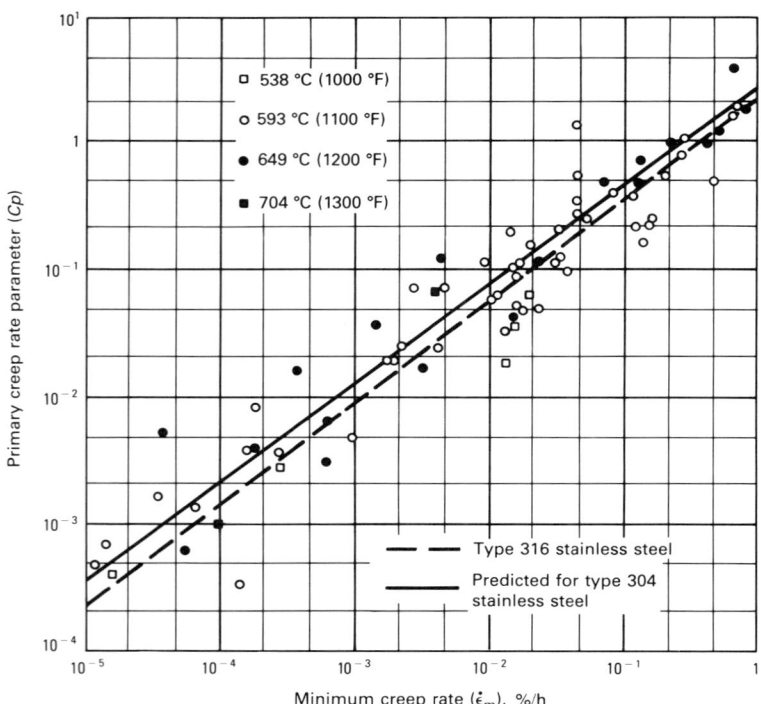

Minimum creep rate ($\dot{\varepsilon}_m$), %/h

vidual lots later used to establish a strength trend versus temperature curve. Parametric analysis is typically performed by using all data as a single population, if only because there are seldom sufficient data to perform such an analysis on each lot separately.

Direct isothermal extrapolation can be implemented analytically, but this has usually been performed by a manual extrapolation on log-log paper. This technique directly addresses the problem of lot-to-lot variations. However, it does have limitations. Graphical extrapolation can require considerable interpretation by the analyst. Uncertainties are greatly increased if log σ-log t_r isotherms are nonlinear. Conversely, assumption of such linearity may sometimes be erroneous, thus introducing additional extrapolation errors.

Because data at only one temperature are considered at any time, information from other temperatures is not considered. Moreover, if data for a given lot are sparse, those data can only be used to draw an approximate isotherm, assuming data are available for higher and/or lower temperatures. Thus, the method does not make efficient use of the available information. Additionally, data at different temperatures may represent different lots. What may appear to be a temperature effect may be due to thermomechanical processing differences.

Although the parametric approach considers all data together, this method has several inherent disadvantages. Lot-to-lot variations are not considered directly. Ignoring this significant effect may result in large errors. For example, a few points for unusually strong or weak heats can significantly distort the shape of the best fit curve. Any given parameter involves very specific and rigid assumptions about behavior. If the wrong parameter is used (i.e., if the assumptions are not met), the results may contain significant errors.

Because hundreds of parameter forms are available, selection of the correct parameter can be formidable. Also, available data are often dominated by tests run at a single temperature. In these cases, it may be difficult to accurately determine temperature dependence by standard parametric techniques.

Several recent advances in the use of computerized analytical techniques for the treatment of creep and creep-rupture data may help solve these problems (Ref 38-40). The digital computer has made possible improved treatment of lot-to-lot variations, selection of model forms, and statistical analysis of results. Further developments in this technology may enable:

- Treatment of lot-to-lot variations as an integral part of the analysis
- Sufficient flexibility to allow fitting a wide

Fig. 11 Relationship between minimum creep rate and primary creep parameter 1/p

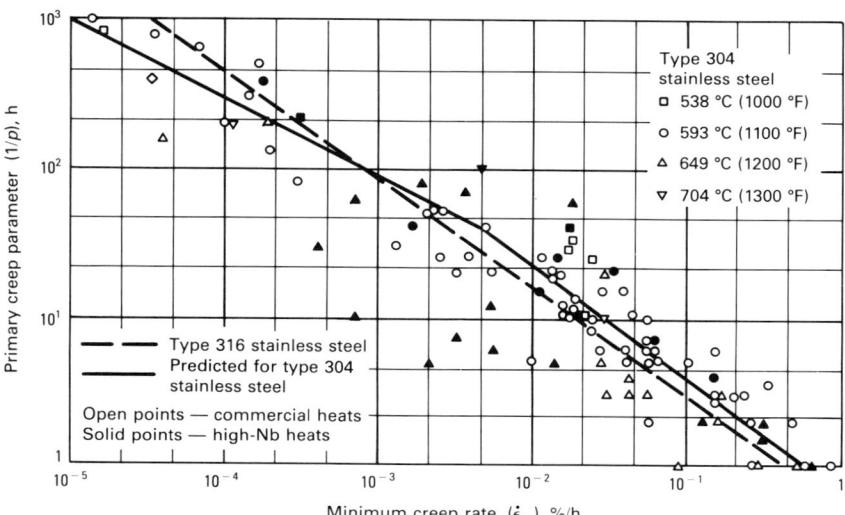

range of behavior by automatic consideration of many models

- Establishment of a statistically viable estimate of average and minimum behavior
- Minimization of vulnerability to "bad" data distributions, such as concentration of the data over a narrow temperature range or availability of only a few data for each of several lots
- Ease of applicability and minimization of manual labor to produce results, particularly for large data sets

Lot-Centered Regression Analysis

Computerized techniques offer a wide range of possible approaches. One such approach is discussed below, although other approaches are also suitable. The general usefulness of this technique has been previously established in a variety of situations (Ref 41-44). The method is essentially a synthesis of the methods used in Ref 38 and 39.

Because this method involves the use of lot-centered data analysis procedures (Ref 38), it provides the maximum protection against poorly distributed databases. Thus, its use is particularly recommended for randomly scattered data distributions. However, it is also useful for well-distributed data.

To implement the method, assume that the logarithm of rupture life (log t_r) has been selected as the dependent variable for the analysis. References 38 to 41 discuss dependent variable selection. Label log t_r as Y. Now assume that Y can be expressed as a

linear function, in the regression sense, of terms involving stress σ and temperature T. Label these terms X_i, which yields:

$$\hat{Y}_K = \sum_{i=0}^{N} a_i X_{iK} \qquad \text{(Eq 20)}$$

where a_i are constants estimated by regression, and $\hat{Y}_K$ is the predicted value of log-rupture life at the Kth set of values of the independent or predictor variables X_{iK}. Note that X_0 is always unity and that a_0 is a constant intercept term.

Then each variable (Y and each X) is lot centered, and the equation becomes:

$$\hat{Y}_{Kh} - \bar{Y}_h = \sum_{i=1}^{N} a_i'(X_{iKh} - \bar{X}_{ih}) \qquad \text{(Eq 21)}$$

where the barred variables represent average values for a given lot, and h represents the index of the lot. The prediction of log-rupture life is then given by:

$$\hat{Y}_{Kh} = \bar{Y}_h - \sum_{i=1}^{N} a_i'\bar{X}_{ih} + \sum_{i=1}^{N} a_i'X_{iKh} \qquad \text{(Eq 22)}$$

The expression $\bar{Y}_h - \sum_{i=1}^{N} a_i'\bar{X}_{ih}$ is a constant for a given lot and replaces the intercept term a_0 in the uncentered analysis. Thus, each lot will have a different intercept term, but all other coefficients a_i' will be common to all lots. There is no separate a_0 term, because it would be superfluous.

Lot centering of the data does not involve complicated mathematics. However, for large data sets, the simple operations can become quite tedious, and centering is best accomplished by computer. Implications of lot centering are also straightforward.

As indicated above, different lots are considered to have different intercept values, but all other equation constants are lot independent. Thus, all lots vary in a similar manner with the independent variables, but any two lots will always be separated by a constant increment in log t_r space. This assumption of parallelism may or may not be sound in any given case, but it has frequently been found to be appropriate (Ref 42-44). Adjustments can be made in case of a lack of parallelism (Ref 45).

If any lot is represented by a single data point, all lot-centered variables will be zero. That lot will not contribute to the establishment of stress and temperature dependence, although it will contribute to the calculation of average and minimum values, as described below. If all data for a given lot occur at a single temperature, all pure temperature variables will be zero, and that lot will not contribute to the estimation of temperature dependence. Thus, lot-to-lot variation is addressed directly, thereby ensuring minimum vulnerability to "bad" data distributions.

Prediction of Average and Minimum Behavior. Fitting a multi-lot set of creep-rupture data by use of lot-centered regression can yield results that accurately portray the stress and temperature dependencies of the material under consideration. Predictions also include different intercept values to yield different strength levels for different lots or heats of material for which data are

Fig. 12 Comparison of predicted and experimental creep strain/time behavior for type 304 stainless steel lot 187-7 at 207 MPa (30 ksi) and 593 °C (1100 °F)

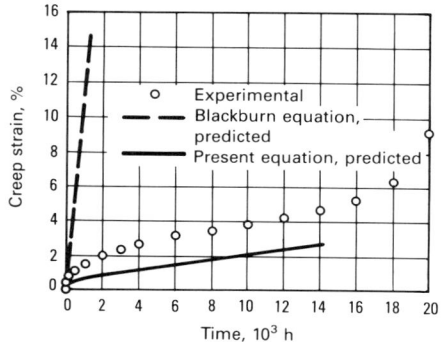

Fig. 13 Comparison of predicted and experimental creep strain/time behavior for type 304 stainless steel lot 544 at 117 MPa (17 ksi) and 593 °C (1100 °F)

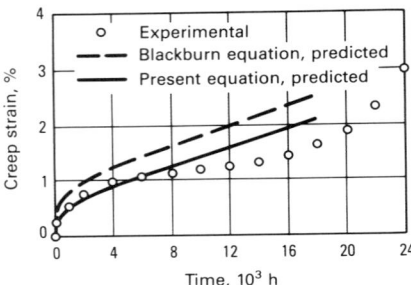

available. An average strength level can also be predicted by the analysis.

Aspects of the method that permit accurate determination of minimum values are discussed, although detailed methods of defining minima are beyond the scope of this article. Although results are discussed in terms of rupture data, all discussions in this article also apply to tensile data or any other data on which this method is used.

Equation 21 is fitted to the data as written, with $Y_{Kh} - \bar{Y}_h$ as the dependent variable, where Y_{Kh} is the experimental value of log t_r. However, because $\bar{Y}_h$ is a constant for a given lot, the error in prediction lies in the estimation of $\hat{Y}_{Kh}$. Thus, when Eq 21 is fitted to data by least squares and the a_i' values are determined, the total "error" in fitting the model can be described by a residual sum of squares (RSS), given by:

$$RSS = \sum_{h=1}^{H} \sum_{K=1}^{n_h} (\hat{Y}_{Kh} - Y_{Kh})^2 \qquad (Eq\ 23)$$

If n equals the total data, RSS has a number of degrees of freedom (df), given by:

$$df = n - N - H \qquad (Eq\ 24)$$

where N is the number of terms in the model, and H is the number of lots and thus the number of lot averages involved in the fitting.

By separating different lots by their different lot constants, this method attempts to describe only within-lot variations in behavior. No between-lot differences have been modeled to date. Thus, the variance defined by the fit is an estimate of the pooled within-lot variance (V_w):

$$V_w = \frac{RSS}{df} \qquad (Eq\ 25)$$

Equation 21 can now be transformed to Eq 23 or:

$$\hat{Y}_{Kh} = C_h + \sum_{i=1}^{N} a_i' X_{iKh} \qquad (Eq\ 26)$$

where the differences in behavior of different lots are not explicitly defined in terms of the lot constants (C_h), where:

$$C_h = \bar{Y}_h - \sum_{i=1}^{N} a_i' X_{ih} \qquad (Eq\ 27)$$

Because C_h is a single constant for a given lot, estimation of average behavior consists only of estimating the average lot constant $\bar{C}_h$. Two methods are immediately apparent. First, $\bar{C}_h$ may be defined as the arithmetic mean of C_h. If the between-lot variability is much larger than the within-lot variability, such an approach would be justified. However, if the amount of within-lot variability is significant, the estimates of C_h will contain some error. Lots with more data will have better estimates of C_h than lots with fewer data. Thus, not all lots should be weighted equally.

Alternatively, each lot should be weighted according to the number of data available for that lot. This approach is correct only if the within-lot variability is much larger than the between-lot variability. If not, this procedure, which weights each test equally, is not valid, because no one lot is necessarily more important in the collection of lots available, even if it is represented by more data.

A possible solution comes from Ref 46, which studies variations in behavior caused by measurement of chemical variables at different laboratories. Based on Ref 38, lab-

to-lab variation results can be extrapolated to the lot-to-lot variation data. Following this approach, C_h for each lot should be given a weight (w_h) of:

$$w_h = \frac{k_h}{k_h \lambda + 1} \qquad (Eq\ 28)$$

where k_h is the number of data for lot h, and $\lambda = V_B/V_w$, where V_B is the between-lot variance. Knowing the appropriate weights, $\bar{C}_h$ can be calculated by:

$$\bar{C}_h = \frac{\sum_{h=1}^{H} C_h w_h}{\sum_{h=1}^{H} w_h} \qquad (Eq\ 29)$$

However, w_h cannot be estimated at this point, because V_B and thus λ are unknown. As a result, one equation exists with two unknowns, and a solution can be obtained only by iterative techniques. However, such techniques are easily implemented by computer.

An iterative technique that results in a solution for both $\bar{C}_h$ and V_B is presented in Ref 46. Typically, results are obtained after only three or four iterations. Reference 38 has reported a similar rapid convergence to a solution. The result represents the most fairly weighted estimate of average behavior that can be obtained by any current technique.

By the direct separation of the variability into its two components, V_B and V_w, this method also yields better estimates of error than can be obtained by estimates of error that are a mixture of within-lot and between-

Fig. 14 Comparison of predicted and experimental creep strain/time behavior for type 304 stainless steel lot 330 at 117 MPa (17 ksi) and 593 °C (1100 °F)

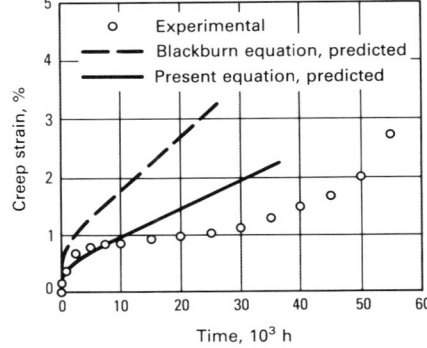

Fig. 15 Comparison of predicted and experimental creep strain/time behavior for type 304 stainless steel lot 380 at 117 MPa (17 ksi) and 593 °C (1100 °F)

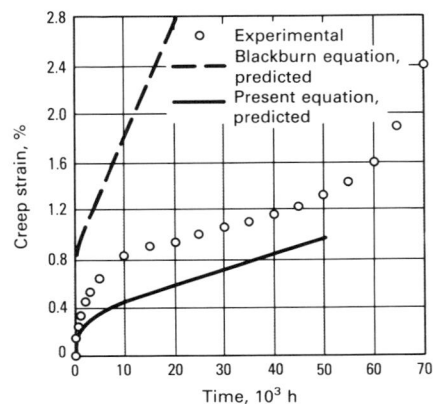

lot variability. Because variance estimation is central to the estimation of any statistical limit, regression on lot-centered data thus may lead to improved techniques of estimating these limits.

Selection of a Regression Model. Equations 20 to 22 encompass thousands of potential model forms. The selection of a particular model form for a given application involves techniques described in Ref 39. Basically selection entails:

• Identification of terms to include in potential models by preliminary graphical assessment

Fig. 16 Comparison of predicted and experimental creep strain/time behavior for type 304 stainless steel lot 796K at 649 °C (1200 °F)

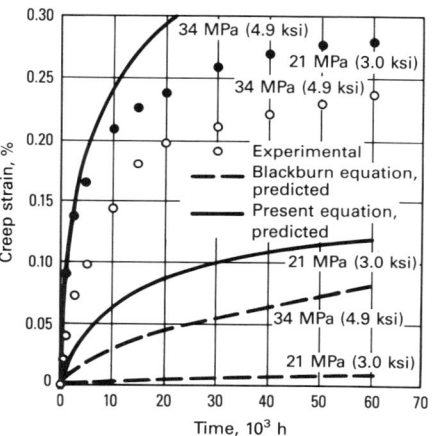

Fig. 17 Comparison of predicted and experimental creep strain/time behavior for type 304 stainless steel lot 796K at 83 MPa (12 ksi) and 538 °C (1000 °F)

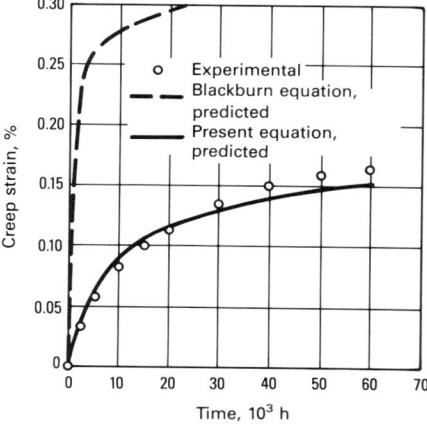

• Use of a computer program to scan the potential models, to identify candidate models for detailed study (usually 5 to 10 out of thousands of potential models)
• Selection of final model on the basis of fit to data, behavior on extrapolation, physical implications, and ease of use in design calculations

Creep ductility data (reduction of area and total elongation to failure) typically show considerably more scatter and variation than data for properties such as rupture life and minimum creep rate. Therefore, for some applications, it may be sufficient to present plots of available data with no analysis.

However, such data often exhibit trends with test temperature and stress, and thus time, particularly when examined on an individual heat basis. In that case, it may be necessary to estimate the trends analytically to allow a reasonable extrapolation to conditions of design relevance.

Generally, such analyses can best be performed by use of the average strain rate concept (Ref 47). For example, let ϵ_t be the total strain to failure. Then, the average strain rate to failure ($\dot{\epsilon}_t$) is defined as:

$$\dot{\epsilon}_t = \frac{\epsilon_t}{t_r} \qquad \text{(Eq 30)}$$

The quantity $\dot{\epsilon}_t$ then exhibits much less scatter than ϵ_t. It can be analyzed by techniques similar to those used for $\dot{\epsilon}_m$ or perhaps related directly to $\dot{\epsilon}_m$. A similar approach can be used for various other ductility quantities, such as reduction of area and strain to the onset of tertiary creep (Ref 48, 49).

REFERENCES

1. R.C. Rice, Ed., "Reference Document for the Analysis of Creep and Stress Rupture Data in MIL-HDBK-5C," AFWAL-TR-81-4087, Air Force Wright Aeronautical Laboratories, Wright-Patterson Air Force Base, OH, Sept 1981
2. D.I. Roberts and S.A. Sterling, A Parametric Method for the Development of Isochronous Stress-Strain Curves, in *The Generation of Isochronous Stress-Strain Curves*, A.O. Schaefer, Ed., American Society of Mechanical Engineers, New York, 1972, p 1-14
3. *Interpretations of the ASME Boiler and Pressure Vessel Code, Case N-47*, American Society of Mechanical Engineers, New York, 1977
4. S.A. Sterling, "A Temperature-Dependent Power Law for Monotonic Creep,"

GA-A13027 (Revision), General Atomics Division, San Diego, CA, June 1974, revised March 1976
5. R.F. Gill and R.M. Goldhoff, Analysis of Long Time Creep Data for Determining Long Term Strength, *Met. Eng. Quart.*, Vol 10, 1970, p 30-39
6. M.C. Murphy, Rating the Creep Behavior of Heat-Resistant Steels for Steam Power Plant, *Met. Eng. Quart.*, Vol 13, 1973, p 41-50
7. R.M. Goldhoff and R.F. Gill, A Method for Predicting Creep Data for Commercial Alloys on a Correlation Between Creep Strength and Rupture Strength, *J. Basic Eng.*, Vol 94, 1972, p 1-6
8. Y.N. Rabotnov, Some Problems on the Theory of Creep, *Vestnik Moskovskoyo Universiteta*, No. 10, 1948; available in English as NACA Tech. Memo. 1353, National Advisory Committee for Aeronautics, Washington, DC, April 1953
9. R.M. Goldhoff, The Application of Rabotnov's Creep Parameter, *Proc. Am. Soc. Test. Mater.*, Vol 61, 1961, p 908-919
10. R.K. Penny and D.L. Marriott, *Design for Creep*, McGraw-Hill, London, 1971, p 192-238
11. L.D. Blackburn, Isochronous Stress-Strain Curves for Austenitic Stainless Steels, in *The Generation of Isochronous Stress-Strain Curves*, A.O. Schaefer, Ed., American Society of Mechanical Engineers, New York, 1972, p 15-48
12. M.K. Booker, Analytical Description of the Effects of Melting Practice and Heat Treatment on the Creep Properties of 2¼Cr-1Mo Steel, in *Effects of Melting and Processing Variables on the Mechanical Properties of Steel*, MPC-6, Metals Properties Council, American Society of Mechanical Engineers, New York, 1977, p 323-343
13. W.E. Stillman, M.K. Booker, and V.K. Sikka, Mathematical Description of the Creep Strain-Time Behavior of Type 316 Stainless Steel, in *Proceedings of the Second International Conference on Mechanical Behavior of Material*, Boston, Aug 16-20, 1976, American Society for Metals, 1976, p 424-428
14. M.K. Booker, An Analytical Representation of the Creep and Creep-Rupture Behavior of Alloy 800H, in *Characterization of Materials for Service at Elevated Temperatures*, MPC-7, American Society of Mechanical Engineers, New York, 1978, p 1-27
15. J.B. Conway, *Numerical Methods for Creep and Rupture Analyses*, Gordon and Breach, New York, 1967
16. F. Garofalo, *Fundamentals of Creep*

and *Creep-Rupture in Metals,* Macmillan, New York, 1965, p 10-20

17. I.A. Oding et al., *Creep and Stress Relaxation in Metals,* transl. by E. Bishop, A.J. Kennedy, Ed., Oliver and Boyd, Edinburgh, 1965

18. A.J. Kennedy, *Processes of Creep and Fatigue in Metals,* Oliver and Boyd, Edinburgh, 1962

19. P.G. McVetty, Factors Affecting Choice of Working Stresses for High-Temperature Service, *Mech. Eng.,* Vol 55, 1933, p 99-104

20. F. Garofalo et al., Strain-Time, Rate-Stress, and Rate-Temperature Relations During Large Deformation in Creep, in *Joint International Conference on Creep,* Sect. 1, The Institution of Mechanical Engineers, London, 1963, p 31-39

21. A. Ahmadieh and A.K. Mukherjee, Stress-Temperature-Time Correlation for High Temperature Creep Curves, *Mat. Sci. Eng.,* Vol 21, 1975, p 115-124

22. A. Ahmadieh and A.K. Mukherjee, Transient and Steady-State Creep Curves in NiFe Alloy System, *Scr. Metall.,* Vol 9, 1975, p 1299-1304

23. B.A. Movchan et al., A Phenomenological Structural Approach to High-Temperature Creep, *Strength Mater.,* Vol 6, 1974, p 1041-1046 (transl. from *Prob. Prochn.*)

24. B.A. Movchan and E.V. Dabizha, A Phenomenological Model for High-Temperature Creep in Pure Nickel, *Strength Mater.,* Vol 7, 1975, p 278-283 (transl. from *Prob. Prochn.*)

25. D. Sidey and B. Wilshire, Mechanisms of Creep and Recovery in Nimonic 80A, *Met. Sci. J.,* Vol 3, 1969, p 56-60

26. B. Wilshire, Some Grain Size Effects in Creep and Fracture, *Scr. Metall.,* Vol 4, 1970, p 361-366

27. W.J. Evans and B. Wilshire, Transient and Steady-State Creep Behavior of a Copper-15 at. % Aluminum Alloy, *Met. Sci. J.,* Vol 4, 1970, p 89-94

28. P.W. Davies et al., An Equation to Represent Strain/Time Relationships During High Temperature Creep, *Scr. Metall.,* Vol 3, 1969, p 671-674

29. W.J. Evans and B. Wilshire, The High Temperature Creep and Fracture Behavior of 70-30 Alpha-Brass, *Met. Trans.,* Vol 1, 1970, p 2133-2139

30. P.L. Threadgill and B. Wilshire, Mechanisms of Transient and Steady-State in a γ′-Hardened Austenitic Steel, in

Creep Strength in Steel and High Temperature Alloys, The Metals Society, London, 1974, p 8-14

31. G.A. Webster, A.P.D. Cox, and J.E. Dorn, A Relationship Between Transient and Steady-State Creep at Elevated Temperature, *Met. Sci. J.,* Vol 3, 1969, p 221-225

32. M. Mejía, R. Gómez-Ramírez, and M.A. Martínez, The Phenomenological Theory of Creep, *Scr. Metall.,* Vol 10, 1976, p 589-591

33. D.O. Hobson and M.K. Booker, "Materials Applications and Mathematical Properties of the Rational Polynomial Creep Equation," ORNL-5202, Oak Ridge National Laboratory, Dec 1976

34. A.M. Freudenthal, Theory of Wide-Span Arches in Concrete and Reinforced Concrete, *Int. Assoc. Bridge Structural Eng.,* Vol 4, 1936, p 249, as cited by Conway (Ref 19)

35. I.A. Oding, *Izv. Akad. Nauk SSSR, Otd. Tekh. Nauk.,* 1948, as cited by Oding et al. (Ref 21)

36. L.A. Bunatyan, Proceedings of a Seminar on the Strength of Engineering Components, *Izv. Akad. Nauk SSSR,* Vol 1, 1953, as cited by Oding et al. (Ref 21)

37. G.V. Smith, Evaluation of Elevated-Temperature Strength Data, *J. Mater.,* Vol 4 (No. 4), Dec 1969, p 878-908

38. L.H. Sjodahl, A Comprehensive Method of Rupture Data Analysis with Simplified Models, in *Characterization of Materials for Service at Elevated Temperatures,* MPC-7, Metal Properties Council, American Society of Mechanical Engineers, New York, 1978, p 501-515

39. M.K. Booker, Use of Generalized Regression Models for the Analysis of Stress-Rupture Data, in *Characterization of Materials for Service at Elevated Temperatures,* MPC-7, Metal Properties Council, American Society of Mechanical Engineers, New York, 1978, p 459-499

40. D.R. Rummler, Stress-Rupture Data Correlation—Generalized Regression Analysis, An Alternative to Parametric Methods, in *Reproducibility and Accuracy of Mechanical Tests,* ASTM, Philadelphia, 1977, p 110-126

41. G.J. Hahn, The Role of Statistics, in *Development of a Standard Methodology for the Correlation and Extrapolation of Elevated Temperature Creep and Rupture Data,* EPRI FP-1062, Vol 1,

Electric Power Research Institute, Palo Alto, CA, 1979, p 4-1—4-45

42. M.K. Booker, V.K. Sikka, and B.L.P. Booker, Comparison of the Mechanical Strength Properties of Several High-Chromium Ferritic Steels, in *Ferritic Steels for High Temperature Applications,* American Society for Metals, 1983, p 257-273

43. M.K. Booker and B.L.P. Booker, New Methods for Analysis of Materials Strength Data for the ASME Boiler and Pressure Vessel Code, in *Use of Computers in Managing Material Property Data,* MPC-14, Metal Properties Council, American Society of Mechanical Engineers, New York, 1980, p 31-64

44. M.K. Booker, B.L.P. Booker, and R.W. Swindeman, *Analysis of Elevated-Temperature Tensile and Creep Properties of Normalized and Tempered 2¼Cr-1Mo Steel,* ORNL/TM-8075, Oak Ridge National Laboratory, Jan 1982

45. M.K. Booker and B.L.P. Booker, Analysis of the Elevated Temperature Tensile and Creep Properties of Wrought Carbon Steels, in *Factors Influencing the Time-Dependent Properties of Carbon Steels for Elevated Temperature Pressure Vessels,* MPC-19, Metal Properties Council, American Society of Mechanical Engineers, New York, 1983

46. J. Mandel and R.C. Paule, Interlaboratory Evaluation of a Material with Unequal Numbers of Replicates, *Anal. Chem.,* Vol 42, 1970, p 1194-1197, corrected in *Anal. Chem.,* Vol 43, 1971, p 1287

47. G.V. Smith, *Properties of Metals at Elevated Temperatures,* McGraw-Hill, New York, 1950, p 151

48. M.K. Booker, C.R. Brinkman, and V.K. Sikka, Correlation and Extrapolation of Creep Ductility Data for Four Elevated-Temperature Structural Materials, in *Structural Materials for Service at Elevated Temperatures in Nuclear Power Generation,* MPC-1, Metal Properties Council, American Society of Mechanical Engineers, New York, 1975, p 108-145

49. M.K. Booker and V.K. Sikka, A Study of Tertiary Creep Instability in Several Elevated-Temperature Structural Materials, in *Ductility and Toughness Considerations in Elevated Temperature Service,* MPC-8, Metal Properties Council, American Society of Mechanical Engineers, New York, 1978, p 325-343

Fatigue Data Analysis

By Richard C. Rice
Manager, Structural Integrity Projects Office
Mechanics Section
Battelle Columbus Laboratories

FATIGUE DATA are subject to considerable scatter or variability. This scatter can cause the results of a fatigue test program to be confusing if the test matrix is not well designed and the results are not evaluated in an appropriate statistical manner.

Fatigue testing usually is performed to estimate the relationship between the amplitude of stress or strain (or load or deflection) and the cycles-to-failure for a particular material or component. Failure may be defined in several ways, but the primary concerns are that the failure criteria be held constant in a given material evaluation and that the criteria be well defined. Fatigue testing is also conducted to compare the fatigue properties of two or more materials or components. In either case, the reliability of any decisions based on the results of a fatigue testing program is directly related to the manner in which the experiments are designed and analyzed.

This article reviews the planning of fatigue experiments, including the structure of a test plan, randomization, identification of realistic control modes and test types, sample size requirements, replication, and nuisance variables. The statistical characterization of the S/N (stress/life) or ε/N (strain/life) response of a single material tested under a single condition is discussed, as well as techniques for defining a mean fatigue curve and evaluating scatter or variability about that mean.

Information on the construction of confidence limits on the mean curve and on the equation parameters is discussed, as is the construction of tolerance limits on the fatigue response of a material. Probability-stress-life curve construction is also reviewed, and current approaches for the treatment of runouts are compared.

Standard techniques for statistical characterization of the fatigue strength or fatigue limit of a single material are presented by use of the Probit method, the up-and-down (staircase) method, and two-point procedures.

Stress-level selection methods are also presented.

Comparison of the fatigue behavior of two or more materials for data generated at a single stress or strain level is presented. Treatments to compare data generated over a range of stress or strain levels are included. Comparison of the fatigue strengths of two or more materials at a given life is also presented.

Consolidation of fatigue data generated at different conditions is discussed. The different conditions include sources or heats, mean stresses or strains, notch concentrations, and temperatures. Related articles of interest in this Volume include "Fatigue Crack Initiation," "Fatigue Crack Propagation," "Environmental Effects on Fatigue Crack Propagation," and "Fatigue Crack Growth Data Analysis."

Planning of Fatigue Experiments

Successful fatigue data analysis depends on proper experiment planning and preparation. If the test data have been generated previously, guidelines cannot be used to change the quality of the data. However, specific statistical procedures can be applied to help evaluate the completeness and adequacy of the data for a given application.

Defining a Structure. A concise organizational structure must be defined when conducting a series of fatigue experiments. The complexity of the plan will vary, depending on the number of key variables or treatments to be considered. For materials tests, these variables may include the effect of different heat treatment procedures, surface treatments, environments, mean stress levels, stress-concentration factors, control modes, or test temperatures. Many other variables may also be considered. The experimentalist should identify major variables or treatments

that are to be studied in a test program before testing begins.

The organizational structure should include tables and figures that illustrate the interrelationship of all variables in the test program. It should also identify an order of testing for the individual specimens, based on an appropriate randomization procedure. Finally, the organizational structure should identify the specific statistical procedures that will be applied to the data, because some issues depend on the analysis procedures to be used and on the ultimate goals of the investigation. The organizational structure will include the number of treatment levels, treatment combinations, the degree of replication, and the appropriate point of termination of a test series.

Defining Treatment Levels. Treatment levels are the distinct levels of each key variable to be examined. The levels may be qualitative or quantitative. A qualitative level might be different surface treatments (chemically milled versus ground and polished, for example) or different specimen geometries (hourglass versus uniform gage). A quantitative level might be different test temperatures ($T = 800$, 900, 1000, and 1100 °F, for example) or different theoretical stress-concentration factors ($K_t = 1.0$, 2.0, 3.0, 5.0).

Whether a treatment level is quantitative or qualitative, each level should be defined as precisely as possible, and the necessary controls should be introduced to ensure that all specimens are prepared within the precise limits defined for each level. The number of treatment levels is sometimes dictated by the specific problem being addressed. In other cases, the number of treatment levels is left to the discretion of the designer of the experiment. Minimizing the number of quantitative treatment levels is both practical and cost-effective.

Each level should represent a condition of interest that is expected to exhibit a significantly different fatigue behavior or response. For example, in the case of designated test temperatures for a fatigue evaluation of a heat-resistant alloy for a turbine blade, investigating temperature levels below 540 °C (1000 °F) may be nonproductive, even if the temperature levels are widely spaced. However, temperature differences as small as 14 °C (25 °F) may lead to significant differences in fatigue response if the temperature is in a range where microstructural changes can be expected to occur.

Several different treatments can apply (or be applied) to the same specimen. A combination of treatments may effectively be viewed as one separate treatment. In some cases, the order of application of the different treatments is critical and must be documented. For example, when investigating the influence of heat treatment and surface preparation, the net effect on the fatigue performance of the samples to be tested may be different, depending on the order of application of these treatments.

Randomization. When a regression analysis is performed on a collection of fatigue data, it is generally assumed that the observations are independent and randomized. This set of assumptions is seldom the case, however, and can lead to confusing or invalid conclusions.

For example, assume that fatigue data are available from two different heats of material. Also assume that it was of interest to establish whether the fatigue resistance of the two heats differed significantly. Because one heat was available before the other, the specimens were machined to somewhat different surface finish requirements, and each heat was tested in a different machine.

The second heat was tested over a somewhat different range of stress, because the first series of tests did not produce sufficiently high life data. Without additional information concerning the influence of extraneous variables, it is impossible to quantify accurately any observed difference in the fatigue resistance of these two heats.

Proper randomization provides the engineer with insurance against unknown effects of extraneous variables, either random during the experiment or systematic effects that may be associated with the experimental material or the structure of the experiment. Further, more precise evaluation of the effect of the key variables or treatments usually result from the use of randomization. Mechanical randomization imposed by the experimenter through use of random number tables is required to obtain unbiased estimates of the unknown differences between any combinations of key variables. In particular, if any statistical inferences are to be made, by the use of analysis of variance or related techniques, about observed differences, then the treatments must be randomized onto the experimental units for such inferences to have statistical validity. Reference 1 provides several examples of randomized test programs, including a completely randomized design, a randomized complete block design, and elementary split plot designs.

Identifying Control Modes and Test Types. In planning fatigue experiments, it is important to keep in mind the probable end use of the data. This should influence the environmental conditions chosen for the experiments and, in many cases, should also influence decisions concerning the nature of the test itself.

A wide range of fatigue testing equipment is available today that provides many options for control mode and test type. Fatigue tests can be performed on simple specimens, components, or complete structures. Conditions of constant-amplitude cyclic stress, strain, displacement, or load can be prescribed. Tests can be performed to induce cyclic bending, tension, torsion, and various other combinations of conditions.

The objective of a typical fatigue testing program is to characterize the fatigue resistance of a material under one or more conditions and at one or more stress or strain levels. Simple unnotched axial fatigue specimens and test methods have been standardized for this purpose (Ref 2). These specimens generally are tested under either load or strain control. Load-control tests are preferred for high-cycle fatigue testing (above 10^5 cycles), because operating frequencies can be higher and, therefore, test times are shorter.

Strain-control tests are preferred for low-cycle fatigue testing (below 10^3 cycles), because the cyclic plasticity induced under severe conditions can lead to cyclic ratcheting when load control is used, causing tensile rather than fatigue failures. High-stress, low-cycle fatigue tests performed under load (stress) control are sometimes subject to high scatter. Some specimens stabilize and fail in fatigue, while others ratchet and fail in tension.

Strain-control test results can display greater variability than load-control test results if the criterion for failure is not well defined. An unnotched axial fatigue specimen tested under load control generally will fail soon after the formation of a visible crack. However, complete failure may not occur in a similar specimen tested using strain control until long after a visible crack has formed. The cycles-to-failure may be several times as large as the cycles-to-crack-initiation. This tendency is particularly evident in ductile materials. Also, the same material tested in the same manner can fail abruptly if the fatigue crack initiates just outside the extensometer probes.

Some fatigue data are still developed in cantilever bending and rotating-beam-type test systems (Ref 3). These data should be considered separate from axial fatigue data, because bending data commonly exhibit higher (unconservative) fatigue data trends (Ref 4).

Sample Sizes and Replication. Because fatigue data exhibit considerable variability, sufficient data should be developed to identify statistically meaningful trends. Because fatigue tests are relatively expensive to run, some laboratories do not generate enough data to permit obtaining the desired estimates from test results.

The recommended number of tests and the distribution required to produce information over the life range of interest depend on the goals of the experimental plan, previous knowledge of the fatigue behavior of the material, and inherent variability in the data. Most fatigue data are developed on an exploratory basis to approximate the S/N or ϵ/N curve. For this approximation, the number of different stress levels that are tested should be based on the anticipated complexity of the underlying S/N response. When stress or strain (or the logarithm of stress or strain) is plotted versus the logarithm of fatigue life, one of several basic curve shapes generally results, as shown in Fig. 1. If a complex curvature is anticipated (Fig. 1a), five or more stress or strain levels may have to be investigated. However, in the majority of cases, three or four levels will be sufficient (away from the fatigue limit) to define the shape of the fatigue curve (Fig. 1b).

After the general curve shape is defined by individual specimens tested at each stress or strain level, the remaining specimens should be used to provide repeat testing (replication) at the same condition. Replication is most important under long-life test conditions, where variability tends to be the highest. Percent replication in a test program is calculated as:

Replication % =

$$100 \left[1 - \frac{\text{total number of different stress levels used in testing}}{\text{total number of specimens tested}} \right]$$

(Eq 1)

Using Eq 1, the guidelines presented in Table 1 have been proposed for the percent replication and the minimum number of specimens per S/N or ϵ/N curve (Ref 1). Note that a replication level of 25% implies repeat testing at only about one third of the chosen

Fig. 1 Typical S/N and ε/N curves with straight and curved regions

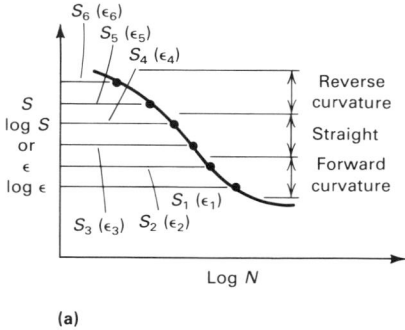

(a)

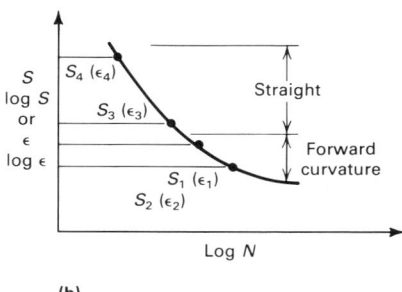

(b)

Type of test	Minimum percent replication	Minimum No. specimens
Exploratory or preliminary ...	17-33	6-12
Research and development ...	33-50	6-12
Design allowables data	50-75	12-24
Reliability data	75-88	12-24

Source: Ref 1

stress or strain levels, while a replication level of 75% implies the need for an average of four replicates at each stress or strain level. A higher number of replicates per stress level becomes more useful if lower bounds on the data—prediction limits, tolerance limits, etc.—are to be defined.

Minimizing Nuisance Variables. Nuisance variables are factors that may have an effect on the fatigue behavior of a material, but that are not of interest in the analysis. Such variables may include test machine response, specimen design and fabrication procedures, residual stresses, failure criteria, test frequency, waveform, and test type (axial versus bending).

Some nuisance variables can be eliminated, some can be minimized, while others cannot be readily controlled. The design of an experimental plan can systematically handle nuisance variables so that the observed variability does not confuse interpretation of other controlled variables. Test samples can be organized into homogeneous blocks so that primary effects can be studied free from biases caused by the nuisance variables. Reference 1 provides further discussion of nuisance variables and includes sample problems to illustrate use of the technique of blocking to minimize the influence of nuisance variables. References 5 and 6 also contain information on handling nuisance variables in fatigue experiments.

Statistical Characterization of the S/N or ε/N Response of a Material

Historically, millions of S/N curves have been generated. Although many analysis procedures have evolved for describing S/N and ε/N curves, none of these procedures is universally accepted and applied. Several of the important statistical issues regarding the analysis of S/N and ε/N data are discussed below.

Selection of the Dependent Variable. An S/N or ε/N curve illustrates the relationship between stress or strain and fatigue life. Conventionally, S/N curves are constructed on semilogarithmic plots of maximum stress versus cycles-to-failure, as shown in Fig. 2. Individual curves are constructed for specific mean stress levels or stress ratios. The most common conditions are a stress ratio of −1.0 (mean stress = 0.0) and a stress ratio of 0.0 (mean stress = one half the maximum stress).

For strain-control fatigue data, strain amplitude is plotted versus cycles-to-failure (or cycles-to-crack-initiation, determined by a specific criterion) on a log-log scale (Fig. 3). As with S/N data, curves sometimes are constructed for different mean strains or strain ratios, although the most common condition

is a strain ratio of −1.0 (mean strain = 0.0).

If an analytical expression is to be defined between stress and life or strain and life, it is important to decide which variable will be selected as the dependent variable. A relationship between two variables can be defined as:

$$y = A_0 + A_1x + \cdots \qquad \text{(Eq 2)}$$

where y is the dependent variable, and x is the independent variable. Equation 2 can take on many complicated forms; in any form, however, y is predicted from specific values of x. The controlled variable is x, and y is the predicted variable. In the case of S/N and ε/N data trends, regression equations take the form:

$$N = A_0 + A_1S + \cdots \qquad \text{(Eq 3)}$$

and

Fig. 2 S/N data and best-fit curves for aluminum alloy 2219-T851 at five different stress ratios

Stresses are based on net section. 1.0 ksi = 6.8948 MPa

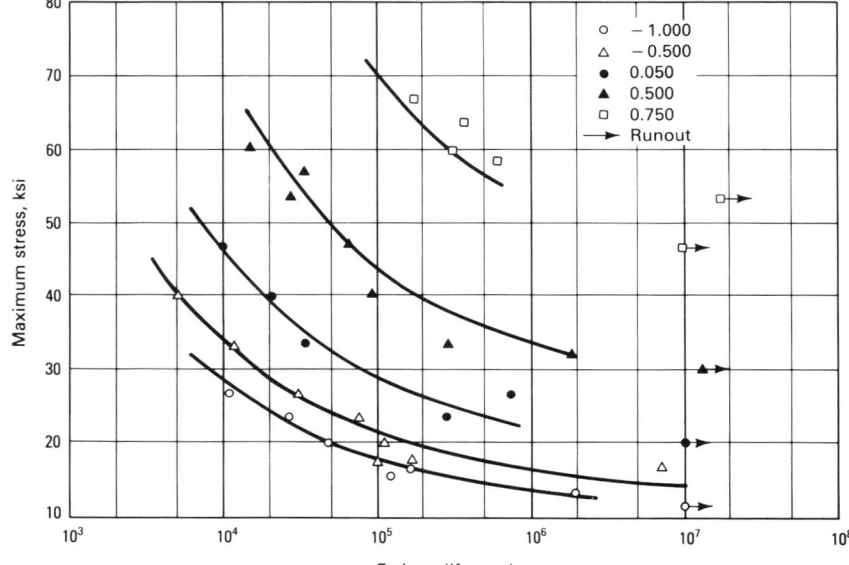

Fig. 3 Typical strain-life curve for a wrought steel

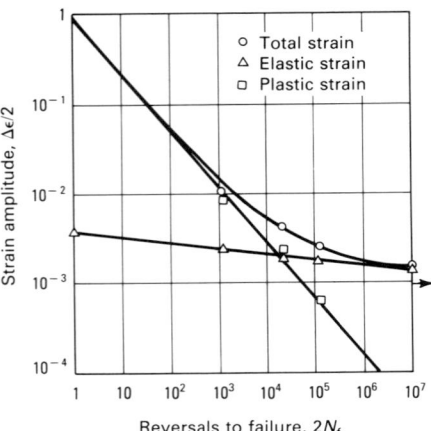

$$N = A_0 + A_1\epsilon + \cdots \qquad \text{(Eq 4)}$$

because stress or strain is the controlled variable and fatigue life is the predicted variable.

Certainly, stress or strain is never controlled completely in a fatigue experiment. However, the uncertainty, or scatter, in life data generally is greater than the uncertainty in maximum and minimum load, which defines the stress cycle, or the uncertainty in the maximum and minimum localized displacement, which defines the strain cycle.

Most test equipment controls stress or strain to within 1.0% of the chosen limits. However, the resultant scatter in fatigue life is at least a factor of two if multiple specimens are run under the same conditions. Thus, the effective uncertainty in life is commonly over 100 times greater than the uncertainty in stress or strain.

Therefore, stress or strain is the true independent variable, and fatigue life is the dependent variable. However, many stress-strain/fatigue-life relationships treat these variables in an opposite fashion. For example, the Basquin-Coffin-Manson (Ref 7-9) strain-life relationship is defined as:

$$\Delta\epsilon = A_1(2N_f)^{n_1} + A_2(2N_f)^{n_2} \qquad \text{(Eq 5)}$$

In this equation, the total strain range, $\Delta\epsilon$, is the dependent variable, and cycles-to-failure, N_f, is treated as the independent variable, because it is possible to "specify" the fatigue life and then compute an expected strain range for that life. This methodology is convenient in design, because determination of a strain level required for a part to survive a particular number of cycles is often of interest.

The modified Langer relation (Ref 10), another commonly used strain-life equation, is formulated as:

$$\Delta\epsilon = A_1 N_f^{n_1} + A_2 \qquad \text{(Eq 6)}$$

A form similar to Eq 6 is also used to analyze load-controlled fatigue data. In this case, the stress amplitude is treated as the dependent variable as follows (Ref 4):

$$\sigma_a = A_1(2N_f)^{n_1} \qquad \text{(Eq 7)}$$

Equations 5, 6, and 7 are similar, because all treat fatigue life as the independent variable.

Another difficulty arises in that fatigue-life data are plotted backward from the normal statistical convention of plotting the dependent variable as the ordinate, or vertical axis, and the independent variable as the abscissa, or horizontal axis. Visually, best-fit mean fatigue curves usually are constructed to minimize deviations in stress rather than life, because the eye tends to average vertical scatter more than horizontal scatter.

As stated previously, fatigue life commonly has been used as an independent variable. The statistically correct choice, however, is the opposite—fatigue life should be treated as the dependent variable. The use of fatigue life as the independent variable makes least squares estimators biased and inconsistent, forcing the model to overestimate and/or underestimate the true dependent variable.

This consequence is illustrated in Fig. 4, which represents a case in which three stress levels were chosen and three replicates were run at each stress level. A factor of ± 2 scatter about the mean was observed at each stress level. If maximum stress is used as the dependent variable, the dashed line results. If fatigue life is used as the dependent variable, the solid line results. Clearly, the dashed line underestimates the high-stress mean life and overestimates the low-stress mean life. Therefore, the expression involving fatigue life as the dependent variable is most realistic.

Mean Curve Definition. Using fatigue life as the dependent variable, a variety of simple linear regression models can be constructed that suggest a relationship between stress or strain and fatigue life. For example:

$$N = A_1 + A_2 (S \text{ or } \epsilon) \qquad \text{(Eq 8a)}$$

or

$$\log N = A_1 + A_2 (S \text{ or } \epsilon) \qquad \text{(Eq 8b)}$$

or

$$N = A_1 + A_2 \log (S \text{ or } \epsilon) \qquad \text{(Eq 8c)}$$

or

Fig. 4 Comparison of best-fit fatigue curves using stress and life as dependent variables

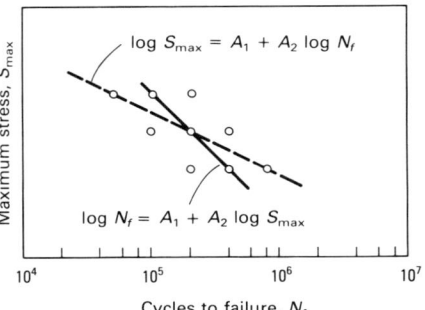

$$\log N = A_1 + A_2 \log (S \text{ or } \epsilon) \qquad \text{(Eq 8d)}$$

are all possibilities, where S and ϵ may refer to (1) the maximum value of stress or strain, (2) the amplitude or range of stress or strain, or (3) some combination of the independent variables. Equations 8(b) and 8(d), which involve the logarithm of fatigue life, are more commonly used than Eq 8(a) and 8(c).

Other more complicated expressions than the general form of Eq 8 have also been developed to include fatigue life as the dependent variable. Some include a fatigue-limit term (A_3) as follows (Ref 11):

$$\log N_f = A_1 + A_2 \log (S - A_3) \qquad \text{(Eq 9)}$$

Others are based on a hyperbolic relation of the form (Ref 12):

$$\log N_f = A_1(S)^{n_1} + A_2(S)^{n_2} \qquad \text{(Eq 10)}$$

Equation 10 is somewhat similar to Eq 5, although the physical interpretation of the coefficients and exponents is less well defined.

Several different techniques can be used to optimize the coefficients in Eq 8. Least squares regression analysis is the most commonly used technique in engineering fields and will therefore be discussed in this article. However, maximum likelihood fitting procedures are also available (Ref 13-16).

If the logarithms of the fatigue lives are assumed to be normally distributed about the mean line, and the scatter in log lives is assumed constant over the range of stress levels being considered, then log N can be considered the dependent variable in the analysis. Stress or strain is the independent variable, as in Eq 8(b). The analysis is most straightforward if there are no runouts in the data set.

Assuming that the fatigue data do conform to a log-linear S/N relationship, as in Eq 8(b), the equation parameters can be defined by the method of least squares as follows:

$$A_1 = \frac{\Sigma \log N \Sigma(S^2) - \Sigma S \Sigma(S \log N)}{n\Sigma(S^2) - (\Sigma S)^2} \quad \text{(Eq 11)}$$

$$A_2 = \frac{n\Sigma(S \log N) - (\Sigma S)(\Sigma \log N)}{n\Sigma(S^2) - (\Sigma S)^2} \quad \text{(Eq 12)}$$

where n is the number of fatigue tests in the sample.

The optimum equation coefficients for the other simple linear models (Eq 8a, 8c, and 8d) can be determined in the same manner. The optimum equation coefficients for other more complicated expressions (such as Eq 9 and 10) must be solved iteratively using a computer.

Data Scatter About the Mean. The distribution of fatigue lives about the mean curve is generally unknown. When a functional relationship is established between stress or strain and fatigue life, it is commonly assumed that the variability in fatigue lives does not change with increasing mean life. However, there is considerable evidence that this is not the case (Ref 17, 18). Fatigue lives tend to be more highly scattered as the mean life increases, as shown in Fig. 5 (Ref 4). The variability remains fairly constant up to about 10^6 cycles-to-failure, and generally increases significantly above this. Figure 6 shows the trend toward increasing variability with increasing mean life, as evidenced by the gradual flattening of the cumulative probability curves.

The distribution of fatigue lives at a given stress (or strain) is well modeled in many cases by log normal distribution; that is, the distribution of log N values at a given stress value tend to follow a normal distribution. When analyzing fatigue data, the residuals in log N as a function of stress should be examined to determine the quality of the fit and the uniformity of the scatter. A residual is the deviation of an individual point from the fitted curve (i.e., predicted minus the observed life). Figure 7 shows a series of hypothetical residual plots. In Fig. 7(a), the residuals display relatively constant variability. In Fig. 7(b), the residuals exhibit a distinct increase in variability for the lower stress levels, but the mean trends appear to be well represented. In Fig 7(c), the tendency is not one of nonuniform variance; rather, there is a problem with the assumed stress-life model. As shown, the mean fatigue lives are overestimated at high and low stresses, while they are underestimated at intermediate stress levels. The latter case should be reexamined with another S/N formula that is more flexi-

ble and/or better represents the mean fatigue trends over the complete range of interest.

If only mean trends are to be estimated, the nonuniformity of variance is not normally of great concern. However, increased variability at long lives does require more replication to define mean trends with equal precision. If the scatter or dispersion in the data increases by a factor of 2, then the number of tests required to define the mean curve with the same precision will also increase by about a factor of 2 (Ref 19).

Nonuniformity of variance becomes particularly important when lower bounds on the data are to be constructed. In this case, both the assumed distribution of fatigue lives for a given stress and the assumed pattern of increasing variability with increasing mean life greatly influence the limits that are defined.

If replicate data are available at individual stress levels and if a log normal distribution of fatigue lives can be assumed, the variability can be quantified in terms of the standard deviation in log N as:

$$S_l = \sqrt{\frac{\Sigma(\log N_i - \log \overline{N_l})^2}{n_l - 1}} \quad \text{(Eq 13)}$$

where n_l is the number of tests at level l, log $\overline{N_l}$ is the average logarithmic life value, log N_i is the ith observation of life at level l, and s_l is the standard deviation in log N at level l.

If the fatigue data have not been replicated, Eq 13 cannot be calculated at a single stress level. However, a measure of variability in fatigue lives can be obtained if several assumptions are made. First, it must be assumed that the underlying model for the mean curve is correct. Second, it must be assumed that the variance is uniform. With these assumptions, it is possible to compute an estimate of the fatigue-life variability (for a simple linear model, as in Eq 8) as follows:

Fig. 5 Schematic S/N diagram showing log normal distribution of lives at various stress levels

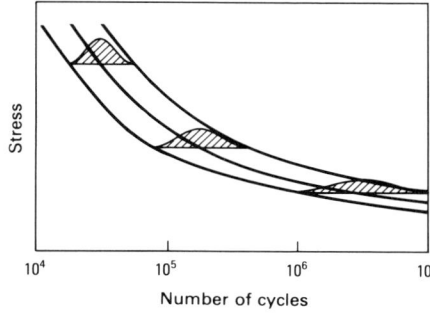

Fig. 6 Probability plot of percent failures versus life for different stress levels

Reversed bending. Square test section 12.7 by 12.7 mm (0.500 by 0.500 in.). Edges: sheared, sawed, milled

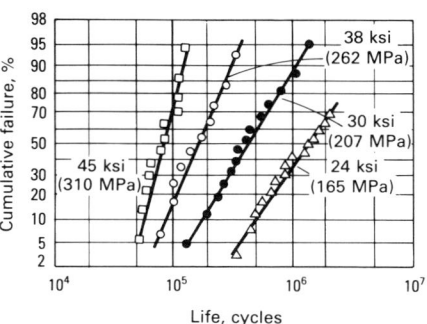

Fig. 7 Residual plots

(a) Uniform scatter. (b) Increasing scatter. (c) Uniform scatter, poor mean trend definition

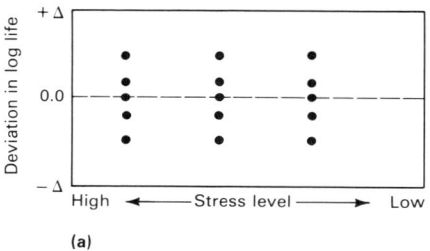

(a)

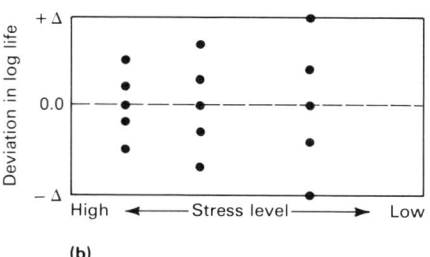

(b)

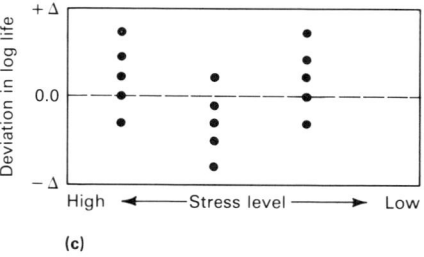

(c)

$$S_{yx} = \sqrt{\frac{\Sigma(\log N_i - \log N_{ic})^2}{n - 2}} \quad \text{(Eq 14)}$$

where n is the total number of test results, log N_{ic} is the calculated values of log N at the stress level for the ith test, and log N_i is the actual value of log N for the ith test.

The quantity computed in Eq 14 is known as the standard error of estimate. This formula can be used for other more complicated stress-life equations, but the denominator under the radical must reflect the proper degrees of freedom. For example, a second-order stress-life equation involving three optimized coefficients would require a denominator of $n - 3$ in the standard error of estimate calculation.

Confidence Limits on the Mean. Once an average curve has been established, the statistical confidence limits on the mean curve, or confidence bounds on the parameters in the equation for the mean curve, should be defined. Equation 8 illustrates the simple linear regression method of determining the regression parameter. Confidence limits can be computed in more complicated cases as well, if the equation is linear in all parameters (i.e., does not require an iterative solution). These procedures also assume that the regression residuals are normally distributed.

Confidence limits for a simple linear regression line (the mean curve), such as Eq 8(b), can be computed as follows:

$$\log N_{ic} \pm t_{\alpha/2,(n-2)}\, S_{yx} \sqrt{\frac{1}{n} + \frac{(S - \bar{S})^2}{(n-1)\, S_x^2}}$$

$$\text{(Eq 15)}$$

where n is the total number of test results, S is the stress value that produces regression estimates of log N_{ic}, $\bar{S}$ is the average stress value for all test results, S_x^2 is the standard deviation of stress values for all test results, S_{yx} is the standard error of estimate (Eq 14), and $t_{\alpha/2,(n-2)}$ is the Student's t value for a confidence level of $1 - \alpha$ and $n - 2$ degrees of freedom.

Confidence limits on the mean curve (log N_f as the dependent variable) for the hypothetical data presented in Fig. 4 are shown in Fig. 8. The limits are hyperbolic and have minimum widths at the mean stress value. The limits become wider as the distance from the mean stress value increases. The limits converge on the mean as the sample size becomes very large. Statistically, these limits define bounds within which the population mean curve could be expected to be located with a stated level of confidence.

Confidence limits can also be constructed on the least squares parameters. Confidence limits on the intercept term in Eq 8 are computed as follows:

$$A_1 \pm t_{\alpha/2,(n-2)}\, S_{yx} \sqrt{\frac{1}{n} + \frac{\bar{S}^2}{(n-1)\, S_x^2}}$$

$$\text{(Eq 16)}$$

Similarly, the confidence limits on the slope term in Eq 8 are computed as follows:

$$A_2 + t_{\alpha/2,(n-2)}\, S_{yx} \sqrt{\frac{1}{\Sigma(S_i - \bar{S})^2}}$$

$$\text{(Eq 17a)}$$

which is equivalent to:

$$A_2 \pm t_{\alpha/2,(n-2)}\, S_{yx} \sqrt{\frac{1}{(n-1)S_x^2}}$$

$$\text{(Eq 17b)}$$

With a simple linear regression equation, the confidence limits on the slope term should be examined to evaluate the significance of the regression equation. If the confidence interval about A_2 does not contain zero, then the regression can be considered significant at a $1 - \alpha$ confidence level. This result depends on the validity of the assumption that the true model for the data is linear.

If the fatigue data have been replicated at each stress level, the validity of the assumption of linearity can be evaluated as described in ASTM E 739 (Ref 20), which involves the computation of a critical parameter that must be less than the appropriate F statistic. This standard practice should be followed when the validity of a linear model is to be investigated.

Tolerance Limits on the Data. A lower tolerance limit can be constructed on a collection of S/N data to define a bound, above which a certain percentage of future observations can be expected to fall at a specified level of confidence. The tolerance limit that is constructed depends significantly on the assumptions that are made concerning the underlying distribution of residuals about the mean curve and the assumptions that are made regarding the uniformity of variance with increasing mean life. If a log normal distribution of residuals and a constant variance are assumed, a one-sided tolerance limit can be constructed on Eq 8(b) as follows:

$$\log N_{ic} - k_{\alpha,(n-2)}\, S_{yx} \sqrt{1 + \frac{1}{n} + \frac{(S - \bar{S})^2}{(n-1)S_x^2}}$$

$$\text{(Eq 18)}$$

where $k_{\alpha,(n-2)}$ is the tolerance limit factor for a confidence level of $1 - \alpha$ and $n - 2$ degrees of freedom (Ref 21).

One-sided lower tolerance limits constructed in accordance with Eq 18 are shown in Fig. 9 for the hypothetical data of Fig. 4. Note the great distance of these data bounds from the mean curve compared to the confidence limits on the mean shown in Fig. 8. The bounds must be significantly farther from the mean, because they are expected to bound a large percentage of future observations at a high level of confidence. By comparison, the confidence limits are only

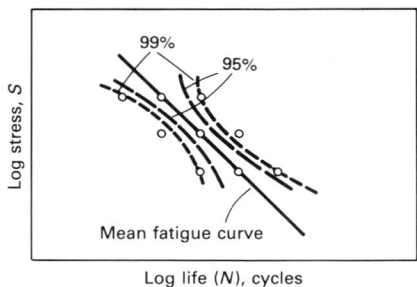

Fig. 8 Confidence limits on the mean fatigue curve

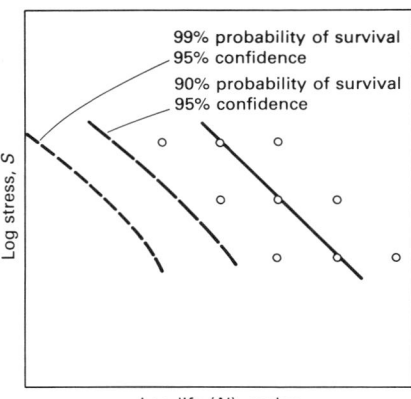

Fig. 9 One-sided tolerance limits on data

expected to contain the true mean curve at the same level of confidence.

The tolerance limit defined in Eq 18 is only valid if the stated assumptions are not violated. In some situations, the underlying distribution cannot be assumed, or evidence may suggest a Weibull, rather than log normal, distribution. Uniformity of variance may also be questionable. References 13 and 22 offer guidelines for the construction of tolerance limits on fatigue data in more complex situations.

Probability-Stress-Life Relationships. If a large number of fatigue experiments are completed at each of a number of different stress levels, it is possible to construct a probability-stress-life plot (Fig. 10). For each value of stress, it is possible to identify a range of fatigue lives that corresponds to different failure probabilities. For each value of life, it is thus possible to identify a range of fatigue strengths that correspond to different failure probabilities. Normally, a log-log presentation of stress versus life is used.

The most appropriate statistical model for the distribution of fatigue lives should be identified. Log normal and Weibull two- or

Fig. 10 Schematic of probability-stress-life plot

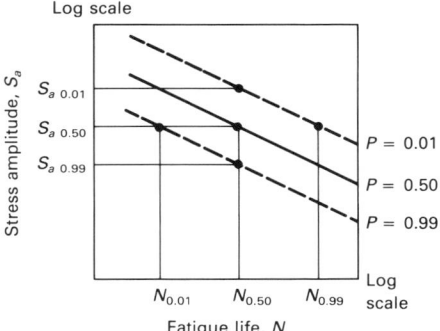

three-parameter distributions are used most commonly. Identification of the most appropriate statistical model for the distribution of fatigue strengths at a specific life is more complicated, because fatigue life is the dependent (uncontrolled) variable. A normal distribution of fatigue strengths has been shown to be a good approximation for cases in which sufficient data were available to perform a meaningful analysis (Ref 17, 23). Reference 24 provides an overview of the procedures to construct probability-stress-life plots and analytical models.

Treatment of Runouts. Runouts, or discontinued tests, occur in most fatigue testing programs. No one technique to accommodate these nonfailures statistically is used universally. It is generally agreed that the treatment procedure affects the fatigue trends that are established. Alternative procedures that have been used for the treatment of runouts include:
1. Disregard of all runouts in defining fatigue trends
2. Treatment of all runouts as failures
3. Conducting a trial analysis without runouts, then rerunning the analysis to include all runouts that fall above the initial mean curve
4. Incorporation of the runouts into the analysis via maximum likelihood methods
5. Incorporation of runouts into the analysis via iterative least squares procedure

If runouts are dispersed through the data collection at different stress levels and different fatigue lives, as is commonly the case, selecting the best procedure is not a straightforward decision. In the hypothetical example of Fig. 11, three tests were run at each of four stress levels, and the tests were continued to failure or stopped at 10^7 cycles. The mean fatigue curve is shown by the solid line.

If Alternative 1 above were used, 5 of the 12 test results would be ignored, and the mean curve would be biased downward

slightly in the long-life regime. The bias for this case would be relatively small, but it could be quite significant in other cases if only limited and/or highly scattered fatigue test failures were available.

If Alternative 2 were taken, the mean curve would be biased even further in the conservative direction at long lives. Alternative 2 can give either a conservative or unconservative estimate, depending on the number and location of the runouts in the sample. The normal bias is toward the conservative side.

Alternative 3 is essentially a hybrid of Alternatives 1 and 2. It tends to produce less conservative results in the long-life regime than either Alternative 1 or 2. Although the results of such an analysis may be more realistic than the trends produced with Alternatives 1 and 2, it is not an exact procedure and should only be used with caution.

Alternatives 4 and 5 are the most statistically rigorous procedures. Alternative 4 is detailed in Ref 13, and the results of many different analyses are presented to support this approach. Alternative 5 is discussed in Ref 25 and 26. These maximum likelihood and iterative least squares procedures are superior to the first three approaches. Although the first three approaches offer simplicity, they cannot provide unbiased estimates of mean fatigue trends on data sets with a significant number of runouts.

Treatment of Overstrain Data. Fatigue data trends for some materials are altered significantly in the long-life regime if the material is subjected to initial or periodic overstrains that are sufficient to cause some cyclic plasticity (Ref 27-29). For some materials, the apparent fatigue limit is virtually eliminated with the introduction of periodic overstrains. In the case illustrated in Fig. 12, the periodic overstrain failures occurred at strain levels well below the level at which runouts were routinely observed in the constant-amplitude experiments. Based on linear damage concepts (Ref 30), the periodic overstrains contributed a negligible fraction of the total damage (Ref 31).

Long-life fatigue data generally shows the most difference between periodic overstrain test conditions and constant-amplitude conditions. The former generally fall significantly below the latter. This effect should be considered in data analyses.

Scatter in periodic overstrain data also tends to be less than in constant-amplitude data. This difference can influence the construction of prediction limits and tolerance limits. It should be noted that variable-amplitude fatigue-life predictions have been shown to be more accurate when periodic overstrain data are used (Ref 30). This trend is particularly evident for stress or strain his-

Fig. 11 Schematic of fatigue data collection, including runouts

Value above symbol indicates number of observations.

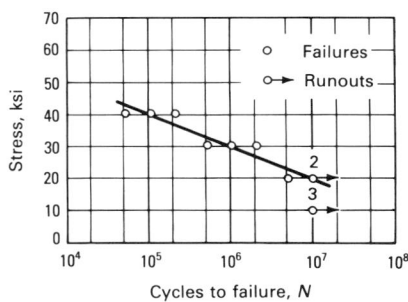

tories involving large numbers of small-amplitude cycles, intermixed with a few larger cycles.

Statistical Characterization of Fatigue Strength or Fatigue Limit

The fatigue strength of a material at a particular life can be characterized statistically using several different techniques. Median fatigue strength values may be identified, as well as lower confidence and/or tolerance limit values, for the fatigue strength of a material. The following procedures can be applied at almost any fatigue life and for any specific stress or strain ratio.

However, these approaches typically are used to characterize the so-called infinite life, fatigue limit, or endurance limit for a material at a stress ratio of -1.0. This approach is based on the concept that a stress amplitude exists below which most materials will not fail, at least on the average, even if they are cycled an infinite number of times.

If the material is prestrained or periodically overstrained, the statistically based fatigue strength should be calculated at a given finite life that is within the range of available

Fig. 12 Periodic overstrain fatigue behavior of SAE 1045 hot rolled bar

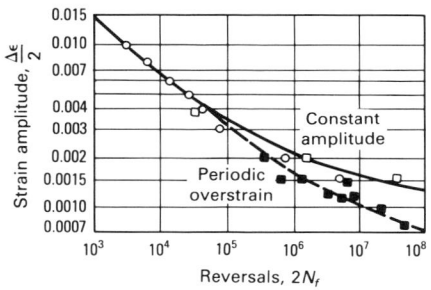

data. Caution should be exercised when determining whether such a fatigue strength still applies at higher fatigue lives.

Probit Method

The Probit method of statistically defining the fatigue strength of a material is based on sensitivity testing. This technique assumes that a certain level of stimulus to a test item will cause either a response or no response. The test typically is destructive. Either the item must be destroyed, or the characteristics of the item must be changed so significantly that further testing is meaningless. In addition, the percentage of items expected to respond (fail) must increase as the severity of the test increases.

For example, in fatigue testing, a certain number of stress or strain cycles of a certain magnitude will cause a material specimen or coupon either to fail or not fail. Failure can mean total separation (as it usually does in a load-control fatigue test), or it can mean initiation of a crack of a prescribed dimension (as it sometimes does in a strain-control fatigue test). The percentage of specimens that will fail after a specific number of cycles for a particular material will increase as the severity of the stress or strain cycle increases.

The exact response level (stress level) required to cause any specific specimen to fail in a particular number of cycles cannot be predicted exactly. More than one specimen can be tested at a given stress level for a specific number of cycles, however, and inferences can be made about the distribution of critical stress levels for a collection of specimens made from a particular material.

In most cases, the distribution of critical stress levels for fatigue data is assumed to be normal. However, three other types of distributions are applicable to sensitivity data: logistic (Ref 32), angular (Ref 33), and extreme value (Ref 34). Selection of a distribution is not of great concern, unless failure or survival probability levels of less than 10% are of concern.

Applied to fatigue data, the Probit technique involves a group of specimens tested at each of several uniformly spaced stress levels. Each test is continued until specimen failure occurs, or until the prescribed cycle limit is reached. When fatigue-limit or endurance-limit values are desired for a material, the cycle limit is usually set between 2×10^6 and 1×10^8 cycles. The exact values chosen depend on the material, test frequencies, and the application for which the test data are generated.

Typical fatigue test Probit response data are shown in Table 2. The stress levels were separated by 1.5 ksi, and a total of five different levels were considered. Replication was lowest at the stress level at which close to 50% failures were obtained, and replication was highest at the stress levels at which the lowest percentages of failures and non-failures (runouts) were observed.

This distribution of test samples produced a non-zero, non-unity fraction of failures at each level, which facilitated plotting of the data on a response curve (Fig. 13). Because a normal distribution of failure percentages was assumed, normal probability paper was used. The assumption of normality appears to be reasonable, because the failure percentages tend to fall in a straight line.

Typically, at least five specimens are tested at stress levels near the median fatigue strength value, or 50% failure stress (which in this case is about 43 ksi, or 296 MPa). Proportionally more samples are required at lower and higher failure percentages to define the response curve away from the median value with equal precision. About twice as many samples are necessary at the 10 and 90% failure stress levels as are required at the 50% failure stress level, and about three times as many are needed at the 5 and 95% failure stress levels.

Estimation of the specimen requirements at adjacent stress levels is facilitated by successive selection of the stress levels, beginning with the levels near the median fatigue strength value. Established fatigue data are helpful in identifying the stress levels near the median fatigue strength, at which testing should begin. If no previously generated data are available, exploratory tests may be necessary.

The Probit method requires testing of a large number of specimens. About 50 samples typically are required to develop a response curve to the 5 and 95% survival limits. Some savings in test time and number of samples can be achieved if the focus of the testing is directed toward the low-failure probability part of the response curve in the regime of the 50% survival limit and above. Development of the response curve in this region helps define safe operating stress levels for a particular material. The primary drawback in focusing on the lower end of the response curve is that the assumption of normality in the response curve may not be well validated. This approach also eliminates most of the failures in the experimental program, which take somewhat less time to run than nonfailures.

Table 2 Typical Probit fatigue test response data

Applied stress, ksi	No. of specimens tested	No. of specimens surviving 10^7 cycles	Survival (p), %
40.0	15	14	93.33
41.5	8	6	75.00
43.0	5	3	60.00
44.5	8	2	25.00
46.0	15	1	6.67

Note: 1.0 ksi = 6.8948 MPa
Source: Ref 34

Fig. 13 Typical response curve (normal probability) fitted to the example data of Table 2
Source: Ref 34

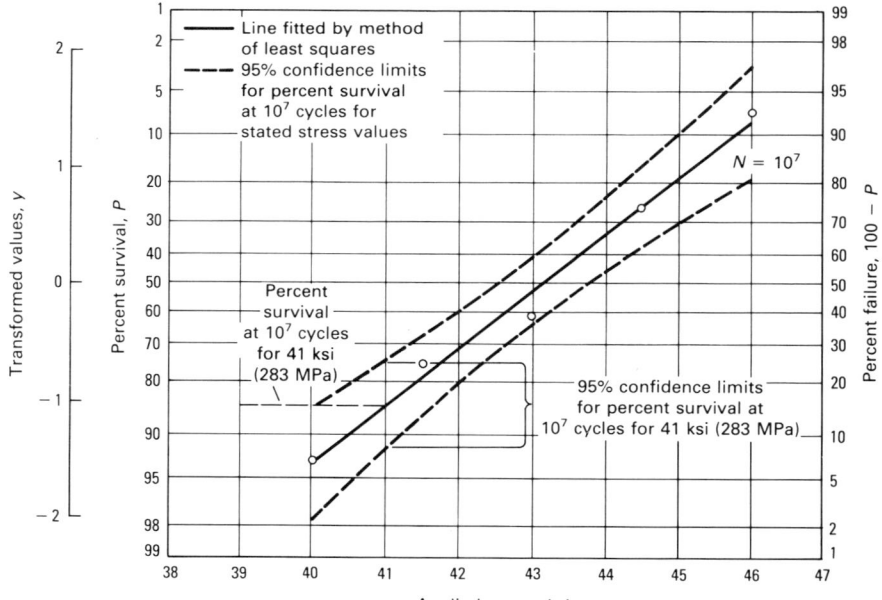

c

The advantage of the Probit test procedure is that it can be used to develop accurate estimates of the survival percentage at points away from the median stress level on the response curve. An equation for the response curve can be defined, and confidence limits can be constructed about that curve. The equation for the fitted line is normally defined as a simple first-order linear regression expression:

$$y = A_1 + A_2 (x - \bar{x}) \qquad \text{(Eq 19)}$$

where $y = z$, the normal deviate, or the transformed value of the failure percentage obtained from a table of areas under the normal curve; x is the stress or strain level; $\bar{x}$ is the average stress or strain level; and A_1 and A_2 are optimized regression coefficients, where A_2 is the slope term and A_1 is the value of y for $x = \bar{x}$, i.e.:

$$A_1 = \bar{y}$$

$$A_2 = \frac{\Sigma xy - k\bar{x}\bar{y}}{\Sigma x^2 - k\bar{x}^2}$$

where k is the number of groups tested.

Sample response curve computations for data analyzed by the Probit method shown in Table 2 are presented in Table 3. The predicted probabilities of survival correspond closely to the actual survival percentages.

If confidence limits are desired about the mean response curve, as shown in Fig. 13, they can be computed as follows (assuming that the normal transformation of the percent survival values is suitable):

$$y_f \pm S_{yx}\sqrt{2F_{2,k-2}} \sqrt{\frac{1}{k} + \frac{(x' - \bar{x})^2}{\Sigma (x - \bar{x})^2}}$$

$$\text{(Eq 20)}$$

where x' is the value of applied stress, x represents the individual stress values used to fit the response curve, $\bar{x}$ is the average of the stress values used to fit the response curve, k is the number of stress values, $F_{2,k-2}$ is the F statistic for 2 degrees of freedom for the numerator and $k - 2$ degrees of freedom for the denominator, and S_{yx} is the standard error of estimate:

$$S_{yx} = \sqrt{\frac{\Sigma (y - y_f)^2}{k - 2}}$$

Confidence limits can be computed for any stress levels within the range of the test data. To illustrate the process, 95% confidence limit calculations will be presented for the response data in Fig. 13. First, the standard error of estimate, S_{yx}, is computed to be 0.1427. Next, the F statistic can be found in

standard tables, for 95% confidence, to be 9.55. The value of $\bar{x}$ is 43.0 from Table 3, and k equals 5. The quantity $\Sigma (x - \bar{x})^2$ is 22.5. With this information, the confidence limits can be computed for 1-ksi increments in stress, as shown in Table 4. The statistical interpretation of the confidence limits shown in Table 4 and Fig. 13 is that they represent bounds within which the unknown response curve can be expected to lie with 95% confidence. At specific stress levels—for example, the lowest level of 40 ksi (276 MPa)—the probability of survival is likely to be in the range of about 85 to 98%.

Staircase Method

If only a limited number of specimens are available and if only the median fatigue resistance is required to be estimated, the staircase procedure should be applied rather than the Probit method. The following discussion applies to load-control tests on components and strain-control tests on single specimens—the approach is the same, but load or strain values rather than stress values are selected in the staircase procedure.

The staircase, or up-and-down, procedure is an abbreviated form of the Probit method. Specimens are tested sequentially, one at a time, with the first specimen being tested at a stress level equal to the estimated median fatigue limit, using preliminary S/N data to define this initial stress level.

If the first specimen survives the chosen fatigue life value, the next specimen is tested at a stress level that is an increment higher. If it fails before reaching the chosen fatigue-life value, the next specimen should be tested at a stress level that is an increment lower. Specimen by specimen, the stress levels are incremented up or down, depending on whether the current test specimen fails or is a runout (does not fail).

A hypothetical sequence is shown in Fig. 14. Specimens that do not fail are denoted by open circles, and those that fail are designated by closed circles. The sequential chart indicates the necessary level for the next test in a series. This procedure determines the median fatigue strength value range more quickly than the Probit method, but it provides little information concerning the probabilities of survival at stress levels above and below the median level.

The increments between stress levels should be correctly identified; if too far apart, sequential specimens are likely to "bounce" between failures and nonfailures,

Table 3 Computation of least squares response curve data presented in Table 2

	Applied stress (x), ksi	Survival (p), %	Transformed values of y	xy	x^2	Predicted y_f	p_f
	40.0	93.33	−1.501	−60.04	1600	−1.527	93.66
	41.5	75.00	−0.674	−27.97	1722.25	−0.789	78.49
	43.0	60.00	−0.254	−10.92	1849	−0.051	52.03
	44.5	25.00	0.674	29.99	1980.25	0.687	24.63
	46.0	6.67	1.501	69.05	2116	1.425	7.71
Sum	215	...	−0.254	0.11	9267.5	...	...

Note: 1.0 ksi = 6.8948 MPa
$\bar{x} = 43.0$
$\bar{y} = -0.051$
$A_1 = \bar{y} = -0.051$
$A_2 = \dfrac{0.11 - 5(43.0)(-0.051)}{9267.5 - 5(43)^2} = \dfrac{11.07}{22.50} = 0.492$

$y = -0.051 + 0.492 (x - 43.0)$

Table 4 Computation of 95% confidence limits on the response curve shown in Fig. 13

Applied stress (x'), ksi	y_f	y_c(a)	Confidence limits y values(b) Lower	Upper	p values(c) Lower	Upper
40.0	−1.527	0.483	−2.010	−1.044	97.8	85.2
41.0	−1.035	0.383	−1.418	−0.652	92.2	74.3
42.0	−0.543	0.308	−0.851	−0.235	80.3	59.3
43.0	−0.051	0.279	−0.330	−0.228	62.9	41.0
44.0	0.441	0.308	0.133	0.749	44.7	22.7
45.0	0.933	0.383	0.550	1.316	29.1	9.4
46.0	1.425	0.483	0.942	1.908	17.3	2.8

Note: 1.0 ksi = 6.8948 MPa

(a) $y_c = S_{yx}\sqrt{2F_{(2, k-2)}} \sqrt{\dfrac{1}{k} + \dfrac{(x' - \bar{x})^2}{\Sigma(x - \bar{x})^2}} = 0.1427(4.37)\sqrt{\dfrac{1}{5} + \dfrac{(x' - 43.0)^2}{22.5}}$.

(b) $y_{lower} = y_f - y_c$. $y_{upper} = y_f + y_c$. (c) Percent survival from a table of normal deviates for specific y values

Fig. 14 Schematic of staircase test method

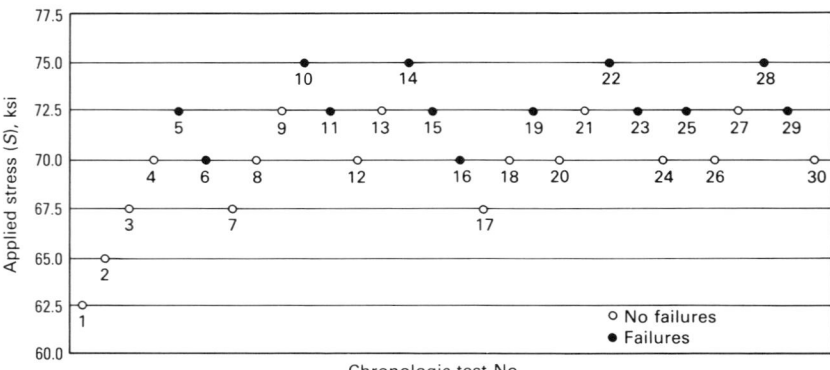

Chronologic test No.

○ No failures
● Failures

Table 5 Tabulation of staircase data used in computation of mean fatigue strength

Stress level, ksi	i(a)	N_i(b)	$i(N_i)$
75.0 2		4	6
72.5 1		7	7
70.0 0		2	0
Sums 13			13

Note: 1.0 ksi = 6.8948 MPa
(a) Numerical value of 0 stress level was assigned to lowest stress level where a failure occurred, and increasing numbers were assigned to succeeding higher stress levels. (b) Number of failures at the ith stress level

with the higher level producing almost all failures and the lower level producing almost all runouts. Conversely, if the stress levels are too close together, a large number of specimens may be required to complete a series of steps in the staircase.

Statistical analysis of staircase data is straightforward for producing estimates of the mean fatigue strength at a specific fatigue-life value. To complete the analysis, the following procedures should be followed:

1. Determine whether fewer failures or runouts have occurred in the staircase test series. The set with fewer occurrences becomes the area of interest. For example, in Fig. 14, there are 17 runouts and 13 failures. Failures should be considered in the following steps.
2. The stress levels are numbered, and a table of failures/runouts is constructed, as shown in Table 5.
3. The mean fatigue strength is then estimated by:

$$m = S_o + d\left(\frac{\Sigma_i N_i}{n} \pm \frac{1}{2}\right) \qquad \text{(Eq 21)}$$

where n is the total number of less frequent events, d is the stress increment, and S_o is the first stress level. Generally, $\pm\frac{1}{2}$ is the added quantity if the less frequent event is a runout, and it is the subtracted quantity if the less frequent event is a failure.

For the example case shown in Fig. 14 and Table 5, the mean fatigue strength estimate is:

$$m = 70.0 + 2.5\left(\frac{13}{13} - \frac{1}{2}\right)$$

$$= 70.0 + 1.25 = 71.25$$

No exact estimate of the confidence limits on the mean value is obtained when computed in this manner, although the procedure described below may be applicable if there are a sufficient number of replicates ($n \geq 6$) at stress levels above and below the mean

fatigue strength estimate and if non-unity, non-zero percentages of failure are available at these stress levels.

Two-Point Strategy

The two-point strategy is more widely used than the straight staircase procedure, because it requires fewer specimens. Both methods begin in the same manner, and the allocation of specimens to specific stress levels is the same until two stress levels are identified that have produced a non-unity and non-zero percentage of failures.

Typically this involves testing until a stress level that has produced all runouts produces a failure and a higher stress level that has produced all failures produces a runout. In Fig. 14, Test 6 produced the first non-zero, non-unity failure percentage, and Test 9 produced the second. After this point is reached, all further testing is concentrated at these two levels. At least six replicates at these two levels should then be generated.

Use of the two-point method, rather than the straight staircase procedure, would have eliminated several of the tests in Fig. 14—specifically, Tests 10, 14, 17, 22, and 28. The number of tests would have been reduced by five, which represents a 15 to 20% savings in testing time. In addition, if the number of replicates at each of these two levels had been limited to 6, only 16 tests would have been required. This modification would have reduced the original testing time by nearly 50%.

Development of median fatigue strength estimates for the two-point method is straightforward if certain assumptions are made. The first step is to compute the fractions failed at the two levels. In Fig. 14, the fraction failed at 72.5 ksi (500 MPa) was 64% (7 out of 11), while the fraction failed at 70.0 ksi (483 MPa) was 20% (2 out of 10). These fractions are then plotted on normal probability paper, assuming the distribution of fatigue strengths is believed to be normal.

If experience dictates the use of some other distribution of fatigue strengths (e.g., logistic, extreme value), then the appropriate alternative probability paper should be used. Regardless of the distribution selected, failure probabilities are assumed to plot as a straight line on the appropriate probability paper.

The fractions failed at 70.0 and 72.5 ksi (483 and 500 MPa) from Fig. 14 are plotted on normal probability paper in Fig. 15. The stress value associated with 50% failure is the median fatigue strength estimate. In this case, the median fatigue strength estimate is approximately 71.7 ksi (494 MPa) by graphical interpolation. This value can be determined with greater accuracy if the failure percentages are converted to fractiles of the normal distribution and the resultant fractiles are used in an analytical interpolation. For example, a failure percentage of 2 out of 10 corresponds to a 0.20 fractile, and for an assumed normal distribution this represents a normal deviate of $\hat{Y}_1 = -0.842$.

Similarly, a failure percentage of 7 out of 11 corresponds to a 0.636 fractile, which is equivalent to a normal deviate of $\hat{Y}_2 = +0.349$. Using these normal deviate values, it is possible to analytically compute a median fatigue strength as follows:

$$\hat{S}_{50} = 70 + 2.50\left[\frac{0.842}{(0.842 + 0.349)}\right]$$

$$= 71.77$$

(Eq 22)

Uncertainty in the median fatigue strength value can be identified through the calculation of the asymptotic variance, which is given by:

$$\hat{\sigma}_{50}^2 = \frac{1}{(S_2 - S_1)^2} \times \left[\frac{(S_2 - \hat{S}_{50})^2}{n_1 \hat{W}_1}\right.$$
$$\left. + \frac{(S_1 - \hat{S}_{50})^2}{n_2 \hat{W}_2}\right] \qquad \text{(Eq 23)}$$

where S_1 and S_2 are the replicated stress levels, $\hat{S}_{50}$ is the median fatigue strength, n_1 and

Fig. 15 Analysis of data developed by the two-point strategy

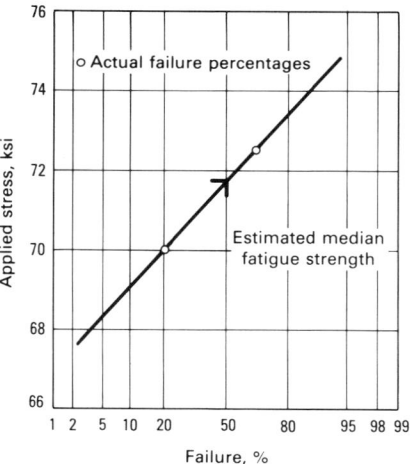

n_2 are the numbers of replicates at S_1 and S_2, and $\hat{W}_1$ and $\hat{W}_2$ are the weight factors associated with failure percentages at S_1 and S_2.

The weight factors used in the above equation can be estimated from Fig. 16 (Ref 1) by intersecting the abscissa at the failure percentage for level 1 to find $\hat{W}_1$ on the ordinate. The same procedure can be followed to find $\hat{W}_2$ from the failure percentage for level 2. For this example, $\hat{W}_1$ is approximately 0.49, and $\hat{W}_2$ is approximately 0.61 (because a normal distribution of failure percentages was assumed). Using Eq 23, it is then possible to compute the variance as follows:

$$\hat{\sigma}_{50}^2 = \frac{1}{(2.5)^2} \left[\frac{(0.73)^2}{10(0.49)} + \frac{(1.77)^2}{11(0.61)} \right]$$

$$= 0.092$$

Therefore, the standard deviation, $\hat{\sigma}$, on the median fatigue strength is approximately $0.303(\sqrt{\hat{\sigma}_{50}^2})$. Because $\pm 2\hat{\sigma}$ limits are approximately equivalent to a 95% confidence interval (assuming normality), the unknown median fatigue limit is expected to lie in the interval from about 70.5 to 73.0 with 95% confidence. These limiting values were computed as follows:

Approximate 95%

confidence limits $= \hat{S}_{50} \pm 2\sigma(B)$

where:

$$B = \frac{S_2 - S_1}{\hat{Y}_2 - \hat{Y}_1} = \frac{2.5}{1.19} = 2.10$$

so that:

95% confidence limits

$= 71.77 \pm 0.606(2.10)$

$= 70.50, 73.04$

The confidence limits tend to "close-in" on the unknown median fatigue strength value as the number of replicates increases, but the convergence is slow. This is illustrated in Table 6, in which the trends in the median fatigue strength value and the width of the confidence interval are shown for different points in the test series. The first computation of the median strength and 95% confidence limits was based only on test results at the 70.0 and 72.5 stress level up through Test 18. The next computation was based on results up through Test 19 and so on. Table 6 illustrates that even though the failure percentages at both levels varied significantly between Tests 18 and 30, the median fatigue strength estimate varied by no more than 0.725% from the final estimate.

The 95% confidence limits on the median value did narrow by almost 1 ksi. These results illustrate why a sample size of six at each of the levels used in the two-point strategy is sufficient to define statistically median fatigue strength (if a non-unity, non-zero percentage of failures is achieved after six tests at both levels).

Selecting Stress Levels

Application of the staircase or two-point strategies depends on a reasonable initial estimate of the median fatigue strength and the proper selection of the stress interval to be used. Preliminary or exploratory testing can be valuable in this regard. If sufficient specimens are available, exploratory testing can be started at a stress level significantly above the estimated fatigue strength. Testing can then work down in stress level so that several points on the S/N curve can be identified in the process, as shown in Fig. 17.

If only a limited number of specimens are available, specimens will be conserved if

Fig. 16 Plot of statistical weight W (per specimen) versus P, the true probability of failure at the given stress level used in testing

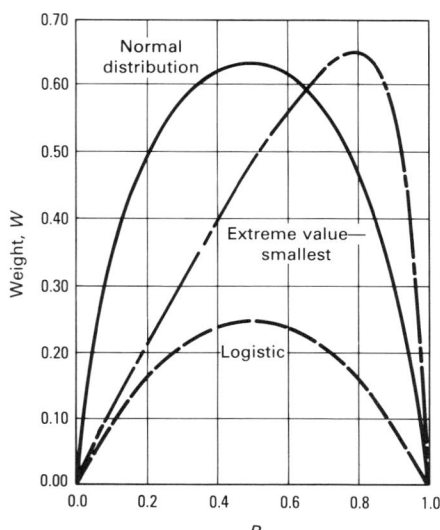

exploratory testing is started at a stress level below the estimated fatigue strength. If the specimen does not fail, it may be retested at a higher stress level, as shown in Fig. 18, assuming that it is a material that does not "coax." Coaxing is the process of artificially increasing the fatigue strength of a material susceptible to strain aging. It is achieved by fatigue cycling below the normal fatigue limit and subsequently stepping up the stress level in small increments. In most cases such a practice should be avoided, because it produces unconservative estimates of the fatigue resistance of a material. A dis-

Fig. 17 S-N data generated as preliminary test for subsequent two-point program beginning with up-and-down strategy

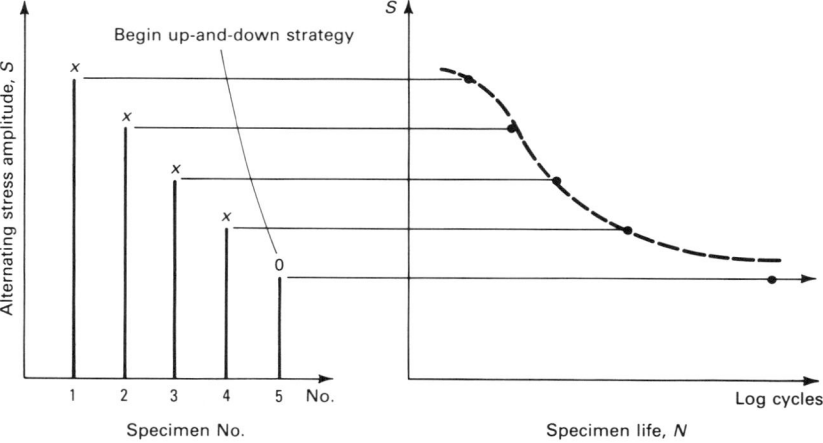

advantage of using the single-specimen approach on noncoaxing materials is that a considerable amount of testing time may be required to fail the first specimen, particularly if the steps increasing toward the fatigue limit are too small.

The most advantageous spacing of the stress levels is in the range of ⅔ to ½ of the standard deviation of the underlying distribution (Ref 35-37). The spacing in the example in Fig. 14 was approximately 1.2 standard deviations. This spacing corresponds to approximately 1 to 3% of the ultimate strength of most wrought materials.

Larger spacings, up to as high as 6 to 8% of the ultimate strength of the material, may be necessary for some materials with inconsistent fatigue properties. Initial staircase testing helps to define the appropriate spacing of the stress levels. The primary goal of the two-point strategy is to reach two stress levels that will produce significantly different non-zero and non-unity failure percentages.

Small-Sample Procedures

Recently, a method of obtaining valid median fatigue-limit estimates based on as few as six test points has emerged. This procedure is based on the staircase test method, but it requires fewer specimens after the first pair of failure and nonfailure results is obtained. This method depends greatly on prior information regarding the stress levels at which specimens are likely to fail for the fatigue life of interest. This method also depends on an assumed distribution of fatigue strengths, as is generally the case with the more traditional staircase and two-point methods (Ref 36, 37).

Fig. 18 Preliminary testing in which a single test specimen is used

This procedure is not recommended for low-carbon steels and other materials exhibiting marked coaxing behavior.

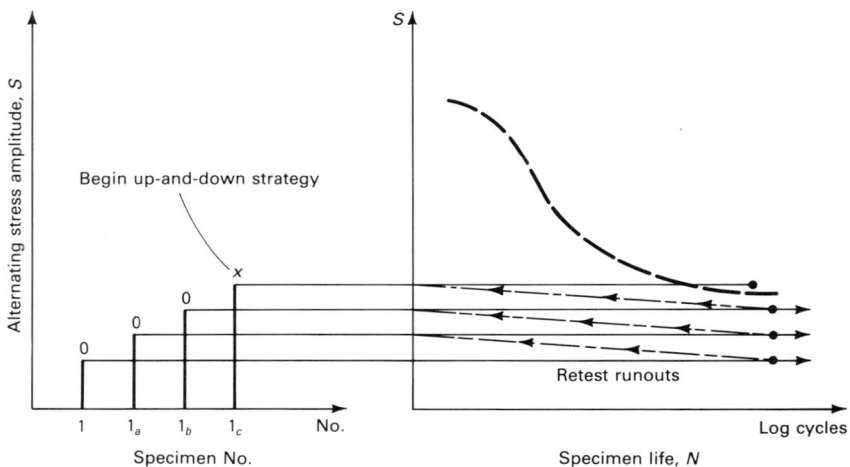

Table 6 Trends in median fatigue strength value and width of the confidence interval with increasing sample size

Test No.	Failure percentage Level 1 at 72.5 ksi	Level 2 at 70.0 ksi	Median fatigue strength estimate (S_{50}), ksi	Width of 95% confidence interval about median, ksi
18	⅗ = 0.600	²⁄₆ = 0.333	71.57	3.51
19	⁴⁄₆ = 0.667	...	71.25	3.28
20	...	²⁄₇ = 0.286	71.42	3.17
21	⁴⁄₇ = 0.571	...	71.90	3.04
23	⅝ = 0.625	...	71.60	2.93
24	...	²⁄₈ = 0.250	71.70	2.92
25	⁶⁄₉ = 0.667	...	71.52	2.74
26	...	²⁄₉ = 0.222	71.60	2.70
27	⁶⁄₁₀ = 0.600	...	71.88	2.61
29	⁷⁄₁₁ = 0.636	...	71.72	2.56
30	...	²⁄₁₀ = 0.200	71.77	2.54

Note: 1.0 ksi = 6.8948 MPa

Comparison of Fatigue Behavior of Different Materials

Material selection often must be based on compromises in mechanical properties and cost. Good fatigue properties are important in many design applications. An intelligent choice can only be made after it has been decided whether the fatigue properties of one material are better for a specific application than the fatigue properties of another material. Several factors must be taken into consideration when comparing the fatigue properties of two materials:

- Are average fatigue properties of greatest interest, or are superior lower bound failure probabilities of greatest concern?
- Is maximum fatigue life at a specific stress level or a combination of stress levels most important, or is maximum fatigue strength for a specific service period and number of stress or strain cycles most important?
- Does the application involve potential material failure sites at which conditions are essentially displacement limited, or load limited, or some combination of the two conditions?
- Is the application primarily a low-cycle fatigue situation in which ductility is most important, or is the application a high-cycle fatigue situation in which strength is more important than ductility?
- Is failure based on a durability approach to design, in which crack initiation is equivalent to failure, or on a damage-tolerant approach to design, in which failure is not predicted until initiated cracks have propagated to the point of component failure?
- Are time-dependent factors such as creep effects and corrosion effects important?

These are some of the background questions that should be considered carefully before comparing the fatigue behavior of two materials (Ref 38).

Fatigue Resistance at a Single Stress or Strain Level

If two or more groups of specimens are tested, the observed differences in the values may be due to chance or to differences in the populations from which the groups were drawn. For example, the observed differences could arise because of differences in material lots or differences in the characteristics of the testing machines. Determination of the origin of these differences is crucial to valid fatigue data analysis (Ref 38).

Nonparametric Evaluation of Group Medians. The rank tests described below assume that the several groups are drawn independently and randomly from populations of the same shape, but that the populations may differ with respect to their medians. All observed values in one group are assumed to come from one population. Because the populations are assumed to be of the same (although unknown) shape, only those groups that are tested at the same stress level should be compared, because the form of the distribution tends to change with changes in stress level.

Rank Test for Two Groups. In the rank test for two groups, the rank of each observation in the combined two groups is determined. The lowest value is given the rank of 1; the next higher observed value is given the rank of 2, etc. If one value appears several times—i.e., if there is a tie—the average of the ranks for those numbers is assigned to each one. For example, if the 11th, 12th, 13th, and 14th values are all equal, they are each given the rank of $(11 + 12 + 13 + 14)/4 = 12.5$.

The ranks for the two groups are totaled separately, and the total for one of the groups (the one with the smaller number of observations if the group sizes are unequal) is compared with the critical values given in Table 7 for sample sizes equal to the group sizes. If the observed value falls within the range of values given in the table for the chosen significance level (5 or 1% are common choices), the groups may be considered to have come from one population. If the observed value falls outside the range of values given in the table, the two groups are significantly different, i.e., come from two populations with different medians. The possibility of claiming that the groups can be considered significantly different when they are actually drawn from one population is reduced by use of a low (perhaps 1%) significance level.

Example 1. To compare two machines, the rank test was applied to the data given in Table 8, which represents 27 specimens randomly assigned to two testing machines. According to Table 7, the rank total for Machine A, which has the smaller number of measurements, should be between 101 and 179 ($N_1 = 10$, $N_2 = 17$) for the 5% level of significance and between 89 and 191 for the 1% level of significance. Thus, it would be expected that the actual total (87) would not occur as often as once in 100 samples due to chance alone if the two machines were completely interchangeable. Consequently, the conclusion is that, on the average, the machines are giving significantly different fatigue life values.

Rank Test for More Than Two Groups. The method of assigning ranks to more than two groups is the same as for the two-group test—that is, ranking the observations for all the groups combined. The ranks are totaled separately for each group, and the following test statistic, H, is computed from the rank totals (Ref 39):

$$H = \left[\frac{12}{N(N + 1)} \sum_{i=1}^{k} \frac{R_i^2}{n_i} \right] - 3(N + 1)$$

$$\text{(Eq 24)}$$

where k is the number of groups; n_i is the number of observations in the ith group; $N = \Sigma n_i$, the number of observations in all groups combined; and R_i is the sum of the ranks in the ith group.

The test statistic H is distributed approximately as χ^2 with $k - 1$ degrees of freedom, if each n_i is at least 5 (for a discussion of χ^2, see Ref 40). Thus, the value of H calculated from the observed data may be compared with the values of χ^2 given in Table 9 to determine whether there is a significant difference among the populations from which the groups were drawn. If H is greater than the χ^2 value for $k - 1$ degrees of freedom and the chosen significance level, the populations are stated to be different—that is, the groups may be said to have been drawn from two or more populations. Inspection of the rank totals usually indicates which groups are different if the difference is significant.

Example 2. To compare five machines, the rank test was applied to the data in Table 10, which represents 25 specimens randomly assigned to the five machines.

$$H = \frac{12}{25 \times 26} \times 4363.8 - 3 \times 26$$

$$= \frac{12 \times 4363.8}{650} - 78$$

$$= 80.56 - 78 = 2.56$$

In Table 9, with 4 degrees of freedom (one less than the number of groups), $\chi^2 = 9.49$, corresponding to a 5% significance level, or a percentile of 95. Because the computed value of H is 2.56, which is much smaller than 9.49, the observed values of fatigue life are considered to be from one population; i.e., the machines are considered interchangeable.

Nonparametric Evaluation of Percent Survival Values. The test statistic used to test the significance of the differences among percentage values computed from observed data is χ^2. The formula for χ^2 may be written in two ways; the second is preferred for computation purposes.

When the sample sizes are unequal:

$$\chi^2 = \frac{1}{\bar{p}(1 - \bar{p})} \Sigma n_i (p_i - \bar{p})^2$$

$$= \frac{\Sigma \dfrac{x_i^2}{n_i} - \dfrac{(\Sigma n_i)^2}{\Sigma n_i}}{\dfrac{\Sigma x_i}{\Sigma n_i} - \dfrac{(\Sigma x_i)^2}{\Sigma n_i}}$$

$$\text{(Eq 25)}$$

where k is the number of samples; $\Sigma = \sum_{i=1}^{k}$ is the sum over k samples; n_i is the size of ith sample ($i = 1, 2, \ldots, k$); x_i is the observed number of events in the ith sample; p_i is the observed fraction for the ith sample, $p_i = \chi_i/n_i$; and $\bar{p} = \Sigma x_i/\Sigma n_i$ is the average fraction for all samples combined. When the sample sizes are equal (Ref 15):

$$\chi^2 = \frac{n}{\bar{p}(1 - \bar{p})} \Sigma (p_i - \bar{p})^2$$

$$= \frac{\Sigma x_i^2 - k\bar{x}^2}{\bar{x}\left(1 - \dfrac{\bar{x}}{n}\right)}$$

$$\text{(Eq 26)}$$

where n is the sample size, and $\bar{x} = \Sigma x_i/k$.

The computed value of χ^2 may be compared with the tabular values given in Table 9 for $k - 1$ degrees of freedom (d.f. = $k - 1$). If the computed value of χ^2 is larger than the tabular value corresponding to a percentile equal to 100 minus the chosen significance level (which is commonly 5 or 10%), the percentages are said to be significantly different; i.e., the samples were drawn from different populations. If the computed value of χ^2 is smaller than the tabular value, the samples are considered to have come from one population.

Another use of the χ^2 test is to test whether the observed percentage values differ significantly from an arbitrary value. The method of computation is the same as that given previously, except that the first variation of the formula for χ^2 is used for the computations, the arbitrary value p' replaces $\bar{p}$, and the degrees of freedom equals k.

Example 3. To compare six lots of phosphor bronze strip, the χ^2 test was applied to the data given in Table 11, using a significance level of 10%. The tabular value of χ^2 corresponding to d.f. = 5 and percentile = 90% is 9.24. Because the computed value is 2.28, which is much smaller than 9.24, no significant difference among samples is indicated by these data, and the samples may be considered to have come from one population. If the computed value of χ^2 were larger than 9.24, the lots from which the samples were drawn would be considered to be significantly different.

Parametric Evaluation of Differences Between Two Groups of Fatigue Data. Before testing the averages of two samples to determine whether the parent populations are significantly different, the standard deviations of the two samples must be determined and checked for homogeneity. In the following example, the fatigue lives are assumed normal. Thus, let s_1 be the standard deviation for a sample of size n_1 from the first population, and let s_2 and n_2 be similarly

Table 7 Unpaired rank test

Critical lower and upper rank totals for the 5 and 1% levels of significance. Values in the body of the table refer to the group with the smaller number of measurements (N_1).

N_2		$N_1=4$	5	6	7	8	9	10	11	12	13	14	15	16	17	18	19	20	21	22	23	24	25
4	0.05%	11 25																					
	0.01%																						
5	0.05%	12 28	18 37																				
	0.01%		15 40																				
6	0.05%	13 31	19 41	26 52																			
	0.01%	10 34	16 44	23 55																			
7	0.05%	13 35	20 45	28 56	37 68																		
	0.01%	11 37	17 48	24 60	33 72																		
8	0.05%	14 38	21 49	29 61	39 73	49 87																	
	0.01%	11 41	18 52	25 65	34 78	44 92																	
9	0.05%	15 41	22 53	31 65	41 78	51 93	63 108																
	0.01%	12 44	19 56	27 69	36 83	46 98	57 114																
10	0.05%	16 44	24 56	33 69	43 83	54 98	66 114	79 131															
	0.01%	12 48	19 61	28 74	37 89	47 105	59 121	71 139															
11	0.05%	16 48	25 60	34 74	44 89	56 104	68 121	82 138	96 157														
	0.01%	13 51	20 65	29 79	38 95	49 111	61 128	74 146	88 165														
12	0.05%	17 51	26 64	36 78	46 94	58 110	71 127	85 145	100 164	116 184													
	0.01%	13 55	21 69	30 84	40 100	51 117	63 135	76 154	90 174	106 194													
13	0.05%	18 54	27 68	37 83	48 99	61 115	74 133	88 152	104 171	120 192	137 214												
	0.01%	14 58	22 73	31 89	41 106	53 123	65 142	79 161	93 182	109 203	126 226												
14	0.05%	19 57	28 72	39 87	50 104	63 121	77 139	91 159	107 179	124 200	142 222	160 246											
	0.01%	14 62	23 77	32 94	43 111	55 129	67 149	81 169	96 190	112 212	129 235	147 259											
15	0.05%		30 75	41 91	52 109	65 127	79 146	95 165	111 186	128 208	146 231	165 255	185 280										
	0.01%		23 82	33 99	44 117	56 136	70 155	84 176	99 198	116 220	133 244	151 269	171 294										
16	0.05%			42 96	54 114	68 132	82 152	98 172	114 194	132 216	150 240	170 264	190 290	212 316									
	0.01%			35 103	46 122	58 142	72 162	87 183	102 206	119 229	137 253	156 278	175 305	196 332									
17	0.05%				56 119	70 138	85 158	101 179	118 201	136 224	155 248	175 273	196 299	218 326	241 354								
	0.01%				47 128	60 148	74 169	89 191	105 214	122 238	141 262	159 289	179 316	200 344	223 372								
18	0.05%					73 143	88 164	104 186	121 209	140 232	159 257	179 283	201 309	223 337	247 365	271 395							
	0.01%					62 154	76 179	92 198	108 222	126 246	144 272	163 299	184 326	205 355	228 384	251 415							
19	0.05%						91 170	107 193	125 216	144 240	163 266	184 292	206 319	229 347	253 376	278 407	303 438						
	0.01%						79 182	94 206	111 230	129 255	147 282	167 309	188 337	210 366	233 396	257 427	282 459						
20	0.05%							111 199	129 223	148 248	168 274	189 301	211 329	234 358	259 387	284 418	310 450	338 482					
	0.01%							97 213	114 238	133 263	151 291	171 319	193 347	215 377	238 408	263 439	288 472	315 505					
21	0.05%								132 231	152 256	172 283	194 310	216 339	240 368	265 398	290 430	317 462	345 495	374 529				
	0.01%								117 246	136 272	155 300	175 329	197 358	220 388	244 419	268 452	294 485	321 519	349 554				
22	0.05%									156 264	177 291	199 319	222 348	246 378	271 409	297 441	324 474	352 508	381 543	412 579			
	0.01%									139 281	158 310	179 339	202 368	225 399	249 431	274 464	300 498	328 532	356 568	385 605			
23	0.05%										181 300	203 329	227 358	251 389	277 420	303 453	331 486	360 521	389 556	420 592	451 630		
	0.01%										162 319	184 348	206 379	230 410	254 443	280 476	306 511	334 546	363 582	392 620	423 658		
24	0.05%											208 338	232 368	257 399	283 431	310 464	338 498	367 533	397 569	428 606	460 644	493 683	
	0.01%											188 358	211 389	235 421	260 455	286 489	313 524	341 559	370 596	400 634	431 673	463 713	
25	0.05%												237 378	263 409	289 442	316 476	345 510	374 546	405 582	436 620	469 658	502 698	536
	0.01%												215 400	239 433	265 466	291 501	319 536	347 573	377 611	407 649	438 689	471 729	505
26	0.05%													268 420	295 453	323 487	352 522	382 558	412 596	444 634	477 673	511 713	546
	0.01%													244 444	270 478	297 513	325 549	354 586	383 625	414 664	446 704	479 745	513
27	0.05%														301 464	329 499	359 534	389 571	420 609	452 648	486 687	520 728	555
	0.01%														275 490	303 525	331 562	360 600	390 639	422 678	454 719	487 761	522
28	0.05%															336 510	366 546	396 584	428 622	461 661	494 702	529 743	565
	0.01%															308 538	337 575	367 613	397 653	429 693	462 734	495 777	530
29	0.05%																373 558	404 596	436 635	469 675	503 716	538 758	575
	0.01%																343 588	373 627	404 667	436 708	469 750	504 792	539
30	0.05%																	411 609	444 648	477 689	512 730	547 773	584
	0.01%																	380 640	411 681	444 722	477 765	512 808	547

Source: Ref 38

defined for the second. The value of the test statistic is thus computed as:

$$V = \frac{s_1^2}{s_2^2}$$

To determine whether the difference is significant, a significance level must have been selected previously. If the chosen risk (as a proportion) is denoted by α, compute $\beta = 1 - \alpha$, and for $\beta = 0.95$ or 0.975 obtain the value of F_β corresponding to $n_1 - 1$ degrees of freedom for the numerator and $n_2 - 1$ degrees of freedom for the denominator (Table 12). If $1/F < V < F$, then the sample variances (and therefore the sample standard deviations) are not considered to be significantly different. If $V < 1/F$, or if $V > F$, the sample standard deviations are considered to be significantly different.

Differences Between Two Means When the Sample Standard Deviations Are Not Significantly Different. The hypothesis that the population means are equal can be tested by computing an estimate of the common variance, s^2:

$$s^2 = \frac{(n_1 - 1)\, s_1^2 + (n_2 - 1)\, s_2^2}{n_1 + n_2 - 2} \qquad \text{(Eq 27)}$$

Next, the test statistic is computed:

$$t = \frac{\bar{x}_1 - \bar{x}_2}{s\sqrt{\dfrac{1}{n_1} + \dfrac{1}{n_2}}} \qquad \text{(Eq 28)}$$

where $\bar{x}_1$ and $\bar{x}_2$ are the sample means for the first and second samples, respectively. With a preassigned significance level—α, for example—$\beta = 1 - \alpha$ is computed, and t_β is found in Table 13 corresponding to d.f. $= (n_1 + n_2) - 2$. If $|t| > t_\beta$, the populations can be considered to be different in mean and/or in variance. On the average,

Table 9   Percentiles of the χ^2 distribution

Degrees of freedom	0.5	1	2.5	5	10	90	95	97.5	99	99.5
1	0.000039	0.00016	0.00098	0.0039	0.0158	2.71	3.84	5.02	6.63	7.88
2	0.0100	0.0201	0.0506	0.1026	0.2107	4.61	5.99	7.38	9.21	10.60
3	0.0717	0.115	0.216	0.352	0.584	6.25	7.81	9.35	11.34	12.84
4	0.207	0.297	0.484	0.711	1.064	7.78	9.49	11.14	13.28	14.86
5	0.412	0.554	0.831	1.15	1.61	9.24	11.07	12.83	15.09	16.75
6	0.676	0.872	1.24	1.64	2.20	10.64	12.59	14.45	16.81	18.55
7	0.989	1.24	1.69	2.17	2.83	12.02	14.07	16.01	18.48	20.28
8	1.34	1.65	2.18	2.73	3.49	13.36	15.51	17.53	20.09	21.96
9	1.73	2.09	2.70	3.33	4.17	14.68	16.92	19.02	21.67	23.59
10	2.16	2.56	3.25	3.94	4.87	15.99	18.31	20.48	23.21	25.19
11	2.60	3.05	3.82	4.57	5.58	17.28	19.68	21.92	24.73	26.76
12	3.07	3.57	4.40	5.23	6.30	18.55	21.03	23.34	26.22	28.30
13	3.57	4.11	5.01	5.89	7.04	19.81	22.36	24.74	27.69	29.82
14	4.07	4.66	5.63	6.57	7.79	21.06	23.68	26.12	29.14	31.32
15	4.60	5.23	6.26	7.26	8.55	22.31	25.00	27.49	30.58	32.80
16	5.14	5.81	6.91	7.96	9.31	23.54	26.30	28.85	32.00	34.27
18	6.26	7.01	8.23	9.39	10.86	25.99	28.87	31.53	34.81	37.16
20	7.43	8.26	9.59	10.85	12.44	28.41	31.41	34.17	37.57	40.00
24	9.89	10.86	12.40	13.85	15.66	33.20	36.42	39.36	42.98	45.56
30	13.79	14.95	16.79	18.49	20.60	40.26	43.77	46.98	50.89	53.67
40	20.71	22.16	24.43	26.51	29.05	51.81	55.76	59.34	63.69	66.77
60	35.53	37.48	40.48	43.19	46.46	74.40	79.08	83.30	88.38	91.95
120	83.85	86.92	91.58	95.70	100.62	140.23	146.57	152.21	158.95	163.64

Note: For large values of degrees of freedom, the approximate formula

$$\chi_\alpha^2 = \frac{1}{2}(z_\alpha + \sqrt{2n-1})^2$$

where z_α is the normal deviate, and n is the number of degrees of freedom, may be used. For example, $\chi_{0.99}^2 = \frac{1}{2}(2.326 + 10.909)^2 = 87.6$ for the 99th percentile for 60 degrees of freedom.
Source: Ref 40

Table 8   Fatigue test data for Example 1

Rank	Kilocycles
Machine A	
1	624
2	662
4	681
5	688
6	699
8	732
9	774
10	781
18	865
24	948
87 total	
Machine B	
3	667
7	715
11	811
12	822
13	833
14	841
15.5	842
15.5	842
17	849
19	869
20	892
21	903
22	944
23	946
25	1032
26	1067
27	1092
291 total	

Table 10   Fatigue test data for Example 2

		A	B	C	D	E	
		(5) 596	(6) 599	(3) 539	(2) 530	(1) 477	
		(10) 640	(13) 661	(12) 651	(8) 624	(4) 568	
		(11) 646	(21) 760	(14) 662	(9) 638	(7) 607	
		(18) 733	(22) 774	(15) 675	(16) 684	(17) 719	
		(24) 807	(23) 781	(19) 744	(25) 889	(20) 757	
Sum of ranks	R_i	68	85	63	60	49	
	$[R_i^2]$	4624	7225	3969	3600	2401	
							Total
	$\left[\dfrac{R_i^2}{n_i}\right]$	924.8	1445.0	793.8	720.0	480.2	4363.8

Table 11   Percentages surviving 10^8 cycles to illustrate nonparametric evaluation of percent survival values
Stress $= \pm 25$ ksi (± 172 MPa)

Lot No.	Sample size, n_i	Percent surviving, p_i	No. surviving, x_i	$\dfrac{x_i^2}{n_i}$
1	15	60.0	9	5.40
2	20	40.0	8	3.20
3	17	58.8	10	5.88
4	25	48.0	12	5.76
5	19	57.9	11	6.37
6	14	50.0	7	3.50
Total	$110 = \Sigma n_i$		$57 = \Sigma x_i$	$30.11 = \Sigma\dfrac{x_i^2}{n_i}$

$$\chi^2 = \frac{30.11 - \dfrac{3249}{110}}{0.52 - 0.27} = \frac{30.11 - 29.54}{0.25} = 2.28$$

Table 12 F distributions

Upper 5% points ($F_{0.95}$)

df denom	Degrees of freedom for numerator																		
	1	2	3	4	5	6	7	8	9	10	12	15	20	24	30	40	60	120	∞
1	161	200	216	225	230	234	237	239	241	242	244	246	248	249	250	251	252	253	254
2	18.5	19.0	19.2	19.2	19.3	19.3	19.4	19.4	19.4	19.4	19.4	19.4	19.4	19.5	19.5	19.5	19.5	19.5	19.5
3	10.1	9.55	9.28	9.12	9.01	8.94	8.89	8.85	8.81	8.79	8.74	8.70	8.66	8.64	8.62	8.59	8.57	8.55	8.53
4	7.71	6.94	6.59	6.39	6.26	6.16	6.09	6.04	6.00	5.96	5.91	5.86	5.80	5.77	5.75	5.72	5.69	5.66	5.63
5	6.61	5.79	5.41	5.19	5.05	4.95	4.88	4.82	4.77	4.74	4.68	4.62	4.56	4.53	4.50	4.46	4.43	4.40	4.37
6	5.99	5.14	4.76	4.53	4.39	4.28	4.21	4.15	4.10	4.06	4.00	3.94	3.87	3.84	3.81	3.77	3.74	3.70	3.67
7	5.59	4.74	4.35	4.12	3.97	3.87	3.79	3.73	3.68	3.64	3.57	3.51	3.44	3.41	3.38	3.34	3.30	3.27	3.23
8	5.32	4.46	4.07	3.84	3.69	3.58	3.50	3.44	3.39	3.35	3.28	3.22	3.15	3.12	3.08	3.04	3.01	2.97	2.93
9	5.12	4.26	3.86	3.63	3.48	3.37	3.29	3.23	3.18	3.14	3.07	3.01	2.94	2.90	2.86	2.83	2.79	2.75	2.71
10	4.96	4.10	3.71	3.48	3.33	3.22	3.14	3.07	3.02	2.98	2.91	2.85	2.77	2.74	2.70	2.66	2.62	2.58	2.54
11	4.84	3.98	3.59	3.36	3.20	3.09	3.01	2.95	2.90	2.85	2.79	2.72	2.65	2.61	2.57	2.53	2.49	2.45	2.40
12	4.75	3.89	3.49	3.26	3.11	3.00	2.91	2.85	2.80	2.75	2.69	2.62	2.54	2.51	2.47	2.43	2.38	2.34	2.30
13	4.67	3.81	3.41	3.18	3.03	2.92	2.83	2.77	2.71	2.67	2.60	2.53	2.46	2.42	2.38	2.34	2.30	2.25	2.21
14	4.60	3.74	3.34	3.11	2.96	2.85	2.76	2.70	2.65	2.60	2.53	2.46	2.39	2.35	2.31	2.27	2.22	2.18	2.13
15	4.54	3.68	3.29	3.06	2.90	2.79	2.71	2.64	2.59	2.54	2.48	2.40	2.33	2.29	2.25	2.20	2.16	2.11	2.07
16	4.49	3.63	3.24	3.01	2.85	2.74	2.66	2.59	2.54	2.49	2.42	2.35	2.28	2.24	2.19	2.15	2.11	2.06	2.01
17	4.45	3.59	3.20	2.96	2.81	2.70	2.61	2.55	2.49	2.45	2.38	2.31	2.23	2.19	2.15	2.10	2.06	2.01	1.96
18	4.41	3.55	3.16	2.93	2.77	2.66	2.58	2.51	2.46	2.41	2.34	2.27	2.19	2.15	2.11	2.06	2.02	1.97	1.92
19	4.38	3.52	3.13	2.90	2.74	2.63	2.54	2.48	2.42	2.38	2.31	2.23	2.16	2.11	2.07	2.03	1.98	1.93	1.88
20	4.35	3.49	3.10	2.87	2.71	2.60	2.51	2.45	2.39	2.35	2.28	2.20	2.12	2.08	2.04	1.99	1.95	1.90	1.84
21	4.32	3.47	3.07	2.83	2.68	2.57	2.49	2.42	2.37	2.32	2.25	2.18	2.10	2.05	2.01	1.96	1.92	1.87	1.81
22	4.30	3.44	3.05	2.82	2.66	2.55	2.46	2.40	2.34	2.30	2.23	2.15	2.07	2.03	1.98	1.94	1.89	1.84	1.78
23	4.28	3.42	3.03	2.80	2.64	2.53	2.44	2.37	2.32	2.27	2.20	2.13	2.05	2.01	1.96	1.91	1.86	1.81	1.76
24	4.26	3.40	3.01	2.78	2.62	2.51	2.42	2.36	2.30	2.25	2.18	2.11	2.03	1.98	1.94	1.89	1.84	1.79	1.73
25	4.24	3.39	2.99	2.76	2.60	2.49	2.40	2.34	2.28	2.24	2.16	2.09	2.01	1.96	1.92	1.87	1.82	1.77	1.71
30	4.17	3.32	2.92	2.69	2.53	2.42	2.33	2.27	2.21	2.16	2.09	2.01	1.93	1.89	1.84	1.79	1.74	1.68	1.62
40	4.08	3.23	2.84	2.61	2.45	2.34	2.25	2.18	2.12	2.08	2.00	1.92	1.84	1.79	1.74	1.69	1.64	1.58	1.51
60	4.00	3.15	2.76	2.53	2.37	2.25	2.17	2.10	2.04	1.99	1.92	1.84	1.75	1.70	1.65	1.59	1.53	1.47	1.39
120	3.92	3.07	2.68	2.45	2.29	2.18	2.09	2.02	1.96	1.91	1.83	1.75	1.66	1.61	1.55	1.50	1.43	1.35	1.25
∞	3.84	3.00	2.60	2.37	2.21	2.10	2.01	1.94	1.88	1.83	1.75	1.67	1.57	1.52	1.46	1.39	1.32	1.22	1.00

Upper 2.5% points ($F_{0.975}$)

df denom	Degrees of freedom for numerator																		
	1	2	3	4	5	6	7	8	9	10	12	15	20	24	30	40	60	120	∞
1	648	800	864	900	922	937	948	957	963	969	977	985	993	997	1,001	1,006	1,010	1,014	1,018
2	38.5	39.0	39.2	39.2	39.3	39.3	39.4	39.4	39.4	39.4	39.4	39.4	39.4	39.5	39.5	39.5	39.5	39.5	39.5
3	17.4	16.0	15.4	15.1	14.9	14.7	14.6	14.5	14.5	14.4	14.3	14.3	14.2	14.1	14.1	14.0	14.0	13.9	13.9
4	12.2	10.6	9.98	9.60	9.36	9.20	9.07	8.98	8.90	8.84	8.75	8.66	8.56	8.51	8.46	8.41	8.36	8.31	8.26
5	10.0	8.43	7.76	7.39	7.15	6.98	6.85	6.76	6.68	6.62	6.52	6.43	6.33	6.28	6.23	6.18	6.12	6.07	6.02
6	8.81	7.26	6.60	6.23	5.99	5.82	5.70	5.60	5.52	5.46	5.37	5.27	5.17	5.12	5.07	5.01	4.96	4.90	4.85
7	8.07	6.54	5.89	5.52	5.29	5.12	4.99	4.90	4.82	4.76	4.67	4.57	4.47	4.42	4.36	4.31	4.25	4.20	4.14
8	7.57	6.06	5.42	5.05	4.82	4.65	4.53	4.43	4.36	4.30	4.20	4.10	4.00	3.95	3.89	3.84	3.78	3.73	3.67
9	7.21	5.71	5.08	4.72	4.48	4.32	4.20	4.10	4.03	3.96	3.87	3.77	3.67	3.61	3.56	3.51	3.45	3.39	3.33
10	6.94	5.46	4.83	4.74	4.24	4.07	3.95	3.85	3.78	3.72	3.62	3.52	3.42	3.37	3.31	3.26	3.20	3.14	3.08
11	6.72	5.26	4.63	4.28	4.04	3.88	3.76	3.66	3.59	3.53	3.43	3.33	3.23	3.17	3.12	3.06	3.00	2.94	2.88
12	6.55	5.10	4.47	4.12	3.89	3.73	3.61	3.51	3.44	3.37	3.28	3.18	3.07	3.02	2.96	2.91	2.85	2.79	2.72
13	6.41	4.97	4.35	4.00	3.77	3.60	3.48	3.39	3.31	3.25	3.15	3.05	2.95	2.89	2.84	2.78	2.72	2.66	2.60
14	6.30	4.86	4.24	3.89	3.66	3.50	3.38	3.28	3.21	3.15	3.05	2.95	2.84	2.79	2.73	2.67	2.61	2.55	2.49
15	6.20	4.77	4.15	3.80	3.58	3.41	3.29	3.20	3.12	3.06	2.96	2.86	2.76	2.70	2.64	2.59	2.52	2.46	2.40
16	6.12	4.69	4.08	3.73	3.50	3.34	3.22	3.12	3.05	2.99	2.89	2.79	2.68	2.63	2.57	2.51	2.45	2.38	2.32
17	6.04	4.62	4.01	3.66	3.44	3.28	3.16	3.06	2.98	2.92	2.82	2.72	2.62	2.56	2.50	2.44	2.38	2.32	2.25
18	5.98	4.56	3.95	3.61	3.38	3.22	3.10	3.01	2.93	2.87	2.77	2.67	2.56	2.50	2.44	2.38	2.32	2.26	2.19
19	5.92	4.51	3.90	3.56	3.33	3.17	3.05	2.96	2.88	2.82	2.72	2.62	2.51	2.45	2.39	2.33	2.27	2.20	2.13
20	5.87	4.46	3.86	3.51	3.29	3.13	3.01	2.91	2.84	2.77	2.68	2.57	2.46	2.41	2.35	2.29	2.22	2.16	2.09
21	5.83	4.42	3.82	3.48	3.25	3.09	2.97	2.87	2.80	2.73	2.64	2.53	2.42	2.37	2.31	2.25	2.18	2.11	2.04
22	5.79	4.38	3.78	3.44	3.22	3.05	2.93	2.84	2.76	2.70	2.60	2.50	2.39	2.33	2.27	2.21	2.14	2.08	2.00
23	5.75	4.35	3.75	3.41	3.18	3.02	2.90	2.81	2.73	2.67	2.57	2.47	2.36	2.30	2.24	2.18	2.11	2.04	1.97
24	5.72	4.32	3.72	3.38	3.15	2.99	2.87	2.78	2.70	2.64	2.54	2.44	2.33	2.27	2.21	2.15	2.08	2.01	1.94
25	5.69	4.29	3.69	3.35	3.13	2.97	2.85	2.75	2.68	2.61	2.51	2.41	2.30	2.24	2.18	2.12	2.05	1.98	1.91
30	5.57	4.18	3.59	3.25	3.03	2.87	2.75	2.65	2.57	2.51	2.41	2.31	2.20	2.14	2.07	2.01	1.94	1.87	1.79
40	5.42	4.05	3.46	3.13	2.90	2.74	2.62	2.53	2.45	2.39	2.29	2.18	2.07	2.01	1.94	1.88	1.80	1.72	1.64
60	5.29	3.93	3.34	3.01	2.79	2.63	2.51	2.41	2.33	2.27	2.17	2.06	1.94	1.88	1.82	1.74	1.67	1.58	1.48
120	5.15	3.80	3.23	2.89	2.67	2.52	2.39	2.30	2.22	2.16	2.05	1.95	1.82	1.76	1.69	1.61	1.53	1.43	1.31
∞	5.02	3.69	3.12	2.79	2.57	2.41	2.29	2.19	2.11	2.05	1.94	1.83	1.71	1.64	1.57	1.48	1.39	1.27	1.00

Degrees of freedom for denominator

Source: Ref 41

Table 13 Values of the *t* statistic

Degrees of freedom	$t_{0.95}$	$t_{0.975}$	$t_{0.9875}$	$t_{0.995}$	$t_{0.9975}$
1	6.31	12.7	25.5	63.7	127
2	2.92	4.30	6.21	9.92	14.1
3	2.35	3.18	4.18	5.84	7.45
4	2.13	2.78	3.50	4.60	5.60
5	2.01	2.57	3.16	4.03	4.77
6	1.94	2.45	2.97	3.71	4.32
7	1.89	2.36	2.84	3.50	4.03
8	1.86	2.31	2.75	3.36	3.83
9	1.83	2.26	2.69	3.25	3.69
10	1.81	2.23	2.63	3.17	3.58
11	1.80	2.20	2.59	3.11	3.50
12	1.78	2.18	2.56	3.05	3.43
13	1.77	2.16	2.53	3.01	3.37
14	1.76	2.14	2.51	2.98	3.33
15	1.75	2.13	2.49	2.95	3.29
16	1.75	2.12	2.47	2.92	3.25
17	1.74	2.11	2.46	2.90	3.22
18	1.73	2.10	2.45	2.88	3.20
19	1.73	2.09	2.43	2.86	3.17
20	1.72	2.09	2.42	2.85	3.15
21	1.72	2.08	2.41	2.83	3.14
22	1.72	2.07	2.41	2.82	3.12
23	1.71	2.07	2.40	2.81	3.10
24	1.71	2.06	2.39	2.80	3.09
25	1.71	2.06	2.38	2.79	3.08
26	1.71	2.06	2.38	2.78	3.07
27	1.70	2.05	2.37	2.77	3.06
28	1.70	2.05	2.37	2.76	3.05
29	1.70	2.05	2.36	2.76	3.04
30	1.70	2.04	2.36	2.75	3.03
40	1.68	2.02	2.33	2.70	2.97
60	1.67	2.00	2.30	2.66	2.91
120	1.66	1.98	2.27	2.62	2.86
∞	1.64	1.96	2.24	2.58	2.81

Degrees of freedom	$t_{0.05}$	$t_{0.025}$	$t_{0.0125}$	$t_{0.005}$	$t_{0.0025}$

Note: When the table is read from the bottom, the values are to be prefixed with a negative sign. Interpolation should be performed using the reciprocals of the degrees of freedom.
Source: Ref 40

identical populations will be erroneously judged different about 100 $\alpha\%$ of the time.

If the samples are large enough so that any important difference in the variances would have been detected, a value of $|t| > t_\beta$ can be attributed to different population means. With the same reservation about sample size, if $|t| < t_\beta$, the population means may be considered equal.

Difference Between Two Means When the Sample Standard Deviations Are Significantly Different. If the sample standard deviations are significantly different, the hypothesis that the population means are equal can be tested by the following computations (Ref 16):

$$t' = \frac{\bar{x}_1 - \bar{x}_2}{\sqrt{\dfrac{s_1^2}{n_1} + \dfrac{s_2^2}{n_2}}} \qquad \text{(Eq 29)}$$

and

$$c = \frac{\dfrac{s_1^2}{n_1}}{\dfrac{s_1^2}{n_1} + \dfrac{s_2^2}{n_2}} \qquad \text{(Eq 30)}$$

From Table 13, a value of t_β is read corresponding to:

$$\text{d.f.} = \left[\frac{c^2}{n_1 - 1} + \frac{(1 - c)^2}{n_2 - 1}\right]^{-1} \qquad \text{(Eq 31)}$$

If this value is not an integer, the nearest smaller integer is used. If $|t'| < t_\beta$, the population means are judged to be equal, and if $|t'| > t_\beta$, they are determined to be unequal; $\alpha = 1 - \beta$ is the approximate proportion of the time when the means are actually equal, but will be incorrectly judged unequal.

Example 4. After testing specimens with one surface finish, another lot is fabricated with a different finish and tested at the same stress level. The tests are to be analyzed to determine whether the change in surface finish significantly affects the fatigue life at the stress level used for the tests (see Table 14). The distribution of fatigue lives at this stress level is almost log normal; therefore, the computations are in terms of log N rather than N.

The mean of each sample and the variance (s^2) of each sample have been obtained. To test for significant differences, the F ratio test is used and then the Student's t-test, using a significance level of 0.05. The computed F ratio is:

$$F = \frac{0.00527}{0.00276} = 1.91$$

From Table 12, $F_{0.95}$ (d.f. numerator $= 7$)(d.f. denominator $= 9$) $= 3.29$. Therefore, the variances of the two samples are not considered to be significantly different. Thus, the test given in Eq 27 and 28 may be used:

$$s^2 = \frac{(7 \times 0.00527) + (9 \times 0.00276)}{8 + 10 - 2}$$

$$= 0.00386$$

$$t = \frac{4.9661 - 4.6495}{\sqrt{0.00386}\sqrt{\dfrac{1}{8} + \dfrac{1}{10}}}$$

$$= \frac{0.3166}{0.0621\sqrt{0.225}} = \frac{0.3166}{0.0294} = 10.8$$

From Table 13, $t_{0.95}$(d.f. $= 16$) $= 1.75$. Because $t = 10.8$ is larger than $t_{0.95} = 1.75$, the mean of the first sample can be considered to be significantly larger than the mean of the second sample. In terms of the fatigue life at the test stress level, the second surface finish appears to be inferior to the first finish.

Differences Among Multiple Means. If $k(k > 2)$ sets of data have been obtained, each of which is a random sample from a normal population, and these populations are assumed to have a common standard deviation, the hypothesis that these populations have a common mean can be tested. Let x_{ij} be the jth observation from the ith group, let $\bar{x}_i$ be the mean of the ith group, and let n_i be the size of the ith group. s_w^2 and s_b^2 can be defined by:

$$s_w^2 = \frac{\displaystyle\sum_{i=1}^{k}\sum_{j=1}^{n_i}(x_{ij} - \bar{x}_i)^2}{\displaystyle\sum_{i=1}^{k}n_i - k} \qquad \text{(Eq 32)}$$

$$s_b^2 = \frac{\displaystyle\sum_{i=1}^{k}n_i(\bar{x}_i - \bar{x})^2}{k - 1} \qquad \text{(Eq 33)}$$

Table 14 Computations for significance tests

First surface finish			Second surface finish		
$x_i = \log N_i$	$x_i - \bar{x}$	$(x_i - \bar{x})^2$	$x_i = \log N_i$	$x_i - \bar{x}$	$(x_i - \bar{x})^2$
4.8388	−0.1273	162.1×10^{-4}	4.5315	−0.1180	139.2×10^{-4}
4.9243	−0.0418	17.5	4.6232	−0.0263	6.9
4.9445	−0.0216	4.7	4.6232	−0.0263	6.9
4.9542	−0.0119	1.4	4.6435	−0.0060	0.4
4.9731	+0.0070	0.5	4.6435	−0.0060	0.4
4.9777	+0.0116	1.3	4.6532	+0.0037	0.1
5.0334	+0.0673	45.3	4.6721	+0.0226	5.1
5.0828	+0.1167	136.2	4.6902	+0.0407	16.6
			4.6902	+0.0407	16.6
			4.7243	+0.0748	56.0
39.7288		369.0×10^{-4}	46.4949		248.2×10^{-4}

$n_1 = 8$ d.f. $= 7$
$\bar{x}_1 = 39.7288/8 = 4.9661$
$s_1^2 = 369.0 \times 10^{-4}/7 = 0.00527$

$n_2 = 10$ d.f. $= 9$
$\bar{x}_2 = 46.4949/10 = 4.6495$
$s_2^2 = 248.2 \times 10^{-4}/9 = 0.00276$

where $\bar{x}$ is the mean of the numbers $\bar{x}_i$. Compute the ratio:

$$V = \frac{s_b^2}{s_w^2} \qquad \text{(Eq 34)}$$

From Table 12, the value of F_β ($\beta = 1 - \alpha$) corresponding to $k - 1$ degrees of freedom in the numerator and $\sum_{i=1}^{k} n_i - k$ degrees of freedom in the denominator is read. The hypothesis that the means of the k populations are equal is accepted if and only if $V < F$.

Fatigue Resistance Over a Range of Stress or Strain Levels

In many fatigue applications, both the high- and low-cycle fatigue resistance of a material must be considered. Comparisons are inevitably made between materials in terms of their overall S/N or ϵ/N fatigue resistance. The choice of material usually is based on which provides the highest predicted fatigue resistance. Predicted fatigue resistance in a variable-amplitude loading situation depends on several factors, including the assumed stress history and the damage-accumulation model, which may embody mean stress effects, history effects, multiaxial effects, etc.

Fatigue resistance for materials of the same class, but with different strength levels, commonly exhibit trends such as those shown in Fig. 19. The "rocking chair" effect illustrates the tendency of high-strength materials to exhibit poor low-cycle fatigue resistance and superior high-cycle fatigue resistance. However, lower strength materials of the same basic alloy group tend to exhibit superior low-cycle fatigue properties and lower high-cycle fatigue properties.

In some situations, the problem can be simplified by comparison of low-cycle fatigue resistance or high-cycle fatigue strength. In other cases, the relative importance of the two regimes must be weighed through a fatigue-damage analysis, and selection of a material should be based on overall fatigue-damage resistance.

If low-cycle fatigue properties are important in a comparison of two materials, it may be sufficient to consider data below the transition fatigue life and to compare fatigue lives over the applicable range of strain amplitudes. If the data are replicated, data trends can be evaluated to see if they are parallel in log life. It also can be determined whether the two materials provide significantly different fatigue-life trends.

Standard analysis of variance techniques can be used for this purpose. If the data are not replicated, visual inspection may determine whether parallelism can be assumed. If parallelism can be assumed, a "dummy variable" analysis can be performed in which the data for both materials are analyzed using a regression model such as the following:

$$\log N = A_1 + A_2 y + A_3 \log \epsilon' \qquad \text{(Eq 35)}$$

where ϵ' is an appropriate form of strain parameter, such as $\Delta\epsilon$ (strain range) or $\Delta\epsilon p/2$ (plastic-strain amplitude), and $y = 1$ for Material A and 0 for Material B. Examination of the significance of the A_2 term in Eq 35 reveals the significance of a difference in fatigue performance of the two materials.

If parallelism cannot be assumed and if the data have not been replicated, an extension of Eq 35 can be considered as follows:

$$\log N = A_1 + A_2 y + A_3 y \log \epsilon' + A_4 \log \epsilon' \qquad \text{(Eq 36)}$$

where $y = 1$ for Material A and 0 for Material B. In Eq 36, the significance of A_2 and the difference between A_3 and A_4 should be examined.

This approach can be extended to multiple materials if additional dummy variables are added for each new material and if statistical significance tests are performed on the family of dummy variable coefficients. The linear model dummy variable approach can be applied to load-control data and to long-life strain-control data, as long as the comparisons are made over a range of the independent variable that produces a linear relationship with fatigue life (or the logarithm of fatigue life).

Consolidation of Fatigue Data

In some situations, the available quantity of fatigue data at a given condition is so limited that it is difficult to define even a mean S/N or ϵ/N curve with confidence. In other situations, projection of the fatigue response of a material at conditions intermediate to those for which data are available is of interest. In these situations, appropriate analytical models can be used to consolidate available fatigue data.

None of the techniques for consolidating fatigue data is universally applicable, and all should be used with careful consideration. Care should be taken to ensure that the apparent trends in the data are not distorted. Extrapolation of the trends beyond the range of the data should only be undertaken with caution.

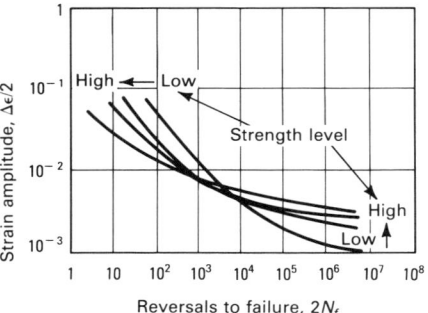

Fig. 19 Schematic of rocking chair effect

Increasing strength typically improves long-life fatigue strength and degrades low-cycle fatigue resistance.

Different Sources or Heats. Fatigue data are sometimes available that are similar in all respects except the testing source and/or the material heat. These data may be replicated test results at several stress levels or nonreplicated S/N curves. Statistical decisions must be made regarding whether it is appropriate to combine fatigue data from different heats or sources.

If data are available at a single stress or strain level, the comparison can be based on a nonparametric ranking procedure or on a parametric basis using the F and t procedures, as described earlier. If the data were replicated at multiple stress levels, standard analysis of variance techniques can be used to discern the significance of differences in sources or heats.

If multiple heats are to be included in the definition of the fatigue response of a material, the use of heat-centering techniques should be considered (Ref 42, 43). Techniques for treatment of heats involving unequal sample sizes (Ref 44) can also be used in such analyses.

Different Mean Stresses or Strains. Fatigue loading on actual components generally involves variable-amplitude conditions, which produce local stress and strain cycles covering a range of amplitudes and mean levels. Both the stress or strain amplitude and the mean stress or strain have a major impact on the fatigue resistance of a material.

This interrelationship between stress amplitude and mean stress with fatigue life has been represented (Ref 45, 46) in constant-life diagrams, known as Goodman diagrams (Fig. 20). Each line represents the combinations of mean stress and stress amplitude that will produce an average fatigue life for a specified number of cycles.

The combinations of stress amplitude and mean stress defining conditions of constant severity in terms of fatigue can also be rep-

Fig. 20 Goodman diagram for fatigue data

$R = S_{min}/S_{max}$

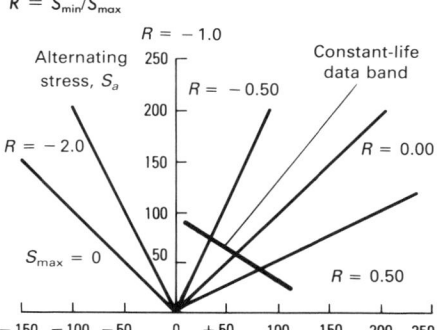

Fig. 21(a) Goodman diagram plot of the Stulen equivalent stress parameter (Eq 37)

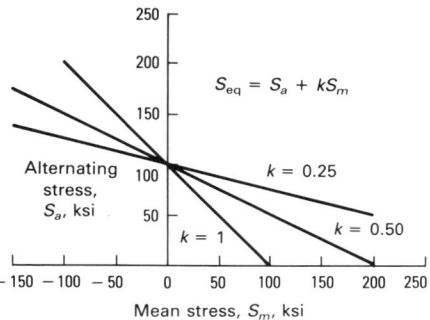

Fig. 21(b) Goodman diagram of the Topper-Sandor equivalent stress parameter (Eq 38)

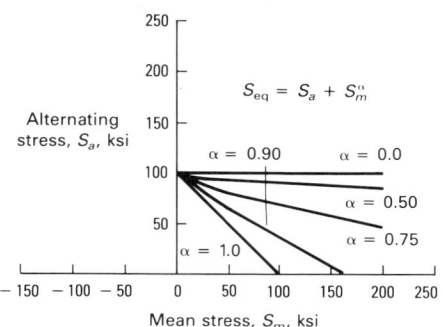

Fig. 21(c) Goodman diagram plot of the Walker equivalent stress parameter (Eq 39)

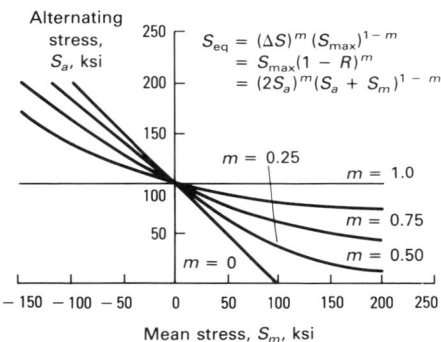

Fig. 21(d) Goodman diagram plot of the Leis equivalent stress parameter (Eq 40)

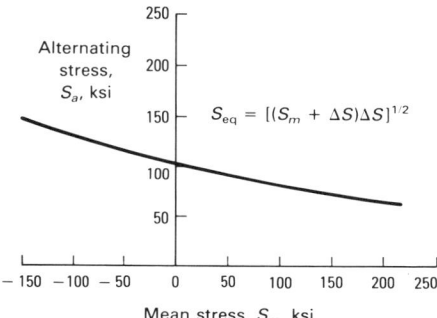

resented in terms of equivalent stress parameters. Some of the more commonly used equivalent stress parameters (Ref 47-50) are:

$$S_{eq} = S_a + kS_m \qquad (Eq\ 37)$$

$$S_{eq} = S_a + S_m^{\alpha} \qquad (Eq\ 38)$$

$$S_{eq} = (\Delta S)^m (S_{max})^{1-m} = S_{max}(1-R)^m \qquad (Eq\ 39)$$

$$S_{eq} = [(S_m + \Delta S)\Delta S]^{1/2} \qquad (Eq\ 40)$$

where S_a is alternating stress; S_m is mean stress; S_{max} is maximum stress; ΔS is stress range; k, α, and m are optimized constants; and R is stress ratio, or minimum stress/maximum stress.

The above equations are not applicable to all materials and all stress ratios. Stress ratio effects are best modeled in the range of about -1.0 to $+0.6$, which is the range of practical importance in many engineering applications. However, each of these models can produce unrealistic trends for very low and high stress ratios.

The behavior of each material system should be described by the model that is applicable to that system. The possible range of values for the optimized constants in these four expressions is shown in Fig. 21(a) to 21(d). In each case, different values for these optimized constants affect the shape of the predicted constant-life line on a Goodman diagram.

Models for consolidating data generated at different mean stress-strain conditions in strain control have also been formulated. Each equivalent stress model has an equivalent strain parameter counterpart. The equivalent strain forms of Eq 37 to 40 are expressed as:

$$\epsilon_{eq} = \epsilon_a + \frac{kS_m}{E} \qquad (Eq\ 41)$$

$$\epsilon_{eq} = \epsilon_a + \frac{S_m^{\alpha}}{E} \qquad (Eq\ 42)$$

$$\epsilon_{eq} = (\Delta\epsilon)^m \left(\frac{S_{max}}{E}\right)^{1-m} \qquad (Eq\ 43)$$

$$\epsilon_{eq} = \left[\left(\frac{S_m}{E} + \Delta\epsilon\right)\Delta\epsilon\right]^{1/2} \qquad (Eq\ 44)$$

where ϵ_a is the alternating strain; $\Delta\epsilon$ is the strain range; S_m is the stable mean stress; S_{max} is the stable maximum stress; E is the elastic modulus; and k, α, and m are optimized constants.

In the above equations, the mean and maximum stress values are obtained from stable hysteresis loops developed in the course of the strain-controlled test, normally at the half-life point, or are computed through an analytically defined cyclic stress-strain curve (Ref 51).

Application of Laboratory Fatigue Data to Real Component Life Assessments. Many factors can influence the fatigue response of a material and should therefore be considered in the application of laboratory data to real components. Typical factors include notch concentrations, elevated or reduced temperatures, surface finish variations, aggressive environments, multiaxial stress conditions, and residual stresses. Detailed procedures for accommodating these factors can be found in Ref 52 to 56.

Special Analysis Requirements of Composite Materials

By Won J. Park
Professor of Mathematics and Statistics
Wright State University

and

Ran Y. Kim
Senior Research Engineer
University of Dayton Research Institute

FIBER-REINFORCED COMPOSITE MATERIALS, such as graphite/epoxy, boron/epoxy, and glass/epoxy exhibit complex failure mechanisms under static and fatigue loading because of anisotropic characteristics in their strength and stiffness. Instead of a predominant single crack, often observed in most isotropic, brittle materials, extensive damage throughout the specimen usually accompanies fatigue failure in composites. Basic failure mechanisms are matrix cracking, delamination, fiber breakage, and interfacial debonding. Any combination of these causes fatigue damage, which may result in reduced fatigue strength and stiffness. Damage varies widely, depending on material properties, laminations (including stacking sequence), type of fatigue loading, etc. Damage development under fatigue and static loading is similar, except that fatigue causes additional damage as a function of cycles at a given stress level.

Failure Mechanisms

Matrix Cracking. To illustrate matrix cracking, consider the matrix of a $(0/90/\pm45)_s$ graphite/epoxy laminate subjected to uniaxial tension (see Fig. 22). Successive transverse cracks in the respective off-axis layer are expected as the load applied to the laminate increases. The first crack occurs in the 90° layers; with greater load, the number of cracks increases and is confined to the 90° layers. As the load increases, new cracks occur in the adjacent 45° layers, appearing at the tip of the 90° cracks and extending to the interface of the $+45/-45$ layers. Subsequently, the number of cracks increases with the load level until final laminate failure (Ref 57). However, some laminates reach a crack-density limit, after which no new cracks occur before final failure, despite additional loading (Ref 58). The crack-density limit for a given layer varies with its thickness, but appears to be independent of laminate type (Ref 57-61).

During fatigue loading, more cracks occur in each layer and reach a crack-density limit than during static loading. Many cracks in the $-45°$ layers of the $(0/90/\pm45)_s$ laminate occur during fatigue loading, whereas few or none occurs during static loading. In addition, many axial cracks initiate at the tip of the transverse crack and extend along the axial direction as fatigue cycles increase. Cracks in the 90° layers of the $(0/90/\pm45)_s$ laminate are found at less than 10^6 cycles at a fatigue stress level of 25 ksi, which is less than the 40-ksi stress level at which the first layer fails under static loading.

The 0° layers are also susceptible to cracking in the fiber direction, due to transverse stress in the 0° layer of a multidirectional laminate. Because transverse stress usually is slight, axial cracking in 0° layers may not occur under static loading. However, under fatigue loading, axial cracking in cross laminates appears.

Delamination. In composites, free-edge delamination is caused by interlaminar stresses localized around the free edge under in-plane loading (Ref 62, 63). Among the interlaminar stresses, tensile normal stress is a significant factor in most delaminations (Ref 3, 59, 64). The laminate stacking sequence determines whether interlaminar normal stress produces tension or compression at the free edges. For example, a quasi-isotropic laminate with two stacking sequences, $(0/90/\pm45)_s$ and $(0/\pm45/90)_s$, is subjected to uniaxial tensile loading. The latter stacking sequence produces tensile normal stress along the free edges of a coupon, whereas the former produces compression. Consequently, the $(0/\pm45/90)_s$ laminate shows extensive delamination under tension before final failure, but shows no delamination under compression.

During fatigue loading, the delamination that occurs at stress levels that are lower than during static loading at the onset of delami-

nation also occurs in early fatigue life and rapidly propagates toward the middle of the specimen width as the number of cycles increases. The point at which delamination begins to occur under fatigue varies, depending on the type of laminate.

In addition to interlaminar tensile stress, other mechanisms such as transverse cracking and interlaminar shearing appear to be significant in the onset and growth of delamination under fatigue loading (Ref 65). However, delamination grows more slowly from these causes than from interlaminar tension.

Fiber break and interface debonding differ widely, depending on the properties of the constituent materials and fiber defects. In most advanced composites, such as boron and graphite fiber with polymer matrix, the resistance to failure is greater in the matrix than in the fibers. Therefore, fibers can break, because of defects or weakness, before the interface fails. The crack created by a fiber break grows into the matrix as load increases along a path varying with matrix and interface properties. If bonding is strong, the crack grows into the matrix, resulting in a fairly smooth surface across the section. With a weak bond, the crack is more likely to lead to interfacial debonding and extensive fiber pullout. An intermediate bond shows irregular surface failure, with some fiber pullout. These failure mechanisms occur under static and fatigue loading, although fatigue failure depends on the sensitivity of the matrix, interface, and fiber. Because in most advanced composites the matrix remains elastic until the composite fails, fatigue damage in the interface is negligible, except at the site of fiber breaks. Consequently, the modulus and strength of a unidirectional laminate fatigue loaded along the fiber direction does not decrease until fracture is imminent.

Weibull Parameters Estimation

For static-strength or fatigue testing of composite materials, the two-parameter Weibull distribution is used widely for failure estimation. A random variable, X (static strength or fatigue life) has a Weibull distribution with shape parameter α and scale parameter β if the cumulative distribution function, $F_X(x)$, is:

$$F_X(x) = P_r\{X \le x\}$$

$$= 1 - \exp\left[-\left(\frac{x}{\beta}\right)^{\alpha}\right], x > 0$$

or its density function, $f_X(x)$, is given by:

Fig. 22 Schematic of the laminate code $(0/90/\pm45)_s$ for a composite material

The laminate code follows an ascending order from the bottom ply. The numerals are ply (or fiber) orientation with respect to the x-axis. The subscript s denotes that the laminate is symmetric with respect to the midplane ($z = 0$).

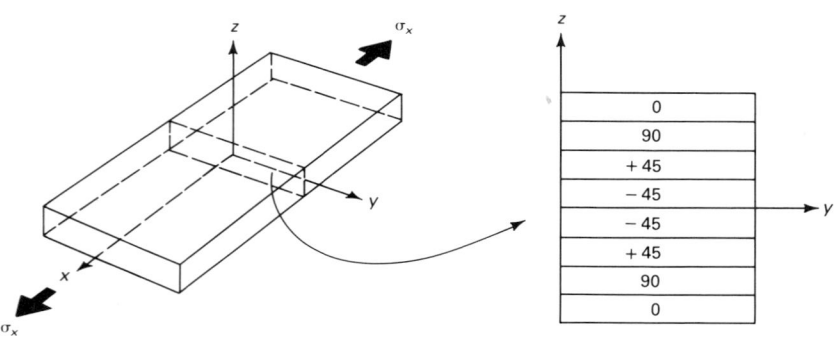

$$f_X(x) = \left(\frac{\alpha}{\beta}\right)\left(\frac{x}{\beta}\right)^{\alpha-1}\exp\left[-\left(\frac{x}{\beta}\right)^{\alpha}\right], x > 0$$

A maximum likelihood estimation method is applied for the parameter estimation. If X_1, X_2, ..., X_n is a random sample of size n (data) from the $W(\alpha, \beta)$ (Weibull distribution), the likelihood function, $L(\alpha, \beta)$, is given by:

$$L(\alpha, \beta) = \prod_{i=1}^{n} f(x_i, \alpha, \beta)$$

$$= \left(\frac{\alpha}{\beta}\right)^n \prod_{i=1}^{n}\left(\frac{x_i}{\beta}\right)^{\alpha-1}\exp\left[-\sum_{i=1}^{n}\left(\frac{x_i^{\alpha}}{\beta^{\alpha}}\right)\right]$$

(Eq 45)

The maximum likelihood estimations of α and β maximize $L(\alpha, \beta)$ and therefore should solve the equations:

$$\frac{\partial}{\partial\alpha}L(\alpha, \beta) = 0$$

and

$$\frac{\partial}{\partial\beta}L(\alpha, \beta) = 0$$

That is, α and β are solutions of the maximum likelihood equation:

$$\frac{\sum_{i=1}^{n}x_i^{\hat\alpha}\ln x_i}{\sum_{i=1}^{n}x_i^{\hat\alpha}} - \frac{1}{\hat\alpha} - \frac{\sum_{i=1}^{n}\ln x_i}{n} = 0$$

$$\hat\beta = \left[\frac{1}{n}\sum_{i=1}^{n}x_i^{\hat\alpha}\right]^{1/\hat\alpha}$$

(Eq 46)

An efficient iterative technique for obtaining $\hat\alpha$, the solution of $f(\hat\alpha) = 0$ (the first part of Eq 46), is the Newton-Raphson method (see Ref 66), in which the $(j + 1)$st successive approximation, $\hat\alpha_{j+1}$ to $\hat\alpha_j$, is given by:

$$\hat\alpha_{j+1} = \hat\alpha_j - \frac{f(\hat\alpha_j)}{f'(\hat\alpha_j)}$$

The maximum likelihood estimation method has advantages, in that the confidence intervals for α and β can be computed based on the tables given in Ref 67, and in that it can be applied to a life-test model in which censoring is progressive (a portion of the survivors is withdrawn several times during the test).

Example 5. The data in Table 15 are analyzed below, and the following estimates are obtained using the Newton-Raphson approximation:

$$\hat\alpha = 18.01$$
$$\hat\beta = 86.78$$

To find the 95% confidence intervals for α and β, let $l_{0.95} = 1.342$ and let $l_{0.95}^* = 0.34$ ($n = 29$) from Tables 16 and 17; therefore, 13.42 and 85.16 are 95% confidence intervals for α and β, respectively, because:

$$\frac{\hat\alpha}{l_r} = \frac{18.01}{1.342} = 13.42 < \alpha$$

$$\frac{\hat\beta^{l_r/\hat\alpha}}{e} = \frac{86.78}{\left(\frac{e^{0.34}}{18.01}\right)} = 85.16 < \beta$$

In fatigue testing, m levels of stress commonly are used, with n specimens tested at each level of stress. The model for the underlying distribution of stress cycles-to-failure at each level (ith level) often is a two-parameter Weibull distribution with a common shape parameter, α:

$$F_i(x) = 1 - \exp\left[-\left(\frac{x}{\beta_i}\right)^{\alpha}\right], x > 0$$

where β_i are scale parameters. From the pooled fatigue data, $x_{i1}, x_{i2}, \dots x_{in}$ (under ith level of stress), $i = 1, 2, \dots m$, the unknown parameters, α and β_i, can be estimated by the maximum likelihood method.

The maximum likelihood equation for estimating α and β_i is:

$$\sum_{i=1}^{m}\left[\frac{\sum_{i=1}^{n}x_{ij}^{\hat\alpha}\ln x_{ij}}{\sum_{j=1}^{n}x_{ij}^{\hat\alpha}}\right] - \frac{m}{\hat\alpha} - \frac{\sum_{i=1}^{m}\sum_{j=1}^{n}\ln x_{ij}}{n} = 0$$

$$\hat\beta_i = \left[\frac{1}{n}\sum_{j=1}^{n}x_{ij}^{\hat\alpha}\right]^{1/\hat\alpha}$$

$i = 1, 2, \cdots, m$

(Eq 47)

where $\hat\alpha$ is the pooled estimate of α. This pooled estimate is more efficient than the one introduced in Eq 46.

As before, the solution $\hat\alpha$ of Eq 47 can be obtained by the iterative technique of the Newton-Raphson method. Tables from Ref 68 are useful in computing the confidence intervals of α and β_i, respectively.

Example 6. The fatigue data given in Table 18 are analyzed below. According to Eq 47, $\hat\alpha = 1.46$, and the 95% confidence interval for α is $(0.86/\infty)$, because $\hat\alpha/l_\gamma = 1.46/1.704 = 0.86$. Here $l_{0.95} = 1.704$, from

Tables 19 and 20. Also, $l_\gamma^* = 0.849$; therefore, 95% confidence intervals for β_i for stress levels 70, 63.64, 53.85, and 50 ksi are $(9776/\infty)$, $(25\,536/\infty)$, $(931\,659/\infty)$, and $(1\,832\,160/\infty)$, respectively. For more information on the Weibull distribution, see the article "Statistical Distributions" in this Section.

Table 15 Ultimate static strength data, ksi

Graphite/epoxy composite material T300/5208 (0/90/±45)$_s$

72.04	72.44	76.61	77.19	77.36
79.36	80.48	81.52	81.97	82.20
82.22	82.51	83.35	83.90	84.99
85.70	85.79	86.43	86.61	86.77
87.98	88.54	88.86	89.60	90.17
91.44	91.73	92.37	93.33	

Table 16 Percentage points l_γ so that $P_r\{\hat\alpha/\alpha < l_\gamma\} = \gamma$

n \ γ	0.90	0.95	0.98
5	2.277	2.779	3.518
6	2.030	2.436	3.067
7	1.861	2.183	2.640
8	1.747	2.015	2.377
9	1.665	1.896	2.199
10	1.602	1.807	2.070
12	1.513	1.682	1.894
14	1.452	1.597	1.777
16	1.406	1.535	1.693
18	1.356	1.487	1.630
20	1.343	1.449	1.579
22	1.320	1.418	1.538
24	1.301	1.392	1.504
26	1.284	1.370	1.475
28	1.269	1.351	1.450
30	1.257	1.334	1.429
40	1.211	1.273	1.351
50	1.182	1.235	1.301

Table 17 Percentage points l_γ^* so that $P_r\{\hat\alpha\ln(\hat\beta/\beta) < l_\gamma^*\} = \gamma$

n \ γ	0.90	0.95	0.98
5	0.772	1.107	1.582
6	0.666	0.939	1.291
7	0.598	0.829	1.120
8	0.547	0.751	1.003
9	0.507	0.691	0.917
10	0.475	0.644	0.851
12	0.425	0.572	0.752
14	0.389	0.520	0.681
16	0.360	0.480	0.627
18	0.338	0.447	0.584
20	0.318	0.421	0.549
22	0.302	0.398	0.519
24	0.288	0.379	0.494
26	0.276	0.362	0.472
28	0.265	0.347	0.453
30	0.256	0.334	0.435
40	0.222	0.285	0.371
50	0.195	0.253	0.328

Source: Ref 67

Table 18 Tension-tension failure cycles

Graphite/epoxy composite material T300/5208 (0/90/±45)_s

Number of cycles, n	Maximum stress, ksi			
	70.0	63.64	53.85	50.0
4	5 490	21 320	360 270	1 907 860
	5 830	31 880	373 770	1 913 620
5	20 710	34 990	421 530	3 295 850
	22 570	40 340	489 900	3 569 000
	26 950	87 080	4 643 180	5 175 030
$\hat{\beta}_i$	17 490	45 683	1 666 737	3 277 735

Table 19 Percentage points, l_γ, so that $P_r\{\alpha/\alpha < l_\gamma\} = \gamma$

m	γ^n	5	6	7	8	9	10
3	0.90	1.642	1.534	1.457	1.391	1.360	1.329
	0.95	1.811	1.677	1.569	1.497	1.450	1.406
	0.98	2.049	1.852	1.723	1.620	1.561	1.515
4	0.90	1.554	1.457	1.393	1.343	1.313	1.288
	0.95	1.704	1.557	1.490	1.430	1.388	1.357
	0.98	1.872	1.704	1.610	1.530	1.480	1.442
5	0.90	1.505	1.412	1.353	1.313	1.280	1.258
	0.95	1.602	1.502	1.438	1.385	1.348	1.321
	0.98	1.763	1.625	1.536	1.480	1.430	1.399

Source: Ref 68

Table 20 Percentage points, l_γ^*, so that $P_r\{\alpha \ln(\beta/\beta) < l_\gamma^*\} = \gamma$

m	γ^n	5	6	7	8	9	10
3	0.90	0.655	0.578	0.533	0.488	0.453	0.420
	0.95	0.875	0.768	0.701	0.639	0.589	0.548
	0.98	1.116	0.997	0.905	0.816	0.751	0.710
4	0.90	0.656	0.571	0.521	0.480	0.445	0.422
	0.95	0.849	0.746	0.679	0.628	0.581	0.555
	0.98	1.094	0.955	0.858	0.781	0.736	0.700
5	0.90	0.648	0.570	0.513	0.466	0.436	0.416
	0.95	0.836	0.737	0.660	0.606	0.569	0.537
	0.98	1.063	0.940	0.845	0.764	0.710	0.671

Source: Ref 68

Strength Degradation Model

The residual strength degradation model is important in fatigue design, because it can predict the statistical distribution of fatigue life and the residual strength. In this section, the degradation model is presented under tension-tension constant-amplitude fatigue loading. The probability of surviving failure stress, X_o, in static tension, referred to in this article as static-strength distribution, is given by a two-parameter Weibull distribution:

$$R_o(x) = P_r\{X_o \geq x\} = \exp\left[-\left(\frac{x}{\beta_o}\right)^{\alpha_o}\right]$$

$$\text{(Eq 48)}$$

where α_o is the shape parameter, and β_o is the scale parameter (characteristic strength).

Because composite failure is characterized by many cracks, rather than a single dominant crack, residual strength, rather than crack length, describes the seriousness of the damage. Nevertheless, the change in residual strength is postulated, analogous to crack-growth laws in metals.

In terms of material age, L, the change in residual strength, $X(n)$, after n fatigue cycles, is assumed to be (Ref 69, 70):

$$\frac{dX(n)}{dL} = -\frac{1}{c+1}X^{-c}(n) \qquad \text{(Eq 49)}$$

Upon integration, Eq 49 yields the relationship between static and residual strength, $X(n)$:

$$X^c(n) = X^c(o) - L \qquad \text{(Eq 50)}$$

where $X(o)$ is static strength.

The material age, L, is related to the maximum stress level, stress ratio, frequency, and number of cycles. Consequently, Eq 50 can be expressed as:

$$X^c(n) = X^c(o) - f(S, \omega, R) \cdot n \qquad \text{(Eq 51)}$$

For simplicity, the loading frequency, ω, and the stress ratio, R, are fixed, so that $f(S, \omega, R) = f(S)$, where S is the applied stress level.

Crack-growth law is deterministic; that is, crack length at any time can be related to initial length, which determines the static strength. Thus, a one-to-one relationship can be established between static strength and fatigue life under a similar fatigue-damage process (Ref 70). In other words, a specimen that is strong under static loading is also strong under fatigue loading. Therefore, if the static-strength distribution is given by Eq 48, the residual-strength distribution, $R_n(X)$, follows by substituting Eq 51 into Eq 48:

$$R_n(x) = P\{X(n) \geq x\}$$

$$= \exp\left[-\left\{\left(\frac{x}{\beta_o}\right)^c + \frac{n}{\frac{\beta_o^c}{f(S)}}\right\}^{\alpha_o/c}\right]$$

$$\text{(Eq 52)}$$

Let N denote the number of cycles at which fatigue failure occurs under applied stress level, S. At the moment of fatigue fracture, the relationship $X(n) = S$ if and only if $N = n$ holds that the distribution of fatigue life, N, can be obtained from Eq 48 and 51 as:

$$R_n(n) = P_r\{N \geq n\}$$

$$= \exp\left[-\left\{\frac{n + \left(\frac{S^c}{f(S)}\right)}{\frac{\beta_o^c}{f(S)}}\right\}^{\alpha_o/c}\right] \qquad \text{(Eq 53)}$$

In general, the stress level is low, so that $(S/\beta_o)^c \ll 1$, and Eq 53 is reduced to:

$$R_N(n) = \exp\left[-\left\{\frac{n}{\frac{\beta_o^c}{f(S)}}\right\}^{\alpha_o/c}\right] \qquad \text{(Eq 54)}$$

The fatigue life, N, has a two-parameter Weibull distribution with the characteristic life $\beta_o^c/f(S)$, and the shape α_o/c, which is independent of the stress level, S.

From a classical S-N curve, $KS^bN = 1$, an expression for $f(S)$ is obtained as $f(S) = \beta_o^c KS^b$, because the characteristic life is $N = 1/KS^b = \beta_o^c/f(S)$, where K and b are constants. Note the Eq 51, 53, and 54 can be expressed as:

$$X(n)^c = X(o)^c - \beta_o^c KS^b n \qquad \text{(Eq 55)}$$

$$R_n(x) = \exp\left[-\left\{\left(\frac{x}{\beta_o}\right)^c + \frac{n}{\frac{1}{KS^b}}\right\}^{\alpha_o/c}\right]$$

$$\text{(Eq 56)}$$

and

$$R_N(n) = \exp\left[-\left(\frac{n}{\frac{1}{KS^b}}\right)^{\alpha_o/c}\right] \quad \text{(Eq 57)}$$

The term $1/KS^b$ in Eq 57 is the power law relating the characteristic life N to S_{max}. Therefore, the slope b of the log N-log S_{max} relationship is a measure of material aging. The steeper the slope, the more rapidly the material ages.

Example 7. For the graphite/epoxy composite material T300/5208 $(0/90/\pm45)_s$, the parameter estimations were obtained from previous examples as follows:

$\hat{\alpha}_o = 18.01$ and $\hat{\beta}_o = 86.78$
  (see Example 5)
$b = 16.09$ and $K = 1.874 \times 10^{-34}$
  (see Example 6)
$\hat{\alpha}_f = 1.46$ (see Example 6)

Therefore:

$$c = \frac{\hat{\alpha}_o}{\hat{\alpha}_f} = 12.34$$

The static-strength distribution is:

$$R(x_o) = \exp\left[-\left(\frac{x_o}{86.78}\right)^{18.01}\right]$$

The residual strength distribution after 10^4 cycles at 60 ksi is:

$R_n(x) =$

$$\exp\left\{-\left[\left(\frac{x}{86.78}\right)^{12.34} + 0.0764\right]^{1.46}\right\}$$

Note that $KS_{max}^b n$ is 0.0764.

Values for nine replicas of specimens tested experimentally for residual strength after undergoing 10^4 fatigue cycles are given in Table 21. Comparisons of experimental data (Table 15 and 21) with predicted distributions for static and residual strength are shown in Fig. 23. The median rank, $P_j = (j - 0.3)/(n + 0.4)$, of each experimental data point $x_{(j)}$ was used to represent the strength distribution.

Proof Testing and Data Analysis

Proof testing is an alternative to nondestructive inspection for control of fatigue failure of composite components if the damage incurred during the proof test is considered in reporting results (Ref 69, 71-73). In proof testing, specimens are loaded to the desired proof-stress level, unloaded, then fatigued until final failure. This procedure removes weak specimens and guarantees a minimum fatigue life for specimens that survive the proof test.

The relationship between static strength, x_o, and fatigue cycle, n—a specimen strong under static loading is also strong under fatigue loading—can be obtained from Eq 48 and 57 by equating R_o and R_n:

$$\frac{x_o}{\beta_o} = \left(\frac{n}{N}\right)^{\alpha_f/\alpha_o} \quad \text{(Eq 58)}$$

If proof loading, σ_p, is applied to a specimen, the corresponding fatigue cycle in Eq 58 is the minimum fatigue cycle. Thus, the minimum fatigue cycle, n_{min}, for a specimen that survives the proof loading, σ_p, is obtained by:

$$n_{min} = N\left(\frac{\sigma_p}{\beta_o}\right)^{\alpha_o/\alpha_f} \quad \text{(Eq 59)}$$

The relationship is valid if severe damage is not incurred during proof loading.

The static-strength distribution of specimens that survive proof loading, σ_p, is a truncated Weibull distribution:

$$R_{s,p}(x) = \exp\left[-\left(\frac{x}{\beta_o}\right)^{\alpha_o} + \left(\frac{\sigma_p}{\beta_o}\right)^{\alpha_o}\right]$$
$$\text{(Eq 60)}$$

The corresponding fatigue-life distribution is:

$$R_{f,p}(n) = \exp\left[-\left(\frac{n}{N}\right)^{\alpha_f} + \left(\frac{n_{min}}{N}\right)^{\alpha_f}\right]$$
$$\text{(Eq 61)}$$

where n_{min} satisfies Eq 58.

Table 21   Residual strength data

Graphite/epoxy composite material T300/5208 $(0/90/\pm45)_s$, after $n = 10\ 000$ fatigue cycles at $S = 60$ ksi

68.63	82.79	85.36
76.83	82.82	86.03
81.45	83.50	89.00

Example 8. To illustrate the concept of proof testing, the graphite/epoxy T300/5208 $(0/90/\pm45)_s$ laminate is used in this example. Figure 24 shows three types of tests performed on a closed-loop electrohydraulic testing machine with friction-type jaw grips. In a fatigue test, two stress levels, 50 ksi and 60 ksi, were employed, with a ratio of maximum to minimum stress of 0.1 and a frequency of 10 Hz.

The experimental data on static strength and fatigue life for two stress levels, 50 ksi and 60 ksi, were obtained by testing 29 specimens at both levels. Both static-strength and fatigue-life data were fitted to two-parameter Weibull distributions, and the parameters were estimated by the maximum likelihood method:

$\hat{\alpha}_o = 18.00$, $\hat{\beta}_o = 86.78$ ksi (static data)
$\hat{\alpha}_f = 1.58$, $\hat{N} = 1\ 009\ 350$ (under 50 ksi)
$\hat{\alpha}_f = 1.71$, $\hat{N} = 39\ 650$ (under 50 ksi)

Fatigue data points fitted to the two-parameter Weibull distribution are shown in Fig. 25.

To show the relationship between static strength and fatigue life, each pair of normalized static-strength and fatigue-life points of

Fig. 23 Distributions of static and residual strength

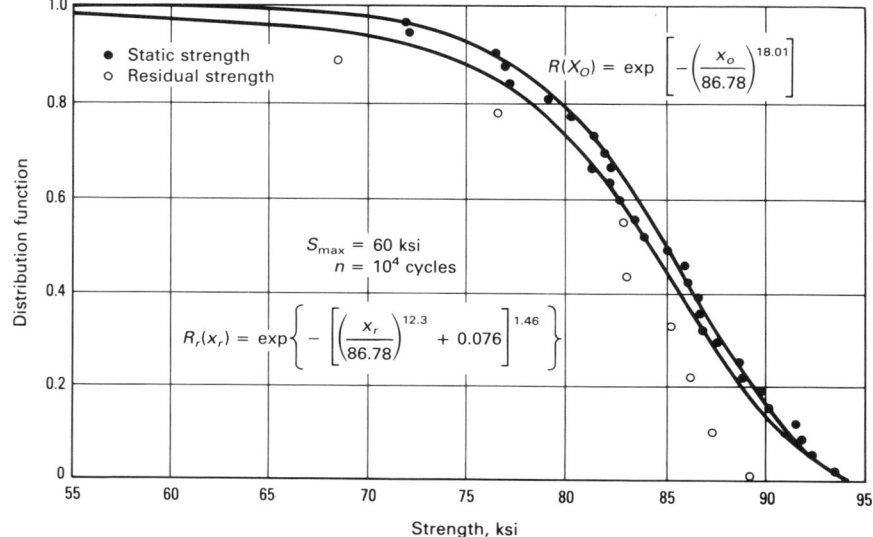

Fig. 24 Three types of loading

(a) Static tension. (b) Cyclic tension-tension fatigue. (c) Proof test followed by fatigue

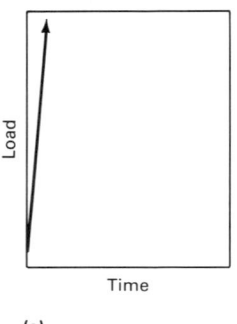
(a)

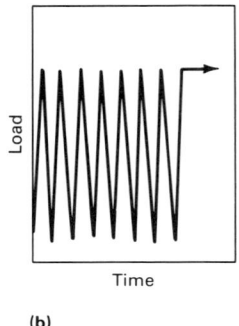
(b)

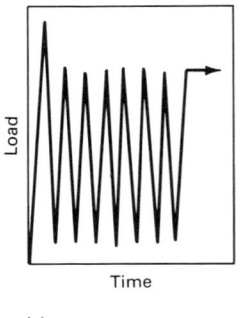
(c)

Fig. 25 Statistical distribution of fatigue life for graphite/epoxy T300/5208 (0/90/±45)$_s$

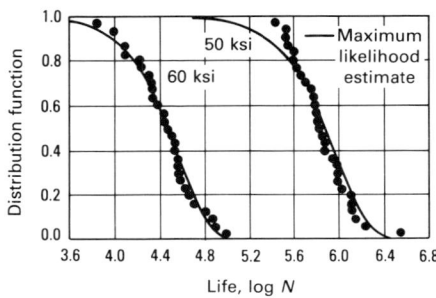

Fig. 26 Relationship between static strength and fatigue life

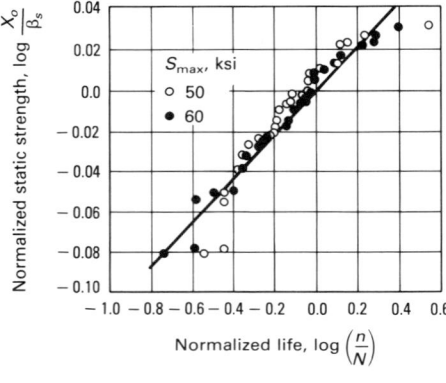

Fig. 27 Results of proof testing

(a) At 60 ksi. (b) At 50 ksi

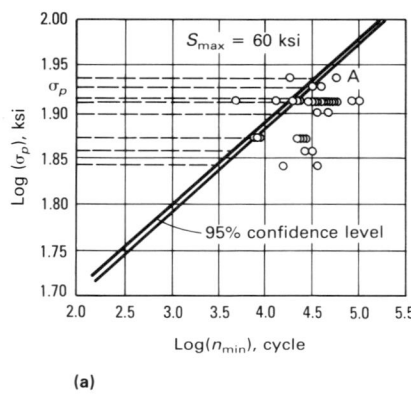
(a)

(b)

Fig. 28 Truncated fatigue life obtained by static proof test

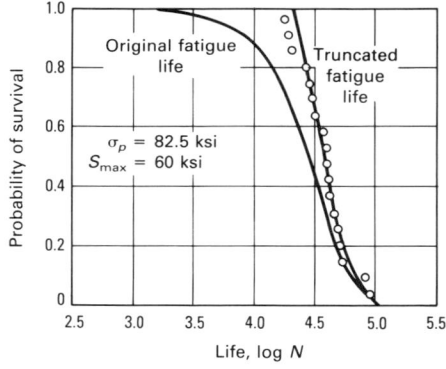

the same rank in both distributions is plotted on a rectangular coordinate system (Fig. 26).

The relationship between minimum fatigue cycle, n_{min}, and proof loading, σ_p, given in Eq 59 is plotted on a rectangular coordinate, with a logarithmic scale for stress levels, S_{max}, 60 and 50 ksi in Fig. 27. The validity of Eq 59 is checked through several proof stress levels. For example, specimen A was subjected to a proof stress, σ_p, unloaded, then fatigued at 60 ksi until failure. This specimen is represented by open circle A in Fig. 27(a). Because the specimen did not fail at proof stress, σ_p, its static strength is greater than σ_p. The minimum expected life is given by the intercept between the solid line and the dotted line at σ_p.

A total of 52 specimens were subjected to proof loading and then to fatigue. Seven specimens were subjected to residual-strength testing after exceeding minimum expected life, of which 5 specimens failed during proof loading, and 40 specimens were subjected to fatigue failure testing, of which 9 failed prior to minimum expected life, as indicated in Fig. 27. The premature failure

points in Fig. 27 are due to the high proof stress level, compared with the static strength of each specimen.

Predicting minimum fatigue life by proof testing (Eq 59) is satisfactory, as long as the proof stress level does not exceed a critical point. The effect of proof stress is also shown by the truncated life distribution, Eq 61. Because 18 specimens exceeded their minimum expected life at the proof stress of 82.26 ksi, with a maximum fatigue stress of 60 ksi, the proof stress levels were converted to a corresponding fatigue life (22 530 cycles), using Eq 59. The truncated life distribution at 22 530 cycles is shown, along with experimental data, in Fig. 28. The experimental data compare favorably, with the lower-tail distribution eliminated up to proof load level, thus increasing reliability.

REFERENCES

1. R.E. Little, *Manual on Statistical Planning and Analysis for Fatigue Experiments*, STP 588, ASTM, Philadelphia, 1975
2. "Standard Recommended Practice for

Constant-Amplitude Low-Cycle Fatigue Testing," E 606, *Annual Book of ASTM Standards*, Vol. 03.01, ASTM, Philadelphia, 1984, p 653-670
3. P.G. Forrest, *Fatigue of Metals*, Pergamon Press, Oxford, 1962, p 24-39
4. N.E. Frost, K.J. Marsh, and L.P. Pook, *Metal Fatigue*, Clarendon Press, Ox-

ford, 1974, p 54-61

5. R.E. Little and E.H. Jebe, *Statistical Design of Fatigue Experiments*, Applied Science, London, 1975

6. D.R. Cox, *Planning of Experiments*, John Wiley & Sons, New York, 1958

7. O.H. Basquin, "The Exponential Law of Endurance Tests," in *Proceedings*, Vol 10, Part II, ASTM, Philadelphia, 1910, p 625-630

8. L.F. Coffin, Jr. and J.F. Tavernelli, The Cyclic Straining and Fatigue of Metals, *Trans. Met. Soc. AIME*, Vol 215, Oct 1959, p 794-806

9. S.S. Manson, and M.H. Hirschberg, "Fatigue Behavior in Strain Cycling in the Low and Intermediate Cycle Range," Tenth Sagamore Army Materials Research Conference, Sagamore, NY, Aug 13-16, 1963

10. B.F. Langer, *Trans. ASME*, Vol 84, 1962, p 389

11. *Metallic Materials and Elements for Aerospace Vehicle Structures*, MIL-HDBK-5D, June 1983, p 9-53 to 9-65

12. R.C. Rice, K.B. Davies, C.E. Jaske, and C.E. Feddersen, "Consolidation of Fatigue and Fatigue-Crack-Propagation Data for Design Use," Battelle Columbus Laboratories, NASA Contractor Report, NASA CR-2586, Oct 1975

13. W. Nelson, Fitting of Fatigue Curves with Nonconstant Standard Deviation to Data with Runouts, *J. Test. Eval.*, Vol 12 (No. 2), March 1984, p 69-77

14. W. Nelson, *Applied Life Data Analysis*, John Wiley & Sons, New York, 1982

15. R.J. Carroll and D. Ruppert, "A Comparison Between Maximum Likelihood and Generalized Least Squares in a Heteroscedastic Linear Model," Institute of Statistics Mimeo Series 1334, Department of Statistics, University of North Carolina, Chapel Hill, 1981

16. A. Hald, *Statistical Theory with Engineering Applications*, Wiley Publications in Statistics, New York, 1957, p 204-208

17. F.A. Bastenaire, New Method for the Statistical Evaluation of Constant Stress Amplitude Fatigue-Test Results, in *Probabilistic Aspects of Fatigue*, STP 511, ASTM, Philadelphia, 1972, p 3-28

18. S. Nishijima, Statistical Fatigue Properties of Some Heat-Treated Steels for Machine Structural Use, in *Statistical Analysis of Fatigue Data*, STP 744, R.E. Little and J.C. Ekvall, Ed., ASTM, Philadelphia, 1981, p 75-88

19. C. Lipson and N.J. Sheth, *Statistical Design and Analysis of Engineering Experiments*, McGraw-Hill, New York, 1973, p 264-270

20. "Standard Practice for Statistical Analysis of Linear or Linearized Stress-Life (S-N) and Strain Life (ε-N) Fatigue Data," E 739, *Annual Book of ASTM Standards*, Vol. 03.01, ASTM, Philadelphia, 1984, p 732-740

21. D.B. Owen, "Tables of Factors for One-Sided Tolerance Limits for a Normal Distribution," Sandia Corporation, Albuquerque, April 1958

22. R.E. Little, Review of Statistical Analyses of Fatigue Life Data Using One-Sided Lower Statistical Tolerance Limits, in *Statistical Analysis of Fatigue Data*, STP 744, R.E. Little and J.C. Ekvall, Ed., ASTM, Philadelphia, 1981, p 3-23

23. P.R. Abelkis, "Statistical Aspects of Fatigue and Analysis of Fatigue Test Results," Lecture Notes on Design for Long Life Aircraft Structures, presented by Continuing Education in Engineering and Science University Extension, UCLA, April 1973, p 19

24. P.H. Wirsching, "Statistical Summaries of Fatigue Data for Design Purposes," University of Arizona, Tucson, NASA Contractor Report 3697, 1983

25. D.T. Raske, "Statistical Treatment of Fatigue Test Data," Argonne National Laboratory, prepared for USDOE/UKAEA Exchange Meeting on Mechanical Properties for FBR Structural Materials, Warrington, England, Sept 22-24, 1980

26. J. Schmee and G.J. Hahn, A Simple Method for Regression Analysis with Censored Data, *Technometrics*, Vol 21 (No. 4), 1979, p 417-432

27. W.R. Brose, N.E. Dowling, and J. Morrow, "Effect of Periodic Large Strain Cycles on the Fatigue Behavior of Steels," SAE Paper 740221, Society of Automotive Engineers, Warrendale, PA, 1974

28. A. Conle and T.H. Topper, Overstrain Effects During Variable Amplitude Service History Testing, *Int. J. Fatigue*, July 1980, p 130-136

29. R.C. Rice and D. Broek, "Fatigue Crack Initiation Properties of Rail Steels," Battelle Columbus Laboratories, Contractor Report DOT/FRA/ORD-82/05, U.S. Department of Transportation, Federal Railroad Administration, March 1982, p 45-56

30. M.A. Miner, Cumulative Damage in Fatigue, *J. Appl. Mech.*, Vol 12, 1945, p A159

31. R.C. Rice, private communication, Battelle Columbus Laboratories, Columbus, OH, Feb 24, 1984

32. J.L. Hodges, Jr., Fitting the Logistic by Maximum Likelihood, *Biometrics*, Vol 14 (No. 4), Dec 1958, p 453-461

33. L.F. Knudson and J.M. Curtis, The Use of the Angular Transformation in Biological Assays, *J. Amer. Stat. Assoc.*, Vol 42, 1947, p 282-296

34. R.E. Little, The Up-and-Down Method for Small Samples with Extreme Value Response Distributions, *J. Amer. Stat. Assoc.*, Vol 69, 1974, p 202-206

35. R.E. Little, A Reliability Analysis of Fatigue Limits Based on Large Sample Quantal Response Data, in *Mechanical Behavior of Materials*, Vol V, The Society of Materials Science, Tokyo, 1972, p 471-479

36. R.E. Little, *Tables for Estimating Median Fatigue Limits*, STP 731, ASTM, Philadelphia, 1980

37. R.E. Little, Estimating the Median Fatigue Limit for Very Small Up-and-Down Quantal Response Tests and for S-N Data with Runouts, in *Probabilistic Aspects of Fatigue*, STP 511, ASTM, Philadelphia, 1972, p 29-42

38. *A Tentative Guide for Fatigue Testing and the Statistical Analysis of Fatigue Data*, STP 91-A, ASTM, Philadelphia, 1958

39. W.H. Kruskal and W.A. Wallis, Use of Ranks in One Criterion Variance Analysis, *J. Amer. Stat. Assoc.*, Vol 47, 1952, p 583-621

40. W.J. Dixon and F.J. Massey, Jr., *Introduction to Statistical Analysis*, McGraw-Hill, New York, 1957

41. E.S. Pearson and H.O. Hartley, Ed., *Biometrika Tables for Statisticians*, Table 18, Vol 1, Biometrika Trustees, 1954

42. L.H. Sjodahl, A Comprehensive Method of Rupture Data Analysis with Simplified Models, in *Characterization of Materials for Service at Elevated Temperatures*, MPC-7, Metal Properties Council, American Society of Mechanical Engineers, New York, 1978, p 501-515

43. D.R. Rummler, Stress-Rupture Data Correlation—Generalized Regression Analysis, An Alternative to Parametric Methods, in *Reproducibility and Accuracy of Mechanical Tests*, ASTM, Philadelphia, 1977, p 110-126

44. J. Mandel and R. Paule, Interlaboratory Evaluation of a Material with Unequal Numbers of Replicates, *Anal. Chem.*, Vol 42 (No. 11), Sept 1970, p 1194-1197, correction in Vol 43 (No. 10), Aug 1971

45. J. Goodman, *Mechanics Applied to Engineering*, Longman, Green and Company, London, 1899

46. W.Z. Gerber, Bestimmung der Zulossigne Spannugen in Eisen Construc-

tionen, *Bayer. Archit. Ing. Ver.*, Vol 6, 1874, p 101 (in German)

47. F.B. Stulen, "Fatigue Life Data Displayed by a Single Quantity Relating Alternating and Mean Stress," Technical Report AFML-TR-65-121, Air Force Materials Laboratory, Wright-Patterson Air Force Base, 1965

48. T.H. Topper and B.I. Sandor, Effects of Mean Stress and Prestrain on Fatigue-Damage Summation, in *Effects of Environment and Complex Load History on Fatigue Life*, STP 462, ASTM, Philadelphia, 1970, p 93-104

49. K. Walker, The Effect of Stress Ratio During Crack Propagation and Fatigue for 2024-T3 and 7075-T6 Aluminum, in *Effects of Environment and Complex Load History on Fatigue Life*, STP 462, ASTM, Philadelphia, 1970, p 1-14

50. B.N. Leis, An Energy-Based Fatigue and Creep-Fatigue Damage Parameter, *J. Pressure Vessel Technol.*, Vol 99 (No. 4), Nov 1977, p 524-533

51. R.W. Landgraf, J. Morrow, and T. Endo, Determination of the Cyclic Stress-Strain Curve, *J. Mat.*, Vol 4 (No. 1), March 1969, p 176-188

52. J.A. Graham, Ed., *Fatigue Design Handbook*, AE-4, Society of Automotive Engineers, Warrendale, PA, 1968

53. P. Watson and R.G. Rebbeck, Modern Methods of Fatigue Assessment, *Railway Eng. J.*, Nov 1975, p 10-23

54. J. Lankford, D.L. Davidson, W.L. Morris, and R.P. Wei, Ed., *Fatigue Mechanisms: Advances in Quantitative Measurement of Physical Damage*, STP 811, ASTM, Philadelphia, 1983

55. C. Amyallag, B.N. Leis, and P. Rabbe, Ed., *Low-Cycle Fatigue and Life Prediction*, STP 770, ASTM, Philadelphia, 1982

56. P.R. Abelkis and C.M. Hudson, Ed., *Design of Fatigue and Fracture Resistant Structures*, STP 761, ASTM, Philadelphia, 1982

57. R.Y. Kim, Experimental Assessment of Static and Fatigue Damage of Graphite/Epoxy Laminates, in *Advances in Composite Materials*, Vol 2, A.R. Bunsell, Ed., Third International Conference on Composite Materials, Paris, 1980

58. K.L. Reifsnider, E.G. Henneke, II, and W.W. Stinchcomb, "Defect-Property Relationships in Composite Materials," Technical Report AFML-TR-76-81 Part IV, Air Force Materials Laboratory, Wright-Patterson Air Force Base, 1979

59. A.S.D. Wang and F.W. Crossman, Initiation and Growth of Transverse Cracks and Edge Delamination in Composite Laminates, Part 1, An Energy Method, *J. Compos. Mater. Suppl.*, 1980, p 71

60. K.W. Garrett and J.E. Bailey, Multiple Transverse Fracture in 90° Cross-ply Laminates of a Glass Fiber-Reinforced Polyester, *J. Mater. Sci.*, Vol 12, 1977, p 157

61. A. Parvizi and J.E. Bailey, On Multiple Transverse Cracking in Glass Fiber Epoxy Cross-ply Laminates, *J. Mater. Sci.*, Vol 13, 1978, p 2131

62. N.J. Pagano and R.B. Pipes, Some Observations on the Interlaminar Strength of Composite Laminates, *Int. J. Mech. Sci.*, Vol 15, 1973, p 679

63. N.J. Pagano, Stress Fields in Composite Laminates, *Int. J. Solids Struct.*, Vol 14, 1978, p 385

64. R.Y. Kim and S.R. Soni, Experimental and Analytical Studies on the Onset of

Delamination in Laminated Composites, *J. Compos. Mater.*, Vol 18, Jan 1984, p 70

65. S.R. Soni and R.Y. Kim, "Delamination of Composite Laminates Stimulated by Interlaminar Shear," presented at ASTM 7th Symposium on Composite Materials: Design and Testing, ASTM, Philadelphia, April 1984

66. J.F. Lawless, *Statistical Models and Methods for Lifetime Data*, John Wiley & Sons, New York, 1981, p 528

67. D.R. Thoman, L.J. Bain, and C.E. Antle, Inferences on the Parameters of the Weibull Distribution, *Technometrics*, Vol 11, 1969, p 445-450

68. W.J. Park, Percentiles of Pooled Estimates of Weibull Parameters, *IEEE Trans. Reliability*, April 1983, p 91-94

69. H.T. Hahn and R.Y. Kim, Proof Testing of Composite Materials, *J. Compos. Mater.*, Vol 9, 1975, p 297-311

70. J.N. Yang and M.D. Liu, Residual Strength Degradation and Theory of Periodic Proof Tests for Graphite/Epoxy Laminate, *J. Compos. Mater.*, Vol 11, 1977, p 193-203

71. J.N. Yang, *Reliability Prediction of Composites Under Proof Tests in Service*, STP 617, ASTM, Philadelphia, 1977, p 272-295

72. R.Y. Kim and W.J. Park, Proof Testing under Cyclic Tension-Tension Fatigue, *J. Compos. Mater.*, Vol 14, Jan 1980, p 70-79

73. P.C. Chou and A.S.D. Wang, "Statistical Analysis of Fatigue of Composite Materials," Technical Report AFML-TR-78-96, Air Force Materials Laboratory, Wright-Patterson Air Force Base, 1978

Appendix:
Metric and Conversion
Data for Mechanical Testing

This Appendix is intended as a guide for expressing weights and measures in the Système International d'Unités (SI) for use in mechanical testing. The purpose of SI units, developed and maintained by the General Conference of Weights and Measures, is to provide a basis for world-wide standardization of units and measure. For more information on metric conversions, the reader should consult the following references:

- "Standard for Metric Practice," E 380, 1984, ASTM, 1916 Race Street, Philadelphia, PA 19103
- "Metric Practice," ANSI/IEEE 268-1982, American National Standards Institute, 1430 Broadway, New York, NY 10018
- *Metric Practice Guide—Units and Conversion Factors for the Steel Industry*, 1978, American Iron and Steel Institute, 1000 16th Street NW, Washington, DC 20036
- *The International System of Units*, SP 330, 1981, National Bureau of Standards. Order from Superintendent of Documents, U.S. Government Printing Office, Washington, DC 20402
- *Metric Editorial Guide*, 4th ed., 1984, American National Metric Council, 1625 Massachusetts Ave. NW, Washington, DC 20036
- *ASME Orientation and Guide for Use of SI (Metric) Units*, ASME Guide SI 1, 9th ed., 1982, The American Society of Mechanical Engineers, 345 East 47th Street, New York, NY 10017

Base, supplementary, and derived SI units

Measure	Unit	Symbol	Measure	Unit	Symbol
Base units			Force	newton	N
Amount of substance	mole	mol	Frequency	hertz	Hz
Electric current	ampere	A	Heat capacity	joule per kelvin	J/K
Length	meter	m	Heat flux density	watt per square meter	W/m^2
Luminous intensity	candela	cd	Illuminance	lux	lx
Mass	kilogram	kg	Inductance	henry	H
Thermodynamic temperature	kelvin	K	Irradiance	watt per square meter	W/m^2
Time	second	s	Luminance	candela per square meter	cd/m^2
			Luminous flux	lumen	lm
Supplementary units			Magnetic field strength	ampere per meter	A/m
Plane angle	radian	rad	Magnetic flux	weber	WB
Solid angle	steradian	sr	Magnetic flux density	tesla	T
			Molar energy	joule per mole	J/mol
			Molar entropy	joule per mole kelvin	J/mol · K
Derived units			Molar heat capacity	joule per mole kelvin	J/mol · K
Absorbed dose	gray	Gy	Moment of force	newton meter	N · m
Acceleration	meter per second squared	m/s^2	Permeability	henry per meter	H/m
Activity (of radionuclides)	becquerel	Bq	Permittivity	farad per meter	F/m
Angular acceleration	radian per second squared	rad/s^2	Power, radiant flux	watt	W
Angular velocity	radian per second	rad/s	Pressure, stress	pascal	Pa
Area	square meter	m^2	Quantity of electricity,		
Capacitance	farad	F	electric charge	coulomb	C
Concentration (of amount			Radiance	watt per square meter steradian	W/m^2 · sr
of substance)	mole per cubic meter	mol/m^3			
Conductance	siemens	S	Radiant intensity	watt per steradian	W/sr
Current density	ampere per square meter	A/m^2	Specific heat capacity	joule per kilogram kelvin	J/kg · K
Density, mass	kilogram per cubic meter	kg/m^3	Specific energy	joule per kilogram	J/kg
Electric charge density	coulomb per cubic meter	C/m^3	Specific entropy	joule per kilogram kelvin	J/kg · K
Electric field strength	volt per meter	V/m	Specific volume	cubic meter per kilogram	m^3/kg
Electric flux density	coulomb per square meter	C/m^2	Surface tension	newton per meter	N/m
Electric potential, potential			Thermal conductivity	watt per meter kelvin	W/m · K
difference, electromotive force	volt	V	Velocity	meter per second	m/s
Electric resistance	ohm	Ω	Viscosity, dynamic	pascal second	Pa · s
Energy, work, quantity of heat	joule	J	Viscosity, kinematic	square meter per second	m^2/s
Energy density	joule per cubic meter	J/m^3	Volume	cubic meter	m^3
Entropy	joule per kelvin	J/K	Wavenumber	1 per meter	1/m

SI prefixes—names and symbols

Exponential expression	Multiplication factor	Prefix	Symbol
10^{18}	1 000 000 000 000 000 000	exa	E
10^{15}	1 000 000 000 000 000	peta	P
10^{12}	1 000 000 000 000	tera	T
10^{9}	1 000 000 000	giga	G
10^{6}	1 000 000	mega	M
10^{3}	1 000	kilo	K
10^{2}	100	hecto(a)	h
10^{1}	10	deka(a)	da
10^{0}	1	BASE UNIT	
10^{-1}	0.1	deci(a)	d
10^{-2}	0.01	centi(a)	c
10^{-3}	0.001	milli	m
10^{-6}	0.000 001	micro	μ
10^{-9}	0.000 000 001	nano	n
10^{-12}	0.000 000 000 001	pico	p
10^{-15}	0.000 000 000 000 001	femto	f
10^{-18}	0.000 000 000 000 000 001	atto	a

(a) Nonpreferred. Prefixes should be selected in steps of 10^3 so that the resultant number before the prefix is between 0.1 and 1000. These prefixes should not be used for units of linear measurement, but may be used for higher order units. For example, the linear measurement, decimeter, is nonpreferred, but square decimeter is acceptable.

Conversion factors

To convert from	to	multiply by
Angle		
degree	rad	1.745 329 E − 02
Area		
in.^{2}	mm^2	6.451 600 E + 02
in.^{2}	cm^2	6.451 600 E + 00
in.^{2}	m^2	6.451 600 E − 04
ft^2	m^2	9.290 304 E − 02
Bending moment or torque		
lbf · in.	N · m	1.129 848 E − 01
lbf · ft	N · m	1.355 818 E + 00
kgf · m	N · m	9.806 650 E + 00
ozf · in.	N · m	7.061 552 E − 03
Bending moment or torque per unit length		
lbf · in./in.	N · m/m	4.448 222 E + 00
lbf · ft/in.	N · m/m	5.337 866 E + 01
Current density		
A/in.^{2}	A/mm^2	1.550 003 E − 03
A/ft^2	A/m^2	1.076 400 E + 01
Electricity and magnetism		
gauss	T	1.000 000 E − 04
maxwell	μWb	1.000 000 E − 02
mho	S	1.000 000 E + 00
Oersted	A/m	7.957 700 E + 01
Ω · cm	Ω · m	1.000 000 E − 02
Ω circular-mil/ft	μΩ · m	1.662 426 E − 03
Energy (impact, other)		
ft · lbf	J	1.355 818 E + 00
Btu (thermochemical)	J	1.054 350 E + 03
cal (thermochemical)	J	4.184 000 E + 00
kW · h	J	3.600 000 E + 06
W · h	J	3.600 000 E + 03
Flow rate		
ft^3/h	L/min	4.719 475 E − 01
ft^3/min	L/min	2.831 000 E + 01
gal/h	L/min	6.309 020 E − 02
gal/min	L/min	3.785 412 E + 00

To convert from	to	multiply by
Force		
lbf	N	4.448 222 E + 00
kip (1000 lbf)	N	4.448 222 E + 03
tonf	kN	8.896 443 E + 00
kgf	N	9.806 650 E + 00
Force per unit length		
lbf/ft	N/m	1.459 390 E + 01
lbf/in.	N/m	1.751 268 E + 02
Fracture toughness		
ksi$\sqrt{\text{in.}}$	MPa$\sqrt{m}$	1.098 800 E + 00
Heat content		
Btu/lb	kJ/kg	2.326 000 E + 00
cal/g	kJ/kg	4.186 800 E + 00
Heat input		
J/in.	J/m	3.937 008 E + 01
kJ/in.	kJ/m	3.937 008 E + 01
Length		
Å	nm	1.000 000 E − 01
μin.	μm	2.540 000 E − 02
mil	μm	2.540 000 E + 01
in.	mm	2.540 000 E + 01
in.	cm	2.540 000 E + 00
ft	m	3.048 000 E − 01
yd	m	9.144 000 E − 01
mile	km	1.609 300 E + 00
Mass		
oz	kg	2.834 952 E − 02
lb	kg	4.535 924 E − 01
ton (short, 2000 lb)	kg	9.071 847 E + 02
ton (short, 2000 lb)	kg × 10^3(a)	9.071 847 E − 01
ton (long, 2240 lb)	kg	1.016 047 E + 03
Mass per unit area		
oz/in.^{2}	kg/m^2	4.395 000 E + 01
oz/ft^2	kg/m^2	3.051 517 E − 01
oz/yd^2	kg/m^2	3.390 575 E − 02
lb/ft^2	kg/m^2	4.882 428 E + 00

To convert from	to	multiply by
Mass per unit length		
lb/ft	kg/m	1.488 164 E + 00
lb/in.	kg/m	1.785 797 E + 01
Mass per unit time		
lb/h	kg/s	1.259 979 E − 04
lb/min	kg/s	7.559 873 E − 03
lb/s	kg/s	4.535 924 E − 01
Mass per unit volume (includes density)		
g/cm^3	kg/m^3	1.000 000 E + 03
lb/ft^3	g/cm^3	1.601 846 E − 02
lb/ft^3	kg/m^3	1.601 846 E + 01
lb/in.^{3}	g/cm^3	2.767 990 E + 01
lb/in.^{3}	kg/m^3	2.767 990 E + 04
Power		
Btu/s	kW	1.055 056 E + 00
Btu/min	kW	1.758 426 E − 02
Btu/h	W	2.928 751 E − 01
erg/s	W	1.000 000 E − 07
ft · lbf/s	W	1.355 818 E + 00
ft · lbf/min	W	2.259 697 E − 02
ft · lbf/h	W	3.766 161 E − 04
hp (550 ft · lbf/s)	kW	7.456 999 E − 01
hp (electric)	kW	7.460 000 E − 01
Power density		
W/in.^{2}	W/m^2	1.550 003 E + 03
Pressure (fluid)		
atm (standard)	Pa	1.013 250 E + 05
bar	Pa	1.000 000 E + 05
in.Hg (32 °F)	Pa	3.386 380 E + 03
in.Hg (60 °F)	Pa	3.376 850 E + 03
lbf/in.^{2} (psi)	Pa	6.894 757 E + 03
torr (mmHg, 0 °C)	Pa	1.333 220 E + 02
Specific heat		
Btu/lb · °F	J/kg · K	4.186 800 E + 03
cal/g · °C	J/kg · K	4.186 800 E + 03
Stress (force per unit area)		
tonf/in.^{2} (tsi)	MPa	1.378 951 E + 01
kgf/mm^2	MPa	9.806 650 E + 00

(continued)

Conversion factors (continued)

To convert from	to	multiply by	To convert from	to	multiply by	To convert from	to	multiply by
Stress (force per unit area) (continued)			**Thermal expansion**			strokes	m²/s	1.000 000 E − 04
ksi	MPa	6.894 757 E + 00	in./in. · °C	m/m · K	1.000 000 E + 00	ft²/s	m²/s	9.290 304 E − 02
lbf/in.² (psi)	MPa	6.894 757 E − 03	in./in. · °F	m/m · K	1.800 000 E + 00	in.²/s	mm²/s	6.451 600 E + 02
MN/m²	MPa	1.000 000 E + 00						
			Velocity			**Volume**		
Temperature			ft/h	m/s	8.466 667 E − 05	in.³	m³	1.638 706 E − 05
°F	°C	5/9 · (°F − 32)	ft/min	m/s	5.080 000 E − 03	ft³	m³	2.831 685 E − 02
°R	°K	5/9	ft/s	m/s	3.048 000 E − 01	fluid oz	m³	2.957 353 E − 05
			in./s	m/s	2.540 000 E − 02	gal (U.S. liquid)	m³	3.785 412 E − 03
Temperature interval			km/h	m/s	2.777 778 E − 01			
°F	°C	5/9	mph	km/h	1.609 344 E + 00	**Volume per unit time**		
						ft³/min	m³/s	4.719 474 E − 04
Thermal conductivity			**Velocity of rotation**			ft³/s	m³/s	2.831 685 E − 02
Btu · in./s · ft² · °F	W/m · K	5.192 204 E + 02	rev/min (rpm)	rad/s	1.047 164 E − 01	in.³/min	m³/s	2.731 177 E − 07
Btu/ft · h · °F	W/m · K	1.730 735 E + 00	rev/s	rad/s	6.283 185 E + 00			
Btu · in./h · ft² · °F	W/m · K	1.442 279 E − 01				**Wavelength**		
cal/cm · s · °C	W/m · K	4.184 000 E + 02	**Viscosity**			Å	nm	1.000 000 E − 01
			poise	Pa · s	1.000 000 E + 01			

(a) kg × 10³ = 1 metric ton

Abbreviations and Symbols

a crack length, or crack size, or crystal lattice length along *a* axis

A ampere

A area of cross section, or stress ratio

Å angstrom

ac alternating current

Ac₁ temperature at which austenite begins to form upon heating

Ac_1 temperature at which austenite begins to form upon heating

Ac_3 temperature at which transformation of ferrite to austenite is completed upon heating

Ac_{cm} in hypereutectoid steel, temperature at which cementite completes solution in austenite

AFWAL Air Force Wright Aeronautical Laboratories

AISC American Institute of Steel Construction

AISI American Iron and Steel Institute

AMMRC Army Materials and Mechanics Research Center

AMS Aerospace Material Specification (of SAE)

ANSI American National Standards Institute, Inc.

APC active path corrosion

API American Petroleum Institute

Ar_1 temperature at which transformation to ferrite or to ferrite plus cementite is completed upon cooling

Ar_3 temperature at which transformation of austenite to ferrite begins upon cooling

Ar_{cm} temperature at which cementite begins to precipitate from austenite on cooling

ASLE American Society of Lubrication Engineers

ASM American Society for Metals

ASME American Society of Mechanical Engineers

ASTM American Society for Testing and Materials

at.% atomic percent

atm atmosphere (pressure)

b width

B thickness

bal balance

bcc body-centered cubic

BS British Standards Institution

Btu British thermal unit

cal calorie

CCT center-cracked tension (specimen), or continuous cooling transformation

CDA Copper Development Association

CE carbon equivalent

cm centimeter

CMOD crack mouth opening displacement

COD crack opening displacement

cos cosine

cot cotangent

cpm cycles per minute

CRO cathode-ray oscilloscope

CRT cathode-ray tube

CT compact type (specimen)

CTOA crack tip opening angle

CTOD crack tip opening displacement

CVN Charpy V-notch

d used in mathematical expressions involving a derivative (denotes rate of change), or diameter

D diameter

dB decibel

da/dN fatigue crack growth rate

DBTT ductile brittle transition temperature

dc direct current

DCB double-cantilever beam

d.f. degrees of freedom

diam diameter

DIN German Industrial Standard

DOD Department of Defense

DPH diamond pyramid hardness

DS dispersion strengthened

DT dynamic tear

DWTT drop-weight tear test

e natural logarithm base, 2.71828...

e engineering strain (see also ϵ)

E modulus of elasticity for axial loading (Young's modulus)

e.g. for example

EPFM elastic-plastic fracture mechanics

EPRI Electric Power Research Institute

Eq equation

et al. and others

ETP electrolytic tough pitch (copper)

exp exponent, exponential

F farad

F force

fcc face-centered cubic

FDA Food and Drug Administration

Fig. figure

ft foot

g gram

G gauss

G shear modulus

gf gram-force

GJ gigajoule

GPa gigapascal

h hour

H height

HAZ heat-affected zone

HB Brinell hardness

hcp hexagonal close-packed

HEC hydrogen embrittlement cracking

HEL Hugoniot elastic limit

HEM hydrogen embrittlement

HFRSc Scleroscope hardness (Model C)

HFRSd Scleroscope hardness (Model D)

HK Knoop hardness

hp horsepower

HR Rockwell hardness (requires scale designation such as HRC for Rockwell "C" hardness)

HSLA high-strength low-alloy

HV Vickers hardness

Hz hertz

I moment of inertia

i subscript denoting the ith term

IACS International Annealed Copper Standard

ID inside diameter

i.e. that is

in. inch

ISO International Organization for Standardization

J joule

J polar moment of inertia

JIS Japanese Industrial Standard

K Kelvin

K stress-intensity factor

$\dot{K}$ stress-intensification rate

K_a crack arrest toughness

K_c plane-stress fracture toughness

K_f fatigue notch factor

K_{Ia} plane-strain crack arrest toughness

K_{Ic} plane-strain fracture toughness

K_{Id} dynamic fracture toughness

K_{Iscc} threshold stress intensity for stress-corrosion cracking

K_o crack initiation toughness

K_t theoretical stress-concentration factor

K_{th} threshold stress-intensity factor

kbar kilobar (pressure)

kcal kilocalorie

kg kilogram

kgf kilogram-force

kHz kilohertz

kN kilonewton

kPa kilopascal

ksi 1000 pounds per square inch

kV kilovolt

kW kilowatt

L liter, or longitudinal (direction)

L length

lb pound

lbf pound-force

LDH limiting dome height

LEFM linear elastic fracture mechanics

LMP Larson-Miller parameter

ln natural logarithm (base e)

log common logarithm (base 10)

LT long transverse (direction)

LVDT linear variable differential transformer

m meter

m strain rate sensitivity

M molar solution

M bending moment

max maximum

MCR minimum creep rate

mg milligram

MIL military

MIL-STD Military Standard

min minimum, or minute

mL milliliter

mm millimeter

MN meganewton

mPa millipascal

MPa megapascal

MPC Metal Properties Council

ms millisecond

N newton

N fatigue life (number of cycles), or normal solution

n strain-hardening exponent, or number of stress cycles endured, or sample size

NACA National Advisory Committee for Aeronautics

NACE National Association of Corrosion Engineers

NASA National Aeronautics and Space Administration

NBS National Bureau of Standards

NDTT nil ductility transition temperature

N_f fatigue life (number of cycles to failure)

nm nanometer

No. number

NRC Nuclear Regulatory Commission, or National Research Council

ns nanosecond

NSR notch strength ratio

OD outside diameter

OFHC oxygen-free high conductivity

ORNL Oak Ridge National Laboratory

oz ounce

p page

p probability

P applied (concentrated) load

Pa pascal

PCB printed circuit board

pH negative logarithm of hydrogen-ion activity

PH precipitation hardenable

P/M powder metallurgy

ppb parts per billion

ppm parts per million

PROM programmable read-only memory

psi pounds per square inch

PVRC Pressure Vessel Research Committee

PZT lead-zirconate-titanate

q fatigue notch sensitivity

R stress (load) ratio, or radius of curvature, or gas constant

RA reduction in area

rad radian

RAM random access memory

Ref reference

rem remainder

RF radio frequency

rms root mean square

rpm revolutions per minute

RSS residual sum of squares

s second

s sample standard deviation

s^2 sample variance

S siemens

S nominal engineering stress or normal engineering stress

S_a alternating stress amplitude

SAE Society of Automotive Engineers

SCC stress-corrosion cracking

SCE saturated calomel electrode

SEM scanning electron microscopy

SI Système International d' Unités

sin sine

sinh hyperbolic sine

SLC sustained-load cracking

S_m mean stress

SOA state-of-the-art

SSRT slow strain rate testing

ST short transverse (direction)

SWAP stress wave analyzing program

t time, or thickness

T tesla, or transverse (direction)

T temperature

tan tangent

TEM transmission electron microscopy

T_M or T_m melting temperature

TTT time-temperature transformation

u or *U* displacement

UHF ultra-high frequency

UNS Unified Numbering System (ASTM-SAE)

UTS ultimate tensile strength

v or *V* velocity

V volt

VHF very-high frequency

vol volume

W watt

Wb weber

WOL wedge-opening load

wt% weight percent

yr year

YS yield strength

℄ centerline

° degree, angular measure

°C degree Celsius (centigrade)

°F degree Fahrenheit

⇌ direction of reaction

÷ divided by

= equals

≈ approximately equals

≠ not equal to

≡ identical with

> greater than

≫ much greater than

≥ greater than or equal to

∫ integral of

∞ infinity

∝ varies as, is proportional to

< less than

≪ much less than

≤ less than or equal to

± maximum deviation

− minus, or negative ion charge

× multiplied by, diameters (magnification)

· multiplied by

/ per

% percent

+ plus, in addition to, including, positive ion charge

√ surface roughness

√ square root of

~ similar to, approximately

α angle

Γ or γ shear strain

δ crack (tip) opening displacement

Δ change in quantity, an increment, a range

ε general symbol for strain (natural or true strain)

$\dot{\varepsilon}$ strain rate

θ twist or angular movement

$\dot{\theta}$ twist rate (in torsion)

μ coefficient of friction

μin. micro-inch

μm micron

μs micro-second

ν Poisson's ratio

π pi (3.141592)

ρ density

σ normal or true stress

Σ summation of

τ shear stress

Ω ohm

Index

Z